## SHOWCASE EXAMPLES

These special examples stretch across the entire page and have a tan background that makes them easy to find. *Showcase Examples* introduce key topics and provide "how-to" instruction by walking step-by-step through the problem-solving process. With this format, the left and middle "columns" can be thought of as the instructor's voice offering an explanation (left column) and then summarizing information (middle column) during a classroom lecture.

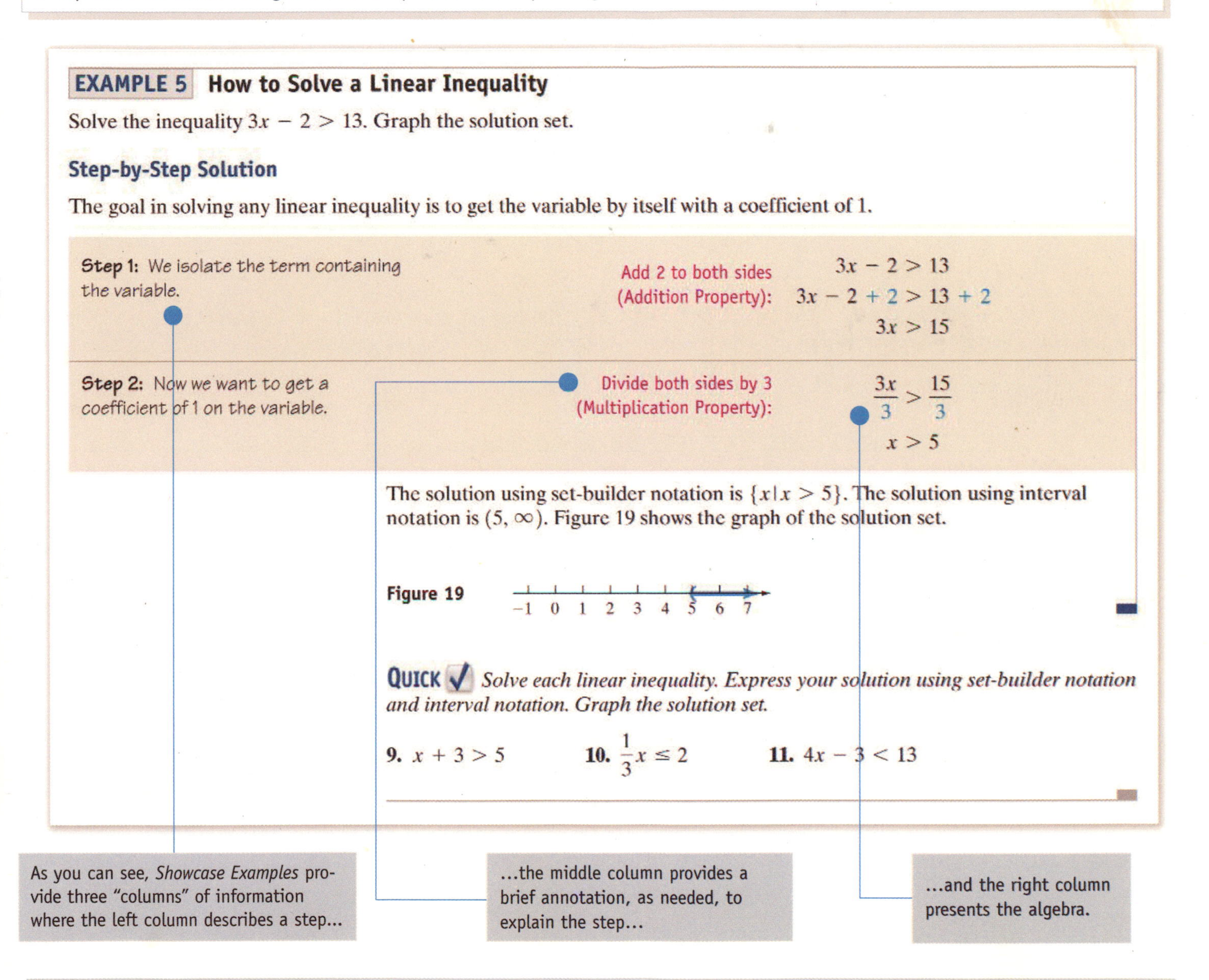

**EXAMPLE 5 How to Solve a Linear Inequality**

Solve the inequality $3x - 2 > 13$. Graph the solution set.

**Step-by-Step Solution**

The goal in solving any linear inequality is to get the variable by itself with a coefficient of 1.

| | | |
|---|---|---|
| **Step 1:** We isolate the term containing the variable. | Add 2 to both sides (Addition Property): | $3x - 2 > 13$<br>$3x - 2 + 2 > 13 + 2$<br>$3x > 15$ |
| **Step 2:** Now we want to get a coefficient of 1 on the variable. | Divide both sides by 3 (Multiplication Property): | $\frac{3x}{3} > \frac{15}{3}$<br>$x > 5$ |

The solution using set-builder notation is $\{x \mid x > 5\}$. The solution using interval notation is $(5, \infty)$. Figure 19 shows the graph of the solution set.

**Figure 19**

−1 0 1 2 3 4 5 6 7

**QUICK ✓** *Solve each linear inequality. Express your solution using set-builder notation and interval notation. Graph the solution set.*

**9.** $x + 3 > 5$ **10.** $\frac{1}{3}x \le 2$ **11.** $4x - 3 < 13$

As you can see, *Showcase Examples* provide three "columns" of information where the left column describes a step...

...the middle column provides a brief annotation, as needed, to explain the step...

...and the right column presents the algebra.

# Prep for Exams with the CHAPTER TEST PREP VIDEO CD

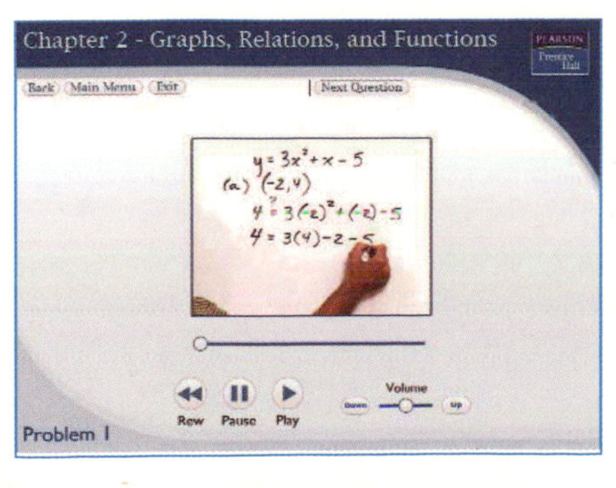


This video CD found at the back of your text contains worked-out solutions to every exercise in each Chapter Test in the text. To make the most of valuable study time when preparing for exams, follow these three steps:

1. Take the Chapter Test found at the end of the textbook chapter.
2. Check your answers in the back of the text.
3. Use the Chapter Test Prep Video CD to review every step of the worked-out solution to the specific questions you need to review.

# Additional Resources to Help You Succeed

## Student Study Pack

A single, easy-to-use package, available bundled with your textbook or by itself, for purchase through your bookstore. This package contains the following resources to help you succeed:

### Student Solutions Manual

- Solutions to the odd-numbered section exercises
- Solutions to the Quick Check exercises
- Solutions to the Preparing for This Section, Putting the Concepts Together (mid-chapter review), Chapter Review, Chapter Test, and Cumulative Review exercises

### Prentice Hall Math Tutor Center

- Staffed by qualified math instructors who provide students with tutoring on examples and odd-numbered exercises from the textbook. Tutoring is available via toll-free telephone, toll-free fax, email, or the Internet.

### CD Lecture Series

- Perfect for review of a section or a specific topic, these mini-lectures cover the key concepts from each section of the text in approximately 10-15 minutes.
- Includes fully worked-out solutions to exercises marked with a CD video icon.

## Online Homework and Tutorial Resources

### MyMathLab MyMathLab

MyMathLab is a series of text specific, easily customizable, online courses for Prentice Hall textbooks in mathematics and statistics. MyMathLab is powered by CourseCompass™—Pearson Education's online teaching and learning environment—and by MathXL®—our online homework, tutorial, and assessment system. MyMathLab gives instructors the tools they need to deliver all or a portion of their course online, whether students are in a lab setting or working from home. MyMathLab provides a rich and flexible set of course materials, featuring free-response exercises that are algorithmically generated for unlimited practice and mastery. Students can also use online tools, such as video lectures, animations, and a multimedia textbook, to independently improve their understanding and performance. MyMathLab is available to qualified adopters. For more information, visit our Web site at *www.mymathlab.com* or contact your Prentice Hall sales representative. (MyMathLab must be set up and assigned by your instructor.)

### MathXL® www.mathxl.com MathXL

MathXL is a powerful online homework, tutorial, and assessment system that accompanies the text. With MathXL, instructors can create, edit, and assign online homework and tests using algorithmically generated exercises correlated to your textbook. All student work is tracked in MathXL's online gradebook. Students can take chapter tests in MathXL and receive personalized study plans based on their test results. The study plan diagnoses weaknesses and links students directly to tutorial exercises for the objectives they need to study and retest. Students can also access supplemental animations and video clips directly from selected exercises. MathXL is available to qualified adopters. For more information, visit our Web site at *www.mathxl.com* or contact your Prentice Hall sales representative for a product demonstration. (MathXL must be set up and assigned by your instructor.)

# Intermediate Algebra

Michael Sullivan, III
Joliet Junior College

Katherine R. Struve
Columbus State Community College

Upper Saddle River, New Jersey 07458

**Library of Congress Cataloging-in-Publication Data**

Sullivan, Michael.
Intermediate algebra/Michael Sullivan, III, Katherine R. Struve.
p. cm.
Includes indexes.
ISBN 0-13-146773-5
1. Algebra—Textbooks. I. Title.
CIP data available.

Executive Editor: Paul Murphy
Editor in Chief: Christine Hoag
Executive Project Manager: Ann Heath
Production Editor: Barbara Mack
Senior Managing Editor: Linda Mihatov Behrens
Executive Managing Editor: Kathleen Schiaparelli
Media Project Manager: Audra J. Walsh
Media Production Editor: Jenelle J. Woodrup
Assistant Managing Editor, Science and Math Print Supplements: Karen Bosch
Manufacturing Buyer: Maura Zaldivar
Manufacturing Manager: Alexis Heydt-Long
Director of Marketing: Patrice Jones
Senior Marketing Manager: Kate Valentine
Marketing Assistant: Jennifer de Leeuwerk
Development Editor: Don Gecewicz
Editor in Chief, Development: Carol Trueheart
Editorial Assistant: Abigail Rethore
Project Manager/Class Testing: Dawn Nuttall
Art Director: Jonathan Boylan
Interior Designers: Wanda España, Mary Siener
Cover Designer: Wanda España
Art Editor: Thomas Benfatti
Creative Director: Juan R. López
Director of Creative Services: Paul Belfanti
Director, Image Resource Center: Melinda Reo
Manager, Rights and Permissions: Zina Arabia
Manager, Visual Research: Beth Brenzel
Manager, Cover Visual Research & Permissions: Karen Sanatar
Image Permission Coordinator: Richard Rodrigues
Photo Researcher: Melinda Alexander
Cover Photo: © Celia Pearson/Pearson Photography
Art Studios: Precision Graphics, Laserwords
Compositor: Interactive Composition Corporation

Pearson Prentice Hall
Pearson Education, Inc.
Upper Saddle River, New Jersey 07458


Printed in the United States of America

10 9 8 7 6 5 4 3 2 1

ISBN 0-13-146773-5

Pearson Education LTD., *London*
Pearson Education Australia PTY, Limited, *Sydney*
Pearson Education Singapore, Pte. Ltd
Pearson Education North Asia Ltd, *Hong Kong*
Pearson Education Canada, Ltd., *Toronto*
Pearson Educación de Mexico, S.A. de C.V.
Pearson Education—Japan, *Tokyo*
Pearson Education Malaysia, Pte. Ltd

*To Michael, Kevin, and Marissa, who patiently waited until Dad could come out to play.*

*- Michael Sullivan*

*In memory of my parents, Paul and Dorothy, who encouraged me to accept any challenge.*

*- Katherine R. Struve*

# About the Authors

With training in mathematics, statistics, and economics, Michael Sullivan, III has a varied teaching background that includes 15 years of instruction in both high school and college-level mathematics. He is currently a full-time professor of mathematics at Joliet Junior College. Michael has numerous textbooks in publication, including an Introductory Statistics series, and a Precalculus series, which he writes with his father, Michael Sullivan.

Michael believes that his experiences writing texts for college-level math and statistics courses give him a unique perspective as to where students are headed once they leave the developmental mathematics tract. This experience is reflected in the philosophy and presentation of his developmental text series. When not in the classroom or writing, Michael enjoys spending time with his three children, Michael, Kevin, and Marissa, and playing golf. Now that his two sons are getting older, he has the opportunity to do both at the same time!

Kathy Struve has been a classroom teacher for nearly 25 years, first at the high school level, and, for the past 13 years, at Columbus State Community College. Kathy emphasizes classroom diversity: diversity of age, learning styles, and previous learning success. She is aware of the challenges of teaching mathematics at a large, urban community college, where students have varied mathematics backgrounds, and may enter college with a high level of mathematics anxiety.

Kathy served as Lead Instructor of the Developmental Algebra sequence at Columbus State where she developed curriculum and provided leadership to adjunct faculty in implementing graphing calculator technology in the classroom. She has authored classroom activities at the Elementary Algebra, Intermediate Algebra, and College Algebra levels and conducted workshops at local, state, and national conferences. In her spare time Kathy enjoys biking, hiking, traveling, and reading British detective mysteries.

# Contents

## A Word about Textbook Design and Student Success

As students and instructors have related in Prentice Hall focus groups and market research surveys, developmental math textbooks should not look "cluttered" or "busy." A busy design can distract a student from what is most important in the text. It can also heighten math anxiety.

As a result of this research, the design of this text is understated and focused on the most important pedagogical elements. Students and instructors helped us to identify the primary elements of this text, which are central to student success. They include:

- Exercise Sets
- Examples and *Quick Check* exercises (practice problems)
- Rules, Property, and Definition boxes
- Study Aids: *In Words*, *Work Smart*, and *Work Smart: Study Skills*

As you will notice, these primary features are the most prominent elements in the design. We have made every attempt to ensure that these components are the features to which the eye is drawn. The remaining features, the secondary elements, blend into the "fabric" or "grain" of the overall design.

Our thanks go to all of the students and instructors who helped us develop the design of this text. Their feedback proved invaluable in helping us to make the right decisions. We are confident the design of this text will be both practical and engaging as it serves its educational and learning purposes.

Sincerely,

Paul Murphy

Executive Editor
Developmental Mathematics
Prentice Hall

Kate Valentine

Senior Marketing Manager
Developmental Mathematics
Prentice Hall

# Preface

Intermediate Algebra is a gateway course to other college-level mathematics courses. The goal of the course is to provide students with the mathematical skills that are prerequisites for courses such as College Algebra, Elementary Statistics, Liberal-Arts Math, Mathematics for Teachers, and so on. In addition, Intermediate Algebra must expose students to a variety of mathematical concepts that build on each other and that range from the basics such as linear equations to sophisticated concepts such as exponential functions.

Of particular importance in this course are rigor and mathematical thinking. It is imperative that this course have sufficient rigor to teach students how to study math successfully. At the same time, the course must develop students' ability to think mathematically. The rigor in the course exists both in the material presented (such as a more thorough development of functions) as well as in the array of problems and examples. As a result, it should be clear to students that this course is not simply a rehash of Elementary Algebra.

Most students have seen the content of this course at some point in their high-school careers or in other college coursework. For some students, success at studying and facility with math concepts did not develop during their previous contact with the material, and they need a fresh start. In addition, the number of nontraditional students who have lost some of their math skills over the course of time continues to grow, especially at community colleges. Nontraditional students often are highly motivated to succeed because of their life experiences, yet they may be rusty at the business of "going to school." For nontraditional students, this course refreshes and reinforces their study skills as well as their mathematical skills.

To address the many needs and the diversity of today's Intermediate Algebra students, we have been guided by the following ideas as broad goals for this text:

- Provide the student with a strong conceptual foundation in mathematics through a clear, comprehensive presentation of topics and a special emphasis on functions.
- Present a variety of pedagogical features, tools, study tips, and easy-to-use aids to help students see the value of the text as an important resource and guide that will increase their success in the course.
- Provide comprehensive exercise sets with paired exercises that build problem-solving skills, show a variety of applications of mathematics, and reinforce mathematical concepts for students.
- Streamline the Intermediate Algebra course through the strategic placement of topics that will provide instructors with the flexibility to review material as needed as opposed to reteaching it.

## A Strong Foundation Through a Functions Approach

The approach that we take in Intermediate Algebra is that the function is the overriding theme of the text. The reason for this stress on functions is twofold. First, Intermediate Algebra is not a terminal course but rather a gateway to the future, and functions form the basis for much study in mathematics. The introduction of functions helps make the "jump" from Intermediate Algebra to College Algebra less severe because students feel more comfortable with functions and function notation. Second, today's students like to learn in context so that they can see the relevancy of the material. The function provides a great way to present the usefulness of the material we are teaching.

## Developing an Effective Text for Use In and Out of the Classroom

Given the hectic lives led by most students, coupled with the anxiety and trepidation with which they approach this course, an outstanding developmental mathematics text must provide pedagogical support that makes the text valuable to students as they study and do assignments. Pedagogy must be presented within a framework that teaches students how to study math; pedagogical devices must also address what students see as the "mystery" of mathematics—and solve that mystery.

To encourage students and to clarify the material, we developed a set of pedagogical features that help students develop good study skills, garner an understanding of the connections between topics, and work smarter in the process. The pedagogy used in this text is based upon the more than 40 years of classroom teaching experience that the authors bring to this text.

**Examples** are often the determining factor in how valuable a textbook is to a student. Students look to Examples to provide them with guidance and instruction when they need it most—the times when they are away from the instructor and the classroom. We have developed several Example formats in an attempt to provide superior guidance and instruction for the students. The formats include:

### Innovative *Sullivan/Struve Examples*

The innovative *Sullivan/Struve Example* has a two-column format in which annotations are provided to the **left** of the algebra, rather than the right, as is the practice in most texts. Because we read from **left-to-right,** placing the annotation on the left will make more sense to the student. It becomes clear that the annotation describes what we are about to do instead of what was just done. The annotations may be thought of as the teacher's voice offering clarification immediately before writing the solution on the board. Consider the following:

**EXAMPLE 3 Solving a Linear Equation by Combining Like Terms**

Solve the linear equation $3y - 2 + 5y = 2y + 5 + 4y + 3$.

**Solution**

| | |
|---|---|
| | $3y - 2 + 5y = 2y + 5 + 4y + 3$ |
| Combine like terms: | $8y - 2 = 6y + 8$ |
| Subtract $6y$ from both sides: | $8y - 2 - 6y = 6y + 8 - 6y$ |
| | $2y - 2 = 8$ |
| Add 2 to both sides: | $2y - 2 + 2 = 8 + 2$ |
| | $2y = 10$ |
| Divide both sides by 2: | $\frac{2y}{2} = \frac{10}{2}$ |
| | $y = 5$ |

### Showcase Examples

*Showcase Examples* are used strategically to introduce key topics or important problem-solving techniques. These examples provide "how-to" instruction by offering a guided, step-by-step approach to solving a problem. Students can then immediately see how each of the steps is employed. We remind students that the *Showcase Example* is meant to provide "how-to" instruction by including the words "how to" in the example title.

The *Showcase Example* has a three-column format in which the left column describes a step, the middle column provides a brief annotation, as needed, to explain the step, and the right column presents the algebra. With this format, the left and middle columns can be thought of as the instructor's voice offering an explanation (left) and then summarizing information (middle) during a lecture.

**EXAMPLE 5 How to Solve a Linear Inequality**

Solve the inequality $3x - 2 > 13$. Graph the solution set.

**Step-by-Step Solution**

The goal in solving any linear inequality is to get the variable by itself with a coefficient of 1.

**Step 1:** We isolate the term containing the variable.

Add 2 to both sides (Addition Property):

$$3x - 2 > 13$$
$$3x - 2 + 2 > 13 + 2$$
$$3x > 15$$

**Step 2:** Now we want to get a coefficient of 1 on the variable.

Divide both sides by 3 (Multiplication Property):

$$\frac{3x}{3} > \frac{15}{3}$$
$$x > 5$$

### *Quick Check* Exercises

Placed at the conclusion of every Example, the *Quick Check* exercises provide the students with an opportunity for immediate reinforcement. By working the problems that mirror the example just presented, students get instant feedback and gain confidence in their understanding of the concept. All the answers to *Quick Check* exercises are provided in the back of the text. We think that the *Quick Check* exercises will make the text more accessible and encourage students to read, consult, and use the text regularly.

## Study Skills and Student Success

We have included study skills and student success as regular themes throughout this text starting with *Section R.1, Success in Mathematics*. In addition to this dedicated section that covers many of the basics that are essential to success in any math course, we have included several recurring study aids that appear in the margin. These features were designed to anticipate the student's needs and to provide immediate help—as if the teacher were looking over his or her shoulder. These margin features include *In Words; Work Smart;* and *Work Smart: Study Skills.*

***Section R.1: Success in Mathematics*** focuses the student on the basics of study skills, including what to do during the first week of the semester; what to do before, during, and after class; how to use the text effectively; and how to prepare for an exam.

***In Words*** helps to address the difficulty that students have in reading mathematically precise definitions and theorems by explaining them in plain English.

***Work Smart*** provides "tricks of the trade" hints, tips, reminders, and alerts. It also identifies some common errors to avoid and helps students work more efficiently.

***Work Smart: Study Skills*** reminds students of study skills that will help them to succeed at various points in the course. Attention to these practices will help them to become better, more proficient, learners.

## Test Preparation and Student Success

The Chapter Tests in this text and the companion Chapter Test Prep Video CD have been designed to help students make the most of their valuable study time.

***Chapter Test***

In preparation for their classroom test, students should take the practice test to make sure they understand the key topics in the chapter. The exercises in the Chapter Tests have been crafted to reflect the level and types of exercises a student is likely to see on a classroom test.

***Chapter Test Prep Video CD***

Packaged with each new copy of the text, the Chapter Test Prep Video CD provides students with help at the critical juncture when they are studying for a test. The CD Video presents step-by-step solutions to the exact exercises found in each of the book's Chapter Tests. Easy video navigation allows students to access instantly the worked-out solutions to the exercises they want to study or review.

## Superior Exercise Sets: Paired with Purpose

Students learn algebra by doing algebra. The superior end-of-section exercise sets in this text provide students with ample practice of both procedures and concepts. The exercises are paired and present problem types with every possible derivative.

The exercises also present a gradual increase in difficulty level. The early, basic exercises keep the student's focus on as few "levels of understanding" as possible. The later or higher-numbered exercises are "multi-task" (or Mixed Practice) exercises where students are required to utilize multiple skills, concepts, or problem-solving techniques.

Throughout the textbook, the exercise sets are grouped into seven categories—some of which appear only as needed:

1. **Concepts and Vocabulary** exercises are fill-in-the-blank, true/false, and open-ended questions that test a student's understanding of the vocabulary and concepts presented within the section. We have found that students do not succeed if they do not become familiar with the basic vocabulary of mathematics.
2. **Building Skills** exercises are drill problems that develop the student's understanding of the procedures and skills in working with the methods presented in the section. Often these exercises can be linked back to a single example in the section.
3. **Mixed Practice** exercises are also drill problems, but they offer a comprehensive assessment of the skills learned in the section by asking problems that relate to more than one concept or objective.
4. **Applying the Concepts** exercises are problems that allow students to see the relevancy of the material learned within the section. Problems in this category are either situational problems that use material learned in the section to solve "real-world" problems or they are problems that ask a series of questions to enhance a student's conceptual understanding of the mathematics presented in the section.
5. **Extending the Concepts Exercises** can be thought of as problems that go beyond the basics. Within this block of problems an instructor will find a variety of problems to sharpen students' critical-thinking skills.
6. Starting with Chapter 6, we provide **Synthesis Review** exercises to help students grasp the "big picture" of algebra—once they have a sufficient conceptual foundation to build upon from their work in Chapters 1 through 5. Synthesis review exercises ask students to perform a single operation (adding, solving, and so on) on several objects (polynomials, rational expressions, and so on). The student is then asked to discuss the similarities and differences in performing the same operation on the different objects.
7. Finally, we also include coverage of the **graphing calculator.** Instructors' philosophies about the use of graphing devices vary considerably. Because instructors disagree about the value of this tool, we have made an effort to make graphing technology entirely optional. When appropriate, technology exercises are included at the close of a section's exercise set.

## How It All Fits Together: The Big Picture

Another important role of the pedagogy in this text is to help students see and understand the connection between the mathematical topics being presented. Several section-opening and margin features help to reinforce connections:

***The Big Picture: Putting It Together (Chapter Opener)***

This feature is based on how we start each chapter in the classroom—with a quick sketch of what we plan to cover. Before tackling a chapter, we tie concepts and techniques together by summarizing material covered previously and then relate these ideas to material we are about to discuss. It is important for students to understand that content truly builds from one chapter to the next. We find that students need to be reminded that the familiar operations of addition, subtraction, multiplication, and division are being applied to different or more complex objects.

***Preparing for This Section***

As part of this building process, we think it is important to remind students of specific material that they will need from earlier in the course to be successful within a given section. The *Preparing for*... feature that begins each section not only provides a list of prerequisite skills that a student should understand before tackling the content of a new section but also offers a short quiz to test students' preparedness. Answers to the quiz are provided as a footnote on the same page, and a cross-reference to the material in the text is provided so that the student can remediate when necessary.

***Putting the Concepts Together (Mid-Chapter Review)***

Each chapter has a group of exercises at the appropriate point in the chapter, entitled Putting the Concepts Together. These exercises serve as a review—synthesizing material introduced up to that point in the chapter. The exercises in these mid-chapter reviews are carefully chosen to assist students in seeing the "big picture."

***Synthesis Review Exercises***

Starting with Chapter 6, we provide Synthesis Review exercises to help students grasp the "big picture" of algebra—once they have a sufficient conceptual foundation to build upon from their work in Chapters 1 through 5.

***Cumulative Review***

Learning algebra is a building process and building involves considerable reinforcement. The cumulative review exercises at the end of each odd-numbered chapter, starting with Chapter 1, help students to reinforce and solidify their knowledge by revisiting concepts and using them in context. This way, studying for the final exam should be fairly easy.

## Streamlining Intermediate Algebra: Getting Ready for Chapter . . . Review Sections

To maintain the pace of the course, we created several ***Getting Ready*** sections that review material taught in Elementary Algebra courses. The *Getting Ready* sections are designed to allow students to brush up on topics and skills as needed before beginning the chapters in the Intermediate Algebra text where the skills will be used or further developed. These optional, yet integrated, sections provide the student with timely review. They also streamline the Intermediate Algebra course by providing the instructors with the flexibility to decide if the *Getting Ready* sections should be covered in their entirety, briefly reviewed, or skipped, depending upon the needs of their students. *Getting Ready* review sections have been placed before Chapters 5, 6, and 7 in the text.

## In Closing

When we started writing this textbook, we discussed what improvement we could make in coverage; in staples such as examples and problems; and in any pedagogical features that we found truly useful. After writing and rewriting, and reading many thoughtful reviews from instructors, we focused on the following features of the text to set it apart.

- **Functions** are introduced early and revisited often throughout the course. This integration helps to prepare students for the quantitative courses that they will take after Intermediate Algebra.
- The **innovative *Sullivan/Struve Examples*** and ***Showcase Examples*** provide students with superior guidance and instruction when they need it most—when they are away from the instructor and the classroom.

- The ***Quick Check*** exercises provide students with immediate reinforcement and instant feedback to determine their understanding of the concepts presented in the examples.
- We developed each of the margin features such as ***In Words, Work Smart,*** and ***Work Smart: Study Skills*** with the goals of improving study skills, making the textbook easier to navigate, and increasing student success.
- ***Exercise Sets: Paired with Purpose***—The exercise sets are structured to assess student understanding of vocabulary, concepts, drill, problem solving, and applications. The exercise sets are graded in difficulty level to build confidence and to enhance students' mathematical thinking.
- ***Putting the Concepts Together*** and ***Synthesis Review*** help students see the big picture and provide a structure for learning each new concept and skill in the course.
- The text is written to streamline Intermediate Algebra (and distinguish it from Elementary Algebra) through the strategic placement of ***Getting Ready*** review sections that provide instructors with the flexibility to review material instead of reteaching it.

## Instructor and Student Resources

The following resources are available to help instructors and students use this text more effectively.

### Instructor Resources

#### Annotated Instructor's Edition (0-13-146775-1)

#### Instructor Solutions Manual (0-13-146776-X)

#### Instructor's Resource Manual with Tests (0-13-146774-3)

#### CD Lecture Series—Lab Pack (0-13-227686-0)

#### TestGen (0-13-146779-4)

- Enables instructors to build, edit, print, and administer tests.
- Features a computerized bank of questions developed to cover all text objectives.
- Available on dual-platform Windows/Macintosh CD-Rom.

#### MyMathLab Instructor Version (0-13-147898-2)

MyMathLab is a series of text specific, easily customizable, online courses for Prentice Hall textbooks in mathematics and statistics. MyMathLab is powered by Course Compass™—Pearson Education's online teaching and learning environment—and by MathXL®—our online homework, tutorial, and assessment system.

#### MathXL® Instructor Version (013-147895-8) www.mathxl.com

MathXL is a powerful online homework, tutorial, and assessment system that accompanies the text. With MathXL, instructors can create, edit, and assign online homework and tests using algorithmically generated exercises correlated to your textbook.

### Student Resources

#### Student Solutions Manual (0-13-146781-6)

- Solutions to the odd-numbered section exercises.
- Solutions to the Quick Check exercises.
- Solutions to the Preparing for This Section, Putting the Concepts Together (mid-chapter review), Chapter Review, Chapter Test, Cumulative Review, and Math for the Future exercises.

### Prentice Hall Math Tutor Center (0-13-064604-0)

- Staffed by qualified math instructors who provide students with tutoring on examples and odd-numbered exercises from the textbook.
- Tutoring is available via toll-free telephone, toll-free fax, e-mail, or the Internet.
- White board technology allows tutors and students to see problems worked while they "talk" in real time over the Internet during tutoring sessions.

### Intermediate Algebra Student Study Pack (0-13-199265-1)

The Student Study Pack includes:

- CD Lecture Series
- Student Solutions Manual
- Prentice Hall Math Tutor Center access code

### Chapter Test Prep Video CD—Standalone (0-13-219677-8)

- Includes fully worked-out solutions to every problem from each Chapter Test in the text.

### MathXL® Tutorial on CD (0-13-134606-7)

- Provides algorithmically generated practice exercises that correlate to exercises at the end of sections.
- Every exercise is accompanied by an example and a guided solution; selected exercises include a video clip.
- The software recognizes student errors and provides feedback. It can also generate printed summaries of students' progress.

### Interact Math® Tutorial Web Site www.interactmath.com

Get practice and tutorial help online! This interactive tutorial Web site provides algorithmically generated practice exercises that correlate directly to the exercises in your textbook.

## Acknowledgments

Textbooks are written by authors but evolve through the efforts of many people. We would like to extend our thanks to the following individuals for their important contributions to the project. From Prentice Hall: Paul Murphy, who saw the vision of this text from its inception and made it happen; Kate Valentine and Patrice Jones for their innovative marketing ideas; Ann Heath for her dedication, enthusiasm, and attention to detail (quite honestly, Ann was the cement of the project); Chris Hoag for her support and encouragement; Dawn Nuttall for her perseverance, publishing acumen, and attention to detail with the class testing effort; Barbara Mack for her attention to detail throughout production; Jonathan Boylan and the design team for the attractive and functional design; Thomas Benfatti for his attentive eye in overseeing the creation of literally thousands of pieces of art; Maura Zaldivar for coordinating the scheduling of this project with the compositor and the printer; Linda Behrens for her watchful eye and management over countless production details; and finally, the Prentice Hall sales team, for their confidence and support of our books.

We would like to thank two people who contributed greatly to the development of this book. Thanks to Don Gecewicz for his ability to manage and maintain a single voice as well as his talent to make us think about each sentence. Thanks also go to Janet Mazzarella for her perspective, teaching philosophies, and numerous contributions to the text and its supplement package, including the selection of the MathXL® exercises and the creation of the mini-lectures in the resource manual.

We would like to offer special thanks to a number of instructors who helped us with this project, including: Kevin Bodden and Randy Gallaher for their many contributions to the text and the creation of its accompanying solutions manuals; John Close for his review of the exercises to make them complete; Darren Wiberg for his thorough accuracy review of the final manuscript and participation in the Chapter Test Prep Video CD; Kimberly Neuburger for lending her teaching style and caring to the CD Lecture Series; Katalin Szucs and Pat Foard for lending their expertise to the resource manual; and Andreana Grimaldo and Denise Robichaud for their creativity with Chapter Activities. We would also like to thank Rafiq Ladhani, Jon Stockdale, and Joshua Gay for their dedication to accuracy in checking the art, examples, and answers; and Sarah Streett and Jenny Crawford for attention to detail and consistency in accuracy checking the solutions manuals. We also thank Jenny for her careful creation of the answer manuscript.

We offer many thanks to all the instructors from across the country who participated in reviewer conferences and focus groups, and reviewed and/or class-tested some aspect of the manuscript. Their insights and ideas form the backbone of this text. Hundreds of instructors contributed their time, energy, and ideas to help us shape this text. We will attempt to thank them all here. We apologize for any omissions. *Note:* At the time this book went to press, more class tests were being secured. Our thanks also go to those instructors who tested the manuscript with their students after the printing deadline.

## Class Testers

Marwan Abu-Sawwa, *Florida Community College—Jacksonville*
Mary Lou Baker, *Columbia State Community College*
Donna Beatty, *Ventura College*
Becky Bradshaw, *Lake Superior College*
Tim Britt, *Jackson State Community College*
Beverly Broomell, *SUNY Suffolk*
Hien Bui, *Hillsborough Community College—Dale Mabry*
Elena Catoiu, *Joliet Junior College*
John Close, *Salt Lake Community College*
Shirley Davis, *South Plains College*
Erica Egizio, *Joliet Junior College*
Sanford Geraci, *Broward Community College*
Susan Grody, *Broward Community College*
Pete Herrera, *Southwestern College*
Becky Hubiak, *Tidewater Community College—Virginia Beach*
Sally Jackman, *Richland College*
Nancy Johnson, *Broward Community College*
Mike Kirby, *Tidewater Community College—Virginia Beach*
Carla Kulinsky, *Salt Lake Community College*
Lynn Marecek, *Santa Ana College*
Janet Mazzarella, *Southwestern College*
Michael McComas, *Marshall University*
Judy Meckley, *Joliet Junior College*
Ron Moore, *Florida Community College—Jacksonville*
Hossein Navid-Tabrizi, *Houston Community College*
Charlotte Newsom, *Tidewater Community College—Virginia Beach*
Charles Odion, *Houston Community College*
Eugenia Peterson, *Daley College*
Elise Price, *Tarrant County Community College*
RB Pruitt, *South Plains College*
William Radulovich, *Florida Community College—Jacksonville*
Pavlov Rameau, *Miami Dade Community College—Wolfson*
Nancy Ressler, *Oakton Community College*
George Rhys, *College of the Canyons*
Togba Sapolucia, *Houston Community College*
Gisela Spieler-Persad, *Rio Hondo Community College*
Patrick Stevens, *Joliet Junior College*
Jennifer Strehler, *Oakton Community College*
Katalin Szucs, *East Carolina University*
Jo Tucker, *Tarrant County Community College*
Richard Watkins, *Tidewater Community College*

## Reviewers

Darla Aguilar, *Pima State University*
Grant Alexander, *Joliet Junior College*
Philip Anderson, *South Plains College*
Mary Lou Baker, *Columbia State Community College*
Bill Bales, *Rogers State*
Tony Barcellos, *American River College*
John Beachy, *Northern Illinois University*
David Bell, *Florida Community College—Jacksonville*
Sandy Berry, *Hinds Community College*
Lori Braselton, *Georgia Southern University*
Beverly Broomell, *Suffolk Community College*
Joanne Brunner, *Joliet Junior College*
Connie Buller, *Metropolitan Community College*
Annette Burden, *Youngstown State University*
James Butterbach, *Joliet Junior College*
Marc Campbell, *Daytona Beach Community College*
Elena Catoiu, *Joliet Junior College*
Nancy Chell, *Anne Arundel Community College*
John Close, *Salt Lake Community College*
Bobbi Cook, *Indian River Community College*
Carlos Corona, *San Antonio College*
Faye Dang, *Joliet Junior College*
Vivian Dennis-Monzingo, *Eastfield College*
Alvio Dominguez, *Miami Dade Community College—Wolfson*
Karen Driskell, *South Plains College*
Brenda Dugas, *McNeese State University*
Doug Dunbar, *Okaloosa-Walton Junior College*
Laura Dyer, *Southwestern Illinois State University*
Bill Echols, *Houston Community College—Northwest*
Erica Egizio, *Joliet Junior College*
Jason Eltrevoog, *Joliet Junior College*
Nancy Eschen, *Florida Community College—Jacksonville*
Mike Everett, *Santa Ana College*
Phil Everett, *Ohio State University*
Scott Fallstrom, *Shoreline Community College*
Betsy Farber, *Bucks County Community College*
Fitzroy Farqharson, *Valencia Community College—West*

Dorothy French, *Community College of Philadelphia*
Donna Gerken, *Miami Dade Community College—Kendall*
Adrienne Goldstein, *Miami Dade Community College—Kendall*
Marion Graziano, *Montgomery County Community College*
Susan Grody, *Broward Community College*
Tom Grogan, *Cincinnati State University*
Barbara Grover, *Salt Lake Community College*
Shawna Haider, *Salt Lake Community College*
Margaret Harris, *Milwaukee Area Technical College*
Teresa Hasenauer, *Indian River Community College*
Mary Henderson, *Okaloosa-Walton Junior College*
Celeste Hernandez, *Richland College*
Bob Hervey, *Hillsborough Community College—Dale Mabry*
Teresa Hodge, *Broward Community College*
Sandee House, *Georgia Perimeter College*
Becky Hubiak, *Tidewater Community College—Virginia Beach*
John Jarvis, *Utah Valley State College*
Steven Kahn, *Anne Arundel Community College*
Linda Kass, *Bergen Community College*
Donna Katula, *Joliet Junior College*
Mohammed Kazemi, *University of North Carolina—Charlotte*
Doreen Kelly, *Mesa Community College*
Mike Kirby, *Tidewater Community College—Virginia Beach*
Keith Kuchar, *College of Dupage*
Carla Kulinsky, *Salt Lake Community College*
Julie Labbiento, *Leigh Carbon Community College*
Kathy Lavelle, *Westchester Community College*
Deanna Li, *North Seattle Community College*
Brian Macon, *Valencia Community College—West*
Jim Matovina, *Community College of Southern Nevada*
Jean McArthur, *Joliet Junior College*
Mikal McDowell, *Cedar Valley College*
Lee McEwen, *Ohio State University*
Angela McNulty, *Joliet Junior College*
Debbie McQueen, *Fullerton College*
Judy Meckley, *Joliet Junior College*
Lynette Meslinsky, *Erie Community College—City Campus*
Kausha Miller, *Lexington Community College*
Chris Mizell, *Okaloosa Walton Junior College*
Jim Moore, *Madison Area Technical College*
Ronald Moore, *Florida Community College—Jacksonville*
Elizabeth Morrison, *Valencia Community College—West*
Roya Namavar, *Rogers State*
Hossein Navid-Tabrizi, *Houston Community College*
Carol Nessmith, *Georgia Southern University*
Kim Neuburger, *Portland Community College*
Larry Newberry, *Glendale Community College*
Elsie Newman, *Owens Community College*
Charlotte Newsome, *Tidewater Community College*
Charles Odion, *Houston Community College*
Viann Olson, *Rochester Community and Technical College*
Linda Padilla, *Joliet Junior College*
Carol Perry, *Marshall Community and Technical College*
Faith Peters, *Miami Dade Community College—Wolfson*
Philip Pina, *Florida Atlantic University*
Carol Poos, *Southwestern Illinois University*
William Radulovich, *Florida Community College—Jacksonville*
David Ray, *University of Tennessee—Martin*
Michael Reynolds, *Valencia Community College—West*
George Rhys, *College of the Canyons*
Jorge Romero, *Hillsborough Community College—Dale Mabry*
David Ruffato, *Joliet Junior College*
Carol Rychly, *Augusta State University*
David Santos, *Community College of Philadelphia*
Doug Smith, *Tarrant Community College*
Gisela Spieler-Persad, *Rio Hondo Community College*
Raju Sriram, *Okaloosa-Walton Junior College*
Patrick Stevens, *Joliet Junior College*
Bryan Stewart, *Tarrant Community College*
Elizabeth Suco, *Miami Dade Community College—Wolfson*
Katalin Szucs, *East Carolina University*
KD Taylor, *Utah Valley State College*
Suzanne Topp, *Salt Lake Community College*
Suzanne Trabucco, *Nassau Community College*
Jo Tucker, *Tarrant Community College*
Bob Tuskey, *Joliet Junior College*
Mary Vachon, *San Joaquin Delta College*
Carol Walker, *Hinds Community College*
Kim Ward, *Eastern Connecticut State University*
Natalie Weaver, *Daytona Beach Community College*
Darren Wiberg, *Utah Valley State College*
Rachel Wieland, *Bergen Community College*
Christine Wilson, *Western Virginia University*
Brad Wind, *Miami Dade Community College—North*
Roberta Yellott, *McNeese State University*
Steve Zuro, *Joliet Junior College*

## Additional Acknowledgments

We also would like to extend thanks to our colleagues at Joliet Junior College and Columbus State Community College, who provided encouragement, support, and the teaching environment where the ideas and teaching philosophies in this text were developed.

Michael Sullivan, III
Katherine R. Struve

CHAPTER

# R Real Numbers and Algebraic Expressions

The image to the right is that of the Fibonacci Coil #6. The art is generated using a sequence of numbers called the Fibonacci sequence. See Problem 95 in Section R.3.

## OUTLINE



## The Big Picture: Putting It Together

As the "R" in the title implies, this chapter is a review. The purpose of the chapter is to help you recall facts that you learned in earlier courses. The topics chosen for inclusion here are important building blocks that will help you succeed in this course.

Your instructor may decide to cover this chapter or not, depending on the course syllabus. Regardless, as you proceed through the book, references will be made to Chapter R so that you can use it as a "just in time" review.

# R.1 Success in Mathematics

**OBJECTIVES**

1. What to Do the First Week of the Semester
2. What to Do Before, During, and After Class
3. How to Use the Text Effectively
4. How to Prepare for an Exam

Let's start by having a frank discussion about the "big picture" goals of the course and how this book can help you to be successful at mathematics. The first "big picture" goal of the class is to develop algebraic skills and gain an appreciation for the power of algebra and mathematics. But there is also a second "big picture" goal. By studying mathematics, we develop our sense of logic and exercise the part of our brain that deals with logical thinking. The examples and problems that appear throughout the text are like the crunches that we do in a gym to exercise our body. The goal of running or walking is to get from point A to point B, so doing fifty crunches on a mat does not accomplish this goal, but crunches do make our upper bodies, backs, and heart stronger when we need to run or walk.

**In Words**
Doing crunches doesn't solve the problem of running a race, but they are truly effective at conditioning your body.

Logical thinking can assist us in solving difficult everyday problems, so solving algebra problems "builds the muscles" in the part of our brain that performs logical thinking. So, when you are studying algebra and getting frustrated with the amount of work that needs to be done, and you say, "My brain hurts," remember the phrase that we all use in the gym, "No pain, no gain."

Another phrase to keep in mind is "Success Breeds Success." Mathematics is everywhere. You already are successful at doing some everyday mathematics. With practice, you can take your initial successes and become even more successful. Have you ever done any of the following everyday activities?

- Compare the price per ounce of different sizes of jars of peanut butter or jam.
- Leave a tip at a restaurant.
- Figure out how many calories your bowl of breakfast cereal gives you.
- Take an opinion survey along with many other people.
- Measure the distances between cities as you plan your summer vacation.
- Order the appropriate number of gallons of paint to cover the walls of a room that you are renovating.
- Buy a car and take out a car loan with interest.
- Double a cookie recipe.
- Change American dollars for Canadian dollars.
- Fill up a basketball or soccer ball with air (balls are spheres, after all).
- Coach a Little League team (scores, statistics, catching, and throwing all involve math).
- Check the percentages of saturated and unsaturated fats in a chocolate bar.

We just listed twelve of many everyday mathematical activities, and you may do five or ten in a single day! The everyday mathematics that you already know is the foundation for your success in this course.

## 1 What to Do the First Week of the Semester

You have enrolled in an intermediate algebra course. The first week of the semester gives you the opportunity to prepare your road to success. Here are the things that you should do:

1. **Pick a good seat.** As you enter the classroom for the first time, choose a seat that gives you a good view of the room. Sit close enough to the front so that you can easily see the board and hear the professor.
2. **Read the syllabus to learn about your instructor and the course.** Be sure to take note of your instructor's name, office location, e-mail address, telephone number, and office hours. Also, pay attention to any additional help that can be found on campus such as tutoring centers, videos in the library, software, online tutorials,

and so on. Make sure that you fully understand all of the instructor's policies for the class. This includes the policy on absences, missed exams or quizzes, and homework. Ask questions.

3. **Learn the names of some of your classmates and exchange contact information.** One of the best ways to learn math is through group study sessions. Try to create time each week to study with your classmates. Knowing how to get in contact with classmates is also useful if you ever miss class because you can obtain the assignment for the day.

**Work Smart: Study Skills**
Plan on studying two hours outside of class for each hour in class every week.

4. **Budget your time.** Most students have a tendency to "bite off more than they can chew." To help with time management, consider the following general rule for studying mathematics: You should plan on studying *at least* two hours outside of class for each hour in class. So, if you enrolled in a four-hour math class, you should set aside at least eight hours each week to study for the course. If this is not your only course, you will have to set aside time for other courses as well. Consider your work schedule and personal life when creating your budget as well.

## 2 What to Do Before, During, and After Class

**Work Smart: Study Skills**
Take a few minutes to plan out the upcoming academic term.

Now that the semester is underway, we present the following ideas for what to do before, during, and after each class meeting. While these suggestions may sound overwhelming, we guarantee that by following them, you will be successful in mathematics (and other courses). Also, you will find that studying for exams becomes much easier by following this plan.

### Before Class Begins

1. Make sure you are mentally prepared for class. This means that your mind should be alert and ready to concentrate for the entire class period. (Invest in a cup of coffee at breakfast and eats lots of protein!)
2. Read the section or sections that will be covered in the upcoming class meeting.
3. Based upon your reading, prepare a list of questions. Jot them down. In many cases, your questions will be answered through the lecture. You can then ask any that are not answered completely.

### During Class

1. Arrive early enough to prepare your mind and material for the lecture.
2. Stay alert. Do not doze off or daydream during class. It will be very difficult to understand the lecture when you "return to class."
3. Take thorough notes. It is normal not to get certain topics the first time that you hear them through the lecture. However, this does not mean that you throw your hands up in despair. Rather, continue to write your class notes.

**Work Smart: Study Skills**
Be sure to ask questions during class.

4. You can ask questions when appropriate. Do not be afraid to ask questions. In fact, instructors love when students ask questions, for two reasons. First, we know as teachers that if one student has a question that there are many more in class with the same question. Second, by asking questions, you are teaching the teacher what topics cause difficulty.

### After Class

1. Reread (and possibly rewrite) your class notes. In our experience as students, we were amazed how often confusion that existed during class went away after studying our in-class notes later when we had more time to absorb the material.
2. Reread the section. This is an especially important step. Once you have heard the lecture, the section will make more sense and you will understand much more.

**Work Smart: Study Skills**
The reason for doing homework is to build your skill and confidence. Don't skip assignments.

3. Do your homework. **Homework is not optional.** There is an old Chinese proverb that says,

   I hear . . . and I forget
   I see . . . and I remember
   I do . . . and I understand

   This proverb applies to any situation in life in which you want to succeed. Would a pianist expect to be the best if she didn't practice? The only way you are going to learn algebra is by doing algebra. Remember: Success breeds success.
4. And don't forget, when you get a problem wrong, try to figure out why you got the problem wrong. If you can't discover your error, be sure to ask for help.
5. If you have questions, visit your professor during office hours. You can also ask someone in your study group or go to the tutoring center on campus, if available.

### Math Courses: No Brain Freezes Here!

Learning algebra is a building process. Learning is the art of making connections between thousands of neurons (specialized cells) in the brain. Memory is the ability to reactivate these neural networks—it is a conversation among neurons.

Math isn't a mystery. You already know some math. But you do have to practice what you know and expand your knowledge. Why? The brain contains thousands of neurons. Through repeated practice, a special coating forms that allows the signals to travel faster and reduces interference. The cells "fire" more quickly and connections are made faster and with less effort. Practice forms the pathways that allow us to retrieve concepts and facts at test time. Remember those crunches, which are a way of making your body more robust and nimble—learning does the same to your brain.

### Have We Mentioned Asking Questions?

To move information from short-term memory to long-term memory, we need to think about the information, comprehend its meaning, and ask questions about it.

## 3 How to Use the Text Effectively

When we sat down to write this text, we knew based upon experience from teaching our own students that students typically do not read their mathematics text. Rather than saying to you, "Ah, but our book is different—it can be read!," we decided to accept how students study math.

Students usually go through the following steps:

1. Attend the lecture and watch the instructor do some problems on the board. Perhaps work some problems in class.
2. Go home and work on the homework assignment.
3. After each problem, check the answer in the back of the text. If right, move on, but if wrong, go back and see where the solution went wrong.
4. Maybe, the mistake can be identified, but if not, go to the class notes or try to find a similar example in the text. With a little luck, a similar example can be found and you can determine where the solution went wrong in the problem.
5. If not, mark the problem and ask about it in the next class meeting, which leads us back to step 1.

**Work Smart: Study Skills**
Learn what the different features of this book are designed to do. Decide which ones you may need the most.

So with this model in mind, we started to develop this text so that there is more than one way to extract the information you need from it.

All of the features have been included in the text to help you succeed. We list each feature in the order they appear and briefly explain its purpose and how it can be used to help you succeed in this course:

## Preparing for This Section: Warming Up

Immediately after the title of the section, each section (after Chapter R) begins with a short "readiness quiz." The readiness quiz presents questions about material that was presented earlier in the course that is needed for the upcoming section. You should take the readiness quiz to be sure that you understand the material that the new section will be based on. Answers to the readiness quiz appear as footnotes on the page of the quiz. Check your answers. If you get a problem wrong, or don't know how to do a problem, go back to the section listed and review the material.

## Objectives: A "Road Map" through the Course

To the left of the readiness quiz, we present a list of objectives to be covered in the section. If you follow the objectives, you will get a good idea of the book's "big picture"—the important concepts, techniques, and procedures.

The objectives are numbered. When we begin discussing a particular objective, the number appears in the left-hand column of the text along with the stated objective.

## Examples: Where to Look for Information

You look to examples to provide you with guidance and instruction when you need it most—when you are away from the instructor and the classroom. With this in mind, we have developed two example formats.

*Step-by-Step Examples* have a three-column format where the left column describes a step, the middle column provides a brief explanation of the step, and the right column presents the algebra. With this format, the left and middle columns can be thought of as your instructor's voice during a lecture. *Step-by-Step Examples* are used to introduce key topics or important problem-solving strategies. They are meant to provide easy-to-understand, practical instructions by including the words "how to" in the examples' headline.

*Annotated Examples* have a two-column format in which explanations are provided to the left of the algebra. Because we read from left to right, placing the explanation on the left clearly describes what we are about to do. Again, the annotations can be thought of as your instructor's voice right before he or she writes the solution on the board.

## Quick Check: Practice for the Examples

After a concept has been presented, we provide between 2 and 4 (occasionally up to 8 for some concepts) problems to solve. The answers are in the back of the book so you can compare your answer to the correct answer providing immediate feedback about your understanding of the concepts. The Quick Checks usually follow each example in the text. This feature came from our belief that students use the examples as a template for solving problems, so we decided to place the Quick Checks just after an example. If you can do the Quick Check problems, you will be prepared for the end-of-section exercises, which are a comprehensive review of the material in the section.

## In Words: Math in Everyday Language

Have you ever been given a math definition in class and said, "What in the world does that mean?" As teachers, we have heard that from our students. So we added the "In Words" feature, which takes definitions that are in their mathematical form and restates them in everyday language. These boxes will help you to understand the language of mathematics better.

## Work Smart

These are "tricks of the trade" that can be used to help you solve problems. They also show alternative approaches to solving problems. Yes, there is more than one way to solve a math problem!

### Work Smart—Study Skills

Working smart also means studying smart. We provide tips throughout the text to help you understand the study skills required for success in this and other mathematics courses.

### Chapter Review

The chapter review is arranged section by section. For each section, we list key concepts, key terms, and objectives. For each objective, we provide the examples from the text that illustrate the objective, along with page references. Also, for each objective, we list the problems in the review exercises that test your understanding. If you get a problem wrong, use this feature to help you to identify how to work the problem. Cumulative Review exercises are also provided at the end of every odd-numbered chapter to refresh your memory on important skills and concepts.

### Chapter Test

We have included a chapter test. Once you think that you are prepared for the exam, take the chapter test. If you do well on the chapter test, chances are you will do well on your in-class exam. Be sure to take the chapter test under the conditions that you will face in class.

### Chapter Test Prep Video CD

Packaged with each new copy of the text, the Chapter Test Prep Video CD provides students with help at the critical juncture when they are studying for a test. The CD video presents step-by-step solutions to the exact exercises found in each of the book's Chapter Tests. Easy video navigation allows students to access instantly the worked-out solutions to the exercises they want to study or review.

## 4 How to Prepare for an Exam

The following steps are time-tested suggestions to help you prepare for an exam.

**Step 1: Revisit your homework and the chapter review problems:** Beginning about one week before your exam, start to redo your homework assignments. If you don't understand a topic, be sure to seek out help. You should also work the problems given in the chapter review. The problems are keyed to the objectives in the course. If you get a problem wrong, identify the objective and examples that illustrate the objective. Then review this material and try the problem in the chapter review again. If you get the problem wrong again, seek out help.

**Step 2: Test yourself:** A day or two before the exam, take the chapter test under test conditions. Be sure to check your answers. If you got any problems wrong, determine why you got them wrong and remedy the situation.

**Step 3: The Chapter Test Prep Video CD** provides you with step-by-step solutions to the exact exercises found in each of the book's Chapter Tests. To get the most from this valuable resource, follow the worked-out solutions to any of the exercises on the Chapter Test that you want to study or review.

**Step 4: Follow these rules as you train:** Be sure to arrive early at the location of the exam. Prepare your mind for the exam. Also, be sure that you are well rested. Don't try to pull "all-nighters." If you need to study all night long for an exam, then your time management is poor and you should rethink how you are using your time or whether you have enough time set aside for the course.

**Work Smart: Study Skills**
Do not "cram" for an exam by pulling an "all nighter."

# R.1 Exercises

1. Why do you want to be successful in mathematics? Are your goals positive or negative? If you stated your goal negatively ("Just get me out of this course!), can you restate it positively?

2. Name three activities in your daily life that involve the use of math (for instance, scoring a game of poker, bridge, or pinochle, operating your computer, or reading a credit-card bill).

3. What is your instructor's name?

4. What are your instructor's office hours? Where is your instructor's office?

5. Does your instructor have an e-mail address? If so, what is it?

6. Does your class have a Web site? Do you know how to access it? What information is located on the Web site?

7. Are there tutors available for this course? If so, where are they located? When are they available?

8. Name two other students in your class. What is their contact information? When can you meet with them to study?

9. List some of the things that you should do before class begins.

10. List some of the things that you should do during class.

11. List some of the things that you should do after class.

12. What is the point of the Chinese proverb on page 4?

13. What is the "readiness quiz"? How should it be used?

14. Name three features that appear in the margins. What is the purpose of each of them?

15. How should the chapter review material be used?

16. How should the chapter test be used?

17. What is the Chapter Test Prep Video CD?

18. List the four steps that should be followed when preparing for an exam.

19. Use the chart on the following page to help manage your time. Be sure to fill in time allocated to various activities in your life including school, work, and leisure.

| | Monday | Tuesday | Wednesday | Thursday | Friday | Saturday | Sunday |
|---|---|---|---|---|---|---|---|
| 7 am | | | | | | | |
| 8 am | | | | | | | |
| 9 am | | | | | | | |
| 10 am | | | | | | | |
| 11 am | | | | | | | |
| Noon | | | | | | | |
| 1 pm | | | | | | | |
| 2 pm | | | | | | | |
| 3 pm | | | | | | | |
| 4 pm | | | | | | | |
| 5 pm | | | | | | | |
| 6 pm | | | | | | | |
| 7 pm | | | | | | | |
| 8 pm | | | | | | | |
| 9 pm | | | | | | | |

**20.** How is mathematics like doing crunches at the gym?

# R.2 Sets and Classification of Numbers

**OBJECTIVES**

1. Use Set Notation
2. Know the Classification of Numbers
3. Approximate Decimals by Rounding or Truncating
4. Plot Points on the Real Number Line
5. Use Inequalities to Order Real Numbers

## 1 Use Set Notation

A **set** is a collection of well-defined objects. The word "well-defined" means that there is a rule for determining whether a given object is in the set. For example, we could treat the students enrolled in Intermediate Algebra at your college as a set. As another example, consider the set of numbers 0, 1, 2, 3, 4, 5, 6, 7, 8, and 9. If we let $D$ represent this set, then we can write

$$D = \{0, 1, 2, 3, 4, 5, 6, 7, 8, 9\}$$

In this notation, the braces { } are used to enclose the objects, or **elements,** in the set. This method of representing a set is called the **roster method.**

### EXAMPLE 1 Using the Roster Method

Write the set that represents the vowels.

**Solution**

The vowels are $a$, $e$, $i$, $o$, and $u$. If we let $V$ represent the set, then

$$V = \{a, e, i, o, u\}$$

**In Words**
We read the set to the right as "$D$ is the set of all $x$ such that $x$ is a digit."

Another way to denote a set is to use **set-builder notation.** The numbers in the set $D = \{0, 1, 2, 3, 4, 5, 6, 7, 8, 9\}$ are called digits. Using set-builder notation the set $D$ of digits is written as

$$D = \{x | x \text{ is a digit}\}$$

In algebra, we use letters such as $x, y, a, b$, and $c$ to represent numbers. When the letter can be any number in a set of numbers, it is called a **variable.** In the set $D$, the letter $x$ is used to represent any digit, so that $x$ is a variable that can take on the value 0, 1, 2, 3, 4, 5, 6, 7, 8, or 9.

### EXAMPLE 2 Using Set-Builder Notation

Use set-builder notation to represent the following sets.

**(a)** The set of all even digits **(b)** The set of all odd digits

**Solution**

**(a)** We will let $E$ represent the set of all even digits, so that

$$E = \{x | x \text{ is an even digit}\}$$

**(b)** We will let $O$ represent the set of all odd digits, so that

$$O = \{x | x \text{ is an odd digit}\}$$

**QUICK ✓** *Use set-builder notation and the roster method to represent the following sets.*

**1.** The set of all digits less than 5.

**2.** The set of all digits greater than or equal to 6.

---

**Work Smart**
Never list elements in a set more than once.

When we list the elements in a set, we never list an element more than once because each element is distinct. For example, we would never write $\{1, 2, 3, 2\}$; the correct listing is $\{1, 2, 3\}$. Also, the order in which the elements are listed does not matter. For example, $\{2, 3\}$ and $\{3, 2\}$ both represent the same set.

We now introduce some notation that is common when describing sets.

**Work Smart**
The empty set is represented as either $\emptyset$ or $\{\ \}$. We never write $\{\emptyset\}$ to represent the empty set because this notation means a set that contains the set called the empty set.

**SET NOTATION**

- If two sets $A$ and $B$ have the same elements, then we say that $A$ **equals** $B$ and write $A = B$.
- If every element of a set $A$ is also an element of a set $B$, then we say that $A$ is a **subset** of $B$ and write $A \subseteq B$.
- If $A \subseteq B$ and $A \neq B$, then we say that $A$ is a **proper subset** of $B$ and write $A \subset B$. Put another way, $A$ is a proper subset of $B$ if all elements in $A$ are also in $B$ and there are elements in $B$ that are not in $A$.
- If a set $A$ has no elements, it is called the **empty set,** or **null set,** and is denoted by the symbol $\emptyset$ or $\{\ \}$. The empty set is a subset of every set; that is $\emptyset \subseteq A$ for any set $A$.

When working with sets, we usually designate a **universal set,** which is the set of all elements of interest to us. For example, in Example 2, we were interested in the set of all digits, so the universal set is the set of all digits.

It is often helpful to draw pictures of sets because the pictures help us to visualize relations among sets. We call pictures of sets **Venn diagrams,** in honor of John Venn (1834–1923). In Venn diagrams we represent sets as circles enclosed in a rectangle.

The rectangle represents the universal set. For example, suppose that $A = \{1, 2\}$, $B = \{1, 2, 3, 4, 5\}$ and the universal set is $U = \{0, 1, 2, 3, 4, 5, 6, 7, 8, 9\}$, then $A \subset B$. Figure 1 illustrates the relation between sets $A$ and $B$ in a Venn diagram.

**Figure 1**
Venn diagram with $A \subset B$.

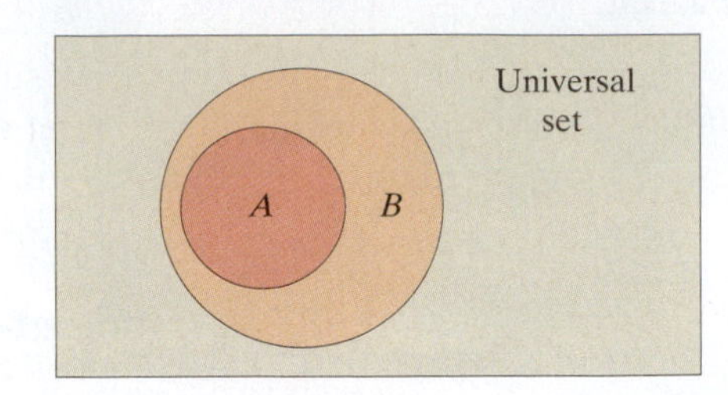


## EXAMPLE 3 Using Set Notation

Let $A = \{0, 1, 2, 3, 4, 5\}$, $B = \{3, 4, 5\}$, $C = \{5, 4, 3\}$, and $D = \{3, 4, 5, 6\}$. Write True or False to each statement.

**(a)** $B \subseteq A$ **(b)** $D \subseteq A$ **(c)** $B = C$
**(d)** $C = D$ **(e)** $B \subset C$ **(f)** $\emptyset \subseteq C$

### Solution

**(a)** The statement $B \subseteq A$ is True because all the elements that are in $B$ are also elements in $A$.

**(b)** The statement $D \subseteq A$ is False because there is an element that is in $D$, 6, that is not in $A$.

**(c)** The statement $B = C$ is True because sets $B$ and $C$ have the same elements.

**(d)** The statement $C = D$ is False because there is an element in $D$, 6, that is not in $C$.

**(e)** In order for $B$ to be a proper subset of $C$, it must be the case that all the elements that are in $B$ are also elements in $C$. In addition, there must be elements that are in $C$ that are not in $B$. Because $B = C$, the statement $B \subset C$ is False. Note, however, that $B \subseteq C$ is True.

**(f)** The statement $\emptyset \subseteq C$ is True because the empty set is a subset of every set.

**QUICK ✓** *Let $A = \{a, b, c, d, e, f, g\}$, $B = \{a, b, c\}$, $C = \{c, d\}$, and $D = \{c, b, a\}$. Write True or False to each statement. Be sure to justify your answer.*

**3.** $B \subseteq A$ **4.** $B = C$ **5.** $B \subset D$ **6.** $\emptyset \subseteq A$

---

We use the symbol $\in$ (read "is an element of") to denote that a particular element is in a set. For example, $7 \in \{1, 3, 5, 7, 9\}$ means "7 is an element of the set $\{1, 3, 5, 7, 9\}$." If an element is not in a set, we use the symbol $\notin$ (read "is not an element of"). For example, we can use mathematical notation to represent the expression "$b$ is not a vowel" as $b \notin \{a, e, i, o, u\}$.

## EXAMPLE 4 Using Set Notation

Write True or False to each statement.

**(a)** $3 \in \{x \mid x \text{ is a digit}\}$

**(b)** $\frac{1}{2} \notin \left\{x \mid x = \frac{p}{q}, \text{ where } p \text{ and } q \text{ are digits}, q \neq 0\right\}$

**(c)** $a \in \{a, e, i, o, u\}$

**Solution**

**(a)** The statement $3 \in \{x \mid x \text{ is a digit}\}$ is True because 3 is a digit.

**(b)** The statement $\frac{1}{2} \notin \left\{x \mid x = \frac{p}{q} \text{ where } p \text{ and } q \text{ are digits, } q \neq 0\right\}$ is False because $\frac{1}{2}$ is of the form $\frac{p}{q}$, where $p = 1$ and $q = 2$.

**(c)** The statement $a \in \{a, e, i, o, u\}$ is True because $a$ is an element of the set $\{a, e, i, o, u\}$.

**QUICK ✓** *Answer True or False for each statement.*

**7.** $5 \in \{0, 1, 2, 3, 4, 5\}$ **8.** Michigan $\notin$ {Illinois, Indiana, Michigan, Wisconsin}

**9.** $\frac{8}{3} \in \left\{x \mid x = \frac{p}{q}, \text{ where } p \text{ and } q \text{ are digits, } q \neq 0\right\}$

---

**SUMMARY: Set Notation**

Below we provide a quick overview of the notation used when describing sets.

| | | |
|---|---|---|
| $A = B$ | means | all the elements in set $A$ are also elements in set $B$ and all the elements in set $B$ are also elements in set $A$. |
| $A \subseteq B$ | means | all the elements that are in set $A$ are also elements in set $B$ |
| $A \subset B$ | means | all the elements that are in set $A$ are also elements in set $B$, but there are elements in set $B$ that are not in set $A$. |
| $\emptyset$ or $\{\ \}$ | means | the empty set. The set has no elements. |
| $5 \in A$ | means | 5 is an element in set $A$. |
| $5 \notin A$ | means | 5 is not an element in set $A$. |

## 2 Know the Classification of Numbers

The reason that we discussed sets is because it is helpful to classify the various kinds of numbers that we deal with as sets.

**DEFINITION**

The **natural numbers,** or **counting numbers,** are the numbers in the set $\mathbb{N} = \{1, 2, 3, \ldots\}$. The three dots, called *ellipsis,* indicate that the pattern continues indefinitely.

As their name implies, the counting numbers are often used to count things. For example, we can count the number of cars that arrive at a McDonald's drive-through between 12 noon and 1:00 P.M. or we can count the letters in the alphabet.

Suppose you had \$7 in your pocket and purchase a drink and hot dog at a baseball game for \$7. Can we use the counting numbers to describe the amount of money you have in your pocket after the purchase? No! We need a new number system to describe the amount in your pocket after the purchase. For this, we use the *whole number system.*

**DEFINITION**

The **whole numbers** are the numbers in the set $W = \{0, 1, 2, 3, \ldots\}$.

We can see that the whole numbers consists of the set of counting numbers together with the number 0, so that $\mathbb{N} \subset W$.

Now suppose you have a balance of \$100 in your checking account and you write a check for \$120. Can the whole numbers be used to describe your new balance? No! To describe situations such as this, we need the *integers*.

**DEFINITION**

The **integers** are the numbers in the set $\mathbb{Z} = \{\ldots, -3, -2, -1, 0, 1, 2, 3, \ldots\}$.

**In Words**

The set of natural numbers is a subset of the set of whole numbers. The set of whole numbers is a subset of the set of integers.

You should notice that the $\mathbb{N} \subset W$ and $W \subset \mathbb{Z}$. As we expand a number system, we are able to discuss new, and usually more complicated, problems. For example, the whole numbers allow us to discuss the absence of something because they include zero and the counting numbers do not. The integers allow us to deal with problems involving both negative and positive quantities. For example, we could not discuss profit (positive counting numbers) and loss (negative counting numbers) without integers.

A question that you may be asking yourself is, "How might I represent parts of a whole?" For example, how can I represent a portion of a dollar or a portion of a whole pie? To address this problem, we enlarge our number system to include *rational numbers*.

**DEFINITION**

A **rational number** is a number that can be expressed as a quotient $\frac{p}{q}$ of two integers. The integer $p$ is called the **numerator,** and the integer $q$, which cannot be 0, is called the **denominator.** The set of rational numbers are the numbers $\mathbb{Q} = \left\{x \middle| x = \frac{p}{q}, \text{ where } p, q \text{ are integers and } q \neq 0\right\}$.

Examples of rational numbers are $\frac{3}{4}, \frac{4}{3}, \frac{0}{6}, -\frac{4}{5}$, and $\frac{33}{5}$. In addition, because $\frac{p}{1} = p$ for any integer $p$, it follows that the set of integers is a subset of the set of rational numbers ($\mathbb{Z} \subset \mathbb{Q}$). For example, 5 is a rational number because it can be written as $\frac{5}{1}$, but more specifically, it is an integer. More specifically than that, it is a counting number.

We can also represent rational numbers as *decimals*. The **decimal** representation of a number is found carrying out the division indicated. For example, $\frac{4}{5}, \frac{7}{2}, -\frac{2}{3}$, and $\frac{2}{11}$ may be represented as follows:

$$\frac{4}{5} = 0.8 \qquad \frac{7}{2} = 3.5 \qquad -\frac{2}{3} = -0.6666\ldots = -0.\overline{6} \qquad \frac{2}{11} = 0.181818\ldots = 0.\overline{18}$$

**Work Smart**

$\frac{2}{11}$ can be expressed as a decimal by writing $11\overline{)2}$. We carry out this division as follows:

$$\begin{array}{r} 0.181 \\ 11\overline{)2.000} \\ \underline{11}\phantom{00} \\ 90\phantom{0} \\ \underline{88}\phantom{0} \\ 20 \\ \underline{11} \\ 9 \end{array}$$

and so on.

Notice the line above the 6 in the decimal representation of $-\frac{2}{3}$. The line, called the **repeat bar,** is another way of representing the fact that the pattern continues. So, in the decimal representation of $\frac{2}{11}$ we know that the block of numbers 18 will continue indefinitely to the right of the decimal point.

In looking at the decimal representation of rational numbers, we notice that the decimals either **terminate** (as in the case of $\frac{4}{5}$ and $\frac{7}{2}$) or do not terminate (as in the case of $-\frac{2}{3}$ and $\frac{2}{11}$). When the decimal representation of a rational number does not terminate, it will have a block of decimals that repeat. For example, $\frac{2}{3} = 0.6666\ldots$ or $0.\overline{6}$. Every rational number may be represented by a decimal that either terminates or is **nonterminating** with a repeating block of decimals, and vice versa.

What if a decimal neither terminates, nor has a block of decimals that repeat? These types of decimals represent a set of numbers called *irrational numbers*. Every **irrational number** may be represented by a decimal that neither terminates nor repeats. This means that irrational numbers cannot be written in the form $\frac{p}{q}$, where $p$ and $q$ are integers and $q \neq 0$. An example of an irrational number is the symbol $\pi$, whose value is approximately 3.14159265359. Another example of an irrational number is the symbol $\sqrt{2}$, whose value is approximately 1.414213562.

**Work Smart**

While it is true that $\sqrt{2}$ is an irrational number, not all numbers under the $\sqrt{\ }$ symbol are irrational. For example, $\sqrt{4} = 2$, a positive integer.

**DEFINITION**

Together, the set of rational numbers and the set of irrational numbers form the set of **real numbers.** The set of real numbers is denoted using the symbol $\mathbb{R}$.

Figure 2 shows the relationship among the various types of numbers. The universal set in Figure 2 is the set of all real numbers. As you continue in your study of mathematics, it will be extremely important that you are able to distinguish among the various number systems—in particular, it will be important for you to know the distinction between a rational number and an irrational number. Notice that if a number is rational, it cannot be irrational and vice-versa.

**Figure 2**

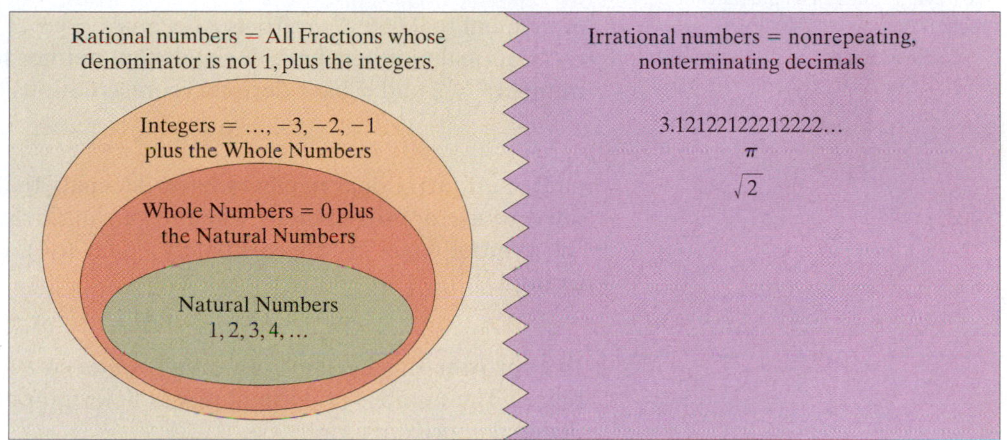


## EXAMPLE 5 Classifying Numbers in a Set

List the numbers in the set

$$\left\{3, -\frac{3}{5}, -1, 0, \sqrt{2}, 5.\overline{94}, 4.2122122212222\ldots, -\frac{8}{2}, \sqrt{-1}, \frac{\pi}{2}\right\}$$

that are

**(a)** Natural numbers **(b)** Whole numbers **(c)** Integers
**(d)** Rational numbers **(e)** Irrational numbers **(f)** Real numbers

### Solution

**(a)** 3 is the only natural number

**(b)** 0 and 3 are the whole numbers

**(c)** 3, $-1$, 0, and $-\frac{8}{2}$ are the integers ($-\frac{8}{2}$ is an integer because it simplifies to $-4$)

**(d)** 3, $-\frac{3}{5}$, $-1$, 0, $5.\overline{94}$, and $-\frac{8}{2}$ are the rational numbers

**(e)** $\sqrt{2}$, 4.2122122212222..., and $\frac{\pi}{2}$ are the irrational numbers. Note that 4.2122122212222... is irrational because there are infinitely many nonrepeating decimals.

**(f)** All the numbers except $\sqrt{-1}$ listed are real numbers. There is no real number whose square is $-1$.

**QUICK ✓** *List the numbers in the set*

$$\left\{\frac{7}{3}, -9, 10, 4.\overline{56}, 5.7377377737777\ldots, \frac{0}{3}, \pi, -\frac{4}{7}, \frac{12}{4}, \sqrt{-4}\right\}$$

*that are*

**10.** Natural numbers **11.** Whole numbers **12.** Integers

**13.** Rational numbers **14.** Irrational numbers **15.** Real numbers

## 3 Approximate Decimals by Rounding or Truncating

**Work Smart**

Rational numbers have decimals that either terminate or do not terminate, but have a block of numbers that repeat.

Every number written in decimal form is a real number that may either be rational or irrational. In addition, every real number can be represented by a decimal. For example, the rational number $\frac{3}{4}$ may be written in decimal form as 0.75. The rational number $\frac{2}{3}$ is equivalent to 0.666... or $0.\overline{6}$ as a decimal.

Irrational numbers have decimals that neither terminate, nor repeat. The irrational numbers $\sqrt{2}$ and $\pi$ have decimal representations that begin as follows:

$$\sqrt{2} = 1.414213\ldots \qquad \pi = 3.14159\ldots$$

Because irrational numbers have decimals that neither terminate, nor repeat, we need to use approximations when displaying irrational numbers as decimals. We use the symbol $\approx$ (read "approximately equal to"), to write approximate decimals. For example,

$$\sqrt{2} \approx 1.4142 \qquad \pi \approx 3.1416$$

In approximating decimals, we either *round* or *truncate* to a given number of decimal places. The number of decimal places determines the location of the *final digit* in the decimal approximation.

**DEFINITION**

**Truncation:** Drop all the digits that follow the specified final digit in the decimal.

**Rounding:** Identify the specified final digit in the decimal. If the next digit is 5 or more, add 1 to the final digit; if the next digit is 4 or less, leave the final digit as it is. Then truncate all digits to the right of the final digit.

### EXAMPLE 6 Approximating a Decimal by Truncating and Rounding

Approximate 13.9463 to two decimal places by

**(a)** Truncating **(b)** Rounding

**Solution**

First, we notice that we want to approximate the decimal to two decimal places. Therefore, the final digit is 4 as indicated in red (13.9463).

**(a)** To truncate, we remove all digits to the right of the final digit, 4. So, 13.9463 truncated to two decimal places is 13.94.

**(b)** To round to two decimal places, we determine the value of the digit to the right of the final digit, 4, is 6. Because 6 is 5 or more, we add 1 to the final digit 4 and then truncate everything to the right of the final digit. So, 13.9463 rounded to two decimal places is 13.95.

**QUICK ✓** *Approximate each number by (a) truncating and (b) rounding to the indicated number of decimal places.*

**16.** 5.694392; 3 decimal places **17.** −4.9369102; 2 decimal places

**THE GRAPHING CALCULATOR: DOES YOUR CALCULATOR TRUNCATE OR ROUND?**

Calculators are only capable of displaying a certain number of decimals. For example, most scientific calculators display 8 digits. Most graphing calculators display 10–12 digits. When a number has more digits than the calculator can display, the calculator will either round or truncate.

To see whether your calculator rounds or truncates divide 2 by 3. How many digits do you see? Is the last digit a 6 or a 7? If it is a 6, your calculator truncates; if it is a 7, your calculator rounds. Figure 3 shows the results on a TI-84 Plus graphing calculator. Does the TI-84 Plus round or truncate? Because the last digit displayed is a 7, the TI-84 Plus rounds.

Figure 3

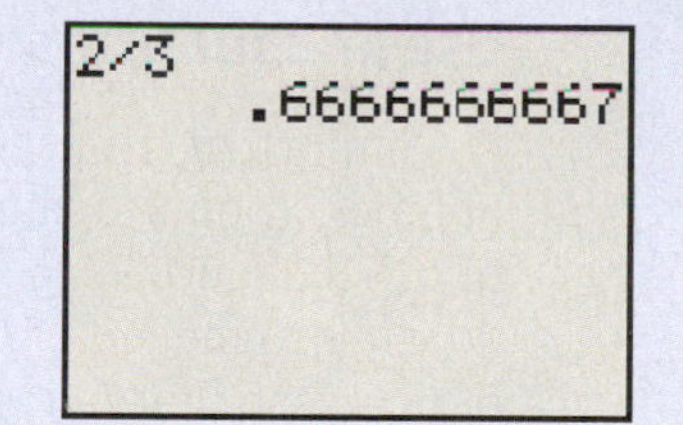


## 4 Plot Points on the Real Number Line

The real numbers can be represented by points on a line called the **real number line.** Every real number corresponds to a point on a the line, and each point on the line has a unique real number associated with it.

To construct a real number line, pick a point on a line somewhere in the center, and label it $O$. This point, called the **origin,** corresponds to the real number 0. See Figure 4. The point 1 unit to the right of $O$ corresponds to the real number 1. The distance between 0 and 1 determines the **scale** of the number line. For example, the point associated with the number 2 is twice as far from $O$ as 1 is. Notice that an arrowhead on the right end of the line indicates the direction in which the numbers increase. Points to the left of the origin correspond to the real numbers −1, −2, and so on.

**Figure 4**
The real number line.

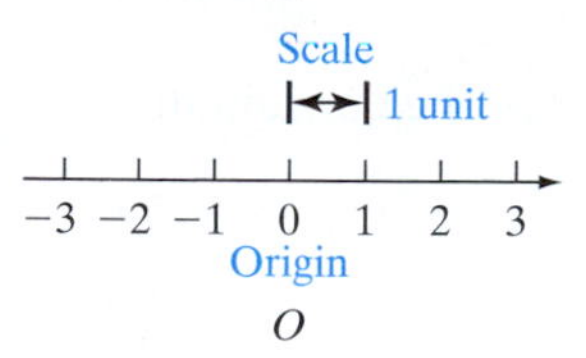


**In Words**
We can think of the real number line as the graph of the set of all real numbers.

**DEFINITION**

The real number associated with a point $P$ is called the **coordinate** of $P$. The **real number line** is the set of all points that have been assigned coordinates.

### EXAMPLE 7 Plotting Points on the Real Number Line

On the real number line, label the points with coordinates $0, 5, -1, 1.5, -\frac{1}{2}$.

**Solution**

We draw a real number line with a scale of 1 and then plot the points. See Figure 5.

**Figure 5**

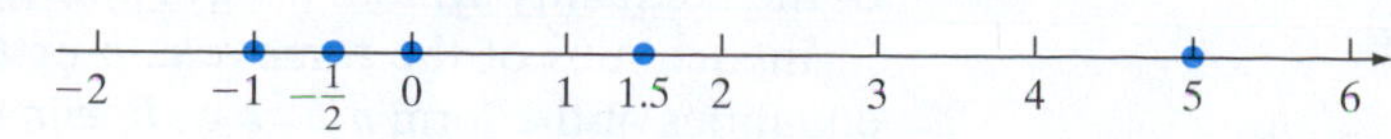

**QUICK ✓**

**18.** On the real number line, label the points with coordinates 0, 3, −2, $\frac{1}{2}$, and 3.5.

The real number line consists of three classes (or categories) of real numbers, as shown in Figure 6.

**Figure 6**

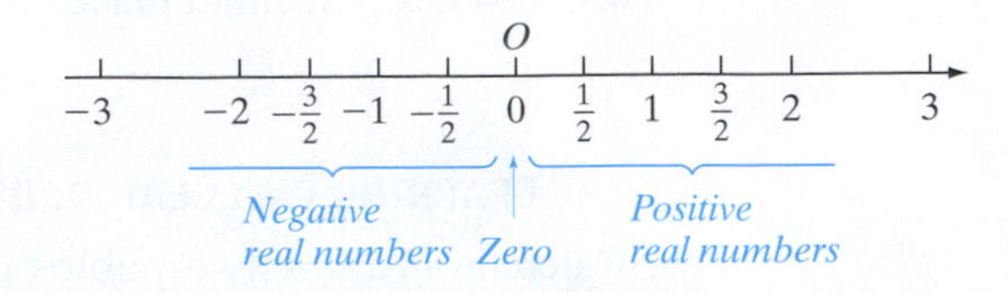


- The **negative real numbers** are the coordinates of points to the left of the origin $O$.
- The real number **zero** is the coordinate of the origin $O$.
- The **positive real numbers** are the coordinates of points to the right of the origin $O$.

## 5 Use Inequalities to Order Real Numbers

**Figure 7**

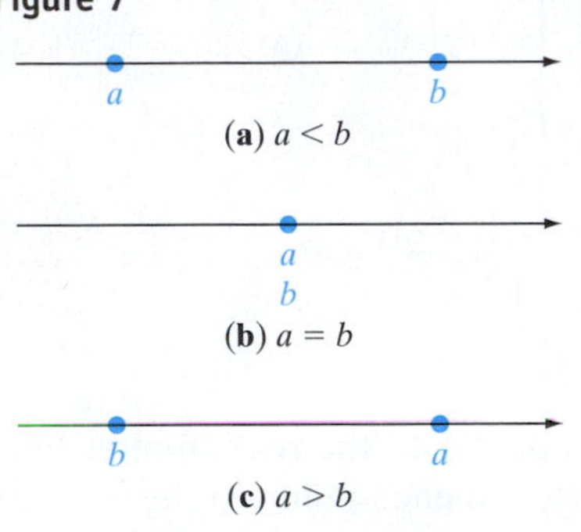


An important property of the real number line follows from the fact that given two numbers (points) $a$ and $b$, either $a$ is to the left of $b$, $a$ is the same as $b$, or $a$ is to the right of $b$. See Figure 7.

If $a$ is to the left of $b$, we say that "$a$ is less than $b$" and write $a < b$. If $a$ is to the right of $b$, we say that "$a$ is greater than $b$" and write $a > b$. If $a$ is at the same location as $b$, then we say that "$a$ is equal to $b$" and write $a = b$. If $a$ is either less than or equal to $b$, we write $a \leq b$. Similarly, $a \geq b$ means that $a$ is either greater than or equal to $b$. Collectively, the symbols $<$, $>$, $\leq$, and $\geq$ are called **inequality symbols.**

Note that $a < b$ and $b > a$ mean the same thing. For example, it does not matter whether we write $2 < 3$ (2 is to the left of 3) or $3 > 2$ (3 is to the right of 2).

### EXAMPLE 8 Using Inequality Symbols

**(a)** $2 < 5$ because the coordinate 2 lies to the left of the coordinate 5 on the real number line.

**(b)** $-1 > -3$ because the coordinate $-1$ lies to the right of the coordinate $-3$ on the real number line.

**(c)** $3.5 \leq \frac{7}{2}$ because $3.5 = \frac{7}{2}$.

**(d)** $\frac{5}{6} > \frac{4}{5}$ because $\frac{5}{6} = 0.8\overline{3}$ and $\frac{4}{5} = 0.8$, so the coordinate $\frac{5}{6}$ lies to the right of the coordinate $\frac{4}{5}$ on the real number line.

 *Replace the question mark by <, > or =, whichever is correct.*

**19.** 3 ? 6 **20.** −3 ? −2 **21.** $\frac{2}{3}$ ? $\frac{1}{2}$ **22.** $\frac{5}{7}$ ? 0.7 **23.** $\frac{2}{3}$ ? $\frac{10}{15}$

If you look carefully at the results of Example 8, you should notice that the direction of the inequality symbol always points to the smaller number.

Inequalities of the form $a < b$ or $b > a$ are called **strict inequalities,** whereas inequalities of the form $a \leq b$ or $b \geq a$ are called **nonstrict** (or **weak**) **inequalities.**

Based upon the discussion so far, we conclude that

| | | |
|---|---|---|
| $a > 0$ | is equivalent to | $a$ is positive |
| $a < 0$ | is equivalent to | $a$ is negative |

We read $a > 0$ by saying that "$a$ is positive" or "$a$ is greater than 0." If $a \geq 0$, then either $a > 0$ or $a = 0$, and we may read this as "$a$ is nonnegative" or "$a$ is greater than or equal to 0."

# R.2 Exercises

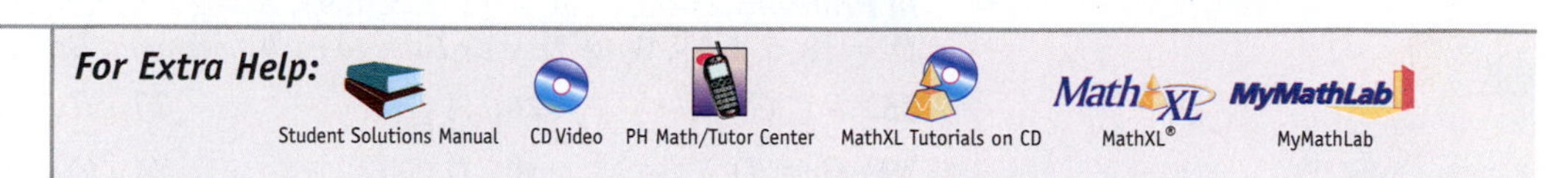


## Concepts and Vocabulary

*In Problems 1–4, fill in the blanks.*

**1.** A _____ is a collection of well-defined objects.

**2.** If every element of a set $A$ is also an element of a set $B$, then we say that $A$ is a _____ of $B$ and write $A$ _____ $B$.

**3.** The numbers in the set $\left\{ x \middle| x = \dfrac{p}{q}, \text{ where } p, q \text{ are integers and } q \neq 0 \right\}$ are called _____ numbers.

**4.** The set of _____ _____ have decimal representations that neither terminate, nor repeat.

*In Problems 5–10, answer True or False to each statement.*

**5.** The order in which elements are listed in a set does not matter.

**6.** If a set has no elements, it is called the empty set and is denoted $\{\emptyset\}$.

**7.** The rational numbers are a subset of the set of irrational numbers.

**8.** If a number is expressed as a decimal, then it is rational.

**9.** Every rational number is a real number.

**10.** If a number is rational, it cannot be irrational.

**11.** Are there any real numbers that are both rational and irrational? Are there any real numbers that are neither? Explain your reasoning.

**12.** Explain why the sum of a rational number and an irrational number must be irrational.

**13.** In your own words, explain what a set is. Give an example of a set.

**14.** Explain why it is impossible to list the set of rational numbers using the roster method.

**15.** Explain the difference between a subset and a proper subset.

**16.** Describe the difference between 0.45 and $0.\overline{45}$. Are both rational?

**17.** Explain the circumstances under which rounding and truncating will both result in the same decimal approximation.

**18.** Is there a positive real number "closest" to 0?

## Building Skills

*In Problems 19–24, write each set using the roster method.*

**19.** $\{x|x$ is a whole number less than $6\}$

**20.** $\{x|x$ is a natural number less than $4\}$

**21.** $\{x|x$ is an integer between $-3$ and $5\}$

**22.** $\{x|x$ is an integer between $-4$ and $6\}$

**23.** $\{x|x$ is a natural number less than $1\}$

**24.** $\{x|x$ is a whole number less than $0\}$

*In Problems 25–32, let* $A = \{1, 3, 5, 7, 9\}$, $B = \{2, 4, 6, 8\}$, $C = \{1, 2, 3, 4, 5, 6, 7, 8, 9\}$, *and* $D = \{8, 6, 4, 2\}$. *Write True or False to each statement. Be sure to justify your answer.*

**25.** $B \subseteq C$ **26.** $A \subseteq C$ **27.** $B \subset D$ **28.** $A \subset C$

**29.** $B = D$ **30.** $B \subset D$ **31.** $\emptyset \subset C$ **32.** $\emptyset \subset B$

*In Problems 33–36, fill in the blank with the appropriate symbol,* $\in$ or $\notin$.

**33.** $\frac{1}{2}$ _____ $\{x|x$ is an integer$\}$

**34.** $4.\overline{5}$ _____ $\{x|x$ is a rational number$\}$

**35.** $\pi$ _____ $\{x|x$ is a real number$\}$

**36.** $0$ _____ $\{x|x$ is a natural number$\}$

*In Problems 37–40, list numbers in each set that are (a) Natural numbers, (b) Integers, (c) Rational numbers, (d) Irrational numbers, (e) Real numbers.*

**37.** $A = \left\{-5, 4, \frac{4}{3}, -\frac{7}{5}, 5.\overline{1}, \pi\right\}$

**38.** $B = \left\{13, 0, -4.5656\ldots, 2.43, \sqrt{2}\right\}$

**39.** $C = \left\{100, -5.423, \frac{8}{7}, \sqrt{2} + 4, -64, \sqrt{-9}\right\}$

**40.** $D = \left\{15, -\frac{6}{1}, 7.3, \sqrt{2} + \pi, \sqrt{-100}\right\}$

*In Problems 41–44, approximate each number by (a) truncating and (b) rounding to the indicated number of decimal places.*

**41.** 19.93483; 4 decimal places

**42.** $-93.432101$; 2 decimal places

**43.** 0.06345; 1 decimal place

**44.** 9.9999; 2 decimal places

**45.** On the real number line, label the points with coordinates $2, 0, -3, 1.5, -\frac{3}{2}$.

**46.** On the real number line, label the points with coordinates $4, -5, 2.5, \frac{5}{3}, -\frac{1}{2}$.

*In Problems 47–52, replace the question mark by* $<$, $>$, *or* $=$, *whichever is correct.*

**47.** $-5 \; ? \; -3$ **48.** $4 \; ? \; 2$ **49.** $\frac{3}{2} \; ? \; 1.5$

**50.** $\frac{2}{3} \; ? \; \frac{2}{5}$ **51.** $\frac{1}{3} \; ? \; 0.3$ **52.** $-\frac{8}{3} \; ? \; -\frac{8}{5}$

## Applying the Concepts

**53. Death Valley** Death Valley in California is the lowest point in the United States with an elevation that is 282 feet below sea level. Express this elevation as an integer. (*Source: Information Please* Almanac, Web site: *www.Infoplease.com*)

**54. Dead Sea** The Dead Sea, Israel–Jordan, is the lowest point in the world with an elevation that is 1349 feet below sea level. Express this elevation as an integer. (*Source: Information Please* Almanac)

**55. Tellabs** Tellabs is a telecommunications company that lost \$0.04 per share of common stock in the third quarter of 2003. Express this loss as a rational number. (*Source:* Yahoo!Finance)

**56. Sun Microsystems** Sun Microsystems reported a loss of \$0.08 per share of common stock in the first quarter of 2004. Express this loss as a rational number. (*Source:* Yahoo!Finance)

**57. Golf** In the game of golf, your score is often given in relation to par. For example, if par is 72 and a player shoots a 66, then he is 6 under par. Express this score as an integer.

**58. It's a Cold One!** The normal high temperature in Las Vegas, NV on April 23 is 80°F. On April 23, 2003, the temperature was 23°F below normal. Express the departure from normal as an integer. (*Source: USA Today*)

### Extending the Concepts

**59.** Research the history of the set of irrational numbers. Your research should concentrate on the Greek cult called the Pythagoreans.

**60.** Research the origins of the number 0. Is there any single person who can claim its discovery?

**61.** The first known computation of the decimal approximation of the number $\pi$ is attributed to Archimedes around 200 B.C. Research Archimedes and find out his approximation.

**62.** The irrational number $e$ is attributed to Euler. Research Euler and find out the decimal approximation of $e$.

### Graphing Calculator Exercises

**63.** Use your calculator to express $\frac{8}{7}$ rounded to three decimal places. Express $\frac{8}{7}$ truncated to three decimal places.

**64.** Use your calculator to express $\frac{19}{7}$ rounded to four decimal places. Express $\frac{19}{7}$ truncated to four decimal places.

# R.3 Operations on Signed Numbers; Properties of Real Numbers

**OBJECTIVES**

1. Compute the Absolute Value of a Real Number
2. Add and Subtract Signed Numbers
3. Multiply and Divide Signed Numbers
4. Perform Operations on Fractions
5. Know the Associative and Distributive Properties of Real Numbers

## Operations

The symbols used in algebra for the operations of addition, subtraction, multiplication, and division are $+$, $-$, $\cdot$, and /, respectively. The words used to describe the results of these operations are **sum, difference, product,** and **quotient.** Table 1 summarizes these ideas.

**Table 1**

| Operation | Symbol | Words |
|---|---|---|
| Addition | $a + b$ | Sum: $a$ plus $b$ |
| Subtraction | $a - b$ | Difference: $a$ minus $b$<br>$b$ is subtracted from $a$ |
| Multiplication | $a \cdot b$, $(a) \cdot b$, $a \cdot (b)$, $(a) \cdot (b)$,<br>$ab$, $(a)b$, $a(b)$, $(a)(b)$ | Product: $a$ times $b$ |
| Division | $a/b$ or $\frac{a}{b}$ | Quotient: $a$ divided by $b$ |

In algebra, we avoid using the multiplication sign $\times$ (to avoid confusion with the often used $x$) and the division sign $\div$ (to avoid confusion with $+$). Also, when two expressions are placed next to each other without an operation symbol, as in $ab$, or in parentheses, as in $(a)(b)$, it is understood that the expressions, called **factors,** are to be multiplied.

**Work Smart**
Do not use mixed numbers in algebra.

We also do not use mixed numbers in algebra. When mixed numbers are used, addition is understood. For example, $2\frac{3}{4}$ means $2 + \frac{3}{4}$. In algebra, use of a mixed number may be confusing because the absence of an operation symbol between two terms is taken to mean multiplication. To avoid this confusion, $2\frac{3}{4}$ is written as 2.75 or as $\frac{11}{4}$.

The symbol $=$, called an **equal sign** and read as "equals" or "is," is used to express the idea that the expression on the left side of the equal sign is equivalent to the expression on the right.

## 1 Compute the Absolute Value of a Real Number

In Section R.2, we introduced the real number line. The real number line can be used to describe the concept of *absolute value.*

**DEFINITION**
The **absolute value** of a number $a$, written $|a|$, is the distance from 0 to $a$ on the real number line.

**Work Smart**
The absolute value of a number can never be negative because it represents a distance.

**Figure 8**

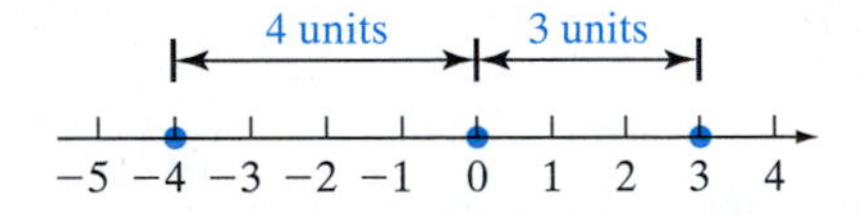


This definition of absolute value is sometimes called the geometric definition because it provides a geometric interpretation. For example, because the distance from 0 to 3 on the real number line is 3, the absolute value of 3, $|3|$, is 3. Because the distance from 0 to $-4$ on the real number line is 4, the absolute value of $-4$, $|-4|$, is 4. See Figure 8.

### EXAMPLE 1 Computing Absolute Value

**(a)** $|8| = 8$ because the distance from 0 to 8 on the real number line is 8.

**(b)** $|-5| = 5$ because the distance from 0 to $-5$ on the real number line is 5.

**(c)** $|0| = 0$ because the distance from 0 to 0 on the real number line is 0.

**QUICK ✓** *Evaluate each expression.*

**1.** $|6|$ **2.** $|-10|$

## 2 Add and Subtract Signed Numbers

**In Words**
To add two real numbers that are the same sign, add their absolute values and then keep the sign of the original numbers. To add two real numbers that are different signs, subtract the absolute value of the smaller number from the absolute value of the larger number. The sign will be the sign of the number whose absolute value is larger.

The following rules are used to add two real numbers.

**ADDING TWO NONZERO REAL NUMBERS**
The approach to adding two real numbers depends upon the sign of the two numbers.

1. *Both Positive:* Add the numbers. This is the same as the techniques used in arithmetic.
2. *Both Negative:* Add the absolute values of each number. The sum will be negative.

**3.** *One Positive, One Negative:* Determine the absolute value of each number. Subtract the smaller absolute value from the larger absolute value.

- If the larger absolute value was originally the positive number, then the sum is positive.
- If the larger absolute value was originally the negative number, then the sum is negative.
- If the two absolute values are equal, then the sum is 0.

## EXAMPLE 2 Adding Two Real Numbers

Perform the indicated operation.

**(a)** $-11 + (-3)$ **(b)** $9.3 + (-6.4)$

### Solution

**(a)** Both numbers are negative, so we first determine their absolute value: $|-11| = 11$ and $|-3| = 3$. We now add the absolute values and obtain $11 + 3 = 14$. Because both numbers to be added are negative, the sum is negative. So

$$-11 + (-3) = -14$$

**(b)** One number is positive, while the other is negative. We determine the absolute value of each number: $|9.3| = 9.3$ and $|-6.4| = 6.4$. We now subtract the smaller absolute value from the larger absolute value and obtain $9.3 - 6.4 = 2.9$. Because the larger absolute value was originally positive, the sum will be positive. So

$$9.3 + (-6.4) = 2.9$$

**Work Smart**

Another approach to adding two real numbers is through a number line. In Example 2(a), we place a point at $-11$ on the real number line. Then, move 3 places to the left (since we are adding $-3$). See Figure 9.

**Figure 9**

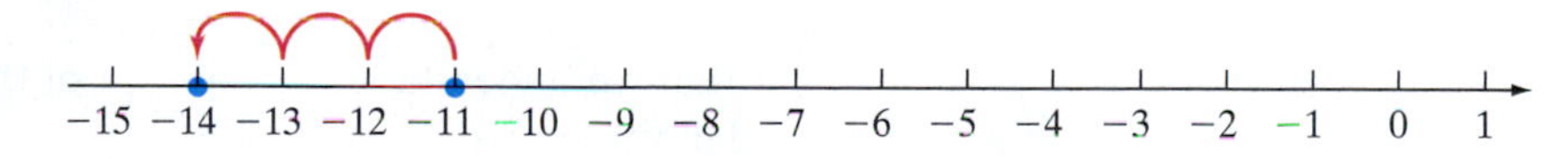


**QUICK ✓** *Add.*

**3.** $18 + (-6)$ **4.** $-21 + 10$ **5.** $-5.4 + (-1.2)$

**6.** $-6.5 + 4.3$ **7.** $-9 + 9$

**Work Smart**

The additive inverse of $a$, that is, $-a$, is sometimes called the *negative* of $a$ or the *opposite* of $a$. Be careful when using these terms because they suggest that the additive inverse is a negative number, which may not be true! For example, the additive inverse of $-3$ is 3, a positive number.

Look back at Problem 7 in the Quick Check above. What do you notice? The result of Problem 7 is true in general.

**ADDITIVE INVERSE PROPERTY**

For any real number $a$ other than 0, there is a real number $-a$, called the **additive inverse,** or **opposite,** of $a$, having the following property:

$$a + (-a) = -a + a = 0$$

The additive inverse of $a$ is sometimes called the *opposite of a*. For example, $-3$ is the additive inverse of 3 so that $-3 + 3 = 0$.

**EXAMPLE 3** **Finding an Additive Inverse**

**(a)** The additive inverse of 6 is $-6$ because $6 + (-6) = 0$.

**(b)** The additive inverse of $-10$ is $-(-10) = 10$ because $-10 + 10 = 0$.

**QUICK ✓** *Determine the additive inverse of the given real number.*

**8.** 5 **9.** $\frac{4}{5}$ **10.** $-12$ **11.** $-\frac{5}{3}$

The real number 0 has another interesting property that is related to the Additive Inverse Property. It turns out that 0 is the only number that can be added to any real number $a$ and result in the same number $a$. We call this the *Identity Property of Addition.*

**IDENTITY PROPERTY OF ADDITION**

For any real number $a$,

$$0 + a = a + 0 = a$$

That is, the sum of any number and 0 is that number. We call 0 the **additive identity.**

We will use the Identity Property of Addition throughout the entire course (and future courses) to create a new expression that is equivalent to a previous expression. For example, $0 + (-0.5) = -0.5$ or $\frac{1}{2} + 0 = \frac{1}{2}$.

Addition is also *commutative*. That is to say, we get the same result whether we compute $a + b$ or $b + a$.

**COMMUTATIVE PROPERTY**

If $a$ and $b$ are real numbers, then

$$a + b = b + a$$

Now that we understand the concept of the additive inverse, we can present a more formal definition of absolute value.

**In Words**
The absolute value of a number greater than or equal to 0 is the number itself. To find the absolute value of a number less than zero, determine the additive inverse of the number.

**DEFINITION**

The **absolute value** of a real number $a$, denoted by the symbol $|a|$, is defined by the rules

$$|a| = a \quad \text{if } a \geq 0 \qquad \text{and} \qquad |a| = -a \quad \text{if } a < 0$$

This definition of absolute value is sometimes called the algebraic definition. For example $|12| = 12$ and $|-13| = -(-13) = 13$.

We can use the idea behind the additive inverse to define subtraction between two real numbers.

**In Words**
To subtract $b$ from $a$, add the "opposite" of $b$ to $a$.

**DEFINITION**

If $a$ and $b$ are real numbers, then the **difference** $a - b$, also read "$a$ less $b$" or "$a$ minus $b$," is defined as

$$a - b = a + (-b)$$

Based on this definition, we should see that subtracting $b$ from $a$ is really just adding the additive inverse of $b$ to $a$.

### EXAMPLE 4 Working with Differences

Evaluate each expression.

**(a)** $10 - 4$ **(b)** $-7.3 - (-4.2)$

**Solution**

**(a)** $10 - 4 = 10 + (-4) = 6$

**(b)** Notice that we are subtracting a negative number. Subtracting $-4.2$ is the same as adding 4.2, so

$$-7.3 - (-4.2) = -7.3 + 4.2$$
$$= -3.1$$

**QUICK ✓** *Evaluate each expression.*

**12.** $6 - 2$ **13.** $4 - 13$ **14.** $-3 - 8$

**15.** $12.5 - 3.4$ **16.** $-8.5 - (-3.4)$ **17.** $-6.9 - 9.2$

## 3 Multiply and Divide Signed Numbers

When we first learned how to multiply the counting numbers, we were told that we can think of multiplication as repeated addition. For example, $3 \cdot 5$ is equivalent to adding 5 three times. That is,

$$3 \cdot 5 = \underbrace{5 + 5 + 5}_{\text{Add 5 three times}} = 15$$

Also, because

$$5 \cdot 3 = 3 + 3 + 3 + 3 + 3 = 15$$

we see that multiplication of two real numbers $a$ and $b$ is commutative, just like addition.

**COMMUTATIVE PROPERTY OF MULTIPLICATION**

If $a$ and $b$ are real numbers, then

$$a \cdot b = b \cdot a$$

When we multiply real numbers, we must be concerned with rules for determining the sign of the product. We present a summary for the rules of signs below.

**RULES OF SIGNS FOR MULTIPLYING TWO REAL NUMBERS**

1. If we multiply two positive real numbers, the product is positive.
2. If we multiply one positive real number and one negative real number, the product is negative.
3. If we multiply two negative real numbers, the product is positive.

### EXAMPLE 5 Multiplying Signed Numbers

**(a)** $3(-5) = -(3 \cdot 5) = -15$ **(b)** $-7 \cdot 3 = -(7 \cdot 3) = -21$

**(c)** $(-9) \cdot (-4) = 9 \cdot 4 = 36$ **(d)** $-1.5 \cdot (2.6) = -3.9$

**Work Smart**

In Example 5(a), $3 \cdot (-5)$ means to add $-5$ three times. That is, $3 \cdot (-5)$ means

$$-5 + (-5) + (-5)$$

which is $-15$. In Example 5(b),

$$\begin{aligned} -7 \cdot 3 &= 3 \cdot (-7) \\ &= -7 + (-7) + (-7) \\ &= -21 \end{aligned}$$

Do you see why a positive times a negative is negative?

**QUICK ✓** *Multiply.*

**18.** $-6 \cdot (8)$ **19.** $12 \cdot (-5)$ **20.** $4 \cdot 14$

**21.** $-7 \cdot (-15)$ **22.** $-1.9 \cdot (-2.7)$

The real number 1 has an interesting property. Recall that the expression $5 \cdot 3$ is equivalent to $5 + 5 + 5$. Therefore, $5 \cdot 1$ means to add 5 one time, so, $5 \cdot 1 = 5$. This result is true in general.

**IDENTITY PROPERTY OF MULTIPLICATION**

For any real number $a$,

$$a \cdot 1 = 1 \cdot a = a$$

That is, the product of any number and 1 is that number. We call 1 the **multiplicative identity.**

We use the Identity Property of Multiplication throughout our math careers to create new expressions that are equivalent to previous expressions. For example, the expressions

$$\frac{4}{5} \quad \text{and} \quad \frac{4}{5} \cdot \frac{3}{3}$$

are equivalent because $\frac{3}{3} = 1$.

For each *nonzero* real number $a$, there is a real number $\frac{1}{a}$, called the *multiplicative inverse* of $a$, having the following property:

**MULTIPLICATIVE INVERSE PROPERTY**

For each *nonzero* real number $a$, there is a real number $\frac{1}{a}$, called the **multiplicative inverse** or **reciprocal** of $a$, having the following property:

$$a \cdot \frac{1}{a} = \frac{1}{a} \cdot a = 1 \qquad a \neq 0$$

The multiplicative inverse of a nonzero real number $a$, $\frac{1}{a}$, is also referred to as the **reciprocal** of $a$.

## EXAMPLE 6 Finding the Multiplicative Inverse or Reciprocal

**(a)** The multiplicative inverse or reciprocal of 5 is $\frac{1}{5}$.

**(b)** The multiplicative inverse or reciprocal of $-4$ is $-\frac{1}{4}$.

**(c)** The multiplicative inverse of $\frac{2}{3}$ is $\frac{3}{2}$ because $\frac{2}{3} \cdot \frac{3}{2} = 1$.

**QUICK ✓** *Find the multiplicative inverse or reciprocal of the given real number.*

**23.** 10 **24.** $-8$ **25.** $\frac{2}{5}$ **26.** $-\frac{1}{5}$

We now use the idea behind the multiplicative inverse to define division of real numbers.

**In Words**
Quotients can be thought of as rational numbers.

**DEFINITION**

If $a$ is a real number and $b$ is a nonzero real number, the **quotient** $\frac{a}{b}$, read as "$a$ divided by $b$" or "the ratio of $a$ to $b$," is defined as

$$\frac{a}{b} = a \cdot \frac{1}{b} \quad \text{if } b \neq 0$$

For example, $\frac{5}{8} = 5 \cdot \frac{1}{8}$. Because division of real numbers can be represented as multiplication, the same rules of signs that apply to multiplication also apply to division.

**RULES OF SIGNS FOR DIVIDING TWO REAL NUMBERS**

1. If we divide two positive numbers, the quotient is positive.
2. If we divide one positive real number and one negative real number, the quotient is negative.
3. If we divide two negative real numbers, the quotient is positive.

Put another way, if $a$ and $b$ are real numbers and $b \neq 0$, then

$$-\frac{a}{b} = \frac{-a}{b} = \frac{a}{-b} \quad \text{and} \quad \frac{-a}{-b} = \frac{a}{b}$$

We now introduce additional properties of the numbers 0 and 1.

**MULTIPLICATION BY ZERO**

For any real number $a$, the product of $a$ and 0 is always 0; that is,

$$a \cdot 0 = 0 \cdot a = 0$$

**Work Smart**
Division by 0 is not defined! One reason is to avoid the following problem: If $\frac{2}{0} = x$, then it must be the case that $2 = 0 \cdot x$. But, according to the Multiplication by Zero Property, $0 \cdot x = 0$ for any real number $x$, so there is no number $x$ such that $\frac{2}{0} = x$.

**DIVISION PROPERTIES**

For any nonzero real number $a$,

$$\frac{0}{a} = 0 \qquad \frac{a}{a} = 1 \qquad \frac{a}{0} \text{ is undefined if } a \neq 0$$

Perhaps you are wondering what $\frac{0}{0}$ equals. The answer, which may surprise you, is that the value of $\frac{0}{0}$ cannot be determined! Why? Remember that $\frac{a}{b} = c$ means that $b \cdot c = a$. So, $\frac{0}{0}$ could equal 0 since $0 \cdot 0 = 0$, but $\frac{0}{0}$ could also equal 1 since $0 \cdot 1 = 0$. In addition, $\frac{0}{0}$ could equal 10 since $0 \cdot 10 = 0$. For this reason, we say that $\frac{0}{0}$ is **indeterminate.**

## 4 Perform Operations on Fractions

We now discuss addition, subtraction, multiplication, and division of fractions. First, we must discuss a property used to write a fraction in *lowest terms*. A fraction is said to be in **lowest** terms if there are no common factors between the numerator and the denominator other than 1.

**REDUCTION PROPERTY**

If $a$, $b$, and $c$ are real numbers, then

$$\frac{ac}{bc} = \frac{a}{b} \quad \text{if } b \neq 0, c \neq 0$$

### EXAMPLE 7 Using the Reduction Property

**(a)**
$$\frac{5 \cdot 3}{5 \cdot 4} = \frac{\cancel{5} \cdot 3}{\cancel{5} \cdot 4}$$

Divide out the 5's: $= \frac{3}{4}$

**(b)** We need to factor the numerator and denominator so that we can divide out any common factors.

$$\frac{18}{12} = \frac{\cancel{6} \cdot 3}{\cancel{6} \cdot 2}$$

Divide out the 6's: $= \frac{3}{2}$

**QUICK ✓** *Use the Reduction Property to simplify the expression.*

**27.** $\frac{2 \cdot 6}{2 \cdot 5}$ **28.** $\frac{-5 \cdot 9}{-2 \cdot 5}$ **29.** $\frac{25}{15}$ **30.** $\frac{-24}{20}$

We now have all the tools that we need in order to perform arithmetic operations on rational numbers.

**In Words**

To multiply two rational numbers, multiply the numerators and then multiply the denominators. To divide two rational numbers, multiply the rational number in the numerator by the reciprocal of the rational number in the denominator. To add two rational numbers, write the rational numbers over a common denominator and then add the numerators.

**ARITHMETIC OF RATIONAL NUMBERS**

$$\frac{a}{b} \cdot \frac{c}{d} = \frac{ac}{bd} \quad \text{if } b \neq 0, d \neq 0$$

$$\frac{a}{b} \div \frac{c}{d} = \frac{\frac{a}{b}}{\frac{c}{d}} = \frac{a}{b} \cdot \frac{d}{c} = \frac{ad}{bc} \quad \text{if } b \neq 0, c \neq 0, d \neq 0$$

$$\frac{a}{c} + \frac{b}{c} = \frac{a+b}{c} \quad \text{if } c \neq 0$$

$$\frac{a}{b} + \frac{c}{d} = \frac{ad}{bd} + \frac{bc}{bd} = \frac{ad+bc}{bd} \quad \text{if } b \neq 0, d \neq 0$$

### EXAMPLE 8 Multiplying and Dividing Rational Numbers

Perform the indicated operation. Be sure to express the answer in lowest terms.

**(a)** $\frac{8}{3} \cdot \frac{15}{4}$ **(b)** $-\frac{3}{5} \div \frac{6}{7}$

**Solution**

**(a)** Here, we multiply the numerators and then multiply the denominators using $\frac{a}{b} \cdot \frac{c}{d} = \frac{ac}{bd}$.

$$\frac{8}{3} \cdot \frac{15}{4} = \frac{8 \cdot 15}{3 \cdot 4}$$

Factor the numerator: $= \frac{4 \cdot 2 \cdot 5 \cdot 3}{3 \cdot 4}$

Divide out like factors: $= \frac{\cancel{4} \cdot 2 \cdot 5 \cdot \cancel{3}}{\cancel{3} \cdot \cancel{4}}$

$$= \frac{2 \cdot 5}{1}$$

$$= \frac{10}{1}$$

$$= 10$$

**Work Smart**

Notice when all the factors in the denominator of Example 8(a) divide out, the denominator is 1.

**(b)** First, we need to rewrite the division problem as an equivalent multiplication problem using $\frac{a}{b} \div \frac{c}{d} = \frac{a}{b} \cdot \frac{d}{c}$.

$$-\frac{3}{5} \div \frac{6}{7} = -\frac{3}{5} \cdot \frac{7}{6}$$

$$= -\frac{\cancel{3} \cdot 7}{5 \cdot \cancel{3} \cdot 2}$$

$$= -\frac{7}{10}$$

**QUICK ✓** *Perform the indicated operation. Express your answer in lowest terms.*

**31.** $\frac{2}{3} \cdot \left(-\frac{5}{4}\right)$ **32.** $\frac{5}{3} \cdot \frac{12}{25}$ **33.** $\frac{4}{3} \div \frac{8}{3}$ **34.** $\frac{\frac{10}{3}}{\frac{5}{12}}$

## EXAMPLE 9 Adding and Subtracting Rational Numbers With the Same Denominator

Perform the indicated operation. Be sure to express the answer in lowest terms.

**(a)** $\frac{5}{12} + \frac{11}{12}$ **(b)** $\frac{4}{15} - \frac{7}{15}$

**Solution**

**(a)** Because the denominators are the same, we add the numerators and write the sum over the common denominator.

$$\frac{5}{12} + \frac{11}{12} = \frac{5 + 11}{12}$$

$$= \frac{16}{12}$$

Factor the numerator and denominator. Divide out like factors: $= \frac{\cancel{4} \cdot 4}{\cancel{4} \cdot 3}$

$$= \frac{4}{3}$$

**(b)** Because the denominators are the same, we subtract the numerators and write the difference over the common denominator.

$$\frac{4}{15} - \frac{7}{15} = \frac{4-7}{15} = \frac{-3}{15} = \frac{-1 \cdot \cancel{3}}{\cancel{3} \cdot 5} = -\frac{1}{5}$$

**QUICK ✓** *Perform the indicated operation. Express your answer in lowest terms.*

**35.** $\frac{3}{11} + \frac{2}{11}$ **36.** $\frac{8}{15} - \frac{13}{15}$ **37.** $\frac{3}{7} + \frac{8}{7}$ **38.** $\frac{8}{5} - \frac{3}{5}$

When we need to add two rational numbers with different denominators, we first find the *least common denominator*. The **least common denominator (LCD)** is the smallest number that each denominator has as a common multiple. The next example illustrates the idea.

## EXAMPLE 10 How to Find the Least Common Denominator

Find the least common denominator of the rational numbers $\frac{8}{15}$ and $\frac{5}{12}$. Then rewrite each rational number with the least common denominator.

### Step-by-Step Solution

| | |
|---|---|
| **Step 1:** Factor each denominator. | $15 = 5 \cdot 3$<br>$12 = 4 \cdot 3 = 2 \cdot 2 \cdot 3$ |
| **Step 2:** Write down the common factor(s) between each denominator. Then copy the remaining factors. | The common factor is 3. The remaining factors are 5, 2, and 2. |
| **Step 3:** Multiply the factors listed in Step 2. The product is the least common denominator (LCD). | $\text{LCD} = 3 \cdot 5 \cdot 2 \cdot 2 = 60$ |
| **Step 4:** Rewrite $\frac{8}{15}$ and $\frac{5}{12}$ with a denominator of 60. Multiply $\frac{8}{15}$ by $\frac{4}{4}$. Multiply $\frac{5}{12}$ by $\frac{5}{5}$. | $\frac{8}{15} = \frac{8}{15} \cdot \frac{4}{4} = \frac{8 \cdot 4}{15 \cdot 4} = \frac{32}{60}$<br>$\frac{5}{12} = \frac{5}{12} \cdot \frac{5}{5} = \frac{5 \cdot 5}{12 \cdot 5} = \frac{25}{60}$ |

**QUICK ✓** *Find the least common denominator (LCD) of each pair of rational numbers. Then rewrite each rational number with the LCD.*

**39.** $\frac{3}{20}$ and $\frac{2}{15}$ **40.** $\frac{5}{18}$ and $-\frac{1}{45}$

Let's do an example where we add two rational numbers using the LCD.

### EXAMPLE 11 Adding Rational Numbers Using the Least Common Denominator

Perform the indicated operation:

**(a)** $\frac{8}{15} + \frac{5}{12}$ **(b)** $-\frac{3}{10} - \frac{2}{15}$

**Solution**

**(a)** From Example 10, we know the least common denominator between 15 and 12 is 60. We rewrite each fraction with a denominator of 60.

$$\frac{8}{15} + \frac{5}{12} = \frac{8}{15}\cdot\frac{4}{4} + \frac{5}{12}\cdot\frac{5}{5}$$
$$= \frac{32}{60} + \frac{25}{60}$$
$$\frac{a}{c} + \frac{b}{c} = \frac{a+b}{c}: \quad = \frac{32 + 25}{60}$$
$$= \frac{57}{60}$$
$$= \frac{19}{20}$$

**(b)** First, write the subtraction problem as an equivalent addition problem using $a - b = a + (-b)$.

$$-\frac{3}{10} - \frac{2}{15} = \frac{-3}{10} + \left(\frac{-2}{15}\right)$$
$$\text{LCD} = 30: \quad = \frac{-3}{10}\cdot\frac{3}{3} + \left(\frac{-2}{15}\cdot\frac{2}{2}\right)$$
$$= \frac{-9}{30} + \left(\frac{-4}{30}\right)$$
$$= \frac{-9 + (-4)}{30}$$
$$= \frac{-13}{30}$$
$$\frac{-a}{b} = -\frac{a}{b}: \quad = -\frac{13}{30}$$

**QUICK ✓** *Perform the indicated operation. Express your answers in lowest terms.*

**41.** $\frac{3}{20} + \frac{2}{15}$ **42.** $\frac{5}{14} - \frac{11}{21}$ **43.** $\frac{-4}{25} - \frac{7}{30}$ **44.** $-\frac{5}{18} - \frac{1}{45}$

## 5 Know the Associative and Distributive Properties of Real Numbers

Now let's illustrate another property of real numbers. Example 12 illustrates that the order in which we add or multiply three real numbers does not affect the final result. This property is called the *Associative Property*.

### EXAMPLE 12 Illustrating the Associative Property

**(a)** $17 + (3 + 5) = 17 + 8 = 25$
$(17 + 3) + 5 = 20 + 5 = 25$
$17 + (3 + 5) = (17 + 3) + 5$

**(b)** $13\cdot(2\cdot 5) = 13\cdot 10 = 130$
$(13\cdot 2)\cdot 5 = 26\cdot 5 = 130$
$13\cdot(2\cdot 5) = (13\cdot 2)\cdot 5$

**ASSOCIATIVE PROPERTY OF ADDITION AND MULTIPLICATION**

If $a$, $b$, and $c$ are real numbers, then

$$a + (b + c) = (a + b) + c = a + b + c$$
$$a \cdot (b \cdot c) = (a \cdot b) \cdot c = a \cdot b \cdot c$$

The next property of real numbers will be used throughout the course and in future courses.

**THE DISTRIBUTIVE PROPERTY**

If $a$, $b$, and $c$ are real numbers, then

$$a \cdot (b + c) = a \cdot b + a \cdot c$$
$$(a + b) \cdot c = a \cdot c + b \cdot c$$

### EXAMPLE 13 Using the Distributive Property

Use the Distributive Property to remove the parentheses.

**(a)** $2(x + 3)$ **(b)** $-3(2y + 1)$ **(c)** $(z - 4) \cdot 3$ **(d)** $\frac{1}{2}(4x - 10)$

**Solution**

**(a)** $2(x + 3) = 2 \cdot x + 2 \cdot 3 = 2x + 6$

**(b)** $-3(2y + 1) = -3 \cdot 2y + (-3) \cdot 1 = -6y - 3$

**(c)** $(z - 4) \cdot 3 = z \cdot 3 - 4 \cdot 3 = 3z - 12$

**(d)** $\frac{1}{2}(4x - 10) = \frac{1}{2} \cdot 4x - \frac{1}{2} \cdot 10 = \frac{4x}{2} - \frac{10}{2} = 2x - 5$

**QUICK ✓** *Use the Distributive Property to remove the parentheses.*

**45.** $5(x + 3)$ **46.** $-6(x + 1)$ **47.** $-4(z - 8)$ **48.** $\frac{1}{3}(6x + 9)$

**SUMMARY: Properties of the Real Number System**

If $a$, $b$, and $c$ are real numbers,

- Identity Properties: $a + 0 = 0 + a = a$; $1 \cdot a = a \cdot 1 = a$
- Inverse Properties: $a + (-a) = (-a) + a = 0$; $a \cdot \frac{1}{a} = \frac{1}{a} \cdot a = 1$ $(a \neq 0)$
- Double Negative Property: $-(-a) = a$
- Commutative Property: $a + b = b + a$; $a \cdot b = b \cdot a$
- Multiplication Property of 0: $a \cdot 0 = 0 \cdot a = 0$
- Division Properties: $\frac{0}{a} = 0, a \neq 0$; $\frac{a}{a} = 1, a \neq 0$; $\frac{a}{0}$ is undefined, $a \neq 0$
- Reduction Property: $\frac{ac}{bc} = \frac{a}{b}, b \neq 0, c \neq 0$
- Associative Property: $(a + b) + c = a + (b + c)$; $(ab)c = a(bc)$
- Distributive Property: $a(b + c) = ab + ac$

# R.3 Exercises

**For Extra Help:**  Student Solutions Manual  CD Video  PH Math/Tutor Center  MathXL Tutorials on CD  MathXL® MyMathLab

## Concepts and Vocabulary

*In Problems 1–3, fill in the blanks.*

**1.** In the expression $a \cdot b$, the expressions $a$ and $b$ are called _____.

**2.** The _____ _____ states that $a \cdot (b + c) = a \cdot b + a \cdot c$.

**3.** The additive inverse of $a$, $-a$ is also called the _____ of $a$. The multiplicative inverse of $a$, $\frac{1}{a}$, is also called the _____ of $a$.

*In Problems 4–6, answer True or False to each statement.*

**4.** The absolute value of a number is always positive.

**5.** Addition and multiplication are commutative.

**6.** We call 0 the additive identity and we call 1 the multiplicative identity.

**7.** What is the additive inverse of 0?

**8.** In your own words, explain why 0 does not have a multiplicative inverse.

**9.** Why does $2(4 \cdot 5)$ not equal $(2 \cdot 4) \cdot (2 \cdot 5)$?

**10.** Explain the flaw in the following reasoning:

$$\frac{4+3}{2+5} = \frac{2 \cdot 2 + 3}{2 + 5} = \frac{\cancel{2} \cdot 2 + 3}{\cancel{2} + 5} = \frac{2+3}{1+5} = \frac{5}{6}.$$

**11.** Is subtraction commutative? Support your conclusion with an example.

**12.** Is subtraction associative? Support your conclusion with an example.

**13.** Is division commutative? Support your conclusion with an example.

**14.** Is division associative? Support your conclusion with an example.

## Building Skills

*In Problems 15–26, determine (a) the additive inverse, (b) the multiplicative inverse, and (c) the reciprocal of the given number.*

**15.** $\frac{9}{5}$ **16.** $\frac{16}{5}$ **17.** $-15$ **18.** $-73$ **19.** $\frac{4}{3}$ **20.** $-\frac{1}{5}$

**21.** $8$ **22.** $10$ **23.** $-4$ **24.** $-6$ **25.** $\frac{2}{5}$ **26.** $-\frac{5}{4}$

*In Problems 27–34, use the Distributive Property to remove the parentheses.*

**27.** $2(x + 4)$ **28.** $3(y - 5)$ **29.** $3(z + 2)$ **30.** $5(x + 4)$

**31.** $(x - 10) \cdot 3$ **32.** $(3x + y) \cdot 2$ **33.** $\frac{3}{4}(8x - 10)$ **34.** $-\frac{2}{3}(3x + 15)$

*In Problems 35–38, use the Reduction Property to simplify each expression.*

**35.** $\frac{4 \cdot z}{20}$ **36.** $\frac{7 \cdot y}{35}$ **37.** $\frac{30}{25}$ **38.** $\frac{40}{16}$

*In Problems 39–78, perform the indicated operation. Express all rational numbers in lowest terms.*

**39.** $8 + 7$ **40.** $-6 + 10$ **41.** $12 - 5$ **42.** $9 + (-3)$

**43.** $|13 - 16|$ **44.** $|-4| + 12$ **45.** $4.3 - 6.8$ **46.** $-8.2 - 4.5$

**47.** $4 \cdot (-8)$ **48.** $-5 \cdot (-15)$ **49.** $-6 \cdot (-14)$ **50.** $7 \cdot (-15)$

**51.** $\frac{3}{4} \cdot \frac{20}{9}$ **52.** $\frac{2}{5} \cdot \frac{15}{8}$ **53.** $\frac{2}{3} \cdot \left(-\frac{9}{14}\right)$ **54.** $-\frac{7}{3} \cdot \left(-\frac{12}{35}\right)$

**55.** $\frac{7}{4} \div \frac{21}{8}$ **56.** $-\frac{10}{3} \div \frac{15}{21}$ **57.** $\frac{\frac{2}{5}}{\frac{8}{25}}$ **58.** $\frac{\frac{18}{7}}{\frac{3}{14}}$

**59.** $-\frac{5}{12} + \frac{7}{12}$ **60.** $\frac{8}{5} - \frac{18}{5}$ **61.** $\frac{8}{15} + \frac{9}{14}$ **62.** $\frac{3}{8} - \frac{7}{18}$

**63.** $\frac{41}{42} - \frac{8}{35}$ **64.** $-\frac{17}{45} - \frac{23}{24}$ **65.** $5 \div \frac{15}{4}$ **66.** $\frac{2}{3} \div 8$

**67.** $|6.2 - 9.5|$ **68.** $|-5.4 + 10.5|$ **69.** $-|-8 \cdot (4)|$ **70.** $-|-5 \cdot 9|$

**71.** $\left|-\frac{1}{2} - \frac{4}{5}\right|$ **72.** $\left|\frac{4}{3} - \frac{8}{7}\right|$ **73.** $\left|-\frac{5}{6} - \frac{3}{10}\right|$ **74.** $\left|\frac{4}{15} - \frac{1}{6}\right|$

**75.** $\frac{21}{32} + (-5)$ **76.** $-\frac{10}{21} + 6$ **77.** $\frac{\frac{2}{3}}{6}$ **78.** $\frac{20}{\frac{5}{4}}$

*In Problems 79–86, state the property that is being illustrated.*

**79.** $5 \cdot 3 = 3 \cdot 5$ **80.** $9 + (-9) = 0$ **81.** $5 \cdot \frac{1}{5} = 1$

**82.** $\frac{a}{a} = 1, a \neq 0$ **83.** $\frac{42}{10} = \frac{21}{5}$ **84.** $3(x - 4) = 3x - 12$

**85.** $3 + (4 + 5) = (3 + 4) + 5$ **86.** $\frac{0}{6} = 0$

## Applying the Concepts

**87. Age of Presidents** The youngest president at the time of inauguration was Theodore Roosevelt (42 years of age). The oldest president at the time of inauguration was Ronald Reagan (69 years of age). What is the difference in age of the oldest and youngest president at the time of inauguration?

**88. Life Expectancy** In South Korea, the life expectancy for females born in 1950 was 49 years of age. The life expectancy for females born in 1998 was 78 years of age. Compute the difference in life expectancy from 1950 to 1998.

**89. Football** The Chicago Bears obtained the following yardages for each of the first three plays of a game: 4, −3, 8. How many total yards did they gain for the first three plays? If 10 yards are required for a first down, did the Bears obtain a first down?

**90. Balancing a Checkbook** At the beginning of the month, Paul had \$400 in his checking account. During the month he wrote four checks for \$20, \$45, \$60, and \$105. He also made a deposit in the amount of \$150. What is Paul's balance at the end of the month?

**91. Peaks and Valleys** In the United States, the highest elevation is Mount McKinley in Alaska (20,320 feet above sea level); the lowest elevation is Death Valley in California (282 feet below sea level). What is the difference between the highest and lowest elevation?

**92. More Peaks and Valleys** In Louisiana, the highest elevation is Driskill Mountain (535 feet above sea level); the lowest elevation is New Orleans (8 feet below sea level). What is the difference between the highest and lowest elevation?

## Extending the Concepts

**93.** Illustrate why the product of two positive numbers is positive.

**94.** Illustrate why the product of a positive number and a negative number is negative.

**95. The Fibonacci Sequence** The Fibonacci Sequence is a famous sequence of numbers that were discovered by Leonardo Fibonacci of Pisa. The numbers in the sequence are $1, 1, 2, 3, 5, 8, 13, 21, 34, 55, \ldots$, where each term after the second term is the sum of the two preceding terms.

**(a)** Compute the ratio of consecutive terms in the sequence. That is compute $\frac{1}{1}, \frac{2}{1}, \frac{3}{2}, \frac{5}{3}$, and so on.

**(b)** What number does the ratio approach? This number is called the **golden ratio** and has application in many different areas.

**(c)** Research Fibonacci numbers and cite three different applications.

*Problems 96–101 use the following definition.*

If $P$ and $Q$ are two points on a real number line with coordinates $a$ and $b$, respectively, the **distance between $P$ and $Q$,** denoted by $d(P, Q)$, is

$$d(P, Q) = |b - a|$$

Since $|b - a| = |a - b|$, it follows that $d(P, Q) = d(Q, P)$.

**96.** Plot the points $P = -4$ and $Q = 10$ on the real number line and then find $d(P, Q)$.

**97.** Plot the points $P = -2$ and $Q = 6$ on the real number line and then find $d(P, Q)$.

**98.** Plot the points $P = -3.2$ and $Q = 7.2$ on the real number line and then find $d(P, Q)$.

**99.** Plot the points $P = -9.3$ and $Q = 1.6$ on the real number line and then find $d(P, Q)$.

**100.** Plot the points $P = -\frac{10}{3}$ and $Q = \frac{6}{5}$ on the real number line and then find $d(P, Q)$.

**101.** Plot the points $P = -\frac{7}{5}$ and $Q = 6$ on the real number line and then find $d(P, Q)$.

**102. Proof that the Product of Two Negatives Is Positive** In this problem, we use the Distributive Property to prove that the product of two negative real numbers is positive.

**(a)** Express the product of any real number $a$ and 0.
**(b)** Use the Additive Inverse Property to write 0 as $b + (-b)$.
**(c)** Use the Distributive Property to distribute the $a$ into the expression in parentheses.
**(d)** Suppose that $a < 0$ and $b > 0$. What can be said about the product $ab$? Now, what must be true regarding the product $a(-b)$ in order for the sum to be zero?

*In Problems 103 and 104, find the set of all ratios $\frac{p}{q}$ such that $p \in A$ and $q \in B$.*

**103.** $A = \{-6, -3, -2, -1, 1, 2, 3, 6\}, B = \{-2, -1, 1, 2\}$

**104.** $A = \{-9, -3, -1, 1, 3, 9\}, B = \{-4, -2, -1, 1, 2, 4\}$

## The Graphing Calculator

Many calculators have the ability to compute absolute value. Figure 10 shows the results of Example 1 using a TI-84 Plus graphing calculator. Calculators have the ability to add, subtract, multiply, and divide rational numbers and write the solution in lowest terms! Figure 11 shows the results of Examples 8(b) and 11(b).

**Figure 10**

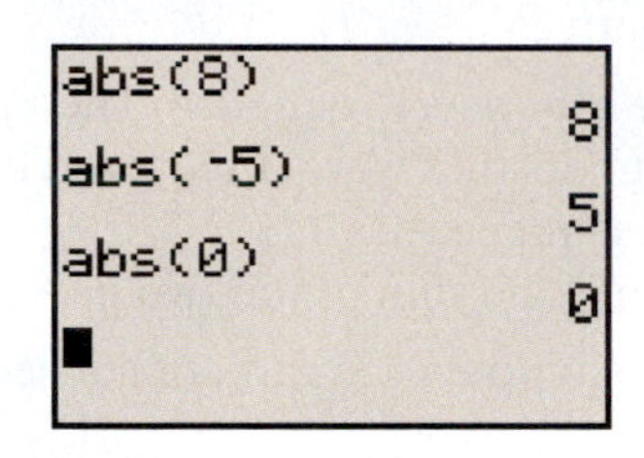


**Figure 11**

*In Problems 105–114, use a calculator to perform the indicated operation. Express all rational numbers in lowest terms.*

**105.** $5.4 - 9.2$ **106.** $-2.9 + (-6.3)$ **107.** $-3(6.4)$ **108.** $-5.4(-4.8)$

**109.** $\frac{3}{10} - \frac{4}{15}$ **110.** $-\frac{3}{8} + \frac{1}{10}$ **111.** $|4.5(-3.2)|$ **112.** $|-3.65| \cdot |5.4|$

**113.** $\frac{6}{5} \cdot \frac{45}{12}$ **114.** $-\frac{3}{4} \div \frac{9}{20}$

# R.4 Order of Operations

**OBJECTIVES**

1. Evaluate Real Numbers with Exponents
2. Use the Order of Operations to Evaluate Expressions

## 1 Evaluate Real Numbers with Exponents

Integer exponents provide a shorthand device for representing repeated multiplications of a real number. For example,

$$3^4 = 3 \cdot 3 \cdot 3 \cdot 3 = 81$$

In the expression $3^4$, we call 3 the **base** and 4 the **exponent** or **power.** The power indicates the number of times we should multiply the base to itself.

**In Words**
The notation $a^n$ means to multiply $a$ by itself $n$ times.

**DEFINITION**

If $a$ is a real number and $n$ is a positive integer, then the symbol $a^n$ represents the product of $n$ factors of $a$. That is,

$$a^n = \underbrace{a \cdot a \cdot \cdots \cdot a}_{n \text{ factors}}$$

Here, it is understood that $a^1 = a$.

In particular, we have

$$a^1 = a$$
$$a^2 = a \cdot a$$
$$a^3 = a \cdot a \cdot a$$

and so on.

We read $a^n$ as "$a$ raised to the power $n$" or as "$a$ to the $n$th power." We usually read $a^2$ as "$a$ squared" and $a^3$ as "$a$ cubed."

### EXAMPLE 1 Evaluating Expressions Containing Exponents

Evaluate each expression.

**(a)** $2^3$ **(b)** $\left(\frac{1}{2}\right)^2$ **(c)** $12^1$

**Solution**

**(a)** The expression $2^3$ means that we should multiply 2 by itself 3 times, so

$$2^3 = 2 \cdot 2 \cdot 2 = 8$$

**(b)** $\left(\frac{1}{2}\right)^2 = \frac{1}{2} \cdot \frac{1}{2} = \frac{1}{4}$

**(c)** $12^1 = 12$

### EXAMPLE 2 Evaluating Expressions Containing Exponents

Evaluate each expression.

**(a)** $(-3)^4$ **(b)** $(-3)^5$ **(c)** $-3^4$ **(d)** $-(-3)^4$

**Solution**

**(a)** $(-3)^4 = (-3) \cdot (-3) \cdot (-3) \cdot (-3) = 81$

**(b)** $(-3)^5 = (-3) \cdot (-3) \cdot (-3) \cdot (-3) \cdot (-3) = -243$

**(c)** The expression $-3^4$ means that we should determine the additive inverse of $3^4$. That is, $-3^4 = -(3) \cdot (3) \cdot (3) \cdot (3) = -81$.

**(d)** $-(-3)^4 = -(-3) \cdot (-3) \cdot (-3) \cdot (-3) = -81$

**Work Smart**

The expressions $(-3)^4$ and $-3^4$ are different. $(-3)^4$ means to multiply $(-3)$ by itself four times, while $-3^4$ means to determine the additive inverse of $3^4$. That is,

$$(-3)^4 = (-3)(-3)(-3)(-3) = 81$$

and

$$-3^4 = -(3)(3)(3)(3) = -81$$

Based upon the results of Examples 2(a) and 2(b), you might conclude that raising a negative number to a positive, even integer exponent results in a positive number, while raising a negative number to a positive, odd integer exponent results in a negative number. This conclusion is correct in general.

**QUICK ✓** *Evaluate each expression.*

**1.** $4^3$ **2.** $(-7)^2$ **3.** $(-10)^3$ **4.** $\left(\frac{2}{3}\right)^3$ **5.** $-8^2$ **6.** $-(-5)^3$

## 2 Use the Order of Operations to Evaluate Expressions

When we are asked to evaluate an expression that contains both multiplication and addition, such as $2 + 3 \cdot 7$, we agree to the following.

> Whenever the two operations of addition and multiplication separate three numbers, the multiplication operation is always performed first, followed by the addition operation.

Therefore, $2 + 3 \cdot 7 = 2 + 21 = 23$.

### EXAMPLE 3 Finding the Value of an Expression

Evaluate each expression.

**(a)** $3 + 4 \cdot 5$ **(b)** $8 \cdot 2 + 1$

**Solution**

**(a)** Don't forget! We do multiplication before addition.

$$\begin{aligned} 3 + 4 \cdot 5 &= 3 + 20 \\ &= 23 \end{aligned}$$

**(b)** $$\begin{aligned} 8 \cdot 2 + 1 &= 16 + 1 \\ &= 17 \end{aligned}$$

**QUICK ✓** *Evaluate each expression.*

**7.** $5 \cdot 2 + 6$ **8.** $3 \cdot 2 + 5 \cdot 6$

If we want to add two numbers first and then multiply, we use parentheses and write $(2 + 3) \cdot 7$. In other words, any operation in parentheses is always done prior to multiplication.

### EXAMPLE 4 Finding the Value of an Expression

**(a)** $$\begin{aligned} (6 + 2) \cdot 4 &= 8 \cdot 4 \\ &= 32 \end{aligned}$$

**(b)** $$\begin{aligned} (8 - 3) \cdot (7 + 3) &= 5 \cdot 10 \\ &= 50 \end{aligned}$$

**QUICK ✓** *Evaluate each expression.*

**9.** $4 \cdot (5 + 3)$ **10.** $8 \cdot (9 - 3)$

**11.** $(12 - 4) \cdot (18 - 13)$ **12.** $(4 + 9) \cdot (6 - 4)$

When we divide two expressions, as in

$$\frac{4 + 5}{6 + 12}$$

it is understood that the division bar acts like parentheses. This means that we must evaluate the expressions in the numerator and denominator FIRST, and then perform the division. That is,

$$\frac{4 + 5}{6 + 12} = \frac{(4 + 5)}{(6 + 12)} = \frac{9}{18} = \frac{\cancel{9} \cdot 1}{\cancel{9} \cdot 2} = \frac{1}{2}$$

It will be extremely important that you remember this as you progress in the course.

**Work Smart**

Never use the Reduction Property across addition. We CANNOT simplify $\frac{4+5}{6+12}$ as follows:

$$\frac{4+5}{6+12} = \frac{\cancel{2}\cdot 2+5}{\cancel{2}\cdot 3+12} = \frac{2+5}{3+12} = \frac{7}{15}$$

So $\frac{4+5}{6+12} \neq \frac{7}{15}$.

If an expression contains a set of grouping symbols embedded within a second set of grouping symbols as in $2 + [3 + 2(9 - 4)]$, we agree to simplify the innermost grouping symbols first.

## EXAMPLE 5 Finding the Value of an Expression

Evaluate each expression.

**(a)** $\frac{3+7}{6+4\cdot 6}$ **(b)** $9\cdot[8+4(10-7)]$

### Solution

**(a)**

$$\frac{3+7}{6+4\cdot 6} = \frac{3+7}{6+24} = \frac{10}{30} = \frac{1}{3}$$

**(b)** Evaluate the expression in the innermost parentheses first.

$$\begin{aligned} 9\cdot[8+4(10-7)] &= 9\cdot[8+4(3)] \\ &= 9\cdot[8+12] \\ &= 9\cdot 20 \\ &= 180 \end{aligned}$$

**Work Smart**

When an expression contains more than one set of parentheses, work inside out.

**QUICK ✓** *Evaluate each expression.*

**13.** $\frac{3+7}{4+9}$ **14.** $1 - 4 + 8\cdot 2 + 5$

**15.** $25\cdot[2(8-3)-8]$ **16.** $\frac{3+5}{2\cdot(9-4)}$

When do we evaluate exponents in the order of operations? Let's consider the expression $3\cdot 2^4$. Do we multiply 3 and 2 first and then raise this product to the 4th power to get 1296, or do we first raise 2 to the 4th power to get 16 and multiply this by 3 and obtain 48? Because $2^4 = 2\cdot 2\cdot 2\cdot 2$, we have $3\cdot 2^4 = 3\cdot 2\cdot 2\cdot 2\cdot 2 = 48$, which implies that we evaluate exponents before multiplication.

Because parentheses always mean "Do the operation in parentheses first!", we have the following order of operations.

**ORDER OF OPERATIONS**

1. Evaluate expressions within **parentheses** first. When an expression has multiple parentheses, begin with the innermost parentheses and work outward.
2. Evaluate expressions containing **exponents,** working from left to right.
3. Perform **multiplication** and **division,** working from left to right.
4. Perform **addition** and **subtraction,** working from left to right.

**QUICK ✓** *Evaluate each expression.*

**17.** $6 + 5 \cdot 2$ **18.** $(3 + 9) \cdot 4$ **19.** $\dfrac{3 + 7}{4 + 5}$

**20.** $\dfrac{7 + 5}{4 + 10}$ **21.** $4 + [(8 - 3) \cdot 2]$ **22.** $[4 \cdot (6 - 2) - 9]$

## EXAMPLE 6 Evaluating Expressions Using the Order of Operations

Evaluate each expression.

**(a)** $6 + 4 \cdot 3^2$ **(b)** $\dfrac{1}{4} \cdot 2^3 + 9 \cdot 4 - 2 \cdot 5^2$

### Solution

**(a)** Evaluate exponents first.

$$6 + 4 \cdot 3^2 = 6 + 4 \cdot 9$$

Multiply: $= 6 + 36$

$= 42$

**(b)** Evaluate exponents first.

$$\frac{1}{4} \cdot 2^3 + 9 \cdot 4 - 2 \cdot 5^2 = \frac{1}{4} \cdot 8 + 9 \cdot 4 - 2 \cdot 25$$

Multiply: $= 2 + 36 - 50$

Add from left to right: $= 38 - 50$

$= -12$

## EXAMPLE 7 Evaluating Expressions Using the Order of Operations

**(a)** $\dfrac{2 + 3 \cdot 4^2}{5 \cdot (6 - 2)}$ **(b)** $-2 \cdot |-(-2)^3 - 4 \cdot (10 - 7)^2|$

### Solution

**(a)**

$$\frac{2 + 3 \cdot 4^2}{5 \cdot (6 - 2)} = \frac{2 + 3 \cdot 16}{5 \cdot 4}$$

$$= \frac{2 + 48}{20}$$

$$= \frac{50}{20}$$

$$= \frac{5 \cdot 10}{2 \cdot 10}$$

Divide out common factor: $= \dfrac{5}{2}$

**(b)** The absolute value bars act just like parentheses. So, we simplify everything in the absolute bars first.

Evaluate inner parentheses first

$$
\begin{aligned}
-2\cdot|-(-2)^3 - 4\cdot(10-7)^2| &= -2\cdot|-(-2)^3 - 4\cdot(3)^2| \\
\text{Exponents:} \quad &= -2\cdot|-(-8) - 4\cdot 9| \\
&= -2\cdot|8 - 36| \\
&= -2\cdot|-28| \\
&= -2\cdot 28 \\
&= -56
\end{aligned}
$$

**QUICK ✓** *Evaluate each expression.*

**23.** $-8 + 2\cdot 5^2$ **24.** $5\cdot 3 - 3\cdot 2^3$ **25.** $5\cdot(10-8)^2$

**26.** $3\cdot(-2)^2 + 6\cdot 3 - 3\cdot 4^2$ **27.** $\dfrac{4+6^2}{2\cdot 3 + 2}$ **28.** $-3\cdot|7^2 - 2\cdot(8-5)^3|$

# R.4 Exercises

**For Extra Help:**  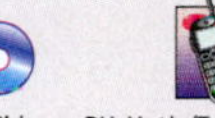  
Student Solutions Manual | CD Video | PH Math/Tutor Center | MathXL Tutorials on CD | MathXL® | MyMathLab

## Concepts and Vocabulary

*In Problems 1 and 2, fill in the blanks.*

**1.** In the expression $5^6$, the number 5 is called the _____ and 6 is called the _____ or _____.

**2.** The order of operations is (1) _____, (2) _____, (3) _____ and _____, (4) _____ and _____.

*In Problems 3 and 4, answer True or False to each statement.*

**3.** $-7^4 = (-7)\cdot(-7)\cdot(-7)\cdot(-7)$

**4.** In the order of operations, we evaluate exponents before multiplication or division.

**5.** Explain why $2 + 3\cdot 4 = 14$, whereas $(2+3)\cdot 4 = 20$.

**6.** Explain the difference between $-4^3$ and $(-4)^3$.

**7.** In your own words, write a few sentences that justify performing multiplication before addition.

**8.** In your own words, write a few sentences that justify evaluating exponents before multiplication.

## Building Skills

*In Problems 9–44, evaluate each expression.*

**9.** $4^2 - 3^2$ **10.** $(-3)^2 + (-2)^2$ **11.** $3\cdot 2 + 9$

**12.** $-5\cdot 3 + 12$ **13.** $4 + 2\cdot(6-2)$ **14.** $2 + 5\cdot(8-5)$

**15.** $-2[10 - (3-7)]$ **16.** $3[15 - (7-3)]$ **17.** $\dfrac{5-(-7)}{4}$

**18.** $\dfrac{12-4}{-2}$ **19.** $|3\cdot 2-4\cdot 5|$ **20.** $|6\cdot 2-5\cdot 3|$

**21.** $12\cdot\dfrac{2}{3}-5\cdot 2$ **22.** $15\cdot\dfrac{3}{5}+4\cdot 3$ **23.** $3\cdot[2+3\cdot(1+5)]$

**24.** $2\cdot[25-2\cdot(10-4)]$ **25.** $(3^2-3)\cdot(3-(-3)^3)$ **26.** $-2\cdot(5-2)-(-5)^2$

**27.** $|3\cdot(6-3^2)|$ **28.** $-2\cdot(4+|2\cdot 3-5^2|)$ **29.** $|4\cdot[2\cdot 5+(-3)\cdot 4]|$

**30.** $|6\cdot(3\cdot 2-10)|$ **31.** $\dfrac{2\cdot 5+15}{2^2+3\cdot 2}$ **32.** $\dfrac{2\cdot 3^2}{4^2-4}$

**33.** $\dfrac{2\cdot(4+8)}{3+3^2}$ **34.** $\dfrac{3\cdot(5+2^2)}{2\cdot 3^3}$ **35.** $\dfrac{6\cdot[12-3\cdot(5-2)]}{5\cdot[21-2\cdot(4+5)]}$

**36.** $\dfrac{4\cdot[3+2\cdot(8-6)]}{5\cdot[14-2\cdot(2+3)]}$ **37.** $\left(\dfrac{2}{3}\right)^2\cdot\left(\dfrac{1+2^3}{2^3-2}\right)$ **38.** $\left(\dfrac{3^2}{29-3\cdot 2^3}\right)\cdot\dfrac{5}{4+5}$

**39.** $\dfrac{3^3-2^4\cdot 3}{4\cdot(3^2-2\cdot 3)}$ **40.** $2^5+4\cdot(-5)+4^2$

**41.** $\dfrac{2\cdot 4-5}{4^2+(-2)^3}+\dfrac{3^2}{2^3}$ **42.** $\dfrac{5\cdot(37-6^2)}{6\cdot 2-3^2}+\dfrac{7\cdot 2-4^2}{5+4}$

**43.** $\dfrac{\dfrac{2\cdot 3+3^2}{2\cdot 5-8}+\dfrac{4}{3}}{\dfrac{7\cdot(5-3)}{14-2^3}}$ **44.** $\dfrac{\dfrac{4}{4^2-1}-\dfrac{3}{5\cdot(7-5)}}{\dfrac{-(-2)^2}{4\cdot 7+2}}$

## Applying the Concepts

*In Problems 45–48, insert parentheses in order to make the statement true.*

**45.** $3\cdot 7-2=15$ **46.** $-2\cdot 3-5=4$

**47.** $3+5\cdot 6-3=18$ **48.** $3+5\cdot 6-3=24$

△ **49. Geometry** The surface area of a right circular cylinder whose radius is 5 inches and height is 12 inches is given approximately by $2\cdot 3.1416\cdot 5^2+2\cdot 3.1416\cdot 5\cdot 12$. Evaluate this expression rounded to two decimal places.

△ **50. Geometry** The surface area of a sphere whose radius is 3 centimeters is given approximately by $4\cdot 3.1416\cdot 3^2$. Evaluate this expression rounded to two decimal places.

**51. Hitting a Golf Ball** The height (in feet) of a golf ball hit with an initial speed of 100 feet per second after 3 seconds is given by $-16\cdot 3^2+50\cdot 3$. Evaluate this expression in order to determine the height of the golf ball after 3 seconds.

**52. Horsepower** The horsepower rating of an engine is $\dfrac{10^2\cdot 8}{2.5}$. Evaluate this expression in order to determine the horsepower rating of the engine.

## Extending the Concepts

**53.** Explain why $\frac{2 + 7}{2 + 9} \neq \frac{7}{9}$.

**54.** Develop an example that explains why we perform multiplication before addition when there are no grouping symbols.

**55.** Develop an example that explains why we evaluate exponents before multiplication when there are no grouping symbols.

*Problems 56–58 show some computations required in statistics. Completely simplify each expression.*

**56.** $\frac{105 + 80 + 115 + 95 + 105}{5}$

**57.** $\frac{(105 - 100)^2 + (80 - 100)^2 + (115 - 100)^2 + (95 - 100)^2 + (105 - 100)^2}{4}$

**58.** $\frac{65 - 50}{10}$

## The Graphing Calculator

Calculators can be used to evaluate exponential expressions. Figure 12 shows the results of Examples 2(a), (b), and (c) using a TI-84 Plus graphing calculator.

Graphing calculators know the order of operations. Figure 13(a) shows the result of Example 6(b) and Figure 13(b) shows the result of Example 7(a) using a TI-84 Plus graphing calculator. Be careful with the placement of parentheses when using the calculator!

**Figure 12**

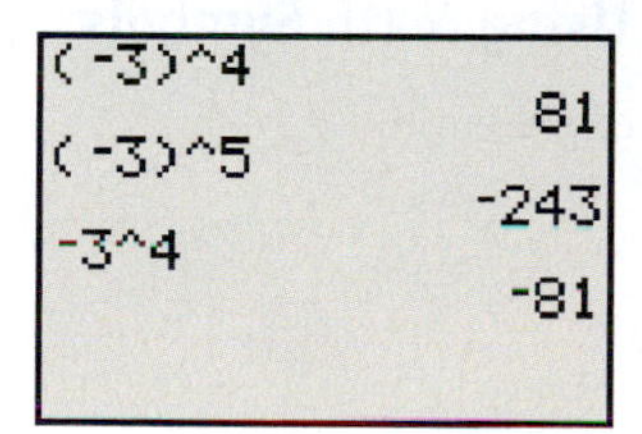


**Figure 13**

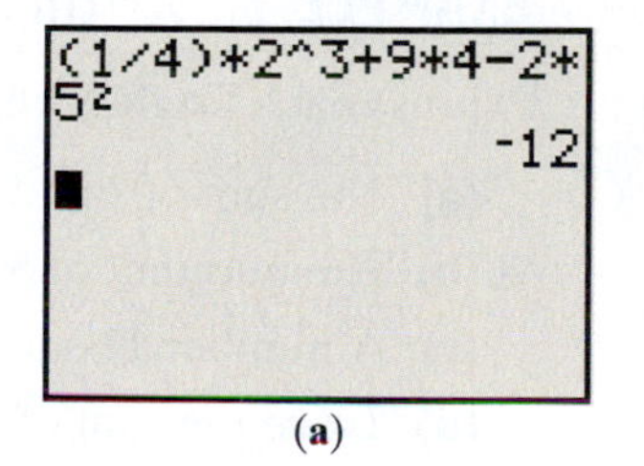


(a)

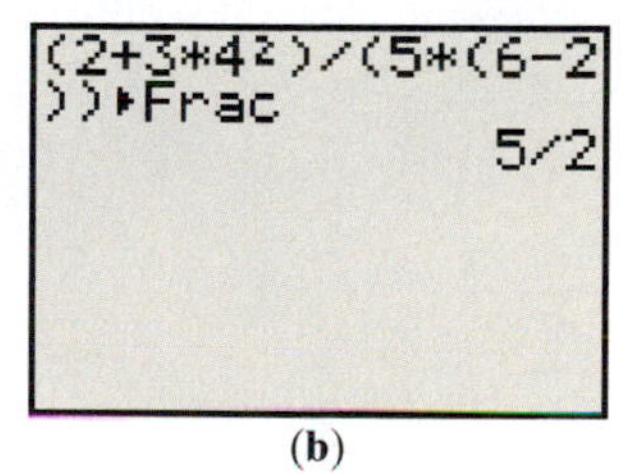


(b)

*In Problems 59–66, use a calculator to evaluate each expression. When necessary, express answers rounded to two decimal places.*

**59.** $\frac{4}{5} - \left(\frac{2}{3}\right)^2$

**60.** $3 - \left(\frac{6}{5}\right)^3$

**61.** $\frac{4^2 + 1}{13}$

**62.** $\frac{3^2 - 2^3}{1 + 3 \cdot 2}$

**63.** $\frac{3.5^3}{1.3^2} - 6.2^3$

**64.** $2.3^4 \cdot \frac{4}{11} - (3.7)^2 \cdot \frac{8}{3}$

**65.** $4.3[9.3^2 - 4(34.2 + 18.5)]$

**66.** $6.3^2 + 4.2^2$

# R.5 Algebraic Expressions

**OBJECTIVES**

1. Translate English Expressions into the Language of Mathematics
2. Evaluate Algebraic Expressions
3. Simplify Algebraic Expressions by Combining Like Terms
4. Determine the Domain of a Variable

We mentioned earlier that in algebra we use letters such as $x$, $y$, $a$, $b$, and $c$ to represent numbers. If a letter is used to represent *any* number from a given set of numbers, it is called a **variable.** A **constant** is either a fixed number, such as 5 or $\sqrt{2}$, or a letter that represents a fixed (possibly unspecified) number. For example, in Einstein's Theory of Relativity, $E = mc^2$, $E$ and $m$ are variables that represent total energy and mass, respectively, while $c$ is a constant that represents the speed of light (299,792,458 meters per second).

An **algebraic expression** is any combination of variables, constants, grouping symbols such as parentheses ( ) or brackets [ ], and mathematical operations such as

**In Words**
In algebra, letters of the alphabet are used to represent numbers.

addition, subtraction, multiplication, division, or exponents. The following are examples of algebraic expressions.

$$3x + 4 \qquad 2y^2 - 5y - 12z \qquad \frac{5v^4 - v}{4 - v}$$

## 1 Translate English Expressions into the Language of Mathematics

**Work Smart**
An algebraic expression is not the same as an algebraic statement. An expression might be $x + 3$ while a statement might be $x + 3 = 7$. Do you see the difference?

One of the neat features of mathematics is that the symbols we use allow us to express English phrases briefly and consistently. There are certain words or phrases in English that easily translate into math symbols. Table 2 lists various English words or phrases and their corresponding math symbol.

**Table 2 Math Symbols and the Words They Represent**

| Add (+) | Subtract (−) | Multiply (·) | Divide ( / ) |
|---|---|---|---|
| sum | difference | product | quotient |
| plus | minus | times | divided by |
| more than | subtracted from | of | per |
| exceeds by | less | twice | ratio |
| in excess of | less than | | |
| added to | decreased by | | |
| increased by | | | |

### EXAMPLE 1 Writing English Phrases Using Math Symbols

Express each English phrase as an algebraic expression.

**(a)** The sum of 3 and 8
**(b)** The quotient of 50 and some number $y$
**(c)** A number 11 subtracted from $z$
**(d)** Twice the sum of a number $x$ and 5

**Solution**

**(a)** Because we are talking about a sum, we know to use the + symbol, so "The sum of 3 and 8" is represented mathematically as $3 + 8$.

**(b)** "The quotient of 50 and some number $y$" is represented mathematically as $\frac{50}{y}$.

**(c)** "A number 11 subtracted from $z$" is represented mathematically as $z - 11$.

**(d)** "Twice the sum of a number $x$ and 5" is represented as an algebraic expression as $2(x + 5)$. We know the mathematical representation of the phrase is $2(x + 5)$ rather than $2x + 5$ because the phrase said "twice the sum," which means to multiply the sum of the two numbers by 2. The English phrase that would result in $2x + 5$ might be "the sum of twice a number and 5." Do you see the difference?

**QUICK ✓** *Express each English phrase using an algebraic expression.*

**1.** The sum of 3 and 11
**2.** The product of 6 and 7
**3.** The quotient of $y$ and 4
**4.** The difference of 3 and $z$
**5.** Twice the difference of $x$ and 3
**6.** Five plus the ratio of $z$ and 2

## 2 Evaluate Algebraic Expressions

To **evaluate an algebraic expression,** substitute or replace each variable with its numerical value.

### EXAMPLE 2 Evaluating an Algebraic Expression

Evaluate each expression for the given value of the variable.

**(a)** $4x + 3$ for $x = 5$ **(b)** $\dfrac{-z^2 + 4z}{z + 1}$ for $z = 9$ **(c)** $|10y - 8|$ for $y = \dfrac{1}{2}$

**Solution**

**(a)** We substitute 5 for $x$ in the expression $4x + 3$.

$$4(5) + 3 = 20 + 3 = 23$$

**(b)** We substitute 9 for $z$ in the expression $\dfrac{-z^2 + 4z}{z + 1}$.

$$\frac{-(9)^2 + 4 \cdot 9}{9 + 1} = \frac{-81 + 36}{10} = \frac{-45}{10} = \frac{-9 \cdot \cancel{5}}{\cancel{5} \cdot 2} = \frac{-9}{2} = -\frac{9}{2}$$

**(c)** We substitute $\frac{1}{2}$ for $y$ in the expression $|10y - 8|$.

$$\left|10 \cdot \frac{1}{2} - 8\right| = |5 - 8| = |-3| = 3$$

**QUICK ✓** *Evaluate each expression for the given value of the variable.*

**7.** $-5x + 3$ for $x = 2$

**8.** $y^2 - 6y + 1$ for $y = -4$

**9.** $\dfrac{w + 8}{3w}$ for $w = 4$

**10.** $|4x - 5|$ for $x = \dfrac{1}{2}$

### EXAMPLE 3 Evaluating an Algebraic Expression

The algebraic expression $2x + 2(x + 10)$ represents the perimeter of a rectangular field whose width is 10 yards more than its length. See Figure 14. Evaluate the algebraic expression for $x = 4$, 8, and 10.

Figure 14

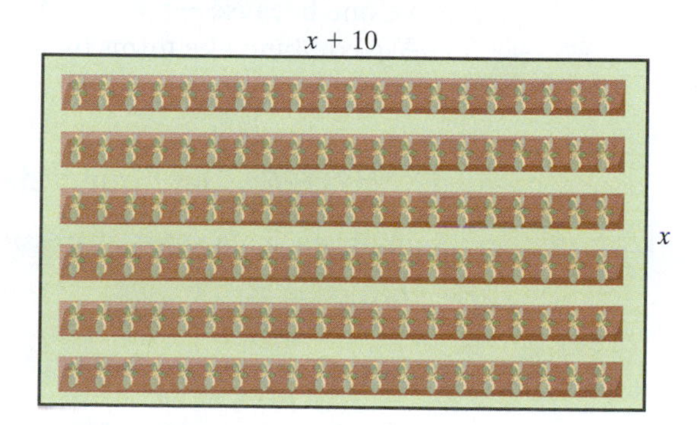


**Solution**

We evaluate the algebraic expression $2x + 2(x + 10)$ for each value of $x$.

$x = 4$: $2(4) + 2(4 + 10) = 8 + 2(14) = 8 + 28 = 36$ yards

$x = 8$: $2(8) + 2(8 + 10) = 16 + 2(18) = 16 + 36 = 52$ yards

$x = 10$: $2(10) + 2(10 + 10) = 20 + 2(20) = 20 + 40 = 60$ yards

**QUICK ✓**

**11.** The algebraic expression $118x$ represents the number of Japanese Yen that you could purchase for $x$ dollars. Evaluate the algebraic expression for $x = 100, 1000,$ and 10,000 dollars. (*Source:* Yahoo! Finance)

**12.** The algebraic expression $\frac{5}{9}(x - 32)$ represents the equivalent temperature in degrees Celsius for $x$ degrees Fahrenheit. Determine the equivalent temperature for $x = 32, 86$ and 212 degrees Fahrenheit.

## 3 Simplify Algebraic Expressions by Combining Like Terms

Often, we can simplify complicated algebraic expressions into simpler algebraic expressions. One technique for doing this is through the elimination of parentheses and combining *like terms*. A **term** is a number or the product of a number and one or more variables raised to a power. In algebraic expressions, terms are separated by addition signs. For example, consider Table 3, where various algebraic expressions are given and the terms of the expression identified.

**Table 3**

| Algebraic Expression | Terms |
|---|---|
| $5x + 4$ | $5x, 4$ |
| $7x^2 - 8x + 3$ | $7x^2, -8x, 3$ |
| $3x^2 + 7y^2$ | $3x^2, 7y^2$ |

Notice in the algebraic expression $7x^2 - 8x + 3$, we first rewrite it with only addition signs as in $7x^2 + (-8x) + 3$ to identify the terms.

Terms that have the same variable(s) and the same exponent(s) on the variable(s) are called **like terms.** For example $3x$ and $8x$ are like terms because both terms have the variable $x$ raised to the 1st power. Also, $-4x^3y$ and $10x^3y$ are like terms because both have the variable $x$ raised to the 3rd power along with $y$ raised to the 1st power. The number in front of the variable expression is called the **coefficient.** For example, the coefficient of $3x$ is 3; the coefficient of $-4x^3y$ is $-4$. For terms that have no coefficient, such as $xy$, the implied coefficient is one. The implied coefficient for $-z$ would be negative one because $-z = -1 \cdot z$.

We combine like terms by using the Distributive Property "in reverse."

### EXAMPLE 4 Combining Like Terms

Simplify each algebraic expression by combining like terms.

**(a)** $5x + 3x$ **(b)** $z + 7z - 5$

**Solution**

**(a)** $5x + 3x = (5 + 3)x = 8x$

**(b)** We should note that $z = 1 \cdot z$ because of the multiplicative identity. So

$$\begin{aligned} z + 7z - 5 &= 1z + 7z - 5 \\ &= (1 + 7)z - 5 \\ &= 8z - 5 \end{aligned}$$

**QUICK ✓** *Simplify each expression by combining like terms.*

**13.** $4x - 9x$

**14.** $-2x^2 + 13x^2$

**15.** $-5x - 3x + 6 - 3$

**16.** $6x - 10x - 4y + 12y$

---

Often, we must rearrange the terms in an algebraic expression using the commutative properties of real numbers.

## EXAMPLE 5 Combining Like Terms Using the Associative and Distributive Properties

Simplify each algebraic expression.

**(a)** $13y^2 + 8 - 4y^2 + 3$

**(b)** $12z + 5 + 5z - 8z - 2$

### Solution

**(a)** We rearrange the terms so that we can use the Distributive Property.

$$13y^2 + 8 - 4y^2 + 3 = 13y^2 - 4y^2 + 8 + 3$$

Use the Distributive Property "in reverse" $= (13 - 4)y^2 + 8 + 3$

Combine like terms: $= 9y^2 + 11$

**(b)** Again, we rearrange terms before using the Distributive Property.

$$12z + 5 + 5z - 8z - 2 = 12z + 5z - 8z + 5 - 2$$

Use the Distributive Property "in reverse" $= (12 + 5 - 8)z + 5 - 2$

Combine like terms: $= 9z + 3$

**QUICK ✓** *Simplify each expression by combining like terms.*

**17.** $10y - 3 + 5y + 2$

**18.** $0.5x^2 + 1.3 + 1.8x^2 - 0.4$

**19.** $4z + 6 - 8z - 3 - 2z$

---

Often, we need to first remove parentheses by using the Distributive Property before we can collect like terms. This is consistent with the Rule for Order of Operations as expressions must be simplified by multiplying before adding or subtracting.

## EXAMPLE 6 Combining Like Terms Using the Distributive Property

Simplify each algebraic expression.

**(a)** $5(x - 3) - 2x$

**(b)** $x + 4 - 2(x + 3)$

### Solution

**(a)** We first must use the Distributive Property to remove the parentheses.

$$5(x - 3) - 2x = 5x - 15 - 2x$$

Rearrange terms: $= 5x - 2x - 15$

Combine like terms: $= 3x - 15$

**(b)** Distribute $-2$ to remove the parentheses.

$$x + 4 - 2(x + 3) = x + 4 - 2x - 6$$

Rearrange terms: $= x - 2x + 4 - 6$

Combine like terms: $= -x - 2$

## EXAMPLE 7 Combining Like Terms Using the Distributive Property

Simplify each algebraic expression.

**(a)** $5(x - 3) - (7x - 4)$ **(b)** $\frac{1}{2}(4x + 3) + \frac{8x - 3}{5}$

### Solution

**(a)** When there is a subtraction sign in front of parentheses, we treat the subtraction sign as $-1$.

$$5(x - 3) - (7x - 4) = 5(x - 3) - 1 \cdot (7x - 4)$$

Use the Distributive Property to remove the parentheses: $= 5x - 15 - 7x + 4$

Rearrange terms: $= 5x - 7x - 15 + 4$

Combine like terms: $= -2x - 11$

**(b)** We use the fact that $\frac{a}{b} = \frac{1}{b} \cdot a$ and rewrite $\frac{8x - 3}{5} = \frac{1}{5}(8x - 3)$.

$$\frac{1}{2}(4x + 3) + \frac{8x - 3}{5} = \frac{1}{2}(4x + 3) + \frac{1}{5}(8x - 3)$$

Use the Distributive Property to remove parentheses: $= \frac{1}{2} \cdot 4x + \frac{1}{2} \cdot 3 + \frac{1}{5} \cdot 8x - \frac{1}{5} \cdot 3$

Simplify: $= 2x + \frac{3}{2} + \frac{8}{5}x - \frac{3}{5}$

Rearrange terms; rewrite fractions with least common denominator: $= \frac{10}{5}x + \frac{8}{5}x + \frac{15}{10} - \frac{6}{10}$

Combine like terms: $= \frac{18}{5}x + \frac{9}{10}$

**QUICK ✓** *Simplify each expression by combining like terms.*

**20.** $3(x - 2) + x$

**21.** $5(y + 3) - 10y - 4$

**22.** $3(z + 4) - 2(3z + 1)$

**23.** $-4(x - 2) - (2x + 4)$

**24.** $\frac{1}{2}(6x + 4) - \frac{15x + 5}{5}$

**25.** $\frac{5x - 1}{3} + \frac{5x + 9}{2}$

## 4 Determine the Domain of a Variable

When working with an algebraic expression, the variable may only be allowed to take on values from a certain set of numbers. For example, because division by zero is not defined, any value of the variable that causes division by zero must be excluded from the set of numbers that the variable can take on. So, in the expression $\frac{1}{x}$, the variable $x$ cannot take on the value 0, because this would cause division by 0 which is not defined.

**DEFINITION**

The set of values that a variable may assume is called the **domain of the variable.**

### EXAMPLE 8 Determining the Domain of a Variable

Determine which of the following numbers are in the domain of the variable $x$ for the expression $\frac{4}{x+3}$.

**(a)** $x = 3$ **(b)** $x = 0$ **(c)** $x = -3$

#### Solution

We need to determine whether the value of the variable causes division by 0. That is, we need to determine if the value of the variable causes $x + 3$ to equal 0. If it does, we exclude it from the domain.

**(a)** When $x = 3$, we have that $x + 3 = 3 + 3 = 6$, so 3 is in the domain of the variable.

**(b)** When $x = 0$, we have that $x + 3 = 0 + 3 = 3$, so 0 is in the domain of the variable.

**(c)** When $x = -3$, we have that $x + 3 = -3 + 3 = 0$, so $-3$ is NOT in the domain of the variable.

**QUICK ✓** *Determine which of the following numbers are in the domain of the variable.*

**(a)** $x = 2$ **(b)** $x = 0$ **(c)** $x = 4$ **(d)** $x = -3$

**26.** $\frac{2}{x-4}$ **27.** $\frac{x}{x+3}$ **28.** $\frac{x+3}{x^2+x-6}$

# R.5 Exercises

**For Extra Help:** Student Solutions Manual, CD Video, PH Math/Tutor Center, MathXL Tutorials on CD, MathXL®, MyMathLab

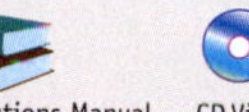

## Concepts and Vocabulary

*In Problems 1–3, fill in the blanks.*

**1.** A _____ is a letter used to represent any number from a given set of numbers.

**2.** A _____ is a letter used to represent a fixed (possibly unspecified) number.

**3.** A _____ is any algebraic expression separated by addition signs.

*In Problems 4 and 5, answer True or False to each statement.*

**4.** The terms $4x^2$ and $-10y^2$ are like terms.

**5.** There are three terms in the algebraic expression $5x^2 + 4x + 6$.

**6.** Explain the difference between a variable and a constant.

**7.** Explain the difference between a term and a factor.

**8.** In your own words, explain what "like terms" are. Explain how the Distributive Property is used to combine like terms.

## Building Skills

*In Problems 9–24, evaluate each expression for the given value of the variable.*

**9.** $4x + 3$ for $x = 2$

**10.** $-5x + 1$ for $x = 3$

**11.** $x^2 + 5x - 3$ for $x = -2$

**12.** $y^2 - 4y + 5$ for $y = 3$

**13.** $-z^2 + 4$ for $z = 5$

**14.** $-2z^2 + z + 3$ for $z = -4$

**15.** $\dfrac{2w}{w^2 + 2w + 1}$ for $w = 3$

**16.** $\dfrac{4z + 3}{z^2 - 4}$ for $z = 3$

**17.** $\dfrac{v^2 + 2v + 1}{v^2 + 3v + 2}$ for $v = 5$

**18.** $\dfrac{2x^2 + 5x + 2}{x^2 + 5x + 6}$ for $x = 3$

**19.** $|5x - 4|$ for $x = -5$

**20.** $|x^2 - 6x + 1|$ for $x = 2$

**21.** $|15y + 10|$ for $y = -\dfrac{3}{5}$

**22.** $|4z - 1|$ for $z = -\dfrac{5}{2}$

**23.** $\dfrac{(x + 2)^2}{|4x - 10|}$ for $x = 1$

**24.** $\dfrac{|3 - 5z|}{(z - 4)^2}$ for $z = 4$

*In Problems 25–54, simplify each expression by combining like terms.*

**25.** $3x - 2x$

**26.** $5y + 2y$

**27.** $-4z - 2z + 3$

**28.** $8x - 9x + 1$

**29.** $13z + 2 - 14z - 7$

**30.** $-10x + 6 + 4x - x + 1$

**31.** $\dfrac{3}{4}x + \dfrac{1}{6}x$

**32.** $\dfrac{3}{10}y + \dfrac{4}{15}y$

**33.** $2x + 3x^2 - 5x + x^2$

**34.** $-x - 3x^2 + 4x - x^2$

**35.** $-1.3x - 3.4 + 2.9x + 3.4$

**36.** $2.5y - 1.8 - 1.4y + 0.4$

**37.** $3x - 2 - x + 3 - 5x$

**38.** $10y + 3 - 2y + 6 + y$

**39.** $-2(5x - 4) - (4x + 1)$

**40.** $3(2y + 5) - 6(y + 2)$

**41.** $5(z + 2) - 6z$

**42.** $\dfrac{1}{2}(20x - 14) + \dfrac{1}{3}(6x + 9)$

**43.** $\dfrac{2}{5}(5x - 10) + \dfrac{1}{4}(8x + 4)$

**44.** $4(w + 2) + 3(4w + 3)$

**45.** $2(v - 3) + 5(2v - 1)$

**46.** $-4(w - 3) - (2w + 1)$

**47.** $\dfrac{3}{5}(10x + 4) - \dfrac{8x + 3}{2}$

**48.** $\dfrac{4}{3}(5y + 1) - \dfrac{2}{5}(3y - 4)$

**49.** $\dfrac{5}{6}\left(\dfrac{3}{10}x - \dfrac{2}{5}\right) + \dfrac{2}{3}\left(\dfrac{1}{6}x + \dfrac{1}{2}\right)$

**50.** $\dfrac{1}{4}\left(\dfrac{2}{3}x - \dfrac{1}{2}\right) + \dfrac{1}{10}\left(\dfrac{5}{2}x - \dfrac{15}{4}\right)$

**51.** $4.3(1.2x - 2.3) + 9.3x - 5.6$

**52.** $0.4(2.9x - 1.6) - 2.7(0.3x + 6.2)$

**53.** $6.2(x - 1.4) - 5.4(3.2x - 0.6)$

**54.** $9.3(0.2x - 0.8) + 3.8(1.3x + 6.3)$

*In Problems 55–66, express each English phrase using an algebraic expression.*

**55.** The sum of 5 and a number $x$

**56.** The difference of 10 and a number $y$

**57.** The product of 4 and a number $z$

**58.** The ratio of a number $x$ and 5

**59.** A number $y$ decreased by 7

**60.** A number $z$ increased by 30

**61.** Twice the sum of a number $t$ and 4

**62.** The sum of a number $x$ and 5 divided by 10

**63.** Three less than five times a number $x$

**64.** Three times a number $z$ increased by the quotient of $z$ and 8

**65.** The quotient of some number $y$ and 3 increased by the product of 6 and some number $x$

**66.** Twice some number $x$ decreased by the ratio of a number $y$ and 3

*In Problems 67–72, determine which of the following numbers are in the domain of the variable.*

**(a)** $x = 5$ **(b)** $x = -1$ **(c)** $x = -4$ **(d)** $x = 0$

**67.** $\dfrac{3}{x - 5}$

**68.** $\dfrac{7}{x + 1}$

**69.** $\dfrac{x + 1}{x + 5}$

**70.** $\dfrac{x + 4}{x - 1}$

**71.** $\dfrac{x}{x^2 - 5x}$

**72.** $\dfrac{x + 1}{x^2 + 5x + 4}$

## Applying the Concepts

△ **73. Volume of a Cube** The algebraic expression $s^3$ represents the volume of a cube whose sides are length $s$. See the figure. Evaluate the algebraic expression for $s = 1, 2, 3$, and 4 inches.

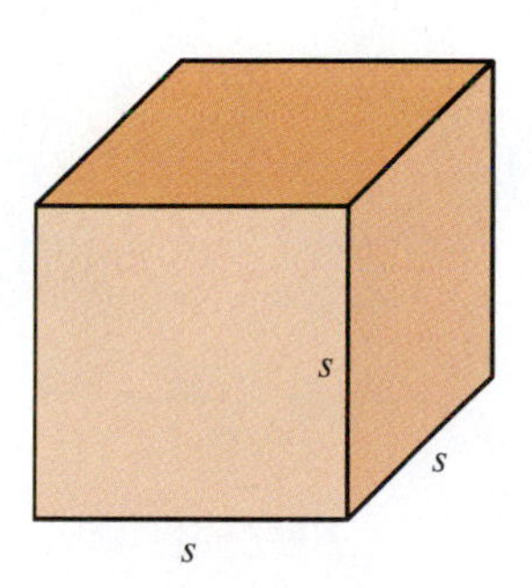


△ **74. Area of a Triangle** The algebraic expression $\frac{1}{2}(h + 2)h$ represents the area of a triangle whose base is two centimeters longer than its height $h$. See the figure. Evaluate the algebraic expression for $h = 2, 5$, and 10 centimeters.

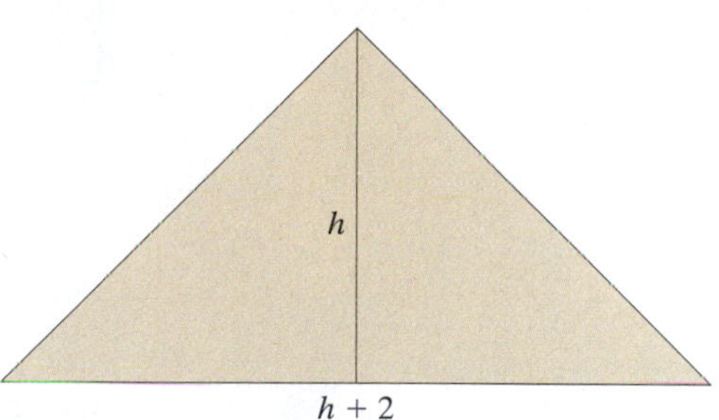


**75. Projectile Motion** The algebraic expression $-16t^2 + 75t$ represent the height (in feet) of a golf ball hit at an angle of 30° to the horizontal after $t$ seconds.

**(a)** Evaluate the algebraic expression for $t = 0, 1, 2, 3$, and 4 seconds.

**(b)** Use the results of part (a) to describe what happens to the golf ball as time passes.

**76. Cost of Production** The algebraic expression $30x + 1000$ represents the cost of manufacturing $x$ watches in a day. Evaluate the algebraic expression for $x = 20, 30$, and 40 watches.

**77. How Old Is Tony?** Suppose that Bob is $x$ years of age. Write a mathematical expression for the following: "Tony is 5 years older than Bob." Evaluate the expression if Bob is $x = 13$ years of age.

**78. Getting a Discount** Suppose the regular price of a computer is $p$ dollars. Write a mathematical expression for the following: "I'll give you \$50 off regular price." Evaluate the expression if the regular price of the computer is \$890.

**79. Getting a Big Discount** Suppose the regular price of a computer is $p$ dollars. Write a mathematical expression for the following: "I'll give you half off regular price." Evaluate the expression if the regular price of the computer is \$900.

**80. How Old Is Marissa's Mother?** Suppose that Marissa is $x$ years of age. Write a mathematical expression for the following: "Marissa's mother is twice the sum of Marissa's age and 3." How old is Marissa's mother if Marissa is $x = 18$ years of age?

*For Problems 81 and 82, use the formula* $Z = \frac{X - \mu}{\sigma}$ *to evaluate each expression for the given values.*

**81.** $X = 120, \mu = 100, \sigma = 15$

**82.** $X = 40, \mu = 50, \sigma = 10$

## Extending the Concepts

*In Problems 83–88, write an English phrase that would translate into the given mathematical expression.*

**83.** $2z - 5$

**84.** $5x + 3$

**85.** $2(z - 5)$

**86.** $5(x + 3)$

**87.** $\frac{z + 3}{2}$

**88.** $\frac{t}{3} - 2t$

## The Graphing Calculator

A graphing calculator can be used to evaluate an algebraic expression. Figure 15 shows the results of Example 2 using a TI-84 Plus graphing calculator.

**Figure 15**

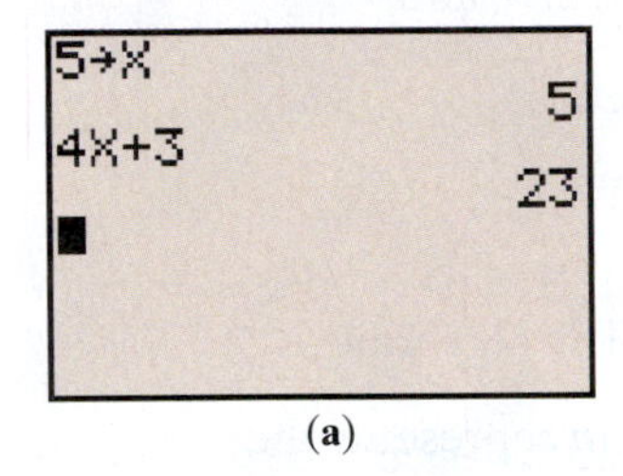

(a)

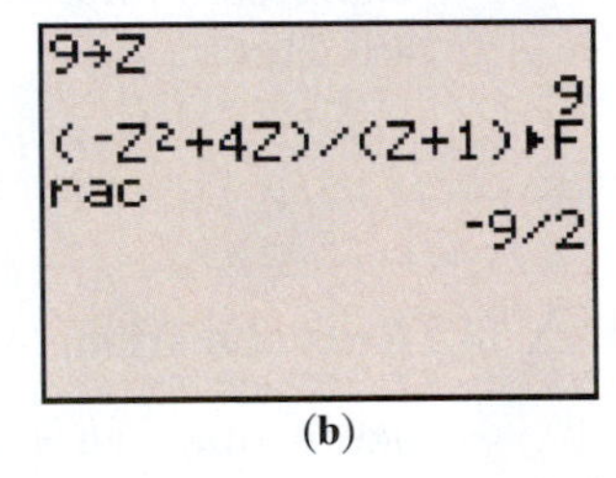

(b)

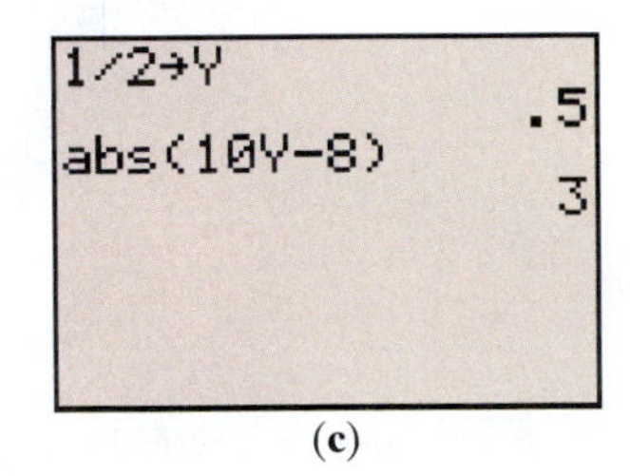

(c)

*In Problems 89–96, use a graphing calculator to evaluate each algebraic expression for the given values of the variable.*

**89.** $-4x + 3$ for (a) $x = 0$ (b) $x = -3$

**90.** $-5x + 9$ for (a) $x = 4$ (b) $x = -3$

**91.** $4x^2 - 8x + 3$ for (a) $x = 5$ (b) $x = -2$

**92.** $-9x^2 + x - 5$ for (a) $x = 3$ (b) $x = -4$

**93.** $\frac{3z - 1}{z^2 + 1}$ for (a) $z = -2$ (b) $z = 8$

**94.** $\frac{2y^2 + 5}{3y - 1}$ for (a) $y = 3$ (b) $y = -8$

**95.** $|-9x + 5|$ for (a) $x = 8$ (b) $x = -3$

**96.** $|-3x^2 + 5x - 2|$ for (a) $x = 6$ (b) $x = -4$

**97.** Use your graphing calculator to evaluate $\frac{2x + 1}{x - 5}$ when $x = 5$. What result does the calculator display? Why?

**98.** Use your graphing calculator to evaluate $\frac{x + 2}{(x + 2)(x - 2)}$ when $x = -2$. What result does your calculator display? Why?

CHAPTER

# 1 Linear Equations and Inequalities

Did you know that the amount of income tax that we pay to the federal government can be found by solving a linear equation? See Problems 81 and 82 in Section 1.1. Linear inequalities also come up in our tax code. See Example 9 in Section 1.5.

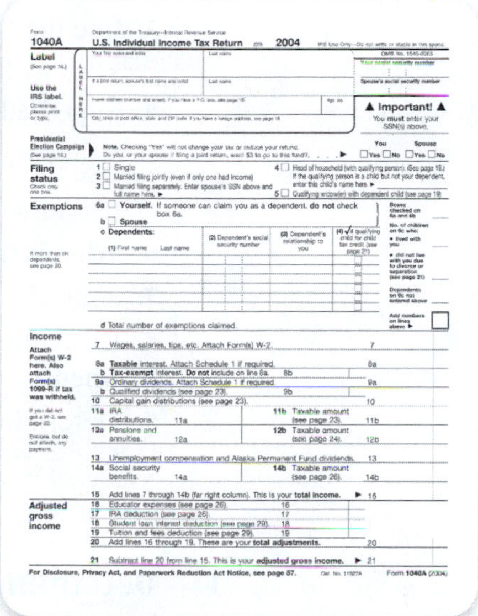

## OUTLINE



## The Big Picture: Putting It Together

In Chapter R, we reviewed skills learned in earlier courses. These skills will be used throughout the text and should always be kept fresh in your mind.

With these skills in hand, we are ready to begin our discussion of *algebra*. The word "algebra" is derived from the Arabic word *al-jabr*. This word is part of a title of a ninth-century work, "Hisâb al-jabr w'al-muqâbalah," written during the golden age of Islamic science and mathematics by Muhammad ibn Mûsâ al-Khowârizmî.

The word *al-jabr* means "restoration," a reference to the fact that, if a number is added to one side of an equation, then it must also be added to the other side in order to "restore" the equality. The title of the work "Hisâb al-jabr w'al-muqâbalah" means "the science of restoring and canceling." Today, algebra has come to mean a great deal more, but for now we are going to concentrate on the "restoration" part.

# 1.1 Linear Equations

**OBJECTIVES**

1. Determine Whether a Number Is a Solution to an Equation
2. Solve Linear Equations
3. Determine Whether an Equation Is a Conditional Equation, Identity, or Contradiction

## *Preparing for Linear Equations*

*Before getting started, take this readiness quiz. If you get a problem wrong, go back to the section cited and review the material.*

**1.** Determine the additive inverse of 5. [Section R.3, pp. 21–22]

**2.** Determine the multiplicative inverse of $-3$. [Section R.3, p. 24]

**3.** Use the Reduction Property to simplify $\frac{1}{5} \cdot 5x$. [Section R.3, p. 26]

**4.** Find the Least Common Denominator of $\frac{3}{8}$ and $\frac{5}{12}$. [Section R.3, p. 28]

**5.** Use the Distributive Property to remove the parentheses: $6(z - 2)$. [Section R.3, p. 30]

**6.** What is the coefficient of $-4x$? [Section R.5, p. 44]

**7.** Simplify by combining like terms: $4(y - 2) - y + 5$. [Section R.5, pp. 44–46]

**8.** Evaluate the expression $-5(x + 3) - 8$ when $x = -2$. [Section R.5, pp. 43–44]

**9.** Determine whether $x = 3$ is in the domain of $\frac{2}{x + 3}$. Is $x = -3$ in the domain? [Section R.5, pp. 46–47]

## 1 Determine Whether a Number Is a Solution to an Equation

An **equation in one variable** is a statement made up of two expressions that are equal, and in the statement, at least one of the expressions contains the variable. The expressions are called the **sides** of the equation. Examples of equations in one variable are

$$2y + 5 = 0 \qquad 4x + 5 = -2x + 10 \qquad \frac{3}{z + 2} = 9$$

**In Words**

In the equation $2y + 5 = 0$, the expression $2y + 5$ is the *left side* of the equation and 0 is the *right side*. In this equation only the left side contains the variable, $y$. In the equation $4x + 5 = -2x + 10$, the expression $4x + 5$ is the *left side* of the equation and $-2x + 10$ is the right side. In this equation both sides have expressions that contain the variable.

In this section, we will concentrate on solving *linear equations in one variable*.

**DEFINITION**

A **linear equation in one variable** is an equation that has one unknown and the unknown is written to the first power. Linear equations in one variable can be written in the form

$$ax + b = 0$$

where $a$ and $b$ are real numbers and $a \neq 0$.

The following are all examples of linear equations in one variable because they can be written in the form $ax + b = 0$ with a little algebraic manipulation.

$$4x - 3 = 12 \qquad \frac{2}{3}y + \frac{1}{5} = \frac{2}{15} \qquad -0.73p + 1.23 = 1.34p + 8.05$$

Because an equation is a statement, it can be either true or false, depending upon the value of the variable. Any value of the variable that results in a true statement is called a **solution** of the equation. When a value of the variable results in a true statement, we say that the value **satisfies** the equation. To determine whether a number satisfies an equation, we replace the variable with the number and determine whether the left side of the equation equals the right side of the equation—if it does, then we have a true statement and the number substituted is a solution.

---

*Preparing for...Answers* **1.** $-5$ **2.** $-\frac{1}{3}$ **3.** $x$ **4.** 24 **5.** $6z - 12$ **6.** $-4$ **7.** $3y - 3$ **8.** $-13$ **9.** Yes; No

### EXAMPLE 1 Determining Whether a Number Is a Solution to a Linear Equation

Determine if the following numbers are solutions to the equation

$$3(x - 1) = -2x + 12$$

**(a)** $x = 5$ **(b)** $x = 3$

**Solution**

**In Words**
The symbol $\stackrel{?}{=}$ is used to indicate that we are unsure whether the left side of the equation equals the right side of the equation.

**(a)** Let $x = 5$ in the equation and simplify.

$$\begin{aligned} 3(x - 1) &= -2x + 12 \\ 3(5 - 1) &\stackrel{?}{=} -2(5) + 12 \\ \text{Simplify: } 3(4) &\stackrel{?}{=} -10 + 12 \\ 12 &\neq 2 \end{aligned}$$

Because the left side of the equation does not equal the right side of the equation, we do not have a true statement. Therefore, $x = 5$ is not a solution.

**(b)** Let $x = 3$ in the equation and simplify.

$$\begin{aligned} 3(x - 1) &= -2x + 12 \\ 3(3 - 1) &\stackrel{?}{=} -2(3) + 12 \\ \text{Simplify: } 3(2) &\stackrel{?}{=} -6 + 12 \\ 6 &= 6 \quad \text{True} \end{aligned}$$

Because the left side of the equation equals the right side of the equation, we have a true statement. Therefore, $x = 3$ is a solution to the equation.

**QUICK ✓** *Determine which of the given numbers are solutions to the equation.*

1. $-5x + 3 = -2; x = -2, x = 1, x = 3$
2. $3x + 2 = 2x - 5; x = 0, x = 6, x = -7$
3. $-3(z + 2) = 4z + 1; z = -3, z = -1, z = 2$

## 2 Solve Linear Equations

**Work Smart**
The directions solve, simplify, and evaluate are different! We *solve* equations. We *simplify* algebraic expressions to form equivalent algebraic expressions. We *evaluate* algebraic expressions to find the value of the expression for a specific value of the variable.

To **solve an equation** means to find ALL the solutions of the equation. The set of all solutions to the equation is called the **solution set** of the equation.

One method for solving equations algebraically requires that a series of *equivalent equations* be developed from the original equation until a solution results.

**DEFINITION**
Two or more equations that have precisely the same solutions are called **equivalent equations.**

But how do we obtain equivalent equations? The first method we introduce for obtaining an equivalent equation is called the *Addition Property of Equality*.

**In Words**
The Addition Property says that whatever you add to one side of an equation, you must also add to the other side.

**ADDITION PROPERTY OF EQUALITY**
The **Addition Property of Equality** states that for real numbers $a$, $b$, and $c$,

$$\text{if } a = b, \quad \text{then} \quad a + c = b + c$$

For example, if $x = 3$, then $x + 2 = 3 + 2$ (we added 2 to both sides of the equation). Because $a - b$ is equivalent to $a + (-b)$, the Addition Property can be used to add a real number to each side of an equation or subtract a real number from each side of an equation. You will use this handy property a great deal in algebra.

A second method that results in an equivalent equation is called the *Multiplication Property of Equality*.

**In Words**

The Multiplication Property says that whenever you multiply one side of an equation by a nonzero expression, you must also multiply the other side by the same nonzero expression.

**MULTIPLICATION PROPERTY OF EQUALITY**

The **Multiplication Property of Equality** states that for real numbers $a$, $b$, and $c$ where $c \neq 0$,

$$\text{if } a = b, \quad \text{then} \quad ac = bc$$

For example, if $5x = 30$, then $\frac{1}{5} \cdot 5x = \frac{1}{5} \cdot 30$. Remember, the quotient $\frac{a}{b}$ is equivalent to the product $a \cdot \frac{1}{b}$, so dividing by some number $b$ is really multiplying by the multiplicative inverse of $b$, $\frac{1}{b}$. So, the Multiplication Property can be used to multiply or divide each side of the equation by some nonzero quantity.

## EXAMPLE 2 Using the Addition and Multiplication Properties to Solve a Linear Equation

Solve the linear equation $2x + 9 = 15$.

### Solution

**Work Smart**

The number in front of the variable expression is the coefficient. For example, the coefficient in the expression $2x$ is 2.

The goal in solving any linear equation is to get the variable by itself with a coefficient of 1, that is, to isolate the variable.

$$2x + 9 = 15$$

Subtract 9 from both sides: $2x + 9 - 9 = 15 - 9$

$$2x = 6$$

Divide both sides by 2: $\frac{2x}{2} = \frac{6}{2}$

$$x = 3$$

**Check** $2x + 9 = 15$

Let $x = 3$ in the original equation: $2(3) + 9 \stackrel{?}{=} 15$

$$6 + 9 \stackrel{?}{=} 15$$

$$15 = 15 \quad \text{True}$$

Because $x = 3$ satisfies the equation, the solution of the equation is 3, or the solution set is $\{3\}$.

**QUICK ✓** *Solve each equation and verify your solution.*

**4.** $3x + 8 = 17$ **5.** $-4a - 7 = 1$ **6.** $5y + 1 = 2$

Often, we must combine like terms or use the Distributive Property to eliminate parentheses before we can use the Addition or Multiplication Properties. Remember, when solving linear equations, our goal is to get all terms involving the variable on one side of the equation and all constants on the other side.

### EXAMPLE 3 Solving a Linear Equation by Combining Like Terms

Solve the linear equation $3y - 2 + 5y = 2y + 5 + 4y + 3$.

**Solution**

$$3y - 2 + 5y = 2y + 5 + 4y + 3$$

Combine like terms: $8y - 2 = 6y + 8$

Subtract $6y$ from both sides: $8y - 2 - 6y = 6y + 8 - 6y$

$$2y - 2 = 8$$

Add 2 to both sides: $2y - 2 + 2 = 8 + 2$

$$2y = 10$$

Divide both sides by 2: $\dfrac{2y}{2} = \dfrac{10}{2}$

$$y = 5$$

**Check**

$$3y - 2 + 5y = 2y + 5 + 4y + 3$$

Let $y = 5$ in the original equation: $3(5) - 2 + 5(5) \stackrel{?}{=} 2(5) + 5 + 4(5) + 3$

$$15 - 2 + 25 \stackrel{?}{=} 10 + 5 + 20 + 3$$

$38 = 38$ True

Because $y = 5$ satisfies the equation, the solution of the equation is 5, or the solution set is $\{5\}$.

**QUICK ✓** *Solve each linear equation. Be sure to verify your solution.*

**7.** $2x + 3 + 5x + 1 = 4x + 10$

**8.** $4b + 3 - b - 8 - 5b = 2b - 1 - b - 1$

**9.** $2w + 8 - 7w + 1 = 3w - 1 + 2w - 5$

### EXAMPLE 4 Solving a Linear Equation Using the Distributive Property

Solve the linear equation $4(x + 3) = x - 3(x - 2)$.

**Solution**

$$4(x + 3) = x - 3(x - 2)$$

Use the Distributive Property to remove parentheses: $4x + 12 = x - 3x + 6$

Combine like terms: $4x + 12 = -2x + 6$

Add $2x$ to both sides: $4x + 12 + 2x = -2x + 6 + 2x$

$$6x + 12 = 6$$

Subtract 12 from both sides: $6x + 12 - 12 = 6 - 12$

$$6x = -6$$

Divide both sides by 6: $\dfrac{6x}{6} = \dfrac{-6}{6}$

$$x = -1$$

**Check**

$$4(x + 3) = x - 3(x - 2)$$

Let $x = -1$ in the original equation: $4(-1 + 3) \stackrel{?}{=} -1 - 3(-1 - 2)$

$$4(2) \stackrel{?}{=} -1 - 3(-3)$$

$$8 \stackrel{?}{=} -1 + 9$$

$8 = 8$ True

Because $x = -1$ satisfies the equation, the solution of the equation is $-1$, or the solution set is $\{-1\}$.

**QUICK ✓** *Solve each linear equation. Be sure to verify your solution.*

**10.** $4(x - 1) = 12$

**11.** $-2(x - 4) - 6 = 3(x + 6) + 4$

**12.** $4(x + 3) - 8x = 3(x + 2) + x$

**13.** $5(x - 3) + 3(x + 3) = 2x - 3$

---

We now summarize the steps for solving a linear equation. Bear in mind that it is possible that one or more of these steps may not be necessary when solving a linear equation.

### SUMMARY: STEPS FOR SOLVING A LINEAR EQUATION

**Step 1:** Remove any parentheses using the Distributive Property.

**Step 2:** Combine like terms on each side of the equation.

**Step 3:** Use the Addition Property of Equality to get all variables on one side of the equation and all constants on the other side.

**Step 4:** Use the Multiplication Property of Equality to get the coefficient of the variable to equal 1.

**Step 5:** Check your answer to be sure that it satisfies the original equation.

## Linear Equations with Fractions or Decimals

Fractions in a linear equation can be removed by multiplying both sides of the equation by the Least Common Denominator (LCD) of all the fractions in the equation.

### EXAMPLE 5 How to Solve a Linear Equation That Contains Fractions

Solve the linear equation $\frac{y+1}{4} + \frac{y-2}{10} = \frac{y+7}{20}$.

#### Step-by-Step Solution

Before we follow the summary steps, we need to eliminate the fractions by multiplying both sides of the equation by the Least Common Denominator (LCD). The LCD is 20, so we multiply both sides of the equation by 20 and obtain

$$20 \cdot \left(\frac{y+1}{4} + \frac{y-2}{10}\right) = 20 \cdot \left(\frac{y+7}{20}\right)$$

Now we can follow Steps 1–5 for solving a linear equation.

**Step 1:** Remove all parentheses using the Distributive Property.

$$20 \cdot \left(\frac{y+1}{4} + \frac{y-2}{10}\right) = 20 \cdot \left(\frac{y+7}{20}\right)$$

Use the Distributive Property: $20 \cdot \frac{y+1}{4} + 20 \cdot \frac{y-2}{10} = 20 \cdot \frac{y+7}{20}$

Divide out common factors: $5(y + 1) + 2(y - 2) = y + 7$

Use the Distributive Property: $5y + 5 + 2y - 4 = y + 7$

**Step 2:** Combine like terms on each side of the equation.

$$7y + 1 = y + 7$$

**Step 3:** Use the Addition Property of Equality to get all variables on one side of the equation and all constants on the other side.

Subtract $y$ from both sides:

$$7y + 1 - y = y + 7 - y$$
$$6y + 1 = 7$$

Subtract 1 from both sides:

$$6y + 1 - 1 = 7 - 1$$
$$6y = 6$$

**Step 4:** Use the Multiplication Property of Equality to get the coefficient on the variable to equal 1.

Divide both sides by 6:

$$\frac{6y}{6} = \frac{6}{6}$$
$$y = 1$$

**Step 5: Check** Verify the solution.

$$\frac{y+1}{4} + \frac{y-2}{10} = \frac{y+7}{20}$$

Let $y = 1$ in the original equation:

$$\frac{1+1}{4} + \frac{1-2}{10} \stackrel{?}{=} \frac{1+7}{20}$$
$$\frac{2}{4} + \frac{-1}{10} \stackrel{?}{=} \frac{8}{20}$$

Rewrite each rational number with LCD = 20:

$$\frac{2}{4}\cdot\frac{5}{5} + \frac{-1}{10}\cdot\frac{2}{2} \stackrel{?}{=} \frac{8}{20}$$
$$\frac{10}{20} + \frac{-2}{20} \stackrel{?}{=} \frac{8}{20}$$
$$\frac{8}{20} = \frac{8}{20} \quad \text{True}$$

Because $y = 1$ satisfies the equation, the solution of the equation is 1, or the solution set is $\{1\}$.

**Work Smart: Study Skills**

It is a good idea to clear an equation of fractions before following Steps 1–5.

**QUICK ✓** *Solve each linear equation. Be sure to verify your solution.*

**14.** $\frac{3y}{2} + \frac{y}{6} = \frac{10}{3}$

**15.** $\frac{3x}{4} - \frac{5}{12} = \frac{5x}{6}$

**16.** $\frac{x+2}{6} + 2 = \frac{5}{3}$

**17.** $\frac{4x+3}{9} - \frac{2x+1}{2} = \frac{1}{6}$

When decimals occur in linear equations, we can eliminate the decimal using the same techniques that we used in eliminating fractions. The idea behind the procedure is to multiply both sides of the equation by a power of 10 so that the decimal is removed. For example, because $0.7 = \frac{7}{10}$, multiplying 0.7 by 10 eliminates the decimal since $10(0.7) = 10\cdot\frac{7}{10} = 7$. Because $0.03 = \frac{3}{100}$, multiplying 0.03 by 100 eliminates the decimal.

### EXAMPLE 6 Solving a Linear Equation That Contains Decimals

Solve the linear equation $0.5x - 0.4 = 0.3x + 0.2$.

**Solution**

We want to eliminate the decimal from the equation. This is done by multiplying both sides of the equation by 10. Do you see why? Each of the decimals is written to the tenths position, so multiplying by 10 will eliminate the decimal.

$$10(0.5x - 0.4) = 10(0.3x + 0.2)$$

Use the Distributive Property: $$10(0.5x) - 10(0.4) = 10(0.3x) + 10(0.2)$$

$$5x - 4 = 3x + 2$$

Subtract $3x$ from both sides: $$5x - 4 - 3x = 3x + 2 - 3x$$

$$2x - 4 = 2$$

Add 4 to both sides: $$2x - 4 + 4 = 2 + 4$$

$$2x = 6$$

Divide both sides by 2: $$\frac{2x}{2} = \frac{6}{2}$$

$$x = 3$$

**Check**

$$0.5x - 0.4 = 0.3x + 0.2$$

Let $x = 3$ in the original equation: $$0.5(3) - 0.4 \stackrel{?}{=} 0.3(3) + 0.2$$

$$1.5 - 0.4 \stackrel{?}{=} 0.9 + 0.2$$

$$1.1 = 1.1 \quad \text{True}$$

Because $x = 3$ satisfies the equation, the solution of the equation is 3, or the solution set is $\{3\}$.

**QUICK ✓** *Solve each linear equation. Be sure to verify your solution.*

**18.** $0.2t + 1.4 = 0.8$

**19.** $0.07x - 1.3 = 0.05x - 1.1$

**20.** $0.4(y + 3) = 0.5(y - 4)$

## 3 Determine Whether an Equation Is a Conditional Equation, Identity, or Contradiction

All of the linear equations that we have studied thus far have had one solution. While it is tempting to say that all linear equations must have one solution, this statement is not true in general. In fact, linear equations may have either one solution, no solution, or infinitely many solutions. We give names to the type of equation depending upon the number of solutions that the linear equation has.

The equations that we have solved thus far are called *conditional equations.*

**DEFINITION**

A **conditional equation** is an equation that is true for some values of the variable and false for other values of the variable.

For example, the equation

$$x + 7 = 10$$

is a conditional equation because it is true when $x = 3$ and false for every other real number $x$.

**DEFINITION**

An equation that is false for every value of the variable is called a **contradiction.**

For example, the equation

$$3x + 8 = 3x + 6$$

is a contradiction because it is false for any value of $x$. Contradictions are identified through the process of creating equivalent equations. For example, if we subtract $3x$ from both sides of $3x + 8 = 3x + 6$, we obtain $8 = 6$, which is clearly false. Contradictions have no solution and therefore the solution set is empty. We express the solution set of contradictions as either

$$\emptyset \text{ or } \{\ \}.$$

**DEFINITION**

An equation that is satisfied for every choice of the variable for which both sides of the equation are defined is called an **identity.**

**In Words**

Conditional equations are true for some values of the variable and false for others. Contradictions are false for all values of the variable. Identities are true for all allowed values of the variable.

For example,

$$2x + 3 + x + 8 = 3x + 11$$

is an example of an identity because any real number $x$ satisfies the equation. Just as with contradictions, identities are recognized through the process of creating equivalent equations. For example, if we combine like terms in the equation $2x + 3 + x + 8 = 3x + 11$ we obtain $3x + 11 = 3x + 11$, which is true no matter what value of $x$ we choose. Therefore, the solution set of linear identities is the set of all real numbers. We express the solution of linear identities as either

$$\{x \mid x \text{ is a real number}\} \quad \text{or} \quad \mathbb{R}$$

## EXAMPLE 7 Categorizing a Linear Equation

Solve the linear equation

$$3(x + 3) - 6x = 5(x + 1) - 8x$$

State whether the equation is an identity, contradiction or conditional equation.

### Solution

As with any linear equation, our goal is to get the variable by itself with a coefficient of 1.

$$3(x + 3) - 6x = 5(x + 1) - 8x$$

Use the Distributive Property: $3x + 9 - 6x = 5x + 5 - 8x$

Combine like terms: $-3x + 9 = -3x + 5$

Add $3x$ to both sides: $-3x + 9 + 3x = -3x + 5 + 3x$

$$9 = 5$$

**Work Smart**

In the solution to Example 7 we obtained the equation

$$-3x + 9 = -3x + 5$$

You may recognize at this point that the equation is a contradiction and state the solution set as $\emptyset$ or $\{\ \}$.

The last statement states that $9 = 5$. This is a false statement, so the equation is a contradiction. The original equation is a contradiction and has no solution. The solution set is $\emptyset$ or $\{\ \}$.

### EXAMPLE 8 Categorizing a Linear Equation

Solve the linear equation

$$-4(x - 2) + 3(4x + 2) = 2(4x + 7)$$

State whether the equation is an identity, contradiction, or conditional equation.

**Solution**

$$-4(x - 2) + 3(4x + 2) = 2(4x + 7)$$

Use the Distributive Property: $-4x + 8 + 12x + 6 = 8x + 14$

Combine like terms: $8x + 14 = 8x + 14$

At this point it should be clear that the equation $8x + 14 = 8x + 14$ is true for all real numbers $x$. So, the original equation is an identity and its solution set is all real numbers or

$$\{x \mid x \text{ is any real number}\} \quad \text{or} \quad \mathbb{R}$$

Had we continued to solve the equation in Example 8 by subtracting $8x$ from both sides of the equation, we would obtain

$$8x + 14 - 8x = 8x + 14 - 8x$$
$$14 = 14$$

The statement $14 = 14$ is true for all real numbers $x$, so the solution set of the original equation is all real numbers.

**QUICK ✓** *Solve the equation and state whether it is an identity, contradiction, or conditional equation.*

**21.** $4(x + 2) = 4x + 2$

**22.** $3(x - 2) = 2x - 6 + x$

**23.** $-4x + 2 + x + 1 = -4(x + 2) + 11$

**24.** $-3(z + 1) + 2(z - 3) = z + 6 - 2z - 15$

# 1.1 Exercises

*For Extra Help:*

## Concepts and Vocabulary

*In Problems 1–3, fill in the blanks.*

**1.** Two or more equations that have precisely the same solutions are called __________ __________.

**2.** The values of the variable that result in a true statement are called __________, or __________ the equation.

**3.** A __________ __________ is an equation that is true for some values of the variable and false for other values of the variable.

*In Problems 4–6, answer True or False to each statement.*

**4.** A linear equation can have one solution, no solution, or infinitely many solutions.

**5.** An equation that is satisfied for every choice of the variable for which both sides of the equation are defined is called a contradiction.

**6.** The Multiplication Property of Equality states that for real numbers $a$, $b$, and $c$, where $c \neq 0$, if $a = b$, then $ac = bc$.

**7.** Explain the difference between $4(x + 1) - 2$ and $4(x + 1) = 2$. In general, what is the difference between an algebraic expression and an equation?

**8.** Explain the difference between the directions "solve" and "simplify."

## Building Skills

*In Problems 9–14, determine which of the numbers are solutions to the given equation.*

**9.** $8x - 10 = 6; x = -2, x = 1, x = 2$

**10.** $-4x - 3 = -15; x = -2, x = 1, x = 3$

**11.** $5m - 3 = -3m + 5; m = -2, m = 1, m = 3$

**12.** $6x + 1 = -2x + 9; x = -2, x = 1, x = 4$

**13.** $4(x - 1) = 3x + 1; x = -1, x = 2, x = 5$

**14.** $3(t + 1) - t = 4t + 9; t = -3, t = -1, t = 2$

*In Problems 15–52, solve each linear equation. Identify each equation as being an identity, contradiction, or conditional equation. Be sure to verify your solution.*

**15.** $3x + 1 = 7$

**16.** $8x - 6 = 18$

**17.** $5x + 4 = 14$

**18.** $-6x - 5 = 13$

**19.** $4z + 3 = 2$

**20.** $8y + 3 = 5$

**21.** $-3w + 2w + 5 = -4$

**22.** $-7t - 3 + 5t = 11$

**23.** $3m + 4 = 2m - 5$

**24.** $-5z + 3 = -3z + 1$

**25.** $5x + 2 - 2x + 3 = 7x + 2 - x + 5$

**26.** $-6x + 2 + 2x + 9 + x = 5x + 10 - 6x + 11$

**27.** $3(x + 2) = -6$

**28.** $4(z - 2) = 12$

**29.** $4(x + 1) = 4x$

**30.** $5(s + 3) = 3s + 2s$

**31.** $4m + 1 - 6m = 2(m + 3) - 4m$

**32.** $10(x - 1) - 4x = 2x - 1 + 4(x + 1)$

**33.** $2(y + 1) - 3(y - 2) = 5y + 8 - 6y$

**34.** $8(w + 2) - 3w = 7(w + 2) + 2(1 - w)$

**35.** $\dfrac{4y}{5} - \dfrac{14}{15} = \dfrac{y}{3}$

**36.** $\dfrac{3x}{2} + \dfrac{x}{6} = -\dfrac{5}{3}$

**37.** $\dfrac{x}{4} + \dfrac{3x}{10} = -\dfrac{33}{20}$

**38.** $\dfrac{z - 2}{4} + \dfrac{2z - 3}{6} = 7$

**39.** $3p - \dfrac{p}{4} = \dfrac{11p}{4} + 1$

**40.** $\dfrac{r}{2} + 2(r - 1) = \dfrac{5r}{2} + 4$

**41.** $\dfrac{4x + 3}{9} - \dfrac{2x + 1}{2} = \dfrac{1}{6}$

**42.** $\dfrac{2x + 1}{3} - \dfrac{6x - 1}{4} = -\dfrac{5}{12}$

**43.** $\dfrac{2x + 1}{2} - \dfrac{x + 1}{5} = \dfrac{23}{10}$

**44.** $\dfrac{3x + 1}{4} - \dfrac{7x - 4}{2} = \dfrac{26}{3}$

**45.** $0.5x - 3.2 = -1.7$

**46.** $0.3z + 0.8 = -0.1$

**47.** $0.14x + 2.23 = 0.09x + 1.98$

**48.** $0.12y - 5.26 = 0.05y + 1.25$

**49.** $0.4(z + 1) - 0.7z = -0.1z + 0.7 - 0.2z - 0.3$

**50.** $0.9(z - 3) - 0.2(z - 5) = 0.4(z + 1) + 0.3z - 2.1$

**51.** $\dfrac{1}{3}(2x - 3) + 2 = \dfrac{5}{6}(x + 3) - \dfrac{11}{12}$

**52.** $\dfrac{4}{5}(y - 4) + 3 = \dfrac{2}{3}(y + 1) + \dfrac{4}{15}$

## Mixed Practice

*In Problems 53–68, solve each linear equation and verify the solution.*

**53.** $7y - 8 = -7$

**54.** $4y + 5 = 7$

**55.** $4a + 3 - 2a + 4 = 5a - 7 + a$

**56.** $-5x + 5 + 3x + 7 = 5x - 6 + x + 12$

**57.** $4(p + 3) = 3(p - 2) + p + 18$

**58.** $7(x + 2) = 5(x - 2) + 2(x + 12)$

**59.** $4b - 3(b + 1) - b = 5(b - 1) - 5b$

**60.** $13z - 8(z + 1) = 2(z - 3) + 3z$

**61.** $\frac{m + 1}{4} + \frac{5}{6} = \frac{2m - 1}{12}$

**62.** $\frac{z - 4}{6} - \frac{2z + 1}{9} = \frac{1}{3}$

**63.** $0.3x - 1.3 = 0.5x - 0.7$

**64.** $-0.8y + 0.3 = 0.2y - 3.7$

**65.** $-0.8(x + 1) = 0.2(x + 4)$

**66.** $0.5(x + 3) = 0.2(x - 6)$

**67.** $\frac{1}{4}(x - 4) + 3 = \frac{1}{3}(2x + 6) - \frac{5}{6}$

**68.** $\frac{1}{5}(2a - 5) - 4 = \frac{1}{2}(a + 4) - \frac{7}{10}$

## Applying the Concepts

**69.** Find $a$ such that the solution set of $ax + 3 = 15$ is $\{-3\}$.

**70.** Find $a$ such that the solution set of $ax + 6 = 20$ is $\{7\}$.

**71.** Find $a$ such that the solution set of $a(x - 1) = 3(x - 1)$ is the set of all real numbers.

**72.** Find $a$ such that the solution set of $ax + 3 = 2x + 3(x + 1)$ is the set of all real numbers.

*In Section R.5 we introduced the domain of a variable. The domain of a variable is the set of all values that a variable may assume. Recall that division by zero is not defined, so any value of the variable that results in division by zero must be excluded from the domain. In Problems 73–78, determine which values of the variable must be excluded from the domain.*

**73.** $\frac{5}{2x + 1}$

**74.** $\frac{-3}{5x + 8}$

**75.** $\frac{3x + 7}{4x - 3}$

**76.** $\frac{2x}{3x + 1}$

**77.** $\frac{6x - 2}{3(x + 1) - 6}$

**78.** $\frac{-2x + 7}{4(x - 3) + 2}$

**79. Interest** Suppose you have a credit card debt of \$2000. Last month, the bank charged you \$25 interest on the debt. The solution to the equation $25 = \frac{2000}{12} \cdot r$ represents the annual interest rate on the credit card, $r$. Find the annual interest rate on the credit card.

**80. How Much Do I Make?** Last week, before taxes, you earned \$539 after working 26 hours at your regular hourly rate and 6 hours at time-and-a-half. The solution to the equation $26x + 9x = 539$ represents your regular hourly rate, $x$. Determine your regular hourly rate.

**81. Paying Your Taxes** You are single and just determined that you paid \$2335 in federal income taxes in 2005. The solution to the equation $2335 = 0.15(x - 7300) + 730$ represents the amount $x$ that you earned in 2005. Determine how much you earned in 2005. (*Source:* Internal Revenue Service)

**82. Paying Your Taxes** You are married and just determined that you paid \$6020 in federal income taxes in 2005. The solution to the equation $6020 = 0.15(x - 14{,}600) + 1460$ represents the amount $x$ that you and your spouse earned in 2005. Determine how much you and your spouse earned in 2005. (*Source:* Internal Revenue Service)

### Extending the Concepts

**83.** Make up a linear equation that has one solution. Make up a linear equation that has no solution. Make up a linear equation that is an identity. Comment on the differences and similarities in making up each equation.

# 1.2 An Introduction to Problem Solving

**OBJECTIVES**

1. Translate English Sentences into Mathematical Statements
2. Model and Solve Direct Translation Problems
3. Model and Solve Mixture Problems
4. Model and Solve Uniform Motion Problems

### *Preparing for Problem Solving*

*Before getting started, take the following readiness quiz. If you get a problem wrong, go back to the section cited and review the material.*

*In Problems 1–6, express each English phrase as an algebraic expression.* *[Section R.5, p. 42].*

**1.** The sum of 12 and a number $z$.

**2.** The product of 4 and $x$.

**3.** A number $y$ decreased by 87.

**4.** The quotient of $z$ and 12.

**5.** Four times the sum of $x$ and 7.

**6.** The sum of four times a number $x$ and 7.

## 1 Translate English Sentences into Mathematical Statements

**Work Smart**

We learned in the last section that an equation is a statement in which two algebraic expressions, separated by an equal sign, are equal.

In English, a complete sentence must contain a subject and a verb, so expressions or "phrases" are not complete sentences. For example, "Beats me!" is an expression, not a complete sentence, because it does not contain a subject. The expression "5 more than a number $x$" does not contain a verb and therefore is not a complete sentence either. The statement "5 more than a number $x$ is 18" is a complete sentence because it contains a subject and a verb. Because this is a complete sentence, we can translate it into a mathematical statement. In mathematics, statements can be represented symbolically through equations. In English, statements can be true or false. For example, "The moon is made of green cheese" is a false statement, while "Humans are mammals" is a true statement. Mathematical statements can be true or false as well—we called them conditional equations.

You may want to look back at page 42 in Section R.5 for a review of key words that translate into mathematical symbols. Table 1 provides a summary of words that typically translate into an equal sign.

**Table 1 Words That Translate into an Equal Sign**

| | | | |
|---|---|---|---|
| is | yields | are | equals |
| was | gives | results in | is equal to |
| is equivalent to | | | |

Let's look at some examples where we translate English sentences into mathematical statements.

***Preparing for...Answers*** **1.** $12 + z$ **2.** $4x$ **3.** $y - 87$ **4.** $\frac{z}{12}$ **5.** $4(x + 7)$ **6.** $4x + 7$

### EXAMPLE 1 Translating English Sentences into Mathematical Statements

Translate each of the following sentences into a mathematical statement. Do not solve the equation.

**(a)** Five more than a number $x$ is 20.
**(b)** Four times the sum of a number $z$ and 3 is 15.
**(c)** The difference of $x$ and 5 equals the quotient of $x$ and 2.

**Solution**

**(a)** $\underbrace{\text{5 more than a number } x}_{x + 5}\ \underbrace{\text{is}}_{=}\ \underbrace{20}_{20}$

$$x + 5 = 20$$

**Work Smart**

The English statement "The sum of four times a number $z$ and 3 is 15" would be expressed mathematically as $4z + 3 = 15$. Do you see the difference between this statement and the one in Example 1(b)?

**(b)** The expression that reads "Four times the sum of" tells us we need to find the sum first and then multiply this result by 4.

Four times the sum of a number $z$ and 3 | is | 15

$$4(z + 3) = 15$$

**(c)** The difference of $x$ and 5 | equals | the quotient of $x$ and 2

$$x - 5 = \frac{x}{2}$$

**QUICK ✓** *Translate each English statement into a mathematical statement. Do not solve the equation.*

1. The product of 3 and $y$ is equal to 21.
2. Two times the sum of 3 and $x$ is equivalent to the product of 5 and $x$.
3. The difference of $x$ and 10 equals the quotient of $x$ and 2.
4. Three less than a number $y$ is five times $y$.

## An Introduction to Problem Solving and Mathematical Models

Every day we encounter various types of problems that must be solved. **Problem solving** is the ability to use information, tools, and our own skills to achieve a goal. For example, suppose Kevin wants a glass of water, but he is too short to reach the sink. Kevin has a problem. To solve the problem, he finds a step stool and pulls it over to the sink. He uses the step stool to climb on the counter, opens the kitchen cabinet and pulls out a cup. He then crawls along the counter top, turns on the faucet, and proceeds to fill his cup with water. Problem solved!

Of course, this is not the only way that Kevin could solve the problem. Can you think of any other solutions? Just as there are various approaches to solving life's everyday problems, there are many ways to solve problems using mathematics. However, regardless of the approach, there are always some common aspects in solving any problem. For example, regardless of how Kevin ultimately ends up with his cup of water, someone must get a cup from the cabinet and someone must turn on the faucet.

One of the purposes of learning algebra is to be able to solve certain types of problems. To solve these problems, we will need techniques that can help us translate the verbal description of the problem into an equation that can be solved. The process of taking a verbal description of a problem and developing a mathematical equation that can be used to solve the problem is **mathematical modeling.**

Mathematical modeling begins with a problem. The problem is then summarized as a verbal description. The verbal description is then translated into the language of mathematics. This translation results in an equation that can be solved (the mathematical problem). The solution must be checked against the mathematical problem (the equation) and the verbal description. This entire process is called the **modeling process.** We call the equation that is developed the **mathematical model.** See Figure 1.

**Figure 1**

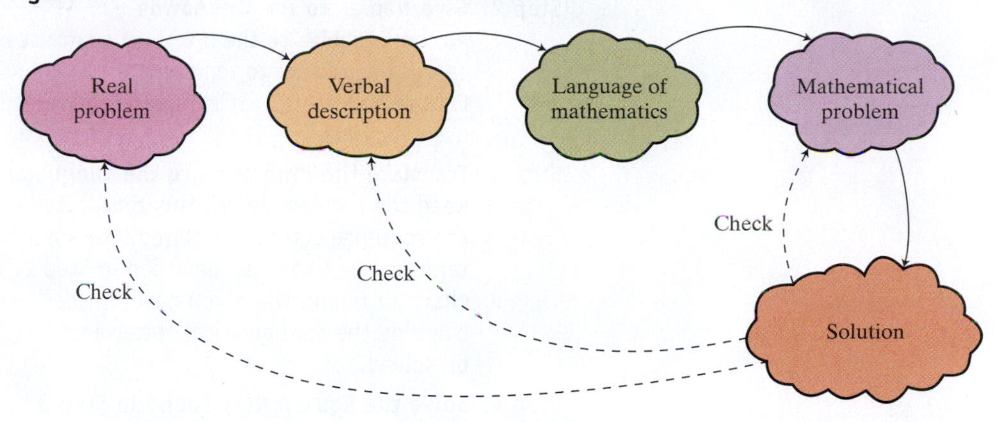


Not all models are mathematical. In general, a **model** is a way of representing reality through graphs, pictures, small-scale reproductions, equations, or even verbal descriptions. Because the world is an extremely complex place, we often need to simplify information when we develop a model. For example, a map is a model of our road system. Maps don't show all the details of the system such as trees, buildings, or potholes, but they do a good job of describing how to get from point $A$ to point $B$. Mathematical models are similar in that we often make assumptions regarding our world in order to make the mathematics more manageable. The fewer assumptions that are made, the more complicated the mathematics becomes in the model.

It is difficult to give a step-by-step approach for solving problems because each problem is unique in some way. However, because there are common links to many types of problems, we can categorize problems. In this text, we will present five categories of problems.

**Five Categories of Problems**

1. **Direct Translation**—problems where we must translate from English into Mathematics by using key words in the verbal description
2. **Mixture**—problems where two or more quantities are combined in some fashion
3. **Geometry**—problems where the unknown quantities are related through geometric formulas
4. **Uniform Motion**—problems where an object travels at a constant speed
5. **Work Problems**—problems where two or more entities join forces to complete a job

We will present strategies for solving these categories of problems throughout the text. In this section, we will concentrate on direct translation problems, mixture problems, and uniform motion problems.

Regardless of the type of problem we must solve, there are certain steps that can be used to assist in solving the problem. Below we provide you with a series of steps that should be followed when developing any mathematical model. As we proceed through this course, and in future courses, you will use the techniques that you have studied in this course to solve more complicated problems, but the approach remains the same.

## Steps for Solving Problems with Mathematical Models

**Step 1: Identify What You Are Looking For**

Read the problem very carefully, perhaps two or three times. Identify the type of problem and the information that we wish to learn from the problem. It is fairly typical that the last sentence in the problem indicates what it is we wish to solve for.

**Step 2: Give Names to the Unknowns**
Assign variables to the unknown quantities in the problem. Remember you can use any letter to represent the unknown(s) when you make your model. Choose a variable that is representative of the unknown quantity. For example, use $t$ for time.

**Step 3: Translate the Problem into the Language of Mathematics**
Read the problem again. This time, after each sentence is read, determine if the sentence can be translated into a mathematical statement or expression in terms of the variables identified in Step 2. It is often helpful to create a table, chart, or figure. When you have finished reading the problem, if necessary, combine the mathematical statements or expressions into an equation that can be solved.

**Step 4: Solve the Equation(s) Found in Step 3**
Solve the equation for the variable and then answer the question posed by the original problem.

**Step 5: Check the Reasonableness of Your Answer**
Check your answer to be sure that it makes sense. If it does not, go back and try again.

**Step 6: Answer the Question**
Write your answer in a complete sentence.

**Work Smart**

When you solve a problem by making a mathematical model, check your work to make sure you have the right answer. Typically, errors can happen in two ways.

- One type of error occurs if you correctly translate the problem into a model but then make an error solving the equation. This type of error is usually easy to find.
- However, if you misinterpret the problem and develop an incorrect model, then the solution you obtain may still satisfy your model, but it probably will not be the correct solution to the original problem. We can check for this type of error by determining whether the solution is reasonable. Does your answer make sense? Always be sure that you are answering the question that is being asked.

## 2 Model and Solve Direct Translation Problems

Let's look at solving **direct translation** problems, which are problems that can be set up by reading the problem and using everyday language to translate the verbal description into a mathematical equation.

### EXAMPLE 2 Consecutive Integers

The sum of three consecutive odd integers results in 45. Find the integers.

**Solution**

**Step 1: Identify** This is a direct translation problem. We are looking for three odd integers. The odd integers are 1, 3, 5, and so on.

**Step 2: Name** If we let $x$ represent the first odd integer, then $x + 2$ is the next odd integer, and $x + 4$ is the third odd integer.

**Step 3: Translate** Since we know that their sum is 45, we have

$$\underbrace{x}_{\text{First Integer}} + \underbrace{x + 2}_{\text{Second Integer}} + \underbrace{x + 4}_{\text{Third Integer}} = 45 \quad \text{The Model}$$

**Step 4: Solve** We now proceed to solve the equation.

$$x + x + 2 + x + 4 = 45$$

Combine like terms: $3x + 6 = 45$

Subtract 6 from both sides: $3x = 39$

Divide both sides by 3: $x = 13$

**Step 5: Check** Since $x$ represents the first odd integer, the remaining two odd integers are $13 + 2 = 15$ and $13 + 4 = 17$. It is always a good idea to make sure your answer is reasonable. Since $13 + 15 + 17 = 45$, we know we have the right answer!

**Step 6: Answer the Question** The three consecutive odd integers whose sum is 45 are 13, 15, and 17.

**QUICK ✓** *Translate each English statement into an equation and solve the equation.*

**5.** The sum of three consecutive even integers is 60. Find the integers.

**6.** The sum of three consecutive integers is 78. Find the integers.

## EXAMPLE 3 How Much Do I Make per Hour?

Before taxes, Marissa earned \$725 one week after working 52 hours. Her employer pays time-and-a-half for all hours worked in excess of 40 hours. What is Marissa's hourly wage?

### Solution

**Step 1: Identify** This is a direct translation problem. We are looking for Marissa's hourly wage.

**Step 2: Name** Let $w$ represent Marissa's hourly wage.

**Step 3: Translate** We know that Marissa earned \$725 by working 40 hours at her regular wage and 12 hours earning 1.5 times her regular wage. For each hour that Marissa works for the first 40 hours, she earns $w$ dollars and for each hour she works for the next 12 hours, she earns $1.5w$ dollars. Therefore, her total salary is

$$\overbrace{40w}^{\text{Regular Earnings}} + \overbrace{12(1.5w)}^{\text{Overtime Earnings}} = \overbrace{725}^{\text{Total Earnings}} \quad \text{The Model}$$

**Step 4: Solve**

$$40w + 12(1.5w) = 725$$

Simplify: $40w + 18w = 725$

Combine like terms: $58w = 725$

Divide both sides by 58: $\dfrac{58w}{58} = \dfrac{725}{58}$

$$w = 12.50$$

**Step 5: Check** We believe that Marissa's hourly wage is \$12.50. So, for the first 40 hours she earned $\$12.50(40) = \$500$ and for the next 12 hours she earned $1.5(\$12.50)(12) = \$225$. Her total salary was $\$500 + \$225 = \$725$. This checks with the information presented in the problem.

**Step 6: Answer the Question** Marissa's hourly wage is \$12.50.

**QUICK ✓**

**7.** Before taxes, Melody earned \$735 one week after working 46 hours. Her employer pays time-and-a-half for all hours worked in excess of 40 hours. What is Melody's hourly wage?

**8.** Before taxes, Jim earned \$564 one week after working 30 hours at his regular wage, 6 hours at time-and-a-half on Saturday, and 4 hours at double-time on Sunday. What is Jim's hourly wage?

---

## EXAMPLE 4 Choosing a Long-Distance Carrier

MCI has a long-distance phone plan that charges \$2.00 a month plus \$0.09 per minute of usage. Sprint has a long-distance phone plan that charges \$3.50 a month plus \$0.07 per minute of usage. For how many minutes of long-distance calls will the costs for the two plans be the same? (*Source:* MCI and Sprint)

### Solution

**Step 1: Identify** This is a direct translation problem. We are looking for the number of minutes for which the two plans cost the same.

**Step 2: Name** Let $m$ represent the number of long-distance minutes used in the month.

**Step 3: Translate** The monthly fee for MCI is \$2.00 plus \$0.09 for each minute used. So, if one minute is used, the fee is $2.00 + 0.09(1) = 2.09$ dollars. If two minutes are used, the fee is $2.00 + 0.09(2) = 2.18$ dollars. In general, if $m$ minutes are used, the monthly fee is $2.00 + 0.09m$ dollars. Similar logic results in the monthly fee for Sprint being $3.50 + 0.07m$ dollars. We want to know the number of minutes for which the cost for the two plans will be the same, which means we need to solve

$$\text{Cost for MCI} = \text{Cost for Sprint}$$

$$2.00 + 0.09m = 3.50 + 0.07m \quad \text{The Model}$$

**Step 4: Solve**

$$2.00 + 0.09m = 3.50 + 0.07m$$

Subtract 2.00 from both sides: $0.09m = 1.50 + 0.07m$

Subtract $0.07m$ from both sides: $0.02m = 1.50$

Divide both sides by 0.02: $m = 75$

**Step 5: Check** We believe the cost of the two plans will be the same if 75 minutes are used. The cost of MCI's plan will be $2.00 + 0.09(75) = \$8.75$. The cost of Sprint's plan will be $3.50 + 0.07(75) = \$8.75$. They are the same!

**Step 6: Answer the Question** The cost of the two plans will be the same if 75 minutes are used.

**QUICK ✓**

**9.** You need to rent a moving truck. You have identified two companies that rent trucks. EZ-Rental charges \$35 per day plus \$0.15 per mile. Do It Yourself Rental charges \$20 per day plus \$0.25 per mile. For how many miles will the cost of renting be the same?

**10.** You need a new cell phone for emergencies only. Company A charges \$12 per month plus \$0.10 per minute, while Company B charges \$0.15 per minute with no monthly service charge. For how many minutes will the monthly cost be the same?

There are many types of direct translation problems. One specific type of direct translation problem is a "percent problem." Typically, these problems involve discounts or markups that businesses use in determining their prices.

Percent means divided by 100 or per hundred. We use the symbol % to denote percent, so 45% means 45 out of 100 or $\frac{45}{100}$ or 0.45. In applications involving percents, we often encounter the word "of," as in 20% of 100. Remember that the word "of" translates into multiplication in mathematics so 20% of 100 means

$$20\% \cdot 100 \quad \text{or} \quad 0.20 \cdot 100$$

So, 20% of 100 is 20.

When dealing with percents and the price of goods, it is helpful to remember the following:

$$\text{Original Price} - \text{Discount} = \text{Sale Price}$$
$$\text{Net Price} + \text{Markup} = \text{Gross Price}$$

## EXAMPLE 5 Discounted Price

Suppose that you have just entered your favorite clothing store and find that everything in the store is marked at a discount of 40% off. If the sale price of a suit is $144, what was the original price?

### Solution

**Step 1: Identify** This is a direct translation problem involving percents. We wish to find the original price of the suit.

**Step 2: Name** Let $p$ represent the original price.

**Step 3: Translate** We know that the original price minus the discount will give us the sale price. We also know that the sale price was $144, so

$$p - \text{discount} = 144$$

The discount was 40% off of the original price so that the discount is $0.40p$. Substituting into the equation $p - \text{discount} = 144$, we obtain

$$p - 0.40p = 144 \quad \text{The Model}$$

**Work Smart**

You could eliminate the decimals by multiplying both sides of the equation by 10.

**Step 4: Solve** We now solve for $p$, the original price.

$$p - 0.40p = 144$$

Combine like terms: $0.60p = 144$

Divide both sides by 0.60: $\frac{0.60p}{0.60} = \frac{144}{0.60}$

Cancel and simplify: $p = 240$

**Step 5: Check** If the original price of the suit was $240, then the discount would be $0.4(\$240) = \$96$. Subtracting $96 from the original price results in a sale price of $144. This agrees with the information in the problem.

**Step 6: Answer the Question** The original price of the suit was $240.

### QUICK ✓

**11.** What is 40% of 100?

**12.** 8 is 5% of what?

**13.** 15 is what percent of 20?

**14.** Suppose that you have just entered your favorite clothing store and a find that everything in the store is marked at a discount of 30% off. If the sale price of a shirt is $21, what was the original price?

(continued)

**15.** A Milex Tune-Up automotive facility marks up its parts 35%. Suppose that Milex's charges its customers \$1.62 for each spark plug it installs. What is Milex's cost for each spark plug?

---

**Interest** is money paid for the use of money. The total amount borrowed is called the **principal.** The principal can be in the form of a loan (an individual borrows from the bank) or a deposit (the bank borrows from the individual). The **rate of interest,** expressed as a percent, is the amount charged for the use of the principal for a given period of time, usually on yearly (that is, per annum) basis.

**SIMPLE INTEREST FORMULA**

If a principal of $P$ dollars is borrowed for a period of $t$ years at an annual interest rate $r$, expressed as a decimal, the interest $I$ charged is

$$I = Prt$$

Interest charged according to this formula is called **simple interest.**

### EXAMPLE 6 Computing Credit Card Interest

Suppose that Yolanda has a credit card balance of \$2800. Each month, the credit card charges 14% annual simple interest on any outstanding balances. What is the interest that Yolanda will be charged on this loan after one month? What is Yolanda's credit card balance after one month?

**Solution**

We wish to know the interest $I$ charged on the loan. Because the interest rate is given as an annual rate, the length of time that the money is borrowed must be expressed in years. Since we are talking about 1 month, the length of time is $\frac{1}{12}$ year so that $t = \frac{1}{12}$. The outstanding balance, or principal, is $P = \$2800$. The annual interest rate is $r = 14\% = 0.14$. Substituting into $I = Prt$, we obtain

$$I = (\$2800)(0.14)\left(\frac{1}{12}\right) = \$32.67$$

Yolanda will owe the amount borrowed, \$2800, plus accrued interest, \$32.67, for a total of $\$2800 + \$32.67 = \$2832.67$.

**QUICK ✓**

**16.** Suppose that Dave has a car loan of \$6500. The bank charges 6% annual simple interest. What is the interest charge on Dave's car loan after 1 month?

**17.** Suppose that you have \$1400 in a savings account. The bank pays 1.5% annual simple interest. What would be the interest paid after 6 months? What is the balance in the account?

---

## 3 Model and Solve Mixture Problems

**Mixture problems** are problems in which two or more items are combined to form a third item. There are a number of different types of mixture problems, but they all follow a basic approach to solving the problem. In solving mixture problems remember the following idea:

Portion from Item A + Portion from Item B = Whole or Total

One type of mixture problem is the so-called interest problem.

### EXAMPLE 7 Financial Planning

Kevin has $15,000 to invest. His goal is to obtain an overall annual rate of return of 9% or $1350 annually. His financial advisor recommends that he invest some of the money in corporate bonds that pay 12% and the rest in government-backed Treasury bonds paying 4%. How much should be placed in each investment in order for Kevin to achieve his goal?

**Solution**

**Step 1: Identify** This is a mixture problem involving simple interest. Kevin needs to know how much to place into corporate bonds and how much to place into Treasury bonds in order to earn $1350 in interest.

**Step 2: Name** Let $b$ represent the amount invested in corporate bonds, so that $\$15{,}000 - b$ represents the amount invested in Treasury bonds.

**Step 3: Translate** We organize the information given in Table 2.

**Table 2**

| | Principal $ | Rate % | Time Yr | Interest $ |
|---|---|---|---|---|
| Corporate Bond | $b$ | 0.12 | 1 | $0.12b$ |
| Treasury Bond | $15{,}000 - b$ | 0.04 | 1 | $0.04(15{,}000 - b)$ |
| Total | 15,000 | 0.09 | 1 | $0.09(15{,}000) = \$1350$ |

Kevin wants to earn 9% each year on his principal of $15,000. Using the simple interest formula, Kevin wants to earn $\$15{,}000(0.09)(1) = \$1350$ each year in interest. The total interest will be the sum of the interest from the corporate bonds and the Treasury bonds.

$$\text{Interest from corporate bonds} + \text{Interest from Treasury bonds} = \$1350$$
$$0.12b + 0.04(15{,}000 - b) = 1350 \quad \text{The Model}$$

**Step 4: Solve**

Use the Distributive Property to remove parentheses:
$$0.12b + 0.04(15{,}000 - b) = 1350$$
$$0.12b + 600 - 0.04b = 1350$$
Combine like terms:
$$0.08b + 600 = 1350$$
Subtract 600 from both sides:
$$0.08b = 750$$
Divide both sides by 0.08:
$$b = 9375$$

**Step 5: Check** It appears that Kevin should invest $9375 in corporate bonds and $\$15{,}000 - \$9375 = \$5625$ in Treasury bonds. The simple interest earned each year on the corporate bonds is $(\$9375)(0.12)(1) = \$1125$. The simple interest earned each year on the Treasury bonds is $(\$5625)(0.04)(1) = \$225$. The total interest earned is $\$1125 + \$225 = \$1350$. Kevin wanted to earn 9% annually and $(\$15{,}000)(0.09)(1) = \$1350$. This agrees with the information presented in the problem.

**Step 6: Answer the Question** Kevin will invest $9375 in corporate bonds and $5625 in Treasury bonds.

**QUICK ✓**

**18.** Sophia has recently retired and requires an extra $5400 per year in income. She has $90,000 to invest and can invest in either an Aaa-rated bond that pays 5% per annum or a B-rated bond paying 9% per annum. How much should be placed in each investment in order for Sophia to achieve her goal?

**19.** Steve has $25,000 to invest and wishes to earn an overall annual rate of return of 8%. His financial advisor recommends that he invest some of the money in a 5 year CD paying 4% per annum and the rest in a corporate bond paying 9% per annum. How much should be placed in each investment in order for Steve to achieve his goal?

**Work Smart**

Remember that mixtures can include interest (money), solids (nuts), liquids (chocolate milk), and even gases (Earth's atmosphere). "Mixture" is a simplifying assumption.

Often, new blends are created by mixing two quantities. For example, a chef might mix buckwheat flour with wheat flour to make buckwheat pancakes. Or a coffee shop might mix two different types of coffee to create a new coffee blend.

### EXAMPLE 8 Blending Coffees

The manager of a coffee shop wishes to form a new blend of coffee. She wants to mix Sumatra beans, known for their strong, distinctive taste, that sell for $12 per pound with milder Brazilian beans that sell for $8 per pound to get 50 pounds of the new blend. The new blend will sell for $9 per pound and there will be no difference in revenue from selling the new blend versus selling the beans separately. See Figure 2. How many pounds of the Sumatra and Brazilian beans are required?

**Figure 2**

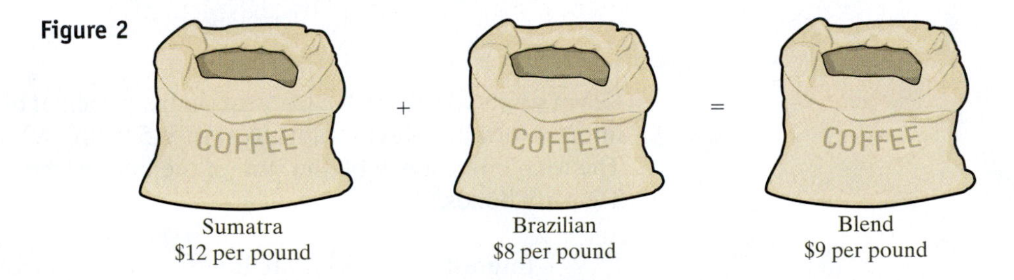


**Solution**

**Step 1: Identify** This is a mixture problem. We want to know the number of pounds of Sumatra beans and the number of pounds of Brazilian beans that are required in the new blend.

**Step 2: Name** Let $s$ represent the number of pounds of Sumatra beans that are required. So, $50 - s$ will be the number of pounds of Brazilian coffee required.

**Step 3: Translate** We are told that there is to be no difference in revenue between selling the Sumatra and Brazilian separately versus the blend. This means that if the blend contains one pound of Sumatra and one pound of Brazilian, we should collect $\$12(1) + \$8(1) = \$20$ because that is how much we would collect if we sold the beans separately.

We set up Table 3.

**Table 3**

| | Price $/Pound | Number of Pounds | Revenue |
|---|---|---|---|
| **Sumatra** | 12 | $s$ | $12s$ |
| **Brazilian** | 8 | $50 - s$ | $8(50 - s)$ |
| **Blend** | 9 | 50 | $9(50) = 450$ |

In general, if the mixture contains $s$ pounds of Sumatra beans, we should collect $\$12s$. If the mixture contains $50 - s$ pounds of Brazilian beans, we should collect $\$8(50 - s)$. If the mixture sells for $9 per pound and we make 50 pounds of the blend, then we should collect $\$9(50) = \$450$ for the blend.

$$\begin{pmatrix}\text{Price per pound}\\ \text{of Sumatra}\end{pmatrix}\begin{pmatrix}\text{Pounds of}\\ \text{Sumatra}\end{pmatrix} + \begin{pmatrix}\text{Price per pound}\\ \text{of Brazilian}\end{pmatrix}\begin{pmatrix}\text{Pounds of}\\ \text{Brazilian}\end{pmatrix} = \begin{pmatrix}\text{Price per pound}\\ \text{of blend}\end{pmatrix}\begin{pmatrix}\text{Pounds of}\\ \text{blend}\end{pmatrix}$$

$$\underbrace{12 \cdot s}_{\text{Revenue from Sumatra}} + \underbrace{8 \cdot (50 - s)}_{\text{Revenue from Brazilian}} = \underbrace{\$9 \cdot 50}_{\text{Revenue from Mixture}}$$

We have the equation

$$12s + 8(50 - s) = 450 \quad \text{The Model}$$

**Step 4: Solve** Solve the equation for $s$.

$$12s + 8(50 - s) = 450$$

Use the Distributive Property to remove the parentheses: $12s + 400 - 8s = 450$

Combine like terms: $4s + 400 = 450$

Subtract 400 from both sides: $4s = 50$

Divide both sides by 4: $s = 12.5$

**Step 5: Check** It appears that we should mix 12.5 pounds of Sumatra with $50 - 12.5 = 37.5$ pounds of Brazilian. The 12.5 pounds of Sumatra beans would sell for $\$12(12.5) = \$150$ and the 37.5 pounds of Brazilian beans would sell for $\$8(37.5) = \$300$; the total revenue would be $\$150 + \$300 = \$450$, which equals the revenue obtained from selling the blend. This checks with the information presented in the problem.

**Step 6: Answer the Question** The manager should mix 12.5 pounds of Sumatra beans with 37.5 pounds of Brazilian beans to make the blend.

**QUICK ✓**

**20.** Suppose that you want to blend two teas in order to obtain 10 pounds of the new blend that sells for $3.50 per pound. Tea A sells for $4.00 per pound and Tea B sells for $2.75 per pound. Assuming that there will be no difference in revenue from selling the new blend versus selling the tea separately, determine the number of pounds of each tea that will be required in the blend.

**21.** "We're Nuts!" sells cashews for $6.00 per pound and peanuts for $1.50 per pound. The manager has decided to make a "trail mix" that combines the cashews and peanuts. She wants the trail mix to sell for $3.00 per pound and there should be no loss in revenue from selling the trail mix versus selling the nuts alone. How many pounds of cashews and peanuts are required to create 30 pounds of trail mix?

## 4 Model and Solve Uniform Motion Problems

Objects that move at a constant velocity are said to be in **uniform motion.** When the average velocity of an object is known, it can be interpreted as its constant velocity. For example, a car traveling at an average velocity of 40 miles per hour is in uniform motion.

**In Words**
The uniform motion formula states that distance equals rate times time.

**UNIFORM MOTION FORMULA**

If an object moves at an average speed $r$, the distance $d$ covered in time $t$ is given by the formula

$$d = rt$$

## EXAMPLE 9 Uniform Motion

Roger and Bill decide to have a 10-mile race. Roger can run at an average speed of 12 miles per hour while Bill can run at an average speed of 10 miles per hour. To "even things up," Roger agrees to give Bill a head start of 0.15 hour. When will Roger catch up to Bill?

### Solution

**Step 1: Identify** This is a uniform motion problem. We wish to know the number of hours it will take for Roger to catch up to Bill if each man runs at his average speed.

**Step 2: Name** Let $t$ represent the number of hours it takes for Roger to catch up to Bill. Since Bill is given a head start of 0.15 hour, Bill will have been running for $t + 0.15$ hours.

**Step 3: Translate** Figure 3 illustrates the situation. We will set up Table 4.

**Figure 3**

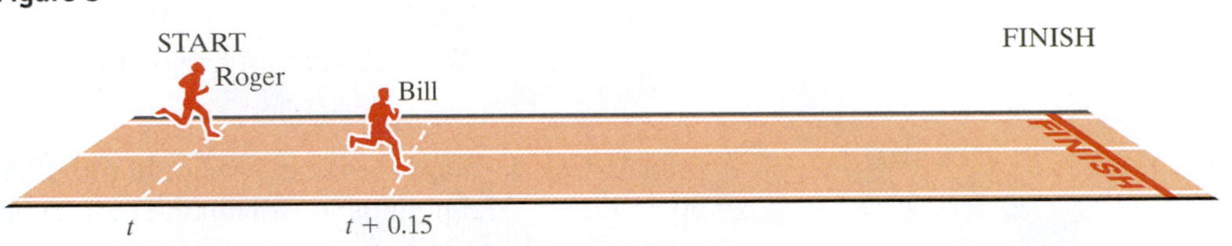


**Table 4**

| | Rate, mph | Time, hours | Distance, miles |
|---|---|---|---|
| Roger | 12 | $t$ | $12t$ |
| Bill | 10 | $t + 0.15$ | $10(t + 0.15)$ |

The distance that Roger will travel is $12t$, since Roger's average velocity is 12 miles per hour and he runs for $t$ hours. The distance that Bill will travel is $10(t + 0.15)$ since Bill's average velocity is 10 miles per hour and he runs for $t + 0.15$ hours. We want the distance that each travels to be equal, so we set up the model so that

$$\text{Distance Roger runs} = \text{Distance Bill runs}$$
$$12t = 10(t + 0.15) \quad \text{The Model}$$

**Step 4: Solve** We wish to solve for $t$:

$$12t = 10(t + 0.15)$$

Use the Distributive Property: $12t = 10t + 1.5$

Subtract $10t$ from both sides: $2t = 1.5$

Divide both sides by 2: $t = \frac{1.5}{2} = 0.75$

**Step 5: Check** It appears that it will take Roger 0.75 hour to catch up to Bill. After 0.75 hour, Roger will have traveled $12(0.75) = 9$ miles and Bill will have traveled $10(0.75 + 0.15) = 9$ miles. It checks!

**Step 6: Answer the Question** It will take Roger 0.75 hour or 45 minutes to catch up to Bill.

**QUICK ✓**

**22.** A Chevrolet Cavalier left Los Angeles traveling at an average velocity of 40 miles per hour. Two hours later, a BMW 540i left Los Angeles traveling on the same road at an average velocity of 60 miles per hour. When will the BMW catch up to the Cavalier? How far will each car have traveled?

**23.** A train leaves Union Station traveling at an average velocity of 50 miles per hour. Four hours later, a helicopter follows along the train tracks traveling at an average velocity of 90 miles per hour. When will the helicopter catch up to the train? How far will each have traveled?

# 1.2 Exercises

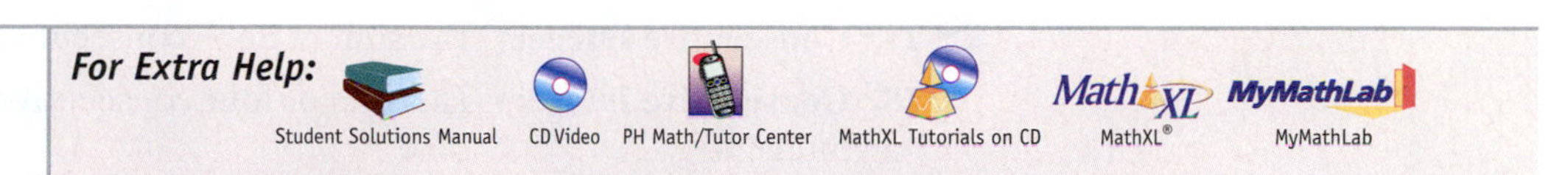


## Concepts and Vocabulary

*In Problems 1 and 2, fill in the blanks.*

**1.** Letting variables represent unknown quantities and then expressing relationships among the variables in the form of equations is called __________ __________.

**2.** Objects that move at a constant velocity are said to be in __________ __________.

*In Problems 3 and 4, answer True or False to each statement.*

**3.** Interest is the money paid for the use of money.

**4.** Consider the phrase "... there will be no difference in revenue from selling the new blend versus selling the beans separately" presented in Example 8. This means that the sum of the revenue from each item, if sold separately, would equal the revenue of the blend.

**5.** How is mathematical modeling related to problem solving?

**6.** Why do we make assumptions when creating mathematical models?

**7.** Name the different categories of problems presented in this section. Name two different kinds of mixture problems.

**8.** Think of two models that you use in your everyday life. Are any of them mathematical models? Describe your models to members of your class.

## Building Skills

**9.** What is 25% of 40?

**10.** What is 150% of 70?

**11.** 12 is 30% of what?

**12.** 50 is 90% of what?

**13.** 30 is what percent of 80?

**14.** 90 is what percent of 120?

*In Problems 15–24, translate each of the following English statements into a mathematical statement. Then solve the equation.*

**15.** The sum of a number $x$ and 12 is 20.

**16.** The difference between 10 and a number $z$ is 6.

**17.** Twice the sum of $y$ and 3 is 16.

**18.** The sum of two times $y$ and 3 is 16.

**19.** The difference between $w$ and 22 equals three times $w$.

**20.** The sum of $x$ and 4 results in twice $x$.

**21.** Four times a number $x$ is equivalent to the sum of two times $x$ and 14.

**22.** Five times a number $x$ is equivalent to the difference of three times $x$ and 10.

**23.** 80% of a number is equivalent to the sum of the number and 5.

**24.** 40% of a number equals the difference between the number and 10.

## Applying the Concepts

**25. Number Sense** Grant is thinking of two numbers. He says that one of the numbers is twice the other number and the sum of the numbers is 39. What are the numbers?

**26. Number Sense** Pattie is thinking of two numbers. She says that one of the numbers is 8 more than the other number and the sum of the numbers is 56. What are the numbers?

**27. Consecutive Integers** The sum of three consecutive integers is 75. Find the integers.

**28. Consecutive Integers** The sum of four consecutive odd integers is 104. Find the integers.

**29. Computing Grades** Going into the final exam, which counts as two grades, Kendra has test scores of 84, 78, 64, and 88. What score does Kendra need on the final exam in order to have an average of 80?

**30. Computing Grades** Going into the final exam, which counts as three grades, Mark has test scores of 65, 79, 83, and 68. What score does Mark need on the final exam in order to have an average of 70?

**31. Comparing Printers** Jacob is trying to decide between two laser printers, one manufactured by Hewlett-Packard, the other by Brother. Both have similar features and warranties, so price is the determining factor. The Hewlett-Packard costs $180 and printing costs are approximately $0.03 per page. The Brother costs $230 and printing costs are approximately $0.01 per page. How many pages need to be printed for the cost of the two printers to be the same?

**32. Comparing Job Offers** Maria has just been offered two sales jobs. The first job offer is a base monthly salary of $2500 plus a commission of 3% of total sales. The second job offer is a base monthly salary of $1500 plus a commission of 3.5% of total sales. For what level of monthly sales is the salary of the two jobs equivalent?

**33. Finance** An inheritance of $800,000 is to be divided among Avery, Connor, and Olivia in the following manner: Olivia is to receive $^3/_4$ of what Connor gets, while Avery gets $^1/_4$ of what Connor gets. How much does each receive?

**34. Sharing the Cost of a Pizza** Judy and Linda agree to share the cost of a $21 pizza based on how much each ate. If Judy ate the $^3/_4$ of the amount that Linda ate, how much should each pay?

**35. Sales Tax** In the state of Colorado there is a sales tax of 2.9% on all goods purchased. If Jan buys a television for $599, what will be the final bill, including sales tax?

**36. Sales Tax** In the state of Texas there is a sales tax of 6.25% on all goods purchased. If Megan buys 4 compact disks for $39.83, what will be the final bill, including sales tax?

**37. Markups** A new 2005 Honda Accord has a list price of $25,800. Suppose that the dealer markup on this car is 15%. What is the dealer's cost?

**38. Markups** Suppose that the price of a new Intermediate Algebra text is $95. The bookstore has a policy of marking texts up 30%. What is the cost of the text to the bookstore?

**39. Discount Pricing** Suppose that you just received an email alert from buy.com indicating that 256 megabyte memory cards have just been discounted by 20%. If the sale price of the memory card is now $13.60, what was the original price?

**40. Discount Pricing** Suppose that you just received an email alert from Kohls indicating that the fall line of clothing has just been discounted by 30% and knit polo shirts are now $28. What was the original price of a polo shirt?

**41. Cars** A Mazda 6s weighs 20 pounds more than a Nissan Altima. A Honda Accord EX weighs 70 pounds more than a Nissan Altima. The total weight of all three cars is 10,050 pounds. How much does each car weigh? (*Source: road and track* magazine)

**42. Cars** A Mazda 6s requires 18 more feet to stop from 80 miles per hour than a Nissan Altima. A Honda Accord EX requires 26 more feet to stop from 80 miles per hour than a Nissan Altima. The combined distance that each car requires to stop from 80 miles per hour is 734 feet. What is the stopping distance required for each car? (*Source: road and track* magazine)

**43. 2003 Super Bowl** The 2003 Super Bowl was played between the Tampa Bay Buccaneers and the Oakland Raiders. The Buccaneers rushed for 90 more yards than the Raiders. The total yards rushed by both teams was 560. How many yards did the Raiders rush for?

**44. 2003 Pebble Beach Pro-Am** The 2003 Pebble Beach Pro-Am is a golf tournament played at Pebble Beach Golf Links on the Monterey Peninsula in California. Davis Love III, the tournament winner, had a score that was 22 strokes less than Phil Mickelson. Combined, the two players required 570 strokes. How many strokes were required by Phil Mickelson?

**45. Finance** A total of $20,000 is going to be split between Adam and Krissy with Adam receiving $3000 less than Krissy. How much will each get?

**46. Finance** A total of $40,000 is going to be invested in stocks and bonds. A financial advisor recommends that $6000 more should be invested in stocks than bonds. How much is invested in stocks? How much is invested in bonds?

**47. Investments** Suppose that your long lost Aunt Sara has left you an unexpected inheritance of $24,000. You have decided to invest the money rather than spend it on frivolous purchases. Your financial advisor has recommended that you diversify by placing some of the money in stocks and some in bonds. Based upon current market conditions, she has recommended that the amount in bonds should equal three-fifths of the amount invested in stocks. How much should be invested in stocks? How much should be invested in bonds?

**48. Investments** Jack and Diane have $60,000 to invest. Their financial advisor has recommended that they diversify by placing some of the money in stocks and some in bonds. Based upon current market conditions, he has recommended that the amount in bonds should equal two-thirds of the amount invested in stocks. How much should be invested in stocks? How much should be invested in bonds?

**49. Simple Interest** Elena has a credit card balance of $2500. The credit card company charges 14% per annum simple interest. What is the interest charge on Elena's credit card after 1 month?

**50. Simple Interest** Faye has a home equity loan of $70,000. The bank charges Faye 6% per annum simple interest. What is the interest charge on Faye's loan after 1 month?

**51. Banking** A bank has loaned out $500,000, part of it at 6% per annum and the rest of it at 11% per annum. If the bank receives $43,750 in interest each year, how much was loaned at 6%?

52. **Banking** Patrick is a loan officer at a bank. He has $2,000,000 to lend out and has two loan programs. His home equity loan is currently priced at 6% per annum, while his unsecured personal loan is priced at 14%. The bank president wants Patrick to earn a rate of return of 12% on the $2,000,000 available. How much should Patrick lend out at 6%?

53. **Investments** Pedro wants to invest his $25,000 bonus check. His investment advisor has recommended that he put some of the money in a bond fund that yields 5% per annum and the rest in a stock fund that yields 9% per annum. If Pedro wants to earn $1875 each year from his investments, how much should he place in each investment?

54. **Investments** Johnny is a shrewd 8-year-old. For Christmas, his grandparents gave him $10,000. Johnny decides to invest some of the money in a savings account that pays 2% per annum and the rest in a stock fund paying 10% per annum. Johnny wants his investments to yield 7% per annum. How much should Johnny put into each account?

55. **Making Coffee** Suppose that you want to blend two coffees in order to obtain 50 pounds of a new blend. If $x$ represents the number of pounds of coffee A, write an algebraic expression that represents the number of pounds of coffee B.

56. **Candy** "Sweet Tooth!" candy store sells chocolate-covered almonds for $6.50 per pound and chocolate-covered peanuts for $4.00 per pound. The manager decides to make a bridge mix that combines the almonds with the peanuts. She wants the bridge mix to sell for $5.00 per pound, and there should be no loss in revenue from selling the bridge mix versus the almonds and peanuts alone. How many pounds of chocolate-covered almonds and chocolate-covered peanuts are required to create 50 pounds of bridge mix?

57. **Coins** Bobby has been saving quarters and dimes. He opened up his piggy bank and determined that it contained 47 coins worth $9.50. Determine how many dimes and quarters were in the piggy bank.

58. **More Coins** Diana has been saving nickels and dimes. She opened up her piggy bank and determined that it contained 48 coins worth $4.50. Determine how many nickels and dimes were in the piggy bank.

59. **Gold** The purity of gold is measured in karats, with pure gold being 24 karats. Other purities of gold are expressed as proportional parts of pure gold. For example, 18 karat gold is $\frac{18}{24}$, or 75%, pure gold; 12 karat gold is $\frac{12}{24}$, or 50%, pure gold; and so on. How much pure gold should be mixed with 12 karat gold to obtain 72 grams of 18 karat gold?

60. **Antifreeze** The cooling system of a car has a capacity of 15 liters. If the system is currently filled with a mixture that is 40% antifreeze, how much of this mixture should be drained and replaced with pure antifreeze so that the system is filled with a solution that is 50% antifreeze?

61. **A Biathlon** Suppose that you have entered a 62-mile biathlon that consists of a run and a bicycle race. During your run, your average velocity is 8 miles per hour and during your bicycle race, your average velocity is 20 miles per hour. You finish the race in 4 hours. What is the distance of the run? What is the distance of the bicycle race?

62. **A Biathlon** Suppose that you have entered a 15-mile biathlon that consists of a run and swim. During your run, your average velocity is 7 miles per hour and during your swim, your average velocity is 2 miles per hour. You finish the race in 2.5 hours. What is the length of the swim? What is the distance of the run?

63. **Collision Course** Two cars that are traveling toward each other are 455 miles apart. One car is traveling 10 miles per hour faster than the other car. The cars pass after 3.5 hours. How fast is each car traveling?

64. **Collision Course** Two planes that are traveling toward each other are 720 miles apart. One plane is traveling 40 miles per hour faster than the other. The planes pass after 0.75 hour. How fast is each plane traveling?

65. **Boats** Two boats leave a port at the same time, one traveling north and the other traveling south. The north bound boat travels at 12 miles per hour (mph) faster than the south bound boat. If the south bound boat is traveling at 25 mph, how long before they are 155 miles apart?

66. **Cyclists** Two cyclists leave a city at the same time, one going east and the other going west. The west bound cyclist bikes at 3 mph faster than the east bound cyclist. If after 6 hours they are 162 miles apart, how fast is each cyclist riding?

67. **Walking** At 10:00 A.M. two people leave their homes that are 15 miles apart, walking toward each other. If one person walks at a rate that is 2 mph faster than the other and they meet after 1.5 hours, how fast was each person walking?

68. **Trains** At 9:00 A.M., two trains are 715 miles apart traveling toward each other on parallel tracks. If one train is traveling at a rate that is 10 miles per hour (mph) faster than the other train and they meet after 5.5 hours, how fast is each train traveling?

### Extending the Concepts

69. **Computing Average Speed** On a recent trip to Florida, we averaged 50 miles per hour. On the return trip we averaged 60 miles per hour. What do you think the average speed of the trip to Florida and back was? Defend your position. Algebraically, determine the average speed of the trip to Florida and back.

70. **Discount Pricing** Suppose that you are the manager of a clothing store and have just purchased 100 shirts for $15 each. After 1 month of selling the shirts at the regular price, you plan to have a sale giving 30% off the original selling price. However, you still want to make a profit of $6 on each shirt at the sale price. What should you price the shirts at initially to ensure this?

71. **Critical Thinking** Make up an applied problem that would result in the equation $x - 0.05x = 60$.

72. **Critical Thinking** Make up an applied problem that would result in the equation $10 + 0.14x = 50$.

73. **Uniform Motion** Suppose that you are walking along the side of train tracks at 2 miles per hour. A train that is traveling 20 miles per hour in the same direction requires 1 minute to pass you. How long is the train?

# 1.3 Using Formulas to Solve Problems

**OBJECTIVES**

1. Solve for a Variable in a Formula
2. Use Formulas to Solve Problems

### *Preparing for Formulas*

*Before getting started, take this readiness quiz. If you get a problem wrong, go back to the section cited and review the material.*

*In Problems 1 and 2, (a) round, (b) truncate each decimal to the indicated number of places. [Section R.2, pp. 14–15]*

1. 3.00343; 3 decimal places
2. 14.957; 2 decimal places

*Preparing for...Answers* **1. (a)** 3.003 **(b)** 3.003 **2. (a)** 14.96 **(b)** 14.95

A known relation that exists between two or more variables can be used to solve certain types of problems. A **formula** is an equation that describes how two or more variables are related.

**Work Smart**
Think of a formula as another kind of mathematical model.

## EXAMPLE 1 Answering Questions with Formulas

The information in Table 5 gives descriptions in words of known relations and the corresponding formula.

**Table 5**

| Verbal Description | Formula |
|---|---|
| *Geometry:* What is the area of a rectangle? The area $A$ of a rectangle is the product of its length $l$ and width $w$. | $A = lw$ |
| *Physics:* How do we measure the energy of an object in motion? Kinetic energy $K$ is one-half the product of the mass $m$ and the square of the velocity $v$. | $K = \frac{1}{2}mv^2$ |
| *Economics:* What proportion of the population is in the labor force? The participation rate $R$ is the sum of the number of employed $E$ and the number of unemployed $U$ divided by the adult population $P$. | $R = \frac{E + U}{P}$ |
| *Finance:* How much money will I have next year? The future value $A$ is the product of the present value $P$ and 1 plus the annual interest rate $r$. | $A = P(1 + r)$ |

**QUICK ✓** *Translate the verbal description into a mathematical formula.*

1. The area $A$ of a circle is the product of the number $\pi$ and the square of its radius $r$.
2. The volume $V$ of a right circular cylinder is the product of the number $\pi$, the square of its radius $r$, and its height $h$.
3. The daily cost $C$ of manufacturing computers is \$175 times the number of computers manufactured $x$ plus \$7000.
4. The distance $s$ that an object free-falls is one-half the product of acceleration due to gravity $g$ and the square of time $t$.

## 1 Solve for a Variable in a Formula

The expression "solve for the variable" means to get the variable by itself on one side of the equation with all other variables and constants, if any, on the other side by forming equivalent equations. For example, in the formula for the area of a rectangle, $A = lw$, the formula is solved for $A$ because $A$ is by itself on one side of the equation while all other variables are on the other side.

When solving certain problems, it becomes important for us to be able to solve formulas for certain variables. The steps that we follow when solving a formula for a certain variable are identical to those that we followed when solving an equation.

**Figure 4**

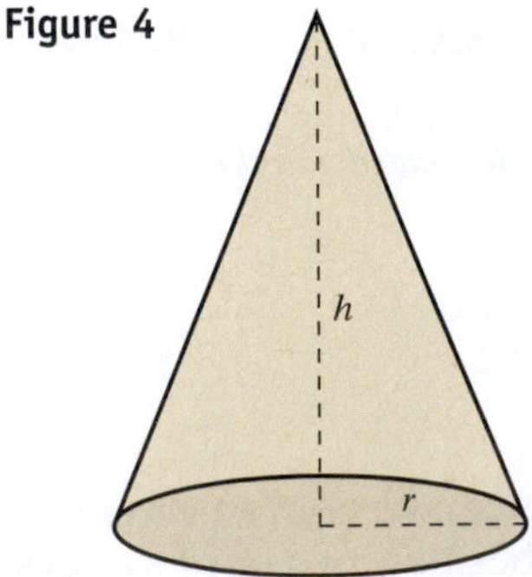


## EXAMPLE 2 Solving for a Variable in a Formula

The volume $V$ of a cone is given by the formula $V = \frac{1}{3}\pi r^2 h$, where $r$ is the radius and $h$ is the height of the cone. See Figure 4.

**(a)** Solve the formula for $h$.

**(b)** Use the result from part (a) to find the height of a cone if its volume is $50\pi$ cubic feet and its radius is 5 feet.

**Solution**

**Work Smart**

Solving for a variable is just like solving an equation with one unknown. When solving for a variable, treat all the other variables as constants.

**(a)** Because we want to solve the formula for $h$, we want to get $h$ by itself and all other variables and constants on the other side of the equation.

$$V = \frac{1}{3}\pi r^2 h$$

Multiply both sides by 3: $$3 \cdot V = \left(\frac{1}{3}\pi r^2 h\right) \cdot 3$$

Divide out common factors: $$3V = \pi r^2 h$$

Divide both sides by $\pi r^2$: $$\frac{3V}{\pi r^2} = \frac{\pi r^2 h}{\pi r^2}$$

Divide out common factors: $$\frac{3V}{\pi r^2} = h$$

If $a = b$, then $b = a$: $$h = \frac{3V}{\pi r^2}$$

**Work Smart**

When working with formulas, keep track of the units. Verify the units in your answer are reasonable.

**(b)** Substituting $V = 50\pi \text{ ft}^3$ and $r = 5$ feet into $h = \frac{3V}{\pi r^2}$, we obtain

$$h = \frac{3(50\pi \text{ ft}^3)}{\pi(5 \text{ ft})^2}$$

$$h = \frac{150\pi \text{ ft}^3}{25\pi \text{ ft}^2}$$

$$h = 6 \text{ feet}$$

Example 2 presents a formula from geometry. Formulas from geometry are useful in solving many types of problems. We list some of these formulas in Table 6.

**Table 6**

| Figure | Formulas |
|---|---|
| Square (sides $s$, $s$) | **Area:** $A = s^2$ <br> **Perimeter:** $P = 4s$ |
| Rectangle (length $l$, width $w$) | **Area:** $A = lw$ <br> **Perimeter:** $P = 2l + 2w$ |
| Triangle (sides $a$, $c$, base $b$, height $h$) | **Area:** $A = \frac{1}{2}bh$ <br> **Perimeter:** $P = a + b + c$ |
| Trapezoid (sides $a$, $c$, bases $b$, $B$, height $h$) | **Area:** $A = \frac{1}{2}h(B + b)$ <br> **Perimeter:** $P = a + b + c + B$ |

(continued)

| Figure | Formulas |
|---|---|
| Parallelogram | **Area:** $A = ah$<br>**Perimeter:** $P = 2a + 2b$ |
| Circle | **Area:** $A = \pi r^2$<br>**Circumference:** $C = 2\pi r = \pi d$ |
| Cube | **Volume:** $V = s^3$<br>**Surface Area:** $S = 6s^2$ |
| Rectangular Solid | **Volume:** $V = lwh$<br>**Surface Area:** $S = 2lw + 2lh + 2wh$ |
| Sphere | **Volume:** $V = \frac{4}{3}\pi r^3$<br>**Surface Area:** $S = 4\pi r^2$ |
| Right Circular Cylinder | **Volume:** $V = \pi r^2 h$<br>**Surface Area:** $S = 2\pi r^2 + 2\pi rh$ |
| Cone | **Volume:** $V = \frac{1}{3}\pi r^2 h$ |

**QUICK ✓**

**5.** The area $A$ of a triangle is given by the formula $A = \frac{1}{2}bh$, where $b$ is the base of the triangle and $h$ is the height.

**(a)** Solve the formula for $h$.

**(b)** Find the height of the triangle whose area is 10 square inches and whose base is 4 inches.

**6.** The perimeter $P$ of a parallelogram is given by the formula $P = 2a + 2b$, where $a$ is the length of one side of the parallelogram and $b$ is the length of the adjacent side.

**(a)** Solve the formula for $b$.

**(b)** Find the length of one side of a parallelogram whose perimeter is 60 cm and whose adjacent length is 20 cm.

### EXAMPLE 3 Solving for a Variable in a Formula

The formula $Y = C + bY + I + G + N$ is a model used in economics to describe the total income of an economy. In the model, $Y$ is income, $C$ is consumption, $I$ is investment in capital, $G$ is government spending, $N$ is net exports, and $b$ is a constant. Solve the formula for $Y$.

#### Solution

We want to get all terms with $Y$ on the same side of the equal sign.

$$Y = C + bY + I + G + N$$

Subtract $bY$ from both sides:
$$Y - bY = C + bY + I + G + N - bY$$

Combine like terms:
$$Y - bY = C + I + G + N$$

Use the Distributive Property "in reverse" to isolate $Y$:
$$Y(1 - b) = C + I + G + N$$

Divide both sides by $1 - b$:
$$\frac{Y(1 - b)}{1 - b} = \frac{C + I + G + N}{1 - b}$$

Simplify:
$$Y = \frac{C + I + G + N}{1 - b}$$

**QUICK ✓** *Solve for the indicated variable.*

**7.** $I = Prt$ for $P$

**8.** $Ax + By = C$ for $y$

**9.** $2xh - 4x = 3h - 3$ for $h$

**10.** $S = na + (n - 1)d$ for $n$

## 2 Use Formulas to Solve Problems

Formulas are often needed in order to solve certain types of word problems. We follow the same steps to solving these problems as was presented in Section 1.2 on page 65.

### EXAMPLE 4 The Perimeter of a Window

The perimeter of a rectangular picture window is 466 inches. The length of the window is 55 inches more than the width. See Figure 5. Find the dimensions of the window.

**Figure 5**

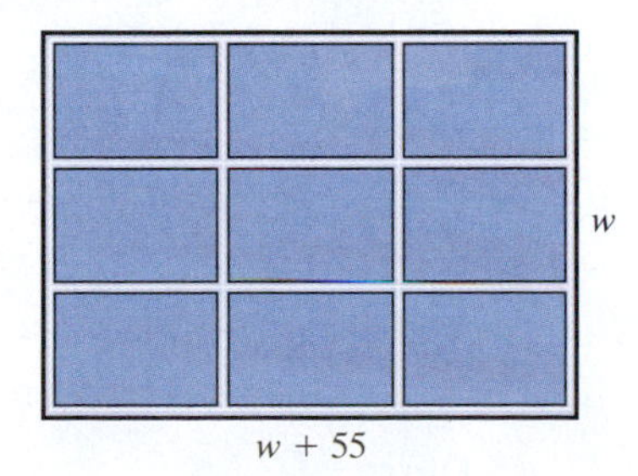


#### Solution

**Step 1: Identify** This is a geometry problem that requires the formula for the perimeter of a rectangle. We want to determine the dimensions of the window, which is in the shape of a rectangle. That is, we want to determine the length and width of the window.

**Step 2: Name** Let $w$ represent the width. Since the length is 55 inches more than the width, we know that $l = w + 55$.

**Step 3: Translate** The perimeter of a rectangle is $P = 2l + 2w$. We substitute the known values into the formula for the perimeter of a rectangle.

$$P = 2l + 2w$$

$P = 466; l = w + 55$: $\quad 466 = 2(w + 55) + 2w$

**Step 4: Solve**

$$466 = 2(w + 55) + 2w$$

Distribute the 2: $\quad 466 = 2w + 110 + 2w$

Combine like terms: $\quad 466 = 4w + 110$

Subtract 110 from both sides: $\quad 356 = 4w$

Divide both sides by 4: $\quad 89 = w$

**Step 5: Check** It appears that the width of the window is 89 inches, so the length is $89 + 55 = 144$ inches. The perimeter of the window is $2(144) + 2(89) = 466$ inches. It checks!

**Step 6: Answer the Question** The width of the window is 89 inches and the length is 144 inches.

There is an interesting side note to the result of Example 4. If we compute the ratio of the length of the window to the width, we obtain $\frac{144}{89} \approx 1.618$. Rectangles whose dimensions form this ratio are called **golden rectangles.** Golden rectangles are said to have dimensions that are "pleasing to the eye." The golden rectangle was first constructed by the Greek philosopher Pythagoras in the sixth century B.C. These rectangles are used in architecture (The Parthenon) and in art (the Mona Lisa). See Figure 6.

**Figure 6**

**QUICK ✓**

**11.** The perimeter of a rectangular pool is 180 feet. If the length of the pool is to be 10 feet more than the width, find the dimensions of the pool.

**12.** The opening of a rectangular bookcase has a perimeter of 224 inches. If the height of the bookcase is 32 inches more than the width, determine the dimensions of the opening of the bookcase.

## EXAMPLE 5 Constructing a Soup Can

A can of Campbell's soup has a surface area of 46.5 square inches. See Figure 7. The surface area $A$ of a right circular cylinder is $A = 2\pi r^2 + 2\pi rh$, where $r$ is the radius of the can and $h$ is the height of the can. Find the height of a can of Campbell's soup if its radius is 1.375 inches. Round your answer to two decimal places.

Figure 7

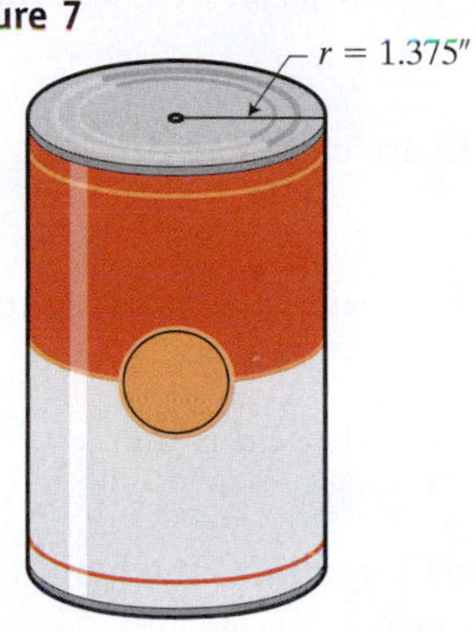


### Solution

**Step 1: Identify** This is a geometry problem that requires the formula for the surface area of a cylinder. We wish to find the height of the can of soup.

**Step 2: Name** Let $h$ represent the height of the can of soup.

**Step 3: Translate** We know that the surface area $A$ of the can of soup is 46.5 square inches. The radius of the can of soup is 1.375 inches. We substitute these values into the formula for the surface area of the can.

$$A = 2\pi r^2 + 2\pi rh$$

$$A = 46.5;\ r = 1.375:\quad 46.5 = 2\pi(1.375)^2 + 2\pi(1.375)h \quad \text{The Model}$$

**Step 4: Solve** Solve for $h$. We will not compute any of the values until the last calculation. This is done to avoid round-off error.

$$46.5 = 2\pi(1.375)^2 + 2\pi(1.375)h$$

Subtract $2\pi(1.375)^2$ from both sides: $$46.5 - 2\pi(1.375)^2 = 2\pi(1.375)h$$

Divide both sides by $2\pi(1.375)$: $$\frac{46.5 - 2\pi(1.375)^2}{2\pi(1.375)} = h$$

Figure 8

We will use a calculator to evaluate this expression. Figure 8 shows the output from a TI-84 Plus graphing calculator.

Rounded to two decimal places, we obtain $h = 4.01$ inches.

**Step 5: Check** The surface area $A$ of the can is

$$\begin{aligned} A &= 2\pi r^2 + 2\pi rh \\ &= 2\pi(1.375)^2 + 2\pi(1.375)(4.01) \\ &= 46.5 \text{ square inches.} \end{aligned}$$

**Step 6: Answer the Question** The height of the can is 4.01 inches rounded to two decimal places.

**Work Smart**

Round-off error occurs when decimals are continually rounded during the course of solving a problem. The more times we round, the more inaccurate the results may be. So, do not do any arithmetic until the last step.

### QUICK ✓

**13.** A can of peaches has a surface area of 51.8 square inches. The surface area $A$ of a right circular cylinder is $A = 2\pi r^2 + 2\pi rh$, where $r$ is the radius of the can and $h$ is the height of the can. Find the height of a can of peaches if its radius is 1.5 inches. Round your answer to two decimal places.

**14.** A can of baked beans has a surface area of 72.7 square inches. The surface area $A$ of a right circular cylinder is $A = 2\pi r^2 + 2\pi rh$, where $r$ is the radius of the can and $h$ is the height of the can. Find the height of a can of baked beans if its radius is 1.625 inches. Round your answer to two decimal places.

# 1.3 Exercises

**For Extra Help:**  Student Solutions Manual  CD Video  PH Math/Tutor Center MathXL Tutorials on CD  MathXL®  MyMathLab

## Concepts and Vocabulary

*In Problems 1–3, fill in the blank.*

1. A _______ is an equation that describes how two or more variables are related.
2. The formula for the area of a circle is _______.
3. Rectangles whose dimensions form a ratio of about 1.618 are called _______ _______.

*In Problems 4 and 5, answer True or False to each statement.*

4. The formula for the perimeter $P$ of a rectangle is $P = l + w$ where $l$ is the length and $w$ is the width of the rectangle.
5. The formula for the circumference $C$ of a circle is $C = \pi d$ where $d$ is the diameter.
6. If the radius of a circle is doubled, does the area double? Explain. If the length of the side of a cube is doubled, what happens to the volume?

## Building Skills

*In Problems 7–10, translate the verbal description into a mathematical formula.*

7. Force $F$ equals the product of mass $m$ and acceleration $a$.
8. The area $A$ of a triangle is one-half the product of its base $b$ and its height $h$.
9. The volume $V$ of a sphere is four-thirds the product of the number $\pi$ and the cube of its radius $r$.
10. The revenue $R$ of selling computers is \$800 times the number of computers sold $x$.

*In Problems 11–22, solve the formula for the indicated variable.*

11. **Uniform Motion** Solve $d = rt$ for $r$.
12. **Direct Variation** Solve $y = kx$ for $k$.
13. **Algebra** Solve $y - y_1 = m(x - x_1)$ for $m$.
14. **Algebra** Solve $y = mx + b$ for $m$.
15. **Statistics** Solve $Z = \dfrac{x - \mu}{\sigma}$ for $x$.
16. **Statistics** Solve $E = \dfrac{Z \cdot \sigma}{\sqrt{n}}$ for $\sqrt{n}$.
17. **Newton's Law of Gravitation** Solve $F = G\dfrac{m_1 m_2}{r^2}$ for $m_1$.
18. **Sequences** Solve $S - rS = a - ar^5$ for $S$.
19. **Finance** Solve $A = P + Prt$ for $P$.
20. **Bernoulli's Equation** Solve $p + \dfrac{1}{2}\rho v^2 + \rho g y = a$ for $\rho$.
21. **Temperature Conversion** Solve $C = \dfrac{5}{9}(F - 32)$ for $F$.
22. **Trapezoid** Solve $A = \dfrac{1}{2}h(B + b)$ for $b$.

*In Problems 23–30, solve for y.*

23. $2x + y = 13$
24. $-4x + y = 12$
25. $9x - 3y = 15$
26. $4x + 2y = 20$

**27.** $4x + 3y = 13$ **28.** $5x - 6y = 18$ **29.** $\frac{1}{2}x + \frac{1}{6}y = 2$ **30.** $\frac{2}{3}x - \frac{5}{2}y = 5$

## Applying the Concepts

**31. Cylinders** The volume $V$ of a right circular cylinder is given by the formula $V = \pi r^2 h$, where $r$ is the radius and $h$ is the height.

**(a)** Solve the formula for $h$.
**(b)** Find the height of a right circular cylinder whose volume is $32\pi$ cubic inches and whose radius is 2 inches.

**32. Cylinders** The surface area $A$ of a right circular cylinder is given by the formula $A = 2\pi rh + 2\pi r^2$, where $r$ is the radius and $h$ is the height.

**(a)** Solve the formula for $h$.
**(b)** Determine the height of a right circular cylinder whose surface area is $72\pi$ square centimeters and whose radius is 4 centimeters.

**33. Maximum Heart Rate** The model $M = -0.711A + 206.3$ was developed by Londeree and Moeschberger to determine the maximum heart rate $M$ of an individual who is age $A$. (*Source:* Londeree and Moeschberger, "Effect of Age and Other Factors on HR max," *Research Quarterly for Exercise and Sport,* 53(4), 297–304)

**(a)** Solve the model for $A$.
**(b)** According to this model, what is the age of an individual whose maximum heart rate is 160?

**34. Maximum Heart Rate** The model $M = -0.85A + 217$ was developed by Miller to determine the maximum heart rate $M$ of an individual who is age $A$. (*Source:* Miller et al., "Predicting max HR," *Medicine and Science in Sports and Exercise,* 25(9), 1077–1081)

**(a)** Solve the model for $A$.
**(b)** According to this model, what is the age of an individual whose maximum heart rate is 160?

**35. Finance** The formula $A = P(1 + r)^t$ can be used to relate the future value $A$ of a deposit of $P$ dollars in an account that earns an annual interest rate $r$ (expressed as a decimal) after $t$ years.

**(a)** Solve the formula for $P$.
**(b)** How much would you have to deposit today in order to have \$5000 in 5 years in a bank account that pays 4% annual interest?

**36. Federal Taxes** According to the tax code in 2004, a married couple filing a joint income tax return that earns over \$142,700 per year is subject to having their itemized deductions reduced. The formula $P = D - 0.03(I - 142{,}700)$ can be used to determine the permitted deductions $P$ where $D$ represents the amount of deductions from Schedule A and $I$ represents the couples adjusted gross income.

**(a)** Solve the formula for $I$.
**(b)** Determine the adjusted gross income of a married couple filing a joint return whose allowed deductions were \$16,941 and Schedule A deductions were \$18,000.

**37. Supplementary Angles** Two angles are **supplementary** if the sum of the measures of the angles is 180°. If one angle is 30° more than its supplement, find the measures of the two angles.

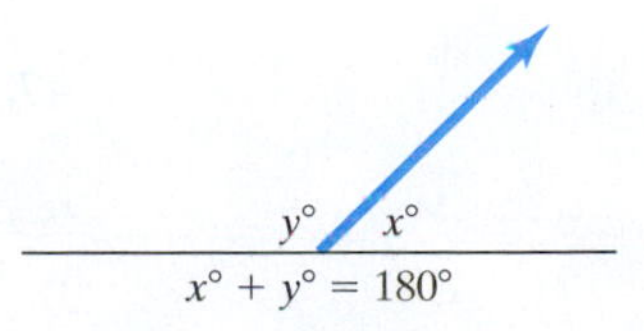

**38. Supplementary Angles** See Problem 37. If one angle is twice the measure of its supplement, find the measures of the two angles.

**39. Complementary Angles** Two angles are **complementary** if the sum of the measures of the angles is 90°. If one angle is ten more than 3 times its complement, find the measures of the two angles.

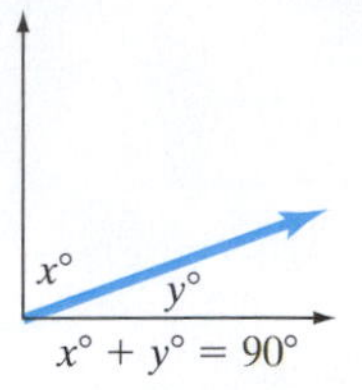


**40. Complementary Angles** See Problem 39. If one angle is 30° less than twice its complement, find the measures of the two angles.

**41. Dimensions of a Window** The perimeter of a rectangular window is 26 feet. The width of the window is 3 more than the length. What are the dimensions of the window?

**42. Dimensions of a Window** The perimeter of a rectangular window is 120 inches. The length of the window is twice its width. What are the dimensions of the window?

**43. Art** An artist wants to place a piece of round stained glass into a square that is made of copper wire. See the figure. If the perimeter of the square is 40 inches, what is the area of the largest circular piece of stained glass that can fit into the copper square?

**44. Art** An artist wants to place two circular pieces of stained glass into a rectangle that is made of copper wire. See the figure. The perimeter of the rectangle is 36 centimeters and the length is twice the width. Find the area of each circle assuming they are to be as large as possible and still fit inside the rectangle.

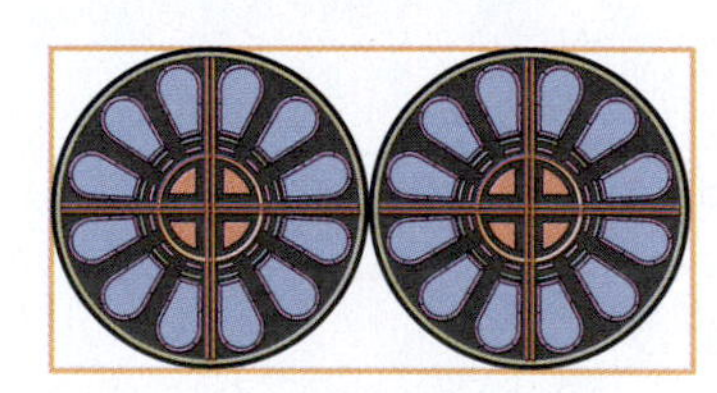

**45. Angles in a Triangle** The sum of the measure of the interior angles in a triangle is 180°. The measure of the second angle is 15° more than the measure of the first angle. The measure of the third angle is 45° more than the measure of the first angle. Find the measures of the interior angles in the triangle.

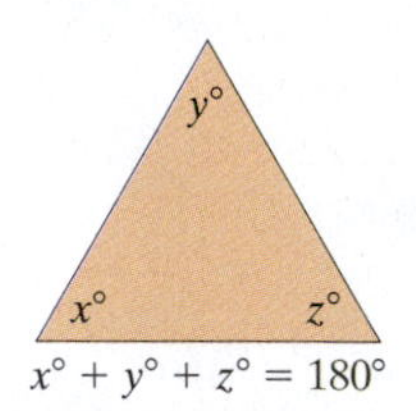


**46. Angles in a Triangle** See Problem 45. The measure of the second angle is 3 times the measure of the first angle. The measure of the third angle is 20° more than the measure of the first angle. Find the measure of the interior angles in the triangle.

**47. Designing a Patio** Suppose that you wish to build a rectangular cement patio that is to have a perimeter of 80 feet. The length is to be 5 feet more than the width.

**(a)** Find the dimensions of the patio.

**(b)** If building code requires that the cement be 4 inches deep, how much cement do you have to purchase?

**48. Designing a Foundation** You have just purchased a circular gazebo whose diameter is 12 feet. Before you have the gazebo delivered, you must lay a cement foundation to place the gazebo on.

**(a)** Find the area of the base of the gazebo.

**(b)** If the building code requires that the cement be 4 inches deep, how much cement do you have to purchase?

### Extending the Concepts

**49. Critical Thinking** Suppose that you have just purchased a swimming pool whose diameter is 25 feet. You decide to build a deck around the pool that is 3 feet wide.

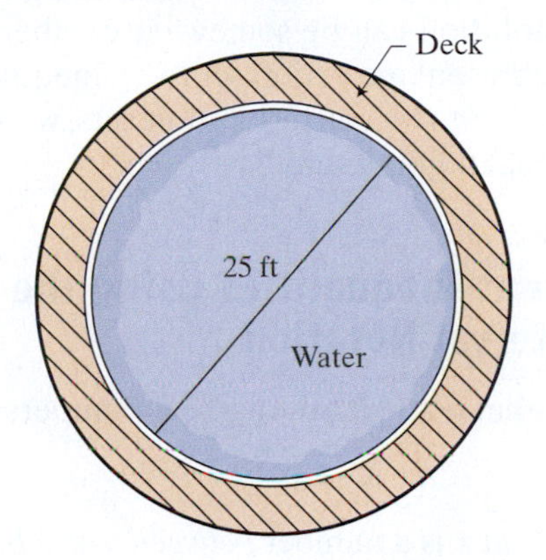


**(a)** What is the area of the deck?

**(b)** Building code requires that the pool must be enclosed in a fence. How much fence is required to encircle the pool?

**(c)** If the fence costs \$25 per linear foot to install, how much will the fence cost?

**50. Critical Thinking** Suppose that you wish to install a window whose dimensions are given in the figure.

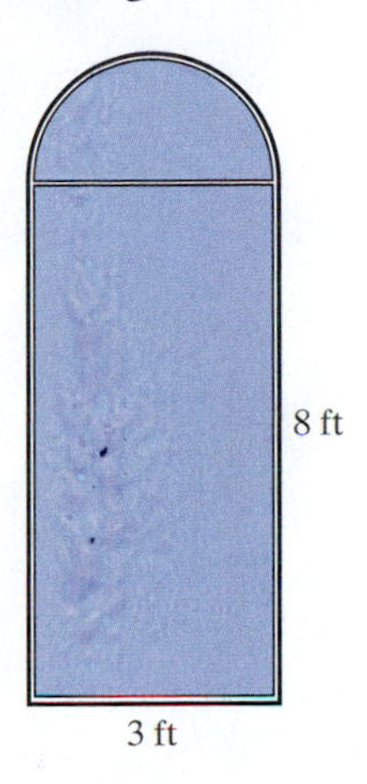


**(a)** What is the area of the opening of the window?

**(b)** What is the perimeter of the window?

**(c)** If glass costs \$8.25 per square foot, what is the cost of the glass for the window?

# 1.4 Linear Inequalities

## OBJECTIVES

1. Represent Inequalities Using the Real Number Line and Interval Notation
2. Understand the Properties of Inequalities
3. Solve Linear Inequalities
4. Solve Problems Involving Linear Inequalities

### *Preparing for Linear Inequalities*

*Before getting started, take the following readiness quiz. If you get a problem wrong, go back to the section cited and review the material.*

*In Problems 1–4, replace the question mark by $<$, $>$, or $=$ to make the statement true. [Section R.2, pp. 16–17]*

**1.** $3 \; ? \; 6$ **2.** $-3 \; ? \; -6$ **3.** $\frac{1}{2} \; ? \; 0.5$ **4.** $\frac{2}{3} \; ? \; \frac{3}{5}$

*In Problem 5, answer True or False.*

**5.** The inequality $\geq$ is called a strict inequality. [Section R.2, p. 16]

**6.** Use set builder notation to represent the set of all digits that are divisible by 3. [Section R.2, pp. 8–11]

An **inequality in one variable** is a statement involving two expressions, at least one containing the variable, separated by one of the inequality symbols $<$, $\leq$, $>$, or $\geq$. To **solve an inequality** means to find all values of the variable for which the statement is true. These values are called **solutions** of the inequality. The set of all solutions is called the **solution set.**

***Preparing for...Answers*** **1.** $<$ **2.** $>$ **3.** $=$ **4.** $>$ **5.** False **6.** $\{x \mid x \text{ is a digit that is divisible by } 3\}$

**DEFINITION**

A **linear inequality in one variable** is an inequality that can be written in the form

$$ax + b < c \quad \text{or} \quad ax + b \leq c \quad \text{or} \quad ax + b > c \quad \text{or} \quad ax + b \geq c$$

where $a$, $b$, and $c$ are real numbers and $a \neq 0$.

For example, the following are all linear inequalities involving one variable:

$$x - 4 > 9 \qquad 5x - 1 \leq 14 \qquad 8z < 0 \qquad 5x - 1 \geq 3x + 8$$

Before we discuss methods for solving linear inequalities, we will present three ways of representing the solution set. One of the methods for representing the solution set is through set-builder notation—something we are already familiar with. However, set-builder notation can be somewhat cumbersome, so we introduce a more streamlined way to represent a solution set to an inequality using *interval notation.* Finally, because we often like to visualize solution sets, we present a method for graphing the solution set on a real number line.

## 1 Represent Inequalities Using the Real Number Line and Interval Notation

Suppose that $a$ and $b$ are two real numbers and $a < b$. We shall use the notation

$$a < x < b$$

to mean that $x$ is a number *between* $a$ and $b$. So, the expression $a < x < b$ is equivalent to the two inequalities $a < x$ and $x < b$. Similarly, the expression $a \leq x \leq b$ is equivalent to the two inequalities $a \leq x$ and $x \leq b$. We define $a \leq x < b$ and $a < x \leq b$ similarly. Expressions such as $-2 < x < 5$ or $x \geq 5$ are said to be in **inequality notation.**

**Work Smart**

Remember that the inqualities $<$ and $>$ are called strict inequalities, while $\leq$ and $\geq$ are called nonstrict inequalities.

While the expression $3 \geq x \geq 2$ is technically correct, it is not the preferred way to write inequalities. For ease of reading, we prefer that the numbers in the inequality go from smaller values to larger values. So, we would write $3 \geq x \geq 2$ as $2 \leq x \leq 3$.

A statement such as $3 \leq x \leq 1$ is false because there is no number $x$ for which $3 \leq x$ and $x \leq 1$. We also never mix inequalities as in $2 \leq x \geq 3$.

In addition to representing inequalities using inequality notation, we can use *interval notation.*

**DEFINITION: INTERVAL NOTATION**

Let $a$ and $b$ represent two real numbers with $a < b$.

A **closed interval,** denoted by $[a, b]$, consists of all real numbers $x$ for which $a \leq x \leq b$.

An **open interval,** denoted by $(a, b)$, consists of all real numbers $x$ for which $a < x < b$.

The **half-open,** or **half-closed, intervals** are $(a, b]$, consisting of all real numbers $x$ for which $a < x \leq b$, and $[a, b)$, consisting of all real numbers $x$ for which $a \leq x < b$.

In each of these definitions, $a$ is called the **left endpoint** and $b$ is called the **right endpoint** of the interval.

The symbol $\infty$ (read as "infinity") is not a real number, but a notational device used to indicate unboundedness in the positive direction. In other words, the symbol $\infty$ means that there is no right endpoint on the inequality. The symbol $-\infty$ (read as "minus infinity" or "negative infinity") also is not a real number, but a notational device used to indicate unboundedness in the negative direction. The symbol $-\infty$ means that there is no left endpoint on the inequality. Using the symbols $\infty$ and $-\infty$, we can define five other kinds of intervals.

**Work Smart**

The symbols $\infty$ and $-\infty$ are never included as endpoints because they are not real numbers. So, we use parentheses when $-\infty$ or $\infty$ are endpoints.

**INTERVALS INCLUDING $\infty$**

$[a, \infty)$ consists of all real numbers $x$ for which $x \geq a$

$(a, \infty)$ consists of all real numbers $x$ for which $x > a$

$(-\infty, a]$ consists of all real numbers $x$ for which $x \leq a$

$(-\infty, a)$ consists of all real numbers $x$ for which $x < a$

$(-\infty, \infty)$ consists of all real numbers $x$ (or $-\infty < x < \infty$)

In addition to representing inequalities using interval notation, we can represent inequalities using a graph on the real number line. The inequality $x > 3$ or the interval $(3, \infty)$ consists of all numbers $x$ that lie to the right of 3 on the real number line. We can represent these values by shading the real number line to the right of 3. To indicate that 3 is not included in the set, we will agree to use a parenthesis on the endpoint. See Figure 9.

**Figure 9**

$x > 3$

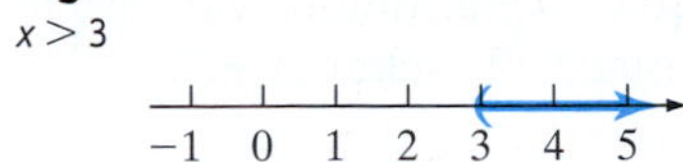


To represent the inequality $x \geq 3$ or the interval $[3, \infty)$ graphically, we also shade to the right of 3, but this time we use a bracket on the endpoint to indicate that 3 is included in the set. See Figure 10.

**Figure 10**

$x \geq 3$

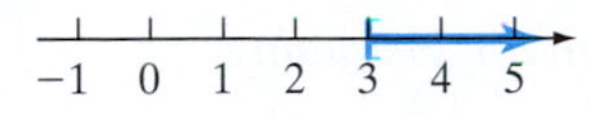


Table 7 summarizes interval notation, inequality notation, and their graphs.

**Table 7**

| Interval | Inequality Notation | Graph |
|---|---|---|
| The open interval $(a, b)$ | $\{x \mid a < x < b\}$ | |
| The closed interval $[a, b]$ | $\{x \mid a \leq x \leq b\}$ | |
| The half-open interval $[a, b)$ | $\{x \mid a \leq x < b\}$ | |
| The half-open interval $(a, b]$ | $\{x \mid a < x \leq b\}$ | |
| The interval $[a, \infty)$ | $\{x \mid x \geq a\}$ | |
| The interval $(a, \infty)$ | $\{x \mid x > a\}$ | |
| The interval $(-\infty, a]$ | $\{x \mid x \leq a\}$ | |
| The interval $(-\infty, a)$ | $\{x \mid x < a\}$ | |
| The interval $(-\infty, \infty)$ | $\{x \mid x \text{ is a real number}\}$ | |

## EXAMPLE 1 Using Interval Notation and Graphing Inequalities

Write each inequality using interval notation. Graph the inequality.

**(a)** $-2 \leq x \leq 4$

**(b)** $1 < x \leq 5$

**Solution**

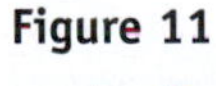

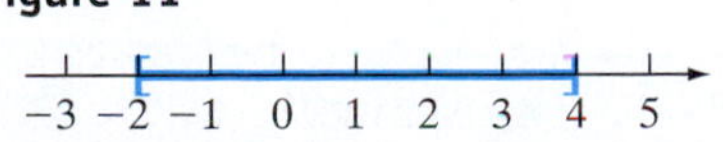


**(a)** $-2 \leq x \leq 4$ describes all numbers $x$ between $-2$ and 4, inclusive. In interval notation, we write $[-2, 4]$. To graph $-2 \leq x \leq 4$, we place brackets at $-2$ and 4 and shade in between. See Figure 11.

**Figure 12**

−1 0 1 2 3 4 5 6

**(b)** $1 < x \leq 5$ describes all numbers $x$ greater than 1 and less than or equal to 5. In interval notation, we write $(1, 5]$. To graph $1 < x \leq 5$, we place a parenthesis at 1 and a bracket at 5 and shade in between. See Figure 12.

## EXAMPLE 2 Using Interval Notation and Graphing Inequalities

Write each inequality using interval notation. Graph the inequality.

**(a)** $x < 2$ **(b)** $x \geq -3$

### Solution

**(a)** $x < 2$ describes all numbers $x$ less than 2. In interval notation, we write $(-\infty, 2)$. To graph $x < 2$, we place a parenthesis at 2 and then shade to the left. See Figure 13.

**Figure 13**

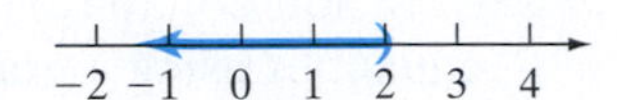


**(b)** $x \geq -3$ describes all numbers $x$ greater than or equal to $-3$. In interval notation, we write $[-3, \infty)$. To graph $x \geq -3$, we place a bracket at $-3$ and then shade to the right. See Figure 14.

**Figure 14**

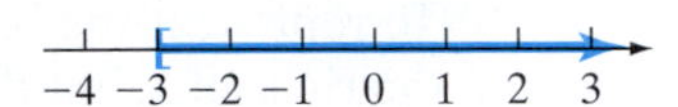


**QUICK ✓** *Write each inequality in interval notation. Graph the inequality.*

**1.** $-3 \leq x \leq 2$ **2.** $3 \leq x < 6$ **3.** $x \leq 3$ **4.** $\frac{1}{2} < x < \frac{7}{2}$

---

## EXAMPLE 3 Using Inequality Notation and Graphing Inequalities

Write each interval in inequality notation involving $x$. Graph the inequality.

**(a)** $[-2, 4)$ **(b)** $(1, 5)$

### Solution

**(a)** The interval $[-2, 4)$ consists of all numbers $x$ for which $-2 \leq x < 4$. See Figure 15 for the graph.

**Figure 15**

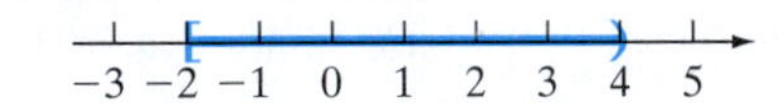


**(b)** The interval $(1, 5)$ consists of all numbers $x$ for which $1 < x < 5$. See Figure 16 for the graph.

**Figure 16**

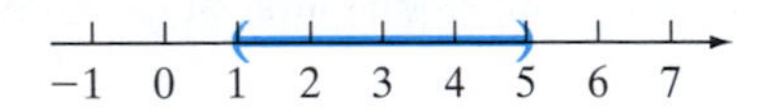


## EXAMPLE 4 Using Inequality Notation and Graphing Inequalities

Write each interval in inequality notation involving $x$. Graph the inequality.

**(a)** $\left[\frac{3}{2}, \infty\right)$ **(b)** $(-\infty, 1)$

### Solution

**(a)** The interval $\left[\frac{3}{2}, \infty\right)$ consists of all numbers $x$ for which $x \geq \frac{3}{2}$. See Figure 17 for the graph.

**Figure 17**

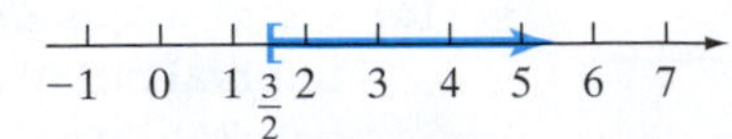


**(b)** The interval $(-\infty, 1)$ consists of all numbers $x$ for which $x < 1$. See Figure 18 for the graph.

**Figure 18**

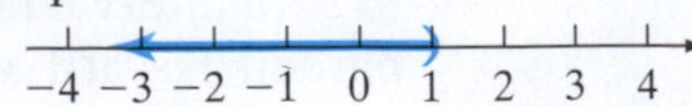

**QUICK ✓** *Write each interval as an inequality. Graph the inequality.*

**5.** $(0, 5]$ **6.** $(-6, 0)$ **7.** $(5, \infty)$ **8.** $\left(-\infty, \frac{8}{3}\right]$

## 2 Understand the Properties of Inequalities

Consider the inequality $2 < 5$. If we add 3 to both sides of the inequality, the expression on the left becomes 5 and the expression on the right becomes 8. Since $5 < 8$, we can see that adding the same quantity to both sides of an inequality does not change the sense, or direction, of the inequality. This result is called the *Addition Property of Inequalities.*

**In Words**
The Addition Property states that the direction of the inequality does not change when the same quantity is added to each side of the inequality.

**ADDITION PROPERTY OF INEQUALITIES**

For real numbers $a$, $b$, and $c$,

$$\text{If} \quad a < b, \quad \text{then} \quad a + c < b + c$$
$$\text{If} \quad a > b, \quad \text{then} \quad a + c > b + c$$

For example, since $2 < 5$, we have that $2 + 4 < 5 + 4$ or $6 < 9$. In addition, since $3 > -1$, we have that $3 + (-2) > -1 + (-2)$ or $1 > -3$.

We've seen what happens when we add a real number to both sides of an inequality. What happens when we multiply both sides by a non-zero constant? Let's see.

Consider the inequality $3 < 5$. Multiply both sides of the inequality by 2. The expression on the left side of the inequality becomes $2(3) = 6$ and the expression on the right becomes $2(5) = 10$. Certainly $6 < 10$, so the direction of the inequality did not change.

Again consider the inequality $3 < 5$. Now multiply both sides of the inequality by $-2$. The expression on the left side of the inequality becomes $-2(3) = -6$ and the expression on the right becomes $-2(5) = -10$. Because $-6 > -10$ we see that the direction of the inequality is reversed.

These results are true in general and lead us to the *Multiplication Properties for Inequalities.*

**In Words**
The Multiplication Property states that if both sides of an inequality are multiplied by a positive real number, the direction of the inequality is unchanged. If both sides of an inequality are multiplied by a negative real number, the direction of the inequality is reversed.

**MULTIPLICATION PROPERTIES FOR INEQUALITIES**

Let $a$, $b$, and $c$ be real numbers.

$$\text{If } a < b \text{ and if } c > 0, \text{ then } ac < bc$$
$$\text{If } a > b \text{ and if } c > 0, \text{ then } ac > bc$$

$$\text{If } a < b \text{ and if } c < 0, \text{ then } ac > bc$$
$$\text{If } a > b \text{ and if } c < 0, \text{ then } ac < bc$$

## 3 Solve Linear Inequalities

**In Words**
If the sides of an inequality are interchanged, the direction of the inequality reverses.

Two inequalities that have exactly the same solution set are called **equivalent inequalities.** As with equations, one method for solving a linear inequality is to replace it by a series of equivalent inequalities until an inequality with an obvious solution, such as $x > 2$, is obtained. We obtain equivalent inequalities by applying some of the same operations as those used to find equivalent equations. The Addition Property and Multiplication Properties form the basis for the procedures.

Although not essential, it is easier to read an inequality if the variable is placed on the left side and the constant on the right. If the variable does end up on the right side of the inequality, we can rewrite it with the variable on the left side using the fact that

$$a < x \quad \text{is equivalent to} \quad x > a$$
and
$$a > x \quad \text{is equivalent to} \quad x < a$$

## EXAMPLE 5 How to Solve a Linear Inequality

Solve the inequality $3x - 2 > 13$. Graph the solution set.

### Step-by-Step Solution

The goal in solving any linear inequality is to get the variable by itself with a coefficient of 1.

**Step 1:** We isolate the term containing the variable.

Add 2 to both sides (Addition Property):

$$3x - 2 > 13$$
$$3x - 2 + 2 > 13 + 2$$
$$3x > 15$$

**Step 2:** Now we want to get a coefficient of 1 on the variable.

Divide both sides by 3 (Multiplication Property):

$$\frac{3x}{3} > \frac{15}{3}$$
$$x > 5$$

The solution using set-builder notation is $\{x | x > 5\}$. The solution using interval notation is $(5, \infty)$. Figure 19 shows the graph of the solution set.

**Figure 19**

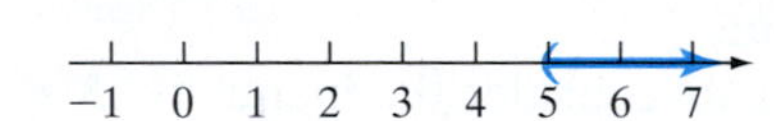

**QUICK ✓** *Solve each linear inequality. Express your solution using set-builder notation and interval notation. Graph the solution set.*

**9.** $x + 3 > 5$  **10.** $\frac{1}{3}x \leq 2$  **11.** $4x - 3 < 13$

## EXAMPLE 6 Solving Linear Inequalities

Solve the inequality $x - 4 \geq 5x + 12$.

### Solution

$$x - 4 \geq 5x + 12$$

Add 4 to both sides:

$$x - 4 + 4 \geq 5x + 12 + 4$$
$$x \geq 5x + 16$$

Subtract $5x$ from both sides:

$$x - 5x \geq 5x + 16 - 5x$$
$$-4x \geq 16$$

Divide both sides by $-4$. Don't forget to change the direction of the inequality:

$$\frac{-4x}{-4} \leq \frac{16}{-4}$$
$$x \leq -4$$

The solution using set-builder notation is $\{x | x \leq -4\}$. The solution using interval notation is $(-\infty, -4]$. See Figure 20 for the graph of the solution set.

**Figure 20**

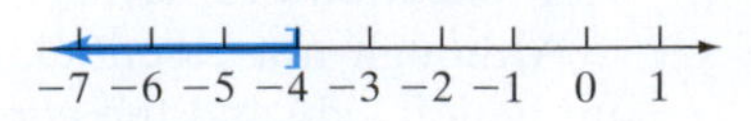

**QUICK ✓** *Solve each linear inequality. Express your solution using set-builder notation and interval notation. Graph the solution set.*

**12.** $3x + 1 > x - 5$ **13.** $-2x + 1 \le 3x + 11$ **14.** $-5x + 12 < x - 3$

---

### EXAMPLE 7 Solving Linear Inequalities

Solve the inequality $3(x - 1) + 2x < 6x + 3$.

**Solution**

$$3(x - 1) + 2x < 6x + 3$$

Distribute the 3: $$3x - 3 + 2x < 6x + 3$$

Combine like terms: $$5x - 3 < 6x + 3$$

Add 3 to both sides: $$5x - 3 + 3 < 6x + 3 + 3$$

$$5x < 6x + 6$$

Subtract $6x$ from both sides: $$5x - 6x < 6x + 6 - 6x$$

$$-x < 6$$

Multiply both sides of the inequality by $-1$: $$(-1)(-x) > (-1)6$$

$$x > -6$$

The solution using set-builder notation is $\{x | x > -6\}$. The solution using interval notation is $(-6, \infty)$. See Figure 21 for the graph of the solution set.

**Figure 21**

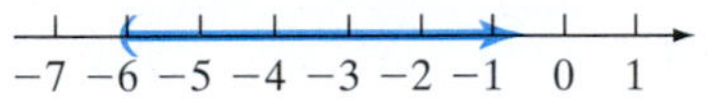

$-7 \quad -6 \quad -5 \quad -4 \quad -3 \quad -2 \quad -1 \quad 0 \quad 1$

---

**QUICK ✓** *Solve each linear inequality. Express your solution using set-builder notation and interval notation. Graph the solution set.*

**15.** $4(x - 2) < 3x - 4$ **16.** $-2(x + 1) \ge 4(x + 3)$

**17.** $7 - 2(x + 1) \le 3(x - 5)$

---

### EXAMPLE 8 Solving Linear Inequalities Involving Fractions

Solve the inequality $\dfrac{2x + 1}{3} > \dfrac{x - 2}{2}$.

**Solution**

We begin by clearing the linear inequality of fractions by multiplying both sides of the inequality by 6, the Least Common Denominator.

$$6 \cdot \left(\frac{2x + 1}{3}\right) > 6 \cdot \left(\frac{x - 2}{2}\right)$$

$$2(2x + 1) > 3(x - 2)$$

Distribute: $$4x + 2 > 3x - 6$$

Subtract 2 from both sides: $$4x + 2 - 2 > 3x - 6 - 2$$

$$4x > 3x - 8$$

Subtract $3x$ from both sides: $$4x - 3x > 3x - 8 - 3x$$

$$x > -8$$

The solution using set-builder notation is $\{x | x > -8\}$. The solution using interval notation is $(-8, \infty)$. See Figure 22 for the graph of the solution set.

**Figure 22**

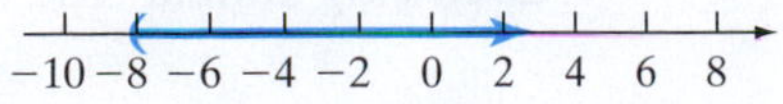

$-10 \quad -8 \quad -6 \quad -4 \quad -2 \quad 0 \quad 2 \quad 4 \quad 6 \quad 8$

**QUICK ✓** *Solve each linear inequality. Express your solution using set-builder notation and interval notation. Graph the solution set.*

**18.** $\frac{3x + 1}{5} \geq 2$ **19.** $\frac{2}{5}x + \frac{3}{10} < \frac{1}{2}$ **20.** $\frac{1}{2}(x + 3) > \frac{1}{3}(x - 4)$

## 4 Solve Problems Involving Linear Inequalities

When you are confronted with a word problem, one of the first things that you need to do is look for key words that tip you off as to the type of word problem that it is. There are certain phrases that frequently occur in problems that lead to linear inequalities. We list some of these phrases for you in Table 7.

**Table 7**

| Phrase | Inequality |
|---|---|
| At least | $\geq$ |
| No less than | $\geq$ |
| More than | $>$ |
| Greater than | $>$ |
| No more than | $\leq$ |
| At most | $\leq$ |
| Fewer than | $<$ |
| Less than | $<$ |

When solving applications involving linear inequalities, we use the same steps for setting up applied problems that we introduced in Section 1.2 on pages 65–66.

### EXAMPLE 9 Comparing Credit Cards

BankOne has offered you two different credit card options. The Southwest rewards card charges an annual fee of \$39 plus 12.90% simple interest on all outstanding balances. The Marriott rewards card charges an annual fee of \$30 plus 14.15% simple interest on all outstanding balances. What annual balance results in the Southwest card costing less than the Marriott card? (*Source:* BankOne.com)

#### Solution

**Step 1: Identify** We want to know the credit card balance for which the Southwest credit card costs less than the Marriott rewards card. The phrase "costs less" implies that this is an inequality problem.

**Step 2: Name** Let $b$ represent the credit card balance.

**Step 3: Translate** Each card charges an annual fee plus simple interest. So, for each card the cost will be "annual fee + interest."

Recall from Section 1.2 that simple interest is found using the formula

$$I = Prt$$

where $I$ is the interest charged, $P$ is the balance on the credit card, $r$ is the annual interest rate, and $t$ is time.

In this problem, we let $b$ represent the outstanding balance. The annual interest rate $r$ will either be 0.129 (for Southwest) or 0.1415 for (Marriott). Because we are discussing annual cost, we have that $t = 1$.

Since we want to know what balance results in Southwest costing less than Marriott, we have the following inequality:

Southwest Credit Card Cost < Marriott Credit Card Cost

$$\underbrace{39}_{\text{Annual Fee for Southwest}} + \underbrace{0.129b}_{\text{Interest Charged by Southwest}} < \underbrace{30}_{\text{Annual Fee for Marriott}} + \underbrace{0.1415b}_{\text{Interest Charged by Marriott}} \quad \text{The Model}$$

**Step 4: Solve** Solve the inequality for $b$.

$$39 + 0.129b < 30 + 0.1415b$$

Subtract 39 from both sides: $0.129b < -9 + 0.1415b$

Subtract $0.1415b$ from both sides: $-0.0125b < -9$

Divide both sides by $-0.0125$. Don't forget to reverse the inequality symbol: $b > 720$

**Step 5: Check** If the balance is \$720, then the annual cost for Southwest is $39 + 0.129(720) = \$131.88$. The annual cost for Marriott is $30 + 0.1415(720) = \$131.88$. For a balance greater than \$720, say \$750, the annual cost for Southwest is $39 + 0.129(750) = \$135.75$. The annual cost for Marriott is $30 + 0.1415(750) = \$136.13$.

**Step 6: Answer the Question** If the annual balance is greater than \$720, then Southwest offers a better deal than Marriott.

**21.** You have just received two credit card applications in the mail. The card from Bank A has an annual fee of \$25 and charges 9.9% simple interest. The card from Bank B has no annual fee, but charges 14.9% simple interest. For what annual balance will the card from Bank A cost less than the card from Bank B?

**22.** Suppose the daily revenue from selling $x$ boxes of candy is given by the equation $R = 12x$. The daily cost of operating the store and making the candy is given by the equation $C = 8x + 96$. For how many boxes of candy will revenue exceed costs? That is, solve $R > C$.

# 1.4 Exercises

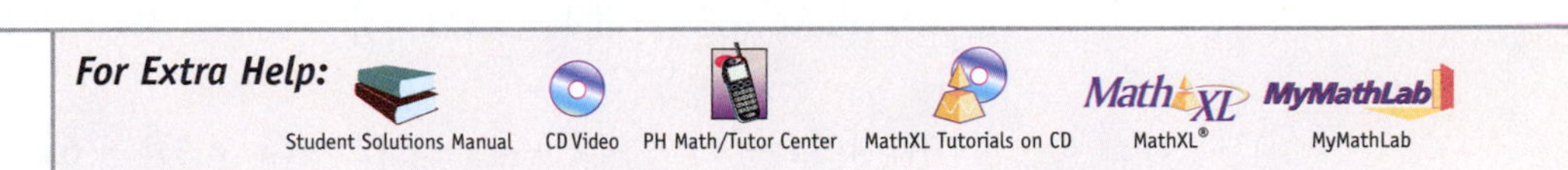


## Concepts and Vocabulary

*In Problems 1–3, fill in the blanks.*

**1.** A(n) _____ _____, denoted $[a, b]$, consists of all real numbers $x$ for which $a \le x \le b$.

**2.** The _____ _____ states that the direction, or sense, of an inequality remains the same if each side is multiplied by a positive number, while the direction is reversed is each side is multiplied by a negative number.

**3.** In the interval $[a, b]$, $a$ is called the _____ _____ and $b$ is called the _____ _____ of the interval.

*In Problems 4–6, answer True or False to each statement. In each statement, assume that $a < b$ and $c < 0$.*

**4.** $a \pm c < b \pm c$ **5.** $ac > bc$ **6.** $\dfrac{a}{c} < \dfrac{b}{c}$

**7.** Explain why it is incorrect to write $x \geq 4$ in interval notation as $[4, \infty]$.

**8.** Explain why it is incorrect to write $x < 4$ as $(4, -\infty)$.

## Building Skills

*In Problems 9–46, solve each linear inequality. Express your solution using set-builder notation and interval notation. Graph the solution set.*

**9.** $x - 4 \leq 2$ **10.** $x + 6 < 9$ **11.** $6x < 24$ **12.** $4x \geq 20$

**13.** $-7x < 21$ **14.** $-8x > 32$ **15.** $\frac{4}{15}x > \frac{8}{5}$ **16.** $\frac{3}{8}x < \frac{9}{16}$

**17.** $3x + 2 > 11$ **18.** $5x - 4 \leq 16$ **19.** $-3x + 1 > 13$ **20.** $-6x - 5 < 13$

**21.** $6x + 5 \leq 3x + 2$ **22.** $8x + 3 \geq 5x - 9$

**23.** $-3x + 1 < 2x + 11$ **24.** $3x + 4 \geq 5x - 8$

**25.** $3(x - 3) < 2(x + 4)$ **26.** $3(x - 2) + 5 > 4(x + 1) + x$

**27.** $4(x + 1) - 2x \geq 5(x - 2) + 2$ **28.** $-3(x + 4) + 5x < 4(x + 3) - 14$

**29.** $0.5x + 4 \leq 0.2x - 5$ **30.** $2.3x - 1.2 > 1.8x + 0.4$

**31.** $\frac{3x + 1}{4} < \frac{1}{2}$ **32.** $\frac{2x - 3}{3} > \frac{4}{3}$

**33.** $\frac{1}{2}(x - 4) > \frac{3}{4}(2x + 1)$ **34.** $\frac{1}{3}(3x + 5) < \frac{1}{6}(x + 4)$

**35.** $\frac{3}{5} - x > \frac{5}{3}$ **36.** $\frac{2}{3} - \frac{5}{6}x > 2$

**37.** $-5(x - 3) \geq 3[4 - (x + 4)]$ **38.** $-3(2x + 1) \leq 2[3x - 2(x - 5)]$

**39.** $4(3x - 1) - 5(x + 4) \geq 3[2 - (x + 3)] - 6x$

**40.** $7(x + 2) - 4(2x + 3) < -2[5x - 2(x + 3)] + 7x$

**41.** $\frac{2}{3}(4x - 1) - \frac{4}{9}(x - 4) > \frac{5}{12}(2x + 3)$

**42.** $\frac{5}{6}(3x - 2) - \frac{2}{3}(4x - 1) < -\frac{2}{9}(2x + 5)$

**43.** $\frac{4x - 3}{3} < 3$ **44.** $\frac{2}{5}x + \frac{3}{10} < \frac{1}{2}$

**45.** $\frac{2}{3}x < \frac{1}{4}(2x + 3)$ **46.** $\frac{x}{12} \geq \frac{x}{2} - \frac{2x + 1}{4}$

## Mixed Practice

*In Problems 47–60, solve each linear inequality. Express your solution using set-builder notation and interval notation. Graph the solution set.*

**47.** $y + 8 > -7$

**48.** $y - 5 \geq 7$

**49.** $-3a \leq -21$

**50.** $-5x < 30$

**51.** $13x - 5 > 10x - 6$

**52.** $4x + 3 \geq -6x - 2$

**53.** $4(x + 2) \leq 3(x - 2)$

**54.** $5(y + 7) < 6(y + 4)$

**55.** $3(4 - 3x) > 6 - 5x$

**56.** $2(5 - x) - 3 \leq 4 - 5x$

**57.** $2[4 - 3(x + 1)] \leq -4x + 8$

**58.** $3[1 + 2(x - 4)] \geq 3x + 3$

**59.** $\frac{x}{2} + \frac{3}{4} \geq \frac{3}{8}$

**60.** $\frac{b}{3} + \frac{5}{6} < \frac{11}{12}$

## Applying the Concepts

**61.** Find the set of all $x$ such that the sum of twice $x$ and 5 is at least 13.

**62.** Find the set of all $x$ such that the difference between 3 times $x$ and 2 is less than 7.

**63.** Find the set of all $z$ such that the product of 4 and $z$ minus 3 is no more than 9.

**64.** Find the set of all $y$ such that the sum of twice $y$ and 3 is greater than 13.

**65. Computing Grades** In order to earn an $A$ in Mr. Ruffatto's Intermediate Algebra course, Jackie must earn at least 540 points. Thus far, Jackie has earned 90, 83, 95, and 90 points on her four exams. The final exam, which counts as 200 points, is rapidly approaching. How many points does Jackie need to earn on the final to earn an $A$ in Mr. Ruffatto's class?

**66. Computing Grades** In order to earn an $A$ in Mrs. Padilla's Intermediate Algebra course, Mark must obtain an average score of at least 90. On his first four exams Mark scored 94, 83, 88, and 92. The final exam counts as two test scores. What score does Mark need on the final to earn an $A$ in Mrs. Padilla's class?

**67. McDonald's** Suppose that you have ordered one medium order of French Fries and one 16 ounce triple thick chocolate shake from McDonald's. The fries have 22 grams of fat and the shake has 17 grams of fat. Each McDonald's hamburger has 10 grams of fat. How many hamburgers can you order and still keep the total fat content of the meal no more than 69 grams? (*Source:* McDonald's Corporation)

**68. Burger King** Suppose that you have ordered one medium order of onion rings and one 16 ounce chocolate shake from Burger King. The onion rings have 16 grams of fat and the shake has 8 grams of fat. Each cheeseburger has 21 grams of fat. How many cheeseburgers can you order and still keep the total fat content of the meal no more than 87 grams? (*Source:* Burger King)

69. **Payload Restrictions** An Airbus A320 has a maximum payload of 45,686 pounds. Suppose that on a flight from Chicago to New Orleans, United Airlines sold out the flight at 179 seats. Assuming that the average passenger weighs 150 pounds, determine the maximum weight of the luggage and other cargo that the plane can carry. (*Source:* Airbus)

70. **Moving Trucks** A 15-foot moving truck from Budget costs \$39.95 per day plus \$0.65 per mile. If your budget only allows for you to spend \$125.75, what is the maximum number of miles you can drive? (*Source:* Budget)

71. **Health Benefits** The average monthly benefit $B$, in dollars, for individuals on disability is given by the equation $B = 19.25t + 585.72$, where $t$ is the number of years since 1990. In what year will the monthly benefit $B$ exceed \$1000? That is, solve $B > 1000$.

72. **Health Expenditures** Total private health expenditures $H$, in billions of dollars, are given by the equation $H = 26t + 411$, where $t$ is the number of years since 1990. In what year will total private health expenditures exceed \$1 trillion (\$1000 billion)? That is, solve $26t + 411 > 1000$.

73. **Commissions** Susan sells computer systems. Her annual base salary is \$34,000. She also earns a commission of 1.2% on the sale price of all computer systems that she sells. For what value of the computer systems sold will Susan's annual salary be at least \$100,000?

74. **Commissions** Al sells used cars for a Chevy dealer. His annual base salary is \$24,300. He also earns a commission of 3% on the sale price of the cars that he sells. For what value of the cars sold will Al's annual salary be more than \$60,000? If used cars at this particular dealership sell for an average of \$15,000, how many cars does Al have to sell to meet his salary goal?

75. **Supply and Demand** The quantity demanded of custom monogrammed shirts is given by the equation $D = 1000 - 20p$. The quantity supplied of custom monogrammed shirts is given by the equation $S = -200 + 10p$, where $p$ is the price of a shirt. For what prices will quantity supplied exceed quantity demanded, thereby resulting in a surplus of shirts? That is, solve $S > D$.

76. **Supply and Demand** The quantity supplied of digital cameras is given by the equation $S = -2800 + 13p$. The quantity demanded of digital cameras is given by the equation $D = 1800 - 12p$, where $p$ is the price of a camera. For what prices will quantity demanded exceed quantity supplied, thereby resulting in a shortage of cameras? That is, solve $D > S$.

## Extending the Concepts

77. Solve the linear inequality $3(x + 2) + 2x > 5(x + 1)$.

78. Solve the linear inequality $-3(x - 2) + 7x > 2(2x + 5)$.

79. Write a brief paragraph that explains the circumstances under which the direction, or sense, of an inequality changes.

## PUTTING THE CONCEPTS TOGETHER (SECTIONS 1.1–1.4)

*These problems cover important concepts from Sections 1.1 to 1.4. We designed these problems so that you can review the chapter so far and show your mastery of the concepts. Take time to work these problems before proceeding with the next section. The answers to these problems are located at the back of the text starting on page AN-3.*

**1.** Determine which, if any, of the following are solutions to

$$5(2x - 3) + 1 = 2x - 6$$

**(a)** $x = -3$ **(b)** $x = 1$

*In Problems 2 and 3, solve the equation.*

**2.** $3(2x - 1) + 6 = 5x - 2$ **3.** $\frac{7}{3}x + \frac{4}{5} = \frac{5x + 12}{15}$

*Determine if the equation is an identity, a contradiction, or a conditional equation.*

**4.** $5 - 2(x + 1) + 4x = 6(x + 1) - (3 + 4x)$

*In Problems 5 and 6, translate the English statement into a mathematical statement. Do not solve the equation.*

**5.** The difference of a number and 3 is two more than half the number.

**6.** The quotient of a number and 2 is less than the number increased by 5.

**7. Mixture** Two acid solutions are available to a chemist. One is a 20% nitric acid solution and the other is a 40% nitric acid solution. How much of each type of solution should be mixed together to form 16 liters of a 35% nitric acid solution?

**8. Travel** Two cars leave from the same location and travel in opposite directions along a straight road. One car travels 30 miles per hour while the other travels at 45 miles per hour. How long will it take the two cars to be 255 miles apart?

**9.** Solve $3x - 2y = 4$ for $y$.

**10.** Solve the formula $A = P + Prt$ for $r$.

**11.** The volume of a right circular cylinder is given by the formula $V = \pi r^2 h$, where $r$ is the radius of the cylinder and $h$ is the height of the cylinder.

**(a)** Solve the formula for $h$.

**(b)** Use the result from part (a) to find the height of a right circular cylinder with volume $V = 294\pi$ in.$^3$ and radius $r = 7$ in.

**12.** Write the following inequalities in interval notation and graph on a real number line.

**(a)** $x > -3$ **(b)** $2 < x \leq 5$

**13.** Write the interval in inequality notation and graph the inequality.

**(a)** $(-\infty, -1.5]$ **(b)** $(-3, 1]$

*In Problems 14–16, solve the inequality and graph the solution set on a real number line.*

**14.** $2x + 3 \leq 4x - 9$ **15.** $-3 > 3x - (x + 5)$ **16.** $x - 9 \leq x + 3(2 - x)$

**17. Birthday Party** A recreational center offers a children's birthday party for $75 plus $5 for each child. How many children can Logan invite to his birthday party if the budget for the party is no more than $125?

# 1.5 Compound Inequalities

## OBJECTIVES

1. Determine the Intersection or Union of Two Sets.
2. Solve Compound Inequalities Involving "and"
3. Solve Compound Inequalities Involving "or"
4. Solve Problems Using Compound Inequalities

### *Preparing for Compound Inequalities*

*Before getting started, take the following readiness quiz. If you get a problem wrong, go back to the section cited and review the material.*

**1.** Use set-builder notation and the roster method to represent the set of all integers between $-2$ and 4, inclusive. [Section R.2, pp. 8–11]

**2.** Use set-builder notation and the roster method to represent the set of all positive integers less than 3. [Section R.2, pp. 8–11]

## 1 Determine the Intersection or Union of Two Sets

Consider the information presented in Table 8 regarding students enrolled in an Intermediate Algebra course.

**Table 8**

| Student | Age | Gender |
|---|---|---|
| Grace | 19 | Female |
| Sophia | 23 | Female |
| Kevin | 20 | Male |
| Robert | 32 | Male |
| Jack | 19 | Male |
| Mary | 35 | Female |
| Nancy | 40 | Female |
| George | 22 | Male |
| Teresa | 20 | Female |

We can classify the people in the course in a set using the roster method. For example, suppose we define set $A$ as the set of all students whose age is less than 25. Then

$$A = \{\text{Grace, Sophia, Kevin, Jack, George, Teresa}\}$$

Suppose we define set $B$ as the set of all students who are female. Then

$$B = \{\text{Grace, Sophia, Mary, Nancy, Teresa}\}$$

Now list all the students that are in set $A$ and set $B$. That is, list all the students who are less than 25 years of age and female.

$$A \text{ and } B = \{\text{Grace, Sophia, Teresa}\}$$

Now list all the students that are either in set $A$ or set $B$ or both.

$$A \text{ or } B = \{\text{Grace, Sophia, Kevin, Jack, George, Teresa, Mary, Nancy}\}$$

Figure 23 shows a Venn diagram illustrating the relation among $A$, $B$, $A$ and $B$, and $A$ or $B$. Notice that Grace, Sophia, and Teresa are in both $A$ and $B$, while Robert is neither in $A$ nor $B$.

**Figure 23**

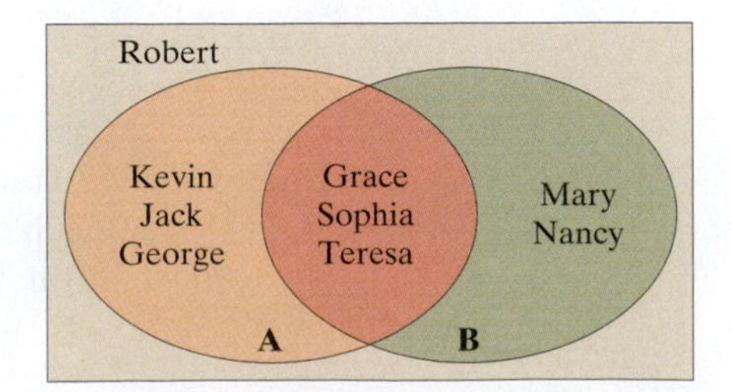


*Preparing for...Answers*

**1.** $\{x \mid -2 \le x \le 4, x \text{ is an integer}\}$; $\{-2, -1, 0, 1, 2, 3, 4\}$

**2.** $\{x \mid 0 < x < 3, x \text{ is an integer}\}$; $\{1, 2\}$

When we used the word *and* to obtain the set, we listed elements that were common to both set $A$ and set $B$. When we used the word *or* to obtain the set, we listed elements that were in either set $A$ or set $B$ or both. These results lead us to the following definitions.

**DEFINITIONS:**

- The **intersection** of two sets $A$ and $B$, denoted $A \cap B$, is the set of all elements that belong to both set $A$ and set $B$.
- The **union** of two sets $A$ and $B$, denoted $A \cup B$, is the set of all elements that are in the set $A$ or in the set $B$ or in both $A$ and $B$.
- The word **and** implies intersection, while the word **or** implies union.

## EXAMPLE 1 Finding the Intersection and Union of Sets

Let $A = \{1, 3, 5, 7, 9\}$ and let $B = \{1, 2, 3, 4, 5\}$. Find

**(a)** $A \cap B$ **(b)** $A \cup B$

### Solution

**(a)** $A \cap B$ is the set of all elements that are in both $A$ and $B$. So, $A \cap B = \{1, 3, 5\}$.

**(b)** $A \cup B$ is the set of all elements that are in $A$ or $B$, or both. So, $A \cup B = \{1, 2, 3, 4, 5, 7, 9\}$.

**Work Smart**
When finding the union of two sets, we only list each element once, even if it occurs in both sets.

**QUICK ✓** *Let* $A = \{1, 2, 3, 4, 5, 6\}$, $B = \{1, 3, 5, 7\}$, *and* $C = \{2, 4, 6, 8\}$.

**1.** Find $A \cap B$. **2.** Find $A \cap C$. **3.** Find $A \cup B$.
**4.** Find $A \cup C$. **5.** Find $B \cap C$. **6.** Find $B \cup C$.

Let's look at the intersection and union of two sets from the point of view of inequalities.

## EXAMPLE 2 Finding the Intersection and Union of Two Sets

Suppose $A = \{x | x \leq 5\}$, $B = \{x | x \geq 1\}$, and $C = \{x | x < -2\}$.

**(a)** Determine $A \cap B$. Graph the set on a real number line. Write the set $A \cap B$ using both set-builder notation and interval notation.

**(b)** Determine $B \cup C$. Graph the set on a real number line. Write the set $B \cup C$ using both set-builder notation and interval notation.

### Solution

**(a)** $A \cap B$ is the set of all real numbers that are less than or equal to 5 and greater than or equal to 1. We can identify this set by determining where the graphs of the inequalities overlap. See Figure 24.

**Figure 24**

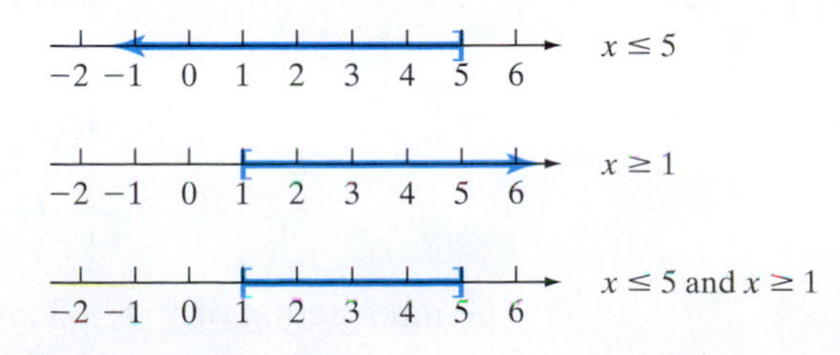


We can represent the set $A \cap B$ using set-builder notation as $\{x | 1 \leq x \leq 5\}$ or interval notation as $[1, 5]$.

**Work Smart**

Throughout the text, we will use the word "or" when using set-builder notation and we will use the union symbol, $\cup$, when using interval notation.

**(b)** $B \cup C$ is the set of all real numbers that are greater than or equal to 1 or less than $-2$. The union of these two sets would be all real numbers less than $-2$ or greater than or equal to 1. See Figure 25.

**Figure 25**

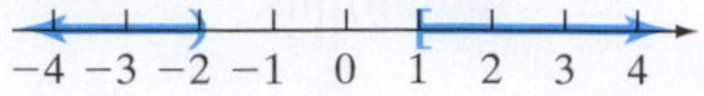

$-4\ -3\ -2\ -1\ 0\ 1\ 2\ 3\ 4$

We can represent the set $B \cup C$ using set-builder notation as $\{x \mid x < -2 \text{ or } x \geq 1\}$ or interval notation as $(-\infty, -2) \cup [1, \infty)$.

**QUICK ✓** *Let* $A = \{x \mid x > 2\}$, $B = \{x \mid x < 7\}$, *and* $C = \{x \mid x \leq -3\}$.

**7.** Determine $A \cap B$. Graph the set on a real number line. Write the set $A \cap B$ using both set-builder notation and interval notation.

**8.** Determine $A \cup C$. Graph the set on a real number line. Write the set $A \cup C$ using both set-builder notation and interval notation.

## 2 Solve Compound Inequalities Involving "and"

A **compound inequality** is formed by joining two inequalities with the word "and" or "or." For example,

$$3x + 1 > 4 \text{ and } 2x - 3 < 7$$
$$5x - 2 \leq 13 \text{ or } 2x - 5 > 3$$

are examples of compound inequalities. To **solve a compound inequality** means to find all possible values of the variable such that the compound inequality results in a true statement. For example, the compound inequality

$$3x + 1 > 4 \text{ and } 2x - 3 < 7$$

is true for $x = 2$, but false for $x = 0$.

Let's look at an example that illustrates how to solve compound inequalities involving the word "and."

### EXAMPLE 3 How to Solve a Compound Inequality Involving "and"

Solve $3x + 2 > -7$ and $4x + 1 \leq 9$. Graph the solution set.

**Step-by-Step Solution**

**Step 1:** Solve each inequality separately.

$$3x + 2 > -7 \qquad\qquad 4x + 1 \leq 9$$

Subtract 2 from both sides: $3x > -9$ Subtract 1 from both sides: $4x \leq 8$

Divide both sides by 3: $x > -3$ Divide both sides by 4: $x \leq 2$

**Step 2:** Find the intersection of the solution sets, which will represent the solution set to the compound inequality.

To find the intersection of the two solution sets, we graph each inequality separately. See Figure 26.

**Figure 26**

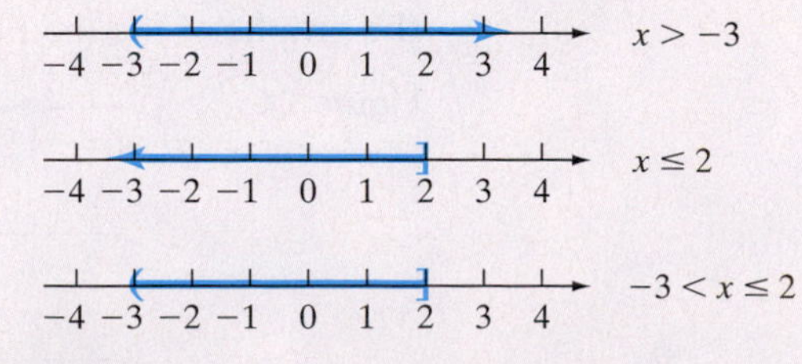


The intersection of $x > -3$ and $x \leq 2$ is $-3 < x \leq 2$.

The solution set is $\{x \mid -3 < x \leq 2\}$ or, using interval notation, $(-3, 2]$.

The steps below summarize the procedure for solving compound inequalities involving "and."

### Steps for Solving Compound Inequalities Involving "and"

**Step 1:** Solve each inequality separately.

**Step 2:** Find the INTERSECTION of the solution sets of each inequality.

### EXAMPLE 4 Solving a Compound Inequality with "and"

Solve $-2x + 5 > -1$ and $5x + 6 \le -4$. Graph the solution set.

**Solution**

We solve each inequality separately

| | | | |
|---|---|---|---|
| | $-2x + 5 > -1$ | | $5x + 6 \le -4$ |
| Subtract 5 from both sides: | $-2x > -6$ | Subtract 6 from both sides: | $5x \le -10$ |
| Divide both sides by $-2$; don't forget to reverse the direction of the inequality! | $x < 3$ | Divide both sides by 5: | $x \le -2$ |

Find the intersection of the solution sets, which will represent the solution set to the compound inequality. To find the intersection of the two solution sets, we graph each inequality separately. See Figure 27.

**Figure 27**

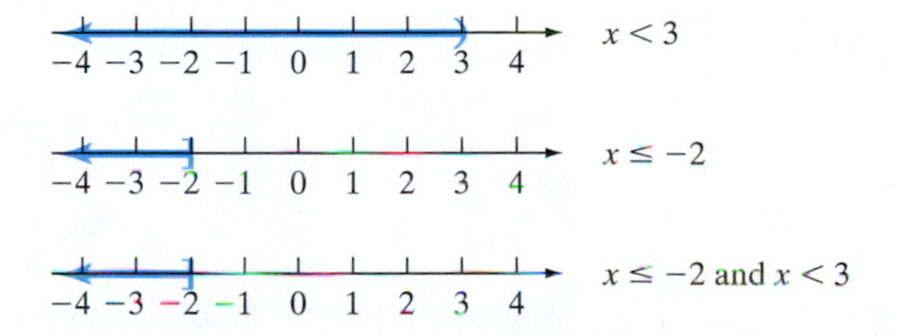


The intersection of $x < 3$ and $x \le -2$ is $x \le -2$. The solution set is $\{x | x \le -2\}$ or, using interval notation, $(-\infty, -2]$.

**QUICK ✓** *Solve each compound inequality. Express your solution using set-builder notation and interval notation. Graph the solution set.*

**9.** $2x + 1 \ge 5$ and $-3x + 2 < 5$

**10.** $4x - 5 < 7$ and $3x - 1 > -10$

**11.** $-8x + 3 < -5$ and $\frac{2}{3}x + 1 < 3$

### EXAMPLE 5 Solving a Compound Inequality with "and"

Solve $x - 5 > -1$ and $2x - 3 \le -5$. Graph the solution set.

**Solution**

We solve each inequality separately

| | | | |
|---|---|---|---|
| | $x - 5 > -1$ | | $2x - 3 \le -5$ |
| Add 5 to both sides: | $x > 4$ | Add 3 to both sides: | $2x \le -2$ |
| | | Divide both sides by 2: | $x \le -1$ |

Find the intersection of the solution sets, which will represent the solution set to the compound inequality. To find the intersection of the two solution sets, we graph each inequality separately. See Figure 28.

**Figure 28**

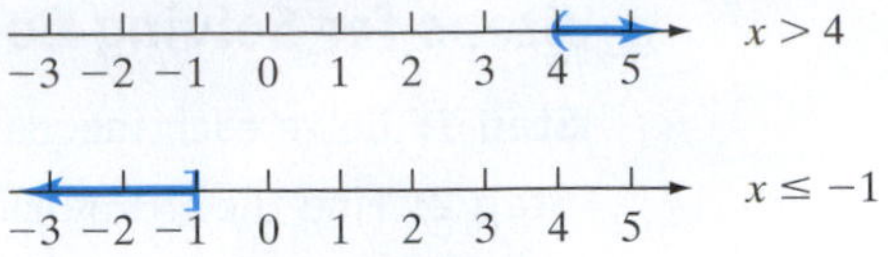

−3 −2 −1 0 1 2 3 4 5 $x > 4$

−3 −2 −1 0 1 2 3 4 5 $x \leq -1$

**Work Smart**
The braces are empty in an empty set.

The intersection of $x > 4$ and $x \leq -1$ is the empty set. The solution set is $\{\ \}$ or $\emptyset$.

**QUICK ✓** *Solve each compound inequality. Express your solution using set-builder notation and interval notation. Graph the solution set.*

**12.** $3x - 5 < -8$ and $2x + 1 > 5$ **13.** $5x + 1 \leq 6$ and $3x + 2 \geq 5$

Sometimes, we can combine "and" inequalities into a more streamlined notation.

**WRITING INEQUALITIES INVOLVING "AND" COMPACTLY**

If $a < b$, then we can write

$$a < x \quad \text{and} \quad x < b$$

more compactly as

$$a < x < b$$

For example, we can write

$$-3 < -4x + 1 \quad \text{and} \quad -4x + 1 < 13$$

as

$$-3 < -4x + 1 < 13$$

When compound inequalities come in this form, we solve the inequality by getting the variable by itself in the "middle" with a coefficient of 1.

## EXAMPLE 6 Solving a Compound Inequality

Solve $-3 < -4x + 1 < 13$ and graph the solution set.

### Solution

Our goal is to get the variable by itself in the "middle" with a coefficient of 1.

$$-3 < -4x + 1 < 13$$

Subtract 1 from all three parts (Addition Property): $-3 - 1 < -4x + 1 - 1 < 13 - 1$

$$-4 < -4x < 12$$

Divide all three parts by −4. Don't forget to reverse the direction of the inequalities. $\dfrac{-4}{-4} > \dfrac{-4x}{-4} > \dfrac{12}{-4}$

$$1 > x > -3$$

If $b > x > a$, then $a < x < b$: $-3 < x < 1$

The solution using set-builder notation is $\{x \mid -3 < x < 1\}$. The solution using interval notation is $(-3, 1)$. Figure 29 shows the graph of the solution set.

**Figure 29**

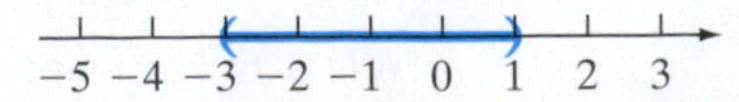

**QUICK ✓** *Solve each compound inequality. Express your solution using set-builder notation and interval notation. Graph the solution set.*

**14.** $-2 < 3x + 1 < 10$ **15.** $0 < 4x - 5 \le 3$

**16.** $3 \le -2x - 1 \le 11$

## 3 Solve Compound Inequalities Involving "or"

We now address compound inequalities involving the word "or." The solution to these types of inequalities is the union of the solutions to each inequality.

### EXAMPLE 7 How to Solve a Compound Inequality Involving "or"

Solve $3x - 5 < -2$ or $4 - 5x \le -16$. Graph the solution set.

#### Step-by-Step Solution

**Step 1:** Solve each inequality separately.

| | | | |
|---|---|---|---|
| | $3x - 5 < -2$ | | $4 - 5x \le -16$ |
| Add 5 to each side: | $3x < 3$ | Subtract 4 from both sides: | $-5x \le -20$ |
| Divide both sides by 3: | $x < 1$ | Divide both sides by $-5$: Don't forget to reverse the direction of the inequality. | $x \ge 4$ |

**Step 2:** Find the union of the solution sets, which will represent the solution set to the compound inequality.

The union of the two solution sets is $x < 1$ or $x \ge 4$. The solution set using set-builder notation is $\{x | x < 1 \text{ or } x \ge 4\}$. The solution set using interval notation is $(-\infty, 1) \cup [4, \infty)$. Figure 30 shows the graph of the solution set.

**Figure 30**

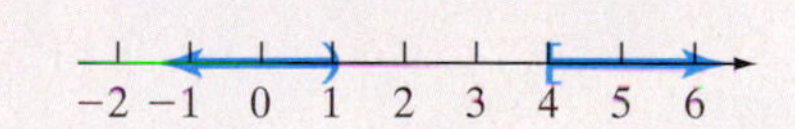


**Work Smart**

Remember, the union of sets $A$ and $B$ is the set of all elements that are in the set $A$ or in the set $B$.

Below is a summary of the steps for solving compound inequalities involving "or."

### Steps for Solving Compound Inequalities Involving "or"

**Step 1:** Solve each inequality separately.

**Step 2:** Find the UNION of the solution sets of each inequality.

**Work Smart**

A common error to avoid is to write the solution $x < 1$ or $x > 4$ as $1 > x > 4$, which is incorrect. There are no real numbers that are less than 1 *and* greater than 4. Another common error is to "mix" symbols as in $1 < x > 4$. This makes no sense!

**QUICK ✓** *Solve each compound inequality. Express your solution using set-builder notation and interval notation. Graph the solution set.*

**17.** $x + 3 < 1$ or $x - 2 > 3$ **18.** $3x + 1 \le 7$ or $2x - 3 > 9$

**19.** $2x - 3 \ge 1$ or $6x - 5 \ge 1$ **20.** $\frac{3}{4}(x + 4) < 6$ or $\frac{3}{2}(x + 1) > 15$

### EXAMPLE 8 Solving Compound Inequalities Involving "or"

Solve $\frac{1}{2}x - 1 < 1$ or $\frac{2x - 1}{3} \geq -1$. Graph the solution set.

**Solution**

First, we solve each inequality separately.

$$\frac{1}{2}x - 1 < 1 \qquad\qquad \frac{2x - 1}{3} \geq -1$$

Add 1 to each side: $\frac{1}{2}x < 2$ Multiply both sides by 3: $2x - 1 \geq -3$

Multiply both sides by 2: $x < 4$ Add 1 to both sides: $2x \geq -2$

Divide both sides by 2: $x \geq -1$

Find the union of the solution sets of each inequality. If we graph the solution set of each inequality separately, we notice that the union of the two solutions sets is the set of all real numbers. See Figure 31.

**Figure 31**

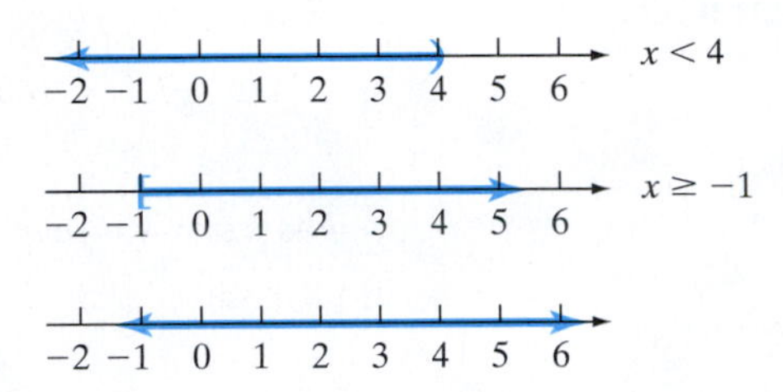


The solution set using set-builder notation is $\{x \mid x \text{ is any real number}\}$. The solution set using interval notation is $(-\infty, \infty)$.

**QUICK ✓** *Solve each compound inequality. Express your solution using set-builder notation and interval notation. Graph the solution set.*

**21.** $3x - 2 > -5$ or $2x - 5 \leq 1$ **22.** $-5x - 2 \leq 3$ or $7x - 9 > 5$

## 4 Solve Problems Using Compound Inequalities

We now look at an application involving compound inequalities.

### EXAMPLE 9 Federal Income Taxes

In 2005, a married couple filing a joint federal tax return whose income places them in the 25% tax bracket will pay federal income taxes between \$8180 and \$23,317.50, inclusive. The couple must pay federal income taxes equal to \$8180 plus 25% of the amount over \$59,400. Find the range of taxable income in order for a married couple to be in the 25% tax bracket. (*Source:* Internal Revenue Service)

**Solution**

**Step 1: Identify** We want to find the range of the taxable income for a married couple in the 25% tax bracket. This is a direct translation problem involving an inequality.

**Step 2: Name** We let $t$ represent the taxable income.

**Step 3: Translate** The federal tax bill equals \$8180 plus 25% of the taxable income over \$59,400. If the couple has taxable income equal to \$60,400, their tax bill will be

**Work Smart**

The word "range" tells us that an inequality is to be solved.

\$8180 plus 25% of \$1000 (\$1000 is the amount over \$59,400). In general, if the couple has taxable income $t$, then their tax bill will be

$$\underbrace{\$8180}_{8180} \underbrace{+}_{\text{plus}} \underbrace{0.25}_{25\%} \cdot \underbrace{(t - \$59,400)}_{\text{of the amount over } \$59,400}$$

Because the tax bill is between \$8180 and \$23,317.50, we have

$$8180 \leq 8180 + 0.25(t - 59,400) \leq 23,317.50 \quad \text{The Model}$$

**Step 4: Solve**

$$8180 \leq 8180 + 0.25(t - 59,400) \leq 23,317.50$$

Remove the parentheses by distributing 0.25: $8180 \leq 8180 + 0.25t - 14,850 \leq 23,317.50$

Combine like terms: $8180 \leq -6670 + 0.25t \leq 23,317.50$

Add 6670 to all three parts: $14,850 \leq 0.25t \leq 29,987.50$

Divide all three parts by 0.25: $59,400 \leq t \leq 119,950$

**Step 5: Check** If a married couple has taxable income of \$59,400, then their tax bill will be \$8180 + 0.25(\$59,400 − \$59,400) = \$8180. If a married couple has taxable income of \$119,950, then their tax bill will be \$8180 + 0.25(\$119,950 − \$59,400) = \$23,317.50.

**Step 6: Answer the Question** A married couple who files a joint tax return with a tax bill between \$8180 and \$23,317.50 has taxable income between \$59,400 and \$119,950.

**QUICK ✓**

**23.** In 2005, an individual filing a federal tax return whose income places them in the 25% tax bracket will pay federal income taxes between \$4090 and \$14,652.50. The individual must pay federal income taxes equal to \$4090 plus 25% of the amount over \$29,700. Find the range of taxable income in order for an individual to be in the 25% tax bracket. (*Source:* Internal Revenue Service)

**24.** AT&T offers a long-distance phone plan that charges \$4.95 per month plus \$0.07 per minute. During the course of a year, Sophia's long-distance phone bill ranges from \$13.00 to \$22.80. What was the range of monthly minutes?

# 1.5 Exercises

***For Extra Help:*** Student Solutions Manual | CD Video | PH Math/Tutor Center | MathXL Tutorials on CD | MathXL® | MyMathLab

## Concepts and Vocabulary

*In Problems 1–3, fill in the blanks.*

**1.** The _____ of two sets $A$ and $B$, denoted $A \cap B$, is the set of all elements that belong to both set $A$ and set $B$.

**2.** The word _____ implies intersection. The word _____ implies union.

**3.** A _____ _____ is formed by joining two inequalities with the word "and" or "or."

*In Problems 4–6, answer True or False to each statement.*

**4.** The intersection of two sets can be the empty set.

**5.** The symbol for the union of two sets is $\cap$.

**6.** The inequalities $5 > x > -2$ and $-2 < x < 5$ are equivalent.

**7.** Explain why the inequality $4 < x < 2$ makes no sense.

**8.** Explain why it is incorrect to write $-3 < x > 2$.

**9.** Is $x = 3$ a solution of $3x + 1 > 4$ and $2x - 3 < 7$?

**10.** Is $x = -1$ a solution of $5x - 2 \le 13$ or $2x - 5 > 3$?

## Building Skills

*In Problems 11–16, use $A = \{4, 5, 6, 7, 8, 9\}$, $B = \{1, 5, 7, 9\}$, and $C = \{2, 3, 4, 6\}$ to find each set.*

**11.** $A \cup B$ **12.** $A \cup C$ **13.** $A \cap B$

**14.** $A \cap C$ **15.** $B \cap C$ **16.** $B \cup C$

*In Problems 17–20, use the graph of the inequality to find each set.*

**17.** $A = \{x | x \le 5\}$; $B = \{x | x > -2\}$. Find (a) $A \cap B$ and (b) $A \cup B$.

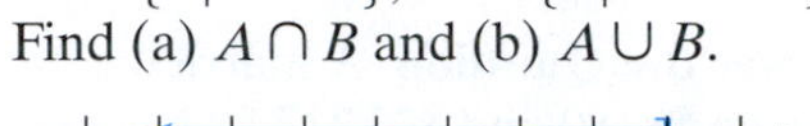

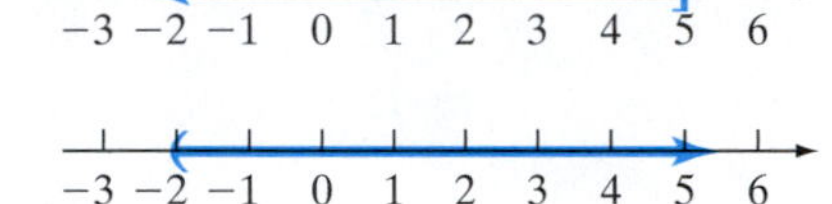

**18.** $A = \{x | x \ge 4\}$; $B = \{x | x < 1\}$. Find (a) $A \cap B$ and (b) $A \cup B$.

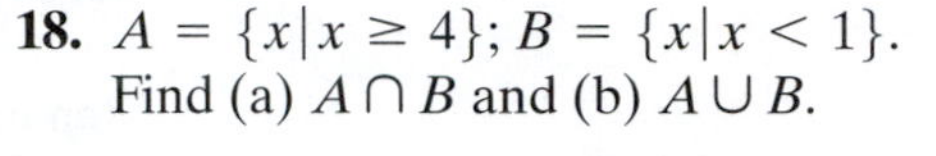

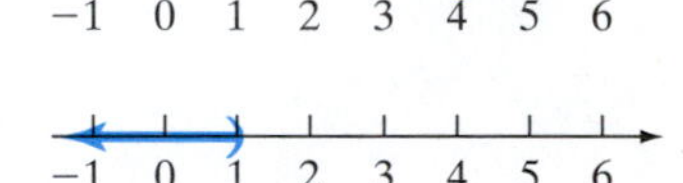

**19.** $E = \{x | x > 3\}$; $F = \{x | x < -1\}$. Find (a) $E \cap F$ and (b) $E \cup F$.

**20.** $E = \{x | x \le 2\}$; $F = \{x | x \ge -2\}$. Find (a) $E \cap F$ and (b) $E \cup F$.

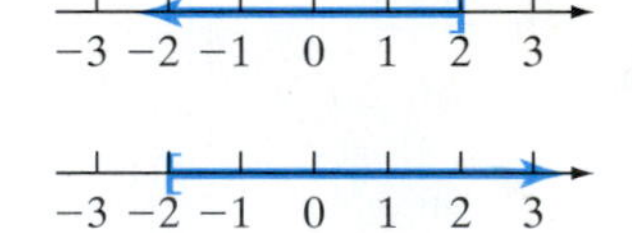

*In Problems 21–58, solve each compound inequality. Graph the solution set.*

**21.** $x < 3$ and $x \ge -2$

**22.** $x \le 5$ and $x > 0$

**23.** $x < -2$ or $x > 3$

**24.** $x < 0$ or $x \ge 6$

**25.** $4x - 4 < 0$ and $-5x + 1 \le -9$

**26.** $6x - 2 \le 10$ and $10x > -20$

**27.** $4x - 3 < 5$ and $-5x + 3 > 13$

**28.** $x - 3 \le 2$ and $6x + 5 \ge -1$

**29.** $-4x - 1 < 3$ and $-x - 2 > 3$

**30.** $7x + 2 \ge 9$ and $4x + 3 \le 7$

**31.** $x - 2 < -4$ or $x + 3 > 8$

**32.** $x + 3 \le 5$ or $x - 2 \ge 3$

**33.** $6(x - 2) < 12$ or $4(x + 3) > 12$

**34.** $4x + 3 > -5$ or $8x - 5 < 3$

**35.** $-8x + 6x - 2 > 0$ or $5x > 3x + 8$

**36.** $3x \ge 7x + 8$ or $x < 4x - 9$

**37.** $-3 \le 5x + 2 < 17$

**38.** $-10 < 6x + 8 \le -4$

**39.** $-3 \le 6x + 1 \le 10$

**40.** $-12 < 7x + 2 \le 6$

**41.** $2x + 5 \le -1$ or $\frac{4}{3}x - 3 > 5$

**42.** $-\frac{4}{5}x - 5 > 3$ or $7x - 3 > 4$

**43.** $3 \le -5x + 7 < 12$

**44.** $-6 < -3x + 6 \le 4$

**45.** $-1 \le \frac{1}{2}x - 1 \le 3$

**46.** $0 < \frac{3}{2}x - 3 \le 3$

**47.** $3 \le -2x - 1 \le 11$

**48.** $-3 < -4x + 1 < 17$

**49.** $\frac{1}{2}x < 3$ or $\frac{3x - 1}{2} > 4$

**50.** $\frac{2}{3}x + 2 \le 4$ or $\frac{5x - 3}{3} \ge 4$

**51.** $\frac{2}{3}x + \frac{1}{2} < \frac{5}{6}$ and $-\frac{1}{5}x + 1 < \frac{3}{10}$

**52.** $x - \frac{3}{2} \le \frac{5}{4}$ and $-\frac{2}{3}x - \frac{2}{9} < \frac{8}{9}$

**53.** $-2 < \frac{3x + 1}{2} \le 8$

**54.** $-4 \le \frac{4x - 3}{3} < 3$

**55.** $-8 \le -2(x + 1) < 6$

**56.** $-6 < -3(x - 2) < 15$

**57.** $3(x - 1) + 5 < 2$ or $-2(x - 3) < 1$

**58.** $2(x + 1) - 5 \le 4$ or $-(x + 3) \le -2$

## Mixed Practice

*In Problems 59–72, solve each compound inequality. Graph the solution set.*

**59.** $3a + 5 < 5$ and $-2a + 1 \le 7$

**60.** $5x - 1 < 9$ and $5x > -20$

**61.** $5(x + 2) < 20$ or $4(x - 4) > -20$

**62.** $3(x + 7) < 24$ or $6(x - 4) > -30$

**63.** $-4 \le 3x + 2 \le 10$

**64.** $-8 \le 5x - 3 \le 4$

**65.** $2x + 7 < -13$ or $5x - 3 > 7$

**66.** $3x - 8 < -14$ or $4x - 5 > 7$

**67.** $5 < 3x - 1 < 14$

**68.** $-5 < 2x + 7 \le 5$

**69.** $\frac{x}{3} \le -1$ or $\frac{4x - 1}{2} > 7$

**70.** $\frac{x}{2} \le -4$ or $\frac{2x - 1}{3} \ge 2$

**71.** $-3 \le -2(x + 1) < 8$

**72.** $-15 < -3(x + 2) \le 1$

## Applying the Concepts

*In Problems 73–78, use the Addition Property and/or Multiplication Properties to find a and b.*

**73.** If $-3 < x < 4$, then $a < x + 4 < b$.

**74.** If $-2 < x < 3$, then $a < x - 3 < b$.

**75.** If $4 < x < 10$, then $a < 3x < b$.

**76.** If $2 < x < 12$, then $a < \frac{1}{2}x < b$.

**77.** If $-2 < x < 6$, then $a < 3x + 5 < b$.

**78.** If $-4 < x < 3$, then $a < 2x - 7 < b$.

**79. Systolic Blood Pressure** Blood pressure is measured using two numbers. One of the numbers measures systolic blood pressure. The systolic blood pressure represents the pressure while the heart is beating. In a healthy person, the systolic blood pressure should be greater than 90 and less than 140. If we let the variable $x$ represent a person's systolic blood pressure, express the systolic blood pressure of a healthy person using a compound inequality.

**80. Diastolic Blood Pressure** Blood pressure is measured using two numbers. One of the numbers measures diastolic blood pressure. The diastolic blood pressure represents the pressure while the heart is resting between beats. In a healthy person, the diastolic blood pressure should be greater than 60 and less than 90. If we let the variable $x$ represent a person's diastolic blood pressure, express the diastolic blood pressure of a healthy person using a compound inequality.

**81. Computing Grades** Joanna desperately wants to earn a $B$ in her History class. Her current test scores are 74, 86, 77, and 89. Her final exam is worth 2 test scores. In order to earn a $B$, Joanna's average must lie between 80 and 89, inclusive. What range of scores can Joanna receive on the final and earn a $B$ in the course?

**82. Computing Grades** Jack needs to earn a $C$ in his Sociology class. His current test scores are 67, 72, 81, and 75. His final exam is worth 3 test scores. In order to earn a $C$, Jack's average must lie between 70 and 79, inclusive. What range of scores can Jack receive on the final exam and earn a $C$ in the course?

**83. Federal Tax Withholding** The percentage method of withholding for federal income tax (2002) states that a single person whose weekly wages, after subtracting withholding allowances, are over \$517, but not over \$1105, shall have \$69.60 plus 28% of the excess over \$517 withheld. Over what range does the amount withheld vary if the weekly wages vary from \$600 to \$700, inclusive? (*Source:* Internal Revenue Service)

**84. Federal Tax Withholding** Rework Problem 83 if the weekly wages vary from \$800 to \$900, inclusive.

**85. Gas Bill** Pacific Gas and Electric Company charges \$65.05 plus \$1.15855 per therm for gas usage in excess of 70 therms. In the winter of 2004/2005, one homeowner's bill ranged from a low of \$157.73 to a high of \$175.11. Over what range did gas usage vary (in therms)? (*Source:* Pacific Gas and Electric Company)

**86. Electric Bills** In North Carolina, Duke Energy charges \$31.52 plus \$0.075895 for each additional kilowatt hour (kwh) used during the months from November through June for usage in excess of 350 kwh. Suppose one homeowner's electric bill ranged from a high of \$69.47 to a low of \$39.11 during this time period. Over what range did the usage vary (in kwh)? (*Source:* Duke Energy)

**87. The Arithmetic Mean** If $a < b$, show that $a < \frac{a+b}{2} < b$. We call $\frac{a+b}{2}$ the **arithmetic mean** of $a$ and $b$.

**88. Identifying Triangles** A triangle is one such that the length of the longest side is greater than the difference of the other sides and the length of the longest side is less than the sum of the other sides. That is, if $a$, $b$, and $c$ are sides such that $a \le b \le c$, then $b - a < c < b + a$. Determine which of the following could be lengths of the sides of a triangle.

**(a)** 3, 4, 5 **(b)** 4, 7, 12 **(c)** 3, 3, 5 **(d)** 1, 9, 10

## Extending the Concepts

**89.** Solve $2x + 1 \le 5x + 7 \le x - 5$.

**90.** Solve $x - 3 \le 3x + 1 \le x + 11$.

**91.** Solve $4x + 1 > 2(2x + 1)$. Provide an explanation that generalizes the result.

**92.** Solve $4x + 1 > 2(2x - 1)$. Provide an explanation that generalizes the result.

**93.** Consider the following analysis assuming that $x < 2$.

$$5 > 2$$
$$5(x - 2) > 2(x - 2)$$
$$5x - 10 > 2x - 4$$
$$3x > 6$$
$$x > 2$$

How can it be that the final line in the analysis states that $x > 2$, when the original assumption stated that $x < 2$?

# 1.6 Absolute Value Equations and Inequalities

**OBJECTIVES**

1. Solve Absolute Value Equations
2. Solve Absolute Value Inequalities Involving $<$ or $\leq$
3. Solve Absolute Value Inequalities Involving $>$ or $\geq$
4. Solve Applied Problems Involving Absolute Value

### *Preparing for Absolute Value Equations and Inequalities*

*Before getting started, take the following readiness quiz. If you get a problem wrong, go back to the section cited and review the material.*

*In Problems 1–4, evaluate each expression.* [Section R.3, p. 20]

**1.** $|3|$ **2.** $|-4|$ **3.** $|-1.6|$ **4.** $|0|$

**5.** Express the distance between the origin, 0, and 5 as an absolute value. [Section R.3, p. 20]

**6.** Express the distance between the origin, 0, and $-8$ as an absolute value. [Section R.3, p. 20]

Recall from Section R.3 that we defined the absolute value of a number as the distance between the number and the origin on the real number line. For example, $|-5| = 5$ because the distance on the real number from 0 to $-5$ is 5 units. See Figure 32 for a geometric interpretation of absolute value.

**Figure 32**
$|-5| = 5$

This interpretation of absolute value forms the basis for solving absolute value equations.

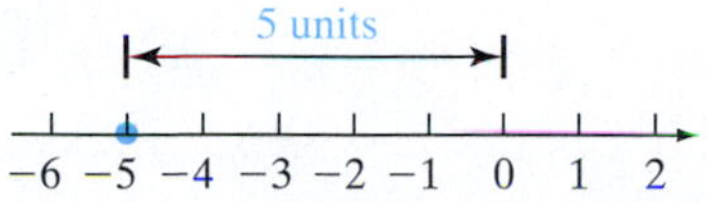

## 1 Solve Absolute Value Equations

We begin with an example.

### EXAMPLE 1 Solving an Absolute Value Equation

Solve the equation $|x| = 4$.

**Solution**

The equation $|x| = 4$ is asking, "Tell me all real numbers $x$ such that the distance from the origin to $x$ on the real number line is 4 units." There are two such numbers as indicated in Figure 33, $-4$ and 4. The solution set is $\{-4, 4\}$.

**Figure 33**

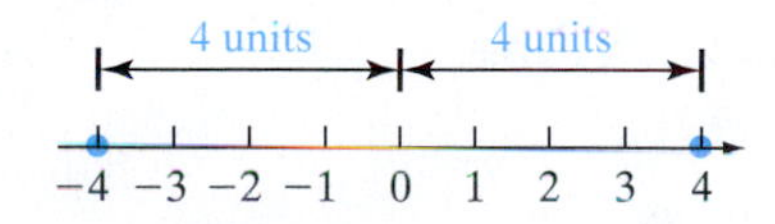


**QUICK ✓** *Solve the equation.*

**1.** $|x| = 7$ **2.** $|z| = 1$

The results of Example 1 lead us to the following result.

**EQUATIONS INVOLVING ABSOLUTE VALUE**

If $a$ is a positive real number and if $u$ is any algebraic expression, then

$$|u| = a \quad \text{is equivalent to} \quad u = a \quad \text{or} \quad u = -a$$

**Note:** If $a = 0$, the equation $|u| = 0$ is equivalent to $u = 0$. If $a < 0$, the equation $|u| = a$ has no real solution.

Example 2 shows us how to solve an equation involving absolute value.

*Preparing for...Answers* **1.** 3 **2.** 4 **3.** 1.6 **4.** 0 **5.** $|5|$ **6.** $|-8|$

## EXAMPLE 2 How to Solve an Equation Involving Absolute Value

Solve the equation $|2x - 1| + 3 = 12$.

### Step-by-Step Solution

**Step 1:** Isolate the expression containing the absolute value.

$$|2x - 1| + 3 = 12$$

Subtract 3 from both sides: $|2x - 1| = 9$

**Step 2:** Rewrite the absolute value equation as two equations: $u = a$ and $u = -a$, where $u$ is the algebraic expression in the absolute value symbol. Here $u = 2x - 1$ and $a = 9$.

$$2x - 1 = 9 \quad \text{or} \quad 2x - 1 = -9$$

**Step 3:** Solve each equation.

$2x - 1 = 9$

Add 1 to each side: $2x = 10$

Divide both sides by 2: $x = 5$

$2x - 1 = -9$

Add 1 to each side: $2x = -8$

Divide both sides by 2: $x = -4$

**Step 4: Check** Verify each solution.

Let $x = 5$:

$$|2x - 1| + 3 = 12$$
$$|2(5) - 1| + 3 \stackrel{?}{=} 12$$
$$|10 - 1| + 3 \stackrel{?}{=} 12$$
$$9 + 3 \stackrel{?}{=} 12$$
$$12 = 12 \quad \text{True}$$

Let $x = -4$:

$$|2x - 1| + 3 = 12$$
$$|2(-4) - 1| + 3 \stackrel{?}{=} 12$$
$$|-8 - 1| + 3 \stackrel{?}{=} 12$$
$$9 + 3 \stackrel{?}{=} 12$$
$$12 = 12 \quad \text{True}$$

Both solutions check, so the solution set is $\{-4, 5\}$.

The following steps can be used to solve an absolute value equation.

### Steps for Solving Absolute Value Equations with One Absolute Value

**Step 1:** Isolate the expression containing the absolute value.

**Step 2:** Rewrite the absolute value equation as two equations: $u = a$ and $u = -a$, where $u$ is the algebraic expression in the absolute value symbol.

**Step 3:** Solve each equation.

**Step 4:** Verify your solution.

**QUICK ✓** *Solve each equation.*

**3.** $|2x - 3| = 7$ **4.** $|3x - 2| + 3 = 10$ **5.** $|-5x + 2| - 2 = 5$ **6.** $3|x + 2| - 4 = 5$

## EXAMPLE 3 Solving an Equation Involving Absolute Value with No Solution

Solve the equation $|-x + 5| + 7 = 5$.

### Solution

$$|-x + 5| + 7 = 5$$

Subtract 7 from both sides: $|-x + 5| = -2$

**Work Smart**

The equation $|u| = a$, where $a$ is a negative real number has no real solution.

Since the absolute value of any real number is always nonnegative (greater than or equal to zero), the equation has no real solution. The solution set is { } or ∅. ▪

 *Solve each equation.*

**7.** $|5x + 3| = -2$ **8.** $|2x + 5| + 7 = 3$ **9.** $|x + 1| + 3 = 3$

---

What if an absolute value equation has two absolute values as in $|3x - 1| = |x + 5|$? How do we handle this situation? Well, there are four possibilities for the algebraic expressions in the absolute value symbol:

1. both algebraic expressions are positive,
2. both are negative,
3. the left is positive, and the right is negative, or
4. the left is negative and the right is positive.

To see how the solution works, we need to remember the definition for absolute value given in Section R.3 on page 22. This definition states that $|a| = a$, if $a \geq 0$ and $|a| = -a$ if $a < 0$.

So, if $3x - 1 \geq 0$, then $|3x - 1| = 3x - 1$. However, if $3x - 1 < 0$, then $|3x - 1| = -(3x - 1)$. This leads us to a method for solving absolute value equations with two absolute values.

| Case 1: Both Algebraic Expressions Are Positive | Case 2: Both Algebraic Expressions Are Negative | Case 3: The Algebraic Expression on the Right Is Positive, and the Left Is Negative | Case 4: The Algebraic Expression on the Left Is Negative, and the Right Is Positive |
|---|---|---|---|
| $\|3x - 1\| = \|x + 5\|$ | $\|3x - 1\| = \|x + 5\|$ | $\|3x - 1\| = \|x + 5\|$ | $\|3x - 1\| = \|x + 5\|$ |
| $3x - 1 = x + 5$ | $-(3x - 1) = -(x + 5)$ | $3x - 1 = -(x + 5)$ | $-(3x - 1) = x + 5$ |
| | $3x - 1 = x + 5$ | | |

Whether both algebraic expressions are positive, or both negative, we end up with equivalent equations. Also, if one side is positive and the other is negative, we end up with equivalent equations. So, the four possibilities reduce to two possibilities.

**EQUATIONS INVOLVING TWO ABSOLUTE VALUES**

If $u$ and $v$ are any algebraic expression, then

$$|u| = |v| \quad \text{is equivalent to} \quad u = v \quad \text{or} \quad u = -v$$

## EXAMPLE 4 Solving an Absolute Value Equation Involving Two Absolute Values

Solve the equation $|2x - 3| = |x + 6|$.

### Solution

The equation is in the form $|u| = |v|$, where $u = 2x - 3$ and $v = x + 6$. We rewrite the equation as two equations that do not involve absolute value:

$$2x - 3 = x + 6 \quad \text{or} \quad 2x - 3 = -(x + 6)$$

Now, we solve each equation.

| | | | |
|---|---|---|---|
| | $2x - 3 = x + 6$ | | $2x - 3 = -(x + 6)$ |
| | | Distribute the $-1$: | $2x - 3 = -x - 6$ |
| Add 3 to each side: | $2x = x + 9$ | Add 3 to both sides: | $2x = -x - 3$ |
| | | Add $x$ to both sides: | $3x = -3$ |
| Subtract $x$ from both sides: | $x = 9$ | Divide both sides by 3: | $x = -1$ |

**Check**

$x = 9$: $|2(9) - 3| \stackrel{?}{=} |9 + 6|$
$|18 - 3| \stackrel{?}{=} |15|$
$|15| \stackrel{?}{=} 15$
$15 = 15$ True

$x = -1$: $|2(-1) - 3| \stackrel{?}{=} |-1 + 6|$
$|-2 - 3| \stackrel{?}{=} |5|$
$|-5| \stackrel{?}{=} 5$
$5 = 5$ True

Both solutions check, so the solution set is $\{-1, 9\}$.

**QUICK ✓** *Solve each equation.*

**10.** $|x - 3| = |2x + 5|$

**11.** $|8z + 11| = |6z + 17|$

**12.** $|3 - 2y| = |4y + 3|$

**13.** $|2x - 3| = |5 - 2x|$

## 2 Solve Absolute Value Inequalities Involving $<$ or $\leq$

The method for solving absolute value equations relies on the geometric interpretation of absolute value. Namely, the absolute value of a real number $x$ is the distance from the origin to $x$ on the real number line. We use this same interpretation to solve absolute value inequalities.

### EXAMPLE 5 Solving an Absolute Value Inequality

Solve the inequality $|x| < 4$. Graph the solution set.

**Solution**

The inequality $|x| < 4$ is asking, "Tell me all real numbers $x$ such that the distance from the origin to $x$ on the real number line is less than 4." Figure 34 illustrates the situation. We can see from the figure that any number between $-4$ and 4 satisfies the inequality. The solution set consists of all real numbers $x$ for which $-4 < x < 4$ or, using interval notation, $(-4, 4)$.

**Figure 34**

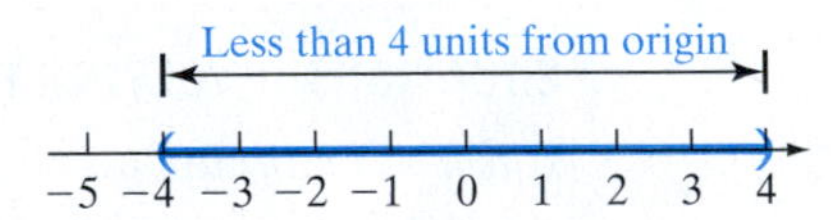


**QUICK ✓** *Solve each inequality. Graph the solution set.*

**14.** $|x| \leq 5$

**15.** $|x| < \frac{3}{2}$

The results of Example 5 lead to the following results.

**INEQUALITIES OF THE FORM $<$ OR $\leq$ INVOLVING ABSOLUTE VALUE**

If $a$ is a positive real number and if $u$ is an algebraic expression, then

$|u| < a$ is equivalent to $-a < u < a$

$|u| \leq a$ is equivalent to $-a \leq u \leq a$

**Note:** If $a = 0$, $|u| < 0$ has no real solution, $|u| \leq 0$ is equivalent to $u = 0$. If $a < 0$, the inequality has no real solution.

## EXAMPLE 6 How to Solve an Absolute Value Inequality Involving ≤

Solve the inequality $|2x + 3| \leq 5$. Graph the solution set.

### Step-by-Step Solution

**Step 1:** The inequality is in the form $|u| \leq a$, where $u = 2x + 3$ and $a = 5$. We rewrite the inequality as a compound inequality that does not involve absolute value.

$$|2x + 3| \leq 5$$

Use the fact that $|u| \leq a$ means $-a \leq u \leq a$: $\quad -5 \leq 2x + 3 \leq 5$

**Step 2:** Solve the resulting compound inequality.

Subtract 3 from all three parts: $\quad -5 - 3 \leq 2x + 3 - 3 \leq 5 - 3$

$$-8 \leq 2x \leq 2$$

Divide all three parts of the inequality by 2: $\quad \frac{-8}{2} \leq \frac{2x}{2} \leq \frac{2}{2}$

$$-4 \leq x \leq 1$$

The solution using set-builder notation is $\{x \mid -4 \leq x \leq 1\}$. The solution using interval notation is $[-4, 1]$. Figure 35 shows the graph of the solution set.

**Figure 35**

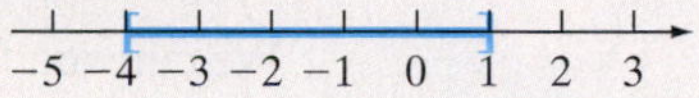

−5 −4 −3 −2 −1 0 1 2 3

### Work Smart

Although not a complete check of the solution of Example 6, we can choose a number in the interval and see if it works. Let's try $x = -3$.

$$|2(-3) + 3| \stackrel{?}{\leq} 5$$
$$|-6 + 3| \stackrel{?}{\leq} 5$$
$$|3| \stackrel{?}{\leq} 5$$
$$3 \leq 5 \checkmark$$

**QUICK ✓** *Solve each inequality. Graph the solution set.*

**16.** $|x + 3| < 5$ **17.** $|2x - 3| \leq 7$ **18.** $|7x + 2| < -3$

## EXAMPLE 7 Solving an Absolute Value Inequality Involving <

Solve the inequality $|-3x + 2| + 4 < 14$. Graph the solution set.

### Solution

First, we want to isolate the absolute value by subtracting 4 from both sides of the inequality.

$$|-3x + 2| + 4 < 14$$

Subtract 4 from both sides: $\quad |-3x + 2| < 10$

Use $|u| < a$ means $-a < u < a$: $\quad -10 < -3x + 2 < 10$

Subtract 2 from all three parts: $\quad -10 - 2 < -3x + 2 - 2 < 10 - 2$

$$-12 < -3x < 8$$

Divide all three parts by −3. Be sure to reverse the direction of the inequalities. $\quad \frac{-12}{-3} > \frac{-3x}{-3} > \frac{8}{-3}$

$$4 > x > -\frac{8}{3}$$

Use $b > x > a$ is equivalent to $a < x < b$: $\quad -\frac{8}{3} < x < 4$

The solution using set-builder notation is $\left\{x \middle| -\frac{8}{3} < x < 4\right\}$. The solution using interval notation is $\left(-\frac{8}{3}, 4\right)$. Figure 36 shows the graph of the solution set.

**Figure 36**

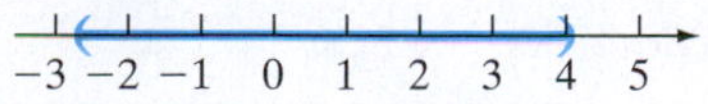

−3 −2 −1 0 1 2 3 4 5

**QUICK ✓** *Solve each inequality. Graph the solution set.*

**19.** $|x| + 4 < 6$ **20.** $|x - 3| + 4 \leq 8$ **21.** $3|2x + 1| \leq 9$ **22.** $|-3x + 1| - 5 < 3$

## 3 Solve Absolute Value Inequalities Involving > or ≥

Now let's look at absolute value inequalities involving $>$ or $\geq$.

### EXAMPLE 8 Solving an Absolute Value Inequality Involving >

Solve the inequality $|x| > 3$. Graph the solution set.

**Solution**

The inequality $|x| > 3$ is asking, "Tell me all real numbers $x$ such that the distance from the origin to $x$ on the real number line is more than 3 units." Figure 37 illustrates the situation.

**Figure 37**

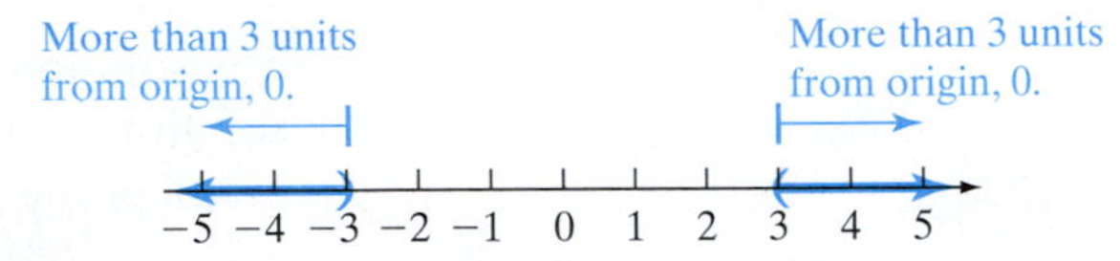


We can see from the figure that any number less than $-3$ or greater than 3 satisfies the inequality. The solution set consists of all real numbers $x$ for which $x < -3$ or $x > 3$ or, using interval notation, $(-\infty, -3) \cup (3, \infty)$.

**QUICK ✓** *Solve each inequality. Graph the solution set.*

**23.** $|x| \geq 6$ **24.** $|x| > \dfrac{5}{2}$

Based upon Example 8, we are led to the following results.

**INEQUALITIES OF THE FORM > OR ≥ INVOLVING ABSOLUTE VALUE**

If $a$ is a positive real number and $u$ is an algebraic expression, then

$|u| > a$ is equivalent to $u < -a$ or $u > a$

$|u| \geq a$ is equivalent to $u \leq -a$ or $u \geq a$

### EXAMPLE 9 How to Solve an Inequality Involving >

Solve the inequality $|2x - 5| > 3$. Graph the solution set.

**Step-by-Step Solution**

**Step 1:** The inequality is in the form $|u| > a$, where $u = 2x - 5$ and $a = 3$. We rewrite the inequality as a compound inequality that does not involve absolute value:

$$|2x - 5| > 3$$

$$2x - 5 < -3 \quad \text{or} \quad 2x - 5 > 3$$

**Step 2:** Solve each inequality separately.

| | | | |
|---|---|---|---|
| | $2x - 5 < -3$ | | $2x - 5 > 3$ |
| Add 5 to both sides: | $2x < 2$ | Add 5 to both sides: | $2x > 8$ |
| Divide both sides by 2: | $x < 1$ | Divide both sides by 2: | $x > 4$ |

**Step 3:** Find the union of the solution sets of each inequality.

The solution set is $\{x \mid x < 1 \text{ or } x > 4\}$, or using interval notation, $(-\infty, 1) \cup (4, \infty)$. See Figure 38 for the graph of the solution set.

**Figure 38**

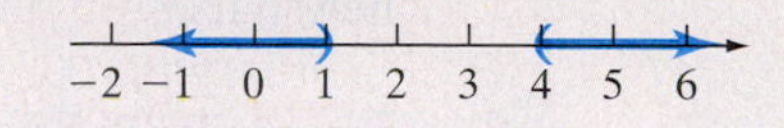


**Work Smart**

$|u| > a$

CANNOT be written as

$-a > u > a$

**QUICK ✓** *Solve each inequality. Graph the solution set.*

**25.** $|x + 3| > 4$ **26.** $|4x - 3| \geq 5$ **27.** $|-3x + 2| > 7$

**28.** $|2x + 5| - 2 > -2$ **29.** $|6x - 5| \geq 0$ **30.** $|2x + 1| > -3$

Below, we summarize the techniques for solving absolute value equations and inequalities.

## SUMMARY: Solving Absolute Value Equations and Inequalities

| Absolute Value Form | Equation/Inequality Form | Example |
|---|---|---|
| $\lvert u\rvert = a$ | $u = -a$ or $u = a$ | $\lvert 2x - 3\rvert = 4$<br>$2x - 3 = -4$ or $2x - 3 = 4$<br>$2x = -1$ or $2x = 7$<br>$x = -\frac{1}{2}$ or $x = \frac{7}{2}$<br>The solution set is $\left\{-\frac{1}{2}, \frac{7}{2}\right\}$. |
| $\lvert u\rvert = \lvert v\rvert$ | $u = v$ or $u = -v$ | $\lvert x + 6\rvert = \lvert 3x - 4\rvert$<br>$x + 6 = 3x - 4$ or $x + 6 = -(3x - 4)$<br>$x = 3x - 10$ or $x + 6 = -3x + 4$<br>$-2x = -10$ or $4x = -2$<br>$x = 5$ or $x = -\frac{1}{2}$<br>The solution set is $\left\{-\frac{1}{2}, 5\right\}$. |
| $\lvert u\rvert < a$<br>$\lvert u\rvert \leq a$ | $-a < u < a$<br>$-a \leq u \leq a$ | $2\lvert x - 1\rvert + 3 \leq 11$<br>$2\lvert x - 1\rvert \leq 8$<br>$\lvert x - 1\rvert \leq 4$<br>$-4 \leq x - 1 \leq 4$<br>$-3 \leq x \leq 5$<br>The solution set is $\{x \mid -3 \leq x \leq 5\}$ or $[-3, 5]$. |
| $\lvert u\rvert > a$<br>$\lvert u\rvert \geq a$ | $u < -a$ or $u > a$<br>$u \leq -a$ or $u \geq a$ | $\lvert 3x + 2\rvert > 8$<br>$3x + 2 < -8$ or $3x + 2 > 8$<br>$3x < -10$ or $3x > 6$<br>$x < -\frac{10}{3}$ or $x > 2$<br>The solution set is<br>$\left\{x \mid x < -\frac{10}{3} \text{ or } x > 2\right\}$ or<br>$\left(-\infty, -\frac{10}{3}\right) \cup (2, \infty)$. |

## 4 Solve Applied Problems Involving Absolute Value Inequalities

You may frequently read phrases such as "margin of error" and "tolerance" in the newspaper or on the Internet. For example, according to a Gallup poll conducted February 6, 2003, 57% of Americans felt that the United States rates favorably internationally. The poll had a margin of error of 3%. The 57% reported is an estimate of the true percentage of U.S. residents who believe that the United States rates favorably internationally. If we let $p$ represent the true percentage of U.S. residents that believe that the United States rates favorably internationally, then we can represent the poll's margin of error mathematically as

$$|p - 57| \leq 3$$

As another example, the tolerance of a belt whose width is 6 inches is $\frac{1}{16}$ inch. If $x$ represents the actual width of the belt, then we can represent the acceptable belt widths as

$$|x - 6| \leq \frac{1}{16}$$

### EXAMPLE 10 Analyzing the Margin of Error in a Poll

The inequality

$$|p - 57| \leq 3$$

represents the percentage of Americans who feel that the United States rates favorably internationally. Solve the inequality and interpret the results.

**Solution**

$$|p - 57| \leq 3$$

Use $|u| \leq a$ means $-a \leq u \leq a$: $\quad -3 \leq p - 57 \leq 3$

Add 57 to all three parts of the inequality: $\quad 54 \leq p \leq 60$

The percentage of Americans who feel that the United States rates favorably internationally is between 54% and 60%, inclusive.

**QUICK ✓**

31. The inequality $|x - 4| \leq \frac{1}{32}$ represents the acceptable belt widths $x$ (in inches) for a belt that is manufactured for a pulley system. Determine the acceptable belt widths.

32. In a recent poll conducted by ABC News, 9% of respondents stated that they have been shot at. The margin of error in the poll was 1.7%. If we let $p$ represent the true percentage of people who have been shot at, we can represent the margin of error as

$$|p - 9| \leq 1.7$$

Solve the inequality and interpret the results.

# 1.6 Exercises

**For Extra Help:**  Student Solutions Manual  CD Video PH Math/Tutor Center  MathXL Tutorials on CD 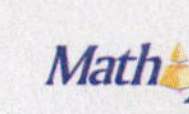 MathXL® MyMathLab

## Concepts and Vocabulary

*In Problems 1–3, fill in the blanks.*

**1.** $|u| = a$ is equivalent to $u =$ _____ or $u =$ _____.

**2.** $|u| < a$ is equivalent to _____.

**3.** $|u| < a$ will have no solution if $a$ _____ 0.

*In Problems 4–6, answer True or False to each statement.*

**4.** $|x| = -4$ has no real solution.

**5.** $|x| > -2$ has no real solution.

**6.** $|u| > a$ is equivalent to $-a < u < a$.

## Building Skills

*In Problems 7–28, solve each absolute value equation.*

**7.** $|x| = 10$

**8.** $|z| = 9$

**9.** $|y - 3| = 4$

**10.** $|x + 3| = 5$

**11.** $|-3x + 5| = 8$

**12.** $|-4y + 3| = 9$

**13.** $|y| - 7 = -2$

**14.** $|x| + 3 = 5$

**15.** $|2x + 3| - 5 = 3$

**16.** $|3y + 1| - 5 = -3$

**17.** $-2|x - 3| + 10 = -4$

**18.** $3|y - 4| + 4 = 16$

**19.** $|-3x| - 5 = -5$

**20.** $|-2x| + 9 = 9$

**21.** $\left|\frac{3x - 1}{4}\right| = 2$

**22.** $\left|\frac{2x - 3}{5}\right| = 2$

**23.** $|3x + 2| = |2x - 5|$

**24.** $|5y - 2| = |4y + 7|$

**25.** $|8 - 3x| = |2x - 7|$

**26.** $|5x + 3| = |12 - 4x|$

**27.** $|4y - 7| = |9 - 4y|$

**28.** $|5x - 1| = |9 - 5x|$

*In Problems 29–54, solve each absolute value inequality. Graph the solution set on a real number line.*

**29.** $|x| < 9$

**30.** $|x| \le \frac{5}{4}$

**31.** $|x - 4| \le 7$

**32.** $|y + 4| < 6$

**33.** $|3x + 1| < 8$

**34.** $|4x - 3| \le 9$

**35.** $|6x + 5| < -1$

**36.** $|4x + 3| \le 0$

**37.** $|y - 5| > 2$

**38.** $|x + 4| \ge 7$

**39.** $|-4x - 3| \ge 5$

**40.** $|-5y + 3| > 7$

**41.** $2|y| + 3 > 1$

**42.** $3|z| + 8 > 2$

**43.** $2|x - 3| + 3 < 9$

**44.** $3|y + 2| - 2 > 7$

**45.** $|-5x - 3| > -3$

**46.** $|-9x + 2| \geq -1$

**47.** $|2 - 5x| + 3 < 10$

**48.** $|-3x + 2| - 7 \leq -2$

**49.** $|-2x + 1| > 1$

**50.** $|8x + 3| \geq 3$

**51.** $|1 - 2x| \geq |-5|$

**52.** $|3 - 5x| < |-7|$

**53.** $|(2x - 3) - 1| < 0.01$

**54.** $|(3x + 2) - 8| < 0.01$

## Mixed Practice

*In Problems 55–74, solve each absolute value equation or inequality. For absolute value inequalities, graph the solution set on a real number line.*

**55.** $|x| > 5$

**56.** $|x| \geq \frac{8}{3}$

**57.** $|2x + 5| = 3$

**58.** $|4x + 3| = 1$

**59.** $7|x| = 35$

**60.** $8|y| = 32$

**61.** $|5x + 2| \leq 8$

**62.** $|7y - 3| < 11$

**63.** $|-2x + 3| = -4$

**64.** $|3x - 4| = -9$

**65.** $|3x + 2| \geq 5$

**66.** $|5y + 3| > 2$

**67.** $|3x - 2| + 7 > 9$

**68.** $|4y + 3| - 8 \geq -3$

**69.** $|5x + 3| = |3x + 5|$

**70.** $|3z - 2| = |z + 6|$

**71.** $|4x + 7| + 6 < 5$

**72.** $|4x + 1| > 0$

**73.** $\left|\frac{x - 2}{4}\right| = \left|\frac{2x + 1}{6}\right|$

**74.** $\left|\frac{1}{2}x - 3\right| = \left|\frac{2}{3}x + 1\right|$

## Applying the Concepts

**75.** Express the fact that $x$ differs from 5 by less than 3 as an inequality involving absolute value. Solve for $x$.

**76.** Express the fact that $x$ differs from $-4$ by less than 2 as an inequality involving absolute value. Solve for $x$.

**77.** Express the fact that twice $x$ differs from $-6$ by more than 3 as an inequality involving absolute value. Solve for $x$.

**78.** Express the fact that twice $x$ differs from 7 by more than 3 as an inequality involving absolute value. Solve for $x$.

**79. Tolerance** A certain rod in an internal combustion engine is supposed to be 5.7 inches. The tolerance on the rod is 0.0005 inches. If $x$ represents the length of a rod, the acceptable lengths of a rod can be expressed as $|x - 5.7| \leq 0.0005$. Determine the acceptable lengths of the rod. (*Source:* WiseCo Piston)

**80. Tolerance** A certain rod in an internal combustion engine is supposed to be 6.125 inches. The tolerance on the rod is 0.0005 inches. If $x$ represents the length of a rod, the acceptable lengths of a rod can be expressed as $|x - 6.125| \leq 0.0005$. Determine the acceptable lengths of the rod.

**81. IQ Scores** According to the Stanford-Binet IQ test, a normal IQ score is 100. It can be shown that anyone with an IQ $x$ that satisfies the inequality $\left|\frac{x - 100}{15}\right| > 1.96$ has an unusual IQ score. Determine the IQ scores that would be considered unusual.

**82. Gestation Period** The length of human pregnancy is about 266 days. It can be shown that a mother whose gestation period $x$ satisfies the inequality $\left|\frac{x - 266}{16}\right| > 1.96$ has an unusual length of pregnancy. Determine the length of pregnancy that would be considered unusual.

## Extending the Concepts

**83.** Explain why $|2x - 3| + 1 = 0$ has no solution.

**84.** Explain why the solution set of $|5x - 3| > -5$ is the set of all real numbers.

**85.** Explain why $|4x + 3| + 3 < 0$ has the empty set as the solution set.

**86.** Solve $|x - 5| = |5 - x|$. Explain why the result is reasonable. What do we call this type of equation?

*In Problems 87–94, solve each equation.*

**87.** $|x| - x = 5$

**88.** $|y| + y = 3$

**89.** $z + |-z| = 4$

**90.** $y - |-y| = 12$

**91.** $|4x + 1| = x - 2$

**92.** $|2x + 1| = x - 3$

**93.** $|x + 5| = -(x + 5)$

**94.** $|y - 4| = y - 4$

# CHAPTER 1 ACTIVITY: PASS THE PAPER

**Focus:** Solving equations and inequalities

**Time:** 15 minutes

**Group size:** 4

**Materials needed:** One blank piece of notebook paper per group member

Below are three equations and one inequality. In this activity you will work together to solve these problems by following the procedure below. Be sure to read through the entire procedure together before beginning the activity so all group members will understand the procedure.

**Procedure**

1. Write one of the problems shown below at the top of your paper. Be sure that each group member chooses a different problem.

2. Two lines below the original problem, write out the first step for solving the problem.
3. Fold the top of the paper down to cover the original problem, leaving your first step visible.
4. Pass your paper to a different group member. You might want to arrange your seats so that the papers can be passed around in a circle.
5. Continue solving the problems one step at a time, covering the step above yours, and passing the paper to the next group member, until all problems are solved.
6. As a group, discuss the solutions and decide whether or not they are correct. If any of the solutions are incorrect, solve them correctly together.

**Problems**

**(a)** $4 - (x - 3) = -10 + 5(x + 1)$ **(b)** $3(x - 2) + 8 = 5x - 2(x - 1)$

**(c)** $3(x - 4) - 5x > 2x + 12$ **(d)** $-2|x - 4| + 4 = -6$

# CHAPTER 1 REVIEW

## Section 1.1 Linear Equations

| KEY CONCEPTS | KEY TERMS |
|---|---|
| • **Linear Equation in One Variable**<br>An equation equivalent to one of the form $ax + b = 0$, where $a$ and $b$ are real numbers with $a \neq 0$.<br>• **Addition Property of Equality**<br>For real numbers $a$, $b$, and $c$, if $a = b$, then $a + c = b + c$.<br>• **Multiplication Property of Equality**<br>For real numbers $a$, $b$, and $c$ where $c \neq 0$, if $a = b$, then $ac = bc$. | Equation in one variable<br>Sides of the equation<br>Solution<br>Satisfies<br>Solve an equation<br>Solution set<br>Equivalent equations<br>Conditional equation<br>Contradiction<br>Identity |

| YOU SHOULD BE ABLE TO... | EXAMPLE | REVIEW EXERCISES |
|---|---|---|
| 1 Determine whether a number is a solution to an equation (p. 52) | Example 1 | 1–4 |
| 2 Solve linear equations (p. 53) | Examples 2 through 6 | 5–18 |
| 3 Determine whether an equation is a conditional equation, identity, or contradiction (p. 58) | Examples 7 and 8 | 5–14 |

*In Problems 1–4, determine which of the numbers, if any, are solutions to the given equation.*

**1.** $3x - 4 = 6 + x;\ x = 5, x = 6$

**2.** $-1 - 4x = 2(3 - 2x) - 7;\ x = -2, x = -1$

**3.** $4y - (1 - y) + 5 = -6 - 2(3y - 5) - 2y;\ y = -2, y = 0$

**4.** $\frac{w - 7}{3} - \frac{w}{4} = -\frac{7}{6};\ w = -14, w = 7$

*In Problems 5–14, solve the linear equation. State whether the equation is an identity, contradiction, or conditional equation.*

**5.** $2w + 9 = 15$

**6.** $-4 = 8 - 3y$

**7.** $2x + 5x - 1 = 20$

**8.** $7x + 5 - 8x = 13$

**9.** $-2(x - 4) = 8 - 2x$

**10.** $3(2r + 1) - 5 = 9(r - 1) - 3r$

**11.** $\dfrac{2y + 3}{4} - \dfrac{y}{2} = 5$

**12.** $\dfrac{x}{3} + \dfrac{2x}{5} = \dfrac{x - 20}{15}$

**13.** $0.2(x - 6) + 1.75 = 4.25 + 0.1(3x + 10)$

**14.** $2.1w - 3(2.4 - 0.2w) = 0.9(3w - 5) - 2.7$

*In Problems 15 and 16, determine which values of the variable must be excluded from the domain.*

**15.** $\dfrac{8}{2x + 3}$

**16.** $\dfrac{6x - 5}{6(x - 1) + 3}$

**17. State Income Tax** A resident of Missouri completes her state tax return and determines that she paid \$2370 in state income tax in 2003. The solution to the equation $2370 = 0.06(x - 9000) + 315$ represents her Missouri taxable income $x$ in 2003. Solve the equation to determine her Missouri taxable income. (*Source:* Missouri Department of Revenue)

**18. Movie Club** The DVD club to which you belong offers unlimited DVDs at \$10 off the regular price if you buy 1 at the regular price. You purchase 5 DVDs through this offer and spend \$69.75 (not including tax and shipping). The solution to the equation $x + 4(x - 10) = 69.75$ represents the regular club price $x$ for a DVD. Solve the equation to determine the regular club price for a DVD.

## Section 1.2 An Introduction to Problem Solving

| KEY CONCEPTS | KEY TERMS |
|---|---|
| • **Simple Interest Formula**<br>$I = Prt$, where $I$ is interest, $P$ is principal, $r$ is the per annum interest rate expressed as a decimal, $t$ is time in years<br>• **Uniform Motion Formula**<br>$d = rt$, where $d$ is distance, $r$ is average speed, $t$ is time | Problem solving<br>Mathematical modeling<br>Modeling process<br>Mathematical model<br>Direct translation<br>Interest<br>Principal<br>Rate of interest<br>Simple interest<br>Mixture problems<br>Uniform motion |

| YOU SHOULD BE ABLE TO . . . | EXAMPLE | REVIEW EXERCISES |
|---|---|---|
| 1 Translate English sentences into mathematical statements (p. 63) | Example 1 | 19–22 |
| 2 Model and solve direct translation problems (p. 66) | Examples 2 through 6 | 23–28 |
| 3 Model and solve mixture problems (p. 70) | Examples 7 and 8 | 29–32 |
| 4 Model and solve uniform motion problems (p. 73) | Example 9 | 33–34 |

*In Problems 19–22, translate each of the following English statements into a mathematical statement. Do not solve the equation.*

**19.** The sum of three times a number and 7 is 22.

**20.** The difference of a number and 3 is equivalent to the quotient of the number and 2.

21. 20% of a number equals the difference of the number and 12.

22. The product of six and a number is the same as 4 less than twice the number.

*For Problems 23 and 24, translate each English statement into a mathematical statement. Then solve the equation.*

23. Shawn is 8 years older than Payton and the sum of their ages is 18. What are their ages?

24. The sum of five consecutive odd integers is 125. Find the integers.

25. **Computing Grades** Logan is in an elementary statistics course and has test scores of 85, 81, 84, and 77. If the final exam counts the same as two tests, what score does Logan need on the final to have an average of 80?

26. **Home Equity Loans** On December 18, 2003, Bank of America offered a home equity line of credit at a rate of 4.25% annual simple interest. If Cherie has such a credit line with a balance of $3200, how much interest will she accrue at the end of 1 month?

27. **Discounted Price** Suppose that REI sells a 0° sleeping bag at the discounted price of $94.50. If this price represents a discount of 30% off the original selling price, find the original price.

28. **Minimum Wage** On November 4, 2003, the city of San Francisco passed a law raising its minimum wage to $8.50 per hour (an amount that was 65% higher than the federal minimum wage). Determine the federal minimum wage at that time.

29. **Making a Mixture** CoffeeAM sells chocolate covered blueberries for $10.95 per pound and chocolate covered strawberries for $13.95 per pound. The company wants to sell a mix of the two that would sell for $12.95 per pound with no loss in revenue. How many pounds of each treat should be used to make 12 pounds of the mix?

30. **A Sports Mix** The Candy Depot sells baseball gumballs for $3.50 per pound and soccer gumballs for $4.50 per pound. The company wants to sell a "sports mix" that sells for $3.75 per pound with no loss in revenue. How many pounds of each gumball type should be included to make 10 pounds of the mix?

31. **Investments** Angie received an $8000 bonus and wants to invest the money. She can invest part of the money at 8% simple interest with a moderate risk and the rest at 18% simple interest with a high risk. She wants an overall annual return of 12% but does not want to risk losing any more than necessary. How much should she invest at 18% to reach her goal?

32. **Antifreeze** A 2004 Pontiac Montana has an engine coolant system capacity of 10.1 liters. If the system is currently filled with a mixture that is 30% antifreeze, how much of this mixture should be drained and replaced with pure antifreeze so that the system is filled with a mixture that is 50% antifreeze?

33. **Road Trip** On a 300-mile trip to Chicago, Josh drove part of the time at 60 miles per hour and the remainder of the trip at 70 miles per hour. If the total trip took 4.5 hours, for how many miles did Josh drive at a rate of 60 miles per hour?

34. **Uniform Motion** An F15 Strike Eagle near New York City and an F14 Tomcat near San Diego are about 2200 miles apart and traveling towards each other. The F15 is traveling 200 miles per hour faster than the F14 and the planes pass each other after 50 minutes. How fast is each plane traveling?

## Section 1.3 Using Formulas to Solve Problems

| KEY CONCEPTS | KEY TERMS |
|---|---|
| • **Geometry Formulas (see pp. 81–82 for the figures).** | Formula<br>Golden Rectangle<br>Supplementary Angles<br>Complementary Angles |

| Shape | Formulas |
|---|---|
| Square | **Area:** $A = s^2$<br>**Perimeter:** $P = 4s$ |
| Rectangle | **Area:** $A = lw$<br>**Perimeter:** $P = 2l + 2w$ |
| Triangle | **Area:** $A = \frac{1}{2}bh$<br>**Perimeter:** $P = a + b + c$ |
| Trapezoid | **Area:** $A = \frac{1}{2}h(B + b)$<br>**Perimeter:** $P = a + b + c + B$ |
| Parallelogram | **Area:** $A = bh$<br>**Perimeter:** $P = 2a + 2b$ |
| Circle | **Area:** $A = \pi r^2$<br>**Circumference:** $C = 2\pi r = \pi d$ |
| Cube | **Volume:** $V = s^3$<br>**Surface Area:** $S = 6s^2$ |
| Rectangular Box | **Volume:** $V = lwh$<br>**Surface Area:** $S = 2lw + 2lh + 2wh$ |
| Sphere | **Volume:** $V = \frac{4}{3}\pi r^3$<br>**Surface Area:** $S = 4\pi r^2$ |
| Right Circular Cylinder | **Volume:** $V = \pi r^2 h$<br>**Surface Area:** $S = 2\pi r^2 + 2\pi rh$ |
| Cone | **Volume:** $V = \frac{1}{3}\pi r^2 h$ |

| YOU SHOULD BE ABLE TO . . . | EXAMPLE | REVIEW EXERCISES |
|---|---|---|
| 1 Solve for a variable in a formula (p. 80) | Examples 2 and 3 | 35–44 |
| 2 Use formulas to solve problems (p. 83) | Examples 4 and 5 | 45–52 |

*In Problems 35–40, solve for the indicated variable.*

**35.** Solve $y = \frac{k}{x}$ for $x$.

**36.** Solve $F = \frac{9}{5}C + 32$ for $C$.

**37.** Solve $P = 2L + 2W$ for $W$.

**38.** Solve $\rho = m_1 v_1 + m_2 v_2$ for $m_2$.

**39.** Solve $PV = nRT$ for $T$.

**40.** Solve $S = 2LW + 2LH + 2WH$ for $W$.

*In Problems 41–44, solve for y.*

**41.** $3x + 4y = 2$

**42.** $-5x + 4y = 10$

**43.** $4.8x - 1.2y = 6$

**44.** $\frac{2}{5}x + \frac{1}{3}y = 8$

**45. Temperature Conversions** To convert temperatures from Fahrenheit to Celsius, we can use the formula. $C = \frac{5}{9}(F - 32)$. If the melting point for platinum is 3221.6°F, convert this temperature to degrees Celsius.

**46. Angles in a Triangle** The measure of each congruent angle in an isosceles triangle is 30 degrees larger than the measure of the remaining angle. Determine the measures of all three angles.

**47. Window Dimensions** The perimeter of a rectangular window is 76 feet. The window is 8 feet longer than it is wide. Find the dimensions of the window.

**48. Long-Distance Phone Calls** A long-distance telephone company charges a monthly fee of \$2.95 and a per minute charge of \$0.04. The monthly cost for long distance is given by $C = 2.95 + 0.04x$ where $x$ is the number of minutes used.

**(a)** Solve the equation for $x$.

**(b)** How many full minutes can Debbie use in one month on this plan if she does not want to spend more than \$20 in long distance in one month?

**49. Concrete** Rick has 80 cubic feet of concrete to pour for his new patio. If the patio is rectangular and Rick wants it to be 12 ft by 18 ft, how thick will the patio be if he uses all the concrete?

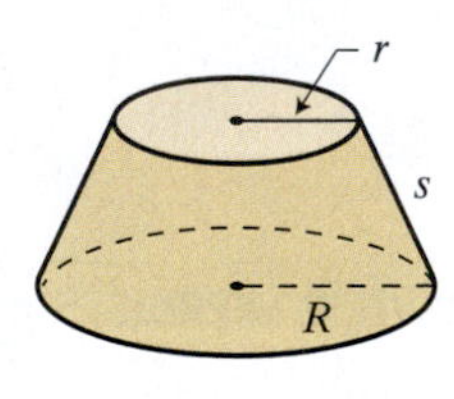


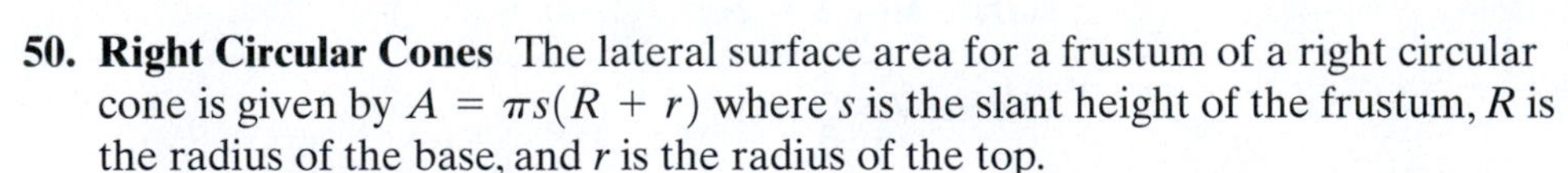

**50. Right Circular Cones** The lateral surface area for a frustum of a right circular cone is given by $A = \pi s(R + r)$ where $s$ is the slant height of the frustum, $R$ is the radius of the base, and $r$ is the radius of the top.

**(a)** Solve the equation for $r$.

**(b)** If the frustum of a right circular cone has a lateral surface area of $10\pi$ square feet, a slant height of 2 feet, and a base whose radius is 3 feet, what is the radius of the top of the frustum?

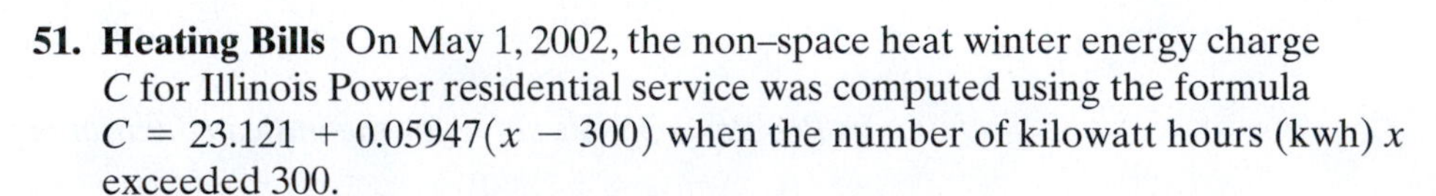

**51. Heating Bills** On May 1, 2002, the non–space heat winter energy charge $C$ for Illinois Power residential service was computed using the formula $C = 23.121 + 0.05947(x - 300)$ when the number of kilowatt hours (kwh) $x$ exceeded 300.

**(a)** Solve the equation for $x$.

**(b)** How many kwh were used if the non–space heat winter energy charge was \$115.30? Round to the nearest whole number.

**52. Supplementary and Complementary Angles** The supplement of an angle and the complement of the same angle sum to 150°. What is the measure of the angle?

## Section 1.4 Linear Inequalities

### KEY CONCEPTS

- **Linear Inequality in One Variable**
  An inequality of the form $ax + b < c, ax + b \leq c, ax + b > c$, or $ax + b \geq c$, where $a$, $b$, and $c$ are real numbers with $a \neq 0$.
- **Interval Notation versus Inequality Notation**

| Interval | Inequality Notation | Graph |
|---|---|---|
| The open interval $(a, b)$ | $\{x \mid a < x < b\}$ | |
| The closed interval $[a, b]$ | $\{x \mid a \leq x \leq b\}$ | |
| The half-open interval $[a, b)$ | $\{x \mid a \leq x < b\}$ | |
| The half-open interval $(a, b]$ | $\{x \mid a < x \leq b\}$ | |
| The interval $[a, \infty)$ | $\{x \mid x \geq a\}$ | |
| The interval $(a, \infty)$ | $\{x \mid x > a\}$ | |
| The interval $(-\infty, a]$ | $\{x \mid x \leq a\}$ | |
| The interval $(-\infty, a)$ | $\{x \mid x < a\}$ | |
| The interval $(-\infty, \infty)$ | $\{x \mid x$ is a real number$\}$ | |

- **Nonnegative Property of Inequalities**
  For any real number $a$, $a^2 > 0$
- **Addition Property of Inequalities**
  For real numbers $a$, $b$, and $c$
  If $a < b$, then $a + c < b + c$.
  If $a > b$, then $a + b > b + c$.
- **Multiplication Properties of Inequalities**
  For real numbers $a$, $b$, and $c$
  If $a < b$ and if $c > 0$, then $ac < bc$.
  If $a > b$ and if $c > 0$, then $ac > bc$.
  If $a < b$ and if $c < 0$, then $ac > bc$.
  If $a > b$ and if $c < 0$, then $ac < bc$.

### KEY TERMS

Solve an inequality
Solutions
Solution set
Interval Notation
Closed interval
Open interval
Half-open or half-closed interval
Left endpoint
Right endpoint
Equivalent inequalities

| YOU SHOULD BE ABLE TO . . . | EXAMPLE | REVIEW EXERCISES |
|---|---|---|
| 1 Represent inequalities using the real number line and interval notation (p. 90) | Examples 1 through 4 | 53–56 |
| 2 Understand the properties of inequalities (p. 93) | | 57–58 |
| 3 Solve linear inequalities (p. 93) | Examples 5 through 8 | 59–68 |
| 4 Solve problems involving linear inequalities (p. 96) | Example 9 | 69–72 |

*In Problems 53 and 54, write each inequality using interval notation and graph the inequality.*

**53.** $2 < x \leq 7$

**54.** $x > -2$

*In Problems 55 and 56, write each interval in inequality notation and graph the inequality.*

**55.** $(-\infty, 4]$

**56.** $[-1, 3)$

*In Problems 57 and 58, use the Addition Property and/or Multiplication Property to find a and b.*

**57.** If $5 \leq x \leq 9$, then $a \leq 2x - 3 \leq b$.

**58.** If $-2 < x < 0$, then $a < 3x + 5 < b$.

*In Problems 59–68, solve each linear inequality. Express your solution using set-builder notation and interval notation. Graph the solution set.*

**59.** $3x + 12 \leq 0$

**60.** $2 < 1 - 3x$

**61.** $-7 \leq 3(h + 1) - 8$

**62.** $-7x - 8 < -22$

**63.** $3(p - 2) + (5 - p) > 2 - (p - 3)$

**64.** $2(x + 1) + 1 > 2(x - 2)$

**65.** $5(x - 1) - 7x > 2(2 - x)$

**66.** $0.03x + 0.10 > 0.52 - 0.07x$

**67.** $-\frac{4}{9}w + \frac{7}{12} < \frac{5}{36}$

**68.** $\frac{2}{5}y - 20 > \frac{2}{3}y + 12$

*In Problems 69–72, write a linear inequality and solve.*

**69. Octoberfest** The German Club plans to rent a hall for their annual Octoberfest banquet. The hall costs \$150 to rent plus \$7.50 for each person who attends. If the club does not want to spend more than \$600 for the event, how many people can attend the banquet?

**70. Car Rentals** A Ford Taurus at Enterprise Rent-a-Car rents for \$43.46 per day. You receive 150 free miles per day but are charged \$0.25 per mile for any additional miles. How many miles can you drive per day, on average, and not exceed your daily budget of \$60.00?

**71. Fund Raising** A middle school band sells \$1 candy bars at a carnival to raise money for new instruments. The band pays \$50.00 to rent a booth and must pay the candy company \$0.60 for each bar sold. How many bars must the band sell to be making a profit?

**72. Movie Club** A DVD club offers unlimited DVDs for \$9.95 if you purchase one for \$24.95. How many DVDs can you purchase without spending more than \$72.00?

## Section 1.5 Compound Inequalities

| KEY CONCEPTS | KEY TERMS |
|---|---|
| • If $a < b$, then we can write $a < x$ and $x < b$ as $a < x < b$. | Intersection<br>Union<br>Compound inequality<br>Solve a compound inequality |

| YOU SHOULD BE ABLE TO . . . | EXAMPLE | REVIEW EXERCISES |
|---|---|---|
| 1 Determine the intersection or union of two sets (p. 102) | Examples 1 and 2 | 73–78 |
| 2 Solve compound inequalities involving "and" (p. 104) | Examples 3 through 6 | 79, 80, 83, 84, 88 |
| 3 Solve compound inequalities involving "or" (p. 107) | Examples 7 and 8 | 81, 82, 85, 86, 87 |
| 4 Solve problems using compound inequalities (p. 108) | Example 9 | 89–90 |

*In Problems 73–76, use* $A = \{2, 4, 6, 8\}$, $B = \{-1, 0, 1, 2, 3, 4\}$, *and* $C = \{1, 2, 3, 4\}$ *to find each set.*

**73.** $A \cup B$ **74.** $A \cap C$ **75.** $B \cap C$ **76.** $A \cup C$

*In Problems 77 and 78, use the graph of the inequality to find each set.*

**77.** $A = \{x \mid x \le 4\}$; $B = \{x \mid x > 2\}$. Find (a) $A \cap B$ and (b) $A \cup B$.

**78.** $E = \{x \mid x \ge 3\}$; $F = \{x \mid x < -2\}$. Find (a) $E \cap F$ and (b) $E \cup F$.

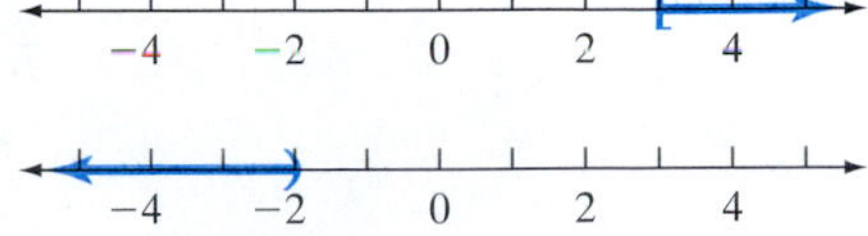

*In Problems 79–88, solve each compound inequality. Graph the solution set.*

**79.** $x < 4$ and $x + 3 > 2$

**80.** $3 < 2 - x < 7$

**81.** $x + 3 < 1$ or $x > 2$

**82.** $x + 6 \ge 10$ or $x \le 0$

**83.** $3x + 2 \le 5$ and $-4x + 2 \le -10$

**84.** $1 \le 2x + 5 < 13$

**85.** $x - 3 \le -5$ or $2x + 1 > 7$

**86.** $3x + 4 > -2$ or $4 - 2x \ge -6$

**87.** $\frac{1}{3}x > 2$ or $\frac{2}{5}x < -4$

**88.** $x + \frac{3}{2} \ge 0$ and $-2x + \frac{3}{2} > \frac{1}{4}$

**89. Heart Rates** The normal heart rate for healthy adults between the ages of 21 and 60 should be between 70 and 75 beats per minute (inclusive). If we let $x$ represent the heart rate of an adult between the ages of 21 and 60, express the normal range of values using a compound inequality.

**90. Heating Bills** For usage above 300 kilowatt hours, the non–space heat winter energy charge for Illinois Power residential service was \$23.12 plus \$0.05947 per kilowatt hour over 300. During one winter, a customer's charge ranged from a low of \$50.28 to a high of \$121.43. Over what range of values did electric usage vary (in kilowatt hours)?

## Section 1.6 Absolute Value Equations and Inequalities

### KEY CONCEPTS

- **Equations Involving Absolute Value**
  If $a$ is a positive real number and if $u$ is any algebraic expression, then $|u| = a$ is equivalent to $u = a$ or $u = -a$.
- **Equations Involving Two Absolute Values**
  If $u$ and $v$ are any algebraic expression, then $|u| = |v|$ is equivalent to $u = v$ or $u = -v$.
- **Inequalities of the Form < or ≤ Involving Absolute Value**
  If $a$ is a positive real number and if $u$ is any algebraic expression, then $|u| < a$ is equivalent to $-a < u < a$ and $|u| \le a$ is equivalent to $-a \le u \le a$.
- **Inequalities of the Form > or ≥ Involving Absolute Value**
  If $a$ is a positive real number and if $u$ is any algebraic expression, then $|u| > a$ is equivalent to $u < -a$ or $u > a$ and $|u| \ge a$ is equivalent to $u \le -a$ or $u \ge a$.

| YOU SHOULD BE ABLE TO . . . | EXAMPLE | REVIEW EXERCISES |
|---|---|---|
| 1 Solve absolute value equations (p. 113) | Examples 1 through 4 | 91–96 |
| 2 Solve absolute value inequalities involving $<$ or $\le$ (p. 116) | Examples 5 through 7 | 97, 99, 102, 103 |
| 3 Solve absolute value inequalities involving $>$ or $\ge$ (p. 118) | Examples 8 and 9 | 98, 100, 101, 104 |
| 4 Solve applied problems involving absolute value (p. 120) | Example 10 | 105–106 |

*In Problems 91–96, solve the absolute value equation.*

**91.** $|x| = 4$

**92.** $|3x - 5| = 4$

**93.** $|-y + 4| = 9$

**94.** $-3|x + 2| - 5 = -8$

**95.** $|2w - 7| = -3$

**96.** $|x + 3| = |3x - 1|$

*In Problems 97–104, solve each absolute value inequality. Graph the solution set on a real number line.*

**97.** $|x| < 2$

**98.** $|x| \ge \frac{7}{2}$

**99.** $|x + 2| \le 3$

**100.** $|4x - 3| \ge 1$

**101.** $3|x| + 6 \ge 1$

**102.** $|7x + 5| + 4 < 3$

**103.** $|(x - 3) - 2| \le 0.01$

**104.** $\left|\frac{2x - 3}{4}\right| > 1$

**105. Tolerance** The diameter of a certain ball bearing is required to be 0.503 inches. The tolerance on the bearing is 0.001 inches. If $x$ represents the diameter of a bearing, the acceptable diameters of the bearing can be expressed as $|x - 0.503| \le 0.001$. Determine the acceptable diameters of the bearing.

**106. Tensile Strength** The tensile strength of paper used to make grocery bags is about 40 lb/in.$^2$ A paper grocery bag whose tensile strength satisfies the inequality $\left|\frac{x - 40}{2}\right| > 1.96$ has an unusual tensile strength. Determine the tensile strengths that would be considered unusual.

## CHAPTER 1 TEST

*Remember to use your Chapter Test Prep Video CD to see fully worked-out solutions to any problems you would like to review.*

**1.** Determine which, if any, of the following are solutions to $3(x - 7) + 5 = x - 4$.

**(a)** $x = 6$ **(b)** $x = -2$

**2.** Write the following inequalities in interval notation and graph on a real number line.

**(a)** $x > -4$ **(b)** $3 < x \le 7$

*In Problems 3 and 4, translate the English statement into a mathematical statement. Do not attempt to solve.*

**3.** Three times a number, decreased by 8, is 4 more than the number.

**4.** Two-thirds of a number, increased by twice the difference of the number and 5, is more than 7.

*In Problems 5–7, solve the equation. Determine if the equation is an identity, a contradiction, or a conditional equation.*

**5.** $5x - (x - 2) = 6 + 2x$

**6.** $|2x + 5| - 3 = 0$

**7.** $7 + (x - 3) = 3(x + 1) - 2x$

*In Problems 8–12, solve the inequality and graph the solution set on a real number line.*

**8.** $x + 2 \leq 3x - 4$

**9.** $4x + 7 > 2x - 3(x - 2)$

**10.** $-x + 4 \leq x + 3$

**11.** $x + 2 < 8$ and $2x + 5 \geq 1$

**12.** $x > 4$ or $2(x - 1) + 3 < -2$

**13.** Solve $7x + 4y = 3$ for $y$.

**14.** For the sets $A = \{3, 6, 9, 12\}$ and $B = \{1, 3, 5, 7, 9\}$, find each of the following.

**(a)** $A \cup B$

**(b)** $A \cap B$

**15. Computer Sales** Glen works as a computer salesman and earns $400 weekly plus 8% commission on his weekly sales. If he wants to make at least $750 in a week, how much must his sales be?

**16. Shower** The Crescent Rod is a curved shower rod designed to provide more room in the shower area than a standard rod while keeping water from leaking out of the sides. The Crescent Rod fits in a 60-inch opening but has an adjustability of $\pm 1$ inch. Let $x$ represent the width, in inches, of the shower area. Write an absolute value inequality that represents the possible widths that the Crescent Rod can fit. Solve the inequality.

**17. Party Costs** A recreational center offers a children's birthday party for $75 plus $5 for each child. How many children were at Payton's birthday party if the total cost for the party was $145?

**18. Sandbox** Rick is building a rectangular sandbox for his daughter. He wants the length of the sandbox to be 2 feet more than the width and he has 20 feet of lumber to build the frame. Find the dimensions of the sandbox.

**19. Mixture** Two acid solutions are available to a chemist. One is a 10% nitric acid solution and the other is a 40% nitric acid solution. How much of each type of solution should be mixed together to form 12 liters of a 20% nitric acid solution?

**20. Ironman Race** The last leg of the Ironman competition is a 26.2 mile run. Contestant A runs at a constant rate of 8 miles per hour. If Contestant B starts the run 30 minutes after Contestant A and runs at a constant rate of 10 miles per hour, how long will it take Contestant B to catch up to Contestant A?

## CUMULATIVE REVIEW CHAPTERS R-1

**1.** Approximate each number by (i) truncating and (ii) rounding to the indicated number of decimal places.

**(a)** 27.2357; 3 decimal places.

**(b)** 1.0729; 1 decimal place.

**2.** Plot the points $-4$, $-\frac{5}{2}$, 0, and $\frac{7}{2}$ on a real number line.

*In Problems 3–8, evaluate the expressions.*

**3.** $-|-14|$

**4.** $(-3) + 4 - 7$

**5.** $\dfrac{-3(12)}{-6}$

**6.** $(-3)^4$

**7.** $5 - 2(1 - 4)^3 + 5 \cdot 3$

**8.** $\frac{2}{3} + \frac{1}{2} - \frac{1}{4}$

**9.** Evaluate $3x^2 + 2x - 7$ when $x = 2$.

**10.** Simplify: $4a^2 - 6a + a^2 - 12 + 2a - 1$.

**11.** Determine if the given values are in the domain of $x$ for the expression $\frac{x + 2}{x^2 + x - 2}$.

**(a)** $x = -2$ **(b)** $x = 0$

**12.** Use the Distributive Property to remove parentheses and then simplify.
$3(x + 2) - 4(2x - 1) + 8$

**13.** Determine whether $x = 3$ is a solution to the equation $x - (2x + 3) = 5x - 1$.

*In Problems 14–16, solve the equation.*

**14.** $4x - 3 = 2(3x - 2) - 7$

**15.** $\frac{x + 1}{3} = x - 4$

**16.** $2|x - 1| + 2 = 9$

**17.** Solve $2x - 5y = 6$ for $y$.

*In Problems 18–20, solve the inequality and graph the solution set on a real number line.*

**18.** $\frac{x + 3}{2} \leq \frac{3x - 1}{4}$

**19.** $|x - 4| \leq 3$

**20.** $|x + 7| > 2$

**21. Computing Grades** Shawn really wants an A in his geometry class. His four exam scores are 94, 95, 90, and 97. The final exam is worth two exam scores. To have an A, his average must be between 93 and 100, inclusive. For what range of scores on the final exam will Shawn be able to earn an A in the course?

**22. Body Mass Index** The body mass index (BMI) of a person 62 inches tall and weighing $x$ pounds is given by $0.2x - 2$. A BMI of 30 or more is considered to be obese. For what weights would a person 62 inches tall be considered obese?

**23. Supplementary Angles** Two angles are supplementary. The measure of the larger angle is 15 degrees more than twice the measure of the smaller angle. Find the angle measures.

**24. Cylinders** Max has 100 square inches of aluminum with which to make a closed cylinder. If the radius of the cylinder must be 2 inches, how tall will the cylinder be? (Round to the nearest hundredth of an inch.)

**25. Consecutive Integers** Find three consecutive even integers such that the sum of the first two is 22 more than the third.

CHAPTER

# 2 Graphs, Relations, and Functions

Graphs come in many different forms. For example, the graph shown to the right was drawn by Florence Nightingale and is called a *polar area graph*. The area of each shaded region is proportional to the number of deaths represented by the shaded region. Nightingale used the graph to illustrate that more soldiers died because of unsanitary conditions in hospitals than on the battlefield during the Crimean War. This graph helped to increase the focus on the sanitary conditions in hospitals and shows how math can be used to save lives.

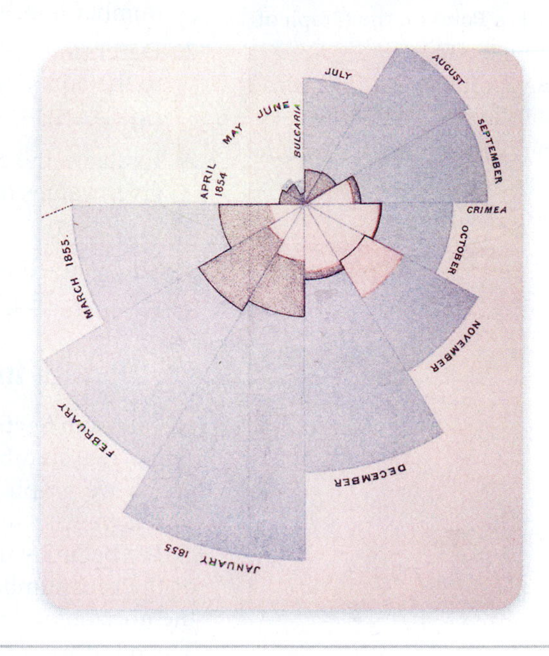


## OUTLINE



## The Big Picture: Putting It Together

In Chapter R, we learned how to plot points on the real number line. We can think of this as plotting in one dimension. Then, in Chapter 1, we solved linear equations in one variable. We learned that a linear equation in one variable may have no solution, one solution, or infinitely many solutions. When the equation has one solution, it can be represented as a single point on the real number line. When the equation has infinitely many solutions, the solution is the entire real number line.

In this chapter, we introduce the *rectangular coordinate system*. This system allows us to plot points in two dimensions, which gives us a method for representing solutions to equations in two variables graphically. The rectangular coordinate system provides the connection between algebra and geometry. Prior to the introduction of the rectangular coordinate system, algebra and geometry were thought to be separate subjects. We also introduce the concept of a function in this chapter. The function is arguably the single most important concept of algebra.

# 2.1 Rectangular Coordinates and Graphs of Equations

**OBJECTIVES**

1. Plot Points in the Rectangular Coordinate System
2. Determine Whether an Ordered Pair Is a Point on the Graph of an Equation
3. Graph an Equation Using the Point-Plotting Method
4. Identify the Intercepts from the Graph of an Equation
5. Interpret Graphs

***Preparing for Rectangular Coordinates and Graphs of Equations***

*Before getting started, take this readiness quiz. If you get a problem wrong, go back to the section cited and review the material.*

**1.** Plot the following points on the real number line: $-2, 4, 0, \frac{1}{2}$. [Section R.2, pp. 15–16]

**2.** Determine which of the following are solutions to the equation $3x - 5(x + 2) = 4$.
**(a)** $x = 0$ **(b)** $x = -3$ **(c)** $x = -7$ [Section 1.1, pp. 52–53]

**3.** Evaluate the expression $2x^2 - 3x + 1$ for the given values of the variable.
**(a)** $x = 0$ **(b)** $x = 2$ **(c)** $x = -3$ [Section R.5, pp. 43–44]

**4.** Solve the equation $3x + 2y = 8$ for $y$. [Section 1.3, pp. 81–84]

**5.** Evaluate $|-4|$. [Section R.3, p. 20]

## 1 Plot Points in the Rectangular Coordinate System

Recall from Section R.1 that we locate a point on the real number line by assigning it a single real number, called the *coordinate of the point*. See *Preparing for* Problem 1. When we graph a point on the real number line, we are working in one dimension. When we wish to work in two dimensions, we locate a point using two real numbers.

We begin by drawing two real number lines that intersect at right (90°) angles. One of the real number lines is drawn horizontal, while the other is drawn vertical. We call the horizontal real number line the ***x*-axis,** and the vertical real number line is the ***y*-axis.** The point where the $x$-axis and $y$-axis intersect is called the **origin, *O*.** See Figure 1.

**Figure 1**

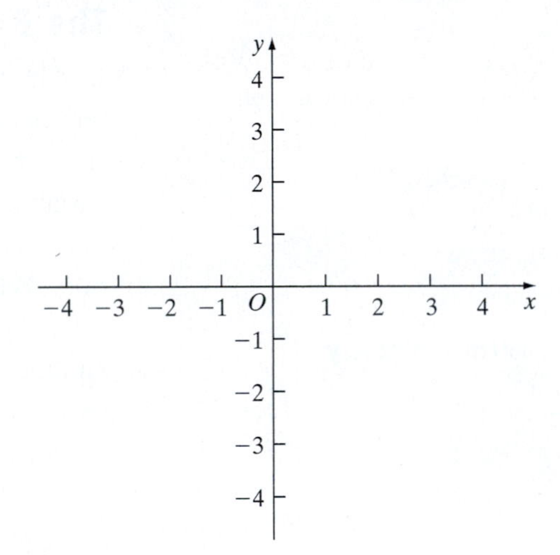


The origin $O$ has a value of 0 on the $x$-axis and the $y$-axis. Points on the $x$-axis to the right of $O$ are positive real numbers; points on the $x$-axis to the left of $O$ are negative real numbers. Points on the $y$-axis that are above $O$ are positive real numbers; points on the $y$-axis that are below $O$ are negative real numbers. In Figure 1 we label the $x$-axis "$x$" and the $y$-axis "$y$." Notice that an arrow is used at the end of each axis to denote the positive direction. We do not use an arrow to denote the negative direction.

The coordinate system presented in Figure 1 is called a **rectangular** or **Cartesian coordinate system,** named after René Descartes (1596–1650), a French mathematician, philosopher, and theologian. The plane formed by the $x$-axis and $y$-axis is often referred to as the ***xy*-plane,** and the $x$-axis and $y$-axis are called the **coordinate axes.**

We can represent any point $P$ in the rectangular coordinate system by using an **ordered pair $(x, y)$** of real numbers. If $x > 0$, we travel $x$ units to the right of the $y$-axis; if $x < 0$, we travel $|x|$ units to the left of the $y$-axis. If $y > 0$, we travel $y$ units above the $x$-axis; if $y < 0$, we travel $|y|$ units below the $x$-axis. The ordered pair $(x, y)$ is also called the **coordinates** of $P$.

***Preparing for...Answers***

**1.**

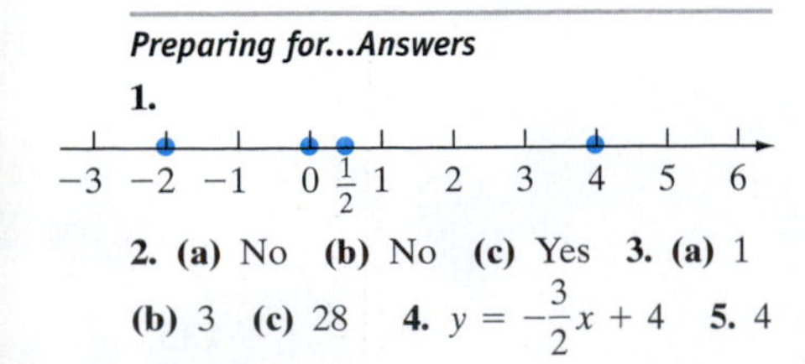


**2. (a)** No **(b)** No **(c)** Yes **3. (a)** 1 **(b)** 3 **(c)** 28 **4.** $y = -\frac{3}{2}x + 4$ **5.** 4

The origin $O$ has coordinates $(0, 0)$. Any point on the $x$-axis has coordinates of the form $(x, 0)$, and any point on the $y$-axis has coordinates of the form $(0, y)$.

If $(x, y)$ are the coordinates of a point $P$, then $x$ is called the **$x$-coordinate** or **abscissa,** of $P$ and $y$ is called the **$y$-coordinate** or **ordinate,** of $P$.

If you look back at Figure 1, you should notice that the $x$- and $y$-axes divide the plane into four separate regions or **quadrants.** In quadrant I, both the $x$- and $y$-coordinate are positive; in quadrant II, $x$ is negative and $y$ is positive; in quadrant III, both $x$ and $y$ are negative; and in quadrant IV, $x$ is positive and $y$ is negative. Points on the coordinate axes do not belong to a quadrant. See Figure 2.

**Figure 2**

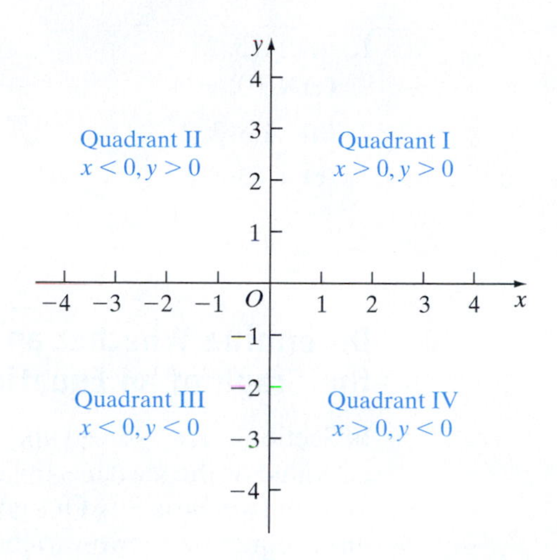


## EXAMPLE 1 Plotting Points in the Rectangular Coordinate System and Determining the Quadrant in which the Point Lies

Plot the points in the $xy$-plane. Tell which quadrant each point is in.

**(a)** $A(3, 2)$ **(b)** $B(-2, 4)$ **(c)** $C(-1, -3)$
**(d)** $D(3, -4)$ **(e)** $E(-2, 0)$

### Solution

Before we plot the points, we draw a rectangular or Cartesian coordinate system. See Figure 3(a). We now plot the points.

**(a)** To plot point $A(3, 2)$, from the origin $O$, we travel 3 units to the right and then 2 units up. Label the point $A$. See Figure 3(b). Point $A$ is in quadrant I because both $x$ and $y$ are positive.

**Figure 3**

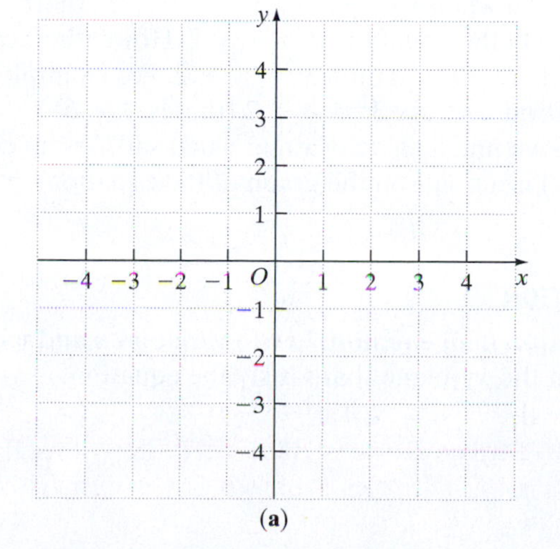

(a)

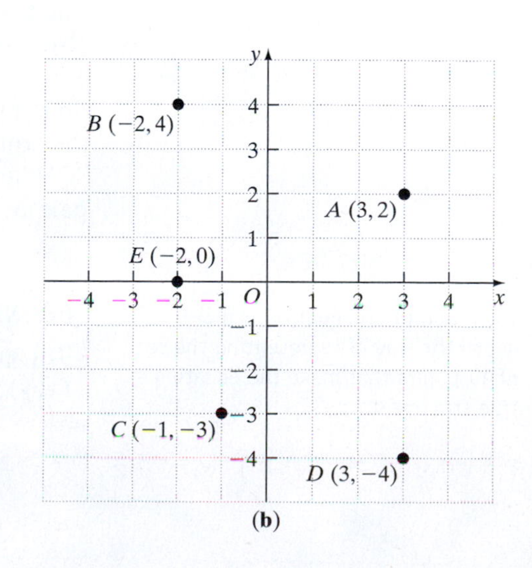


(b)

**(b)** To plot point $B(-2, 4)$, from the origin $O$, we travel 2 units to the left and then 4 units up. Label the point $B$. See Figure 3(b). Point $B$ is in quadrant II.

**(c)** See Figure 3(b). Point $C$ is in quadrant III.

**(d)** See Figure 3(b). Point $D$ is in quadrant IV.

**(e)** See Figure 3(b). Point $E$ is not in a quadrant because it lies on the $x$-axis.

**QUICK ✓** *Plot each point in the xy-plane. Tell in which quadrant or on what coordinate axis each point lies.*

**1. (a)** $A(5, 2)$ **(b)** $B(4, -2)$
**(c)** $C(0, -3)$ **(d)** $D(-4, -3)$

**2. (a)** $A(-3, 2)$ **(b)** $B(-4, 0)$
**(c)** $C(3, -2)$ **(d)** $D(6, 1)$

## 2 Determine Whether an Ordered Pair Is a Point on the Graph of an Equation

In Section 1.1, we solved linear equations in one variable. The solution was either a single value of the variable, the empty set, or all real numbers. We will now look at equations in two variables. Our goal is to learn a method for representing the solution to a linear equation in two variables.

**DEFINITION**

An **equation in two variables,** say $x$ and $y$, is a statement in which the algebraic expressions involving $x$ and $y$ are equal. The expressions are called **sides** of the equation.

Since an equation is a statement, it may be true or false, depending upon the values of the variables. Any values of the variable that make the equation a true statement are said to **satisfy** the equation.

For example, the following are all equations in two variables.

$$x^2 = y + 2 \qquad 3x + 2y = 6 \qquad y = -4x + 5$$

The first equation $x^2 = y + 2$ is satisfied when $x = 3$ and $y = 7$ since $3^2 = 7 + 2$. It is also satisfied when $x = -2$ and $y = 2$. In fact, there are infinitely many choices of $x$ and $y$ that satisfy the equation $x^2 = y + 2$. However, there are some choices of $x$ and $y$ that do not satisfy the equation $x^2 = y + 2$. For example, $x = 3$ and $y = 4$ does not satisfy the equation because $3^2 \neq 4 + 2$ (that is, $9 \neq 6$).

When we find a value of $x$ and $y$ that satisfies an equation, it means that the ordered pair $(x, y)$ is a point on the graph of the equation.

**In Words**
The graph of an equation is a geometric way of representing the set of all points that make the equation a true statement.

**DEFINITION**

The **graph of an equation in two variables** $x$ and $y$ is the set of all ordered pairs $(x, y)$ in the $xy$-plane that satisfy the equation.

### EXAMPLE 2 Determining Whether a Point Is on the Graph of an Equation

Determine if the following points are on the graph of $3x - y = 6$.

**(a)** $(2, 0)$ **(b)** $(1, -2)$ **(c)** $\left(\frac{1}{2}, -\frac{9}{2}\right)$

### Solution

**(a)** For the point $(2, 0)$, we check to see if $x = 2$, $y = 0$ satisfies the equation $3x - y = 6$.

$$\begin{aligned} & 3x - y = 6 \\ \text{Let } x = 2,\ y = 0{:}\quad & 3(2) - 0 \stackrel{?}{=} 6 \\ & 6 = 6 \quad \text{True} \end{aligned}$$

The statement is true when $x = 2$ and $y = 0$, so the point $(2, 0)$ is on the graph.

**(b)** For the point $(1, -2)$, we have

$$\begin{aligned} & 3x - y = 6 \\ \text{Let } x = 1,\ y = -2{:}\quad & 3(1) - (-2) \stackrel{?}{=} 6 \\ & 3 + 2 \stackrel{?}{=} 6 \\ & 5 = 6 \quad \text{False} \end{aligned}$$

The statement $5 = 6$ is false when $x = 1$ and $y = -2$, so the point $(1, -2)$ is not on the graph.

**(c)** For the point $\left(\frac{1}{2}, -\frac{9}{2}\right)$, we have

$$\begin{aligned} & 3x - y = 6 \\ \text{Let } x = \tfrac{1}{2},\ y = -\tfrac{9}{2}{:}\quad & 3\left(\frac{1}{2}\right) - \left(-\frac{9}{2}\right) \stackrel{?}{=} 6 \\ & \frac{3}{2} + \frac{9}{2} \stackrel{?}{=} 6 \\ & \frac{12}{2} \stackrel{?}{=} 6 \\ & 6 = 6 \quad \text{True} \end{aligned}$$

The statement is true when $x = \frac{1}{2}$ and $y = -\frac{9}{2}$, so the point $\left(\frac{1}{2}, -\frac{9}{2}\right)$ is on the graph.

**QUICK ✓**

**3.** Determine if the following points are on the graph of $2x - 4y = 12$.

**(a)** $(2, -3)$ **(b)** $(2, -2)$ **(c)** $\left(\frac{3}{2}, -\frac{9}{4}\right)$

**4.** Determine if the following points are on the graph of $y = x^2 + 3$.

**(a)** $(1, 4)$ **(b)** $(-2, -1)$ **(c)** $(-3, 12)$

## 3 Graph an Equation Using the Point-Plotting Method

One of the most elementary methods for graphing an equation is the **point-plotting method.** With this method, we choose values for one of the variables and use the equation to determine the corresponding values of the remaining variable. If $x$ and $y$ are the variables in the equation, it does not matter whether we choose values of $x$ and use the equation to find the corresponding $y$ or choose $y$ and find $x$. Convenience will determine which way we go.

### EXAMPLE 3 How to Graph an Equation by Plotting Points

Graph the equation $y = -2x + 4$ by plotting points.

#### Step-by-Step Solution

**Step 1:** We want to find all points $(x, y)$ that satisfy the equation. To determine these points we choose values of $x$ (do you see why?) and use the equation to determine the corresponding values of $y$. See Table 1.

Table 1

| $x$ | $y = -2x + 4$ | $(x, y)$ |
|---|---|---|
| $-3$ | $-2(-3) + 4 = 10$ | $(-3, 10)$ |
| $-2$ | $-2(-2) + 4 = 8$ | $(-2, 8)$ |
| $-1$ | $-2(-1) + 4 = 6$ | $(-1, 6)$ |
| $0$ | $-2(0) + 4 = 4$ | $(0, 4)$ |
| $1$ | $-2(1) + 4 = 2$ | $(1, 2)$ |
| $2$ | $-2(2) + 4 = 0$ | $(2, 0)$ |
| $3$ | $-2(3) + 4 = -2$ | $(3, -2)$ |

**Step 2:** We plot the points listed in the third column of Table 1 as shown in Figure 4(a). Now connect the points to obtain the graph of the equation (a line) as shown in Figure 4(b).

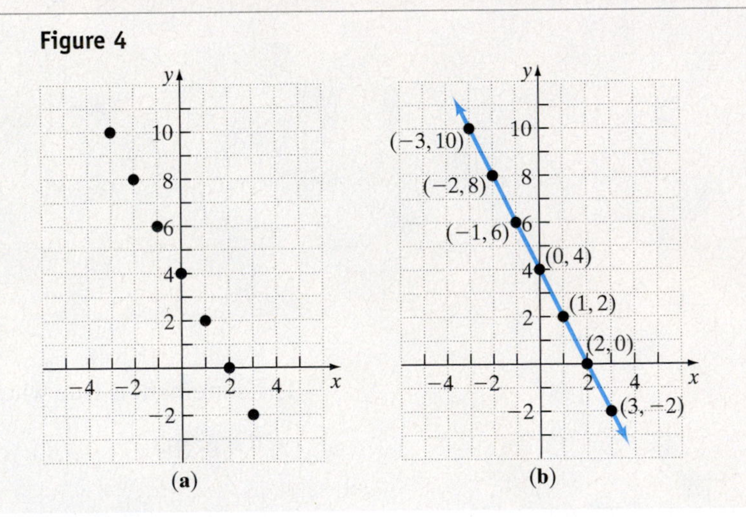

Figure 4

The graph of the equation shown in Figure 4(b) does not show all the points that satisfy $y = -2x + 4$. For example, in Figure 4(b) the point $(8, -12)$ is part of the graph of $y = -2x + 4$, but it is not shown. Since the graph of $y = -2x + 4$ can be extended as far as we please, we use arrows on the ends of the graph to indicate that the pattern shown continues. It is important to show enough of the graph so that anyone who is looking at it will "see" the rest of it as an obvious continuation of what is there. This is called a **complete graph.**

## EXAMPLE 4 Graphing an Equation by Plotting Points

Graph the equation $y = x^2$ by plotting points.

### Solution

Table 2 shows several points on the graph.

**Table 2**

| $x$ | $y = x^2$ | $(x, y)$ |
|---|---|---|
| $-4$ | $y = (-4)^2 = 16$ | $(-4, 16)$ |
| $-3$ | $y = (-3)^2 = 9$ | $(-3, 9)$ |
| $-2$ | $y = (-2)^2 = 4$ | $(-2, 4)$ |
| $-1$ | $y = (-1)^2 = 1$ | $(-1, 1)$ |
| $0$ | $y = (0)^2 = 0$ | $(0, 0)$ |
| $1$ | $y = (1)^2 = 1$ | $(1, 1)$ |
| $2$ | $y = (2)^2 = 4$ | $(2, 4)$ |
| $3$ | $y = (3)^2 = 9$ | $(3, 9)$ |
| $4$ | $y = (4)^2 = 16$ | $(4, 16)$ |

In Figure 5(a), we plot the ordered pairs listed in Table 2. In Figure 5(b), we connect the points in a smooth curve.

**Figure 5**

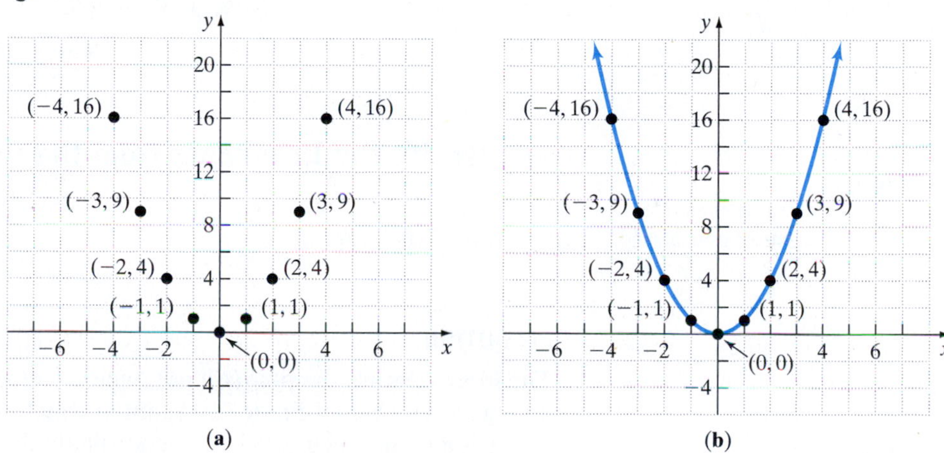


**Work Smart**
Notice we use a different scale on the $x$- and $y$-axis in Figure 5.

**Work Smart**
Experience will play a huge role in determining which $x$-values to choose in creating a table of values. For the time being, start by choosing values of $x$ around $x = 0$ as in Table 2.

Two questions that you might be asking yourself right now are "How do I know how many points are sufficient?" and "How do I know which $x$-values (or $y$-values) I should choose in order to obtain points on the graph?" Often, the type of equation we wish to graph indicates the number of points that are necessary. For example, we will learn later that if the equation is of the form $y = mx + b$, then its graph is a line and only two points are required to obtain the graph (as in Example 3). Other times, more points are required. At this stage in your math career, you will need to plot quite a few points to obtain a complete graph. However, as your experience and knowledge grows, you will learn to be more efficient in obtaining complete graphs.

**QUICK ✓** *Graph each equation using the point-plotting method.*

**5.** $y = 3x + 1$ **6.** $2x + 3y = 8$ **7.** $y = x^2 + 3$

## EXAMPLE 5 Graphing the Equation $x = y^2$

Graph the equation $x = y^2$ by plotting points.

### Solution

Because the equation is solved for $x$, we will choose values of $y$ and use the equation to find the corresponding values of $x$. See Table 3. We plot the ordered pairs listed in Table 3 and connect the points in a smooth curve. See Figure 6.

**Table 3**

| $y$ | $x = y^2$ | $(x, y)$ |
|---|---|---|
| $-3$ | $(-3)^2 = 9$ | $(9, -3)$ |
| $-2$ | $(-2)^2 = 4$ | $(4, -2)$ |
| $-1$ | $(-1)^2 = 1$ | $(1, -1)$ |
| 0 | $0^2 = 0$ | $(0, 0)$ |
| 1 | $1^2 = 1$ | $(1, 1)$ |
| 2 | $2^2 = 4$ | $(4, 2)$ |
| 3 | $3^2 = 9$ | $(9, 3)$ |

Figure 6

**QUICK ✓** *Graph each equation using the point-plotting method.*

**8.** $x = y^2 + 2$  **9.** $x = (y - 1)^2$

## 4 Identify the Intercepts from the Graph of an Equation

**Work Smart**
In order for a graph to be complete, all of its intercepts must be displayed.

One of the key components that should be displayed in a complete graph is the *intercepts* of the graph.

**DEFINITION**

The **intercepts** are the points, if any, where a graph crosses or touches the coordinate axes. The $x$-coordinate of a point at which the graph crosses or touches the $x$-axis is an **$x$-intercept,** and the $y$-coordinate of a point at which the graph crosses or touches the $y$-axis is a **$y$-intercept.**

See Figure 7 for an illustration. Notice that an $x$-intercept exists when $y = 0$ and a $y$-intercept exists when $x = 0$.

Figure 7

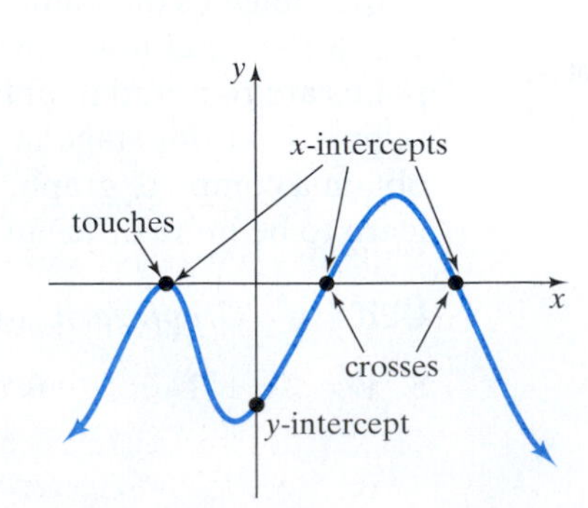

## EXAMPLE 6 Finding Intercepts from a Graph

Find the intercepts of the graph shown in Figure 8. What are the $x$-intercepts? What are the $y$-intercepts?

**Figure 8**

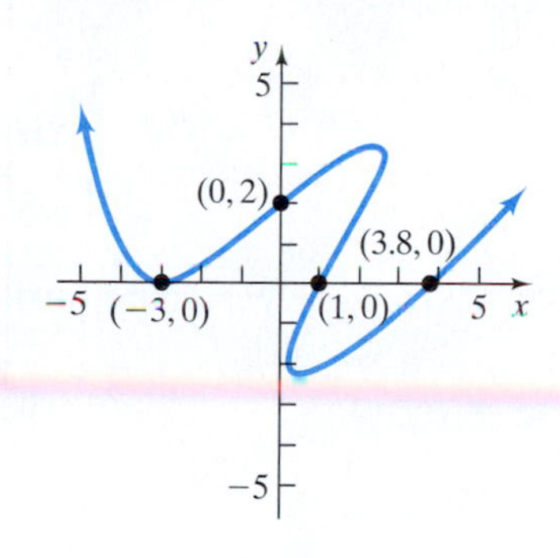


### Solution

The intercepts of the graph are the points

$$(-3, 0), \quad (0, 2), \quad (1, 0), \quad \text{and} \quad (3.8, 0)$$

The $x$-intercepts are $-3$, 1, and 3.8. The $y$-intercept is 2.

In Example 6, you should notice the following: If we do not specify the type of intercept ($x$- versus $y$-), then we report the intercept as an ordered pair. However, if we specify the type of intercept, then we only need report the coordinate of the intercept. For $x$-intercepts, we report the $x$-coordinate of the intercept; for $y$-intercepts, we report the $y$-coordinate of the intercept.

10. Find the intercepts of the graph shown in the figure. What are the $x$-intercepts? What are the $y$-intercepts?

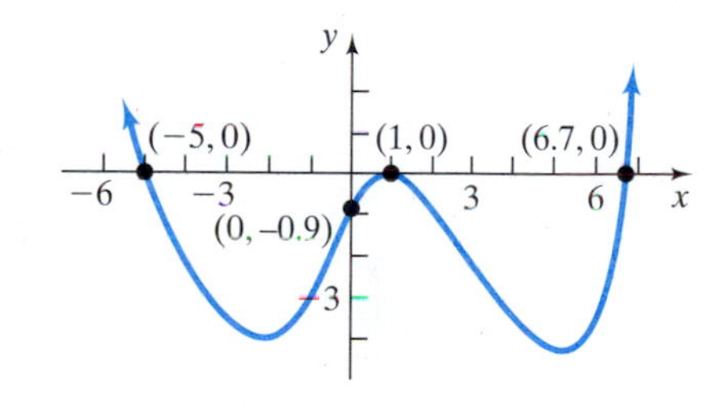


## 5 Interpret Graphs

Graphs play an important role in helping us to visualize relationships that exist between two variables or quantities. We have all heard the expression "A picture is worth a thousand words." A graph is a "picture" that illustrates the relationship between two variables. By visualizing this relationship, we are able to see important information and draw conclusions regarding the relationship between the two variables.

## EXAMPLE 7 Interpret a Graph

The graph in Figure 9 shows the revenue $R$ for selling $x$ gallons of gasoline in an hour at a gas station. The vertical axis represents the revenue and the horizontal axis represents the number of gallons of gasoline sold.

**(a)** What is the revenue if 150 gallons of gasoline are sold?

**(b)** How many gallons of gasoline are sold when revenue is highest? What is the highest revenue?

**(c)** Identify and interpret the intercepts.

Figure 9

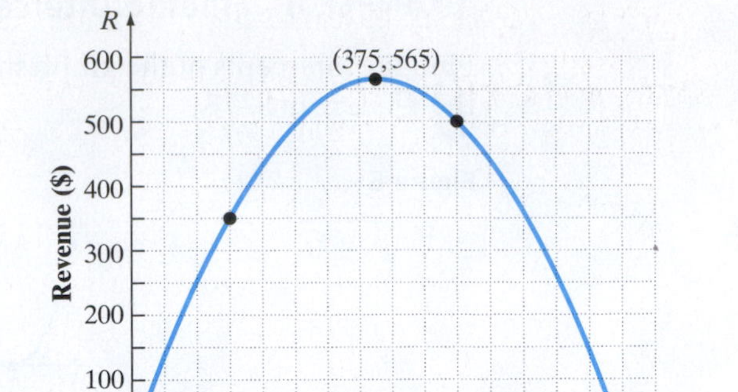


## Solution

**(a)** Draw a vertical line up from 150 on the horizontal axis until we reach the point on the graph. Then draw a horizontal line from this point to the vertical axis. The point where the horizontal line intersects the vertical axis is the revenue when 150 gallons of gasoline are sold. The revenue from selling 150 gallons of gasoline is \$350.

**(b)** The revenue is highest when 375 gallons of gasoline are sold. The highest revenue is \$565.

**(c)** The intercepts are $(0, 0)$ and $(750, 0)$. If the price of gasoline is too high, demand for gasoline (in theory) will be 0 gallons. This is the explanation for selling 0 gallons of gasoline. If the price of gasoline is \$0, then revenue will also be 0. The 750 gallons "sold" at a price of \$0 per gallon represents the maximum number of gallons that can be pumped at the station. In other words, it is the station's capacity.

### QUICK ✓

**11.** The graph shown below represents the cost $C$ (in thousands) of refining $x$ gallons of gasoline per hour (in thousands). The vertical axis represents the cost and the horizontal axis represents the number of gallons of gasoline refined.

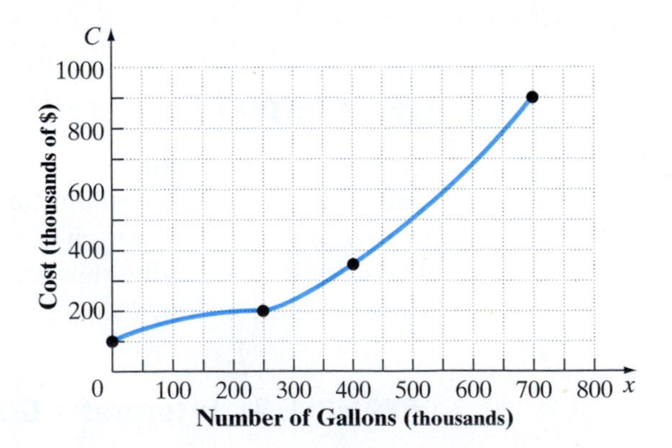


**(a)** What is the cost of refining 250 thousand gallons of gasoline per hour?

**(b)** What is the cost of refining 400 thousand gallons of gasoline per hour?

**(c)** In the context of the problem, explain the meaning of the graph ending at 700 thousand gallons of gasoline.

**(d)** Identify and interpret the intercept.

# 2.1 Exercises

**For Extra Help:** Student Solutions Manual · CD Video · PH Math/Tutor Center · MathXL Tutorials on CD · MathXL® · MyMathLab

## Concepts and Vocabulary

*In Problems 1–3, fill in the blanks.*

1. The point where the $x$-axis and $y$-axis intersect in the Cartesian coordinate system is called the _____.
2. If a point lies in quadrant II of the Cartesian coordinate system, then $x$ _____ $(>, <)$ 0 and $y$ _____ $(>, <)$ 0.
3. The points, if any, at which a graph crosses or touches a coordinate axis are called _____.

*In Problems 4–6, answer True or False to each statement.*

4. If a point lies in quadrant III of the Cartesian coordinate system, then both $x$ and $y$ are negative.
5. The graph of an equation must have at least one $x$-intercept.
6. The graph of an equation in two variables $x$ and $y$ is the set of all ordered pairs $(x, y)$ in the $xy$-plane that satisfy the equation.
7. Explain what is meant by a complete graph.
8. Explain what the graph of an equation represents.
9. What is the point-plotting method for graphing an equation?
10. What is the $y$-coordinate of a point that is an $x$-intercept? What is the $x$-coordinate of a point that is a $y$-intercept?

## Building Skills

11. Determine the coordinates of each of the points plotted. Tell in which quadrant or on what coordinate axis each point lies.

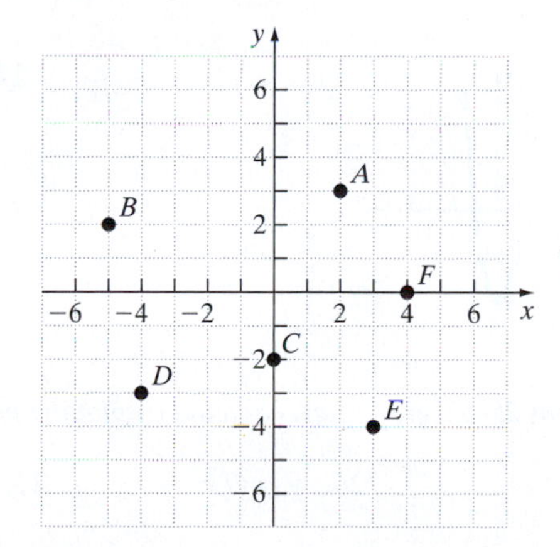


12. Determine the coordinates of each of the points plotted. Tell in which quadrant or on what coordinate axis each point lies.

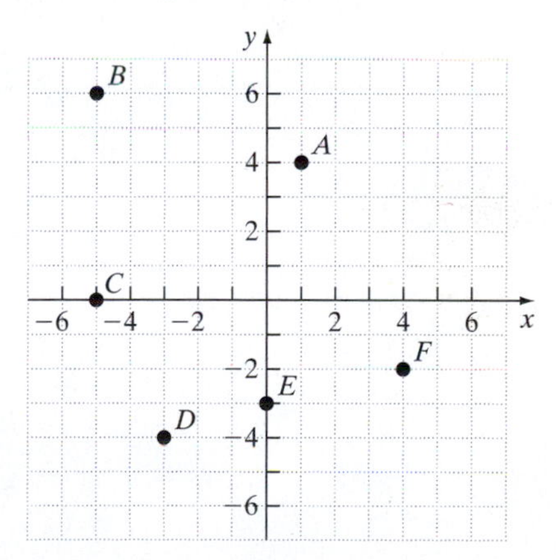


*In Problems 13 and 14, plot each point in the xy-plane. Tell in which quadrant or on what coordinate axis each point lies.*

13. $A(3, 5)$
$B(-2, -6)$
$C(5, 0)$
$D(1, -6)$
$E(0, 3)$
$F(-4, 1)$

14. $A(-3, 1)$
$B(-6, 0)$
$C(2, -5)$
$D(-6, -2)$
$E(1, 2)$
$F(0, -5)$

*In Problems 15–20, determine whether the given points are on the graph of the equation.*

**15.** $2x + 5y = 12$

**(a)** $(1, 2)$
**(b)** $(-2, 3)$
**(c)** $(-4, 4)$
**(d)** $\left(-\frac{3}{2}, 3\right)$

**16.** $-4x + 3y = 18$

**(a)** $(1, 7)$
**(b)** $(0, 6)$
**(c)** $(-3, 10)$
**(d)** $\left(\frac{3}{2}, 4\right)$

**17.** $y = -2x^2 + 3x - 1$

**(a)** $(-2, -15)$
**(b)** $(3, 10)$
**(c)** $(0, 1)$
**(d)** $(2, -3)$

**18.** $y = x^3 - 3x$

**(a)** $(2, 2)$
**(b)** $(3, 8)$
**(c)** $(-3, -18)$
**(d)** $(0, 0)$

**19.** $y = |x - 3|$

**(a)** $(1, 4)$
**(b)** $(4, 1)$
**(c)** $(-6, 9)$
**(d)** $(0, 3)$

**20.** $x^2 + y^2 = 1$

**(a)** $(0, 1)$
**(b)** $(1, 1)$
**(c)** $\left(\frac{1}{2}, \frac{1}{2}\right)$
**(d)** $\left(\frac{\sqrt{3}}{2}, \frac{1}{2}\right)$

*In Problems 21–24, the graph of an equation is given. List the intercepts of the graph.*

**21.**

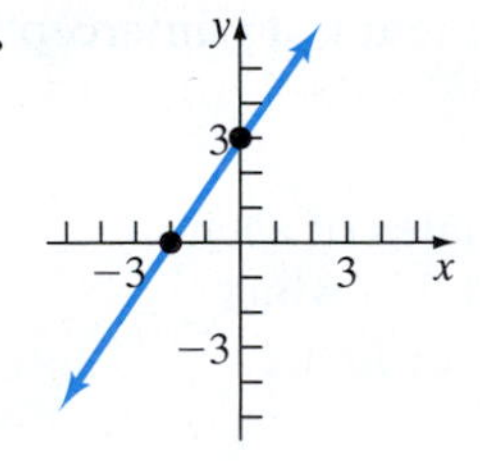


**22.**

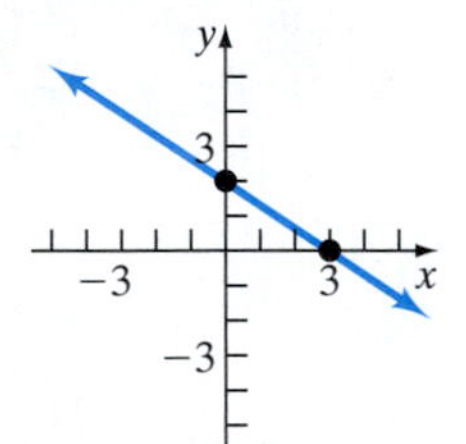


**23.**

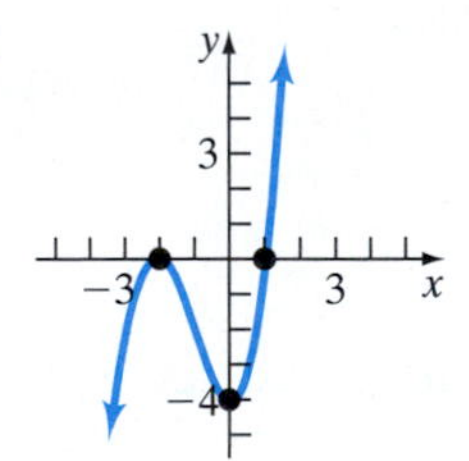


**24.**

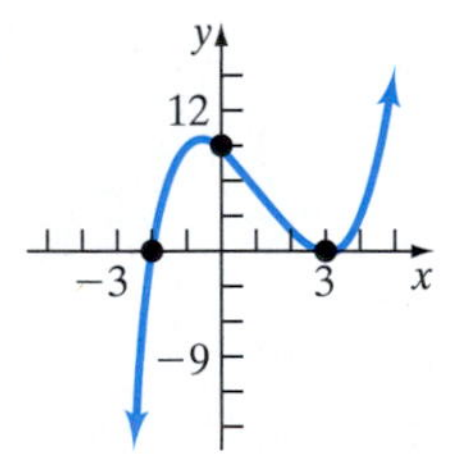


*In Problems 25–52, graph each equation by plotting points.*

**25.** $y = 4x$ **26.** $y = 2x$ **27.** $y = -\frac{1}{2}x$ **28.** $y = -\frac{1}{3}x$

**29.** $y = x + 3$ **30.** $y = x - 2$ **31.** $y = -3x + 1$ **32.** $y = -4x + 2$

**33.** $y = \frac{1}{2}x - 4$ **34.** $y = -\frac{1}{2}x + 2$ **35.** $2x + y = 7$ **36.** $3x + y = 9$

**37.** $y = -x^2$ **38.** $y = x^2 - 2$ **39.** $y = 2x^2 - 8$ **40.** $y = -2x^2 + 8$

**41.** $y = |x|$ **42.** $y = |x| - 2$ **43.** $y = |x - 1|$ **44.** $y = -|x|$

**45.** $y = x^3$ **46.** $y = -x^3$ **47.** $y = x^3 + 1$ **48.** $y = x^3 - 2$

**49.** $x^2 - y = 4$ **50.** $x^2 + y = 5$ **51.** $x = y^2 - 1$ **52.** $x = y^2 + 2$

## Applying the Concepts

**53.** If $(a, 4)$ is a point on the graph of $y = 4x - 3$, what is $a$?

**54.** If $(a, -2)$ is a point on the graph of $y = -3x + 5$, what is $a$?

**55.** If $(3, b)$ is a point on the graph of $y = x^2 - 2x + 1$, what is $b$?

**56.** If $(-2, b)$ is a point on the graph of $y = -2x^2 + 3x + 1$, what is $b$?

**57. Area of a Window** Bob Villa wishes to put a new window in his home. He wants the perimeter of the window to be 100 feet. The graph below shows the relation between the width, $x$, of the opening and the area of the opening.

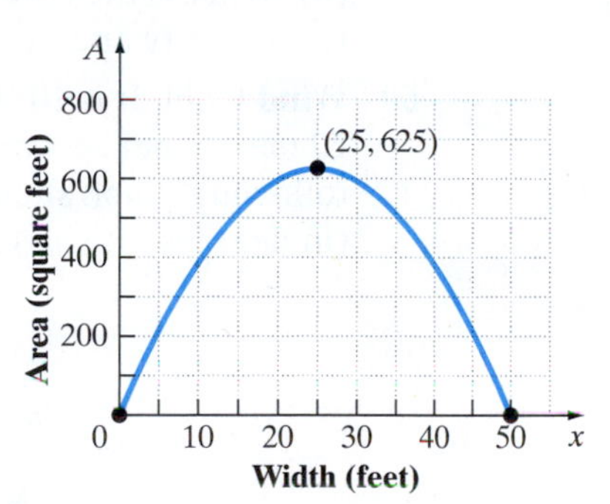


**(a)** What is the area of the opening if the width is 10 feet?
**(b)** What is the width of the opening in order for area to be a maximum? What is the maximum area of the opening?
**(c)** Identify and interpret the intercepts.

**58. Projectile Motion** The graph below shows the height, in feet, of a ball thrown straight up with an initial speed of 80 feet per second from an initial height of 96 feet after $t$ seconds.

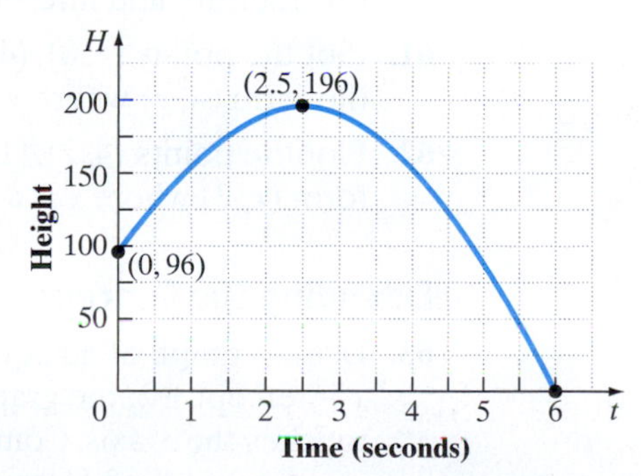


**(a)** What is the height of the object after 1.5 seconds?
**(b)** At what time is the height a maximum? What is the maximum height?
**(c)** Identify and interpret the intercepts.

**59. Cell Phones** We all struggle with selecting a cellular phone provider. The graph below shows the relation between the monthly cost of a cellular phone and the number of minutes used, $m$, when using the Sprint PCS 500-minute plan. (*Source: SprintPCS.com*)

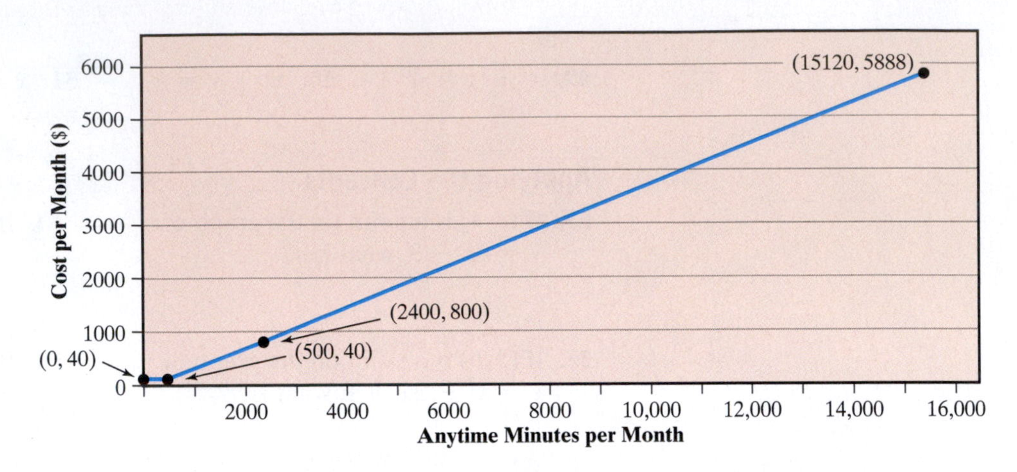


**(a)** What is the cost of talking for 200 minutes in a month? 500 minutes?
**(b)** What is the cost of talking 2400 minutes in a month?
**(c)** Identify and interpret the intercept.

**60. Wind Chill** It is 10° Celsius outside. The wind is calm but then gusts up to 20 meters per second. You feel the chill go right through your bones. The following graph shows the relation between the wind chill temperature (in degrees Celsius) and wind speed (in meters per second).

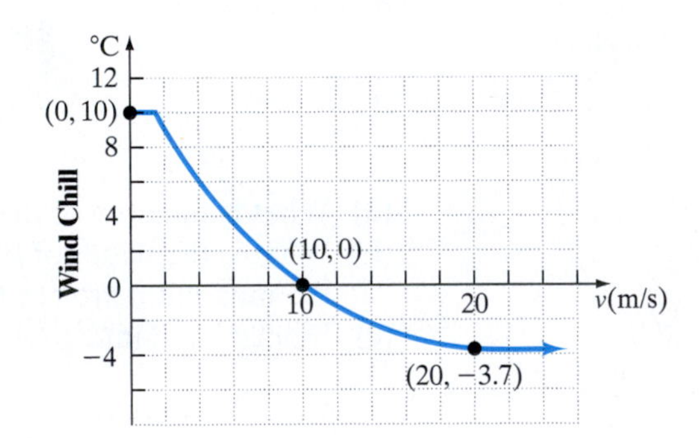


**(a)** What is the wind chill if the wind is blowing 4 meters per second?
**(b)** What is the wind chill if the wind is blowing 20 meters per second?
**(c)** Identify and interpret the intercepts.

**61.** Plot the points $(4, 0)$, $(4, 2)$, $(4, -3)$, and $(4, -6)$. Describe the set of all points of the form $(4, y)$ where $y$ is a real number.

**62.** Plot the points $(4, 2)$, $(1, 2)$, $(0, 2)$, and $(-3, 2)$. Describe the set of all points of the form $(x, 2)$ where $x$ is a real number.

## Extending the Concepts

**63.** Draw a graph of an equation that contains two $x$-intercepts, $-2$ and $3$. At the $x$-intercept $-2$, the graph crosses the $x$-axis; at the $x$-intercept $3$, the graph touches the $x$-axis. Compare your graph with those of your classmates. How are they similar? How are they different?

**64.** Draw a graph that contains the points $(-3, -1)$, $(-1, 1)$, $(0, 3)$, and $(1, 5)$. Compare your graph with those of your classmates. How many of the graphs are straight lines? How many are "curved"?

**65.** Make up an equation that contains the points $(2, 0)$, $(4, 0)$, and $(1, 0)$. Compare your equation with those of your classmates. How are they similar? How are they different?

**66.** Make up an equation that contains the points $(0, 3)$, $(1, 3)$, and $(-4, 3)$. Compare your equation with those of your classmates. How many are the same?

### The Graphing Calculator

Just as we have graphed equations using point-plotting, the graphing calculator also graphs equations by plotting points. Figure 10 shows the graph of $y = x^2$ and Table 4 shows points on the graph of $y = x^2$ using a TI-84 Plus graphing calculator.

Figure 10

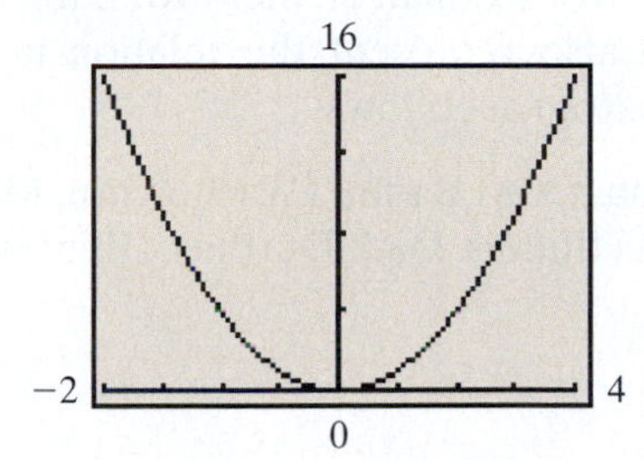


Table 4

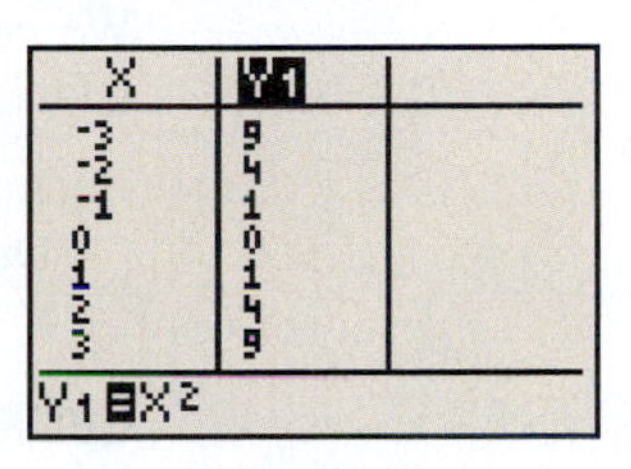

| X | Y1 |
|---|---|
| -3 | 9 |
| -2 | 4 |
| -1 | 1 |
| 0 | 0 |
| 1 | 1 |
| 2 | 4 |
| 3 | 9 |

Y1 ■ X²

*In Problems 67–74, use a graphing calculator to draw a complete graph of each equation. Use the TABLE feature to assist in selecting an appropriate viewing window.*

**67.** $y = 3x - 9$ **68.** $y = -5x + 8$ **69.** $y = -x^2 + 8$ **70.** $y = 2x^2 - 4$

**71.** $y + 2x^2 = 13$ **72.** $y - x^2 = -15$ **73.** $y = x^3 - 6x + 1$ **74.** $y = -x^3 + 3x$

# 2.2 Relations

**OBJECTIVES**

1. Understand Relations
2. Find the Domain and the Range of a Relation
3. Graph a Relation Defined by an Equation

***Preparing for Relations***

*Before getting started, take this readiness quiz. If you get a problem wrong, go back to the section cited and review the material.*

1. Write the inequality $-4 \le x \le 4$ in interval notation. [Section 1.4, pp. 91–94]
2. Write the interval $[2, \infty)$ using an inequality. [Section 1.4, pp. 91–94]

## 1 Understand Relations

We often see situations where one variable is somehow linked to the value of some other variable. For example, an individual's level of education is linked to annual income. Engine size is linked to gas mileage. When the value of one variable is related to the value of a second variable, we have a *relation.*

**DEFINITION**

When the elements in one set are linked to elements in a second set, we have a **relation.** If $x$ and $y$ are two elements in these sets and if a relation exists between $x$ and $y$, then we say that $x$ **corresponds** to $y$ or that $y$ **depends on** $x$, and we write $x \rightarrow y$. We may also write a relation where $y$ depends on $x$ as an ordered pair $(x, y)$.

***Preparing for...Answers*** **1.** $[-4, 4]$ **2.** $x \ge 2$

### EXAMPLE 1 Illustrating a Relation

Consider the data presented in Figure 11, where a correspondence between states and senators in 2005 is shown for randomly selected senators. We might name the relation "is represented by." So, we would say "Indiana is represented by Evan Bayh."

Figure 11

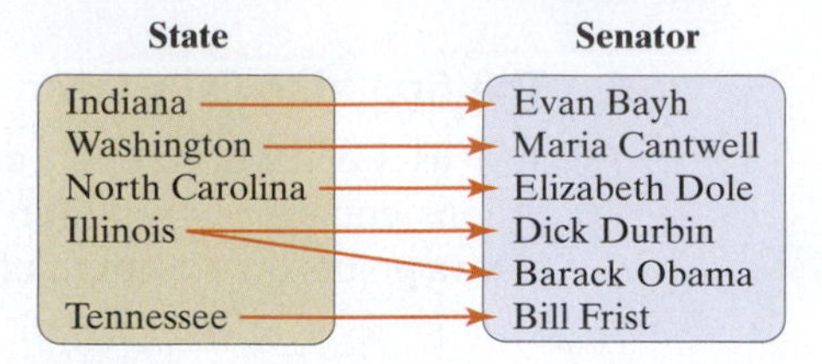


By representing the relation as in Figure 11, we are using **mapping,** in which we draw an arrow from an element from the set "state" to an element in the set "senator." We could also represent the relation in Figure 11 using ordered pairs in the form (state, senator) as follows:

{(Indiana, Evan Bayh), (Washington, Maria Cantwell), (North Carolina, Elizabeth Dole), (Illinois, Dick Durbin), (Illinois, Barack Obama), (Tennessee, Bill Frist)}

QUICK ✓

1. Use the map to represent the relation as a set of ordered pairs.

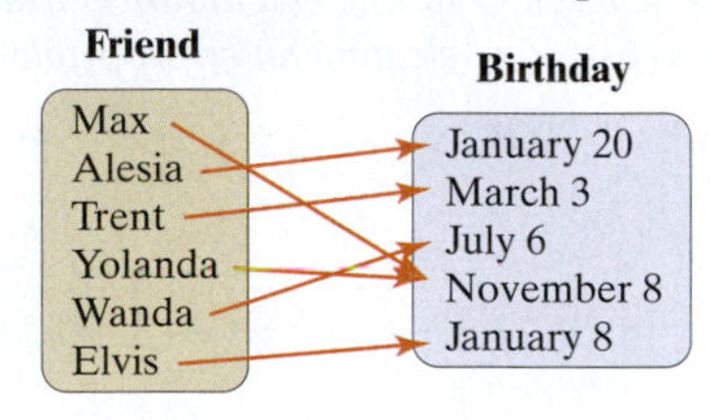


2. Use the set of ordered pairs to represent the relation as a map.

$$\{(1, 3), (5, 4), (8, 4), (10, 13)\}$$

## 2 Find the Domain and the Range of a Relation

In a relation we say that $y$ depends on $x$ and could write the relation as a set of ordered pairs $(x, y)$. We can think of the set of all $x$ as the **inputs** of the relation. The set of all $y$ can be thought of as the **outputs** of the relation. We use this interpretation of a relation to define *domain* and *range*.

**DEFINITION**

The **domain** of a relation is the set of all inputs of the relation. The **range** is the set of all outputs of the relation.

### EXAMPLE 2 Finding the Domain and the Range of a Relation

Find the domain and the range of the relation presented in Figure 11 from Example 1.

**Solution**

The domain is the set of all inputs and the range is the set of all outputs. The inputs, and therefore the domain, of the relation are

{Indiana, Washington, North Carolina, Illinois, Tennessee}

The outputs, and therefore the range, of the relation are

{Evan Bayh, Maria Cantwell, Elizabeth Dole, Dick Durbin, Barack Obama, Bill Frist}

**Work Smart**
Never list elements in the domain or range more than once.

The careful reader will notice that we did not list Illinois twice in the domain because the domain and the range are sets and we never list elements in a set more than once.

**QUICK ✓**

3. State the domain and the range of the relation.

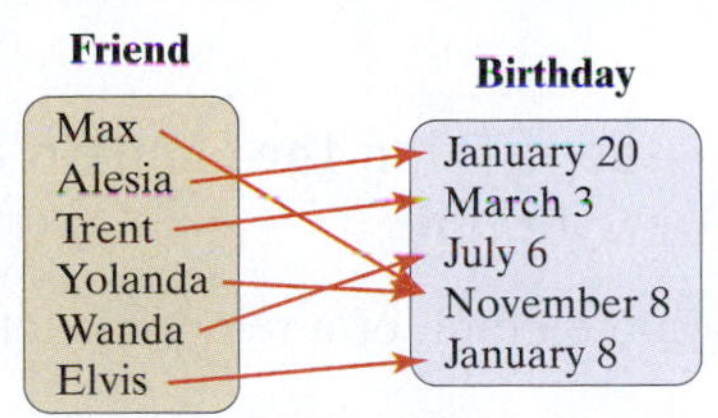


4. State the domain and the range of the relation.

$$\{(1, 3), (5, 4), (8, 4), (10, 13)\}$$

Relations can also be represented by plotting a set of ordered pairs. The set of all $x$-coordinates represents the domain of the relation and the set of all $y$-coordinates represents the range of the relation.

## EXAMPLE 3 Finding the Domain and the Range of a Relation

Figure 12 shows the graph of a relation. Identify the domain and the range of the relation.

**Figure 12**

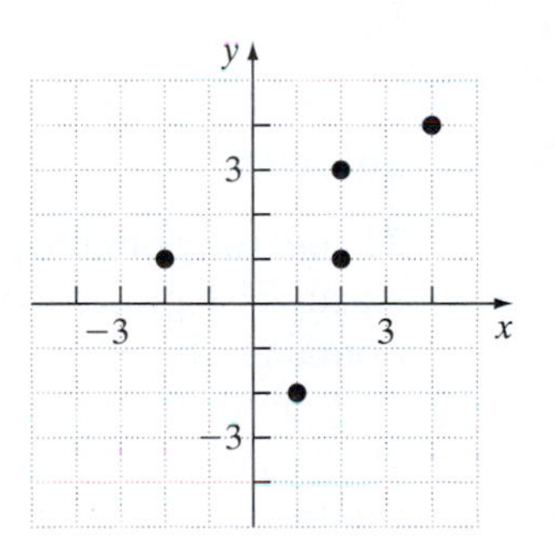


### Solution

**Work Smart**
First write the points as ordered pairs to assist in finding the domain and range.

First, we notice that the ordered pairs in the graph are $(-2, 1)$, $(1, -2)$, $(2, 1)$, $(2, 3)$, and $(4, 4)$. The domain is the set of all $x$-coordinates: $\{-2, 1, 2, 4\}$. The range is the set of all $y$-coordinates: $\{-2, 1, 3, 4\}$.

**QUICK ✓**

5. Identify the domain and the range of the relation shown in the figure.

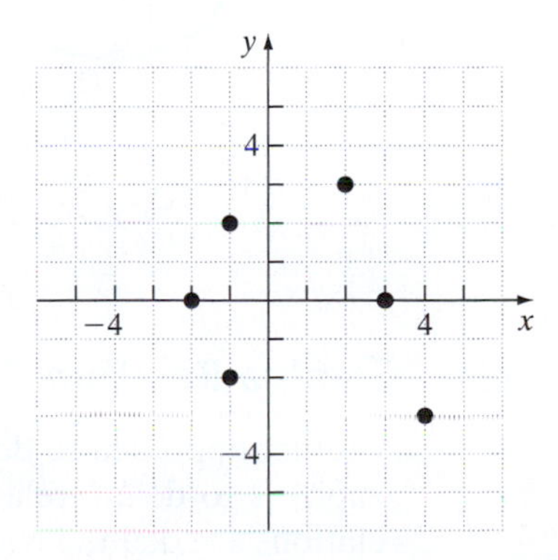


We have learned that a relation can be defined by a map or by a set of ordered pairs. A relation can also be defined by a graph. Remember that a graph is the set of all ordered pairs $(x, y)$ such that the equation is a true statement. If a graph exists for some ordered pair $(x, y)$, then the $x$-coordinate is in the domain and the $y$-coordinate is in the range. Think of it this way: When a graph of a relation is given, its domain may be viewed as the shadow created by the graph on the $x$-axis by vertical beams of light. Its range can be viewed as the shadow created by the graph on the $y$-axis by horizontal beams of light.

### EXAMPLE 4 Identifying the Domain and the Range of a Relation from Its Graph

Figure 13 shows the graph of a relation. Determine the domain and the range of the relation.

**Figure 13**

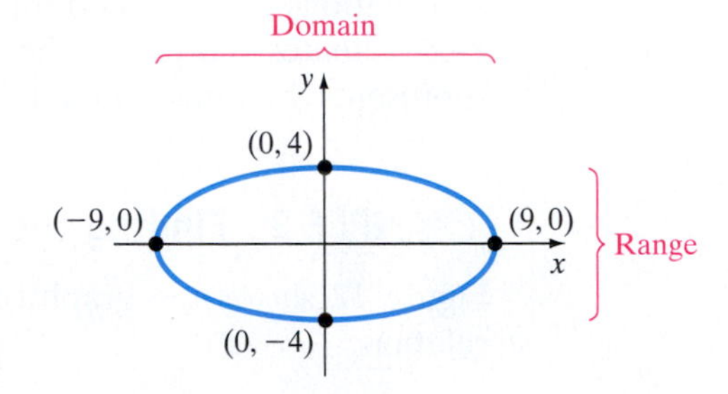


**Solution**

To find the domain of the relation, we determine the $x$-coordinates for which the graph exists. The graph exists for all $x$-values between $-9$ and 9, inclusive. Therefore, the domain is $\{x \mid -9 \le x \le 9\}$ or, using interval notation, $[-9, 9]$.

To find the range of the relation, we determine the $y$-coordinates for which the graph exists. The graph exists for all $y$-values between $-4$ and 4, inclusive. Therefore, the range is $\{y \mid -4 \le y \le 4\}$ or, using interval notation $[-4, 4]$.

**QUICK ✓** *Identify the domain and the range of the relation from its graph.*

**6.**

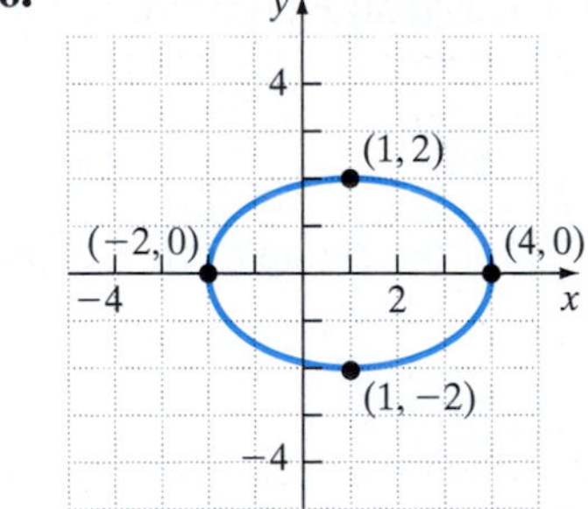


**7.**

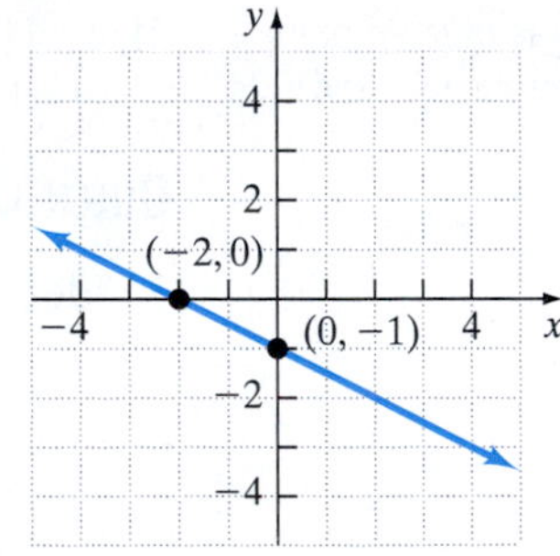


## 3 Graph a Relation Defined by an Equation

**Work Smart**

We can define relations by
1. Mapping
2. Sets of ordered pairs
3. Graphs
4. Equations

Another approach to defining a relation (instead of a map, a set of ordered pairs, or a graph) is to define relations through equations such as $x + y = 4$ or $x = y^2$. When relations are defined by equations, we typically graph the relation so that we can visualize how $y$ depends upon $x$. As was seen in Example 4, the graph of the relation is also useful for helping us to identify the domain and the range of the relation.

**EXAMPLE 5 Relations Defined by Equations**

Graph the relation $y = -x^2 + 4$. Use the graph to determine the domain and the range of the relation.

### Solution

The relation says to take the input $x$, square it, multiply this result by $-1$, and then add 4 to get the output $y$. We use the point-plotting method to graph the relation. Table 5 shows some points on the graph. Figure 14 shows a graph of the relation.

Table 5

| $x$ | $y = -x^2 + 4$ | $(x, y)$ |
|---|---|---|
| $-3$ | $-(-3)^2 + 4 = -5$ | $(-3, -5)$ |
| $-2$ | $-(-2)^2 + 4 = 0$ | $(-2, 0)$ |
| $-1$ | 3 | $(-1, 3)$ |
| 0 | 4 | $(0, 4)$ |
| 1 | 3 | $(1, 3)$ |
| 2 | 0 | $(2, 0)$ |
| 3 | $-5$ | $(3, -5)$ |

Figure 14

From the graph, we can see that the graph extends indefinitely to the left and to the right (that is, the graph exists for all $x$-values). Therefore, the domain of the relation is the set of all real numbers, or $\{x \mid x \text{ is a real number}\}$, or using interval notation, $(-\infty, \infty)$. We also notice from the graph that there are no $y$-values greater than 4, but the graph exists everywhere for $y$-values less than or equal to 4. The range of the relation is $\{y \mid y \leq 4\}$, or using interval notation $(-\infty, 4]$.

**QUICK ✓** *Graph each relation. Use the graph to identify the domain and the range.*

**8.** $y = 3x - 8$ **9.** $y = x^2 - 8$ **10.** $x = y^2 + 1$

## 2.2 Exercises

**For Extra Help:**

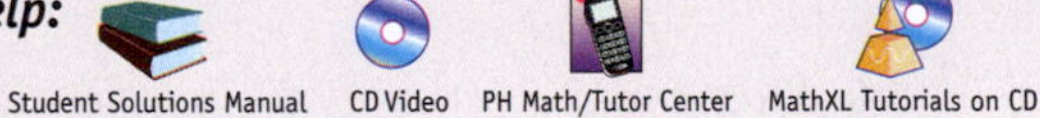

Student Solutions Manual | CD Video | PH Math/Tutor Center | MathXL Tutorials on CD | MathXL® | MyMathLab

### Concepts and Vocabulary

*In Problems 1 and 2, fill in the blanks.*

**1.** If a relation exists between $x$ and $y$, then we say that $y$ _____ to $x$ or that $y$ _____ on $x$, and we write $x \rightarrow y$.

**2.** The _____ of a relation is the set of all inputs to the relation. The _____ is the set of all outputs of the relation.

*In Problems 3 and 4, answer True or False to each statement.*

**3.** If the graph of a relation does not exist at $x = 3$, then 3 is not in the domain of the relation.

**4.** The range of a relation is always the set of all real numbers.

**5.** In your own words, explain what a relation is. Be sure to include an explanation of domain and range.

6. State the four methods for describing a relation presented in this section. When is using ordered pairs most appropriate? When is using a graph most appropriate? Support your opinion.

## Building Skills

*In Problems 7–10, write each relation as a set of ordered pairs. Then identify the domain and the range of the relation.*

**7.**

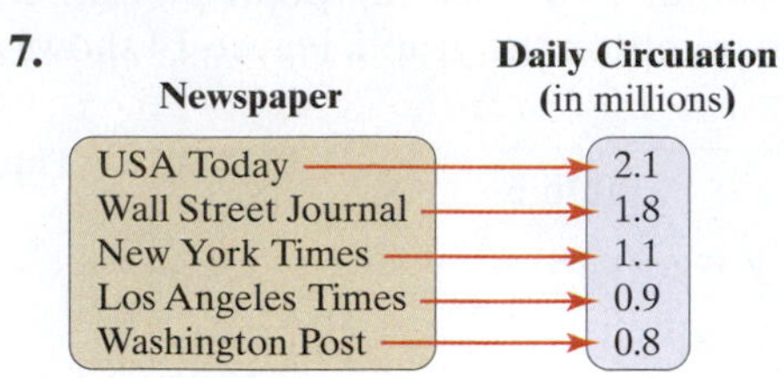


SOURCE: *Information Please Almanac*

**8.**

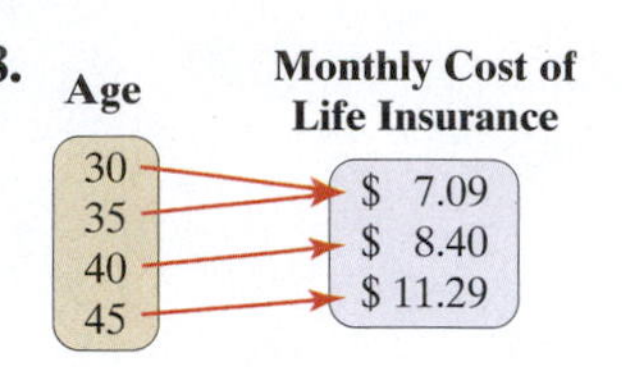


SOURCE: *eterm.com*

**9.**

| Level of Education | Average Annual Income |
|---|---|
| Less than 9th Grade | $ 17,261 |
| 9th – 12th Grade, No Diploma | $ 21,737 |
| High School Graduate | $ 35,744 |
| Associate's Degree | $ 49,279 |
| Bachelor's Degree or Higher | $ 69,804 |

SOURCE: *United States Census Bureau*

**10.**

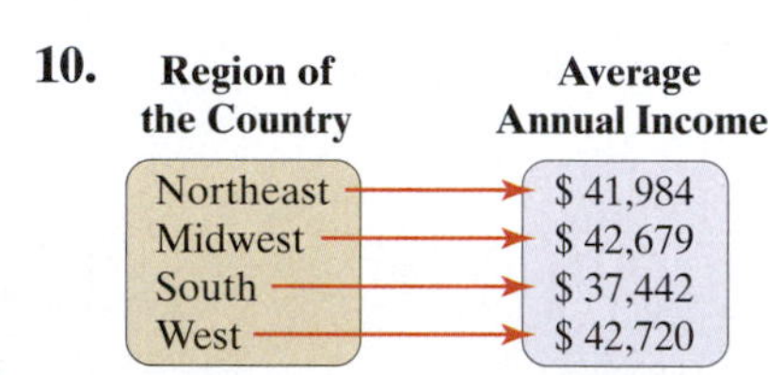


SOURCE: *United States Census Bureau*

*In Problems 11–16, write each relation as a map. Then identify the domain and the range of the relation.*

**11.** $\{(-3, 4), (-2, 6), (-1, 8), (0, 10), (1, 12)\}$

**12.** $\{(-2, 6), (-1, 3), (0, 0), (1, -3), (2, 6)\}$

**13.** $\{(-2, 4), (-1, 2), (0, 0), (1, 2), (2, 4)\}$

**14.** $\{(-2, -8), (-1, -1), (0, 0), (1, 1), (2, 8)\}$

**15.** $\{(0, -4), (-1, -1), (-2, 0), (-1, 1), (0, 4)\}$

**16.** $\{(-3, 0), (0, 3), (3, 0), (0, -3)\}$

*In Problems 17–24, identify the domain and the range of the relation from the graph.*

**17.**

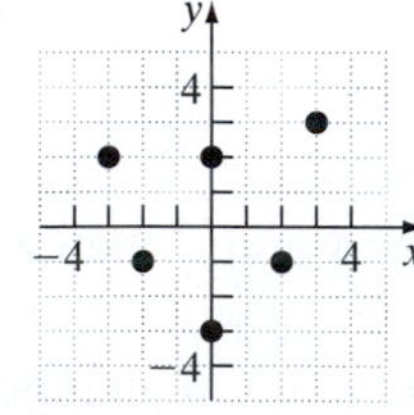


**18.**

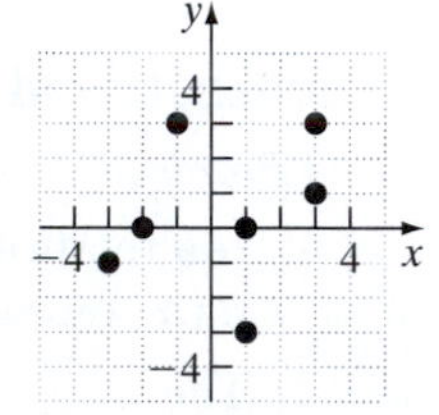


**19.**

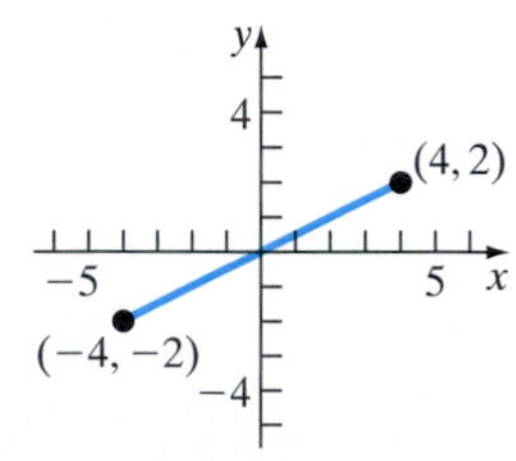


**20.**

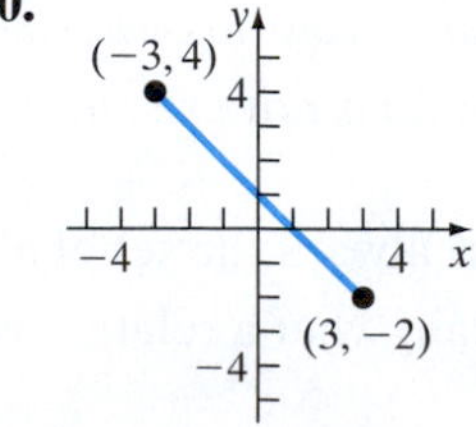


**21.**

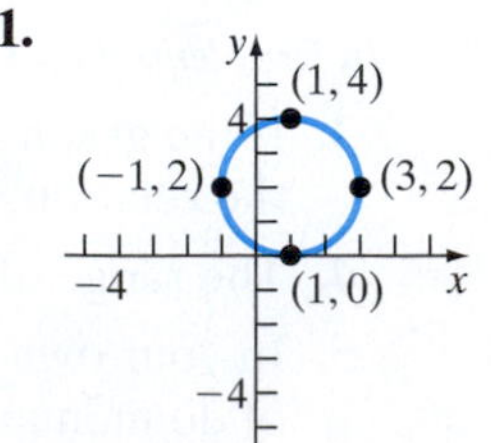


**22.**

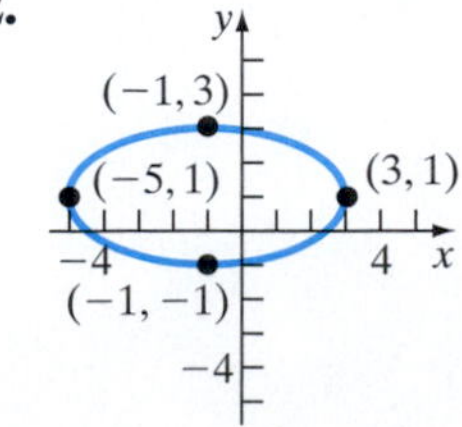

**23.**

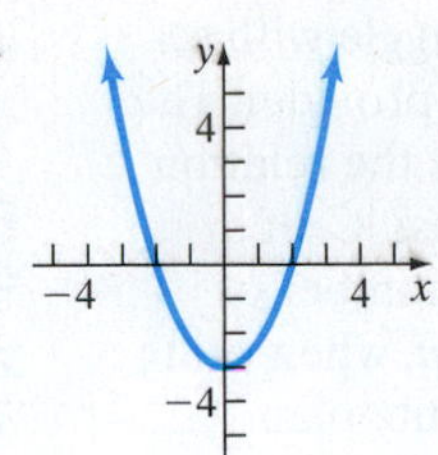


**24.**

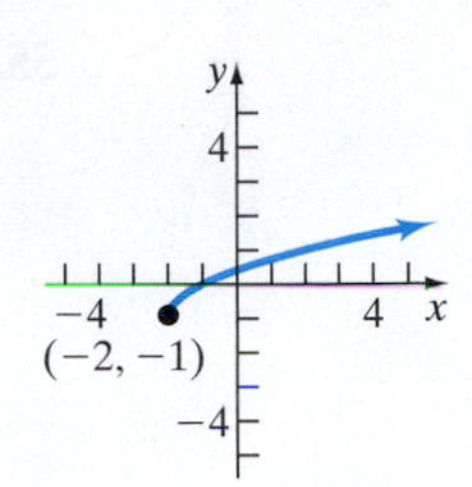


*In Problems 25–52, use the graph the relation obtained in Problems 25–52 from Section 2.1 to identify the domain and the range of the relation.*

**25.** $y = 4x$

**26.** $y = 2x$

**27.** $y = -\frac{1}{2}x$

**28.** $y = -\frac{1}{3}x$

**29.** $y = x + 3$

**30.** $y = x - 2$

**31.** $y = -3x + 1$

**32.** $y = -4x + 2$

**33.** $y = \frac{1}{2}x - 4$

**34.** $y = -\frac{1}{2}x + 2$

**35.** $2x + y = 7$

**36.** $3x + y = 9$

**37.** $y = -x^2$

**38.** $y = x^2 - 2$

**39.** $y = 2x^2 - 8$

**40.** $y = -2x^2 + 8$

**41.** $y = |x|$

**42.** $y = |x| - 2$

**43.** $y = |x - 1|$

**44.** $y = -|x|$

**45.** $y = x^3$

**46.** $y = -x^3$

**47.** $y = x^3 + 1$

**48.** $y = x^3 - 2$

**49.** $x^2 - y = 4$

**50.** $x^2 + y = 5$

**51.** $x = y^2 - 1$

**52.** $x = y^2 + 2$

## Applying the Concepts

**53. Area of a Window** Bob Villa wishes to put a new window in his home. He wants the perimeter of the window to be 100 feet. The graph below shows the relation between the width, $x$, of the opening and the area of the opening.

**(a)** What is the domain and the range of the relation?

**(b)** Provide an explanation as to why the domain obtained in part (a) is reasonable.

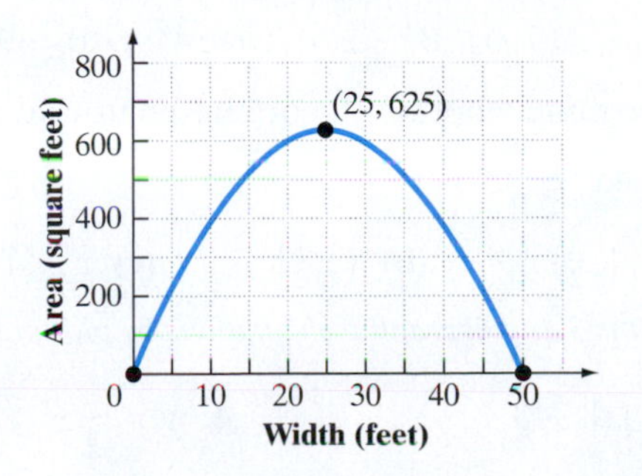


**54. Projectile Motion** The graph below shows the height, in feet, of a ball thrown straight up with an initial speed of 80 feet per second from an initial height of 96 feet after $t$ seconds. Determine the domain and the range of the relation.

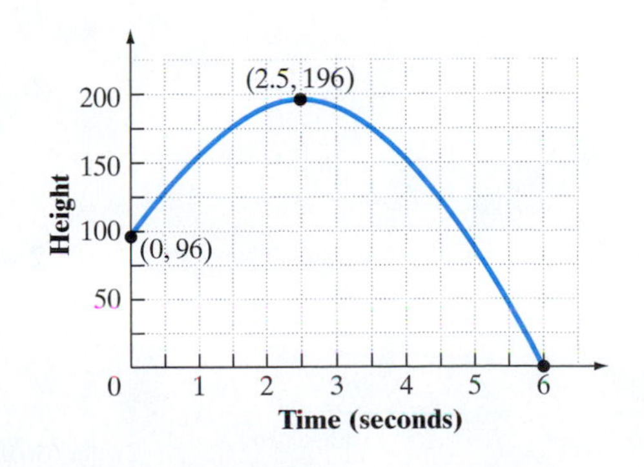

**55. Cell Phones** We all struggle with selecting a cellular phone provider. The graph to the right shows the relation between the monthly cost, $C$, of a cellular phone and the number of anytime minutes used, $m$, when using the Sprint PCS 500-minute plan. (*Source: SprintPCS.com*)

**(a)** Determine the domain and the range of the relation.

**(b)** If anytime minutes are from 7:00 A.M. to 7:00 P.M. Monday through Friday, provide an explanation as to why the domain obtained in part (a) is reasonable assuming there are 21 non-weekend days.

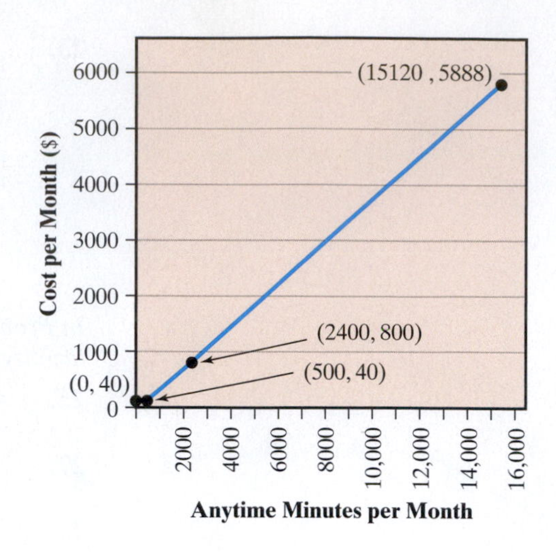


**56. Wind Chill** It is 10° Celsius outside. The wind is calm but then gusts up to 20 meters per second. You feel the chill go right through your bones. The graph to the right shows the relation between the wind chill temperature (in degrees Celsius) and wind speed (in meters per second). Determine the domain and the range of the relation.

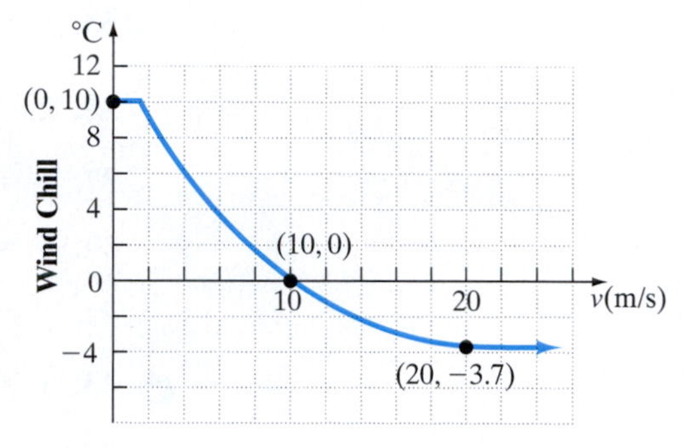


## Extending the Concepts

**57.** Draw the graph of a relation whose domain is all real numbers, but whose range is a single real number. Compare your graph with those of your classmates. How are they similar?

**58.** Draw the graph of a relation whose domain is a single real number, but whose range is all real numbers. Compare your graph with those of your classmates. How are they similar?

# PUTTING THE CONCEPTS TOGETHER (SECTIONS 2.1–2.2)

*These problems cover important concepts from Sections 2.1 to 2.2. We designed these problems so that you can review the chapter so far and show your mastery of the concepts. Take time to work these problems before proceeding with the next section. The answers to these problems are located at the back of the text starting on page AN-8.*

**1.** Plot the following ordered pairs in the same $xy$-plane. Tell in which quadrant or on what coordinate axis each point lies.

$$A(7, 0), B(-2, 6), C(8, 4), D(0, -9), E(-5, -10), F(6, -3)$$

**2.** Determine whether the ordered pair is a point on the graph of the equation $y = 4x - \frac{3}{2}$.

**(a)** $\left(1, \frac{5}{2}\right)$ **(b)** $\left(\frac{1}{2}, \frac{1}{2}\right)$ **(c)** $\left(\frac{1}{4}, \frac{1}{4}\right)$

*In Problems 3 and 4, graph the equations by plotting points.*

**3.** $y = |x| + 3$ **4.** $y = \frac{1}{2}x^2 - 1$

**5.** Identify the intercepts from the graph below.

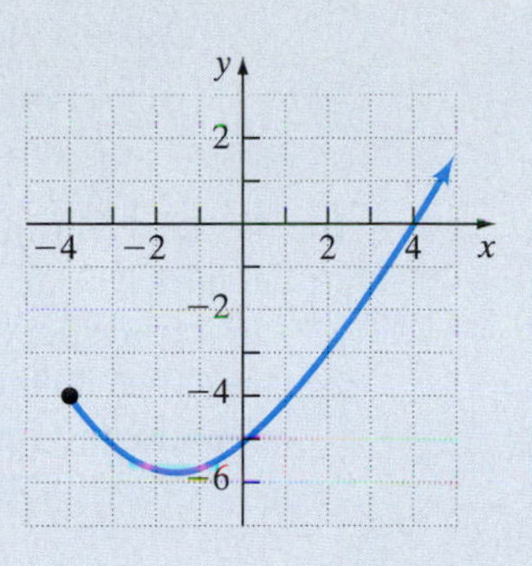


**6.** The graph below shows the average selling price of a new home from August 2002 to July 2003. The vertical axis represents the selling price, in thousands of dollars, and the horizontal axis represents the month.

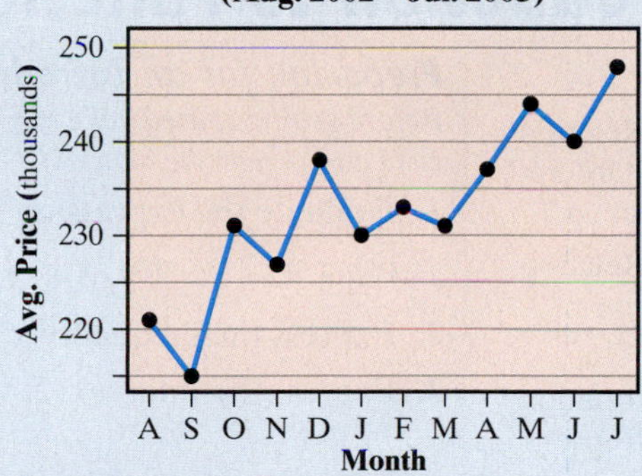


**(a)** What was the average selling price of a new home in January of 2003?
**(b)** In what month was the average selling price the highest? What was the approximate average selling price?
**(c)** Between what two consecutive months did the average selling price increase the most? What was the approximate increase?

**7.** Write the following relation as a set of ordered pairs.

| Domain | Range |
|---|---|
| −2 | −1 |
| −1 | 0 |
| 0 | 1 |
| 1 | 2 |
| 2 | 3 |

**8.** Identify the domain and range of the relation from each graph.

**(a)**

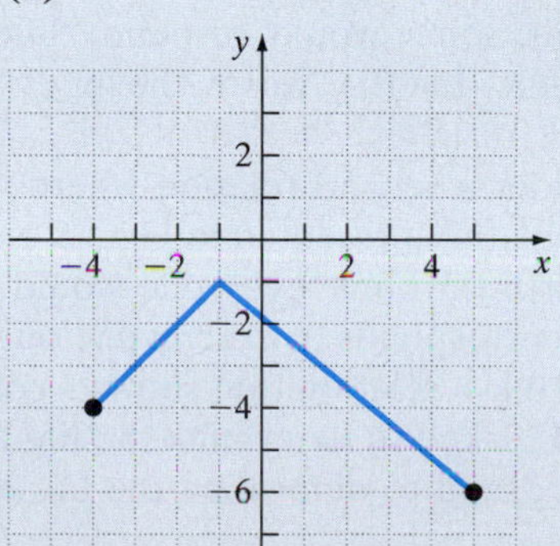


**(b)**

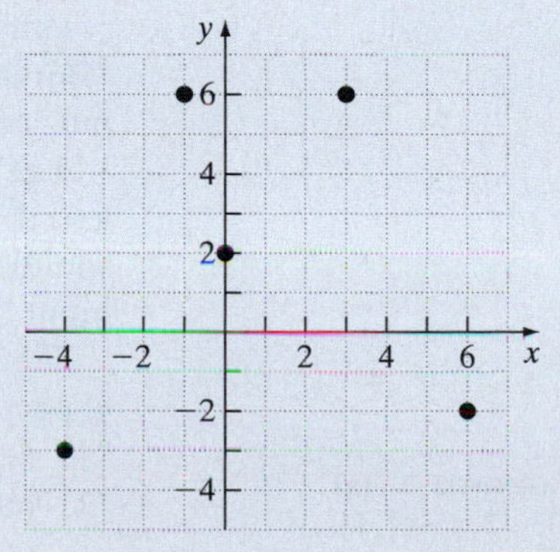

**9.** Graph each relation and use the graph to identify the domain and the range of the relation.

**(a)** $y = |x - 2| - 3$ **(b)** $y = \frac{1}{2}x^2 + 1$

**10. Vertical Motion** The graph to the right shows the height, $h$, in feet, of a ball thrown straight up with an initial speed of 40 feet per second from an initial height of 80 feet after $t$ seconds. What are the domain and the range of the relation?

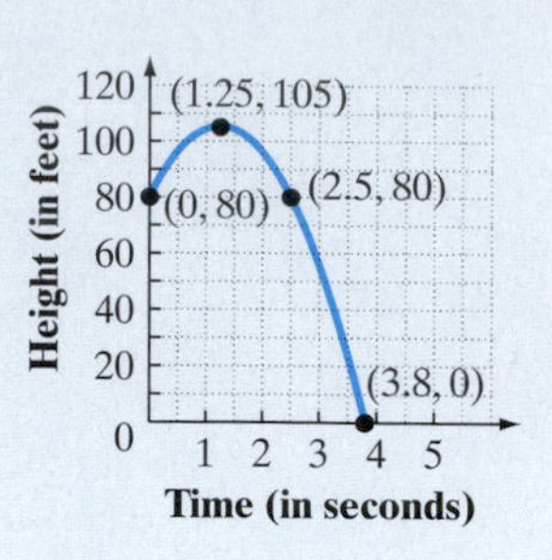


# 2.3 An Introduction to Functions

## OBJECTIVES

1. Determine Whether a Relation Expressed as a Map or Ordered Pairs Represents a Function
2. Determine Whether a Relation Expressed as an Equation Represents a Function
3. Determine Whether a Relation Expressed as a Graph Represents a Function
4. Find the Value of a Function
5. Graph a Function
6. Work with Applications of Functions

### *Preparing for an Introduction to Functions*

*Before getting started, take this readiness quiz. If you get a problem wrong, go back to the section cited and review the material.*

**1.** Evaluate the expression $2x^2 - 5x$ for

**(a)** $x = 1$ **(b)** $x = 4$ **(c)** $x = -3$ [Section R.5, pp. 43–44]

**2.** Express the inequality $x \leq 5$ using interval notation. [Section 1.4, pp. 91–94]

**3.** Express the interval $(2, \infty)$ using set-builder notation. [Section 1.4, pp. 91–94]

## 1 Determine Whether a Relation Expressed as a Map or Ordered Pairs Represents a Function

We now present what is one of the most important concepts in algebra—the *function.* A function is a special type of relation. To understand the idea behind a function, let's revisit the relation presented in Example 1 from Section 2.2 shown again in Figure 15. Recall, this is a correspondence between states and their senators. We named the relation "is represented by."

**Figure 15**

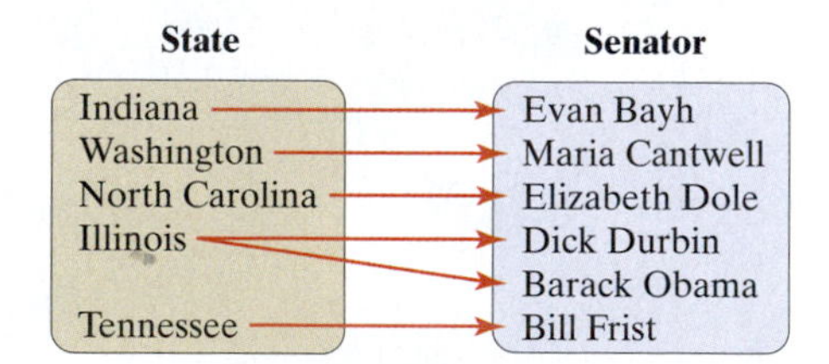


For the relation in Figure 15, if someone were asked to name the senator who represents Illinois, some would respond "Dick Durbin," while others would respond "Barack Obama." In other words, the input "state" does not correspond to a single output "senator."

Let's consider a second relation where we have a correspondence between states and their population presented in Figure 16(a). If asked for the population that corresponds to North Carolina, everyone would respond "8,320,146." In other words, each input "state" corresponds to exactly one output "population."

Figure 16(b) is a relation that shows a correspondence between "animals" and "life expectancy." If asked to determine the life expectancy of a dog, we would all respond "11 years." If asked to determine the life expectancy of a cat, we would all respond "11 years."

*Preparing for...Answers* **1. (a)** $-3$ **(b)** 12 **(c)** 33 **2.** $(-\infty, 5]$ **3.** $\{x | x > 2\}$

Figure 16

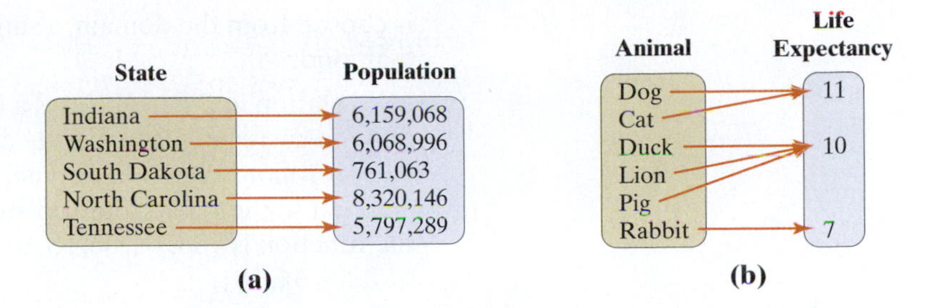


Looking carefully at the three relations, we should notice that the relations presented in Figures 16(a) and 16(b) have something in common, while the relation in Figure 15 is different. What is it? The common link between the relations in Figures 16(a) and (b) is that each input corresponds to only one output. However, in Figure 15 the input Illinois corresponds to two outputs—Dick Durbin and Barack Obama. This leads to the definition of a *function*.

**In Words**

For a relation to be classified as a function, each input may have only one output.

**DEFINITION**

A **function** is a relation in which each element in the domain (the inputs) of the relation corresponds to exactly one element in the range (the outputs) of the relation.

## EXAMPLE 1 Determining Whether a Relation Represents a Function

Determine whether the following relations represent functions. If the relation is a function, then state its domain and range.

**(a)** See Figure 17(a). For this relation, the domain represents the length (mm) of the right humerus and the range represents the length (mm) of the right tibia for each of five rats sent to space. The lengths were measured once the rats returned from their trip.

**(b)** See Figure 17(b). For this relation, the domain represents the weight of pear-cut diamonds and the range represents their price.

**(c)** See Figure 17(c). For this relation, the domain represents the age of 5 males and the range represents their HDL (good) cholesterol (mg/dL).

Figure 17

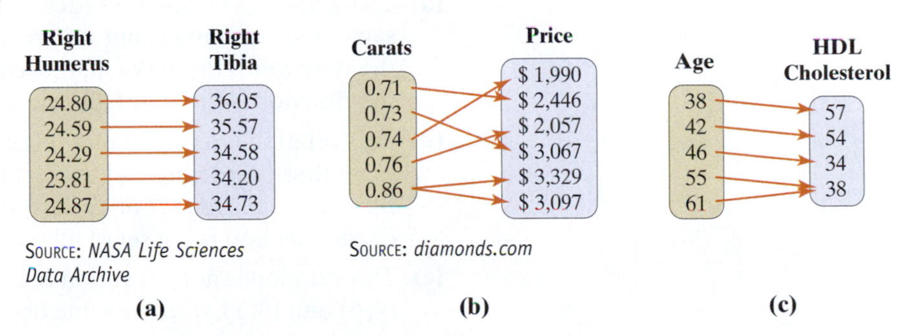


SOURCE: *NASA Life Sciences Data Archive*

SOURCE: *diamonds.com*

(a) (b) (c)

### Solution

**(a)** The relation in Figure 17(a) is a function because each element in the domain corresponds to exactly one element in the range. The domain of the function is $\{24.80, 24.59, 24.29, 23.81, 24.87\}$. The range of the function is $\{36.05, 35.57, 34.58, 34.20, 34.73\}$.

**(b)** The relation in Figure 17(b) is not a function because there is an element in the domain, 0.86, that corresponds to two elements in the range. If 0.86

is chosen from the domain, a single price cannot be determined for the diamond.

**(c)** The relation in Figure 17(c) is a function because each element in the domain corresponds to exactly one element in the range. Notice that it is okay for more than one element in the domain to correspond to the same element in the range (both 55 and 61 correspond to 38). The domain of the function is $\{38, 42, 46, 55, 61\}$. The range of the function is $\{57, 54, 34, 38\}$.

**Work Smart**

All functions are relations, but not all relations are functions!

**QUICK ✓** *Determine whether the relation represents a function. If the relation is a function, state its domain and range.*

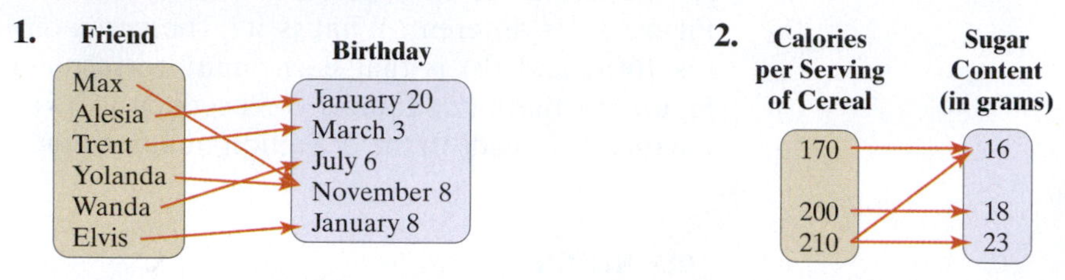


We may also think of a function as a set of ordered pairs $(x, y)$ in which no two ordered pairs have the same first coordinate, but different second coordinates.

## EXAMPLE 2 Determining Whether a Relation Represents a Function

Determine whether each relation represents a function. If the relation is a function, then state its domain and range.

**(a)** $\{(1, 3), (-1, 4), (0, 6), (2, 8)\}$

**(b)** $\{(-2, 6), (-1, 3), (0, 2), (1, 3), (2, 6)\}$

**(c)** $\{(0, 3), (1, 4), (4, 5), (9, 5), (4, 1)\}$

### Solution

**(a)** This relation is a function because there are no ordered pairs with the same first coordinate, but different second coordinates. The domain of the function is the set of all first coordinates, $\{-1, 0, 1, 2\}$. The range of the function is the set of all second coordinates, $\{3, 4, 6, 8\}$.

**(b)** This relation is a function because there are no ordered pairs with the same first coordinate, but different second coordinates. The domain of the function is the set of all first coordinates, $\{-2, -1, 0, 1, 2\}$. The range of the function is the set of all second coordinates, $\{2, 3, 6\}$.

**(c)** This relation is not a function because there are two ordered pairs, $(4, 5)$ and $(4, 1)$, with the same first coordinate, but different second coordinates.

**Work Smart**

In a function, two different inputs cannot correspond to the same output, but two different outputs can be the result of a single input.

In Example 2(b), notice that $-2$ and 2 in the domain each correspond to 6 in the range. This does not violate the definition of a function—two different first coordinates can have the same second coordinate. A violation of the definition occurs when two ordered pairs have the same first coordinate and different second coordinates as in Example 2(c).

**QUICK ✓** *Determine whether each relation represents a function. If the relation is a function, then state its domain and range.*

**3.** $\{(-3,3),(-2,2),(-1,1),(0,0),(1,1)\}$ **4.** $\{(-3,2),(-2,5),(-1,8),(-3,6)\}$

## 2 Determine Whether a Relation Expressed as an Equation Represents a Function

At this point, we have shown how to identify when a relation defined by a map or ordered pairs is a function. In Section 2.2, we also learned how to express relations as equations and graphs. We will now address the circumstances under which equations are functions.

To determine whether an equation, where $y$ depends upon $x$, is a function, it is often easiest to solve the equation for $y$. If a value of $x$ corresponds to exactly one $y$, the equation defines a function; otherwise it does not define a function.

### EXAMPLE 3 Determining Whether an Equation Represents a Function

Determine whether the equation $y = 3x + 5$ shows $y$ as a function of $x$.

#### Solution

The rule for getting from $x$ to $y$ is to multiply $x$ by 3 and then add 5. Since there is only one output $y$ that can result by performing these operations on any given input $x$, the equation is a function.

### EXAMPLE 4 Determining Whether an Equation Represents a Function

**In Words**
The symbol $\pm$ is a shorthand device and is read "plus or minus." For example, $\pm 4$ means "negative four or positive four."

Determine whether the equation $y = \pm x^2$ shows $y$ as a function of $x$.

#### Solution

Notice that for any single value of $x$ (other than 0), two values of $y$ result. For example, if $x = 2$, then $y = \pm 4$ ($-4$ or $+4$). Since a single $x$ corresponds to more than one $y$, the equation is not that of a function.

**QUICK ✓** *Determine whether each equation shows y as a function of x.*

**5.** $y = -2x + 5$ **6.** $y = \pm 3x$ **7.** $y = x^2 + 5x$ **8.** $x + y^2 = 9$

## 3 Determine Whether a Relation Expressed as a Graph Represents a Function

Remember that the graph of an equation is the set of all ordered pairs $(x, y)$ that satisfy the equation. For a relation to be a function, each number $x$ in the domain can correspond to only one $y$ in the range. This means that the graph of an equation will *not* represent a function if two points with the same $x$-coordinate have different $y$-coordinates.

**VERTICAL LINE TEST**

A set of points in the $xy$-plane is the graph of a function if and only if every vertical line intersects the graph in at most one point.

### EXAMPLE 5 Using the Vertical Line Test to Identify Graphs of Functions

Which of the graphs in Figure 18 are graphs of functions?

Figure 18

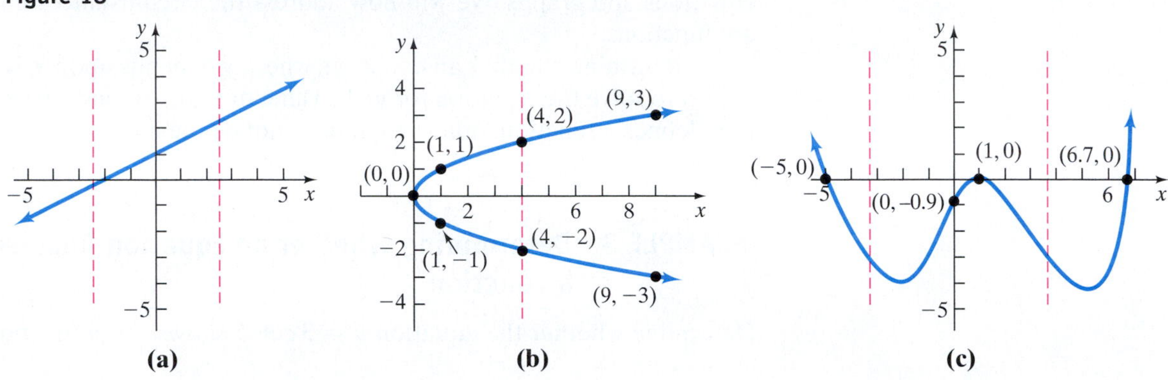


**Solution**

The graph in Figure 18(a) is a function, because every vertical line intersects the graph in at most one point. The graph in Figure 18(b) is not a function, because a vertical line intersects the graph in more than one point. The graph in Figure 18(c) is a function, because a vertical line intersects the graph in at most one point.

Based on the results of Example 5, do you see why the vertical line test works? If a vertical line intersects the graph of an equation in two or more points, then the same $x$-coordinate corresponds to two or more different $y$-coordinates and we have violated the definition of a function.

**QUICK ✓** *Use the vertical line test to determine whether the graph is that of a function.*

**9.**

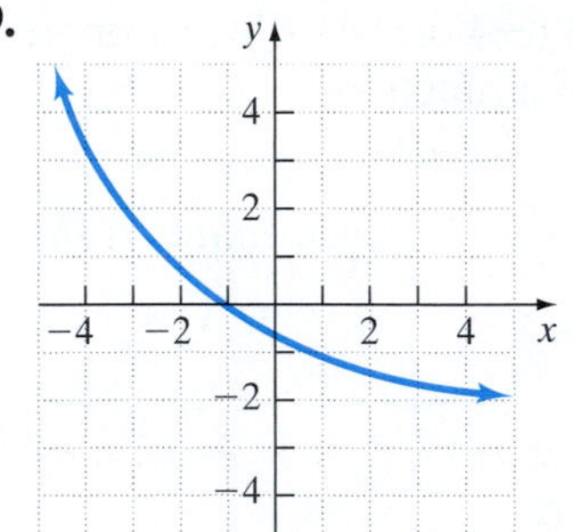

**10.**

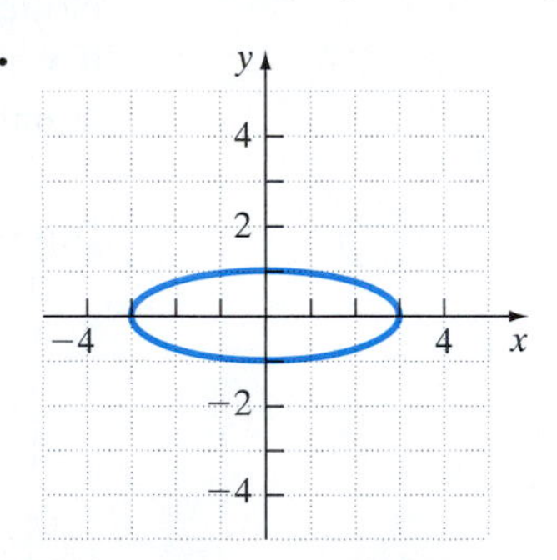

## 4 Find the Value of a Function

Functions are often denoted by letters such as $f$, $F$, $g$, $G$, and so on. If $f$ is a function, then for each number $x$ in its domain, the corresponding value in the range is denoted $f(x)$, read as "$f$ of $x$" or as "$f$ at $x$." We call $f(x)$ the **value of $f$ at the number $x$**; $f(x)$ is the number that results when the function is applied to $x$; $f(x)$ does not mean "$f$ times $x$." For example, the function $y = 3x + 5$ given in Example 3 may be written as $f(x) = 3x + 5$.

**Work Smart**

Be careful with function notation. In the expression $y = f(x)$, $y$ is the dependent variable, $x$ is the independent variable, and $f$ is the name given to a rule that relates the input $x$ to the output $y$.

For a function $y = f(x)$, the variable $x$ is called the **independent variable,** because it can be assigned any of the numbers in the domain. The variable $y$ is called the **dependent variable,** because its value depends on $x$.

Any symbol can be used to represent the independent variable. For example, if $f$ is the *square function,* then $f$ can be defined by $f(x) = x^2$, $f(t) = t^2$, or $f(z) = z^2$. All three functions are the same: Each tells us to square the independent variable.

In practice, the symbols used for the independent and dependent variables should remind us what they represent. For example, in economics, we use $C$ for cost and $q$ for quantity, so that $C(q)$ represents the cost of manufacturing $q$ units of a good. Here, $C$ is the dependent variable, $q$ is the independent variable, and $C(q)$ is the rule that tells us how to get the output $C$ from the input $q$.

The independent variable is also called the **argument** of the function. Thinking of the independent variable as an argument can sometimes make it easier to find the value of a function. For example, if $f$ is the function defined by $f(x) = x^2$, then $f$ tells us to square the argument. So, $f(2)$ means to square 2, $f(a)$ means to square $a$, and $f(x + h)$ means to square the quantity $x + h$.

## EXAMPLE 6 Finding Values of a Function

For the function defined by $f(x) = x^2 + 6x$, evaluate:

**(a)** $f(3)$ **(b)** $f(-2)$

### Solution

**(a)** Wherever we see an $x$ in the equation defining the function $f$ we substitute 3 to get

$$\begin{aligned} f(3) &= 3^2 + 6(3) \\ &= 9 + 18 \\ &= 27 \end{aligned}$$

**(b)** We substitute $-2$ for $x$ in the expression $x^2 + 6x$ to get

$$\begin{aligned} f(-2) &= (-2)^2 + 6(-2) \\ &= 4 + (-12) \\ &= -8 \end{aligned}$$

The notation $f(x)$ plays a dual role—it represents the rule for getting from the input to the output and its value represents the output $y$ of the function. For example, in Example 6(a), the rule for getting from the input to the output is given by $f(x) = x^2 + 6x$. In words, the function says to "take some input $x$, square it, and add the result to six times the input $x$." If the input is 3, then $f(3)$ represents the output, 27.

## EXAMPLE 7 Finding Values of a Function

For the function $F(z) = 4z + 7$, evaluate:

**(a)** $F(z + 3)$ **(b)** $F(z) + F(3)$

### Solution

**(a)** Wherever we see a $z$ in the equation defining $F$, we substitute $z + 3$ to get

$$\begin{aligned} F(z + 3) &= 4(z + 3) + 7 \\ &= 4z + 12 + 7 \\ &= 4z + 19 \end{aligned}$$

**(b)** $F(z) + F(3) = \underbrace{4z + 7}_{F(z)} + \underbrace{4 \cdot 3 + 7}_{F(3)}$

$= 4z + 7 + 12 + 7$

$= 4z + 26$

**QUICK ✓** *Let $f(x) = 3x + 2$ and $g(x) = -2x^2 + x - 3$ to evaluate each function.*

**11.** $f(4)$ **12.** $g(-2)$ **13.** $f(x - 2)$ **14.** $f(x) - f(2)$

---

**SUMMARY: Important Facts about Functions**

1. For each $x$ in the domain there corresponds exactly one $y$ in the range.
2. $f$ is a symbol that we use to denote the function. It represents the equation that we use to get from an $x$ in the domain to $f(x)$ in the range.
3. If $y = f(x)$, then $x$ is called the independent variable or argument of $f$, and $y$ is called the dependent variable or the value of $f$ at $x$.

Figure 19

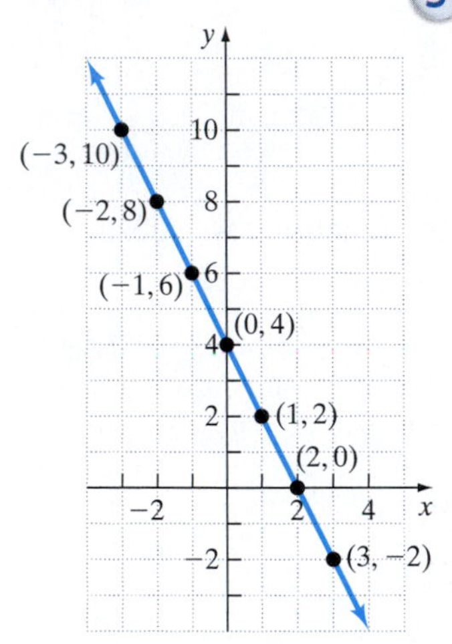


## 5 Graph a Function

In Example 3 from Section 2.1, we graphed the equation $y = -2x + 4$. We reproduce its graph in Figure 19. The graph passes the vertical line test so the equation $y = -2x + 4$ is that of a function. Therefore, we can express the relation using function notation as $f(x) = -2x + 4$. The graph of an equation is the same as the graph of the function where the horizontal axis represents the independent variable and the vertical axis represents the dependent variable. When we graph functions, we label the vertical axis either by $y$ or by the name of the function.

**DEFINITION**

When a function is defined by an equation in $x$ and $y$, the **graph of the function** is the set of *all* ordered pairs $(x, y)$ such that $y = f(x)$.

### EXAMPLE 8 Graphing a Function

Graph the function $f(x) = |x|$.

#### Solution

To graph the function $f(x) = |x|$, we first determine some ordered pairs $(x, f(x)) = (x, y)$ such that $y = |x|$. See Table 6. We now plot the ordered pairs $(x, y)$ in Table 6 in an $xy$-plane and connect the points as shown in Figure 20.

Table 6

| $x$ | $f(x)$ | $(x, f(x))$ |
|---|---|---|
| −3 | $\lvert -3 \rvert = 3$ | (−3, 3) |
| −2 | $\lvert -2 \rvert = 2$ | (−2, 2) |
| −1 | $\lvert -1 \rvert = 1$ | (−1, 1) |
| 0 | $\lvert 0 \rvert = 0$ | (0, 0) |
| 1 | $\lvert 1 \rvert = 1$ | (1, 1) |
| 2 | $\lvert 2 \rvert = 2$ | (2, 2) |
| 3 | $\lvert 3 \rvert = 3$ | (3, 3) |

Figure 20

$f(x) = |x|$

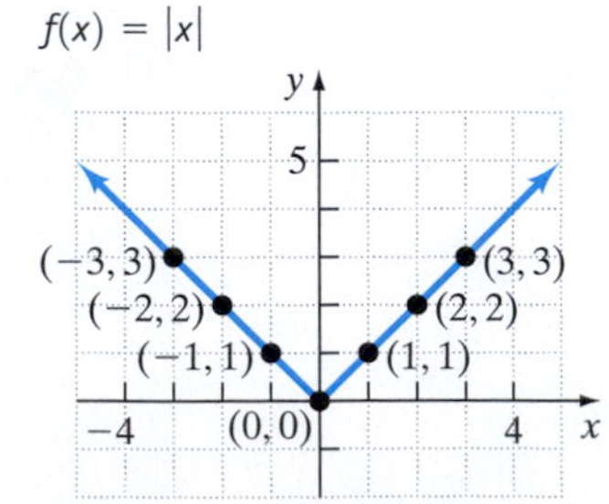

**QUICK ✓** *Graph each function.*

**15.** $f(x) = -2x + 9$ **16.** $f(x) = x^2 + 2$ **17.** $f(x) = |x - 2|$

## 6 Work with Applications of Functions

### EXAMPLE 9 Life Cycle Hypothesis

The Life Cycle Hypothesis from Economics was presented by Franco Modigliani in 1954. It states that income is a function of age. The function $I(a) = -55a^2 + 5119a - 54{,}448$ represents the relation between average annual income $I$ and age $a$.

**(a)** Identify the dependent and independent variables.

**(b)** Evaluate $I(20)$. Provide a verbal explanation of the meaning of $I(20)$.

**Solution**

**(a)** Because income depends upon age, we have that the dependent variable is income, $I$, and the independent variable is age, $a$.

**(b)** We let $a = 20$ in the function.

$$I(20) = -55(20)^2 + 5119(20) - 54{,}448$$
$$= 25{,}932$$

The average annual income of an individual who is 20 years of age is $25,932.

**QUICK ✓**

**18.** In 2002, the Prestige oil tanker sank and started leaking oil off the coast of Spain. The oil slick takes the shape of a circle. Suppose that the area $A$ (in square miles) of the circle contaminated with oil can be determined using the function $A(t) = 0.25\pi t^2$, where $t$ represents the number of days since the tanker sprung a leak.

**(a)** Identify the dependent and independent variable.

**(b)** Evaluate $A(30)$. Provide a verbal explanation of the meaning of $A(30)$.

# 2.3 Exercises

**For Extra Help:** Student Solutions Manual | CD Video | PH Math/Tutor Center | MathXL Tutorials on CD | MathXL® | MyMathLab

## Concepts and Vocabulary

*In Problems 1–3, fill in the blanks.*

**1.** A _____ is a relation in which each element in the domain of the relation corresponds to exactly one element in the range of the relation.

**2.** In the function $H(q) = 2q^2 - 5q + 1$, $H$ is called the _____ variable and $q$ is called the _____ variable.

**3.** The independent variable is also called the _____ of the function.

*In Problems 4 and 5, answer True or False to each statement.*

**4.** Every relation is a function.

**5.** In order for a graph to be that of a function, any vertical line can intersect the graph in at most one point.

6. In your own words, explain why the vertical line test can be used to identify the graph of a function.

7. What are the four forms of a function presented in this section?

8. Explain why the term independent variable for $x$ and dependent variable for $y$ make sense in the function $y = f(x)$.

## Building Skills

*In Problems 9–18, determine whether each relation represents a function. State the domain and the range of each relation.*

**9.**

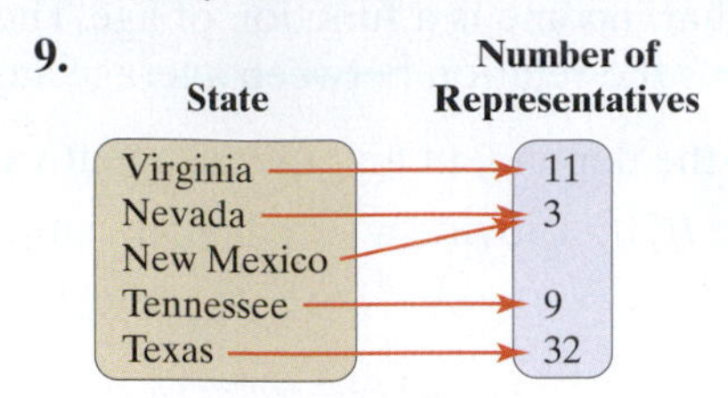


**10.**

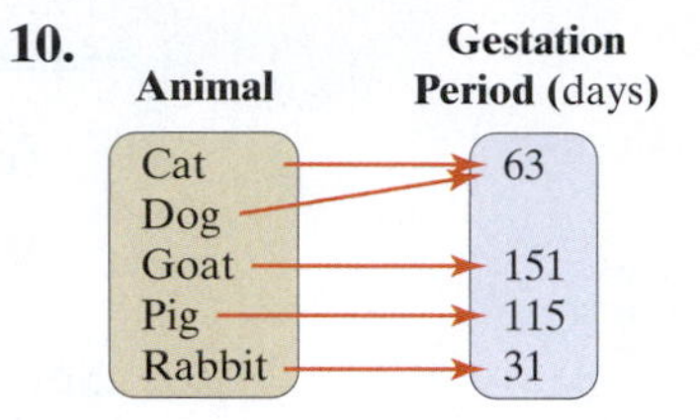


**11.**

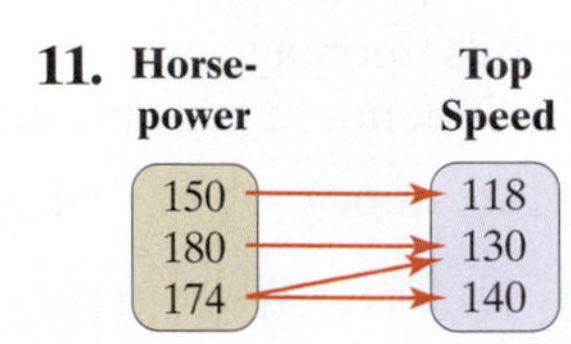


**12.**

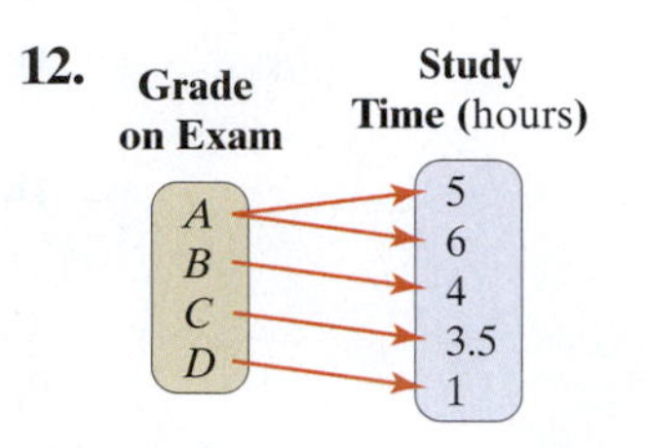


**13.** $\{(0, 3), (1, 4), (2, 5), (3, 6)\}$

**14.** $\{(-1, 4), (0, 1), (1, -2), (2, -5)\}$

**15.** $\{(-3, 5), (1, 5), (4, 5), (7, 5)\}$

**16.** $\{(-2, 3), (-2, 1), (-2, -3), (-2, 9)\}$

**17.** $\{(-10, 1), (-5, 4), (0, 3), (-5, 2)\}$

**18.** $\{(-5, 3), (-2, 1), (5, 1), (7, -3)\}$

*In Problems 19–28, determine whether each equation shows y as a function of x.*

**19.** $y = 2x + 9$

**20.** $y = -6x + 3$

**21.** $2x + y = 10$

**22.** $6x - 3y = 12$

**23.** $y = \pm 5x$

**24.** $y = \pm 2x^2$

**25.** $y = x^2 + 2$

**26.** $y = x^3 - 3$

**27.** $x + y^2 = 10$

**28.** $y^2 = x$

*In Problems 29–36, determine whether the graph is that of a function.*

**29.**

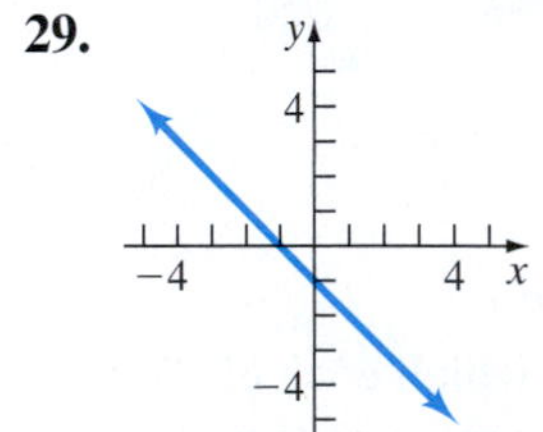

**30.**

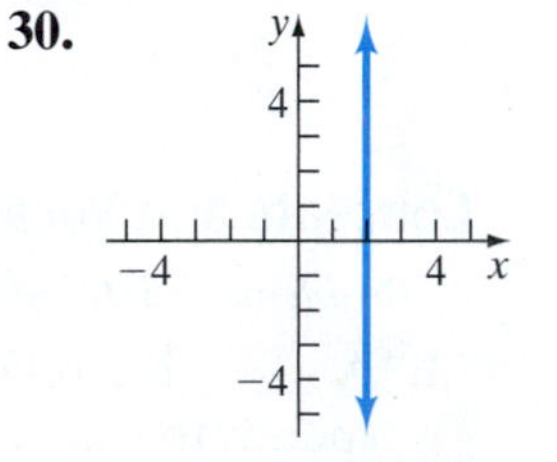

**31.**

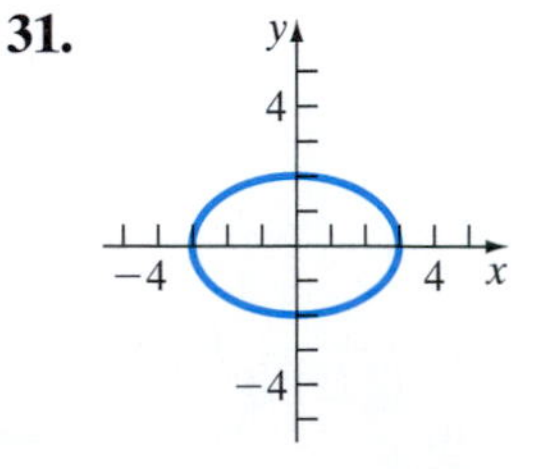

**32.**

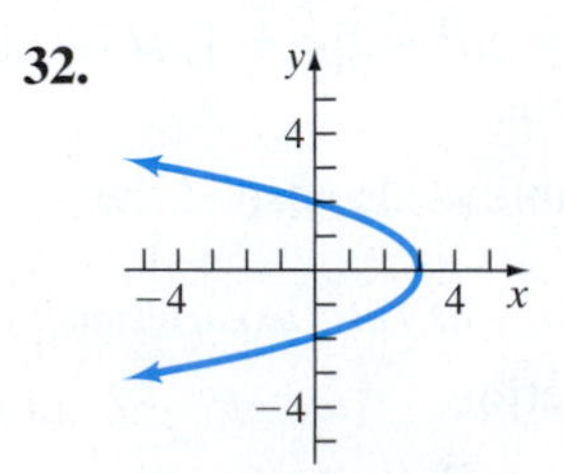

**33.**

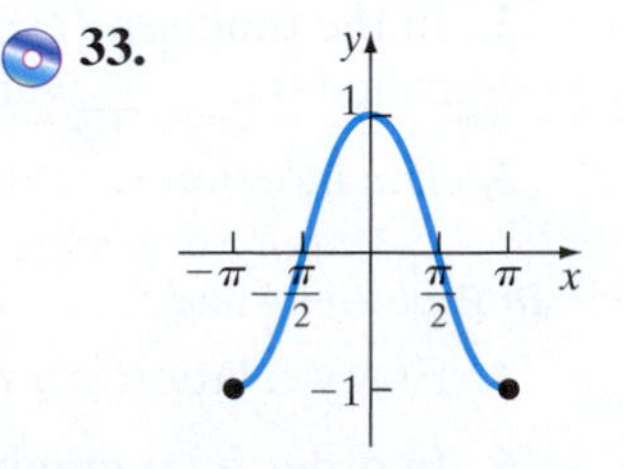

**34.**

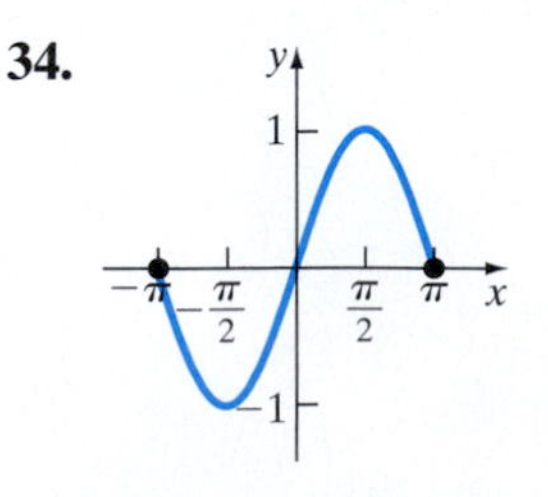

**35.**

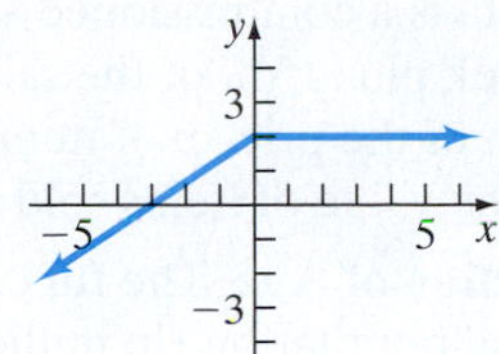


**36.**

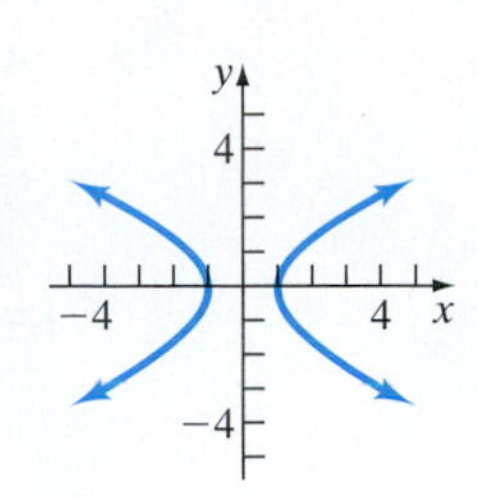


*In Problems 37–40, find the following values for each function:*

**(a)** $f(0)$ **(b)** $f(3)$ **(c)** $f(-2)$ **(d)** $f(-x)$
**(e)** $-f(x)$ **(f)** $f(x + 2)$ **(g)** $f(2x)$ **(h)** $f(x + h)$

**37.** $f(x) = 2x + 3$

**38.** $f(x) = 3x + 1$

**39.** $f(x) = -5x + 2$

**40.** $f(x) = -2x - 3$

*In Problems 41–48, find the value of each function.*

**41.** $f(x) = x^2 + 3; f(2)$

**42.** $f(x) = -2x^2 + x + 1; f(-3)$

**43.** $s(t) = -t^3 - 4t; s(-2)$

**44.** $g(h) = -h^2 + 5h - 1; g(4)$

**45.** $F(x) = |x - 2|; F(-3)$

**46.** $G(z) = 2|z + 5|; G(-6)$

**47.** $F(z) = \frac{z + 2}{z - 5}; F(4)$

**48.** $h(q) = \frac{3q^2}{q + 2}; h(2)$

*In Problems 49–56, graph each function.*

**49.** $f(x) = 4x - 6$

**50.** $g(x) = -3x + 5$

**51.** $h(x) = x^2 - 2$

**52.** $F(x) = x^2 + 1$

**53.** $G(x) = |x - 1|$

**54.** $H(x) = |x + 1|$

**55.** $g(x) = x^3$

**56.** $h(x) = x^3 - 3$

## Applying the Concepts

**57.** If $f(x) = 3x^2 - x + C$ and $f(3) = 18$, what is the value of $C$?

**58.** If $f(x) = -2x^2 + 5x + C$ and $f(-2) = -15$, what is the value of $C$?

**59.** If $f(x) = \frac{2x + 5}{x - A}$ and $f(0) = -1$, what is the value of $A$?

**60.** If $f(x) = \frac{-x + B}{x - 5}$ and $f(3) = -1$, what is the value of $B$?

△ **61. Geometry** Express the area $A$ of a circle as a function of its radius, $r$. Determine the area of circle whose radius is 4 inches. That is, find $A(4)$.

△ **62. Geometry** Express the area $A$ of a triangle as a function of its height $h$ assuming that the length of the base is 8 centimeters. Determine the area of this triangle if its height is 5 centimeters. That is, find $A(5)$.

**63. Salary** Express the gross salary $G$ of Jackie, who earns \$15 per hour as a function of the number of hours worked, $h$. Determine the gross salary of Jackie if she works 25 hours. That is, find $G(25)$.

**64. Commissions** Roberta is a commissioned salesperson. She earns a base weekly salary of \$250 per week plus 15% of the sales price of items sold. Express her gross salary $G$ as a function of the price $p$ of items sold. Determine the weekly gross salary of Roberta if the value of items sold is \$10,000. That is, find $G(10{,}000)$.

**65. Population as a Function of Age** The function $P(a) = 0.025a^2 - 5.633a + 300.517$ represents the population (in millions) of Americans in 2005, $P$, that are $a$ years of age or older. (*Source:* United States Census Bureau)

**(a)** Identify the dependent and independent variable.
**(b)** Evaluate $P(20)$. Provide a verbal explanation of the meaning of $P(20)$.
**(c)** Evaluate $P(0)$. Provide a verbal explanation of the meaning of $P(0)$.

**66. Number of Rooms** The function $N(r) = -1.33r^2 + 14.68r - 17.09$ represents the number of housing units (in millions), $N$, in 2005 that have $r$ rooms, where $1 \le r \le 9$. (*Source*: United States Census Bureau)

**(a)** Identify the dependent and independent variable.
**(b)** Evaluate $N(3)$. Provide a verbal explanation of the meaning of $N(3)$.
**(c)** Why is it unreasonable to evaluate $N(0)$?

**67. Revenue Function** The function $R(p) = -p^2 + 200p$ represents the daily revenue $R$ earned from selling personal digital assistants (PDAs) at $p$ dollars for $0 \le p \le 200$.

**(a)** Identify the dependent and independent variable.
**(b)** Evaluate $R(50)$. Provide a verbal explanation of the meaning of $R(50)$.
**(c)** Evaluate $R(120)$. Provide a verbal explanation of the meaning of $R(120)$.

**68. Average Trip Length** The function $T(x) = 0.01x^2 - 0.12x + 8.89$ represents the average vehicle trip length $T$ (in miles) $x$ years since 1969.

**(a)** Identify the dependent and independent variable.
**(b)** Evaluate $T(35)$. Provide a verbal explanation of the meaning of $T(35)$.
**(c)** Evaluate $T(0)$. Provide a verbal explanation of the meaning of $T(0)$.

## Extending the Concepts

**69.** Investigate when the use of function notation $y = f(x)$ first appeared. Start by researching Lejeune Dirichlet.

**70.** Are all relations functions? Are all functions relations? Explain your answers.

## The Graphing Calculator

Graphing calculators have the ability to evaluate any function you wish. Figure 21 shows the results obtained in Example 6 using a TI-84 Plus graphing calculator.

**Figure 21**

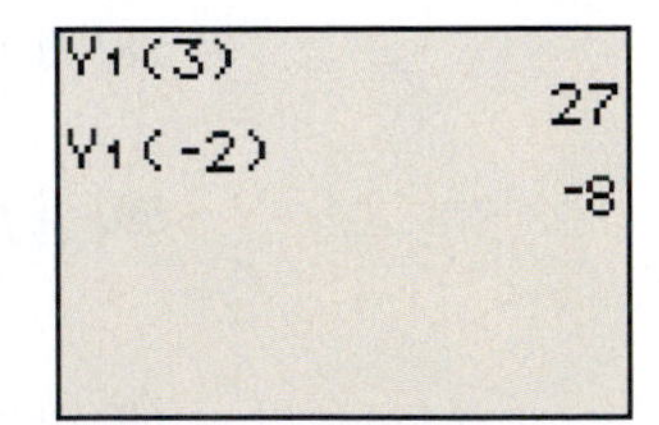


*In Problems 71–78, use a graphing calculator to find the value of each function.*

**71.** $f(x) = x^2 + 3; f(2)$

**72.** $f(x) = -2x^2 + x + 1; f(-3)$

**73.** $F(x) = |x - 2|; F(-3)$

**74.** $g(h) = \sqrt{2h + 1}; g(4)$

**75.** $H(x) = \sqrt{4x - 3}; H(7)$

**76.** $G(z) = 2|z + 5|; G(-6)$

**77.** $F(z) = \dfrac{z + 2}{z - 5}; F(4)$

**78.** $h(q) = \dfrac{3q^2}{q + 2}; h(2)$

# 2.4 Functions and Their Graphs

**OBJECTIVES**

1. Find the Domain of a Function
2. Obtain Information from the Graph of a Function
3. Interpret Graphs of Functions

## *Preparing for Functions and Their Graphs*

*Before getting started, take this readiness quiz. If you get a problem wrong, go back to the section cited, and review the material.*

**1.** Determine whether each of the following numbers is in the domain of $x$ for the algebraic expression $\frac{x+1}{x-5}$

**(a)** $x = -1$ **(b)** $x = 5$ **(c)** $x = 0$ [Section R.5, pp. 46–47]

**2.** Solve the equation $3x - 12 = 0$. [Section 1.1, pp. 53–56]

## 1 Find the Domain of a Function

When working with functions, we need to determine the set of inputs for which a function makes sense. Often the set of inputs for which the function makes sense is not specified; instead, only the equation defining the function is given.

**In Words**

The domain of a function is the set of all inputs for which the function gives an output that is a real number or makes sense.

**DEFINITION**

When only the equation of a function is given, we agree that the **domain of $f$** is the largest set of real numbers for which $f(x)$ is a real number.

The domain of $f$ is the same as the domain of the variable $x$ in the expression $f(x)$. When identifying the domain of a function don't forget that division by zero is undefined, so exclude values of the variable that cause division by zero.

### EXAMPLE 1 Finding the Domain of a Function

Find the domain of each of the following functions:

**(a)** $G(x) = x^2 + 1$

**(b)** $g(z) = \frac{z-3}{z+1}$

**Solution**

**(a)** The function $G$ tells us to square a number $x$ and then add 1 to that number. These operations can be performed on any real number, so the domain of $G$ is the set of all real numbers. We can express the domain as $\{x \mid x \text{ is a real number}\}$ or, using interval notation, $(-\infty, \infty)$.

**(b)** The function $g$ tells us to divide $z - 3$ by $z + 1$. Since division by 0 is not defined, the denominator $z + 1$ can never be 0. Therefore, $z$ can never equal $-1$. The domain of $g$ is $\{z \mid z \neq -1\}$, or using interval notation, $(-\infty, -1) \cup (-1, \infty)$.

**QUICK ✓** *Find the domain of each function.*

**1.** $f(x) = 3x^2 + 2$

**2.** $h(x) = \frac{x+1}{x-3}$

When we use functions in applications, the domain may be restricted by physical or geometric considerations, rather than by pure mathematical restrictions. For example, the domain of the function defined by $f(x) = x^2$ is the set of all real numbers.

*Preparing for...Answers* **1. (a)** Yes **(b)** No **(c)** Yes **2.** $\{4\}$

However, if $f$ is used to obtain the area of a square when the length $x$ of a side is known, then we must restrict the domain of $f$ to the positive real numbers, since the length of a side can never be 0 or negative.

### EXAMPLE 2 Finding the Domain of a Function

The number $N$ of computers produced at one of Dell Computers manufacturing facilities in one day after $t$ hours is given by the function, $N(t) = 336t - 7t^2$. What is the domain of this function?

**Solution**

The independent variable in this function is $t$, where $t$ represents the number of hours in the day. Therefore, the domain of the function is $\{t \mid 0 \leq t \leq 24\}$, or the interval $[0, 24]$.

**QUICK ✓**

**3.** The function $A(r) = \pi r^2$ gives the area of a circle $A$ as a function of the radius $r$. What is the domain of the function?

## 2 Obtain Information from the Graph of a Function

We can find the domain and the range of a function from its graph. The approach to finding the domain and the range of a function from its graph is identical to the approach taken to find the domain and the range of a relation from its graph.

### EXAMPLE 3 Determining the Domain and the Range of a Function from Its Graph

Figure 22 shows the graph of a function.

**(a)** Determine the domain and the range of the function.

**(b)** Identify the intercepts.

**Figure 22**

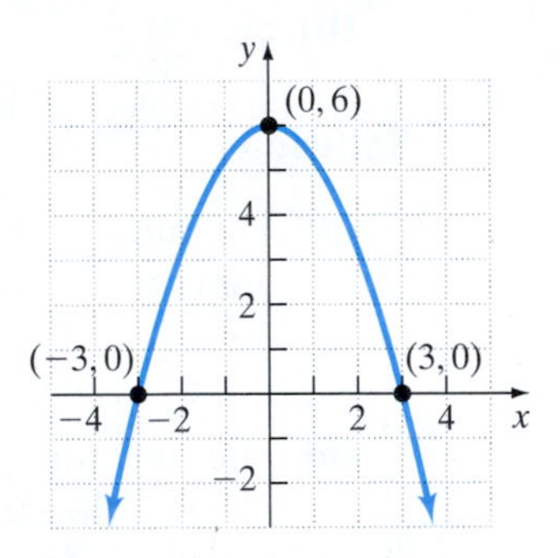


**In Words**

When the graph of a function is given, its domain may be viewed as the shadow created by the graph on the $x$-axis by vertical beams of light. Its range can be viewed as the shadow created by the graph on the $y$-axis by horizontal beams of light.

**Solution**

**(a)** To find the domain of the function, we determine the $x$-coordinates for which the graph of the function exists. Because the graph exists for all real numbers $x$, the domain is $\{x \mid x \text{ is a real number}\}$, or using interval notation, $(-\infty, \infty)$.

To find the range of the function, we determine the $y$-coordinates for which the graph of the function exists. Because the graph exists for all real numbers $y$ less than or equal to 6, the range is $\{y \mid y \leq 6\}$, or using interval notation, $(-\infty, 6]$.

**(b)** The intercepts are the points $(-3, 0)$, $(0, 6)$, and $(3, 0)$. The $x$-intercepts are $-3$ and $3$. The $y$-intercept is 6.

QUICK ✓

**4.** Use the graph of the function to answer parts (a) and (b).

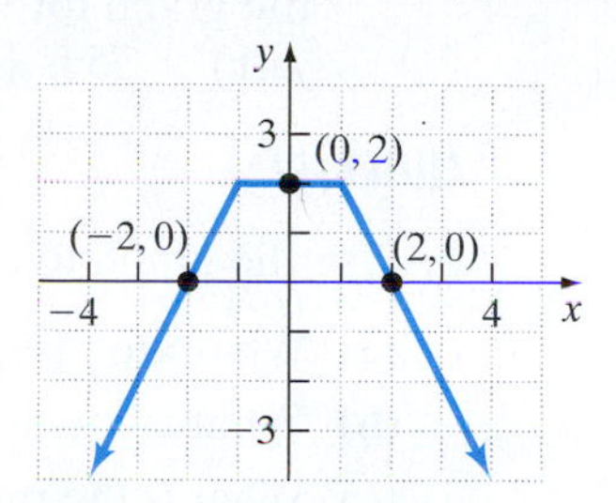


**(a)** Determine the domain and the range of the function.

**(b)** Identify the intercepts.

---

Remember, if $(x, y)$ is a point on the graph of a function $f$, then $y$ is the value of $f$ at $x$, that is, $y = f(x)$. So, if $(1, 5)$ is a point on a function $f$, then $f(1) = 5$. The next example illustrates how to obtain information about a function if its graph is known.

## EXAMPLE 4 Obtaining Information from the Graph of a Function

The Wonder Wheel is a Ferris wheel located in Coney Island. See Figure 23. Let $f$ be the distance above the ground of a person riding on the Wonder Wheel as a function of time $x$ (in minutes). Figure 24 represents the graph of the function $f$. Use the graph to answer the following questions.

Figure 23

Figure 24

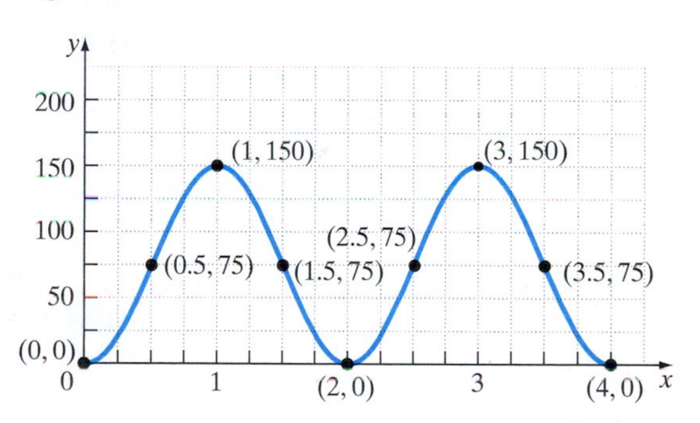


**(a)** What are $f(1.5)$ and $f(3)$? Interpret these values.

**(b)** What is the domain of $f$?

**(c)** What is the range of $f$?

**(d)** List the intercepts.

**(e)** For what values of $x$ does $f(x) = 75$? That is, solve $f(x) = 75$.

### Solution

**(a)** Since $(1.5, 75)$ is on the graph of $f$, then $f(1.5) = 75$. After 1.5 minutes, an individual on the Wonder Wheel is 75 feet in the air. Similarly, we find that since $(3, 150)$ is on the graph we have that $f(3) = 150$. After 3 minutes, an individual on the Wonder Wheel is 150 feet in the air.

**(b)** To determine the domain of $f$, we notice that for each number $x$ between 0 and 4 inclusive, there are points $(x, f(x))$ on the graph of $f$. Therefore, the domain of $f$ is $\{x | 0 \le x \le 4\}$, or the interval $[0, 4]$.

**(c)** The points on the graph have $y$-coordinates between 0 and 150, inclusive. Therefore, the range of $f$ is $\{y | 0 \le y \le 150\}$, or the interval $[0, 150]$.

**(d)** The intercepts are $(0, 0)$, $(2, 0)$, and $(4, 0)$.

**(e)** Since $(0.5, 75)$, $(1.5, 75)$, $(2.5, 75)$, and $(3.5, 75)$ are the only points on the graph for which $y = f(x) = 75$, the solution set to the equation $f(x) = 75$ is $\{0.5, 1.5, 2.5, 3.5\}$.

**QUICK ✓**

**5.** Use the graph to answer the following questions.

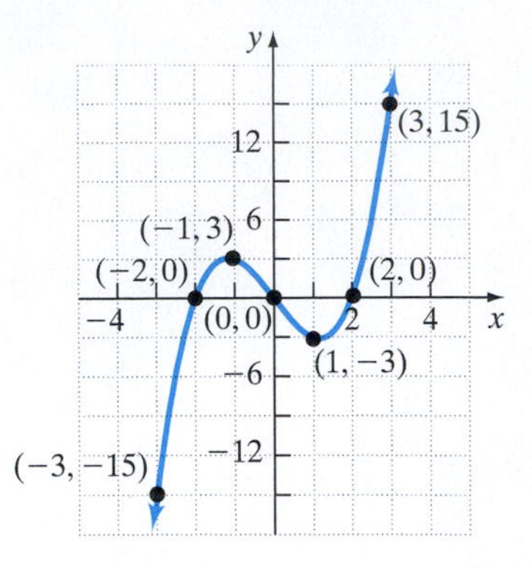


**(a)** What are $f(-3)$ and $f(1)$?

**(b)** What is the domain of $f$?

**(c)** What is the range of $f$?

**(d)** List the intercepts.

**(e)** For what value of $x$ does $f(x) = 15$? That is, solve $f(x) = 15$.

## EXAMPLE 5 Obtaining Information about the Graph of a Function

Consider the function $f(x) = 2x - 5$.

**(a)** Is the point $(3, -1)$ on the graph of the function?

**(b)** If $x = 1$, what is $f(x)$? What point is on the graph of the function?

**(c)** If $f(x) = 3$, what is $x$? What point is on the graph of $f$?

### Solution

**(a)** When $x = 3$, then

$$f(x) = 2x - 5$$
$$f(3) = 2(3) - 5 = 6 - 5 = 1$$

Since $f(3) = 1$, the point $(3, 1)$ is on the graph; the point $(3, -1)$ is not on the graph.

**(b)** If $x = 1$, then

$$f(x) = 2x - 5$$
$$f(1) = 2(1) - 5 = 2 - 5 = -3$$

The point $(1, -3)$ is on the graph of $f$.

**(c)** If $f(x) = 3$, then

$$f(x) = 3$$
$$2x - 5 = 3$$

Add 5 to both sides: $2x = 8$

Divide both sides by 2: $x = 4$

If $f(x) = 3$, then $x = 4$. The point $(4, 3)$ is on the graph of $f$.

**Work Smart: Study Skills**

Do not confuse the directions "Find $f(3)$" with "If $f(x) = 3$, what is $x$?" Write down and study errors that you commonly make so that you can avoid them.

**QUICK ✓**

**6.** Consider the function $f(x) = -3x + 7$.

**(a)** Is the point $(-2, 1)$ on the graph of the function?

**(b)** If $x = 3$, what is $f(x)$? What point is on the graph of the function?

**(c)** If $f(x) = -8$, what is $x$? What point is on the graph of $f$?

## 3 Interpret Graphs of Functions

We can use the graph of a function to give a visual description of many different scenarios. Consider the following example.

### EXAMPLE 6 Graphing a Verbal Description

Maria decides to take a walk. She leaves her house and walks 3 blocks in 2 minutes at a constant speed. She realizes that she left her front door unlocked, so she runs home in 1 minute. It takes Maria 1 minute to find her keys and lock the door. She then decides to run 10 blocks in 3 minutes. She is a little tired now, so she rests for 1 minute and then walks an additional 4 blocks in 10 minutes. She hitches a ride home with her neighbor who happens to drive by and gets home in 2 minutes. Draw a graph of Maria's distance from home (in blocks) as a function of time.

#### Solution

First, we recognize that distance from home is a function of time. Therefore, we draw a Cartesian Plane with the horizontal axis representing the independent variable, time, and the vertical axis representing the dependent variable, distance from home.

The ordered pair $(2, 3)$ corresponds to being 3 blocks from home after 2 minutes. We start the graph at the origin and then draw a straight line from $(0, 0)$ to $(2, 3)$. From the point $(2, 3)$, we draw a straight line to $(3, 0)$, which represents the return trip home to lock the door. Draw a line segment from $(3, 0)$ to $(4, 0)$ to represent the time it takes to lock the door. Draw a line segment from $(4, 0)$ to $(7, 10)$, which represents the 10 block run in 3 minutes. Now we draw a horizontal line from $(7, 10)$ to $(8, 10)$. This represents the resting period. Draw a line from $(8, 10)$ to $(18, 14)$ to represent the 4 block walk in 10 minutes. Finally, draw a line segment from $(18, 14)$ to $(20, 0)$ to represent the ride home. See Figure 25.

**Figure 25**

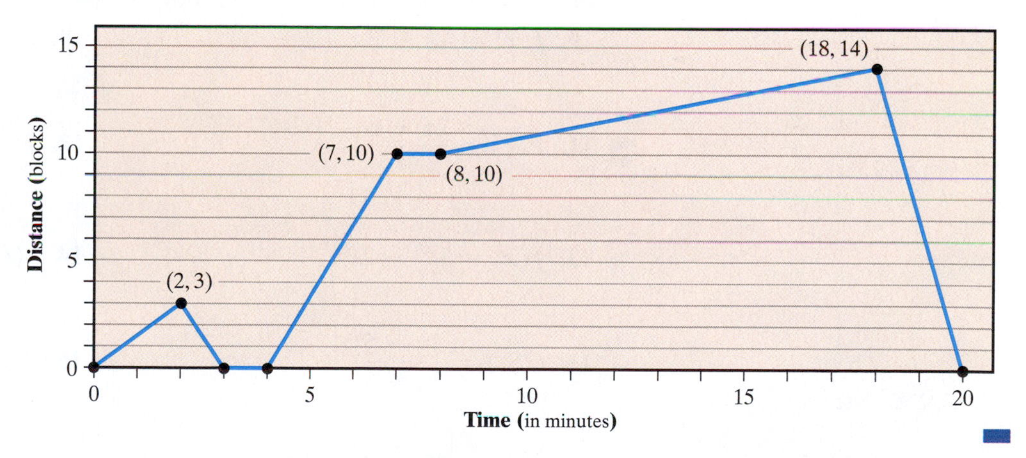


**QUICK ✓**

7. Maria decides to take a walk. She leaves her house and walks 5 blocks in 5 minutes at a constant speed. She realizes that she left her front door unlocked, so she runs home in 2 minutes. It takes Maria 1 minute to find her keys and lock the door. She then decides to jog 8 blocks in 5 minutes. She then runs 3 blocks in 1 minute. She is a little tired now, so she rests for 2 minutes and then walks home in 10 minutes. Draw a graph of Maria's distance from home (in blocks) as a function of time.

# 2.4 Exercises

**For Extra Help:** Student Solutions Manual | CD Video | PH Math/Tutor Center | MathXL Tutorials on CD | MathXL® | MyMathLab

## Concepts and Vocabulary

*In Problems 1–3, fill in the blanks.*

**1.** If the point (3, 8) is on the graph of a function $f$, then $f($_____$)$ = _____.

**2.** If $g(-2) = 4$, then (_____, _____) is a point on the graph of $g$.

**3.** When only the equation of a function is given, we agree that the _____ of $f$ is the largest set of real numbers for which $f(x)$ is a real number.

*In Problems 4 and 5, answer True or False to each statement.*

**4.** A function can have more than one $y$-intercept.

**5.** A function can have more than one $x$-intercept.

**6.** Using the definition of a function, explain why the graph of a function can have at most one $y$-intercept.

**7.** In your own words, explain what the domain of a function is. In your explanation, provide a discussion as to how domains are determined in applications.

**8.** In your own words, explain what the range of a function is.

## Building Skills

*In Problems 9–18, find the domain of each function.*

**9.** $f(x) = 4x + 7$

**10.** $G(x) = -8x + 3$

**11.** $F(z) = \dfrac{2z + 1}{z - 5}$

**12.** $H(x) = \dfrac{x + 5}{2x + 1}$

**13.** $f(x) = 3x^4 - 2x^2$

**14.** $s(t) = 2t^2 - 5t + 1$

**15.** $G(x) = \dfrac{3x - 5}{3x + 1}$

**16.** $H(q) = \dfrac{1}{6q + 5}$

**17.** $f(x) = \dfrac{10x + 7}{3}$

**18.** $f(x) = \dfrac{4x - 9}{7}$

*In Problems 19–28, for each graph of a function, find (a) the domain and the range, and (b) the intercepts, if any.*

**19.**

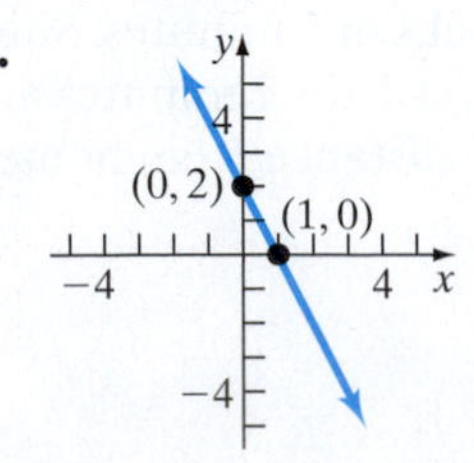


**20.**

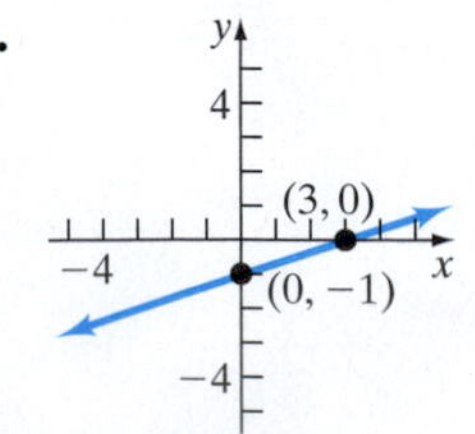

**21.**

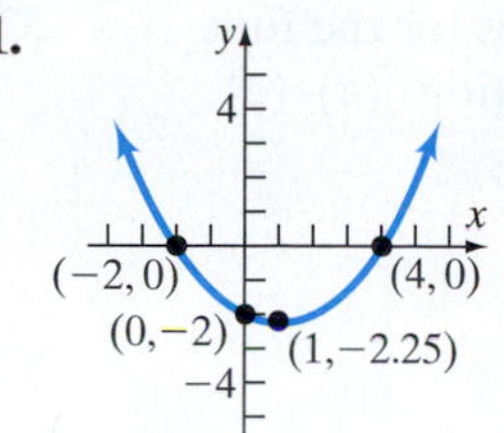


**22.**

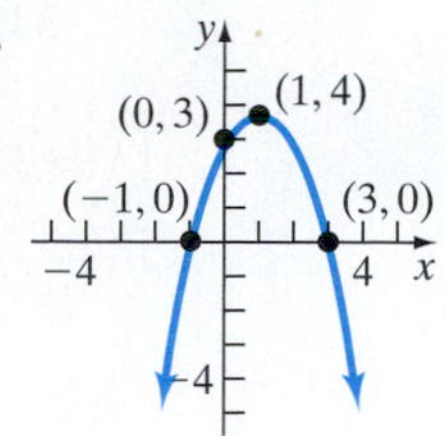


**23.**

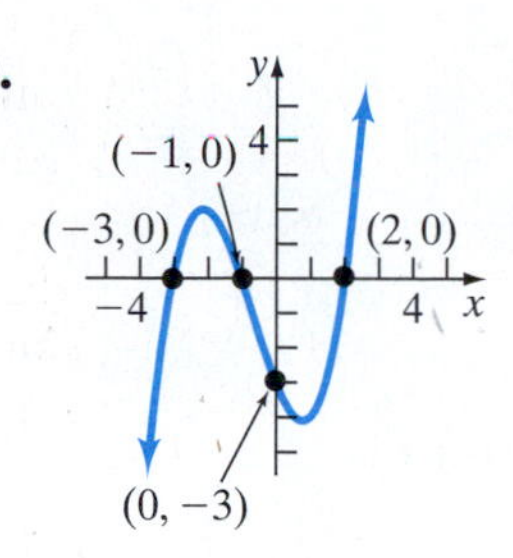


**24.**

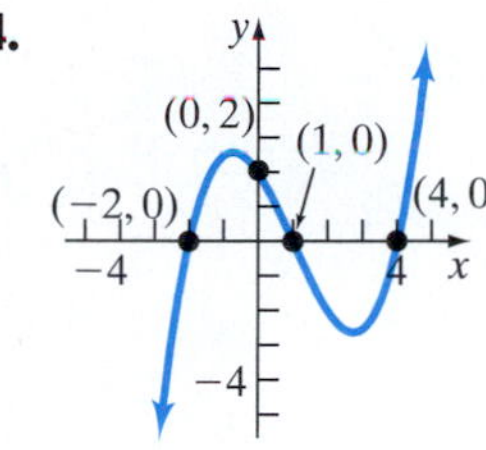


**25.**

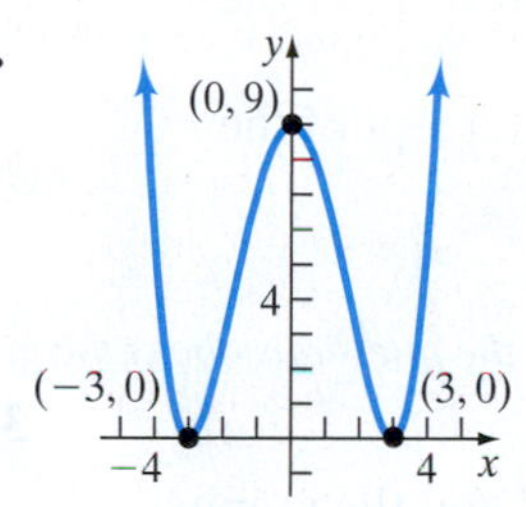


**26.**

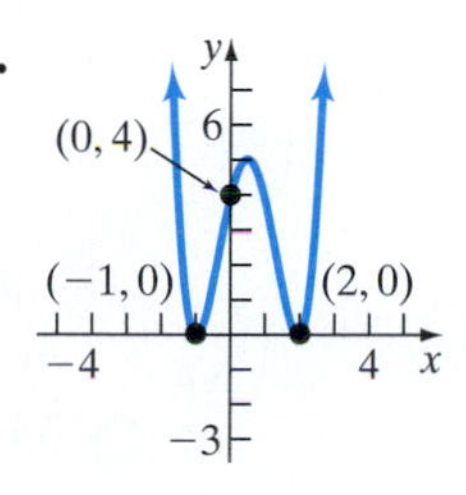


**27.**

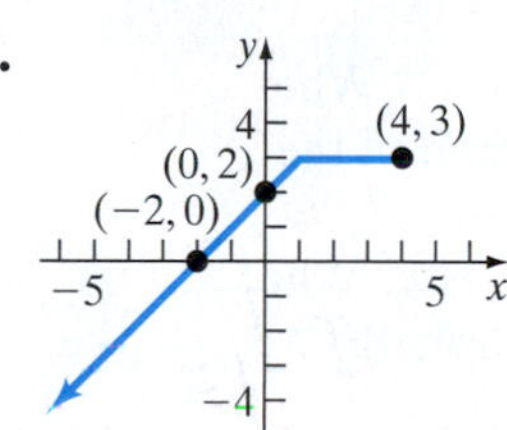


**28.**

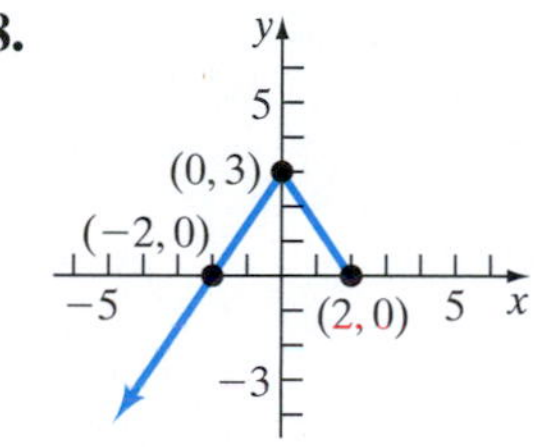


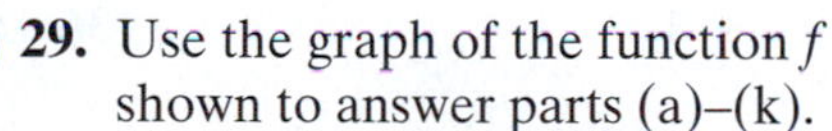

**29.** Use the graph of the function $f$ shown to answer parts (a)–(k).

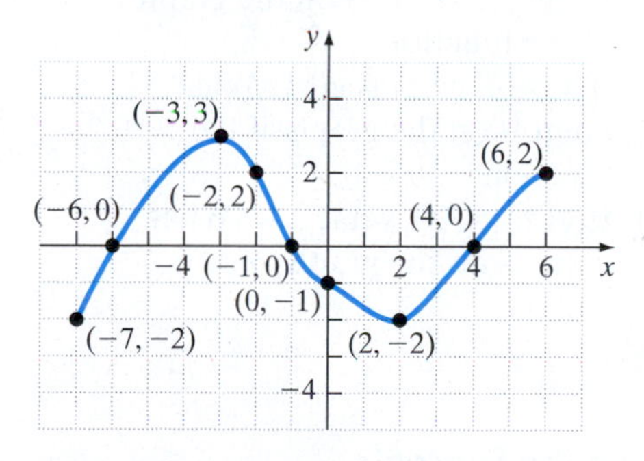


**(a)** Find $f(-7)$.
**(b)** Find $f(-3)$.
**(c)** Find $f(6)$.
**(d)** Is $f(2)$ positive or negative?
**(e)** For what numbers $x$ is $f(x) = 0$?
**(f)** What is the domain of $f$?
**(g)** What is the range of $f$?
**(h)** What are the $x$-intercepts?
**(i)** What is the $y$-intercept?
**(j)** For what numbers $x$ is $f(x) = -2$?
**(k)** For what number $x$ is $f(x) = 3$?

**30.** Use the graph of the function $g$ shown to answer parts (a)–(k).

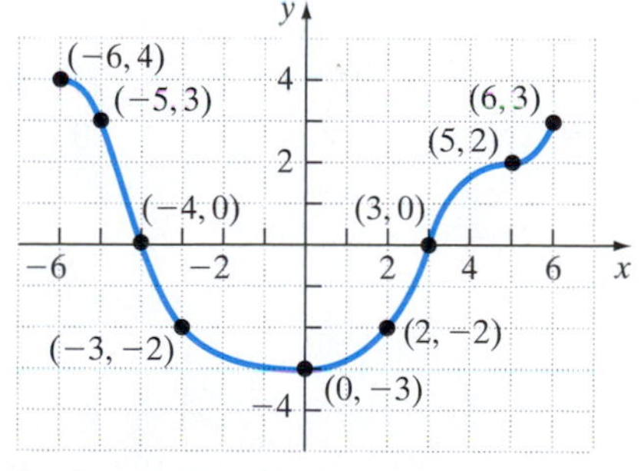


**(a)** Find $g(-3)$.
**(b)** Find $g(5)$.
**(c)** Find $g(6)$.
**(d)** Is $g(-5)$ positive or negative?
**(e)** For what numbers $x$ is $g(x) = 0$?
**(f)** What is the domain of $g$?
**(g)** What is the range of $g$?
**(h)** What are the $x$-intercepts?
**(i)** What is the $y$-intercept?
**(j)** For what numbers $x$ is $g(x) = -2$?
**(k)** For what number $x$ is $g(x) = 3$?

**31.** Use the table of values for the function $F$ to answer questions (a)–(e).

| $x$ | $F(x)$ |
|---|---|
| −4 | 0 |
| −2 | 3 |
| −1 | 5 |
| 0 | 2 |
| 3 | −6 |

**(a)** What is $F(-2)$?
**(b)** What is $F(3)$?
**(c)** For what number(s) $x$ is $F(x) = 5$?
**(d)** What is the $x$-intercept of the graph of $F$?
**(e)** What is the $y$-intercept of the graph of $F$?

**32.** Use the table of values for the function $G$ to answer questions (a)–(e).

| $x$ | $G(x)$ |
|---|---|
| −5 | −3 |
| −4 | 0 |
| 0 | 5 |
| 3 | 8 |
| 7 | 5 |

**(a)** What is $G(3)$?
**(b)** What is $G(7)$?
**(c)** For what number(s) $x$ is $G(x) = 5$?
**(d)** What is the $x$-intercept of the graph of $G$?
**(e)** What is the $y$-intercept of the graph of $G$?

*In Problems 33–36, answer the questions about the given function.*

**33.** $f(x) = 4x - 9$
**(a)** Is the point $(2, 1)$ on the graph of the function?
**(b)** If $x = 3$, what is $f(x)$? What point is on the graph of the function?
**(c)** If $f(x) = 7$, what is $x$? What point is on the graph of $f$?

**34.** $f(x) = 3x + 5$
**(a)** Is the point $(-2, 1)$ on the graph of the function?
**(b)** If $x = 4$, what is $f(x)$? What point is on the graph of the function?
**(c)** If $f(x) = -4$, what is $x$? What point is on the graph of $f$?

**35.** $g(x) = -\frac{1}{2}x + 4$
**(a)** Is the point $(4, 2)$ on the graph of the function?
**(b)** If $x = 6$, what is $g(x)$? What point is on the graph of the function?
**(c)** If $g(x) = 10$, what is $x$? What point is on the graph of $g$?

**36.** $H(x) = \frac{2}{3}x - 4$
**(a)** Is the point $(3, -2)$ on the graph of the function?
**(b)** If $x = 6$, what is $H(x)$? What point is on the graph of the function?
**(c)** If $H(x) = -4$, what is $x$? What point is on the graph of $f$?

## Applying the Concepts

△ **37. Geometry** The volume $V$ of a sphere as a function of its radius $r$ is given by $V(r) = \frac{4}{3}\pi r^3$. What is the domain of this function?

△ **38. Geometry** The area $A$ of a triangle as a function of its height $h$ assuming that the length of the base is 5 centimeters is $A = \frac{5}{2}h$. What is the domain of the function?

**39. Salary** The gross salary $G$ of Jackie as a function of the number of hours worked, $h$ is given by $G(h) = 22.5h$. What is the domain of the function if she can work up to 60 hours per week?

**40. Commissions** Roberta is a commissioned salesperson. She earns a base weekly salary of \$350 per week plus 12% of the sales price of items sold. Her gross salary $G$ as a function of the price $p$ of items sold is given by $G(p) = 350 + 0.12p$. What is the domain of the function?

**41. Demand for Hot Dogs** Suppose the function $D(p) = 1200 - 10p$ represents the demand for hot dogs, whose price is $p$, at a baseball game. Find the domain of the function.

**42. Revenue Function** The function $R(p) = -p^2 + 200p$ represents the daily revenue earned from selling personal digital assistants (PDAs) at $p$ dollars for $0 \leq p \leq 200$. Explain why any $p$ greater than \$200 is not in the domain of the function.

**43.** Match each of the following functions with the graph that best describes the situation.

**(a)** The distance from ground level of a person who is jumping on a trampoline as a function of time
**(b)** The cost of a telephone call as a function of time
**(c)** The height of a human as a function of time
**(d)** The revenue earned from selling cars as a function of price
**(e)** The book value of a machine that is depreciated by equal amounts each year as a function of the year

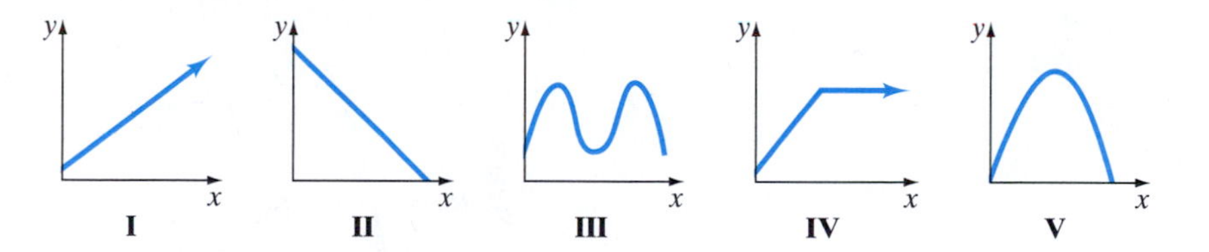


**44.** Match each of the following functions with the graph that best describes the situation.

**(a)** The average high temperature each day as a function of the day of the year
**(b)** The number of bacteria in a Petri dish as a function of time
**(c)** The distance that a person rides her bicycle at a constant speed as a function of time
**(d)** The temperature of a pizza after it is removed from the oven as a function of time
**(e)** The value of car as a function of time

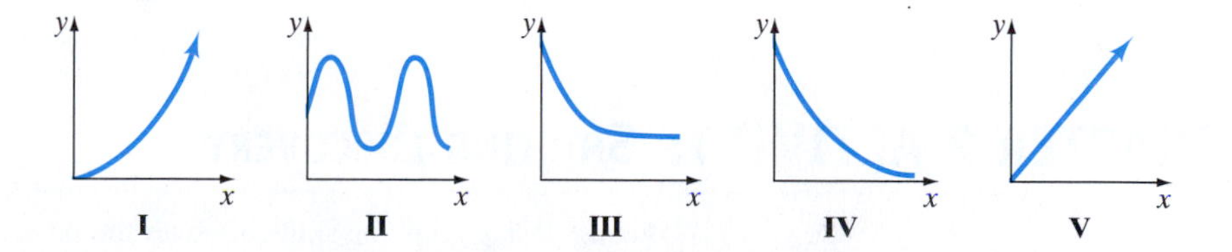


**45. Pulse Rate** Consider the following scenario: Zach starts jogging on a treadmill. His resting pulse rate is 70. As he continues to jog on the treadmill, his pulse increases at a constant rate until, after 10 minutes, his pulse is 120. He then starts jogging faster and his pulse increases at a constant rate, until after 2 minutes, at which time his pulse is up to 150. He then begins a cooling off period for 7 minutes until his pulse backs down to 110. He then gets off the treadmill and his pulse returns to 70 after 12 minutes. Draw a graph of Zach's pulse as a function of time.

46. **Altitude of an Airplane** Suppose that a plane is flying from Chicago to New Orleans. The plane leaves the gate and taxis for 5 minutes. The plane takes off and quickly gets up to 10,000 feet after 5 minutes. The plane continues to ascend at constant rate until it reaches its cruising altitude of 35,000 feet after another 25 minutes. For the next 80 minutes, the plane maintains a constant height of 35,000 feet. The plane then descends at a constant rate until it lands after 20 minutes. It requires 5 minutes to taxi to the gate. Draw a graph of the height of the plane as a function of time.

47. **Height of a Swing** An 8-year-old girl gets on a swing and starts swinging for 10 minutes. Draw a graph that represents the height of the child from the ground as a function of time.

48. **Temperature of Pizza** Marissa is hungry and would like a pizza. Her mother pulls a frozen pizza out of the freezer and puts it in the oven. After 12 minutes the pizza is done, but Mom lets the pizza cool for 5 minutes before serving it to Marissa. Draw a graph that represents the temperature of the pizza as a function of time.

49. The graph below shows the weight of a person as a function of his age. Describe the weight of the individual over the course of his life.

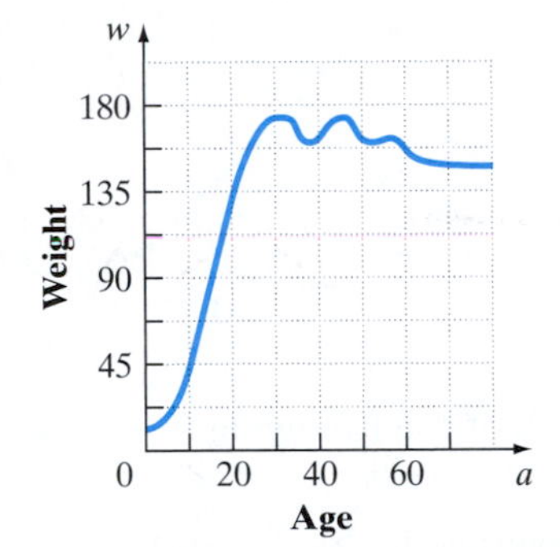


50. The graph below shows the depth of a lake (in feet) as a function of time (in days). Describe the depth of the lake over the course of the year.

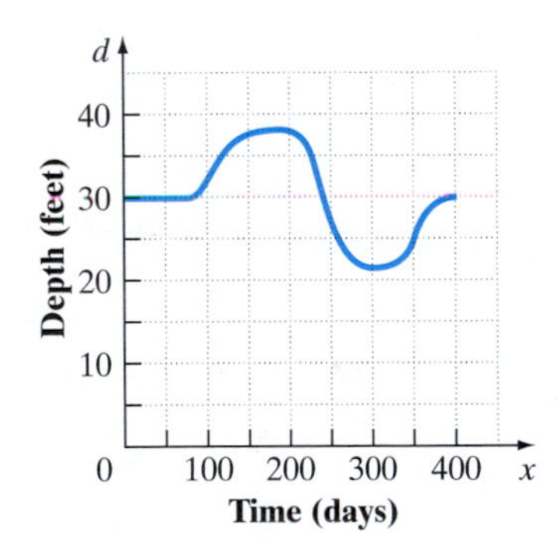


## Extending the Concepts

51. Draw a graph of a function $f$ with the following characteristics: $x$-intercepts: $-4, -1, 2$; $y$-intercept: $-2$; $f(-3) = 7$ and $f(3) = 8$.

52. Draw a graph of a function $f$ with the following characteristics: $x$-intercepts: $-3, 2$, and $5$; $y$-intercept: $3$; $f(3) = -2$.

# CHAPTER 2 ACTIVITY: SHIFTING DISCOVERY

**Focus:** Using graphing skills, discover the possible "rules" for graphing functions.

**Time:** 30–35 minutes

**Group size:** 4

**Materials Needed:** Graph paper (2–3 pieces)

*Each member of the group needs to*

1. Draw a coordinate plane and label the $x$-axis and $y$-axis.
2. By plotting points, graph the primary function: $f(x) = x^2$.

3. On the same coordinate plane, each group member graphs *one* of the following functions by plotting points. Be sure your graphs of the primary function and one of the functions (a)–(d) are on the same coordinate plane.

**(a)** $f(x) = x^2 + 3$ **(b)** $f(x) = (x - 3)^2$

**(c)** $f(x) = x^2 - 3$ **(d)** $f(x) = (x + 3)^2$

*As a group, discuss the following:*

4. What shape are the graphs?
5. Each member of the group, share the difference between the graph of your primary function and the other function you chose.
6. As a group, can you develop a possible rule for these differences?

*With this possible rule in mind, each member of the group needs to*

7. Draw a coordinate plane and label $x$-axis, $y$-axis, and $-10$ to 10 on each axis.
8. Graph the primary function by plotting points: $f(x) = |x|$.
9. On the same coordinate plane, each group member graphs *one* of the following functions by plotting points. Be sure your graphs of the primary function and one of the functions (a)–(d) are on the same coordinate plane.

**(a)** $f(x) = |x| + 4$ **(b)** $f(x) = |x - 4|$

**(c)** $f(x) = |x| - 4$ **(d)** $f(x) = |x + 4|$

10. Did your possible rules developed in Problem 6 hold true? Discuss.

# Chapter 2 Review

## Section 2.1 Rectangular Coordinates and Graphs of Equations

| KEY CONCEPTS | KEY TERMS | |
|---|---|---|
| • **Graph of an Equation in Two Variables** The set of all ordered pairs $(x, y)$ in the $xy$-plane that satisfy the equation<br>• **Intercepts** The points, if any, where a graph crosses or touches the coordinate axes | $x$-axis<br>$y$-axis<br>Origin<br>Rectangular or Cartesian coordinate system<br>$xy$-plane<br>Coordinate axes<br>Ordered pair<br>Coordinates<br>$x$-coordinate<br>$y$-coordinate<br>Abscissa | Ordinate<br>Quadrants<br>Equation in two variables<br>Sides<br>Satisfy<br>Graph of an equation in two variables<br>Point-plotting method<br>Complete graph<br>Intercept<br>$x$-intercept<br>$y$-intercept |

| YOU SHOULD BE ABLE TO . . . | EXAMPLE | REVIEW EXERCISES |
|---|---|---|
| 1 Plot points in the rectangular coordinate system (p. 136) | Example 1 | 1, 2 |
| 2 Determine whether an ordered pair is a point on the graph of an equation (p. 138) | Example 2 | 3, 4 |
| 3 Graph an equation using the point-plotting method (p. 140) | Examples 3 through 5 | 5–14 |
| 4 Identify the intercepts from the graph of an equation (p. 142) | Example 6 | 15–16 |
| 5 Interpret graphs (p. 143) | Example 7 | 17–20 |

*In Problems 1 and 2, plot each point in the same xy-plane. Tell in which quadrant or on what coordinate axis each point lies.*

**1.** $A(2, -4)$
$B(-1, -3)$
$C(0, 4)$
$D(-5, 1)$
$E(1, 0)$
$F(4, 3)$

**2.** $A(3, 0)$
$B(1, 5)$
$C(-3, -5)$
$D(-1, 4)$
$E(5, -2)$
$F(0, -5)$

*In Problems 3 and 4, determine whether the given points are on the graph of the equation.*

**3.** $3x - 2y = 7$
**(a)** $(3, 1)$
**(b)** $(2, -1)$
**(c)** $(4, 0)$
**(d)** $\left(\frac{1}{3}, -3\right)$

**4.** $y = 2x^2 - 3x + 2$
**(a)** $(-1, 3)$
**(b)** $(1, 1)$
**(c)** $(-2, 16)$
**(d)** $\left(\frac{1}{2}, \frac{3}{2}\right)$

*In Problems 5–14, graph each equation by plotting points.*

**5.** $y = x + 2$ **6.** $2x + y = 3$ **7.** $y = 2x^2 - 3$ **8.** $y = -x^2 + 4$

**9.** $y = -|x| - 2$ **10.** $y = |x + 2| - 1$ **11.** $y = x^3 + 2$ **12.** $y = -x^3 + 1$

**13.** $x = y^2 + 1$ **14.** $y = \dfrac{1}{x - 2}$

*In Problems 15 and 16, the graph of an equation is given. List the intercepts of the graph.*

**15.**

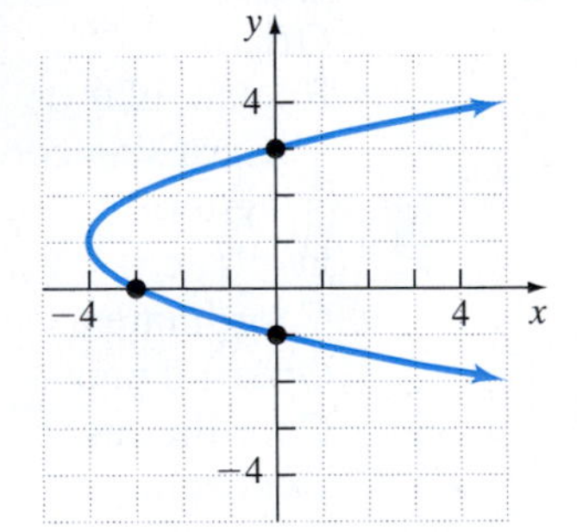


**16.**

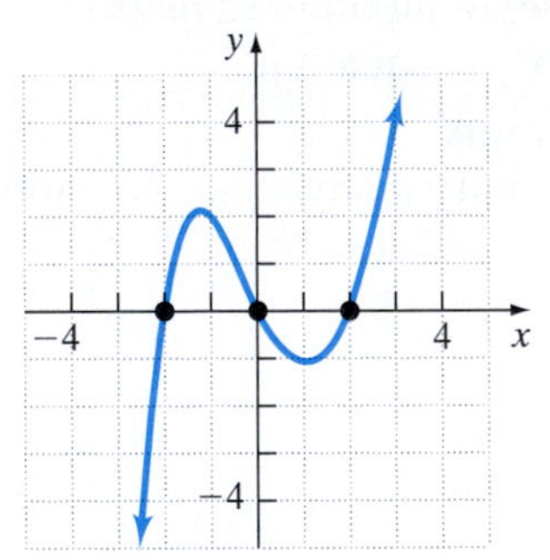


**17. Cell Phones** A cellular phone company offers a plan for \$40 per month for 3000 minutes with additional minutes costing \$0.05 per minute. The graph to the right shows the monthly cost, in dollars, when $x$ minutes are used.

**(a)** If you talk for 2250 minutes in a month, how much is your monthly bill?

**(b)** Use the graph to estimate your monthly bill if you talk for 12 thousand minutes.

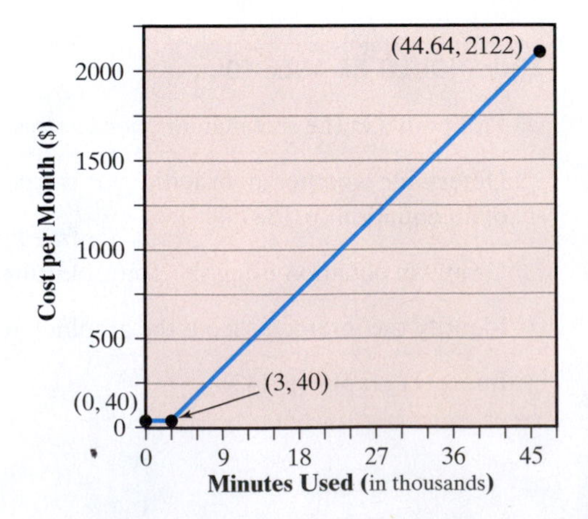

**18. Kentucky Derby** The graph to the right shows the winning times (to the nearest second) in the Kentucky Derby for the years 1995–2004. The vertical axis represents the winning time in seconds over 2 minutes and the horizontal axis represents the year. (*Source:* Churchill Downs Simulcast Network)

**(a)** Use the graph to determine the winning time of the Kentucky Derby in 1999.

**(b)** Use the graph to determine which year between 1995 and 2004 had the fastest winning time for the Kentucky Derby.

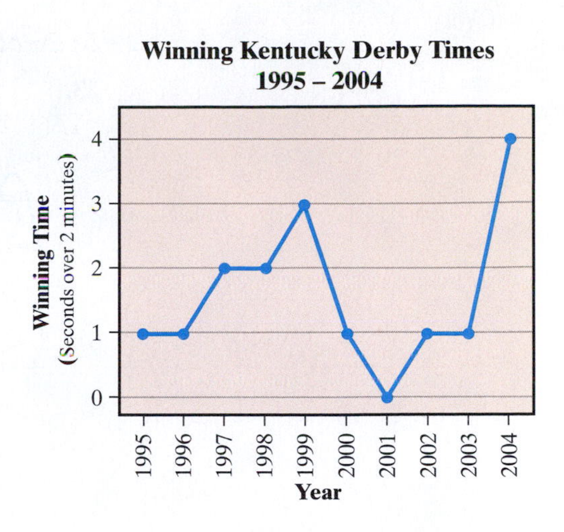


## Section 2.2 Relations

| KEY CONCEPTS | KEY TERMS |
|---|---|
| • **Relation** A correspondence between two variables $x$ and $y$ where $y$ depends on $x$. Relations can be represented through maps, sets of ordered pairs, equations, or graphs. | Relation<br>Corresponds<br>Depends on<br>Mapping<br>Inputs<br>Outputs<br>Domain<br>Range |

| YOU SHOULD BE ABLE TO . . . | EXAMPLE | REVIEW EXERCISES |
|---|---|---|
| 1 Understand relations (p. 149) | Example 1 | 19–22 |
| 2 Find the domain and the range of a relation (p. 150) | Examples 2 through 4 | 19–36, 37, 38 |
| 3 Graph a relation defined by an equation (p. 152) | Example 5 | 27–36 |

*In Problems 19 and 20, write each relation as a set of ordered pairs. Then identify the domain and range of the relation.*

**19.** **20.**

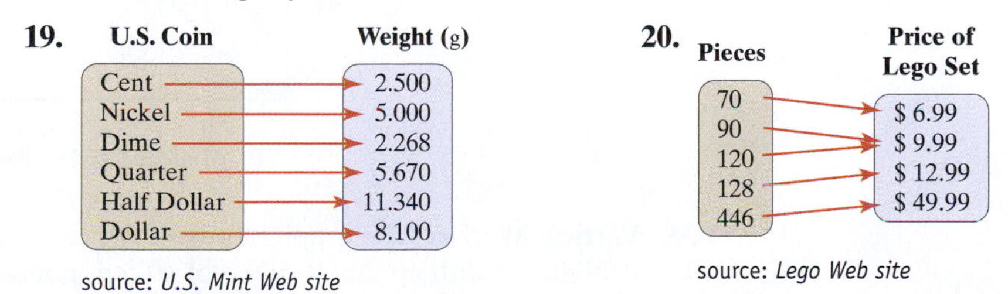


*In Problems 21 and 22, write each relation as a map. Then identify the domain and the range of the relation.*

**21.** $\{(2, 7), (-4, 8), (3, 5), (6, -1), (-2, -9)\}$

**22.** $\{(3, 1), (3, 7), (5, 1), (-2, 8), (1, 4)\}$

*In Problems 23–26, identify the domain and range of the relation from the graph.*

**23.**

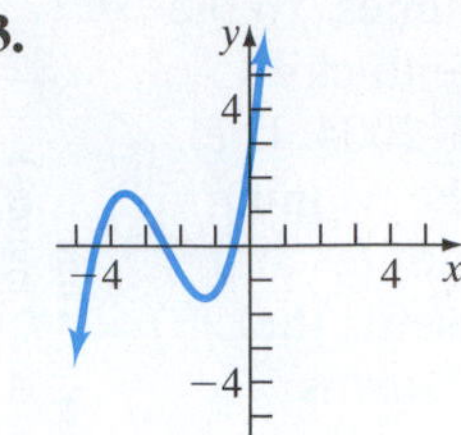


**24.**

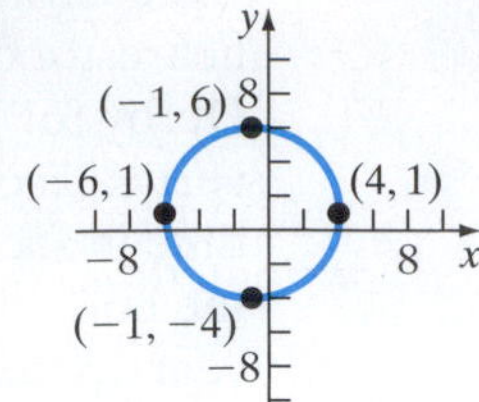


**25.**

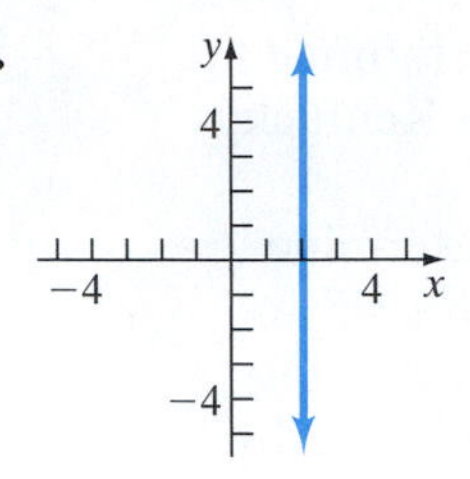


**26.**

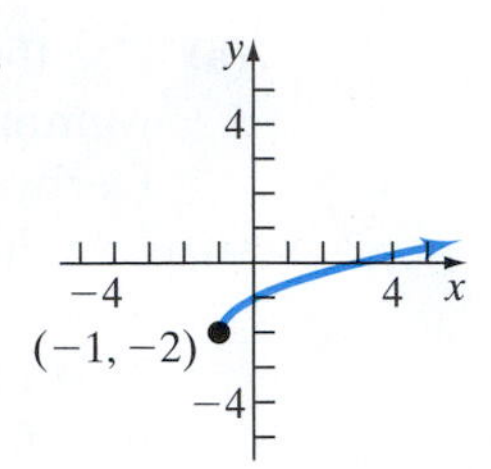


*In Problems 27–36, graph the relation. Use the graph of the relation to identify the domain and the range of the relation. (See Problems 5–14.)*

**27.** $y = x + 2$ **28.** $2x + y = 3$ **29.** $y = 2x^2 - 3$ **30.** $y = -x^2 + 4$

**31.** $y = -|x| - 2$ **32.** $y = |x + 2| - 1$ **33.** $y = x^3 + 2$ **34.** $y = -x^3 + 1$

**35.** $x = y^2 + 1$ **36.** $y = \dfrac{1}{x - 2}$

**37. Cell Phones** A cellular phone company offers a plan for \$40 per month for 3000 minutes with additional minutes costing \$0.05 per minute. The graph below shows the monthly cost, in dollars, when $x$ minutes are used.

**(a)** What is the domain and the range of the relation?

**(b)** Explain why the domain obtained in part (a) is reasonable.

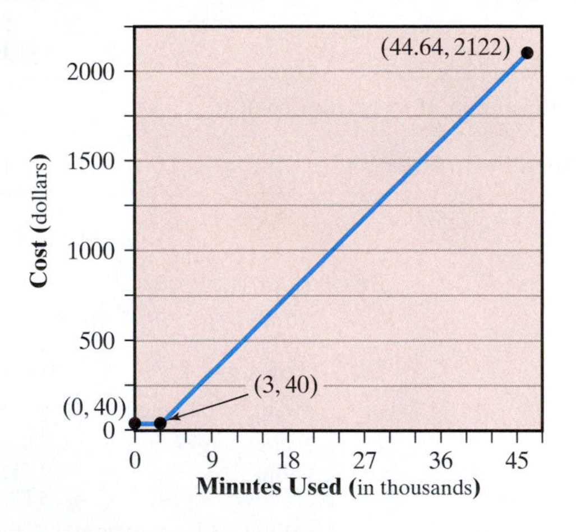


**38. Vertical Motion** The graph below shows the height, in feet, of a ball thrown straight up with an initial speed of 40 feet per second from an initial height of 96 feet after $t$ seconds. What is the domain and the range of the relation?

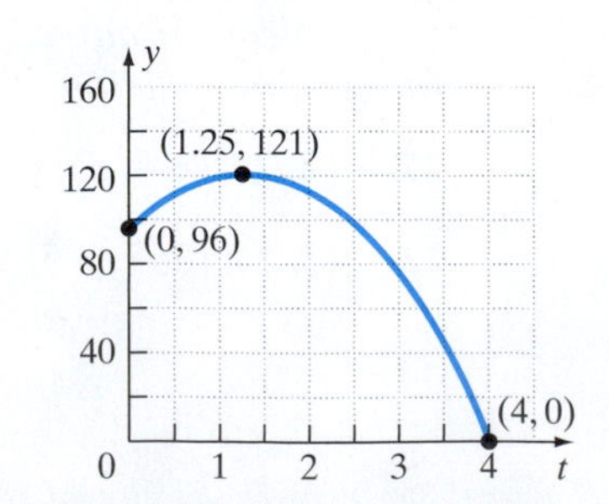

## Section 2.3 An Introduction to Functions

| KEY CONCEPTS | KEY TERMS |
|---|---|
| • **Functions**<br>A special type of relation where any given input, $x$, corresponds to only one output $y$. Functions can be represented through maps, sets of ordered pairs, equations, or graphs.<br>• **Vertical Line Test**<br>A set of points in the $xy$-plane is the graph of a function if and only if every vertical line intersects the graph in at most one point.<br>• **Graph of a Function**<br>The graph of a function, $f$, is the set of all ordered pairs $(x, f(x))$. | Function<br>Vertical Line Test<br>Value of $f$ at the number $x$<br>Independent variable<br>Dependent variable<br>Argument<br>Graph of the function |

| YOU SHOULD BE ABLE TO . . . | EXAMPLE | REVIEW EXERCISES |
|---|---|---|
| 1 Determine whether a relation expressed as a map or ordered pairs represents a function (p. 158) | Examples 1 and 2 | 39, 40 |
| 2 Determine whether a relation expressed as an equation represents a function (p. 161) | Examples 3 and 4 | 41–44 |
| 3 Determine whether a relation expressed as a graph represents a function (p. 161) | Example 5 | 45–48 |
| 4 Find the value of a function (p. 162) | Examples 6 and 7 | 49–52 |
| 5 Graph a function (p. 164) | Example 8 | 53–56 |
| 6 Work with applications of functions (p. 165) | Example 9 | 57–58 |

*In Problems 39 and 40, determine whether the given relation represents a function. State the domain and the range of each relation.*

**39. (a)** $\{(-1, -2), (-1, 3), (5, 0), (7, 2), (9, 4)\}$

**(b)**

| Animal | Typical Lifespan (years) |
|---|---|
| Camel | 50 |
| Macaw | 35 |
| Deer | 14 |
| Fox | 22 |
| Tiger | 45 |
| Crocodile | |

**40. (a)** $\{(-2, 4), (2, 3), (-3, -1), (5, 7), (4, 7)\}$

**(b)**

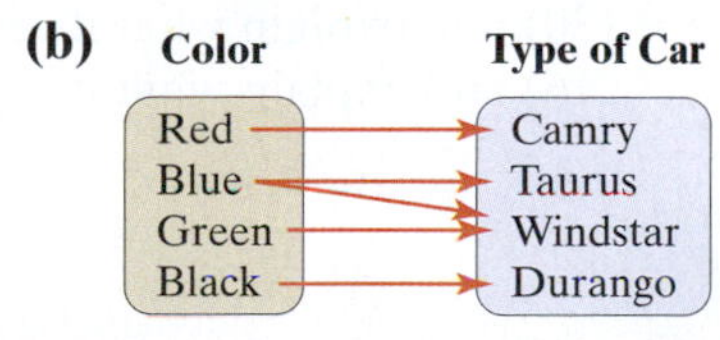


*In Problems 41–44, determine whether each relation shows y as a function of x.*

**41.** $3x - 5y = 18$

**42.** $x^2 + y^2 = 81$

**43.** $y = \pm 10x$

**44.** $y = x^2 - 14$

*In Problems 45–48, determine whether the graph is that of a function.*

**45.**

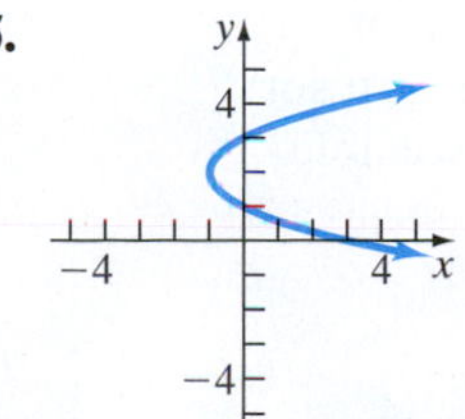

**46.**

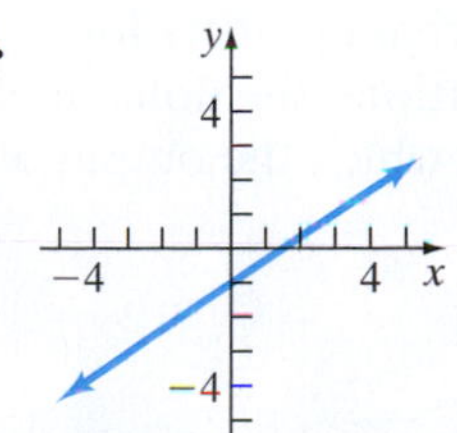

**47.**

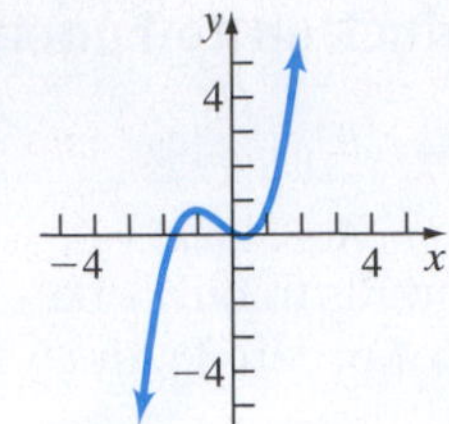


**48.**

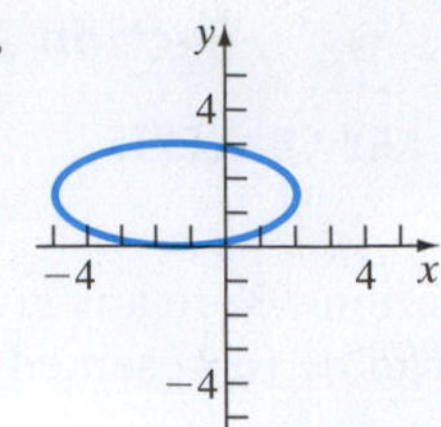


*In Problems 49–52, find the indicated values for the given functions.*

**49.** $f(x) = x^2 + 2x - 5$

**(a)** $f(-2)$ **(b)** $f(3)$

**50.** $g(z) = \dfrac{2z + 1}{z - 3}$

**(a)** $g(0)$ **(b)** $g(2)$

**51.** $F(x) = -2x + 7$

**(a)** $F(5)$ **(b)** $F(-x)$

**52.** $G(x) = 2x + 1$

**(a)** $G(7)$ **(b)** $G(x + h)$

*In Problems 53–56, graph each function.*

**53.** $f(x) = 2x - 5$

**54.** $g(x) = x^2 - 3x + 2$

**55.** $h(x) = (x - 1)^3 - 3$

**56.** $f(x) = |x + 1| - 4$

**57. Population** Using census data from 1900 to 2000, the function $P(t) = 0.144t^2 - 6.613t + 104.448$ represents the population, $P$, of Orange County in Florida (in thousands) $t$ years after 1900.

**(a)** Identify the dependent and independent variables.
**(b)** Evaluate $P(110)$ and explain what it represents.
**(c)** Evaluate $P(-70)$ and explain what it represents. Is the result reasonable? Explain.

**58. Wages** The function $W(a) = -0.058a^2 + 5.410a - 73.839$ represents the 2000 average annual wage, $W$, (in thousands) of a Wyoming resident in the mining industry who is $a$ years old.

**(a)** Identify the dependent and independent variables.
**(b)** Evaluate $W(30)$ and explain what it represents.
**(c)** Evaluate $W(16)$ and explain what it represents. Is this result reasonable? Explain.

## Section 2.4 Functions and Their Graphs

| KEY CONCEPTS | KEY TERMS |
|---|---|
| • **Domain of a Function**<br>When only an equation of a function is given, the domain of the function is the largest set of real numbers for which $f(x)$ is a real number. However, in applications, the domain of a function is the largest set of real numbers for which the output of the function is reasonable. | Domain of $f$ |

| YOU SHOULD BE ABLE TO . . . | EXAMPLE | REVIEW EXERCISES |
|---|---|---|
| 1 Find the domain of a function (p. 169) | Examples 1 and 2 | 59–70 |
| 2 Obtain information from the graph of a function (p. 170) | Examples 3 through 5 | 71–72 |
| 3 Interpret graphs of functions (p. 173) | Example 6 | 73, 74 |

*For Problems 59–64, find the domain of each function.*

**59.** $f(x) = -\frac{3}{2}x + 5$

**60.** $g(w) = \frac{w - 9}{2w + 5}$

**61.** $h(t) = \frac{t + 2}{t - 5}$

**62.** $F(x) = \frac{3}{x - 2}$

**63.** $G(t) = 3t^2 + 4t - 9$

**64.** $H(x) = x^4 - 2x^3 + 7$

*For Problems 65–70, find (a) the domain and the range, and (b) the intercepts, if any.*

**65.**

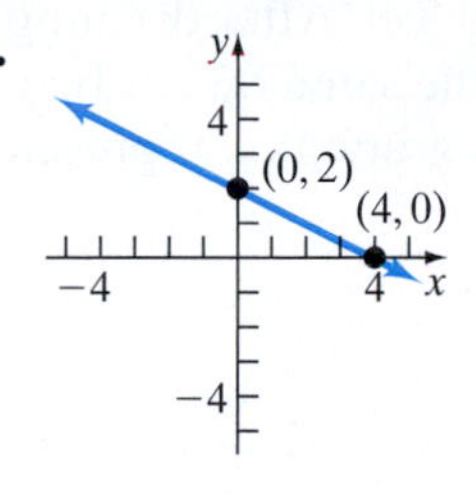


**66.**

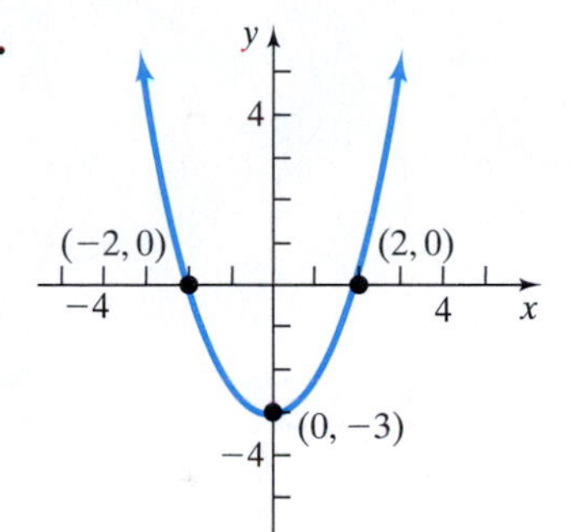


**67.**

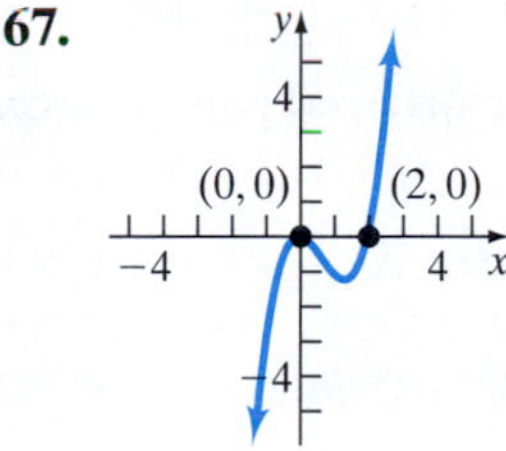


**68.**

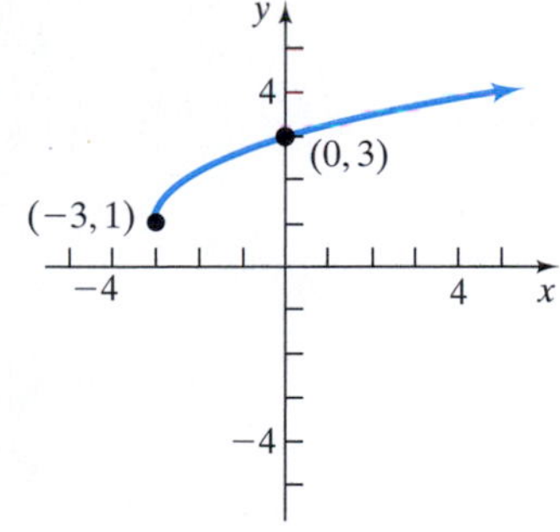


**69.**

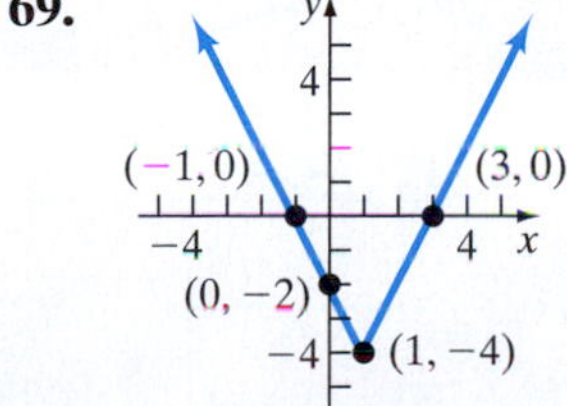


**70.**

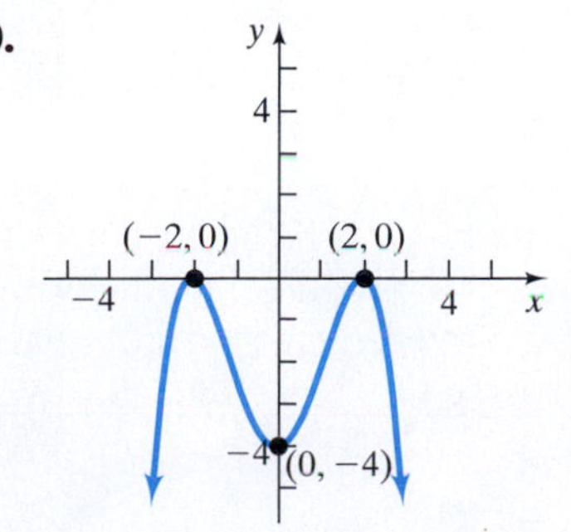

*In Problems 71 and 72, answer the questions about the given function.*

**71.** $h(x) = 2x - 7$

**(a)** Is the point $(3, -1)$ on the graph of the function?
**(b)** If $x = -2$, what is $h(x)$? What point is on the graph of the function?
**(c)** If $h(x) = 4$, what is $x$? What point is on the graph of $h$?

**72.** $g(x) = \frac{3}{5}x + 4$

**(a)** Is the point $(-5, 2)$ on the graph of the function?
**(b)** If $x = 3$, what is $g(x)$? What point is on the graph of the function?
**(c)** If $g(x) = -2$, what is $x$? What point is on the graph of $g$?

**73. Travel by Train** A Metrolink train leaves L.A. Union Station and travels 6 miles at a constant speed for 10 minutes arriving at the Glendale station where it waits 1 minute for passengers to board and depart. The train continues traveling at the same speed for 5 more minutes to reach downtown Burbank, which is 3 miles from Glendale. Sketch a graph that represents the distance of the train as a function of time until it reaches downtown Burbank.

**74. Filling a Tub** With the faucet running at a constant rate, it takes Angie 7 minutes to fill her bathtub. She turns off the faucet when the tub is full and realizes the water is too hot. She then opens the drain letting water out at a constant rate that is half the rate of the faucet. After draining for 2 minutes, she stops the drain and turns on the faucet at the same rate as before (but at a cooler temperature) until the tub is full. Sketch a graph that represents the amount of water in the tub as a function of time.

## CHAPTER 2 TEST

*Remember to use your Chapter Test Prep Video CD to see fully worked-out solutions to any of these problems you would like to review.*

**1.** Plot the following ordered pairs in the same $xy$-plane. Tell in which quadrant or on what coordinate axis each point lies.

$$A(3, -4), B(0, 2), C(3, 0), D(2, 1), E(-1, -4), F(-3, 5)$$

**2.** Determine whether the ordered pair is a point on the graph of the equation $y = 3x^2 + x - 5$.

**(a)** $(-2, 4)$ **(b)** $(-1, -3)$ **(c)** $(2, 9)$

*In Problems 3 and 4, graph the equations by plotting points.*

**3.** $y = 4x - 1$ **4.** $y = 4x^2$

**5.** Identify the intercepts from the graph below.

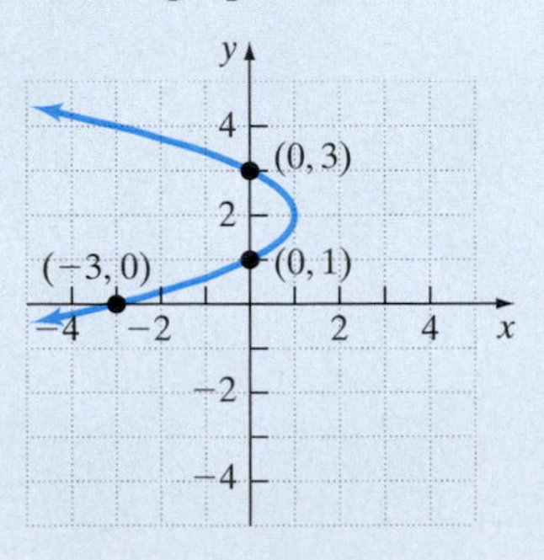

**6.** The graph below shows the monthly unemployment rate for South Bend, Indiana during 2002. The vertical axis represents the unemployment rate (as a percent) and the horizontal axis represents the month. (*Source:* U.S. Bureau of Labor Statistics)

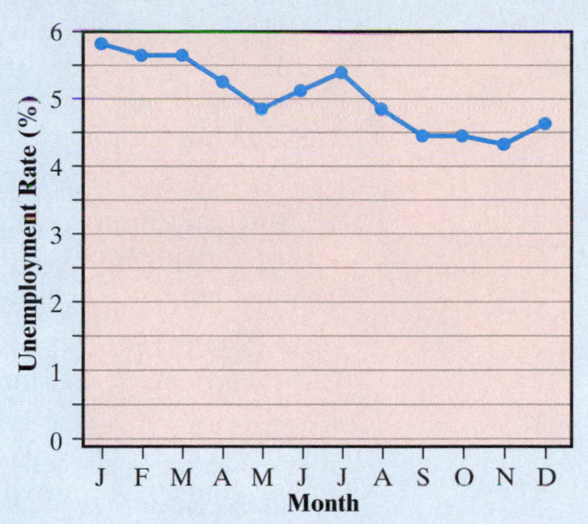


SOURCE: *U.S. Bureau of Labor Statistics*

**(a)** What was the approximate unemployment rate in South Bend for the month of May?
**(b)** In what month was the unemployment rate the highest? What was the approximate rate?
**(c)** In what month was the unemployment rate the lowest? What was the approximate rate?
**(d)** Describe the unemployment rate trend for South Bend during 2002.

**7.** Write the relation as a map. Then identify the domain and the range of the relation.

$$\{(2, 8), (5, -2), (7, 12), (-4, -7), (7, 3), (5, -1)\}$$

**8.** Identify the domain and range of the relation from the graph.

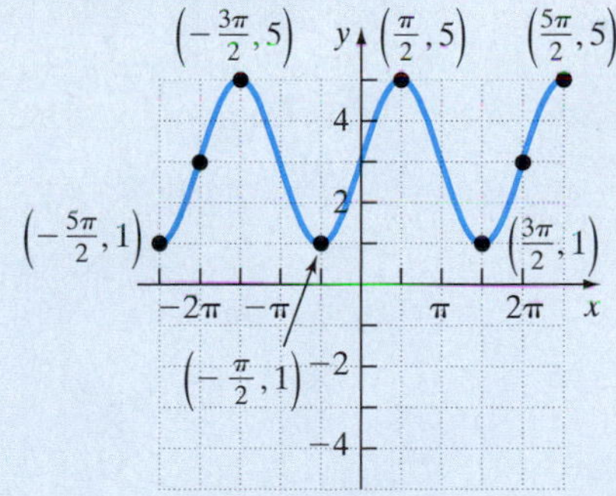


**9.** Graph the relation $y = x^2 - 3$ by plotting points. Use the graph of the relation to identify the domain and range.

*In Problems 10 and 11, determine whether the relations represent functions. Identify the domain and the range of each relation.*

**10.**

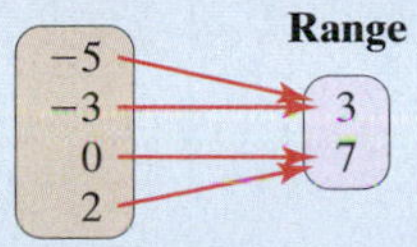


**11.**

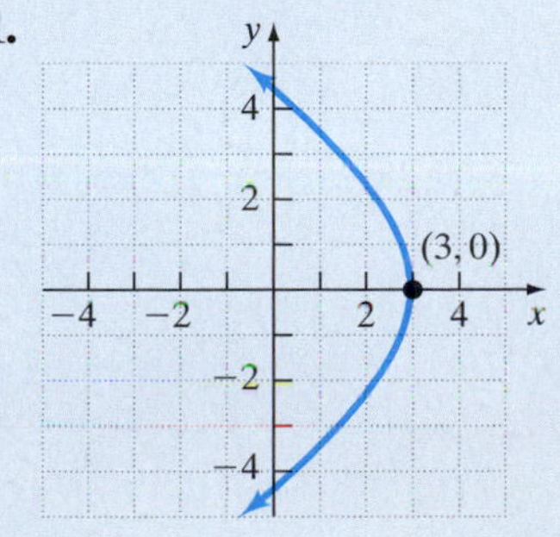

**12.** Does the equation $y = \pm 5x$ represent a function? Why or why not?

**13.** For $f(x) = -3x + 11$, find $f(x + h)$.

**14.** For $g(x) = 2x^2 + x - 1$, find the indicated values.

**(a)** $g(-2)$ **(b)** $g(0)$ **(c)** $g(3)$

**15.** Sketch the graph of $f(x) = x^2 + 3$.

**16.** Using data from 1989 to 2002, the function $P(x) = 0.13x + 3.76$ approximates the average movie ticket price (in dollars) $x$ years after 1989. (*Source:* National Association of Theater Owners)

**(a)** Identify the dependent and independent variables.
**(b)** Evaluate $P(15)$ and explain what it represents.

**17.** Using data from 1960 to 2000, the function $N(x) = 271.40x + 836.83$ represents the approximate number of registered climbers at Mt. Rainier $x$ years after 1960. (*Source:* National Parks Service, U.S. Dept. of the Interior)

**(a)** Identify the dependent and independent variables.
**(b)** Evaluate $N(43)$ and explain what it represents.
**(c)** There were 9714 registered climbers on Mt. Rainier in 2003. Compare this value to your result in part (b) and comment on any differences.

**18.** Find the domain of $f(x) = \dfrac{-15}{x + 2}$.

**19.** $h(x) = -5x + 12$

**(a)** Is the point $(2, 2)$ on the graph of the function?
**(b)** If $x = 3$, what is $h(x)$? What point is on the graph of the function?
**(c)** If $h(x) = 0$, what is $x$? What point is on the graph of $h$?

**20.** The following graph represents the speed of a car as a function of time.

**(a)** When does the car stop accelerating?
**(b)** For how long does the car maintain a constant speed?

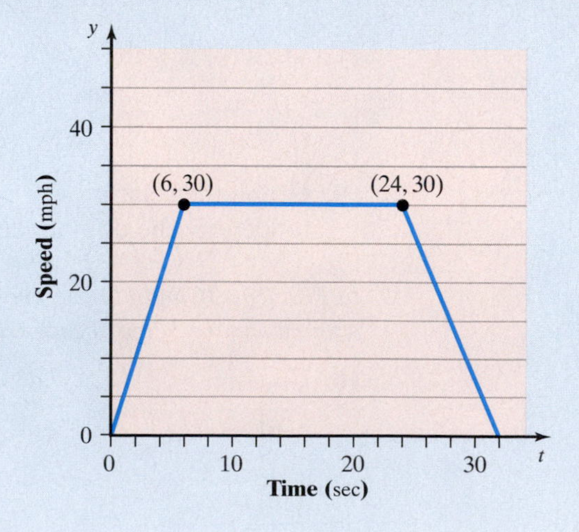

CHAPTER

# 3 Linear Functions and Their Graphs

Change is all around us, and mathematics is perfectly suited for explaining this change.

For example, mathematics can be used to explain the declining population of swordfish. The following quote is from http://www.bigmarinefish.com/swordfish.html: "For a long time, the north Atlantic swordfish stock has been declining more rapidly than any other marine species. It has been steadily doing so at the same rate each year for the past 20 years!" By stating that the rate of change is the same, we know that lines can be used to describe the declining population of swordfish. See Problem 103 in Section 3.2.

## OUTLINE



## The Big Picture: Putting It Together

In Chapter 2, we learned that the graph of an equation is the set of all ordered pairs $(x, y)$ such that the equation is a true statement. We also learned how to identify certain properties of the equation from its graph such as domain, range, and intercepts. We then went on to talk about the idea of a function. The presentation was general because it did not concentrate on any particular type of function.

In this chapter we introduce you to a specific type of function—the *linear function*. We are going to graph linear functions and learn some properties of linear functions such as the domain and the range. We also will learn how to determine the intercepts of the graph of a linear function and how to build linear functions from points in the Cartesian plane. Lastly, we talk about some of the interesting applications that use linear functions. These linear functions allow us to describe the world we live in and help us to make informed decisions.

# 3.1 Linear Equations and Linear Functions

**OBJECTIVES**

1. Graph Linear Equations Using Point Plotting
2. Graph Linear Equations Using Intercepts
3. Graph Vertical and Horizontal Lines
4. Work with Linear Functions

## *Preparing for Linear Equations and Linear Functions*

*Before getting started, take this readiness quiz. If you get a problem wrong, go back to the section cited and review the material.*

1. Solve for $x$: $3x + 12 = 0$ [Section 1.1, pp. 53–56]
2. Graph the equation $-2x + y = -4$ using the point-plotting method. [Section 2.1, pp. 140–142]
3. Based upon the graph found in Problem 2, determine whether the equation $-2x + y = -4$ is a function. [Section 2.3, pp. 161–162]
4. Given the function $f(x) = 2x^2 - 5$, determine $f(3)$. [Section 2.3, pp. 162–164]
5. Find the domain and range of the function whose graph is shown. [Section 2.4, pp. 170–172]

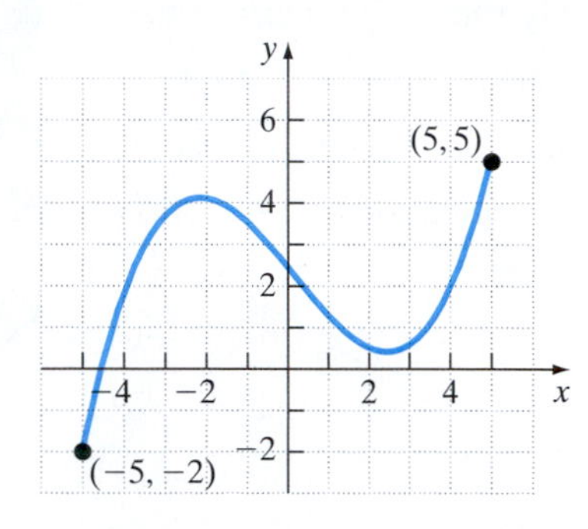


6. If $f(3) = 5$, what point is on the graph of the function? [Section 2.4, p. 172]

## 1 Graph Linear Equations Using Point Plotting

In Section 2.1 we discussed how to graph any equation using the point-plotting method. Remember, the graph of an equation is the set of all ordered pairs $(x, y)$ such that the equation is a true statement.

We are now going to learn methods for graphing a specific type of equation, called a *linear equation.*

**DEFINITION**

A **linear equation** in two variables is an equation of the form

$$Ax + By = C$$

where $A$, $B$, and $C$ are real numbers. $A$ and $B$ cannot both be 0.

When a linear equation is written in the form $Ax + By = C$, we say that the linear equation is in **standard form.**

Some examples of linear equations in standard form are

$$3x - 4y = 9 \qquad \frac{1}{2}x + \frac{2}{3}y = 4 \qquad 3x = 9 \qquad -2y = 5$$

The graph of a linear equation is a **line.** Let's graph a linear equation using the point-plotting method.

***Preparing for...Answers*** **1.** $\{-4\}$

**2.**

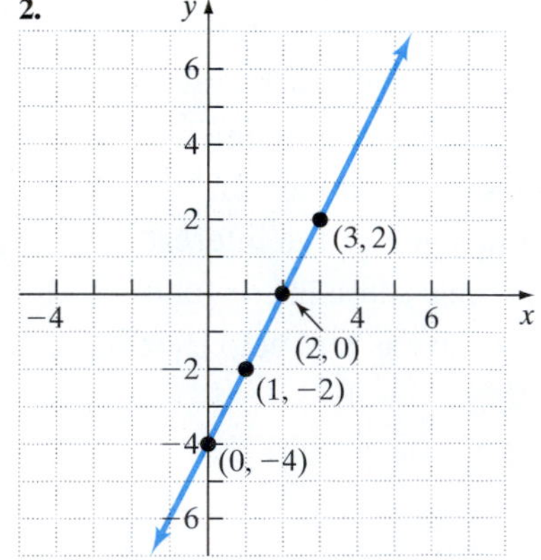


**3.** A function **4.** 13
**5.** Domain: $\{x \mid -5 \le x \le 5\}$ or $[-5, 5]$; Range: $\{y \mid -2 \le y \le 5\}$ or $[-2, 5]$
**6.** $(3, 5)$

## EXAMPLE 1 Graphing a Linear Equation Using the Point-Plotting Method

Graph the linear equation $4x + 2y = 6$.

### Solution

To graph the linear equation, we choose various values for $x$ and then use the equation to find the corresponding values of $y$. For this equation, we will let $x = -2, -1, 0,$ and 1.

**Work Smart**

In Example 1, we chose to pick $x$-values and use the equation to find the corresponding $y$-values, however we could also have chosen a $y$-value and used the equation to find the corresponding $x$-value.

$x = -2$: $4(-2) + 2y = 6$
$-8 + 2y = 6$
Add 8 to both sides: $2y = 14$
Divide both sides by 2: $y = 7$
$(-2, 7)$ is on the graph

$x = -1$: $4(-1) + 2y = 6$
$-4 + 2y = 6$
Add 4 to both sides: $2y = 10$
Divide both sides by 2: $y = 5$
$(-1, 5)$ is on the graph

$x = 0$: $4(0) + 2y = 6$
$0 + 2y = 6$
$2y = 6$
Divide both sides by 2: $y = 3$
$(0, 3)$ is on the graph

$x = 1$: $4(1) + 2y = 6$
$4 + 2y = 6$
Subtract 4 from both sides: $2y = 2$
Divide both sides by 2: $y = 1$
$(1, 1)$ is on the graph

Table 1 shows the points that are on the graph of $4x + 2y = 6$. We plot the ordered pairs $(-2, 7)$, $(-1, 5)$, $(0, 3)$, and $(1, 1)$ in the Cartesian plane and connect the points in a straight line. See Figure 1.

**Table 1**

| $x$ | $y$ | $(x, y)$ |
|---|---|---|
| −2 | 7 | (−2, 7) |
| −1 | 5 | (−1, 5) |
| 0 | 3 | (0, 3) |
| 1 | 1 | (1, 1) |

**Figure 1**



**Work Smart**

We recommend that you find at least three points to be sure your graph is correct. Also, remember that a complete graph is a graph that shows all the interesting features of the graph, such as its intercepts.

One of the problems with using the point-plotting method to graph equations is determining how many points need to be plotted before we obtain a complete graph. Based on the results of Example 1, we can see that only two points are required to obtain a complete graph of a linear equation. To guard against making an error, however, you should plot at least three points.

**QUICK ✓** *Graph each linear equation using the point-plotting method.*

**1.** $y = 2x - 3$ **2.** $\frac{1}{2}x + y = 2$ **3.** $-6x + 3y = 12$

## 2 Graph Linear Equations Using Intercepts

In Section 2.1, we said any complete graph should display the intercepts, if any. Recall, the intercepts of the graph of an equation are the points, if any, where the graph crosses or touches the coordinate axes. See Figure 2.

**Figure 2**

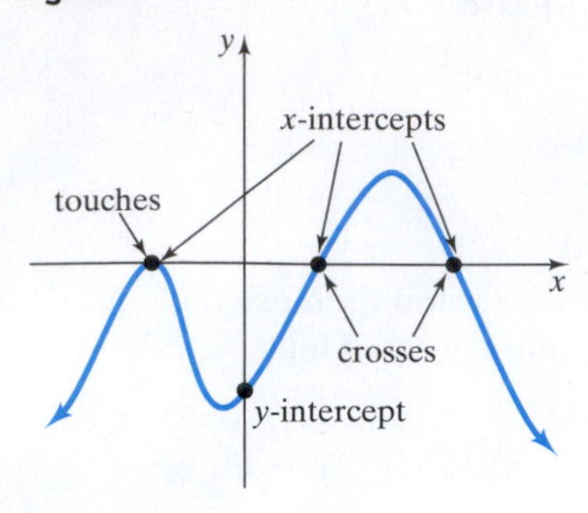


Now we will explain how to find the intercepts algebraically. From Figure 2 it is apparent that an $x$-intercept exists when the value of $y$ is 0 and that a $y$-intercept exists when the value of $x$ is 0. This leads to the following procedure for finding intercepts.

### Procedure for Finding Intercepts

- To find the $x$-intercept(s), if any, of the graph of an equation, let $y = 0$ in the equation and solve for $x$.
- To find the $y$-intercept(s), if any, of the graph of an equation, let $x = 0$ in the equation and solve for $y$.

The procedure given above can be used to find the intercepts of any type of equation. Let's use this procedure to find the intercepts of a linear equation.

## EXAMPLE 2 Graphing a Linear Equation by Finding Its Intercepts

Graph the linear equation $3x + 2y = 12$ by finding its intercepts.

### Solution

To find the $y$-intercept, we let $x = 0$ and solve the equation $3x + 2y = 12$ for $y$.

Let $x = 0$: $3(0) + 2y = 12$

$0 + 2y = 12$

$2y = 12$

Divide both sides by 2: $y = 6$

**Work Smart**

Linear equations in one variable have no solution, one solution, or infinitely many solutions. Because the procedure for finding intercepts of linear equations in two variables results in a linear equation in one variable, linear equations can have no $x$-intercepts, one $x$-intercept, or infinitely many $x$-intercepts. The same applies to $y$-intercepts.

The $y$-intercept is 6, so the point $(0, 6)$ is on the graph of the equation.

To find the $x$-intercept, we let $y = 0$ and solve the equation $3x + 2y = 12$ for $x$.

Let $y = 0$: $3x + 2(0) = 12$

$3x + 0 = 12$

$3x = 12$

Divide both sides by 3: $x = 4$

The $x$-intercept is 4, so the point $(4, 0)$ is on the graph of the equation. We obtain one additional point on the graph by letting $x = 2$ (or any other value of $x$ besides 0 or 4), and find $y$ to be 3. Table 2 shows the points that are on the graph of $3x + 2y = 12$. We plot the points $(0, 6)$, $(4, 0)$, and $(2, 3)$. Connect the points in a straight line and obtain the graph in Figure 3.

**Table 2**

| $x$ | $y$ | $(x, y)$ |
|---|---|---|
| 0 | 6 | (0, 6) |
| 4 | 0 | (4, 0) |
| 2 | 3 | (2, 3) |

**Figure 3**

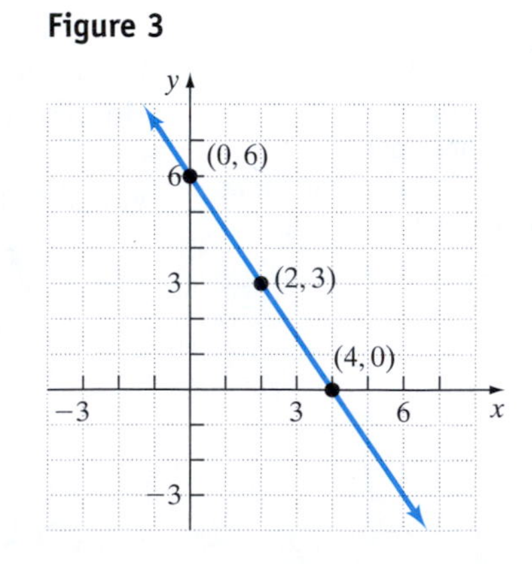


**QUICK ✓** *Graph each linear equation by finding its intercepts.*

**4.** $x + y = 4$ **5.** $4x - 5y = 20$

## EXAMPLE 3 Graphing a Linear Equation by Finding Its Intercepts

Graph the linear equation $x + 3y = 0$ by finding its intercepts.

### Solution

To find the $y$-intercept, we let $x = 0$ and solve the equation $x + 3y = 0$ for $y$.

$$\begin{aligned} \text{Let } x = 0: \quad 0 + 3y &= 0 \\ 3y &= 0 \\ \text{Divide both sides by 3:} \quad y &= 0 \end{aligned}$$

The $y$-intercept is 0, so the point $(0, 0)$ is on the graph of the equation.

To find the $x$-intercept, we let $y = 0$ and solve the equation $x + 3y = 0$ for $x$.

$$\begin{aligned} \text{Let } y = 0: \quad x + 3(0) &= 0 \\ x + 0 &= 0 \\ x &= 0 \end{aligned}$$

The $x$-intercept is 0, so the point $(0, 0)$ is on the graph of the equation.

**Work Smart**

We chose $x = -3$ and $x = 3$, to avoid fractions. This makes plotting the points easier.

Because both the $x$- and $y$-intercepts are 0, we will find *two* additional points on the graph of the equation. By letting $x = 3$, we find that $y = -1$. By letting $x = -3$, we find that $y = 1$. We plot the points $(0, 0)$, $(-3, 1)$, and $(3, -1)$, connect the points in a straight line and obtain the graph in Figure 4.

**Figure 4**

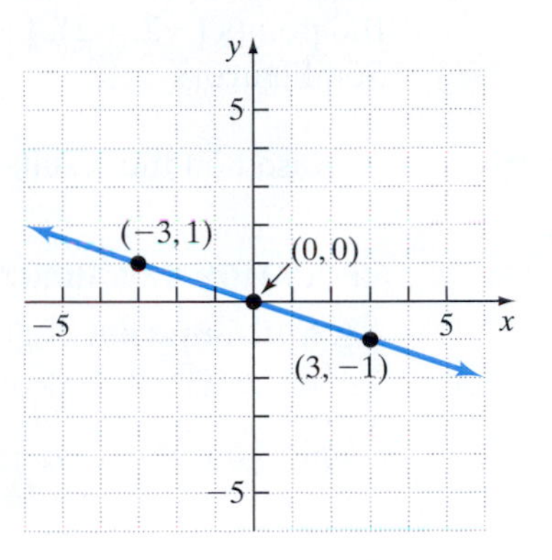


**Work Smart**

Linear equations of the form $Ax + By = 0$, where $A \neq 0$ and $B \neq 0$, have only one intercept at (0, 0).

From Example 3, we learn that any equation of the form $Ax + By = 0$, where $A \neq 0$ and $B \neq 0$, has only one intercept at $(0, 0)$. Therefore, to graph equations of this form, we find two additional points on the graph.

**QUICK ✓** *Graph the equation by finding its intercepts.*

**6.** $3x - 2y = 0$

## 3 Graph Vertical and Horizontal Lines

In the equation of a line, $Ax + By = C$, we said that $A$ and $B$ cannot both be zero. But what if $A = 0$ or $B = 0$?

## EXAMPLE 4 Graphing a Vertical Line

Graph the equation $x = 3$ using the point-plotting method.

### Solution

Because the equation $x = 3$ can be written as $1x + 0y = 3$, we know that the graph is a line. When you look at the equation $x = 3$, it should be clear to you that no matter what value of $y$ we choose, the corresponding value of $x$ is going to be 3. Therefore, the points $(3, -2)$, $(3, -1)$, $(3, 0)$, $(3, 1)$, and $(3, 2)$ are all points on the line. See Figure 5.

**Figure 5**
$x = 3$

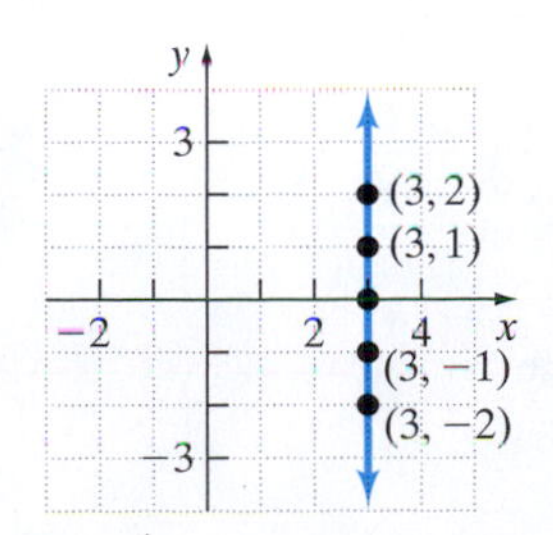

Based on the results of Example 4, we can write the equation of a vertical line:

**EQUATION OF A VERTICAL LINE**

A **vertical line** is given by an equation of the form

$$x = a$$

where $a$ is the $x$-intercept.

Now let's look at equations that lead to graphs that are horizontal lines.

### EXAMPLE 5 Graphing a Horizontal Line

Graph the equation $y = -2$ using the point-plotting method.

**Solution**

Because the equation $y = -2$ can be written as $0x + 1y = -2$, we know that the graph is a line. In looking at the equation $y = -2$, it should be clear that no matter what value of $x$ we choose, the corresponding value of $y$ is going to be $-2$. Therefore, the points $(-2, -2)$, $(-1, -2)$, $(0, -2)$, $(1, -2)$, and $(2, -2)$ are all points on the line. See Figure 6.

**Figure 6**
$y = -2$

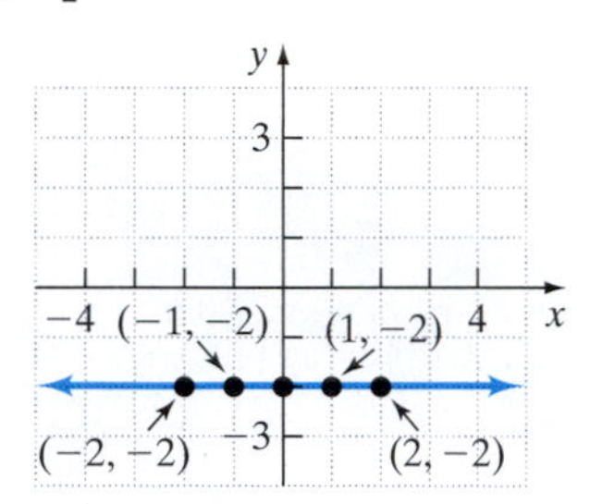


Based on the results of Example 5, we can generalize:

**EQUATION OF A HORIZONTAL LINE**

A **horizontal line** is given by an equation of the form

$$y = b$$

where $b$ is the $y$-intercept.

**QUICK ✓** *Graph each equation.*

**7.** $x = 5$ **8.** $y = -4$

## 4 Work with Linear Functions

If you look at all the linear equations that we graphed in Examples 1–5, you should notice that all except one pass the vertical line test for identifying the graph of a function. The exception is a vertical line itself. We conclude the following: **All linear equations except for equations of the form $x = a$ are functions.**

Because all linear equations except for those of the form $x = a$ are functions, we can write any linear equation that is in the form $Ax + By = C$ using function notation provided that $B \neq 0$ as follows:

$$Ax + By = C \qquad B \neq 0$$

Subtract $Ax$ from both sides: $$By = -Ax + C$$

Divide both sides by $B$: $$\frac{By}{B} = \frac{-Ax + C}{B}$$

Simplify: $$y = -\frac{A}{B}x + \frac{C}{B}$$

$$\updownarrow \quad \updownarrow \quad \updownarrow$$

$$f(x) = mx + b$$

This leads to the following definition.

**DEFINITION**

A **linear function** is a function of the form

$$f(x) = mx + b$$

where $m$ and $b$ are real numbers. The graph of a linear function is called a **line.**

There are many applications of linear functions. For example, the cost of cab fare, sales commissions, or the cost of breakfast as a function of the number of eggs ordered can all be modeled by a linear function.

## EXAMPLE 6 Sales Commissions

Tony's weekly salary at Apple Chevrolet is 0.75% of his weekly sales plus \$450. The linear function $S(x) = 0.0075x + 450$ describes Tony's weekly salary $S$ as a linear function of his weekly sales $x$.

**(a)** What is the implied domain of the function?

**(b)** Draw a graph of the function found in part (a).

**(c)** If Tony sells cars worth a total of \$55,000 one week, what is his salary?

**(d)** If Tony earned \$723.75 one week, what was the value of the cars that he sold?

### Solution

**(a)** The independent variable is weekly sales, $x$. Because it does not make sense to talk about negative weekly sales, we have that the domain of the function is $\{x | x \geq 0\}$ or, using interval notation, $[0, \infty)$.

**(b)** We plot the independent variable, *weekly sales,* on the horizontal axis and the dependent variable, *salary,* on the vertical axis. We graph the equation by plotting points. To obtain points on the graph of the function, evaluate the function for $x = \$0, \$10,000$, and \$20,000.

$$S(0) = 0.0075(0) + 450 = 450$$
(0, 450) is on the graph

$$S(10{,}000) = 0.0075(10{,}000) + 450 = \$525$$
(10000, 525) is on the graph

$$S(20{,}000) = 0.0075(20{,}000) + 450 = \$600$$
(20000, 600) is on the graph

We plot these points and obtain the graph of the linear function shown in Figure 7.

**Work Smart**

Always label the coordinate axes when graphing.

**Work Smart**

The independent variable is always graphed on the horizontal axis and the dependent variable is always graphed on the vertical axis.

Figure 7

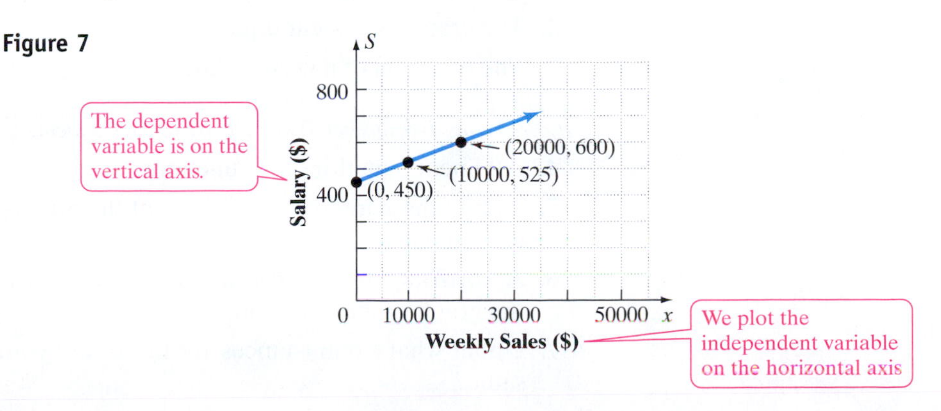

**(c)** We evaluate the function at $x = \$55{,}000$ to obtain

$$S(55000) = 0.0075(55000) + 450$$
$$= \$862.50$$

Tony will earn \$862.50 for the week if he sells \$55,000 worth of cars.

**(d)** Here, we need to solve the equation $S(x) = 723.75$.

$$0.0075x + 450 = 723.75$$

Subtract 450 from both sides: $0.0075x = 273.75$

Divide both sides by 0.0075: $x = \$36{,}500$

If Tony sells \$36,500 worth of cars in a week, he will earn \$723.75.

Notice a few details regarding the graph of the linear function in Figure 7. First, we only graph the function over its domain, $[0, \infty)$, so that in this case we only graph in quadrant I. Also notice that we labeled the horizontal axis $x$ for the independent variable, *weekly sales,* and we labeled the vertical axis $S$ for the dependent variable, *salary.* We also indicated what the independent and dependent variable represent on each coordinate axis. It is always a good practice to label your axes.

**QUICK ✓**

**9.** The cost, $C$, of renting a 12-foot moving truck for a day is \$40 plus \$0.35 times the number of miles driven. The linear function $C(x) = 0.35x + 40$ describes the cost $C$ of driving the truck $x$ miles.

**(a)** What is the implied domain of this linear function?

**(b)** Determine the $y$-intercept of the graph of the linear function.

**(c)** What is the rental cost if the truck is driven 80 miles?

**(d)** Graph the linear function.

**(e)** How many miles was the truck driven if the rental cost is \$85.50? (**Hint:** Solve the equation $C(x) = 85.5$.)

# 3.1 Exercises

**For Extra Help:** Student Solutions Manual · CD Video · PH Math/Tutor Center · MathXL Tutorials on CD · MathXL® · MyMathLab

## Concepts and Vocabulary

*In Problems 1–3, fill in the blanks.*

**1.** A(n) __________ __________ is an equation of the form $Ax + By = C$, where $A$, $B$, and $C$ are real numbers. $A$ and $B$ cannot both be 0.

**2.** The graph of a linear equation is called a __________.

**3.** The equation of a vertical line is __________ where __________ is the $x$-intercept.

*In Problems 4–6, answer True or False to each statement.*

**4.** All linear equations are functions.

**5.** To find the $x$-intercept(s), if any, of the graph of an equation, let $y = 0$ in the equation and solve for $x$.

**6.** A horizontal line is an equation of the form $y = b$ where $b$ is the $x$-intercept of the graph of the equation.

**7.** Under what circumstances are the $x$- and $y$-intercepts of a linear function the same?

**8.** Can there be a linear function that has an $x$-intercept, but no $y$-intercept?

## Building Skills

*In Problems 9–18, graph each linear equation by plotting points.*

**9.** $x + y = 5$

**10.** $x + y = -3$

**11.** $x - 2y = 6$

**12.** $2x - y = -8$

**13.** $3x + 2y = 12$

**14.** $-5x + y = 10$

**15.** $\frac{2}{3}x + y = 6$

**16.** $2x - \frac{3}{2}y = 10$

**17.** $5x - 3y = 6$

**18.** $-7x + 3y = 9$

*In Problems 19–34, graph each linear equation by finding its intercepts.*

**19.** $x + y = 3$

**20.** $x + y = -2$

**21.** $3x + y = 6$

**22.** $-2x + y = 4$

**23.** $5x - 3y = 15$

**24.** $-4x + 3y = 24$

**25.** $-3x + 2y = 7$

**26.** $7x - 2y = 10$

**27.** $\frac{1}{3}x - \frac{1}{2}y = 1$

**28.** $\frac{1}{4}x + \frac{1}{5}y = 2$

**29.** $2x + y = 0$

**30.** $4x + 3y = 0$

**31.** $-3x - y = 0$

**32.** $5x - 3y = 0$

**33.** $\frac{2}{3}x - \frac{1}{2}y = 0$

**34.** $-\frac{3}{2}x + \frac{3}{4}y = 0$

*In Problems 35–42, graph each linear equation.*

**35.** $x = 3$

**36.** $x = 5$

**37.** $y = 1$

**38.** $y = 6$

**39.** $x = -5$

**40.** $x = -8$

**41.** $2y + 8 = -6$

**42.** $3y + 20 = -10$

## Applying the Concepts

**43. Taxes** The function $T(x) = 0.15(x - 7300) + 730$ represents the tax bill $T$ of a single person whose adjusted gross income in 2005 is $x$ dollars for income between \$7300 and \$29,700, inclusive. (*Source:* Internal Revenue Service)

**(a)** What is the implied domain of this linear function?
**(b)** What is a single filer's tax bill if adjusted gross income is \$20,000?
**(c)** Which variable is independent and which is dependent?
**(d)** Graph the linear function over the domain specified in part (a).
**(e)** What is a single filer's adjusted gross income if their tax bill is \$3385? (**Hint:** Solve the equation $T(x) = 1650$.)

**44. Sales Commissions** Tanya works for Prentice Hall as a book representative. The linear function $I(s) = 0.01s + 20{,}000$ describes the annual income $I$ of Tanya when she has total sales $s$.

**(a)** What is the implied domain of this linear function?
**(b)** What is $I(0)$? Explain what this result means.
**(c)** What is Tanya's salary if she sells \$500,000 in books for the year?
**(d)** Graph the linear function.
**(e)** At what level of sales will Tanya's income be \$45,000? (**Hint:** Solve the equation $I(s) = 45{,}000$.)

**45. Cab Fare** The linear function $C(m) = 1.5m + 2$ describes the cab fare $C$ for a ride of $m$ miles.

**(a)** What is the implied domain of this linear function?
**(b)** What is $C(0)$? Explain what this result means.
**(c)** What is cab fare for a 5-mile ride?
**(d)** Graph the linear function.
**(e)** How many miles can you ride in a cab if you have \$13.25?

**46. Luxury Tax** In 2002, Major League Baseball signed a labor agreement with the players. In this agreement, any team whose payroll exceeds \$128 million in 2005 will have to pay a luxury tax of 22.5% (for first time offenses). The linear function $T(p) = 0.225(p - 128)$ describes the luxury tax $T$ of a team whose payroll is $p$ (in millions).

**(a)** What is the implied domain of this linear function?
**(b)** What is the luxury tax for a team whose payroll is \$160 million?
**(c)** Graph the linear function.
**(d)** What is the payroll of a team that pays luxury tax of \$11.7 million?

**47. Health Costs** The annual cost of health insurance $H$ as a function of age $a$ is given by the function $H(a) = 22.8a - 117.5$ for $15 \le a \le 90$. (*Source: Statistical Abstract,* 2005)

**(a)** What are the independent and dependent variables?
**(b)** What is the domain of this linear function?
**(c)** What is the health insurance premium of a 30 year old?
**(d)** Graph the linear function over its domain.
**(e)** What is the age of an individual whose health insurance premium is \$976.90?

**48. Birth Rate** A multiple birth is any birth with 2 or more children born. The birth rate is the number of births per 1000 women. The birth rate $B$ of multiple births as a function of age $a$ is given by the function $B(a) = 1.73a - 14.56$ for $15 \le a \le 44$. (*Source:* Centers for Disease Control)

**(a)** What are the independent and dependent variables?
**(b)** What is the domain of this linear function?
**(c)** What is the multiple birth rate of women who are 22 years of age according to the model?
**(d)** Graph the linear function over its domain.
**(e)** What is the age of women whose multiple birth rate is 49.45?

## Extending the Concepts

**49.** In parts (a)–(e), use the figure shown to the right.

**(a)** Solve $f(x) = 1$.
**(b)** Solve $f(x) = -3$.
**(c)** Solve $f(x) = 2$.
**(d)** What are the intercepts of the function $y = f(x)$?
**(e)** The **zero** of a function is any value of the independent variable that causes the value of the function to be zero. What is the zero of the function?

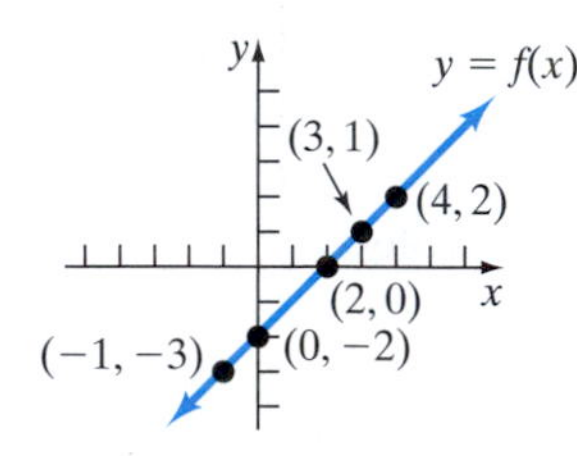


**50.** In parts (a)–(e), use the figure shown to the right.

**(a)** Solve $g(x) = 1$.
**(b)** Solve $g(x) = -1$.
**(c)** Solve $g(x) = 4$.
**(d)** What are the intercepts of the function $y = g(x)$?
**(e)** See Problem 49(e). What is the zero of the function?

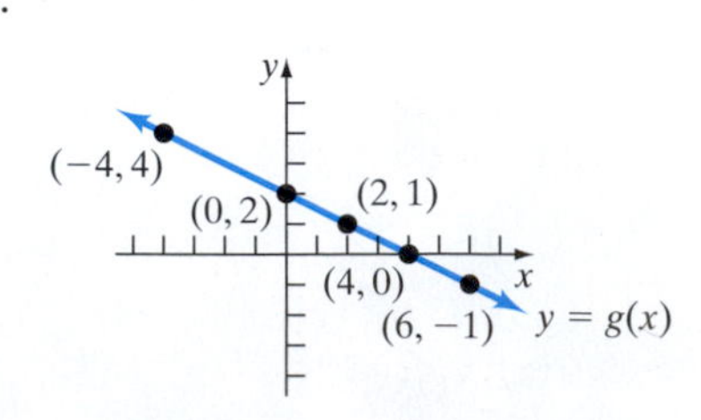

**51. Exploration** Graph $y = 2x$, $y = 2x + 3$, $y = 2x + 7$, and $y = 2x - 4$ on the same Cartesian plane. What pattern do you observe? In general, describe the graph of $y = 2x + b$.

**52. Exploration** Graph $y = \frac{1}{2}x$, $y = x$, and $y = 2x$ on the same Cartesian plane. What pattern do you observe? In general, describe the graph of $y = ax$ with $a > 0$.

**53. Exploration** Graph $y = -\frac{1}{2}x$, $y = -x$, and $y = -2x$ on the same Cartesian plane. What pattern do you observe? In general, describe the graph of $y = ax$ with $a < 0$.

**54. Cab Fare Revisited** In Problem 45, we used the linear function $C(m) = 1.5m + 2$ to model the cab fare, $C$, for a ride of $m$ miles. It is recommended that the cabbie is given 15% of the fare as a tip. The total cost of the trip $T$ as a function of miles driven would be the cost $C$ plus 15% of $C$. Find a linear function relating total cost $T$ to the number of miles driven, $m$.

### The Graphing Calculator

*In Problems 55–66, graph each linear equation using a graphing calculator.*

**55.** $x + y = 3$ **56.** $x + y = -2$ **57.** $3x + y = 6$ **58.** $-2x + y = 4$

**59.** $5x - 3y = 15$ **60.** $-4x + 3y = 24$ **61.** $-3x + 2y = 7$ **62.** $7x - 2y = 10$

**63.** $\frac{1}{3}x - \frac{1}{2}y = 1$ **64.** $\frac{1}{4}x + \frac{1}{5}y = 2$ **65.** $2x + y = 0$ **66.** $4x + 3y = 0$

# 3.2 Slope and Equations of Lines

**OBJECTIVES**

1. Find the Slope of a Line Given Two Points
2. Interpret Slope as an Average Rate of Change
3. Graph a Line Given a Point and Its Slope
4. Use the Point-Slope Form of a Line
5. Identify the Slope and $y$-Intercept of a Line from Its Equation
6. Find the Equation of a Line Given Two Points
7. Build Linear Models Using the Point-Slope Form of a Line

### *Preparing for Slope and Equations of Lines*

*Before getting started, take the following readiness quiz. If you get a problem wrong, go back to the section cited and review the material.*

1. Evaluate: $\dfrac{5 - 2}{-2 - 4}$ [Section R.4, pp. 36–37]
2. Distribute: $-2(x + 3)$ [Section R.3, p. 30]
3. Solve for $y$: $4x + 3y = 15$ [Section 1.3, pp. 80–83]
4. In the function $F(z) = 4z - 3$, what is the independent variable? What is the dependent variable? [Section 2.3, p. 163]

In Section 3.1, we learned that two points are required to graph a line (although we usually find three points for accuracy). In this section, we learn how to determine an equation of a line given two points. The first step in doing this is learning about *slope*.

## 1 Find the Slope of a Line Given Two Points

Consider the staircase drawn in Figure 8(a) on the following page. If we draw a line at the top of each riser on the staircase (in blue), we can see that each step contains exactly the same horizontal **run** and the same vertical **rise.** We call the ratio of the

*Preparing for...Answers* **1.** $-\frac{1}{2}$ **2.** $-2x - 6$ **3.** $y = -\frac{4}{3}x + 5$ **4.** $z$; $F$

rise to the run the *slope* of the line. It is a numerical measure of the steepness of the line. For example, suppose that the staircase in Figure 8(a) has a run of 7 inches and a rise of 6 inches. Then the slope of the line is $\frac{\text{rise}}{\text{run}} = \frac{6 \text{ inches}}{7 \text{ inches}}$. If the run of the stair is increased to 10 inches, while the rise remains the same, then the slope of the line is $\frac{\text{rise}}{\text{run}} = \frac{6 \text{ inches}}{10 \text{ inches}}$. See Figure 8(b). If the run is decreased to 4 inches and the rise remains the same, then the slope of the line is $\frac{\text{rise}}{\text{run}} = \frac{6 \text{ inches}}{4 \text{ inches}}$. See Figure 8(c).

**Figure 8**

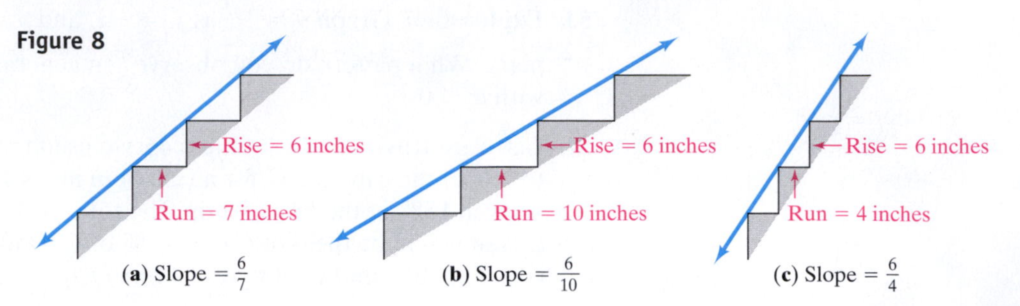


We can define the slope of a line using rectangular coordinates.

**DEFINITION**

Let $P = (x_1, y_1)$ and $Q = (x_2, y_2)$ be two distinct points. If $x_1 \neq x_2$, the **slope *m*** of the nonvertical line $L$ containing $P$ and $Q$ is defined by the formula

$$m = \frac{y_2 - y_1}{x_2 - x_1}, \quad x_1 \neq x_2$$

If $x_1 = x_2$, then $L$ is a vertical line and the slope $m$ of $L$ is **undefined** (since this results in division by 0).

The accepted symbol for the slope of a line is $m$. It comes from the French word *monter,* which means to ascend or climb. Figure 9(a) provides an illustration of the slope of a nonvertical line; Figure 9(b) illustrates a vertical line.

**Figure 9**

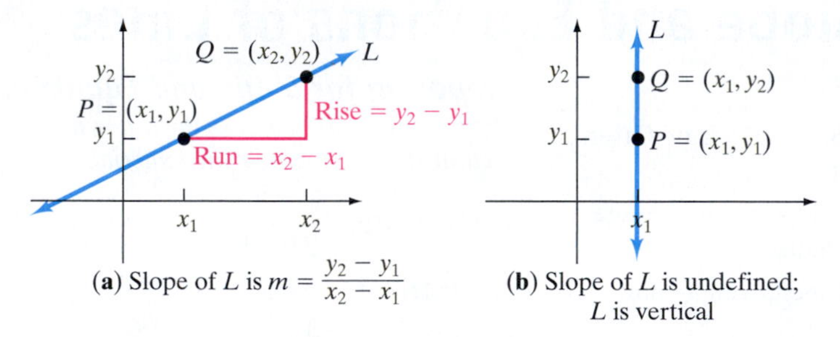


From Figure 9(a) we can see that the slope $m$ of a nonvertical line may be viewed as

$$m = \frac{y_2 - y_1}{x_2 - x_1} = \frac{\text{Rise}}{\text{Run}}$$

**In Words**
Slope is the change in $y$ divided by the change in $x$.

We can also write the slope $m$ of a nonvertical line as

$$m = \frac{y_2 - y_1}{x_2 - x_1} = \frac{\text{Change in } y}{\text{Change in } x} = \frac{\Delta y}{\Delta x}$$

**Work Smart**
The symbol $\Delta$ comes from the Greek word *dunamis*, which means "change" or "power."

The symbol $\Delta$ is the Greek letter delta. In mathematics, we read the symbol $\Delta$ as "change in." So the notation $\frac{\Delta y}{\Delta x}$ is read "change in $y$ divided by change in $x$."

The slope $m$ of a nonvertical line measures the amount that $y$ changes (the vertical change) as $x$ changes from $x_1$ to $x_2$ (the horizontal change). The slope $m$ of a vertical line is undefined since it results in division by zero.

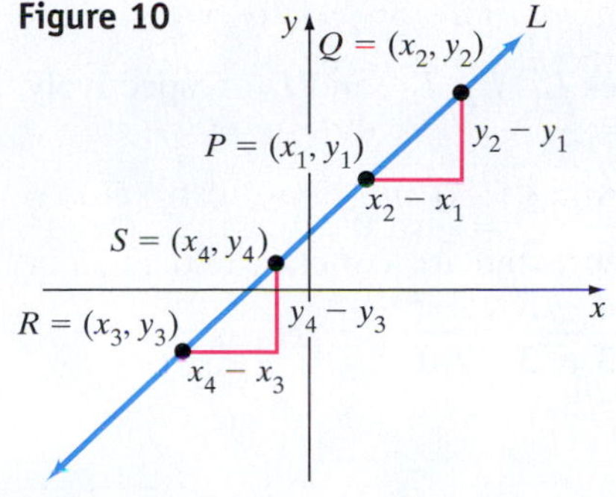

Figure 10

**Work Smart**

It doesn't matter whether we compute the slope of the line from point $P$ to $Q$ or from point $Q$ to $P$.

## Comments Regarding the Slope of a Nonvertical Line

**1.** Any two different points on the line can be used to compute the slope of the line shown in Figure 10. The slope $m$ of the line $L$ is given by

$$m = \frac{y_2 - y_1}{x_2 - x_1} \quad \text{or} \quad m = \frac{y_4 - y_3}{x_4 - x_3}$$

This result is due to the fact that the two triangles formed in Figure 10 are similar (the measure of the angles is the same in both triangles). Therefore, the ratios of the sides are proportional.

**2.** The slope of a line may be computed from $P = (x_1, y_1)$ to $Q = (x_2, y_2)$ or from $Q$ to $P$ because

$$m = \frac{y_2 - y_1}{x_2 - x_1} = \frac{-(y_1 - y_2)}{-(x_1 - x_2)} = \frac{y_1 - y_2}{x_1 - x_2}$$

### EXAMPLE 1 Finding and Interpreting the Slope of a Line

Find and interpret the slope of the line containing the points (3, 6) and (−2, 2).

**Solution**

We plot the points $P = (x_1, y_1) = (3, 6)$ and $Q = (x_2, y_2) = (-2, 2)$ and draw a line through the points as shown in Figure 11. The slope of the line drawn in Figure 11 is

$$m = \frac{y_2 - y_1}{x_2 - x_1} = \frac{2 - 6}{-2 - 3} = \frac{-4}{-5} = \frac{4}{5}$$

We could also compute the slope as

$$m = \frac{y_1 - y_2}{x_1 - x_2} = \frac{6 - 2}{3 - (-2)} = \frac{4}{5}$$

Figure 11

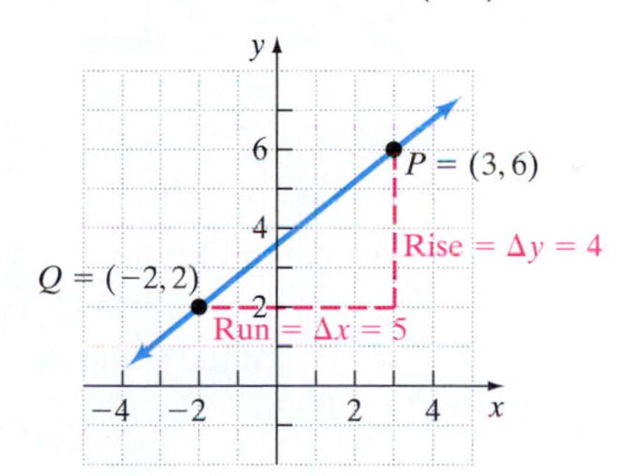


A slope of $\frac{4}{5}$ can be interpreted as: For every 5-unit increase in $x$, $y$ will increase by 4 units. Or for every 5-unit decrease in $x$, $y$ will decrease by 4 units.

**QUICK ✓** *Find and interpret the slope of the line containing the points.*

**1.** (0, 3); (3, 12) **2.** (−1, 3); (3, −4) **3.** (3, 2); (−3, 2) **4.** (−2, 4); (−2, −1)

### EXAMPLE 2 Finding Slopes of Different Lines Each of Which Contain (3, 5)

Find the slope of the lines $L_1$, $L_2$, $L_3$, and $L_4$ containing the following pairs of points. Graph the lines in the Cartesian plane.

| | | |
|---|---|---|
| $L_1$: | $P(3, 5)$ | $Q_1(5, 8)$ |
| $L_2$: | $P(3, 5)$ | $Q_2(6, 5)$ |
| $L_3$: | $P(3, 5)$ | $Q_3(5, -2)$ |
| $L_4$: | $P(3, 5)$ | $Q_4(3, 0)$ |

### Solution

Let $m_1$, $m_2$, $m_3$, and $m_4$ denote the slopes of the lines $L_1$, $L_2$, $L_3$, and $L_4$, respectively. Then

$$m_1 = \frac{8-5}{5-3} = \frac{3}{2} \qquad m_2 = \frac{5-5}{6-3} = \frac{0}{3} = 0$$

$$m_3 = \frac{-2-5}{5-3} = \frac{-7}{2} = -\frac{7}{2} \qquad m_4 = \frac{0-5}{3-3} = \frac{-5}{0} \quad \text{undefined}$$

The graphs of the four lines are shown in Figure 12.

Figure 12

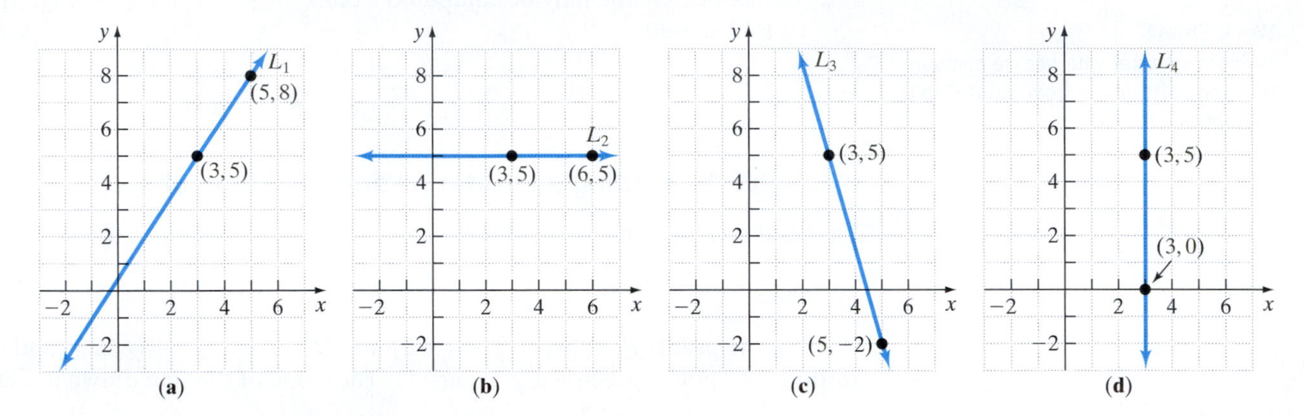


Relying on Figure 12, we have the following properties of slope:

**PROPERTIES OF SLOPE**

- When the slope of a line is positive, the line slants upward from left to right, as shown by $L_1$ in Figure 12(a).
- When the slope of a line is zero, the line is horizontal, as shown by $L_2$ in Figure 12(b).
- When the slope of a line is negative, the line slants downward from left to right, as shown by $L_3$ in Figure 12(c).
- When the slope of a line is undefined, the line is vertical, as shown by $L_4$ in Figure 12(d).

**QUICK ✓**

**5.** Find the slope of the lines $L_1$, $L_2$, $L_3$, and $L_4$ containing the following pairs of points. Graph all four lines on the same Cartesian plane.

$L_1$: $P(1, 3)$ $Q_1(6, 4)$ $\qquad$ $L_2$: $P(1, 3)$ $Q_2(1, 8)$

$L_3$: $P(1, 3)$ $Q_3(-3, 7)$ $\qquad$ $L_4$: $P(1, 3)$ $Q_4(-4, 3)$

## 2 Interpret Slope as an Average Rate of Change

The slope $m$ of a nonvertical line measures the amount that $y$ changes as $x$ changes from $x_1$ to $x_2$. The slope of a line is also called the **average rate of change** of $y$ with respect to $x$.

In applications, we are often interested in knowing how the change in one variable might impact some other variable. For example, if your income increases by $1000, how much will your spending (on average) change? Or, if the speed of your car increases by 10 miles per hour, how much (on average) will your car's gas mileage change?

## EXAMPLE 3 Slope as an Average Rate of Change

A strain of *E. coli* Beu 397-recA441 is placed into a Petri dish at 30° Celsius and allowed to grow. The data shown in Table 3 are collected. The population is measured in grams and the time in hours. The population growth is shown in Figure 13.

**Table 3**

| Time (hours), $x$ | Population (grams), $y$ |
|---|---|
| 0 | 0.09 |
| 1 | 0.12 |
| 2 | 0.16 |
| 3 | 0.22 |
| 4 | 0.29 |
| 5 | 0.39 |

SOURCE: *Dr. Polly Lavery, Joliet Junior College*

**Figure 13**

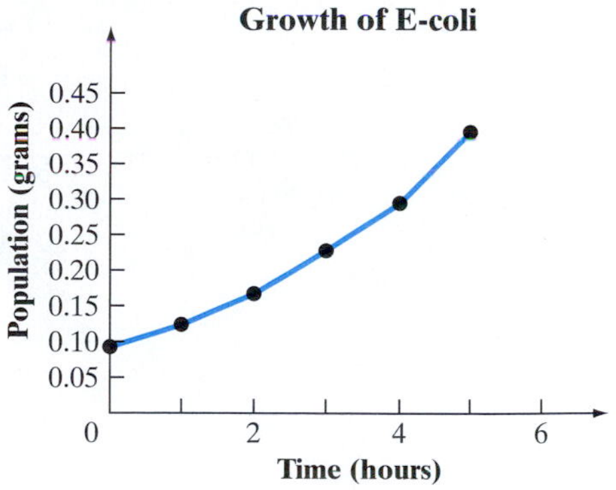

**(a)** Compute and interpret the average rate of change in the population between 0 and 1 hour.

**(b)** Compute and interpret the average rate of change in the population between 3 and 4 hours.

**(c)** Based upon your results to parts (a) and (b), do you think that the population grows linearly? Why?

### Solution

**(a)** To find the average rate of change, we compute the slope of the line between the points $(0, 0.09)$ and $(1, 0.12)$.

$$m = \text{average rate of change} = \frac{0.12 - 0.09}{1 - 0} = 0.03 \text{ gram per hour}$$

The population of *E. coli* was growing at the rate of 0.03 gram per hour between 0 and 1 hour.

**(b)** We want to compute the slope of the line between the points $(3, 0.22)$ and $(4, 0.29)$.

$$m = \text{average rate of change} = \frac{0.29 - 0.22}{4 - 3} = 0.07 \text{ gram per hour}$$

The population of *E. coli* was growing at the rate of 0.07 gram per hour between 3 and 4 hours.

**(c)** The population is not growing linearly because the average rate of change (slope) is not constant. In fact, because the average rate of change is increasing as time passes, the population is growing more rapidly over time. ■

### QUICK ✓

| Number of Bicycles, $x$ | Total Revenue, $y$ |
|---|---|
| 0 | 0 |
| 25 | 28,000 |
| 60 | 45,000 |
| 102 | 53,400 |
| 150 | 59,160 |

**6.** The data to the left represent the total revenue that would be received from selling $x$ bicycles at Gibson's Bicycle Shop.

**(a)** Plot the ordered pairs $(x, y)$ on a graph and connect the points with straight lines.

**(b)** Compute and interpret the average rate of change in the revenue between 0 and 25 bicycles sold.

**(c)** Compute and interpret the average rate of change in the revenue between 102 and 150 bicycles sold.

**(d)** Based upon your results to parts (a), (b), and (c), do you think that the revenue grows linearly? Why?

## 3 Graph a Line Given a Point and Its Slope

We now illustrate how to use the slope of a line to graph lines.

### EXAMPLE 4 Graphing a Line Given a Point and Its Slope

Draw a graph of the line that contains the point $(1, 2)$ and has a slope of

**(a)** 3 **(b)** $-\frac{3}{2}$

**Solution**

**(a)** Because the slope $= \frac{\text{Rise}}{\text{Run}} = \frac{\Delta y}{\Delta x}$, we have that $3 = \frac{3}{1} = \frac{\Delta y}{\Delta x}$. This means that if $x$ increases by 1 unit, then $y$ will increase by 3 units. So, if we start at $(1, 2)$ and move 1 unit to the right and then 3 units up, we end up at the point $(2, 5)$. We then draw a line through the points $(1, 2)$ and $(2, 5)$ to obtain the graph of the line. See Figure 14.

**(b)** Because the slope $= \frac{\text{Rise}}{\text{Run}} = \frac{\Delta y}{\Delta x}$, we have that $-\frac{3}{2} = \frac{-3}{2} = \frac{\Delta y}{\Delta x}$. This means that if $x$ increases by 2 units, then $y$ will decrease by 3 units. So, if we start at $(1, 2)$ and move 2 units to the right and then 3 units down, we end up at the point $(3, -1)$. We then draw a line through the points $(1, 2)$ and $(3, -1)$ to obtain the graph of the line. See Figure 15.

It is perfectly acceptable to set $\frac{\Delta y}{\Delta x} = -\frac{3}{2} = \frac{3}{-2}$ so that we move left 2 units from $(1, 2)$ and then up 3 units. We would then end up at $(-1, 5)$, which is also on the graph of the line as indicated in Figure 15.

**Figure 14**

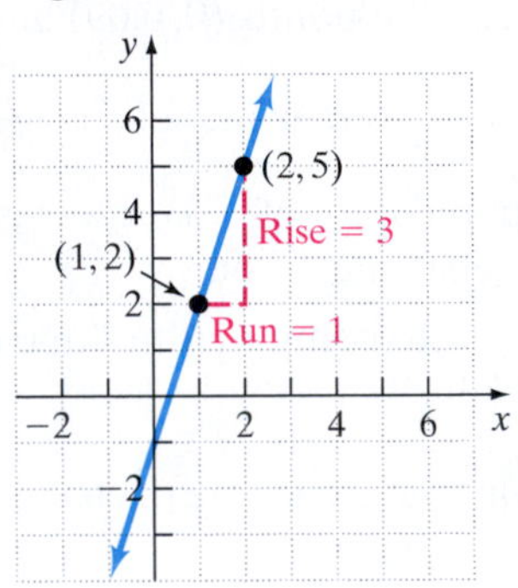


**Figure 15**

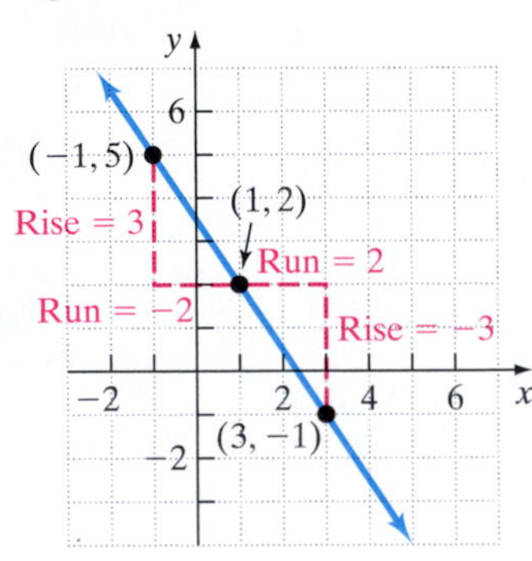


**QUICK ✓**

**7.** Draw a graph of the line that contains the point $(-1, 3)$ and has a slope of

**(a)** $\frac{1}{3}$ **(b)** $-4$ **(c)** 0

---

## 4 Use the Point-Slope Form of a Line

Suppose that $L$ is a nonvertical line with slope $m$ containing the point $(x_1, y_1)$. See Figure 16.

**Figure 16**

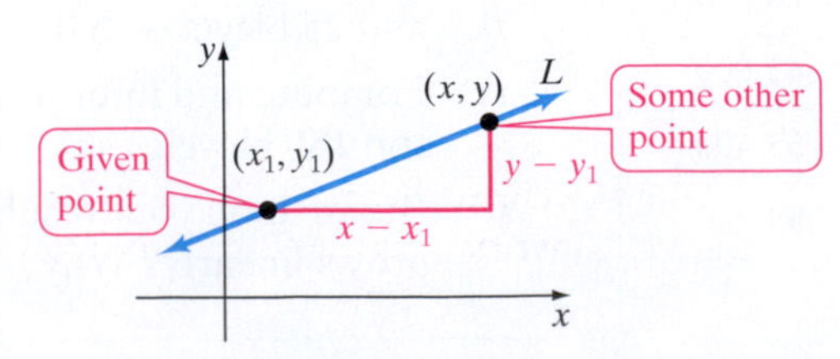

For any other point $(x, y)$ on $L$, we know from the formula for the slope of a line that

$$m = \frac{y - y_1}{x - x_1}$$

Multiplying both sides by $x - x_1$, we can rewrite this expression as

$$y - y_1 = m(x - x_1)$$

**POINT-SLOPE FORM OF AN EQUATION OF A LINE**

An equation of a nonvertical line with slope $m$ that contains the point $(x_1, y_1)$ is

Slope ↓

$$y - y_1 = m(x - x_1)$$

↑ Given Point ↑

## EXAMPLE 5 Using the Point-Slope Form of an Equation of a Line

Find the equation of a line whose slope is 3 and contains the point $(-2, 5)$. Graph the line.

### Solution

Because we are given the slope and a point containing the line, we use the point-slope form of a line with $m = 3$ and $(x_1, y_1) = (-2, 5)$.

$$y - y_1 = m(x - x_1)$$

$m = 3,\ x_1 = -2,\ y_1 = 5$: $\quad y - 5 = 3(x - (-2))$

$$y - 5 = 3(x + 2)$$

See Figure 17 for a graph of the line.

Figure 17

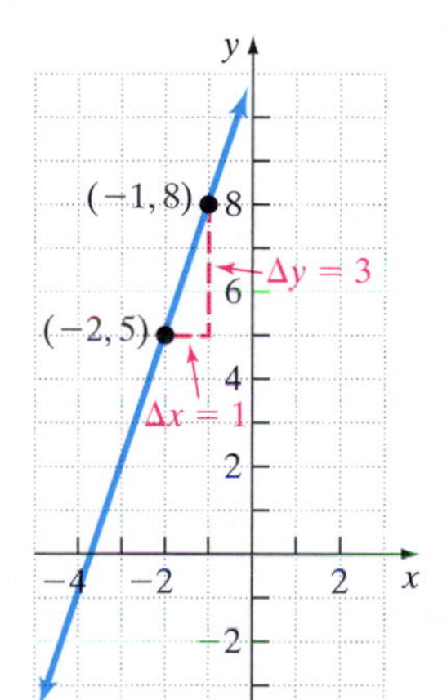


**QUICK ✓** *Find an equation of the line with the given properties. Graph the line.*

**8.** $m = 2, (x_1, y_1) = (3, 5)$

**9.** $m = -4, (x_1, y_1) = (-2, 3)$

**10.** $m = \frac{1}{3}, (x_1, y_1) = (3, -4)$

**11.** $m = 0, (x_1, y_1) = (4, -2)$

## 5 Identify the Slope and y-Intercept of a Line from Its Equation

Because nonvertical lines are functions, we prefer to write the equation of the line using function notation. That is, in the form $y = f(x)$. This requires that we solve the equation for $y$. For example, if we solve the equation in Example 5 for $y$, we obtain the following:

$$y - 5 = 3(x + 2)$$

Distribute the 3 to remove parentheses: $\quad y - 5 = 3x + 6$

Add 5 to both sides of the equation: $\quad y = 3x + 11$

If we call the function $f$, then we can write the equation using function notation as

$$f(x) = 3x + 11$$

The coefficient of $x$, 3, is the slope and the $y$-intercept is $f(0) = 11$. When an equation is written in the form $y = f(x) = mx + b$, we say the equation is in *slope-intercept form*.

**SLOPE-INTERCEPT FORM OF AN EQUATION OF A LINE**

An equation of a line $L$ with slope $m$ and $y$-intercept $b$ is

$$y = f(x) = mx + b$$

### EXAMPLE 6 Finding the Slope and $y$-Intercept of a Line from Its Equation

Write the equation $x - 3y = 9$ in slope-intercept form. Find the slope $m$ and $y$-intercept $b$ of the line. Graph the line.

**Solution**

To put the equation in slope-intercept form, we solve the equation for $y$.

$$x - 3y = 9$$

Subtract $x$ from both sides of the equation: $$-3y = -x + 9$$

Divide both sides of the equation by $-3$: $$y = \frac{-x + 9}{-3}$$

Divide $-3$ into both terms in the numerator: $$y = f(x) = \frac{1}{3}x - 3$$

Comparing $f(x) = \frac{1}{3}x - 3$ to $f(x) = mx + b$, we see that the coefficient of $x$, $\frac{1}{3}$, is the slope, and the $y$-intercept is $-3$.

We graph the line by plotting a point at $(0, -3)$. We then use the slope to find an additional point on the graph by moving right 3 units and up 1 unit from the point $(0, -3)$ to the point $(3, -2)$. Draw a line through the two points and obtain the graph shown in Figure 18.

**Figure 18**

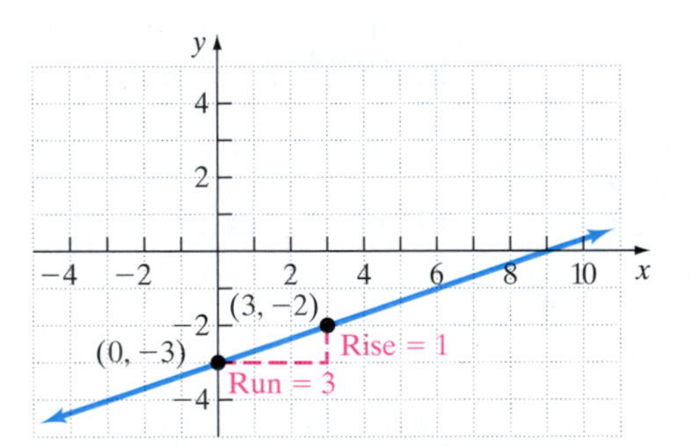


**QUICK ✓** *Find the slope and y-intercept of each line. Graph the line.*

**12.** $3x - y = 2$

**13.** $6x + 2y = 8$

**14.** $3x - 2y = 7$

**15.** $7x + 3y = 0$

## 6 Find the Equation of a Line Given Two Points

From Section 3.1, we know that two points are all that is needed to graph a line. If we are given two points, we can find an equation of the line through the points by first finding the slope of the line and then using the point-slope form of a line.

## EXAMPLE 7 How to Find an Equation of a Line from Two Points

Find the equation of a line through the points $(-1, 4)$ and $(2, -5)$. If possible, write the equation in slope-intercept form. Graph the line.

### Step-by-Step Solution

**Step 1:** Find the slope of the line containing the points.

Let $(x_1, y_1) = (-1, 4)$ and $(x_2, y_2) = (2, -5)$. Substitute these values into the formula for the slope of a line.

$$m = \frac{y_2 - y_1}{x_2 - x_1} = \frac{-5 - 4}{2 - (-1)} = \frac{-9}{3} = -3$$

**Step 2:** Use the point-slope form of a line to find the equation.

With $m = -3$, $x_1 = -1$, and $y_1 = 4$, we have

$$y - y_1 = m(x - x_1)$$
$$y - 4 = -3(x - (-1))$$
$$y - 4 = -3(x + 1)$$

**Step 3:** Solve the equation for $y$.

Distribute the $-3$: $y - 4 = -3x - 3$

Add 4 to both sides: $y = f(x) = -3x + 1$

The slope of the line is $-3$ and the $y$-intercept is 1. See Figure 19 for the graph.

**Figure 19**

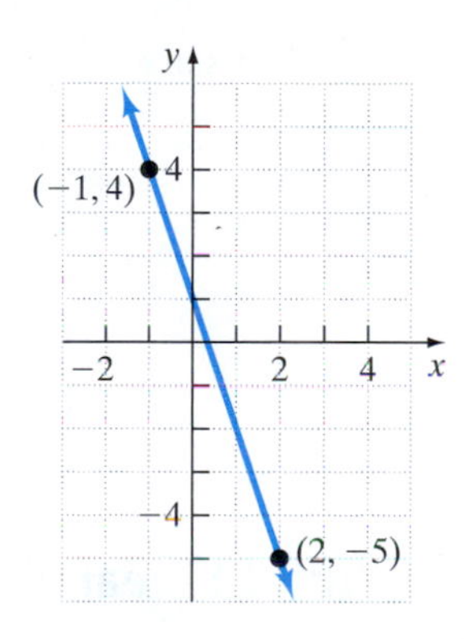


**Work Smart**

In Step 2 of Example 7, we chose to use $x_1 = -1$ and $y_1 = 4$, but we could also have used $x_1 = 2$ and $y_1 = -5$. Choose the values of $x$ and $y$ that make the algebra easiest.

**QUICK ✓** *Find the equation of the line containing the given points. If possible, write the equation in slope-intercept form. Graph the line.*

**16.** $(1, 3); (4, 9)$ **17.** $(-2, 4); (2, 2)$

## EXAMPLE 8 Finding an Equation of a Line from Two Points

Find the equation of a line through the points $(-3, 2)$ and $(-3, -2)$. If possible, write the equation in slope-intercept form. Graph the line.

### Solution

Let $(x_1, y_1) = (-3, 2)$ and $(x_2, y_2) = (-3, -2)$. Substitute these values into the formula for the slope of a line.

$$m = \frac{y_2 - y_1}{x_2 - x_1} = \frac{-2 - 2}{-3 - (-3)} = \frac{-4}{0}$$

The slope is undefined, so the line is vertical. The equation of the line is $x = -3$. See Figure 20 for the graph.

**Work Smart**
If you plot the points first, it will be clear that the line through the points is vertical.

**Figure 20**

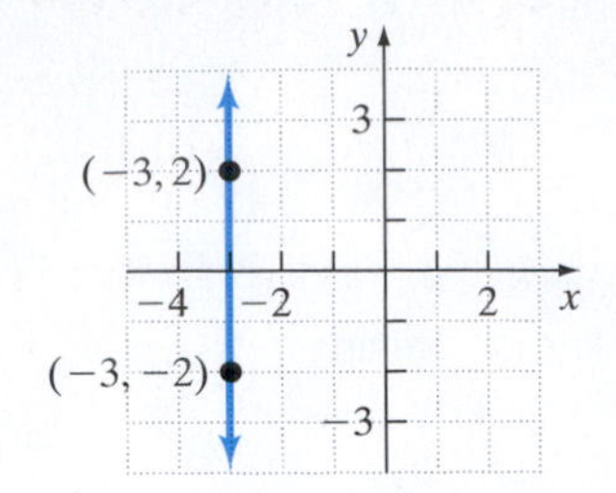


**QUICK ✓**

**18.** Find an equation of the line containing the points $(3, 2)$ and $(3, -4)$. If possible, write the answer in slope-intercept form. Graph the line.

**Work Smart: Study Smart**
To determine which equation of a line to use, ask yourself "What information do I know?"
If you know (1) the slope and a point that isn't the $y$-intercept, then use the point-slope form (2) the slope and the $y$-intercept, then use the slope-intercept form (3) two points, then use the slope formula with the point-slope form (4) If the slope is undefined, use the vertical line.

### SUMMARY: Equations of Lines

| Form of Line | Formula | Comments |
|---|---|---|
| Horizontal Line | $y = b$ | Graph is a horizontal line (slope is 0) with $y$-intercept $b$. |
| Vertical Line | $x = a$ | Graph is a vertical line (undefined slope) with $x$-intercept $a$. |
| Point-slope | $y - y_1 = m(x - x_1)$ | Useful for finding the equation of a line given a point and a slope or two points. |
| Slope-intercept | $y = f(x) = mx + b$ | This is the form of a line expressed in function notation. Useful for quickly determining the slope and $y$-intercept of the line. |
| Standard | $Ax + By = C$ | Straight forward to find the $x$- and $y$-intercepts. |

## 7 Build Linear Models Using the Point-Slope Form of a Line

We can use the point-slope form of a line to build linear models from data.

### EXAMPLE 9 Building a Linear Model from Data

The enrollment $E$ for children 3 to 4 years old in preschool has been increasing at a constant rate for the past 8 years in School District 205. In 1990 the number of 3- to 4-year-olds enrolled in the district was 203. In 2000, the number of 3- to 4-year-olds enrolled in the district was 278. A school administrator wishes to have an idea what the enrollment of 3- to 4-year-olds will be in 2008. In addition, she wishes to be able to tell the school board how much enrollment is increasing each year.

**(a)** Find a linear function that relates school enrollment $E$ to the year $x$ treating year as the independent variable.

**(b)** Predict the enrollment in 2008.

**(c)** Interpret the slope.

**(d)** Assuming the linear trend continues, in what year will enrollment be 368?

## Solution

**(a)** For ease of computation, we let $x$ represent the number of years since 1990, so $x = 0$ represents 1990 and $x = 10$ represents 2000. When $x = 0$, $E = 203$; when $x = 10$, $E = 278$. We have the ordered pairs $(0, 203)$ and $(10, 278)$ and wish to find the equation of the line through these points. To help visualize enrollment growth, we plot these points and draw a line through the points as shown in Figure 21.*

**Work Smart**
When graphing, we use the horizontal axis for the independent variable and the vertical axis for the dependent variable.

**Figure 21**

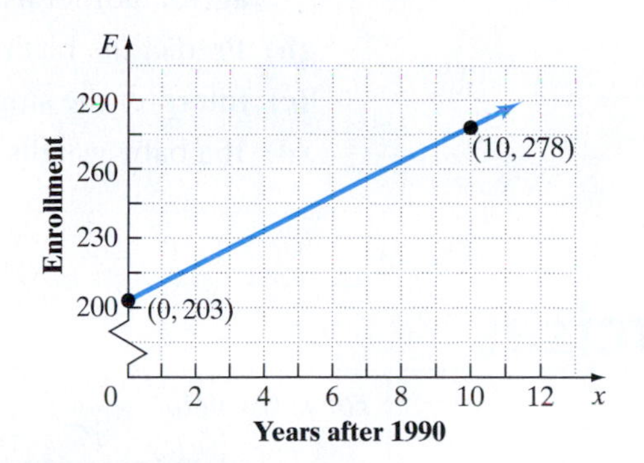


Because we have two points, we will use the point-slope form of a line to find the equation of the line.

First, we must find the slope:

**Work Smart**
Rather than thinking of the slope as the change in $y$ divided by the change in $x$, think of slope as the change in the dependent variable divided by the change in the independent variable.

$$m = \frac{E_2 - E_1}{x_2 - x_1} = \frac{278 - 203}{10 - 0} = \frac{75}{10} = 7.5$$

We use the point-slope form of a line with $m = 7.5$, $x_1 = 0$, and $E_1 = 203$:

$$E - E_1 = m(x - x_1)$$

$m = 7.5$, $x_1 = 0$, and $E_1 = 203$: $\quad E - 203 = 7.5(x - 0)$

$$E - 203 = 7.5x$$

Add 203 to both sides of the equation: $\quad E = 7.5x + 203$

We can express the equation using function notation as $E(x) = 7.5x + 203$.

**(b)** The year 2008 corresponds with $x = 18$, so the predicted enrollment in 2008 would be $E(18)$:

$$E(18) = 7.5(18) + 203$$
$$= 338$$

We predict that enrollment will be 338 students in 2008.

**(c)** The slope is 7.5, so that each year, enrollment is increasing by about 7.5 students. We are allowed to have an average rate of change of 7.5 students per year because averages do not have to be integers.

**(d)** To determine the year in which enrollment will be 368, we let $E = 368$ and solve the resulting equation.

$$E = 7.5x + 203$$
$$368 = 7.5x + 203$$

Subtract 203 from both sides: $\quad 165 = 7.5x$

Divide both sides by 7.5: $\quad 22 = x$

Since $x$ represents the number of years since 1990, we predict that enrollment will be 368 in 2012. ■

*The symbol ⋛ indicates that a portion of the graph has been removed. We typically do this when a large amount of "white space" would result in the graph.

**QUICK ✓**

**19.** According to the National Center for Health Statistics, the average birth weight of babies born to 22-year-old mothers is 3280 grams. The average birth weight of babies born to 32-year-old mothers is 3370 grams. Suppose that the relation between age of mother and birth weight is linear.

**(a)** Find a linear function that relates age of mother $a$ to birth weight $W$ treating age of mother as the independent variable.

**(b)** Predict the birth weight of a mother who is 30 years old.

**(c)** Interpret the slope.

**(d)** If a baby weighs 3310 grams, how old do you expect the mother to be?

## 3.2 Exercises

### Concepts and Vocabulary

*In Problems 1–3, fill in the blanks.*

**1.** If a line is vertical, then its slope is _______.

**2.** On a line, for every 10-foot run there is a 4-foot rise. The slope of the line is _______.

**3.** For the graph of the function $G(z) = 5z + 1$, the slope is _______ and the $y$-intercept is _______.

*In Problems 4–6, answer True or False to each statement.*

**4.** If $P = (x_1, y_1)$ and $Q = (x_2, y_2)$ are two distinct points with $x_1 \neq x_2$, the slope $m$ of the nonvertical line $L$ containing $P$ and $Q$ is defined by the formula

$$m = \frac{x_2 - x_1}{y_2 - y_1}, \quad y_1 \neq y_2.$$

**5.** When the slope of a line is negative, the line slants downward from left to right.

**6.** If the slope of a line is $\frac{1}{2}$, then if $x$ increases by 2, $y$ will increase by 1.

**7.** What is the only linear equation that is not a function?

**8.** Name the five forms of equations of lines given in this section.

**9.** What type of line has one $x$-intercept, but no $y$-intercept?

**10.** What type of line has one $y$-intercept, but no $x$-intercept?

**11.** What type of line has one $x$-intercept and one $y$-intercept?

**12.** Are there any lines that have no intercepts? Explain your answer.

### Building Skills

*In Problems 13–16, (a) find the slope of the line and (b) interpret the slope.*

**13.** **14.** **15.** **16.**

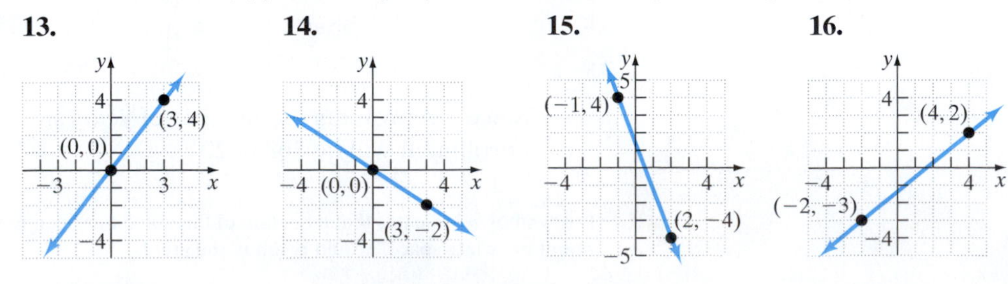

*In Problems 17–32, plot each pair of points and determine the slope of the line containing them. Graph the line.*

**17.** $(0, 0); (1, 5)$
**18.** $(0, 0); (-2, 5)$
**19.** $(2, 1); (4, 9)$
**20.** $(4, 1); (7, 10)$
**21.** $(-2, 3); (1, -6)$
**22.** $(3, -1); (-2, 11)$
**23.** $(-2, 3); (3, 7)$
**24.** $(1, -4); (-1, 3)$
**25.** $(-3, 2); (4, 2)$
**26.** $(-3, 1); (2, 1)$
**27.** $(10, 2); (10, -3)$
**28.** $(4, 1); (4, -3)$
**29.** $\left(\frac{1}{2}, \frac{5}{3}\right); \left(\frac{9}{4}, \frac{11}{6}\right)$
**30.** $\left(\frac{7}{3}, \frac{5}{2}\right); \left(\frac{13}{9}, \frac{13}{4}\right)$
**31.** $\left(-\frac{5}{4}, \frac{14}{3}\right); \left(\frac{1}{5}, -\frac{11}{2}\right)$
**32.** $\left(\frac{1}{3}, -\frac{15}{4}\right); \left(-\frac{12}{5}, \frac{3}{7}\right)$

*In Problems 33–42, graph the line containing the point P and having slope m. Do not find the equation of the line.*

**33.** $m = 3; (1, 2)$
**34.** $m = 2; (-1, 4)$
**35.** $m = -2; (-3, 1)$
**36.** $m = -4; (-1, 5)$
**37.** $m = \frac{1}{3}; (-3, 4)$
**38.** $m = \frac{4}{3}; (-2, -5)$
**39.** $m = -\frac{3}{2}; (2, 5)$
**40.** $m = -\frac{1}{2}; (3, 3)$
**41.** $m = 0; (1, 2)$
**42.** $m$ is undefined; $(-5, 2)$

*In Problems 43–46, the slope and a point on a line are given. Use the information to find three additional points on the line. Answers may vary.*

**43.** $m = 1; (-3, 4)$
**44.** $m = -2; (1, 4)$
**45.** $m = \frac{5}{2}; (-2, 3)$
**46.** $m = -\frac{2}{3}; (1, -3)$

*In Problems 47–52, find an equation of the line. Express your answer in slope-intercept form.*

**47.**

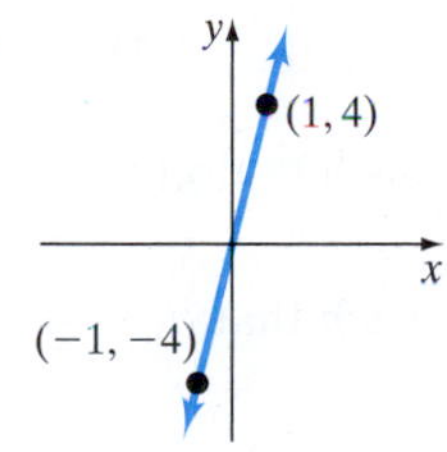


**48.**

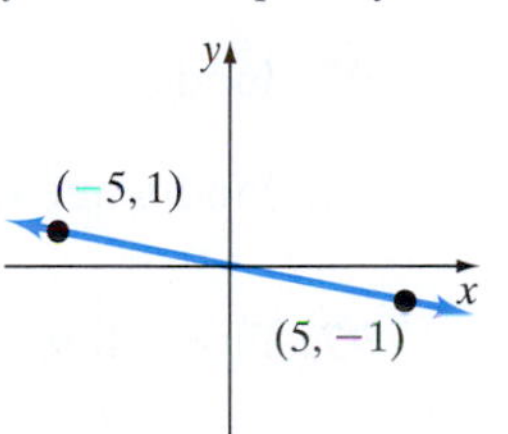


**49.**

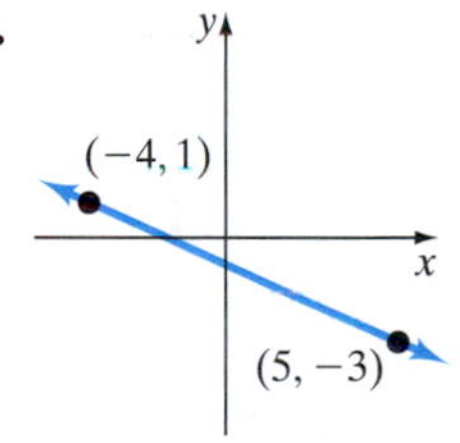


**50.**

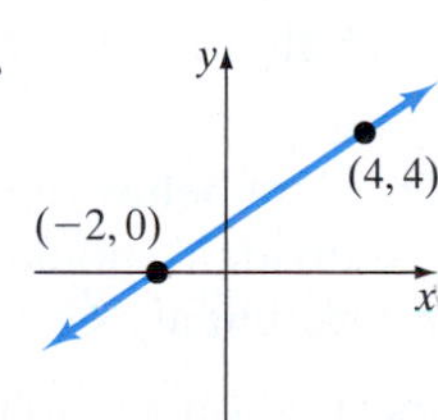


**51.**

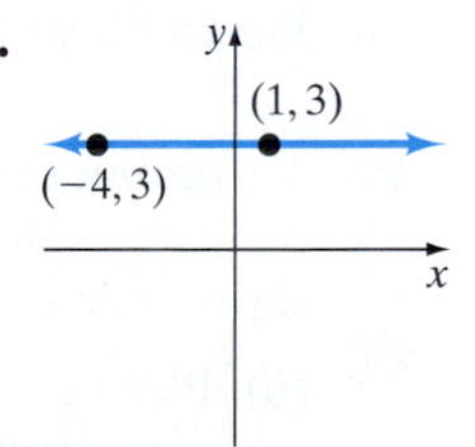


**52.**

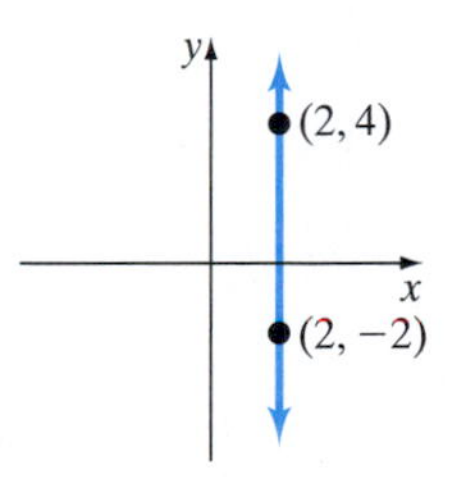


*In Problems 53–62, find an equation of the line with the given slope and containing the given point. Express your answer in slope-intercept form.*

**53.** $m = 2; (0, 0)$
**54.** $m = -1; (0, 0)$
**55.** $m = -3; (-1, 1)$
**56.** $m = 4; (2, -1)$
**57.** $m = \frac{4}{3}; (3, 2)$
**58.** $m = \frac{1}{2}; (2, 1)$
**59.** $m = -\frac{5}{4}; (-2, 4)$
**60.** $m = -\frac{4}{3}; (1, -3)$
**61.** $m$ undefined; $(6, 1)$
**62.** $m = 0; (3, -2)$

*In Problems 63–76, find an equation of the line containing the given points. Express your answer in slope-intercept form.*

**63.** $(0, 0); (5, 7)$
**64.** $(0, 0); (4, -3)$
**65.** $(3, 2); (4, 7)$
**66.** $(1, 3); (3, 7)$
**67.** $(-2, 1); (5, -2)$
**68.** $(-3, 1); (1, 6)$
**69.** $(-1, -3); (-1, 5)$
**70.** $(-3, -4); (1, -4)$
**71.** $(1, 3); (-3, -7)$
**72.** $(-5, 1); (1, -1)$
**73.** $(2, 4); (-4, 4)$
**74.** $(3, 1); (3, -4)$
**75.** $\left(\frac{5}{2}, \frac{7}{3}\right); \left(\frac{3}{4}, -\frac{1}{6}\right)$
**76.** $\left(\frac{2}{3}, \frac{9}{2}\right); \left(-\frac{4}{9}, \frac{11}{4}\right)$

*In Problems 77–88, find the slope and y-intercept of each line. Graph the line.*

**77.** $y = 2x - 1$
**78.** $y = 3x + 2$
**79.** $y = -4x$
**80.** $y = -7x$
**81.** $2x + y = 3$
**82.** $-3x + y = 1$
**83.** $4x + 2y = 8$
**84.** $3x + 6y = 12$
**85.** $x - 4y - 2 = 0$
**86.** $2x - 5y - 10 = 0$
**87.** $x = 3$
**88.** $y = -4$

*In Problems 89–92, graph each function using the slope and y-intercept.*

**89.** $f(x) = -3x + 7$
**90.** $g(x) = -6x + 2$
**91.** $h(x) = -\frac{5}{3}x + 4$
**92.** $G(x) = \frac{1}{2}x - 5$

## Applying the Concepts

**93.** Find an equation for the $x$-axis.

**94.** Find an equation for the $y$-axis.

**95.** Find a linear function $f$ such that $f(2) = 6$ and $f(5) = 12$. What is $f(-2)$?

**96.** Find a linear function $g$ such that $g(1) = 5$ and $g(5) = 17$. What is $g(-3)$?

**97.** Find a linear function $h$ such that $h(3) = 7$ and $h(-1) = 14$. What is $h\left(\frac{1}{2}\right)$?

**98.** Find a linear function $F$ such that $F(2) = 5$ and $F(-3) = 9$. What is $F\left(-\frac{3}{2}\right)$?

**99. Maximum Heart Rate** The data below represent the maximum number of heartbeats that a healthy individual should have during a 15-second interval of time while exercising for different ages.

**(a)** Plot the ordered pairs $(x, y)$ on a graph and connect the points with straight lines.
**(b)** Compute and interpret the average rate of change in the maximum number of heartbeats between 20 and 30 years of age.
**(c)** Compute and interpret the average rate of change in the maximum number of heartbeats between 50 and 60 years of age.
**(d)** Based upon your results to parts (a), (b), and (c), do you think that the maximum number of heartbeats is linearly related to age? Why?

| Age, $x$ | Maximum Number of Heartbeats, $y$ |
|---|---|
| 20 | 50 |
| 30 | 47.5 |
| 40 | 45 |
| 50 | 42.5 |
| 60 | 40 |
| 70 | 37.5 |

SOURCE: *American Heart Association*

**100. Raisins** The following data represent the weight (in grams) of a box of raisins and the number of raisins in the box.

**(a)** Plot the ordered pairs $(x, y)$ on a graph and connect the points with straight lines.
**(b)** Compute and interpret the average rate of change in the number of raisins between 42.3 and 42.5 grams.
**(c)** Compute and interpret the average rate of change in the number of raisins between 42.7 and 42.8 grams.
**(d)** Based upon your results to parts (a), (b), and (c), do you think that the number of raisins is linearly related to weight? Why?

| Weight (in grams), $x$ | Number of Raisins, $y$ |
|---|---|
| 42.3 | 82 |
| 42.5 | 86 |
| 42.6 | 89 |
| 42.7 | 91 |
| 42.8 | 93 |

SOURCE: *Jennifer Maxwell, student at Joliet Junior College*

**101. Average Income** An individual's income varies with age. The following data show the average income of individuals of different ages in the United States for 2000.

**(a)** Plot the ordered pairs $(x, y)$ on a graph and connect the points with straight lines.
**(b)** Compute and interpret the average rate of change in average income between 20 and 30 years of age.
**(c)** Compute and interpret the average rate of change in average income between 50 and 60 years of age.
**(d)** Based upon your results to parts (a), (b), and (c), do you think that average income is linearly related to age? Why?

| Age, $x$ | Average Income, $y$ |
|---|---|
| 20 | $27,711 |
| 30 | $44,477 |
| 40 | $53,243 |
| 50 | $58,217 |
| 60 | $44,993 |
| 70 | $23,047 |

SOURCE: *Statistical Abstract, 2002*

**102. U.S. Population** The following data represent the population of the United States between 1930 and 2000.

**(a)** Plot the ordered pairs $(x, y)$ on a graph and connect the points with straight lines.
**(b)** Compute and interpret the average rate of change in population between 1930 and 1940.
**(c)** Compute and interpret the average rate of change in population between 1990 and 2000.
**(d)** Based upon your results to parts (a), (b), and (c), do you think that population is linearly related to the year? Why?

| Year, $x$ | Population, $y$ |
|---|---|
| 1930 | 123,202,624 |
| 1940 | 132,164,569 |
| 1950 | 151,325,798 |
| 1960 | 179,323,175 |
| 1970 | 203,302,031 |
| 1980 | 226,542,203 |
| 1990 | 248,709,873 |
| 2000 | 281,421,906 |

SOURCE: *U.S. Census Bureau*

**103. The Swordfish Population** The term "landing" refers to the amount of fish caught. Landings of swordfish on the east coast of the United States have followed the same trend as the population of swordfish. In 1988, swordfish landings were 10.23 million pounds, while in 1995, swordfish landings were 6.3 million pounds. (*Source:* Natural Resources Defense Council)

**(a)** Assuming a linear relation between year and swordfish landings, find a linear function that relates swordfish landings to year treating year $x$ as the independent variable. Round your answers to two decimal places. (**Hint:** Let $x$ represent the number of years since 1988 so that $x = 0$ represents 1988 and $x = 7$ represents 1995.)

**(b)** Predict the number of landings of swordfish in 2000 assuming the linear trend continues.

**(c)** Interpret the slope.

**(d)** Assuming that the linear trend continues, in what year will landings of swordfish be 2.39 million pounds?

**104. Catching Fish** The year 1993 was a record year for catching fish with a total of 10,467 million pounds of fish caught. In that year, the average price per pound of fish caught was 33.2 cents. In 2000, there were 9069 million pounds of fish caught at an average price per pound of 39.1 cents.

**(a)** Find a linear function that relates price per pound to number of pounds of fish caught treating number of pounds of fish caught as the independent variable.

**(b)** Predict the price per pound if 9830 million pounds of fish are caught.

**(c)** Interpret the slope.

**(d)** If the average price per pound of fish one year was 38.4 cents, approximate the number of pounds of fish that were caught.

**105. Diamonds** The relation between the cost of a diamond and its weight is linear. In looking at two diamonds, we find that one of the diamonds weighs 0.7 Carat and costs $3543, while the other diamond weighs 0.8 Carat and costs $4378. (*Source: diamonds.com*)

**(a)** Find a linear function that relates the price of a diamond to its weight treating weight as the independent variable.

**(b)** Predict the price of a diamond that weighs 0.77 Carat.

**(c)** Interpret the slope.

**(d)** If a diamond costs $5300, what do you think it should weigh?

**106. Apartments** In the North Chicago area, an 820 square foot apartment rents for $1507 per month. A 970 square foot apartment rents for $1660. (*Source: apartments.com*) Suppose that the relation between area and rent is linear.

**(a)** Find a linear function that relates the rent of a North Chicago apartment to its area treating area as the independent variable.

**(b)** Predict the rent of a 900 square foot apartment in North Chicago.

**(c)** Interpret the slope.

**(d)** If the rent of a North Chicago apartment is $1300 per month, how big would you expect it to be?

**107. The Consumption Function** A famous theory in economics developed by John Maynard Keynes states that personal consumption expenditures are a linear function of disposable income. An economist wishes to develop a model that relates income and consumption and obtains the following information from the United States Bureau of Economic Analysis. In 1990, personal disposable income was $4293.6 billion and personal consumption expenditures were $3831.5 billion. In 2000, personal disposable income was $7031.0 billion and personal consumption expenditures were $6728.4 billion.

**(a)** Find a linear function that relates personal consumption expenditures to disposable income treating disposable income as the independent variable.

**(b)** In 2001, personal disposable income was $7417.3 billion. Use this information to predict personal consumption expenditures in 2001.

**(c)** Interpret the slope. In economics, this slope is called the **marginal propensity to consume.**

**(d)** If personal consumption expenditures were $7210 billion, what do you think that disposable income was?

**108. Measuring Temperature** The relationship between Celsius (°C) and Fahrenheit (°F) degrees for measuring temperature is linear. Find an equation relating °C and °F if 0°C corresponds to 32°F and 100°C corresponds to 212°F. Use the equation to find the Celsius measure of 60°F.

## Extending the Concepts

**109.** Which of the following equations might have the graph shown? (More than one answer is possible.)

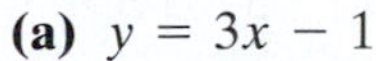

**(a)** $y = 3x - 1$
**(b)** $y = -2x + 3$
**(c)** $y = 2x + 3$
**(d)** $3x - 2y = 4$
**(e)** $-3x + 2y = -4$

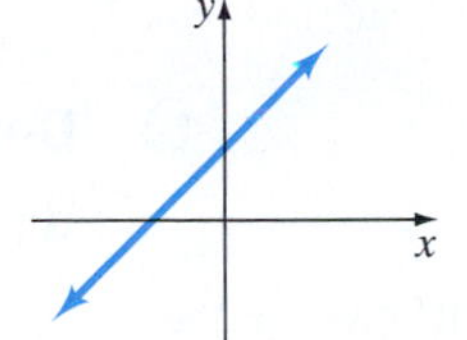


**110.** Which of the following equations might have the graph shown? (More than one answer is possible.)

**(a)** $y = 2x - 5$
**(b)** $y = -x + 2$
**(c)** $y = -\frac{2}{3}x - 3$
**(d)** $4x + 3y = -5$
**(e)** $-2x + y = -4$

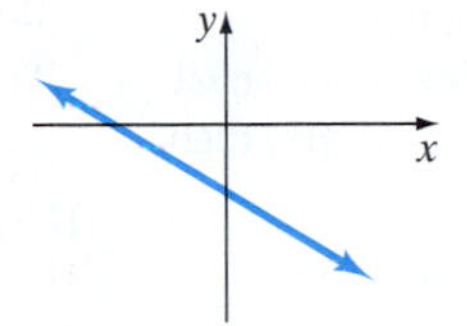


**111.** The equation $2x + y = C$ defines a **family of functions,** one line for each value of $C$. On one Cartesian plane, graph the members of the family when $C = -3$, $C = 0$, and $C = 3$. Can you draw a conclusion about each member of the family?

**112.** Rework Problem 111 for the family of lines $Cx + y = 1$.

**113. Building Codes** As a result of the Americans with Disabilities Act (ADA, 1990), the building code states that access ramps must have a slope not steeper than $\frac{1}{12}$. Interpret what this result means.

**114. Forestry** According to the Forestry Harvesting Information System when building roads while foresting, the grade should be about 15% if the distance traveled is less than 200 feet. Research the idea of grade as it pertains to roads and explain this recommendation.

## The Graphing Calculator

**115.** To see the role that the slope $m$ plays in the graph of a linear equation $y = mx + b$, graph the following lines on the same screen.

$$Y_1 = 0x + 2 \qquad Y_2 = \frac{1}{2}x + 2 \qquad Y_3 = 2x + 2 \qquad Y_4 = 6x + 2$$

State some general conclusions about the graph of $y = mx + b$ for $m \geq 0$. Now graph

$$Y_1 = -\frac{1}{2}x + 2 \qquad Y_2 = -2x + 2 \qquad Y_3 = -6x + 2$$

State some general conclusions about the graph of $y = mx + b$ for $m < 0$.

**116.** To see the role that the $y$-intercept $b$ plays in the graph of a linear equation $y = mx + b$, graph the following lines on the same screen.

$$Y_1 = 2x \qquad Y_2 = 2x + 2 \qquad Y_3 = 2x + 5 \qquad Y_4 = 2x - 4$$

State some general conclusions about the graph of $y = 2x + b$.

# 3.3 Parallel and Perpendicular Lines

**OBJECTIVES**

1. Define Parallel Lines
2. Find Equations of Parallel Lines
3. Define Perpendicular Lines
4. Find Equations of Perpendicular Lines

***Preparing for Parallel and Perpendicular Lines***

*Before getting started, take this readiness quiz. If you get a problem wrong, go back to the section cited and review the material.*

**1.** Determine the reciprocal of 3. [Section R.3, p. 24]

**2.** Determine the reciprocal of $-\frac{3}{5}$. [Section R.3, p. 24]

## 1 Define Parallel Lines

When two lines (in the Cartesian plane) do not intersect (that is, they have no points in common), they are said to be *parallel.*

**Work Smart**

The words "if and only if" given in the definition mean that there are two statements being made.

If two nonvertical lines are parallel, then their slopes are equal and they have different $y$-intercepts.

If two nonvertical lines have equal slopes and different $y$-intercepts, then they are parallel.

**DEFINITION**

Two nonvertical lines are **parallel** if and only if their slopes are equal and they have different $y$-intercepts. Vertical lines are parallel if they have different $x$-intercepts.

Figure 22(a) shows nonvertical parallel lines. Figure 22(b) shows vertical parallel lines.

**Figure 22**
Parallel lines

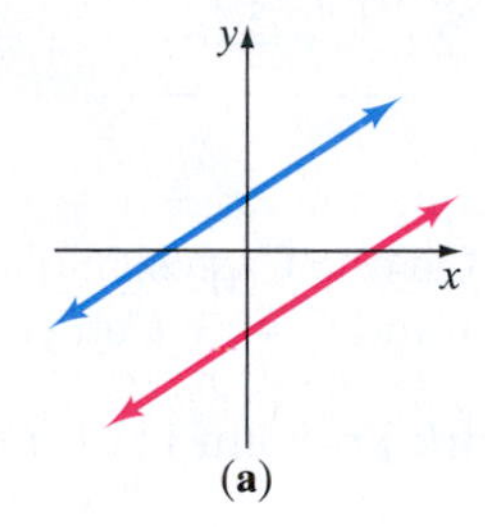

(a)

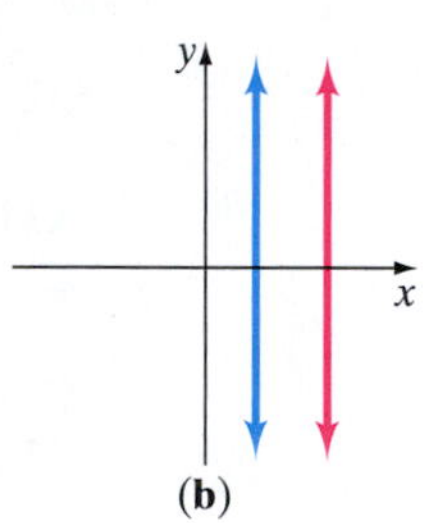

(b)

### EXAMPLE 1 Determining Whether Two Lines Are Parallel

Determine whether the given lines are parallel.

**(a)** $L_1: 4x + y = 8$
$L_2: 6x + 2y = 12$

**(b)** $L_1: -3x + 2y = 6$
$L_2: 6x - 4y = 8$

**Solution**

To determine whether two lines are parallel, we determine the slope and $y$-intercept of each line by putting the equation of the line in slope-intercept form. If the slopes are the same, but the $y$-intercepts are different, then the lines are parallel.

**(a)**

Solve $L_1$ for $y$: $4x + y = 8$

Subtract $4x$ from both sides: $y = -4x + 8$

The slope of $L_1$ is $-4$ and the $y$-intercept is 8.

Solve $L_2$ for $y$: $6x + 2y = 12$

Subtract $6x$ from both sides: $2y = -6x + 12$

Divide both sides by 2: $y = \frac{-6x + 12}{2}$

Divide each term in the numerator by 2: $y = -3x + 6$

The slope of $L_2$ is $-3$ and the $y$-intercept is 6.

Because the lines have different slopes, they are not parallel.

*Preparing for...Answers* **1.** $\frac{1}{3}$ **2.** $-\frac{5}{3}$

**(b)**

| | | | |
|---|---|---|---|
| Solve $L_1$ for $y$: | $-3x + 2y = 6$ | Solve $L_2$ for $y$: | $6x - 4y = 8$ |
| Add $3x$ to both sides: | $2y = 3x + 6$ | Subtract $6x$ from both sides: | $-4y = -6x + 8$ |
| Divide both sides by 2: | $y = \frac{3x + 6}{2}$ | Divide both sides by $-4$: | $y = \frac{-6x + 8}{-4}$ |
| Divide each term in the numerator by 2: | $y = \frac{3}{2}x + 3$ | Divide each term in the numerator by $-4$: | $y = \frac{3}{2}x - 2$ |

The slope of $L_1$ is $\frac{3}{2}$ and the $y$-intercept is 3.

The slope of $L_2$ is $\frac{3}{2}$ and the $y$-intercept is $-2$.

Because the lines have the same slope, $\frac{3}{2}$, but different $y$-intercepts, the lines are parallel.

**QUICK ✓** *Determine whether the two lines are parallel.*

1. $L_1$: $y = 3x + 1$
   $L_2$: $y = -3x - 3$
2. $L_1$: $6x + 3y = 3$
   $L_2$: $-8x - 4y = 12$
3. $L_1$: $-3x + 5y = 10$
   $L_2$: $6x + 10y = 10$

## 2 Find Equations of Parallel Lines

Now that we know how to identify parallel lines, let's discuss how to find the equation of a line that is parallel to a given line.

### EXAMPLE 2 How to Find the Equation of a Line That Is Parallel to a Given Line

Find an equation for the line that is parallel to $4x + 2y = 2$ and contains the point $(-2, 3)$. Graph the lines in the Cartesian plane.

**Step-by-Step Solution**

**Step 1:** Find the slope of the given line by putting the equation in slope-intercept form.

$$4x + 2y = 2$$

Subtract $4x$ from both sides: $2y = -4x + 2$

Divide both sides by 2: $y = -2x + 1$

The slope of the line is $-2$.

**Step 2:** Use the point-slope form of a line with the given point and the slope found in Step 1 to find the equation of the parallel line.

$$y - y_1 = m(x - x_1)$$

$m = -2, x_1 = -2, y_1 = 3$: $y - 3 = -2(x - (-2))$

**Step 3:** Put the equation in slope-intercept form by solving for $y$.

$$y - 3 = -2(x + 2)$$

Distribute the $-2$: $y - 3 = -2x - 4$

Add 3 to both sides: $y = -2x - 1$

(continued)

The line parallel to $4x + 2y = 2$ containing $(-2, 3)$ is $y = -2x - 1$. Notice that the slopes of the two lines are the same, but the $y$-intercepts are different. Figure 23 shows the graph of the parallel lines.

**Figure 23**

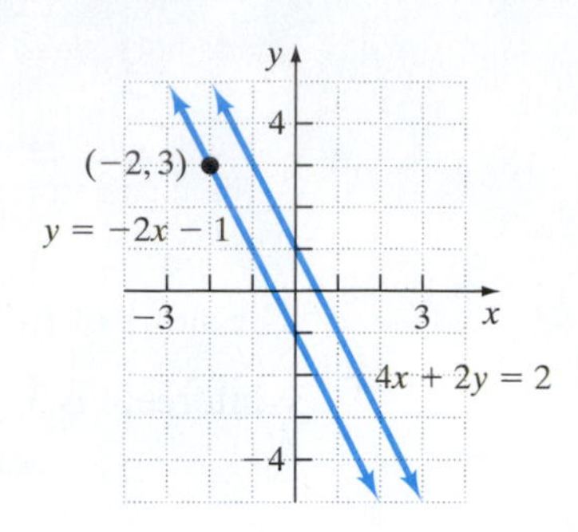


**QUICK ✓** *Find the equation of the line that contains the given point and is parallel to the given line. Write the line in slope-intercept form. Graph the lines.*

**4.** $(5, 8)$; $y = 3x + 1$

**5.** $(-2, 4)$; $3x + 2y = 10$

## 3 Define Perpendicular Lines

When two lines intersect at a right angle (90°), they are said to be **perpendicular.** See Figure 24.

**Figure 24**
Perpendicular lines.

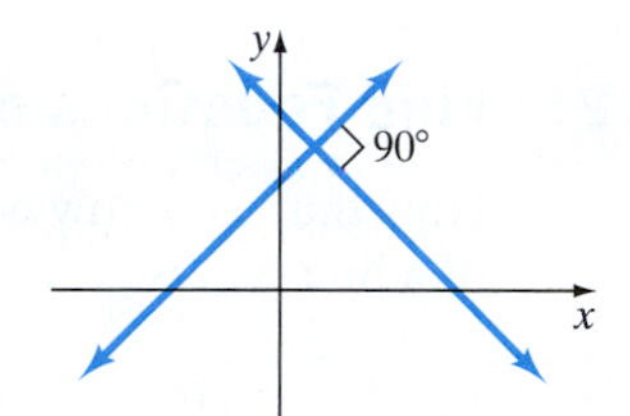


We use the slopes of the lines to determine whether two lines are perpendicular.

**Work Smart**
If $m_1$ and $m_2$ are negative reciprocals of each other, then $m_1 = \frac{-1}{m_2}$.

**DEFINITION**
Two nonvertical lines are **perpendicular** if and only if the product of their slopes is $-1$. Alternatively, two nonvertical lines are perpendicular if their slopes are negative reciprocals of each other. Any vertical line is perpendicular to any horizontal line.

### EXAMPLE 3 Finding the Slope of a Line Perpendicular to a Given Line

Find the slope of the line perpendicular to a line whose slope is $\frac{5}{4}$.

**Solution**

The negative reciprocal of $\frac{5}{4}$ is $\frac{-1}{\frac{5}{4}} = -1 \cdot \frac{4}{5} = -\frac{4}{5}$. Any line whose slope is $-\frac{4}{5}$ will be perpendicular to the line whose slope is $\frac{5}{4}$ because $-\frac{4}{5} \cdot \left(\frac{5}{4}\right) = -1$.

**QUICK ✓**

**6.** Find the slope of the line perpendicular to a line whose slope is $-3$.

**EXAMPLE 4** **Determining Whether Two Lines Are Perpendicular**

Determine whether the given lines are perpendicular.

**(a)** $L_1$: $y = 4x + 1$
$L_2$: $y = -4x - 3$

**(b)** $L_1$: $y = \frac{2}{3}x - 5$
$L_2$: $y = -\frac{3}{2}x + 2$

**Solution**

**(a)** The slope of $L_1$ is $m_1 = 4$. The slope of $L_2$ is $m_2 = -4$. Because the product of the slopes, $m_1 m_2 = 4 \cdot (-4) = -16 \neq -1$, the lines are not perpendicular. Notice that the slopes are not negative reciprocals of each other.

**(b)** The slope of $L_1$ is $m_1 = \frac{2}{3}$. The slope of $L_2$ is $m_2 = -\frac{3}{2}$. Because the product of the slopes is $-1$, the lines are perpendicular. Notice that the slopes are negative reciprocals of each other.

**QUICK ✓** *Determine whether the given lines are perpendicular.*

**7.** $L_1$: $y = 5x - 3$
$L_2$: $y = -\frac{1}{5}x - 4$

**8.** $L_1$: $4x - y = 3$
$L_2$: $x - 4y = 2$

## 4 Find Equations of Perpendicular Lines

Now that we know how to find the slope of a line perpendicular to a second line, we can find the equation of a line perpendicular a second line.

**EXAMPLE 5** **How to Find the Equation of a Line Perpendicular to a Given Line**

Find an equation of the line that is perpendicular to the line $2x + 5y = 10$ and contains the point $(4, -1)$. Write the equation in slope-intercept form. Graph the two lines.

**Step-by-Step Solution**

**Step 1:** Find the slope of the given line by putting the equation in slope-intercept form.

$$2x + 5y = 10$$

Subtract $2x$ from both sides:
$$5y = -2x + 10$$

Divide both sides by 5:
$$y = -\frac{2}{5}x + 2$$

The slope of the line is $-\frac{2}{5}$.

**Step 2:** Find the slope of the perpendicular line.

The slope of the perpendicular line is the negative reciprocal of $-\frac{2}{5}$, which is $\frac{5}{2}$.

**Step 3:** Use the point-slope form of a line with the given point and the slope found in Step 2 to find the equation of the perpendicular line.

$$y - y_1 = m(x - x_1)$$

$m = \frac{5}{2}, x_1 = 4, y_1 = -1$:
$$y - (-1) = \frac{5}{2}(x - 4)$$

(continued)

**Step 4:** Put the equation in slope-intercept form by solving for y.

$$y + 1 = \frac{5}{2}(x - 4)$$

Distribute the $\frac{5}{2}$:
$$y + 1 = \frac{5}{2}x - 10$$

Subtract 1 from both sides:
$$y = \frac{5}{2}x - 11$$

The equation of the line perpendicular to $2x + 5y = 10$ through $(4, -1)$ is $y = \frac{5}{2}x - 11$. Figure 25 shows the graphs of the two lines.

**Figure 25**

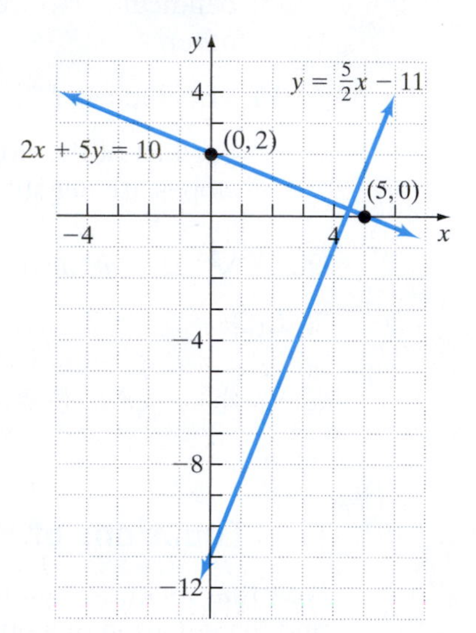


**QUICK ✓** *Find the equation of the line that contains the given point and is perpendicular to the given line. Write the line in slope-intercept form. Graph the lines.*

**9.** $(-4, 2); y = 2x + 1$ **10.** $(-3, -4); 3x - 4y = 8$

# 3.3 Exercises

***For Extra Help:***

Student Solutions Manual

CD Video

PH Math/Tutor Center

MathXL Tutorials on CD

MathXL®

MyMathLab

## Concepts and Vocabulary

*In Problems 1 and 2, fill in the blank.*

**1.** Two lines are parallel if and only if they have the same __________ and different __________.

**2.** Two lines are perpendicular if and only if the product of their slopes is __________.

*In Problems 3 and 4, answer True or False to each statement.*

**3.** Perpendicular lines have slopes that are reciprocals of each other.

**4.** The lines $y = 3x - 1$ and $y = 3x + 1$ are parallel.

**5.** If two nonvertical lines have the same $x$-intercept, but different $y$-intercepts can they be parallel? Explain your answer.

**6.** Why don't we say that a horizontal line is perpendicular to a vertical line if they have slopes that are negative reciprocals of each other?

## Building Skills

*In Problems 7–10, a slope of a line is given. Determine (a) the slope of the line parallel to the line whose slope is given and (b) the slope of the line perpendicular to the line whose slope is given.*

**7.** $m = 5$ **8.** $m = -\frac{8}{5}$ **9.** $-\frac{5}{6}$ **10.** $0$

*In Problems 11–18, determine whether the given linear equations are parallel, perpendicular, or neither.*

**11.** $y = 5x + 4$; $y = 5x - 7$

**12.** $y = 3x - 1$; $y = -\frac{1}{3}x - 5$

**13.** $8x + y = 12$; $2x - 8y = 3$

**14.** $-3x - y = 3$; $6x + 2y = 9$

**15.** $-4x + 2y = 12$; $x + 2y = 6$

**16.** $10x - 3y = 5$; $5x + 6y = 3$

**17.** $-x + \frac{1}{3}y = \frac{1}{3}$; $x - \frac{1}{3}y = \frac{5}{3}$

**18.** $\frac{1}{2}x - \frac{3}{2}y = 3$; $2x + \frac{2}{3}y = 1$

*In Problems 19–24, two points on $L_1$ and two points on $L_2$ are given. Plot the points in the Cartesian plane and draw a line through the points. Compute the slope of the line containing these points and determine whether the lines are parallel, perpendicular, or neither.*

**19.** $L_1$: $(1, 2)$; $(6, 5)$ $L_2$: $(-2, 3)$; $(1, -2)$

**20.** $L_1$: $(1, 1)$; $(4, 3)$ $L_2$: $(-1, 3)$; $(3, -3)$

**21.** $L_1$: $(-2, 4)$; $(1, -3)$ $L_2$: $(-1, -6)$; $(-4, 1)$

**22.** $L_1$: $(-3, 0)$, $(0, 2)$ $L_2$: $(4, 8)$; $(2, 5)$

**23.** $L_1$: $(0, 5)$; $(1, 3)$ $L_2$: $(-4, 6)$; $(0, 14)$

**24.** $L_1$: $(1, -3)$; $(5, -4)$ $L_2$: $(0, 4)$; $(8, 2)$

*In Problems 25–30, find an equation of the line L. Express your answer in slope-intercept form.*

**25.**

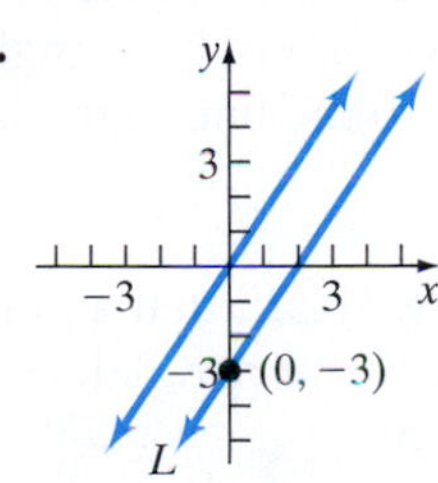


$L$ is parallel to $y = \frac{3}{2}x$

**26.**

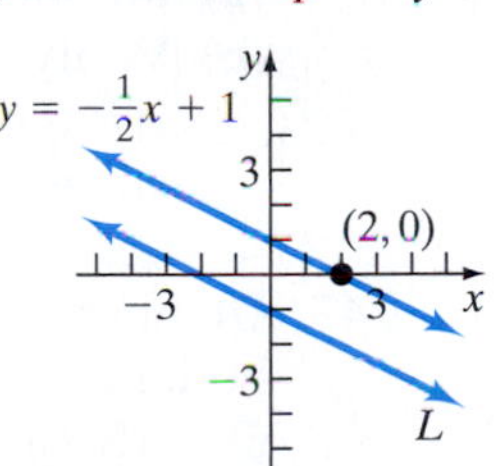


$L$ is parallel to $y = -\frac{1}{2}x + 1$

**27.**

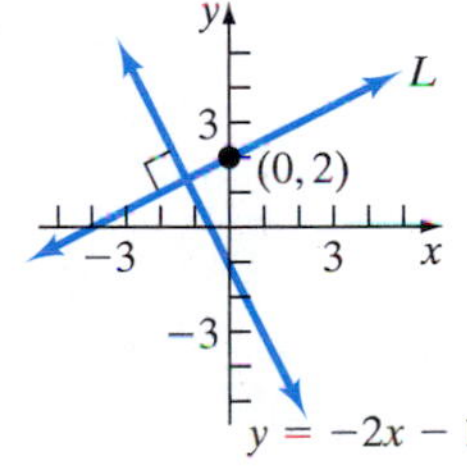


$L$ is perpendicular to $y = -2x - 1$

**28.**

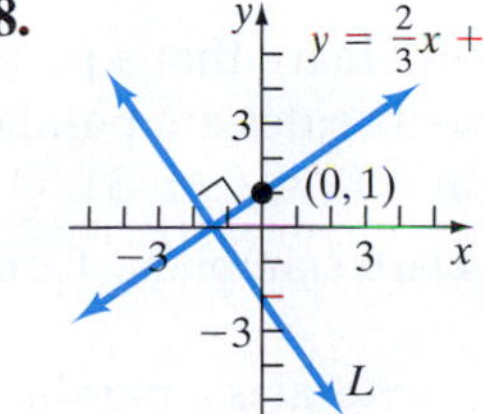


$L$ is perpendicular to $y = \frac{2}{3}x + 1$

**29.**

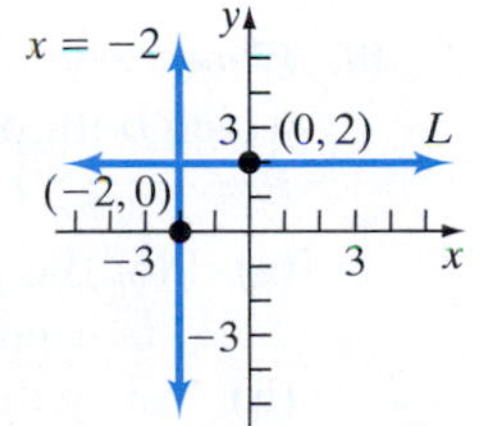


$L$ is perpendicular to $x = -2$

**30.**

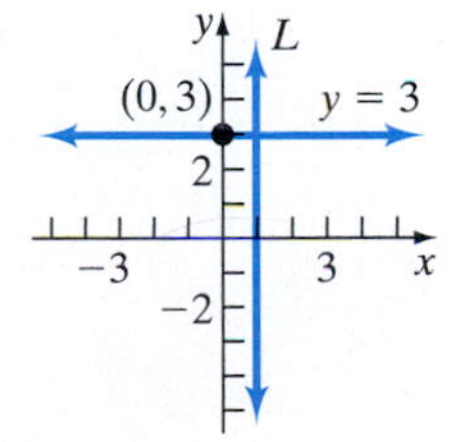


$L$ is perpendicular to $y = 3$

*In Problems 31–44, find an equation of the line with the given properties. Express your answer in slope-intercept form. Graph the lines.*

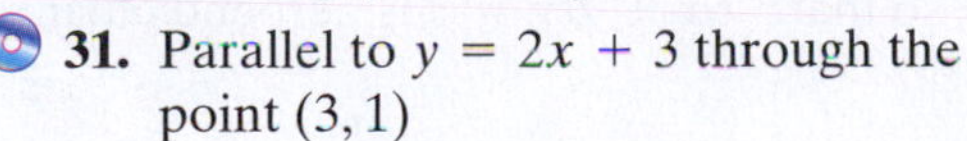

**31.** Parallel to $y = 2x + 3$ through the point $(3, 1)$

**32.** Parallel to $y = -3x + 1$ through the point $(2, 5)$

**33.** Perpendicular to $y = -2x + 1$ through the point $(2, 3)$

**34.** Perpendicular to $y = 4x + 3$ through the point $(4, 1)$

**35.** Parallel to $y = 1$ through the point $(-1, -3)$

**36.** Parallel to $x = -2$ through the point $(2, 5)$

**37.** Perpendicular to $x = 1$ through the point $(1, 3)$

**38.** Perpendicular to $y = 8$ through the point $(2, -4)$

**39.** Parallel to $3x - y = 2$ through the point $(1, 5)$

**40.** Parallel to $2x + y = 5$ through the point $(-4, 3)$

**41.** Perpendicular to $4x + 3y - 1 = 0$ through the point $(-4, 1)$

**42.** Perpendicular to $-2x + 5y - 3 = 0$ through the point $(2, -3)$

**43.** Parallel to $5x + 2y = 1$ through the point $(-2, -3)$

**44.** Perpendicular to $3x + y = 1$ through the point $(3, -1)$

## Applying the Concepts

△ **45. Geometry** Given the points $A = (1, 1)$, $B = (4, 3)$, and $C = (2, 6)$,

**(a)** Plot the points in a Cartesian plane. Connect the points to form a triangle.

**(b)** Verify that the triangle is a right triangle by showing that the line segment $\overline{AB}$ is perpendicular to the line segment $\overline{BC}$ and therefore forms a right angle.

△ **46. Geometry** Given the points $A = (-2, -2)$, $B = (3, 1)$, and $C = (-5, 3)$,

**(a)** Plot the points in a Cartesian plane. Connect the points to form a triangle.

**(b)** Verify that the triangle is a right triangle by showing that the line segment $\overline{AB}$ is perpendicular to the line segment $\overline{AC}$ and therefore forms a right angle.

△ **47. Geometry** In geometry, we learn that a **parallelogram** is a quadrilateral in which both pairs of opposite sides are parallel. Given the points $A = (2, 2)$, $B = (7, 3)$, $C = (8, 6)$, and $D = (3, 5)$,

**(a)** Plot the points in a Cartesian plane. Connect the points to form a quadrilateral.

**(b)** Verify that the quadrilateral is a parallelogram by showing that the opposite sides are parallel.

△ **48. Geometry** In geometry, we learn that a parallelogram is a quadrilateral in which both pairs of opposite sides are parallel. Given the points $A = (-2, -1)$, $B = (4, 1)$, $C = (5, 5)$, and $D = (-1, 3)$,

**(a)** Plot the points in a Cartesian plane. Connect the points to form a quadrilateral.

**(b)** Verify that the quadrilateral is a parallelogram by showing that the opposite sides are parallel.

## Extending the Concepts

**49.** Find $A$ so that $Ax + 4y = 12$ is perpendicular to $4x + y = 3$.

**50.** Find $B$ so that $-6x + By = 3$ is perpendicular to $2x - 3y = 8$.

**51.** The figure shows the graph of two parallel lines. Which of the following pairs of equations might have such a graph?

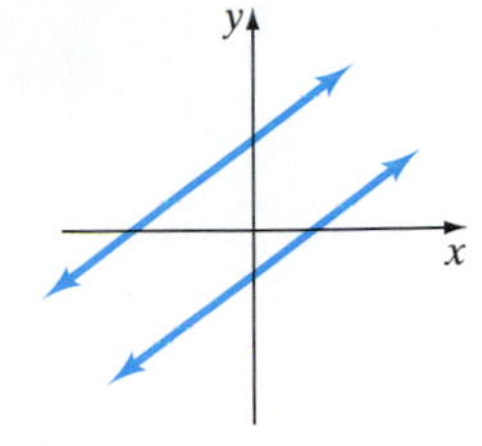


**(a)** $y = x + 3$
$y = -x - 1$

**(b)** $y = 2x + 3$
$y = 2x + 1$

**(c)** $x - 2y = 4$
$x - 2y = -3$

**(d)** $-2x + y = 5$
$-2x + y = 2$

**(e)** $x - y = 3$
$3x - 3y = 9$

**52.** The figure shows the graph of two perpendicular lines. Which of the following pairs of equations might have such a graph?

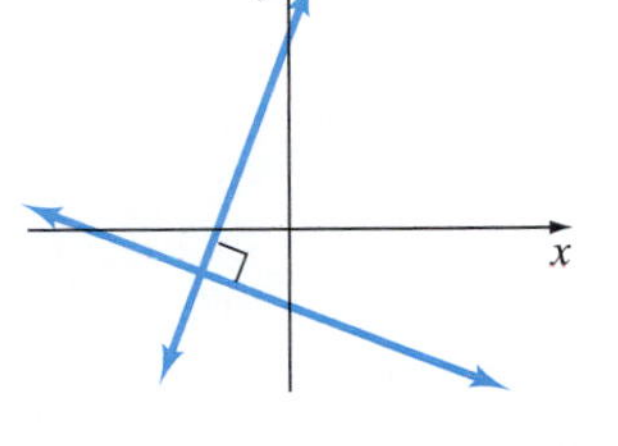


**(a)** $y = 3x + 4$
$y = -\frac{1}{3}x - 2$

**(b)** $-2x + y = 3$
$x + 2y = 1$

**(c)** $2x + 3y = -2$
$3x - 2y = 5$

**(d)** $3x + 4y = 5$
$-3x + 4y = -2$

**(e)** $x - 2y = 6$
$2y + x = 2$

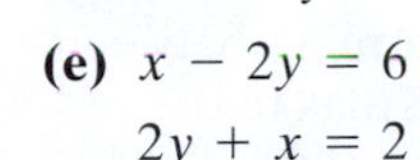

## PUTTING THE CONCEPTS TOGETHER (SECTIONS 3.1–3.3)

*These problems cover important concepts from Sections 3.1 to 3.3. We designed these problems so that you can review the chapter so far and show your mastery of the concepts. Take time to work these problems before proceeding with the next section. The answers to these problems are located at the back of the text starting on page AN-14.*

*In Problems 1–6, graph each linear equation or function, using any method you prefer.*

**1.** $y = -x + 6$ **2.** $3x - 4y = 0$ **3.** $2x - 5y = 20$

**4.** $\frac{1}{2}x - \frac{2}{3}y = 1$ **5.** $y = 6$ **6.** $g(x) = -\frac{3}{4}x + 1$

**7.** Find and interpret the slope of the line containing the points $(-1, 5)$ and $(3, 7)$.

**8.** Find the slope and the $y$-intercept of the line with equation $4x - 5y = 20$.

**9.** Draw the graph of the line that contains the point $(-2, -4)$ and has a slope of $\frac{3}{4}$.

**10.** Determine whether the graphs of the following pair of linear equations are parallel, perpendicular, or neither.

$$12x - 4y = 1$$
$$x - 3y = -12$$

*In Problems 11–14, find the equation of the line with the given properties. Express your answer in slope-intercept form.*

**11.** Through the point $(6, -4)$ and having a slope of $\frac{3}{2}$

**12.** Through the points $(-1, 7)$ and $(4, -3)$

**13.** Parallel to $2x - 3y = 12$ and through the point $(-3, 4)$

**14.** Perpendicular to $4x - 3y = 3$ and through the point $(4, -1)$

**15. Cargo Van Weight** An empty cargo van weighs 5700 pounds. The van is used to transport television sets weighing 40 pounds each.

**(a)** Write a linear function that expresses the total weight $W$ as a function of the number of television sets $x$ loaded on the van.
**(b)** What is the implied domain of this linear function?
**(c)** What is the total weight if 62 television sets are loaded on the van?
**(d)** Graph the linear function.
**(e)** If the total weight is 8580 pounds, how many television sets are on the van?

**16. Commission Earnings** The following chart shows the monthly income that a sales associate will earn, depending on the sales made that month.

**(a)** Plot the ordered pairs $(x, y)$ on a graph and connect the points with straight lines.
**(b)** Compute and interpret the average rate of change in earnings between sales of $5000 and $10,000.
**(c)** Compute and interpret the average rate of change in earnings between sales of $20,000 and $25,000.
**(d)** Based upon the results to parts (a), (b), and (c), do you think that the monthly earnings are linearly related to sales? Why?

| Sales, $x$ | Earnings, $y$ |
|---|---|
| $5000 | $750 |
| $10,000 | $900 |
| $15,000 | $1050 |
| $20,000 | $1200 |
| $25,000 | $1350 |

# 3.4 Linear Inequalities in Two Variables

**OBJECTIVES**

1. Determine Whether an Ordered Pair Is a Solution to a Linear Inequality
2. Graph Linear Inequalities
3. Solve Problems Involving Linear Inequalities

### *Preparing for Linear Inequalities in Two Variables*

*Before getting started, take this readiness quiz. If you get a problem wrong, go back to the section cited and review the material.*

**1.** Determine whether $x = 4$ satisfies the inequality $3x + 1 \geq 7$. [Section 1.4, pp. 93–96]

**2.** Solve the inequality $-4x - 3 > 9$. [Section 1.4, pp. 93–96]

In Chapter 1, we solved inequalities in one variable. In this section, we discuss linear inequalities in two variables.

## 1 Determine Whether an Ordered Pair Is a Solution to a Linear Inequality

Linear inequalities in two variables are inequalities in one of the forms

$$Ax + By < C \qquad Ax + By > C \qquad Ax + By \leq C \qquad Ax + By \geq C$$

where $A$, $B$, and $C$ are real numbers and $A$ and $B$ are not both zero.

If we replace the inequality symbol with an equal sign, we obtain the equation of a line, $Ax + By = C$. The line separates the $xy$-plane into two regions, called **half-planes.** See Figure 26.

*Preparing for...Answers* **1.** Satisfies **2.** $\{x \mid x < -3\}$ or $(-\infty, -3)$

**Figure 26**

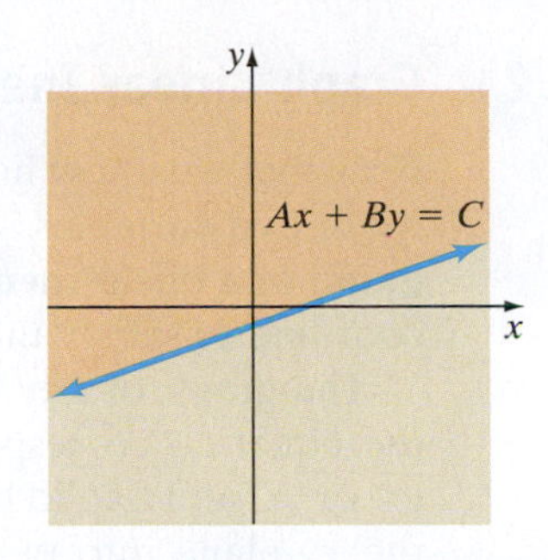


A linear inequality in two variables $x$ and $y$ is **satisfied** by an ordered pair $(a, b)$ if, when $x$ is replaced by $a$ and $y$ is replaced by $b$, a true statement results.

### EXAMPLE 1 Determining Whether an Ordered Pair Is a Solution to a Linear Inequality In Two Variables

Determine which of the following ordered pairs are solutions to the linear inequality $3x + y < 7$.

**(a)** $(2, 4)$ **(b)** $(-3, 1)$ **(c)** $(1, 3)$

**Solution**

**(a)** Let $x = 2$ and $y = 4$ in the inequality. If a true statement results, then $(2, 4)$ is a solution to the inequality.

$$\begin{aligned} 3x + y &< 7 \\ x = 2, y = 4: \quad 3(2) + 4 &\stackrel{?}{<} 7 \\ 6 + 4 &\stackrel{?}{<} 7 \\ 10 &\stackrel{?}{<} 7 \quad \text{False} \end{aligned}$$

Because 10 is not less than 7, the statement is false, so $(2, 4)$ is not a solution to the inequality.

**(b)** Let $x = -3$ and $y = 1$ in the inequality. If a true statement results, then $(-3, 1)$ is a solution to the inequality.

$$\begin{aligned} 3x + y &< 7 \\ x = -3, y = 1: \quad 3(-3) + 1 &\stackrel{?}{<} 7 \\ -9 + 1 &\stackrel{?}{<} 7 \\ -8 &\stackrel{?}{<} 7 \quad \text{True} \end{aligned}$$

Because $-8$ is less than 7, the statement is true, so $(-3, 1)$ is a solution to the inequality.

**(c)** Let $x = 1$ and $y = 3$ in the inequality.

$$\begin{aligned} 3x + y &< 7 \\ x = 1, y = 3: \quad 3(1) + 3 &\stackrel{?}{<} 7 \\ 3 + 3 &\stackrel{?}{<} 7 \\ 6 &\stackrel{?}{<} 7 \quad \text{True} \end{aligned}$$

Because 6 is less than 7, the statement is true, so $(1, 3)$ is a solution to the inequality.

**QUICK ✓**

1. Determine which of the following points are solutions to the linear inequality $-2x + 3y \geq 3$.

**(a)** $(4, 1)$ **(b)** $(-1, 2)$ **(c)** $(2, 3)$ **(d)** $(0, 1)$

## 2 Graph Linear Inequalities

**Work Smart**

To review the distinction between strict and nonstrict inequalities, turn back to page 16.

Now that we know how to determine whether a point is a solution to a linear inequality in two variables, we are prepared to graph linear inequalities in two variables. A **graph of a linear inequality in two variables** $x$ and $y$ consists of all points $(x, y)$ whose coordinates satisfy the inequality.

The graph of any linear inequality in two variables may be obtained by graphing the equation corresponding to the inequality, using dashes if the inequality is strict ($<$ or $>$) and a solid line if the inequality is nonstrict ($\leq$ or $\geq$). This graph will separate the $xy$-plane into two half planes. In each half-plane either all points satisfy the inequality or no points satisfy the inequality. So the use of a single test point is all that is required to obtain the graph of a linear inequality in two variables.

### EXAMPLE 2 How to Graph a Linear Inequality in Two Variables

Graph the linear inequality $3x + y < 7$.

#### Step-by-Step Solution

**Step 1:** We replace the inequality symbol with an equal sign and graph the corresponding line. If the inequality is strict ($<$ or $>$), graph the line as a dashed line. If the inequality is nonstrict ($\leq$ or $\geq$), graph the line as a solid line.

We replace $<$ with $=$ to obtain $3x + y = 7$. We graph the line $3x + y = 7$ ($y = -3x + 7$) using a dashed line because the inequality is strict. See Figure 27(a).

**Step 2:** We select any test point that is not on the line and determine whether the test point satisfies the inequality. When the line does not contain the origin, it is usually easiest to choose the origin, (0, 0), as the test point.

$$3x + y < 7$$

Test Point: (0, 0): $3(0) + 0 \overset{?}{<} 7$

$0 \overset{?}{<} 7$ True

Because 0 is less than 7, the point $(0, 0)$ satisfies the inequality. Therefore, we shade the half-plane containing the point $(0, 0)$. See Figure 27(b). The shaded region represents the solution to the linear inequality.

**Figure 27**

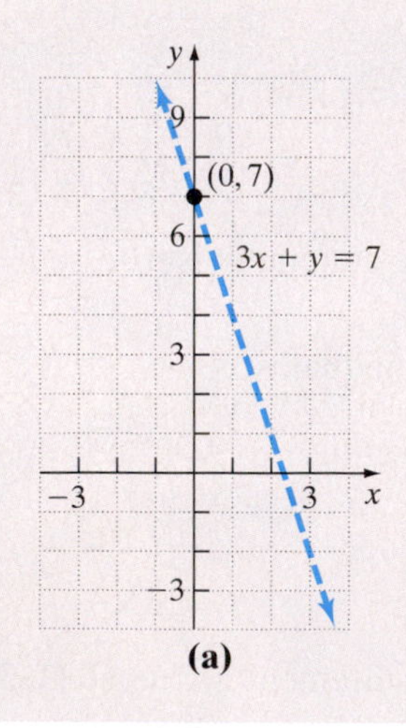

(a)

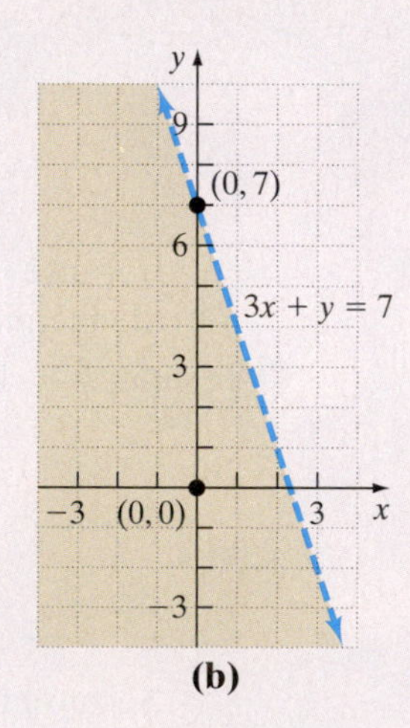

(b)

Only one test point is needed to obtain the graph of the inequality. Why? Consider Example 1. We notice that $A(2, 4)$ does not satisfy the inequality, while $B(-3, 1)$ and $C(1, 3)$ do satisfy the inequality. Notice that point $A$ is not in the shaded region of Figure 27(b), while points $B$ and $C$ are in the shaded region. So, if a point does not satisfy the inequality, then none of the points in the half-plane containing that point satisfy the inequality. If a point does satisfy the inequality, then all the points in the half-plane containing the point satisfy the inequality.

Below we summarize the steps for graphing a linear inequality in two variables.

**Work Smart**
An alternative to using test points is to solve the inequality for $y$. If the inequality is of the form $y >$ or $y \geq$, shade above the line. If the inequality is of the form $y <$ or $y \leq$, shade below the line.

### Steps for Graphing a Linear Inequality in Two Variables

**Step 1:** Replace the inequality symbol with an equal sign and graph the resulting equation. If the inequality is strict ($<$ or $>$), use dashes to graph the line; if the inequality is nonstrict ($\leq$ or $\geq$) use a solid line. The graph separates the $xy$-plane into two half-planes.

**Step 2:** Select a test point $P$ that is not on the line (that is, select a test point in one of the half-planes).

**(a)** If the coordinates of $P$ satisfy the inequality, then shade the half-plane containing $P$.

**(b)** If the coordinates of $P$ do not satisfy the inequality, then shade the half-plane that does not contain $P$.

**QUICK ✓** *Graph each linear inequality.*

**2.** $y < -2x + 3$ **3.** $6x - 3y \leq 15$

---

**EXAMPLE 3** **Graphing a Linear Inequality in Two Variables**

Graph the linear inequality $y \geq \frac{1}{2}x$.

**Solution**

We replace the inequality symbol with an equal sign to obtain $y = \frac{1}{2}x$. We graph the line $y = \frac{1}{2}x$ using a solid line because the inequality is non-strict. See Figure 28(a).

Now, we select any test point that is not on the line and determine whether the test point satisfies the inequality. Because the line contains the origin, we choose (2, 0) as the test point.

**Work Smart**
Do not use (0, 0) as a test point for equations of the form $Ax + By = 0$ because the graph of this equation contains the origin.

$$y \geq \frac{1}{2}x$$

Test Point: (2, 0): $$0 \overset{?}{\geq} \frac{1}{2} \cdot 2$$

$$0 \overset{?}{\geq} 1 \quad \text{False}$$

Because 0 is not greater than 1, the point (2, 0) does not satisfy the inequality. Therefore, we shade the half-plane that does not contain (2, 0). See Figure 28(b). The shaded region represents the solution to the linear inequality.

**Figure 28**

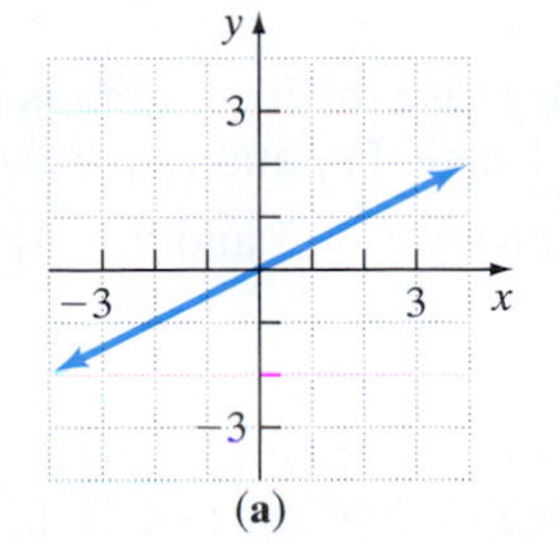

(a)

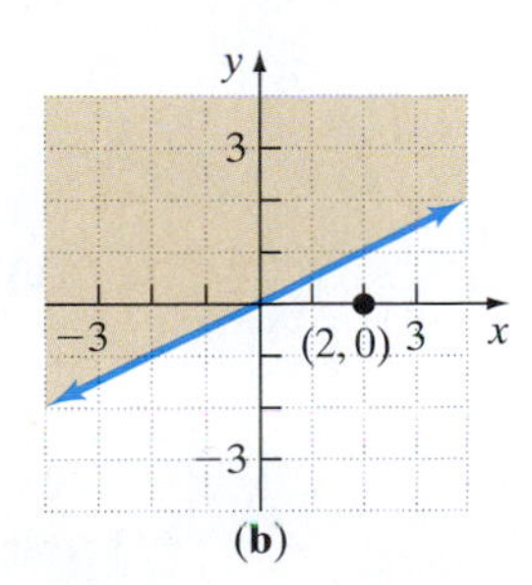

(b)

**QUICK ✓**

**4.** Graph the linear inequality $2x + y < 0$.

## 3 Solve Problems Involving Linear Inequalities

There are many applications of linear inequalities involving two variables that can be used to solve problems in areas such as nutrition, manufacturing, or sales. Let's look at an application of linear inequalities from nutrition.

### EXAMPLE 4 Saturated Fat Intake

Randy really enjoys Wendy's Junior Cheeseburger and Biggie French Fries. However, he knows that his intake of saturated fat during lunch should not exceed 16 grams. Each Junior Cheeseburger contains 6 grams of saturated fat and each Biggie Fries contains 3 grams of saturated fat. (*Source: wendys.com*)

**(a)** Write a linear inequality that describes Randy's options for eating at Wendy's. That is, write an inequality that represents all the combinations of Junior Cheeseburgers and Biggie Fries that Randy can order.

**(b)** Can Randy eat 2 Junior Cheeseburgers and 1 Biggie Fry during lunch and stay within his allotment of saturated fat?

**(c)** Can Randy eat 3 Junior Cheeseburgers and 1 Biggie Fry during lunch and stay within his allotment of saturated fat?

#### Solution

**(a)** We are going to use the first three steps in the problem solving strategy given in Chapter 1 on pages 65–66 to help us develop the linear inequality.

**Step 1: Identify** We want to determine the number of Junior Cheeseburgers and Biggie Fries Randy can eat while not exceeding 16 grams of saturated fat.

**Step 2: Name the Unknowns** Let $x$ represent the number of Junior Cheeseburgers that Randy eats and let $y$ represent the number of Biggie Fries Randy eats.

**Step 3: Translate** If Randy eats one Junior Cheeseburger, then he will consume 6 grams of saturated fat. If he eats two, then he will consume 12 grams of saturated fat. In general, if he eats $x$ Junior Cheeseburgers, he will consume $6x$ grams of saturated fat. Similar logic for the Biggie Fries tells us that if Randy eats $y$ Biggie Fries, he will consume $3y$ grams of saturated fat. The words "cannot exceed" imply a $\leq$ inequality. Therefore, a linear inequality that describes Randy's options for eating at Wendy's is

$$6x + 3y \leq 16$$

**(b)** Letting $x = 2$ and $y = 1$, we obtain

$$6(2) + 3(1) \overset{?}{\leq} 16$$

$$15 \overset{?}{\leq} 16 \quad \text{True}$$

Because the inequality is true, Randy can eat 2 Junior Cheeseburgers and 1 Biggie Fry and remain within the allotment of 16 grams of saturated fat.

**(c)** Letting $x = 3$ and $y = 1$, we obtain

$$6(3) + 3(1) \overset{?}{\leq} 16$$

$$21 \overset{?}{\leq} 16 \quad \text{False}$$

Because the inequality is false, Randy cannot eat 3 Junior Cheeseburgers and 1 Biggie Fry and remain within the allotment of 16 grams of saturated fat.

**QUICK ✓**

5. Avery is on a diet that requires that he consume no more than 800 calories for lunch. He really enjoys Wendy's Chicken Breast filet and Frosties. Each Chicken Breast filet contains 430 calories and each Frosty contains 330 calories.
   **(a)** Write a linear inequality that describes Avery's options for eating at Wendy's.
   **(b)** Can Avery eat 1 Chicken Breast filet and 1 Frosty and stay within his allotment of calories?
   **(c)** Can Avery eat 2 Chicken Breast filets and 1 Frosty and stay within his allotment of calories?

# 3.4 Exercises

**For Extra Help:** Student Solutions Manual | CD Video | PH Math/Tutor Center | MathXL Tutorials on CD | MathXL® | MyMathLab

## Concepts and Vocabulary

*In Problems 1 and 2, fill in the blanks.*

**1.** If we replace the inequality symbol in $Ax + By > C$ with an equal sign, we obtain the equation of a line, $Ax + By = C$. The line separates the $xy$-plane into two regions, called ________.

**2.** When we graph the line corresponding to $Ax + By \geq C$, we use a _____ line.

*In Problems 3 and 4, answer True or False to each statement.*

**3.** The graph of a linear inequality is a line.

**4.** In a graph of a linear inequality in two variables with a strict inequality, the line separating the two half-planes should be dashed.

**5.** If an ordered pair does not satisfy a linear inequality, then we shade the side opposite the point. Explain why this works.

**6.** Explain why we cannot use a test point that lies on the line separating the two half-planes when determining the solution to a linear inequality involving two variables.

## Building Skills

*In Problems 7–10, determine whether the given points are solutions to the linear inequality.*

**7.** $x + 3y < 6$
(a) $(0, 1)$
(b) $(-2, 4)$
(c) $(8, -1)$

**8.** $2x + y > -3$
(a) $(2, -1)$
(b) $(1, -3)$
(c) $(-5, 4)$

**9.** $-3x + 4y \geq 12$
(a) $(-4, 2)$
(b) $(0, 2)$
(c) $(0, 3)$

**10.** $2x - 5y \leq 2$
(a) $(1, -1)$
(b) $(3, 0)$
(c) $(-3, -2)$

*In Problems 11–30, graph each inequality.*

**11.** $y > 3$ **12.** $y < -2$ **13.** $x \geq -2$ **14.** $x < 7$

**15.** $y < 5x$ **16.** $y \geq \frac{2}{3}x$ **17.** $y > 2x + 3$ **18.** $y < -3x + 1$

**19.** $y \leq \frac{1}{2}x - 5$ **20.** $y \geq -\frac{4}{3}x + 5$ **21.** $3x + y \leq 4$ **22.** $-4x + y \geq -5$

**23.** $2x + 5y \leq -10$ **24.** $3x + 4y \geq 12$ **25.** $-4x + 6y > 24$ **26.** $-5x + 3y < 30$

**27.** $\frac{x}{2} + \frac{y}{3} < 1$ **28.** $\frac{x}{3} - \frac{y}{4} \leq 1$ **29.** $\frac{2}{3}x - \frac{3}{2}y \geq 2$ **30.** $\frac{5}{4}x - \frac{3}{5}y \leq 2$

## Applying the Concepts

**31. Nutrition** Sammy goes to McDonald's for lunch. He is on a diet that requires that he consume no more than 150 grams of carbohydrates for lunch. Sammy enjoys McDonald's Filet-o-Fish sandwiches and French fries. Each Filet-o-Fish has 45 grams of carbohydrates and each small order of French fries has 40 grams of carbohydrates. (*Source:* McDonald's Corp.)

**(a)** Write a linear inequality that describes Sammy's options for eating at McDonald's.
**(b)** Can Sammy eat two Filet-o-Fish and one order of fries and meet his carbohydrate requirement?
**(c)** Can Sammy eat three Filet-o-Fish and one order of fries and meet his carbohydrate requirement?

**32. Salesperson** Juanita sells two different computer models. For each Model A computer sold she makes \$45 and for each Model B computer sold she makes \$65. Juanita set a monthly goal of earning at least \$4000.

**(a)** Write a linear inequality that describes Juanita's options for making her sales goal.
**(b)** Will Juanita make her sales goal if she sells 50 Model A and 28 Model B computers?
**(c)** Will Juanita make her sales goal if she sells 41 Model A and 33 Model B computers?

**33. Production Planning** Acme Switch Company is a small manufacturing firm that makes two different styles of microwave switches. Switch A requires 2 hours to assemble, while Switch B requires 1.5 hours to assemble. Suppose that there are at most 80 hours of assembly time available each week.

**(a)** Write a linear inequality that describes the Acme's options for making the microwave switches.
**(b)** Can Acme manufacture 24 of Switch A and 41 of Switch B in a week?
**(c)** Can Acme manufacture 16 of Switch A and 45 of Switch B in a week?

**34. Budget Constraints** Johnny can spend no more than \$3.00 that he got from his grandparents. He goes to the candy store and wants to buy gummy bears that cost \$0.10 each and suckers that cost \$0.25 each.

**(a)** Write a linear inequality that describes Johnny's options for buying candy.
**(b)** Can Johnny buy 18 gummy bears and 5 suckers?
**(c)** Can Johnny buy 19 gummy bears and 4 suckers?

## Extending the Concepts

*In Problems 35–38, determine the linear inequality whose graph is given.*

**35.**

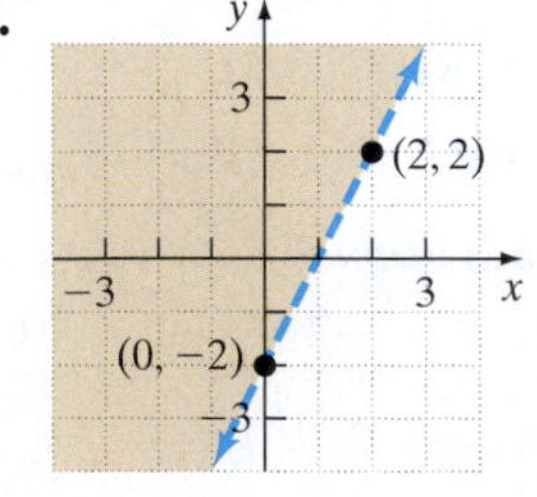


**36.**

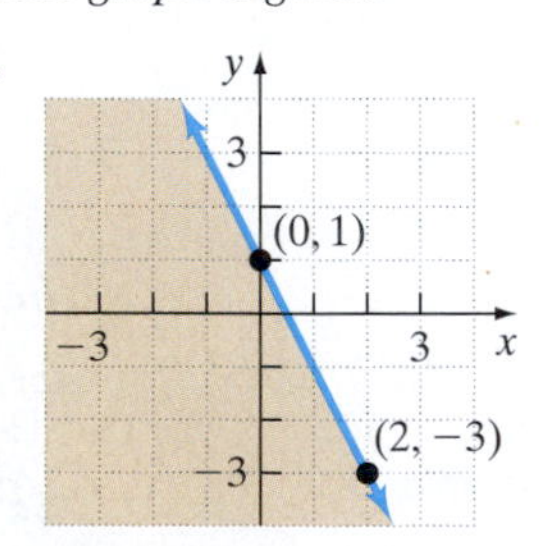


**37.**

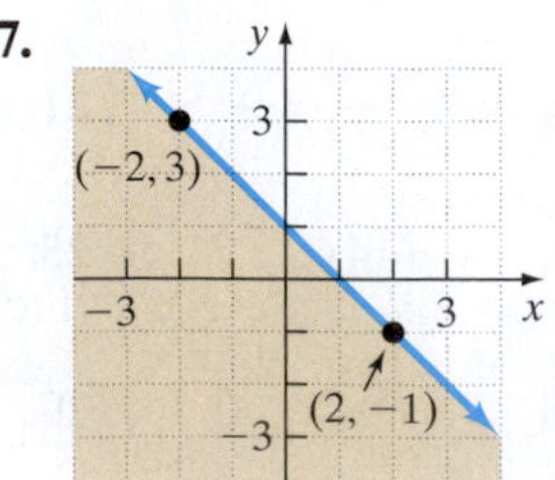


**38.**

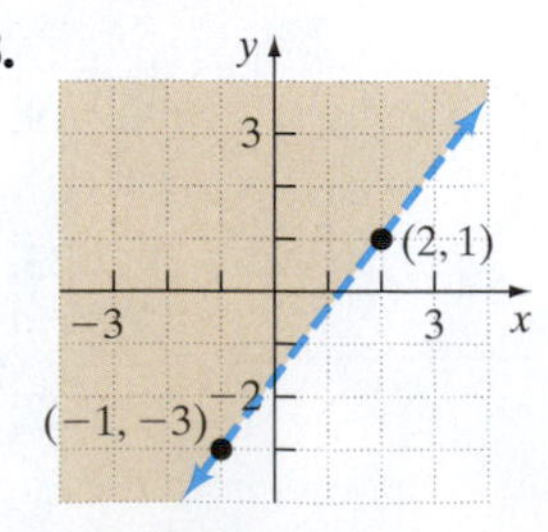

### The Graphing Calculator

Graphing calculators can be used to graph linear inequalities in two variables. Figure 29 shows the result of Example 2 using a TI-84 Plus graphing calculator. Consult your owner's manual for specific keystrokes.

**Figure 29**

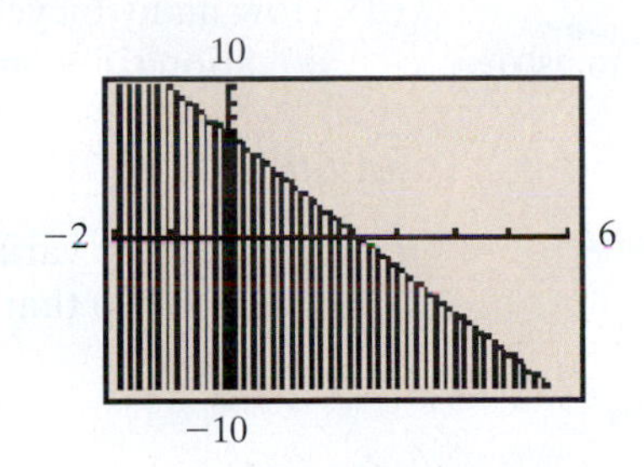


*In Problems 39–50, graph the inequalities using a graphing calculator.*

**39.** $y > 3$ **40.** $y < -2$ **41.** $y < 5x$ **42.** $y \geq \frac{2}{3}x$

**43.** $y > 2x + 3$ **44.** $y < -3x + 1$ **45.** $y \leq \frac{1}{2}x - 5$ **46.** $y \geq -\frac{4}{3}x + 5$

**47.** $3x + y \leq 4$ **48.** $-4x + y \geq -5$ **49.** $2x + 5y \leq -10$ **50.** $3x + 4y \geq 12$

# 3.5 Building Linear Models

**OBJECTIVES**

1. Build Linear Models from Verbal Descriptions
2. Build Linear Models Involving Direct Variation
3. Build Linear Models from Data

### *Preparing for Building Linear Models*

*Before getting started, take the following readiness quiz. If you get a problem wrong, go back to the section cited and review the material.*

1. What are the steps in our problem-solving approach? [Section 1.2, pp. 65–66]
2. Plot the points (0, 2), (3, 5), (1, 4), (5, 8) in the Cartesian plane. [Section 2.1, pp. 136–138]

## 1 Build Linear Models from Verbal Descriptions

In Section 3.1, we said that a linear function is of the form $f(x) = mx + b$, where $m$ and $b$ are real numbers. We now know that $m$ is the slope of the linear function and $b$ is its $y$-intercept. In Section 3.2, we said that the slope $m$ can be thought of as an average rate of change. The slope describes by how much a dependent variable changes for a given change in the independent variable. For example, in the linear function $f(x) = 4x + 3$, the slope is $4 = \frac{4}{1} = \frac{\Delta y}{\Delta x}$ so that if $x$ (the independent variable) increases by 1, then the dependent variable will increase by 4. When the average rate of change of a function is constant, then we can use linear functions to model the situation. For example, if your phone company charges you \$0.05 per minute to talk regardless of the number of minutes on the phone, then we can use a linear function to model the cost of talking with slope $m = \frac{0.05 \text{ dollars}}{1 \text{ minute}}$.

*Preparing for...Answers*

1. See pages 65–66.

2.

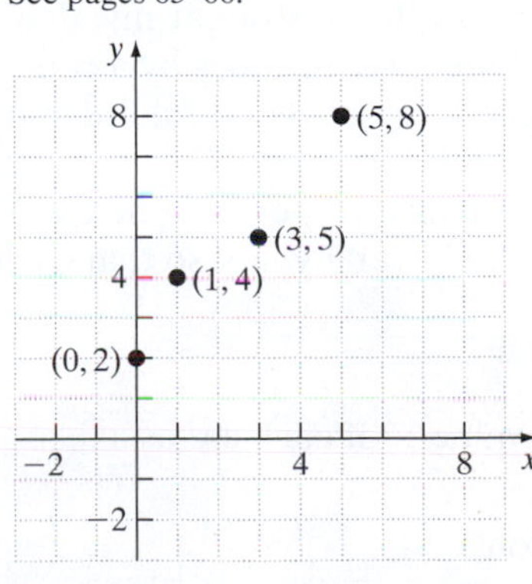


### EXAMPLE 1 Cost Function

The simplest cost function is the linear cost function $C(x) = ax + b$, where $b$ represents the fixed costs of operating a business and $a$ represents the variable costs (the cost of manufacturing one additional item). Suppose that a small bicycle manufacturer has daily fixed costs of \$2000 and each bicycle costs \$80 to manufacture.

Figure 30

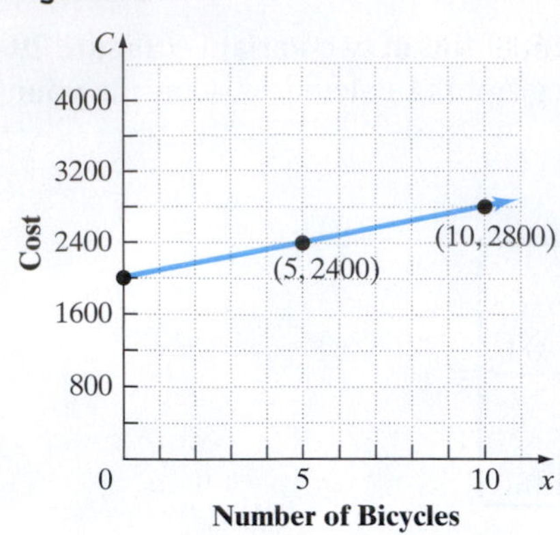


**(a)** Write a linear function that expresses the cost of manufacturing $x$ bicycles in a day.

**(b)** Graph the linear function.

**(c)** What is the cost of manufacturing 18 bicycles in a day?

**(d)** How many bicycles can be manufactured for $4080? (**Hint:** Solve the equation $C(x) = 4080$.)

## Solution

**(a)** Because the variable costs are $80, we have that $a = 80$. The fixed costs are $2000 so that $b = 2000$. Therefore, the cost function is

$$C(x) = 80x + 2000$$

**(b)** Label the horizontal axis $x$ and the vertical axis $C$. Figure 30 shows the graph of the cost function.

**(c)** We evaluate the function for $x = 18$ and obtain

$$\begin{aligned} C(18) &= 80(18) + 2000 \\ &= \$3440 \end{aligned}$$

It will cost $3440 to manufacture 18 bicycles.

**(d)** We solve $C(x) = 4080$.

$$\begin{aligned} C(x) &= 4080 \\ 80x + 2000 &= 4080 \\ \text{Subtract 2000 from both sides:} \quad 80x &= 2080 \\ \text{Divide both sides by 80:} \quad x &= 26 \end{aligned}$$

So 26 bicycles can be manufactured for a cost of $4080.

**QUICK ✓**

1. Suppose that the government imposes a tax of $1 per bicycle manufactured for the business presented in Example 1.

   **(a)** Write a linear function that expresses the cost $C$ of manufacturing $x$ bicycles in a day.

   **(b)** Graph the linear function.

   **(c)** What is the cost of manufacturing 18 bicycles in a day?

   **(d)** How many bicycles can be manufactured for $4025?

## EXAMPLE 2 Straight-Line Depreciation

*Book value* is the value of an asset such as a building or piece of machinery that a company uses to create its balance sheet. Some companies will use straight-line depreciation to depreciate their assets so that the value of the asset declines by a constant amount each year. The amount of the decline depends upon the useful life that the company places on the asset. Suppose that Prentice Hall Publishing Company just purchased a new fleet of cars for its sales force at a cost of $29,400 per car. The company will depreciate the cars using the straight-line method over 7 years, so that each car depreciates by $\dfrac{\$29{,}400}{7} = \$4200$ per year.

**(a)** Write a linear function that expresses the book value $V$ of each car as a function of its age, $x$.

**(b)** What is the implied domain of this linear function?

**(c)** What is the book value of each car after 3 years?

**(d)** When will the book value of each car be \$12,600? (**Hint:** Solve the equation $V(x) = 12{,}600$.)

**(e)** Graph the linear function.

## Solution

**(a)** We let $V(x)$ represent the book value of each car after $x$ years, so $V(x) = mx + b$. The original value of the car is \$29,400, so $V(0) = 29{,}400$. The $V$-intercept of the linear function is \$29,400. Because each car depreciates by \$4200 per year, the slope of the linear function is $-4200$. The linear function that represents the book value of the car after $x$ years is given by

$$V(x) = -4200x + 29{,}400$$

**(b)** Because the car cannot have a negative age, we know that the age, $x$, must be greater than or equal to zero. In addition, the car is depreciated over 7 years. After 7 years the book value of the car is $V(7) = 0$. Therefore, the implied domain of the function is $\{x \mid 0 \leq x \leq 7\}$, or using interval notation $[0, 7]$.

**(c)** The book value of the car after $x = 3$ years is given by $V(3)$.

$$\begin{aligned} V(3) &= -4200(3) + 29{,}400 \\ &= \$16{,}800 \end{aligned}$$

**(d)** To find when the book value is \$12,600, we solve the equation

$$\begin{aligned} V(x) &= 12{,}600 \\ -4200x + 29{,}400 &= 12{,}600 \\ \text{Subtract 29,400 from both sides:} \quad -4200x &= -16{,}800 \\ \text{Divide both sides by 4:} \quad x &= 4 \end{aligned}$$

Each car will have a book value of \$12,600 after 4 years.

**(e)** Label the horizontal axis $x$ and the vertical axis $V$. Since $V(0) = 29{,}400$, we know that $(0, 29400)$ is on the graph. Since $V(7) = 0$, we know that $(7, 0)$ is on the graph. To graph the function, we use these points (the intercepts), along with points $(3, 16800)$ and $(4, 12600)$ found in parts (c) and (d). See Figure 31.

**Figure 31**

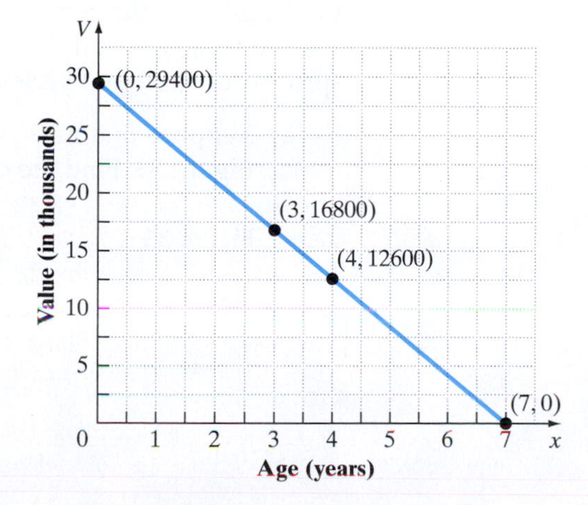

QUICK ✓

2. Roberta just purchased a new car. Her monthly payments are \$250 per month. She estimates that maintenance and gas cost her \$0.18 per mile.
   **(a)** Write a linear function that relates the monthly cost $C$ of operating the car as a function of miles driven, $x$.
   **(b)** What is the implied domain of this linear function?
   **(c)** What is the monthly cost of driving 320 miles?
   **(d)** Graph the linear function.
   **(e)** How many miles can Roberta drive each month if she can afford the monthly cost to be \$282.40?

## 2 Build Linear Models Involving Direct Variation

Often two variables are related in terms of proportions. For example, we say "Revenue is proportional to sales" or "Force is proportional to acceleration." When we say that one variable is proportional to another variable, we are really talking about *variation*. **Variation** refers to how one quantity varies in relation to some other quantity. Quantities may vary *directly, inversely,* or *jointly*. We will discuss direct variation now.

**DEFINITION**

Suppose we let $x$ and $y$ represent two quantities. We say that $y$ **varies directly** with $x$, or $y$ is **directly proportional to** $x$, if there is a nonzero number $k$ such that

$$y = kx$$

The number $k$ is called the **constant of proportionality.**

Figure 32
$y = kx, k > 0, x \geq 0$

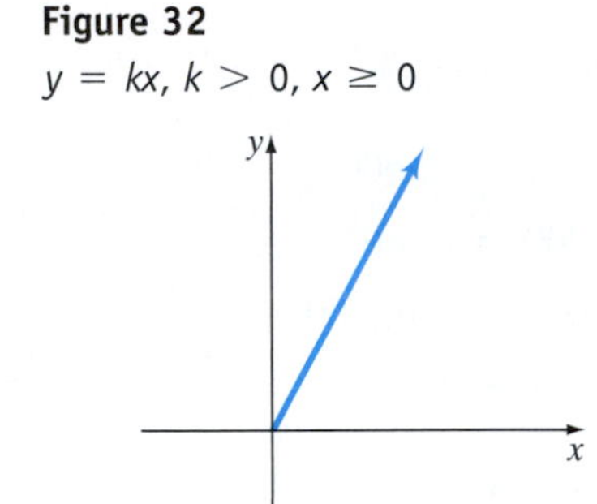


If $y$ varies directly with $x$, then $y$ is a linear function of $x$. The graph in Figure 32 illustrates the relationship between $y$ and $x$ if $y$ varies directly with $x$ and $k > 0$, $x \geq 0$. Notice that the constant of proportionality is the slope of the line and the $y$-intercept is 0.

If we know that two quantities vary directly, then knowing the value of each quantity in one instance allows us to write a formula that is true in all cases.

### EXAMPLE 3 Hooke's Law

Hooke's Law states that the force (or weight) on a spring is directly proportional to the length that the spring stretches from its "at rest" position. That is, $F = kx$, where $F$ is the force exerted, $x$ is the extension of the spring, and $k$ is the proportionality (or spring) constant that varies from spring to spring.

**(a)** Suppose that a 20-pound weight causes a spring to stretch 10 inches. See Figure 33. Find the constant of proportionality, $k$.

Figure 33

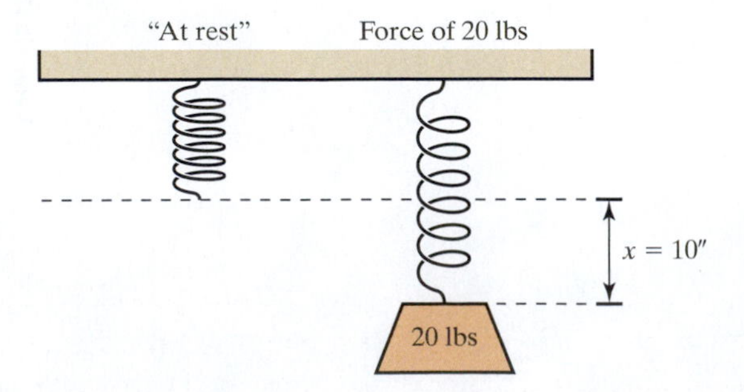

**(b)** Write the relation between force and stretch length using function notation.

**(c)** Suppose a weight is attached to the spring and it stretches 8 inches. What is the weight attached to the spring?

**(d)** Graph the relation between force and length that the spring stretches.

### Solution

**(a)** We know that $F = kx$ and that $F = 20$ pounds when $x = 10$ inches. Substituting, we have that

$$F = kx$$
$$20 = k(10)$$
$$k = 2$$

**(b)** With $k = 2$, we have that $F = 2x$. We write this using function notation as

$$F(x) = 2x$$

**(c)** If the spring stretches $x = 8$ inches, the force exerted is

$$F(8) = 2(8) = 16 \text{ pounds}$$

**(d)** Figure 34 shows the relation between force and length of the spring.

Figure 34

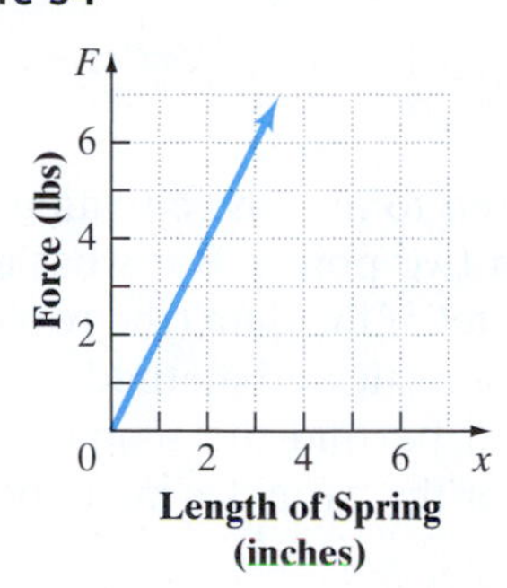


## EXAMPLE 4 Car Payments

Suppose that Dulce just purchased a used car for $10,000. She decides to put $1000 down on the car and borrow the remaining $9000. The bank lends Dulce $9000 at 4.9% interest for 48 months. Her payments are $206.86. The monthly payment $p$ on a car varies directly with the amount borrowed $b$.

**(a)** Find a function that relates the monthly payment $p$ to the amount borrowed $b$ for any car loan with the same terms.

**(b)** Suppose that Dulce put $2000 down on the car instead. What would her monthly payment be?

**(c)** Graph the relation between monthly payment and amount borrowed.

### Solution

**(a)** Because $p$ varies directly with $b$, we know that

$$p = kb$$

for some constant $k$. Because $p = \$206.86$ when $b = \$9000$, it follows that

$$206.86 = k(9000).$$

Solving this equation for $k$ by dividing both sides of the equation by 9000, we find that

$$k = 0.022984$$

So, we have that

$$p = 0.022984b$$

We can write this as a linear function:

$$p(b) = 0.022984b$$

**(b)** When $b = \$8000$, we have that

$$p(8000) = 0.022984(8000)$$
$$= \$183.87$$

If Dulce put $2000 down, her payment would be $183.87.

**(c)** Figure 35 shows the relationship between the monthly payment $p$ and the amount borrowed $b$.

**Work Smart**

To avoid round-off error, do not round the value of $k$ to fewer than 5 decimal places.

Figure 35

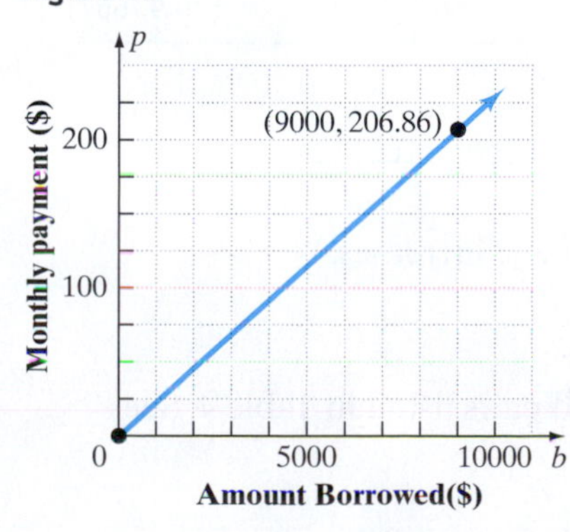

QUICK ✓

3. The cost of gas $C$ varies directly with the number of gallons pumped, $g$. Suppose that the cost of pumping 8 gallons of gas is \$22.39.
   **(a)** Find a function that relates the cost of gas $C$ to the number of gallons pumped $g$.
   **(b)** Suppose that 5.5 gallons are pumped into your car. What would the cost be?
   **(c)** Graph the relation between cost and number of gallons pumped.

## 3 Build Linear Models from Data

In Section 3.2, we learned that only two points are required to find the equation of a line. We also learned how to build a linear function from two points. But what if we have a set of data with more than two points? How can we tell if the data (the two variables) are related linearly? While there are some rather sophisticated methods for determining whether two variables are linearly related (and beyond the scope of this course), we can draw a picture of the data and learn whether the variables might be linearly related. This picture is called a *scatter diagram.*

### Scatter Diagrams

Often, we are interested in finding an equation that can describe the relation between two variables. The first step in determining the type of equation that should be used to describe the relation is to plot the ordered pairs that make up the relation in the Cartesian plane. The graph that results is called a **scatter diagram.**

### EXAMPLE 5 Drawing a Scatter Diagram

In baseball, the on-base percentage for a team represents the percentage of time that the team safely reaches base. The data given in Table 4 represent the number of runs scored and the on-base percentage for various teams during the 2002 baseball season.

**Table 4**

| Team | On-Base Percentage, $x$ | Runs Scored, $y$ | $(x, y)$ |
|---|---|---|---|
| NY Yankees | 35.4 | 897 | (35.4, 897) |
| Anaheim Angels | 34.1 | 851 | (34.1, 851) |
| Texas Rangers | 33.8 | 843 | (33.8, 843) |
| Toronto Blue Jays | 32.7 | 813 | (32.7, 813) |
| Minnesota Twins | 33.2 | 768 | (33.2, 768) |
| Oakland A's | 33.9 | 800 | (33.9, 800) |
| Kansas City Royals | 32.3 | 737 | (32.3, 737) |
| Baltimore Orioles | 30.9 | 667 | (30.9, 667) |

SOURCE: *espn.com*

**(a)** Draw a scatter diagram of the data treating on-base percentage as the independent variable.

**(b)** Describe what happens as the on-base percentage increases.

**Solution**

**(a)** To draw a scatter diagram we plot the ordered pairs listed in Table 4. See Figure 36.

**Figure 36**

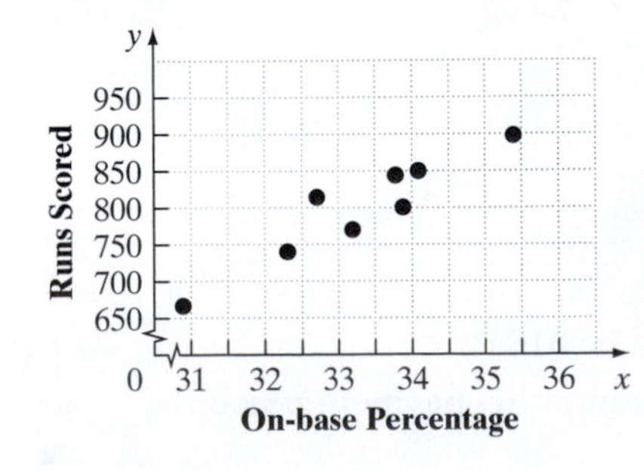

**(b)** From the scatter diagram, we can see that as the on-base percentage increases, the number of runs scored also increases. While the relation between on-base percentage and number of runs scored does not follow a perfect linear relation (because the points don't all fall on a straight line), we can agree that the pattern of the data is linear.

**QUICK ✓**

**4.** The data listed below represent the total cholesterol (in mg/dL) and age of males.

| Age | Total Cholesterol | Age | Total Cholesterol |
|---|---|---|---|
| 25 | 180 | 38 | 239 |
| 25 | 195 | 48 | 204 |
| 28 | 186 | 51 | 243 |
| 32 | 180 | 62 | 228 |
| 32 | 197 | 65 | 269 |

**(a)** Draw a scatter diagram treating age as the independent variable.
**(b)** Describe the relation between age and total cholesterol.

## Recognizing the Type of Relation That Appears to Exist between Two Variables

We use scatter diagrams to help us see the type of relation that exists between two variables. In this text, we will look at a few different types of relations between two variables. For now, however, our only goal is to distinguish between linear and nonlinear relations. See Figure 37.

**Figure 37**

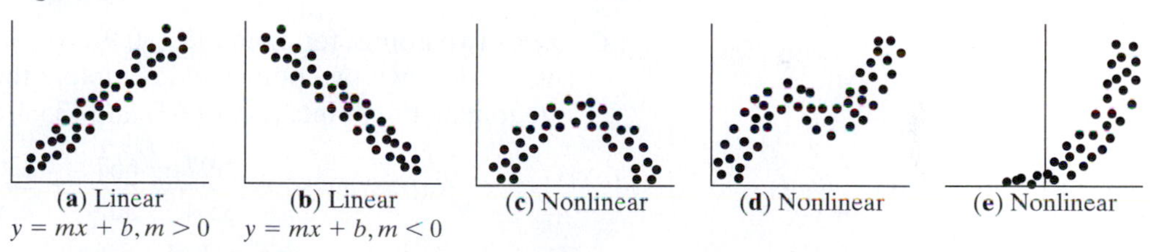

(a) Linear $y = mx + b, m > 0$ (b) Linear $y = mx + b, m < 0$ (c) Nonlinear (d) Nonlinear (e) Nonlinear

## EXAMPLE 6 Distinguishing between Linear and Nonlinear Relations

Determine whether the relation between the two variables in Figure 38 is linear or nonlinear. If the relation is linear, indicate whether the slope is positive or negative.

**Figure 38**



(a) (b) (c)

**Solution**

**(a)** Linear with negative slope **(b)** Nonlinear **(c)** Nonlinear

**QUICK ✓** *Determine whether the relation between the two variables is linear or nonlinear. If it is linear, determine whether the slope is positive or negative.*

**5.**

**6.**

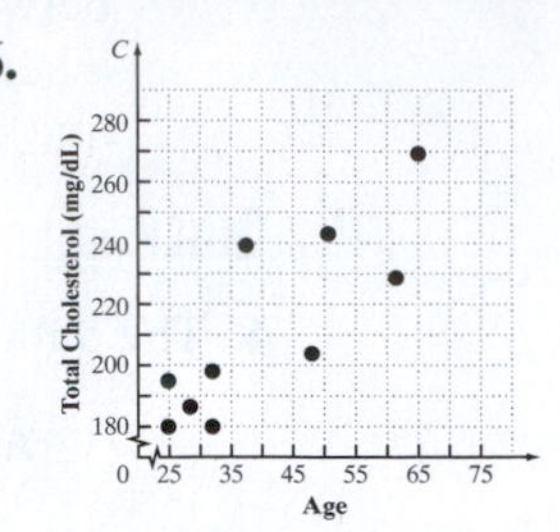


## Fitting a Line to Data

Suppose that the scatter diagram of a set of data appears to be linearly related as in Figure 37(a) or (b). We might wish to find an equation of a line that relates the two variables. One way to obtain an equation for data that appears to follow a linear pattern is to draw a line through two points on the scatter diagram and determine the equation of the line.

### EXAMPLE 7 Finding an Equation for Linearly Related Data

Using the data in Table 4 from Example 5,

**(a)** Select two points and find an equation of the line containing the points.

**(b)** Graph the line on the scatter diagram obtained in Example 5(a).

**(c)** Use the line found in part (a) to predict the number of runs scored by a team whose on-base percentage is 34.6%.

**(d)** Interpret the slope. Does it make sense to interpret the $y$-intercept?

**Solution**

**(a)** Select two points, for example, (30.9, 667) and (35.4, 897). (You should select your own two points and complete the solution.) The slope of the line joining the points (30.9, 667) and (35.4, 897) is

$$m = \frac{897 - 667}{35.4 - 30.9} = \frac{230}{4.5} = 51.1$$

The equation of the line with slope 51.1 and passing through (30.9, 667) is found using the point-slope form with $m = 51.1$, $x_1 = 30.9$, and $y_1 = 667$.

Point-slope form: $y - y_1 = m(x - x_1)$

$m = 51.1; x_1 = 30.9, y_1 = 667$: $y - 667 = 51.1(x - 30.9)$

Distribute 51.1: $y - 667 = 51.1x - 1578.99$

Add 667 to both sides: $y = f(x) = 51.1x - 911.99$

**(b)** Figure 39 shows the scatter diagram with the graph of the line found in part (a). We obtain the graph of the line by drawing the line through the two points selected in part (a).

**Figure 39**

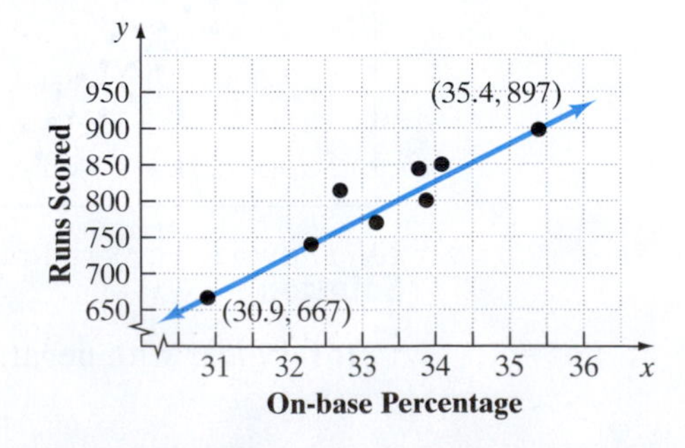

**(c)** We evaluate $f(x) = 51.1x - 911.99$ at $x = 34.6$.

$$\begin{aligned} f(34.6) &= 51.1(34.6) - 911.99 \\ &= 856.07 \end{aligned}$$

We round this to the nearest whole number. We predict that a team whose on-base percentage is 34.6% will score 856 runs.

**(d)** The slope of the linear function is 51.1. This means that if the on-base percentage increases by 1%, then the number of runs scored will increase by about 51 runs. The $y$-intercept, $-911.99$, represents the runs scored when on-base percentage is 0. Since negative runs scored does not make sense and we do not have any observations near zero, it does not make sense to interpret the $y$-intercept.

**7.** Using the data from Quick Check Problem 4 on page 237,

**(a)** Select two points and find an equation of the line containing the points.

**(b)** Graph the line on the scatter diagram obtained in Quick Check Problem 4 (page 237).

**(c)** Predict the total cholesterol of a 39-year-old male.

**(d)** Interpret the slope. Does it make sense to interpret the $y$-intercept?

# 3.5 Exercises

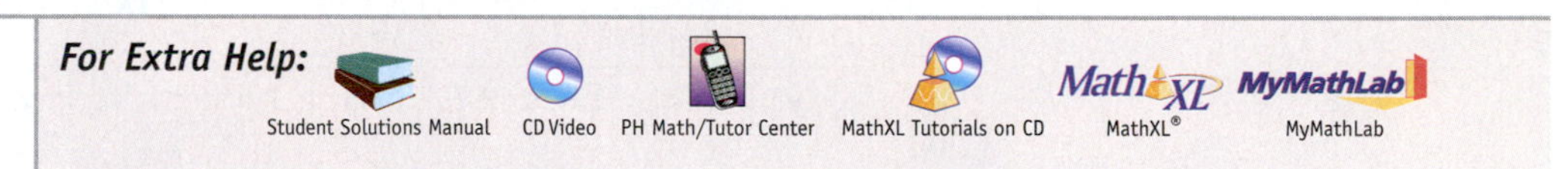


## Concepts and Vocabulary

*In Problems 1–3, fill in the blanks.*

**1.** A _____ _____ is used to help us to see the type of relation, if any, that may exist between two variables.

**2.** _____ refers to how one quantity varies in relation to some other quantity.

**3.** If $x$ and $y$ are two quantities, then $y$ is directly proportional to $x$ if there is a nonzero number $k$ such that _____.

*In Problem 4, answer True or False to the statement.*

**4.** If $y$ varies directly with $x$, then $y = kx$. The number $k$ is called the constant of proportionality.

**5.** What does it mean if we say that a scatter diagram indicates that two variables are linearly related with a positive slope?

**6.** What would it mean if a teacher states that a student's average in the class is directly proportional to the amount of time spent studying for the course?

## Building Skills

*In Problems 7–12, (a) find the constant of proportionality k, (b) write the linear function relating the two variables, and (c) find the quantity indicated.*

**7.** Suppose that $y$ varies directly with $x$. When $x = 5$, then $y = 30$. Find $y$ when $x = 7$.

**8.** Suppose that $y$ varies directly with $x$. When $x = 3$, then $y = 15$. Find $y$ when $x = 5$.

**9.** Suppose that $y$ is directly proportional to $x$. When $x = 7$, then $y = 3$. Find $y$ when $x = 28$.

**10.** Suppose that $y$ is directly proportional to $x$. When $x = 20$, then $y = 4$. Find $y$ when $x = 35$.

**11.** Suppose that $y$ is directly proportional to $x$. When $x = 8$, then $y = 4$. Find $y$ when $x = 30$.

**12.** Suppose that $y$ is directly proportional to $x$. When $x = 12$, then $y = 8$. Find $y$ when $x = 20$.

*In Problems 13–16, determine whether the scatter diagram indicates that a linear relation may exist between the two variables. If a linear relation does exist, indicate whether the slope is positive or negative.*

**13.**

**14.**

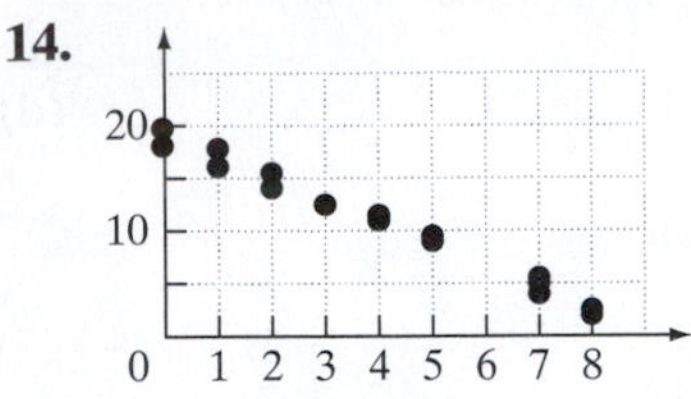


**15.**

**16.**

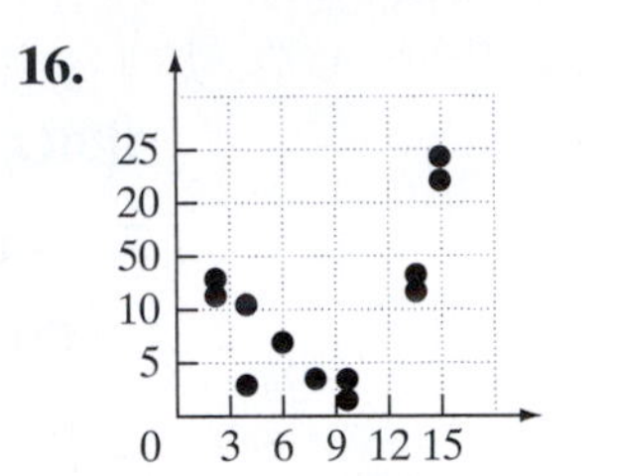


*In Problems 17–20,*

*(a) Draw a scatter diagram of the data.*

*(b) Select two points from the scatter diagram and find the equation of the line containing the points selected.**

*(c) Graph the line found in part (b) on the scatter diagram.*

**17.**

| $x$ | 2 | 4 | 5 | 8 | 9 |
|---|---|---|---|---|---|
| $y$ | 1.4 | 1.8 | 2.1 | 2.3 | 2.6 |

**18.**

| $x$ | 2 | 3 | 5 | 6 | 7 |
|---|---|---|---|---|---|
| $y$ | 5.7 | 5.2 | 2.8 | 1.9 | 1.8 |

**19.**

| $x$ | 1.2 | 1.8 | 2.3 | 3.5 | 4.1 |
|---|---|---|---|---|---|
| $y$ | 8.4 | 7.0 | 7.3 | 4.5 | 2.4 |

**20.**

| $x$ | 0 | 0.5 | 1.4 | 2.1 | 3.9 |
|---|---|---|---|---|---|
| $y$ | 0.8 | 1.3 | 1.9 | 2.5 | 5.0 |

## Applying the Concepts

**21. Phone Charges** Sprint has a long-distance phone plan that charges a monthly fee of \$8.95 plus \$0.05 per minute. (*Source: Sprint.com*)

**(a)** Find a linear function that expresses the monthly bill $B$ as a function of minutes used $m$.
**(b)** What are the independent and dependent variables?
**(c)** What is the implied domain of this linear function?
**(d)** What is the monthly bill if 300 minutes are used for long-distance phone calls?
**(e)** How many minutes were used for long distance if the long-distance phone bill was \$20.95?
**(f)** Graph the linear function.

**22. RV Rental** The rental cost $R$ of a class C 20-foot recreational vehicle is \$129.50 plus \$0.15 per mile. (*Source: westernrv.com*)

**(a)** Find a linear function that expresses the cost $R$ as a function of miles driven $m$.
**(b)** What are the independent and dependent variables?
**(c)** What is the implied domain of this linear function?
**(d)** What is the rental cost if 860 miles are driven?
**(e)** How many miles were driven if the rental cost is \$213.80?
**(f)** Graph the linear function.

*Answers will vary.

**23. Depreciation** Suppose that a company has just purchased a new computer for \$2700. The company chooses to depreciate the computer using the straight-line method over 3 years.

**(a)** Find a linear function that expresses the book value $V$ of the computer as a function of its age $x$.
**(b)** What is the implied domain of this linear function?
**(c)** What is the book value of the computer after the first year?
**(d)** What are the intercepts of the graph of the linear function?
**(e)** When will the book value of the computer be \$900?
**(f)** Graph the linear function.

**24. Depreciation** Suppose that a company just purchased a new machine for its manufacturing facility for \$1,200,000. The company chooses to depreciate the machine using the straight-line method over 20 years.

**(a)** Find a linear function that expresses the book value $V$ of the machine as a function of its age $x$.
**(b)** What is the implied domain of this linear function?
**(c)** What is the book value of the machine after three years?
**(d)** What are the intercepts of the graph of the linear function?
**(e)** When will the book value of the machine be \$480,000?
**(f)** Graph the linear function.

**25. Mortgage Payments** The monthly payment $p$ on a mortgage varies directly with the amount borrowed $b$. Suppose that you decide to borrow \$120,000 using a 30-year mortgage at 5.75% interest. You are told that your payment is \$700.29.

**(a)** Write a linear function that relates the monthly payment $p$ to the amount borrowed $b$ for a mortgage with the same terms.
**(b)** Assume that you have decided to buy a more expensive home that requires you to borrow \$140,000. What will your monthly payment be?
**(c)** Graph the relation between monthly payment and amount borrowed.

**26. Mortgage Payments** The monthly payment $p$ on a mortgage varies directly with the amount borrowed $b$. Suppose that you decide to borrow \$120,000 using a 15-year mortgage at 5.5% interest. You are told that your payment is \$980.50.

**(a)** Write a linear function that relates the monthly payment $p$ to the amount borrowed $b$ for a mortgage with the same terms.
**(b)** Assume that you have decided to buy a more expensive home that requires you to borrow \$150,000. What will your monthly payment be?
**(c)** Graph the relation between monthly payment and amount borrowed.

**27. Revenue Function** The cost $C$ of purchasing chocolate-covered almonds varies directly with the weight $w$ in pounds. Suppose that the cost of purchasing 5 pounds of chocolate-covered almonds is \$28.

**(a)** Write a linear function that relates the cost $C$ to the number of pounds of chocolate-covered almonds purchased $w$.
**(b)** What would it cost to purchase 3.5 pounds of chocolate-covered almonds?
**(c)** Graph the relation between cost and weight.

**28. Conversion** Suppose that you are planning a trip to Europe so you need to obtain some euros. The amount received in euros varies directly with the amount in U.S. dollars. Your friend just converted \$600 into 520 euros.

**(a)** Write a linear function that relates the number of euros $E$ to the number of U.S. dollars $d$.
**(b)** If you wish to convert \$700 into euros, how many euros would you receive?
**(c)** Graph the relation between euros and U.S. dollars.

**29. Falling Objects** The velocity of a falling object (ignoring air resistance) $v$ is directly proportional to the time $t$ of the fall. If, after 2 seconds, the velocity of the object is 64 feet per second, what will its velocity be after 3 seconds?

**30. Circumference of a Circle** The circumference of a circle $C$ is directly proportional to its radius $r$. If the circumference of a circle whose radius is 5 inches is $10\pi$ inches, what is the circumference of a circle whose radius is 8 inches?

**31. Concrete** As concrete cures, it gains strength. The following data represent the 7-day and 28-day strength (in pounds per square inch) of a certain type of concrete.

| 7-day Strength, *x* | 28-day Strength, *y* | 7-day Strength, *x* | 28-day Strength, *y* |
|---|---|---|---|
| 2300 | 4070 | 2480 | 4120 |
| 3390 | 5220 | 3380 | 5020 |
| 2430 | 4640 | 2660 | 4890 |
| 2890 | 4620 | 2620 | 4190 |
| 3330 | 4850 | 3340 | 4630 |

**(a)** Draw a scatter diagram of the data treating 7-day strength as the independent variable.
**(b)** What type of relation appears to exist between 7-day strength and 28-day strength?
**(c)** Select two points and find an equation of the line containing the points.
**(d)** Graph the line on the scatter diagram drawn in part (a).
**(e)** Predict the 28-day strength of a slab of concrete if its 7-day strength is 3000 psi.
**(f)** Interpret the slope of the line found in part (c).

**32. Candy** The following data represent the weight (in grams) of various candy bars and the corresponding number of calories.

| Candy Bar | Weight, *x* | Calories, *y* |
|---|---|---|
| Hershey's Milk Chocolate | 44.28 | 230 |
| Nestle Crunch | 44.84 | 230 |
| Butterfinger | 61.30 | 270 |
| Baby Ruth | 66.45 | 280 |
| Almond Joy | 47.33 | 220 |
| Twix (with Caramel) | 58.00 | 280 |
| Snickers | 61.12 | 280 |
| Heath | 39.52 | 210 |

SOURCE: *Megan Pocius, student at Joliet Junior College*

**(a)** Draw a scatter diagram of the data treating weight as the independent variable.
**(b)** What type of relation appears to exist between the weight of a candy bar and the number of calories?
**(c)** Select two points and find an equation of the line containing the points.
**(d)** Graph the line on the scatter diagram drawn in part (a).
**(e)** Predict the number of calories in a candy bar that weighs 62.3 grams.
**(f)** Interpret the slope of the line found in part (c).

**33. Raisins** The following data represent the weight (in grams) of a box of raisins and the number of raisins in the box.

| Weight, *w* | Number of Raisins, *N* | Weight, *w* | Number of Raisins, *N* |
|---|---|---|---|
| 42.3 | 87 | 42.4 | 90 |
| 42.7 | 91 | 42.3 | 82 |
| 42.8 | 93 | 42.5 | 86 |
| 42.4 | 87 | 42.7 | 86 |
| 42.6 | 89 | 42.5 | 86 |

SOURCE: *Jennifer Maxwell, student at Joliet Junior College*

**(a)** Does the relation defined by the set of ordered pairs $(w, N)$ represent a function?
**(b)** Draw a scatter diagram of the data treating weight as the independent variable.
**(c)** Select two points and find the equation of the line containing the points.
**(d)** Graph the line on the scatter diagram drawn in part (b).
**(e)** Express the relationship found in part (c) using function notation.
**(f)** Predict the number of raisins in a box that weighs 42.5 grams.
**(g)** Interpret the slope of the line found in part (c).

**34. Height versus Head Circumference** The following data represent the height (in inches) and head circumference (in inches) of 9 randomly selected children.

**(a)** Does the relation defined by the set of ordered pairs $(h, C)$ represent a function?
**(b)** Draw a scatter diagram of the data treating height as the independent variable.
**(c)** Select two points and find the equation of the line containing the points.
**(d)** Graph the line on the scatter diagram drawn in part (b).
**(e)** Express the relationship found in part (c) using function notation.
**(f)** Predict the head circumference of a child who is 26.5 inches tall.
**(g)** Interpret the slope of the line found in part (c).

| Height, $h$ | Head Circumference, $C$ |
|---|---|
| 25.25 | 16.4 |
| 25.75 | 16.9 |
| 25 | 16.9 |
| 27.75 | 17.6 |
| 26.50 | 17.3 |
| 27.00 | 17.5 |
| 26.75 | 17.3 |
| 26.75 | 17.5 |
| 27.5 | 17.5 |

SOURCE: *Denise Slucki, student at Joliet Junior College*

## Extending the Concepts

**35.** A strain of *E. coli* Beu 397-recA441 is placed into a Petri dish at 30° Celsius and allowed to grow. The population is estimated by means of an optical device in which the amount of light that passes through the Petri dish is measured. The following data are collected. Do you think that a linear function could be used to describe the relation between the two variables? Why or why not?

| Time, $x$ | Population, $y$ |
|---|---|
| 0 | 0.09 |
| 2.5 | 0.18 |
| 3.5 | 0.26 |
| 4.5 | 0.35 |
| 6 | 0.50 |

SOURCE: *Dr. Polly Lavery, Joliet Junior College*

**36.** Suppose that $y$ is directly proportional to $x^2$. If $x$ is doubled, what happens to the value of $y$?

## The Graphing Calculator

The line obtained in Example 7 depends on the points selected, which will vary from person to person. So the line we found might be different from the line that you found. Although the line that we found in Example 7 fits the data well, there may be a line that "fits it better." Do you think that your line fits the data better? Is there a line of *best fit*? As it turns out, there is a method for finding the line that best fits linearly related data (called the *line of best fit*).*

*We shall not discuss in this book the underlying mathematics of lines of best fit. Books in elementary statistics discuss this topic.

Graphing utilities can be used to draw scatter diagrams and find the line of best fit. Figure 40(a) shows a scatter diagram of the data presented in Table 4 from Example 5 drawn on a TI-84 Plus graphing calculator. Figure 40(b) shows the line of best fit from a TI-84 Plus graphing calculator.

The line of best fit is $y = 50.3x - 877.4$.

**Figure 40**

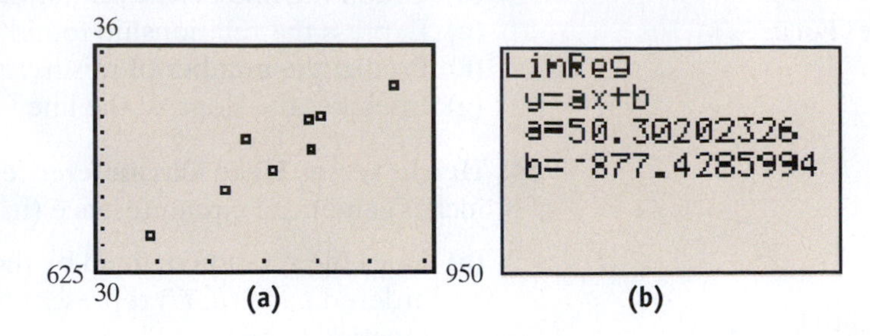


*In Problems 37–40,*

**(a)** *Draw a scatter diagram using a graphing calculator*

**(b)** *Find the line of best fit using a graphing calculator*

*for the data in the problem specified.*

**37.** Problem 31

**38.** Problem 32

**39.** Problem 33

**40.** Problem 34

## CHAPTER 3 ACTIVITY: DOES A COMMON THREAD EXIST?

**Focus:** Discover the common theme (thread) that exists between different linear equations and inequalities.

**Time:** 15–20 minutes

**Group size:** 2–4

Each member, on your own, examine the expressions in each row. Does a common thread exist? Be ready to support your opinion with mathematical vocabulary and properties.

| | A | B | C |
|---|---|---|---|
| 1 | $2y - x = x - y - 15$ | $8(y + 6) - 5(x - 1) = 2(x + 4) + 5$ | $2(y + 5) - 3x = -2x$ |
| 2 | $x(x + 5) - 8y = x^2 + 3$ | $\frac{5}{2}\left(x - \frac{2}{5}\right) - 4y + \frac{1}{3}(x + 1) = \frac{1}{2} + \frac{1}{3}(x + 1)$ | $2(x + 2) + 3\left(x - \frac{8}{3}y\right) = 7$ |
| 3 | $\frac{1}{2}x(x + 1) = \frac{1}{2}(x^2 - 3)$ | $y$-axis | A line perpendicular to $-3y + 9 = -3$ |
| 4 | $3\left(x - \frac{1}{3}\right) \leq -(4y + 1)$ | $3(x + 2) \geq y + 6$ | $x = 0$ |
| 5 | A rental car that costs \$100 per week and \$0.13 for each mile driven. | A cell phone plan that charges \$19.99 per month and \$0.10 per minute used. | Joe's weekly salary is \$250 plus \$10 for every new account he registers. |

# CHAPTER 3 REVIEW

## Section 3.1 Linear Equations and Linear Functions

| KEY CONCEPTS | KEY TERMS |
|---|---|
| • **General Form of a Line** $Ax + By = C$, where $A$, $B$, and $C$ are real numbers. $A$ and $B$ are not both 0.<br>• **Finding Intercepts** $x$-intercept(s): Let $y = 0$ in the equation and solve for $x$; $y$-intercept(s): Let $x = 0$ in the equation and solve for $y$<br>• **Equation of a Vertical Line** $x = a$, where $a$ is the $x$-intercept<br>• **Equation of a Horizontal Line** $y = b$, where $b$ is the $y$-intercept<br>• **Linear Function** $f(x) = mx + b$, where $m$ and $b$ are real numbers. Its graph is a line. | Linear equation<br>Standard form<br>Line<br>Vertical line<br>Horizontal line<br>Linear function |

| YOU SHOULD BE ABLE TO . . . | EXAMPLE | REVIEW EXERCISES |
|---|---|---|
| 1 Graph linear equations using point plotting (p. 190) | Example 1 | 1–4 |
| 2 Graph linear equations using intercepts (p. 191) | Examples 2 and 3 | 5–8 |
| 3 Graph vertical and horizontal lines (p. 193) | Examples 4 and 5 | 9–12 |
| 4 Work with linear functions (p. 194) | Example 6 | 13–14 |

*In Problems 1–4, graph each linear equation by plotting points.*

**1.** $x + y = 7$ **2.** $x - y = -4$ **3.** $5x - 2y = 6$ **4.** $-3x + 2y = 8$

*In Problems 5–8, graph each linear equation by finding its intercepts.*

**5.** $5x + 3y = 30$ **6.** $4x + 3y = 0$

**7.** $\frac{3}{4}x - \frac{1}{2}y = 1$ **8.** $4x + y = 8$

*In Problems 9–12, graph each linear equation.*

**9.** $x = 4$ **10.** $y = -8$ **11.** $x = -2$ **12.** $y = 5$

**13. Long Distance** A phone company offers a plan for long-distance calls that charges \$5.00 per month plus 7¢ per minute. The monthly long-distance cost $C$ for talking $x$ minutes is given by the linear function $C(x) = 0.07x + 5$.

**(a)** What is the implied domain of this linear function?
**(b)** What is the cost if a person made 235 minutes worth of long-distance calls during one month?
**(c)** Graph the linear function.
**(d)** In one month, how many minutes of long-distance can be purchased for \$75?

**14. Straight-Line Depreciation** Using straight-line depreciation, the value $V$ of a particular computer $x$ years after purchase is given by the linear function $V(x) = 1800 - 360x$ for $0 \le x \le 5$.

**(a)** What are the independent and dependent variables?
**(b)** What is the domain of this linear function?
**(c)** What is the initial value of the computer?
**(d)** What is the value of the computer 2 years after purchase?
**(e)** Graph the linear function over its domain.
**(f)** After how long will the value of the computer be \$0?

## Section 3.2 Slope and Equations of Lines

| KEY CONCEPTS | KEY TERMS |
|---|---|
| • **Slope of a Line**<br>Let $P = (x_1, y_1)$ and $Q = (x_2, y_2)$ be two distinct points. If $x_1 \neq x_2$, the slope $m$ of the nonvertical line $L$ containing $P$ and $Q$ is defined by the formula<br>$m = \frac{y_2 - y_1}{x_2 - x_1}, \quad x_1 \neq x_2$<br>If $x_1 = x_2$, then $L$ is a vertical line and the slope $m$ of $L$ is undefined (since this results in division by 0).<br>• When the slope of a line is positive, the line slants upward from left to right.<br>• When the slope of a line is negative, the line slants downward from left to right.<br>• When the slope of a line is zero, the line is horizontal.<br>• When the slope of a line is undefined, the line is vertical.<br>• **Point-slope form of a line**<br>An equation of a nonvertical line of slope $m$ that contains the point $(x_1, y_1)$ is $y - y_1 = m(x - x_1)$.<br>• **Slope-intercept form of a line**<br>An equation of a nonvertical line with slope $m$ and $y$-intercept $b$ is $y = f(x) = mx + b$. | Run<br>Rise<br>Slope<br>Undefined slope<br>Average rate of change |

| YOU SHOULD BE ABLE TO . . . | EXAMPLE | REVIEW EXERCISES |
|---|---|---|
| 1 Find the slope of a line given two points (p. 199) | Examples 1 and 2 | 15–18 |
| 2 Interpret slope as an average rate of change (p. 202) | Example 3 | 19–20 |
| 3 Graph a line given a point and its slope (p. 204) | Example 4 | 21–24 |
| 4 Use the point-slope form of a line (p. 204) | Example 5 | 25–28 |
| 5 Identify the slope and $y$-intercept of a line from its equation (p. 205) | Example 6 | 33–40 |
| 6 Find the equation of a line given two points (p. 206) | Examples 7 and 8 | 29–32 |
| 7 Build linear models using the point-slope form of a line (p. 208) | Example 9 | 41–42 |

*In Problems 15 and 16, (a) find the slope of the line and (b) interpret the slope.*

**15.** **16.**

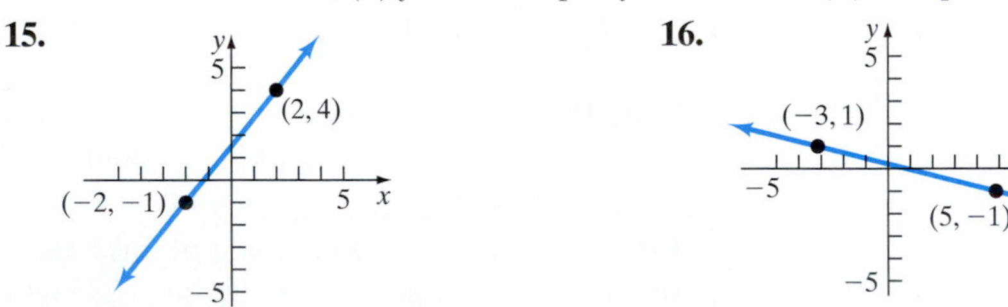


*In Problems 17 and 18, plot each pair of points and determine the slope of the line containing them. Graph the line.*

**17.** $(-1, 5); (2, -1)$

**18.** $(4, 5); (0, -1)$

**19. Illinois's Population** The following data represent the population of Illinois between 1940 and 2000.

**(a)** Plot the ordered pairs $(x, y)$ on a graph and connect the points with straight lines.

**(b)** Compute and interpret the average rate of change in population between 1940 and 1950.

**(c)** Compute and interpret the average rate of change in population between 1980 and 1990.

**(d)** Compute and interpret the average rate of change in population between 1990 and 2000.

**(e)** Based upon the results to parts (a), (b), (c), and (d), do you think that population is linearly related to the year? Why?

| Year, $x$ | Population, $y$ |
|---|---|
| 1940 | 7,897,241 |
| 1950 | 8,712,176 |
| 1960 | 10,081,158 |
| 1970 | 11,110,285 |
| 1980 | 11,427,409 |
| 1990 | 11,430,602 |
| 2000 | 12,419,293 |

SOURCE: *U.S. Census Bureau*

**20. U.S. Community Colleges** The following data represent the number of public community colleges in the United States for selected years from 1920 to 1990.

**(a)** Plot the ordered pairs $(x, y)$ on a graph and connect the points with straight lines.

**(b)** Compute and interpret the average rate of change in the number of public community colleges between 1920 and 1930.

**(c)** Compute and interpret the average rate of change in the number of public community colleges between 1960 and 1970.

**(d)** Compute and interpret the average rate of change in the number of public community colleges between 1980 and 1990.

**(e)** Based upon the results to parts (a), (b), (c), and (d), do you think that the number of public community colleges is linearly related to the year? Why?

| Year, $x$ | Public Community Colleges, $y$ |
|---|---|
| 1920 | 70 |
| 1930 | 178 |
| 1940 | 258 |
| 1950 | 337 |
| 1960 | 390 |
| 1970 | 847 |
| 1980 | 1049 |
| 1990 | 1282 |

SOURCE: *National Profile of Community Colleges: Trends and Statistics, 3rd ed., Community College Press, 2000*

*In Problems 21–24, graph the line containing the point P and having slope m. Do not find the equation of the line.*

**21.** $m = 4$; $P(-1, -5)$

**22.** $m = -\dfrac{2}{3}$; $P(3, 2)$

**23.** $m$ is undefined; $P(2, -4)$

**24.** $m = 0$; $P(-3, 1)$

*In Problems 25 and 26, find an equation of the line. Express your answer in either slope-intercept or standard form, whichever you prefer.*

**25.** 

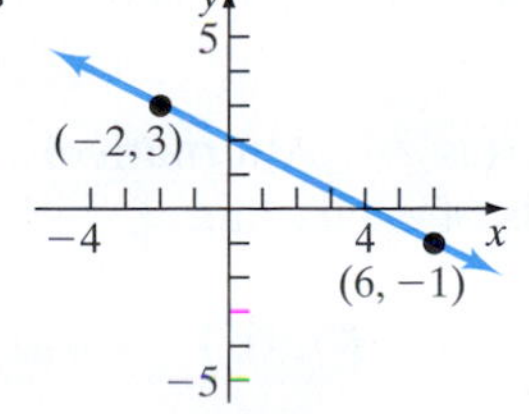


**26.**

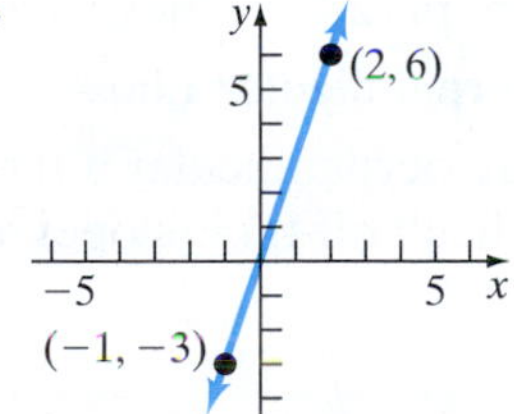


*In Problems 27 and 28, find an equation of the line with the given slope and containing the given point. Express your answer in either slope-intercept or standard form, whichever you prefer.*

**27.** $m = -1$; $(3, 2)$

**28.** $m = \dfrac{3}{5}$; $(-10, -4)$

*In Problems 29–32, find an equation of the line containing the given points. Express your answer in either slope-intercept or standard form, whichever you prefer.*

**29.** $(6, 2); (-3, 5)$ **30.** $(-2, 3); (4, 3)$ **31.** $(4, -1); (1, -7)$ **32.** $(-1, 2); (8, -1)$

*In Problems 33–36, find the slope and y-intercept of each line. Graph the line.*

**33.** $y = 4x - 6$ **34.** $2x + 3y = 12$ **35.** $x - y = 4$ **36.** $2x + 6y = 3$

*In Problems 37–40, graph each linear function.*

**37.** $g(x) = 2x - 6$

**38.** $H(x) = -\frac{4}{3}x + 5$

**39.** $F(x) = -x - 3$

**40.** $f(x) = \frac{3}{4}x - 3$

**41. Federal Tax Returns** In 1996, approximately 12.6% of U.S. Federal Tax Returns were filed electronically. In 2001, approximately 30.7% were filed electronically. (*Source:* Internal Revenue Service)

**(a)** Assuming a linear relation between year and percentage of electronic filings, find a linear function that relates the percentage of electronic filings to year, treating the number of years since 1996 as the independent variable, $x$.

**(b)** Predict the percentage of electronic filings in 2004 if the linear trend continues.

**(c)** Interpret the slope.

**(d)** Assuming the linear trend continues, in what year will the percentage of electronic returns be 48.8%?

**42. Heart Rates** According to the American Geriatric Society, the maximum recommended heart rate for a 20-year-old man under stress is 200 beats per minute. The maximum recommended heart rate for a 60-year-old man under stress is 160 beats per minute.

**(a)** Find a linear function that relates the maximum recommended heart rate for men to age.

**(b)** Predict the maximum recommended heart rate for a 45-year-old man under stress.

**(c)** Interpret the slope.

**(d)** For what age would the maximum recommended heart rate under stress be 168 beats per minute?

## Section 3.3 Parallel and Perpendicular Lines

| KEY CONCEPTS | KEY TERMS |
|---|---|
| • **Parallel Lines** Two lines are parallel if they have the same slope, but different $y$-intercepts. | Parallel lines |
| • **Slopes of Perpendicular Lines** Two lines are perpendicular if the product of their slopes is $-1$. Alternatively, two lines are perpendicular if their slopes are negative reciprocals of each other. | Perpendicular lines |

| YOU SHOULD BE ABLE TO . . . | EXAMPLE | REVIEW EXERCISES |
|---|---|---|
| 1 Define parallel lines (p. 216) | Example 1 | 43, 45–48 |
| 2 Find equations of parallel lines (p. 217) | Example 2 | 49–51 |
| 3 Define perpendicular lines (p. 218) | Examples 3 and 4 | 44, 45–48 |
| 4 Find equations of perpendicular lines (p. 219) | Example 5 | 52–54 |

*In Problems 43–44, the slope of a line L is* $m = -\frac{3}{8}$.

**43.** Determine the slope of a line that is parallel to $L$.

**44.** Determine the slope of a line that is perpendicular to $L$.

*In Problems 45–48, determine whether the given pairs of linear equation are parallel, perpendicular, or neither.*

**45.** $x - 3y = 9$
$9x + 3y = -3$

**46.** $6x - 8y = 16$
$3x + 4y = 28$

**47.** $2x - y = 3$
$-6x + 3y = 0$

**48.** $x = 2$
$y = 2$

*In Problems 49–54, find an equation of the line with the given properties. Express your answer in slope-intercept form. Graph the lines.*

**49.** Parallel to $y = -2x - 5$ through $(1, 2)$

**50.** Parallel to $5x - 2y = 8$ through $(4, 3)$

**51.** Parallel to $x = -3$ through $(1, -4)$

**52.** Perpendicular to $y = 3x + 7$ through $(6, 2)$

**53.** Perpendicular to $3x + 4y = 6$ through $(-3, -2)$

**54.** Perpendicular to $x = 2$ through $(5, -4)$

## Section 3.4 Linear Inequalities in Two Variables

| KEY CONCEPT | KEY TERMS |
|---|---|
| • Linear inequalities in two variables are inequalities in one of the forms $Ax + By < C \quad Ax + By > C \quad Ax + By \le C \quad Ax + By \ge C$ where $A$, $B$, and $C$ are real numbers and $A$ and $B$ are not both zero. | Half-planes<br>Satisfied<br>Graph of a linear inequality in two variables |

| YOU SHOULD BE ABLE TO . . . | EXAMPLE | REVIEW EXERCISES |
|---|---|---|
| 1 Determine whether an ordered pair is a solution to a linear inequality (p. 224) | Example 1 | 55–56 |
| 2 Graph linear inequalities (p. 226) | Examples 2 and 3 | 57–62 |
| 3 Solve problems involving linear inequalities (p. 228) | Example 4 | 63–64 |

*In Problems 55 and 56, determine whether the given points are solutions to the linear inequality.*

**55.** $5x + 3y \le 15$
**(a)** $(4, -2)$
**(b)** $(-6, 15)$
**(c)** $(5, -1)$

**56.** $x - 2y > -4$
**(a)** $(2, 3)$
**(b)** $(5, -2)$
**(c)** $(-1, 3)$

*In Problems 57–62, graph each inequality.*

**57.** $y < 3x - 2$

**58.** $2x - 4y \le 12$

**59.** $3x + 4y > 20$

**60.** $y \ge 5$

**61.** $2x + 3y < 0$

**62.** $x > -8$

**63. Entertainment Budget** Ethan's entertainment budget permits him to spend a maximum of \$60 per month on movie tickets and music CDs. Movie tickets cost on average \$7.50 each. Music CDs average \$15.00 each.
**(a)** Write a linear inequality that describes Ethan's options for spending the \$60 maximum budget.
**(b)** Can Ethan buy 5 movie tickets and 2 music CDs?
**(c)** Can Ethan buy 2 movie tickets and 2 music CDs?

**64. Fund Raising** For a fund raiser, the math club agrees to sell candy bars and candles. The club's profit will be 50¢ for each candy bar and \$2.00 for each candle it sells. The club needs to earn at least \$1000 in order to pay for an upcoming field trip.
**(a)** Write a linear inequality that describes the combination of candy bars and candles that must be sold.
**(b)** Will selling 500 candy bars and 350 candles earn enough for the trip?
**(c)** Will selling 600 candy bars and 400 candles earn enough for the trip?

## Section 3.5 Building Linear Models

| KEY CONCEPT | KEY TERMS |
|---|---|
| • **Direct Variation**<br>Suppose we let $x$ and $y$ represent two quantities. We say that $y$ varies directly with $x$, or $y$ is directly proportional to $x$, if there is a nonzero number $k$ such that $y = kx$. The number $k$ is called the constant of proportionality. | Variation<br>Directly proportional<br>Varies directly<br>Constant of proportionality<br>Scatter diagrams |

| YOU SHOULD BE ABLE TO . . . | EXAMPLE | REVIEW EXERCISES |
|---|---|---|
| 1 Build linear models from verbal descriptions (p. 231) | Examples 1 and 2 | 65–67 |
| 2 Build linear models involving direct variation (p. 234) | Examples 3 and 4 | 68–71 |
| 3 Build linear models from data (p. 236) | Examples 5 through 7 | 72–75 |

**65. Car Rental** The daily rental charge for a particular car is \$35 plus 12¢ per mile.
**(a)** Find a linear function that expresses the rental cost $C$ as a function of the miles driven $m$.
**(b)** What are the independent and dependent variables?
**(c)** What is the implied domain of this linear function?
**(d)** For a one-day rental, what is the rental cost if 124 miles are driven?
**(e)** For a one-day rental, how many miles were driven if the rental cost was \$67.16?
**(f)** Graph the linear function.

**66. Satellite Television Bill** A satellite television company charges \$33.99 per month for a 100-channel package, plus \$3.50 for each pay-per-view movie watched that month.
**(a)** Find a linear function that expresses the monthly bill $B$ as a function of $x$, the number of pay-per-view movies watched that month.
**(b)** What are the independent and dependent variables?
**(c)** What is the implied domain of this linear function?
**(d)** What is the monthly bill if 5 pay-per-view movies are watched that month?
**(e)** For one month, how many pay-per-view movies were watched if the bill was \$58.49?
**(f)** Graph the linear function.

**67. Depreciation** A farmer just bought a new tractor for \$84,600. The farmer will depreciate the tractor using the straight-line method over 12 years.

**(a)** Find a linear function that expresses the book value $V$ of the tractor as a function of its age $x$.
**(b)** What is the implied domain?
**(c)** What is the book value of the tractor after 5 years?
**(d)** What are the intercepts of the graph of the linear function?
**(e)** When will the book value of the tractor be \$28,200?
**(f)** Graph the linear function.

*In Problems 68 and 69, (a) find the constant of proportionality k, (b) write the linear function relating the two variables, and (c) find the quantity indicated.*

**68.** Suppose that $y$ varies directly with $x$. When $x = 9$, then $y = 108$. Find $y$ when $x = 4$.

**69.** Suppose the $y$ is directly proportional to $x$. When $x = 6$, the $y = 11$. Find $y$ when $x = 24$.

**70. Drug Dosage** The recommended dosage $d$ of an antibiotic drug is directly proportional to a person's weight $w$. If a 168-pound person is given 3024 milligrams of the drug, find the recommended dosage for a 146-pound person.

**71. Perimeter of a Square** The perimeter $P$ of a square varies directly with the length of its side $s$. If the perimeter of a square with side length 8 feet is 32 feet, what is the perimeter of a square whose side length is 6 feet?

*In Problems 72 and 73,*

**(a)** *Draw a scatter diagram of the data.*
**(b)** *Select two points from the scatter diagram and find the equation of the line containing the points selected.**
**(c)** *Graph the line found in part* (b) *on the scatter diagram.*

**72.**

| $x$ | 2 | 5 | 8 | 11 | 14 |
|---|---|---|---|---|---|
| $y$ | 13.3 | 11.6 | 8.4 | 7.2 | 4.6 |

**73.**

| $x$ | 0 | 0.4 | 1.5 | 2.3 | 4.2 |
|---|---|---|---|---|---|
| $y$ | 0.6 | 1.1 | 1.3 | 1.8 | 3.0 |

**74.** The table below gives the number of calories and the total carbohydrates (in grams) for a one-cup serving of seven name-brand cereals (not including milk).

| Cereal | Calories, $x$ | Total Carbohydrates (in grams), $y$ |
|---|---|---|
| Rice Krispies® | 96 | 23.2 |
| Life® | 160 | 33.3 |
| Lucky Charms® | 120 | 25.0 |
| Kellogg's Complete® | 120 | 30.7 |
| Wheaties® | 110 | 24.0 |
| Cheerios® | 110 | 22.0 |
| Honey Nut Chex® | 160 | 34.7 |

SOURCE: *Quaker Oats, General Mills, and Kellogg*

**(a)** Draw a scatter diagram of the data treating calories as the independent variable.
**(b)** What type of relation appears to exist between calories and total carbohydrates in a one-cup serving of cereal?

*Answers will vary.

**(c)** Select two points and find an equation of the line containing the points.*
**(d)** Graph the line on the scatter diagram drawn in part (a).
**(e)** Predict the total carbohydrates in a one-cup serving of cereal that has 140 calories.
**(f)** Interpret the slope of the line found in part (c).

**75. Second-Day Delivery Costs** The table below lists some selected prices charged by Federal Express for FedEx 2Day delivery, depending on the weight of the package.

| Weight (in pounds), $x$ | FedEx 2Day® Delivery Charge, $y$ |
|---|---|
| 1 | $9.25 |
| 3 | $12.00 |
| 6 | $17.25 |
| 8 | $21.25 |
| 9 | $23.00 |
| 11 | $26.50 |

SOURCE: *Federal Express Corporation*

**(a)** Draw a scatter diagram of the data treating weight as the independent variable.
**(b)** What type of relation appears to exist between the weight of the package and the FedEx 2Day delivery charge?
**(c)** Select two points and find an equation of the line containing the points.*
**(d)** Graph the line on the scatter diagram drawn in part (a).
**(e)** Predict the FedEx 2Day delivery charge for shipping a 5-pound package.
**(f)** Interpret the slope of the line found in part (c).

*Answers will vary.

## CHAPTER 3 TEST

*Remember to use your Chapter Test Prep Video CD to see fully worked-out solutions to any of these problems you would like to review.*

*In Problems 1–6, graph each linear equation or function, using the method you prefer.*

**1.** $x - y = 8$ **2.** $3x + 5y = 0$ **3.** $3x + 2y = 12$

**4.** $\frac{3}{2}x - \frac{1}{4}y = 1$ **5.** $x = -7$ **6.** $f(x) = -\frac{2}{3}x + 6$

**7.** Find and interpret the slope of the line containing the points $(5, -2)$ and $(-1, 6)$.

**8.** Draw a graph of the line that contains the point $(2, -4)$ and has a slope of $-\frac{3}{5}$. Do not find the equation of the line.

**9.** Determine whether the graphs of the following pair of linear equations are parallel, perpendicular, or neither.

$$8x - 2y = 1$$
$$x + 4y = -2$$

*In Problems 10–13, find the equation of the line with the given properties. Express your answer in either slope intercept or standard form, which ever you prefer.*

**10.** Through the point $(-3, 1)$ and having a slope of 4

**11.** Through the points $(6, 1)$ and $(-3, 7)$

**12.** Parallel to $x - 5y = 15$ and through the point $(10, -1)$

**13.** Perpendicular to $3x - y = 4$ and through the point $(6, 2)$

**14.** Determine whether the given points are solutions to the linear inequality $3x - y > 10$.

**(a)** $(3, -1)$ **(b)** $(4, 5)$ **(c)** $(5, 3)$

*In Problems 15 and 16, graph each linear inequality.*

**15.** $y \leq -2x + 1$

**16.** $5x - 2y < 0$

**17. Crafts Fair Sales** Henry plans to sell small wooden shelves at a crafts fair for $30 each. A booth at the fair costs $100 to rent. Henry estimates his expenses for producing the shelves to be $12 each, so his profit will be $18 per shelf.

**(a)** Write a function that expresses Henry's profit $P$ as a function of the number of shelves $x$ sold.
**(b)** What is the implied domain of this linear function?
**(c)** What is the profit if Henry sells 34 shelves?
**(d)** Graph the linear function.
**(e)** If Henry's profit is $764, how many shelves did he sell?

**18. Area of a Circle** The following data show the relationship between the diameter of a circle and the area of that circle.

**(a)** Plot the ordered pairs $(x, y)$ on a graph and connect the points with straight lines.
**(b)** Compute and interpret the average rate of change in area between diameters lengths of 1 and 3 feet.
**(c)** Compute and interpret the average rate of change in area between diameter lengths of 10 and 13 feet.
**(d)** Based upon the results to parts (a), (b), and (c), do you think that the area of circles is linearly related to the diameter? Why?

| Diameter (feet), $x$ | Area (square feet), $y$ |
|---|---|
| 1 | 0.79 |
| 3 | 7.07 |
| 6 | 28.27 |
| 7 | 38.48 |
| 8 | 50.27 |
| 10 | 78.54 |
| 13 | 132.73 |

**19. Earnings** The amount of money earned $E$ varies directly with the amount of time worked $t$. For one job, Benjamin earns $296.40 for 24 hours of work. How much will Benjamin earn for 10 hours of work?

**20. Shetland Pony Weights** The table below lists the average weight of a Shetland pony, depending on the age of the pony.

(a) Draw a scatter diagram of the data treating weight as the independent variable.
(b) What type of relation appears to exist between the age and the weight of the Shetland pony?
(c) Select two points and find an equation of the line containing the points.*
(d) Graph the line on the scatter diagram drawn in part (a).
(e) Predict the weight of a 9 month old Shetland pony.
(f) Interpret the slope of the line found in part (c).

| Age (months), $x$ | Average weight (kilograms), $y$ |
|---|---|
| 3 | 60 |
| 6 | 95 |
| 12 | 140 |
| 18 | 170 |
| 24 | 185 |

SOURCE: *The Merck Veterinary Manual*

*Answers will vary.

# CUMULATIVE REVIEW CHAPTERS R–3

*In Problems 1–3, evaluate each expression.*

**1.** $200 \div 25 \cdot (-2)$ **2.** $\frac{3}{4} + \frac{1}{6} - \frac{2}{3}$ **3.** $\frac{8 - 3(5 - 3^2)}{7 - 2 \cdot 6}$

**4.** Evaluate $x^3 + 3x^2 - 5x - 7$ for $x = -3$.

**5.** Simplify: $8m - 5m^2 - 3 + 9m^2 - 3m - 6$

*In Problems 6–8, solve each equation.*

**6.** $8(n + 2) - 7 = 6n - 5$ **7.** $\frac{2}{5}x + \frac{1}{6} = -\frac{2}{3}$ **8.** $\frac{|2x - 5|}{3} + 1 = 4$

**9.** Solve $A = \frac{1}{2}h(b + B)$ for $B$.

*In Problems 10–12, solve each linear inequality. Express the solution using set-builder notation and interval notation. Graph the solution set.*

**10.** $6x - 7 > -31$ **11.** $5(x - 3) \geq 7(x - 4) + 3$ **12.** $|2x + 5| < 9$

**13.** Plot the following ordered pairs in the same Cartesian plane.

$A(-3, 0)$, $B(4, -2)$, $C(1, 5)$, $D(0, 3)$, $E(-4, -5)$, $F(-5, 2)$

*In Problems 14–16, determine whether the relation represents a function. State the domain and range of the relation.*

**14.** $\{(-1, 3), (0, 4), (-1, 6), (1, -2), (2, -5)\}$

**15.**

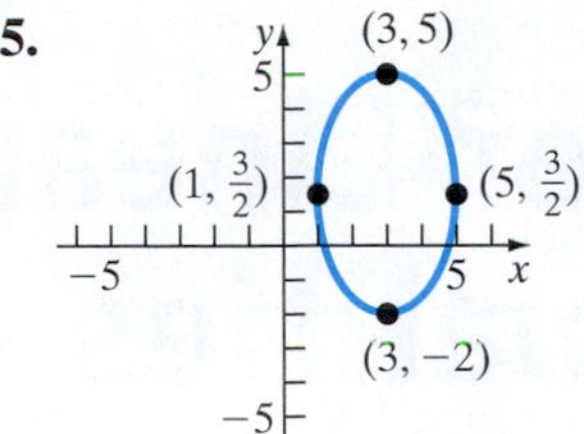


**16.**

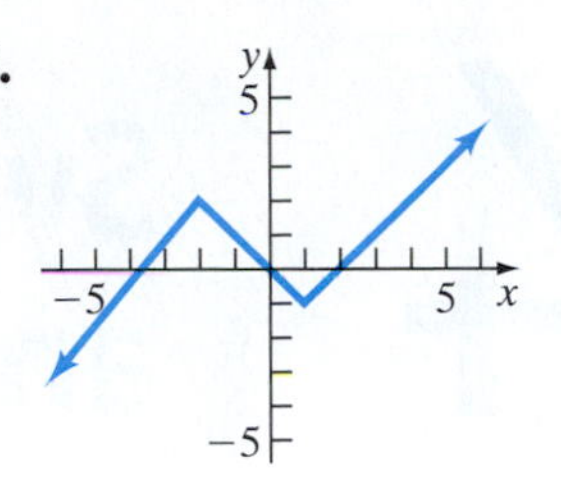


*In Problems 17 and 18, find the indicated values for the given function.*

**17.** $H(x) = \dfrac{3x + 5}{x - 2}$ **(a)** $H(1)$ **(b)** $H\left(\dfrac{1}{2}\right)$

**18.** $g(x) = 2x - 3$ **(a)** $g(-4)$ **(b)** $g(x + h)$

*In Problems 19 and 20, graph the linear equation using the method you prefer.*

**19.** $y = -\dfrac{1}{2}x + 4$

**20.** $4x - 5y = 15$

*In Problems 21 and 22, find the equation of the line with the given properties. Express your answer in either slope-intercept or standard form, whichever you prefer.*

**21.** Through the points $(3, -2)$ and $(-6, 10)$

**22.** Parallel to $y = -3x + 10$ and through the point $(-5, 7)$

**23.** Graph $x - 3y > 12$.

**24. Cell Phone Use** Using data from Forrester Research, the function $C(t) = 14.27t + 106.39$ represents the projected number of cell phones (in millions) that will be in use $t$ years after 2000.

**(a)** Identify the dependent and independent variables.
**(b)** Evaluate $C(5)$ and explain what it represents.
**(c)** Evaluate $C(-8)$ and explain what it represents. Is the result reasonable? Why?

**25. Total Carbohydrates** The amount of total carbohydrates $C$ ingested while eating is directly proportional to the volume $V$ of the cereal that is consumed. According to the nutrition facts provided on the box of Cap'n Crunch's Peanut Butter Crunch® cereal, a $\frac{3}{4}$-cup serving of the cereal contains 21 grams of carbohydrates. How many grams of carbohydrates will be in 2 cups of Peanut Butter Crunch? (*Source:* The Quaker Oats Company)

CHAPTER

# 4 Systems of Linear Equations and Inequalities

Why does it take longer to fly west than east? The answer is the jet stream. But what impact does the jet stream have on flight time? Is it the same for all airplanes? See Example 6 in Section 4.2 and Quick Check Problem 6 on page 278.

## OUTLINE



## The Big Picture: Putting It Together

In Chapter 1, we solved equations and inequalities in one variable. Recall that linear equations in one variable can have no solution (a contradiction), one solution, or infinitely many solutions (an identity). In Chapter 3, we learned how to graph both linear equations and linear inequalities in two variables. The graph of a linear equation will be used extensively in this chapter to help us visualize results.

In this chapter, we will discuss solving two or more linear equations involving two or more variables (called *systems of equations*). We are going to learn a variety of techniques that can be used to solve these systems. We also will learn that these systems can have no solution, one solution, or infinitely many solutions, just like linear equations. We conclude the chapter by looking at systems of linear inequalities. These systems require us to determine the region of a Cartesian plane that satisfies two or more linear inequalities simultaneously.

# 4.1 Systems of Linear Equations in Two Variables

**OBJECTIVES**

1. Determine Whether an Ordered Pair Is a Solution to a System of Linear Equations
2. Solve a System of Two Linear Equations Containing Two Unknowns by Graphing
3. Solve a System of Two Linear Equations Containing Two Unknowns by Substitution
4. Solve a System of Two Linear Equations Containing Two Unknowns by Elimination
5. Identify Inconsistent Systems
6. Express the Solution of a System of Dependent Equations

### *Preparing for Systems of Linear Equations in Two Variables*

*Before getting started, take the following readiness quiz. If you get a problem wrong, go back to the section cited and review the material.*

**1.** Evaluate $2x - 3y$ for $x = 5$, $y = 4$. [Section R.5, pp. 43–44]

**2.** Determine whether the point $(4, -1)$ is on the graph of the equation $2x - 3y = 11$. [Section 2.1, pp. 138–139]

**3.** Graph $y = 3x - 7$. [Section 3.2, p. 206]

**4.** Find the equation of the line parallel to $y = -3x + 1$ containing the point $(2, 3)$. [Section 3.3, pp. 217–218]

**5.** Determine the slope and $y$-intercept of $4x - 3y = 15$. [Section 3.2, p. 206]

**6.** What is the additive inverse of 4? [Section R.3, pp. 21–22]

**7.** Solve: $2x - 3(-3x + 1) = -36$ [Section 1.1, p. 53–56]

Recall, from Section 3.1, that an equation in two variables is linear provided that it can be written in the form $Ax + By = C$, where $A$, $B$, and $C$ are real numbers and $A$ and $B$ are not both zero. However, linear equations can have more than two variables.

Some examples of linear equations are

$$\underbrace{4x - 3y = 9}_{\text{Linear equation in two variables, } x \text{ and } y} \qquad \underbrace{-2x + y - 5z = -3}_{\text{Linear equation in three variables, } x,\ y, \text{ and } z} \qquad \underbrace{3w - x + 5y - 2z = 12}_{\text{Linear equation in four variables, } w,\ x,\ y, \text{ and } z}$$

A **system of linear equations** is a grouping of two or more linear equations, each of which contains one or more variables.

### EXAMPLE 1 Examples of Systems of Linear Equations

**(a)** $\begin{cases} 2x + y = 5 \\ x - 5y = -10 \end{cases}$ Two equations containing two variables, $x$ and $y$

**(b)** $\begin{cases} x + 3y + z = 8 \\ 3x - y + 6z = 12 \\ -4x - y + 2z = -1 \end{cases}$ Three equations containing three variables, $x$, $y$, and $z$

We use a brace, as shown in the systems in Example 1, to remind us that we are dealing with a system of equations. In this section, we concentrate on systems of two linear equations containing two variables such as the system in Example 1(a).

## 1 Determine Whether an Ordered Pair Is a Solution to a System of Linear Equations

A **solution** of a system of equations consists of values for the variables that are solutions of each equation of the system. When we are solving systems of two linear equations containing two unknowns, we represent the solution as an ordered pair, $(x, y)$.

### EXAMPLE 2 Determining Whether Values Are a Solution to a System of Linear Equations

Determine whether the given ordered pairs are solutions to the system of equations.

$$\begin{cases} 2x + 3y = 9 \\ -5x - 3y = 0 \end{cases}$$

**(a)** $(6, -1)$ **(b)** $(-3, 5)$

*Preparing for...Answers* **1.** $-2$ **2.** Yes
**3.**

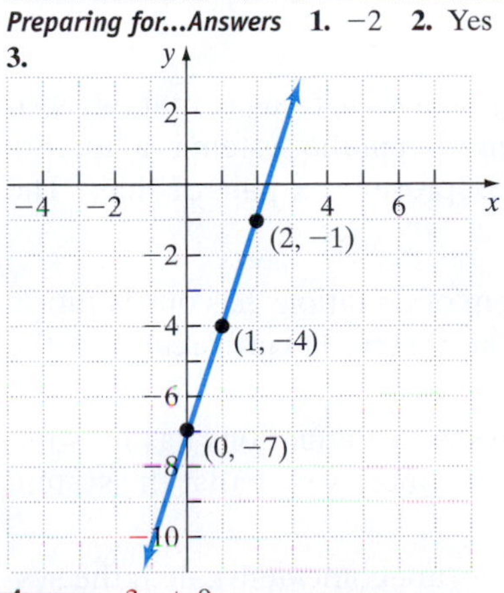


**4.** $y = -3x + 9$
**5.** slope $= 4/3$; $y$-intercept $= -5$
**6.** $-4$ **7.** $\{-3\}$

**Solution**

To help us organize our thoughts, we name $2x + 3y = 9$ equation (1) and $-5x - 3y = 0$ equation (2).

$$\begin{cases} 2x + 3y = 9 & (1) \\ -5x - 3y = 0 & (2) \end{cases}$$

**(a)** Let $x = 6$ and $y = -1$ in both equations (1) and (2). If both equations are true, then $(6, -1)$ is a solution.

Equation (1): $2x + 3y = 9$

$x = 6, y = -1$: $2(6) + 3(-1) \stackrel{?}{=} 9$

$12 - 3 \stackrel{?}{=} 9$

$9 = 9$ True

Equation (2): $-5x - 3y = 0$

$x = 6, y = -1$: $-5(6) - 3(-1) \stackrel{?}{=} 0$

$-30 + 3 \stackrel{?}{=} 0$

$-27 = 0$ False

**In Words**

A solution to a system must satisfy all of the equations in the system.

Although $x = 6$, $y = -1$ satisfy equation (1), they do not satisfy equation (2); therefore, $(6, -1)$ is not a solution of the system of equations.

**(b)** Let $x = -3$ and $y = 5$ in both equations (1) and (2). If both equations are true, then $(-3, 5)$ is a solution.

Equation (1): $2x + 3y = 9$

$x = -3, y = 5$: $2(-3) + 3(5) \stackrel{?}{=} 9$

$-6 + 15 \stackrel{?}{=} 9$

$9 = 9$ True

Equation (2): $-5x - 3y = 0$

$x = -3, y = 5$: $-5(-3) - 3(5) \stackrel{?}{=} 0$

$15 - 15 \stackrel{?}{=} 0$

$0 = 0$ True

Because the values $x = -3$ and $y = 5$ satisfy both equations (1) and (2), the ordered pair $(-3, 5)$ is a solution of the system of equations.

**Work Smart**

It is a good idea to number the equations in a system so that it is easier to keep track of your work.

For the remainder of the chapter, we shall number each equation as we did in Example 1. When solving homework problems, you should do the same.

**QUICK ✓**

**1.** Which of the following points is a solution to the system of equations?

$$\begin{cases} 2x + 3y = 7 \\ 3x + \phantom{3}y = -7 \end{cases}$$

**(a)** $(3, 1)$ **(b)** $(-4, 5)$ **(c)** $(-2, -1)$

## Visualizing the Solutions in a System of Two Linear Equations Containing Two Unknowns

We can view the problem of solving a system of two linear equations containing two variables as a geometry problem. The graph of each equation in the system is a line. So, a system of two equations containing two variables represents a pair of lines. The graphs of the two lines can appear in one of three ways:

1. **INTERSECT:** If the lines intersect, then the system of equations has one solution given by the point of intersection. We say that the system is **consistent** and the equations are **independent.** See Figure 1(a).
2. **PARALLEL:** If the lines are parallel, then the system of equations has no solution because the lines never intersect. In this circumstance, we say that the system is **inconsistent.** See Figure 1(b).
3. **COINCIDENT:** If the lines lie on top of each other (are coincident), then the system of equations has infinitely many solutions. The solution set is the set of all points on the line. The system is **consistent** and the equations are **dependent.** See Figure 1(c).

Figure 1

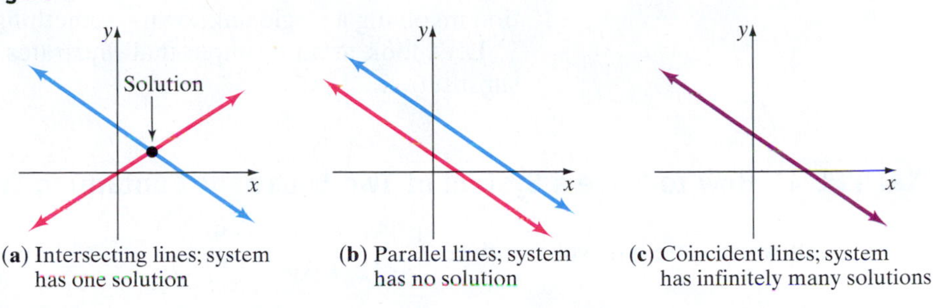


(a) Intersecting lines; system has one solution (b) Parallel lines; system has no solution (c) Coincident lines; system has infinitely many solutions

For now, we will concentrate on solving systems for which there is a single solution.

## 2 Solve a System of Two Linear Equations Containing Two Unknowns by Graphing

Let's look at an example where we use graphing to solve a system.

### EXAMPLE 3 Solving a System of Two Linear Equations Using Graphing

Solve the following system by graphing: $\begin{cases} x + y = -1 \\ -2x + y = -7 \end{cases}$

**Solution**

First, we name $x + y = -1$ equation (1) and $-2x + y = -7$ equation (2).

$$\begin{cases} x + y = -1 & (1) \\ -2x + y = -7 & (2) \end{cases}$$

Figure 2

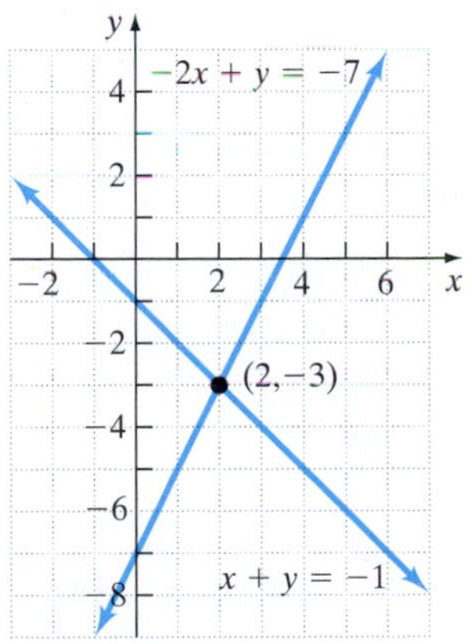

In order to graph each equation, we put them in slope-intercept form. Equation (1) in slope-intercept form is $y = -x - 1$, which has slope $-1$ and $y$-intercept $-1$. Equation (2) in slope-intercept form is $y = 2x - 7$, which has slope 2 and $y$-intercept $-7$. Figure 2 shows their graphs (Note that we could also have graphed the lines using the intercepts). The lines appear to intersect at $(2, -3)$, so we believe that the ordered pair $(2, -3)$ is the solution to the system.

**Check** Let $x = 2$ and $y = -3$ in both equations in the system:

Equation (1): $x + y = -1$

$x = 2, y = -3$: $2 + (-3) \stackrel{?}{=} -1$

$2 - 3 \stackrel{?}{=} -1$

$-1 = -1$ True

Equation (2): $-2x + y = -7$

$x = 2, y = -3$: $-2(2) + (-3) \stackrel{?}{=} -7$

$-4 - 3 \stackrel{?}{=} -7$

$-7 = -7$ True

Both equations are true, so the solution is the ordered pair $(2, -3)$.

**QUICK ✓** *Solve the system by graphing.*

**2.** $\begin{cases} y = -3x + 10 \\ y = 2x - 5 \end{cases}$ **3.** $\begin{cases} 2x + y = -1 \\ -2x + 2y = 10 \end{cases}$

## 3 Solve a System of Two Linear Equations Containing Two Unknowns by Substitution

**Work Smart**

Obtaining exact solutions using graphical methods can be difficult. Therefore, algebraic methods should be used.

If the $x$- and $y$-coordinates of the point of intersection between two lines are not integers, then obtaining an exact result graphically can be difficult. Therefore, rather than using graphical methods to obtain solutions to systems of two linear equations, we prefer to use algebraic methods. The first algebraic method that we present is the *method of substitution*. The goal of the method of substitution is to obtain a single linear equation involving a single unknown—something we already know how to solve.

Let's look at an example that illustrates how to solve a system of equations using substitution.

## EXAMPLE 4 How to Solve a System of Two Equations Containing Two Unknowns by Substitution

Solve the following system by substitution: $\begin{cases} 3x + y = -9 & (1) \\ -2x + 3y = 17 & (2) \end{cases}$

### Step-by-Step Solution

| Step | | |
|---|---|---|
| **Step 1:** Solve one of the equations for one of the unknowns. It is easiest to solve equation (1) for y since the coefficient of y is 1. | | $3x + y = -9$ |
| | Subtract 3x from both sides: | $y = -3x - 9$ |
| **Step 2:** Substitute $-3x - 9$ for y in equation (2). | Equation (2): | $-2x + 3y = 17$ |
| | | $-2x + 3(-3x - 9) = 17$ |
| **Step 3:** Solve the equation for x. | Distribute the 3: | $-2x - 9x - 27 = 17$ |
| | Combine like terms: | $-11x - 27 = 17$ |
| | Add 27 to both sides: | $-11x = 44$ |
| | Divide both sides by −11: | $x = -4$ |
| **Step 4:** We substitute −4 for x into the equation from Step 1. | | $y = -3x + 9$ |
| | | $y = -3(-4) - 9$ |
| | | $y = 12 - 9$ |
| | | $y = 3$ |

**Step 5: Check** We check our answer that $x = -4$ and $y = 3$.

| | Equation (1): | | Equation (2): |
|---|---|---|---|
| | $3x + y = -9$ | | $-2x + 3y = 17$ |
| $x = -4, y = 3$: | $3(-4) + 3 \stackrel{?}{=} -9$ | | $-2(-4) + 3(3) \stackrel{?}{=} 17$ |
| | $-12 + 3 \stackrel{?}{=} -9$ | | $8 + 9 \stackrel{?}{=} 17$ |
| | $-9 = -9$ True | | $17 = 17$ True |

Both equations are satisfied so the solution is the ordered pair $(-4, 3)$.

**Work Smart**

When using substitution solve for the variable whose coefficient is 1 or −1 in order to simplify the algebra.

### Steps for Solving a System of Two Linear Equations Containing Two Unknowns by Substitution

**Step 1:** Solve one of the equations for one of the unknowns. For example, we might solve equation (1) for $y$ in terms of $x$. Choose the equation that is easiest to solve for a variable. Typically, this would be the equation that has a variable whose coefficient is 1 or −1.

**Step 2:** Substitute the expression that equals the variable solved for in Step 1 into the other equation. The result will be a single linear equation in one unknown. For example, if we solved equation (1) for $y$ in terms of $x$ in Step 1, then we would replace $y$ in equation (2) with the algebraic expression in $x$.

**Step 3:** Solve the linear equation in one unknown found in Step 2.

**Step 4:** Substitute the value of the variable into the expression found in Step 1 to find the value of the other variable.

**Step 5:** Check your answer.

## EXAMPLE 5 Solving a System of Two Equations Containing Two Unknowns by Substitution

Solve the following system by substitution: $\begin{cases} 2x - 3y = -6 & (1) \\ -8x + 3y = 3 & (2) \end{cases}$

### Solution

We choose to solve equation (1) for $x$ because it seems easiest.

Equation (1): $2x - 3y = -6$

Add $3y$ to both sides: $2x = 3y - 6$

Divide both sides by 2: $x = \frac{3y - 6}{2}$

Divide 2 into both terms in the numerator: $x = \frac{3}{2}y - 3$

Now substitute $\frac{3}{2}y - 3$ for $x$ in equation (2) and then solve for $y$.

**Work Smart**

Notice how we use substitution to reduce a system of two linear equations involving two unknowns down to one linear equation involving one unknown. Again, we use algebraic techniques to reduce a problem down to one we already know how to solve!

Equation (2): $-8x + 3y = 3$

$$-8\left(\frac{3}{2}y - 3\right) + 3y = 3$$

Distribute the $-8$: $-8 \cdot \frac{3}{2}y - (-8) \cdot 3 + 3y = 3$

Multiply: $-12y + 24 + 3y = 3$

Combine like terms: $-9y + 24 = 3$

Subtract 24 from both sides: $-9y = -21$

Divide both sides by $-9$: $y = \frac{-21}{-9}$

$$y = \frac{7}{3}$$

Now we substitute $\frac{7}{3}$ for $y$ into $x = \frac{3}{2}y - 3$ to find the value of $x$.

$$x = \frac{3}{2}\left(\frac{7}{3}\right) - 3$$

Multiply: $x = \frac{7}{2} - 3$

$$x = \frac{1}{2}$$

**Check** We check our answer that $x = \frac{1}{2}$ and $y = \frac{7}{3}$.

Equation (1): $2x - 3y = -6$

$x = \frac{1}{2}, y = \frac{7}{3}$: $2\left(\frac{1}{2}\right) - 3\left(\frac{7}{3}\right) \stackrel{?}{=} -6$

$1 - 7 \stackrel{?}{=} -6$

$-6 = -6$ True

Equation (2): $-8x + 3y = 3$

$-8\left(\frac{1}{2}\right) + 3\left(\frac{7}{3}\right) \stackrel{?}{=} 3$

$-4 + 7 \stackrel{?}{=} 3$

$3 = 3$ True

Both equations are satisfied so the solution is the ordered pair $\left(\frac{1}{2}, \frac{7}{3}\right)$.

**QUICK ✓** *Solve the system using substitution.*

4. $\begin{cases} y = -3x - 5 \\ 5x + 3y = 1 \end{cases}$

5. $\begin{cases} 2x + y = -2 \\ -3x - 2y = -2 \end{cases}$

**Work Smart**

Substitution is a method to use if one of the variables has a coefficient of 1 or if one of the variables is already solved for; otherwise use elimination.

## 4 Solve a System of Two Linear Equations Containing Two Unknowns by Elimination

Using substitution to solve the system in Example 5 led to some rather complicated equations containing fractions. A second algebraic method for solving a system of linear equations is the *method of elimination*. This method is usually preferred over the method of substitution if substitution leads to fractions.

The basic idea in using elimination is to get the coefficients of one of the variables to be additive inverses, such as 5 and $-5$, so that we can add the equations together and get a single linear equation involving one unknown. Remember that this is the same goal we had when using the method of substitution.

Let's go over an example to illustrate how to solve a system of linear equations by elimination.

### EXAMPLE 6 How to Solve a System of Linear Equations by Elimination

Solve: $\begin{cases} 5x + 2y = -5 & (1) \\ -2x - 4y = -14 & (2) \end{cases}$

#### Step-by-Step Solution

**Step 1:** Our first goal is to get the coefficients on one of the variables to be additive inverses. In looking at this system, we can make the coefficients of $y$ to be additive inverses by multiplying equation (1) by 2.

$$\begin{cases} 5x + 2y = -5 & (1) \\ -2x - 4y = -14 & (2) \end{cases}$$

Multiply both sides of (1) by 2: $\begin{cases} 2(5x + 2y) = 2(-5) & (1) \\ -2x - 4y = -14 & (2) \end{cases}$

Use the Distributive Property: $\begin{cases} 10x + 4y = -10 & (1) \\ -2x - 4y = -14 & (2) \end{cases}$

**Step 2:** We now add equations (1) and (2) to eliminate the variable $y$ and then solve for $x$.

$$\begin{cases} 10x + 4y = -10 & (1) \\ -2x - 4y = -14 & (2) \end{cases}$$

Add (1) and (2): $8x = -24$

Divide both sides by 8: $x = -3$

**Step 3:** Back-substitute $-3$ for $x$ into either equation (1) or (2) and solve for $y$. We will back-substitute $-3$ for $x$ into equation (1).

Equation (1): $5x + 2y = -5$

$x = -3$: $5(-3) + 2y = -5$

$-15 + 2y = -5$

Add 15 to both sides: $2y = 10$

Divide both sides by 2: $y = 5$

We have that $x = -3$ and $y = 5$.

**Step 4: Check**

Equation (1): $5x + 2y = -5$

$x = -3, y = 5$: $5(-3) + 2(5) \stackrel{?}{=} -5$

$-15 + 10 \stackrel{?}{=} -5$

$-5 = -5$ True

Equation (2): $-2x - 4y = -14$

$-2(-3) - 4(5) \stackrel{?}{=} -14$

$6 - 20 \stackrel{?}{=} -14$

$-14 = -14$ True

The solution is the ordered pair $(-3, 5)$.

The following steps can be used to solve a system of linear equations by elimination.

**Steps for Solving a System of Linear Equations by Elimination**

**Step 1:** Multiply both sides of one or both equations by a nonzero constant so that the coefficients of one of the variables are additive inverses.

**Step 2:** Add equations (1) and (2) to eliminate the variable whose coefficients are now additive inverses. Solve the resulting equation for the unknown.

**Step 3:** Back-substitute the value of the variable found in Step 2 into one of the original equations to find the value of the remaining variable.

**Step 4:** Check your answer.

**In Words**
"Back-substitute" means to "plug in" the known value of the variable into one of the equations in the system.

What allows us to add two equations and use the result to replace an equation? Remember, an equation is a statement that the left side equals the right side. When we add equation (2) to equation (1), we are adding the same quantity to both sides of equation (1).

## EXAMPLE 7 Solving a System of Linear Equations by Elimination

Solve: $\begin{cases} \dfrac{5}{2}x + 2y = 5 & (1) \\ \dfrac{3}{2}x + \dfrac{3}{2}y = \dfrac{9}{4} & (2) \end{cases}$

### Solution

Because both equations (1) and (2) have fractions, our first goal is to get rid of the fractions by multiplying both sides of equation (1) by 2 and both sides of equation (2) by 4.

**Work Smart**
If you are not afraid of fractions, you could eliminate $x$ by multiplying both sides of equation (1) by $-3$ and both sides of equation (2) by 5. Try it!

$$\begin{cases} \dfrac{5}{2}x + 2y = 5 & (1) \\ \dfrac{3}{2}x + \dfrac{3}{2}y = \dfrac{9}{4} & (2) \end{cases}$$

Multiply both sides of (1) by 2:
Multiply both sides of (2) by 4:
$$\begin{cases} 2\left(\dfrac{5}{2}x + 2y\right) = 2 \cdot 5 & (1) \\ 4\left(\dfrac{3}{2}x + \dfrac{3}{2}y\right) = 4 \cdot \dfrac{9}{4} & (2) \end{cases}$$

$$\begin{cases} 5x + 4y = 10 & (1) \\ 6x + 6y = 9 & (2) \end{cases}$$

Divide both sides of equation (2) by 3:
$$\begin{cases} 5x + 4y = 10 & (1) \\ 2x + 2y = 3 & (2) \end{cases}$$

Multiply both sides of equation (2) by $-2$:
Add (1) and (2):
$$\begin{cases} 5x + 4y = 10 & (1) \\ -4x - 4y = -6 & (2) \end{cases}$$
$$x = 4$$

**Work Smart**
Although it is not necessary to divide both sides of equation (2) by 3, it makes solving the problem easier. Do you see why?

Back-substitute 4 for $x$ into the original equation (1).

Equation (1): $\frac{5}{2}x + 2y = 5$

$x = 4$: $\frac{5}{2}(4) + 2y = 5$

$10 + 2y = 5$

Subtract 10 from both sides: $2y = -5$

Divide both sides by 2: $y = -\frac{5}{2}$

We have that $x = 4$ and $y = -\frac{5}{2}$.

We leave the check to you. Both equations are satisfied.

The solution is the ordered pair $\left(4, -\frac{5}{2}\right)$.

**QUICK ✓** *Solve the system using elimination.*

6. $\begin{cases} -2x + y = 4 \\ -5x + 3y = 7 \end{cases}$

7. $\begin{cases} -3x + 2y = 3 \\ 4x - 3y = -6 \end{cases}$

**SUMMARY: Which Method Should I Use?**

We have presented three methods for solving systems of two linear equations containing two unknowns. A question that remains unanswered is "When should I use each method?" Below, we present a summary of the methods, the advantages of each method, and when each method should be used.

| Method | Advantages/Disadvantages | When Should I Use It? |
|---|---|---|
| Graphical | Allows us to "see" the answer but if the solutions are not integers, it can be difficult to determine the solution. | When a visual solution is required. |
| Substitution | Method gives exact solutions. The algebra can be easy provided one of the variables has a coefficient of 1. If none of the coefficients are one, the algebra can get messy. | If one of the coefficients of the variables is 1 or one of the variables is already solved for (as in $x =$ or $y =$). |
| Elimination | Method gives exact solutions. It is easy to use when none of the variables has a coefficient of 1. | If both equations are in standard form ($Ax + By = C$). |

**Work Smart**

The figure below shows a consistent and independent system.

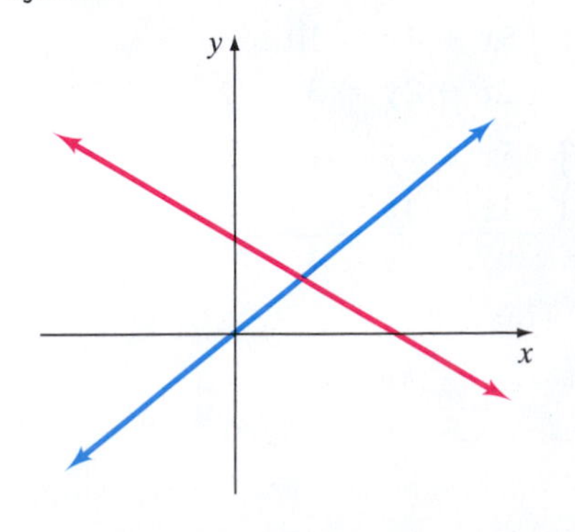


## 5 Identify Inconsistent Systems

Examples 2–7 dealt only with consistent and independent systems of equations. That is, we only discussed systems of equations with a single solution. Remember that there are two other possibilities for the solution of a system of linear equations: (1) The system could be inconsistent, which means that the lines in the system are parallel, or (2) the system could be consistent, but dependent, which means that the lines in the system are coincident (the same line).

In this next example, we look at an inconsistent system of equations.

### EXAMPLE 8 An Inconsistent System

Solve: $\begin{cases} 3x + 2y = 2 & (1) \\ -6x - 4y = 8 & (2) \end{cases}$

**Solution**

It seems easiest to use the method of elimination to solve this system because none of the variables have a coefficient of 1.

In looking at this system, we can make the coefficients on the variable $x$ additive inverses by multiplying equation (1) by 2.

$$\begin{cases} 3x + 2y = 2 & (1) \\ -6x - 4y = 8 & (2) \end{cases}$$

Multiply (1) by 2: $\begin{cases} 2(3x + 2y) = 2(2) & (1) \\ -6x - 4y = 8 & (2) \end{cases}$

Use the Distributive Property: $\begin{cases} 6x + 4y = 4 & (1) \\ -6x - 4y = 8 & (2) \end{cases}$

Add (1) and (2): $0 = 12$

**Work Smart**

When solving a system of equations, if you end up with a statement "0 = some nonzero constant," the system is inconsistent. Graphically, the lines in the system are parallel.

The equation $0 = 12$ is false. We conclude that the system has no solution, so the solution set is $\emptyset$ or $\{\ \}$. The system is inconsistent.

Figure 3 shows the pair of lines whose equations form the system in Example 8. Notice that the graphs of the two equations are lines, each with slope $-\frac{3}{2}$. Equation (1) has a $y$-intercept of 1, while equation (2) has a $y$-intercept of $-2$. Therefore, the lines are parallel and do not intersect. This geometric statement is equivalent to the algebraic statement that the system has no solution.

**Figure 3**

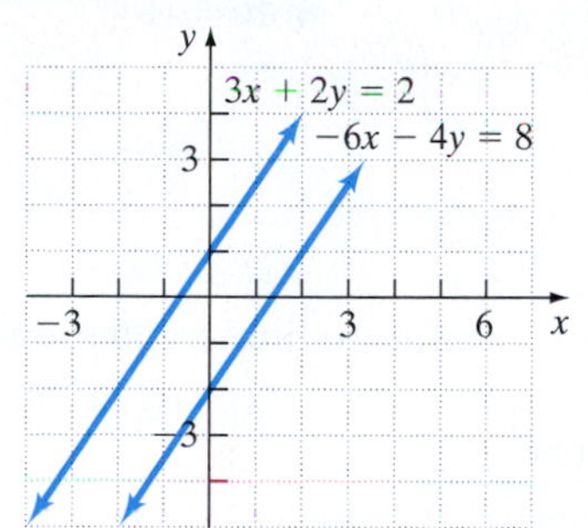


**QUICK ✓** *Show that the system is inconsistent. Draw a graph to support your result.*

**8.** $\begin{cases} -3x + \quad y = 2 \\ 6x - 2y = 1 \end{cases}$

## 6 Express the Solution of a System of Dependent Equations

The next example illustrates a system with infinitely many solutions.

### EXAMPLE 9 Solving a System of Dependent Equations

Solve: $\begin{cases} 3x + \quad y = 1 & (1) \\ -6x - 2y = -2 & (2) \end{cases}$

**Solution**

We choose to use the substitution method because solving equation (1) for $y$ is straightforward.

Equation (1): $3x + y = 1$

Subtract $3x$ from both sides: $y = -3x + 1$

Substitute $-3x + 1$ for $y$ in equation (2).

Equation (2): $-6x - 2y = -2$

$-6x - 2(-3x + 1) = -2$

Distribute the $-2$: $-6x + 6x - 2 = -2$

$-2 = -2$

The equation $-2 = -2$ is true. This means that as long as $y$ is chosen so that it equals $-3x + 1$, we will have a solution to the system. For example, if $x = 0$, then $y = -3(0) + 1 = 1$; if $x = 1$, then $y = -3(1) + 1 = -2$; if $x = 2$, then

**In Words**
The solution to a dependent system is "the set of all ordered pairs such that one of the equations in the system is true."

$y = -3(2) + 1 = -5$. The system of equations is consistent, but dependent (the value of $y$ that makes the equation true *depends* on the value of $x$) so that there are infinitely many solutions. We will write the solution as

$$\{(x, y) | 3x + y = 1\}$$

**Figure 4**

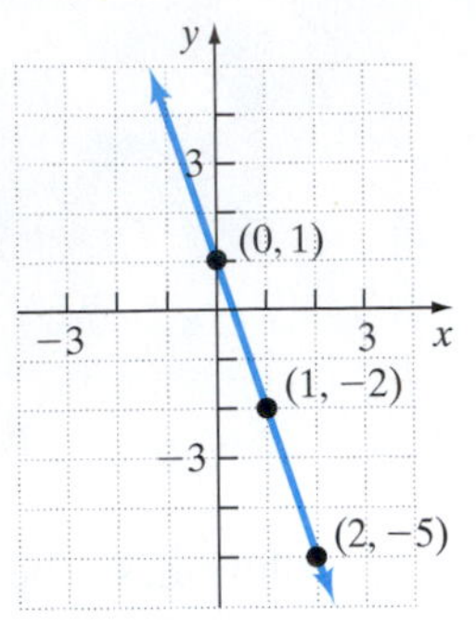


Figure 4 illustrates the situation presented in Example 9. The graphs of the two equations are lines, each with slope $-3$ and each with $y$-intercept 1. The lines are coincident.

Look back at the equations in Example 9. Notice that the terms in equation (2) are two times the terms in equation (1). This is another way to identify dependent systems when you have two equations with two unknowns.

**QUICK ✓** *Solve the system. Draw a graph to support your result.*

9. $\begin{cases} -3x + 2y = 8 \\ 6x - 4y = -16 \end{cases}$

# 4.1 Exercises

***For Extra Help:***

Student Solutions Manual

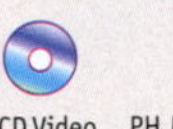
CD Video

PH Math/Tutor Center

MathXL Tutorials on CD

MathXL®

MyMathLab

## Concepts and Vocabulary

*In Problems 1–3, fill in the blanks.*

1. A _______ _______ _______ _______ is a grouping of two or more linear equations, each of which contains one or more variables.
2. If a system of equations has no solution, it is said to be ___________.
3. If a system of equations has infinitely many solutions, the system is said to be _______ and the equations are _______.

*In Problems 4–6, answer True or False to each statement.*

4. A system of two linear equations containing two variables always has at least one solution.
5. When the lines in a system of equations are parallel, then the system is inconsistent and has no solution.
6. A method for obtaining an equivalent system of equations is to multiply both sides of an equation in the system by a nonzero constant.
7. In this section, we presented two algebraic methods for solving a system of linear equations. Are there any circumstances where one method is preferable to the other? What are these circumstances?
8. Describe geometrically the three possibilities for a solution to a system of two linear equations containing two variables.
9. The solution to a system of two linear equations in two unknowns is $x = 3$, $y = -2$. Where do the lines in the system intersect?
10. In the process of solving a system of linear equations, what tips you off that the system is consistent but dependent? What tips you off that the system is inconsistent?

## Building Skills

*In Problems 11–14, determine whether the given ordered pairs listed are solutions of the system of linear equations.*

11. $\begin{cases} 2x + y = 13 \\ -5x + 3y = 6 \end{cases}$

**(a)** $(5, 3)$ **(b)** $(3, 7)$

12. $\begin{cases} x - 2y = -11 \\ 3x + 2y = -1 \end{cases}$

**(a)** $(-5, 3)$ **(b)** $(-3, 4)$

**13.** $\begin{cases} 5x + 2y = 9 \\ -10x - 4y = -18 \end{cases}$

**(a)** $(1, 2)$ **(b)** $\left(2, -\frac{1}{2}\right)$

**14.** $\begin{cases} -3x + y = 5 \\ 6x - 2y = 6 \end{cases}$

**(a)** $(-2, -1)$ **(b)** $(2, 0)$

*In Problems 15–18, use the graph of the system to determine the solution.*

**15.** $\begin{cases} x + y = 1 \\ x + 2y = 0 \end{cases}$

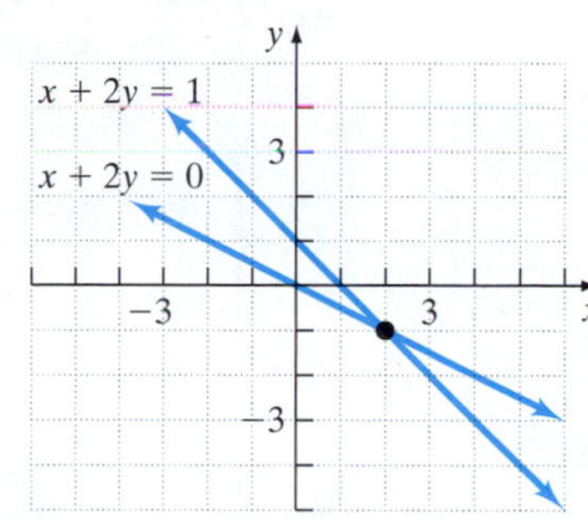


**16.** $\begin{cases} -2x + y = 4 \\ 2x + y = 0 \end{cases}$

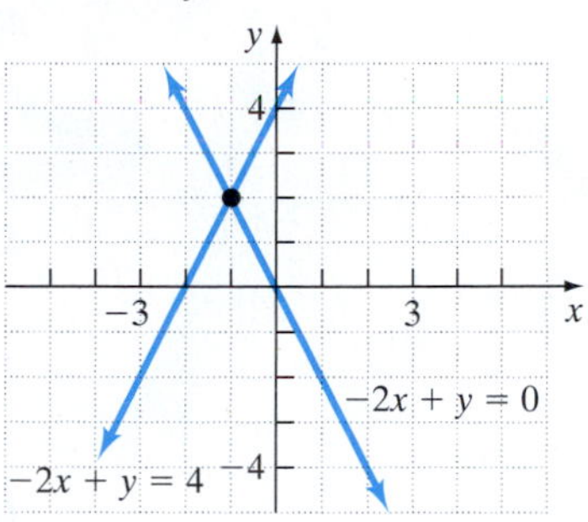


**17.** $\begin{cases} x - 2y = -2 \\ x - 2y = 2 \end{cases}$

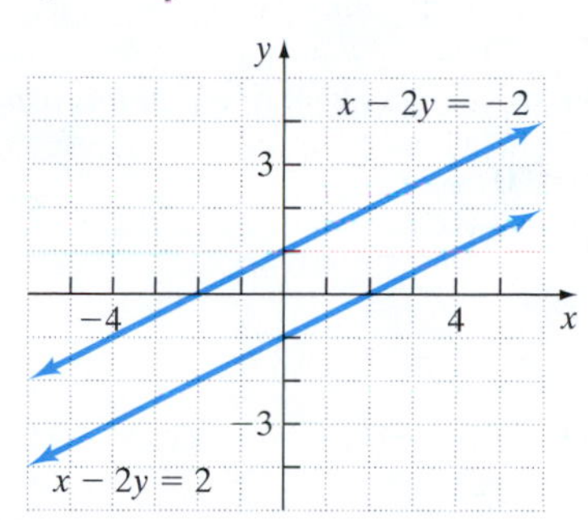


**18.** $\begin{cases} 3x + y = 1 \\ -6x - 2y = -2 \end{cases}$

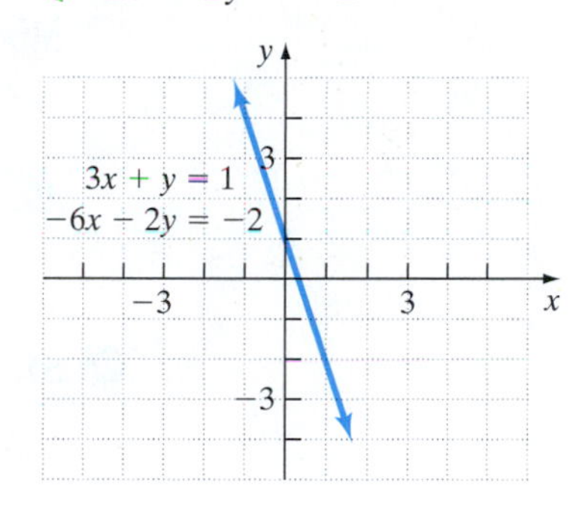


*In Problems 19–24, solve the system of equations by graphing.*

**19.** $\begin{cases} y = 3x \\ y = -2x + 5 \end{cases}$

**20.** $\begin{cases} y = -2x + 4 \\ y = 2x - 4 \end{cases}$

**21.** $\begin{cases} y = \dfrac{1}{2}x + 1 \\ 2x - 4y = -4 \end{cases}$

**22.** $\begin{cases} y = -\dfrac{2}{3}x + 3 \\ 2x + 3y = 9 \end{cases}$

**23.** $\begin{cases} 2x + y = 2 \\ x + 3y = -9 \end{cases}$

**24.** $\begin{cases} -x + 2y = -9 \\ 2x + y = -2 \end{cases}$

*In Problems 25–36, solve the system of equations using substitution.*

**25.** $\begin{cases} y = -\dfrac{1}{2}x + 1 \\ y = -2x + 10 \end{cases}$

**26.** $\begin{cases} y = -3x - 4 \\ y = 4x + 17 \end{cases}$

**27.** $\begin{cases} x = \dfrac{2}{3}y \\ 3x - y = -3 \end{cases}$

**28.** $\begin{cases} y = \dfrac{1}{2}x \\ x - 4y = -4 \end{cases}$

**29.** $\begin{cases} 3x + y = 1 \\ -6x - 2y = -4 \end{cases}$

**30.** $\begin{cases} -2x + 4y = 9 \\ x - 2y = -3 \end{cases}$

**31.** $\begin{cases} 2x - 4y = 2 \\ x + 2y = 0 \end{cases}$

**32.** $\begin{cases} 3x + 2y = 0 \\ 6x + 2y = 5 \end{cases}$

**33.** $\begin{cases} x + 3y = 6 \\ -\dfrac{x}{3} - y = -2 \end{cases}$

**34.** $\begin{cases} -4x + y = 8 \\ x - \dfrac{y}{4} = -2 \end{cases}$

**35.** $\begin{cases} x + y = 10{,}000 \\ 0.05x + 0.07y = 650 \end{cases}$

**36.** $\begin{cases} x + y = 5000 \\ 0.04x + 0.08y = 340 \end{cases}$

*In Problems 37–48, solve the system of equations using elimination.*

**37.** $\begin{cases} x + y = -5 \\ -x + 2y = 14 \end{cases}$

**38.** $\begin{cases} x + y = -6 \\ -2x - y = 0 \end{cases}$

**39.** $\begin{cases} x + 2y = -5 \\ 3x + 3y = 9 \end{cases}$

**40.** $\begin{cases} -3x + 2y = -5 \\ 2x - y = 10 \end{cases}$

**41.** $\begin{cases} 5x - 2y = 2 \\ -10x + 4y = 3 \end{cases}$

**42.** $\begin{cases} 6x - 4y = 6 \\ -3x + 2y = 3 \end{cases}$

**43.** $\begin{cases} 2x + 5y = -3 \\ x + \frac{5}{4}y = -\frac{1}{2} \end{cases}$

**44.** $\begin{cases} x + 2y = -\frac{8}{3} \\ 3x - 3y = 5 \end{cases}$

**45.** $\begin{cases} \frac{1}{3}x - 2y = 6 \\ -\frac{1}{2}x + 3y = -9 \end{cases}$

**46.** $\begin{cases} \frac{5}{4}x - \frac{1}{2}y = 6 \\ -\frac{5}{3}x + \frac{2}{3}y = -8 \end{cases}$

**47.** $\begin{cases} 0.05x + 0.1y = 5.25 \\ 0.08x - 0.02y = 1.2 \end{cases}$

**48.** $\begin{cases} 0.04x + 0.06y = 2.1 \\ 0.06x - 0.03y = 0.15 \end{cases}$

## Mixed Practice

*In Problems 49–56, solve the system of equations using either substitution or elimination.*

**49.** $\begin{cases} x + 3y = 0 \\ -2x + 4y = 30 \end{cases}$

**50.** $\begin{cases} 2x + y = -1 \\ -3x - 2y = 7 \end{cases}$

**51.** $\begin{cases} x = 5y - 3 \\ -3x + 15y = 9 \end{cases}$

**52.** $\begin{cases} y = \frac{1}{2}x + 2 \\ x - 2y = -4 \end{cases}$

**53.** $\begin{cases} 2x - 4y = 18 \\ 3x + 5y = -3 \end{cases}$

**54.** $\begin{cases} 12x + 45y = 0 \\ 8x + 6y = 24 \end{cases}$

**55.** $\begin{cases} \frac{5}{6}x - \frac{1}{3}y = -5 \\ -x + \frac{2}{5}y = 1 \end{cases}$

**56.** $\begin{cases} \frac{1}{3}x - \frac{1}{2}y = -5 \\ -\frac{4}{5}x + \frac{6}{5}y = 1 \end{cases}$

## Applying the Concepts

*In Problems 57–60, write each equation in the system of equations in slope-intercept form. Use the slope-intercept form to determine the number of solutions the system has.*

**57.** $\begin{cases} 2x + y = -5 \\ 5x + 3y = 1 \end{cases}$

**58.** $\begin{cases} 4x - 2y = 8 \\ -10x + 5y = 5 \end{cases}$

**59.** $\begin{cases} 3x - 2y = -2 \\ -6x + 4y = 4 \end{cases}$

**60.** $\begin{cases} 2x - y = -5 \\ -4x + 3y = 9 \end{cases}$

△ **61. Parallelogram** Use the parallelogram shown to the right to answer parts (a) and (b).

**(a)** Find the equation of the line for the diagonal through the points $(-1, 3)$ and $(3, 1)$. Find the equation for the line through the diagonal through $(-2, -1)$ and $(4, 5)$.

**(b)** Find the point of intersection of the diagonals.

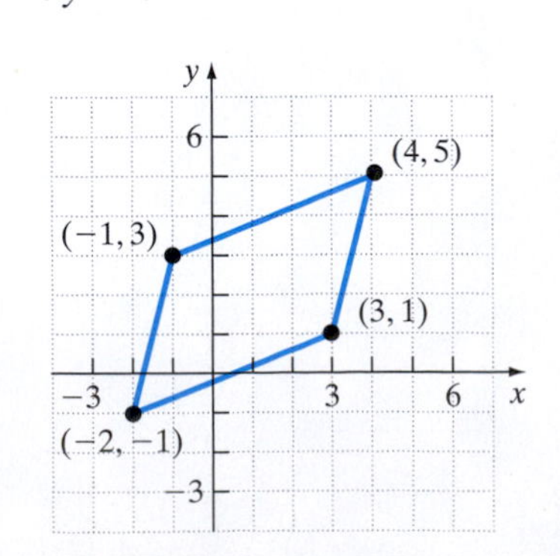

△ **62. Rhombus** A rhombus is a parallelogram whose adjacent sides are congruent. Use the rhombus to the right to answer parts (a), (b), and (c).

**(a)** Find the equation of the line for the diagonal through the points $(-1, 3)$ and $(3, 1)$. Find the equation for the line through the diagonal through $(-1, -2)$ and $(3, 6)$.

**(b)** Find the point of intersection of the diagonals.

**(c)** Compare the slopes of the diagonals. What can be said about the diagonals?

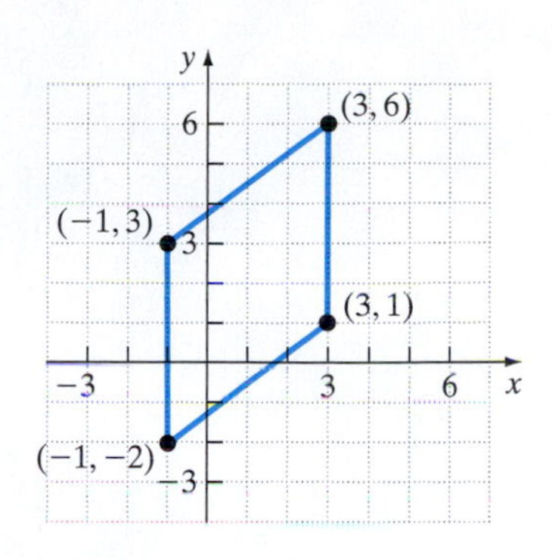


## Extending the Concepts

**63.** Which of the following ordered pairs could be a solution to the system graphed to the right?

**(a)** $(2, 4)$
**(b)** $(-2, 0)$
**(c)** $(-3, 1)$
**(d)** $(5, -2)$
**(e)** $(-1, -3)$
**(f)** $(-1, 3)$

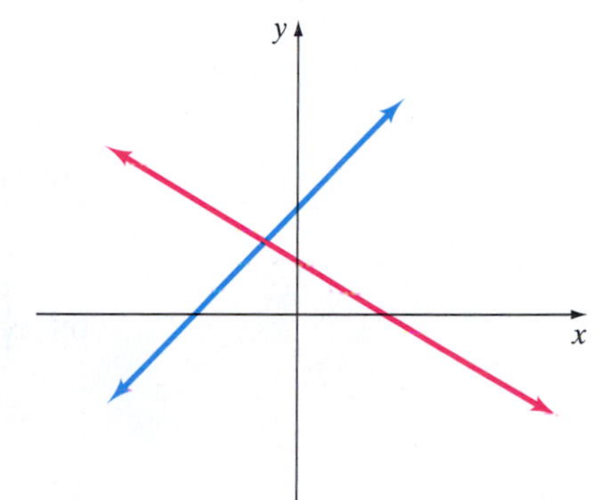


**64.** Which of the following systems of equations could have the graph to the right?

**(a)** $\begin{cases} 2x + 3y = 12 \\ 2x + \phantom{3}y = -2 \end{cases}$

**(b)** $\begin{cases} 2x + 3y = 3 \\ -2x + y = 2 \end{cases}$

**(c)** $\begin{cases} 2x - 3y = 12 \\ x + 2y = 2 \end{cases}$

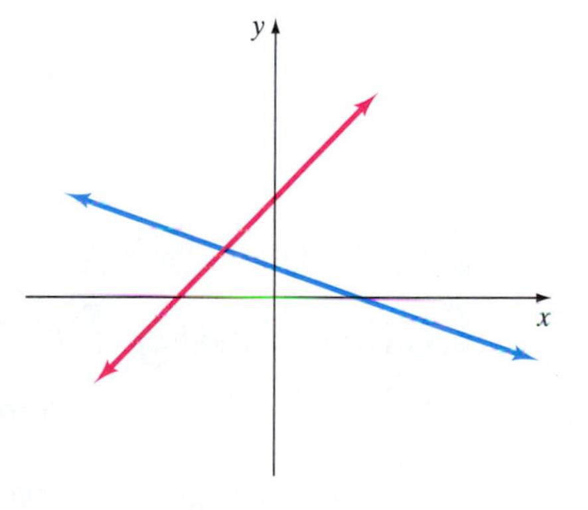


**65.** For the system $\begin{cases} Ax + 3By = 2 \\ -3Ax + By = -11 \end{cases}$, find $A$ and $B$ such that $x = 3$, $y = 1$ is a solution.

**66.** Write a system of equations that has $(3, 5)$ as a solution.

**67.** Write a system of equations that has $(-1, 4)$ as a solution.

△ **68. Centroid** The medians of a triangle are the line segments from each vertex to the midpoint of the opposite side. The centroid of a triangle is the point where the medians of the triangle intersect. Use the information given in the figure of the triangle to find its centroid.

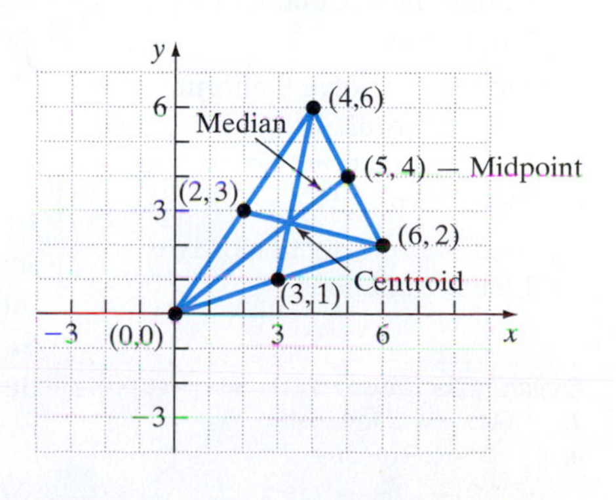

### The Graphing Calculator

A graphing calculator can be used to approximate the point of intersection between two equations using its INTERSECT command. We illustrate this feature of the graphing calculator by doing Example 3. Start by graphing each equation in the system as shown in Figure 5(a). Then use the INTERSECT command and find that the lines intersect at $x = 2$, $y = -3$. See Figure 5(b). The solution is the ordered pair $(2, -3)$.

**Figure 5**

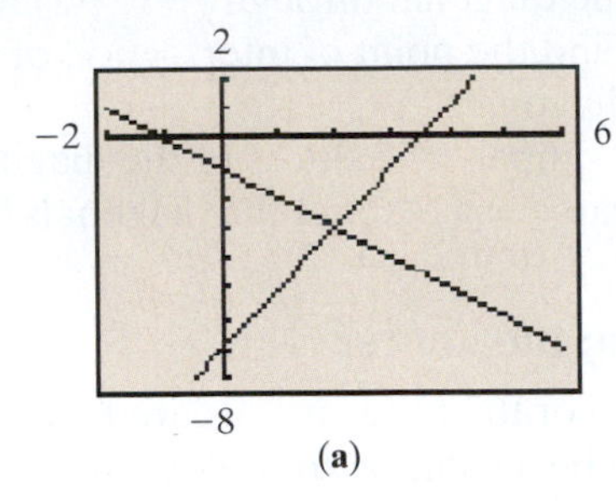


(a)

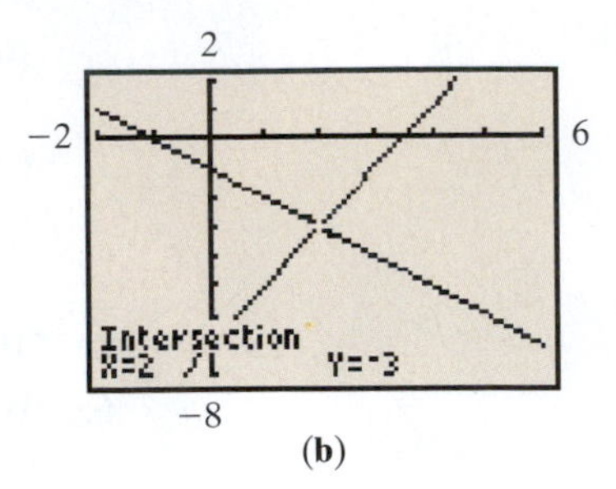


(b)

*In Problems 69–76, use a graphing calculator to solve each system of equations. If necessary, express your solution rounded to two decimal places.*

**69.** $\begin{cases} y = 3x - 1 \\ y = -2x + 5 \end{cases}$

**70.** $\begin{cases} y = \dfrac{3}{2}x - 4 \\ y = -\dfrac{1}{4}x + 3 \end{cases}$

**71.** $\begin{cases} 3x - y = -1 \\ -4x + y = -3 \end{cases}$

**72.** $\begin{cases} -6x - 2y = 4 \\ 5x + 3y = -2 \end{cases}$

**73.** $\begin{cases} 4x - 3y = 1 \\ -8x + 6y = -2 \end{cases}$

**74.** $\begin{cases} -2x + 5y = -2 \\ 4x - 10y = 1 \end{cases}$

**75.** $\begin{cases} 2x - 3y = 12 \\ 5x + y = -2 \end{cases}$

**76.** $\begin{cases} x - 3y = 21 \\ x + 6y = -2 \end{cases}$

# 4.2 Problem Solving: Systems of Two Linear Equations Containing Two Unknowns

## OBJECTIVES

1. Model and Solve Direct Translation Problems Involving Two Linear Equations Containing Two Unknowns
2. Model and Solve Geometry Problems Involving Two Linear Equations Containing Two Unknowns
3. Model and Solve Mixture Problems Involving Two Linear Equations Containing Two Unknowns
4. Model and Solve Uniform Motion Problems Involving Two Linear Equations Containing Two Unknowns
5. Find the Intersection of Two Linear Functions

***Preparing for...Answers*** **1.** See page 65. **2.** $25{,}000 - s$ **3.** \$36.46 **4.** $C(x) = 15x + 500$

### *Preparing for Problem Solving*

*Before getting started, take the following readiness quiz. If you get a problem wrong, go back to the section cited and review the material.*

1. List the problem solving steps given in Section 1.2. [Section 1.2, pp. 65–66]
2. If a total of \$25,000 is to be invested in stocks and bonds with $s$ representing the amount in stocks, write an algebraic expression for the amount invested in bonds. [Section 1.2, pp. 71–72]
3. Suppose that you have a credit card balance of \$3500 and the credit card company charges 12.5% annual interest on outstanding balances. How much interest will you have to pay after 1 month? [Section 1.2, pp. 70–71]
4. Write a linear cost function if the fixed costs are \$500 and the variable costs are \$15 per unit. [Section 3.5, pp. 231–232]

In Sections 1.2 and 1.3, we modeled and solved problems using a single variable. In this section, we learn how to model problems using two variables. The variables will be related through a system of equations. This approach is different from the problems in Section 1.2 because the Section 1.2 problems had a single unknown whose value was determined from a single equation. If you haven't already, go back to Section 1.2 and review the problem-solving strategy given on pages 64–66.

## 1 Model and Solve Direct Translation Problems Involving Two Linear Equations Containing Two Unknowns

One classification of problems that we discussed in Chapter 1 was direct translation problems. Let's look at a couple of examples where direct translation results in a system of two linear equations containing two unknowns.

### EXAMPLE 1 Take Me Out to the Ball Game

The Cummings family and the Freese family recently went to a baseball game. In the second inning, Roger Cummings bought 4 hot dogs and 3 large Cokes for \$22.25. In the seventh inning, Dave Freese bought 5 hot dogs and 4 large Cokes for \$28.50. How much does each hot dog cost? How much does each large Coke cost?

**Solution**

**Step 1: Identify** We want to know the cost of each hot dog and the cost of each large Coke.

**Step 2: Name** Let $h$ represent the cost of a hot dog. Let $c$ represent the cost of a large Coke.

**Step 3: Translate** One hot dog costs $h$ dollars. Two hot dogs would cost $2h$ dollars, so four hot dogs would cost $4h$ dollars. Similar logic tells us that 3 large Cokes would cost $3c$ dollars. We add these costs together to get \$22.25. This leads to equation (1):

Roger's purchase, equation (1): $4h + 3c = 22.25$

Using the same notation, we obtain equation (2):

Dave's purchase, equation (2): $5h + 4c = 28.50$

We combine equations (1) and (2) to form the following system:

$$\begin{cases} 4h + 3c = 22.25 & (1) \\ 5h + 4c = 28.50 & (2) \end{cases} \quad \text{The Model}$$

**Step 4: Solve**

We will eliminate $h$ by multiplying equation (1) by 5 and equation (2) by $-4$:
$$\begin{cases} 5(4h + 3c) = 5(22.25) & (1) \\ -4(5h + 4c) = -4(28.50) & (2) \end{cases}$$

Distribute:
$$\begin{cases} 20h + 15c = 111.25 & (1) \\ -20h - 16c = -114 & (2) \end{cases}$$

Add equations (1) and (2): $-c = -2.75$

Divide both sides by $-1$: $c = 2.75$

Substitute 2.75 for $c$ in equation (1) and solve for $h$.

$$4h + 3(2.75) = 22.25$$
$$4h + 8.25 = 22.25$$

Subtract 8.25 from both sides: $4h = 14$

Divide both sides by 4: $h = 3.50$

**Step 5: Check** If each hot dog costs \$3.50 and Roger buys 4 hot dogs, then he'll spend $4(3.50) = \$14.00$. If each Coke costs \$2.75 and Roger buys 3 Cokes, then he'll spend $3(\$2.75) = \$8.25$. His total bill will be $\$14.00 + \$8.25 = \$22.25$. If Dave buys 5 hot dogs, then he'll spend $5(3.50) = \$17.50$. If each Coke costs \$2.75 and Dave buys 4 Cokes, then he'll spend $4(\$2.75) = \$11.00$. His total bill will be $\$17.50 + \$11.00 = \$28.50$. It checks.

**Step 6: Answer** Each hot dog costs \$3.50 and each large Coke costs \$2.75.

QUICK ✓

1. At a fast food joint, 4 cheeseburgers and 2 medium shakes cost \$10.10. At the same fast food joint, 3 cheeseburgers and 3 medium shakes cost \$10.35. What is the cost of a cheeseburger? What is the cost of a medium shake?

## 2 Model and Solve Geometry Problems Involving Two Linear Equations Containing Two Unknowns

Formulas from geometry are often needed in order to solve certain types of problems. The type of formula that you will need is dictated by the problem. Remember that a formula is a model—and in geometry, formulas describe relationships and shapes.

### EXAMPLE 2 Enclosing a Yard with a Fence

The Fitzgeralds just bought a dog, so they need to enclose their back yard with a fence. Jim Fitzgerald measured the perimeter of the rectangular yard to be 180 feet. He also knows that the difference between the length and the width of the yard is 25 feet with the width being larger. What are the dimensions of the back yard?

**Solution**

**Step 1: Identify** We are looking for the length and the width of the back yard.

**Step 2: Name** Let $l$ represent the length and $w$ represent the width of the back yard.

**Step 3: Translate** The perimeter $P$ of a rectangle is $P = 2w + 2l$, where $l$ is the length and $w$ is the width. So we know that

Equation (1): $2w + 2l = 180$

In addition, the difference between the length and width is 25 feet with the width being larger, so

Equation (2): $w - l = 25$

Equations (1) and (2) are combined to form the following system:

$$\begin{cases} 2w + 2l = 180 & (1) \\ w - l = 25 & (2) \end{cases} \quad \text{The Model}$$

**Step 4: Solve** We will use the method of substitution by solving equation (2) for $w$.

| | |
|---|---|
| Equation (2): | $w - l = 25$ |
| Add $l$ to both sides: | $w = l + 25$ |
| Let $w = l + 25$ in equation (1): | $2(l + 25) + 2l = 180$ |
| Distribute: | $2l + 50 + 2l = 180$ |
| Combine like terms: | $4l + 50 = 180$ |
| Subtract 50 from both sides: | $4l = 130$ |
| Divide both sides by 4: | $l = 32.5$ |

Substitute 32.5 for $l$ in $w = l + 25$.

$$w = 32.5 + 25$$
$$w = 57.5$$

**Step 5: Check** With $l = 32.5$ and $w = 57.5$, the perimeter would be $2(32.5) + 2(57.5) = 180$ feet. The difference between the length and width is $57.5 - 32.5 = 25$ feet.

**Step 6: Answer** The length of the back yard is 32.5 feet and the width is 57.5 feet.

**QUICK ✓**

**2.** A rectangular field has a perimeter of 360 yards. The length of the field is twice the width. What is the length of the field? What is the width of the field?

## EXAMPLE 3 Solving a Geometry Problem

Two lines that are not parallel are shown in Figure 6. Suppose that we know that the measure of angle 1 is $(x + 3y)^\circ$, the measure of angle 2 is $(x + y)^\circ$ and the measure of angle 3 is $(3x - 2y - 5)^\circ$. Find $x$ and $y$.

**Figure 6**

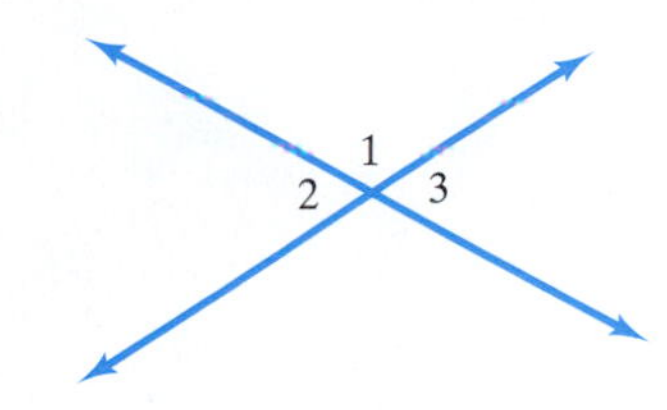


### Solution

**Step 1: Identify** This is a geometry problem involving supplementary angles. From geometry, we know that angles 1 and 2 are supplementary, which means that the measure of angle 1 plus the measure of angle 2 must equal 180°. In addition, we know that angle 2 and angle 3 are vertical angles. Therefore, the measure of angle 2 must equal the measure of angle 3.

**Work Smart**

Two angles are supplementary if the sum of the angles is 180°. When two lines intersect, the nonadjacent angles are vertical.

**Step 2: Name** The names of the unknowns, $x$ and $y$, were given in the problem.

**Step 3: Translate** Because the measure of angle 1 plus the measure of angle 2 equals 180°, we have that

$$\underbrace{x + 3y}_{\text{Measure of Angle 1}} + \underbrace{x + y}_{\text{Measure of Angle 2}} = 180$$

This simplifies to

$$\text{Equation (1):} \quad 2x + 4y = 180$$

In addition, we know that the measure of angle 2 must equal the measure of angle 3, so that

$$\underbrace{x + y}_{\text{Measure of Angle 2}} = \underbrace{3x - 2y - 5}_{\text{Measure of Angle 3}}$$

This simplifies to

$$\text{Equation (2):} \quad -2x + 3y = -5$$

We put equations (1) and (2) into the following system of equations:

$$\begin{cases} 2x + 4y = 180 & (1) \\ -2x + 3y = -5 & (2) \end{cases} \quad \text{The Model}$$

**Step 4: Solve** We solve the system using the method of elimination by adding equations (1) and (2):

$$\begin{cases} 2x + 4y = 180 & (1) \\ -2x + 3y = -5 & (2) \end{cases}$$

$$\text{Add:} \quad 7y = 175$$

$$\text{Divide both sides by 7:} \quad y = 25$$

Let $y = 25$ in equation (1) and find that $x = 40$.

**Step 5: Check** If $x = 25$ and $y = 40$, then the measure of angle 1 is $(x + 3y)^\circ = (40 + 3(25))^\circ = 115^\circ$. The measure of angle 2 is $(x + y)^\circ = (40 + 25)^\circ = 65^\circ$. The measure of angle 3 is $(3x - 2y - 5)^\circ = (3(40) - 2(25) - 5)^\circ = 65^\circ$. We add the measure of angle 1 to the measure of angle 2 and obtain $115° + 65° = 180°$, so that angle 1 and angle 2 are supplementary. In addition, we find that the measure of angle 2 equals the measure of angle 3.

**Step 6: Answer** The value of $x$ is 40 and the value of $y$ is 25.

**QUICK ✓**

3. Suppose that lines $m$ and $n$ are parallel. Line $p$ transverses lines $m$ and $n$. Suppose that it is known that the measure of angle 1 is $(x + 3y)°$, the measure of angle 3 is $(3x + y)°$, and the measure of angle 5 is $(5x + y)°$. Find $x$ and $y$.

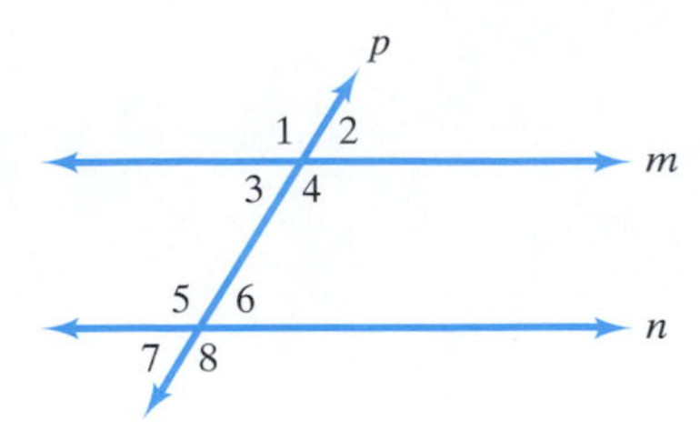


## 3 Model and Solve Mixture Problems Involving Two Linear Equations Containing Two Unknowns

In Section 1.2, we solved a variety of mixture problems. These problems included mixing investments and mixing quantities to form a new quantity. When we solved these problems in Section 1.2, we developed a model that required a single equation with a single unknown. An alternative approach to modeling these problems is to develop a system of equations. This is one of the beauties of modeling: different models can be developed to solve the exact same problem! Let's revisit a mixture problem and solve it by developing a model that involves systems of equations.

### EXAMPLE 4 Financial Planning

Kevin has \$15,000 to invest. His goal is to obtain an overall annual rate of return of 9% or \$1350 annually. His financial advisor recommends that he invest some of the money in corporate bonds that pay 12% and the rest in government-backed Treasury notes paying 4%. How much should be placed in each investment in order for Kevin to achieve his goal?

#### Solution

**Step 1: Identify** This is a mixture problem involving simple interest. Kevin needs to know how much to place into corporate bonds and how much to place into Treasury notes to earn \$1350 in interest.

**Step 2: Name** Let $c$ represent the amount invested in corporate bonds and let $t$ represent the amount invested in Treasury notes.

**Step 3: Translate** We organize the information given in Table 1.

Table 1

| | Principal \$ | Rate % | Time Yr | Interest \$ |
|---|---|---|---|---|
| **Corporate Bond** | $c$ | 0.12 | 1 | $0.12c$ |
| **Treasury Notes** | $t$ | 0.04 | 1 | $0.04t$ |
| **Total** | 15,000 | 0.09 | 1 | $0.09(15,000) = \$1350$ |

Kevin wants to earn 9% each year on his principal of \$15,000. Using the simple interest formula, Kevin wants to earn $\$15{,}000(0.09)(1) = \$1350$ each year in interest. The total interest will be the sum of the interest from the corporate bonds and the Treasury notes.

Since $t = 1$ year, we have

Interest from corporate bonds + Interest from Treasury notes = \$1350

Equation (1): $0.12c + 0.04t = 1350$

In addition, the total investment is to be \$15,000 so that the amount invested in corporate bonds plus the amount invested in Treasury notes must equal \$15,000:

Equation (2): $c + t = 15{,}000$

We use equations (1) and (2) to form a system of equations.

$$\begin{cases} 0.12c + 0.04t = 1350 & (1) \\ c + t = 15{,}000 & (2) \end{cases} \quad \text{The Model}$$

**Step 4: Solve**

Use the substitution method by solving equation (2) for $t$: $t = 15{,}000 - c$

Substitute $15{,}000 - c$ for $t$ in equation (1): $0.12c + 0.04(15{,}000 - c) = 1350$

Use the Distributive Property to remove parentheses: $0.12c + 600 - 0.04c = 1350$

Combine like terms: $0.08c + 600 = 1350$

Subtract 600 from both sides: $0.08c = 750$

Divide both sides by 0.08: $c = 9375$

Since $c = 9375$, we have that $t = 15{,}000 - c = 15{,}000 - 9375 = 5625$.

**Step 5: Check** The simple interest earned each year on the corporate bonds is $(\$9375)(0.12)(1) = \$1125$. The simple interest earned each year on the Treasury bonds is $(\$5625)(0.04)(1) = \$225$. The total interest earned is $\$1125 + \$225 = \$1350$. Kevin wanted to earn 9% annually and $(\$15{,}000)(0.09)(1) = \$1350$. This agrees with the information presented in the problem.

**Step 6: Answer the Question** Kevin will invest \$9375 in corporate bonds and \$5625 in Treasury bonds.

**QUICK ✓**

4. Maria has recently retired and requires an extra \$7200 per year in income. She has \$120,000 to invest and can invest in either an Aaa-rated bond that pays 5% per annum or a B-rated bond paying 10% per annum. How much should be placed in each investment in order for Maria to achieve her goal?

## EXAMPLE 5 Blending Coffees

The manager of a coffee shop wishes to form a new blend of coffee. She wants to mix Sumatra beans, known for their strong, distinctive taste, that sell for \$12 per pound with milder Brazilian beans that sell for \$8 per pound to get 50 pounds of the new blend. The new blend will sell for \$9 per pound and there will be no difference in revenue from selling the new blend versus selling the beans separately. How many pounds of the Sumatra and Brazilian beans are required?

## Solution

**Step 1: Identify** This is a mixture problem. We want to know the number of pounds of Sumatra beans and the number of pounds of Brazilian beans that are required in the new blend.

**Step 2: Name** Let $s$ represent the number of pounds of Sumatra beans and $b$ represent the number of pounds of Brazilian beans that are required.

**Step 3: Translate** We are told that there is to be no difference in revenue between selling the Sumatra and Brazilian separately versus the blend. This means that if the blend contains one pound of Sumatra and one pound of Brazilian, we should collect $\$12(1) + \$8(1) = \$20$ because that is how much we would collect if we sold the beans separately.

Table 2 summarizes the information and Figure 7 illustrates the idea.

**Table 2**

| | Price $/Pound | Number of Pounds | Revenue |
|---|---|---|---|
| **Sumatra** | 12 | $s$ | $12s$ |
| **Brazilian** | 8 | $b$ | $8b$ |
| **Blend** | 9 | 50 | $9(50) = 450$ |

**Figure 7**

In general, if the mixture contains $s$ pounds of Sumatra beans, we should collect $\$12s$. If the mixture contains $b$ pounds of Brazilian beans, we should collect $\$8b$. If the mixture sells for \$9 per pound and we make 50 pounds of the blend, then we should collect $\$9(50) = \$450$ for the blend.

$$\begin{pmatrix}\text{Price per pound}\\ \text{of Sumatra}\end{pmatrix}\begin{pmatrix}\text{Pounds of}\\ \text{Sumatra}\end{pmatrix} + \begin{pmatrix}\text{Price per pound}\\ \text{of Brazilian}\end{pmatrix}\begin{pmatrix}\text{Pounds of}\\ \text{Brazilian}\end{pmatrix} = \begin{pmatrix}\text{Price per pound}\\ \text{of blend}\end{pmatrix}\begin{pmatrix}\text{Pounds of}\\ \text{blend}\end{pmatrix}$$

$$\$12 \cdot s + \$8 \cdot b = \$9 \cdot 50$$

We have the equation

Equation (1): $12s + 8b = 450$

The number of pounds of Sumatra beans plus the number of pounds of Brazilian beans should equal 50 pounds. We have the equation

Equation (2): $s + b = 50$

We use equations (1) and (2) to form a system of equations.

$$\begin{cases} 12s + 8b = 450 & (1) \\ s + b = 50 & (2) \end{cases}$$ The Model

**Step 4: Solve**

| | |
|---|---|
| Use the substitution method by solving equation (2) for $b$: | $b = 50 - s$ |
| Substitute $50 - s$ for $b$ in equation (1): | $12s + 8(50 - s) = 450$ |
| Use the Distributive Property: | $12s + 400 - 8s = 450$ |
| Combine like terms: | $4s + 400 = 450$ |
| Subtract 400 from both sides: | $4s = 50$ |
| Divide both sides by 4: | $s = 12.5$ |

With $s = 12.5$ pounds, we have that $b = 50 - s = 50 - 12.5 = 37.5$ pounds.

**Step 5: Check** It appears that we should mix 12.5 pounds of Sumatra with 37.5 pounds of Brazilian. The 12.5 pounds of Sumatra beans would sell for $12(12.5) = $150 and the 37.5 pounds of Brazilian beans would sell for $8(37.5) = $300; the total revenue would be $150 + $300 = $450 which equals the revenue obtained from selling the blend. This checks with the information presented in the problem.

**Step 6: Answer the Question** The manager should mix 12.5 pounds of Sumatra beans with 37.5 pounds of Brazilian beans to make the blend.

**QUICK ✓**

**Work Smart**
Remember that mixtures can include interest (money), solids (nuts), liquids (chocolate milk), and even gases (Earth's atmosphere).

5. "We're Nuts!" sells cashews for $7.00 per pound and peanuts for $2.50 per pound. The manager has decided to make a "trail mix" that combines the cashews and peanuts. She wants the trail mix to sell for $4.00 per pound and there should be no loss in revenue from selling the trail mix versus selling the nuts alone. How many pounds of cashews and peanuts are required to create 30 pounds of trail mix?

## 4 Model and Solve Uniform Motion Problems Involving Two Linear Equations Containing Two Unknowns

Let's now look at a problem involving uniform motion. Remember, these problems use the fact that distance equals rate times time ($d = rt$).

### EXAMPLE 6 Uniform Motion

The airspeed of a plane is its speed through the air. This speed contrasts with the plane's groundspeed—its speed relative to the ground. The groundspeed of an airplane is affected by the speed of the wind. Suppose that a Boeing 767 flying west from Washington, D.C. to San Francisco, a distance of 2400 miles, takes 6 hours. The trip east from San Francisco to Washington, D.C. takes 4 hours. Find the airspeed of the plane and the effect wind resistance has on the plane.

**Solution**

**Step 1: Identify** This is a uniform motion problem. We want to determine the airspeed of the plane.

**Step 2: Name** There are two unknowns in the problem—the airspeed of the plane and the impact of wind resistance on the plane. We will let $a$ represent the airspeed of the plane and $w$ represent the impact of wind resistance.

**Step 3: Translate** Going west, the plane is flying into the jet stream, so that the plane is hindered by the wind. Therefore, the groundspeed of the plane will be

$a - w$. Going east, the jet stream is helping the plane. Therefore, the groundspeed of the plane will be $a + w$. We set up Table 3.

**Table 3**

| | Distance = | Rate | · Time |
|---|---|---|---|
| **Against Wind (West)** | 2400 | $a - w$ | 6 |
| **With Wind (East)** | 2400 | $a + w$ | 4 |

Going against the wind, we have the equation

Equation (1): $2400 = 6(a - w)$ or $2400 = 6a - 6w$

**Work Smart**

In equation (1), we could have divided both sides by 6 rather than distributing. What could we have done with equation (2)?

Going with the wind, we have the equation

Equation (2): $2400 = 4(a + w)$ or $2400 = 4a + 4w$

Combining equations (1) and (2), we form a system of two linear equations containing two unknowns.

$$\begin{cases} 6a - 6w = 2400 & (1) \\ 4a + 4w = 2400 & (2) \end{cases} \quad \text{The Model}$$

**Step 4: Solve** We will use the elimination method by multiplying equation (1) by 3 and equation (2) by 2 and then adding equations (1) and (2).

$$\begin{cases} 2(6a - 6w) = 2(2400) & (1) \\ 3(4a + 4w) = 3(2400) & (2) \end{cases}$$

$$\text{Distribute:} \quad \begin{cases} 12a - 12w = 4800 & (1) \\ 12a + 12w = 7200 & (2) \end{cases}$$

$$\text{Add:} \quad 24a = 12{,}000$$

$$\text{Divide both sides by 24:} \quad a = 500$$

We use equation (1) with $a = 500$ to find that the effect of wind resistance is $w = 100$.

**Work Smart**

When checking your work make sure that the airspeed of the plane is greater than the windspeed.

**Step 5: Check** Flying west, the groundspeed of the plane is $500 - 100 = 400$ miles per hour. Flying east, the groundspeed of the plane is $500 + 100 = 600$ miles per hour.

**Step 6: Answer** The airspeed of the plane is 500 miles per hour. The impact of wind resistance on the plane is 100 miles per hour.

**QUICK ✓**

6. The airspeed of a plane is its speed through the air. The groundspeed of a plane is its speed relative to the ground. The groundspeed of an airplane is affected by the speed of the wind. Suppose that a plane flying 1200 miles west requires 4 hours and flying 1200 miles east requires 3 hours. Find the airspeed of the plane and the effect wind resistance has on the plane.

## 5 Find the Intersection of Two Linear Functions

We have seen that linear functions can be used to model a variety of situations such as costs of phone plans. Often, we are interested in knowing when two linear functions are equal. For example, we might want to know the value of $x$ such that $f(x) = 3x + 1$ will equal $g(x) = -2x + 11$. This would require that we solve the equation $f(x) = g(x)$ or $3x + 1 = -2x + 11$. The solution to this equation is represented geometrically as the point of intersection of the two functions as shown in Figure 8.

**Figure 8**

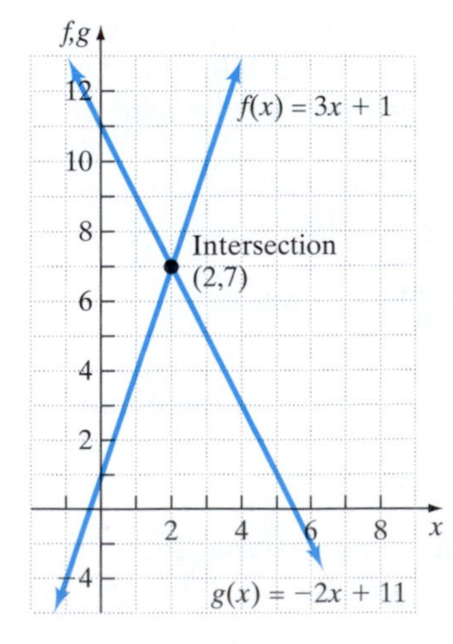


Let's look at an application of intersecting functions from business.

## EXAMPLE 7 Break-Even Analysis

Business is motivated by profit. A company's profit is the difference between revenues and cost. It is key that a company understand the number of units of their product they must manufacture and sell in order to be profitable. Suppose that a gas grill company sells their entry level grill for \$130. The variable cost of manufacturing the grill is \$80 per grill and the fixed costs per month are \$8500.

**(a)** Write revenue $R$ as a function of the number of grills sold $x$.

**(b)** Write cost $C$ as a function of the number of grills manufactured $x$.

**(c)** Graph the revenue function and cost function on the same Cartesian plane.

**(d)** The **break-even point** is the point where revenue equals cost. Find the break-even number of grills to be manufactured and sold. What is the revenue when this number of grills is sold? Label the break-even point on the graph drawn in part (c).

### Solution

**(a)** If one grill is sold, revenue will be \$130. If two grills are sold, revenue will be \$130(2) = \$260. In general, if $x$ grills are sold, revenue will be \$130$x$. The revenue function is $R(x) = 130x$.

**(b)** The cost function is $C(x) = ax + b$, where $a$ is the variable cost and $b$ is the fixed cost. With $a = 80$ and $b = 8500$, we have $C(x) = 80x + 8500$.

**(c)** Figure 9 shows the graphs of the revenue and cost functions.

**Figure 9**

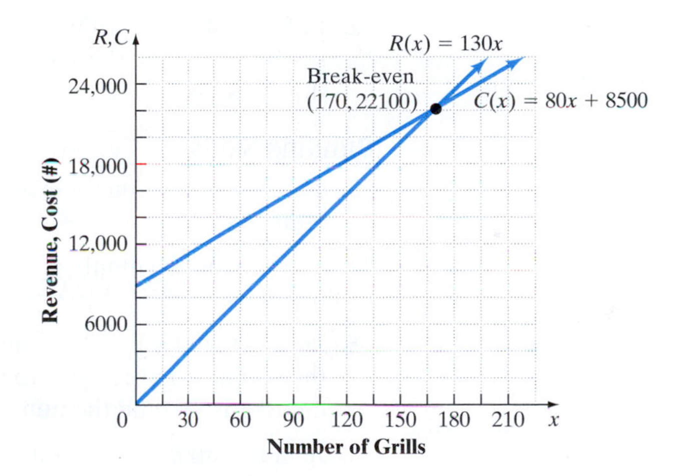


**(d)** To find the break-even point, we need to solve the equation $R(x) = C(x)$.

$$R(x) = C(x)$$
$$130x = 80x + 8500$$

Subtract $80x$ from both sides: $$50x = 8500$$

Divide both sides by 50: $$x = 170$$

It appears that the company needs to sell 170 grills each month to break even. We verify this by determining the revenue and cost when $x = 170$.

$$R(170) = 130(170) = \$22,100$$
$$C(170) = 80(170) + 8500 = \$22,100$$

Since the revenue equals the cost when $x = 170$ grills are manufactured and sold, we know that the break-even point is 170 grills per month. We label the break-even point on the graph in Figure 9. Notice that if more than 170 grills are sold, the company will make a profit.

**7.** Suppose that a nursery sells 8-foot Austrian Pine trees for $230. The nursery can buy the trees from a tree farm for $160. The fixed costs at the nursery amount to $2100 per month.

**(a)** Write revenue $R$ as a function of the number of trees sold $x$.

**(b)** Write cost $C$ as a function of the number of trees purchased from the tree farm $x$.

**(c)** Graph the revenue function and cost function on the same Cartesian plane.

**(d)** Find the break-even number of trees to be sold. What is the revenue when this number of trees is sold? Label the break-even point on the graph drawn in part (c).

# 4.2 Exercises

*For Extra Help:*

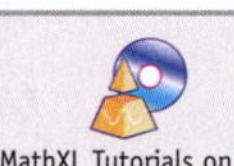

Student Solutions Manual | CD Video | PH Math/Tutor Center | MathXL Tutorials on CD | MathXL® | MyMathLab

## Concepts and Vocabulary

**1.** Do you prefer to use one variable or a system of equations to model mixture problems? Support your opinion.

**2.** In Example 6, we found the effect of wind resistance on the plane was 100 miles per hour. Do you think that the actual wind speed (the jet stream) is more or less than this? Why?

## Building Skills

**3.** The sum of two numbers is 18. The difference of the two numbers is $-2$. Find the numbers.

**4.** The sum of two numbers is 25. The difference of two numbers is 3. Find the numbers.

**5.** Janice is thinking of two numbers. She says that two times the first number plus the second number is 47. In addition, the first number plus three times the second number is 81. Find the numbers.

**6.** Juan is thinking of two numbers. He says the 3 times the first number minus the second number is 118. In addition, two times the first number plus the second number is 147. Find the numbers.

**7.** The sum of twice a first number and three times a second number is 81. If the second number is subtracted from three times the first number the result is 17. Find the numbers.

**8.** The sum of four times a first number and a second number is 68. If the first number is decreased by twice the second number the result is $-1$. Find the numbers.

*In Problems 9 and 10, let R represent a company's revenue, let C represent the company's cost, and let x represent the number of units produced and sold each day.*

**(a)** *Graph the revenue function and cost function on the same Cartesian plane.*

**(b)** *Find the company's break-even point; that is, find x so that $R = C$. Label this point on the graph drawn in part (a).*

**9.** $R(x) = 12x$
$C(x) = 5.5x + 9880$

**10.** $R(x) = 16x$
$C(x) = 7x + 3645$

## Applying the Concepts

**11. Rental Costs** Gina rented a moon-walk for 5 hours at a total cost of \$235. Lori rented the same moon-walk for 3 hours at a total cost of \$165. The cost of renting is based upon a flat set-up fee plus a rental rate per hour. How much is the set-up fee? What is the hourly rental fee?

**12. Making Change** Johnny has \$6.75 in dimes and quarters. He has 8 more dimes than quarters. How many quarters does Johnny have? How many dimes does Johnny have?

**13. Counting Calories** Suppose that Kristin ate two McDonald's hamburgers and drank one medium Coke, for a total of 770 calories. Kristin's friend, Jack, ate three hamburgers and drank two medium Cokes (Jack takes advantage of free refills) for a total of 1260 calories. How many calories are in a McDonald's hamburger? How many calories are in a medium Coke?

**14. Carbs** Yvette and José go to McDonald's for breakfast. Yvette orders two sausage biscuits and one orange juice. The entire meal had 98 grams of carbohydrates. José ordered three sausage biscuits and two orange juices and his meal had 168 grams of carbohydrates. How many grams of carbohydrates are in a sausage biscuit? How many grams of carbohydrates are in an orange juice?

△ **15. Perimeter** The perimeter of a rectangle is 120 meters. If the length of the rectangle is 20 meters more than the width, what are the dimensions of the rectangle?

△ **16. Perimeter** The perimeter of a rectangle is 260 centimeters. If the width of the rectangle is 15 centimeters less than the length, what are the dimensions of the rectangle?

△ **17. Triangle** An isosceles triangle is one in which two of the sides are the same length (congruent). The perimeter of an isosceles triangle is 35 cm. If the length of each of the congruent sides is 3 times the length of the third side, find the dimensions of the triangle.

△ **18. Trapezoid** A trapezoid is a quadrilateral in which two of the sides are parallel. A trapezoid is an isosceles trapezoid if two sides are congruent (the same length). See the figure. Suppose that the length of each of the congruent sides of an isosceles trapezoid is 12 cm. The perimeter of the trapezoid is 100 cm. Of the remaining two sides, one length is 14 cm shorter than two times the other. Find the length of each of the two remaining sides.

△ **19. Angles** Two lines that are not parallel are shown in the figure to the right. Suppose that it is known that the measure of angle 1 is $(10x + 6y)^\circ$, the measure of angle 2 is $(4y)^\circ$ and the measure of angle 3 is $(7x + 2y)^\circ$. Find $x$ and $y$.

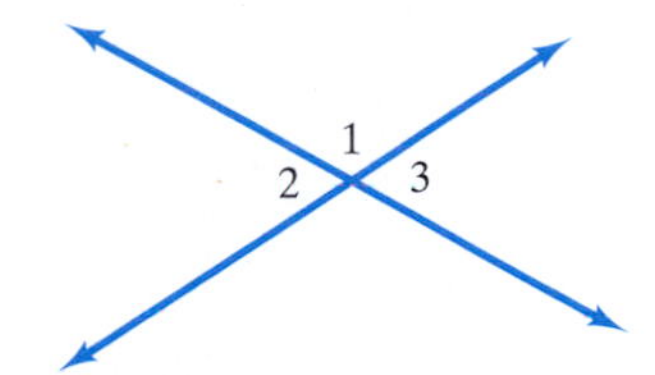


△ **20. Angles** Suppose that lines $m$ and $n$ are parallel. Line $p$ transverses lines $m$ and $n$. Suppose that it is known that the measure of angle 1 is $(2x + 3y)^\circ$, the measure of angle 2 is $(4x)^\circ$, and the measure of angle 3 is $(5x + y + 1)^\circ$. Find $x$ and $y$.

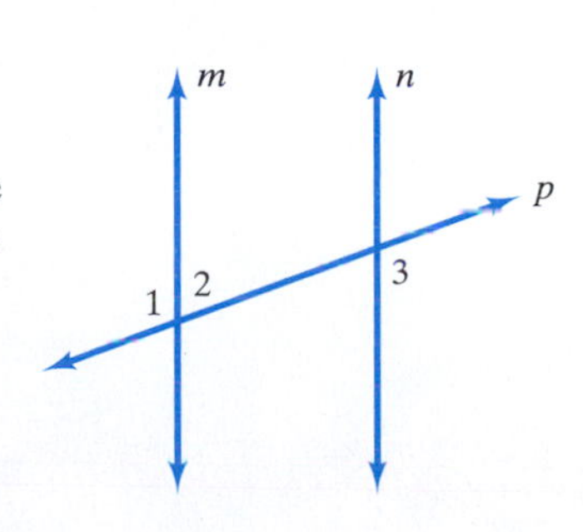

△ **21. Angles** Find the values of $x$ and $y$ based upon the values for the angles in the following triangle. The measure of angle 1 is $(4x + 2y)°$, the measure of angle 2 is $(10x + 5y)°$, the measure of angle 3 is $(7x + 5y)°$, and the measure of angle 4 is $(9x + 8y)°$.

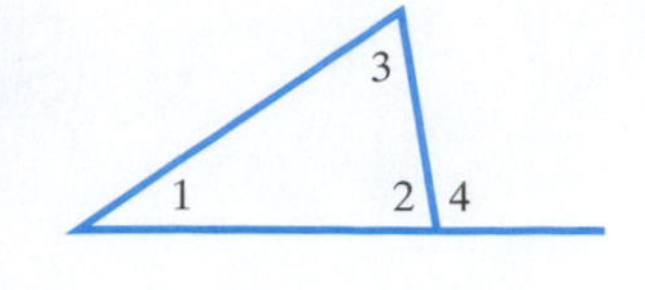


△ **22. Angles** Find the values of $x$ and $y$ based upon the values for the angles in the following parallelogram. The measure of angle 1 is $(5x + 7y)°$, the measure of angle 2 is $(10x + 5y)°$, and the measure of angle 3 is $(15x - 9y)°$.

**23. Investments** Suppose that you received an unexpected inheritance of $36,000. You have decided to invest the money by placing some of the money in stocks and some in bonds. To diversify, you decided that five times the amount in bonds should equal three times the amount invested in stocks. How much should be invested in stocks? How much should be invested in bonds?

**24. Investments** Marge and Homer have $80,000 to invest. Their financial advisor has recommended that they diversify by placing some of the money in stocks and some in bonds. Based upon current market conditions, he has recommended that three times the amount in bonds should equal two times the amount invested in stocks. How much should be invested in stocks? How much should be invested in bonds?

**25. Banking** A bank has loaned out $750,000, part of it at 5% per annum and the rest of it at 8% per annum. If the bank receives $52,500 in interest each year, how much was loaned at 5%?

**26. One Bad Investment** Horace invested $70,000 in two stocks. After 1 year, one of the stocks increased by 13%, while the other stock declined by 5%. His total gain for the year was $2800. How much was invested in the stock that earned 13%? How much was invested in the stock that lost 5%?

**27. Making Coffee** Suppose that you want to blend two coffees in order to obtain a new blend. The blend will be made with the best Arabica beans that sell for $9.00 per pound and select African Robustas that sell for $11.50 per pound to obtain 100 pounds of the new blend. The new blend will sell for $10.00 per pound and there will be no difference in revenue from selling the new blend versus selling the beans separately. How many pounds of the Arabica and Robusta beans are required?

**28. Candy** A candy store sells chocolate-covered almonds for $6.50 per pound and chocolate-covered peanuts for $4.00 per pound. The manager decides to make a bridge mix that combines the almonds with the peanuts. She wants the bridge mix to sell for $6.00 per pound, and there should be no loss in revenue from selling the bridge mix versus the almonds and peanuts alone. How many pounds of chocolate-covered almonds and chocolate-covered peanuts are required to create 50 pounds of bridge mix?

**29. Pharmacy** A doctor's prescription calls for a daily intake of liquid containing 40 mg of vitamin C and 30 mg of vitamin D. Your pharmacy stocks two liquids that can be used: One contains 20% vitamin C and 30% vitamin D, the other 50% vitamin C and 25% vitamin D. How many milligrams of each liquid should be mixed to fill the prescription?

**30. Pharmacy** A doctor's prescription calls for the creation of pills that contain 20 units of vitamin $B_{12}$ and 12 units of vitamin E. Your pharmacy stocks two powders that can be used to make these pills: One contains 20% vitamin $B_{12}$ and 40% vitamin E, the other 50% vitamin $B_{12}$ and 30% vitamin E. How many units of each powder should be mixed in each pill?

**31. Canoeing in the River** Jonathon and Samantha row their canoe 26 miles downstream in 2 hours. After a picnic, they row back upstream. After 3 hours they only travel 9 miles. Assuming that they canoe at a constant rate and the river's current is constant, find the speed at which Jonathon and Samantha can row in still water.

**32. Against the Wind** A Piper Arrow can fly 510 miles in 3 hours with a tail wind. Against this same wind, the plane can fly 390 miles in 3 hours. Find the airspeed of the plane. What is the impact of the wind on the plane?

**33. A Car Ride** An Infiniti G35 travels at an average speed of 50 miles per hour. A Lincoln Aviator travels at an average speed of 40 miles per hour. In the time it takes the Lincoln to travel a certain distance $d$, the Infiniti travels 100 miles farther. Find the distance that each car travels and the time of the trip.

**34. Runners** Enrique leaves his house and starts to run at an average speed of 6 miles per hour. Half an hour later, Enrique's younger (and faster) brother leaves the house to catch up to Enrique running at an average speed of 8 miles per hour. How long will it take for Enrique's brother to run half the distance that Enrique has run?

**35. Supply and Demand** Suppose that the quantity supplied $S$ and quantity demanded $D$ of hot dogs at a baseball game are given by the following functions:

$$S(p) = -2000 + 3000p$$
$$D(p) = 10{,}000 - 1000p$$

where $p$ is the price. The **equilibrium price** of a market is defined as the price at which quantity supplied equals quantity demanded ($S = D$).

**(a)** Graph each of the two functions on the same Cartesian plane.
**(b)** Find the equilibrium price for hot dogs at the baseball game. What is the equilibrium quantity? Label the equilibrium point on the graph drawn in part (a).

**36. Supply and Demand** Suppose that the quantity supplied $S$ and quantity demanded $D$ of baseball hats at a baseball game are given by the following functions:

$$S(p) = 9p - 17$$
$$D(p) = -35p + 995$$

where $p$ is the price. The **equilibrium price** of a market is defined as the price at which quantity supplied equals quantity demanded ($S = D$).

**(a)** Graph each of the two functions on the same Cartesian plane.
**(b)** Find the equilibrium price for baseball hats at the baseball game. What is the equilibrium quantity? Label the equilibrium point on the graph drawn in part (a).

**37. College Grads** The percent of adult males with at least 4 years of college $M$ as a function of the year $t$ is $M(t) = 0.327t + 24.24$, where $t$ is the number of years since 1990. The percent of adult females with at least 4 years of college $F$ as a function of the year $t$ is $F(t) = 0.532t + 18.12$ where $t$ is the number of years since 1990. (*Source:* Statistical Abstract)

**(a)** Graph each of the two functions on the same Cartesian plane.
**(b)** Assuming that this linear trend continues, find the year in which the percent of male college grads and female college grads is equal. That is, solve $M(t) = F(t)$. What is the percent of college grads in this year? Label this point on the graph.

**38. Weekly Earnings** The average weekly earnings of 16- to 24-year-old males $M$ as a function of the year $t$ is $M(t) = 8.92t + 272.11$, where $t$ is the number of years since 1990. The average weekly earnings of 16 – 24 year old females $F$ as a function of the year $t$ is $F(t) = 9.44t + 244.92$, where $t$ is the number of years since 1990.

**(a)** Graph each of the two functions on the same Cartesian plane.
**(b)** Assuming that this linear trend continues, find the year in which the average weekly earnings of 16- to 24-year-old males will equal the average weekly earnings of 16- to 24-year-old females. That is, solve $M(t) = F(t)$. What is the average weekly earnings in this year? Label this point on the graph.

**39. Phone Charges** Sprint has two different long-distance phone plans. Plan $A$ charges a monthly fee of \$8.95 plus \$0.05 per minute. Plan $B$ charges a monthly fee of \$5.95 plus \$0.07 per minute. (*Source: Sprint.com*)

**(a)** Find a function for each long-distance plan treating number of minutes used as the independent variable and cost as the dependent variable.
**(b)** Graph each of the two functions on the same Cartesian plane.
**(c)** Find the number of minutes for which the cost of plan $A$ equals the cost of plan $B$. What is the cost of each plan at this number of minutes? Label this point on the graph.

**40. Salary** Suppose that you are offered a sales position for a pharmaceutical company. They offer you two salary options. Option A would pay you an annual base salary of \$15,000 plus a commission of 1% on sales. Option B would pay you an annual base salary of \$25,000 plus a commission of 0.75% on sales.

**(a)** Find a function for each salary option.
**(b)** Graph each function found in part (a) on the same Cartesian plane.
**(c)** Determine the annual sales required for the options to result in the same annual salary. What would the annual salary be? Label this point on the graph.

**41. Break-Even** A wood craftsman makes children's desks. He sells the desks for \$60 each. His monthly fixed costs of operating the business are \$3500. Each desk costs \$35 in material.

**(a)** Find the revenue function $R$ treating the number of desks $x$ as the independent variable.
**(b)** Find the cost function $C$ treating the number of desks $x$ as the independent variable.
**(c)** Graph the revenue and cost function on the same Cartesian plane.
**(d)** Find the break-even number of desks that must be manufactured and sold. What is the revenue and cost at this number of desks? Label the point on the graph drawn in part (c).

**42. Break-Even** Audra wants to establish a lemonade stand on her corner. Her father buys some lumber and other materials and makes Audra a lemonade stand for \$40. Audra goes to the store and buys lemonade. She determines that each cup of lemonade will cost her \$0.03 to make, but the actual cup will cost her an additional \$0.07. She decides to sell the lemonade for \$0.30 a cup.

**(a)** Find the revenue function $R$ treating the number of cups of lemonade $x$ as the independent variable.
**(b)** Find the cost function $C$ treating the number of cups of lemonade $x$ as the independent variable.
**(c)** Graph the revenue and cost function on the same Cartesian plane.
**(d)** Find the break-even number of cups of lemonade that Audra must sell. What is the revenue and cost at this number of cups of lemonade? Label the point on the graph drawn in part (c).

## Extending the Concepts

**43. The Olympics** The data in the following table represent the winning times in the 200-meter run in the finals of the Olympics.

**(a)** Draw a scatter diagram of the men's winning time treating the year as the independent variable.

**(b)** On the same graph, draw a scatter diagram of the women's winning time treating the year as the independent variable. Be sure to use a different plotting symbol to label the points (such as a □ and a ○).

**(c)** Draw a line through any two points for the men's winning time. Use these points to find the equation of the line.

**(d)** Draw a line through any two points for the women's winning time. Use these points to find the equation of the line.

**(e)** Use the equations found in parts (c) and (d) to find the year in which the winning time for men will equal the winning time for women. What is the winning time?

**(f)** Do you think that your answer to part (e) is reasonable? Why or why not?

| Year | Men's Time (in seconds) | Women's Time (in seconds) |
|---|---|---|
| 1968 | 19.83 | 22.50 |
| 1972 | 20.00 | 22.40 |
| 1976 | 20.23 | 22.37 |
| 1980 | 20.19 | 22.03 |
| 1984 | 19.80 | 21.81 |
| 1988 | 19.75 | 21.34 |
| 1992 | 19.73 | 21.72 |
| 1996 | 19.32 | 22.12 |
| 2000 | 20.09 | 21.84 |
| 2004 | 19.79 | 22.05 |

# 4.3 Systems of Linear Equations in Three Variables

## OBJECTIVES

1. Solve Systems of Three Linear Equations Containing Three Variables
2. Identify Inconsistent Systems
3. Express the Solution of a System of Dependent Equations
4. Model and Solve Problems Involving Three Linear Equations Containing Three Unknowns

### *Preparing for Systems of Linear Equations in Three Variables*

**1.** Evaluate the expression $3x - 2y + 4z$ for $x = 1$, $y = -2$, and $z = 3$ [Section R.5, pp. 43–44]

## 1 Solve Systems of Three Linear Equations Containing Three Variables

An example of a linear equation in three variables is $2x - y + z = 8$. An example of a system of three linear equations containing three variables is

Three equations containing three variables, $x$, $y$, and $z$

$$\begin{cases} x + 3y + z = 8 \\ 3x - y + 6z = 12 \\ -4x - y + 2z = -1 \end{cases}$$

Systems of three linear equations containing three variables have the same possible solutions as a system of two linear equations containing two variables:

1. **Exactly one solution**—A consistent system with independent equations
2. **No solution**—An inconsistent system
3. **Infinitely many solutions**—A consistent system with dependent equations

We can view the problem of solving a system of three linear equations containing three variables as a geometry problem. The graph of each equation in a system of linear equations containing three variables is a plane in space. A system of three linear equations containing three variables represents three planes in space. Figure 10 illustrates some of the possibilities.

*Preparing for...Answers* **1.** 19

**Figure 10**

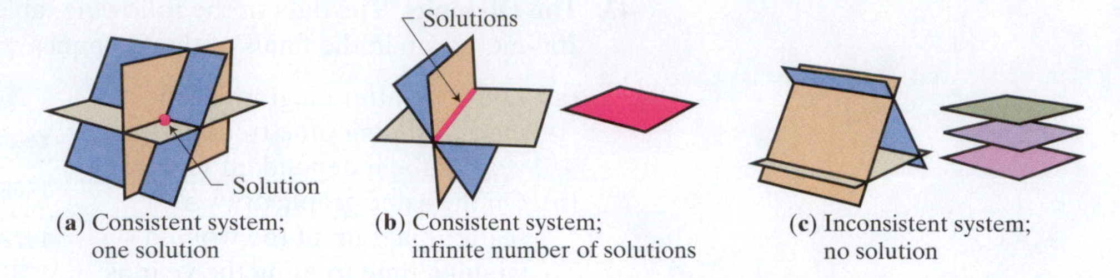


(a) Consistent system; one solution (b) Consistent system; infinite number of solutions (c) Inconsistent system; no solution

Recall that a **solution** to a system of equations consists of values for the variables that are solutions of each equation of the system. We write the solution to a system of three equations containing three unknowns as an **ordered triple** $(x, y, z)$.

## EXAMPLE 1 Determining Whether Values Are a Solution to a System of Linear Equations

Determine which of the following ordered triples are solutions to the system of equations.

$$\begin{cases} x + y + z = 0 \\ 2x - y + 3z = 17 \\ -3x + 2y - z = -21 \end{cases}$$

**(a)** $(1, 3, -4)$ **(b)** $(3, -5, 2)$

### Solution

First, we name $x + y + z = 0$ equation (1), $2x - y + 3z = 17$ equation (2), and $-3x + 2y - z = -21$ equation (3) as follows:

$$\begin{cases} x + y + z = 0 & (1) \\ 2x - y + 3z = 17 & (2) \\ -3x + 2y - z = -21 & (3) \end{cases}$$

**(a)** Let $x = 1$, $y = 3$, and $z = -4$ in equations (1), (2), and (3). If all three equations are true, then $(1, 3, -4)$ is a solution.

Equation (1):
$$\begin{aligned} x + y + z &= 0 \\ 1 + 3 + (-4) &\stackrel{?}{=} 0 \\ 0 &= 0 \quad \text{True} \end{aligned}$$

Equation (2):
$$\begin{aligned} 2x - y + 3z &= 17 \\ 2(1) - 3 + 3(-4) &\stackrel{?}{=} 17 \\ 2 - 3 - 12 &\stackrel{?}{=} 17 \\ -13 &= 17 \quad \text{False} \end{aligned}$$

Equation (3):
$$\begin{aligned} -3x + 2y - z &= -21 \\ -3(1) + 2(3) - (-4) &\stackrel{?}{=} -21 \\ -3 + 6 + 4 &\stackrel{?}{=} -21 \\ 7 &= -21 \quad \text{False} \end{aligned}$$

Although these values satisfy equation (1), they do not satisfy equations (2) or (3). Because the values $x = 1$, $y = 3$, and $z = -4$ do not satisfy equations (2) or (3), $(1, 3, -4)$ is not a solution.

**(b)** Let $x = 3$, $y = -5$, and $z = 2$ in equations (1), (2), and (3). If all three equations are true, then $(3, -5, 2)$ is a solution.

Equation (1):
$$\begin{aligned} x + y + z &= 0 \\ 3 + (-5) + (2) &\stackrel{?}{=} 0 \\ 0 &= 0 \quad \text{True} \end{aligned}$$

Equation (2):
$$\begin{aligned} 2x - y + 3z &= 17 \\ 2(3) - (-5) + 3(2) &\stackrel{?}{=} 17 \\ 6 + 5 + 6 &\stackrel{?}{=} 17 \\ 17 &= 17 \quad \text{True} \end{aligned}$$

Equation (3):
$$-3x + 2y - z = -21$$
$$-3(3) + 2(-5) - (2) \stackrel{?}{=} -21$$
$$-9 - 10 - 2 \stackrel{?}{=} -21$$
$$-21 = -21 \quad \text{True}$$

Because the values $x = 3$, $y = -5$, and $z = 2$ satisfy all three equations, the ordered triple $(3, -5, 2)$ is a solution.

**QUICK ✓**

**1.** Determine which of the following ordered triples are solutions to the system of equations.

$$\begin{cases} x + y + z = 3 \\ 3x + y - 2z = -23 \\ -2x - 3y + 2z = 17 \end{cases}$$

**(a)** $(3, 2, -2)$ **(b)** $(-4, 1, 6)$

---

Typically, when solving a system of three linear equations containing three variables, we use the method of elimination. Recall that the idea behind the method of elimination is to eliminate variables from the system. We eliminate variables by multiplying equations by nonzero constants in order to get the coefficients of the variables to be additive inverses. We then add the equations to remove the variable that has coefficients that are additive inverses. Other methods that we can use are to interchange any two equations or multiply (or divide) each side of an equation by the same non-zero constant.

The process for solving a system of three linear equations containing three unknowns is an extension of the elimination method presented in Section 4.1. Let's look at an example to see how to solve a system of three linear equations with three variables.

## EXAMPLE 2 How to Solve a System of Three Linear Equations with Three Variables

Use the method of elimination to solve the system:
$$\begin{cases} x + y - z = -1 & (1) \\ 2x - y + 2z = 8 & (2) \\ -3x + 2y + z = -9 & (3) \end{cases}$$

### Step-by-Step Solution

**Step 1:** Our goal is to eliminate the same variable from two of the equations. In looking at the system, we notice that we can use equation (1) to eliminate the variable x from equations (2) and (3). We can do this by multiplying equation (1) by $-2$ and adding the result to equation (2). The equation that results becomes equation (4). Why do we do this? Because the coefficients on x are now additive inverses and adding the equations eliminates the variable x. We also multiply equation (1) by 3 and add the result to equation (3). The equation that results becomes equation (5).

$$x + y - z = -1 \quad (1)$$
$$2x - y + 2z = 8 \quad (2)$$

Multiply (1) by $-2$:
$$-2x - 2y + 2z = 2 \quad (1)$$
$$2x - y + 2z = 8 \quad (2)$$
Add:
$$-3y + 4z = 10$$

$$x + y - z = -1 \quad (1)$$
$$-3x + 2y + z = -9 \quad (3)$$

Multiply (1) by 3:
$$3x + 3y - 3z = -3 \quad (1)$$
$$-3x + 2y + z = -9 \quad (3)$$
Add:
$$5y - 2z = -12$$

$$\begin{cases} x + y - z = 1 & (1) \\ -3y + 4z = 10 & (4) \\ 5y - 2z = -12 & (5) \end{cases}$$

*(continued)*

**Step 2:** We now concentrate on equations (4) and (5), treating them as a system of two equations containing two variables. It is easiest to eliminate the variable z by multiplying equation (5) by 2 and then add equations (4) and (5). The result will be equation (6).

$$-3y + 4z = 10 \quad (4)$$
$$5y - 2z = -12 \quad (5)$$

Multiply (5) by 2:
$$\begin{aligned} -3y + 4z &= 10 \quad (4) \\ 10y - 4z &= -24 \quad (5) \end{aligned}$$

Add: $7y = -14$

$$\begin{cases} x + y - z = -1 & (1) \\ -3y + 4z = 10 & (4) \\ 7y = -14 & (6) \end{cases}$$

**Step 3:** We solve equation (6) for y by dividing both sides of the equation by 7.

$$\begin{cases} x + y - z = -1 & (1) \\ -3y + 4z = 10 & (4) \\ y = -2 & (6) \end{cases}$$

**Step 4:** Back-substitute −2 for y in equation (4) and solve for z.

$$\begin{aligned} -3y + 4z &= 10 \\ -3(-2) + 4z &= 10 \\ 6 + 4z &= 10 \\ 4z &= 4 \\ z &= 1 \end{aligned}$$

**Step 5:** Back-substitute −2 for y and 1 for z into equation (1) and solve for x.

$$\begin{aligned} x + y - z &= -1 \\ x + (-2) - 1 &= -1 \\ x - 3 &= -1 \\ x &= 2 \end{aligned}$$

The solution appears to be $x = 2$, $y = -2$, $z = 1$.

**Step 6:** *Check* Verify that $x = 2$, $y = -2$, and $z = 1$ is the solution.

Equation (1): $x + y - z = -1$

$x = 2, y = -2, z = 1$: $2 + (-2) - 1 \stackrel{?}{=} -1$

$-1 = -1$ True

Equation (2): $2x - y + 2z = 8$

$2(2) - (-2) + 2(1) \stackrel{?}{=} 8$

$4 + 2 + 2 \stackrel{?}{=} 8$

$8 = 8$ True

Equation (3): $-3x + 2y + z = -9$

$-3(2) + 2(-2) + 1 \stackrel{?}{=} -9$

$-6 - 4 + 1 \stackrel{?}{=} -9$

$-9 = -9$ True

The solution is the ordered triple $(2, -2, 1)$.

We now summarize the steps used in Example 1.

## Steps for Solving a System of Three Linear Equations Containing Three Unknowns by Elimination

**Step 1:** Select two of the equations and eliminate one of the variables from one of the equations. Select any two other equations and eliminate the *same variable* from one of the equations.

**Step 2:** You will have two equations that have only two unknowns. Eliminate a second variable from the two linear equations in two unknowns.

**Step 3:** Solve for the remaining variable.

**Work Smart**

By eliminating a variable in two of the equations in Step 1, we are creating a system of 2 equations with 2 unknowns that can be solved. Remember, this is the goal in mathematics—reduce the problem to one you already know how to solve.

**Step 4:** Use the value of the variable found in Step 3 to find the value of a second variable.

**Step 5:** Use the two known values of the variables identified in Steps 3 and 4 to find the value of the third variable.

**Step 6:** Check your answer.

**QUICK ✓**

2. Use the method of elimination to solve the system: $\begin{cases} x + y + z = -3 \\ 2x - 2y - z = -7 \\ -3x + y + 5z = 5 \end{cases}$

## EXAMPLE 3 Solving a System of Three Linear Equations with Three Variables

Use the method of elimination to solve the system: $\begin{cases} 4x + z = 4 & (1) \\ 2x + 3y = -4 & (2) \\ 2y - 4z = -15 & (3) \end{cases}$

### Solution

We eliminate $z$ from equation (3) by multiplying equation (1) by 4 and adding the result to equation (3). The equation that results becomes equation (4).

$$\begin{array}{ll} 4x + z = 4 & (1) \\ 2y - 4z = -15 & (3) \end{array} \quad \text{Multiply (1) by 4:} \quad \begin{array}{ll} 16x + 4z = 16 & (1) \\ 2y - 4z = -15 & (3) \\ \hline 16x + 2y = 1 & \end{array} \rightarrow \begin{cases} 4x + z = 4 & (1) \\ 2x + 3y = -4 & (2) \\ 16x + 2y = 1 & (4) \end{cases}$$

We now concentrate on equations (2) and (4), treating them as a system of two equations containing two variables. It is easiest to eliminate the variable $x$ by multiplying equation (2) by $-8$ and then add equations (2) and (4). The result will be equation (5).

$$\begin{array}{ll} 2x + 3y = -4 & (2) \\ 16x + 2y = 1 & (4) \end{array} \quad \text{Multiply (2) by } -8\text{:} \quad \begin{array}{ll} -16x - 24y = 32 & (2) \\ 16x + 2y = 1 & (4) \\ \hline -22y = 33 & \end{array} \rightarrow \begin{cases} 4x + z = 4 & (1) \\ 2x + 3y = -4 & (2) \\ -22y = 33 & (5) \end{cases}$$

**Work Smart**

We chose to eliminate $z$ from equation (3) because the coefficient of $z$ in equation (1) is 1. We could also have eliminated $x$ from equation (2) by multiplying both sides of equation (2) by $-2$ and then added (1) and (2). There is more than one approach to solving the system!

We solve equation (5) for $y$ by dividing both sides of the equation by $-22$.

Equation (5): $-22y = 33$

$$y = \frac{33}{-22} = -\frac{3}{2}$$

Now we back-substitute $-\frac{3}{2}$ for $y$ into equation (2) and solve for $x$.

Equation (2): $2x + 3y = -4$

$$2x + 3\left(-\frac{3}{2}\right) = -4$$

$$2x - \frac{9}{2} = -4$$

$$2x = \frac{1}{2}$$

$$x = \frac{1}{4}$$

Now back-substitute $\frac{1}{4}$ for $x$ into equation (1) and solve for $z$.

$$\begin{aligned} \text{Equation (1):} \quad 4x + z &= 4 \\ 4\left(\frac{1}{4}\right) + z &= 4 \\ 1 + z &= 4 \\ z &= 3 \end{aligned}$$

The solution is the ordered triple $\left(\frac{1}{4}, -\frac{3}{2}, 3\right)$.

**QUICK ✓**

3. Use the method of elimination to solve the system: $\begin{cases} 2x \quad\quad - 4z = -7 \\ x + 6y \quad\quad = 5 \\ \quad\quad 2y - z = 2 \end{cases}$

---

**Work Smart**

The figures below geometrically illustrate inconsistent systems.

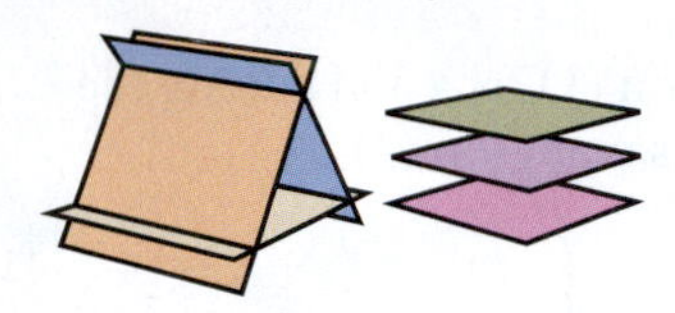

## 2 Identify Inconsistent Systems

Examples 2 and 3 were consistent and independent systems resulting in a single solution. We now look at an inconsistent system.

### EXAMPLE 4 An Inconsistent System of Linear Equations

Use the method of elimination to solve the system: $\begin{cases} x + 2y - z = 4 & (1) \\ -2x + 3y + z = -4 & (2) \\ x + 9y - 2z = 1 & (3) \end{cases}$

**Solution**

In looking at the system, notice that we can use equation (1) to eliminate the variable $x$ from equations (2) and (3). We can do this by multiplying equation (1) by 2 and adding the result to equation (2). The equation that results becomes equation (4). We also multiply equation (1) by $-1$ and add the result to equation (3). The equation that results becomes equation (5).

$$\begin{aligned} x + 2y - z &= 4 \quad (1) & \text{Multiply by 2:} \quad 2x + 4y - 2z &= 8 \quad (1) \\ -2x + 3y + z &= -4 \quad (2) & -2x + 3y + z &= -4 \quad (2) \\ & & 7y - z &= 4 \end{aligned}$$

$$\begin{aligned} x + 2y - z &= 4 \quad (1) & \text{Multiply by } -1\text{:} \quad -x - 2y + z &= -4 \quad (1) \\ x + 9y - 2z &= 1 \quad (3) & x + 9y - 2z &= 1 \quad (3) \\ & & 7y - z &= -3 \end{aligned}$$

$$\begin{cases} x + 2y - z = 4 & (1) \\ 7y - z = 4 & (4) \\ 7y - z = -3 & (5) \end{cases}$$

We now concentrate on equations (4) and (5), treating them as a system of two equations containing two variables. We multiply equation (4) by $-1$ and then add equations (4) and (5). The result will be equation (6).

$$\begin{aligned} 7y - z &= 4 \quad (4) & \text{Multiply by } -1\text{:} \quad -7y + z &= -4 \quad (4) \\ 7y - z &= -3 \quad (5) & 7y - z &= -3 \quad (5) \\ & & 0 &= -7 \end{aligned}$$

$$\begin{cases} x + 2y - z = 4 & (1) \\ 7y - z = 4 & (4) \\ 0 = -7 & (6) \text{ False} \end{cases}$$

**Work Smart**

Whenever you end up with a false statement such as $0 = -7$, you have an inconsistent system.

Equation (6) now states that $0 = -7$, which is a false statement. Therefore, the system is inconsistent. The solution set is $\emptyset$ or $\{ \}$.

**Work Smart**

The figures below illustrate dependent systems.

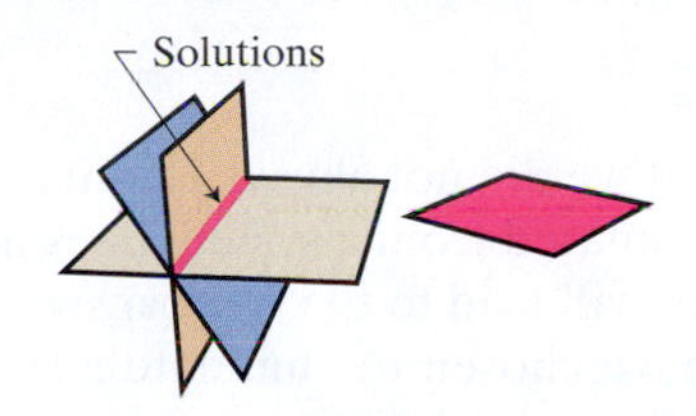


**QUICK ✓**

4. Use the method of elimination to solve the system: $\begin{cases} x - y + 2z = -7 \\ -2x + y - 3z = 5 \\ x - 2y + 3z = 2 \end{cases}$

## 3 Express the Solution of a System of Dependent Equations

Now we look at a system of dependent equations.

### EXAMPLE 5 Solving a System of Dependent Equations

Use the method of elimination to solve the system: $\begin{cases} x - 3y - z = 4 & (1) \\ x - 2y + 2z = 5 & (2) \\ 2x - 5y + z = 9 & (3) \end{cases}$

**Solution**

We can use equation (1) to eliminate the variable $x$ from equations (2) and (3). We can do this by multiplying equation (1) by $-1$ and adding the result to equation (2). The equation that results becomes equation (4). We also multiply equation (1) by $-2$ and add the result to equation (3). The equation that results becomes equation (5).

$$\begin{aligned} x - 3y - z &= 4 \quad (1) \\ x - 2y + 2z &= 5 \quad (2) \end{aligned} \qquad \text{Multiply by } -1: \quad \begin{aligned} -x + 3y + z &= -4 \quad (1) \\ x - 2y + 2z &= 5 \quad (2) \\ \hline y + 3z &= 1 \end{aligned}$$

$$\begin{aligned} x - 3y - z &= 4 \quad (1) \\ 2x - 5y + z &= 9 \quad (3) \end{aligned} \qquad \text{Multiply by } -2: \quad \begin{aligned} -2x + 6y + 2z &= -8 \quad (1) \\ 2x - 5y + z &= 9 \quad (3) \\ \hline y + 3z &= 1 \end{aligned}$$

$$\begin{cases} x - 3y - z = 4 & (1) \\ y + 3z = 1 & (4) \\ y + 3z = 1 & (5) \end{cases}$$

We now concentrate on equations (4) and (5), treating them as a system of two equations containing two variables. We multiply equation (4) by $-1$ and then add equations (4) and (5). The result will be equation (6).

**Work Smart**

Whenever you end up with a true statement such as $3 = 3$ or $0 = 0$, you have a dependent system.

$$\begin{aligned} y + 3z &= 1 \quad (4) \\ y + 3z &= 1 \quad (5) \end{aligned} \qquad \text{Multiply by } -1: \quad \begin{aligned} -y - 3z &= -1 \quad (4) \\ y + 3z &= 1 \quad (5) \\ \hline 0 &= 0 \end{aligned} \qquad \begin{cases} x - 3y - z = 4 & (1) \\ y + 3z = 1 & (4) \\ 0 = 0 & (6) \end{cases}$$

The statement $0 = 0$ in equation (6) indicates that we have a dependent system, so the system has infinitely many solutions. We can express how the values for $x$ and $y$ *depend* on the value of $z$ by letting $z$ represent any real number. Then, we can solve equation (4) for $y$ and obtain $y$ in terms of $z$.

$$\begin{aligned} \text{Equation (4):} \quad & y + 3z = 1 \\ \text{Subtract } 3z \text{ from both sides:} \quad & y = -3z + 1 \end{aligned}$$

Let $y = -3z + 1$ in equation (1) and solve for $x$ in terms of $z$.

$$\begin{aligned} \text{Equation (1):} \quad & x - 3y - z = 4 \\ \text{Let } y = -3z + 1 \text{ in (1):} \quad & x - 3(-3z + 1) - z = 4 \\ \text{Distribute:} \quad & x + 9z - 3 - z = 4 \\ \text{Combine like terms:} \quad & x + 8z - 3 = 4 \\ \text{Subtract } 8z \text{ from both sides; Add 3 to both sides:} \quad & x = -8z + 7 \end{aligned}$$

The solution to the system is $\{(x, y, z) | x = -8z + 7, y = -3z + 1, z \text{ is any real number}\}$. To find specific solutions to the system, choose any value of $z$ and use the equations $x = -8z + 7$ and $y = -3z + 1$ to determine $x$ and $y$. Some specific solutions to the system are $x = 7, y = 1, z = 0$; $x = -1, y = -2, z = 1$; or $x = 15, y = 4, z = -1$.

**QUICK ✓**

**5.** Use the method of elimination to solve the system: $\begin{cases} x - y + 3z = 2 \\ -x + 2y - 5z = -3 \\ 2x - y + 4z = 3 \end{cases}$

---

If you look back at Examples 2–5, you will notice that we did not always eliminate $x$ first and $y$ second. The order in which variables are eliminated from a system does not matter and different approaches to solving the problem will lead to the right answer if done correctly. For example, in Example 4 we could have chosen to eliminate $z$ from equations (1) and (3). We would have ended up with the same solution.

## 4 Model and Solve Problems Involving Three Linear Equations Containing Three Unknowns

Now let's look at problems that can be solved using three equations containing three variables.

### EXAMPLE 6 Production

A swing-set manufacturer has three different models of swing sets. The Monkey requires 2 hours to cut the wood, 2 hours to stain, and 3 hours to assemble. The Gorilla requires 3 hours to cut the wood, 4 hours to stain, and 4 hours to assemble. The King Kong requires 4 hours to cut the wood, 5 hours to stain, and 5 hours to assemble. The company has 61 hours available to cut the wood, 73 hours available to stain, and 83 hours available to assemble each day. How many of each type of swing set can be manufactured each day?

**Solution**

**Step 1: Identify** We want to determine the number of Monkey swing sets, Gorilla swing sets, and King Kong swing sets that can be manufactured each day.

**Step 2: Name** Let $m$ represent the number of Monkey swing sets, $g$ represent the number of Gorilla swing sets, and $k$ represent the number of King Kong swing sets.

**Step 3: Translate** We organize the information given in Table 4.

**Table 4**

| | Monkey | Gorilla | King Kong | Total Hours Available |
|---|---|---|---|---|
| Cut Wood | 2 | 3 | 4 | 61 |
| Stain | 2 | 4 | 5 | 73 |
| Assemble | 3 | 4 | 5 | 83 |

If we manufacture $m$ Monkey swing sets, then we need $2m$ hours to cut wood. If we manufacture $g$ Gorilla swing sets, then we need $3g$ hours to cut wood. If we manufacture $k$ King Kong swing sets, then we need $4k$ hours to cut wood. There are a total of 61 hours available, so

$$2m + 3g + 4k = 61 \quad \text{Equation (1)}$$

If we manufacture $m$ Monkey swing sets, then we need $2m$ hours to stain. If we manufacture $g$ Gorilla swing sets, then we need $4g$ hours to stain. If we manufacture $k$ King Kong swing sets, then we need $5k$ hours to stain. There are a total of 73 hours available, so

$$2m + 4g + 5k = 73 \quad \text{Equation (2)}$$

If we manufacture $m$ Monkey swing sets, then we need $3m$ hours to assemble. If we manufacture $g$ Gorilla swing sets, then we need $4g$ hours to assemble. If we manufacture $k$ King Kong swing sets, then we need $5k$ hours to assemble. There are a total of 83 hours available, so

$$3m + 4g + 5k = 83 \quad \text{Equation (3)}$$

We combine equations (1), (2), and (3) to form the following system:

$$\begin{cases} 2m + 3g + 4k = 61 & (1) \\ 2m + 4g + 5k = 73 & (2) \\ 3m + 4g + 5k = 83 & (3) \end{cases} \quad \text{The Model}$$

**Step 4: Solve** If we solve the system of equations found in Step 3, we find that $m = 10$, $g = 7$, and $k = 5$.

**Step 5: Check** By manufacturing 10 Monkeys, 7 Gorillas, and 5 King Kongs, we need $2(10) + 3(7) + 4(5) = 61$ hours to cut wood; $2(10) + 4(7) + 5(5) = 73$ hours to stain; and $3(10) + 4(7) + 5(5) = 83$ hours to assemble.

**Step 6: Answer** The company should manufacture 10 Monkey swing sets, 7 Gorilla swing sets, and 5 King Kong swing sets.

6. The Mowing 'Em Down lawn mower company manufactures three styles of lawn mower. The 21-inch model requires 2 hours to mold, 3 hours for engine manufacturing, and 1 hour to assemble. The 24-inch model requires 3 hours to mold, 3 hours for engine manufacturing, and 1 hour to assemble. The 40-inch riding mower requires 4 hours to mold, 4 hours for engine manufacturing, and 2 hours to assemble. The company has 81 hours available to mold, 95 hours available for engine manufacturing, and 35 hours available to assemble each day. How many of each type of mower can be manufactured each day?

## 4.3 Exercises

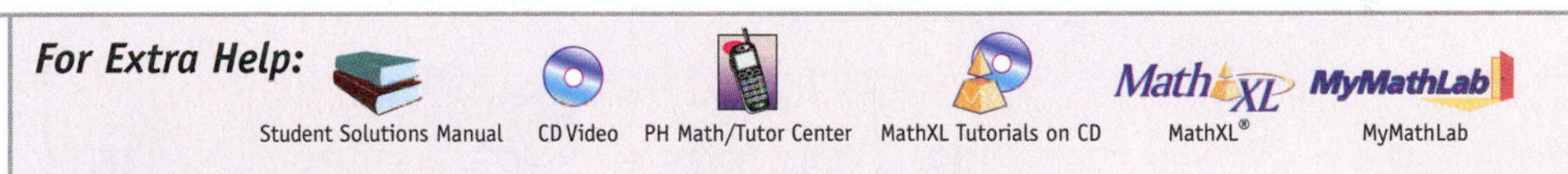


### Concepts and Vocabulary

*In Problems 1–3, fill in the blanks.*

1. If a system of equations has no solution, it is said to be _______.

2. If a system of equations has infinitely many solutions, the system is said to be _______ and the equations are _______.

3. A _______ to a system of equations consists of values for the variables that are solutions of each equation of the system.

*In Problems 4–6, answer True or False to each statement.*

4. A system of three linear equations containing three variables always has at least one solution.

5. When the planes in a system of equations are parallel, then the system is inconsistent and has no solution.

6. One of the rules for obtaining an equivalent system of equations allows us to replace any equation in the system by the sum (or difference) of that equation and a nonzero multiple of any other equation in the system.

**7.** Suppose that (3, 2, 5) is the only solution to a system of three linear equations containing three variables. What does this mean geometrically?

**8.** Why is it necessary to eliminate the same variable in the first step of the "Steps for solving a system of three linear equations containing three unknowns by elimination" (p. 288)?

## Building Skills

*In Problems 9 and 10, determine whether the given ordered triples listed are solutions of the system of linear equations.*

**9.** $\begin{cases} x + y + 2z = 6 \\ -2x - 3y + 5z = 1 \\ 2x + y + 3z = 5 \end{cases}$ **(a)** $(6, 2, -1)$ **(b)** $(-3, 5, 2)$

**10.** $\begin{cases} 2x + y - 2z = 6 \\ -2x + y + 5z = 1 \\ 2x + 3y + z = 13 \end{cases}$ **(a)** $(3, 2, 1)$ **(b)** $(10, -4, 5)$

*In Problems 11–22, solve each system of three linear equations containing three unknowns.* (**Hint:** *Each system has exactly one solution.*)

**11.** $\begin{cases} x + y + z = 5 \\ -2x - 3y + 2z = 8 \\ 3x - y - 2z = 3 \end{cases}$

**12.** $\begin{cases} x + 2y - z = 4 \\ 2x - y + 3z = 8 \\ -2x + 3y - 2z = 10 \end{cases}$

**13.** $\begin{cases} x - 3y + z = 13 \\ 3x + y - 4z = 13 \\ -4x - 4y + 2z = 0 \end{cases}$

**14.** $\begin{cases} x + 2y - 3z = -19 \\ 3x + 2y - z = -9 \\ -2x - y + 3z = 26 \end{cases}$

**15.** $\begin{cases} 2x - y + 2z = 1 \\ -2x + 3y - 2z = 3 \\ 4x - y + 6z = 7 \end{cases}$

**16.** $\begin{cases} x - y + 3z = 2 \\ -2x + 3y - 8z = -1 \\ 2x - 2y + 4z = 7 \end{cases}$

**17.** $\begin{cases} x - 4y + z = 5 \\ 4x + 2y + z = 2 \\ -4x + y - 3z = -8 \end{cases}$

**18.** $\begin{cases} 2x + 2y - z = -7 \\ x + 2y - 3z = -8 \\ 4x - 2y + z = -11 \end{cases}$

**19.** $\begin{cases} x - 3y = 12 \\ 2y - 3z = -9 \\ 2x + z = 7 \end{cases}$

**20.** $\begin{cases} 2x + z = -7 \\ 3y - 2z = 17 \\ -4x - y = 7 \end{cases}$

**21.** $\begin{cases} 2y - z = -3 \\ -2x + 3y = 10 \\ 4x + 3z = -11 \end{cases}$

**22.** $\begin{cases} x - 3z = -3 \\ 3y + 4z = -5 \\ 3x - 2y = 6 \end{cases}$

## Mixed Practice

*In Problems 23–32, solve each system of equations.*

**23.** $\begin{cases} x - y + z = 5 \\ -2x + y - z = 2 \\ x - 2y + 2z = 1 \end{cases}$

**24.** $\begin{cases} x - y + 2z = 3 \\ 2x + y - 2z = 1 \\ 4x - y + 2z = 0 \end{cases}$

**25.** $\begin{cases} x - 2y + z = 5 \\ -2x + y - z = 2 \\ x - 5y - 4z = 8 \end{cases}$

**26.** $\begin{cases} x + 2y - z = -4 \\ -2x + 4y - z = 6 \\ 2x + 2y + 3z = 1 \end{cases}$

**27.** $\begin{cases} x + 2y - z = 1 \\ 2x + 7y + 4z = 11 \\ x + 3y + z = 4 \end{cases}$

**28.** $\begin{cases} x + y - 2z = 3 \\ -2x - 3y + z = -7 \\ x + 2y + z = 4 \end{cases}$

**29.** $\begin{cases} x + y + z = 5 \\ 3x + 4y + z = 16 \\ -x - 4y + z = -6 \end{cases}$

**30.** $\begin{cases} x + y + z = 4 \\ 2x + 3y - z = 8 \\ x + y - z = 3 \end{cases}$

**31.** $\begin{cases} x + y + z = 3 \\ -x + \frac{1}{2}y + z = \frac{1}{2} \\ -x + 2y + 3z = 4 \end{cases}$

**32.** $\begin{cases} x + \frac{1}{2}y + \frac{1}{2}z = \frac{3}{2} \\ -x + 2y + 3z = 1 \\ 3x + 4y + 5z = 7 \end{cases}$

## Applying the Concepts

**33. Role Reversal** Write a system of three linear equations containing three unknowns that has the solution $(2, -1, 3)$.

**34. Role Reversal** Write a system of three linear equations containing three unknowns that has the solution $(-4, 1, -3)$.

**35. Curve Fitting** The function $f(x) = ax^2 + bx + c$ is a quadratic function, where $a$, $b$, and $c$ are constants.

**(a)** If $f(1) = 4$, then $4 = a(1)^2 + b(1) + c$ or $a + b + c = 4$. Find two additional linear equations if $f(-1) = -6$, and $f(2) = 3$.

**(b)** Use the three linear equations found in part (a) to determine $a$, $b$, and $c$. What is the quadratic function that contains the points $(-1, -6)$, $(1, 4)$, and $(2, 3)$?

**36. Curve Fitting** The function $f(x) = ax^2 + bx + c$ is a quadratic function, where $a$, $b$, and $c$ are constants.

**(a)** If $f(-1) = 6$, then $6 = a(-1)^2 + b(-1) + c$ or $a - b + c = 6$. Find two additional linear equations if $f(1) = 2$, and $f(2) = 9$.

**(b)** Use the three linear equations found in part (a) to determine $a$, $b$, and $c$. What is the quadratic function that contains the points $(-1, 6)$, $(1, 2)$, and $(2, 9)$?

**37. Electricity: Kirchhoff's Rules** An application of Kirchhoff's Rule to the circuit shown results in the following system of equations:

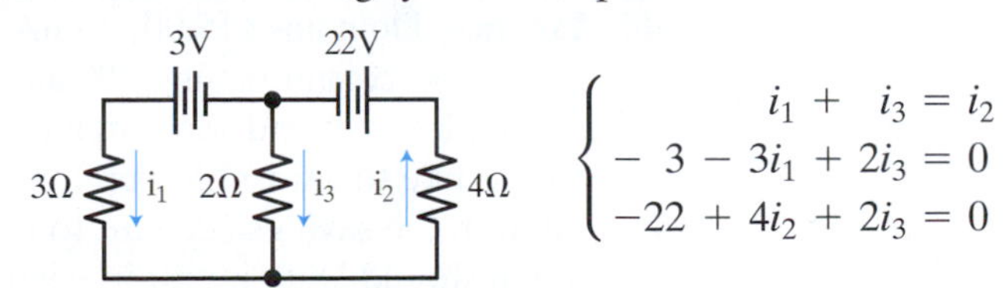


$$\begin{cases} i_1 + i_3 = i_2 \\ -3 - 3i_1 + 2i_3 = 0 \\ -22 + 4i_2 + 2i_3 = 0 \end{cases}$$

In the system circuit $V$ is the voltage, $\Omega$ is resistance, and $i$ is the current. Find the currents $i_1$, $i_2$, and $i_3$.

**38. Electricity: Kirchhoff's Rules** An application of Kirchhoff's Rule to the circuit shown results in the following system of equations:

$$\begin{cases} i_1 + i_3 = i_2 \\ -8 - 5i_1 + 8i_3 = 0 \\ -48 + 6i_2 + 8i_3 = 0 \end{cases}$$

In the system circuit $V$ is the voltage, $\Omega$ is resistance, and $i$ is the current. Find the currents $i_1$, $i_2$, and $i_3$.

**39. Minor League Baseball** In the Joliet Jackhammers baseball stadium, there are three types of seats available. Box seats are $9, reserved seats are $7, and lawn seats are $5. The stadium capacity is 4100. If all the seats are sold, the total revenue to the club is $28,400. If $\frac{1}{2}$ of the box seats are sold, $\frac{1}{2}$ of the reserved seats are sold, and all the lawn seats are sold, the total revenue is $18,300. How many are there of each kind of seat?

**40. Theater Revenues** A theater has 600 seats, divided into orchestra, main floor, and balcony seating. Orchestra seats sell for $80, main floor seats for $60, and balcony seats for $25. If all the seats are sold, the total revenue to the theater is $33,500. One evening, all the orchestra seats were sold, $\frac{3}{5}$ of the main seats were sold, and only $\frac{4}{5}$ of the balcony seats were sold. The total revenue collected was $24,640. How many are there of each kind of seat?

**41. Nutrition** Nancy's dietitian wants her to consume 470 mg of sodium, 89 g of carbohydrates, and 20 g of protein for breakfast. This morning, Nancy wants to have Chex® cereal, 2% milk, and orange juice for breakfast. Each serving of Chex® cereal contains 220 mg of sodium, 26 g of carbohydrates, and 1 g of protein. Each serving of 2% milk contains 125 mg of sodium, 12 g of carbohydrates, and 8 g of protein. Each serving of orange juice contains 0 mg of sodium, 26 mg of carbohydrates, and 2 g of protein. How many servings of each does Nancy need?

**42. Nutrition** Antonio is on a special diet that requires he consume 1325 calories, 172 grams of carbohydrates, and 63 grams of protein for lunch. He goes to Wendy's and wishes to have their Broccoli and Cheese Baked Potato, Chicken BLT Salad, and a medium Coke. Each Broccoli and Cheese Baked Potato has 480 calories, 80 g of carbohydrates, and 9 g of protein. Each Chicken BLT Salad has 310 calories, 10 g of carbohydrates, and 33 g of protein. Each Coke has 140 calories, 37 g of carbohydrates, and 0 g of protein. How many servings of each does Antonio need?

**43. Finance** Sachi has $25,000 to invest. Her financial planner suggests that she diversify her investment into three investment categories: Treasury bills that yield 3% simple interest annually, municipal bonds that yield 5% simple interest annually, and corporate bonds that yield 9% simple interest annually. Sachi would like to earn $1210 per year in income. In addition, Sachi wants her investment in Treasury bills to be $7000 more than her investment in corporate bonds. How much should Sachi invest in each investment category?

**44. Finance** Delu has $15,000 to invest. She decides to place some of the money into a savings account paying 2% annual interest, some in Treasury bonds paying 5% annual interest and some in a mutual fund paying 10% annual interest. Delu would like to earn $720 per year in income. In addition, Delu wants her investment in the savings account to be twice the amount in the mutual fund. How much should Delu invest in each investment category?

△ **45. Geometry** A circle is inscribed in $\triangle ABC$ as shown in the figure. Suppose that $AB = 6$, $AC = 14$, and $BC = 12$. Find the length of $\overline{AM}$, $\overline{BN}$, and $\overline{CO}$. (**Hint:** $\overline{AM} \cong \overline{AO}$; $\overline{BM} \cong \overline{BN}$; $\overline{NC} \cong \overline{OC}$)

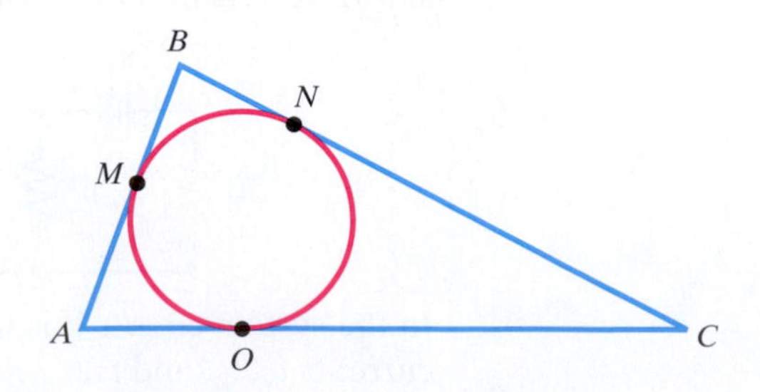

## Extending the Concepts

*In Problems 46–49, solve each system of equations.*

**46.** $\begin{cases} \frac{1}{4}x + \frac{1}{4}y + \frac{1}{2}z = 6 \\ -\frac{1}{8}x + \frac{1}{2}y - \frac{1}{5}z = -5 \\ \frac{1}{2}x + \frac{1}{2}y - \frac{1}{2}z = -3 \end{cases}$

**47.** $\begin{cases} \frac{2}{5}x + \frac{1}{2}y - \frac{1}{3}z = 0 \\ \frac{3}{5}x - \frac{1}{4}y + \frac{1}{2}z = 10 \\ -\frac{1}{5}x + \frac{1}{4}y - \frac{1}{6}z = -4 \end{cases}$

**48.** $\begin{cases} x + y + z + w = 0 \\ 2x - 3y - z + w = -17 \\ 3x + y + 2z - w = 8 \\ -x + 2y - 3z + 2w = -7 \end{cases}$

**49.** $\begin{cases} x + y + z + w = 3 \\ -2x - y + 3z - w = -1 \\ 2x + 2y - 2z + w = 2 \\ -x + 2y - 3z + 2w = 12 \end{cases}$

# PUTTING THE CONCEPTS TOGETHER (SECTIONS 4.1–4.3)

*These problems cover important concepts from Sections 4.1 to 4.3. We designed these problems so that you can review the chapter so far and show your mastery of the concepts. Take time to work these problems before proceeding with the next section. The answers to these problems are located at the back of the text starting on page AN-20.*

**1.** Use the graph of the system of linear equations to determine the solution.

$$\begin{cases} 5x - 2y = -25 \\ x + y = 2 \end{cases}$$

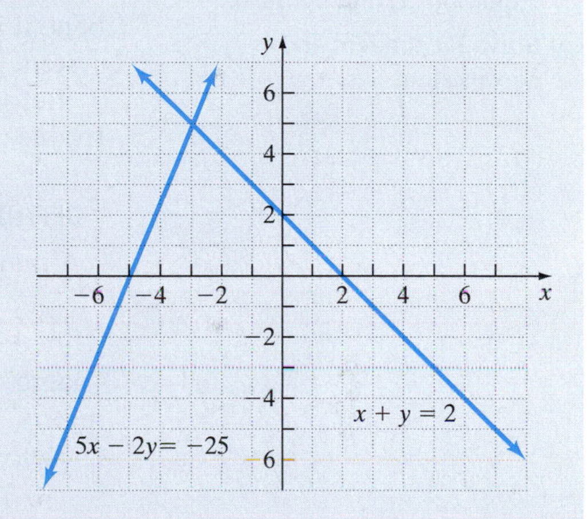


**2.** Solve the system of linear equations by graphing.

$$\begin{cases} 3x + y = 7 \\ -2x + 3y = -12 \end{cases}$$

*In Problems 3–6, solve the system of linear equations using either substitution or elimination.*

**3.** $\begin{cases} x = 2 - 3y \\ 3x + 10y = 5 \end{cases}$

**4.** $\begin{cases} 4x + 3y = -1 \\ 2x - y = 3 \end{cases}$

**5.** $\begin{cases} x + 3y = 8 \\ \frac{1}{5}x + \frac{1}{2}y = 1 \end{cases}$

**6.** $\begin{cases} 8x - 4y = 12 \\ -10x + 5y = -15 \end{cases}$

*In Problems 7 and 8, solve each system of three linear equations containing three unknowns.*

**7.** $\begin{cases} 2x + y + 3z = 10 \\ x - 2y + z = 10 \\ -4x + 3y + 2z = 5 \end{cases}$

**8.** $\begin{cases} x + 2y - 2z = 3 \\ x + 3y - 4z = 6 \\ 4x + 5y - 2z = 6 \end{cases}$

**9. Museum Tickets** A museum sells adult tickets for $9.00 and youth tickets for $5.00. On Saturday, the museum collected a total of $5925. It is also known that 825 people passed through the turnstiles. How many adult tickets were sold? How many youth tickets were sold?

**10. Theater Seating** A theater has 1200 seats that are divided into three sections: orchestra, mezzanine, and balcony. Each seat sells for $65 in the orchestra section, $48 each in the mezzanine section, and $35 each in the balcony section. There are 150 more balcony seats than mezzanine seats. If all of the seats are sold out, the theater will bring in $55,640 in revenue. Find the number of seats in each section of the theater.

# 4.4 Using Matrices to Solve Systems

## OBJECTIVES

1. Write the Augmented Matrix of a System of Linear Equations
2. Write the System from the Augmented Matrix
3. Perform Row Operations on a Matrix
4. Solve Systems of Linear Equations Using Matrices
5. Solve Dependent and Inconsistent Systems

### *Preparing for Using Matrices to Solve Systems*

*Before getting started, take the following readiness quiz. If you get a problem wrong, go back to the section cited and review the material.*

**1.** Determine the coefficients of the expression $4x - 2y + z$. [Section R. 5, p. 44]

**2.** Solve $x - 4y = 3$ for $x$. [Section 1.3, pp. 81–84]

**3.** Evaluate $3x - 2y + z$ when $x = 1$, $y = -3$, and $z = 2$. [Section R.5, pp. 43–44]

We now present an alternative approach to solving systems of linear equations. The benefit in this approach is that it streamlines the notation and makes working with the system more manageable.

Before we discuss this method for solving systems of linear equations, we need to introduce a new concept, called the *matrix*.

**DEFINITION**

A **matrix** is a rectangular array of numbers.

A matrix has rows and columns. The number of rows and columns are used to "name" the matrix. For example, a matrix with 2 rows and 3 columns is called a "2 by 3 matrix" and is denoted "2 × 3."

Below are some examples of matrices.

**2 × 3 matrix**

$$\begin{bmatrix} 3 & -1 & 4 \\ 8 & 0 & -5 \end{bmatrix}$$

**3 × 3 matrix**

$$\begin{bmatrix} 2 & -8 & 12 \\ 0 & 7 & -2 \\ 5 & -2 & 1 \end{bmatrix}$$

**Work Smart**

A spreadsheet such as Microsoft Excel is a matrix!

## 1 Write the Augmented Matrix of a System of Linear Equations

We will use matrix notation to represent a system of linear equations. The matrix used to represent a system of linear equations is called an **augmented matrix.** In order to write the augmented matrix of a system, the terms containing the variables of each equation must be on the left side of the equal sign in the same order ($x$, $y$, and $z$, for example) and the constants on the right side. A variable that does not appear in an equation has a coefficient of 0. For example, the system of two linear equations containing two unknowns,

$$\begin{cases} a_1x + b_1y = c_1 \\ a_2x + b_2y = c_2 \end{cases} \quad \text{is written as an augmented matrix as} \quad \left[\begin{array}{cc|c} a_1 & b_1 & c_1 \\ a_2 & b_2 & c_2 \end{array}\right]$$

Notice that the first column represents the coefficients on the variable $x$, the second column represents the coefficients on the variable $y$, the vertical bar represents the equal signs, and the constants are to the right of the vertical bar. The first row in the matrix is equation (1) and the second row in the matrix is equation (2).

*Preparing for...Answers* **1.** 4, −2, 1 **2.** $x = 4y + 3$ **3.** 11

### EXAMPLE 1 Writing the Augmented Matrix of a System of Linear Equations

Write each system of linear equations as an augmented matrix.

**(a)** $\begin{cases} x - 3y = 2 \\ -2x + 5y = 7 \end{cases}$ **(b)** $\begin{cases} 2x + 3y - z = 1 \\ x - 2z + 1 = 0 \\ -4x - y + 3z = 5 \end{cases}$

**Solution**

**(a)** In the augmented matrix, we let the first column represent the coefficients on the variable $x$. The second column represents the coefficients on the variable $y$. The vertical line signifies the equal signs. The third column represents the constants to the right of the equal sign.

$$\begin{array}{cc} x & y \\ \end{array}$$
$$\left[\begin{array}{cc|c} 1 & -3 & 2 \\ -2 & 5 & 7 \end{array}\right] \quad \begin{array}{r} x - 3y = 2 \\ -2x + 5y = 7 \end{array}$$

**(b)** We must be careful to write the system of equations so that the variables are all on the left side of the equal sign and the constants are on the right. Also, if a variable is missing from an equation, then its coefficient is understood to be 0. The system

$$\begin{cases} 2x + 3y - z = 1 \\ x - 2z + 1 = 0 \\ -4x - y + 3z = 5 \end{cases} \quad \text{gets rearranged as} \quad \begin{cases} 2x + 3y - z = 1 \\ x + 0y - 2z = -1 \\ -4x - y + 3z = 5 \end{cases}$$

The augmented matrix is

$$\left[\begin{array}{ccc|c} 2 & 3 & -1 & 1 \\ 1 & 0 & -2 & -1 \\ -4 & -1 & 3 & 5 \end{array}\right]$$

**QUICK ✓** *Write the augmented matrix of each system of equations.*

**1.** $\begin{cases} 3x - y = -10 \\ -5x + 2y = 0 \end{cases}$ **2.** $\begin{cases} x + 2y - 2z = 11 \\ -x - 2z = 4 \\ 4x - y + z - 3 = 0 \end{cases}$

## 2 Write the System from the Augmented Matrix

We are now going to write the system of linear equations that corresponds to a given augmented matrix.

### EXAMPLE 2 Writing the System of Linear Equations from the Augmented Matrix

Write the system of linear equations corresponding to each augmented matrix.

**(a)** $\left[\begin{array}{cc|c} 2 & 1 & -5 \\ -1 & 3 & 2 \end{array}\right]$ **(b)** $\left[\begin{array}{ccc|c} 1 & -3 & 2 & 5 \\ -2 & 0 & 4 & -3 \\ -1 & 4 & 1 & 0 \end{array}\right]$

**Solution**

**(a)** The augmented matrix has two rows and so represents a system of two equations. Because there are two columns to the left of the vertical bar, the system has two variables. If we call these two variables $x$ and $y$, the system of equations is

$$\begin{cases} 2x + y = -5 \\ -x + 3y = 2 \end{cases}$$

**(b)** Since the augmented matrix has three rows, it represents a system of three equations. Since there are three columns to the left of the vertical bar, the system contains three variables. If we call these three variables $x$, $y$, and $z$, the system of equations is

$$\begin{cases} x - 3y + 2z = 5 \\ -2x \qquad\; + 4z = -3 \\ -x + 4y + \;\; z = 0 \end{cases}$$

**QUICK ✓** *Write the system of linear equations corresponding to the given augmented matrix.*

**3.** $\left[\begin{array}{cc|c} 1 & -3 & 7 \\ -2 & 5 & -3 \end{array}\right]$ **4.** $\left[\begin{array}{ccc|c} 1 & -3 & 2 & 4 \\ 3 & 0 & -1 & -1 \\ -1 & 4 & 0 & 0 \end{array}\right]$

## 3 Perform Row Operations on a Matrix

Row operations are used on an augmented matrix to solve the corresponding system of equations. There are three basic row operations.

**In Words**
The first row operation is like "flip-flopping" two equations in a system. The second row operation is like multiplying both sides of the equation by a nonzero constant. The third row operation is like adding two equations and replacing an equation with the sum.

**ROW OPERATIONS**

1. Interchange any two rows.
2. Replace a row by a nonzero multiple of that row.
3. Replace a row by the sum of that row and a nonzero multiple of some other row.

If you look carefully, these are the same types of operations that we can perform on a system of equations from Section 4.3. The main reason for using a matrix to solve a system of equations is that it is more efficient notation and helps us to organize the mathematics. To see how row operations work, consider the following augmented matrix.

$$\left[\begin{array}{cc|c} 1 & 3 & -1 \\ -2 & 1 & 3 \end{array}\right]$$

Suppose that we want to apply a row operation to this matrix that results in a matrix whose entry in row 2, column 1 is a 0. The current entry in row 2, column 1 is $-2$. The row operation to use is

**Multiply each entry in row 1 by 2 and then add the result to the corresponding entry in row 2. Have this result replace the current row 2.**

That's a whole lot of words! To streamline this a little, we introduce some notation. If we use $R_2$ to represent the new entries in row 2 and we use $r_1$ and $r_2$ to represent the original entries in rows 1 and 2 respectively, then we can represent the row operation given above by

Multiply each entry in row 1 by 2 and add the result to the corresponding entry in row 2 . . .

$$R_2 = 2r_1 + r_2$$

. . . to obtain the "new" row 2.

We demonstrate this row operation below.

$$\left[\begin{array}{cc|c} 1 & 3 & -1 \\ -2 & 1 & 3 \end{array}\right] \xrightarrow{R_2 = 2r_1 + r_2} \left[\begin{array}{cc|c} 1 & 3 & -1 \\ 2(1) + (-2) & 2(3) + 1 & 2(-1) + 3 \end{array}\right] = \left[\begin{array}{cc|c} 1 & 3 & -1 \\ 0 & 7 & 1 \end{array}\right]$$

Notice that we now have a 0 in row 2, column 1, as desired.

### EXAMPLE 3 Applying a Row Operation to an Augmented Matrix

Apply the row operation $R_2 = -3r_1 + r_2$ to the augmented matrix

$$\left[\begin{array}{cc|c} 1 & 2 & -3 \\ 3 & 4 & 5 \end{array}\right]$$

**Solution**

The row operation $R_2 = -3r_1 + r_2$ tells us to multiply the entries in row 1 by $-3$ and then add the result to the entries in row 2. The result should replace the current row 2.

$$\left[\begin{array}{cc|c} 1 & 2 & -3 \\ 3 & 4 & 5 \end{array}\right] \xrightarrow[R_2 = -3r_1 + r_2]{} \left[\begin{array}{cc|c} 1 & 2 & -3 \\ -3(1) + 3 & -3(2) + 4 & -3(-3) + 5 \end{array}\right] = \left[\begin{array}{cc|c} 1 & 2 & -3 \\ 0 & -2 & 14 \end{array}\right]$$

**QUICK ✓**

5. Apply the row operation $R_2 = 4r_1 + r_2$ to the augmented matrix $\left[\begin{array}{cc|c} 1 & -2 & 5 \\ -4 & 5 & -11 \end{array}\right]$.

### EXAMPLE 4 Finding a Particular Row Operation

For the augmented matrix

$$\left[\begin{array}{cc|c} 1 & -3 & -12 \\ 0 & 1 & 5 \end{array}\right]$$

find a row operation that will result in the entry in row 1, column 2 becoming a 0 and perform the row operation.

**Solution**

We want a 0 in row 1, column 2. We can accomplish this by multiplying row 2 by 3 and adding the result to row 1. That is, we apply the row operation $R_1 = 3r_2 + r_1$.

$$\left[\begin{array}{cc|c} 1 & -3 & -12 \\ 0 & 1 & 5 \end{array}\right] \xrightarrow[R_1 = 3r_2 + r_1]{} \left[\begin{array}{cc|c} 3(0) + 1 & 3(1) + (-3) & 3(5) + (-12) \\ 0 & 1 & 5 \end{array}\right] = \left[\begin{array}{cc|c} 1 & 0 & 3 \\ 0 & 1 & 5 \end{array}\right]$$

**Work Smart**

If you want to change the entries in row 1, then you should multiply the entries in some other row and then add this result to the entries in row 1.

**QUICK ✓**

6. For the augmented matrix $\left[\begin{array}{cc|c} 1 & 5 & 13 \\ 0 & 1 & 2 \end{array}\right]$, find a row operation that will result in the entry in row 1, column 2 becoming a 0 and perform the row operation.

## 4 Solve Systems of Linear Equations Using Matrices

To solve a system of linear equations using matrices, we use row operations on the augmented matrix of the system to obtain a matrix that is in *row echelon form*.

**DEFINITION**

A matrix is in **row echelon form** when

1. The entry in row 1, column 1 is a 1, and 0s appear below it.
2. The first nonzero entry in each row after the first row is a 1, 0s appear below it, and it appears to the right of the first nonzero entry in any row above.
3. Any rows that contain all 0s to the left of the vertical bar appear at the bottom.

For a system of two equations containing two variables with a single solution (that is, a system that is consistent and independent), the augmented matrix is in row echelon form if it is of the form

$$\left[\begin{array}{cc|c} 1 & a & b \\ 0 & 1 & c \end{array}\right]$$

where $a$, $b$, and $c$ are real numbers. A system of three linear equations containing three variables that is consistent and independent is in row echelon form if it is of the form

$$\left[\begin{array}{ccc|c} 1 & a & b & d \\ 0 & 1 & c & e \\ 0 & 0 & 1 & f \end{array}\right]$$

where $a$, $b$, $c$, $d$, $e$, and $f$ are real numbers.

The augmented matrix $\left[\begin{array}{cc|c} 1 & a & b \\ 0 & 1 & c \end{array}\right]$, is equivalent to the system $\begin{cases} x + ay = b & (1) \\ y = c & (2) \end{cases}$.

In other words, because $c$ is a known number, we know $y$. We can then back-substitute to find $x$. For example, the augmented matrix $\left[\begin{array}{cc|c} 1 & 3 & -5 \\ 0 & 1 & -3 \end{array}\right]$ is in row echelon form and is equivalent to the system $\begin{cases} x + 3y = -5 & (1) \\ y = -3 & (2) \end{cases}$. From equation (2), we know that $y = -3$. Using back-substitution, we find that $x = 4$. So, the solution is $(4, -3)$.

## EXAMPLE 5 How to Solve a System of Two Linear Equations Containing Two Variables Using Matrices

Solve: $\begin{cases} 3x - 2y = -19 \\ x + 2y = 7 \end{cases}$

### Step-by-Step Solution

**Step 1:** Write the augmented matrix of the system.

$$\left[\begin{array}{cc|c} 3 & -2 & -19 \\ 1 & 2 & 7 \end{array}\right]$$

**Step 2:** We want the entry in row 1, column 1 to be 1. We can interchange rows 1 and 2 in the augmented matrix.

$$\left[\begin{array}{cc|c} 1 & 2 & 7 \\ 3 & -2 & -19 \end{array}\right]$$

**Step 3:** We want the entry in row 2, column 1 to be 0. We use the row operation $R_2 = -3r_1 + r_2$ to accomplish this. The entries in row 1 remain unchanged.

$$\left[\begin{array}{cc|c} 1 & 2 & 7 \\ 0 & -8 & -40 \end{array}\right]$$

**Step 4:** Now we want the entry in row 2, column 2 to be 1. This is accomplished by multiplying row 2 by $-\frac{1}{8}$. We use the row operation $R_2 = -\frac{1}{8}r_2$.

$$\left[\begin{array}{cc|c} 1 & 2 & 7 \\ 0 & 1 & 5 \end{array}\right]$$

**Step 5:** From row 2 we have that $y = 5$. Row 1 represents the equation $x + 2y = 7$. Back-substitute 5 for $y$ and solve for $x$.

$$x + 2y = 7$$
$$x + 2(5) = 7$$
$$x + 10 = 7$$
$$x = -3$$

**Step 6: Check** We let $x = -3$ and $y = 5$ in both equations (1) and (2) to verify our solution.

Equation (1): $3x - 2y = -19$

$x = -3, y = 5$: $3(-3) - 2(5) \stackrel{?}{=} -19$

$-9 - 10 \stackrel{?}{=} -19$

$-19 = -19$ True

Equation (2): $x + 2y = 7$

$x = -3, y = 5$: $-3 + 2(5) \stackrel{?}{=} 7$

$-3 + 10 \stackrel{?}{=} 7$

$7 = 7$ True

The solution is the ordered pair $(-3, 5)$.

**QUICK ✓**

**7.** Solve the following system using matrices. $\begin{cases} 2x - 4y = 20 \\ 3x + y = 16 \end{cases}$

## EXAMPLE 6 How to Solve a System of Three Linear Equations Containing Three Variables Using Matrices

Solve: $\begin{cases} x + y + z = 3 \\ -2x - 3y + 2z = 13 \\ 4x + 5y + z = -3 \end{cases}$

### Step-by-Step Solution

**Step 1:** Write the augmented matrix of the system.

$$\left[\begin{array}{ccc|c} 1 & 1 & 1 & 3 \\ -2 & -3 & 2 & 13 \\ 4 & 5 & 1 & -3 \end{array}\right]$$

**Step 2:** We want the entry in row 1, column 1 to be 1. This is already done.

$$\left[\begin{array}{ccc|c} 1 & 1 & 1 & 3 \\ -2 & -3 & 2 & 13 \\ 4 & 5 & 1 & -3 \end{array}\right]$$

**Step 3:** We want the entry in row 2, column 1 to be 0. We use the row operation $R_2 = 2r_1 + r_2$ to accomplish this. We also want the entry in row 3, column 1 to be 0. We use the row operation $R_3 = -4r_1 + r_3$ to accomplish this. The entries in row 1 remain unchanged.

$$\begin{array}{r} \\ R_2 = 2r_1 + r_2: \\ R_3 = -4r_1 + r_3: \end{array} \left[\begin{array}{ccc|c} 1 & 1 & 1 & 3 \\ 0 & -1 & 4 & 19 \\ 0 & 1 & -3 & -15 \end{array}\right]$$

**Step 4:** Now we want the entry in row 2, column 2 to be 1. This is accomplished by interchanging rows 2 and 3.

$$\left[\begin{array}{ccc|c} 1 & 1 & 1 & 3 \\ 0 & 1 & -3 & -15 \\ 0 & -1 & 4 & 19 \end{array}\right]$$

**Step 5:** We need the entry in row 3, column 2 to be 0. We use the row operation $R_3 = r_2 + r_3$.

$$\begin{array}{r} \\ \\ R_3 = r_2 + r_3: \end{array} \left[\begin{array}{ccc|c} 1 & 1 & 1 & 3 \\ 0 & 1 & -3 & -15 \\ 0 & 0 & 1 & 4 \end{array}\right]$$

*(continued)*

**Step 6:** We want the entry in row 3, column 3 to be 1. This is already done.

$$\left[\begin{array}{ccc|c} 1 & 1 & 1 & 3 \\ 0 & 1 & -3 & -15 \\ 0 & 0 & 1 & 4 \end{array}\right]$$

**Step 7:** The augmented matrix is in row echelon form. Write the system of equations corresponding to the augmented matrix and solve.

$$\begin{cases} x + y + z = 3 & (1) \\ y - 3z = -15 & (2) \\ z = 4 & (3) \end{cases}$$

From equation (3), we have that $z = 4$. Letting $z = 4$ in equation (2), we can determine $y$.

Equation (2): $y - 3z = -15$

$z = 4$: $y - 3(4) = -15$

$y - 12 = -15$

$y = -3$

Let $y = -3$ and $z = 4$ in equation (1) to find that $x = 2$.

**Step 8: Check** We let $x = 2$, $y = -3$, and $z = 4$ in each equation in the original system to verify our solution.

We leave it to you to verify the solution.

The solution is the ordered triple $(2, -3, 4)$.

**Work Smart**

Look back at Step 4 in Example 6. There is more than one option for obtaining a 1 in row 2, column 2. We could also have used the row operation $R_2 = 2r_3 + r_2$ or $R_2 = -r_2$.

**8.** Solve: $\begin{cases} x - y + 2z = 7 \\ 2x - 2y + z = 11 \\ -3x + y - 3z = -14 \end{cases}$

## SUMMARY: Steps for Solving a System of Linear Equations Using Matrices

**Step 1:** Write the augmented matrix of the system.

**Step 2:** Perform row operations so that the entry in row 1, column 1 is 1.

**Step 3:** Perform row operations so that all the entries below the 1 in row 1, column 1 are 0's.

**Step 4:** Perform row operations so that the entry in row 2, column 2 is 1. Make sure that the entries in column 1 remain unchanged. If it is impossible to place a 1 in row 2, column 2, then proceed to use operations to place a 1 in row 2, column 3. (*Note:* If a row with all 0's is obtained, then it should be placed in the last row of the matrix.)

**Step 5:** Once a 1 is in place, perform row operations to place 0s below it.

**Step 6:** Repeat Steps 4 and 5 until you have the augmented matrix in echelon form.

**Step 7:** With the augmented matrix in echelon form, write the corresponding system of equations. Back-substitute to find the values of the variables.

**Step 8:** Check your answer.

## 5 Solve Dependent and Inconsistent Systems

The matrix method for solving a system of linear equations also identifies systems that have infinitely many solutions (dependent systems) and systems with no solution (inconsistent).

### EXAMPLE 7 Solving a Dependent System Using Matrices

$$\text{Solve: } \begin{cases} x + y + 3z = 3 \\ -2x - y - 8z = -5 \\ 3x + 2y + 11z = 8 \end{cases}$$

**Solution**

We will write the augmented matrix of the system and perform row operations to get the matrix into row echelon form.

$$\left[\begin{array}{rrr|r} 1 & 1 & 3 & 3 \\ -2 & -1 & -8 & -5 \\ 3 & 2 & 11 & 8 \end{array}\right] \xrightarrow[R_3 = -3r_1 + r_3]{R_2 = 2r_1 + r_2} \left[\begin{array}{rrr|r} 1 & 1 & 3 & 3 \\ 0 & 1 & -2 & 1 \\ 0 & -1 & 2 & -1 \end{array}\right] \xrightarrow{R_3 = r_2 + r_3} \left[\begin{array}{rrr|r} 1 & 1 & 3 & 3 \\ 0 & 1 & -2 & 1 \\ 0 & 0 & 0 & 0 \end{array}\right]$$

Notice that the last row is all 0s. The augmented matrix is in row echelon form. The system of equations corresponding to this matrix is

$$\begin{cases} x + y + 3z = 3 & (1) \\ y - 2z = 1 & (2) \\ 0 = 0 & (3) \end{cases}$$

The statement $0 = 0$ in equation (3) indicates that we have a dependent system. If we let $z$ represent any real number, then we can solve equation (2) for $y$ in terms of $z$.

Equation (2): $y - 2z = 1$

Add $2z$ to both sides: $y = 2z + 1$

Let $y = 2z + 1$ in equation (1) and solve for $x$ in terms of $z$.

Equation (1): $x + y + 3z = 3$

Let $y = 2z + 1$: $x + (2z + 1) + 3z = 3$

Combine like terms: $x + 5z + 1 = 3$

Subtract $5z + 1$ from both sides: $x = -5z + 2$

The solution to the system is $\{(x, y, z) | x = -5z + 2, y = 2z + 1, z \text{ is any real number}\}$. To find specific solutions to the system, choose any value of $z$ and use the equations $x = -5z + 2$ and $y = 2z + 1$ to determine $x$ and $y$. Some specific solutions to the system are $x = 2, y = 1, z = 0$; $x = -3, y = 3, z = 1$; or $x = 7, y = -1, z = -1$.

If we evaluate the system at some of the specific solutions, we see that all of them satisfy the original system. This leads us to believe that our solution is correct.

**QUICK ✓**

**9.** $\text{Solve: } \begin{cases} x + y - 3z = 8 \\ 2x + 3y - 10z = 19 \\ -x - 2y + 7z = -11 \end{cases}$

**EXAMPLE 8** **Solving an Inconsistent System Using Matrices**

$$\text{Solve: } \begin{cases} 2x + y - z = 5 \\ -x + 2y + z = 1 \\ 3x + 4y - z = -2 \end{cases}$$

**Solution**

We will write the augmented matrix of the system and perform row operations to get the matrix into row echelon form.

$$\left[\begin{array}{rrr|r} 2 & 1 & -1 & 5 \\ -1 & 2 & 1 & 1 \\ 3 & 4 & -1 & -2 \end{array}\right] \xrightarrow{R_1 = r_2 + r_1} \left[\begin{array}{rrr|r} 1 & 3 & 0 & 6 \\ -1 & 2 & 1 & 1 \\ 3 & 4 & -1 & -2 \end{array}\right] \xrightarrow[R_3 = -3r_1 + r_3]{R_2 = r_1 + r_2} \left[\begin{array}{rrr|r} 1 & 3 & 0 & 6 \\ 0 & 5 & 1 & 7 \\ 0 & -5 & -1 & -20 \end{array}\right]$$

$$\xrightarrow[R_2 = \frac{1}{5}r_2]{} \left[\begin{array}{rrr|r} 1 & 3 & 0 & 6 \\ 0 & 1 & \frac{1}{5} & \frac{7}{5} \\ 0 & -5 & -1 & -20 \end{array}\right] \xrightarrow{R_3 = 5r_2 + r_3} \left[\begin{array}{rrr|r} 1 & 3 & 0 & 6 \\ 0 & 1 & \frac{1}{5} & \frac{7}{5} \\ 0 & 0 & 0 & -13 \end{array}\right]$$

We want the entry in row 3, column 3 to be 1. This cannot be done. The augmented matrix is in row echelon form. The system of equations corresponding to this matrix is

$$\begin{cases} x + 3y = 6 & (1) \\ y + \frac{1}{5}z = \frac{7}{5} & (2) \\ 0 = -13 & (3) \end{cases}$$

The statement $0 = -13$ in equation (3) indicates that we have an inconsistent system. The solution is $\emptyset$ or $\{\ \}$.

**QUICK ✓**

10. Solve: $\begin{cases} -x + 2y - z = 5 \\ 2x + y + 4z = 3 \\ 3x - y + 5z = 0 \end{cases}$

# 4.4 Exercises

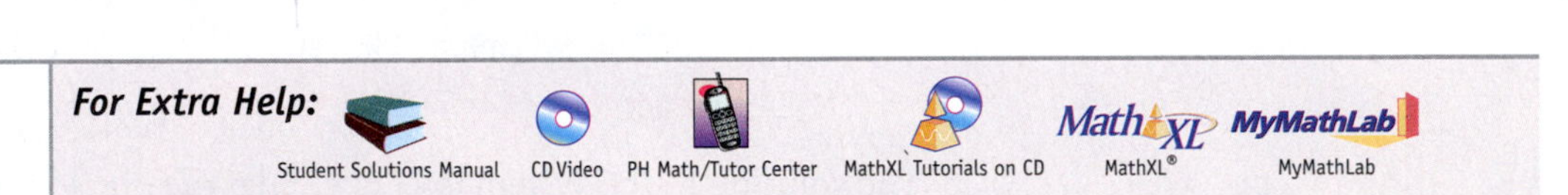


## Concepts and Vocabulary

*In Problems 1–3, fill in the blanks.*

1. An $m$ by $n$ rectangular array of numbers is called a(n) _______.
2. The matrix used to represent a system of linear equations is called a(n) _______ matrix.
3. A $4 \times 3$ matrix has _______ rows and _______ columns.

*In Problems 4–6, answer True or False to each statement.*

4. The augmented matrix of a system of two equations containing 2 unknowns has 2 rows and 2 columns.
5. The matrix $\left[\begin{array}{rr|r} 1 & 3 & 2 \\ 0 & 1 & -5 \end{array}\right]$ is in row echelon form.

**6.** A $2 \times 5$ matrix has 2 rows and 5 columns.

**7.** Write a paragraph that outlines the strategy for putting an augmented matrix in row echelon form.

**8.** When solving a system of linear equations using matrices, how do you know that the system is inconsistent?

**9.** What would be the next row operation on the given augmented matrix?

$$\left[\begin{array}{cc|c} 1 & 3 & 8 \\ 0 & 5 & 10 \end{array}\right]$$

**10.** What would your recommend as the next row operation on the given augmented matrix? Why?

$$\left[\begin{array}{ccc|c} 1 & 3 & -2 & 4 \\ 0 & 5 & 3 & 2 \\ 0 & -4 & 6 & -5 \end{array}\right]$$

## Building Skills

*In Problems 11–18, write the augmented matrix of the given system of equations.*

**11.** $\begin{cases} x - 3y = 2 \\ 2x + 5y = 1 \end{cases}$

**12.** $\begin{cases} -x + y = 6 \\ 5x - y = -3 \end{cases}$

**13.** $\begin{cases} x + y + z = 3 \\ 2x - y + 3z = 1 \\ -4x + 2y - 5z = -3 \end{cases}$

**14.** $\begin{cases} x + y - z = 2 \\ -2x + y - 4z = 13 \\ 3x - y - 2z = -4 \end{cases}$

**15.** $\begin{cases} -x + y - 2 = 0 \\ 5x + y + 5 = 0 \end{cases}$

**16.** $\begin{cases} 6x + 4y + 2 = 0 \\ -x - y + 1 = 0 \end{cases}$

**17.** $\begin{cases} x + z = 2 \\ 2x + y = 13 \\ x - y + 4z + 4 = 0 \end{cases}$

**18.** $\begin{cases} 2x + 7 = 1 \\ -x - 6z = 5 \\ 5x + 2y - 4z + 1 = 0 \end{cases}$

*In Problems 19–24, perform each row operation on the given augmented matrix.*

**19.** $\left[\begin{array}{cc|c} 1 & -3 & 2 \\ -2 & 5 & 1 \end{array}\right]$

**(a)** $R_2 = 2r_1 + r_2$ followed by
**(b)** $R_2 = -r_2$

**20.** $\left[\begin{array}{cc|c} 1 & 5 & 7 \\ 3 & 11 & 13 \end{array}\right]$

**(a)** $R_2 = -3r_1 + r_2$ followed by
**(b)** $R_2 = -\frac{1}{4}r_2$

**21.** $\left[\begin{array}{ccc|c} 1 & 1 & -1 & 4 \\ 2 & 5 & 3 & -3 \\ -1 & -3 & 2 & 1 \end{array}\right]$

**(a)** $R_2 = -2r_1 + r_2$ followed by
**(b)** $R_3 = r_1 + r_3$

**22.** $\left[\begin{array}{ccc|c} 1 & -1 & 1 & 6 \\ -2 & 1 & -3 & 3 \\ 3 & 2 & -2 & -5 \end{array}\right]$

**(a)** $R_2 = 2r_1 + r_2$ followed by
**(b)** $R_3 = -3r_1 + r_3$

**23.** $\left[\begin{array}{ccc|c} 1 & 1 & 1 & 4 \\ 0 & 5 & 3 & -3 \\ 0 & -4 & 2 & 8 \end{array}\right]$

**(a)** $R_2 = r_3 + r_2$ followed by
**(b)** $R_3 = \frac{1}{2}r_3$

**24.** $\left[\begin{array}{ccc|c} 1 & -3 & 4 & 11 \\ 0 & 3 & 6 & -12 \\ 0 & -2 & -3 & 8 \end{array}\right]$

**(a)** $R_2 = \frac{1}{3}r_2$ followed by
**(b)** $R_3 = 2r_2 + r_3$

*In Problems 25–30, the reduced row echelon form of a system of linear equations is given. Write the system of equations corresponding to the given matrix. Use x, y; or x, y, z as variables. Determine whether the system is consistent and independent, consistent and dependent, or inconsistent. If it is consistent and independent, give the solution.*

**25.** $\left[\begin{array}{cc|c} 1 & 4 & -5 \\ 0 & 1 & -2 \end{array}\right]$

**26.** $\left[\begin{array}{cc|c} 1 & -2 & 3 \\ 0 & 1 & -5 \end{array}\right]$

**27.** $\left[\begin{array}{ccc|c} 1 & 3 & -2 & 6 \\ 0 & 1 & 5 & -2 \\ 0 & 0 & 0 & 4 \end{array}\right]$

**28.** $\left[\begin{array}{ccc|c} 1 & -2 & -4 & 6 \\ 0 & 1 & -5 & -3 \\ 0 & 0 & 0 & 0 \end{array}\right]$

**29.** $\left[\begin{array}{ccc|c} 1 & -2 & -1 & 3 \\ 0 & 1 & -2 & -8 \\ 0 & 0 & 1 & 5 \end{array}\right]$

**30.** $\left[\begin{array}{ccc|c} 1 & 2 & -1 & -7 \\ 0 & 1 & 2 & -4 \\ 0 & 0 & 1 & -3 \end{array}\right]$

*In Problems 31–52, solve each system of equations using matrices. If the system has no solution, say that it is inconsistent.*

**31.** $\begin{cases} x - 3y = 18 \\ 2x + y = 1 \end{cases}$

**32.** $\begin{cases} x + 5y = 2 \\ -2x + 3y = 9 \end{cases}$

**33.** $\begin{cases} 2x + 4y = 10 \\ x + 2y = 3 \end{cases}$

**34.** $\begin{cases} 5x - 2y = 3 \\ -15x + 6y = -9 \end{cases}$

**35.** $\begin{cases} x - 6y = 8 \\ 2x + 8y = -9 \end{cases}$

**36.** $\begin{cases} 3x + 3y = -1 \\ 2x + y = 1 \end{cases}$

**37.** $\begin{cases} 4x - y = 8 \\ 2x - \frac{1}{2}y = 4 \end{cases}$

**38.** $\begin{cases} 5x - 2y = 10 \\ 2x - \frac{4}{5}y = 4 \end{cases}$

**39.** $\begin{cases} x + y + z = 0 \\ 2x - 3y + z = 19 \\ -3x + y - 2z = -15 \end{cases}$

**40.** $\begin{cases} x + y + z = 5 \\ 2x - y + 3z = -3 \\ -x + 2y - z = 10 \end{cases}$

**41.** $\begin{cases} 2x + y - z = 13 \\ -x - 3y + 2z = -14 \\ -3x + 2y - 3z = 3 \end{cases}$

**42.** $\begin{cases} 2x - y + 2z = 13 \\ -x + 2y - z = -14 \\ 3x + y - 2z = -13 \end{cases}$

**43.** $\begin{cases} 2x - y + 3z = 1 \\ -x + 3y + z = -4 \\ 3x + y + 7z = -2 \end{cases}$

**44.** $\begin{cases} -x + 2y + z = 1 \\ 2x - y + 3z = -3 \\ -x + 5y + 6z = 2 \end{cases}$

**45.** $\begin{cases} 3x + y - 4z = 0 \\ -2x - 3y + z = 5 \\ -x - 5y - 2z = 3 \end{cases}$

**46.** $\begin{cases} -x + 4y - 3z = 1 \\ 3x + y - z = -3 \\ x + 9y - 7z = -1 \end{cases}$

**47.** $\begin{cases} 2x - y + 3z = -1 \\ 3x + y - 4z = 3 \\ x + 7y - 2z = 2 \end{cases}$

**48.** $\begin{cases} x + y + z = 8 \\ 2x + 3y + z = 19 \\ 2x + 2y + 4z = 21 \end{cases}$

**49.** $\begin{cases} 3x + 5y + 2z = 6 \\ 10y - 2z = 5 \\ 6x + 4z = 8 \end{cases}$

**50.** $\begin{cases} 2x + y + 3z = 3 \\ 2x - 3y = 7 \\ 4y + 6z = -2 \end{cases}$

**51.** $\begin{cases} x - z = 3 \\ 2x + y = -3 \\ 2y - z = 7 \end{cases}$

**52.** $\begin{cases} x + 2y - z = 3 \\ y + z = 1 \\ x - 3z = 2 \end{cases}$

## Applying the Concepts

**53. Curve Fitting** The function $f(x) = ax^2 + bx + c$ is a quadratic function, where $a$, $b$, and $c$ are constants.

**(a)** If $f(-1) = 6$, then $6 = a(-1)^2 + b(-1) + c$ or $a - b + c = 6$. Find two additional linear equations if $f(1) = 0$, and $f(2) = 3$.

**(b)** Use the three linear equations found in part (a) to determine $a$, $b$, and $c$. What is the quadratic function that contains the points $(-1, 6)$, $(1, 0)$, and $(2, 3)$?

**54. Curve Fitting** The function $f(x) = ax^2 + bx + c$ is a quadratic function, where $a$, $b$, and $c$ are constants.

**(a)** If $f(-1) = -6$, then $-6 = a(-1)^2 + b(-1) + c$ or $a - b + c = -6$. Find two additional linear equations if $f(1) = 0$, and $f(2) = -3$.

**(b)** Use the three linear equations found in part (a) to determine $a$, $b$, and $c$. What is the quadratic function that contains the points $(-1, -6)$, $(1, 0)$, and $(2, -3)$?

**55. Finance** Carissa has $20,000 to invest. Her financial planner suggests that she diversify her investment into three investment categories: Treasury bills that yield 4% simple interest, municipal bonds that yield 5% simple interest, and corporate bonds that yield 8% simple interest. Carissa would like to earn $1070 per year in income. In addition, she wants her investment in Treasury bills to be $3000 more than her investment in corporate bonds. How much should Carissa invest in each investment category?

**56. Finance** Marlon has $12,000 to invest. He decides to place some of the money into a savings account paying 2% interest, some in Treasury bonds paying 4% interest, and some in a mutual fund paying 9% interest. Marlon would like to earn $440 per year in income. In addition, Marlon wants his investment in the savings account to be $4000 more than the amount in Treasury bonds. How much should Marlon invest in each investment category?

## Extending the Concepts

*Sometimes it is advantageous to write a matrix in* ***reduced row echelon form.*** *In this form, row operations are used to obtain entries that are 0 above and below the leading 1 in a row. Augmented matrices for systems in reduced row echelon form with 2 equations containing 2 variables and 3 equations containing 3 variables are shown below.*

$$\left[\begin{array}{cc|c} 1 & 0 & a \\ 0 & 1 & b \end{array}\right] \qquad \left[\begin{array}{ccc|c} 1 & 0 & 0 & a \\ 0 & 1 & 0 & b \\ 0 & 0 & 1 & c \end{array}\right]$$

*The obvious advantage to writing an augmented matrix in reduced row echelon form is that the solution to the system is readily seen. In the system with 2 equations containing 2 variables, the solution is $x = a$ and $y = b$. In the system with 3 equations containing 3 variables, the solution is $x = a$, $y = b$, and $z = c$.*

*In Problems 57–60, solve the system of equations by writing the augmented matrix in reduced row echelon form.*

**57.** $\begin{cases} 2x + y = 1 \\ -3x - 2y = -5 \end{cases}$

**58.** $\begin{cases} 2x + y = -1 \\ -3x - 2y = -3 \end{cases}$

**59.** $\begin{cases} x + y + z = 3 \\ 2x + y - 4z = 25 \\ -3x + 2y + z = 0 \end{cases}$

**60.** $\begin{cases} x + y + z = 3 \\ 3x - 2y + 2z = 38 \\ -2x - 3z = -19 \end{cases}$

## The Graphing Calculator

Graphing calculators have the ability to solve systems of linear equations by writing the equation in echelon form. We enter the augmented matrix into the graphing calculator and name it $A$. See Figure 11(a). Figures 11(b) shows the results of Example 6 using a TI-84 Plus graphing calculator. Since the entire matrix does not fit on the screen, we need to scroll right to the see the rest of it. See Figure 11(c).

**Figure 11**

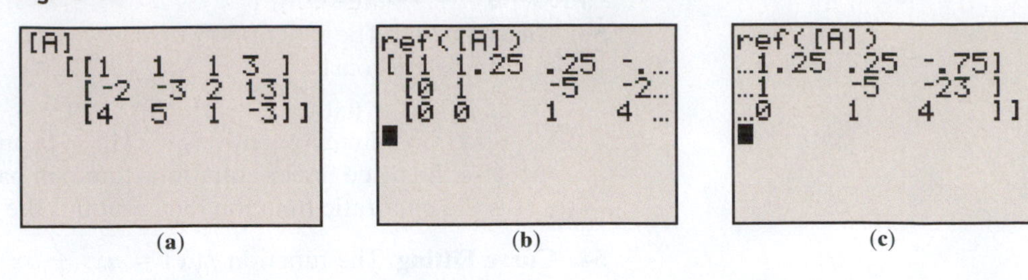


(a) (b) (c)

Notice that the row echelon form of the augmented matrix using the graphing calculator differs from the row echelon form in the algebraic solution presented in Example 6, yet both matrices provide the same solution! This is because the two solutions use different row operations to obtain the row echelon form.

*In Problems 61–66, solve each system of equations using a graphing calculator.*

**61.** $\begin{cases} 2x + 3y = 1 \\ -3x - 4y = -3 \end{cases}$

**62.** $\begin{cases} 3x + 2y = 4 \\ -5x - 3y = -4 \end{cases}$

**63.** $\begin{cases} 2x + 3y - 2z = -12 \\ -3x + y + 2z = 0 \\ 4x + 3y - z = 3 \end{cases}$

**64.** $\begin{cases} 2x - 3y - 4z = 16 \\ -3x + y + 2z = -23 \\ 4x + 3y - z = 13 \end{cases}$

# 4.5 Determinants and Cramer's Rule

**OBJECTIVES**

1. Evaluate the Determinant of a $2 \times 2$ Matrix
2. Use Cramer's Rule to Solve a System of Two Equations Containing Two Variables
3. Evaluate the Determinant of a $3 \times 3$ Matrix
4. Use Cramer's Rule to Solve a System of Three Equations Containing Three Variables

### *Preparing for Deteminants and Cramer's Rule*

*Before getting started, take the following readiness quiz. If you get a problem wrong, go back to the section cited and review the material.*

**1.** Evaluate $4 \cdot 2 - 3 \cdot (-3)$ [Section R.4, pp. 35–36]

**2.** Simplify $\dfrac{18}{6}$ [Section R.3, p. 26]

Up to this point, we have learned how to solve systems of linear equations using substitution, elimination and row operations on augmented matrices. This section presents another method for solving a system of linear equations. However, to use this method, the number of equations must equal the number of variables. The method for solving these systems is called *Cramer's Rule* and is based on the concept of a *determinant*.

## 1 Evaluate the Determinant of a 2 × 2 Matrix

A matrix is **square** provided that the number of rows equals the number of columns. For any square matrix, we can compute its *determinant*.

**Work Smart**

The methods for solving systems presented in this section only work when the number of equations equals the number of variables.

**DEFINITION**

Suppose that $a$, $b$, $c$, and $d$ are four real numbers. The **determinant** of a $2 \times 2$ matrix $\begin{bmatrix} a & b \\ c & d \end{bmatrix}$, denoted $\begin{vmatrix} a & b \\ c & d \end{vmatrix}$, is

$$\begin{vmatrix} a & b \\ c & d \end{vmatrix} = ad - bc$$

*Preparing for...Answers* **1.** 17 **2.** 3

### EXAMPLE 1 Evaluating a 2 × 2 Determinant

Evaluate each determinant:

**(a)** $\begin{vmatrix} 5 & 2 \\ 3 & 4 \end{vmatrix}$ **(b)** $\begin{vmatrix} -1 & 3 \\ -4 & 5 \end{vmatrix}$

**Solution**

**(a)**
$$\begin{vmatrix} 5 & 2 \\ 3 & 4 \end{vmatrix} = 5(4) - 3(2) = 20 - 6 = 14$$

**(b)**
$$\begin{vmatrix} -1 & 3 \\ -4 & 5 \end{vmatrix} = -1(5) - (-4)(3) = -5 - (-12) = -5 + 12 = 7$$

**QUICK ✓** *Evaluate each determinant.*

**1.** $\begin{vmatrix} 5 & 3 \\ 4 & 6 \end{vmatrix}$ **2.** $\begin{vmatrix} -2 & -5 \\ 1 & 7 \end{vmatrix}$

## 2 Use Cramer's Rule to Solve a System of Two Equations Containing Two Variables

We can use 2 × 2 determinants to solve a system of two equations containing two variables.

**CRAMER'S RULE FOR TWO EQUATIONS CONTAINING TWO VARIABLES**

The solution to the system of equations

$$\begin{cases} ax + by = s & (1) \\ cx + dy = t & (2) \end{cases}$$

is given by

$$x = \frac{\begin{vmatrix} s & b \\ t & d \end{vmatrix}}{\begin{vmatrix} a & b \\ c & d \end{vmatrix}} = \frac{D_x}{D}, \quad y = \frac{\begin{vmatrix} a & s \\ c & t \end{vmatrix}}{\begin{vmatrix} a & b \\ c & d \end{vmatrix}} = \frac{D_y}{D}$$

provided that

$$D = \begin{vmatrix} a & b \\ c & d \end{vmatrix} = ad - bc \neq 0$$

Look very carefully at the pattern in Cramer's Rule. The denominator in the solution is the determinant of the coefficients of the variables.

$$\begin{cases} ax + by = s \\ cx + dy = t \end{cases}, \quad D = \begin{vmatrix} a & b \\ c & d \end{vmatrix}$$

In the solution for $x$, the numerator is the determinant formed by replacing the entries in the first column (the coefficients of $x$) in $D$ by the constants on the right side of the equal sign. That is,

$$D_x = \begin{vmatrix} s & b \\ t & d \end{vmatrix}$$

In the solution for $y$, the numerator is the determinant formed by replacing the entries in the second column (the coefficients of $y$) in $D$ by the constants on the right side of the equal sign. That is,

$$D_y = \begin{vmatrix} a & s \\ c & t \end{vmatrix}$$

**EXAMPLE 2** **How to Solve a System of Two Linear Equations Containing Two Variables Using Cramer's Rule**

Use Cramer's Rule to solve the system $\begin{cases} x + 2y = 0 \\ -2x - 8y = -9 \end{cases}$

**Step-by-Step Solution**

**Step 1:** Determine the determinant of the coefficients of the variables, $D$.

$$D = \begin{vmatrix} 1 & 2 \\ -2 & -8 \end{vmatrix} = 1(-8) - (-2)(2) = -8 - (-4) = -8 + 4 = -4$$

**Step 2:** Because $D \neq 0$, we continue by determining $D_x$ by replacing the first column in $D$ with the constants on the right side of the equal sign in the system. We determine $D_y$ by replacing the second column in $D$ with the constants on the right side of the equal sign in the system.

$$D_x = \begin{vmatrix} 0 & 2 \\ -9 & -8 \end{vmatrix} = 0(-8) - (-9)(2) = 0 - (-18) = 18$$

$$D_y = \begin{vmatrix} 1 & 0 \\ -2 & -9 \end{vmatrix} = 1(-9) - (-2)(0) = -9 - 0 = -9$$

**Step 3:** Find $x = \dfrac{D_x}{D}$ and $y = \dfrac{D_y}{D}$

$$x = \frac{D_x}{D} = \frac{18}{-4} = -\frac{9}{2} \qquad y = \frac{D_y}{D} = \frac{-9}{-4} = \frac{9}{4}$$

**Step 4:** Check $x = -\dfrac{9}{2}$ and $y = \dfrac{9}{4}$.

We leave the check to you.

The solution is the ordered pair $\left(-\dfrac{9}{2}, \dfrac{9}{4}\right)$.

If the determinant of the coefficients of the variables, $D$, is found to be 0 when using Cramer's Rule, then the system of equations is either consistent, but dependent, or inconsistent. We will learn how to deal with these possibilities shortly.

**QUICK ✓** *Use Cramer's Rule to solve the system, if possible.*

**3.** $\begin{cases} 3x + 2y = 1 \\ -2x - y = 1 \end{cases}$

**4.** $\begin{cases} 4x - 2y = 8 \\ -6x + 3y = 3 \end{cases}$

## 3 Evaluate the Determinant of a 3 × 3 Matrix

To use Cramer's Rule to solve a system of three equations containing three variables, we need to define a $3 \times 3$ determinant.

The **determinant of a 3 × 3 matrix** is symbolized by

$$\begin{vmatrix} a_{1,1} & a_{1,2} & a_{1,3} \\ a_{2,1} & a_{2,2} & a_{2,3} \\ a_{3,1} & a_{3,2} & a_{3,3} \end{vmatrix}$$

where $a_{1,1}, a_{1,2}, \ldots, a_{3,3}$ are real numbers.

As with matrices, the subscript is used to identify the row and column of an entry. For example, $a_{2,3}$ is the entry in row 2, column 3.

**DEFINITION**

The **value of a determinant of a 3 × 3 matrix** may be defined in terms of 2 × 2 determinants as follows:

$$\begin{vmatrix} a_{1,1} & a_{1,2} & a_{1,3} \\ a_{2,1} & a_{2,2} & a_{2,3} \\ a_{3,1} & a_{3,2} & a_{3,3} \end{vmatrix} = a_{1,1}\begin{vmatrix} a_{2,2} & a_{2,3} \\ a_{3,2} & a_{3,3} \end{vmatrix} - a_{1,2}\begin{vmatrix} a_{2,1} & a_{2,3} \\ a_{3,1} & a_{3,3} \end{vmatrix} + a_{1,3}\begin{vmatrix} a_{2,1} & a_{2,2} \\ a_{3,1} & a_{3,2} \end{vmatrix}$$

Minus

Plus

2 × 2 determinant left after removing the row and column containing $a_{1,1}$.

2 × 2 determinant left after removing the row and column containing $a_{1,2}$.

2 × 2 determinant left after removing the row and column containing $a_{1,3}$.

The 2 × 2 determinants shown in the definition above are called **minors** of the 3 × 3 determinant. Notice that once again we have reduced a problem into something we already know how to do—here we reduced the 3 × 3 determinant into 3 different 2 × 2 determinants.

The formula given in the definition above is easiest to remember by noting that each entry in row 1 is multiplied by the 2 × 2 determinant that remains after the row and column containing the entry have been removed.

## EXAMPLE 3 Evaluating a 3 × 3 Determinant

Evaluate: $\begin{vmatrix} 3 & 2 & -4 \\ 1 & 7 & -3 \\ 0 & 2 & -5 \end{vmatrix}$

### Solution

$$\begin{aligned} \begin{vmatrix} 3 & 2 & -4 \\ 1 & 7 & -3 \\ 0 & 2 & -5 \end{vmatrix} &= 3\begin{vmatrix} 7 & -3 \\ 2 & -5 \end{vmatrix} - 2\begin{vmatrix} 1 & -3 \\ 0 & -5 \end{vmatrix} + (-4)\begin{vmatrix} 1 & 7 \\ 0 & 2 \end{vmatrix} \\ &= 3(-35 - (-6)) - 2(-5 - 0) - 4(2 - 0) \\ &= 3(-29) - 2(-5) - 4(2) \\ &= -87 + 10 - 8 \\ &= -85 \end{aligned}$$

The definition demonstrates one way to find the value of a 3 × 3 determinant—by expanding across row 1. In fact, the expansion can take place across any row or down any column. The terms to be added or subtracted consist of the row (or column) entry times the value of the $2 \times 2$ determinant that remains after removing the row and column containing the entry. There is only one glitch – the signs of the terms in the expansion change depending upon the row or column that is expanded on. The signs of the terms obey the following scheme:

$$\begin{matrix} + & - & + \\ - & + & - \\ + & - & + \end{matrix}$$

For example, if we choose to expand down column 2, we obtain

Minus Plus Minus

$$\begin{vmatrix} a_{1,1} & a_{1,2} & a_{1,3} \\ a_{2,1} & a_{2,2} & a_{2,3} \\ a_{3,1} & a_{3,2} & a_{3,3} \end{vmatrix} = -a_{1,2}\begin{vmatrix} a_{2,1} & a_{2,3} \\ a_{3,1} & a_{3,3} \end{vmatrix} + a_{2,2}\begin{vmatrix} a_{1,1} & a_{1,3} \\ a_{3,1} & a_{3,3} \end{vmatrix} - a_{3,2}\begin{vmatrix} a_{1,1} & a_{1,3} \\ a_{2,1} & a_{2,3} \end{vmatrix}$$

## EXAMPLE 4 Evaluating a $3 \times 3$ Determinant

Redo Example 3 by expanding down column 1. That is, evaluate $\begin{vmatrix} 3 & 2 & -4 \\ 1 & 7 & -3 \\ 0 & 2 & -5 \end{vmatrix}$ by expanding down column 1.

### Solution

We choose to expand down column 1 because it has a 0 in it.

$$\begin{aligned} \begin{vmatrix} 3 & 2 & -4 \\ 1 & 7 & -3 \\ 0 & 2 & -5 \end{vmatrix} &= 3\begin{vmatrix} 7 & -3 \\ 2 & -5 \end{vmatrix} - 1\begin{vmatrix} 2 & -4 \\ 2 & -5 \end{vmatrix} + 0\begin{vmatrix} 2 & -4 \\ 7 & -3 \end{vmatrix} \\ &= 3(-35 - (-6)) - 1(-10 - (-8)) + 0(-6 - (-28)) \\ &= 3(-29) - 1(-2) + 0 \\ &= -87 + 2 \\ &= -85 \end{aligned}$$

**Work Smart**

Expand across the row or down the column with the most 0s.

Notice that the results of Example 3 and 4 are the same! However, the computation in Example 4 was a little easier because we expanded down the column that contains a 0. To make the computation easier, expand across the row or down the column that contains the most 0s.

**QUICK ✓**

**5.** Evaluate $\begin{vmatrix} 2 & -3 & 5 \\ 0 & 4 & -1 \\ 3 & 8 & -7 \end{vmatrix}$

## 4 Use Cramer's Rule to Solve a System of Three Equations Containing Three Variables

Cramer's Rule can also be applied to a system of three linear equations containing three unknowns.

**CRAMER'S RULE FOR THREE EQUATIONS CONTAINING THREE VARIABLES**

For the system of three equations containing three variables

$$\begin{cases} a_1x + b_1y + c_1z = d_1 \\ a_2x + b_2y + c_2z = d_2 \\ a_3x + b_3y + c_3z = d_3 \end{cases}$$

with

$$D = \begin{vmatrix} a_1 & b_1 & c_1 \\ a_2 & b_2 & c_2 \\ a_3 & b_3 & c_3 \end{vmatrix} \neq 0 \quad D_x = \begin{vmatrix} d_1 & b_1 & c_1 \\ d_2 & b_2 & c_2 \\ d_3 & b_3 & c_3 \end{vmatrix} \quad D_y = \begin{vmatrix} a_1 & d_1 & c_1 \\ a_2 & d_2 & c_2 \\ a_3 & d_3 & c_3 \end{vmatrix} \quad D_z = \begin{vmatrix} a_1 & b_1 & d_1 \\ a_2 & b_2 & d_2 \\ a_3 & b_3 & d_3 \end{vmatrix}$$

then

$$x = \frac{D_x}{D} \qquad y = \frac{D_y}{D} \qquad z = \frac{D_z}{D}$$

## EXAMPLE 5 How to Use Cramer's Rule

Use Cramer's Rule, if applicable, to solve the following system:

$$\begin{cases} x - 2y + z = -9 \\ -3x + y - 2z = 5 \\ 4x + 3z = 1 \end{cases}$$

### Step-by-Step Solution

**Step 1:** Find the determinant of the coefficients of the variables, $D$. We choose to expand down column 2. Do you see why?

$$D = \begin{vmatrix} 1 & -2 & 1 \\ -3 & 1 & -2 \\ 4 & 0 & 3 \end{vmatrix} = -(-2)\begin{vmatrix} -3 & -2 \\ 4 & 3 \end{vmatrix} + 1\begin{vmatrix} 1 & 1 \\ 4 & 3 \end{vmatrix} - 0\begin{vmatrix} 1 & 1 \\ -3 & -2 \end{vmatrix}$$

$$= 2[-3(3) - 4(-2)] + 1[1(3) - 4(1)] - 0$$

$$= 2(-1) + 1(-1) - 0$$

$$= -2 - 1$$

$$= -3$$

**Step 2:** Because $D \neq 0$, we continue by determining $D_x$ by replacing the first column in $D$ with the constants on the right side of the equal sign in the system. We determine $D_y$ by replacing the second column in $D$ with the constants on the right side of the equal sign in the system. We determine $D_z$ by replacing the third column in $D$ with the constants on the right side of the equal sign in the system.

Expand down column 1

$$D_x = \begin{vmatrix} -9 & -2 & 1 \\ 5 & 1 & -2 \\ 1 & 0 & 3 \end{vmatrix} = -(-2)\begin{vmatrix} 5 & -2 \\ 1 & 3 \end{vmatrix} + 1\begin{vmatrix} -9 & 1 \\ 1 & 3 \end{vmatrix} - 0\begin{vmatrix} -9 & 1 \\ 5 & -2 \end{vmatrix}$$

$$= 2[15 - (-2)] + 1[-27 - 1] - 0$$

$$= 2(17) - 28$$

$$= 6$$

Expand down column 1

$$D_y = \begin{vmatrix} 1 & -9 & 1 \\ -3 & 5 & -2 \\ 4 & 1 & 3 \end{vmatrix} = 1\begin{vmatrix} 5 & -2 \\ 1 & 3 \end{vmatrix} - (-3)\begin{vmatrix} -9 & 1 \\ 1 & 3 \end{vmatrix} - 4\begin{vmatrix} -9 & 1 \\ 5 & -2 \end{vmatrix}$$

$$= -15$$

Expand down column 2

$$D_z = \begin{vmatrix} 1 & -2 & -9 \\ -3 & 1 & 5 \\ 4 & 0 & 1 \end{vmatrix} = -(-2)\begin{vmatrix} -3 & 5 \\ 4 & 1 \end{vmatrix} + 1\begin{vmatrix} 1 & -9 \\ 4 & 1 \end{vmatrix} - 0\begin{vmatrix} 1 & -9 \\ -3 & 5 \end{vmatrix}$$

$$= -9$$

(continued)

**Step 3:** Find $x = \frac{D_x}{D}$, $y = \frac{D_y}{D}$, and $z = \frac{D_z}{D}$.

$$x = \frac{D_x}{D} = \frac{6}{-3} = -2 \qquad y = \frac{D_y}{D} = \frac{-15}{-3} = 5 \qquad z = \frac{D_z}{D} = \frac{-9}{-3} = 3$$

**Step 4:** Check your answer.

$$\begin{aligned} x - 2y + z &= -9 \\ -2 - 2(5) + 3 &\stackrel{?}{=} -9 \\ -2 - 10 + 3 &\stackrel{?}{=} -9 \\ -9 &= -9 \quad \text{True} \end{aligned}$$

$$\begin{aligned} -3x + y - 2z &= 5 \\ -3(-2) + 5 - 2(3) &\stackrel{?}{=} 5 \\ 6 + 5 - 6 &\stackrel{?}{=} 5 \\ 5 &= 5 \quad \text{True} \end{aligned}$$

$$\begin{aligned} 4x + 3z &= 1 \\ 4(-2) + 3(3) &\stackrel{?}{=} 1 \\ -8 + 9 &\stackrel{?}{=} 1 \\ 1 &= 1 \quad \text{True} \end{aligned}$$

The solution is the ordered triple $(-2, 5, 3)$.

**Quick ✓**

6. Use Cramer's Rule to solve the system $\begin{cases} x - y + 3z = -2 \\ 4x + 3y + z = 9 \\ -2x + 5z = 7 \end{cases}$

We already know that Cramer's Rule does not apply when the determinant of the coefficients on the variables, $D$, is 0. But can we learn anything about the system other than it is not a consistent and independent system if $D = 0$? The answer is yes!

**Work Smart: Study Skills**

At the end of Section 4.1, we summarized the methods that can be used to solve a system of linear equations in two variables. Make up a summary table of your own for solving systems of linear equations in three variables. When is elimination appropriate? When would you use matrices? When is Cramer's Rule best?

**CRAMER'S RULE WITH INCONSISTENT OR DEPENDENT SYSTEMS**

- If $D = 0$ and at least one of the determinants $D_x$, $D_y$, or $D_z$ is different from 0, then the system is inconsistent and the solution set is $\emptyset$ or $\{\ \}$.
- If $D = 0$ and all the determinants $D_x$, $D_y$, or $D_z$ equal 0, then the system is consistent and dependent so that there are infinitely many solutions.

# 4.5 Exercises

**For Extra Help:**

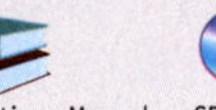
Student Solutions Manual

CD Video

PH Math/Tutor Center

MathXL Tutorials on CD

MathXL®

MyMathLab

## Concepts and Vocabulary

*In Problems 1–3, fill in the blanks.*

1. Cramer's Rule uses _________ to solve a system of linear equations.

2. $D = \begin{vmatrix} a & b \\ c & d \end{vmatrix} =$ _________.

3. A matrix is _________ provided that the number of rows equals the number of columns.

*In Problems 4 and 5, answer True or False to each statement.*

4. A $3 \times 3$ determinant can never equal 0.

5. If $D = 0$, then the system of equations is inconsistent.

**6.** Why can't Cramer's Rule be used if the determinant of the coefficient matrix is zero? What are the possibilities for the solution to the system of equations if the coefficient matrix is zero?

**7.** Suppose that you wish to solve a system of equations using Cramer's Rule and find that $D = 4$, $D_x = -4$, $D_y = 8$, and $D_z = 0$. What is the solution to the system?

**8.** Suppose that you wish to solve a system of equations using Cramer's Rule and find that $D = 0$, $D_x = 3$, $D_y = 7$, and $D_z = -13$. What is the solution to the system?

## Building Skills

*In Problems 9–18, find the value of each determinant.*

**9.** $\begin{vmatrix} 4 & 2 \\ 1 & 3 \end{vmatrix}$ **10.** $\begin{vmatrix} 5 & 3 \\ 2 & 4 \end{vmatrix}$ **11.** $\begin{vmatrix} -2 & -4 \\ 1 & 3 \end{vmatrix}$ **12.** $\begin{vmatrix} -8 & 5 \\ -4 & 3 \end{vmatrix}$

**13.** $\begin{vmatrix} 2 & 0 & -1 \\ 3 & 8 & -3 \\ 1 & 5 & -2 \end{vmatrix}$ **14.** $\begin{vmatrix} -2 & 1 & 6 \\ -3 & 2 & 5 \\ 1 & 0 & -2 \end{vmatrix}$ **15.** $\begin{vmatrix} -3 & 2 & 3 \\ 0 & 5 & -2 \\ 1 & 4 & 8 \end{vmatrix}$ **16.** $\begin{vmatrix} 8 & 4 & -1 \\ 2 & -7 & 1 \\ 0 & 5 & -3 \end{vmatrix}$

**17.** $\begin{vmatrix} 0 & 2 & 1 \\ 1 & -6 & -4 \\ -3 & 4 & 5 \end{vmatrix}$ **18.** $\begin{vmatrix} -3 & 4 & -2 \\ 1 & -2 & 0 \\ 0 & 6 & 6 \end{vmatrix}$

*In Problems 19–26, solve each system of equations using Cramer's Rule, if applicable.*

**19.** $\begin{cases} x + y = -4 \\ x - y = -12 \end{cases}$ **20.** $\begin{cases} x + y = 6 \\ x - y = 4 \end{cases}$ **21.** $\begin{cases} 2x + 3y = 3 \\ -3x + y = -10 \end{cases}$

**22.** $\begin{cases} 2x + 4y = -6 \\ 3x + 2y = 7 \end{cases}$ **23.** $\begin{cases} 3x + 4y = 1 \\ -6x + 8y = 4 \end{cases}$ **24.** $\begin{cases} 2x + 4y = 6 \\ 3x + 6y = 1 \end{cases}$

**25.** $\begin{cases} 2x - 6y - 12 = 0 \\ 3x - 5y - 11 = 0 \end{cases}$ **26.** $\begin{cases} 3x - 6y - 2 = 0 \\ x + 2y - 4 = 0 \end{cases}$

*In Problems 27–38, solve each system of equations using Cramer's Rule, if applicable.*

**27.** $\begin{cases} x - y + z = -4 \\ x + 2y - z = 1 \\ 2x + y + 2z = -5 \end{cases}$ **28.** $\begin{cases} x + y - z = 6 \\ x + 2y + z = 6 \\ -x - y + 2z = -7 \end{cases}$

**29.** $\begin{cases} x + y + z = 4 \\ 5x + 2y - 3z = 7 \\ 2x - y - z = 5 \end{cases}$ **30.** $\begin{cases} x + y + z = -3 \\ -2x - 3y - z = 1 \\ 2x - y - 3z = -5 \end{cases}$

**31.** $\begin{cases} 2x + y - z = 4 \\ -x + 2y + 2z = -6 \\ 5x + 5y - z = 6 \end{cases}$ **32.** $\begin{cases} -x + 2y - z = 2 \\ 2x + y + 2z = -6 \\ -x + 7y - z = 0 \end{cases}$

**33.** $\begin{cases} 3x + y + z = 5 \\ x + y - 3z = 9 \\ 4x + 3y + z = 11 \end{cases}$ **34.** $\begin{cases} x - 2y - z = 1 \\ 2x + 2y + z = 3 \\ x - 4y + 3z = 14 \end{cases}$

**35.** $\begin{cases} 2x + z = 27 \\ -x - 3y = 6 \\ x - 2y + z = 27 \end{cases}$ **36.** $\begin{cases} x + 2y + z = 0 \\ -x - 3y - 2z = 3 \\ 2x - 3z = 7 \end{cases}$

**37.** $\begin{cases} 5x + 3y = 2 \\ -10x + 3z = -3 \\ y - 2z = -9 \end{cases}$

**38.** $\begin{cases} 2x + 4y = 0 \\ -2x + z = -5 \\ -4y - 3z = 1 \end{cases}$

## Applying the Concepts

*In Problems 39–42, solve for x.*

**39.** $\begin{vmatrix} x & 3 \\ 1 & 2 \end{vmatrix} = 7$

**40.** $\begin{vmatrix} -2 & x \\ 3 & 4 \end{vmatrix} = 1$

**41.** $\begin{vmatrix} x & -1 & -2 \\ 1 & 0 & 4 \\ 3 & 2 & 5 \end{vmatrix} = 5$

**42.** $\begin{vmatrix} 2 & x & -1 \\ 3 & 5 & 0 \\ -4 & 1 & 2 \end{vmatrix} = 0$

*Problems 43–46, use the following result. Determinants can be used to find the area of a triangle. If $(a_1, b_1)$, $(a_2, b_2)$, and $(a_3, b_3)$ are the vertices of a triangle, the area of the triangle is $|D|$, where*

$$D = \frac{1}{2}\begin{vmatrix} a_1 & a_2 & a_3 \\ b_1 & b_2 & b_3 \\ 1 & 1 & 1 \end{vmatrix}$$

△ **43. Geometry: Area of a Triangle** Given the points $A = (1, 1)$, $B = (5, 1)$, and $C = (5, 6)$,

**(a)** Plot the points in the Cartesian plane and form triangle $ABC$.
**(b)** Find the area of the triangle $ABC$.

△ **44. Geometry: Area of a Triangle** Given the points $A = (-1, -1)$, $B = (3, 2)$, and $C = (0, 6)$,

**(a)** Plot the points in the Cartesian plane and form triangle $ABC$.
**(b)** Find the area of the triangle $ABC$.

△ **45. Geometry: Area of a Parallelogram** Find the area of a parallelogram by doing the following.

**(a)** Plot the points $A = (2, 1)$, $B = (7, 2)$, $C = (8, 4)$, and $D = (3, 3)$ in the Cartesian plane.
**(b)** Form triangle $ABC$ and find the area of triangle $ABC$.
**(c)** Find the area of the triangle $ADC$.
**(d)** Conclude that the diagonal of the parallelogram forms two triangles of equal area. Use this result to find the area of the parallelogram.

△ **46. Geometry: Area of a Parallelogram** Find the area of a parallelogram by doing the following.

**(a)** Plot the points $A = (-3, -2)$, $B = (3, 1)$, $C = (4, 4)$, and $D = (-2, 1)$ in the Cartesian plane.
**(b)** Form triangle $ABC$ and find the area of triangle $ABC$.
**(c)** Find the area of the triangle $ADC$.
**(d)** Conclude that the diagonal of the parallelogram forms two triangles of equal area. Use this result to find the area of the parallelogram.

△ **47. Geometry: Equation of a Line** An equation of the line containing the two points $(x_1, y_1)$ and $(x_2, y_2)$ may be expressed as the determinant

$$\begin{vmatrix} x & y & 1 \\ x_1 & y_1 & 1 \\ x_2 & y_2 & 1 \end{vmatrix} = 0$$

**(a)** Find the equation of the line containing $(3, 2)$ and $(5, 1)$ using the determinant.
**(b)** Verify your result by using the slope formula and the point-slope formula.

△ **48. Geometry: Collinear Points** The distinct points $(x_1, y_1)$, $(x_2, y_2)$, and $(x_3, y_3)$ are collinear if and only if

$$\begin{vmatrix} x_1 & y_1 & 1 \\ x_2 & y_2 & 1 \\ x_3 & y_3 & 1 \end{vmatrix} = 0$$

**(a)** Plot the points $(-3, -2)$, $(1, 2)$, and $(7, 8)$ in the Cartesian plane.
**(b)** Show that the points are collinear using the determinant.
**(c)** Show that the points are collinear using the idea of slopes.

## Extending the Concepts

**49.** Evaluate the determinant $\begin{vmatrix} 3 & -2 \\ 1 & 4 \end{vmatrix}$. Now interchange rows 1 and 2 and recompute the determinant. What do you notice? Do you think that this result is true in general?

**50.** Evaluate the determinant $\begin{vmatrix} -3 & 1 \\ 6 & 5 \end{vmatrix}$. Multiply the entries in column 2 by 3 and recompute the determinant. What do you notice? Do you think that this result is true in general?

## The Graphing Calculator

Graphing calculators have the ability to evaluate $2 \times 2$ determinants. First, we enter the matrix into the graphing calculator and name it $A$. Then, we compute the determinant of matrix $A$. Figure 12 shows the results of Example 1(a) using a TI-84 Plus graphing calculator.

Because graphing calculators have the ability to compute determinants of matrices, they can be used to solve systems of equations using Cramer's Rule. Figure 13 shows the results of Example 2 using Cramer's Rule, where $A = D$, $B = D_x$, and $C = D_y$.

**Figure 12**

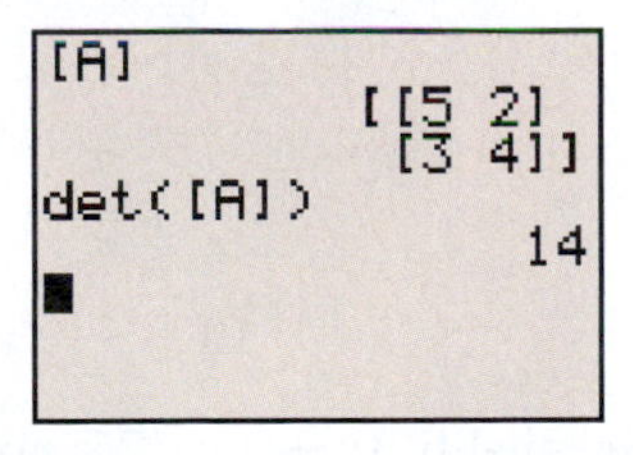


**Figure 13**

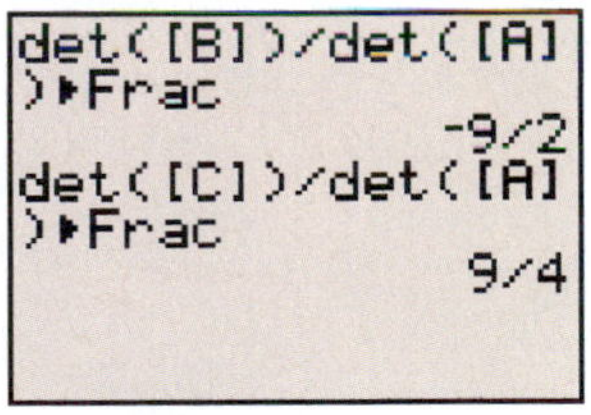

det([B])/det([A]) ▸Frac
-9/2
det([C])/det([A]) ▸Frac
9/4

*In Problems 51–56, use a graphing calculator to solve the system of equations.*

**51.** $\begin{cases} x + y = -4 \\ x - y = -12 \end{cases}$

**52.** $\begin{cases} x + y = 6 \\ x - y = 4 \end{cases}$

**53.** $\begin{cases} 2x + 3y = 3 \\ -3x + y = -10 \end{cases}$

**54.** $\begin{cases} 2x + 4y = -6 \\ 3x + 2y = 7 \end{cases}$

**55.** $\begin{cases} x - y + z = -4 \\ x + 2y - z = 1 \\ 2x + y + 2z = -5 \end{cases}$

**56.** $\begin{cases} x + y - z = 6 \\ x + 2y + z = 6 \\ -x - y + 2z = -7 \end{cases}$

# 4.6 Systems of Linear Inequalities

**OBJECTIVES**

1. Determine Whether an Ordered Pair Is a Solution to a System of Linear Inequalities
2. Graph a System of Linear Inequalities
3. Solve Problems Involving Systems of Linear Inequalities

***Preparing for Systems of Linear Inequalities***

*Before getting started, take the following readiness quiz. If you get a problem wrong, go back to the section cited and review the material.*

**1.** Determine whether $x = -2$ satisfies the inequality $-3x + 2 \geq 7$. [Section 1.4, p. 95]

**2.** Solve the inequality $-2x + 1 > 7$. [Section 1.4, p. 95]

**3.** Graph the inequality $3x + 2y > -6$. [Section 3.4, pp. 226–227]

In Section 3.4, we graphed a single linear inequality in two variables. In this section, we discuss how to graph a system of linear inequalities in two variables.

## 1 Determine Whether an Ordered Pair Is a Solution to a System of Linear Inequalities

An ordered pair **satisfies** a system of linear inequalities if it makes each inequality in the system a true statement.

### EXAMPLE 1 Determining Whether an Ordered Pair Is a Solution to a System of Linear Inequalities

Which of the following points, if any, is a solution to the system of linear inequalities?

$$\begin{cases} 3x + y \leq 7 \\ 4x - 2y \leq 8 \end{cases}$$

**(a)** $(2, 4)$ **(b)** $(-3, 1)$

**Solution**

**(a)** Let $x = 2$ and $y = 4$ in each inequality in the system. If $3(2) + 4 \leq 7$ and $4(2) - 2(4) \leq 8$, then $(2, 4)$ is a solution to the system of inequalities.

$$\begin{aligned} 3x + y &\leq 7 \\ 3(2) + 4 &\stackrel{?}{\leq} 7 \\ 6 + 4 &\stackrel{?}{\leq} 7 \\ 10 &\leq 7 \quad \text{False} \end{aligned} \qquad \begin{aligned} 4x - 2y &\leq 8 \\ 4(2) - 2(4) &\stackrel{?}{\leq} 8 \\ 8 - 8 &\stackrel{?}{\leq} 8 \\ 0 &\leq 8 \quad \text{True} \end{aligned}$$

The inequality $3x + y \leq 7$ is not true when $x = 2$ and $y = 4$, so $(2, 4)$ is not a solution to the system.

**(b)** Let $x = -3$ and $y = 1$ in each inequality in the system. If each inequality is true, then $(-3, 1)$ is a solution to the system of inequalities.

$$\begin{aligned} 3x + y &\leq 7 \\ 3(-3) + 1 &\stackrel{?}{\leq} 7 \\ -9 + 1 &\stackrel{?}{\leq} 7 \\ -8 &\leq 7 \quad \text{True} \end{aligned} \qquad \begin{aligned} 4x - 2y &\leq 8 \\ 4(-3) - 2(1) &\stackrel{?}{\leq} 8 \\ -12 - 2 &\stackrel{?}{\leq} 8 \\ -14 &\leq 8 \quad \text{True} \end{aligned}$$

Both inequalities are true when $x = -3$ and $y = 1$, so $(-3, 1)$ is a solution to the system.

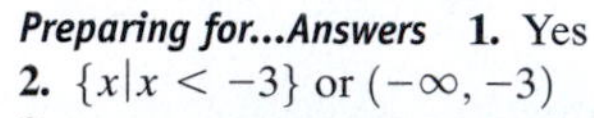

***Preparing for...Answers*** **1.** Yes **2.** $\{x | x < -3\}$ or $(-\infty, -3)$ **3.**

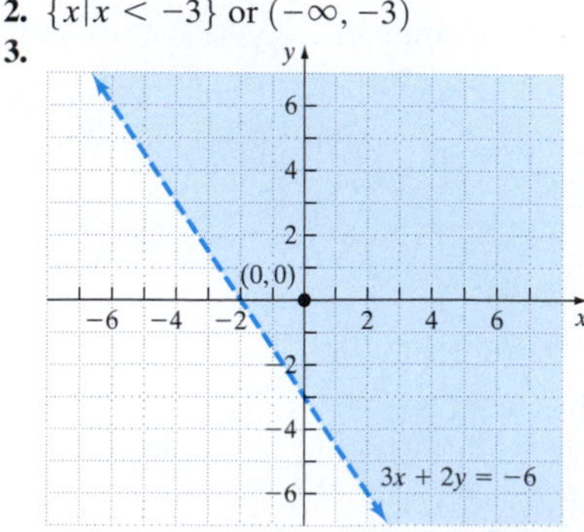

**QUICK ✓**

1. Determine which of the following points is a solution to the system of linear inequalities.

$$\begin{cases} -4x + y < -5 \\ 2x - 5y < 10 \end{cases}$$

**(a)** $(1, 2)$ **(b)** $(3, 1)$

## 2 Graph a System of Linear Inequalities

**Work Smart**

Don't forget that we graph with a solid line when the inequality is nonstrict ($\leq$ or $\geq$) and a dashed line when the inequality is strict ($<$ or $>$).

The graph of a system of inequalities in two variables $x$ and $y$ is the set of all points $(x, y)$ that simultaneously satisfy *each* of the inequalities in the system. The graph of a system of linear inequalities can be obtained by graphing each linear inequality individually and then determining where, if at all, they intersect. The ONLY way we show the solution of a system of linear inequalities is by its graphical representation.

### EXAMPLE 2 Graphing a System of Linear Inequalities

Graph the system: $\begin{cases} 3x + y \leq 7 \\ 4x - 2y \leq 8 \end{cases}$

**Solution**

First, we graph the inequality $3x + y \leq 7$. We replace the inequality symbol with an equal sign to obtain $3x + y = 7$ and graph the solid boundary line $y = -3x + 7$. The test point $(0, 0)$ satisfies the inequality $3x + y \leq 7$ so we shade below the boundary line. See Figure 14(a). We then graph the inequality $4x - 2y \leq 8$ by first graphing the solid boundary line $4x - 2y = 8$ (or $y = 2x - 4$). The test point $(0, 0)$ satisfies the inequality $4x - 2y \leq 8$ so we shade above the boundary line. See Figure 14(b).

**Work Smart**

To obtain the boundary line $y = 2x - 4$, solve the equation $4x - 2y = 8$ for $y$.

Now combine the graphs in Figures 14(a) and (b). The intersection of the shaded regions is the solution to the system of linear inequalities. See Figure 14(c). Notice that the point $(-3, 1)$, which was verified to be a solution in Example 1(b), is in the shaded region.

**Figure 14**

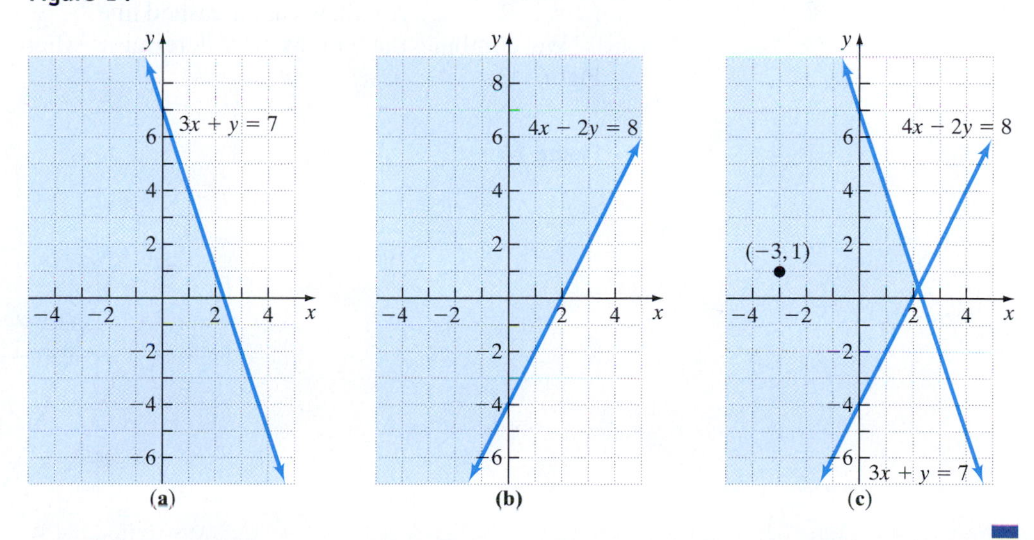


Have you noticed that if we write the inequality in the form $y >$ or $y \geq$, we shade above the line? If the inequality is of the form $y <$ or $y \leq$, shade below the line. For example, $3x + y \leq 7$ can be written as $y \leq -3x + 7$, and we shade below the boundary line. Instead of using a test point to determine shading, we will use this approach in future examples.

Rather than use multiple graphs to obtain the solution set to the system of linear inequalities, we can use a single graph to determine the overlapping region.

## EXAMPLE 3 Graphing a System of Linear Inequalities

Graph the system: $\begin{cases} -2x + \quad y > 5 \\ \;\;3x + 2y \geq -4 \end{cases}$

### Solution

We graph the inequality $-2x + y > 5$ by adding $2x$ to both sides and obtaining $y > 2x + 5$. We graph $y = 2x + 5$ as a dashed line and shade above (since the inequality is of the form $y >$). On the same Cartesian plane, we graph the inequality $3x + 2y \geq -4$. We solve for $y$ and obtain $y \geq -\frac{3}{2}x - 2$. We graph $y = -\frac{3}{2}x - 2$ as a solid line and shade above (since the inequality is of the form $y>$).

The shaded region represents the solution. Figure 15 shows the solution.

Figure 15

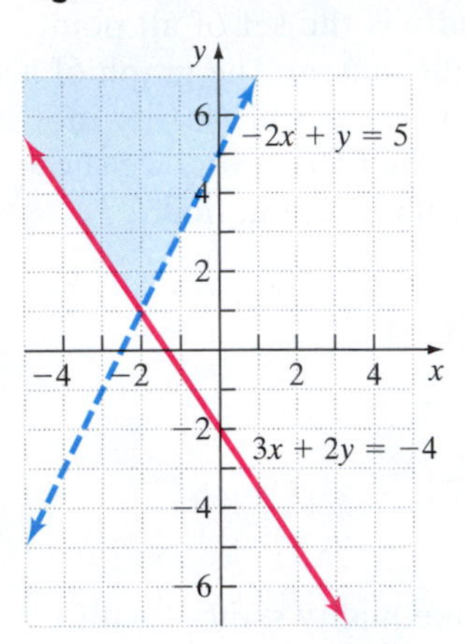


**QUICK ✓** *Graph the system of linear inequalities.*

**2.** $\begin{cases} 2x + y \leq 5 \\ -x + y \geq -4 \end{cases}$ **3.** $\begin{cases} 3x + \quad y > -2 \\ 2x + 3y < 3 \end{cases}$

## EXAMPLE 4 Graphing a System of Linear Inequalities with No Solution

Graph the system: $\begin{cases} 2x + y > 2 \\ 2x + y < -2 \end{cases}$

### Solution

We graph the inequality $2x + y > 2$ $(y > -2x + 2)$ using a dashed line because the inequality is strict. On the same Cartesian plane, we graph the inequality $2x + y < -2$ $(y < -2x - 2)$. Again, we use a dashed line.

We combine the graphs and determine where the shaded regions overlap. See Figure 16.

Figure 16

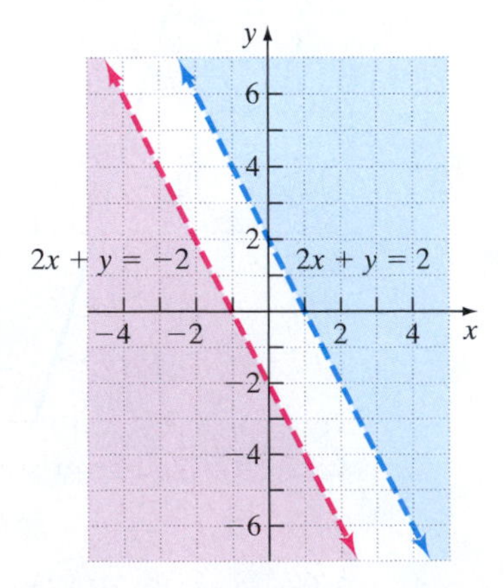


Because no overlapping region results, there are no points in the Cartesian plane that satisfy both inequalities. The system has no solution, so the solution set is $\emptyset$ or $\{ \; \}$.

**QUICK ✓** *Graph the system of linear inequalities.*

**4.** $\begin{cases} -2x + 3y \geq 9 \\ 6x - 9y \geq 9 \end{cases}$

Often, we encounter systems with more than two linear inequalities.

### EXAMPLE 5 Graphing a System of Four Linear Inequalities

Graph the system: $\begin{cases} x + y \leq 5 \\ 2x + y \leq 7 \\ x \geq 0 \\ y \geq 0 \end{cases}$

**Solution**

The two inequalities $x \geq 0$ and $y \geq 0$ require that the graph be in quadrant I. The inequality $x + y \leq 5$ requires that we shade below the line $x + y = 5$. The inequality $2x + y \leq 7$ requires that we shade below the line $2x + y = 7$. Figure 17 shows the graph.

Figure 17

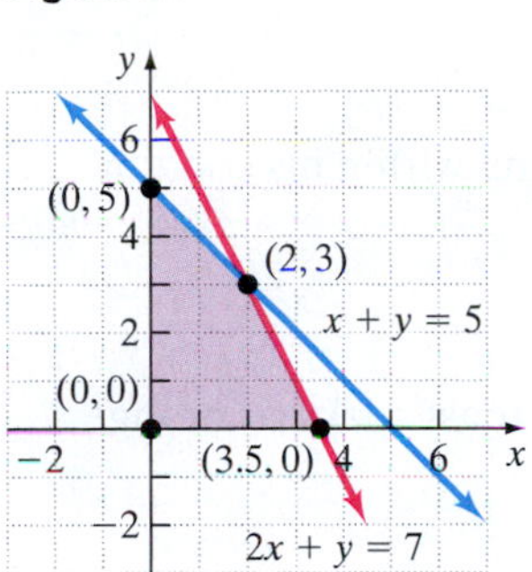


In looking carefully at Figure 17, you should notice that we have labeled specific points on the graph. These points belong to the graph of the system and represent points of intersection between two boundary lines in the system and are called **corner points.** The corner points in Figure 17 are (0, 0), (0, 5), (2, 3), and (3.5, 0) and are found by solving the system of equations formed by two boundary lines. The corner points in Figure 17 are found by solving the following systems:

$\begin{cases} x = 0 \\ y = 0 \end{cases}$ Corner point: (0, 0)

$\begin{cases} x + y = 5 \\ x = 0 \end{cases}$ Corner point: (0, 5)

$\begin{cases} x + y = 5 \\ 2x + y = 7 \end{cases}$ Corner point: (2, 3)

$\begin{cases} 2x + y = 7 \\ y = 0 \end{cases}$ Corner point: (3.5, 0)

The graph of the system of linear inequalities discussed in Example 5 is said to be **bounded** because it can be contained within a circle. Look at the systems graphed in Figures 14 or 15. They are **unbounded** because the solution set cannot be contained within a circle. The graphs extend indefinitely.

**QUICK ✓** *Graph the system of linear inequalities. Tell whether the graph is bounded or unbounded, and label the corner points.*

**5.** $\begin{cases} x + y \leq 6 \\ 2x + y \leq 10 \\ x \geq 0 \\ y \geq 0 \end{cases}$

## 3 Solve Problems Involving Systems of Linear Inequalities

Now let's look at some problems that lead to systems of linear inequalities. As always, we use the problem-solving strategy presented in Section 1.2.

### EXAMPLE 6 Breakfast at Burger King

Aman decides to eat breakfast at Burger King. He is a big fan of their French toast sticks and likes orange juice. He wants to eat no more than 500 calories and consume no more than 425 mg of sodium. Each French toast stick (with syrup) has 90 calories and 100 mg of sodium. Each small orange juice has 140 calories and 25 mg of sodium. Write a system of linear inequalities that represents the possible combination of French toast sticks and orange juice that Aman can consume. Graph the system.

### Solution

**Step 1: Identify** We want to know the number of French toast sticks and orange juices Aman can consume while staying within his diet restrictions.

**Step 2: Name** Let $x$ represent the number of French toast sticks and $y$ represent the number of orange juices that Aman consumes.

**Step 3: Translate** If Aman eats one French toast stick, then he will consume 90 calories. If he eats two, then he will consume 180 calories. In general, if he eats $x$ French toast sticks, he will consume $90x$ calories. Similar logic for orange juice tells us that if Aman drinks $y$ orange juices, he will consume $140y$ calories. The words "no more than" imply a $\leq$ inequality. A linear inequality that describes Aman's options while staying within his calorie restriction is

$$90x + 140y \leq 500$$

A linear inequality that describes Aman's options while staying within his sodium restriction is

$$100x + 25y \leq 425$$

**In Words**
A nonnegativity constraint means that the value of the variable has to be greater than or equal to zero.

Finally, Aman cannot consume negative quantities of French toast sticks or orange juice, so we have the **nonnegativity constraints**

$$x \geq 0 \quad \text{and} \quad y \geq 0$$

If we combine these four linear inequalities, we obtain the system

$$\begin{cases} 90x + 140y \leq 500 \\ 100x + 25y \leq 425 \\ x \geq 0 \\ y \geq 0 \end{cases}$$

**Step 4: Solve** Figure 18 shows the graph of the system.

**Figure 18**

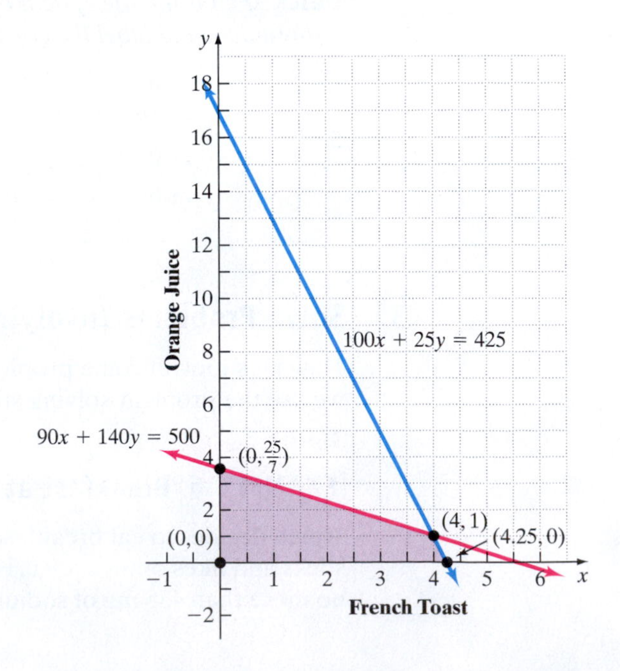

**Step 5: Check** Although not a check, we can gather evidence that our graph is correct by choosing a point in the shaded region and determining if all the inequalities in the system are satisfied. For example, $x = 3$ French toast sticks and $y = 1$ orange juice satisfies all three inequalities.

QUICK ✓

6. Jack and Mary recently retired and they have $25,000 to invest. Their financial advisor has recommended that they place at least $10,000 in Treasury notes yielding 5% and no more than $15,000 in corporate bonds yielding 8%. Write a system of linear inequalities that represents the possible combination of investments. Graph the system.

# 4.6 Exercises

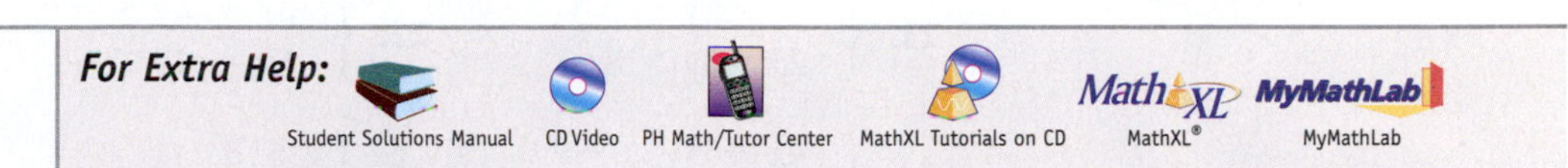


## Concepts and Vocabulary

*In Problems 1 and 2, fill in the blanks.*

1. An ordered pair ________ a system of linear inequalities if it makes each inequality in the system a true statement.

2. The points of intersection between two boundary lines in a system of linear inequalities are called ________ ________.

*In Problems 3 and 4, answer True or False to each statement.*

3. The graph of the system of linear inequalities is sometimes unbounded.

4. A system of linear inequalities must always have a solution.

5. Explain how to graph a system of linear inequalities in two variables.

6. If one inequality in a system of two linear inequalities contains a strict inequality and the other inequality contains a nonstrict inequality, is the corner point a solution to the system? Explain your answer.

7. Is it possible for a system of linear inequalities to have the entire Cartesian plane as the solution set? Explain your answer, and if your answer is yes, then provide an example.

8. Is it possible for a system of linear inequalities to have a straight line as the solution set? Explain your answer, and if your answer is yes, then provide an example.

## Building Skills

*In Problems 9–14, determine which of the points, if any, satisfies the system.*

9. $\begin{cases} x + y \leq 4 \\ 3x + y \geq 10 \end{cases}$
   (a) $(3, 2)$
   (b) $(5, -1)$

10. $\begin{cases} x + y \geq 2 \\ -3x + y \leq 10 \end{cases}$
   (a) $(-3, 6)$
   (b) $(4, 1)$

11. $\begin{cases} 3x + 2y < 12 \\ -2x + 3y > 12 \end{cases}$
   (a) $(3, 2)$
   (b) $(1, 4)$

**12.** $\begin{cases} 5x + 2y < 10 \\ 4x - 3y < 24 \end{cases}$

**(a)** $(1, 3)$
**(b)** $(1, 1)$

**13.** $\begin{cases} x + y \leq 8 \\ 3x + y \leq 12 \\ x \geq 0 \\ y \geq 0 \end{cases}$

**(a)** $(4, 2)$
**(b)** $(2, 5)$

**14.** $\begin{cases} x + y \geq 6 \\ 2x + y \geq 10 \\ x \geq 0 \\ y \geq 0 \end{cases}$

**(a)** $(4, 2)$
**(b)** $(2, 5)$

*In Problems 15–26, graph each system of linear inequalities.*

**15.** $\begin{cases} x + y \leq 4 \\ 3x + y \geq 10 \end{cases}$

**16.** $\begin{cases} x + y \geq 2 \\ -3x + y \leq 10 \end{cases}$

**17.** $\begin{cases} 3x + 2y < 12 \\ -2x + 3y > 12 \end{cases}$

**18.** $\begin{cases} 5x + 2y < 10 \\ 4x - 3y < 24 \end{cases}$

**19.** $\begin{cases} x - \frac{1}{2}y \leq 3 \\ \frac{3}{2}x + y > 3 \end{cases}$

**20.** $\begin{cases} -x + \frac{1}{3}y < 3 \\ \frac{4}{3}x + y \geq 4 \end{cases}$

**21.** $\begin{cases} -2x + y < -8 \\ 2x - y > 8 \end{cases}$

**22.** $\begin{cases} 3x - 2y < -6 \\ -6x + 4y > 12 \end{cases}$

**23.** $\begin{cases} -5x + 3y < 12 \\ 5x - 3y < 9 \end{cases}$

**24.** $\begin{cases} 4x + 3y > -9 \\ -8x - 6y > 12 \end{cases}$

**25.** $\begin{cases} y \leq 8 \\ x \geq 3 \end{cases}$

**26.** $\begin{cases} y \leq 4 \\ x \geq -1 \end{cases}$

*In Problems 27–34, graph each system of linear inequalities. Tell whether the graph is bounded or unbounded, and label the corner points.*

**27.** $\begin{cases} x + y \leq 6 \\ 3x + y \leq 12 \\ x \geq 0 \\ y \geq 0 \end{cases}$

**28.** $\begin{cases} x + y \leq 9 \\ 3x + 2y \leq 24 \\ x \geq 0 \\ y \geq 0 \end{cases}$

**29.** $\begin{cases} x + y \geq 8 \\ 4x + 2y \geq 28 \\ x \geq 0 \\ y \geq 0 \end{cases}$

**30.** $\begin{cases} x + y \geq 8 \\ x + 3y \geq 12 \\ x \geq 0 \\ y \geq 0 \end{cases}$

**31.** $\begin{cases} 2x + 3y \leq 30 \\ 3x + 2y \leq 25 \\ 5x + 2y \leq 35 \\ x \geq 0 \\ y \geq 0 \end{cases}$

**32.** $\begin{cases} 2x + 3y \leq 36 \\ x + y \leq 14 \\ 3x + y \leq 30 \\ x \geq 0 \\ y \geq 0 \end{cases}$

**33.** $\begin{cases} 2x + y \geq 13 \\ x + 2y \geq 11 \\ x \geq 4 \\ y \geq 0 \end{cases}$

**34.** $\begin{cases} 7x + 3y \geq 45 \\ 5x + 3y \geq 39 \\ x \geq 0 \\ y \geq 3 \end{cases}$

## Applying the Concepts

**35. Breakfast** Daria is on a special diet that requires she consume at least 500 mg of potassium and 14 grams of protein for breakfast. One morning, Daria decides to have orange juice, which contains 450 mg of potassium and 2 g of protein per serving, and Apple Cinnamon Cheerios® (with 2% milk), which contains 50 mg of potassium and 6 g of protein per serving.

**(a)** Let $x$ denote the number of servings of orange juice and $y$ denote the number of servings of cereal. Write a system of linear inequalities that represents the possible combination of foods.

**(b)** Graph the system and label the corner points.

**36. Manufacturing Printers** Bob's Printer Emporium manufactures two printers, a color ink jet printer, and a black-and-white laser printer. Each color printer requires 2 hours for molding and 3 hours for assembly; each laser printer requires 3 hours for molding and 4 hours for assembly. There are a total of 80 hours per week available in the work schedule for molding and 120 hours per week available for assembly.

**(a)** Using $x$ to denote the number of color printers and $y$ to denote the number of laser printers, write a system of linear inequalities that describes the possible number of each model printer that can be manufactured in a week (it is possible to manufacturer a portion of a printer).

**(b)** Graph the system and label the corner points.

**37. Financial Planning** Jack and Mary recently retired, and they have \$25,000 to invest. Their financial advisor has recommended that they place at least \$5000 in Treasury Notes yielding 5% and no more than \$15,000 in corporate bonds yielding 8%. In addition, they want to earn at least \$1400 each year.

**(a)** Let $x$ denote the amount invested in Treasury Notes and $y$ denote the amount invested in corporate bonds. Write a system of linear inequalities that represents the possible combination of investments.

**(b)** Graph the system and label the corner points.

**38. Mixing Nuts** You've Got to Be Nuts is a store that specializes in selling nuts. The owner finds that she has excess inventory of 100 pounds (1600 ounces) of cashews and 120 pounds (1920 ounces) of peanuts. She decides to make two types of 1 pound nut mixes from the excess inventory. A premium mix will contain 12 ounces of cashews and 4 ounces of peanuts while the standard mix will contain 6 ounces of cashews and 6 ounces of peanuts.

**(a)** Use $x$ to denote the number of premium mixes and $y$ denote the number of standard mixes. Write a system of linear inequalities that describe the possible number of each kind of mix.

**(b)** Graph the system and label the corner points.

## Extending the Concepts

*In Problems 39–42, write a system of linear inequalities that has the given graph.*

**39.**

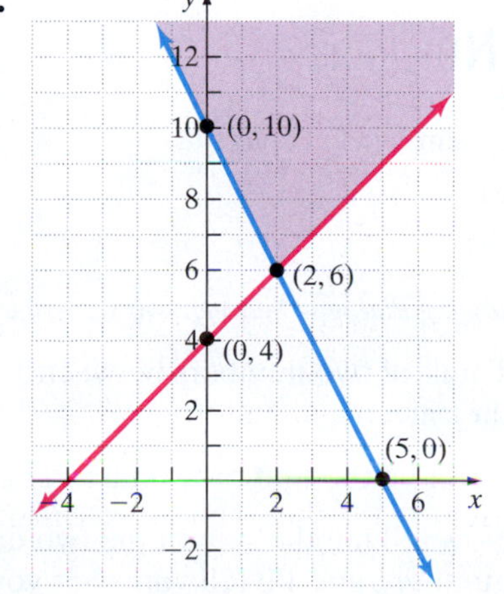

**40.**

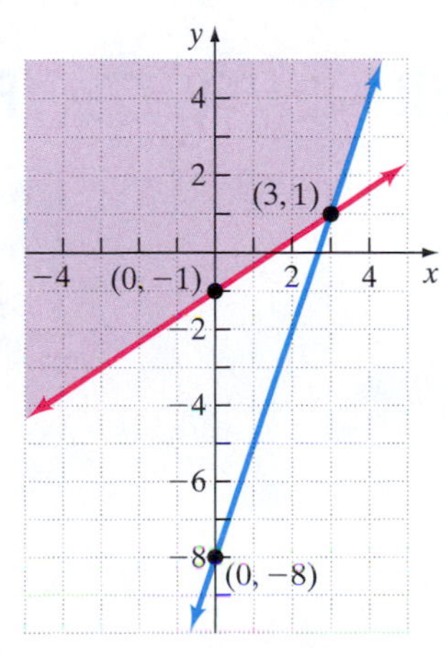

**41.**

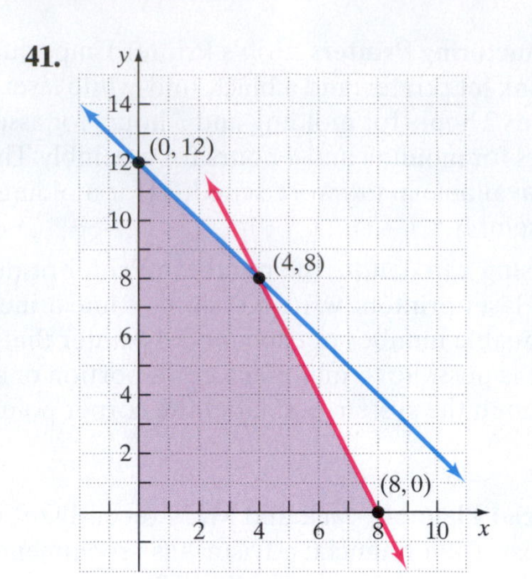


**42.**

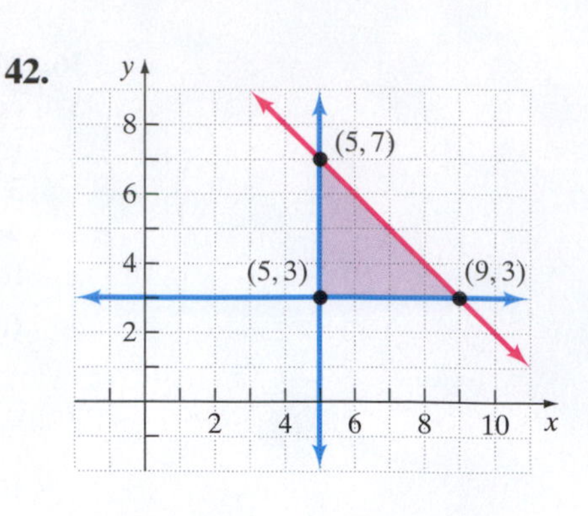


### The Graphing Calculator

Graphing calculators have the ability to graph systems of linear inequalities. Figure 19 shows the results of Example 3 using a TI-84 Plus graphing calculator.

**Figure 19**

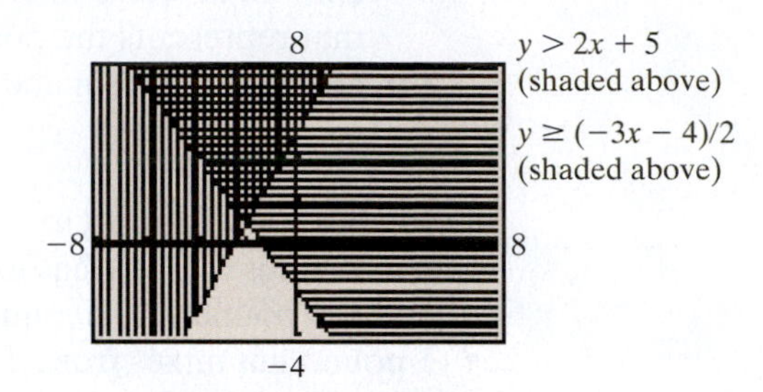


*In Problems 43–48, graph each system of linear inequalities using a graphing calculator.*

**43.** $\begin{cases} x + y \leq 4 \\ 3x + y \geq 10 \end{cases}$

**44.** $\begin{cases} x + y \geq 2 \\ -3x + y \leq 10 \end{cases}$

**45.** $\begin{cases} 3x + 2y < 12 \\ -2x + 3y > 12 \end{cases}$

**46.** $\begin{cases} 5x + 2y < 10 \\ 4x - 3y < 24 \end{cases}$

**47.** $\begin{cases} x + y \leq 6 \\ 3x + y \leq 12 \\ x \geq 0 \\ y \geq 0 \end{cases}$

**48.** $\begin{cases} x + y \leq 9 \\ 3x + 2y \leq 24 \\ x \geq 0 \\ y \geq 0 \end{cases}$

## CHAPTER 4 ACTIVITY: Find the Numbers

**Focus:** Solving systems of equations

**Time:** 15 minutes

**Group size:** 2

*Consider the following dialogue between two students.*

*Michele:* Think of two numbers between 1 and 10, and don't tell me what they are.

*Rafael:* OK, I've thought of two numbers.

*Michele:* Now tell me the sum of the two numbers and the difference of the two numbers, and I'll tell you what your two numbers are.

*Rafael:* Their sum is 14 and their difference is 6.

*Michele:* Your numbers are 10 and 4.

*Rafael:* That's right! How did you do that?

*Michele:* I set up a system of equations using the sum and difference that you gave me, along with the variables $x$ and $y$.

**1.** Each group member should set up and solve the system of equations described by Michele.
Discuss your results and be sure that you both arrive at the solutions 10 and 4.

**2.** Now each of you will think of two new numbers and the other will try to find the numbers by solving a system of equations. But this time the system will be a bit trickier! Give each other the following information about your two numbers, and then figure out each other's numbers:

**five more than twice the sum of the numbers**
**four less than three times the difference of the numbers**

**3.** Would the systems in this activity work if negative numbers were used? Try it and see!

# Chapter 4 Review

## Section 4.1 Systems of Linear Equations in Two Variables

| KEY CONCEPTS | KEY TERMS |
|---|---|
| • **Recognizing Solutions to Systems of Two Linear Equations with Two Unknowns**<br>• If the lines in a system of two linear equations containing two unknowns intersect, then the point of intersection is the solution and the system is consistent and independent.<br>• If the lines in a system of two linear equations containing two unknowns are parallel, then the system has no solution and the system is inconsistent.<br>• If the lines in a system of two linear equations containing two unknowns lie on top of each other, then the system has infinitely many solutions. The solution set is the set of all points on the line and the system is consistent, but dependent. | System of linear equations<br>Solution<br>Consistent and independent<br>Inconsistent<br>Consistent and dependent |

| YOU SHOULD BE ABLE TO . . . | EXAMPLE | REVIEW EXERCISES |
|---|---|---|
| (1) Determine whether an ordered pair is a solution to a system of linear equations (p. 257) | Example 2 | 1–4 |
| (2) Solve a system of two linear equations containing two unknowns by graphing (p. 259) | Example 3 | 5–10 |
| (3) Solve a system of two linear equations containing two unknowns by substitution (p. 259) | Examples 4 and 5 | 11–14, 19–24 |
| (4) Solve a system of two linear equations containing two unknowns by elimination (p. 262) | Examples 6 and 7 | 15–18, 19–24 |
| (5) Identify inconsistent systems (p. 264) | Example 8 | 9, 23 |
| (6) Express the solution of a system of dependent equations (p. 265) | Example 9 | 12, 17 |

*In Problems 1–4, determine whether the given values of the variables listed are solutions to the system of equations.*

**1.** $\begin{cases} x + 3y = -2 \\ 2x - y = 10 \end{cases}$

**(a)** $x = 3, y = -1$
**(b)** $x = 4, y = -2$

**2.** $\begin{cases} -2x + 5y = 4 \\ 4x - 5y = -3 \end{cases}$

**(a)** $x = \frac{1}{2}, y = 1$
**(b)** $x = -1, y = \frac{1}{3}$

**3.** $\begin{cases} 6x - 5y = -12 \\ x - y = -3 \end{cases}$

**(a)** $x = 3, y = 6$
**(b)** $x = 1, y = 4$

**4.** $\begin{cases} 3x - y = 9 \\ 8x + 3y = 7 \end{cases}$

**(a)** $x = 3, y = 0$
**(b)** $x = 2, y = -3$

*In Problems 5 and 6, use the graph of the system to determine the solution.*

**5.** $\begin{cases} x + 4y = 12 \\ 2x - y = 6 \end{cases}$

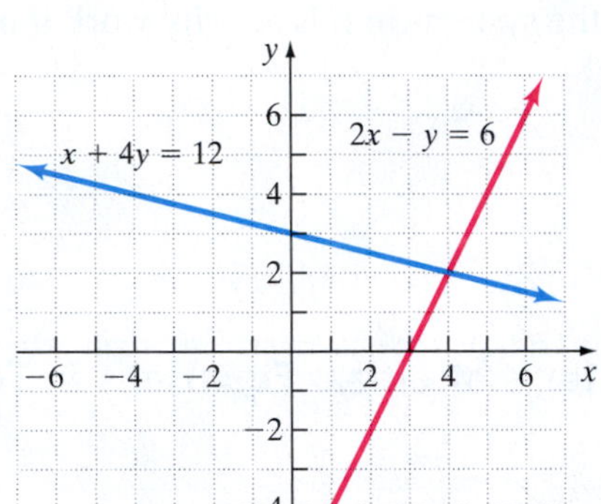


**6.** $\begin{cases} 3x + 2y = 4 \\ x - 2y = 4 \end{cases}$

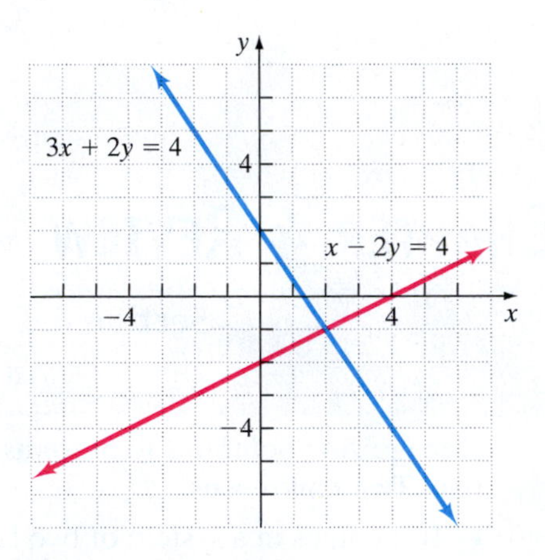


*In Problems 7–10, solve each system of linear equations by graphing.*

**7.** $\begin{cases} y = -3x + 1 \\ y = \frac{1}{2}x - 6 \end{cases}$

**8.** $\begin{cases} -2x + 3y = -9 \\ 3x + y = 8 \end{cases}$

**9.** $\begin{cases} y = -\frac{1}{3}x + 2 \\ x + 3y = -9 \end{cases}$

**10.** $\begin{cases} 2x - 3y = 0 \\ 2x - y = -4 \end{cases}$

*In Problems 11–14, solve each system of linear equation using substitution.*

**11.** $\begin{cases} y = -\frac{1}{4}x + 2 \\ y = 4x - 32 \end{cases}$

**12.** $\begin{cases} y = -\frac{3}{4}x + 2 \\ 3x + 4y = 8 \end{cases}$

**13.** $\begin{cases} y = 3x - 9 \\ 4x + 3y = -1 \end{cases}$

**14.** $\begin{cases} x - 2y = 7 \\ 3x - y = -4 \end{cases}$

*In Problems 15–18, solve each system of linear equation using elimination.*

**15.** $\begin{cases} 2x - y = 9 \\ 3x + y = 11 \end{cases}$

**16.** $\begin{cases} -x + 3y = 4 \\ 3x - 4y = -2 \end{cases}$

**17.** $\begin{cases} 2x - 4y = 8 \\ -3x + 6y = -12 \end{cases}$

**18.** $\begin{cases} 3x - 4y = -11 \\ 2x - 3y = -7 \end{cases}$

*In Problems 19–24, solve each system of linear equation using either substitution or elimination.*

**19.** $\begin{cases} x + y = -4 \\ 2x - 3y = 12 \end{cases}$ **20.** $\begin{cases} 5x - 3y = 2 \\ x + 2y = -10 \end{cases}$ **21.** $\begin{cases} 3x - 2y = 5 \\ 4x - 5y = 9 \end{cases}$

**22.** $\begin{cases} 12x + 20y = 21 \\ 3x - 2y = 0 \end{cases}$ **23.** $\begin{cases} 6x + 9y = -3 \\ 8x + 12y = 7 \end{cases}$ **24.** $\begin{cases} 6x + 11y = 2 \\ 5x + 8y = -3 \end{cases}$

## Section 4.2 Problem Solving: Systems of Two Linear Equations Containing Two Unknowns

**KEY TERM**

Break-even point

| YOU SHOULD BE ABLE TO . . . | EXAMPLE | REVIEW EXERCISES |
|---|---|---|
| 1 Model and solve direct translation problems involving two linear equations containing two unknowns (p. 271) | Example 1 | 25–27 |
| 2 Model and solve geometry problems involving two linear equations containing two unknowns (p. 272) | Examples 2 and 3 | 28–30 |
| 3 Model and solve mixture problems involving two linear equations containing two unknowns (p. 274) | Examples 4 and 5 | 31–33 |
| 4 Model and solve uniform motion problems involving two linear equations containing two unknowns (p. 277) | Example 6 | 34–36 |
| 5 Find the intersection of two linear functions (p. 278) | Example 7 | 37–38 |

**25. Numbers** The sum of two numbers is 56. The difference of the two numbers is 14. Find the two numbers.

**26. Phi Theta Kappa** The local chapter of the Phi Theta Kappa International Honor Society for Two-Year Colleges will induct 73 new members this semester. There are 11 more female inductees than male inductees. How many males and how many females will be inducted?

**27. Counting Calories** Shawn and Randy ate lunch at a Pizza Hut® buffet. Shawn ate 3 slices of pepperoni pizza and 5 slices of Italian sausage pizza for a total of 2600 calories. Randy ate 4 slices of pepperoni pizza and 2 slices of Italian sausage pizza for a total of 1880 calories. Assuming the pizza slices are exactly the same size, how many calories are in a slice of pepperoni pizza? How many calories are in a slice of Italian sausage pizza? (*Source:* Pizza Hut, Inc., August 2003)

△ **28. Rectangle** The length of a rectangle is 5 inches less than twice the width. The perimeter of the rectangle is 68 inches. Find the dimensions of the rectangle.

△ **29. Angles** In the right triangle shown in the figure to the right, the measure of angle $y$ is twice the measure of angle $x$. Find the measures of angles $x$ and $y$.

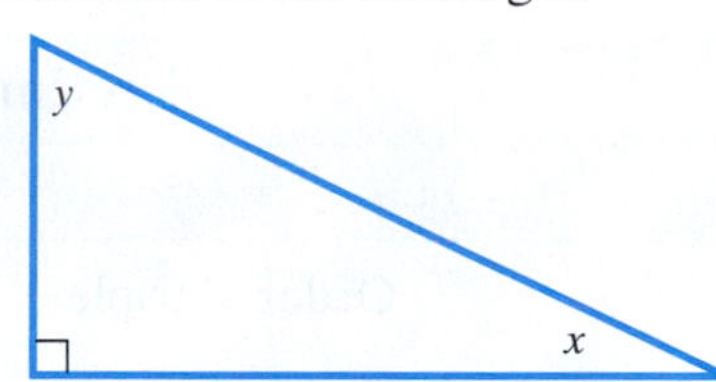


△ **30. Angles** Two intersecting lines are shown in the figure to the right. Suppose that the measure of angle 1 is $(x + 4y)°$, the measure of angle 2 is $(8x + 5y)°$, and the measure of angle 3 is $(2x + y)°$. Find $x$ and $y$.

**31. Coins** Jerome's piggy bank contains a total of 40 nickels and quarters. The value of the coins in the bank is $4.40. How many nickels and how many quarters are in the bank?

**32. Hydrochloric Acid Solution** For an experiment, a chemist needs 12 liters of a solution that is 30% hydrochloric acid. However, she only has two solutions that are 25% hydrochloric acid and 40% hydrochloric acid. How many liters of each should she mix in order to obtain the needed solution?

**33. Investments** Verna invested part of a $10,000 inheritance in stocks paying 6.5% simple interest annually. She invested the rest in bonds paying 4.25% simple interest annually. If Verna will receive $582.50 in interest after one year, how much did she invest in stocks? How much did she invest in bonds?

**34. Airplane Speed** An airplane averaged 160 miles per hour with the wind and 112 miles per hour against the wind. Determine the speed of the plane and the speed of the wind.

**35. Catching Up** A Pontiac Grand Am enters the Will Rodgers Turnpike traveling at 50 miles per hour. One-half hour later, a Ford Mustang enters the turnpike at the same location and travels in the same direction at 80 miles per hour. How long will it take for the Mustang to catch up to the Grand Am? When it catches up, how far will the two cars have traveled on the turnpike?

**36. Boat Speed** A boat traveled 30 miles upstream in 1.5 hours. The return trip downstream took only 1 hour. Assuming the boat travels at a constant rate and that the river's current is constant, find the speed of the boat in still water and the speed of the current.

**37. High School Sport Participation** The number of male students (in millions) who participate in high school sports can be estimated by the function $M(t) = 0.056t + 3.354$, where $t$ is the number of years since 1990. The number of female students (in millions) who participate in high school sports can be estimated by $F(t) = 0.103t + 1.842$. (*Source:* National Federation of State High School Associations)

**(a)** Graph each of the two functions on the same Cartesian plane.
**(b)** Assuming these linear trends continue, find the year in which participation by male and female students will be the same.

**38. Break-Even** Karen bakes pies for a living. She sells pies for $15 each. Her monthly fixed costs are $1200. Each pie costs $2.50 in materials.

**(a)** Find the revenue function $R$ treating the number of pies $x$ as the independent variable.
**(b)** Find the cost function $C$ treating the number of pies $x$ as the independent variable.
**(c)** Graph the revenue and cost functions on the same Cartesian plane.
**(d)** Find the break-even number of pies that Karen must make and sell each month. What is the revenue and cost at this number of pies? Label the point on the graph in part (c).

## Section 4.3 Systems of Linear Equations in Three Variables

**KEY TERMS**

Ordered triple | Exactly one solution | No solution | Infinitely many solutions

| YOU SHOULD BE ABLE TO . . . | EXAMPLE | REVIEW EXERCISES |
|---|---|---|
| 1 Solve systems of three linear equations containing three variables (p. 285) | Examples 1 through 3 | 39–46 |
| 2 Identify inconsistent systems (p. 290) | Example 4 | 42 |
| 3 Express the solution of a system of dependent equations (p. 291) | Example 5 | 43 |
| 4 Model and solve problems involving three linear equations containing three unknowns (p. 292) | Example 6 | 47, 48 |

*In Problems 39–46, solve each system of three linear equations containing three unknowns.*

**39.** $\begin{cases} x + y - z = 1 \\ x - y + z = 7 \\ x + 2y + z = 1 \end{cases}$

**40.** $\begin{cases} 2x - 2y + z = -10 \\ 3x + y - 2z = 4 \\ 5x + 2y - 3z = 7 \end{cases}$

**41.** $\begin{cases} x + 2y = -1 \\ 3y + 4z = 7 \\ 2x - z = 6 \end{cases}$

**42.** $\begin{cases} x + 2y - 3z = -4 \\ x + y + 3z = 5 \\ 3x + 4y + 3z = 7 \end{cases}$

**43.** $\begin{cases} 3x + y - 2z = 6 \\ x + y - z = -2 \\ -x - 3y + 2z = 14 \end{cases}$

**44.** $\begin{cases} 9x - y + 2z = -5 \\ -3x - 4y + 4z = 3 \\ 15x + 3y - 2z = -10 \end{cases}$

**45.** $\begin{cases} 4x - 5y + 2z = -8 \\ 3x + 7y - 3z = 21 \\ 7x - 4y + 2z = -5 \end{cases}$

**46.** $\begin{cases} 3x - 2y + 5z = -7 \\ 4x + y + 3z = -2 \\ 2x - 3y + 7z = -4 \end{cases}$

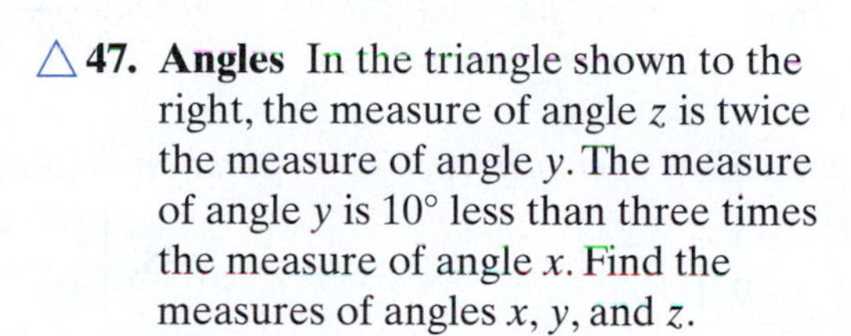

△ **47. Angles** In the triangle shown to the right, the measure of angle $z$ is twice the measure of angle $y$. The measure of angle $y$ is 10° less than three times the measure of angle $x$. Find the measures of angles $x$, $y$, and $z$.

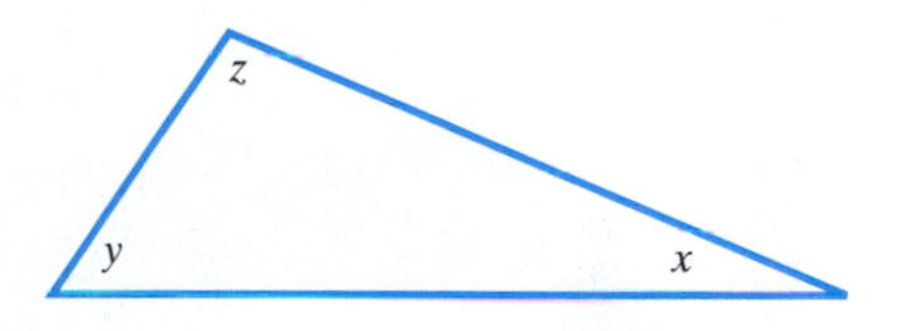


**48. Counting Calories** Mike, Clint, and Charlie decide to stop by Burger King for a snack. Mike consumed 1040 calories by eating a cheeseburger, a medium order of fries, and a medium Coke. Clint consumed 1400 calories by eating two cheeseburgers, a medium order of fries, and a medium Coke. Charlie consumed 1680 calories by eating two cheeseburgers, a medium order of fries, and two medium Cokes. Find the number of calories each in a Burger King cheeseburger, in a medium order of fries, and in a medium Coke. (*Source:* Burger King Corporation, January 1999)

## Section 4.4 Using Matrices to Solve Systems

| KEY CONCEPTS | KEY TERMS |
|---|---|
| • **Augmented Matrix** $\begin{cases} a_1x + b_1y = c_1 \\ a_2x + b_2y = c_2 \end{cases}$ is written as an augmented matrix as $\left[\begin{array}{cc\|c} a_1 & b_1 & c_1 \\ a_2 & b_2 & c_2 \end{array}\right]$ <br> • **Row Operations** <br> **1.** Interchange any two rows. <br> **2.** Replace a row by a nonzero multiple of that row. <br> **3.** Replace a row by the sum of that row and a nonzero multiple of some other row. | Matrix <br> Augmented matrix <br> Row echelon form |

| YOU SHOULD BE ABLE TO . . . | EXAMPLE | REVIEW EXERCISES |
|---|---|---|
| ① Write the augmented matrix of a system of linear equations (p. 298) | Example 1 | 49–52 |
| ② Write the system from the augmented matrix (p. 299) | Example 2 | 53–56 |
| ③ Perform row operations on a matrix (p. 300) | Examples 3 and 4 | 57–60 |
| ④ Solve systems of linear equations using matrices (p. 301) | Examples 5 and 6 | 61–70 |
| ⑤ Solve dependent and inconsistent systems (p. 305) | Examples 7 and 8 | 64, 65, 68, 69 |

*In Problems 49–52, write the augmented matrix of the given system of linear equations.*

**49.** $\begin{cases} 3x + y = 7 \\ 2x + 5y = 9 \end{cases}$

**50.** $\begin{cases} x - 5y = 14 \\ -x + y = -3 \end{cases}$

**51.** $\begin{cases} 5x - y + 4z = 6 \\ -3x - 3z = -1 \\ x - 2y = 0 \end{cases}$

**52.** $\begin{cases} 8x - y + 3z = 14 \\ -3x + 5y - 6z = -18 \\ 7x - 4y + 5z = 21 \end{cases}$

*In Problems 53–56, write the system of linear equations corresponding to the given augmented matrix.*

**53.** $\left[\begin{array}{cc|c} 1 & 2 & 12 \\ 0 & 3 & 15 \end{array}\right]$

**54.** $\left[\begin{array}{cc|c} 3 & -4 & -5 \\ -1 & 2 & 7 \end{array}\right]$

**55.** $\left[\begin{array}{ccc|c} 1 & 3 & 4 & 20 \\ 0 & 1 & -2 & -16 \\ 0 & 0 & 1 & 7 \end{array}\right]$

**56.** $\left[\begin{array}{ccc|c} -3 & 7 & 9 & 1 \\ 4 & 10 & 7 & 5 \\ 2 & -5 & -6 & -8 \end{array}\right]$

*In Problems 57–60, perform each row operation on the given augmented matrix.*

**57.** $\left[\begin{array}{cc|c} 1 & -5 & 22 \\ -2 & 9 & -40 \end{array}\right]$

**(a)** $R_2 = 2r_1 + r_2$ followed by
**(b)** $R_2 = -r_2$

**58.** $\left[\begin{array}{cc|c} 1 & -4 & 7 \\ 3 & -7 & 6 \end{array}\right]$

**(a)** $R_2 = -3r_1 + r_2$ followed by
**(b)** $R_2 = \frac{1}{5}r_2$

**59.** $\left[\begin{array}{ccc|c} -1 & 2 & 1 & 1 \\ 2 & -1 & 3 & -3 \\ -1 & 5 & 6 & 2 \end{array}\right]$

**(a)** $R_2 = 2r_1 + r_2$ followed by
**(b)** $R_3 = -r_1 + r_3$

**60.** $\left[\begin{array}{ccc|c} 1 & 3 & 4 & 4 \\ 0 & 5 & 10 & -15 \\ 0 & -4 & -7 & 7 \end{array}\right]$

**(a)** $R_2 = \frac{1}{5}r_2$ followed by
**(b)** $R_3 = 4r_2 + r_3$

*In Problems 61–70, solve each system of linear equations using matrices. If the system has no solution, say it is inconsistent.*

**61.** $\begin{cases} x + 2y = 1 \\ x + 3y = -2 \end{cases}$

**62.** $\begin{cases} 6x - 2y = -7 \\ 4x + 3y = 17 \end{cases}$

**63.** $\begin{cases} 3x + 2y = -10 \\ 2x - y = -9 \end{cases}$

**64.** $\begin{cases} -3x + 9y = 15 \\ 5x - 15y = 11 \end{cases}$

**65.** $\begin{cases} 4x - 2y = 6 \\ 6x - 3y = 9 \end{cases}$

**66.** $\begin{cases} x + 4y = 4 \\ 4x - 8y = 7 \end{cases}$

**67.** $\begin{cases} x + 2y - 2z = -11 \\ 2x + z = -6 \\ 5y - 3z = -7 \end{cases}$

**68.** $\begin{cases} 2x - 5y + 2z = 9 \\ -x + y - 2z = -2 \\ -x - 2y - 4z = 8 \end{cases}$

**69.** $\begin{cases} 2x - 7y + 11z = -5 \\ 4x - 2y + 6z = 2 \\ -2x + 19y - 27z = 17 \end{cases}$

**70.** $\begin{cases} 5x + 3y + 7z = 9 \\ 3x + 5y + 4z = 8 \\ x + 3y + 3z = 9 \end{cases}$

## Section 4.5 Determinants and Cramer's Rule

### KEY CONCEPTS

- **Evaluate a 2 × 2 Determinant**

$$\begin{vmatrix} a & b \\ c & d \end{vmatrix} = ad - bc$$

- **Evaluate a 3 × 3 Determinant**

$$\begin{vmatrix} a_{1,1} & a_{1,2} & a_{1,3} \\ a_{2,1} & a_{2,2} & a_{2,3} \\ a_{3,1} & a_{3,2} & a_{3,3} \end{vmatrix} = a_{1,1}\begin{vmatrix} a_{2,2} & a_{2,3} \\ a_{3,2} & a_{3,3} \end{vmatrix} - a_{1,2}\begin{vmatrix} a_{2,1} & a_{2,3} \\ a_{3,1} & a_{3,3} \end{vmatrix} + a_{1,3}\begin{vmatrix} a_{2,1} & a_{2,2} \\ a_{3,1} & a_{3,2} \end{vmatrix}$$

- **Cramer's Rule:**

The solution to the system of equations $\begin{cases} ax + by = s \\ cx + dy = t \end{cases}$

is given by $x = \dfrac{\begin{vmatrix} s & b \\ t & d \end{vmatrix}}{\begin{vmatrix} a & b \\ c & d \end{vmatrix}} = \dfrac{D_x}{D}$, $y = \dfrac{\begin{vmatrix} a & s \\ c & t \end{vmatrix}}{\begin{vmatrix} a & b \\ c & d \end{vmatrix}} = \dfrac{D_y}{D}$, with $D \neq 0$

The solution to the system of equations $\begin{cases} a_1x + b_1y + c_1z = d_1 \\ a_2x + b_2y + c_2z = d_2 \\ a_3x + b_3y + c_3z = d_3 \end{cases}$

with $D = \begin{vmatrix} a_1 & b_1 & c_1 \\ a_2 & b_2 & c_2 \\ a_3 & b_3 & c_3 \end{vmatrix} \neq 0$

$$D_x = \begin{vmatrix} d_1 & b_1 & c_1 \\ d_2 & b_2 & c_2 \\ d_3 & b_3 & c_3 \end{vmatrix} \quad D_y = \begin{vmatrix} a_1 & d_1 & c_1 \\ a_2 & d_2 & c_2 \\ a_3 & d_3 & c_3 \end{vmatrix} \quad D_z = \begin{vmatrix} a_1 & b_1 & d_1 \\ a_2 & b_2 & d_2 \\ a_3 & b_3 & d_3 \end{vmatrix}$$

is given by

$$x = \frac{D_x}{D} \quad y = \frac{D_y}{D} \quad z = \frac{D_z}{D}$$

### KEY TERMS

Square matrix
2 × 2 determinant
3 × 3 determinant
Minors

| YOU SHOULD BE ABLE TO . . . | EXAMPLE | REVIEW EXERCISES |
|---|---|---|
| 1 Evaluate the determinant of a 2 × 2 matrix (p. 310) | Example 1 | 71–74 |
| 2 Use Cramer's Rule to solve a system of two equations containing two variables (p. 311) | Example 2 | 79–84 |
| 3 Evaluate the determinant of a 3 × 3 matrix (p. 312) | Examples 3 and 4 | 75–78 |
| 4 Use Cramer's Rule to solve a system of three equations containing three variables (p. 314) | Example 5 | 85–90 |

*In Problems 71–78, find the value of each determinant.*

**71.** $\begin{vmatrix} 3 & 4 \\ -1 & 2 \end{vmatrix}$ **72.** $\begin{vmatrix} 2 & -3 \\ -6 & 9 \end{vmatrix}$ **73.** $\begin{vmatrix} -5 & 7 \\ -4 & 6 \end{vmatrix}$ **74.** $\begin{vmatrix} -7 & 2 \\ 6 & -3 \end{vmatrix}$

**75.** $\begin{vmatrix} 5 & 0 & 1 \\ 2 & -3 & -1 \\ 3 & 6 & -4 \end{vmatrix}$ **76.** $\begin{vmatrix} 1 & -4 & 5 \\ 0 & 1 & -3 \\ 2 & -6 & 4 \end{vmatrix}$ **77.** $\begin{vmatrix} 3 & 0 & -1 \\ 2 & 6 & 7 \\ 2 & 5 & 4 \end{vmatrix}$ **78.** $\begin{vmatrix} 2 & 3 & 1 \\ 1 & -3 & -7 \\ -5 & 4 & 8 \end{vmatrix}$

*In Problems 79–90, solve each system of linear equations using Cramer's Rule, if applicable.*

**79.** $\begin{cases} x + 2y = -1 \\ 2x + 3y = 1 \end{cases}$

**80.** $\begin{cases} 4x + 9y = -13 \\ -5x + 6y = 22 \end{cases}$

**81.** $\begin{cases} 4x - y = -6 \\ 3x + 5y = 7 \end{cases}$

**82.** $\begin{cases} x - y = 2 \\ 2x + y = 5 \end{cases}$

**83.** $\begin{cases} 6x + 2y = 5 \\ 15x + 5y = 8 \end{cases}$

**84.** $\begin{cases} 12x + y = 6 \\ -6x + 7y = 15 \end{cases}$

**85.** $\begin{cases} x - y + 2z = 9 \\ 3x + 2y - 4z = 7 \\ 3y + 5z = -1 \end{cases}$

**86.** $\begin{cases} x - 3y - 3z = -5 \\ 7x + y - 2z = 24 \\ 6x - 5y - 4z = -9 \end{cases}$

**87.** $\begin{cases} x + y + z = 1 \\ 3x + 2y + 4z = -1 \\ 2x + 2y + 3z = 0 \end{cases}$

**88.** $\begin{cases} 4x - 3y + z = -6 \\ 4x - 2y + 3z = -3 \\ 8x - 5y - 2z = -12 \end{cases}$

**89.** $\begin{cases} x - y - 4z = 7 \\ 4x - 3y - 3z = 4 \\ 3x - 2y + z = -3 \end{cases}$

**90.** $\begin{cases} x + y = -3 \\ 2y - z = -1 \\ 5x + z = 1 \end{cases}$

## Section 4.6 Systems of Linear Inequalities

**KEY TERMS**

| | |
|---|---|
| Satisfies | Bounded |
| Corner points | Unbounded |

| YOU SHOULD BE ABLE TO . . . | EXAMPLE | REVIEW EXERCISES |
|---|---|---|
| 1 Determine whether an ordered pair is a solution to a system of linear inequalities (p. 320) | Example 1 | 91–94 |
| 2 Graph a system of linear inequalities (p. 321) | Examples 2 through 5 | 95–102 |
| 3 Solve problems involving a system of linear inequalities (p. 323) | Example 6 | 103, 104 |

*In Problems 91–94, determine which of the points, if any, satisfies the system of linear inequalities.*

**91.** $\begin{cases} x - y > 2 \\ 2x + 3y > 8 \end{cases}$

**(a)** $(5, 2)$
**(b)** $(-3, 5)$

**92.** $\begin{cases} 7x + 3y \le 21 \\ -2x + y \ge 5 \end{cases}$

**(a)** $(-1, 2)$
**(b)** $(-3, 4)$

**93.** $\begin{cases} x + 2y \le 6 \\ 3x - y \ge 2 \\ x \ge 0 \\ y \ge 0 \end{cases}$

**(a)** $(2, 1)$
**(b)** $(1, 2)$

**94.** $\begin{cases} 3x - 2y \le 12 \\ 2x + y \le 15 \\ x \ge 0 \\ y \ge 0 \end{cases}$

**(a)** $(-2, 3)$
**(b)** $(4, 1)$

*In Problems 95–100, graph each system of linear inequalities.*

**95.** $\begin{cases} x + y \ge 7 \\ 2x - y \ge 5 \end{cases}$

**96.** $\begin{cases} 2x + 3y > 9 \\ x - 3y > -18 \end{cases}$

**97.** $\begin{cases} x - 4y > -4 \\ x + 2y \le 8 \end{cases}$

**98.** $\begin{cases} x - y < -2 \\ 2x - 2y > 6 \end{cases}$

**99.** $\begin{cases} 2x - y \ge -2 \\ 2x - 3y \le 6 \end{cases}$

**100.** $\begin{cases} x \ge 2 \\ y < -3 \end{cases}$

*In Problems 101 and 102, graph each system of linear inequalities. Tell whether the graph is bounded or unbounded, and label the corner points.*

**101.** $\begin{cases} 3x + 5y \le 30 \\ 4x - 5y \ge 5 \\ x \le 8 \\ y \ge 0 \end{cases}$

**102.** $\begin{cases} 3x + 2y \ge 10 \\ x + 2y \ge 6 \\ x \ge 0 \\ y \ge 0 \end{cases}$

**103. Investments Options** Anna has a maximum of \$4000 to invest. One investment option pays 6% simple interest annually, while a second investment option pays 8% simple interest annually. Anna's financial advisor recommends that she invest at least \$500 at 6% and at least \$2500 at 8%. Anna wants to earn at least \$275 annually in interest.

**(a)** Let $x$ denote the amount invested at 6% and $y$ denote the amount invested at 8%. Write a system of linear inequalities that represents Anna's possible combinations of investments.

**(b)** Graph the system and label the corner points.

**104. Purchasing Tulip Bulbs** Jordan has a maximum of \$144 to purchase tulip bulbs for her yard. Red tulip bulbs cost \$6 per dozen, and yellow tulip bulbs cost \$4 per dozen. Jordan wants at least 4 more dozens of red tulip bulbs than yellow.

**(a)** Let $x$ denote the number of dozens of red tulip bulbs and $y$ denote the number of dozens of yellow tulip bulbs. Write a system of linear inequalities that represents Jordan's possible combinations of purchases.

**(b)** Graph the system and label the corner points.

## CHAPTER 4 TEST

*Remember to use your Chapter Test Prep Video CD to see fully worked-out solutions to any of these problems you would like to review.*

**1.** Solve the system of linear equations by graphing.

$$\begin{cases} 2x - y = 0 \\ 4x - 5y = 12 \end{cases}$$

*In Problems 2–5, solve the system of linear equations using either substitution or elimination.*

**2.** $\begin{cases} 5x + 2y = -3 \\ y = 2x - 6 \end{cases}$

**3.** $\begin{cases} 9x + 3y = 1 \\ x - 2y = 4 \end{cases}$

**4.** $\begin{cases} 6x - 9y = 5 \\ 8x - 12y = 7 \end{cases}$

**5.** $\begin{cases} 2x + y = -4 \\ \frac{1}{3}x + \frac{1}{2}y = 2 \end{cases}$

*In Problems 6 and 7, solve each system of three linear equations containing three unknowns.*

**6.** $\begin{cases} x - 2y + 3z = 1 \\ x + y - 3z = 7 \\ 3x - 4y + 5z = 7 \end{cases}$

**7.** $\begin{cases} 2x + 4y + 3z = 5 \\ 3x - y + 2z = 8 \\ x + y + 2z = 0 \end{cases}$

*In Problems 8 and 9, perform each row operation on the given augmented matrix.*

**8.** $\left[\begin{array}{cc|c} 1 & -3 & -2 \\ 2 & -4 & 8 \end{array}\right]$

**(a)** $R_2 = -2r_1 + r_2$ followed by

**(b)** $R_2 = \frac{1}{2}r_2$

**9.** $\left[\begin{array}{ccc|c} 1 & -2 & 1 & -2 \\ 3 & -5 & 2 & 1 \\ 0 & -4 & 5 & -32 \end{array}\right]$

**(a)** $R_2 = -3r_1 + r_2$ followed by

**(b)** $R_3 = 4r_2 + r_3$

*In Problems 10 and 11, write the augmented matrix for each system of linear equations. Then, use the matrix to solve the system.*

**10.** $\begin{cases} x - 5y = 2 \\ 2x + y = 4 \end{cases}$

**11.** $\begin{cases} x + 2y + z = 3 \\ 4y + 3z = 5 \\ 2x + 3y = 1 \end{cases}$

*In Problems 12 and 13, find the value of each determinant.*

**12.** $\begin{vmatrix} 3 & -5 \\ 4 & -8 \end{vmatrix}$

**13.** $\begin{vmatrix} 0 & 1 & 2 \\ 3 & 3 & -1 \\ -2 & 1 & 2 \end{vmatrix}$

*In Problems 14 and 15, solve each system of linear equations using Cramer's Rule, if applicable.*

**14.** $\begin{cases} x - y = -2 \\ 5x + 3y = -8 \end{cases}$

**15.** $\begin{cases} x + y + z = -2 \\ x + y - 2z = 1 \\ 4x + 2y + 3z = -15 \end{cases}$

*In Problems 16 and 17, graph each system of linear inequalities. Label the corner points.*

**16.** $\begin{cases} 2x - y > -2 \\ x - 3y < 9 \end{cases}$

**17.** $\begin{cases} 3x + 2y \leq 12 \\ x - 2y \geq -4 \\ x \geq 0 \\ y \geq 0 \end{cases}$

**18. Grain Storage** A grain-storage warehouse has a total of 50 bins. Some of the bins hold 25 tons of grain each, and the rest hold 20 tons each. How many of each type of bin are there if the capacity of the warehouse is 1160 tons?

△ **19. Triangle** In the triangle shown below, the measure of angle $z$ is 10° larger than the measure of angle $y$. Five times the measure of angle $x$ equals the sum of the measures of angles $y$ and $z$. Find the measures of angles $x$, $y$, and $z$.

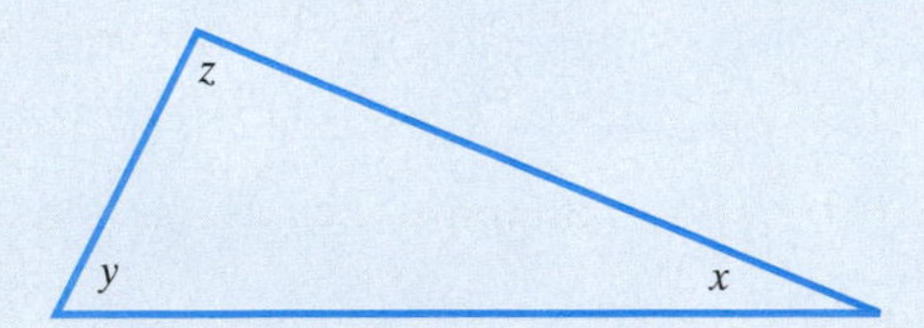


**20. Shopping** It's Margaret's lucky day. While shopping, she happened onto a sale where designer blouses were being sold for \$12 each and designer sweaters were being sold for \$18 each. Unfortunately, Margaret has no more than \$180 to spend. She also decides that she can buy at most 13 items.

**(a)** Let $x$ denote the number of blouses and $y$ denote the number of sweaters that Margaret will buy. Write a system of linear inequalities that represents the possible combinations of purchases.

**(b)** Graph the system and label the corner points.

# Getting Ready for Chapter 5: Polynomials and Polynomial Functions

## OBJECTIVES

1. Simplify Exponential Expressions Using the Product Rule
2. Simplify Exponential Expressions Using the Quotient Rule
3. Evaluate Exponential Expressions with a Zero or Negative Exponent
4. Simplify Exponential Expressions Using the Power Rule
5. Simplify Exponential Expressions Containing Products or Quotients
6. Simplify Exponential Expressions Using the Laws of Exponents
7. Convert Between Scientific Notation and Decimal Notation
8. Use Scientific Notation to Multiply and Divide

A quick review for dealing with positive integer exponents is in order. Recall that if $a$ is a real number and $n$ is a positive integer, then the symbol $a^n$ means that we should multiply $a$ by itself $n$ times. That is,

$$a^n = \underbrace{a \cdot a \cdot \cdots \cdot a}_{n \text{ factors}}$$

For example,

$$4^3 = \underbrace{4 \cdot 4 \cdot 4}_{3 \text{ factors}}$$

In the notation $a^n$, we call $a$ the **base** and $n$ the **power** or **exponent.**

### Work Smart

We read $a^n$ as "$a$ raised to the power $n$" or "$a$ raised to the $n$th power." We usually read $a^2$ as "$a$ squared" and $a^3$ as "$a$ cubed."

## 1 Simplify Exponential Expressions Using the Product Rule

Several general rules can be discovered for simplifying expressions involving *positive* integer exponents. The first rule that we introduce is used when multiplying two exponential expressions that have the same base. Consider the following:

$$\underset{\uparrow \text{ Same base}}{x^2 \cdot x^4} = \underbrace{(x \cdot x)}_{2 \text{ factors}} \underbrace{(x \cdot x \cdot x \cdot x)}_{4 \text{ factors}} = \underbrace{x \cdot x \cdot x \cdot x \cdot x \cdot x}_{6 \text{ factors}} = x^6$$

Sum of powers 2 and 4 ($x^6$); Same base ($x$)

Based on the above, we have the following result:

### In Words

When multiplying two exponential expressions with the same base, add the exponents. Then write the common base to the power of this sum.

**PRODUCT RULE FOR EXPONENTS (POSITIVE INTEGER EXPONENTS)**

If $a$ is a real number and $m$ and $n$ are positive integers, then

$$a^m \cdot a^n = a^{m+n}$$

### EXAMPLE 1 Using the Product Rule to Simplify Expressions Involving Exponents

Simplify each expression. All answers should contain only positive integer exponents.

**(a)** $2^2 \cdot 2^3$ **(b)** $3z^2 \cdot 4z^4$

**Solution**

**(a)** $2^2 \cdot 2^3 = 2^{2+3}$
$= 2^5$
$= 32$

**(b)** $3z^2 \cdot 4z^4 = 3 \cdot 4 \cdot z^2 \cdot z^4$
$= 12z^{2+4}$
$= 12z^6$

**QUICK ✓** *Simplify each expression. All answers should contain only positive integer exponents.*

**1.** $5^2 \cdot 5$ **2.** $(-3)^2 \cdot (-3)^3$ **3.** $y^4 \cdot y^3$

**4.** $(5x^2) \cdot (-2x^5)$ **5.** $(6y^3) \cdot (-y^2)$

## 2 Simplify Exponential Expressions Using the Quotient Rule

To find a general rule for the quotient of two exponential expressions with *positive* integer exponents, we use the Reduction Property by dividing out common factors. Consider the following:

$$\frac{y^6}{y^2} = \frac{\overbrace{y \cdot y \cdot y \cdot y \cdot \cancel{y} \cdot \cancel{y}}^{6 \text{ factors}}}{\underbrace{\cancel{y} \cdot \cancel{y}}_{2 \text{ factors}}} = \underbrace{y \cdot y \cdot y \cdot y}_{4 \text{ factors}} = y^4$$

Difference of powers 6 and 2

Divide out common factors

We conclude from this result that

$$\frac{y^6}{y^2} = y^{6-2} = y^4$$

This result is true in general.

**In Words**

When dividing two exponential expressions with a common base, subtract the exponent in the denominator from the exponent in the numerator. Then write the common base to the power of this difference.

**QUOTIENT RULE FOR EXPONENTS (POSITIVE INTEGER EXPONENTS)**

If $a$ is a real number and if $m$ and $n$ are positive integers, then

$$\frac{a^m}{a^n} = a^{m-n} \quad \text{if } a \neq 0$$

**EXAMPLE 2 Using the Quotient Rule to Simplify Expressions Involving Exponents**

Simplify each expression. Answers should contain only positive integer exponents.

**(a)** $\frac{8^5}{8^3}$ **(b)** $\frac{27z^9}{12z^4}$

**Solution**

**(a)** $\frac{8^5}{8^3} = 8^{5-3}$

$= 8^2$

$= 64$

**(b)** $\frac{27z^9}{12z^4} = \frac{9 \cdot 3}{4 \cdot 3} z^{9-4}$

$= \frac{9}{4} z^5$

**QUICK ✓** *Simplify each expression. All answers should contain only positive integer exponents.*

**6.** $\frac{5^6}{5^4}$ **7.** $\frac{y^8}{y^6}$ **8.** $\frac{16a^6}{10a^5}$ **9.** $\frac{-24b^5}{16b^3}$

## 3 Evaluate Exponential Expressions with a Zero or Negative Exponent

To this point, we only dealt with exponential expressions with positive integer exponents. We now wish to extend the definition of exponential expressions to *all* integer exponents. That is, we wish to evaluate exponential expressions where the exponent can be a positive integer, zero, or a negative integer. We begin with raising a real number to the 0 power.

**DEFINITION: ZERO-EXPONENT RULE**

If $a$ is a nonzero real number (that is, if $a \neq 0$), we define

$$a^0 = 1 \qquad \text{if } a \neq 0$$

The reason that $a^0 = 1$ is easy to see based upon the Product Rule and the Identity Property of Multiplication. From the Product Rule for Exponents we have that

$$\begin{aligned} a^0 a^n &= a^{0+n} \\ &= a^n \end{aligned}$$

From the Identity Property of Multiplication, it must be that $a^0 = 1$.

Suppose that we wanted to simplify $\frac{z^3}{z^5}$. If we use the Quotient Rule for Exponents, we obtain

$$\frac{z^3}{z^5} = z^{3-5} = z^{-2}$$

We could also simplify this expression directly using the Reduction Property.

$$\frac{z^3}{z^5} = \frac{\cancel{z} \cdot \cancel{z} \cdot \cancel{z}}{\cancel{z} \cdot \cancel{z} \cdot \cancel{z} \cdot z \cdot z} = \frac{1}{z^2}$$

This implies that $z^{-2} = \frac{1}{z^2}$. Based upon this result, we define $a$ raised to a negative power as follows:

**DEFINITION: NEGATIVE-EXPONENT RULE**

If $n$ is a positive integer and if $a$ is a nonzero real number (that is, if $a \neq 0$), then we define

$$a^{-n} = \frac{1}{a^n} \quad \text{or} \quad \frac{1}{a^{-n}} = a^n \qquad \text{if } a \neq 0$$

## EXAMPLE 3 Evaluating Exponential Expressions Containing Integer Exponents

Simplify each expression. All exponents should be positive integers.

**(a)** $3^{-4}$ **(b)** $4x^{-5}$ **(c)** $5x^0$ **(d)** $\frac{1}{3^{-2}}$

### Solution

**(a)** $3^{-4} = \frac{1}{3^4} = \frac{1}{81}$ **(b)** $4x^{-5} = \frac{4}{x^5}$ **(c)** $5x^0 = 5 \cdot 1 = 5$ **(d)** $\frac{1}{3^{-2}} = 3^2 = 9$

**QUICK ✓** *Simplify each expression. All exponents should be positive integers.*

**10.** $5^{-3}$ **11.** $5z^{-7}$ **12.** $\frac{1}{x^{-4}}$ **13.** $\frac{5}{y^{-3}}$ **14.** $-4^0$ **15.** $(-10)^0$

## EXAMPLE 4 Evaluating Exponential Expressions Containing Integer Exponents

Simplify each expression. All exponents should be positive integers.

**(a)** $\left(\frac{2}{3}\right)^{-3}$ **(b)** $\left(\frac{1}{7}\right)^{-2}$

**Solution**

**(a)** $\left(\frac{2}{3}\right)^{-3} = \frac{1}{\left(\frac{2}{3}\right)^3} = \frac{1}{\frac{2}{3}\cdot\frac{2}{3}\cdot\frac{2}{3}} = \frac{1}{\frac{8}{27}} = \frac{27}{8}$

**(b)** $\left(\frac{1}{7}\right)^{-2} = \frac{1}{\left(\frac{1}{7}\right)^2} = \frac{1}{\frac{1}{7}\cdot\frac{1}{7}} = \frac{1}{\frac{1}{49}} = 49$

The following shortcut is based upon the results of Example 4:

**In Words**

To evaluate $\left(\frac{a}{b}\right)^{-n}$, determine the reciprocal of the base and then raise it to the $n$th power.

If $a$ and $b$ are real numbers and $n$ is an integer, then

$$\left(\frac{a}{b}\right)^{-n} = \left(\frac{b}{a}\right)^{n} \quad \text{if } a \neq 0, b \neq 0$$

**QUICK ✓** *Simplify each expression. All exponents should be positive integers.*

**16.** $\left(\frac{4}{3}\right)^{-2}$ **17.** $\left(-\frac{1}{4}\right)^{-3}$ **18.** $\left(\frac{3}{x}\right)^{-2}$ **19.** $\frac{5}{2^{-2}}$

Now that we have definitions for 0 as an exponent and negative exponents, we restate the Product Rule and Quotient Rule for Exponents assuming that the exponent is any integer (positive, negative, or zero).

**PRODUCT RULE FOR EXPONENTS**

If $a$ is a real number and $m$ and $n$ are integers, then

$$a^m \cdot a^n = a^{m+n}$$

If $m$, $n$, or $m + n$ is 0 or negative, then $a$ cannot be 0.

**QUOTIENT RULE FOR EXPONENTS**

If $a$ is a real number and if $m$ and $n$ are integers, then

$$\frac{a^m}{a^n} = a^{m-n} \quad \text{if } a \neq 0$$

You should notice that allowing the exponents to be any integer (instead of just positive integers) requires that we include restrictions on the value of the base.

## EXAMPLE 5 Using the Product Rule to Simplify Expressions Involving Exponents

Simplify each expression. All exponents should be positive integers.

**(a)** $(-3)^2(-3)^{-4}$ **(b)** $\frac{3}{4}y^5 \cdot \frac{20}{9}y^{-2}$

**Solution**

**(a)** $(-3)^2(-3)^{-4} = (-3)^{2+(-4)}$
$= (-3)^{-2}$
$= \dfrac{1}{(-3)^2}$
$= \dfrac{1}{9}$

**(b)** $\dfrac{3}{4}y^5 \cdot \dfrac{20}{9}y^{-2} = \dfrac{3}{4}\cdot\dfrac{20}{9}y^{5+(-2)}$
$= \dfrac{5}{3}y^3$

### EXAMPLE 6 Using the Quotient Rule to Simplify Expressions Involving Exponents

**(a)** $\dfrac{w^{-2}}{w^{-5}}$ **(b)** $\dfrac{20a^3b}{4ab^4}$

**Solution**

**(a)** $\dfrac{w^{-2}}{w^{-5}} = w^{-2-(-5)}$
$= w^{-2+5}$
$= w^3$

**(b)** $\dfrac{20a^3b}{4ab^4} = 5a^{3-1}b^{1-4}$
$= 5a^2b^{-3}$
$a^{-n} = \dfrac{1}{a^n}$: $= \dfrac{5a^2}{b^3}$

**QUICK ✓** *Simplify each expression. All exponents should be positive integers.*

**20.** $6^3 \cdot 6^{-5}$ **21.** $\dfrac{10^{-3}}{10^{-5}}$ **22.** $(4x^2y^3)\cdot(5xy^{-4})$

**23.** $\left(\dfrac{3}{4}a^3b\right)\cdot\left(\dfrac{8}{9}a^{-2}b^3\right)$ **24.** $\dfrac{-24b^5}{16b^{-3}}$ **25.** $\dfrac{50s^2t}{15s^5t^{-4}}$

## 4 Simplify Exponential Expressions Using the Power Rule

Another law of exponents applies when an exponential expression containing a power is itself raised to a power.

$$(3^2)^4 = \underbrace{3^2\cdot 3^2\cdot 3^2\cdot 3^2}_{\text{4 factors}} = \underbrace{\underbrace{(3\cdot 3)}_{\text{2 factors}}\cdot\underbrace{(3\cdot 3)}_{\text{2 factors}}\cdot\underbrace{(3\cdot 3)}_{\text{2 factors}}\cdot\underbrace{(3\cdot 3)}_{\text{2 factors}}}_{2\cdot 4 = 8\text{ factors}} = 3^8$$

We have the following result:

**In Words**
If an exponential expression contains a power raised to a power, keep the base and multiply the powers.

**POWER RULE FOR EXPONENTIAL EXPRESSIONS**

If $a$ is a real number and $m$ and $n$ are integers, then

$$(a^m)^n = a^{m\cdot n}$$

If $m$ or $n$ is 0 or negative, then $a$ must not be 0.

### EXAMPLE 7 Using the Power Rule to Simplify Exponential Expressions

Simplify each expression. All exponents should be positive integers.

**(a)** $(4^3)^5$ **(b)** $[(-3)^3]^2$ **(c)** $(6^3)^0$

**Solution**

**(a)** $(4^3)^5 = 4^{3 \cdot 5}$
$= 4^{15}$

**(b)** $[(-3)^3]^2 = (-3)^{3 \cdot 2}$
$= (-3)^6$
$= 729$

**(c)** $(6^3)^0 = 6^{3 \cdot 0}$
$= 6^0$
$= 1$

**QUICK ✓** *Simplify each expression. All exponents should be positive integers.*

**26.** $(2^2)^3$ **27.** $(5^8)^0$ **28.** $[(-4)^3]^2$
**29.** $(a^3)^5$ **30.** $(z^3)^{-6}$ **31.** $(s^{-3})^{-7}$

## 5 Simplify Exponential Expressions Containing Products or Quotients

We will now look at two additional laws of exponents. The first deals with raising a product to a power, while the second deals with raising a quotient to a power. Consider the following where we have a product to a power:

$$(x \cdot y)^3 = (x \cdot y) \cdot (x \cdot y) \cdot (x \cdot y)$$
$$= (x \cdot x \cdot x) \cdot (y \cdot y \cdot y)$$
$$= x^3 \cdot y^3$$

We have the following result:

**Work Smart**

Do not use this rule to try and simplify $(a + b)^2$ as $a^2 + b^2$ or $(a + b)^3$ as $a^3 + b^3$. To use this rule the base must be the product of two numbers—not a sum.

**PRODUCT TO A POWER RULE**

If $a$, $b$ are real numbers and $n$ is an integer, then

$$(a \cdot b)^n = a^n \cdot b^n$$

If $n$ is 0 or negative, neither $a$ nor $b$ can be 0.

### EXAMPLE 8 Using the Product to a Power Rule to Simplify Exponential Expressions

Simplify each expression. All exponents should be positive integers.

**(a)** $(3z)^4$ **(b)** $(3y^{-2})^{-3}$ **(c)** $(-3a^2)^{-2}$

**Solution**

**(a)** $(3z)^4 = 3^4 z^4$
$= 81z^4$

**(b)** $(3y^{-2})^{-3} = 3^{-3}(y^{-2})^{-3}$
$= \dfrac{y^{-2(-3)}}{3^3}$
$= \dfrac{y^6}{27}$

**(c)** $(-3a^2)^{-2} = \dfrac{1}{(-3a^2)^2}$
$= \dfrac{1}{(-3)^2(a^2)^2}$
$= \dfrac{1}{9a^4}$

**QUICK ✓** *Simplify each expression. All exponents should be positive integers.*

**32.** $(5y)^3$ **33.** $(6y)^0$ **34.** $(3x^2)^4$ **35.** $(4a^3)^{-2}$

Now let's look at a quotient raised to a power:

$$\left(\frac{2}{3}\right)^4 = \left(\frac{2}{3}\right) \cdot \left(\frac{2}{3}\right) \cdot \left(\frac{2}{3}\right) \cdot \left(\frac{2}{3}\right) = \frac{2^4}{3^4}$$

We have the following result:

**QUOTIENT TO A POWER RULE**

If $a$ and $b$ are real numbers and $n$ is an integer, then

$$\left(\frac{a}{b}\right)^n = \frac{a^n}{b^n} \qquad \text{if } b \neq 0$$

If $n$ is negative or 0, then $a$ cannot be 0.

### EXAMPLE 9 Using the Quotient to a Power Rule to Simplify Exponential Expressions

Simplify each expression. All exponents should be positive integers.

**(a)** $\left(\frac{w}{4}\right)^3$ **(b)** $\left(\frac{2x^2}{y^3}\right)^4$

**Solution**

**(a)** $\left(\frac{w}{4}\right)^3 = \frac{w^3}{4^3}$

$= \frac{w^3}{64}$

**(b)** $\left(\frac{2x^2}{y^3}\right)^4 = \frac{(2x^2)^4}{(y^3)^4}$

$= \frac{2^4(x^2)^4}{(y^3)^4}$

$= \frac{16x^{2\cdot 4}}{y^{3\cdot 4}}$

$= \frac{16x^8}{y^{12}}$

**QUICK ✓** *Simplify each expression. All exponents should be positive integers.*

**36.** $\left(\frac{z}{3}\right)^4$ **37.** $\left(\frac{x}{2}\right)^{-5}$ **38.** $\left(\frac{x^2}{y^3}\right)^4$ **39.** $\left(\frac{3a^{-2}}{b^4}\right)^3$

## 6 Simplify Exponential Expressions Using the Laws of Exponents

We now summarize the Laws of Exponents.

**THE LAWS OF EXPONENTS**

If $a$ and $b$ are real numbers and if $m$ and $n$ are integers, then assuming the expression is defined,

**Zero Exponent Rule:** $a^0 = 1$ if $a \neq 0$

**Negative Exponent Rule:** $a^{-n} = \frac{1}{a^n}$ if $a \neq 0$

**Product Rule:** $a^m \cdot a^n = a^{m+n}$

**Quotient Rule:** $\frac{a^m}{a^n} = a^{m-n}$ if $a \neq 0$

**Power Rule:** $(a^m)^n = a^{m\cdot n}$

*(continued)*

**THE LAWS OF EXPONENTS (*continued*)**

**Product to Power Rule:** $(a \cdot b)^n = a^n \cdot b^n$

**Quotient to Power Rule:** $\left(\frac{a}{b}\right)^n = \frac{a^n}{b^n}$ if $b \neq 0$

**Quotient to a Negative Power Rule:** $\left(\frac{a}{b}\right)^{-n} = \left(\frac{b}{a}\right)^n$ if $a \neq 0, b \neq 0$

Now let's do some examples where we use one or more of the rules listed above.

## EXAMPLE 10 Using the Laws of Exponents

Simplify each expression. All exponents should be positive integers. None of the variables are zero.

**(a)** $\frac{a^3b^{-1}}{(a^2b)^3}$ **(b)** $\left(\frac{3xy}{x^2y^{-2}}\right)^2 \cdot \left(\frac{9x^2y^{-3}}{x^3y^2}\right)^{-1}$

### Solution

**(a)**

$$\frac{a^3b^{-1}}{(a^2b)^3} = \frac{a^3b^{-1}}{(a^2)^3b^3} \quad (a \cdot b)^n = a^n \cdot b^n$$

$$(a^m)^n = a^{m \cdot n}: \quad = \frac{a^3b^{-1}}{a^6b^3}$$

$$\frac{a^m}{a^n} = a^{m-n}: \quad = a^{3-6}b^{-1-3}$$

$$= a^{-3}b^{-4}$$

$$a^{-n} = \frac{1}{a^n}: \quad = \frac{1}{a^3b^4}$$

**(b)**

$$\left(\frac{3xy}{x^2y^{-2}}\right)^2 \cdot \left(\frac{9x^2y^{-3}}{x^3y^2}\right)^{-1} = (3x^{1-2}y^{1-(-2)})^2 \cdot (9x^{2-3}y^{-3-2})^{-1} \quad \frac{a^m}{a^n} = a^{m-n}$$

$$= (3x^{-1}y^3)^2 \cdot (9x^{-1}y^{-5})^{-1}$$

$$(a \cdot b)^n = a^n \cdot b^n: \quad = 3^2 \cdot (x^{-1})^2(y^3)^2 \cdot 9^{-1} \cdot (x^{-1})^{-1}(y^{-5})^{-1}$$

$$(a^m)^n = a^{m \cdot n}: \quad = 9x^{-2}y^6 \cdot \frac{1}{9} \cdot xy^5$$

$$a^ma^n = a^{m+n}: \quad = x^{-2+1}y^{6+5}$$

$$a^{-n} = \frac{1}{a^n}: \quad = x^{-1}y^{11} = \frac{y^{11}}{x}$$

**Work Smart: Study Skills**

Many different approaches may be taken to simplify exponential expressions. In Example 10(b), we could have used the Quotient to Power Rule first and then continued to simplify, for example. Try working a problem one way and then working it again a second way to see if you obtain the same answer.

**QUICK ✓** *Simplify each expression. All exponents should be positive integers.*

**40.** $\frac{(3x^2y)^2}{12xy^{-2}}$ **41.** $(3ab^3)^3 \cdot (6a^2b^2)^{-2}$ **42.** $\left(\frac{2x^2y^{-1}}{x^{-2}y^2}\right)^2 \cdot \left(\frac{4x^3y^2}{xy^{-2}}\right)^{-1}$

## 7 Convert Between Scientific Notation and Decimal Notation

Figure 1

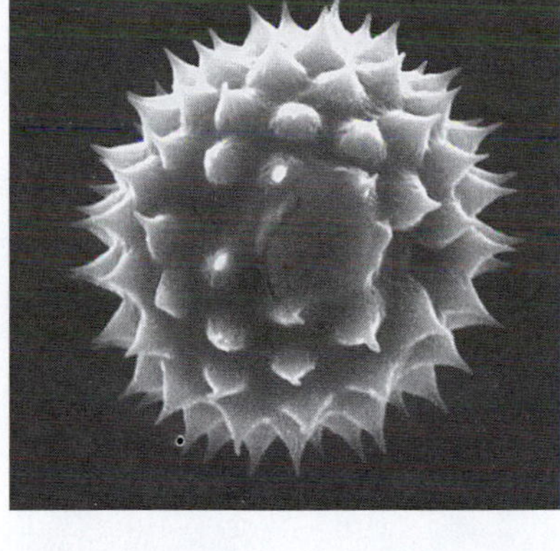

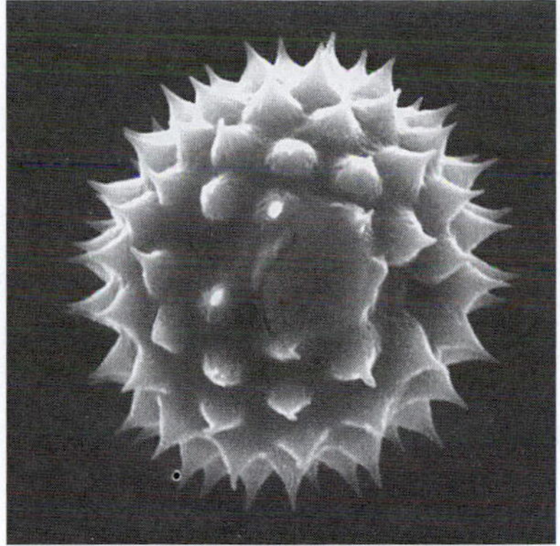

Figure 2

Measurements of physical quantities can range from very small to very large. For example, the mass of an electron (Figure 1) is approximately 0.000000000000000000000000000911 gram and the mass of Earth (Figure 2) is about 5,980,000,000,000,000,000,000,000 kilograms. These numbers are difficult to write and difficult to read, so we use exponents to write them.

**DEFINITION: SCIENTIFIC NOTATION**

When a number has been written as the product of a number $a$, where $1 \leq |a| < 10$, and a power of 10, it is said to be written in **scientific notation.** That is, a number is written in scientific notation when it is of the form

$$a \times 10^N$$

where

$$1 \leq |a| < 10$$

$$N \text{ is an integer}$$

For example, in scientific notation,

$$\text{Mass of an Electron} = 9.11 \times 10^{-28} \text{ gram}$$
$$\text{Mass of Earth} = 5.98 \times 10^{24} \text{ kilograms}$$

### Converting a Decimal to Scientific Notation

To change a positive number into scientific notation

**Step 1:** Count the number $N$ of decimal places that the decimal point must be moved in order to arrive at a number $a$, where $1 \leq |a| < 10$.

**Step 2:** If the absolute value of the original number is greater than or equal to 1, the scientific notation is $a \times 10^N$. If the absolute value of the original number is between 0 and 1, the scientific notation is $a \times 10^{-N}$.

### EXAMPLE 11 How to Convert from Decimal Notation to Scientific Notation

Write each number in scientific notation.

**(a)** 94,873 **(b)** 0.042

#### Step-by-Step Solution

For a number to be in scientific notation, the decimal must be moved so there is a single non-zero digit to the left of the decimal point. All remaining digits must appear to the right of the decimal point.

**(a)**

| | |
|---|---|
| **Step 1:** The decimal point in 94,873 follows the 3. Therefore, we will move the decimal to the left until it is between the 9 and the 4. Do you see why? This requires that we move the decimal $N = 4$ places. | 9 4 8 7 3 . |
| **Step 2:** The original number is greater than 1, so we write 94,873 in scientific notation as | $9.4873 \times 10^4$ |

**Work Smart**

We move the decimal four places to the left because, when the number is written in scientific notation, we are multiplying 9.4873 by $10^4 = 10{,}000$—which, if simplified, would move the decimal four places to the right.

**(b)**

| | |
|---|---|
| **Step 1:** Because 0.042 is less than 1, we move the decimal point to the right until it is between the 4 and the 2. This requires that we move the decimal $N = 2$ places. | 0 . 0 4 2 |
| **Step 2:** The original number is between 0 and 1, so we write 0.042 in scientific notation as | $4.2 \times 10^{-2}$ |

## EXAMPLE 12 Converting a Negative Number from Decimal Notation to Scientific Notation

Write $-520{,}000{,}000$ in scientific notation.

### Solution

Because the absolute value of the number is greater than 1, we move the decimal to the left $N = 8$ places. Therefore,

$$-520{,}000{,}000 = -5.2 \times 10^8$$

**QUICK ✓** *Write each number in scientific notation.*

**43.** 532 **44.** $-1{,}230{,}000$ **45.** 0.034 **46.** $-0.0000845$

Now we are going to convert a number from scientific notation to decimal notation.

**CONVERTING A NUMBER FROM SCIENTIFIC NOTATION TO DECIMAL NOTATION**

Determine the exponent on the number 10. If the exponent is negative, then move the decimal $|N|$ decimal places to the left. If the exponent is positive, then move the decimal $N$ decimal places to the right.

## EXAMPLE 13 Converting from Scientific Notation to Decimal Notation

Write each number in decimal notation.

**(a)** $3.2 \times 10^3$ **(b)** $7.54 \times 10^{-5}$

### Solution

**(a)** The exponent on the 10 is 3 so we will move the decimal three places to the right.

$$3.2 \times 10^3 = 3.200 \times 10^3 = 3200$$

**(b)** The exponent on the 10 is $-5$ so we will move the decimal five places to the left.

$$7.54 \times 10^{-5} = 000007.54 \times 10^{-5} = 0.0000754$$

**Work Smart**

We move the decimal three places to the right in Example 13(a) because $10^3 = 1000$. So, we are really multiplying 3.2 by 1000. We move the decimal five places to the left in Example 13(b) because

$$10^{-5} = \frac{1}{10^5} = \frac{1}{100{,}000} = 0.00001$$

**QUICK ✓** *Write each number in decimal notation.*

**47.** $5 \times 10^2$ **48.** $9.1 \times 10^5$ **49.** $1.8 \times 10^{-4}$ **50.** $1 \times 10^{-6}$

## 8 Use Scientific Notation to Multiply and Divide

Once a number is presented in scientific notation, the Laws of Exponents make it relatively straightforward to multiply and divide the numbers. The two Laws of Exponents that we make use of are

$$a^m \cdot a^n = a^{m \cdot n} \quad \text{and} \quad \frac{a^m}{a^n} = a^{m-n}$$

We will use these laws where the base is 10.

### EXAMPLE 14 Multiplying Using Scientific Notation

Perform the indicated operation. Express the answer in scientific notation.

**(a)** $(5 \times 10^2) \cdot (3 \times 10^8)$
**(b)** $(2.5 \times 10^{-3}) \cdot (4.3 \times 10^{-4})$

**Solution**

**(a)**
$$\begin{aligned}(5 \times 10^2) \cdot (3 \times 10^8) &= (5 \cdot 3) \times (10^2 \cdot 10^8)\\ &= 15 \times 10^{10}\\ &= (1.5 \times 10^1) \times 10^{10}\\ &= 1.5 \times 10^{11}\end{aligned}$$

**(b)**
$$\begin{aligned}(2.5 \times 10^{-3}) \cdot (4.3 \times 10^{-4}) &= (2.5 \cdot 4.3) \times (10^{-3} \cdot 10^{-4})\\ &= 10.75 \times 10^{-7}\\ &= (1.075 \times 10^1) \times 10^{-7}\\ &= 1.075 \times 10^{-6}\end{aligned}$$

**QUICK ✓** *Perform the indicated operation. Express the solution in scientific notation.*

**51.** $(3 \times 10^3) \cdot (2 \times 10^5)$ **52.** $(2 \times 10^{-4}) \cdot (4 \times 10^{-7})$ **53.** $(6 \times 10^{-5}) \cdot (4 \times 10^8)$

### EXAMPLE 15 Dividing Using Scientific Notation

Perform the indicated operation. Express the answer in scientific notation.

**(a)** $\dfrac{8 \times 10^5}{2 \times 10^3}$ **(b)** $\dfrac{2.5 \times 10^5}{5 \times 10^{-3}}$

**Solution**

**(a)**
$$\frac{8 \times 10^5}{2 \times 10^3} = \frac{8}{2} \times \frac{10^5}{10^3} = 4 \times 10^2$$

**(b)**
$$\begin{aligned}\frac{2.5 \times 10^5}{5 \times 10^{-3}} &= \frac{2.5}{5} \times \frac{10^5}{10^{-3}}\\ &= 0.5 \times 10^{5-(-3)}\\ &= 0.5 \times 10^8\\ &= (5 \times 10^{-1}) \times 10^8\\ &= 5 \times 10^7\end{aligned}$$

**QUICK ✓** *Perform the indicated operation. Express the solution in scientific notation.*

**54.** $\dfrac{6 \times 10^8}{3 \times 10^6}$ **55.** $\dfrac{6.8 \times 10^{-8}}{3.4 \times 10^{-5}}$ **56.** $\dfrac{4.8 \times 10^7}{9.6 \times 10^3}$ **57.** $\dfrac{3 \times 10^{-5}}{8 \times 10^7}$

### EXAMPLE 16 Using Scientific Notation to Multiply and Divide

Perform the indicated operation. Express the answer in decimal notation.

**(a)** $(3{,}000{,}000) \cdot (90{,}000)$ **(b)** $\dfrac{0.00000075}{0.00015}$

#### Solution

In each of these problems, we will first write the numbers in scientific notation and then perform the indicated operation.

**(a)**
$$\begin{aligned}(3{,}000{,}000) \cdot (90{,}000) &= (3 \times 10^6) \cdot (9 \times 10^4)\\ &= 27 \times 10^{10}\\ &= 2.7 \times 10^{11}\\ &= 270{,}000{,}000{,}000\end{aligned}$$

**(b)**
$$\begin{aligned}\frac{0.00000075}{0.00015} &= \frac{7.5 \times 10^{-7}}{1.5 \times 10^{-4}}\\ &= \frac{7.5}{1.5} \times \frac{10^{-7}}{10^{-4}}\\ &= 5 \times 10^{-7-(-4)}\\ &= 5 \times 10^{-3}\\ &= 0.005\end{aligned}$$

**QUICK ✓** *Perform the indicated operation. Express the solution in decimal notation.*

**58.** $(8{,}000{,}000) \cdot (30{,}000)$ **59.** $\dfrac{0.000000012}{0.000004}$

**60.** $(25{,}000{,}000) \cdot (0.00003)$ **61.** $\dfrac{0.000039}{13{,}000{,}000}$

# Getting Ready for Chapter 5 Exercises

**For Extra Help:**

Student Solutions Manual | CD Video | PH Math/Tutor Center | MathXL Tutorials on CD | MathXL® | MyMathLab

## Concepts and Vocabulary

*In Problems 1–4, fill in the blanks.*

**1.** In the notation $a^n$ we call $a$ the _______ and $n$ the _______ or _______.

**2.** Using a Law of Exponents, $a^m \cdot a^n =$ _______.

**3.** $a^0 = 1$, provided $a \neq$ _______.

**4.** To put the number 1234.567 into scientific notation, move the decimal _______ places to the _______.

*In Problems 5–8, answer True or False to each statement.*

**5.** $(3a)^{-2} = \dfrac{-9}{a^2}$

**6.** To divide two exponential expressions having the same base, keep the base and subtract the exponents.

**7.** When a number is expressed in scientific notation, it is expressed as the product of number $x$, $0 \leq x < 1$, and a power of 10.

**8.** The number 0.0001 in scientific notation is $1 \times 10^4$.

**9.** Explain why $a$ cannot be 0 when $m$, $n$, or $m + n$ is negative or 0 in the expression $a^{m+n}$. Use examples to support your explanation.

**10.** Explain why $a$ cannot be 0 when $n$ is negative or 0 in the expression $(a^m)^n$. Use examples to support your explanation.

**11.** Explain why neither $a$ nor $b$ can be 0 when $n$ is 0 or negative in the expression $(a \cdot b)^n = a^n \cdot b^n$.

**12.** Write a paragraph to justify the definition given in the text that $a^0 = 1, a \neq 0$.

**13.** Provide a justification for the product rule for exponential expressions.

**14.** Provide a justification for the quotient rule for exponential expressions.

**15.** Provide a justification for the power rule for exponential expressions.

**16.** Provide a justification for the product to a power rule for exponential expressions.

**17.** In your own words, explain how to convert a decimal number to scientific notation. In your own words, explain how to convert a number in scientific notation to decimal notation.

**18.** Explain the benefits of using scientific notation to multiply or divide very large or very small numbers.

## Building Skills

*In Problems 19–36, simplify each expression.*

**19.** $5^2$ **20.** $(-5)^2$ **21.** $-5^2$ **22.** $5^{-2}$

**23.** $-5^{-2}$ **24.** $-5^0$ **25.** $8^2 \cdot 8^{-3}$ **26.** $\dfrac{8^7}{8^5}$

**27.** $\left(\dfrac{4}{9}\right)^{-2}$ **28.** $\left(\dfrac{3}{4}\right)^{-3}$ **29.** $(-2)^3 \cdot 4^0$ **30.** $(-3)^2 \cdot 2^0$

**31.** $(-3)^2 \cdot (-3)^{-5}$ **32.** $(-4)^{-5} \cdot (-4)^3$ **33.** $\dfrac{(-4)^2}{(-4)^{-1}}$ **34.** $\dfrac{(-3)^3}{(-3)^{-2}}$

**35.** $\dfrac{2^3 \cdot 3^{-2}}{2^{-2} \cdot 3^{-4}}$ **36.** $\dfrac{3^{-2} \cdot 5^3}{3^2 \cdot 5}$

*In Problems 37–82, simplify each expression. All exponents should be positive integers.*

**37.** $z^6 \cdot z^{-2}$ **38.** $y^{-3} \cdot y^7$ **39.** $\dfrac{y^8}{y^6}$ **40.** $\dfrac{z^5}{z^{-2}}$

**41.** $(6x)^0$ **42.** $(5a^2)^0$ **43.** $\dfrac{1}{6^{-2}}$ **44.** $10^{-1}$

**45.** $(2s^{-2}t^4)(-5s^2t)$ **46.** $(6ab) \cdot (3a^3b^{-4})$ **47.** $\left(\dfrac{1}{4}xy\right) \cdot (20xy^{-2})$

**48.** $(3xy^3) \cdot \left(\dfrac{1}{9}x^2y\right)$ **49.** $\dfrac{16a^6}{10a^5}$ **50.** $\dfrac{12b^6}{18b^2}$

**51.** $\dfrac{36x^7y^3}{9x^5y^2}$ **52.** $\dfrac{25a^2b^3}{5ab^6}$ **53.** $\dfrac{21a^2b}{14a^3b^{-2}}$ **54.** $\dfrac{25x^{-2}y}{10xy^3}$

**55.** $\dfrac{6x^2y^5}{2x^3y^3}$ **56.** $\dfrac{25xy^{-4}}{10x^{-2}y^2}$ **57.** $(x^{-2})^4$ **58.** $(z^2)^{-6}$

**59.** $(3x^2y)^3$ **60.** $(5a^2b^{-1})^2$ **61.** $\left(\dfrac{z}{4}\right)^3$ **62.** $\left(\dfrac{x}{y}\right)^{-8}$

**63.** $(3a^{-3})^{-2}$ **64.** $(2y^{-2})^{-4}$ **65.** $(-2a^2b^3)^{-4}$ **66.** $(-4a^{-2}b^2)^{-2}$

**67.** $\dfrac{x^2yz^{-3}}{xy^{-4}z^2}$ **68.** $\dfrac{5x^{-3}y^5z}{x^{-2}y^6z^{-2}}$ **69.** $\dfrac{2^3 \cdot xy^{-2}}{12(x^2)^{-2}y}$ **70.** $\dfrac{3^2 \cdot x^{-3}(y^2)^3}{15x^2y^8}$

**71.** $\left(\dfrac{15a^2b^3}{3a^{-4}b^5}\right)^{-2}$ **72.** $\left(\dfrac{15x^4y^7}{18x^{-3}y}\right)^{-1}$ **73.** $(4x^4y^{-2})^{-1} \cdot (2x^2y^{-1})^2$

**74.** $(9a^2b^{-4})^{-1} \cdot (3ab^{-2})^2$ **75.** $(-4x^2y) \cdot (2xy^{-3})^{-2}$ **76.** $(-5a^2b)^2 \cdot (10a^3b^2)^{-1}$

**77.** $\dfrac{(-2)^2x^3(yz)^2}{-4xy^{-2}z}$ **78.** $\dfrac{(-3)^3a^3(ab)^{-2}}{9ab^4}$ **79.** $\dfrac{(3x^{-1}yz^2)^2}{(xy^{-2}z)^3}$

**80.** $\dfrac{(2ab^2c)^{-1}}{(a^{-1}b^3c^2)^{-2}}$ **81.** $\dfrac{(6a^3b^{-2})^{-1}}{(2a^{-2}b)^{-2}} \cdot \left(\dfrac{3ab^3}{2a^2b^{-3}}\right)^2$ **82.** $\left(\dfrac{a^{-3}b^{-1}}{2a^4b^{-2}}\right)^2 \cdot \dfrac{(4a^2b)^2}{(2a^{-2}b)^3}$

*In Problems 83–90, write each number in scientific notation.*

**83.** 4,500,000 **84.** 94,000,000 **85.** −230,000

**86.** −567,000 **87.** 0.00034 **88.** 0.000123

**89.** −0.0000001 **90.** −0.000004

*In Problems 91–106, perform the indicated operation. Express the solution in scientific notation.*

**91.** $(3.4 \times 10^5) \cdot (2 \times 10^{-8})$ **92.** $(1.8 \times 10^{-7}) \cdot (3 \times 10^3)$

**93.** $(-5.3 \times 10^{-4}) \cdot (2.8 \times 10^{-3})$ **94.** $(6.2 \times 10^3) \cdot (-3.8 \times 10^5)$

**95.** $\dfrac{3.6 \times 10^{12}}{1.2 \times 10^7}$ **96.** $\dfrac{8.2 \times 10^{-5}}{4.1 \times 10^4}$

**97.** $(4 \times 10^6)^3$ **98.** $(5 \times 10^8)^2$

**99.** $\dfrac{5 \times 10^{-6}}{8 \times 10^{-4}}$ **100.** $\dfrac{1 \times 10^7}{5 \times 10^{-4}}$

**101.** $\dfrac{(4 \times 10^3)(6 \times 10^7)}{3 \times 10^4}$ **102.** $\dfrac{(3 \times 10^9)(8 \times 10^{-6})}{4 \times 10^4}$

**103.** $\dfrac{(2 \times 10^{-4}) \cdot (-4.8 \times 10^{-6})}{6 \times 10^3}$ **104.** $\dfrac{(-1.5 \times 10^4) \cdot (3.2 \times 10^2)}{5 \times 10^5}$

**105.** $\dfrac{6.2 \times 10^{-3}}{(3.1 \times 10^4) \cdot (2 \times 10^{-7})}$ **106.** $\dfrac{1.5 \times 10^{13}}{(3 \times 10^4) \cdot (5 \times 10^8)}$

*In Problems 107–118, perform the indicated operation by changing to scientific notation first. Express the solution in decimal notation.*

**107.** (4,000,000) · (3,000,000) **108.** (15,000) · (3,000,000)

**109.** (6000) · (2,000,000) **110.** (12,000,000) · (80,000)

**111.** $\dfrac{0.00008}{0.002}$ **112.** $\dfrac{0.00012}{0.0000002}$

**113.** $\dfrac{0.000036}{0.0009}$ **114.** $\dfrac{0.00000048}{0.00016}$

**115.** $\dfrac{-60{,}000 \cdot 4{,}000{,}000}{0.008}$ **116.** $\dfrac{-5{,}000{,}000 \cdot 0.0015}{25{,}000}$

**117.** $\dfrac{(0.000004) \cdot 1{,}600{,}000}{(0.0008) \cdot (0.002)}$ **118.** $\dfrac{(0.0001) \cdot (3{,}500{,}000)}{(0.0005) \cdot (1{,}400{,}000)}$

## Applying the Concepts

△ **119. Cubes** Suppose the length of a side of a cube is $x^2$. Find the volume of the cube in terms of $x$.

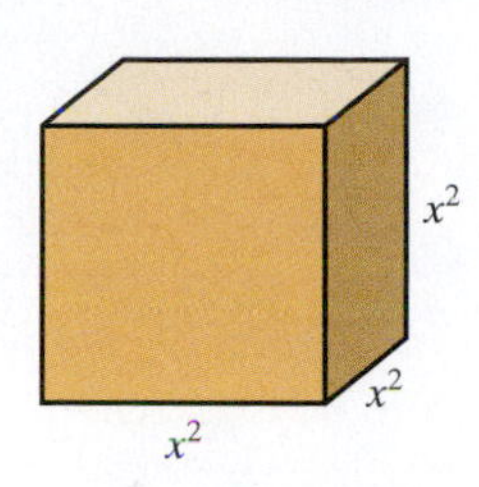


△ **120. Circles** The radius $r$ of a circle is $\frac{d}{2}$, where $d$ is the diameter. The area of a circle is given by the formula $A = \pi r^2$. Find the area of a circle in terms of its diameter $d$.

**121. Diameter of a Plant Cell** The diameter of a plant cell is 0.00001276 meter. Express this number in scientific notation.

**122. Diameter of an Atom** The diameter of an atom is about 0.0000000001 meter. Express this diameter as a number in scientific notation.

**123. Diameter of Earth** The diameter of Earth is 12,760,000 meters. Express this number in scientific notation.

**124. National Debt** As of May 25, 2005, the debt of the United States federal government was $7,772,000,000,000. Express this number in scientific notation.

**125. A Nanometer** One nanometer is $1 \times 10^{-9}$ meters. Express this number in decimal notation.

**126. A Lightyear** One lightyear is $1 \times 10^{16}$ meters. Express this number in decimal notation.

**127. Rubik Cube** The Rubik's Cube has $4 \times 10^{19}$ possible states. Express this number in decimal notation.

**128. Hair Growth** The human hair grows at a rate of $1 \times 10^{-8}$ miles per hour. Express this number in decimal notation. What is the growth rate in miles per year?

**129. Per Capita GDP** The per capita Gross Domestic Product (GDP) is the GDP divided by the United States population. It represents the average output of each resident of the United States and is measured in dollars per person. In 2000, the Gross Domestic Product of the United States was $\$9.96 \times 10^{12}$. In 2000, the United States population was $2.8 \times 10^8$ people. Determine the per capita GDP of the United States in 2000. Express your answer in decimal notation.

**130. Taxes** In 2000, the United States government collected $\$9.9 \times 10^{11}$ in personal income tax revenue. In 2000, the United States population was $2.8 \times 10^8$ people. The per capita tax bill is defined as the total personal income tax revenue divided by the size of the United States population. Determine the per capita tax bill in the United States in 2000. Express your answer in decimal notation.

**131. Astronomy** The speed of light is 186,000 miles per second. How long does it take a beam of light to reach Earth from the Sun when the Sun is 93,000,000 miles from Earth? Express your answer in seconds, using scientific notation.

**132. Arts and Entertainment Industry** According to the United States Census, in 2000 the total payroll of the arts, entertainment and recreation industry was $\$43.2 \times 10^9$. Also in 2000, there were $1.74 \times 10^6$ employees in the industry. What was the salary per employee in 2000? Express your answer in decimal notation to the nearest dollar.

**133. Population Density** Population density is the number of people living per some unit of land. In other words, Population Density $= \frac{\text{Population}}{\text{Area of Land}}$. The United States is $3.54 \times 10^6$ square miles. In 1990, the United States population was $2.49 \times 10^8$ people. In 2000, the United States population was $2.81 \times 10^8$ people.

**(a)** Determine the population density of the United States in 1990. Interpret the result.

**(b)** Determine the population density of the United States in 2000. Interpret the result.

**(c)** By how much did the population density increase? What does this result mean?

**134. Fossil Fuels** In 2003, the United States imported $3.52 \times 10^9$ barrels of oil. The United States population was $2.91 \times 10^8$ people in 2003.

**(a)** Determine the per capita number of barrels of oil imported into the United States in 2003.

**(b)** One barrel of oil contains 42 gallons. Determine the number of gallons of oil imported into the United States in 2003.

**(c)** Use the result from part (b) to compute the per capita number of gallons of oil imported into the United States in 2003.

*In Problems 135–142, simplify each algebraic expression by rewriting each factor with a common base. (***Hint:** *Consider that* $8 = 2^3$.*)*

**135.** $\dfrac{2^{x+3}}{4}$

**136.** $\dfrac{3^{2x}}{27}$

**137.** $3^x \cdot 27^{3x+1}$

**138.** $9^{-x} \cdot 3^{x+1}$

**139.** If $3^x = 5$, what does $3^{4x}$ equal?

**140.** If $4^x = 6$, what does $4^{5x}$ equal?

**141.** If $2^x = 7$, what does $2^{-4x}$ equal?

**142.** If $5^x = 3$, what does $5^{-3x}$ equal?

**143.** A friend of yours has a homework problem in which he must simplify $(x^4)^3$. He tells you that he thinks the answer is $x^7$. Is he right? If not, explain where he went wrong.

**144.** A friend of yours is convinced that $x^0$ must equal 0. Write an explanation that details why $x^0 = 1$. Include any restrictions that must be placed on $x$.

CHAPTER

# 5 Polynomials and Polynomial Functions

The City of Chicago offers boat tours of Chicago's historical architecture along the Chicago River and lakefront. Often companies will offer high volume discounts to customers. Did you know that the equations that we present in this chapter can be used to model various pricing schemes? See Example 8 on page 430 in Section 5.8.

## OUTLINE



## The Big Picture: Putting It Together

In Chapter 3, we presented a discussion of linear functions. We learned properties of linear functions and how to graph them.

We are now going to generalize our discussion of functions a little more. It turns out that linear functions belong to a general class of functions called polynomial functions. Polynomial functions are arguably the simplest of all functions because they require only addition and repeated multiplication to evaluate. In this chapter, we learn how to add, subtract, multiply, and divide polynomials and polynomial functions. We then learn techniques for "undoing" polynomial multiplication, called factoring. You will find that the Distributive Property comes in handy in many of these operations. Finally, we learn how to use factoring to solve equations that have polynomial expressions or equations that involve polynomial functions.

# 5.1 Adding and Subtracting Polynomials

**OBJECTIVES**

1. Define Monomial and Determine the Coefficient and Degree of a Monomial
2. Define Polynomial and Determine the Degree of a Polynomial
3. Simplify Polynomials by Combining Like Terms
4. Evaluate Polynomial Functions
5. Add and Subtract Polynomial Functions

### *Preparing for Adding and Subtracting Polynomials*

*Before getting started, take the following readiness quiz. If you get a problem wrong, go back to the section cited and review the material.*

**1.** What is the coefficient of $-4x^5$? [Section R.5, p. 44]

**2.** Combine like terms: $5x^2 - 3x + 1 - 2x^2 - 6x + 3$ [Section R.5, pp. 44–46]

**3.** Use the Distributive Property to remove the parentheses: $-4(x - 3)$ [Section R.3, p. 30]

**4.** Given $f(x) = -4x + 3$, find $f(3)$. [Section 2.3, pp. 162–164]

## 1 Define Monomial and Determine the Coefficient and Degree of a Monomial

Recall from Section R.5 that a term is a number or the product of a number and one or more variables raised to a power. The numerical factor of a term is the coefficient. For example, consider Table 1, where various algebraic expressions are given and the terms of the expression identified.

**Table 1**

| Algebraic Expression | Terms |
|---|---|
| $5x + 4$ | $5x$, $4$ |
| $7x^2 - 8x + 3 = 7x^2 + (-8x) + 3$ | $7x^2$, $-8x$, $3$ |
| $3x^2 + 7y^{-1}$ | $3x^2$, $7y^{-1}$ |

In this chapter, we study *polynomials*. Polynomials have terms that are *monomials*.

**Work Smart**

The nonnegative integers are 0, 1, 2, 3, . . . .

**DEFINITION**

A **monomial in one variable** is the product of a constant and a variable raised to a nonnegative integer power. A monomial in one variable is of the form

$$ax^k$$

where $a$ is a constant, $x$ is a variable, and $k \geq 0$ is an integer. The constant $a$ is called the **coefficient** of the monomial. If $a \neq 0$, then $k$ is called the **degree** of the monomial.

### EXAMPLE 1 Identifying the Coefficient and Degree of Monomials

| | MONOMIAL | COEFFICIENT | DEGREE |
|---|---|---|---|
| **(a)** | $5x^3$ | $5$ | $3$ |
| **(b)** | $-\frac{2}{3}x^6$ | $-\frac{2}{3}$ | $6$ |
| **(c)** | $8 = 8x^0$ | $8$ | $0$ |
| **(d)** | $x^2 = 1x^2$ | $1$ | $2$ |
| **(e)** | $-x = -1 \cdot x$ | $-1$ | $1$ |

Now let's look at some expressions that are not monomials.

### EXAMPLE 2 Expressions That Are Not Monomials

**(a)** $4x^{\frac{1}{2}}$ is not a monomial because the exponent of the variable $x$ is $\frac{1}{2}$ and $\frac{1}{2}$ is not a nonnegative integer.

**(b)** $5x^{-3}$ is not a monomial because the exponent of the variable $x$ is $-3$ and $-3$ is not a nonnegative integer.

*Preparing for...Answers* **1.** $-4$ **2.** $3x^2 - 9x + 4$ **3.** $-4x + 12$ **4.** $-9$

**QUICK ✓** *Determine whether the expression is a monomial. For those that are monomials, name the coefficient and give the degree.*

**1.** $8x^5$ **2.** $5x^{-2}$ **3.** 12 **4.** $x^{\frac{1}{3}}$

A monomial may contain more than one variable factor, such as $ax^m y^n$, where $a$ is a constant (called the coefficient), $x$ and $y$ are variables, and $m$ and $n$ are whole numbers. The **degree of the monomial** $ax^m y^n$ is the sum of the exponents, $m + n$.

### EXAMPLE 3 Monomials in More than One Variable

**(a)** $-4x^3y^4$ is a monomial in $x$ and $y$ of degree $3 + 4 = 7$. The coefficient is $-4$.

**(b)** $10ab^5$ is a monomial in $a$ and $b$ of degree $1 + 5 = 6$. The coefficient is 10.

**QUICK ✓** *Determine whether the expression is a monomial. For those that are monomials, determine the coefficient and degree.*

**5.** $3x^5y^2$ **6.** $-2m^3n$ **7.** $4ab^{\frac{1}{2}}$ **8.** $-xy$

## 2 Define Polynomial and Determine the Degree of a Polynomial

We begin with a definition.

**DEFINITION**

A **polynomial** is a monomial or the sum of monomials.

A polynomial is in **standard form** if it is written with the terms in descending order according to degree. The **degree of a polynomial** is the highest degree of all the terms of the polynomial. Remember, the degree of a nonzero constant is 0 and the number 0 has no degree.

### EXAMPLE 4 Examples of Polynomials

| | POLYNOMIAL | DEGREE |
|---|---|---|
| **(a)** | $7x^3 - 2x^2 + 6x + 4$ | 3 |
| **(b)** | $3 - 8x + x^2 = x^2 - 8x + 3$ | 2 |
| **(c)** | $-7x^4 + 24$ | 4 |
| **(d)** | $x^3y^4 - 3x^3y^2 + 2x^3y$ | 7 |
| **(e)** | $p^2q - 8p^3q^2 + 3 = -8p^3q^2 + p^2q + 3$ | 5 |
| **(f)** | 6 | 0 |
| **(g)** | 0 | No Degree |

### EXAMPLE 5 Is the Algebraic Expression a Polynomial?

**(a)** $4x^{-2} - 5x + 1$ is not a polynomial because the exponent on the first term, $-2$, is negative.

**(b)** $\dfrac{4}{x^3}$ is not a polynomial because it can be written as $4x^{-3}$ and $-3$ is less than 0. Remember, the exponents on polynomials must be integers greater than or equal to 0.

**(c)** $\dfrac{8x^2 + 16}{2}$ is a polynomial of degree 2 because it can be written as $4x^2 + 8$ after dividing 2 into each term in the numerator.

**(d)** $\dfrac{3xy + 1}{xy - 2}$ is not a polynomial because it is the quotient of two polynomials, and the expression cannot be simplified to a polynomial.

**QUICK ✓** *Determine whether the algebraic expression is a polynomial. For those that are polynomials, determine the degree.*

**9.** $-3x^3 + 7x^2 - x + 5$ **10.** $5z^{-1} + 3$ **11.** $\dfrac{x - 1}{x + 1}$

**12.** $\dfrac{3x^2 - 9x + 27}{3}$ **13.** $5p^3q - 8pq^2 + pq$

**Work Smart**

The prefix "bi" means "two" as in bicycle. The prefix "tri" means "three" as in tricycle. The prefix "poly" means "many."

A polynomial with exactly one term is a monomial; a polynomial that has two monomials that are not like terms is called a **binomial;** and a polynomial that contains three monomials that are not like terms is called a **trinomial.** So

| | | |
|---|---|---|
| $-14x$ is a polynomial | but more specifically | $-14x$ is a monomial |
| $2x^3 - 5x$ is a polynomial | but more specifically | $2x^3 - 5x$ is a binomial |
| $-x^3 - 4x + 11$ is a polynomial | but more specifically | $-x^3 - 4x + 11$ is a trinomial |
| $3x^2 + 6xy - 2y^2$ is a polynomial | but more specifically | $3x^2 + 6xy - 2y^2$ is a trinomial |

We use the term *monomial* to describe a polynomial with a *single* term, and the term *polynomial* to describe the sum of two or more monomials.

## 3 Simplify Polynomials by Combining Like Terms

In Section R.5, we learned how to combine like terms. To simplify a polynomial, means to perform all indicated operations and combine like terms. One operation we perform on polynomials is addition. **To add polynomials, combine the like terms of the polynomials.**

### EXAMPLE 6 Simplifying Polynomials: Addition

Simplify: $(-4x^3 + 9x^2 + x - 3) + (2x^3 + 6x + 5)$

**Solution**

We can find the sum using either horizontal addition or vertical addition.

*Horizontal Addition:*

The idea here is to combine like terms.

**Work Smart**

Remember, like terms have the same variable and the same exponent on the variable.

$$
\begin{aligned}
& (-4x^3 + 9x^2 + x - 3) + (2x^3 + 6x + 5) \\
\text{Remove parentheses: } & = -4x^3 + 9x^2 + x - 3 + 2x^3 + 6x + 5 \\
\text{Rearrange terms: } & = -4x^3 + 2x^3 + 9x^2 + x + 6x - 3 + 5 \\
\text{Distributive Property: } & = (-4 + 2)x^3 + 9x^2 + (1 + 6)x + (-3 + 5) \\
\text{Simplify: } & = -2x^3 + 9x^2 + 7x + 2
\end{aligned}
$$

*Vertical Addition:*

The idea here is to line up like terms in each polynomial vertically and then add the coefficients.

$$
\begin{array}{r}
-4x^3 + 9x^2 + \phantom{6}x - 3 \\
2x^3 \phantom{+ 9x^2} + 6x + 5 \\
\hline
-2x^3 + 9x^2 + 7x + 2
\end{array}
$$

**QUICK ✓** *Simplify by adding the polynomials.*

**14.** $(2x^2 - 3x + 1) + (4x^2 + 5x - 3)$

**15.** $(5w^4 - 3w^3 + w - 8) + (-2w^4 + w^3 - 7w^2 + 3)$

---

## EXAMPLE 7 Simplifying Polynomials in Two Variables: Addition

Simplify: $(5a^2b - 3ab + 2ab^2) + (a^2b + 5ab - 4ab^2)$

### Solution

To save space, we only present the horizontal format. The first step is to remove the parentheses.

$$(5a^2b - 3ab + 2ab^2) + (a^2b + 5ab - 4ab^2) = 5a^2b - 3ab + 2ab^2 + a^2b + 5ab - 4ab^2$$

Rearrange terms: $= 5a^2b + a^2b - 3ab + 5ab + 2ab^2 - 4ab^2$

Distributive Property: $= (5 + 1)a^2b + (-3 + 5)ab + (2 - 4)ab^2$

Simplify: $= 6a^2b + 2ab - 2ab^2$

**QUICK ✓** *Simplify by adding the polynomials.*

**16.** $(8x^2y + 2x^2y^2 - 7xy^2) + (-3x^2y + 5x^2y^2 + 3xy^2)$

---

We can subtract polynomials using either the horizontal or vertical approach as well. Remember,

$$a - b = a + (-b)$$

So, to subtract one polynomial from another, we add the opposite of each term in the polynomial following the subtraction sign and then combine like terms.

## EXAMPLE 8 Simplifying Polynomials: Subtraction

Simplify: $(5z^3 + 3z^2 - 3) - (-2z^3 + 7z^2 - z + 2)$

### Solution

*Horizontal Subtraction:*

Recall that $a - b = a + (-1) \cdot b$.

$$(5z^3 + 3z^2 - 3) - (-2z^3 + 7z^2 - z + 2) = 5z^3 + 3z^2 - 3 + (-1)(-2z^3 + 7z^2 - z + 2)$$

Distribute the $-1$: $= 5z^3 + 3z^2 - 3 + 2z^3 - 7z^2 + z - 2$

Rearrange terms: $= 5z^3 + 2z^3 + 3z^2 - 7z^2 + z - 3 - 2$

Combine like terms: $= 7z^3 - 4z^2 + z - 5$

*Vertical Subtraction:*

We line up like terms, change the sign of each coefficient of the second polynomial, and add.

$$\begin{array}{rrrr} z^3 & z^2 & z^1 & z^0 \\ 5z^3 & + 3z^2 & & - 3 \\ \hline -(-2z^3 & + 7z^2 & - z & + 2) \end{array} \qquad \begin{array}{rrrr} z^3 & z^2 & z^1 & z^0 \\ 5z^3 & + 3z^2 & & - 3 \\ + 2z^3 & - 7z^2 & + z & - 2 \\ \hline 7z^3 & - 4z^2 & + z & - 5 \end{array}$$

**QUICK ✓** *Simplify by subtracting the polynomials.*

**17.** $(5x^3 - 6x^2 + x + 9) - (4x^3 + 10x^2 - 6x + 7)$

**18.** $(8y^3 - 5y^2 + 3y + 1) - (-3y^3 + 6y + 8)$

**19.** $(8x^2y + 2x^2y^2 - 7xy^2) - (-3x^2y + 5x^2y^2 + 3xy^2)$

## 4 Evaluate Polynomial Functions

Up to now, we have only discussed polynomial expressions, such as $4x^3 + x^2 - 7x + 1$. If we write $f(x) = 4x^3 + x^2 - 7x + 1$, then we have a *polynomial function*.

**DEFINITION**

A **polynomial function** is a function whose rule is a polynomial. The domain of all polynomial functions is the set of all real numbers. The **degree** of a polynomial function is the value of the largest exponent on the variable.

In Chapter 3, we studied linear functions. Recall, a linear function is a function of the form $f(x) = mx + b$. Table 2 illustrates some specific types of polynomial functions, their functional forms and specific examples.

**Table 2**

| Polynomial | Functional Form | Examples | Degree |
|---|---|---|---|
| Linear | $f(x) = a_1x + a_0$ where $a_1$ and $a_0$ are not both 0, $a_1$ is the slope, and $a_0$ is the $y$-intercept | $f(x) = 4x + 5$<br>$f(x) = -6x$<br>$f(x) = 3$ | 1<br>1<br>0 |
| Quadratic | $f(x) = a_2x^2 + a_1x + a_0$, $a_2 \neq 0$ | $f(x) = 3x^2 - 5x + 1$ | 2 |
| Cubic | $f(x) = a_3x^3 + a_2x^2 + a_1x + a_0$, $a_3 \neq 0$ | $f(x) = -2x^3 + 4x - 1$ | 3 |

To **evaluate** a polynomial function, we substitute for the value of the variable and simplify, just as we did in Section 2.3.

### EXAMPLE 9 Evaluating a Polynomial Function

For the polynomial function $P(x) = 2x^3 - 5x^2 + x - 3$, find

**(a)** $P(3)$ **(b)** $P(-1)$

### Solution

**(a)** We substitute 3 for $x$ in the expression $2x^3 - 5x^2 + x - 3$ to get

$$\begin{aligned} P(3) &= 2(3)^3 - 5(3)^2 + 3 - 3 \\ &= 2 \cdot 27 - 5 \cdot 9 + 3 - 3 \\ &= 54 - 45 + 3 - 3 \\ &= 9 \end{aligned}$$

**(b)** We substitute $-1$ for $x$ in the expression $2x^3 - 5x^2 + x - 3$ to get

$$\begin{aligned} P(-1) &= 2(-1)^3 - 5(-1)^2 + (-1) - 3 \\ &= 2 \cdot (-1) - 5 \cdot 1 + (-1) - 3 \\ &= -2 - 5 - 1 - 3 \\ &= -11 \end{aligned}$$

**QUICK ✓**

**20.** Find the following values of the polynomial function $g(x) = -2x^3 + 7x + 1$.

**(a)** $g(0)$ **(b)** $g(2)$ **(c)** $g(-3)$

---

Polynomial functions can be used to model a variety of situations.

## EXAMPLE 10 A Polynomial Model for the Murder Rate

The bar graph shown in Figure 1 represents the murder rate (murders per 100,000 population) in the United States for the years 1993–2002. The polynomial function

$$M(t) = 0.055t^2 - 0.964t + 9.742$$

can be used to approximate the murder rate $M$, where $t$ is the number of years since 1993.

**(a)** Use the function to estimate the murder rate in 1993.

**(b)** Use the function to estimate the murder rate in 2008.

**Figure 1**

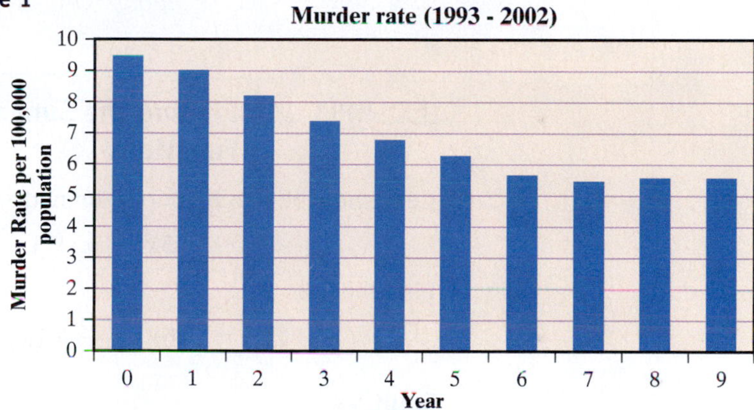


SOURCE: *Statistical Abstract*, 2004

### Solution

**(a)** The independent variable $t$ represents the number of years since 1993, so $t = 0$ corresponds to 1993. We evaluate $M(0)$.

$$\begin{aligned} M(t) &= 0.055t^2 - 0.964t + 9.742 \\ M(0) &= 0.055(0)^2 - 0.964(0) + 9.742 \\ &= 9.742 \end{aligned}$$

We estimate that the murder rate in 1993 was approximately 9.74 murders per 100,000 population. This result is pretty close to the actual murder rate of 9.5 murders per 100,000 population.

**Work Smart**

Always answer applied problems with complete sentences.

**(b)** Since $2008 - 1993 = 15$, $t = 15$ corresponds to 2008. So we evaluate $M(15)$.

$$\begin{aligned} M(t) &= 0.055t^2 - 0.964t + 9.742 \\ M(15) &= 0.055(15)^2 - 0.964(15) + 9.742 \\ &= 12.375 - 14.46 + 9.742 \\ &= 7.66 \end{aligned}$$

We estimate that the murder rate in 2008 will be 7.66 murders per 100,000 population.

**Quick ✓**

**21.** The polynomial function $F(t) = 0.13t^2 - 2.48t + 58.93$ can be used to approximate the fertility rate $F$ for women 15–19 years of age in the United States, where $t$ is the number of years since 1994.

**(a)** Use the function to estimate the fertility rate in 1994.

**(b)** Use the function to estimate the fertility rate in 2008.

## 5 Add and Subtract Polynomial Functions

Functions, just like numbers, can be added, subtracted, multiplied, and divided. In this section, we concentrate on adding and subtracting polynomial functions.

**DEFINITION**

If $f$ and $g$ are two functions,

The **sum** $\boldsymbol{f + g}$ is the function defined by

$$(f + g)(x) = f(x) + g(x)$$

The **difference** $\boldsymbol{f - g}$ is the function defined by

$$(f - g)(x) = f(x) - g(x)$$

### EXAMPLE 11 Finding the Sum and Difference of Two Polynomial Functions

Let $f$ and $g$ be two functions defined as

$$f(x) = 3x^2 - x + 1 \quad \text{and} \quad g(x) = -x^2 + 5x - 6$$

Find the following:

**(a)** $(f + g)(x)$ **(b)** $(f - g)(x)$ **(c)** $(f + g)(2)$ **(d)** $(f - g)(-1)$

**Solution**

**(a)**
$$\begin{aligned}(f + g)(x) &= f(x) + g(x)\\ &= (3x^2 - x + 1) + (-x^2 + 5x - 6)\\ &= 3x^2 - x + 1 - x^2 + 5x - 6\\ &= 3x^2 - x^2 - x + 5x + 1 - 6\\ &= 2x^2 + 4x - 5\end{aligned}$$

**(b)**
$$\begin{aligned}(f - g)(x) &= f(x) - g(x)\\ &= (3x^2 - x + 1) - (-x^2 + 5x - 6)\\ &= 3x^2 - x + 1 + x^2 - 5x + 6\\ &= 3x^2 + x^2 - x - 5x + 1 + 6\\ &= 4x^2 - 6x + 7\end{aligned}$$

**(c)** Because $(f + g)(x) = 2x^2 + 4x - 5$, we have

$$\begin{aligned}(f + g)(2) &= 2(2)^2 + 4(2) - 5\\ &= 8 + 8 - 5\\ &= 11\end{aligned}$$

**Work Smart**

We could also have evaluated $(f + g)(2)$ as follows:

$$\begin{aligned}(f + g)(2) &= f(2) + g(2)\\ &= 11 + 0\\ &= 11\end{aligned}$$

**(d)** Because $(f - g)(x) = 4x^2 - 6x + 7$, we have

$$\begin{aligned}(f - g)(-1) &= 4(-1)^2 - 6(-1) + 7\\ &= 4 + 6 + 7\\ &= 17\end{aligned}$$

**QUICK ✓**

**22.** Let $f$ and $g$ be two functions defined as

$$f(x) = 3x^2 - x + 1 \qquad \text{and} \qquad g(x) = -x^2 + 5x - 6$$

Find the following:

**(a)** $(f + g)(x)$ **(b)** $(f - g)(x)$ **(c)** $(f + g)(1)$ **(d)** $(f - g)(-2)$

Let's look at an application of the difference of two functions.

## EXAMPLE 12 The Profit Function

Profit is defined as total revenue minus total cost. The profit function of a company that manufactures and sells $x$ units of a product is given by

$$P(x) = R(x) - C(x)$$

where $P$ represents the company's profits
$R$ represents the company's revenue
$C$ represents the company's cost

**(a)** If a company sells a scientific calculator for \$12, its revenue function is $R(x) = 12x$. If the company's variable cost is \$7 per calculator and fixed costs are \$1200 per week, its cost is given by $C(x) = 7x + 1200$. Find the company's profit function, $P(x)$.

**(b)** Determine and interpret $P(800)$.

### Solution

**(a)** The profit function is

$$\begin{aligned} P(x) &= R(x) - C(x) \\ &= 12x - (7x + 1200) \\ &= 12x - 7x - 1200 \\ &= 5x - 1200 \end{aligned}$$

**(b)** $$\begin{aligned} P(800) &= 5(800) - 1200 \\ &= 4000 - 1200 \\ &= 2800 \end{aligned}$$

If the company manufactures and sells 800 calculators in a week, its profit will be \$2800.

**QUICK ✓**

**23.** The calculator company presented in Example 12 just gave its employees a raise that increases the variable cost to \$8 per calculator. In addition, it renewed its lease on the plant so that weekly fixed costs increased to \$1250.

**(a)** Determine the new profit function.

**(b)** Determine and interpret $P(800)$.

# 5.1 Exercises

**For Extra Help:** Student Solutions Manual  CD Video 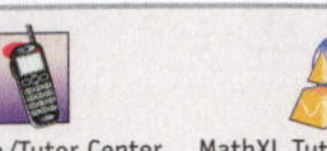 PH Math/Tutor Center  MathXL Tutorials on CD 

MathXL® MyMathLab

## Concepts and Vocabulary

*In Problems 1–3, fill in the blanks.*

**1.** Terms that have the same variable(s) and the same exponent(s) on the variable(s) are called _______ _______.

**2.** The degree of the polynomial 5 is _______.

**3.** The _______ of a polynomial function is the value of the largest exponent on the variable.

*In Problems 4–6, answer True or False to each statement.*

**4.** A monomial in one variable is the product of a constant and a variable raised to an integer power.

**5.** The sum or difference of two monomials that are not like terms is called a binomial.

**6.** The degree of $-3x^3 + 5x^2 - x + 10$ is 3.

**7.** Explain the difference between a term and a monomial term.

**8.** What is the degree of a polynomial that is linear?

**9.** How many terms are in a binomial? How many terms are in a trinomial?

**10.** Give a definition of *polynomial* using your own words. Provide examples of polynomials that are monomials, binomials, and trinomials. In addition, give examples of polynomials that are linear.

**11.** What is the difference between a polynomial and a polynomial function?

**12.** Explain why the degree of the sum of two polynomials is equal to the degree of the polynomial of highest degree.

## Building Skills

*In Problems 13–20, determine the coefficient and degree of each monomial.*

**13.** $3x^2$ **14.** $5x^4$ **15.** $-8x^2y^3$ **16.** $-12xy$

**17.** $\frac{4}{3}x^6$ **18.** $-\frac{5}{3}z^5$ **19.** 2 **20.** $-7$

*In Problems 21–24, state why each of the following is not a polynomial.*

**21.** $2x^{-1} + 3x$ **22.** $6p^{-3} - p^{-2} + 3p^{-1}$ **23.** $\dfrac{4}{z - 1}$ **24.** $\dfrac{x^2 + 2}{x}$

*In Problems 25–40, determine whether the algebraic expression is a polynomial (Yes or No). If it is a polynomial, write the polynomial in standard form, determine the degree and state if it is a monomial, binomial or trinomial. If it is a polynomial with more than 3 terms, identify the expression as a polynomial.*

**25.** $5x^2 - 9x + 1$ **26.** $-3y^2 + 8y + 1$ **27.** $\dfrac{-20}{n}$ **28.** $\dfrac{1}{x}$

**29.** $3y^{\frac{1}{3}} + 2$ **30.** $8m - 4m^{\frac{1}{2}}$ **31.** $\dfrac{5}{8}$ **32.** $-12$

**33.** $5 - 8y + 2y^2$ **34.** $7 - 5p + 2p^2 - p^3$

**35.** $7x^{-1} + 4$

**36.** $4y^{-2} + 6y - 1$

**37.** $3x^2y^2 + 2xy^4 + 4$

**38.** $4mn^3 - 2m^2n^3 + mn^8$

**39.** $4pqr + 2p^2q + 3pq^{\frac{1}{4}}$

**40.** $-2xyz^2 + 7x^3z - 8y^{\frac{1}{2}}z$

*In Problems 41–60, simplify each polynomial by adding or subtracting, as indicated. Express your answer as a single polynomial in standard form.*

**41.** $5z^3 + 8z^3$

**42.** $10y^4 - 6y^4$

**43.** $(x^2 + 5x + 1) + (3x^2 - 2x - 3)$

**44.** $(x^2 - 4x + 1) + (5x^2 + 2x + 7)$

**45.** $(6p^3 - p^2 + 3p - 4) + (2p^3 - 7p + 3)$

**46.** $(2w^3 - w^2 + 6w - 5) + (-3w^3 + 5w^2 + 9)$

**47.** $(5x^2 + 9x + 4) - (3x^2 + 5x + 1)$

**48.** $(7y^2 + 9y + 12) - (4y^2 + 8y - 3)$

**49.** $(7s^2t^3 + st^2 - 5t - 8) - (4s^2t^3 + 5st^2 - 7)$

**50.** $(-2x^3y^3 + 7xy - 3) - (x^3y^3 + 5y^2 + xy - 3)$

**51.** $(3 - 5x + x^2) + (-2 + 3x - 5x^2)$

**52.** $(-3 - 5z + 3z^2) + (1 + 2z + z^2)$

**53.** $(6 - 2y + y^3) - (-2 + y^2 - 2y^3)$

**54.** $(8 - t^3) - (1 + 3t + 3t^2 + t^3)$

**55.** $\left(\frac{1}{4}x^2 + \frac{3}{2}x + 3\right) + \left(\frac{1}{2}x^2 - \frac{1}{4}x - 2\right)$

**56.** $\left(\frac{3}{4}y^3 - \frac{1}{8}y + \frac{2}{3}\right) + \left(\frac{1}{2}y^3 + \frac{5}{12}y - \frac{5}{6}\right)$

**57.** $(5x^2y^2 - 8x^2y + xy^2) + (3x^2y^2 + x^2y - 4xy^2)$

**58.** $(7a^3b + 9ab^2 - 4a^2b) + (-4a^3b + 3a^2b - 8ab^2)$

**59.** $(3x^2y + 7xy^2 + xy) - (2x^2y - 4xy^2 - xy)$

**60.** $(-5xy^2 + 3xy - 9y^2) - (5xy^2 + 7xy - 8y^2)$

*In Problems 61–66, find the following values for each polynomial function.*

**(a)** $f(0)$ **(b)** $f(2)$ **(c)** $f(-3)$

**61.** $f(x) = x^2 - 4x + 1$

**62.** $f(x) = x^2 + 5x - 3$

**63.** $f(x) = 2x^3 - 7x + 3$

**64.** $f(x) = -2x^3 + 3x - 1$

**65.** $f(x) = -x^3 + 3x^2 - 2x + 3$

**66.** $f(x) = -2x^3 + x^2 + 5x - 3$

*In Problems 67–72, for the given functions f and g, find the following functions.*

**(a)** $(f + g)(x)$ **(b)** $(f - g)(x)$ **(c)** $(f + g)(2)$ **(d)** $(f - g)(1)$

**67.** $f(x) = 2x + 5;\ g(x) = -5x + 1$

**68.** $f(x) = 4x + 3;\ g(x) = 2x - 3$

**69.** $f(x) = x^2 - 5x + 3;\ g(x) = 2x^2 + 3$

**70.** $f(x) = 3x^2 + x + 2;\ g(x) = x^2 - 3x - 1$

**71.** $f(x) = x^3 + 6x^2 + 12x + 2;\ g(x) = x^3 - 8$

**72.** $f(x) = 8x^3 + 1;\ g(x) = x^3 + 3x^2 + 3x + 1$

## Applying the Concepts

**73.** Add $5x^3 - 5x + 3$ to $-4x^3 + x^2 - 2x + 1$.

**74.** Add $2x^3 - 3x^2 - 5x + 7$ to $x^3 + 3x^2 - 6x - 4$.

**75.** Subtract $4b^3 - b^2 + 3b - 1$ from $2b^3 + 5b^2 - b + 3$.

**76.** Subtract $2q^3 - 3q^2 + 7q - 2$ from $-5q^3 + q^2 + 2q - 1$.

△ **77. Area** A rectangle has one corner on the graph of $y = 6 - 2x$, another at the origin, a third on the positive $y$-axis, and the fourth on the positive $x$-axis (see the figure).

The polynomial function $A(x) = -2x^2 + 6x$ can be used to find the area $A$ of the rectangle whose vertex is at $(x, y)$.

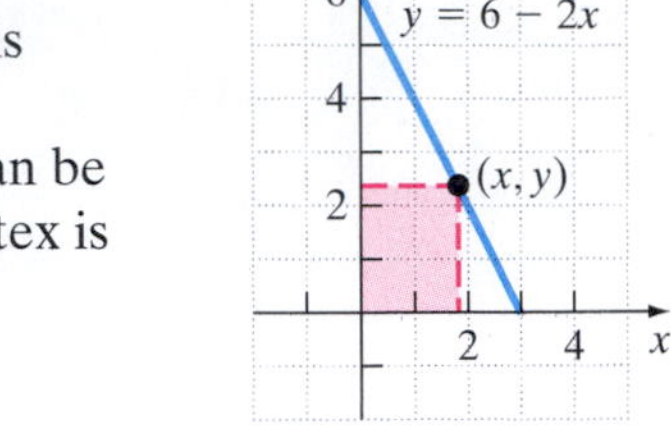


**(a)** Find the area of the rectangle whose vertex is at $(2, 2)$.

**(b)** Find the area of the rectangle whose vertex is at $(1, 4)$.

**(c)** How can the coordinates of the vertex be used to find the area? What is the area?

△ **78. Area** A rectangle has one corner on the graph of $y = 10 - 2x$, another at the origin, a third on the positive $y$-axis, and the fourth on the positive $x$-axis (see the figure).

The polynomial function $A(x) = -2x^2 + 10x$ can be used to find the area $A$ of the rectangle whose vertex is at $(x, y)$.

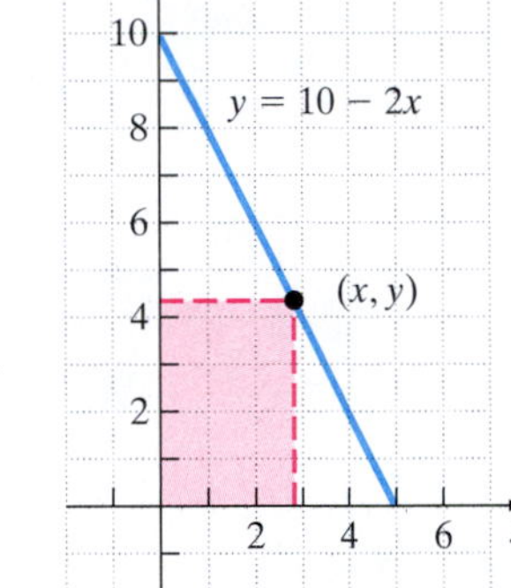


**(a)** Find the area of the rectangle whose vertex is at $(3, 4)$.

**(b)** Find the area of the rectangle whose vertex is at $(4, 2)$.

**79. Price of a New Home** The bar graph shown represents the average price (in thousands of dollars) for a new single-family home in the United States for the years 1991–2003.

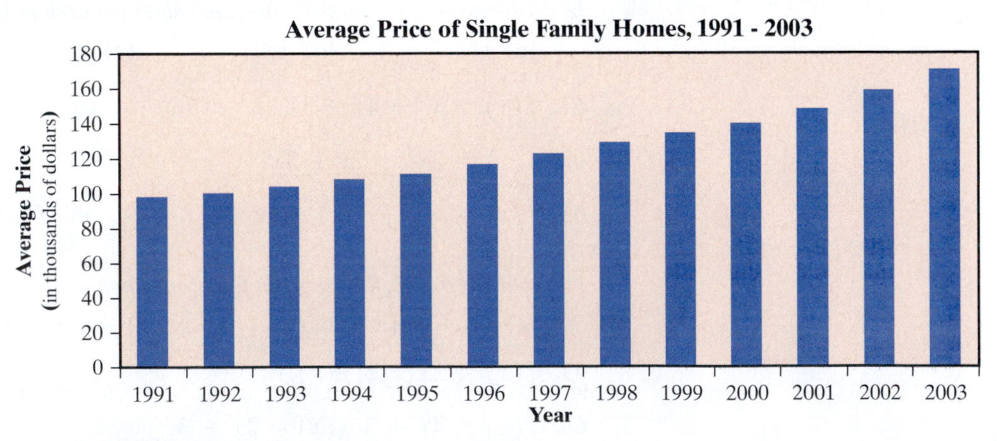


SOURCE: *Statistical Abstract*, 2004

The polynomial function $P(t) = 0.332t^2 + 1.848t + 97.826$ can be used to approximate the average price $P$, where $t$ is the number of years since 1991.

**(a)** Use the function to estimate the average price in 1991.

**(b)** Use the function to estimate the average price in 2010.

**80. Income** The bar graph shown represents the average per-capita income for male residents of the United States by age in 2002.

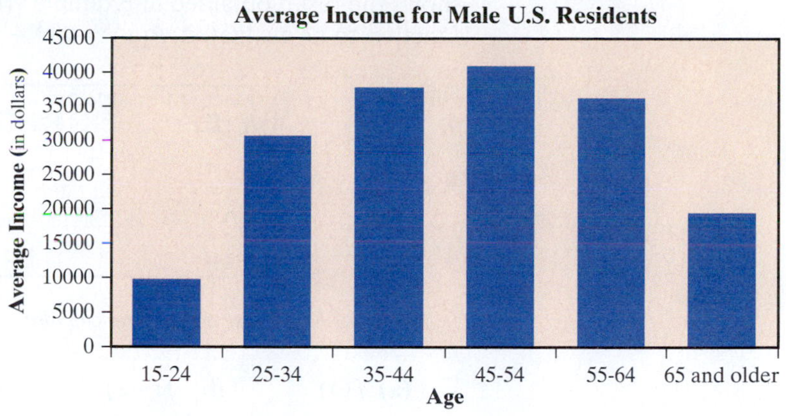


SOURCE: *Statistical Abstract,* 2004

The polynomial function $I(a) = -42.32a^2 + 4005.76a - 53062.54$ can be used to approximate the average income $I$, where $a$ is the age of the individual.

**(a)** Use the function to estimate the average income of a 20-year-old in 2002.
**(b)** Use the function to estimate the average income of a 55-year-old in 2002.

**81. Profit Function** Suppose that the revenue $R$ from selling $x$ cell phones is $R(x) = -1.2x^2 + 220x$. The cost $C$ of selling $x$ cell phones is $C(x) = 0.05x^3 - 2x^2 + 65x + 500$.

**(a)** Find the profit function, $P(x)$.
**(b)** Find the profit if $x = 15$ cell phones are sold.
**(c)** Determine and interpret $P(100)$.

**82. Profit Function** Suppose that the revenue $R$ from selling $x$ clocks is $R(x) = -0.3x^2 + 30x$. The cost $C$ of selling $x$ clocks is $C(x) = 0.1x^2 + 7x + 400$.

**(a)** Find the profit function, $P(x)$.
**(b)** Find the profit if $x = 15$ clocks are sold.
**(c)** Determine and interpret $P(40)$.

## Extending the Concepts

**83.** If $f(x) = 3x + 2$ and $g(x) = ax - 5$, find $a$ such that $(f + g)(3) = 12$.

**84.** If $f(x) = -5x + 1$ and $g(x) = ax - 3$, find $a$ such that $(f - g)(2) = 10$.

**85.** The graph of two functions, $f$ and $g$, is shown to the right. Use the graph to answer parts (a)–(d).

**(a)** $(f + g)(2)$
**(b)** $(f + g)(4)$
**(c)** $(f - g)(6)$
**(d)** $(g - f)(6)$

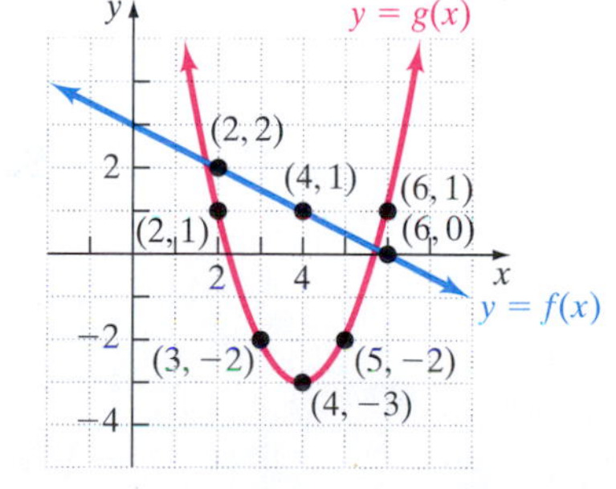


**86. Work-Related Disability** Suppose that $M(a)$ represents the number of American males with a work-related disability who are $a$ years of age. Suppose that $F(a)$ represents the number of American females with a work-related disability who are $a$ years of age. Determine a function $T$ that represents the total number of Americans with work-related disabilities who are $a$ years of age.

**87. Taxes** Let $T(x)$ represent the total taxes paid by everyone who worked in the United States in year $x$. Let $F(x)$ represent the taxes paid to the federal government in year $x$. Write a function $S$ that represents the total taxes paid other than federal (state, property, sales, and so on) in year $x$.

### The Graphing Calculator

Graphing calculators can be used to evaluate any function that you wish. Figure 2 shows the result obtained in Example 9(b) on a TI-84 Plus graphing calculator with the function to be evaluated, $P(x) = 2x^3 - 5x^2 + x - 3$ in $Y_1$.

**Figure 2**

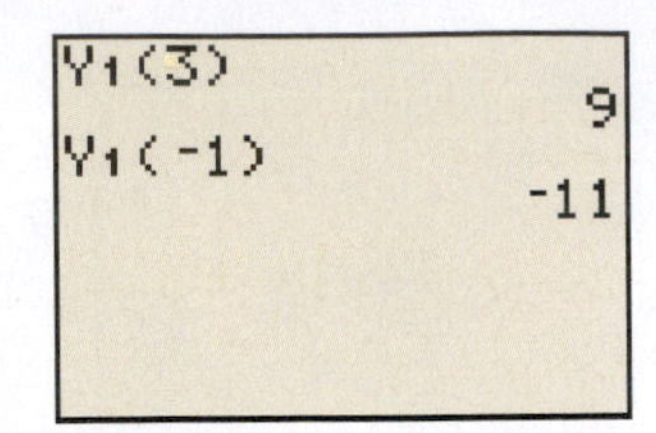


*In Problems 88–91, use a graphing calculator to find the following values for each polynomial function.*

**(a)** $f(4)$ **(b)** $f(-2)$ **(c)** $f(6)$

**88.** $f(x) = -2x^2 + 5x + 3$

**89.** $f(x) = 4x^2 - 7x + 1$

**90.** $f(x) = -3x^3 + 5x^2 - 8x + 1$

**91.** $f(x) = 2x^3 - 5x^2 + x + 5$

# 5.2 Multiplying Polynomials

**OBJECTIVES**

1. Multiply a Monomial and a Polynomial
2. Multiply a Binomial by a Binomial
3. Multiply a Polynomial by a Polynomial
4. Multiply Special Products
5. Multiply Polynomial Functions

### *Preparing for Multiplying Polynomials*

*Before getting started, take the following readiness quiz. If you get a problem wrong, go back to the section cited and review the material.*

**1.** Simplify: $4x^2 \cdot 3x^3$ [Getting Ready, p. 339]

**2.** Simplify: $(-3x)^2$ [Getting Ready, p. 344]

**3.** Use the Distributive Property to remove the parentheses: $4(x - 5)$ [Section R.3, p. 30]

Multiplying polynomials is based on two concepts learned earlier: the Product Rule for Exponents (Getting Ready, p. 339) and the Distributive Property (Section R.3, p. 30). The following example will help you to review these concepts.

**EXAMPLE 1 Review the Product Rule for Exponents and the Distributive Property**

**(a)** Simplify: $(2a^3b)(-6a^2b^4)$

**(b)** Remove the parentheses: $2(z - 3)$

**Solution**

**(a)**

$$(2a^3b)(-6a^2b^4) = (2\cdot(-6))(a^3\cdot a^2)(b\cdot b^4)$$

Product Rule: $a^m \cdot a^n = a^{m+n}$:

$$= -12a^{3+2}b^{1+4}$$

$$= -12a^5b^5$$

**(b)** The Distributive Property is used to remove parentheses:

$$2(z - 3) = 2\cdot z - 2\cdot 3 = 2z - 6$$

**QUICK ✓** *Simplify each expression completely. All exponents should be positive.*

**1.** $(3x^5)(2x^2)$

**2.** $(-7a^3b^2)(3ab^4)$

**3.** $\left(\frac{2}{3}x^4\right)\left(\frac{15}{8}x\right)$

**4.** $-3(x + 2)$

*Preparing for...Answers* **1.** $12x^5$ **2.** $9x^2$ **3.** $4x - 20$

## 1 Multiply a Monomial and a Polynomial

**In Words**

The Extended Form of the Distributive Property says to multiply each term in parentheses by $a$.

From Example 1(b), we can see that when we multiply a binomial by a monomial, we use the Distributive Property to remove the parentheses. In general, when we multiply a polynomial by a monomial, we use the *Extended Form of the Distributive Property.*

**EXTENDED FORM OF THE DISTRIBUTIVE PROPERTY**

$$a(b_1 + b_2 + \cdots + b_n) = a \cdot b_1 + a \cdot b_2 + \cdots + a \cdot b_n$$

where $a, b_1, b_2, \ldots, b_n$ are real numbers.

### EXAMPLE 2 Using the Extended Form of the Distributive Property

Multiply and simplify each of the following expressions:

**(a)** $3x^2(x^2 + 4x + 2)$ **(b)** $\frac{1}{2}xy^3\left(\frac{2}{3}xy^2 + \frac{6}{5}y + \frac{3}{4}\right)$

**Solution**

**(a)** $3x^2(x^2 + 4x + 2) = 3x^2 \cdot x^2 + 3x^2 \cdot 4x + 3x^2 \cdot 2$
$= 3x^4 + 12x^3 + 6x^2$

**(b)** $\frac{1}{2}xy^3\left(\frac{2}{3}xy^2 + \frac{6}{5}y + \frac{3}{4}\right) = \frac{1}{2}xy^3 \cdot \frac{2}{3}xy^2 + \frac{1}{2}xy^3 \cdot \frac{6}{5}y + \frac{1}{2}xy^3 \cdot \frac{3}{4}$
$= \frac{1}{3}x^2y^5 + \frac{3}{5}xy^4 + \frac{3}{8}xy^3$

**QUICK ✓** *Multiply and simplify each of the expressions.*

**5.** $5x(x^2 + 3x + 2)$ **6.** $2xy(3x^2 - 5xy + 2y^2)$ **7.** $\frac{3}{4}y^2\left(\frac{4}{3}y^2 + \frac{2}{9}y + \frac{16}{3}\right)$

## 2 Multiply a Binomial by a Binomial

When we find the product of two binomials, we use the Distributive Property by distributing the first binomial to each term in the second binomial. Just as with addition and subtraction of polynomials, we can use either a horizontal or vertical format.

### EXAMPLE 3 Multiplying Two Binomials

Find the product: $(3x + 4)(2x - 5)$

**Solution**

VERTICAL MULTIPLICATION

$$\begin{array}{r} 3x + 4 \\ \times\ 2x - 5 \\ \hline -15x - 20 \\ 6x^2 + 8x \quad \\ \hline 6x^2 - 7x - 20 \end{array}$$

$-15x - 20 \leftarrow -5(3x + 4)$

$6x^2 + 8x \leftarrow 2x(3x + 4)$

HORIZONTAL MULTIPLICATION

We distribute the first binomial to each term in the second binomial.

$(3x + 4)(2x - 5) = (3x + 4)(2x) + (3x + 4)(-5)$
$= 3x \cdot 2x + 4 \cdot 2x + 3x \cdot (-5) + 4 \cdot (-5)$
$= 6x^2 + 8x - 15x - 20$

Combine like terms: $= 6x^2 - 7x - 20$

In either case, $(3x + 4)(2x - 5) = 6x^2 - 7x - 20$.

## The FOIL Method

**Work Smart**

The FOIL method is a form of the Distributive Property.

When multiplying two binomials, we can use a method referred to as the **FOIL method.** The acronym FOIL stands for **F**irst, **O**uter, **I**nner, **L**ast and is illustrated below.

$$(ax + b)(cx + d) = \overset{F}{ax \cdot cx} + \overset{O}{ax \cdot d} + \overset{I}{b \cdot cx} + \overset{L}{b \cdot d}$$

First Last Inner Outer

### EXAMPLE 4 Using the FOIL Method to Multiply Two Binomials

Find the product:

**(a)** $(4y - 3)(2y + 5)$ **(b)** $(2m + n)(m - 3n)$

**Solution**

**(a)**
$$\begin{aligned}(4y - 3)(2y + 5) &= \overset{F}{4y \cdot 2y} + \overset{O}{4y \cdot 5} - \overset{I}{3 \cdot 2y} - \overset{L}{3 \cdot 5}\\ &= 8y^2 + 20y - 6y - 15\\ &= 8y^2 + 14y - 15\end{aligned}$$

First Last Inner Outer

**(b)**
$$\begin{aligned}(2m + n)(m - 3n) &= \overset{F}{2m \cdot m} - \overset{O}{2m \cdot 3n} + \overset{I}{n \cdot m} - \overset{L}{n \cdot 3n}\\ &= 2m^2 - 6mn + mn - 3n^2\\ &= 2m^2 - 5mn - 3n^2\end{aligned}$$

**QUICK ✓** *Find the product.*

**8.** $(x + 4)(x + 1)$ **9.** $(3v + 5)(2v - 3)$ **10.** $(2a - b)(a + 5b)$

## 3 Multiply a Polynomial by a Polynomial

**Work Smart**

When multiplying polynomials, it is a good idea to write the polynomials in descending order of degree.

When multiplying a trinomial by a binomial or a trinomial by a trinomial, we make repeated use of the Distributive Property. The approach is similar to that taken in Example 3 presented earlier. It is a good idea to write each polynomial in standard form. Although you also have the choice of the horizontal or vertical format, we only present the horizontal format.

### EXAMPLE 5 Multiplying Two Polynomials

Find the product:

**(a)** $(3x + 1)(x^2 + 4x - 3)$ **(b)** $(x^2 + 5x + 2)(2x^2 - x + 3)$

**Solution**

**(a)** We begin by distributing the binomial $3x + 1$ to each term in the trinomial.

$$(3x + 1)(x^2 + 4x - 3) = (3x + 1) \cdot x^2 + (3x + 1) \cdot 4x + (3x + 1) \cdot (-3)$$

Distributive Property: $= 3x^3 + x^2 + 12x^2 + 4x - 9x - 3$

Combine like terms: $= 3x^3 + 13x^2 - 5x - 3$

**(b)** Distribute $x^2 + 5x + 2$ to each term in the second trinomial.

$$(x^2 + 5x + 2)(2x^2 - x + 3) = (x^2 + 5x + 2) \cdot 2x^2 + (x^2 + 5x + 2) \cdot (-x) + (x^2 + 5x + 2) \cdot 3$$

Distributive Property: $$= x^2 \cdot 2x^2 + 5x \cdot 2x^2 + 2 \cdot 2x^2 + x^2 \cdot (-x) + 5x \cdot (-x) + 2 \cdot (-x) + x^2 \cdot 3 + 5x \cdot 3 + 2 \cdot 3$$

$$= 2x^4 + 10x^3 + 4x^2 - x^3 - 5x^2 - 2x + 3x^2 + 15x + 6$$

Combine like terms: $$= 2x^4 + 9x^3 + 2x^2 + 13x + 6$$

**QUICK ✓** *Find the product.*

**11.** $(2y - 3)(y^2 + 4y + 5)$

**12.** $(z^2 - 3z + 2)(2z^2 + z + 6)$

## 4 Multiply Special Products

There are certain products, called **special products,** which occur quite frequently. Because these products are special, we give them names.

### EXAMPLE 6 Products of the Form $(A - B)(A + B)$

Find the product: $(x - 7)(x + 7)$

**Solution**

We use FOIL and obtain

$$(x - 7)(x + 7) = \overset{F}{x \cdot x} + \overset{O}{x \cdot 7} - \overset{I}{7 \cdot x} - \overset{L}{7 \cdot 7}$$
$$= x^2 + 7x - 7x - 49$$
$$= x^2 - 49$$

**Work Smart**

When multiplying products of the form $(A - B)(A + B)$, the "middle terms" will always be "opposites" and therefore sum to 0.

We have the following based upon the results of Example 6.

**DIFFERENCE OF TWO SQUARES**

$$(A - B)(A + B) = A^2 - B^2$$

### EXAMPLE 7 Using the Difference of Two Squares Formula

**(a)** $(A + B)\ (A - B) = A^2 \quad - B^2$

$$(3x + 2)(3x - 2) = (3x)^2 - 2^2$$
$$= 9x^2 - 4$$

**(b)** In this next example, $A = 2m$ and $B = 5n^2$.

$$(2m - 5n^2)(2m + 5n^2) = (2m)^2 - (5n^2)^2$$
$$= 4m^2 - 25n^4$$

**QUICK ✓** *Find each product.*

**13.** $(5y + 2)(5y - 2)$

**14.** $(7y + 2z^3)(7y - 2z^3)$

### EXAMPLE 8 Products of the Form $(A + B)^2$ or $(A - B)^2$

Find each product:

**(a)** $(4x + 3)^2$ **(b)** $(3z - 5)^2$

**Solution**

**(a)**
$$(4x + 3)^2 = (4x + 3)(4x + 3)$$
$$= (4x)^2 + \underbrace{(4x)(3) + (4x)(3)}_{= 2(4x)(3)} + 3^2$$
$$= 16x^2 + 24x + 9$$

**(b)**
$$(3z - 5)^2 = (3z - 5)(3z - 5)$$
$$= (3z)^2 - \underbrace{(3z)(5) - (3z)(5)}_{= -2(3z)(5)} + 5^2$$
$$= 9z^2 - 30z + 25$$

**Work Smart**

$$(A + B)^2 \neq A^2 + B^2$$
$$(A - B)^2 \neq A^2 - B^2$$

Whenever you feel the urge to perform an operation that you're not quite sure about, try it with actual numbers. For example, does

$$(3 + 2)^2 = 3^2 + 2^2?$$

NO! So

$$(x + y)^2 \neq x^2 + y^2$$

Example 8 leads to some general results.

**SQUARES OF BINOMIALS, OR PERFECT SQUARE TRINOMIALS**

$$(A + B)^2 = A^2 + 2AB + B^2$$
$$(A - B)^2 = A^2 - 2AB + B^2$$

### EXAMPLE 9 Using the Perfect Square Trinomial Formulas

$$(A + B)^2 = A^2 + 2\ A\ B + B^2$$

**(a)**
$$(w + 5)^2 = w^2 + 2 \cdot w \cdot 5 + 5^2$$
$$= w^2 + 10w + 25$$

$$(A - B)^2 = A^2 - 2\ A\ B + B^2$$

**(b)**
$$(6p - 5)^2 = (6p)^2 - 2 \cdot 6p \cdot 5 + 5^2$$
$$= 36p^2 - 60p + 25$$

**(c)**
$$(3x + 5y^2)^2 = (3x)^2 + 2 \cdot 3x \cdot 5y^2 + (5y^2)^2$$
$$= 9x^2 + 30xy^2 + 25y^4$$

**Work Smart**

If you can't remember the formulas for a perfect square, don't panic! Use the fact that

$$(A + B)^2 = (A + B)(A + B)$$

and then FOIL. The same logic applies to perfect squares of the form $(A - B)^2$.

**QUICK ✓** *Find each product.*

**15.** $(z - 8)^2$ **16.** $(6p + 5)^2$ **17.** $(4a - 3b)^2$

## 5 Multiply Polynomial Functions

In Section 5.1, we introduced adding and subtracting functions. We now present the multiplication of functions.

**DEFINITION**

Let $f$ and $g$ be two functions. The **product $f \cdot g$** is the function defined by

$$(f \cdot g)(x) = f(x) \cdot g(x)$$

## EXAMPLE 10 Finding the Product of Two Functions

Suppose that $f(x) = 2x + 5$ and $g(x) = x^2 - 7x + 5$.

**(a)** Find $f(4) \cdot g(4)$.

**(b)** Find $(f \cdot g)(x)$.

**(c)** Use the result from part (b) to determine $(f \cdot g)(4)$.

### Solution

**(a)** We will evaluate the functions at $x = 4$ separately.

$$f(4) = 2(4) + 5 = 13 \qquad g(4) = (4)^2 - 7(4) + 5 = 16 - 28 + 5 = -7$$

$$f(4) \cdot g(4) = (13)(-7) = -91$$

**(b)**

$$\begin{aligned} (f \cdot g)(x) &= f(x) \cdot g(x) \\ &= (2x + 5)(x^2 - 7x + 5) \\ \text{Distribute (twice):} \quad &= 2x \cdot x^2 - 2x \cdot 7x + 2x \cdot 5 + 5 \cdot x^2 - 5 \cdot 7x + 5 \cdot 5 \\ \text{Simplify:} \quad &= 2x^3 - 14x^2 + 10x + 5x^2 - 35x + 25 \\ \text{Combine like terms:} \quad &= 2x^3 - 9x^2 - 25x + 25 \end{aligned}$$

**(c)**

$$\begin{aligned} (f \cdot g)(4) &= 2(4)^3 - 9(4)^2 - 25(4) + 25 \\ &= 2(64) - 9(16) - 100 + 25 \\ &= 128 - 144 - 100 + 25 \\ &= -91 \end{aligned}$$

Notice that $(f \cdot g)(4) = f(4) \cdot g(4)$, as we would expect.

**QUICK ✓**

**18.** Suppose that $f(x) = 5x - 3$ and $g(x) = x^2 + 3x + 1$.

**(a)** Find $f(2) \cdot g(2)$.

**(b)** Find $(f \cdot g)(x)$.

**(c)** Use the result from part (b) to determine $(f \cdot g)(2)$.

---

Back in Section 2.3, we learned to evaluate functions when the argument, the value at which we are evaluating the function, was, itself, an algebraic expression.

## EXAMPLE 11 Evaluating Functions

For the function $f(x) = x^2 + 5x$, find

**(a)** $f(x + 3)$ **(b)** $f(x + h) - f(x)$

### Solution

$$f(\text{input}) = (\text{input})^2 + 5(\text{input})$$

**(a)**

$$\begin{aligned} f(x + 3) &= (x + 3)^2 + 5(x + 3) \\ &= x^2 + 6x + 9 + 5x + 15 \\ &= x^2 + 11x + 24 \end{aligned}$$

**(b)** To evaluate $f(x + h) - f(x)$, we first evaluate $f(x + h)$ by replacing $x$ in $f(x)$ with $x + h$. We then subtract $f(x)$ from this result.

$$f(x+h) - f(x) = \overbrace{(x+h)^2 + 5(x+h)}^{f(x+h)} - \overbrace{(x^2 + 5x)}^{f(x)}$$

$$= x^2 + 2xh + h^2 + 5x + 5h - x^2 - 5x$$

Combine like terms: $\quad = 2xh + h^2 + 5h$

**QUICK ✓**

**19.** For the function $f(x) = x^2 - 2x$, find (a) $f(x - 3)$ (b) $f(x + h) - f(x)$.

# 5.2 Exercises

**For Extra Help:** Student Solutions Manual  CD Video  PH Math/Tutor Center  MathXL Tutorials on CD · MathXL® · MyMathLab

## Concepts and Vocabulary

*In Problems 1–3, fill in the blanks.*

**1.** $(A - B)(A + B) =$ ______.

**2.** $(A - B)^2 =$ ______.

**3.** $(f \cdot g)(x) =$ ______ $\cdot$ ______.

*In Problems 4–6, answer True or False to each statement.*

**4.** $(x + a)^2 = x^2 + a^2$.

**5.** The acronym FOIL stands for First, Outer, Inner, Last.

**6.** The product of a binomial and a binomial is always a trinomial.

**7.** The product of a binomial of degree 2 and a trinomial of degree 3 will be a polynomial of degree 5. Explain why.

**8.** Do you prefer the horizontal or vertical format when multiplying polynomials? Justify your position.

## Building Skills

*In Problems 9–20, find the product.*

**9.** $(5xy^2)(-3x^2y^3)$

**10.** $(9a^3b^2)(-3a^2b^5)$

**11.** $\left(\frac{3}{4}yz^3\right)\left(\frac{20}{9}y^3z^2\right)$

**12.** $\left(\frac{12}{5}x^2y\right)\left(\frac{15}{4}x^4y^3\right)$

**13.** $5x(x^2 + 4x + 2)$

**14.** $6y(y^2 - 4y + 3)$

**15.** $-4a^2b(3a^2 + 2ab - b^2)$

**16.** $-3mn^3(4m^2 - mn + 5n^2)$

**17.** $\frac{2}{3}ab\left(\frac{3}{4}a^2b - \frac{9}{8}ab^3 + 6ab\right)$

**18.** $\frac{5}{2}xy\left(\frac{4}{15}x^2y - \frac{6}{5}xy + \frac{3}{10}xy^2\right)$

**19.** $0.4x^2(1.2x^2 - 0.8x + 1.5)$

**20.** $0.8y(0.4y^2 + 1.1y - 2.5)$

*In Problems 21–34, find the product of the two binomials. You may use any method you wish.*

**21.** $(x + 3)(x + 5)$

**22.** $(y - 2)(y - 6)$

**23.** $(a + 5)(a - 3)$

**24.** $(z - 8)(z + 3)$

**25.** $(4a + 3)(3a - 1)$

**26.** $(5x - 3)(x + 4)$

**27.** $(-3x + 1)(2x + 7)$

**28.** $(-x + 4)(6x + 1)$

**29.** $(4 - 5x)(3 + 2x)$

**30.** $(2 - 7y)(5 + 2y)$ **31.** $\left(\frac{2}{3}x + 2\right)\left(\frac{1}{2}x - 4\right)$ **32.** $\left(\frac{3}{2}y + 4\right)\left(\frac{4}{3}y - 1\right)$

**33.** $(4a + 3b)(a - 5b)$ **34.** $(3m - 5n)(m + 2n)$

*In Problems 35–50, find the product of the polynomials. You may use any method you wish.*

**35.** $(x + 1)(x^2 + 4x + 2)$ **36.** $(y - 2)(y^2 + 5y - 3)$

**37.** $(3a - 2)(2a^2 + a - 5)$ **38.** $(2b + 3)(3b^2 - 2b + 1)$

**39.** $(5z^2 + 3z + 2)(4z + 3)$ **40.** $(3p^2 - 5p + 3)(7p - 2)$

**41.** $(x - 3)(x^3 - 4x^2 + 2x - 7)$ **42.** $(w + 2)(w^3 - 6w^2 - 3w + 2)$

**43.** $(4 + y)(2y^2 - 3 + 5y)$ **44.** $(3 + 2z)(z^2 + 5 - 3z)$

**45.** $(w^2 + 2w + 1)(2w^2 - 3w + 1)$ **46.** $(a^2 + 4a + 4)(3a^2 - a - 2)$

**47.** $(x + y)(3x^2 - 2xy + 4y^2)$ **48.** $(a + 2b)(2a^2 - 5ab + 3b^2)$

**49.** $(2ab + 5)(4a^2 - 2ab + b^2)$ **50.** $(xy - 2)(x^2 + 2xy + 4y^2)$

*In Problems 51–68, find the special product.*

**51.** $(x - 6)(x + 6)$ **52.** $(y + 9)(y - 9)$ **53.** $(a + 8)^2$

**54.** $(b + 3)^2$ **55.** $(3y - 1)^2$ **56.** $(4z - 5)^2$

**57.** $(5a + 3b)(5a - 3b)$ **58.** $(8y + 3z)(8y - 3z)$ **59.** $(8z + y)^2$

**60.** $(4a + 7b)^2$ **61.** $(10x - y)^2$ **62.** $(7p - 3q)^2$

**63.** $(a^3 + 2b)(a^3 - 2b)$ **64.** $(m^2 - 2n^3)(m^2 + 2n^3)$

**65.** $[3x - (y + 1)][3x + (y + 1)]$ **66.** $[5 - (a + b)][5 + (a + b)]$

**67.** $[2a + (b - 3)]^2$ **68.** $[(m + 4) - n]^2$

*In Problems 69–74, for the given functions find*

**(a)** $(f \cdot g)(x)$ **(b)** $(f \cdot g)(3)$

**69.** $f(x) = x + 4; g(x) = x - 1$ **70.** $f(x) = x + 5; g(x) = 2x + 1$

**71.** $f(x) = 4x - 3; g(x) = 2x + 5$ **72.** $f(x) = 5x - 1; g(x) = 4x + 5$

**73.** $f(x) = (x - 2); g(x) = x^2 + 5x - 3$ **74.** $f(x) = (x + 5); g(x) = x^2 - 2x + 3$

*In Problems 75–80, for the given functions, find*

**(a)** $f(x + 2)$ **(b)** $f(x + h) - f(x)$

**75.** $f(x) = x^2 + 1$ **76.** $f(x) = x^2 - 4$ **77.** $f(x) = x^2 + 5x - 2$

**78.** $f(x) = x^2 - 2x + 3$ **79.** $f(x) = 3x^2 - x + 1$ **80.** $f(x) = -2x^2 + x - 5$

## Mixed Practice

*In Problems 81–100, simplify the expression.*

**81.** $5ab(a - b)^2$ **82.** $-3x(x - 3)^2$ **83.** $(5y + 1)(5y - 1)$

**84.** $(9b + 2)^2$ **85.** $(8z + 3)(3z - 1)$ **86.** $(4x + 3)(3x - 7)$

**87.** $(2m - 3n)(4m + n) - (m - 2n)^2$ **88.** $(6p + q)(5p - 2q) - (p + 3q)^2$

**89.** $(x + 3)(x^2 - 3x + 9)$ **90.** $(2y + 3)(4y^2 - 6y + 9)$

**91.** $\left(2x - \frac{1}{2}\right)^2$

**92.** $\left(3x + \frac{1}{3}\right)^2$

**93.** $(p + 2)^3$

**94.** $(z - 3)^3$

**95.** $(7x - 5y + 2)(3x - 2y + 1)$

**96.** $(2a + b - 5)(4a - 2b + 1)$

**97.** $(2p - 1)(p + 3) + (p - 3)(p + 3)$

**98.** $(3z + 2)(z - 2) + (z + 2)(z - 2)$

**99.** $(x + 3)(x - 3)(x^2 - 9) - (x + 1)(x^2 - 3)$

**100.** $(a + 2)(a - 2)(a^2 - 4) - (a + 3)(a^2 - 3)$

## Applying the Concepts

△ *We can visualize the product of polynomials by using area of rectangles. In Problems 101–104, find a polynomial expression for the total area of each of the following figures.*

**101.** **102.** **103.** **104.**

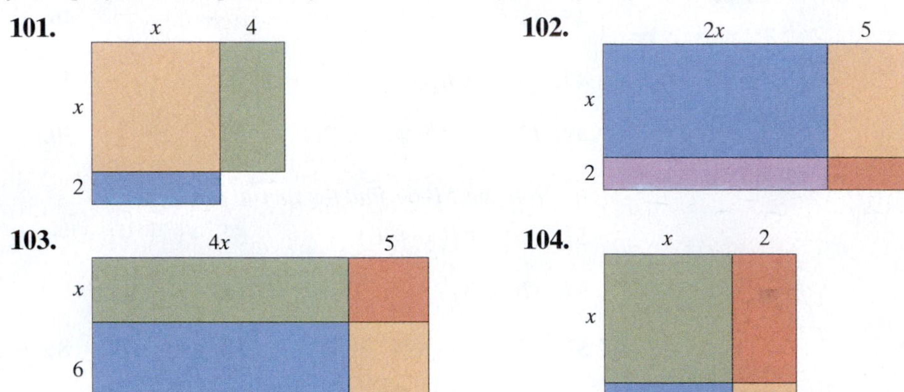


△ *In Problems 105 and 106, write a polynomial expression for the area of the shaded region of the figure.*

**105.**

$2x + 4$

$x + 2$

$3x - 1$

$2x - 5$

**106.**

$3x + 4$

$x + 2$

$2x + 1$

$x - 1$

△ **107. Perfect Square** Why is the expression $(a + b)^2$ called a perfect square? Consider the figure to the right.

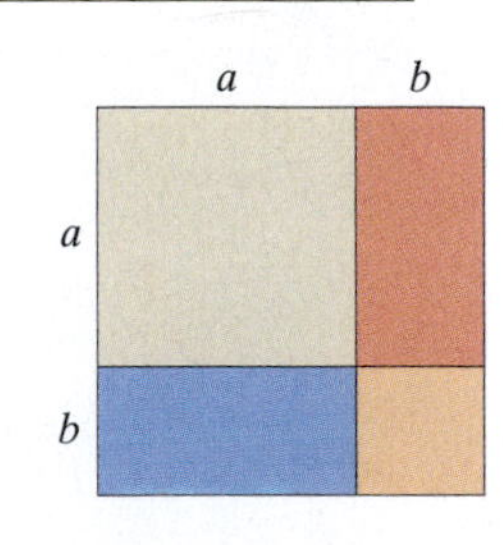


**(a)** Find the area of each of the four quadrilaterals.

**(b)** Use the result from part (a) to find the area of the entire region.

**(c)** Find the length and width of the entire region in terms of $a$ and $b$. Use this result to find the area of the entire region. What do you notice?

△ **108. Area** Express the area of the shaded region in the figure shown as a polynomial in standard form.

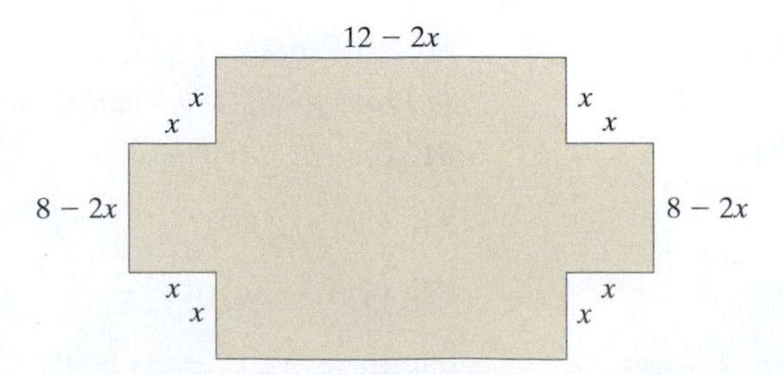

### Extending the Concepts

*In Problems 109–112, find the product.*

**109.** $(2^x + 3)(2^x - 4)$

**110.** $(3^x - 1)(3^x - 9)$

**111.** $(5^y - 1)^2$

**112.** $(2^z - 4)^2$

# 5.3 Dividing Polynomials; Synthetic Division

**OBJECTIVES**

1. Divide a Polynomial by a Monomial
2. Divide Polynomials Using Long Division
3. Divide Polynomials Using Synthetic Division
4. Divide Polynomial Functions
5. Use the Remainder and Factor Theorems

### *Preparing for Dividing Polynomials and Synthetic Division*

*Before getting started, take the following readiness quiz. If you get a problem wrong, go back to the section cited and review the material.*

**1.** Simplify: $\dfrac{15x^5}{12x^3}$. [Getting Ready, p. 340]

**2.** Add: $\dfrac{2}{7} + \dfrac{5}{7}$. [Section R.3, pp. 27–28]

We have now presented a discussion of adding, subtracting, and multiplying polynomials. All that's left is to discuss polynomial division! We begin with dividing a polynomial by a monomial.

When we added, subtracted, and multiplied polynomials, the result was also a polynomial. This is not true for polynomial division—a polynomial divided by a polynomial may not be a polynomial.

## 1 Divide a Polynomial by a Monomial

Dividing a polynomial by a monomial is based on the Quotient Rule for Exponents (see Getting Ready, p. 340). For example,

$$\frac{24z^5}{18z^2} = \frac{6 \cdot 4}{6 \cdot 3}z^{5-2}$$
$$= \frac{4}{3}z^3$$

**In Words**
To divide a polynomial by a monomial, divide the monomial into each term in the polynomial.

Recall that

$$\frac{A}{C} + \frac{B}{C} = \frac{A + B}{C}$$

We reverse this process when dividing a polynomial by a monomial.

For example, if $A$, $B$, and $C$ are monomials, we can write $\dfrac{A + B}{C}$ as shown below.

$$\frac{A + B}{C} = \frac{A}{C} + \frac{B}{C}$$

We can extend this result to polynomials with three or more terms.

### EXAMPLE 1 Dividing a Polynomial by a Monomial

Divide and simplify:

**(a)** $\dfrac{12p^4 + 15p^3 + 3p^2}{3p^2}$

**(b)** $\dfrac{6a^2b^2 - 4a^2b + 3ab^2}{2a^2b^2}$

*Preparing for...Answers* **1.** $\frac{5}{4}x^2$ **2.** 1

### Solution

For each of the problems, we divide the monomial in the denominator into each term in the numerator.

**(a)** $$\frac{12p^4 + 15p^3 + 3p^2}{3p^2} = \frac{12p^4}{3p^2} + \frac{15p^3}{3p^2} + \frac{3p^2}{3p^2}$$

$\frac{a^m}{a^n} = a^{m-n}$: $$= \frac{12}{3}p^{4-2} + \frac{15}{3}p^{3-2} + \frac{3}{3}\cdot p^{2-2}$$

Simplify: $$= 4p^2 + 5p + 1$$

**(b)** $$\frac{6a^2b^2 - 4a^2b + 3ab^2}{2a^2b^2} = \frac{6a^2b^2}{2a^2b^2} - \frac{4a^2b}{2a^2b^2} + \frac{3ab^2}{2a^2b^2}$$

$\frac{a^m}{a^n} = a^{m-n}$: $$= \frac{6}{2}a^{2-2}b^{2-2} - \frac{4}{2}a^{2-2}b^{1-2} + \frac{3}{2}a^{1-2}b^{2-2}$$

Simplify: $$= 3a^0b^0 - 2a^0b^{-1} + \frac{3}{2}a^{-1}b^0$$

$a^0 = 1;\ b^0 = 1;\ a^{-1} = \frac{1}{a};\ b^{-1} = \frac{1}{b}$: $$= 3 - \frac{2}{b} + \frac{3}{2a}$$

**QUICK ✓** *Find the quotient.*

**1.** $\frac{9p^4 - 12p^3 + 3p^2}{3p}$ **2.** $\frac{14m^4 - 10m^3 + 2m^2}{2m^2}$ **3.** $\frac{x^4y^4 + 8x^2y^2 - 4xy}{4x^3y}$

## 2 Divide Polynomials Using Long Division

The procedure for dividing two polynomials is similar to the procedure for dividing two integers. Although this procedure should be familiar to you, we review it in the following example.

### EXAMPLE 2 Dividing an Integer by an Integer

Divide 645 by 14.

### Solution

```
             46   ← Quotient
Divisor → 14)645  ← Dividend
            -56   ← 4·14 (Subtract)
             85
            -84   ← 6·14 (Subtract)
              1   ← Remainder
```

So $\frac{645}{14} = 46 + \frac{1}{14}$.

In Example 2, the number 14 is called the **divisor,** the number 645 is called the **dividend,** the number 46 is called the **quotient,** and the number 1 is called the **remainder.**

We can always check our work after completing a division problem by multiplying the quotient by the divisor and adding this product to the remainder. The result should be the dividend. That is,

$$(\text{Quotient})(\text{Divisor}) + \text{Remainder} = \text{Dividend}$$

For example, we can check the results of Example 2 as follows:

$$(46)(14) + 1 = 644 + 1 = 645$$

To divide two polynomials, we must first write each polynomial in standard form (descending order of degree). Use the pattern in Example 2 to guide your work.

## EXAMPLE 3 How to Divide Two Polynomials

Find the quotient and remainder when

$$4x^2 + 7x + 3 \text{ is divided by } x + 3$$

### Step-by-Step Solution

Each polynomial is in standard form. The dividend is $4x^2 + 7x + 3$ and the divisor is $x + 3$.

**Step 1:** Divide the leading term of the dividend, $4x^2$, by the leading term of the divisor, $x$. Enter the result over the term $4x^2$.

$$\frac{4x^2}{x} = 4x$$

$$\begin{array}{r} 4x \phantom{+ 7x + 3} \\ x + 3\overline{)4x^2 + 7x + 3} \end{array}$$

**Step 2:** Multiply $4x$ by $x + 3$. Be sure to vertically align like terms.

$$\begin{array}{r} 4x \phantom{+ 7x + 3} \\ x + 3\overline{)4x^2 + 7x + 3} \\ \underline{4x^2 + 12x} \phantom{+ 3} \end{array} \quad \leftarrow 4x(x + 3) = 4x^2 + 12x$$

**Step 3:** Subtract $4x^2 + 12x$ from $4x^2 + 7x + 3$.

$$\begin{array}{r} 4x \phantom{+ 7x + 3} \\ x + 3\overline{)4x^2 + 7x + 3} \\ \underline{-(4x^2 + 12x)} \phantom{+ 3} \\ -5x + 3 \end{array}$$

**Step 4:** Repeat Steps 1–3 treating $-5x + 3$ as the dividend by dividing $x$ into $-5x$ to obtain $-5$.

$$\begin{array}{r} 4x - 5 \phantom{+ 3} \\ x + 3\overline{)4x^2 + 7x + 3} \\ \underline{-(4x^2 + 12x)} \phantom{+ 3} \\ -5x + 3 \\ \underline{-(-5x - 15)} \\ 18 \end{array} \quad \begin{array}{l} \leftarrow \dfrac{-5x}{x} = -5 \\ \\ \\ \\ \leftarrow -5(x + 3) = -5x - 15 \\ \end{array}$$

Because 18 is a lower degree than the divisor, $x + 3$, the process ends. The quotient is $4x - 5$ and the remainder is 18.

**Step 5: Check** We verify that (Quotient)(Divisor) + Remainder = Dividend.

$$(4x - 5)(x + 3) + 18 = 4x^2 + 12x - 5x - 15 + 18$$

Combine like terms: $= 4x^2 + 7x + 3$

The answer checks, so

$$\frac{4x^2 + 7x + 3}{x + 3} = 4x - 5 + \frac{18}{x + 3}$$

**Work Smart**
When the degree of the remainder is less than the degree of the divisor, you are finished dividing.

Always write the results of polynomial division as follows:

$$\frac{\text{Dividend}}{\text{Divisor}} = \text{Quotient} + \frac{\text{Remainder}}{\text{Divisor}}$$

### EXAMPLE 4 Dividing Two Polynomials

Simplify by performing long division: $\dfrac{6x^3 - 11x^2 - 7x + 2}{3x + 2}$

**Solution**

$$\frac{6x^3}{3x} = 2x^2; \quad \frac{-15x^2}{3x} = -5x; \quad \frac{3x}{3x} = 1$$

$$\begin{array}{rll}
 & 2x^2 - \phantom{1}5x + 1 & \\
3x + 2 \,\big) & 6x^3 - 11x^2 - 7x + 2 & \\
 & -(6x^3 + \phantom{1}4x^2) & \leftarrow 2x^2(3x + 2) \\
 & -15x^2 - 7x + 2 & \\
 & -(-15x^2 - 10x) & \leftarrow -5x(3x + 2) \\
 & 3x + 2 & \\
 & -(3x + 2) & \leftarrow 1(3x + 2) \\
 & 0 & \leftarrow \text{Remainder}
\end{array}$$

The quotient is $2x^2 - 5x + 1$ and the remainder is 0. We now check our work.

**Check** (Quotient)(Divisor) + Remainder = Dividend

$$(2x^2 - 5x + 1)(3x + 2) + 0 = 6x^3 - 11x^2 - 7x + 2$$

Our answer checks, so

$$\frac{6x^3 - 11x^2 - 7x + 2}{3x + 2} = 2x^2 - 5x + 1$$

In Example 4, the remainder was 0. Therefore,

$$6x^3 - 11x^2 - 7x + 2 = (2x^2 - 5x + 1)(3x + 2)$$

which means that $2x^2 - 5x + 1$ and $3x + 2$ are *factors* of $6x^3 - 11x^2 - 7x + 2$. This result is true in general: If the remainder is zero, then the divisor and quotient are factors of the dividend.

### EXAMPLE 5 Dividing Two Polynomials

Simplify by performing long division: $\dfrac{8 - 15x + x^2 + 4x^3 + 3x^5}{3 + x^2}$

**Solution**

When setting up this division problem, you should write the dividend and divisor in standard form. Write the missing $x^4$ term in the dividend as $0x^4$.

$$\begin{array}{rll}
 & 3x^3 \phantom{+ 0x^4 + 4x^3 + x^2} - 5x + 1 & \\
x^2 + 3 \,\big) & 3x^5 + 0x^4 + 4x^3 + x^2 - 15x + 8 & \\
 & -(3x^5 \phantom{+ 0x^4} + 9x^3) & \leftarrow 3x^3(x^2 + 3) \\
 & -5x^3 + x^2 - 15x + 8 & \\
 & -(-5x^3 \phantom{+ x^2} - 15x) & \leftarrow -5x(x^2 + 3) \\
 & x^2 \phantom{- 15x} + 8 & \\
 & -(x^2 \phantom{- 15x} + 3) & \leftarrow 1(x^2 + 3) \\
 & 5 & \leftarrow \text{Remainder}
\end{array}$$

The quotient is $3x^3 - 5x + 1$ and the remainder is 5. We now check our work.

**Check** (Quotient)(Divisor) + Remainder = Dividend

$$(3x^3 - 5x + 1)(x^2 + 3) + 5 = 3x^5 + 4x^3 + x^2 - 15x + 8$$

Our answer checks, so

$$\frac{8 - 15x + x^2 + 4x^3 + 3x^5}{3 + x^2} = 3x^3 - 5x + 1 + \frac{5}{3 + x^2}$$

**QUICK ✓** *Simplify by performing long division.*

4. $\dfrac{x^3 + 3x^2 - 31x + 21}{x - 4}$ 5. $\dfrac{2x^3 + 7x^2 - 7x - 12}{2x - 3}$ 6. $\dfrac{2 + 12x^2 - 2x^3 - 5x^4 + x^5}{x^2 - 2}$

## 3 Divide Polynomials Using Synthetic Division

**Work Smart**

Synthetic division can be used only when the divisor is of the form $x - c$ or $x + c$.

To find the quotient and remainder when a polynomial of degree 1 or higher is divided by $x - c$, a shortened version of long division called **synthetic division** makes the task easier.

To see how synthetic division works, we will use long division to divide the polynomial $2x^3 - 5x^2 - 7x + 20$ by $x - 3$. Synthetic division comes from rewriting the long division in a more compact form. For example, in the long division below, the terms in red ink are not really necessary because they are identical to the terms directly above them. The subtraction signs are not necessary, because subtraction is understood. With these items removed, we have the division shown on the right.

$$\begin{array}{r} 2x^2 + x - 4 \quad \leftarrow \text{Quotient} \\ x - 3\overline{)2x^3 - 5x^2 - 7x + 20} \\ -(2x^3 - 6x^2) \qquad\qquad \\ x^2 - 7x \qquad\quad \\ -(x^2 - 3x) \qquad\quad \\ -4x + 20 \\ -(-4x + 12) \\ 8 \quad \leftarrow \text{Remainder} \end{array} \qquad \begin{array}{r} 2x^2 + x - 4 \\ x - 3\overline{)2x^3 - 5x^2 - 7x + 20} \\ -6x^2 \qquad\qquad \\ x^2 \qquad\qquad \\ -3x \qquad \\ -4x \qquad \\ 12 \\ 8 \end{array}$$

The $x$'s that appear in the division on the right are not necessary if we are careful about positioning each coefficient. As long as the right-most number under the division symbol is the constant, the number to its left is the coefficient of $x$ and so on, we can remove the $x$'s. Now we have

$$\begin{array}{r} 2x^2 + x - 4 \\ x - 3\overline{)2 \quad -5 \quad -7 \quad 20} \\ -6 \qquad\qquad \\ \boxed{1} \qquad\qquad \\ -3 \qquad \\ \boxed{-4} \qquad \\ 12 \\ \boxed{8} \end{array}$$

We can make this display more compact by moving the lines up until the "boxed" numbers align horizontally.

$$\begin{array}{r} 2x^2 + x - 4 \\ x - 3\overline{)2 \quad -5 \quad -7 \quad 20} \\ -6 \quad -3 \quad 12 \\ \hline \square \quad 1 \quad -4 \quad 8 \end{array}$$

Because the leading coefficient of the divisor is always 1, we know that the leading coefficient of the dividend will always be the leading coefficient of the quotient. So, we place the leading coefficient of the quotient, 2, in the boxed position.

$$\begin{array}{rrrrrl} & 2x^2 & +\ x & -\ 4 & & \text{Row 1} \\ x-3\,) & 2 & -5 & -7 & 20 & \\ & & -6 & -3 & 12 & \\ \hline & 2 & 1 & -4 & 8 & \text{Row 4} \end{array}$$

The first three numbers in Row 4 are the coefficients of the quotient. The last number in row 4 is the remainder. Now, Row 1 above is not needed.

$$\begin{array}{rrrrrl} x-3\,) & 2 & -5 & -7 & 20 & \text{Row 1} \\ & & -6 & -3 & 12 & \text{Row 2} \\ \hline & 2 & 1 & -4 & 8 & \text{Row 3} \end{array}$$

Remember, the entries in Row 3 are obtained by subtracting the entries in Row 2 from the entries in Row 1. Rather than subtracting the entries in Row 2, we can change the sign of each entry and then add. With this modification, our display becomes

$$\begin{array}{rrrrrl} x-3\,) & 2 & -5 & -7 & 20 & \text{Row 1} \\ & & 6 & 3 & -12 & \text{Row 2 (add)} \\ \hline & 2 & 1 & -4 & 8 & \text{Row 3} \end{array}$$

**Work Smart**

If there are any missing powers of $x$ in the dividend, you must insert a coefficient of 0 for the missing term when doing synthetic division.

Notice that the entries in Row 2 are three times the entries one column to the left in Row 3 (for example, the 6 in Row 2 is 3 times 2; the 3 in Row 2 is 3 times 1, and so on). We remove the $x - 3$ and replace it with 3. The entries in Row 3 give us the quotient and remainder.

$$\begin{array}{rrrrrl} 3\,) & 2 & -5 & -7 & 20 & \text{Row 1} \\ & & 6 & 3 & -12 & \text{Row 2 (add)} \\ \hline & 2 & 1 & -4 & 8 & \text{Row 3} \end{array}$$

$$\underbrace{2x^2 + x - 4}_{\text{Quotient}} \quad \underbrace{8}_{\text{Remainder}}$$

Let's go over an example step by step.

## EXAMPLE 6 How to Use Synthetic Division to Divide Polynomials

Use synthetic division to find the quotient and remainder when $3x^3 + 11x^2 + 14$ is divided by $x + 4$.

### Step-by-Step Solution

**Step 1:** Write the dividend in descending powers of x. Then copy the coefficients of the dividend. Remember to insert a 0 for any missing powers of x.

$$3x^3 + 11x^2 + 14 = 3x^3 + 11x^2 + 0x + 14$$

$$\begin{array}{rrrrl} 3 & 11 & 0 & 14 & \text{Row 1} \end{array}$$

**Step 2:** Insert the division symbol. Rewrite the divisor in the form $x - c$ and insert the value of $c$ to the left of the division symbol.

$$x + 4 = x - (-4)$$

$$\begin{array}{rrrrrl} -4\,) & 3 & 11 & 0 & 14 & \text{Row 1} \end{array}$$

**Step 3:** Bring the 3 down two rows and enter it in Row 3.

$$\begin{array}{rrrrrl} -4\,) & 3 & 11 & 0 & 14 & \text{Row 1} \\ & \downarrow & & & & \text{Row 2} \\ \hline & 3 & & & & \text{Row 3} \end{array}$$

**Step 4:** Multiply the latest entry in Row 3 by −4 and place the result in Row 2, one column over to the right.

$$\begin{array}{r|rrrrl} -4 & 3 & 11 & 0 & 14 & \text{Row 1} \\ & \downarrow & -12 & & & \text{Row 2} \\ \hline & 3^{-4(3)\nearrow} & & & & \text{Row 3} \end{array}$$

**Step 5:** Add the entry in Row 2 to the entry above it in Row 1. Enter the sum in Row 3.

$$\begin{array}{r|rrrrl} -4 & 3 & 11 & 0 & 14 & \text{Row 1} \\ & \downarrow & -12 & & & \text{Row 2} \\ \hline & 3^{-4(3)\nearrow} & -1 & & & \text{Row 3} \end{array}$$

**Step 6:** Repeat Steps 4 and 5 until no more entries are available in Row 1.

$$\begin{array}{r|rrrrl} -4 & 3 & 11 & 0 & 14 & \text{Row 1} \\ & \downarrow & -12 & 4 & -16 & \text{Row 2} \\ \hline & 3^{-4(3)\nearrow} & -1^{-4(-1)\nearrow} & 4^{-4(4)\nearrow} & -2 & \text{Row 3} \end{array}$$

**Step 7:** The final entry in Row 3, −2, is the remainder; the other entries in Row 3, 3, −1, and 4, are the coefficients of the quotient, in descending order. The quotient is a polynomial whose degree is one less than the degree of the dividend.

Quotient: $3x^2 - x + 4$

Remainder: $-2$

**Step 8:** Check

(Quotient)(Divisor) + Remainder = Dividend

$$\begin{aligned} &(3x^2 - x + 4)(x + 4) - 2 \\ &= 3x^3 + 12x^2 - x^2 - 4x + 4x + 16 - 2 \\ &= 3x^3 + 11x^2 + 14 \end{aligned}$$

So $\dfrac{3x^3 + 11x^2 + 14}{x + 4} = 3x^2 - x + 4 - \dfrac{2}{x + 4}$.

**Work Smart**

The number opposite the number following the "$x$" in $x - c$ is "$c$." For example, in $x + 3$, $c = -3$. In $x - 5$, $c = 5$.

Let's do one more example where we consolidate all the steps given in Example 6.

## EXAMPLE 7 Dividing Two Polynomials Using Synthetic Division

Use synthetic division to find the quotient and remainder when $x^4 - 5x^3 - 6x^2 + 33x - 15$ is divided by $x - 5$.

### Solution

The divisor is $x - 5$ so that $c = 5$.

$$\begin{array}{r|rrrrr} 5 & 1 & -5 & -6 & 33 & -15 \\ & \downarrow & 5 & 0 & -30 & 15 \\ \hline & 1 & 0 & -6 & 3 & 0 \end{array}$$

The dividend is a fourth-degree polynomial, so the quotient is a third-degree polynomial. The quotient is $x^3 + 0x^2 - 6x + 3 = x^3 - 6x + 3$ and the remainder is 0. So

$$\frac{x^4 - 5x^3 - 6x^2 + 33x - 15}{x - 5} = x^3 - 6x + 3$$

**Work Smart: Study Skills**

Knowing when a method **does not** apply is as essential as knowing when the method **does** apply. Identify when synthetic division can and cannot be used to divide polynomials.

In Example 7, because $\dfrac{x^4 - 5x^3 - 6x^2 + 33x - 15}{x - 5} = x^3 - 6x + 3$, we know that $x - 5$ and $x^3 - 6x + 3$ are factors of $x^4 - 5x^3 - 6x^2 + 33x - 15$. Therefore, we can write

$$x^4 - 5x^3 - 6x^2 + 33x - 15 = (x - 5)(x^3 - 6x + 3)$$

**QUICK ✓** *Use synthetic division to find the quotient.*

**7.** $\dfrac{2x^3 + x^2 - 7x - 13}{x - 2}$

**8.** $\dfrac{x^4 + 8x^3 + 15x^2 - 2x - 6}{x + 3}$

## 4 Divide Polynomial Functions

We have discussed adding, subtracting, and multiplying polynomial functions. All that is left is dividing polynomial functions.

**DEFINITION**

If $f$ and $g$ are functions, then the **quotient** $\dfrac{f}{g}$ is the function defined by

$$\left(\frac{f}{g}\right)(x) = \frac{f(x)}{g(x)}, \quad g(x) \neq 0$$

### EXAMPLE 8 Dividing Two Polynomial Functions

If $f(x) = 2x^4 - x^3 - 3x^2 + 13x - 4$ and $g(x) = x^2 - 2$, find

**(a)** $\left(\dfrac{f}{g}\right)(x)$ **(b)** $\left(\dfrac{f}{g}\right)(2)$

**Solution**

**(a)** $\left(\dfrac{f}{g}\right)(x) = \dfrac{f(x)}{g(x)} = \dfrac{2x^4 - x^3 - 3x^2 + 13x - 4}{x^2 - 2}$

We use long division to find the quotient because the divisor is not of the form $x - c$.

$$\begin{array}{r} 2x^2 - x + 1 \phantom{)} \\ x^2 - 2\,\overline{)\,2x^4 - x^3 - 3x^2 + 13x - 4} \\ -(2x^4 \phantom{- x^3} - 4x^2) \phantom{+ 13x - 4} \\ \hline -x^3 + x^2 + 13x - 4 \\ -(-x^3 \phantom{+ x^2} + 2x) \phantom{- 4} \\ \hline x^2 + 11x - 4 \\ -(x^2 \phantom{+ 11x} - 2) \\ \hline 11x - 2 \end{array}$$

So $\left(\dfrac{f}{g}\right)(x) = \dfrac{f(x)}{g(x)} = 2x^2 - x + 1 + \dfrac{11x - 2}{x^2 - 2}$.

**(b)** $\left(\dfrac{f}{g}\right)(2) = 2(2)^2 - 2 + 1 + \dfrac{11(2) - 2}{(2)^2 - 2}$

$= 8 - 2 + 1 + \dfrac{20}{2}$

$= 17$

**QUICK ✓** *Find (a)* $\left(\dfrac{f}{g}\right)(x)$ *(b)* $\left(\dfrac{f}{g}\right)(3)$.

**9.** $f(x) = 3x^4 - 4x^3 - 3x^2 + 10x - 5;\ g(x) = x^2 - 2$

## 5 Use the Remainder and Factor Theorems

Look back at Example 6 where we used synthetic division to find the quotient and remainder when $3x^3 + 11x^2 + 14$ is divided by $x + 4$. If we let $f(x) = 3x^3 + 11x^2 + 14$, we find that $f(-4) = -2$. Looking back at Example 6, we find that the remainder when $3x^3 + 11x^2 + 14$ is divided by $x + 4$ is $-2$. The value of the function $f$ at $x = -4$ is the same as the remainder when $f$ is divided by $x + 4 = x - (-4)$. This result is not a coincidence and is true in general! It is called the *Remainder Theorem.*

**In Words**

A theorem is a big idea that can be shown to be true in general. The word comes from a Greek verb meaning "to view."

**THE REMAINDER THEOREM**

Let $f$ be a polynomial function. If $f(x)$ is divided by $x - c$, then the remainder is $f(c)$.

### EXAMPLE 9 Using the Remainder Theorem

Use the Remainder Theorem to find the remainder if $f(x) = 2x^3 - 3x + 8$ is divided by $x + 3$.

**Solution**

The divisor is $x + 3 = x - (-3)$, so the Remainder Theorem says that the remainder is $f(-3)$.

$$\begin{aligned} f(x) &= 2x^3 - 3x + 8 \\ f(-3) &= 2(-3)^3 - 3(-3) + 8 \\ &= 2(-27) + 9 + 8 \\ &= -54 + 9 + 8 \\ &= -37 \end{aligned}$$

When $f(x) = 2x^3 - 3x + 8$ is divided by $x + 3$, the remainder is $-37$.

**Check** Using synthetic division, we find that the remainder is, in fact, $-37$.

$$\begin{array}{r|rrrr} -3 & 2 & 0 & -3 & 8 \\ & & -6 & 18 & -45 \\ \hline & 2 & -6 & 15 & -37 \end{array} \leftarrow \text{Remainder}$$

**QUICK ✓**

**10.** Use the Remainder Theorem to find the remainder if $f(x) = 3x^3 + 10x^2 - 9x - 4$ is divided by

**(a)** $x - 2$ **(b)** $x + 4$

We saw from Example 7 that when the remainder is 0, then the quotient and divisor are factors of the dividend. The Remainder Theorem can be used to determine whether an expression of the form $x - c$ is a factor of the dividend. This result is called the *Factor Theorem.*

**Work Smart**

"If and only if" statements are used to compress two statements into one. The Factor Theorem is two statements:

1. If $f(c) = 0$, then $x - c$ is a factor of $f$.
2. If $x - c$ is a factor of $f$, then $f(x) = 0$.

**THE FACTOR THEOREM**

Let $f$ be a polynomial function. Then $x - c$ is a factor of $f(x)$ if and only if $f(c) = 0$.

We can use the Factor Theorem to determine whether a polynomial has a particular factor.

### EXAMPLE 10 Using the Factor Theorem

Use the Factor Theorem to determine whether the function $f(x) = 2x^3 - 3x^2 - 18x - 8$ has the factor

**(a)** $x - 3$ **(b)** $x + 2$

### Solution

The Factor Theorem states that if $f(c) = 0$, then $x - c$ is a factor of $f$.

**(a)** Because $x - 3$ is of the form $x - c$ with $c = 3$, we find the value of $f(3)$.

$$f(3) = 2(3)^3 - 3(3)^2 - 18(3) - 8 = -35 \neq 0$$

Since $f(3) \neq 0$, we know that $x - 3$ is not a factor of $f$.

**(b)** Because $x + 2 = x - (-2)$ is of the form $x - c$ with $c = -2$, we find the value of $f(-2)$. Rather than evaluating the function using substitution, we use synthetic division.

$$\begin{array}{r|rrrr} -2 & 2 & -3 & -18 & -8 \\ & & -4 & 14 & 8 \\ \hline & 2 & -7 & -4 & 0 \end{array} \quad \leftarrow \text{Remainder}$$

**In Words**

If $f(c) = 0$, then $f(x)$ can be written in factored form as

$f(x) = (x - c)(\text{quotient})$

The remainder is 0, so that $f(-2) = 0$. Because $f(-2) = 0$, we know that $x + 2$ is a factor of $f$. This means the dividend can be written as the product of the quotient and divisor. The quotient is $2x^2 - 7x - 4$ and the divisor is $x + 2$, so

$$2x^3 - 3x^2 - 18x - 8 = (x + 2)(2x^2 - 7x - 4)$$

**QUICK ✓**

**11.** Use the Factor Theorem to determine whether $x - c$ is a factor of $f(x) = 2x^3 - 9x^2 - 6x + 5$ for the given values of $c$. If $x - c$ is a factor, then write $f$ in factored form. That is, write $f(x) = (x - c)(\text{quotient})$.

**(a)** $c = -2$ **(b)** $c = 5$

# 5.3 Exercises

***For Extra Help:***

Student Solutions Manual

CD Video

PH Math/Tutor Center

MathXL Tutorials on CD

MathXL® MyMathLab

## Concepts and Vocabulary

*In Problems 1–3, fill in the blanks.*

**1.** Because $\dfrac{3x^2 + 2x - 1}{x + 1} = 3x - 1$, the remainder when dividing $3x^2 + 2x - 1$ by $x + 1$ is ______ and $3x^2 + 2x - 1 =$ ______ · ______.

**2.** Given that $\dfrac{6x^3 - x^2 - 9x + 8}{3x + 1} = 2x^2 - 3x + 1 + \dfrac{4}{3x + 1}$, we call $6x^3 - x^2 - 9x + 8$ the ______, $3x + 1$ the ______ and 4 the ______.

**3.** To find the quotient and remainder when a polynomial of degree 1 or higher is divided by $x - c$, a shortened version of long division, called ______ ______ can be used.

*In Problems 4–6, answer True or False to each statement.*

**4.** We can divide $-4x^3 + 5x^2 + 10x - 3$ by $x^2 - 2$ using synthetic division.

**5.** To check division, we compute (Quotient)(Divisor) + Remainder and determine whether it equals the Dividend.

**6.** If a polynomial function $f$ is divided by $x - c$, then the remainder is $f(c)$.

**7.** If $f$ is a polynomial of degree $n$ and it is divided by $x + 4$, the quotient will be a polynomial of degree $n - 1$. Explain why.

**8.** Explain the Remainder Theorem in your own words. Explain the Factor Theorem in your own words.

**9.** Given that $\dfrac{6x^3 - x^2 - 9x + 4}{3x + 4} = 2x^2 - 3x + 1$, is $3x + 4$ a factor of $6x^3 - x^2 - 9x + 4$? If so, write $6x^3 - x^2 - 9x + 4$ in factored form.

**10.** Suppose that you were asked to divide $8x^3 - 3x + 1$ by $x + 3$. Would you use long division or synthetic division? Why?

## Building Skills

*In Problems 11–18, divide and simplify.*

**11.** $\dfrac{8x^2 + 12x}{4x}$

**12.** $\dfrac{6z^3 + 9z^2}{3z^2}$

**13.** $\dfrac{2a^3 - 15a^2 + 10a}{5a}$

**14.** $\dfrac{4b^3 + 12b^2 + 24b}{6b}$

**15.** $\dfrac{2y^3 + 6y}{4y^2}$

**16.** $\dfrac{3z^4 + 12z^2}{6z^3}$

**17.** $\dfrac{4m^2n^2 + 6m^2n - 18mn^2}{4m^2n^2}$

**18.** $\dfrac{2x^2y^3 - 9xy^3 + 16x^2y}{2x^2y^2}$

*In Problems 19–36, divide using long division.*

**19.** $\dfrac{x^2 + 5x + 6}{x + 2}$

**20.** $\dfrac{x^2 - 4x - 21}{x + 3}$

**21.** $\dfrac{2x^2 + x - 4}{x - 2}$

**22.** $\dfrac{3z^2 - 7z - 28}{z - 4}$

**23.** $\dfrac{2w^2 + 5w - 49}{2w - 7}$

**24.** $\dfrac{4x^2 - 17x - 33}{4x + 7}$

**25.** $\dfrac{x^3 + 8x^2 + x - 42}{x + 3}$

**26.** $\dfrac{x^3 + x^2 - 22x - 40}{x + 2}$

**27.** $\dfrac{w^3 - 21w - 20}{w + 4}$

**28.** $\dfrac{a^3 - 49a + 120}{a + 8}$

**29.** $\dfrac{6x^3 - 13x^2 - 80x - 25}{2x + 5}$

**30.** $\dfrac{4p^3 + 20p^2 + 19p - 15}{2p + 5}$

**31.** $\dfrac{x^3 - 7x^2 + 3x - 21}{x^2 + 3}$

**32.** $\dfrac{x^3 - 5x^2 - 2x + 10}{x^2 - 2}$

**33.** $\dfrac{3z^3 + 21z^2 + 5z + 2}{3z^2 + 1}$

**34.** $\dfrac{2k^3 + 10k^2 - 6k - 8}{2k^2 - 3}$

**35.** $\dfrac{2x^4 - 11x^3 + 8x^2 - 22x + 144}{x^2 + 2x + 5}$

**36.** $\dfrac{2x^4 - 7x^3 - 50x^2 - 10x + 96}{x^2 + x - 3}$

*In Problems 37–50, divide using synthetic division.*

**37.** $\dfrac{x^2 - 3x - 10}{x - 5}$

**38.** $\dfrac{x^2 + 4x - 12}{x - 2}$

**39.** $\dfrac{2x^2 + 11x + 12}{x + 4}$

**40.** $\dfrac{3x^2 + 19x - 40}{x + 8}$

**41.** $\dfrac{x^2 - 3x - 14}{x - 6}$

**42.** $\dfrac{x^2 + 2x - 17}{x - 4}$

**43.** $\dfrac{x^3 - 19x - 15}{x - 5}$

**44.** $\dfrac{x^3 - 13x - 17}{x + 3}$

**45.** $\dfrac{3x^4 - 5x^3 - 21x^2 + 17x + 25}{x - 3}$

**46.** $\dfrac{2x^4 - x^3 - 38x^2 + 16x + 103}{x + 4}$

**47.** $\dfrac{x^4 - 40x^2 + 109}{x + 6}$

**48.** $\dfrac{a^4 - 65a^2 + 55}{a - 8}$

**49.** $\dfrac{2x^3 + 3x^2 - 14x - 15}{x - \frac{5}{2}}$

**50.** $\dfrac{3x^3 + 13x^2 + 8x - 12}{x - \frac{2}{3}}$

*In Problems 51–60, find (a)* $\left(\frac{f}{g}\right)(x)$, *(b)* $\left(\frac{f}{g}\right)(2)$.

**51.** $f(x) = 4x^3 - 8x^2 + 12x;\ g(x) = 4x$

**52.** $f(x) = 3x^3 - 9x^2 + 12x;\ g(x) = 3x$

**53.** $f(x) = x^2 - x - 12;\ g(x) = x - 4$

**54.** $f(x) = x^2 + 3x - 4;\ g(x) = x + 4$

**55.** $f(x) = 2x^2 + 5x - 1;\ g(x) = x + 3$

**56.** $f(x) = 3x^2 - 6x + 5;\ g(x) = 2x + 1$

**57.** $f(x) = 2x^3 + 9x^2 + x - 12;\ g(x) = 2x + 3$

**58.** $f(x) = 3x^3 - 2x^2 - 19x - 6;\ g(x) = 3x + 1$

**59.** $f(x) = x^3 - 13x - 12;\ g(x) = x^2 - 9$

**60.** $f(x) = x^3 - 19x + 30;\ g(x) = x^2 - x - 6$

*In Problems 61–68, use the Remainder Theorem to find the remainder.*

**61.** $f(x) = x^2 - 5x + 1$ is divided by $x - 2$

**62.** $f(x) = x^2 + 4x - 5$ is divided by $x + 2$

**63.** $f(x) = x^3 - 2x^2 + 5x - 3$ is divided by $x + 4$

**64.** $f(x) = x^3 + 3x^2 - x + 1$ is divided by $x - 3$

**65.** $f(x) = 2x^3 - 4x + 1$ is divided by $x - 5$

**66.** $f(x) = 3x^3 + 2x^2 - 5$ is divided by $x + 3$

**67.** $f(x) = x^4 + 1$ is divided by $x - 1$

**68.** $f(x) = x^4 - 1$ is divided by $x - 1$

*In Problems 69–76, use the Factor Theorem to determine whether $x - c$ is a factor of the given function for the given values of c. If $x - c$ is a factor, then write f in factored form. That is, write $f(x) = (x - c)(\text{quotient})$.*

**69.** $f(x) = x^2 - 3x + 2;\ c = 2$

**70.** $f(x) = x^2 + 5x + 6;\ c = 3$

**71.** $f(x) = 2x^2 + 5x + 2;\ c = -2$

**72.** $f(x) = 3x^2 + x - 2;\ c = 2$

**73.** $f(x) = 4x^3 - 9x^2 - 49x - 30;\ c = 3$

**74.** $f(x) = 2x^3 - 9x^2 - 2x + 24;\ c = 1$

**75.** $f(x) = 4x^3 - 7x^2 - 5x + 6;\ c = -1$

**76.** $f(x) = 5x^3 + 8x^2 - 7x - 6;\ c = -2$

## Mixed Practice

*In Problems 77–90, divide using any appropriate method.*

**77.** $\dfrac{3a^3b^2 - 9a^2b + 18ab}{3ab}$

**78.** $\dfrac{5s^4t^3 - 15s^3t^2 + 50s^2t}{5s^2t}$

**79.** $\dfrac{3y^2 + 11y + 6}{3y + 2}$

**80.** $\dfrac{4a^2 + 23a + 15}{4a + 3}$

**81.** $\dfrac{x^3 + 6x^2 + 8x + 32}{x^2 + 5}$

**82.** $\dfrac{x^3 - 3x^2 + 5x - 12}{x^2 + 3}$

**83.** $\dfrac{8x^3 + 6x}{12x^2}$ **84.** $\dfrac{3x^4 + 6x^2}{9x^3}$ **85.** $\dfrac{x^3 + 7x^2 + 2x - 46}{x + 4}$

**86.** $\dfrac{x^3 + 5x^2 - 29x - 97}{x - 5}$ **87.** $\dfrac{3x^3 + 5x^2 - 24x - 40}{3x + 5}$ **88.** $\dfrac{4b^3 + 11b^2 - 28b - 17}{4b + 3}$

**89.** $\dfrac{8x^3 - 27}{2x - 3}$ **90.** $\dfrac{8a^3 + 125}{2a + 5}$

## Applying the Concepts

△ **91. Area** The area of a rectangle is $15x^2 + x - 2$ square feet. If the width of the rectangle is $3x - 1$ feet, find the length.

△ **92. Area** The area of a rectangle is $10x^2 + 9x - 9$ square centimeters. If the width of the rectangle is $2x + 3$ centimeters, find the length.

△ **93. Volume** The volume of the box shown is $2x^3 + 9x^2 - 20x - 75$ cubic centimeters. Find the length if the width is $x + 5$ centimeters and the height is $x - 3$ centimeters.

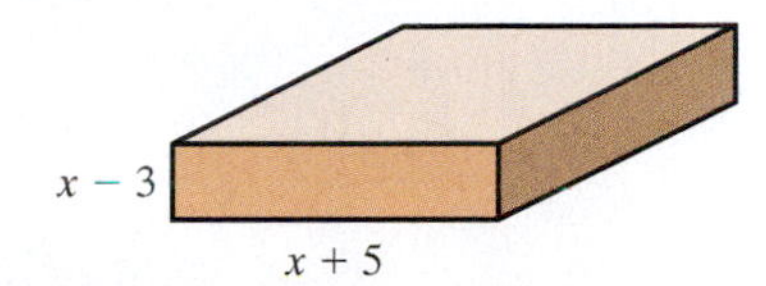


△ **94. Volume** The volume of the box shown is $4x^3 + 35x^2 + 52x + 21$ cubic feet. Find the height if the width is $4x + 3$ feet and the length is $x + 1$ feet.

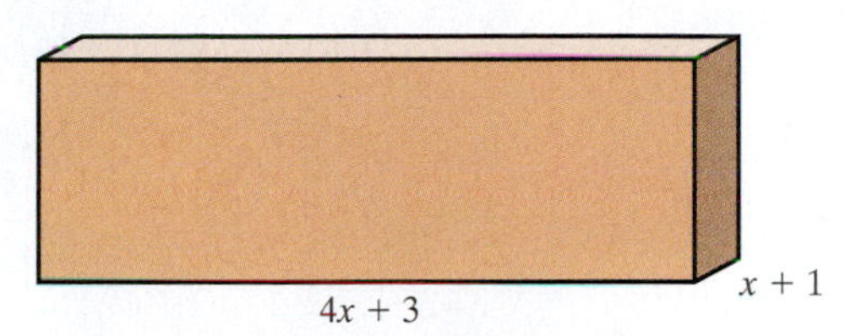


**95. Average Cost** The average cost function is defined in economics as $\overline{C}(x) = \dfrac{C(x)}{x}$, where $C$ is the cost of manufacturing $x$ units of a good. Suppose that $C(x) = 0.01x^3 - 0.4x^2 + 13x + 400$, where $x$ is the number of wristwatches manufactured in a day.

**(a)** What is $\overline{C}(x) = \dfrac{C(x)}{x}$?

**(b)** What is the average cost of manufacturing $x = 50$ wristwatches in a day?

**96. Average Cost** The average cost function is defined in economics as $\overline{C}(x) = \dfrac{C(x)}{x}$, where $C$ is the cost of manufacturing $x$ units of a good. Suppose that $C(x) = 0.01x^3 - 0.45x^2 + 16.5x + 600$, where $x$ is the number of wagons manufactured in a day.

**(a)** What is $\overline{C}(x) = \dfrac{C(x)}{x}$?

**(b)** What is the average cost of manufacturing $x = 50$ wagons in a day?

**97.** If $\dfrac{f(x)}{x - 5} = 3x + 5$, find $f(x)$. **98.** If $\dfrac{f(x)}{x + 3} = 2x + 7$, find $f(x)$.

**99.** If $\dfrac{f(x)}{x - 3} = x + 8 + \dfrac{4}{x - 3}$, find $f(x)$.

**100.** If $\dfrac{f(x)}{x - 3} = x^2 + 2 + \dfrac{7}{x - 3}$, find $f(x)$.

## Extending the Concepts

**101.** Find the sum of $a, b, c,$ and $d$ if $\dfrac{2x^3 - 3x^2 - 26x - 37}{x + 2} = ax^2 + bx + c + \dfrac{d}{x + 2}$.

**102.** What is the remainder when $f(x) = 2x^{30} - 3x^{20} + 4x^{10} - 2$ is divided by $x - 1$?

## PUTTING THE CONCEPTS TOGETHER (SECTIONS 5.1–5.3)

*These problems cover important concepts from Sections 5.1 to 5.3. We designed these problems so that you can review the chapter so far and show your mastery of the concepts. Take time to work these problems before proceeding with the next section. The answers to these problems are located at the back of the text starting on page AN-25.*

**1.** Write the polynomial in standard form and determine its degree:

$$3m + 5m^4 - 2m^3 + 8$$

**2.** Add: $(7a^2 - 4a^3 + 7a - 1) + (2a^2 - 6a - 7)$

**3.** Subtract: $\left(\frac{1}{5}y^2 + 2y - 6\right) - (4y^2 - y + 2)$

**4.** For $f(x) = 2x^3 - x^2 + 4x + 9$, find $f(2)$.

**5.** For $f(x) = 6x + 5$ and $g(x) = -x^2 + 2x + 3$, find $(f + g)(-3)$.

**6.** For $f(x) = 2x^2 + 7$ and $g(x) = x^2 - 4x - 3$, find $(f - g)(x)$.

**7.** Multiply: $2mn^3(m^2n - 4mn + 6)$

**8.** Multiply: $(3a - 5b)^2$

**9.** Multiply: $(7n^2 + 3)(7n^2 - 3)$

**10.** Multiply: $(3a + 2b)(6a^2 - 2ab + b^2)$

**11.** For $f(x) = x + 2$ and $g(x) = x^2 - 4x + 11$, find $(f \cdot g)(x)$.

**12.** Divide using long division: $\dfrac{10z^3 + 41z^2 + 7z - 49}{2z + 7}$

**13.** Divide using synthetic division: $\dfrac{2x^3 + 25x^2 + 62x - 6}{x + 9}$

**14.** For $f(x) = x^3 + 2x^2 - 4x + 5$ and $g(x) = x - 1$, find $\left(\dfrac{f}{g}\right)(x)$.

**15.** Use the Factor Theorem to determine if $x + 5$ is a factor of $f(x) = 3x^3 + 8x^2 - 23x + 60$. If so, write the function in factored form—that is, in the form $f(x) = (x + 5)(\text{quotient})$.

# 5.4 Greatest Common Factor; Factoring by Grouping

**OBJECTIVES**

1. Factor the Greatest Common Factor
2. Factor by Grouping

### *Preparing for Greatest Common Factor; Factoring by Grouping*

*Before getting started, take the following readiness quiz. If you get a problem wrong, go back to the section cited and review the material.*

**1.** Write 24 as the product of prime factors. [Section R.3, pp. 19–29]

**2.** Distribute: $4(3x - 5)$ [Section R.3, pp. 29–30]

Consider the following products:

$$5y(y^2 - 2y + 5) = 5y^3 - 10y^2 + 25y$$
$$(3x + 1)(x - 5) = 3x^2 - 14x - 5$$

The polynomials on the left side are called **factors** of the polynomial on the right side. Expressing a polynomial with integer coefficients as the product of two or more other polynomials with integer coefficients is called **factoring over the integers.**

**In Words**
Factoring is "undoing" multiplication.

A polynomial with integer coefficients is **prime** if it cannot be written as the product of two other polynomials with integer coefficients (excluding 1 and $-1$). When a polynomial has been written as a product consisting only of prime factors, it is said to be **factored completely.** The word *prime* has the same meaning as it does for integers. For example, 3, 7, and 13 are prime numbers while $4\ (= 2 \cdot 2)$, $12\ (= 2 \cdot 6)$, and

*Preparing for...Answers* **1.** $24 = 2 \cdot 2 \cdot 2 \cdot 3$ **2.** $12x - 20$

35 ($= 5 \cdot 7$) are not prime. If we write 4 as $2 \cdot 2$, we have factored 4 completely. If we write 12 as $2 \cdot 6$, we have not factored completely, because 6 can be further factored as $2 \cdot 3$. So, 12 factored completely would be written $2 \cdot 2 \cdot 3$.

## 1 Factor the Greatest Common Factor

**Work Smart**
The first step in factoring is to look for the greatest common factor.

The first step in factoring any polynomial is to look for the *greatest common factor*. The **greatest common factor (GCF)** of a polynomial is the largest polynomial that is a factor of all the terms in the polynomial. To find the GCF of a polynomial, find the largest polynomial that is a factor of (or divides evenly into) each term in the polynomial.

### EXAMPLE 1 Finding the Greatest Common Factor

Find the greatest common factor (GCF) of the terms.

**(a)** $4x, 12$ **(b)** $6x^3, 12x^2, 15x$ **(c)** $4x^3y^4, 8x^2y^3, 12xy^2$

**Solution**

In each case, we look for the largest polynomial that is a factor of each polynomial.

**(a)** Look at the coefficients first. The largest number that divides into 4 and 12 evenly is 4, so 4 is part of the GCF. Because 12 does not have a variable factor, the GCF is 4.

**(b)** The largest number that divides evenly into 6, 12, and 15 is 3, so 3 is part of the GCF. Now, look at the variable expressions, $x^3$, $x^2$, and $x$. We choose the variable expression with the smallest exponent, $x$ ($= x^1$). The GCF is $3x$.

**(c)** The largest number that divides into 4, 8, and 12 is 4, so 4 is part of the GCF. Look at the expression involving $x$. The smallest exponent involving $x$ is 1, so $x$ is part of the GCF. The smallest exponent involving $y$ is $y^2$, so $y^2$ is part of the GCF. The GCF is $4xy^2$.

**Work Smart**
The Distributive Property comes in handy again for factoring out the GCF. Note how it is used in the check step, too.

**QUICK ✓** *Find the greatest common factor (GCF) of the terms.*

**1.** $5y, 15$ **2.** $4z^3, 10z^2, 12z$ **3.** $3x^3y^5, 9x^2y^3, 12xy^4$

Once the GCF is identified, we use the Distributive Property to factor the GCF out.

### EXAMPLE 2 How to Factor Out the Greatest Common Factor

Factor out the greatest common factor: $4a^2b^2 - 10ab^3 + 18a^3b^4$

**Step-by-Step Solution**

**Step 1:** Find the GCF.

$$\text{GCF} = 2ab^2$$

**Step 2:** Rewrite each term as the product of the GCF and remaining factor.

$$4a^2b^2 - 10ab^3 + 18a^3b^4 = 2ab^2 \cdot 2a - 2ab^2 \cdot 5b + 2ab^2 \cdot 9a^2b^2$$

**Step 3:** Factor out GCF.

$$= 2ab^2(2a - 5b + 9a^2b^2)$$

**Step 4:** Check

$$2ab^2(2a - 5b + 9a^2b^2) = 2ab^2 \cdot 2a - 2ab^2 \cdot 5b + 2ab^2 \cdot 9a^2b^2$$
$$= 4a^2b^2 - 10ab^3 + 18a^3b^4$$

So $4a^2b^2 - 10ab^3 + 18a^3b^4 = 2ab^2(2a - 5b + 9a^2b^2)$.

### Factoring the Greatest Common Factor

**Step 1:** Identify the greatest common factor (GCF) of each term.

**Step 2:** Rewrite each term as the product of the GCF and remaining factor.

**Step 3:** Use the Distributive Property to factor out the GCF.

**Step 4:** Use Distributive Property to verify that the factorization is correct.

**QUICK ✓** *Factor out the greatest common factor.*

**4.** $7z^2 - 14z$ **5.** $6y^3 - 14y^2 + 10y$ **6.** $2m^4n^2 + 8m^3n^4 - 6m^2n^5$

---

When the coefficient of the term of highest degree is negative, we often prefer to factor the negative out of the polynomial as part of the GCF.

### EXAMPLE 3 Factoring Out a Negative Number as Part of the GCF

Factor out the greatest common factor.

**(a)** $-8a + 16$ **(b)** $-2b^3 + 10b^2 + 8b$

**Solution**

GCF = −8

**(a)** $-8a + 16 = -8 \cdot a + (-8) \cdot (-2)$

$= -8(a - 2)$

**Check** $-8(a - 2) = -8 \cdot a + (-8) \cdot (-2)$

$= -8a + 16$

GCF = −2b

**(b)** $-2b^3 + 10b^2 + 8b = -2b \cdot b^2 + (-2b) \cdot (-5b) + (-2b) \cdot (-4)$

$= -2b(b^2 - 5b - 4)$

**Check** $-2b(b^2 - 5b - 4) = -2b \cdot b^2 + (-2b) \cdot (-5b) + (-2b) \cdot (-4)$

$= -2b^3 + 10b^2 + 8b$

**QUICK ✓** *Factor out the greatest common factor.*

**7.** $-5y^2 + 10y$ **8.** $-3a^3 + 6a^2 - 12a$

---

Sometimes the greatest common factor is a binomial.

### EXAMPLE 4 Factoring Out a Binomial as the Greatest Common Factor

Factor out the greatest common binomial factor.

**(a)** $4x(x - 3) + 5(x - 3)$

**(b)** $3y(2y + 1) - 5(2y + 1)^2$

**(c)** $(c + 4)(c - 1) + (5c - 2)(c - 1)$

**Solution**

GCF = (x − 3)

**(a)** $4x(x - 3) + 5(x - 3) = (x - 3) \cdot 4x + (x - 3) \cdot 5$

$= (x - 3)(4x + 5)$

**(b)** $3y(2y + 1) - 5(2y + 1)^2 = (2y + 1)\cdot 3y - (2y + 1)\cdot 5(2y + 1)$

$= (2y + 1)(3y - 5(2y + 1))$

$= (2y + 1)(3y - 10y - 5)$

$= (2y + 1)(-7y - 5)$

Factor out $-1$: $= -1(2y + 1)(7y + 5)$

**(c)** $(c + 4)(c - 1) + (5c - 2)(c - 1) = (c - 1)\cdot(c + 4) + (c - 1)\cdot(5c - 2)$

$= (c - 1)(c + 4 + 5c - 2)$

$= (c - 1)(6c + 2)$

GCF = 2 in $6c + 2$: $= 2(c - 1)(3c + 1)$

**QUICK ✓** *Factor out the greatest common factor.*

**9.** $4a(a - 3) + 3(a - 3)$ **10.** $(w + 2)(w - 5) + (2w + 1)(w - 5)$

## 2 Factor by Grouping

Sometimes a common factor does not occur in every term of the polynomial, but a common factor does occur in some of the terms of the polynomial (one group of polynomials) and a second common factor occurs in the remaining terms of the polynomial (a second group of polynomials). When this happens, the common factor can be factored out of each group using the Distributive Property. This technique is called **factoring by grouping** and is used often when a polynomial contains four terms.

### EXAMPLE 5 How to Factor by Grouping

Factor by grouping: $4x - 4y + ax - ay$

#### Step-by-Step Solution

**Step 1:** Group terms with common factors. In this problem the first two terms have a common factor of 4 and the last two terms have a common factor of $a$.

$$4x - 4y + ax - ay = (4x - 4y) + (ax - ay)$$

**Step 2:** In each grouping, factor out the common factor.

$$= 4(x - y) + a(x - y)$$

**Step 3:** Factor out the common factor that remains.

$$= (x - y)(4 + a)$$

**Step 4:** Check

FOIL

$$(x - y)(4 + a) = 4x + ax - 4y - ay$$

Rearrange terms: $= 4x - 4y + ax - ay$

So $4x - 4y + ax - ay = (x - y)(4 + a)$.

Based on Example 5, the following steps should be followed to factor by grouping.

### Steps to Factor by Grouping

**Step 1:** Group the terms with common factors. Sometimes it will be necessary to rearrange the terms.

**Step 2:** In each grouping, factor out the common factor.

**Step 3:** Factor out the common factor that remains.

**Step 4:** Check your work.

### EXAMPLE 6 Factoring by Grouping

Factor by grouping:

**(a)** $x^3 + 3x^2 + 2x + 6$ **(b)** $6x^2 + 9x - 10x - 15$

**Solution**

**(a)** First, we group terms with common factors. Notice the first two terms, $x^3$ and $3x^2$, have a common factor of $x^2$; the last two terms have a common factor of 2.

**Work Smart**

In Example 6(a), grouping the first three terms would result in a common factor of $x$, but the second grouping would not have any common factor.

$$x^3 + 3x^2 + 2x + 6 = (x^3 + 3x^2) + (2x + 6)$$

Factor out common factor: $= x^2(x + 3) + 2(x + 3)$

$= (x + 3)(x^2 + 2)$

**Check** $(x + 3)(x^2 + 2) = x^3 + 2x + 3x^2 + 6$

Rearrange terms: $= x^3 + 3x^2 + 2x + 6$

So $x^3 + 3x^2 + 2x + 6 = (x + 3)(x^2 + 2)$.

**(b)** $6x^2 + 9x - 10x - 15 = (6x^2 + 9x) + (-10x - 15)$

Factor out $-5$ in the second grouping: $= 3x(2x + 3) + (-5)(2x + 3)$

$= (2x + 3)(3x - 5)$

We leave the check to you.

**QUICK ✓** *Factor by grouping.*

**11.** $5x + 5y + bx + by$

**12.** $w^3 - 3w^2 + 4w - 12$

**13.** $6z^2 + 2z + 9z + 3$

**14.** $2x^2 + x - 10x - 5$

# 5.4 Exercises

*For Extra Help:*

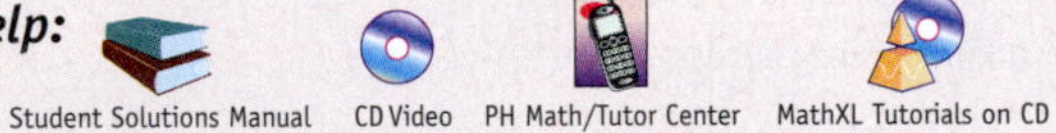

Student Solutions Manual CD Video PH Math/Tutor Center MathXL Tutorials on CD MathXL® MyMathLab

## Concepts and Vocabulary

*In Problems 1–3, fill in the blanks.*

**1.** In $(3x + 1)(x - 5) = 3x^2 - 14x - 5$, the polynomials on the left side are called _______ of the polynomial on the right side.

**2.** If a polynomial cannot be written as the product of two other polynomials (excluding 1 and $-1$), then the polynomial is said to be _______.

**3.** The _______ _______ _______ of a polynomial is the largest polynomial that is a factor of all the terms in the polynomial.

*In Problems 4–6, answer True or False to each statement.*

**4.** The polynomial $x + 1$ is prime.

**5.** The polynomial $3x + 12$ is prime.

**6.** The first step in any factoring problem is to look for the greatest common factor.

**7.** Explain how to find the greatest common factor.

**8.** Explain the steps in factoring by grouping.

**9.** Determine the greatest common factor of the following terms: $4x^2$, $10x$, $12x^3$

**10.** Determine the greatest common factor of the following terms: $3a^2b^3$, $6ab^4$, $12a^3b^2$

## Building Skills

*In Problems 11–28, factor out the greatest common factor.*

**11.** $5a + 35$

**12.** $8z + 48$

**13.** $-3y + 21$

**14.** $-4b + 32$

**15.** $14x^2 - 21x$

**16.** $12a^2 + 45a$

**17.** $3z^3 - 6z^2 + 18z$

**18.** $2w^3 + 10w^2 - 14$

**19.** $-5p^4 + 10p^3 - 25p^2$

**20.** $-6q^3 + 36q^2 - 48q$

**21.** $49m^3n + 84mn^3 - 35m^4n^2$

**22.** $64x^4y^2 - 40x^3y^4 + 96xy^5$

**23.** $-18z^3 + 14z^2 + 4z$

**24.** $-18b^3 + 10b^2 + 6b$

**25.** $5c(3c - 2) - 3(3c - 2)$

**26.** $6z(5z + 3) + 5(5z + 3)$

**27.** $(4a + 3)(a - 3) + (2a - 7)(a - 3)$

**28.** $(5b + 3)(b + 4) + (3b + 1)(b + 4)$

*In Problems 29–40, factor by grouping.*

**29.** $5x + 5y + ax + ay$

**30.** $8x - 8y + bx - by$

**31.** $2z^3 + 10z^2 - 5z - 25$

**32.** $3y^3 + 9y^2 - 5y - 15$

**33.** $w^2 - 5w + 3w - 15$

**34.** $p^2 - 3p + 8p - 24$

**35.** $2x^2 - 8x - 4x + 16$

**36.** $3a^2 - 15a - 9a + 45$

**37.** $3x^3 + 15x^2 - 12x^2 - 60x$

**38.** $2y^3 + 14y^2 - 4y^2 - 28y$

**39.** $2ax - 2ay - bx + by$

**40.** $15x^2 - 5xy + 18xy - 6y^2$

## Mixed Practice

*In Problems 41–50, factor each polynomial completely.*

**41.** $(w + 3)(w - 3) - (w - 2)(w - 3)$

**42.** $(x + 5)(x - 3) - (x - 1)(x - 3)$

**43.** $2y^2 + 5y - 4y - 10$

**44.** $3q^2 + 5q - 12q - 20$

**45.** $6x^3y^3 + 9x^2y - 21x^3y^2$

**46.** $8a^4b^2 + 12a^3b^3 - 36ab^4$

**47.** $x^3 + x^2 + 3x + 3$

**48.** $c^3 - c^2 + 5c - 5$

**49.** $x(x - 2) + 3(x - 2)^2$

**50.** $2y(y + 4) + 3(y + 4)^2$

## Applying the Concepts

△ **51. Area** Write the area of the shaded region in factored form.

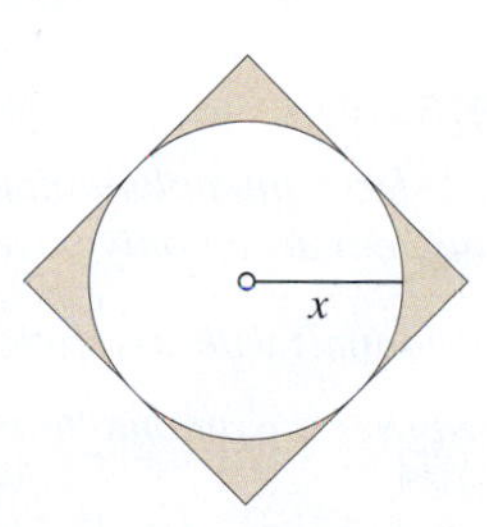


△ **52. Area** Write the area of the shaded region in factored form.

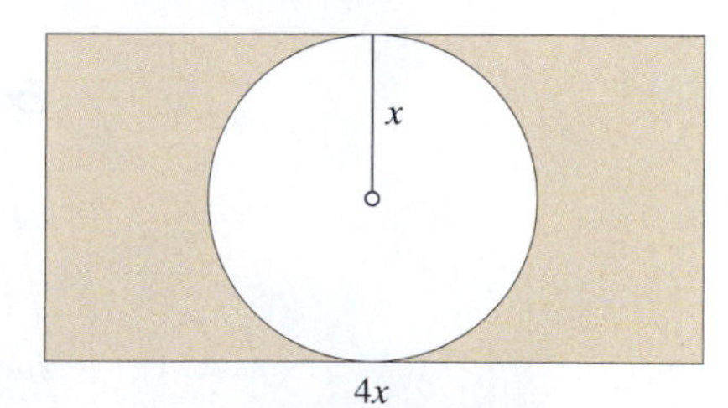


**53. Surface Area** The surface area of a cylindrical can whose radius is $r$ inches and height is 4 inches is given by $S = 2\pi r^2 + 8\pi r$. Express the surface area in factored form.

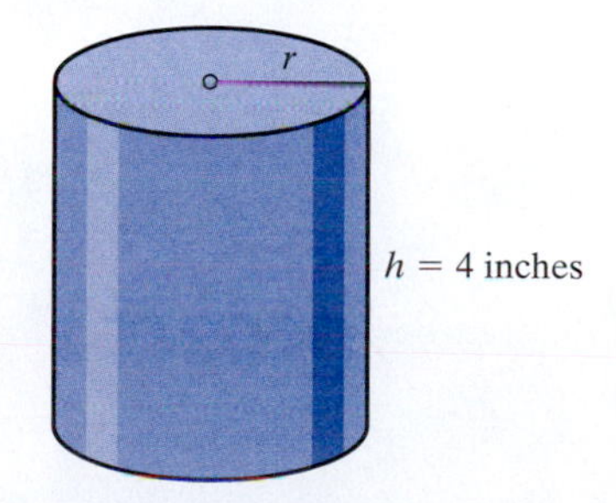

**54. Volume** The volume of a right circular cylinder of height $h$ and radius $r$ inscribed in a sphere of fixed radius $R$ is given by $V = \pi h R^2 - \frac{\pi h^3}{4}$. Express the volume of the cylinder in factored form.

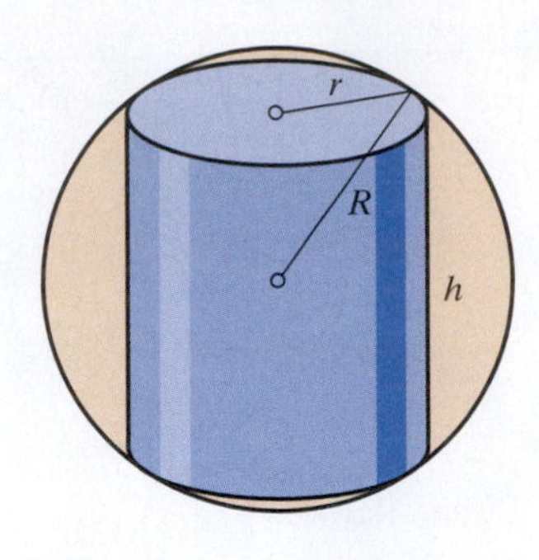


**55. Markups and Discounts** Suppose that a clothing store marks up its clothes 40% when it buys them from the supplier.

**(a)** Let $x$ represent the cost of designer shirts purchased from the supplier. Write an algebraic expression representing the selling price of the shirt.
**(b)** After 1 month, the manager of the clothing store discounts the shirts by 40%. Write an algebraic expression representing the sale price of the shirt in terms of $x$.
**(c)** Write the algebraic expression in factored form.
**(d)** Based on your answer to part (c), is the store selling the shirt for the same price that it paid for it?

**56. Summer Clearance** Suppose that an electronics store decides to sell last year's model 35-inch televisions for a 20% discount.

**(a)** Let $x$ represent the original cost of the television. Write an algebraic expression representing the selling price of the television.
**(b)** After 1 month, the manager of the store discounts the TVs by another 15%. Write an algebraic expression representing the sale price of the televisions in terms of $x$, the original selling price.
**(c)** Write the algebraic expression in factored form.
**(d)** If the original price of the television was $650, what is the sale price after the second discount?

**57. Stock Prices** Suppose that Christina purchased a stock for $x$ dollars. During the first year, the stock's price rose 15%.

**(a)** Write an algebraic expression for the price of the stock after the first year in terms of $x$.
**(b)** During the second year, the stock's price rose 10%. Write an algebraic expression for the price of the stock after the second year in terms of $x$.
**(c)** Write the algebraic expression found in part (b) in factored form.
**(d)** If the stock was originally purchased for $x = \$20$, what is the value of the stock after two years?

## Extending the Concepts

*In this section, we discussed factoring polynomials with integer coefficients over the integers. This means that factors of a polynomial can only have polynomials with integer coefficients. However, we could also discuss polynomials with, say, rational coefficients such as* $\frac{1}{3}x + \frac{2}{3}$. *A polynomial such as this can be factored over the rational numbers by factoring out the greatest common factor* $\frac{1}{3}$, *so that* $\frac{1}{3}x + \frac{2}{3} = \frac{1}{3}(x + 2)$. *Use this idea to factor out the greatest common factor in Problems 58–61.*

**58.** $\frac{1}{2}x + \frac{3}{2}$ **59.** $\frac{1}{4}x - \frac{7}{4}$ **60.** $\frac{2}{3}x^2 + \frac{4}{9}x$ **61.** $\frac{1}{5}b^3 + \frac{8}{25}b$

**62. The Better Deal** Which is the better deal: (a) Receiving a 30% discount or (b) receiving a 15% discount and then another 15% discount after the first 15% discount was applied? Prove it!

# 5.5 Factoring Trinomials

## OBJECTIVES

1. Factor Trinomials of the Form $x^2 + bx + c$
2. Factor Trinomials of the Form $ax^2 + bx + c, a \neq 1$
3. Factor Trinomials Using Substitution

### *Preparing for Factoring Trinomials*

*Before getting started, take the following readiness quiz. If you get a problem wrong, go back to the section cited and review the material.*

**1.** List the factors of 18 whose sum is 11.

**2.** List the factors of $-24$ whose sum is $-2$.

**3.** Determine the coefficients of $4x^2 - 9x + 2$ [Section R.5, p. 44]

In this section, we are going to factor trinomials of the form $ax^2 + bx + c$, where $a$, $b$, and $c$ are integers. We begin by looking at trinomials where the leading coefficient is 1.

## 1 Factor Trinomials of the Form $x^2 + bx + c$

The idea behind factoring a second-degree polynomial of the form $x^2 + bx + c$ is to see whether it can be written as the product of two first-degree polynomials.

For example,

Multiplication →

$$\text{Factored Form} \rightarrow (x-5)(x+2) = x^2 - 3x - 10 \leftarrow \text{Product}$$

← Factoring

The factors of $x^2 - 3x - 10$ are $x - 5$ and $x + 2$. Notice the following:

$$x^2 - 3x - 10 = (x-5)(x+2)$$

The sum of $-5$ and $2$ is $-3$

The product of $-5$ and $2$ is $-10$

In general, if $x^2 + bx + c = (x+m)(x+n)$, then $mn = c$ and $m + n = b$.

*Preparing for...Answers* **1.** 9 and 2 **2.** $-6$ and 4 **3.** $4, -9, 2$

### EXAMPLE 1 How to Factor a Trinomial of the Form $x^2 + bx + c$

Factor: $x^2 + 8x + 12$

#### Step-by-Step Solution

**Step 1:** We are looking for factors of $c = 12$ whose sum is $b = 8$. We begin by listing all factors of 12 and computing the sum of these factors.

| **Integers Whose Product Is 12** | 1, 12 | 2, 6 | 3, 4 | −1, −12 | −2, −6 | −3, −4 |
|---|---|---|---|---|---|---|
| **Sum** | 13 | 8 | 7 | −13 | −8 | −7 |

We can see that $2 \cdot 6 = 12$ and $2 + 6 = 8$, so $m = 2$ and $n = 6$.

**Step 2:** We write the trinomial in the form $(x + m)(x + n)$.

$$x^2 + 8x + 12 = (x+2)(x+6)$$

**Step 3: Check** We FOIL $(x + 2)(x + 6)$ to verify our solution.

$$(x+2)(x+6) = x^2 + 6x + 2x + 2(6) = x^2 + 8x + 12$$

So $x^2 + 8x + 12 = (x+2)(x+6)$. Because multiplication is commutative, the order in which we list the factors does not matter. So $x^2 + 8x + 12$ is also equal to $(x+6)(x+2)$.

We summarize the steps used in Example 1 below.

**Factoring a Trinomial of the Form $x^2 + bx + c$**

**Step 1:** Find the pair of integers whose product is $c$ and whose sum is $b$. That is, determine $m$ and $n$ such that $mn = c$ and $m + n = b$.

**Step 2:** Write $x^2 + bx + c = (x + m)(x + n)$.

**Step 3:** Check your work by multiplying out the factored form.

**Work Smart**

In a trinomial $x^2 + bx + c$, if both $b$ and $c$ are positive, then $m$ and $n$ must both be positive.

In Example 1, notice that the coefficient of the middle term is positive and the constant is positive. If the coefficient of the middle term and constant are both positive, then $m$ and $n$ must both be positive.

**QUICK ✓** *Factor each trinomial.*

**1.** $y^2 + 9y + 18$ **2.** $p^2 + 14p + 24$

**EXAMPLE 2 Factoring a Trinomial of the Form $x^2 + bx + c$**

Factor: $p^2 - 10p + 24$

**Solution**

We are looking for factors of $c = 24$ whose sum is $b = -10$. We begin by listing all factors of 24 and computing the sum of these factors.

| Integers Whose Product Is 24 | 1, 24 | 2, 12 | 3, 8 | 4, 6 | −1, −24 | −2, −12 | −3, −8 | −4, −6 |
|---|---|---|---|---|---|---|---|---|
| Sum | 25 | 14 | 11 | 10 | −25 | −14 | −11 | −10 |

We can see that $-4 \cdot (-6) = 24$ and $-4 + (-6) = -10$, so $m = -4$ and $n = -6$. We write the trinomial in the form $(x + m)(x + n)$.

$$\begin{aligned} x^2 - 10x + 24 &= (x + (-4))(x + (-6)) \\ &= (x - 4)(x - 6) \end{aligned}$$

**Check** We FOIL $(x - 4)(x - 6)$ to verify our solution.

$$\begin{aligned} (x - 4)(x - 6) &= x^2 - 6x - 4x + (-4)(-6) \\ &= x^2 - 10x + 24 \end{aligned}$$

**Work Smart**

In a trinomial $x^2 + bx + c$, if $b$ is negative and $c$ is positive, then $m$ and $n$ must both be negative.

In Example 2, notice that the coefficient of the middle term is negative and the constant is positive. If the coefficient of the middle term is negative and constant is positive, then $m$ and $n$ must both be negative.

**QUICK ✓** *Factor each trinomial.*

**3.** $q^2 - 6q + 8$ **4.** $x^2 - 8x + 12$

## EXAMPLE 3 Factoring a Trinomial of the Form $x^2 + bx + c$, $c < 0$

Factor: $z^2 - 3z - 28$

### Solution

We are looking for factors of $c = -28$ whose sum is $b = -3$. We begin by listing all factors of $-28$ and computing the sum of these factors.

| Integers Whose Product Is $-28$ | $-1, 28$ | $-2, 14$ | $-4, 7$ | $1, -28$ | $2, -14$ | $4, -7$ |
|---|---|---|---|---|---|---|
| Sum | 27 | 12 | 3 | $-27$ | $-12$ | $-3$ |

We can see that $4 \cdot (-7) = -28$ and $4 + (-7) = -3$, so $m = 4$ and $n = -7$. We write the trinomial in the form $(x + m)(x + n)$.

$$\begin{aligned} x^2 - 3x - 28 &= (x + 4)(x + (-7)) \\ &= (x + 4)(x - 7) \end{aligned}$$

**Check** We FOIL $(x + 4)(x - 7)$ to verify our solution.

$$(x + 4)(x - 7) = x^2 - 7x + 4x + 4(-7) = x^2 - 3x - 28$$

**Work Smart**

In a trinomial $x^2 + bx + c$, if both $b$ and $c$ are negative, then $m$ and $n$ must have opposite signs.

In Example 3, notice that the coefficient of the middle term is negative and the constant is also negative. If the coefficient of the middle term is negative and constant is negative, then $m$ and $n$ must be opposite in sign. The factor with the larger absolute value must be negative.

**QUICK ✓** *Factor each trinomial.*

**5.** $w^2 - 4w - 21$ **6.** $q^2 - 9q - 36$

Remember, a polynomial whose coefficients are integers that cannot be written as the product of two other polynomials whose coefficients are integers (other than 1 or $-1$) is said to be prime.

## EXAMPLE 4 Identifying a Prime Trinomial

Factor: $y^2 + 4y + 12$

### Solution

We are looking for factors of $c = 12$ whose sum is $b = 4$. Because both $b$ and $c$ are positive, we know that $m$ and $n$ must both be positive, so we only list positive factors of 12 and compute the sum of these factors.

| Integers Whose Product Is 12 | 1, 12 | 2, 6 | 3, 4 |
|---|---|---|---|
| Sum | 13 | 8 | 7 |

There are no factors of 12 whose sum is 4. Therefore, $y^2 + 4y + 12$ is prime.

**QUICK ✓**

**7.** Factor: $t^2 - 5t + 8$

Sometimes we need to factor out a common factor before factoring the trinomial.

## EXAMPLE 5 Factoring Trinomials with a Common Factor

Factor: $3u^3 - 9u^2 - 120u$

### Solution

**Work Smart**

What's the first thing to look for when factoring any polynomial? A greatest common factor!

We start by noticing that there is a common factor of $3u$ in the trinomial. We factor the $3u$ out:

$$3u^3 - 9u^2 - 120u = 3u(u^2 - 3u - 40)$$

We now concentrate on factoring the trinomial in parentheses, $u^2 - 3u - 40$. We are looking for factors of $c = -40$ whose sum is $b = -3$. We begin by listing all factors of $-40$. Because $c = -40$ and $b = -3$, we know the larger factor of 40 will be negative.

| Integers Whose Product Is $-40$ | $1, -40$ | $2, -20$ | $4, -10$ | $5, -8$ |
|---|---|---|---|---|
| Sum | $-39$ | $-18$ | $-6$ | $-3$ |

We can see that $5 \cdot (-8) = -40$ and $5 + (-8) = -3$, so $m = 5$ and $n = -8$. We write the trinomial in the form $(x + m)(x + n)$.

$$\begin{aligned} u^2 - 3u - 40 &= (u + 5)(u + (-8)) \\ &= (u + 5)(u - 8) \end{aligned}$$

Remember, we already factored out a common factor, so we have that $3u^3 - 9u^2 - 120u = 3u(u^2 - 3u - 40) = 3u(u + 5)(u - 8)$.

**Check**

$$\begin{aligned} 3u(u + 5)(u - 8) &= 3u(u^2 - 8u + 5u - 40) \\ &= 3u(u^2 - 3u - 40) \\ &= 3u^3 - 9u^2 - 120u \end{aligned}$$

**QUICK ✓** *Factor each trinomial completely.*

**8.** $2x^3 - 12x^2 - 54x$ **9.** $-3z^2 - 21z - 30$

If a trinomial has more than one variable, we take the same approach as that used for trinomials with one variable. So trinomials of the form

$$x^2 + bxy + cy^2$$

will factor as

$$(x + \_\,y)(x + \_\,y)$$

where the blanks need to be determined.

## EXAMPLE 6 Factoring Trinomials with Two Variables

Factor: $p^2 + 6pq - 16q^2$

### Solution

The trinomial $p^2 + 6pq - 16q^2$ will factor as $(p + mq)(p + nq)$ where $mn = -16$ and $m + n = 6$. We are looking for factors of $c = -16$ whose sum is $b = 6$. We list factors of $-16$, but because $c = -16$ and $b = 6$, we know the larger factor of 16 will be positive.

| Integers Whose Product Is $-16$ | $-1, 16$ | $-2, 8$ | $-4, 4$ |
|---|---|---|---|
| Sum | 15 | 6 | 0 |

We can see that $-2 \cdot 8 = -16$ and $-2 + 8 = 6$, so $m = -2$ and $n = 8$. We write the trinomial in the form $(p + mq)(p + nq)$.

$$p^2 + 6pq - 16q^2 = (p + (-2)q)(p + 8q)$$
$$= (p - 2q)(p + 8q)$$

**Check** $$(p - 2q)(p + 8q) = p^2 + 8pq - 2pq - 16q^2$$
$$= p^2 + 6pq - 16q^2$$

**QUICK ✓** *Factor completely each trinomial.*

**10.** $x^2 + 8xy + 15y^2$ **11.** $m^2 + mn - 20n^2$

## 2 Factor Trinomials of the Form $ax^2 + bx + c, a \neq 1$

When it comes to factoring trinomials of the form $ax^2 + bx + c$, where $a$ is not 1, we have two methods that can be used:

1. Factoring by grouping
2. Trial and error

There are pros and cons to both methods. We will point out these pros and cons as we proceed. We start with factoring by grouping.

### Factoring $ax^2 + bx + c, a \neq 1$ by Grouping

Example 7 illustrates how to factor $ax^2 + bx + c, a \neq 1$ by grouping.

**EXAMPLE 7 How to Factor $ax^2 + bx + c, a \neq 1$ by Grouping**

Factor: $3x^2 + 14x + 8$

**Step-by-Step Solution**

First, we notice that $3x^2 + 14x + 8$ has no common factors and that $a = 3$, $b = 14$, and $c = 8$.

**Step 1:** Find the value of $ac$.

The value of $a \cdot c = 3 \cdot 8 = 24$.

**Step 2:** We want to determine the integers whose product is 24 and whose sum is 14. Because both 24 and 14 are positive, we only list the positive factors of 24.

| Integers Whose Product Is 24 | 1, 24 | 2, 12 | 3, 8 | 4, 6 |
|---|---|---|---|---|
| Sum | 25 | 14 | 11 | 10 |

The integers whose product is 24 and sum is 14 are 2 and 12.

**Step 3:** Write

$$ax^2 + bx + c = ax^2 + mx + nx + c.$$

Write $3x^2 + 14x + 8$ as $3x^2 + 2x + 12x + 8$

$14x = 2x + 12x$

**Step 4:** Factor the expression in Step 3 by grouping.

$$3x^2 + 2x + 12x + 8 = (3x^2 + 2x) + (12x + 8)$$
$$= x(3x + 2) + 4(3x + 2)$$

Factor out $3x + 2$: $$= (3x + 2)(x + 4)$$

**Step 5:** Check

$$(3x + 2)(x + 4) = 3x^2 + 12x + 2x + 8$$
$$= 3x^2 + 14x + 8$$

So $3x^2 + 14x + 8 = (3x + 2)(x + 4)$.

To factor a second-degree polynomial $ax^2 + bx + c$ when $a \neq 1$ and there are no common factors in the polynomial, we use the following steps.

**Factoring $ax^2 + bx + c$, $a \neq 1$ by Grouping: $a$, $b$, and $c$ Have No Common Factors**

**Step 1:** Find the value of $ac$.

**Step 2:** Find the pair of integers whose product is $ac$ and whose sum is $b$. That is, find $m$ and $n$ so that $mn = ac$ and $m + n = b$.

**Step 3:** Write $ax^2 + bx + c = ax^2 + mx + nx + c$.

**Step 4:** Factor the expression in Step 3 by grouping.

**Step 5:** Multiply out the factored form to verify your answer.

## EXAMPLE 8 Factoring $ax^2 + bx + c$, $a \neq 1$ by Grouping

Factor: $12x^2 - x - 6$

### Solution

First, we notice that $12x^2 - x - 6$ has no common factors and that $a = 12$, $b = -1$, and $c = -6$. The value of $a \cdot c = 12 \cdot (-6) = -72$.

We want to determine the integers whose product is $-72$ and whose sum is $-1$. Because $-72 < 0$, we know that one integer will be positive and the other negative. Because $-1 < 0$, we know the larger factor of $-72$ will be negative.

| **Integers Whose Product Is −72** | 1, −72 | 2, −36 | 3, −24 | 4, −18 | 6, −12 | 8, −9 |
|---|---|---|---|---|---|---|
| **Sum** | −71 | −34 | −21 | −14 | −6 | −1 |

The integers whose product is $-72$ and sum is $-1$ are 8 and $-9$.

**Work Smart**

Be careful with the negative sign on $-9x$.

$-x = 8x - 9x$

$$\begin{aligned} 12x^2 - x - 6 &= 12x^2 + 8x - 9x - 6 \\ &= (12x^2 + 8x) + (-9x - 6) \\ &= 4x(3x + 2) - 3(3x + 2) \\ \text{Factor out } 3x + 2\text{:} \quad &= (3x + 2)(4x - 3) \end{aligned}$$

**Check**
$$\begin{aligned} (3x + 2)(4x - 3) &= 12x^2 - 9x + 8x - 6 \\ &= 12x^2 - x - 6 \end{aligned}$$

So $12x^2 - x - 6 = (3x + 2)(4x - 3)$.

**QUICK ✓** *Factor each trinomial completely.*

**12.** $6x^2 + 11x + 3$ **13.** $2b^2 + 7b - 15$

The advantage of factoring by grouping to factor trinomials of the form $ax^2 + bx + c$, $a \neq 1$ is that it is algorithmic (that is, step by step). However, if the product $a \cdot c$ gets large, then there are a lot of factors of $ac$ whose sum must be determined. This can get overwhelming. Under these circumstances, it may be better to try the second method, trial and error.

## Using Trial and Error to Factor $ax^2 + bx + c$, $a \neq 1$

The idea behind using trial and error is to list various binomials and use FOIL to find their product until the combination of binomials that results in the original trinomial is found. While this method may sound haphazard, experience and logic play a role in minimizing the number of possibilities that must be tried before a factored form is found.

### EXAMPLE 9 How to Factor $ax^2 + bx + c$, $a \neq 1$ Using Trial and Error

Factor: $10x^2 + 19x + 6$

**Step-by-Step Solution**

There are no common factors in $10x^2 + 19x + 6$.

**Step 1:** List the possibilities for the first terms of each binomial whose product is $ax^2$.

We list all possible ways of representing the first term, $10x^2$.

$$(10x + \_\_)(x + \_\_)$$
$$(5x + \_\_)(2x + \_\_)$$

**Step 2:** List the possibilities for the last terms of each binomial whose product is $c$.

The last term, 6, has the factors:

$$1 \cdot 6,\ 2 \cdot 3,\ -1 \cdot (-6), \text{ or } -2 \cdot (-3).$$

**Step 3:** Write out all the combinations of factors found in Steps 1 and 2. Multiply the binomials out until a product is found that equals the trinomial.

| Possible Factorization | Product | Possible Factorization | Product |
|---|---|---|---|
| $(10x + 1)(x + 6)$ | $10x^2 + 61x + 6$ | $(5x + 1)(2x + 6)$ | $10x^2 + 32x + 6$ |
| $(10x + 6)(x + 1)$ | $10x^2 + 16x + 6$ | $(5x + 6)(2x + 1)$ | $10x^2 + 17x + 6$ |
| $(10x + 2)(x + 3)$ | $10x^2 + 32x + 6$ | $(5x + 2)(2x + 3)$ | $10x^2 + 19x + 6$ |
| $(10x + 3)(x + 2)$ | $10x^2 + 23x + 6$ | $(5x + 3)(2x + 2)$ | $10x^2 + 16x + 6$ |
| $(10x - 1)(x - 6)$ | $10x^2 - 61x + 6$ | $(5x - 1)(2x - 6)$ | $10x^2 - 32x + 6$ |
| $(10x - 6)(x - 1)$ | $10x^2 - 16x + 6$ | $(5x - 6)(2x - 1)$ | $10x^2 - 17x + 6$ |
| $(10x - 2)(x - 3)$ | $10x^2 - 32x + 6$ | $(5x - 2)(2x - 3)$ | $10x^2 - 19x + 6$ |
| $(10x - 3)(x - 2)$ | $10x^2 - 23x + 6$ | $(5x - 3)(2x - 2)$ | $10x^2 - 16x + 6$ |

The highlighted row is the factorization that works, so $10x^2 + 19x + 6 = (5x + 2)(2x + 3)$.

In looking at the solution to Example 9, you may feel a little overwhelmed—so many possibilities! However, many of the possibilities could have been eliminated with a little thought. For example, we notice that the middle and last terms are both positive. This means that the factors of $c$ must be positive—this alone would eliminate half of the combinations listed in Example 9. Further, because the original polynomial has no common factors, the binomials in the factored form cannot have common factors either. This would eliminate three additional possibilities.

If we had employed these hints, the list in Example 9 would become

| Possible Factorization | Product |
|---|---|
| $(10x + 1)(x + 6)$ | $10x^2 + 61x + 6$ |
| $(10x + 3)(x + 2)$ | $10x^2 + 23x + 6$ |
| $(5x + 6)(2x + 1)$ | $10x^2 + 17x + 6$ |
| $(5x + 2)(2x + 3)$ | $10x^2 + 19x + 6$ |

Not too bad! There are only four possibilities. We summarize the steps to factor using trial and error and some helpful hints below.

### Steps to Factor $ax^2 + bx + c$, $a \neq 1$ Using Trial and Error: $a$, $b$, and $c$ Have No Common Factors

**Step 1:** List the possibilities for the first terms of each binomial whose product is $ax^2$.

$$(\_x + \quad)(\_x + \quad) = ax^2 + bx + c$$

**Step 2:** List the possibilities for the last terms of each binomial whose product is $c$.

$$(\_x + \square)(\_x + \square) = ax^2 + bx + c$$

**Step 3:** Write out all the combinations of factors found in Steps 1 and 2. Multiply the binomials out until a product is found that equals the trinomial.

**HELPFUL HINTS IN USING TRIAL AND ERROR TO FACTOR $ax^2 + bx + c$, $a \neq 1$**

- If the constant $c$ is positive, then the factors of $c$ must be the same sign as $b$.
- If $ax^2 + bx + c$ has no common factor, then the binomials in the factored form cannot have common factors either.
- If the value of $b$ is small, then choose factors of $ac$ that are close to each other. If the value of $b$ is large, then choose factors of $ac$ that are far from each other.
- If the value of $b$ is correct, but is the wrong sign, then interchange the signs in the binomial factors.

### EXAMPLE 10 Factoring $ax^2 + bx + c$, $a \neq 1$ Using Trial and Error

Factor: $18x^2 + 3x - 10$

#### Solution

There are no common factors in $18x^2 + 3x - 10$. We list all possible ways of representing the first term, $18x^2$.

$$(18x + \_)(x + \_)$$
$$(9x + \_)(2x + \_)$$
$$(6x + \_)(3x + \_)$$

The last term, $-10$, has the factors: $1 \cdot (-10)$, $2 \cdot (-5)$, $-1 \cdot 10$, or $-2 \cdot 5$.

We list some possible combinations of factors found in Steps 1 and 2. We should notice that the middle term is $+3x$. Because this middle term is positive and small, the binomial factors we list should have outer and inner products that sum to a positive, small number. Therefore, we will start with $(6x + \_)(3x + \_)$ and the factors

$2 \cdot (-5)$ and $-2 \cdot 5$. We do not use $6x + 2$ or $6x - 2$ as a possible factor because there is a common factor in these binomials.

Let's try $(6x - 5)(3x + 2)$.

$$\begin{aligned}(6x - 5)(3x + 2) &= 18x^2 + 12x - 15x - 10 \\ &= 18x^2 - 3x - 10\end{aligned}$$

Close! The only problem is that the middle term is the opposite sign that we want. Therefore, let's change the signs on $-5$ and 2 in the binomial and try $(6x + 5)(3x - 2)$.

$$\begin{aligned}(6x + 5)(3x - 2) &= 18x^2 - 12x + 15x - 10 \\ &= 18x^2 + 3x - 10\end{aligned}$$

So $18x^2 + 3x - 10 = (6x + 5)(3x - 2)$.

The moral of the story in Example 10 is that the name *trial and error* is a bit misleading. You won't have to haphazardly choose binomial factors "until the cows come home" provided you use the helpful hints and some careful thought.

**QUICK ✓** *Factor each trinomial completely.*

**14.** $8x^2 + 14x + 5$ **15.** $12y^2 + 32y - 35$

---

## EXAMPLE 11 Factoring Trinomials with Two Variables

Factor: $24x^2 + 13xy - 2y^2$

### Solution

There are no common factors in $24x^2 + 13xy - 2y^2$. The trinomial will factor in the form $24x^2 + 13xy - 2y^2 = (\_x + \_y)(\_x + \_y)$.

We list all possible ways of representing the first term, $24x^2$.

$$\begin{aligned}&(24x + \_y)(x + \_y) \\ &(12x + \_y)(2x + \_y) \\ &(8x + \_y)(3x + \_y) \\ &(6x + \_y)(4x + \_y)\end{aligned}$$

The last term, $-2$, has the factors: $1 \cdot (-2)$ or $2 \cdot (-1)$.

We should notice that the middle term is $+13x$. Because this middle term is positive and neither large nor small, the binomial factors we list should have outer and inner products that sum to a positive, midsize number. The only coefficients on $y$ in the factored form are 1 and 2 (ignoring their sign for a second). We cannot have 2 as a factor in the forms $(12x + \_y)(2x + \_y)$ or $(6x + \_y)(4x + \_y)$ because 2 creates a common factor and a common factor does not exist in the original trinomial. Therefore, we will start with $(8x + \_y)(3x + \_y)$ and the factors $1 \cdot (-2)$ and $2 \cdot (-1)$.

Let's try $(8x + 1y)(3x - 2y)$.

$$\begin{aligned}(8x + 1y)(3x - 2y) &= 24x^2 - 16xy + 3xy - 2y^2 \\ &= 24x^2 - 13xy - 2y^2\end{aligned}$$

Close! The only problem is that the middle term is the opposite sign that we want. Therefore, let's change the signs on 1 and $-2$ in the binomial and try $(8x - 1y)(3x + 2y)$.

$$\begin{aligned}(8x - 1y)(3x + 2y) &= 24x^2 + 16xy - 3xy - 2y^2 \\ &= 24x^2 + 13xy - 2y^2\end{aligned}$$

So $24x^2 + 13xy - 2y^2 = (8x - y)(3x + 2y)$.

**QUICK ✓** *Completely factor the trinomial.*

**16.** $30x^2 + 7xy - 2y^2$

---

### EXAMPLE 12 Factoring Trinomials with a Negative Leading Coefficient

Factor: $-14x^2 + 29x + 15$

#### Solution

While there are no common factors in $-14x^2 + 29x + 15$, we do notice that the coefficient of the square term is negative. It is easier to factor trinomials when the leading coefficient is positive. Therefore, we can factor $-1$ out of the trinomial to obtain

$$-14x^2 + 29x + 15 = -1(14x^2 - 29x - 15)$$

Now we factor the expression in parentheses using either the grouping technique or trial and error and obtain

$$-14x^2 + 29x + 15 = -1(14x^2 - 29x - 15)$$

Remember the $-1$ that was factored out: $= -1(7x + 3)(2x - 5)$

**QUICK ✓** *Factor each trinomial completely.*

**17.** $-6y^2 + 23y + 4$

**18.** $-9x^2 - 21xy - 10y^2$

---

## 3 Factor Trinomials Using Substitution

Sometimes it is possible to factor a complicated-looking polynomial through a substitution of one variable for another. When this approach is used, we say that we are **factoring by substitution.**

### EXAMPLE 13 Factoring by Substitution

Factor: $2z^4 - z^2 - 15$

#### Solution

Notice that $2z^4 - z^2 - 15$ can be written as $2(z^2)^2 - z^2 - 15$ so that the trinomial is in the form $au^2 + bu + c$, where $u = z^2$. If we substitute $u$ for $z^2$, we obtain

**Work Smart**

To see if a trinomial can be factored using substitution, check to see if the trinomial can be written in the form

$a(☺)^2 + b(☺) + c$

where ☺ is some algebraic expression.

$$2z^4 - z^2 - 15 = 2(z^2)^2 - z^2 - 15$$

Let $z^2 = u$: $= 2u^2 - u - 15$

Factor: $= (2u + 5)(u - 3)$

Let $u = z^2$: $= (2z^2 + 5)(z^2 - 3)$

So $2z^4 - z^2 - 15 = (2z^2 + 5)(z^2 - 3)$.

### EXAMPLE 14 Factoring by Substitution

Factor: $3(x - 3)^2 + 11(x - 3) - 4$

#### Solution

Notice that $3(x - 3)^2 + 11(x - 3) - 4$ can be written in the form $au^2 + bu + c$ where $u = x - 3$. If we substitute $u$ for $x - 3$, we obtain

$$3(x - 3)^2 + 11(x - 3) - 4 = 3u^2 + 11u - 4$$

Factor: $= (3u - 1)(u + 4)$

We do not want to factor an expression in $u$, we want to factor an expression in $x$. So, we substitute $x - 3$ for $u$ and obtain

$$\begin{aligned} 3u^2 + 11u - 4 &= (3u - 1)(u + 4) \\ \text{Let } u = x - 3: \quad &= [3(x - 3) - 1][(x - 3) + 4] \\ &= (3x - 9 - 1)(x + 1) \\ &= (3x - 10)(x + 1) \end{aligned}$$

So $3(x - 3)^2 + 11(x - 3) - 4 = (3x - 10)(x + 1)$.

**QUICK ✓** *Factor each trinomial by substitution.*

**19.** $y^4 - 2y^2 - 24$ **20.** $4(x - 3)^2 + 5(x - 3) - 6$

# 5.5 Exercises

**For Extra Help:**

## Concepts and Vocabulary

*In Problems 1 and 2, fill in the blanks.*

**1.** When factoring a trinomial of the form $x^2 + bx + c$, we need to find pairs of integers $m$ and $n$ such that _____ $= c$ and _____ $= b$.

**2.** When factoring $3(2x - 3)^2 + 5(2x - 3) + 2$ by substitution, we would let $u =$ _____.

*In Problems 3 and 4, answer True or False to each statement.*

**3.** $5x^2 - 6x + 1 = (5x - 1)(x - 1)$

**4.** $4(2x + 1)^2 - 3(2x + 1) - 1 = 2x(8x + 5)$

**5.** State the circumstances for which trial and error is a better approach than grouping when factoring $ax^2 + bx + c$, $a \neq 1$.

**6.** When factoring any polynomial, what is always the first step?

**7.** A student factored the trinomial $3p^2 - 9p - 30$ as $(3p + 6)(p - 5)$ on an exam, but only received partial credit. Can you explain why?

**8.** How can you tell if a trinomial is not factorable? Make up an example to demonstrate your reasoning.

## Building Skills

*In Problems 9–54, factor each trinomial completely.*

**9.** $x^2 + 8x + 15$ **10.** $x^2 + 8x + 12$ **11.** $p^2 + 3p - 18$

**12.** $z^2 + 3z - 28$ **13.** $r^2 + 10r + 25$ **14.** $y^2 - 12y + 36$

**15.** $s^2 + 7s - 60$ **16.** $q^2 + 2q - 80$ **17.** $x^2 - 15x + 56$

**18.** $z^2 - 16z + 48$ **19.** $-w^2 - 2w + 24$ **20.** $-p^2 + 3p + 54$

**21.** $x^2 + 7xy + 12y^2$ **22.** $m^2 + 7mn + 10n^2$ **23.** $p^2 + 2pq - 24q^2$

**24.** $x^2 - 4xy - 21y^2$ **25.** $2x^2 + 12x - 54$ **26.** $3y^2 - 6y - 189$

**27.** $-3r^2 + 39r - 120$ **28.** $-4s^2 - 32s - 48$ **29.** $2p^2 - 15p - 8$

**30.** $3z^2 - 13z - 10$ **31.** $4y^2 - 11y + 6$ **32.** $6x^2 - 37x + 6$

**33.** $8s^2 + 2s - 3$ **34.** $12r^2 + 11r - 15$ **35.** $16z^2 + 8z - 15$

**36.** $18y^2 + 43y - 5$ **37.** $18y^2 + 17y + 4$ **38.** $20r^2 + 23r + 6$

**39.** $-16m^2 + 12m + 70$ **40.** $-24y^2 - 39y + 18$ **41.** $48z^2 + 124z + 28$

**42.** $48w^2 + 20w - 42$ **43.** $2x^2 + 11xy - 21y^2$ **44.** $3m^2 + 7mn - 6n^2$

**45.** $4r^2 - 23rs + 15s^2$ **46.** $6r^2 - 25rs + 4s^2$ **47.** $24r^2 + 23rs - 12s^2$

**48.** $18x^2 + 37xy - 20y^2$ **49.** $3x^3 - 6x^2 - 240x$ **50.** $4x^3 - 52x^2 + 144x$

**51.** $8x^3y^2 - 76x^2y^2 + 140xy^2$ **52.** $-24m^3n - 18m^2n + 27mn$

**53.** $70r^4s - 36r^3s - 16r^2s$ **54.** $54x^3y + 33x^2y - 72xy$

*In Problems 55–64, factor each trinomial completely.*

**55.** $x^4 - 3x^2 + 2$ **56.** $y^4 + 5y^2 + 6$

**57.** $m^2n^2 + 5mn - 14$ **58.** $r^2s^2 + 8rs - 48$

**59.** $(x + 1)^2 - 6(x + 1) - 16$ **60.** $(y - 3)^2 + 3(y - 3) + 2$

**61.** $(3r - 1)^2 - 9(3r - 1) + 20$ **62.** $(5z - 3)^2 - 12(5z - 3) + 32$

**63.** $2(y - 3)^2 + 13(y - 3) + 15$ **64.** $3(z + 3)^2 + 14(z + 3) + 8$

## Mixed Practice

*In Problems 65–82, factor each polynomial completely. If the polynomial cannot be factored, say it is prime.*

**65.** $10w^2 + 41w + 21$ **66.** $8q^2 + 26q + 11$

**67.** $4(2y + 1)^2 - 3(2y + 1) - 1$ **68.** $2(3z - 1)^2 + 3(3z - 1) + 1$

**69.** $12x^2 + 23xy + 10y^2$ **70.** $24m^2 + 58mn + 9n^2$

**71.** $y^2 + 2y - 27$ **72.** $t^2 - 5t + 8$

**73.** $x^4 + 6x^2 + 8$ **74.** $a^6 + 7a^3 + 12$

**75.** $z^6 + 9z^3 + 20$ **76.** $r^6 - 6r^3 + 8$

**77.** $r^2 - 12rs + 32s^2$ **78.** $p^2 - 14pq + 45q^2$

**79.** $8(z + 1)^2 + 2(z + 1) - 1$ **80.** $9(a + 2)^2 - 10(a + 2) + 1$

**81.** $3x^2 - 7x - 12$ **82.** $5w^2 - 10w + 12$

## Applying the Concepts

**83. How to Make an Open Box** The volume of an open box with a rectangular base is to be made from a piece of cardboard that is 24 by 30 inches by cutting a square piece of cardboard from each corner and turning up the sides. See the illustration. The volume $V$ of the box as a function of the length $x$ of the side of the square cut from each corner is

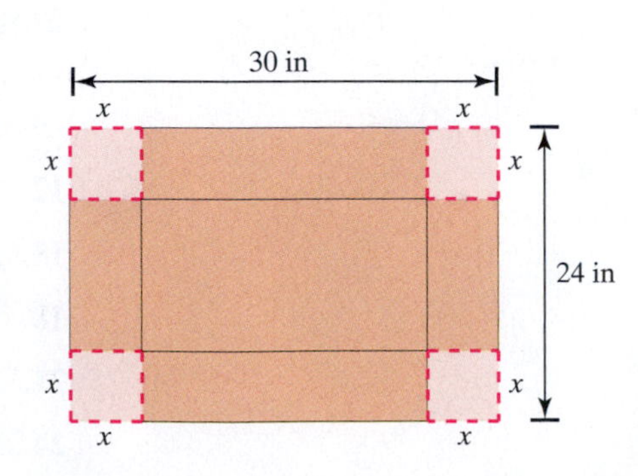


$$V(x) = 4x^3 - 108x^2 + 720x$$

**(a)** Find and interpret $V(3)$.
**(b)** Completely factor the function $V(x) = 4x^3 - 108x^2 + 720x$.
**(c)** Find $V(3)$ using the factored form found in part (b).
**(d)** Did you find it easier to evaluate $V(3)$ in part (a) or in part (c)?

**84. Projectile Motion** A boy is standing on a cliff that is 240 feet high and throws a rock out toward the ocean. The height of the rock above the sea can be described by the function in which $t$ is in seconds and $s$ is in feet:

$$s(t) = -16t^2 + 32t + 240$$

**(a)** Find and interpret $s(3)$.
**(b)** Completely factor the function $s(t) = -16t^2 + 32t + 240$.
**(c)** Find $s(3)$ using the factored form found in part (b).
**(d)** Did you find it easier to evaluate $s(3)$ in part (a) or in part (c)?

**85.** Suppose that we know one factor of $6x^2 - 11x - 10$ is $3x + 2$. What is the other factor?

**86.** Suppose that we know one factor of $8x^2 + 22x - 21$ is $2x + 7$. What is the other factor?

## Extending the Concepts

*In this section, we discussed factoring polynomials whose coefficients are integers (factoring over the integers). We can also factor polynomials whose coefficients are rational numbers in which case we factor out the rational numbers. One technique for doing this is to first factor out the coefficient of the highest degree term as a greatest common factor as follows:*

$$\begin{aligned}\frac{1}{3}x^2 + \frac{4}{3}x + 1 &= \frac{1}{3}(x^2 + 4x + 3)\\ &= \frac{1}{3}(x + 3)(x + 1)\end{aligned}$$

*Use this technique in Problems 87–92.*

**87.** $\frac{1}{2}x^2 + 3x + 4$ **88.** $\frac{1}{4}x^2 + 2x + 3$ **89.** $\frac{1}{3}p^2 - \frac{2}{3}p - 1$

**90.** $\frac{1}{4}z^2 + \frac{1}{2}z - \frac{15}{4}$ **91.** $\frac{4}{3}a^2 - \frac{8}{3}a - 32$ **92.** $\frac{3}{8}b^2 - \frac{15}{8}b - 9$

# 5.6 Factoring Special Products

**OBJECTIVES**

1. Factor Perfect Square Trinomials
2. Factor the Difference of Two Squares
3. Factor the Sum or Difference of Two Cubes

### *Preparing for Factoring Special Products*

*Before getting started, take the following readiness quiz. If you get a problem wrong, go back to the section cited and review the material.*

**1.** What is $1^2$? $2^2$? $3^2$? $4^2$? $5^2$? [Section R.4, pp. 34–35]

**2.** What is $\left(-\frac{3}{2}\right)^2$? [Section R.4, pp. 34–35]

In this section, we look at polynomials that can be categorized as having "special formulas" for factoring. We begin with perfect square trinomials.

## 1 Factor Perfect Square Trinomials

Recall from Section 5.2 that

$$\begin{aligned}(A + B)^2 &= A^2 + 2AB + B^2\\ (A - B)^2 &= A^2 - 2AB + B^2\end{aligned}$$

We call these perfect square trinomials. Reversing the formulas, we obtain a method for factoring perfect square trinomials.

***Preparing for...Answers*** **1.** 1, 4, 9, 16, 25 **2.** $\frac{9}{4}$

**PERFECT SQUARE TRINOMIALS**

$$A^2 + 2AB + B^2 = (A + B)^2$$
$$A^2 - 2AB + B^2 = (A - B)^2$$

**Work Smart**

The perfect squares are $1^2 = 1$, $2^2 = 4$, $3^2 = 9$, and so on. Any variable raised to an even exponent is a perfect square. So $x^2$, $x^4 = (x^2)^2$, $x^6 = (x^3)^2$ are all perfect squares.

In order for a polynomial to be a perfect square trinomial, two conditions must be satisfied.

**1.** The first and last terms must be perfect squares. Examples of perfect squares are

| | | |
|---|---|---|
| 81 | because | $9^2 = 81$ |
| 144 | because | $12^2 = 144$ |
| $4x^2$ | because | $(2x)^2 = 4x^2$ |
| $25a^4$ | because | $(5a^2)^2 = 25a^4$ |

**2.** The "middle term" must equal 2 or $-2$ times the product of the expressions being squared in the first and last term.

Perfect square trinomials can be factored using the methods introduced in the last section, however, they can be factored much quicker using the above formulas when they are recognized.

## EXAMPLE 1 Factoring Perfect Square Trinomials

Factor:

**(a)** $z^2 - 10z + 25$ **(b)** $9x^2 + 48xy + 64y^2$ **(c)** $32m^4 - 48m^2 + 18$

**Solution**

**(a)** The first term, $z^2$, and the third term, $25 = 5^2$, are perfect squares. Because the middle term, $-10z$, is $-2$ times the product of $z$ and 5, we have a perfect square. So

$$z^2 - 10z + 25 = z^2 - 2 \cdot z \cdot 5 + 5^2$$

$A = z$, $B = 5$, $A^2 - 2AB + B^2 = (A - B)^2$: $\quad = (z - 5)^2$

**(b)** The first term, $9x^2 = (3x)^2$, and the third term, $64y^2 = (8y)^2$, are perfect squares. Because the middle term is 2 times the product of $3x$ and $8y$, we have a perfect square. So

$$9x^2 + 48xy + 64y^2 = (3x)^2 + 2 \cdot 3x \cdot 8y + (8y)^2$$

$A = 3x$, $B = 8y$, $A^2 + 2AB + B^2 = (A + B)^2$: $\quad = (3x + 8y)^2$

**(c)** Remember, the first step in any factoring problem is to look for a common factor.

$$32m^4 - 48m^2 + 18 = 2(16m^4 - 24m^2 + 9)$$

Look at the expression in parentheses. The first term, $16m^4 = (4m^2)^2$, and the third term, $9 = 3^2$, are perfect squares. Because the middle term is $-2$ times the product of $4m^2$ and 3, we have a perfect square. So

$$16m^4 - 24m^2 + 9 = (4m^2)^2 - 2 \cdot 4m^2 \cdot 3 + 3^2$$
$$= (4m^2 - 3)^2$$

Therefore,

$$32m^4 - 48m^2 + 18 = 2(16m^4 - 24m^2 + 9)$$
$$= 2(4m^2 - 3)^2$$

**QUICK ✓** *Factor each trinomial completely.*

**1.** $x^2 - 18x + 81$ **2.** $4x^2 + 20xy + 25y^2$ **3.** $18p^4 - 84p^2 + 98$

## 2 Factor the Difference of Two Squares

Another "special product" introduced in Section 5.2 was

$$(A - B)(A + B) = A^2 - B^2$$

We call this product the difference of two squares.

**DIFFERENCE OF TWO SQUARES**

$$A^2 - B^2 = (A - B)(A + B)$$

### EXAMPLE 2 Factoring the Difference of Two Squares

Factor:

**(a)** $y^2 - 100$ **(b)** $9x^2 - 16y^4$

**Solution**

**(a)** We notice that $y^2 - 100$ is the difference of two squares, $y^2$ and $100 = 10^2$. So

$$y^2 - 100 = y^2 - 10^2$$
$$= (y - 10)(y + 10)$$

**Work Smart**
Don't forget that you can always check your answer by multiplying out the factored form.

**(b)** We notice that $9x^2 - 16y^4$ is the difference of two squares, $9x^2 = (3x)^2$ and $16y^4 = (4y^2)^2$. So

$$9x^2 - 16y^4 = (3x)^2 - (4y^2)^2$$
$$= (3x - 4y^2)(3x + 4y^2)$$

**QUICK ✓** *Factor completely.*

**4.** $z^2 - 16$ **5.** $16m^2 - 81n^2$ **6.** $4a^2 - 9b^4$

### EXAMPLE 3 Factoring the Difference of Two Squares

Factor:

**(a)** $32x^4 - 2$ **(b)** $x^2 + 10x + 25 - y^2$

**Solution**

**(a)** Remember, the first step in any factoring problem is to look for a common factor. In this polynomial, we have a common factor of 2.

$$32x^4 - 2 = 2(16x^4 - 1)$$

Difference of Two Squares, $16x^4 = (4x^2)^2$, $1 = 1^2$: $= 2((4x^2)^2 - 1^2)$

$A^2 - B^2 = (A - B)(A + B)$, where $A = 4x^2$ and $B = 1$: $= 2(4x^2 - 1)(4x^2 + 1)$

**Work Smart**
$4x^2 + 1$ is the **sum** of two squares. Remember that the sum of two squares does not factor over the integers.

We are not quite finished. Notice that $4x^2 - 1$ is the difference of two squares, $4x^2 = (2x)^2$ and $1 = 1^2$: $= 2((2x)^2 - 1^2)(4x^2 + 1)$

$A^2 - B^2 = (A - B)(A + B)$, where $A = 2x$ and $B = 1$: $= 2(2x - 1)(2x + 1)(4x^2 + 1)$

**Check** $2(2x - 1)(2x + 1)(4x^2 + 1) = 2(4x^2 - 1)(4x^2 + 1)$
$= 2(16x^4 + 4x^2 - 4x^2 - 1)$
$= 2(16x^4 - 1)$
$= 32x^4 - 2$

So $32x^4 - 2 = 2(2x - 1)(2x + 1)(4x^2 + 1)$.

**Work Smart**

Perfect square trinomials are trinomials of the form

$$A^2 + 2AB + B^2 = (A + B)^2$$
$$A^2 - 2AB + B^2 = (A - B)^2$$

**(b)** Remember, when we have four or more terms, we should try to factor by grouping. In Section 5.4, we grouped two terms in each group. Any attempt to group two terms in each group in this problem will fail because we won't have a common factor in each group (try it!). However, if we look closely, we should notice that the first three terms, $x^2 + 10x + 25$, form a perfect square trinomial (which factors into a perfect square) and the last term, $y^2$, is a perfect square.

$$x^2 + 10x + 25 - y^2 = (x^2 + 10x + 25) - y^2$$

$x^2 + 10x + 25 = x^2 + 2(5)x + 5^2 = (x + 5)^2$: $\quad = (x + 5)^2 - y^2$

$(x + 5)^2 - y^2$ is the difference of two squares, where $A = (x + 5)$ and $B = y$: $\quad = (\underbrace{x + 5}_{A} - \underbrace{y}_{B})(\underbrace{x + 5}_{A} + \underbrace{y}_{B})$

**Check**

$$(x + 5 - y)(x + 5 + y) = x^2 + 5x + xy + 5x + 25 + 5y - xy - 5y - y^2$$
$$= x^2 + 10x + 25 - y^2$$

So $x^2 + 10x + 25 - y^2 = (x + 5 - y)(x + 5 + y)$.

**QUICK ✓** *Factor completely.*

**7.** $3b^4 - 48$ **8.** $p^2 - 8p + 16 - q^2$

You may be asking yourself, "What about the sum of two squares—how does it factor?" The answer is that **the sum of two squares, $a^2 + b^2$, is prime** and does not factor over the integers. In fact, the sum of two squares cannot be factored for any real numbers! So binomials such as $x^2 + 4$ or $4y^2 + 81$ are prime.

## 3 Factor the Sum or Difference of Two Cubes

**Work Smart**

The perfect cubes are $1^3 = 1$, $2^3 = 8$, $3^3 = 27$, and so on. Any variable raised to a multiple of 3 is a perfect cube. So $x^3$, $x^6 = (x^2)^3$, $x^9 = (x^3)^3$ are all perfect cubes.

Consider the following products:

$$(A + B)(A^2 - AB + B^2) = A^3 - A^2B + AB^2 + A^2B - AB^2 + B^3$$
$$= A^3 + B^3$$
$$(A - B)(A^2 + AB + B^2) = A^3 + A^2B + AB^2 - A^2B - AB^2 - B^3$$
$$= A^3 - B^3$$

These products show us that we can factor the sum or difference of two cubes as follows:

**THE SUM OF TWO CUBES**

$$A^3 + B^3 = (A + B)(A^2 - AB + B^2)$$

**THE DIFFERENCE OF TWO CUBES**

$$A^3 - B^3 = (A - B)(A^2 + AB + B^2)$$

## EXAMPLE 4 Factoring the Sum or Difference of Two Cubes

Factor:

**(a)** $x^3 - 27$ **(b)** $8m^3 + 125n^6$

### Solution

**(a)** We notice that we have the difference of two cubes, $x^3$ and $27 = 3^3$. We let $A = x$ and $B = 3$ in the factoring formula for the difference of two cubes.

$$A^3 - B^3 = (A - B)(A^2 + AB + B^2)$$

$$x^3 - 27 = x^3 - 3^3 = (x - 3)(x^2 + x(3) + 3^2)$$
$$= (x - 3)(x^2 + 3x + 9)$$

So $x^3 - 27 = (x - 3)(x^2 + 3x + 9)$.

**Work Smart**
The trinomial in the factored form of the sum or difference of two cubes will always be prime.

**(b)** We notice that we have the sum of two cubes, $8m^3 = (2m)^3$ and $125n^6 = (5n^2)^3$. We let $A = 2m$ and $B = 5n^2$ in the factoring formula for the sum of two cubes.

$$8m^3 + 125n^6 = ((2m)^3 + (5n^2)^3)$$
$$= (2m + 5n^2)((2m)^2 - (2m)(5n^2) + (5n^2)^2)$$
$$= (2m + 5n^2)(4m^2 - 10mn^2 + 25n^4)$$

So $8m^3 + 125n^6 = (2m + 5n^2)(4m^2 - 10mn^2 + 25n^4)$.

**QUICK ✓** *Completely factor.*

**9.** $z^3 + 64$ **10.** $125p^3 - 216q^6$

## EXAMPLE 5 Factoring the Sum or Difference of Two Cubes

Factor:

**(a)** $27x^3 - 216y^6$ **(b)** $(x - 4)^3 + 8x^3$

### Solution

**(a)** Although $27x^3 = (3x)^3$ and $216y^6 = (6y^2)^3$ is the difference of two cubes, we should notice that there is a common factor of 27 in the binomial $27x^3 - 216y^6$. We factor out the 27.

$$27x^3 - 216y^6 = 27(x^3 - 8y^6)$$
$$= 27(x^3 - (2y^2)^3)$$

Let $A = x$ and $B = 2y^2$ in the factoring formula for the difference of two cubes: $= 27(x - 2y^2)(x^2 + 2xy^2 + 4y^4)$

**(b)** We notice that we have the sum of two cubes, $(x - 4)^3$ and $8x^3 = (2x)^3$. We let $A = x - 4$ and $B = 2x$ in the factoring formula for the sum of two cubes.

$$(x - 4)^3 + 8x^3 = (x - 4)^3 + (2x)^3$$
$$= (x - 4 + 2x)((x - 4)^2 - (x - 4)\cdot 2x + (2x)^2)$$

Combine like terms; multiply: $= (3x - 4)(x^2 - 8x + 16 - 2x^2 + 8x + 4x^2)$

Combine like terms: $= (3x - 4)(3x^2 + 16)$

So $(x - 4)^3 + 8x^3 = (3x - 4)(3x^2 + 16)$.

**QUICK ✓** *Factor completely.*

**11.** $32m^3 + 500n^6$ **12.** $(x + 1)^3 - 27x^3$

# 5.6 Exercises

**For Extra Help:** Student Solutions Manual  CD Video  PH Math/Tutor Center  MathXL Tutorials on CD MathXL® MyMathLab

## Concepts and Vocabulary

*In Problems 1 and 2, fill in the blanks.*

**1.** A trinomial of the form $A^2 + 2AB + B^2$ or $A^2 - 2AB + B^2$ is called a ______ ______ ______.

**2.** $A^3 + B^3 = (\_\_\_\_\_\_)(\_\_\_\_\_\_)$.

*In Problems 3 and 4, answer True or False to each statement.*

**3.** $4x^2 + 6x + 9$ is a perfect square trinomial.

**4.** The binomial $x^2 + 25$ can be factored over the real numbers.

**5.** Show that the binomial $x^2 + 4$ is prime.

**6.** Give an example of a polynomial of degree 2 that is a perfect square trinomial, and then factor it.

## Building Skills

*In Problems 7–26, factor each perfect square trinomial completely.*

**7.** $x^2 + 4x + 4$
**8.** $y^2 + 6y + 9$
**9.** $36 + 12w + w^2$
**10.** $49 - 14d + d^2$
**11.** $4x^2 + 4x + 1$
**12.** $9z^2 - 6z + 1$
**13.** $9p^2 - 30p + 25$
**14.** $16y^2 - 24y + 9$
**15.** $25a^2 + 90a + 81$
**16.** $36b^2 + 84b + 49$
**17.** $9x^2 + 24xy + 16y^2$
**18.** $4a^2 + 20ab + 25b^2$
**19.** $3w^2 - 30w + 75$
**20.** $4c^2 - 24c + 36$
**21.** $-5t^2 - 70t - 245$
**22.** $-2a^2 - 32a - 128$
**23.** $32a^2 - 80ab + 50b^2$
**24.** $12x^2 - 84xy + 147y^2$
**25.** $z^4 - 6z^2 + 9$
**26.** $b^4 + 8b^2 + 16$

*In Problems 27–42, factor the difference of two squares completely.*

**27.** $x^2 - 9$
**28.** $z^2 - 64$
**29.** $4 - y^2$
**30.** $81 - a^2$
**31.** $4z^2 - 9$
**32.** $16y^2 - 81$
**33.** $100m^2 - 81n^2$
**34.** $x^4 - 9y^2$
**35.** $m^4 - 36n^2$
**36.** $x^4 - 100y^2$
**37.** $8p^2 - 18q^2$
**38.** $12m^2 - 75n^2$
**39.** $80p^2r - 245b^2r$
**40.** $36x^2z - 64y^2z$
**41.** $(x + y)^2 - 9$
**42.** $16 - (x - y)^2$

*In Problems 43–60, factor the sum or difference of two cubes completely.*

**43.** $x^3 - 8$
**44.** $z^3 + 64$
**45.** $125 + m^3$
**46.** $216 - n^3$
**47.** $x^6 - 64y^3$
**48.** $m^6 - 27n^3$
**49.** $24x^3 - 375y^3$
**50.** $16m^3 + 54n^3$
**51.** $(p + 1)^3 - 27$

**52.** $(y - 2)^3 - 8$

**53.** $(3y + 1)^3 + 8y^3$

**54.** $(2z + 3)^3 + 27z^3$

**55.** $(x + 3)^3 - (x - 3)^3$

**56.** $(y + 5)^3 + (y - 5)^3$

**57.** $y^6 + z^9$

**58.** $m^9 + n^{12}$

**59.** $y^9 - 1$

**60.** $x^{12} - 1$

## Mixed Practice

*In Problems 61–82, factor each polynomial completely.*

**61.** $25x^2 - y^2$

**62.** $9a^2 - b^2$

**63.** $8x^3 + 27$

**64.** $64x^3 - 125$

**65.** $z^2 - 8z + 16$

**66.** $p^2 - 20p + 100$

**67.** $5x^4 - 40xy^3$

**68.** $3m^4 - 81mn^3$

**69.** $49m^2 - 42mn + 9n^2$

**70.** $81p^2 - 72pq + 16q^2$

**71.** $y^4 - 8y^2 + 16$

**72.** $p^4 - 18p^2 + 81$

**73.** $4a^2b^2 + 12ab + 9$

**74.** $9m^2n^2 - 30mn + 25$

**75.** $x^2 - 4x + 4 - y^2$

**76.** $p^2 + 8p + 16 - q^2$

**77.** $2n^2 - 2m^2 + 40m - 200$

**78.** $2a^2 - 2b^2 - 24b - 72$

**79.** $3y^3 - 24$

**80.** $-5a^3 - 40$

**81.** $16x^2 + 24xy + 9y^2 - 100$

**82.** $36m^2 + 12mn + n^2 - 81$

## Applying the Concepts

△ *In Problems 83–86, find an expression in factored form for the area of the shaded region.*

**83.**

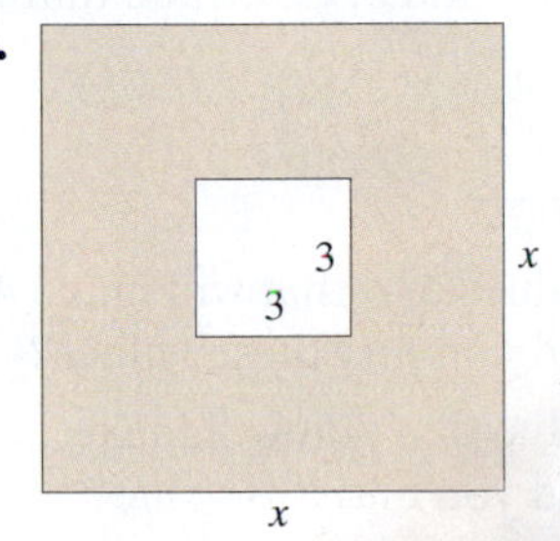


**84.**

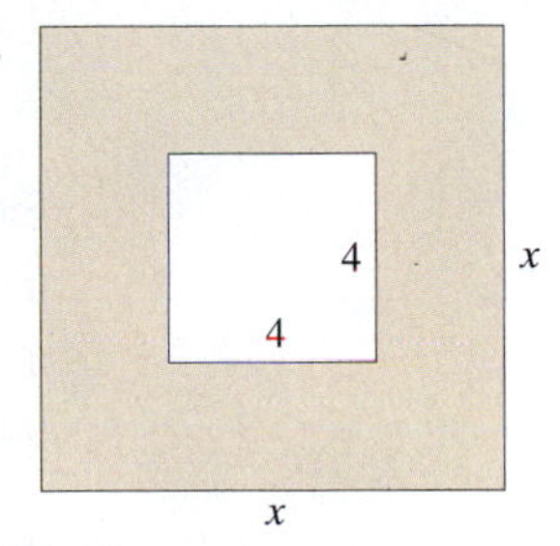


**85.**

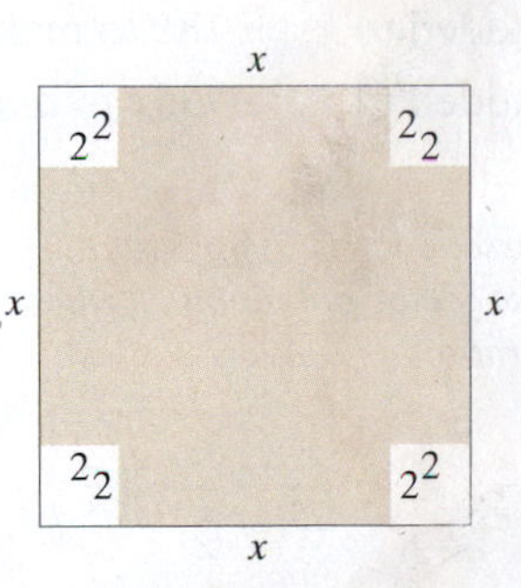


**86.**

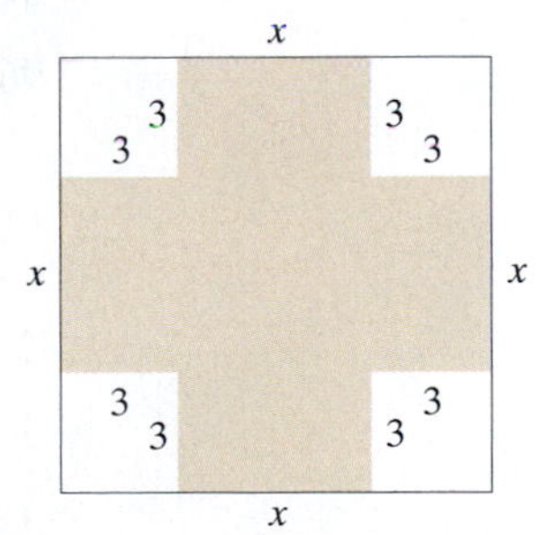


△ *In Problems 87–90, find an expression in factored form for the area or volume of the shaded region.*

**87.** Circle: *Area* $= \pi r^2$

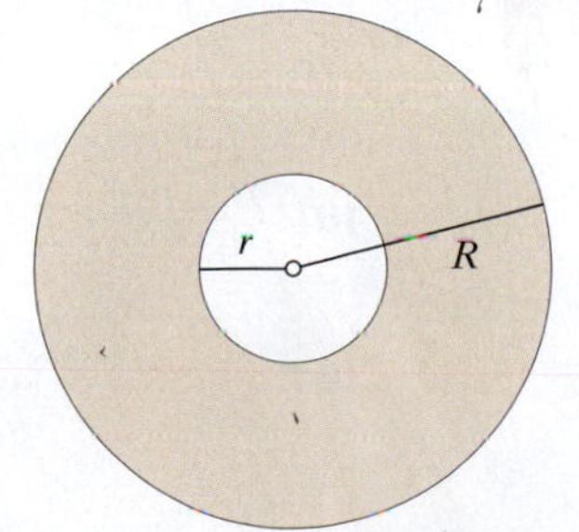


**88.** Cylinder: *Volume* $= \pi r^2 h$

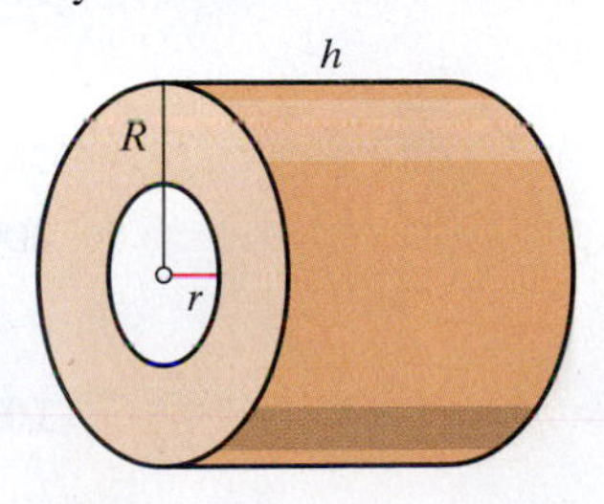

**89.** Rectangular solid: *Volume* $= lwh$

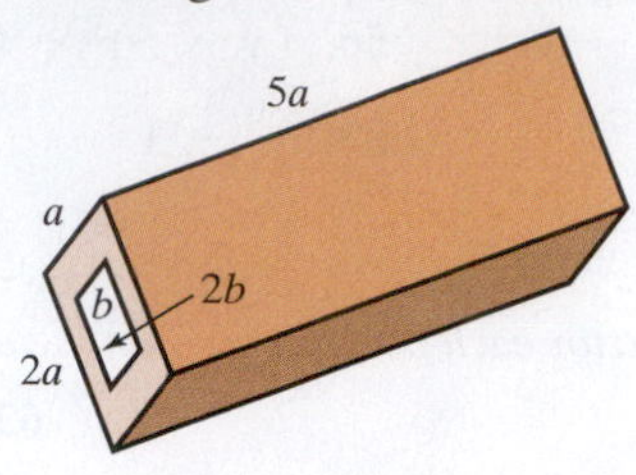


**90.** Sphere: *Volume* $= \frac{4}{3}\pi r^3$

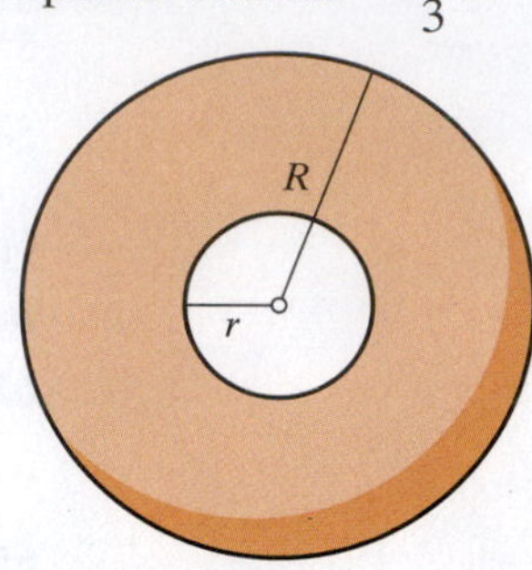


△ **91. A Perfect Square** What is a perfect square? The figure to the right shows a square whose dimensions are $(a + b)$ by $(a + b)$. The area of the square is therefore $(a + b)^2$. Write the area of the square as the sum of the four quadrilaterals in the figure. Then show that this sum is equal to $(a + b)^2$.

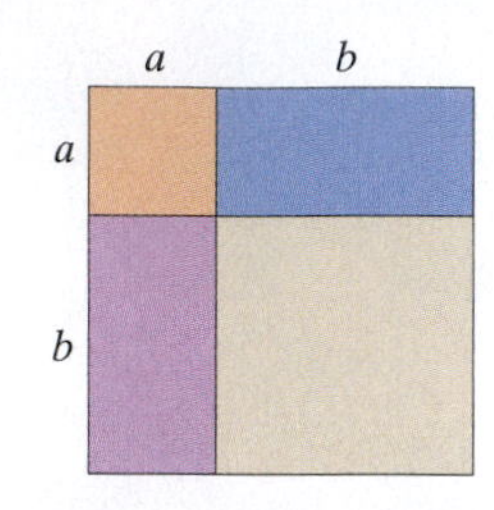


△ **92. Difference of Two Squares** What is the difference of two squares? The figure to the right shows two squares. The length of the sides on the "outer" square is $a$ and the length of the sides on the "inner" square is $b$. The area of the shaded region is $a^2 - b^2$. Express the area of the shaded region as the sum of the two shaded regions in terms of $a$ and $b$. Conclude that $a^2 - b^2 = (a + b)(a - b)$.

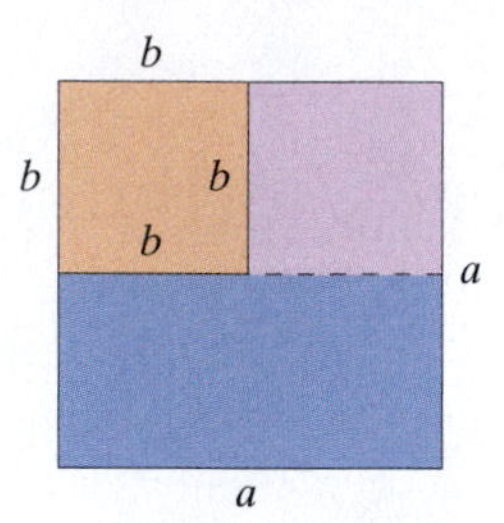


## Extending the Concepts

**93.** Determine two values of $b$ that will make $4x^2 + bx + 81$ a perfect square trinomial. How did you find these values?

**94.** Determine the value of $c$ that will make $16y^2 + 24y + c$ a perfect square trinomial. How did you find this value?

**95.** What has to be added to $x^2 + 18x$ to make it a perfect square trinomial?

**96.** What has to be added to $4x^2 + 36x$ to make it a perfect square trinomial?

*In this section, we discussed factoring polynomials whose coefficients are integers (factoring over the integers). We can also factor polynomials whose coefficients are rational numbers in which case we factor over the rational numbers. For example,*

$$\frac{x^2}{25} - \frac{1}{4} = \left(\frac{x}{5}\right)^2 - \left(\frac{1}{2}\right)^2$$
$$= \left(\frac{x}{5} - \frac{1}{2}\right)\left(\frac{x}{5} + \frac{1}{2}\right)$$

*Use this technique to factor Problems 97–104.*

**97.** $b^2 - 0.4b + 0.04$

**98.** $x^2 + 0.6x + 0.09$

**99.** $9b^2 - \frac{1}{25}$

**100.** $100x^2 - \frac{1}{81}$

**101.** $\frac{x^2}{9} - \frac{y^2}{25}$

**102.** $\frac{a^2}{36} - \frac{b^2}{49}$

**103.** $\frac{x^3}{8} - \frac{y^3}{27}$

**104.** $\frac{a^3}{27} + \frac{b^3}{64}$

# 5.7 Factoring: A General Strategy

**OBJECTIVE**

1 Factor Polynomials Completely

## 1 Factor Polynomials Completely

We have one objective in this special section—to put together all the various factoring techniques we have discussed in Sections 5.4–5.6. The following steps should be followed for any factoring problem.

### Steps for Factoring

**Step 1:** Factor out the greatest common factor (GCF), if any exists.

**Step 2:** Count the number of terms.

**Step 3:** **(a)** 2 terms

- Is it the difference of two squares? If so,

$$A^2 - B^2 = (A - B)(A + B)$$

- Is it the difference of two cubes? If so,

$$A^3 - B^3 = (A - B)(A^2 + AB + B^2)$$

- Is it the sum of two cubes? If so,

$$A^3 + B^3 = (A + B)(A^2 - AB + B^2)$$

**(b)** 3 terms

- Is it a perfect square trinomial? If so,

$$A^2 + 2AB + B^2 = (A + B)^2 \quad \text{or} \quad A^2 - 2AB + B^2 = (A - B)^2$$

- Is the coefficient of the square term 1? If so,

$$x^2 + bx + c = (x + m)(x + n) \text{ where } mn = c \text{ and } m + n = b$$

- Is the coefficient of the square term different from 1? If so,
  **a.** Use factoring by grouping
  **b.** Use trial and error

**(c)** 4 terms

- Use factoring by grouping

**Step 4:** Check your work by multiplying out the factored form.

**Work Smart**

Remember, the sum of two squares, $a^2 + b^2$, does not factor over the real numbers.

**EXAMPLE 1 How to Factor Completely**

Factor: $8x^2 - 16x - 42$

**Step-by-Step Solution**

| | | |
|---|---|---|
| **Step 1:** Factor out the greatest common factor (GCF), if any exists. | The GCF is 2, so we factor 2 out of the expression. | $8x^2 - 16x - 42 = 2(4x^2 - 8x - 21)$ |
| **Step 2:** Count the number of terms. | There are three terms in the polynomial in parentheses. | |

**Step 3:** We concentrate on the trinomial in parentheses, $4x^2 - 8x - 21$. It is not a perfect square trinomial. Because the coefficient on the square term, 4, and the constant, $-21$, aren't that big, we choose to factor by grouping.

$$ac = 4(-21) = -84.$$

Because $6 \cdot (-14) = -84$ and $6 + (-14) = -8$, we rewrite $4x^2 - 8x - 21$ as

$$4x^2 + 6x - 14x - 21 = (4x^2 + 6x) + (-14x - 21)$$

GCF in 1st grouping: $2x$; GCF in 2nd grouping: $-7$

$$= 2x(2x + 3) - 7(2x + 3)$$

Factor out $2x + 3$:

$$= (2x + 3)(2x - 7)$$

**Step 4: Check**

$$2(2x + 3)(2x - 7) = 2(4x^2 - 14x + 6x - 21)$$

$$= 2(4x^2 - 8x - 21)$$

Distribute:

$$= 8x^2 - 16x - 42$$

The answer checks, so $8x^2 - 16x - 42 = 2(2x + 3)(2x - 7)$.

**QUICK ✓** *Factor the polynomial completely.*

**1.** $2p^2q - 8pq^2 - 90q^3$ **2.** $-45x^2y + 66xy + 27y$

## EXAMPLE 2 How to Factor Completely

Factor: $9p^2 - 25q^2$

### Step-by-Step Solution

**Step 1:** Factor out the greatest common factor (GCF), if any exists.

There is no GCF.

**Step 2:** Count the number of terms.

There are two terms.

**Step 3:** Because the first term, $9p^2 = (3p)^2$, and the second term, $25q^2 = (5q)^2$, are both perfect squares, we have the difference of two squares.

$$9p^2 - 25q^2 = (\overbrace{(3p)}^{A})^2 - (\overbrace{(5q)}^{B})^2$$

$A^2 - B^2 = (A - B)(A + B)$:

$$= (3p - 5q)(3p + 5q)$$

**Step 4: Check**

$$(3p - 5q)(3p + 5q) = 9p^2 + 15pq - 15pq - 25q^2$$

Combine like terms:

$$= 9p^2 - 25q^2$$

So $9p^2 - 25q^2 = (3p - 5q)(3p + 5q)$.

**QUICK ✓** *Factor the polynomial completely.*

**3.** $81x^2 - 100y^2$ **4.** $-3m^2n + 147n$

## EXAMPLE 3 Factoring Completely

Factor: $16x^2 + 112xy + 196y^2$

### Solution

We first look for a common factor. Each term has a factor of 4, so the GCF is 4.

$$16x^2 + 112xy + 196y^2 = 4(4x^2 + 28xy + 49y^2)$$

We concentrate on the polynomial in parentheses. There are three terms in the parentheses. Notice that the first term is a perfect square, $4x^2 = (2x)^2$. The third term is also a perfect square, $49y^2 = (7y)^2$. The middle term is 2 times the product of $2x$ and $7y$. The polynomial in parentheses is a perfect square trinomial.

$$4(4x^2 + 28xy + 49y^2) = 4((2x)^2 + 2(2x)(7y) + (7y)^2)$$

$A = 2x$; $B = 7y$;

$A^2 + 2AB + B^2 = (A + B)^2$: $\quad = 4(2x + 7y)^2$

**Check** $\quad 4(2x + 7y)^2 = 4(2x + 7y)(2x + 7y)$

FOIL: $\quad = 4(4x^2 + 14xy + 14xy + 49y^2)$

Combine like terms: $\quad = 4(4x^2 + 28xy + 49y^2)$

Distribute: $\quad = 16x^2 + 112xy + 196y^2$

So $16x^2 + 112xy + 196y^2 = 4(2x + 7y)^2$.

**QUICK ✓** *Factor each polynomial completely.*

**5.** $p^2 - 16pq + 64q^2$ **6.** $20x^2 + 60x + 45$

## EXAMPLE 4 Factoring Completely

Factor: $8m^3 + 27n^6$

### Solution

We notice that the polynomial does not have a common factor and it has two terms. Because the first term is a perfect cube, $8m^3 = (2m)^3$, and the second term is a perfect cube, $27n^6 = (3n^2)^3$, we have the sum of two cubes.

$$8m^3 + 27n^6 = (2m)^3 + (3n^2)^3$$

$A = 2m$; $B = 3n^2$;

$A^3 + B^3 = (A + B)(A^2 - AB + B^2)$: $\quad = (2m + 3n^2)((2m)^2 - (2m)(3n^2) + (3n^2)^2)$

$\quad = (2m + 3n^2)(4m^2 - 6mn^2 + 9n^4)$

**Check** $\quad (2m + 3n^2)(4m^2 - 6mn^2 + 9n^4) = 8m^3 - 12m^2n^2 + 18mn^4 + 12m^2n^2 - 18mn^4 + 27n^6$

Combine like terms: $\quad = 8m^3 + 27n^6$

So $8m^3 + 27n^6 = (2m + 3n^2)(4m^2 - 6mn^2 + 9n^4)$.

**QUICK ✓** *Factor each polynomial completely.*

**7.** $64y^3 - 125$ **8.** $-16m^3 - 2n^3$

### EXAMPLE 5 Factoring Completely

Factor: $-4xy^2 + 12xy + 132x$

**Solution**

We notice that each term has a common factor of $-4x$. Factor out the GCF of $-4x$.

$$-4xy^2 + 12xy + 132x = -4x(y^2 - 3y - 33)$$

There are three terms in the polynomial in parentheses, $y^2 - 3y - 33$. Because 33 is not a perfect square, $y^2 - 3y - 33$ is not a perfect square trinomial. We need to find two factors of $-33$ whose sum is $-3$. There are no such factors. Therefore, $y^2 - 3y - 33$ is prime. So

$$-4xy^2 + 12xy + 132x = -4x(y^2 - 3y - 33)$$

**QUICK ✓** *Factor each polynomial completely.*

**9.** $10z^2 - 15z + 35$ **10.** $6xy^2 + 81x^3$

### EXAMPLE 6 Factoring Completely

Factor: $6x^3 - 4x^2 - 24x + 16$

**Solution**

Each term has a common factor of 2. Factor out the GCF of 2.

$$6x^3 - 4x^2 - 24x + 16 = 2(3x^3 - 2x^2 - 12x + 8)$$

Because there are four terms remaining in the parentheses, we attempt to factor by grouping. Group the first two terms and the last two terms. Watch out for the subtraction sign in front of the third term!

$$2(3x^3 - 2x^2 - 12x + 8) = 2[(3x^3 - 2x^2) + (-12x + 8)]$$

Common factor of $x^2$ in 1st grouping;
Common factor of $-4$ in 2nd grouping: $= 2[x^2(3x - 2) - 4(3x - 2)]$

Factor out $3x - 2$ as a common factor: $= 2(3x - 2)(x^2 - 4)$

$x^2 - 4$ is the difference of two squares: $= 2(3x - 2)(x - 2)(x + 2)$

**Check** $2(3x - 2)(x - 2)(x + 2) = 2(3x - 2)(x^2 - 4)$

FOIL: $= 2(3x^3 - 12x - 2x^2 + 8)$

Distribute; Rearrange terms: $= 6x^3 - 4x^2 - 24x + 16$

It checks, so $6x^3 - 4x^2 - 24x + 16 = 2(3x - 2)(x - 2)(x + 2)$.

**QUICK ✓** *Factor each polynomial completely.*

**11.** $2x^3 + 5x^2 + 4x + 10$ **12.** $9x^3 + 3x^2 - 9x - 3$

### EXAMPLE 7 Factoring Completely

Factor: $x^2 - 6xy + 9y^2 - 25$

#### Solution

There are no common factors and there are four terms, so we try factoring by grouping. Attempts to form two groups of two terms fail. Before thinking that the polynomial is prime, consider that the first three terms form a perfect square trinomial that factors to a perfect square. And the last term is a perfect square. We group the first three terms.

$$\begin{aligned} x^2 - 6xy + 9y^2 - 25 &= (x^2 - 6xy + 9y^2) - 25 \\ &= [x^2 - 2x(3y) + (3y)^2] - 25 \\ \text{A = x; B = 3y; } A^2 - 2AB + B^2 = (A - B)^2\text{:} \quad &= (x - 3y)^2 - 5^2 \\ \text{A = x − 3y; B = 5; } A^2 - B^2 = (A - B)(A + B)\text{:} \quad &= (x - 3y - 5)(x - 3y + 5) \end{aligned}$$

**Check** $(x - 3y - 5)(x - 3y + 5) = x^2 - 3xy + 5x - 3xy + 9y^2 - 15y - 5x + 15y - 25$
$= x^2 - 6xy + 9y^2 - 25$

**QUICK ✓** *Factor each polynomial completely.*

**13.** $4x^2 + 4xy + y^2 - 81$ **14.** $16 - m^2 - 8mn - 16n^2$

# 5.7 Exercises

**For Extra Help:** Student Solutions Manual, CD Video, PH Math/Tutor Center, MathXL Tutorials on CD, MathXL®, MyMathLab

## Concepts and Vocabulary

**1.** Write out the steps for factoring any polynomial.

**2.** What does factored completely mean?

## Mixed Practice

*In Problems 3–50, factor each polynomial completely.*

**3.** $2x^2 - 12x - 144$

**4.** $3x^2 + 6x - 105$

**5.** $-3y^2 + 27$

**6.** $-5a^2 + 80$

**7.** $4b^2 + 20b + 25$

**8.** $8m^2 - 42m + 49$

**9.** $16w^3 + 2y^6$

**10.** $54p^6 - 2q^3$

**11.** $-3z^2 + 12z - 18$

**12.** $-4c^3 + 16c^2 - 28c$

**13.** $20y^2 - 9y - 18$

**14.** $18t^2 - 9t - 20$

**15.** $x^3 - 4x^2 + 5x - 20$

**16.** $p^3 + 7p^2 - 3p - 21$

**17.** $200x^2 + 18y^2$

**18.** $12p^2 + 50q^2$

**19.** $x^4 - 81$

**20.** $16w^4 - 1$

**21.** $3x^2 - 7x - 16$

**22.** $4w^2 - 3w - 6$

**23.** $36q^3 + 24q^2 + 4q$

**24.** $20k^3 - 60k^2 + 45k$

**25.** $24m^3n - 66m^2n - 63mn$

**26.** $20p^3q - 2p^2q - 4pq$

**27.** $3r^5 - 24r^2s^3$

**28.** $54p^5 + 16p^2q^3$

**29.** $2x^3 + 8x^2 - 18x - 72$

**30.** $3x^3 - 6x^2 - 48x + 96$

**31.** $9x^4 - 1$

**32.** $4z^4 - 25$

**33.** $3w^4 + 4w^2 - 15$

**34.** $4b^4 + 4b^2 - 15$

**35.** $(2y + 3)^2 - 5(2y + 3) + 6$

**36.** $(3x + 5)^2 + 4(3x + 5) - 21$

**37.** $p^2 - 10p + 25 - 36q^2$

**38.** $a^2 + 12a + 36 - 4b^2$

**39.** $y^6 + 6y^3 - 16$

**40.** $w^6 + 4w^3 - 5$

**41.** $p^6 - 1$

**42.** $q^6 + 1$

**43.** $-3x^3 - 15x^2 + 27x + 135$

**44.** $-2y^3 - 4y^2 + 32y + 64$

**45.** $3a - 27a^3$

**46.** $-5z - 20z^3$

**47.** $8t^5 + 14t^3 - 72t$

**48.** $18h^5 + 154h^3 - 72h$

**49.** $2x^4y + 10x^3y - 18x^2y - 90xy$

**50.** $2p^4q + 14p^3q - 32p^2q - 224pq$

## Applying the Concepts

△ *In Problems 51–54, write an algebraic expression in completely factored form for the area that is shaded.*

**51.**

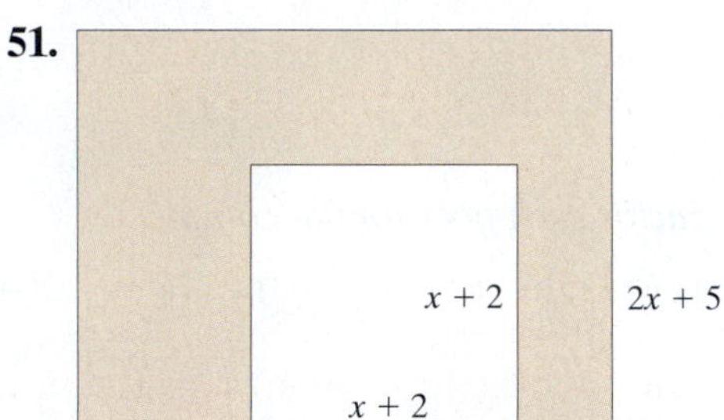


**52.**

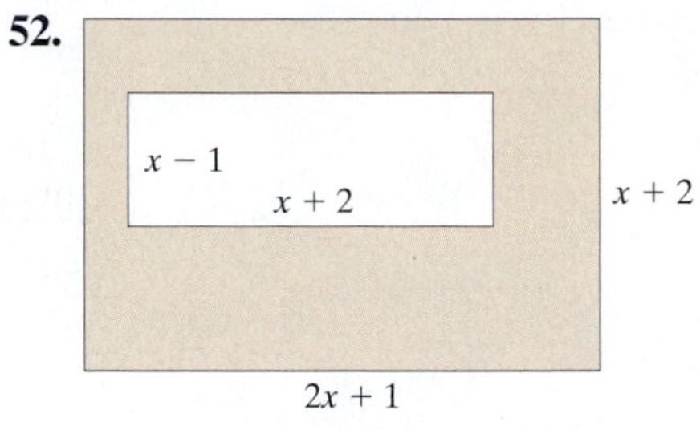


**53.**

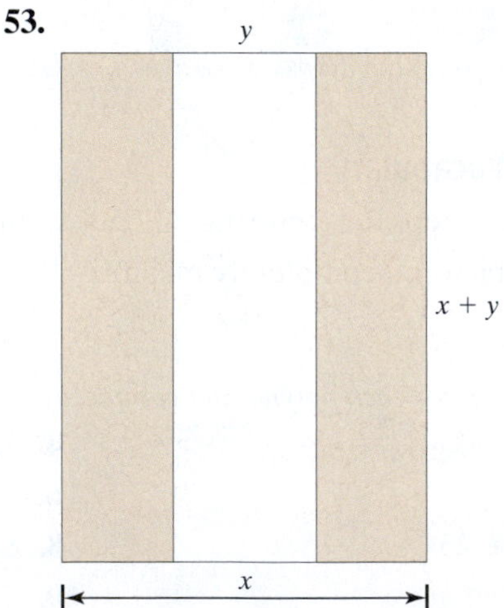


**54.**

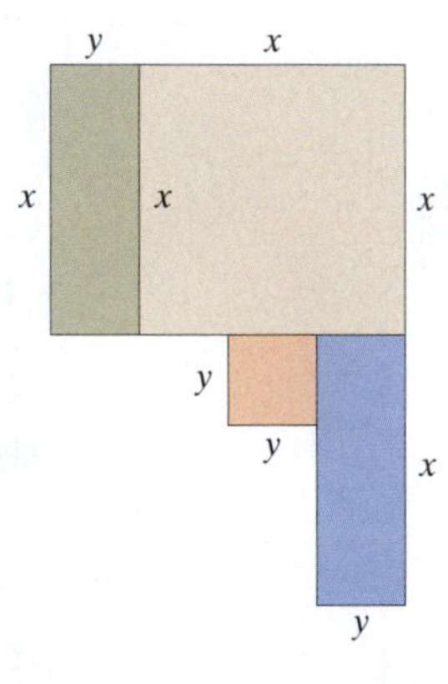


△ *In Problems 55 and 56, find an algebraic expression in factored form for the difference in the volumes of the two cubes shown.*

**55.**

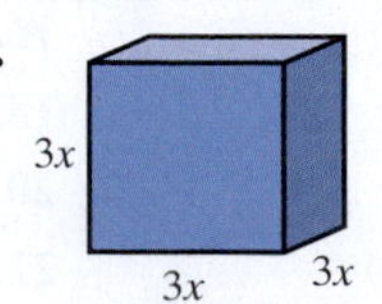


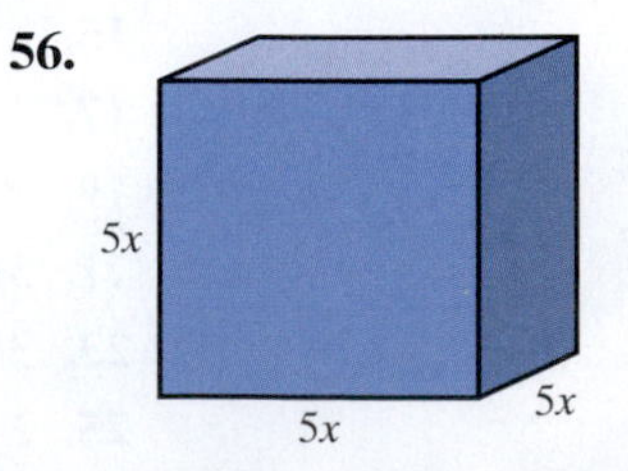


**56.**



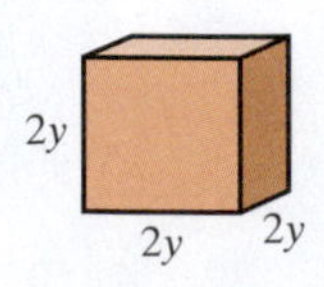

### Extending the Concepts

**57.** Show that $x^2 + 9$ is prime.

**58.** Show that $x^2 - 4x - 8$ is prime.

*In Problems 59–62, factor out the indicated common factor of the expression. Then completely factor the expression.*

**59.** $x^{\frac{5}{2}} - 9x^{\frac{1}{2}}; x^{\frac{1}{2}}$

**60.** $x^{\frac{3}{2}} - 4x^{-\frac{1}{2}}; x^{-\frac{1}{2}}$

**61.** $1 + 6x^{-1} + 8x^{-2}; x^{-2}$

**62.** $3 - 2x^{-1} - x^{-2}; x^{-2}$

# 5.8 Polynomial Equations

**OBJECTIVES**

1. Solve Polynomial Equations Using the Zero-Product Property
2. Solve Equations Involving Polynomial Functions
3. Model and Solve Problems Involving Polynomials

## *Preparing for Polynomial Equations*

*Before getting started, take the following readiness quiz. If you get a problem wrong, go back to the cited section and review the material.*

**1.** Solve: $x + 4 = 0$ [Section 1.1, pp. 53–54]

**2.** Solve: $3(x - 2) - 12 = 0$ [Section 1.1, pp. 55–56]

**3.** Evaluate $2x^2 + 3x + 1$ when (a) $x = 2$ (b) $x = -1$. [Section R.5, p. 43]

**4.** If $f(x) = 4x + 3$, solve $f(x) = 11$. What point is on the graph of $f$? [Section 2.4, p. 172]

**5.** If $f(x) = -2x + 8$, find $f(4)$. What point is on the graph of $f$? [Section 2.4, p. 172]

In Sections 5.4–5.7, we learned how to factor polynomial expressions. Why is factoring an important skill? As we progress through the course you will see that there are many uses of factoring. We can present one important use now. It turns out that factoring is really handy for solving *polynomial equations.* A **polynomial equation** is any equation that contains a polynomial expression. The **degree of a polynomial equation** is the degree of the polynomial expression in the equation. Some examples of polynomial equations are

| $4x + 5 = 17$ | $2x^2 - 5x - 3 = 0$ | $y^3 + 4y^2 = 3y + 18$ |
|---|---|---|
| Polynomial equation of degree 1 | Polynomial equation of degree 2 | Polynomial equation of degree 3 |

**Work Smart**

Remember, the degree of a polynomial is the value of the largest exponent on the variable. For example, the degree of $4x^3 - 9x^2 + 1$ is 3.

The polynomial equation $4x + 5 = 17$ is, more specifically, a linear or first-degree equation. Remember, we studied linear equations back in Section 1.1. In this section, we will learn methods for solving polynomial equations when the polynomial expression is of degree 2 or higher and factorable over the integers.

## 1 Solve Polynomial Equations Using the Zero-Product Property

If a polynomial equation can be solved using factoring, then we make use of the following property.

**THE ZERO-PRODUCT PROPERTY**

If the product of two numbers is zero, then at least one of the numbers is 0. That is, if $ab = 0$, then $a = 0$ or $b = 0$ or both $a$ and $b$ are 0.

For example, if $2x = 0$, then either $2 = 0$ or $x = 0$. Since $2 \neq 0$, it must be that $x = 0$.

*Preparing for...Answers* **1.** $\{-4\}$ **2.** $\{6\}$ **3. (a)** 15 **(b)** 0 **4.** $\{2\}$; $(2, 11)$ **5.** 0; $(4, 0)$

### EXAMPLE 1 Using the Zero-Product Property

Solve: $(x + 5)(2x - 3) = 0$

**Solution**

We have the product of two numbers, $x + 5$ and $2x - 3$, set equal to 0. By the Zero-Product Property, at least one of the numbers must equal 0. Therefore, we set each of the expressions to 0 and solve each equation separately:

$$x + 5 = 0 \quad \text{or} \quad 2x - 3 = 0$$

Subtract 5 from both sides: $x = -5$

Add 3 to both sides: $2x = 3$

Divide both sides by 2: $x = \frac{3}{2}$

**Check**

$x = -5$: $(x + 5)(2x - 3) = 0$

$$(-5 + 5)(2(-5) - 3) \stackrel{?}{=} 0$$

$$0(-13) \stackrel{?}{=} 0$$

$$0 = 0 \quad \text{True}$$

$x = \frac{3}{2}$: $(x + 5)(2x - 3) = 0$

$$\left(\frac{3}{2} + 5\right)\left(2 \cdot \left(\frac{3}{2}\right) - 3\right) \stackrel{?}{=} 0$$

$$\left(\frac{13}{2}\right) \cdot (3 - 3) \stackrel{?}{=} 0$$

$$\frac{13}{2} \cdot 0 \stackrel{?}{=} 0$$

$$0 = 0 \quad \text{True}$$

The solution set is $\left\{-5, \frac{3}{2}\right\}$.

**QUICK ✓** *Use the Zero-Product Property to solve the equation.*

**1.** $x(x + 7) = 0$  **2.** $(x - 3)(4x + 3) = 0$

## Using the Zero-Product Property to Solve Quadratic Equations

The Zero-Product Property can be used to solve *quadratic equations*.

**Work Smart**

Why can't $a$ equal 0? If $a$ is equal to zero, the equation would be $bx + c = 0$, a linear equation.

**DEFINITION**

A **quadratic equation** is an equation equivalent to one of the form

$$ax^2 + bx + c = 0$$

where $a$, $b$, and $c$ are real numbers and $a \neq 0$.

Quadratic equations are equations such as

$$3x^2 + 5x + 2 = 0 \qquad -7z^2 + 14z = 0 \qquad y^2 - 16 = 0 \qquad p^2 + 8p = 16$$

A quadratic equation written in the form $ax^2 + bx + c = 0$ is said to be in **standard form.** The first three equations listed above are in standard form; the equation $p^2 + 8p = 16$ is not in standard form.

Sometimes, a quadratic equation is called a **second-degree equation** because the left side is a polynomial of degree 2.

When a quadratic equation is written in standard form, $ax^2 + bx + c = 0$, it may be possible to factor the expression $ax^2 + bx + c$ as the product of two first-degree polynomials. We will present methods for solving $ax^2 + bx + c = 0$ when we cannot factor the expression on the left side in Sections 8.2 and 8.3.

**EXAMPLE 2** **How to Solve a Quadratic Equation by Factoring**

Solve: $2x^2 - 5x = 3$

**Step-by-Step Solution**

**Step 1:** Write the quadratic equation in standard form, $ax^2 + bx + c = 0$.

$$2x^2 - 5x = 3$$

Subtract 3 from both sides: $2x^2 - 5x - 3 = 0$

**Step 2:** Factor the expression on the left side of the equation.

$$(2x + 1)(x - 3) = 0$$

**Step 3:** Set each factor to 0.

$$2x + 1 = 0 \quad \text{or} \quad x - 3 = 0$$

**Step 4:** Solve each first-degree equation.

$$2x = -1 \quad \text{or} \quad x = 3$$

$$x = -\frac{1}{2}$$

**Step 5: Check:** Substitute $-\frac{1}{2}$ for $x$ and 3 for $x$ into the original equation.

$x = -\frac{1}{2}$: $2x^2 - 5x = 3$

$$2\left(-\frac{1}{2}\right)^2 - 5\left(-\frac{1}{2}\right) \stackrel{?}{=} 3$$

$$2\left(\frac{1}{4}\right) + \frac{5}{2} \stackrel{?}{=} 3$$

$$\frac{1}{2} + \frac{5}{2} \stackrel{?}{=} 3$$

$$3 = 3 \quad \text{True}$$

$x = 3$: $2x^2 - 5x = 3$

$$2(3)^2 - 5(3) \stackrel{?}{=} 3$$

$$18 - 15 \stackrel{?}{=} 3$$

$$3 = 3 \quad \text{True}$$

The solution set is $\left\{-\frac{1}{2}, 3\right\}$.

We summarize the steps to solve a quadratic equation by factoring below.

**Steps for Solving a Quadratic Equation by Factoring**

**Step 1:** Write the quadratic equation in standard form, $ax^2 + bx + c = 0$.

**Step 2:** Factor the expression on the left side of the equation.

**Step 3:** Set each factor found in Step 2 equal to zero using the Zero-Product Property.

**Step 4:** Solve each first-degree equation for the variable.

**Step 5:** Be sure to check your answers by substituting into the *original* equation.

**QUICK ✓** *Solve each quadratic equation by factoring.*

**3.** $p^2 - 5p + 6 = 0$ **4.** $3t^2 - 14t = 5$ **5.** $4y^2 + 8y + 3 = y^2 - 1$

### EXAMPLE 3 Solving a Quadratic Equation by Factoring

Solve: $(m - 1)(2m + 3) = 6m$

**Solution**

First, we need to write the equation in standard form.

**Work Smart**

Do not attempt to solve the equation $(m - 1)(2m + 3) = 2m$ by setting each factor equal to $2m$.

$$(m - 1)(2m + 3) = 6m$$

FOIL: $$2m^2 + m - 3 = 6m$$

Subtract $6m$ from both sides: $$2m^2 - 5m - 3 = 0$$

Factor: $$(2m + 1)(m - 3) = 0$$

Set each factor to 0: $$2m + 1 = 0 \quad \text{or} \quad m - 3 = 0$$

$$2m = -1 \quad \text{or} \quad m = 3$$

$$m = -\frac{1}{2}$$

**Check** Substitute $m = -\frac{1}{2}$ and $m = 3$ into the original equation.

$m = -\frac{1}{2}$: $$(m - 1)(2m + 3) = 6m$$

$$\left(-\frac{1}{2} - 1\right)\left(2\cdot\left(-\frac{1}{2}\right) + 3\right) \stackrel{?}{=} 6\cdot\left(-\frac{1}{2}\right)$$

$$\left(-\frac{3}{2}\right)(-1 + 3) \stackrel{?}{=} -3$$

$$-3 = -3 \quad \text{True}$$

$m = 3$: $$(m - 1)(2m + 3) = 6m$$

$$(3 - 1)(2(3) + 3) \stackrel{?}{=} 6(3)$$

$$2(9) \stackrel{?}{=} 18$$

$$18 = 18 \quad \text{True}$$

The solution set is $\left\{-\frac{1}{2}, 3\right\}$.

**QUICK ✓** *Solve each quadratic equation by factoring.*

**6.** $x(x + 3) = -2$ **7.** $(x - 3)(x + 5) = 9$

### Using the Zero-Product Property to Solve Higher-Degree Equations

Up to now, we have solved only second-degree equations. We can use an extended form of the Zero-Product Property to solve polynomial equations of degree three or higher. The basic idea is to first write the equation so that the right hand side is zero (that is, write the equation in standard form), factor the expression that equals zero, and then set each factor to zero and solve.

**EXAMPLE 4 How to Solve an Equation Using the Zero-Product Property**

Solve: $w^3 + 5w^2 - 4w = 20$

**Step-by-Step Solution**

**Step 1:** We put the equation in standard form by subtracting 20 from both sides of the equation.

$$w^3 + 5w^2 - 4w = 20$$
$$w^3 + 5w^2 - 4w - 20 = 0$$

**Step 2:** Factor the expression on the left side of the equation. Because there are four terms, we factor by grouping.

Group the 1st two terms; group the last two terms: $(w^3 + 5w^2) + (-4w - 20) = 0$

Factor out the common factor in each group: $w^2(w + 5) - 4(w + 5) = 0$

Factor out $w + 5$: $(w + 5)(w^2 - 4) = 0$

Factor $w^2 - 4$: $(w + 5)(w + 2)(w - 2) = 0$

**Step 3:** Set each factor to 0.

$w + 5 = 0$ or $w + 2 = 0$ or $w - 2 = 0$

**Step 4:** Solve each first-degree equation.

$w = -5$ or $w = -2$ or $w = 2$

**Step 5: Check** Substitute $w = -5$, $w = -2$, and $w = 2$ into the original equation.

$w = -5$:
$$w^3 + 5w^2 - 4w = 20$$
$$(-5)^3 + 5(-5)^2 - 4(-5) \stackrel{?}{=} 20$$
$$-125 + 125 + 20 \stackrel{?}{=} 20$$
$$20 = 20$$
True

$w = -2$:
$$w^3 + 5w^2 - 4w = 20$$
$$(-2)^3 + 5(-2)^2 - 4(-2) \stackrel{?}{=} 20$$
$$-8 + 20 + 8 \stackrel{?}{=} 20$$
$$20 = 20$$
True

$w = 2$:
$$w^3 + 5w^2 - 4w = 20$$
$$(2)^3 + 5(2)^2 - 4(2) \stackrel{?}{=} 20$$
$$8 + 20 - 8 \stackrel{?}{=} 20$$
$$20 = 20$$
True

The solution set is $\{-5, -2, 2\}$.

**QUICK ✓** *Solve the polynomial equation.*

**8.** $y^3 - y^2 - 9y + 9 = 0$

## 2 Solve Equations Involving Polynomial Functions

Suppose that we were given the function $f(x) = x^2 - 5x + 4$ and wanted to know the values of $x$ such that $f(x) = 4$. This requires solving the equation

$$\overbrace{x^2 - 5x + 4}^{f(x)} = 4$$

**EXAMPLE 5 Solving an Equation Involving a Polynomial Function**

Suppose that $f(x) = x^2 + 6x - 3$. Find the values of $x$ such that $f(x) = 4$. What points are on the graph of $f$?

### Solution

We want to solve $f(x) = 4$. That is, we want to solve $x^2 + 6x - 3 = 4$. We start by putting the equation in standard form.

$$x^2 + 6x - 3 = 4$$

Subtract 4 from both sides: $$x^2 + 6x - 7 = 0$$

Factor: $$(x + 7)(x - 1) = 0$$

Set each factor to 0: $$x + 7 = 0 \quad \text{or} \quad x - 1 = 0$$

$$x = -7 \quad \text{or} \quad x = 1$$

**Check**

$x = -7$: $x^2 + 6x - 3 = 4$

$(-7)^2 + 6(-7) - 3 \stackrel{?}{=} 4$

$49 - 42 - 3 \stackrel{?}{=} 4$

$4 = 4$ True

$x = 1$: $x^2 + 6x - 3 = 4$

$(1)^2 + 6(1) - 3 \stackrel{?}{=} 4$

$1 + 6 - 3 \stackrel{?}{=} 4$

$4 = 4$ True

The values of $x$ such that $f(x) = 4$ are $-7$ and 1. So, the points $(-7, 4)$ and $(1, 4)$ are on the graph of $f$.

QUICK ✓ *Solve the equations.*

**9.** Suppose that $g(x) = x^2 - 8x + 3$. Find the values of $x$ such that

**(a)** $g(x) = 12$ **(b)** $g(x) = -4$

What points are on the graph of $g$?

---

**Work Smart**

If $f(r) = 0$, then $r$ is a zero of $f$. If $r$ is a zero of $f$, then $r$ is an $x$-intercept of the graph of $f$.

A **zero** of a function $f$ is any number $r$ such that $f(r) = 0$. In addition, if $r$ is a zero of a function, then $r$ is also an $x$-intercept of the graph of the function. In the next example, we find the zeros of a quadratic function.

## EXAMPLE 6 Finding the Zeros of a Quadratic Function

Find the zeros of $f(x) = 4x^2 - 5x - 6$. What are the $x$-intercepts of the graph of the function?

### Solution

The zeros are found by solving the equation $f(x) = 0$ or $4x^2 - 5x - 6 = 0$. The solutions to this equation are $-\frac{3}{4}$ and 2, so the zeros of $f(x) = 4x^2 - 5x - 6$ are $-\frac{3}{4}$ and 2. Because the zeros are $-\frac{3}{4}$ and 2, the $x$-intercepts of the graph of the function are also $-\frac{3}{4}$ and 2.

QUICK ✓

**10.** Find the zeros of $h(x) = 2x^2 + 3x - 20$. What are the $x$-intercepts of the graph of the function?

---

## 3 Model and Solve Problems Involving Polynomials

Many applied problems require solving polynomial equations by factoring. For example, the height of a projectile over time can be described by a polynomial equation. We can use the equation to determine the time at which the projectile is a certain height. As always, we shall employ the problem-solving strategy first presented in Section 1.2.

### EXAMPLE 7 Geometry: Area of a Rectangle

The length of a rectangle is 8 feet more than its width. If the area of the rectangle is 84 square feet, what are the dimensions of the rectangle? See Figure 3.

Figure 3

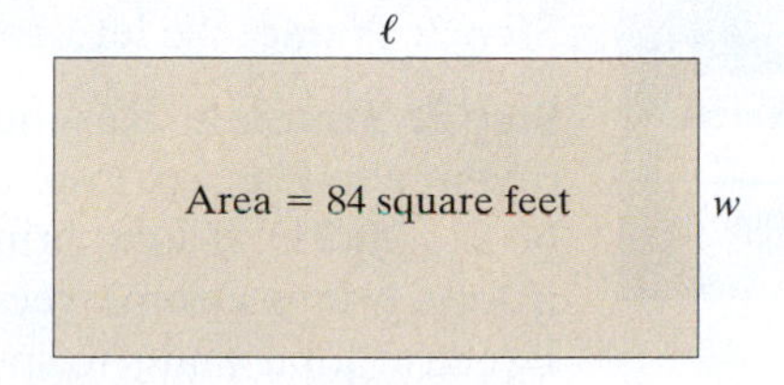


#### Solution

**Step 1: Identify** This is a geometry problem involving the area of a rectangle.

**Step 2: Name** We let $w$ represent the width of the rectangle and $l$ represent the length of the rectangle.

**Step 3: Translate** Because the length is 8 feet more than the width, we know that $l = w + 8$. In addition, we are given that the area of the rectangle is 84 square feet.

$$\text{Area} = (\text{length})(\text{width})$$

$$\text{Area} = lw$$

$$84 = (w + 8)w \quad \text{The Model}$$

**Step 4: Solve** We now proceed to solve the equation.

$$w(w + 8) = 84$$

Distribute: $w^2 + 8w = 84$

Subtract 84 from both sides: $w^2 + 8w - 84 = 0$

Factor: $(w + 14)(w - 6) = 0$

Set each factor to 0: $w + 14 = 0$ or $w - 6 = 0$

Solve: $w = -14$ or $w = 6$

**Step 5: Check** Since $w$ represents the width of the rectangle, we discard the solution $w = -14$. If the width of the rectangle is 6 feet, then the length would be $6 + 8 = 14$ feet. The area of a rectangle that is 6 feet by 14 feet would be $6(14) = 84$ square feet. We have the right answer!

**Step 6: Answer** The dimensions of the rectangle are 6 feet by 14 feet.

**QUICK ✓**

11. The width of a rectangular plot of land is 6 miles less than its length. If the area of the land is 135 square miles, what are the dimensions of the land?

## EXAMPLE 8 Pricing a Charter

Chicago Tours offers boat charters along the Chicago coastline on Lake Michigan. John Alferivich decides to take his company on a tour as a thank-you to his employees. He strikes a deal with Chicago Tours. Normally, a ticket costs \$20 per person, but for each person John brings in excess of 30 people, Chicago Tours will lower the price of the ticket by \$0.10. Assuming that John knows more than 30 employees will go on the trip and that the capacity of the boat is 120 passengers, how many employees can attend if John is willing to spend \$900 for the tour?

### Solution

**Step 1: Identify** This is a direct translation problem involving revenue. Remember, revenue is price times quantity.

**Step 2: Name** We let $x$ represent the number of employees in excess of 30 that attend.

**Step 3: Translate** Revenue is price times quantity. If John brings 30 employees, the revenue to Chicago Tours will be \$20(30). If John brings 31 employees, revenue will be \$19.90(31). If John brings 32 employees, revenue will be \$19.80(32) In general, if John brings $x$ employees in excess of 30, revenue will be $(20 - 0.1x)(x + 30)$. Because John wants to spend \$900, we have

$$(20 - 0.1x)(x + 30) = 900 \quad \text{The Model}$$

**Step 4: Solve** We now proceed to solve the equation.

$$(20 - 0.1x)(x + 30) = 900$$

FOIL: $20x + 600 - 0.1x^2 - 3x = 900$

Combine like terms; rearrange terms: $-0.1x^2 + 17x + 600 = 900$

Subtract 900 from both sides: $-0.1x^2 + 17x - 300 = 0$

Multiply both sides by $-10$ to make the coefficient of $x^2$ equal to 1: $x^2 - 170x + 3000 = 0$

Factor: $(x - 150)(x - 20) = 0$

Set each factor to 0: $x - 150 = 0$ or $x - 20 = 0$

Solve: $x = 150$ or $x = 20$

**Step 5: Check** Remember that $x$ represents the number of passengers in excess of 30. We discard the solution $x = 150$ because it causes us to exceed the capacity of 120 passengers. Therefore, $30 + 20 = 50$ passengers can go on the trip. The cost per ticket would be $\$20 - 0.1(20) = \$20 - \$2 = \$18$. Multiplying the cost per ticket by the number of passengers we obtain $\$18(50) = \$900$. We have the right answer!

**Step 6: Answer** A total of 50 passengers can attend.

12. A compact disk manufacturer charges \$100 for each box of CDs ordered. However, for orders in excess of 30 boxes, but less than 65 boxes, it reduces the price by \$1 per box. If a customer placed an order that qualified for the discount pricing and the bill was \$4200, how many boxes of CDs were ordered?

## EXAMPLE 9 Projectile Motion

Figure 4

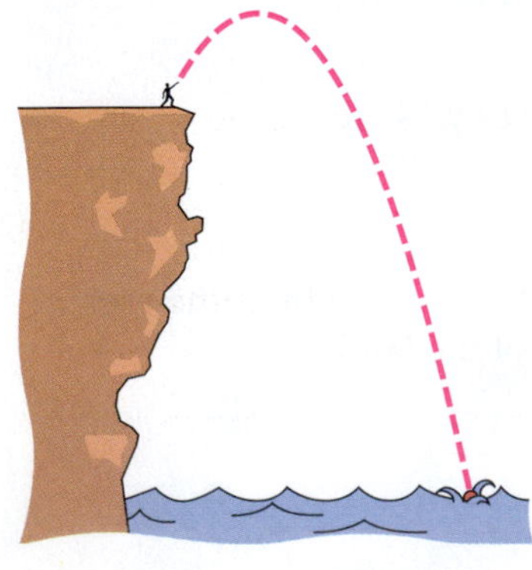

A ball is thrown off a cliff from a height of 240 feet above sea level, as pictured in Figure 4. The height $s$ of the ball above the water (in feet) as a function of time (in seconds) can be modeled by the function

$$s(t) = -16t^2 + 32t + 240$$

**(a)** When will the height of the ball be 240 feet?

**(b)** When will the ball strike the water?

### Solution

**(a)** To determine when the height of the ball will be 240 feet, we solve the equation $s(t) = 240$.

$$s(t) = 240$$
$$-16t^2 + 32t + 240 = 240$$

Subtract 240 from both sides: $-16t^2 + 32t = 0$

Factor out $-16t$: $-16t(t - 2) = 0$

Set each factor to 0: $-16t = 0 \quad \text{or} \quad t - 2 = 0$

Solve each equation: $t = 0 \quad \text{or} \quad t = 2$

The ball will be at a height of 240 feet the instant the ball leaves the child's hand and after 2 seconds of flight.

**(b)** The ball will strike the water at the instant its height is 0. So, we need to solve the equation $s(t) = 0$.

$$s(t) = 0$$
$$-16t^2 + 32t + 240 = 0$$

Factor out $-16$: $-16(t^2 - 2t - 15) = 0$

Factor: $-16(t - 5)(t + 3) = 0$

Set each factor to 0: $-16 = 0 \quad \text{or} \quad t - 5 = 0 \quad \text{or} \quad t + 3 = 0$

Solve each equation: $t = 5 \quad \text{or} \quad t = -3$

The equation $-16 = 0$ is false and $t = -3$ makes no sense. Therefore, the ball will strike the water after 5 seconds.

**QUICK ✓**

**13.** A model rocket is fired straight up from the ground. The height $s$ of the rocket (in feet) as a function of time (in seconds) can be modeled by the function $s(t) = -16t^2 + 160t$.

**(a)** When will the height of the rocket be 384 feet from the ground?

**(b)** When will the rocket strike the ground?

## 5.8 Exercises

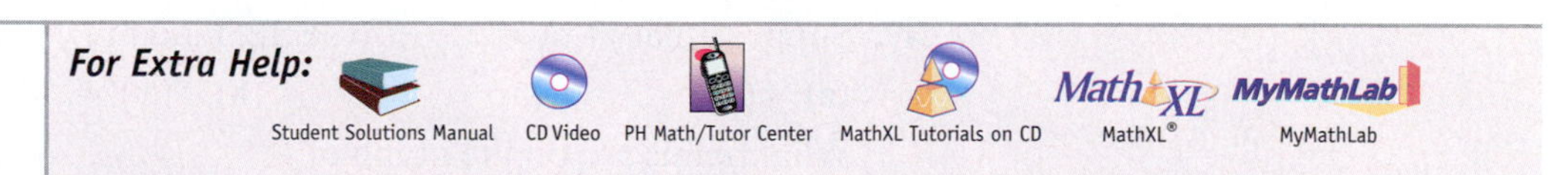


### Concepts and Vocabulary

*In Problems 1–3, fill in the blanks.*

**1.** A _____ _____ is any equation that contains a polynomial expression.

**2.** A _____ _____ is an equation equivalent to one of the form $ax^2 + bx + c = 0$ where $a$, $b$, and $c$ are real numbers and $a \neq 0$.

**3.** The equation $x^3 - 7x^2 + 4x + 2 = 0$ could be called a _____-degree equation.

*In Problems 4–6, answer True or False to each statement.*

**4.** The degree of a polynomial equation is the degree of the polynomial expression in the equation.

**5.** The Zero-Product Property states that if $ab = 0$, then $a = 0$ and $b = 0$.

**6.** If $f(5) = 0$, then 5 is a zero of the function $f$.

**7.** Explain how to determine the degree of a polynomial equation.

**8.** What role, if any, does the degree of a polynomial equation play in the number of solutions the equation has?

## Building Skills

*In Problems 9–48, solve each equation.*

**9.** $(x - 3)(x + 1) = 0$

**10.** $(x + 3)(x - 8) = 0$

**11.** $2x(3x + 4) = 0$

**12.** $4x(2x - 3) = 0$

**13.** $y(y - 5)(y + 3) = 0$

**14.** $3a(a - 9)(a + 11) = 0$

**15.** $3p^2 - 12p = 0$

**16.** $5c^2 + 15c = 0$

**17.** $2w^2 = 16w$

**18.** $4t^2 = -20t$

**19.** $m^2 + 2m - 15 = 0$

**20.** $x^2 + 3x - 40 = 0$

**21.** $w^2 - 13w = -36$

**22.** $y^2 + 13y = -42$

**23.** $p^2 - 6p + 9 = 0$

**24.** $a^2 + 12a + 36 = 0$

**25.** $5x^2 = 2x + 3$

**26.** $4c^2 + 6 = 25c$

**27.** $6m^2 = 23m - 15$

**28.** $6z^2 + 17z = -5$

**29.** $3p^2 + 9p - 120 = 0$

**30.** $4y^2 - 20y - 56 = 0$

**31.** $-4b^2 - 14b + 60 = 0$

**32.** $-6n^2 - 9n + 60 = 0$

**33.** $\frac{1}{2}x^2 + 2x - 6 = 0$

**34.** $\frac{1}{2}t^2 - 3t - 8 = 0$

**35.** $\frac{2}{3}x^2 + x = \frac{14}{3}$

**36.** $\frac{2}{3}x^2 + \frac{7}{3}x = 5$

**37.** $x(x + 8) = 33$

**38.** $y(y + 4) = 45$

**39.** $(x - 2)(x + 1)(x + 5) = 0$

**40.** $(y - 4)(y - 1)(3y + 2) = 0$

**41.** $2z^3 - 5z^2 = 3z$

**42.** $7q^3 + 31q^2 = -12q$

**43.** $2p^3 + 5p^2 - 8p - 20 = 0$

**44.** $w^3 + 5w^2 - 16w - 80 = 0$

**45.** $-30b^3 - 38b^2 = 12b$

**46.** $-24b^3 + 27b = 18b^2$

**47.** $(x - 2)^3 = x^3 - 2x$

**48.** $(x + 2)^3 = x^3 - 2x$

**49.** Suppose that $f(x) = x^2 + 7x + 12$. Find the values of $x$ such that
**(a)** $f(x) = 2$ **(b)** $f(x) = 20$
What points are on the graph of $f$?

**50.** Suppose that $f(x) = x^2 + 5x + 3$. Find the values of $x$ such that
**(a)** $f(x) = 3$ **(b)** $f(x) = 17$
What points are on the graph of $f$?

**51.** Suppose that $g(x) = 2x^2 - 6x - 5$. Find the values of $x$ such that
**(a)** $g(x) = 3$ **(b)** $g(x) = 15$
What points are on the graph of $g$?

**52.** Suppose that $h(x) = 3x^2 - 9x - 8$. Find the values of $x$ such that
**(a)** $h(x) = -8$ **(b)** $h(x) = 22$
What points are on the graph of $h$?

**53.** Suppose that $F(x) = -3x^2 + 12x + 5$. Find the values of $x$ such that

**(a)** $F(x) = 5$ **(b)** $F(x) = -10$

What points are on the graph of $F$?

**54.** Suppose that $G(x) = -x^2 + 4x + 6$. Find the values of $x$ such that

**(a)** $G(x) = 1$ **(b)** $G(x) = 9$

What points are on the graph of $G$?

*In Problems 55–60, find the zeros of the function. What are the x-intercepts of the graph of the function?*

**55.** $f(x) = x^2 + 9x + 14$

**56.** $f(x) = x^2 - 13x + 42$

**57.** $g(t) = 6t^2 - 25t - 9$

**58.** $h(p) = 8p^2 - 18p - 35$

**59.** $s(d) = 2d^3 + 2d^2 - 40d$

**60.** $f(a) = 3a^3 - 15a^2 - 42a$

## Mixed Practice

*In Problems 61–70, solve each equation.*

**61.** $(x + 3)(x - 5) = 9$

**62.** $(x + 7)(x - 3) = 11$

**63.** $2q^2 + 3q - 14 = 0$

**64.** $3t^2 + 7t - 20 = 0$

**65.** $-3b^2 + 21b = 0$

**66.** $-7z^2 + 42z = 0$

**67.** $(x + 2)(x + 3) = 0$

**68.** $(x + 7)(x - 6) = 0$

**69.** $x^3 + 5x^2 - 4x - 20 = 0$

**70.** $2c^3 + 3c^2 - 8c - 12 = 0$

## Applying the Concepts

△ **71. Area** The length of a rectangle is 8 centimeters less than its width. What are the dimensions of the rectangle if its area is 128 square centimeters?

△ **72. Area** The length of a rectangle is twice the sum of its width and 3. What are the dimensions of the rectangle if its area is 216 square inches?

△ **73. Area** The height of a triangle is 12 feet more than its base. What are the height and base of the triangle if its area is 110 square feet?

△ **74. Area** The width of a triangle is 4 meters shorter than its height. What are the height and width of the triangle if its area is 48 square meters?

△ **75. Convex Polygons** A **convex polygon** is a polygon whose interior angles are between 0° and 180°. The number of diagonals $D$ in a convex polygon with $n$ sides is given by the formula $D = \frac{n(n-3)}{2}$. Determine the number of sides $n$ in a convex polygon that has 20 diagonals.

**76. Consecutive Integers** The sum $S$ of the consecutive integers $1, 2, 3, \ldots, n$ is given by the formula $S = \frac{n(n+1)}{2}$. That is, $1 + 2 + 3 + \cdots + n = \frac{n(n+1)}{2}$. How many consecutive integers must be added together to obtain a sum of 36?

**77. Enclosing an Area with a Fence** A farmer has 100 meters of fencing and wants to enclose a rectangular plot that borders a river. If the farmer does not fence the side along the river, what are the dimensions of the land enclosed if the area enclosed is 800 square meters?

**78. Enclosing an Area with a Fence** A farmer has 300 feet of fencing and wants to enclose a rectangular corral that borders his barn on one side and then divide it into two plots with a fence parallel to one of the sides (see the figure). Assuming that the farmer will not fence the side along the barn, what are the lengths of the parts of the fence if the total area enclosed is 4800 square feet?

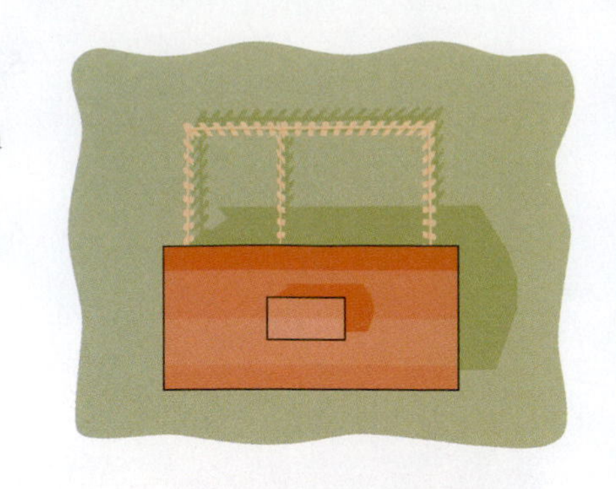

**79. Landscape Design** Robert Boehm just designed a cloister (a rectangular garden surrounded by a covered walkway on all four sides). The outside dimensions of the garden are 12 feet by 8 feet, and the area of the garden and the walkway together are 252 square feet. What is the width of the walkway?

**80. Picture Frame** The outside dimensions of a picture frame are 40 inches by 32 inches. The area of the picture within the frame is 1008 square inches. Find the width of the frame.

**81. Making a Box** A box is to be made from a rectangular piece of corrugated cardboard where the length is 5 more inches than the width by cutting a square piece 2 inches on side from each corner. The volume of the box is to be 168 cubic inches. Find the dimensions of the rectangular piece of cardboard.

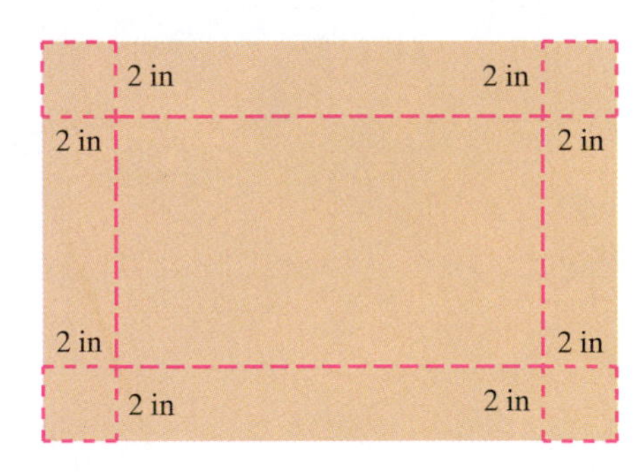


**82. Making a Box** A box is to be made from a rectangular piece of corrugated cardboard where the length is 8 inches more than the width by cutting a square piece 3 inches on side from each corner. The volume of the box is to be 315 cubic inches. Find the dimensions of the rectangular piece of cardboard.

**83. Marginal Cost** Marginal cost can be thought of as the cost of producing one additional unit of output. For example, if the marginal cost of producing the 30th unit of output is \$9.30, then it cost \$9.30 to increase production from 29 to 30 units. The marginal cost $C$ (in dollars) to produce $x$ bicycles is given by $C(x) = x^2 - 40x + 600$.

**(a)** Find the marginal cost of producing 30 bicycles.
**(b)** How many bicycles can be manufactured so that marginal cost equals $200? That is, solve $C(x) = 200$.
**(c)** Economic theory states that, to maximize profit, production should continue until marginal revenue equals marginal cost. Marginal revenue is the additional revenue received for each bicycle sold. In certain situations, the marginal revenue is the price of the product. Assuming that marginal revenue equals $225, how many bicycles should be manufactured?

**84. Marginal Cost** (See Problem 83.) Suppose that the marginal cost of manufacturing $x$ cellular telephones $C(x) = \frac{1}{2}x^2 - 30x + 475$.

**(a)** Find the marginal cost of producing 30 cell phones.
**(b)** How many cell phones can be manufactured so that marginal cost equals $75? That is, solve $C(x) = 75$.
**(c)** Economic theory states that, to maximize profit, production should continue until marginal revenue equals marginal cost. Assuming that marginal revenue equals $97, how many cell phones should be manufactured?

**85. Projectile Motion** Tiger Woods hits a golf ball with an initial speed of 240 feet per second. The height $s$ of the ball (in feet) as a function of time (in seconds) can be modeled by the function

$$s(t) = -16t^2 + 120t$$

**(a)** When will the height of the ball be 200 feet?
**(b)** When will the ball hit the ground?

**86. Projectile Motion** A cannonball is fired from a cliff that is 260 feet high with an initial speed of 128 feet per second. The height $s$ of the cannonball (in feet) as a function of time (in seconds) can be modeled by the function

$$s(t) = -16t^2 + 64t + 260$$

**(a)** When will the height of the cannonball be 320 feet?
**(b)** When will the cannonball hit the ground?

## Extending the Concepts

*In Problems 87–90, solve each equation.*

**87.** $4x^4 - 17x^2 + 4 = 0$

**88.** $9z^4 - 13z^2 + 4 = 0$

**89.** $(a + 3)^2 - 5(a + 3) = -6$

**90.** $(2b + 1)^2 + 7(2b + 1) = -12$

*In Problems 91–94, find the domain of each function.*

**91.** $f(x) = \dfrac{5}{x^2 - 4}$

**92.** $f(x) = \dfrac{-9}{x^2 + 6x + 5}$

**93.** $g(x) = \dfrac{4x + 3}{2x^2 - 3x + 1}$

**94.** $h(x) = \dfrac{x + 4}{3x^2 - 7x - 6}$

## The Graphing Calculator

The graphing calculator can be used to find solutions to any equation. There are two different methods that can be used. The ZERO (or ROOT) feature of a graphing calculator can be used to find solutions of an equation when one side of the equation is 0. Solving an equation for $x$ when one side of the equation is 0 is equivalent to finding where the graph of the corresponding equation crosses or touches the $x$-axis. For example, to solve the equation in Example 2, $2x^2 - 5x - 3 = 0$, we would graph $Y_1 = 2x^2 - 5x - 3$ and then use ZERO (or ROOT) to determine where $Y_1 = 0$. See Figures 5(a) and (b).

**Figure 5**

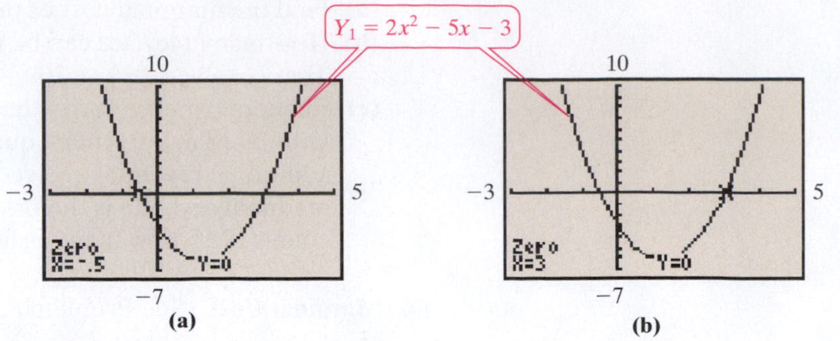

(a) (b)

*In Problems 95–100, solve each equation using a graphing calculator. Round your answers to two decimal places, if necessary.*

**95.** $2x^2 - 7x - 5 = 0$

**96.** $2x^2 - x - 10 = 0$

**97.** $0.2x^2 - 5.1x + 3 = 0$

**98.** $0.4x^2 - 2.7x + 1 = 0$

**99.** $-x^2 + 0.6x = -2$

**100.** $-3.1x^2 - 0.4x = -3$

## CHAPTER 5 ACTIVITY: WHAT IS THE QUESTION?

**Focus:** Performing operations and solving equations with polynomials

**Time:** 15–20 minutes

**Group size:** 2 or 4

In this activity you will work as a team to solve eight multiple choice questions. However, these questions are different from most multiple choice questions. You are given the answer to a problem and must determine which of the multiple choice options has the correct question for the given answer.

Before beginning the activity, decide how you will approach this task as a team. For example:

- If there are 2 members on your team . . . one member will always examine choices (a) and (b) and the other will always examine choices (c) and (d).
- If there are 4 members on your team . . . one member will always examine choice (a), another member will always examine choice (b), . . . etc.

**1.** The answer is $-3x^2 - 10x$. What is the question?

**(a)** Simplify: $2x - 3x(x^2 + 4)$
**(b)** Find the quotient: $(6x^3 - 20x^2) \div (-2x)$
**(c)** Simplify: $-3x(x^2 + 3) + 1$
**(d)** Find the quotient: $(-6x^4 - 20x^3) \div (2x^2)$

**2.** The answer is $(f + g)(-1) = 5$. What is the question?

**(a)** $f(x) = 2x + 2, g(x) = -3x - 1$
**(b)** $f(x) = 2x + 3, g(x) = -3x + 1$
**(c)** $f(x) = -2x + 5, g(x) = -3x + 4$
**(d)** $f(x) = -2x - 2, g(x) = 3x - 4$

**3.** The answer is $(6x + 1)(2x - 3)$. What is the question?

**(a)** Factor: $12x^2 - 20x + 3$
**(b)** Factor: $12x^2 + 16x - 3$
**(c)** Factor: $12x^2 - 16x - 3$
**(d)** Factor: $12x^2 + 20x + 3$

**4.** The answer is $x^2 + 5x + 6$. What is the question?

**(a)** Find the product: $(x + 6)(x - 1)$
**(b)** Simplify: $2x^2 + 7x + 9 - (x^2 - 2x - 3)$
**(c)** Find the product: $(x + 2)(x + 3)$
**(d)** Simplify: $(x + 6)^2$

**5.** The answer is 3. What is the question?

**(a)** What is the name of the variable in $16z^2 + 3z - 5$?
**(b)** What is the degree of the polynomial $2mn + 6m - 3$?
**(c)** How many terms are in the polynomial $2mn + 6m - 3$?
**(d)** What is the coefficient of $b$ in the polynomial $3a^2b - 9a + 5b$?

**6.** The answer is $x = -5$ or $x = 3$. What is the question?

**(a)** Solve: $x(x + 2) = 15$
**(b)** Find the values of $x$ such that $f(x) = 8$ if $f(x) = x^2 + 4x + 3$
**(c)** Solve: $\frac{2}{3}x + 5 = \frac{1}{3}x^2$
**(d)** Find the $x$-intercepts of the graph of the function $f(x) = x^2 - 2x - 15$

**7.** The answer is $8x^2$. What is the question?

**(a)** Simplify: $9x^2(x + 1) - 3x^2(3x + 5)$
**(b)** Find the greatest common factor: $16x^2y^2 - 8x^3y - 24x^2$
**(c)** Find the quotient: $(-8x^3 - 8x^2) \div (x + 1)$
**(d)** Factor by grouping: $8x^3 + 8x^2 - x - 1$

**8.** The answer is $x + 2$. What is the question?

**(a)** Find the quotient: $(x^3 + x^2 - 7x - 2) \div (x^2 + 3x - 1)$
**(b)** Find the binomial factor: $x^3 - 8$
**(c)** Find the quotient: $(x^2 + 2x - 3) \div (x - 1)$
**(d)** Find the binomial factor: $x^3 + 8$

# CHAPTER 5 REVIEW

## Section 5.1 Adding and Subtracting Polynomials

| KEY CONCEPTS | KEY TERMS |
|---|---|
| • **Monomial**<br>A monomial in one variable is the product of a constant and a variable raised to a nonnegative integer power. A monomial in one variable is of the form $ax^k$ where $a$ is a constant, $x$ is a variable, and $k \geq 0$ is an integer. The constant $a$ is called the **coefficient** of the monomial. If $a \neq 0$, then $k$ is called the **degree** of the monomial.<br>• **Polynomial**<br>A polynomial is a monomial or the sum of monomials.<br>• **Polynomial Function**<br>A polynomial function is a function whose rule is a polynomial. The domain of all polynomial functions is the set of all real numbers.<br>• **Sum or Difference of Two Functions**<br>If $f$ and $g$ are two functions:<br>The sum $f + g$ is the function defined by $(f + g)(x) = f(x) + g(x)$<br>The difference $f - g$ is the function defined by $(f - g)(x) = f(x) - g(x)$ | Monomial<br>Coefficient<br>Degree<br>Polynomial<br>Standard form<br>Binomial<br>Trinomial<br>Polynomial function<br>Evaluate<br>Sum of two functions<br>Difference of two functions |

| YOU SHOULD BE ABLE TO . . . | EXAMPLE | REVIEW EXERCISES |
|---|---|---|
| 1 Define monomial and determine the coefficient and degree of a monomial (p. 356) | Examples 1 through 3 | 1–2 |
| 2 Define polynomial and determine the degree of a polynomial (p. 357) | Examples 4 and 5 | 3–4 |
| 3 Simplify polynomials by combining like terms (p. 358) | Examples 6 through 8 | 5–10 |
| 4 Evaluate polynomial functions (p. 360) | Examples 9 and 10 | 11–12, 13(b), 14(b), 15(b), 16 |
| 5 Add and subtract polynomial functions (p. 362) | Examples 11 and 12 | 13(a), 14(a), 15(a) |

*In Problems 1 and 2, determine the coefficient and degree of each monomial.*

**1.** $-7x^4$

**2.** $\frac{1}{9}w^3$

*In Problems 3 and 4, write each polynomial in standard form. Then determine the degree of each polynomial.*

**3.** $x + 7x^3 - 8 - 2x^2$

**4.** $3 + 2y - 3y^2 + y^4$

*In Problems 5–10, add or subtract as indicated. Express your answer as a single polynomial in standard form.*

**5.** $(x^2 + 2x - 7) + (3x^2 - x - 4)$

**6.** $(4x^3 - 3x^2 + x - 5) - (x^4 + 2x^2 - 7x + 1)$

**7.** $\left(\frac{1}{4}x^2 - \frac{1}{2}x\right) - \left(4x - \frac{1}{6}\right)$

**8.** $\left(\frac{1}{2}x^2 - x + \frac{1}{4}\right) + \left(\frac{1}{3}x^2 + \frac{2}{5}\right)$

**9.** $(x^3y^2 + 6x^2y^2 - xy) + (-x^3y^2 + 4x^2y^2 + xy)$

**10.** $(a^2b - 4ab^2 + 3) - (2a^2b + 2ab^2 + 7)$

*In Problems 11–14, find the indicated function or function value.*

**11.** $f(x) = -3x^2 + 2x - 8$
**(a)** $f(-2)$ **(b)** $f(0)$ **(c)** $f(3)$

**12.** $f(x) = x^3 - 5x^2 + 3x - 1$
**(a)** $f(-3)$ **(b)** $f(0)$ **(c)** $f(2)$

**13.** $f(x) = 4x - 3$; $g(x) = x^2 + 3x + 2$
**(a)** $(f + g)(x)$ **(b)** $(f + g)(3)$

**14.** $f(x) = 2x^3 + x^2 - 7$;
$g(x) = 3x^2 - x + 5$
**(a)** $(f - g)(x)$ **(b)** $(f - g)(2)$

**15. Profit** Suppose that the revenue function $R$ from selling $x$ graphing calculators is $R(x) = -1.5x^2 + 180x$. The cost $C$ of selling $x$ graphing calculators is $C(x) = x^2 - 100x + 3290$.

**(a)** Find the profit function.
**(b)** Find the profit if $x = 25$ calculators are sold.

**16. Area** The area $A$ of the region shown in quadrant I is given by $A(x) = -x^2 + 5x$ where $(x, y)$ is a point in quadrant I on the graph of the line $y = -2x + 5$.

**(a)** Find the area of the region when the given point is (2, 1).
**(b)** Find the area of the region when the given point is (1, 3).

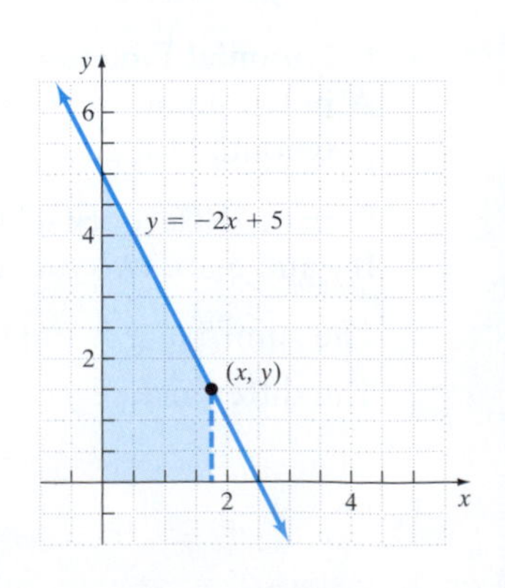

## Section 5.2 Multiplication of Polynomials

| KEY CONCEPTS | KEY TERMS |
|---|---|
| • **Product Rule for Exponents** If $a$ is a real number and $m$ and $n$ are integers, then $a^m \cdot a^n = a^{m+n}$. If $m, n$, or $m + n$ is 0 or negative, then $a$ cannot be 0.<br>• **Distributive Property** If $a$, $b$, and $c$ are real numbers, then $a \cdot (b + c) = a \cdot b + a \cdot c$ and $(a + b) \cdot c = a \cdot c + b \cdot c$.<br>• **Extended Form of the Distributive Property** $a(b_1 + b_2 + \cdots + b_n) = a \cdot b_1 + a \cdot b_2 + \cdots + a \cdot b_n$ where $a, b_1, b_2, \ldots, b_n$ are real numbers.<br>• **Difference of Two Squares** $(A - B)(A + B) = A^2 - B^2$<br>• **Squares of Binomials or Perfect Squares** $(A + B)^2 = A^2 + 2AB + B^2$ $(A - B)^2 = A^2 - 2AB + B^2$<br>• **Product of Two Functions** Let $f$ and $g$ be two functions. The product $f \cdot g$ is the function defined by $(f \cdot g)(x) = f(x) \cdot g(x)$. | FOIL<br>Special products |

| YOU SHOULD BE ABLE TO . . . | EXAMPLE | REVIEW EXERCISES |
|---|---|---|
| 1 Multiply a monomial and a polynomial (p. 369) | Example 2 | 17–20 |
| 2 Multiply a binomial by a binomial (p. 369) | Examples 3 and 4 | 21–24 |
| 3 Multiply a polynomial by a polynomial (p. 370) | Example 5 | 25–28 |
| 4 Multiply special products (p. 371) | Examples 6 through 9 | 29–34 |
| 5 Multiply polynomial functions (p. 372) | Example 10 | 35–36 |

*In Problems 17–20, find the product.*

**17.** $(-3x^3y)(4xy^2)$

**18.** $\left(\frac{1}{3}mn^4\right)(18m^3n^3)$

**19.** $5ab(-2a^2b + ab^2 - 3ab)$

**20.** $0.5c(1.7c^2 + 4.3c + 8.9)$

*In Problems 21–24, find the product of the two binomials.*

**21.** $(x + 2)(x - 9)$

**22.** $(-3x + 1)(2x - 8)$

**23.** $(m - 4n)(2m + n)$

**24.** $(2a + 15)(-a + 3)$

*In Problems 25–28, find the product of the polynomials.*

**25.** $(x + 2)(3x^2 - 5x + 1)$

**26.** $(w - 4)(w^2 + w - 8)$

**27.** $(m^2 - 2m + 3)(2m^2 + 5m - 7)$

**28.** $(2p - 3q)(p^2 + 7pq - 4q^2)$

*In Problems 29–34, find the special products.*

**29.** $(3w + 1)(3w - 1)$

**30.** $(2x - 5y)(2x + 5y)$

**31.** $(6k - 5)^2$

**32.** $(3a + 2b)^2$

**33.** $(x + 2)(x^2 - 2x + 4)$

**34.** $(2x - 3)(4x^2 + 6x + 9)$

*In Problems 35–38, find the indicated function or value.*

**35.** $f(x) = 3x - 7; g(x) = 6x + 5$
**(a)** $(f \cdot g)(x)$ **(b)** $(f \cdot g)(-2)$

**36.** $f(x) = x + 2; g(x) = 3x^2 - x + 1$
**(a)** $(f \cdot g)(x)$ **(b)** $(f \cdot g)(4)$

**37.** $f(x - 3)$ when $f(x) = 5x^2 + 8$.

**38.** $f(x + h) - f(x)$ when $f(x) = -x^2 + 3x - 5$.

## Section 5.3 Division of Polynomials and Synthetic Division

| KEY CONCEPTS | KEY TERMS |
|---|---|
| • **Quotient Rule for Exponents**<br>If $a$ is a real number and $m$ and $n$ are integers, then $\frac{a^m}{a^n} = a^{m-n}, a \neq 0$.<br>• **Quotient of Two Functions**<br>Let $f$ and $g$ be two functions. The quotient $\frac{f}{g}$ is the function defined by<br>$\left(\frac{f}{g}\right)(x) = \frac{f(x)}{g(x)}, g(x) \neq 0$.<br>• **The Remainder Theorem**<br>Let $f$ be a polynomial function. If $f(x)$ is divided by $x - c$, then the remainder is $f(c)$.<br>• **The Factor Theorem**<br>Let $f$ be a polynomial function. Then $x - c$ is a factor of $f(x)$ if and only if $f(c) = 0$. | Divisor<br>Dividend<br>Quotient<br>Remainder<br>Synthetic division |

| YOU SHOULD BE ABLE TO . . . | EXAMPLE | REVIEW EXERCISES |
|---|---|---|
| 1 Divide a polynomial by a monomial (p. 377) | Example 1 | 39–42 |
| 2 Divide polynomials using long division (p. 378) | Examples 3 through 5 | 43–48, 63 |
| 3 Divide polynomials using synthetic division (p. 381) | Examples 6 and 7 | 49–54, 64 |
| 4 Divide polynomial functions (p. 384) | Example 8 | 55–58 |
| 5 Use the remainder and factor theorems (p. 385) | Examples 9 and 10 | 59–62 |

*In Problems 39–42, divide and simplify.*

**39.** $\frac{12x^3 - 6x^2}{3x}$

**40.** $\frac{15w^5 - 5w^3 + 25w^2 + 10w}{5w}$

**41.** $\frac{7y^3 + 12y^2 - 6y}{2y}$

**42.** $\frac{2m^3n^2 + 8m^2n^2 - 14mn^3}{4m^2n^3}$

*In Problems 43–48, divide using long division.*

**43.** $\frac{3x^2 - 2x - 8}{x - 2}$

**44.** $\frac{-2x^2 - 3x + 40}{x + 5}$

**45.** $\frac{6z^3 + 9z^2 + 4z - 6}{2z + 3}$

**46.** $\frac{12k^3 - 29k^2 - 14k + 16}{3k - 8}$

**47.** $\frac{16x^4 - 81}{2x - 3}$

**48.** $\frac{2x^4 - 11x^3 + 35x^2 - 54x + 55}{x^2 - 3x + 4}$

*In Problems 49–54, divide using synthetic division.*

**49.** $\dfrac{5x^2 + 11x + 8}{x + 2}$

**50.** $\dfrac{9a^2 - 14a - 8}{a - 2}$

**51.** $\dfrac{3m^3 + 11m^2 - 5m - 33}{m + 3}$

**52.** $\dfrac{n^3 + 2n^2 - 39n + 67}{n - 4}$

**53.** $\dfrac{x^4 + 6x^2 - 7}{x + 1}$

**54.** $\dfrac{2x^3 + 5x - 8}{x + 2}$

*In Problems 55–58, find the indicated function or value.*

**55.** $f(x) = 5x^3 + 25x^2 - 15x$; $g(x) = 5x$

**(a)** $\left(\dfrac{f}{g}\right)(x)$ **(b)** $\left(\dfrac{f}{g}\right)(2)$

**56.** $f(x) = 9x^2 + 54x - 31$; $g(x) = 3x - 2$

**(a)** $\left(\dfrac{f}{g}\right)(x)$ **(b)** $\left(\dfrac{f}{g}\right)(-3)$

**57.** $f(x) = 2x^3 + 12x^2 + 9x - 28$; $g(x) = x + 4$

**(a)** $\left(\dfrac{f}{g}\right)(x)$ **(b)** $\left(\dfrac{f}{g}\right)(-2)$

**58.** $f(x) = 3x^4 - 14x^3 + 31x^2 - 58x + 22$; $g(x) = x^2 - x + 5$

**(a)** $\left(\dfrac{f}{g}\right)(x)$ **(b)** $\left(\dfrac{f}{g}\right)(4)$

*In Problems 59 and 60, use the Remainder Theorem to find the remainder.*

**59.** $f(x) = 4x^2 - 7x + 23$ is divided by $x - 4$.

**60.** $f(x) = x^3 - 2x^2 + 12x - 5$ is divided by $x + 2$.

*In Problems 61 and 62, use the Factor Theorem to determine whether $x - c$ is a factor of the given function for the given value of c. If $x - c$ is a factor, then write the function in factored form.*

**61.** $f(x) = 3x^2 + x - 14$; $c = 2$

**62.** $f(x) = 2x^2 + 13x + 22$; $c = -4$

**63.** The area of a rectangle is $20x^2 - 11x - 3$ square meters. If the width of the rectangle is $4x - 3$ meters, find an expression for the length.

**64.** The volume of a rectangular box is $2x^3 + x^2 - 7x - 6$ cubic centimeters. If the height of the box is $x - 2$ centimeters, find an expression for the area of the top of the box.

## Section 5.4 Greatest Common Factor and Factoring by Grouping

| KEY CONCEPTS | KEY TERMS |
|---|---|
| • **Factoring out the greatest common factor** Identify the greatest common factor (GCF) of each term. Rewrite each term as the product of the GCF and remaining factor. Use the Distributive Property to factor out the GCF. Check your work using the Distributive Property.<br>• **Factoring by grouping** Group the terms with common factors. Sometimes it will be necessary to rearrange the terms. In each grouping, factor out the common factor. Factor out the common factor that remains. Check your work. | Factors<br>Factoring over the integers<br>Prime<br>Factored completely<br>Greatest common factor (GCF)<br>Factoring by grouping |

| YOU SHOULD BE ABLE TO . . . | EXAMPLE | REVIEW EXERCISES |
|---|---|---|
| 1 Factor the greatest common factor (p. 391) | Examples 1 through 4 | 65–72, 79, 80 |
| 2 Factor by grouping (p. 393) | Examples 5 and 6 | 73–78 |

*In Problems 65–72, factor out the greatest common factor.*

**65.** $4z + 24$

**66.** $-7y^2 + 91y$

**67.** $14x^3y^2 + 2xy^2 - 8x^2y$

**68.** $30a^4b^3 + 15a^3b - 25a^2b^2$

**69.** $3x(x + 5) - 4(x + 5)$

**70.** $-4c(2c + 9) + 3(2c + 9)$

**71.** $(5x + 3)(x - 5y) + (x + 2)(x - 5y)$

**72.** $(3a - b)(a + 7) - (a + 1)(a + 7)$

*In Problems 73–78, factor by grouping.*

**73.** $x^2 + 6x - 3x - 18$

**74.** $c^2 + 2c - 5c - 10$

**75.** $14z^2 + 16z - 21z - 24$

**76.** $21w^2 - 28w + 6w - 8$

**77.** $2x^3 + 2x^2 - 18x^2 - 18x$

**78.** $10a^4 + 15a^3 + 70a^3 + 105a^2$

**79. Integers** The sum of the first $n$ positive integers is given by $\frac{1}{2}n^2 + \frac{1}{2}n$.

**(a)** Write this expression in factored form.
**(b)** Use the factored form to determine the sum of the first 32 positive integers.

**80. Revenue** A computer manufacturer estimates that its revenue for selling $x$ computer systems can be approximated by the function $R(x) = 5200x - 2x^3$. Express the revenue function in factored form.

## Section 5.5 Factoring Trinomials

| KEY CONCEPTS | KEY TERM |
|---|---|
| • **Factoring $x^2 + bx + c$**<br>$x^2 + bx + c = (x + m)(x + n)$, where $mn = c$ and $m + n = b$<br>• **Factoring $ax^2 + bx + c$ by grouping**<br>See page 402<br>• **Factoring $ax^2 + bx + c$ by trial and error**<br>See page 404 | Factoring by substitution |

| YOU SHOULD BE ABLE TO . . . | EXAMPLE | REVIEW EXERCISES |
|---|---|---|
| 1 Factor trinomials of the form $x^2 + bx + c$ (p. 397) | Examples 1 through 6 | 81–86 |
| 2 Factor trinomials of the form $ax^2 + bx + c, a \neq 1$ (p. 401) | Examples 7 through 12 | 87–94 |
| 3 Factor trinomials using substitution (p. 406) | Examples 13 and 14 | 95–98 |

*In Problems 81–98, factor each trinomial completely. If the polynomial cannot be factored, say it is prime.*

**81.** $w^2 - 11w - 26$

**82.** $x^2 - 9x + 15$

**83.** $-t^2 + 6t + 72$

**84.** $m^2 + 10m + 21$

**85.** $x^2 + 4xy - 320y^2$

**86.** $r^2 - 5rs + 6s^2$

**87.** $5x^2 + 13x - 6$

**88.** $6m^2 + 41m + 44$

**89.** $4y^2 - 5y + 7$

**90.** $8t^2 + 22t - 6$

**91.** $6x^2 - 13x + 5$

**92.** $21r^2 - rs - 2s^2$

**93.** $20x^2 - 57xy + 27y^2$

**94.** $-2s^2 + 12s + 14$

**95.** $x^4 - 10x^2 - 11$

**96.** $10x^2y^2 + 41xy + 4$

**97.** $(a + 4)^2 - 9(a + 4) - 36$

**98.** $2(w - 1)^2 + 11(w - 1) + 9$

## Section 5.6 Factoring Special Products

**KEY CONCEPTS**

- **Squares of Binomials or Perfect Square Trinomials**

  $A^2 + 2AB + B^2 = (A + B)^2$

  $A^2 - 2AB + B^2 = (A - B)^2$

- **Difference of Two Squares**

  $A^2 - B^2 = (A - B)(A + B)$

- **Sum or Difference of Two Cubes**

  $A^3 + B^3 = (A + B)(A^2 - AB + B^2)$

  $A^3 - B^3 = (A - B)(A^2 + AB + B^2)$

**KEY TERM**

Sum of two squares

| YOU SHOULD BE ABLE TO . . . | EXAMPLE | REVIEW EXERCISES |
|---|---|---|
| 1 Factor perfect square trinomials (p. 409) | Example 1 | 99–104 |
| 2 Factor the difference of two squares (p. 411) | Examples 2 and 3 | 105–110 |
| 3 Factor the sum or difference of two cubes (p. 412) | Examples 4 and 5 | 111–116 |

*In Problems 99–116, factor completely.*

**99.** $x^2 + 22x + 121$ **100.** $w^2 - 34w + 289$ **101.** $144 - 24c + c^2$

**102.** $x^2 - 8x + 16$ **103.** $64y^2 + 80y + 25$ **104.** $12z^2 + 48z + 48$

**105.** $x^2 - 196$ **106.** $49 - y^2$ **107.** $t^2 - 225$

**108.** $4w^2 - 81$ **109.** $36x^4 - 25y^2$ **110.** $80mn^2 - 20m$

**111.** $x^3 - 343$ **112.** $729 - y^3$ **113.** $27x^3 - 125y^3$

**114.** $8m^6 + 27n^3$ **115.** $2a^6 - 2b^6$ **116.** $(y - 1)^3 + 64$

## Section 5.7 Factoring: A General Strategy

**KEY CONCEPT**

**Steps for Factoring**

**Step 1:** Factor out the greatest common factor (GCF), if any exists.

**Step 2:** Count the number of terms.

**Step 3:** **(a)** 2 terms

- Is it the difference of two squares? If so,

  $A^2 - B^2 = (A - B)(A + B)$

- Is it the difference of two cubes? If so,

  $A^3 - B^3 = (A - B)(A^2 + AB + B^2)$

- Is it the sum of two cubes? If so,

  $A^3 + B^3 = (A + B)(A^2 - AB + B^2)$

**(b)** 3 terms

- Is it a perfect square trinomial? If so,

  $A^2 + 2AB + B^2 = (A + B)^2$ or $A^2 - 2AB + B^2 = (A - B)^2$

- Is the coefficient of the square term 1? If so,

  $x^2 + bx + c = (x + m)(x + n)$ where $mn = c$ and $m + n = b$

- Is the coefficient of the square term different from 1? If so,
  **a.** Use factoring by grouping
  **b.** Use trial and error

**(c)** 4 terms
- Use factoring by grouping

**Step 4:** Check your work by multiplying out the factored form.

| YOU SHOULD BE ABLE TO . . . | EXAMPLE | REVIEW EXERCISES |
|---|---|---|
| 1 Factor polynomials completely (p. 417) | Examples 1 through 7 | 117–138 |

*In Problems 117–136, factor each polynomial completely.*

**117.** $x^2 + 7x + 6$ **118.** $c^2 - 24c + 144$ **119.** $z^2 - 9z - 112$

**120.** $-8x^2y^3 + 12xy^3$ **121.** $7x^3 - 28x^2 + 63x$ **122.** $3x^2 - 3x - 18$

**123.** $4z^2 - 60z + 225$ **124.** $12x^2 + 7x - 49$ **125.** $45 + 6x - 3x^2$

**126.** $10n^2 - 33n - 7$ **127.** $8 - 2y - y^2$ **128.** $2x^3 - 10x^2 + 6x - 30$

**129.** $(w - 3z)^2 - (z + 2)^2$ **130.** $(3h + 2)^3 + 64$ **131.** $5p^3q^2 - 80p$

**132.** $36a^2 - 20a - 27a + 15$ **133.** $m^4 - 5m^2 + 4$

**134.** $686 - 16m^6$ **135.** $h^3 + 2h^2 - h - 2$ **136.** $108x^3 + 4y^3$

*In Problems 137 and 138, write an expression for the shaded area in factored form.*

**137.**

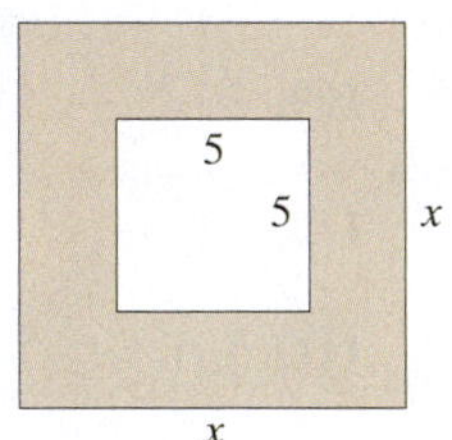


**138.**

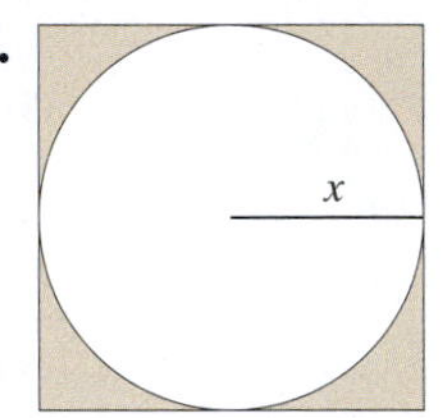


## Section 5.8 Polynomial Equations

| KEY CONCEPTS | KEY TERMS |
|---|---|
| • **The Zero-Product Property** If the product of two numbers is zero, then at least one of the numbers is 0. That is, if $ab = 0$, then $a = 0$ or $b = 0$ or both $a$ and $b$ are 0.<br>• **Zero of a Function** A zero of a function $f$ is any number $r$ such that $f(r) = 0$. If $r$ is a zero of a function, then $r$ is also an $x$-intercept of the graph of the function. | Polynomial equation<br>Degree of a polynomial equation<br>Zero-product property<br>Quadratic equation<br>Standard form<br>Second-degree equation<br>Zero |

| YOU SHOULD BE ABLE TO . . . | EXAMPLE | REVIEW EXERCISES |
|---|---|---|
| 1 Solve polynomial equations using the Zero-Product Property (p. 423) | Examples 1 through 4 | 139–148 |
| 2 Solve equations involving polynomial functions (p. 427) | Examples 5 and 6 | 149–152 |
| 3 Model and solve problems involving polynomials (p. 429) | Examples 7 through 9 | 153–154 |

*In Problems 139–148, solve each equation.*

**139.** $(w + 5)(w - 13) = 0$

**140.** $x(2x + 1)(3x - 5) = 0$

**141.** $5a^2 = -20a$

**142.** $y^2 + 2y = 15$

**143.** $x^2 + 21x + 54 = 0$

**144.** $15x^2 + 29x - 14 = 0$

**145.** $x(x + 1) = 110$

**146.** $\frac{1}{2}x^2 + 5x + 12 = 0$

**147.** $(b + 1)(b - 3) = 5$

**148.** $(x + 7)^3 = x^3 + 133$

**149.** Suppose that $f(x) = x^2 + 5x - 18$. Find values of $x$ such that

**(a)** $f(x) = 6$ **(b)** $f(x) = -4$

What points are on the graph of $f$?

**150.** Suppose that $f(x) = 5x^2 - 4x + 3$. Find values of $x$ such that

**(a)** $f(x) = 3$ **(b)** $f(x) = 4$

What points are on the graph of $f$?

*In Problems 151 and 152, find the zeros of the function. What are the x-intercepts of the graph of the function?*

**151.** $f(x) = 3x^3 + 18x^2 + 24x$

**152.** $f(x) = -4x^2 + 22x + 42$

**153. Falling Object** At one point, a four-foot flagpole on top of the KXJB-TV mast in Galesburg, North Dakota made it the world's tallest structure standing 2064 feet tall. If an object is dropped from the top of this mast, the height $s$ of the object (in feet) as a function of time (in seconds) can be modeled by the function $s(t) = -16t^2 + 2064$. When will the object be 1280 feet above the ground?

**154. Reliability** A simple parallel system with two identical components has a reliability given by $R = 1 - (1 - r)^2$, where $r$ is the reliability of the individual components.

**(a)** What is the reliability of the individual components if the system reliability is $R = 0.96$?

**(b)** What is the reliability of the individual components if the system reliability is $R = 0.99$?

## CHAPTER 5 TEST

*Remember to use your Chapter Test Prep Video CD to see fully worked-out solutions to any of these problems you would like to review.*

**1.** Write the polynomial in standard form and determine its degree.

$$7x^2 + x^4 - 5x^7 + 1 - x$$

**2.** Add: $(-2a^3b^2 + 5a^2b + ab + 1) + \left(\frac{1}{3}a^3b^2 + 4a^2b - 6ab - 5\right)$.

**3.** For $f(x) = x^3 + 3x^2 - x + 1$, find $f(-2)$.

**4.** For $f(x) = 7x^3 - 1$ and $g(x) = 4x^2 + 3x - 2$, find $(f - g)(x)$.

*In Problems 5–7, perform the indicated multiplication.*

**5.** $\frac{1}{2}a^2b(4ab^2 - 6ab + 8)$

**6.** $(3x - 1)(4x + 17)$

**7.** $(2m - n)^2$

**8.** Divide using long division: $\dfrac{6z^3 - 14z^2 + z + 4}{2z^2 + 1}$

**9.** Divide using synthetic division: $\dfrac{5x^2 - 27x - 18}{x - 6}$

**10.** For $f(x) = 6x^2 + x - 12$ and $g(x) = 2x + 3$, find $\left(\dfrac{f}{g}\right)(2)$.

**11.** Use the Remainder Theorem to determine the remainder when $f(x) = 2x^3 - 3x^2 - 4x + 7$ is divided by $x - 3$.

**12.** Factor out the greatest common factor: $12a^3b^2 + 8a^2b^2 - 16ab^3$

*In Problems 13–18, factor completely.*

**13.** $6c^2 + 21c - 4c - 14$

**14.** $x^2 - 13x - 48$

**15.** $-14p^2 - 17p + 6$

**16.** $5(z - 1)^2 + 17(z - 1) - 12$

**17.** $-98x^2 + 112x - 32$

**18.** $16x^2 - 196$

**19.** Solve: $3m^2 - 5m = 5m - 7$.

**20.** One side of a rectangular patio is 3 meters longer than the other. If the area of the patio is 108 square meters, what are the dimensions of the patio?

# CUMULATIVE REVIEW CHAPTERS R–5

**1.** Evaluate: $(-3)^2 + 4 - 16 \div 2$.

**2.** Evaluate: $|2 - 3^2| + 7$.

**3.** Simplify: $2(5x + 1) - 4(x - 3)$.

*In Problems 4 and 5, solve the indicated equations.*

**4.** $3(x + 2) - 10 = 4x$

**5.** $2|x - 5| + 3 = 5$

**6.** Solve $4x - 5y = 30$ for $y$.

**7.** Solve the inequality. Write your answer in interval notation and graph the solution set on a number line.

$$4x - 3(x + 1) \geq 7 - 4x$$

**8.** **Acid Solution** How many liters of a 70% sulfuric acid solution must be mixed with 12 liters of a 30% sulfuric acid solution to obtain a 40% sulfuric acid solution?

**9.** Determine the intercepts of the graph.

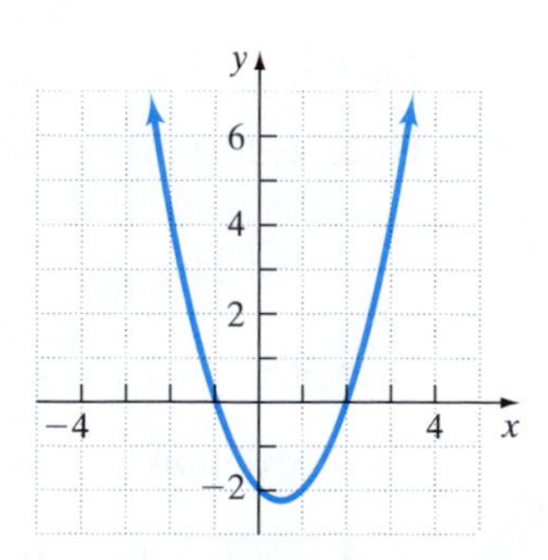


**10.** Graph the relation $y = 2x^2 + 4x - 1$ by plotting points. Use the graph to determine the domain and range.

**11.** Determine if each relation is a function. In each case, determine the domain and range.

**(a)** $\{(-2, 1), (3, 4), (5, -3), (7, 4), (10, 13)\}$

**(b)**

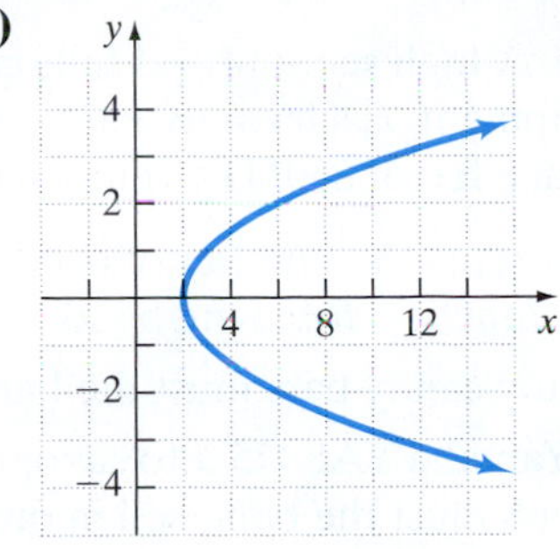


**12.** Sketch the graph of $3x - 2y = 6$ by plotting points.

**13.** Find the equation of the line with slope $\frac{2}{5}$ that passes through the point $(10, -4)$.

**14. Gas Prices** The following graph shows the average price per gallon in the United States for regular gasoline over a 6 month period beginning in August of 2003. (*Source: GasBuddy.com*)

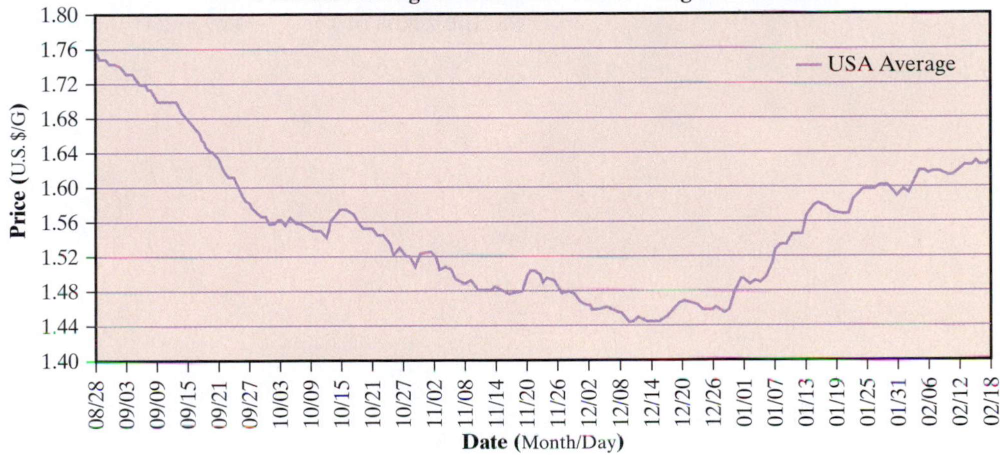


Source: *GasBuddy.com, 2004*

**(a)** Describe the trend in gas prices over the indicated time period.

**(b)** Approximately when was gas the cheapest during this time period?

**(c)** What was the average price per gallon on January 1, 2004?

**15.** Graph the inequality $4x - 2y < 5$.

**16.** Solve using substitution or elimination.

$$\begin{cases} y = x - 3 \\ 3x - 2y = 4 \end{cases}$$

**17.** Solve using matrices.

$$\begin{cases} 6x + 2y = 2 \\ 3x - y = 11 \end{cases}$$

**18. Super Bowl Ads** The average cost of a 30 second ad during the Super Bowl was \$900,000 in 1994 and \$2.3 million in 2004. Determine the average rate of change in Super Bowl ad price over this time period.

**19. Pizza Sales** At a recent fundraiser, Brandon made \$77.50 by selling 4 cheese pizzas, 5 sausage pizzas, and 2 pepperoni pizzas. Matt made \$110.50 by selling 8 cheese, 3 sausage, and 5 pepperoni. Ethan made \$92.00 by selling 1 cheese, 4 sausage, and 7 pepperoni. Determine the price of each type of pizza.

**20. Candy Sales** A high school band sells candy bars during lunch to raise money for new equipment. Each candy bar costs the band \$0.22 and they must pay a one-time usage fee of \$50. The candy bars sell for \$0.75.

**(a)** If $x$ represents the number of candy bars sold, write a cost function $C$ and a revenue function $R$ for the band.

**(b)** How many candy bars must the band sell to begin making a profit?

**(c)** A new Yamaha YAS475 alto saxophone is priced at \$1396.00. How many candy bars must the band sell in order to purchase this saxophone?

*For Problems 21–23, perform the indicated operation.*

**21.** $(12x^3 + 5x^2 - 3x + 1) - (2x^3 - 4x + 8)$

**22.** $(2x + 1)(x^2 - 3x + 5)$ **23.** $\dfrac{3x^3 + 10x^2 - 23x + 1}{x + 5}$

**24.** Factor completely.

**(a)** $\frac{1}{3}x^2 + 2x + 3$ **(b)** $w^2 - 7w - 60$ **(c)** $32a^4 - 128a^2$

**25.** Find the zeros of $f(x) = 4x^2 - 2x - 30$.

# Getting Ready for Chapter 6: Rational Expressions and Rational Functions

## OBJECTIVES

1. Write Rational Numbers in Lowest Terms
2. Multiply and Divide Rational Numbers
3. Add and Subtract Rational Numbers

The purpose of this "Getting Ready" section is to provide a review of operations on rational numbers. As you go through the section pay attention to the methods used to perform each operation because these same methods will be used in Chapter 6 when we discuss operations on rational expressions.

## 1 Write Rational Numbers in Lowest Terms

We prefer to write rational numbers in **lowest terms;** that is, we write rational numbers so that there are not any common factors in the numerator and the denominator of the rational number. We obtain rational numbers in lowest terms using the *Reduction Property*.

**REDUCTION PROPERTY**

If $a$, $b$, and $c$ are real numbers, then

$$\frac{ac}{bc} = \frac{a}{b} \quad \text{if } b \neq 0, c \neq 0$$

**EXAMPLE 1 Writing Rational Numbers in Lowest Terms**

Divide out the 9's

$$\frac{45}{18} = \frac{\cancel{9} \cdot 5}{\cancel{9} \cdot 2} = \frac{5}{2}$$

**QUICK ✓** *Write each rational number in lowest terms.*

**1.** $\frac{13 \cdot 5}{13 \cdot 6}$  **2.** $\frac{80}{12}$

## 2 Multiply and Divide Rational Numbers

We now review the methods for multiplying and dividing rational numbers.

**Multiplying Rational Numbers**

**Step 1:** Completely factor each integer in the numerator and denominator.

**Step 2:** Divide out common factors in the numerator and denominator.

**Step 3:** Use the fact that if $\frac{a}{b}$ and $\frac{c}{d}$, $b \neq 0$, $d \neq 0$, are two rational numbers, then $\frac{a}{b} \cdot \frac{c}{d} = \frac{ac}{bd}$ to multiply the rational numbers.

**Dividing Rational Numbers**

To divide rational numbers, use the fact that if $\frac{a}{b}$ and $\frac{c}{d}$, $b \neq 0$, $c \neq 0$, $d \neq 0$, are two rational numbers, then $\frac{\frac{a}{b}}{\frac{c}{d}} = \frac{a}{b} \div \frac{c}{d} = \frac{a}{b} \cdot \frac{d}{c} = \frac{ad}{bc}$, then follow the steps for multiplying two rational numbers.

**In Words**

To multiply two rational numbers, multiply the numerators and then multiply the denominators. To divide two rational numbers, multiply the rational number in the numerator by the reciprocal of the rational number in the denominator.

**EXAMPLE 2 Multiplying and Dividing Rational Numbers**

Perform the indicated operation. Be sure to express the result in lowest terms.

**(a)** $\frac{10}{3} \cdot \frac{18}{25}$ **(b)** $\frac{\frac{14}{5}}{\frac{21}{10}}$

**Solution**

$10 = 2 \cdot 5;\ 18 = 9 \cdot 2 = 3 \cdot 3 \cdot 2;\ 25 = 5 \cdot 5$

**(a)**
$$\frac{10}{3} \cdot \frac{18}{25} = \frac{2 \cdot 5}{3} \cdot \frac{3 \cdot 3 \cdot 2}{5 \cdot 5}$$

Use the Reduction Property to divide out common factors:
$$= \frac{2 \cdot \cancel{5}}{\cancel{3}} \cdot \frac{\cancel{3} \cdot 3 \cdot 2}{\cancel{5} \cdot 5}$$
$$= \frac{2}{1} \cdot \frac{3 \cdot 2}{5}$$

Multiply:
$$= \frac{12}{5}$$

**Work Smart**

In Example 2(a), rather than factoring each numerator and denominator individually and then dividing out like factors, we might proceed as follows:

$$\frac{10}{3} \cdot \frac{18}{25} = \frac{\overset{2}{\cancel{10}}}{\underset{1}{\cancel{3}}} \cdot \frac{\overset{6}{\cancel{18}}}{\underset{5}{\cancel{25}}}$$
$$= \frac{2 \cdot 6}{1 \cdot 5}$$

Multiply numerators: Multiply denominators:
$$= \frac{12}{5}$$

Notice that we use slashes in the same direction when writing in lowest terms and put the remaining factor next to the reduced factor.

**(b)** We rewrite the division problem as a multiplication problem by multiplying the numerator, $\frac{14}{5}$, by the reciprocal of the denominator, $\frac{21}{10}$. The reciprocal of $\frac{21}{10}$ is $\frac{10}{21}$.

$$\frac{\frac{14}{5}}{\frac{21}{10}} = \frac{14}{5} \cdot \frac{10}{21}$$
$$= \frac{7 \cdot 2}{5} \cdot \frac{5 \cdot 2}{7 \cdot 3}$$

Use the Reduction Property to divide out common factors:
$$= \frac{\cancel{7} \cdot 2}{\cancel{5}} \cdot \frac{\cancel{5} \cdot 2}{\cancel{7} \cdot 3}$$
$$= \frac{2}{1} \cdot \frac{2}{3} = \frac{4}{3}$$

**QUICK ✓** *Perform the indicated operation. Express your answer in lowest terms.*

**3.** $\frac{5}{7} \cdot \left(-\frac{21}{10}\right)$ **4.** $\frac{35}{15} \cdot \frac{3}{14}$ **5.** $\frac{\frac{4}{5}}{\frac{12}{25}}$ **6.** $\frac{24}{35} \div \left(-\frac{8}{7}\right)$

## 3 Add and Subtract Rational Numbers

We now review addition and subtraction of rational numbers.

**In Words**

To add two rational numbers with a common denominator, add the numerators and write the sum over the common denominator.

**Adding or Subtracting Rational Numbers**

**Step 1:** If $\frac{a}{c}$ and $\frac{b}{c}$, $c \neq 0$, are two rational numbers, then $\frac{a}{c} + \frac{b}{c} = \frac{a+b}{c}$ and $\frac{a}{c} - \frac{b}{c} = \frac{a-b}{c}$.

**Step 2:** Write the result in lowest terms.

**EXAMPLE 3** **Adding or Subtracting Rational Numbers with Common Denominators**

Perform the indicated operation. Be sure to express the result in lowest terms.

**(a)** $\frac{7}{24} + \frac{11}{24}$ **(b)** $\frac{2}{15} - \frac{7}{15}$

**Solution**

**(a)**

$$\frac{7}{24} + \frac{11}{24} = \frac{7 + 11}{24}$$

$$= \frac{18}{24}$$

Factor numerator and denominator: $= \frac{6 \cdot 3}{6 \cdot 4}$

Divide out common factors: $= \frac{3}{4}$

**(b)**

$$\frac{2}{15} - \frac{7}{15} = \frac{2 - 7}{15}$$

$$= \frac{-5}{15}$$

Factor numerator and denominator: $= \frac{-1 \cdot 5}{3 \cdot 5}$

Divide out common factors: $= \frac{-1}{3} = -\frac{1}{3}$

**QUICK ✓** *Perform the indicated operation. Express your answer in lowest terms.*

**7.** $\frac{11}{12} + \frac{5}{12}$ **8.** $\frac{3}{18} - \frac{13}{18}$

What if the denominators of the rational numbers to be added or subtracted are not the same? In this case, we must rewrite each rational number over a *least common denominator.* The **least common denominator (LCD)** is the smallest integer that is a multiple of each denominator in the rational numbers to be added or subtracted.

**EXAMPLE 4** **How to Find the Least Common Denominator**

Find the least common denominator of the rational numbers $\frac{7}{30}$ and $\frac{5}{12}$. Then rewrite each rational number with the least common denominator.

**Step-by-Step Solution**

**Step 1:** Factor each denominator.

It is easier to see the common factors by listing them directly over each other.

$$30 = 5 \cdot 6 = 5 \cdot 3 \cdot 2$$

$$12 = 3 \cdot 4 = \quad 3 \cdot 2 \cdot 2$$

**Work Smart**

Line up the factors vertically to find the LCD.

**Step 2:** Write down the common factor(s) between each denominator. Then copy the remaining factors. The product of these factors is the least common denominator (LCD).

The common factors are 3 and 2. The remaining factors are 5 and 2.

$$\text{LCD} = 3 \cdot 2 \cdot 5 \cdot 2 = 60$$

We rewrite $\frac{7}{30}$ with a denominator of 60 by multiplying the numerator and denominator by 2. We rewrite $\frac{5}{12}$ with a denominator of 60 by multiplying the numerator and denominator by 5.

**Work Smart**

When we multiply $\frac{7}{30}$ by $\frac{2}{2}$, we are using the Multiplicative Identity Property of real numbers where $1 = \frac{2}{2}$.

$$\frac{7}{30} = \frac{7}{30} \cdot \frac{2}{2} = \frac{7 \cdot 2}{30 \cdot 2} = \frac{14}{60}$$

$$\frac{5}{12} = \frac{5}{12} \cdot \frac{5}{5} = \frac{5 \cdot 5}{12 \cdot 5} = \frac{25}{60}$$

**QUICK ✓** *Find the least common denominator (LCD) of each pair of rational numbers. Then rewrite each rational number with the LCD.*

**9.** $\frac{3}{25}$ and $\frac{2}{15}$ **10.** $\frac{5}{18}$ and $-\frac{1}{63}$

Now that we know how to obtain the least common denominator, we can discuss how to add or subtract rational numbers that have unlike denominators.

**EXAMPLE 5** **How to Add or Subtract Fractions Using the Least Common Denominator**

Perform the indicated operation:

**(a)** $\frac{5}{2} - \frac{4}{3}$ **(b)** $\frac{5}{28} - \frac{5}{12}$

**Step-by-Step Solution**

**(a)** $\frac{5}{2} - \frac{4}{3}$

| | |
|---|---|
| **Step 1:** Find the least common denominator. | Each denominator is prime, so the LCD $= 2 \cdot 3 = 6$. |
| **Step 2:** Rewrite each rational number with the common denominator. | $\frac{5}{2} - \frac{4}{3} = \frac{5}{2} \cdot \frac{3}{3} - \frac{4}{3} \cdot \frac{2}{2}$ <br> $= \frac{15}{6} - \frac{8}{6}$ |
| **Step 3:** Add or subtract the rational numbers found in Step 2. | $= \frac{15 - 8}{6}$ <br> $= \frac{7}{6}$ |
| **Step 4:** Write the result in lowest terms, if necessary. | The rational number is already in lowest terms. |

**(b)** $\dfrac{5}{28} - \dfrac{5}{12}$

| | |
|---|---|
| **Step 1:** Find the least common denominator. | $28 = 7 \cdot 4$<br>$12 = \quad 4 \cdot 3$<br>The LCD $= 4 \cdot 7 \cdot 3 = 84$ |
| **Step 2:** Rewrite each rational number with the common denominator. | $\dfrac{5}{28} - \dfrac{5}{12} = \dfrac{5}{28} \cdot \dfrac{3}{3} - \dfrac{5}{12} \cdot \dfrac{7}{7}$<br>$= \dfrac{15}{84} - \dfrac{35}{84}$ |
| **Step 3:** Add or subtract the rational numbers found in Step 2. | $= \dfrac{15 - 35}{84}$<br>$= \dfrac{-20}{84}$ |
| **Step 4:** Write the result in lowest terms, if necessary. | $= \dfrac{-5 \cdot 4}{21 \cdot 4}$<br>$= \dfrac{-5}{21} = -\dfrac{5}{21}$ |

We summarize the steps used in Example 5 below.

### Adding or Subtracting Rational Numbers with Unlike Denominators

**Step 1:** Find the least common denominator.

**Step 2:** Rewrite each rational number with the common denominator.

**Step 3:** Add or subtract the rational numbers found in Step 2.

**Step 4:** Write the result in lowest terms, if necessary.

**QUICK ✓** *Perform the indicated operation. Express your answers in lowest terms.*

**11.** $\dfrac{3}{4} + \dfrac{1}{5}$ **12.** $\dfrac{3}{20} + \dfrac{2}{15}$ **13.** $\dfrac{5}{14} - \dfrac{11}{21}$

# Getting Ready for Chapter 6 Exercises

***For Extra Help:***

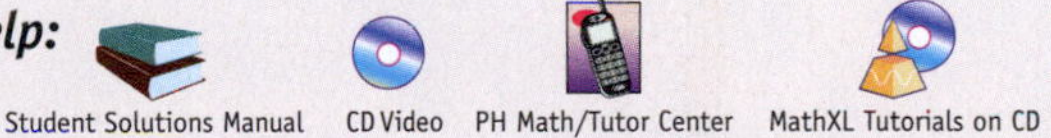

Student Solutions Manual · CD Video · PH Math/Tutor Center · MathXL Tutorials on CD · MathXL® · MyMathLab

## Concepts and Vocabulary

*In Problems 1 and 2, fill in the blanks.*

**1.** When a rational number is written so that there are not any common factors in the numerator and the denominator, we say that the rational number is in _____ _____.

**2.** The _____ _____ _____ is the smallest integer that is a multiple of each denominator in the rational number to be added or subtracted.

3. Explain how to reduce a rational number to lowest terms.
4. Explain how to find the least common denominator between two rational numbers.
5. Explain how to add two rational numbers that do not have a common denominator.
6. Explain how to divide two rational numbers.

## Building Skills

*In Problems 7–12, write each rational number in lowest terms.*

7. $\frac{4}{12}$ 8. $\frac{6}{18}$ 9. $\frac{-15}{35}$ 10. $\frac{-12}{28}$ 11. $\frac{-50}{-10}$ 12. $\frac{81}{-27}$

*In Problems 13–24, multiply or divide the rational numbers. Express each product or quotient as a rational number in lowest terms.*

13. $\frac{3}{4} \cdot \frac{20}{9}$ 14. $\frac{2}{3} \cdot \frac{15}{6}$ 15. $-\frac{5}{6} \cdot \frac{18}{5}$ 16. $-\frac{9}{8} \cdot \frac{16}{3}$

17. $\frac{5}{8} \cdot \frac{2}{15}$ 18. $\frac{3}{14} \cdot \frac{7}{12}$ 19. $\frac{5}{2} \div \frac{25}{4}$ 20. $\frac{2}{3} \div \frac{8}{9}$

21. $\dfrac{\frac{-6}{5}}{\frac{8}{15}}$ 22. $\dfrac{\frac{12}{7}}{-\frac{18}{21}}$ 23. $\dfrac{-\frac{9}{2}}{\frac{-3}{4}}$ 24. $\dfrac{\frac{10}{-7}}{\frac{-5}{14}}$

*In Problems 25–43, add or subtract the rational numbers. Express each sum or difference as a rational number in lowest terms.*

25. $\frac{5}{3} + \frac{1}{3}$ 26. $\frac{3}{4} + \frac{9}{4}$ 27. $\frac{11}{6} - \frac{1}{6}$ 28. $\frac{19}{6} - \frac{5}{6}$

29. $\frac{3}{4} + \frac{1}{5}$ 30. $\frac{-1}{7} + \frac{4}{9}$ 31. $\frac{1}{4} + \frac{5}{6}$ 32. $\frac{5}{8} + \frac{5}{12}$

33. $-\frac{3}{10} - \frac{7}{15}$ 34. $\frac{7}{16} - \frac{9}{20}$ 35. $\frac{5}{18} - \frac{11}{15}$ 36. $\frac{5}{24} + \frac{7}{32}$

37. $-\frac{7}{8} + \frac{3}{10}$ 38. $-\frac{7}{12} + \frac{2}{15}$ 39. $\frac{7}{24} - \frac{3}{20}$ 40. $\frac{3}{28} + \frac{5}{12}$

41. $-\frac{7}{9} - \frac{2}{15}$ 42. $-\frac{5}{18} - \frac{1}{45}$ 43. $\frac{-4}{25} - \frac{7}{30}$

CHAPTER

# 6 Rational Expressions and Rational Functions

We all enjoy clean lakes and rivers, but at some point the additional cost of cleaning becomes prohibitive. Environmentalists use rational expressions to describe the costs versus the benefits of cleaning the environment. See Problem 57 in Section 6.4.

## OUTLINE



## The Big Picture: Putting It Together

In Chapter 5, we learned how to add, subtract, multiply, and divide polynomials. We then learned how to factor polynomial expressions and use the result (along with the Zero-Product Property) to solve some polynomial equations.

We will do the same thing with *rational expressions*. A rational expression is a polynomial divided by another polynomial. Factoring comes in handy when working with rational expressions. The methods for reducing, adding, subtracting, multiplying, and dividing rational expressions are identical to the methods used to do these operations on rational numbers. The point is this—the algebra of rational expressions can be thought of as a generic arithmetic of rational numbers.

Beginning with this chapter, we introduce a new category of problems called Synthesis Review. The synthesis review problems are created to assist you in seeing the "big picture." For example, we might ask you to add various objects (polynomials, rational expressions, and so on). We then ask for you to discuss similarities and differences in the performing the operation on the objects.

# 6.1 Multiplying and Dividing Rational Expressions

**OBJECTIVES**

1. Determine the Domain of a Rational Expression
2. Simplify Rational Expressions
3. Multiply Rational Expressions
4. Divide Rational Expressions
5. Work with Rational Functions

***Preparing for Multiplying and Dividing Rational Expressions***

*Before getting started, take the following readiness quiz. If you get a problem wrong, go back to the section cited and review the material.*

**1.** Factor: $2x^2 - 11x - 21$ [Section 5.5, pp. 401–406]

**2.** Solve: $q^2 - 16 = 0$ [Section 5.7, pp. 423–426]

**3.** Determine the reciprocal of $\frac{5}{2}$. [Section R.3, p. 24]

**4.** Explain what *domain* means. [Section 2.2, pp. 150–152]

If we form the quotient of two polynomials, then we have a **rational expression.** Some examples of rational expressions are

**(a)** $\frac{x-5}{2x+1}$ **(b)** $\frac{x^2-7x-18}{x^2-4}$ **(c)** $\frac{2a^2+5ab+2b^2}{a^2-6ab+8b^2}$ **(d)** $\frac{1}{x-3}$

Expressions (a), (b), and (d) are rational expressions in one variable, $x$, while expression (c) is a rational expression in two variables, $a$ and $b$.

Rational expressions are described the same way as rational numbers. In the expression $\frac{x-5}{2x+1}$, we call $x - 5$ the **numerator** and $2x + 1$ the **denominator.** When the numerator and denominator have no common factors (except 1 and $-1$), we say that the rational expression is expressed in **lowest terms,** or **simplified.**

## 1 Determine the Domain of a Rational Expression

To find the domain of a rational expression, we determine all values of the variable that cause the denominator to equal 0 and exclude these values from the domain because division by 0 is not defined.

### EXAMPLE 1 Determining the Domain of a Rational Expression

For each of the following rational expressions, determine the domain.

**(a)** $\frac{2x}{x+3}$ **(b)** $\frac{p^2+5p+6}{p^2-4}$

**Solution**

**(a)** We want to find all values of $x$ that cause $x + 3$ to equal 0, so we solve

$$x + 3 = 0$$

Subtract 3 from both sides: $x = -3$

So $-3$ causes the denominator, $x + 3$, to equal 0. Therefore, the domain of $\frac{2x}{x+3}$ is $\{x|x \neq -3\}$.

**Work Smart**

Because we are working with the set of real numbers, the domain of a variable is understood to be all real numbers except those listed. For example, the notation $\{x|x \neq -3\}$ means $x$ is any real number except $-3$.

**(b)** We want to find all values of $p$ that cause $p^2 - 4$ to equal 0.

$$p^2 - 4 = 0$$

Factor the difference of two squares: $(p - 2)(p + 2) = 0$

Zero-Product Property: $p - 2 = 0$ or $p + 2 = 0$

$p = 2$ or $p = -2$

The domain of $\frac{p^2+5p+6}{p^2-4}$ is $\{p|p \neq -2, p \neq 2\}$.

***Preparing for...Answers***

**1.** $(2x + 3)(x - 7)$ **2.** $\{-4, 4\}$ **3.** $\frac{2}{5}$ **4.** The set of all inputs for which an algebraic expression is defined.

**QUICK ✓** *Determine the domain of the rational expression.*

**1.** $\dfrac{x-4}{x+6}$ **2.** $\dfrac{z^2-9}{z^2+3z-28}$

## 2 Simplify Rational Expressions

A rational expression is simplified by completely factoring the numerator and the denominator and dividing out any common factors using the Reduction Property.

$$\frac{ac}{bc} = \frac{a\cancel{c}}{b\cancel{c}} = \frac{a}{b} \qquad \text{if } b \neq 0, c \neq 0$$

We simplify rational expressions in the same way that we write rational numbers in lowest terms. For example, the rational number $\frac{12}{20}$ is not written in lowest terms because there is a common factor of 4. We write $\frac{12}{20}$ in lowest terms as follows: $\frac{12}{20} = \frac{4 \cdot 3}{4 \cdot 5} = \frac{3}{5}$.

### EXAMPLE 2 Simplifying a Rational Expression

Simplify each rational expression:

**(a)** $\dfrac{x^2+2x-15}{2x^2-3x-9}$ **(b)** $\dfrac{q^3-8}{3q^2-6q}$

**Solution**

**(a)** Factor the numerator and denominator and divide out common factors using the Reduction Property.

$$\frac{x^2+2x-15}{2x^2-3x-9} = \frac{(x+5)\cancel{(x-3)}}{(2x+3)\cancel{(x-3)}}$$

$$= \frac{x+5}{2x+3} \qquad x \neq -\frac{3}{2}, x \neq 3$$

**(b)** Factor the numerator and denominator. Then divide out common factors.

$$\frac{q^3-8}{3q^2-6q} = \frac{\cancel{(q-2)}(q^2+2q+4)}{3q\cancel{(q-2)}}$$

$$= \frac{q^2+2q+4}{3q} \qquad q \neq 0, q \neq 2$$

**Work Smart**

When we divide out like factors, the quotient is 1. For example,

$$\frac{\overset{1}{\cancel{(x+3)}}}{2x\underset{1}{\cancel{(x+3)}}} = \frac{1}{2x}$$

To keep the rational expressions equivalent, we must restrict from the domain all values of the variable that are not in the domain of the *original* rational expression. Consider Example 2(a). We include the restriction $x \neq -\frac{3}{2}, x \neq 3$ for two reasons:

1. To remind us of the restrictions on the variable $x$.
2. To keep the rational expressions equal. Without the restriction $x \neq -\frac{3}{2}, x \neq 3$, the expression $\frac{x^2+2x-15}{2x^2-3x-9}$ is not equal to $\frac{x+5}{2x+3}$ because in $\frac{x^2+2x-15}{2x^2-3x-9}$, $x$ cannot take on the value 3, while in $\frac{x+5}{2x+3}$, $x$ can take on the value 3. By not allowing $x$ to equal 3 in both instances, the expressions remain equal.

**Work Smart**

$$\frac{x^2+2x-15}{2x^2-3x-9} = \frac{x+5}{2x+3}$$

is not true unless we include the restrictions

$$x \neq -\frac{3}{2}, x \neq 3$$

because $x$ **cannot** take on the value 3 in $\frac{x^2+2x-15}{2x^2-3x-9}$ while $x$ **can** take on the value 3 in $\frac{x+5}{2x+3}$.

The same logic is used to justify the restrictions on $q$ in Example 2(b). For the remainder of the text, we shall not include the restrictions on the variable, but you should be aware that the restrictions are necessary to maintain equality.

Sometimes, we can obtain common factors in the numerator and denominator of a rational expression by factoring $-1$ out of one of the factors. Consider the expressions $3 - 4x$ and $4x - 3$. If we factor $-1$ out of $3 - 4x$, we obtain $-1(-3 + 4x)$ or $-1(4x - 3)$ so that we can now divide out the common factor, $4x - 3$.

### EXAMPLE 3 Simplifying a Rational Expression

Simplify: $\dfrac{3x^2 + 11x - 4}{1 - 3x}$

**Solution**

$$\frac{3x^2 + 11x - 4}{1 - 3x} = \frac{(3x - 1)(x + 4)}{1 - 3x}$$

Factor $-1$ from $1 - 3x$; Divide out common factors:

$$= \frac{\cancel{(3x - 1)}(x + 4)}{-1\cancel{(3x - 1)}}$$

$$= \frac{(x + 4)}{-1}$$

$\dfrac{a}{-1} = -a$:

$$= -(x + 4)$$

**Work Smart**

A common error for students to make is to divide out terms rather than dividing out factors. **When simplifying, we can only divide out common factors, not common terms!**

**WRONG!:** $\dfrac{x + 1}{x} = \dfrac{\cancel{x} + 1}{\cancel{x}} = 1$ **WRONG!:** $\dfrac{x^2 + x + 2}{x + 2} = \dfrac{x^2 + \cancel{x} + \cancel{2}}{\cancel{x} + \cancel{2}} = x^2$

If you aren't quite sure whether you can simplify, try the computation with actual numbers and see if it works. For example, does $\dfrac{4}{3} = \dfrac{3 + 1}{3} = \dfrac{\cancel{3} + 1}{\cancel{3}} = 1$? NO! So, $\dfrac{x + 1}{x} \neq 1$.

**QUICK ✓** *Simplify each rational expression.*

**3.** $\dfrac{x^2 - 7x + 12}{x^2 + 4x - 21}$ **4.** $\dfrac{z^3 - 64}{2z^2 - 3z - 20}$ **5.** $\dfrac{3w^2 + 13w - 10}{2 - 3w}$

## 3 Multiply Rational Expressions

We show how to multiply rational expressions in the next example.

### EXAMPLE 4 How to Multiply Rational Expressions

Multiply $\dfrac{x^2 + 2x - 15}{x + 1} \cdot \dfrac{x^2 + 7x}{x^2 + 4x - 21}$. Simplify the product.

**Step-by-Step Solution**

**Step 1:** Completely factor each polynomial in the numerator and denominator.

$$\frac{x^2 + 2x - 15}{x + 1} \cdot \frac{x^2 + 7x}{x^2 + 4x - 21} = \frac{(x + 5)(x - 3)}{x + 1} \cdot \frac{x(x + 7)}{(x + 7)(x - 3)}$$

**Step 2:** Divide out common factors in the numerator and denominator.

$$= \frac{(x+5)\cancel{(x-3)}}{x+1} \cdot \frac{\cancel{x(x+7)}}{\cancel{(x+7)}\,\cancel{(x-3)}}$$

**Step 3:** Multiply.

$$= \frac{x(x+5)}{x+1}$$

The steps for multiplying rational expressions are the same as the steps for multiplying rational numbers.

**In Words**
To multiply two rational expressions, factor each polynomial, divide out common factors, and then write the expression as a single fraction in factored form.

### Multiplying Rational Expressions

**Step 1:** Completely factor each polynomial in the numerator and denominator.

**Step 2:** Divide out common factors in the numerator and denominator.

**Step 3:** Use the fact that if $\frac{a}{b}$ and $\frac{c}{d}$, $b \neq 0$, $d \neq 0$, are two rational expressions, then $\frac{a}{b} \cdot \frac{c}{d} = \frac{ac}{bd}$ to multiply the rational expressions.

When using the steps given above, always leave your answer in factored form because the factored form is required when solving rational equations (Section 6.4) and rational inequalities (Section 6.5).

**QUICK ✓** *Multiply the rational expressions. Simplify the product if possible.*

**6.** $\frac{p^2 - 9}{p^2 + 5p + 6} \cdot \frac{3p^2 - p - 2}{2p - 6}$

## EXAMPLE 5 Multiplying Rational Expressions

Multiply each of the following rational expressions. Simplify the product if possible.

**(a)** $\frac{y^2 + 2y + 1}{3y^2 + y - 2} \cdot \frac{2 - 3y}{y^2 + 5y + 4}$ **(b)** $\frac{p^2 + 5pq + 6q^2}{2p^2 + 7pq + 3q^2} \cdot \frac{2p + q}{3p + 6q}$

**Solution**

**(a)**

$$\frac{y^2 + 2y + 1}{3y^2 + y - 2} \cdot \frac{2 - 3y}{y^2 + 5y + 4} = \frac{(y+1)(y+1)}{(3y-2)(y+1)} \cdot \frac{-1(3y-2)}{(y+4)(y+1)}$$

Factor the numerator and denominator:

$$= \frac{\cancel{(y+1)}\,\cancel{(y+1)}}{\cancel{(3y-2)}\,\cancel{(y+1)}} \cdot \frac{-1\cancel{(3y-2)}}{(y+4)\cancel{(y+1)}}$$

Divide out common factors:

$$= -\frac{1}{y+4}$$

**Work Smart: Study Skills**
It is a good idea to use a different symbol for each factor that divides out. Or use colored pencils to highlight the like factors that divide out.

**(b)**

$$\frac{p^2 + 5pq + 6q^2}{2p^2 + 7pq + 3q^2} \cdot \frac{2p + q}{3p + 6q} = \frac{(p+2q)(p+3q)}{(2p+q)(p+3q)} \cdot \frac{2p+q}{3(p+2q)}$$

Factor the numerator and denominator:

$$= \frac{\cancel{(p+2q)}\,\cancel{(p+3q)}}{\cancel{2p+q}\,\cancel{(p+3q)}} \cdot \frac{\cancel{2p+q}}{3\cancel{(p+2q)}}$$

Divide out common factors:

$$= \frac{1}{3}$$

Did you notice in Example 5(b) that all the factors in the numerator divide out? This means that a factor of 1 remains in the numerator.

**QUICK ✓** *Multiply the rational expressions. Simplify the product, if possible.*

**7.** $\dfrac{2x+8}{2x^2+11x+12}\cdot\dfrac{2x^2-3x-9}{6-2x}$ **8.** $\dfrac{m^2+2mn+n^2}{2m^2+3mn+n^2}\cdot\dfrac{2m^2-5mn-3n^2}{3n-m}$

## 4 Divide Rational Expressions

The rule for dividing rational expressions is the same as the rule for dividing rational numbers.

**In Words**

To divide two rational expressions, multiply the rational expression in the numerator by the reciprocal of the rational expression in the denominator.

**DIVIDING RATIONAL EXPRESSIONS**

To divide rational expressions, use the fact that if $\dfrac{a}{b}$ and $\dfrac{c}{d}$, $b \neq 0$, $c \neq 0$, $d \neq 0$, are two rational expressions, then $\dfrac{\frac{a}{b}}{\frac{c}{d}} = \dfrac{a}{b}\cdot\dfrac{d}{c} = \dfrac{ad}{bc}$, and then follow the steps for multiplying two rational expressions.

### EXAMPLE 6 Dividing Rational Expressions

Divide each of the following rational expressions. Simplify the quotient, if possible.

**(a)** $\dfrac{\frac{20x^5}{3y}}{\frac{4x^2}{15y^5}}$ **(b)** $\dfrac{\frac{x^2-4x-12}{4x^3-6x^2}}{\frac{x^3+8}{2x^3-4x^2+8x}}$

**Solution**

For parts (a) and (b), rewrite the division problem as a multiplication problem by multiplying the numerator by the reciprocal of the denominator.

**(a)**

$$\dfrac{\frac{20x^5}{3y}}{\frac{4x^2}{15y^5}} = \dfrac{20x^5}{3y}\cdot\dfrac{15y^5}{4x^2}$$

$$= \dfrac{20\cdot 15\cdot x^5\cdot y^5}{3\cdot 4\cdot y\cdot x^2}$$

Divide out like factors: $$= \dfrac{\overset{5}{\cancel{20}}\cdot\overset{5}{\cancel{15}}}{\underset{1}{\cancel{3}}\cdot\underset{1}{\cancel{4}}}x^{5-2}y^{5-1}$$

Simplify: $$= 25x^3y^4$$

**Work Smart**

When you see $\dfrac{\frac{20x^5}{3y}}{\frac{4x^2}{15y^5}}$, you may find it helpful to think

$$\frac{20x^5}{3y} \div \frac{4x^2}{15y^5}$$

**(b)**

$$\dfrac{\frac{x^2-4x-12}{4x^3-6x^2}}{\frac{x^3+8}{2x^3-4x^2+8x}} = \dfrac{x^2-4x-12}{4x^3-6x^2}\cdot\dfrac{2x^3-4x^2+8x}{x^3+8}$$

Factor the numerator and denominator: $$= \dfrac{(x-6)(x+2)}{2x^2(2x-3)}\cdot\dfrac{2x(x^2-2x+4)}{(x+2)(x^2-2x+4)}$$

Divide out common factors: $$= \dfrac{(x-6)\cancel{(x+2)}}{\cancel{2}x^{\cancel{2}}(2x-3)}\cdot\dfrac{\cancel{2x}\cancel{(x^2-2x+4)}}{\cancel{(x+2)}\cancel{(x^2-2x+4)}}$$

Simplify: $$= \dfrac{x-6}{x(2x-3)}$$

**QUICK ✓** *Divide the rational expressions. Simplify the quotient, if possible.*

**9.** $\dfrac{\dfrac{12a^4}{5b^2}}{\dfrac{4a^2}{15b^5}}$

**10.** $\dfrac{\dfrac{m^2 - 5m}{m - 7}}{\dfrac{2m}{m^2 - 6m - 7}}$

## 5 Work with Rational Functions

Ratios of integers such, as $\frac{3}{5}$, are called *rational numbers* and ratios of polynomial expressions are called *rational expressions*. Similarly, ratios of polynomial functions are called *rational functions*.

**DEFINITION**

A **rational function** is a function of the form

$$R(x) = \frac{p(x)}{q(x)}$$

where $p$ and $q$ are polynomial functions and $q$ is not the zero polynomial. The domain consists of all real numbers except those for which the denominator $q$ is 0.

### EXAMPLE 7 Finding the Domain of a Rational Function

Find the domain of $R(x) = \dfrac{x + 2}{x^2 - 5x - 14}$.

**Solution**

The domain is the set of all real numbers such that the denominator, $x^2 - 5x - 14$, does not equal zero. We will find the values of $x$ such that $x^2 - 5x - 14 = 0$ and exclude these values from the domain.

$$x^2 - 5x - 14 = 0$$

Factor: $(x - 7)(x + 2) = 0$

Set each factor to 0: $x - 7 = 0$ or $x + 2 = 0$

Solve: $x = 7$ or $x = -2$

The values $x = 7$ and $x = -2$ cause the denominator to equal 0, so the domain of $R$ is $\{x | x \neq 7, x \neq -2\}$.

**QUICK ✓** *Find the domain of the rational function.*

**11.** $R(x) = \dfrac{2x}{x^2 + x - 30}$

We can multiply and divide rational functions just as we multiplied and divided polynomial functions.

### EXAMPLE 8 Multiplying and Dividing Rational Functions

Given that $f(x) = \dfrac{x^2 - 4}{3x^2 + 9x}$, $g(x) = \dfrac{x + 3}{x^2 - 2x - 8}$, and $h(x) = \dfrac{2x^2 + 7x + 6}{x^2 + 5x}$, find

**(a)** $R(x) = f(x) \cdot g(x)$ **(b)** $H(x) = \dfrac{f(x)}{h(x)}$

and state the domain of each function.

**Solution**

**(a)**

$$R(x) = f(x) \cdot g(x) = \frac{x^2 - 4}{3x^2 + 9x} \cdot \frac{x + 3}{x^2 - 2x - 8}$$

Factor the numerator and denominator:
$$= \frac{(x + 2)(x - 2)}{3x(x + 3)} \cdot \frac{x + 3}{(x - 4)(x + 2)}$$

Divide out common factors:
$$= \frac{\cancel{(x + 2)}(x - 2)}{3x\cancel{(x + 3)}} \cdot \frac{\cancel{x + 3}}{(x - 4)\cancel{(x + 2)}}$$

Simplify:
$$= \frac{x - 2}{3x(x - 4)}$$

The domain of $f(x)$ is $\{x \mid x \neq -3, x \neq 0\}$. The domain of $g(x)$ is $\{x \mid x \neq -2, x \neq 4\}$. Therefore, the domain of $R(x)$ is $\{x \mid x \neq -3, x \neq -2, x \neq 0, x \neq 4\}$.

**Work Smart**

In Example 8(b), $3x^2 + 9x$ cannot equal 0. Why? In addition, $2x^2 + 7x + 6$ cannot equal 0. Why? $x^2 + 5x$ cannot equal 0. Why?

**(b)**

$$H(x) = \frac{f(x)}{h(x)} = \frac{\dfrac{x^2 - 4}{3x^2 + 9x}}{\dfrac{2x^2 + 7x + 6}{x^2 + 5x}}$$

$$= \frac{x^2 - 4}{3x^2 + 9x} \cdot \frac{x^2 + 5x}{2x^2 + 7x + 6}$$

Factor numerator and denominator:
$$= \frac{(x - 2)(x + 2)}{3x(x + 3)} \cdot \frac{x(x + 5)}{(2x + 3)(x + 2)}$$

Divide out common factors:
$$= \frac{(x - 2)\cancel{(x + 2)}}{3\cancel{x}(x + 3)} \cdot \frac{\cancel{x}(x + 5)}{(2x + 3)\cancel{(x + 2)}}$$

Simplify:
$$= \frac{(x - 2)(x + 5)}{3(x + 3)(2x + 3)}$$

The domain of $f(x)$ is $\{x \mid x \neq -3, x \neq 0\}$. The domain of $h(x)$ is $\{x \mid x \neq -5, x \neq 0\}$. Because the denominator of $H(x)$ cannot equal 0, we must exclude those values of $x$ such that $2x^2 + 7x + 6 = 0$. These values are $x = -2$ and $x = -\frac{3}{2}$. Therefore, the domain of $H(x)$ is $\{x \mid x \neq -5, x \neq -3, x \neq -2, x \neq -\frac{3}{2}, x \neq 0\}$.

**QUICK ✓**

12. Given that $f(x) = \dfrac{x^2 - 4x - 5}{3x - 5}$, $g(x) = \dfrac{3x^2 + 4x - 15}{x^2 - 2x - 15}$, and $h(x) = \dfrac{4x^2 + 7x + 3}{9x^2 - 15x}$, find

**(a)** $R(x) = f(x) \cdot g(x)$ **(b)** $H(x) = \dfrac{f(x)}{h(x)}$

and state the domain of each function.

# 6.1 Exercises

**For Extra Help:**

Student Solutions Manual

CD Video

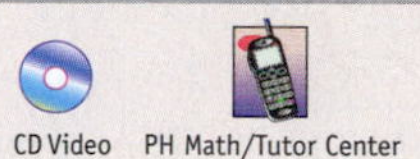
PH Math/Tutor Center

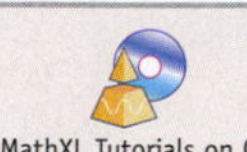
MathXL Tutorials on CD

MathXL®   MyMathLab

## Concepts and Vocabulary

*In Problems 1 and 2, fill in the blanks.*

**1.** The quotient of two polynomials is called a _____ _____.

**2.** In the expression $\frac{x+3}{3x-5}$, we call $x + 3$ the _____, and $3x - 5$ is called the _____.

*In Problems 3 and 4, answer True or False to each statement.*

**3.** When the numerator and denominator have no common factors (except 1 and $-1$), we say that the rational expression is simplified.

**4.** The domain of all rational functions is the set of all real numbers.

**5.** Define rational expression. Explain how a rational expression is related to a rational number.

**6.** What does it mean when we say that a rational expression is simplified?

**7.** Explain why we can only divide out common factors, not common terms.

**8.** Why is $\frac{\sqrt{x}}{x+1}$ not a rational expression?

## Building Skills

*In Problems 9–18, state the domain of each rational expression.*

**9.** $\frac{3}{x+5}$ **10.** $\frac{4}{x-7}$ **11.** $\frac{x-1}{x^2-6x-16}$ **12.** $\frac{2x+1}{x^2+4x-45}$

**13.** $\frac{p^2-4}{2p^2+p-10}$ **14.** $\frac{m^2+5m+6}{3m^2+4m-4}$ **15.** $\frac{x+1}{x^2+1}$ **16.** $\frac{x-2}{x^2+4}$

**17.** $\frac{3x-2}{(x-1)^2}$ **18.** $\frac{x+5}{x^2+8x+16}$

*In Problems 19–38, simplify each rational expression.*

**19.** $\frac{2x+8}{x^2-16}$ **20.** $\frac{x^2-3x}{x^2-9}$ **21.** $\frac{p^2+4p+3}{p+1}$

**22.** $\frac{a^2-2a-24}{a+4}$ **23.** $\frac{5x+25}{x^3+5x^2}$ **24.** $\frac{6x-42}{x^3-7x^2}$

**25.** $\frac{q^2-3q-18}{q^2-8q+12}$ **26.** $\frac{w^2+5w-14}{w^2+6w-16}$ **27.** $\frac{2y^2-3y-20}{2y^2+15y+25}$

**28.** $\frac{3n^2+n-2}{3n^2-20n+12}$ **29.** $\frac{9-x^2}{x^2+2x-15}$ **30.** $\frac{25-k^2}{k^2+2k-35}$

**31.** $\frac{x^3+2x^2-8x}{2x^4-32x^2}$ **32.** $\frac{2z^2-10z-28}{4z^3-32z^2+28z}$ **33.** $\frac{x^2-xy-6y^2}{x^2-4y^2}$

**34.** $\frac{a^2+5ab+4b^2}{a^2+8ab+16b^2}$ **35.** $\frac{x^3-5x^2+3x-15}{x^2-10x+25}$ **36.** $\frac{v^3+3v^2-5v-15}{v^2+6v+9}$

**37.** $\dfrac{x^3 + 8}{x^2 - 5x - 14}$

**38.** $\dfrac{27q^3 + 1}{6q^2 - 7q - 3}$

*In Problems 39–50, multiply each rational expression. Simplify the product, if possible.*

**39.** $\dfrac{3x}{x^2 - x - 12} \cdot \dfrac{x - 4}{12x^2}$

**40.** $\dfrac{5x^2}{x + 3} \cdot \dfrac{x^2 + 7x + 12}{20x}$

**41.** $\dfrac{2x^2 - x - 6}{x^2 + 3x - 4} \cdot \dfrac{x^2 - x - 20}{2x^2 - 7x - 15}$

**42.** $\dfrac{3x^2 + 14x - 5}{x^2 + x - 30} \cdot \dfrac{x^2 - 2x - 15}{3x^2 + 8x - 3}$

**43.** $\dfrac{x^2 - 9}{x^2 - 25} \cdot \dfrac{x^2 - 2x - 15}{x^2 + 4x - 21}$

**44.** $\dfrac{p^2 - 16}{p^2 - 25} \cdot \dfrac{p^2 + 2p - 24}{p^2 + 3p - 4}$

**45.** $\dfrac{2q^2 - 5q - 3}{3q^2 + 19q + 6} \cdot \dfrac{3q^2 + 7q + 2}{3 - q}$

**46.** $\dfrac{2y^2 - 5y - 12}{2y^2 - y - 6} \cdot \dfrac{4y^2 - 5y - 6}{4 - y}$

**47.** $\dfrac{x^2 - 5x + 6}{x^2 + 2x - 8} \cdot (x + 4)$

**48.** $\dfrac{p^2 - 4p - 5}{p^2 - 5p - 6} \cdot (p - 6)$

**49.** $\dfrac{m^2 - n^2}{5m - 5n} \cdot \dfrac{10m + 5n}{2m^2 + 3mn + n^2}$

**50.** $\dfrac{a^2 + 2ab + b^2}{3a + 3b} \cdot \dfrac{b - a}{a^2 - b^2}$

*In Problems 51–58, divide each rational expression. Simplify the quotient, if possible.*

**51.** $\dfrac{\dfrac{x + 3}{2x - 8}}{\dfrac{4x}{9}}$

**52.** $\dfrac{\dfrac{x - 2}{3x}}{\dfrac{5x - 10}{x}}$

**53.** $\dfrac{\dfrac{4a}{b^2}}{\dfrac{2a^2}{b}}$

**54.** $\dfrac{\dfrac{9m^3}{2n^2}}{\dfrac{3m}{8n^4}}$

**55.** $\dfrac{\dfrac{p^2 - 4p - 5}{2p^2 - 3p - 2}}{\dfrac{p^2 + p}{p^2 + p - 6}}$

**56.** $\dfrac{\dfrac{y^2 - 9}{2y^2 - y - 15}}{\dfrac{3y^2 + 10y + 3}{2y^2 + y - 10}}$

**57.** $\dfrac{\dfrac{x^3 - 1}{x^2 - 1}}{\dfrac{3x^2 + 3x + 3}{x^2 + 3x + 1}}$

**58.** $\dfrac{\dfrac{8x^3 + 1}{2x}}{\dfrac{x^3 + 2x^2 - 15x}{2x^2 - 5x - 3}}$

*In Problems 59–68, determine the domain of each rational function.*

**59.** $R(x) = \dfrac{2}{x - 1}$

**60.** $R(x) = \dfrac{5}{x + 3}$

**61.** $R(x) = \dfrac{x - 2}{(2x + 1)(x - 4)}$

**62.** $R(x) = \dfrac{3x + 2}{(4x - 1)(x + 5)}$

**63.** $R(x) = \dfrac{x + 9}{x^2 + 6x + 5}$

**64.** $R(x) = \dfrac{5x - 2}{x^2 - 6x - 16}$

**65.** $R(x) = \dfrac{x - 2}{2x^2 - 9x + 10}$

**66.** $R(x) = \dfrac{x + 3}{3x^2 + 7x - 6}$

**67.** $R(x) = \dfrac{x - 1}{x^2 + 1}$

**68.** $R(x) = \dfrac{4x}{4x^2 + 1}$

**69.** If $f(x) = \dfrac{x^2 - 2x - 15}{x + 6}$, $g(x) = \dfrac{x^2 + 5x - 6}{2x^2 - 7x - 15}$, and $h(x) = \dfrac{x + 3}{3x^2 + 17x - 6}$, find (a) $R(x) = f(x) \cdot g(x)$ and state its domain. (b) $R(x) = \dfrac{f(x)}{h(x)}$ and state its domain.

**70.** If $f(x) = \dfrac{x^2 - 7x - 8}{2x - 5}$, $g(x) = \dfrac{2x^2 + 3x - 20}{x^2 - 10x + 16}$, and $h(x) = \dfrac{x^2 - 3x - 40}{x + 9}$, find (a) $R(x) = f(x) \cdot g(x)$ and state its domain. (b) $R(x) = \dfrac{f(x)}{h(x)}$ and state its domain.

**71.** If $f(x) = \dfrac{3x^2 - x - 10}{x^3 - 1}$, $g(x) = \dfrac{x^2 + 5x - 6}{2x^2 + 3x - 14}$, and $h(x) = \dfrac{3x^2 + 8x + 5}{x^2 - 1}$, find (a) $R(x) = (f \cdot g)(x)$ and state its domain. (b) $R(x) = \left(\dfrac{f}{h}\right)(x)$ and state its domain.

**72.** If $f(x) = \dfrac{4x^2 - 9x - 9}{x^3 - 8}$, $g(x) = \dfrac{x^2 + 7x - 18}{5x^2 - 14x - 3}$, and $h(x) = \dfrac{x^2 - 6x + 9}{x^2 - 4}$, find (a) $R(x) = (f \cdot g)(x)$ and state its domain. (b) $R(x) = \left(\dfrac{f}{h}\right)(x)$ and state its domain.

## Mixed Practice

*In Problems 73–80, multiply or divide each rational expression, as indicated. Simplify the product or quotient, if possible.*

**73.** $\dfrac{z^3 + 8}{z^2 - 3z - 10} \cdot \dfrac{z^2 - 2z - 15}{2z^2 - 4z + 8}$

**74.** $\dfrac{x^3 - 27}{2x^2 + 5x - 25} \cdot \dfrac{x^2 + 2x - 15}{x^3 + 3x^2 + 9x}$

**75.** $\dfrac{\dfrac{m^2 - 4n^2}{m^3 - n^3}}{\dfrac{2m + 4n}{m^2 + mn + n^2}}$

**76.** $\dfrac{\dfrac{x^2 + 2xy + y^2}{x^2 + 3xy + 2y^2}}{\dfrac{x^2 - y^2}{x + 2y}}$

**77.** $\dfrac{4w + 8}{w^2 - 4w} \cdot \dfrac{w^2 - 3w - 4}{w^2 + 3w + 2}$

**78.** $\dfrac{5m - 5}{m^2 + 6m} \cdot \dfrac{m^2 + 2m - 24}{m^2 + 3m - 4}$

**79.** $\dfrac{\dfrac{2x - 6}{x^2 + x}}{\dfrac{x^2 - 4x + 3}{x^2}}$

**80.** $\dfrac{\dfrac{3x + 15}{2x + 4}}{\dfrac{x + 5}{x^2 - 4}}$

## Applying the Concepts

**81.** Make up a rational expression that is undefined at $x = 3$.

**82.** Make up a rational expression that is undefined at $x = -2$.

**83.** Make up a rational expression that is undefined at $x = -4$ and at $x = 5$.

**84.** Make up a rational expression that is undefined at $x = -6$ and at $x = 0$.

**85.** Develop a rational function $R$ that is undefined at $x = -2$ and at $x = 1$ such that $R(-3) = 1$.

**86.** Develop a rational function $R$ that is undefined at $x = -4$ and at $x = 3$ such that $R(4) = 1$.

**87. Gravity** In physics, it is established that the acceleration due to gravity $g$ at a height $h$ meters above sea level is given by

$$g(h) = \frac{3.99 \times 10^{14}}{(6.374 \times 10^6 + h)^2}$$

where $6.374 \times 10^6$ is the radius of Earth in meters.

**(a)** What is the acceleration due to gravity at sea level?

**(b)** What is the acceleration due to gravity in Denver, Colorado, elevation 1600 meters?

**(c)** What is the acceleration due to gravity on the peak of Mount Everest, elevation 8848 meters?

**88. Economics** The Gross Domestic Product (GDP) is the total value of all goods and services manufactured within the United States. A model for determining the change in GDP of a government spending plan is given by the function

$$G(s) = \frac{s}{1 - b}$$

where $G$ is the change in Gross Domestic Product, $s$ is the amount spent by the government, and $b$ is the marginal propensity to consume with $0 < b < 1$. The marginal propensity to consume can be thought of as the amount an individual would spend for each additional dollar of income earned. For example, if $b = 0.9$, then individuals will spend \$0.90 for each additional dollar earned. If $b = 0.95$, then individuals will spend \$0.95 for each additional dollar earned.

**(a)** Suppose the government spends \$100 million on highway infrastructure and the marginal propensity to consume in the United States is $b = 0.9$. What will be the change in GDP?

**(b)** Suppose the government spends \$100 million on highway infrastructure and the marginal propensity to consume in the United States is $b = 0.95$. What will be the change in GDP?

## Extending the Concepts

**89.** $f(x) = \dfrac{1}{x - 2}$

**(a)** Determine the domain of $f$.

**(b)** Fill in the following table. What happens to the values of $f$ as $x$ approaches 2, but remains greater than 2?

| $x$ | 3 | 2.5 | 2.1 | 2.01 | 2.001 | 2.0001 |
|---|---|---|---|---|---|---|
| $f(x)$ | | | | | | |

**(c)** Fill in the following table. What happens to the values of $f$ as $x$ approaches 2, but remains less than 2?

| $x$ | 1 | 1.5 | 1.9 | 1.99 | 1.999 | 1.9999 |
|---|---|---|---|---|---|---|
| $f(x)$ | | | | | | |

**(d)** To the right is the graph of $f(x) = \dfrac{1}{x - 2}$. What happens to the graph of the function as $x$ approaches 2 for values of $x$ larger than 2? What happens to the graph of the function as $x$ approaches 2 for values of $x$ smaller than 2? Compare your results to the results obtained in parts (b) and (c).

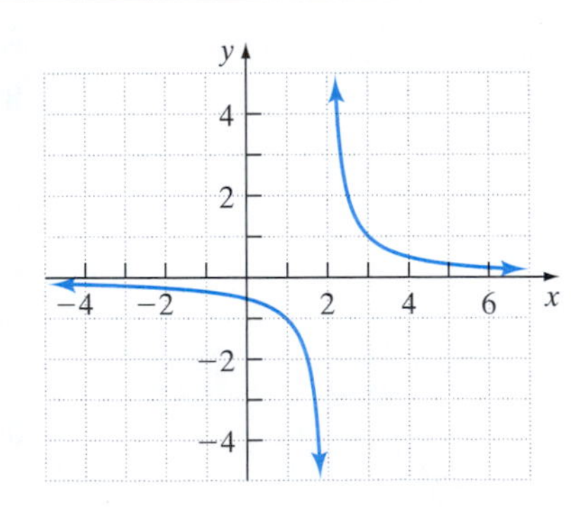


**90.** $R(x) = \dfrac{2x + 1}{x - 2}$

**(a)** Fill in the following table. What happens to the values of $R$ as $x$ gets larger in the positive direction?

| $x$ | 5 | 10 | 50 | 100 | 1000 |
|---|---|---|---|---|---|
| $R(x)$ | | | | | |

**(b)** Fill in the following table. What happens to the values of $R$ as $x$ gets larger in the negative direction?

| $x$ | −5 | −10 | −50 | −100 | −1000 |
|---|---|---|---|---|---|
| $R(x)$ | | | | | |

**(c)** What is the term of highest degree in the numerator? What is the term of highest degree in the denominator? What is the ratio of the coefficients on the terms of highest degree in the numerator and denominator? Compare this result to your results in parts (a) and (b).

**(d)** To the right is the graph of $R(x) = \frac{2x + 1}{x - 2}$. What happens to the graph of the function as $x$ gets larger? That is, what happens to the graph of the function as $x$ approaches $\infty$? What happens to the graph of the function as $x$ gets smaller? That is, what happens to the graph of the function as $x$ approaches $-\infty$? Compare your results to the results obtained in parts (b) and (c).

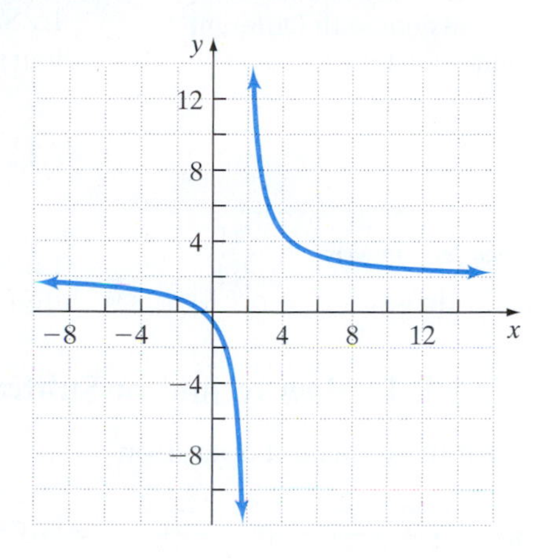


*If $R(x) = \frac{p(x)}{q(x)}$ is a rational function and if p and q have no common factors, then the rational function R is said to be in* **lowest terms.** *For a rational function $R(x) = \frac{p(x)}{q(x)}$ in lowest terms, the zeros, if any, of the numerator are the x-intercepts of the graph of R. In Problems 91 and 92, (a) write the rational function in lowest terms, (b) find the x-intercepts of the graph of R. That is, solve $R(x) = 0$.*

**91.** $R(x) = \frac{x^2 - 2x - 35}{x^2 - x - 42}$

**92.** $R(x) = \frac{2x^2 - 3x - 20}{x^2 + 4x - 32}$

**93.** Write $\frac{x^{-1}}{x + 1}$ as a rational expression.

**94.** What is the difference between $f(x) = \frac{x^2 - 3x + 2}{x - 1}$ and $g(x) = x - 2$?

## Synthesis Review

*In Problems 95–100, graph each function.*

**95.** $f(x) = 2x - 3$

**96.** $g(x) = 3x - 6$

**97.** $F(x) = -4x + 8$

**98.** $G(x) = -5x + 10$

**99.** $h(x) = x^2 - 4$

**100.** $H(x) = -x^2 + 4$

**101.** Discuss the methods that you used to graph each function. What methods were the same? What was different?

# 6.2 Adding and Subtracting Rational Expressions

**OBJECTIVES**

1. Add or Subtract Rational Expressions with a Common Denominator
2. Find the Least Common Denominator of Two or More Rational Expressions
3. Add or Subtract Rational Expressions with Different Denominators

### *Preparing for Adding and Subtracting Rational Expressions*

*Before getting started, take the following readiness quiz. If you get a problem wrong, go back to the section cited and review the material.*

**1.** $\frac{1}{6} - \frac{5}{8}$. [Getting Ready, pp. 450–453]

**2.** Determine the additive inverse of (a) 5 (b) $x - 2$ [Section R.3, pp. 21–22]

In Section 6.1, we learned how to multiply and divide rational expressions. We now learn to add and subtract rational expressions.

*Preparing for...Answers* **1.** $-\frac{11}{24}$ **2.** **(a)** $-5$ **(b)** $-(x-2)$ or $2 - x$

## 1 Add or Subtract Rational Expressions with a Common Denominator

Let's go over an example that illustrates how to add or subtract rational expressions that have a common denominator.

### EXAMPLE 1 How to Add or Subtract Rational Expressions

Perform the indicated operation.

**(a)** $\frac{x^2 - 3x + 6}{x + 3} + \frac{7x - 3}{x + 3}, x \neq -3$ **(b)** $\frac{3x - 5}{x + 1} - \frac{x + 3}{x + 1}, x \neq -1$

**Step-by-Step Solution**

**(a)** $\frac{x^2 - 3x + 6}{x + 3} + \frac{7x - 3}{x + 3}, x \neq -3$. Notice that the rational expressions have a common denominator, $x + 3$.

**Step 1:** Add the numerators and write the result over the common denominator.

$$\frac{x^2 - 3x + 6}{x + 3} + \frac{7x - 3}{x + 3} = \frac{x^2 - 3x + 6 + (7x - 3)}{x + 3}$$

Combine like terms in the numerator: $= \frac{x^2 + 4x + 3}{x + 3}$

**Step 2:** Simplify the rational expression.

Factor the numerator: $= \frac{(x + 3)(x + 1)}{x + 3}$

Divide out like factors: $= x + 1$

**(b)** $\frac{3x - 5}{x + 1} - \frac{x + 3}{x + 1}, x \neq -1$. The rational expressions have a common denominator, $x + 1$.

**Step 1:** Subtract the numerators and write the result over the common denominator.

$$\frac{3x - 5}{x + 1} - \frac{x + 3}{x + 1} = \frac{3x - 5 - (x + 3)}{x + 1}$$

Distribute the $-1$: $= \frac{3x - 5 - x - 3}{x + 1}$

Combine like terms in the numerator: $= \frac{2x - 8}{x + 1}$

**Step 2:** Simplify the rational expression.

Factor out the common factor: $= \frac{2(x - 4)}{x + 1}$

The rules for adding and subtracting rational expressions are the same as the rules for adding and subtracting rational numbers.

**In Words**
To add or subtract rational expressions with the same denominator, we add or subtract the numerators and write the result over the common denominator.

### Adding or Subtracting Rational Expressions

**Step 1:** If $\frac{a}{c}$ and $\frac{b}{c}$, $c \neq 0$, are two rational expressions, then $\frac{a}{c} + \frac{b}{c} = \frac{a+b}{c}$ and $\frac{a}{c} - \frac{b}{c} = \frac{a-b}{c}$.

**Step 2:** Simplify the result.

**QUICK ✓** *Perform the indicated operation. Be sure to simplify the result.*

**1.** $\frac{x^2 - 3x - 1}{x - 2} + \frac{x^2 - 2x + 3}{x - 2}$

**2.** $\frac{4x + 3}{x + 5} - \frac{x - 6}{x + 5}$

---

**EXAMPLE 2** **Adding Rational Expressions with Denominators That Are Additive Inverses**

Perform the indicated operation and simplify the result.

$$\frac{3x}{x - 2} + \frac{2}{2 - x}, \ x \neq 2$$

**Solution**

Although the denominators of the two rational expressions are different, we should notice that

$$2 - x = -x + 2 = -1(x - 2)$$

So

$$\frac{3x}{x - 2} + \frac{2}{2 - x} = \frac{3x}{x - 2} + \frac{2}{-1(x - 2)}$$

Use $\frac{a}{-b} = \frac{-a}{b}$:

$$= \frac{3x}{x - 2} + \frac{-2}{x - 2}$$

$$= \frac{3x - 2}{x - 2}$$

**QUICK ✓** *Perform the indicated operation. Be sure to simplify the result.*

**3.** $\frac{4x}{x - 5} + \frac{3}{5 - x}$

---

## 2 Find the Least Common Denominator of Two or More Rational Expressions

What if the denominators of the rational expressions to be added or subtracted are not the same? In this case, we must rewrite each rational expression over a *least common denominator*. The **least common denominator (LCD)** is the smallest polynomial that is a multiple of each denominator in the rational expressions to be added or subtracted. The idea is exactly the same as that used to add rational numbers that do not have common denominators. Let's see how it's done.

## EXAMPLE 3 How to Find the Least Common Denominator

Find the least common denominator of each expression.

**(a)** $\dfrac{4}{3x^2y^2}$ and $\dfrac{5}{6xy^3}$ **(b)** $\dfrac{x-1}{x^2+4x+3}$ and $\dfrac{3x-5}{x^3+2x^2+x}$

### Step-by-Step Solution

**(a)** $\dfrac{4}{3x^2y^2}$ and $\dfrac{5}{6xy^3}$

**Step 1:** Factor each denominator.

$3x^2y^2$ is factored completely.

$6xy^3 = 2 \cdot 3 \cdot xy^3$

**Step 2:** List the factors that are common to all denominators. Then list the uncommon factors.

- 3 is common to each denominator.
- We list $x^2$ because 2 is the highest exponent on the factor $x$.
- We list $y^3$ because 3 is the highest exponent on the factor $y$.
- The factor that is not common is 2.

$$\begin{aligned}\text{LCD} &= 2 \cdot 3 \cdot x^2 \cdot y^3 \\ &= 6x^2y^3\end{aligned}$$

**(b)** $\dfrac{x-1}{x^2+4x+3}$ and $\dfrac{3x-5}{x^3+2x^2+x}$

**Step 1:** Factor each denominator.

$$\begin{aligned}x^2+4x+3 &= (x+3)(x+1) \\ x^3+2x^2+x &= x(x^2+2x+1) \\ &= x(x+1)^2\end{aligned}$$

**Step 2:** List the factors that are common to all denominators. Then list the uncommon factors.

- We list $(x+1)^2$ because 2 is the highest exponent on the factor $x+1$.
- The factors that are not common are $x$ and $x+3$.

$$\text{LCD} = x(x+3)(x+1)^2$$

### Finding the Least Common Denominator

**Step 1:** Factor each denominator completely. When factoring, write the factored form using powers. For example, write $x^2 + 4x + 4$ as $(x+2)^2$.

**Step 2:** List the common factors. If factors are common except for their power, then list the factor with the highest power. Then list the factors that are not common.

**QUICK ✓** *Find the least common denominator of each expression.*

**4.** $\dfrac{5}{8x^2y}$ and $\dfrac{1}{12xy^3}$ **5.** $\dfrac{4x-3}{x^2-5x-14}$ and $\dfrac{x+1}{x^2+4x+4}$

## 3 Add or Subtract Rational Expressions with Different Denominators

Now that we know how to obtain the least common denominator, we can discuss how to add or subtract rational expressions that have unlike denominators.

### EXAMPLE 4 How to Add Rational Expressions with Unlike Denominators

Add $\frac{3}{8x^2} + \frac{1}{12x}$. Simplify the result.

**Step-by-Step Solution**

**Step 1:** Find the least common denominator.

$8x^2 = 4 \cdot 2 \cdot x^2$

$12x = 4 \cdot 3 \cdot x$

The LCD is $4 \cdot 2 \cdot 3 \cdot x^2 = 24x^2$.

**Step 2:** Rewrite each rational expression with the common denominator.

Multiply $\frac{3}{8x^2}$ by $\frac{3}{3}$; Multiply $\frac{1}{12x}$ by $\frac{2x}{2x}$

$$\frac{3}{8x^2} + \frac{1}{12x} = \frac{3}{8x^2} \cdot \frac{3}{3} + \frac{1}{12x} \cdot \frac{2x}{2x}$$

$$= \frac{9}{24x^2} + \frac{2x}{24x^2}$$

**Step 3:** Add the rational expressions found in Step 2.

$\frac{a}{c} + \frac{b}{c} = \frac{a+b}{c}$: $$= \frac{9 + 2x}{24x^2} = \frac{2x + 9}{24x^2}$$

**Step 4:** The rational expression is simplified.

So $\frac{3}{8x^2} + \frac{1}{12x} = \frac{2x + 9}{24x^2}$.

**Work Smart**

Notice in Step 2 of Example 4, when we rewrite the rational expression with the common denominator, we are simply multiplying by a "disguised" 1 such as $1 = \frac{3}{3}$ or $1 = \frac{2x}{2x}$.

**Adding or Subtracting Rational Expressions with Unlike Denominators**

**Step 1:** Find the least common denominator.

**Step 2:** Rewrite each rational expression with the common denominator. You will need to multiply out the numerator, but leave the denominator in factored form.

**Step 3:** Add or subtract the rational expression found in Step 2.

**Step 4:** Simplify the result, if necessary.

**QUICK ✓** *Perform the indicated operation and simplify the result.*

**6.** $\frac{3}{10a} + \frac{4}{15a^2}$

### EXAMPLE 5 Adding Rational Expressions with Unlike Denominators

Perform the indicated operation and simplify the result.

**(a)** $\frac{x-1}{x+3} + \frac{x}{x+2}$ **(b)** $\frac{x-1}{x^2+2x-8} + \frac{x-1}{x^2-16}$

**Solution**

**(a)** $\dfrac{x-1}{x+3} + \dfrac{x}{x+2}$

The LCD is $(x + 3)(x + 2)$.

Multiply $\dfrac{x-1}{x+3}$ by $\dfrac{x+2}{x+2}$; Multiply $\dfrac{x}{x+2}$ by $\dfrac{x+3}{x+3}$

$$\frac{x-1}{x+3} + \frac{x}{x+2} = \frac{x-1}{x+3}\cdot\frac{x+2}{x+2} + \frac{x}{x+2}\cdot\frac{x+3}{x+3}$$

Multiply out the numerator: $$= \frac{x^2+x-2}{(x+3)(x+2)} + \frac{x^2+3x}{(x+3)(x+2)}$$

Use $\frac{a}{c} + \frac{b}{c} = \frac{a+b}{c}$: $$= \frac{x^2+x-2+(x^2+3x)}{(x+3)(x+2)}$$

Combine like terms: $$= \frac{2x^2+4x-2}{(x+3)(x+2)} = \frac{2(x^2+2x-1)}{(x+3)(x+2)}$$

So $\dfrac{x-1}{x+3} + \dfrac{x}{x+2} = \dfrac{2(x^2+2x-1)}{(x+3)(x+2)}$.

**(b)** $\dfrac{x-1}{x^2+2x-8} + \dfrac{x-1}{x^2-16}$

First, we factor the denominators to find the LCD.

$$x^2 + 2x - 8 = (x+4)(x-2)$$
$$x^2 - 16 = (x+4)(x-4)$$
$$\text{LCD} = (x+4)(x-2)(x-4)$$

Multiply $\dfrac{x-1}{(x+4)(x-2)}$ by $\dfrac{x-4}{x-4}$; Multiply $\dfrac{x-1}{(x+4)(x-4)}$ by $\dfrac{x-2}{x-2}$

$$\frac{x-1}{x^2+2x-8} + \frac{x-1}{x^2-16} = \frac{x-1}{(x+4)(x-2)}\cdot\frac{x-4}{x-4} + \frac{x-1}{(x+4)(x-4)}\cdot\frac{x-2}{x-2}$$

Multiply out the numerator: $$= \frac{x^2-5x+4}{(x+4)(x-2)(x-4)} + \frac{x^2-3x+2}{(x+4)(x-2)(x-4)}$$

Use $\frac{a}{c} + \frac{b}{c} = \frac{a+b}{c}$: $$= \frac{x^2-5x+4+(x^2-3x+2)}{(x+4)(x-2)(x-4)}$$

Combine like terms: $$= \frac{2x^2-8x+6}{(x+4)(x-2)(x-4)}$$

Factor the numerator: $$= \frac{2(x-3)(x-1)}{(x+4)(x-2)(x-4)}$$

So $\dfrac{x-1}{x^2+2x-8} + \dfrac{x-1}{x^2-16} = \dfrac{2(x-3)(x-1)}{(x+4)(x-2)(x-4)}$.

**QUICK ✓** *Perform the indicated operation and simplify the result.*

**7.** $\dfrac{3x}{x-1} + \dfrac{x+5}{x+2}$ **8.** $\dfrac{x-1}{2x^2+7x+6} + \dfrac{x-1}{x^2+6x+8}$

## EXAMPLE 6 How to Subtract Rational Expressions with Unlike Denominators

Perform the indicated operation and simplify the result.

$$\frac{2x-1}{2x^2-7x-4}-\frac{x-1}{2x^2+3x+1}$$

### Step-by-Step Solution

**Step 1:** Find the least common denominator.

$$2x^2-7x-4=(2x+1)(x-4)$$
$$2x^2+3x+1=(2x+1)(x+1)$$
$$\text{LCD}=(2x+1)(x-4)(x+1)$$

**Step 2:** Rewrite each rational expression with the common denominator.

Multiply $\frac{2x-1}{(2x+1)(x-4)}$ by $\frac{x+1}{x+1}$; Multiply $\frac{x-1}{(2x+1)(x+1)}$ by $\frac{x-4}{x-4}$

$$\frac{2x-1}{2x^2-7x-4}-\frac{x-1}{2x^2+3x+1}=\frac{2x-1}{(2x+1)(x-4)}\cdot\frac{x+1}{x+1}-\frac{x-1}{(2x+1)(x+1)}\cdot\frac{x-4}{x-4}$$

Multiply out the numerator: $$=\frac{2x^2+x-1}{(2x+1)(x-4)(x+1)}-\frac{x^2-5x+4}{(2x+1)(x-4)(x+1)}$$

**Step 3:** Subtract the rational expressions found in Step 2.

Use $\frac{a}{c}-\frac{b}{c}=\frac{a-b}{c}$: $$=\frac{2x^2+x-1-(x^2-5x+4)}{(2x+1)(x-4)(x+1)}$$

Distribute the $-1$: $$=\frac{2x^2+x-1-x^2+5x-4}{(2x+1)(x-4)(x+1)}$$

Combine like terms: $$=\frac{x^2+6x-5}{(2x+1)(x-4)(x+1)}$$

**Step 4:** Simplify the rational expression.

The rational expression is simplified.

So $$\frac{2x-1}{2x^2-7x-4}-\frac{x-1}{2x^2+3x+1}=\frac{x^2+6x-5}{(2x+1)(x-4)(x+1)}.$$

**QUICK ✓** *Perform the indicated operation and simplify the result.*

**9.** $\frac{3x+4}{2x^2+x-6}-\frac{x-1}{x^2+4x+4}$

## EXAMPLE 7 Adding and Subtracting Three Rational Expressions

Perform the indicated operations and simplify the result.

$$\frac{6}{x^2-9}+\frac{x+1}{x+3}-\frac{x-2}{x-3}$$

### Solution

We factor each denominator.

$$x^2-9=(x+3)(x-3)$$
$$x+3=x+3$$
$$x-3=x-3$$

The LCD $= (x + 3)(x - 3)$.

Multiply $\frac{x+1}{x+3}$ by $\frac{x-3}{x-3}$; Multiply $\frac{x-2}{x-3}$ by $\frac{x+3}{x+3}$

$$\frac{6}{x^2-9} + \frac{x+1}{x+3} - \frac{x-2}{x-3} = \frac{6}{(x+3)(x-3)} + \frac{x+1}{x+3}\cdot\frac{x-3}{x-3} - \frac{x-2}{x-3}\cdot\frac{x+3}{x+3}$$

Multiply out the numerator: $= \frac{6}{(x+3)(x-3)} + \frac{x^2-2x-3}{(x+3)(x-3)} - \frac{x^2+x-6}{(x+3)(x-3)}$

Use $\frac{a}{c} + \frac{b}{c} = \frac{a+b}{c}$; $\frac{a}{c} - \frac{b}{c} = \frac{a-b}{c}$: $= \frac{6 + x^2 - 2x - 3 - (x^2 + x - 6)}{(x+3)(x-3)}$

Distribute the $-1$: $= \frac{6 + x^2 - 2x - 3 - x^2 - x + 6}{(x+3)(x-3)}$

Combine like terms: $= \frac{-3x+9}{(x+3)(x-3)}$

Factor: $= \frac{-3(x-3)}{(x+3)(x-3)}$

Divide out like factors: $= \frac{-3}{x+3}$

So $\frac{6}{x^2-9} + \frac{x+1}{x+3} - \frac{x-2}{x-3} = \frac{-3}{x+3}$.

**QUICK ✓** *Perform the indicated operations and simplify the result.*

**10.** $\frac{4}{x^2-4} - \frac{x+3}{x-2} + \frac{x+3}{x+2}$

# 6.2 Exercises

**For Extra Help:**

Student Solutions Manual | CD Video | PH Math/Tutor Center | MathXL Tutorials on CD | MathXL® | MyMathLab

## Concepts and Vocabulary

*In Problems 1 and 2, fill in the blanks.*

**1.** The _____ _____ _____ is the smallest polynomial that is a multiple of each denominator in the rational expressions to be added or subtracted.

**2.** $\frac{a}{c} - \frac{b}{c} =$ _____ provided that $c \neq 0$.

*In Problems 3 and 4, answer True or False to each statement.*

**3.** To add or subtract rational expressions, we write the rational expressions over a common denominator and then add or subtract the numerators.

**4.** The least common denominator of $\frac{3}{(x+2)^2(x-1)}$ and $\frac{2x-7}{(x+2)(x-1)^2}$ is $(x+2)^2(x-1)^2$.

**5.** In your own words, explain how to find the least common denominator when adding or subtracting two rational expressions with unlike denominators.

**6.** In your own words, explain how to add or subtract rational expressions with unlike denominators.

## Building Skills

*In Problems 7–18, perform the indicated operation and simplify the result.*

**7.** $\frac{3x}{x+1} + \frac{5}{x+1}$

**8.** $\frac{5x}{x-3} + \frac{2}{x-3}$

**9.** $\frac{2x}{2x+5} - \frac{1}{2x+5}$

**10.** $\frac{9x}{6x-5} - \frac{2}{6x-5}$

**11.** $\frac{2x}{2x^2-7x-15} + \frac{3}{2x^2-7x-15}$

**12.** $\frac{x}{3x^2+8x-3} + \frac{3}{3x^2+8x-3}$

**13.** $\frac{2x^2-5x+7}{x^2-2x-15} - \frac{x^2+3x-8}{x^2-2x-15}$

**14.** $\frac{3x^2+8x-1}{x^2-3x-28} - \frac{2x^2+2x-9}{x^2-3x-28}$

**15.** $\frac{3x}{x-5} + \frac{1}{5-x}$

**16.** $\frac{3x}{x-6} + \frac{2}{6-x}$

**17.** $\frac{2x^2-4x-1}{x-3} - \frac{x^2-3x+4}{3-x}$

**18.** $\frac{x^2+2x-5}{x-4} - \frac{x^2-5x-15}{4-x}$

*In Problems 19–28, find the least common denominator.*

**19.** $\frac{3}{4x^3}$ and $\frac{9}{8x}$

**20.** $\frac{5}{3a^3}$ and $\frac{2}{9a^2}$

**21.** $\frac{1}{15xy^2}$ and $\frac{7}{18x^3y}$

**22.** $\frac{1}{8a^3b}$ and $\frac{5}{12ab^2}$

**23.** $\frac{5x}{x-4}$ and $\frac{3}{x+2}$

**24.** $\frac{x-3}{x+2}$ and $\frac{x+7}{x-5}$

**25.** $\frac{x-4}{x^2-x-12}$ and $\frac{2x+1}{x^2-9x+20}$

**26.** $\frac{2m-7}{m^2+3m-18}$ and $\frac{5m+1}{m^2-7m+12}$

**27.** $\frac{p+1}{2p^2+3p-2}$ and $\frac{4p-1}{p^3+2p^2}$

**28.** $\frac{x-6}{x^2-9}$ and $\frac{3x}{x^3-3x^2}$

*In Problems 29–54, add or subtract, as indicated, and simplify the result.*

**29.** $\frac{3}{4x^2} + \frac{5}{8x}$

**30.** $\frac{2}{9x} + \frac{5}{3x^2}$

**31.** $\frac{5}{12a^2b} - \frac{4}{15ab^2}$

**32.** $\frac{3}{14mn^3} - \frac{2}{21m^2n}$

**33.** $\frac{y+2}{y-5} - \frac{y-4}{y+3}$

**34.** $\frac{x+2}{x-3} - \frac{x+2}{x+1}$

**35.** $\frac{a+5}{a-2} - \frac{5a+18}{a^2-4}$

**36.** $\frac{z+1}{z+3} - \frac{z+17}{z^2-z-12}$

**37.** $\frac{3}{(x-2)(x+3)} - \frac{5}{(x+3)(x+4)}$

**38.** $\frac{1}{(x-1)(x+3)} + \frac{5}{(x+1)(x-1)}$

**39.** $\frac{x-3}{x^2+3x+2} + \frac{x-1}{x^2-4}$

**40.** $\frac{x-5}{x^2+4x+3} + \frac{x-2}{x^2-1}$

**41.** $\frac{w-4}{2w^2+3w+1} - \frac{w+3}{2w^2-5w-3}$

**42.** $\frac{y+4}{3y^2-y-2} - \frac{1}{3y^2+14y+8}$

**43.** $\frac{x+y}{x^2-6xy+9y^2} + \frac{x+2y}{x^2-2xy-3y^2}$

**44.** $\frac{m-2n}{m^2+4mn+4n^2} + \frac{m-n}{m^2-mn-6n^2}$

**45.** $\frac{3}{x^2-4x-5} - \frac{2}{x^2-6x+5}$

**46.** $\frac{3}{x^2+7x+10} - \frac{4}{x^2+6x+5}$

**47.** $\dfrac{p^2 - 3p - 10}{p^2 - 16} + \dfrac{p^2 - 3p - 10}{16 - p^2}$

**48.** $\dfrac{y^2 + 4y + 4}{y^2 - 9} + \dfrac{y^2 + 4y + 4}{9 - y^2}$

**49.** $\dfrac{2}{w + 2} - \dfrac{3}{w} + \dfrac{w + 10}{w^2 - 4}$

**50.** $\dfrac{7}{m - 3} - \dfrac{5}{m} - \dfrac{2m + 6}{m^2 - 9}$

**51.** $\dfrac{p - 2}{p^2 + 6p + 9} + \dfrac{1}{p + 3} - \dfrac{2p + 1}{2p^2 + p - 15}$

**52.** $\dfrac{x - 1}{x^2 - 16} + \dfrac{1}{x + 4} - \dfrac{4x + 1}{3x^2 - 7x - 20}$

**53.** $\dfrac{2}{x} - \dfrac{2}{x - 1} + \dfrac{3}{(x - 1)^2}$

**54.** $\dfrac{2}{x} - \dfrac{2}{x + 2} + \dfrac{2}{(x + 2)^2}$

**55.** Given that $f(x) = \dfrac{3}{x - 2}$ and $g(x) = \dfrac{2}{x + 1}$,
**(a)** find $R(x) = f(x) + g(x)$,
**(b)** state the domain of $R(x)$.

**56.** Given that $f(x) = \dfrac{5}{x + 2}$ and $g(x) = \dfrac{3}{x - 1}$,
**(a)** find $R(x) = f(x) + g(x)$,
**(b)** state the domain of $R(x)$.

**57.** Given that $f(x) = \dfrac{x + 1}{x^2 - 3x - 4}$ and $g(x) = \dfrac{x + 4}{x^2 - x - 12}$,
**(a)** find $R(x) = f(x) + g(x)$,
**(b)** state the domain of $R(x)$,
**(c)** find $H(x) = f(x) - g(x)$,
**(d)** state the domain of $H(x)$.

**58.** Given that $f(x) = \dfrac{x + 5}{x^2 - 5x + 6}$ and $g(x) = \dfrac{x + 1}{x^2 - 4x - 12}$,
**(a)** find $R(x) = f(x) + g(x)$,
**(b)** state the domain of $R(x)$,
**(c)** find $H(x) = f(x) - g(x)$,
**(d)** state the domain of $H(x)$.

## Mixed Practice

*In Problems 59–70, perform the indicated operation and simplify the result.*

**59.** $\dfrac{1}{x - 3} - \dfrac{x^2 + 18}{x^3 - 27}$

**60.** $\dfrac{1}{x + 2} + \dfrac{x - 10}{x^3 + 8}$

**61.** $\dfrac{2x^2 - x}{x + 3} + \dfrac{3x}{x + 3} - \dfrac{x^2 + 3}{x + 3}$

**62.** $\dfrac{2x^2}{x - 1} - \dfrac{x^2 - 2x}{x - 1} + \dfrac{x - 4}{x - 1}$

**63.** $6 + \dfrac{x - 3}{x + 3}$

**64.** $3 + \dfrac{x + 4}{x - 4}$

**65.** $\dfrac{b + 3}{b^2 + 2b - 8} - \dfrac{b + 2}{b^2 - 4}$

**66.** $\dfrac{a + 3}{a^2 - 8a + 15} + \dfrac{a + 3}{a^2 - 9}$

**67.** $\dfrac{y - 1}{y + 4} + \dfrac{y - 2}{y + 3} - \dfrac{y^2 + 3y + 1}{y^2 + 7y + 12}$

**68.** $\dfrac{z + 3}{z - 6} + \dfrac{z - 1}{z - 2} - \dfrac{6z}{z^2 - 8z + 12}$

**69.** $\dfrac{x + 4}{x^2 - 5x + 6} + \dfrac{x - 1}{x^2 - 2x - 3} - \dfrac{2x + 1}{x^2 - x - 2}$

**70.** $\dfrac{x - 1}{x^2 + 4x - 5} + \dfrac{3x - 1}{x^2 + 3x - 10} - \dfrac{4x + 1}{x^2 - 3x + 2}$

## Applying the Concepts

△ **71. Surface Area of a Box** The volume of a closed box with a square base is 2000 cubic inches. Its surface area $S$ as a function of the length of the base $x$ is given by the function

$$S(x) = 2x^2 + \frac{8000}{x}$$

**(a)** Write $S$ over a common denominator. That is, write $S$ so that the rule is a single rational expression.
**(b)** Find and interpret $S(10)$.

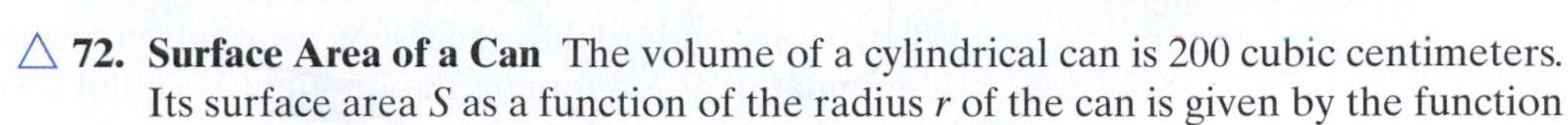

△ **72. Surface Area of a Can** The volume of a cylindrical can is 200 cubic centimeters. Its surface area $S$ as a function of the radius $r$ of the can is given by the function

$$S(r) = 2\pi r^2 + \frac{400}{r}$$

**(a)** Write $S$ over a common denominator. That is write $S$ so that the rule is a single rational expression.
**(b)** Find and interpret $S(4)$. Round your answer to two decimal places.

**73. Road Trip** Suppose you and a group of your friends decide to go on a road trip to a neighboring university that is 200 miles away. Your average speed for the first 50 miles of the trip is 10 miles an hour slower than the average speed for the remaining 150 miles of the trip. If we let $s$ represent your average speed for the first 50 miles of the trip, the function

$$T(s) = \frac{50}{s} + \frac{150}{s + 10}$$

represents the amount of time $T$ it will take to get to the neighboring university.

**(a)** Write $T$ over a common denominator. That is, write $T$ so that the rule is a single rational expression.
**(b)** Find and interpret $T(50)$.

**74. Vacation** The distance from Chicago, Illinois to Naples, Florida is approximately 1200 miles. Atlanta, Georgia is approximately the midpoint between Chicago and Naples. On a recent trip to Naples, the Sullivan family averaged $s$ miles per hour between Chicago and Atlanta and they averaged 5 miles per hour faster between Atlanta and Naples. The time $T$ of their trip as a function of their average speed $s$ between Chicago and Atlanta is given by the function

$$T(s) = \frac{600}{s} + \frac{600}{s + 5}$$

**(a)** Write $T$ over a common denominator. That is, write $T$ so that the rule is a single rational expression.
**(b)** Find and interpret $T(50)$. Round your answer to two decimal places.
**(c)** Using the result from part (b), compute the average speed of the Sullivans for the entire trip. Are you surprised by the result?

## Extending the Concepts

**75.** Write $x^{-1} + y^{-1}$ as a single rational expression with no negative exponents.

**76.** Write $\left(\frac{a}{b}\right)^{-1} + \left(\frac{b}{a}\right)^{-1}$ as a single rational expression with no negative exponents.

## Synthesis Review

*In Problems 77–82, multiply the expressions.*

**77.** $4a(a - 3)$ **78.** $-5z(z + 4)$ **79.** $(p - 3)(p + 3)$

**80.** $(3q + 1)(3q - 1)$ **81.** $(w - 2)(w^2 + 2w + 4)$ **82.** $(2v + 1)(4v^2 - 2v + 1)$

# 6.3 Complex Rational Expressions

**OBJECTIVES**

1. Simplify a Complex Rational Expression by Simplifying the Numerator and Denominator Separately
2. Simplify a Complex Rational Expression Using the Least Common Denominator

***Preparing for Complex Rational Expressions***

*Before getting started, take the following readiness quiz. If you get a problem wrong, go back to the section cited and review the material.*

**1.** Factor $6y^2 - 5y - 6$ [Section 5.5, pp. 401–406]

**2.** Simplify $\left(\dfrac{3ab^2}{2a^{-1}b^5}\right)^{-2}$ [Getting Ready, pp. 345–346]

**In Words**

A complex rational expression is simplified when it is of the form *polynomial over polynomial* and the polynomials have no common factors.

When sums and/or differences of rational expressions occur in the numerator or denominator of a quotient, the quotient is called a **complex rational expression.** The following are examples of complex rational expressions.

$$\frac{3 - \dfrac{1}{x}}{1 + \dfrac{1}{x}} \quad \text{and} \quad \frac{\dfrac{x+1}{x-2} - \dfrac{3}{x+2}}{\dfrac{2x+3}{x-2} + 1}$$

To **simplify** a complex rational expression means to write it as a rational expression in simplest form. This can be done using one of two methods: (1) Simplify the numerator and the denominator separately, or (2) simplify by using the least common denominator. We'll start by using method (1).

## Simplify a Complex Rational Expression by Simplifying the Numerator and Denominator Separately

Let's look at an example that illustrates how to simplify a complex rational expression.

*Preparing for...Answers*

**1.** $(3y + 2)(2y - 3)$ **2.** $\dfrac{4b^6}{9a^4}$

### EXAMPLE 1 How to Simplify a Complex Rational Expression Using Method I

Simplify: $\dfrac{\dfrac{1}{3} + \dfrac{1}{x}}{\dfrac{x+3}{2}}, x \neq -3, 0$

**Step-by-Step Solution**

Notice that $x$ cannot equal $-3$ because it will cause the denominator to equal zero; $x$ cannot equal 0 because it will cause division by 0 in $\dfrac{1}{x}$ in the numerator.

**Step 1:** Write the numerator of the complex rational expression as a single rational expression.

The least common denominator is $3x$.

$$\frac{1}{3} + \frac{1}{x} = \frac{1}{3} \cdot \frac{x}{x} + \frac{1}{x} \cdot \frac{3}{3}$$

$$= \frac{x}{3x} + \frac{3}{3x}$$

$\dfrac{a}{c} + \dfrac{b}{c} = \dfrac{a+b}{c}$: $= \dfrac{x+3}{3x}$

**Step 2:** Write the denominator of the complex rational expression as a single rational expression.

This is already done.

**Step 3:** Rewrite the complex rational expression using the rational expressions determined in Steps 1 and 2.

$$\frac{\frac{1}{3} + \frac{1}{x}}{\frac{x + 3}{2}} = \frac{\frac{x + 3}{3x}}{\frac{x + 3}{2}}$$

**Step 4:** Simplify the rational expression using the techniques for dividing rational expressions from Section 6.1.

Rewrite the division problem as a multiplication problem: $= \frac{x + 3}{3x} \cdot \frac{2}{x + 3}$

Divide like factors: $= \frac{\cancel{x + 3}}{3x} \cdot \frac{2}{\cancel{x + 3}}$

$$= \frac{2}{3x}$$

So $\dfrac{\frac{1}{3} + \frac{1}{x}}{\frac{x + 3}{2}} = \dfrac{2}{3x}$.

We summarize the steps to simplify a complex rational expression.

### Simplifying a Complex Rational Expression by Simplifying the Numerator and Denominator Separately (Method I)

**Step 1:** Write the numerator of the complex rational expression as a single rational expression.

**Step 2:** Write the denominator of the complex rational expression as a single rational expression.

**Step 3:** Rewrite the complex rational expression using the rational expressions determined in Steps 1 and 2.

**Step 4:** Simplify the rational expression using the techniques for dividing rational expressions from Section 6.1.

**QUICK ✓**

**1.** Simplify: $\dfrac{\frac{z}{4} - \frac{4}{z}}{\frac{z + 4}{16}}, z \neq 0, -4$

We will not state the domain restrictions for the remaining examples, but you should be aware that the restrictions are needed to maintain equality.

### EXAMPLE 2 Simplifying a Complex Rational Expression Using Method I

Simplify: $\dfrac{\frac{2x}{x + 4} - \frac{x - 7}{x^2 - 16}}{x - \frac{x^2 + 4}{x + 4}}$

### Solution

Write the numerator of the complex rational expression as a single rational expression.

LCD $= (x-4)(x+4)$

$$\frac{2x}{x+4} - \frac{x-7}{x^2-16} = \frac{2x}{x+4}\cdot\frac{x-4}{x-4} - \frac{x-7}{x^2-16}$$

$$= \frac{2x^2-8x}{x^2-16} - \frac{x-7}{x^2-16}$$

Use $\frac{a}{c} - \frac{b}{c} = \frac{a-b}{c}$: $$= \frac{2x^2-9x+7}{x^2-16}$$

Write the denominator of the complex rational expression as a single rational expression.

LCD $= x+4$

$$x - \frac{x^2+4}{x+4} = \frac{x}{1}\cdot\frac{x+4}{x+4} - \frac{x^2+4}{x+4}$$

$$= \frac{x^2+4x}{x+4} - \frac{x^2+4}{x+4}$$

Use $\frac{a}{c} - \frac{b}{c} = \frac{a-b}{c}$: $$= \frac{4x-4}{x+4}$$

Rewrite the complex rational expression using the numerator and denominator just found, and then simplify.

$$\frac{\frac{2x}{x+4} - \frac{x-7}{x^2-16}}{x - \frac{x^2+4}{x+4}} = \frac{\frac{2x^2-9x+7}{x^2-16}}{\frac{4x-4}{x+4}}$$

Rewrite the division problem as a multiplication problem: $$= \frac{2x^2-9x+7}{x^2-16}\cdot\frac{x+4}{4x-4}$$

Factor and divide out like factors: $$= \frac{(2x-7)\cancel{(x-1)}}{(x-4)\cancel{(x+4)}}\cdot\frac{\cancel{(x+4)}}{4\cancel{(x-1)}}$$

$$= \frac{2x-7}{4(x-4)}$$

So $$\frac{\frac{2x}{x+4} - \frac{x-7}{x^2-16}}{x - \frac{x^2+4}{x+4}} = \frac{2x-7}{4(x-4)}.$$

**QUICK ✓**

**2.** Simplify: $$\frac{\frac{2x}{x+1} - \frac{x^2-3}{x^2+3x+2}}{4 + \frac{4}{x+2}}$$

## 2 Simplify a Complex Rational Expression Using the Least Common Denominator

We now simplify complex rational expressions using the least common denominator. We will redo Example 1 using this second method so that you can compare the two methods.

## EXAMPLE 3 How to Simplify a Complex Rational Expression Using Method II

Simplify: $\dfrac{\dfrac{1}{3}+\dfrac{1}{x}}{\dfrac{x+3}{2}}$

### Step-by-Step Solution

**Step 1:** Find the least common denominator among all the denominators in the complex rational expression.

The denominators of the complex rational expression are 3, $x$, and 2. The least common denominator is $2 \cdot 3 \cdot x = 6x$.

**Step 2:** Multiply both the numerator and denominator of the complex rational expression by the least common denominator found in Step 1.

$$\frac{\dfrac{1}{3}+\dfrac{1}{x}}{\dfrac{x+3}{2}} \cdot \frac{6x}{6x} = \frac{\left(\dfrac{1}{3}+\dfrac{1}{x}\right)\cdot 6x}{\left(\dfrac{x+3}{2}\right)\cdot 6x}$$

Distribute $6x$ to each term:
$$= \frac{\dfrac{1}{3}\cdot 6x + \dfrac{1}{x}\cdot 6x}{\dfrac{x+3}{2}\cdot 6x}$$

**Step 3:** Simplify the rational expression.

Divide out like factors:
$$= \frac{2x+6}{3x(x+3)}$$

Factor and divide out like factors:
$$= \frac{2\cancel{(x+3)}}{3x\cancel{(x+3)}} = \frac{2}{3x}$$

So $\dfrac{\dfrac{1}{3}+\dfrac{1}{x}}{\dfrac{x+3}{2}} = \dfrac{2}{3x}$, the same result as we obtained in Example 1!

### Simplifying a Complex Rational Expression Using the Least Common Denominator (Method II)

**Step 1:** Find the least common denominator among all the denominators in the complex rational expression.

**Step 2:** Multiply both the numerator and denominator of the complex rational expression by the least common denominator found in Step 1.

**Step 3:** Simplify the rational expression.

**QUICK ✓**

**3.** Simplify $\dfrac{\dfrac{z}{4}-\dfrac{4}{z}}{\dfrac{z+4}{16}}$ using Method II.

**EXAMPLE 4 Simplifying Complex Rational Expressions Using Method II**

Simplify: $\dfrac{\dfrac{1}{x} + \dfrac{1}{x-2}}{\dfrac{x}{x^2-4} + \dfrac{1}{x-2}}$

**Solution**

Since $x^2 - 4 = (x-2)(x+2)$, the least common denominator is $x(x-2)(x+2)$, so we multiply the numerator and denominator by $x(x-2)(x+2)$.

$$\frac{\dfrac{1}{x} + \dfrac{1}{x-2}}{\dfrac{x}{x^2-4} + \dfrac{1}{x-2}} = \frac{\dfrac{1}{x} + \dfrac{1}{x-2}}{\dfrac{x}{x^2-4} + \dfrac{1}{x-2}} \cdot \frac{x(x-2)(x+2)}{x(x-2)(x+2)}$$

Distribute the LCD to each term:
$$= \frac{\dfrac{1}{x} \cdot x(x-2)(x+2) + \dfrac{1}{x-2} \cdot x(x-2)(x+2)}{\dfrac{x}{x^2-4} \cdot x(x-2)(x+2) + \dfrac{1}{x-2} \cdot x(x-2)(x+2)}$$

Factor and divide out like factors:
$$= \frac{\dfrac{1}{\cancel{x}} \cdot \cancel{x}(x-2)(x+2) + \dfrac{1}{\cancel{x-2}} \cdot x\cancel{(x-2)}(x+2)}{\dfrac{x}{\cancel{(x-2)}\cancel{(x+2)}} \cdot x\cancel{(x-2)}\cancel{(x+2)} + \dfrac{1}{\cancel{x-2}} \cdot x\cancel{(x-2)}(x+2)}$$

$$= \frac{(x-2)(x+2) + x(x+2)}{x^2 + x(x+2)}$$

Factor out $(x + 2)$:
$$= \frac{(x+2)[x - 2 + x]}{x^2 + x^2 + 2x}$$

$$= \frac{(x+2)(2x-2)}{2x^2 + 2x}$$

Factor and divide out like factors:
$$= \frac{\cancel{2}(x+2)(x-1)}{\cancel{2}x(x+1)}$$

$$= \frac{(x+2)(x-1)}{x(x+1)}$$

**QUICK ✓**

**4.** Simplify: $\dfrac{\dfrac{x+2}{x+5} - \dfrac{x+2}{x+1}}{\dfrac{2x+1}{x+1} - 1}$

## Comparing Methods

We will work through the next example using both methods. As you work through problems in the exercise set, be sure to start developing a sense as to when Method I might be preferred over Method II, and vice versa.

## EXAMPLE 5 Comparing the Methods I and II

Simplify $\dfrac{x^{-1} + y^{-1}}{x^{-3} + y^{-3}}$ as a rational expression that contains no negative exponents.

### Solution

First, we rewrite the expression so that it does not contain any negative exponents.

**Work Smart**

Remember, $x^{-1} = \dfrac{1}{x}$ and $y^{-1} = \dfrac{1}{y}$ but

$x^{-1} + y^{-1} \neq \dfrac{1}{x + y}$

$x^{-1} + y^{-1} = \dfrac{1}{x} + \dfrac{1}{y}$

$$\frac{x^{-1} + y^{-1}}{x^{-3} + y^{-3}} = \frac{\dfrac{1}{x} + \dfrac{1}{y}}{\dfrac{1}{x^3} + \dfrac{1}{y^3}}$$

**Method I**

We write the numerator of the complex rational expression as a single quotient. The LCD is $xy$.

$$\begin{aligned}\frac{1}{x} + \frac{1}{y} &= \frac{1}{x} \cdot \frac{y}{y} + \frac{1}{y} \cdot \frac{x}{x}\\ &= \frac{y}{xy} + \frac{x}{xy}\\ &= \frac{y + x}{xy}\\ &= \frac{x + y}{xy}\end{aligned}$$

Write the denominator of the complex rational expression as a single quotient. The LCD is $x^3y^3$.

$$\begin{aligned}\frac{1}{x^3} + \frac{1}{y^3} &= \frac{1}{x^3} \cdot \frac{y^3}{y^3} + \frac{1}{y^3} \cdot \frac{x^3}{x^3}\\ &= \frac{y^3}{x^3y^3} + \frac{x^3}{x^3y^3}\\ &= \frac{y^3 + x^3}{x^3y^3}\\ &= \frac{x^3 + y^3}{x^3y^3}\end{aligned}$$

Now we rewrite the complex rational expression using the numerator and denominator just found and then simplify.

$$\frac{x^{-1} + y^{-1}}{x^{-3} + y^{-3}} = \frac{\dfrac{1}{x} + \dfrac{1}{y}}{\dfrac{1}{x^3} + \dfrac{1}{y^3}} = \frac{\dfrac{x + y}{xy}}{\dfrac{x^3 + y^3}{x^3y^3}}$$

Multiply the rational expression in the numerator by the reciprocal of the rational expression in the denominator: $= \dfrac{x + y}{xy} \cdot \dfrac{x^3y^3}{x^3 + y^3}$

Factor: $= \dfrac{x + y}{xy} \cdot \dfrac{x^3y^3}{(x + y)(x^2 - xy + y^2)}$

Divide out like factors: $= \dfrac{\cancel{x + y}}{\cancel{xy}} \cdot \dfrac{x^{\cancel{3}2}y^{\cancel{3}2}}{\cancel{(x + y)}(x^2 - xy + y^2)}$

$$= \frac{x^2y^2}{x^2 - xy + y^2}$$

**Method II** The least common denominator of all denominators is $x^3y^3$. We multiply the numerator and denominator of the complex rational expression by $x^3y^3$.

$$\frac{\frac{1}{x}+\frac{1}{y}}{\frac{1}{x^3}+\frac{1}{y^3}} = \frac{\frac{1}{x}+\frac{1}{y}}{\frac{1}{x^3}+\frac{1}{y^3}}\cdot\frac{x^3y^3}{x^3y^3}$$

Distribute the LCD to each term:
$$= \frac{\frac{1}{x}\cdot x^3y^3 + \frac{1}{y}\cdot x^3y^3}{\frac{1}{x^3}\cdot x^3y^3 + \frac{1}{y^3}\cdot x^3y^3}$$

$$= \frac{x^2y^3 + x^3y^2}{y^3 + x^3}$$

Factor out common factor in numerator:
Factor the sum of two cubes in the denominator:
$$= \frac{x^2y^2(x + y)}{(x + y)(x^2 - xy + y^2)}$$

Divide out and simplify:
$$= \frac{x^2y^2}{x^2 - xy + y^2}$$

QUICK ✓ *Simplify the expression so that it does not contain any negative exponents. Use both methods and decide which method you prefer for this problem.*

5. $\dfrac{3a^{-1} + b^{-1}}{9a^{-2} - b^{-2}}$

# 6.3 Exercises

**For Extra Help:**

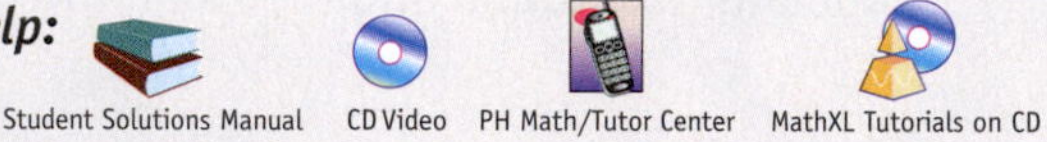

## Concepts and Vocabulary

*In Problem 1, fill in the blank.*

1. When sums and/or differences of rational expressions occur in the numerator or denominator of a quotient (or fraction), the quotient is called a ______ ______ ______.

*In Problem 2, answer True or False.*

2. To simplify a complex rational expression means to write it as a rational expression in simplest form.

3. In your own words, provide a definition for a complex rational expression.

4. Which of the two methods for simplifying complex rational expressions do you prefer? Write a paragraph supporting your opinion.

## Building Skills

*In Problems 5–14, simplify the complex rational expression using both Method I and Method II. State which method you prefer for the problem.*

5. $\dfrac{1 + \frac{1}{x}}{1 - \frac{1}{x}}$

6. $\dfrac{1 + \frac{1}{x^2}}{1 - \frac{1}{x^2}}$

7. $\dfrac{w - \frac{1}{w}}{w + \frac{1}{w}}$

8. $\dfrac{\frac{7}{w} + \frac{9}{x}}{\frac{9}{w} - \frac{7}{x}}$

**9.** $\dfrac{1 - \dfrac{a}{a-2}}{2 - \dfrac{a+2}{a}}$ **10.** $\dfrac{\dfrac{a}{a+1} - 1}{\dfrac{a+3}{a} - 2}$ **11.** $\dfrac{\dfrac{x+2}{x-1} - \dfrac{x+5}{x+3}}{x+11}$ **12.** $\dfrac{\dfrac{x+5}{x-2} - \dfrac{x+3}{x-1}}{3x+1}$

**13.** $\dfrac{\dfrac{x-1}{x-4} - \dfrac{x}{x-2}}{1 - \dfrac{3}{x-4}}$ **14.** $\dfrac{\dfrac{x-4}{x-1} - \dfrac{x}{x-3}}{3 + \dfrac{12}{x-3}}$

## Mixed Practice

*In Problems 15–36, simplify the complex rational expression using either Method I or Method II.*

**15.** $\dfrac{\dfrac{3}{y} - 1}{\dfrac{9}{y} - y}$ **16.** $\dfrac{1 - \dfrac{4}{z}}{z - \dfrac{16}{z}}$ **17.** $\dfrac{\dfrac{4n}{m} - \dfrac{4m}{n}}{\dfrac{2}{m} + \dfrac{2}{n}}$ **18.** $\dfrac{\dfrac{n^2}{m} - \dfrac{m^2}{n}}{\dfrac{1}{m} - \dfrac{1}{n}}$

**19.** $\dfrac{2 + \dfrac{3}{x}}{\dfrac{2x^2}{x+3} - 3}$ **20.** $\dfrac{1 + \dfrac{5}{x}}{1 + \dfrac{1}{x+4}}$ **21.** $\dfrac{\dfrac{x}{3} - \dfrac{3}{x}}{\dfrac{3}{x^2} - \dfrac{1}{3}}$ **22.** $\dfrac{\dfrac{5}{x} - \dfrac{x}{5}}{\dfrac{1}{5} - \dfrac{5}{x^2}}$

**23.** $\dfrac{\dfrac{2x+1}{x-1} - \dfrac{x-1}{x-3}}{x^2 - 3x - 4}$ **24.** $\dfrac{\dfrac{x+5}{x-3} - \dfrac{x}{x+4}}{3x^2 - 4x - 15}$ **25.** $\dfrac{\dfrac{x-4}{x+4} + \dfrac{x-4}{x-2}}{1 - \dfrac{2}{x-2}}$

**26.** $\dfrac{\dfrac{x-3}{x+3} + \dfrac{x-3}{x-4}}{1 + \dfrac{x+3}{x-4}}$ **27.** $\dfrac{\dfrac{z+2}{z-2} + \dfrac{z-2}{z+2}}{\dfrac{z+2}{z-2} - \dfrac{z-2}{z+2}}$ **28.** $\dfrac{\dfrac{m+3}{m-3} - \dfrac{m-3}{m+3}}{\dfrac{m+3}{m-3} + \dfrac{m-3}{m+3}}$

**29.** $\dfrac{\dfrac{b^2}{b^2-25} - \dfrac{b}{b+5}}{\dfrac{b}{b^2-25} - \dfrac{1}{b-5}}$ **30.** $\dfrac{\dfrac{-6}{x^2+5x+6}}{\dfrac{2}{x+3} - \dfrac{3}{x+2}}$ **31.** $\dfrac{3x^{-1} + 3y^{-1}}{x^{-2} - y^{-2}}$

**32.** $\dfrac{2x^{-1} + 2y^{-1}}{xy^{-1} - x^{-1}y}$ **33.** $\dfrac{(m+n)^{-1}}{m^{-1} + n^{-1}}$ **34.** $\dfrac{(x-y)^{-1}}{x^{-1} - y^{-1}}$

**35.** $\dfrac{a^{-2}b^{-1} - a^{-1}b^{-2}}{4a^{-2} - 4b^{-2}}$ **36.** $\dfrac{a^{-3} + 8b^{-3}}{a^{-2} - 4b^{-2}}$

## Applying the Concepts

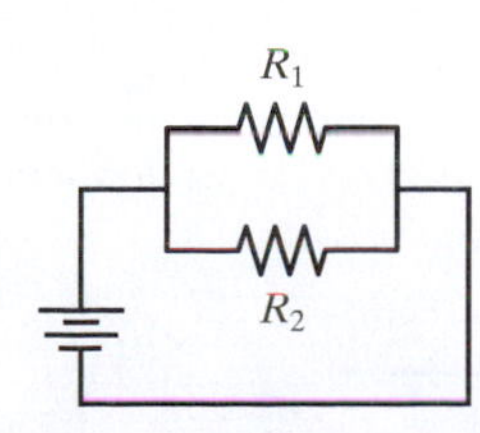


**37. Electric Circuits** An electrical circuit contains two resistors connected in parallel, as shown in the figure. If the resistance of each is $R_1$ and $R_2$ ohms, respectively, then their combined resistance $R$ is given by the formula

$$R = \frac{1}{\dfrac{1}{R_1} + \dfrac{1}{R_2}}$$

**(a)** Express $R$ as a simplified rational expression.
**(b)** Evaluate the rational expression if $R_1 = 4$ ohms and $R_2 = 10$ ohms.

**38. Electric Circuits** An electrical circuit contains three resistors connected in parallel. If the resistance of each is $R_1$, $R_2$, and $R_3$ ohms, respectively, then their combined resistance is given by the formula

$$R = \frac{1}{\frac{1}{R_1} + \frac{1}{R_2} + \frac{1}{R_3}}$$

**(a)** Express $R$ as a simplified rational expression.
**(b)** Evaluate the rational expression if $R_1 = 4$ ohms, $R_2 = 6$ ohms and $R_3 = 10$ ohms.

**39. Future Value of Money** The value of an account $V$ in which $P$ dollars is deposited every year for the next 5 years paying an interest rate $i$ (expressed as a decimal) is given by the formula

$$V = \frac{Pi}{1 - \frac{1}{(1 + i)^5}}$$

**(a)** Express $V$ as a simplified rational expression.
**(b)** Determine the value of an account paying 5% when the annual deposit is \$1000. Express your answer to the nearest penny.

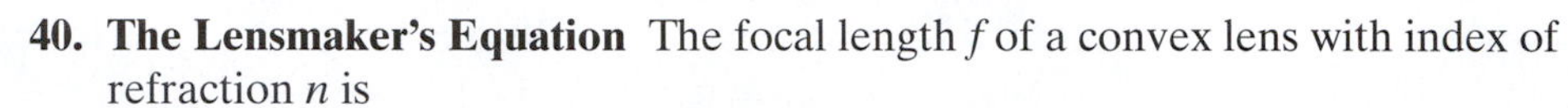

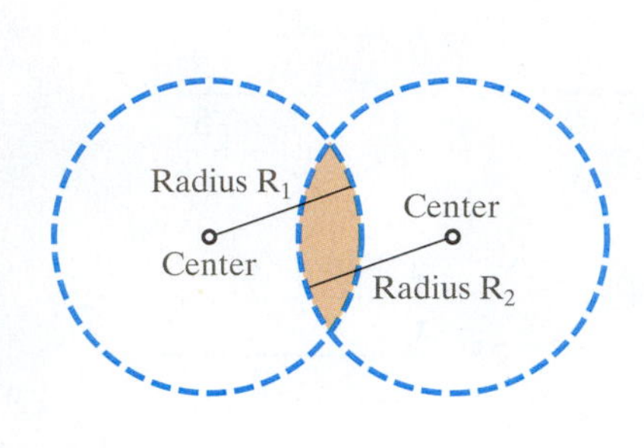


**40. The Lensmaker's Equation** The focal length $f$ of a convex lens with index of refraction $n$ is

$$f = \frac{1}{(n - 1)\left[\frac{1}{R_1} + \frac{1}{R_2}\right]}$$

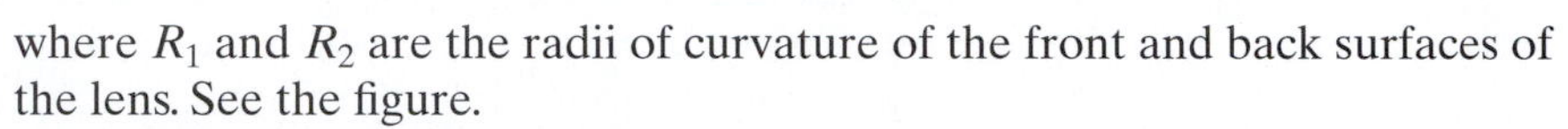

where $R_1$ and $R_2$ are the radii of curvature of the front and back surfaces of the lens. See the figure.

**(a)** Express $f$ as a simplified rational expression.
**(b)** Determine the focal length of a lens for $n = 1.5$, $R_1 = 0.5$ meter and $R_2 = 0.3$ meter.

**41. Harmonic Mean** The harmonic mean is used to determine an average value when the data are measured as a rate of change such as 50 miles per hour. The harmonic mean is found using the formula

$$H = \frac{n}{\frac{1}{x_1} + \frac{1}{x_2} + \cdots + \frac{1}{x_n}}$$

where $x_1, x_2, \ldots, x_n$ are the $n$ rates of change.

**(a)** Suppose that a family drove from Chicago, Illinois to Naples, Florida. The distance each way is about 1200 miles. The trip from Chicago to Naples resulted in an average speed of 48 miles per hour, while the return trip resulted in an average speed of 52 miles per hour. Compute the average speed of the entire trip.
**(b)** A 300-Kb file is downloaded four separate times resulting in download speeds of 4.3 Kb/s, 4.1 Kb/s, 3.8 Kb/s, and 4.3 Kb/s. Compute the average download time.

## Extending the Concepts

**42. Fibonacci Strikes Again!** Write each of the following expressions in the form $\frac{ax + b}{bx + c}$. (**Hint:** Use Method I.)

**(a)** $1 + \frac{1}{1 + \frac{1}{x}}$

**(b)** $1 + \frac{1}{1 + \frac{1}{1 + \frac{1}{x}}}$

**(c)** $1 + \cfrac{1}{1 + \cfrac{1}{1 + \cfrac{1}{1 + \cfrac{1}{x}}}}$

**(d)** $1 + \cfrac{1}{1 + \cfrac{1}{1 + \cfrac{1}{1 + \cfrac{1}{1 + \cfrac{1}{x}}}}}$

**(e)** Write down the values of $a$, $b$, and $c$ from part (a). Now write down the values of $a$, $b$, and $c$ from part (b), followed by the values of $a$, $b$, and $c$ from part (c), followed by the values of $a$, $b$, and $c$ from part (d). What is the pattern? Write the sequence of numbers in increasing order. This sequence of numbers forms the first six numbers in the **Fibonacci sequence.**

### Synthesis Review

*In Problems 43–48, solve each equation.*

**43.** $4x + 3 = 15$

**44.** $-5a + 2 = 22$

**45.** $\frac{1}{2}w + 3 = 5$

**46.** $\frac{2}{3}x - \frac{5}{7}(x + 21) = \frac{11}{21}x + \frac{4}{7}$

**47.** $y^2 - 5y = 50$

**48.** $3p^2 + 19p = 14$

## PUTTING THE CONCEPTS TOGETHER (SECTIONS 6.1–6.3)

*These problems cover important concepts from Sections 6.1 to 6.3. We designed these problems so that you can review the chapter so far and show your mastery of the concepts. Take time to work these problems before proceeding with the next section. The answers to these problems are located at the back of the text starting on page AN-29.*

**1.** Determine the domain of the rational function $g(x) = \frac{3x + 1}{3x^2 - 17x - 6}$.

*In Problems 2 and 3, write each rational expression in lowest terms.*

**2.** $\frac{24n - 4n^2}{2n^2 - 9n - 18}$

**3.** $\frac{2p^2 - pq - 10q^2}{3p^2 + 2pq - 8q^2}$

*In Problems 4–8, perform the indicated operations.*

**4.** $\frac{a^2 - 16}{12a^2 + 48a} \cdot \frac{6a^3 - 30a^2}{a^2 + 2a - 24}$

**5.** $\dfrac{\dfrac{x^2 + x - 2}{3x^2 - 5x - 2}}{\dfrac{3x^2 - 2x - 1}{x^2 - 9x + 14}}$

**6.** $\frac{x^2 - 10}{x^2 - 4} - \frac{3x}{x^2 - 4}$

**7.** $\frac{3n}{n^2 - 7n + 10} - \frac{2n}{n^2 - 8n + 15}$

**8.** $\frac{3y + 2}{y^2 + 5y - 24} + \frac{7}{y^2 + 4y - 32}$

*In Problems 9 and 10, use the functions $f(x) = \frac{2x + 1}{x^2 - 11x + 28}$, $g(x) = \frac{3x - 12}{4x^2 + 4x + 1}$, and $h(x) = \frac{3x}{x - 7}$ to find each difference or product.*

**9.** $P(x) = f(x) \cdot g(x)$

**10.** $D(x) = h(x) - f(x)$

*In Problems 11 and 12, simplify each complex rational expression using the method you wish.*

**11.** $\dfrac{\dfrac{1}{m^2} - \dfrac{1}{n^2}}{\dfrac{1}{m} - \dfrac{1}{n}}$

**12.** $\dfrac{\dfrac{z^2 - 2}{z^2 - 4} + \dfrac{7}{z - 2}}{\dfrac{z^2 + z - 24}{z^2 - 4} - \dfrac{2}{z + 2}}$

# 6.4 Rational Equations

**OBJECTIVES**

1. Solve Equations Containing Rational Expressions
2. Solve Equations Involving Rational Functions

### *Preparing for Rational Equations*

*Before getting started, take the following readiness quiz. If you get a problem wrong, go back to the section cited and review the material.*

**1.** Solve: $\frac{2}{3}x + \frac{1}{2} = \frac{3}{4}$ [Section 1.1, pp. 53–57]

**2.** Factor: $3z^2 + 11z - 4$ [Section 5.5, pp. 401–407]

**3.** Solve: $6y^2 - y - 12 = 0$ [Section 5.8, pp. 423–426]

**4.** Determine which of the following is in the domain of $\frac{x+4}{x^2 - 5x - 24}$

**(a)** $x = -4$ **(b)** $x = 8$ [Section R.5, pp. 46–47]

**5.** If $f(x) = x^2 - 3x - 15$, solve $f(x) = 3$ [Section 5.8, pp. 427–428]

**6.** If $g(4) = 3$, what point is on the graph of $g$? [Section 2.4, p. 172]

*Preparing for...Answers* **1.** $\left\{\frac{3}{8}\right\}$ **2.** $(3z - 1)(z + 4)$ **3.** $\left\{-\frac{4}{3}, \frac{3}{2}\right\}$ **4. (a)** Yes **(b)** No **5.** $\{-3, 6\}$ **6.** $(4, 3)$

## 1 Solve Equations Containing Rational Expressions

Up to this point, we have learned how to solve linear equations (Section 1.1), quadratic equations (Section 5.8), and equations that contain polynomial expressions that can be factored (Section 5.8). We now introduce another type of equation, the *rational equation*. A **rational equation** is an equation that contains a rational expression. Examples of rational equations are

$$\frac{3}{x+4} = \frac{5}{x-1} + \frac{1}{x^2 + 3x - 4} \quad \text{and} \quad \frac{x-5}{x^2 + 4x - 12} = 3$$

Remember, the domain of a variable is the set of all values that the variable can take on. Because division by zero is not defined, we exclude from the domain all values of the variable that result in division by zero.

**EXAMPLE 1 How to Solve a Rational Equation**

Solve: $\frac{2x-1}{x-3} = \frac{2(x+1)}{x-2}$

### Step-by-Step Solution

**Step 1:** Determine the domain of the variable in the rational equation.

Because $x = 2$ and $x = 3$ result in division by zero, the domain of $x$ is

$$\{x \mid x \neq 2, x \neq 3\}$$

**Step 2:** Determine the least common denominator (LCD) of all the denominators.

The LCD is $(x - 3)(x - 2)$.

**Step 3:** Multiply both sides of the equation by the LCD and simplify the expression on each side of the equation.

$$\frac{2x-1}{x-3} = \frac{2(x+1)}{x-2}$$

Multiply both sides by the LCD, $(x-3)(x-2)$: $(x-3)(x-2) \cdot \frac{2x-1}{x-3} = (x-3)(x-2) \cdot \frac{2(x+1)}{x-2}$

Divide out like factors: $(x-2)(2x-1) = 2(x-3)(x+1)$

FOIL: $2x^2 - 5x + 2 = 2(x^2 - 2x - 3)$

Distribute the 2: $2x^2 - 5x + 2 = 2x^2 - 4x - 6$

| | | |
|---|---|---|
| **Step 4:** Solve the resulting equation. | Subtract $2x^2$ from both sides: | $-5x + 2 = -4x - 6$ |
| | Add $4x$ to both sides; Subtract 2 from both sides: | $-x = -8$ |
| | Divide both sides by $-1$: | $x = 8$ |
| **Step 5:** Verify your solution using the original equation. | Let $x = 8$ in the original equation: | $\frac{2(8) - 1}{8 - 3} \stackrel{?}{=} \frac{2(8 + 1)}{8 - 2}$ |
| | | $\frac{16 - 1}{5} \stackrel{?}{=} \frac{2(9)}{6}$ |
| | | $\frac{15}{5} \stackrel{?}{=} \frac{18}{6}$ |
| | | $3 = 3$ True |

The solution checks, so the solution set is $\{8\}$.

We summarize the steps that can be used to solve any rational equation.

### Solving a Rational Equation

**Step 1:** Determine the domain of the variable in the rational equation.

**Step 2:** Determine the least common denominator (LCD) of all the denominators.

**Step 3:** Multiply both sides of the equation by the LCD and simplify the expression on each side of the equation.

**Step 4:** Solve the resulting equation.

**Step 5:** Verify your solution using the original equation.

**In Words**

The purpose of Step 3 is to "clear the fractions" so that we transform the equation into one that we already know how to solve, such as a linear equation.

**QUICK ✓**

**1.** Solve: $\frac{x - 4}{x^2 + 4} = \frac{3}{3x + 2}$

## EXAMPLE 2 Solving a Rational Equation

Solve: $\frac{2}{x} - \frac{1}{6} = \frac{5}{2x} - \frac{1}{3}$

### Solution

The domain of the variable is $\{x | x \neq 0\}$. The LCD of all denominators is $6x$, so we multiply both sides of the equation by $6x$.

$$\frac{2}{x} - \frac{1}{6} = \frac{5}{2x} - \frac{1}{3}$$

$$6x \cdot \left(\frac{2}{x} - \frac{1}{6}\right) = 6x \cdot \left(\frac{5}{2x} - \frac{1}{3}\right)$$

Distribute the $6x$: $$6x \cdot \frac{2}{x} - 6x \cdot \frac{1}{6} = 6x \cdot \frac{5}{2x} - 6x \cdot \frac{1}{3}$$

Simplify: $$12 - x = 15 - 2x$$

Add $2x$ to both sides: $$12 + x = 15$$

Subtract 12 from both sides: $$x = 3$$

**Check** Let $x = 3$ in the original equation:
$$\frac{2}{3} - \frac{1}{6} \stackrel{?}{=} \frac{5}{2 \cdot 3} - \frac{1}{3}$$
$$\frac{4}{6} - \frac{1}{6} \stackrel{?}{=} \frac{5}{6} - \frac{2}{6}$$
$$\frac{3}{6} = \frac{3}{6} \quad \text{True}$$

The solution checks, so the solution set is $\{3\}$.

**QUICK ✓**

**2.** Solve: $\frac{5}{x} + \frac{1}{4} = \frac{3}{2x} - \frac{3}{2}$

## EXAMPLE 3 Solving a Rational Equation

Solve: $\frac{3}{p^2 - 4p + 3} + \frac{6}{p^2 - 2p - 3} = \frac{5}{p^2 - 1}$

### Solution

First, we find the domain of the variable, $p$.

$p^2 - 4p + 3 = (p - 3)(p - 1)$, so $p \neq 3, p \neq 1$ in the first term.

$p^2 - 2p - 3 = (p - 3)(p + 1)$, so $p \neq 3, p \neq -1$ in the second term.

$p^2 - 1 = (p - 1)(p + 1)$, so $p \neq 1, p \neq -1$ in the third term.

The domain of the variable $p$ is $\{p | p \neq -1, p \neq 1, p \neq 3\}$. The factored form of each denominator found above allows us to determine that the LCD of all denominators is $(p - 1)(p + 1)(p - 3)$, so we multiply both sides of the equation by $(p - 1)(p + 1)(p - 3)$.

$$\frac{3}{p^2 - 4p + 3} + \frac{6}{p^2 - 2p - 3} = \frac{5}{p^2 - 1}$$

$$(p - 1)(p + 1)(p - 3) \cdot \left(\frac{3}{(p - 3)(p - 1)} + \frac{6}{(p - 3)(p + 1)}\right) = (p - 1)(p + 1)(p - 3) \cdot \frac{5}{(p - 1)(p + 1)}$$

Distribute the LCD and simplify: $3(p + 1) + 6(p - 1) = 5(p - 3)$

Distribute: $3p + 3 + 6p - 6 = 5p - 15$

Combine like terms: $9p - 3 = 5p - 15$

Subtract $5p$ from both sides; add 3 to both sides: $4p = -12$

Divide both sides by 4: $p = -3$

We leave the check to you. The solution set is $\{-3\}$.

**QUICK ✓**

**3.** Solve: $\frac{3}{x^2 + 5x + 4} + \frac{2}{x^2 - 3x - 4} = \frac{4}{x^2 - 16}$

## EXAMPLE 4 A Rational Equation with No Solution

Solve: $\dfrac{3}{y^2 - 5y + 4} + \dfrac{2}{y^2 - 10y + 24} = \dfrac{2}{y^2 - 7y + 6}$

### Solution

First, we find the domain of the variable, $y$.

$y^2 - 5y + 4 = (y - 4)(y - 1)$, so $y \neq 4$, $y \neq 1$ in the first term.

$y^2 - 10y + 24 = (y - 6)(y - 4)$, so $y \neq 6$, $y \neq 4$ in the second term.

$y^2 - 7y + 6 = (y - 6)(y - 1)$, so $y \neq 6$, $y \neq 1$ in the third term.

The domain of the variable $y$ is $\{y \mid y \neq 1, y \neq 4, y \neq 6\}$. The LCD of all denominators is $(y - 1)(y - 4)(y - 6)$. Multiply both sides of the equation by the LCD.

$$\frac{3}{y^2 - 5y + 4} + \frac{2}{y^2 - 10y + 24} = \frac{2}{y^2 - 7y + 6}$$

$$(y - 1)(y - 4)(y - 6) \cdot \left(\frac{3}{(y - 1)(y - 4)} + \frac{2}{(y - 4)(y - 6)}\right) = (y - 1)(y - 4)(y - 6) \cdot \frac{2}{(y - 6)(y - 1)}$$

Distribute the LCD and simplify: $3(y - 6) + 2(y - 1) = 2(y - 4)$

Distribute: $3y - 18 + 2y - 2 = 2y - 8$

Combine like terms: $5y - 20 = 2y - 8$

Subtract $2y$ from both sides; Add 20 to both sides: $3y = 12$

Divide both sides by 3: $y = 4$

Notice that $y = 4$ is not in the domain of the variable $y$, so there is no solution to the equation. The solution set is $\{\ \}$ or $\emptyset$.

**In Words**

The word *extraneous* means "not constituting a vital part."

We call $y = 4$ an *extraneous solution*. **Extraneous solutions** are results that develop through the solution process but do not satisfy the original equation.

**Quick ✓**

**4.** Solve: $\dfrac{5}{z^2 + 2z - 3} - \dfrac{3}{z^2 + z - 2} = \dfrac{1}{z^2 + 5z + 6}$

## EXAMPLE 5 Solving a Rational Equation That Leads to a Quadratic Equation

Solve: $\dfrac{w + 3}{w - 1} + \dfrac{w + 5}{w} = \dfrac{3w + 1}{w - 1}$

### Solution

The domain of the variable $w$ is $\{w \mid w \neq 0, w \neq 1\}$. The LCD of all denominators is $w(w - 1)$, so we multiply both sides of the equation by $w(w - 1)$.

$$\frac{w + 3}{w - 1} + \frac{w + 5}{w} = \frac{3w + 1}{w - 1}$$

$$w(w - 1) \cdot \left(\frac{w + 3}{w - 1} + \frac{w + 5}{w}\right) = w(w - 1) \cdot \frac{3w + 1}{w - 1}$$

**Work Smart**
Notice that this equation has a term with a square in it. So, we know the equation is a quadratic equation—this is why we rewrite the equation with 0 on one side (standard form) rather than isolating the variable (as we do in solving linear equations).

Distribute the LCD: $w(w + 3) + (w - 1)(w + 5) = w(3w + 1)$

$w^2 + 3w + w^2 + 5w - w - 5 = 3w^2 + w$

Combine like terms: $2w^2 + 7w - 5 = 3w^2 + w$

Put equation in standard form: $0 = w^2 - 6w + 5$

If $a = b$, then $b = a$: $w^2 - 6w + 5 = 0$

Factor: $(w - 5)(w - 1) = 0$

Zero-Product Property: $w = 5$ or $w = 1$

Since $w = 1$ is not in the domain of the variable, it is an extraneous solution. The only potential solution is 5.

**Check** Let $w = 5$ in the original equation.

$$\frac{5 + 3}{5 - 1} + \frac{5 + 5}{5} \stackrel{?}{=} \frac{3(5) + 1}{5 - 1}$$

$$\frac{8}{4} + \frac{10}{5} \stackrel{?}{=} \frac{16}{4}$$

$$2 + 2 = 4 \quad \text{True}$$

The solution set is $\{5\}$.

**QUICK ✓**

**5.** Solve: $\dfrac{z + 1}{z + 4} + \dfrac{z + 1}{z - 3} = \dfrac{z^2 + z + 16}{z^2 + z - 12}$

## 2 Solve Equations Involving Rational Functions

Now let's look at a problem involving a rational function that leads to a rational equation.

### EXAMPLE 6 Working with Rational Functions

For the function $f(x) = x + \dfrac{4}{x}$, solve $f(x) = 5$. What point(s) are on the graph of $f$?

**Solution**

We wish to solve the equation $x + \dfrac{4}{x} = 5$. The domain of the variable is $\{x | x \neq 0\}$.

The LCD of all denominators is $x$, so we multiply both sides of the equation by $x$.

$$x \cdot \left(x + \frac{4}{x}\right) = 5 \cdot x$$

Distribute: $x^2 + 4 = 5x$

Subtract $5x$ from both sides: $x^2 - 5x + 4 = 0$

Factor: $(x - 4)(x - 1) = 0$

Zero-Product Property: $x - 4 = 0$ or $x - 1 = 0$

$x = 4$ or $x = 1$

We leave the check to you. We have that $f(1) = 5$, so the point $(1, 5)$ is on the graph of $f$. We have that $f(4) = 5$ so the point $(4, 5)$ is on the graph of $f$.

**QUICK ✓**

**6.** For the function $f(x) = 2x - \dfrac{3}{x}$, solve $f(x) = 1$. What point(s) are on the graph of $f$?

### EXAMPLE 7 An Application of Rational Functions: Drug Concentration

The concentration $C$ of a drug in a patient's bloodstream in milligrams per liter $t$ hours after ingestion is modeled by

$$C(t) = \frac{40t}{t^2 + 9}$$

When will the concentration of the drug be 4 milligrams per liter?

**Solution**

Since we want to know when the concentration of the drug is 4, we wish to solve the equation $C(t) = 4$.

$$\frac{40t}{t^2 + 9} = 4$$

Multiply both sides by $t^2 + 9$: $40t = 4(t^2 + 9)$

Divide both sides by 4: $10t = t^2 + 9$

Subtract $10t$ from both sides: $0 = t^2 - 10t + 9$

Factor: $0 = (t - 1)(t - 9)$

Zero-product property: $t = 1$ or $t = 9$

The concentration of the drug will be 4 milligrams per liter after 1 hour and after 9 hours.

**QUICK ✓**

7. The concentration $C$ of a drug in a patient's bloodstream in milligrams per liter $t$ hours after ingestion is modeled by $C(t) = \dfrac{50t}{t^2 + 6}$. When will the concentration of the drug be 4 milligrams per liter?

## 6.4 Exercises

*For Extra Help:*     

Student Solutions Manual | CD Video | PH Math/Tutor Center | MathXL Tutorials on CD | MathXL® | MyMathLab

### Concepts and Vocabulary

*In Problems 1 and 2, fill in the blanks.*

1. _____ _____ are results that develop through the solution process but do not satisfy the original equation.

2. To eliminate the denominator from all the expressions in a rational equation, we multiply both sides of the equation by the _____ _____ _____.

*In Problems 3 and 4, answer True or False to each statement.*

3. A rational equation is an equation that contains a rational expression.

4. Some rational equations have no solution.

5. Explain the role that domain plays in solving a rational equation.

6. Is the solution set to the equation $\dfrac{x - 6}{x - 6} = 1$ the set of all real numbers? Explain.

## Building Skills

*In Problems 7–36, solve each equation. Be sure to verify your results.*

**7.** $\frac{3}{z} - \frac{1}{2z} = -\frac{5}{8}$

**8.** $\frac{8}{p} + \frac{1}{4p} = \frac{11}{8}$

**9.** $\frac{4}{x+2} = \frac{7}{x+4}$

**10.** $\frac{3}{x-5} = \frac{-2}{x+2}$

**11.** $\frac{y+2}{y-5} = \frac{y+6}{y+1}$

**12.** $\frac{w-4}{w+1} = \frac{w-3}{w+3}$

**13.** $\frac{x+8}{x+4} = \frac{x+2}{x-2}$

**14.** $\frac{2x+1}{x+3} = \frac{4(x-1)}{2x+3}$

**15.** $a - \frac{5}{a} = 4$

**16.** $m + \frac{8}{m} = 6$

**17.** $6p - \frac{3}{p} = 7$

**18.** $8b - \frac{3}{b} = 2$

**19.** $1 + \frac{10}{w-3} = \frac{2}{w}$

**20.** $2 - \frac{3}{p+2} = \frac{6}{p}$

**21.** $\frac{5-p}{p-5} + 2 = \frac{1}{p}$

**22.** $\frac{3-y}{y-3} + 2 = \frac{2}{y}$

**23.** $\frac{3}{2} + \frac{5}{x-3} = \frac{x+9}{2x-6}$

**24.** $\frac{4}{3} + \frac{7}{x-4} = \frac{x-1}{3x-12}$

**25.** $1 + \frac{3}{x+3} = \frac{4}{x-3}$

**26.** $\frac{5}{x+2} = 1 - \frac{3}{x-2}$

**27.** $\frac{4}{x-5} + \frac{3}{x-2} = \frac{x+1}{x^2-7x+10}$

**28.** $\frac{4}{x+3} + \frac{5}{x-6} = \frac{4x+1}{x^2-3x-18}$

**29.** $\frac{2}{x+5} = \frac{2x}{x^2-25} - \frac{3}{x-5}$

**30.** $\frac{3}{x-4} = \frac{5x+4}{x^2-16} - \frac{4}{x+4}$

**31.** $\frac{3}{x-1} - \frac{2}{x+4} = \frac{x^2+8x+6}{x^2+3x-4}$

**32.** $\frac{5}{z-4} + \frac{3}{z-2} = \frac{z^2-z-2}{z^2-6z+8}$

**33.** $\frac{7}{y^2+y-12} - \frac{4y}{y^2+7y+12} = \frac{6}{y^2-9}$

**34.** $\frac{3}{x^2-5x-6} + \frac{3}{x^2-7x+6} = \frac{6}{x^2-1}$

**35.** $\frac{2x+3}{x-3} + \frac{x+6}{x-4} = \frac{x+6}{x-3}$

**36.** $\frac{x+5}{x+1} + 1 = \frac{x-5}{x-2}$

**37.** For the function $f(x) = x + \frac{9}{x}$, solve $f(x) = 10$. What point(s) are on the graph of $f$?

**38.** For the function $f(x) = x + \frac{7}{x}$, solve $f(x) = 8$. What point(s) are on the graph of $f$?

**39.** For the function $f(x) = 2x + \frac{4}{x}$, solve $f(x) = -9$. What point(s) are on the graph of $f$?

**40.** For the function $f(x) = 2x + \frac{8}{x}$, solve $f(x) = -10$. What point(s) are on the graph of $f$?

**41.** For the function $f(x) = \frac{x+3}{x-4}$, solve $f(x) = \frac{9}{2}$. What point(s) are on the graph of $f$?

**42.** For the function $f(x) = \frac{x+5}{x-3}$, solve $f(x) = \frac{1}{5}$. What point(s) are on the graph of $f$?

**43.** Let $f(x) = \dfrac{x + 2}{2x + 9}$ and $g(x) = \dfrac{x - 1}{x + 3}$. For what value(s) of $x$ does $f(x) = g(x)$? What are the point(s) of intersection of the graphs of $f$ and $g$?

**44.** Let $f(x) = \dfrac{4x + 1}{8x + 5}$ and $g(x) = \dfrac{x - 4}{2x - 7}$. For what value(s) of $x$ does $f(x) = g(x)$? What are the point(s) of intersection of the graphs of $f$ and $g$?

## Mixed Practice

*In Problems 45–54, solve each equation. Be sure to verify your results.*

**45.** $\dfrac{4}{z + 4} - \dfrac{3}{4} = \dfrac{5z + 2}{4z + 16}$

**46.** $\dfrac{2b - 1}{b + 5} - \dfrac{2}{3} = \dfrac{1}{3b + 15}$

**47.** $x + \dfrac{9}{x} = 6$

**48.** $p + \dfrac{25}{p} = 10$

**49.** $\dfrac{2}{z^2 + 2z - 3} + \dfrac{3}{z^2 + 4z + 3} = \dfrac{6}{z^2 - 1}$

**50.** $\dfrac{3}{a^2 + 3a - 10} + \dfrac{2}{a^2 + 7a + 10} = \dfrac{4}{a^2 - 4}$

**51.** $\dfrac{4}{x} - \dfrac{5}{2x} = \dfrac{3}{4}$

**52.** $\dfrac{9}{b} + \dfrac{4}{5b} = \dfrac{7}{10}$

**53.** $\dfrac{3y + 1}{y - 1} + 3 = \dfrac{y + 2}{y + 1}$

**54.** $\dfrac{x + 3}{x - 2} + 4 = \dfrac{x + 2}{x + 1}$

## Applying the Concepts

**55. Average Cost** Suppose that the average daily cost $\overline{C}$ of manufacturing $x$ bicycles is given by the function

$$\overline{C}(x) = \frac{x^2 + 75x + 5000}{x}$$

Determine the level of production for which the average daily cost will be \$225.

**56. Population** When loggers began cutting in a region in the Amazon rain forest, a rare insect species was discovered. To protect the species, government scientists declared the insect endangered and moved them into a protected area. The population $P$ of the insect $t$ months after being transplanted is modeled by

$$P(t) = \frac{200(1 + 0.4t)}{2(1 + 0.01t)}$$

Predict when the population will be 1350 insects. Round your answer to two decimal places.

**57. Cost-Benefit Model** Environmental scientists often use cost-benefit models to estimate the cost of removing a pollutant from the environment as a function of the percentage of pollutant removed. Suppose a cost-benefit function for the cost $C$ (in millions of dollars) of removing $x$ percent of the pollutants from Maple Lake is given by

$$C(x) = \frac{25x}{100 - x}$$

**(a)** If the federal government budgets \$100 million to clean up the lake, what percent of the pollutants can be removed?
**(b)** If the federal government budgets \$225 million to clean up the lake, what percent of the pollutants can be removed?

**58. The Learning Curve** Suppose that a student is given 500 vocabulary words to learn. The function

$$P(x) = \frac{0.8x - 0.8}{0.8x + 0.1}$$

models the proportion $P$ of words learned after $x$ hours of studying.

**(a)** How long would a student need to study to learn 70% (0.7) of the words?

**(b)** How long would a student need to study to learn 400 words?

**59. Runs in Baseball** In his book *Moneyball,* author Michael Lewis cites a formula for predicting the number of runs a team will score in a season. According to the formula, the number of runs $R$ a team will score is given by

$$R = \frac{(h + w)t}{b + w}$$

where $h$ is the number of hits, $w$ is the number of walks, $t$ is the total number of bases, and $b$ is the number of official at-bats. Suppose that the Oakland Athletics scored 750 runs in a season, had 1400 hits, 2250 total bases, and 5500 total at-bats. Use the formula to predict the number of walks that the Athletics had.

△ **60. Regular Polygons** A regular polygon is a polygon that is both equilateral and equiangular. The measure $I$ of each interior angle of a regular polygon with $n$ sides is $I = \dfrac{180°(n - 2)}{n}$. Find the number of sides of a regular polygon whose interior angles measure 135°.

## Extending the Concepts

**61.** Make up a rational equation that has one real solution.

**62.** Make up a rational equation that has no real solution.

**63.** Solve: $\left(\dfrac{4}{x + 3}\right)^2 - 5\left(\dfrac{4}{x + 3}\right) + 6 = 0$

**64.** Solve: $2 + 11a^{-1} = -12a^{-2}$

## Synthesis Review

*In Problems 65–69, simplify or solve.*

**65.** $\dfrac{2a^6}{(a^3)^2} - \dfrac{5a^2}{a^3} = 3\dfrac{a}{a^3}$

**66.** $\left(\dfrac{z^{-2}}{2z^{-3}}\right)^{-1} + 3(z - 1)^{-1}$

**67.** $\dfrac{3}{x - 2} - \dfrac{2x + 1}{x + 1}$

**68.** $\dfrac{5}{x - 6} + \dfrac{2}{x + 2} = \dfrac{1}{x^2 - 4x - 12}$

**69.** $\dfrac{x + 1}{2x + 3} - \dfrac{3}{x - 4} = \dfrac{-3}{2x^2 - 5x - 12}$

**70.** Write a sentence or two explaining the difference between "simplify" and "solve."

## The Graphing Calculator

We can use a graphing calculator to approximate solutions to rational equations using the INTERSECT or ZERO (or ROOT) feature of the graphing calculator. We use the ZERO or ROOT feature of the graphing calculator when one side of the equation is 0; we use the INTERSECT feature when neither side of the equation is 0. For example, to solve $\dfrac{x - 4}{x + 1} = 4$, we would graph $Y_1 = \dfrac{x - 4}{x + 1}$ and $Y_2 = 4$. The $x$-coordinates of the point(s) of intersection represent the solution set as shown in Figure 1.

**Figure 1**

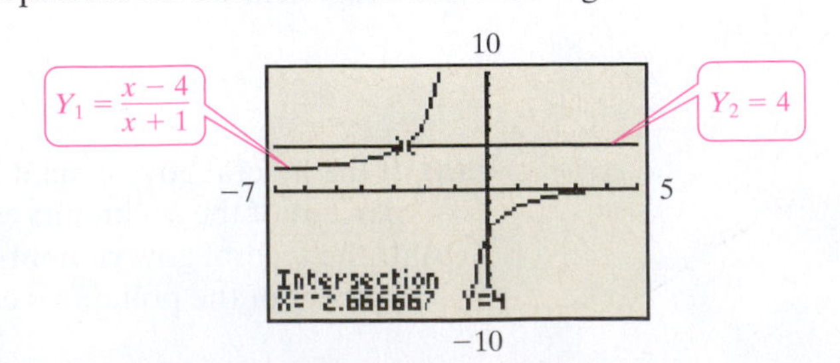

When using a graphing calculator to approximate solutions to equations, we will express the solution as a decimal rounded to two decimal places. The solution to the equation $\frac{x-4}{x+1} = 4$ is $x = -2.67$ rounded to two decimal places.

*In Problems 71–76, use a graphing utility to solve the equation by graphing each side of the equation and finding the point(s) of intersection.*

**71.** $\frac{x-4}{x+4} = \frac{1}{2}$ **72.** $\frac{x-6}{x+1} = \frac{2}{3}$ **73.** $\frac{3}{5} + \frac{4}{x+6} = \frac{x+12}{5x+30}$

**74.** $\frac{4}{3} + \frac{7}{x-4} = \frac{-7}{3x-12}$ **75.** $\frac{2x^2+11x+12}{x+4} = -5$ **76.** $\frac{3x^2+10x+3}{x+3} = -8$

# 6.5 Rational Inequalities

**OBJECTIVE**

1 Solve a Rational Inequality

### *Preparing for Rational Inequalities*

*Before getting started, take the following readiness quiz. If you get a problem wrong, go back to the section cited and review the material.*

**1.** Write $-1 < x \le 8$ in interval notation. [Section 1.4, pp. 90–93]

**2.** Solve: $2x + 3 > 4x - 9$ [Section 1.4, pp. 93–96]

Back in Section 1.4, we solved linear inequalities in one variable such as $2x - 3 > 4x + 5$. We were able to solve these inequalities using methods that were similar to solving linear equations. We also learned to represent the solution set to such an inequality using either set-builder notation or interval notation.

Although the approach to solving inequalities involving rational expressions is not a simple extension of solving rational equations, we will use the skills developed in solving equations to solve rational inequalities.

## 1 Solve a Rational Inequality

A **rational inequality** is an inequality that contains a rational expression. Examples of rational inequalities include

$$\frac{1}{x} > 1 \qquad \frac{x-1}{x+5} \le 0 \qquad \frac{x^2+3x+2}{x-5} > 0 \qquad \frac{3}{x-5} < \frac{4x}{2x-1} + \frac{1}{x}$$

*Preparing for...Answers* **1.** $(-1, 8]$ **2.** $\{x|x < 6\}$ or $(-\infty, 6)$

There are two keys to solving rational inequalities:

1. The quotient of two positive numbers is positive; the quotient of a positive and negative number is negative; and the quotient of two negative numbers is positive.
2. A rational expression may change signs (positive to negative or negative to positive) on either side of a value of the variable that makes the rational expression equal to 0 or for values for which the rational expression is undefined.

### EXAMPLE 1 How to Solve a Rational Inequality

Solve $\frac{x+3}{x-4} \ge 0$. Graph the solution set.

**Step-by-Step Solution**

**Step 1:** Write the inequality so that a rational expression is on one side of the inequality and 0 is on the other. Be sure to write the rational expression as a single quotient.

The rational expression is already in the form that we need.

$$\frac{x+3}{x-4} \ge 0$$

(continued)

**Step 2:** Determine the numbers for which the rational expression equals 0 or is undefined.

The rational expression will equal 0 when $x = -3$. The rational expression is undefined when $x = 4$.

**Step 3:** Use the numbers found in Step 2 to separate the real number line into intervals.

We separate the real number line into the following intervals:

$(-\infty, -3)$ $(-3, 4)$ $(4, \infty)$

−5 −4 −3 −2 −1 0 1 2 3 4 5 6 7

Because the rational expression is undefined at $x = 4$, we plot an open circle at 4.

**Step 4:** Choose a test point within each interval formed in Step 3 to determine the sign of $x + 3$ and $x - 4$. Then determine the sign of the quotient.

In the interval $(-\infty, -3)$ we choose a test point of $-4$. The expression $x + 3$ equals $-1$ when $x = -4$. The expression $x - 4$ equals $-8$ when $x = -4$. Since the quotient of two negatives is positive, the expression $\frac{x+3}{x-4}$ will be positive when $x = -4$. So the expression $\frac{x+3}{x-4}$ will be positive for all $x$ in the interval $(-\infty, -3)$. In the interval $(-3, 4)$ we choose a test point of 0. For $x = 0$, $x + 3$ is positive, while $x - 4$ is negative, so $\frac{x+3}{x-4}$ will be negative when $x = 0$. In the interval $(4, \infty)$ we choose a test point of 5. For $x = 5$, both $x + 3$ and $x - 4$ are positive, so $\frac{x+3}{x+4}$ will be positive. Table 1 shows these results, the sign of $\frac{x+3}{x-4}$ in each interval and the value of $\frac{x+3}{x-4}$ at $x = -3$ and $x = 4$. We want to know where $\frac{x+3}{x-4}$ is greater than or equal to zero, so we include the value of $x$ where $\frac{x+3}{x-4}$ is equal to zero. The solution set is $\{x | x \le -3 \text{ or } x > 4\}$ using set-builder notation; the solution is $(-\infty, -3] \cup (4, \infty)$ using interval notation. Notice that $-3$ is part of the solution set since $x = -3$ causes $\frac{x+3}{x-4}$ to equal zero, but 4 is not part of the solution set because it is not in the domain of $\frac{x+3}{x-4}$. Figure 2 shows the graph of the solution set.

**Figure 2**

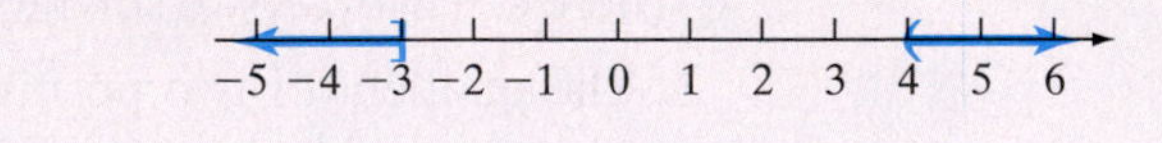


**Table 1**

| | | −3 | 0 | 4 | |
|---|---|---|---|---|---|
| **Interval** | $(-\infty, -3)$ | | $(-3, 4)$ | | $(4, \infty)$ |
| **Test Point** | −4 | −3 | 0 | 4 | 5 |
| **Sign of $x + 3$** | Negative | 0 | Positive | Positive | Positive |
| **Sign of $x - 4$** | Negative | Negative | Negative | 0 | Positive |
| **Sign of $\frac{x+3}{x-4}$** | Positive | 0 | Negative | Undefined | Positive |
| **Conclusion** | $\frac{x+3}{x-4}$ is positive, so $(-\infty, -3)$ is part of the solution set. | Because the inequality is nonstrict, −3 is part of the solution. | $\frac{x+3}{x-4}$ is negative, so $(-3, 4)$ is not part of the solution set. | 4 cannot be part of the solution set because it causes division by 0. | $\frac{x+3}{x-4}$ is positive, so $(4, \infty)$ is part of the solution set. |

### Solving Rational Inequalities

**Step 1:** Write the inequality so that a rational expression is on one side of the inequality and 0 is on the other. Be sure to write the rational expression as a single quotient in factored form.

**Step 2:** Determine the numbers for which the rational expression equals 0 or is undefined.

**Step 3:** Use the numbers found in Step 2 to separate the real number line into intervals.

**Step 4:** Choose a test point within each interval formed in Step 3 to determine the sign of each factor in the numerator and denominator. Then determine the sign of the quotient.

- If the quotient is positive, then the rational expression is positive for all numbers $x$ in the interval.
- If the quotient is negative, then the rational expression is negative for all numbers $x$ in the interval.

Also determine the value of the rational expression at each value found in Step 2. If the inequality is not strict ($\leq$ or $\geq$), include the values of the variable for which the rational expression equals 0 in the solution set, but do not include the values for which the rational expression is undefined!

**QUICK ✓**

**1.** Solve $\dfrac{x-7}{x+3} \geq 0$. Graph the solution set.

## EXAMPLE 2 Solving a Rational Inequality

Solve $\dfrac{x+3}{x-1} > 2$. Graph the solution set.

### Solution

First, we write the inequality so that a rational expression is on one side of the inequality and 0 is on the other.

$$\frac{x+3}{x-1} > 2$$

Subtract 2 from both sides:
$$\frac{x+3}{x-1} - 2 > 0$$

LCD $= x - 1$; multiply $-2$ by $\frac{x-1}{x-1}$:
$$\frac{x+3}{x-1} - 2 \cdot \frac{x-1}{x-1} > 0$$

Write rational expression over common denominator:
$$\frac{x+3-2(x-1)}{x-1} > 0$$

Distribute $-2$:
$$\frac{x+3-2x+2}{x-1} > 0$$

Combine like terms in numerator:
$$\frac{-x+5}{x-1} > 0$$

We can see that the rational expression will equal 0 when $x = 5$. The rational ession is undefined when $x = 1$. We separate the real number line into the followir ervals:

$(-\infty, 1)$ $(1, 5)$ $(5, \infty)$

−2 −1 0 1 2 3 4 5 6 7 8 9 10

Table 2 shows the sign of $-x + 5$, $x - 1$, and $\dfrac{-x+5}{x-1}$ in each interval. In addition, it shows the value of $\dfrac{-x+5}{x-1}$ at $x = 1$ and $x = 5$.

**Table 2**

| | | | | | |
|---|---|---|---|---|---|
| | | 1 | | 5 | |
| **Interval** | $(-\infty, 1)$ | | $(1, 5)$ | | $(5, \infty)$ |
| **Test Point** | 0 | 1 | 3 | 5 | 6 |
| **Sign of $-x + 5$** | Positive | Positive | Positive | 0 | Negative |
| **Sign of $x - 1$** | Negative | 0 | Positive | Positive | Positive |
| **Sign of $\dfrac{-x+5}{x-1}$** | Negative | Undefined | Positive | 0 | Negative |
| **Conclusion** | $\dfrac{-x+5}{x-1}$ is negative, so $(-\infty, 1)$ is not part of the solution set. | Because $\dfrac{-x+5}{x-1}$ is undefined at $x = 1$, it is not part of the solution set. | $\dfrac{-x+5}{x-1}$ is positive, so $(1, 5)$ is part of the solution set. | Because the inequality is strict, 5 is not part of the solution set. | $\dfrac{-x+5}{x-1}$ is negative, so $(5, \infty)$ is not part of the solution set. |

We want to know the values of $x$ such that $\dfrac{x+3}{x-1}$ is greater than 2. This is equivalent to determining where $\dfrac{-x+5}{x-1}$ is greater than zero. So the solution set is $\{x \mid 1 < x < 5\}$ using set-builder notation. The solution is $(1, 5)$ using interval notation. Notice that the endpoints of the interval are not part of the solution because the inequality in the original problem is strict. Figure 3 shows the graph of the solution set.

**Figure 3**

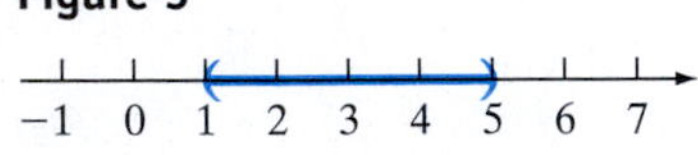


**QUICK ✓**

**2.** Solve $\dfrac{4x+5}{x+2} < 3$. Graph the solution set.

# 6.5 Exercises

## Concepts and Vocabulary

*In Problem 1, fill in the blank.*

**1.** The inequality $\dfrac{2x-3}{x+6} > 1$ is an example of a(n) _____ inequality.

*In Problem 2, answer True or False to the statement.*

**2.** A solution to the inequality $\dfrac{x+1}{x-1} > 0$ is 2.

**3.** In solving the rational inequality $\dfrac{x-4}{x+1} \le 0$, a student determines that the only interval that makes the inequality true is $(-1, 4)$. He states that the solution set is $\{x \mid -1 \le x \le 4\}$. What is wrong with this solution?

**4.** In Step 2 of the steps for solving a rational inequality, we determine the numbers for which the rational expression equals 0 or is undefined. We then use these numbers to form intervals on the real number line. Explain why this guarantees that there is not a change in the sign of the rational expression within any given interval.

## Building Skills

*In Problems 5–26, solve each rational inequality. Graph the solution set.*

**5.** $\frac{x-4}{x+1} > 0$ **6.** $\frac{x+5}{x-2} > 0$ **7.** $\frac{x+9}{x-3} < 0$ **8.** $\frac{x+8}{x+2} > 0$

**9.** $\frac{x+10}{x-4} \geq 0$ **10.** $\frac{x+12}{x-2} \geq 0$ **11.** $\frac{x+7}{x-8} \leq 0$ **12.** $\frac{x-10}{x+5} \leq 0$

**13.** $\frac{(2x-1)(x+3)}{x-5} > 0$ **14.** $\frac{(5x-2)(x+4)}{x-5} < 0$

**15.** $\frac{(3x+5)(x+8)}{x-2} \leq 0$ **16.** $\frac{(3x-2)(x-6)}{x+1} \geq 0$

**17.** $\frac{x-5}{x+1} < 1$ **18.** $\frac{x+3}{x-4} > 1$ **19.** $\frac{3x-1}{x+4} \geq 2$ **20.** $\frac{3x-7}{x+2} \leq 2$

**21.** $\frac{2x-9}{x-3} > 4$ **22.** $\frac{3x+20}{x+6} < 5$ **23.** $\frac{3}{x-4} + \frac{1}{x} \geq 0$ **24.** $\frac{2}{x+3} + \frac{2}{x} \leq 0$

**25.** $\frac{3}{x-2} \leq \frac{4}{x+5}$ **26.** $\frac{1}{x-4} \geq \frac{3}{2x+1}$

*In Problems 27–30, for each function find the values of x that satisfy the given condition. Graph the solution set.*

**27.** Solve $R(x) \leq 0$ if $R(x) = \frac{x-6}{x+1}$ **28.** Solve $R(x) \geq 0$ if $R(x) = \frac{x+3}{x-8}$

**29.** Solve $R(x) < 0$ if $R(x) = \frac{2x-5}{x+2}$ **30.** Solve $R(x) < 0$ if $R(x) = \frac{3x+2}{x-4}$

## Applying the Concepts

**31. Average Cost** Suppose that the daily cost $C$ of manufacturing $x$ bicycles is given by $C(x) = 80x + 5000$. Then the average daily cost $\overline{C}$ is given by $\overline{C}(x) = \frac{80x + 5000}{x}$. How many bicycles must be produced each day in order for the average cost to be no more than $130?

**32. Average Cost** See Problem 31. Suppose that the government imposes a $10 tax on each bicycle manufactured so that the daily cost $C$ of manufacturing $x$ bicycles is now given by $C(x) = 90x + 5000$. Now the average daily cost $\overline{C}$ is given by $\overline{C}(x) = \frac{90x + 5000}{x}$. How many bicycles must be produced each day in order for the average cost to be no more than $130?

## Extending the Concepts

**33.** Write a rational inequality that has $(2, \infty)$ as the solution set.

**34.** Write a rational inequality that has $(-2, 5]$ as the solution set.

## Synthesis Review

*In Problems 35–40, find the x-intercepts of the graph of each function.*

**35.** $F(x) = 6x - 12$ **36.** $G(x) = 5x + 30$

**37.** $f(x) = 2x^2 + 3x - 14$ **38.** $h(x) = -3x^2 - 7x + 20$

**39.** $R(x) = \dfrac{3x - 2}{x + 4}$

**40.** $R(x) = \dfrac{x^2 + 5x + 6}{x + 2}$

## The Graphing Calculator

We can use a graphing calculator to approximate solutions to rational inequalities using the INTERSECT or ZERO (or ROOT) feature of the graphing calculator. We use the ZERO or ROOT feature of the graphing calculator when one side of the inequality is 0; we use the INTERSECT feature when neither side of the inequality is 0. For example, to solve $\dfrac{x - 4}{x + 1} > \dfrac{7}{2}$, we would graph $Y_1 = \dfrac{x - 4}{x + 1}$ and $Y_2 = \dfrac{7}{2}$. To determine the $x$-values such that the graph of $Y_1$ is above that of $Y_2$, we find $x$-coordinates of the point(s) of intersection. The graph of $Y_1$ is above that of $Y_2$ between $x = -3$ and $x = -1$. See Figure 4.

**Figure 4**

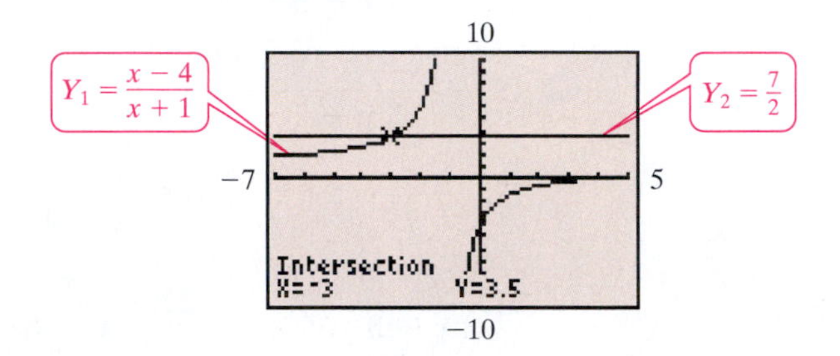


When using a graphing calculator to approximate solutions to inequalities, we typically express the solution as a decimal rounded to two decimal places, if exact answers cannot be found. The solution to the inequality $\dfrac{x - 4}{x + 1} > \dfrac{7}{2}$ is $\{x \mid -3 \leq x < -1\}$, or $[-3, -1)$ using interval notation.

*In Problems 41–44, solve each inequality using a graphing calculator.*

**41.** $\dfrac{x - 5}{x + 1} \leq 3$

**42.** $\dfrac{x + 2}{x - 5} > -2$

**43.** $\dfrac{2x + 5}{x - 7} > 3$

**44.** $\dfrac{2x - 1}{x + 5} \leq 4$

# 6.6 Models Involving Rational Expressions

## OBJECTIVES

1. Solve for a Variable in a Rational Expression
2. Model and Solve Ratio and Proportion Problems
3. Model and Solve Work Problems
4. Model and Solve Uniform Motion Problems
5. Model and Solve Problems Involving Inverse Variation
6. Model and Solve Problems Involving Joint or Combined Variation

### *Preparing for Models Involving Rational Expressions*

*Before getting started, take the following readiness quiz. If you get a problem wrong, go back to the section cited and review the material.*

**1.** Solve for $y$: $4x - 2y = 10$ [Section 1.3, pp. 80–83]

**2.** $y$ varies directly with $x$ and when $y = 4$, $x = 12$. Find $y$ when $x = 24$. [Section 3.5, pp. 234–236]

## 1 Solve for a Variable in a Rational Expression

The expression "solve for the variable" means to get the variable by itself on one side of the equation with all other variables and constants, if any, on the other side. The steps that we follow when solving formulas for a certain variable are identical to those that we followed when solving rational equations.

***Preparing for...Answers*** **1.** $y = 2x - 5$ **2.** 8

## EXAMPLE 1 Solving for a Variable in a Lens Construction Formula

The formula $\frac{1}{f} = \frac{1}{p} + \frac{1}{q}$ is used in telescope and camera construction, where $f$ is the focal length of the lens. In general, the larger $f$, the more power the telescope has. The variable $p$ is the distance from the object we wish to see and the lens; the variable $q$ is the distance from the lens to point of focus (such as the film or your eye). See Figure 5.

**Figure 5**

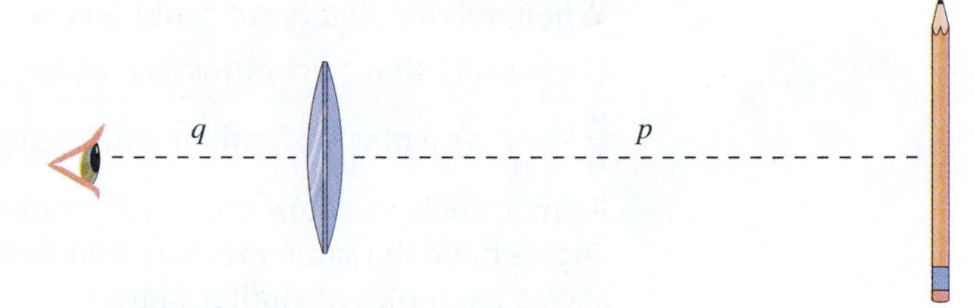


**(a)** Solve the formula for $q$.

**(b)** Suppose that a camera has a focal length of 100 mm and the camera is focusing on an object 5000 mm away. What is $q$, the distance from the lens to the point of focus?

### Solution

**(a)** Our goal is to get $q$ by itself. We follow the same steps that we used to solve a rational equation. First, we note that none of the variables can equal 0. The LCD of all denominators is $pqf$, so we multiply both sides of the equation by $pqf$.

$$\frac{1}{f} = \frac{1}{p} + \frac{1}{q}$$

$$pqf \cdot \frac{1}{f} = pqf \cdot \left(\frac{1}{p} + \frac{1}{q}\right)$$

Distribute the $pqf$: $$pqf \cdot \frac{1}{f} = pqf \cdot \frac{1}{p} + pqf \cdot \frac{1}{q}$$

Simplify: $$pq = qf + pf$$

Subtract $qf$ from both sides: $$pq - qf = qf + pf - qf$$

$$pq - qf = pf$$

Factor out $q$: $$q(p - f) = pf$$

Divide both sides by $p - f$: $$q = \frac{pf}{p - f}$$

**(b)** Substitute $f = 100$ mm and $p = 5000$ mm in $q = \frac{pf}{p - f}$.

$$q = \frac{5000 \cdot 100}{5000 - 100}$$

$$\approx 102 \text{ mm}$$

The distance from the lens to the point of focus is approximately 102 mm. ■

### QUICK ✓

**1.** The formula $Y = \frac{G}{1 - b}$ is used in economics to determine the impact on Gross Domestic Product (GDP) $Y$ by increasing government spending by $G$ dollars if the proportion of additional income that people spend is $b$.

**(a)** Solve the formula for $b$.

**(b)** Find $b$ if the government increased spending by \$100 billion and GDP increased by \$1000 billion.

## 2 Model and Solve Ratio and Proportion Problems

The problems in this objective focus on the idea of ratio and proportion. The **ratio** of two numbers $a$ and $b$ can be written as

$$a{:}b \quad \text{or} \quad \frac{a}{b}$$

When solving algebraic problems, we write ratios as $\frac{a}{b}$. A **proportion** is a statement (equation) that two ratios are equal. That is, proportions are equations of the form $\frac{a}{b} = \frac{c}{d}$. You may be familiar with using proportions to solve problems involving similar figures, such as triangles, from geometry. Recall, that two figures are **similar** if their angles have the same measure and their corresponding sides are proportional. Figure 6 shows examples of similar figures.

Figure 6

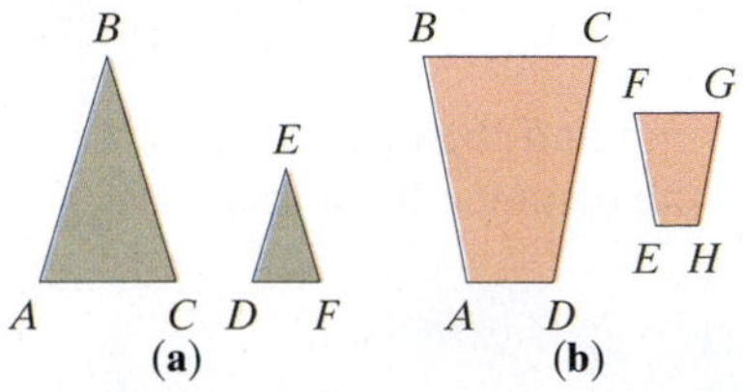

(a) (b)

In Figure 6(a), $\triangle ABC$ is similar to $\triangle DEF$ and in Figure 6(b), quadrilateral $ABCD$ is similar to quadrilateral $EFGH$. Because $\triangle ABC$ is similar to $\triangle DEF$, we know that the ratio of $AB$ to $AC$ equals the ratio of $DE$ to $DF$. That is,

$$\frac{AB}{AC} = \frac{DE}{DF}$$

The principle that the ratios of corresponding sides are equal can be used to find unknown lengths in similar figures.

### EXAMPLE 2 Similar Figures

Suppose that you are standing next to a tall building and wish to know the building's height. A light post that is 20 feet tall casts a shadow that is 12 feet long. The shadow from the building is measured to be 660 feet. Determine the height of the building.

#### Solution

The ratio of the length of the building's shadow to the height of the building equals the ratio of the length of the light post's shadow to the height of the light post because the building and its shadow form a triangle that is similar to the light post and its shadow. See Figure 7.

Figure 7

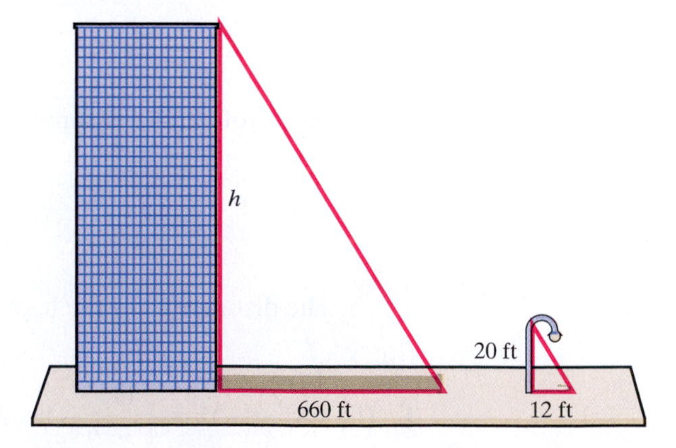


We set up the proportion problem as

$$\frac{660}{h} = \frac{12}{20}$$

and solve the equation. The domain of $h$ is $\{h \mid h > 0\}$ since $h$ represents the height of the building. The LCD of all denominators is $20h$, so we multiply both sides of the equation by $20h$.

$$\frac{660}{h} = \frac{12}{20}$$

$$20h \cdot \frac{660}{h} = 20h \cdot \frac{12}{20}$$

$$13{,}200 = 12h$$

Divide both sides by 12: $$1100 = h$$

The height of the building is 1100 feet. ■

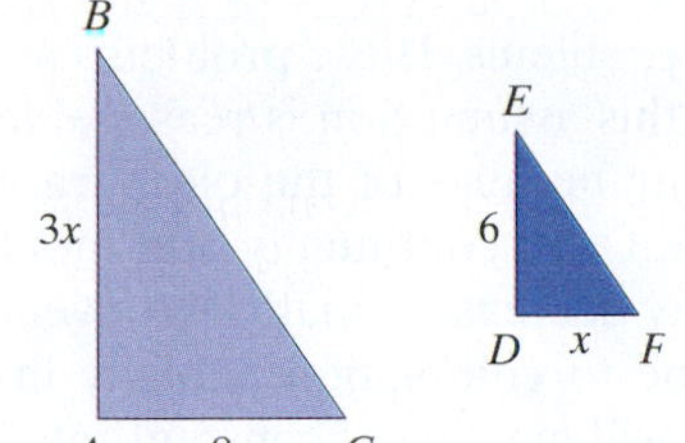

**2.** Suppose that $\triangle ABC$ is similar to $\triangle DEF$ as shown in the figure to the left. Find the length of $AB$ and $DF$.

---

Proportion problems also come up in direct translation problems as well.

## EXAMPLE 3 A Proportion Problem

According to the National Vital Statistics Report dated December 17, 2003, the birth rate for unmarried women (15–44 years of age) was 44 live births per 1000 population in the United States in 2002. In 2002, there were 1,366,000 births to unmarried women. Determine the population of unmarried women between 15 and 44 years of age in 2002.

### Solution

**Step 1: Identify** This is a direct translation problem involving proportions. We are looking for the population of unmarried women in 2002.

**Step 2: Name** We let $p$ represent the population of unmarried women in 2002.

**Step 3: Translate** Since we know that the rate of births was 44 per 1000 and that there were 1,366,000 births, we have the model

$$\frac{44}{1000} = \frac{1{,}366{,}000}{p}$$

**Step 4: Solve** We now proceed to solve the equation.

$$\frac{44}{1000} = \frac{1{,}366{,}000}{p}$$

Multiply both sides by the LCD, $1000p$: $$1000p \cdot \frac{44}{1000} = 1000p \cdot \frac{1{,}366{,}000}{p}$$

Divide out common factors: $$44p = 1{,}366{,}000{,}000$$

Divide both sides by 44: $$p = 31{,}045{,}455$$

**Step 5: Check** It is always a good idea to make sure your answer is reasonable. Since there were over 60 million women 15–44 years of age in the United States in 2002, the answer seems reasonable.

**Step 6: Answer** The population of unmarried women 15–44 years of age in the United States in 2002 was 31,045,455. ■

**QUICK ✓**

3. According to the American Cancer Society, the incidence rate of melanoma of the skin in 2005 was 20 per 100,000 population in the United States. There were approximately 59,850 reported cases of melanoma in the United States in 2005. Determine the population of the United States in 2005.

## 3 Model and Solve Work Problems

We are now going to solve work or "constant rate jobs" problems. These problems assume that jobs are performed at a **constant rate.** While this assumption is reasonable for machines, it is not likely to be true for people simply because of the old phrase "too many chefs spoil the broth." Think of it this way—if you continually add more people to paint a room, the time to complete the job may decrease initially, but eventually the painters get in each other's way and the time to completion actually increases. While we could model situations such as this, we will make the "constant rate" assumption for humans as well in order to keep the mathematics manageable.

**Work Smart**
Remember, when we model we make simplifying assumptions to make the math easier to deal with.

The constant rate assumption states that if it takes $t$ units of time to complete a job, then $\frac{1}{t}$ of the job is done in 1 unit of time. For example, if it takes 5 hours to paint a room, then $\frac{1}{5}$ of the room should be painted in 1 hour.

### EXAMPLE 4 Working Together on a Job

It's Saturday and Kevin needs to cut and edge the grass. At 9 A.M., Michael asks Kevin to go golfing at 11:00 A.M. Typically, it takes Kevin 3 hours to cut and edge the grass. When Michael cuts and edges the grass, it takes 4 hours. If they worked together, would they be able to finish the lawn and still make the golf date?

#### Solution

**Step 1: Identify** We want to know how long it will take for Michael and Kevin to finish the lawn.

**Step 2: Name** We let $t$ represent the time (in hours) that it takes to finish the lawn working together. Then, in 1 hour they will complete $\frac{1}{t}$ of the job.

**Step 3: Translate** Since we know that Kevin can finish the job in 3 hours, Kevin will finish $\frac{1}{3}$ of the job in 1 hour. We know that Michael can finish the job in 4 hours, so Michael will finish $\frac{1}{4}$ of the job in 1 hour. We set up the model using the following logic:

$$\begin{pmatrix}\text{Part done by Kevin}\\ \text{in 1 hour}\end{pmatrix} + \begin{pmatrix}\text{Part done by Michael}\\ \text{in 1 hour}\end{pmatrix} = \begin{pmatrix}\text{Part done together}\\ \text{in 1 hour}\end{pmatrix}$$

$$\frac{1}{3} + \frac{1}{4} = \frac{1}{t}$$

**Step 4: Solve** We now proceed to solve the equation.

$$\frac{1}{3} + \frac{1}{4} = \frac{1}{t}$$

Multiply both sides by the LCD, $12t$: $$12t \cdot \left(\frac{1}{3} + \frac{1}{4}\right) = 12t \cdot \frac{1}{t}$$

Distribute: $$12t \cdot \frac{1}{3} + 12t \cdot \frac{1}{4} = 12t \cdot \frac{1}{t}$$

Divide like factors: $4t + 3t = 12$

Combine like terms: $7t = 12$

Divide both sides by 7: $t = \frac{12}{7} \approx 1.714$

**Work Smart**

We convert 0.714 hours to minutes by multiplying 0.714 by 60 minutes and obtain 43 minutes.

**Step 5: Check** It is always a good idea to make sure your answer is reasonable. We expect our answer to be greater than 0 but less than 3 (because it takes Kevin 3 hours working by himself). Our answer of 1.714 hours or 1 hour, 43 minutes seems reasonable.

**Step 6: Answer** If they start right away, they should finish at 10:43 A.M. As long as they can get to the course in 17 minutes, they can make the tee time.

**QUICK ✓**

**4.** Juan and Maria have a pool in their backyard. If they use their hose alone to fill the pool, the pool can be filled in 30 hours. Their neighbor has the same pool and was able to fill it in 24 hours. Suppose their neighbor agrees to let them use his hose. How long will it take to fill the pool with both hoses?

## EXAMPLE 5 The Kitchen Sink

Suppose that the kitchen sink can be filled in 5 minutes. If the sink is full, it takes 8 minutes to drain the sink when the drain is left partially open. If the sink's drain is accidentally left partially open, how long will it take to fill the sink?

### Solution

**Step 1: Identify** We want to know how long it will take to fill the sink with the drain open.

**Step 2: Name** We let $t$ represent the time (in minutes) that it takes to fill the sink. Then, in 1 minute $\frac{1}{t}$ of the sink will be full.

**Step 3: Translate** We know that the sink can be filled in 5 minutes when the drain is closed, so after 1 minute, $\frac{1}{5}$ of the sink is full. It takes 8 minutes to drain, so after 1 minute, $\frac{1}{8}$ of the sink is drained. We set up the model using the following logic:

**Work Smart**

Notice we subtract the portion of sink drained after 1 minute since it is "working against us."

$$\left(\begin{array}{c}\text{Portion of sink filled}\\ \text{in 1 minute with}\\ \text{closed drain}\end{array}\right) - \left(\begin{array}{c}\text{Portion of sink drained}\\ \text{after 1 minute}\end{array}\right) = \left(\begin{array}{c}\text{Portion of sink filled}\\ \text{after 1 minute with}\\ \text{open drain}\end{array}\right)$$

$$\frac{1}{5} - \frac{1}{8} = \frac{1}{t}$$

**Step 4: Solve** We now proceed to solve the equation.

$$\frac{1}{5} - \frac{1}{8} = \frac{1}{t}$$

Multiply both sides by the LCD: $40t$: $40t \cdot \left(\frac{1}{5} - \frac{1}{8}\right) = 40t \cdot \frac{1}{t}$

Distribute: $40t \cdot \frac{1}{5} - 40t \cdot \frac{1}{8} = 40t \cdot \frac{1}{t}$

Simplify: $8t - 5t = 40$

Combine like terms: $3t = 40$

Divide both sides by 3: $t = \frac{40}{3} \approx 13.3$ minutes

**Step 5: Check** It is always a good idea to make sure your answer is reasonable. We expect our answer to be greater than 5 since this is the time it takes to fill the sink with the drain in. Our answer of 13.3 minutes or 13 minutes, 20 seconds seems reasonable.

**Step 6: Answer** It will take 13.3 minutes or 13 minutes, 20 seconds to fill the sink.

**QUICK ✓**

**5.** A children's inflatable pool takes 20 minutes to fill with an electric air pump. It takes 50 minutes to let the air out of the pool. If the pool's valve is accidentally left open, how long will it take to fill the pool?

## 4 Model and Solve Uniform Motion Problems

We first introduced uniform motion problems back in Section 1.2. Recall that uniform motion problems use the fact that distance equals rate times time, that is, $d = rt$. When modeling uniform motion problems that lead to rational equations, we usually end up using an alternative form of this model, $t = \frac{d}{r}$.

### EXAMPLE 6 A Round-Trip Flight

A plane flies 990 miles west (into the wind) and makes the return trip following the same flight path. The effect of the wind on the plane is 20 miles per hour. The round trip takes 10 hours. What is the speed of the plane in still air?

#### Solution

**Step 1: Identify** This is a uniform motion problem. We wish to know the speed of the plane in still air.

**Step 2: Name** Let $r$ represent the speed of the plane in still air.

**Step 3: Translate** Going west, the plane is flying into the wind. The speed of the plane is its rate in still air less the impact of the wind, so $r - 20$ represents the speed of the plane going west. Similar logic tells us that the speed of the plane going east is $r + 20$. We set up Table 3. Remember, $d = r \cdot t$, so that $t = \frac{d}{r}$.

**Table 3**

| | Distance (miles) | Rate (miles per hour) | Time (hours) |
|---|---|---|---|
| **West** | 990 | $r - 20$ | $\frac{990}{r - 20}$ |
| **East** | 990 | $r + 20$ | $\frac{990}{r + 20}$ |

The round trip takes 10 hours, so that we have the following model:

$$\text{Time going west} + \text{Time going east} = 10$$

$$\frac{990}{r - 20} + \frac{990}{r + 20} = 10 \quad \text{The Model}$$

**Step 4: Solve** We wish to solve for $r$:

$$\frac{990}{r-20}+\frac{990}{r+20}=10$$

Multiply both sides by the LCD, $(r-20)(r+20)$: $$(r-20)(r+20)\left(\frac{990}{r-20}+\frac{990}{r+20}\right)=10(r-20)(r+20)$$

Distribute: $$(r-20)(r+20)\frac{990}{r-20}+(r-20)(r+20)\frac{990}{r+20}=10(r-20)(r+20)$$

Divide out common factors: $$990(r+20)+990(r-20)=10(r-20)(r+20)$$

Distribute: $$990r+19{,}800+990r-19{,}800=10(r^2-400)$$

Combine like terms: $$1980r=10(r^2-400)$$

Divide both sides by 10: $$198r=r^2-400$$

Set equal to 0: $$0=r^2-198r-400$$

Factor: $$0=(r-200)(r+2)$$

Zero-Product Property: $$r=200 \text{ or } r=-2$$

**Step 5: Check** We disregard the $-2$ because the rate of the plane must be positive. So it appears that plane will travel at a rate of 200 miles per hour in still air. Flying west, the plane travels at 180 miles per hour, so it takes $\frac{990 \text{ miles}}{180 \text{ miles per hour}}=$ 5.5 hours to fly west. Flying east, the plane travels at 220 miles per hour, so the trip takes $\frac{990 \text{ miles}}{220 \text{ miles per hour}}=4.5$ hours. The total trip is $5.5+4.5=10$ hours. It checks!

**Step 6: Answer the Question** The plane travels at 200 miles per hour in still air.

**QUICK ✓**

6. A canoe travels on a river whose current is running at 2 miles per hour. After traveling 12 miles upstream, the canoe turns around and makes the 12 mile trip back downstream. The trip up and back takes 8 hours. What is the speed of the canoe in still water?

## 5 Model and Solve Problems Involving Inverse Variation

Back in Section 3.5, we solved problems that involved direct variation. Recall that if $x$ and $y$ are two quantities, then $y$ varies directly with $x$ or $y$ is directly proportional to $x$ if there is a nonzero number $k$ such that $y=kx$. We call the number $k$ the constant of proportionality. The graph of the relationship between $x$ and $y$ is linear.

We now turn our attention to another kind of variation.

**DEFINITION**

Suppose $x$ and $y$ represent two quantities. We say that $y$ **varies inversely** with $x$, or $y$ is **inversely proportional to** $x$, if there is a nonzero number $k$ such that

$$y=\frac{k}{x}$$

**Figure 8**

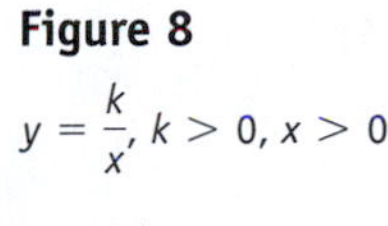

$y=\frac{k}{x}, k>0, x>0$

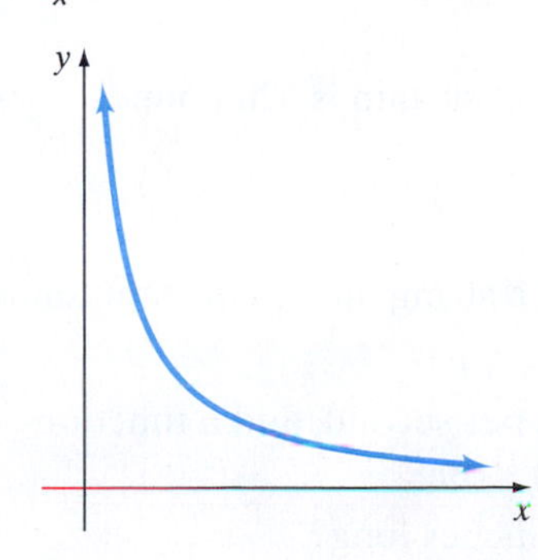


The graph in Figure 8 illustrates the relationship between $y$ and $x$ if $y$ varies inversely with $x$ with $k>0$ and $x>0$. Notice from the graph that as $x$ increases the value of $y$ decreases.

### EXAMPLE 7 Weight That Can Be Supported by a Beam

The weight $W$ that can be safely supported by a 2-inch by 4-inch (2-by-4) piece of lumber varies inversely with its length $l$. See Figure 9.

**Figure 9**

**(a)** Experiments indicate that the maximum weight a 12-foot pine 2-by-4 can support is 400 pounds. Find a function that relates the maximum weight to the length $l$ for any pine 2-by-4.

**(b)** Determine the maximum weight that a 15-foot pine 2-by-4 can sustain.

**Solution**

**(a)** Because $W$ varies inversely with $l$, we know that

$$W = \frac{k}{l}$$

for some constant $k$. Because $W = 400$ when $l = 12$, it follows that

$$400 = \frac{k}{12}$$

Solving this equation for $k$ by multiplying both sides of the equation by 12, we find that

$$k = 4800$$

So we have that

$$W = \frac{4800}{l}$$

We can write this as a function:

$$W(l) = \frac{4800}{l}$$

**(b)** When $l = 15$ feet, we have that

$$W(15) = \frac{4800}{15}$$
$$= 320 \text{ pounds}$$

The maximum weight that a 15-foot pine 2-by-4 can sustain is 320 pounds. ■

**QUICK ✓**

**7.** The rate of vibration (in oscillations per second) $V$ of a string under constant tension varies inversely with the length $l$.

**(a)** If a string is 30 inches long and vibrates 500 times per second, find a function that relates the rate of vibration to the length of a string.

**(b)** What is the rate of vibration of a string that is 50 inches long?

## 6 Model and Solve Problems Involving Combined or Joint Variation

When a variable quantity $Q$ is proportional to the product of two or more other variables, we say that $Q$ **varies jointly** with these quantities. For example, the equation $y = kxz$ can be read as "$y$ varies jointly with $x$ and $z$." When direct and inverse variation occur at the same time, we have **combined variation.** For example, the equation $y = \frac{kx}{z}$ can be read as "$y$ varies directly with $x$ and inversely with $z$." The equation $y = \frac{kmn}{p}$ can be read "$y$ varies jointly with $m$ and $n$ and inversely with $p$."

### EXAMPLE 8 Force of the Wind—Joint Variation

The force $F$ of the wind on a flat surface positioned at a right angle to the direction of the wind varies jointly with the area $A$ of the surface and the square of the speed $v$ of the wind. A wind of 30 miles per hour blowing on a window measuring 4 feet by 5 feet has a force of 150 pounds. What is the force on a window measuring 2 feet by 6 feet caused by a hurricane-force wind of 100 miles per hour?

#### Solution

Because $F$ varies jointly with $A$ and the square of $v$, we know that

$$F = kAv^2$$

for some constant $k$. The dimensions of the window are 4 feet by 5 feet, so the area $A$ of the window is (4 feet)(5 feet) = 20 square feet. Because $F = 150$ when $A = 20$ and $v = 30$, it follows that

$$150 = k(20)(30^2)$$

We solve this equation for $k$.

$$150 = 18{,}000k$$

$$k = \frac{1}{120}$$

So we have that

$$F = \frac{1}{120}Av^2$$

For a wind of 100 miles per hour blowing on a window whose area is $A =$ (2 feet)(6 feet) = 12 square feet, the force $F$ is

$$F = \frac{1}{120}(12)(100)^2$$

$$= 1000 \text{ pounds}$$

**QUICK ✓**

**8.** The kinetic energy $K$ of an object varies jointly with its mass and the square of its velocity. The kinetic energy of a 110 kg linebacker running at 9 meters per second is 4455 joules. Determine the kinetic energy of a 140-kg linebacker running at 5 meters per second.

### EXAMPLE 9 Centripetal Force—Combined Variation

The force required to keep an object traveling in a circular motion is called the centripetal force. The centripetal force $F$ required to keep an object of a fixed mass in circular motion varies directly to the square of the velocity $v$ of the object and inversely to the radius $r$ of the circle. See Figure 10 on page 511.

**Figure 10**

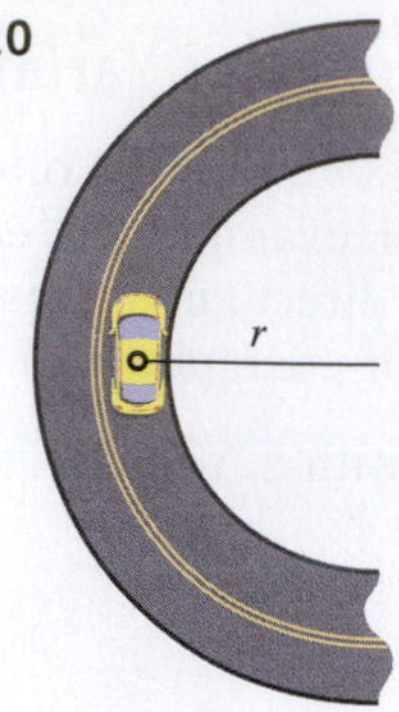


The force required to keep a car traveling 30 meters per second on a circular road with radius 50 meters is 21,600 newtons. Determine the force required to keep the same car on a circular road if it is traveling 40 meters per second on a road whose radius is 30 meters.

### Solution

Because the force varies directly to the square of the velocity of the object and inversely to the radius of the circle, we have combined variation with

$$F = \frac{kv^2}{r}$$

for some constant $k$. Because $F = 21{,}600$ when $v = 30$ and $r = 50$, it follows that

$$21{,}600 = \frac{k \cdot 30^2}{50}$$

Solving for $k$, we find that $k = 1200$. So we have that

$$F = \frac{1200v^2}{r}$$

For a car traveling 40 meters per second on a circular road with radius 30 meters, the force required to keep the car on the road is

$$F = \frac{1200 \cdot 40^2}{30}$$

$$= 64{,}000 \text{ newtons}$$

**QUICK ✓**

9. The electrical resistance of a wire $R$ varies directly with the length of the wire $l$ and inversely with the square of the diameter of the wire $d$. If a wire 432 feet long and 4 millimeters in diameter has a resistance of 1.24 ohms, find the resistance in a wire that is 282 feet long with a diameter of 3 millimeters. Round your answer to two decimal places.

# 6.6 Exercises

**For Extra Help:**

Student Solutions Manual

CD Video

PH Math/Tutor Center

MathXL Tutorials on CD

MathXL® MyMathLab

## Concepts and Vocabulary

*In Problems 1 and 2, fill in the blanks.*

1. A(n) _____ is a statement (equation) that two ratios are equal.

2. Two figures are _____ if their angles have the same measure and their corresponding sides are proportional.

*In Problems 3 and 4, answer True or False to each statement.*

3. When solving work problems, we assume that individuals work at a constant rate.

4. Suppose we let $x$ and $y$ represent two quantities. We say that $y$ varies inversely with $x$, or $y$ is inversely proportional to $x$, if there is a nonzero number $k$ such that $y = \frac{k}{x}$.

5. Explain what it means when a variable $Q$ varies jointly with $x$ and $y$. Explain what is meant by combined variation.

**6.** When solving work problems, we assume that each individual works at a constant rate and that there is no gain or loss of efficiency when additional individuals are added to the job. Explain what "no gain or loss of efficiency" means. Do you think this assumption is reasonable? If not, then why do we make the assumption?

## Building Skills

*In Problems 7–16, solve each formula for the indicated variable.*

**7. Chemistry (Gas Laws)** Solve $\frac{V_1}{V_2} = \frac{P_2}{P_1}$ for $P_1$.

**8. Chemistry (Gas Laws)** Solve $\frac{V_1}{V_2} = \frac{P_2}{P_1}$ for $V_2$.

**9. Finance** Solve $R = \frac{r}{1 - t}$ for $t$.

**10. Finance** Solve $P = \frac{A}{1 + r}$ for $r$.

**11. Slope** Solve $m = \frac{y - y_1}{x - x_1}$ for $x$.

**12. Slope** Solve $m = \frac{y - y_1}{x - x_1}$ for $x_1$.

**13. Physics** Solve $\omega = \frac{rmv}{I + mr^2}$ for $v$.

**14. Physics** Solve $\omega = \frac{rmv}{I + mr^2}$ for $I$.

**15. Physics** Solve $V = \frac{mv}{M + m}$ for $m$.

**16. Physics** Solve $v_2 = \frac{2m_1}{m_1 + m_2} v_1$ for $m_1$.

*In Problems 17–24, (a) find the constant of proportionality k, (b) write the function relating the two variables, and (c) find the quantity indicated.*

**17.** Suppose that $y$ varies inversely with $x$. When $x = 10$, $y = 2$. Find $y$ if $x = 5$.

**18.** Suppose that $y$ varies inversely with $x$. When $x = 3$, $y = 15$. Find $y$ if $x = 5$.

**19.** Suppose that $y$ is inversely proportional to $x$. When $x = 7$, $y = 3$. Find $y$ if $x = 28$.

**20.** Suppose that $y$ is inversely proportional to $x$. When $x = 20$, $y = 4$. Find $y$ if $x = 35$.

**21.** Suppose that $y$ varies jointly with $x$ and $z$. When $y = 10$, $x = 8$ and $z = 5$. Find $y$ if $x = 12$ and $z = 9$.

**22.** Suppose that $y$ varies jointly with $x$ and $z$. When $y = 20$, $x = 6$ and $z = 10$. Find $y$ if $x = 8$ and $z = 15$.

**23.** Suppose that $Q$ varies directly with $x$ and inversely with $y$. When $Q = \frac{13}{12}$, $x = 5$ and $y = 6$. Find $Q$ if $x = 9$ and $y = 4$.

**24.** Suppose that $Q$ varies directly with $x$ and inversely with $y$. When $Q = \frac{14}{5}$, $x = 4$ and $y = 3$. Find $Q$ if $x = 8$ and $y = 3$.

## Applying the Concepts

*In Problems 25–30, solve the proportion problem.*

**25.** Suppose that $\triangle ABC$ is similar to $\triangle DEF$ as shown in the figure. Find the length of $AB$ and $DF$.

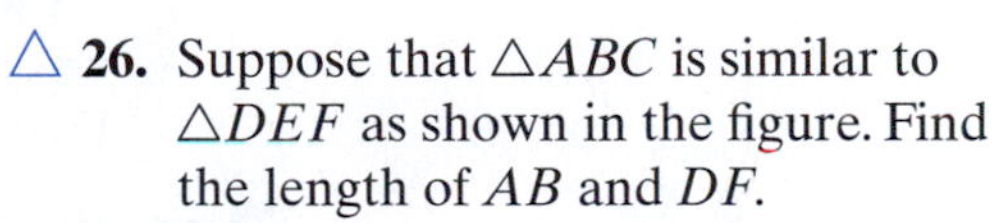

**26.** Suppose that $\triangle ABC$ is similar to $\triangle DEF$ as shown in the figure. Find the length of $AB$ and $DF$.

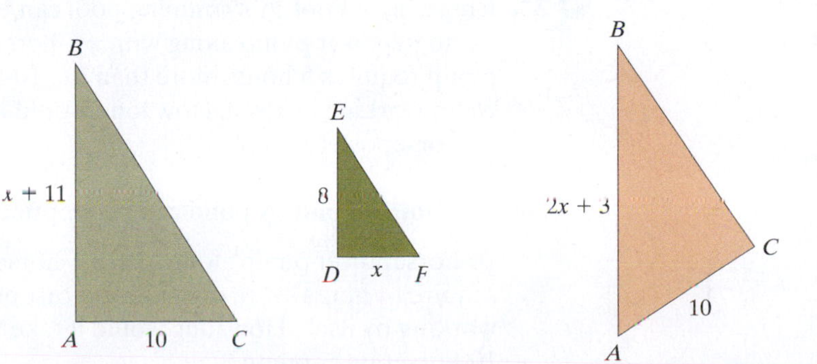

**27. Motor Vehicle Death Rates** According to the Centers for Disease Control, the death rate as a result of a motor vehicle accident is 16.1 per 100,000 population. In 2000, there were 41,821 fatalities in motor vehicle accidents. What was the population in the United States in 2000?

**28. Flight Accidents** According to the *Statistical Abstract of the United States,* in 2001, there were 1.22 fatal airplane accidents per 100,000 flight hours. Also, in 2001, there were a total of 321 fatal accidents. How many flight hours were flown in 2001?

**29. Road Trip** At current prices, Roberta can drive her car 12.8 miles per dollar of gasoline that she buys. Roberta and three of her friends decide to go on a road trip to a neighboring university that is 105 miles away and agree to split the cost evenly. How much money will each have to contribute to get to the university and back?

**30. Car Payments** At current rates, a 60-month term car loan is being offered where the monthly payments are $0.0191 per dollar borrowed. Suppose that Eduardo's car payment is $340 per month. How much did Eduardo borrow?

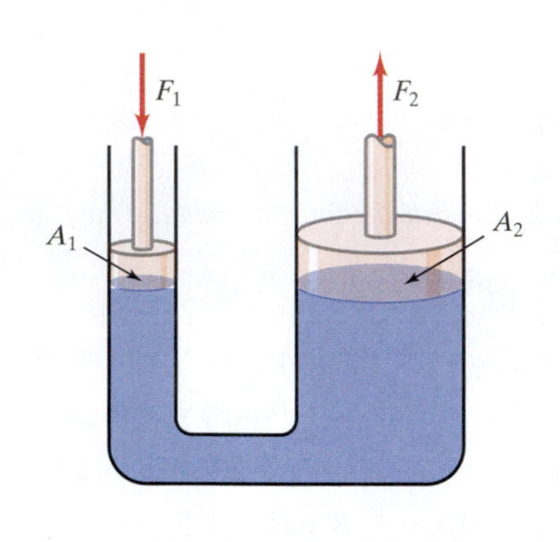


**31. Pascal's Principle** Pascal's Principle applied to a hydraulic lever states that the ratio of the force exerted on an input piston $F_1$ to the area displaced $A_1$ will equal the ratio of force on the output piston $F_2$ to the area displaced $A_2$. See the figure. If a force of 30 pounds is exerted with an area of 12 square feet of water displaced in the right pipe and an area of 5 square feet is displaced in the left pipe, determine the force exerted by the left pipe.

**32. Pascal's Principle** See Problem 31. If a force of 40 pounds is exerted with an area of 15 square feet of water displaced in the right pipe and an area of 8 square feet is displaced in the left pipe, determine the force exerted by the left pipe.

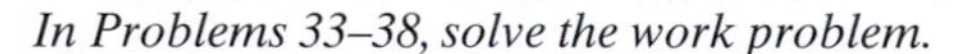

*In Problems 33–38, solve the work problem.*

**33. Sharing a Paper Route** Amiri can deliver his newspapers in 80 minutes. It takes Horus 60 minutes to do the same route. How long would it take them to deliver the newspapers if they work together?

**34. Painting a Room** Latoya can paint five 10-foot-by-14-foot rooms by herself in 14 hours. Lisa can paint five 10-foot-by-14-foot rooms by herself in 10 hours. Working together, how long would it take to paint five 10-foot-by-14-foot rooms?

**35. Cutting the Grass** Avery can cut the grass working by himself in 3 hours. When Avery cuts the grass with his younger brother Connor, it takes 2 hours. How long would it take Connor to cut the grass if he worked by himself?

**36. Assembling a Swing Set** Alexandra and Frank can assemble a King Kong swing set working together in 6 hours. One day, when Frank called in sick, Alexandra was able to assemble a King Kong swing set in 10 hours. How long would it take Frank to assemble a King Kong swing set if he worked by himself?

**37. Emptying a Pool** A swimming pool can be emptied in 6 hours using a 10-horsepower pump along with a 6-horsepower pump. The 6-horsepower pump requires 5 hours more than the 10-horsepower pump to empty the pool when working by itself. How long would it take to empty the pool using just the 10-horsepower pump?

**38. Draining a Pond** A pond can be emptied in 3.75 $\left(=\frac{15}{4}\right)$ hours using a 10-horsepower pump along with a 4-horsepower pump. The 4-horsepower pump requires 4 hours more than the 10-horsepower pump to empty the pond when working by itself. How long would it take to empty the pond using just the 10-horsepower pump?

*In Problems 39–48, solve the uniform motion problem.*

**39. Tough Commute** You have a 20-mile commute into work. Since you leave very early, the trip going to work is easier than the trip home. You can travel to work in the same time that it takes for you to make it 16 miles on the trip back home. Your average speed coming home is 7 miles per hour slower than your average speed going to work. What is your average speed going to work?

**40. Riding Your Bicycle** Every weekend, you ride your bicycle on a forest preserve path. The path is 20 miles long and ends at a waterfall, at which point you relax and then make the trip back to the starting point. One weekend, you find that in the same time it takes you to travel to the waterfall, you are only able to return 12 miles. Your average speed going to the waterfall is 4 miles per hour faster than the return trip. What was your average speed going to the waterfall?

**41. Moving Walkway** In order to access the outer part of Terminal 1 at O'Hare International Airport, you must walk quite some distance in a tunnel that travels under part of the airport. To make the walk less difficult, there is a moving walkway that travels at 2 feet per second. Suppose that Hana can travel 152 feet while walking on the walkway in the same amount of time it takes her to travel 72 feet while walking on the pavement without the aid of the moving sidewalk. How fast does Hana walk?

**42. Escalator** When exiting Terminal 1 at O'Hare International Airport, you can either take an escalator up to the main level or you can take traditional stairs. Suppose that the escalator travels 1.5 feet per second. Karli can walk up the 50 foot escalator in the same amount of time it takes her to walk 30 feet up the stairs. How fast does Karli walk up stairs?

**43. Football** Suppose that Jeremy Shockey of the New York Giants can run 100 yards in 12 seconds. Further suppose that Brian Urlacher of the Chicago Bears can run 100 yards in 9 seconds. Suppose that Shockey catches a pass at his own 20-yard line in stride and starts running away from Urlacher who is at the 15-yard line directly behind Shockey. See the figure. At what yard line will Urlacher catch up to Shockey?

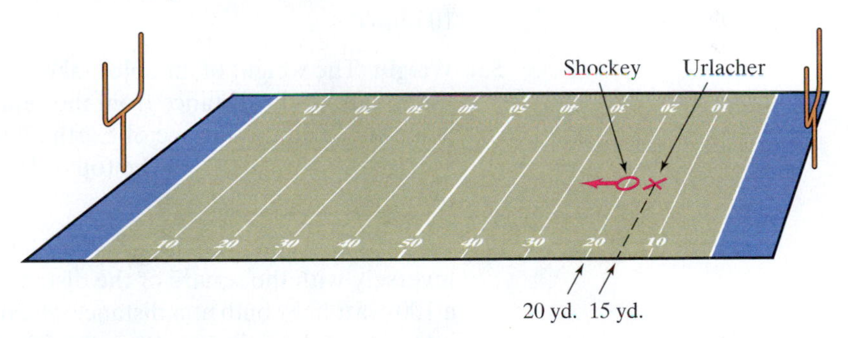


**44. Running a Race** Roger can run one mile in 8 minutes. Jeff can run one mile in 6 minutes. If Jeff gives Roger a 1 minute head start, how long will it take before Jeff catches up to Roger? How far will each have run?

**45. Uphill/Downhill** A bicyclist rides his bicycle 12 miles up a hill and then comes back down. His speed coming downhill is 8 miles per hour faster than going uphill. The roundtrip takes 2 hours and 15 minutes $\left(=\frac{9}{4}\text{hours}\right)$. What was the speed of the bicyclist going uphill?

**46. Round Trip** A plane flies 600 miles west (into the wind) and makes the return trip following the same flight path. The effect of the jet stream on the plane is 15 miles per hour. The round trip takes 9 hours. What is the speed of the plane in still air?

**47. Scenic Drive** Joe and Nancy live in Morro Bay, California right off of Highway 1. They decide to take a trip north to Monterey. The first 50 miles of the drive is pretty easy, while the last 68 miles of the drive is filled with curves. They drove at an average of 9 miles per hour faster for the first 50 miles of the trip. The entire trip took 3 hours. How fast were Joe and Nancy driving for the first 50 miles of the trip?

**48. A Race** Dirk and Garret decide to have a 40-mile bicycle race. During the race, Dirk averages 2 miles per hour faster than Garret and beats Garret by $\frac{2}{3}$ of an hour. What was Dirk's average speed?

*In Problems 49–60, solve the variation problem.*

**49. Demand** Suppose that the demand $D$ for candy at the movie theater is inversely related to the price $p$.

**(a)** When the price of candy is \$2.50 per bag, the theater sells 150 bags of candy. Express the demand of candy as a function of its price.

**(b)** Determine the number of bags of candy that will be sold if the price is raised to \$3 a bag.

**50. Driving to School** The time $t$ that it takes to get to school varies inversely with your average speed $s$.

**(a)** Suppose that it takes you 30 minutes to drive to school when your average speed is 35 miles per hour. Express the driving time to school as a function of average speed.

**(b)** Suppose that your average speed driving to school is 30 miles per hour. How long will it take you to get to school?

**51. Pressure** The volume of a gas $V$ held at a constant temperature in a closed container varies inversely with its pressure $P$. If the volume of a gas is 600 cubic centimeters (cc) when the pressure is 150 millimeters of mercury (mm Hg), find the volume when the pressure is 200 mm Hg.

**52. Resistance** The current $i$ in a circuit is inversely proportional to its resistance $R$ measured in ohms. Suppose that when the current in a circuit is 30 amperes, the resistance is 8 ohms. Find the current in the same circuit when the resistance is 10 ohms.

**53. Weight** The weight of an object above the surface of Earth varies inversely with the square of the distance from the center of Earth. If Maria weighs 120 pounds when she is on the surface of Earth (3960 miles from the center), determine Maria's weight if she is at the top of Mount McKinley (3.8 miles from the surface of Earth).

**54. Intensity of Light** The intensity $I$ of light (measured in foot-candles) varies inversely with the square of the distance from the bulb. Suppose the intensity of a 100-watt light bulb at a distance of 2 meters is 0.075 foot-candles. Determine the intensity of the bulb at a distance of 3 meters.

**55. Drag Force** When an object moves through air, a frictionlike drag force tends to slow the object down. The drag force $D$ on a parachutist free-falling varies jointly with the surface area of the parachutist and the square of his velocity. The drag force on a parachutist with surface area 2 square meters falling at 40 meters per second is 1152 newtons. Find the drag force on a parachutist whose surface area is 2.5 square meters falling at 50 meters per second.

**56. Kinetic Energy** The kinetic energy $K$ (measured in joules) of a moving object varies jointly with the mass of the object and the square of its velocity $v$. The kinetic energy of a linebacker weighing 110 kilograms and running at a speed of 8 meters per second is 3520 joules. Find the kinetic energy of a wide receiver weighing 90 kilograms and running at a speed of 10 meters per second.

57. **Newton's Law of Gravitation** According to Newton's law of universal gravitation, the force $F$ of gravity between any two objects varies jointly with the masses of the objects $m_1$ and $m_2$ and inversely with the square of the distance between the objects $r$. The force of gravity between a 105-kg man and his 80-kg wife when they are separated by a distance of 5 meters is $2.24112 \times 10^{-8}$ newtons. Find the force of gravity between the man and his wife when they are 2 meters apart.

58. **Electrical Resistance** The electrical resistance of a wire varies directly with the length of the wire and inversely with the square of the diameter of the wire. If a wire 50 feet long and 3 millimeters in diameter has a resistance of 0.255 ohms, find the length of a wire of the same material whose resistance is 0.147 ohms and whose diameter is 2.5 millimeters.

59. **Stress of Material** The stress in the material of a pipe subject to internal pressure varies jointly with the internal pressure and internal diameter of the pipe and inversely with the thickness of the pipe. The stress is 100 pounds per square inch when the diameter is 5 inches, the thickness is 0.75 inch, and the internal pressure is 25 pounds per square inch. Find the stress when the internal pressure is 50 pounds per square inch, the diameter is 6 inches, and the thickness is 0.5 inch.

60. **Gas Laws** The volume $V$ of an ideal gas varies directly with the temperature $T$ and inversely with the pressure $P$. If a cylinder contains oxygen at a temperature of 300 kelvin (K) and a pressure of 15 atmospheres in a volume of 100 liters, what is the constant of proportionality $k$? If a piston is lowered into the cylinder, decreasing the volume occupied by the gas to 70 liters and raising the temperature to 315 K, what is the pressure?

## Extending the Concepts

61. **David and Goliath** The force $F$ (in newtons) required to maintain an object in a circular path varies jointly with the mass $m$ (in kilograms) of the object and the square of its speed (measured in meters per second) and inversely with the radius $r$ (in meters) of the circular path. Suppose that David has a rope that is 3 meters long. On the end of the rope he has attached a pouch that holds a 0.5-kilogram stone. Suppose that David is able to spin the rope in a circular motion at the rate of 50 revolutions per minute.

    **(a)** The spinning rate of 50 revolutions per minute can be converted to an *angular velocity* $\omega$ (lowercase Greek letter omega) using the formula $\omega = 2\pi \cdot$ revolutions per minute. Write the spinning rate as an angular velocity rounded to two decimal places.

    **(b)** Use the fact that $v = \omega r$, where $r$ is the radius of the circle to find the linear velocity (in meters per minute) of the stone if it were released. Now convert the linear velocity to meters per second.

    **(c)** Suppose that the force on the rope required to keep the rock in a circular motion is 2.3 newtons. Use the result from part (b) to find the constant of proportionality $k$.

    **(d)** David is fairly certain that he will require more force than this to beat Goliath in a battle, so he increases the circular motion to 80 revolutions per minute and increases the length of the rope to 4 meters. What is the force required to keep the stone in a circular motion?

62. **The Olympics** The current world record holder in the 100-meter dash is Maurice Greene, with a time of 9.79 seconds. In the 1984 Olympics, Carl Lewis won the gold medal in the 100-meter dash with a time of 9.99 seconds. If these two athletes ran in the same race repeating their respective times, by how many meters would Greene beat Lewis?

## Synthesis Review

*In Problems 63–68, use the Laws of Exponents to simplify each expression.*

**63.** $(a^3)^5$ **64.** $a(a^2b^{-3})^3$ **65.** $(ab)^{-2} \cdot \left(\frac{a^3}{b^2}\right)^2$

**66.** $\left(\frac{13a^5b^2}{ab^{-7}}\right)^0$ **67.** $\left(\frac{3m^3n^{-1}}{mn^5}\right)^{-2}$ **68.** $\left(\frac{12pq^{-3}}{3p^4q^{-4}}\right)^2$

# CHAPTER 6 ACTIVITY: CORRECT THE QUIZ

**Focus:** Performing operations and solving equations and inequalities with rational expressions
**Time:** 20 minutes
**Group size:** 2

In this activity you will work as a team to grade the student quiz shown below. One of you will grade the odd questions, and the other will grade the even questions. If an answer is correct, mark it correct. If an answer is wrong, mark it wrong and show the correct answer.

Once all of the quiz questions are graded, explain your results to each other and compute the final score for the quiz. Be prepared to discuss your results with the rest of the class.

| Student Quiz | |
|---|---|
| Name: *Ima Student* | Quiz Score: ______ |
| (1) Multiply: $\frac{2x^3 + 54}{5x^2 + 5x - 30} \cdot \frac{6x + 12}{3x^2 - 9x + 27}$ | Answer: $\frac{4(x + 2)}{5(x - 2)}$ |
| (2) Subtract: $\frac{xy}{x^2 - y^2} - \frac{y}{x + y}$ | Answer: $\frac{y^2}{x^2 - y^2}$ |
| (3) Solve: $\frac{3}{x - 3} + \frac{4}{x} = \frac{-12}{x^2 + 3x}$ | Answer: *no solution* |
| (4) Divide: $\frac{x^2 + x - 6}{5x^2 - 7x - 6} \div \frac{3x^2 + 13x + 12}{6x^2 + 17x + 12}$ | Answer: $\frac{2}{5}$ |
| (5) Solve: Joe can mow his lawn in 2 hrs. Mike can mow the same lawn in 3 hrs. If they work together, how long will it take them to mow the lawn? | Answer: *1 hr, 12 min* |
| (6) Simplify: $\frac{\frac{1}{x + 5} - \frac{2}{x - 7}}{\frac{4}{x - 7} + \frac{1}{x + 5}}$ | Answer: $\frac{x + 17}{5x + 13}$ |
| (7) Add: $\frac{x}{x^2 + 10x + 25} + \frac{4}{x^2 + 6x + 5}$ | Answer: $\frac{x^2 + 5x + 20}{(x + 5)(x + 1)}$ |
| (8) Solve: $\frac{4x}{x - 3} \geq 5$ | Answer: $(3, 15]$ |

# Chapter 6 Review

## Section 6.1 Multiplying and Dividing Rational Expressions

| KEY CONCEPTS | KEY TERMS |
|---|---|
| • **Multiplying Rational Expressions**<br>If $\frac{a}{b}$ and $\frac{c}{d}$, $b \neq 0$, $d \neq 0$, are two rational expressions, then $\frac{a}{b} \cdot \frac{c}{d} = \frac{ac}{bd}$.<br>• **Dividing Rational Expressions**<br>If $\frac{a}{b}$ and $\frac{c}{d}$, $b \neq 0$, $c \neq 0$, $d \neq 0$, are two rational expressions, then $\dfrac{\frac{a}{b}}{\frac{c}{d}} = \frac{a}{b} \cdot \frac{d}{c} = \frac{ad}{bc}$. | Rational expression<br>Numerator<br>Denominator<br>Lowest terms<br>Simplified<br>Rational function |

| YOU SHOULD BE ABLE TO . . . | EXAMPLE | REVIEW EXERCISES |
|---|---|---|
| 1 Determine the domain of a rational expression (p. 456) | Example 1 | 1–4 |
| 2 Simplify rational expressions (p. 457) | Examples 2 and 3 | 5–10 |
| 3 Multiply rational expressions (p. 458) | Examples 4 and 5 | 11–16 |
| 4 Divide rational expressions (p. 460) | Example 6 | 17–22 |
| 5 Work with rational functions (p. 461) | Examples 7 and 8 | 23–26 |

*In Problems 1–4, state the domain of each rational expression.*

**1.** $\dfrac{x - 5}{3x - 2}$ **2.** $\dfrac{a^2 - 16}{a^2 - 3a - 28}$ **3.** $\dfrac{m - 3}{m^2 + 9}$ **4.** $\dfrac{n^2 + 7n + 10}{n^2 - 2n - 8}$

*In Problems 5–10, simplify each rational expression.*

**5.** $\dfrac{6x + 30}{x^2 - 25}$ **6.** $\dfrac{4y^2 - 28y}{2y^5 - 14y^4}$ **7.** $\dfrac{w^2 - 4w - 21}{w^2 + 7w + 12}$

**8.** $\dfrac{6a^2 - 7ab - 3b^2}{10a^2 - 11ab - 6b^2}$ **9.** $\dfrac{7 - m}{3m^2 - 20m - 7}$ **10.** $\dfrac{n^3 - 4n^2 + 3n - 12}{n^2 - 8n + 16}$

*In Problems 11–22, multiply or divide each rational expression, as indicated. Simplify the product or quotient, if necessary.*

**11.** $\dfrac{4p^2}{p^2 - 3p - 18} \cdot \dfrac{p + 3}{8p}$ **12.** $\dfrac{q^2 + 6q}{6q + 12} \cdot \dfrac{4q + 8}{q^2 + q - 30}$

**13.** $\dfrac{x^3 - 4x^2}{x^2 - 4} \cdot \dfrac{x^2 + 4x - 12}{x^3 + 2x^2}$ **14.** $\dfrac{y^2 - 3y - 28}{y^3 + 4y^2} \cdot \dfrac{2y^2 + 10y}{y^2 - 12y + 35}$

**15.** $\dfrac{6a^2 + ab - b^2}{3a^2 + 2ab - b^2} \cdot \dfrac{3a^2 + 4ab + b^2}{4a^2 - b^2}$ **16.** $\dfrac{m^2 + m - 20}{m^3 - 64} \cdot \dfrac{3m^2 + 12m + 48}{m^2 + 3m - 10}$

**17.** $\dfrac{\frac{4c^2}{3d^4}}{\frac{8c}{27d}}$ **18.** $\dfrac{\frac{6z - 24}{7z + 21}}{\frac{z - 4}{z^2 - 9}}$ **19.** $\dfrac{\frac{x^2 - 11x + 30}{x^2 - 8x + 15}}{\frac{x^2 - 5x - 6}{x^2 + 8x + 7}}$

**20.** $\dfrac{\dfrac{m^2 + mn - 12n^2}{m^3 - 27n^3}}{\dfrac{m + 5n}{m^2 + 3mn + 9n^2}}$

**21.** $\dfrac{\dfrac{4p^3 - 4pq^2}{p^2 - 5pq - 24q^2}}{\dfrac{2p^3 + 4p^2q + 2pq^2}{p^2 - 7pq - 8q^2}}$

**22.** $\dfrac{\dfrac{15a^2 + 11a - 14}{25a^2 - 49}}{\dfrac{27a^3 - 8}{10a^2 + 11a - 35}}$

*In Problems 23–26, use the functions* $f(x) = \dfrac{2x^2 + 3x - 2}{x - 5}$, $g(x) = \dfrac{x^2 - 3x - 10}{2x - 1}$, *and* $h(x) = \dfrac{2x - 1}{x^2 + 9x + 14}$ *to find each product or quotient. State the domain of each product or quotient.*

**23.** $P(x) = f(x) \cdot g(x)$

**24.** $R(x) = g(x) \cdot h(x)$

**25.** $Q(x) = \dfrac{g(x)}{f(x)}$

**26.** $T(x) = \dfrac{f(x)}{h(x)}$

## Section 6.2 Adding and Subtracting Rational Expressions

| KEY CONCEPTS | KEY TERM |
|---|---|
| • **Adding/Subtracting Rational Expressions** If $\frac{a}{c}$ and $\frac{b}{c}$, $c \neq 0$, are two rational expressions, then $\frac{a}{c} + \frac{b}{c} = \frac{a + b}{c}$ and $\frac{a}{c} - \frac{b}{c} = \frac{a - b}{c}$. *Note:* If the rational expressions do not have a common denominator, then the least common denominator can be found using the steps on page 470. Then follow the steps listed on page 471 to add or subtract rational expressions with unlike denominators. | Least common denominator |

| YOU SHOULD BE ABLE TO . . . | EXAMPLE | REVIEW EXERCISES |
|---|---|---|
| 1 Add or subtract rational expressions with a common denominator (p. 468) | Examples 1 and 2 | 27–32 |
| 2 Find the least common denominator of two or more rational expressions (p. 469) | Example 3 | 33–36 |
| 3 Add or subtract rational expressions with different denominators (p. 471) | Examples 4 through 7 | 37–50 |

*In Problems 27–32, perform the indicated operation and simplify the result.*

**27.** $\dfrac{4x}{x - 5} + \dfrac{3}{x - 5}$

**28.** $\dfrac{4y}{y - 3} - \dfrac{12}{y - 3}$

**29.** $\dfrac{a^2 - 2a - 4}{a^2 - 6a + 8} + \dfrac{4a - 20}{a^2 - 6a + 8}$

**30.** $\dfrac{3b^2 + 8b - 5}{2b^2 - 5b - 12} - \dfrac{2b^2 + 7b + 15}{2b^2 - 5b - 12}$

**31.** $\dfrac{5c^2 - 8c}{c - 8} + \dfrac{2c^2 + 16c}{8 - c}$

**32.** $\dfrac{2d^2 + d}{d^2 - 1} - \dfrac{d^2 + 1}{d^2 - 1} + \dfrac{d - 2}{d^2 - 1}$

*In Problems 33–36, find the least common denominator.*

**33.** $\dfrac{4}{9x^4}$ and $\dfrac{5}{12x^2}$

**34.** $\dfrac{2y + 1}{y - 9}$ and $\dfrac{3y}{y + 2}$

**35.** $\dfrac{3p + 4}{2p^2 - 3p - 20}$ and $\dfrac{7p^2}{2p^3 + 5p^2}$

**36.** $\dfrac{q - 4}{q^2 + 4q - 5}$ and $\dfrac{q - 6}{q^2 + 2q - 15}$

*In Problems 37–48, perform the indicated operation and simplify the result.*

**37.** $\dfrac{1}{mn^4} + \dfrac{4}{m^3n^2}$

**38.** $\dfrac{3}{2xy^3} - \dfrac{7}{6x^2y}$

**39.** $\dfrac{p}{p - q} - \dfrac{q}{p + q}$

**40.** $\dfrac{x + 8}{x^2 - 10x + 21} - \dfrac{x - 5}{x^2 - 3x - 28}$

**41.** $\dfrac{3}{y^2 - 2y + 1} - \dfrac{2}{y^2 + y - 2}$

**42.** $\dfrac{3a - 5b}{4a^2 - 9b^2} + \dfrac{4}{2a - 3b}$

**43.** $\dfrac{4x^2 - 10x}{x^2 - 9} + \dfrac{8x - 2x^2}{9 - x^2}$

**44.** $\dfrac{1}{n + 5} - \dfrac{n^2 - 10n}{n^3 + 125}$

**45.** $\dfrac{m + n}{m + 3n} - \dfrac{m - 4n}{m - 7n} + \dfrac{7m + n^2}{m^2 - 4mn - 21n^2}$

**46.** $\dfrac{z^2 + 10z + 3}{z^2 - 9} - \dfrac{2z}{z - 3} + \dfrac{z}{z + 3}$

**47.** $\dfrac{y - 1}{y - 2} - \dfrac{y + 1}{y + 2} + \dfrac{y - 6}{y^2 - 4}$

**48.** $\dfrac{2a}{a^2 - 16} - \dfrac{1}{a - 4} - \dfrac{1}{a^2 + 2a - 8}$

**49.** Given that $f(x) = \dfrac{5}{x - 4}$ and $g(x) = \dfrac{x}{x + 2}$,
(a) find $S(x) = f(x) + g(x)$, and (b) state the domain of $S(x)$.

**50.** Given that $f(x) = \dfrac{x + 3}{2x^2 + x - 15}$ and $g(x) = \dfrac{x - 7}{4x^2 - 8x - 5}$,
(a) find $D(x) = f(x) - g(x)$, and (b) state the domain of $D(x)$.

## Section 6.3 Complex Rational Expressions

| KEY CONCEPT | KEY TERMS |
|---|---|
| • There are two methods that can be utilized in simplifying a complex rational expression. The steps for Method I are presented on page 479 while the steps for Method II are presented on page 481. | Complex rational expression<br>Complex fraction<br>Simplify |

| YOU SHOULD BE ABLE TO . . . | EXAMPLE | REVIEW EXERCISES |
|---|---|---|
| 1 Simplify a complex rational expression by simplifying the numerator and denominator separately (p. 478) | Examples 1, 2, and 5 | 51–54, 59–66 |
| 2 Simplify a complex rational expression using the least common denominator (p. 480) | Examples 3 through 5 | 55–58, 59–66 |

*In Problems 51–54, simplify each complex rational expression by using Method I (that is, by simplifying the numerator and denominator separately).*

**51.** $\dfrac{x - \dfrac{1}{x}}{1 - \dfrac{1}{x}}$

**52.** $\dfrac{\dfrac{1}{x} - \dfrac{1}{y}}{\dfrac{1}{x^2} - \dfrac{1}{y^2}}$

**53.** $\dfrac{\dfrac{a}{b} - \dfrac{a - b}{a + b}}{\dfrac{a}{b} + \dfrac{a + b}{a - b}}$

**54.** $\dfrac{\dfrac{2}{a + 2} - 1}{\dfrac{1}{a + 2} + 1}$

*In Problems 55–58, simplify each complex rational expression using Method II (that is, by using the least common denominator).*

**55.** $\dfrac{\frac{3}{t}+\frac{4}{t^2}}{5+\frac{1}{t^2}}$ **56.** $\dfrac{\frac{1}{a}-\frac{1}{b}}{\frac{b}{a}-\frac{a}{b}}$ **57.** $\dfrac{\frac{1}{z-1}-\frac{1}{z}}{\frac{1}{z}-\frac{1}{z+1}}$ **58.** $\dfrac{1+\frac{x}{x+1}}{\frac{2x+1}{x-1}}$

*In Problems 59–66, simplify the complex rational expression using either Method I or Method II.*

**59.** $\dfrac{\frac{x}{y}+1}{\frac{x}{y}-1}$ **60.** $\dfrac{\frac{a}{a-b}-\frac{b}{a+b}}{\frac{b}{a-b}+\frac{a}{a+b}}$ **61.** $\dfrac{\frac{1}{x-2}-\frac{x}{x^2-4}}{1-\frac{2}{x+2}}$ **62.** $\dfrac{z-\frac{5z}{z+5}}{z+\frac{5z}{z-5}}$

**63.** $\dfrac{\frac{m-n}{m+n}+\frac{n}{m}}{\frac{m}{n}-\frac{m-n}{m+n}}$ **64.** $\dfrac{\frac{x+4}{x-2}-\frac{x-3}{x+1}}{5x^2+4x-1}$ **65.** $\dfrac{3x^{-1}-3y^{-1}}{(x+y)^{-1}}$ **66.** $\dfrac{2c^{-1}-(3d)^{-1}}{(6d)^{-1}}$

## Section 6.4 Rational Equations

| KEY CONCEPT | KEY TERMS |
|---|---|
| • The steps for solving any rational equation are given on page 489. | Rational equation<br>Extraneous solution |

| YOU SHOULD BE ABLE TO . . . | EXAMPLE | REVIEW EXERCISES |
|---|---|---|
| 1 Solve equations containing rational expressions (p. 488) | Examples 1 through 5 | 67–78 |
| 2 Solve equations involving rational functions (p. 492) | Examples 6 and 7 | 79, 80 |

*In Problems 67–78, solve each equation. Be sure to verify your results.*

**67.** $\dfrac{2}{z}-\dfrac{1}{3z}=\dfrac{1}{6}$ **68.** $\dfrac{4}{m-4}=\dfrac{-5}{m+2}$ **69.** $m-\dfrac{14}{m}=5$ **70.** $\dfrac{2}{n+3}=\dfrac{1}{n-3}$

**71.** $\dfrac{s}{s-1}=1+\dfrac{2}{s}$ **72.** $\dfrac{3}{x^2-7x+10}+2=\dfrac{x-4}{x-5}$

**73.** $\dfrac{1}{k-1}+\dfrac{1}{k+2}=\dfrac{3}{k^2+k-2}$ **74.** $x+\dfrac{3x}{x-3}=\dfrac{9}{x-3}$

**75.** $\dfrac{2}{a+3}-\dfrac{4}{a^2-4}=\dfrac{a+1}{a^2+5a+6}$

**76.** $\dfrac{2}{z^2+2z-8}=\dfrac{1}{z^2+9z+20}+\dfrac{4}{z^2+3z-10}$

**77.** $\dfrac{x-3}{x+4}=\dfrac{14}{x^2+6x+8}$ **78.** $\dfrac{5}{y-5}+4=\dfrac{3y-10}{y-5}$

**79.** For the function $f(x)=\dfrac{6}{x-2}$, solve $f(x)=2$. What point(s) are on the graph of $f$?

**80.** For the function $g(x)=x-\dfrac{21}{x}$, solve $g(x)=4$. What point(s) are on the graph of $g$?

## Section 6.5 Rational Inequalities

| KEY CONCEPT | KEY TERM |
|---|---|
| • The steps for solving any rational inequality are given on page 499. | Rational inequality |

| YOU SHOULD BE ABLE TO . . . | EXAMPLE | REVIEW EXERCISES |
|---|---|---|
| 1 Solve a rational inequality (p. 497) | Examples 1 and 2 | 81–90 |

*In Problems 81–88, solve each rational inequality. Graph the solution set.*

**81.** $\dfrac{x-4}{x+2} \geq 0$

**82.** $\dfrac{y-5}{y+4} < 0$

**83.** $\dfrac{4}{z^2-9} \leq 0$

**84.** $\dfrac{w^2+5w-14}{w-4} < 0$

**85.** $\dfrac{m-5}{m^2+3m-10} \geq 0$

**86.** $\dfrac{4}{n-2} \leq -2$

**87.** $\dfrac{a+1}{a-2} > 3$

**88.** $\dfrac{4}{c-2} - \dfrac{3}{c} < 0$

*In Problems 89 and 90, for each function, find the values of x that satisfy the given condition. Graph the solution set.*

**89.** Solve $Q(x) < 0$ if $Q(x) = \dfrac{2x+3}{x-4}$.

**90.** Solve $R(x) \geq 0$ if $R(x) = \dfrac{x+5}{x+1}$.

## Section 6.6 Models Involving Rational Expressions

| KEY CONCEPTS | KEY TERMS |
|---|---|
| • Let $x$ and $y$ represent two quantities. We say that $y$ varies inversely with $x$, or $y$ is inversely proportional to $x$, if there is a nonzero number $k$ such that $y = \dfrac{k}{x}$.<br>• When a variable quantity $Q$ is proportional to the product of two or more other variables, we say that $Q$ varies jointly with these quantities.<br>• When direct and inverse variation occur at the same time, we have combined variation. | Ratio<br>Proportion<br>Similar<br>Constant rate<br>Varies inversely<br>Inversely proportional to<br>Joint variation<br>Combined variation |

| YOU SHOULD BE ABLE TO . . . | EXAMPLE | REVIEW EXERCISES |
|---|---|---|
| 1 Solve for a variable in a rational expression (p. 502) | Example 1 | 91–94 |
| 2 Model and solve ratio and proportion problems (p. 504) | Examples 2 and 3 | 95–98 |
| 3 Model and solve work problems (p. 506) | Examples 4 and 5 | 99–102 |
| 4 Model and solve uniform motion problems (p. 508) | Example 6 | 103–106 |
| 5 Model and solve problems involving inverse variation (p. 509) | Example 7 | 107, 109, 111, 112 |
| 6 Model and solve problems involving joint or combined variation (p. 511) | Examples 8 and 9 | 108, 110, 113, 114 |

*In Problems 91–94, solve each formula for the indicated variable.*

**91. Electronics (Capacitance)** Solve $\dfrac{1}{C_1} + \dfrac{1}{C_2} = \dfrac{1}{C}$ for $C$.

**92. Chemistry (Ideal Gas Law)** Solve $\dfrac{P_1V_1}{T_1} = \dfrac{P_2V_2}{T_2}$ for $T_2$.

**93. Physics (Kepler's Third Law)** Solve $T = \dfrac{4\pi^2 a^2}{MG}$ for $G$.

**94. Statistics ($z$-score)** Solve $z = \dfrac{x - \mu}{\sigma}$ for $x$.

*In Problems 95–98, solve each proportion problem.*

**95.** Suppose that $\triangle ABC$ is similar to $\triangle DEF$ as shown in the figure. Find the lengths of $\overline{AB}$ and $\overline{DF}$.

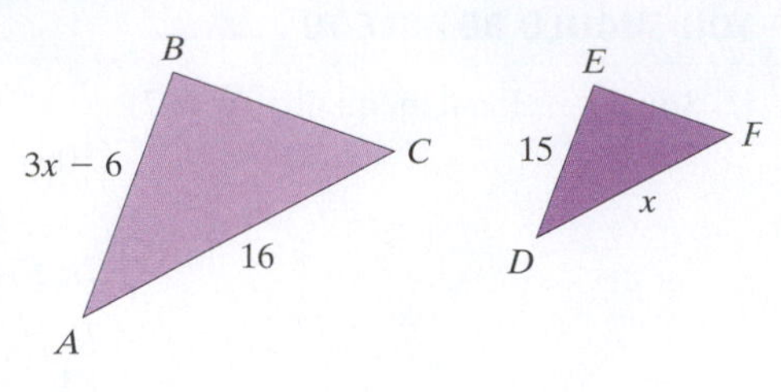


**96.** At a particular time of day, a pine tree casts a 30-foot shadow. At the exact same time, a nearby 5-foot post casts an 8-foot shadow. Find the height of the tree.

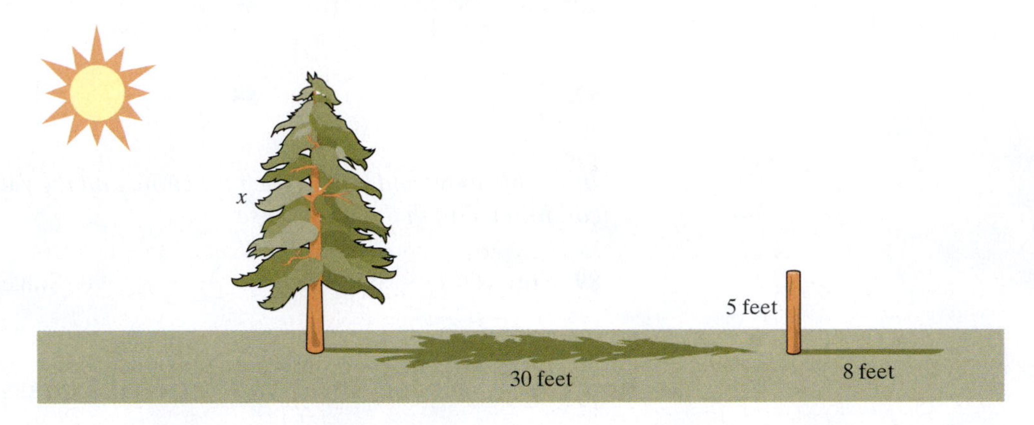


**97.** According to the nutrition facts on a box of Honey Nut Chex® (*Source:* General Mills), a $\frac{3}{4}$-cup serving contains 26 grams of total carbohydrates. How many grams of total carbohydrates are in a 3-cup bowl of the cereal?

**98.** One day, Jeri earns \$48.75 for 5 hours of work. How much will she earn for 8 hours of work?

*In Problems 99–102, solve each work problem.*

**99. Filling a Tank** One pipe can fill a tank in 48 minutes. Another pipe can fill the tank in 1 hour and 12 minutes. If both pipes are used, how long will it take to fill the tank?

**100. Mowing the Lawn** Together, Diane and Craig can mow their lawn in 1 hour and 10 minutes. Working alone, Diane can mow the lawn in 2 hours. How long will it take Craig to mow the lawn when working alone?

**101. Carpeting** Together, Rick and John can carpet a large room in 12 hours. Alone, Rick can carpet the same size room in 7 hours less time than John. How long would it take Rick to carpet the same size room if working alone? How long would it take John if working alone?

**102. Draining a Sink** A faucet can fill a sink in 1 minute when the drain is plugged, but when the drain is unplugged, it takes 1 minute and 30 seconds to fill the sink. With the faucet off, how long would it take the drain to empty a full sink?

*In Problems 103–106, solve each motion problem.*

**103. Pleasure Flight** In his private plane, Nick can fly 180 miles per hour if the wind is not blowing. One day, Nick took a pleasure flight. He flew 100 miles directly against the wind and then returned (flying with the wind) to his point of origin. If the total time of the flight was 1 hour and 15 minutes, what was the speed of the wind?

**104. Boating** In his motorboat, Jesse can travel 20 miles downstream in the same amount of time it takes him to travel 10 miles upstream. If the speed of the current is 5 miles per hour, how fast can Jesse's motorboat travel in still water?

**105. Running/Walking** To stay in shape, Todd first runs 3 miles and then walks 1 mile every morning. Todd's average running speed is 4 times his average walking speed. If Todd spends a total of 35 minutes running and walking each morning, find the average speed at which he walks and the average speed at which he runs.

**106. Road Trip** Because of heavy traffic, Danielle averaged only 30 miles per hour for the first 20 miles of her trip. If she averaged 50 miles per hour for the entire 100-mile trip, what was her average speed for the last 80 miles?

*In Problems 107–110, (a) find the constant of proportionality k, (b) write the function relating the variables, and (c) find the quantity indicated.*

**107.** Suppose that $y$ varies inversely with $x$. If $y = 15$ when $x = 4$, find $y$ when $x = 5$.

**108.** Suppose that $y$ varies jointly with $x$ and $z$. If $y = 45$ when $x = 6$ and $z = 10$, find $y$ when $x = 8$ and $z = 7$.

**109.** Suppose that $s$ varies inversely with the square of $t$. If $s = 18$ when $t = 2$, find $s$ when $t = 3$.

**110.** Suppose that $w$ varies directly with $x$ and inversely with $z$. If $w = \frac{4}{3}$ when $x = 10$ and $z = 12$, find $w$ when $x = 9$ and $z = 16$.

*In Problems 111–114, solve each variation problem.*

**111. Radio Signals** The frequency of a radio signal varies inversely with the wavelength. A signal of 800 kilohertz has a wavelength of 375 meters. What frequency has a signal of wavelength 250 meters?

**112. Ohm's Law** The electrical current flowing through a wire varies inversely with the resistance of the wire. If the current is 8 amperes when the resistance is 15 ohms, for what resistance will the current be 10 amperes?

**113. Volume of a Cylinder** The volume $V$ of a right circular cylinder varies jointly with the height $h$ and the square of the diameter $d$. If the volume of a cylinder is 231 cubic centimeters when the diameter is 7 centimeters and the height is 6 centimeters, find the volume when the diameter is 8 centimeters and the height is 14 centimeters.

**114. Volume of a Pyramid** The volume $V$ of a pyramid varies jointly with the base area $B$ and the height $h$. If the volume of a pyramid is 270 cubic inches when the base area is 81 square inches and the height is 10 inches, find the volume of a pyramid with base area 125 square inches and height 9 inches.

## CHAPTER 6 TEST

*Remember to use your Chapter Test Prep Video CD to see fully worked-out solutions to any of these problems you would like to review.*

**1.** Determine the domain of $f(x) = \dfrac{2x + 1}{2x^2 - 13x - 7}$.

*In Problems 2 and 3, simplify each rational expression.*

**2.** $\dfrac{2m^2 + 5m - 12}{3m^2 + 11m - 4}$

**3.** $\dfrac{2b - 3a}{3a^2 + 10ab - 8b^2}$

*In Problems 4–7, perform the indicated operations.*

**4.** $\dfrac{4x^2 - 12x}{x^2 - 9} \cdot \dfrac{2x^2 + 11x + 15}{8x^3 - 32x^2}$

**5.** $\dfrac{\dfrac{y^2 + 2y - 8}{4y^2 - 5y - 6}}{\dfrac{3y^2 - 14y - 5}{4y^2 - 17y - 15}}$

**6.** $\dfrac{3p^2 + 3pq}{p^2 - q^2} - \dfrac{3p - 2q}{p - q}$

**7.** $\dfrac{9c + 2}{3c^2 - 2c - 8} + \dfrac{7}{3c^2 + c - 4}$

*In Problems 8 and 9, use the functions* $f(x) = \dfrac{3x}{x^2 - 4}$, $g(x) = \dfrac{6}{x^2 + 2x}$, *and* $h(x) = \dfrac{9x^2 - 45x}{x^2 - 2x - 8}$ *to find each sum or quotient. State the domain of each.*

**8.** $Q(x) = \dfrac{f(x)}{h(x)}$

**9.** $S(x) = f(x) + g(x)$

*In Problems 10 and 11, simplify each complex rational expression using either Method I or Method II.*

**10.** $\dfrac{1 - \dfrac{1}{a}}{1 - \dfrac{1}{a^2}}$

**11.** $\dfrac{\dfrac{5}{d + 2} - \dfrac{1}{d - 2}}{\dfrac{3}{d + 2} - \dfrac{6}{d - 2}}$

*In Problems 12 and 13, solve each rational equation. Be sure to verify the results.*

**12.** $\dfrac{1}{6x} - \dfrac{1}{3} = \dfrac{5}{4x} + \dfrac{3}{4}$

**13.** $\dfrac{7n}{n + 3} + \dfrac{21}{n - 3} = \dfrac{126}{n^2 - 9}$

**14.** Solve $\dfrac{x + 5}{x - 2} \geq 3$. Graph the solution set.

**15. Electronics (Coulomb's Law)** Solve $\dfrac{1}{F} = \dfrac{D^2}{kq_1q_2}$ for $k$.

**16. Printing Documents** A particular laser printer can print out a 10-page document in 25 seconds. How long will it take to print out a 48-page document?

**17. Cleaning House** Linnette can clean the house in 4 hours. Her husband Darrell can do the same job in 6 hours. If the two work together, how long will it take them to clean the house?

**18. Kayaking** Chuck kayaked 4 miles upstream in the same time it took him to kayak 10 miles downstream. If Chuck can average 7 miles per hour in still water, what was the rate of the current?

**19. Using a Lever** Using a lever, the force $F$ required to lift a weight is inversely proportional to the length $l$ of the force arm of the lever (assuming all other factors are constant). If a force of 50 pounds is required to lift a granite boulder when the force arm length is 4 feet, how much force will be required to lift the boulder if the force arm length is 10 feet?

**20. Lateral Surface Area of a Cylinder** The lateral surface area $L$ of a right circular cylinder varies jointly with its radius $r$ and height $h$. If the lateral surface area of a cylinder with radius 7 centimeters and height 12 centimeters is 528 square centimeters, what would be the lateral surface area if the radius is 9 centimeters and the height is 14 centimeters?

# Getting Ready for Chapter 7: Radicals and Rational Exponents: Square Roots

**OBJECTIVES**

1. Evaluate Square Roots of Perfect Squares
2. Determine Whether a Square Root Is Rational, Irrational, or Not a Real Number
3. Find Square Roots of Variable Expressions

In Section R.4, we introduced the concept of exponents. Exponents are used to indicate repeated multiplication. For example, $4^2$ means $4 \cdot 4$, so $4^2 = 16$; $(-6)^2$ means $(-6) \cdot (-6)$, so $(-6)^2 = 36$. Now, we will reverse the process of raising a number to the second power and ask questions such as, "What number, or numbers, when squared, give me 16?"

## 1 Evaluate Square Roots of Perfect Squares

**In Words**
Taking the square root of a number is the "inverse" of squaring a number.

A real number is squared when it is raised to the power 2. The inverse of squaring a number is finding the **square root.** For example, since $5^2 = 25$ and $(-5)^2 = 25$, the square roots of 25 are $-5$ and 5. The square roots of $\frac{16}{49}$ are $-\frac{4}{7}$ and $\frac{4}{7}$. If we want only the positive square root of a number, we use the symbol $\sqrt{\phantom{x}}$, called a **radical sign,** to denote the **principal square root,** or nonnegative (zero or positive) square root.

**In Words**
Nonnegative real numbers are positive real numbers or 0.

**DEFINITION**

If $a$ is a nonnegative real number, the nonnegative real number $b$ such that $b^2 = a$, is the **principal square root** of $a$ and is denoted by $b = \sqrt{a}$.

For example, if we want the positive square root of 25, we would write $\sqrt{25} = 5$. We read $\sqrt{25} = 5$ as "the positive square root of 25 is 5." But what if we want the negative square root of a real number? In that case, we use the expression $-\sqrt{25} = -5$ to obtain the negative square root of 25.

**PROPERTIES OF SQUARE ROOTS**

- Every positive real number has two square roots, one positive and one negative.
- The square root of 0 is 0. That is, $\sqrt{0} = 0$.
- We use the symbol $\sqrt{\phantom{x}}$, called a radical, to denote the nonnegative square root of a real number. The nonnegative square root is called the principal square root.
- The number under the radical is called the **radicand.** For example, the radicand in $\sqrt{25}$ is 25.
- For any real number $c$, such that $c \geq 0$, $(\sqrt{c})^2 = c$. For example, $(\sqrt{4})^2 = 4$ and $(\sqrt{8.3})^2 = 8.3$.

To **evaluate** a square root, we ask ourselves, "What is the nonnegative number whose square is equal to the radicand?"

**EXAMPLE 1** **Evaluating Square Roots**

Evaluate each square root.

**(a)** $\sqrt{36}$ **(b)** $\sqrt{\frac{1}{9}}$

**(c)** $\sqrt{0.01}$ **(d)** $(\sqrt{2.3})^2$

### Solution

**(a)** Is there a positive number whose square is 36? Because $6^2 = 36$, $\sqrt{36} = 6$.

**(b)** $\sqrt{\frac{1}{9}} = \frac{1}{3}$ because $\left(\frac{1}{3}\right)^2 = \frac{1}{9}$.

**(c)** $\sqrt{0.01} = 0.1$ because $0.1^2 = 0.01$.

**(d)** $\left(\sqrt{2.3}\right)^2 = 2.3$ because $\left(\sqrt{c}\right)^2 = c$ when $c \geq 0$.

**Figure 1**

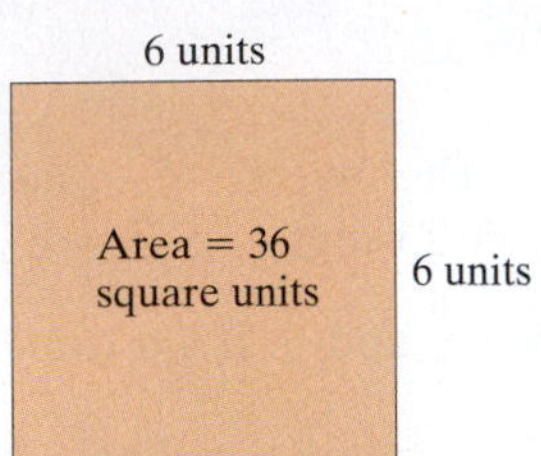


A rational number is a **perfect square** if it is the square of a rational number. Examples 1(a), (b), and (c) are square roots of perfect squares since $6^2 = 36$, $\left(\frac{1}{3}\right)^2 = \frac{1}{9}$, and $0.1^2 = 0.01$. We can think of perfect squares geometrically as shown in Figure 1, where we have a square whose area is 36. The square root of the area, $\sqrt{36}$, gives us the length of each side of the square, 6 units.

**QUICK ✓** *Evaluate each square root.*

**1.** $\sqrt{81}$ **2.** $\sqrt{900}$ **3.** $\sqrt{\frac{1}{4}}$ **4.** $\sqrt{0.16}$ **5.** $\left(\sqrt{13}\right)^2$

### EXAMPLE 2 Evaluating an Expression Containing Square Roots

Evaluate each expression:

**(a)** $-4\sqrt{36}$ **(b)** $\sqrt{9} + \sqrt{16}$ **(c)** $\sqrt{9 + 16}$ **(d)** $\sqrt{64 - 4 \cdot 7 \cdot 1}$

### Solution

**(a)** The expression $-4\sqrt{36}$ is asking us to find $-4$ times the positive square root of 36. So we first find the positive square root of 36 and then multiply this result by $-4$.

$$-4\sqrt{36} = -4 \cdot 6$$
$$= -24$$

**Work Smart**

In Examples 2(b) and (c), notice that

$$\sqrt{9} + \sqrt{16} \neq \sqrt{9 + 16}$$

In general,

$$\sqrt{a} + \sqrt{b} \neq \sqrt{a + b}$$

The radical acts like a grouping symbol, so always simplify the radicand before taking the square root.

**(b)** $\sqrt{9} + \sqrt{16} = 3 + 4$
$= 7$

**(c)** $\sqrt{9 + 16} = \sqrt{25}$
$= 5$

**(d)** $\sqrt{64 - 4 \cdot 7 \cdot 1} = \sqrt{64 - 28}$
$= \sqrt{36}$
$= 6$

**QUICK ✓** *Evaluate each expression.*

**6.** $5\sqrt{9}$ **7.** $\sqrt{36 + 64}$ **8.** $\sqrt{36} + \sqrt{64}$ **9.** $\sqrt{25 - 4 \cdot 3 \cdot (-2)}$

## 2 Determine Whether a Square Root Is Rational, Irrational, or Not a Real Number

Not all radical expressions will simplify to a rational number. For example, because there is no rational number whose square is 5, $\sqrt{5}$ is not a rational number. In fact, $\sqrt{5}$ is an *irrational* number. Remember, an irrational number is a number that cannot be written as the quotient of two integers.

**Work Smart**

The square roots of negative real numbers are not real.

What if we wanted to evaluate $\sqrt{-16}$? Because any positive real number squared is positive, any negative real number squared is also positive, and 0 squared is 0, there is no real number whose square is $-16$. We conclude: **Negative real numbers do not have square roots that are real numbers!**

The following comments regarding square roots are important.

**MORE PROPERTIES OF SQUARE ROOTS**

- The square root of a perfect square is a rational number.
- The square root of a positive rational number that is not a perfect square is an irrational number. For example, $\sqrt{20}$ is an irrational number because 20 is not a perfect square.
- The square root of a negative real number is not a real number. For example, $\sqrt{-2}$ is not a real number.

When a radical has a radicand that is not a perfect square, we can do one of two things:

1. Write a decimal approximation of the radical.
2. Simplify the radical using properties of radicals, if possible (Section 7.2).

## EXAMPLE 3 Writing a Radical as a Decimal Using a Calculator

Write $\sqrt{5}$ as a decimal rounded to two decimal places.

### Solution

Figure 2 shows the results from a TI-84 Plus graphing calculator. From the display, we see that $\sqrt{5} \approx 2.24$.

**Figure 2**

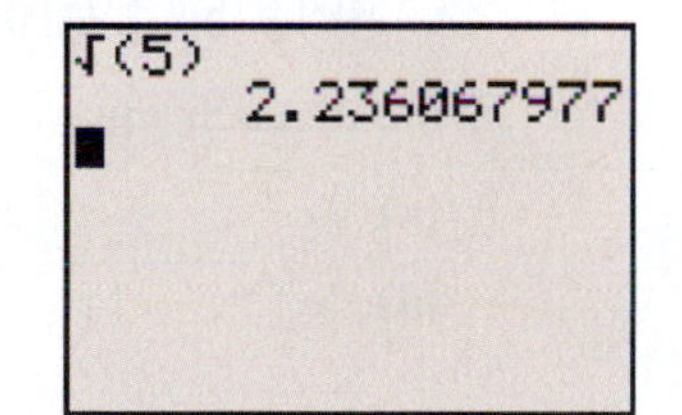


## EXAMPLE 4 Determining Whether a Square Root of an Integer Is Rational, Irrational, or Not a Real Number

Determine if each square root is rational, irrational, or not a real number. Then evaluate each real square root. For each square root that is irrational, express the square root as a decimal rounded to two decimal places.

**(a)** $\sqrt{51}$ **(b)** $\sqrt{169}$ **(c)** $\sqrt{-81}$

### Solution

**Work Smart**

$\sqrt{-81}$ is not a real number, but $-\sqrt{81}$ is a real number because $-\sqrt{81} = -9$. Note the placement of the negative sign!

**(a)** $\sqrt{51}$ is irrational because 51 is not a perfect square. That is, there is no rational number whose square is 51. Using a calculator, we find $\sqrt{51} \approx 7.14$.

**(b)** $\sqrt{169}$ is a rational number because $13^2 = 169$. So, $\sqrt{169} = 13$.

**(c)** $\sqrt{-81}$ is not a real number. There is no real number whose square is $-81$.

**QUICK ✓** *Determine whether each square root is rational, irrational, or not a real number. Evaluate each square root that is rational. For each square root that is irrational, approximate the square root rounded to two decimal places.*

**10.** $\sqrt{400}$ **11.** $\sqrt{40}$ **12.** $\sqrt{-25}$ **13.** $-\sqrt{196}$

## 3 Find Square Roots of Variable Expressions

What is $\sqrt{4^2}$? Because $4^2 = 16$, we have that $\sqrt{4^2} = \sqrt{16} = 4$. Based on this result, we might conclude that $\sqrt{a^2} = a$ for any real number $a$. Before we jump to this conclusion, let's consider $\sqrt{(-4)^2}$. Our "formula" says that $\sqrt{a^2} = a$, so we would think that $\sqrt{(-4)^2} = -4$, right? Wrong! $\sqrt{(-4)^2} = \sqrt{16} = 4$. So $\sqrt{4^2} = 4$ and $\sqrt{(-4)^2} = 4$. Regardless of whether the "$a$" in $\sqrt{a^2}$ is positive or negative, the result ends up being positive. So to say that $\sqrt{a^2} = a$ would not quite be correct. How can we fix our "formula"? In Section R.3, we learned that $|a|$ will be a positive number if $a$ is nonzero. From this, we have the following result:

**In Words**
The square root of a nonzero number squared will always be positive. The absolute value ensures this.

For any **real number** $a$,

$$\sqrt{a^2} = |a|$$

The bottom line is this—if you are taking the square root of some variable expression raised to the second power, the result will be the absolute value of the variable expression.

### EXAMPLE 5 Evaluating Square Roots

Evaluate each square root.

**(a)** $\sqrt{7^2}$ **(b)** $\sqrt{(-15)^2}$ **(c)** $\sqrt{x^2}$
**(d)** $\sqrt{(3x-1)^2}$ **(e)** $\sqrt{x^2 + 6x + 9}$

### Solution

**(a)** $\sqrt{7^2} = 7$

**(b)** $\sqrt{(-15)^2} = |-15| = 15$

**(c)** We don't know whether the real number $x$ is positive, negative, or zero. To ensure that the result is positive or zero, we write $\sqrt{x^2} = |x|$.

**(d)** $\sqrt{(3x-1)^2} = |3x - 1|$

**(e)** Notice that the radicand is a perfect square trinomial, so that $x^2 + 6x + 9$ factors to $(x+3)^2$. Therefore,

$$\sqrt{x^2 + 6x + 9} = \sqrt{(x+3)^2} = |x + 3|$$

**Work Smart**
Note how we use absolute values to get nonnegative answers in these examples.

**QUICK ✓** *Evaluate each square root.*

**14.** $\sqrt{(-14)^2}$ **15.** $\sqrt{z^2}$ **16.** $\sqrt{(2x+3)^2}$ **17.** $\sqrt{p^2 - 12p + 36}$

# Getting Ready for Chapter 7 Exercises

**For Extra Help:**

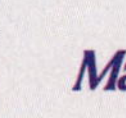


## Concepts and Vocabulary

*In Problems 1–3, fill in the blanks.*

**1.** The symbol $\sqrt{\ }$ is called a _____ _____.

**2.** If $a$ is a nonnegative real number, the nonnegative number $b$ such that $b^2 = a$ is the _____ _____ _____ of $a$ and is denoted by $b = \sqrt{a}$.

**3.** For any real number $a$, $\sqrt{a^2} =$ _____.

*In Problems 4–6, answer True or False to each statement.*

**4.** The square root of 36 is 6.

**5.** Negative numbers do not have square roots in the real number system.

**6.** The square root of a negative real number is a negative real number.

**7.** Give four examples of perfect squares.

**8.** Explain why $\sqrt{a^2} = |a|$. Provide examples to support your explanation.

## Building Skills

*In Problems 9–18, evaluate each square root.*

**9.** $\sqrt{1}$ **10.** $\sqrt{9}$ **11.** $-\sqrt{100}$ **12.** $-\sqrt{144}$ **13.** $\sqrt{\frac{1}{4}}$

**14.** $\sqrt{\frac{4}{81}}$ **15.** $\sqrt{0.36}$ **16.** $\sqrt{0.16}$ **17.** $(\sqrt{1.6})^2$ **18.** $(\sqrt{3.7})^2$

*In Problems 19–30, tell if the square root is rational, irrational, or not a real number. If the square root is rational, find the exact value; if the square root is irrational, write the approximate value rounded to two decimal places.*

**19.** $\sqrt{-14}$ **20.** $\sqrt{-50}$ **21.** $\sqrt{64}$ **22.** $\sqrt{121}$

**23.** $\sqrt{\frac{1}{16}}$ **24.** $\sqrt{\frac{49}{100}}$ **25.** $\sqrt{44}$ **26.** $\sqrt{24}$

**27.** $\sqrt{50}$ **28.** $\sqrt{12}$ **29.** $\sqrt{-16}$ **30.** $\sqrt{-64}$

*In Problems 31–42, simplify each square root.*

**31.** $\sqrt{8^2}$ **32.** $\sqrt{5^2}$ **33.** $\sqrt{(-19)^2}$ **34.** $\sqrt{(-13)^2}$

**35.** $\sqrt{r^2}$ **36.** $\sqrt{w^2}$ **37.** $\sqrt{(x+4)^2}$ **38.** $\sqrt{(x-8)^2}$

**39.** $\sqrt{(4x-3)^2}$ **40.** $\sqrt{(5x+2)^2}$

**41.** $\sqrt{4y^2 + 12y + 9}$ **42.** $\sqrt{9z^2 - 24z + 16}$

## Mixed Practice

*In Problems 43–60, simplify each expression. If necessary, express results that are not rational numbers as a decimal rounded to two decimal places.*

**43.** $\sqrt{25 + 144}$

**44.** $\sqrt{9 + 16}$

**45.** $\sqrt{25} + \sqrt{144}$

**46.** $\sqrt{9} + \sqrt{16}$

**47.** $\sqrt{-144}$

**48.** $\sqrt{-36}$

**49.** $3\sqrt{25}$

**50.** $-10\sqrt{16}$

**51.** $5\sqrt{\dfrac{16}{25}} - \sqrt{144}$

**52.** $2\sqrt{\dfrac{9}{4}} - \sqrt{4}$

**53.** $\sqrt{8^2 - 4 \cdot 1 \cdot 7}$

**54.** $\sqrt{9^2 - 4 \cdot 1 \cdot 20}$

**55.** $\sqrt{(-5)^2 - 4 \cdot 2 \cdot 5}$

**56.** $\sqrt{(-3)^2 - 4 \cdot 3 \cdot 2}$

**57.** $\dfrac{-(-1) + \sqrt{(-1)^2 - 4 \cdot 6 \cdot (-2)}}{2 \cdot (-1)}$

**58.** $\dfrac{-7 + \sqrt{7^2 - 4 \cdot 2 \cdot 6}}{2 \cdot 2}$

**59.** $\sqrt{(6 - 1)^2 + (15 - 3)^2}$

**60.** $\sqrt{(2 - (-1))^2 + (6 - 2)^2}$

**61.** What are the square roots of 36? What is $\sqrt{36}$?

**62.** What are the square roots of 64? What is $\sqrt{64}$?

*For Problems 63 and 64, use the formula* $Z = \dfrac{X - \mu}{\frac{\sigma}{\sqrt{n}}}$ *from statistics (a formula used to determine the relative value of one observation to another) to evaluate each expression for the given values. Write the exact value and then write your answer rounded to two decimal places.*

**63.** $X = 120, \mu = 100, \sigma = 15, n = 13$

**64.** $X = 40, \mu = 50, \sigma = 10, n = 5$

CHAPTER

# 7 Radicals and Rational Exponents

Where would we be without electricity? In 1752, when Ben Franklin first sent his kite into the clouds, the idea of a number whose square is $-1$ was still being developed by mathematicians. Yet it turns out that this concept is extremely useful in describing alternating electric currents. See Problems 113 and 114 from Section 7.8.

## OUTLINE



## The Big Picture: Putting It Together

In Chapter 5 we simplified polynomial expressions by adding, subtracting, multiplying, and dividing. We also factored polynomials. In Chapter 6 we used the skills learned in Chapter 5 to simplify rational expressions and perform operations on rational expressions.

We now present a similar discussion with radical expressions. We will learn how to add, subtract, multiply, and divide radical expressions. In addition, we will use factoring to simplify radical expressions. Throughout this discussion, keep in mind that radicals perform the "inverse" operation from raising a real number to an integer exponent. For example, a square root undoes the "squaring" operation.

# 7.1 nth Roots and Rational Exponents

**OBJECTIVES**

1. Evaluate $n$th Roots
2. Simplify Expressions of the Form $\sqrt[n]{a^n}$
3. Evaluate Expressions of the Form $a^{1/n}$
4. Evaluate Expressions of the Form $a^{m/n}$

### *Preparing for nth Root and Rational Exponents*

*Before getting started, take the following readiness quiz. If you get a problem wrong, go to the section cited and review the material.*

**1.** Simplify: $\left(\dfrac{x^2y}{xy^{-2}}\right)^{-3}$ [Getting Ready: Integer Exponents, pp. 344–345]

**2.** Simplify: $\left(\sqrt{7}\right)^2$ [Getting Ready: Square Roots, pp. 527–528]

**3.** Evaluate: $\sqrt{64}$ [Getting Ready: Square Roots, pp. 527–528]

In the Getting Ready: Square Roots, we reviewed skills for evaluating square roots. We now extend this skill to other types of roots.

## 1 Evaluate nth Roots

A real number is cubed when it is raised to the power 3. The inverse of cubing a number is finding the *cube root*. For example, since $2^3 = 8$, the cube root of 8 is 2 and since $(-2)^3 = -8$, the cube root of $-8$ is $-2$. In general, we can find $n$th roots of numbers.

**In Words**
When you see the notation $\sqrt[n]{a} = b$, think to yourself, "Find a number $b$ such that raising that number to the $n$th power gives me $a$."

**DEFINITION**

The **principal nth root of a number a,** symbolized by $\sqrt[n]{a}$, where $n \geq 2$ is an integer, is defined as follows:

$$\sqrt[n]{a} = b \qquad \text{means} \qquad a = b^n$$

- If $n \geq 2$ and even, then $a$ and $b$ must be greater than or equal to 0.
- If $n \geq 3$ and odd, then $a$ and $b$ can be any real number.

**Work Smart**
If the index is even, then the radicand must be greater than or equal to zero in order for a radical to simplify to a real number. If the index is odd, the radicand can be any real number.

In the notation $\sqrt[n]{a}$, the integer $n$, $n \geq 2$, is called the **index.** If a radical is written without the index, it is understood that we mean the square root, so $\sqrt{a}$ represents the square root of $a$. If the index is 3, we call $\sqrt[3]{a}$ the **cube root** of $a$. If the index is even, then the radicand must be greater than or equal to 0. If the index is odd, then the radicand can be any real number. Do you know why? Since $\sqrt[n]{a} = b$ means $b^n = a$, if the index $n$ is even, then $b^n \geq 0$ so $a \geq 0$. If we have an odd index $n$, then $b^n$ can be any real number, so $a$ can be any real number.

Before we evaluate $n$th roots, we list some "perfect" powers of 2, 3, 4, and 5. Having this list will be a great help in finding roots in the examples that follow. Some perfect cubes are $1^3 = 1$, $(-2)^3 = -8$, $3^3 = 27$, $4^3 = 64$ and $(-5)^3 = -125$, and so on. Some perfect fourths are $1^4 = 1$, $2^4 = 16$, $3^4 = 81$, $4^4 = 256$, $5^4 = 625$, and so on. Notice that perfect cubes can be negative, but perfect fourths cannot. Do you know why?

| Perfect Squares | Perfect Cubes | Perfect Fourths | Perfect Fifths |
|---|---|---|---|
| $1^2 = 1$ | $(-2)^3 = -8$ | $1^4 = 1$ | $(-2)^5 = -32$ |
| $2^2 = 4$ | $(-1)^3 = -1$ | $2^4 = 16$ | $(-1)^5 = -1$ |
| $3^2 = 9$ | $1^3 = 1$ | $3^4 = 81$ | $1^5 = 1$ |
| $4^2 = 16$ and so on | $2^3 = 8$ and so on | $4^4 = 256$ and so on | $2^5 = 32$ and so on |

### EXAMPLE 1 Evaluating nth Roots of Real Numbers

Evaluate:

**(a)** $\sqrt[3]{64}$ **(b)** $\sqrt[4]{16}$ **(c)** $\sqrt[3]{-8}$ **(d)** $\sqrt[4]{-81}$

*Preparing for...Answers*
**1.** $\dfrac{1}{x^3y^9}$ **2.** 7 **3.** 8

**Solution**

**(a)** $\sqrt[3]{64} = 4$ since $4^3 = 64$.
**(b)** $\sqrt[4]{16} = 2$ since $2^4 = 16$.
**(c)** $\sqrt[3]{-8} = -2$ since $(-2)^3 = -8$.
**(d)** Because there is no real number $b$ such that $b^4 = -81$, $\sqrt[4]{-81}$ is not a real number.

**QUICK ✓** *Evaluate each root.*

**1.** $\sqrt[3]{125}$ **2.** $\sqrt[4]{81}$ **3.** $\sqrt[3]{-216}$ **4.** $\sqrt[4]{-32}$ **5.** $\sqrt[5]{\frac{1}{32}}$

The $n$th roots in Examples 1(a)–(c) were all rational numbers. This is not always the case. Just as we can approximate square roots using a calculator, we can also approximate $n$th roots.

### EXAMPLE 2 Approximating an $n$th Root Using a Calculator

**(a)** Write $\sqrt[3]{25}$ as a decimal rounded to two decimal places.
**(b)** Write $\sqrt[4]{18}$ as a decimal rounded to two decimal places.

**Solution**

**(a)** Because $\sqrt[3]{8} = 2$ and $\sqrt[3]{27} = 3$, we expect $\sqrt[3]{25}$ is between 2 and 3 (closer to 3). Figure 1(a) shows the approximate value of $\sqrt[3]{25}$ obtained from a TI-84 Plus graphing calculator. So $\sqrt[3]{25} \approx 2.92$.
**(b)** Because $\sqrt[4]{16} = 2$ and $\sqrt[4]{81} = 3$, we expect $\sqrt[4]{18}$ is between 2 and 3 (closer to 2). Figure 1(b) shows the approximate value of $\sqrt[4]{18}$ obtained from a TI-84 Plus graphing calculator. So $\sqrt[4]{18} \approx 2.06$.

**Figure 1**

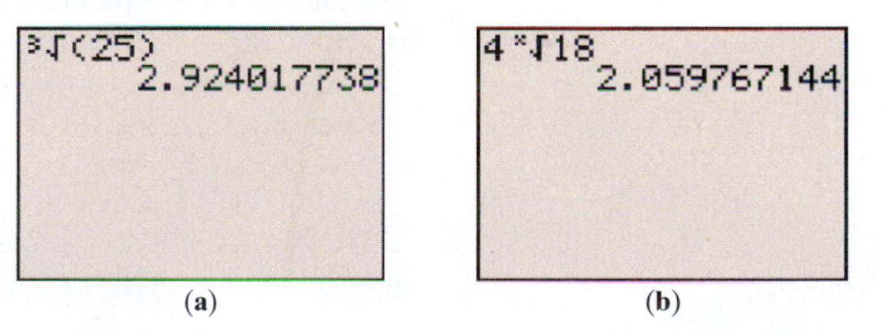


(a) (b)

**QUICK ✓** *Use a calculator to write the approximate value of each radical rounded to two decimal places.*

**6.** $\sqrt[3]{50}$ **7.** $\sqrt[4]{80}$ **8.** $\sqrt[5]{40}$

## 2 Simplify Expressions of the Form $\sqrt[n]{a^n}$

We have already seen that $\sqrt{a^2} = |a|$. But what about $\sqrt[3]{a^3}$ or $\sqrt[4]{a^4}$? Recall that the definition of the principal $n$th root, $\sqrt[n]{a}$, requires that $a \geq 0$ when $n$ is even and $a$ can be any real number when $n$ is odd.

**SIMPLIFYING $\sqrt[n]{a^n}$**

If $n \geq 2$ is a positive integer and $a$ is a real number, then

$$\sqrt[n]{a^n} = a \quad \text{if } n \geq 3 \text{ is odd}$$
$$\sqrt[n]{a^n} = |a| \quad \text{if } n \geq 2 \text{ is even}$$

**EXAMPLE 3 Simplifying Radicals**

Simplify:

**(a)** $\sqrt[3]{x^3}$ **(b)** $\sqrt[4]{(x-7)^4}$ **(c)** $-\sqrt[6]{(-3)^6}$ **(d)** $\sqrt[3]{-\frac{8}{125}}$

**Solution**

**(a)** Because the index, 3, is odd, we have that $\sqrt[3]{x^3} = x$.

**(b)** Because the index, 4, is even, we have that $\sqrt[4]{(x-7)^4} = |x-7|$.

**(c)** $-\sqrt[6]{(-3)^6} = -|-3| = -3$

**(d)** $\sqrt[3]{-\frac{8}{125}} = \sqrt[3]{\frac{-8}{125}} = \sqrt[3]{\frac{(-2)^3}{5^3}} = \sqrt[3]{\left(\frac{-2}{5}\right)^3} = -\frac{2}{5}$

**QUICK ✓** *Simplify each radical.*

**9.** $\sqrt[4]{5^4}$ **10.** $\sqrt[6]{z^6}$ **11.** $\sqrt[7]{(3x-2)^7}$ **12.** $\sqrt[8]{(-2)^8}$ **13.** $\sqrt[5]{\frac{-32}{243}}$

## 3 Evaluate Expressions of the Form $a^{1/n}$

**Work Smart**

Remember, a rational number is a number of the form $\frac{p}{q}$, where $p$ and $q$ are integers and $q \neq 0$.

In the "Getting Ready: Integer Exponents," we carefully developed methods for simplifying algebraic expressions that contained integer exponents. This development started with methods for simplifying algebraic expressions containing only positive integer exponents. We then presented a definition for raising a nonzero real number to the power of 0. With this material in hand, we were able to develop rules for simplifying algebraic expressions involving *all* integer exponents. Of course, the world cannot easily be described using only integers, so it is logical that we would want to extend the rules of exponents to rational exponents.

We start by providing a definition for "$a$ raised to the power $\frac{1}{n}$," where $a$ is a real number and $n$ is a positive integer. This definition needs to be written so that the laws of exponents apply. For example, we know that $a^2 = a \cdot a$, so

$$\begin{aligned}\left(5^{\frac{1}{2}}\right)^2 &= 5^{\frac{1}{2}} \cdot 5^{\frac{1}{2}} \\ a^m \cdot a^n = a^{m+n}: \quad &= 5^{\frac{1}{2}+\frac{1}{2}} \\ &= 5^1 \\ &= 5\end{aligned}$$

We also know that $\left(\sqrt{5}\right)^2 = 5$, so it is reasonable to conclude that

$$5^{\frac{1}{2}} = \sqrt{5}$$

This suggests the following definition:

**DEFINITION OF $a^{1/n}$**

If $a$ is a real number and $n$ is an integer with $n \geq 2$, then

$$a^{\frac{1}{n}} = \sqrt[n]{a}$$

provided that $\sqrt[n]{a}$ exists.

**EXAMPLE 4 Evaluating Expressions Containing Exponents of the Form $1/n$**

Write each of the following expressions as a radical and simplify, if possible.

**(a)** $9^{\frac{1}{2}}$ **(b)** $(-64)^{\frac{1}{3}}$ **(c)** $-100^{\frac{1}{2}}$ **(d)** $(-100)^{\frac{1}{2}}$ **(e)** $z^{\frac{1}{2}}$

**Work Smart**

Notice the use of parentheses in Examples 4(c) and (d). In Example 4(c), $-100^{1/2}$ means that we should evaluate $100^{1/2}$ first and then multiply the result by $-1$. Remember, exponents before multiplication.

**Solution**

**(a)** $9^{\frac{1}{2}} = \sqrt{9} = 3$

**(b)** $(-64)^{\frac{1}{3}} = \sqrt[3]{-64} = -4$

**(c)** $-100^{\frac{1}{2}} = -1 \cdot 100^{\frac{1}{2}} = -\sqrt{100} = -10$

**(d)** $(-100)^{\frac{1}{2}} = \sqrt{-100}$ is not a real number because there is no real number whose square is $-100$.

**(e)** $z^{\frac{1}{2}} = \sqrt{z}$

**QUICK ✓** *Write each of the following expressions and simplify, if possible.*

**14.** $25^{\frac{1}{2}}$ **15.** $(-27)^{\frac{1}{3}}$ **16.** $-64^{\frac{1}{2}}$ **17.** $(-64)^{\frac{1}{2}}$ **18.** $b^{\frac{1}{2}}$

## EXAMPLE 5 Writing Radicals with Rational Exponents

Rewrite each of the following radicals with a rational exponent.

**(a)** $\sqrt[4]{7a}$ **(b)** $\sqrt[5]{\dfrac{xy^3}{4}}$

**Solution**

**(a)** The index on the radical is 4, so this becomes the denominator of the rational exponent. The parentheses are necessary because the radicand is $7a$ and the exponent $\frac{1}{4}$ is applied to both the 7 and the $a$.

$$\sqrt[4]{7a} = (7a)^{\frac{1}{4}}$$

**(b)** The index on the radical is 5, so this becomes the denominator of the rational exponent.

$$\sqrt[5]{\frac{xy^3}{4}} = \left(\frac{xy^3}{4}\right)^{\frac{1}{5}}$$

**QUICK ✓** *Rewrite each of the following radicals with a rational exponent.*

**19.** $\sqrt[5]{8b}$ **20.** $\sqrt[8]{\dfrac{mn^5}{3}}$

## 4 Evaluate Expressions of the Form $a^{\frac{m}{n}}$

We now look for a definition for $a^{\frac{m}{n}}$, where $m$ and $n$ are integers, $\frac{m}{n}$ is reduced to lowest terms, and $n \geq 2$. The definition we provide should obey all the laws of exponents presented earlier. For example,

$$a^{\frac{m}{n}} = a^{m \cdot \frac{1}{n}} = (a^m)^{\frac{1}{n}} = \sqrt[n]{a^m}$$

and

$$a^{\frac{m}{n}} = a^{\frac{1}{n} \cdot m} = \left(a^{\frac{1}{n}}\right)^m = \left(\sqrt[n]{a}\right)^m$$

This suggests the following definition:

**In Words**

The expression $a^{\frac{m}{n}} = \sqrt[n]{a^m}$ means that we will raise $a$ to the $m$th power first, and then take the $n$th root. The expression $a^{\frac{m}{n}} = \left(\sqrt[n]{a}\right)^m$ means that we will take the $n$th root of $a$ first, and then raise to the power of $m$.

**DEFINITION OF $a^{\frac{m}{n}}$**

If $a$ is a real number, $m/n$ is a rational number in lowest terms with $n \geq 2$, then

$$a^{\frac{m}{n}} = \sqrt[n]{a^m} = \left(\sqrt[n]{a}\right)^m$$

provided that $\sqrt[n]{a}$ exists.

When simplifying $a^{\frac{m}{n}}$ either $\sqrt[n]{a^m}$ or $\left(\sqrt[n]{a}\right)^m$ may be used. Use the one that makes simplifying the expression easier. Generally, taking the root first, as in $\left(\sqrt[n]{a}\right)^m$, is easier.

## EXAMPLE 6 Evaluating Expressions of the Form $a^{m/n}$

Evaluate each of the following expressions, if possible.

**(a)** $25^{\frac{3}{2}}$ **(b)** $64^{\frac{2}{3}}$ **(c)** $-9^{\frac{5}{2}}$ **(d)** $(-8)^{\frac{4}{3}}$ **(e)** $(-81)^{\frac{7}{2}}$

### Solution

**(a)** $25^{\frac{3}{2}} = \left(\sqrt{25}\right)^3 = 5^3 = 125$

**(b)** $64^{\frac{2}{3}} = \left(\sqrt[3]{64}\right)^2 = 4^2 = 16$

**(c)** $-9^{\frac{5}{2}} = -1 \cdot 9^{\frac{5}{2}} = -1 \cdot \left(\sqrt{9}\right)^5 = -1 \cdot 3^5 = -1 \cdot 243 = -243$

**(d)** $(-8)^{\frac{4}{3}} = \left(\sqrt[3]{-8}\right)^4 = (-2)^4 = 16$

**(e)** $(-81)^{\frac{7}{2}}$ is not a real number because $(-81)^{\frac{7}{2}} = \left(\sqrt{-81}\right)^7$ and $\sqrt{-81}$ is not a real number.

**Work Smart**

When simplifying expressions of the form $a^{\frac{m}{n}}$ it is typically easier to evaluate the radical first.

**QUICK ✓** *Evaluate each expression, if possible.*

**21.** $16^{\frac{3}{2}}$ **22.** $27^{\frac{2}{3}}$ **23.** $-16^{\frac{3}{4}}$ **24.** $(-64)^{\frac{2}{3}}$ **25.** $(-25)^{\frac{5}{2}}$

---

The expressions in Examples 6(a)–(d) were all rational numbers. Not all expressions involving rational exponents will simplify to rational numbers.

## EXAMPLE 7 Approximating Expressions Involving Rational Exponents

Write $35^{\frac{3}{4}}$ as a decimal rounded to two decimal places.

### Solution

Figure 2

Figure 2 shows the results obtained from a TI-84 Plus graphing calculator. So $35^{\frac{3}{4}} \approx 14.39$.

**QUICK ✓** *Approximate the expression rounded to two decimal places.*

**26.** $50^{\frac{2}{3}}$ **27.** $40^{0.15}$

---

## EXAMPLE 8 Writing Radicals with Rational Exponents

Rewrite each of the following radicals with a rational exponent.

**(a)** $\sqrt[3]{x^2}$ **(b)** $\left(\sqrt[5]{10a^2b}\right)^4$

### Solution

**(a)** The index, 3, is the denominator of the rational exponent and the power on the radicand, 2, is the numerator of the rational exponent.

$$\sqrt[3]{x^2} = x^{\frac{2}{3}}$$

**(b)** The index, 5, is the denominator of the rational exponent and the power, 4, is the numerator of the rational exponent.

$$\left(\sqrt[5]{10a^2b}\right)^4 = (10a^2b)^{\frac{4}{5}}$$

**QUICK ✓** *Write each radical with a rational exponent.*

**28.** $\sqrt[8]{a^3}$ **29.** $\left(\sqrt[4]{12ab^3}\right)^9$

If a rational exponent is negative, then we can use the rule for negative rational exponents given below.

**DEFINITION: NEGATIVE-EXPONENT RULE**

If $\frac{m}{n}$ is a rational number, and if $a$ is a nonzero real number (that is, if $a \neq 0$), then we define

$$a^{-\frac{m}{n}} = \frac{1}{a^{\frac{m}{n}}} \quad \text{and} \quad \frac{1}{a^{-\frac{m}{n}}} = a^{\frac{m}{n}} \quad \text{if } a \neq 0$$

**EXAMPLE 9 Evaluating Expressions with Negative Rational Exponents**

Rewrite each of the following with positive exponents, and completely simplify, if possible.

**(a)** $36^{-\frac{1}{2}}$ **(b)** $\frac{1}{27^{-\frac{2}{3}}}$ **(c)** $(6a)^{-\frac{5}{4}}$

**Solution**

**(a)** $36^{-\frac{1}{2}} = \frac{1}{36^{\frac{1}{2}}} = \frac{1}{\sqrt{36}} = \frac{1}{6}$

**(b)** Because the negative exponent is in the exponential expression in the denominator, we use $\frac{1}{a^{-\frac{m}{n}}} = a^{\frac{m}{n}}$ to simplify:

$$\frac{1}{27^{-\frac{2}{3}}} = 27^{\frac{2}{3}} = \left(\sqrt[3]{27}\right)^2 = 3^2 = 9$$

**(c)** $(6a)^{-\frac{5}{4}} = \frac{1}{(6a)^{\frac{5}{4}}}$

**QUICK ✓** *Rewrite each of the following with positive exponents, and completely simplify, if possible.*

**30.** $81^{-\frac{1}{2}}$ **31.** $\frac{1}{8^{-\frac{2}{3}}}$ **32.** $(13x)^{-\frac{3}{2}}$

# 7.1 Exercises

**For Extra Help:**

Student Solutions Manual

CD Video

PH Math/Tutor Center

MathXL Tutorials on CD

MathXL®
MyMathLab

## Concepts and Vocabulary

*In Problems 1–3, fill in the blanks.*

**1.** If $a$ is a nonnegative real number and $n \geq 2$ is an integer, then $a^{\frac{1}{n}} =$ _____.

**2.** If $\frac{m}{n}$ is a rational number and if $a$ is a nonzero real number (that is, if $a \neq 0$), then we define $a^{-\frac{m}{n}} =$ _____.

**3.** In the notation $\sqrt[n]{a}$, the integer $n$, $n \geq 2$, is called the _____.

*In Problems 4 and 5, answer True or False to each statement.*

**4.** If $a$ is a real number, and $m/n$ is a rational number in lowest terms with $n \geq 2$, then $a^{\frac{m}{n}} = \sqrt[n]{a^m} = \left(\sqrt[n]{a}\right)^m$.

**5.** If $a < 0$ is a real number and $n \geq 2$ is an integer, then $a^{\frac{1}{n}}$ exists.

6. Explain why $(-9)^{\frac{1}{2}}$ is not a real number, but $-9^{\frac{1}{2}}$ is a real number.
7. In your own words, provide a justification for why $a^{\frac{1}{n}} = \sqrt[n]{a}$.
8. Under what conditions is $a^{\frac{m}{n}}$ a real number?

## Building Skills

*In Problems 9–26, simplify each radical.*

9. $\sqrt[3]{125}$ 10. $\sqrt[3]{216}$ 11. $\sqrt[3]{-27}$ 12. $\sqrt[3]{-64}$
13. $-\sqrt[4]{625}$ 14. $-\sqrt[4]{256}$ 15. $\sqrt[3]{-\frac{1}{8}}$ 16. $\sqrt[3]{\frac{8}{125}}$
17. $-\sqrt[5]{-243}$ 18. $-\sqrt[5]{-1024}$ 19. $\sqrt[3]{5^3}$ 20. $\sqrt[4]{6^4}$
21. $\sqrt[4]{m^4}$ 22. $\sqrt[5]{n^5}$ 23. $\sqrt[9]{(x-3)^9}$ 24. $\sqrt[6]{(2x-3)^6}$
25. $-\sqrt[4]{(3p+1)^4}$ 26. $-\sqrt[3]{(6z-5)^3}$

*In Problems 27–56, evaluate each expression, if possible.*

27. $4^{\frac{1}{2}}$ 28. $16^{\frac{1}{2}}$ 29. $-36^{\frac{1}{2}}$ 30. $-25^{\frac{1}{2}}$ 31. $8^{\frac{1}{3}}$
32. $27^{\frac{1}{3}}$ 33. $-16^{\frac{1}{4}}$ 34. $-81^{\frac{1}{4}}$ 35. $\left(\frac{4}{25}\right)^{\frac{1}{2}}$ 36. $\left(\frac{8}{27}\right)^{\frac{1}{3}}$
37. $(-125)^{\frac{1}{3}}$ 38. $(-216)^{\frac{1}{3}}$ 39. $(-4)^{\frac{1}{2}}$ 40. $(-81)^{\frac{1}{2}}$ 41. $4^{\frac{5}{2}}$
42. $25^{\frac{3}{2}}$ 43. $-16^{\frac{3}{2}}$ 44. $-100^{\frac{5}{2}}$ 45. $8^{\frac{4}{3}}$ 46. $27^{\frac{4}{3}}$
47. $(-64)^{\frac{2}{3}}$ 48. $(-125)^{\frac{2}{3}}$ 49. $-(-32)^{\frac{3}{5}}$ 50. $-(-216)^{\frac{2}{3}}$ 51. $144^{-\frac{1}{2}}$
52. $121^{-\frac{1}{2}}$ 53. $\frac{1}{25^{-\frac{3}{2}}}$ 54. $\frac{1}{49^{-\frac{3}{2}}}$ 55. $\frac{1}{8^{-\frac{5}{3}}}$ 56. $27^{-\frac{4}{3}}$

*In Problems 57–68, rewrite each of the following radicals with a rational exponent.*

57. $\sqrt[3]{3x}$ 58. $\sqrt[5]{2y}$ 59. $\sqrt[4]{\frac{x}{3}}$ 60. $\sqrt{\frac{w}{2}}$ 61. $\sqrt[4]{x^3}$
62. $\sqrt[3]{p^5}$ 63. $(\sqrt[5]{3x})^2$ 64. $(\sqrt[4]{6z})^3$ 65. $\sqrt{\left(\frac{5x}{y}\right)^3}$ 66. $\sqrt[6]{\left(\frac{2a}{b}\right)^5}$
67. $\sqrt[3]{(9ab)^4}$ 68. $\sqrt[4]{(3pq)^7}$

*In Problems 69–78, use a calculator to write each expression as a decimal rounded to two decimal places.*

69. $\sqrt[3]{25}$ 70. $\sqrt[3]{85}$ 71. $\sqrt[4]{12}$ 72. $\sqrt[4]{2}$ 73. $20^{\frac{1}{2}}$
74. $5^{\frac{1}{2}}$ 75. $4^{\frac{5}{3}}$ 76. $100^{\frac{3}{4}}$ 77. $10^{0.1}$ 78. $100^{0.25}$

## Mixed Practice

*In Problems 79–92, evaluate each expression, if possible.*

79. $\sqrt[3]{512}$ 80. $\sqrt[3]{-125}$ 81. $9^{\frac{5}{2}}$ 82. $100^{\frac{3}{2}}$ 83. $\sqrt[4]{-16}$
84. $\sqrt[4]{-1}$ 85. $144^{-\frac{1}{2}}$ 86. $125^{-\frac{1}{3}}$ 87. $\sqrt[3]{0.008}$ 88. $\sqrt[4]{0.0081}$
89. $4^{\frac{1}{2}} + 25^{\frac{3}{2}}$ 90. $100^{\frac{1}{2}} - 4^{\frac{3}{2}}$ 91. $(-25)^{\frac{5}{2}}$ 92. $(-125)^{-\frac{1}{3}}$

## Applying the Concepts

93. What are the cube roots of 1,000? What is $\sqrt[3]{1000}$?
94. What are the cube roots of 729? What is $\sqrt[3]{729}$?

**95. Wind Chill** According to the National Weather Service, the wind chill temperature is how cold people and animals feel when outside. Wind chill is based on the rate of heat loss from exposed skin caused by wind and cold. The formula for computing wind chill $W$ is

$$W = 35.74 + 0.6215T - 35.75v^{0.16} + 0.4275Tv^{0.16}$$

where $T$ is the air temperature in degrees Fahrenheit and $v$ is the wind speed in miles per hour.

**(a)** What is the wind chill if it is 30°F and the wind speed is 10 miles per hour?
**(b)** What is the wind chill if it is 30°F and the wind speed is 20 miles per hour?
**(c)** What is the wind chill if it is 0°F and the wind speed is 10 miles per hour?

**96. Money** The annual rate of interest $r$ (expressed as a decimal) required to have $A$ dollars after $t$ years from an initial deposit of $P$ dollars is given by

$$r = \left(\frac{A}{P}\right)^{\frac{1}{t}} - 1$$

**(a)** If you deposit \$100 in a mutual fund today and have \$144 in the account in 2 years, what was your annual rate of interest earned?
**(b)** If you deposit \$100 in a mutual fund today and have \$337.50 in 3 years, what was your annual rate of interest earned?
**(c)** The Rule of 72 states that your money will double in $\frac{72}{100r}$ years where $r$ is the rate of interest earned (expressed as a decimal). Suppose that you deposit \$1000 in a mutual fund today and have \$2000 in 8 years. What rate of interest did you earn? Compute $\frac{72}{100r}$ for this rate of interest. Is it close?

**97. Terminal Velocity** Terminal speed is the maximum speed that a body falling through air can reach. The speed is limited by air resistance. Terminal velocity is given by the formula $v_t = \sqrt{\frac{2mg}{C\rho A}}$, where $m$ is the mass of the falling object, $g$ is acceleration due to gravity ($\approx$ 9.81 meters per second$^2$), $C$ is a drag coefficient with $0.5 \le C \le 1.0$, $\rho$ is the density of air ($\approx$ 1.2 kg/m$^3$), and $A$ is the cross-sectional area of the object. Suppose that a raindrop whose radius is 1.5 mm falls from the sky. The mass of the raindrop is given by $m = \frac{4}{3}\pi r^3 \rho_w$ where $r$ is its radius and $\rho_w = 1000$ kg/m$^3$. The cross-sectional area of the raindrop is $A = \pi r^2$.

**(a)** Substitute the formulas for the mass and area of a raindrop into the formula for terminal speed and simplify the expression.
**(b)** Determine the terminal velocity of a raindrop whose radius is 0.0015 m with $C = 0.6$.

**98. Kepler's Law** Early in the seventeenth century, Johannes Kepler (1571–1630) discovered that the square of the period $T$ of a planet varies directly with the cube of its mean distance $r$ from the Sun. The period of a planet is the amount of time (in years) for the planet to complete one orbit around the Sun. Kepler's Law can be expressed using rational exponents as $T = kr^{\frac{3}{2}}$, where $k$ is the constant of proportionality.

**(a)** The period of Mercury is 0.241 years and its mean distance from the Sun is $5.79 \times 10^{10}$ meters. Use this information to state Kepler's Law (find the value of $k$).
**(b)** The mean distance of Mars to the Sun is $2.28 \times 10^{11}$ m. Use this information along with the result of part (a) to find the amount of time it takes Mars to complete one orbit around the Sun.

### Extending the Concepts

*In Problems 99–102, evaluate each function.*

**99.** $f(x) = x^{\frac{3}{2}}$; find $f(4)$

**100.** $g(x) = x^{-\frac{3}{2}}$; find $g(16)$

**101.** $F(z) = z^{\frac{4}{3}}$; find $F(-8)$

**102.** $G(a) = a^{\frac{5}{3}}$; find $G(-8)$

### Synthesis Review

*In Problems 103–107, simplify completely each expression.*

**103.** $\left(\dfrac{x^2y}{y^{-2}}\right)^3$

**104.** $\dfrac{(x+2)^2(x-1)^4}{(x+2)(x-1)}$

**105.** $\dfrac{(x-1)^2(x^2+5x+6)}{(x+2)\sqrt{x-1}}$

**106.** $\dfrac{(3a^2+5a-3)-(a^2-2a-9)}{4a^2+12a+9}$

**107.** $\dfrac{(4z^2-7z+3)+(-3z^2-z+9)}{(4z^2-2z-7)+(-3z^2-z+9)}$

# 7.2 Simplify Expressions Using the Laws of Exponents

**OBJECTIVES**

1. Use the Laws of Exponents to Simplify Expressions Involving Rational Exponents
2. Use the Laws of Exponents to Simplify Radical Expressions
3. Factor Expressions Containing Rational Exponents

***Preparing for Simplifying Expressions Using the Laws of Exponents***

*Before getting started, take the following readiness quiz. If you get a problem wrong, go to the section cited and review the material.*

**1.** Simplify: $z^{-3}$ [Getting Ready: Integer Exponents, pp. 340–342]

**2.** Simplify: $x^{-2} \cdot x^5$ [Getting Ready: Integer Exponents, pp. 342–343]

**3.** Simplify: $\left(\dfrac{2a^2}{b^{-1}}\right)^3$ [Getting Ready: Integer Exponents, pp. 344–345]

**4.** Evaluate: $\sqrt{64}$ [Getting Ready: Square Roots, pp. 527–528]

## 1 Use the Laws of Exponents to Simplify Expressions Involving Rational Exponents

The Laws of Exponents that were presented in Getting Ready: Integer Exponents on pages 339–346 applied to integer exponents. These same laws apply to rational exponents as well.

**THE LAWS OF EXPONENTS**

If $a$ and $b$ are real numbers and if $r$ and $s$ are rational numbers, then assuming the expression is defined,

| | | |
|---|---|---|
| Zero-Exponent Rule: | $a^0 = 1$ | if $a \neq 0$ |
| Negative-Exponent Rule: | $a^{-r} = \dfrac{1}{a^r}$ | if $a \neq 0$ |
| Product Rule: | $a^r \cdot a^s = a^{r+s}$ | |
| Quotient Rule: | $\dfrac{a^r}{a^s} = a^{r-s} = \dfrac{1}{a^{s-r}}$ | if $a \neq 0$ |
| Power Rule: | $(a^r)^s = a^{r \cdot s}$ | |
| Product to Power Rule: | $(a \cdot b)^r = a^r \cdot b^r$ | |
| Quotient to Power Rule: | $\left(\dfrac{a}{b}\right)^r = \dfrac{a^r}{b^r}$ | if $b \neq 0$ |
| Quotient to a Negative Power Rule: | $\left(\dfrac{a}{b}\right)^{-r} = \left(\dfrac{b}{a}\right)^r$ | if $a \neq 0, b \neq 0$ |

***Preparing for...Answers*** **1.** $\dfrac{1}{z^3}$ **2.** $x^3$ **3.** $8a^6b^3$ **4.** 8

**Work Smart**
We *simplify* expressions (no equal sign) and *solve* equations.

The direction **simplify** shall mean the following:

- All the exponents are positive.
- Each base only occurs once.
- There are no parentheses in the expression.
- There are no powers written to powers.

## EXAMPLE 1 Simplifying Expressions Involving Rational Exponents

Simplify each of the following:

**(a)** $27^{\frac{1}{2}} \cdot 27^{\frac{5}{6}}$ **(b)** $\dfrac{8^{\frac{1}{3}}}{8^{\frac{5}{3}}}$

**Solution**

**(a)** $a^r \cdot a^s = a^{r+s}$

$$\begin{aligned} 27^{\frac{1}{2}} \cdot 27^{\frac{5}{6}} &= 27^{\frac{1}{2}+\frac{5}{6}} \\ &= 27^{\frac{3}{6}+\frac{5}{6}} \\ &= 27^{\frac{8}{6}} \\ &= 27^{\frac{4}{3}} \\ a^{\frac{m}{n}} = \left(\sqrt[n]{a}\right)^m: \quad &= \left(\sqrt[3]{27}\right)^4 \\ &= 3^4 \\ &= 81 \end{aligned}$$

**(b)** $\dfrac{a^r}{a^s} = a^{r-s}$

$$\begin{aligned} \frac{8^{\frac{1}{3}}}{8^{\frac{5}{3}}} &= 8^{\frac{1}{3}-\frac{5}{3}} \\ &= 8^{-\frac{4}{3}} \\ &= \frac{1}{8^{\frac{4}{3}}} \\ a^{\frac{m}{n}} = \left(\sqrt[n]{a}\right)^m: \quad &= \frac{1}{\left(\sqrt[3]{8}\right)^4} \\ &= \frac{1}{2^4} \\ &= \frac{1}{16} \end{aligned}$$

## EXAMPLE 2 Simplifying Expressions Involving Rational Exponents

Simplify each of the following:

**(a)** $\left(36^{\frac{2}{5}}\right)^{\frac{5}{4}}$ **(b)** $\left(x^{\frac{1}{2}} \cdot y^{\frac{2}{3}}\right)^{\frac{3}{2}}$

**Solution**

**(a)** $(a^r)^s = a^{r \cdot s}$

$$\begin{aligned} \left(36^{\frac{2}{5}}\right)^{\frac{5}{4}} &= 36^{\frac{2}{5}\cdot\frac{5}{4}} \\ &= 36^{\frac{10}{20}} \\ &= 36^{\frac{1}{2}} \\ &= 6 \end{aligned}$$

**(b)** $(ab)^r = a^r \cdot b^r$

$$\begin{aligned} \left(x^{\frac{1}{2}} \cdot y^{\frac{2}{3}}\right)^{\frac{3}{2}} &= \left(x^{\frac{1}{2}}\right)^{\frac{3}{2}} \cdot \left(y^{\frac{2}{3}}\right)^{\frac{3}{2}} \\ &= x^{\frac{1}{2}\cdot\frac{3}{2}} \cdot y^{\frac{2}{3}\cdot\frac{3}{2}} \\ (a^r)^s = a^{r \cdot s}: \quad &= x^{\frac{3}{4}}y \end{aligned}$$

**QUICK ✓** *Simplify each expression.*

**1.** $5^{\frac{3}{4}} \cdot 5^{\frac{1}{6}}$

**2.** $\dfrac{32^{\frac{6}{5}}}{32^{\frac{3}{5}}}$

**3.** $\left(100^{\frac{3}{8}}\right)^{\frac{4}{3}}$

**4.** $\left(a^{\frac{3}{2}} \cdot b^{\frac{5}{4}}\right)^{\frac{2}{3}}$

### EXAMPLE 3 Simplifying Expressions Involving Rational Exponents

Simplify each of the following:

**(a)** $\left(x^{\frac{2}{3}}y^{-1}\right)\cdot\left(x^{-1}y^{\frac{1}{2}}\right)^{\frac{2}{3}}$ **(b)** $\left(\dfrac{9xy^{\frac{4}{3}}}{x^{\frac{5}{6}}y^{-\frac{2}{3}}}\right)^{\frac{1}{2}}$

**Solution**

**(a)**

Product to Power Rule: $(ab)^r = a^r b^r$

$$\left(x^{\frac{2}{3}}y^{-1}\right)\cdot\left(x^{-1}y^{\frac{1}{2}}\right)^{\frac{2}{3}} = x^{\frac{2}{3}}y^{-1}(x^{-1})^{\frac{2}{3}}\left(y^{\frac{1}{2}}\right)^{\frac{2}{3}}$$

Power Rule: $(a^r)^s = a^{rs}$: $= x^{\frac{2}{3}}y^{-1}x^{-\frac{2}{3}}y^{\frac{1}{3}}$

Product Rule: $a^r \cdot a^s = a^{r+s}$: $= x^{\frac{2}{3}+\left(-\frac{2}{3}\right)}y^{-1+\frac{1}{3}}$

$= x^0 y^{-\frac{2}{3}}$

$a^0 = 1$; Negative Exponent Rule: $a^{-r} = \dfrac{1}{a^r}$: $= \dfrac{1}{y^{\frac{2}{3}}}$

**(b)**

Quotient Rule: $\dfrac{a^r}{a^s} = a^{r-s}$

$$\left(\frac{9xy^{\frac{4}{3}}}{x^{\frac{5}{6}}y^{-\frac{2}{3}}}\right)^{\frac{1}{2}} = \left(9x^{1-\frac{5}{6}}y^{\frac{4}{3}-\left(-\frac{2}{3}\right)}\right)^{\frac{1}{2}}$$

$x^{1-\frac{5}{6}} = x^{\frac{6}{6}-\frac{5}{6}} = x^{\frac{1}{6}}$;
$y^{\frac{4}{3}-\left(-\frac{2}{3}\right)} = y^{\frac{4}{3}+\frac{2}{3}} = y^{\frac{6}{3}} = y^2$: $= \left(9x^{\frac{1}{6}}y^2\right)^{\frac{1}{2}}$

Power Rule: $(a^r)^s = a^{r\cdot s}$: $= 9^{\frac{1}{2}}\cdot\left(x^{\frac{1}{6}}\right)^{\frac{1}{2}}\cdot(y^2)^{\frac{1}{2}}$

$9^{\frac{1}{2}} = \sqrt{9} = 3$; Power Rule: $(a^r)^s = a^{r\cdot s}$: $= 3x^{\frac{1}{12}}y$

**QUICK ✓** *Simplify each expression.*

**5.** $\left(8x^{\frac{3}{4}}y^{-1}\right)^{\frac{2}{3}}$ **6.** $\left(\dfrac{25x^{\frac{1}{2}}y^{\frac{3}{4}}}{x^{-\frac{3}{4}}y}\right)^{\frac{1}{2}}$ **7.** $8\left(125a^{\frac{3}{4}}b^{-1}\right)^{\frac{2}{3}}$

## 2 Use the Laws of Exponents to Simplify Radical Expressions

Rational exponents can be used to simplify radicals.

### EXAMPLE 4 Simplifying Radicals Using Rational Exponents

Use rational exponents to simplify the radicals.

**(a)** $\sqrt[8]{16^4}$ **(b)** $\sqrt[3]{64x^6y^3}$ **(c)** $\dfrac{\sqrt{x}}{\sqrt[3]{x^2}}$ **(d)** $\sqrt{\sqrt[3]{z}}$

**Solution**

The idea in all these problems is to rewrite the radical as an expression involving a rational exponent. Then use the Laws of Exponents to simplify the expression. Finally, write the simplified expression as a radical.

**(a)**

Write radical as a rational exponent using $\sqrt[n]{a^m} = a^{\frac{m}{n}}$.

$$\sqrt[8]{16^4} = 16^{\frac{4}{8}}$$

Simplify the exponent: $= 16^{\frac{1}{2}}$

Write rational exponent as a radical: $= \sqrt{16} = 4$

**(b)**

$\sqrt[n]{a^m} = a^{\frac{m}{n}}$

$$\sqrt[3]{64x^6y^3} = (64x^6y^3)^{\frac{1}{3}}$$

$(ab)^r = a^r \cdot b^r$: $= 64^{\frac{1}{3}} \cdot (x^6)^{\frac{1}{3}} \cdot (y^3)^{\frac{1}{3}}$

$(a^r)^s = a^{rs}$: $= 64^{\frac{1}{3}} \cdot x^{6 \cdot \frac{1}{3}} \cdot y^{3 \cdot \frac{1}{3}}$

$64^{\frac{1}{3}} = \sqrt[3]{64} = 4$: $= 4x^2y$

**(c)**

$\sqrt{a} = a^{\frac{1}{2}}$; $\sqrt[n]{a^m} = a^{\frac{m}{n}}$

$$\frac{\sqrt{x}}{\sqrt[3]{x^2}} = \frac{x^{\frac{1}{2}}}{x^{\frac{2}{3}}}$$

$\frac{a^r}{a^s} = a^{r-s}$: $= x^{\frac{1}{2} - \frac{2}{3}}$

LCD $= 6$; $\frac{1}{2} - \frac{2}{3} = \frac{3}{6} - \frac{4}{6} = -\frac{1}{6}$: $= x^{-\frac{1}{6}}$

$a^{-r} = \frac{1}{a^r}$: $= \frac{1}{x^{\frac{1}{6}}}$

Write rational exponent as a radical: $= \frac{1}{\sqrt[6]{x}}$

**(d)**

Write radicand with a rational exponent.

$$\sqrt{\sqrt[3]{z}} = \sqrt{z^{\frac{1}{3}}}$$

$\sqrt{a} = a^{\frac{1}{2}}$: $= \left(z^{\frac{1}{3}}\right)^{\frac{1}{2}}$

$(a^r)^s = a^{rs}$: $= z^{\frac{1}{3} \cdot \frac{1}{2}}$

$= z^{\frac{1}{6}}$

Write rational exponent as a radical: $= \sqrt[6]{z}$

**QUICK ✓** *Use rational exponents to simplify each radical.*

**8.** $\sqrt[10]{36^5}$ **9.** $\sqrt[4]{16a^8b^{12}}$ **10.** $\frac{\sqrt[3]{x^2}}{\sqrt[4]{x}}$ **11.** $\sqrt[4]{\sqrt[3]{a^2}}$

## 3 Factor Expressions Containing Rational Exponents

Often, expressions involving rational exponents contain a common factor. When this occurs, we want to factor out the common factor to write the expression in simplified form. The goal of these types of problems is to write the expression as either a single product or a single quotient. We present two examples to illustrate the idea.

### EXAMPLE 5 Writing an Expression Containing Rational Exponents as a Single Product

Simplify $9x^{\frac{4}{3}} + 4x^{\frac{1}{3}}(3x + 5)$ by factoring out $x^{\frac{1}{3}}$.

**Solution**

Clearly, $x^{\frac{1}{3}}$ is a factor of the second term, $4x^{\frac{1}{3}}(3x + 5)$. It is also a factor of the first term, $9x^{\frac{4}{3}}$. We can see this by rewriting $9x^{\frac{4}{3}}$ as $9x^{\frac{3}{3}+\frac{1}{3}} = 9x^{\frac{3}{3}} \cdot x^{\frac{1}{3}} = 9x \cdot x^{\frac{1}{3}}$. Now we

**Work Smart**

When factoring out the greatest common factor, factor out the variable expression raised to the smallest exponent that the expressions have in common. For example,

$3x^5 + 12x^2 = 3x^2(x^3 + 4)$

proceed to factor out $x^{\frac{1}{3}}$.

$$9x^{\frac{4}{3}} + 4x^{\frac{1}{3}}(3x + 5) = 9x \cdot x^{\frac{1}{3}} + 4x^{\frac{1}{3}}(3x + 5)$$

Factor out $x^{1/3}$: $= x^{\frac{1}{3}}(9x + 4(3x + 5))$

Distribute the 4: $= x^{\frac{1}{3}}(9x + 12x + 20)$

Combine like terms: $= x^{\frac{1}{3}}(21x + 20)$

**QUICK ✓**

**12.** Simplify $8x^{\frac{3}{2}} + 3x^{\frac{1}{2}}(4x + 3)$ by factoring out $x^{\frac{1}{2}}$.

### EXAMPLE 6 Writing an Expression Containing Rational Exponents as a Single Quotient

Simplify $4x^{\frac{1}{2}} + x^{-\frac{1}{2}}(2x + 1)$ by factoring out $x^{-\frac{1}{2}}$.

#### Solution

Clearly, $x^{-\frac{1}{2}}$ is a factor of the second term, $x^{-\frac{1}{2}}(2x + 1)$. It is also a factor of the first term, $4x^{\frac{1}{2}}$. We can see this by rewriting $4x^{\frac{1}{2}}$ as $4x^{\frac{2}{2}-\frac{1}{2}} = 4x^{\frac{2}{2}} \cdot x^{-\frac{1}{2}} = 4x \cdot x^{-\frac{1}{2}}$. Now we proceed to factor out $x^{-\frac{1}{2}}$.

$$4x^{\frac{1}{2}} + x^{-\frac{1}{2}}(2x + 1) = 4x \cdot x^{-\frac{1}{2}} + x^{-\frac{1}{2}}(2x + 1)$$

Factor out $x^{-1/2}$: $= x^{-\frac{1}{2}}(4x + (2x + 1))$

Combine like terms: $= x^{-\frac{1}{2}}(6x + 1)$

Rewrite without negative exponents: $= \dfrac{6x + 1}{x^{\frac{1}{2}}}$

**QUICK ✓**

**13.** Simplify $9x^{\frac{1}{3}} + x^{-\frac{2}{3}}(3x + 1)$ by factoring out $x^{-\frac{2}{3}}$.

# 7.2 Exercises

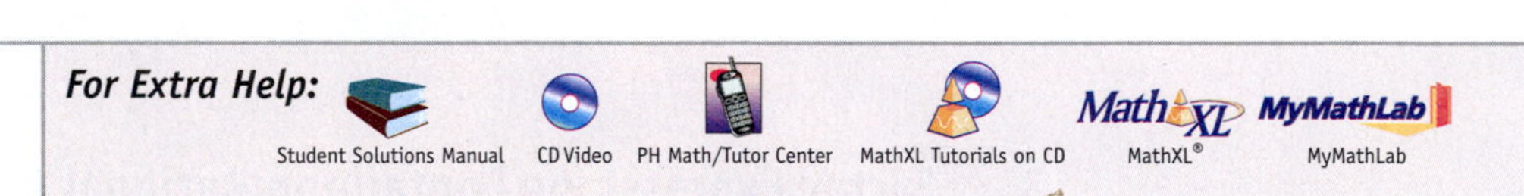


## Concepts and Vocabulary

*In Problems 1 and 2, fill in the blanks.*

**1.** If $a$ and $b$ are real numbers and if $r$ and $s$ are rational numbers, then assuming the expression is defined, $(ab)^r =$ _____.

**2.** If $a$ and $b$ are real numbers and if $r$ and $s$ are rational numbers, then assuming the expression is defined, $a^r \cdot a^s =$ _____.

## Building Skills

*In Problems 3–24, simplify each of the following expressions.*

**3.** $5^{\frac{1}{2}} \cdot 5^{\frac{3}{2}}$

**4.** $3^{\frac{1}{3}} \cdot 3^{\frac{5}{3}}$

**5.** $\dfrac{8^{\frac{5}{4}}}{8^{\frac{1}{4}}}$

**6.** $\dfrac{10^{\frac{7}{5}}}{10^{\frac{2}{5}}}$

**7.** $2^{\frac{1}{3}} \cdot 2^{-\frac{3}{2}}$ **8.** $9^{-\frac{5}{4}} \cdot 9^{\frac{1}{3}}$ **9.** $\dfrac{x^{\frac{1}{4}}}{x^{\frac{5}{6}}}$ **10.** $\dfrac{y^{\frac{1}{5}}}{y^{\frac{9}{10}}}$

**11.** $\left(4^{\frac{4}{3}}\right)^{\frac{3}{8}}$ **12.** $\left(9^{\frac{3}{5}}\right)^{\frac{5}{6}}$ **13.** $\left(25^{\frac{3}{4}} \cdot 4^{-\frac{3}{4}}\right)^2$ **14.** $\left(36^{-\frac{1}{4}} \cdot 9^{\frac{3}{4}}\right)^{-2}$

**15.** $\left(x^{\frac{3}{4}} \cdot y^{\frac{1}{3}}\right)^{\frac{2}{3}}$ **16.** $\left(a^{\frac{5}{4}} \cdot b^{\frac{3}{2}}\right)^{\frac{2}{5}}$

**17.** $\left(x^{-\frac{1}{3}} \cdot y\right)\left(x^{\frac{1}{2}} \cdot y^{-\frac{4}{3}}\right)$ **18.** $\left(a^{\frac{4}{3}} \cdot b^{-\frac{1}{2}}\right)\left(a^{-2} \cdot b^{\frac{5}{2}}\right)$

**19.** $\left(4a^2b^{-\frac{3}{2}}\right)^{\frac{1}{2}}$ **20.** $\left(25p^{\frac{2}{5}}q^{-1}\right)^{\frac{1}{2}}$

**21.** $\left(\dfrac{x^{\frac{2}{3}}y^{-\frac{1}{3}}}{8x^{\frac{1}{2}}y}\right)^{\frac{1}{3}}$ **22.** $\left(\dfrac{64m^{\frac{1}{2}}n}{m^{-2}n^{\frac{4}{3}}}\right)^{\frac{1}{2}}$

**23.** $\left(\dfrac{50x^{\frac{3}{4}}y}{2x^{\frac{1}{2}}}\right)^{\frac{1}{2}} + \left(\dfrac{x^{\frac{1}{2}}y^{\frac{1}{2}}}{9x^{\frac{3}{4}}y^{\frac{3}{2}}}\right)^{-\frac{1}{2}}$ **24.** $\left(\dfrac{27x^{\frac{1}{2}}y^{-1}}{y^{-\frac{2}{3}}x^{-\frac{1}{2}}}\right)^{1/3} - \left(\dfrac{4x^{\frac{1}{3}}y^{\frac{4}{9}}}{x^{-\frac{1}{3}}y^{\frac{2}{3}}}\right)^{\frac{1}{2}}$

*In Problems 25–30, distribute and simplify.*

**25.** $x^{\frac{1}{2}}\left(x^{\frac{3}{2}} - 2\right)$ **26.** $x^{\frac{1}{3}}\left(x^{\frac{5}{3}} + 4\right)$ **27.** $2y^{-\frac{1}{3}}(1 + 3y)$

**28.** $3a^{-\frac{1}{2}}(2 - a)$ **29.** $4z^{\frac{3}{2}}\left(z^{\frac{3}{2}} - 8z^{-\frac{3}{2}}\right)$ **30.** $8p^{\frac{2}{3}}\left(p^{\frac{4}{3}} - 4p^{-\frac{2}{3}}\right)$

*In Problems 31–46, use rational exponents to simplify each radical. Assume all variables are positive.*

**31.** $\sqrt{x^8}$ **32.** $\sqrt[3]{x^6}$ **33.** $\sqrt[12]{8^4}$ **34.** $\sqrt[9]{125^6}$

**35.** $\sqrt[3]{8a^3b^{12}}$ **36.** $\sqrt{25x^4y^6}$ **37.** $\dfrac{\sqrt{x}}{\sqrt[4]{x}}$ **38.** $\dfrac{\sqrt[3]{y^2}}{\sqrt{y}}$

**39.** $\sqrt{x} \cdot \sqrt[3]{x}$ **40.** $\sqrt[4]{p^3} \cdot \sqrt[3]{p}$ **41.** $\sqrt{\sqrt[4]{x^3}}$ **42.** $\sqrt[3]{\sqrt{x^3}}$

**43.** $\sqrt{3} \cdot \sqrt[3]{9}$ **44.** $\sqrt{5} \cdot \sqrt[3]{25}$ **45.** $\dfrac{\sqrt{6}}{\sqrt[4]{36}}$ **46.** $\dfrac{\sqrt[4]{49}}{\sqrt{7}}$

**47.** Simplify $2x^{\frac{3}{2}} + 3x^{\frac{1}{2}}(x + 5)$ by factoring out $x^{\frac{1}{2}}$.

**48.** Simplify $6x^{\frac{4}{3}} + 4x^{\frac{1}{3}}(2x - 3)$ by factoring out $x^{\frac{1}{3}}$.

**49.** Simplify $5(x + 2)^{\frac{2}{3}}(3x - 2) + 9(x + 2)^{\frac{5}{3}}$ by factoring out $(x + 2)^{\frac{2}{3}}$.

**50.** Simplify $3(x - 5)^{\frac{1}{2}}(3x + 1) + 6(x - 5)^{\frac{3}{2}}$ by factoring out $(x - 5)^{\frac{1}{2}}$.

**51.** Simplify $x^{-\frac{1}{2}}(2x + 5) + 4x^{\frac{1}{2}}$ by factoring out $x^{-\frac{1}{2}}$.

**52.** Simplify $x^{-\frac{2}{3}}(3x + 2) + 9x^{\frac{1}{3}}$ by factoring out $x^{-\frac{2}{3}}$.

**53.** Simplify $2(x - 4)^{-\frac{1}{3}}(4x - 3) + 12(x - 4)^{\frac{2}{3}}$ by factoring out $2(x - 4)^{-\frac{1}{3}}$.

**54.** Simplify $4(x + 3)^{\frac{1}{2}} + (x + 3)^{-\frac{1}{2}}(2x + 1)$ by factoring out $(x + 3)^{-\frac{1}{2}}$.

**55.** Simplify $15x(x^2 + 4)^{\frac{1}{2}} + 5(x^2 + 4)^{\frac{3}{2}}$.

**56.** Simplify $24x(x^2 - 1)^{\frac{1}{3}} + 9(x^2 - 1)^{\frac{4}{3}}$.

### Mixed Practice

*In Problems 57–68, simplify each expression.*

**57.** $\sqrt[8]{4^4}$ **58.** $\sqrt[6]{27^2}$ **59.** $(-2)^{\frac{1}{2}} \cdot (-2)^{\frac{3}{2}}$ **60.** $25^{\frac{3}{4}} \cdot 25^{\frac{3}{4}}$

**61.** $\left(100^{\frac{1}{3}}\right)^{\frac{3}{2}}$ **62.** $(8^4)^{\frac{5}{12}}$ **63.** $\left(\sqrt[4]{25}\right)^2$ **64.** $\left(\sqrt[6]{27}\right)^2$

**65.** $\sqrt[4]{x^2} - \dfrac{\sqrt[4]{x^6}}{x}$ **66.** $\sqrt[9]{a^6} - \dfrac{\sqrt[6]{a^5}}{\sqrt[6]{a}}$ **67.** $\left(4 \cdot 9^{\frac{1}{4}}\right)^{-2}$ **68.** $\left(4^{-1} \cdot 81^{\frac{1}{2}}\right)^{\frac{1}{2}}$

### Applying the Concepts

**69.** If $3^x = 25$, what does $3^{\frac{x}{2}}$ equal?

**70.** If $5^x = 64$, what does $5^{\frac{x}{3}}$ equal?

**71.** If $7^x = 9$, what does $\sqrt{7^x}$ equal?

**72.** If $5^x = 27$, what does $\sqrt[3]{5^x}$ equal?

### Extending the Concepts

*In Problems 73 and 74, simplify the expression using rational exponents.*

**73.** $\sqrt[4]{\sqrt[3]{\sqrt{x}}}$

**74.** $\sqrt[5]{\sqrt[3]{\sqrt{x^2}}}$

**75.** Without using a calculator, determine the value of $\left(6^{\sqrt{2}}\right)^{\sqrt{2}}$.

**76.** Determine the domain of $g(x) = (x - 3)^{\frac{1}{2}}(x - 1)^{-\frac{1}{2}}$.

**77.** Determine the domain of $f(x) = (x + 3)^{\frac{1}{2}}(x + 1)^{-\frac{1}{2}}$.

### Synthesis Review

*In Problems 78–81, simplify each expression.*

**78.** $(2x - 1)(x + 4) - (x + 1)(x - 1)$

**79.** $3a(a - 3) + (a + 3)(a - 2)$

**80.** $\dfrac{\sqrt{x^2 + 4x + 4}}{x + 2}, x + 2 > 0$

**81.** $\dfrac{x^2 - 4}{x + 2} \cdot (x + 5) - (x + 4)(x - 1)$

# 7.3 Simplifying Radical Expressions

**OBJECTIVES**

1. Use the Product Property to Multiply Radical Expressions
2. Use the Product Property to Simplify Radical Expressions
3. Use the Quotient Property to Simplify Radical Expressions
4. Multiply Radicals with Unlike Indices

## *Preparing for Simplifying Radical Expressions*

*Before getting started, take the following readiness quiz. If you get a problem wrong, go back to the section cited and review the material.*

**1.** List the positive integer powers of 2 that are less than 200.

**2.** List the positive integer powers of 3 that are less than 200.

**3.** Simplify: (a) $\sqrt{16}$ (b) $\sqrt{p^2}$ [Getting Ready: Square Roots, pp. 527–530]

## 1 Use the Product Property to Multiply Radical Expressions

Perhaps you are noticing a trend at this point. When we introduce a new algebraic expression, we then learn how to multiply, divide, add and subtract the algebraic expression. Well, here we go again! First, we are going to learn how to multiply radical expressions when they have the same index.

Consider the following:

$$\sqrt{4 \cdot 25} = \sqrt{100} = 10 \quad \text{and} \quad \sqrt{4} \cdot \sqrt{25} = 2 \cdot 5 = 10$$

***Preparing for...Answers*** **1.** 1, 4, 9, 16, 25, 36, 49, 64, 81, 100, 121, 144, 169, 196 **2.** 1, 8, 27, 64, 125 **3. (a)** 4 **(b)** $|p|$

This suggests the following result:

**In Words**
$\sqrt[n]{a} \cdot \sqrt[n]{b} = \sqrt[n]{ab}$ means "the product of the roots equals the root of the product provided the index is the same."

**PRODUCT PROPERTY OF RADICALS**

If $\sqrt[n]{a}$ and $\sqrt[n]{b}$ are real numbers, and $n \geq 2$ is an integer, then

$$\sqrt[n]{a} \cdot \sqrt[n]{b} = \sqrt[n]{ab}$$

We can justify this formula using rational exponents.

$$\sqrt[n]{a} \cdot \sqrt[n]{b} = a^{\frac{1}{n}} \cdot b^{\frac{1}{n}}$$

Product to a Power Rule: $= (a \cdot b)^{\frac{1}{n}}$

$a^{\frac{1}{n}} = \sqrt[n]{a}$: $= \sqrt[n]{a \cdot b}$

## EXAMPLE 1 Using the Product Property to Multiply Radicals

Multiply.

**(a)** $\sqrt{5} \cdot \sqrt{3}$
**(b)** $\sqrt[3]{2} \cdot \sqrt[3]{13}$
**(c)** $\sqrt{x-3} \cdot \sqrt{x+3}$
**(d)** $\sqrt[5]{6c} \cdot \sqrt[5]{7c^2}$

### Solution

**(a)** $\sqrt{5} \cdot \sqrt{3} = \sqrt{5 \cdot 3} = \sqrt{15}$
**(b)** $\sqrt[3]{2} \cdot \sqrt[3]{13} = \sqrt[3]{2 \cdot 13} = \sqrt[3]{26}$
**(c)** $\sqrt{x-3} \cdot \sqrt{x+3} = \sqrt{(x-3)(x+3)} = \sqrt{x^2-9}$
**(d)** $\sqrt[5]{6c} \cdot \sqrt[5]{7c^2} = \sqrt[5]{6c \cdot 7c^2} = \sqrt[5]{42c^3}$

**Work Smart**
In Example 1(c), notice that $\sqrt{x^2-9}$ does not equal $\sqrt{x^2} - \sqrt{9}$.

**QUICK ✓** *Multiply each radical expression.*

**1.** $\sqrt{11} \cdot \sqrt{7}$ **2.** $\sqrt[4]{6} \cdot \sqrt[4]{7}$ **3.** $\sqrt{x-5} \cdot \sqrt{x+5}$ **4.** $\sqrt[7]{5p} \cdot \sqrt[7]{4p^3}$

## 2 Use the Product Property to Simplify Radical Expressions

Up to now, we have simplified radicals only when the radicand simplified to a perfect square, such as $\sqrt{81} = 9$ or $\sqrt[3]{\frac{1}{8}} = \frac{1}{2}$. When a radical does not simplify to a rational number, we can do one of two things:

1. Write a decimal approximation of the radical.
2. Simplify the radical using properties of radicals, if possible.

We learned how to approximate radicals using a calculator in the Getting Ready: Square Roots and Section 7.1. Now we are going to learn how to use properties of radicals to write the radical in simplified form.

Recall that a number that is the square of a rational number is called a perfect square. So $1^2 = 1, 2^2 = 4, 3^2 = 9$, and so on are perfect squares. A number that is the cube of a rational number is called a perfect cube. So $1^3 = 1, 2^3 = 8, 3^3 = 27$, and so on are perfect cubes. In general, if $n$ is the index of a radical, then $a^n$ is a perfect power of index where $a$ is a rational number.

**Work Smart**
index $\rightarrow \sqrt[n]{a} \leftarrow$ radicand

We say that a radical expression is **simplified** provided that the radicand does not contain any factors that are perfect powers of the index. For example, $\sqrt{50}$ is not simplified because 25 is a factor of 50 and 25 is a perfect square or $\sqrt[3]{16}$ is not simplified because 8 is a factor of 16 and 8 is a perfect cube. When the radicand contains variables, the exponent on the variable must be less than the index in order for the radical to be simplified.

To simplify radicals that contain perfect square factors, we use the Product Property of Radicals "in reverse." That is, we use $\sqrt[n]{ab} = \sqrt[n]{a} \cdot \sqrt[n]{b}$.

**EXAMPLE 2 How to Use the Product Property to Simplify a Radical**

Simplify: $\sqrt{18}$

### Step-by-Step Solution

**Step 1:** What is the index on the radical? Since the index is 2, we write each factor of the radicand as the product of two factors, one of which is a perfect square.

The perfect squares are 1, 4, 9, 16, 25, .... Because 9 is a factor of 18 and 9 is a perfect square, we write 18 as $9 \cdot 2$. $\sqrt{18} = \sqrt{9 \cdot 2}$

**Step 2:** Write the radicand as the product of two radicals, one of which contains a perfect square. $= \sqrt{9} \cdot \sqrt{2}$

**Step 3:** Take the square root of each perfect power. $= 3\sqrt{2}$

We summarize the steps used in Example 2 below.

**Work Smart**

When performing Step 1 with real numbers, we want to look for the *largest* factor of the radicand that is a perfect power of the index.

**Simplifying a Radical Expression**

**Step 1:** Write each factor of the radicand as the product of two factors, one of which is a perfect power of the index.

**Step 2:** Write the radicand as the product of two radicals, one of which contains perfect squares using the Product Property of Radicals.

**Step 3:** Take the *n*th root of each perfect power.

**EXAMPLE 3 Using the Product Property to Simplify a Radical**

Simplify each of the following:

**(a)** $5\sqrt[3]{24}$ **(b)** $\sqrt{128x^2}$ **(c)** $\sqrt[4]{20}$

### Solution

**(a)** We are looking for the largest factor of 24 that is a perfect cube. The perfect cubes are 1, 8, 27, .... Because 8 is a factor of 24 and 8 is a perfect cube we write 24 as $8 \cdot 3$.

$$5\sqrt[3]{24} = 5 \cdot \sqrt[3]{8 \cdot 3}$$

$$\sqrt[n]{ab} = \sqrt[n]{a} \cdot \sqrt[n]{b}: \quad = 5 \cdot \sqrt[3]{8} \cdot \sqrt[3]{3}$$

$$= 5 \cdot 2 \cdot \sqrt[3]{3}$$

$$= 10\sqrt[3]{3}$$

**(b)** Because 64 is a factor of 128 and 64 is a perfect square, we write 128 as $64 \cdot 2$, $x^2$ is a perfect square.

$$\sqrt{128x^2} = \sqrt{64x^2 \cdot 2}$$

$$\sqrt[n]{ab} = \sqrt[n]{a} \cdot \sqrt[n]{b}: \quad = \sqrt{64x^2} \cdot \sqrt{2}$$

$$\sqrt[n]{ab} = \sqrt[n]{a} \cdot \sqrt[n]{b}: \quad = \sqrt{64} \cdot \sqrt{x^2} \cdot \sqrt{2}$$

$$\sqrt{64} = 8,\ \sqrt{x^2} = |x|: \quad = 8|x|\sqrt{2}$$

**(c)** In $\sqrt[4]{20}$, the index is 4. The fourth powers (or perfect fourths) are 1, 16, 81, .... There are no factors of 20 that are fourth powers, so the radical $\sqrt[4]{20}$ cannot be simplified any further.

**QUICK ✓** *Simplify each of the radical expressions.*

**5.** $\sqrt{48}$ **6.** $4\sqrt[3]{54}$ **7.** $\sqrt{200a^2}$ **8.** $\sqrt[4]{40}$

## EXAMPLE 4 Simplifying an Expression Involving a Square Root

Simplify: $\dfrac{4-\sqrt{20}}{2}$

### Solution

We can simplify this expression using two different approaches.

**Method 1:**

4 is the largest perfect square factor of 20

$$\frac{4-\sqrt{20}}{2} = \frac{4-\sqrt{4\cdot 5}}{2}$$

Use $\sqrt{a\cdot b} = \sqrt{a}\cdot\sqrt{b}$: $= \dfrac{4-\sqrt{4}\cdot\sqrt{5}}{2}$

$$= \frac{4-2\cdot\sqrt{5}}{2}$$

Factor out the 2 in the numerator: $= \dfrac{2(2-\sqrt{5})}{2}$

Divide out common factor: $= 2-\sqrt{5}$

**Method 2:**

4 is the largest perfect square factor of 20

$$\frac{4-\sqrt{20}}{2} = \frac{4-\sqrt{4\cdot 5}}{2}$$

Use $\sqrt{a\cdot b} = \sqrt{a}\cdot\sqrt{b}$: $= \dfrac{4-\sqrt{4}\cdot\sqrt{5}}{2}$

$$= \frac{4-2\cdot\sqrt{5}}{2}$$

Use $\dfrac{A+B}{C} = \dfrac{A}{C}+\dfrac{B}{C}$: $= \dfrac{4}{2} - \dfrac{2\cdot\sqrt{5}}{2}$

Divide out common factor: $= 2-\sqrt{5}$

**QUICK ✓** *Simplify the expression.*

**9.** $\dfrac{6+\sqrt{45}}{3}$ **10.** $\dfrac{-2+\sqrt{32}}{4}$

Recall how to simplify $\sqrt[n]{a^n}$ from Section 7.1.

**SIMPLIFYING $\sqrt[n]{a^n}$**

If $n \geq 2$ is a positive integer and $a$ is a real number, then

$$\sqrt[n]{a^n} = a \quad \text{if } n \geq 3 \text{ is odd}$$

$$\sqrt[n]{a^n} = |a| \quad \text{if } n \geq 2 \text{ is even}$$

This means that

$$\sqrt{a^2} = |a| \qquad \sqrt[3]{a^3} = a \qquad \sqrt[4]{a^4} = |a| \qquad \sqrt[5]{a^5} = a \qquad \text{and so on}$$

**In order to make our mathematical lives a little easier, for the remainder of the text, we shall assume that all variables that appear in the radicand are greater than or equal to zero (nonnegative).** So

$$\sqrt{a^2} = a \qquad \sqrt[3]{a^3} = a \qquad \sqrt[4]{a^4} = a \qquad \sqrt[5]{a^5} = a \qquad \text{and so on}$$

What if the exponent on the radicand is greater than the index as in $\sqrt{x^3}$ or $\sqrt[3]{x^6}$? We could use the Laws of Exponents along with the rule for simplifying $\sqrt[n]{a^n}$ or we could use rational exponents.

$$\sqrt{x^6} = \sqrt{(x^3)^2} = x^3 \qquad \text{or} \qquad \sqrt{x^6} = x^{\frac{6}{2}} = x^3$$
$$\sqrt[3]{x^{12}} = \sqrt[3]{(x^4)^3} = x^4 \qquad \text{or} \qquad \sqrt[3]{x^{12}} = x^{\frac{12}{3}} = x^4$$

## EXAMPLE 5 Simplifying a Radical with a Variable Radicand

Simplify $\sqrt{20x^{10}}$. Assume $x \geq 0$.

### Solution

Because 4 is a factor of 20 and 4 is a perfect square, we write 20 as $4 \cdot 5$.

$$\sqrt{20x^{10}} = \sqrt{4 \cdot 5 \cdot x^{10}}$$
$$= \sqrt{4x^{10}} \cdot \sqrt{5}$$

$\sqrt{4} = 2$, $\sqrt{x^{10}} = \sqrt{(x^5)^2} = x^5$ or $\sqrt{x^{10}} = x^{\frac{10}{2}} = x^5$: $\quad = 2x^5\sqrt{5}$

**QUICK ✓**

**11.** Simplify $\sqrt{75a^6}$. Assume $a \geq 0$.

---

What if the index does not divide evenly into the exponent on the variable in the radicand as in $\sqrt[3]{x^8}$? Under these circumstances, we rewrite the variable expression as the product of two variable expressions where one of the factors has an exponent that is a multiple of the index. For example, we can write

$$\sqrt[3]{x^8} \qquad \text{as} \qquad \sqrt[3]{x^6 \cdot x^2}$$

so that

$$\sqrt[3]{x^8} = \sqrt[3]{x^6 \cdot x^2} = \sqrt[3]{x^6} \cdot \sqrt[3]{x^2} = x^2\sqrt[3]{x^2}$$

## EXAMPLE 6 Simplifying Radicals

Simplify:

**(a)** $\sqrt{80a^3}$ **(b)** $\sqrt[3]{27m^4n^{14}}$

Assume all variables are greater than or equal to zero.

### Solution

**(a)**

$80 = 16 \cdot 5$; $a^3 = a^2 \cdot a$

$$\sqrt{80a^3} = \sqrt{16 \cdot 5 \cdot a^2 \cdot a}$$
$$= \sqrt{16a^2 \cdot 5a}$$

$\sqrt[n]{ab} = \sqrt[n]{a} \cdot \sqrt[n]{b}$: $\quad = \sqrt{16a^2} \cdot \sqrt{5a}$

$\sqrt{a^2} = a$ assuming $a \geq 0$: $\quad = 4a\sqrt{5a}$

**(b)**

$m^4 = m^3 \cdot m;\ n^{14} = n^{12} \cdot n^2$

$$\sqrt[3]{27m^4n^{14}} = \sqrt[3]{27 \cdot m^3 \cdot m \cdot n^{12} \cdot n^2}$$
$$= \sqrt[3]{27 \cdot m^3 \cdot n^{12} \cdot m \cdot n^2}$$
$\sqrt[n]{ab} = \sqrt[n]{a} \cdot \sqrt[n]{b}$:
$$= \sqrt[3]{27m^3n^{12}} \cdot \sqrt[3]{mn^2}$$
$\sqrt[3]{27} = 3;\ \sqrt[3]{m^3} = m;\ \sqrt[3]{n^{12}} = n^{\frac{12}{3}} = n^4$:
$$= 3mn^4\sqrt[3]{mn^2}$$

**QUICK ✓** *Simplify each radical. Assume all variables are greater than or equal to zero.*

**12.** $\sqrt{18a^5}$ **13.** $\sqrt[3]{128x^6y^{10}}$ **14.** $\sqrt[4]{16a^5b^{11}}$

In this next example, we first multiply radical expressions and then simplify the product.

## EXAMPLE 7 Multiplying and Simplifying Radicals

Multiply and simplify:

**(a)** $\sqrt{3} \cdot \sqrt{15}$ **(b)** $3\sqrt[3]{4x} \cdot \sqrt[3]{2x^4}$ **(c)** $\sqrt[4]{27a^2b^5} \cdot \sqrt[4]{6a^3b^6}$

Assume all variables are greater than or equal to zero.

### Solution

Remember, to multiply two radicals the index must be the same. When you have the same index, we multiply the radicands and then we simplify the product.

**(a)** We start by looking to see if the index is the same. The index on both radicals is 2, so we multiply the radicands.

$$\sqrt{3} \cdot \sqrt{15} = \sqrt{3 \cdot 15}$$
$$= \sqrt{45}$$
9 is the largest factor of 45 that is a perfect square:
$$= \sqrt{9 \cdot 5}$$
Product Property of Radicals:
$$= \sqrt{9} \cdot \sqrt{5}$$
$$= 3\sqrt{5}$$

**(b)** The index on both radicals is 3, so we multiply the radicands.

$$3\sqrt[3]{4x} \cdot \sqrt[3]{2x^4} = 3\sqrt[3]{4x \cdot 2x^4}$$
$$= 3\sqrt[3]{8x^5}$$
$x^3$ is a perfect cube; 8 is a perfect cube:
$$= 3\sqrt[3]{8x^3 \cdot x^2}$$
Product Property of Radicals:
$$= 3\sqrt[3]{8x^3} \cdot \sqrt[3]{x^2}$$
$\sqrt[3]{8} = 2;\ \sqrt[3]{x^3} = x$:
$$= 3 \cdot 2 \cdot x \cdot \sqrt[3]{x^2}$$
$$= 6x\sqrt[3]{x^2}$$

**Work Smart**

Notice that $27 = 3^3$ and that $6 = 3 \cdot 2$, so that $27 \cdot 6 = 3^3 \cdot 3 \cdot 2 = 3^4 \cdot 2$. This makes finding the perfect power of 4 a lot easier!

**(c)** The index on both radicals is 4, so we multiply the radicands.

$$\sqrt[4]{27a^2b^5} \cdot \sqrt[4]{6a^3b^6} = \sqrt[4]{3^3 \cdot a^2b^5 \cdot 3 \cdot 2 \cdot a^3b^6}$$
$$= \sqrt[4]{3^4 \cdot 2 \cdot a^5b^{11}}$$
$$= \sqrt[4]{3^4 \cdot 2 \cdot a^4 \cdot a \cdot b^8 \cdot b^3}$$
Product Property of Radicals:
$$= \sqrt[4]{3^4a^4b^8} \cdot \sqrt[4]{2ab^3}$$
$$= 3ab^2\sqrt[4]{2ab^3}$$

**QUICK ✓** *Multiply and simplify the radicals. Assume all variables are greater than or equal to zero.*

**15.** $\sqrt{6} \cdot \sqrt{8}$ **16.** $\sqrt[3]{12a^2} \cdot \sqrt[3]{10a^4}$ **17.** $4\sqrt[3]{8a^2b^5} \cdot \sqrt[3]{6a^2b^4}$

## 3 Use the Quotient Property to Simplify Radical Expressions

Now consider the following:

$$\sqrt{\frac{64}{4}} = \sqrt{\frac{4 \cdot 16}{4}} = \sqrt{16} = 4 \quad \text{and} \quad \frac{\sqrt{64}}{\sqrt{4}} = \frac{8}{2} = 4$$

This suggests the following result:

**In Words**

$$\sqrt[n]{\frac{a}{b}} = \frac{\sqrt[n]{a}}{\sqrt[n]{b}}$$

means "the root of the quotient equals the quotient of the roots" provided that the radicals have the same index.

**QUOTIENT PROPERTY OF RADICALS**

If $\sqrt[n]{a}$ and $\sqrt[n]{b}$ are real numbers, $b \neq 0$, and $n \geq 2$ is an integer, then

$$\frac{\sqrt[n]{a}}{\sqrt[n]{b}} = \sqrt[n]{\frac{a}{b}}$$

We can justify this formula using rational exponents.

$$\frac{\sqrt[n]{a}}{\sqrt[n]{b}} = \frac{a^{\frac{1}{n}}}{b^{\frac{1}{n}}} = \left(\frac{a}{b}\right)^{\frac{1}{n}} = \sqrt[n]{\frac{a}{b}}$$

### EXAMPLE 8 Using the Quotient Property to Simplify Radicals

Simplify:

**(a)** $\sqrt{\frac{18}{25}}$ **(b)** $\sqrt[3]{\frac{6z^3}{125}}$ **(c)** $\sqrt[4]{\frac{10a^2}{81b^4}}, b \neq 0$

Assume all variables are greater than or equal to zero.

#### Solution

In all three of these problems, you should notice that the expression in the denominator is a perfect power of the index. Therefore, we are going to use the Quotient Rule "in reverse" to simplify the expressions. That is, we use $\sqrt[n]{\frac{a}{b}} = \frac{\sqrt[n]{a}}{\sqrt[n]{b}}$.

**(a)** $\sqrt{\frac{18}{25}} = \frac{\sqrt{18}}{\sqrt{25}} = \frac{3\sqrt{2}}{5}$

**(b)** $\sqrt[3]{\frac{6z^3}{125}} = \frac{\sqrt[3]{6z^3}}{\sqrt[3]{125}} = \frac{z\sqrt[3]{6}}{5}$

**(c)** $\sqrt[4]{\frac{10a^2}{81b^4}} = \frac{\sqrt[4]{10a^2}}{\sqrt[4]{81b^4}} = \frac{\sqrt[4]{10a^2}}{3b}$

**QUICK ✓** *Simplify the radicals. Assume all variables are greater than or equal to zero.*

**18.** $\sqrt{\frac{13}{49}}$ **19.** $\sqrt[3]{\frac{27p^3}{8}}$ **20.** $\sqrt[4]{\frac{3q^4}{16}}$

## EXAMPLE 9 Using the Quotient Property to Simplify Radicals

Simplify:

**(a)** $\dfrac{\sqrt{24a^3}}{\sqrt{6a}}$ **(b)** $\dfrac{-2\sqrt[3]{54a}}{\sqrt[3]{2a^4}}$ **(c)** $\dfrac{\sqrt[3]{-375x^2y}}{\sqrt[3]{3x^{-1}y^7}}$

Assume all variables are greater than zero.

### Solution

In these problems, we notice that the radical expression in the denominator cannot be simplified. However, the index on the numerator and denominator of each expression is the same, so we can write each expression as a single radical.

**(a)**
$$\frac{\sqrt{24a^3}}{\sqrt{6a}} = \sqrt{\frac{24a^3}{6a}}$$
$$= \sqrt{4a^2}$$
$$= 2a$$

**(b)**
$$\frac{-2\sqrt[3]{54a}}{\sqrt[3]{2a^4}} = -2\cdot\sqrt[3]{\frac{54a}{2a^4}}$$
$$= -2\cdot\sqrt[3]{\frac{27}{a^3}}$$
$$= -2\cdot\frac{3}{a}$$
$$= -\frac{6}{a}$$

**(c)**
$$\frac{\sqrt[3]{-375x^2y}}{\sqrt[3]{3x^{-1}y^7}} = \sqrt[3]{\frac{-375x^2y}{3x^{-1}y^7}}$$
$$= \sqrt[3]{-125x^{2-(-1)}y^{1-7}}$$
$$= \sqrt[3]{-125x^3y^{-6}}$$
$$= \sqrt[3]{\frac{-125x^3}{y^6}}$$
$$= \frac{-5x}{y^2}$$

**QUICK ✓** *Simplify the radicals. Assume all variables are greater than zero.*

**21.** $\dfrac{\sqrt{12a^5}}{\sqrt{3a}}$ **22.** $\dfrac{\sqrt[3]{-24x^2}}{\sqrt[3]{3x^{-1}}}$ **23.** $\dfrac{\sqrt[3]{250a^5b^{-2}}}{\sqrt[3]{2ab}}$

## 4 Multiply Radicals with Unlike Indices

To multiply radicals we use the fact that $\sqrt[n]{a}\cdot\sqrt[n]{b} = \sqrt[n]{ab}$. This rule only works when the index on each radical is the same. What if the index on each radical is different? Can we still simplify the product? The answer is yes! To perform the multiplication we use the fact that $\sqrt[n]{a} = a^{\frac{1}{n}}$. Let's go over an example.

## EXAMPLE 10 Multiplying Radicals with Unlike Indices

Multiply and simplify:

$$\sqrt[4]{8}\cdot\sqrt[3]{5}$$

### Solution

Notice that the index is not the same, so $\sqrt[n]{a}\cdot\sqrt[n]{b} = \sqrt[n]{ab}$ cannot be used to find the product. We will use rational exponents along with $\sqrt[n]{a} = a^{\frac{1}{n}}$ instead.

$$\sqrt[4]{8}\cdot\sqrt[3]{5} = 8^{\frac{1}{4}}\cdot 5^{\frac{1}{3}}$$

$$\text{LCD} = 12: \quad = 8^{\frac{3}{12}}\cdot 5^{\frac{4}{12}}$$

$$a^{\frac{r}{s}} = (a^r)^{\frac{1}{s}}: \quad = \left[(8^3)^{\frac{1}{12}}\cdot(5^4)^{\frac{1}{12}}\right]$$

$$a^r\cdot b^r = (ab)^r: \quad = [(8^3)(5^4)]^{\frac{1}{12}}$$

$$= (320{,}000)^{\frac{1}{12}}$$

$$a^{\frac{1}{n}} = \sqrt[n]{a}: \quad = \sqrt[12]{320{,}000}$$

**QUICK ✓** *Multiply and simplify.*

**24.** $\sqrt[4]{5}\cdot\sqrt[3]{3}$ **25.** $\sqrt{10}\cdot\sqrt[3]{12}$

# 7.3 Exercises

**For Extra Help:** Student Solutions Manual  CD Video  PH Math/Tutor Center 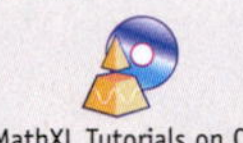 MathXL Tutorials on CD MathXL® MyMathLab

## Concepts and Vocabulary

*In Problems 1 and 2, fill in the blanks.*

**1.** If $\sqrt[n]{a}$ and $\sqrt[n]{b}$ are real numbers and $n \geq 2$ is an integer, then $\sqrt[n]{a}\cdot\sqrt[n]{b} =$ _____.

**2.** A number that is the square of a rational number is called a _____ _____.
A number that is the cube of a rational number is called a _____ _____.

*In Problems 3 and 4, answer True or False to each statement.*

**3.** The radical expression $\sqrt{8x}$ is simplified.

**4.** If $\sqrt[n]{a}$ and $\sqrt[n]{b}$ are real numbers, $b \neq 0$, and $n \geq 2$ is an integer, then $\dfrac{\sqrt[n]{a}}{\sqrt[n]{b}} = \sqrt[n]{\dfrac{a}{b}}$.

**5.** List the first six integers that are perfect squares.

**6.** List the first six positive integers that are perfect cubes.

**7.** In your own words, explain how you would simplify $\sqrt[3]{16a^5}$.

**8.** In order to use the Product Property to multiply radicals, what must be true about the index in each radical?

## Building Skills

*In Problems 9–18, use the Product Property to multiply. Assume that all variables can be any real number.*

**9.** $\sqrt{2}\cdot\sqrt{7}$ **10.** $\sqrt{3}\cdot\sqrt{10}$ **11.** $\sqrt[3]{6}\cdot\sqrt[3]{10}$

**12.** $\sqrt[3]{-5}\cdot\sqrt[3]{7}$ **13.** $\sqrt{3a}\cdot\sqrt{5b}$ **14.** $\sqrt[4]{6a^2}\cdot\sqrt[4]{7b^2}$

**15.** $\sqrt{x-7}\cdot\sqrt{x+7}$ **16.** $\sqrt{p-5}\cdot\sqrt{p+5}$ **17.** $\sqrt{\dfrac{5x}{3}}\cdot\sqrt{\dfrac{3}{x}}$

**18.** $\sqrt[3]{\dfrac{-9x^2}{4}}\cdot\sqrt[3]{\dfrac{4}{3x}}$

*In Problems 19–52, simplify each radical using the Product Property. Assume that all variables can be any real number.*

**19.** $\sqrt{50}$ **20.** $\sqrt{32}$ **21.** $\sqrt[3]{54}$

**22.** $\sqrt[4]{162}$ **23.** $\sqrt{48x^2}$ **24.** $\sqrt{20a^2}$

**25.** $\sqrt[3]{-27x^3}$ **26.** $\sqrt[3]{-64p^3}$ **27.** $\sqrt[4]{32m^4}$

**28.** $\sqrt[4]{48z^4}$ **29.** $\sqrt{12p^2q}$ **30.** $\sqrt{45m^2n}$

**31.** $3\sqrt{24a}$ **32.** $4\sqrt{27b}$ **33.** $\sqrt{162m^4}$

**34.** $\sqrt{98w^8}$ **35.** $\sqrt{y^{13}}$ **36.** $\sqrt{s^9}$

**37.** $\sqrt[3]{c^8}$ **38.** $\sqrt[5]{x^{12}}$ **39.** $\sqrt{m^5n^3}$

**40.** $\sqrt{x^7y^2}$ **41.** $\sqrt{125p^3q^4}$ **42.** $\sqrt{243ab^5}$

**43.** $\sqrt[3]{-16x^9}$ **44.** $\sqrt[3]{-54q^{12}}$ **45.** $\sqrt{p^3q^5r}$

**46.** $\sqrt[4]{x^6y^9z^4}$ **47.** $\sqrt[5]{-16m^8n^2}$ **48.** $\sqrt{75x^6y}$

**49.** $\sqrt[4]{(x-y)^5}, x > y$ **50.** $\sqrt[3]{(a+b)^5}$ **51.** $\sqrt[3]{8x^3 - 8y^3}$

**52.** $\sqrt[3]{8a^3 + 8b^3}$

*In Problems 53–60, simplify each expression.*

**53.** $\dfrac{4+\sqrt{36}}{2}$ **54.** $\dfrac{5-\sqrt{100}}{5}$ **55.** $\dfrac{9+\sqrt{18}}{3}$ **56.** $\dfrac{10-\sqrt{75}}{5}$

**57.** $\dfrac{-4-\sqrt{162}}{6}$ **58.** $\dfrac{-6+\sqrt{48}}{8}$ **59.** $\dfrac{7-\sqrt{98}}{14}$ **60.** $\dfrac{-6+\sqrt{108}}{6}$

*In Problems 61–78, multiply and simplify. Assume that all variables are greater than or equal to zero.*

**61.** $\sqrt{5}\cdot\sqrt{5}$ **62.** $\sqrt{6}\cdot\sqrt{6}$

**63.** $\sqrt{2}\cdot\sqrt{8}$ **64.** $\sqrt{3}\cdot\sqrt{12}$

**65.** $\sqrt[3]{4}\cdot\sqrt[3]{2}$ **66.** $\sqrt[3]{9}\cdot\sqrt[3]{3}$

**67.** $\sqrt{5x}\cdot\sqrt{15x}$ **68.** $\sqrt{6x}\cdot\sqrt{30x}$

**69.** $\sqrt[3]{4b^2}\cdot\sqrt[3]{6b^2}$ **70.** $\sqrt[3]{9a}\cdot\sqrt[3]{6a^2}$

**71.** $2\sqrt{6ab}\cdot3\sqrt{15ab^3}$ **72.** $3\sqrt{14pq^3}\cdot2\sqrt{7pq}$

**73.** $\sqrt[4]{27p^3q^2}\cdot\sqrt[4]{12p^2q^2}$ **74.** $\sqrt[3]{16m^2n}\cdot\sqrt[3]{27m^2n}$

**75.** $\sqrt[5]{-8a^3b^4}\cdot\sqrt[5]{12a^3b}$ **76.** $\sqrt[5]{-27x^4y^2}\cdot\sqrt[5]{18x^3y^4}$

**77.** $\sqrt[4]{8(x-y)^2}\cdot\sqrt[4]{6(x-y)^3}, x > y$ **78.** $\sqrt[3]{9(a+b)^2}\cdot\sqrt[3]{6(a+b)^5}$

*In Problems 79–88, simplify. Assume that all variables are greater than zero.*

**79.** $\sqrt{\dfrac{3}{16}}$ **80.** $\sqrt{\dfrac{5}{36}}$ **81.** $\sqrt{\dfrac{121}{100}}$ **82.** $\sqrt{\dfrac{81}{64}}$

**83.** $\sqrt[4]{\dfrac{5x^4}{16}}$ **84.** $\sqrt[4]{\dfrac{2a^8}{81}}$ **85.** $\sqrt{\dfrac{9y^2}{25x^2}}$ **86.** $\sqrt{\dfrac{4a^4}{81b^2}}$

**87.** $\sqrt[3]{\dfrac{-27x^9}{64y^{12}}}$ **88.** $\sqrt[5]{\dfrac{-32a^{15}}{243b^{10}}}$

*In Problems 89–102, divide and simplify. Assume that all variables are greater than zero.*

**89.** $\dfrac{\sqrt{8}}{\sqrt{2}}$ **90.** $\dfrac{\sqrt{27}}{\sqrt{3}}$ **91.** $\dfrac{\sqrt[3]{128}}{\sqrt[3]{2}}$ **92.** $\dfrac{\sqrt[4]{64}}{\sqrt[4]{4}}$

**93.** $\dfrac{\sqrt{48a^3}}{\sqrt{6a}}$ **94.** $\dfrac{\sqrt{54y^5}}{\sqrt{3y}}$ **95.** $\dfrac{\sqrt{24a^5b}}{\sqrt{3ab^3}}$ **96.** $\dfrac{\sqrt{360m^7n^3}}{\sqrt{5mn^5}}$

**97.** $\dfrac{\sqrt{512a^7b}}{3\sqrt{2ab^3}}$ **98.** $\dfrac{\sqrt{375x^2y^7}}{10\sqrt{3y}}$ **99.** $\dfrac{\sqrt[3]{104a^5}}{\sqrt[3]{4a^{-1}}}$ **100.** $\dfrac{\sqrt[3]{-128x^8}}{\sqrt[3]{2x^{-1}}}$

**101.** $\dfrac{\sqrt{90x^3y^{-1}}}{\sqrt{2x^{-3}y}}$ **102.** $\dfrac{\sqrt{96a^5b^{-3}}}{\sqrt{3a^{-5}b}}$

*In Problems 103–110, multiply and simplify.*

**103.** $\sqrt{3}\cdot\sqrt[3]{4}$ **104.** $\sqrt{2}\cdot\sqrt[3]{7}$ **105.** $\sqrt[3]{2}\cdot\sqrt[6]{3}$ **106.** $\sqrt[4]{3}\cdot\sqrt[8]{5}$

**107.** $\sqrt{3}\cdot\sqrt[3]{18}$ **108.** $\sqrt{6}\cdot\sqrt[3]{9}$ **109.** $\sqrt[4]{9}\cdot\sqrt[6]{12}$ **110.** $\sqrt[5]{8}\cdot\sqrt[10]{16}$

## Mixed Practice

*In Problems 111–126, perform any indicated operation and simplify. Assume all variables are greater than zero.*

**111.** $\sqrt[3]{\dfrac{5x}{8}}$ **112.** $\sqrt[3]{\dfrac{7a^2}{64}}$ **113.** $\sqrt[3]{5a}\cdot\sqrt[3]{9a}$

**114.** $\sqrt[5]{8b^2}\cdot\sqrt[5]{3b}$ **115.** $\sqrt{72a^4}$ **116.** $\sqrt{24b^6}$

**117.** $\sqrt[3]{6a^2b}\cdot\sqrt[3]{9ab}$ **118.** $\sqrt[4]{8x^3y^2}\cdot\sqrt[4]{4x^2y^3}$ **119.** $\dfrac{\sqrt[3]{-32a}}{\sqrt[3]{2a^4}}$

**120.** $\dfrac{\sqrt[3]{-250p^2}}{\sqrt[3]{2p^5}}$ **121.** $-5\sqrt[3]{32m^3}$ **122.** $-7\sqrt[3]{250p^3}$

**123.** $\sqrt[3]{81a^4b^7}$ **124.** $\sqrt[5]{32p^7q^{11}}$ **125.** $\sqrt[3]{12}\cdot\sqrt[3]{18}$

**126.** $\sqrt[4]{8}\cdot\sqrt[4]{18}$

## Applying the Concepts

△ **127. Length of a Line Segment** The length of the line segment joining the points $(2, 5)$ and $(-1, -1)$ is given by

$$\sqrt{(5-(-1))^2+(2-(-1))^2}$$

**(a)** Plot the points in the Cartesian plane and draw a line segment connecting the points.
**(b)** Express the length of the line segment as a radical in simplified form.

△ **128. Length of a Line Segment** The length of the line segment joining the points $(4, 2)$ and $(-2, 4)$ is given by

$$\sqrt{(4-2)^2+(-2-4)^2}$$

**(a)** Plot the points in the Cartesian plane and draw a line segment connecting the points.
**(b)** Express the length of the line segment as a radical in simplified form.

**129. Revenue Growth** Suppose that the annual revenue $R$ (in millions of dollars) of a company after $t$ years of operating is modeled by the function

$$R(t) = \sqrt[3]{\frac{t}{2}}$$

**(a)** Predict the revenue of the company after 8 years of operation.
**(b)** Predict the revenue of the company after 27 years of operation.

△ **130. Sphere** The radius $r$ of a sphere whose volume is $V$ is given by

$$r = \sqrt[3]{\frac{3V}{4\pi}}$$

**(a)** Write the radius of a sphere whose volume is 9 cubic centimeters as a radical in simplified form.
**(b)** Write the radius of a sphere whose volume is $32\pi$ cubic meters as a radical in simplified form.

## Extending the Concepts

**131.** Suppose that $f(x) = \sqrt{2x}$ and $g(x) = \sqrt{8x^3}$.
**(a)** Find $(f \cdot g)(x)$.
**(b)** Evaluate $(f \cdot g)(3)$.

*In Problems 132–135, evaluate the formula*

$$x = \frac{-b \pm \sqrt{b^2 - 4ac}}{2a}$$

*for the given values of a, b, and c. Note that the symbol* $\pm$ *is shorthand notation to indicate that there are two solutions. One solution is obtained when you add the quantity after the* $\pm$ *symbol, and another is obtained when you subtract. This formula can be used to solve any equation of the form* $ax^2 + bx + c = 0$.

**132.** $a = 1, b = 4, c = 1$

**133.** $a = 1, b = 6, c = 3$

**134.** $a = 2, b = 1, c = -1$

**135.** $a = 3, b = 4, c = -1$

## Synthesis Review

*In Problems 136–139, solve the following.*

**136.** $4x + 3 = 13$

**137.** $2|7x - 1| + 4 = 16$

**138.** $\frac{3}{5}x + 2 \le 28$

**139.** $\frac{5}{2}|x + 1| + 1 \le 11$

**140.** How are the solutions in Problems 136 and 138 similar? How are the solutions in Problems 137 and 139 similar?

## The Graphing Calculator

**141. Exploration** To understand the circumstances under which absolute value symbols are required when simplifying radicals, do the following.
**(a)** Graph $Y_1 = \sqrt{x^2}$ and $Y_2 = x$. Do you think that $\sqrt{x^2} = x$? Now graph $Y_1 = \sqrt{x^2}$ and $Y_2 = |x|$. Do you think that $\sqrt{x^2} = |x|$?
**(b)** Graph $Y_1 = \sqrt[3]{x^3}$ and $Y_2 = x$. Do you think that $\sqrt[3]{x^3} = x$? Now graph $Y_1 = \sqrt[3]{x^3}$ and $Y_2 = |x|$. Do you think that $\sqrt[3]{x^3} = |x|$?
**(c)** Graph $Y_1 = \sqrt[4]{x^4}$ and $Y_2 = x$. Do you think that $\sqrt[4]{x^4} = x$? Now graph $Y_1 = \sqrt[4]{x^4}$ and $Y_2 = |x|$. Do you think that $\sqrt[4]{x^4} = |x|$?
**(d)** In your own words, make a generalization about $\sqrt[n]{x^n}$.

# 7.4 Adding, Subtracting, and Multiplying Radical Expressions

**OBJECTIVES**

1. Add or Subtract Radical Expressions
2. Multiply Radical Expressions

***Preparing for Adding, Subtracting, and Multiplying Radical Expressions***

*Before getting started, take the following readiness quiz. If you get a problem wrong, go back to the section cited and review the material.*

**1.** Add: $4y^3 - 2y^2 + 8y - 1 + (-2y^3 + 7y^2 - 3y + 9)$ [Section 5.1, pp. 358–359]

**2.** Subtract: $5z^2 + 6 - (3z^2 - 8z - 3)$ [Section 5.1, pp. 359–360]

**3.** Multiply: $(4x + 3)(x - 5)$ [Section 5.2, pp. 369–370]

**4.** Multiply: $(2y - 3)(2y + 3)$ [Section 5.2, p. 371]

## 1 Add or Subtract Radical Expressions

Recall, a radical expression is an algebraic expression that contains a radical. We say that two radicals are **like radicals** if each radical has the same index and the same radicand. For example,

$$4\sqrt[3]{x - 4} \quad \text{and} \quad 10\sqrt[3]{x - 4}$$

are like radicals because each has the same index, 3, and the same radicand, $x - 4$.

To add or subtract radical expressions we combine like radicals using the Distributive Property as follows:

**Work Smart**

You will be doing three of the four basic operations in this section. Division of radical expressions is just a bit more involved, so we discuss it in the next section by itself.

Factor out $\sqrt[3]{x-4}$

$$4\sqrt[3]{x - 4} + 10\sqrt[3]{x - 4} = (4 + 10)\sqrt[3]{x - 4}$$

$4 + 10 = 14$: $\quad = 14\sqrt[3]{x - 4}$

### EXAMPLE 1 Adding and Subtracting Radical Expressions

Add or subtract, as indicated.

**(a)** $5\sqrt{2x} + 9\sqrt{2x}$

**(b)** $3\sqrt[3]{10} + 7\sqrt[3]{10} - 5\sqrt[3]{10}$

**Solution**

**(a)** Both radicals have the same index, 2, and the same radicand, $2x$.

**Work Smart**

Remember that to add or subtract radicals both the index and the radicand must be the same.

$$5\sqrt{2x} + 9\sqrt{2x} = (5 + 9)\sqrt{2x}$$

$5 + 9 = 14$: $\quad = 14\sqrt{2x}$

**(b)** All three radicals have the same index, 3, and the same radicand, 10.

$$3\sqrt[3]{10} + 7\sqrt[3]{10} - 5\sqrt[3]{10} = (3 + 7 - 5)\sqrt[3]{10}$$

$3 + 7 - 5 = 5$: $\quad = 5\sqrt[3]{10}$

***Preparing for...Answers***
**1.** $2y^3 + 5y^2 + 5y + 8$ **2.** $2z^2 + 8z + 9$
**3.** $4x^2 - 17x - 15$ **4.** $4y^2 - 9$

**QUICK ✓** *Add or subtract, as indicated.*

1. $9\sqrt{13y} + 4\sqrt{13y}$ 2. $\sqrt[4]{5} + 9\sqrt[4]{5} - 3\sqrt[4]{5}$

Sometimes we have to simplify the radical so that the radicands are the same before adding or subtracting.

### EXAMPLE 2 Adding and Subtracting Radical Expressions

Add or subtract, as indicated. Assume all variables are real numbers greater than or equal to zero.

**(a)** $3\sqrt{12} + 7\sqrt{3}$ **(b)** $3x\sqrt{20x} - 7\sqrt{5x^3}$ **(c)** $3\sqrt{5} + 7\sqrt{13}$

**Solution**

**(a)** The index on each radical is the same, but the radicands are different. However, we can simplify the radicals to make the radicands the same.

$$\sqrt{12} = \sqrt{4}\cdot\sqrt{3}$$

$$\begin{aligned}
3\sqrt{12} + 7\sqrt{3} &= 3\sqrt{4}\cdot\sqrt{3} + 7\sqrt{3} \\
\sqrt{4} = 2{:}\quad &= 3\cdot 2\sqrt{3} + 7\sqrt{3} \\
&= 6\sqrt{3} + 7\sqrt{3} \\
\text{Factor out } \sqrt{3}{:}\quad &= (6 + 7)\sqrt{3} \\
&= 13\sqrt{3}
\end{aligned}$$

**(b)** The index on each radical is the same, but the radicands are different. However, we can simplify the radicals to make the radicands the same.

$$\sqrt{20x} = \sqrt{4}\cdot\sqrt{5x};\ \sqrt{5x^3} = \sqrt{x^2}\cdot\sqrt{5x}$$

$$\begin{aligned}
3x\sqrt{20x} - 7\sqrt{5x^3} &= 3x\cdot\sqrt{4}\cdot\sqrt{5x} - 7\cdot\sqrt{x^2}\cdot\sqrt{5x} \\
\text{Simplify radicals:}\quad &= 3x\cdot 2\cdot\sqrt{5x} - 7\cdot x\cdot\sqrt{5x} \\
\text{Multiply:}\quad &= 6x\sqrt{5x} - 7x\sqrt{5x} \\
\text{Factor out } \sqrt{5x}{:}\quad &= (6x - 7x)\sqrt{5x} \\
6x - 7x = -x{:}\quad &= -x\sqrt{5x}
\end{aligned}$$

**(c)** For $3\sqrt{5} + 7\sqrt{13}$ the index on each radical is the same, but the radicands are different and we cannot simplify the radicals. ■

**QUICK ✓** *Add or subtract, as indicated. Assume all variables are real numbers greater than or equal to zero.*

3. $4\sqrt{18} - 3\sqrt{8}$ 4. $-5x\sqrt[3]{54x} + 7\sqrt[3]{2x^4}$ 5. $7\sqrt{10} - 6\sqrt{3}$

### EXAMPLE 3 Adding or Subtracting Radical Expressions

Add or subtract, as indicated. Assume that all variables are real numbers greater than or equal to zero.

**(a)** $\sqrt[3]{16x^4} - 7x\sqrt[3]{-2x} + \sqrt[3]{54x}$ **(b)** $3\sqrt[4]{m^4n} - 5m\sqrt[8]{n^2}$

**Solution**

**(a)** The index on each radical is the same, but the radicands are different. However, we can simplify the radicals to make the radicands the same.

$\sqrt[3]{16x^4} = \sqrt[3]{8x^3} \cdot \sqrt[3]{2x}$; $\sqrt[3]{-2x} = \sqrt[3]{-1} \cdot \sqrt[3]{2x}$; $\sqrt[3]{54x} = \sqrt[3]{27} \cdot \sqrt[3]{2x}$

$$\sqrt[3]{16x^4} - 7x\sqrt[3]{-2x} + \sqrt[3]{54x} = \sqrt[3]{8x^3} \cdot \sqrt[3]{2x} - 7x\sqrt[3]{-1} \cdot \sqrt[3]{2x} + \sqrt[3]{27} \cdot \sqrt[3]{2x}$$

$\sqrt[3]{8x^3} = 2x$; $\sqrt[3]{-1} = -1$; $\sqrt[3]{27} = 3$: $= 2x\sqrt[3]{2x} - 7x(-1)\sqrt[3]{2x} + 3\sqrt[3]{2x}$

$-7x(-1) = 7x$: $= 2x\sqrt[3]{2x} + 7x\sqrt[3]{2x} + 3\sqrt[3]{2x}$

Factor out $\sqrt[3]{2x}$: $= (2x + 7x + 3)\sqrt[3]{2x}$

Simplify: $= (9x + 3)\sqrt[3]{2x}$

Factor out 3: $= 3(3x + 1)\sqrt[3]{2x}$

**Work Smart: Study Skills**

Contrast adding radicals with multiplying radicals:

Add: $3\sqrt{5} + 8\sqrt{5} = (3 + 8)\sqrt{5}$
$= 11\sqrt{5}$

Multiply:
$3\sqrt{5} \cdot 8\sqrt{5} = 3 \cdot 8 \cdot \sqrt{5} \cdot \sqrt{5}$
$= 24\sqrt{25}$
$= 24 \cdot 5$
$= 120$

Ask yourself these questions:

How must the radicals be "like" to be added?

How must the radicals be "like" to be multiplied?

**(b)** Here, the index on the radical and the radicand are different. We start by dealing with the index using rational exponents.

$\sqrt[n]{a^m} = a^{\frac{m}{n}}$

$$3\sqrt[4]{m^4n} - 5m\sqrt[8]{n^2} = 3\sqrt[4]{m^4n} - 5m \cdot n^{\frac{2}{8}}$$

Reduce rational exponent: $= 3\sqrt[4]{m^4n} - 5m \cdot n^{\frac{1}{4}}$

Rewrite as radical: $= 3\sqrt[4]{m^4n} - 5m \cdot \sqrt[4]{n}$

Now the index is the same, but the radicands are different. We can deal with this issue as well.

$\sqrt[4]{m^4n} = \sqrt[4]{m^4} \cdot \sqrt[4]{n}$: $= 3\sqrt[4]{m^4} \cdot \sqrt[4]{n} - 5m \cdot \sqrt[4]{n}$

Simplify: $= 3m \cdot \sqrt[4]{n} - 5m \cdot \sqrt[4]{n}$

Factor out $\sqrt[4]{n}$: $= (3m - 5m) \cdot \sqrt[4]{n}$

Simplify: $= -2m \cdot \sqrt[4]{n}$

**QUICK ✓** *Add or subtract, as indicated. Assume all variables are real numbers greater than or equal to zero.*

**6.** $\sqrt[3]{8z^4} - 2z\sqrt[3]{-27z} + \sqrt[3]{125z}$ **7.** $\sqrt{25m} - 3\sqrt[4]{m^2}$

## 2 Multiply Radical Expressions

We have already multiplied radical expressions in which a single radical was multiplied by a second single radical. Now we concentrate on multiplying radical expressions involving more than one radical. These expressions are multiplied in the same way that we multiplied polynomials.

### EXAMPLE 4 Multiplying Radical Expressions

Multiply and simplify:

**(a)** $\sqrt{5}(3 - 4\sqrt{5})$ **(b)** $\sqrt[3]{2}(3 + \sqrt[3]{4})$ **(c)** $(3 + 2\sqrt{7})(2 - 3\sqrt{7})$

**Solution**

**(a)** We use the Distributive Property and multiply $\sqrt{5}$ by each term in the parentheses.

$$\sqrt{5}(3 - 4\sqrt{5}) = \sqrt{5}\cdot 3 - \sqrt{5}\cdot 4\sqrt{5}$$

Multiply radicals: $= 3\sqrt{5} - 4\cdot\sqrt{25}$

$\sqrt{25} = 5$: $= 3\sqrt{5} - 4\cdot 5$

Simplify: $= 3\sqrt{5} - 20$

**(b)** We use the Distributive Property and multiply $\sqrt[3]{2}$ by each term in parentheses.

$$\sqrt[3]{2}(3 + \sqrt[3]{4}) = \sqrt[3]{2}\cdot 3 + \sqrt[3]{2}\cdot\sqrt[3]{4}$$

Multiply radicals: $= 3\sqrt[3]{2} + \sqrt[3]{8}$

$\sqrt[3]{8} = 2$: $= 3\sqrt[3]{2} + 2$

**(c)** We treat this just like the product of two binomials and use the FOIL method to multiply.

First, Last, Inner, Outer

$$(3 + 2\sqrt{7})(2 - 3\sqrt{7}) = \overset{F}{3\cdot 2} - \overset{O}{3\cdot 3\sqrt{7}} + \overset{I}{2\sqrt{7}\cdot 2} - \overset{L}{2\sqrt{7}\cdot 3\sqrt{7}}$$

Multiply: $= 6 - 9\sqrt{7} + 4\sqrt{7} - 6\sqrt{49}$

$\sqrt{49} = 7$: $= 6 - 9\sqrt{7} + 4\sqrt{7} - 6\cdot 7$

$= 6 - 9\sqrt{7} + 4\sqrt{7} - 42$

Simplify: $= -36 - 5\sqrt{7}$

**Work Smart**

$-36 - 5\sqrt{7} \neq -41\sqrt{7}$.
Do you know why?

**QUICK ✓** *Multiply and simplify.*

**8.** $\sqrt{6}(3 - 5\sqrt{6})$ **9.** $\sqrt[3]{12}(3 - \sqrt[3]{2})$ **10.** $(2 - 7\sqrt{3})(5 + 4\sqrt{3})$

We can use our special products formulas (Section 5.6) to multiply radicals as well. In particular, we are going to use the formulas for perfect squares, $(a + b)^2 = a^2 + 2ab + b^2$ and $(a - b)^2 = a^2 - 2ab + b^2$, as well as the formula for the difference of two squares, $(a + b)(a - b) = a^2 - b^2$.

## EXAMPLE 5 Multiplying Radical Expressions Involving Special Products

Multiply and simplify.

**(a)** $(2\sqrt{3} + \sqrt{5})^2$ **(b)** $(3 + \sqrt{7})(3 - \sqrt{7})$

**Solution**

$(a + b)^2 = a^2 + 2ab + b^2$

**(a)** $(2\sqrt{3} + \sqrt{5})^2 = (2\sqrt{3})^2 + 2\cdot 2\sqrt{3}\cdot\sqrt{5} + (\sqrt{5})^2$

Multiply: $= 4\sqrt{9} + 4\sqrt{15} + \sqrt{25}$

Simplify: $= 4\cdot 3 + 4\sqrt{15} + 5$

Combine like terms: $= 17 + 4\sqrt{15}$

**Work Smart**

Notice in Example 5(a),
$(2\sqrt{3} + \sqrt{5})^2 \neq (2\sqrt{3})^2 + (\sqrt{5})^2$

**(b)** We should notice that $(3 + \sqrt{7})(3 - \sqrt{7})$ is in the form $(a + b)(a - b)$, so that

$(a + b)(a - b) = a^2 - b^2$

$$(3 + \sqrt{7})(3 - \sqrt{7}) = 3^2 - (\sqrt{7})^2$$

$= 9 - 7$

$= 2$

Notice that the product found in Example 5(b) is an integer. That is, there are no radicals in the product. Radical expressions such as $3 + \sqrt{7}$ and $3 - \sqrt{7}$ are called **conjugates** of each other. When we multiply radical expressions involving square roots that are conjugates, the result will never contain a radical. This result plays a huge role in the next section.

**QUICK ✓** *Multiply and simplify.*

**11.** $(5\sqrt{2} + \sqrt{3})^2$ **12.** $(\sqrt{7} - 3\sqrt{2})^2$ **13.** $(\sqrt{3} + \sqrt{2})(\sqrt{3} - \sqrt{2})$

# 7.4 Exercises

***For Extra Help:***

Student Solutions Manual

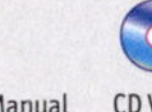

CD Video

PH Math/Tutor Center

MathXL Tutorials on CD

MathXL®

MyMathLab

## Concepts and Vocabulary

*In Problems 1 and 2, fill in the blanks.*

**1.** Two radicals are _________ _________ if each radical has the same index and the same radicand.

**2.** The radical expressions $4 + \sqrt{5}$ and $4 - \sqrt{5}$ are examples of _________.

*In Problems 3 and 4, answer True or False to each statement.*

**3.** To add or subtract radicals, the index and radicand must be the same.

**4.** The conjugate of $-5 + \sqrt{2}$ is $5 - \sqrt{2}$.

**5.** In your own words, explain how to add or subtract radicals.

**6.** Multiply $(\sqrt{a} - \sqrt{b})(\sqrt{a} + \sqrt{b})$ and provide a general result regarding the product of conjugates involving square roots.

## Building Skills

*In Problems 7–14, add or subtract as indicated.*

**7.** $3\sqrt{2} + 7\sqrt{2}$

**8.** $6\sqrt{3} + 8\sqrt{3}$

**9.** $5\sqrt[3]{x} - 3\sqrt[3]{x}$

**10.** $12\sqrt[4]{z} - 5\sqrt[4]{z}$

**11.** $8\sqrt{5x} - 3\sqrt{5x} + 9\sqrt{5x}$

**12.** $4\sqrt[3]{3y} + 8\sqrt[3]{3y} - 10\sqrt[3]{3y}$

**13.** $4\sqrt[3]{5} - 3\sqrt{5} + 7\sqrt[3]{5} - 8\sqrt{5}$

**14.** $12\sqrt{7} + 5\sqrt[4]{7} - 5\sqrt{7} + 6\sqrt[4]{7}$

*In Problems 15–36, add or subtract as indicated. Assume all variables are positive or zero.*

**15.** $\sqrt{8} + 6\sqrt{2}$

**16.** $6\sqrt{3} + \sqrt{12}$

**17.** $\sqrt[3]{24} - 4\sqrt[3]{3}$

**18.** $\sqrt[3]{32} - 5\sqrt[3]{4}$

**19.** $\sqrt[3]{54} - 7\sqrt[3]{128}$

**20.** $7\sqrt[4]{48} - 4\sqrt[4]{243}$

**21.** $5\sqrt{54x} - 3\sqrt{24x}$

**22.** $2\sqrt{48z} - \sqrt{75z}$

**23.** $2\sqrt{8} + 3\sqrt{10}$

**24.** $4\sqrt{12} + 2\sqrt{20}$

**25.** $\sqrt{12x^3} + 5x\sqrt{108x}$

**26.** $3\sqrt{63z^3} + 2z\sqrt{28z}$

**27.** $\sqrt{12x^2} + 3x\sqrt{2} - 2\sqrt{98x^2}$

**28.** $\sqrt{48y^2} - 4y\sqrt{12} + \sqrt{108y^2}$

**29.** $\sqrt[3]{-54x^3} + 3x\sqrt[3]{16} - 2\sqrt[3]{128}$

**30.** $2\sqrt[3]{-5x^3} + 4x\sqrt[3]{40} - \sqrt[3]{135}$

**31.** $\sqrt{9x - 9} + \sqrt{4x - 4}$

**32.** $\sqrt{4x + 12} - \sqrt{9x + 27}$

**33.** $\sqrt{16x} - \sqrt[6]{x^3}$

**34.** $\sqrt{25x} - \sqrt[4]{x^2}$

**35.** $\sqrt[3]{27x} + 2\sqrt[9]{x^3}$

**36.** $\sqrt[4]{16y} + \sqrt[8]{y^2}$

*In Problems 37–70, multiply and simplify. Assume all variables are positive or zero.*

**37.** $\sqrt{3}(2 - 3\sqrt{2})$

**38.** $\sqrt{5}(5 + 3\sqrt{3})$

**39.** $\sqrt{3}(\sqrt{2} + \sqrt{6})$

**40.** $\sqrt{2}(\sqrt{5} - 2\sqrt{10})$

**41.** $\sqrt[3]{4}(\sqrt[3]{3} - \sqrt[3]{6})$

**42.** $\sqrt[3]{6}(\sqrt[3]{2} + \sqrt[3]{12})$

**43.** $\sqrt{2x}(3 - \sqrt{10x})$

**44.** $\sqrt{5x}(6 + \sqrt{15x})$

**45.** $(3 + \sqrt{2})(4 + \sqrt{3})$

**46.** $(5 + \sqrt{5})(3 + \sqrt{6})$

**47.** $(6 + \sqrt{3})(2 - \sqrt{7})$

**48.** $(7 - \sqrt{3})(6 + \sqrt{5})$

**49.** $(4 - 2\sqrt{7})(3 + 3\sqrt{7})$

**50.** $(9 + 5\sqrt{10})(1 - 3\sqrt{10})$

**51.** $(\sqrt{2} + 3\sqrt{6})(\sqrt{3} - 2\sqrt{2})$

**52.** $(2\sqrt{3} + \sqrt{10})(\sqrt{5} - 2\sqrt{2})$

**53.** $(2\sqrt{5} + \sqrt{3})(4\sqrt{5} - 3\sqrt{3})$

**54.** $(\sqrt{6} - 2\sqrt{2})(2\sqrt{6} + 3\sqrt{2})$

**55.** $(1 + \sqrt{3})^2$

**56.** $(2 - \sqrt{3})^2$

**57.** $(\sqrt{2} - \sqrt{5})^2$

**58.** $(\sqrt{7} - \sqrt{3})^2$

**59.** $(\sqrt{x} - \sqrt{2})^2$

**60.** $(\sqrt{z} + \sqrt{5})^2$

**61.** $(\sqrt{2} - 1)(\sqrt{2} + 1)$

**62.** $(\sqrt{3} - 1)(\sqrt{3} + 1)$

**63.** $(3 - 2\sqrt{5})(3 + 2\sqrt{5})$

**64.** $(6 + 3\sqrt{2})(6 - 3\sqrt{2})$

**65.** $(\sqrt{2x} + \sqrt{3y})(\sqrt{2x} - \sqrt{3y})$

**66.** $(\sqrt{5a} + \sqrt{7b})(\sqrt{5a} - \sqrt{7b})$

**67.** $(\sqrt[3]{x} + 4)(\sqrt[3]{x} - 3)$

**68.** $(\sqrt[3]{y} - 6)(\sqrt[3]{y} + 3)$

**69.** $(\sqrt[3]{2a} - 5)(\sqrt[3]{2a} + 5)$

**70.** $(\sqrt[3]{4p} - 1)(\sqrt[3]{4p} + 3)$

## Mixed Practice

*In Problems 71–90, perform the indicated operation and simplify. Assume all variables are positive or zero.*

**71.** $\sqrt{5}(\sqrt{3} + \sqrt{10})$

**72.** $\sqrt{7}(\sqrt{14} + \sqrt{3})$

**73.** $\sqrt{28x^5} - x\sqrt{7x^3} + 5\sqrt{175x^5}$

**74.** $\sqrt{180a^5} + a^2\sqrt{20} - a\sqrt{80a^3}$

**75.** $(2\sqrt{3} + 5)(2\sqrt{3} - 5)$

**76.** $(4\sqrt{2} - 2)(4\sqrt{2} + 2)$

**77.** $\sqrt[3]{7}(2 + \sqrt[3]{4})$

**78.** $\sqrt[3]{9}(5 + 2\sqrt[3]{2})$

**79.** $(2\sqrt{2} + 5)(4\sqrt{2} - 4)$

**80.** $(5\sqrt{5} - 3)(3\sqrt{5} - 4)$

**81.** $4\sqrt{18} + 2\sqrt{32}$

**82.** $5\sqrt{20} + 2\sqrt{80}$

**83.** $(\sqrt{5} - \sqrt{3})^2$

**84.** $(\sqrt{2} - \sqrt{7})^2$

**85.** $3\sqrt[3]{5x^3y} + \sqrt[3]{40y}$

**86.** $5\sqrt[3]{3m^3n} + \sqrt[3]{81n}$

**87.** $(\sqrt{2x} - \sqrt{7y})(\sqrt{2x} + \sqrt{7y})$

**88.** $(\sqrt{3a} - \sqrt{4b})(\sqrt{3a} + \sqrt{4b})$

**89.** $-\frac{3}{5} \cdot \left(-\frac{\sqrt{5}}{5}\right) - \frac{4}{5} \cdot \left(-\frac{2\sqrt{5}}{5}\right)$

**90.** $\frac{4}{5} \cdot \left(-\frac{\sqrt{5}}{5}\right) + \left(-\frac{3}{5}\right) \cdot \left(-\frac{2\sqrt{5}}{5}\right)$

## Applying the Concepts

*In Problems 91 and 92, find the perimeter and area of the figures shown. Express your answer as a radical in simplified form.*

**91.**

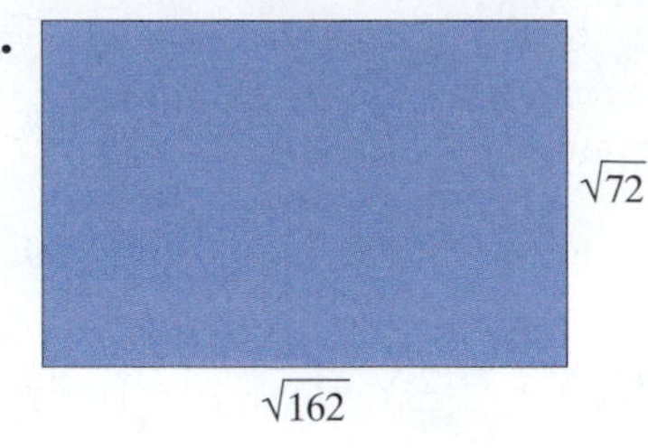


**92.**

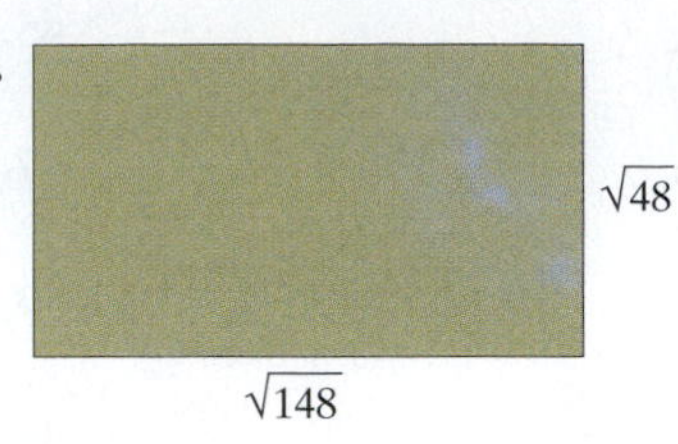


*Problems 93 and 94, use* ***Heron's Formula*** *for finding the area of a triangle whose sides are known. Heron's Formula states that the area A of a triangle with sides a, b, and c is*

$$A = \sqrt{s(s - a)(s - b)(s - c)}$$

*where*

$$s = \frac{1}{2}(a + b + c)$$

*Find the area of the shaded region by computing the difference in the areas of each triangle. That is, compute "area of larger triangle minus area of smaller triangle." Write your answer as a radical in simplified form.*

**93.**

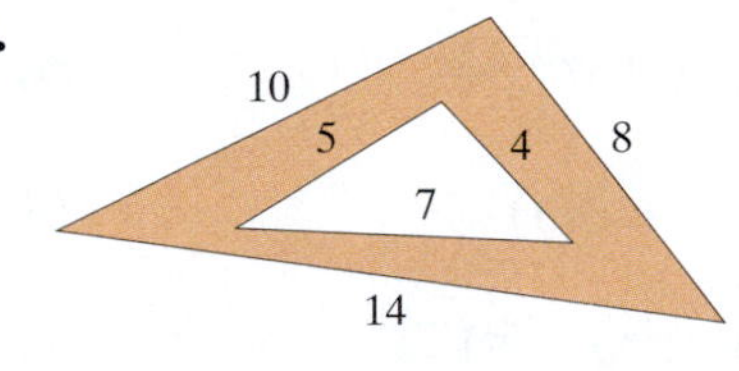


**94.**

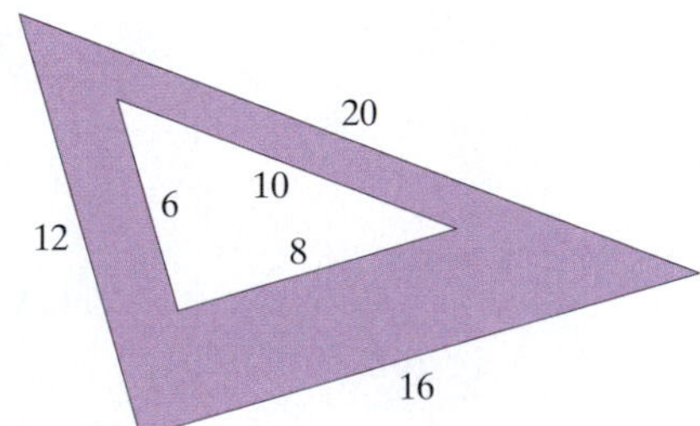


## Extending the Concepts

**95.** Suppose that $f(x) = \sqrt{3x}$ and $g(x) = \sqrt{12x}$; find
**(a)** $(f + g)(x)$ **(b)** $(f + g)(4)$ **(c)** $(f \cdot g)(x)$

**96.** Suppose that $f(x) = \sqrt{4x - 4}$ and $g(x) = \sqrt{25x - 25}$; find
**(a)** $(f + g)(x)$ **(b)** $(f + g)(10)$ **(c)** $(f \cdot g)(x)$

**97.** Show that $-2 + \sqrt{5}$ is a solution to the equation $x^2 + 4x - 1 = 0$. Show that $-2 - \sqrt{5}$ is also a solution.

## Synthesis Review

*In Problems 98–103, multiply each of the following.*

**98.** $(3a^3b)(4a^2b^4)$

**99.** $(3p - 1)(2p + 3)$

**100.** $(3y + 2)(2y - 1)$

**101.** $(m - 4)(m + 4)$

**102.** $(5w + 2)(5w - 2)$

**103.** $(\sqrt{x} + 2)(\sqrt{x} - 2)$

# 7.5 Rationalizing Radical Expressions

**OBJECTIVES**

1. Rationalize a Denominator Containing One Term
2. Rationalize a Denominator Containing Two Terms

## *Preparing for Rationalizing Radical Expressions*

*Before getting started, take the following readiness quiz. If you get a problem wrong, go back to the section cited and review the material.*

**1.** What would we need to multiply 12 by in order to make it the smallest perfect square that is a multiple of 12?

**2.** Simplify: $\sqrt{25x^2}$, $x > 0$ [Section 7.3, pp. 549–551]

**In Words**
We call the process "rationalizing the denominator" because we are making the denominator a rational number (no radicals).

When radical expressions appear in the denominator of a quotient, it is customary to rewrite the quotient so that the denominator does not contain any radicals. The process is referred to as **rationalizing the denominator.** In this section, we shall concentrate on rationalizing denominators that contain a single term and denominators that contain two terms.

## 1 Rationalize a Denominator Containing One Term

To rationalize a denominator containing a single square root, we multiply the numerator and denominator of the quotient by a square root so that the radicand in the denominator becomes a perfect square. For example, if the denominator of a quotient contains $\sqrt{5}$, we would multiply the numerator and denominator by $\sqrt{5}$ because $5 \cdot 5 = 25$, which is a perfect square. If the denominator of a quotient contains $\sqrt{8}$, we would multiply the numerator and denominator by $\sqrt{2}$ because $8 \cdot 2 = 16$, which is a perfect square.

### EXAMPLE 1 Rationalizing a Denominator Containing a Square Root

Rationalize the denominator of each expression:

**(a)** $\dfrac{1}{\sqrt{7}}$ **(b)** $\dfrac{\sqrt{5}}{\sqrt{12}}$ **(c)** $\dfrac{2}{3\sqrt{2x}}$

**Solution**

**(a)** We have $\sqrt{7}$ in the denominator. So, we ask ourselves, "What can I multiply by 7 and obtain a perfect square?" Because $7 \cdot 7 = 49$ and 49 is a perfect square, we multiply the numerator and denominator by $\sqrt{7}$.

$$\frac{1}{\sqrt{7}} = \frac{1}{\sqrt{7}} \cdot \frac{\sqrt{7}}{\sqrt{7}}$$

Multiply numerators; multiply denominators: $= \dfrac{\sqrt{7}}{\sqrt{49}}$

$\sqrt{49} = 7$: $= \dfrac{\sqrt{7}}{7}$

**Work Smart**
Remember that $a = 1 \cdot a$ and that 1 can take many forms. In Example 1(a), $1 = \dfrac{\sqrt{7}}{\sqrt{7}}$.

**(b)** We have $\sqrt{12}$ in the denominator. Again, we ask, "What can I multiply by 12 and obtain a perfect square?" Certainly, multiplying 12 by itself would result in a perfect square, but is this the best choice? No, you want to find the smallest integer that makes the radicand a perfect square. So, multiply 12 by 3 to obtain 36, a perfect square.

$$\frac{\sqrt{5}}{\sqrt{12}} = \frac{\sqrt{5}}{\sqrt{12}} \cdot \frac{\sqrt{3}}{\sqrt{3}}$$

Multiply numerators; multiply denominators: $= \dfrac{\sqrt{15}}{\sqrt{36}}$

$\sqrt{36} = 6$: $= \dfrac{\sqrt{15}}{6}$

**Work Smart**
An alternative approach to Example 1(b) would be to simplify $\sqrt{12}$ first, as follows:

$$\frac{\sqrt{5}}{\sqrt{12}} = \frac{\sqrt{5}}{\sqrt{4 \cdot 3}} = \frac{\sqrt{5}}{2\sqrt{3}} = \frac{\sqrt{5}}{2\sqrt{3}} \cdot \frac{\sqrt{3}}{\sqrt{3}} = \frac{\sqrt{15}}{6}$$

*Preparing for...Answers* **1.** 3 **2.** $5x$

**(c)** $$\frac{2}{3\sqrt{2x}} = \frac{2}{3\sqrt{2x}} \cdot \frac{\sqrt{2x}}{\sqrt{2x}}$$

Multiply numerators; multiply denominators: $$= \frac{2\sqrt{2x}}{3\sqrt{4x^2}}$$

$\sqrt{4x^2} = 2x$: $$= \frac{2\sqrt{2x}}{3 \cdot 2x}$$

Divide out common factor, 2: $$= \frac{\sqrt{2x}}{3x}$$

**QUICK ✓** *Rationalize each denominator.*

**1.** $\frac{1}{\sqrt{3}}$ **2.** $\frac{\sqrt{5}}{\sqrt{8}}$ **3.** $\frac{5}{\sqrt{10x}}$

---

In general, to rationalize a denominator when the denominator contains a single radical, we multiply the numerator and denominator of the quotient by a radical so that the product in the denominator has a radicand that is a perfect power of the index $n$. So, if the denominator contains a radical whose index is 3, we multiply the numerator and denominator by a cube root so that the radicand in the denominator becomes a perfect cube. For example, if the denominator of a quotient contains $\sqrt[3]{4}$, we would multiply the numerator and denominator of the quotient by $\sqrt[3]{2}$ because $4 \cdot 2 = 8$ and 8 is a perfect cube.

### EXAMPLE 2 Rationalizing a Denominator Containing Cube Roots and Fourth Roots

Rationalize the denominator of each expression:

**(a)** $\frac{1}{\sqrt[3]{6}}$ **(b)** $\sqrt[3]{\frac{5}{18}}$ **(c)** $\frac{6}{\sqrt[4]{4z^3}}$

Assume all variables represent positive real numbers.

**Work Smart**

The radicand in $\sqrt[3]{6}$ is 6. Think of 6 as $6^1$. To make it a perfect cube, $6^3$, we need to multiply by $6^2$ or 36. Then $6 \cdot 36 = 6^1 \cdot 6^2 = 6^3$. The cube root of $6^3$ or 216 is 6.

### Solution

**(a)** We have $\sqrt[3]{6}$ in the denominator. We want the radicand in the denominator to be a perfect cube (since the index is 3), so we multiply the numerator and denominator by $\sqrt[3]{6^2} = \sqrt[3]{36}$ since $6 \cdot 36 = 216$ and 216 is a perfect cube.

$$\frac{1}{\sqrt[3]{6}} = \frac{1}{\sqrt[3]{6}} \cdot \frac{\sqrt[3]{6^2}}{\sqrt[3]{6^2}}$$

Multiply numerators; multiply denominators: $$= \frac{\sqrt[3]{36}}{\sqrt[3]{6^3}}$$

$\sqrt[3]{6^3} = 6$: $$= \frac{\sqrt[3]{36}}{6}$$

**(b)** First, we use the Quotient Property $\left(\sqrt[n]{\frac{a}{b}} = \frac{\sqrt[n]{a}}{\sqrt[n]{b}}\right)$ to rewrite the radical as the quotient of two radicals.

$$\sqrt[3]{\frac{5}{18}} = \frac{\sqrt[3]{5}}{\sqrt[3]{18}}$$

What do we need to multiply $\sqrt[3]{18}$ by in order to make it a perfect cube? We will write 18 as $9 \cdot 2 = 3^2 \cdot 2^1$. Therefore, if we multiply $18 = 3^2 \cdot 2^1$ by $3^1 \cdot 2^2 = 12$, we will have a perfect cube as the radicand in the denominator.

$$\sqrt[3]{\frac{5}{18}} = \frac{\sqrt[3]{5}}{\sqrt[3]{18}} = \frac{\sqrt[3]{5}}{\sqrt[3]{3^2 \cdot 2}} \cdot \frac{\sqrt[3]{3 \cdot 2^2}}{\sqrt[3]{3 \cdot 2^2}}$$

Multiply numerators; multiply denominators:
$$= \frac{\sqrt[3]{60}}{\sqrt[3]{3^3 \cdot 2^3}}$$

$$= \frac{\sqrt[3]{60}}{6}$$

**(c)** We rewrite the denominator as $\sqrt[4]{2^2 \cdot z^3}$. To make the radicand a perfect power of 4, we need to multiply $\sqrt[4]{2^2 \cdot z^3}$ by $\sqrt[4]{2^2 \cdot z} = \sqrt[4]{4z}$ to obtain $\sqrt[4]{2^4 z^4}$ in the denominator.

$$\frac{6}{\sqrt[4]{4z^3}} = \frac{6}{\sqrt[4]{2^2 \cdot z^3}} \cdot \frac{\sqrt[4]{2^2 \cdot z}}{\sqrt[4]{2^2 \cdot z}}$$

Multiply numerators; multiply denominators:
$$= \frac{6\sqrt[4]{4z}}{\sqrt[4]{2^4 \cdot z^4}}$$

$$= \frac{6\sqrt[4]{4z}}{2z}$$

Simplify:
$$= \frac{3\sqrt[4]{4z}}{z}$$

**QUICK ✓** *Rationalize each denominator. Assume all variables are positive.*

**4.** $\frac{4}{\sqrt[3]{3}}$ **5.** $\sqrt[3]{\frac{3}{20}}$ **6.** $\frac{3}{\sqrt[4]{p}}$

## 2 Rationalize a Denominator Containing Two Terms

To rationalize a denominator containing two terms involving square roots, we use the fact that

$$(a + b)(a - b) = a^2 - b^2$$

and multiply both the numerator and denominator of the quotient by the conjugate of the denominator. For example, if the quotient is $\frac{3}{\sqrt{3} + 2}$, we would multiply both the numerator and the denominator by the conjugate of $\sqrt{3} + 2$, which is $\sqrt{3} - 2$. We know from the last section that the product $(\sqrt{3} + 2)(\sqrt{3} - 2)$ will not contain a radical.

### EXAMPLE 3 Rationalizing a Denominator Containing Two Terms

Rationalize the denominator: $\frac{\sqrt{2}}{\sqrt{6} + 2}$

### Solution

We see that $\sqrt{6} + 2$ is in the denominator of the quotient, so we multiply the numerator and denominator by the conjugate of $\sqrt{6} + 2$, $\sqrt{6} - 2$.

$$\frac{\sqrt{2}}{\sqrt{6}+2} = \frac{\sqrt{2}}{\sqrt{6}+2} \cdot \frac{\sqrt{6}-2}{\sqrt{6}-2}$$

Multiply the numerators and denominators:
$$= \frac{\sqrt{2}(\sqrt{6}-2)}{(\sqrt{6}+2)(\sqrt{6}-2)}$$

Distribute in the numerator; $(a+b)(a-b) = a^2 - b^2$ in the denominator:
$$= \frac{\sqrt{12} - 2\sqrt{2}}{(\sqrt{6})^2 - 2^2}$$

$\sqrt{12} = 2\sqrt{3}$:
$$= \frac{2\sqrt{3} - 2\sqrt{2}}{6 - 4}$$

Factor out common factor of 2 in numerator:
$$= \frac{2(\sqrt{3} - \sqrt{2})}{2}$$

Divide out the 2s:
$$= \sqrt{3} - \sqrt{2}$$

**QUICK ✓** *Rationalize the denominator.*

**7.** $\dfrac{4}{\sqrt{3}+1}$

**8.** $\dfrac{\sqrt{2}}{\sqrt{6}-\sqrt{2}}$

---

## EXAMPLE 4 Rationalizing a Denominator Containing Two Terms

Rationalize the denominator: $\dfrac{\sqrt{6}-3}{\sqrt{10}-\sqrt{6}}$

### Solution

We see that $\sqrt{10} - \sqrt{6}$ is in the denominator of the quotient, so we multiply the numerator and denominator by the conjugate of the denominator, $\sqrt{10} + \sqrt{6}$.

$$\frac{\sqrt{6}-3}{\sqrt{10}-\sqrt{6}} = \frac{\sqrt{6}-3}{\sqrt{10}-\sqrt{6}} \cdot \frac{\sqrt{10}+\sqrt{6}}{\sqrt{10}+\sqrt{6}}$$

Multiply the numerators and denominators:
$$= \frac{(\sqrt{6}-3)(\sqrt{10}+\sqrt{6})}{(\sqrt{10}-\sqrt{6})(\sqrt{10}+\sqrt{6})}$$

FOIL the numerator; $(a+b)(a-b) = a^2 - b^2$ in the denominator:
$$= \frac{\sqrt{60} + \sqrt{36} - 3\sqrt{10} - 3\sqrt{6}}{(\sqrt{10})^2 - (\sqrt{6})^2}$$

Simplify radicals:
$$= \frac{2\sqrt{15} + 6 - 3\sqrt{10} - 3\sqrt{6}}{10 - 6}$$

$$= \frac{2\sqrt{15} + 6 - 3\sqrt{10} - 3\sqrt{6}}{4}$$

**QUICK ✓** *Rationalize the denominator.*

**9.** $\dfrac{\sqrt{5}+4}{\sqrt{5}-\sqrt{2}}$

# 7.5 Exercises

**For Extra Help:**    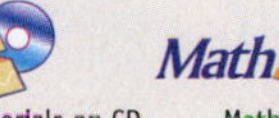 

Student Solutions Manual · CD Video · PH Math/Tutor Center · MathXL Tutorials on CD · MathXL® · MyMathLab

## Concepts and Vocabulary

*In Problems 1 and 2, fill in the blanks.*

1. Rewriting a quotient to remove radicals from the denominator is called _______ _______ _______.

2. To rationalize the denominator of $\frac{\sqrt{5}}{\sqrt{7}}$, we would multiply the numerator and denominator by _______.

*In Problem 3, answer True or False.*

3. To rationalize the denominator of $\frac{4-\sqrt{3}}{-2+\sqrt{7}}$, we would multiply the numerator and denominator by $2-\sqrt{7}$.

4. Explain why it is necessary to multiply the numerator and denominator by the conjugate of the denominator when rationalizing a denominator containing two terms.

## Building Skills

*In Problems 5–48, rationalize each denominator. Assume all variables are positive.*

5. $\frac{1}{\sqrt{2}}$
6. $\frac{2}{\sqrt{3}}$
7. $-\frac{6}{5\sqrt{3}}$
8. $-\frac{3}{2\sqrt{3}}$
9. $\frac{3}{\sqrt{12}}$
10. $\frac{5}{\sqrt{20}}$
11. $\frac{\sqrt{2}}{\sqrt{6}}$
12. $\frac{\sqrt{3}}{\sqrt{11}}$
13. $\sqrt{\frac{2}{p}}$
14. $\sqrt{\frac{5}{z}}$
15. $\frac{\sqrt{8}}{\sqrt{y^3}}$
16. $\frac{\sqrt{32}}{\sqrt{a^5}}$
17. $\frac{2}{\sqrt[3]{2}}$
18. $\frac{5}{\sqrt[3]{3}}$
19. $\sqrt[3]{\frac{7}{q}}$
20. $\sqrt[3]{\frac{-4}{p}}$
21. $\sqrt[3]{\frac{-3}{50}}$
22. $\sqrt[3]{\frac{-5}{72}}$
23. $\frac{2}{\sqrt[3]{20y}}$
24. $\frac{8}{\sqrt[3]{36z^2}}$
25. $\frac{-4}{\sqrt[4]{3x^3}}$
26. $\frac{6}{\sqrt[4]{9b^2}}$
27. $\frac{12}{\sqrt[5]{m^3n^2}}$
28. $\frac{-3}{\sqrt[5]{ab^3}}$
29. $\frac{4}{\sqrt{6}-2}$
30. $\frac{6}{\sqrt{7}-2}$
31. $\frac{5}{\sqrt{5}+2}$
32. $\frac{10}{\sqrt{10}+3}$
33. $\frac{8}{\sqrt{7}-\sqrt{3}}$
34. $\frac{12}{\sqrt{11}-\sqrt{7}}$
35. $\frac{\sqrt{2}}{\sqrt{10}-\sqrt{6}}$
36. $\frac{\sqrt{3}}{\sqrt{15}-\sqrt{6}}$
37. $\frac{\sqrt{p}}{\sqrt{p}+\sqrt{q}}$
38. $\frac{\sqrt{a}}{\sqrt{a}+\sqrt{b}}$
39. $\frac{18}{2\sqrt{3}+3\sqrt{2}}$
40. $\frac{15}{3\sqrt{5}+4\sqrt{3}}$
41. $\frac{\sqrt{7}+3}{\sqrt{7}-3}$
42. $\frac{\sqrt{5}+3}{\sqrt{5}-3}$
43. $\frac{\sqrt{3}-4\sqrt{2}}{2\sqrt{3}+5\sqrt{2}}$
44. $\frac{3\sqrt{6}+5\sqrt{7}}{2\sqrt{6}-3\sqrt{7}}$
45. $\frac{\sqrt{p}+2}{\sqrt{p}-2}$
46. $\frac{\sqrt{x}-4}{\sqrt{x}+4}$
47. $\frac{\sqrt{2}-3}{\sqrt{8}-\sqrt{2}}$
48. $\frac{2\sqrt{3}+3}{\sqrt{12}-\sqrt{3}}$

## Mixed Practice

*In Problems 49–56, perform the indicated operation and simplify.*

**49.** $\sqrt{3} + \dfrac{1}{\sqrt{3}}$

**50.** $\sqrt{5} - \dfrac{1}{\sqrt{5}}$

**51.** $\dfrac{\sqrt{10}}{2} - \dfrac{1}{\sqrt{2}}$

**52.** $\dfrac{\sqrt{5}}{2} + \dfrac{3}{\sqrt{5}}$

**53.** $\sqrt{\dfrac{1}{3}} + \sqrt{12} + \sqrt{75}$

**54.** $\sqrt{\dfrac{2}{5}} + \sqrt{20} - \sqrt{45}$

**55.** $\dfrac{3}{\sqrt{18}} - \sqrt{\dfrac{1}{2}}$

**56.** $\sqrt{\dfrac{4}{3}} + \dfrac{4}{\sqrt{48}}$

*In Problems 57–68, simplify each expression so that the denominator does not contain a radical. Work smart because in some of the problems, it will be easier if you divide the radicands before attempting to rationalize the denominator.*

**57.** $\dfrac{\sqrt{3}}{\sqrt{12}}$

**58.** $\dfrac{\sqrt{2}}{\sqrt{18}}$

**59.** $\dfrac{3}{\sqrt{72}}$

**60.** $\dfrac{7}{\sqrt{98}}$

**61.** $\sqrt{\dfrac{4}{3}}$

**62.** $\sqrt{\dfrac{9}{5}}$

**63.** $\dfrac{\sqrt{3} - 3}{\sqrt{3} + 3}$

**64.** $\dfrac{\sqrt{2} - 5}{\sqrt{2} + 5}$

**65.** $\dfrac{2}{\sqrt{5} + 2}$

**66.** $\dfrac{5}{\sqrt{6} + 4}$

**67.** $\dfrac{\sqrt{8}}{\sqrt{2}}$

**68.** $\dfrac{\sqrt{75}}{\sqrt{3}}$

## Applying the Concepts

*In Problems 69–74, find the reciprocal of the given number. Be sure to rationalize the denominator.*

**69.** $\sqrt{3}$

**70.** $\sqrt{7}$

**71.** $\sqrt[3]{12}$

**72.** $\sqrt[3]{18}$

**73.** $\sqrt{3} + 5$

**74.** $7 - \sqrt{2}$

*Problems 75 and 76 contain expressions that are seen in a course in Trigonometry. Simplify each expression completely.*

**75.** $\dfrac{1}{\sqrt{2}} \cdot \dfrac{\sqrt{3}}{2} - \dfrac{1}{\sqrt{2}} \cdot \dfrac{1}{2}$

**76.** $-\sqrt{\dfrac{2}{3}} \cdot \left(-\dfrac{2}{\sqrt{5}}\right) + \dfrac{1}{\sqrt{3}} \cdot \dfrac{1}{\sqrt{5}}$

*Sometimes we are asked to rationalize a numerator. In Problems 77–80, rationalize each expression by multiplying the numerator and denominator by the conjugate of the numerator.*

**77.** $\dfrac{\sqrt{2} + 1}{3}$

**78.** $\dfrac{\sqrt{3} + 2}{2}$

**79.** $\dfrac{\sqrt{x} - \sqrt{h}}{\sqrt{x}}$

**80.** $\dfrac{\sqrt{a} - \sqrt{b}}{\sqrt{2}}$

## Extending the Concepts

**81.** When two quantities $a$ and $b$ are positive, we can verify that $a = b$, by showing that $a^2 = b^2$. Verify that $\dfrac{\sqrt{6} + \sqrt{2}}{4} = \dfrac{\sqrt{2 + \sqrt{3}}}{2}$ by squaring each side.

**82.** Rationalize the denominator: $\dfrac{2}{\sqrt{2} + \sqrt{3} - \sqrt{9}}$

## Synthesis Review

*In Problems 83–86, graph each of the following functions using point plotting.*

**83.** $f(x) = 5x - 3$

**84.** $g(x) = -3x + 9$

**85.** $G(x) = x^2$

**86.** $F(x) = x^3$

# PUTTING THE CONCEPTS TOGETHER (SECTIONS 7.1–7.5)

*These problems cover important concepts from Sections 7.1–7.5. We designed these problems so that you can review the chapter so far and show your mastery of the concepts. Take time to work these problems before proceeding with the next section. The answers to these problems are located at the back of the text starting on page AN-33.*

**1.** Evaluate: $-25^{\frac{1}{2}}$ **2.** Evaluate: $(-64)^{-2/3}$

**3.** Write the expression $\sqrt[4]{3x^3}$ with a rational exponent.

**4.** Write the expression $7z^{4/5}$ as a radical expression.

**5.** Evaluate using rational exponents: $\sqrt[3]{\sqrt{64x^3}}$

**6.** Distribute and simplify: $c^{1/2}(c^{3/2} + c^{5/2})$

*In Problems 7–9, use Laws of Exponents to simplify each expression. Assume all variables in the radicand are greater than or equal to zero. Express answers with positive exponents.*

**7.** $(a^{2/3}b^{-1/3})(a^{4/3}b^{-5/3})$ **8.** $\dfrac{x^{3/4}}{x^{1/8}}$ **9.** $(x^{3/4}y^{-1/8})^8$

*In Problems 10–19, perform the indicated operation and simplify. Assume all variables in the radicand are greater than or equal to zero.*

**10.** $\sqrt{15a} \cdot \sqrt{2b}$ **11.** $\sqrt{10m^3n^2} \cdot \sqrt{20mn}$

**12.** $\sqrt[3]{\dfrac{-32xy^4}{4x^{-2}y}}$ **13.** $2\sqrt{108} - 3\sqrt{75} + \sqrt{48}$

**14.** $-5b\sqrt{8b} + 7\sqrt{18b^3}$ **15.** $\sqrt[3]{16y^4} - y\sqrt[3]{2y}$ **16.** $(3\sqrt{x})(4\sqrt{x})$

**17.** $3\sqrt{x} + 4\sqrt{x}$ **18.** $(2 - 3\sqrt{2})(10 + \sqrt{2})$ **19.** $(4\sqrt{2} - 3)^2$

*In Problems 20 and 21, rationalize the denominator.*

**20.** $\dfrac{3}{2\sqrt{32}}$ **21.** $\dfrac{4}{\sqrt{3} - 8}$

# 7.6 Functions Involving Radicals

**OBJECTIVES**

1. Evaluate Functions Whose Rule Is a Radical Expression
2. Find the Domain of a Function Whose Rule Contains a Radical
3. Graph Functions Involving Square Roots
4. Graph Functions Involving Cube Roots

### *Preparing for Functions Involving Radicals*

*Before getting started, take the following readiness quiz. If you get a problem wrong, go back to the section cited and review the material.*

**1.** Simplify: $\sqrt{121}$ [Getting Ready: Square Roots, pp. 527–528]

**2.** Simplify: $\sqrt{p^2}$ [Getting Ready: Square Roots, p. 530]

**3.** Given $f(x) = x^2 - 4$, find $f(3)$. [Section 2.3, pp. 162–164]

**4.** Solve: $-2x + 3 \geq 0$ [Section 1.4, pp. 93–96]

**5.** Graph $f(x) = x^2 + 1$ using point plotting. [Section 2.3, pp. 164–165]

***Preparing for...Answers*** **1.** 11 **2.** $|p|$ **3.** 5 **4.** $\left\{x \middle| x \leq \dfrac{3}{2}\right\}$ or $\left(-\infty, \dfrac{3}{2}\right]$

**5.**

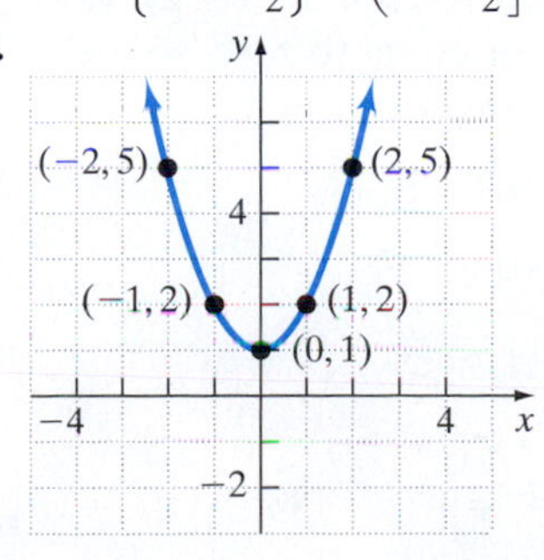


## 1 Evaluate Functions Whose Rule Is a Radical Expression

We can evaluate functions whose rule contains a radical by substituting the value of the independent variable into the rule, just as we did in Section 2.3.

**EXAMPLE 1 Evaluating Functions for Which the Rule Is a Radical Expression**

For the functions $f(x) = \sqrt{x + 2}$ and $g(x) = \sqrt[3]{3x + 1}$, find

**(a)** $f(7)$ **(b)** $f(10)$ **(c)** $g(-3)$

**Solution**

**(a)** $f(x) = \sqrt{x+2}$
$f(7) = \sqrt{7+2}$
$= \sqrt{9}$
$= 3$

**(b)** $f(x) = \sqrt{x+2}$
$f(10) = \sqrt{10+2}$
$= \sqrt{12}$
$= 2\sqrt{3}$

**(c)** $g(x) = \sqrt[3]{3x+1}$
$g(-3) = \sqrt[3]{3(-3)+1}$
$= \sqrt[3]{-8}$
$= -2$

**QUICK ✓** *Find the following values for each function.*

**1.** $f(x) = \sqrt{3x+7}$
**(a)** $f(3)$ **(b)** $f(7)$

**2.** $g(x) = \sqrt[3]{2x+7}$
**(a)** $g(-4)$ **(b)** $g(10)$

## 2 Find the Domain of a Function Whose Rule Contains a Radical

Recall the definition of the principal $n$th root of a number $a$, $\sqrt[n]{a}$. This definition states that $\sqrt[n]{a} = b$ means $a = b^n$. From this definition we learned that for $n \geq 2$ and even, the radicand, $a$, must be greater than or equal to 0. For $n \geq 3$ and odd, the radicand, $a$, can be any real number. This leads to a procedure for finding the domain of a function whose rule contains a radical.

**FINDING THE DOMAIN OF A FUNCTION FOR WHICH THE RULE CONTAINS A RADICAL**

- If the index on a radical is even, then the radicand must be greater than or equal to zero.
- If the index on a radical is odd, then the radicand can be any real number.

### EXAMPLE 2 Finding the Domain of a Radical Function

Find the domain of each of the following functions:

**(a)** $f(x) = \sqrt{x-5}$ **(b)** $G(x) = \sqrt[3]{2x+1}$ **(c)** $h(t) = \sqrt[4]{5-2t}$

**Solution**

**(a)** First, we take a look at the rule given in the function and interpret it. The function $f(x) = \sqrt{x-5}$ tells us to take the square root of $x - 5$. We can only take square roots of numbers greater than or equal to zero, so the radicand, $x - 5$, must be greater than or equal to zero. This requires that

$$x - 5 \geq 0$$

Add 5 to both sides: $x \geq 5$

The domain of $f$ is $\{x | x \geq 5\}$ or the interval $[5, \infty)$.

**(b)** The function $G(x) = \sqrt[3]{2x+1}$ tells us to take the cube root of $2x + 1$. We can take the cube root of any real number, so the domain of $G$ is any real number.

**(c)** The function $h(t) = \sqrt[4]{5-2t}$ tells us to take the fourth root of $5 - 2t$. We can only take fourth roots of numbers greater than or equal to zero, so the radicand, $5 - 2t$, must be greater than or equal to zero. This requires that

$$5 - 2t \geq 0$$

Subtract 5 from both sides: $-2t \geq -5$

Divide both sides by $-2$ (Don't forget to change the direction of the inequality!): $t \leq \frac{5}{2}$

The domain of $h$ is $\left\{t \middle| t \leq \frac{5}{2}\right\}$ or the interval $\left(-\infty, \frac{5}{2}\right]$.

**QUICK ✓** *Find the domain of each function.*

**3.** $H(x) = \sqrt{x+6}$ **4.** $g(t) = \sqrt[5]{3t-1}$ **5.** $F(m) = \sqrt[4]{6-3m}$

## 3 Graph Functions Involving Square Roots

The **square root function** is given by $f(x) = \sqrt{x}$. The domain of the square root function is $\{x|x \geq 0\}$ or using interval notation $[0, \infty)$. We can obtain the graph of $f(x) = \sqrt{x}$ by determining some ordered pairs $(x, y)$ such that $y = \sqrt{x}$. We then plot the ordered pairs in the $xy$-plane and connect the points. To make life easy, we choose values of $x$ that are perfect squares (0, 1, 4, 9, and so on). Table 1 shows some points on the graph of $f(x) = \sqrt{x}$. Figure 3 shows the graph of $f(x) = \sqrt{x}$. From the graph of $f(x) = \sqrt{x}$ given in Figure 3, we can see that the range of $f(x) = \sqrt{x}$ is $[0, \infty)$.

**Table 1**

| $x$ | $f(x) = \sqrt{x}$ | $(x, y)$ or $(x, f(x))$ |
|---|---|---|
| 0 | $f(0) = 0$ | (0, 0) |
| 1 | $f(1) = 1$ | (1, 1) |
| 4 | $f(4) = 2$ | (4, 2) |
| 9 | $f(9) = 3$ | (9, 3) |
| 16 | $f(16) = 4$ | (16, 4) |

**Figure 3**
$f(x) = \sqrt{x}$

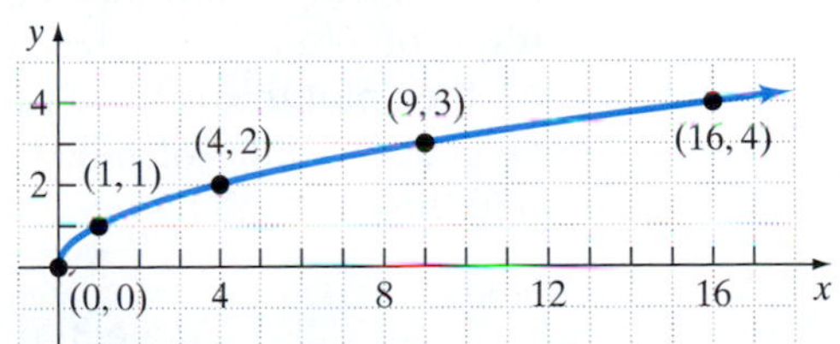


The point-plotting method can be used to graph a variety of functions involving square roots.

### EXAMPLE 3 Graphing a Function Involving a Square Root

For the function $f(x) = \sqrt{x-2}$,

**(a)** Find the domain.

**(b)** Graph the function using point plotting.

**(c)** Based on the graph, determine the range.

**Solution**

**(a)** The function $f(x) = \sqrt{x-2}$ tells us to take the square root of $x - 2$. We can only take square roots of numbers greater than or equal to zero, so the radicand, $x - 2$, must be greater than or equal to zero. This requires that

$$x - 2 \geq 0$$

Add 2 to both sides: $x \geq 2$

The domain of $f$ is $\{x|x \geq 2\}$ or the interval $[2, \infty)$.

**(b)** We choose values of $x$ that are greater than or equal 2. Again, to make life easy, we choose values of $x$ that will make the radicand a perfect square. See Table 2. Figure 4 shows the graph of $f(x) = \sqrt{x - 2}$.

**Table 2**

| $x$ | $f(x) = \sqrt{x-2}$ | $(x, y)$ or $(x, f(x))$ |
|---|---|---|
| 2 | $f(2) = 0$ | (2, 0) |
| 3 | $f(3) = 1$ | (3, 1) |
| 6 | $f(6) = 2$ | (6, 2) |
| 11 | $f(11) = 3$ | (11, 3) |
| 18 | $f(18) = 4$ | (18, 4) |

**Figure 4**

**(c)** From the graph of $f(x) = \sqrt{x - 2}$ given in Figure 4, we can see that the range of $f(x) = \sqrt{x - 2}$ is $[0, \infty)$.

**QUICK ✓**

**6.** For the function $f(x) = \sqrt{x + 3}$,

**(a)** Find the domain. **(b)** Graph the function using point plotting.

**(c)** Based on the graph, determine the range.

## 4 Graph Functions Involving Cube Roots

The **cube root function** is given by $f(x) = \sqrt[3]{x}$. The domain of the cube root function is $\{x | x \text{ is any real number}\}$ or using interval notation $(-\infty, \infty)$. We can obtain the graph of $f(x) = \sqrt[3]{x}$ by determining some ordered pairs $(x, y)$ such that $y = \sqrt[3]{x}$. We then plot the ordered pairs in the $xy$-plane and connect the points. To make life easy, we choose values of $x$ that are perfect cubes ($-8, -1, 0, 1, 8$, and so on). See Table 3. Figure 5 shows the graph of $f(x) = \sqrt[3]{x}$.

**Table 3**

| $x$ | $f(x) = \sqrt[3]{x}$ | $(x, y)$ or $(x, f(x))$ |
|---|---|---|
| −8 | $f(-8) = -2$ | (−8, −2) |
| −1 | $f(-1) = -1$ | (−1, −1) |
| 0 | $f(0) = 0$ | (0, 0) |
| 1 | $f(1) = 1$ | (1, 1) |
| 8 | $f(8) = 2$ | (8, 2) |

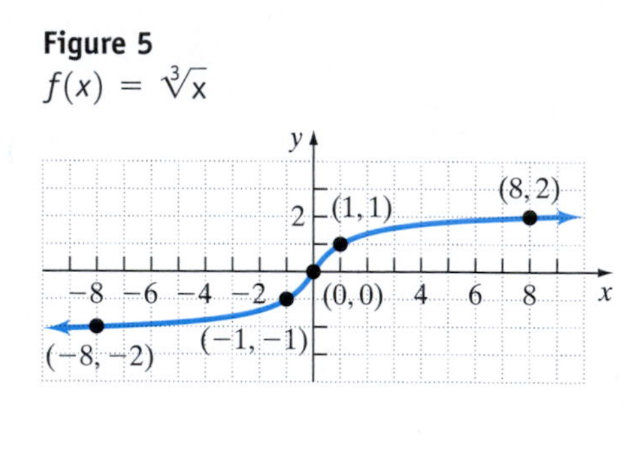


**Figure 5**
$f(x) = \sqrt[3]{x}$

From the graph of $f(x) = \sqrt[3]{x}$ given in Figure 5, we can see that the range of $f(x) = \sqrt[3]{x}$ is $(-\infty, \infty)$.

The point-plotting method can be used to graph a variety of functions involving cube roots.

### EXAMPLE 4 Graphing a Function Involving a Cube Root

For the function $g(x) = \sqrt[3]{x} + 2$,

**(a)** Find the domain. **(b)** Graph the function using point plotting.

**(c)** Based on the graph, determine the range.

### Solution

**(a)** The function $g(x) = \sqrt[3]{x} + 2$ tells us to take the cube root of $x$ and then add 2. We can take a cube root of any real number, so the domain of $g$ is $\{x | x \text{ is any real number}\}$ or the interval $(-\infty, \infty)$.

**(b)** We choose values of $x$ that make the radicand a perfect cube. See Table 4. Figure 6 shows the graph of $g(x) = \sqrt[3]{x} + 2$.

**Table 4**

| $x$ | $g(x) = \sqrt[3]{x} + 2$ | $(x, y)$ or $(x, g(x))$ |
|---|---|---|
| $-8$ | $g(-8) = \sqrt[3]{-8} + 2 = 0$ | $(-8, 0)$ |
| $-1$ | $g(-1) = 1$ | $(-1, 1)$ |
| $0$ | $g(0) = 2$ | $(0, 2)$ |
| $1$ | $g(1) = 3$ | $(1, 3)$ |
| $8$ | $g(8) = 4$ | $(8, 4)$ |

**Figure 6**

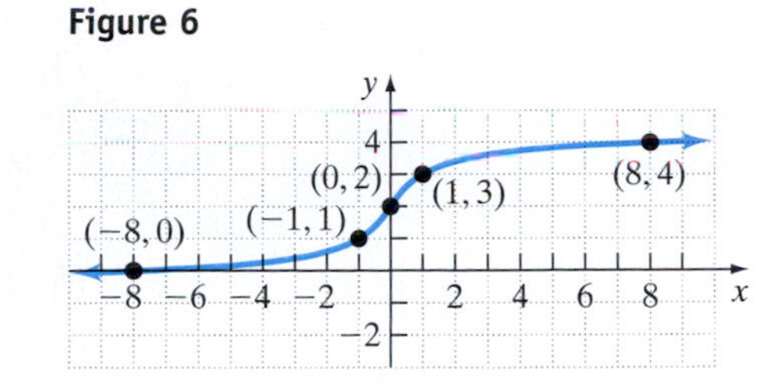


**(c)** From the graph of $g(x) = \sqrt[3]{x} + 2$ given in Figure 6 we can see that the range of $g(x) = \sqrt[3]{x} + 2$ is $(-\infty, \infty)$.

**QUICK ✓**

**7.** For the function $G(x) = \sqrt[3]{x} - 1$,

**(a)** Find the domain.

**(b)** Graph the function using point plotting.

**(c)** Based on the graph, determine the range.

# 7.6 Exercises

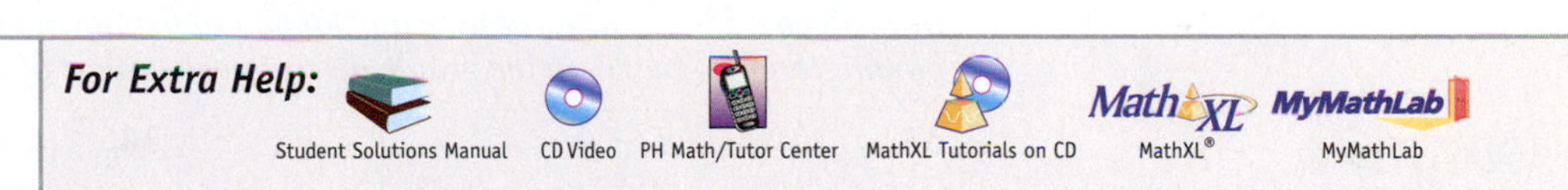


## Concepts and Vocabulary

*In Problems 1 and 2, fill in the blanks.*

**1.** If the index on a radical expression is _____, then the radicand must be greater than or equal to zero.

**2.** If the index on a radical expression is _____, then the radicand can be any real number.

*In Problem 3 and 4, answer True or False.*

**3.** The domain and range of $f(x) = \sqrt{x}$ is $[0, \infty)$.

**4.** The domain and range of $f(x) = \sqrt[3]{x}$ is $(-\infty, \infty)$.

## Building Skills

*In Problems 5–16, evaluate each radical function at the indicated values.*

**5.** $f(x) = \sqrt{x + 6}$
**(a)** $f(3)$ **(b)** $f(8)$ **(c)** $f(-2)$

**6.** $f(x) = \sqrt{x + 10}$
**(a)** $f(6)$ **(b)** $f(2)$ **(c)** $f(-6)$

**7.** $g(x) = -\sqrt{2x + 3}$
**(a)** $g(11)$ **(b)** $g(-1)$ **(c)** $g\left(\frac{1}{8}\right)$

**8.** $g(x) = -\sqrt{4x + 5}$
**(a)** $g(1)$ **(b)** $g(10)$ **(c)** $g\left(\frac{1}{8}\right)$

**9.** $G(m) = 2\sqrt{5m - 1}$

**(a)** $G(1)$ **(b)** $G(5)$ **(c)** $G\left(\frac{1}{2}\right)$

**10.** $G(p) = 3\sqrt{4p + 1}$

**(a)** $G(2)$ **(b)** $G(11)$ **(c)** $G\left(\frac{1}{8}\right)$

**11.** $H(z) = \sqrt[3]{z + 4}$

**(a)** $H(4)$ **(b)** $H(-12)$ **(c)** $H(-20)$

**12.** $G(t) = \sqrt[3]{t - 6}$

**(a)** $G(7)$ **(b)** $G(-21)$ **(c)** $G(22)$

**13.** $f(x) = \sqrt{\frac{x - 2}{x + 2}}$

**(a)** $f(7)$ **(b)** $f(6)$ **(c)** $f(10)$

**14.** $f(x) = \sqrt{\frac{x - 4}{x + 4}}$

**(a)** $f(5)$ **(b)** $f(8)$ **(c)** $f(12)$

**15.** $g(z) = \sqrt[3]{\frac{2z}{z - 4}}$

**(a)** $g(-4)$ **(b)** $g(8)$ **(c)** $g(12)$

**16.** $H(z) = \sqrt[3]{\frac{3z}{z + 5}}$

**(a)** $H(3)$ **(b)** $H(4)$ **(c)** $H(-1)$

*In Problems 17–32, find the domain of the radical function.*

**17.** $f(x) = \sqrt{x - 7}$

**18.** $f(x) = \sqrt{x + 4}$

**19.** $g(x) = \sqrt{2x + 7}$

**20.** $g(x) = \sqrt{3x + 7}$

**21.** $F(x) = \sqrt{4 - 3x}$

**22.** $G(x) = \sqrt{5 - 2x}$

**23.** $H(z) = \sqrt[3]{2z + 1}$

**24.** $G(z) = \sqrt[3]{5z - 3}$

**25.** $W(p) = \sqrt[4]{7p - 2}$

**26.** $C(y) = \sqrt[4]{3y - 2}$

**27.** $g(x) = \sqrt[5]{x - 3}$

**28.** $g(x) = \sqrt[5]{x + 9}$

**29.** $f(x) = \sqrt{\frac{3}{x + 5}}$

**30.** $f(x) = \sqrt{\frac{3}{x - 3}}$

**31.** $H(x) = \sqrt{\frac{x + 3}{x - 3}}$

**32.** $H(x) = \sqrt{\frac{x - 5}{x}}$

*In Problems 33–54, (a) determine the domain of the function; (b) graph the function using point plotting; and (c) based on the graph, determine the range of the function.*

**33.** $f(x) = \sqrt{x - 4}$

**34.** $f(x) = \sqrt{x - 1}$

**35.** $g(x) = \sqrt{x + 2}$

**36.** $g(x) = \sqrt{x + 5}$

**37.** $G(x) = \sqrt{2 - x}$

**38.** $F(x) = \sqrt{4 - x}$

**39.** $f(x) = \sqrt{x} + 3$

**40.** $f(x) = \sqrt{x} + 1$

**41.** $g(x) = \sqrt{x} - 4$

**42.** $g(x) = \sqrt{x} - 2$

**43.** $H(x) = 2\sqrt{x}$

**44.** $h(x) = 3\sqrt{x}$

**45.** $f(x) = \frac{1}{2}\sqrt{x}$

**46.** $g(x) = \frac{1}{4}\sqrt{x}$

**47.** $G(x) = -\sqrt{x}$

**48.** $F(x) = \sqrt{-x}$

**49.** $h(x) = \sqrt[3]{x + 2}$

**50.** $g(x) = \sqrt[3]{x - 4}$

**51.** $f(x) = \sqrt[3]{x} - 3$

**52.** $H(x) = \sqrt[3]{x} + 3$

**53.** $G(x) = 2\sqrt[3]{x}$

**54.** $F(x) = 3\sqrt[3]{x}$

## Applying the Concepts

**55. Distance to a Point on a Graph** Suppose that $P = (x, y)$ is a point on the graph of $y = x^2 - 4$. The distance from $P$ to $(0, 1)$ is given by the function

$$d(x) = \sqrt{x^4 - 9x^2 + 25}$$

See the figure.

**(a)** What is the distance from $P = (0, -4)$ to $(0, 1)$? That is, what is $d(0)$?

**(b)** What is the distance from $P = (1, -3)$ to $(0, 1)$? That is, what is $d(1)$?

**(c)** What is the distance from $P = (5, 21)$ to $(0, 1)$? That is, what is $d(5)$?

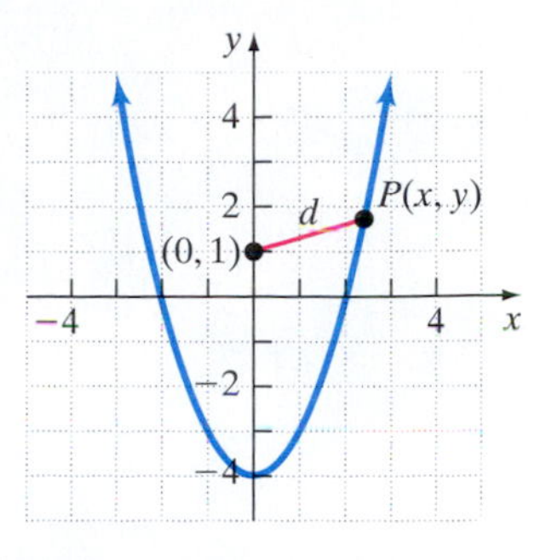


**56. Distance to a Point on a Graph** Suppose that $P = (x, y)$ is a point on the graph of $y = x^2 - 2$. The distance from $P$ to $(0, 2)$ is given by the function

$$d(x) = \sqrt{x^4 - 7x^2 + 16}$$

See the figure.

**(a)** What is the distance from $P = (0, -2)$ to $(0, 2)$? That is, what is $d(0)$?

**(b)** What is the distance from $P = (1, -1)$ to $(0, 2)$? That is, what is $d(1)$?

**(c)** What is the distance from $P = (4, 14)$ to $(0, 2)$? That is, what is $d(4)$?

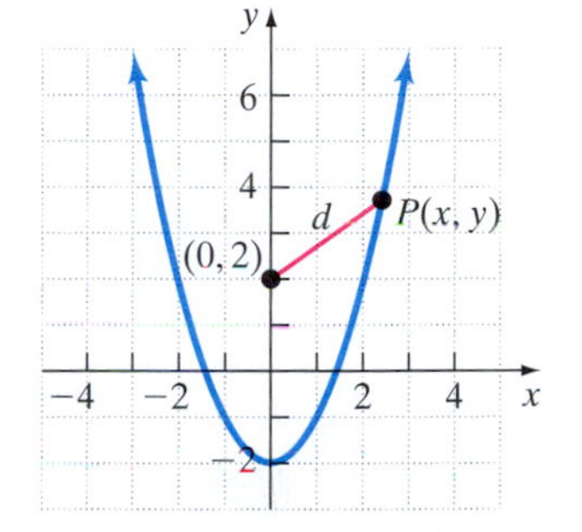


△ **57. Area** A rectangle is inscribed in a semicircle of radius 3 as shown in the figure. Let $P = (x, y)$ be the point in Quadrant I that is a vertex of the rectangle and is on the circle.

The area $A$ of the rectangle as a function of $x$ is given by

$$A(x) = 2x\sqrt{9 - x^2}$$

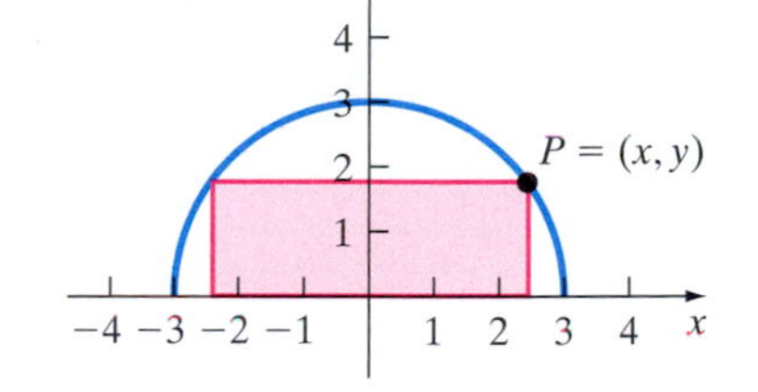


**(a)** What is the area of the rectangle whose vertex is at $\left(1, 2\sqrt{2}\right)$?

**(b)** What is the area of the rectangle whose vertex is at $\left(2, \sqrt{5}\right)$?

**(c)** What is the area of the rectangle whose vertex is at $\left(\sqrt{2}, \sqrt{7}\right)$?

△ **58. Area** A rectangle is inscribed in a semicircle of radius 4 as shown in the figure. Let $P = (x, y)$ be the point in Quadrant I that is a vertex of the rectangle and is on the circle.

The area $A$ of the rectangle as a function of $x$ is given by

$$A(x) = 2x\sqrt{16 - x^2}$$

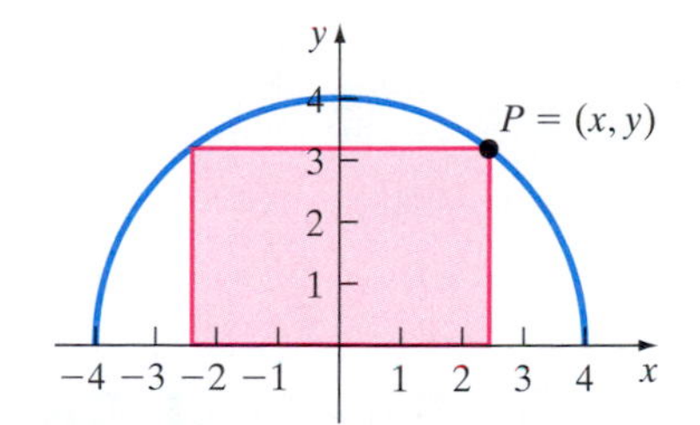


**(a)** What is the area of the rectangle whose vertex is at $\left(1, \sqrt{15}\right)$?

**(b)** What is the area of the rectangle whose vertex is at $\left(2, 2\sqrt{3}\right)$?

**(c)** What is the area of the rectangle whose vertex is at $\left(2\sqrt{2}, 2\sqrt{2}\right)$?

## Extending the Concepts

**59.** Use the results of Problems 33–36 to make a generalization about how to obtain the graph of $g(x) = \sqrt{x + c}$ from the graph of $f(x) = \sqrt{x}$.

**60.** Use the results of Problems 39–42 to make a generalization about how to obtain the graph of $g(x) = \sqrt{x} + c$ from the graph of $f(x) = \sqrt{x}$.

## Synthesis Review

*In Problems 61–65, add each of the following.*

**61.** $\frac{1}{3} + \frac{1}{2}$

**62.** $\frac{1}{5} + \frac{3}{4}$

**63.** $\frac{1}{x} + \frac{3}{x+1}$

**64.** $\frac{5}{x-3} + \frac{2}{x+1}$

**65.** $\frac{4}{x-1} + \frac{3}{x+1}$

**66.** Explain how adding rational numbers that do not have denominators with any common factors is similar to adding rational expressions that do not have denominators with any common factors.

## The Graphing Calculator

The graphing calculator can graph square root and cube root functions. The figure to the right shows the graph of $f(x) = \sqrt{x-2}$ using a TI-84 Plus graphing calculator. Note how the graph shown only exists for $x \geq 2$. This is useful in verifying the domain that we found algebraically.

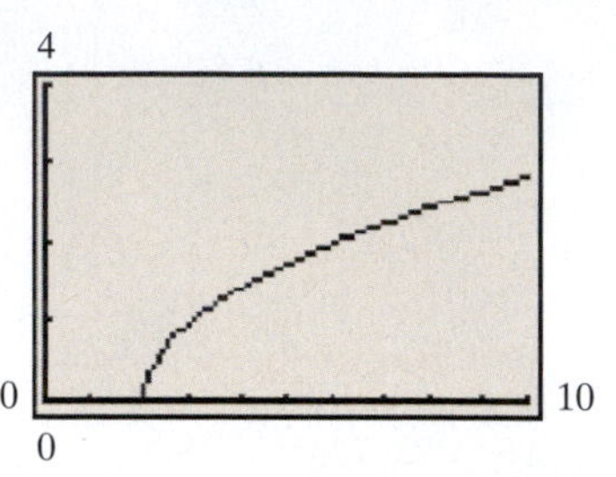


*In Problems 67–88, graph the function using a graphing calculator. Compare the graphs obtained on the calculator to the hand drawn graphs in Problems 33–54.*

**67.** $f(x) = \sqrt{x-4}$

**68.** $f(x) = \sqrt{x-1}$

**69.** $g(x) = \sqrt{x+2}$

**70.** $g(x) = \sqrt{x+5}$

**71.** $G(x) = \sqrt{2-x}$

**72.** $F(x) = \sqrt{4-x}$

**73.** $f(x) = \sqrt{x} + 3$

**74.** $f(x) = \sqrt{x} + 1$

**75.** $g(x) = \sqrt{x} - 4$

**76.** $g(x) = \sqrt{x} - 2$

**77.** $H(x) = 2\sqrt{x}$

**78.** $h(x) = 3\sqrt{x}$

**79.** $f(x) = \frac{1}{2}\sqrt{x}$

**80.** $g(x) = \frac{1}{4}\sqrt{x}$

**81.** $G(x) = -\sqrt{x}$

**82.** $F(x) = \sqrt{-x}$

**83.** $h(x) = \sqrt[3]{x+2}$

**84.** $g(x) = \sqrt[3]{x-4}$

**85.** $f(x) = \sqrt[3]{x} - 3$

**86.** $H(x) = \sqrt[3]{x} + 3$

**87.** $G(x) = 2\sqrt[3]{x}$

**88.** $F(x) = 3\sqrt[3]{x}$

# 7.7 Radical Equations and Their Applications

### OBJECTIVES

1. Solve Radical Equations Containing One Radical
2. Solve Radical Equations Containing Two Radicals
3. Solve for a Variable in a Radical Equation

### *Preparing for Radical Equations and Their Applications*

*Before getting started, take the following readiness quiz. If you get a problem wrong, go back to the section cited and review the material.*

**1.** Solve: $3x - 5 = 0$ [Section 1.1, pp. 53–56]

**2.** Solve: $2p^2 + 4p - 6 = 0$ [Section 5.7, pp. 423–426]

**3.** Simplify: $\left(\sqrt[3]{x-5}\right)^3$ [Section 7.1, pp. 535–536]

***Preparing for...Answers***

**1.** $\left\{\frac{5}{3}\right\}$ **2.** $\{-3, 1\}$ **3.** $x - 5$

When the variable in an equation occurs in a radicand, the equation is called a **radical equation.** Examples of radical equations are

$$\sqrt{3x+1} = 5 \qquad \sqrt[3]{x-5} - 5 = 12 \qquad \sqrt{x-2} - \sqrt{2x+5} = 2$$

In this section, we are going to solve radical equations involving one or two radicals. We start with radical equations containing one radical.

## 1 Solve Radical Equations Containing One Radical

Example 1 illustrates how to solve a radical equation.

### EXAMPLE 1 How to Solve a Radical Equation Containing One Radical

Solve: $\sqrt{2x-3} - 5 = 0$

**Step-by-Step Solution**

**Step 1:** Isolate the radical.

$$\sqrt{2x-3} - 5 = 0$$

Add 5 to both sides: $\sqrt{2x-3} = 5$

**Step 2:** Raise both sides to the power of the index.

The index is 2, so we square both sides: $\left(\sqrt{2x-3}\right)^2 = 5^2$

$$2x - 3 = 25$$

**Step 3:** Solve the equation that results.

Add 3 to both sides: $2x - 3 + 3 = 25 + 3$

$$2x = 28$$

Divide both sides by 2: $\dfrac{2x}{2} = \dfrac{28}{2}$

$$x = 14$$

**Step 4:** Check

$$\sqrt{2x-3} - 5 = 0$$

Let $x = 14$ in the original equation: $\sqrt{2 \cdot 14 - 3} - 5 \stackrel{?}{=} 0$

$$\sqrt{28-3} - 5 \stackrel{?}{=} 0$$

$$5 - 5 = 0 \quad \text{True}$$

The solution set is $\{14\}$.

Below we summarize the steps to solve a radical equation that contains a single radical.

### Solving a Radical Equation Containing One Radical

**Step 1:** Isolate the radical. That is, get the radical by itself on one side of the equation and everything else on the other side.

**Step 2:** Raise both sides of the equation to the power of the index. This will eliminate the radical from the equation.

**Step 3:** Solve the equation that results.

**Step 4:** Check your answer. When solving radical equations containing an even index, apparent solutions that are not solutions to the original equation creep in. These solutions are called **extraneous solutions.**

**QUICK ✓**

**1.** Solve: $\sqrt{3x+1} - 4 = 0$

## EXAMPLE 2 Solving a Radical Equation Containing One Radical

Solve:

**(a)** $\sqrt{4x + 1} - 2 = 1$ **(b)** $\sqrt{5x - 1} + 7 = 5$

### Solution

**(a)**

$$\sqrt{4x + 1} - 2 = 1$$

Add 2 to both sides: $\sqrt{4x + 1} - 2 + 2 = 1 + 2$

$$\sqrt{4x + 1} = 3$$

The index is 2, so we square both sides: $\left(\sqrt{4x + 1}\right)^2 = 3^2$

$$4x + 1 = 9$$

Subtract 1 from both sides: $4x + 1 - 1 = 9 - 1$

$$4x = 8$$

Divide both sides by 4: $\dfrac{4x}{4} = \dfrac{8}{4}$

$$x = 2$$

**Check**

$$\sqrt{4x + 1} - 2 = 1$$

Let $x = 2$ in the original equation: $\sqrt{4 \cdot 2 + 1} - 2 \stackrel{?}{=} 1$

$$\sqrt{8 + 1} - 2 \stackrel{?}{=} 1$$

$$\sqrt{9} - 2 \stackrel{?}{=} 1$$

$$3 - 2 \stackrel{?}{=} 1$$

$$1 = 1 \quad \text{True}$$

The solution set is $\{2\}$.

**(b)**

$$\sqrt{5x - 1} + 7 = 5$$

Subtract 7 from both sides: $\sqrt{5x - 1} + 7 - 7 = 5 - 7$

$$\sqrt{5x - 1} = -2$$

The equation has no real solution because the principal square root of a number cannot be less than 0. Put another way, there is no real number whose square root is $-2$, so the equation has no real solution. The solution set is $\varnothing$ or $\{\ \}$.

### QUICK ✓

**2.** Solve: $\sqrt{6x + 4} - 5 = 3$ **3.** Solve: $\sqrt{2x + 3} + 8 = 6$

## EXAMPLE 3 Solving a Radical Equation Containing One Radical

Solve: $\sqrt{x + 5} = x - 1$

### Solution

$$\sqrt{x + 5} = x - 1$$

The index is 2, so we square both sides: $\left(\sqrt{x + 5}\right)^2 = (x - 1)^2$

$$x + 5 = x^2 - 2x + 1$$

Subtract $x$ and 5 from both sides: $x + 5 - x - 5 = x^2 - 2x + 1 - x - 5$

$$0 = x^2 - 3x - 4$$

Factor: $0 = (x - 4)(x + 1)$

Zero-Product Property: $x - 4 = 0$ or $x + 1 = 0$

$$x = 4 \quad \text{or} \quad x = -1$$

**Work Smart**

$(x - 1)^2 = x^2 - 2x + 1$

Do not write

$(x - 1)^2 = x^2 + 1$

**Check** $\sqrt{x+5} = x - 1$

$x = 4$: $\sqrt{4+5} \stackrel{?}{=} 4 - 1$; $\sqrt{9} \stackrel{?}{=} 3$; $3 = 3$ True

$x = -1$: $\sqrt{-1+5} \stackrel{?}{=} -1 - 1$; $\sqrt{4} \stackrel{?}{=} -2$; $2 = -2$ False

The apparent solution $x = -1$ does not check, so it is an extraneous solution. The solution set is $\{4\}$.

**QUICK ✓**

**4.** Solve: $\sqrt{2x+1} = x - 1$

## EXAMPLE 4 Solving a Radical Equation Containing One Radical

Solve: $\sqrt[3]{3x+2} - 1 = 1$

### Solution

$$\sqrt[3]{3x+2} - 1 = 1$$

Add 1 to both sides: $\sqrt[3]{3x+2} - 1 + 1 = 1 + 1$

$$\sqrt[3]{3x+2} = 2$$

The index is 3, so we cube both sides: $\left(\sqrt[3]{3x+2}\right)^3 = 2^3$

$$3x + 2 = 8$$

Subtract 2 from both sides: $3x + 2 - 2 = 8 - 2$

$$3x = 6$$

Divide both sides by 3: $\dfrac{3x}{3} = \dfrac{6}{3}$

$$x = 2$$

**Check** $\sqrt[3]{3x+2} - 1 = 1$

Let $x = 2$ in the original equation: $\sqrt[3]{3 \cdot 2 + 2} - 1 \stackrel{?}{=} 1$

$$\sqrt[3]{6+2} - 1 \stackrel{?}{=} 1$$

$$\sqrt[3]{8} - 1 \stackrel{?}{=} 1$$

$$2 - 1 \stackrel{?}{=} 1$$

$$1 = 1 \quad \text{True}$$

The solution set is $\{2\}$.

**QUICK ✓**

**5.** Solve: $\sqrt[3]{6x+3} - 4 = -1$

Sometimes, rather than an equation containing radicals, it will contain rational exponents. When solving these problems, we can rewrite the equation with a radical or we use the fact that $(a^r)^s = a^{r \cdot s}$.

## EXAMPLE 5 Solving an Equation Containing a Rational Exponent

Solve: $(5x - 1)^{\frac{1}{2}} + 3 = 10$

**Solution**

$$(5x-1)^{\frac{1}{2}}+3=10$$

Subtract 3 from both sides: $(5x-1)^{\frac{1}{2}}=7$

Square both sides: $\left((5x-1)^{\frac{1}{2}}\right)^2=7^2$

Use $(a^r)^s=a^{r\cdot s}$: $(5x-1)^{\frac{1}{2}\cdot 2}=49$

$5x-1=49$

Add 1 to both sides: $5x=50$

Divide both sides by 5: $x=10$

**Check** $(5x-1)^{\frac{1}{2}}+3=10$

Let $x=10$ in the original equation: $(5\cdot 10-1)^{\frac{1}{2}}+3\stackrel{?}{=}10$

$(49)^{\frac{1}{2}}+3\stackrel{?}{=}10$

$7+3=10$

$10=10$ True

The solution set is $\{10\}$.

**QUICK ✓**

**6.** Solve: $(2x-3)^{\frac{1}{3}}-7=-4$

## 2 Solve Radical Equations Containing Two Radicals

Example 6 illustrates how to solve a radical equation containing two radicals.

### EXAMPLE 6 How to Solve a Radical Equation Containing Two Radicals

Solve: $\sqrt[3]{p^2-4p-4}=\sqrt[3]{-3p+2}$

**Step-by-Step-Solution**

**Step 1:** Isolate one of the radicals.

The radical on the left side of the equation is isolated: $\sqrt[3]{p^2-4p-4}=\sqrt[3]{-3p+2}$

**Step 2:** Raise both sides to the power of the index.

The index is 3, so we cube both sides: $\left(\sqrt[3]{p^2-4p-4}\right)^3=\left(\sqrt[3]{-3p+2}\right)^3$

$p^2-4p-4=-3p+2$

**Step 3:** Because there is no radical, we solve the equation that results.

Add $3p$ to both sides; subtract 2 from both sides: $p^2-4p-4+3p-2=-3p+2+3p-2$

Combine like terms: $p^2-p-6=0$

Factor: $(p-3)(p+2)=0$

Zero-Product Property: $p-3=0$ or $p+2=0$

$p=3$ or $p=-2$

**Step 4:** Check

$$\sqrt[3]{p^2-4p-4}=\sqrt[3]{-3p+2}$$

$p=-2$:

$\sqrt[3]{(-2)^2-4(-2)-4}\stackrel{?}{=}\sqrt[3]{-3(-2)+2}$

$\sqrt[3]{4+8-4}\stackrel{?}{=}\sqrt[3]{6+2}$

$\sqrt[3]{8}=\sqrt[3]{8}$ True

$p=3$:

$\sqrt[3]{(3)^2-4(3)-4}\stackrel{?}{=}\sqrt[3]{-3(3)+2}$

$\sqrt[3]{9-12-4}\stackrel{?}{=}\sqrt[3]{-9+2}$

$\sqrt[3]{-7}=\sqrt[3]{-7}$ True

Both apparent solutions check. The solution set is $\{-2, 3\}$.

When a radical equation contains two radicals, the following steps should be used to solve the equation for the variable.

### Solving a Radical Equation Containing Two Radicals

**Step 1:** Isolate one of the radicals. That is, get one of the radicals by itself on one side of the equation and everything else on the other side.

**Step 2:** Raise both sides of the equation to the power of the index. This will eliminate one radical or both radicals from the equation.

**Step 3:** If a radical remains in the equation, then follow the steps for solving a radical equation containing one radical. Otherwise, solve the equation that results.

**Step 4:** Check your answer. When solving radical equations, apparent solutions that, in fact, are not solutions to the original equation may creep in. Remember, these solutions are called extraneous solutions.

**QUICK ✓**

**7.** Solve: $\sqrt[3]{m^2 + 4m + 4} = \sqrt[3]{2m + 7}$

## EXAMPLE 7 Solving a Radical Equation Containing Two Radicals

Solve: $\sqrt{3x + 6} - \sqrt{x + 6} = 2$

### Solution

**Work Smart**
When there is more than one radical, it is best to isolate the radical with the more complicated radicand.

$$\sqrt{3x+6} - \sqrt{x+6} = 2$$

Add $\sqrt{x + 6}$ to both sides: $\sqrt{3x+6} = 2 + \sqrt{x+6}$

Square both sides: $\left(\sqrt{3x+6}\right)^2 = \left(2 + \sqrt{x+6}\right)^2$

Use $(a + b)^2 = a^2 + 2ab + b^2$: $3x + 6 = 4 + 4\sqrt{x+6} + \left(\sqrt{x+6}\right)^2$

$\left(\sqrt{x+6}\right)^2 = x + 6$: $3x + 6 = 4 + 4\sqrt{x+6} + x + 6$

Isolate the radical: $2x - 4 = 4\sqrt{x+6}$

Factor out 2: $2(x - 2) = 4\sqrt{x+6}$

Divide both sides by 2: $x - 2 = 2\sqrt{x+6}$

Square both sides: $(x-2)^2 = \left(2\sqrt{x+6}\right)^2$

$$x^2 - 4x + 4 = 4(x + 6)$$

Distribute: $x^2 - 4x + 4 = 4x + 24$

Subtract $4x$ and 24 from both sides: $x^2 - 8x - 20 = 0$

Factor: $(x - 10)(x + 2) = 0$

Zero-Product Property: $x - 10 = 0$ or $x + 2 = 0$

$$x = 10 \quad \text{or} \quad x = -2$$

**Check** $x = -2$:

$$\sqrt{3\cdot(-2) + 6} - \sqrt{-2 + 6} \stackrel{?}{=} 2$$
$$\sqrt{0} - \sqrt{4} \stackrel{?}{=} 2$$
$$0 - 2 \stackrel{?}{=} 2$$
$$-2 = 2 \quad \text{False}$$

$x = 10$:

$$\sqrt{3\cdot 10 + 6} - \sqrt{10 + 6} \stackrel{?}{=} 2$$
$$\sqrt{36} - \sqrt{16} \stackrel{?}{=} 2$$
$$6 - 4 \stackrel{?}{=} 2$$
$$2 = 2 \quad \text{True}$$

The apparent solution $x = -2$ does not check, so it is an extraneous solution. The solution set is $\{10\}$.

**QUICK ✓**

**8.** Solve: $\sqrt{2x+1} - \sqrt{x+4} = 1$

## 3 Solve for a Variable in a Radical Equation

In many situations, you will be required to solve for a variable in a formula. For instance, in Example 8, we are assessing how much error there is in estimates based upon a statistical study. This commonly used formula from statistics contains a radical.

### EXAMPLE 8 Solving for a Variable

A formula from statistics for finding the margin of error in estimating a population mean is given by

$$E = z\cdot\frac{\sigma}{\sqrt{n}}$$

**(a)** Solve this equation for $n$.

**(b)** Find $n$ when $\sigma = 12$, $z = 2$, and $E = 3$.

**Solution**

**(a)**
$$E = z\cdot\frac{\sigma}{\sqrt{n}}$$

Multiply both sides by $\sqrt{n}$: $\sqrt{n}\cdot E = z\sigma$

Divide both sides by $E$: $\sqrt{n} = \frac{z\sigma}{E}$

Square both sides: $n = \left(\frac{z\sigma}{E}\right)^2$

**In Words**
The symbol $\sigma$ is pronounced "sigma."

**(b)** $n = \left(\frac{z\sigma}{E}\right)^2 = \left(\frac{2\cdot 12}{3}\right)^2$
$= 64$

**QUICK ✓**

**9.** The period of a pendulum is the time it takes to complete one trip back and forth. The period $T$, in seconds, of a pendulum of length $L$, in feet, may be approximated using the formula $T = 2\pi\sqrt{\frac{L}{32}}$.

**(a)** Solve the equation for $L$.

**(b)** Determine the length of a pendulum whose period is $2\pi$ seconds.

# 7.7 Exercises

***For Extra Help:***

Student Solutions Manual

CD Video

PH Math/Tutor Center

MathXL Tutorials on CD

MathXL®

MyMathLab

## Concepts and Vocabulary

*In Problems 1 and 2, fill in the blanks.*

**1.** When the variable in an equation occurs in a radical, the equation is called a _______ _______.

**2.** When an apparent solution is not a solution of the original equation, we say the apparent solution is an _______ solution.

*In Problems 3 and 4, answer True or False to each statement.*

**3.** The first step in solving $x + \sqrt{x - 3} = 5$ is to square both sides of the equation.

**4.** When solving radical equations, extraneous solutions only occur when the index on the radical is even.

**5.** Why is it always necessary to check apparent solutions when solving radical equations?

**6.** How can you tell by inspection that the equation $\sqrt{x - 2} + 5 = 0$ will have no real solution?

## Building Skills

*In Problems 7–38, solve each equation.*

**7.** $\sqrt{x} = 4$ **8.** $\sqrt{p} = 6$ **9.** $\sqrt{x - 3} = 2$

**10.** $\sqrt{y - 5} = 3$ **11.** $\sqrt{2t + 3} = 5$ **12.** $\sqrt{3w - 2} = 4$

**13.** $\sqrt{4x + 3} = -2$ **14.** $\sqrt{6p - 5} = -5$ **15.** $\sqrt[3]{4t} = 2$

**16.** $\sqrt[3]{9w} = 3$ **17.** $\sqrt[3]{5q + 4} = 4$ **18.** $\sqrt[3]{7m + 20} = 5$

**19.** $\sqrt{y} + 3 = 8$ **20.** $\sqrt{q} - 5 = 2$ **21.** $\sqrt{x + 5} - 3 = 1$

**22.** $\sqrt{x - 4} + 4 = 7$ **23.** $\sqrt{2x - 1} + 5 = 8$ **24.** $\sqrt{4x + 1} - 2 = 3$

**25.** $3\sqrt{x} + 5 = 8$ **26.** $4\sqrt{t} - 2 = 10$ **27.** $\sqrt{4 - x} - 3 = 0$

**28.** $\sqrt{6 - w} - 3 = 1$ **29.** $\sqrt{p} = 2p$ **30.** $\sqrt{q} = 3q$

**31.** $\sqrt{x + 6} = x$ **32.** $\sqrt{2p + 8} = p$ **33.** $\sqrt{w} = w - 6$

**34.** $\sqrt{m} = m - 12$ **35.** $\sqrt{17 - 2x} + 1 = x$ **36.** $\sqrt{1 - 4x} - 5 = x$

**37.** $\sqrt{w^2 - 11} + 5 = w + 4$ **38.** $\sqrt{z^2 - z - 7} + 3 = z + 2$

*In Problems 39–52, solve each equation.*

**39.** $\sqrt{x + 9} = \sqrt{2x + 5}$ **40.** $\sqrt{3x + 1} = \sqrt{2x + 7}$

**41.** $\sqrt[3]{4x - 3} = \sqrt[3]{2x - 9}$ **42.** $\sqrt[3]{3y - 2} = \sqrt[3]{5y + 8}$

**43.** $\sqrt{2w^2 - 3w - 4} = \sqrt{w^2 + 6w + 6}$ **44.** $\sqrt{2x^2 + 7x - 10} = \sqrt{x^2 + 4x + 8}$

**45.** $\sqrt{3w + 4} = 2 + \sqrt{w}$ **46.** $\sqrt{3y - 2} = 2 + \sqrt{y}$

**47.** $\sqrt{x + 1} - \sqrt{x - 2} = 1$ **48.** $\sqrt{2x - 1} - \sqrt{x - 1} = 1$

**49.** $\sqrt{2x + 6} - \sqrt{x - 1} = 2$ **50.** $\sqrt{2x + 6} - \sqrt{x - 6} = 3$

**51.** $\sqrt{2x + 5} - \sqrt{x - 1} = 2$ **52.** $\sqrt{4x + 1} - \sqrt{2x + 1} = 2$

*In Problems 53–58, solve each equation.*

**53.** $(2x + 3)^{\frac{1}{2}} = 3$ **54.** $(4x + 1)^{\frac{1}{2}} = 5$

**55.** $(6x - 1)^{\frac{1}{4}} = (2x + 15)^{\frac{1}{4}}$ **56.** $(6p + 3)^{\frac{1}{5}} = (4p - 9)^{\frac{1}{5}}$

**57.** $(x + 3)^{\frac{1}{2}} - (x - 5)^{\frac{1}{2}} = 2$ **58.** $(3x + 1)^{\frac{1}{2}} - (x - 1)^{\frac{1}{2}} = 2$

**59. Finance** Solve $A = P\sqrt{1 + r}$ for $r$.

**60. Centripetal Acceleration** Solve $v = \sqrt{ar}$ for $a$.

**61. Volume of a Sphere** Solve $r = \sqrt[3]{\frac{3V}{4\pi}}$ for $V$.

**62. Surface Area of a Sphere** Solve $r = \sqrt{\frac{S}{4\pi}}$ for $S$.

**63. Coulomb's Law** Solve $r = \sqrt{\frac{4F\pi\varepsilon_0}{q_1q_2}}$ for $F$.

**64. Potential Energy** Solve $V = \sqrt{\frac{2U}{C}}$ for $U$.

## Mixed Practice

*In Problems 65–80, solve each equation.*

**65.** $\sqrt{5p-3}+7=3$

**66.** $\sqrt{3b-2}+8=5$

**67.** $\sqrt{x+12}=x$

**68.** $\sqrt{x+20}=x$

**69.** $\sqrt{2p+12}=4$

**70.** $\sqrt{3a-5}=2$

**71.** $\sqrt[4]{x+7}=2$

**72.** $\sqrt[5]{x+23}=2$

**73.** $(3x+1)^{\frac{1}{3}}+2=0$

**74.** $(5x-2)^{\frac{1}{3}}+3=0$

**75.** $\sqrt{x}+5=7$

**76.** $\sqrt{a}-5=-2$

**77.** $\sqrt{2x+5}=\sqrt{3x-4}$

**78.** $\sqrt{4c-5}=\sqrt{3c+1}$

**79.** $\sqrt{x-1}+\sqrt{x+4}=5$

**80.** $\sqrt{x-3}+\sqrt{x+4}=7$

## Applying the Concepts

**81.** Suppose that $f(x)=\sqrt{x-2}$.

**(a)** Solve $f(x)=0$. What point is on the graph of $f$?
**(b)** Solve $f(x)=1$. What point is on the graph of $f$?
**(c)** Solve $f(x)=2$. What point is on the graph of $f$?
**(d)** Use the information obtained in parts (a)–(c) to graph $f(x)=\sqrt{x-2}$.
**(e)** Use the graph and the concept of the range of a function to explain why the equation $f(x)=-1$ has no solution.

**82.** Suppose that $g(x)=\sqrt{x+3}$.

**(a)** Solve $g(x)=0$. What point is on the graph of $g$?
**(b)** Solve $g(x)=1$. What point is on the graph of $g$?
**(c)** Solve $g(x)=2$. What point is on the graph of $g$?
**(d)** Use the information obtained in parts (a)–(c) to graph $g(x)=\sqrt{x+3}$.
**(e)** Use the graph and the concept of the range of a function to explain why the equation $g(x)=-1$ has no solution.

△ **83. Finding a *y*-Coordinate** The solutions to the equation

$$\sqrt{4^2+(y-2)^2}=5$$

represent the $y$-coordinates such that the distance from the point $(3, 2)$ to $(-1, y)$ in the Cartesian plane is 5 units.

**(a)** Solve the equation for $y$.
**(b)** Plot the points in the Cartesian plane and label the lengths of the sides of the figure formed.

△ **84. Finding an *x*-Coordinate** The solutions to the equation

$$\sqrt{x^2+4^2}=5$$

represent the $x$-coordinates such that the distance from the point $(0, 3)$ to $(x, -1)$ in the Cartesian plane is 5 units.

**(a)** Solve the equation for $x$.
**(b)** Plot the points in the Cartesian plane and label the lengths of the sides of the figure formed.

**85. Revenue Growth** Suppose that the annual revenue $R$ (in millions of dollars) of a company after $t$ years of operating is modeled by the function

$$R(t)=\sqrt[3]{\frac{t}{2}}$$

**(a)** After how many years can the company expect to have annual revenue of \$1 million?
**(b)** After how many years can the company expect to have annual revenue of \$2 million?

△ **86. Sphere** The radius $r$ of a sphere whose volume is $V$ is given by

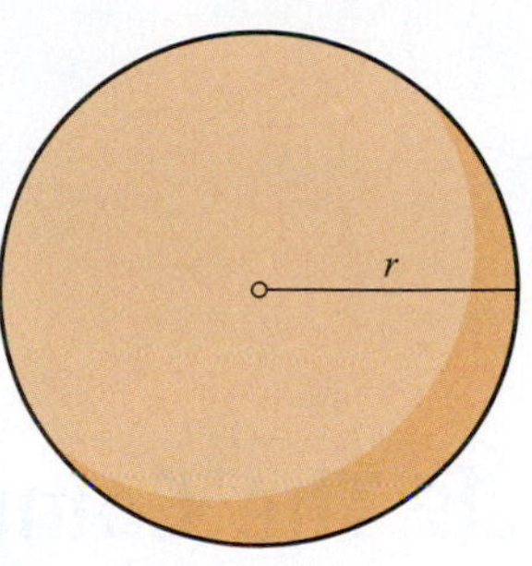


$$r = \sqrt[3]{\frac{3V}{4\pi}}$$

**(a)** Find the volume of a sphere whose radius is 3 meters.
**(b)** Find the volume of a sphere whose radius is 2 meters.

**87. Birth Rates** A plural birth is a live birth to twins, triplets, and so forth. The function $R(t) = 26 \cdot \sqrt[10]{t}$ models the plural birth rate $R$ (live births per 1,000 live births), where $t$ is the number of years since 1995.

**(a)** Use the model to predict the year in which the plural birth rate will be 39.
**(b)** Use the model to predict the year in which the plural birth rate will be 36.

**88. Money** The annual rate of interest $r$ (expressed as a decimal) required to have $A$ dollars after $t$ years from an initial deposit of $P$ dollars is given by

$$r = \sqrt[t]{\frac{A}{P}} - 1$$

**(a)** Suppose that you deposit \$1,000 in an account that pays 5% annual interest so that $r = 0.05$. How much will you have after $t = 2$ years?
**(b)** Suppose that you deposit \$1,000 in an account that pays 5% annual interest so that $r = 0.05$. How much will you have after $t = 3$ years?

## Extending the Concepts

**89.** Solve: $\sqrt{3\sqrt{x+1}} = \sqrt{2x+3}$ **90.** Solve: $\sqrt[3]{2\sqrt{x-2}} = \sqrt[3]{x-1}$

**91.** Which step in the process of solving a radical equation leads to the possibility of extraneous solutions?

## Synthesis Review

*In Problems 92–95, identify which of the numbers in the set*

$$\left\{0, -4, 12, \frac{2}{3}, 1.\overline{56}, \sqrt{2^3}, \pi, \sqrt{-5}, \sqrt[3]{-4}\right\}$$

*are . . .*

**92.** Integers **93.** Rational numbers **94.** Irrational numbers **95.** Real numbers

**96.** State the difference between a rational number and an irrational number. Why is $\sqrt{-1}$ not real?

## The Graphing Calculator

A graphing calculator can be used to verify solutions obtained algebraically. To solve $\sqrt{2x-3} = 5$ presented in Example 1, we graph $Y_1 = \sqrt{2x-3}$ and $Y_2 = 5$ and determine the $x$-coordinate of the point of intersection.

The $x$-coordinate of the point of intersection is 14, so the solution set is $\{14\}$.

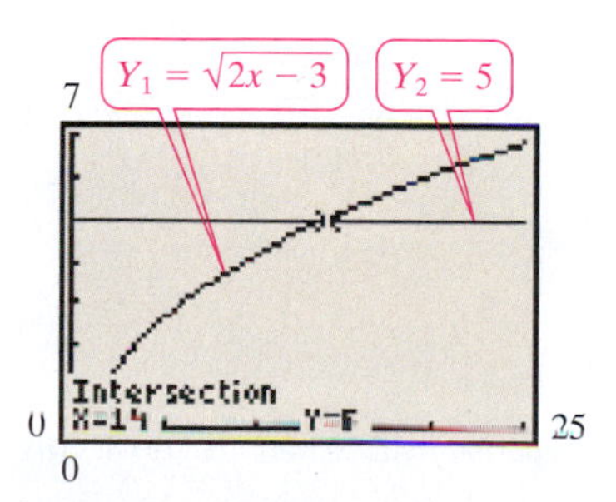


**97.** Verify your solution to Problem 11 by graphing $Y_1 = \sqrt{2x+3}$ and $Y_2 = 5$ and then finding the $x$-coordinate of their point of intersection.

**98.** Verify your solution to Problem 12 by graphing $Y_1 = \sqrt{3x-2}$ and $Y_2 = 4$ and then finding the $x$-coordinate of their point of intersection.

**99.** When you solved Problem 13 algebraically, you should have determined that the equation has no real solution. Verify this result by graphing $Y_1 = \sqrt{4x + 3}$ and $Y_2 = -2$. Explain what the algebraic solution means graphically.

**100.** When you solved Problem 14 algebraically, you should have determined that the equation has no real solution. Verify this result by graphing $Y_1 = \sqrt{6x - 5}$ and $Y_2 = -5$. Explain what the algebraic solution means graphically.

# 7.8 The Complex Number System

## OBJECTIVES

1. Evaluate the Square Root of Negative Real Numbers
2. Add or Subtract Complex Numbers
3. Multiply Complex Numbers
4. Divide Complex Numbers
5. Evaluate the Powers of $i$

## *Preparing for The Complex Number System*

*Before getting started, take the following readiness quiz. If you get a problem wrong, go back to the section cited and review the material.*

**1.** List the numbers in the set $\left\{8, -\frac{1}{3}, -23, 0, \sqrt{2}, 1.\overline{26}, -\frac{12}{3}, \sqrt{-5}\right\}$ that are

- **(a)** Natural numbers
- **(b)** Whole numbers
- **(c)** Integers
- **(d)** Rational numbers
- **(e)** Irrational numbers
- **(f)** Real numbers [Section R.2, pp. 11–14]

**2.** Distribute: $3x(4x - 3)$ [Section 5.2, p. 369]

**3.** Multiply: $(z + 4)(3z - 2)$ [Section 5.2, pp. 369–370]

**4.** Multiply: $(2y + 5)(2y - 5)$ [Section 5.2, p. 371]

If you look back at Section R.2, where we introduced the various number systems, you should notice that each time we encounter a situation where a number system can't handle a problem, we expand the number system. For example, if we only considered the whole numbers, we could not describe a negative balance in a checking account, so we introduced integers. If the world could only be described by integers, then we could not talk about parts of a whole as in $\frac{1}{2}$ a pizza or $\frac{3}{4}$ of a dollar, so we introduced rational numbers. If we only considered rational numbers, then we wouldn't be able to find a number whose square is 2, so we introduced the irrational numbers, so that $(\sqrt{2})^2 = 2$. By combining the rational numbers with the irrational numbers, we created the real number system. The real number system is usually sufficient for solving most problems in mathematics, but not for all problems.

For example, suppose we wanted to determine a number whose square is $-1$. We know that when we square any real number, the result is never negative. We call this property of real numbers, the *Nonnegativity Property*.

**NONNEGATIVITY PROPERTY OF REAL NUMBERS**

For any real number $a$, $a^2 \geq 0$.

Because the square of any real number is never negative, there is no real number $x$ for which

$$x^2 = -1$$

To remedy this situation, we introduce a new number.

**DEFINITION**

The **imaginary unit,** denoted by $i$, is the number whose square is $-1$. That is,

$$i^2 = -1$$

If we take the square root of both sides of $i^2 = -1$, we find that

$$i = \sqrt{-1}$$

*Preparing for...Answers* **1. (a)** 8 **(b)** 8, 0 **(c)** 8, −23, 0, −12/3 **(d)** 8, −1/3, −23, 0, $1.\overline{26}$, −12/3 **(e)** $\sqrt{2}$ **(f)** $8, -\frac{1}{3}, -23, 0, \sqrt{2}, 1.\overline{26}, -\frac{12}{3}$ **2.** $12x^2 - 9x$ **3.** $3z^2 + 10z - 8$ **4.** $4y^2 - 25$

In looking at the development of the real number system, each new number system contained the earlier number system as a subset. By introducing the number $i$, we now have a new number system called the **complex number system.**

**DEFINITION**

**Complex numbers** are numbers of the form $\boldsymbol{a + bi}$, where $a$ and $b$ are real numbers. The real number $a$ is called the **real part** of the number $a + bi$; the real number $b$ is called the **imaginary part** of $a + bi$.

**In Words**

The real number system is a subset of the complex number system. This means that all real numbers are, more generally, complex numbers.

For example, the complex number $6 + 2i$ has the real part 6 and the imaginary part 2. The complex number $4 - 3i = 4 + (-3)i$ has the real part 4 and the imaginary part $-3$.

When a complex number is written in the form $a + bi$, where $a$ and $b$ are real numbers, we say that it is in **standard form.** The complex number $a + 0i$ is typically written as $a$. This serves as a reminder that the real number system is a subset of the complex number system. The complex number $0 + bi$ is usually written as $bi$. Any number of the form $bi$ is called a **pure imaginary number.** Figure 7 shows the relation between the number systems.

**Figure 7**
The Complex Number System

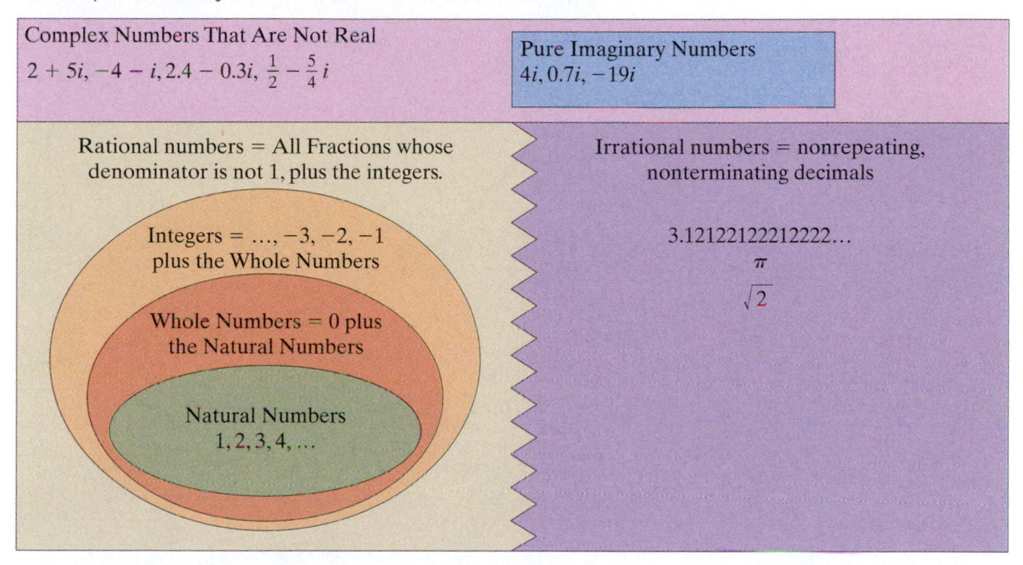


## 1 Evaluate the Square Root of Negative Real Numbers

Using the definition of $i$ along with the fact that $\sqrt{ab} = \sqrt{a} \cdot \sqrt{b}$ for real numbers $a$ and $b$, we now have the ability to simplify square roots of negative numbers.

**EVALUATING SQUARE ROOTS OF NEGATIVE NUMBERS**

If $N$ is a positive real number, we define the **principal square root of $-N$,** denoted by $\sqrt{-N}$, as

$$\sqrt{-N} = \sqrt{N}i$$

where $i = \sqrt{-1}$.

### EXAMPLE 1 Evaluating the Square Root of a Negative Number

Write each of the following as a pure imaginary number.

**(a)** $\sqrt{-16}$ **(b)** $\sqrt{-3}$ **(c)** $\sqrt{-18}$

**Work Smart**

When writing complex numbers whose imaginary part is a radical as in $\sqrt{3}i$ be sure that the "$i$" is not written under the radical.

**Solution**

(a) $\sqrt{-16} = \sqrt{16\cdot(-1)}$

$\sqrt{ab} = \sqrt{a}\cdot\sqrt{b}$: $= \sqrt{16}\cdot\sqrt{-1}$

$= 4i$

(b) $\sqrt{-3} = \sqrt{3\cdot(-1)}$

$= \sqrt{3}\cdot\sqrt{-1}$

$= \sqrt{3}i$

(c) $\sqrt{-18} = \sqrt{18\cdot(-1)}$

$= \sqrt{18}\cdot\sqrt{-1}$

$= 3\sqrt{2}i$

**QUICK ✓** *Write each radical as a pure imaginary number.*

**1.** $\sqrt{-36}$ **2.** $\sqrt{-5}$ **3.** $\sqrt{-12}$

## EXAMPLE 2 Writing Complex Numbers in Standard Form

Write each of the following in standard form.

**(a)** $2 - \sqrt{-25}$ **(b)** $3 + \sqrt{-50}$ **(c)** $\dfrac{4 - \sqrt{-12}}{2}$

**Solution**

The standard form of a complex number is $a + bi$.

**(a)** $2 - \sqrt{-25} = 2 - \sqrt{25\cdot(-1)}$

$\sqrt{ab} = \sqrt{a}\cdot\sqrt{b}$: $= 2 - \sqrt{25}\cdot\sqrt{-1}$

$= 2 - 5i$

**(b)** $3 + \sqrt{-50} = 3 + \sqrt{50\cdot(-1)}$

$\sqrt{ab} = \sqrt{a}\cdot\sqrt{b}$: $= 3 + \sqrt{50}\cdot\sqrt{-1}$

$\sqrt{50} = \sqrt{25}\cdot\sqrt{2} = 5\sqrt{2}$: $= 3 + 5\sqrt{2}i$

**(c)** $\dfrac{4 - \sqrt{-12}}{2} = \dfrac{4 - \sqrt{12\cdot(-1)}}{2}$

$= \dfrac{4 - \sqrt{12}\cdot\sqrt{-1}}{2}$

$\sqrt{12} = \sqrt{4}\cdot\sqrt{3} = 2\sqrt{3}$: $= \dfrac{4 - 2\sqrt{3}i}{2}$

Factor out 2: $= \dfrac{2(2 - \sqrt{3}i)}{2}$

Divide out the 2's: $= 2 - \sqrt{3}i$

**QUICK ✓** *Write each expression in the standard form of a complex number, $a + bi$.*

**4.** $4 + \sqrt{-100}$ **5.** $-2 - \sqrt{-8}$ **6.** $\dfrac{6 - \sqrt{-72}}{3}$

## 2 Add or Subtract Complex Numbers

Throughout your math career, whenever you were introduced to a new number system, you learned how to perform the four binary operations of addition, subtraction, multiplication, and division using this number system. Therefore, we will now show how to add and subtract complex numbers. Then we will discuss multiplying and dividing complex numbers.

Two complex numbers are added by adding the real parts and then adding the imaginary parts.

**In Words**

To add two complex numbers, add the real parts, then add the imaginary parts. To subtract two complex numbers, subtract the real parts, then subtract the imaginary parts.

**SUM OF COMPLEX NUMBERS**

$$(a + bi) + (c + di) = (a + c) + (b + d)i$$

To subtract two complex numbers, we use this rule:

**DIFFERENCE OF COMPLEX NUMBERS**

$$(a + bi) - (c + di) = (a - c) + (b - d)i$$

### EXAMPLE 3 Adding Complex Numbers

Add:

**(a)** $(4 - 3i) + (-2 + 5i)$ **(b)** $\left(4 + \sqrt{-25}\right) + \left(6 - \sqrt{-16}\right)$.

**Solution**

**(a)**
$$\begin{aligned}(4 - 3i) + (-2 + 5i) &= [4 + (-2)] + (-3 + 5)i \\ &= 2 + 2i\end{aligned}$$

**(b)** $\sqrt{-25} = 5i$; $\sqrt{-16} = 4i$

$$\begin{aligned}\left(4 + \sqrt{-25}\right) + \left(6 - \sqrt{-16}\right) &= (4 + 5i) + (6 - 4i) \\ &= (4 + 6) + (5 - 4)i \\ &= 10 + 1i \\ &= 10 + i\end{aligned}$$

**Work Smart**

Adding or subtracting complex numbers is just like combining like terms. For example,

$$\begin{aligned}&(4 - 3x) + (-2 + 5x) \\ &\quad = 4 + (-2) - 3x + 5x \\ &\quad = 2 + 2x\end{aligned}$$

so

$$\begin{aligned}&(4 - 3i) + (-2 + 5i) \\ &\quad = 4 + (-2) - 3i + 5i \\ &\quad = 2 + 2i\end{aligned}$$

### EXAMPLE 4 Subtracting Complex Numbers

Subtract:

**(a)** $(-3 + 7i) - (5 - 4i)$ **(b)** $\left(3 + \sqrt{-12}\right) - \left(-2 - \sqrt{-27}\right)$.

**Solution**

**(a)**
$$\begin{aligned}(-3 + 7i) - (5 - 4i) &= (-3 - 5) + (7 - (-4))i \\ &= -8 + 11i\end{aligned}$$

**(b)** $\sqrt{-12} = 2\sqrt{3}i$; $\sqrt{-27} = 3\sqrt{3}i$

$$\begin{aligned}\left(3 + \sqrt{-12}\right) - \left(-2 - \sqrt{-27}\right) &= \left(3 + 2\sqrt{3}i\right) - \left(-2 - 3\sqrt{3}i\right) \\ \text{Distribute the minus:} \quad &= 3 + 2\sqrt{3}i + 2 + 3\sqrt{3}i \\ &= [3 + 2] + \left[2\sqrt{3} + 3\sqrt{3}\right]i \\ &= 5 + 5\sqrt{3}i\end{aligned}$$

**QUICK ✓** *Add or subtract as indicated.*

7. $(4 + 6i) + (-3 + 5i)$
8. $(4 - 2i) - (-2 + 7i)$
9. $\left(4 - \sqrt{-4}\right) + \left(-7 + \sqrt{-9}\right)$

## 3 Multiply Complex Numbers

We multiply complex numbers using the Distributive Property. The methods are almost the same methods that we used to multiply polynomials.

### EXAMPLE 5 Multiplying Complex Numbers

Multiply:

**(a)** $4i(3 - 6i)$ **(b)** $(-2 + 4i)(3 - i)$

**Solution**

**(a)** We distribute the $4i$ into each term in the parentheses.

$$\begin{aligned} 4i(3 - 6i) &= 4i \cdot 3 - 4i \cdot 6i \\ &= 12i - 24i^2 \\ i^2 = -1: \quad &= 12i - 24 \cdot (-1) \\ &= 24 + 12i \end{aligned}$$

**(b)**

$$\begin{aligned} (-2 + 4i)(3 - i) &= -2 \cdot 3 - 2 \cdot (-i) + 4i \cdot 3 + 4i \cdot (-i) \\ &= -6 + 2i + 12i - 4i^2 \\ \text{Combine like terms; } i^2 = -1: \quad &= -6 + 14i - 4(-1) \\ &= -6 + 14i + 4 \\ &= -2 + 14i \end{aligned}$$

**QUICK ✓** *Multiply.*

**10.** $3i(5 - 4i)$ **11.** $(-2 + 5i)(4 - 2i)$

Look back at the Product Property for Radicals on page 589. You should notice that the property only applies when $\sqrt[n]{a}$ and $\sqrt[n]{b}$ are real numbers. This means that

$$\sqrt{a} \cdot \sqrt{b} \neq \sqrt{ab} \quad \text{if } a < 0 \text{ or } b < 0$$

So how do we perform this multiplication? Well, we first need to write the radical as a complex number using the fact that $\sqrt{-N} = \sqrt{N}i$ and then perform the multiplication.

### EXAMPLE 6 Multiplying Square Roots of Negative Numbers

Multiply:

**(a)** $\sqrt{-25} \cdot \sqrt{-4}$ **(b)** $\left(2 + \sqrt{-16}\right)\left(1 - \sqrt{-4}\right)$

**Solution**

**(a)** We cannot use the Product Property of Radicals to multiply these radicals because neither $\sqrt{-25}$ nor $\sqrt{-4}$ are real numbers. Therefore, we express the radicals as pure imaginary numbers and then multiply.

$$\begin{aligned} \sqrt{-25} \cdot \sqrt{-4} &= 5i \cdot 2i \\ &= 10i^2 \\ i^2 = -1: \quad &= -10 \end{aligned}$$

**Work Smart**

$\sqrt{-25} \cdot \sqrt{-4} \neq \sqrt{(-25)(-4)}$ because the Product Property of Radicals only applies when the radical is a real number.

**(b)** First, we rewrite the expression as a complex number in standard form.

$$(2+\sqrt{-16})(1-\sqrt{-4}) = (2+4i)(1-2i)$$

$$\begin{aligned}
\text{FOIL:} \quad &= 2\cdot 1 + 2\cdot(-2i) + 4i\cdot 1 + 4i\cdot(-2i) \\
&= 2 - 4i + 4i - 8i^2 \\
\text{Combine like terms; } i^2 = -1\text{:} \quad &= 2 - 8(-1) \\
&= 2 + 8 \\
&= 10
\end{aligned}$$

 *Multiply.*

**12.** $\sqrt{-9}\cdot\sqrt{-36}$

**13.** $(2+\sqrt{-36})(4-\sqrt{-25})$

## Complex Conjugates

We now introduce a special product that involves the *conjugate* of a complex number.

**In Words**

To find the complex conjugate of $a + bi$, simply change the sign from "+" to "−" or "−" to "+" between the "$a$" and "$b$" in the complex number.

**COMPLEX CONJUGATE**

If $a + bi$ is a complex number, then its **conjugate** is defined as $a - bi$.

For example,

| Complex Number | Conjugate |
|---|---|
| $3 + 5i$ | $3 - 5i$ |
| $-10 - 3i$ | $-10 + 3i$ |

Notice what happens when we multiply a complex number and its conjugate.

### EXAMPLE 7 Multiplying a Complex Number by Its Conjugate

Find the product of $4 + 3i$ and its conjugate, $4 - 3i$.

**Solution**

$$\begin{aligned}
(4+3i)(4-3i) &= 4\cdot 4 + 4\cdot(-3i) + 3i\cdot 4 + 3i\cdot(-3i) \\
&= 16 - 12i + 12i - 9i^2 \\
&= 16 - 9(-1) \\
&= 16 + 9 \\
&= 25
\end{aligned}$$

Wow! The product of the complex number $4 + 3i$ and its conjugate $4 - 3i$ is 25—a real number! In fact, the results of Example 7 are true in general.

**PRODUCT OF A COMPLEX NUMBER AND ITS CONJUGATE**

The product of a complex number and its conjugate is a nonnegative real number. That is,

$$(a+bi)(a-bi) = a^2 + b^2$$

Perhaps you noticed that multiplying a complex number and its conjugate is akin to multiplying $(a+b)(a-b) = a^2 - b^2$.

**QUICK ✓** *Multiply.*

**14.** $(3 - 8i)(3 + 8i)$ **15.** $(-2 + 5i)(-2 - 5i)$

## 4 Divide Complex Numbers

Now that we understand the product of a complex number and its conjugate we can proceed to divide complex numbers.

### EXAMPLE 8 How to Divide Complex Numbers

Divide: $\dfrac{-3 + i}{5 + 3i}$

**Step-by-Step Solution**

**Step 1:** Write the numerator and denominator in standard form, $a + bi$.

The numerator and denominator are already in standard form.

**Step 2:** Multiply the numerator and denominator by the complex conjugate of the denominator.

The complex conjugate of $5 + 3i$ is $5 - 3i$:

$$\frac{-3 + i}{5 + 3i} = \frac{-3 + i}{5 + 3i} \cdot \frac{5 - 3i}{5 - 3i}$$

$$= \frac{(-3 + i)(5 - 3i)}{(5 + 3i)(5 - 3i)}$$

**Step 3:** Simplify by writing the quotient in standard form, $a + bi$.

Multiply numerator and denominator; $(a + bi)(a - bi) = a^2 + b^2$:

$$= \frac{-3 \cdot 5 - 3 \cdot (-3i) + i \cdot 5 + i \cdot (-3i)}{5^2 + 3^2}$$

Combine like terms:

$$= \frac{-15 + 9i + 5i - 3i^2}{25 + 9}$$

$i^2 = -1$:

$$= \frac{-15 + 14i + 3}{34}$$

$$= \frac{-12 + 14i}{34}$$

Divide 34 into each term in the numerator to write in standard form:

$$= \frac{-12}{34} + \frac{14}{34}i$$

Write each fraction in lowest terms:

$$= -\frac{6}{17} + \frac{7}{17}i$$

We summarize the steps used in Example 8 below.

### Dividing Complex Numbers

**Step 1:** Write the numerator and denominator in standard form, $a + bi$.

**Step 2:** Multiply the numerator and denominator by the complex conjugate of the denominator.

**Step 3:** Simplify by writing the quotient in standard form, $a + bi$.

## EXAMPLE 9 Dividing Complex Numbers

Divide: $\dfrac{3 + 4i}{2i}$

**Solution**

$$\frac{3 + 4i}{2i} = \frac{3 + 4i}{0 + 2i}$$

The complex conjugate of $0 + 2i$ is $0 - 2i$: $= \dfrac{3 + 4i}{0 + 2i} \cdot \dfrac{0 - 2i}{0 - 2i}$

The 0s are not necessary: $= \dfrac{3 + 4i}{2i} \cdot \dfrac{-2i}{-2i}$

Multiply numerator; multiply denominator: $= \dfrac{(3 + 4i)(-2i)}{(2i)(-2i)}$

Distribute in numerator: $= \dfrac{-6i - 8i^2}{-4i^2}$

$i^2 = -1$: $= \dfrac{-6i + 8}{4}$

Divide 4 into each term in the numerator to write in standard form: $= \dfrac{8}{4} - \dfrac{6}{4}i$

Reduce each fraction to lowest terms: $= 2 - \dfrac{3}{2}i$

**Work Smart**

We could also have multiplied the numerator and denominator in Example 9 by $i$. Do you know why? Try it yourself! Which approach do you prefer?

Look back to Example 5(b), where we found $(-2 + 4i)(3 - i) = -2 + 14i$. Now find $\dfrac{-2 + 14i}{3 - i}$. What result do you expect? Verify that $\dfrac{-2 + 14i}{3 - i} = -2 + 4i$.

**QUICK ✓** *Divide.*

**16.** $\dfrac{-4 + i}{3i}$ **17.** $\dfrac{4 + 3i}{1 - 3i}$

## 5 Evaluate the Powers of *i*

The **powers of *i*** follow a pattern.

| | | | |
|---|---|---|---|
| $i^1 = i$ | $i^2 = -1$ | $i^3 = i^2 \cdot i^1 = -1 \cdot i = -i$ | $i^4 = i^2 \cdot i^2 = (-1)(-1) = 1$ |
| $i^5 = i^4 \cdot i = 1 \cdot i = i$ | $i^6 = i^4 \cdot i^2 = 1 \cdot (-1) = -1$ | $i^7 = i^4 \cdot i^3 = 1 \cdot (-i) = -i$ | $i^8 = i^4 \cdot i^4 = 1 \cdot 1 = 1$ |
| $i^9 = i^8 \cdot i = 1 \cdot i = i$ | $i^{10} = i^8 \cdot i^2 = 1 \cdot (-1) = -1$ | $i^{11} = i^8 \cdot i^3 = 1 \cdot (-i) = -i$ | $i^{12} = (i^4)^3 = (1)^3 = 1$ |

Do you see the pattern? In the first column, the expressions all simplify to $i$; in the second column, the expressions all simplify to $-1$; in the third column, the expressions all simplify to $-i$; in the fourth column, the expressions all simplify to 1. That is, the powers of $i$ repeat with every fourth power. Figure 8 shows the pattern.

**Figure 8**

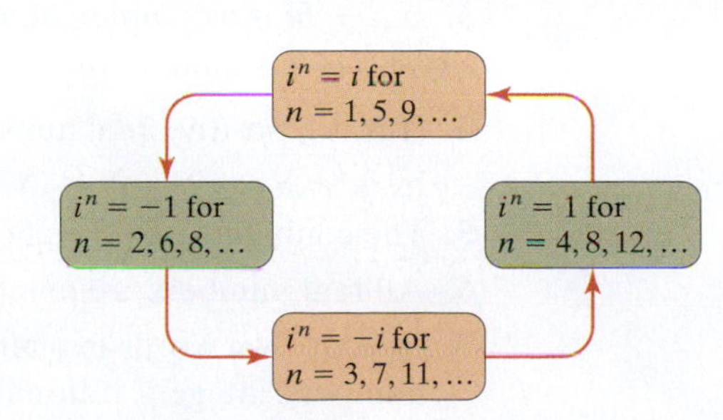

For this reason any power of $i$ can be expressed in terms of 1, $-1$, $i$, or $-i$. The following steps can be used to simplify any power of $i$.

**Simplifying the Powers of $i$**

**Step 1:** Divide the exponent of $i$ by 4. Rewrite $i^n$ as $(i^4)^q \cdot i^r$, where $q$ is the quotient and $r$ is the remainder of the division.

**Step 2:** Simplify the product in Step 1 to $i^r$ since $i^4 = 1$.

**EXAMPLE 10 Simplifying Powers of $i$**

Simplify:

**(a)** $i^{34}$ **(b)** $i^{101}$

**Solution**

**(a)** We divide 34 by 4 and obtain of quotient of $q = 8$ and a remainder of $r = 2$. So

$$\begin{aligned} i^{34} &= (i^4)^8 \cdot i^2 \\ &= (1)^8 \cdot (-1) \\ &= -1 \end{aligned}$$

**(b)** We divide 101 by 4 and obtain a quotient of $q = 25$ and a remainder of $r = 1$. So

$$\begin{aligned} i^{101} &= (i^4)^{25} \cdot i^1 \\ &= (1)^{25} \cdot i \\ &= i \end{aligned}$$

**QUICK ✓** *Simplify the power of i.*

**18.** $i^{43}$ **19.** $i^{98}$

# 7.8 Exercises

***For Extra Help:***

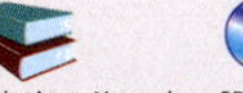
Student Solutions Manual

CD Video

PH Math/Tutor Center

MathXL Tutorials on CD

MathXL® MyMathLab

## Concepts and Vocabulary

*In Problems 1–3, fill in the blanks.*

**1.** The _____ _____, denoted by $i$, is the number whose square is $-1$.

**2.** Any number of the form $bi$ is called a _____ _____ _____.

**3.** If $a + bi$ is a complex number, then its _____ is defined as $a - bi$.

*In Problems 4–6, answer True or False to each statement.*

**4.** If $N$ is a positive real number, we define the principal square root of $-N$, denoted by $\sqrt{-N}$, as $\sqrt{-N} = \sqrt{N}i$.

**5.** The conjugate of $-3 + 5i$ is $3 - 5i$.

**6.** All real numbers are complex numbers.

**7.** In your own words explain the relation between the natural numbers, whole numbers, integers, rational numbers, real numbers and complex number system.

**8.** In your own words, explain why the product of a complex number and its conjugate is a nonnegative real number. That is, explain why $(a + bi)(a - bi) = a^2 + b^2$.

**9.** How is multiplying two complex numbers related to multiplying two binomials?

**10.** How is the method used to rationalize denominators related to the method used to write the quotient of two complex numbers in standard form?

## Building Skills

*In Problems 11–20, write each expression as a pure imaginary number.*

**11.** $\sqrt{-4}$ **12.** $\sqrt{-25}$ **13.** $-\sqrt{-81}$ **14.** $-\sqrt{-100}$

**15.** $\sqrt{-45}$ **16.** $\sqrt{-48}$ **17.** $\sqrt{-300}$ **18.** $\sqrt{-162}$

**19.** $\sqrt{-7}$ **20.** $\sqrt{-13}$

*In Problems 21–28, write each expression as a complex number in standard form.*

**21.** $5 + \sqrt{-49}$ **22.** $4 - \sqrt{-36}$ **23.** $-2 - \sqrt{-28}$ **24.** $10 + \sqrt{-32}$

**25.** $\dfrac{4 + \sqrt{-4}}{2}$ **26.** $\dfrac{10 - \sqrt{-25}}{5}$ **27.** $\dfrac{4 + \sqrt{-8}}{12}$ **28.** $\dfrac{15 - \sqrt{-50}}{5}$

*In Problems 29–36, add or subtract as indicated.*

**29.** $(4 + 5i) + (2 - 7i)$ **30.** $(-6 + 2i) + (3 + 12i)$

**31.** $(4 + i) - (8 - 5i)$ **32.** $(-7 + 3i) - (-3 + 2i)$

**33.** $\left(4 - \sqrt{-4}\right) - \left(2 + \sqrt{-9}\right)$ **34.** $\left(-4 + \sqrt{-25}\right) + \left(1 - \sqrt{-16}\right)$

**35.** $\left(-2 + \sqrt{-18}\right) + \left(5 - \sqrt{-50}\right)$ **36.** $\left(-10 + \sqrt{-20}\right) - \left(-6 + \sqrt{-45}\right)$

*In Problems 37–60, multiply.*

**37.** $6i(2 - 4i)$ **38.** $3i(-2 - 6i)$

**39.** $-\dfrac{1}{2}i(4 - 10i)$ **40.** $\dfrac{1}{3}i(12 + 15i)$

**41.** $(2 + i)(4 + 3i)$ **42.** $(3 - i)(1 + 2i)$

**43.** $(-3 - 5i)(2 + 4i)$ **44.** $(5 - 2i)(-1 + 2i)$

**45.** $(-3 + 5i)(5 - 3i)$ **46.** $(2 + 8i)(-3 - i)$

**47.** $\left(3 - \sqrt{2}i\right)\left(-2 + \sqrt{2}i\right)$ **48.** $\left(1 + \sqrt{3}i\right)\left(-4 - \sqrt{3}i\right)$

**49.** $\left(\dfrac{1}{2} - \dfrac{1}{4}i\right)\left(\dfrac{2}{3} + \dfrac{3}{4}i\right)$ **50.** $\left(-\dfrac{2}{3} + \dfrac{4}{3}i\right)\left(\dfrac{1}{2} - \dfrac{3}{2}i\right)$

**51.** $(3 + 2i)^2$ **52.** $(2 + 5i)^2$

**53.** $(-4 - 5i)^2$ **54.** $(2 - 7i)^2$

**55.** $\sqrt{-9} \cdot \sqrt{-4}$ **56.** $\sqrt{-36} \cdot \sqrt{-4}$

**57.** $\sqrt{-8} \cdot \sqrt{-10}$ **58.** $\sqrt{-12} \cdot \sqrt{-15}$

**59.** $\left(2 + \sqrt{-81}\right)\left(-3 - \sqrt{-100}\right)$ **60.** $\left(1 - \sqrt{-64}\right)\left(-2 + \sqrt{-49}\right)$

*In Problems 61–66, (a) find the conjugate of the complex number, and (b) multiply the complex number by its conjugate.*

**61.** $3 + 5i$ **62.** $5 + 2i$ **63.** $2 - 7i$

**64.** $9 - i$ **65.** $-7 + 2i$ **66.** $-1 - 4i$

*In Problems 67–80, divide.*

**67.** $\dfrac{1 + i}{3i}$ **68.** $\dfrac{2 - i}{2i}$ **69.** $\dfrac{-5 + 2i}{5i}$ **70.** $\dfrac{-4 + 5i}{6i}$

**71.** $\dfrac{3}{2 + i}$ **72.** $\dfrac{2}{4 + i}$ **73.** $\dfrac{-2}{-3 - 7i}$ **74.** $\dfrac{-4}{-5 - 3i}$

**75.** $\dfrac{2+3i}{3-2i}$ **76.** $\dfrac{2+5i}{5-2i}$ **77.** $\dfrac{4+2i}{1-i}$ **78.** $\dfrac{-6+2i}{1+i}$

**79.** $\dfrac{4-2i}{1+3i}$ **80.** $\dfrac{5-3i}{2+4i}$

*In Problems 81–88, simplify.*

**81.** $i^{53}$ **82.** $i^{72}$ **83.** $i^{43}$ **84.** $i^{110}$

**85.** $i^{153}$ **86.** $i^{131}$ **87.** $i^{-45}$ **88.** $i^{-26}$

## Mixed Practice

*In Problems 89–102, perform the indicated operation.*

**89.** $(-4-i)(4+i)$ **90.** $(-5+2i)(5-2i)$ **91.** $(3+2i)^2$

**92.** $(-3+2i)^2$ **93.** $\dfrac{-3+2i}{3i}$ **94.** $\dfrac{5-3i}{4i}$

**95.** $\dfrac{-4+i}{-5-3i}$ **96.** $\dfrac{-4+6i}{-5-i}$ **97.** $(10-3i)+(2+3i)$

**98.** $(-4+5i)+(4-2i)$ **99.** $5i(-4+3i)$ **100.** $2i(3-4i)$

**101.** $\sqrt{-10}\cdot\sqrt{-15}$ **102.** $\sqrt{-8}\cdot\sqrt{-12}$

## Applying the Concepts

*In Problems 103–108, find the reciprocal of the complex number. Write each number in standard form.*

**103.** $5i$ **104.** $7i$ **105.** $2-i$

**106.** $3-5i$ **107.** $-4+5i$ **108.** $-6+2i$

**109.** Suppose that $f(x)=x^2$; find (a) $f(i)$ (b) $f(1+i)$.

**110.** Suppose that $f(x)=x^2+x$; find (a) $f(i)$ (b) $f(1+i)$.

**111.** Suppose that $f(x)=x^2+2x+2$; find (a) $f(3i)$ (b) $f(1-i)$.

**112.** Suppose that $f(x)=x^2+x-1$; find (a) $f(2i)$ (b) $f(2+i)$.

**113. Impedance (Series Circuit)** The total impedance, $Z$, of an ac circuit containing components in series is equivalent to the sum of the individual impedances. Impedance is measured in ohms ($\Omega$) and is expressed as an imaginary number of the form $Z=R+i\cdot X$. Here, $R$ represents resistance and $X$ represents reactance.

**(a)** If the impedance in one part of a series circuit is $7+3i$ ohms and the impedance of the remainder of the circuit is $3-4i$ ohms, find the total impedance of the circuit.

**(b)** What is the total resistance of the circuit?

**(c)** What is the total reactance of the circuit?

**114. Impedance (Parallel Circuit)** The total impedance, $Z$, of an ac circuit consisting of two parallel pathways is given by the formula $\dfrac{1}{Z}=\dfrac{1}{Z_1}+\dfrac{1}{Z_2}$, where $Z_1$ and $Z_2$ are the impedances of each pathway. If the impedances of the individual pathways are $Z_1=5$ ohms and $Z_2=1-2i$ ohms, find the total impedance of the circuit.

## Extending the Concepts

**115.** For the function $f(x)=x^2+4x+5$, find (a) $f(-2+i)$ (b) $f(-2-i)$.

**116.** For the function $f(x)=x^2-2x+2$, find (a) $f(1+i)$ (b) $f(1-i)$.

**117.** For the function $f(x)=x^3+1$, find (a) $f(-1)$ (b) $f\left(\dfrac{1}{2}+\dfrac{\sqrt{3}}{2}i\right)$ (c) $f\left(\dfrac{1}{2}-\dfrac{\sqrt{3}}{2}i\right)$.

**118.** For the function $f(x) = x^3 - 1$, find (a) $f(1)$ (b) $f\left(-\frac{1}{2} + \frac{\sqrt{3}}{2}i\right)$ (c) $f\left(-\frac{1}{2} - \frac{\sqrt{3}}{2}i\right)$.

**119.** Any complex number $z$ such that $f(z) = 0$ is called a **complex zero** of $f$. Look at the complex zeros in Problems 115–118. Conjecture a general result regarding the complex zeros of polynomials that have real coefficients.

## Synthesis Review

**120.** Expand: $(x + 2)^3$

**121.** Expand: $(y - 4)^2$

**122.** Evaluate: $(3 + i)^3$

**123.** Evaluate: $(4 - 3i)^2$

**124.** How is raising a complex number to a positive integer power related to raising a binomial to a positive integer power?

## The Graphing Calculator

Graphing calculators have the ability to add, subtract, multiply and divide complex numbers. First, put the calculator into complex mode as shown in Figure 9(a). Figure 9(b) shows the results of Examples 3(a) and 3(b). Figure 9(c) shows the results of Examples 5(b) and 8.

**Figure 9**

Normal Sci Eng
Float 0123456789
Radian Degree
Func Par Pol Seq
Connected Dot
Sequential Simul
Real a+bi re^θi
Full Horiz G-T

(a)

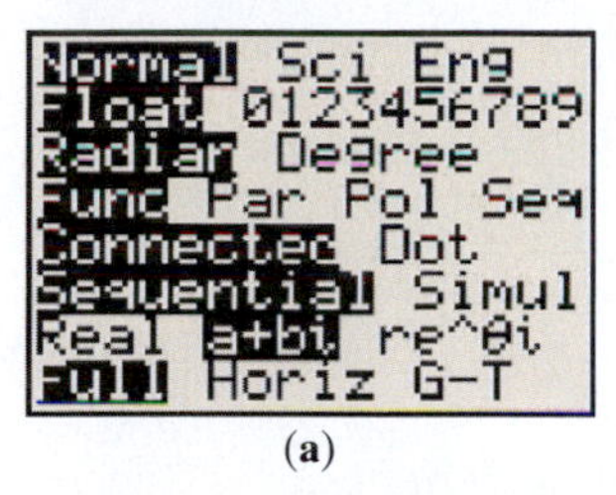


(b)

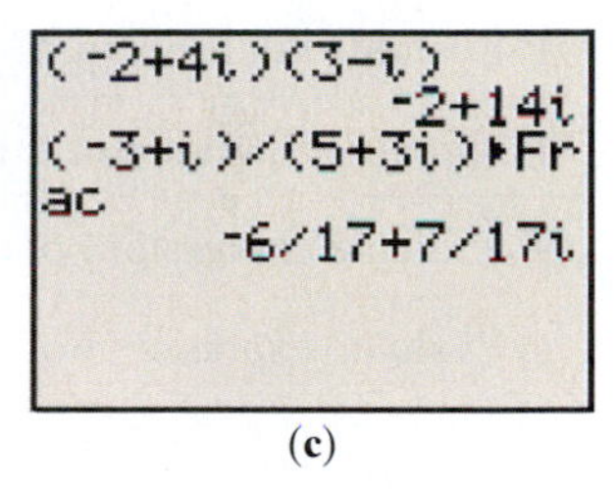


(c)

*In Problems 125–132, use a graphing calculator to simplify each expression. Write your answers in standard form, a + bi.*

**125.** $(1.3 - 4.3i) + (-5.3 + 0.7i)$

**126.** $(-3.4 + 1.9i) - (6.5 - 5.3i)$

**127.** $(0.3 - 5.2i)(1.2 + 3.9i)$

**128.** $(-4.3 + 0.2i)(7.2 - 0.5i)$

**129.** $\dfrac{4}{3 - 8i}$

**130.** $\dfrac{1 - 7i}{-4 + i}$

**131.** $3i^6 + (5 - 0.3i)(3 + 2i)$

**132.** $-6i^{14} + 5i^6 - (2 + i)(3 + 11i)$

# CHAPTER 7 ACTIVITY: WHICH ONE DOES NOT BELONG?

**Focus:** Simplifying Radicals and Rational Exponents

**Time:** 15–25 minutes

**Group Size:** 2–4

**5–10 Minutes:** Individually, evaluate each row below. Decide which item does not belong and why. Be creative with your reasons!

**10–15 Minutes:** Discuss each row and make a list of each member's response. Is there any example that everyone agrees with?

| | A | B | C | D |
|---|---|---|---|---|
| 1 | $16^{\frac{3}{2}}$ | $-16^{\frac{3}{2}}$ | $-16^{\frac{2}{3}}$ | $-16^{-\frac{2}{3}}$ |
| 2 | $\sqrt[2]{(a-b)^3}$, $a > b$ | $\sqrt[2]{a^3 - b^3}$ | $\sqrt[2]{(a-b)^3}$, $a < b$ | $\sqrt[2]{a - b}$, $a > b$ |
| 3 | The range of $x^2$ | The range of $\sqrt[3]{x}$ | The range of $\lvert x\rvert$ | The range of $x^3$ |
| 4 | $i^{212}$ | $i^0$ | $i^{-108}$ | $i^1$ |

# Chapter 7 Review

## Section 7.1 nth Roots and Rational Exponents

| KEY CONCEPTS | KEY TERMS |
|---|---|
| • **Simplifying $\sqrt[n]{a^n}$** If $n \geq 2$ is a positive integer and $a$ is a real number, then $\sqrt[n]{a^n} = a$ if $n \geq 3$ is odd; $\sqrt[n]{a^n} = \lvert a \rvert$ if $n \geq 2$ is even <br> • If $a$ is a real number and $n \geq 2$ is an integer, then $a^{\frac{1}{n}} = \sqrt[n]{a}$ provided that $\sqrt[n]{a}$ exists. <br> • If $a$ is a real number, $m/n$ is a rational number in lowest terms with $n \geq 2$, the $a^{\frac{m}{n}} = \sqrt[n]{a^m} = \left(\sqrt[n]{a}\right)^m$ provided that $\sqrt[n]{a}$ exists. <br> • If $m/n$ is a rational number and if $a$ is a nonzero real number, then $a^{-\frac{m}{n}} = \frac{1}{a^{\frac{m}{n}}}$ or $\frac{1}{a^{-\frac{m}{n}}} = a^{\frac{m}{n}}$. | Principal $n$th root of a number $a$ <br> Index <br> Cube root |

| YOU SHOULD BE ABLE TO . . . | EXAMPLE | REVIEW EXERCISES |
|---|---|---|
| 1 Evaluate $n$th roots (p. 534) | Examples 1 and 2 | 1–5, 19, 20 |
| 2 Simplify expressions of the form $\sqrt[n]{a^n}$ (p. 535) | Example 3 | 6–8 |
| 3 Evaluate expressions of the form $a^{\frac{1}{n}}$ (p. 536) | Examples 4 and 5 | 9–12, 17, 21 |
| 4 Evaluate expressions of the form $a^{\frac{m}{n}}$ (p. 537) | Examples 6 through 9 | 13–16, 18, 22–24 |

*In Problems 1–8, simplify each radical.*

**1.** $\sqrt[3]{343}$ **2.** $\sqrt[3]{-125}$ **3.** $\sqrt[3]{\frac{8}{27}}$ **4.** $\sqrt[4]{81}$

**5.** $-\sqrt[5]{-243}$ **6.** $\sqrt[3]{10^3}$ **7.** $\sqrt[5]{z^5}$ **8.** $\sqrt[4]{(5p-3)^4}$

*In Problems 9–16, evaluate each of the following expressions.*

**9.** $81^{1/2}$ **10.** $(-256)^{1/4}$ **11.** $-4^{1/2}$ **12.** $729^{1/3}$

**13.** $16^{7/4}$ **14.** $-(-27)^{2/3}$ **15.** $-121^{3/2}$ **16.** $\frac{1}{36^{-1/2}}$

*In Problems 17–20, use a calculator to write each expression rounded to two decimal places.*

**17.** $(-65)^{1/3}$ **18.** $4^{3/5}$ **19.** $\sqrt[3]{100}$ **20.** $\sqrt[4]{10}$

*In Problems 21–24, rewrite each of the following radicals with a rational exponent.*

**21.** $\sqrt[3]{5a}$ **22.** $\sqrt[5]{p^7}$ **23.** $\left(\sqrt[4]{10z}\right)^3$ **24.** $\sqrt[6]{(2ab)^5}$

## Section 7.2 Simplify Expressions Using the Laws of Exponents

| KEY CONCEPTS | KEY TERM |
|---|---|
| • **If $a$ and $b$ are real numbers and if r and s are rational numbers, then assuming the expression is defined,** <br> Zero-Exponent Rule: $a^0 = 1$ if $a \neq 0$ <br> Negative-Exponent Rule: $a^{-r} = \frac{1}{a^r}$ if $a \neq 0$ <br> Product Rule: $a^r \cdot a^s = a^{r+s}$ <br> Quotient Rule: $\frac{a^r}{a^s} = a^{r-s} = \frac{1}{a^{s-r}}$ if $a \neq 0$ | Simplify |

Power Rule: $(a^r)^s = a^{r \cdot s}$

Product to Power Rule: $(a \cdot b)^r = a^r \cdot b^r$

Quotient to Power Rule: $\left(\frac{a}{b}\right)^r = \frac{a^r}{b^r}$ if $b \neq 0$

Quotient to a Negative Power Rule: $\left(\frac{a}{b}\right)^{-r} = \left(\frac{b}{a}\right)^r$ if $a \neq 0, b \neq 0$

| YOU SHOULD BE ABLE TO . . . | EXAMPLE | REVIEW EXERCISES |
|---|---|---|
| 1 Use the Laws of Exponents to simplify expressions involving rational exponents (p. 542) | Examples 1 through 3 | 25–30 |
| 2 Use the Laws of Exponents to simplify radical expressions (p. 544) | Example 4 | 31–34 |
| 3 Factor expressions containing rational exponents (p. 545) | Examples 5 and 6 | 35, 36 |

*In Problems 25–36, simplify the expression.*

**25.** $4^{2/3} \cdot 4^{7/3}$
**26.** $\frac{k^{1/2}}{k^{3/4}}$
**27.** $(p^{4/3} \cdot q^4)^{3/2}$
**28.** $(32a^{-3/2} \cdot b^{1/4})^{1/5}$

**29.** $5m^{-2/3}(2m + m^{-1/3})$
**30.** $\left(\frac{16x^{1/3}}{x^{-1/3}}\right)^{-1/2} + \left(\frac{x^{-3/2}}{64x^{-1/2}}\right)^{1/3}$

**31.** $\sqrt[8]{x^6}$
**32.** $\sqrt{121x^4y^{10}}$
**33.** $\sqrt[3]{m^2} \cdot \sqrt{m^3}$
**34.** $\frac{\sqrt[3]{c}}{\sqrt[6]{c^4}}$

**35.** $2(3m - 1)^{1/4} + (m - 7)(3m - 1)^{5/4}$
**36.** $3(x^2 - 5)^{1/3} - 4x(x^2 - 5)^{-2/3}$

## Section 7.3 Simplifying Radical Expressions

| KEY CONCEPTS | KEY TERM |
|---|---|
| • **Product Property of Radicals** If $\sqrt[n]{a}$ and $\sqrt[n]{b}$ are real numbers and $n \geq 2$ is an integer, then $\sqrt[n]{a} \cdot \sqrt[n]{b} = \sqrt[n]{ab}$. <br> • **Quotient Property of Radicals** If $\sqrt[n]{a}$ and $\sqrt[n]{b}$ are real numbers, $b \neq 0$ and $n \geq 2$ is an integer, then $\frac{\sqrt[n]{a}}{\sqrt[n]{b}} = \sqrt[n]{\frac{a}{b}}$. | Simplify |

| YOU SHOULD BE ABLE TO . . . | EXAMPLE | REVIEW EXERCISES |
|---|---|---|
| 1 Use the Product Property to multiply radical expressions (p. 548) | Example 1 | 37, 38, 49–52 |
| 2 Use the Product Property to simplify radical expressions (p. 549) | Examples 2 through 7 | 39–52 |
| 3 Use the Quotient Property to simplify radical expressions (p. 554) | Examples 8 and 9 | 53–60 |
| 4 Multiply radicals with unlike indices (p. 555) | Example 10 | 61, 62 |

*In Problems 37 and 38, use the Product Property to multiply. Assume that all variables can be any real number.*

**37.** $\sqrt{15} \cdot \sqrt{7}$
**38.** $\sqrt[4]{2ab^2} \cdot \sqrt[4]{6a^2b}$

*In Problems 39–44, simplify each radical using the Product Property. Assume that all variables can be any real number.*

**39.** $\sqrt{80}$
**40.** $\sqrt[3]{-500}$
**41.** $\sqrt[3]{162m^6n^4}$

**42.** $\sqrt[4]{50p^8q^4}$
**43.** $2\sqrt{16x^6y}$
**44.** $\sqrt{(2x + 1)^3}$

*In Problems 45–48, simplify each radical using the Product Property. Assume that all variables are greater than or equal to zero.*

**45.** $\sqrt{w^3z^2}$ **46.** $\sqrt{45x^4yz^3}$ **47.** $\sqrt[3]{16a^{12}b^5}$ **48.** $\sqrt{4x^2 + 8x + 4}$

*In Problems 49–52, multiply and simplify. Assume that all variables are greater than or equal to zero.*

**49.** $\sqrt{15} \cdot \sqrt{18}$ **50.** $\sqrt[3]{20} \cdot \sqrt[3]{30}$

**51.** $\sqrt[3]{-3x^4y^7} \cdot \sqrt[3]{24x^3y^2}$ **52.** $3\sqrt{4xy^2} \cdot 5\sqrt{3x^2y}$

*In Problems 53–60, simplify. Assume that all variables are greater than zero.*

**53.** $\sqrt{\dfrac{121}{25}}$ **54.** $\sqrt{\dfrac{5a^4}{64b^2}}$ **55.** $\sqrt[3]{\dfrac{54k^2}{9k^5}}$ **56.** $\sqrt[3]{\dfrac{-160w^{11}}{343w^{-4}}}$

**57.** $\dfrac{\sqrt{12h^3}}{\sqrt{3h}}$ **58.** $\dfrac{\sqrt{50a^3b^3}}{\sqrt{8a^5b^{-3}}}$ **59.** $\dfrac{\sqrt[3]{-8x^7y}}{\sqrt[3]{27xy^4}}$ **60.** $\dfrac{\sqrt[4]{48m^2n^7}}{\sqrt[4]{3m^6n}}$

*In Problems 61 and 62, multiply and simplify.*

**61.** $\sqrt{5} \cdot \sqrt[3]{2}$ **62.** $\sqrt[4]{8} \cdot \sqrt[6]{4}$

| **Section 7.4** | **Adding, Subtracting, and Multiplying Radical Expressions** | |
|---|---|---|
| **KEY CONCEPTS** | | **KEY TERM** |
| • To add or subtract radicals, the index and the radicand must be the same.<br>• When we multiply radicals, the index on each radical must be the same. Then we multiply the radicands. | | Like radicals |
| **YOU SHOULD BE ABLE TO . . .** | **EXAMPLE** | **REVIEW EXERCISES** |
| 1 Add or subtract radical expressions (p. 560) | Examples 1 through 3 | 63–72 |
| 2 Multiply radical expressions (p. 562) | Examples 4 and 5 | 73–82 |

*In Problems 63–72, add or subtract as indicated. Assume all variables are positive or zero.*

**63.** $2\sqrt[4]{x} + 6\sqrt[4]{x}$ **64.** $7\sqrt[3]{4y} + 2\sqrt[3]{4y} - 3\sqrt[3]{4y}$

**65.** $5\sqrt{2} - 2\sqrt{12}$ **66.** $\sqrt{18} + 2\sqrt{50}$

**67.** $\sqrt[3]{-16z} + \sqrt[3]{54z}$ **68.** $7\sqrt[3]{8x^2} - \sqrt[3]{-27x^2}$

**69.** $\sqrt{16a} + \sqrt[6]{729a^3}$ **70.** $\sqrt{27x^2} - x\sqrt{48} + 2\sqrt{75x^2}$

**71.** $5\sqrt[3]{4m^5y^2} - \sqrt[6]{16m^{10}y^4}$ **72.** $\sqrt{y^3 - 4y^2} - 2\sqrt{y - 4} + \sqrt[4]{y^2 - 8y + 16}$

*In Problems 73–82, multiply and simplify.*

**73.** $\sqrt{3}(\sqrt{5} - \sqrt{15})$ **74.** $\sqrt[3]{5}(3 + \sqrt[3]{4})$

**75.** $(3 + \sqrt{5})(4 - \sqrt{5})$ **76.** $(7 + \sqrt{3})(6 + \sqrt{2})$

**77.** $(1 - 3\sqrt{5})(1 + 3\sqrt{5})$ **78.** $(\sqrt[3]{x} + 1)(9\sqrt[3]{x} - 4)$

**79.** $(\sqrt{x} - \sqrt{5})^2$

**80.** $(11\sqrt{2} + \sqrt{5})^2$

**81.** $(\sqrt{2a} - b)(\sqrt{2a} + b)$

**82.** $(\sqrt[3]{6s} + 2)(\sqrt[3]{6s} - 7)$

## Section 7.5 Rationalizing Radical Expressions

| KEY CONCEPTS | KEY TERM |
|---|---|
| • To rationalize the denominator when the denominator contains a single radical, multiply the numerator and denominator by a radical such that the radicand in the denominator is a perfect power of the index, $n$.<br>• To rationalize a denominator containing two terms, use the fact that $(a + b)(a - b) = a^2 - b^2$. | Rationalizing the denominator |

| YOU SHOULD BE ABLE TO . . . | EXAMPLE | REVIEW EXERCISES |
|---|---|---|
| 1 Rationalize a denominator containing one term (p. 567) | Examples 1 and 2 | 83–90, 99 |
| 2 Rationalize a denominator containing two terms (p. 569) | Examples 3 and 4 | 91–98, 100 |

*In Problems 83–98, rationalize the denominator.*

**83.** $\dfrac{2}{\sqrt{6}}$

**84.** $\dfrac{6}{\sqrt{3}}$

**85.** $\dfrac{\sqrt{48}}{\sqrt{p^3}}$

**86.** $\dfrac{5}{\sqrt{2a}}$

**87.** $\dfrac{-2}{\sqrt{6y^3}}$

**88.** $\dfrac{3}{\sqrt[3]{5}}$

**89.** $\sqrt[3]{\dfrac{-4}{45}}$

**90.** $\dfrac{27}{\sqrt[5]{8p^3q^4}}$

**91.** $\dfrac{6}{7 - \sqrt{6}}$

**92.** $\dfrac{3}{\sqrt{3} - 9}$

**93.** $\dfrac{\sqrt{3}}{3 + \sqrt{2}}$

**94.** $\dfrac{\sqrt{k}}{\sqrt{k} - \sqrt{m}}$

**95.** $\dfrac{\sqrt{10} + 2}{\sqrt{10} - 2}$

**96.** $\dfrac{3 - \sqrt{y}}{3 + \sqrt{y}}$

**97.** $\dfrac{4}{2\sqrt{3} + 5\sqrt{2}}$

**98.** $\dfrac{\sqrt{5} - \sqrt{6}}{\sqrt{10} + \sqrt{3}}$

**99.** Simplify: $\dfrac{\sqrt{7}}{3} + \dfrac{6}{\sqrt{7}}$

**100.** Find the reciprocal: $4 - \sqrt{7}$

## Section 7.6 Functions Involving Radicals

| KEY CONCEPT | KEY TERMS |
|---|---|
| • **Finding the Domain of a Function Whose Rule Is a Radical Expression**<br>**1.** If the index on a radical expression is even, then the radicand must be greater than or equal to zero.<br>**2.** If the index on a radical expression is odd, then the radicand can be any real number. | Square root function<br>Cube root function |

| YOU SHOULD BE ABLE TO . . . | EXAMPLE | REVIEW EXERCISES |
|---|---|---|
| 1 Evaluate functions whose rule is a radical expression (p. 573) | Example 1 | 101–104 |
| 2 Find the domain of a function whose rule is a radical (p. 574) | Example 2 | 105–110, 111(a), 112(a), 113(a), 114(a) |
| 3 Graph functions involving square roots (p. 575) | Example 3 | 111–113 |
| 4 Graph functions involving cube roots (p. 576) | Example 4 | 114 |

*In Problems 101–104, evaluate each radical function at the indicated values.*

**101.** $f(x) = \sqrt{x + 4}$

**(a)** $f(-3)$ **(b)** $f(0)$ **(c)** $f(5)$

**102.** $g(x) = \sqrt{3x - 2}$

**(a)** $g\left(\frac{2}{3}\right)$ **(b)** $g(2)$ **(c)** $g(6)$

**103.** $H(t) = \sqrt[3]{t + 3}$

**(a)** $H(-2)$ **(b)** $H(-4)$ **(c)** $H(5)$

**104.** $G(z) = \sqrt{\frac{z - 1}{z + 2}}$

**(a)** $G(1)$ **(b)** $G(-3)$ **(c)** $G(2)$

*In Problems 105–110, find the domain of the radical function.*

**105.** $f(x) = \sqrt{3x - 5}$

**106.** $g(x) = \sqrt[3]{2x - 7}$

**107.** $h(x) = \sqrt[4]{6x + 1}$

**108.** $F(x) = \sqrt[5]{2x - 9}$

**109.** $G(x) = \sqrt{\frac{4}{x - 2}}$

**110.** $H(x) = \sqrt{\frac{x - 3}{x}}$

*In Problems 111–114, (a) determine the domain of the function; (b) graph the function using point-plotting; and (c) based on the graph, determine the range of the function.*

**111.** $f(x) = \frac{1}{2}\sqrt{1 - x}$

**112.** $g(x) = \sqrt{x + 1} - 2$

**113.** $h(x) = -\sqrt{x + 3}$

**114.** $F(x) = \sqrt[3]{x + 1}$

| **Section 7.7** | **Radical Equations and Their Applications** | |
|---|---|---|
| **KEY TERMS** | | |
| Radical equation | Extraneous solution | |
| **YOU SHOULD BE ABLE TO . . .** | **EXAMPLE** | **REVIEW EXERCISES** |
| 1 Solve radical equations containing one radical (p. 581) | Examples 1 through 5 | 115–124, 129, 130 |
| 2 Solve radical equations containing two radicals (p. 584) | Examples 6 and 7 | 125–128 |
| 3 Solve for a variable in a radical equation (p. 586) | Example 8 | 131, 132 |

*In Problems 115–130, solve each equation.*

**115.** $\sqrt{m} = 13$

**116.** $\sqrt[3]{3t + 1} = -2$

**117.** $\sqrt[4]{3x - 8} = 3$

**118.** $\sqrt{2x + 5} + 4 = 2$

**119.** $\sqrt{k + 4} - 3 = -1$

**120.** $3\sqrt{t} - 4 = 11$

**121.** $2\sqrt[3]{m} + 5 = -11$

**122.** $\sqrt{q + 2} = q$

**123.** $\sqrt{w + 11} + 3 = w + 2$

**124.** $\sqrt{p^2 - 2p + 9} = p + 1$

**125.** $\sqrt{a + 10} = \sqrt{2a - 1}$

**126.** $\sqrt{5x + 9} = \sqrt{7x - 3}$

**127.** $\sqrt{c - 8} + \sqrt{c} = 4$

**128.** $\sqrt{x + 2} - \sqrt{x + 9} = 7$

**129.** $(4x - 3)^{1/3} - 3 = 0$

**130.** $(x^2 - 9)^{1/4} = 2$

**131. Height of a Cone** Solve $r = \sqrt{\frac{3V}{\pi h}}$ for $h$.

**132. Ball Slide Speed Factor** Solve $f_s = \sqrt[3]{\frac{30}{v}}$ for $v$.

## Section 7.8 The Complex Number System

| KEY CONCEPTS | KEY TERMS |
|---|---|
| • **Nonnegativity Property of Real Numbers** For any real number $a$, $a^2 \geq 0$<br>• **Imaginary Unit** The imaginary unit, denoted by $i$, is the number whose square is $-1$. That is, $i^2 = -1$.<br>• **Complex numbers** Complex numbers are numbers of the form $a + bi$, where $a$ and $b$ are real numbers. The real number $a$ is called the real part of the number $a + bi$; the real number $b$ is called the imaginary part of $a + bi$.<br>• **Square Roots of Negative Numbers** If $N$ is a positive real number, the principal square root of $-N$, denoted by $\sqrt{-N}$, is $\sqrt{-N} = \sqrt{N}\,i$, where $i = \sqrt{-1}$.<br>• **Sum of Complex Numbers** $(a + bi) + (c + di) = (a + c) + (b + d)i$<br>• **Difference of Complex Numbers** $(a + bi) - (c + di) = (a - c) + (b - d)i$<br>• **Complex Conjugate** If $a + bi$ is a complex number, then its conjugate is $a - bi$.<br>• **Product of a Complex Number and Its Conjugate** $(a + bi)(a - bi) = a^2 + b^2$ | Imaginary unit<br>Complex number system<br>Complex number<br>Real part<br>Imaginary part<br>Standard form<br>Pure imaginary number<br>Principal square root<br>Conjugate<br>Powers of $i$ |

| YOU SHOULD BE ABLE TO . . . | EXAMPLE | REVIEW EXERCISES |
|---|---|---|
| 1 Evaluate the square root of negative real numbers (p. 591) | Examples 1 and 2 | 133–136 |
| 2 Add or subtract complex numbers (p. 592) | Examples 3 and 4 | 137–140 |
| 3 Multiply complex numbers (p. 594) | Examples 5 through 7 | 141–146 |
| 4 Divide complex numbers (p. 596) | Examples 8 and 9 | 147–150 |
| 5 Evaluate the powers of $i$ (p. 597) | Example 10 | 151–152 |

*In Problems 133 and 134, write each expression as a pure imaginary number.*

**133.** $\sqrt{-29}$

**134.** $\sqrt{-54}$

*In Problems 135 and 136, write each expression as a complex number in standard form.*

**135.** $14 - \sqrt{-162}$

**136.** $\dfrac{6 + \sqrt{-45}}{3}$

*In Problems 137–150, perform the indicated operation.*

**137.** $(3 - 7i) + (-2 + 5i)$

**138.** $(4 + 2i) - (9 - 8i)$

**139.** $\left(8 - \sqrt{-45}\right) - \left(3 + \sqrt{-80}\right)$

**140.** $\left(1 + \sqrt{-9}\right) + \left(-6 + \sqrt{-16}\right)$

**141.** $(4 - 5i)(3 + 7i)$

**142.** $\left(\frac{1}{2} + \frac{2}{3}i\right)(4 - 9i)$

**143.** $\sqrt{-3} \cdot \sqrt{-27}$

**144.** $\left(1 + \sqrt{-36}\right)\left(-5 - \sqrt{-144}\right)$

**145.** $(1 + 12i)(1 - 12i)$

**146.** $(7 + 2i)(5 + 4i)$

**147.** $\dfrac{4}{3+5i}$

**148.** $\dfrac{-3}{7-2i}$

**149.** $\dfrac{2-3i}{5+2i}$

**150.** $\dfrac{4+3i}{1-i}$

*In Problems 151 and 152, simplify.*

**151.** $i^{59}$

**152.** $i^{173}$

## CHAPTER 7 TEST

*Remember to use your Chapter Test Prep Video CD to see fully worked-out solutions to any of these problems you would like to review.*

**1.** Evaluate: $49^{-1/2}$

*In Problems 2 and 3, simplify using rational exponents.*

**2.** $\sqrt[3]{8x^{1/2}y^3} \cdot \sqrt{9xy^{1/2}}$

**3.** $\sqrt[5]{(2a^4b^3)^7}$

*In Problems 4–9, perform the indicated operation and simplify. Assume all variables in the radicand are greater than or equal to zero.*

**4.** $\sqrt{3m} \cdot \sqrt{13n}$

**5.** $\sqrt{32x^7y^4}$

**6.** $\dfrac{\sqrt{9a^3b^{-3}}}{\sqrt{4ab}}$

**7.** $\sqrt{5x^3} + 2\sqrt{45x}$

**8.** $\sqrt{9a^2b} - \sqrt[4]{16a^4b^2}$

**9.** $(11 + 2\sqrt{x})(3 - \sqrt{x})$

*In Problems 10 and 11, rationalize the denominator.*

**10.** $\dfrac{-2}{3\sqrt{72}}$

**11.** $\dfrac{\sqrt{5}}{\sqrt{5}+2}$

**12.** For $f(x) = \sqrt{-2x+3}$, find the following:

**(a)** $f(1)$ **(b)** $f(-3)$

**13.** Determine the domain of the function $g(x) = \sqrt{-3x+5}$.

**14.** For $f(x) = \sqrt{x} - 3$, do the following:

**(a)** Determine the domain of the function.
**(b)** Graph the function using point plotting.
**(c)** From the graph, determine the range of the function.

*In Problems 15–17, solve the given equations.*

**15.** $\sqrt{x+3} = 4$

**16.** $\sqrt{x+13} - 4 = x - 3$

**17.** $\sqrt{x-1} + \sqrt{x+2} = 3$

*In Problems 18–20, perform the indicated operation.*

**18.** $(13+2i) + (4-15i)$

**19.** $(4-7i)(2+3i)$

**20.** $\dfrac{7-i}{12+11i}$

## CUMULATIVE REVIEW CHAPTERS R–7

**1.** Evaluate: $6 - 3^2 \div (9-3)$

**2.** Simplify: $(3x+2y) - (2x-5y+3) + 9$

**3.** Solve: $(3x+5) - 2 = 7x - 13$

**4.** Solve and write your answer in interval notation: $6x + \dfrac{1}{2}(4x-2) \leq 3x + 9$

**5.** For $f(x) = 3x^2 - x + 5$, find the following:

**(a)** $f(-2)$ **(b)** $f(3)$

**6.** Find the domain of $g(x) = \dfrac{x^2 - 9}{x^2 - 2x - 8}$.

**7.** The annual sales for a certain computer game is given by $n(p) = -50p + 6000$, where $n$ is the number of games sold and $p$ is the price in dollars.

**(a)** How many games will be sold in one year if the price were $50?

**(b)** At what price will there be no sales during a given year?

**8.** Find the equation of the line that passes through the points $(-1, 6)$ and $(3, -2)$.

**9.** Graph the inequality: $6x + 3y > 24$

**10.** Solve the system using substitution:

$$\begin{cases} 4x - \phantom{6}y = 17 \\ 5x + 6y = 14 \end{cases}$$

**11.** Dried fruit is to be mixed with nuts to create a trail mix blend. The fruit costs $3.45 per pound and the nuts cost $2.10 per pound. How much of each should be used to make 10 pounds of trail mix that will sell for $2.64 per pound if the total revenue should remain the same?

**12.** Find the determinant:

$$\begin{vmatrix} 5 & -2 & 3 \\ 3 & -3 & 4 \\ -2 & 4 & 1 \end{vmatrix}$$

**13.** Add: $(8x^3 - 4x^2 + 5x + 3) + (2x^2 - 8x + 7)$

**14.** Multiply: $(2x - 1)(4x^2 + 2x - 9)$

**15.** Divide: $\dfrac{6x^4 + 13x^3 - 21x^2 - 28x + 37}{2x^2 + 3x - 5}$

**16.** Factor completely: $8x^2 - 44x - 84$

**17.** Subtract: $\dfrac{2x}{x - 3} - \dfrac{x + 1}{x + 2}$

**18.** Solve: $\dfrac{9}{k - 2} = \dfrac{6}{k} + 3$

**19.** Solve and graph the solution set on a number line: $\dfrac{x + 8}{x - 4} \le 3$

**20.** Shawn can paint a certain room in 4 hours. Payton can paint the same room in 6 hours. How long will it take them to paint the room if they work together?

**21.** Simplify: $\dfrac{\sqrt{50a^3b}}{\sqrt{2a^{-1}b^3}}$

**22.** Find the domain of the function $f(x) = \sqrt[4]{8 - 3x}$.

**23.** Solve: $\sqrt{x + 7} - 8 = x - 7$

**24.** Divide: $\dfrac{3i}{1 - 7i}$

**25.** Rationalize the denominator: $\dfrac{2}{4 - \sqrt{11}}$

CHAPTER

# 8 Quadratic Equations and Functions

One of the more unusual sports is found in Millsboro, Delaware—Punkin Chunkin. Participants catapult or fire pumpkins to see who can toss them the farthest (and most accurately). Interestingly, this bizarre ritual is an application of a quadratic function at work. See Problems 73–76 in Section 8.5.

## OUTLINE



## The Big Picture: Putting It Together

In Chapter 1, we reviewed solving linear equations and inequalities. In Chapter 3, we completed our discussion of "everything linear" after covering linear functions and their graphs.

This chapter is dedicated to completing our discussion of "everything quadratic" that began in Section 5.7, when we solved quadratic equations $ax^2 + bx + c = 0$, where the expression $ax^2 + bx + c$ was factorable. Remember, if the expression $ax^2 + bx + c$ was not factorable, we did not have a method for solving the equation. The missing piece for solving this type of quadratic equation was the idea of a radical. Having learned how to work with radicals in Chapter 7, we now have the tools necessary to expand our understanding of solving all quadratic equations $ax^2 + bx + c = 0$, where the expression $ax^2 + bx + c$ may or may not be factorable. The theme of the chapter will be to solve quadratic equations, graph quadratic functions, and solve quadratic inequalities.

# 8.1 Solving Quadratic Equations by Completing the Square

## OBJECTIVES

1. Solve Quadratic Equations Using the Square Root Property
2. Complete the Square in One Variable
3. Solve Quadratic Equations by Completing the Square
4. Solve Problems Using the Pythagorean Theorem

### *Preparing for Solving Quadratic Equations by Completing the Square*

*Before getting started, take the following readiness quiz. If you get a problem wrong, go back to the section cited and review the material.*

**1.** Multiply: $(2p + 3)^2$ [Section 5.2, p. 372]

**2.** Factor: $y^2 - 8y + 16$ [Section 5.6, pp. 409–411]

**3.** Solve: $x^2 + 5x - 14 = 0$ [Section 5.7, pp. 423–426]

**4.** Solve: $x^2 - 16 = 0$ [Section 5.7, pp. 423–426]

**5.** Simplify: (a) $\sqrt{36}$ (b) $\sqrt{45}$ (c) $\sqrt{-12}$ [Getting Ready, pp. 527–528; Section 7.3, pp. 549–551; Section 7.8, pp. 591–592]

**6.** Find the complex conjugate of $-3 + 2i$. [Section 7.8, pp. 595–596]

## 1 Solve Quadratic Equations Using the Square Root Property

Suppose that we wanted to solve the quadratic equation

$$x^2 = p$$

where $p$ is any real number. In words, this equation is saying, "give me all numbers whose square is $p$." So, if $p = 16$, then we would have the equation

$$x^2 = 16$$

which means we want "all numbers whose square is 16." There are two numbers whose square is 16, $-4$ and 4, so the solution set to the equation $x^2 = 16$ is $\{-4, 4\}$.

In general, we have the following method for solving equations of the form $x^2 = p$.

**THE SQUARE ROOT PROPERTY**

If $x^2 = p$, then $x = \sqrt{p}$ or $x = -\sqrt{p}$.

**Work Smart**

The Square Root Property is useful for solving equations of the form "some unknown squared equals a real number." To solve this equation, we take the square root of both sides of the equation, but don't forget the $\pm$ symbol to obtain the positive and negative square root.

When using the Square Root Property to solve an equation such as $x^2 = p$, we usually abbreviate the solutions as $x = \pm\sqrt{p}$, read "$x$ equals plus or minus the square root of $p$." For example, the two solutions of the equation

$$x^2 = 16$$

are

$$x = \pm\sqrt{16}$$

and since $\sqrt{16} = 4$, we have

$$x = \pm 4$$

Let's look at an example that discusses how to solve a quadratic equation using the Square Root Property.

***Preparing for...Answers***

**1.** $4p^2 + 12p + 9$ **2.** $(y - 4)^2$ **3.** $\{-7, 2\}$ **4.** $\{-4, 4\}$ **5. (a)** 6 **(b)** $3\sqrt{5}$ **(c)** $2\sqrt{3}i$ **6.** $-3 - 2i$

**EXAMPLE 1 How to Solve a Quadratic Equation Using the Square Root Property**

Solve: $p^2 - 9 = 0$

**Step-by-Step Solution**

**Step 1:** Isolate the expression containing the square term.

$$p^2 - 9 = 0$$

Add 9 to both sides: $p^2 = 9$

**Step 2:** Use the Square Root Property. Don't forget the $\pm$ symbol.

$$p = \pm\sqrt{9}$$

Simplify the radical: $= \pm 3$

(continued)

| | | |
|---|---|---|
| **Step 3:** Isolate the variable, if necessary. | The variable is already isolated. | |
| **Step 4:** Verify your solution(s). | $p = -3$: $(-3)^2 - 9 \stackrel{?}{=} 0$<br>$9 - 9 = 0$ True | $p = 3$: $3^2 - 9 \stackrel{?}{=} 0$<br>$9 - 9 = 0$ True |

The solution set is $\{-3, 3\}$.

We summarize the steps used to solve quadratic equations using the Square Root Property.

### Solving Quadratic Equations Using the Square Root Property

**Step 1:** Isolate the expression containing the square term.

**Step 2:** Use the Square Root Property, which states if $x^2 = p$, then $x = \pm\sqrt{p}$. Don't forget the $\pm$ symbol.

**Step 3:** Isolate the variable, if necessary.

**Step 4:** Verify your solution(s).

You could also solve the equation in Example 1(a) by factoring the difference of two squares:

$$p^2 - 9 = 0$$
$$(p - 3)(p + 3) = 0$$
Zero-Product Property: $p - 3 = 0$ or $p + 3 = 0$
$$p = 3 \quad \text{or} \quad p = -3$$

**Work Smart**

As our mathematical knowledge develops, we will find there is more than one way to solve a problem.

So there is more than one way to obtain the solution! However, factoring only works nicely when solving equations of the form $x^2 = p$ when $p$ is a perfect square. It doesn't work nicely when $p$ is not a perfect square as Example 2 illustrates.

## EXAMPLE 2 Solving a Quadratic Equation Using the Square Root Property

Solve: $3x^2 - 60 = 0$

### Solution

$$3x^2 - 60 = 0$$
Add 60 to both sides of the equation: $3x^2 = 60$

Divide both sides by 3: $x^2 = 20$

Use the Square Root Property: $x = \pm\sqrt{20}$

Simplify the radical: $= \pm 2\sqrt{5}$

**Check**

$x = -2\sqrt{5}$: $3(-2\sqrt{5})^2 - 60 \stackrel{?}{=} 0$

$(ab)^2 = a^2b^2$: $3(-2)^2(\sqrt{5})^2 - 60 \stackrel{?}{=} 0$

$3 \cdot 4 \cdot 5 - 60 \stackrel{?}{=} 0$

$60 - 60 = 0$ True

$x = 2\sqrt{5}$: $3(2\sqrt{5})^2 - 60 \stackrel{?}{=} 0$

$3 \cdot (2)^2(\sqrt{5})^2 - 60 \stackrel{?}{=} 0$

$3 \cdot 4 \cdot 5 - 60 \stackrel{?}{=} 0$

$60 - 60 = 0$ True

The solution set is $\{-2\sqrt{5}, 2\sqrt{5}\}$.

**QUICK ✓** *Solve the quadratic equation using the Square Root Method.*

**1.** $p^2 = 48$ **2.** $3b^2 = 75$ **3.** $s^2 - 81 = 0$

There is no reason that the solution to a quadratic equation must be real. The next example illustrates a solution to a quadratic equation that is a non-real complex number.

## EXAMPLE 3 Solving a Quadratic Equation Using the Square Root Property

Solve: $y^2 + 14 = 2$

### Solution

$$y^2 + 14 = 2$$

Subtract 14 from both sides of the equation: $y^2 = -12$

Use the Square Root Property: $y = \pm\sqrt{-12}$

Simplify the radical: $= \pm 2\sqrt{3}i$

**Check**

$y = -2\sqrt{3}i$:

$$(-2\sqrt{3}i)^2 + 14 \stackrel{?}{=} 2$$
$$(-2\sqrt{3})^2 i^2 + 14 \stackrel{?}{=} 2$$
$$4 \cdot 3 \cdot (-1) + 14 \stackrel{?}{=} 2$$
$$-12 + 14 \stackrel{?}{=} 2$$
$$2 = 2 \quad \text{True}$$

$y = 2\sqrt{3}i$:

$$(2\sqrt{3}i)^2 + 14 \stackrel{?}{=} 2$$
$$(2\sqrt{3})^2 i^2 + 14 \stackrel{?}{=} 2$$
$$4 \cdot 3 \cdot (-1) + 14 \stackrel{?}{=} 2$$
$$-12 + 14 \stackrel{?}{=} 2$$
$$2 = 2 \quad \text{True}$$

The solution set is $\{-2\sqrt{3}i, 2\sqrt{3}i\}$.

**QUICK ✓** *Solve the quadratic equation using the Square Root Property.*

**4.** $d^2 = -72$

**5.** $3q^2 + 27 = 0$

## EXAMPLE 4 Solving Quadratic Equations Using the Square Root Property

Solve:

**(a)** $(x - 2)^2 = 25$

**(b)** $(y + 5)^2 + 24 = 0$

### Solution

**(a)**

$$(x - 2)^2 = 25$$

Use the Square Root Property: $x - 2 = \pm\sqrt{25}$

Simplify the radical: $x - 2 = \pm 5$

Add 2 to each side: $x = 2 \pm 5$

$2 \pm 5$ means $2 - 5$ or $2 + 5$: $x = 2 - 5 = -3$ or $x = 2 + 5 = 7$

**Check**

$x = -3$:

$$(-3 - 2)^2 \stackrel{?}{=} 25$$
$$(-5)^2 \stackrel{?}{=} 25$$
$$25 = 25 \quad \text{True}$$

$x = 7$:

$$(7 - 2)^2 \stackrel{?}{=} 25$$
$$5^2 \stackrel{?}{=} 25$$
$$25 = 25 \quad \text{True}$$

The solution set is $\{-3, 7\}$.

**(b)**

$$(y + 5)^2 + 24 = 0$$

Subtract 24 from both sides: $(y + 5)^2 = -24$

Use the Square Root Property: $y + 5 = \pm\sqrt{-24}$

$\sqrt{-24} = \sqrt{24} \cdot \sqrt{-1} = \sqrt{4 \cdot 6}\,i = 2\sqrt{6}\,i$: $y + 5 = \pm 2\sqrt{6}\,i$

Subtract 5 from each side: $y = -5 \pm 2\sqrt{6}\,i$

$$y = -5 - 2\sqrt{6}\,i \quad \text{or} \quad y = -5 + 2\sqrt{6}\,i$$

**Check**

$y = -5 - 2\sqrt{6}\,i$:

$$(-5 - 2\sqrt{6}\,i + 5)^2 + 24 \stackrel{?}{=} 0$$
$$(-2\sqrt{6}\,i)^2 + 24 \stackrel{?}{=} 0$$
$$4 \cdot 6 \cdot i^2 + 24 \stackrel{?}{=} 0$$
$$-24 + 24 = 0 \quad \text{True}$$

$y = -5 + 2\sqrt{6}\,i$:

$$(-5 + 2\sqrt{6}\,i + 5)^2 + 24 \stackrel{?}{=} 0$$
$$(2\sqrt{6}\,i)^2 + 24 \stackrel{?}{=} 0$$
$$4 \cdot 6 \cdot i^2 + 24 \stackrel{?}{=} 0$$
$$-24 + 24 = 0 \quad \text{True}$$

The solution set is $\{-5 - 2\sqrt{6}\,i, -5 + 2\sqrt{6}\,i\}$.

**QUICK ✓** *Solve the quadratic equation using the Square Root Property.*

**6.** $(y + 3)^2 = 100$

**7.** $(q - 5)^2 + 20 = 4$

## 2 Complete the Square in One Variable

We now introduce the method of **completing the square.** The idea behind completing the square in one variable is to "adjust" the left side of a quadratic equation of the form $x^2 + bx + c$ in order to make it a perfect square trinomial. Recall that perfect square trinomials are trinomials of the form

$$A^2 + 2AB + B^2 = (A + B)^2$$

or

$$A^2 - 2AB + B^2 = (A - B)^2$$

For example, $x^2 + 6x + 9$ is a perfect square trinomial because $x^2 + 6x + 9 = (x + 3)^2$. Or $p^2 - 12p + 36$ is a perfect square trinomial because $p^2 - 12p + 36 = (p - 6)^2$.

We "adjust" the left side of $x^2 + bx + c$ by adding a number to make it a perfect square trinomial. For example, to make $x^2 + 6x$ a perfect square we would add 9. But where does this 9 come from? If we divide the coefficient on the first-degree term, 6, by 2, and then square the result, we obtain 9. This approach works in general.

**Work Smart**

To complete the square, the coefficient of $x^2$ must be 1.

**OBTAINING A PERFECT SQUARE TRINOMIAL**

Identify the coefficient of the first-degree term. Multiply this coefficient by $\frac{1}{2}$ and then square the result. That is, determine the value of $b$ in $x^2 + bx + c$ and compute $\left(\frac{1}{2}b\right)^2$.

**EXAMPLE 5 Obtaining a Perfect Square Trinomial**

Determine the number that must be added to each expression in order to make it a perfect square trinomial. Then factor the expression.

| Start | Add | Result | Factored Form |
|---|---|---|---|
| $y^2 + 8y$ | $\left(\frac{1}{2}\cdot 8\right)^2 = 16$ | $y^2 + 8y + 16$ | $(y + 4)^2$ |
| $x^2 + 12x$ | $\left(\frac{1}{2}\cdot 12\right)^2 = 36$ | $x^2 + 12x + 36$ | $(x + 6)^2$ |
| $a^2 - 20a$ | $\left(\frac{1}{2}\cdot(-20)\right)^2 = 100$ | $a^2 - 20a + 100$ | $(a - 10)^2$ |
| $p^2 - 5p$ | $\left(\frac{1}{2}\cdot(-5)\right)^2 = \frac{25}{4}$ | $p^2 - 5p + \frac{25}{4}$ | $\left(p - \frac{5}{2}\right)^2$ |

**Work Smart**

It is common to write the value $\left(\frac{1}{2}b\right)^2$ as a fraction, not a decimal.

Did you notice in the factored form that the perfect square trinomial always factors so that

$$x^2 + bx + \left(\frac{b}{2}\right)^2 = \left(x + \frac{b}{2}\right)^2 \quad \text{or} \quad x^2 - bx + \left(\frac{b}{2}\right)^2 = \left(x - \frac{b}{2}\right)^2$$

That is, the perfect square trinomial will always factor as $\left(x \pm \frac{b}{2}\right)^2$, where we use the + if the coefficient of the first-degree term is positive and we use the − if the coefficient of the first-degree term is negative. The $\frac{b}{2}$ represents $\frac{1}{2}$ the value of the coefficient of the first-degree term.

**QUICK ✓** *Determine the number that must be added to the expression to make it a perfect square trinomial. Then factor the expression.*

**8.** $p^2 + 14p$ **9.** $w^2 + 3w$

**Figure 1**

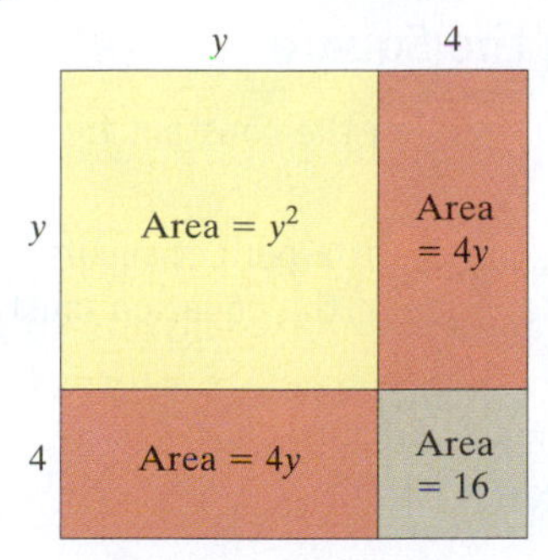


Are you wondering why we call making an expression a perfect square trinomial "completing the square"? Consider the expression $y^2 + 8y$ given in Example 5. We can geometrically represent this algebraic expression as shown in Figure 1. The yellow area is $y^2$ and each orange area is $4y$ (for a total area of $8y$). But what is the area of the green region in order to make the square complete? The dimensions of the green region must be 4 by 4, so the area of the green region is 16. The area of the entire square region, $(y + 4)^2$, equals the sum of the area of the regions that make up the square: $y^2 + 4y + 4y + 16 = y^2 + 8y + 16$.

## 3 Solve Quadratic Equations by Completing the Square

Up to this point, we have only been able to solve quadratic equations of the form $ax^2 + bx + c = 0$ when $ax^2 + bx + c$ was factorable. This raises the question, Is there a method to solve $ax^2 + bx + c = 0$ when $ax^2 + bx + c$ is not factorable? The answer is yes! We begin by presenting a method for solving quadratic equations of the form $x^2 + bx + c = 0$. That is, the coefficient of the square term is 1.

**EXAMPLE 6** **How to Solve a Quadratic Equation by Completing the Square**

Solve: $x^2 + 6x + 1 = 0$

### Step-by-Step Solution

**Step 1:** Rewrite $x^2 + bx + c = 0$ as $x^2 + bx = -c$ by subtracting the constant from both sides of the equation.

$$x^2 + 6x + 1 = 0$$

Subtract 1 from both sides:
$$x^2 + 6x = -1$$

**Step 2:** Complete the square in the expression $x^2 + bx$ by making it a perfect square trinomial.

$\left(\frac{1}{2}\cdot 6\right)^2 = 9$; Add 9 to both sides:
$$x^2 + 6x + 9 = -1 + 9$$
$$x^2 + 6x + 9 = 8$$

**Step 3:** Factor the perfect square trinomial on the left side of the equation.

$x^2 + 6x + 9 = (x + 3)^2$:
$$(x + 3)^2 = 8$$

**Step 4:** Solve the equation using the Square Root Property.

$$x + 3 = \pm\sqrt{8}$$

$\sqrt{8} = 2\sqrt{2}$:
$$x + 3 = \pm 2\sqrt{2}$$

Subtract 3 from both sides:
$$x = -3 \pm 2\sqrt{2}$$

$a \pm b$ means $a - b$ or $a + b$:
$$x = -3 - 2\sqrt{2} \quad \text{or} \quad x = -3 + 2\sqrt{2}$$

**Step 5:** Verify your solution(s).

$$x^2 + 6x + 1 = 0$$

$x = -3 - 2\sqrt{2}$:
$$(-3 - 2\sqrt{2})^2 + 6(-3 - 2\sqrt{2}) + 1 \stackrel{?}{=} 0$$
$$9 + 12\sqrt{2} + 8 - 18 - 12\sqrt{2} + 1 \stackrel{?}{=} 0$$
$$0 = 0 \quad \text{True}$$

$x = -3 + 2\sqrt{2}$:
$$(-3 + 2\sqrt{2})^2 + 6(-3 + 2\sqrt{2}) + 1 \stackrel{?}{=} 0$$
$$9 - 12\sqrt{2} + 8 - 18 + 12\sqrt{2} + 1 \stackrel{?}{=} 0$$
$$0 = 0 \quad \text{True}$$

The solution set is $\{-3 - 2\sqrt{2}, -3 + 2\sqrt{2}\}$.

The following is a summary of the steps used to solve a quadratic equation by completing the square.

### Solving a Quadratic Equation by Completing the Square

**Step 1:** Rewrite $x^2 + bx + c = 0$ as $x^2 + bx = -c$ by subtracting the constant from both sides of the equation.

**Step 2:** Complete the square in the expression $x^2 + bx$ by making it a perfect square trinomial. Don't forget, whatever you add to the left side of the equation must also be added to the right side.

**Step 3:** Factor the perfect square trinomial on the left side of the equation.

**Step 4:** Solve the equation using the Square Root Property.

**Step 5:** Verify your solutions.

**QUICK ✓** *Solve the equation by completing the square.*

**10.** $b^2 + 2b - 8 = 0$

**Work Smart**

If the coefficient of the square term is not 1, we must divide each side of the equation by the coefficient of the square term so that it becomes 1 before using the method of completing the square.

Up to this point we have only looked at quadratic equations where the coefficient of the square term is 1. When the coefficient of the square term is not 1, we multiply or divide both sides of the equation by a nonzero constant so this coefficient becomes 1. The next example demonstrates this method.

### EXAMPLE 7 Solving a Quadratic Equation by Completing the Square When the Coefficient of the Square Term Is Not 1

Solve: $2x^2 + 4x + 3 = 0$

#### Solution

First, we notice that the coefficient of the square term is 2, so we divide both sides of the equation by 2 in order to make the coefficient of the square term equal to 1.

$$2x^2 + 4x + 3 = 0$$

Divide both sides of the equation by 2: $$\frac{2x^2 + 4x + 3}{2} = \frac{0}{2}$$

Simplify: $$x^2 + 2x + \frac{3}{2} = 0$$

Now we are ready to solve the equation by completing the square.

Subtract $\frac{3}{2}$ from both sides: $$x^2 + 2x = -\frac{3}{2}$$

$\left(\frac{1}{2} \cdot 2\right)^2 = 1$; Add 1 to both sides: $$x^2 + 2x + 1 = -\frac{3}{2} + 1$$

Simplify: $$x^2 + 2x + 1 = -\frac{1}{2}$$

Factor expression on left: $$(x + 1)^2 = -\frac{1}{2}$$

Use Square Root Property: $$x + 1 = \pm\sqrt{-\frac{1}{2}}$$

$\sqrt{\frac{a}{b}} = \frac{\sqrt{a}}{\sqrt{b}}$: $$x + 1 = \pm\frac{\sqrt{-1}}{\sqrt{2}}$$

$\frac{\sqrt{-1}}{\sqrt{2}} = \frac{i}{\sqrt{2}} = \frac{\sqrt{2}}{2}i$: $$x + 1 = \pm\frac{\sqrt{2}}{2}i$$

Subtract 1 from both sides: $$x = -1 \pm \frac{\sqrt{2}}{2}i$$

$a \pm b$ means $a - b$ or $a + b$: $$x = -1 - \frac{\sqrt{2}}{2}i \quad \text{or} \quad x = -1 + \frac{\sqrt{2}}{2}i$$

**Work Smart**

Notice the solutions in Example 7 are conjugates of each other.

We leave it to you to verify the solutions. The solution set is $\left\{-1 - \frac{\sqrt{2}}{2}i, -1 + \frac{\sqrt{2}}{2}i\right\}$.

**QUICK ✓** *Solve the quadratic equation by completing the square.*

**11.** $2q^2 + 6q - 1 = 0$

## 4 Solve Problems Using the Pythagorean Theorem

**Figure 2**

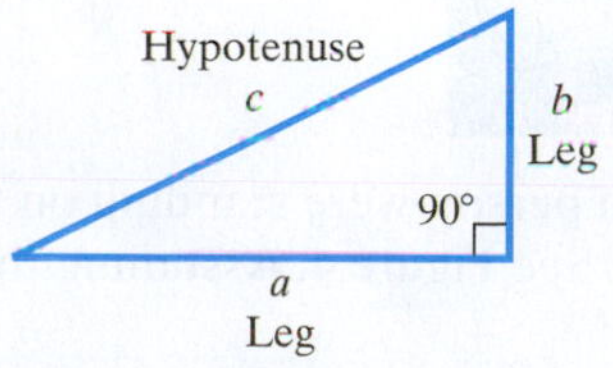

The Pythagorean Theorem is a statement about *right triangles.* A **right triangle** is one that contains a **right angle,** that is, an angle of 90°. The side of the triangle opposite the 90° angle is called the **hypotenuse;** the remaining two sides are called **legs.** In Figure 2 we use $c$ to represent the length of the hypotenuse and $a$ and $b$ to represent the lengths of the legs. Notice the use of the symbol ⌐ to show the 90° angle.

We now state the Pythagorean Theorem.

**THE PYTHAGOREAN THEOREM**

In a right triangle, the square of the length of the hypotenuse is equal to the sum of the squares of the lengths of the legs. That is, in the right triangle shown in Figure 2,

$$c^2 = a^2 + b^2$$

## EXAMPLE 8 Finding the Hypotenuse of a Right Triangle

In a right triangle, one leg is of length 5 inches and the other is of length 12 inches. What is the length of the hypotenuse?

### Solution

Since the triangle is a right triangle, we use the Pythagorean Theorem with $a = 5$ and $b = 12$ to find the length $c$ of the hypotenuse.

$$\begin{aligned} c^2 &= a^2 + b^2 \\ c^2 &= 5^2 + 12^2 \\ &= 25 + 144 \\ &= 169 \end{aligned}$$

We now use the Square Root Property to find $c$, the length of the hypotenuse.

$$c = \sqrt{169} = 13$$

The length of the hypotenuse is 13 inches. Notice that we only find the positive square root of 169 since $c$ represents the length of a side of a triangle and a negative length does not make sense.

**QUICK ✓** *The lengths of the legs of a right triangle are given. Find the length of the hypotenuse.*

**12.** $a = 3, b = 4$  **13.** $a = 6, b = 6$

## EXAMPLE 9 How Far Can You See?

The Currituck Lighthouse is located in Corolla, North Carolina. As part of North Carolina's Outer Banks, the lighthouse was completed in 1875. It stands 162 feet tall with the observation deck located 158 feet above the ground. See Figure 3.

**Figure 3**

The Web site for the Currituck Lighthouse states that if a person were standing on the observation deck, they could see approximately 18 miles. See Figure 4. Assuming that the radius of the Earth is 3960 miles, verify this claim.

**Figure 4**

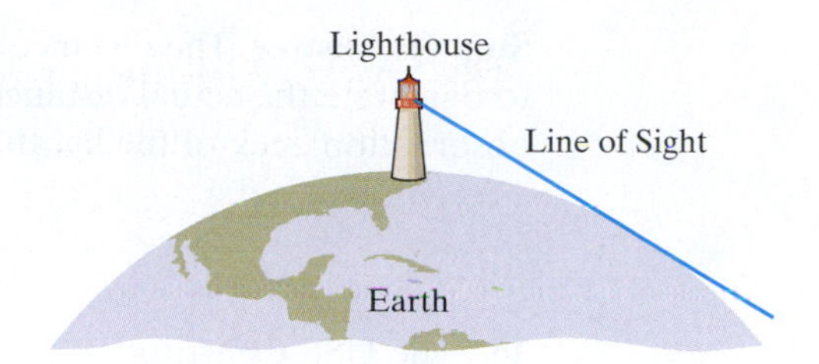


## Solution

**Step 1: Identify** We want to know how far a person can see from the lighthouse.

**Step 2: Name** We will call this unknown distance, $d$.

**Step 3: Translate** To help with the translation, we draw a picture. From the Center of Earth, draw two lines: one through the lighthouse and the other to the farthest point a person can see from the lighthouse. See Figure 5.

**Figure 5**

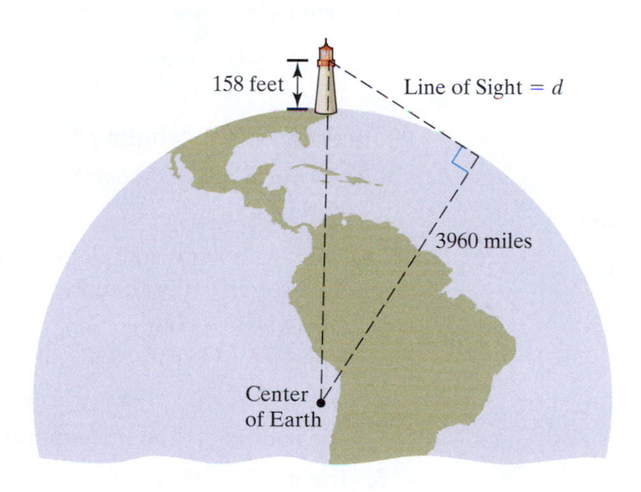


The line of sight and the two lines drawn from the center of Earth form a right triangle. So the angle where the line of sight touches the horizon measures 90°. From the Pythagorean Theorem, we know that

$$\text{Hypotenuse}^2 = \text{Leg}^2 + \text{Leg}^2$$

The length of the hypotenuse is 3960 miles plus 158 feet. We can't add 3960 miles to 158 feet without first converting the height of the tower to miles. Since 158 feet $= 158 \text{ feet} \cdot \dfrac{1 \text{ mile}}{5280 \text{ feet}} = \dfrac{158}{5280}$ mile, we have that the hypotenuse is $\left(3960 + \dfrac{158}{5280}\right)$ miles. One of the legs is 3960 miles. The length of the other leg is our unknown, $d$. So we have

$$3960^2 + d^2 = \left(3960 + \frac{158}{5280}\right)^2 \quad \text{The Model}$$

**Step 4: Solve**

$$3960^2 + d^2 = \left(3960 + \frac{158}{5280}\right)^2$$

Subtract $3960^2$ from both sides: $$d^2 = \left(3960 + \frac{158}{5280}\right)^2 - 3960^2$$

Use a calculator: $$d^2 \approx 237.000895$$

Square Root Property: $$d \approx \sqrt{237.000895}$$
$$\approx 15.39 \text{ miles}$$

**Step 5: Check** Our answer is less than the distance given on the Web site.

**Step 6: Answer** The distance given on the Currituck Lighthouse Web site appears to overstate the actual distance a person could see. Someone standing on the observation deck of the lighthouse could see about 15.39 miles.

QUICK ✓

**14.** The USS Constitution (aka *Old Ironsides*) is perhaps the most famous ship from United States Naval history. The mainmast of the Constitution is 220 feet high. Suppose that a sailor climbs the mainmast to a height of 200 feet in order to look for enemy vessels. How far could the sailor see? Assume the radius of the Earth is 3960 miles.

## 8.1 Exercises

*For Extra Help:*

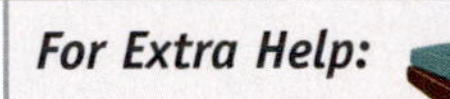

Student Solutions Manual | CD Video | PH Math/Tutor Center | MathXL Tutorials on CD | MathXL® | MyMathLab

### Concepts and Vocabulary

*In Problems 1–3, fill in the blanks.*

**1.** If $x^2 = p$, then $x =$ _____ or $x =$ _____.

**2.** The idea behind completing the square is to adjust the left side of a quadratic equation in order to make it a _____ _____ _____.

**3.** The side of a right triangle opposite the 90° angle is called the _____; the remaining two sides are called _____.

*In Problems 4–6, answer True or False to each statement.*

**4.** The solution set of the equation $(z - 2)^2 - 25 = 0$ is $\{-3, 7\}$.

**5.** The method of completing the square to solve a quadratic equation only works when the coefficient on the square term is 1.

**6.** The Pythagorean Theorem states that for any triangle, the length of the hypotenuse is equal to the sum of the squares of the lengths of the legs.

**7.** Explain in your own words what the expression "completing the square" means. You may want to use a figure similar to Figure 1 to assist in your explanation.

**8.** What would be your first step in solving the quadratic equation $3x^2 - 6x + 12 = 0$? Why?

### Building Skills

*In Problems 9–36, solve each equation using the Square Root Property.*

**9.** $y^2 = 100$

**10.** $x^2 = 81$

**11.** $p^2 = 50$

**12.** $z^2 = 48$

**13.** $m^2 = -25$

**14.** $n^2 = -49$

**15.** $w^2 = \frac{5}{4}$

**16.** $z^2 = \frac{8}{9}$

**17.** $x^2 + 5 = 13$

**18.** $w^2 - 6 = 14$

**19.** $3z^2 = 48$

**20.** $4y^2 = 100$

**21.** $3x^2 = 8$

**22.** $5y^2 = 32$

**23.** $2p^2 + 23 = 15$

**24.** $-3x^2 - 5 = 22$

**25.** $(z + 3)^2 = 64$

**26.** $(y - 2)^2 = 9$

**27.** $(d - 1)^2 = -18$

**28.** $(z + 4)^2 = -24$

**29.** $3(q + 5)^2 - 1 = 8$

**30.** $5(x - 3)^2 + 2 = 27$

**31.** $(3q + 1)^2 = 9$

**32.** $(2p + 3)^2 = 16$

**33.** $\left(x - \frac{2}{3}\right)^2 = \frac{5}{9}$

**34.** $\left(y + \frac{3}{2}\right)^2 = \frac{3}{4}$

**35.** $x^2 + 8x + 16 = 81$

**36.** $q^2 - 6q + 9 = 16$

*In Problems 37–46, complete the square in each expression. Then factor the perfect square trinomial.*

**37.** $x^2 + 10x$ **38.** $y^2 + 16y$ **39.** $z^2 - 18z$ **40.** $p^2 - 4p$

**41.** $y^2 + 7y$ **42.** $x^2 + x$ **43.** $w^2 + \frac{1}{2}w$ **44.** $z^2 - \frac{1}{3}z$

**45.** $q^2 - \frac{3}{5}q$ **46.** $m^2 + \frac{5}{2}m$

*In Problems 47–74, solve each quadratic equation by completing the square.*

**47.** $w^2 - 5w = 14$ **48.** $z^2 - 6z = 7$ **49.** $x^2 + 4x - 12 = 0$

**50.** $y^2 + 3y - 18 = 0$ **51.** $x^2 - 4x + 1 = 0$ **52.** $p^2 - 6p + 4 = 0$

**53.** $z^2 + 8z + 9 = 0$ **54.** $b^2 + 10b + 19 = 0$ **55.** $a^2 - 4a + 5 = 0$

**56.** $m^2 - 2m + 5 = 0$ **57.** $b^2 + 5b - 2 = 0$ **58.** $q^2 + 7q + 7 = 0$

**59.** $p^2 - 3p - 2 = 0$ **60.** $x^2 - 5x - 3 = 0$ **61.** $m^2 = 8m + 3$

**62.** $n^2 = 10n + 5$ **63.** $p^2 - p + 3 = 0$ **64.** $z^2 - 3z + 5 = 0$

**65.** $2y^2 - 5y - 12 = 0$ **66.** $3a^2 - 4a - 4 = 0$ **67.** $3y^2 - 6y + 2 = 0$

**68.** $2y^2 - 2y - 1 = 0$ **69.** $2z^2 - 5z + 1 = 0$ **70.** $2x^2 - 7x + 2 = 0$

**71.** $2x^2 + 4x + 5 = 0$ **72.** $2z^2 + 6z + 5 = 0$ **73.** $3b^2 + b - 5 = 0$

**74.** $3m^2 + 2m - 7 = 0$

*In Problems 75–84, the lengths of the legs of a right triangle are given. Find the hypotenuse. Give exact answers and decimal approximations rounded to two decimal places.*

**75.** $a = 6, b = 8$ **76.** $a = 7, b = 24$ **77.** $a = 12, b = 16$

**78.** $a = 15, b = 8$ **79.** $a = 5, b = 5$ **80.** $a = 3, b = 3$

**81.** $a = 1, b = \sqrt{3}$ **82.** $a = 2, b = \sqrt{5}$ **83.** $a = 6, b = 10$

**84.** $a = 8, b = 10$

*In Problems 85–88, use the right triangle shown to the right and find the missing length. Give exact answers and decimal approximations rounded to two decimal places.*

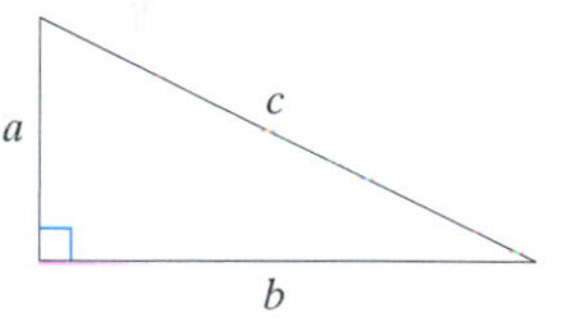


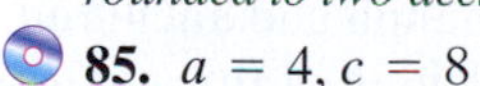

**85.** $a = 4, c = 8$ **86.** $a = 4, c = 10$

**87.** $b = 8, c = 12$ **88.** $b = 2, c = 10$

**89.** Given that $f(x) = (x - 3)^2$, find all $x$ such that $f(x) = 36$.

**90.** Given that $f(x) = (x - 5)^2$, find all $x$ such that $f(x) = 49$.

**91.** Given that $g(x) = (x + 2)^2$, find all $x$ such that $g(x) = 18$.

**92.** Given that $h(x) = (x + 1)^2$, find all $x$ such that $h(x) = 32$.

## Applying the Concepts

*In Problems 93 and 94, find the length of the diagonal in each figure.*

**93.**

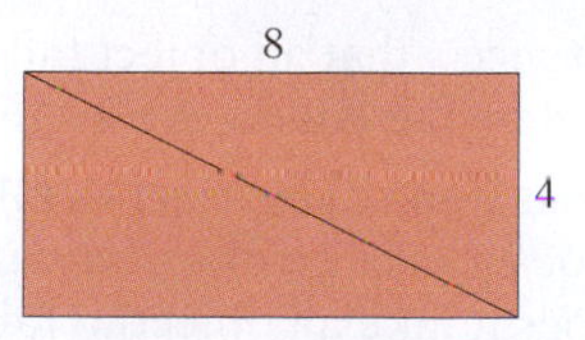


**94.**

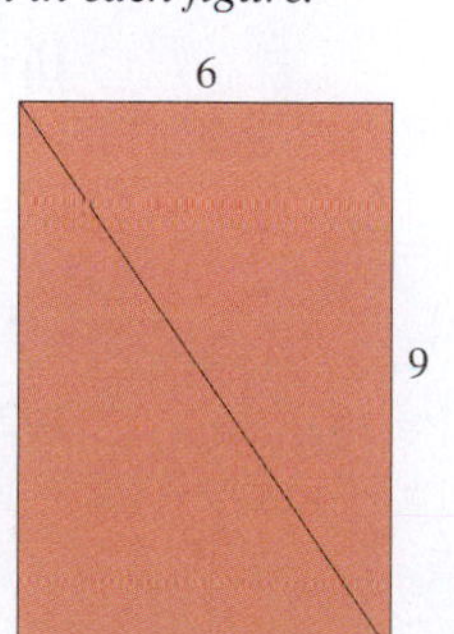

**95. Golf** A golfer hits an errant tee shot that lands in the rough. The golfer finds that the ball is exactly 30 yards to the right of the 100 yard marker, which indicates the distance to the center of the green as shown in the figure. How far is the ball from the center of the green?

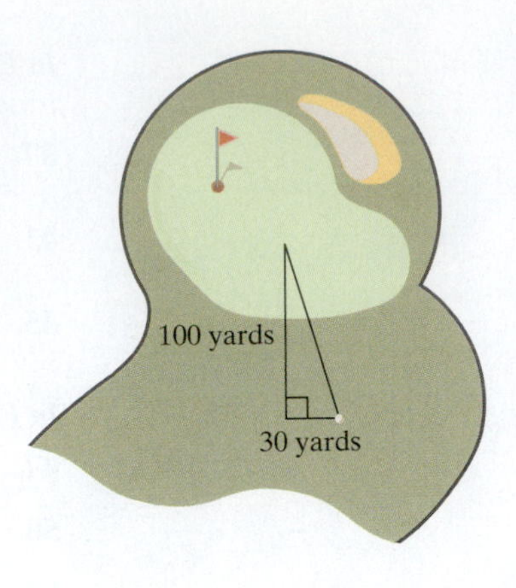


**96. Baseball** Jermaine Dye plays right field for the Chicago White Sox. He catches a fly ball 40 feet from the right field foul line, as indicated in the figure. How far is it to home plate?

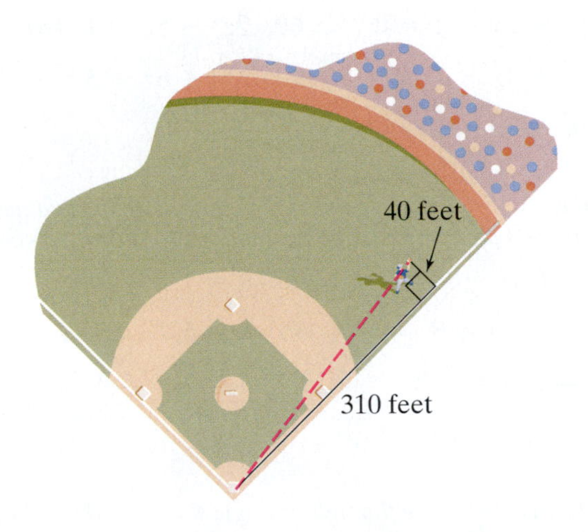


**97. Guy Wire** A guy wire is a wire used to support telephone poles. Suppose that a guy wire is located 30 feet up a telephone pole and is anchored to the ground 10 feet from the base of the pole. How long is the guy wire?

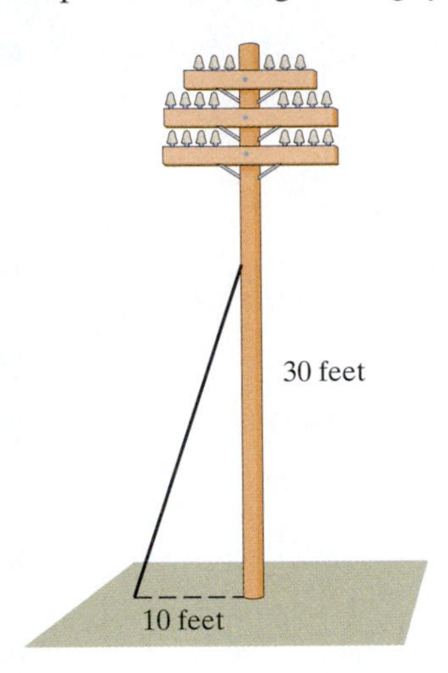


**98. Guy Wire** A guy wire is used to support an antenna on a roof top. The wire is located 40 feet up on the antenna and anchored to the roof 8 feet from the base of the antenna. What is the length of the guy wire?

**99. Ladder** Bob needs to wash the windows on his house. He has a 25-foot ladder and places the base of the ladder 10 feet from the wall on the house.

**(a)** How far up the wall will the ladder reach?

**(b)** If his windows are 20 feet above the ground, what is the farthest distance the base of the ladder can be from the wall?

**100. Fire Truck Ladder** A fire truck has a 75-foot ladder. If the truck can safely park 20 feet from a building, how far up the building can the ladder reach assuming that the top of the base of the ladder is resting on top of the truck and the truck is 10 feet tall?

**101. Gravity** The distance $s$ that an object falls (in feet) after $t$ seconds ignoring air resistance is given by the equation $s = 16t^2$.

**(a)** How long does it take an object to fall 16 feet?

**(b)** How long does it take an object to fall 48 feet?

**(c)** How long does it take an object to fall 64 feet?

△ **102. Equilateral Triangles** An equilateral triangle is one whose sides are all the same length. The area $A$ of an equilateral triangle whose sides are each length $x$ is given by $A = \frac{\sqrt{3}}{4}x^2$.

**(a)** What is the length of each side of an equilateral triangle whose area is $\frac{8\sqrt{3}}{9}$ square feet?

**(b)** What is the length of each side of an equilateral triangle whose area is $\frac{25\sqrt{3}}{4}$ square meters?

*Problems 103 and 104 are based on the following discussion. If P dollars are invested today at an annual interest rate r compounded once a year, then the value of the account A after 2 years is given by the formula* $A = P(1 + r)^2$.

**103. Value of Money** Find the rate of interest required to turn an investment of \$1000 into \$1200 after 2 years. Express your answer as a percent rounded to two decimal places.

**104. Value of Money** Find the rate of interest required to turn an investment of \$5000 into \$6200 after 2 years. Express your answer as a percent rounded to two decimal places.

## Extending the Concepts

*If you look carefully at the Pythagorean Theorem, it states, "If we have a right triangle, then* $c^2 = a^2 + b^2$, *where c is the length of the hypotenuse." In this theorem "a right triangle" represents the hypothesis, while "*$c^2 = a^2 + b^2$*" represents the conclusion. The* ***converse*** *of a theorem interchanges the hypothesis and conclusion. The converse of the Pythagorean Theorem is true.*

**CONVERSE OF THE PYTHAGOREAN THEOREM**

In a triangle, if the square of the length of one side equals the sum of the squares of the lengths of the other two sides, then the triangle is a right triangle. The 90° angle is opposite the longest side.

*In Problems 105–108, the lengths of the sides of a triangle are given. Determine if the triangle is a right triangle. If it is, identify the hypotenuse.*

**105.** 8, 15, 17

**106.** 4, 6, 8

**107.** 14, 18, 20

**108.** 20, 48, 52

**109. Pythagorean Triples** Suppose that $m$ and $n$ are positive integers with $m > n$. If $a = m^2 - n^2$, $b = 2mn$, and $c = m^2 + n^2$, show that $a$, $b$, and $c$ are the lengths of the sides of a right triangle using the Converse of the Pythagorean Theorem. We call any numbers $a$, $b$, and $c$ found from the above formulas **Pythagorean Triples.**

**110.** Solve $ax^2 + bx + c = 0$ for $x$ by completing the square.

## Synthesis Review

*In Problems 111–114, solve each equation.*

**111.** $a^2 - 5a - 36 = 0$

**112.** $p^2 + 4p = 32$

**113.** $|4q + 1| = 3$

**114.** $\left|\frac{3}{4}w - \frac{2}{3}\right| = \frac{5}{2}$

**115.** In Problems 111–114, you are asked to solve quadratic and absolute value equations. In solving both types of equations, we reduce the equation to a simpler equation. What is the simpler equation?

# 8.2 Solving Quadratic Equations by the Quadratic Formula

**OBJECTIVES**

1. Solve Quadratic Equations Using the Quadratic Formula
2. Use the Discriminant to Determine the Nature of Solutions in a Quadratic Equation
3. Model and Solve Problems Involving Quadratic Equations

## *Preparing for Solving Quadratic Equations by the Quadratic Formula*

*Before getting started, take the following readiness quiz. If you get a problem wrong, go back to the section cited and review the material.*

**1.** Simplify: (a) $\sqrt{54}$ (b) $\sqrt{121}$ [Getting Ready, pp. 527–528; Section 7.3, pp. 549–551]

**2.** Simplify: (a) $\sqrt{-9}$ (b) $\sqrt{-72}$ [Section 7.8, pp. 591–592]

**3.** Simplify: $\dfrac{3 + \sqrt{18}}{6}$ [Section 7.3, p. 551]

At this stage of the course, we know three methods for solving quadratic equations: (1) factoring, (2) The Square Root Property, and (3) completing the square. Why do we need three methods? Well, each method provides the "quickest" route to the solution when used appropriately. For example, if the quadratic expression is easy to factor, the method of factoring will get you to the solution faster than the other two. If the equation is in the form $x^2 = p$, using the Square Root Property is fastest. If the quadratic expression is not factorable, we have no choice but to complete the square. But the method of completing the square is tedious. So you may be asking yourself if there is an alternative to this method. The answer is yes!

## 1 Solve Quadratic Equations Using the Quadratic Formula

We can use the method of completing the square to obtain a general formula for solving the quadratic equation

$$ax^2 + bx + c = 0, \qquad a \neq 0$$

To complete the square, we first must get the constant $c$ on the right-hand side of the equation.

$$ax^2 + bx = -c, \qquad a \neq 0$$

Since $a \neq 0$, we can divide both sides of the equation by $a$ to get

$$x^2 + \frac{b}{a}x = -\frac{c}{a}$$

Now that the coefficient of $x^2$ is 1, we complete the square on the left side by adding the square of $\frac{1}{2}$ of the coefficient of $x$ to both sides of the equation. That is, we add

$$\left(\frac{1}{2}\cdot\frac{b}{a}\right)^2 = \frac{b^2}{4a^2}$$

to both sides of the equation. We now have

$$x^2 + \frac{b}{a}x + \frac{b^2}{4a^2} = -\frac{c}{a} + \frac{b^2}{4a^2}$$

or

$$x^2 + \frac{b}{a}x + \frac{b^2}{4a^2} = \frac{b^2}{4a^2} - \frac{c}{a}$$

We combine like terms on the right-hand side by writing the right-hand side over a common denominator. The least common denominator on the right-hand side is $4a^2$. So we multiply $-\frac{c}{a}$ by $\frac{4a}{4a}$:

$$x^2 + \frac{b}{a}x + \frac{b^2}{4a^2} = \frac{b^2}{4a^2} - \frac{c}{a}\cdot\frac{4a}{4a}$$

$$x^2 + \frac{b}{a}x + \frac{b^2}{4a^2} = \frac{b^2}{4a^2} - \frac{4ac}{4a^2}$$

$$x^2 + \frac{b}{a}x + \frac{b^2}{4a^2} = \frac{b^2 - 4ac}{4a^2}$$

*Preparing for...Answers* **1. (a)** $3\sqrt{6}$ **(b)** 11 **2. (a)** $3i$ **(b)** $6\sqrt{2}i$ **3.** $\frac{1}{2} + \frac{\sqrt{2}}{2}$ or $\frac{1 + \sqrt{2}}{2}$

**Work Smart**

To factor any perfect square trinomial of the form $x^2 + bx + c$, we write $\left(x + \frac{b}{2}\right)^2$.

The expression on the left-hand side is a perfect square trinomial. We factor the left-hand side and obtain

$$\left(x + \frac{b}{2a}\right)^2 = \frac{b^2 - 4ac}{4a^2}$$

At this point we will assume that $a > 0$ (you'll see why in a little while). This assumption does not compromise the results because if $a < 0$, we could multiply both sides of the equation $ax^2 + bx + c = 0$ by $-1$ to make it positive. With this assumption, we use the Square Root Property and get

$$x + \frac{b}{2a} = \pm\sqrt{\frac{b^2 - 4ac}{4a^2}}$$

$\sqrt{\frac{a}{b}} = \frac{\sqrt{a}}{\sqrt{b}}$: $$x + \frac{b}{2a} = \pm\frac{\sqrt{b^2 - 4ac}}{\sqrt{4a^2}}$$

$\sqrt{4a^2} = 2a$ since $a > 0$: $$x + \frac{b}{2a} = \pm\frac{\sqrt{b^2 - 4ac}}{2a}$$

Subtract $\frac{b}{2a}$ from both sides: $$x = -\frac{b}{2a} \pm \frac{\sqrt{b^2 - 4ac}}{2a}$$

Write over a common denominator: $$x = \frac{-b \pm \sqrt{b^2 - 4ac}}{2a}$$

**In Words**

The quadratic formula says that the solution(s) to the equation $ax^2 + bx + c = 0$ is (are) "the opposite of $b$ plus or minus the square root of $b$ squared minus $4ac$ all over $2a$."

This gives us the *quadratic formula.*

**THE QUADRATIC FORMULA**

The solution(s) to the quadratic equation $ax^2 + bx + c = 0$, $a \neq 0$, are given by the **quadratic formula**

$$x = \frac{-b \pm \sqrt{b^2 - 4ac}}{2a}$$

**Work Smart: Study Skill**

When solving homework problems always write the quadratic formula as part of the solution so you "accidentally" memorize the formula.

## EXAMPLE 1 How to Solve a Quadratic Equation Using the Quadratic Formula

Solve: $12x^2 + 5x - 3 = 0$

### Step-by-Step Solution

**Step 1:** Write the equation in standard form $ax^2 + bx + c = 0$ and identify the values of $a$, $b$, and $c$.

$a = 12$, $b = 5$, $c = -3$

$$12x^2 + 5x - 3 = 0$$

**Step 2:** Substitute the values of $a$, $b$, and $c$ into the quadratic formula.

$$x = \frac{-b \pm \sqrt{b^2 - 4ac}}{2a}$$

$$x = \frac{-5 \pm \sqrt{5^2 - 4(12)(-3)}}{2(12)}$$

(continued)

**Step 3:** Simplify the expression found in Step 2.

$$= \frac{-5 \pm \sqrt{25 + 144}}{24}$$

$$= \frac{-5 \pm \sqrt{169}}{24}$$

$\sqrt{169} = 13$: $$= \frac{-5 \pm 13}{24}$$

$a \pm b$ means $a - b$ or $a + b$: $$x = \frac{-5 - 13}{24} \quad \text{or} \quad x = \frac{-5 + 13}{24}$$

$$= \frac{-18}{24} \quad \text{or} \quad = \frac{8}{24}$$

Reduce: $$= -\frac{3}{4} \quad \text{or} \quad = \frac{1}{3}$$

**Step 4:** Check

$$12x^2 + 5x - 3 = 0$$

$x = -\frac{3}{4}$: $$12\left(-\frac{3}{4}\right)^2 + 5\left(-\frac{3}{4}\right) - 3 \stackrel{?}{=} 0$$

$$12 \cdot \frac{9}{16} - \frac{15}{4} - 3 \stackrel{?}{=} 0$$

$$\frac{27}{4} - \frac{15}{4} - 3 \stackrel{?}{=} 0$$

$$\frac{12}{4} - 3 \stackrel{?}{=} 0$$

$$0 = 0 \quad \text{True}$$

$x = \frac{1}{3}$: $$12\left(\frac{1}{3}\right)^2 + 5\left(\frac{1}{3}\right) - 3 \stackrel{?}{=} 0$$

$$12 \cdot \frac{1}{9} + \frac{5}{3} - 3 \stackrel{?}{=} 0$$

$$\frac{4}{3} + \frac{5}{3} - 3 \stackrel{?}{=} 0$$

$$\frac{9}{3} - 3 \stackrel{?}{=} 0$$

$$0 = 0 \quad \text{True}$$

The solution set is $\left\{-\frac{3}{4}, \frac{1}{3}\right\}$.

**Work Smart**

If $b^2 - 4ac$ is a perfect square, then the quadratic equation can be solved by factoring.

Notice that the solutions to the equation in Example 1 are rational numbers and the expression $b^2 - 4ac$ under the radical in the quadratic formula, 169, is a perfect square. This leads to a generalization. Whenever the expression $b^2 - 4ac$ is a perfect square, then the quadratic equation will have rational solutions and the quadratic equation can be solved by factoring.

We summarize the steps used to solve a quadratic equation using the quadratic formula.

### Solving a Quadratic Equation Using the Quadratic Formula

**Step 1:** Write the equation in standard form $ax^2 + bx + c = 0$ and identify the values of $a$, $b$, and $c$.

**Step 2:** Substitute the values of $a$, $b$, and $c$ into the quadratic formula.

**Step 3:** Simplify the expression found in Step 2.

**Step 4:** Verify your solution(s).

**QUICK ✓** *Solve each equation using the quadratic formula.*

**1.** $2x^2 - 3x - 9 = 0$

**2.** $2x^2 + 7x = 4$

## EXAMPLE 2 Solving a Quadratic Equation Using the Quadratic Formula

Solve: $3p^2 = 6p - 1$

### Solution

First, we must write the equation in standard form to identify $a$, $b$, and $c$.

$$3p^2 = 6p - 1$$

Subtract $6p$ from both sides; Add 1 to both sides:
$$3p^2 - 6p + 1 = 0$$

The variable in the equation is $p$, so write "$p =$":
$$p = \frac{-b \pm \sqrt{b^2 - 4ac}}{2a}$$

$a = 3$, $b = -6$, $c = 1$:
$$p = \frac{-(-6) \pm \sqrt{(-6)^2 - 4(3)(1)}}{2(3)}$$

$$= \frac{6 \pm \sqrt{36 - 12}}{6}$$

$$= \frac{6 \pm \sqrt{24}}{6}$$

$\sqrt{24} = \sqrt{4 \cdot 6} = 2\sqrt{6}$:
$$= \frac{6 \pm 2\sqrt{6}}{6}$$

$\frac{a+b}{c} = \frac{a}{c} + \frac{b}{c}$:
$$= \frac{6}{6} \pm \frac{2\sqrt{6}}{6}$$

Simplify:
$$= 1 \pm \frac{\sqrt{6}}{3}$$

$a \pm b$ means $a - b$ or $a + b$:
$$p = 1 - \frac{\sqrt{6}}{3} \quad \text{or} \quad p = 1 + \frac{\sqrt{6}}{3}$$

We leave it to you to verify the solutions. The solution set is $\left\{1 - \frac{\sqrt{6}}{3}, 1 + \frac{\sqrt{6}}{3}\right\}$.

**Work Smart**

We could also simplify $\frac{6 \pm 2\sqrt{6}}{6}$ by factoring:

$$\frac{6 \pm 2\sqrt{6}}{6} = \frac{2(3 \pm \sqrt{2})}{6} = \frac{3 \pm \sqrt{2}}{3}$$

This is equivalent to $1 \pm \frac{\sqrt{6}}{3}$. Ask your instructor which form of the solution is preferred, if any.

Notice in Example 2 that the value of $b^2 - 4ac$ is positive, but not a perfect square. There are two solutions to the quadratic equation and they are irrational.

**QUICK ✓**

**3.** Solve: $4z^2 + 1 = 8z$

## EXAMPLE 3 Solving a Rational Equation That Leads to a Quadratic Equation

Solve: $9m + \frac{4}{m} = 12$

### Solution

We first note that $m$ cannot equal 0. To clear the equation of rational expressions, we multiply both sides of the equation by the LCD, $m$.

$$m\left(9m + \frac{4}{m}\right) = 12 \cdot m$$

**Work Smart**

Notice that the entire expression $-b \pm \sqrt{b^2 - 4ac}$ is in the numerator of the quadratic formula.

Distribute the $m$: $9m^2 + 4 = 12m$

Subtract $12m$ from both sides: $9m^2 - 12m + 4 = 0$

$$m = \frac{-b \pm \sqrt{b^2 - 4ac}}{2a}$$

$a = 9, b = -12, c = 4$: $$m = \frac{-(-12) \pm \sqrt{(-12)^2 - 4(9)(4)}}{2(9)}$$

$$= \frac{12 \pm \sqrt{144 - 144}}{18}$$

$$= \frac{12 \pm \sqrt{0}}{18}$$

$$= \frac{12}{18} = \frac{2}{3}$$

We leave it to you to verify the solution. The solution set is $\left\{\frac{2}{3}\right\}$.

In Example 3, we had one solution rather than two (as in Examples 1 and 2). In fact, the solution of $\frac{2}{3}$ is called a **repeated root** because it actually occurs twice! To see why, we solve the equation given in Example 3 by factoring.

$$9m^2 + 4 = 12m$$

Subtract $12m$ from both sides: $9m^2 - 12m + 4 = 0$

$$(3m - 2)(3m - 2) = 0$$

Zero-Product Property: $3m - 2 = 0$ or $3m - 2 = 0$

$$m = \frac{2}{3} \quad \text{or} \quad m = \frac{2}{3}$$

So we obtain two solutions because the expression $9m^2 - 12m + 4$ is a perfect square trinomial. We know that the equation $9m^2 - 12m + 4 = 0$ has a repeated root because the value $b^2 - 4ac$ equals 0. We will have more to say about this soon.

**QUICK ✓**

4. Solve: $4w + \frac{25}{w} = 20$

---

## EXAMPLE 4 Solving a Quadratic Equation Using the Quadratic Formula

Solve: $y^2 - 4y + 13 = 0$

### Solution

$$y^2 - 4y + 13 = 0$$

$$1y^2 - 4y + 13 = 0$$

$$y = \frac{-b \pm \sqrt{b^2 - 4ac}}{2a}$$

$a = 1, b = -4, c = 13$: $$y = \frac{-(-4) \pm \sqrt{(-4)^2 - 4(1)(13)}}{2(1)}$$

$$= \frac{4 \pm \sqrt{16 - 52}}{2}$$

$\sqrt{-36} = 6i$: $\quad = \dfrac{4 \pm \sqrt{-36}}{2} = \dfrac{4 \pm 6i}{2}$

$\dfrac{a+b}{c} = \dfrac{a}{c} + \dfrac{b}{c}$: $\quad = \dfrac{4}{2} \pm \dfrac{6}{2}i = 2 \pm 3i$

$a \pm b$ means $a - b$ or $a + b$: $\quad x = 2 - 3i \quad$ or $\quad x = 2 + 3i$

We leave it to you to verify the solution. The solution set is $\{2 - 3i, 2 + 3i\}$.

Notice in Example 4 that the value of $b^2 - 4ac$ is negative and the equation has two complex solutions that are not real.

**QUICK ✓**

**5.** Solve: $z^2 + 2z + 26 = 0$

---

## 2 Use the Discriminant to Determine the Nature of Solutions in a Quadratic Equation

In the quadratic formula $x = \dfrac{-b \pm \sqrt{b^2 - 4ac}}{2a}$, the quantity $\boldsymbol{b^2 - 4ac}$ is called the **discriminant** of the quadratic equation, because its value tells us the number of solutions and the type of solution to expect from the quadratic formula.

**THE DISCRIMINANT AND THE NATURE OF THE SOLUTION OF A QUADRATIC EQUATION**

For a quadratic equation $ax^2 + bx + c = 0$, the discriminant $b^2 - 4ac$ can be used to describe the nature of the solution as shown:

| Discriminant | Number of Solutions | Type of Solution | Example |
|---|---|---|---|
| Positive and a perfect square | 2 | Rational | 1 |
| Positive and not a perfect square | 2 | Irrational | 2 |
| Zero | 1 (repeated root) | Rational | 3 |
| Negative | 2 | Complex, nonreal | 4 |

If you look back at the results of Example 4, you should notice that the solutions are complex conjugates of each other. In general, for any quadratic equation of the form $ax^2 + bx + c = 0$, where $a$, $b$, and $c$ are real numbers and $b^2 - 4ac < 0$, the equation will have two complex solutions that are not real and are complex conjugates of each other.

This result is a consequence of the quadratic formula. Suppose that $b^2 - 4ac = -N < 0$. Then, by the quadratic formula, the solutions are

$$x = \frac{-b + \sqrt{b^2 - 4ac}}{2a} = \frac{-b + \sqrt{-N}}{2a}$$

$$= \frac{-b + \sqrt{N}i}{2a} = \frac{-b}{2a} + \frac{\sqrt{N}}{2a}i$$

and

$$x = \frac{-b - \sqrt{b^2 - 4ac}}{2a} = \frac{-b - \sqrt{-N}}{2a}$$

$$= \frac{-b - \sqrt{N}i}{2a} = \frac{-b}{2a} - \frac{\sqrt{N}}{2a}i$$

which are conjugates of each other.

### EXAMPLE 5 Determining the Nature of the Solutions of a Quadratic Equation

For each quadratic equation, determine the discriminant. Use the value of the discriminant to determine whether the quadratic equation has two unequal rational solutions, two irrational solutions, one repeated real solution, or two complex solutions that are not real.

**(a)** $x^2 - 5x + 2 = 0$ **(b)** $9y^2 + 6y + 1 = 0$ **(c)** $3p^2 - p = -5$

**Solution**

**(a)** We compare $x^2 - 5x + 2 = 0$ to the standard form $ax^2 + bx + c = 0$.

$$x^2 - 5x + 2 = 0$$

$a = 1$ $b = -5$ $c = 2$

We have that $a = 1$, $b = -5$, and $c = 2$. Substituting these values into the formula for the discriminant, $b^2 - 4ac$, we obtain

$$b^2 - 4ac = (-5)^2 - 4(1)(2) = 25 - 8 = 17$$

Because $b^2 - 4ac = 17$ and 17 is positive, but not a perfect square, the quadratic equation will have two irrational solutions.

**(b)** For the quadratic equation $9y^2 + 6y + 1 = 0$, we have that $a = 9$, $b = 6$, and $c = 1$. Substituting these values into the formula for the discriminant, $b^2 - 4ac$, we obtain

$$b^2 - 4ac = 6^2 - 4(9)(1) = 36 - 36 = 0$$

Because $b^2 - 4ac = 0$, the quadratic equation will have one repeated real solution.

**(c)** Is the quadratic equation $3p^2 - p = -5$ in standard form? No! We add 5 to both sides of the equation and write the equation as $3p^2 - p + 5 = 0$. So we have that $a = 3$, $b = -1$, and $c = 5$. Substituting these values into the formula for the discriminant, $b^2 - 4ac$, we obtain

$$b^2 - 4ac = (-1)^2 - 4(3)(5) = 1 - 60 = -59$$

Because $b^2 - 4ac = -59 < 0$, the quadratic equation will have two complex solutions that are not real. The solutions will be complex conjugates of each other.

**QUICK ✓** *Use the value of the discriminant to determine whether the quadratic equation has two unequal real solutions, one repeated real solution, or two complex solutions that are not real.*

**6.** $2z^2 + 5z + 4 = 0$ **7.** $4y^2 + 12y = -9$ **8.** $2x^2 - 4x + 1 = 0$

## Which Method Should I Use?

We have now introduced four methods for solving quadratic equations:

1. Factoring
2. Square Root Property
3. Completing the Square
4. The Quadratic Formula

You are probably asking yourself, "Which method should I use?" and "Does it matter which method I use?" The answer to the second question is that it does not matter which method you use, but one method may be more efficient than the others. Table 1 contains guidelines to help you solve any quadratic equation. Notice how the value of the discriminant can be used to guide us in choosing the most efficient method.

**Table 1**

| Form of the Quadratic Equation | Method | Example |
|---|---|---|
| $x^2 = p$, where $p$ is any real number | Square Root Property | $x^2 = 45$<br>Square Root Property: $x = \pm\sqrt{45}$<br>$= \pm 3\sqrt{5}$ |
| $ax^2 + c = 0$ | Square Root Property | $3p^2 + 12 = 0$<br>Subtract 12 from both sides: $3p^2 = -12$<br>Divide both sides by 3: $p^2 = -4$<br>Square Root Property: $p = \pm\sqrt{-4}$<br>$= \pm 2i$ |
| $ax^2 + bx + c = 0$, where $b^2 - 4ac$ is a perfect square. That is, $b^2 - 4ac$ is 1, 4, 9, 16, 25, ... | Factoring or the Quadratic Formula | $a = 2, b = 1, c = -10$: $2m^2 + m - 10 = 0$<br>$b^2 - 4ac = 1^2 - 4(2)(-10) = 1 + 80 = 81$<br>81 is a perfect square, so we can use factoring:<br>$2m^2 + m - 10 = 0$<br>$(2m + 5)(m - 2) = 0$<br>$2m + 5 = 0$ or $m - 2 = 0$<br>$m = -\frac{5}{2}$ or $m = 2$ |
| $ax^2 + bx + c = 0$, where $b^2 - 4ac$ is not a perfect square. | Quadratic Formula or Completing the Square | $a = 2, b = 4, c = -1$: $2x^2 + 4x - 1 = 0$<br>$b^2 - 4ac = 4^2 - 4(2)(-1) = 16 + 8 = 24$<br>24 is not a perfect square, so we use the quadratic formula (since it's easier than completing the square):<br>$x = \frac{-b \pm \sqrt{b^2 - 4ac}}{2a}$<br>$= \frac{-4 \pm \sqrt{24}}{2(2)}$<br>$= \frac{-4 \pm 2\sqrt{6}}{4}$<br>$= -1 \pm \frac{\sqrt{6}}{2}$<br>$x = -1 - \frac{\sqrt{6}}{2}$ or $x = -1 + \frac{\sqrt{6}}{2}$ |

Notice if the value of the discriminant is a perfect square, we can either factor or use the quadratic formula to solve the equation. We should factor if the quadratic expression is easy to factor, otherwise use the quadratic formula.

Also, you may have noticed that we did not recommend completing the square as one of the methods to use in solving a quadratic equation. This is because the quadratic formula was developed by completing the square of $ax^2 + bx + c = 0$. Besides, completing the square is a cumbersome task, whereas the quadratic formula is fairly straightforward to use. We did not waste your time by discussing completing the square, however, because it is needed to present a discussion of the quadratic formula. In addition, completing the square is a skill that you will need later in this course and in future math courses.

**QUICK ✓** *Solve each quadratic equation using any method you wish.*

**9.** $5n^2 - 45 = 0$ **10.** $-2y^2 + 5y - 6 = 0$ **11.** $3w^2 + 2w = 5$

## 3 Model and Solve Problems Involving Quadratic Equations

Many applied problems require solving quadratic equations. In the example below, we use a quadratic equation to determine the number of units that a company must sell in order to earn a certain amount of revenue. As always, we shall employ the problem-solving strategy first presented in Section 1.2.

### EXAMPLE 6 Revenue

The revenue $R$ received by a company selling $x$ specialty T-shirts per week is given by the function $R(x) = -0.005x^2 + 30x$.

**(a)** How many T-shirts must be sold in order for revenue to be \$25,000 per week?

**(b)** How many T-shirts must be sold in order for revenue to be \$45,000 per week?

#### Solution

**(a)** **Step 1: Identify** Here, we are looking to determine the number of T-shirts $x$ required so that $R = \$25{,}000$.

**Step 2: Name** We know that $x$ represents the number of T-shirts sold.

**Step 3: Translate** We need to solve the equation $R(x) = 25{,}000$.

$$\begin{aligned} R(x) &= 25{,}000 \\ -0.005x^2 + 30x &= 25{,}000 \\ -0.005x^2 + 30x - 25{,}000 &= 0 \end{aligned}$$

**Step 4: Solve** $a = -0.005, b = 30, c = -25{,}000$

$$\begin{aligned} b^2 - 4ac &= 30^2 - 4(-0.005)(-25{,}000) \\ &= 400 \end{aligned}$$

Because 400 is a perfect square, we can solve the equation by factoring or using the quadratic formula. It is not obvious how to factor $-0.005x^2 + 30x - 25000$, so we will use the quadratic formula to solve the equation.

$b^2 - 4ac = 400$

$$\begin{aligned} x &= \frac{-30 \pm \sqrt{400}}{2(-0.005)} \\ &= \frac{-30 \pm 20}{-0.01} \end{aligned}$$

$$x = \frac{-30 - 20}{-0.01} = 5000 \quad \text{or} \quad x = \frac{-30 + 20}{-0.01} = 1000$$

**Step 5: Check** If 1000 T-shirts are sold, then revenue is $R(1000) = -0.005(1000)^2 + 30(1000) = \$25{,}000$. If 5000 T-shirts are sold, then revenue is $R(5000) = -0.005(5000)^2 + 30(5000) = \$25{,}000$.

**Step 6: Answer** The company needs to sell either 1000 or 5000 T-shirts each week to earn \$25,000 in revenue.

**(b)** **Step 1: Identify** Here, we are looking to determine the number of T-shirts $x$ required so that $R = \$45{,}000$. That is, we wish to solve the equation $R(x) = 45{,}000$.

**Step 2: Name** We know that $x$ represents the number of T-shirts sold.

**Step 3: Translate** We need to solve the equation $R(x) = 45{,}000$.

$$R(x) = 45{,}000$$
$$-0.005x^2 + 30x = 45{,}000$$
$$-0.005x^2 + 30x - 45{,}000 = 0$$

**Step 4: Solve** $a = -0.005, b = 30, c = -45{,}000$

$$b^2 - 4ac = 30^2 - 4(-0.005)(-45{,}000) = 0$$

Because the discriminant is 0, the quadratic equation will have a single real solution. In addition, because the discriminant is 0, we can solve the equation by factoring or using the quadratic formula. It is not obvious how to factor $-0.005x^2 + 30x - 45{,}000$, so we will use the quadratic formula to solve the equation.

$$x = \frac{-30 \pm \sqrt{0}}{2(-0.005)} \quad b^2 - 4ac = 0$$
$$= \frac{-30 \pm 0}{-0.01}$$
$$= \frac{-30}{-0.01} = 3000$$

**Step 5: Check** If 3000 T-shirts are sold, then revenue is $R(3000) = -0.005(3000)^2 + 30(3000) = \$45{,}000$.

**Step 6: Answer** The company needs to sell 3,000 T-shirts each week to earn $45,000 in revenue.

**QUICK ✓**

12. The revenue $R$ received by a video store renting $x$ DVDs per day is given by the function $R(x) = -0.005x^2 + 4x$.
   **(a)** How many DVDs must be rented in order for revenue to be $600 per day?
   **(b)** How many DVDs must be rented in order for revenue to be $800 per day?

## EXAMPLE 7 Designing a Window

A window designer wishes to design a window so that the diagonal is 20 feet. In addition, the length of the window needs to be 4 feet more than the height. What are the dimensions of the window?

### Solution

**Step 1: Identify** We wish to know the dimensions of the window. That is, we want to know the length and height of the window.

**Step 2: Name** Let $h$ represent the height of the window so that $h + 4$ is the length (since the length is 4 feet more than the height).

**Step 3: Translate** Figure 6 illustrates the situation. From the figure, we can see that the three sides form a right triangle. We can express the relation among the three sides using the Pythagorean Theorem.

Figure 6

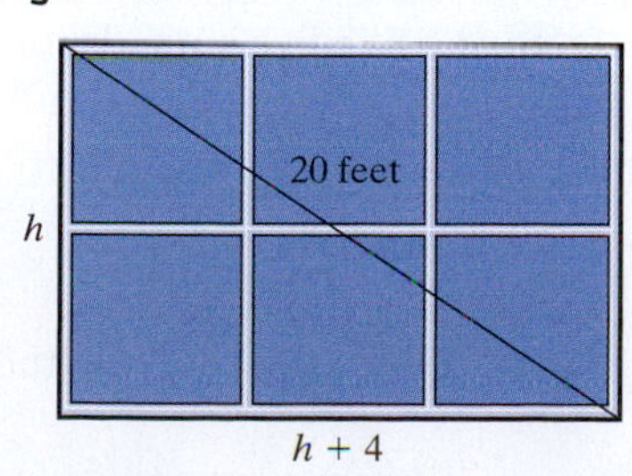


| | |
|---|---|
| leg² + leg² = hypotenuse²: | $h^2 + (h + 4)^2 = 20^2$ |
| FOIL: | $h^2 + h^2 + 8h + 16 = 400$ |
| Combine like terms: | $2h^2 + 8h + 16 = 400$ |
| Subtract 400 from both sides: | $2h^2 + 8h - 384 = 0$ |
| Divide both sides by 2: | $h^2 + 4h - 192 = 0$ |

**Step 4: Solve** In the model, we have that $a = 1, b = 4, c = -192$. The discriminant is $b^2 - 4ac = 4^2 - 4(1)(-192) = 784$ and $784 = 28^2$. So we can solve the equation by factoring.

$$h^2 + 4h - 192 = 0$$
$$(h + 16)(h - 12) = 0$$
$$h + 16 = 0 \quad \text{or} \quad h - 12 = 0$$
$$h = -16 \quad \text{or} \quad h = 12$$

**Step 5: Check** We disregard the solution $h = -16$ because $h$ represents the height of the window. We see if a window whose dimensions are 12 feet by $12 + 4 = 16$ feet has a diagonal that is 20 feet by verifying that $12^2 + 16^2 = 20^2$.

$$12^2 + 16^2 \stackrel{?}{=} 20^2$$
$$144 + 256 \stackrel{?}{=} 400$$
$$400 = 400$$

**Step 6: Answer** The dimensions of the window are 12 feet by 16 feet.

**13.** A rectangular plot of land is designed so that its length is 14 meters more than its width. The diagonal of the land is known to be 34 meters. What are the dimensions of the land?

# 8.2 Exercises

## Concepts and Vocabulary

*In Problems 1–3, fill in the blanks.*

**1.** The solution(s) to the quadratic equation $ax^2 + bx + c = 0, a \neq 0$, are given by the quadratic formula $x =$ ________.

**2.** In the quadratic formula, the quantity $b^2 - 4ac$ is called the ________ of the quadratic equation.

**3.** If the discriminant of a quadratic equation is ________, then the quadratic equation has two complex solutions that are not real.

*In Problems 4–6, answer True or False to each statement.*

**4.** If the discriminant of a quadratic equation is a perfect square, then the equation can be solved by factoring.

**5.** If the discriminant of a quadratic equation is zero, then the equation has no solution.

**6.** When solving a quadratic equation in which the solutions are complex numbers that are not real, the solutions will be complex conjugates of each other.

**7.** Explain the circumstances for which you would use factoring to solve a quadratic equation.

**8.** Explain the circumstances for which you would use the Square Root Property to solve a quadratic equation.

**9.** State the quadratic formula.

**10.** If you were to use the quadratic formula to solve $3x^2 - x = 5$, what would be the values of $a$, $b$, and $c$?

## Building Skills

*In Problems 11–28, solve each equation using the quadratic formula.*

**11.** $x^2 - 4x - 12 = 0$ **12.** $p^2 - 4p - 32 = 0$ **13.** $6y^2 - y - 15 = 0$

**14.** $10x^2 + x - 2 = 0$ **15.** $4m^2 - 8m + 1 = 0$ **16.** $2q^2 - 4q + 1 = 0$

**17.** $3w - 6 = \frac{1}{w}$ **18.** $x + \frac{1}{x} = 3$ **19.** $3p^2 = -2p + 4$

**20.** $5w^2 = -3w + 1$ **21.** $x^2 - 2x + 7 = 0$ **22.** $y^2 - 4y + 5 = 0$

**23.** $2z^2 + 7 = 2z$ **24.** $2z^2 + 7 = 4z$ **25.** $4x^2 = 2x + 1$

**26.** $6p^2 = 4p + 1$ **27.** $1 = 3q^2 + 4q$ **28.** $1 = 5w^2 + 6w$

*In Problems 29–38, determine the discriminant of each quadratic equation. Use the value of the discriminant to determine whether the quadratic equation has two rational solutions, two irrational solutions, one repeated real solution, or two complex solutions that are not real.*

**29.** $x^2 - 5x + 1 = 0$ **30.** $p^2 + 4p - 2 = 0$ **31.** $3z^2 + 2z + 5 = 0$

**32.** $2y^2 - 3y + 5 = 0$ **33.** $9q^2 - 6q + 1 = 0$ **34.** $16x^2 + 24x + 9 = 0$

**35.** $3w^2 = 4w - 2$ **36.** $6x^2 - x = -4$ **37.** $6x = 2x^2 - 1$

**38.** $10w^2 = 3$

## Mixed Practice

*In Problems 39–64, solve each equation using any method you wish.*

**39.** $w^2 - 5w + 5 = 0$ **40.** $q^2 - 7q + 7 = 0$ **41.** $3x^2 + 5x = 8$

**42.** $4p^2 + 5p = 9$ **43.** $2x^2 = 3x + 35$ **44.** $3x^2 + 5x = 2$

**45.** $q^2 + 2q + 8 = 0$ **46.** $w^2 + 4w + 9 = 0$ **47.** $5z^2 = 2z + 3$

**48.** $6x^2 = 2x + 4$ **49.** $7q - 2 = \frac{4}{q}$ **50.** $5m - 4 = \frac{5}{m}$

**51.** $5a^2 - 80 = 0$ **52.** $4p^2 - 100 = 0$ **53.** $8n^2 + 1 = 4n$

**54.** $4q^2 + 1 = 2q$ **55.** $27x^2 + 36x + 12 = 0$ **56.** $8p^2 - 40p + 50 = 0$

**57.** $\frac{1}{3}x^2 + \frac{2}{9}x - 1 = 0$ **58.** $\frac{1}{2}x^2 + \frac{3}{4}x - 1 = 0$ **59.** $(x - 5)(x + 1) = 4$

**60.** $(a - 3)(a + 1) = 2$ **61.** $\frac{x - 2}{x + 2} = x - 3$ **62.** $\frac{x - 5}{x + 3} = x - 3$

**63.** $\frac{x - 4}{x^2 + 2} = 2$ **64.** $\frac{x - 1}{x^2 + 4} = 1$

**65.** Suppose that $f(x) = x^2 + 4x - 21$.
**(a)** Solve $f(x) = 0$ for $x$.
**(b)** Solve $f(x) = -21$ for $x$.

**66.** Suppose that $f(x) = x^2 + 2x - 8$.
**(a)** Solve $f(x) = 0$ for $x$.
**(b)** Solve $f(x) = -8$ for $x$.

**67.** Suppose that $H(x) = -2x^2 - 4x + 1$.
**(a)** Solve $H(x) = 0$ for $x$.
**(b)** Solve $H(x) = 2$ for $x$.

**68.** Suppose that $g(x) = 3x^2 + x - 1$.
**(a)** Solve $g(x) = 0$ for $x$.
**(b)** Solve $g(x) = 4$ for $x$.

**69.** Suppose that $G(x) = 3x^2 + 2x + 2$.
**(a)** Solve $G(x) = 0$ for $x$.
**(b)** Solve $G(x) = 4$ for $x$.

**70.** Suppose that $F(x) = -x^2 + 3x - 3$.
**(a)** Solve $F(x) = 0$ for $x$.
**(b)** Solve $F(x) = -2$ for $x$.

## Applying the Concepts

*In Problems 71–74, use the Pythagorean Theorem to determine the value of x for the given measurements of each right triangle.*

**71.**

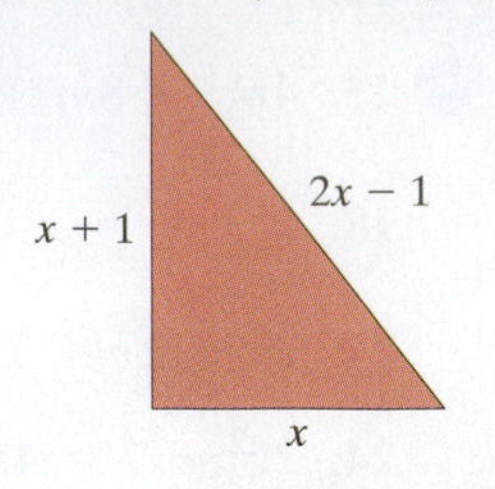


**72.**

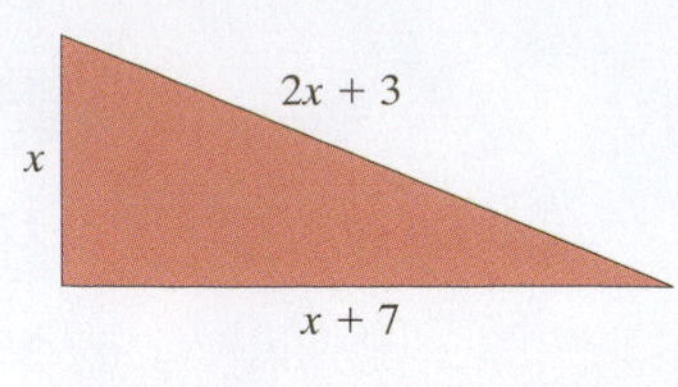


**73.**

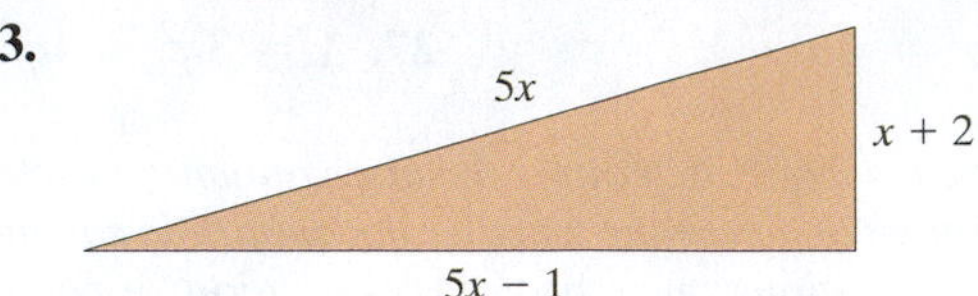


**74.**

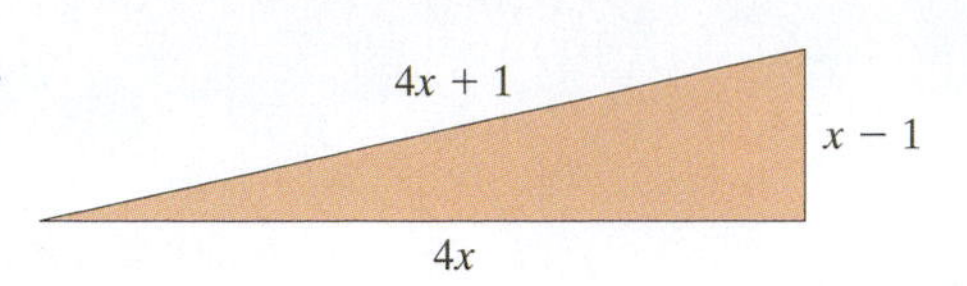


△ **75. Area** The area of a rectangle is 40 square inches. The width of the rectangle is 4 inches more than the length. What are the dimensions of the rectangle?

△ **76. Area** The area of a rectangle is 60 square inches. The width of the rectangle is 6 inches more than the length. What are the dimensions of the rectangle?

△ **77. Area** The area of a triangle is 25 square inches. The height of the triangle is 3 inches less than the base. What are the base and height of the triangle?

△ **78. Area** The area of a triangle is 35 square inches. The height of the triangle is 2 inches less than the base. What are the base and height of the triangle?

**79. Revenue** The revenue $R$ received by a company selling $x$ pairs of sunglasses per week is given by the function $R(x) = -0.1x^2 + 70x$.

**(a)** Find and interpret the values of $R(17)$ and $R(25)$.
**(b)** How many pairs of sunglasses must be sold in order for revenue to be $10,000 per week?
**(c)** How many pairs of sunglasses must be sold in order for revenue to be $12,250 per week?

**80. Revenue** The revenue $R$ received by a company selling $x$ "all day passes" to a small amusement park per day is given by the function $R(x) = -0.02x^2 + 24x$.

**(a)** Find and interpret the values of $R(300)$ and $R(800)$.
**(b)** How many tickets must be sold in order for revenue to be $4000 per day?
**(c)** How many tickets must be sold in order for revenue to be $7200 per day?

**81. Projectile Motion** The height $s$ of a ball after $t$ seconds when thrown straight up with an initial speed of 70 feet per second from an initial height of 5 feet can be modeled by the function

$$s(t) = -16t^2 + 70t + 5$$

**(a)** When will the height of the ball be 40 feet? Round your answer to the nearest tenth of a second.
**(b)** When will the height of the ball be 70 feet? Round your answer to the nearest tenth of a second.
**(c)** Will the ball ever reach a height of 150 feet? How does the result of the equation tell you this?

**82. Projectile Motion** The height $s$ of a toy rocket after $t$ seconds when fired straight up with an initial speed of 150 feet per second from an initial height of 2 feet can be modeled by the function

$$s(t) = -16t^2 + 150t + 2$$

**(a)** When will the height of the rocket be 200 feet? Round your answer to the nearest tenth of a second.
**(b)** When will the height of the rocket be 300 feet? Round your answer to the nearest tenth of a second.
**(c)** Will the rocket ever reach a height of 500 feet?

△ **83. Similar Triangles** Consult the figure. Suppose that $\triangle ABC \sim \triangle DEC$. The length of $\overline{BC}$ is 24 inches and the length of $\overline{DE}$ is 6 inches. If the length of $\overline{AB}$ equals the length of $\overline{CE}$, which we call $x$, find $x$.

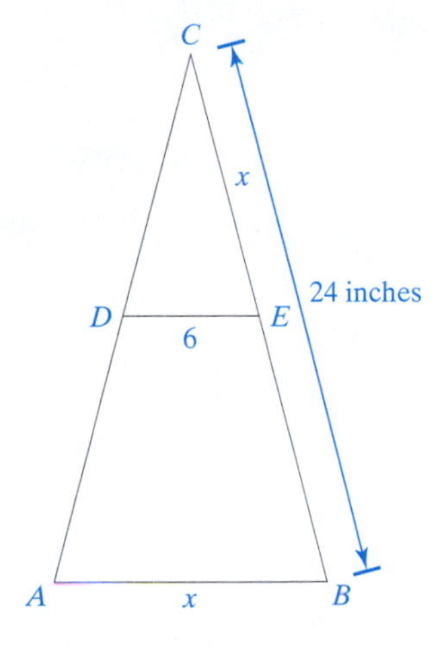


**84. Number Sense** Three times the square of a number equals the sum of two times the number and 5. Find the number(s).

**85. Life Cycle Hypothesis** The Life Cycle Hypothesis from Economics was presented by Franco Modigliani in 1954. One of its components states that income is a function of age. The function $I(a) = -55a^2 + 5119a - 54{,}448$ represents the relation between average annual income $I$ and age $a$.
**(a)** For what age does average income $I$ equal \$40,000? Round your answer to the nearest year.
**(b)** For what age does average income $I$ equal \$50,000? Round your answer to the nearest year.

**86. Population** The function $P(a) = 0.015a^2 - 4.962a + 290.580$ represents the population (in millions) of Americans in 2001, $P$, that are $a$ years of age or older. (*Source:* United States Census Bureau)
**(a)** For what age range is the population 200 million? Round your answer to the nearest year.
**(b)** For what age range is the population 50 million? Round your answer to the nearest year.

**87. Upstream and Back** Zene decides to canoe 4 miles upstream on a river to a waterfall and then canoe back. The total trip (excluding the time spent at the waterfall) takes 6 hours. Zene knows she can canoe at an average speed of 5 miles per hour in still water. What is the speed of the current?

**88. Round Trip** A Cessna aircraft flies 200 miles due west into the jet stream and flies back home on the same route. The total time of the trip (excluding the time on the ground) takes 4 hours. The Cessna aircraft can fly 120 miles per hour in still air. What is the net effect of the jet stream on the aircraft?

**89. Work** Robert and Susan have a newspaper route. When they work the route together, it takes 2 hours to deliver all the newspapers. One morning Robert told Susan he was too sick to deliver the papers. Susan doesn't remember how long it takes for her to deliver the newspapers working alone, but she does remember that Robert can finish the route one hour sooner than Susan can when working alone. How long will it take Susan to finish the route?

**90. Work** Demitrius needs to fill up his pool. When he rents a water tanker to fill the pool with the help of the hose from his house, it takes 5 hours to fill the pool. One year, money is tight and he can't afford to rent the water tanker to fill the pool. He doesn't remember how long it takes for his house hose to fill the pool, but does remember that the tanker hose filling the pool alone can finish the job in 8 fewer hours than using his house hose alone. How long will it take Demitrius to fill his pool using only his house hose?

## Extending the Concepts

**91.** Show that the sum of the solutions to a quadratic equation is $-\frac{b}{a}$.

**92.** Show that the product of the solutions to a quadratic equation is $\frac{c}{a}$.

**93.** Show that the real solutions of the equation $ax^2 + bx + c = 0$ are the negatives of the real solutions of the equation $ax^2 - bx + c = 0$. Assume that $b^2 - 4ac \geq 0$.

**94.** Show that the real solutions of the equation $ax^2 + bx + c = 0$ are the reciprocals of the real solutions of the equation $cx^2 + bx + a = 0$. Assume that $b^2 - 4ac \geq 0$.

## Synthesis Review

**95.** **(a)** Graph $f(x) = x^2 + 3x + 2$ by plotting points.
**(b)** Solve the equation $x^2 + 3x + 2 = 0$.
**(c)** Compare the solutions to the equation in part (b) to the $x$-intercepts of the graph drawn in part (a). What do you notice?

**96.** **(a)** Graph $f(x) = x^2 - x - 6$ by plotting points.
**(b)** Solve the equation $x^2 - x - 6 = 0$.
**(c)** Compare the solutions to the equation in part (b) to the $x$-intercepts of the graph drawn in part (a). What do you notice?

**97.** **(a)** Graph $g(x) = x^2 - 2x + 1$ by plotting points.
**(b)** Solve the equation $x^2 - 2x + 1 = 0$.
**(c)** Compare the solutions to the equation in part (b) to the $x$-intercepts of the graph drawn in part (a). What do you notice?

**98.** **(a)** Graph $g(x) = x^2 + 4x + 4$ by plotting points.
**(b)** Solve the equation $x^2 + 4x + 4 = 0$.
**(c)** Compare the solutions to the equation in part (b) to the $x$-intercepts of the graph drawn in part (a). What do you notice?

## Graphing Calculator

*In Problems 99–102, the graph of the quadratic function f is given. For each function determine the discriminant of the equation $f(x) = 0$ in order to determine the nature of the solutions the equation $f(x) = 0$ has. Compare the nature of solutions based on the discriminant to the graph of the function.*

**99.** $f(x) = x^2 - 7x + 3$

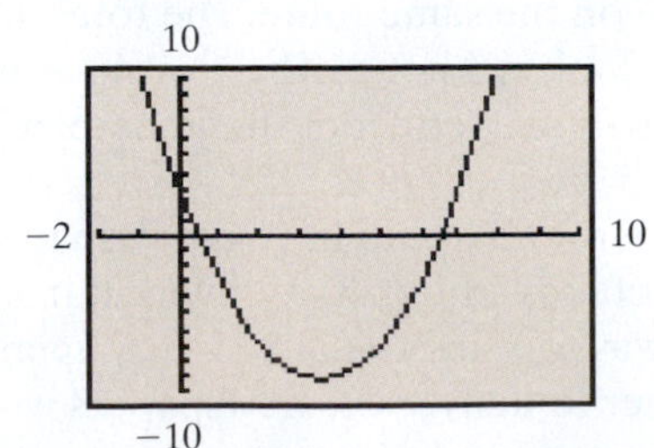


**100.** $f(x) = -x^2 - 5x + 1$

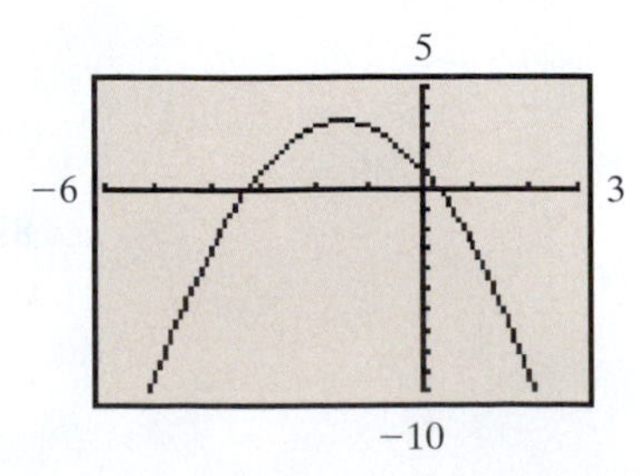

**101.** $f(x) = -x^2 - 3x - 4$

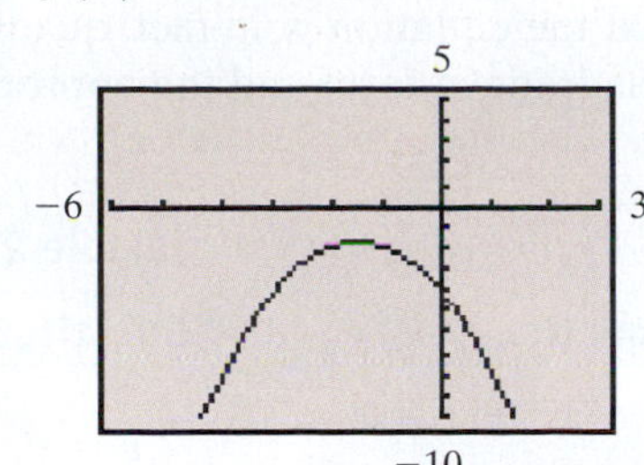


**102.** $f(x) = x^2 - 6x + 9$

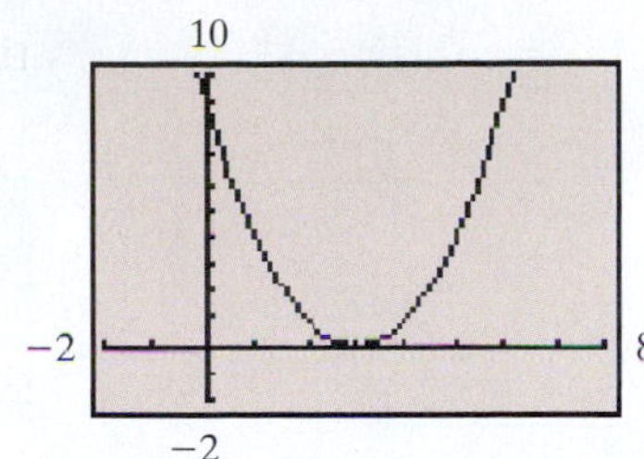


**103.** **(a)** Solve the equation $x^2 - 5x - 24 = 0$ algebraically.
**(b)** Graph $Y_1 = x^2 - 5x - 24$. Compare the $x$-intercepts of the graph to the solutions found in part (a).

**104.** **(a)** Solve the equation $x^2 - 4x - 45 = 0$ algebraically.
**(b)** Graph $Y_1 = x^2 - 4x - 45$. Compare the $x$-intercepts of the graph to the solutions found in part (a).

**105.** **(a)** Solve the equation $x^2 - 6x + 9 = 0$ algebraically.
**(b)** Graph $Y_1 = x^2 - 6x + 9$. Compare the $x$-intercepts of the graph to the solutions found in part (a).

**106.** **(a)** Solve the equation $x^2 + 10x + 25 = 0$ algebraically.
**(b)** Graph $Y_1 = x^2 + 10x + 25$. Compare the $x$-intercepts of the graph to the solutions found in part (a).

**107.** **(a)** Solve the equation $x^2 + 5x + 8 = 0$ algebraically.
**(b)** Graph $Y_1 = x^2 + 5x + 8$. How does the result of part (a) relate to the graph?

**108.** **(a)** Solve the equation $x^2 + 2x + 5 = 0$ algebraically.
**(b)** Graph $Y_1 = x^2 + 2x + 5$. How does the result of part (a) relate to the graph?

# 8.3 Solving Equations Quadratic in Form

**OBJECTIVE**

1 Solve Equations That Are Quadratic in Form

### *Preparing for Solving Equations Quadratic in Form*

*Before getting started, take the following readiness quiz. If you get a problem wrong, go back to the section cited and review the material.*

**1.** Factor: $x^4 - 5x^2 - 6$ [Section 5.5, pp. 406–407]

**2.** Factor: $2(p + 3)^2 + 3(p + 3) - 5$ [Section 5.5, pp. 406–407]

**3.** Simplify: (a) $(x^2)^2$ (b) $(p^{-1})^2$ [Getting Ready: Integer Exponents, pp. 345–346]

## 1 Solve Equations That Are Quadratic in Form

Consider the equation $x^4 - 4x^2 - 12 = 0$. While this equation is not of the form $ax^2 + bx + c = 0$, we can write the equation as $(x^2)^2 - 4x^2 - 12 = 0$. Then, if we let $u = x^2$ in the equation, we would obtain $u^2 - 4u - 12 = 0$, which is of the form $ax^2 + bx + c = 0$. Now we can solve $u^2 - 4u - 12 = 0$ for $u$ by factoring. Then, using the fact that $u = x^2$, we could find $x$—which was our goal in the first place.

In general, if a substitution $u$ transforms an equation into one of the form

$$au^2 + bu + c = 0$$

then the original equation is called an **equation quadratic in form.**

***Preparing for...Answers***
**1.** $(x^2 - 6)(x^2 + 1)$
**2.** $(2p + 11)(p + 2)$
**3.** **(a)** $x^4$ **(b)** $p^{-2} = \frac{1}{p^2}$

The difficulty in solving equations that are quadratic in form is that it is often hard to determine that the equation is, in fact, quadratic in form. Table 2 shows some equations that are quadratic in form and the appropriate substitution.

**Table 2**

| Original Equation | Substitution | Equation with Substitution |
|---|---|---|
| $2x^4 - 3x^2 + 5 = 0$<br>$2(x^2)^2 - 3x^2 + 5 = 0$ | $u = x^2$ | $2u^2 - 3u + 5 = 0$ |
| $3(z-5)^2 + 4(z-5) + 1 = 0$ | $u = z - 5$ | $3u^2 + 4u + 1 = 0$ |
| $-2y + 5\sqrt{y} - 2 = 0$<br>$-2(\sqrt{y})^2 + 5\sqrt{y} - 2 = 0$ | $u = \sqrt{y}$ | $-2u^2 + 5u - 2 = 0$ |

## EXAMPLE 1 How to Solve Equations That Are Quadratic in Form

Solve: $x^4 + x^2 - 12 = 0$

### Step-by-Step Solution

**Step 1:** Determine the appropriate substitution and write the equation in the form $au^2 + bu + c = 0$.

$$x^4 + x^2 - 12 = 0$$

Let $u = x^2$:
$$(x^2)^2 + x^2 - 12 = 0$$
$$u^2 + u - 12 = 0$$

**Step 2:** Solve the equation $au^2 + bu + c = 0$.

$$(u + 4)(u - 3) = 0$$
$$u + 4 = 0 \quad \text{or} \quad u - 3 = 0$$
$$u = -4 \quad \text{or} \quad u = 3$$

**Step 3:** Solve for the variable in the original equation using the value of $u$ found in Step 2.

We want to know $x$, so replace $u$ with $x^2$:
$$x^2 = -4 \quad \text{or} \quad x^2 = 3$$

Square Root Property:
$$x = \pm\sqrt{-4} \quad \text{or} \quad x = \pm\sqrt{3}$$
$$= \pm 2i$$

**Step 4:** Verify your solutions.

$x = 2i$:
$$(2i)^4 + (2i)^2 - 12 \stackrel{?}{=} 0$$
$$2^4i^4 + 2^2i^2 - 12 \stackrel{?}{=} 0$$
$$16(1) + 4(-1) - 12 \stackrel{?}{=} 0$$
$$16 - 4 - 12 \stackrel{?}{=} 0$$
$$0 = 0 \quad \text{True}$$

$x = -2i$:
$$(-2i)^4 + (-2i)^2 - 12 \stackrel{?}{=} 0$$
$$(-2)^4i^4 + (-2)^2i^2 - 12 \stackrel{?}{=} 0$$
$$16(1) + 4(-1) - 12 \stackrel{?}{=} 0$$
$$16 - 4 - 12 \stackrel{?}{=} 0$$
$$0 = 0 \quad \text{True}$$

$x = \sqrt{3}$:
$$(\sqrt{3})^4 + (\sqrt{3})^2 - 12 \stackrel{?}{=} 0$$
$$\sqrt{3^4} + \sqrt{3^2} - 12 \stackrel{?}{=} 0$$
$$\sqrt{81} + 3 - 12 \stackrel{?}{=} 0$$
$$9 + 3 - 12 \stackrel{?}{=} 0$$
$$0 = 0 \quad \text{True}$$

$x = -\sqrt{3}$:
$$(-\sqrt{3})^4 + (-\sqrt{3})^2 - 12 \stackrel{?}{=} 0$$
$$(-1)^4\sqrt{3^4} + (-1)^2\sqrt{3^2} - 12 \stackrel{?}{=} 0$$
$$\sqrt{81} + 3 - 12 \stackrel{?}{=} 0$$
$$9 + 3 - 12 \stackrel{?}{=} 0$$
$$0 = 0 \quad \text{True}$$

The solution set is $\{2i, -2i, \sqrt{3}, -\sqrt{3}\}$.

We summarize the steps can be used to solve an equation that is quadratic in form.

### Solving Equations Quadratic in Form

**Step 1:** Determine the appropriate substitution and write the equation in the form $au^2 + bu + c = 0$.

**Step 2:** Solve the equation $au^2 + bu + c = 0$.

**Step 3:** Solve for the variable in the original equation using the value of $u$ found in Step 2.

**Step 4:** Verify your solutions.

**QUICK ✓** *Solve each equation.*

**1.** $x^4 - 13x^2 + 36 = 0$ **2.** $p^4 - 7p^2 = 18$

---

### EXAMPLE 2 Solving Equations That Are Quadratic in Form

Solve: $(z^2 - 5)^2 - 3(z^2 - 5) - 4 = 0$

**Solution**

$$(z^2 - 5)^2 - 3(z^2 - 5) - 4 = 0$$

Let $u = z^2 - 5$: $u^2 - 3u - 4 = 0$

$$(u - 4)(u + 1) = 0$$

$$u - 4 = 0 \quad \text{or} \quad u + 1 = 0$$

$$u = 4 \quad \text{or} \quad u = -1$$

Replace $u$ with $z^2 - 5$ and solve for $z$: $z^2 - 5 = 4 \quad \text{or} \quad z^2 - 5 = -1$

$$z^2 = 9 \quad \text{or} \quad z^2 = 4$$

Square Root Property: $z = \pm 3 \quad \text{or} \quad z = \pm 2$

**Check**

$z = -3$:

$$((-3)^2 - 5)^2 - 3((-3)^2 - 5) - 4 \stackrel{?}{=} 0$$
$$(9 - 5)^2 - 3(9 - 5) - 4 \stackrel{?}{=} 0$$
$$4^2 - 3(4) - 4 \stackrel{?}{=} 0$$
$$16 - 12 - 4 \stackrel{?}{=} 0$$
$$0 = 0 \quad \text{True}$$

$z = 3$:

$$((3)^2 - 5)^2 - 3((3)^2 - 5) - 4 \stackrel{?}{=} 0$$
$$(9 - 5)^2 - 3(9 - 5) - 4 \stackrel{?}{=} 0$$
$$4^2 - 3(4) - 4 \stackrel{?}{=} 0$$
$$16 - 12 - 4 \stackrel{?}{=} 0$$
$$0 = 0 \quad \text{True}$$

$z = -2$:

$$((-2)^2 - 5)^2 - 3((-2)^2 - 5) - 4 \stackrel{?}{=} 0$$
$$(4 - 5)^2 - 3(4 - 5) - 4 \stackrel{?}{=} 0$$
$$(-1)^2 - 3(-1) - 4 \stackrel{?}{=} 0$$
$$1 + 3 - 4 \stackrel{?}{=} 0$$
$$0 = 0 \quad \text{True}$$

$z = 2$:

$$((2)^2 - 5)^2 - 3((2)^2 - 5) - 4 \stackrel{?}{=} 0$$
$$(4 - 5)^2 - 3(4 - 5) - 4 \stackrel{?}{=} 0$$
$$(-1)^2 - 3(-1) - 4 \stackrel{?}{=} 0$$
$$1 + 3 - 4 \stackrel{?}{=} 0$$
$$0 = 0 \quad \text{True}$$

The solution set is $\{-3, -2, 2, 3\}$.

**QUICK ✓** *Solve each equation.*

**3.** $(p^2 - 2)^2 - 9(p^2 - 2) + 14 = 0$ **4.** $2(2z^2 - 1)^2 + 5(2z^2 - 1) - 3 = 0$

---

When we have to raise both sides of an equation to an even power (such as squaring both sides of the equation), there is a possibility that we will introduce extraneous solutions to the equation. Under these circumstances, it is imperative that we verify our solutions.

### EXAMPLE 3 Solving Equations That Are Quadratic in Form

Solve: $3x - 5\sqrt{x} - 2 = 0$

**Solution**

$$3x - 5\sqrt{x} - 2 = 0$$

$$3(\sqrt{x})^2 - 5\sqrt{x} - 2 = 0$$

Let $u = \sqrt{x}$: $$3u^2 - 5u - 2 = 0$$

$$(3u + 1)(u - 2) = 0$$

$$3u + 1 = 0 \quad \text{or} \quad u - 2 = 0$$

$$3u = -1 \quad \text{or} \quad u = 2$$

$$u = \frac{-1}{3}$$

Replace $u$ with $\sqrt{x}$ and solve for $x$: $$\sqrt{x} = \frac{-1}{3} \quad \text{or} \quad \sqrt{x} = 2$$

Square both sides: $$x = \frac{1}{9} \quad \text{or} \quad x = 4$$

**Check**

$x = \frac{1}{9}$:

$$3 \cdot \frac{1}{9} - 5\sqrt{\frac{1}{9}} - 2 = 0$$

$$\frac{1}{3} - 5 \cdot \frac{1}{3} - 2 = 0$$

$$\frac{1}{3} - \frac{5}{3} + \frac{6}{3} = 0$$

$$\frac{2}{3} = 0 \quad \text{False}$$

$x = 4$:

$$3 \cdot 4 - 5\sqrt{4} - 2 = 0$$

$$12 - 5 \cdot 2 - 2 = 0$$

$$12 - 10 - 2 = 0$$

$$0 = 0 \quad \text{True}$$

The apparent solution $x = \frac{1}{9}$ is extraneous. The solution set is $\{4\}$.

We could also have solved the equation in Example 3 using the methods introduced in Section 7.6 by isolating the radical and squaring both sides.

**QUICK ✓** *Solve the equation.*

**5.** $3w - 14\sqrt{w} + 8 = 0$ **6.** $2q - 9\sqrt{q} - 5 = 0$

---

## EXAMPLE 4 Solving Equations That Are Quadratic in Form

Solve: $4x^{-2} + 13x^{-1} - 12 = 0$

### Solution

$$4x^{-2} + 13x^{-1} - 12 = 0$$
$$4(x^{-1})^2 + 13x^{-1} - 12 = 0$$

Let $u = x^{-1}$: $4u^2 + 13u - 12 = 0$

Factor: $(4u - 3)(u + 4) = 0$

$$4u - 3 = 0 \quad \text{or} \quad u + 4 = 0$$
$$4u = 3 \quad \text{or} \quad u = -4$$
$$u = \frac{3}{4}$$

Replace $u$ with $x^{-1}$ and solve for $x$: $x^{-1} = \frac{3}{4}$ or $x^{-1} = -4$

$x^{-1} = \frac{1}{x}$: $\frac{1}{x} = \frac{3}{4}$ or $\frac{1}{x} = -4$

Take the reciprocal of both sides of the equation: $x = \frac{4}{3}$ or $x = \frac{1}{-4} = -\frac{1}{4}$

**Check** $x = \frac{4}{3}$:

$$4\left(\frac{4}{3}\right)^{-2} + 13\cdot\left(\frac{4}{3}\right)^{-1} - 12 \stackrel{?}{=} 0$$
$$4\left(\frac{3}{4}\right)^{2} + 13\cdot\frac{3}{4} - 12 \stackrel{?}{=} 0$$
$$4\cdot\frac{9}{16} + \frac{39}{4} - 12 \stackrel{?}{=} 0$$
$$\frac{9}{4} + \frac{39}{4} - \frac{48}{4} \stackrel{?}{=} 0$$
$$0 = 0 \quad \text{True}$$

$x = -\frac{1}{4}$:

$$4\left(-\frac{1}{4}\right)^{-2} + 13\cdot\left(-\frac{1}{4}\right)^{-1} - 12 \stackrel{?}{=} 0$$
$$4(-4)^2 + 13\cdot(-4) - 12 \stackrel{?}{=} 0$$
$$4\cdot 16 - 52 - 12 \stackrel{?}{=} 0$$
$$64 - 52 - 12 \stackrel{?}{=} 0$$
$$0 = 0 \quad \text{True}$$

The solution set is $\left\{-\frac{1}{4}, \frac{4}{3}\right\}$.

**QUICK ✓** *Solve the equation.*

**7.** $5x^{-2} + 12x^{-1} + 4 = 0$

## EXAMPLE 5 Solving Equations Quadratic in Form

Solve: $a^{2/3} + 3a^{1/3} - 28 = 0$

### Solution

$$a^{2/3} + 3a^{1/3} - 28 = 0$$
$$(a^{1/3})^2 + 3a^{1/3} - 28 = 0$$

Let $u = a^{1/3}$: $u^2 + 3u - 28 = 0$

$$(u + 7)(u - 4) = 0$$
$$u + 7 = 0 \quad \text{or} \quad u - 4 = 0$$
$$u = -7 \quad \text{or} \quad u = 4$$
$$a^{1/3} = -7 \quad \text{or} \quad a^{1/3} = 4$$

Replace $u$ with $a^{1/3}$ and solve for $a$: $(a^{1/3})^3 = (-7)^3$ or $(a^{1/3})^3 = 4^3$

Cube both sides of the equation: $a = -343$ or $a = 64$

**Check** $a = -343$:

$$\begin{aligned}(-343)^{2/3} + 3(-343)^{1/3} - 28 &= 0\\ \left(\sqrt[3]{-343}\right)^2 + 3\cdot\sqrt[3]{-343} - 28 &\stackrel{?}{=} 0\\ (-7)^2 + 3\cdot(-7) - 28 &\stackrel{?}{=} 0\\ 49 - 21 - 28 &\stackrel{?}{=} 0\\ 0 &= 0 \quad \text{True}\end{aligned}$$

$a = 64$:

$$\begin{aligned}(64)^{2/3} + 3(64)^{1/3} - 28 &= 0\\ \left(\sqrt[3]{64}\right)^2 + 3\cdot\sqrt[3]{64} - 28 &\stackrel{?}{=} 0\\ 4^2 + 3\cdot 4 - 28 &\stackrel{?}{=} 0\\ 16 + 12 - 28 &\stackrel{?}{=} 0\\ 0 &= 0 \quad \text{True}\end{aligned}$$

The solution set is $\{-343, 64\}$.

**QUICK ✓** *Solve the equation.*

**8.** $p^{2/3} - 4p^{1/3} - 5 = 0$

## 8.3 Exercises

**For Extra Help:**  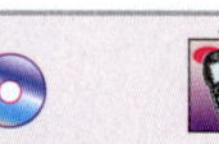 
Student Solutions Manual · CD Video · PH Math/Tutor Center · MathXL Tutorials on CD · MathXL® · MyMathLab

### Concepts and Vocabulary

*In Problems 1 and 2, fill in the blanks.*

**1.** If a substitution $u$ transforms an equation into one of the form $au^2 + bu + c = 0$, then the original equation is called a(n) _____ _____ _____ _____.

**2.** For the equation $2(3x + 1)^2 - 5(3x + 1) + 2 = 0$, an appropriate substitution would be $u =$ _____.

*In Problems 3 and 4, answer True or False to each statement.*

**3.** The equation $3\left(\frac{x}{x-2}\right)^2 - \frac{5x}{x-2} + 3 = 0$ is quadratic in form.

**4.** The equation $x - 5\sqrt{x} - 6 = 0$ can be solved either using the methods of this section or by isolating the radical and squaring both sides.

**5.** In your own words, explain the steps required to solve an equation quadratic in form. Be sure to include an explanation as to how to identify the appropriate substitution.

**6.** Under what circumstances might extraneous solutions occur when solving equations quadratic in form?

**7.** What is the appropriate choice for $u$ when solving the equation $4(3x - 2)^2 + 7(3x - 2) + 2 = 0$?

**8.** What is the appropriate choice for $u$ when solving the equation $2\cdot\frac{1}{x^2} - 6\cdot\frac{1}{x} + 3 = 0$?

### Building Skills

*In Problems 9–44, solve each equation.*

**9.** $x^4 - 5x^2 + 4 = 0$

**10.** $x^4 - 10x^2 + 9 = 0$

**11.** $q^4 + 13q^2 + 36 = 0$

**12.** $z^4 + 10z^2 + 9 = 0$

**13.** $4a^4 - 17a^2 + 4 = 0$

**14.** $4b^4 - 5b^2 + 1 = 0$

**15.** $p^4 + 6 = 5p^2$

**16.** $q^4 + 15 = 8q^2$

**17.** $(x-3)^2 - 6(x-3) - 7 = 0$

**18.** $(x+2)^2 - 3(x+2) - 10 = 0$

**19.** $(x^2-1)^2 - 11(x^2-1) + 24 = 0$

**20.** $(p^2-2)^2 - 8(p^2-2) + 12 = 0$

**21.** $(y^2+2)^2 + 7(y^2+2) + 10 = 0$

**22.** $(q^2+4)^2 + 3(q^2+4) - 4 = 0$

**23.** $x - 3\sqrt{x} - 4 = 0$

**24.** $x - 5\sqrt{x} - 6 = 0$

**25.** $w + 5\sqrt{w} + 6 = 0$

**26.** $z + 7\sqrt{z} + 6 = 0$

**27.** $2x + 5\sqrt{x} = 3$

**28.** $3x = 11\sqrt{x} + 4$

**29.** $x^{-2} + 3x^{-1} = 28$

**30.** $q^{-2} + 2q^{-1} = 15$

**31.** $10z^{-2} + 11z^{-1} = 6$

**32.** $10a^{-2} + 23a^{-1} = 5$

**33.** $x^{2/3} + 3x^{1/3} - 4 = 0$

**34.** $y^{2/3} - 2y^{1/3} - 3 = 0$

**35.** $z^{2/3} - z^{1/3} = 2$

**36.** $w^{2/3} + 2w^{1/3} = 3$

**37.** $a + a^{1/2} = 30$

**38.** $b + 3b^{1/2} = 28$

**39.** $\frac{1}{x^2} - \frac{5}{x} + 6 = 0$

**40.** $\frac{1}{x^2} - \frac{7}{x} + 12 = 0$

**41.** $\left(\frac{1}{x+2}\right)^2 + \frac{4}{x+2} = 5$

**42.** $\left(\frac{1}{x+2}\right)^2 + \frac{6}{x+2} = 7$

**43.** $p^6 - 28p^3 + 27 = 0$

**44.** $y^6 - 7y^3 - 8 = 0$

## Mixed Practice

*In Problems 45–58, solve each equation.*

**45.** $8a^{-2} + 2a^{-1} = 1$

**46.** $6b^{-2} - b^{-1} = 1$

**47.** $z^4 = 4z^2 + 32$

**48.** $x^4 + 3x^2 = 4$

**49.** $x^{1/2} + x^{1/4} - 6 = 0$

**50.** $c^{1/2} + c^{1/4} - 12 = 0$

**51.** $w^4 - 5w^2 - 36 = 0$

**52.** $p^4 - 15p^2 - 16 = 0$

**53.** $\left(\frac{1}{x+3}\right)^2 + \frac{2}{x+3} = 3$

**54.** $\left(\frac{1}{x-1}\right)^2 + \frac{7}{x-1} = 8$

**55.** $x - 7\sqrt{x} + 12 = 0$

**56.** $x - 8\sqrt{x} + 12 = 0$

**57.** $2(x-1)^2 - 7(x-1) = 4$

**58.** $3(y-2)^2 - 4(y-2) = 4$

**59.** Suppose that $f(x) = x^4 + 7x^2 + 12$. Find the values of $x$ such that (a) $f(x) = 12$ (b) $f(x) = 6$.

**60.** Suppose that $f(x) = x^4 + 5x^2 + 3$. Find the values of $x$ such that (a) $f(x) = 3$ (b) $f(x) = 17$.

**61.** Suppose that $g(x) = 2x^4 - 6x^2 - 5$. Find the values of $x$ such that (a) $g(x) = -5$ (b) $g(x) = 15$.

**62.** Suppose that $h(x) = 3x^4 - 9x^2 - 8$. Find the values of $x$ such that (a) $h(x) = -8$ (b) $h(x) = 22$.

**63.** Suppose that $F(x) = x^{-2} - 5x^{-1}$. Find the values of $x$ such that (a) $F(x) = 6$ (b) $F(x) = 14$.

**64.** Suppose that $f(x) = x^{-2} - 3x^{-1}$. Find the values of $x$ such that (a) $f(x) = 4$ (b) $f(x) = 18$.

*In Problems 65–70, find the zeros of the function.* [**Hint:** *Remember, r is a zero if* $f(r) = 0$.]

**65.** $f(x) = x^4 + 9x^2 + 14$

**66.** $f(x) = x^4 - 13x^2 + 42$

**67.** $g(t) = 6t - 25\sqrt{t} - 9$

**68.** $h(p) = 8p - 18\sqrt{p} - 35$

**69.** $s(d) = \frac{1}{(d+3)^2} - \frac{4}{d+3} + 3$

**70.** $f(a) = \frac{1}{(a-2)^2} + \frac{3}{a-2} - 4$

## Applying the Concepts

**71.** **(a)** Solve $x^2 - 5x + 6 = 0$.
**(b)** Solve $(x - 3)^2 - 5(x - 3) + 6 = 0$. Compare the solutions to part (a).
**(c)** Solve $(x + 2)^2 - 5(x + 2) + 6 = 0$. Compare the solutions to part (a).
**(d)** Solve $(x - 5)^2 - 5(x - 5) + 6 = 0$. Compare the solutions to part (a).
**(e)** Conjecture a generalization for the solution of $(x - a)^2 - 5(x - a) + 6 = 0$.

**72.** **(a)** Solve $x^2 + 3x - 18 = 0$.
**(b)** Solve $(x - 1)^2 + 3(x - 1) - 18 = 0$. Compare the solutions to part (a).
**(c)** Solve $(x + 5)^2 + 3(x + 5) - 18 = 0$. Compare the solutions to part (a).
**(d)** Solve $(x - 3)^2 + 3(x - 3) - 18 = 0$. Compare the solutions to part (a).
**(e)** Conjecture a generalization for the solution of $(x - a)^2 + 3(x - a) - 18 = 0$.

**73.** For the function $f(x) = 2x^2 - 3x + 1$,
**(a)** Solve $f(x) = 0$.
**(b)** Solve $f(x - 2) = 0$. Compare the solutions to part (a).
**(c)** Solve $f(x - 5) = 0$. Compare the solutions to part (a).
**(d)** Conjecture a generalization for the zeros of $f(x - a)$.

**74.** For the function $f(x) = 3x^2 - 5x - 2$,
**(a)** Solve $f(x) = 0$.
**(b)** Solve $f(x - 1) = 0$. Compare the solutions to part (a).
**(c)** Solve $f(x - 4) = 0$. Compare the solutions to part (a).
**(d)** Conjecture a generalization for the zeros of $f(x - a)$.

**75.** **Revenue** The function $R(x) = \dfrac{(x - 1990)^2}{2} + \dfrac{3(x - 1990)}{2} + 3000$ models the revenue $R$ (in thousands of dollars) of a start-up computer consulting firm in year $x$, where $x \geq 1990$.
**(a)** Determine and interpret $R(1990)$.
**(b)** Solve and interpret $R(x) = 3065$.
**(c)** According to the model, in what year can the firm expect to receive \$3350 thousand in revenue?

**76.** **Revenue** The function $R(x) = \dfrac{(x - 2000)^2}{3} + \dfrac{5(x - 2000)}{3} + 2000$ models the revenue $R$ (in thousands of dollars) of a start-up computer software firm in year $x$, where $x \geq 2000$.
**(a)** Determine and interpret $R(2000)$.
**(b)** Solve and interpret $R(x) = 2250$.
**(c)** According to the model, in what year can the firm expect to receive \$2350 thousand in revenue?

## Extending the Concepts

*All of the problems given in this section resulted in equations quadratic in form that could be factored after the appropriate substitution. However, this is not a necessary requirement to solving equations quadratic in form. In Problems 77–80, determine the appropriate substitution, and then use the quadratic formula to find the value of u. Finally determine the value of the variable in the equation.*

**77.** $x^4 + 5x^2 + 2 = 0$

**78.** $x^4 + 7x^2 + 4 = 0$

**79.** $2(x - 2)^2 + 8(x - 2) - 1 = 0$

**80.** $3(x + 1)^2 + 6(x + 1) - 1 = 0$

## Synthesis Review

*In Problems 81–85, add or subtract the expressions.*

**81.** $(4x^2 - 3x - 1) + (-3x^2 + x + 5)$

**82.** $(5y^3 - 2y^2 + y + 4) - (2y^3 + 6y^2 - 3)$

**83.** $(3p^{-2} - 4p^{-1} + 8) - (2p^{-2} - 8p^{-1} - 1)$

**84.** $3\sqrt{2x} - \sqrt{8x} + \sqrt{50x}$

**85.** $\sqrt[3]{16a} + \sqrt[3]{54a} - \sqrt[3]{128a^4}$

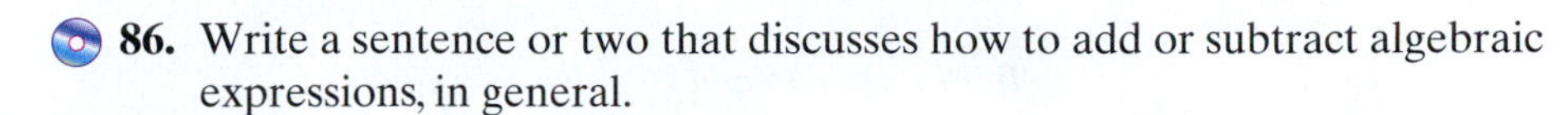

**86.** Write a sentence or two that discusses how to add or subtract algebraic expressions, in general.

## The Graphing Calculator

*In Problems 87–92, use a graphing calculator to find the real solutions to the equations using either the ZERO or INTERSECT feature. Round your answers to two decimal places, if necessary.*

**87.** $x^4 + 5x^2 - 14 = 0$

**88.** $x^4 - 4x^2 - 12 = 0$

**89.** $2(x - 2)^2 = 5(x - 2) + 1$

**90.** $3(x + 3)^2 = 2(x + 3) + 6$

**91.** $x - 5\sqrt{x} = -3$

**92.** $x + 4\sqrt{x} = 5$

**93.** **(a)** Graph $Y_1 = x^2 - 5x - 6$. Find the $x$-intercepts of the graph.
**(b)** Graph $Y_1 = (x + 2)^2 - 5(x + 2) - 6$. Find the $x$-intercepts of the graph.
**(c)** Graph $Y_1 = (x + 5)^2 - 5(x + 5) - 6$. Find the $x$-intercepts of the graph.
**(d)** Make a generalization based upon the results of parts (a), (b), and (c).

**94.** **(a)** Graph $Y_1 = x^2 + 4x + 3$. Find the $x$-intercepts of the graph.
**(b)** Graph $Y_1 = (x - 3)^2 + 4(x - 3) + 3$. Find the $x$-intercepts of the graph.
**(c)** Graph $Y_1 = (x - 6)^2 + 4(x - 6) + 3$. Find the $x$-intercepts of the graph.
**(d)** Make a generalization based upon the results of parts (a), (b), and (c).

# PUTTING THE CONCEPTS TOGETHER (SECTIONS 8.1–8.3)

*These problems cover important concepts from Sections 8.1 to 8.3. We designed these problems so that you can review the chapter so far and show your mastery of the concepts. Take time to work these problems before proceeding with the next section. The answers to these problems are located at the back of the text on page AN-39.*

*In Problems 1–3, complete the square in the given expression. Then factor the perfect square trinomial.*

**1.** $z^2 + 10z$

**2.** $x^2 + 7x$

**3.** $n^2 - \frac{1}{4}n$

*In Problems 4–6, solve each quadratic equation using the stated method.*

**4.** $(2x - 3)^2 - 5 = -1$; square root method

**5.** $x^2 + 8x + 4 = 0$; completing the square

**6.** $x(x - 6) = -7$; quadratic formula

*In Problems 7–10, solve each equation using the method you prefer.*

**7.** $49x^2 - 80 = 0$

**8.** $p^2 - 8p + 6 = 0$

**9.** $3y^2 + 6y + 4 = 0$

**10.** $\frac{1}{4}n^2 + n = \frac{1}{6}$

*In Problems 11–13, determine the discriminant of each quadratic equation. Use the value of the discriminant to determine whether the equation has two rational solutions, two irrational solutions, one repeated real solution, or two complex solutions that are not real.*

**11.** $9x^2 + 12x + 4 = 0$ **12.** $3x^2 + 6x - 2 = 0$ **13.** $2x^2 + 6x + 5 = 0$

**14.** Find the missing length in the right triangle shown below.

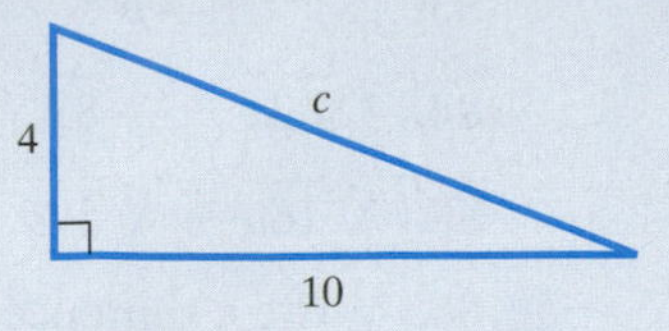


*In Problems 15 and 16, solve each equation.*

**15.** $2m + 7\sqrt{m} - 15 = 0$ **16.** $p^{-2} - 3p^{-1} - 18 = 0$

**17. Revenue** The revenue $R$ received by a company selling $x$ microwave ovens per day is given by the function $R(x) = -0.4x^2 + 140x$. How many microwave ovens must be sold in order for revenue to be \$12,000 per day?

**18. Airplane Ride** An airplane flies 300 miles into the wind and then flies home against the wind. The total time of the trip (excluding time on the ground) is 5 hours. If the plane can fly 140 miles per hour in still air, what was the speed of the wind? Round your answer to the nearest tenth.

# 8.4 Graphing Quadratic Functions Using Transformations

## OBJECTIVES

1. Graph Quadratic Functions of the Form $f(x) = x^2 + k$
2. Graph Quadratic Functions of the Form $f(x) = (x - h)^2$
3. Graph Quadratic Functions of the Form $f(x) = ax^2$
4. Graph Quadratic Functions of the Form $f(x) = ax^2 + bx + c$
5. Find a Quadratic Function from Its Graph

***Preparing for...Answers***

1.

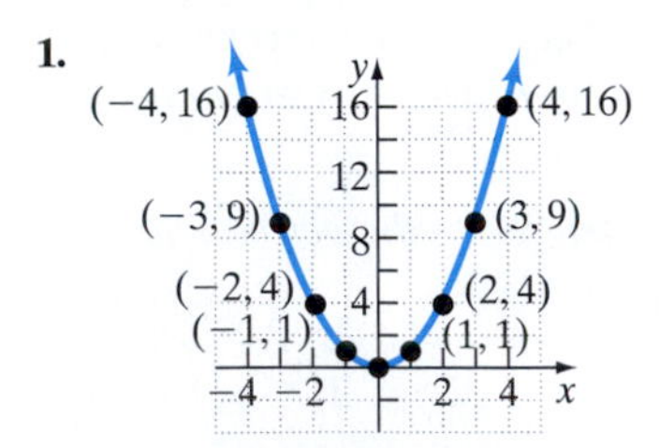


2.

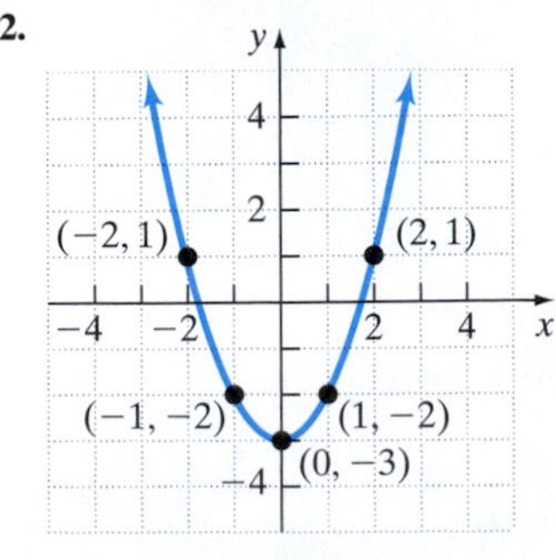


3. The set of all real numbers

## *Preparing for Graphing Quadratic Functions Using Transformations*

*Before getting started, take the following readiness quiz. If you get a problem wrong, go back to the section cited and review the material.*

1. Graph $y = x^2$ using point plotting. [Section 2.1, pp. 140–142]
2. Using the point-plotting method, graph $y = x^2 - 3$. [Section 2.1, pp. 140–142]
3. What is the domain of $f(x) = 2x^2 + 5x + 1$? [Section 5.1, p. 360]

We begin with a definition.

**DEFINITION**

A **quadratic function** is a function of the form

$$f(x) = ax^2 + bx + c$$

where $a$, $b$, and $c$ are real numbers and $a \neq 0$. The domain of a quadratic function consists of all real numbers.

Many situations can be modeled using quadratic functions. For example, we saw in Example 9 of Section 2.3 that Franco Modigliani used the quadratic function $I(a) = -55a^2 + 5119a - 54{,}448$ to model the relation between average annual income, $I$, and age, $a$.

A second situation in which a quadratic function appears involves the motion of a projectile. If we ignore the effect of air resistance on a projectile, the height $H$ of the projectile as a function of horizontal distance traveled, $x$, can be modeled using a quadratic function. See Figure 7.

**Figure 7**

The major goal of this and the next section is to learn methods that will allow us to obtain the graph of a quadratic function. Back in Section 2.1 we learned how to graph virtually any type of equation using point plotting. However, we also discovered that this method is inefficient and could lead to incomplete graphs. Remember, a graph is complete if it shows all of the "interesting features" of the graph. Some of the interesting features that must be included are the intercepts and the high and low points of the graph. We will present two methods that can be used to graph quadratic functions that are superior to the point-plotting method. The first method utilizes a technique called *transformations*. This will be the subject of this section. The second method uses properties of quadratic functions, which is discussed in the next section.

**Work Smart**

Consider the quadratic function $f(x) = ax^2 + bx + c$, where $a = 1$, $b = 0$, and $c$ is any real number. This is a function of the form $f(x) = x^2 + k$.

## 1 Graph Quadratic Functions of the Form $f(x) = x^2 + k$

We begin by looking at the graph of any quadratic function of the form $f(x) = x^2 + k$ such as $f(x) = x^2 + 3$ or $f(x) = x^2 - 4$.

In Example 4 from Section 2.1, we graph the equation $y = x^2$. For convenience, we provide the graph of $y = f(x) = x^2$ in Figure 8.

**Figure 8**

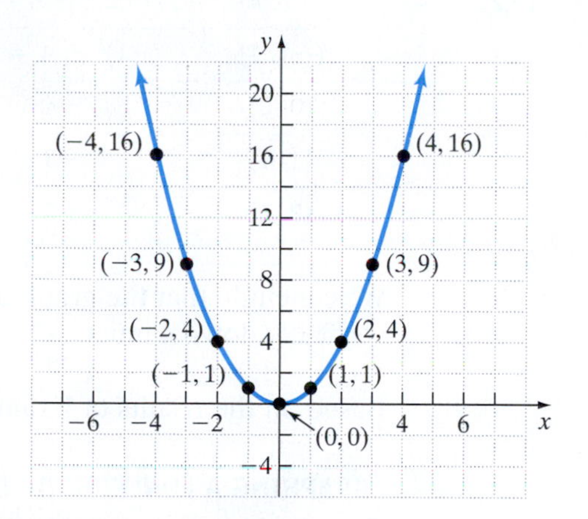


What effect does adding a real number $k$ to the function $f(x) = x^2$ have on its graph? Let's see!

### EXAMPLE 1 Graphing a Quadratic Function of the Form $f(x) = x^2 + k$

On the same Cartesian plane, graph $g(x) = x^2$ and $f(x) = x^2 + 3$.

### Solution

We begin by obtaining some points on the graphs of $g$ and $f$. For example, when $x = 0$, then $y = g(0) = 0$ and $y = f(0) = 0^2 + 3 = 3$. When $x = 1$, then $y = g(1) = 1$ and $y = f(1) = 1^2 + 3 = 4$. Table 3 lists these points along with a few others. Notice that the $y$-coordinates on the graph of $f(x) = x^2 + 3$ are exactly 3 units larger than the corresponding $y$-coordinates on the graph of $g(x) = x^2$. Figure 9 shows the graphs of $f$ and $g$.

**Table 3**

| $x$ | $g(x) = x^2$ | $(x, g(x))$ | $f(x) = x^2 + 3$ | $(x, f(x))$ |
|---|---|---|---|---|
| −2 | $(-2)^2 = 4$ | (−2, 4) | $(-2)^2 + 3 = 7$ | (−2, 7) |
| −1 | $(-1)^2 = 1$ | (−1, 1) | $(-1)^2 + 3 = 4$ | (−1, 4) |
| 0 | 0 | (0, 0) | 3 | (0, 3) |
| 1 | 1 | (1, 1) | 4 | (1, 4) |
| 2 | 4 | (2, 4) | 7 | (2, 7) |

**Figure 9**

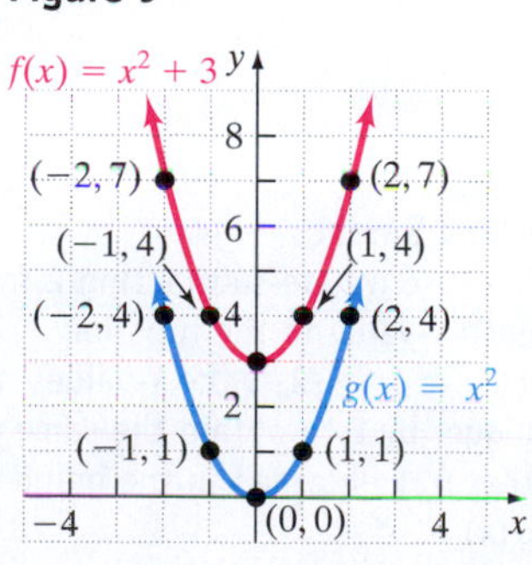

We conclude that the graph of $f$ is identical to the graph of $g$ except that it is shifted vertically up 3 units.

Let's look at another example.

**EXAMPLE 2 Graphing a Quadratic Function of the Form $f(x) = x^2 + k$**

On the same Cartesian plane, graph $g(x) = x^2$ and $f(x) = x^2 - 4$.

**Solution**

Table 4 lists some points on the graphs of $g$ and $f$. Notice that the $y$-coordinates on the graph of $f(x) = x^2 - 4$ are exactly 4 units smaller than the corresponding $y$-coordinates on the graph of $g(x) = x^2$. Figure 10 shows the graphs of $f$ and $g$.

**Table 4**

| $x$ | $g(x) = x^2$ | $(x, g(x))$ | $f(x) = x^2 - 4$ | $(x, f(x))$ |
|---|---|---|---|---|
| $-2$ | $(-2)^2 = 4$ | $(-2, 4)$ | $(-2)^2 - 4 = 0$ | $(-2, 0)$ |
| $-1$ | $(-1)^2 = 1$ | $(-1, 1)$ | $(-1)^2 - 4 = -3$ | $(-1, -3)$ |
| $0$ | $0$ | $(0, 0)$ | $-4$ | $(0, -4)$ |
| $1$ | $1$ | $(1, 1)$ | $-3$ | $(1, -3)$ |
| $2$ | $4$ | $(2, 4)$ | $0$ | $(2, 0)$ |

**Figure 10**

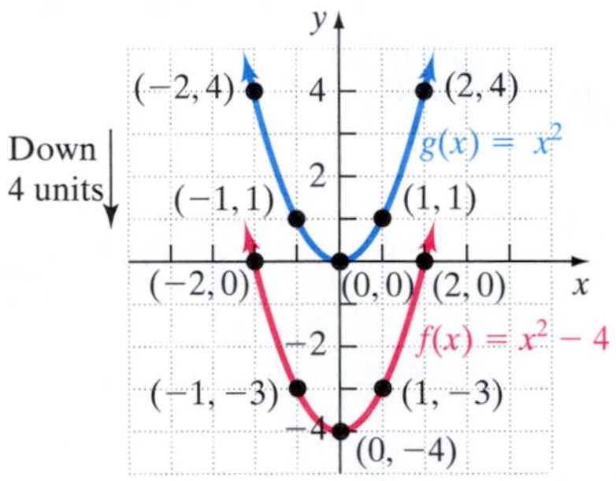

We conclude that the graph of $g$ is identical to the graph of $f$ except that it is shifted vertically down 4 units.

Based on the results of Examples 1 and 2, we are led to the following conclusion.

**GRAPHING A FUNCTION OF THE FORM $f(x) = x^2 + k$**

To obtain the graph of $f(x) = x^2 + k$, $k > 0$, from the graph of $y = x^2$, shift the graph of $y = x^2$ vertically up $k$ units. To obtain the graph of $f(x) = x^2 - k$, $k > 0$, from the graph of $y = x^2$ shift the graph of $y = x^2$ vertically down $k$ units.

**QUICK ✓** *Use the graph of $g(x) = x^2$ to obtain graph the quadratic function.*

**1.** $f(x) = x^2 + 5$ **2.** $f(x) = x^2 - 2$

## 2 Graph Quadratic Functions of the Form $f(x) = (x - h)^2$

We now look at the graph of any quadratic function of the form $f(x) = (x - h)^2$ such as $f(x) = (x + 3)^2$ or $f(x) = (x - 2)^2$. Our goal here is to determine the effect subtracting a real positive number $h$ from $x$ has on the graph of the function $f(x) = x^2$.

**EXAMPLE 3 Graphing a Quadratic Function of the Form $f(x) = (x - h)^2$**

On the same Cartesian plane, graph $g(x) = x^2$ and $f(x) = (x - 2)^2$.

**Work Smart**

Because we are subtracting 2 from each $x$-value in the function $f(x) = (x - 2)^2$, the $x$-values must be bigger by 2 to obtain the same $y$-value that was obtained in the graph of $g(x) = x^2$.

**Solution**

Again, we use the point-plotting method. Table 5 lists some points on the graphs of $g$ and $f$. Notice when $g(x) = 0$, $x = 0$, and when $f(x) = 0$, $x = 2$. Also, when $g(x) = 4$, $x = -2$ or 2, and when $f(x) = 4$, $x = 0$ or 4. We conclude that the graph of $f$ is identical to that of $g$, except that it is shifted 2 units to the right. See Figure 11.

**Table 5**

| $x$ | $g(x) = x^2$ | $(x, g(x))$ | $f(x) = (x - 2)^2$ | $(x, f(x))$ |
|---|---|---|---|---|
| $-2$ | $(-2)^2 = 4$ | $(-2, 4)$ | $(-2 - 2)^2 = 16$ | $(-2, 16)$ |
| $-1$ | $(-1)^2 = 1$ | $(-1, 1)$ | $(-1 - 2)^2 = 9$ | $(-1, 9)$ |
| 0 | 0 | $(0, 0)$ | 4 | $(0, 4)$ |
| 1 | 1 | $(1, 1)$ | 1 | $(1, 1)$ |
| 2 | 4 | $(2, 4)$ | 0 | $(2, 0)$ |
| 3 | 9 | $(3, 9)$ | 1 | $(3, 1)$ |
| 4 | 16 | $(4, 16)$ | 4 | $(4, 4)$ |

**Figure 11**

What if we add a positive number $h$ to $x$?

## EXAMPLE 4 Graphing a Quadratic Function of the Form $f(x) = (x + h)^2$

On the same Cartesian plane, graph $g(x) = x^2$ and $f(x) = (x + 3)^2$.

### Solution

Table 6 lists some points on the graphs of $g$ and $f$. Notice that when $g(x) = 0$, then $x = 0$, and when $f(x) = 0$, then $x = -3$. Also, when $g(x) = 4$, then $x = -2$ or 2, and when $f(x) = 4$, then $x = -5$ or $-1$. We conclude that the graph of $f$ is identical to that of $g$, except that it is shifted 3 units to the left. See Figure 12.

**Table 6**

| $x$ | $g(x) = x^2$ | $(x, g(x))$ | $f(x) = (x + 3)^2$ | $(x, f(x))$ |
|---|---|---|---|---|
| $-5$ | $(-5)^2 = 25$ | $(-5, 25)$ | $(-5 + 3)^2 = 4$ | $(-5, 4)$ |
| $-4$ | $(-4)^2 = 16$ | $(-4, 16)$ | $(-4 + 3)^2 = 1$ | $(-4, 1)$ |
| $-3$ | $(-3)^2 = 9$ | $(-3, 9)$ | $(-3 + 3)^2 = 0$ | $(-3, 0)$ |
| $-2$ | 4 | $(-2, 4)$ | 1 | $(-2, 1)$ |
| $-1$ | 1 | $(-1, 1)$ | 4 | $(-1, 4)$ |
| 0 | 0 | $(0, 0)$ | 9 | $(0, 9)$ |
| 1 | 1 | $(1, 1)$ | 16 | $(1, 16)$ |
| 2 | 4 | $(2, 4)$ | 25 | $(2, 25)$ |

**Figure 12**

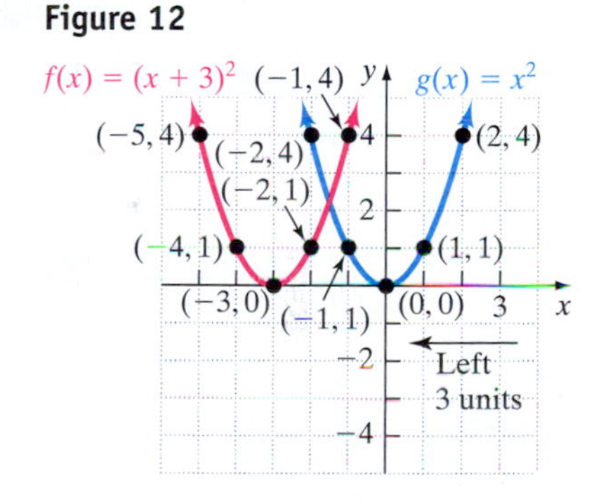


**Work Smart**

Because we are adding 3 to each $x$-value in the function $f(x) = (x + 3)^2$, the $x$-values must be smaller by 3 to obtain the same $y$-value that was obtained in the graph of $g(x) = x^2$.

**Work Smart**

If the function is of the form $f(x) = (x - h)^2$, shift the graph of $y = x^2$ right $h$ units, if the function is of the form $f(x) = (x + h)^2$, shift the graph of $y = x^2$ left $h$ units.

Based upon the results of Examples 3 and 4, we are led to the following conclusion.

**GRAPHING A FUNCTION OF THE FORM $f(x) = (x - h)^2$**

To obtain the graph of $f(x) = (x - h)^2$, $h > 0$, from the graph of $y = x^2$, shift the graph of $y = x^2$ horizontally to the right $h$ units. To obtain the graph of $f(x) = (x + h)^2$ from the graph of $y = x^2$, $h > 0$, shift the graph of $y = x^2$ horizontally to the left $h$ units.

**QUICK ✓** *Use the graph of $y = x^2$ to obtain graph the quadratic function.*

**3.** $f(x) = (x + 5)^2$

**4.** $f(x) = (x - 1)^2$

Let's do an example where we combine a horizontal shift with a vertical shift.

### EXAMPLE 5 Combining Horizontal and Vertical Shifts

Graph the function $f(x) = (x + 2)^2 - 3$.

#### Solution

We will graph $f$ in steps. We begin with the graph of $y = x^2$ as shown in Figure 13(a). We shift the graph of $y = x^2$ horizontally 2 units to the left to get the graph of $y = (x + 2)^2$. See Figure 13(b). Then, we shift the graph of $y = (x + 2)^2$ vertically down 3 units to get the graph of $y = (x + 2)^2 - 3$. See Figure 13(c). Notice that we keep track of key points plotted on each graph.

**Figure 13**

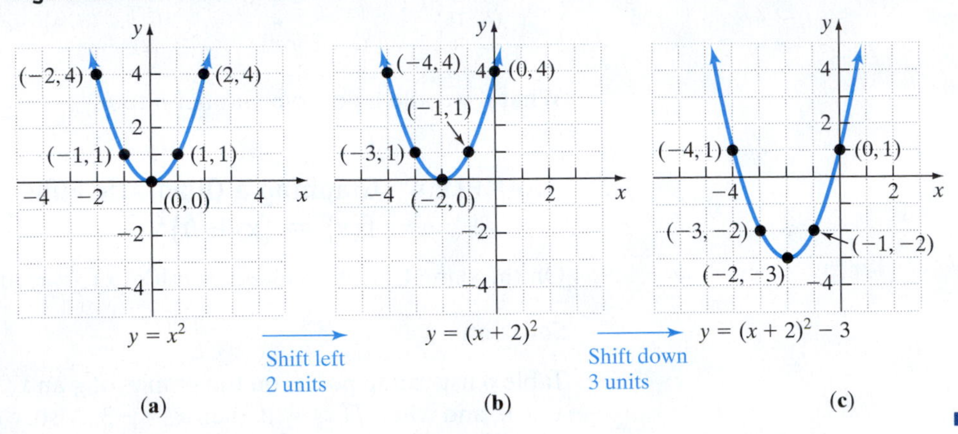


**Note:** The order in which the transformations take place does not matter. We could just as easily have shifted down 3 units first and then shifted left 2 units.

**QUICK ✓** *Graph each quadratic function using horizontal and vertical shifts.*

**5.** $f(x) = (x - 3)^2 + 2$ **6.** $f(x) = (x + 1)^2 - 4$

## 3 Graph Quadratic Functions of the Form $f(x) = ax^2$

In the examples presented thus far, the coefficient of the square term has been equal to 1. We now discuss the impact that the value of $a$ has on the graph of $f(x) = ax^2 + bx + c$. To make the discussion a little easier, we will only consider quadratic functions of the form $f(x) = ax^2, a \neq 0$.

First, let's consider situations in which the value of $a$ is positive. Table 7 shows points on the graphs of $f(x) = x^2$, $g(x) = \frac{1}{2}x^2$ and $h(x) = 2x^2$. Figure 14 shows the graphs of

**Table 7**

| $x$ | $f(x) = x^2$ | $g(x) = \frac{1}{2}x^2$ | $h(x) = 2x^2$ |
|---|---|---|---|
| −2 | $(-2)^2 = 4$ | $\frac{1}{2}(-2)^2 = 2$ | $2(-2)^2 = 8$ |
| −1 | $(-1)^2 = 1$ | $\frac{1}{2}(-1)^2 = \frac{1}{2}$ | $2(-1)^2 = 2$ |
| 0 | 0 | 0 | 0 |
| 1 | 1 | $\frac{1}{2}$ | 2 |
| 2 | 4 | 2 | 8 |

**Figure 14**
$f(x) = ax^2$. Since $a > 0$, the graphs open up.

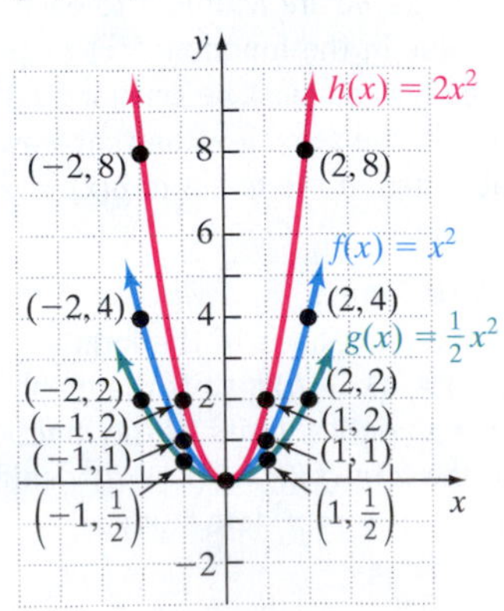

$f(x) = x^2$, $g(x) = \frac{1}{2}x^2$ and $h(x) = 2x^2$. Notice that the $y$-coordinates on the graph of $g$ are exactly $\frac{1}{2}$ of the values of the $y$-coordinates on the graph of $f$. The $y$-coordinates on the graph of $h$ are exactly 2 times the values of the $y$-coordinates on the graph of $f$. Put another way, the larger the value of $a$, the "taller" the graph is, and the smaller the value of $a$, the "shorter" the graph is. Also, notice that all three graphs open "up."

Now let's consider what happens when $a$ is negative. Table 8 shows points on the graphs of $f(x) = -x^2$, $g(x) = -\frac{1}{2}x^2$ and $h(x) = -2x^2$. Figure 15 shows the graphs of $f(x) = -x^2$, $g(x) = -\frac{1}{2}x^2$ and $h(x) = -2x^2$. Notice that the $y$-coordinates on the graph of $g$ are exactly $-\frac{1}{2}$ times the values of the $y$-coordinates on the graph of $f$. The $y$-coordinates on the graph of $h$ are exactly $-2$ times the values of the $y$-coordinates on the graph of $f$. Put another way, the larger the value of $|a|$, the "taller" the graph is, and the smaller the value of $|a|$, the "shorter" the graph is. Also, notice that all three graphs open "down."

**Table 8**

| $x$ | $f(x) = -x^2$ | $g(x) = -\frac{1}{2}x^2$ | $h(x) = -2x^2$ |
|---|---|---|---|
| $-2$ | $-(-2)^2 = -4$ | $-\frac{1}{2}(-2)^2 = -2$ | $-2(-2)^2 = -8$ |
| $-1$ | $-(-1)^2 = -1$ | $-\frac{1}{2}(-1)^2 = -\frac{1}{2}$ | $-2(-1)^2 = -2$ |
| 0 | 0 | 0 | 0 |
| 1 | $-1$ | $-\frac{1}{2}$ | $-2$ |
| 2 | $-4$ | $-2$ | $-8$ |

**Figure 15**

$f(x) = ax^2$. Since $a < 0$, the graphs open down.

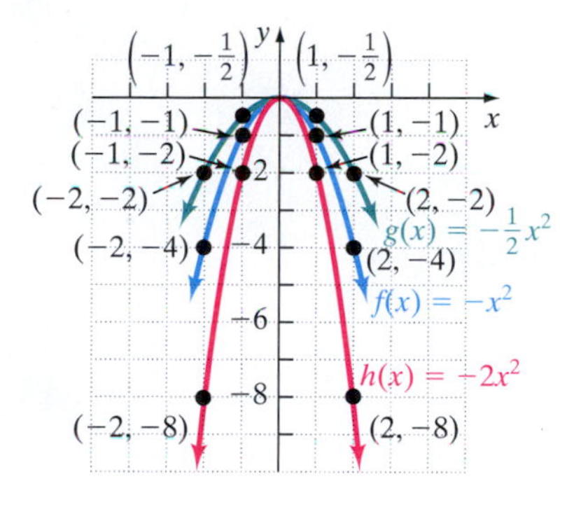


We summarize these conclusions below.

**PROPERTIES OF THE GRAPH OF $f(x) = ax^2$**

- If $a > 0$, the graph of $f(x) = ax^2$ will open upward. In addition, if $0 < a < 1$ ($a$ is between 0 and 1), the opening in the graph will be "wider" than that of $y = x^2$. If $a > 1$, the opening in the graph will be "narrower" than that of $y = x^2$.
- If $a < 0$, the graph of $f(x) = ax^2$ will open downward. In addition, if $0 < |a| < 1$, the opening in the graph will be "wider" than that of $y = x^2$. If $|a| > 1$, the opening in the graph will be "narrower" than that of $y = x^2$.
- When $|a| > 1$, we say that the graph is **vertically stretched** by a factor of $|a|$. When $0 < |a| < 1$, we say that the graph is **vertically compressed** by a factor of $|a|$.

**GRAPHING A FUNCTION OF THE FORM $f(x) = ax^2$**

To obtain the graph of $f(x) = ax^2$ from the graph of $y = x^2$, multiply each $y$-coordinate on the graph of $y = x^2$ by $a$.

**EXAMPLE 6** **Graphing a Quadratic Function of the Form $f(x) = ax^2$**

Use the graph of $y = x^2$ to obtain the graph of $f(x) = -2x^2$.

### Solution

To obtain the graph of $f(x) = -2x^2$ from the graph of $y = x^2$, we multiply each $y$-coordinate on the graph of $y = x^2$ by $-2$ (the value of $a$). See Figure 16.

**Figure 16**

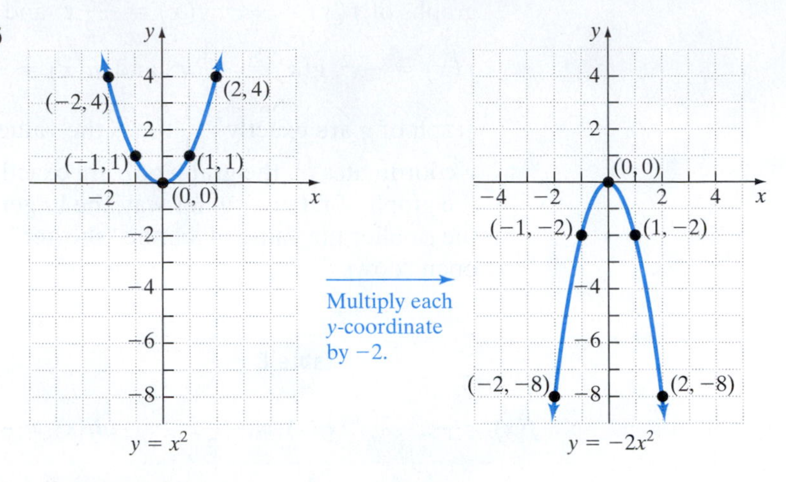


**QUICK ✓** *Use the graph of $y = x^2$ to graph each quadratic function.*

**7.** $f(x) = 3x^2$

**8.** $f(x) = -\frac{1}{4}x^2$

## 4 Graph Quadratic Functions of the Form $f(x) = ax^2 + bx + c$

The graphs obtained in Examples 1–6 are typical of graphs of all quadratic functions. We call the graph of a quadratic function a **parabola** (pronounced puh-ráb-ō-luh). Refer to Figure 17, where two parabolas are shown.

**Figure 17**

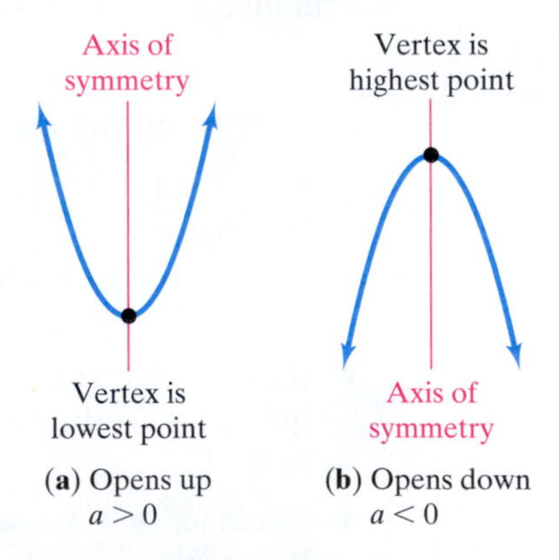


(**a**) Opens up $a > 0$ (**b**) Opens down $a < 0$

**Work Smart**

The axis of symmetry is useful in graphing a quadratic function by hand.

The parabola in Figure 17(a) **opens up** (since $a > 0$) and has a lowest point; the parabola in Figure 17(b) **opens down** (since $a < 0$) and has a highest point. The lowest or highest point of a parabola is called the **vertex.** The vertical line passing through the vertex in each parabola in Figure 17 is called the **axis of symmetry** of the parabola. If we were to take the portion of the parabola to the right of the vertex and fold it over the axis of symmetry, it would lie directly on top of the portion of the parabola to the left of the vertex. Therefore, we say that the parabola is symmetric about its axis of symmetry. It is important to note that the axis of symmetry is not part of the graph of the quadratic function, but it will be useful when graphing quadratic functions using the methods we present in the next section.

**Work Smart**
When graphing parabolas using transformations obtain the graph that results from the vertical compression or stretch first, followed by the horizontal shift, followed by the vertical shift.

Our goal right now is to combine the techniques learned from Examples 1–6 to graph any quadratic function. The techniques of shifting horizontally, shifting vertically, stretching, and compressing are collectively referred to as **transformations.** To graph any quadratic function of the form $f(x) = ax^2 + bx + c$ using transformations, we use the following steps.

### Steps for Graphing Quadratic Functions Using Transformations

**Step 1:** Write the function $f(x) = ax^2 + bx + c$ as $f(x) = a(x - h)^2 + k$ by completing the square in $x$.

**Step 2:** Graph the function $f(x) = a(x - h)^2 + k$ using transformations.

Notice that we must first write the quadratic function $f(x) = ax^2 + bx + c$ as $f(x) = a(x - h)^2 + k$. This is necessary so we can determine the horizontal and vertical shifts.

## EXAMPLE 7 How to Graph a Quadratic Function of the Form $f(x) = ax^2 + bx + c$ Using Transformations

Graph $f(x) = x^2 + 4x + 3$ using transformations. Identify the vertex and axis of symmetry of the parabola. Based on the graph, determine the domain and range of the quadratic function.

### Step-by-Step Solution

**Step 1:** Write the function $f(x) = ax^2 + bx + c$ as $f(x) = a(x - h)^2 + k$ by completing the square in $x$.

$$f(x) = x^2 + 4x + 3$$

Group the terms involving $x$: $= (x^2 + 4x) + 3$

Complete the square in $x$ by taking $\frac{1}{2}$ the coefficient on $x$ and squaring the result: $\left(\frac{1}{2} \cdot 4\right)^2 = 4$.

Because we added 4, we must also subtract 4: $= (x^2 + 4x + 4) + 3 - 4$

$= (x^2 + 4x + 4) - 1$

Factor the perfect squared trinomial in parentheses: $= (x + 2)^2 - 1$

**Step 2:** Graph the function $f(x) = a(x - h)^2 + k$ using transformations.

The graph of $f(x) = (x + 2)^2 - 1$ is the graph of $y = x^2$ shifted 2 units left and 1 unit down. See Figure 18.

**Figure 18**

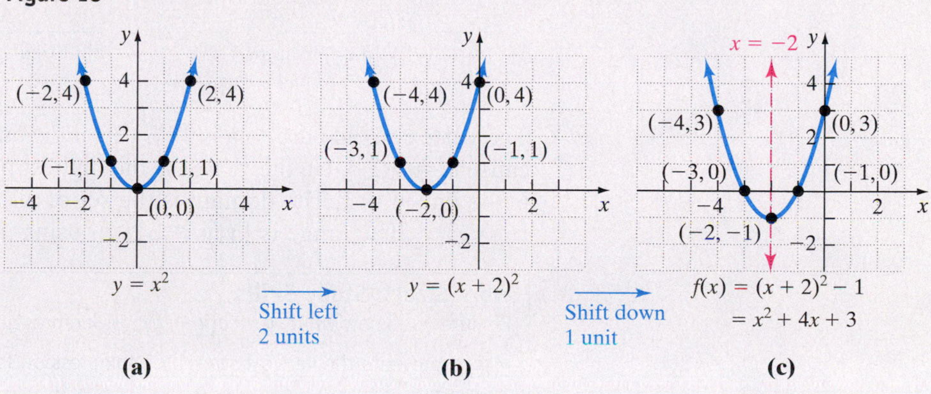


From the graph in Figure 18(c), we can see that the vertex of the parabola is $(-2, -1)$. In addition, because the parabola opens up (because $a = 1 > 0$), the vertex is the lowest point on the graph. The axis of symmetry is the line $x = -2$. The domain is the set of all real numbers, or using interval notation $(-\infty, \infty)$. The range is $\{y \mid y \geq -1\}$, or using interval notation $[-1, \infty)$.

## EXAMPLE 8 Graphing a Quadratic Function of the Form $f(x) = ax^2 + bx + c$ Using Transformations

Graph $f(x) = -2x^2 + 4x + 1$ using transformations. Identify the vertex and axis of symmetry of the parabola. Based on the graph, determine the domain and range of the quadratic function.

### Solution

Write the function $f(x) = ax^2 + bx + c$ as $f(x) = a(x - h)^2 + k$ by completing the square in $x$.

$$f(x) = -2x^2 + 4x + 1$$

Group the terms involving $x$: $= (-2x^2 + 4x) + 1$

Factor out the coefficient of the square term, $-2$, from the parentheses: $= -2(x^2 - 2x) + 1$

Complete the square in $x$ by taking $\frac{1}{2}$ the coefficient on $x$ and squaring the result: $(\frac{1}{2} \cdot 2)^2 = 1$. We add 1 inside the parentheses. Because everything in the parentheses is multiplied by $-2$, so we really added $-2$, so we must add 2 to offset this: $= -2(x^2 - 2x + 1) + 1 + 2$

Factor the perfect square trinomial in parentheses: $= -2(x - 1)^2 + 3$

Now, graph the function $f(x) = a(x - h)^2 + k$ using transformations. Since $a = -2$, the parabola will open down and stretch by a factor of 2. Since $h = 1$ the parabola shifts 1 unit to the right; since $k = 3$, the parabola shifts 3 units up. See Figure 19.

**Figure 19**

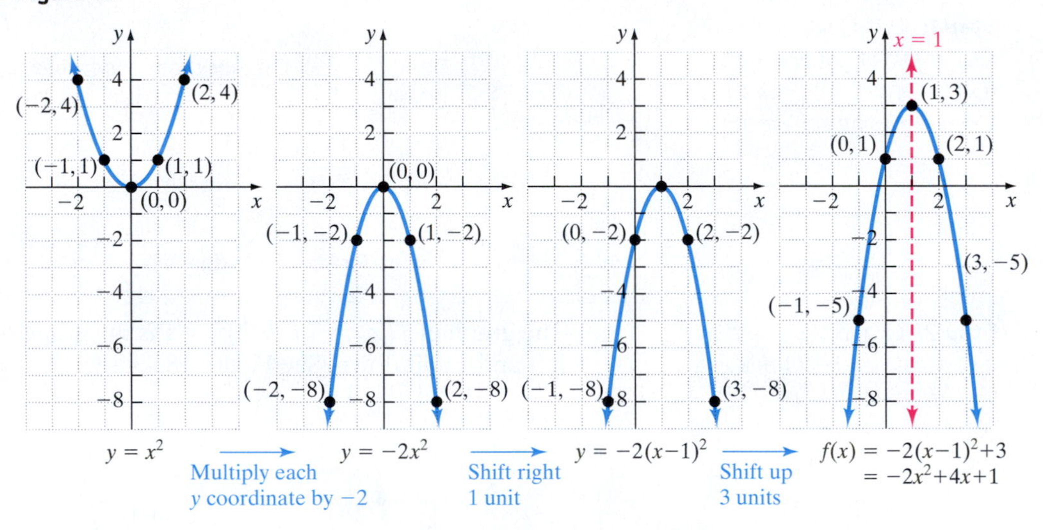


The vertex of the parabola is $(1, 3)$. In addition, because the parabola opens down (because $a = -2 < 0$), the vertex is the highest point on the graph. The axis of symmetry is the line $x = 1$. The domain is the set of all real numbers, or using interval notation $(-\infty, \infty)$. The range is $\{y|y \le 3\}$, or using interval notation $(-\infty, 3]$.

**Work Smart: Study Skills**

Be sure you know what $y = a(x - h)^2 + k$ represents in reference to the graph of $y = x^2$:

- $|a|$ represents the vertical stretch or compression factor: how wide or narrow the graph appears.
- The sign of $a$ determines whether the parabola opens up or down.
- $h$ represents the number of units the graph is shifted horizontally.
- $k$ represents the number of units the graph is shifted vertically.

For example, $y = 4(x + 2)^2 - 3$ means that the graph of $y = x^2$ is stretched vertically by a factor of 4, is shifted 2 units horizontally to the left, and is shifted vertically down 3 units. The vertex of $y = 4(x + 2)^2 - 3$ is at $(-2, -3)$ and represents the low point of the graph since $a > 0$.

**QUICK ✓** *Graph each quadratic function using transformations. Based on the graph, determine the domain and range of each function.*

**9.** $f(x) = -3(x+2)^2 + 1$ **10.** $f(x) = 2x^2 - 8x + 5$

## 5 Find a Quadratic Function from Its Graph

If we are given the vertex, $(h, k)$, and one additional point on the graph of a quadratic function, we can find the quadratic function $f(x) = a(x-h)^2 + k$ that results in the given graph.

### EXAMPLE 9 Finding the Quadratic Function Given Its Vertex and One Other Point

Determine the quadratic function whose graph is given in Figure 20. Write the function in the form $f(x) = a(x-h)^2 + k$.

**Solution**

The vertex is $(2, 3)$, so $h = 2$ and $k = 3$. Substitute these values into $f(x) = a(x-h)^2 + k$.

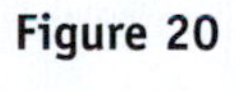
Figure 20

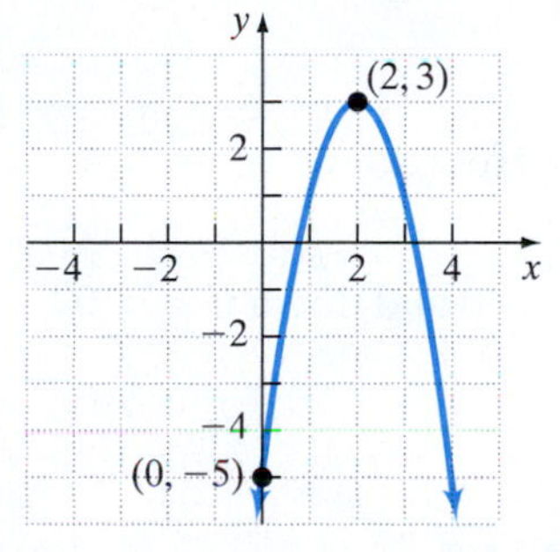


$$f(x) = a(x-h)^2 + k$$

$$h = 2,\ k = 3{:}\quad = a(x-2)^2 + 3$$

To determine the value of $a$, we use the fact that $f(0) = -5$ (the $y$-intercept).

$$f(x) = a(x-2)^2 + 3$$

$$x = 0,\ y = f(0) = -5{:}\quad -5 = a(0-2)^2 + 3$$

$$-5 = a(4) + 3$$

$$-5 = 4a + 3$$

$$\text{Subtract 3 from both sides:}\quad -8 = 4a$$

$$a = -2$$

The quadratic function whose graph is shown in Figure 20 is $f(x) = -2(x-2)^2 + 3$.

**QUICK ✓** *Find the quadratic function whose graph is given. Write the function in the form* $f(x) = a(x-h)^2 + k$.

**11.**

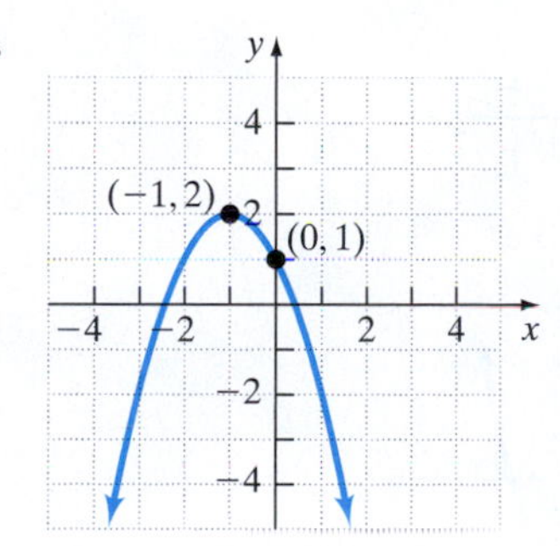

# 8.4 Exercises

**For Extra Help:**

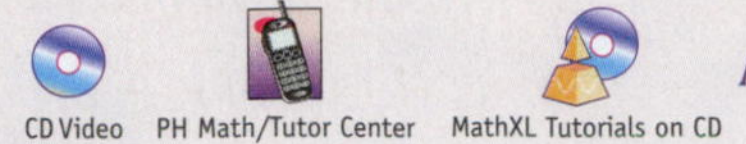

Student Solutions Manual | CD Video | PH Math/Tutor Center | MathXL Tutorials on CD | MathXL® | MyMathLab

## Concepts and Vocabulary

*In Problems 1–3, fill in the blanks.*

**1.** A _________ _________ is a function of the form $f(x) = ax^2 + bx + c$ where $a$, $b$, and $c$ are real numbers and $a \neq 0$.

**2.** To graph $f(x) = x^2 + k$ using the graph of $y = x^2$, shift the graph of $y = x^2$ vertically _____ $k$ units if $k > 0$ and shift the graph of $y = x^2$ vertically _________ $k$ units if $k < 0$.

**3.** When obtaining the graph of $f(x) = ax^2$ from the graph of $y = x^2$, we multiply each __-coordinate on the graph of $y = x^2$ by ___. If $|a| > 1$, we say that the graph is _________ _________ by a factor of $|a|$. If $0 < |a| < 1$, we say that the graph is _________ _________ by a factor of $|a|$.

*In Problems 4–6, answer True or False to each statement.*

**4.** To obtain the graph of $f(x) = x^2 + 5$ from the graph of $y = x^2$, shift the graph of $y = x^2$ vertically up 5 units.

**5.** To obtain the graph of $f(x) = (x + 5)^2$ from the graph of $y = x^2$, shift the graph of $y = x^2$ horizontally to the right 5 units.

**6.** The graph of $f(x) = -3x^2 + x + 6$ opens down.

**7.** What is the lowest or highest point on a parabola called? How do we know whether this point is a high point or a low point?

**8.** Why does the graph of a quadratic function open up if $a > 0$ and down if $a < 0$?

## Building Skills

**9.** Match each quadratic function to its graph.

**(I)** $f(x) = x^2 + 3$ **(II)** $f(x) = (x + 3)^2$

**(III)** $f(x) = x^2 - 3$ **(IV)** $f(x) = (x - 3)^2$

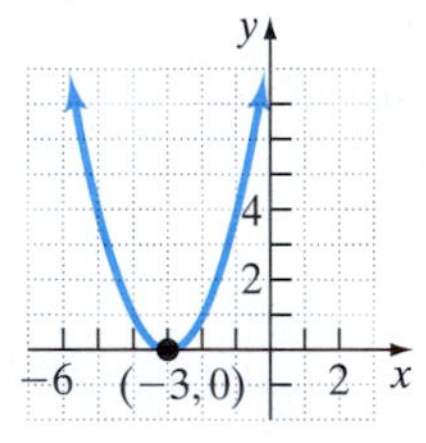

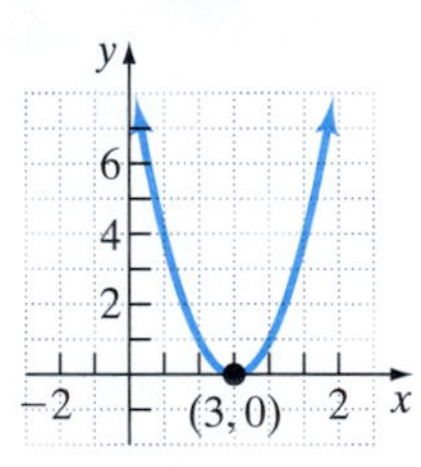

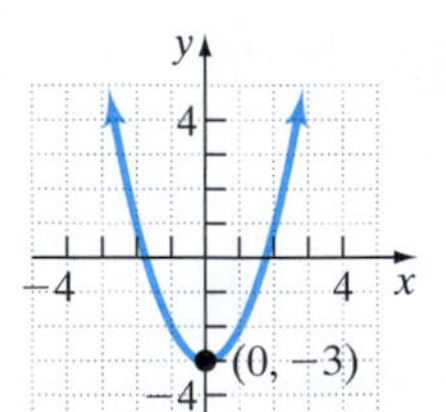


**(D)**

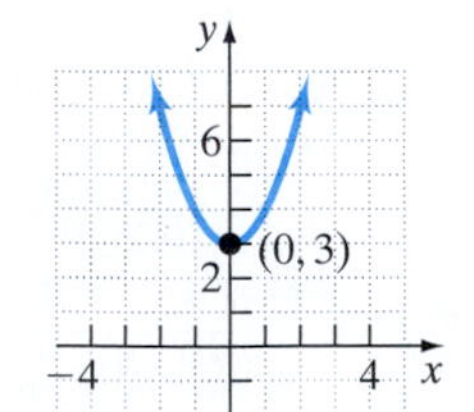

**10.** Match each quadratic function to its graph.

**(I)** $f(x) = (x - 2)^2 - 4$ **(II)** $f(x) = -(x - 2)^2 + 4$

**(III)** $f(x) = -(x + 2)^2 + 4$ **(IV)** $f(x) = 2(x - 2)^2 - 4$

**(A)**

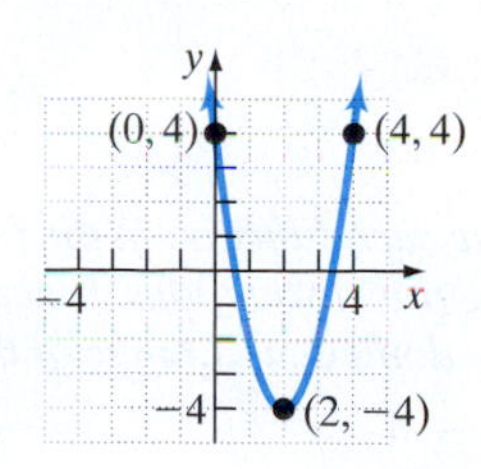


**(B)**

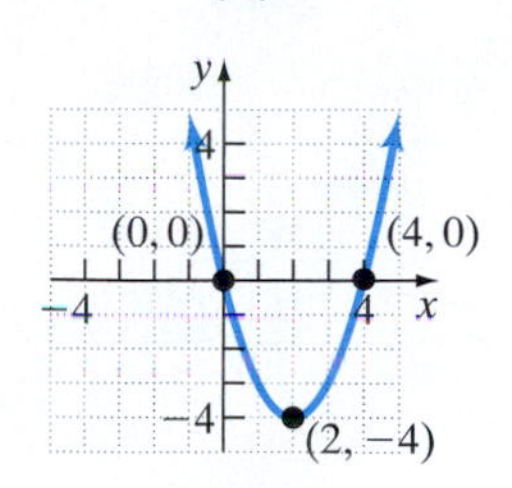


**(C)**

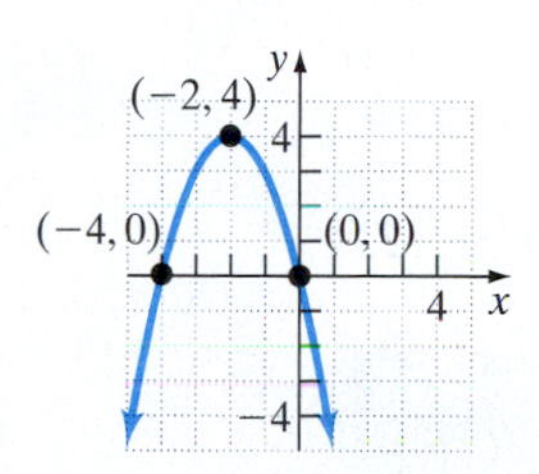


**(D)**

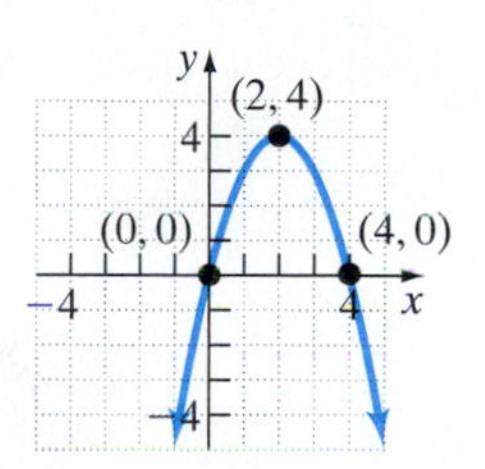


*In Problems 11–18, verbally explain how to obtain the graph of the given quadratic function from the graph of $y = x^2$. For example, to obtain the graph of $f(x) = (x - 3)^2 - 6$ from the graph of $y = x^2$, we take the graph of $y = x^2$ and shift it 3 units to the right and 6 units down.*

**11.** $f(x) = (x + 10)^2$

**12.** $G(x) = (x - 9)^2$

**13.** $F(x) = x^2 + 12$

**14.** $g(x) = x^2 - 8$

**15.** $H(x) = 2(x - 5)^2$

**16.** $h(x) = 4(x + 7)^2$

**17.** $f(x) = -3(x + 5)^2 + 8$

**18.** $F(x) = -\frac{1}{2}(x - 3)^2 - 5$

*In Problems 19–42, use the graph of $y = x^2$ to graph the quadratic function.*

**19.** $f(x) = x^2 + 1$

**20.** $f(x) = x^2 - 1$

**21.** $h(x) = x^2 + 6$

**22.** $g(x) = x^2 - 7$

**23.** $F(x) = (x - 3)^2$

**24.** $F(x) = (x - 2)^2$

**25.** $h(x) = (x + 2)^2$

**26.** $f(x) = (x + 4)^2$

**27.** $g(x) = 4x^2$

**28.** $G(x) = 5x^2$

**29.** $H(x) = \frac{1}{3}x^2$

**30.** $h(x) = \frac{3}{2}x^2$

**31.** $p(x) = -x^2$

**32.** $P(x) = -3x^2$

**33.** $f(x) = (x - 1)^2 - 3$

**34.** $g(x) = (x + 2)^2 - 1$

**35.** $F(x) = (x + 3)^2 + 1$

**36.** $G(x) = (x - 4)^2 + 2$

**37.** $h(x) = -(x + 3)^2 + 2$

**38.** $H(x) = -(x - 3)^2 + 5$

**39.** $G(x) = 2(x + 1)^2 - 2$

**40.** $F(x) = 3(x - 2)^2 - 1$

**41.** $H(x) = -\frac{1}{2}(x + 5)^2 + 3$

**42.** $f(x) = -\frac{1}{2}(x + 6)^2 + 2$

## Mixed Practice

*In Problems 43–64, write each function in the form $f(x) = a(x - h)^2 + k$. Then graph each quadratic function using transformations. Determine the vertex and axis of symmetry. Based on the graph, determine the domain and range of the quadratic function.*

**43.** $f(x) = x^2 + 2x - 4$

**44.** $f(x) = x^2 + 4x - 1$

**45.** $g(x) = x^2 - 4x + 8$

**46.** $G(x) = x^2 - 2x + 7$

**47.** $f(x) = x^2 + 6x - 16$

**48.** $f(x) = x^2 + 4x + 5$

**49.** $F(x) = x^2 + x - 12$

**50.** $h(x) = x^2 - 7x + 10$

**51.** $H(x) = 2x^2 - 4x - 1$

**52.** $g(x) = 2x^2 + 4x - 3$

**53.** $P(x) = 3x^2 + 12x + 13$

**54.** $f(x) = 3x^2 + 18x + 25$

**55.** $F(x) = -x^2 - 10x - 21$

**56.** $g(x) = -x^2 - 8x - 14$

**57.** $g(x) = -x^2 + 6x - 1$

**58.** $f(x) = -x^2 + 10x - 17$

**59.** $H(x) = -2x^2 + 8x - 4$

**60.** $h(x) = -2x^2 + 12x - 17$

**61.** $f(x) = \frac{1}{3}x^2 - 2x + 4$

**62.** $f(x) = \frac{1}{2}x^2 + 2x - 1$

**63.** $G(x) = -12x^2 - 12x + 1$

**64.** $h(x) = -4x^2 + 4x$

## Applying the Concepts

*In Problems 65–74, write a quadratic function in the form $f(x) = a(x - h)^2 + k$ with the properties given.*

**65.** Opens up; vertex at $(3, 0)$.

**66.** Opens up; vertex at $(0, 2)$.

**67.** Opens up; vertex at $(-3, 1)$

**68.** Opens up; vertex at $(4, -2)$

**69.** Opens down; vertex at $(5, -1)$

**70.** Opens down; vertex at $(-4, -7)$

**71.** Opens up; vertically stretched by a factor of 4; vertex at $(9, -6)$.

**72.** Opens up; vertically compressed by a factor of $\frac{1}{2}$; vertex at $(-5, 0)$.

**73.** Opens down; vertically compressed by a factor of $\frac{1}{3}$; vertex at $(0, 6)$.

**74.** Opens down; vertically stretched by a factor of 5; vertex at $(5, 8)$.

*In Problems 75–80, determine the quadratic function whose graph is given.*

**75.**

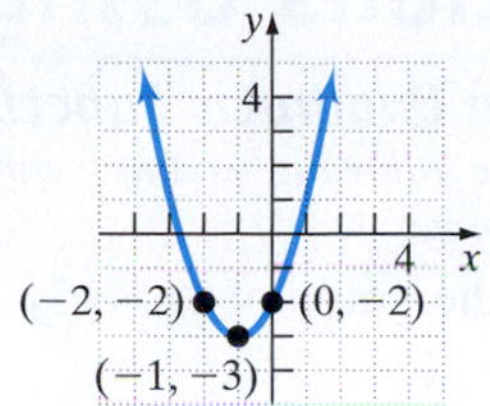


**76.**

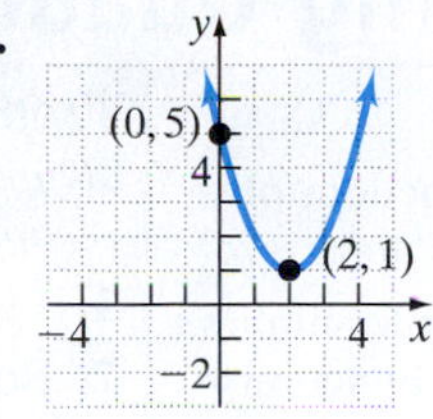


**77.**

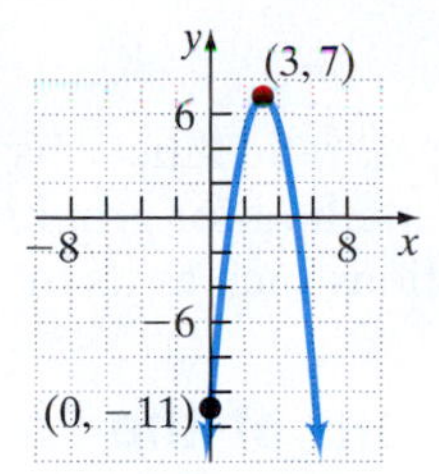


**78.**

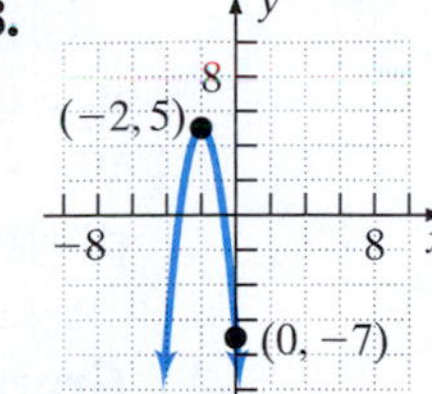


**79.**

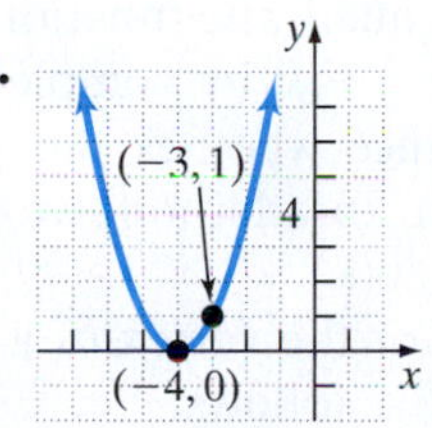


**80.**

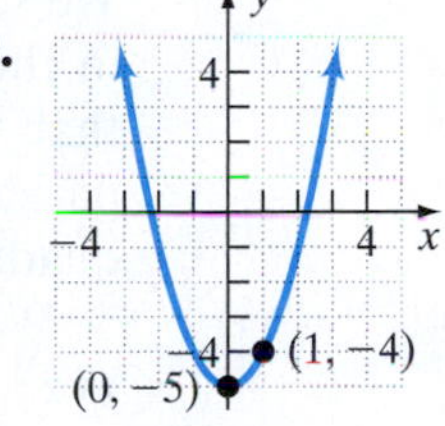


## Extending the Concepts

**81.** Can a quadratic function have a range of $(-\infty, \infty)$? Justify your answer.

**82.** Can the graph of a quadratic function have more than one $y$-intercept? Justify your answer.

## Synthesis Review

*In Problems 83–85, divide.*

**83.** $\dfrac{349}{12}$ **84.** $\dfrac{4x^2 + 19x - 1}{x + 5}$ **85.** $\dfrac{2x^4 - 11x^3 + 13x^2 - 8x}{2x - 1}$

**86.** Explain how division of real numbers is related to division of polynomials.

## The Graphing Calculator

*In Problems 87–94, graph each quadratic function. Determine the vertex and axis of symmetry. Based on the graph, determine the range of the function.*

**87.** $f(x) = x^2 + 1.3$

**88.** $f(x) = x^2 - 3.5$

**89.** $g(x) = (x - 2.5)^2$

**90.** $G(x) = (x + 4.5)^2$

**91.** $h(x) = 2.3(x - 1.4)^2 + 0.5$

**92.** $H(x) = 1.2(x + 0.4)^2 - 1.3$

**93.** $F(x) = -3.4(x - 2.8)^2 + 5.9$

**94.** $f(x) = 0.3(x + 3.8)^2 - 8.9$

# 8.5 Graphing Quadratic Functions Using Properties

**OBJECTIVES**

1. Graph Quadratic Functions of the Form $f(x) = ax^2 + bx + c$
2. Find the Maximum or Minimum Value of a Quadratic Function
3. Model and Solve Optimization Problems Involving Quadratic Functions

### *Preparing for Graphing Quadratic Functions Using Properties*

*Before getting started, take the following readiness quiz. If you get a problem wrong, go back to the section cited and review the material.*

**1.** Find the intercepts of the graph of $2x + 5y = 20$. [Section 3.1, pp. 191–193]

**2.** Solve: $2x^2 - 3x - 20 = 0$ [Section 5.8, pp. 423–426]

**3.** Find the zeros of $f(x) = x^2 - 3x - 4$. [Section 5.8, p. 428]

In Section 8.4, we graphed quadratic functions using a method called transformations. We now introduce a second method for graphing quadratic equations that utilizes the properties of quadratic functions such as its intercepts, axis of symmetry, and vertex.

## 1 Graph Quadratic Functions of the Form $f(x) = ax^2 + bx + c$

We saw in Section 8.4 that a quadratic function $f(x) = ax^2 + bx + c$ can be written in the form $f(x) = a(x - h)^2 + k$ by completing the square in $x$. We also learned that the value of $a$ determines whether the graph of the quadratic function (the parabola) opens up or down. In addition, we know that $(h, k)$ is the vertex of the quadratic function.

We can obtain a formula for the vertex of a parabola by completing the square in $f(x) = ax^2 + bx + c, a \neq 0$, as follows:

$$f(x) = ax^2 + bx + c$$

Group terms involving $x$:
$$= (ax^2 + bx) + c$$

Factor out $a$:
$$= a\left(x^2 + \frac{b}{a}x\right) + c$$

Complete the square in $x$ by taking $\frac{1}{2}$ the coefficient of $x$ and squaring the result: $\left(\frac{1}{2} \cdot \frac{b}{a}\right)^2 = \frac{b^2}{4a^2}$. Because we added $\frac{b^2}{4a^2}$ inside the parentheses, we subtract $a \cdot \frac{b^2}{4a^2} = \frac{b^2}{4a}$ outside the parentheses:
$$= a\left(x^2 + \frac{b}{a}x + \frac{b^2}{4a^2}\right) + c - \frac{b^2}{4a}$$

Factor the perfect square trinomial; multiply $c$ by $\frac{4a}{4a}$ to get a common denominator:
$$= a\left(x + \frac{b}{2a}\right)^2 + c \cdot \frac{4a}{4a} - \frac{b^2}{4a}$$

Write expression in the form $f(x) = a(x - h)^2 + k$:
$$= a\left(x - \left(-\frac{b}{2a}\right)\right)^2 + \frac{4ac - b^2}{4a}$$

If we compare $f(x) = a\left(x - \left(-\frac{b}{2a}\right)\right)^2 + \frac{4ac - b^2}{4a}$ to $f(x) = a(x - h)^2 + k$, we come to the following conclusion:

**THE VERTEX OF A PARABOLA**

Any quadratic function $f(x) = ax^2 + bx + c, a \neq 0$, will have vertex

$$\left(-\frac{b}{2a}, \frac{4ac - b^2}{4a}\right)$$

Because the $y$-coordinate on the graph of any function can be found by evaluating the function at the corresponding $x$-coordinate, we can restate the coordinates of the vertex as

$$\left(-\frac{b}{2a}, f\left(-\frac{b}{2a}\right)\right)$$

**Work Smart**

Another formula for the vertex is

$$\left(-\frac{b}{2a}, \frac{-D}{4a}\right)$$

where $D = b^2 - 4ac$, the discriminant.

***Preparing for...Answers*** **1.** $(0, 4), (10, 0)$ **2.** $\left\{-\frac{5}{2}, 4\right\}$ **3.** $-1$ and $4$

Because the axis of symmetry intersects the vertex, we have that the axis of symmetry for any parabola is $x = -\frac{b}{2a}$. In addition, we know that the parabola will open up if $a > 0$ and down if $a < 0$. Using this information along with the intercepts of the graph of the quadratic function, we can obtain a complete graph.

The $y$-intercept is the value of the quadratic function $f(x) = ax^2 + bx + c$ at $x = 0$, that is, $f(0) = c$.

The $x$-intercepts, if there are any, are found by solving the quadratic equation

$$f(x) = ax^2 + bx + c = 0$$

As we learned in Section 8.2, this equation has two, one, or no real solutions, depending on the value of the discriminant $b^2 - 4ac$. We use the value of the discriminant to determine the number of $x$-intercepts the graph of the quadratic function will have.

**THE $x$-INTERCEPTS OF THE GRAPH OF A QUADRATIC FUNCTION**

1. If the discriminant $b^2 - 4ac > 0$, the graph of $f(x) = ax^2 + bx + c$ has two different $x$-intercepts. The graph will cross the $x$-axis at the solutions to the equation $ax^2 + bx + c = 0$.
2. If the discriminant $b^2 - 4ac = 0$, the graph of $f(x) = ax^2 + bx + c$ has one $x$-intercept. The graph will touch the $x$-axis at the solution to the equation $ax^2 + bx + c = 0$.
3. If the discriminant $b^2 - 4ac < 0$, the graph of $f(x) = ax^2 + bx + c$ has no $x$-intercepts. The graph will not cross or touch the $x$-axis.

Figure 21 illustrates these possibilities for parabolas that open up.

**Figure 21**

$f(x) = ax^2 + bx + c,\ a > 0$

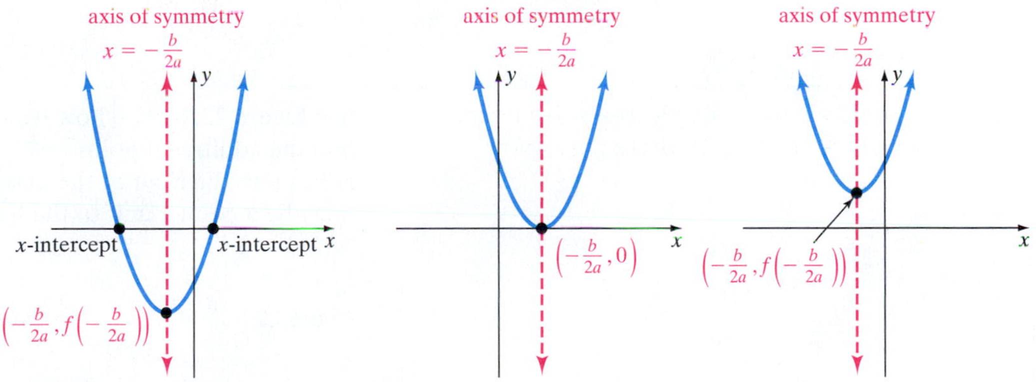

axis of symmetry $x = -\frac{b}{2a}$; $x$-intercept; $x$-intercept; $\left(-\frac{b}{2a}, f\left(-\frac{b}{2a}\right)\right)$

axis of symmetry $x = -\frac{b}{2a}$; $\left(-\frac{b}{2a}, 0\right)$

axis of symmetry $x = -\frac{b}{2a}$; $\left(-\frac{b}{2a}, f\left(-\frac{b}{2a}\right)\right)$

## EXAMPLE 1 How to Graph a Quadratic Function Using Its Properties

Graph $f(x) = x^2 + 2x - 15$ using its properties.

### Step-by-Step Solution

We compare $f(x) = x^2 + 2x - 15$ to $f(x) = ax^2 + bx + c$ and see that $a = 1$, $b = 2$, and $c = -15$.

**Step 1:** *Determine whether the parabola opens up or down.*

The parabola opens up because $a = 1 > 0$.

(continued)

**Step 2:** Determine the vertex and axis of symmetry.

The $x$-coordinate of the vertex is

$$x = -\frac{b}{2a} = -\frac{2}{2(1)} = -1.$$

The $y$-coordinate of the vertex is

$$\begin{aligned} f\left(-\frac{b}{2a}\right) &= f(-1) \\ &= (-1)^2 + 2(-1) - 15 \\ &= 1 - 2 - 15 \\ &= -16 \end{aligned}$$

The vertex is $(-1, -16)$.
The axis of symmetry is the line

$$x = -\frac{b}{2a} = -1$$

**Step 3:** Determine the $y$-intercept, $f(0)$.

$$\begin{aligned} f(0) &= 0^2 + 2(0) - 15 \\ &= -15 \end{aligned}$$

**Step 4:** Find the discriminant, $b^2 - 4ac$, to determine the number (and nature) of the $x$-intercepts. Then determine the $x$-intercepts, if any.

We have that $a = 1, b = 2$, and $c = -15$, so $b^2 - 4ac = (2)^2 - 4(1)(-15) = 64 > 0$. The parabola will have two different $x$-intercepts. We find the $x$-intercepts by solving

$$f(x) = 0$$

$$x^2 + 2x - 15 = 0$$

Factor: $(x + 5)(x - 3) = 0$

Zero-Product Property: $x + 5 = 0$ or $x - 3 = 0$

$x = -5$ or $x = 3$

**Step 5:** Plot the points. Use the axis of symmetry to find an additional point. Draw the graph of the quadratic function.

See Figure 22. Notice how we use the axis of symmetry to find the additional point $(-2, -15)$. The $y$-intercept, $(0, -15)$ is 1 unit to the right of the axis of symmetry, therefore, there must be a point 1 unit to the left of the axis of symmetry, $(-2, -15)$.

**Figure 22**

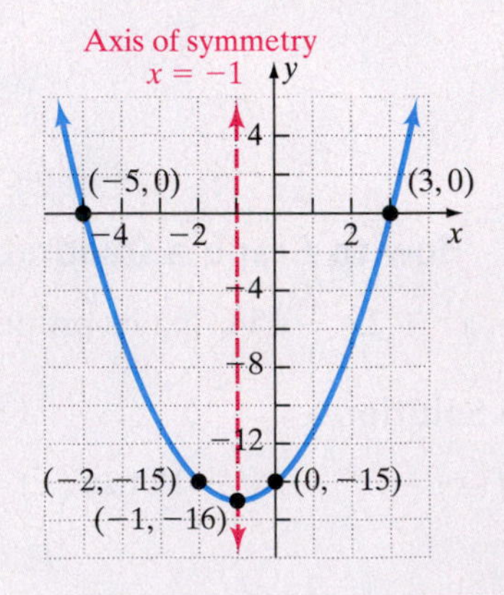

The following steps summarize how to graph any quadratic function using its properties.

## Graphing a Quadratic Function Using Its Properties

To graph any quadratic function of the form $f(x) = ax^2 + bx + c$, $a \neq 0$, we use the following steps:

**Step 1:** Determine whether the parabola opens up or down.

**Step 2:** Determine the vertex and axis of symmetry.

**Step 3:** Determine the $y$-intercept, $f(0)$.

**Step 4:** Determine the discriminant, $b^2 - 4ac$.

- If $b^2 - 4ac > 0$, then the parabola has two $x$-intercepts, which are found by solving $f(x) = 0$ $(ax^2 + bx + c = 0)$.
- If $b^2 - 4ac = 0$, the vertex is the $x$-intercept.
- If $b^2 - 4ac < 0$, there are no $x$-intercepts.

**Step 5:** Plot the points. Use the axis of symmetry to find an additional point. Draw the graph of the quadratic function.

**QUICK ✓** *Graph the quadratic function using its properties.*

**1.** $f(x) = x^2 - 4x - 12$

In Example 1, the function was factorable, so the $x$-intercepts were rational numbers. When the quadratic function cannot be factored, we can use the quadratic formula to find the $x$-intercepts. For the purpose of graphing the quadratic function, we will approximate the $x$-intercepts rounded to two decimal places.

### EXAMPLE 2 Graphing a Quadratic Function Using Its Properties

Graph $f(x) = -2x^2 + 12x - 5$ using its properties.

#### Solution

We compare $f(x) = -2x^2 + 12x - 5$ to $f(x) = ax^2 + bx + c$ and see that $a = -2$, $b = 12$, and $c = -5$. The parabola opens down because $a = -2 < 0$. The $x$-coordinate of the vertex is

$$x = -\frac{b}{2a} = -\frac{12}{2(-2)} = 3$$

The $y$-coordinate of the vertex is

$$\begin{aligned} f\left(-\frac{b}{2a}\right) &= f(3) \\ &= -2(3)^2 + 12(3) - 5 \\ &= -18 + 36 - 5 \\ &= 13 \end{aligned}$$

The vertex is $(3, 13)$. The axis of symmetry is the line

$$x = -\frac{b}{2a} = 3$$

The $y$-intercept is $f(0) = -2(0)^2 + 12(0) - 5 = -5$. Now we find the discriminant, $b^2 - 4ac$ to determine the number (and nature) of the $x$-intercepts. We have that

$a = -2$, $b = 12$, and $c = -5$, so $b^2 - 4ac = (12)^2 - 4(-2)(-5) = 104 > 0$. The parabola will have two different $x$-intercepts. We find the $x$-intercepts by solving

$$f(x) = 0$$
$$-2x^2 + 12x - 5 = 0$$

The equation cannot be solved by factoring (since $b^2 - 4ac$ is not a perfect square), so we use the quadratic formula:

$$x = \frac{-b \pm \sqrt{b^2 - 4ac}}{2a}$$

$a = -2,\ b = 12,\ b^2 - 4ac = 104$: $$= \frac{-12 \pm \sqrt{104}}{2(-2)}$$

$\sqrt{104} = \sqrt{4 \cdot 26} = 2\sqrt{26}$: $$= \frac{-12 \pm 2\sqrt{26}}{-4}$$

Divide $-4$ into each term in the numerator and simplify: $$x = 3 \pm \frac{\sqrt{26}}{-2}$$ Exact solution

**Figure 23**

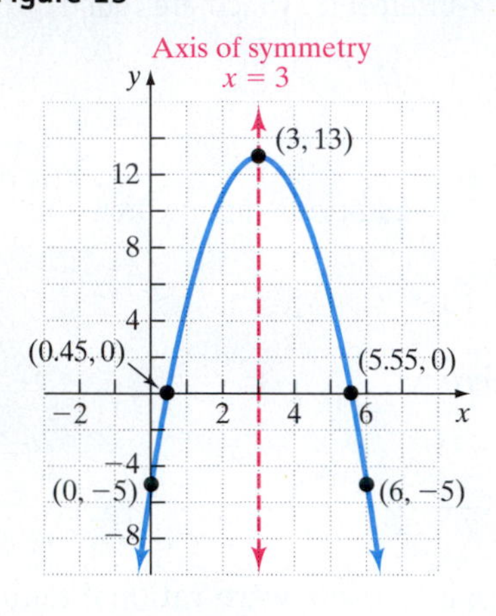


We evaluate the exact solution given by the quadratic formula and find the $x$-intercepts are approximately 0.45 and 5.55. Next, we plot the points and use the axis of symmetry to find an additional point by using the axis of symmetry. Finally, we draw the graph of the quadratic function. See Figure 23. Notice how we use the axis of symmetry to find the additional point $(6, -5)$. The $y$-intercept, $(0, -5)$ is 3 units to the left of the axis of symmetry, therefore, there must be a point 3 units to the right of the axis of symmetry, $(6, -5)$.

**Work Smart**

Notice that the vertex in Example 2 lies in quadrant I (above the $x$-axis) and the graph opens down. This tells us the graph must have two $x$-intercepts.

**QUICK ✓** *Graph the quadratic function using its properties.*

**2.** $f(x) = -3x^2 + 12x - 7$

## EXAMPLE 3 Graphing a Quadratic Function Using Its Properties

Graph $g(x) = x^2 - 8x + 16$ using its properties.

### Solution

We compare $g(x) = x^2 - 8x + 16$ to $g(x) = ax^2 + bx + c$ and see that $a = 1$, $b = -8$, and $c = 16$. The parabola opens up because $a = 1 > 0$. The $x$-coordinate of the vertex is

$$x = -\frac{b}{2a} = -\frac{-8}{2(1)} = 4$$

The $y$-coordinate of the vertex is

$$\begin{aligned} f\left(-\frac{b}{2a}\right) &= f(4) \\ &= 4^2 - 8(4) + 16 \\ &= 16 - 32 + 16 \\ &= 0 \end{aligned}$$

The vertex is $(4, 0)$. The axis of symmetry is the line

$$x = -\frac{b}{2a} = 4$$

**Figure 24**

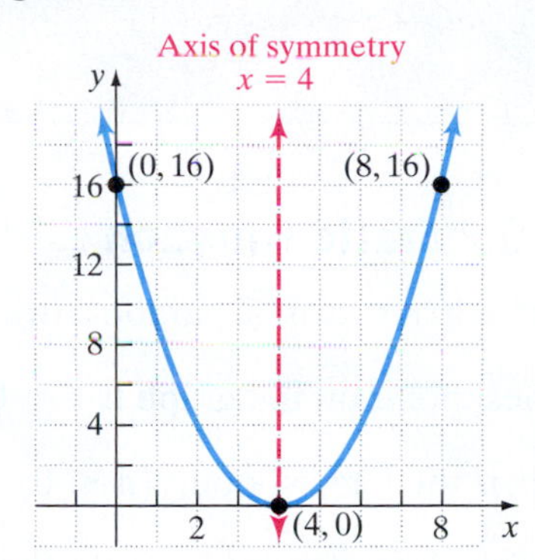


The $y$-intercept is $f(0) = 0^2 - 8(0) + 16 = 16$. We have that $a = 1$, $b = -8$, and $c = 16$, so $b^2 - 4ac = (-8)^2 - 4(1)(16) = 0$. The parabola will have one $x$-intercept, at the vertex. We verify this by solving

$$f(x) = 0$$
$$x^2 - 8x + 16 = 0$$
$$\text{Factor:} \quad (x - 4)^2 = 0$$
$$x = 4$$

Now we graph the parabola. See Figure 24. Notice how we use the axis of symmetry to find the additional point (8, 16). The $y$-intercept, (0, 16) is 4 units to the left of the axis of symmetry, so there must be a point 4 units to the right of the axis of symmetry, (8, 16).

**QUICK ✓** *Graph the quadratic function using its properties.*

**3.** $f(x) = x^2 + 6x + 9$

## EXAMPLE 4 Graphing a Quadratic Function Using Its Properties

Graph $F(x) = 2x^2 + 6x + 5$ using its properties.

### Solution

We compare $F(x) = 2x^2 + 6x + 5$ to $F(x) = ax^2 + bx + c$ and see that $a = 2$, $b = 6$, and $c = 5$. The parabola opens up because $a = 2 > 0$. The $x$-coordinate of the vertex is

$$x = -\frac{b}{2a} = -\frac{6}{2(2)} = -\frac{3}{2}$$

The $y$-coordinate of the vertex is

$$f\left(-\frac{b}{2a}\right) = f\left(-\frac{3}{2}\right)$$
$$= 2\left(-\frac{3}{2}\right)^2 + 6\left(-\frac{3}{2}\right) + 5$$
$$= 2\left(\frac{9}{4}\right) - 9 + 5$$
$$= \frac{1}{2}$$

**Work Smart**

Notice that the vertex lies in quadrant II (above the $x$-axis) and the graph opens up. This tells us the graph will have no $x$-intercepts.

The vertex is $\left(-\frac{3}{2}, \frac{1}{2}\right)$. The axis of symmetry is the line

$$x = -\frac{b}{2a} = -\frac{3}{2}$$

**Figure 25**

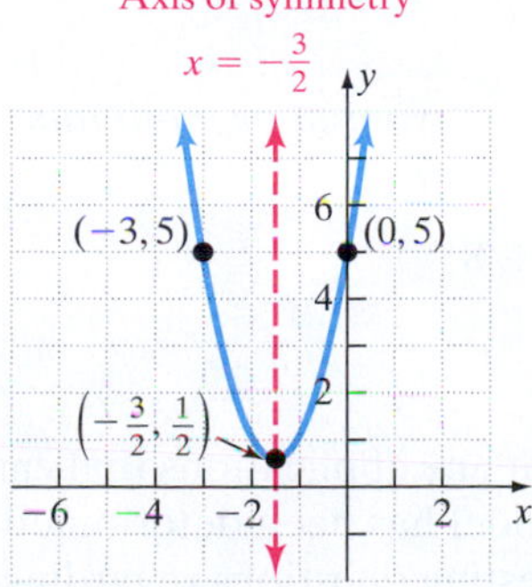


The $y$-intercept is $f(0) = 2(0)^2 + 6(0) + 5 = 5$. Because the parabola opens up and the vertex is in the second quadrant, we expect that the parabola will have no $x$-intercepts. We verify this by computing the discriminant with $a = 2$, $b = 6$, and $c = 5$:

$$b^2 - 4ac = 6^2 - 4(2)(5) = -4 < 0$$

The parabola will have no $x$-intercepts. Now we graph the parabola. See Figure 25. Notice how we use the axis of symmetry to find the additional point $(-3, 5)$. The $y$-intercept, (0, 5) is $\frac{3}{2}(= 1.5)$ units to the right of the axis of symmetry, so there must be a point 1.5 units to the left of the axis of symmetry, $(-3, 5)$.

**QUICK ✓** *Graph the quadratic function using its properties.*

**4.** $G(x) = -3x^2 + 9x - 8$

---

## 2 Find the Maximum or Minimum Value of a Quadratic Function

Recall, the graph of a quadratic function $f(x) = ax^2 + bx + c$ is a parabola with vertex at $\left(-\frac{b}{2a}, f\left(-\frac{b}{2a}\right)\right)$. The vertex will be the highest point on the graph if $a < 0$ and the lowest point on the graph if $a > 0$. If the vertex is the highest point $(a < 0)$, then $f\left(-\frac{b}{2a}\right)$ is the **maximum value** of $f$. If the vertex is the lowest point $(a > 0)$, then $f\left(-\frac{b}{2a}\right)$ is the **minimum value** of $f$.

**Work Smart**

The vertex is not the maximum or minimum value—the $y$-coordinate of the vertex is the maximum or minimum value of the function.

This property of the graph of a quadratic function allows us to answer questions involving *optimization*. **Optimization** is the process whereby we find the maximum or minimum value(s) of a function. In the case of quadratic functions, the maximum $(a < 0)$ or minimum $(a > 0)$ is found through the vertex.

### EXAMPLE 5 Finding the Maximum or Minimum Value of a Quadratic Function

Determine whether the quadratic function

$$f(x) = 3x^2 + 12x - 7$$

has a maximum or minimum value. Then find the maximum or minimum value of the function.

**Solution**

If we compare $f(x) = 3x^2 + 12x - 7$ to $f(x) = ax^2 + bx + c$, we find that $a = 3$, $b = 12$, and $c = -7$. Because $a = 3 > 0$, we know that the graph of the quadratic function will open up, so the function will have a minimum value. The minimum value of the function occurs at

$$x = -\frac{b}{2a} = -\frac{12}{2(3)} = -2$$

The minimum value of the function is

$$\begin{aligned} f\left(-\frac{b}{2a}\right) = f(-2) &= 3(-2)^2 + 12(-2) - 7 \\ &= 3(4) - 24 - 7 \\ &= -19 \end{aligned}$$

So the minimum value of the function is $-19$ and occurs at $x = -2$. ■

**QUICK ✓** *Determine whether the quadratic function has a maximum or minimum value, then find the maximum or minimum value of the function.*

**5.** $f(x) = 2x^2 - 8x + 1$ **6.** $G(x) = -x^2 + 10x + 8$

---

As we stated at the beginning of Section 8.4, there are many applications that can be modeled using a quadratic function. Once a quadratic model has been determined, we can use properties of quadratic functions to answer interesting questions regarding the model.

## EXAMPLE 6 Maximizing Revenue

Suppose that the marketing department of Dell Computer has found that, when a certain model of computer is sold at a price of $p$ dollars, the daily revenue $R$ (in dollars) as a function of the price $p$ is

$$R(p) = -\frac{1}{4}p^2 + 400p$$

**(a)** For what price will the revenue be maximized?

**(b)** What is the maximum daily revenue?

### Solution

**(a)** We notice that the revenue function is a quadratic function whose graph opens down since $a = -\frac{1}{4} < 0$. Therefore, the function will have a maximum at

$$p = -\frac{b}{2a} = -\frac{400}{2 \cdot \left(-\frac{1}{4}\right)} = -\frac{400}{-\frac{1}{2}} = 800$$

Revenue will be maximized when the price is $p = \$800$.

**(b)** The maximum daily revenue is found by letting $p = \$800$ in the revenue function.

$$\begin{aligned} R(800) &= -\frac{1}{4} \cdot (800)^2 + 400 \cdot 800 \\ &= \$160{,}000 \end{aligned}$$

The maximum daily revenue is \$160,000. See Figure 26 for an illustration.

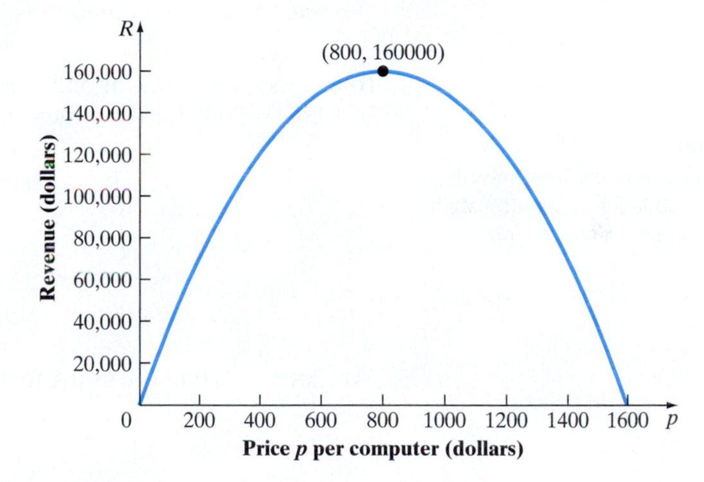


Figure 26

**QUICK ✓**

7. Suppose that the marketing department of Texas Instruments has found that, when a certain model of calculator is sold at a price of $p$ dollars, the daily revenue $R$ (in dollars) as a function of the price $p$ is $R(p) = -0.5p^2 + 75p$.

   **(a)** For what price will the revenue be maximized?

   **(b)** What is the maximum revenue?

## 3 Model and Solve Optimization Problems Involving Quadratic Functions

We now discuss models based on a verbal description that result in quadratic functions. As always, we shall use the problem solving strategy introduced in Section 1.2.

### EXAMPLE 7 Maximizing the Area Enclosed by a Fence

A farmer has 3000 feet of fence to enclose a rectangular field. What is the maximum area that can be enclosed by the fence? What are the dimensions of the rectangle that encloses the most area?

**Solution**

**Step 1: Identify** We wish to determine the dimensions of a rectangle that maximize the area.

**Step 2: Name** We let $w$ represent the width of the rectangle and $l$ will represent the length.

**Step 3: Translate** Figure 27 illustrates the situation.

Figure 27

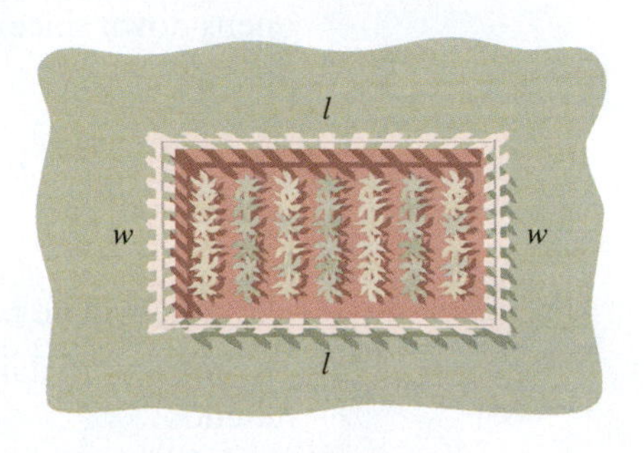


We know that the amount of fence available is 3000 feet. This means that the perimeter of the rectangle will be 3000 feet. Since the perimeter of a rectangle is $2l + 2w$ we have that

$$2l + 2w = 3000$$

The area $A$ of the rectangle is

$$A = lw$$

To express $A$ in terms of only one variable, we solve the equation $2l + 2w = 3000$ for $l$ and then substitute for $l$ in the area formula $A = lw$.

**Work Smart**

We could just as easily have solved $2l + 2w = 3000$ for $w$. We ultimately would obtain the same solution.

$$2l + 2w = 3000$$

Subtract $2w$ from both sides: $$2l = 3000 - 2w$$

Divide both sides by 2: $$l = \frac{3000 - 2w}{2}$$

Simplify: $$l = 1500 - w$$

Now let $l = 1500 - w$ in the formula $A = lw$.

$$\begin{aligned} A &= (1500 - w)w \\ &= -w^2 + 1500w \end{aligned}$$

Now, $A$ is a quadratic function of $w$.

$$A(w) = -w^2 + 1500w$$

**Step 4: Solve** We wish to find the dimensions that result in a maximum area enclosed by the fence. The model we developed in Step 3 is a quadratic function that opens down (because $a = -1 < 0$), so the vertex is a maximum point on the graph of $A$. The maximum value occurs at

$$w = -\frac{b}{2a} = -\frac{1500}{2(-1)} = 750$$

The maximum value of $A$ is

$$\begin{aligned} A\left(-\frac{b}{2a}\right) &= A(750) = -750^2 + 1500(750) \\ &= 562{,}500 \text{ square feet} \end{aligned}$$

We know that $l = 1500 - w$, so if the width is 750 feet, then the length will be $l = 1500 - 750 = 750$ feet.

**Step 5: Check** With a length and width of 750 feet, the perimeter is $2(750) + 2(750) = 3000$ feet. The area is $(750 \text{ feet})(750 \text{ feet}) = 562{,}500$ square feet. Everything checks!

**Step 6: Answer** The largest rectangle that can be enclosed by 3000 feet of fence has an area of 562,500 square feet. Its dimensions are 750 feet by 750 feet.

**QUICK ✓**

**8.** Roberta has 1000 yards of fence to enclose a rectangular field. What is the maximum area that can be enclosed by the fence? What are the dimensions of the rectangle that encloses the most area?

## EXAMPLE 8 Pricing a Charter

Chicago Tours offers boat charters along the Chicago coastline on Lake Michigan. Normally, a ticket costs $20 per person, but for any group, Chicago Tours will lower the price of a ticket by $0.10 per person for each person in excess of 30. Determine the group size that will maximize revenue. What is the maximum revenue that can be earned from a group sale?

### Solution

**Step 1: Identify** This is a direct translation problem involving revenue. Remember, revenue is price times quantity.

**Step 2: Name** We let $x$ represent the number in the group in excess of 30.

**Step 3: Translate** Revenue is price times quantity. If 30 individuals make up a group, the revenue to Chicago Tours will be $20(30). If 31 individuals make up a group, revenue will be $19.90(31). If 32 individuals make up the group, revenue will be $19.80(32) In general, if $x$ individuals make up a group in excess of 30, revenue will be $(20 - 0.1x)(x + 30)$. So the revenue $R$ for a group that has $x + 30$ people in it is given by

$$\begin{aligned} R(x) &= (20 - 0.1x)(x + 30) \\ \text{FOIL:} \quad &= 20x + 600 - 0.1x^2 - 3x \\ \text{This is the model:} \quad &= -0.1x^2 + 17x + 600 \end{aligned}$$

**Step 4: Solve** We wish to know the revenue maximizing number of individuals in a group. The function $R$ is a quadratic function with a graph that opens down (because $a = -0.1 < 0$), so we know that the vertex is a maximum point. The value of $x$ that results in a maximum is given by

$$\begin{aligned} x = -\frac{b}{2a} &= -\frac{17}{2(-0.1)} \\ &= 85 \end{aligned}$$

The maximum revenue is

$$\begin{aligned} R(85) &= -0.1(85)^2 + 17(85) + 600 \\ &= \$1322.50 \end{aligned}$$

**Step 5: Check** Remember that $x$ represents the number of passengers in excess of 30. Therefore, $30 + 85 = 115$ tickets should be sold to maximize revenue. The cost per ticket would be $\$20 - 0.1(85) = \$20 - \$8.50 = \$11.50$. Multiplying the cost per ticket by the number of passengers we obtain $\$11.50(115) = \$1322.50$. We have the right answer!

**Step 6: Answer** A group sale of 115 passengers will maximize revenue. The maximum revenue would be $1322.50.

QUICK ✓

9. A compact disk manufacture charges $100 for each box of CDs ordered. However, it reduces the price by $1 for each box in excess of 30 boxes, but less than 90 boxes. Determine the number of boxes of CDs that should be sold to maximize revenue. What is the maximum revenue?

# 8.5 Exercises

**For Extra Help:** Student Solutions Manual | CD Video | PH Math/Tutor Center | MathXL Tutorials on CD | MathXL® | MyMathLab

## Concepts and Vocabulary

*In Problems 1–3, fill in the blanks.*

1. Any quadratic function $f(x) = ax^2 + bx + c$, $a \neq 0$, will have a vertex whose $x$-coordinate is $x =$ ______.

2. The graph of $f(x) = ax^2 + bx + c$ will have two different $x$-intercepts if $b^2 - 4ac$ ______ 0.

3. If the vertex of a quadratic function $f(x) = ax^2 + bx + c$ is the lowest point $(a > 0)$, then $f\left(-\frac{b}{2a}\right)$ is the ______ ______ of $f$.

*In Problems 4–6, answer True or False to each statement.*

4. The quadratic function $f(x) = -2x^2 - 3x + 6$ will have two $x$-intercepts.

5. The vertex of $f(x) = x^2 + 4x - 3$ is $(2, 29)$.

6. If the vertex of a quadratic function $f(x) = ax^2 + bx + c$ is the highest point, then $f\left(-\frac{b}{2a}\right)$ is the maximum value of $f$.

7. Explain how the discriminant is used to determine the number of $x$-intercepts the graph of a quadratic function will have.

8. Provide two methods for finding the vertex of any quadratic function $f(x) = ax^2 + bx + c$.

## Building Skills

*In Problems 9–16, use the discriminant to determine the number of x-intercepts the graph of each quadratic function will have. Then determine the x-intercepts.*

9. $f(x) = x^2 - 6x - 16$
10. $g(x) = 2x^2 - 7x - 4$
11. $G(x) = -3x^2 + x - 1$
12. $H(x) = x^2 - 3x + 5$
13. $h(x) = 4x^2 + 4x + 1$
14. $f(x) = x^2 - 6x + 9$
15. $F(x) = 4x^2 - x - 1$
16. $P(x) = -2x^2 + 3x + 1$

*In Problems 17–56, graph each quadratic function using its properties by following Steps 1–5 on page 665. Based on the graph, determine the domain and range of the quadratic function.*

17. $f(x) = x^2 - 4x - 5$
18. $f(x) = x^2 - 2x - 8$
19. $G(x) = x^2 + 12x + 32$
20. $g(x) = x^2 - 12x + 27$

**21.** $F(x) = -x^2 + 2x + 8$

**22.** $g(x) = -x^2 + 2x + 15$

**23.** $H(x) = x^2 - 4x + 4$

**24.** $h(x) = x^2 + 6x + 9$

**25.** $g(x) = x^2 + 2x + 5$

**26.** $f(x) = x^2 - 4x + 7$

**27.** $h(x) = -x^2 - 10x - 25$

**28.** $P(x) = -x^2 - 12x - 36$

**29.** $p(x) = -x^2 + 2x - 5$

**30.** $f(x) = -x^2 + 4x - 6$

**31.** $F(x) = 4x^2 - 4x - 3$

**32.** $f(x) = 4x^2 - 8x - 21$

**33.** $G(x) = -9x^2 + 18x + 7$

**34.** $g(x) = -9x^2 - 36x - 20$

**35.** $H(x) = 4x^2 - 4x + 1$

**36.** $h(x) = 9x^2 + 12x + 4$

**37.** $f(x) = -16x^2 - 24x - 9$

**38.** $F(x) = -4x^2 - 20x - 25$

**39.** $f(x) = 2x^2 + 8x + 11$

**40.** $F(x) = 3x^2 + 6x + 7$

**41.** $P(x) = -4x^2 + 6x - 3$

**42.** $p(x) = -2x^2 + 6x + 5$

**43.** $h(x) = x^2 + 5x + 3$

**44.** $H(x) = x^2 + 3x + 1$

**45.** $G(x) = -3x^2 + 8x + 2$

**46.** $F(x) = -2x^2 + 6x + 1$

**47.** $f(x) = 5x^2 - 5x + 2$

**48.** $F(x) = 4x^2 + 4x - 1$

**49.** $H(x) = -3x^2 + 6x$

**50.** $h(x) = -4x^2 + 8x$

**51.** $f(x) = x^2 - \frac{5}{2}x - \frac{3}{2}$

**52.** $g(x) = x^2 + \frac{5}{2}x - 6$

**53.** $G(x) = \frac{1}{2}x^2 + 2x - 6$

**54.** $H(x) = \frac{1}{4}x^2 + x - 8$

**55.** $F(x) = -\frac{1}{4}x^2 + x + 15$

**56.** $G(x) = -\frac{1}{2}x^2 - 8x - 24$

*In Problems 57–68, determine whether the quadratic function has a maximum or minimum value. Then find the maximum or minimum value.*

**57.** $f(x) = x^2 + 8x + 13$

**58.** $f(x) = x^2 - 6x + 3$

**59.** $G(x) = -x^2 - 10x + 3$

**60.** $g(x) = -x^2 + 4x + 12$

**61.** $F(x) = -2x^2 + 12x + 5$

**62.** $H(x) = -3x^2 + 12x - 1$

**63.** $h(x) = 4x^2 + 16x - 3$

**64.** $G(x) = 5x^2 + 10x - 1$

**65.** $f(x) = 2x^2 - 5x + 1$

**66.** $F(x) = 3x^2 + 4x - 3$

**67.** $H(x) = -3x^2 + 4x + 1$

**68.** $h(x) = -4x^2 - 6x + 1$

## Applying the Concepts

**69. Revenue Function** Suppose that the marketing department of Panasonic has found that, when a certain model of DVD player is sold at a price of $p$ dollars, the daily revenue $R$ (in dollars) as a function of the price $p$ is $R(p) = -2.5p^2 + 600p$.

**(a)** For what price will the daily revenue be maximized?

**(b)** What is the maximum daily revenue?

**70. Revenue Function** Suppose that the marketing department of Samsung has found that, when a certain model of cellular telephone is sold at a price of $p$ dollars, the daily revenue $R$ (in dollars) as a function of the price $p$ is $R(p) = -5p^2 + 600p$.

**(a)** For what price will the daily revenue be maximized?
**(b)** What is the maximum daily revenue?

**71. Marginal Cost** The marginal cost of a product can be thought of as the cost of producing one additional unit of output. For example, if the marginal cost of producing the fiftieth product is \$6.30, then it costs \$6.30 to increase production from 49 to 50 units of output. Suppose that the marginal cost $C$ (in dollars) to produce $x$ digital cameras is given by $C(x) = 0.05x^2 - 6x + 215$. How many digital cameras should be produced to minimize marginal cost? What is the minimum marginal cost?

**72. Marginal Cost** (See Problem 71.) The marginal cost $C$ (in dollars) of manufacturing $x$ portable CD players is given by $C(x) = 0.05x^2 - 9x + 435$. How many portable CD players should be manufactured to minimize marginal cost? What is the minimum marginal cost?

**73. Punkin Chunkin** Suppose that an air cannon in the Punkin Chunkin contest whose muzzle is 10 feet above the ground fires a pumpkin at an angle of 45° to the horizontal with a muzzle velocity of 335 feet per second. The model $s(t) = -16t^2 + 240t + 10$ can be used to estimate the height $s$ of an object after $t$ seconds.

**(a)** Determine the time at which the pumpkin is at a maximum height.
**(b)** Determine the maximum height of the pumpkin.
**(c)** After how long will the pumpkin strike the ground?

**74. Punkin Chunkin** Suppose that a catapult in the Punkin Chunkin contest releases a pumpkin 8 feet above the ground at an angle of 45° to the horizontal with an initial speed 220 feet per second. The model $s(t) = -16t^2 + 155t + 8$ can be used to estimate the height $s$ of an object after $t$ seconds.

**(a)** Determine the time at which the pumpkin is at a maximum height.
**(b)** Determine the maximum height of the pumpkin.
**(c)** After how long will the pumpkin strike the ground?

**75. Punkin Chunkin** Suppose that an air cannon in the Punkin Chunkin contest whose muzzle is 10 feet above the ground fires a pumpkin at an angle of 45° to the horizontal with a muzzle velocity of 335 feet per second. The model $h(x) = \frac{-32}{335^2}x^2 + x + 10$ can be used to estimate the height $h$ of an object after the pumpkin has traveled $x$ feet.

**(a)** How far from the cannon will the pumpkin reach a maximum height?
**(b)** What is the maximum height of the pumpkin?
**(c)** How far will the pumpkin travel before it strikes the ground?
**(d)** Compare your answer in part (b) of this problem with the answer found in part (b) of Problem 73. Why might the answers differ?

**76. Punkin Chunkin** Suppose that a catapult in the Punkin Chunkin contest releases a pumpkin 8 feet above the ground at an angle of 45° to the horizontal with an initial speed 220 feet per second. The model $h(x) = \frac{-32}{220^2}x^2 + x + 8$ can be used to estimate the height $h$ of an object after the pumpkin has traveled $x$ feet.

**(a)** How far from the cannon will the pumpkin reach a maximum height?
**(b)** What is the maximum height of the pumpkin?
**(c)** How far will the pumpkin travel before it strikes the ground?
**(d)** Compare your answer in part (b) of this problem with the answer found in part (b) of Problem 74. Why might the answers differ?

**77. Life Cycle Hypothesis** The Life Cycle Hypothesis from Economics was presented by Franco Modigliani in 1954. One of its components states that income is a function of age. The function $I(a) = -55a^2 + 5119a - 54{,}448$ represents the relation between average annual income $I$ and age $a$.

**(a)** According to the model, at what age will average income be a maximum?
**(b)** According to the model, what is the maximum average income?

**78. Advanced Degrees** The function $P(x) = -0.008x^2 + 0.868x - 11.884$ models the percentage of the United States population whose age is given by $x$ that have earned an advanced degree (more than a bachelor's degree) as of March, 2000. (*Source:* Based on data obtained from the U.S. Census Bureau)

**(a)** What is the age for which the highest percentage of Americans have earned an advanced degree?
**(b)** According to the model, what is the percentage of Americans that have earned an advanced degree at the age found in part (a)?

**79. Fun with Numbers** The sum of two numbers is 36. Find the numbers such that their product is a maximum.

**80. Fun with Numbers** The sum of two numbers is 50. Find the numbers such that their product is a maximum.

**81. Fun with Numbers** The difference of two numbers is 18. Find the numbers such that their product is a minimum.

**82. Fun with Numbers** The difference of two numbers is 10. Find the numbers such that their product is a minimum.

**83. Enclosing a Rectangular Field** Maurice has 500 yards of fencing and wishes to enclose a rectangular area. What is the maximum area that can be enclosed by the fence? What are the dimensions of the area enclosed?

**84. Enclosing a Rectangular Field** Maude has 800 yards of fencing and wishes to enclose a rectangular area. What is the maximum area that can be enclosed by the fence? What are the dimensions of the area enclosed?

**85. Maximizing an Enclosed Area** A farmer with 2000 meters of fencing wants to enclose a rectangular plot that borders on a river. If the farmer does not fence the side along the river, what is the largest area that can be enclosed? What is the length of the side of each enclosed area? See the figure.

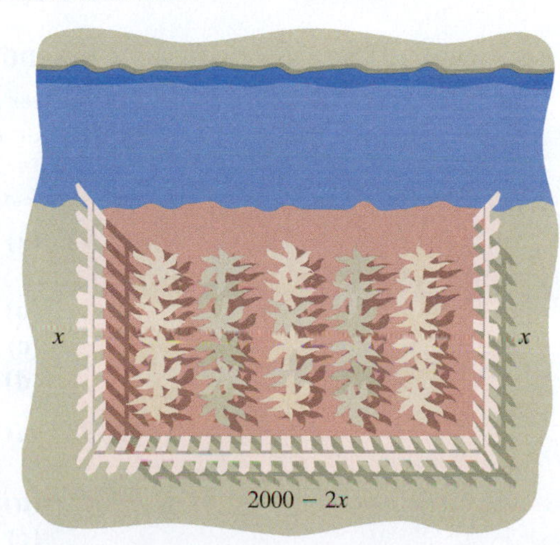

**86. Maximizing an Enclosed Area** A farmer with 8000 meters of fencing wants to enclose a rectangular plot and then divide it into two plots with a fence parallel to one of the sides. See the figure. What is the largest area that can be enclosed? What are the lengths of the sides of each part of the enclosed area?

**87. Constructing Rain Gutters** A rain gutter is to be made of aluminum sheets that are 20 inches wide by turning up the edges 90°. What depth will provide maximum cross-sectional area and hence allow the most water to flow?

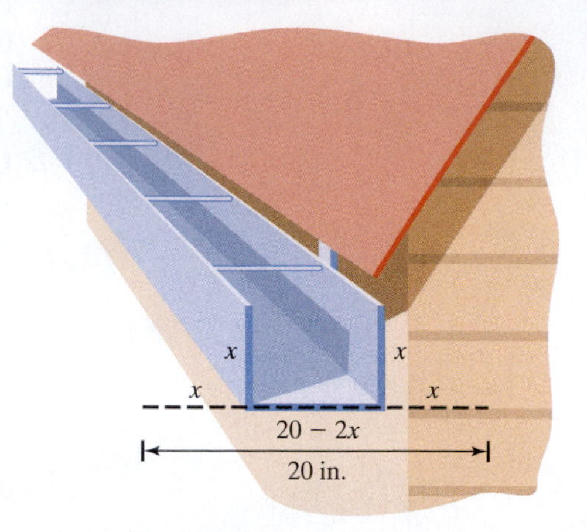


**88. Maximizing the Volume of a Box** A box with a rectangular base is to be constructed such that the perimeter of the base of the box is to be 40 inches. The height of the box must be 15 inches. Find the dimensions of the box such that the volume is maximized. What is the maximum volume?

**89. Revenue Function** Weekly demand for jeans at a department store obeys the demand equation

$$x = -p + 110$$

where $x$ is the quantity demanded and $p$ is the price (in dollars).

**(a)** Express the revenue $R$ as a function of $p$. (**Hint:** $R = xp$)
**(b)** What price $p$ maximizes revenue? What is the maximum revenue?
**(c)** How many pairs of jeans will be sold at the revenue maximizing price?

**90. Revenue Function** Demand for hot dogs at a baseball game obeys the demand equation

$$x = -800p + 8000$$

where $x$ is the quantity demanded and $p$ is the price (in dollars).

**(a)** Express the revenue $R$ as a function of $x$. (**Hint:** $R = xp$)
**(b)** What price $p$ maximizes revenue? What is the maximum revenue?
**(c)** How many hot dogs will be sold at the revenue maximizing price?

## Extending the Concepts

*Answer Problems 91 and 92 using the following information: A quadratic function of the form $f(x) = ax^2 + bx + c$ with $b^2 - 4ac > 0$ may also be written in the form $f(x) = a(x - r_1)(x - r_2)$, where $r_1$ and $r_2$ are the x-intercepts of the graph of the quadratic function.*

**91. (a)** Find a quadratic function whose $x$-intercepts are 2 and 6 with $a = 1$; $a = 2$, and $a = -2$.
**(b)** How does the value of $a$ affect the intercepts?
**(c)** How does the value of $a$ affect the axis of symmetry?
**(d)** How does the value of $a$ affect the vertex?

**92. (a)** Find a quadratic function whose $x$-intercepts are $-1$ and 5 with $a = 1$; $a = 2$, and $a = -2$.
**(b)** How does the value of $a$ affect the intercepts?
**(c)** How does the value of $a$ affect the axis of symmetry?
**(d)** How does the value of $a$ affect the vertex?

**93.** Refer to Example 6 on page 669. Notice that if the price charged for the computer is \$0 or \$1,600 the revenue is \$0. It is easy to explain why revenue would be \$0 if the price charged is \$0, but how can revenue be \$0, if the price charged is \$1600?

## Synthesis Review

*In Problems 94–97, graph each function using point-plotting.*

**94.** $f(x) = -2x + 12$

**95.** $G(x) = \frac{1}{4}x - 2$

**96.** $f(x) = x^2 - 5$

**97.** $f(x) = (x + 2)^2 + 4$

**98.** For each function in Problems 94–97, explain an alternative method for graphing the function. Which method do you prefer? Why?

## The Graphing Calculator: Finding the Vertex

Graphing calculators have a maximum and a minimum feature that allows us to determine the coordinates of the vertex of a parabola. For example, to find the vertex of $f(x) = x^2 + 2x - 15$ using a TI-84 graphing calculator, we use the MINIMUM feature. See Figure 28. The vertex is $(-1, -16)$.

**Figure 28**

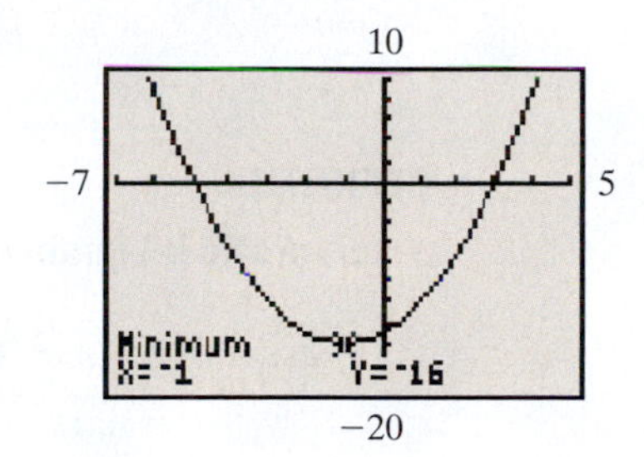


*In Problems 99–106, use a graphing calculator to graph each quadratic function. Using the MAXIMUM or MINIMUM feature on the calculator, determine the vertex. If necessary, round your answers to two decimal places.*

**99.** $f(x) = x^2 - 7x + 3$

**100.** $f(x) = x^2 + 3x + 8$

**101.** $G(x) = -2x^2 + 14x + 13$

**102.** $g(x) = -4x^2 - x + 11$

**103.** $F(x) = 5x^2 + 3x - 20$

**104.** $F(x) = 3x^2 + 2x - 21$

**105.** $H(x) = \frac{1}{2}x^2 - \frac{2}{3}x + 5$

**106.** $h(x) = \frac{3}{4}x^2 + \frac{4}{3}x - 1$

**107.** On the same screen, graph the family of parabolas $f(x) = x^2 + 2x + c$ for $c = -3$, $c = 0$, and $c = 1$. Describe the role that $c$ plays in the graph for this family of functions.

**108.** On the same screen, graph the family of parabolas $f(x) = x^2 + bx + 1$ for $b = -4$, $b = 0$, and $b = 4$. Describe the role that $b$ plays in the graph for this family of functions.

# 8.6 Quadratic Inequalities

**OBJECTIVE**

1 Solve Quadratic Inequalities

### *Preparing for Quadratic Inequalities*

*Before getting started, take the following readiness quiz. If you get a problem wrong, go back to the section cited and review the material.*

**1.** Write $-4 \le x < 5$ in interval notation. [Section 1.4, pp. 90–93]

**2.** Solve: $3x + 5 > 5x - 3$. [Section 1.4, pp. 93–96]

In Section 1.4, we solved linear inequalities in one variable such as $2x - 3 > 4x + 5$. We were able to solve these inequalities using methods that were similar to solving linear equations. We also learned to represent the solution set to such an inequality using either set-builder notation or interval notation.

Unfortunately, the approach to solving inequalities involving quadratic expressions is not a simple extension of solving quadratic equations, but we will use the skills developed in solving quadratic equations to solve quadratic inequalities. The method is similar to the method presented for solving rational inequalities.

## 1 Solve Quadratic Inequalities

We begin with a definition.

**DEFINITION**

A **quadratic inequality** is an inequality of the form

$$ax^2 + bx + c > 0 \quad \text{or} \quad ax^2 + bx + c < 0 \quad \text{or}$$
$$ax^2 + bx + c \ge 0 \quad \text{or} \quad ax^2 + bx + c \le 0$$

where $a \neq 0$.

We will present two methods for solving quadratic inequalities. The first method is a graphical approach to the solution, while the second method is algebraic.

To help understand the logic behind the graphical approach, consider the following. Suppose we were asked to solve the inequality $ax^2 + bx + c > 0$. If we let $f(x) = ax^2 + bx + c$, then we are looking for all $x$-values such that $f(x) > 0$. Since $f$ represents the $y$-values of the graph of the function $f(x) = ax^2 + bx + c$, we are basically looking for all $x$-values such that the graph of $f$ is above the $x$-axis. This occurs when the graph lies in either quadrant I or II of the Cartesian plane. If we were asked to solve $f(x) < 0$, we would look for the $x$-values such that the graph is below the $x$-axis. That is, we would look for all $x$-values such that the graph lies in either quadrant III or IV of the Cartesian plane. See Figure 29 for an illustration of the idea.

**Figure 29**

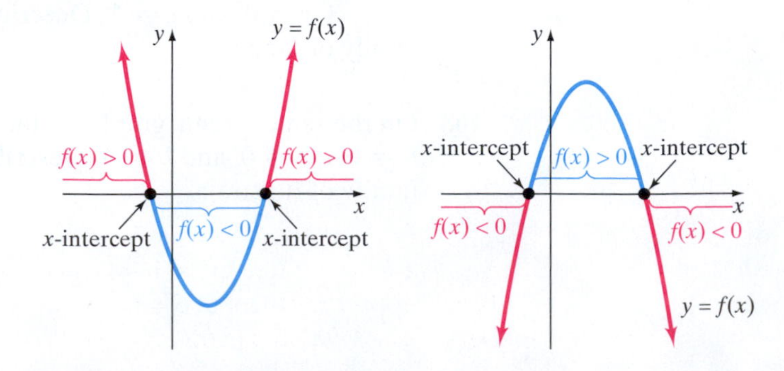


*Preparing for...Answers* **1.** $[-4, 5)$
**2.** $\{x|x < 4\}$ or $(-\infty, 4)$

## EXAMPLE 1 How to Solve a Quadratic Inequality Using the Graphical Method

Solve $x^2 - 4x - 5 \geq 0$ using the graphical method.

### Step-by-Step Solution

**Step 1:** Write the inequality so that $ax^2 + bx + c$ is on one side of the inequality and 0 is on the other.

$$x^2 - 4x - 5 \geq 0$$

**Step 2:** Graph the function $f(x) = ax^2 + bx + c$. Be sure to label the $x$-intercepts of the graph.

We wish to graph $f(x) = x^2 - 4x - 5$.

$x$-intercepts:

$$f(x) = 0$$
$$x^2 - 4x - 5 = 0$$
$$(x - 5)(x + 1) = 0$$
$$x = 5 \quad \text{or} \quad x = -1$$

$y$-intercept: $f(0) = -5$

Vertex: $x = -\dfrac{b}{2a} = -\dfrac{-4}{2(1)} = 2$

$$f\left(-\frac{b}{2a}\right) = f(2) = -9$$

The vertex is at $(2, -9)$. Figure 30 shows the graph.

**Figure 30**

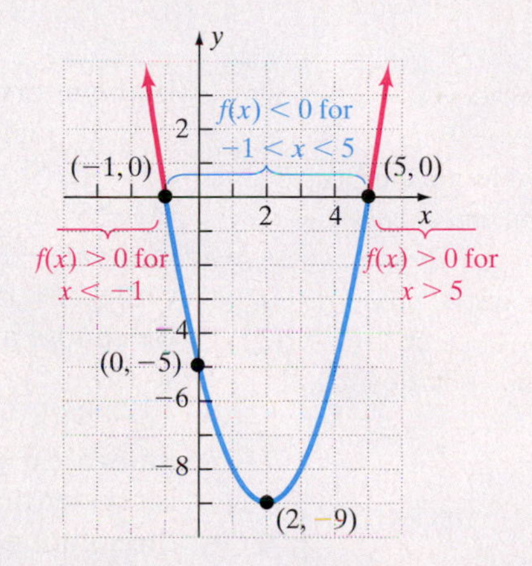


**Step 3:** From the graph, determine where the function is positive and determine where the function is negative. Use the graph to determine the solution set to the inequality.

From the graph shown in Figure 30, we can see that the graph of $f(x) = x^2 - 4x - 5$ is greater than 0 for $x < -1$ or $x > 5$. Because the inequality is nonstrict, we include the $x$-intercepts in the solution. So, the solution is $\{x | x \leq -1 \text{ or } x \geq 5\}$ using set-builder notation; the solution is $(-\infty, -1] \cup [5, \infty)$ using interval notation. See Figure 31 for a graph of the solution set.

**Figure 31**

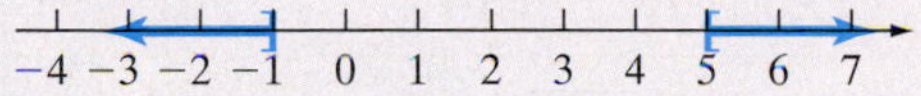

−4 −3 −2 −1 0 1 2 3 4 5 6 7

**QUICK ✓** *Solve the quadratic inequality using the graphical method. Graph the solution set.*

**1.** $x^2 + 3x - 10 \geq 0$

The second method for solving inequalities is an algebraic method. This method is based upon the Rules of Signs when multiplying or dividing two (or more) real numbers. Recall that the product of two positive real numbers is positive, the product of a positive real number and a negative real number is negative, and the product of two negative real numbers is positive.

Let's solve the inequality from Example 1 using the algebraic method.

## EXAMPLE 2 How to Solve a Quadratic Inequality Using the Algebraic Method

Solve $x^2 - 4x - 5 \geq 0$ using the algebraic method.

### Step-by-Step Solution

**Step 1:** Write the inequality so that $ax^2 + bx + c$ is on one side of the inequality and 0 is on the other.

$$x^2 - 4x - 5 \geq 0$$

**Step 2:** Determine the solutions to the equation $ax^2 + bx + c = 0$.

$$x^2 - 4x - 5 = 0$$

Factor: $(x - 5)(x + 1) = 0$

Use Zero-Product Property: $x - 5 = 0$ or $x + 1 = 0$

$x = 5$ or $x = -1$

**Step 3:** Use the solutions to the equation solved in Step 2 to separate the real number line into intervals.

We separate the real number line into the following intervals:

$$(-\infty, -1) \qquad (-1, 5) \qquad (5, \infty)$$

**Step 4:** Write $x^2 - 4x - 5$ in factored form as $(x - 5)(x + 1)$. Within each interval formed in Step 3, choose a test point and determine the sign of each factor. Then determine the sign of the product. Also determine the value of $x^2 - 4x - 5$ at each solution found in Step 2.

In the interval $(-\infty, -1)$, we choose a test point of $-2$. The expression $x - 5$ equals $-7$ when $x = -2$. The expression $x + 1$ equals $-1$ when $x = -2$. Since the product of two negatives is positive, the expression $(x - 5)(x + 1)$ will be positive for all $x$ in the interval $(-\infty, -1)$. In the interval $(-1, 5)$ we choose a test point of 0. For $x = 0$, $x - 5$ is negative, while $x + 1$ is positive, so $(x - 5)(x + 1)$ is negative. In the interval $(5, \infty)$ we choose a test point of 6. For $x = 6$, both $x - 5$ and $x + 1$ are positive, so $(x - 5)(x + 1)$ is positive. Table 9 shows these results and the sign of $(x - 5)(x + 1) = x^2 - 4x - 5$ in each interval. We also list the value of $(x - 5)(x + 1)$ at $x = -1$ and $x = 5$. We want to know where $x^2 - 4x - 5$ is greater than or equal to zero, so we include $-1$ and 5 in the solution, so the solution set is $\{x | x \leq -1 \text{ or } x \geq 5\}$ using set-builder notation; the solution is $(-\infty, -1] \cup [5, \infty)$ using interval notation.

**Table 9**

| | | $-1$ | | 5 | |
|---|---|---|---|---|---|
| **Interval** | $(-\infty, -1)$ | | $(-1, 5)$ | | $(5, \infty)$ |
| **Test Point** | $-2$ | $-1$ | 0 | 5 | 6 |
| **Sign of $(x - 5)$** | Negative | Negative | Negative | 0 | Positive |
| **Sign of $(x + 1)$** | Negative | 0 | Positive | Positive | Positive |
| **Sign of $(x - 5)(x + 1)$** | Positive | 0 | Negative | 0 | Positive |
| **Conclusion** | $(x - 5)(x + 1)$ is positive, so $(-\infty, -1)$ is part of the solution set | Because the inequality is nonstrict, $-1$ is part of the solution | $(x - 5)(x + 1)$ is negative, so $(-1, 5)$ is not part of the solution set | Because the inequality is nonstrict, 5 is part of the solution | $(x - 5)(x + 1)$ is positive, so $(5, \infty)$ is part of the solution set |

**QUICK ✓** *Solve the quadratic inequality using the algebraic method.*

**2.** $x^2 + 3x - 10 \geq 0$

---

**SUMMARY: Solving Quadratic Inequalities**

**Graphical Method**

**Step 1:** Write the inequality so that $ax^2 + bx + c$ is on one side of the inequality and 0 is on the other.

**Step 2:** Graph the function $f(x) = ax^2 + bx + c$. Be sure to label the $x$-intercepts of the graph.

**Step 3:** From the graph, determine where the function is positive and determine where the function is negative. Use the graph to determine the solution set to the inequality.

**Algebraic Method**

**Step 1:** Write the inequality so that $ax^2 + bx + c$ is on one side of the inequality and 0 is on the other.

**Step 2:** Determine the solutions to the equation $ax^2 + bx + c = 0$.

**Step 3:** Use the solutions to the equation solved in Step 2 to separate the real number line into intervals.

**Step 4:** Write $ax^2 + bx + c$ in factored form. Within each interval formed in Step 3, determine the sign of each factor. Then determine the sign of the product. Also determine the value of $ax^2 + bx + c$ at each solution found in Step 2.

**(a)** If the product of the factors is positive, then $ax^2 + bx + c > 0$ for all numbers $x$ in the interval.

**(b)** If the product of the factors is negative, then $ax^2 + bx + c < 0$ for all numbers $x$ in the interval.

If the inequality is not strict ($\leq$ or $\geq$), include the solutions of $ax^2 + bx + c = 0$ in the solution set.

Now let's do an example where we use both methods.

## EXAMPLE 3 Solving a Quadratic Inequality

Solve: $-x^2 + 10 > 3x$

### Solution

**Graphical Method:**

To make our lives a little easier, we will rearrange the inequality so that the coefficient on the square term is positive.

$$-x^2 + 10 > 3x$$

Add $x^2$ to both sides; subtract 10 from both sides: $$0 > x^2 + 3x - 10$$

Use the fact that $0 > b$ is equivalent to $b < 0$: $$x^2 + 3x - 10 < 0$$

We wish to graph $f(x) = x^2 + 3x - 10$. We will graph the function using its properties by finding its intercepts and vertex.

***x*-intercepts:** $f(x) = 0$

$$x^2 + 3x - 10 = 0$$
$$(x + 5)(x - 2) = 0$$
$$x + 5 = 0 \quad \text{or} \quad x - 2 = 0$$
$$x = -5 \quad \text{or} \quad x = 2$$

***y*-intercept:** $f(0) = -10$

**Figure 32**

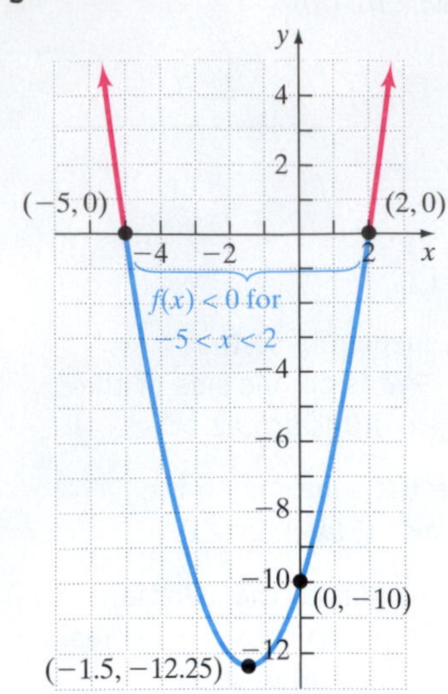


**Vertex:**

$$x = -\frac{b}{2a} = -\frac{3}{2(1)} = -\frac{3}{2} = -1.5$$

$$f\left(-\frac{b}{2a}\right) = f\left(-\frac{3}{2}\right) = -\frac{49}{4} = -12.25$$

The vertex is at $(-1.5, -12.25)$. Figure 32 shows the graph.

From the graph shown in Figure 32, we can see that the graph of $f(x) = x^2 + 3x - 10$ is below the $x$-axis (and therefore $x^2 + 3x - 10 < 0$) for $-5 < x < 2$. Because the inequality is strict, we do not include the $x$-intercepts in the solution. So the solution is $\{x|-5 < x < 2\}$ using set-builder notation; the solution is $(-5, 2)$ using interval notation.

**Algebraic Method:**

Again, to make our lives a little easier, we will rearrange the inequality so that the coefficient on the square term is positive.

$$-x^2 + 10 > 3x$$

Add $x^2$ to both sides; subtract 10 from both sides: $\quad 0 > x^2 + 3x - 10$

Use the fact that $0 > b$ is equivalent to $b < 0$: $\quad x^2 + 3x - 10 < 0$

Now, we solve the equation

$$x^2 + 3x - 10 = 0$$
$$(x + 5)(x - 2) = 0$$
$$x + 5 = 0 \quad \text{or} \quad x - 2 = 0$$
$$x = -5 \quad \text{or} \quad x = 2$$

We use the solutions to separate the real number line into the following intervals:

$$(-\infty, -5) \qquad (-5, 2) \qquad (2, \infty)$$

The factored form of $x^2 + 3x - 10$ is $(x + 5)(x - 2)$. Table 10 shows the sign of each factor and $(x + 5)(x - 2)$ in each interval. We also list the value of $(x + 5)(x - 2)$ at $x = -5$ and $x = 2$.

**Table 10**

| | | −5 | | 2 | |
|---|---|---|---|---|---|
| **Interval** | $(-\infty, -5)$ | | $(-5, 2)$ | | $(2, \infty)$ |
| **Test Point** | −6 | −5 | 0 | 2 | 3 |
| **Sign of $(x + 5)$** | Negative | 0 | Positive | Positive | Positive |
| **Sign of $(x - 2)$** | Negative | Negative | Negative | 0 | Positive |
| **Sign of $(x + 5)(x - 2)$** | Positive | 0 | Negative | 0 | Positive |
| **Conclusion** | $(x + 5)(x - 2)$ is positive in the interval $(-\infty, -5)$, so it is not part of the solution set | The inequality is strict, so −5 is not part of the solution | $(x + 5)(x - 2)$ is negative in the interval $(-5, 2)$, so it is part of the solution set | The inequality is strict, so 2 is not part of the solution | $(x + 5)(x - 2)$ is positive in the interval $(2, \infty)$, so it is not part of the solution set |

We want to know where $(x + 5)(x - 2) = x^2 + 3x - 10$ is negative, so we do not include −5 or 2 in the solution. The solution set is $\{x|-5 < x < 2\}$ using set-builder notation; the solution is $(-5, 2)$ using interval notation.

Figure 33 shows the graph of the solution set.

**Figure 33**

−7 −6 −5 −4 −3 −2 −1 0 1 2 3 4 5

**QUICK ✓** *Solve the quadratic inequality. Graph the solution set.*

**3.** $-x^2 > 2x - 24$

## EXAMPLE 4 Solving a Quadratic Inequality

Solve $2x^2 > 4x - 1$ using both the graphical and algebraic method. Graph the solution set.

### Solution

**Graphical Method:**

$$2x^2 > 4x - 1$$

Subtract $4x$ from both sides; add 1 to both sides: $2x^2 - 4x + 1 > 0$

We wish to graph $f(x) = 2x^2 - 4x + 1$.

**x-intercepts:** $f(x) = 0$ **y-intercept:** $f(0) = 1$

$$2x^2 - 4x + 1 = 0$$

$a = 2; b = -4; c = 1$ $\quad x = \dfrac{-(-4) \pm \sqrt{(-4)^2 - 4(2)(1)}}{2(2)}$

$$= \frac{4 \pm \sqrt{8}}{4}$$

$$= 1 \pm \frac{\sqrt{2}}{2}$$

$$\approx 0.29 \text{ or } 1.71$$

**Vertex:** $x = -\dfrac{b}{2a} = -\dfrac{-4}{2(2)} = 1$

$$f\left(-\frac{b}{2a}\right) = f(1) = -1$$

The vertex is at $(1, -1)$. Figure 34 shows the graph.

**Figure 34**

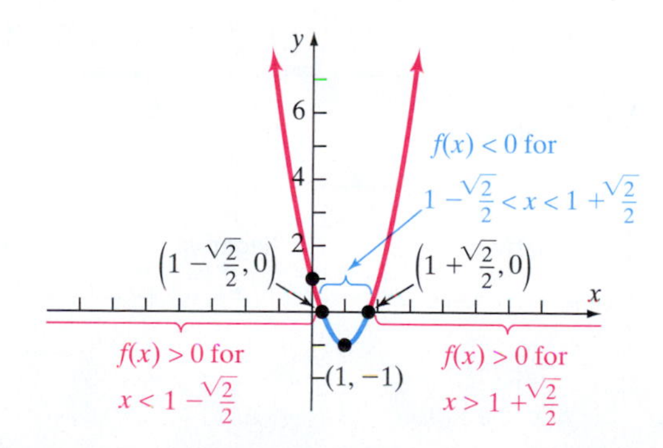


From the graph shown in Figure 34, we can see that the graph of $f(x) = 2x^2 - 4x + 1$ is greater than 0 for $x < 1 - \dfrac{\sqrt{2}}{2}$ or $x > 1 + \dfrac{\sqrt{2}}{2}$. Because the inequality is strict, we do not include the $x$-intercepts in the solution. So the solution is $\left\{x \middle| x < 1 - \dfrac{\sqrt{2}}{2} \text{ or } x > 1 + \dfrac{\sqrt{2}}{2}\right\}$ using set-builder notation; the solution is $\left(-\infty, 1 - \dfrac{\sqrt{2}}{2}\right) \cup \left(1 + \dfrac{\sqrt{2}}{2}, \infty\right)$ using interval notation.

**Algebraic Method:**
First, we put the inequality in the form $ax^2 + bx + c > 0$.

$$2x^2 > 4x - 1$$

Subtract $4x$ from both sides; add 1 to both sides: $2x^2 - 4x + 1 > 0$

Now we determine the solutions to the equation $2x^2 - 4x + 1 = 0$.

$$\begin{aligned} x &= \frac{-(-4) \pm \sqrt{(-4)^2 - 4(2)(1)}}{2(2)} \\ &= \frac{4 \pm \sqrt{8}}{4} \\ &= 1 \pm \frac{\sqrt{2}}{2} \\ &\approx 0.29 \text{ or } 1.71 \end{aligned}$$

We separate the real number line into the following intervals:

$$\left(-\infty, 1 - \frac{\sqrt{2}}{2}\right) \quad \left(1 - \frac{\sqrt{2}}{2}, 1 + \frac{\sqrt{2}}{2}\right) \quad \left(1 + \frac{\sqrt{2}}{2}, \infty\right)$$

**Work Smart**

The factor $\left(x - \left(1 - \frac{\sqrt{2}}{2}\right)\right)$ is close to the factor $x - 0.29$. So we might choose 0 as a test number. Now it's easy to see both factors are negative. Similarly, use $x - 1.71$ as an approximation of $\left(x - \left(1 + \frac{\sqrt{2}}{2}\right)\right)$.

The solutions to the equation $2x^2 - 4x + 1 = 0$ allow us to factor $2x^2 - 4x + 1$ as $\left(x - \left(1 - \frac{\sqrt{2}}{2}\right)\right)\left(x - \left(1 + \frac{\sqrt{2}}{2}\right)\right)$. We set up Table 11 using 0.29 as an approximation of $1 - \frac{\sqrt{2}}{2}$ and 1.71 as an approximation for $1 + \frac{\sqrt{2}}{2}$ as a guide to help us determine the sign of each factor.

**Table 11**

| | | $1 - \frac{\sqrt{2}}{2} \approx 0.29$ | | $1 + \frac{\sqrt{2}}{2} \approx 1.71$ | |
|---|---|---|---|---|---|
| **Interval** | $\left(-\infty, 1 - \frac{\sqrt{2}}{2}\right)$ | | $\left(1 - \frac{\sqrt{2}}{2}, 1 + \frac{\sqrt{2}}{2}\right)$ | | $\left(1 + \frac{\sqrt{2}}{2}, \infty\right)$ |
| **Test Point** | 0 | $1 - \frac{\sqrt{2}}{2}$ | 1 | $1 + \frac{\sqrt{2}}{2}$ | 2 |
| **Sign of** $\left(x - \left(1 - \frac{\sqrt{2}}{2}\right)\right)$ | Negative | 0 | Positive | Positive | Positive |
| **Sign of** $\left(x - \left(1 + \frac{\sqrt{2}}{2}\right)\right)$ | Negative | Negative | Negative | 0 | Positive |
| **Sign of** $\left(x - \left(1 - \frac{\sqrt{2}}{2}\right)\right) \times \left(x - \left(1 + \frac{\sqrt{2}}{2}\right)\right)$ | Positive | 0 | Negative | 0 | Positive |
| **Conclusion** | $2x^2 - 4x + 1$ is positive in the interval $\left(-\infty, 1 - \frac{\sqrt{2}}{2}\right)$, so it is part of the solution set | Since the inequality is strict, $1 - \frac{\sqrt{2}}{2}$ is not part of the solution | $2x^2 - 4x + 1$ is negative in the interval $\left(1 - \frac{\sqrt{2}}{2}, 1 + \frac{\sqrt{2}}{2}\right)$, so it is not part of the solution set | Since the inequality is strict, $1 + \frac{\sqrt{2}}{2}$ is not part of the solution | $2x^2 - 4x + 1$ is positive in the interval $\left(1 + \frac{\sqrt{2}}{2}, \infty\right)$, so it is part of the solution set |

We want to know where $2x^2 - 4x + 1$ is positive, so we do not include $1 - \frac{\sqrt{2}}{2}$ or $1 + \frac{\sqrt{2}}{2}$ in the solution, so the solution set is $\left\{x \middle| x < 1 - \frac{\sqrt{2}}{2} \text{ or } x > 1 + \frac{\sqrt{2}}{2}\right\}$ using set-builder notation; the solution is $\left(-\infty, 1 - \frac{\sqrt{2}}{2}\right) \cup \left(1 + \frac{\sqrt{2}}{2}, \infty\right)$ using interval notation.

Figure 35 shows the graph of the solution set.

**Figure 35**

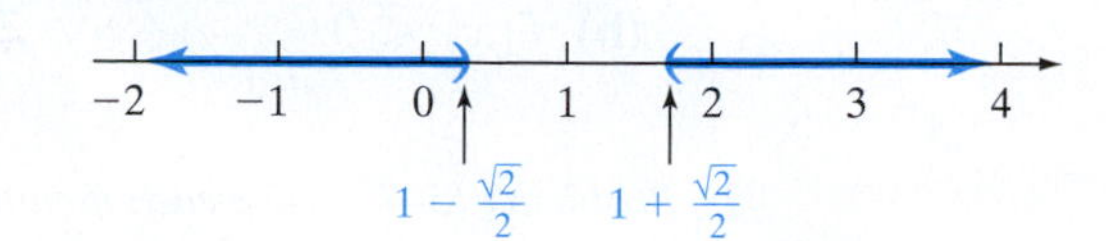


**QUICK ✓**

**4.** Solve $3x^2 > -x + 5$ using both the graphical and algebraic method. Graph the solution set.

# 8.6 Exercises

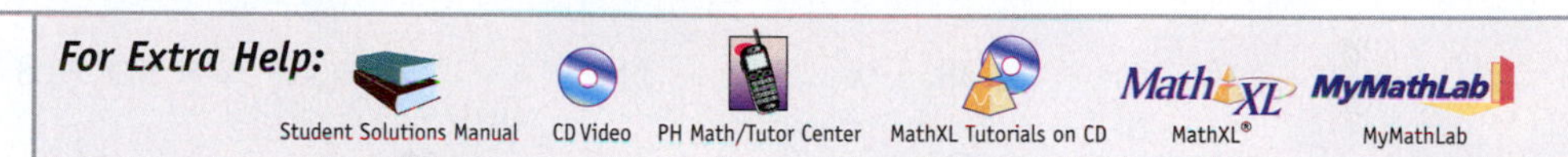


## Concepts and Vocabulary

*In Problems 1 and 2, fill in the blanks.*

**1.** The inequality $3x^2 - 7x + 2 < 0$ is an example of a(n) _______ inequality.

**2.** The sign of $(2x + 1)(x - 3)$ is _______ in the interval $\left(-\frac{1}{2}, 3\right)$.

*In Problems 3 and 4, answer True or False to each statement.*

**3.** A test number for the interval $(-6, -2)$ could be $-1$.

**4.** The inequality $x^2 + 2 > 0$ is true for all real numbers.

**5.** The inequality $x^2 + 3 < -2$ has no solution. Explain why.

**6.** The inequality $x^2 - 1 \geq -1$ has the set of all real numbers as the solution. Explain why.

**7.** Explain when the endpoints of an interval are included in the solution set of a quadratic inequality.

**8.** Is the inequality $x^2 + 1 > 1$ true for all real numbers? Explain.

## Building Skills

*In Problems 9–12, use the graphs of the quadratic function f to determine the solution.*

 **9.**

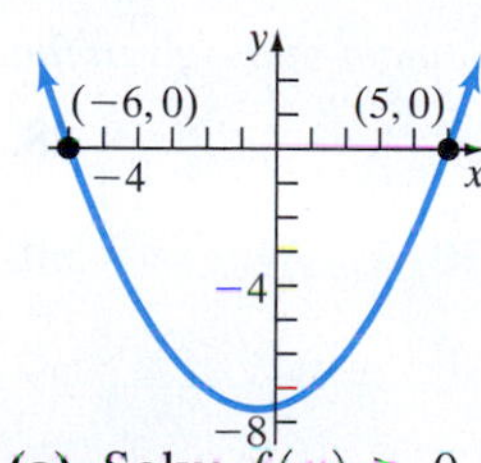


**(a)** Solve $f(x) > 0$.

**(b)** Solve $f(x) \leq 0$.

**10.**

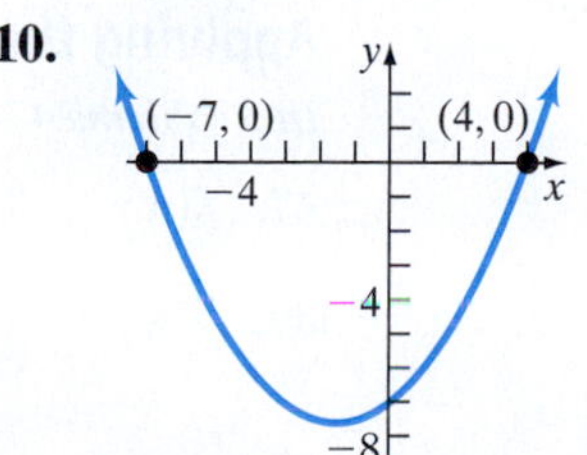


**(a)** Solve $f(x) > 0$.

**(b)** Solve $f(x) \leq 0$.

**11.**

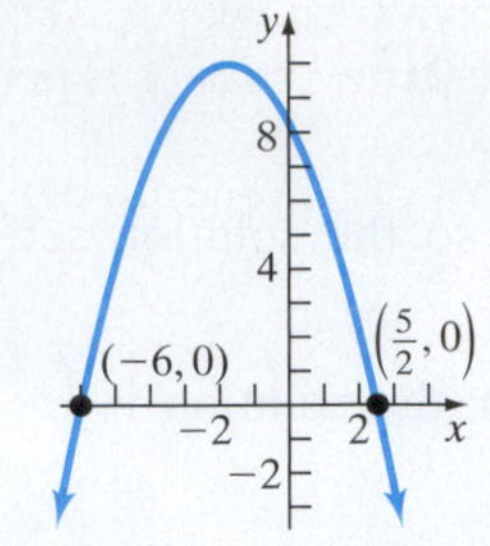

**(a)** $f(x) \geq 0$
**(b)** $f(x) < 0$

**12.**

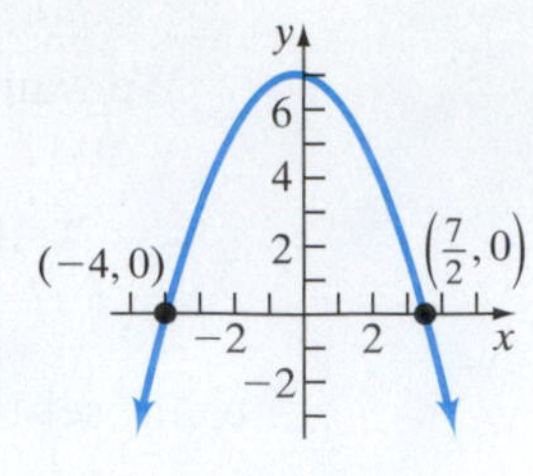


**(a)** $f(x) > 0$
**(b)** $f(x) \leq 0$

*In Problems 13–40, solve each inequality. Graph the solution set.*

**13.** $(x - 5)(x + 2) \geq 0$ **14.** $(x - 8)(x + 1) \leq 0$ **15.** $(x + 3)(x + 7) < 0$

**16.** $(x - 4)(x - 10) > 0$ **17.** $x^2 - 2x - 35 > 0$ **18.** $x^2 + 3x - 18 \geq 0$

**19.** $n^2 - 6n - 8 \leq 0$ **20.** $p^2 + 5p + 4 < 0$ **21.** $m^2 + 5m \geq 14$

**22.** $z^2 > 7z + 8$ **23.** $2q^2 \geq q + 15$ **24.** $2b^2 + 5b < 7$

**25.** $3x + 4 \geq x^2$ **26.** $x + 6 < x^2$ **27.** $-x^2 + 3x < -10$

**28.** $-x^2 > 4x - 21$ **29.** $-3x^2 \leq -10x - 8$ **30.** $-3m^2 \geq 16m + 5$

**31.** $x^2 + 4x + 1 < 0$ **32.** $x^2 - 3x - 5 \geq 0$ **33.** $-2a^2 + 7a \geq -4$

**34.** $-3p^2 < 3p - 5$ **35.** $z^2 + 2z + 3 > 0$ **36.** $y^2 + 3y + 5 \geq 0$

**37.** $2b^2 + 5b \leq -6$ **38.** $3w^2 + w < -2$ **39.** $x^2 + 6x + 9 \geq 0$

**40.** $p^2 - 8p + 16 \leq 0$

*In Problems 41–46, for each function find the values of x that satisfy the given condition. Graph the solution set.*

**41.** Solve $f(x) < 0$ if $f(x) = x^2 - 5x$.

**42.** Solve $f(x) > 0$ if $f(x) = x^2 + 4x$.

**43.** Solve $f(x) \geq 0$ if $f(x) = x^2 - 3x - 28$.

**44.** Solve $f(x) \leq 0$ if $f(x) = x^2 + 2x - 48$.

**45.** Solve $g(x) > 0$ if $g(x) = 2x^2 + x - 10$.

**46.** Solve $F(x) < 0$ if $F(x) = 2x^2 + 7x - 15$.

## Applying the Concepts

*In Problems 47–50, find the domain of the given function.*

**47.** $f(x) = \sqrt{x^2 + 8x}$ **48.** $f(x) = \sqrt{x^2 - 5x}$

**49.** $g(x) = \sqrt{x^2 - x - 30}$ **50.** $G(x) = \sqrt{x^2 + 2x - 63}$

**51. Physics** A ball is thrown vertically upward with an initial speed of 80 feet per second from a cliff 500 feet above the sea level. The height $s$ (in feet) of the ball from the ground after $t$ seconds is $s(t) = -16t^2 + 80t + 500$. For what time $t$ is the ball more than 596 feet above sea level?

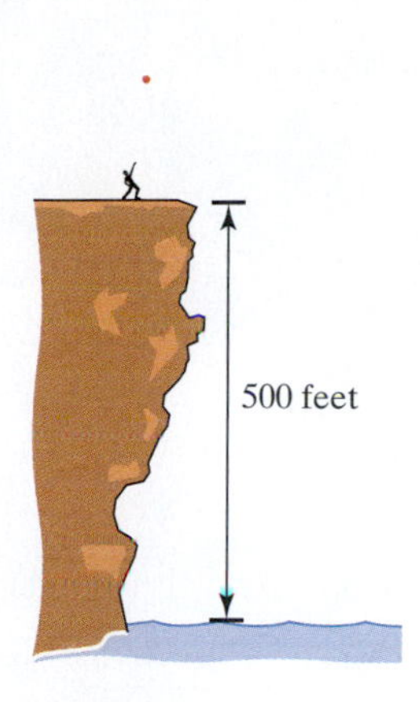


**52. Physics** A water balloon is thrown vertically upward with an initial speed of 64 feet per second from the top of a building 200 feet above the ground. The height $s$ (in feet) of the balloon from the ground after $t$ seconds is $s(t) = -16t^2 + 64t + 200$. For what time $t$ is the balloon more than 248 feet above the ground?

**53. Revenue Function** Suppose that the marketing department of Panasonic has found that, when a certain model of DVD player is sold a price of $p$ dollars, the daily revenue $R$ (in dollars) as a function of the price $p$ is $R(p) = -2.5p^2 + 600p$. Determine the prices for which revenue will exceed \$35,750. That is, solve $R(p) > 35{,}750$.

**54. Revenue Function** Suppose that the marketing department of Samsung has found that, when a certain model of cellular telephone is sold a price of $p$ dollars, the daily revenue $R$ (in dollars) as a function of the price $p$ is $R(p) = -5p^2 + 600p$. Determine the prices for which revenue will exceed \$17,500. That is, solve $R(p) > 17{,}500$.

## Extending the Concepts

*In Problems 55–58, solve each inequality algebraically by inspection. Then provide a verbal explanation of the solution.*

**55.** $(x + 3)^2 \le 0$

**56.** $(x - 4)^2 > 0$

**57.** $(x + 8)^2 > -2$

**58.** $(3x + 1)^2 < -2$

**59.** Write a quadratic inequality that has $[-3, 2]$ as the solution set.

**60.** Write a quadratic inequality that has $(0, 5)$ as the solution set.

**61.** The inequalities $(3x + 2)^2 < 2$ and $(3x + 2)^{-2} > \frac{1}{2}$ have the same solution set. Why?

*In Problems 62–67, use the techniques presented in this section to solve each inequality algebraically.*

**62.** $(x + 1)(x - 2)(x - 5) > 0$

**63.** $(x + 3)(x - 1)(x - 3) < 0$

**64.** $(2x + 1)(x - 4)(x - 9) \le 0$

**65.** $(3x + 4)(x - 2)(x - 6) \ge 0$

**66.** $\dfrac{x^2 + 5x + 6}{x - 2} > 0$

**67.** $\dfrac{x^2 - 3x - 10}{x + 1} < 0$

## Synthesis Review

*In Problems 68–71, simplify each expression.*

**68.** $\dfrac{3a^4b}{12a^{-3}b^5}$

**69.** $(4mn^{-3})(-2m^4n)$

**70.** $\left(\dfrac{3x^4y}{6x^{-2}y^5}\right)^{1/2}$

**71.** $\left(\dfrac{9a^{2/3}b^{1/2}}{a^{-1/9}b^{3/4}}\right)^{-1}$

**72.** Do the Laws of Exponents presented in the "Getting Ready: Integer Exponents" section also apply to the Laws of Exponents for rational exponents presented in Section 7.1?

## The Graphing Calculator

A graphing calculator can be used to solve the quadratic inequality in Example 1 by graphing $Y_1 = x^2 - 4x - 5$. We use the ZERO feature of the calculator to find the $x$-intercepts of the graph. See Figures 36(a) and (b).

**Figure 36**

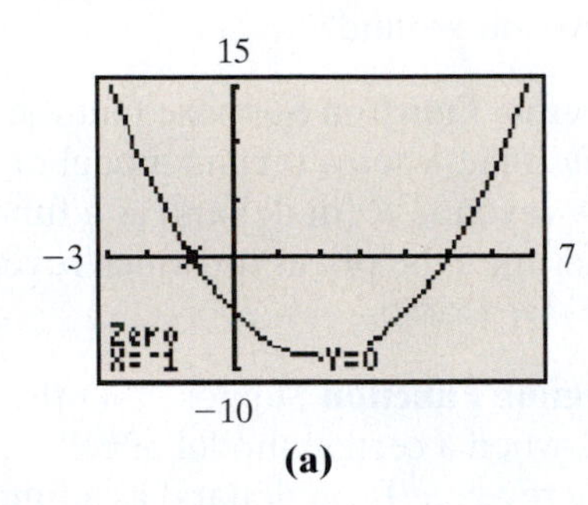


(a)

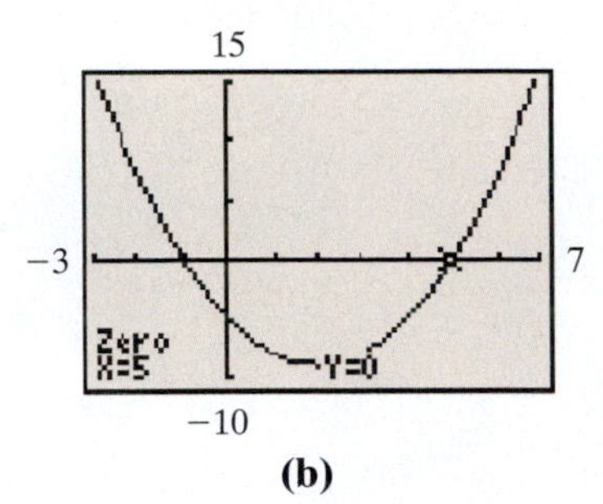


(b)

From Figures 36(a) and (b), we can see that $x^2 - 4x - 5 \geq 0$ for $x \leq -1$ or $x \geq 5$.

*In Problems 73–76, solve each inequality using a graphing calculator.*

**73.** $2x^2 + 7x - 49 > 0$

**74.** $2x^2 + 3x - 27 < 0$

**75.** $6x^2 + x \leq 40$

**76.** $8x^2 + 18x \geq 81$

# CHAPTER 8 ACTIVITY: PRESIDENTIAL DECISION MAKING

**Focus:** Developing quadratic equations.

**Time:** 30–35 minutes

**Group size:** 2–4

**1.** Your boss, Huntington Corporation's President, Gerald Cain, is very concerned about his approval rating with his employees. Last year, January 1, he made some policy changes and he saw his approval rating drop. In fact, his approval rating was 48% just before he made some policy changes. One month later, his rating was at 41%, two months later it was at 40%, and at three months it began to climb and was 45%. He discovered that the following function described his approval rating for that year:

$$R(x) = 3x^2 - 10x + 48$$

**(a)** As a group, use the above function to find when President Cain's approval rating will return to the original rating.

**(b)** If his approval rating continues to climb, when will he reach a 68% approval rating?

2. On January 1 of this year, President Cain surveyed his employees again and found that 68% of them approved of his leadership skills. At this time, President Cain decided to become very strict with his employees and began a series of new policies. He noticed that his approval rating began to steadily slip and reached an all-time low of 38% on March 30 (3 months later). President Cain was not worried because he knows from last year that this drop in popularity will bottom out and eventually rise. He believes his popularity can be modeled by a quadratic function and needs your help. He has more bad news to deliver but does not want to begin the next round of policy changes until his approval rating is back to approximately 50%.
   (a) As a group, write a quadratic function that would model President Cain's approval rating.
   (b) As a group, develop different ways to advise President Cain what date to begin his policy changes. Use graphs and computations, to prove your point.

# CHAPTER 8 REVIEW

## Section 8.1 Solving Quadratic Equations by Completing the Square

| KEY CONCEPTS | KEY TERMS |
|---|---|
| • **Square Root Property** If $x^2 = p$, then $x = \sqrt{p}$ or $x = -\sqrt{p}$. • **Pythagorean Theorem** In a right triangle, the square of the length of the hypotenuse is equal to the sum of the squares of the lengths of the legs. That is, $\text{leg}^2 + \text{leg}^2 = \text{hypotenuse}^2$. | Completing the square<br>Right triangle<br>Right angle<br>Hypotenuse<br>Legs |

| YOU SHOULD BE ABLE TO . . . | EXAMPLE | REVIEW EXERCISES |
|---|---|---|
| 1 Solve quadratic equations using the square root property (p. 611) | Examples 1 through 4 | 1–10 |
| 2 Complete the square in one variable (p. 614) | Example 5 | 11–16 |
| 3 Solve quadratic equations by completing the square (p. 615) | Examples 6 and 7 | 17–26 |
| 4 Solve problems using the Pythagorean Theorem (p. 617) | Examples 8 and 9 | 27–36 |

*In Problems 1–10, solve each equation using the Square Root Property.*

1. $m^2 = 169$
2. $n^2 = 75$
3. $a^2 = -16$
4. $b^2 = \frac{8}{9}$
5. $(x - 8)^2 = 81$
6. $(y - 2)^2 - 62 = 88$
7. $(3z + 5)^2 = 100$
8. $7p^2 = 18$
9. $3q^2 + 251 = 11$
10. $\left(x + \frac{3}{4}\right)^2 = \frac{13}{16}$

*In Problems 11–16, complete the square in each expression. Then factor the perfect square trinomial.*

11. $a^2 + 30a$
12. $b^2 - 14b$
13. $c^2 - 11c$
14. $d^2 + 9d$
15. $m^2 - \frac{1}{4}m$
16. $n^2 + \frac{6}{7}n$

*In Problems 17–26, solve each quadratic equation by completing the square.*

17. $x^2 - 10x + 16 = 0$
18. $y^2 - 3y - 28 = 0$
19. $z^2 - 6z - 3 = 0$
20. $a^2 - 5a - 7 = 0$
21. $b^2 + b + 7 = 0$
22. $c^2 - 6c + 17 = 0$
23. $2d^2 - 7d + 3 = 0$
24. $2w^2 + 2w + 5 = 0$
25. $3x^2 - 9x + 8 = 0$
26. $3x^2 + 4x - 2 = 0$

*In Problems 27–32, the lengths of the legs of a right triangle are given. Find the hypotenuse.*

**27.** $a = 9, b = 12$ **28.** $a = 8, b = 8$ **29.** $a = 3, b = 6$

**30.** $a = 10, b = 24$ **31.** $a = 5, b = \sqrt{11}$ **32.** $a = 6, b = \sqrt{13}$

*In Problems 33–35, use the right triangle shown below and find the missing length.*

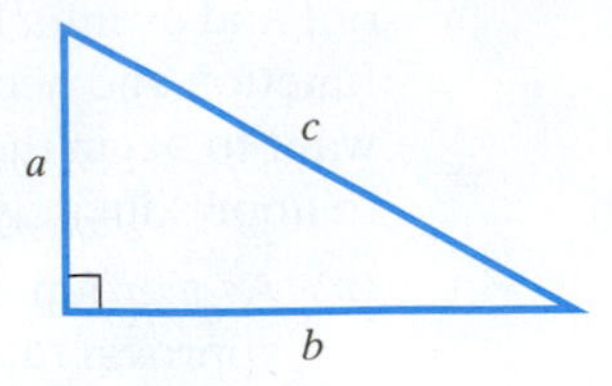


**33.** $a = 9, c = 12$ **34.** $b = 5, c = 10$ **35.** $b = 6, c = 17$

**36. Baseball Diamond** A baseball diamond is really a square that is 90 feet long on each side. (See the figure.) What is the distance between home plate and second base?

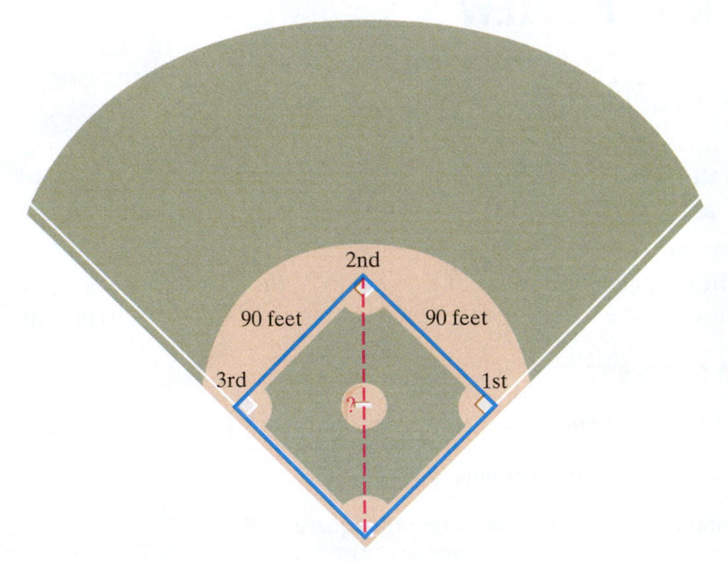


## Section 8.2 Solving Quadratic Equations by the Quadratic Formula

| KEY CONCEPTS | KEY TERM |
|---|---|
| • **The Quadratic Formula**<br>The solutions to the equation $ax^2 + bx + c = 0, a \neq 0$, are given by $x = \dfrac{-b \pm \sqrt{b^2 - 4ac}}{2a}$<br>• **Discriminant**<br>For the quadratic equation $ax^2 + bx + c = 0, a \neq 0$:<br>• If $b^2 - 4ac > 0$, the equation has two unequal real solutions.<br>  • If $b^2 - 4ac$ is a perfect square, the equation has two rational solutions.<br>  • If $b^2 - 4ac$ is not a perfect square, the equation has two irrational solutions.<br>• If $b^2 - 4ac = 0$, the equation has a repeated real solution.<br>• If $b^2 - 4ac < 0$, the equation has two complex solutions that are not real. | Discriminant |

| YOU SHOULD BE ABLE TO . . . | EXAMPLE | REVIEW EXERCISES |
|---|---|---|
| 1 Solve quadratic equations using the quadratic formula (p. 624) | Examples 1 through 4 | 37–46 |
| 2 Use the discriminant to determine the nature of solutions in a quadratic equation (p. 629) | Example 5 | 47–52 |
| 3 Model and solve problems involving quadratic equations (p. 632) | Examples 6 and 7 | 63–68 |

*In Problems 37–46, solve each equation using the quadratic formula.*

**37.** $x^2 - x - 20 = 0$ **38.** $4y^2 = 8y + 21$ **39.** $3p^2 + 8p = -3$

**40.** $2q^2 - 3 = 4q$ **41.** $3w^2 + w = -3$ **42.** $9z^2 + 16 = 24z$

**43.** $m^2 - 4m + 2 = 0$ **44.** $5n^2 + 4n + 1 = 0$ **45.** $5x + 13 = -x^2$

**46.** $-2y^2 = 6y + 7$

*In Problems 47–52, determine the discriminant of each quadratic equation. Use the value of the discriminant to determine whether the quadratic equation has two rational solutions, two irrational solutions, one repeated real solution, or two complex solutions that are not real.*

**47.** $p^2 - 5p - 8 = 0$ **48.** $m^2 + 8m + 16 = 0$ **49.** $3n^2 + n = -4$

**50.** $7w^2 + 3 = 8w$ **51.** $4x^2 + 49 = 28x$ **52.** $11z - 12 = 2z^2$

*In Problems 53–62, solve each equation using any method you wish.*

**53.** $x^2 + 8x - 9 = 0$ **54.** $6p^2 + 13p = 5$ **55.** $n^2 + 13 = -4n$

**56.** $5y^2 - 60 = 0$ **57.** $\frac{1}{4}q^2 - \frac{1}{2}q - \frac{3}{8} = 0$ **58.** $\frac{1}{8}m^2 + m + \frac{5}{2} = 0$

**59.** $(w - 8)(w + 6) = -33$ **60.** $(x - 3)(x + 1) = -2$ **61.** $9z^2 = 16$

**62.** $\frac{1 - 2x}{x^2 + 5} = 1$

**63. Pythagorean Theorem** Use the Pythagorean Theorem to determine the value of $x$ for the given measurements of the right triangle shown below.

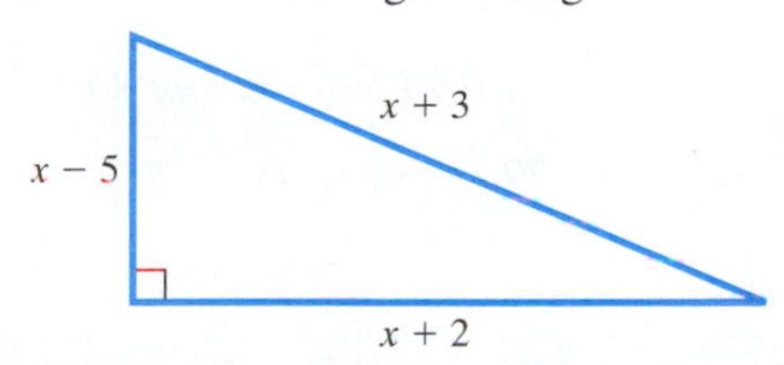


**64. Area** The area of a rectangle is 108 square centimeters. The width of the rectangle is 3 inches less than the length. What are the dimensions of the rectangle?

**65. Revenue** The revenue $R$ received by a company selling $x$ cellular phones per week is given by the function $R(x) = -0.2x^2 + 180x$.

**(a)** How many cellular phones must be sold in order for revenue to be \$36,000 per week?

**(b)** How many cellular phones must be sold in order for revenue to be \$40,500 per week?

**66. Projectile Motion** The height $s$ of a ball after $t$ seconds when thrown straight up with an initial speed of 50 feet per second from an initial height of 180 feet can be modeled by the function $s(t) = -16t^2 + 50t + 180$.

**(a)** When will the height of the ball be 200 feet? Round your answer to the nearest tenth of a second.

**(b)** When will the height of the ball be 100 feet? Round your answer to the nearest tenth of a second.

**(c)** Will the ball ever reach a height of 300 feet? How does the result of the equation tell you this?

**67. Pleasure Boat Ride** A pleasure boat carries passengers 10 miles upstream and then returns to the starting point. The total time of the trip (excluding the time on the ground) takes 2 hours. If the speed of the current is 3 miles per hour, find the speed the boat in still water. Round your answer to the nearest tenth of an hour.

**68. Work** Together, Tom and Beth can wash their car in 30 minutes. By himself, Tom can wash the car in 14 minutes less time than Beth can by herself. How long will it take Beth to wash the car by herself? Round your answer to the nearest tenth of a minute.

## Section 8.3 Solving Equations Quadratic in Form

**KEY TERM**

Equation quadratic in form

| YOU SHOULD BE ABLE TO . . . | EXAMPLE | REVIEW EXERCISES |
|---|---|---|
| 1 Solve equations that are quadratic in form (p. 639) | Examples 1 through 5 | 69–80 |

*In Problems 69–78, solve each equation.*

**69.** $x^4 + 7x^2 - 144 = 0$

**70.** $4w^4 + 5w^2 - 6 = 0$

**71.** $3(a + 4)^2 - 11(a + 4) + 6 = 0$

**72.** $(q^2 - 11)^2 - 2(q^2 - 11) - 15 = 0$

**73.** $y - 13\sqrt{y} + 36 = 0$

**74.** $5z + 2\sqrt{z} - 3 = 0$

**75.** $p^{-2} - 4p^{-1} - 21 = 0$

**76.** $2b^{2/3} + 13b^{1/3} - 7 = 0$

**77.** $m^{1/2} + 2m^{1/4} - 8 = 0$

**78.** $\left(\frac{1}{x + 5}\right)^2 + \frac{3}{x + 5} = 28$

*In Problems 79 and 80, find the zeros of the function.*

**79.** $f(x) = 4x - 20\sqrt{x} + 21$

**80.** $g(x) = x^4 - 17x^2 + 60$

## Section 8.4 Graphing Quadratic Functions Using Transformations

**KEY CONCEPTS**

- **Graphing a Function of the Form $f(x) = x^2 + k$**
  To obtain the graph of $f(x) = x^2 + k$ from the graph of $y = x^2$, shift the graph of $y = x^2$ vertically up $k$ units if $k > 0$ and vertically down $k$ units if $k < 0$.
- **Graphing a Function of the Form $f(x) = (x - h)^2$**
  To obtain the graph of $f(x) = (x - h)^2$ from the graph of $y = x^2$, shift the graph of $y = x^2$ horizontally to the right $h$ units if $h > 0$ and horizontally left $h$ units if $h < 0$.
- **Graphing a Function of the Form $f(x) = ax^2$**
  To obtain the graph of $f(x) = ax^2$ from the graph of $y = x^2$, multiply each $y$-coordinate on the graph of $y = x^2$ by $a$.

**KEY TERMS**

Quadratic function
Vertically stretched
Vertically compressed
Parabola
Opens up
Opens down
Vertex
Axis of symmetry
Transformations

| YOU SHOULD BE ABLE TO . . . | EXAMPLE | REVIEW EXERCISES |
|---|---|---|
| 1 Graph quadratic functions of the form $f(x) = x^2 + k$ (p. 649) | Examples 1 and 2 | 81–82; 87–90 |
| 2 Graph quadratic functions of the form $f(x) = (x - h)^2$ (p. 650) | Examples 3 through 5 | 83–84; 87–90 |
| 3 Graph quadratic functions of the form $f(x) = ax^2$ (p. 652) | Example 6 | 85–86; 89–90 |
| 4 Graph quadratic functions of the form $f(x) = ax^2 + bx + c$ (p. 654) | Examples 7 and 8 | 91–96 |
| 5 Find a quadratic function from its graph (p. 657) | Example 9 | 97–100 |

*In Problems 81–90, use the graph of $y = x^2$ to graph the quadratic function.*

**81.** $f(x) = x^2 + 4$  **82.** $g(x) = x^2 - 5$  **83.** $h(x) = (x + 1)^2$

**84.** $F(x) = (x - 4)^2$  **85.** $G(x) = -4x^2$  **86.** $H(x) = \frac{1}{5}x^2$

**87.** $p(x) = (x - 4)^2 - 3$  **88.** $P(x) = (x + 4)^2 + 2$  **89.** $f(x) = -(x - 1)^2 + 4$

**90.** $F(x) = \frac{1}{2}(x + 2)^2 - 1$

*In Problems 91–96, graph each quadratic function using transformations. Determine the vertex and axis of symmetry. Based on the graph determine the domain and range of each function.*

**91.** $g(x) = x^2 - 6x + 10$  **92.** $G(x) = x^2 + 8x + 11$

**93.** $h(x) = 2x^2 - 4x - 3$  **94.** $H(x) = -x^2 - 6x - 10$

**95.** $p(x) = -3x^2 + 12x - 8$  **96.** $P(x) = \frac{1}{2}x^2 - 2x + 5$

*In Problems 97–100, determine the quadratic function whose graph is given.*

**97.**

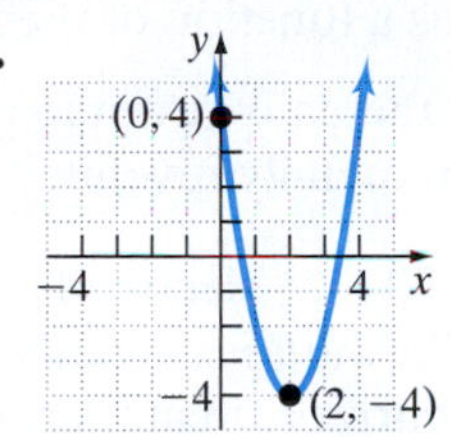


**98.**

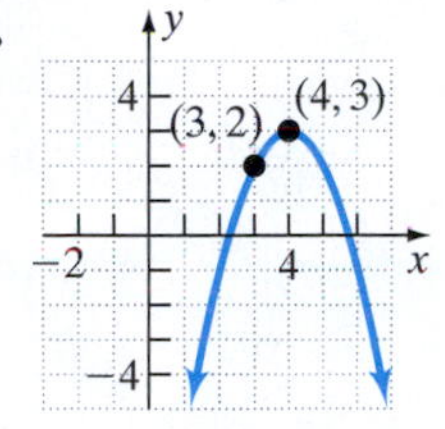


**99.**

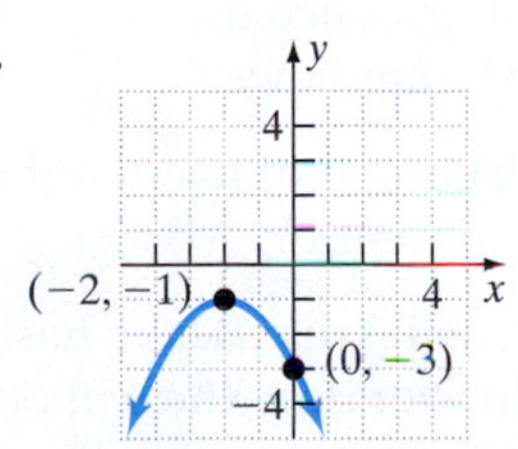


**100.**

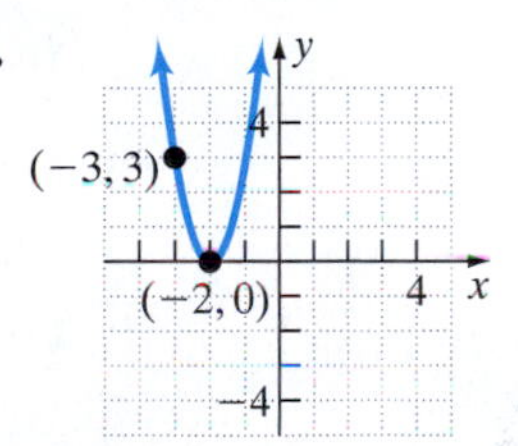


## Section 8.5 Graphing Quadratic Functions Using Properties

**KEY CONCEPTS**

- **Vertex of a Parabola**
  Any quadratic function of the form $f(x) = ax^2 + bx + c, a \neq 0$, will have vertex $\left(-\frac{b}{2a}, f\left(-\frac{b}{2a}\right)\right)$
- **The $x$-Intercepts of the Graph of a Quadratic Function**
  1. If $b^2 - 4ac > 0$, the graph of $f(x) = ax^2 + bx + c$ has two different $x$-intercepts.
  2. If $b^2 - 4ac = 0$, the graph of $f(x) = ax^2 + bx + c$ has one $x$-intercept.
  3. If $b^2 - 4ac < 0$, the graph of $f(x) = ax^2 + bx + c$ has no $x$-intercepts.

**KEY TERMS**

Maximum value
Minimum value
Optimization

| YOU SHOULD BE ABLE TO . . . | EXAMPLE | REVIEW EXERCISES |
|---|---|---|
| 1 Graph quadratic functions of the form $f(x) = ax^2 + bx + c$ (p. 662) | Examples 1 through 4 | 101–108 |
| 2 Find the maximum or minimum value of a quadratic function (p. 668) | Examples 5 and 6 | 109–112 |
| 3 Model and solve optimization problems involving quadratic functions (p. 669) | Examples 7 and 8 | 113–118 |

*In Problems 101–108, graph each quadratic function using its properties by following Steps 1–5 on page 665. Based on the graph determine the domain and range of each function.*

**101.** $f(x) = x^2 + 2x - 8$

**102.** $F(x) = 2x^2 - 5x + 3$

**103.** $g(x) = -x^2 + 6x - 7$

**104.** $G(x) = -2x^2 + 4x + 3$

**105.** $h(x) = 4x^2 - 12x + 9$

**106.** $H(x) = \frac{1}{3}x^2 + 2x + 3$

**107.** $p(x) = \frac{1}{4}x^2 + 3x + 10$

**108.** $P(x) = -x^2 + 4x - 9$

*In Problems 109–112, determine whether the quadratic function has a maximum or minimum value. Then find the maximum or minimum value.*

**109.** $f(x) = -2x^2 + 16x - 10$

**110.** $g(x) = 6x^2 - 3x - 1$

**111.** $h(x) = -4x^2 + 8x + 3$

**112.** $F(x) = -\frac{1}{3}x^2 + 4x - 7$

**113. Revenue** Suppose that the marketing department of Zenith has found that, when a certain model of television is sold for a price of $p$ dollars, the daily revenue $R$ (in dollars) as a function of the price $p$ is $R(p) = -\frac{1}{3}p^2 + 150p$.

**(a)** For what price will the daily revenue be maximized?
**(b)** What is this maximum daily revenue?

**114. Electrical Power** In a 120-volt electrical circuit having a resistance of 16 ohms, the available power $P$ (in watts) is given by the function $P(I) = -16I^2 + 120I$, where $I$ represents the current (in amperes).

**(a)** What current will produce the maximum power in the circuit?
**(b)** What is this maximum power?

**115. Fun with Numbers** The sum of two numbers is 24. Find the numbers such that their product is a maximum.

**116. Maximizing an Enclosed Area** Becky has 15 yards of fencing to make a rectangular kennel for her dog. She will build the kennel next to her garage, so she only needs to enclose three sides. (See the figure.)

**(a)** What dimensions maximize the area of the kennel?
**(b)** What is this maximum area?

**117. Kicking a Football** Ted kicks a football at a 45° angle to the horizontal with an initial velocity of 80 feet per second. The model $h(x) = -0.005x^2 + x$ can be used to estimate the height $h$ of the ball after it has traveled $x$ feet.

**(a)** How far from Ted will the ball reach a maximum height?
**(b)** What is the maximum height of the ball?
**(c)** How far will the ball travel before it strikes the ground?

**118. Revenue** Monthly demand for automobiles at a certain dealership obeys the demand equation $x = -0.002p + 60$, where $x$ is the quantity and $p$ is the price (in dollars).

**(a)** Express the revenue $R$ as a function of $p$. (**Hint:** $R = xp$)
**(b)** What price $p$ maximizes revenue? What is the maximum revenue?
**(c)** How many automobiles will be sold at the revenue-maximizing price?

| Section 8.6 Quadratic Inequalities | | |
|---|---|---|
| KEY TERM | | |
| Quadratic inequality | | |
| **YOU SHOULD BE ABLE TO . . .** | **EXAMPLE** | **REVIEW EXERCISES** |
| 1 Solve quadratic inequalities (p. 678) | Examples 1 through 4 | 119–126 |

*In Problems 119 and 120, use the graphs of the quadratic function f to determine the solution.*

**119.**

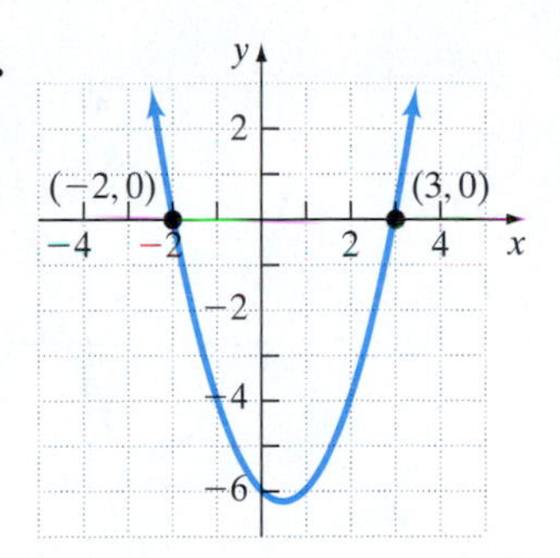


**(a)** $f(x) > 0$ **(b)** $f(x) < 0$

**120.**

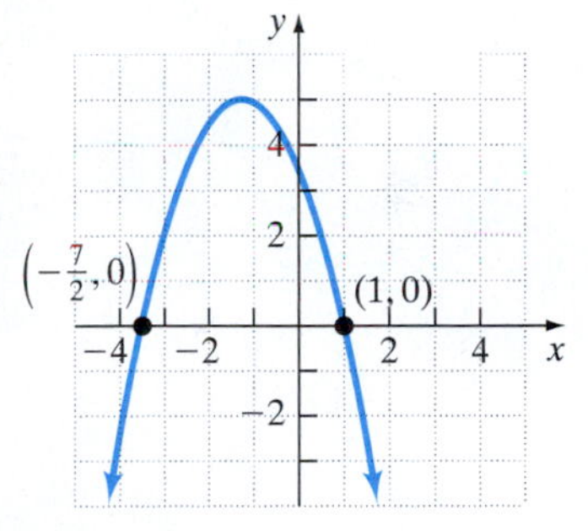


**(a)** $f(x) \geq 0$ **(b)** $f(x) \leq 0$

*In Problems 120–126, solve each inequality. Graph the solution set.*

**121.** $x^2 - 2x - 24 \leq 0$

**122.** $y^2 + 7y - 8 \geq 0$

**123.** $3z^2 - 19z + 20 > 0$

**124.** $p^2 + 4p - 2 < 0$

**125.** $4m^2 - 20m + 25 \geq 0$

**126.** $6w^2 - 19w - 7 \leq 0$

# CHAPTER 8 TEST

*Remember to use your Chapter Test Prep Video CD to see fully worked-out solutions to any of these problems you would like to review.*

*In Problems 1 and 2, complete the square in the given expression. Then factor the perfect square trinomial.*

**1.** $x^2 - 3x$

**2.** $m^2 + \frac{2}{5}m$

*In Problems 3–6, solve each equation using the method you prefer.*

**3.** $9\left(x + \frac{4}{3}\right)^2 = 1$

**4.** $m^2 - 6m + 4 = 0$

**5.** $2w^2 - 4w + 3 = 0$

**6.** $\frac{1}{2}z^2 - \frac{3}{2}z = -\frac{7}{6}$

**7.** Determine the discriminant of $2x^2 + 5x = 4$. Use the value of the discriminant to determine whether the quadratic equation has two rational solutions, two irrational solutions, one repeated real solution, or two complex solutions that are not real.

**8.** Find the missing length in the right triangle shown below.

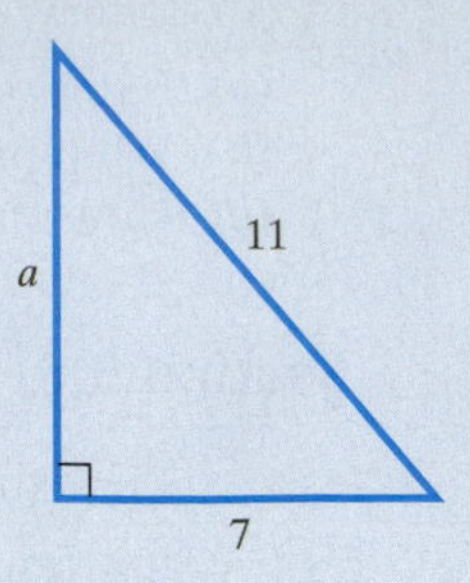


*In Problems 9 and 10, solve each equation.*

**9.** $x^4 - 5x^2 - 36 = 0$

**10.** $6y^{1/2} + 13y^{1/4} - 5 = 0$

*In Problems 11 and 12, graph each quadratic function. Determine the vertex and axis of symmetry. Based on the graph, determine the domain and range of the quadratic function.*

**11.** $f(x) = (x + 2)^2 - 5$

**12.** $g(x) = -2x^2 - 8x - 3$

**13.** Determine the quadratic function whose graph is given.

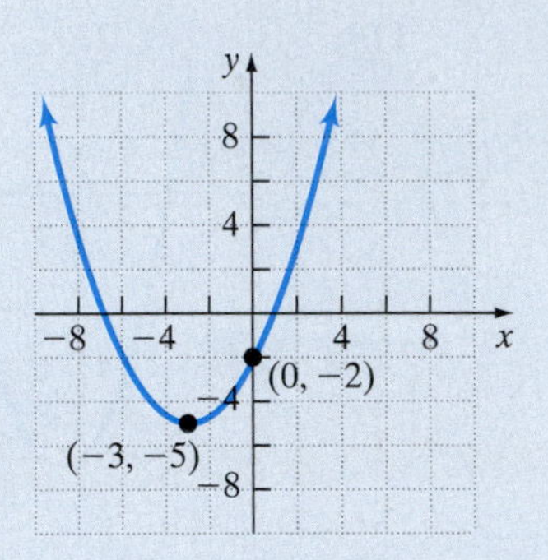


**14.** Determine whether the given quadratic function $h(x) = -\frac{1}{4}x^2 + x + 5$ has a maximum or minimum value. Then find the maximum or minimum value.

*In Problems 15 and 16, solve each inequality. Graph the solution set.*

**15.** $2m^2 + m - 15 > 0$

**16.** $z^2 + 6z - 1 \leq 0$

**17. Projectile Motion** The height $s$ of a rock after $t$ seconds when propelled straight up with an initial speed of 80 feet per second from an initial height of 20 feet can be modeled by the function $s(t) = -16t^2 + 80t + 20$. When will the height of the rock be 50 feet? Round your answer to the nearest tenth of a second.

**18. Work** Together, Lex and Rupert can roof a house in 16 hours. By himself, Rex can roof the house in 4 hours less time than Rupert can by himself. How long will it take Rupert to roof the house by himself? Round your answer to the nearest tenth of an hour.

**19. Revenue** A small company has found that, when their product is sold for a price of $p$ dollars, the weekly revenue (in dollars) as a function of price $p$ is $R(p) = -0.25p^2 + 170p$.

**(a)** For what price will the weekly revenue be maximized?
**(b)** What is this maximum weekly revenue?

**20. Maximizing Volume** A box with a rectangular base is to be constructed such that the perimeter of the base of the box is to be 50 inches. The height of the box must be 12 inches.

**(a)** Find the dimensions that maximize the volume of the box.
**(b)** What is this maximum volume?

CHAPTER

# 9 Exponential and Logarithmic Functions

Earthquakes are one of the most powerful forces in nature. The powerful waves result from a movement of Earth's outer layer that move to the surface. One measure of the strength of an earthquake is the Richter scale. See Problems 111–114 in Section 9.3.

## OUTLINE



## The Big Picture: Putting It Together

Up to now, we have studied polynomial, rational, and radical expressions and functions. We learned how to evaluate, simplify, and solve equations involving these types of algebraic expressions. We also learned how to graph linear functions, quadratic functions and certain types of radical functions.

We now introduce two more types of functions, the *exponential* and *logarithmic functions*. The theme of the text will continue. We will evaluate these functions, graph them, and learn properties of the functions. We also will solve equations that involve the exponential or logarithmic expressions.

# 9.1 Composite Functions and Inverse Functions

**OBJECTIVES**

1. Form the Composite Function
2. Determine Whether or Not a Function Is One-to-One
3. Determine the Inverse of a Function Defined by a Map or Ordered Pair
4. Obtain the Graph of the Inverse Function from the Graph of the Function
5. Find the Inverse of a Function Defined by an Equation

### *Preparing for Composite Functions and Inverse Functions*

*Before getting started, take the following readiness quiz. If you get a problem wrong, go back to the section cited and review the material.*

**1.** Determine the domain of $R(x) = \dfrac{x^2 - 9}{x^2 + 3x - 28}$. [Section 6.1, p. 461]

**2.** If $f(x) = 2x^2 - x + 1$, find (a) $f(-2)$ (b) $f(a + 1)$. [Section 2.3, pp. 162–164]

**3.** The graph of a relation is given. Does the relation represent a function? [Section 2.3, pp. 161–162]

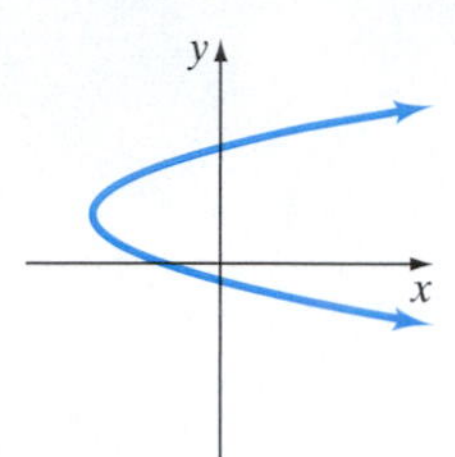


## 1 Form the Composite Function

Consider the function $y = (x - 3)^4$. To evaluate this function, we first evaluate $x - 3$ and then raise the result to the fourth power. So, technically, two different functions form the function $y = (x - 3)^4$. Figure 1 illustrates the idea for $x = 5$.

**Figure 1**

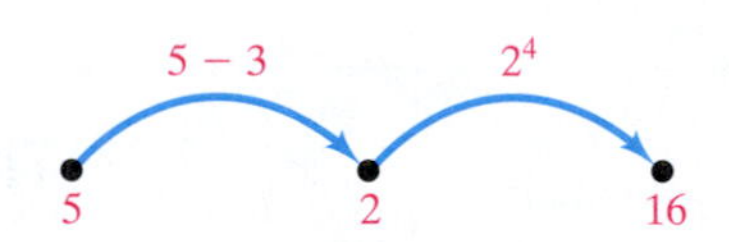


If we let $u = g(x) = x - 3$ and $y = f(u) = u^4$, then $u = g(5) = 5 - 3 = 2$ and $y = f(u) = f(2) = 2^4 = 16$. To shorten the notation, we can express $y = (x - 3)^4$ as $y = f(u) = f(g(x)) = (x - 3)^4$. This process is called **composition.**

In general, suppose that $f$ and $g$ are two functions and that $x$ is a number in the domain of $g$. By evaluating $g$ at $x$, we get $g(x)$. If $g(x)$ is in the domain of $f$, then we may evaluate $f$ at $g(x)$ and obtain the expression $f(g(x))$. The correspondence from $x$ to $f(g(x))$ is called a *composite function* $f \circ g$.

**In Words**

To find $f(g(x))$, first determine $g(x)$. This is the output of $g$. Then evaluate $f$ at the output of $g$. The result is $f(g(x))$. The notation $f(g(x))$ is read "$f$ of $g$ of $x$."

**DEFINITION**

Given two functions $f$ and $g$, the **composite function,** denoted by $f \circ g$ (read as "$f$ composed with $g$"), is defined by

$$(f \circ g)(x) = f(g(x))$$

Figure 2 illustrates the definition. Notice that the "inside" function $g$ in $f(g(x))$ is always evaluated first.

**Figure 2**

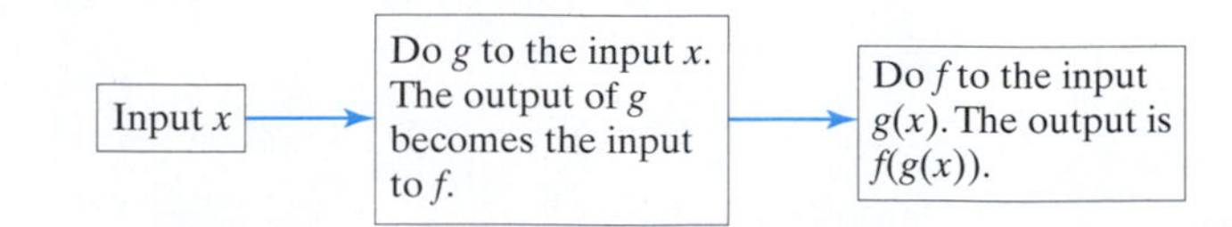


### EXAMPLE 1 Evaluating a Composite Function

Suppose that $f(x) = x^2 - 3$ and $g(x) = 2x + 1$. Find

**(a)** $(f \circ g)(3)$ **(b)** $(g \circ f)(3)$ **(c)** $(f \circ f)(-2)$

*Preparing for...Answers*

**1.** $\{x | x \neq -7, x \neq 4\}$ **2. (a)** 11 **(b)** $2a^2 + 3a + 2$ **3.** No, it fails the vertical line test.

## Solution

**(a)** Using the flowchart in Figure 2, we evaluate $(f \circ g)(3)$ as follows:

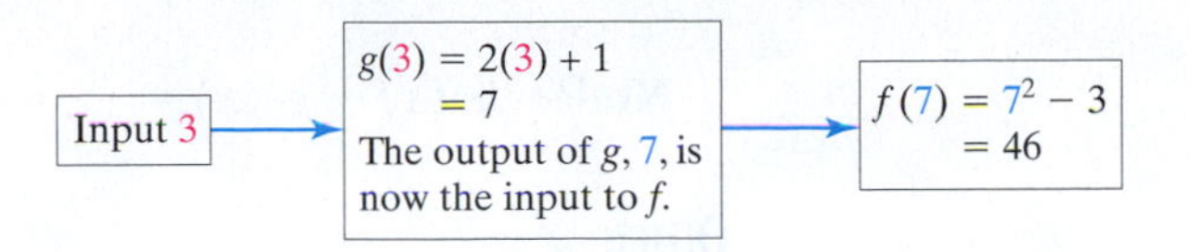


More directly,

$$(f \circ g)(3) = f(g(3)) = f(7) = 7^2 - 3 = 49 - 3 = 46$$

$g(x) = 2x + 1$ $\quad f(x) = x^2 - 3$
$g(3) = 2(3) + 1$
$= 7$

**(b)** $(g \circ f)(3) = g(f(3)) = g(6) = 2(6) + 1 = 13$

$f(x) = x^2 - 3$ $\quad g(x) = 2x + 1$
$f(3) = 3^2 - 3$
$= 6$

**(c)** $(f \circ f)(-2) = f(f(-2)) = f(1) = 1^2 - 3 = -2$

$f(x) = x^2 - 3$ $\quad f(x) = x^2 - 3$
$f(-2) = (-2)^2 - 3$
$= 1$

### QUICK ✓

**1.** Suppose that $f(x) = 4x - 3$ and $g(x) = x^2 + 1$. Find

**(a)** $(f \circ g)(2)$ **(b)** $(g \circ f)(2)$ **(c)** $(f \circ f)(-3)$

Rather than evaluating a composite function at a specific value, we can also form a composite function that is dependent upon the independent variable, $x$.

## EXAMPLE 2 Finding a Composite Function

Suppose that $f(x) = x^2 + 2x$ and $g(x) = 2x - 1$. Find

**(a)** $(f \circ g)(x)$ **(b)** $(g \circ f)(x)$ **(c)** $(f \circ g)(2)$

## Solution

**(a)**

$$(f \circ g)(x) = f(g(x))$$

$g(x) = 2x - 1$: $= f(2x - 1)$

$f(x) = x^2 + 2x$: $= (2x - 1)^2 + 2(2x - 1)$

FOIL; Distribute: $= 4x^2 - 4x + 1 + 4x - 2$

Combine like terms: $= 4x^2 - 1$

**Work Smart**
$(f \circ g)(x)$ does not mean $(f \cdot g)(x)$.

**(b)**

$$(g \circ f)(x) = g(f(x))$$

$f(x) = x^2 + 2x$: $= g(x^2 + 2x)$

$g(x) = 2x - 1$: $= 2(x^2 + 2x) - 1$

Distribute: $= 2x^2 + 4x - 1$

**(c)** Instead of finding $(f \circ g)(2)$ using the approach presented in Example 1, we will find $(f \circ g)(2)$ using the results from part (a).

$$(f \circ g)(2) = f(g(2)) = 4(2)^2 - 1 = 4(4) - 1 = 15$$

Notice that $(f \circ g)(x) \neq (g \circ f)(x)$ in Examples 2(a) and 2(b).

**QUICK ✓**

**2.** Suppose that $f(x) = x^2 - 3x + 1$ and $g(x) = 3x + 2$. Find

**(a)** $(f \circ g)(x)$ **(b)** $(g \circ f)(x)$ **(c)** $(f \circ g)(-2)$

## 2 Determine Whether or Not a Function Is One-to-One

Figures 3 and 4 illustrate two different functions represented as mappings. The function in Figure 3 shows the correspondence between states and their population in millions. The function in Figure 4 shows a correspondence between "animals" and "life expectancy."

**Figure 3**

| State | Population (millions) |
|---|---|
| Indiana | 6.2 |
| Washington | 6.1 |
| South Dakota | 0.8 |
| North Carolina | 8.3 |
| Tennessee | 5.8 |

**Figure 4**

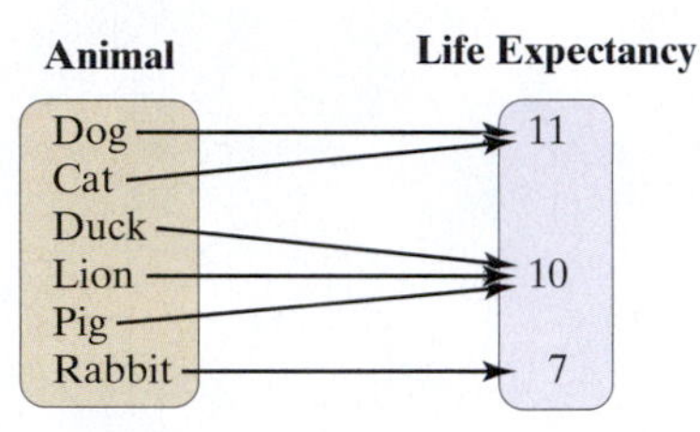


Suppose we asked a group of people to name the state which has a population of 0.8 million based on the function in Figure 3. Everyone in the group would respond "South Dakota." If we asked the same group of people to name the animal whose life expectancy is 11 years based on the function in Figure 4, some would respond "dog," while others would respond "cat." What is the difference between the functions in Figures 3 and 4? In Figure 3, each element in the domain corresponds to one (and only one) element in the range. In Figure 4, this is not the case—there is more than one element in the domain that corresponds to an element in the range. For example, "dog" corresponds to "11," but "cat" also corresponds to "11." We give functions such as the one in Figure 3 a special name.

**In Words**

A function is not one-to-one if two different inputs correspond to the same output.

**DEFINITION**

A function is **one-to-one** if any two different inputs in the domain correspond to two different outputs in the range. That is, if $x_1$ and $x_2$ are two different inputs of a function $f$, then $f(x_1) \neq f(x_2)$.

Put another way, a function is not one-to-one if two different elements in the domain correspond to the same element in the range. So, the function in Figure 4 is not one-to-one because two different elements in the domain, dog and cat, both correspond to 11.

### EXAMPLE 3 Determining Whether a Function Is One-to-One

Determine which of the following functions is one-to-one.

**(a)** For the function to the right, the domain represents the age of 5 males and the range represents their HDL (good) cholesterol (mg/dL).

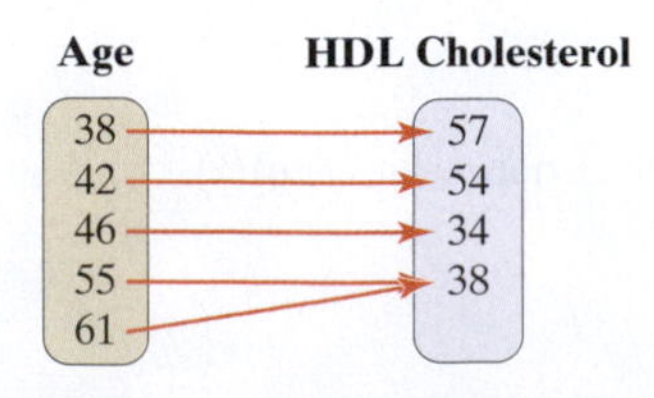


**(b)** $\{(-2, 6), (-1, 3), (0, 2), (1, 4), (2, 8)\}$

## Solution

**(a)** The function is not one-to-one because there are two different inputs, 55 and 61, that correspond to the same output, 38.

**(b)** The function is one-to-one because there are no two distinct inputs that correspond to the same output.

**QUICK ✓** *Determine whether or not the function is one-to-one.*

**3.**

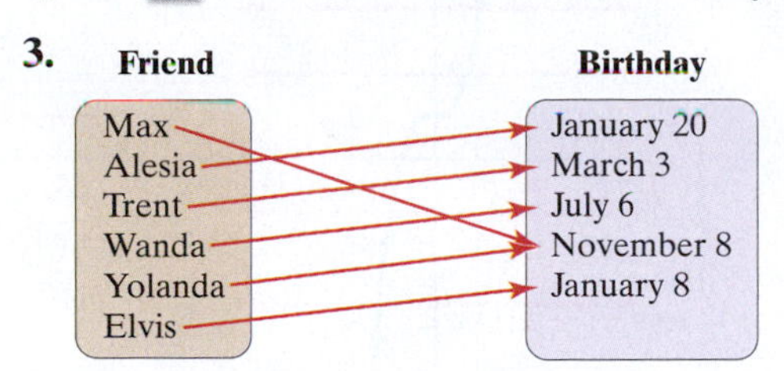


**4.** $\{(-3, 3), (-2, 2), (-1, 1), (0, 0), (1, -1)\}$

Consider the functions $y = 2x - 5$ and $y = x^2 - 4$ shown in Figure 5.

**Figure 5**

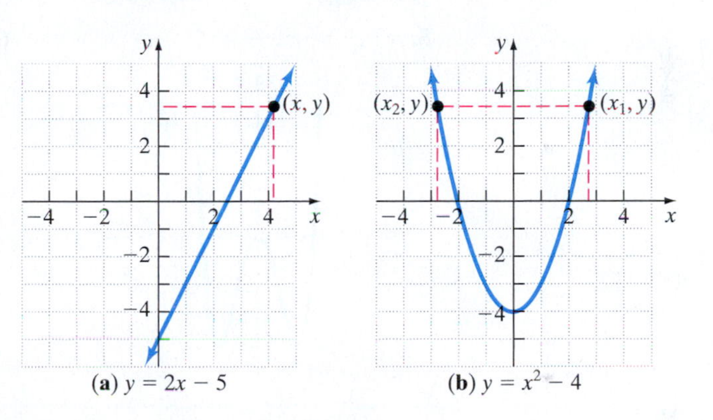


(a) $y = 2x - 5$ (b) $y = x^2 - 4$

Notice for the function $y = 2x - 5$ shown in Figure 5(a) that any output is the result of only one input. However, for the function $y = x^2 - 4$ shown in Figure 5(b) there are instances where a given output is the result of more than one input. For example, the output $-3$ is the image of the two inputs, $-1$ and 1.

If a horizontal line intersects the graph of a function at more than one point, the function cannot be one-to-one because this would mean that two different inputs give the same output. We state this result formally as the *horizontal line test*.

**HORIZONTAL LINE TEST**

If every horizontal line intersects the graph of a function $f$ in at most one point, then $f$ is one-to-one.

## EXAMPLE 4 Using the Horizontal Line Test

For each function, use the graph to determine whether the function is one-to-one.

**(a)** $f(x) = -2x^3 + 1$ **(b)** $g(x) = x^3 - 4x$

## Solution

**(a)** Figure 6(a) illustrates the horizontal line test for $f(x) = -2x^3 + 1$. Because every horizontal line will intersect the graph of $f$ exactly once, it follows that $f$ is one-to-one.

**(b)** Figure 6(b) illustrates the horizontal line test for $g(x) = x^3 - 4x$. The horizontal line $y = 1$ intersects the graph three times, so $g$ is not one-to-one.

**Figure 6**

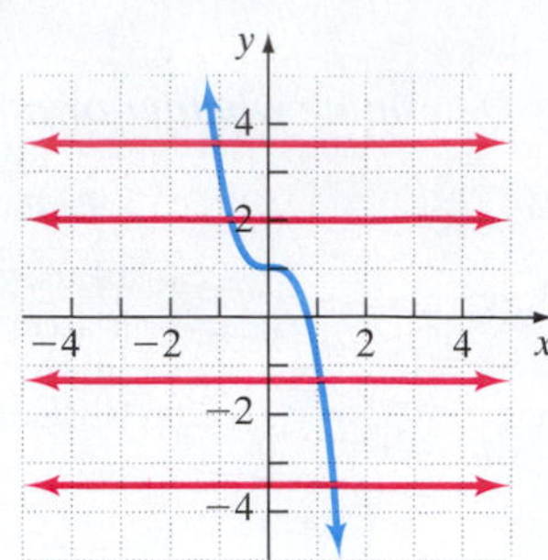


**(a)** Every horizontal line intersects the graph once; $f$ is one-to-one.

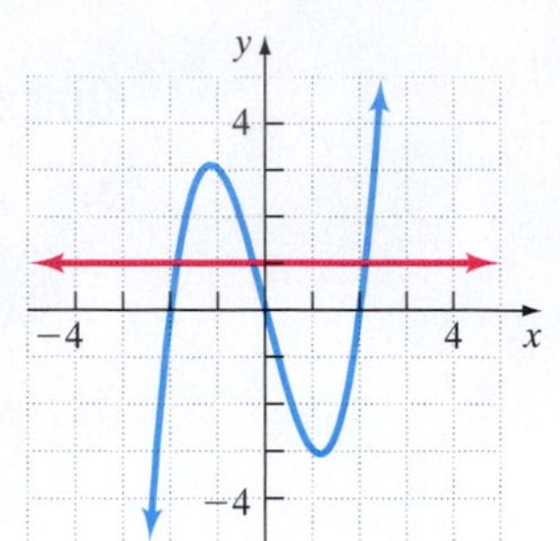


**(b)** A horizontal line intersects the graph three times; $g$ is not one-to-one.

**QUICK ✓** *Use the graph to determine whether the given function is one-to-one.*

**5.** $f(x) = x^4 - 4x^2$

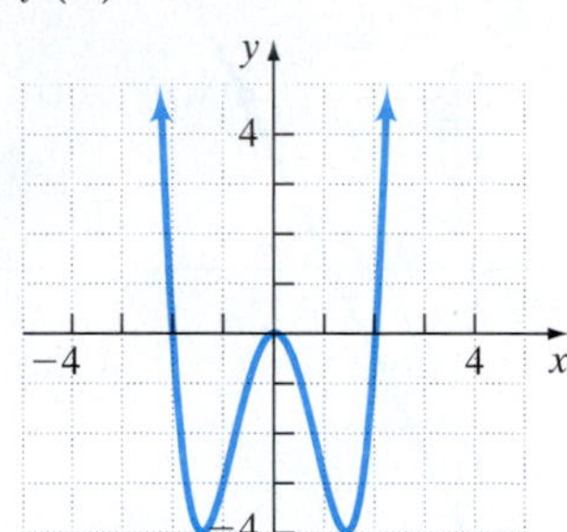


**6.** $f(x) = x^5 + 4x$

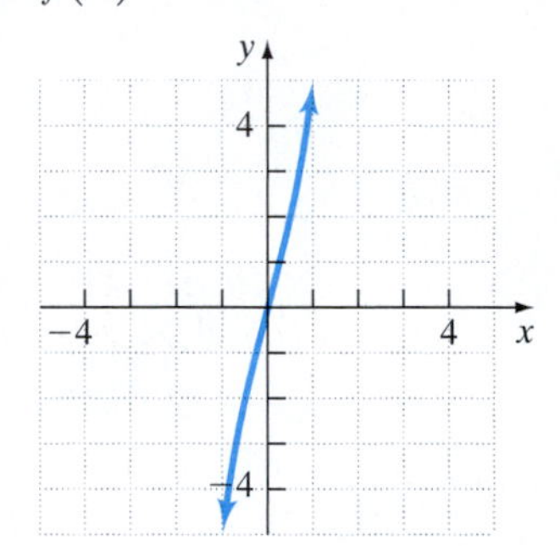


## 3 Determine the Inverse of a Function Defined by a Map or Ordered Pair

Now that we understand the concept of a one-to-one function, we can talk about *inverse functions*.

**DEFINITION**

If $f(x)$ is a one-to-one function with ordered pairs of the form $(a, b)$, then the **inverse** function, denoted $f^{-1}$, is the set of ordered pairs of the form $(b, a)$.

**Work Smart**

The symbol $f^{-1}$ is used to represent the inverse function of $f$. The $-1$ used in $f^{-1}$ is not an exponent. That is, $f^{-1}(x) \neq \dfrac{1}{f(x)}$.

We begin by finding inverses of functions represented by maps and sets of ordered pairs. **In order for the inverse of a function to also be a function, the function must be one-to-one.** To find the inverse of a function defined by a map or a set of ordered pairs, we interchange the inputs and outputs, so for each ordered pair $(a, b)$ that is defined in a function $f$, the ordered pair $(b, a)$ is defined in the inverse function $f^{-1}$. The inverse undoes what the function does. Suppose a function takes the input 11 and gives the output 3. This can be represented as $(3, 11)$. The inverse would take as input 11 and give the output 3, which can be represented as $(11, 3)$.

## EXAMPLE 5 Finding the Inverse of a Function Defined by a Map

Find the inverse of the following function. Let the domain of the function represent certain states and let the range represent the state's population in millions. State the domain and the range of the inverse function.

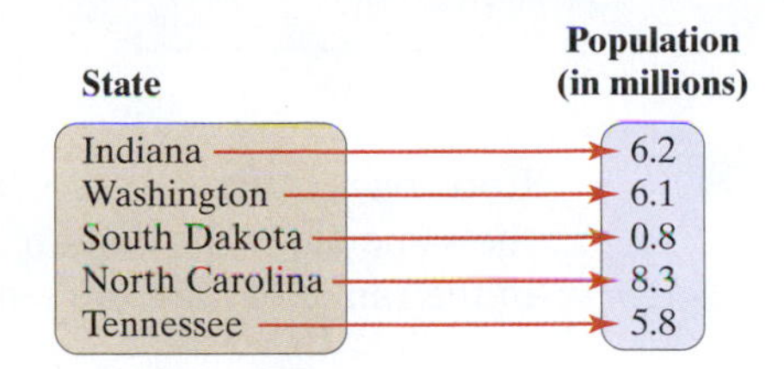


### Solution

The elements in the domain represent the inputs to the function. The elements in the range represent the outputs to the function. The function is one-to-one, so the inverse will be a function. To find the inverse function, we interchange the elements in the domain with the elements in the range. For example, the function receives as input Indiana and outputs 6.2. So, the inverse receives as input 6.2 and outputs Indiana. The inverse function is shown below.

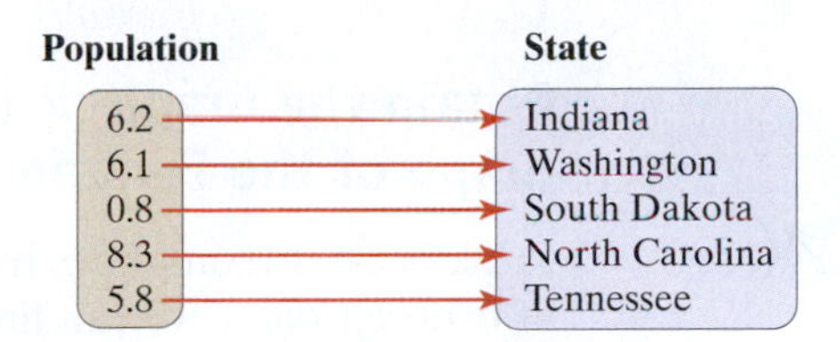


The domain of the inverse function is $\{6.2, 6.1, 0.8, 8.3, 5.8\}$. The range of the inverse function is {Indiana, Washington, South Dakota, North Carolina, Tennessee}.

**QUICK ✓**

7. Find the inverse of the one-to-one function shown to the right. Let the domain of the function represent the length of the right humerus and let the range represent the length of the right tibia of rats sent to space. State the domain and the range of the inverse function.

| Right Humerus | Right Tibia |
|---|---|
| 24.80 | 36.05 |
| 24.59 | 35.57 |
| 24.29 | 34.58 |
| 23.81 | 34.20 |
| 24.87 | 34.73 |

SOURCE: *NASA Life Sciences Data Archive*

## EXAMPLE 6 Finding the Inverse of a Function Defined by an Ordered Pair

Find the inverse of the following function. State the domain and the range of the inverse function.

$$\{(-2, 6), (-1, 3), (0, 2), (1, 5), (2, 8)\}$$

### Solution

The function is one-to-one, so the inverse will be a function. The inverse of a function defined by a set of ordered pairs is found by interchanging the entries in each ordered pair. So the inverse function is given by

$$\{(6, -2), (3, -1), (2, 0), (5, 1), (8, 2)\}$$

The domain of the inverse function is $\{6, 3, 2, 5, 8\}$. The range of the inverse function is $\{-2, -1, 0, 1, 2\}$.

**QUICK ✓**

**8.** Find the inverse of the following function. State the domain and the range of the inverse function.

$$\{(-3, 3), (-2, 2), (-1, 1), (0, 0), (1, -1)\}$$

---

Look back at Examples 5 and 6. Notice that the elements that are in the domain of $f$ are the same elements that are in the range of its inverse. In addition, the elements that are in the range of $f$ are the same elements that are in the domain of its inverse.

**RELATION BETWEEN THE DOMAIN AND RANGE OF A FUNCTION AND ITS INVERSE**

All elements in the domain of a function are also elements in the range of its inverse.

All elements in the range of a function are also elements in the domain of its inverse.

## 4 Obtain the Graph of the Inverse Function from the Graph of the Function

**Figure 7**

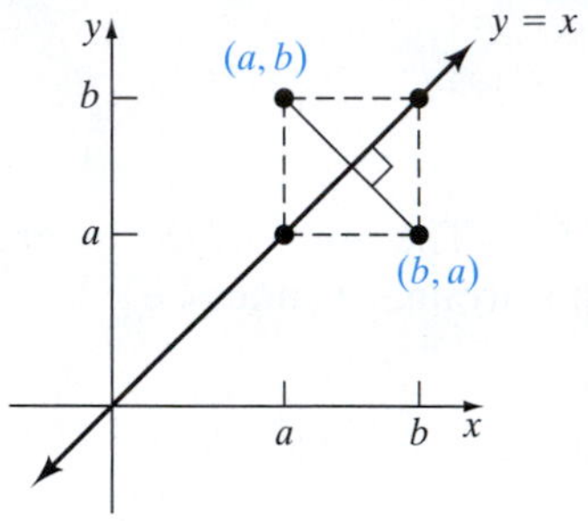

Look back at Example 6. In this example, we found that if a function is defined by a set of ordered pairs, we can find the inverse by interchanging the entries. So if $(a, b)$ is a point on the graph of a function, then $(b, a)$ is a point on the graph of the inverse function. See Figure 7. From the graph, it follows that the point $(b, a)$ on the graph of the inverse function is the reflection about the line $y = x$ of the point $(a, b)$.

**The graph of a function $f$ and the graph of its inverse are symmetric with respect to the line $y = x$.**

### EXAMPLE 7 Graphing the Inverse Function

Figure 8(a) shows the graph of a one-to-one function $y = f(x)$. Draw the graph of its inverse.

### Solution

We begin by adding the graph of $y = x$ to Figure 8(a). Since the points $(-4, -2)$, $(-3, -1)$, $(-1, 0)$, and $(4, 2)$ are on the graph of $f$, we know that the points $(-2, -4)$, $(-1, -3)$, $(0, -1)$, and $(2, 4)$ must be on the graph of the inverse of $f$. Using these points along with the fact that the graph of the inverse of $f$ is a reflection about the line $y = x$ of the graph of $f$, we draw the graph of the inverse. See Figure 8(b).

**Figure 8**

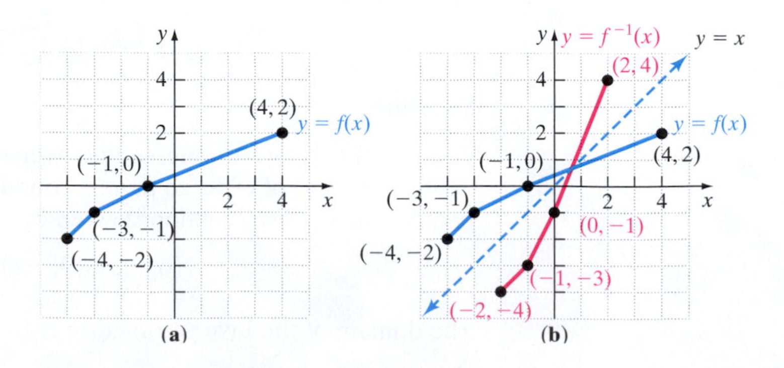

**QUICK ✓**

**9.** Below is the graph of a one-to-one function $y = f(x)$. Draw the graph of its inverse.

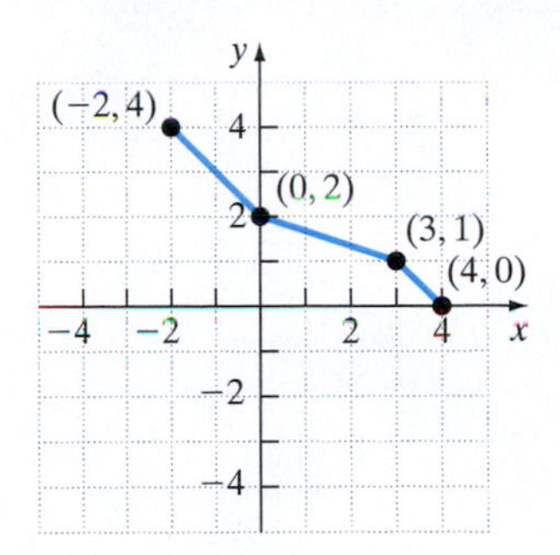


## 5 Find the Inverse of a Function Defined by an Equation

**Work Smart**

Suppose $f(x) = 2x$ so that each input gets doubled. The inverse function would be $f^{-1}(x) = x/2$ so that each input gets cut in half. So, if $x = 10$, then $f(10) = 20$. Now "plug" the output of $f$, 20, into $f^{-1}$, to get $f^{-1}(20) = 10$. Notice we are right back where we started!

We have learned how to find the inverse of a one-to-one function that is defined by a map, a set of ordered pairs, and a graph. We now discuss how to find the inverse of a function defined by an equation.

When a function is defined by an equation we use the notation $y = f(x)$. The notation $y = f^{-1}(x)$ is used to denote the equation whose rule is the inverse function of $f$.

Because the inverse of a function "undoes" what the original function does, we have the following relation between a function $f$ and its inverse, $f^{-1}$.

$$f^{-1}(f(x)) = x \text{ for every } x \text{ in the domain of } f$$
$$f(f^{-1}(x)) = x \text{ for every } x \text{ in the domain of } f^{-1}$$

Let's go over an example to illustrate how to find the inverse of a one-to-one function defined by an equation.

### EXAMPLE 8 How to Find the Inverse of a One-to-One Function

Find the inverse of $f(x) = 2x - 3$.

#### Step-by-Step Solution

Before we find the inverse function, we verify that the function is one-to-one. Because the graph of the function $f$ is a line with $y$-intercept $-3$ and slope 2, we know by the horizontal line test that the function is one-to-one and therefore has an inverse function.

**Step 1:** Replace $f(x)$ with $y$ in the equation for $f(x)$.

$$f(x) = 2x - 3$$
$$y = 2x - 3$$

**Step 2:** In $y = f(x)$, interchange the variables $x$ and $y$ to obtain $x = f(y)$.

$$x - 2y - 3$$

**Step 3:** Solve the equation found in Step 1 for $y$ in terms of $x$.

Add 3 to both sides: $x + 3 = 2y$

Divide both sides by 2: $\dfrac{x + 3}{2} = y$

(Continued)

**Step 4:** Replace $y$ with $f^{-1}(x)$.

$$f^{-1}(x) = \frac{x+3}{2}$$

**Step 5:** Verify your result by showing that

$f^{-1}(f(x)) = x$ and
$f(f^{-1}(x)) = x$.

$$\begin{aligned} f^{-1}(f(x)) &= f^{-1}(2x-3) \\ &= \frac{2x-3+3}{2} \\ &= \frac{2x}{2} \\ &= x \end{aligned}$$

$$\begin{aligned} f(f^{-1}(x)) &= f\left(\frac{x+3}{2}\right) \\ &= 2\left(\frac{x+3}{2}\right) - 3 \\ &= x + 3 - 3 \\ &= x \end{aligned}$$

Everything checks, so $f^{-1}(x) = \dfrac{x+3}{2}$.

Now, let's summarize the steps used in Example 8.

**In Words**

The function $f(x) = 2x - 3$ says we should take an input, double it, then subtract 3. The inverse function $f^{-1}(x) = \dfrac{x+3}{2}$ says we should take an input add 3 and then cut it in half.

### Steps for Finding the Inverse of a One-to-One Function Defined by an Equation

**Step 1:** Replace $f(x)$ with $y$ in the equation for $f(x)$.

**Step 2:** In $y = f(x)$, interchange the variables $x$ and $y$ to obtain $x = f(y)$.

**Step 3:** Solve the equation found in Step 2 for $y$ in terms of $x$.

**Step 4:** Replace $y$ with $f^{-1}(x)$.

**Step 5:** Verify your result by showing that $f^{-1}(f(x)) = x$ and $f(f^{-1}(x)) = x$.

Consider the function $f(x) = 2x - 3$ and its inverse $f^{-1}(x) = \dfrac{x+3}{2}$ from Example 8. Notice that $f(5) = 2(5) - 3 = 7$. What do you think $f^{-1}(7)$ will equal? Notice $f^{-1}(7) = \dfrac{7+3}{2} = \dfrac{10}{2} = 5$. So $f^{-1}$ "undoes" what $f$ did!

**QUICK ✓**

**10.** Find the inverse of $g(x) = 5x - 1$.

## EXAMPLE 9 Finding the Inverse of a One-to-One Function

Find the inverse of $h(x) = x^3 + 4$.

### Solution

$$h(x) = x^3 + 4$$

Replace $h(x)$ with $y$ in the equation for $h(x)$: $y = x^3 + 4$

Interchange the variables $x$ and $y$: $x = y^3 + 4$

Solve the equation for $y$ in terms of $x$: $x - 4 = y^3$

Take the cube root of both sides: $\sqrt[3]{x-4} = y$

Replace $y$ with $h^{-1}(x)$: $h^{-1}(x) = \sqrt[3]{x-4}$

**Check** Verify your result by showing that $h^{-1}(h(x)) = x$ and $h(h^{-1}(x)) = x$.

$$h^{-1}(h(x)) = h^{-1}(x^3 + 4) = \sqrt[3]{x^3 + 4 - 4} = \sqrt[3]{x^3} = x$$

$$h(h^{-1}(x)) = h(\sqrt[3]{x - 4}) = (\sqrt[3]{x - 4})^3 + 4 = x - 4 + 4 = x$$

So $h^{-1}(x) = \sqrt[3]{x - 4}$.

**QUICK ✓**

11. Find the inverse of $f(x) = x^5 + 3$.

# 9.1 Exercises

**For Extra Help:** Student Solutions Manual  CD Video  PH Math/Tutor Center  MathXL Tutorials on CD  MathXL® MyMathLab

## Concepts and Vocabulary

*In Problems 1–3, fill in the blanks.*

1. Given two functions $f$ and $g$, the _____ _____, denoted by $f \circ g$, is defined by $(f \circ g)(x) = f(g(x))$.
2. A function is _____ if any two different inputs in the domain correspond to two different outputs in the range. That is, if $x_1$ and $x_2$ are two different inputs of a function $f$, then $f(x_1) \neq f(x_2)$.
3. If $f$ is some function, the _____ of $f$, denoted $f^{-1}$, receives as input $f(x)$, manipulates it and outputs the value of $x$.

*In Problems 4–6, answer True or False to each statement.*

4. If $f$ is a one-to-one function so that its inverse is $f^{-1}$, then $f^{-1}(f(x)) = x$ for every $x$ in the domain of $f$ and $f(f^{-1}(x)) = x$ for every $x$ in the domain of $f^{-1}$.
5. The notation $f^{-1}(x)$ is equivalent to $\dfrac{1}{f(x)}$.
6. $(f \circ g)(x) = f(x) \cdot g(x)$
7. In your own words, explain what it means for a function to be one-to-one. Why must a function be one-to-one in order for its inverse to be a function?
8. In your own words, explain why domain of $f$ = range of $f^{-1}$ and range of $f$ = domain of $f^{-1}$.
9. State the horizontal line test. Why does it work?
10. If $f(g(x)) = (4x - 3)^5$ and $f(x) = x^5$, what is $g(x)$?

## Building Skills

*In Problems 11–18, for the given functions f and g, find*

**(a)** $(f \circ g)(3)$ **(b)** $(g \circ f)(-2)$ **(c)** $(f \circ f)(1)$ **(d)** $(g \circ g)(-4)$

11. $f(x) = 2x + 5$; $g(x) = x - 4$
12. $f(x) = 4x - 3$; $g(x) = x + 2$
13. $f(x) = x^2 + 4$; $g(x) = 2x + 3$
14. $f(x) = x^2 - 3$; $g(x) = 5x + 1$
15. $f(x) = 2x^3$; $g(x) = -2x^2 + 5$
16. $f(x) = -2x^3$; $g(x) = x^2 + 1$
17. $f(x) = |x - 10|$; $g(x) = \dfrac{12}{x + 3}$
18. $f(x) = \sqrt{x + 8}$; $g(x) = x^2 - 4$

*In Problems 19–30, for the given functions f and g, find*

**(a)** $(f \circ g)(x)$ **(b)** $(g \circ f)(x)$ **(c)** $(f \circ f)(x)$ **(d)** $(g \circ g)(x)$

**19.** $f(x) = x + 1; g(x) = 2x$

**20.** $f(x) = x - 3; g(x) = 4x$

**21.** $f(x) = 2x + 7; g(x) = -4x + 5$

**22.** $f(x) = 3x - 1; g(x) = -2x + 5$

**23.** $f(x) = x^2; g(x) = x - 3$

**24.** $f(x) = x^2 + 1; g(x) = x + 1$

**25.** $f(x) = \sqrt{x}; g(x) = x + 4$

**26.** $f(x) = \sqrt{x + 2}; g(x) = x - 2$

**27.** $f(x) = |x + 4|; g(x) = x^2 - 4$

**28.** $f(x) = |x - 3|; g(x) = x^3 + 3$

**29.** $f(x) = \dfrac{2}{x + 1}; g(x) = \dfrac{1}{x}$

**30.** $f(x) = \dfrac{2}{x - 1}; g(x) = \dfrac{4}{x}$

*In Problems 31–40, determine which of the following functions is one-to-one.*

**31.**

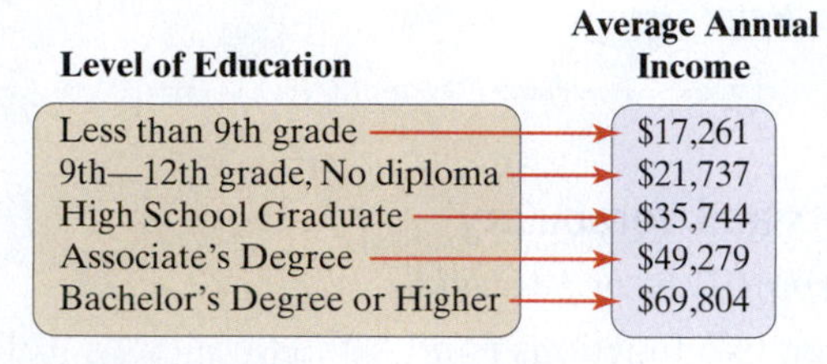


Source: *United States Census Bureau*

**32.**

| Newspaper | Daily Circulation |
|---|---|
| *USA Today* | 2,149,933 |
| *Wall Street Journal* | 1,780, 605 |
| *New York Times* | 1,109,371 |
| *Los Angeles Times* | 944,303 |
| *Washington Post* | 759,864 |

Source: *Information Please Almanac*, September 30, 2001.

**33.**

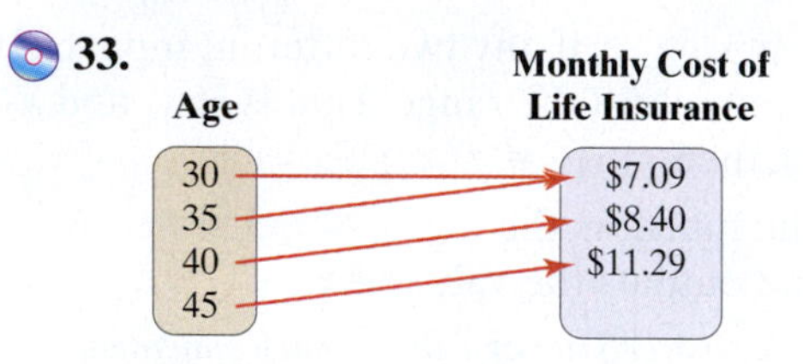


Source: *eterm.com*

**34.**

| State | Number of U.S. Representatives |
|---|---|
| Virginia | 11 |
| Nevada | 3 |
| New Mexico | 3 |
| Tennessee | 9 |
| Texas | 32 |

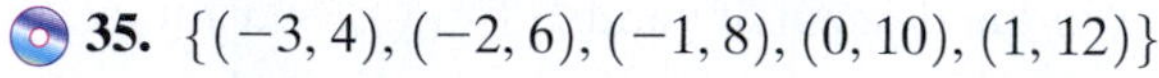

**35.** $\{(-3, 4), (-2, 6), (-1, 8), (0, 10), (1, 12)\}$

**36.** $\{(-2, 6), (-1, 3), (0, 0), (1, -3), (2, 6)\}$

**37.** $\{(-2, 4), (-1, 2), (0, 0), (1, 2), (2, 4)\}$

**38.** $\{(-2, -8), (-1, -1), (0, 0), (1, 1), (2, 8)\}$

**39.** $\{(0, -4), (-1, -1), (-2, 0), (1, 1), (2, 4)\}$

**40.** $\{(-3, 0), (-2, 3), (-1, 0), (0, -3)\}$

*In Problems 41–46, use the horizontal line test to determine whether the function whose graph is given is one-to-one.*

**41.**

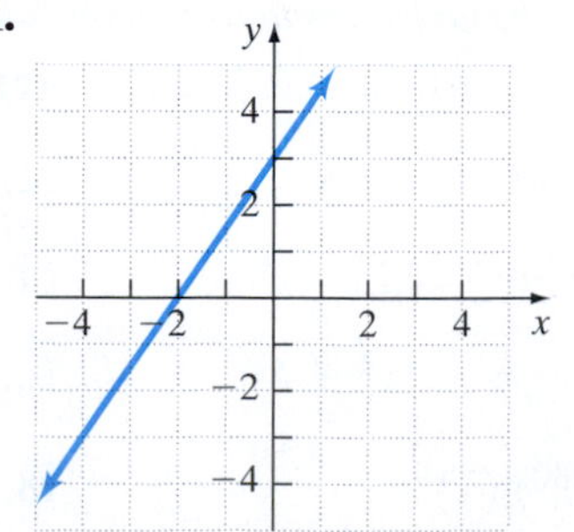


**42.**

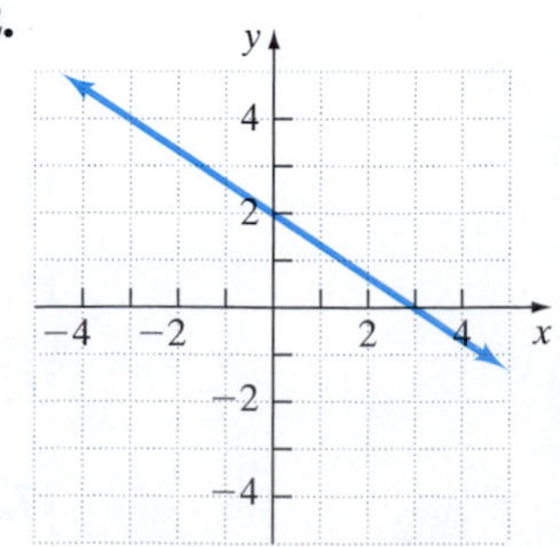

**43.**

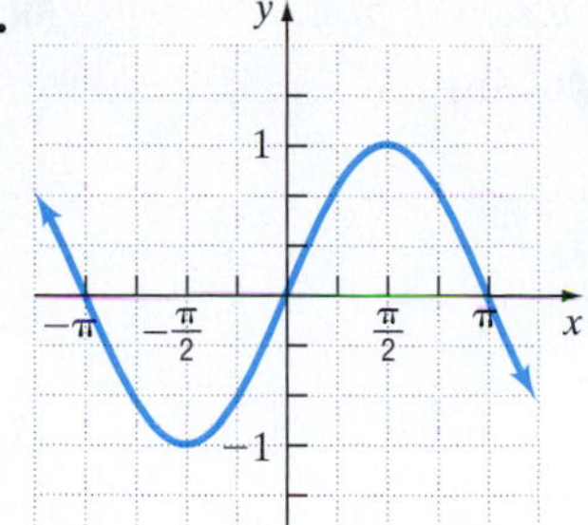


**44.**

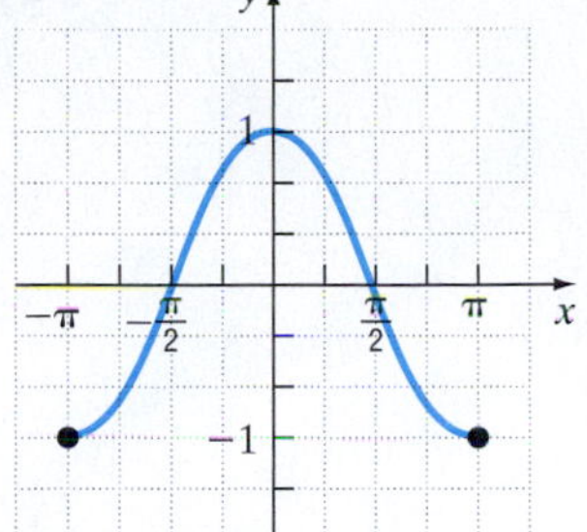


**45.**

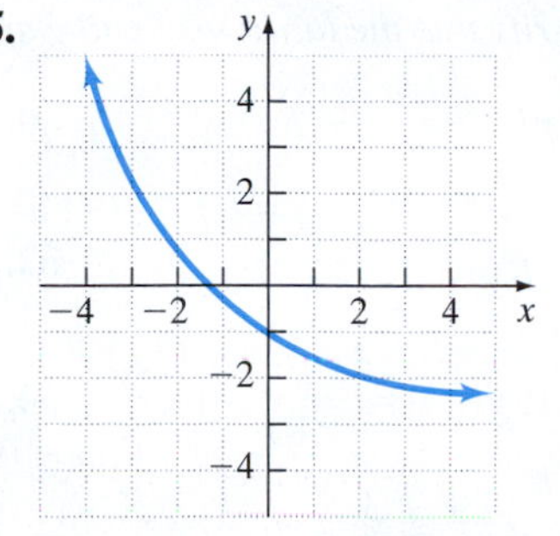


**46.**

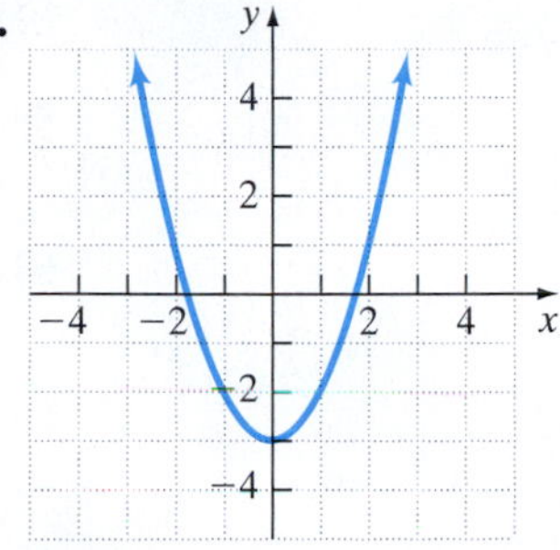


*In Problems 47–52, find the inverse of the following one-to-one functions.*

**47.**

| U.S. Coin | Weight (g) |
|---|---|
| Cent | 2.500 |
| Nickel | 5.000 |
| Dime | 2.268 |
| Quarter | 5.670 |
| Half Dollar | 11.340 |
| Dollar | 8.100 |

SOURCE: *U.S. Mint Web site*

**48.**

| Price ($) | Quantity Demanded |
|---|---|
| 2300 | 152 |
| 2000 | 159 |
| 1700 | 164 |
| 1500 | 171 |
| 1300 | 176 |

**49.** $\{(0, 3), (1, 4), (2, 5), (3, 6)\}$

**50.** $\{(-1, 4), (0, 1), (1, -2), (2, -5)\}$

**51.** $\{(-2, 3), (-2, 1), (-2, -3), (-2, 9)\}$

**52.** $\{(-10, 1), (-5, 4), (0, 3), (-5, 2)\}$

*In Problems 53–58, the graph of a one-to-one function f is given. Draw the graph of the inverse function $f^{-1}$.*

**53.**

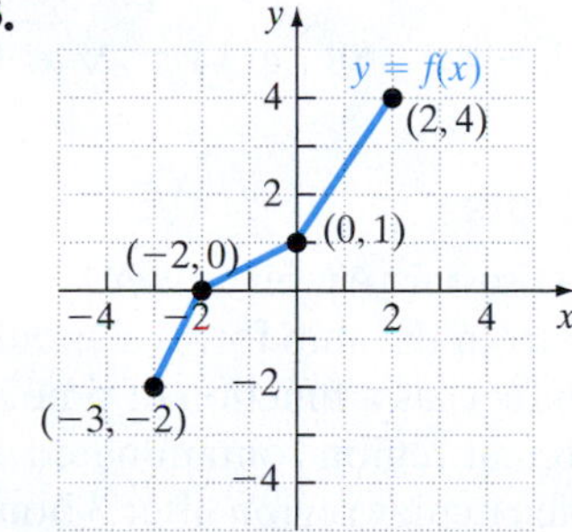


**54.**

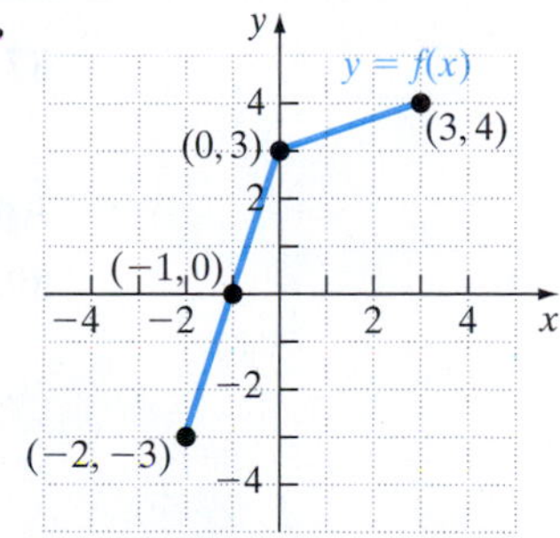


**55.**

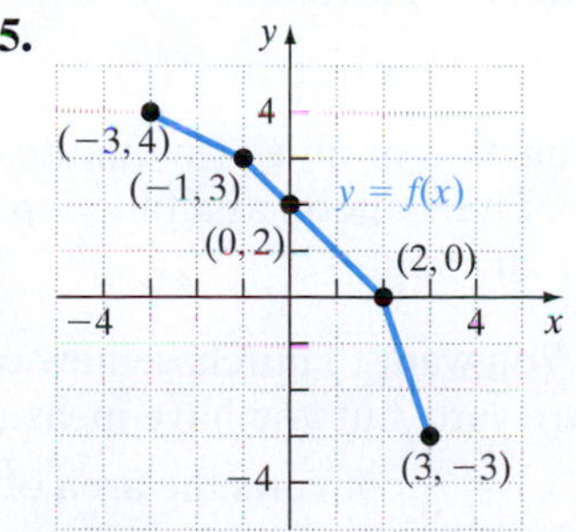


**56.**

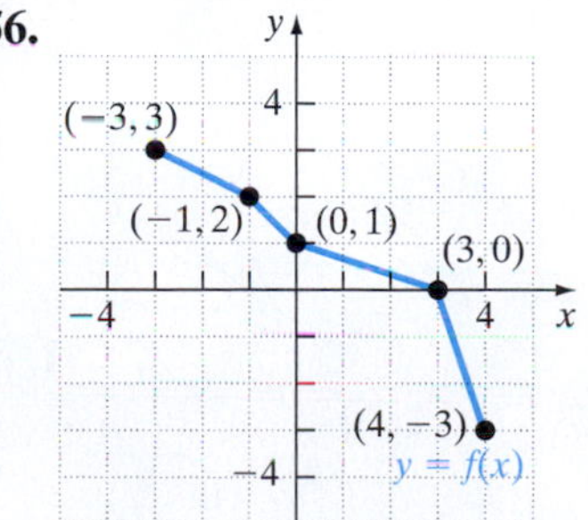

**57.**

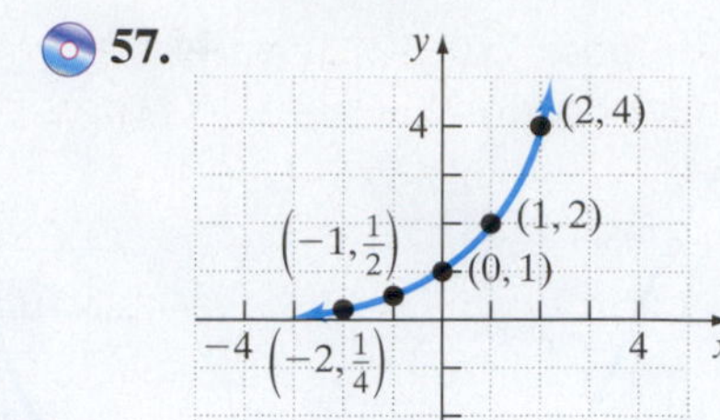


**58.**

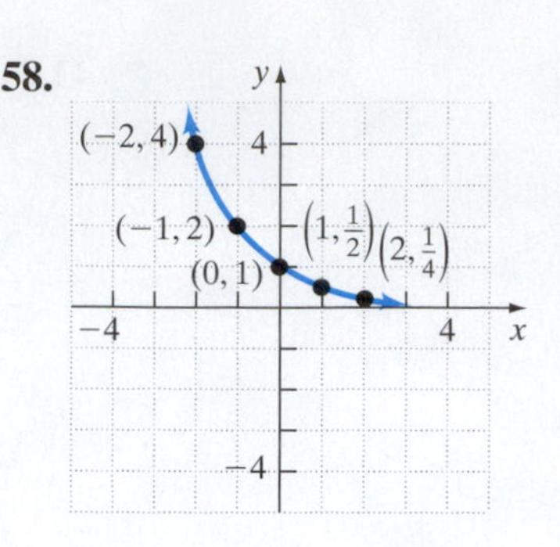


*In Problems 59–68, verify that the functions f and g are inverses of each other.*

**59.** $f(x) = x + 5; g(x) = x - 5$

**60.** $f(x) = 10x; g(x) = \frac{x}{10}$

**61.** $f(x) = 5x + 7; g(x) = \frac{x - 7}{5}$

**62.** $f(x) = 3x - 5; g(x) = \frac{x + 5}{3}$

**63.** $f(x) = 10x - 1; g(x) = \frac{x + 1}{10}$

**64.** $f(x) = 8x + 3; g(x) = \frac{x - 3}{8}$

**65.** $f(x) = \frac{3}{x - 1}; g(x) = \frac{3}{x} + 1$

**66.** $f(x) = \frac{2}{x + 4}; g(x) = \frac{2}{x} - 4$

**67.** $f(x) = \sqrt[3]{x + 4}; g(x) = x^3 - 4$

**68.** $f(x) = \sqrt[3]{2x + 1}; g(x) = \frac{x^3 - 1}{2}$

*In Problems 69–88, find the inverse function of the given one-to-one function.*

**69.** $f(x) = 6x$

**70.** $f(x) = 12x$

**71.** $f(x) = x + 4$

**72.** $g(x) = x + 6$

**73.** $h(x) = 2x - 7$

**74.** $H(x) = 3x + 8$

**75.** $G(x) = 2 - 5x$

**76.** $F(x) = 1 - 6x$

**77.** $g(x) = x^3 + 3$

**78.** $f(x) = x^3 - 2$

**79.** $p(x) = \frac{1}{x + 3}$

**80.** $P(x) = \frac{1}{x + 1}$

**81.** $F(x) = \frac{5}{2 - x}$

**82.** $G(x) = \frac{2}{3 - x}$

**83.** $f(x) = \sqrt[3]{x - 2}$

**84.** $f(x) = \sqrt[5]{x + 5}$

**85.** $R(x) = \frac{x}{x + 2}$

**86.** $R(x) = \frac{2x}{x + 4}$

**87.** $f(x) = \sqrt[3]{x - 1} + 4$

**88.** $g(x) = \sqrt[3]{x + 2} - 3$

## Applying the Concepts

**89. Environmental Disaster** An oil tanker hits a rock that rips a hole in the hull of the ship. Oil leaking from the ship forms a circular region around the ship. If the radius $r$ of the circle (in feet) as a function of time $t$ (in hours) is $r(t) = 20t$, express the area $A$ of the circular region contaminated with oil as a function of time. What will be the area of the circular region after 3 hours? (**Hint:** $A(r) = \pi r^2$)

**90. Volume of a Balloon** The volume $V$ of a hot-air balloon (in cubic meters) as a function of its radius $r$ is given by $V(r) = \frac{4}{3}\pi r^3$. If the radius $r$ of the balloon is increasing as a function of time $t$ (in minutes) according to $r(t) = 3\sqrt[3]{t}$, for $t \geq 0$, find the volume of the balloon as a function of time $t$. What will be the volume of the balloon after 30 minutes?

**91. Buying Carpet** You want to purchase new carpet for your family room. Carpet is sold by the square yard, but you have measured your family room in square feet. The function $A(x) = \frac{x}{9}$ converts the area of a room in square feet to an area in square yards. Suppose that the carpet you have selected is \$18 per square yard installed. Then the function $C(A) = 18A$ represents the cost $C$ of installing carpet in a room that is $A$ square yards.

**(a)** Find the cost $C$ as a function of the square footage of the room $x$.

**(b)** If your family room is 15 feet by 21 feet, what will it cost to install the carpet?

**92. Tax Time** You have a job that pays \$20 per hour. Your gross salary $G$ as a function of hours worked $h$ is given by $G(h) = 20h$. Federal tax withholding $T$ on your paycheck is equal to 18% of gross earnings $G$, so that federal tax withholding as a function of gross pay is given by the function $T(G) = 0.18G$.

**(a)** Find federal tax withholding $T$ as a function of hours worked $h$.

**(b)** Suppose that you worked 28 hours last week. What will be the federal tax withholding on your paycheck?

**93.** If $f(4) = 12$, what is $f^{-1}(12)$?

**94.** If $g(-2) = 7$, what is $g^{-1}(7)$?

**95.** The domain of a function $f$ is $[0, \infty)$, and its range is $[-5, \infty)$. State the domain and the range of $f^{-1}$.

**96.** The domain of a function $f$ is $[5, \infty)$, and its range is $[0, \infty)$. State the domain and the range of $f^{-1}$.

**97.** The domain of a function $g$ is $[-4, 10]$, and its range is $(-6, 12)$. State the domain and the range of $g^{-1}$.

**98.** The domain of a function $g$ is $[0, 15]$, and its range is $(0, 8)$. State the domain and the range of $g^{-1}$.

**99. Taxes** The function $T(x) = 0.15(x - 6000) + 600$ represents the tax bill $T$ of a single person whose adjusted gross income is $x$ dollars for income between \$6,000 and \$28,400, inclusive. *(Source:* Internal Revenue Service) Find the inverse function that expresses adjusted gross income $x$ as a function of taxes $T$. That is, find $x(T)$.

**100. Health Costs** The annual cost of health insurance $H$ as a function of age $a$ is given by the function $H(a) = 22.8a - 117.5$ for $15 \le a \le 90$. *(Source: Statistical Abstract,* 2002) Find the inverse function that expresses age $a$ as a function of health insurance cost $H$. That is, find $a(H)$.

## Extending the Concepts

**101.** If $f(x) = 2x^2 - x + 5$ and $g(x) = x + a$, find $a$ so that the $y$-intercept of the graph of $(f \circ g)(x)$ is 20.

**102.** If $f(x) = x^2 - 3x + 1$ and $g(x) = x - a$, find $a$ so that the $y$-intercept of the graph of $(f \circ g)(x)$ is $-1$.

## The Graphing Calculator

Graphing calculators have the ability to evaluate composite functions. To obtain the results of Example 1, we would let $Y_1 = f(x) = x^2 - 3$ and $Y_2 = g(x) = 2x + 1$. Figure 9 shows the results of Example 1 using a TI-84 Plus graphing calculator.

**Figure 9**

```
Y1(Y2(3))
                46
Y2   1(3))
                13
Y1(Y1(-2))
                -2
```

*In Problems 103–110, use a graphing calculator to evaluate the composite functions. Compare your answers with those found in Problems 11–18.*

**(a)** $(f \circ g)(3)$ **(b)** $(g \circ f)(-2)$ **(c)** $(f \circ f)(1)$ **(d)** $(g \circ g)(-4)$

**103.** $f(x) = 2x + 5;\ g(x) = x - 4$

**104.** $f(x) = 4x - 3;\ g(x) = x + 2$

**105.** $f(x) = x^2 + 4;\ g(x) = 2x + 3$

**106.** $f(x) = x^2 - 3;\ g(x) = 5x + 1$

**107.** $f(x) = 2x^3;\ g(x) = -2x^2 + 5$

**108.** $f(x) = -2x^3;\ g(x) = x^2 + 1$

**109.** $f(x) = |x - 10|;\ g(x) = \dfrac{12}{x + 3}$

**110.** $f(x) = \sqrt{x + 8};\ g(x) = x^2 - 4$

*In Problems 111–116, the functions f and g are inverses. Graph both functions on the same screen along with the line $y = x$ to see the symmetry of the functions about the line $y = x$.*

**111.** $f(x) = x + 5;\ g(x) = x - 5$

**112.** $f(x) = 10x;\ g(x) = \dfrac{x}{10}$

**113.** $f(x) = 5x + 7;\ g(x) = \dfrac{x - 7}{5}$

**114.** $f(x) = 3x - 5;\ g(x) = \dfrac{x + 5}{3}$

**115.** $f(x) = 10x - 1;\ g(x) = \dfrac{x + 1}{10}$

**116.** $f(x) = 8x + 3;\ g(x) = \dfrac{x - 3}{8}$

# 9.2 Exponential Functions

## OBJECTIVES

1. Evaluate Exponential Expressions
2. Graph Exponential Functions
3. Define the Number $e$
4. Solve Exponential Equations
5. Study Exponential Models That Describe Our World

### *Preparing for Exponential Functions*

*Before getting started, take the following readiness quiz. If you get a problem wrong, go back to the section cited and review the material.*

**1.** Evaluate: (a) $2^3$ (b) $2^{-1}$ (c) $3^4$ [Getting Ready, pp. 339–342]

**2.** Graph $f(x) = x^2$. [Section 8.4, p. 649]

**3.** State the definition of a rational number. [Section R.2, p. 12]

**4.** State the definition of an irrational number. [Section R.2, p. 13]

**5.** Write 3.20349193 as a decimal (a) rounded to 4 decimal places (b) truncated to 4 decimal places. [Section R.2, pp. 14–15]

**6.** Simplify: (a) $m^3 \cdot m^5$ (b) $\dfrac{a^7}{a^2}$ (c) $(z^3)^4$ [Getting Ready, pp. 339–344]

**7.** Solve: $x^2 - 5x = 14$ [Section 5.7, pp. 423–426]

***Preparing for...Answers***

**1. (a)** 8 **(b)** $\frac{1}{2}$ **(c)** 81

**2.**

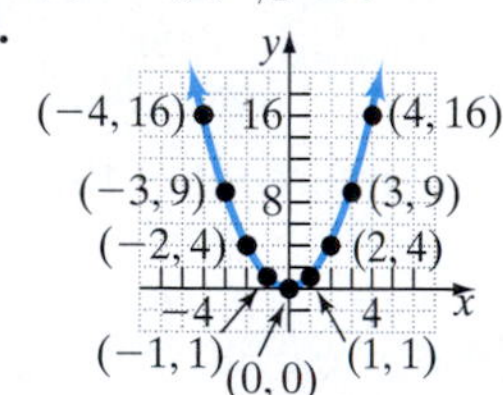


**3.** A rational number is a number that can be expressed as a quotient $\frac{p}{q}$ of two integers. The integer $p$ is called the numerator, and the integer $q$, which cannot be 0, is called the denominator. The set of rational numbers is the numbers $\mathbb{Q} = \left\{x \middle| x = \frac{p}{q}\text{, where } p, q \text{ are integers and } q \neq 0\right\}$.

**4.** An irrational number has a decimal representation that neither repeats nor terminates.

**5. (a)** 3.2035 **(b)** 3.2034

**6. (a)** $m^8$ **(b)** $a^5$ **(c)** $z^{12}$

**7.** $\{-2, 7\}$

Suppose that you have just been hired as a proofreader by Prentice Hall. Paul Murphy, Prentice Hall's editor, offers you two options for getting paid. Option A states that you will get paid \$100 for each error you find in the final page proofs of a text. Option B states that you will get \$2 for the first error you find in the final page proofs and your payment will double for each additional error you find. Based on your experience, you know there are typically about 15–20 errors in the final page proofs of a text. Which option will you go with?

If there is one error, Option A pays \$100, while Option B pays \$2. If there are two errors, Option A pays 2(\$100) = \$200, while Option B pays $\$2^2 = \$4$. If there are three errors, Option A pays 3(\$100) = \$300, while Option B pays $\$2^3 = \$8$. It's looking like Option A is the way to go. To complete the analysis, we set up Table 1, which lists the payment amount as a function of the number of errors in the page proof. Remember, in Option B, the payment amount doubles each time you find an error.

Holy cow! If you find 20 errors, you'll get paid over 1 million dollars! Paul Murphy better reconsider his offer! If we let $x$ represent the number of errors, we can express the salary for Option A as a linear function, $f(x) = 100x$; we can express the salary for Option B as an *exponential function*, $g(x) = 2^x$.

**Table 1**

| Number of Errors | Option A Payment | Option B Payment | Number of Errors | Option A Payment | Option B Payment |
|---|---|---|---|---|---|
| 0 | $0 | $0 | 11 | $1,100 | $2,048 |
| 1 | $100 | $2 | 12 | $1,200 | $4,096 |
| 2 | $200 | $4 | 13 | $1,300 | $8,192 |
| 3 | $300 | $8 | 14 | $1,400 | $16,384 |
| 4 | $400 | $16 | 15 | $1,500 | $32,768 |
| 5 | $500 | $32 | 16 | $1,600 | $65,536 |
| 6 | $600 | $64 | 17 | $1,700 | $131,072 |
| 7 | $700 | $128 | 18 | $1,800 | $262,144 |
| 8 | $800 | $256 | 19 | $1,900 | $524,288 |
| 9 | $900 | $512 | 20 | $2,000 | $1,048,576 |
| 10 | $1,000 | $1,024 | | | |

**DEFINITION**

An **exponential function** is a function of the form

$$f(x) = a^x$$

where $a$ is a positive real number ($a > 0$) and $a \neq 1$. The domain of the exponential function is the set of all real numbers.

We will address the restrictions on the base $a$ shortly. The key point to understand with exponential functions is that the independent variable is in the exponent of the expression. Contrast this idea with polynomial functions (such as $f(x) = x^2 - 4x$ or $g(x) = 2x^3 + x^2 - 5$), where the independent variable is the base of the expression.

## 1 Evaluate Exponential Expressions

In Section 7.1 we gave a definition for raising a real number $a$ to a rational power. From that discussion, we gave meaning to expressions of the form

$$a^{m/n}$$

where the base $a$ is a positive real number and the exponent $m/n$ is a rational number.

However, our world does not only consist of rational numbers, so a logical question to ask is, "What if I want to raise the base $a$ to any real number, rational or irrational?" Although the definition for raising a positive real number to any real number requires advanced mathematics, we can provide a discussion that is reasonable and intuitive.

Suppose that we wanted to determine the value of $3^{\sqrt{2}}$. Using our calculator, we know that $\sqrt{2} \approx 1.414213562$, so it should seem reasonable that we can approximate $3^{\sqrt{2}}$ as $3^{1.4}$, where the 1.4 comes from truncating the decimals to the right of the 4 in the tenths position. A better approximation of $3^{\sqrt{2}}$ would be $3^{1.4142}$, where the digits to the right of the ten-thousandths position have been truncated. The idea is this—the more decimals used in the approximation of an irrational number, the better the approximation of $3^{\sqrt{2}}$.

Fortunately, most calculators can easily evaluate expressions such as $3^{1.4142}$ using the $\boxed{x^y}$ key or a caret $\boxed{\wedge}$ key. To evaluate expressions of the form $a^x$ using a scientific calculator, enter the base $a$, then press the $\boxed{x^y}$ key, enter the exponent $x$, and press $\boxed{=}$.

To evaluate expressions of the form $a^x$ using a graphing calculator, enter the base $a$, then press the caret $\boxed{\wedge}$ key, enter the exponent $x$, and press $\boxed{\text{ENTER}}$.

### EXAMPLE 1 Evaluating Exponential Expressions

Using a calculator, evaluate each of the following expressions. Write as many decimals as your calculator allows.

**(a)** $3^{1.4}$ **(b)** $3^{1.41}$ **(c)** $3^{1.414}$ **(d)** $3^{1.4142}$ **(e)** $3^{\sqrt{2}}$

**Solution**

**(a)** $3^{1.4} \approx 4.655536722$ **(b)** $3^{1.41} \approx 4.706965002$

**(c)** $3^{1.414} \approx 4.727695035$ **(d)** $3^{1.4142} \approx 4.72873393$

**(e)** $3^{\sqrt{2}} \approx 4.728804388$

**QUICK ✓** *Using a calculator, evaluate each of the following expressions. Write as many decimals as your calculator allows.*

**1. (a)** $2^{1.7}$ **(b)** $2^{1.73}$ **(c)** $2^{1.732}$ **(d)** $2^{1.7321}$ **(e)** $2^{\sqrt{3}}$

---

Based on the results of Example 1, it should be clear that we can approximate the value of an exponential expression at any real number. This is why the domain of exponential functions is the set of all real numbers.

When we gave the definition of an exponential function, we excluded the possibility for the base $a$ to equal 1 and we stated that the base $a$ must be positive. We exclude the base $a = 1$ because this function is the constant function $f(x) = 1^x = 1$. We also exclude bases that are negative because we would run into problems for exponents such as $\frac{1}{2}$ or $\frac{3}{4}$. For example, suppose $f(x) = (-2)^x$. In the real number system, we could not evaluate $f\left(\frac{1}{2}\right)$ because $f\left(\frac{1}{2}\right) = (-2)^{1/2} = \sqrt{-2}$, which is not a real number. Finally, we exclude a base of 0, because this function is $f(x) = 0^x$ which equals zero for $x > 0$ and is undefined when $x < 0$. When $x = 0$, $f(x) = 0^x$ is *indeterminate* because its value is not precisely determined.

## 2 Graph Exponential Functions

We use the point-plotting method to learn properties of exponential functions from their graphs.

### EXAMPLE 2 Graphing an Exponential Function

Graph the exponential function $f(x) = 2^x$ using point plotting. From the graph, state the domain and the range of the function.

**Solution**

We begin by locating some points on the graph of $f(x) = 2^x$ as shown in Table 2. We plot the points in Table 2 and connect them in a smooth curve. Figure 10 shows the graph of $f(x) = 2^x$.

The domain of any exponential function is the set of all real numbers. Notice there is no $x$ such that $2^x = 0$ and there is no $x$ such that $2^x$ is negative. Based on this and the graph, we conclude that the range of $f(x) = 2^x$ is the set of all positive real numbers or $\{y | y > 0\}$ or $(0, \infty)$ using interval notation.

**Table 2**

| $x$ | $f(x) = 2^x$ | $(x, f(x))$ |
|---|---|---|
| $-3$ | $f(-3) = 2^{-3} = \frac{1}{2^3} = \frac{1}{8}$ | $\left(-3, \frac{1}{8}\right)$ |
| $-2$ | $f(-2) = 2^{-2} = \frac{1}{2^2} = \frac{1}{4}$ | $\left(-2, \frac{1}{4}\right)$ |
| $-1$ | $f(-1) = 2^{-1} = \frac{1}{2^1} = \frac{1}{2}$ | $\left(-1, \frac{1}{2}\right)$ |
| 0 | $f(0) = 2^0 = 1$ | (0, 1) |
| 1 | $f(1) = 2^1 = 2$ | (1, 2) |
| 2 | $f(2) = 2^2 = 4$ | (2, 4) |
| 3 | $f(3) = 2^3 = 8$ | (3, 8) |

**Figure 10**

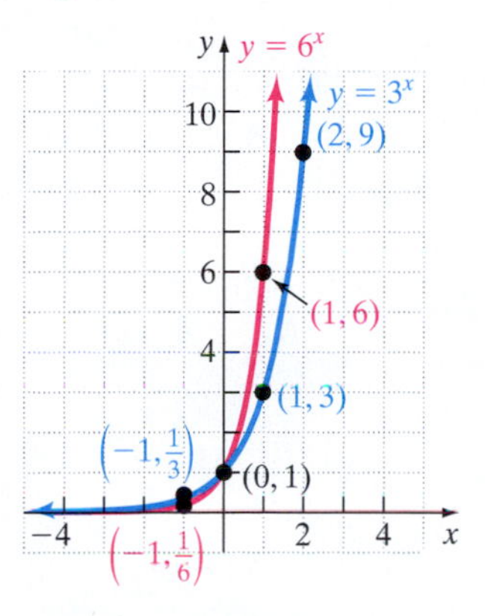


**Figure 11**

The graph of $f(x) = 2^x$ in Figure 10 is typical of all exponential functions that have a base larger than 1. Figure 11 shows the graph of two other exponential functions whose bases are larger than 1. Notice that the larger the base, the steeper the graph is for $x > 0$ and the closer the graph is to the $x$-axis for $x < 0$.

The following display summarizes the information that we have about $f(x) = a^x$ where the base is greater than $1 (a > 1)$.

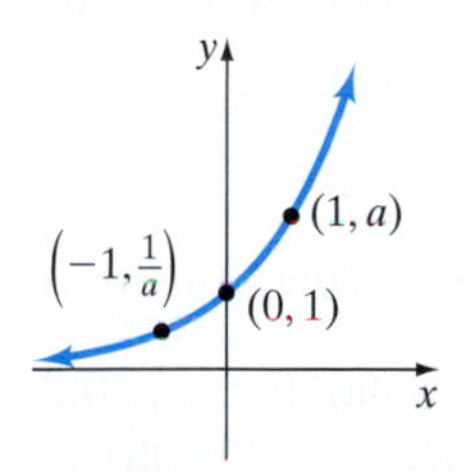


**Figure 12**
$f(x) = a^x, a > 1$

**PROPERTIES OF THE GRAPH OF AN EXPONENTIAL FUNCTION $f(x) = a^x, a > 1$**

1. The domain is the set of all real numbers. The range is the set of all positive real numbers.
2. There are no $x$-intercepts; the $y$-intercept is 1.
3. The graph of $f$ contains the points $\left(-1, \frac{1}{a}\right)$, $(0, 1)$, and $(1, a)$.

See Figure 12.

**QUICK ✓**

2. Graph the exponential function $f(x) = 4^x$ using point plotting. From the graph, state the domain and the range of the function.

Now we consider $f(x) = a^x, 0 < a < 1$.

## EXAMPLE 3 Graphing an Exponential Function

Graph the exponential function $f(x) = \left(\frac{1}{2}\right)^x$ using point plotting. From the graph, state the domain and the range of the function.

### Solution

We begin by locating some points on the graph of $f(x) = \left(\frac{1}{2}\right)^x$ as shown in Table 3 on page 716. We plot the points in Table 3 and connect them in a smooth curve. Figure 13 shows the graph of $f(x) = \left(\frac{1}{2}\right)^x$.

**Table 3**

| $x$ | $f(x) = \left(\frac{1}{2}\right)^x$ | $(x, f(x))$ |
|---|---|---|
| $-3$ | $f(-3) = \left(\frac{1}{2}\right)^{-3} = 2^3 = 8$ | $(-3, 8)$ |
| $-2$ | $f(-2) = \left(\frac{1}{2}\right)^{-2} = 2^2 = 4$ | $(-2, 4)$ |
| $-1$ | $f(-1) = \left(\frac{1}{2}\right)^{-1} = 2^1 = 2$ | $(-1, 2)$ |
| $0$ | $f(0) = \left(\frac{1}{2}\right)^{0} = 1$ | $(0, 1)$ |
| $1$ | $f(1) = \left(\frac{1}{2}\right)^{1} = \frac{1}{2}$ | $\left(1, \frac{1}{2}\right)$ |
| $2$ | $f(2) = \left(\frac{1}{2}\right)^{2} = \frac{1}{4}$ | $\left(2, \frac{1}{4}\right)$ |
| $3$ | $f(3) = \left(\frac{1}{2}\right)^{3} = \frac{1}{8}$ | $\left(3, \frac{1}{8}\right)$ |

Figure 13

Figure 14

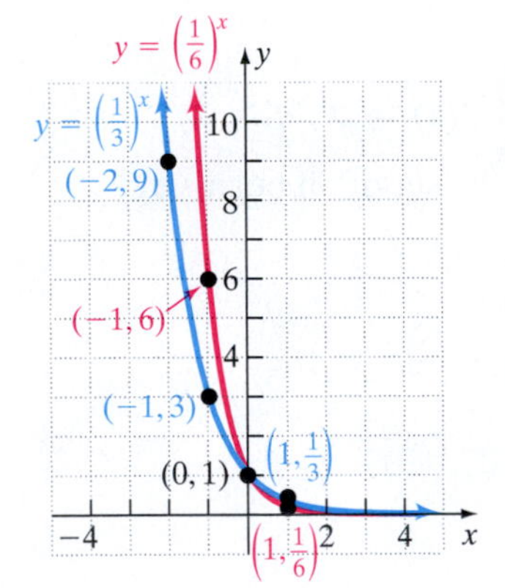


The domain of any exponential function is the set of all real numbers. From the graph, we conclude that the range of $f(x) = \left(\frac{1}{2}\right)^x$ is the set of all positive real numbers or $\{y \mid y > 0\}$ or $(0, \infty)$ using interval notation.

The graph of $f(x) = \left(\frac{1}{2}\right)^x$ in Figure 13 is typical of all exponential functions that have a base between 0 and 1. Figure 14 shows the graph of two additional exponential functions whose bases are between 0 and 1. Notice that the smaller the base, the closer the graph is to the $x$-axis for $x > 0$ and the steeper the graph is for $x < 0$.

The following display summarizes the information that we have about $f(x) = a^x$ where the base is between 0 and 1 $(0 < a < 1)$.

Figure 15

$f(x) = a^x$, $0 < a < 1$

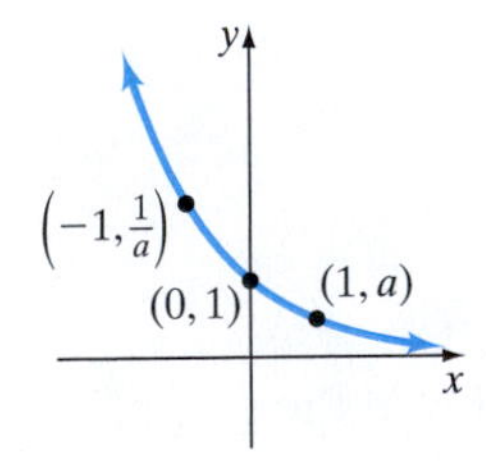


**PROPERTIES OF THE GRAPH OF AN EXPONENTIAL FUNCTION $f(x) = a^x$, $0 < a < 1$**

1. The domain is the set of all real numbers. The range is the set of all positive real numbers.
2. There are no $x$-intercepts; the $y$-intercept is 1.
3. The graph of $f$ contains the points $\left(-1, \frac{1}{a}\right)$, $(0, 1)$, and, $(1, a)$. See Figure 15.

**QUICK ✓**

3. Graph the exponential function $f(x) = \left(\frac{1}{4}\right)^x$ using point plotting. From the graph, state the domain and the range of the function.

## EXAMPLE 4 Graphing an Exponential Function

Use point plotting to graph $f(x) = 3^{x+1}$. From the graph, state the domain and the range of the function.

### Solution

We choose values of $x$ and use the equation that defines the function to find the value of the function. See Table 4. We then plot the ordered pairs and connect them in a smooth curve. See Figure 16.

**Table 4**

| $x$ | $f(x)$ | $(x, f(x))$ |
|---|---|---|
| $-3$ | $f(-3) = 3^{-3+1} = 3^{-2} = \frac{1}{3^2} = \frac{1}{9}$ | $\left(-3, \frac{1}{9}\right)$ |
| $-2$ | $f(-2) = 3^{-2+1} = 3^{-1} = \frac{1}{3^1} = \frac{1}{3}$ | $\left(-2, \frac{1}{3}\right)$ |
| $-1$ | $f(-1) = 3^{-1+1} = 3^0 = 1$ | $(-1, 1)$ |
| $0$ | $f(0) = 3^{0+1} = 3^1 = 3$ | $(0, 3)$ |
| $1$ | $f(1) = 3^{1+1} = 3^2 = 9$ | $(1, 9)$ |

**Figure 16**

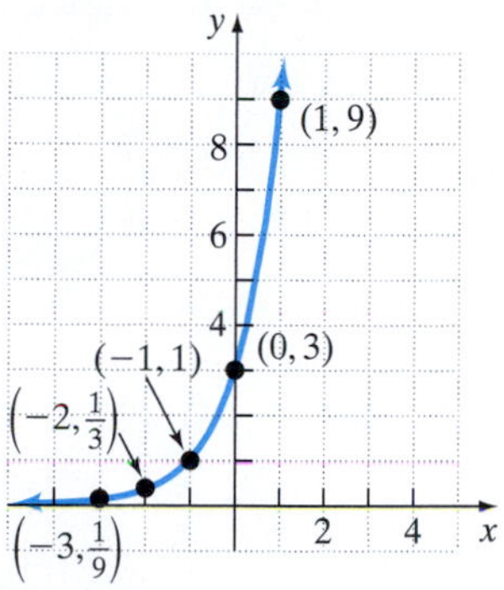

The domain is the set of all real numbers. The range is $\{y \mid y > 0\}$, or using interval notation, $(0, \infty)$.

**QUICK ✓** *Graph each function using point plotting. From the graph, state the domain and the range of each function.*

**4.** $f(x) = 2^{x-1}$

**5.** $f(x) = 3^x + 1$

## 3 Define the Number $e$

Many problems that occur in nature require the use of an exponential function whose base is a certain irrational number, symbolized by the letter $e$. The number $e$ can be used to model the growth of a stock's price or it can be used in a model to estimate the time of death of a carbon-based life form.

**DEFINITION**

The **number** $\boldsymbol{e}$ is defined as the number that the expression

$$\left(1 + \frac{1}{n}\right)^n$$

approaches as $n$ becomes unbounded in the positive direction (that is, as $n$ gets bigger).

Table 5 on page 718 illustrates what happens to the value of $\left(1 + \frac{1}{n}\right)^n$ as $n$ takes on larger and larger values. The last number in the last column in Table 5 is the number $e$ correct to nine decimal places.

**Table 5**

| $n$ | $\frac{1}{n}$ | $1 + \frac{1}{n}$ | $\left(1 + \frac{1}{n}\right)^n$ |
|---|---|---|---|
| 1 | 1 | 2 | 2 |
| 2 | 0.5 | 1.5 | 2.25 |
| 5 | 0.2 | 1.2 | 2.48832 |
| 10 | 0.1 | 1.1 | 2.59374246 |
| 100 | 0.01 | 1.01 | 2.704813829 |
| 1,000 | 0.001 | 1.001 | 2.716923932 |
| 10,000 | 0.0001 | 1.0001 | 2.718145927 |
| 100,000 | 0.00001 | 1.00001 | 2.718268237 |
| 1,000,000 | 0.000001 | 1.000001 | 2.718280469 |
| 1,000,000,000 | $10^{-9}$ | $1 + 10^{-9}$ | 2.718281827 |

The exponential function $f(x) = e^x$, whose base is the number $e$, occurs so often in applications that it is sometimes referred to as *the* exponential function. Most calculators have the key $\boxed{e^x}$ or $\boxed{\exp(x)}$, which may be used to evaluate the exponential function $f(x) = e^x$ for a given value of $x$.

Use your calculator to approximate the values of $f(x) = e^x$ for $x = -1, 0,$ and $1$ as we have done to create Table 6. The graph of the exponential function is shown in Figure 17(a). Since $2 < e < 3$, the graph of $f(x) = e^x$ lies between the graph of $y = 2^x$ and $y = 3^x$. See Figure 17(b).

**Table 6**

| $x$ | $f(x) = e^x$ |
|---|---|
| −2 | $e^{-2} \approx 0.135$ |
| −1 | $e^{-1} \approx 0.368$ |
| 0 | $e^0 = 1$ |
| 1 | $e^1 \approx 2.718$ |
| 2 | $e^2 \approx 7.389$ |

**Figure 17**
$f(x) = e^x$

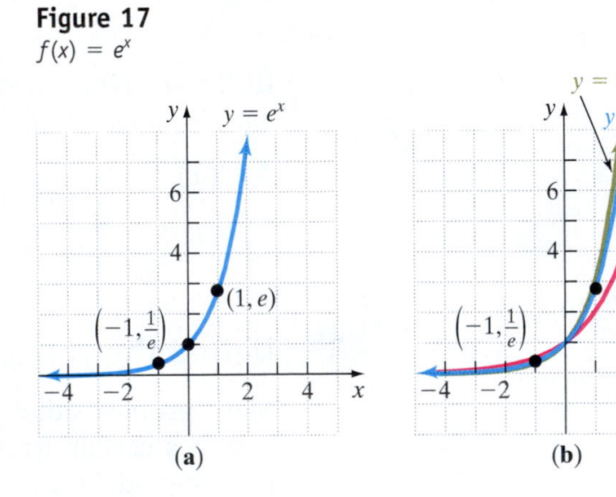

## Solve Exponential Equations

Equations that involve terms of the form $a^x$, $a > 0$, $a \neq 1$, are called **exponential equations.** Sometimes we can solve exponential equations using the Laws of Exponents and the following property.

**Work Smart**
To use the Property for Solving Exponential Equations, both sides of the equation must have the same base.

**PROPERTY FOR SOLVING EXPONENTIAL EQUATIONS**

$$\text{If} \quad a^u = a^v, \quad \text{then} \quad u = v.$$

This results from the fact that exponential functions are one-to-one. This property of exponential functions basically states that any output is the result of one (and only one) input. That is to say, two different inputs cannot yield the same output. For example, $y = x^2$ is not one-to-one because two different inputs, −2 and 2, correspond to the same output, 4.

To use the Property for Solving Exponential Equations, each side of the equality must be written with the same base.

## EXAMPLE 5 How to Solve an Exponential Equation

Solve: $2^{x-3} = 32$

### Step-by-Step Solution

**Step 1:** Use the Laws of Exponents to write both sides of the equation with the same base.

$$2^{x-3} = 32$$

$32 = 2^5$: $$2^{x-3} = 2^5$$

**Step 2:** Set the exponents on each side of the equation equal to each other.

$$x - 3 = 5$$

**Step 3:** Solve the equation resulting from Step 2.

Add 3 to both sides: $$x - 3 + 3 = 5 + 3$$
$$x = 8$$

**Step 4:** Verify your solution(s).

$$2^{x-3} = 32$$

Let $x = 8$: $$2^{8-3} \stackrel{?}{=} 32$$
$$2^5 \stackrel{?}{=} 32$$
$$32 = 32 \quad \text{True}$$

The solution set is $\{8\}$.

Below is a summary of the steps we used in Example 5.

**Work Smart**

Did you notice that we used properties of algebra to reduce the exponential equation to a linear equation?

### Steps for Solving Exponential Equations of the Form $a^u = a^v$

**Step 1:** Use the Laws of Exponents to write both sides of the equation with the same base.

**Step 2:** Set the exponents on each side of the equation equal to each other.

**Step 3:** Solve the equation resulting from Step 2.

**Step 4:** Verify your solution(s).

**QUICK ✓** *Solve each equation.*

**6.** $5^{x-4} = 5^{-1}$

**7.** $3^{x+2} = 81$

## EXAMPLE 6 Solving an Exponential Equation

Solve the following equations:

**(a)** $4^{x^2} = 32$

**(b)** $\dfrac{e^{x^2}}{e^{2x}} = e^8$

## Solution

**(a)**

$$4^{x^2} = 32$$

Rewrite exponential expressions with a common base: $(2^2)^{x^2} = 2^5$

Use $(a^m)^n = a^{m \cdot n}$: $2^{2x^2} = 2^5$

Set the exponents on each side of the equation equal to each other: $2x^2 = 5$

Divide both sides by 5: $x^2 = \dfrac{5}{2}$

Take the square root of both sides: $x = \pm\sqrt{\dfrac{5}{2}}$

Rationalize the denominator: $x = \pm\dfrac{\sqrt{10}}{2}$

**Check**

$x = \dfrac{-\sqrt{10}}{2}$:

$$4^{(-\sqrt{10}/2)^2} \stackrel{?}{=} 32$$
$$4^{10/4} \stackrel{?}{=} 32$$
$$4^{5/2} \stackrel{?}{=} 32$$

$a^{m/n} = \sqrt[n]{a^m}$: $\sqrt{4^5} \stackrel{?}{=} 32$

$$2^5 \stackrel{?}{=} 32$$
$$32 = 32 \quad \text{True}$$

$x = \dfrac{\sqrt{10}}{2}$:

$$4^{(\sqrt{10}/2)^2} \stackrel{?}{=} 32$$
$$4^{10/4} \stackrel{?}{=} 32$$
$$4^{5/2} \stackrel{?}{=} 32$$
$$\sqrt{4^5} \stackrel{?}{=} 32$$
$$2^5 \stackrel{?}{=} 32$$
$$32 = 32 \quad \text{True}$$

The solution set is $\left\{-\dfrac{\sqrt{10}}{2}, \dfrac{\sqrt{10}}{2}\right\}$.

**(b)**

$$\frac{e^{x^2}}{e^{2x}} = e^8$$

Use $\dfrac{a^m}{a^n} = a^{m-n}$: $e^{x^2-2x} = e^8$

Set the exponents on each side of the equation equal to each other: $x^2 - 2x = 8$

Subtract 8 from both sides: $x^2 - 2x - 8 = 0$

Factor: $(x - 4)(x + 2) = 0$

Zero-product property: $x = 4$ or $x = -2$

**Check**

$x = 4$:

$$\frac{e^{4^2}}{e^{2(4)}} \stackrel{?}{=} e^8$$
$$\frac{e^{16}}{e^8} \stackrel{?}{=} e^8$$
$$e^8 = e^8$$

$x = -2$:

$$\frac{e^{(-2)^2}}{e^{2(-2)}} \stackrel{?}{=} e^8$$
$$\frac{e^4}{e^{-4}} \stackrel{?}{=} e^8$$
$$e^8 = e^8$$

The solution set is $\{-2, 4\}$.

**QUICK ✓** *Solve each equation.*

**8.** $e^{x^2} = e^x \cdot e^{4x}$ **9.** $\frac{2^{x^2}}{8} = 2^{2x}$

## 5 Study Exponential Models That Describe Our World

Exponential functions are used in many different disciplines such as biology (half-life), chemistry (carbon-dating), economics (time value of money), and psychology (learning curves). Exponential functions are also used in statistics, as shown in the next example.

### EXAMPLE 7 Exponential Probability

From experience, the manager of a crisis helpline knows that between the hours of 3:00 A.M. and 5:00 A.M., calls occur at the rate of 3 calls per hour (0.05 calls per minute). The following formula from statistics can be used to determine the likelihood that a call will occur within $t$ minutes of 3 A.M.

$$F(t) = 1 - e^{-0.05t}$$

**(a)** Determine the likelihood that a person will call within 5 minutes of 3:00 A.M.

**(b)** Determine the likelihood that a person will call within 20 minutes of 3:00 A.M.

**Solution**

**(a)** The likelihood that a call will occur within 5 minutes of 3:00 A.M. is found by evaluating the function $F(t) = 1 - e^{-0.05t}$ at $t = 5$.

$$F(5) = 1 - e^{-0.05(5)}$$
$$\approx 0.221$$

The likelihood that a call will occur within 5 minutes of 3:00 A.M. is 0.221 = 22.1%.

**(b)** The likelihood that a call will occur within 20 minutes of 3:00 A.M. is found by evaluating the function $F(t) = 1 - e^{-0.05t}$ at $t = 20$.

$$F(20) = 1 - e^{-0.05(20)}$$
$$\approx 0.632$$

The likelihood that a call will occur within 20 minutes of 3:00 A.M. is 0.632 = 63.2%.

**QUICK ✓**

**10.** From experience, the manager of a bank knows that between the hours of 3:00 P.M. and 5:00 P.M., people arrive at the rate of 15 people per hour (0.25 people per minute). The following formula from statistics can be used to determine the likelihood that a person will arrive within $t$ minutes of 3:00 P.M.

$$F(t) = 1 - e^{-0.25t}$$

**(a)** Determine the likelihood that a person will arrive within 10 minutes of 3:00 P.M.

**(b)** Determine the likelihood that a person will arrive within 25 minutes of 3:00 P.M.

### EXAMPLE 8 Radioactive Decay

The radioactive **half-life** for a given radioisotope of an element is the time for half the radioactive nuclei in any sample to decay to some other substance. For example, the half-life of plutonium-239 is 24,360 years. Plutonium-239 is particularly dangerous because it emits alpha particles that are absorbed into bone marrow. The maximum amount of plutonium-239 that an adult can handle without significant injury is 0.13 micrograms (= 0.000000013 grams). Suppose that a researcher possesses a 1-gram sample of Plutonium-239. The amount $A$ (in grams) of Plutonium-239 after $t$ years is given by

$$A(t) = 1 \cdot \left(\frac{1}{2}\right)^{t/24,360}$$

**(a)** How much plutonium-239 is left in the sample after 500 years?

**(b)** How much plutonium-239 is left in the sample after 24,360 years?

**(c)** How much plutonium-239 is left in the sample after 73,080 years?

#### Solution

**(a)** The amount of plutonium-239 left in the sample after 500 years is found by evaluating $A$ at $t = 500$. That is, we determine $A(500)$.

$$A(500) = 1 \cdot \left(\frac{1}{2}\right)^{500/24,360}$$

Use a calculator: $\approx 0.986$ gram

After 500 years, there will be approximately 0.986 gram of Plutonium-239 left in the sample.

**(b)** The amount of plutonium-239 left in the sample after 24,360 years is found by evaluating $A$ at $t = 24{,}360$. That is, we determine $A(24{,}360)$.

$$\begin{aligned} A(24{,}360) &= 1 \cdot \left(\frac{1}{2}\right)^{24,360/24,360} \\ &= 1 \cdot \left(\frac{1}{2}\right)^{1} \\ &= 0.5 \text{ gram} \end{aligned}$$

After 24,360 years, there will be 0.5 gram of plutonium-239 left in the sample.

**(c)** The amount of plutonium-239 left in the sample after 73,080 years is found by evaluating $A$ at $t = 73{,}080$. That is, we determine $A(73{,}080)$.

$$\begin{aligned} A(73{,}080) &= 1 \cdot \left(\frac{1}{2}\right)^{73,080/24,360} \\ &= 1 \cdot \left(\frac{1}{2}\right)^{3} \\ &= \frac{1}{8} \text{ gram} \end{aligned}$$

After 24,360 years, there will be $\frac{1}{8} = 0.125$ gram of plutonium-239 left in the sample.

**QUICK ✓**

**11.** The half-life of thorium-227 is 18.72 days. Suppose that a researcher possesses a 10-gram sample of thorium-227. The amount $A$ (in grams) of thorium-227 after $t$ days is given by

$$A(t) = 10 \cdot \left(\frac{1}{2}\right)^{t/18.72}$$

**(a)** How much thorium-227 is left in the sample after 10 days?

**(b)** How much thorium-227 is left in the sample after 18.72 days?

**(c)** How much thorium-227 is left in the sample after 74.88 days?

**(d)** How much thorium-227 is left in the sample after 100 days?

---

When we deposit money in a bank, the bank pays us interest on the balance in the account. When working with problems involving interest, we use the term **payment period** as shown in Table 7.

**Table 7**

| Payment Period | Number of Times Interest Is Paid |
|---|---|
| Annually | Once per year |
| Semiannually | Twice per year |
| Quarterly | 4 times per year |
| Monthly | 12 times per year |
| Daily | 360 times per year |

When the interest due at the end of a payment period is added to the principal so that the interest computed at the end of the next payment period is based on this new principal amount (old principal + interest), the interest is said to have been **compounded. Compound interest** is interest paid on previously earned interest.

The following formula can be used to determine the value of an account after a certain period of time.

**Work Smart**

When using the compound interest formula, be sure to express the interest as a decimal.

**COMPOUND INTEREST FORMULA**

The amount $A$ after $t$ years due to a principal $P$ invested at an annual interest rate $r$ compounded $n$ times per year is

$$A = P\left(1 + \frac{r}{n}\right)^{nt}$$

For example, if you deposit \$500 into an account paying 3% annual interest compounded monthly, then $P = \$500$, $r = 0.03$, and $n = 12$ (twelve compounding periods per year).

**EXAMPLE 9** **Future Value of Money**

Suppose that you deposit \$3000 into a Roth IRA today. Determine future value $A$ of the deposit if it earns 8% interest compounded quarterly after

**(a)** 1 year **(b)** 10 years **(c)** 35 years, when you plan on retiring

**Solution**

We use the compound interest formula with $P = \$3000$, $r = 0.08$, and $n = 4$, so that

$$A = \$3000\left(1 + \frac{0.08}{4}\right)^{4t}$$
$$= \$3000(1 + 0.02)^{4t}$$
$$= \$3000(1.02)^{4t}$$

**(a)** The value of the account after $t = 1$ year is

$$A = \$3000(1.02)^{4(1)}$$
$$= \$3000(1.02)^{4}$$

Use a calculator: $= \$3000(1.08243216)$

$$= \$3247.30$$

**(b)** The value of the account after $t = 10$ years is

$$A = \$3000(1.02)^{4(10)}$$
$$= \$3000(1.02)^{40}$$

Use a calculator: $= \$3000(2.208039664)$

$$= \$6{,}624.12$$

**(c)** The value of the account after $t = 35$ years is

$$A = \$3000(1.02)^{4(35)}$$
$$= \$3000(1.02)^{140}$$

Use a calculator: $= \$3000(15.99646598)$

$$= \$47{,}989.40$$

**QUICK ✓**

**12.** Suppose that you deposit \$2000 into a Roth IRA today. Determine the future value $A$ of the deposit if it earns 12% interest compounded monthly (12 times per year) after

**(a)** 1 year **(b)** 15 years **(c)** 30 years, when you plan on retiring

# 9.2 Exercises

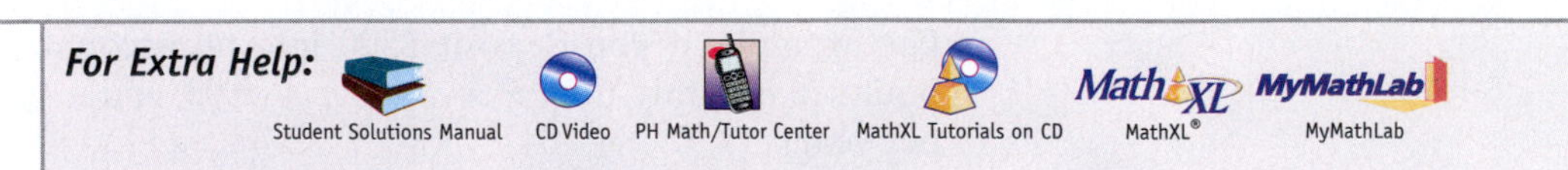


## Concepts and Vocabulary

*In Problems 1–3, fill in the blanks.*

**1.** An exponential function is a function of the form $f(x) = a^x$ where $a$ _____ 0 and $a$ _____ 1.

**2.** The graph of every exponential function $f(x) = a^x$ passes through three points: _____, _____, and _____.

**3.** If $a^u = a^v$, then _____ = _____.

*In Problems 4–6, answer True or False to each statement.*

**4.** The domain of the exponential function $f(x) = a^x$, $a > 0$, $a \neq 1$, is the set of all real numbers.

**5.** The range of the exponential function $f(x) = a^x$, $a > 0$, $a \neq 1$, is the set of all real numbers.

**6.** The number $e$, rounded to five decimal places, is 2.71828.

**7.** As the base $a$ of an exponential function $f(x) = a^x$ increases (for $a > 1$), what happens to the graph of the exponential function for $x > 0$? What happens to the behavior of the graph for $x < 0$?

**8.** The graphs of $f(x) = 2^{-x}$ and $g(x) = \left(\frac{1}{2}\right)^x$ are identical. Why?

**9.** Explain the difference between exponential functions and polynomial functions.

**10.** Can we solve the equation $2^x = 12$ using the fact that if $a^u = a^v$, then $u = v$. Why or why not?

## Building Skills

*In Problems 11–22, approximate each number using a calculator. Express your answer rounded to three decimal places.*

**11.** **(a)** $3^{2.2}$ **(b)** $3^{2.23}$ **(c)** $3^{2.236}$ **(d)** $3^{2.2361}$ **(e)** $3^{\sqrt{5}}$

**12.** **(a)** $5^{1.4}$ **(b)** $5^{1.41}$ **(c)** $5^{1.414}$ **(d)** $5^{1.4142}$ **(e)** $5^{\sqrt{2}}$

**13.** **(a)** $4^{3.1}$ **(b)** $4^{3.14}$ **(c)** $4^{3.142}$ **(d)** $4^{3.1416}$ **(e)** $4^{\pi}$

**14.** **(a)** $10^{2.7}$ **(b)** $10^{2.72}$ **(c)** $10^{2.718}$ **(d)** $10^{2.7183}$ **(e)** $10^{e}$

**15.** **(a)** $3.1^{2.7}$ **(b)** $3.14^{2.72}$ **(c)** $3.142^{2.718}$ **(d)** $3.1416^{2.7183}$ **(e)** $\pi^{e}$

**16.** **(a)** $2.7^{3.1}$ **(b)** $2.72^{3.14}$ **(c)** $2.718^{3.142}$ **(d)** $2.7183^{3.1416}$ **(e)** $e^{\pi}$

**17.** $e^2$ **18.** $e^3$ **19.** $e^{-2}$ **20.** $e^{-3}$ **21.** $e^{2.3}$ **22.** $e^{1.5}$

*In Problems 23–30, the graph of an exponential function is given. Match each graph to one of the following functions. It may prove useful to create a table of values for each function to assist in identifying the correct graph.*

**(a)** $f(x) = 2^x$ **(b)** $f(x) = 2^{-x}$ **(c)** $f(x) = 2^{x+1}$ **(d)** $f(x) = 2^{x-1}$

**(e)** $f(x) = -2^x$ **(f)** $f(x) = 2^x + 1$ **(g)** $f(x) = 2^x - 1$ **(h)** $f(x) = -2^{-x}$

**23.**

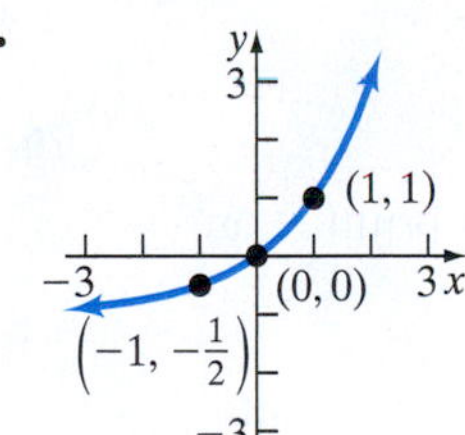


**24.**

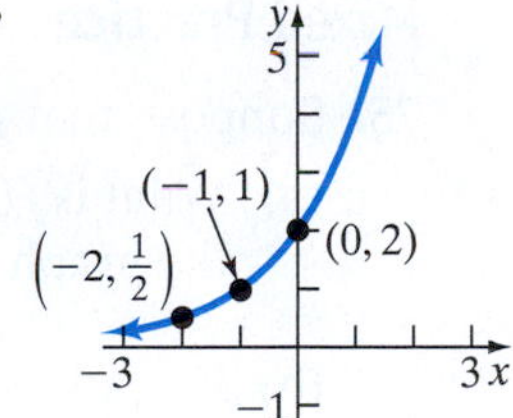


**25.**

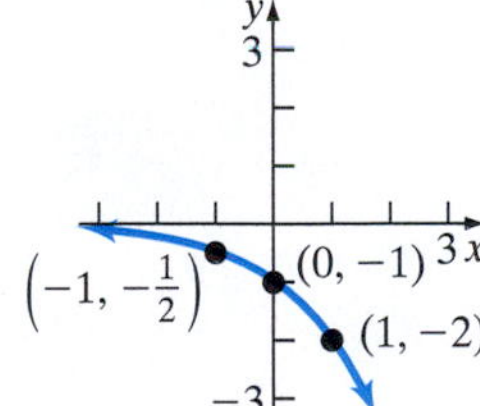


**26.**

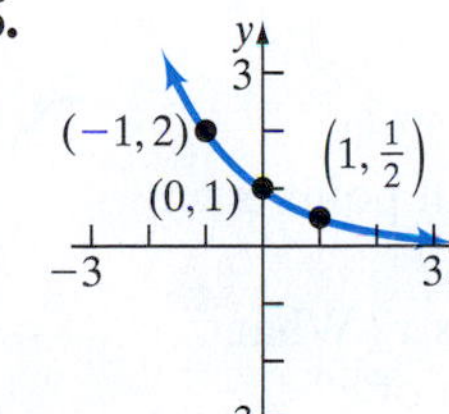


**27.**

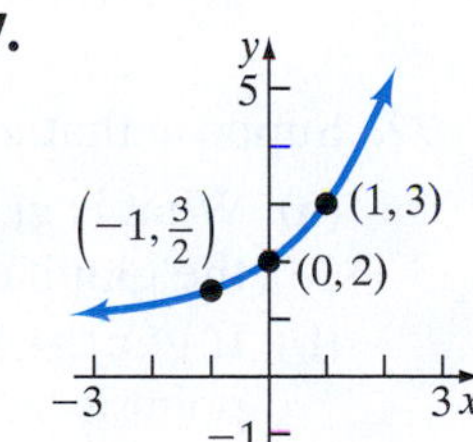


**28.**

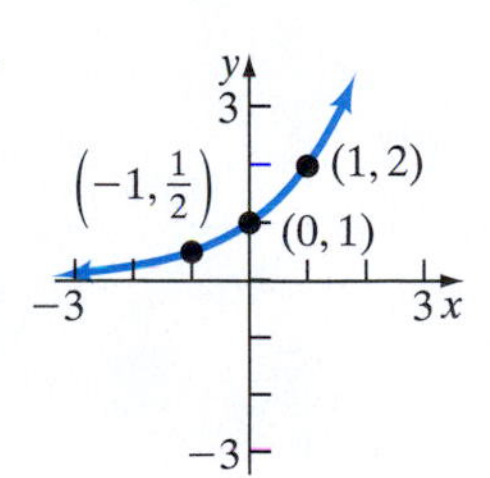


**29.**

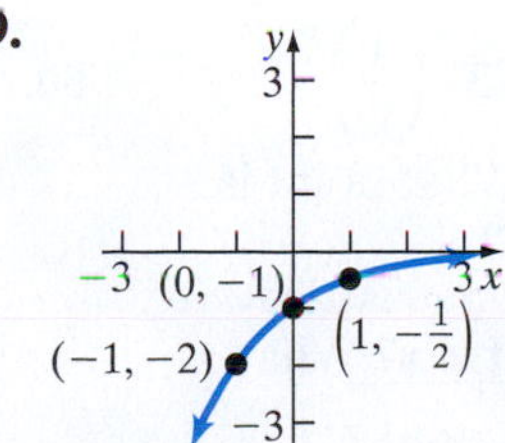


**30.**

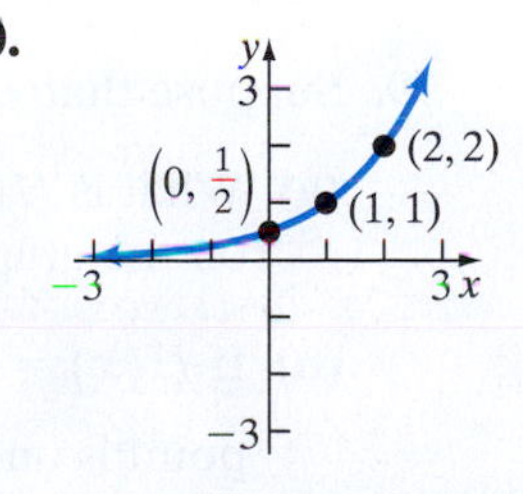

*In Problems 31–48, graph each function. State the domain and the range of the function.*

**31.** $f(x) = 5^x$

**32.** $f(x) = 7^x$

**33.** $g(x) = 10^x$

**34.** $G(x) = 8^x$

**35.** $F(x) = \left(\frac{1}{5}\right)^x$

**36.** $F(x) = \left(\frac{1}{7}\right)^x$

**37.** $G(x) = \left(\frac{1}{10}\right)^x$

**38.** $g(x) = \left(\frac{1}{8}\right)^x$

**39.** $h(x) = 2^{x+2}$

**40.** $H(x) = 2^{x-2}$

**41.** $f(x) = 2^x + 3$

**42.** $F(x) = 2^x - 3$

**43.** $F(x) = \left(\frac{1}{2}\right)^x - 1$

**44.** $G(x) = \left(\frac{1}{2}\right)^x + 2$

**45.** $P(x) = \left(\frac{1}{3}\right)^{x-2}$

**46.** $p(x) = \left(\frac{1}{3}\right)^{x+2}$

**47.** $g(x) = e^{x-1}$

**48.** $f(x) = e^x - 1$

*In Problems 49–74, solve each equation.*

**49.** $2^x = 2^5$

**50.** $3^x = 3^{-2}$

**51.** $3^{-x} = 81$

**52.** $4^{-x} = 64$

**53.** $\left(\frac{1}{2}\right)^x = \frac{1}{32}$

**54.** $\left(\frac{1}{3}\right)^x = \frac{1}{243}$

**55.** $5^{x-2} = 125$

**56.** $2^{x+3} = 128$

**57.** $4^x = 8$

**58.** $9^x = 27$

**59.** $2^{-x+5} = 16^x$

**60.** $3^{-x+4} = 27^x$

**61.** $3^{x^2-4} = 27^x$

**62.** $5^{x^2-10} = 125^x$

**63.** $4^x \cdot 2^{x^2} = 16^2$

**64.** $9^{2x} \cdot 27^{x^2} = 3^{-1}$

**65.** $2^x \cdot 8 = 4^{x-3}$

**66.** $3^x \cdot 9 = 27^x$

**67.** $\left(\frac{1}{5}\right)^x - 25 = 0$

**68.** $\left(\frac{1}{6}\right)^x - 36 = 0$

**69.** $(2^x)^x = 16$

**70.** $(3^x)^x = 81$

**71.** $e^x = e^{3x+4}$

**72.** $e^{3x} = e^2$

**73.** $(e^x)^2 = e^{3x-2}$

**74.** $(e^3)^x = e^2 \cdot e^x$

## Mixed Practice

**75.** Suppose that $f(x) = 2^x$.
- **(a)** What is $f(3)$? What point is on the graph of $f$?
- **(b)** If $f(x) = \frac{1}{8}$, what is $x$? What point is on the graph of $f$?

**76.** Suppose that $f(x) = 3^x$.
- **(a)** What is $f(2)$? What point is on the graph of $f$?
- **(b)** If $f(x) = \frac{1}{81}$, what is $x$? What point is on the graph of $f$?

**77.** Suppose that $g(x) = 4^x - 1$.
- **(a)** What is $g(-1)$? What point is on the graph of $g$?
- **(b)** If $g(x) = 15$, what is $x$? What point is on the graph of $g$?

**78.** Suppose that $g(x) = 5^x + 1$.
- **(a)** What is $g(-1)$? What point is on the graph of $g$?
- **(b)** If $g(x) = 126$, what is $x$? What point is on the graph of $g$?

**79.** Suppose that $H(x) = 3 \cdot \left(\frac{1}{2}\right)^x$.
- **(a)** What is $H(-3)$? What point is on the graph of $H$?
- **(b)** If $H(x) = \frac{3}{4}$, what is $x$? What point is on the graph of $H$?

**80.** Suppose that $F(x) = -2 \cdot \left(\frac{1}{3}\right)^x$.
- **(a)** What is $F(-1)$? What point is on the graph of $F$?
- **(b)** If $F(x) = -18$, what is $x$? What point is on the graph of $F$?

## Applying the Concepts

**81. A Population Model** According to the U.S. Census Bureau, the population of the United States in 2005 was 296 million people. In addition, the population of the United States was growing at a rate of 1.1% per year. Assuming that this growth rate continues, the model $P(t) = 296(1.011)^{t-2005}$ represents the population $P$ (in millions of people) in year $t$.

**(a)** According to this model, what will be the population of the United States in 2008?
**(b)** According to this model, what will be the population of the United States in 2042?
**(c)** The United States Census Bureau predicts that the United States population will be 400 million in 2042. Compare this estimate to the one obtained in part (b). What might account for any differences?

**82. A Population Model** According to the *Statistical Abstract of the United States,* the population of the world in 2005 was 6,448 million people. In addition, the population of the world was growing at a rate of 1.26% per year. Assuming that this growth rate continues, the model $P(t) = 6{,}448(1.0126)^{t-2005}$ represents the population $P$ (in millions of people) in year $t$.

**(a)** According to this model, what will be the population of the world in 2008?
**(b)** According to this model, what will be the population of the world in 2027?
**(c)** The United States Census Bureau predicts that the world population will be 8,000 million (8 billion) in 2027. Compare this estimate to the one obtained in part (b). What might account for any differences?

**83. Time Is Money** Suppose that you deposit $5000 into a Certificate of Deposit (CD) today. Determine the future value $A$ of the deposit if it earns 6% interest compounded monthly after

**(a)** 1 year. **(b)** 3 years. **(c)** 5 years, when the CD comes due.

**84. Time Is Money** Suppose that you deposit $8000 into a Certificate of Deposit (CD) today. Determine the future value $A$ of the deposit if it earns 4% interest compounded quarterly (4 times per year) after

**(a)** 1 year. **(b)** 3 years. **(c)** 5 years, when the CD comes due.

**85. Do the Compounding Periods Matter?** Suppose that you deposit $2000 into an account that pays 3% annual interest. How much will you have after 5 years if interest is compounded

**(a)** annually? **(b)** quarterly? **(c)** monthly? **(d)** daily?
**(e)** Based on the results of parts (a)–(d), what impact does the number of compounding periods have on the future value, all other things equal?

**86. Do the Compounding Periods Matter?** Suppose that you deposit $1000 into an account that pays 6% annual interest. How much will you have after 3 years if interest is compounded

**(a)** annually? **(b)** quarterly? **(c)** monthly? **(d)** daily?
**(e)** Based on the results of parts (a)–(d), what impact does the number of compounding periods have on the future value, all other things equal?

**87. Depreciation** Based on data obtained from the *Kelley Blue Book,* the value $V$ of a Dodge Neon that is $t$ years old can be modeled by $V(t) = 14{,}512(0.82)^t$.

**(a)** According to the model, what is the value of a brand-new Dodge Neon?
**(b)** According to the model, what is the value of a 2-year-old Dodge Neon?
**(c)** According to the model, what is the value of a 5-year-old Dodge Neon?

**88. Depreciation** Based on data obtained from the *Kelley Blue Book,* the value $V$ of a Dodge Stratus that is $t$ years old can be modeled by $V(t) = 19{,}282(0.84)^t$.

**(a)** According to the model, what is the value of a brand-new Dodge Stratus?
**(b)** According to the model, what is the value of a 2-year-old Dodge Stratus?
**(c)** According to the model, what is the value of a 5-year-old Dodge Stratus?

**89. Radioactive Decay** The half-life of beryllium-11 is 13.81 seconds. Suppose that a researcher possesses a 100-gram sample of beryllium-11. The amount $A$ (in grams) of beryllium-11 after $t$ seconds is given by

$$A(t) = 100 \cdot \left(\frac{1}{2}\right)^{t/13.81}$$

**(a)** How much beryllium-11 is left in the sample after 1 second?
**(b)** How much beryllium-11 is left in the sample after 13.81 seconds?
**(c)** How much beryllium-11 is left in the sample after 27.62 seconds?
**(d)** How much beryllium-11 is left in the sample after 100 seconds?

**90. Radioactive Decay** The half-life of carbon-10 is 19.255 seconds. Suppose that a researcher possesses a 200-gram sample of carbon-10. The amount $A$ (in grams) of carbon-10 after $t$ seconds is given by

$$A(t) = 100 \cdot \left(\frac{1}{2}\right)^{t/19.255}$$

**(a)** How much carbon-10 is left in the sample after 1 second?
**(b)** How much carbon-10 is left in the sample after 19.255 seconds?
**(c)** How much carbon-10 is left in the sample after 38.51 seconds?
**(d)** How much carbon-10 is left in the sample after 100 seconds?

**91. Newton's Law of Cooling** Newton's Law of Cooling states that the temperature of a heated object decreases exponentially over time toward the temperature of the surrounding medium. Suppose that a pizza is removed from a 400°F oven and placed in a room whose temperature is 70°F. The temperature $u$ (in °F) of the pizza at time $t$ (in minutes) can be modeled by $u(t) = 70 + 330e^{-0.072t}$.

**(a)** According to the model, what will be the temperature of the pizza after 5 minutes?
**(b)** According to the model, what will be the temperature of the pizza after 10 minutes?
**(c)** If the pizza can be safely consumed when its temperature is 200°F, will it be ready to eat after cooling for 13 minutes?

**92. Newton's Law of Cooling** Newton's Law of Cooling states that the temperature of a heated object decreases exponentially over time toward the temperature of the surrounding medium. Suppose that coffee that is 170°F is poured into a coffee mug and allowed to cool in a room whose temperature is 70°F. The temperature $u$ (in °F) of the coffee at time $t$ (in minutes) can be modeled by $u(t) = 70 + 100e^{-0.045t}$.

**(a)** According to the model, what will be the temperature of the coffee after 5 minutes?
**(b)** According to the model, what will be the temperature of the coffee after 10 minutes?
**(c)** If the coffee doesn't taste good once its temperature reaches 120°F, will it be bad after cooling for 20 minutes?

**93. Learning Curve** Suppose that a student has 200 vocabulary words to learn. If a student learns 20 words in 30 minutes, the function

$$L(t) = 200(1 - e^{-0.0035t})$$

models the number of words $L$ that the student will learn after $t$ minutes.

**(a)** How many words will the student learn after 45 minutes?
**(b)** How many words will the student learn after 60 minutes?

**94. Learning Curve** Suppose that a student has 50 biology terms to learn. If a student learns 10 terms in 30 minutes, the function

$$L(t) = 50(1 - e^{-0.0223t})$$

models the number of terms $L$ that the student will learn after $t$ minutes.

**(a)** How many words will the student learn after 45 minutes?
**(b)** How many words will the student learn after 60 minutes?

**95. Current in an *RL* Circuit** The equation governing the amount of current $I$ (in amperes) after time $t$ (in seconds) in a single $RL$ circuit consisting of a resistance $R$ (in ohms), an inductance $L$ (in henrys), and an electromotive force $E$ (in volts) is

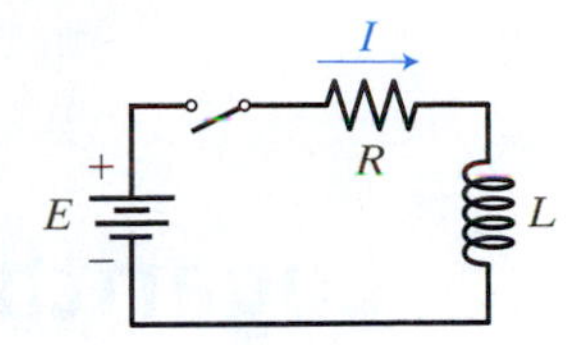


$$I = \frac{E}{R}[1 - e^{-(R/L)t}]$$

**(a)** If $E = 120$ volts, $R = 10$ ohms, and $L = 25$ henrys, how much current $I$ is flowing after 0.05 second?
**(b)** If $E = 240$ volts, $R = 10$ ohms, and $L = 25$ henrys, how much current $I$ is flowing after 0.05 second?

**96. Current in an *RC* Circuit** The equation governing the amount of current $I$ (in amperes) after time $t$ (in microseconds) in a single $RC$ circuit consisting of a resistance $R$ (in ohms), a capacitance $C$ (in microfarads), and an electromotive force $E$ (in volts) is

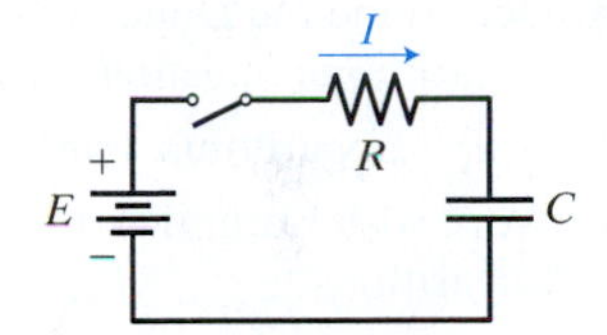


$$I = \frac{E}{R}e^{-t/(RC)}$$

**(a)** If $E = 120$ volts, $R = 2500$ ohms, and $C = 100$ microfarads, how much current $I$ is flowing initially ($t = 0$)? After 50 microseconds?
**(b)** If $E = 240$ volts, $R = 2500$ ohms, and $C = 100$ microfarads, how much current $I$ is flowing initially ($t = 0$)? After 50 microseconds?

## Extending the Concepts

*In Problems 97 and 98, find the exponential function whose graph is given.*

**97.**

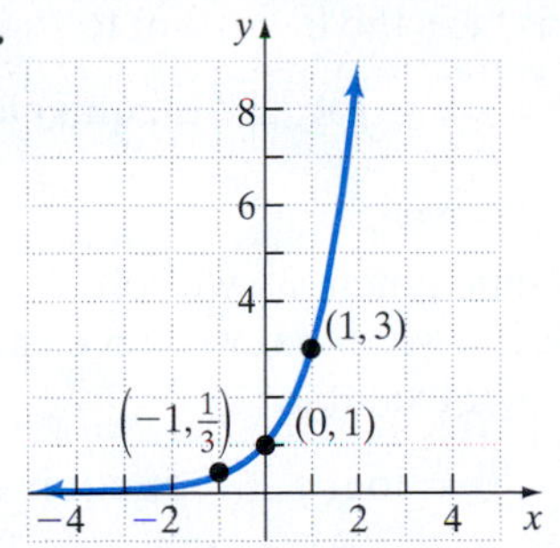

**98.**

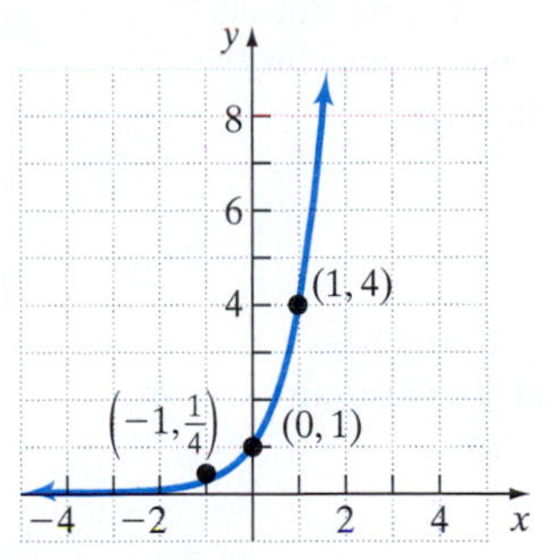

## Synthesis Review

*In Problems 99–104, evaluate each expression.*

**99.** Evaluate $x^2 - 5x + 1$ at (a) $x = -2$ and (b) $x = 3$.

**100.** Evaluate $x^3 - 5x + 2$ at (a) $x = 4$ and (b) $x = 2$.

**101.** Evaluate $\frac{4}{x + 2}$ at (a) $x = -2$ and (b) $x = 3$.

**102.** Evaluate $\frac{2x}{x + 1}$ at (a) $x = -4$ and (b) $x = 5$.

**103.** Evaluate $\sqrt{2x + 5}$ at (a) $x = 2$ and (b) $x = 11$.

**104.** Evaluate $\sqrt[3]{3x - 1}$ at (a) $x = 3$ and (b) $x = 0$.

### The Graphing Calculator

*In Problems 105–112, graph each function using a graphing calculator. State the domain and range of each function.*

**105.** $f(x) = 1.5^x$ **106.** $G(x) = 3.1^x$ **107.** $H(x) = 0.9^x$

**108.** $F(x) = 0.3^x$ **109.** $g(x) = 2.5^x + 3$ **110.** $f(x) = 1.7^x - 2$

**111.** $F(x) = 1.6^{x-3}$ **112.** $g(x) = 0.3^{x+2}$

# 9.3 Logarithmic Functions

**OBJECTIVES**

1. Change Exponential Expressions to Logarithmic Expressions
2. Change Logarithmic Expressions to Exponential Expressions
3. Evaluate Logarithmic Functions
4. Determine the Domain of a Logarithmic Function
5. Graph Logarithmic Functions
6. Work with Natural and Common Logarithms
7. Solve Logarithmic Equations
8. Study Logarithmic Models that Describe Our World

### *Preparing for Logarithmic Functions*

*Before getting started, take the following readiness quiz. If you get a problem wrong, go back to the section cited and review the material.*

**1.** Solve: $3x + 2 > 0$ [Section 1.4, pp. 93–96]

**2.** Solve: $\sqrt{x + 2} = x$ [Section 7.6, pp. 581–583]

**3.** Solve: $x^2 = 6x + 7$ [Section 5.8, pp. 423–426]

We know about the "squaring function" $f(x) = x^2$ and we also know about the square root function, $g(x) = \sqrt{x}$. How are these two functions related? Well, if we evaluate $f(5)$, we obtain 25. If we evaluate $g(25)$, we obtain 5. That is, the input to $f$ is the output of $g$ and the output of $f$ is the input to $g$. Basically, $g$ "undoes" what $f$ does.

In general, whenever we introduce a function in mathematics, we would also like to find a second function that "undoes" the function. For example, we have a square root function to undo the squaring function. We have a cube root function to undo the cubing function. A logical question is, "Do we have a function that "undoes" an exponential function?" The answer is yes. The function is called the *logarithmic function*.

**Work Smart**

We require that $a$ is positive and not equal to 1 for the same reasons that we had these restrictions for the exponential function.

**DEFINITION**

The **logarithmic function to the base $a$,** where $a > 0$ and $a \neq 1$, is denoted by $y = \log_a x$ (read as "$y$ is the logarithm to the base $a$ of $x$") and is defined by

$$y = \log_a x \quad \text{is equivalent to} \quad x = a^y$$

**In Words**

The logarithm to the base $a$ of $x$ is the number $y$ that we must raise $a$ to in order to obtain $x$.

To evaluate logarithmic functions, we convert them to their equivalent exponential expression. Therefore, it is vital that we can easily go from logarithmic form to exponential form, and back. For example,

$$0 = \log_3 1 \quad \text{is equivalent to} \quad 3^0 = 1$$

$$2 = \log_5 25 \quad \text{is equivalent to} \quad 5^2 = 25$$

$$-2 = \log_4 \frac{1}{16} \quad \text{is equivalent to} \quad 4^{-2} = \frac{1}{16}$$

Notice that the base of the logarithm is the base of the exponential; the argument of the logarithm is what the exponential equals; and the value of the logarithm is the exponent of the exponential expression. A logarithm is just a fancy way of writing an exponential expression.

To help see how the logarithmic function "undoes" the exponential function, consider the function $y = 2^x$. If we input $x = 3$, then the output is $y = 2^3 = 8$. To undo this function would require that an input of 8 would give an output of 3. If we compare $2^3 = 8$ to $x = a^y$, we see that $x$ is 8, $a$ is 2, and $y$ is 3, so that $3 = \log_2 8$ using the definition of a logarithm. The input of $\log_2 8$ is 8 and its output is 3.

***Preparing for...Answers***

**1.** $\{x \mid x > -2/3\}$ or $\left(-\frac{2}{3}, \infty\right)$ **2.** $\{2\}$ **3.** $\{-1, 7\}$

## 1 Change Exponential Expressions to Logarithmic Expressions

We can use the definition of a logarithm to change from exponential expressions to logarithmic expressions.

### EXAMPLE 1 Changing Exponential Expressions to Logarithmic Expressions

Rewrite each exponential expression to an equivalent expression involving a logarithm.

**(a)** $5^4 = b$ **(b)** $1.9^c = 12$ **(c)** $z^{1.2} = 7$

#### Solution

In each of these problems, we use the fact that $y = \log_a x$ is equivalent to $a^y = x$ provided that $a > 0$ and $a \neq 1$.

**(a)** If $5^4 = b$, then $4 = \log_5 b$.

**(b)** If $1.9^c = 12$, then $c = \log_{1.9} 12$.

**(c)** If $z^{1.2} = 7$, then $1.2 = \log_z 7$.

**Work Smart: Study Skills**

In doing problems similar to Examples 1 and 2, it is a good idea to say, "$y$ equals the logarithm to the base $a$ of $x$ is equivalent to $a$ to the $y$ equals $x$." So that you memorize the definition of a logarithm.

**QUICK ✓** *Rewrite each exponential expression to an equivalent expression involving a logarithm.*

**1.** $4^3 = w$ **2.** $p^{-2} = 8$ **3.** $5^b = 125$

## 2 Change Logarithmic Expressions to Exponential Expressions

We can also use the definition of a logarithm to change from logarithmic expressions to exponential expressions.

### EXAMPLE 2 Changing Logarithmic Expressions to Exponential Expressions

Change each logarithmic expression to an equivalent expression involving an exponent.

**(a)** $y = \log_3 81$ **(b)** $-3 = \log_a \frac{1}{27}$ **(c)** $2 = \log_4 x$

#### Solution

In each of these problems, we use the fact that $y = \log_a x$ is equivalent to $a^y = x$ provided that $a > 0$ and $a \neq 1$.

**(a)** If $y = \log_3 81$, then $3^y = 81$.

**(b)** If $-3 = \log_a \frac{1}{27}$, then $a^{-3} = \frac{1}{27}$.

**(c)** If $2 = \log_4 x$, then $4^2 = x$.

**QUICK ✓** *Rewrite each logarithmic expression as an equivalent expression involving an exponent.*

**4.** $y = \log_2 16$ **5.** $5 = \log_a 20$ **6.** $-3 = \log_5 z$

## 3 Evaluate Logarithmic Functions

To find the exact value of a logarithm, we write the logarithm in exponential notation and use the fact that if $a^u = a^v$, then $u = v$.

### EXAMPLE 3 Finding the Exact Value of a Logarithmic Expression

Find the exact value of

**(a)** $\log_2 32$ **(b)** $\log_4 \frac{1}{16}$

**Solution**

**(a)** We let $y = \log_2 32$ and convert this expression to an exponential expression.

$$y = \log_2 32$$

Write the logarithm as an exponent: $2^y = 32$

$32 = 2^5$: $2^y = 2^5$

Since we have the same base, we set the exponents equal to each other: $y = 5$

Therefore, $\log_2 32 = 5$.

**(b)** We let $y = \log_4 \frac{1}{16}$ and convert this expression to an exponential expression.

$$y = \log_4 \frac{1}{16}$$

Write the logarithm as an exponent: $4^y = \frac{1}{16}$

$\frac{1}{16} = 4^{-2}$: $4^y = 4^{-2}$

Since we have the same base, we set the exponents equal to each other: $y = -2$

Therefore, $\log_4 \frac{1}{16} = -2$.

**QUICK ✓** *Find the exact value of each logarithmic expression.*

**7.** $\log_5 25$ **8.** $\log_2 \frac{1}{8}$

We could also write $y = \log_a x$ using function notation as $f(x) = \log_a x$. We use this notation in the next example to evaluate logarithmic functions.

### EXAMPLE 4 Evaluating Logarithmic Functions

Find the value of each of the following given that $f(x) = \log_2 x$.

**(a)** $f(2)$ **(b)** $f\left(\frac{1}{4}\right)$

**Solution**

**(a)** $f(2)$ means to evaluate $\log_2 x$ at $x = 2$. So we want to know the value of $\log_2 2$. To determine this value, we follow the approach of Example 3 by letting $y = \log_2 2$ and converting the expression to an exponential expression.

$$y = \log_2 2$$

Write the logarithm as an exponent: $2^y = 2$

$2 = 2^1$: $2^y = 2^1$

Since we have the same base, we set the exponents equal to each other: $y = 1$

Therefore, $f(2) = 1$.

**(b)** $f\left(\frac{1}{4}\right)$ means to evaluate $\log_2 x$ at $x = \frac{1}{4}$. So we want to know the value of $\log_2\left(\frac{1}{4}\right)$. We let $y = \log_2\left(\frac{1}{4}\right)$ and convert this expression to an exponential expression.

$$y = \log_2\left(\frac{1}{4}\right)$$

Write the logarithm as an exponent: $2^y = \frac{1}{4}$

$\frac{1}{4} = 2^{-2}$: $2^y = 2^{-2}$

Since we have the same base, we set the exponents equal to each other: $y = -2$

Therefore, $f\left(\frac{1}{4}\right) = -2$.

**QUICK ✓** *Evaluate the function given that* $g(x) = \log_5 x$.

**9.** $g(25)$ **10.** $g\left(\frac{1}{5}\right)$

## 4 Determine the Domain of a Logarithmic Function

The domain of a function $y = f(x)$ is the set of all $x$ such that the function makes sense and the range is the set of all images of $x$. To find the range of the logarithmic function, we recognize that $y = f(x) = \log_a x$ is equivalent to $x = a^y$. Because we can raise $a$ to any real number (since $a > 0$ and $a \neq 1$), we conclude that $y$ can be any real number in the function $y = f(x) = \log_a x$. In addition, because $a^y$ is positive for any real number, we conclude that $x$ must be positive. Since $x$ represents the input of the logarithmic function, the domain of the logarithmic function is the set of all positive real numbers.

**Work Smart**

Notice that the domain of the logarithmic function is the same as the range of the exponential function. The range of the logarithmic function is the same as the domain of the exponential function.

**DOMAIN AND RANGE OF THE LOGARITHMIC FUNCTION**

Domain of the logarithmic function = $(0, \infty)$
Range of the logarithmic function = $(-\infty, \infty)$

Because the domain of the logarithmic function is the set all positive real numbers, the argument of the logarithmic function must be greater than zero. For example, the logarithmic function $f(x) = \log_{10} x$ is defined for $x = 2$, but is not defined for $x = -1$ or $x = -8$ (or any other $x \leq 0$).

### EXAMPLE 5 Finding the Domain of a Logarithmic Function

Find the domain of each logarithmic function.

**(a)** $f(x) = \log_6(x - 5)$ **(b)** $G(x) = \log_3(3x + 1)$

### Solution

**(a)** The argument of the function $f(x) = \log_6(x - 5)$ is $x - 5$. The domain of $f$ is the set of all real numbers $x$ such that $x - 5 > 0$. We solve this inequality:

$$x - 5 > 0$$

Add 5 to both sides of the inequality: $x > 5$

The domain of $f$ is $\{x | x > 5\}$, or using interval notation, $(5, \infty)$.

**(b)** The argument of the function $G(x) = \log_3(3x + 1)$ is $3x + 1$. The domain of $G$ is the set of all real numbers $x$ such that $3x + 1 > 0$. We solve this inequality:

$$3x + 1 > 0$$

Subtract 1 from both sides of the inequality: $3x > -1$

Divide both sides by 3: $x > -\frac{1}{3}$

The domain of $G$ is $\left\{x \,\middle|\, x > -\frac{1}{3}\right\}$, or using interval notation, $\left(-\frac{1}{3}, \infty\right)$.

**QUICK ✓** *Find the domain of each logarithmic function.*

**11.** $g(x) = \log_8(x + 3)$ **12.** $F(x) = \log_2(5 - 2x)$

## 5 Graph Logarithmic Functions

To graph a logarithmic function $y = \log_a x$, it is helpful to rewrite the function in exponential form as $x = a^y$. We would then choose "nice" values of $y$ and use the expression $x = a^y$ to find the corresponding values of $x$.

### EXAMPLE 6 Graphing a Logarithmic Function

Graph $f(x) = \log_2 x$ using point plotting. From the graph, state the domain and the range of the function.

### Solution

We rewrite $y = f(x) = \log_2 x$, as $x = 2^y$. Table 7 shows various values of $y$, the corresponding values of $x$ and points on the graph of $y = f(x) = \log_2 x$. We plot the ordered pairs in Table 7 and connect them in a smooth curve to obtain the graph of $f(x) = \log_2 x$. See Figure 18. The domain of $f$ is $\{x | x > 0\}$ or, using interval notation, $(0, \infty)$. The range of $f$ is the set of all real numbers or, using interval notation, $(-\infty, \infty)$.

**Table 7**

| $y$ | $x = 2^y$ | $(x, y)$ |
|---|---|---|
| $-2$ | $\frac{1}{4}$ | $\left(\frac{1}{4}, -2\right)$ |
| $-1$ | $\frac{1}{2}$ | $\left(\frac{1}{2}, -1\right)$ |
| 0 | 1 | (1, 0) |
| 1 | 2 | (2, 1) |
| 2 | 4 | (4, 2) |

**Figure 18**

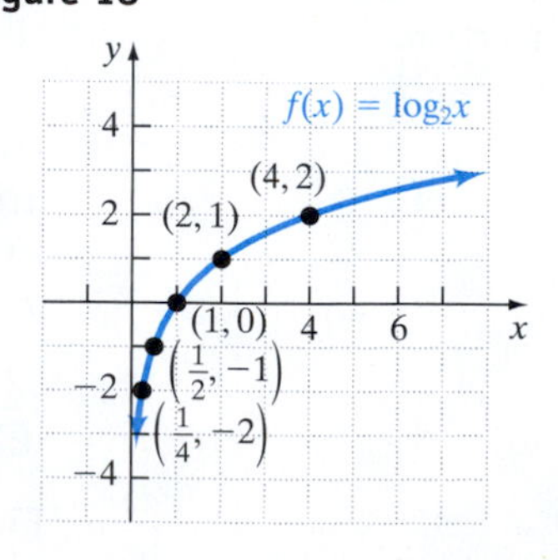

The graph of $f(x) = \log_2 x$ in Figure 18 is typical of all logarithmic functions that have a base larger than 1. Figure 19 shows the graph of two additional logarithmic functions whose bases are larger than 1, $y = \log_3 x$ and $y = \log_6 x$.

**Figure 19**

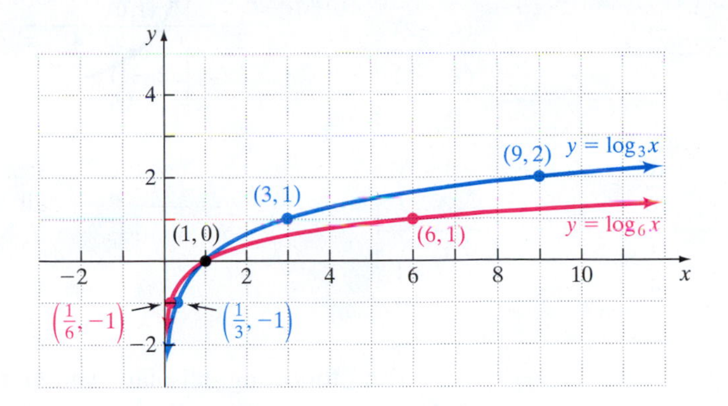


**Figure 20**
$f(x) = \log_a x,\ a > 1$

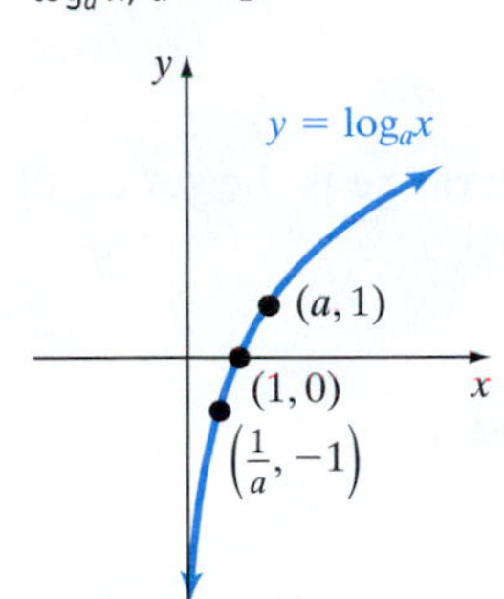


The following display summarizes the information that we have about $f(x) = \log_a x$, where the base is greater than $1 (a > 1)$.

**PROPERTIES OF THE GRAPH OF A LOGARITHMIC FUNCTION $f(x) = \log_a x,\ a > 1$**

1. The domain is the set of all positive real numbers. The range is the set of all real numbers.
2. There are no $y$-intercepts; the $x$-intercept is 1.
3. The graph of $f$ contains the points $\left(\frac{1}{a}, -1\right)$, $(1, 0)$, and $(a, 1)$.

See Figure 20.

**QUICK ✓**

13. Graph the logarithmic function $f(x) = \log_4 x$ using point plotting. From the graph, state the domain and the range of the function.

Now we consider $f(x) = \log_a x,\ 0 < a < 1$.

## EXAMPLE 7 Graphing a Logarithmic Function

Graph $f(x) = \log_{1/2} x$ using point plotting. From the graph, state the domain and the range of the function.

### Solution

We rewrite $y = f(x) = \log_{1/2} x$, as $x = \left(\frac{1}{2}\right)^y$. Table 8 on page 736 shows various values of $y$, the corresponding values of $x$ and points on the graph of $y = f(x) = \log_{1/2} x$. We plot the ordered pairs in Table 8 and connect them in a smooth curve to obtain the graphs of $f(x) = \log_{1/2} x$. See Figure 21 on page 736. The domain of $f$ is $\{x \mid x > 0\}$ or, using interval notation, $(0, \infty)$. The range of $f$ is the set of all real numbers or, using interval notation, $(-\infty, \infty)$. ■

The graph of $f(x) = \log_{1/2} x$ in Figure 21 is typical of all exponential functions that have a base between 0 and 1. Figure 22 shows the graph of two additional logarithmic functions whose bases are larger than 1, $y = \log_{1/3} x$ and $y = \log_{1/6} x$.

**Table 8**

| $y$ | $x = \left(\frac{1}{2}\right)^y$ | $(x, y)$ |
|---|---|---|
| $-2$ | 4 | $(4, -2)$ |
| $-1$ | 2 | $(2, -1)$ |
| 0 | 1 | $(1, 0)$ |
| 1 | $\frac{1}{2}$ | $\left(\frac{1}{2}, 1\right)$ |
| 2 | $\frac{1}{4}$ | $\left(\frac{1}{4}, 2\right)$ |

**Figure 21**

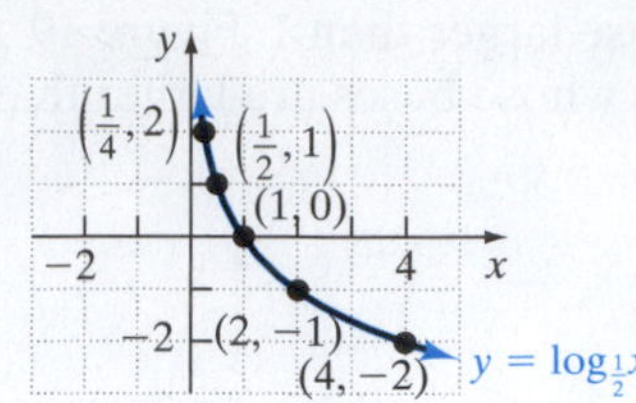


**Figure 22**

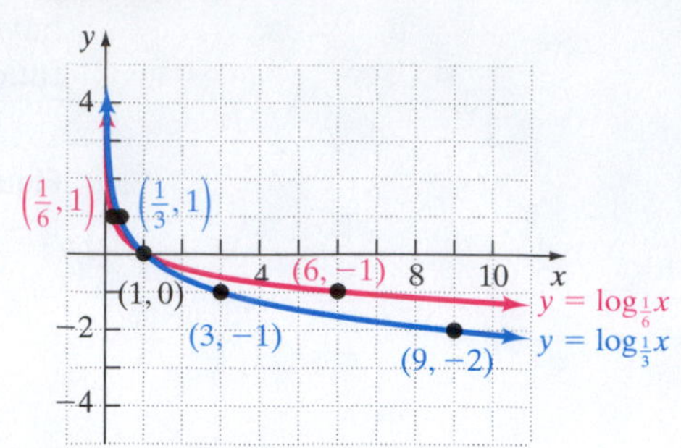


The following display summarizes the information that we have about $f(x) = \log_a x$, where the base $a$ is between 0 and $1(0 < a < 1)$.

**Figure 23**
$f(x) = \log_a x,\ 0 < a < 1$

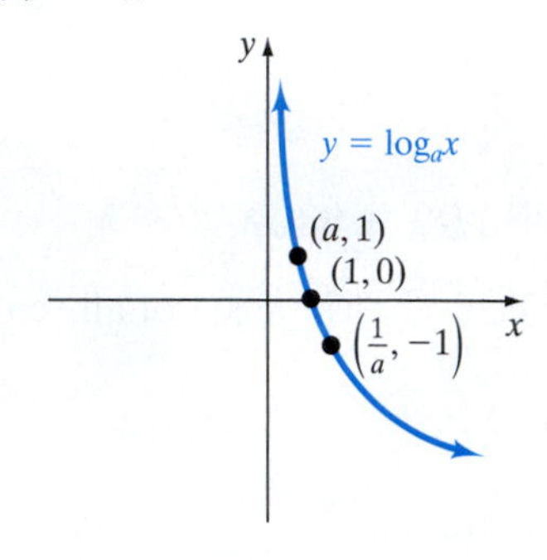


**PROPERTIES OF THE GRAPH OF AN EXPONENTIAL FUNCTION $f(x) = \log_a x,\ 0 < a < 1$**

1. The domain is the set of all positive real numbers. The range is the set of all real numbers.
2. There are no $y$-intercepts; the $x$-intercept is 1.
3. The graph of $f$ contains the points $\left(\frac{1}{a}, -1\right)$, $(1, 0)$, and $(a, 1)$.

See Figure 23.

**QUICK ✓**

**14.** Graph the logarithmic function $f(x) = \log_{\frac{1}{4}} x$ using point plotting. From the graph, state the domain and the range of the function.

## 6 Work with Natural and Common Logarithms

**In Words**
We call a logarithm to the base $e$ the natural logarithm because we can use this function to model many things in nature. In addition,

$\log_e x$ is written $\ln x$

If the base on a logarithmic function is the number $e$, then we have the **natural logarithm function.** This function occurs so frequently in applications, that it is given a special symbol, **ln** (from the Latin *logarithmus naturalis*). So we have the following definition.

**DEFINITION**

The natural logarithm: $y = \ln x$ if and only if $x = e^y$

**Figure 24**

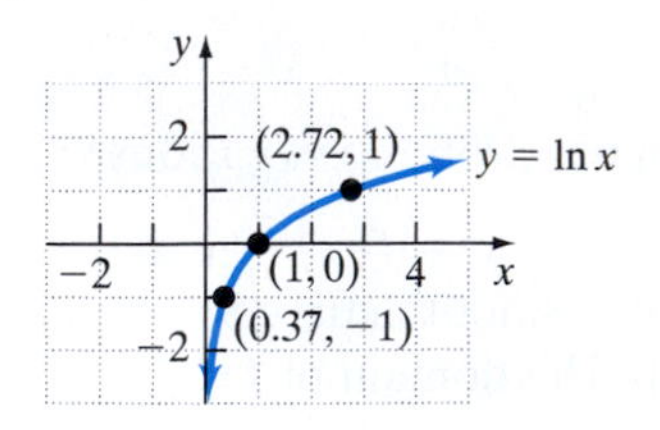


Figure 24 shows the graph of $y = \ln x$.

If the base of a logarithmic function is the number 10, then we have the **common logarithm function.** If the base $a$ of the logarithmic function is not indicated, it is understood to be 10. So, we have the following definition.

**In Words**
When there is no base on a logarithm, then the base is understood to be 10. That is,

$\log x$ is written $\log_{10} x$

**DEFINITION**

The common logarithm: $y = \log x$ if and only if $x = 10^y$

**Figure 25**

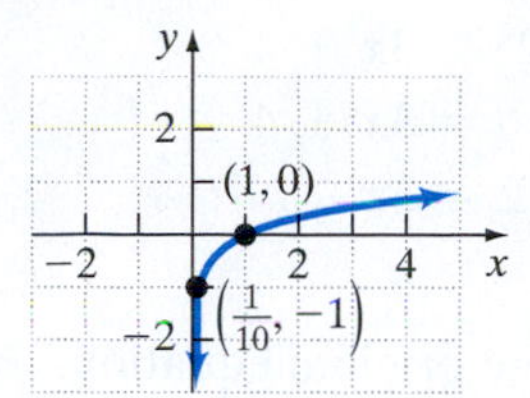


Figure 25 shows the graph of $y = \log x$.

Scientific and graphing calculators have both a natural logarithm button, ln, and a common logarithm button, log. This allows us to use the calculator to approximate the values of logarithms to the base $e$ and base 10, when the results are not exact.

To evaluate logarithmic expressions using a scientific calculator, enter the argument of the logarithm and then press the ln or log button, depending on the base of the logarithm. For example, to evaluate log 80, we would type in 80 and then press the log button. The display should show 1.90308999. Try it!

To evaluate logarithmic expressions using a graphing calculator, press the ln or log button, depending on the base of the logarithm, and then enter the argument of the logarithm. Finally, press ENTER. For example, to evaluate log 80, we would press the log button, type in 80 and then press ENTER. The display should show 1.903089987. Try it!

By the way, it shouldn't be surprising that log 80 is between 1 and 2 since $\log 10 = 1$ (because $10^1 = 10$) and $\log 100 = 2$ (because $10^2 = 100$). It is a good idea to get a sense of the value of the logarithm prior to using your calculator to approximate the value.

### EXAMPLE 8 Evaluating Natural and Common Logarithms on a Calculator

Using a calculator, evaluate each of the following. Round your answers to 3 decimal places.

**(a)** ln 20 **(b)** log 30 **(c)** ln 0.5

**Solution**

**(a)** $\ln 20 \approx 2.996$ **(b)** $\log 30 \approx 1.477$ **(c)** $\ln 0.5 \approx -0.693$

**QUICK ✓** *Evaluate each logarithm using a calculator. Round your answers to 3 decimal places.*

**15.** log 1400 **16.** ln 4.8 **17.** log 0.3

## 7 Solve Logarithmic Equations

Equations that contain logarithms are called **logarithmic equations.** Care must be taken when solving logarithmic equations because there is the possibility that extraneous solutions (an apparent solution that is not actually a solution to the original equation) creep in. This type of solution first appeared when we solved radical equations. To help locate extraneous solutions, we need to remember that in the expression $\log_a M$, both $a$ and $M$ must be positive with $a \neq 1$.

Remember how you learned to change from logarithmic form to exponential form? This skill is important for solving certain types of logarithmic equations!

**Work Smart**
Extraneous solutions can occur while solving logarithmic equations.

### EXAMPLE 9 Solving a Logarithmic Equation

Solve:

**(a)** $\log_2(3x + 4) = 4$ **(b)** $\log_x 25 = 2$

**Solution**

**(a)** We obtain the solution by writing the logarithmic equation as an exponential equation using the fact that if $y = \log_a x$, then $a^y = x$.

$$\log_2(3x + 4) = 4$$

Write as an exponent; If $y = \log_a x$, then $a^y = x$: $2^4 = 3x + 4$

$2^4 = 16$: $16 = 3x + 4$

Subtract 4 from both sides: $12 = 3x$

Divide both sides by 3: $4 = x$

We want to verify our solution by letting $x = 4$ in the original equation:

$$\log_2(3 \cdot 4 + 4) \stackrel{?}{=} 4$$

$$\log_2(16) \stackrel{?}{=} 4$$

Since $2^4 = 16$, $\log_2 16 = 4$. Therefore, the solution set is $\{4\}$.

**(b)** Change the logarithmic equation to an exponential equation.

$$\log_x 25 = 2$$

Write as an exponent; If $y = \log_a x$, then $a^y = x$: $x^2 = 25$

Use the Square Root Property: $x = \pm\sqrt{25} = \pm 5$

Since the base of a logarithm must always be positive, we know that $x = -5$ is extraneous. We leave it to you to verify that $x = 5$ is a solution. The solution set is $\{5\}$.

**QUICK ✓** *Solve each equation. Be sure to verify your solution.*

**18.** $\log_3(5x + 1) = 4$ **19.** $\log_x 16 = 2$

## EXAMPLE 10 Solving Logarithmic Equations

Solve each equation and state the exact solution.

**(a)** $\ln x = 3$ **(b)** $\log(x + 1) = -2$

### Solution

**(a)** We solve the equation by writing the logarithmic equation as an exponential equation.

$$\ln x = 3$$

Write as exponent; If $y = \ln x$, then $e^y = x$: $e^3 = x$

We verify our solution by letting $x = e^3$ in the original equation:

$$\ln e^3 \stackrel{?}{=} 3$$

We know that $\ln e^3 = 3$ can be written as $\log_e e^3 = 3$, which is equivalent to $e^3 = e^3$, so we have a true statement. The solution set is $\{e^3\}$.

**(b)** Write the logarithmic equation as an exponential equation.

$$\log(x + 1) = -2$$

Write as exponent; If $y = \log x$, then $10^y = x$: $10^{-2} = x + 1$

$10^{-2} = 0.01$: $x + 1 = 0.01$

Subtract 1 from both sides: $x = -0.99$

We verify our solution by letting $x = -0.99$ in the original equation:

$$\log(-0.99 + 1) \stackrel{?}{=} -2$$

$$\log(0.01) \stackrel{?}{=} -2$$

$$10^{-2} = 0.01 \quad \text{True}$$

The solution set is $\{-0.99\}$.

**QUICK ✓** *Solve each equation. Be sure to verify your solution.*

**20.** $\ln x = -2$ **21.** $\log(x - 20) = 4$

## 8 Study Logarithmic Models that Describe Our World

Common logarithms often are used when quantities vary from very large to very small numbers. The reason for this is that the common logarithm can "scale down" the measurement. For example, if a certain quantity can vary from $0.00000001 = 10^{-8}$ to $100{,}000{,}000 = 10^8$, the common logarithm of the same quantity would vary from $\log 10^{-8} = -8$ to $\log 10^8 = 8$.

Physicists define the **intensity of a sound wave** as the amount of energy that the sound wave transmits through a given area. For example, the least sound that a human ear can detect at a frequency of 100 hertz is about $10^{-12}$ watt per square meter. The *loudness L* (measured in **decibels** in honor of Alexander Graham Bell) of a sound of intensity $x$ (measured in watts per square meter) is defined as follows.

**DEFINITION**

The **loudness** $L$, measured in decibels, of a sound of intensity $x$, measured in watts per square meter, is

$$L(x) = 10 \log \frac{x}{10^{-12}}$$

The quantity $10^{-12}$ watt per square meter in the definition is the least intense sound that a human ear can detect. So, if we let $x = 10^{-12}$ watt per square meter, we obtain

$$\begin{aligned} L(10^{-12}) &= 10 \log \frac{10^{-12}}{10^{-12}} \\ &= 10 \log 1 \\ &= 10(0) \\ &= 0 \end{aligned}$$

So at the least intense sound a human ear can detect, we measure sound at 0 decibels.

### EXAMPLE 11 Measuring the Loudness of a Sound

Normal conversation has an intensity level of $10^{-6}$ watt per square meter. How many decibels is normal conversation?

#### Solution

We evaluate $L$ at $x = 10^{-6}$.

$$\begin{aligned} L(10^{-6}) &= 10 \log \frac{10^{-6}}{10^{-12}} \\ \text{Laws of exponents; } \frac{a^m}{a^n} = a^{m-n}: \quad &= 10 \log 10^{-6-(-12)} \\ \text{Simplify:} \quad &= 10 \log 10^6 \\ y = \log 10^6 \text{ implies } 10^y = 10^6, \text{ so } y = 6: \quad &= 10(6) \\ &= 60 \text{ decibels} \end{aligned}$$

The loudness of normal conversation is 60 decibels.

**QUICK ✓**

**22.** An MP3 player has an intensity level of $10^{-2}$ watt per square meter when set at its maximum level. How many decibels is the MP3 player on "full blast"?

# 9.3 Exercises

**For Extra Help:**

Student Solutions Manual | CD Video | PH Math/Tutor Center | MathXL Tutorials on CD | MathXL® | MyMathLab

## Concepts and Vocabulary

*In Problems 1–3, fill in the blanks.*

1. The logarithm to the base $a$ of $x$, denoted $y = \log_a x$, can be expressed as an exponent as ______, where $a$ ______ 0 and $a$ ______ 1.
2. The domain of the logarithmic function $f(x) = \log_a x$ is ______.
3. The graph of every logarithmic function $f(x) = \log_a x$, $a > 0$, $a \neq 1$, passes through three points: ______, ______, and ______.

*In Problems 4–6, answer True or False to each statement.*

4. Logarithmic equations never have extraneous solutions.
5. If $y = \log_2 x$, then $y = 2^x$.
6. The base of the natural logarithmic function is $e$.
7. In the definition of the logarithmic function $f(x) = \log_a x$, the base $a$ is not allowed to equal 1. Why?
8. The domain of $f(x) = \log_a(x^2 + 1)$ is the set of all real numbers. Explain why.

## Building Skills

*In Problems 9–20, change each exponential expression to an equivalent expression involving a logarithm.*

**9.** $25 = 5^2$ **10.** $64 = 8^2$ **11.** $64 = 4^3$ **12.** $16 = 2^4$

**13.** $\frac{1}{8} = 2^{-3}$ **14.** $\frac{1}{9} = 3^{-2}$ **15.** $e^x = 12$ **16.** $e^{4.2} = M$

**17.** $a^3 = 19$ **18.** $b^4 = 23$ **19.** $5^{-6} = c$ **20.** $10^{-3} = z$

*In Problems 21–32, change each logarithmic expression to an equivalent expression involving an exponent.*

**21.** $\log_2 16 = 4$ **22.** $\log_3 81 = 4$ **23.** $\log_3 \frac{1}{9} = -2$ **24.** $\log_2 \frac{1}{32} = -5$

**25.** $\ln x = 4$ **26.** $\ln(x - 1) = 3$ **27.** $\log_5 a = -3$ **28.** $\log_6 x = -4$

**29.** $\log_a 4 = 2$ **30.** $\log_a 16 = 2$ **31.** $\log_{1/2} 12 = y$ **32.** $\log_{1/2} 18 = z$

*In Problems 33–40, find the exact value of each logarithm without using a calculator.*

**33.** $\log_3 1$ **34.** $\log_5 5$ **35.** $\log_2 8$ **36.** $\log_4 16$

**37.** $\log_4\left(\frac{1}{16}\right)$ **38.** $\log_5\left(\frac{1}{125}\right)$ **39.** $\log_{\sqrt{2}} 4$ **40.** $\log_{\sqrt{3}} 3$

*In Problems 41–48, evaluate each function given that $f(x) = \log_3 x$, $g(x) = \log_5 x$, $H(x) = \log x$, and $P(x) = \ln x$.*

**41.** $f(81)$ **42.** $f(9)$ **43.** $g(\sqrt{5})$ **44.** $g(\sqrt[3]{5})$

**45.** $H(0.1)$ **46.** $H(100{,}000)$ **47.** $P(e^3)$ **48.** $P(e^{-3})$

*In Problems 49–60, find the domain of each function.*

**49.** $f(x) = \log_2(x - 4)$ **50.** $f(x) = \log_3(x - 2)$ **51.** $G(x) = \log_5(x + 6)$

**52.** $g(x) = \log_4(x + 10)$ **53.** $F(x) = \log_3(2x)$ **54.** $h(x) = \log_4(5x)$

**55.** $f(x) = \log_8(3x - 2)$ **56.** $F(x) = \log_2(4x - 3)$ **57.** $H(x) = \log_7(2x + 1)$

**58.** $f(x) = \log_3(5x + 3)$ **59.** $H(x) = \log_2(1 - 4x)$ **60.** $G(x) = \log_4(3 - 5x)$

*In Problems 61–70, graph each function. From the graph, state the domain and the range of each function.*

**61.** $f(x) = \log_5 x$ **62.** $f(x) = \log_7 x$ **63.** $g(x) = \log_6 x$ **64.** $G(x) = \log_8 x$

**65.** $F(x) = \log_{1/5} x$ **66.** $F(x) = \log_{1/7} x$ **67.** $G(x) = \log_{1/6} x$ **68.** $g(x) = \log_{1/8} x$

**69.** $f(x) = \ln x$ **70.** $f(x) = \log x$

*In Problems 71–82, use a calculator to evaluate each expression. Round your answers to three decimal places.*

**71.** $\log 67$ **72.** $\log 106$ **73.** $\ln 5.4$ **74.** $\ln 10.4$

**75.** $\log 0.35$ **76.** $\log 0.78$ **77.** $\ln 0.2$ **78.** $\ln 0.4$

**79.** $\log\frac{5}{4}$ **80.** $\log\frac{10}{7}$ **81.** $\ln\frac{3}{8}$ **82.** $\ln\frac{1}{2}$

*In Problems 83–102, solve each logarithmic equation.*

**83.** $\log_3(2x + 1) = 2$ **84.** $\log_3(5x - 3) = 3$ **85.** $\log_5(20x - 5) = 3$

**86.** $\log_4(8x + 10) = 3$ **87.** $\log_a 36 = 2$ **88.** $\log_a 81 = 2$

**89.** $\log_a 18 = 2$ **90.** $\log_a 28 = 2$ **91.** $\log_a 1000 = 3$

**92.** $\log_a 243 = 5$ **93.** $\ln x = 5$ **94.** $\ln x = 10$

**95.** $\log(2x - 1) = -1$ **96.** $\log(2x + 3) = 1$ **97.** $\ln e^x = -3$

**98.** $\ln e^{2x} = 8$ **99.** $\log_3 81 = x$ **100.** $\log_4 16 = x + 1$

**101.** $\log_2(x^2 - 1) = 3$ **102.** $\log_3(x^2 + 1) = 2$

## Mixed Practice

**103.** Suppose that $f(x) = \log_2 x$.
- **(a)** What is $f(16)$? What point is on the graph of $f$?
- **(b)** If $f(x) = -3$, what is $x$? What point is on the graph of $f$?

**104.** Suppose that $f(x) = \log_5 x$.
- **(a)** What is $f(5)$? What point is on the graph of $f$?
- **(b)** If $f(x) = -2$, what is $x$? What point is on the graph of $f$?

**105.** Suppose that $G(x) = \log_4(x + 1)$.
- **(a)** What is $G(7)$? What point is on the graph of $G$?
- **(b)** If $G(x) = 2$, what is $x$? What point is on the graph of $G$?

**106.** Suppose that $F(x) = \log_2 x - 3$.
- **(a)** What is $F(8)$? What point is on the graph of $f$?
- **(b)** If $F(x) = -1$, what is $x$? What point is on the graph of $F$?

## Applying the Concepts

**107. Loudness of a Whisper** A whisper has an intensity level of $10^{-10}$ watt per square meter. How many decibels is a whisper?

**108. Loudness of a Concert** If you sit in the front row of a rock concert, you will experience an intensity level of $10^{-1}$ watt per square meter. How many decibels is a rock concert in the front row? If you move back to the 15th row, you will experience intensity of $10^{-2}$ watt per square meter. How many decibels is a rock concert in the 15th row?

**109. Threshold of Pain** The threshold of pain has an intensity level of $10^1$ watt per square meter. How many decibels is the threshold of pain?

**110. Exploding Eardrum** Instant perforation of the eardrum occurs at an intensity level of $10^4$ watt per square meter. At how many decibels will instant perforation of the eardrum occur?

*Problems 111–114 use the following discussion: The* ***Richter scale*** *is one way of converting seismographic readings into numbers that provide an easy reference for measuring the* ***magnitude M*** *of an earthquake. All earthquakes are compared to a* ***zero-level earthquake*** *whose* *seismographic reading measures* 0.001 *millimeter at a distance of* 100 *kilometers from the epicenter. An earthquake whose seismographic reading measures x millimeters has magnitude M given by*

$$M(x) = \log\left(\frac{x}{10^{-3}}\right)$$

*where* $10^{-3}$ *is the reading of a zero-level earthquake* 100 *kilometers from its epicenter.*

**111. San Francisco, 1906** According to the United States Geological Survey, the San Francisco earthquake of 1906 resulted in a seismographic reading of 63,096 millimeters 100 kilometers from its epicenter. What was the magnitude of this earthquake?

**112. Alaska, 1964** According to the United States Geological Survey, an earthquake on March 28, 1964 in Prince William Sound, Alaska resulted in a seismographic reading of 1,584,893 millimeters 100 kilometers from its epicenter. What was the magnitude of this earthquake? This earthquake was the second largest ever recorded, with the largest being the Great Chilean Earthquake of 1960, whose magnitude was 9.5 on the Richter scale.

**113. Ecuador, 1906** According to the United States Geological Survey, an earthquake on January 31, 1906 off the coast of Ecuador had a magnitude of 8.8. What was the seismographic reading 100 kilometers from its epicenter?

**114. South Carolina, 1886** According to the United States Geological Survey, an earthquake on September 1, 1886 in Charleston, South Carolina had a magnitude of 7.3. What was the seismographic reading 100 kilometers from its epicenter?

**115. pH** The pH of a chemical solution is given by the formula

$$\text{pH} = -\log[\text{H}^+]$$

where $[\text{H}^+]$ is the concentration of hydrogen ions in moles per liter. Values of pH range from 0 to 14. A solution whose pH is 7 is considered neutral. The pH of pure water at 25 degrees Celsius is 7. A solution whose pH is less than 7 is considered acidic, while a solution whose pH is greater than 7 is considered basic.

**(a)** What is the pH of household ammonia for which $[\text{H}^+]$ is $10^{-12}$? Is ammonia basic or acidic?

**(b)** What is the pH of black coffee for which $[\text{H}^+]$ is $10^{-5}$? Is black coffee basic or acidic?

**(c)** What is the pH of lemon juice for which $[\text{H}^+]$ is $10^{-2}$? Is lemon juice basic or acidic?

**(d)** What is the concentration of hydrogen ions in human blood ($\text{pH} = 7.4$)?

**116. Energy of an Earthquake** The magnitude and the seismic moment are related to the amount of energy that is given off by an earthquake. The relationship between magnitude and energy is

$$\log E_S = 11.8 + 1.5M$$

where $E_S$ is the energy (in ergs) for an earthquake whose magnitude is $M$. Note that $E_S$ is the amount of energy given off from the earthquake as seismic waves.

**(a)** How much energy is given off by an earthquake that measures 5.8 on the Richter scale?

**(b)** The earthquake on the Rat Islands in Alaska on February 4, 1965 measured 8.7 on the Richter scale. How much energy was given off by this earthquake?

## Extending the Concepts

*In Problems 117–120, find the domain of each function.*

**117.** $f(x) = \log_2(x^2 - 3x - 10)$

**118.** $f(x) = \log_5(x^2 + 2x - 24)$

**119.** $f(x) = \ln\left(\frac{x-3}{x+1}\right)$

**120.** $f(x) = \log\left(\frac{x+4}{x-3}\right)$

**121.** Find $a$ so that the graph of $f(x) = \log_a x$ contains the point $(16, 2)$.

**122.** Find $a$ so that the graph of $f(x) = \log_a x$ contains the point $\left(\frac{1}{4}, -2\right)$.

## Synthesis Review

*In Problems 123–128, add or subtract each expression.*

**123.** $(2x^2 - 6x + 1) - (5x^2 + x - 9)$

**124.** $(x^3 - 2x^2 + 10x + 2) + (3x^3 - 4x - 5)$

**125.** $\dfrac{3x}{x^2 - 1} + \dfrac{x - 3}{x^2 + 3x + 2}$

**126.** $\dfrac{x + 5}{x^2 + 5x + 6} - \dfrac{2}{x^2 + 2x - 3}$

**127.** $\sqrt{8x^3} + x\sqrt{18x}$

**128.** $\sqrt[3]{27a} + 2\sqrt[3]{a} - \sqrt[3]{8a}$

## The Graphing Calculator

*In Problems 129–134, graph each logarithmic function using a graphing calculator. State the domain and the range of each function.*

**129.** $f(x) = \log(x + 1)$

**130.** $g(x) = \log(x - 2)$

**131.** $G(x) = \ln(x) + 1$

**132.** $F(x) = \ln(x) - 4$

**133.** $f(x) = 2\log(x - 3) + 1$

**134.** $G(x) = -\log(x + 1) - 3$

# PUTTING THE CONCEPTS TOGETHER (Sections 9.1–9.3)

*These problems cover important concepts from Sections 9.1 to 9.3. We designed these problems so that you can review the chapter so far and show your mastery of the concepts. Take time to work these problems before proceeding with the next section. The answers to these problems are located at the back of the text starting on page AN–46.*

**1.** Given the functions $f(x) = 2x + 3$ and $g(x) = 2x^2 - 4x$ find

**(a)** $(f \circ g)(x)$ **(b)** $(g \circ f)(x)$ **(c)** $(f \circ g)(3)$

**(d)** $(g \circ f)(-2)$ **(e)** $(f \circ f)(1)$

**2.** Find the inverse of the given one-to-one function.

**(a)** $f(x) = 3x + 4$ **(b)** $g(x) = x^3 - 4$

**3.** Sketch the graph of the inverse of the given one-to-one function.

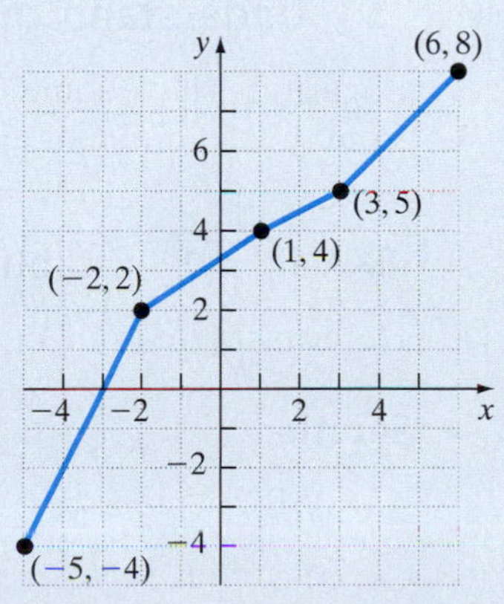


**4.** Approximate each number using a calculator. Express your answer rounded to three decimal places.

**(a)** $2.7^{2.7}$ **(b)** $2.72^{2.72}$ **(c)** $2.718^{2.718}$

**(d)** $2.7183^{2.7183}$ **(e)** $e^e$

**5.** Change each exponential expression to an equivalent expression involving a logarithm.

**(a)** $a^4 = 6.4$ **(b)** $10^x = 278$

**6.** Change each logarithmic expression to an equivalent expression involving an exponent.

**(a)** $\log_2 x = 7$ **(b)** $\ln 16 = M$

**7.** Find the exact value of each expression without using a calculator.

**(a)** $\log_5 625$ **(b)** $\log_{\frac{2}{3}}\left(\frac{9}{4}\right)$

**8.** Determine the domain of the logarithmic function: $f(x) = \log_{13}(2x + 12)$.

*In Problems 9 and 10, graph each function. State the domain and range of the function.*

**9.** $f(x) = \left(\frac{1}{6}\right)^x$ **10.** $g(x) = \log_{\frac{3}{2}} x$

*In Problems 11–14, solve each equation. State the exact solution.*

**11.** $3^{-x+2} = 27$ **12.** $e^x = e^{2x+5}$ **13.** $\log_2(2x + 5) = 4$ **14.** $\ln x = 7$

**15.** Suppose that a student has 150 anatomy and physiology terms to learn. If the student learns 40 terms in 60 minutes, the function $L(t) = 150(1 - e^{-0.0052t})$ models the number of terms $L$ that the student will learn after $t$ minutes. According to the model, how many terms will the student learn after 90 minutes?

# 9.4 Properties of Logarithms

**OBJECTIVES**

1. Understand the Properties of Logarithms
2. Write a Logarithmic Expression as a Sum or Difference of Logarithms
3. Write a Logarithmic Expression as a Single Logarithm
4. Evaluate Logarithms Whose Base Is Neither 10 Nor $e$

### *Preparing for Properties of Logarithms*

*Before getting started, take the following readiness quiz. If you get a problem wrong, go back to the section cited.*

**1.** Round 3.03468 to three decimal points. [Section R.2, pp. 14–15]

**2.** Evaluate $a^0$, $a \neq 0$. [Getting Ready, p. 341]

## 1 Understand the Properties of Logarithms

Logarithms have some very useful properties. These properties can be derived directly from the definition of a logarithm and the laws of exponents.

### EXAMPLE 1 Establishing Properties of Logarithms

Determine the value of the following logarithmic expressions: (a) $\log_a 1$ (b) $\log_a a$.

**Solution**

**(a)** Remember, $y = \log_a x$ is equivalent to $x = a^y$.

$$y = \log_a 1$$

Change to an exponent: $a^y = 1$

Since $a \neq 0$, $a^0 = 1$: $a^y = a^0$

Since the bases are the same, we equate the exponents: $y = 0$

So $\log_a 1 = 0$.

***Preparing for...Answers*** **1.** 3.035 **2.** 1

**(b)** Again, we need to write the logarithm as an exponent in order to evaluate the logarithm. Let $y = \log_a a$, so that

$$y = \log_a a$$

Change to an exponent: $a^y = a$

$a = a^1$: $a^y = a^1$

Since the bases are the same, we equate the exponents: $y = 1$

So $\log_a a = 1$.

We summarize the results of Example 1 in the box below.

$$\log_a 1 = 0 \qquad \log_a a = 1$$

**QUICK ✓** *Evaluate each logarithm.*

**1.** $\log_5 1$ **2.** $\ln 1$ **3.** $\log_4 4$ **4.** $\log 10$

Suppose we were asked to evaluate $3^{\log_3 81}$. If we evaluate the exponent, $\log_3 81$, we obtain 4 (because $y = \log_3 81$ means $3^y = 81$ or $3^y = 3^4$, so $y = 4$). Now $3^{\log_3 81} = 3^4 = 81$. So $3^{\log_3 81} = 81$. This result is true in general.

**In Words**
If we raise the number $a$ to $\log_a M$, we obtain $M$.

**AN INVERSE PROPERTY OF LOGARITHMS**

If $a$ and $M$ are positive real numbers, with $a \neq 1$, then

$$a^{\log_a M} = M$$

This is called an Inverse Property of Logarithms because if we compute the logarithm to the base $a$ of a positive number $M$ and then compute $a$ raised to this power, we end up right back where we started, namely $M$.

## EXAMPLE 2 Using a Property of Logarithms

**(a)** $5^{\log_5 20} = 20$ **(b)** $0.8^{\log_{0.8}\sqrt{23}} = \sqrt{23}$

**QUICK ✓** *Evaluate each logarithm.*

**5.** $12^{\log_{12}\sqrt{2}}$ **6.** $10^{\log 0.2}$

Suppose we were asked to evaluate $\log_5 5^6$. If we let $y = \log_5 5^6$, then $5^y = 5^6$ (since $y = \log_a x$ means $a^y = x$), so that $y = 6$. This result implies that the exponent of the argument is the value of the logarithm if the base of the logarithm and the base of the argument are the same.

**In Words**
The logarithm to the base $a$ of $a$ raised to the power of $r$ is $r$.

**AN INVERSE PROPERTY OF LOGARITHMS**

If $a$ is a positive real numbers, with $a \neq 1$, and $r$ is any real number, then

$$\log_a a^r = r$$

This is also called an Inverse Property of Logarithms because if we compute $a$ raised to some power $r$ and then compute the logarithm to the base $a$ of $a^r$, we end up right back where we started, namely $r$.

**EXAMPLE 3** **Using a Property of Logarithms**

**(a)** $\log_4 4^3 = 3$ **(b)** $\ln e^{-0.5} = -0.5$

**QUICK ✓** *Evaluate each logarithm.*

**7.** $\log_8 8^{1.2}$ **8.** $\log 10^{-4}$

## 2 Write a Logarithmic Expression as a Sum or Difference of Logarithms

The next two properties deal with working with logarithms where the argument is a product or quotient.

If we evaluate $\log_2 8$, we obtain 3. If we evaluate $\log_2 2 + \log_2 4$, we obtain $1 + 2 = 3$. Notice that $\log_2 8 = \log_2 2 + \log_2 4$. This suggests the following result.

**In Words**
The product rule of logarithms states that the log of a product equals the sum of the logs.

**THE PRODUCT RULE OF LOGARITHMS**

If $M$, $N$, and $a$ are positive real numbers, with $a \neq 1$, then

$$\log_a(MN) = \log_a M + \log_a N$$

**EXAMPLE 4** **Using the Product Rule of Logarithms**

Write each of the following logarithms as the sum of logarithms.

**(a)** $\log_2(5 \cdot 3)$ **(b)** $\ln(6z)$

### Solution

Do you notice that each argument contains a product? To simplify, we use the Product Rule of Logarithms.

**(a)** $\log_2(5 \cdot 3) = \log_2 5 + \log_2 3$

**(b)** $\ln(6z) = \ln 6 + \ln z$

**Work Smart**
$\log_a(M + N)$ does not equal $\log_a M + \log_a N$.

**QUICK ✓** *Write each logarithm as the sum of logarithms.*

**9.** $\log_4(9 \cdot 5)$ **10.** $\log(5w)$

If we evaluate $\log_2 8$, we obtain 3. If we evaluate $\log_2 16 - \log_2 2$, we obtain $4 - 1 = 3$. Notice that $\log_2 8 = \log_2 16 - \log_2 2$ and that $8 = \dfrac{16}{2}$. This suggests the following result.

**In Words**
The quotient rule states that the log of a quotient equals the difference of the logs.

**THE QUOTIENT RULE OF LOGARITHMS**

If $M$, $N$, and $a$ are positive real numbers, with $a \neq 1$, then

$$\log_a\left(\frac{M}{N}\right) = \log_a M - \log_a N$$

The results of the product and quotient rules for logarithms are not coincidental. After all, when we multiply exponential expressions with the same base, we add the exponents; when we divide exponential expressions with the same base, we subtract the exponents. The same thing is going on here!

## EXAMPLE 5 Using the Quotient Rule of Logarithms

Write each of the following logarithms as the difference of logarithms.

**(a)** $\log_2\left(\frac{5}{3}\right)$ **(b)** $\log\left(\frac{y}{5}\right)$

### Solution

Do you notice that each argument contains a quotient? To simplify, we use the Quotient Rule of Logarithms.

**(a)** $\log_2\left(\frac{5}{3}\right) = \log_2 5 - \log_2 3$ **(b)** $\log\left(\frac{y}{5}\right) = \log y - \log 5$

**Work Smart**
$\log_a(M - N)$ does not equal $\log_a M - \log_a N$.

**QUICK ✓** *Write each logarithm as the quotient of logarithms.*

**11.** $\log_7\left(\frac{9}{5}\right)$ **12.** $\ln\left(\frac{p}{3}\right)$

We can write a single logarithm as the sum or difference of logs when the argument of the logarithm contains both products and quotients.

## EXAMPLE 6 Writing a Single Logarithm as the Sum or Difference of Logs

Write $\log_3\left(\frac{4x}{y}\right)$ as the sum or difference of logarithms.

### Solution

Notice that the argument of the logarithm contains a product and a quotient. So we will use both the Product and Quotient Rule of Logarithms. When using both rules, it is typically easiest to use the Quotient Rule first.

**Work Smart**
When you need to use both the product and quotient rule, use the quotient rule first.

The Quotient Rule of Logarithms
↓
$$\log_3\left(\frac{4x}{y}\right) = \log_3(4x) - \log_3 y$$

The Product Rule of Logarithms: $= \log_3 4 + \log_3 x - \log_3 y$

**QUICK ✓** *Write each logarithm as the sum or difference of logarithms.*

**13.** $\log_2\left(\frac{3m}{n}\right)$ **14.** $\ln\left(\frac{q}{3p}\right)$

Another useful property of logarithms allows us to express powers on the argument of a logarithm as factors.

**In Words**
If an exponent exists on the quantity we are taking the log of it can be "brought down in front" of the log.

**THE POWER RULE OF LOGARITHMS**

If $M$ and $a$ are positive real numbers, with $a \neq 1$, and $r$ is any real number, then

$$\log_a M^r = r \log_a M$$

## EXAMPLE 7 Using the Power Rule of Logarithms

Use the Power Rule of Logarithms to express all powers as factors.

**(a)** $\log_8 3^5$ **(b)** $\ln x^{\sqrt{3}}$

### Solution

**(a)** $\log_8 3^5 = 5 \log_8 3$ **(b)** $\ln x^{\sqrt{3}} = \sqrt{3} \ln x$

**QUICK ✓** *Write each logarithm so that all powers are factors.*

**15.** $\log_2 5^{1.6}$ **16.** $\log b^5$

**Work Smart**
Whenever you see the word *factor*, you should think product (or multiplication).

We will use the direction *expand the logarithm* to mean to write a logarithm as a sum or difference with exponents written as factors.

## EXAMPLE 8 Expanding a Logarithm

Expand the logarithm. That is, write each logarithm as the sum or difference of logarithms with all exponents written as factors.

**(a)** $\log_2(x^2y^3)$ **(b)** $\log\left(\frac{100x}{\sqrt{y}}\right)$

### Solution

**(a)** Do you see that the argument of the logarithm contains a product? For this reason, we use the Product Rule of Logarithms to write the single log as the sum of two logs.

Product Rule ↓

$$\log_2(x^2y^3) = \log_2 x^2 + \log_2 y^3$$

Write exponents as factors: $= 2\log_2 x + 3\log_2 y$

**(b)** The argument of the logarithm contains a quotient and a product.

**Work Smart**
If the argument of the logarithm contains both a quotient and a product, it is easiest to write the quotient as the difference of two logs first.

Quotient Rule ↓

$$\log\left(\frac{100x}{\sqrt{y}}\right) = \log(100x) - \log\sqrt{y}$$

$\sqrt{y} = y^{1/2}$: $= \log(100x) - \log y^{1/2}$

Product Rule: $= \log 100 + \log x - \log y^{1/2}$

Write exponents as factors: $= \log 100 + \log x - \frac{1}{2}\log y$

$\log 100 = 2$: $= 2 + \log x - \frac{1}{2}\log y$

**QUICK ✓** *Expand each logarithm.*

**17.** $\log_4(a^2b)$ **18.** $\log_3\left(\frac{9m^4}{\sqrt[3]{n}}\right)$

## 3 Write a Logarithmic Expression as a Single Logarithm

**Work Smart**
To write two logs as a single log, the bases on the logs must be the same.

We can also use the Product Rule, Quotient Rule, and Power Rule of Logarithms to write sums and/or differences of logarithms that have the same base as a single logarithm. This skill is particularly useful when solving certain types of logarithmic equations (Section 9.5).

### EXAMPLE 9 Writing Expressions as a Single Logarithm

Write each of the following as a single logarithm.

**(a)** $\log_6 3 + \log_6 12$ **(b)** $\log(x - 2) - \log x$

**Solution**

**(a)** To write the expression as a single logarithm, the base of each log must be the same. Both logarithms are base 6. Since we are adding the logs, we use the Product Rule to write the logs as a single log.

$$\log_6 3 + \log_6 12 = \log_6(3 \cdot 12) \quad \text{Product Rule}$$
$$3 \cdot 12 = 36: \quad = \log_6 36$$
$$6^2 = 36, \text{ so } \log_6 36 = 2: \quad = 2$$

**(b)** The base of each logarithm is 10. Since we are subtracting the logs, we use the Quotient Rule to write the logs as a single log.

$$\log(x - 2) - \log x = \log\left(\frac{x - 2}{x}\right)$$

**QUICK ✓** *Write each expression as a single logarithm.*

**19.** $\log_8 4 + \log_8 16$ **20.** $\log_3(x + 4) - \log_3(x - 1)$

**Work Smart**
When coefficients appear on logarithms, write them as exponents before using the Product or Quotient Rule.

To use the Product or Quotient Rules to write the sum or difference of logs as a single log, the coefficients on the logarithms must be 1. Therefore, when coefficients appear on logarithms, we first use the Power Rule to write the coefficient as a power. For example, we would write $2 \log x$ as $\log x^2$.

### EXAMPLE 10 Writing Expressions as Single Logarithms

Write each of the following as a single logarithm.

**(a)** $2 \log_2(x - 1) + \frac{1}{2}\log_2 x$ **(b)** $\log(x - 1) + \log(x + 1) - 3 \log x$

**Solution**

**(a)** The bases are the same for each logarithm. Because the logarithms have coefficients, we use the Power Rule to write the coefficients as exponents.

$$2 \log_2(x - 1) + \frac{1}{2}\log_2 x = \log_2(x - 1)^2 + \log_2 x^{1/2}$$
$$x^{1/2} = \sqrt{x}: \quad = \log_2(x - 1)^2 + \log_2 \sqrt{x}$$
$$\text{Product Rule:} \quad = \log_2\left[(x - 1)^2\sqrt{x}\right]$$

**(b)** We will work from left to right in order to write the logarithm as a single logarithm. Since the first two logs are being added, we use the Product Rule to write these logs as a single log.

$$\log(x-1)+\log(x+1)-3\log x = \log[(x-1)(x+1)] - 3\log x \quad \text{(Product Rule)}$$

$(x-1)(x+1)=x^2-1$: $\quad = \log(x^2-1) - 3\log x$

Write the coefficient as an exponent: $\quad = \log(x^2-1) - \log x^3$

Quotient Rule: $\quad = \log\dfrac{(x^2-1)}{x^3}$

**QUICK ✓** *Write each expression as a single logarithm.*

**21.** $\log_5 x - 3\log_5 2$ **22.** $\log_2(x+1) + \log_2(x+2) - 2\log_2 x$

## 4 Evaluate Logarithms Whose Base Is Neither 10 Nor *e*

In Section 9.3, we learned how to use a calculator to approximate logarithms whose base was either 10 (the common logarithm) or *e* (the natural logarithm). But what if the base of the logarithm is neither 10 nor *e*? To approximate these types of logarithms, we present the following example.

### EXAMPLE 11 Approximating Logarithms Whose Base Is Neither 10 Nor *e*

Approximate $\log_2 5$. Round the answer to three decimal places.

#### Solution

We let $y = \log_2 5$ and then convert the logarithmic expression to an equivalent exponential expression using the fact that if $y = \log_a x$, then $x = a^y$.

$$y = \log_2 5$$
$$2^y = 5$$

**In Words**
"Taking the logarithm" of both sides of an equation is the same type of approach as squaring both sides of an equation.

Now, we will take the logarithm of both sides. Because we can use a calculator, we will take either the common log or natural log of both sides (it doesn't matter which). Let's take the common log of both sides.

$$\log 2^y = \log 5$$

Use the Power Rule to write the exponent, $y$, as a factor.

$$y\log 2 = \log 5$$

Now divide both sides by log 2 to solve for $y$.

$$y = \frac{\log 5}{\log 2}$$

We approximate $\dfrac{\log 5}{\log 2}$ using a calculator and obtain $\dfrac{\log 5}{\log 2} \approx 2.322$. So $\log_2 5 \approx 2.322$.

Example 11 shows how to approximate a logarithm whose base is 2, but it certainly was a lot of work! The question you may be asking yourself is, "Is there an easier way to do this?" The answer, you'll be happy to hear, is yes!—it is called the **Change-of-Base Formula.**

**CHANGE-OF-BASE FORMULA**

If $a \neq 1$, $b \neq 1$, and $M$ are positive real numbers, then

$$\log_a M = \frac{\log_b M}{\log_b a}$$

Because calculators only have keys for the common logarithm, [log], and natural logarithm, [ln], in practice, the Change-of-Base Formula uses either $b = 10$ or $b = e$, so that

$$\log_a M = \frac{\log M}{\log a} \quad \text{or} \quad \log_a M = \frac{\ln M}{\ln a}$$

## EXAMPLE 12 Using the Change-of-Base Formula

Approximate $\log_4 45$. Round your answer to three decimal places.

### Solution

We use the Change-of-Base Formula.

Using Common Logarithms: $\log_4 45 = \frac{\log 45}{\log 4} \approx 2.746$

Using Natural Logarithms: $\log_4 45 = \frac{\ln 45}{\ln 4} \approx 2.746$

So $\log_4 45 \approx 2.746$.

**QUICK ✓** *Approximate each logarithm. Round your answers to three decimal places.*

**23.** $\log_3 32$

**24.** $\log_{\sqrt{2}} \sqrt{7}$

---

We have presented quite a few properties of logarithms. As a review and for convenience, we provide the following summary of the properties of logarithms.

**SUMMARY: Properties of Logarithms**

In the following properties, $M$, $N$, $a$, and $b$ are positive real numbers, with $a \neq 1$, $b \neq 1$, and $r$ is any real number.

- **Inverse Properties of Logarithms**
  $a^{\log_a M} = M$ and $\log_a a^r = r$
- **The Product Rule of Logarithms**
  $\log_a(MN) = \log_a M + \log_a N$
- **The Quotient Rule of Logarithms**
  $\log_a\left(\frac{M}{N}\right) = \log_a M - \log_a N$
- **The Power Rule of Logarithms**
  $\log_a M^r = r \log_a M$
- **Change-of-Base Formula**
  $\log_a M = \frac{\log_b M}{\log_b a} = \frac{\log M}{\log a} = \frac{\ln M}{\ln a}$

# 9.4 Exercises

**For Extra Help:** 

## Concepts and Vocabulary

*In Problems 1 – 3, fill in the blanks.*

**1.** $\log_a 1 =$ _______; $\log_a a =$ _______.

**2.** $\log_a(xy) =$ ________________.

**3.** $\log_3 10 = \dfrac{\log\_\_\_\_}{\log\_\_\_\_} = \dfrac{\ln\_\_\_\_}{\ln\_\_\_\_}$.

*In Problems 4–6, answer True or False to each statement.*

**4.** $\log(x + 4) = \log x + \log 4$.

**5.** $\log_2(x + 1) + \log_5(x - 2)$ can be written as a single logarithm using the Product Rule.

**6.** $\log_4(4x^2) = 1 + 2\log_4 x$

**7.** State the Product Rule for Logarithms in your own words.

**8.** State the Quotient Rule for Logarithms in your own words.

**9.** Write an example that illustrates why $\log_2(x + y) \neq \log_2 x + \log_2 y$.

**10.** Write an example that illustrates why $(\log_a x)^r \neq r\log_a x$.

## Building Skills

*In Problems 11–24, use properties of logarithms to find the exact value of each expression. Do not use a calculator.*

**11.** $\log_2 2^3$ **12.** $\log_5 5^{-3}$ **13.** $\ln e^{-7}$

**14.** $\ln e^9$ **15.** $3^{\log_3 5}$ **16.** $5^{\log_5 \sqrt{2}}$

**17.** $e^{\ln 2}$ **18.** $e^{\ln 10}$ **19.** $\log 2 + \log 5$

**20.** $\log_6 2 + \log_6 3$ **21.** $\log_3 12 - \log_3 4$ **22.** $\log_4 20 - \log_4 5$

**23.** $10^{\log 8 - \log 2}$ **24.** $e^{\ln 24 - \ln 3}$

*In Problems 25–32, suppose that* $\ln 2 = a$ *and* $\ln 3 = b$. *Use properties of logarithms to write each logarithm in terms of a and b.*

**25.** $\ln 6$ **26.** $\ln\dfrac{3}{2}$ **27.** $\ln 9$ **28.** $\ln 4$

**29.** $\ln 12$ **30.** $\ln 18$ **31.** $\ln\sqrt{2}$ **32.** $\ln\sqrt[4]{3}$

*In Problems 33–54, write each expression as a sum and/or difference of logarithms. Express exponents as factors.*

**33.** $\log(ab)$ **34.** $\log_4\left(\dfrac{a}{b}\right)$ **35.** $\log_5 x^4$ **36.** $\log_3 z^{-2}$

**37.** $\log_2(xy^2)$ **38.** $\log_3(a^3b)$ **39.** $\log_5(25x)$ **40.** $\log_2(8z)$

**41.** $\log_7\left(\dfrac{49}{y}\right)$ **42.** $\log_2\left(\dfrac{16}{p}\right)$ **43.** $\ln(e^2x)$ **44.** $\ln\left(\dfrac{x}{e^3}\right)$

**45.** $\log_3(27\sqrt{x})$ **46.** $\log_2(32\sqrt[4]{z})$ **47.** $\log_5(x^2\sqrt{x^2 + 1})$

**48.** $\log_3(x^3\sqrt{x^2 - 1})$ **49.** $\log\left(\dfrac{x^4}{\sqrt[3]{(x - 1)}}\right)$ **50.** $\ln\left(\dfrac{\sqrt[5]{x}}{(x + 2)^2}\right)$

**51.** $\log_7\sqrt{\dfrac{x + 1}{x}}$ **52.** $\log_6\sqrt[3]{\dfrac{x - 2}{x + 1}}$ **53.** $\log_2\left[\dfrac{x(x - 1)^2}{\sqrt{x + 1}}\right]$

**54.** $\log_4\left[\dfrac{x^3(x - 3)}{\sqrt[3]{x + 1}}\right]$

*In Problems 55–76, write each expression as a single logarithm.*

**55.** $\log 25 + \log 4$
**56.** $\log_4 32 + \log_4 2$
**57.** $\log x + \log 3$
**58.** $\log_2 6 + \log_2 z$
**59.** $\log_3 36 - \log_3 4$
**60.** $\log_2 48 - \log_2 3$
**61.** $3 \log_3 x$
**62.** $8 \log_2 z$
**63.** $\log_4(x + 1) - \log_4 x$
**64.** $\log_5(2y - 1) - \log_5 y$
**65.** $2 \ln x + 3 \ln y$
**66.** $4 \log_2 a + 2 \log_2 b$
**67.** $\frac{1}{2}\log_3 x + 3 \log_3(x - 1)$
**68.** $\frac{1}{3}\log_4 z + 2 \log_4(2z + 1)$
**69.** $\log x^5 - 3 \log x$
**70.** $\log_7 x^4 - 2 \log_7 x$
**71.** $\frac{1}{2}[3 \log x + \log y]$
**72.** $\frac{1}{3}[\ln(x - 1) + \ln(x + 1)]$
**73.** $\log_8(x^2 - 1) - \log_8(x + 1)$
**74.** $\log_5(x^2 + 3x + 2) - \log_5(x + 2)$
**75.** $18 \log\sqrt{x} + 9 \log\sqrt[3]{x} - \log 10$
**76.** $10 \log_4 \sqrt[5]{x} + 4 \log_4 \sqrt{x} - \log_4 16$

*In Problems 77–84, use the Change-of-Base Formula and a calculator to evaluate each logarithm. Round your answer to three decimal places.*

**77.** $\log_2 10$
**78.** $\log_3 18$
**79.** $\log_8 3$
**80.** $\log_7 5$
**81.** $\log_{1/3} 19$
**82.** $\log_{1/4} 3$
**83.** $\log_{\sqrt{2}} 5$
**84.** $\log_{\sqrt{3}} \sqrt{6}$

## Applying the Concepts

**85.** Find the value of $\log_2 3 \cdot \log_3 4 \cdot \log_4 5 \cdot \log_5 6 \cdot \log_6 7 \cdot \log_7 8$.

**86.** Find the value of $\log_2 4 \cdot \log_4 6 \cdot \log_6 8$.

**87.** Find the value of $\log_2 3 \cdot \log_3 4 \cdot \cdots \cdot \log_n(n + 1) \cdot \log_{n+1} 2$.

**88.** Find the value of $\log_3 3 \cdot \log_3 9 \cdot \log_3 27 \cdot \cdots \cdot \log_3 3^n$.

## Extending the Concepts

**89.** Show that $\log_a\left(x + \sqrt{x^2 - 1}\right) + \log_a\left(x - \sqrt{x^2 - 1}\right) = 0$.

**90.** Show that $\log_a\left(\sqrt{x} + \sqrt{x - 1}\right) + \log_a\left(\sqrt{x} - \sqrt{x - 1}\right) = 0$.

**91.** If $f(x) = \log_a x$, show that $f(AB) = f(A) + f(B)$.

**92.** Find the domain of $f(x) = \log_a x^2$ and the domain of $g(x) = 2 \log_a x$. Since $\log_a x^2 = 2 \log_a x$, how can it be that the domains are not equal? Write a brief explanation.

## Synthesis Review

*In Problems 93–98, solve each equation.*

**93.** $4x + 3 = 13$
**94.** $-3x + 10 = 4$
**95.** $x^2 + 4x + 2 = 0$
**96.** $3x^2 = 2x + 1$
**97.** $\sqrt{x + 2} - 3 = 4$
**98.** $\sqrt[3]{2x} - 2 = -5$

## The Graphing Calculator

*We can use the Change-of-Base Formula to graph any logarithmic function on a graphing calculator. For example, to graph $f(x) = \log_2 x$ we would graph $Y_1 = \frac{\log x}{\log 2}$ or $Y_1 = \frac{\ln x}{\ln 2}$.*

*In Problems 99–102, graph each logarithmic function using a graphing calculator. State the domain and the range of each function.*

**99.** $f(x) = \log_3 x$
**100.** $f(x) = \log_5 x$
**101.** $F(x) = \log_{1/2} x$
**102.** $G(x) = \log_{1/3} x$

# 9.5 Exponential and Logarithmic Equations

**OBJECTIVES**

1. Solve Logarithmic Equations Using the Properties of Logarithms
2. Solve Exponential Equations
3. Solve Equations Involving Exponential Models

### *Preparing for Exponential and Logarithmic Equations*

*Before getting started, take the following readiness quiz. If you get a problem wrong, go back to the section cited and review the material.*

**1.** Solve: $2x + 5 = 13$ [Section 1.1, pp. 53–56]

**2.** Solve: $x^2 - 4x = -3$ [Section 5.8, pp. 423–426]

**3.** Solve: $3a^2 = a + 5$ [Section 8.3, pp. 624–629]

**4.** Solve: $(x + 3)^2 + 2(x + 3) - 8 = 0$ [Section 8.4, pp. 639–644]

## 1 Solve Logarithmic Equations Using the Properties of Logarithms

In Section 9.3, we solved logarithmic equations of the form $\log_a x = y$. These equations are solved by changing the logarithmic equation to an equivalent exponential equation. Often, however, logarithmic equations have more than one logarithm in them. In this case, we need to use properties of logarithms to solve the logarithmic equation.

In the last section we learned that if $M = N$, then $\log_a M = \log_a N$. It turns out that the converse of this property is true as well.

**ONE-TO-ONE PROPERTY OF LOGARITHMS**

In the following property, $M$, $N$, and $a$ are positive real numbers, with $a \neq 1$.

$$\text{If } \log_a M = \log_a N, \text{ then } M = N.$$

This property is useful for solving logarithmic equations that have the same base by setting the arguments equal to each other.

### EXAMPLE 1 Solving a Logarithmic Equation

Solve: $2 \log_3 x = \log_3 25$

#### Solution

We notice that both logarithms are to the same base, 3. So if we can write the equation in the form $\log_a M = \log_a N$, we can use the One-to-One Property and set the arguments equal to each other.

$$2 \log_3 x = \log_3 25$$

$r \log_a M = \log_a M^r$: $\quad \log_3 x^2 = \log_3 25$

Set the arguments equal to each other: $\quad x^2 = 25$

Square Root Method: $\quad x = -5 \quad \text{or} \quad x = 5$

The apparent solution $x = -5$ is extraneous because the argument of a logarithm must be greater than zero and $-5$ causes the argument to be negative. We now check the other apparent solution.

**Check** $x = 5$:

$$2 \log_3 5 \stackrel{?}{=} \log_3 25$$

$$2 \log_3 5 \stackrel{?}{=} \log_3 5^2$$

$$2 \log_3 5 = 2 \log_3 5 \quad \text{True}$$

The solution set is $\{5\}$.

*Preparing for...Answers* **1.** $\{4\}$ **2.** $\{1, 3\}$ **3.** $\left\{\frac{1 - \sqrt{61}}{6}, \frac{1 + \sqrt{61}}{6}\right\}$ **4.** $\{-7, -1\}$

**QUICK ✓**

**1.** Solve: $2 \log_4 x = \log_4 9$

If a logarithmic equation contains more than one logarithm on one side of the equation, then we can use properties of logarithms to rewrite the equation as a single logarithm. Once again, we use properties to reduce an equation into a form that is familiar. In this case, we use properties of logarithms to express the sum or difference of logarithms as a single logarithm. We then express the logarithmic equation as an exponential equation and solve for the unknown.

### EXAMPLE 2 Solving a Logarithmic Equation

Solve: $\log_2(x - 2) + \log_2 x = 3$

**Solution**

We use the fact that the sum of two logarithms can be written as the logarithm of the product to write the equation with a single logarithm.

$$\log_2(x - 2) + \log_2 x = 3$$

$\log_a M + \log_a N = \log_a(MN)$:
$$\log_2[x(x - 2)] = 3$$

If $y = \log_a M$, then $a^y = M$:
$$2^3 = x(x - 2)$$

Distribute:
$$8 = x^2 - 2x$$

Write in standard form:
$$x^2 - 2x - 8 = 0$$

Factor:
$$(x - 4)(x + 2) = 0$$

Zero-Product Property:
$$x - 4 = 0 \quad \text{or} \quad x + 2 = 0$$
$$x = 4 \quad \text{or} \quad x = -2$$

**Check** $x = 4$: $\log_2(4 - 2) + \log_2 4 \stackrel{?}{=} 3$
$\log_2(2) + \log_2 4 \stackrel{?}{=} 3$
$1 + 2 = 3$ True

$x = -2$: $\log_2(-2 - 2) + \log_2(-2) \stackrel{?}{=} 3$
$x = -2$ is extraneous because it causes the argument to be negative

**Work Smart**

The apparent solution $x = -2$ in Example 2 is extraneous because it results in us attempting to find the log of a negative number, not because $-2$ is negative.

The solution set is $\{4\}$.

**QUICK ✓**

**2.** Solve: $\log_4(x - 6) + \log_4 x = 2$

## 2 Solve Exponential Equations

In Section 9.2, we solved exponential equations by using the fact that if $a^u = a^v$, then $u = v$. However, in many situations, it is difficult (if not impossible) to write each side of the equation with the same base. For example, to solve the equation $3^x = 5$, using the method presented in Section 9.2, would require that we determine the quantity that we would raise 3 to in order to obtain 5 (but this is the original problem!).

### EXAMPLE 3 Using Logarithms to Solve Exponential Equations

Solve: $3^x = 5$

### Solution

We cannot write 5 so that it is 3 raised to some integer power. Therefore, we write the equation $3^x = 5$ as a logarithm using the fact that $a^y = x$ is equivalent to $y = \log_a x$.

$$3^x = 5$$

Write as a logarithm; If $a^y = x$, then $y = \log_a x$: $\quad x = \log_3 5$ Exact solution

If we want a decimal approximation to the solution, we use the Change-of-Base Formula.

$$x = \log_3 5 = \frac{\log 5}{\log 3}$$

$$\approx 1.465 \quad \text{Approximate solution}$$

An alternative approach to solving the equation would be to take the logarithm of both sides of the equation. Typically, we take either the natural logarithm or common logarithm of both sides of the equation. If we take the natural logarithm of both sides of the equation, we obtain the following:

$$3^x = 5$$

$$\ln 3^x = \ln 5$$

$\log_a M^r = r \log_a M$: $\quad x \ln 3 = \ln 5$

Divide both sides by ln 3: $\quad x = \dfrac{\ln 5}{\ln 3}$ Exact solution

$$\approx 1.465 \quad \text{Approximate solution}$$

The solution set is $\left\{\frac{\ln 5}{\ln 3}\right\}$. If we had taken the common logarithm of both sides, the solution set would be $\left\{\frac{\log 5}{\log 3}\right\}$.

**QUICK ✓** *Solve each equation.*

**3.** $2^x = 11$ **4.** $5^{2x} = 3$

---

## EXAMPLE 4 Using Logarithms to Solve Exponential Equations

Solve: $4e^{3x} = 10$

### Solution

We first need to isolate the exponential expression by dividing both sides of the equation by 4.

$$4e^{3x} = 10$$

$$e^{3x} = \frac{5}{2}$$

We cannot express $\frac{5}{2}$ as $e$ raised to an integer power, so we cannot solve the equation using the fact that if $a^u = a^v$, then $u = v$. However, we can solve the equation by writing the exponential equation as an equivalent logarithmic equation.

$$e^{3x} = \frac{5}{2}$$

Write as a logarithm; If $a^y = x$, then $y = \log_a x$: $\ln\frac{5}{2} = 3x$

Divide both sides by 3: $x = \frac{\ln\left(\frac{5}{2}\right)}{3}$ Exact solution

$x \approx 0.305$ Approximate solution

We leave it to you to verify the solution. The solution set is $\left\{\frac{\ln\left(\frac{5}{2}\right)}{3}\right\}$.

**QUICK ✓** *Solve each equation.*

**5.** $e^{2x} = 5$ **6.** $3e^{-4x} = 20$

## 3 Solve Equations Involving Exponential Models

In Section 9.2, we looked at a variety of models from areas such as statistics, biology, and finance. Now that we have the ability to solve exponential equations, we can look at these models again. This time, rather than evaluating the models at certain values of the independent variable, we will solve equations involving the models.

### EXAMPLE 5 Radioactive Decay

The half-life of plutonium-239 is 24,360 years. The maximum amount of plutonium-239 that an adult can handle without significant injury is 0.13 micrograms (= 0.000000013 grams). Suppose that a researcher possesses a 1-gram sample of plutonium-239. The amount $A$ (in grams) of plutonium-239 after $t$ years is given by

$$A(t) = 1 \cdot \left(\frac{1}{2}\right)^{t/24{,}360}$$

**(a)** How long will it take before 0.9 gram of plutonium-239 is left in the sample?

**(b)** How long will it take before the 1-gram sample is safe? That is, how long will it take before 0.000000013 gram is left?

### Solution

**(a)** We need to determine the time until $A = 0.9$ gram. So we need to solve the equation

$$0.9 = 1 \cdot \left(\frac{1}{2}\right)^{t/24{,}360}$$

for $t$. But how can we get the $t$ out of the exponent? Remember, one of the properties of logarithms that states $\log_a M^r = r \log_a M$. So if we take the logarithm of both sides of the equation, we can "get the variable down in front."

$$\log 0.9 = \log\left(\frac{1}{2}\right)^{t/24{,}360}$$

$$\log 0.9 = \frac{t}{24{,}360}\log\left(\frac{1}{2}\right)$$

Multiply both sides by 24,360: $24{,}360 \log 0.9 = t \log\left(\frac{1}{2}\right)$

Divide both sides by $\log\left(\frac{1}{2}\right)$: $\frac{24{,}360 \log 0.9}{\log\left(\frac{1}{2}\right)} = t$

So $t = \dfrac{24{,}360 \log 0.9}{\log\left(\frac{1}{2}\right)} \approx 3{,}702.8$. After approximately 3703 years, there will be 0.9 gram of plutonium-239.

**(b)** We need to determine the time until $A = 0.000000013$ gram. So we solve the equation

$$0.000000013 = 1 \cdot \left(\frac{1}{2}\right)^{t/24{,}360}$$

Take the logarithm of both sides: $$\log(0.000000013) = \log\left(\frac{1}{2}\right)^{t/24{,}360}$$

$\log_a M^r = r \log_a M$: $$\log(0.000000013) = \frac{t}{24{,}360}\log\left(\frac{1}{2}\right)$$

Multiply both sides by 24,360: $$24{,}360 \log(0.000000013) = t \log\left(\frac{1}{2}\right)$$

Divide both sides by $\log\left(\frac{1}{2}\right)$: $$\frac{24{,}360 \log(0.000000013)}{\log\left(\frac{1}{2}\right)} = t$$

So $t = \dfrac{24{,}360 \log(0.000000013)}{\log\left(\frac{1}{2}\right)} \approx 638{,}156.8$ years. After approximately 638,157 years, the 1-gram sample will be safe!

**QUICK ✓**

**7.** The half-life of thorium-227 is 18.72 days. Suppose that a researcher possesses a 10-gram sample of thorium-227. The amount $A$ (in grams) of thorium-227 after $t$ days is given by

$$A(t) = 10 \cdot \left(\frac{1}{2}\right)^{t/18.72}$$

**(a)** How long will it take before 9 grams of thorium-227 is left in the sample?

**(b)** How long will it take before 3 grams of thorium-227 is left in the sample?

---

Now let's look at an example involving compound interest. Remember, the compound interest formula states that the future value of $P$ dollars invested in an account paying an annual interest rate $r$, compounded $n$ times per year for $t$ years, is given by $A = P\left(1 + \dfrac{r}{n}\right)^{nt}$.

## EXAMPLE 6 Future Value of Money

Suppose that you deposit $3000 into a Roth IRA today. If the deposit earns 8% interest compounded quarterly, how long will it be before the account is worth

**(a)** $4,500?

**(b)** $6,000? That is, how long will you have to wait until your money doubles?

### Solution

First, we write the model with the values of the variables entered. With $P = 3000$, $r = 0.08$ and $n = 4$ (compounded quarterly), we have

$$A = 3000\left(1 + \frac{0.08}{4}\right)^{4t} \quad \text{or} \quad A = 3000(1.02)^{4t}$$

**(a)** We want to know the time $t$ until $A = 4500$. That is, we want to solve

$$4500 = 3000(1.02)^{4t}$$

Divide both sides by 3000: $$1.5 = (1.02)^{4t}$$

Take the logarithm of both sides: $$\log 1.5 = \log(1.02)^{4t}$$

$\log_a M^r = r \log_a M$: $$\log 1.5 = 4t \log(1.02)$$

Divide both sides by 4 log (1.02): $$\frac{\log 1.5}{4\log(1.02)} = t$$

So $t = \dfrac{\log 1.5}{4\log(1.02)} \approx 5.12$. After approximately 5.12 years (5 years, 1.5 months), the account will be worth \$4500.

**(b)** We want to know the time $t$ until $A = 6000$. That is, we want to solve

$$6000 = 3000(1.02)^{4t}$$

Divide both sides by 3000: $$2 = (1.02)^{4t}$$

Take the logarithm of both sides: $$\log 2 = \log(1.02)^{4t}$$

$\log_a M^r = r \log_a M$: $$\log 2 = 4t \log(1.02)$$

Divide both sides by 4 log(1.02): $$\frac{\log 2}{4\log(1.02)} = t$$

So $t = \dfrac{\log 2}{4\log(1.02)} \approx 8.75$. After approximately 8.75 years (8 years, 9 months), the account will be worth \$6000.

**QUICK ✓**

**8.** Suppose that you deposit \$2000 into a Roth IRA today. If the deposit earns 6% interest compounded monthly, how long will it be before the account is worth

**(a)** \$3000?

**(b)** \$4000? That is, how long will you have to wait until your money doubles?

# 9.5 Exercises

**For Extra Help:**

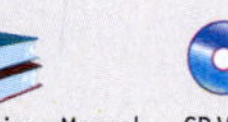

Student Solutions Manual

CD Video

PH Math/Tutor Center

MathXL Tutorials on CD

MathXL®

MyMathLab

## Concepts and Vocabulary

*In Problems 1 and 2, fill in the blanks.*

**1.** If $\log_a M = \log_a N$, then ________.

**2.** To solve $5^x = 12$, we would first take the ________ ________ or ________ ________ of both sides of the equation.

*In Problems 3 and 4, answer True or False to each statement.*

**3.** If a logarithmic equation contains more than one logarithm, we can use the properties of logarithms to write the equation with a single logarithm.

**4.** Logarithmic equations never have extraneous solutions.

**5.** Explain why we can't use the fact that if $a^u = a^v$, then $u = v$ to solve the equation $2^x = 7$.

**6.** Suppose you were solving a logarithmic equation that contains the term $\log_3(x + 3)$ and $x = -2$ was an apparent solution to the equation. Do you think that the solution is extraneous? Why or why not?

## Building Skills

*In Problems 7–44, solve each equation. Express irrational solutions in exact form and as a decimal rounded to three decimal places.*

**7.** $\log_2 x = \log_2 7$

**8.** $\log_5 x = \log_5 13$

**9.** $2 \log_3 x = \log_3 81$

**10.** $2 \log_3 x = \log_3 4$

**11.** $\log_6(3x + 1) = \log_6 10$

**12.** $\log(2x - 3) = \log 11$

**13.** $\frac{1}{2}\ln x = 2 \ln 3$

**14.** $\frac{1}{2}\log_2 x = 2 \log_2 2$

**15.** $\log_2(x + 3) + \log_2 x = 2$

**16.** $\log_2(x - 7) + \log_2 x = 3$

**17.** $\log_2(x - 1) + \log_2(x + 5) = 4$

**18.** $\log_2(x - 2) + \log_2(x + 4) = 4$

**19.** $\log(x + 3) - \log x = 1$

**20.** $\log_3(x + 5) - \log_3 x = 2$

**21.** $\log_4(x + 5) - \log_4(x - 1) = 2$

**22.** $\log_3(x + 2) - \log_3(x - 2) = 4$

**23.** $\log_4(x + 3) + \log_4(x - 6) = \log_4 3$

**24.** $\log_5(x + 3) + \log_5(x - 4) = \log_5 8$

**25.** $2^x = 10$

**26.** $3^x = 8$

**27.** $5^x = 20$

**28.** $4^x = 20$

**29.** $\left(\frac{1}{2}\right)^x = 7$

**30.** $\left(\frac{1}{2}\right)^x = 10$

**31.** $e^x = 5$

**32.** $e^x = 3$

**33.** $10^x = 5$

**34.** $10^x = 0.2$

**35.** $3^{2x} = 13$

**36.** $2^{2x} = 5$

**37.** $\left(\frac{1}{2}\right)^{4x} = 3$

**38.** $\left(\frac{1}{3}\right)^{2x} = 4$

**39.** $4 \cdot 2^x + 3 = 8$

**40.** $3 \cdot 4^x - 5 = 10$

**41.** $-3e^x = -18$

**42.** $\frac{1}{2}e^x = 4$

**43.** $0.2^{x+1} = 3^x$

**44.** $0.4^x = 2^{x-3}$

## Mixed Practice

*In Problems 45–58, solve each equation. Express irrational solutions in exact form and as a decimal rounded to three decimal places.*

**45.** $\log_4 x + \log_4(x - 6) = 2$

**46.** $\log_6 x + \log_6(x + 5) = 2$

**47.** $5^{3x} = 7$

**48.** $3^{2x} = 4$

**49.** $3 \log_2 x = \log_2 8$

**50.** $5 \log_4 x = \log_4 32$

**51.** $\frac{1}{3}e^x = 5$

**52.** $-4e^x = -16$

**53.** $\left(\frac{1}{4}\right)^{x+1} = 8^x$

**54.** $9^x = 27^{x-4}$

**55.** $\log_3 x = \log_3 16$

**56.** $\log_7 x = \log_7 8$

**57.** $\log_2(x + 4) + \log_2(x - 1) = \log_2 6$

**58.** $\log_3(x - 5) + \log_3(x + 1) = \log_3 7$

## Applying the Concepts

**59. A Population Model** According to the U.S. Census Bureau, the population of the United States in 2005 was 296 million people. In addition, the population of the United States was growing at a rate of 1.1% per year. Assuming that this growth rate continues, the model $P(t) = 296(1.011)^{t-2005}$ represents the population $P$ (in millions of people) in year $t$.

**(a)** According to this model, when will the population of the United States be 351 million people?

**(b)** According to this model, when will the population of the United States be 482 million people?

**60. A Population Model** According to the *Statistical Abstract of the United States*, the population of the world in 2005 was 6448 million people. In addition, the population of the world was growing at a rate of 1.26% per year. Assuming that this growth rate continues, the model $P(t) = 6448(1.0126)^{t-2005}$ represents the population $P$ (in millions of people) in year $t$.

**(a)** According to this model, when will the population of the world be 9.65 billion people?
**(b)** According to this model, when will the population of the world be 11.55 billion people?

**61. Time Is Money** Suppose that you deposit $5000 into a Certificate of Deposit (CD) today. If the deposit earns 6% interest compounded monthly, how long will it be before the account is worth

**(a)** $7,000?
**(b)** $10,000? That is, how long will you have to wait until your money doubles?

**62. Time Is Money** Suppose that you deposit $8000 into a Certificate of Deposit (CD) today. If the deposit earns 4% interest compounded quarterly, how long will it be before the account is worth

**(a)** $10,000?
**(b)** $24,000? That is, how long will you have to wait until your money triples?

**63. Depreciation** Based on data obtained from the *Kelley Blue Book,* the value $V$ of a Dodge Neon that is $t$ years old can be modeled by $V(t) = 14{,}512(0.82)^t$.

**(a)** According to the model, when will the car be worth $10,000?
**(b)** According to the model, when will the car be worth $5000?
**(c)** According to the model, when will the car be worth $1000?

**64. Depreciation** Based on data obtained from the *Kelley Blue Book,* the value $V$ of a Dodge Stratus that is $t$ years old can be modeled by $V(t) = 19{,}282(0.84)^t$.

**(a)** According to the model, when will the car be worth $10,000?
**(b)** According to the model, when will the car be worth $5000?
**(c)** According to the model, when will the car be worth $1000?

**65. Radioactive Decay** The half-life of beryllium-11 is 13.81 seconds. Suppose that a researcher possesses a 100-gram sample of beryllium-11. The amount $A$ (in grams) of beryllium-11 after $t$ seconds is given by

$$A(t) = 100 \cdot \left(\frac{1}{2}\right)^{t/13.81}$$

**(a)** When will there be 90 grams of beryllium-11 left in the sample?
**(b)** When will there be 25 grams of beryllium-11 left in the sample?
**(c)** When will there be 10 grams of beryllium-11 left in the sample?

**66. Radioactive Decay** The half-life of carbon-10 is 19.255 seconds. Suppose that a researcher possesses a 100-gram sample of carbon-10. The amount $A$ (in grams) of carbon-10 after $t$ seconds is given by

$$A(t) = 100 \cdot \left(\frac{1}{2}\right)^{t/19.255}$$

**(a)** When will there be 90 grams of carbon-10 left in the sample?
**(b)** When will there be 25 grams of carbon-10 left in the sample?
**(c)** When will there be 10 grams of carbon-10 left in the sample?

**67. Newton's Law of Cooling** Newton's Law of Cooling states that the temperature of a heated object decreases exponentially over time toward the temperature of the surrounding medium. Suppose that a pizza is removed from a 400°F oven and placed in a room whose temperature is 70°F. The temperature $u$ (in °F) of the pizza at time $t$ (in minutes) can be modeled by $u(t) = 70 + 330e^{-0.072t}$.

**(a)** According to the model, when will be the temperature of the pizza be 300°F?
**(b)** According to the model, when will be the temperature of the pizza be 220°F?

**68. Newton's Law of Cooling** Newton's Law of Cooling states that the temperature of a heated object decreases exponentially over time toward the temperature of the surrounding medium. Suppose that coffee that is 170°F is poured into a coffee mug and allowed to cool in a room whose temperature is 70°F. The temperature $u$ (in °F) of the coffee at time $t$ (in minutes) can be modeled by $u(t) = 70 + 100e^{-0.045t}$.

**(a)** According to the model, when will be the temperature of the coffee be 120°F?
**(b)** According to the model, when will be the temperature of the coffee be 100°F?

**69. Learning Curve** Suppose that a student has 200 vocabulary words to learn. If a student learns 20 words in 30 minutes, the function

$$L(t) = 200(1 - e^{-0.0035t})$$

models the number of words $L$ that the student will learn after $t$ minutes.

**(a)** After how long will the student learn 50 words?
**(b)** After how long will the student learn 150 words?

**70. Learning Curve** Suppose that a student has 50 biology terms to learn. If a student learns 10 terms in 30 minutes, the function

$$L(t) = 50(1 - e^{-0.0223t})$$

models the number of terms $L$ that the student will learn after $t$ minutes.

**(a)** After how long will the student learn 10 words?
**(b)** After how long will the student learn 40 words?

## Extending the Concepts

**71. The Rule of 72** The Rule of 72 states that the time for an investment to double in value is approximately given by 72 divided by the annual interest rate. For example, an investment earning 10% annual interest will double in approximately $\frac{72}{10} = 7.2$ years.

**(a)** According to the Rule of 72, approximately how long will it take an investment to double if it earns 8% annual interest?
**(b)** Derive a formula that can be used to find the number of years required for an investment to double. (*Hint:* Let $A = 2P$ in the formula $A = P\left(1 + \frac{r}{n}\right)^{nt}$ and solve for $t$.)
**(c)** Use the formula derived in part (b) to determine the exact amount of time it takes an investment to double that earns 8% interest compounded monthly. Compare the result to the results given by the Rule of 72.

**72. Critical Thinking** Suppose you need to open up a savings account. Bank A offers 4% interest compounded daily, while Bank B offers 4.1% interest compounded quarterly. Which bank offers the better deal? Why?

**73. Critical Thinking** The bacteria in a 2-liter container double every minute. After 30 minutes the container is full. How long did it take to fill half the container?

## Synthesis Review

*In Problems 74–79, find the following values for each function.*

**(a)** $f(3)$ **(b)** $f(-2)$ **(c)** $f(0)$

**74.** $f(x) = 5x + 2$ **75.** $f(x) = -2x + 7$ **76.** $f(x) = \dfrac{x + 3}{x - 2}$

**77.** $f(x) = \dfrac{x}{x - 5}$ **78.** $f(x) = \sqrt{x + 5}$ **79.** $f(x) = 2^x$

## The Graphing Calculator

The techniques for solving equations introduced in this chapter apply only to certain types of exponential or logarithmic equations. Solutions for other types of equations are usually studied in calculus using numerical methods. However, a graphing calculator can be used to approximate solutions using the INTERSECT feature. For example, to solve

$e^x = 3x + 2$, we would graph $Y_1 = e^x$ and $Y_2 = 3x + 2$ on the same screen. We then use the INTERSECT feature to find the $x$-coordinate of the point of intersection. This $x$-coordinate represents the approximate solution as shown in Figure 26.

**Figure 26**

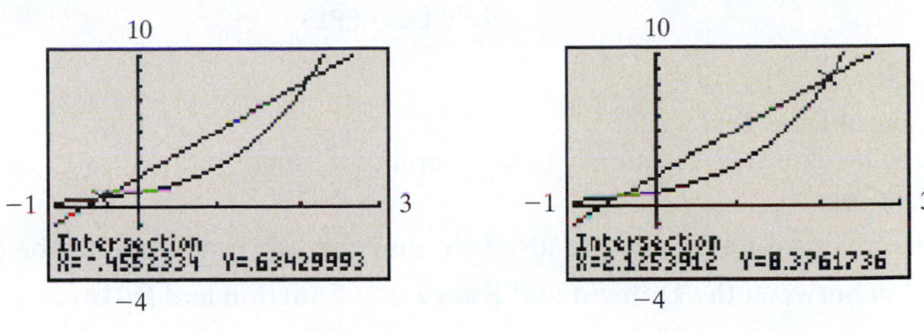


The solution set is $\{-0.46, 2.13\}$ rounded to two decimal places.

*In Problems 80–87, solve each equation using a graphing calculator. Express your answer rounded to two decimal places.*

**80.** $e^x = -3x + 2$ **81.** $e^x = -2x + 5$ **82.** $e^x = x + 2$

**83.** $e^x = x^2$ **84.** $e^x + \ln(x) = 2$ **85.** $e^x - \ln(x) = 4$

**86.** $\ln x = x^2 + 1$ **87.** $\ln x = x^2 - 1$

## CHAPTER 9 ACTIVITY: CORRECT THE QUIZ

**Focus:** Solving exponential and logarithmic equations

**Time:** 10–15 minutes

**Group size:** 2

In this activity you will work as a team to grade the student quiz shown below. One of you will grade the odd questions, and the other will grade the even questions. If an answer is correct, mark it correct. If an answer is wrong, mark it wrong and show the correct answer.

Once all of the quiz questions are graded, explain your results to each other and compute the final score for the quiz. Be prepared to discuss your results with the rest of the class.

| **Student Quiz** | |
|---|---|
| Name: *Ima Student* | Quiz Score: ______ |
| Solve the following equations. Express any irrational answers in exact form. | |
| (1) $\log_2 x = 3$ | Answer: $\{9\}$ |
| (2) $\log_{16} x = \dfrac{3}{4}$ | Answer: $\{8\}$ |
| (3) $\log(2x) - \log 6 = \log(x - 8)$ | Answer: $\{12\}$ |
| (4) $3^x = 27$ | Answer: $\{9\}$ |
| (5) $6^{2x} = 18$ | Answer: $\left\{\dfrac{\log 18}{2 \log 6}\right\}$ |
| (6) $4^{x+9} = 7$ | Answer: $\left\{\dfrac{\log 7}{\log 4} - 9\right\}$ |
| (7) $9^{3x-1} = 27^{4x}$ | Answer: $\left\{-\dfrac{1}{3}\right\}$ |
| (8) $\log_3(2x - 3) = 2$ | Answer: $\left\{\dfrac{11}{2}\right\}$ |

# CHAPTER 9 REVIEW

## Section 9.1 Composite Functions and Inverse Functions

| KEY CONCEPTS | KEY TERMS |
|---|---|
| • $(f \circ g)(x) = f(g(x))$<br>• **Horizontal Line Test**<br>If every horizontal line intersects the graph of a function $f$ in at most one point, then $f$ is one-to-one.<br>• For the inverse of a function to also be a function, the function must be one-to-one.<br>• **Relation between the Domain and Range of a Function and Its Inverse**<br>All elements in the domain of a function are also elements in the range of its inverse<br>All elements in the range of a function are also elements in the domain of its inverse<br>• The graph of a function $f$ and the graph of its inverse are symmetric with respect to the line $y = x$.<br>• $f^{-1}(f(x)) = x$ for every $x$ in the domain of $f$ and<br>$f(f^{-1}(x)) = x$ for every $x$ in the domain of $f^{-1}$ | Composition<br>Composite function<br>One-to-one<br>Inverse function |

| YOU SHOULD BE ABLE TO . . . | EXAMPLE | REVIEW EXERCISES |
|---|---|---|
| 1 Form the composite function (p. 698) | Examples 1 and 2 | 1–8 |
| 2 Determine whether or not a function is one-to-one (p. 700) | Examples 3 and 4 | 9–12 |
| 3 Determine the inverse of a function defined by a map or ordered pair (p. 702) | Examples 5 and 6 | 13–16 |
| 4 Obtain the graph of the inverse function from the graph of a function (p. 704) | Example 7 | 17, 18 |
| 5 Find the inverse of a function defined by an equation (p. 705) | Examples 8 and 9 | 19–22 |

*In Problems 1–4, for the given functions f and g, find:*

**(a)** $(f \circ g)(5)$ **(b)** $(g \circ f)(-3)$ **(c)** $(f \circ f)(-2)$ **(d)** $(g \circ g)(4)$

**1.** $f(x) = 3x + 5;\ g(x) = 2x - 1$

**2.** $f(x) = x - 3;\ g(x) = 5x + 2$

**3.** $f(x) = 2x^2 + 1;\ g(x) = x + 5$

**4.** $f(x) = x - 3;\ g(x) = x^2 + 1$

*In Problems 5–8, for the given functions f and g, find:*

**(a)** $(f \circ g)(x)$ **(b)** $(g \circ f)(x)$ **(c)** $(f \circ f)(x)$ **(d)** $(g \circ g)(x)$

**5.** $f(x) = x + 1;\ g(x) = 5x$

**6.** $f(x) = 2x - 3;\ g(x) = x + 6$

**7.** $f(x) = x^2 + 1;\ g(x) = 2x + 1$

**8.** $f(x) = \dfrac{2}{x + 1};\ g(x) = \dfrac{1}{x}$

*In Problems 9–12, determine which of the following functions is one-to-one.*

**9.** $\{(-5, 8), (-3, 2), (-1, 8), (0, 12), (1, 15)\}$

**10.** $\{(-4, 2), (-2, 1), (0, 0), (1, -1), (2, 8)\}$

**11.**

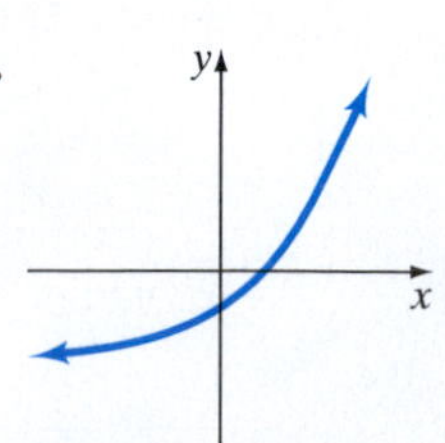


**12.**

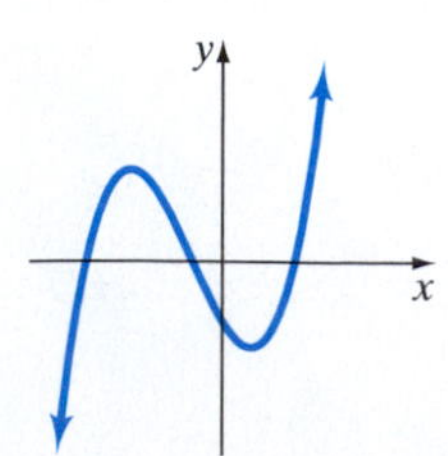

*In Problems 13–16, find the inverse of the following one-to-one functions.*

**13.**

**14.**

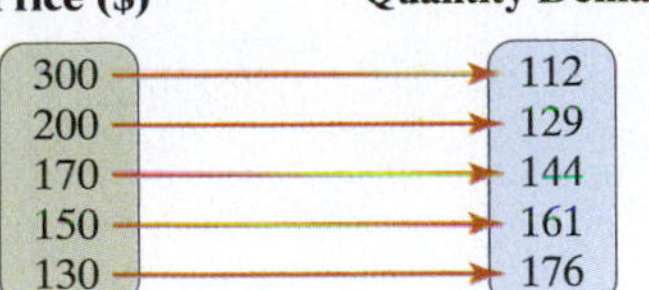


**15.** $\{(-5, 3), (-3, 1), (1, -3), (2, 9)\}$

**16.** $\{(-20, 1), (-15, 4), (5, 3), (25, 2)\}$

*In Problems 17 and 18, the graph of a one-to-one function f is given. Draw the graph of the inverse function $f^{-1}$.*

**17.**

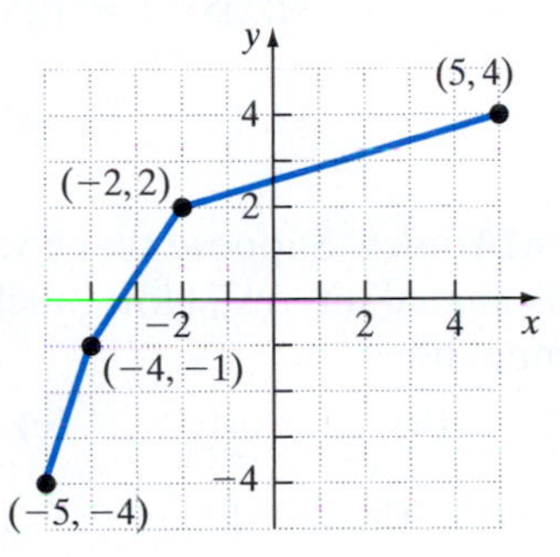


**18.**

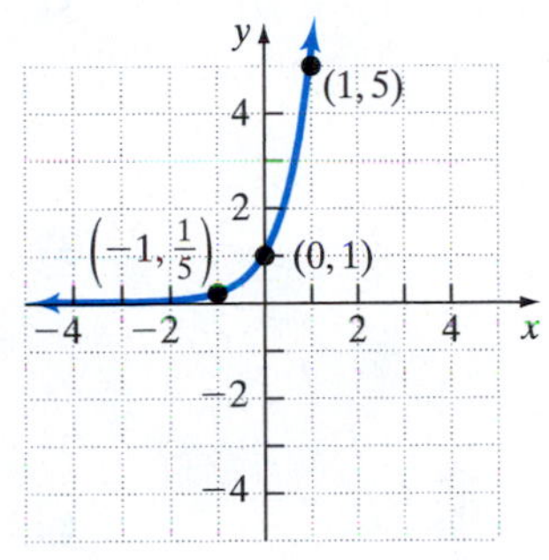


*In Problems 19–22, find the inverse function of the given one-to-one function.*

**19.** $f(x) = 5x$ **20.** $H(x) = 2x + 7$ **21.** $P(x) = \dfrac{4}{x + 2}$ **22.** $g(x) = 2x^3 - 1$

## Section 9.2 Exponential Functions

| KEY CONCEPTS | KEY TERMS |
|---|---|
| • **Properties of Exponential Functions**<br>**1.** The domain is the set of all real numbers. The range is the set of all positive real numbers.<br>**2.** There are no $x$-intercepts. The $y$-intercept is 1.<br>**3.** The graph of an exponential function contains the points $\left(-1, \frac{1}{a}\right)$, $(0, 1)$, and $(1, a)$ if $a > 1$. The graph of an exponential function contains the points $(-1, a)$, $(0, 1)$, and $\left(1, \frac{1}{a}\right)$ if $0 < a < 1$.<br>• **Property for Solving Exponential Equations of the Form $a^u = a^v$**<br>If $a^u = a^v$, then $u = v$. | Exponential function<br>The number $e$<br>Exponential equation<br>Half-life<br>Payment period<br>Compound interest |

| YOU SHOULD BE ABLE TO . . . | EXAMPLE | REVIEW EXERCISES |
|---|---|---|
| 1 Evaluate exponential expressions (p. 713) | Example 1 | 23–25 |
| 2 Graph exponential functions (p. 714) | Examples 2 through 4 | 26–29 |
| 3 Define the number $e$ (p. 717) | | 30 |
| 4 Solve exponential equations (p. 718) | Examples 5 and 6 | 31–36 |
| 5 Study exponential models that describe our world (p. 721) | Examples 7 through 9 | 37–40 |

*In Problems 23–25, approximate each number using a calculator. Express your answer rounded to three decimal places.*

**23.** **(a)** $7^{1.7}$ **(b)** $7^{1.73}$ **(c)** $7^{1.732}$ **(d)** $7^{1.7321}$ **(e)** $7^{\sqrt{3}}$

**24.** **(a)** $10^{3.1}$ **(b)** $10^{3.14}$ **(c)** $10^{3.142}$ **(d)** $10^{3.1416}$ **(e)** $10^{\pi}$

**25.** **(a)** $e^{0.5}$ **(b)** $e^{-1}$ **(c)** $e^{1.5}$ **(d)** $e^{-0.8}$ **(e)** $e^{\sqrt{\pi}}$

*In Problems 26–29, graph each function. State the domain and range of the function.*

**26.** $f(x) = 9^x$ **27.** $g(x) = \left(\frac{1}{9}\right)^x$ **28.** $H(x) = 4^{x-2}$ **29.** $h(x) = 4^x - 2$

**30.** State the definition of the number $e$.

*In Problems 31–36, solve each equation.*

**31.** $2^x = 64$ **32.** $25^{x-2} = 125$ **33.** $27^x \cdot 3^{x^2} = 9^2$

**34.** $\left(\frac{1}{4}\right)^x = 16$ **35.** $(e^2)^{x-1} = e^x \cdot e^7$ **36.** $(2^x)^x = 512$

**37. Future Value of Money** Suppose that you deposit \$2500 into a traditional IRA that pays 4.5% annual interest. How much money will you have after 25 years if interest is compounded

**(a)** annually? **(b)** quarterly? **(c)** monthly? **(d)** daily?

**38. Radioactive Decay** The half-life of the radioactive gas radon is 3.5 days. Suppose a researcher possesses a 100-gram sample of radon gas. The amount $A$ (in grams) of radon after $t$ days is given by $A(t) = 100\left(\frac{1}{2}\right)^{t/3.5}$. How much radon gas is left in the sample after

**(a)** 1 day? **(b)** 3.5 days? **(c)** 7 days? **(d)** 30 days?

**39. A Population Model** According to the U.S. Census Bureau, the population of Nevada in 2000 was 1.998 million people. In addition, the population of Nevada was growing at a rate of 5.2% per year. Assuming that this growth rate continues, the model $P(t) = 1.998(1.052)^{t-2000}$ represents the population (in millions of people) in year $t$. According to this model, what will be the population of Nevada

**(a)** in 2006? **(b)** in 2010?

**40. Newton's Law of Cooling** A baker removes a cake from a 350°F oven and places it in a room whose temperature is 72°F. According to Newton's Law of Cooling, the temperature $u$ (in °F) of the cake at time $t$ (in minutes) can be modeled by $u(t) = 72 + 278e^{-0.0835t}$. According to this model, what will be the temperature of the cake

**(a)** after 15 minutes? **(b)** after 30 minutes?

## Section 9.3 Logarithmic Functions

| KEY CONCEPTS | KEY TERMS |
|---|---|
| • **The Logarithmic Function to the Base $a$**<br>$y = \log_a x$ is equivalent to $x = a^y$<br>• **Properties of the Logarithmic Function**<br>**1.** The domain is the set of all positive real numbers. The range is the set of all real numbers.<br>**2.** There are no $y$-intercepts. The $x$-intercept is 1.<br>**3.** The graph of the logarithmic function contains the points $\left(\frac{1}{a}, -1\right)$, $(1, 0)$, and $(a, 1)$. | Logarithmic function<br>Natural logarithm function<br>Logarithmic equation<br>Intensity of a sound wave<br>Decibels<br>Loudness |

| YOU SHOULD BE ABLE TO . . . | EXAMPLE | REVIEW EXERCISES |
|---|---|---|
| 1 Change exponential expressions to logarithmic expressions (p. 731) | Example 1 | 41–44 |
| 2 Change logarithmic expressions to exponential expressions (p. 731) | Example 2 | 45–48 |
| 3 Evaluate logarithmic functions (p. 732) | Examples 3 and 4 | 49–52 |
| 4 Determine the domain of a logarithmic function (p. 733) | Example 5 | 53–56 |
| 5 Graph logarithmic functions (p. 734) | Examples 6 and 7 | 57, 58 |
| 6 Work with natural and common logarithms (p. 736) | Example 8 | 59–62 |
| 7 Solve logarithmic equations (p. 737) | Examples 9 and 10 | 63–68 |
| 8 Study logarithmic models that describe our world (p. 739) | Example 11 | 69, 70 |

*In Problems 41–44, change each exponential expression to an equivalent expression involving a logarithm.*

**41.** $3^4 = 81$

**42.** $4^{-3} = \frac{1}{64}$

**43.** $b^3 = 5$

**44.** $10^{3.74} = x$

*In Problems 45–48, change each logarithmic expression to an equivalent expression involving an exponent.*

**45.** $\log_8 2 = \frac{1}{3}$

**46.** $\log_5 18 = r$

**47.** $\ln(x + 3) = 2$

**48.** $\log x = -4$

*In Problems 49–52, find the exact value of each logarithm without using a calculator.*

**49.** $\log_8 128$

**50.** $\log_6 1$

**51.** $\log \frac{1}{100}$

**52.** $\log_9 27$

*In Problems 53–56, find the domain of each function.*

**53.** $f(x) = \log_2(x + 5)$

**54.** $g(x) = \log_8(7 - 3x)$

**55.** $h(x) = \ln(3x)$

**56.** $F(x) = \log_{\frac{1}{3}}(4x + 10)$

*In Problems 57 and 58, graph each function.*

**57.** $f(x) = \log_{\frac{5}{2}} x$

**58.** $g(x) = \log_{\frac{2}{5}} x$

*In Problems 59–62, use a calculator to evaluate each expression. Round your answers to three decimal places.*

**59.** $\ln 24$

**60.** $\ln \frac{5}{6}$

**61.** $\log 257$

**62.** $\log 0.124$

*In Problems 63–68, solve each logarithmic equation.*

**63.** $\log_7(4x - 19) = 2$

**64.** $\log_{\frac{1}{3}}(x^2 + 8x) = -2$

**65.** $\log_a \frac{4}{9} = -2$

**66.** $\ln e^{5x} = 30$

**67.** $\log(6 - 7x) = 3$

**68.** $\log_b 75 = 2$

**69. Loudness of a Vacuum Cleaner** A vacuum cleaner has an intensity level of $10^{-4}$ watt per square meter. How many decibels is the vacuum cleaner?

**70. The Great New Madrid Earthquake** According to the United States Geological Survey, an earthquake on December 16, 1811 in New Madrid, Missouri had a magnitude of approximately 8.0. What would have been the seismographic reading 100 kilometers from its epicenter?

## Section 9.4 Properties of Logarithms

### KEY CONCEPTS

- **Properties of Logarithms**

  For the following properties, $a$, $M$, and $N$ are positive real numbers with $a \neq 1$ and $r$ is any real number.

  $a^{\log_a M} = M \qquad \log_a(MN) = \log_a M + \log_a N \qquad \log_a M^r = r \log_a M$

  $\log_a a^r = r \qquad \log_a\left(\frac{M}{N}\right) = \log_a M - \log_a N$

- **Change-of-Base Formula**

  If $a \neq 1$, $b \neq 1$, and $M$ are positive real numbers, then

  $$\log_a M = \frac{\log_b M}{\log_b a} = \frac{\log M}{\log a} = \frac{\ln M}{\ln a}$$

| YOU SHOULD BE ABLE TO . . . | EXAMPLE | REVIEW EXERCISES |
|---|---|---|
| 1 Understand the properties of logarithms (p. 744) | Examples 1 through 3 | 71–76 |
| 2 Write a logarithmic expression as a sum or difference of logarithms (p. 746) | Examples 4 through 8 | 77–80 |
| 3 Write a logarithmic expression as a single logarithm (p. 749) | Examples 9 and 10 | 81–84 |
| 4 Evaluate logarithms whose base is neither 10 nor $e$ (p. 750) | Examples 11 and 12 | 85–88 |

*In Problems 71–76, use properties of logarithms to find the exact value of each expression. Do not use a calculator.*

**71.** $\log_4 4^{21}$ **72.** $7^{\log_7 9.34}$ **73.** $\log_5 5$

**74.** $\log_9 1$ **75.** $\log_4 12 - \log_4 3$ **76.** $12^{\log_{12} 2 + \log_{12} 8}$

*In Problems 77–80, write each expression as a sum and/or difference of logarithms. Write exponents as factors.*

**77.** $\log_7\left(\frac{xy}{z}\right)$ **78.** $\log_3\left(\frac{81}{x^2}\right)$ **79.** $\log 1000r^4$ **80.** $\ln\sqrt{\frac{x-1}{x}}$

*In Problems 81–84, write each expression as a single logarithm.*

**81.** $4 \log_3 x + 2 \log_3 y$ **82.** $\frac{1}{4}\ln x + \ln 7 - 2 \ln 3$

**83.** $\log_2 3 - \log_2 6$ **84.** $\log_6(x^2 - 7x + 12) - \log_6(x - 3)$

*In Problems 85–88, use the Change-of-Base Formula and a calculator to evaluate each logarithm. Round your answer to three decimal places.*

**85.** $\log_6 50$ **86.** $\log_\pi 2$ **87.** $\log_{\frac{2}{3}} 6$ **88.** $\log_{\sqrt{5}} 20$

## Section 9.5 Exponential and Logarithmic Equations

### KEY CONCEPT

- In the following $M$, $N$, and $a$ are positive real numbers with $a \neq 1$.

  If $\log_a M = \log_a N$, then $M = N$

| YOU SHOULD BE ABLE TO . . . | EXAMPLE | REVIEW EXERCISES |
|---|---|---|
| 1 Solve logarithmic equations using properties of logarithms (p. 754) | Examples 1 and 2 | 89–92 |
| 2 Solve exponential equations (p. 755) | Examples 3 and 4 | 93–96 |
| 3 Solve equations involving exponential models (p. 757) | Examples 5 and 6 | 97, 98 |

*In Problems 89–96, solve each equation. Express irrational solutions in exact form and as a decimal rounded to three decimal places.*

**89.** $3 \log_4 x = \log_4 1000$

**90.** $\log_3 x + \log_3(x + 6) = 3$

**91.** $\ln(x + 2) - \ln x = \ln(x + 1)$

**92.** $\frac{1}{3}\log_{12} x = 2 \log_{12} 2$

**93.** $2^x = 15$

**94.** $10^{3x} = 27$

**95.** $\frac{1}{3}e^{7x} = 13$

**96.** $3^x = 2^{x+1}$

**97. Radioactive Decay** The half-life of the radioactive gas radon is 3.5 days. Suppose a researcher possesses a 100-gram sample of radon gas. The amount $A$ (in grams) of radon after $t$ days is given by $A(t) = 100\left(\frac{1}{2}\right)^{t/3.5}$.

**(a)** When will 75 grams of radon gas be left in the sample?
**(b)** When will 1 gram of radon gas be left in the sample?

**98. A Population Model** According to the U.S. Census Bureau, the population of Nevada in 2000 was 1.998 million people. In addition, the population of Nevada was growing at a rate of 5.2% per year. Assuming that this growth rate continues, the model $P(t) = 1.998(1.052)^{t-2000}$ represents the population (in millions of people) in year $t$. According to this model, when will the population of Nevada be

**(a)** 3.0 million people?
**(b)** 4.5 million people?

## CHAPTER 9 TEST

*Remember to use your Chapter Test Prep Video CD to see fully worked-out solutions to any of these problems you would like to review.*

**1.** Determine whether the following function is one-to-one:

$$\{(1, 4), (3, 2), (5, 8), (-1, 4)\}$$

**2.** Find the inverse of $f(x) = 4x - 3$.

**3.** Approximate each number using a calculator. Express your answer rounded to three decimal places.

**(a)** $3.1^{3.1}$ **(b)** $3.14^{3.14}$ **(c)** $3.142^{3.142}$ **(d)** $3.1416^{3.1416}$ **(e)** $\pi^{\pi}$

**4.** Change $4^x = 19$ to an equivalent expression involving a logarithm.

**5.** Change $\log_b x = y$ to an equivalent expression involving an exponent.

**6.** Find the exact value of each expression without using a calculator.

**(a)** $\log_3\left(\frac{1}{27}\right)$ **(b)** $\log 10{,}000$

**7.** Determine the domain of $f(x) = \log_5(7 - 4x)$.

*In Problems 8 and 9, graph each function. State the domain and the range of the function.*

**8.** $f(x) = 6^x$

**9.** $g(x) = \log_{\frac{1}{9}} x$

**10.** Use the properties of logarithms to find the exact value of each expression. Do not use a calculator.

**(a)** $\log_7 7^{10}$ **(b)** $3^{\log_3 15}$

**11.** Write the expression $\log_4 \frac{\sqrt{x}}{y^3}$ as a sum and/or difference of logarithms. Express exponents as factors.

**12.** Write the expression $4 \log M + 3 \log N$ as a single logarithm.

**13.** Use the Change-of-Base Formula and a calculator to evaluate $\log_{\frac{3}{4}} 10$. Round to three decimal places.

*In Problems 14–20, solve each equation. Express irrational solutions in exact form and as a decimal rounded to three places.*

**14.** $4^{x+1} = 2^{3x+1}$

**15.** $5^{x^2} \cdot 125 = 25^{2x}$

**16.** $\log_a 64 = 3$

**17.** $\log_2(x^2 - 33) = 8$

**18.** $2 \log_7(x - 3) = \log_7 3 + \log_7 12$

**19.** $3^{x-1} = 17$

**20.** $\log(x - 2) + \log(x + 2) = 2$

**21.** According to the U.S. Bureau of Census, International Data Base, the population of Canada in 2002 was 31.9 million people. In addition, the population of Canada was growing at a rate of 0.8% annually. Assuming that this growth rate continues, the model $P(t) = 31.9(1.008)^{t-2002}$ represents the population (in millions of people) in year $t$.

**(a)** According to the model, what will be the population of Canada in 2010?
**(b)** According to the model, in what year will the population of Canada be 50 million?

**22.** Rustling leaves have an intensity of $10^{-11}$ watt per square meter. How many decibels are rustling leaves?

## CUMULATIVE REVIEW CHAPTERS R–9

**1.** Solve: $3(5 - 2x) + 8 = 4(x - 7) + 1$

**2.** Solve: $5 - 3|x - 2| \geq -7$

**3.** Determine the domain of $f(x) = \dfrac{9 - x^2}{2x^2 - x - 21}$.

**4.** Graph the linear equation: $4x + 3y = 6$

**5.** Find the equation of the line that passes through the points $(-10, 17)$ and $(5, -4)$. Write your answer in either slope-intercept or standard form, whichever you prefer.

**6.** Write the system of linear equations that corresponds to the following augmented matrix. Then determine the solution to the system.

$$\left[\begin{array}{ccc|c} 1 & -2 & 1 & 4 \\ 0 & 1 & -1 & -2 \\ 0 & 0 & 2 & -6 \end{array}\right]$$

**7.** Evaluate: $\begin{vmatrix} 4 & 0 & -2 \\ 2 & -1 & 1 \\ 1 & 3 & 1 \end{vmatrix}$

**8.** Graph the following system of linear inequalities.

$$\begin{cases} x + 2y \geq 8 \\ 2x - y < 1 \end{cases}$$

*In Problems 9 and 10, add, subtract, multiply, or divide as indicated.*

**9.** $(m^2 - 5m + 13) - (6 - 2m - 3m^2)$

**10.** $(2n + 3)(n^2 - 4n + 6)$

*In Problems 11 and 12, factor completely.*

**11.** $16a^2 + 8ab + b^2$

**12.** $6y^2 - 17y + 7$

*In Problems 13 and 14, perform the indicated operations. Be sure to express the final answer in lowest terms.*

**13.** $\dfrac{2x^2 - 9x - 5}{x^2 - 3x - 10} \cdot \dfrac{3x^2 + 2x - 8}{2x^2 - 13x - 7}$

**14.** $\dfrac{4}{p^2 - 6p + 5} + \dfrac{2}{p^2 - 3p - 10}$

**15.** Solve: $\dfrac{2}{x - 5} = \dfrac{x - 2}{x + 1} + \dfrac{6x - 12}{x^2 - 4x - 5}$

**16.** Simplify: $\sqrt{150} + 4\sqrt{6} - \sqrt{24}$

**17.** Rationalize the denominator: $\dfrac{1 + \sqrt{5}}{3 - \sqrt{5}}$

**18.** Solve: $\sqrt{x - 8} + \sqrt{x} = 4$

**19.** Solve: $3x^2 = 4x + 6$

**20.** Solve: $2a - 7\sqrt{a} + 6 = 0$

**21.** Graph: $f(x) = -x^2 + 6x - 4$

**22.** Solve: $3x^2 + 2x - 8 < 0$

**23.** Graph: $g(x) = 3^x - 4$

**24.** Evaluate: $\log_9\left(\dfrac{1}{27}\right)$

CHAPTER

# 10 Conics

Earth is the center of the universe. While this statement seems ludicrous to us now, it was the common belief held up until the 1500s. A Polish astronomer named Nicolaus Copernicus published a book entitled *De Revolutionibus Orbium Coelestium* (On the Revolutions of the Celestial Spheres), which stated that the Earth (along with the other planets) orbited the Sun in a circular motion. Copernicus's ideas were not readily accepted by the geocentrists, who held on to the belief that Earth is at the center of the universe. Copernicus's model of planetary motion was later improved upon by the German astronomer Johannes Kepler. In 1609, Kepler published Astronomia nova (New Astronomy) in which he proved that the orbit of Mars is an ellipse, with the Sun occupying one of its two foci. See Problems 43–46 on page 807.

## OUTLINE



## The Big Picture: Putting It Together

In Chapter 2, we introduced you to the Cartesian plane or rectangular coordinate system. In that chapter, we stated this system allows us to make connections between algebra and geometry. In this chapter, we develop this connection further by showing how geometric definitions of certain figures lead to algebraic equations.

We start the chapter by showing how to use algebra to find the distance between any two points in the Cartesian plane. The method used to find this distance algebraically is a direct consequence of the Pythagorean Theorem studied in Section 8.1. Knowing this formula allows us to present a complete discussion of the so-called *conic sections.*

If you were asked to tell someone what a circle is, what would you say? In all likelihood, you would draw a picture to illustrate your verbal description. By having the Cartesian plane and the distance formula, we can take the geometric definition of a circle and develop an algebraic equation whose graph would represent a circle. This powerful connection between geometry and algebra allows us to answer all types of interesting questions. The methods that we are about to present form the foundation of an area of mathematics called analytic geometry.

# 10.1 Distance and Midpoint Formulas

**OBJECTIVES**

1. Use the Distance Formula
2. Use the Midpoint Formula

***Preparing for Distance and Midpoint Formulas***

*Before getting started, take the following readiness quiz. If you get a problem wrong, go back to the section cited and review the material.*

**1.** Simplify: (a) $\sqrt{64}$ (b) $\sqrt{24}$ [Section 7.3, pp. 549–551]

**2.** Find the length of the hypotenuse in a right triangle whose legs are 6 and 8. [Section 8.1, pp. 617–620]

## 1 Use the Distance Formula

In Section 2.1, we learned how to plot points in the Cartesian plane. One thing we would like to be able to do is algebraically compute the distance between any two points plotted in the Cartesian plane because this allows us to see a connection between geometry (literally measuring the distance) and algebra. We can find the distance between two points in the Cartesian plane using the Pythagorean Theorem.

### EXAMPLE 1 Finding the Distance between Two Points

Find the distance $d$ between the points $(2, 4)$ and $(5, 8)$.

**Solution**

We first plot the points in the Cartesian plane and connect them with a straight line as shown in Figure 1(a). To find the length $d$, draw a horizontal line through the point $(2, 4)$ and a vertical line through the point $(5, 8)$ and form a right triangle. The right angle of this triangle is at the point $(5, 4)$. Do you see why? If we travel horizontally from the point $(2, 4)$, then there is no "up or down" movement. For this reason, the $y$-coordinate of the point at the right angle must by be 4. Similarly, if we travel vertically straight down from the point $(5, 8)$, we find that the $x$-coordinate of the point at the right angle must be 5. See Figure 1(b).

**Figure 1**

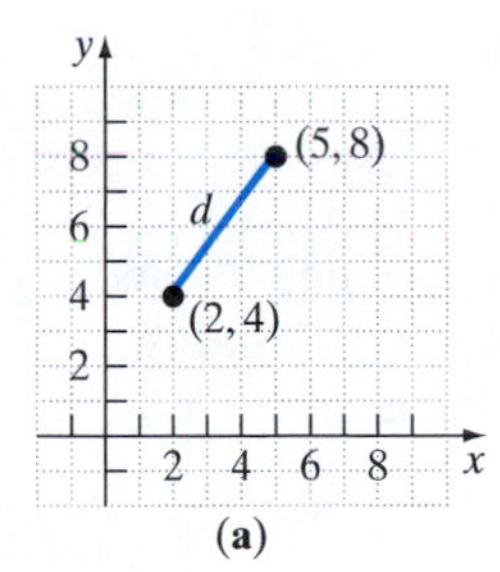

(a)

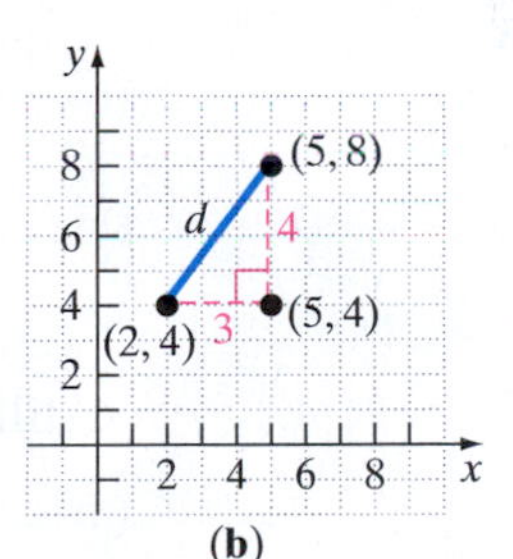

(b)

One leg of the triangle is of length 3 (since $5 - 2 = 3$). The other leg is of length 4 (because $8 - 4 = 4$). By the Pythagorean Theorem, we have that

$$\begin{aligned} d^2 &= 3^2 + 4^2 \\ &= 9 + 16 \\ &= 25 \\ \text{Square Root Property:} \quad d &= \pm 5 \end{aligned}$$

Because $d$ represents the length of the hypotenuse, we discard the solution $d = -5$ and find that the hypotenuse is 5. Therefore, the distance between the points $(2, 4)$ and $(5, 8)$ is 5 units.

***Preparing for...Answers*** **1. (a)** 8 **(b)** $2\sqrt{6}$ **2.** 10

That was quite a bit of work to find the distance between (2,4) and (5,8). The *distance formula* can be used to find the distance between any two points in the Cartesian plane.

**In Words**

To find the distance between two points in the Cartesian plane, find the difference of the $x$-coordinates, square it, and add this to the square of the difference of the $y$-coordinates. The square root of this sum is the distance.

**THE DISTANCE FORMULA**

The distance between two points $P_1 = (x_1, y_1)$ and $P_2 = (x_2, y_2)$, denoted by $d(P_1, P_2)$, is

$$d(P_1, P_2) = \sqrt{(x_2 - x_1)^2 + (y_2 - y_1)^2}$$

**Figure 2**

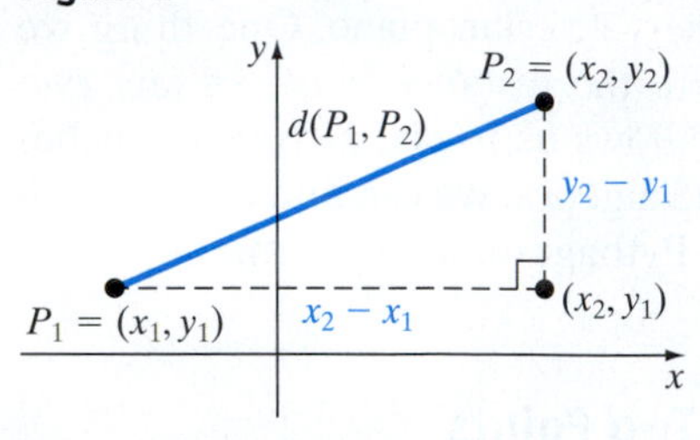


Figure 2 illustrates the theorem.

We can provide a justification for the formula. Let $(x_1, y_1)$ denote the coordinates of a point $P_1$ and let $(x_2, y_2)$ denote the coordinates of point $P_2$. The line joining the points $P_1$ and $P_2$ is neither vertical nor horizontal. Form a right triangle so that the vertex of the right angle is at the point $P_3 = (x_2, y_1)$ as shown in Figure 3(a). The vertical distance from $P_3$ to $P_2$ is the absolute value of the difference of the $y$-coordinates, $|y_2 - y_1|$. The horizontal distance from $P_1$ to $P_3$ is the absolute value of the difference of the $x$-coordinates, $|x_2 - x_1|$. See Figure 3(b).

**Work Smart**

When using the distance formula in applications, be sure that both the $x$-axis and $y$-axis have the same unit of measure, such as inches.

**Figure 3**

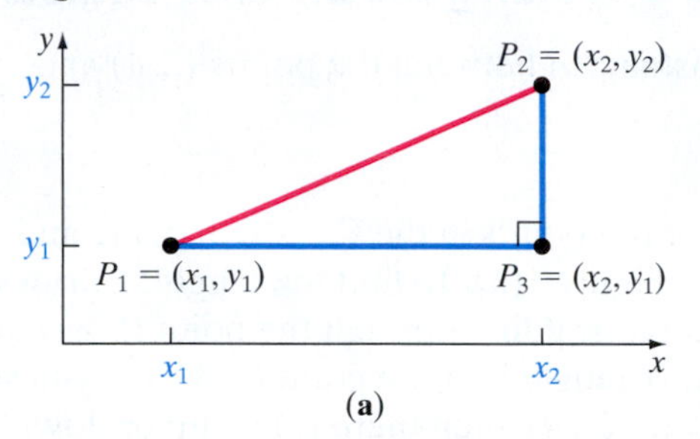


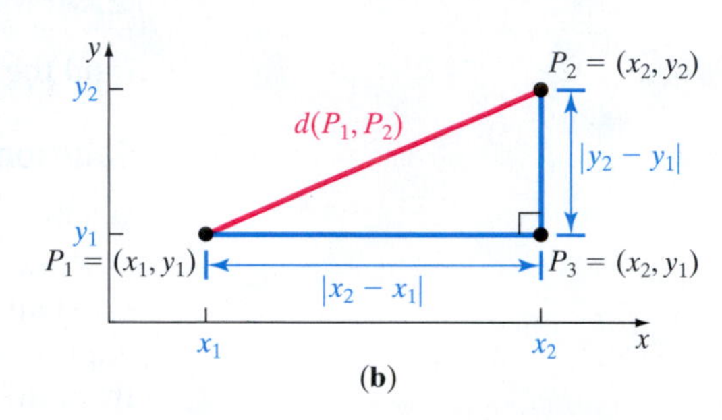


The distance $d(P_1, P_2)$ that we seek is the length of the hypotenuse of the right triangle, so by the Pythagorean Theorem, it follows that

$$[d(P_1, P_2)]^2 = |x_2 - x_1|^2 + |y_2 - y_1|^2$$
$$= (x_2 - x_1)^2 + (y_2 - y_1)^2$$

Take the square root of both sides: $d(P_1, P_2) = \sqrt{(x_2 - x_1)^2 + (y_2 - y_1)^2}$

If the line joining $P_1$ and $P_2$ is horizontal, then the $y$-coordinate of $P_1$ equals the $y$-coordinate of $P_2$; that is, $y_1 = y_2$. See Figure 4(a). In this case, the distance formula still works, because, for $y_1 = y_2$, it becomes

$$d(P_1, P_2) = \sqrt{(x_2 - x_1)^2 + 0^2}$$
$$= \sqrt{(x_2 - x_1)^2} = |x_2 - x_1|$$

A similar argument holds if the line joining $P_1$ and $P_2$ is vertical. See Figure 4(b). The distance formula works in all cases.

**Figure 4**

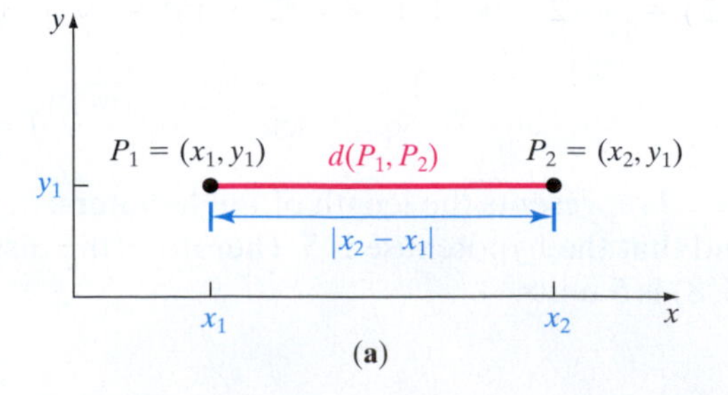


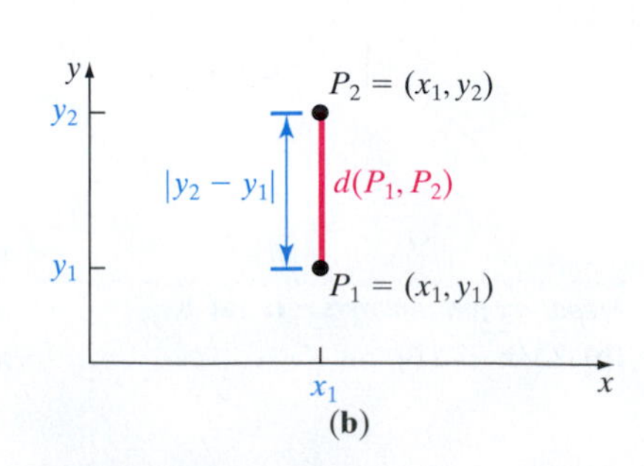

Figure 5

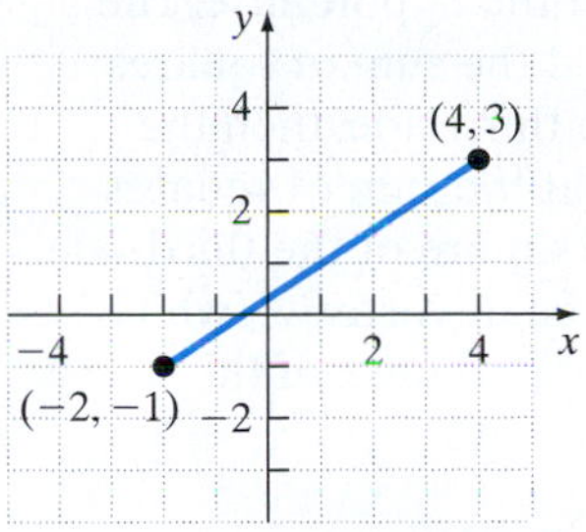

## EXAMPLE 2 Finding the Length of a Line Segment

Find the length of the line segment shown in Figure 5. That is, find the distance between the points in Figure 5.

### Solution

The length of the line segment is the distance between the points $(-2, -1)$ and $(4, 3)$. Using the distance formula with $P_1 = (x_1, y_1) = (-2, -1)$ and $P_2 = (x_2, y_2) = (4, 3)$, the length $d$ is

$$d = \sqrt{(x_2 - x_1)^2 + (y_2 - y_1)^2}$$

$$x_1 = -2, y_1 = -1, x_2 = 4, y_2 = 3: \quad = \sqrt{(4 - (-2))^2 + (3 - (-1))^2}$$

$$= \sqrt{6^2 + 4^2}$$

$$= \sqrt{36 + 16}$$

$$= \sqrt{52}$$

$$= 2\sqrt{13} \approx 7.21$$

**QUICK ✓** *Find the distance between the points.*

**1.** $(3, 8)$ and $(0, 4)$ **2.** $(-2, -5)$ and $(4, 7)$

The distance between two points $P_1 = (x_1, y_1)$ and $P_2 = (x_2, y_2)$ is never a negative number. In addition, the distance between two points is 0 only when the two points are identical, that is, when $x_1 = x_2$ and $y_1 = y_2$. Also, because $(x_2 - x_1)^2 = (x_1 - x_2)^2$ and $(y_2 - y_1)^2 = (y_1 - y_2)^2$, it does not matter whether we compute the distance from $P_1$ to $P_2$ or from $P_2$ to $P_1$. This should seem reasonable since the distance from $P_1$ to $P_2$ equals the distance from $P_2$ to $P_1$.

The next example shows how algebra (the distance formula) can be used to solve geometry problems.

## EXAMPLE 3 Using Algebra to Solve Geometry Problems

Consider the three points $A = (-1, 1)$, $B = (2, -2)$, and $C = (3, 5)$.

**(a)** Plot each point in the Cartesian plane and form the triangle $ABC$.

**(b)** Find the length of each side of the triangle.

**(c)** Verify that the triangle is a right triangle.

**(d)** Find the area of the triangle.

### Solution

**(a)** We plot points $A$, $B$, and $C$ in Figure 6.

**(b)** We use the distance formula to find the length of each side of the triangle.

Figure 6

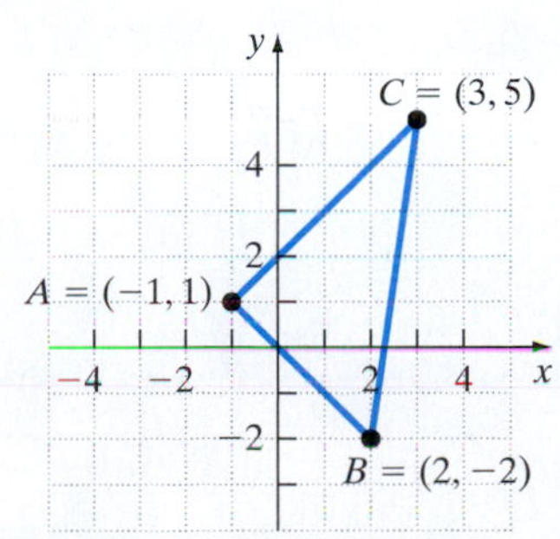


$$d(A, B) = \sqrt{(2 - (-1))^2 + (-2 - 1)^2} = \sqrt{3^2 + (-3)^2}$$

$$= \sqrt{9 + 9} = \sqrt{18} = 3\sqrt{2}$$

$$d(A, C) = \sqrt{(3 - (-1))^2 + (5 - 1)^2} = \sqrt{4^2 + (-4)^2}$$

$$= \sqrt{16 + 16} = \sqrt{32} = 4\sqrt{2}$$

$$d(B, C) = \sqrt{(3 - 2)^2 + (5 - (-2))^2} = \sqrt{1^2 + 7^2} = \sqrt{1 + 49} = \sqrt{50} = 5\sqrt{2}$$

**(c)** The Pythagorean Theorem says that if you have a right triangle, the sum of squares of the two legs will equal the square of the hypotenuse. The converse of this statement is true as well. That is, if the sum of squares of two sides of a triangle equals the square of the third side, then the triangle is a right triangle. So we need to show that the sum of squares of two sides of the triangle in Figure 6 equals the square of the third side. In looking at Figure 6, the right angle appears to be at vertex $A$, which means the side opposite vertex $A$ should be the hypotenuse of the triangle. So we want to show that

$$[d(B, C)]^2 = [d(A, B)]^2 + [d(A, C)]^2$$

We know each of these distances from part (b), so

$$(5\sqrt{2})^2 \stackrel{?}{=} (3\sqrt{2})^2 + (4\sqrt{2})^2$$

$$(ab)^n = a^n \cdot b^n\text{:} \quad 25 \cdot 2 \stackrel{?}{=} 9 \cdot 2 + 16 \cdot 2$$

$$50 = 18 + 32$$

$$50 = 50 \quad \text{True}$$

Since $[d(B, C)]^2 = [d(A, B)]^2 + [d(A, C)]^2$, we know that triangle $ABC$ is a right triangle.

**(d)** The area of a triangle is $\frac{1}{2}$ times the product of the base and the height. From Figure 6, we will say that side $AB$ forms the base and side $AC$ forms the height. The length of side $AB$ is $3\sqrt{2}$ and the length of side $AC$ is $4\sqrt{2}$. So the area of triangle $ABC$ is

$$\text{Area} = \frac{1}{2}(\text{Base})(\text{Height}) = \frac{1}{2}(3\sqrt{2})(4\sqrt{2}) = 12 \text{ square units}$$

**QUICK ✓**

**3.** Consider the three points $A = (-2, -1)$, $B = (4, 2)$, and $C = (0, 10)$.

**(a)** Plot each point in the Cartesian plane and form the triangle $ABC$.

**(b)** Find the length of each side of the triangle.

**(c)** Verify that the triangle is a right triangle.

**(d)** Find the area of the triangle.

## 2 Use the Midpoint Formula

Suppose we had two points $P_1 = (x_1, y_1)$ and $P_2 = (x_2, y_2)$ in the Cartesian plane. Further suppose we wanted to find the point $M = (x, y)$ that is the same distance to each of these two points so that $d(P_1, M) = d(M, P_2)$. We can find this point $M$ using the **midpoint formula.**

**In Words**
To find the midpoint of a line segment, average the $x$-coordinates and average the $y$-coordinates of the endpoints.

**MIDPOINT FORMULA**

The midpoint $M = (x, y)$ of the line segment from $P_1 = (x_1, y_1)$ to $P_2 = (x_2, y_2)$ is

$$M = \left(\frac{x_1 + x_2}{2}, \frac{y_1 + y_2}{2}\right)$$

### EXAMPLE 4 Finding the Midpoint of a Line Segment

Find the midpoint of a line segment joining $P_1 = (-2, 3)$ and $P_2 = (4, 7)$. Plot the points $P_1$ and $P_2$ and their midpoint. Check your answer.

#### Solution

We substitute $x_1 = -2$, $y_1 = 3$, $x_2 = 4$, and $y_2 = 7$ into the midpoint formula. The coordinates $(x, y)$ of the midpoint $M$ are

$$x = \frac{x_1 + x_2}{2} = \frac{-2 + 4}{2} = \frac{2}{2} = 1$$

and

$$y = \frac{y_1 + y_2}{2} = \frac{3 + 7}{2} = \frac{10}{2} = 5$$

Figure 7

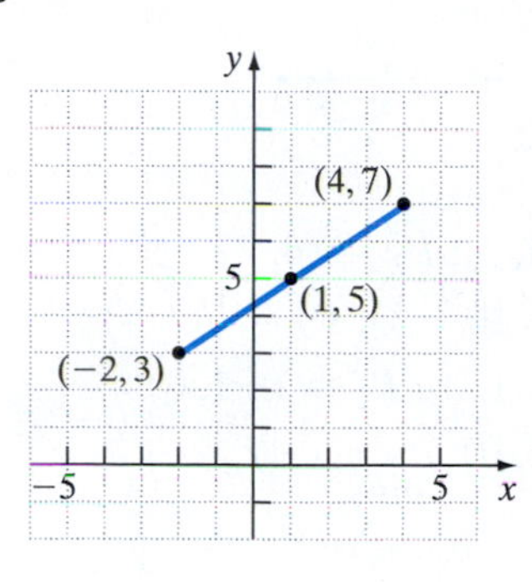


So the midpoint of the line segment joining $P_1 = (-2, 3)$ and $P_2 = (4, 7)$ is $M = (1, 5)$. See Figure 7.

**Check** We can check our solution by verifying that the distance from $P_1$ to $M$ is equal to the distance from $M$ to $P_2$.

$$d(P_1, M) = \sqrt{(1 - (-2))^2 + (5 - 3)^2} = \sqrt{3^2 + 2^2} = \sqrt{13}$$
$$d(M, P_2) = \sqrt{(4 - 1)^2 + (7 - 5)^2} = \sqrt{3^2 + 2^2} = \sqrt{13}$$

It checks!

**QUICK ✓** *Find the midpoint of the line segment joining the points.*

**4.** $(3, 8)$ and $(0, 4)$ **5.** $(-2, -5)$ and $(4, 10)$

# 10.1 Exercises

**For Extra Help:**

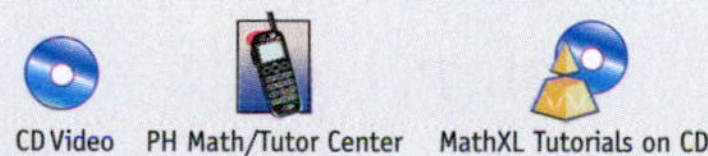

Student Solutions Manual | CD Video | PH Math/Tutor Center | MathXL Tutorials on CD | MathXL® | MyMathLab

## Concepts and Vocabulary

*In Problems 1 and 2, fill in the blanks.*

**1.** The distance between two points $P_1 = (x_1, y_1)$ and $P_2 = (x_2, y_2)$, denoted by $d(P_1, P_2)$, is ____________.

**2.** The midpoint $M = (x, y)$ of the line segment from $P_1 = (x_1, y_1)$ to $P_2 = (x_2, y_2)$ is ____________.

*In Problems 3 and 4, answer True or False to each statement.*

**3.** The distance between two points is sometimes a negative number.

**4.** The midpoint $M$ of a line segment joining $P_1$ and $P_2$ is the point such that $d(P_1, M) = d(M, P_2)$.

**5.** How is the distance formula related to the Pythagorean Theorem?

**6.** How can the distance formula be used to verify that a point is the midpoint of a line segment?

## Building Skills

*In Problems 7–22, find the distance $d(P_1, P_2)$ between the points $P_1$ and $P_2$.*

**7.**

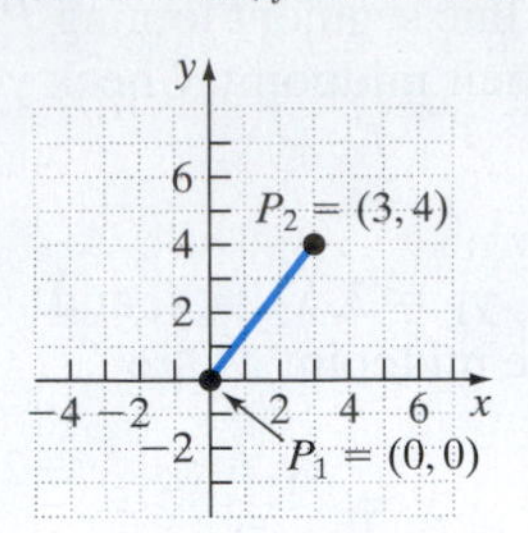


**8.**

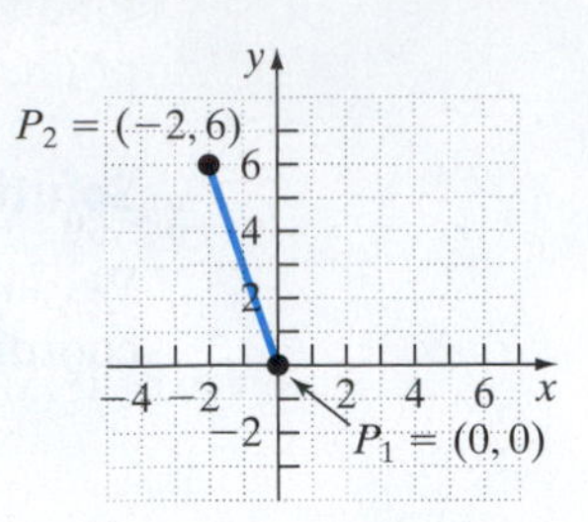


**9.**

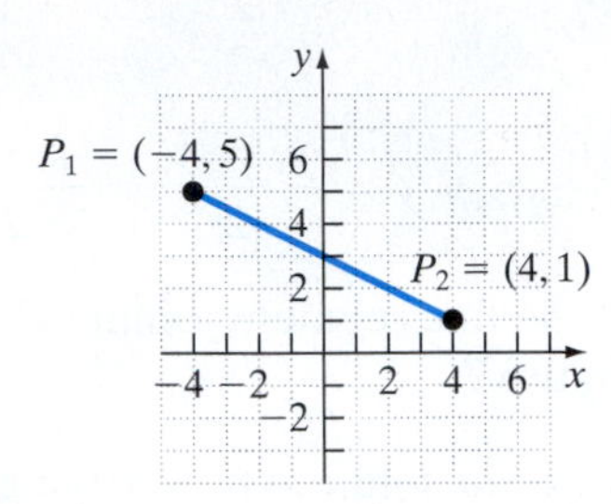


**10.**

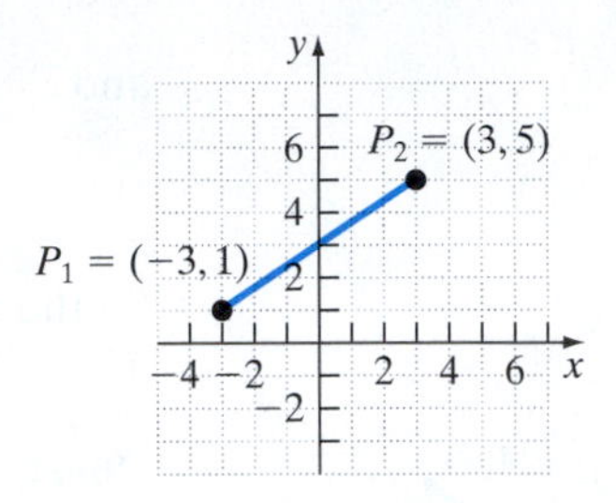


**11.** $P_1 = (2, 1); P_2 = (6, 4)$

**12.** $P_1 = (1, 3); P_2 = (4, 7)$

**13.** $P_1 = (-3, 2); P_2 = (9, -3)$

**14.** $P_1 = (-10, -3); P_2 = (14, 4)$

**15.** $P_1 = (-4, 2); P_2 = (2, 2)$

**16.** $P_1 = (-1, 2); P_2 = (-1, 0)$

**17.** $P_1 = (0, -3); P_2 = (-3, 3)$

**18.** $P_1 = (5, 0); P_2 = (-1, -4)$

**19.** $P_1 = \left(2\sqrt{2}, \sqrt{5}\right); P_2 = \left(5\sqrt{2}, 4\sqrt{5}\right)$

**20.** $P_1 = \left(\sqrt{6}, -2\sqrt{2}\right); P_2 = \left(3\sqrt{6}, 10\sqrt{2}\right)$

**21.** $P_1 = (0.3, -3.3); P_2 = (1.3, 0.1)$

**22.** $P_1 = (-1.7, 1.3); P_2 = (0.3, 2.6)$

*In Problems 23–34, find the midpoint of the line segment formed by joining the points $P_1$ and $P_2$.*

**23.** $P_1 = (2, 2); P_2 = (6, 4)$

**24.** $P_1 = (1, 3); P_2 = (5, 7)$

**25.** $P_1 = (-3, 2); P_2 = (9, -4)$

**26.** $P_1 = (-10, -3); P_2 = (14, 7)$

**27.** $P_1 = (-4, 3); P_2 = (2, 4)$

**28.** $P_1 = (-1, 2); P_2 = (3, 9)$

**29.** $P_1 = (0, -3); P_2 = (-3, 3)$

**30.** $P_1 = (5, 0); P_2 = (-1, -4)$

**31.** $P_1 = \left(2\sqrt{2}, \sqrt{5}\right); P_2 = \left(5\sqrt{2}, 4\sqrt{5}\right)$

**32.** $P_1 = \left(\sqrt{6}, -2\sqrt{2}\right); P_2 = \left(3\sqrt{6}, 10\sqrt{2}\right)$

**33.** $P_1 = (0.3, -3.3); P_2 = (1.3, 0.1)$

**34.** $P_1 = (-1.7, 1.3); P_2 = (0.3, 2.6)$

## Applying the Concepts

**35.** Consider the three points $A = (0, 3)$, $B = (2, 1)$, $C = (6, 5)$.
- **(a)** Plot each point in the Cartesian plane and form the triangle $ABC$.
- **(b)** Find the length of each side of the triangle.
- **(c)** Verify that the triangle is a right triangle.
- **(d)** Find the area of the triangle.

**36.** Consider the three points $A = (0, 2)$, $B = (1, 4)$, $C = (4, 0)$.
- **(a)** Plot each point in the Cartesian plane and form the triangle $ABC$.
- **(b)** Find the length of each side of the triangle.
- **(c)** Verify that the triangle is a right triangle.
- **(d)** Find the area of the triangle.

**37.** Consider the three points $A = (-2, -4)$, $B = (3, 1)$, and $C = (15, -11)$.
- **(a)** Plot each point in the Cartesian plane and form the triangle $ABC$.
- **(b)** Find the length of each side of the triangle.
- **(c)** Verify that the triangle is a right triangle.
- **(d)** Find the area of the triangle.

**38.** Consider the three points $A = (-2, 3)$, $B = (2, 0)$, and $C = (5, 4)$.

**(a)** Plot each point in the Cartesian plane and form the triangle $ABC$.

**(b)** Find the length of each side of the triangle.

**(c)** Verify that the triangle is a right triangle.

**(d)** Find the area of the triangle.

**39.** Find all points having an $x$-coordinate of 2 whose distance from the point $(5, 1)$ is 5.

**40.** Find all points having an $x$-coordinate of 4 whose distance from the point $(0, 3)$ is 5.

**41.** Find all points having a $y$-coordinate of $-3$ whose distance from the point $(2, 3)$ is 10.

**42.** Find all points having a $y$-coordinate of $-3$ whose distance from the point $(-4, 2)$ is 13.

**43. The City of Chicago** The city of Chicago's road system is set up like a Cartesian plane, where streets are indicated by the number of blocks they are from Madison Street and State Street. For example, Wrigley Field in Chicago is located at 1060 West Addison, which is 10 blocks west of State Street and 36 blocks north of Madison Street.

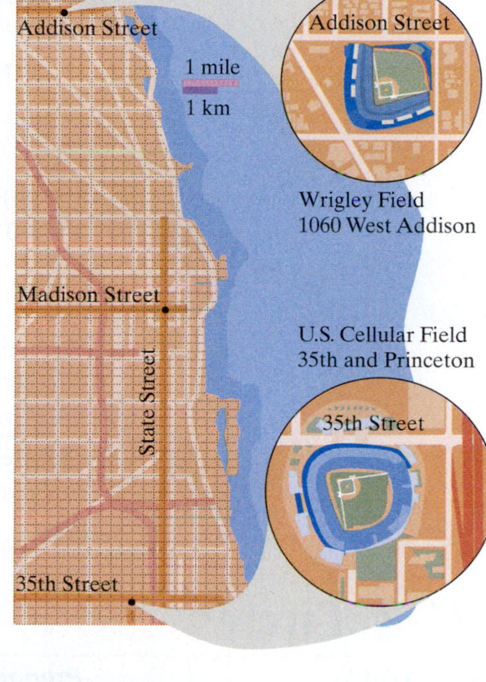


**(a)** Find the distance "as the crow flies" from Madison and State Street to Wrigley Field. Use city blocks as the unit of measurement.

**(b)** U.S. Cellular Field, home of the White Sox, is located at 35th and Princeton, which is 3 blocks west of State Street and 35 blocks south of Madison. Find the distance "as the crow flies" from Madison and State Street to U.S. Cellular Field.

**(c)** Find the distance "as the crow flies" from Wrigley Field to U.S. Cellular Field.

**44. Baseball** A major league baseball "diamond" is actually a square, 90 feet on a side (see the figure). Overlay a Cartesian plane on a major league baseball diamond, so that the origin is at home plate, the positive $x$-axis lies in the direction from home plate to first base, and the positive $y$-axis lies in the direction from home plate to third base.

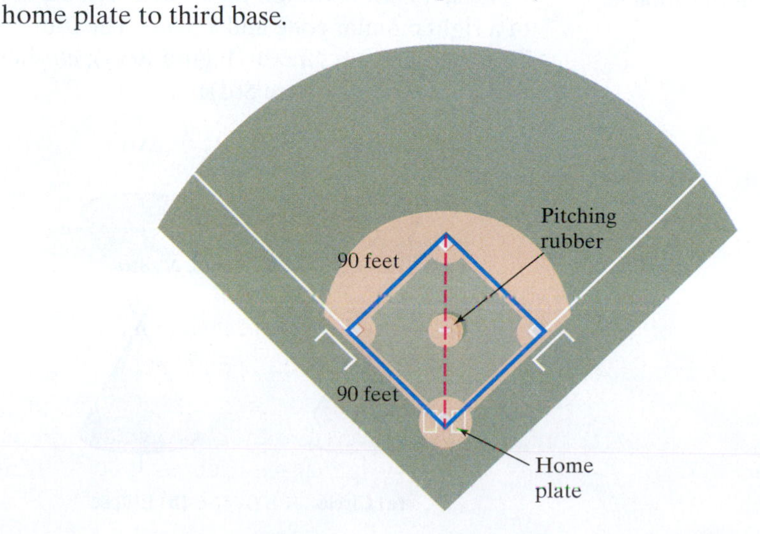

**(a)** What are the coordinates of home plate, first base, second base, and third base? Use feet as the unit of measurement.

**(b)** Suppose the center fielder is located at $(310, 260)$. How far is he from second base?

**(c)** Suppose the shortstop is located at $(60, 100)$. How far is he from second base?

## Extending the Concepts

**45. Baseball** Refer to Problem 44.

**(a)** Suppose the right fielder catches a fly ball at $(320, 20)$. How many seconds will it take to throw the ball to second base if he can throw 130 feet per second? (**Hint:** time = distance divided by speed.)

**(b)** Suppose a runner "tagging up" from first base can run 27 feet per second. Would you "send the runner" as the first base coach if the right fielder requires 0.8 second to catch and throw? Why?

**46.** Let $P = (x, y)$ be a point on the graph of $y = x^2 - 4$.

**(a)** Express the distance $d$ from $P$ to the point $(1, 2)$ as a function of $x$ using the distance formula. See the figure for an illustration of the problem.

**(b)** What is $d$ if $x = 0$?

**(c)** What is $d$ if $x = 3$?

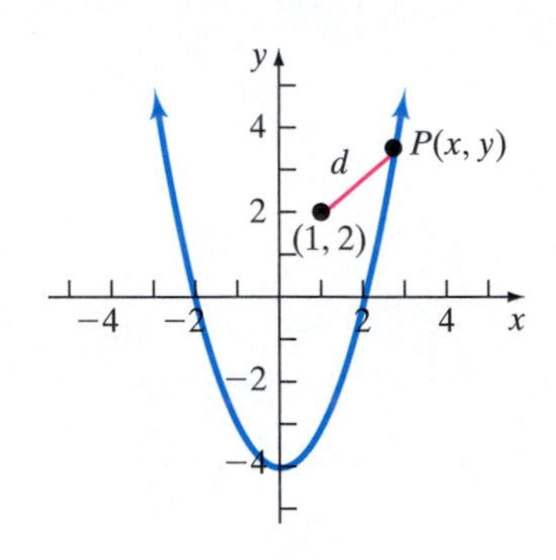


## Synthesis Review

**47.** Evaluate $3^2$. What is $\sqrt{9}$?

**48.** Evaluate $8^2$. What is $\sqrt{64}$?

**49.** Evaluate $(-3)^4$. What is $\sqrt[4]{81}$?

**50.** Evaluate $(-3)^3$. What is $\sqrt[3]{-27}$?

**51.** Describe the relationship between raising a number to a positive integer power, $n$, and the $n$th root of a number.

# 10.2 Circles

## OBJECTIVES

1. Write the Standard Form of the Equation of a Circle
2. Graph a Circle
3. Find the Center and Radius of a Circle From an Equation in General Form

### Preparing for Circles

*Before getting started, take the following readiness quiz. If you get the problem wrong, go back to the section cited and review the material.*

**1.** Complete the square in $x$: $x^2 - 8x$ [Section 8.1, pp. 614–615]

**Conics,** an abbreviation for **conic sections,** are curves that result from the intersection of a right circular cone and a plane. The four conics that we study are shown in Figure 8. These conics are *circles* (Figure 8(a)); *ellipses* (Figure 8(b)); *parabolas* (Figure 8(c)); and *hyperbolas* (Figure 8(d)).

**Figure 8**

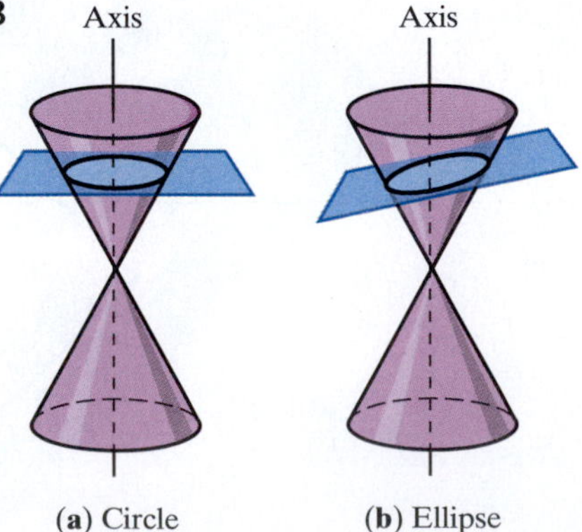


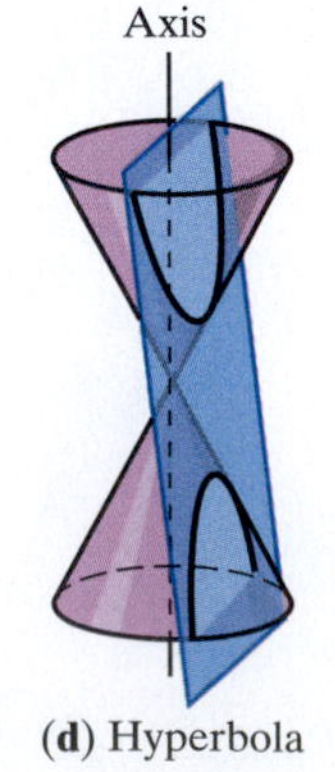


(a) Circle (b) Ellipse (c) Parabola (d) Hyperbola

*Preparing for...Answer*

**1.** $x^2 - 8x + 16$

We study circles in this section, parabolas in Section 10.3, ellipses in Section 10.4, and hyperbolas in Section 10.5.

## 1 Write the Standard Form of the Equation of a Circle

One advantage of a coordinate system (the Cartesian plane) is that it enables us to translate a geometric statement into an algebraic statement, and vice versa. Consider the following geometric statement that defines a circle.

**Figure 9**

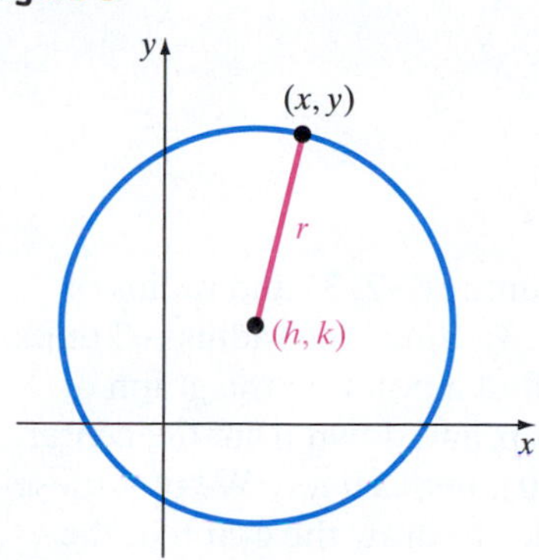


**DEFINITION**

A **circle** is the set of all points in the Cartesian plane that are a fixed distance $r$ from a fixed point $(h, k)$. The fixed distance $r$ is called the **radius,** and the fixed point $(h, k)$ is called the **center** of the circle. See Figure 9.

To find an equation that has this graph, we let $(x, y)$ represent the coordinates of any point on a circle with radius $r$ and center $(h, k)$. Then the distance between the points $(x, y)$ and $(h, k)$ must always equal $r$. That is, by the distance formula,

$$\sqrt{(x-h)^2 + (y-k)^2} = r$$

If we square both sides of the equation, then

$$(x-h)^2 + (y-k)^2 = r^2$$

and we have the equation of a circle.

**DEFINITION**

The **standard form of an equation of a circle** with radius $r$ and center $(h, k)$ is

$$(x-h)^2 + (y-k)^2 = r^2$$

### EXAMPLE 1 Writing the Standard Form of the Equation of a Circle

Write the standard form of the equation of the circle with radius 4 and center $(2, -3)$.

**Solution**

We use the equation $(x-h)^2 + (y-k)^2 = r^2$ with $r = 4, h = 2, k = -3$ and obtain

$$(x-2)^2 + (y-(-3))^2 = 4^2$$
$$(x-2)^2 + (y+3)^2 = 16$$

**QUICK ✓** *Write the standard form of the equation of each circle whose radius is r and center is (h, k).*

**1.** $r = 5; (h, k) = (2, 4)$

**2.** $r = \sqrt{2}; (h, k) = (-2, 0)$

## 2 Graph a Circle

The graph of any equation of the form $(x-h)^2 + (y-k)^2 = r^2$ is that of a circle with radius $r$ and center $(h, k)$.

**EXAMPLE 2** **Graphing a Circle**

Graph the equation: $(x + 2)^2 + (y - 3)^2 = 9$

### Solution

The graph of the equation is a circle because the equation is of the form $(x - h)^2 + (y - k)^2 = r^2$. To graph the equation, we first identify the center and radius of the circle by comparing the given equation to the standard form of the equation of a circle.

$$(x + 2)^2 + (y - 3)^2 = 9$$
$$(x - (-2))^2 + (y - 3)^2 = 3^2$$
$$(x - h)^2 + (y - k)^2 = r^2$$

We see that $h = -2$, $k = 3$, and $r = 3$. The circle has center $(-2, 3)$ and radius of 3 units. To graph this circle, we first plot the center $(-2, 3)$. Since the radius is 3 units, we can go 3 units in any direction from the center and find a point on the graph of the circle. It is easiest to find the four points left, right, up, and down from the center. These four points are $(-5, 3)$, $(1, 3)$, $(-2, 6)$, and $(-2, 0)$, respectively. We plot these points in Figure 10(a). We then use these points as guides to draw the graph of the circle shown in Figure 10(b).

**Figure 10**

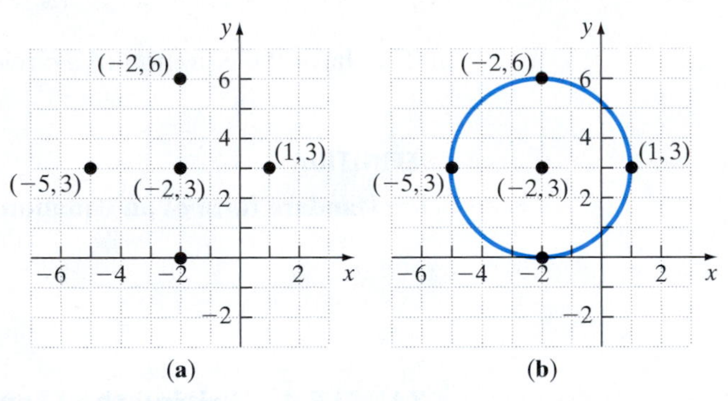


**QUICK ✓** *Graph the circle.*

3. $(x - 3)^2 + (y - 1)^2 = 4$

4. $(x + 5)^2 + y^2 = 16$

## 3 Find the Center and Radius of a Circle from an Equation in General Form

If we eliminate the parentheses from the standard form of the equation of the circle given in Example 2, we get

$$(x + 2)^2 + (y - 3)^2 = 9$$

FOIL: $x^2 + 4x + 4 + y^2 - 6y + 9 = 9$

Subtract 9 from both sides: $x^2 + y^2 + 4x - 6y + 4 = 0$

Any equation of the form

$$x^2 + y^2 + ax + by + c = 0$$

has a graph that is a circle, a point, or no graph at all. For example, the graph of the equation $x^2 + y^2 = 0$ is the single point (0, 0). The equation $x^2 + y^2 + 4 = 0$, or $x^2 + y^2 = -4$ has no graph, because the sum of squares of real numbers are never negative.

**In Words**
The standard form of a circle is $(x - h)^2 + (y - k)^2 = r^2$. The general form is $x^2 + y^2 + ax + by + c = 0$.

**DEFINITION**

The **general form of the equation of a circle** is given by the equation

$$x^2 + y^2 + ax + by + c = 0$$

when the graph exists.

If an equation of a circle is in general form, we use the method of completing the square to put the equation in standard form, $(x - h)^2 + (y - k)^2 = r^2$, so that we can identify its center and radius.

### EXAMPLE 3 Graphing a Circle Whose Equation Is in General Form

Graph the equation: $x^2 + y^2 + 8x - 2y - 8 = 0$

**Solution**

To determine the center and radius of the circle, we first need to put the equation in standard form by completing the square in both $x$ and $y$ (covered in Section 8.1). To do this, we group the terms involving $x$, group the terms involving $y$, and put the constant on the right side of the equation by adding 8 to both sides.

$$(x^2 + 8x) + (y^2 - 2y) = 8$$

Now we complete the square of each expression in parentheses. Remember, any number added to the left side of the equation must also be added to the right.

$$(x^2 + 8x + 16) + (y^2 - 2y + 1) = 8 + 16 + 1$$

$$\left(\frac{1}{2}\cdot 8\right)^2 = 16 \qquad \left(\frac{1}{2}\cdot(-2)\right)^2 = 1$$

Factor: $$(x + 4)^2 + (y - 1)^2 = 25$$

The equation is now in the standard form of a circle. The center of the circle is $(-4, 1)$ and its radius is 5. To graph the equation, we plot the center and the four points to the left, to the right, above, and below the center. Then trace in the circle. See Figure 11.

Figure 11

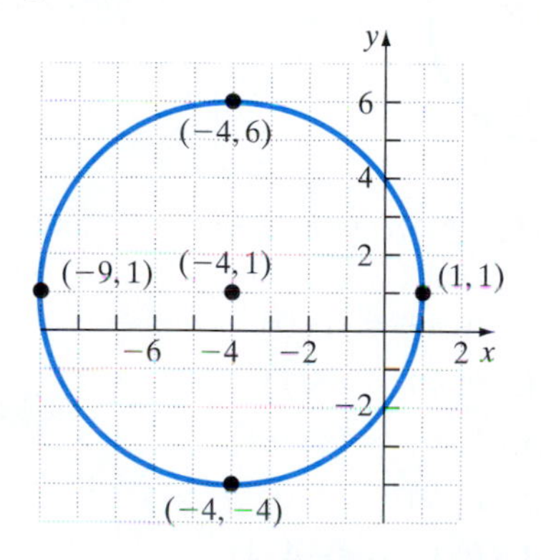


**QUICK ✓** *Graph each circle.*

5. $x^2 + y^2 - 6x - 4y + 4 = 0$

6. $2x^2 + 2y^2 - 16x + 4y - 38 = 0$

# 10.2 Exercises

***For Extra Help:***

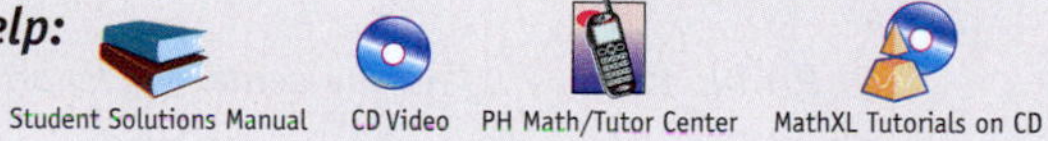

## Concepts and Vocabulary

*In Problems 1 and 2, fill in the blanks.*

1. A __________ is the set of all points in the Cartesian plane that are a fixed distance $r$ from a fixed point $(h, k)$.

2. For a circle, the __________ is the distance from the center to any point on the circle.

*In Problems 3 and 4, answer True or False to each statement.*

**3.** The center of the circle $(x + 1)^2 + (y - 3)^2 = 25$ is $(1, -3)$.

**4.** The center of the circle $x^2 + y^2 = 9$ is $(0, 0)$; its radius is 3.

**5.** How is the distance formula related to the definition of a circle?

**6.** Are circles functions? Why or why not?

**7.** Is $x^2 = 36 - y^2$ the equation of a circle? If so, what is the center and radius?

**8.** Is $3x^2 - 12x + 3y^2 - 15 = 0$ the equation of a circle? If so, what is the center and radius?

## Building Skills

*In Problems 9–12, find the center and radius of each circle. Write the standard form of the equation.*

**9.**

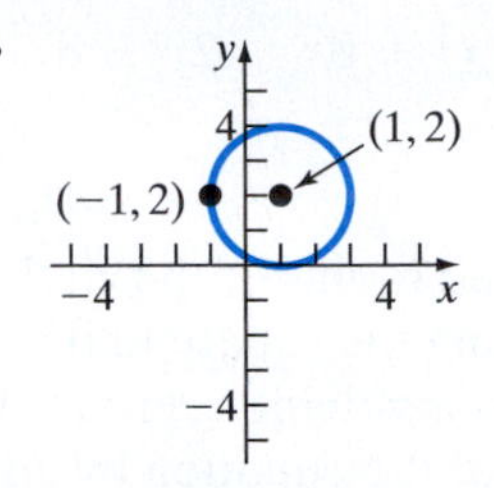


**10.**

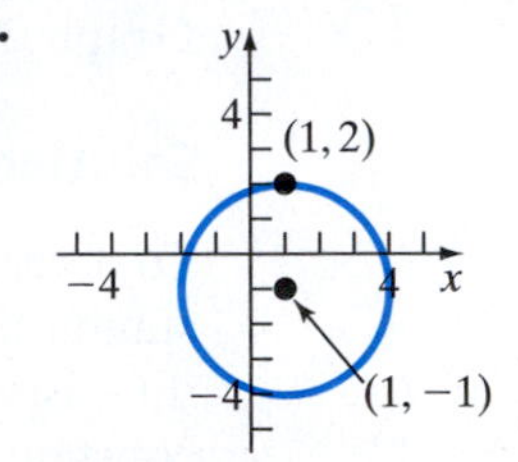


**11.**

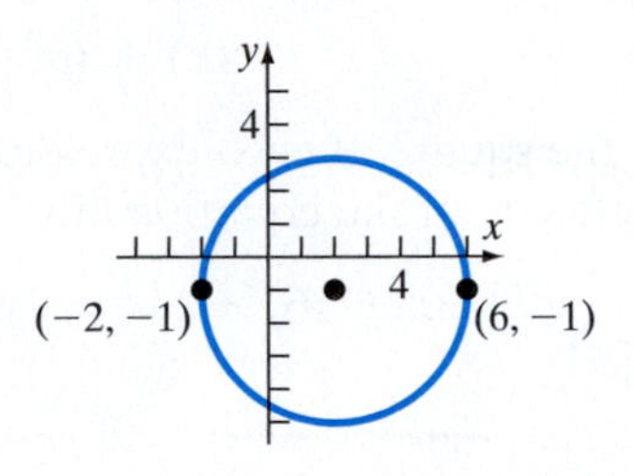


**12.**

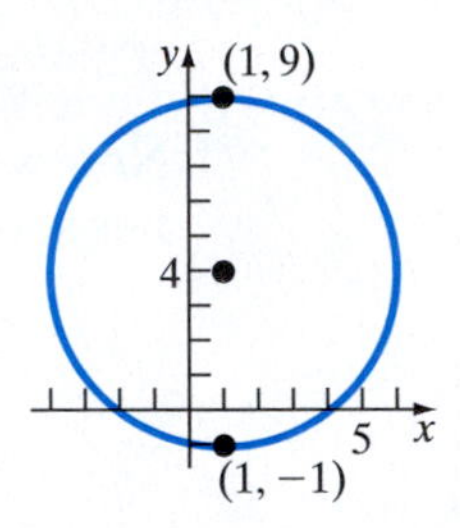


*In Problems 13–24, write the standard form of the equation of each circle whose radius is r and center is (h, k). Graph each circle.*

**13.** $r = 3; (h, k) = (0, 0)$

**14.** $r = 5; (h, k) = (0, 0)$

**15.** $r = 2; (h, k) = (1, 4)$

**16.** $r = 4; (h, k) = (3, 1)$

**17.** $r = 6; (h, k) = (-2, 4)$

**18.** $r = 3; (h, k) = (1, -4)$

**19.** $r = 4; (h, k) = (0, 3)$

**20.** $r = 2; (h, k) = (1, 0)$

**21.** $r = 5; (h, k) = (5, -5)$

**22.** $r = 4; (h, k) = (-4, 4)$

**23.** $r = \sqrt{5}; (h, k) = (1, 2)$

**24.** $r = \sqrt{7}; (h, k) = (5, 2)$

*In Problems 25–34, find the center (h, k) and radius r of each circle. Graph each circle.*

**25.** $x^2 + y^2 = 36$

**26.** $x^2 + y^2 = 144$

**27.** $(x - 4)^2 + (y - 1)^2 = 25$

**28.** $(x - 2)^2 + (y - 3)^2 = 9$

**29.** $(x + 3)^2 + (y - 2)^2 = 81$

**30.** $(x - 5)^2 + (y + 2)^2 = 49$

**31.** $x^2 + (y - 3)^2 = 64$

**32.** $(x - 6)^2 + y^2 = 36$

**33.** $(x - 1)^2 + (y + 1)^2 = \frac{1}{4}$

**34.** $(x - 2)^2 + (y + 2)^2 = \frac{1}{4}$

*In Problems 35–40, find the center (h, k) and radius r of each circle. Graph each circle.*

**35.** $x^2 + y^2 - 6x + 2y + 1 = 0$

**36.** $x^2 + y^2 + 2x - 8y + 8 = 0$

**37.** $x^2 + y^2 + 10x + 4y + 4 = 0$

**38.** $x^2 + y^2 + 4x - 12y + 36 = 0$

**39.** $2x^2 + 2y^2 - 12x + 24y - 72 = 0$

**40.** $2x^2 + 2y^2 - 28x + 20y + 20 = 0$

*In Problems 41–46, find the standard form of the equation of each circle.*

**41.** Center at the origin and containing the point $(4, -2)$.

**42.** Center at $(0, 3)$ and containing the point $(3, 7)$.

**43.** Center at $(-3, 2)$ and tangent to the $y$-axis.

**44.** Center at $(2, -3)$ and tangent to the $x$-axis.

**45.** With endpoints of a diameter at $(2, 3)$ and $(-4, -5)$.

**46.** With endpoints of a diameter at $(-5, -3)$ and $(7, 2)$.

## Applying the Concepts

△ **47.** Find the area and circumference of the circle $(x - 3)^2 + (y - 8)^2 = 64$.

△ **48.** Find the area and circumference of the circle $(x - 1)^2 + (y - 4)^2 = 49$.

△ **49.** Find the area of the square in the figure.

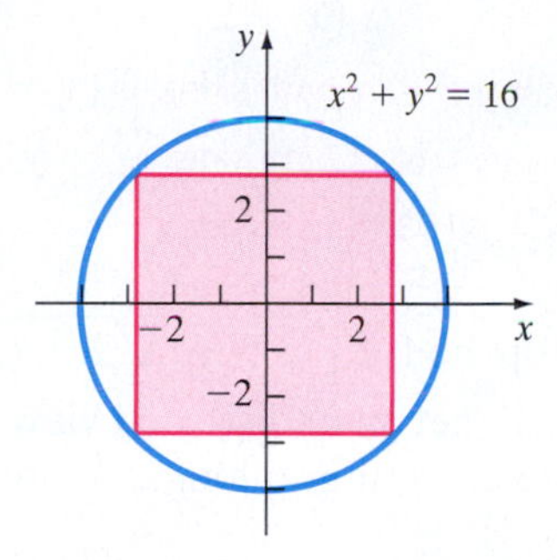


△ **50.** Find the area of the shaded region in the figure.

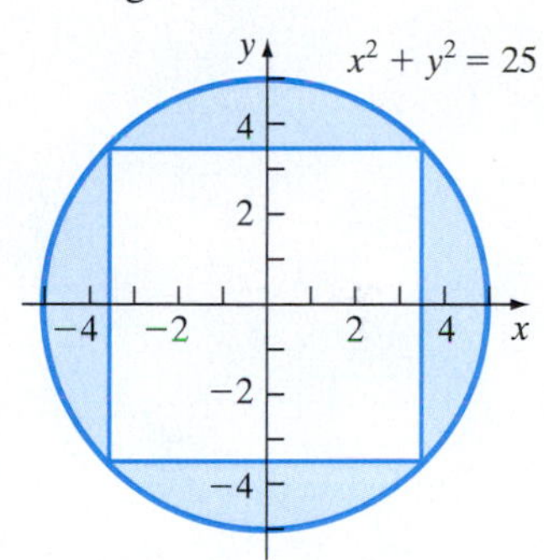


## Extending the Concepts

**51.** Which of the following equations might have the graph shown? (More than one answer is possible.)

**(a)** $(x - 2)^2 + y^2 = 1$
**(b)** $x^2 + (y - 2)^2 = 1$
**(c)** $(x + 4)^2 + y^2 = 9$
**(d)** $(x - 5)^2 + y^2 = 25$
**(e)** $x^2 + y^2 - 8x + 7 = 0$
**(f)** $x^2 + y^2 + 10x + 18 = 0$

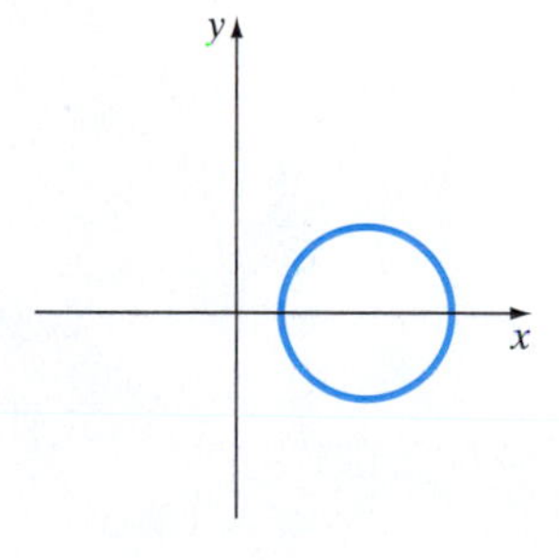


**52.** Which of the following equations might have the graph shown? (More than one answer is possible.)

**(a)** $(x - 2)^2 + (y + 3)^2 = 4$
**(b)** $(x - 3)^2 + (y - 4)^2 = 4$
**(c)** $(x + 3)^2 + (y + 4)^2 = 9$
**(d)** $(x - 5)^2 + (y - 5)^2 = 25$
**(e)** $x^2 + y^2 + 8x + 10y + 32 = 0$
**(f)** $x^2 + y^2 - 4x - 6y - 3 = 0$

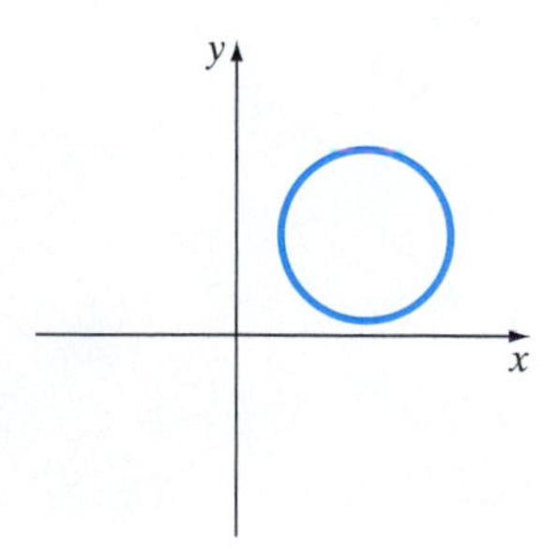

## Synthesis Review

*In Problems 53–57, graph each function using either point plotting or the properties of the function. For example, to graph a line, we would use the slope and y-intercept or the intercepts.*

**53.** $f(x) = 4x - 3$

**54.** $2x + 5y = 20$

**55.** $g(x) = x^2 - 4x - 5$

**56.** $F(x) = -3x^2 + 12x - 12$

**57.** $G(x) = -2(x + 3)^2 - 5$

**58.** Present an argument that supports the approach you took to graphing each function in Problems 53–57. For example, in Problem 53 did you use point plotting? Intercepts? Slope? Which method did you use and why?

## The Graphing Calculator

To graph a circle using a graphing calculator requires that we first solve the equation for $y$. For example, to graph $(x + 2)^2 + (y - 3)^2 = 9$, we solve for $y$ as follows:

$$(x + 2)^2 + (y - 3)^2 = 9$$

Subtract $(x + 2)^2$ from both sides: $$(y - 3)^2 = 9 - (x + 2)^2$$

Take the square root of both sides: $$y - 3 = \pm\sqrt{9 - (x + 2)^2}$$

Add 3 to both sides: $$y = 3 \pm \sqrt{9 - (x + 2)^2}$$

We graph the top half $Y_1 = 3 + \sqrt{9 - (x + 2)^2}$ and the bottom half $Y_2 = 3 - \sqrt{9 - (x + 2)^2}$. To get an undistorted view of the graph, be sure to use the ZOOM SQUARE feature on your graphing calculator. The figure below shows the graph of $(x + 2)^2 + (y - 3)^2 = 9$.

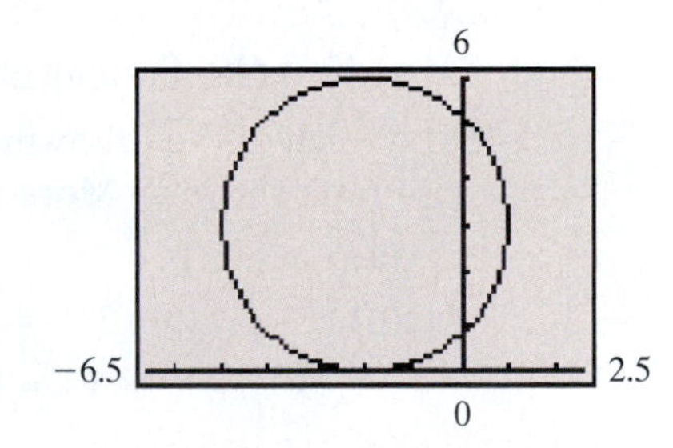


*In Problems 59–68, graph each circle using a graphing calculator. Compare your graphs to the graphs drawn by hand in Problems 25–34.*

**59.** $x^2 + y^2 = 36$

**60.** $x^2 + y^2 = 144$

**61.** $(x - 4)^2 + (y - 1)^2 = 25$

**62.** $(x - 2)^2 + (y - 3)^2 = 9$

**63.** $(x + 3)^2 + (y - 2)^2 = 81$

**64.** $(x - 5)^2 + (y + 2)^2 = 49$

**65.** $x^2 + (y - 3)^2 = 64$

**66.** $(x - 6)^2 + y^2 = 36$

**67.** $(x - 1)^2 + (y + 1)^2 = \frac{1}{4}$

**68.** $(x - 2)^2 + (y + 2)^2 = \frac{1}{4}$

# 10.3 Parabolas

**OBJECTIVES**

1. Graph Parabolas in Which the Vertex Is the Origin
2. Find the Equation of a Parabola
3. Graph Parabolas in Which the Vertex Is Not the Origin
4. Solve Applied Problems Involving Parabolas

## *Preparing for Parabolas*

*Before getting started, take the following readiness quiz. If you get a problem wrong, go back to the section cited and review the material.*

**1.** Identify the vertex and axis of symmetry of $f(x) = -3(x + 4)^2 - 5$. Does the parabola open up or down? Why? [Section 8.4, pp. 654–657]

**2.** Identify the vertex and axis of symmetry of the quadratic function $f(x) = 2x^2 - 8x + 1$. Does the parabola open up or down? Why? [Section 8.5, pp. 662–668]

**3.** Complete the square of $x^2 - 12x$. [Section 8.1, pp. 614–615]

**4.** Solve: $(x - 3)^2 = 25$ [Section 8.1, pp. 611–614]

We began a discussion of parabolas back in Section 8.4 when we studied quadratic functions. To refresh your memory, we include a summary of this information below.

**SUMMARY: Parabolas That Open Up or Down**

The graph of $y = a(x - h)^2 + k$ or $y = ax^2 + bx + c$ is a parabola that

**1.** opens up if $a > 0$ and opens down if $a < 0$.

**2.** has vertex $(h, k)$ if the equation is of the form $y = a(x - h)^2 + k$.

**3.** has a vertex whose $x$-coordinate is $x = -\dfrac{b}{2a}$. The $y$-coordinate is found by evaluating the equation at the $x$-coordinate of the vertex.

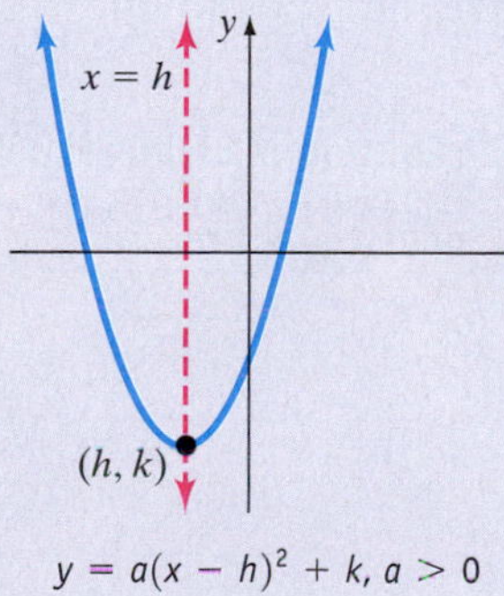


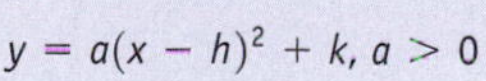

$y = a(x - h)^2 + k,\ a > 0$

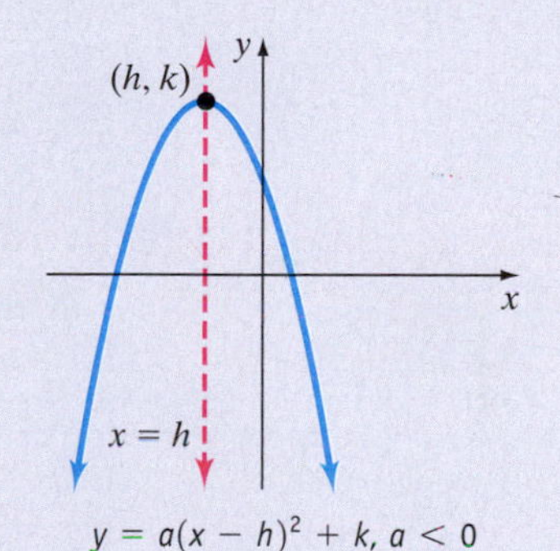


$y = a(x - h)^2 + k,\ a < 0$

The presentation of parabolas given in Sections 8.4 and 8.5 relied more on algebra. In this section, we are going to look at the parabola (and the other conic sections) from a geometric point of view.

**Figure 12**

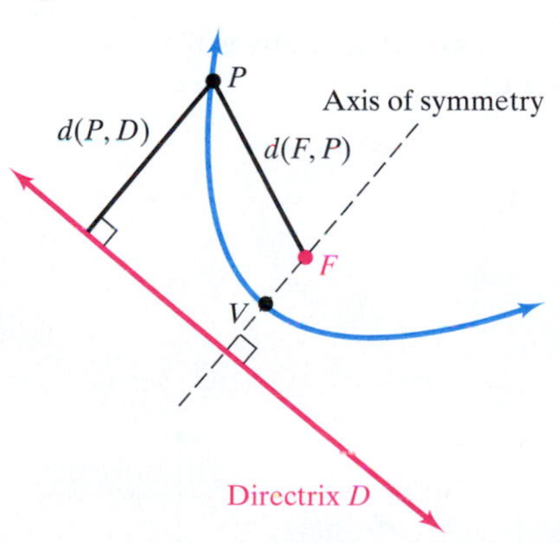


**DEFINITION**

A **parabola** is the collection of all points $P$ in the plane that are the same distance from a fixed point $F$ as they are from a fixed line $D$. The point $F$ is called the **focus** of the parabola, and the line $D$ is its **directrix.** In other words, a parabola is the set of points $P$ for which

$$d(F, P) = d(P, D)$$

Figure 12 shows a parabola. The line through the focus $F$ and perpendicular to the directrix $D$ is called the **axis of symmetry** of the parabola. The point of intersection of the parabola with its axis of symmetry is called the **vertex** $V$.

***Preparing for...Answers*** **1.** Vertex: $(-4, -5)$; axis of symmetry: $x = -4$; opens down since $a = -3 < 0$. **2.** Vertex: $(2, -7)$; axis of symmetry: $x = 2$; opens up since $a = 2 > 0$. **3.** $x^2 - 12x + 36$ **4.** $\{-2, 8\}$

## 1 Graph Parabolas in Which the Vertex Is the Origin

We want to develop an equation for the parabola based on the definition just given. For example, because the vertex is on the graph of the parabola, it must satisfy the definition that the distance from the focus to the vertex will equal the distance from the vertex to the directrix. Let $a$ represent the distance from the focus to the vertex. To develop an equation for a parabola, we start by looking at parabolas whose vertex is at the origin. We consider four possibilities—parabolas that open left, parabolas that open right, parabolas that open up, and parabolas that open down.

Figure 13

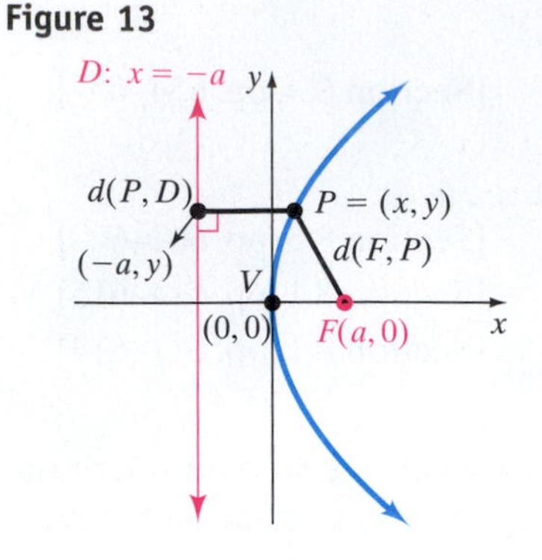


Let's see how to obtain the equation of a parabola whose vertex is at the origin and opens right. To do this, set up a Cartesian plane and position the parabola so that its vertex is at the origin, (0, 0), the focus is on the positive $x$-axis and the directrix is a vertical line in quadrants II and III. Because $a$ represents the distance from the vertex to the focus, we have that the focus is located at $(a, 0)$. Also, because the distance from the vertex to the directrix is $a$ units, the directrix is the line $x = -a$. See Figure 13. If $P = (x, y)$ is any point on the parabola, then the distance from $P$ to the focus $F$, $(a, 0)$, must equal the distance from $P$ to the directrix, $x = -a$. That is,

$$d(F, P) = d(P, D)$$

Use the distance formula: $$\sqrt{(x-a)^2 + (y-0)^2} = \sqrt{(x-(-a))^2 + (y-y)^2}$$

Square both sides; simplify: $$(x-a)^2 + y^2 = (x+a)^2$$

Multiply out binomials: $$x^2 - 2ax + a^2 + y^2 = x^2 + 2ax + a^2$$

Combine like terms: $$y^2 = 4ax$$

So the equation of the parabola whose vertex is at the origin and opens to the right is $y^2 = 4ax$. We obtain the equations of the three other possibilities (opens left, opens up, or opens down) using logic similar to the logic we used to obtain the equation above. We summarize these possibilities in Table 1.

**Table 1 Equations of a Parabola: Vertex at (0, 0); Focus on an Axis; $a > 0$**

| Vertex | Focus | Directrix | Equation | Description |
|---|---|---|---|---|
| (0, 0) | $(a, 0)$ | $x = -a$ | $y^2 = 4ax$ | Parabola, axis of symmetry is the $x$-axis, opens to the right |
| (0, 0) | $(-a, 0)$ | $x = a$ | $y^2 = -4ax$ | Parabola, axis of symmetry is the $x$-axis, opens to the left |
| (0, 0) | $(0, a)$ | $y = -a$ | $x^2 = 4ay$ | Parabola, axis of symmetry is the $y$-axis, opens up |
| (0, 0) | $(0, -a)$ | $y = a$ | $x^2 = -4ay$ | Parabola, axis of symmetry is the $y$-axis, opens down |

The graphs of the four parabolas are given in Figure 14.

Figure 14

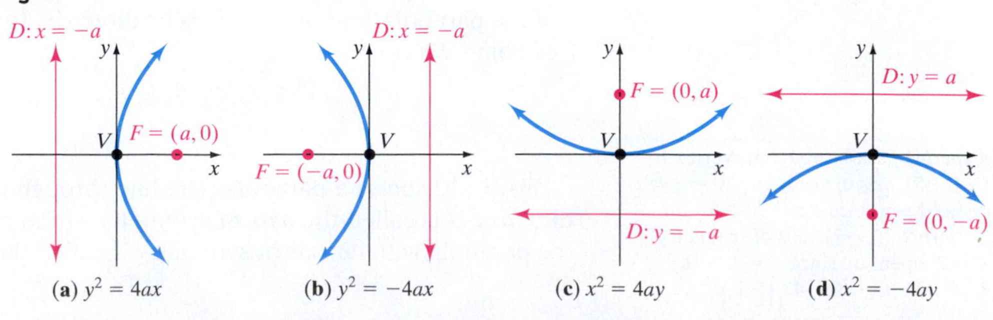

## EXAMPLE 1 Graphing a Parabola That Opens Left

Graph the equation $y^2 = -12x$.

### Solution

The equation $y^2 = -12x$ is of the form $y^2 = -4ax$, where $-4a = -12$, so that $a = 3$. The graph of the equation is a parabola with vertex at $(0, 0)$ and focus at $(-a, 0) = (-3, 0)$ so that the parabola opens to the left. The directrix is the line $x = 3$. To graph the parabola, it is helpful to plot the two points on the graph above and below the focus. Because the points are directly above and below the focus, we let $x = -3$ in the equation $y^2 = -12x$ and solve for $y$.

Figure 15

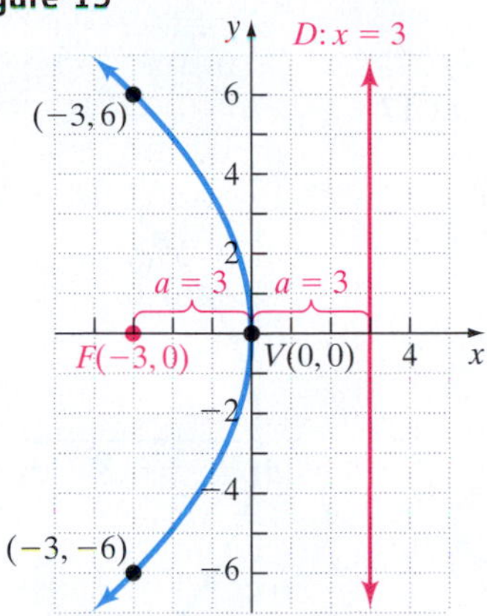


$$y^2 = -12x$$

Let $x = -3$: $\quad = -12(-3)$

$$= 36$$

Take the square root of both sides: $\quad y = \pm 6$

The points on the parabola above and below the focus are $(-3, 6)$ and $(-3, -6)$. These points help in graphing the parabola because they determine the "opening." See Figure 15.

**QUICK ✓** *Graph the equation.*

**1.** $y^2 = 8x$ **2.** $y^2 = -20x$

## EXAMPLE 2 Graphing a Parabola That Opens Up

Graph the equation $x^2 = 8y$.

### Solution

The equation $x^2 = 8y$ is of the form $x^2 = 4ay$, where $4a = 8$, so that $a = 2$. The graph of the equation is a parabola with vertex at $(0, 0)$ and focus at $(0, a) = (0, 2)$ so that the parabola opens up. The directrix is the line $y = -2$. To graph the parabola, it is helpful to plot the two points on the graph to the left and right of the focus. We let $y = 2$ in the equation $x^2 = 8y$ and solve for $x$.

$$x^2 = 8y = 8(2) = 16$$

Take the square root of both sides: $\quad x = \pm 4$

The points on the parabola to the left and right of the focus are $(-4, 2)$ and $(4, 2)$. See Figure 16.

Figure 16

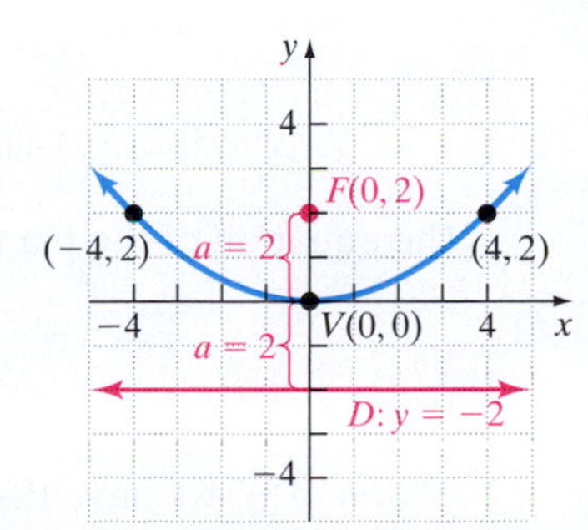


**QUICK ✓** *Graph the equation.*

**3.** $x^2 = 4y$ **4.** $x^2 = -12y$

## 2 Find the Equation of a Parabola

We are now going to change gears and use information regarding the equation of a parabola to obtain its equation.

### EXAMPLE 3 Finding an Equation of a Parabola

Find an equation of the parabola with vertex at $(0, 0)$ and focus at $(5, 0)$. Graph the equation.

**Solution**

**Work Smart**
It is helpful to plot the information given about the parabola before finding its equation. For example, if we plot the vertex and focus of the parabola in Example 3, we can see that it opens to the right.

The distance from the vertex $(0, 0)$ to the focus is $(5, 0)$ is $a = 5$. Because the focus lies on the positive $x$-axis, we know that the parabola will open to the right. This means the equation of the parabola is of the form $y^2 = 4ax$ with $a = 5$:

$$y^2 = 4(5)x = 20x$$

Figure 17 shows the graph of $y^2 = 20x$.

Figure 17

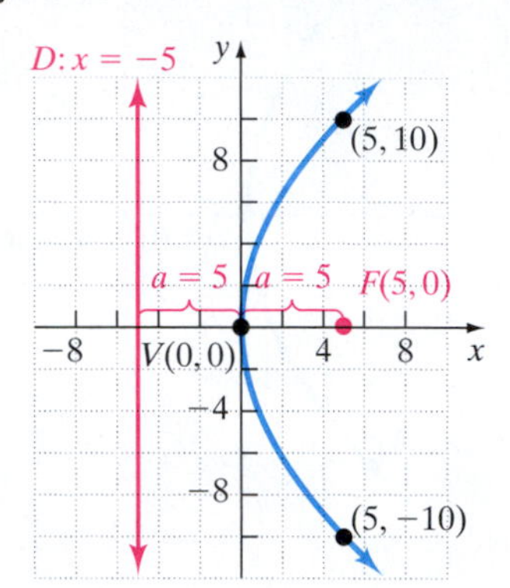


### EXAMPLE 4 Finding the Equation of a Parabola

Find the equation of a parabola with vertex at $(0, 0)$ if its axis of symmetry is the $y$-axis and its graph contains the point $(4, 3)$. Graph the equation.

**Solution**

Which of the four equations of parabolas listed in Table 1 should we use? The vertex is at the origin and the axis of symmetry is the $y$-axis, so the parabola either opens up or down. Because the graph contains the point $(4, 3)$, which is in quadrant I, the parabola must open up. Therefore, the equation of the parabola is of the form

$$x^2 = 4ay$$

**Work Smart**
Again, don't forget to draw a picture of the given information.

Because the point $(4, 3)$ is on the parabola, we let $x = 4$ and $y = 3$ in the equation $x^2 = 4ay$ to determine $a$.

$$x^2 = 4ay$$

$x = 4, y = 3$: $$4^2 = 4a(3)$$

$$16 = 12a$$

Divide both sides by 12: $$a = \frac{16}{12} = \frac{4}{3}$$

The equation of the parabola is

$$x^2 = 4\left(\frac{4}{3}\right)y \quad \text{or} \quad x^2 = \frac{16}{3}y$$

With $a = \frac{4}{3}$, we have that the focus is $\left(0, \frac{4}{3}\right)$ and the directrix is the line $y = -\frac{4}{3}$. We let $y = \frac{4}{3}$ to find points left and right of the focus to determine the "opening" and determine the points $\left(-\frac{8}{3}, \frac{4}{3}\right)$ and $\left(\frac{8}{3}, \frac{4}{3}\right)$ are on the graph. Figure 18 shows the graph of the parabola.

Figure 18

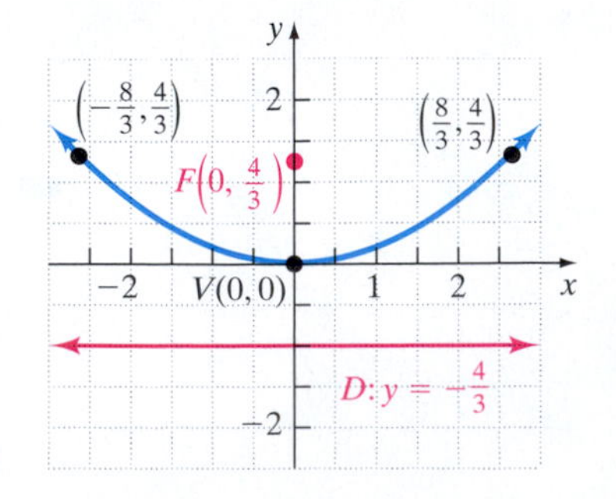

**QUICK ✓** *Find the equation of the parabola described. Graph the equation.*

**5.** Vertex at $(0, 0)$; focus at $(0, -8)$

**6.** Vertex at $(0, 0)$; axis of symmetry the $x$-axis; contains the point $(3, 2)$

## 3 Graph Parabolas in Which the Vertex Is Not the Origin

**Work Smart**

Think back to circles. The equation of a circle whose center is $(h, k)$ is $(x - h)^2 + (y - k)^2 = r^2$. The center shifts horizontally $h$ units and vertically $k$ units.

If a parabola with vertex at the origin and axis of symmetry along a coordinate axis is shifted horizontally $h$ units and then vertically $k$ units, the result is a parabola with vertex at $(h, k)$ and axis of symmetry parallel to either the $x$-axis or $y$-axis. The equations of these parabolas have the same form as those whose vertex is at the origin except that $x$ is replaced with $x - h$ (the horizontal shift) and $y$ is replaced with $y - k$ (the vertical shift). Table 2 givens the equations of the four parabolas. Figure 19(a)–(d) illustrates the graphs for $h > 0$ and $k > 0$.

**Table 2 Parabolas with Vertex at $(h, k)$; Axis of Symmetry Parallel to a Coordinate Axis, $a > 0$**

| Vertex | Focus | Directrix | Equation | Description |
|---|---|---|---|---|
| $(h, k)$ | $(h + a, k)$ | $x = h - a$ | $(y - k)^2 = 4a(x - h)$ | Parabola, axis of symmetry parallel to $x$-axis, opens to the right |
| $(h, k)$ | $(h - a, k)$ | $x = h + a$ | $(y - k)^2 = -4a(x - h)$ | Parabola, axis of symmetry parallel to $x$-axis, opens to the left |
| $(h, k)$ | $(h, k + a)$ | $y = k - a$ | $(x - h)^2 = 4a(y - k)$ | Parabola, axis of symmetry parallel to $y$-axis, opens up |
| $(h, k)$ | $(h, k - a)$ | $y = k + a$ | $(x - h)^2 = -4a(y - k)$ | Parabola, axis of symmetry parallel to $y$-axis, opens down |

**Figure 19**

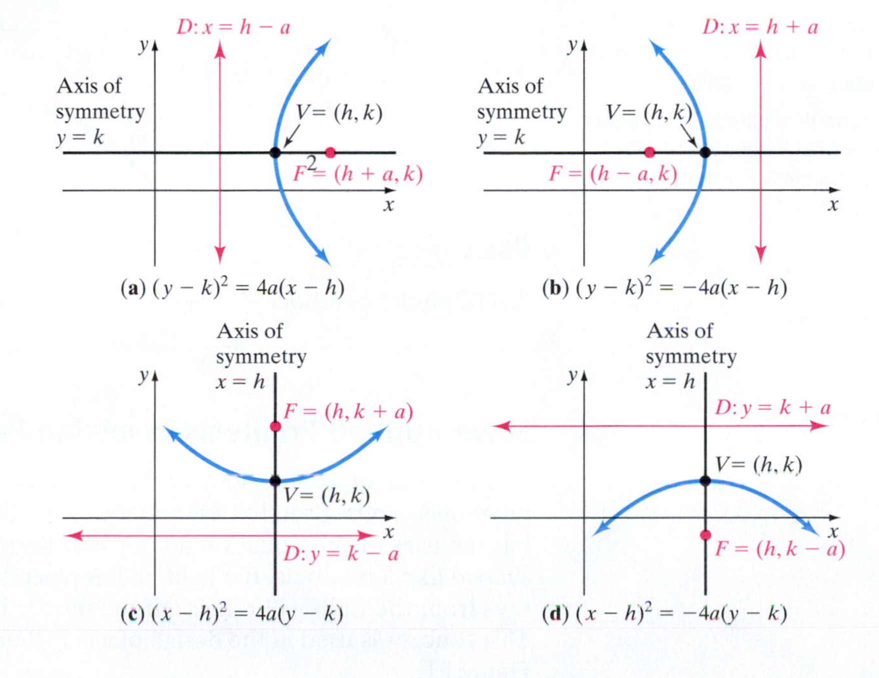


(a) $(y - k)^2 = 4a(x - h)$

(b) $(y - k)^2 = -4a(x - h)$

(c) $(x - h)^2 = 4a(y - k)$

(d) $(x - h)^2 = -4a(y - k)$

## EXAMPLE 5 Graphing a Parabola Whose Vertex Is Not at the Origin

Graph the parabola $x^2 - 2x + 8y + 25 = 0$.

### Solution

Notice the equation is not in any of the forms given in Table 2. We need to complete the square in $x$ to write the equation in standard form.

$$x^2 - 2x + 8y + 25 = 0$$

Isolate the terms involving $x$: $x^2 - 2x = -8y - 25$

Complete the square: $x^2 - 2x + 1 = -8y - 25 + 1$

Simplify: $x^2 - 2x + 1 = -8y - 24$

Factor: $(x - 1)^2 = -8(y + 3)$

The equation is of the form $(x - h)^2 = -4a(y - k)$. This is a parabola that opens down with vertex $(h, k) = (1, -3)$. Since $-4a = -8$, we have that $a = 2$. Because the parabola opens down, the focus will be $a = 2$ units below the vertex at $(1, -5)$. We find two additional points on the graph to the left and right of the focus. To do this, we let $y = -5$ in the equation of the parabola.

Let $y = -5$: $(x - 1)^2 = -8(-5 + 3)$

$$(x - 1)^2 = -8(-2)$$

$$(x - 1)^2 = 16$$

Take the square root of both sides: $x - 1 = \pm 4$

Add 1 to both sides: $x = 1 \pm 4$

$$x = 1 - 4 \quad \text{or} \quad x = 1 + 4$$

$$x = -3 \quad \text{or} \quad x = 5$$

The points $(-3, -5)$ and $(5, -5)$ are on the graph of the parabola. The directrix is $a = 2$ units above the vertex, so $y = -1$ is the directrix. Figure 20 shows the graph of the parabola.

**Work Smart: Study Skills**

Which is the graph of a parabola and which is the graph of a circle?

$x^2 + y^2 - 2y + 3 = 0$

$2y^2 - 12y - x + 22 = 0$

How can you tell the difference between the equation of a circle and the equation of a parabola?

The equation of a parabola has either an $x^2$-term or a $y^2$-term, not both. The equation of a circle has both an $x^2$- and a $y^2$-term.

**Figure 20**

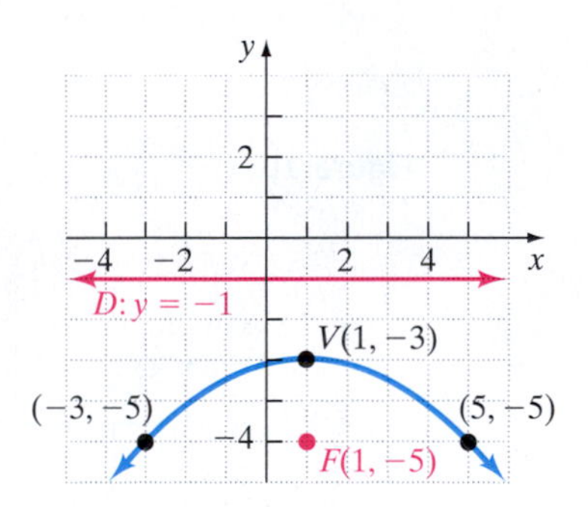


**QUICK ✓**

**7.** Graph the parabola $y^2 - 4y - 12x - 32 = 0$.

## 4 Solve Applied Problems Involving Parabolas

The number of applications of parabolas is astounding. We have already seen how parabolas can be used to describe the shape of cables supporting a suspension bridge, but the uses of this equation do not stop there! For example, suppose that a mirror is shaped like a parabola. If a light bulb is placed at the focus of the parabola, then all the rays from the bulb will reflect off the mirror in lines parallel to the axis of symmetry. This concept is used in the design of a car's headlights, flashlights, and searchlights. See Figure 21.

As another example, suppose that rays of light are received by a parabola. When the rays strike the surface of a parabolic mirror whose axis of symmetry is parallel to these rays, they are all reflected to a single point—the focus. This idea is used in the design of some telescopes and satellite dishes. See Figure 22.

**Figure 21**
Searchlight

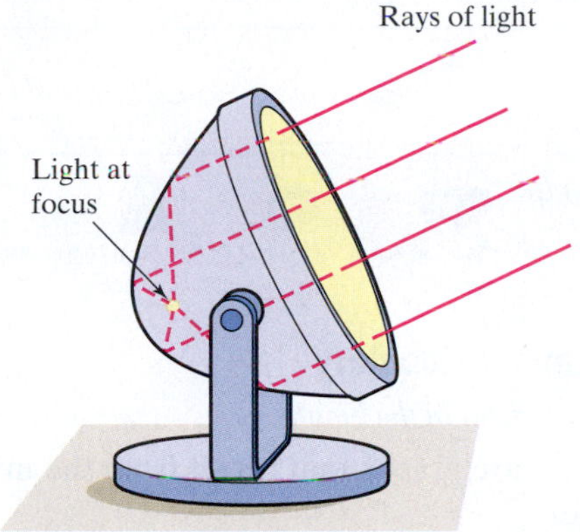

Rays of light

Light at focus

**Figure 22**
Telescope

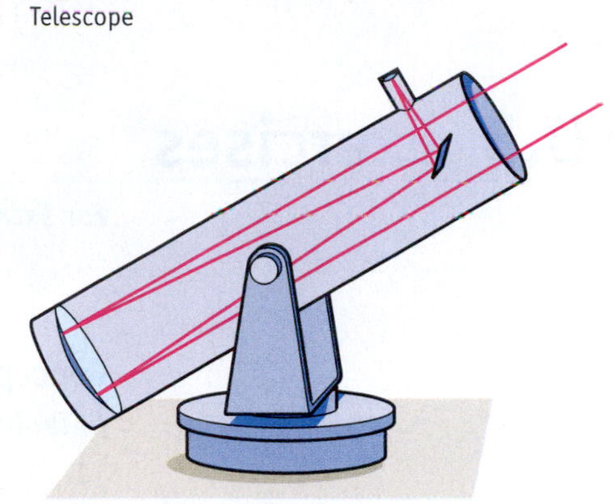

## EXAMPLE 6 A Satellite Dish

A satellite dish is shaped like a parabola. See Figure 23(a). The signals that are received by the dish strike the surface of the dish and are reflected to a single point, where the receiver of the dish is located. If a satellite dish is 2 feet across at its opening and 6 inches deep (0.5 feet) at its center, at what position should the receiver be placed?

### Solution

We want to know where to locate the receiver of the satellite dish. This means we want to know where the focus of the dish is. To solve this problem, we draw a parabola on a Cartesian plane so that the vertex of the parabola is the origin and the focus is on the positive $y$-axis. The width of the dish is 2 feet across and its height is 0.5 feet. So we know two points on the graph of the parabola indicated in Figure 23(b).

**Figure 23**

(a)

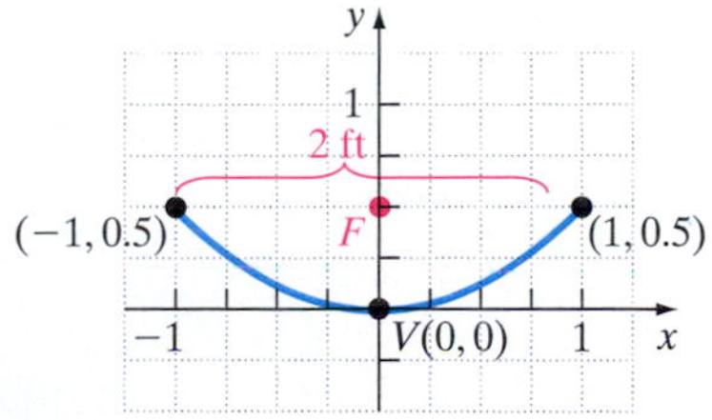

(b)

The parabola is an equation of the form $x^2 = 4ay$. Since $(1, 0.5)$ is a point on the graph, we have

$$x = 1, y = 0.5: \quad 1^2 = 4a(0.5)$$
$$1 = 2a$$
$$\text{Divide both sides by 2:} \quad a = \frac{1}{2}$$

The receiver should be located $\frac{1}{2}$ foot from the base of the dish along its axis of symmetry.

**QUICK ✓**

**8.** A satellite dish is shaped like a parabola. The signals that are received by the dish strike the surface of the dish and are reflected to a single point, where the receiver of the dish is located. If the dish is 4 feet across at its opening and 6 inches deep (0.5 feet) at its center, at what position should the receiver be placed?

# 10.3 Exercises

***For Extra Help:***

Student Solutions Manual | CD Video | PH Math/Tutor Center | MathXL Tutorials on CD | MathXL® | MyMathLab

## Concepts and Vocabulary

*In Problems 1–3, fill in the blanks.*

**1.** ________ are graphs that result from the intersection of a right circular cone and a plane.

**2.** A ________ is the collection of all points $P$ in the plane that are the same distance from a fixed point $F$ as they are from a fixed line $D$.

**3.** The point of intersection of the parabola with its axis of symmetry is called the ________.

*In Problems 4–6, answer True or False to each statement.*

**4.** The line through the focus and perpendicular to the directrix is called the axis of symmetry.

**5.** The parabola $(x + 3)^2 = -14(y - 3)$ opens to the left.

**6.** The vertex of the parabola $(y + 2)^2 = 8(x - 3)$ is $(3, -2)$.

**7.** The distance from a point on a parabola to its focus is 8 units. What is the distance from the same point on the parabola to the directrix?

**8.** Write down the four equations that are parabolas with vertex at $(h, k)$.

**9.** Draw a parabola and label the vertex, axis of symmetry, focus, and directrix.

**10.** Explain the difference between the discussion of parabolas presented in this section and the discussion presented in Sections 8.4 and 8.5.

## Building Skills

*In Problems 11–18, the graph of a parabola is given. Match each graph to its equation.*

**(a)** $y^2 = 8x$ **(b)** $y^2 = -8x$ **(c)** $x^2 = 8y$
**(d)** $x^2 = -8y$ **(e)** $(y - 2)^2 = 8(x + 1)$ **(f)** $(y - 2)^2 = -8(x + 1)$
**(g)** $(x + 1)^2 = 8(y - 2)$ **(h)** $(x + 1)^2 = -8(y - 2)$

**11.**

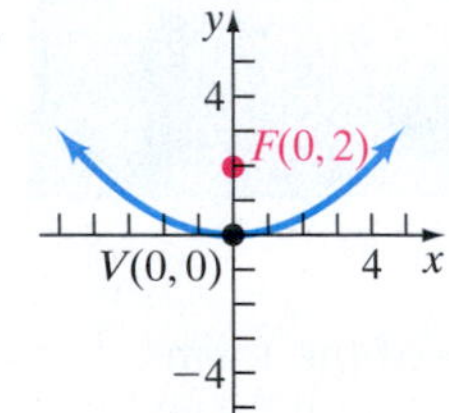


**12.**

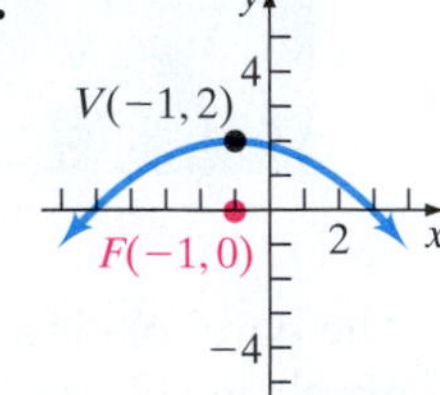


**13.**

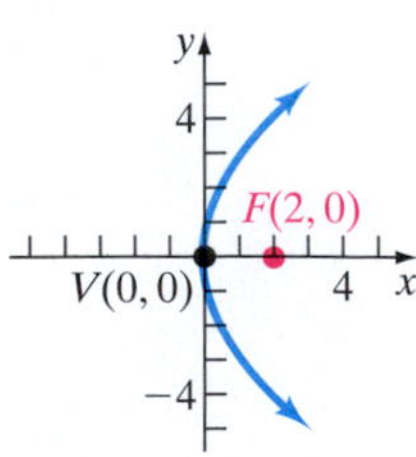


**14.**

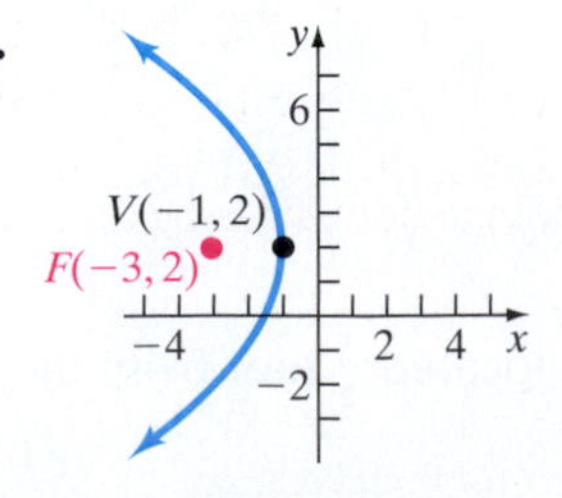


**15.**

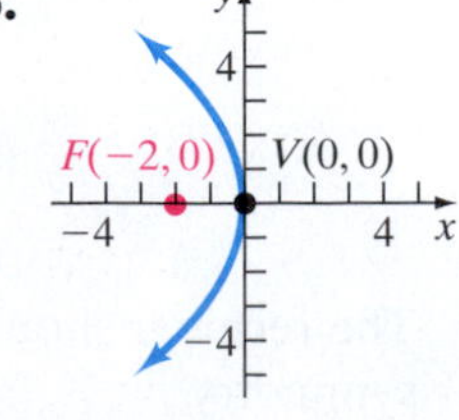


**16.**

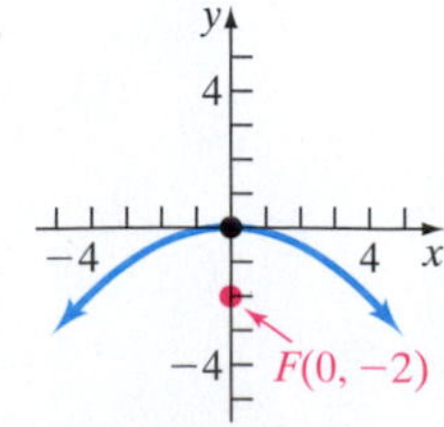

**17.**

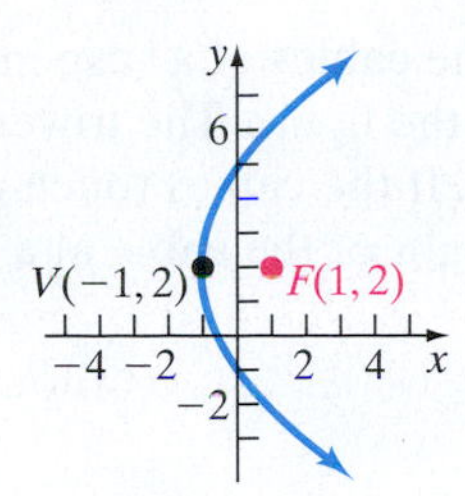


**18.**

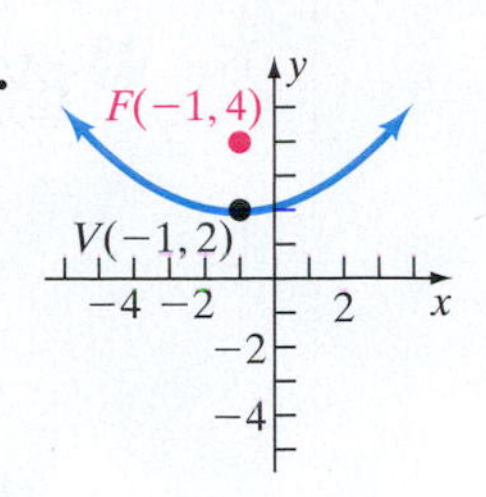


*In Problems 19–28, find the equation of the parabola described. Graph the parabola.*

**19.** Vertex at $(0, 0)$; focus at $(5, 0)$

**20.** Vertex at $(0, 0)$; focus at $(0, 5)$

**21.** Vertex at $(0, 0)$; focus at $(0, -6)$

**22.** Vertex at $(0, 0)$; focus at $(-8, 0)$

**23.** Vertex at $(0, 0)$; contains the point $(6, 6)$; axis of symmetry the $y$-axis

**24.** Vertex at $(0, 0)$; contains the point $(2, 2)$; axis of symmetry the $x$-axis

**25.** Vertex at $(0, 0)$; directrix the line $y = 3$

**26.** Vertex at $(0, 0)$; directrix the line $x = -4$

**27.** Focus at $(-3, 0)$; directrix the line $x = 3$

**28.** Focus at $(0, -2)$; directrix the line $y = 2$.

*In Problems 29–46, find the vertex, focus, and directrix of each parabola. Graph the parabola.*

**29.** $x^2 = 24y$

**30.** $x^2 = 28y$

**31.** $y^2 = -6x$

**32.** $y^2 = 10x$

**33.** $x^2 = -8y$

**34.** $x^2 = -16y$

**35.** $(x - 2)^2 = 4(y - 4)$

**36.** $(x + 4)^2 = -4(y - 1)$

**37.** $(y + 3)^2 = -8(x + 2)$

**38.** $(y - 2)^2 = 12(x + 5)$

**39.** $(x + 5)^2 = -20(y - 1)$

**40.** $(x - 6)^2 = 2(y - 2)$

**41.** $x^2 + 4x + 12y + 16 = 0$

**42.** $x^2 + 2x - 8y + 25 = 0$

**43.** $y^2 - 8y - 4x + 20 = 0$

**44.** $y^2 - 8y + 16x - 16 = 0$

**45.** $x^2 + 10x + 6y + 13 = 0$

**46.** $x^2 - 4x + 10y + 4 = 0$

*In Problems 47 and 48, write an equation for each parabola.*

**47.**

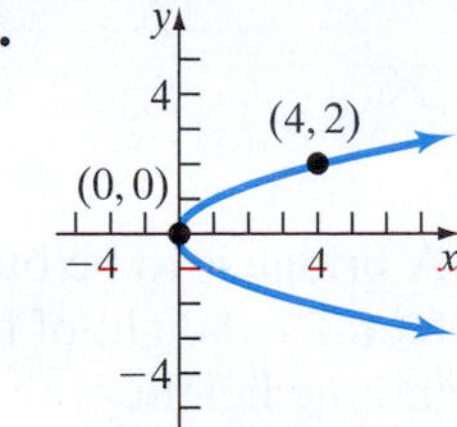


**48.**

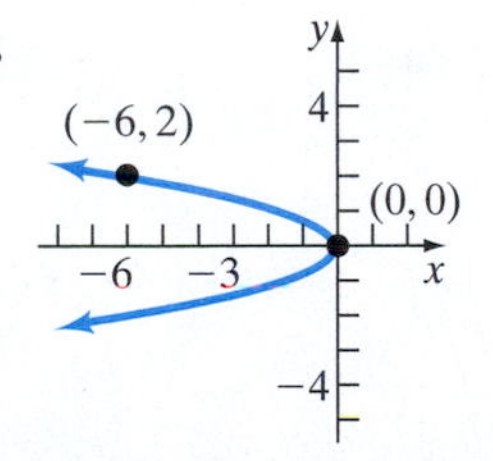


## Applying the Concepts

**49. A Headlight** The headlight of a car is in the shape of a parabola. Its diameter is 4 inches and its depth is 1 inch. How far from the vertex should the light bulb be placed so that the rays will be reflected parallel to the axis?

**50. A Headlight** The headlight of a car is in the shape of a parabola. Suppose the engineers have designed the headlight to be 5 inches in diameter and wish the bulb to be placed at the focus 1 inch from the vertex. What is the depth of the headlight?

**51. Suspension Bridge** The cables of a suspension bridge are in the shape of a parabola, as shown in the figure. The towers supporting the cable are 500 feet apart and 60 feet high. If the cables touch the road surface midway between the towers, what is the height of the cable at a point 150 feet from the center of the bridge?

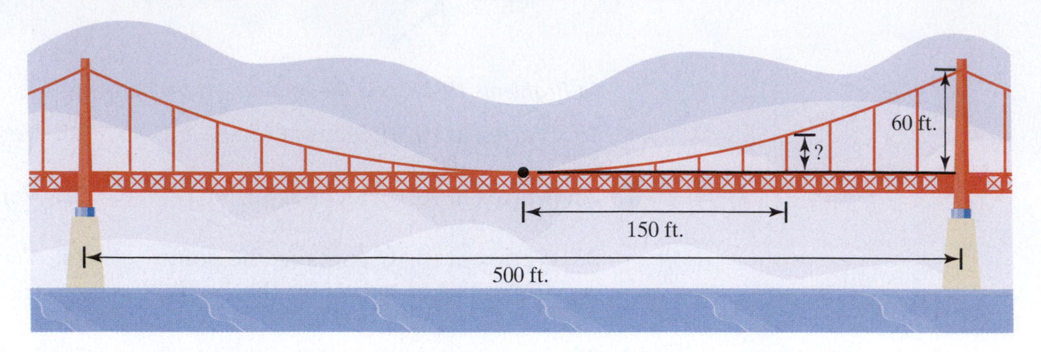


**52. Suspension Bridge** The cables of a suspension bridge are in the shape of a parabola. The towers supporting the cable are 400 feet apart and 80 feet high. If the cables touch the road surface midway between the towers, what is the height of the cable at a point 100 feet from the center of the bridge?

**53. Parabolic Arch Bridge** A bridge is built in the shape of a parabolic arch. The bridge has a span of 100 feet and a maximum height of 30 feet. See the illustration. Choose a suitable rectangular coordinate system and find the height of the arch at distances of 10, 30, and 50 feet from the center.

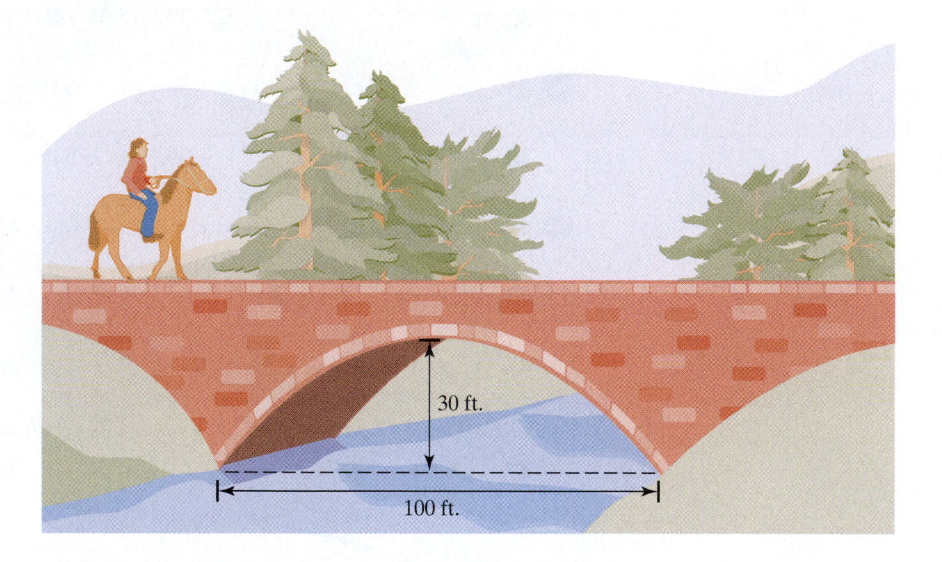


**54. Parabolic Arch Bridge** A bridge is to be built in the shape of a parabolic arch and is to have a span of 120 feet. The height of the arch a distance of 30 feet from the center is to be 15 feet. Find the height of the arch at its center.

## Extending the Concepts

*In Problems 55–58, write an equation for each parabola.*

**55.**

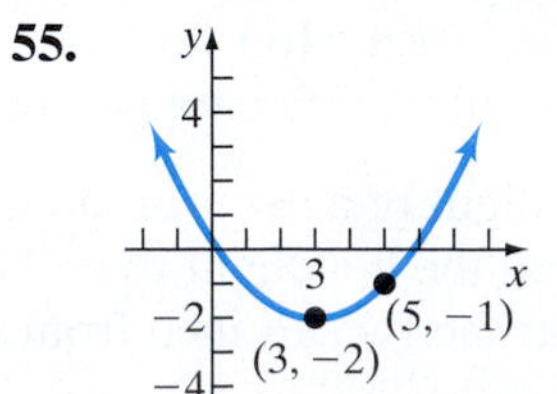


**56.**

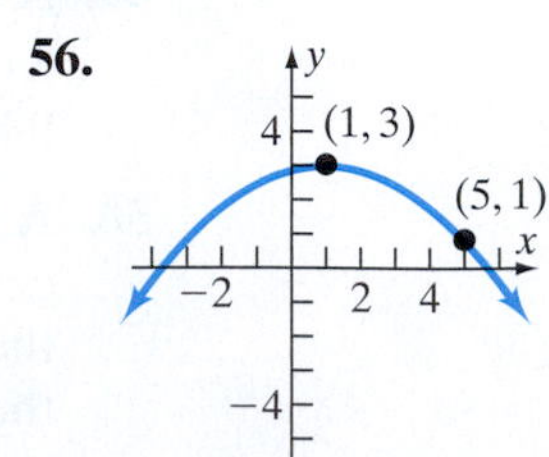

**57.**

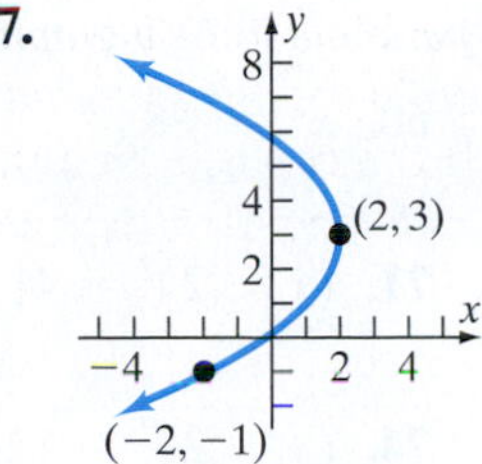


**58.**

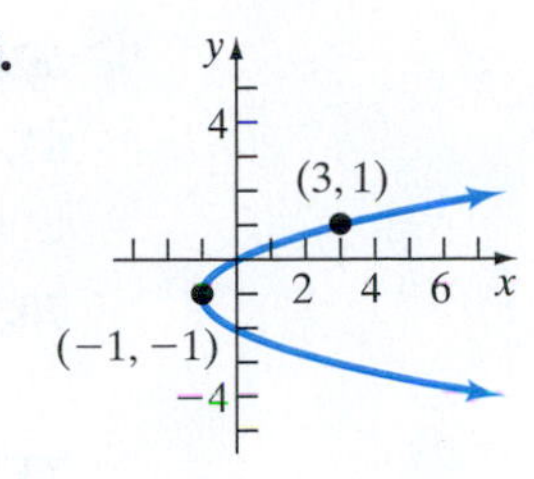


**59.** For the parabola $x^2 = 8y$, (a) verify that $(4, 2)$ is a point on the parabola, and (b) show that the distance from the focus to $(4, 2)$ equals the distance from $(4, 2)$ to the directrix.

**60.** For the parabola $(y - 3)^2 = 12(x + 1)$, (a) verify that $(2, 9)$ is a point on the parabola, and (b) show that the distance from the focus to $(2, 9)$ equals the distance from $(2, 9)$ to the directrix.

## Synthesis Review

**61.** Graph $y = (x + 3)^2$ using the methods of Section 8.4.

**62.** Graph $y = (x + 3)^2 = x^2 + 6x + 9$ using the methods of Section 8.5.

**63.** Graph $y = (x + 3)^2$ using the methods of this section.

**64.** Graph $4(y + 2) = (x - 2)^2$ using the methods of Section 8.4. (**Hint:** Write the equation in the form $y = a(x - h)^2 + k$.)

**65.** Graph $4(y + 2) = (x - 2)^2$ using the methods of this section.

**66.** Compare and contrast the methods of graphing a parabola using the approach in Section 8.4 and the approach in this section. Which do you prefer? Why?

## The Graphing Calculator

To graph a parabola using a graphing calculator, we must solve the equation for $y$, just as we did for circles. This is fairly straightforward if the equation is in the form given in either Table 1 or Table 2. If the equation is not in the form given in Table 1 or Table 2 and it is parabola that opens up or down, then solving for $y$ is also straightforward. However, if it is a parabola that opens left or right, we will need to graph the parabola in two "pieces"—the top half and the bottom half. We can use the quadratic formula to accomplish this task. Consider the equation $y^2 + 8y + x - 4 = 0$. This equation is quadratic in $y$ as shown:

$$y^2 + 8y + x - 4 = 0$$

$a = 1$   $b = 8$   $c = x - 4$

We use the quadratic formula and obtain

$$y = \frac{-8 \pm \sqrt{8^2 - 4(1)(x - 4)}}{2(1)}$$

So we graph

$$Y_1 = \frac{-8 - \sqrt{64 - 4(x - 4)}}{2} \quad \text{and} \quad Y_2 = \frac{-8 + \sqrt{64 - 4(x - 4)}}{2}$$

as shown in Figure 24.

**Figure 24**

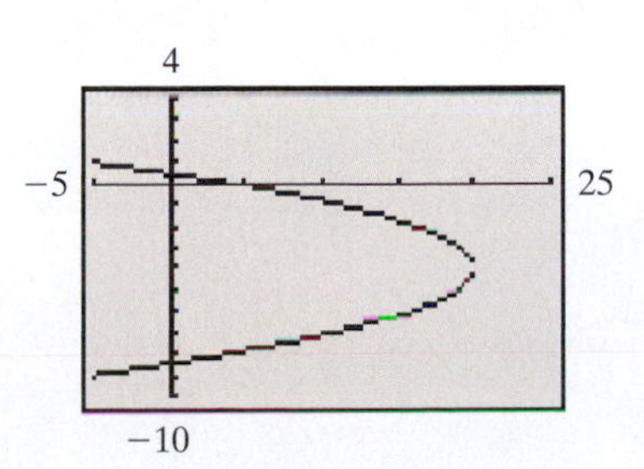

*In Problems 67–78, graph each parabola using a graphing calculator.*

**67.** $x^2 = 24y$ **68.** $x^2 = -8y$ **69.** $y^2 = -6x$

**70.** $y^2 = 10x$ **71.** $(x - 2)^2 = 4(y - 4)$ **72.** $(x + 4)^2 = -4(y - 1)$

**73.** $(y + 3)^2 = -8(x + 2)$ **74.** $(y - 2)^2 = 12(x + 5)$

**75.** $x^2 + 4x + 12y + 16 = 0$ **76.** $x^2 + 2x - 8y + 25 = 0$

**77.** $y^2 - 8y - 4x + 20 = 0$ **78.** $y^2 - 8y + 16x - 16 = 0$

# 10.4 Ellipses

## OBJECTIVES

1. Graph Ellipses in Which the Center Is the Origin
2. Find the Equation of an Ellipse in Which the Center Is the Origin
3. Graph Ellipses in Which the Center Is Not the Origin
4. Solve Applied Problems Involving Ellipses

### *Preparing for Ellipses*

*Before getting started, take the following readiness quiz. If you get a problem wrong, go back to the section cited and review the material.*

**1.** Complete the square of $x^2 + 10x$. [Section 8.1, pp. 614–615]

**2.** Graph $f(x) = (x + 2)^2 - 1$ using transformations. [Section 8.4, pp. 654–657]

## 1 Graph Ellipses in Which the Center Is the Origin

An ellipse is a conic section that is obtained through the intersection of a plane and a cone. See Figure 8(b) on page 780.

**DEFINITION**

An **ellipse** is the collection of points in the plane such that the sum of the distances from two fixed points, called the **foci,** is a constant.

The definition allows us to physically draw an ellipse. To do this, find a piece of string (the length of the string is the constant referred to in the definition). Now take two thumbtacks and stick them on a piece of cardboard so that the distance between them is less than the length of the string. The two thumbtacks represent the foci of the ellipse. Now attach the ends of the string to the thumbtacks and, using the point of a pencil, pull the string taut. Keeping the string taut, rotate the pencil around the two thumbtacks. The pencil traces out an ellipse as shown in Figure 25.

**Figure 25**

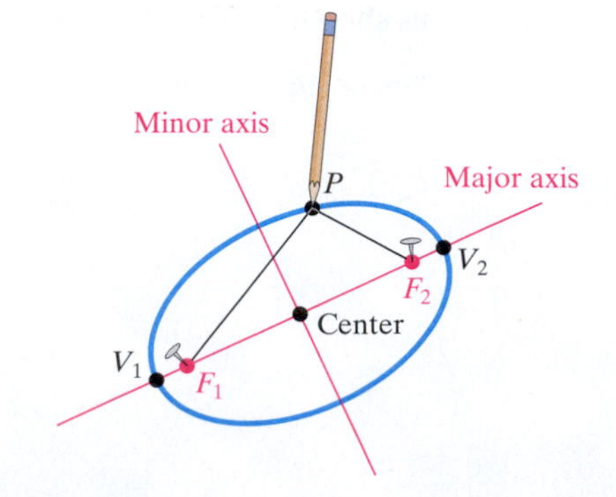


***Preparing for...Answers***

**1.** $x^2 + 10x + 25$

**2.**

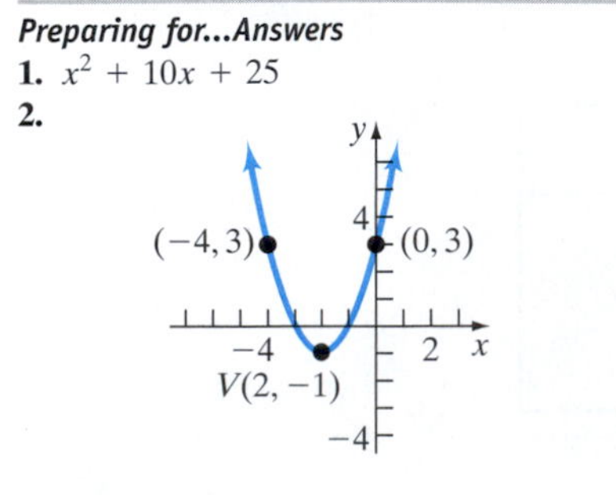

In Figure 25, the foci are labeled $F_1$ and $F_2$. The line containing the foci is called the **major axis.** The midpoint of the line segment joining the foci is called the **center** of the ellipse. The line through the center and perpendicular to the major axis is called the **minor axis.**

The two points of intersection of the ellipse and the major axis are the **vertices,** $V_1$ and $V_2$, of the ellipse. The distance from one vertex to the other is called the **length of the major axis.**

We are now ready to find the equation of an ellipse in a Cartesian plane. First, we place the center of the ellipse at the origin. Second, we position the ellipse so that its major axis coincides with a coordinate axis. Let's have the major axis coincide with the $x$-axis as shown in Figure 26 and call $c$ the distance from the center of the ellipse to a focus, so that one focus is at $F_1 = (-c, 0)$ and the other focus is at $F_2 = (c, 0)$. Let $2a$ represent the constant distance referred to in the definition (the reason for this will be clear shortly).

**Figure 26**

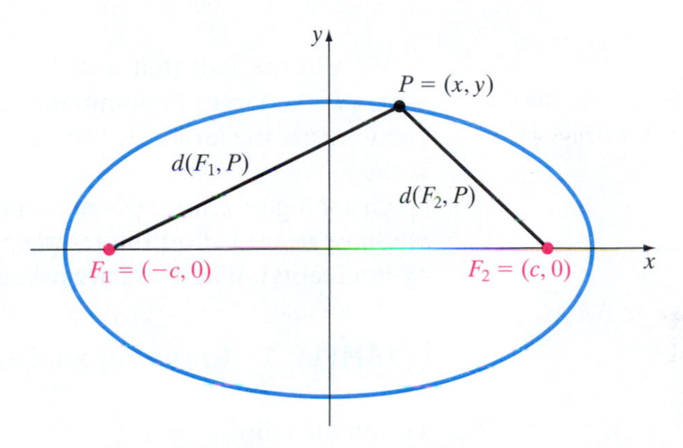


If $P = (x, y)$ is any point on the ellipse, we know from the definition of the ellipse that the sum of the distance from $P$ to the foci equals a constant. Recall, we will let the constant equal $2a$, so that

$$d(F_1, P) + d(F_2, P) = 2a$$

By referring to Figure 26 and using the distance formula, we find that

$$\sqrt{(x-(-c))^2 + (y-0)^2} + \sqrt{(x-c)^2 + (y-0)^2} = 2a$$

This ultimately simplifies to

$$\frac{x^2}{a^2} + \frac{y^2}{b^2} = 1$$

So the equation of an ellipse whose center is at the origin and whose major axis is the $x$-axis is $\frac{x^2}{a^2} + \frac{y^2}{b^2} = 1$. Another possibility is that the major axis is the $y$-axis. We could obtain the equation for this ellipse using logic similar to the logic we used to obtain the equation above. We summarize the equations of ellipses with center at the origin in Table 3.

**Table 3 Ellipses with Center at the Origin**

| Center | Major Axis | Foci | Vertices | Equation |
|---|---|---|---|---|
| (0, 0) | $x$-axis | $(-c, 0)$ and $(c, 0)$ | $(-a, 0)$ and $(a, 0)$ | $\frac{x^2}{a^2} + \frac{y^2}{b^2} = 1$ |
| (0, 0) | $y$-axis | $(0, -c)$ and $(0, c)$ | $(0, -a)$ and $(0, a)$ | $\frac{x^2}{b^2} + \frac{y^2}{a^2} = 1$ |

In both ellipses, $a > b$ and $b^2 = a^2 - c^2$. The graphs of the two ellipses are given in Figure 27.

Figure 27

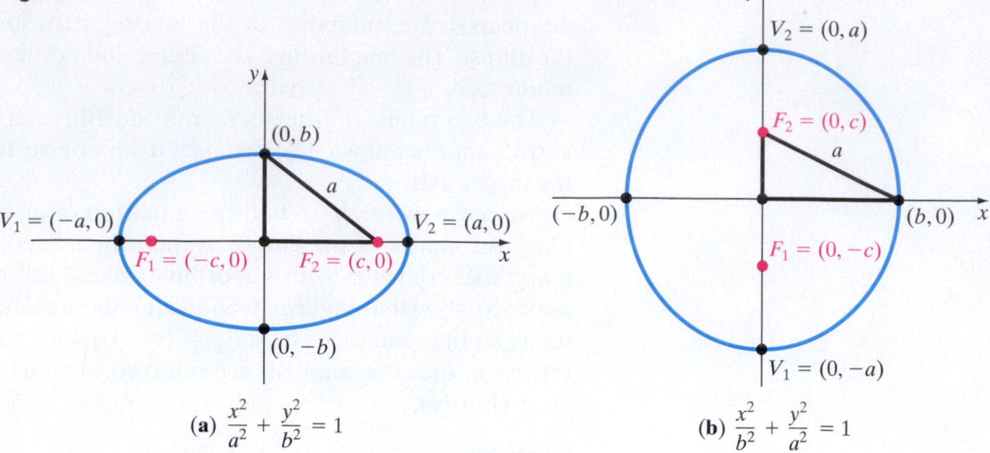


(a) $\frac{x^2}{a^2} + \frac{y^2}{b^2} = 1$  (b) $\frac{x^2}{b^2} + \frac{y^2}{a^2} = 1$

**Work Smart**

The term with the larger denominator tells us which axis is the major axis.

We use the fact that $a > b$ to determine whether the major axis is the $x$-axis or $y$-axis. If the larger denominator is associated with the $x^2$-term, then the major axis is the $x$-axis; if the larger denominator is associated with the $y^2$-term, then the major axis is the $y$-axis.

Graphing an ellipse whose center is at the origin is fairly straightforward because all we have to do is find the intercepts by letting $y = 0$ ($x$-intercepts) and letting $x = 0$ ($y$-intercepts). When you are asked to graph an ellipse, be sure to label the foci.

## EXAMPLE 1 Graphing an Ellipse

Graph the ellipse: $\frac{x^2}{25} + \frac{y^2}{9} = 1$

### Solution

First, we notice that the larger number, 25, is in the denominator of the $x^2$-term. This means that the major axis is the $x$-axis and the equation of the ellipse is of the form $\frac{x^2}{a^2} + \frac{y^2}{b^2} = 1$ so that $a^2 = 25$ and $b^2 = 9$. The center of the ellipse is the origin, $(0, 0)$. Because $b^2 = a^2 - c^2$, or $c^2 = a^2 - b^2$, we have that $c^2 = 25 - 9 = 16$, so that $c = \pm 4$. Since the major axis is the $x$-axis, the foci are $(-4, 0)$ and $(4, 0)$. We now find the intercepts:

$x$-intercepts: Let $y = 0$:

$$\frac{x^2}{25} + \frac{0^2}{9} = 1$$
$$\frac{x^2}{25} = 1$$
$$x^2 = 25$$
$$x = \pm 5$$

$y$-intercepts: Let $x = 0$:

$$\frac{0^2}{25} + \frac{y^2}{9} = 1$$
$$\frac{y^2}{9} = 1$$
$$y^2 = 9$$
$$y = \pm 3$$

The intercepts are $(-5, 0)$, $(5, 0)$, $(0, -3)$, and $(0, 3)$. Figure 28 shows the graph of the ellipse.

Figure 28

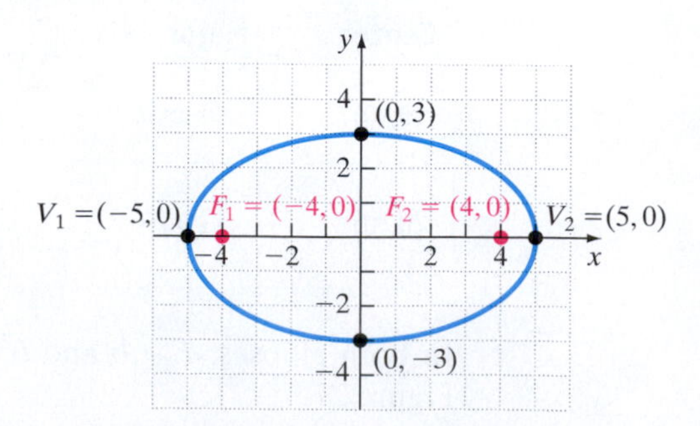

Let's try another one.

## EXAMPLE 2 Graphing an Ellipse

Graph the ellipse: $\frac{x^2}{4} + \frac{y^2}{16} = 1$

### Solution

First, we notice that the larger number, 16, is in the denominator of the $y^2$-term. This means that the major axis is the $y$-axis and the equation of the ellipse is of the form $\frac{x^2}{b^2} + \frac{y^2}{a^2} = 1$, so that $a^2 = 16$ and $b^2 = 4$. The center of the ellipse is the origin, (0, 0). Because $b^2 = a^2 - c^2$, or $c^2 = a^2 - b^2$, we have that $c^2 = 16 - 4 = 12$, so that $c = \pm\sqrt{12} = \pm 2\sqrt{3}$. Since the major axis is the $y$-axis, the foci are $(0, -2\sqrt{3})$ and $(0, 2\sqrt{3})$. We now find the intercepts:

$x$-intercepts: Let $y = 0$:

$$\frac{x^2}{4} + \frac{0^2}{16} = 1$$
$$\frac{x^2}{4} = 1$$
$$x^2 = 4$$
$$x = \pm 2$$

$y$-intercepts: Let $x = 0$:

$$\frac{0^2}{4} + \frac{y^2}{16} = 1$$
$$\frac{y^2}{16} = 1$$
$$y^2 = 16$$
$$y = \pm 4$$

Figure 29

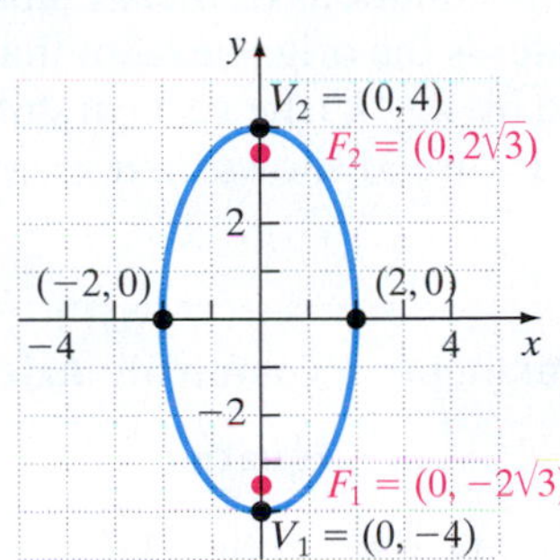


The intercepts are $(-2, 0)$, $(2, 0)$, $(0, -4)$, and $(0, 4)$. Figure 29 shows the graph of the ellipse.

**QUICK ✓** *Graph each ellipse.*

1. $\frac{x^2}{9} + \frac{y^2}{4} = 1$

2. $\frac{x^2}{16} + \frac{y^2}{36} = 1$

## 2 Find the Equation of an Ellipse in Which the Center Is the Origin

Just as we did with parabolas, we are now going to use information about an ellipse to find its equation.

## EXAMPLE 3 Finding the Equation of an Ellipse

Find the equation of the ellipse whose center is the origin with a focus at $(-2, 0)$ and vertex at $(-5, 0)$. Graph the ellipse.

### Solution

By plotting the given focus and vertex, we find that the points lie on the $x$-axis. For this reason, the major axis is the $x$-axis. So the equation of the ellipse is of the form $\frac{x^2}{a^2} + \frac{y^2}{b^2} = 1$. The distance from the center of the ellipse to the vertex is $a = 5$ units. The distance from the center of the ellipse to the focus is $c = 2$ units. Because $b^2 = a^2 - c^2$, we have that $b^2 = 5^2 - 2^2 = 25 - 4 = 21$. So the equation of the ellipse is

$$\frac{x^2}{25} + \frac{y^2}{21} = 1$$

**Work Smart**

*"a" is the distance from the center of an ellipse to one of its vertices. "c" is the distance from the center of an ellipse to one of its foci.*

Figure 30 shows the graph.

**Figure 30**

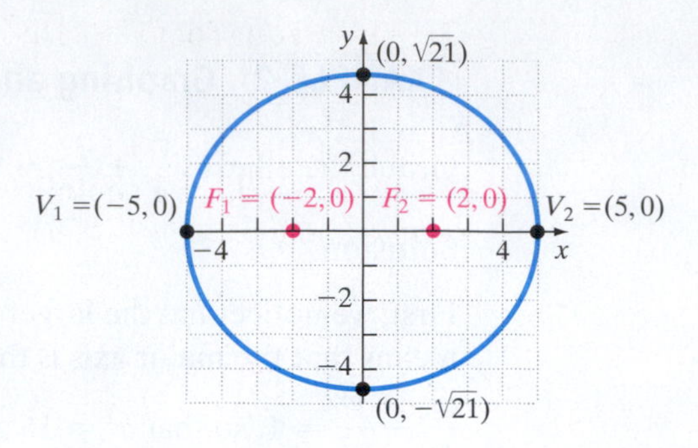


**QUICK ✓**

**3.** Find the equation of the ellipse whose center is the origin with a focus at (0, 3) and vertex at (0, 7). Graph the ellipse.

## 3 Graph Ellipses in Which the Center Is Not the Origin

If an ellipse with center at the origin and major axis coinciding with a coordinate axis is shifted horizontally $h$ units and then vertically $k$ units, the result is an ellipse with center at $(h, k)$ and major axis parallel to a coordinate axis. The equations of these ellipses have the same forms as those given for ellipses whose center is the origin, except that $x$ is replaced by $x - h$ (the horizontal shift) and $y$ is replaced by $y - k$ (the vertical shift). Table 4 gives the forms of the equations for these ellipses. Figure 31 shows their graphs.

**Table 4 Ellipses with Center at (h, k) and Major Axis Parallel to a Coordinate Axis**

| Center | Major Axis | Foci | Vertices | Equation |
|---|---|---|---|---|
| $(h, k)$ | Parallel to $x$-axis | $(h + c, k)$<br>$(h - c, k)$ | $(h + a, k)$<br>$(h - a, k)$ | $\frac{(x-h)^2}{a^2} + \frac{(y-k)^2}{b^2} = 1,$<br>$a > b$ and $b^2 = a^2 - c^2$ |
| $(h, k)$ | Parallel to $y$-axis | $(h, k + c)$<br>$(h, k - c)$ | $(h, k + a)$<br>$(h, k - a)$ | $\frac{(x-h)^2}{b^2} + \frac{(y-k)^2}{a^2} = 1,$<br>$a > b$ and $b^2 = a^2 - c^2$ |

**Work Smart**
Do not attempt to memorize Table 4. Instead, understand the roles of $a$ and $c$ in an ellipse—$a$ is the distance from the center to each vertex and $c$ is the distance from the center to each focus.

**Figure 31**

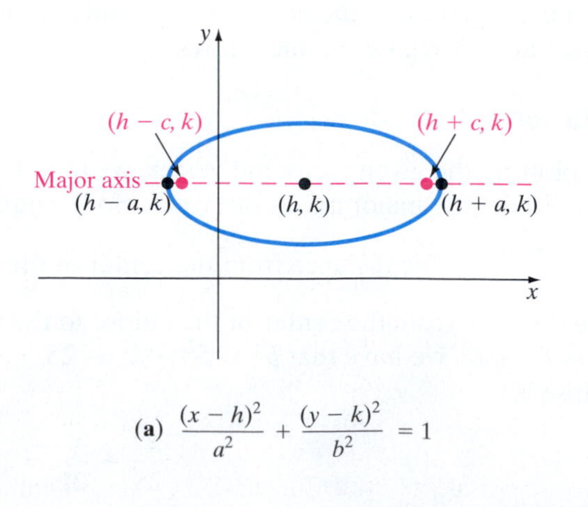


(a) $\frac{(x-h)^2}{a^2} + \frac{(y-k)^2}{b^2} = 1$

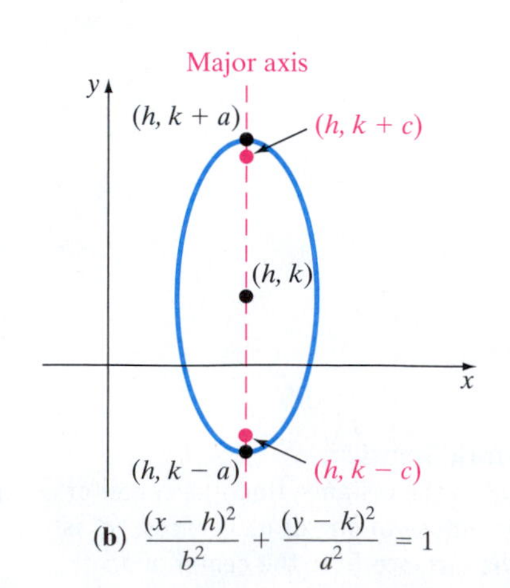


(b) $\frac{(x-h)^2}{b^2} + \frac{(y-k)^2}{a^2} = 1$

## EXAMPLE 4 Graphing an Ellipse

Graph the equation: $x^2 + 16y^2 - 4x + 32y + 4 = 0$

### Solution

We need to write the equation in one of the forms given in Table 4. To do this, we complete the squares in $x$ and in $y$.

$$x^2 + 16y^2 - 4x + 32y + 4 = 0$$

Group like variables; place the constant on the right side:
$$(x^2 - 4x) + (16y^2 + 32y) = -4$$

Factor out 16 from the last two terms:
$$(x^2 - 4x) + 16(y^2 + 2y) = -4$$

Complete each square:
$$(x^2 - 4x + 4) + 16(y^2 + 2y + 1) = -4 + 4 + 16$$

Factor:
$$(x - 2)^2 + 16(y + 1)^2 = 16$$

Divide both sides by 16:
$$\frac{(x - 2)^2}{16} + \frac{(y + 1)^2}{1} = 1$$

**Work Smart**

When we completed the square in $y$, we added a 1, but because it was inside the parentheses with a factor of 16 in front, we must add $16 \cdot 1 = 16$ to the other side.

This is the equation of an ellipse with center at $(2, -1)$. Because 16 is the denominator of the $x^2$-term, we know the major axis is parallel to the $x$-axis. Because $a^2 = 16$ and $b^2 = 1$, we have that $c^2 = a^2 - b^2 = 16 - 1 = 15$. The vertices are $a = 4$ units to the left and right of center at $V_1 = (-2, -1)$ and $V_2 = (6, -1)$. The foci are $c = \sqrt{15}$ units to the left and right of center at $F_1 = \left(2 - \sqrt{15}, -1\right)$ and $F_2 = \left(2 + \sqrt{15}, -1\right)$. We then plot points $b = 1$ unit above and below the center at $(2, 0)$ and $(2, -2)$. Figure 32 shows the graph.

**Figure 32**

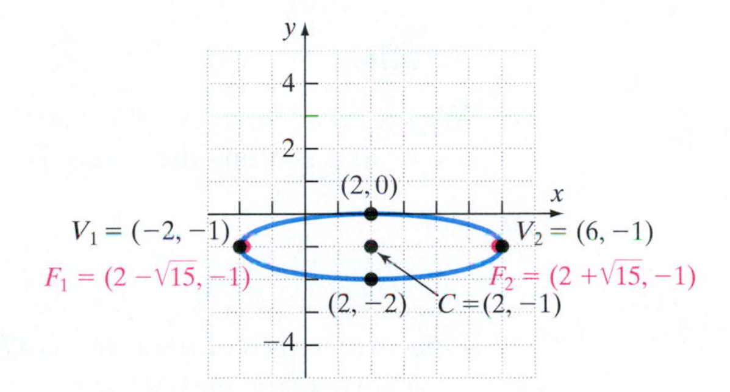


**QUICK ✓**

**4.** Graph the equation: $9x^2 + y^2 + 54x - 2y + 73 = 0$

## 4 Solve Applied Problems Involving Ellipses

Ellipses are found in many applications in science and engineering. For example, the orbits of the planets around the Sun are elliptical, with the Sun's position at a focus. See Figure 33.

Stone and concrete bridges are often the shape of semielliptical arches. Elliptical gears are used in machinery when a variable rate of motion is required.

Ellipses also have an interesting reflection property. If a source of light (or sound) is placed at one focus, the waves transmitted by the source will reflect off the ellipse and concentrate at the other focus. This is the principle behind whispering galleries, which

Figure 33

are rooms designed with elliptical ceilings. A person standing at one focus of the ellipse can whisper and be heard by a person standing at the other focus, because all the sound waves that reach the ceiling are reflected to the other person. National Statuary Hall in the United States Capitol used to be a whispering gallery. The acoustics in the room were such that the design of the room had to be changed.

## EXAMPLE 5 A Whispering Gallery

The whispering gallery in the Museum of Science and Industry in Chicago is 47.3 feet long. The distance from the center of the room to the foci is 20.3 feet. Find an equation that describes the shape of the room. How high is the room at its center?

### Solution

We set up a Cartesian plane so that the center of the ellipse is at the origin and the major axis is along the $x$-axis. The equation of the ellipse is

$$\frac{x^2}{a^2} + \frac{y^2}{b^2} = 1$$

Figure 34

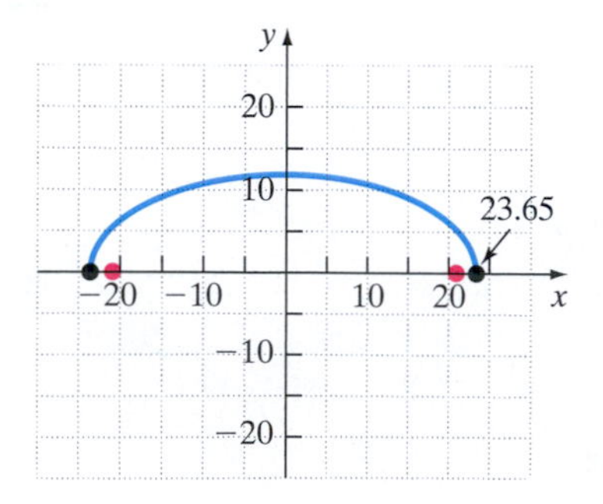


Since the length of the room is 47.3 feet, the distance from the center of the room to each vertex (the end of the room) will be $47.3/2 = 23.65$ feet, so $a = 23.65$ feet. The distance from the center of the room to each focus is $c = 20.3$ feet. See Figure 34. Since $b^2 = a^2 - c^2$, we have that $b^2 = 23.65^2 - 20.3^2 = 147.2325$. An equation that describes the shape of the room is given by

$$\frac{x^2}{23.65^2} + \frac{y^2}{147.2325} = 1$$

The height of the room at its center is $b = b = \sqrt{147.2325} \approx 12.1$ feet.

QUICK ✓

5. A hall 100 feet in length is to be designed as a whispering gallery. If the foci are to be located 30 feet from the center, determine an equation that describes the room. What is the height of the room at its center?

# 10.4 Exercises

**For Extra Help:** Student Solutions Manual · CD Video · PH Math/Tutor Center · MathXL Tutorials on CD · MathXL® · MyMathLab

## Concepts and Vocabulary

*In Problems 1–3, fill in the blanks.*

**1.** An _______ is the collection of points in the plane such that the sum of the distances from two fixed points, called the _______, is a constant.

**2.** For an ellipse, the line containing the foci is called the _______ _______.

**3.** The two points of intersection of the ellipse and the major axis are the _______, $V_1$ and $V_2$, of the ellipse.

*In Problems 4–6, answer True or False to each statement.*

**4.** For any ellipse whose vertex is $(a, 0)$ and focus is $(c, 0)$, then $c^2 = a^2 + b^2$.

**5.** The minor axis of an ellipse is perpendicular to the major axis and contains the center of the ellipse.

**6.** The center of $\dfrac{(x-3)^2}{25} + \dfrac{(y+1)^2}{16} = 1$ is $(-3, 1)$.

**7.** In the ellipse drawn on the right the center is at the origin and the major axis is the $x$-axis. The point $F$ is a focus. In the right triangle (drawn in red), label the lengths of the legs and the hypotenuse using the values of $a$, $b$, and $c$.

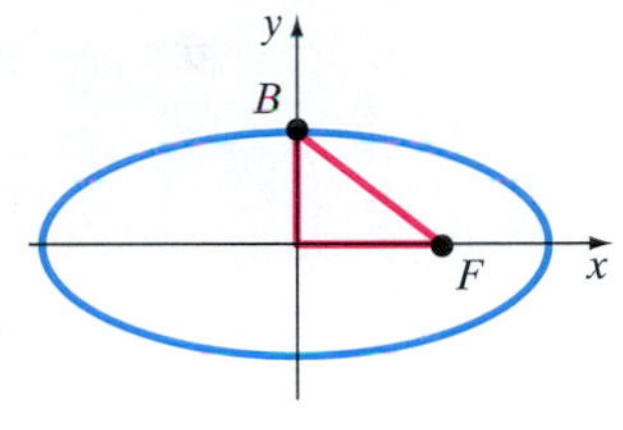


**8.** How does the ellipse given by the equation $\dfrac{x^2}{25} + \dfrac{y^2}{9} = 1$ differ from the ellipse given by the equation $\dfrac{x^2}{9} + \dfrac{y^2}{25} = 1$? How are they the same?

## Building Skills

*In Problems 9–12, the graph of an ellipse is given. Match each graph to its equation.*

**(a)** $x^2 + \dfrac{y^2}{9} = 1$ **(b)** $\dfrac{x^2}{9} + y^2 = 1$ **(c)** $\dfrac{x^2}{16} + \dfrac{y^2}{9} = 1$ **(d)** $\dfrac{x^2}{9} + \dfrac{y^2}{16} = 1$

**9.**

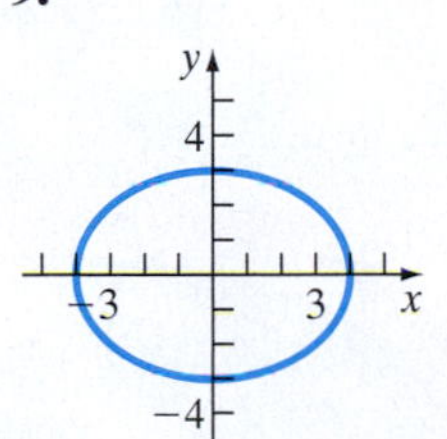


**10.**

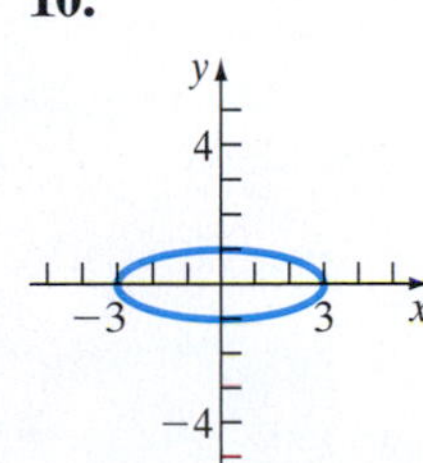


**11.**

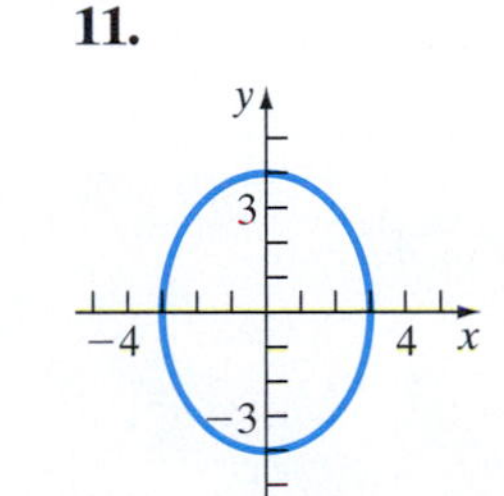


**12.**

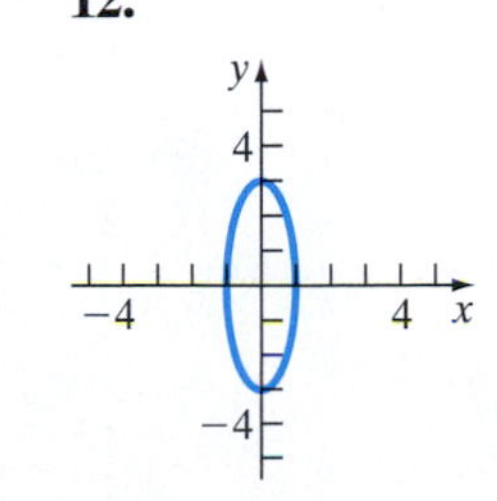

*In Problems 13–22, find the vertices and foci of each ellipse. Graph each ellipse.*

**13.** $\dfrac{x^2}{25} + \dfrac{y^2}{16} = 1$ **14.** $\dfrac{x^2}{25} + \dfrac{y^2}{4} = 1$ **15.** $\dfrac{x^2}{36} + \dfrac{y^2}{100} = 1$ **16.** $\dfrac{x^2}{16} + \dfrac{y^2}{36} = 1$

**17.** $\dfrac{x^2}{49} + \dfrac{y^2}{4} = 1$ **18.** $\dfrac{x^2}{121} + \dfrac{y^2}{100} = 1$ **19.** $x^2 + \dfrac{y^2}{49} = 1$ **20.** $\dfrac{x^2}{64} + y^2 = 1$

**21.** $4x^2 + y^2 = 16$ **22.** $9x^2 + y^2 = 81$

*In Problems 23–30, find an equation for each ellipse. Graph each ellipse.*

**23.** Center at $(0, 0)$; focus at $(4, 0)$; vertex at $(6, 0)$

**24.** Center at $(0, 0)$; focus at $(2, 0)$; vertex at $(5, 0)$

**25.** Center at $(0, 0)$; focus at $(0, -4)$; vertex at $(0, 7)$

**26.** Center at $(0, 0)$; focus at $(0, -1)$; vertex at $(0, 5)$

**27.** Foci at $(\pm 6, 0)$; Vertices at $(\pm 10, 0)$

**28.** Foci at $(0, \pm 2)$; Vertices at $(0, \pm 7)$

**29.** Foci at $(0, \pm 5)$; length of the major axis is 16

**30.** Foci at $(\pm 6, 0)$; length of the major axis is 20

*In Problems 31–40, graph each ellipse.*

**31.** $\dfrac{(x-3)^2}{9} + \dfrac{(y+2)^2}{25} = 1$ **32.** $\dfrac{(x-1)^2}{36} + \dfrac{(y+4)^2}{100} = 1$

**33.** $\dfrac{(x+2)^2}{16} + \dfrac{(y-5)^2}{4} = 1$ **34.** $\dfrac{(x+5)^2}{64} + \dfrac{(y+1)^2}{16} = 1$

**35.** $(x-5)^2 + \dfrac{(y+1)^2}{49} = 1$ **36.** $\dfrac{(x+8)^2}{81} + (y-3)^2 = 1$

**37.** $4(x+2)^2 + 16(y-1)^2 = 64$ **38.** $9(x-3)^2 + (y-4)^2 = 81$

**39.** $4x^2 + y^2 - 24x + 2y - 63 = 0$ **40.** $16x^2 + 9y^2 - 128x + 54y - 239 = 0$

## Applying the Concepts

**41. Semielliptical Arch Bridge** An arch in the shape of the upper half of an ellipse is used to support a bridge that is to span a river 30 meters wide. The center of the arch is 10 meters above the center of the river (see the figure).

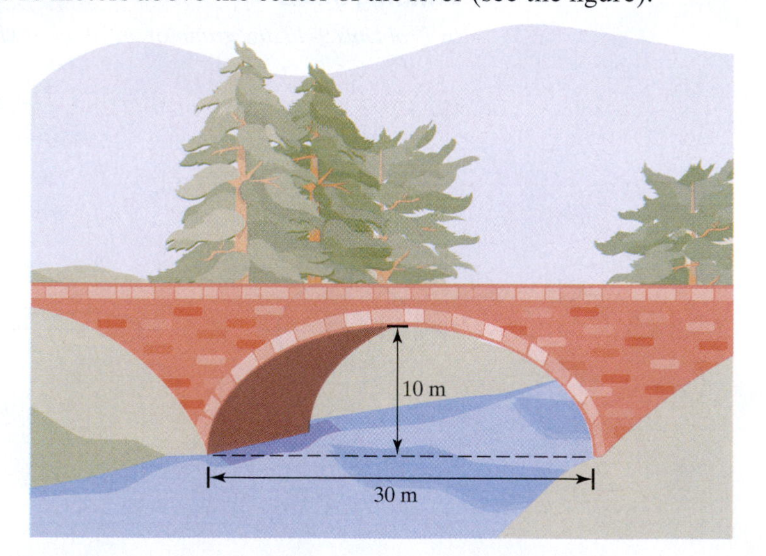

(a) Write the equation for the ellipse in which the $x$-axis coincides with the water and the $y$-axis passes through the center of the arch.

(b) Can a rectangular barge that is 18 meters wide and sits 7 meters above the surface of the water fit through the opening of the bridge?

(c) If heavy rains caused the river's level to increase 1.1 meters, will the barge make it through the opening?

**42. London Bridge** An arch in the shape of the upper half of an ellipse is used to support London Bridge. The main span is 45.6 meters wide. Suppose that the center of the arch is 15 meters above the center of the river.

(a) Write the equation for the ellipse in which the $x$-axis coincides with the water and the $y$-axis passes through the center of the arch.

(b) Can a rectangular barge that is 20 meters wide and sits 12 meters above the surface of the water fit through the opening of the bridge?

(c) If heavy rains caused the river's level to increase 1.5 meters, will the barge make it through the opening?

*In Problems 43–46, use the fact that the orbit of a planet about the Sun is an ellipse, with the Sun at one focus. The* **aphelion** *of a planet is its greatest distance from the Sun, and the* **perihelion** *is its shortest distance. The* **mean distance** *of a planet from the Sun is the length of the semimajor axis of the elliptical orbit. See the illustration.*

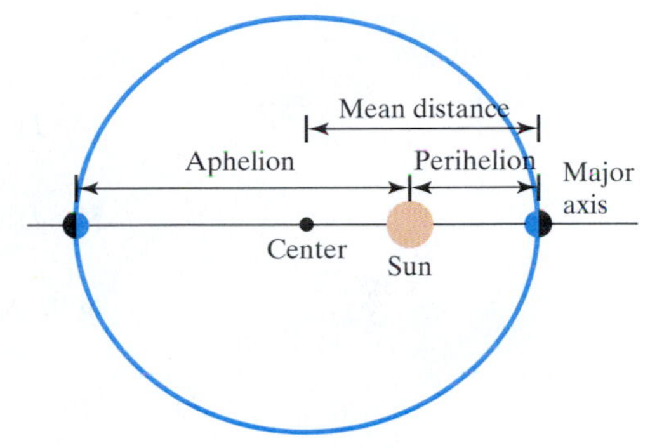


**43. Earth** The mean distance of Earth from the Sun is 93 million miles. If the aphelion of Earth is 94.5 million miles, what is the perihelion? Write an equation for the orbit of Earth around the Sun.

**44. Mars** The mean distance of Mars from the Sun is 142 million miles. If the perihelion of Mars is 128.5 million miles, what is the aphelion? Write an equation for the orbit of Mars about the Sun.

**45. Jupiter** The aphelion of Jupiter is 507 million miles. If the distance from the Sun to the center of its elliptical orbit is 23.2 million miles, what is the perihelion? What is the mean distance? Write an equation for the orbit of Jupiter around the Sun.

**46. Pluto** The perihelion of Pluto is 4551 million miles, and the distance of the Sun from the center of its elliptical orbit is 897.5 million miles. Find the aphelion of Pluto. What is the mean distance of Pluto from the Sun? Write an equation for the orbit of Pluto about the Sun.

## Extending the Concepts

*In Problems 47–50, write an equation for each ellipse.*

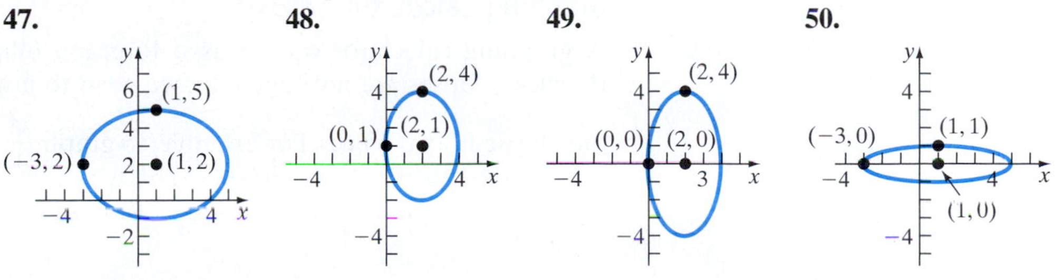

**51.** Show that a circle is a special kind of ellipse by letting $a = b$ in the equation of an ellipse centered at the origin. What is the value of $c$ in a circle? What does this mean regarding the location of the foci?

**52.** The **eccentricity** $e$ of an ellipse is defined as the number $\frac{c}{a}$, where $a$ is the distance from the center of an ellipse to a vertex and $c$ is the distance from the center of an ellipse to a focus. Because $a > c$, it follows that $e < 1$ for an ellipse. Write a paragraph about the general shape of each of the following ellipses. Be sure to justify your conclusion.

**(a)** Eccentricity close to 0 **(b)** Eccentricity = 0.5 **(c)** Eccentricity close to 1

## Synthesis Review

*In Problems 53–56, fill in the following table for each function.*

| $x$ | 5 | 10 | 100 | 1000 |
|---|---|---|---|---|
| $f(x)$ | | | | |

**53.** $f(x) = \dfrac{5}{x + 2}$

**54.** $f(x) = \dfrac{x - 1}{x^2 + 4}$

**55.** $f(x) = \dfrac{2x + 1}{x - 3}$

**56.** $f(x) = \dfrac{3x^2 - x + 1}{x^2 + 1}$

*In Problems 57 and 58, fill in the following table for each function.*

| $x$ | 5 | 10 | 100 | 1000 |
|---|---|---|---|---|
| $f(x)$ | | | | |
| $g(x)$ | | | | |

**57.** $f(x) = \dfrac{x^2 + 3x + 1}{x + 1}$; $g(x) = x + 2$

**58.** $f(x) = \dfrac{x^2 - 3x + 5}{x + 2}$; $g(x) = x - 5$

**59.** For the functions in Problems 53–56, compare the degree of the polynomial in the numerator to the degree of the polynomial in the denominator. Conjecture what happens to a rational function as $x$ increases when the degree of the numerator is less than the degree of the denominator. Conjecture what happens to a rational function as $x$ increases when the degree of the numerator equals the degree of the denominator.

**60.** For the functions in Problems 57 and 58, write $f$ in the form quotient + $\dfrac{\text{remainder}}{\text{dividend}}$. That is, perform the division indicated by the rational function. Now compare the values of $f$ to those of $g$ in the table. What does the function $g$ represent?

## Graphing Calculator

A graphing calculator can be used to graph ellipses by solving the equation for $y$. Because ellipses are not functions, we need to graph the upper half and lower half of the ellipse in two pieces. For example, to graph $\frac{x^2}{4} + \frac{y^2}{9} = 1$, we would graph $Y_1 = 3\sqrt{1 - \frac{x^2}{4}}$ and $Y_2 = -3\sqrt{1 - \frac{x^2}{4}}$. We obtain the graph shown in Figure 35.

Figure 35

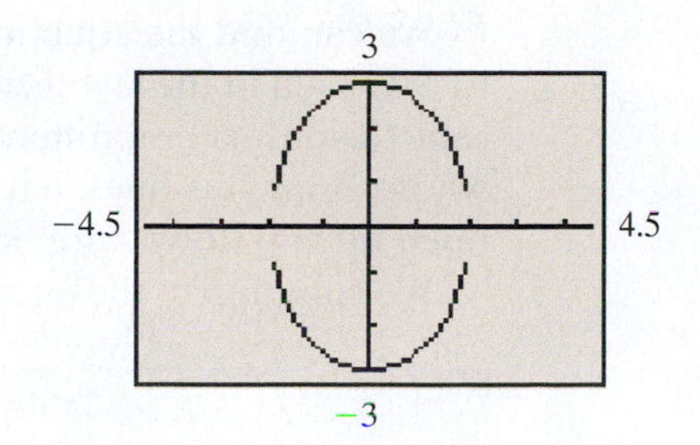


*In Problems 61–68, graph each ellipse using a graphing calculator.*

**61.** $\dfrac{x^2}{25} + \dfrac{y^2}{16} = 1$ **62.** $\dfrac{x^2}{25} + \dfrac{y^2}{4} = 1$ **63.** $4x^2 + y^2 = 16$

**64.** $9x^2 + y^2 = 81$ **65.** $\dfrac{(x-3)^2}{9} + \dfrac{(y+2)^2}{25} = 1$ **66.** $\dfrac{(x-1)^2}{36} + \dfrac{(y+4)^2}{100} = 1$

**67.** $\dfrac{(x+2)^2}{16} + \dfrac{(y-5)^2}{4} = 1$ **68.** $\dfrac{(x+5)^2}{64} + \dfrac{(y+1)^2}{16} = 1$

# 10.5 Hyperbolas

**OBJECTIVES**

1. Graph Hyperbolas in Which the Center Is the Origin
2. Find the Equation of a Hyperbola in Which the Center Is the Origin
3. Find the Asymptotes of a Hyperbola in Which the Center Is the Origin

### *Preparing for Hyperbolas*

*Before getting started, take the following readiness quiz. If you get a problem wrong, go back to the section cited and review the material.*

**1.** Complete the square: $x^2 - 5x$ [Section 8.1, pp. 614–615]

**2.** Solve: $y^2 = 64$ [Section 8.1, pp. 611–614]

## 1 Graph Hyperbolas in Which the Center Is the Origin

Recall from Section 10.3 that a hyperbola is a conic section that is obtained through the intersection of a plane and two cones. See Figure 8(d) on page 780.

A **hyperbola** is the collection of all points in the plane the difference of whose distances from two fixed points, called the **foci,** is a constant.

Figure 36 illustrates a hyperbola with foci $F_1$ and $F_2$. The line containing the foci is called the **transverse axis.** The midpoint of the line segment joining the foci is called the **center** of the hyperbola. The line through the center and perpendicular to the transverse axis is called the **conjugate axis.** The hyperbola consists of two separate curves called **branches.** The two points of intersection of the hyperbola and the transverse axis are the **vertices,** $V_1$ and $V_2$, of the hyperbola.

**In Words**

The distance from $F_1$ to $P$ minus the distance from $F_2$ to $P$ is a constant value for any point $P$ on a hyperbola.

Figure 36

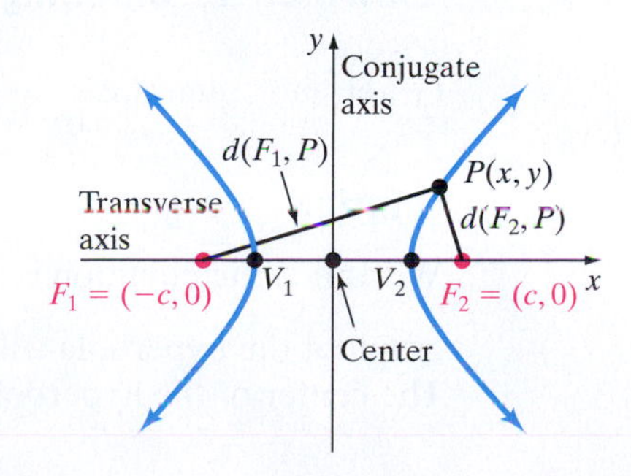


***Preparing for...Answers*** **1.** $x^2 - 5x + \dfrac{25}{4}$ **2.** $\{-8, 8\}$

We can find the equation of a hyperbola using the distance formula in a way similar to that used in the last section to find the equation of an ellipse. However, we won't present this information here. Instead, we will give you the equations of the hyperbolas whose branches open left and right and the equations of hyperbolas whose branches open up and down with both hyperbolas centered at the origin.

We summarize the equations of hyperbolas in Table 5.

**Work Smart**

Notice for hyperbolas that $c^2 = a^2 + b^2$, but for ellipses $c^2 = a^2 - b^2$.

**Table 5 Hyperbolas with Center at the Origin**

| Center | Transverse Axis | Foci | Vertices | Equation |
|---|---|---|---|---|
| (0, 0) | $x$-axis | $(-c, 0)$ and $(c, 0)$ | $(-a, 0)$ and $(a, 0)$ | $\frac{x^2}{a^2} - \frac{y^2}{b^2} = 1$ where $b^2 = c^2 - a^2$ or $c^2 = a^2 + b^2$ |
| (0, 0) | $y$-axis | $(0, -c)$ and $(0, c)$ | $(0, -a)$ and $(0, a)$ | $\frac{y^2}{a^2} - \frac{x^2}{b^2} = 1$ where $b^2 = c^2 - a^2$ or $c^2 = a^2 + b^2$ |

The graphs of the two hyperbolas are given in Figure 37.

**Figure 37**

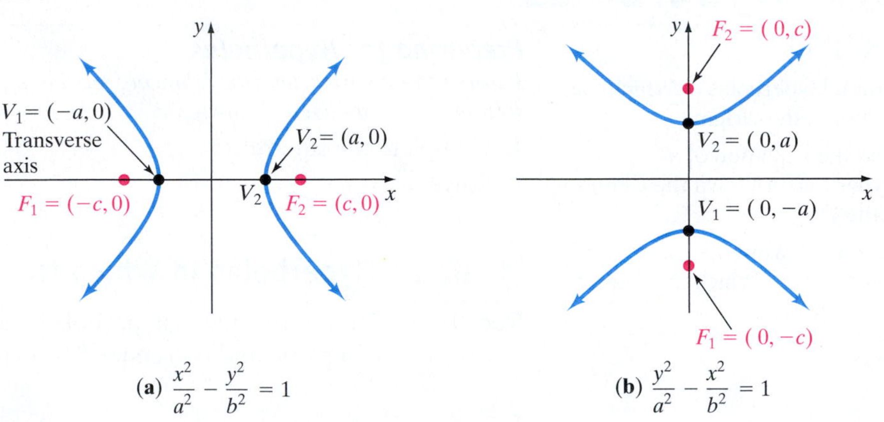


(a) $\frac{x^2}{a^2} - \frac{y^2}{b^2} = 1$ (b) $\frac{y^2}{a^2} - \frac{x^2}{b^2} = 1$

Notice the difference in the two equations given in Table 5. When the $x^2$-term is first, as in Figure 37(a), the transverse axis is the $x$-axis and the hyperbola opens left and right. When the $y^2$-term is first, as in Figure 37(b), the transverse axis is the $y$-axis and the hyperbola opens up and down. In both cases, the value of $a^2$ is in the denominator of the first term.

## EXAMPLE 1 Graphing a Hyperbola

Graph the equation: $\frac{x^2}{16} - \frac{y^2}{4} = 1$

### Solution

We notice the equation is of the form $\frac{x^2}{a^2} - \frac{y^2}{b^2} = 1$. Since the $x^2$-term is first, the graph of the hyperbola will open left and right. We have that $a^2 = 16$ and $b^2 = 4$. The center of the hyperbola is the origin, (0, 0), and the transverse axis is the $x$-axis.

Because $b^2 = c^2 - a^2$ (or $c^2 = a^2 + b^2$), we find that $c^2 = 16 + 4 = 20$, so that $c = 2\sqrt{5}$. The vertices are at $(\pm a, 0) = (\pm 4, 0)$, and the foci are at $(\pm c, 0) = \left(\pm 2\sqrt{5}, 0\right)$.

To obtain the graph we plot the vertices and foci. Then, we locate points above and below the foci (just as we did when graphing parabolas). So we let $x = \pm 2\sqrt{5}$ in the equation $\frac{x^2}{16} - \frac{y^2}{4} = 1$.

$$\frac{\left(\pm 2\sqrt{5}\right)^2}{16} - \frac{y^2}{4} = 1$$

$$\frac{20}{16} - \frac{y^2}{4} = 1$$

Reduce the fraction: $$\frac{5}{4} - \frac{y^2}{4} = 1$$

Subtract $\frac{5}{4}$ from both sides: $$-\frac{y^2}{4} = -\frac{1}{4}$$

Multiply both sides by $-4$: $$y^2 = 1$$

Take the square root of both sides: $$y = \pm 1$$

The points above and below the foci are $\left(\pm 2\sqrt{5}, -1\right)$ and $\left(\pm 2\sqrt{5}, 1\right)$. See Figure 38 for the graph of the hyperbola.

Figure 38

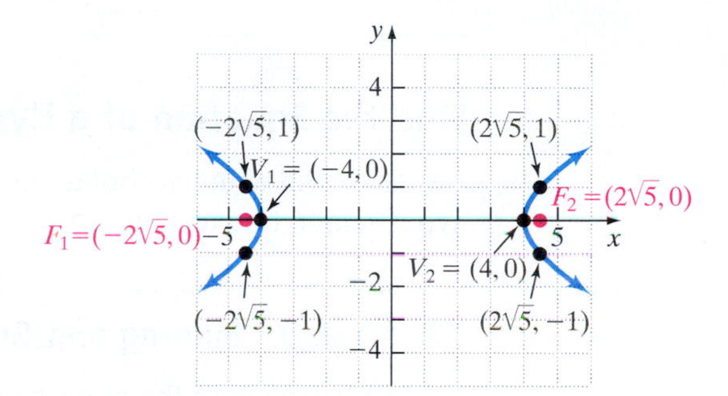


## EXAMPLE 2 Graph a Hyperbola

Graph the equation: $4y^2 - 16x^2 = 16$

### Solution

We wish to put the equation in one of the forms given in Table 5. To do this, we divide both sides of the equation by 16:

$$\frac{y^2}{4} - \frac{x^2}{1} = 1$$

Do you see that the $y^2$-term is first? This means that the hyperbola will open up and down. The center of this hyperbola is the origin, $(0, 0)$, and the transverse axis is the $y$-axis. Comparing the equation to $\frac{y^2}{a^2} - \frac{x^2}{b^2} = 1$, we find that $a^2 = 4$ and $b^2 = 1$. Because $b^2 = c^2 - a^2$ or $c^2 = a^2 + b^2$, we find that $c^2 = 4 + 1 = 5$. The vertices are at $(0, \pm a) = (0, \pm 2)$. The foci are at $(0, \pm c) = \left(0, \pm\sqrt{5}\right)$. Because the hyperbola opens up and down, we locate four additional points on the hyperbola by determining points left and right of each focus. To find these points, we let $y = \pm\sqrt{5}$ in the equation $\dfrac{y^2}{4} - \dfrac{x^2}{1} = 1$ and solve for $x$.

Figure 39

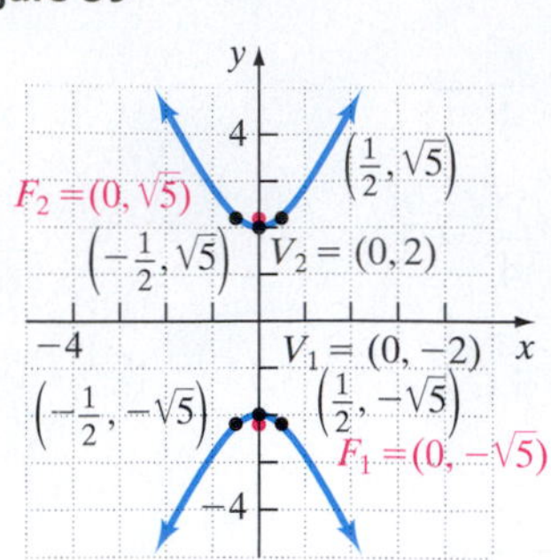


$$\frac{(\pm\sqrt{5})^2}{4} - \frac{x^2}{1} = 1$$

$$\frac{5}{4} - \frac{x^2}{1} = 1$$

Subtract $\frac{5}{4}$ from both sides: $$-x^2 = -\frac{1}{4}$$

Multiply both sides by $-1$: $$x^2 = \frac{1}{4}$$

Take the square root of both sides: $$x = \pm\frac{1}{2}$$

Four additional points on the graph are $\left(-\frac{1}{2}, \pm\sqrt{5}\right)$ and $\left(\frac{1}{2}, \pm\sqrt{5}\right)$. See Figure 39 for the graph of the hyperbola.

**QUICK ✓** *Graph each hyperbola.*

**1.** $\dfrac{x^2}{36} - \dfrac{y^2}{64} = 1$  **2.** $\dfrac{y^2}{9} - \dfrac{x^2}{16} = 1$

---

## 2 Find the Equation of a Hyperbola in Which the Center Is the Origin

Just as we did with parabolas and ellipses, we are now going to use information about a hyperbola to find its equation.

### EXAMPLE 3 Finding and Graphing the Equation of a Hyperbola

Find an equation of the hyperbola with center at the origin, one focus at $(0, -3)$, and vertices at $(0, -2)$ and $(0, 2)$. Graph the equation.

#### Solution

The center of a hyperbola is located at the midpoint of vertices (or foci). Because the vertices are at $(0, -2)$ and $(0, 2)$, the center must be at the origin, $(0, 0)$. If we were to plot the given focus and the vertices, we would notice that they all lie on the $y$-axis. Therefore, the transverse axis is the $y$-axis and the hyperbola opens up and down. Since there is a vertex at $(0, 2)$ and the center is the origin, we have that $a = 2$. Since there is a focus at $(0, -3)$ and the center is the origin, we have that $c = 3$. Since $b^2 = c^2 - a^2$, we have that $b^2 = 3^2 - 2^2 = 9 - 4 = 5$. Since the hyperbola opens up and down and the center is the origin, the equation of the hyperbola is of the form

$$\frac{y^2}{a^2} - \frac{x^2}{b^2} = 1$$

$a^2 = 4;\ b^2 = 5$: $$\frac{y^2}{4} - \frac{x^2}{5} = 1$$

We wish to find points left and right of each focus, so we let $y = \pm 3$ in the equation $\dfrac{y^2}{4} - \dfrac{x^2}{5} = 1$.

$$\frac{(\pm 3)^2}{4} - \frac{x^2}{5} = 1$$

$$\frac{9}{4} - \frac{x^2}{5} = 1$$

Subtract $\frac{9}{4}$ from both sides: $-\frac{x^2}{5} = -\frac{5}{4}$

Multiply both sides by $-5$: $x^2 = \frac{25}{4}$

Take the square root of both sides: $x = \pm\frac{5}{2}$

The points left and right of the foci are $\left(-\frac{5}{2}, \pm 3\right)$ and $\left(\frac{5}{2}, \pm 3\right)$. See Figure 40 for the graph of the hyperbola.

**Figure 40**

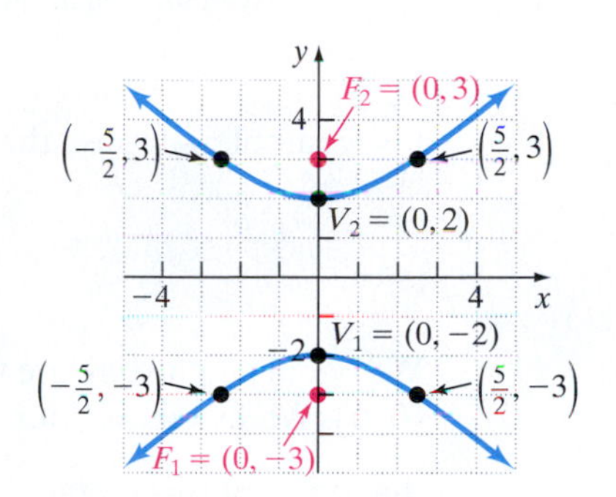


Look at the equations of the hyperbolas in Examples 1 and 3. For the hyperbola in Example 1, $a^2 = 16$ and $b^2 = 4$, so $a > b$; for the hyperbola in Example 3, $a^2 = 4$ and $b^2 = 5$, so $a < b$. We conclude that, for hyperbolas, there are no requirements involving the relative sizes for $a$ and $b$. Contrast this situation to the case of an ellipse, in which the relative sizes of $a$ and $b$ dictate which axis is the major axis.

**QUICK ✓**

**3.** Find the equation of a hyperbola whose vertices are $(-4, 0)$ and $(4, 0)$ and has a focus at $(6, 0)$. Graph the hyperbola.

## 3 Find the Asymptotes of a Hyperbola in Which the Center Is the Origin

**Work Smart**

Asymptotes provide an alternative method for determining the opening of each branch of the hyperbola, rather than finding and plotting four additional points.

As $x$ and $y$ get larger in both the positive and negative direction, the branches of the hyperbola approach two lines, called **asymptotes** of the hyperbola. The asymptotes provide guidance in graphing hyperbolas. Table 6 summarizes the asymptotes of the two hyperbolas discussed in this section.

**Table 6 Asymptotes of a Hyperbola**

| Hyperbola | Asymptotes |
|---|---|
| $\frac{x^2}{a^2} - \frac{y^2}{b^2} = 1$ | $y = -\frac{b}{a}x$ and $y = \frac{b}{a}x$ |
| $\frac{y^2}{a^2} - \frac{x^2}{b^2} = 1$ | $y = -\frac{a}{b}x$ and $y = \frac{a}{b}x$ |

Figure 41 illustrates how the asymptotes can be used to help graph a hyperbola. It is important to remember that the asymptotes are not part of the hyperbola—they only serve as guides in graphing the hyperbola.

**Figure 41**

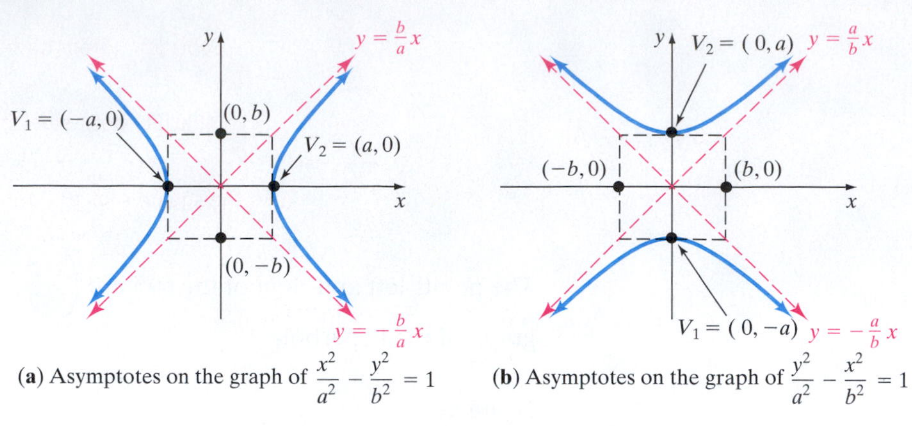


(**a**) Asymptotes on the graph of $\frac{x^2}{a^2} - \frac{y^2}{b^2} = 1$ (**b**) Asymptotes on the graph of $\frac{y^2}{a^2} - \frac{x^2}{b^2} = 1$

For example, suppose that we want to graph the equation

$$\frac{x^2}{a^2} - \frac{y^2}{b^2} = 1$$

We begin by plotting the vertices $(-a, 0)$ and $(a, 0)$. Then we plot the points $(0, -b)$ and $(0, b)$. We use these four points to construct a rectangle as shown in Figure 41(a). The diagonals of this rectangle have slopes $\frac{b}{a}$ and $-\frac{b}{a}$. If we draw a line through the corners of the rectangle, we have the asymptotes of the hyperbola. The equations of these asymptotes are $y = -\frac{b}{a}x$ and $y = \frac{b}{a}x$. Using this technique allows us to avoid plotting additional points, as was done earlier.

**Figure 42**

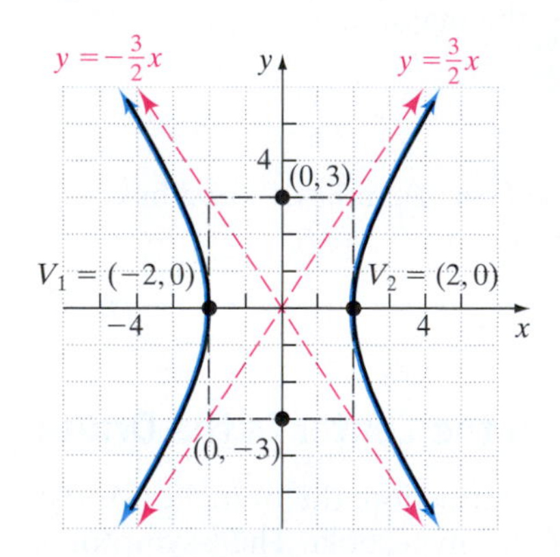


## EXAMPLE 4 Graphing a Hyperbola and Finding Its Asymptotes

Graph the equation $9x^2 - 4y^2 = 36$ using the asymptotes as a guide.

### Solution

We divide both sides of the equation by 36 to put the equation in the form $\frac{x^2}{a^2} - \frac{y^2}{b^2} = 1$.

$$\frac{x^2}{4} - \frac{y^2}{9} = 1$$

The center of the hyperbola is the origin, $(0, 0)$. Because the $x^2$-term is first, the hyperbola opens left and right. The transverse axis is along the $x$-axis. We can see that $a^2 = 4$ and $b^2 = 9$. Because $c^2 = a^2 + b^2$, we have that $c^2 = 4 + 9 = 13$, so that $c = \pm\sqrt{13}$. The vertices are at $(\pm a, 0) = (\pm 2, 0)$; the foci are at $(\pm c, 0) = (\pm\sqrt{13}, 0)$. Since $b^2 = 9$, we have that $b = \pm 3$. The equations of the asymptotes are

$$y = \frac{b}{a}x = \frac{3}{2}x \quad \text{and} \quad y = -\frac{b}{a}x = -\frac{3}{2}x$$

To graph the hyperbola, we form a rectangle using the points $(\pm a, 0) = (\pm 2, 0)$ and $(0, \pm b) = (0, \pm 3)$. The diagonals help us to draw the asymptotes. See Figure 42.

**Work Smart: Study Skills**

Which equation is that of an ellipse and which is the equation of a hyperbola?

$$\frac{x^2}{4} + y^2 = 1$$

$$\frac{x^2}{4} - y^2 = 1$$

Summarize how you can tell the difference between the equation of an ellipse and the equation of a hyperbola.

**QUICK ✓**

4. Graph the equation $x^2 - 9y^2 = 9$ using the asymptotes as a guide.

5. Graph the equation $\frac{y^2}{16} - \frac{x^2}{9} = 1$ using the asymptotes as a guide.

# 10.5 Exercises

**For Extra Help:**

Student Solutions Manual

CD Video

PH Math/Tutor Center

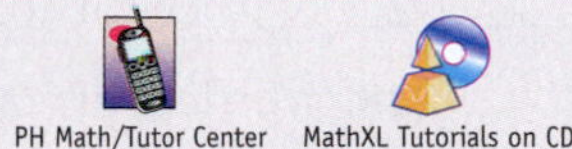

MathXL Tutorials on CD  MathXL®  MyMathLab

## Concepts and Vocabulary

*In Problems 1–3, fill in the blanks.*

1. A(n) _______ is the collection of points in the plane the difference of whose distances from two fixed points is a constant.

2. For a hyperbola, the foci lie on a line called the _______ _______.

3. The asymptotes of the hyperbola $\frac{x^2}{a^2} - \frac{y^2}{b^2} = 1$ are _______ and _______.

*In Problems 4–6, answer True or False to each statement.*

4. The line through the center of a hyperbola that is perpendicular to the transverse axis is called the conjugate axis.

5. Hyperbolas will always have asymptotes.

6. In any hyperbola, it must be the case that $a > b$.

7. Explain how the asymptotes of a hyperbola are helpful in obtaining its graph.

8. How can you tell the difference between the equation of a hyperbola and an ellipse just by looking at its equation?

## Building Skills

*In Problems 9–12, the graph of a hyperbola is given. Match each graph to its equation.*

**(a)** $\frac{x^2}{4} - y^2 = 1$ **(b)** $x^2 - \frac{y^2}{4} = 1$ **(c)** $\frac{y^2}{4} - x^2 = 1$ **(d)** $y^2 - \frac{x^2}{4} = 1$

**9.**

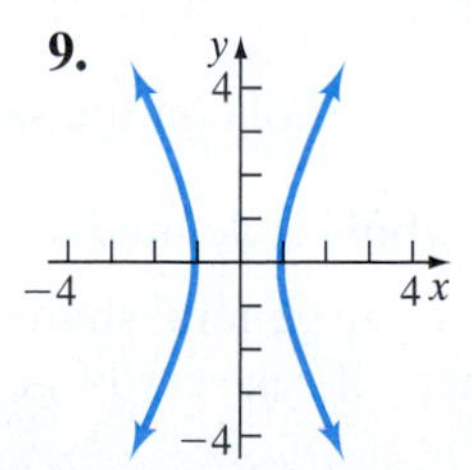


**10.**

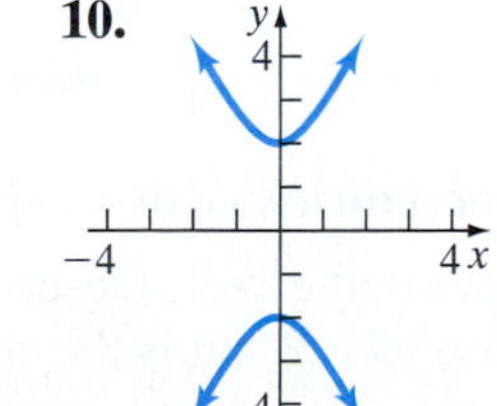


**11.**

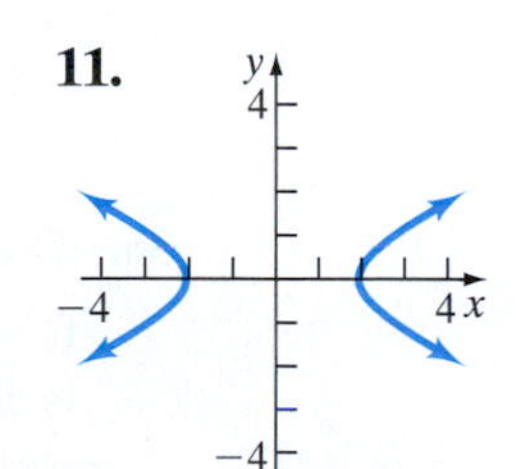


**12.**

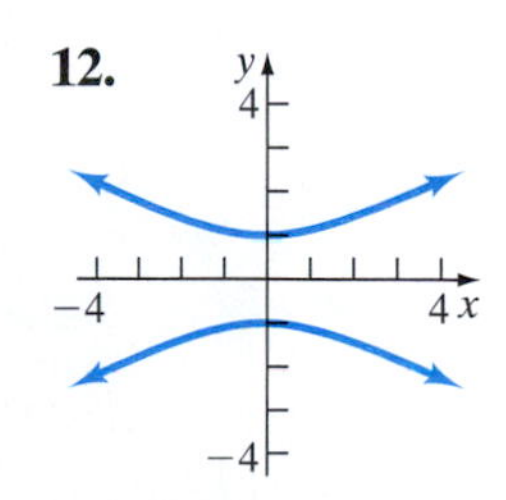


*In Problems 13–20, graph each equation. You may either plot additional points or use the asymptotes to obtain the graph.*

13. $\frac{x^2}{4} - \frac{y^2}{16} = 1$

14. $\frac{x^2}{9} - \frac{y^2}{16} = 1$

15. $\frac{y^2}{25} - \frac{x^2}{36} = 1$

16. $\frac{y^2}{81} - \frac{x^2}{9} = 1$

17. $4x^2 - y^2 = 36$

18. $x^2 - 9y^2 = 36$

19. $25y^2 - x^2 = 100$

20. $4y^2 - 9x^2 = 36$

*In Problems 21–30, find the equation for the hyperbola described. Graph the equation.*

**21.** Center at $(0, 0)$; focus at $(3, 0)$, vertex at $(2, 0)$

**22.** Center at $(0, 0)$; focus at $(-4, 0)$, vertex $(-1, 0)$

**23.** Vertices at $(0, 5)$ and $(0, -5)$; focus at $(0, 7)$

**24.** Vertices at $(0, 6)$ and $(0, -6)$; focus at $(0, 8)$

**25.** Foci at $(-10, 0)$ and $(10, 0)$; vertex at $(-7, 0)$

**26.** Foci at $(-5, 0)$ and $(5, 0)$; vertex at $(-3, 0)$

**27.** Vertices at $(0, -8)$ and $(0, 8)$; asymptote the line $y = 2x$

**28.** Vertices at $(0, -4)$ and $(0, 4)$; asymptote the line $y = 2x$

**29.** Foci at $(-3, 0)$ and $(3, 0)$; asymptote the line $y = x$

**30.** Foci at $(-9, 0)$ and $(9, 0)$; asymptote the line $y = -3x$

## Applying the Concepts

*In Problems 31–34, write the equation of the hyperbola.*

**31.**

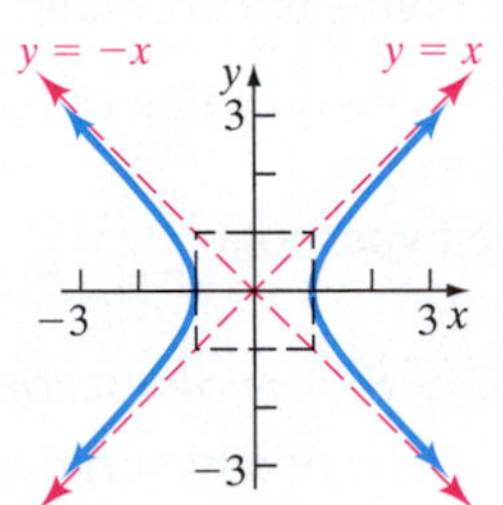


**32.**

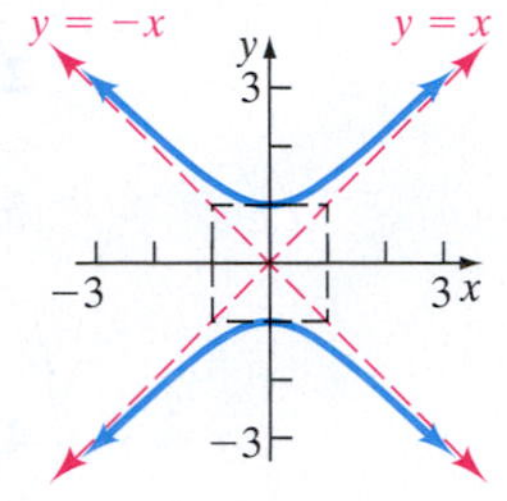


**33.**

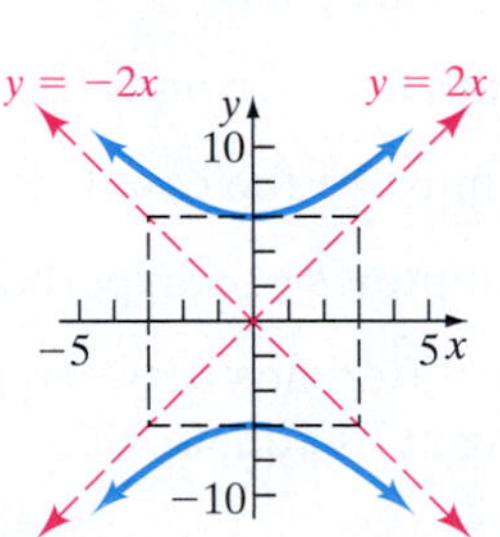


**34.**

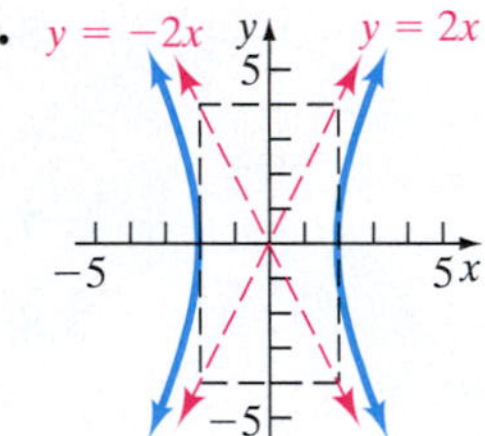


## Extending the Concepts

**35.** Two hyperbolas that have the same set of asymptotes are called **conjugate.** Show that the hyperbolas

$$\frac{x^2}{4} - y^2 = 1 \quad \text{and} \quad y^2 - \frac{x^2}{4} = 1$$

are conjugate. Graph each hyperbola on the same Cartesian plane.

**36.** The **eccentricity** $e$ of a hyperbola is defined as the number $\frac{c}{a}$. Because $c > a$, it follows that $e > 1$. Describe the general shape of a hyperbola whose eccentricity is close to 1. What is the shape if $e$ is very large?

## Synthesis Review

*In Problems 37–41, solve each system of equations using either substitution or elimination.*

**37.** $\begin{cases} 2x - 3y = -9 \\ -x + 5y = 8 \end{cases}$

**38.** $\begin{cases} 3x + 4y = 3 \\ -6x + 2y = -\dfrac{7}{2} \end{cases}$

**39.** $\begin{cases} 2x - 3y = 6 \\ -6x + 9y = -18 \end{cases}$

**40.** $\begin{cases} -2x + y = 8 \\ x - \dfrac{1}{2}y = -4 \end{cases}$

**41.** $\begin{cases} 6x + 3y = 4 \\ -2x - y = -\dfrac{4}{3} \end{cases}$

**42.** Which method did you use more often? Why? Do you think there are situations where substitution is superior to elimination? Are there situations where elimination is superior to substitution? Describe these circumstances.

### The Graphing Calculator

A graphing calculator can be used to graph hyperbolas by solving the equation for $y$. Because hyperbolas are not functions, we need to graph the upper half and lower half of the hyperbola in two pieces. For example, to graph $\frac{x^2}{4} - \frac{y^2}{9} = 1$, we would graph $Y_1 = 3\sqrt{\frac{x^2}{4} - 1}$ and $Y_2 = -3\sqrt{\frac{x^2}{4} - 1}$. We obtain the graph shown in Figure 43.

**Figure 43**

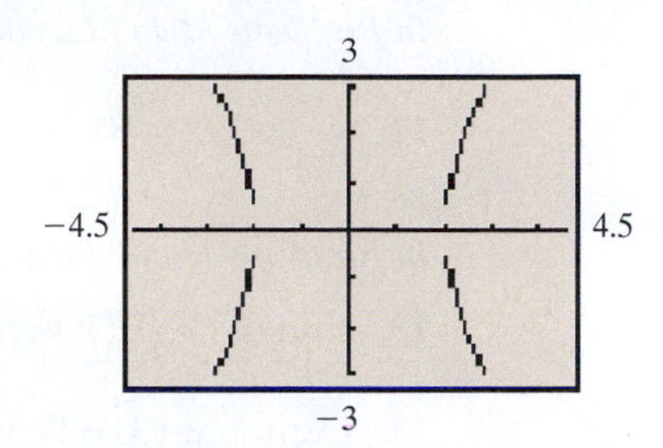


*In Problems 43–50, graph each hyperbola using a graphing calculator.*

**43.** $\frac{x^2}{4} - \frac{y^2}{16} = 1$

**44.** $\frac{x^2}{9} - \frac{y^2}{16} = 1$

**45.** $\frac{y^2}{25} - \frac{x^2}{36} = 1$

**46.** $\frac{y^2}{81} - \frac{x^2}{9} = 1$

**47.** $4x^2 - y^2 = 36$

**48.** $x^2 - 9y^2 = 36$

**49.** $25y^2 - x^2 = 100$

**50.** $4y^2 - 9x^2 = 36$

## PUTTING THE CONCEPTS TOGETHER (SECTIONS 10.1–10.5)

*These problems cover important concepts from Sections 10.1 to 10.5. We designed these problems so that you can review the chapter so far and show your mastery of the concepts. Take time to work these problems before proceeding with the next section. The answers to these problems are located at the back of the text starting on page AN-53.*

**1.** Find the exact distance $d(P_1, P_2)$ between points $P_1 = (-6, 4)$ and $P_2 = (3, -2)$.

**2.** Find the midpoint of the line segment formed by joining the points $P_1 = (-3, 1)$ and $P_2 = (5, -7)$.

*In Problems 3 and 4, find the center (h, k) and the radius r of each circle. Graph each circle.*

**3.** $(x + 2)^2 + (y - 8)^2 = 36$

**4.** $x^2 + y^2 + 6x - 4y - 3 = 0$

*In Problems 5 and 6, find the standard form of the equation of each circle.*

**5.** Center $(0, 0)$; contains the point $(-5, 12)$

**6.** With endpoints of a diameter at $(-1, 5)$ and $(5, -3)$.

*In Problems 7 and 8, find the vertex, focus, and directrix of each parabola. Graph each parabola.*

**7.** $(x + 2)^2 = -4(y - 4)$

**8.** $y^2 + 2y - 8x + 25 = 0$

*In Problems 9 and 10, find an equation for each parabola described.*

**9.** Vertex at $(-1, -2)$; focus at $(-1, -5)$

**10.** Vertex at $(-3, 3)$; contains the point $(-1, 7)$; axis of symmetry parallel to the $x$-axis

*In Problems 11 and 12, find the vertices and foci of each ellipse. Graph each ellipse.*

**11.** $x^2 + 9y^2 = 81$

**12.** $\frac{(x + 1)^2}{36} + \frac{(y - 2)^2}{49} = 1$

*In Problems 13 and 14, find an equation for each ellipse described.*

**13.** Foci at $(0, \pm 6)$; vertices at $(0, \pm 9)$

**14.** Center at $(3, -4)$; vertex at $(7, -4)$; focus at $(6, -4)$

*In Problems 15 and 16, find the vertices, foci, and asymptotes for each hyperbola. Graph each hyperbola using the asymptotes as a guide.*

**15.** $\frac{y^2}{81} - \frac{x^2}{9} = 1$

**16.** $25x^2 - y^2 = 25$

**17.** Find an equation for a hyperbola with center at $(0, 0)$, focus at $(0, -5)$ and vertex at $(0, -2)$.

**18.** A large flood light is in the shape of a parabola. Its diameter is 36 inches and its depth is 12 inches. How far from the vertex should the light bulb be placed so that the rays will be reflected parallel to the axis?

# 10.6 Nonlinear Systems of Equations

## OBJECTIVES

1. Solve a System of Nonlinear Equations Using Substitution
2. Solve a System of Nonlinear Equations Using Elimination

## *Preparing for Nonlinear Systems of Equations*

*Before getting started, take the following readiness quiz. If you get a problem wrong, go back to the section cited and review the material.*

**1.** Solve the system using substitution: $\begin{cases} y = 2x - 5 \\ 2x - 3y = 7 \end{cases}$ [Section 4.1, pp. 259–261]

**2.** Solve the system using elimination: $\begin{cases} 2x - 4y = -11 \\ -x + 5y = 13 \end{cases}$ [Section 4.1, pp. 262–264]

**3.** Solve the system: $\begin{cases} 3x - 5y = 4 \\ -6x + 10y = -8 \end{cases}$ [Section 4.1, pp. 265–266]

Recall from Section 4.1 that a system of equations is a collection of two or more equations, each containing one or more variables. We learned techniques for solving systems of linear equations back in Chapter 4. We now introduce techniques for solving systems of equations where the equations are not linear. We only deal with systems containing two equations with two unknowns. A **system of nonlinear equations** in two variables is a system of equations in which at least one of the equations is not linear. That is, at least

***Preparing for...Answers*** **1.** $(2, -1)$ **2.** $(-1/2, 5/2)$ **3.** $\{(x, y) | 3x - 5y = 4\}$

one of the equations cannot be written in the form $Ax + By = C$. The following are examples of nonlinear systems of equations containing two unknowns.

$$\begin{cases} x + y^2 = 5 & \text{(1) A parabola} \\ 2x + y = 4 & \text{(2) A line} \end{cases} \qquad \begin{cases} x^2 + y^2 = 9 & \text{(1) A circle} \\ -x^2 + y = 9 & \text{(2) A parabola} \end{cases}$$

In Section 4.1, we saw that the solution to a system of linear equations could be found geometrically by determining the point of intersection of the equations in the system. The same idea holds true for nonlinear systems—the point(s) of intersection represent the solution(s) to the system.

In looking back to the problems from Chapter 4, you probably started to get a sense as to when substitution was the best approach for solving a system and when elimination was the best approach. The same deal holds for nonlinear systems—sometimes substitution is best, sometimes elimination is best. Experience and a certain degree of imagination are your friends when solving these problems.

## 1 Solve a System of Nonlinear Equations Using Substitution

The method of substitution for solving a system of nonlinear equations follows much the same approach as that for solving systems of linear equations using substitution.

### EXAMPLE 1 How to Solve a System of Nonlinear Equations Using Substitution

Solve the following system of equations using substitution: $\begin{cases} 3x - y = -2 & \text{(1) A line} \\ 2x^2 - y = 0 & \text{(2) A parabola} \end{cases}$

**Step-by-Step Solution**

**Step 1:** *Graph each equation in the system.*

Figure 44 shows the graphs of $3x - y = -2 (y = 3x + 2)$ and $2x^2 - y = 0 (y = 2x^2)$. The system apparently has two solutions.

**Figure 44**

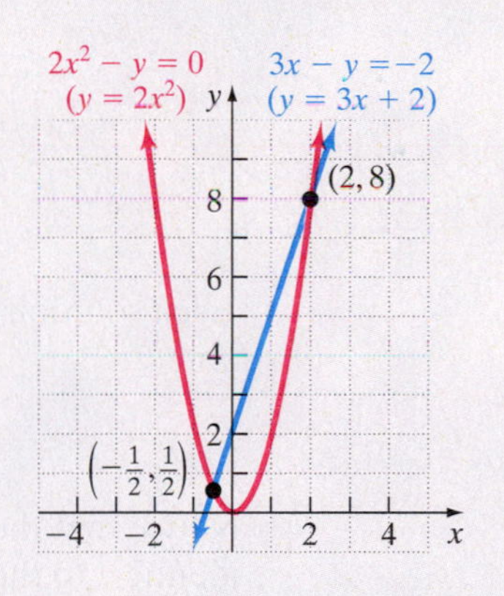


**Step 2:** Solve equation (1) for y.

Add y to both sides; add 2 to both sides:
$$3x - y = -2$$
$$y = 3x + 2$$

**Step 3:** Substitute 3x + 2 for y in equation (2).

Equation (2): $2x^2 - y = 0$

$$2x^2 - (3x + 2) = 0$$

Distribute: $2x^2 - 3x - 2 = 0$

**Step 4:** Solve for x.

Factor: $(2x + 1)(x - 2) = 0$

Zero-Product Property: $2x + 1 = 0$ or $x - 2 = 0$

$$x = -\frac{1}{2} \quad \text{or} \quad x = 2$$

(continued)

**Step 5:** Let $x = -\frac{1}{2}$ and $x = 2$ in equation (1) to determine y.

Equation (1): $3x - y = -2$

$x = -\frac{1}{2}$: $3\left(-\frac{1}{2}\right) - y = -2$

$-\frac{3}{2} - y = -2$

Add $\frac{3}{2}$ to both sides: $-y = -\frac{1}{2}$

Multiply both sides by $-1$: $y = \frac{1}{2}$

$x = 2$: $3(2) - y = -2$

$6 - y = -2$

Subtract 6 from both sides: $-y = -8$

Multiply both sides by $-1$: $y = 8$

The apparent solutions are $\left(-\frac{1}{2}, \frac{1}{2}\right)$ and $(2, 8)$.

**Step 6:** Check

$x = -\frac{1}{2}; y = \frac{1}{2}$:

$$3x - y = -2 \qquad 2x^2 - y = 0$$

$$3\left(-\frac{1}{2}\right) - \frac{1}{2} \stackrel{?}{=} -2 \qquad 2\left(-\frac{1}{2}\right)^2 - \frac{1}{2} \stackrel{?}{=} 0$$

$$-\frac{3}{2} - \frac{1}{2} \stackrel{?}{=} -2 \qquad 2\left(\frac{1}{4}\right) - \frac{1}{2} \stackrel{?}{=} 0$$

$$-\frac{4}{2} = -2 \text{ True} \qquad \frac{1}{2} - \frac{1}{2} = 0 \text{ True}$$

$x = 2; y = 8$:

$$3x - y = -2 \qquad 2x^2 - y = 0$$

$$3(2) - 8 \stackrel{?}{=} -2 \qquad 2(2)^2 - 8 \stackrel{?}{=} 0$$

$$6 - 8 = -2 \text{ True} \qquad 2(4) - 8 = 0 \text{ True}$$

Each solution checks. We now know that the graphs in Figure 44 intersect at $\left(-\frac{1}{2}, \frac{1}{2}\right)$ and $(2, 8)$.

Notice that the steps for solving a system of nonlinear equations using substitution are identical to the steps for solving a system of linear equations using substitution.

**QUICK ✓**

**1.** Solve the following system of equations using substitution: $\begin{cases} 2x + y = -1 \\ x^2 - y = 4 \end{cases}$

## EXAMPLE 2 Solving a System of Nonlinear Equations Using Substitution

Solve the following system of equations using substitution:

$$\begin{cases} x + y = 2 & \text{(1) A line} \\ (x + 2)^2 + (y - 1)^2 = 9 & \text{(2) A circle} \end{cases}$$

Figure 45

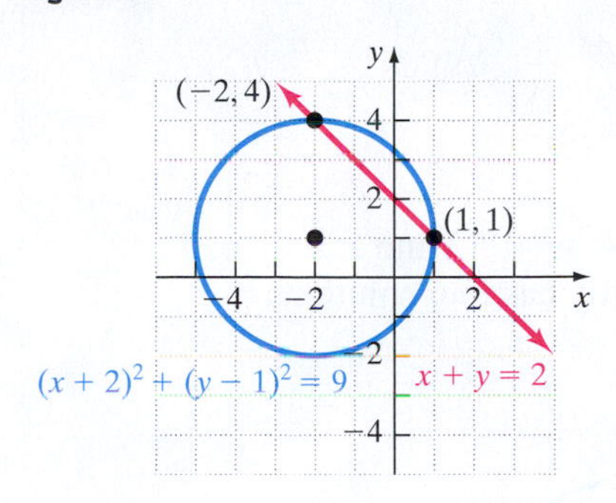


### Solution

We start by graphing each equation in the system. Figure 45 shows the graphs of $x + y = 2 (y = -x + 2)$ and $(x + 2)^2 + (y - 1)^2 = 9$. The system apparently has two solutions.

Solve equation (1) for $y$.

$$x + y = 2$$

Add $x$ to both sides: $y = -x + 2$

Substitute this expression for $y$ into equation (2), $(x + 2)^2 + (y - 1)^2 = 9$, and then solve for $x$.

Substitute $-x + 2$ for $y$ in equation (2): $(x + 2)^2 + (\qquad - 1)^2 = 9$

Combine like terms: $(x + 2)^2 + (1 - x)^2 = 9$

FOIL $x^2 + 4x + 4 + 1 - 2x + x^2 = 9$

Combine like terms: $2x^2 + 2x + 5 = 9$

Put in standard form: $2x^2 + 2x - 4 = 0$

Factor: $2(x^2 + x - 2) = 0$

Divide both sides by 2: $x^2 + x - 2 = 0$

Factor: $(x + 2)(x - 1) = 0$

Zero-Product Property: $x + 2 = 0$ or $x - 1 = 0$

Solve for $x$: $x = -2$ or $x = 1$

Let $x = -2$ and $x = 1$ in equation (1) to determine $y$.

$$x + y = 2$$

$x = -2$: $-2 + y = 2$ $\qquad$ $x = 1$: $1 + y = 2$

Add 2 to both sides: $y = 4$ $\qquad$ Subtract 1 from both sides: $y = 1$

The apparent solutions are $(-2, 4)$ and $(1, 1)$.

**Check**

$x = -2$; $y = 4$: $x + y = 2$ $\qquad (x + 2)^2 + (y - 1)^2 = 9$

$-2 + 4 = 2$ True $\qquad (-2 + 2)^2 + (4 - 1)^2 = 9$

$3^2 = 9$ True

$x = 1$; $y = 1$: $x + y = 2$ $\qquad (x + 2)^2 + (y - 1)^2 = 9$

$1 + 1 = 2$ True $\qquad (1 + 2)^2 + (1 - 1)^2 = 9$

$3^2 = 9$

$9 = 9$ True

Each solution checks. We now know that the graphs in Figure 45 intersect at $(-2, 4)$ and $(1, 1)$.

**QUICK ✓**

2. Solve the following system of equations using substitution:
$$\begin{cases} 2x + y = 0 \\ (x - 4)^2 + (y + 2)^2 = 9 \end{cases}$$

## 2 Solve a System of Nonlinear Equations Using Elimination

Now we discuss the method of elimination. The method of elimination for solving a system of nonlinear equations follows much the same approach as that for solving systems of linear equations using elimination. In fact, the steps are identical.

Recall, the basic idea in using the elimination is to get the coefficients of one of the variables to be additive inverses.

## EXAMPLE 3 How to Solve a System of Nonlinear Equations by Elimination

Solve the following system of equations using elimination: $\begin{cases} x^2 + y^2 = 13 & \text{(1) A circle} \\ x^2 - y = 7 & \text{(2) A parabola} \end{cases}$

### Step-by-Step Solution

**Step 1:** *Graph each equation in the system.*

Figure 46 shows the graphs of $x^2 + y^2 = 13$ and $x^2 - y = 7$ $(y = x^2 - 7)$. The system apparently has four solutions.

**Figure 46**

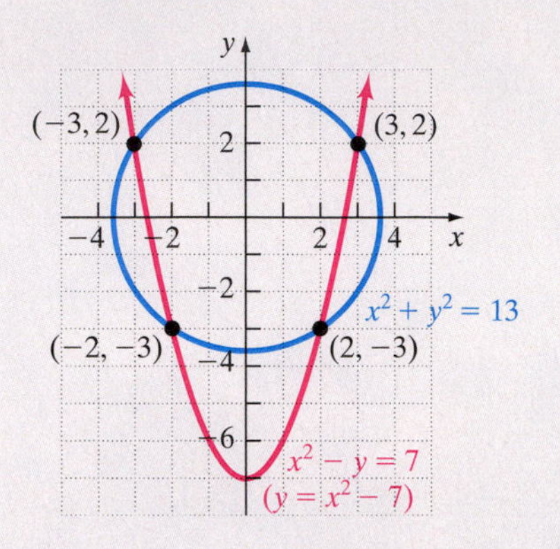


**Step 2:** *Multiply equation (2) by −1 so that the coefficients on $x^2$ become additive inverses.*

$$\begin{cases} x^2 + y^2 = 13 & (1) \\ -x^2 + y = -7 & (2) \end{cases}$$

**Step 3:** *Add equations (1) and (2) to eliminate $x^2$. Solve the resulting equation for y.*

Add:
$$\begin{cases} x^2 + y^2 = 13 & (1) \\ -x^2 + y = -7 & (2) \end{cases}$$

$$y^2 + y = 6$$

Put in standard form: $y^2 + y - 6 = 0$

Factor: $(y + 3)(y - 2) = 0$

Zero-Product Property: $y + 3 = 0$ or $y - 2 = 0$

$y = -3$ or $y = 2$

**Step 4:** *Solve for x using equation (2).*

$$x^2 - y = 7 \quad (2)$$

Using $y = -3$: $x^2 - (-3) = 7$

$x^2 + 3 = 7$

$x^2 = 4$

$x = \pm 2$

Using $y = 2$: $x^2 - 2 = 7$

$x^2 = 9$

$x = \pm 3$

The apparent solutions are $(-2, -3)$, $(2, -3)$, $(-3, 2)$, and $(3, 2)$.

**Work Smart**
We use equation (2) in Step 4 because it's easier to work with.

**Step 5:** *Check*

We leave it to you to verify that all four of the apparent solutions are solutions to the system. The four points $(-2, -3)$, $(2, -3)$, $(-3, 2)$, and $(3, 2)$ are the points of intersection of the graphs. Look again at Figure 46.

### EXAMPLE 4 Solving a System of Nonlinear Equations by Elimination

Solve the following system of equations using elimination:

$$\begin{cases} x^2 - y^2 = 4 & \text{(1) A hyperbola} \\ x^2 - y = 0 & \text{(2) A parabola} \end{cases}$$

### Solution

We graph each equation in the system. Figure 47 shows the graphs of $x^2 - y^2 = 4$ and $x^2 - y = 0 (y = x^2)$. The system apparently has no solution.

Now multiply equation (2) by $-1$ to get the coefficients on $x^2$ to be additive inverses.

$$\begin{cases} x^2 - y^2 = 4 & (1) \\ -x^2 + y = 0 & (2) \end{cases}$$

**Figure 47**

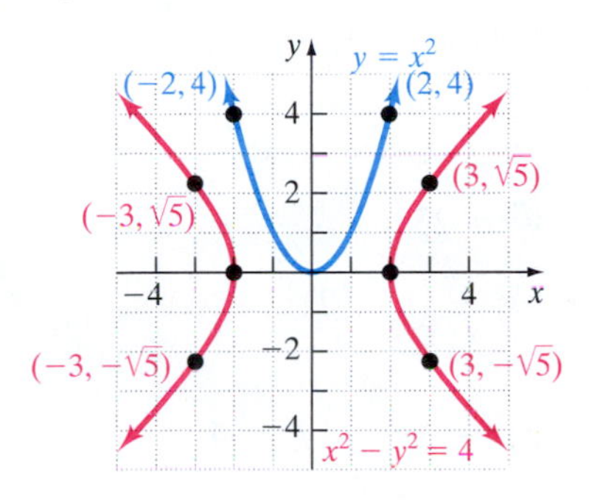


Add equations (1) and (2) to eliminate $x^2$. Solve the resulting equation for $y$.

$$\begin{cases} x^2 - y^2 = 4 & (1) \\ -x^2 + y = 0 & (2) \end{cases}$$

$$-y^2 + y = 4$$

Add $y^2$ to both sides; Add $y$ to both sides: $\quad y^2 - y + 4 = 0$

This is a quadratic equation whose discriminant is $b^2 - 4ac = (-1)^2 - 4(1)(4) = 1 - 16 = -15$. The equation has no real solution. Therefore, the system of equations is inconsistent. The solution set is $\emptyset$ or $\{ \}$. Figure 47 confirms this result.

**QUICK ✓** *Solve each system of nonlinear equations using elimination.*

**3.** $\begin{cases} x^2 + y^2 = 16 \\ x^2 - 2y = 8 \end{cases}$ **4.** $\begin{cases} x^2 - y = -4 \\ x^2 + y^2 = 9 \end{cases}$

# 10.6 Exercises

**For Extra Help:**

Student Solutions Manual | CD Video | PH Math/Tutor Center | MathXL Tutorials on CD | MathXL® | MyMathLab

## Concepts and Vocabulary

*In Problem 1, fill in the blanks.*

**1.** The two methods for solving a system of nonlinear equations in two unknowns discussed in this section are _____ and _____.

*In Problem 2, answer True or False to the statement.*

**2.** A system of nonlinear equations in two unknowns must have at least one solution.

**3.** How does a system of nonlinear equations differ from a system of linear equations?

**4.** Make up a system of nonlinear equations. Then solve the system.

## Building Skills

*In Problems 5–12, solve the system of nonlinear equations using the method of substitution.*

**5.** $\begin{cases} y = x^2 + 4 \\ y = x + 4 \end{cases}$ **6.** $\begin{cases} y = x^3 + 2 \\ y = x + 2 \end{cases}$ **7.** $\begin{cases} y = \sqrt{25 - x^2} \\ x + y = 7 \end{cases}$

**8.** $\begin{cases} y = \sqrt{100 - x^2} \\ x + y = 14 \end{cases}$ **9.** $\begin{cases} x^2 + y^2 = 4 \\ y = x^2 - 2 \end{cases}$ **10.** $\begin{cases} x^2 + y^2 = 16 \\ y = x^2 - 4 \end{cases}$

**11.** $\begin{cases} xy = 4 \\ x^2 + y^2 = 8 \end{cases}$ **12.** $\begin{cases} xy = 1 \\ x^2 - y = 0 \end{cases}$

*In Problems 13–20, solve the system of nonlinear equations using the method of elimination.*

**13.** $\begin{cases} x^2 + y^2 = 4 \\ y^2 - x = 4 \end{cases}$

**14.** $\begin{cases} x^2 + y^2 = 8 \\ x^2 + y^2 + 4y = 0 \end{cases}$

**15.** $\begin{cases} x^2 + y^2 = 7 \\ x^2 - y^2 = 25 \end{cases}$

**16.** $\begin{cases} 4x^2 + 16y^2 = 16 \\ 2x^2 - 2y^2 = 8 \end{cases}$

**17.** $\begin{cases} x^2 + y^2 = 6y \\ x^2 = 3y \end{cases}$

**18.** $\begin{cases} 2x^2 + y^2 = 18 \\ x^2 - y^2 = 9 \end{cases}$

**19.** $\begin{cases} x^2 - 2x - y = 8 \\ 6x + 2y = -4 \end{cases}$

**20.** $\begin{cases} 2x^2 - 5x + y = 12 \\ 14x - 2y = -16 \end{cases}$

## Mixed Practice

*In Problems 21–36, solve the system of nonlinear equations using any method you wish.*

**21.** $\begin{cases} y = x^2 - 6x + 4 \\ 5x + y = 6 \end{cases}$

**22.** $\begin{cases} y = x^2 + 4x + 5 \\ x - y = 9 \end{cases}$

**23.** $\begin{cases} x^2 + y^2 = 16 \\ x^2 - y^2 = 16 \end{cases}$

**24.** $\begin{cases} x^2 + y^2 = 25 \\ x^2 - y^2 = 25 \end{cases}$

**25.** $\begin{cases} (x - 4)^2 + y^2 = 25 \\ x - y = -3 \end{cases}$

**26.** $\begin{cases} (x + 5)^2 + (y - 2)^2 = 100 \\ 8x + y = 18 \end{cases}$

**27.** $\begin{cases} (x - 1)^2 + (y + 2)^2 = 4 \\ y^2 + 4y - x = -1 \end{cases}$

**28.** $\begin{cases} (x + 2)^2 + (y - 1)^2 = 4 \\ y^2 - 2y - x = 5 \end{cases}$

**29.** $\begin{cases} (x + 3)^2 + 4y^2 = 4 \\ x^2 + 6x - y = 13 \end{cases}$

**30.** $\begin{cases} 9x^2 + 4y^2 = 36 \\ x^2 + (y - 7)^2 = 4 \end{cases}$

**31.** $\begin{cases} x^2 - y^2 = 21 \\ x + y = 7 \end{cases}$

**32.** $\begin{cases} y - 2x = 1 \\ 2x^2 + y^2 = 1 \end{cases}$

**33.** $\begin{cases} x^2 + 2y^2 = 16 \\ 4x^2 - y^2 = 24 \end{cases}$

**34.** $\begin{cases} 4x^2 + 3y^2 = 4 \\ 6y^2 - 2x^2 = 3 \end{cases}$

**35.** $\begin{cases} x^2 + y^2 = 25 \\ y = -x^2 + 6x - 5 \end{cases}$

**36.** $\begin{cases} x^2 + y^2 = 65 \\ y = -x^2 + 9 \end{cases}$

## Applying the Concepts

**37. Fun with Numbers** The difference of two numbers is 2. The sum of their squares is 34. Find the numbers.

**38. Fun with Numbers** The sum of two numbers is 8. The sum of their squares is 160. Find the numbers.

△ **39. Perimeter and Area of a Rectangle** The perimeter of a rectangle is 48 feet. The area of the rectangle is 140 square feet. Find the dimensions of the rectangle.

△ **40. Perimeter and Area of a Rectangle** The perimeter of a rectangle is 64 meters. The area of the rectangle is 240 square meters. Find the dimensions of the rectangle.

**41. Constructing a Box** A rectangular piece of cardboard, whose area is 190 square centimeters, is made into an open box by cutting a 2-centimeter square from each corner and turning up the sides. See the figure. If the box is to have a volume of 180 cubic centimeters, what size cardboard should you start with?

**42. Fencing** A farmer has 132 yards of fencing available to enclose a 900 square yard region in the shape of adjoining squares with sides of length $x$ and $y$. See the figure. Find $x$ and $y$.

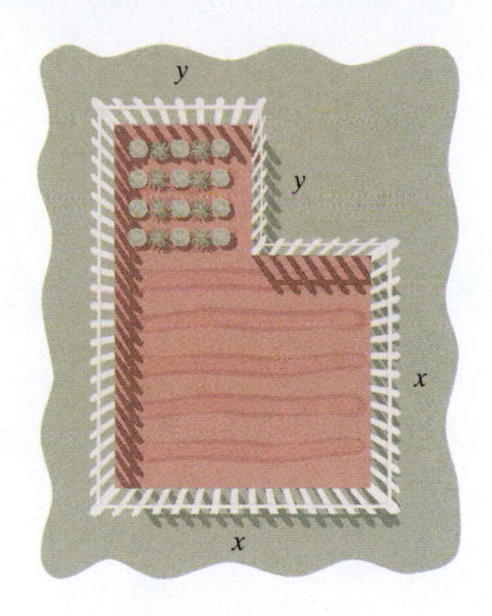


## Extending the Concepts

*In Problems 43–46, solve the system of nonlinear equations. Do not attempt to graph the equations in the system.*

**43.** $\begin{cases} y^2 + y + x^2 - x - 2 = 0 \\ y + 1 + \dfrac{x - 2}{y} = 0 \end{cases}$

**44.** $\begin{cases} x^3 - 2x^2 + y^2 + 3y - 4 = 0 \\ x - 2 + \dfrac{y^2 - y}{x^2} = 0 \end{cases}$

**45.** $\begin{cases} \ln x = 4 \ln y \\ \log_3 x = 2 + 2 \log_3 y \end{cases}$

**46.** $\begin{cases} \ln x = 5 \ln y \\ \log_2 x = 3 + 2 \log_2 y \end{cases}$

**47.** If $r_1$ and $r_2$ are two solutions of a quadratic equation $ax^2 + bx + c = 0$, then it can be shown that

$$r_1 + r_2 = -\frac{b}{a} \quad \text{and} \quad r_1 r_2 = \frac{c}{a}$$

Solve this system of equations for $r_1$ and $r_2$.

**48.** A circle and a line intersect at most twice. A circle and a parabola intersect at most four times. How many times do you think a circle and the graph of a polynomial of degree 3 can intersect? What about a circle and the graph of a polynomial of degree 4? What about a circle and the graph of a polynomial of degree $n$? Explain your conclusions using an algebraic argument.

## Synthesis Review

*In Problems 49–53, evaluate the functions given that $f(x) = 3x + 4$ and $g(x) = 2^x$.*

**49.** **(a)** $f(1)$ **(b)** $g(1)$

**50.** **(a)** $f(2)$ **(b)** $g(2)$

**51.** **(a)** $f(3)$ **(b)** $g(3)$

**52.** **(a)** $f(4)$ **(b)** $g(4)$

**53.** **(a)** $f(5)$ **(b)** $g(5)$

**54.** Use the results of Problems 49–53 to compute

**(a)** $f(2) - f(1)$ **(b)** $f(3) - f(2)$ **(c)** $f(4) - f(3)$ **(d)** $f(5) - f(4)$

**(e)** $\dfrac{g(2)}{g(1)}$ **(f)** $\dfrac{g(3)}{g(2)}$ **(g)** $\dfrac{g(4)}{g(3)}$ **(h)** $\dfrac{g(5)}{g(4)}$

**(i)** Make a generalization about $f(n + 1) - f(n)$ for $n \geq 1$ an integer. Make a generalization about $\dfrac{g(n + 1)}{g(n)}$ for $n \geq 1$ an integer.

### The Graphing Calculator

*In Problems 55–64, use a graphing calculator and the INTERSECT feature to solve the following systems of nonlinear equations.*

**55.** $\begin{cases} y = x^2 - 6x + 4 \\ 5x + y = 6 \end{cases}$

**56.** $\begin{cases} y = x^2 + 4x + 5 \\ x - y = 9 \end{cases}$

**57.** $\begin{cases} x^2 + y^2 = 16 \\ x^2 - y^2 = 16 \end{cases}$

**58.** $\begin{cases} x^2 + y^2 = 25 \\ x^2 - y^2 = 25 \end{cases}$

**59.** $\begin{cases} (x - 4)^2 + y^2 = 25 \\ x - y = -3 \end{cases}$

**60.** $\begin{cases} (x + 5)^2 + (y - 2)^2 = 100 \\ 8x + y = 18 \end{cases}$

**61.** $\begin{cases} (x - 1)^2 + (y + 2)^2 = 4 \\ x^2 + 4y - x = -1 \end{cases}$

**62.** $\begin{cases} (x + 2)^2 + (y - 1)^2 = 4 \\ y^2 - 2y - x = 5 \end{cases}$

**63.** $\begin{cases} x^2 + 4y^2 = 4 \\ x^2 + 6x - y = -13 \end{cases}$

**64.** $\begin{cases} 9x^2 + 4y^2 = 36 \\ x^2 + (y - 7)^2 = 4 \end{cases}$

## CHAPTER 10 ACTIVITY: How Do You Know That . . . ?

**Focus:** A sharing of ideas to identify topics contained in the study of conics

**Time:** 30–35 minutes

**Group size:** 2–4

For each question, every member of the group should spend 1–2 minutes individually listing "How they know . . . ." At the end of the allotted time, the group should convene and conduct a 2-3-minute discussion of the different responses.

### How Do You Know . . .

. . . that a triangle with vertices at $(2, 6)$, $(0, -2)$ and $(5, 1)$ is an isosceles triangle?

. . . the coordinates for the midpoint between two given ordered pairs?

. . . that $x^2 + y^2 + 4x - 8y = 16$ is an equation of a circle and not a parabola?

. . . which way a parabola opens?

. . . whether an ellipse's center is the origin or another point?

. . . that a hyperbola will not intersect its asymptotes?

. . . that a circle and a parabola can have 0, 1, 2, 3, or 4 points of intersection? (If necessary, use a sketch to support your response.)

# CHAPTER 10 REVIEW

## Section 10.1 Distance and Midpoint Formulas

### KEY CONCEPTS

- **The Distance Formula**
  The distance between two points $P_1 = (x_1, y_1)$ and $P_2 = (x_2, y_2)$, denoted by $d(P_1, P_2)$, is $d(P_1, P_2) = \sqrt{(x_2 - x_1)^2 + (y_2 - y_1)^2}$.
- **The Midpoint Formula**
  The midpoint $M = (x, y)$ of the line segment from $P_1 = (x_1, y_1)$ to $P_2 = (x_2, y_2)$ is $M = \left(\frac{x_1 + x_2}{2}, \frac{y_1 + y_2}{2}\right)$.

| YOU SHOULD BE ABLE TO . . . | EXAMPLE | REVIEW EXERCISES |
|---|---|---|
| 1 Use the Distance Formula (p. 773) | Examples 1 through 3 | 1–6, 12 |
| 2 Use the Midpoint Formula (p. 776) | Example 4 | 7–11 |

*In Problems 1–6, find the distance $d(P_1, P_2)$ between points $P_1$ and $P_2$.*

**1.** $P_1 = (0, 0)$ and $P_2 = (-4, -3)$

**2.** $P_1 = (-3, 2)$ and $P_2 = (5, -4)$

**3.** $P_1 = (-1, 1)$ and $P_2 = (5, 3)$

**4.** $P_1 = (6, -7)$ and $P_2 = (6, -1)$

**5.** $P_1 = \left(\sqrt{7}, -\sqrt{3}\right)$ and $P_2 = \left(4\sqrt{7}, 5\sqrt{3}\right)$

**6.** $P_1 = (-0.2, 1.7)$ and $P_2 = (1.3, 3.7)$

*In Problems 7–11, find the midpoint of the line segment formed by joining the points $P_1$ and $P_2$.*

**7.** $P_1 = (-1, 6)$ and $P_2 = (-3, 4)$

**8.** $P_1 = (7, 0)$ and $P_2 = (5, -4)$

**9.** $P_1 = \left(-\sqrt{3}, 2\sqrt{6}\right)$ and $P_2 = \left(-7\sqrt{3}, -8\sqrt{6}\right)$

**10.** $P_1 = (5, -2)$ and $P_2 = (0, 3)$

**11.** $P_1 = \left(\frac{1}{4}, \frac{2}{3}\right)$ and $P_2 = \left(\frac{5}{4}, \frac{1}{3}\right)$

**12.** Consider the three points $A = (-2, 2)$, $B = (1, -1)$, $C = (-1, -3)$.

**(a)** Plot each point in the Cartesian plane and form the triangle $ABC$.
**(b)** Find the length of each side of the triangle.
**(c)** Verify that the triangle is a right triangle.
**(d)** Find the area of the triangle.

## Section 10.2 Circles

| KEY CONCEPTS | KEY TERMS |
|---|---|
| • **Standard Form of an Equation of a Circle** The standard form of a circle with radius $r$ and center $(h, k)$ is $(x - h)^2 + (y - k)^2 = r^2$ • **General Form of the Equation of a Circle** The general form of the equation of a circle is given by the equation $x^2 + y^2 + ax + by + c = 0$ when the graph exists. | Conic sections, Circle, Radius, Center |

| YOU SHOULD BE ABLE TO . . . | EXAMPLE | REVIEW EXERCISES |
|---|---|---|
| 1 Write the standard form of the equation of a circle (p. 781) | Example 1 | 13–20 |
| 2 Graph a circle (p. 781) | Example 2 | 15–18; 21–30 |
| 3 Find the center and radius of a circle from an equation in general form (p. 782) | Example 3 | 27–30 |

*In Problems 13 and 14, find the center and radius of each circle. Write the standard form of the equation.*

**13.**

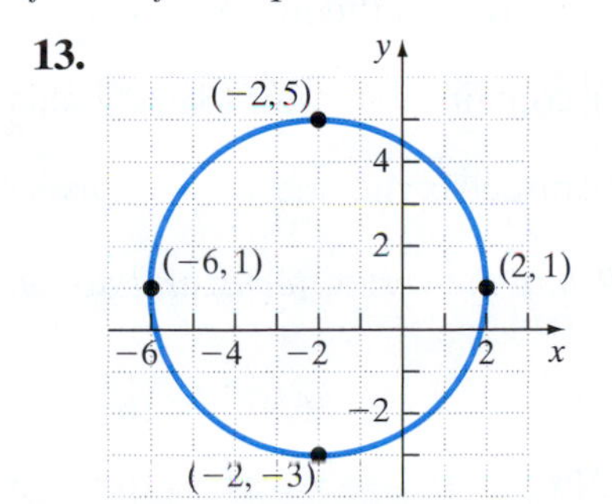


**14.**

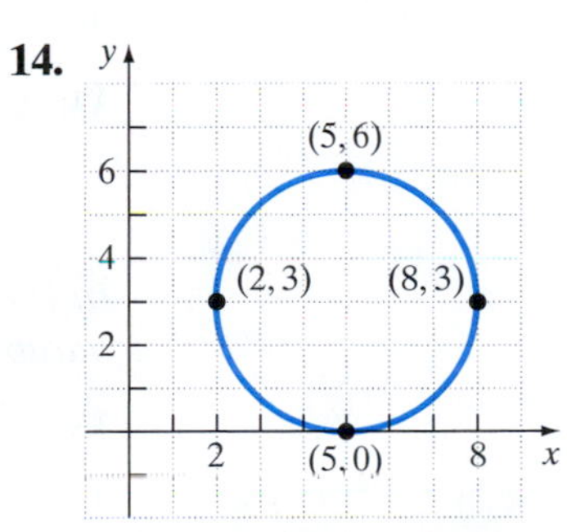


*In Problems 15–18, write the standard form of the equation of each circle whose radius is r and center is (h, k). Graph each circle.*

**15.** $r = 4$; $(h, k) = (0, 0)$

**16.** $r = 3$; $(h, k) = (-3, 1)$

**17.** $r = 1$; $(h, k) = (5, -2)$

**18.** $r = \sqrt{7}$; $(h, k) = (4, 0)$

*In Problems 19 and 20, find the standard form of the equation of each circle.*

**19.** Center at $(2, -1)$ and containing the point $(5, 3)$.

**20.** Endpoints of a diameter at $(1, 7)$ and $(-3, -1)$.

*In Problems 21–26, find the center $(h, k)$ and the radius $r$ of each circle. Graph each circle.*

**21.** $x^2 + y^2 = 25$

**22.** $(x - 1)^2 + (y - 2)^2 = 4$

**23.** $x^2 + (y - 4)^2 = 16$

**24.** $(x + 1)^2 + (y + 6)^2 = 49$

**25.** $(x + 2)^2 + \left(y - \frac{3}{2}\right)^2 = \frac{1}{4}$

**26.** $(x + 3)^2 + (y + 3)^2 = 4$

*In Problems 27–30, find the center $(h, k)$ and the radius $r$ of each circle. Graph each circle.*

**27.** $x^2 + y^2 + 6x + 10y - 2 = 0$

**28.** $x^2 + y^2 - 8x + 4y + 16 = 0$

**29.** $x^2 + y^2 + 2x - 4y - 4 = 0$

**30.** $x^2 + y^2 - 10x - 2y + 17 = 0$

## Section 10.3 Parabolas

| KEY CONCEPTS | KEY TERMS |
|---|---|
| • **Equations of a Parabola: Vertex at (0, 0); $a > 0$** See Table 1 on page 788.<br>• **Equations of a Parabola: Vertex at $(h, k)$; $a > 0$** See Table 2 on page 791. | Parabola<br>Focus<br>Directrix<br>Axis of symmetry<br>Vertex |

| YOU SHOULD BE ABLE TO . . . | EXAMPLE | REVIEW EXERCISES |
|---|---|---|
| 1 Graph parabolas in which the vertex is the origin (p. 788) | Examples 1 and 2 | 31–34; 35, 36 |
| 2 Find the equation of a parabola (p. 790) | Examples 3 and 4 | 31–34 |
| 3 Graph parabolas in which the vertex is not the origin (p. 791) | Example 5 | 37–39 |
| 4 Solve applied problems involving parabolas (p. 792) | Example 6 | 40 |

*In Problems 31–34, find the equation of the parabola described. Graph each parabola.*

**31.** Vertex at $(0, 0)$; focus at $(0, -3)$

**32.** Focus at $(-4, 0)$; directrix the line $x = 4$

**33.** Vertex at $(0, 0)$; contains the point $(8, -2)$; axis of symmetry the $x$-axis

**34.** Vertex at $(0, 0)$; directrix the line $y = -2$; axis of symmetry the $y$-axis

*In Problems 35–39, find the vertex, focus, and directrix of each parabola. Graph each parabola.*

**35.** $x^2 = 2y$

**36.** $y^2 = 16x$

**37.** $(x + 1)^2 = 8(y - 3)$

**38.** $(y - 4)^2 = -2(x + 3)$

**39.** $x^2 - 10x + 3y + 19 = 0$

**40. Radio Telescope** The U.S. Naval Research Laboratory has a giant radio telescope with a dish that is shaped like a parabola. The signals that are received by the dish strike the surface of the dish and are reflected to a single point, where the receiver is located. If the giant dish is 300 feet across and 44 feet deep at its center, at what position should the receiver be placed?

## Section 10.4 Ellipses

| KEY CONCEPTS | KEY TERMS |
|---|---|
| • **Ellipses with Center at (0, 0)** See Table 3 on page 799.<br>• **Ellipses with Center at $(h, k)$** See Table 4 on page 802. | Ellipse<br>Foci<br>Major axis<br>Center<br>Minor axis<br>Vertices |

| YOU SHOULD BE ABLE TO . . . | EXAMPLE | REVIEW EXERCISES |
|---|---|---|
| 1 Graph ellipses in which the center is the origin (p. 798) | Examples 1 and 2 | 41, 42 |
| 2 Find the equation of an ellipse in which the center is the origin (p. 801) | Example 3 | 43–45 |
| 3 Graph ellipses in which the center $(h, k)$ (p. 802) | Example 4 | 46–47 |
| 4 Solve applied problems involving ellipses (p. 803) | Example 5 | 48 |

*In Problems 41 and 42, find the vertices and foci of each ellipse. Graph each ellipse.*

**41.** $\frac{x^2}{9} + y^2 = 1$

**42.** $9x^2 + 4y^2 = 36$

*In Problems 43–45, find an equation for each ellipse. Graph each ellipse.*

**43.** Center at (0, 0); focus at (0, 3); vertex at (0, 5)

**44.** Center at (0, 0); focus at $(-2, 0)$; vertex at $(-6, 0)$

**45.** Foci at $(\pm 8, 0)$; vertices at $(\pm 10, 0)$

*In Problems 46 and 47, find the vertices and foci of each ellipse. Graph each ellipse.*

**46.** $\frac{(x-1)^2}{49} + \frac{(y+2)^2}{25} = 1$

**47.** $25(x+3)^2 + 9(y-4)^2 = 225$

**48. Semielliptical Arch Bridge** An arch in the shape of the upper half of an ellipse is used to support a bridge that is to span a river 60 feet wide. The center of the arch is 16 feet above the center of the river.

**(a)** Write the equation for the ellipse in which the $x$-axis coincides with the water and the $y$-axis passes through the center of the arch.

**(b)** Can a rectangular barge that is 25 feet wide and sits 12 feet above the surface of the water fit through the opening of the bridge?

## Section 10.5 Hyperbolas

| KEY CONCEPTS | KEY TERMS |
|---|---|
| • **Hyperbolas with Center at the Origin** See Table 5 on page 810.<br>• **Asymptotes of a Hyperbola** See Table 6 on page 813. | Hyperbola<br>Foci<br>Transverse axis<br>Center<br>Conjugate axis<br>Branches<br>Vertices |

| YOU SHOULD BE ABLE TO . . . | EXAMPLE | REVIEW EXERCISES |
|---|---|---|
| 1 Graph hyperbolas in which the center is the origin (p. 809) | Examples 1 and 2 | 49–53 |
| 2 Find the equation of a hyperbola in which the center is the origin (p. 812) | Example 3 | 54–56 |
| 3 Find asymptotes of a hyperbola in which the center is the origin (p. 813) | Example 4 | 52, 53 |

*In Problems 49–51, find the vertices and foci of each hyperbola. Graph each hyperbola.*

**49.** $\frac{x^2}{4} - \frac{y^2}{9} = 1$ **50.** $\frac{y^2}{25} - \frac{x^2}{49} = 1$ **51.** $16y^2 - 25x^2 = 400$

*In Problems 52 and 53, graph each hyperbola using the asymptotes as a guide.*

**52.** $\frac{x^2}{36} - \frac{y^2}{36} = 1$ **53.** $\frac{y^2}{25} - \frac{x^2}{4} = 1$

*In Problems 54–56, find an equation for each hyperbola described. Graph each hyperbola.*

**54.** Center at $(0, 0)$; focus at $(-4, 0)$; vertex at $(-3, 0)$

**55.** Vertices at $(0, -3)$ and $(0, 3)$; focus at $(0, 5)$

**56.** Vertices at $(0, \pm 4)$; asymptote the line $y = \frac{4}{3}x$

## Section 10.6 Systems of Nonlinear Equations

| KEY TERM | | |
|---|---|---|
| System of nonlinear equations | | |
| **YOU SHOULD BE ABLE TO . . .** | **EXAMPLE** | **REVIEW EXERCISES** |
| 1 Solve a system of nonlinear equations using substitution (p. 819) | Examples 1 and 2 | 57–60; 65–76 |
| 2 Solve a system of nonlinear equations using elimination (p. 821) | Examples 3 and 4 | 61–64; 65–76 |

*In Problems 57–60, solve the system of nonlinear equations using the method of substitution.*

**57.** $\begin{cases} 4x^2 + y^2 = 10 \\ y = x \end{cases}$ **58.** $\begin{cases} y = 2x^2 + 1 \\ y = x + 2 \end{cases}$

**59.** $\begin{cases} 6x - y = 5 \\ xy = 1 \end{cases}$ **60.** $\begin{cases} x^2 + y^2 = 26 \\ x^2 - 2y^2 = 23 \end{cases}$

*In Problems 61–64, solve the system of nonlinear equations using the method of elimination.*

**61.** $\begin{cases} 4x - y^2 = 0 \\ 2x^2 + y^2 = 16 \end{cases}$ **62** $\begin{cases} x^2 - y = -2 \\ x^2 + y = 4 \end{cases}$

**63.** $\begin{cases} 4x^2 - 2y^2 = 2 \\ -x^2 + y^2 = 2 \end{cases}$ **64.** $\begin{cases} x^2 + y^2 = 8x \\ y^2 = 3x \end{cases}$

*In Problems 65–72, solve the system of nonlinear equations using the method you prefer.*

**65.** $\begin{cases} y = x + 2 \\ y = x^2 \end{cases}$ **66.** $\begin{cases} x^2 + 2y = 9 \\ 5x - 2y = 5 \end{cases}$

**67.** $\begin{cases} x^2 + y^2 = 36 \\ x - y = -6 \end{cases}$ **68.** $\begin{cases} y = 2x - 4 \\ y^2 = 4x \end{cases}$

**69.** $\begin{cases} x^2 + y^2 = 9 \\ x + y = 7 \end{cases}$ **70.** $\begin{cases} 2x^2 + 3y^2 = 14 \\ x^2 - y^2 = -3 \end{cases}$

**71.** $\begin{cases} x^2 + y^2 = 16 \\ x^2 + 4y = 16 \end{cases}$ **72.** $\begin{cases} x = 4 - y^2 \\ x = 2y + 4 \end{cases}$

**73. Fun with Numbers** The sum of two numbers is 12. The difference of their squares is 24. Find the two numbers.

△ **74. Perimeter and Area of a Rectangle** The perimeter of a rectangle is 34 centimeters. The area of the rectangle is 60 square centimeters. Find the dimensions of the rectangle.

△ **75. Dimensions of a Rectangle** The area of a rectangle is 2160 square inches. The diagonal of the rectangle is 78 inches. Find the dimensions of the rectangle.

△ **76. Dimensions of a Triangle** A right triangle has a perimeter of 36 feet and a hypotenuse of 15 feet. Find the lengths of the legs of the right triangle.

## CHAPTER 10 TEST

*Remember to use your Chapter Test Prep Video CD to see fully worked-out solutions to any of these problems you would like to review.*

**1.** Find the distance $d(P_1, P_2)$ between points $P_1 = (-1, 3)$ and $P_2 = (3, -5)$.

**2.** Find the midpoint of the line segment formed by joining the points $P_1 = (-7, 6)$ and $P_2 = (5, -2)$.

*In Problems 3 and 4, find the center (h, k) and the radius r of each circle. Graph each circle.*

**3.** $(x - 4)^2 + (y + 1)^2 = 9$

**4.** $x^2 + y^2 + 10x - 4y + 13 = 0$

*In Problems 5 and 6, find the standard form of the equation of each circle.*

**5.** Radius $r = 6$ and center $(h, k) = (-3, 7)$

**6.** Center at $(-5, 8)$ and containing the point $(3, 2)$.

*In Problems 7 and 8, find the vertex, focus, and directrix of each parabola. Graph each parabola.*

**7.** $(y + 2)^2 = 4(x - 1)$

**8.** $x^2 - 4x + 3y - 8 = 0$

*In Problems 9 and 10, find an equation for each parabola described.*

**9.** Vertex at $(0, 0)$; focus at $(0, -4)$

**10.** Focus at $(3, 4)$; directrix the line $x = -1$

*In Problems 11 and 12, find the vertices and foci of each ellipse. Graph each ellipse.*

**11.** $9x^2 + 25y^2 = 225$

**12.** $\dfrac{(x - 2)^2}{9} + \dfrac{(y + 4)^2}{16} = 1$

*In Problems 13 and 14, find an equation for each ellipse described.*

**13.** Center at $(0, 0)$; focus at $(0, -4)$; vertex at $(0, -5)$

**14.** Vertices at $(-1, 7)$ and $(-1, -3)$; focus at $(-1, -1)$

*In Problems 15 and 16, find the vertices, foci, and asymptotes for each hyperbola. Graph each hyperbola using the asymptotes as a guide.*

**15.** $x^2 - \dfrac{y^2}{4} = 1$

**16.** $16y^2 - 25x^2 = 1600$

**17.** *Find an equation for a hyperbola with foci at $(\pm 8, 0)$ and vertex at $(-3, 0)$.*

*In Problems 18 and 19, solve the system of nonlinear equations using the method you prefer.*

**18.** $\begin{cases} x^2 + y^2 = 17 \\ x + y = -3 \end{cases}$

**19.** $\begin{cases} x^2 + y^2 = 9 \\ 4x^2 - y^2 = 16 \end{cases}$

**20.** An arch in the shape of the upper half of an ellipse is used to support a bridge spanning a creek 30 feet wide. The center of the arch is 10 feet above the center of the creek.

**(a)** Write the equation for the ellipse in which the $x$-axis coincides with the creek and the $y$-axis passes through the center of the arch.

**(b)** What is the height of the arch at a distance 12 feet from the center of the creek?

CHAPTER

# 11 Sequences, Series, and the Binomial Theorem

Population growth has been a topic of debate among scientists for a long time. In fact, over 200 years ago, the English economist and mathematician Thomas Robert Malthus anonymously published a paper predicting that the world's population would overwhelm the Earth's capacity to sustain it—he claimed that food supplies increase arithmetically while population grows geometrically. See Problems 53 and 54 in Section 11.1 and Problem 87 in Section 11.3.

## OUTLINE



## The Big Picture: Putting It Together

In Chapter 2, we defined the domain of a function as the set of real numbers such that the function makes sense. In Sections 11.1–11.3, we introduce a special type of function called a *sequence*. The domain of a sequence is not the set of real numbers, but instead is the set of natural numbers (that is, the positive integers). Functions whose domain is the set all real numbers are useful for modeling situations where we wish to describe what happens to the value of some independent variable that is continuously changing, such as the population of bacteria. Sequences are powerful for modeling situations where we wish to describe what happens to the value of a dependent variable at discrete intervals of time, such as weekly changes in the value of a deposit at a bank.

In Chapter 5, we gave formulas for expanding $(x + a)^2$ and $(x + a)^3$. A logical question to ask yourself is, "Does any formula exist for expanding $(x + a)^n$, where $n$ is an integer greater than 3?" The answer is yes! We discuss this formula in Section 11.4.

# 11.1 Sequences

**OBJECTIVES**

1. Write the First Several Terms of a Sequence
2. Find a Formula for the $n$th Term of a Sequence
3. Use Summation Notation

***Preparing for Sequences***

*Before getting started, take the following readiness quiz. If you get a problem wrong, go back to the section cited and review the material.*

**1.** Evaluate $f(x) = x^2 - 4$ at (a) $x = 3$ (b) $x = -7$. [Section 2.3, pp. 162–164]

**2.** If $g(x) = 2x - 3$, find $g(1) + g(2) + g(3)$. [Section 2.3, pp. 162–164]

**3.** In the function $f(n) = n^2 - 4$, what is the independent variable? [Section 2.3, pp. 162–163]

We begin with a definition.

**DEFINITION**

A **sequence** is a function whose domain is the set of positive integers.

Because a sequence is a function, it will have a graph. In Figure 1(a), we have the graph of the function $f(x) = \frac{1}{x}$ for $x > 0$. If all the points on this graph were removed except those whose $x$-coordinates are positive integers—that is, if all points were removed except $(1, 1)$, $\left(2, \frac{1}{2}\right)$, $\left(3, \frac{1}{3}\right)$—and so on, the remaining points would be the graph of the sequence $f(n) = \frac{1}{n}$, as shown in Figure 1(b). Notice that we use $n$ to represent the independent variable in a sequence. This serves to remind us that $n$ is a positive integer, or natural number.

**Figure 1**

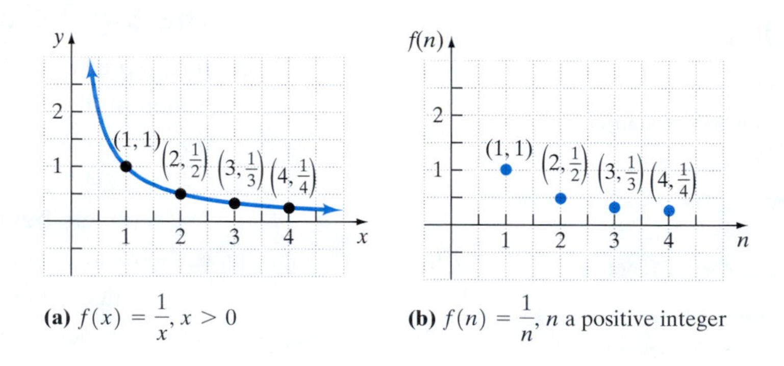


**(a)** $f(x) = \frac{1}{x}, x > 0$ **(b)** $f(n) = \frac{1}{n}$, $n$ a positive integer

A sequence may be represented by listing its values in order. For example, the sequence whose graph is given in Figure 1(b) might be represented as

$$f(1), f(2), f(3), f(4), \ldots \quad \text{or} \quad 1, \frac{1}{2}, \frac{1}{3}, \frac{1}{4}, \ldots$$

The numbers in the ordered list are called the **terms** of the sequence. Notice that the list never ends, as the **ellipsis** (the three dots) indicates. When a sequence does not end it is said to be an **infinite sequence.** We contrast this with **finite sequences** that have a domain that is the first $n$ positive integers. For example,

$$40, 44, 48, 52, 56, 60, 64$$

is a finite sequence because it contains $n = 7$ terms.

**In Words**

The word *infinite* means "without bound," so infinite sequences have no bound. The word *finite* means "bounded in number," so a finite sequence has a certain number of terms.

## 1 Write the First Several Terms of a Sequence

In a sequence we do not use the traditional notation $f(n)$. This is done to distinguish sequences from functions whose domain is the set of all real numbers. Instead, we use the notation $a_n$ to indicate that the function is a sequence. The value of the subscript represents the value of the independent variable. So $a_4$ means to evaluate the function

***Preparing for...Answers*** **1. (a)** 5 **(b)** 45 **2.** 3 **3.** $n$

defined by the sequence $a_n$ at $n = 4$. For the sequence $f(n) = \frac{1}{n}$, we would write the function as $a_n = \frac{1}{n}$ and evaluate the function as follows:

$$a_1 = f(1) = 1, \quad a_2 = f(2) = \frac{1}{2}, \quad a_3 = f(3) = \frac{1}{3}, \quad a_4 = f(4) = \frac{1}{4}, \quad \ldots, \quad a_n = f(n) = \frac{1}{n}$$

In addition, just as we can name functions $f, g, F, G$ and so on, we can name sequences. Typically, we use small letters from the beginning of the alphabet to name a sequence as in $a_n, b_n, c_n$, and so on (although this naming scheme is not necessary).

When a formula for the $n$th term (sometimes called the **general term**) of a sequence is known, rather than write out the terms of the sequence, we may represent the entire sequence by placing braces around the formula for the $n$th term. For example, the sequence whose $n$th term is $b_n = \left(\frac{1}{3}\right)^n$ can be written as

$$\{b_n\} = \left\{\left(\frac{1}{3}\right)^n\right\}$$

or by

$$b_1 = \frac{1}{3}, \quad b_2 = \frac{1}{9}, \quad b_3 = \frac{1}{27}, \quad \ldots, \quad b_n = \left(\frac{1}{3}\right)^n$$

## EXAMPLE 1 Writing the First Five Terms of a Sequence

Write down the first five terms of the sequence $\{a_n\} = \left\{\frac{n}{n+1}\right\}$.

### Solution

To find the first five terms of the sequence, we evaluate the function at $n = 1, 2, 3, 4$, and 5.

$$a_1 = \frac{1}{1+1} = \frac{1}{2}, \quad a_2 = \frac{2}{2+1} = \frac{2}{3}, \quad a_3 = \frac{3}{3+1} = \frac{3}{4}, \quad a_4 = \frac{4}{4+1} = \frac{4}{5}, \quad a_5 = \frac{5}{5+1} = \frac{5}{6}$$

The first five terms of the sequence are $\frac{1}{2}, \frac{2}{3}, \frac{3}{4}, \frac{4}{5}$, and $\frac{5}{6}$.

## EXAMPLE 2 Writing the First Five Terms of a Sequence

Write down the first five terms of the sequence $\{b_n\} = \{(-1)^n \cdot n^2\}$.

### Solution

The first five terms of the sequence are

$$b_1 = (-1)^1 \cdot 1^2 = -1, \quad b_2 = (-1)^2 \cdot 2^2 = 4, \quad b_3 = (-1)^3 \cdot 3^2 = -9,$$
$$b_4 = (-1)^4 \cdot 4^2 = 16, \quad b_5 = (-1)^5 \cdot 5^2 = -25$$

The first five terms of the sequence are $-1, 4, -9, 16$, and $-25$.

Notice in Example 2 that the signs of the terms in the sequence **alternate.** When this occurs, we use factors such as $(-1)^{n+1}$, which equals 1 if $n$ is odd and $-1$ if $n$ is even, or $(-1)^n$, which equals $-1$ if $n$ is odd and 1 if $n$ is even.

**QUICK ✓** *Write the first five terms of the given sequence.*

**1.** $\{a_n\} = \{2n - 3\}$ **2.** $\{b_n\} = \{(-1)^n \cdot 4n\}$

## 2 Find a Formula for the *n*th Term of a Sequence

Sometimes a sequence is indicated by an observed pattern in the first few terms that makes it possible to determine the formula for the $n$th term. In the example that follows, a sufficient number of terms of the sequence are given so that a natural choice for the $n$th term is suggested.

### EXAMPLE 3 Determining a Sequence from a Pattern

Find a formula for the $n$th term of each sequence.

**(a)** $3, 7, 11, 15, \ldots$ **(b)** $2, 4, 8, 16, \ldots$ **(c)** $1, -8, 27, -64, 125, \ldots$

**Solution**

Our goal in all these problems is to find a formula in terms of $n$ such that when $n = 1$ we get the first term, when $n = 2$ we get the second term, and so on.

**(a)** When $n = 1$, we have that $a_1 = 3$; when $n = 2$, we have that $a_2 = 7$. Notice that each subsequent term increases by 4. A formula for the $n$th term is given by $a_n = 4n - 1$.

**(b)** The terms of the sequence are all multiples of 2 with the first term equaling $2^1$, the second term equaling $2^2$, and so on. A formula for the $n$th term is given by $b_n = 2^n$.

**(c)** Notice that the terms alternate with the first term being positive. So $(-1)^{n+1}$ must be part of the formula. Ignoring the sign, we notice that the terms are all perfect cubes. A formula for the $n$th term is given by $c_n = (-1)^{n+1} \cdot n^3$. ■

**QUICK ✓** *Find a formula for the nth term of each sequence.*

**3.** $5, 7, 9, 11, \ldots$ **4.** $\frac{1}{2}, -\frac{1}{3}, \frac{1}{4}, -\frac{1}{5}, \ldots$

---

## 3 Use Summation Notation

In other mathematics courses, such as statistics or calculus, it is important to be able to find the sum of the first $n$ terms of a sequence $\{a_n\}$, that is,

$$a_1 + a_2 + a_3 + \cdots + a_n$$

**In Words**
The symbol $\Sigma$ is read "upper case sigma" or "sigma" for short.

Rather than write down all these terms, mathematicians use a more concise way to express the sum, called **summation notation.** Using summation notation, we would write $a_1 + a_2 + a_3 + \cdots + a_n$ as

$$\sum_{i=1}^{n} a_i = a_1 + a_2 + a_3 + \cdots + a_n$$

The symbol $\Sigma$ is an instruction to sum, or add up, the terms. The integer $i$ is called the **index** of the sum; it tells you where to start the sum and where to end it. The expression

$$\sum_{i=1}^{n} a_i$$

is an instruction to add the terms of the sequence $\{a_i\}$ from $i = 1$ through $i = n$. We read the expression $\sum_{i=1}^{n} a_i$ as "the sum of $a$ sub $i$ from $i = 1$ to $i = n$." When there are a finite number of terms to be added, the sum is called a **partial sum.**

## EXAMPLE 4 Finding a Partial Sum

Write out the sum and determine its value.

**(a)** $\sum_{i=1}^{4}(2i + 5)$ **(b)** $\sum_{i=1}^{5}(i^2 - 5)$

### Solution

**(a)**
$$\sum_{i=1}^{4}(2i + 5) = \underbrace{(2\cdot 1 + 5)}_{i=1} + \underbrace{(2\cdot 2 + 5)}_{i=2} + \underbrace{(2\cdot 3 + 5)}_{i=3} + \underbrace{(2\cdot 4 + 5)}_{i=4}$$
$$= 7 + 9 + 11 + 13$$
$$= 40$$

**(b)**
$$\sum_{i=1}^{5}(i^2 - 5) = \underbrace{(1^2 - 5)}_{i=1} + \underbrace{(2^2 - 5)}_{i=2} + \underbrace{(3^2 - 5)}_{i=3} + \underbrace{(4^2 - 5)}_{i=4} + \underbrace{(5^2 - 5)}_{i=5}$$
$$= -4 + (-1) + 4 + 11 + 20$$
$$= 30$$

**QUICK ✓** *Write out the sum and determine its value.*

**5.** $\sum_{i=1}^{3}(4i - 1)$ **6.** $\sum_{i=1}^{5}(i^3 + 1)$

---

The index of summation does not have to begin with 1. In addition, the index of summation does not have to be $i$. For example, we might have

$$\sum_{k=3}^{6}(2k + 1) = \underbrace{(2\cdot 3 + 1)}_{k=3} + \underbrace{(2\cdot 4 + 1)}_{k=4} + \underbrace{(2\cdot 5 + 1)}_{k=5} + \underbrace{(2\cdot 6 + 1)}_{k=6}$$
$$= 7 + 9 + 11 + 13$$
$$= 40$$

Notice that the terms of this partial sum are identical to those given in Example 4(a). What is the moral of the story? We can make the same sum look entirely different by changing the starting point of the index of summation and changing the variable that represents the index.

Now let's reverse the process. Rather than writing out a sum, we will express a sum using summation notation.

## EXAMPLE 5 Writing a Sum in Summation Notation

Express each sum using summation notation.

**(a)** $1^2 + 2^2 + 3^2 + \cdots + 10^2$ **(b)** $2 + 1 + \frac{2}{3} + \frac{1}{2} + \frac{2}{5} + \cdots + \frac{1}{6}$

### Solution

**(a)** The sum $1^2 + 2^2 + 3^2 + \cdots + 10^2$ has 10 terms, each of the form $i^2$, and starts at $i = 1$ and ends at $i = 10$:

$$1^2 + 2^2 + 3^2 + \cdots + 10^2 = \sum_{i=1}^{10} i^2$$

**(b)** First, we need to figure out the pattern. With a little investigation, we discover that $2 + 1 + \frac{2}{3} + \frac{1}{2} + \frac{2}{5} + \cdots + \frac{1}{6}$ can be written as

$\frac{2}{1} + \frac{2}{2} + \frac{2}{3} + \frac{2}{4} + \frac{2}{5} + \cdots + \frac{2}{12}$, so that the $n$th term of the sum is $\frac{2}{n}$. There are $n = 12$ terms, so

$$2 + 1 + \frac{2}{3} + \frac{1}{2} + \frac{2}{5} + \cdots + \frac{1}{6} = \sum_{n=1}^{12} \frac{2}{n}$$

**Work Smart**

The index of summation can be any variable we desire and can start at any value we desire. Keep this in mind when checking your answers.

**QUICK ✓** *Write each sum using summation notation.*

**7.** $1 + 4 + 9 + \cdots + 144$

**8.** $1 + \frac{1}{2} + \frac{1}{4} + \cdots + \frac{1}{32}$

# 11.1 Exercises

**For Extra Help:**   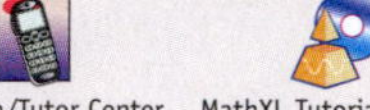 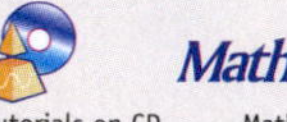

Student Solutions Manual | CD Video | PH Math/Tutor Center | MathXL Tutorials on CD | MathXL® | MyMathLab

## Concepts and Vocabulary

*In Problems 1–3, fill in the blanks.*

**1.** A(n) _________ is a function whose domain is the set of positive integers.

**2.** When a sequence does not end it is said to be a(n) _________ sequence. A(n) _________ sequence has a domain that is the first $n$ positive integers.

**3.** When there is a finite number of terms to be added, the sum is called a _________ _________.

*In Problems 4–6, answer True or False to each statement.*

**4.** A sequence is a function.

**5.** In the expression $\sum_{k=1}^{6} 5k$, the expression $5k$ is called the index of summation.

**6.** We can evaluate the sequence $\{b_n\} = \{5n - 3\}$ at $n = 0$.

**7.** Explain how a sequence and a function are related.

**8.** What does the graph of a sequence look like when compared to the graph of a function? Use the function $f(x) = 3x + 1$ and the sequence $\{a_n\} = \{3n + 1\}$ when doing the comparison.

**9.** Write a sentence that explains the meaning of the symbol $\Sigma$.

**10.** What does it mean when a sequence alternates?

## Building Skills

*In Problems 11–22, write down the first five terms of each sequence.*

**11.** $\{3n + 5\}$ **12.** $\{n - 4\}$ **13.** $\left\{\frac{n}{n + 2}\right\}$ **14.** $\left\{\frac{n + 4}{n}\right\}$

**15.** $\{(-1)^n n\}$ **16.** $\{(-1)^{n+1} n\}$ **17.** $\{2^n + 1\}$ **18.** $\{3^n - 1\}$

**19.** $\left\{\frac{2n}{2^n}\right\}$ **20.** $\left\{\frac{3n}{3^n}\right\}$ **21.** $\left\{\frac{n}{e^n}\right\}$ **22.** $\left\{\frac{n^2}{2}\right\}$

*In Problems 23–30, the given pattern continues. Write down the nth term of each sequence suggested by the pattern.*

**23.** $2, 4, 6, 8, \ldots$ **24.** $5, 10, 15, 20, \ldots$ **25.** $\frac{1}{2}, \frac{2}{3}, \frac{3}{4}, \frac{4}{5}, \ldots$

**26.** $\frac{1}{2}, 1, \frac{3}{2}, 2, \frac{5}{2}, \ldots$ **27.** $3, 6, 11, 18, \ldots$ **28.** $0, 7, 26, 63, \ldots$

**29.** $-1, 4, -9, 16, \ldots$ **30.** $1, -\frac{1}{2}, \frac{1}{4}, -\frac{1}{8}, \ldots$

*In Problems 31–42, write out each sum and determine its value.*

**31.** $\sum_{i=1}^{4}(5i + 1)$ **32.** $\sum_{i=1}^{5}(3i + 2)$ **33.** $\sum_{i=1}^{5}\frac{i^2}{2}$ **34.** $\sum_{i=1}^{4}\frac{i^3}{2}$

**35.** $\sum_{k=1}^{3}2^k$ **36.** $\sum_{k=1}^{4}3^k$ **37.** $\sum_{k=1}^{5}[(-1)^{k+1}\cdot 2k]$ **38.** $\sum_{k=1}^{8}[(-1)^k \cdot k]$

**39.** $\sum_{j=1}^{10}5$ **40.** $\sum_{j=1}^{8}2$ **41.** $\sum_{k=3}^{7}(2k - 1)$ **42.** $\sum_{j=5}^{10}(k + 4)$

*In Problems 43–50, express each sum using summation notation.*

**43.** $1 + 2 + 3 + \cdots + 15$ **44.** $1 + 3 + 5 + \cdots + 17$

**45.** $1 + \frac{1}{2} + \frac{1}{3} + \cdots + \frac{1}{12}$ **46.** $1 + \frac{1}{2} + \frac{1}{4} + \cdots + \frac{1}{2^{15}}$

**47.** $1 - \frac{1}{3} + \frac{1}{9} - \frac{1}{27} + \cdots + (-1)^{9+1}\left(\frac{1}{3^{9-1}}\right)$

**48.** $\frac{2}{3} - \frac{4}{9} + \frac{8}{27} + \cdots + (-1)^{15+1}\left(\frac{2}{3}\right)^{15}$

**49.** $5 + (5 + 2\cdot 1) + (5 + 2\cdot 2) + (5 + 2\cdot 3) + \cdots + (5 + 2\cdot 10)$

**50.** $3 + 3\cdot\frac{1}{2} + 3\cdot\frac{1}{4} + \cdots + 3\cdot\left(\frac{1}{2}\right)^{11}$

## Applying the Concepts

**51. The Future Value of Money** Suppose that you place \$12,000 in your company 401(k) plan that pays 6% interest compounded quarterly. The balance in the account after $n$ quarters is given by

$$a_n = 12{,}000\left(1 + \frac{0.06}{4}\right)^n$$

**(a)** Find the value in the account after 1 quarter.
**(b)** Find the value in the account after 1 year.
**(c)** Find the value in the account after 10 years.

**52. The Future Value of Money** Suppose that you place \$5000 into a company 401(k) plan that pays 8% interest compounded monthly. The balance in the account after $n$ months is given by

$$a_n = 5000\left(1 + \frac{0.08}{12}\right)^n$$

**(a)** Find the value in the account after 1 month.
**(b)** Find the value in the account after 1 year.
**(c)** Find the value in the account after 10 years.

**53. Population Growth** According to the U.S. Census Bureau, the population of the United States in 2005 was 296 million people. In addition, the population of the United States was growing at a rate of 1.1% per year. A model for the population of the United States is given by

$$p_n = 296(1.011)^n$$

where $n$ is the number of years after 2005.

**(a)** Use this model to predict the United States population in 2010 to the nearest million.
**(b)** Use this model to predict the United States population in 2050 to the nearest million.

**54. Population Growth** According to the *Statistical Abstract of the United States,* the population of the world in 2005 was 6448 million people. In addition, the population of the world was growing at a rate of 1.26% per year. A model for the population of the world is given by

$$p_n = 6448(1.0126)^n$$

where $n$ is the number of years after 2005.

**(a)** Use this model to predict the world population in 2010.
**(b)** Use this model to predict the world population in 2050.

**55. Fibonacci Sequence** Let

$$u_n = \frac{\left(1 + \sqrt{5}\right)^n - \left(1 - \sqrt{5}\right)^n}{2^n \cdot \sqrt{5}}$$

define the $n$th term of a sequence. Find the first 10 terms of the sequence. This sequence is called the **Fibonacci sequence.** The terms of the sequence are called **Fibonacci numbers.**

**56. Pascal's Triangle** The triangular array of numbers shown in the right is called Pascal's Triangle. Each number in the triangle is found by adding the entries directly above to the left and right of the number. For example, the 4 in the fourth row is found by adding the 1 and 3 above the 4.

Divide the triangular array using diagonal lines (as shown). Now find the sum of the numbers in each of the highlighted diagonal rows. Do you recognize the sequence? (**Hint:** See Problem 55.)

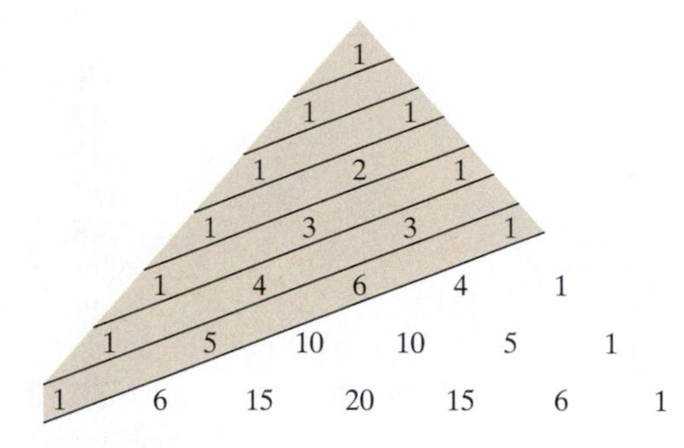


## Extending the Concepts

*A second way of defining a sequence is to assign a value to the first (or first few) term(s) and specify the nth term by a formula or equation that involves one or more of the terms preceding it. Sequences defined this way are said to be defined* ***recursively,*** *and the rule or formula is called a* ***recursive formula.*** *For example* $s_1 = 3$ *and* $s_n = 2s_{n-1}$ *is a recursively defined sequence where* $s_1 = 3, s_2 = 2s_1 = 2(3) = 6, s_3 = 2s_2 = 2(6) = 12, s_4 = 2s_3 = 2(12) = 24$ *and so on.*

*In Problems 57–60, a sequence is defined recursively. Write the first five terms of the sequence.*

**57.** $a_1 = 10, a_n = 1.05a_{n-1}$

**58.** $b_1 = 20, b_n = 3b_{n-1}$

**59.** $b_1 = 8, b_n = n + b_{n-1}$

**60.** $c_1 = 1{,}000, c_n = 1.01c_{n-1} + 100$

**61. Fibonacci Sequence** Use the result of Problem 55 to do the following problems:

**(a)** Compute the ratio $\dfrac{u_{n+1}}{u_n}$ for the first 10 terms.

**(b)** As $n$ gets large, what number does the ratio approach? This number is referred to as the **golden ratio.** Rectangles whose sides are in this ratio were considered pleasing to the eye by the Greeks. For example, the façade of the Parthenon was constructed using the golden ratio.

**(c)** Compute the ratio $\dfrac{u_n}{u_{n+1}}$ for the first 10 terms.

**(d)** As $n$ gets large, what number does the ratio approach? This number is also referred to as the **golden ratio.** This ratio is believed to have been used in the construction of the Great Pyramid in Egypt. The ratio equals the sum of the areas of the four face triangles divided by the total surface area of the Great Pyramid.

**62.** Investigate various applications that lead to a Fibonacci sequence, such as art, architecture, or financial markets. Write an essay on these applications.

## Synthesis Review

*In Problems 63–65, (a) determine the slope of the linear function, (b) Compute $f(1), f(2), f(3)$, and $f(4)$.*

**63.** $f(x) = 4x - 6$ **64.** $f(x) = 2x - 10$ **65.** $f(x) = -5x + 8$

**66.** Compute $f(2) - f(1), f(3) - f(2)$, and $f(4) - f(3)$ for each of the functions in Problems 63–65. How does the difference in the value of the function for consecutive values of the independent variable relate to the slope?

## Graphing Calculator

Graphing calculators can be used to write the terms of a sequence. For example, Figure 2 shows the first five terms of the sequence $\{b_n\} = \{(-1)^n \cdot n^2\}$ that we studied in Example 2. Figure 3 shows how the terms of the sequence can be listed in a table using SEQuence mode.

**Figure 2**

```
seq((-1)^X*X²,X,
1,5,1)
{-1 4 -9 16 -25}
```

**Figure 3**

```
Plot1 Plot2 Plot3
nMin=1
\u(n)■(-1)^n*n²
 u(nMin)■-1■
\v(n)=
 v(nMin)=
\w(n)=
 w(nMin)=
```

| n | u(n) |
|---|---|
| 1 | -1 |
| 2 | 4 |
| 3 | -9 |
| 4 | 16 |
| 5 | -25 |
| 6 | 36 |
| 7 | -49 |

```
u(n)■(-1)^n*n²
```

*In Problems 67–74, use a graphing calculator to find the first five terms of the sequence.*

**67.** $\{3n + 5\}$ **68.** $\{n - 4\}$ **69.** $\left\{\dfrac{n}{n+2}\right\}$ **70.** $\left\{\dfrac{n+4}{n}\right\}$

**71.** $\{(-1)^n n\}$ **72.** $\{(-1)^{n+1} n\}$ **73.** $\{2^n + 1\}$ **74.** $\{3^n - 1\}$

Graphing calculators can also be used to find the sum of a sequence. For example, Figure 4 shows the result of the sum $\sum_{i=1}^{5}(i^2 - 5)$ that we studied in Example 4(b).

**Figure 4**

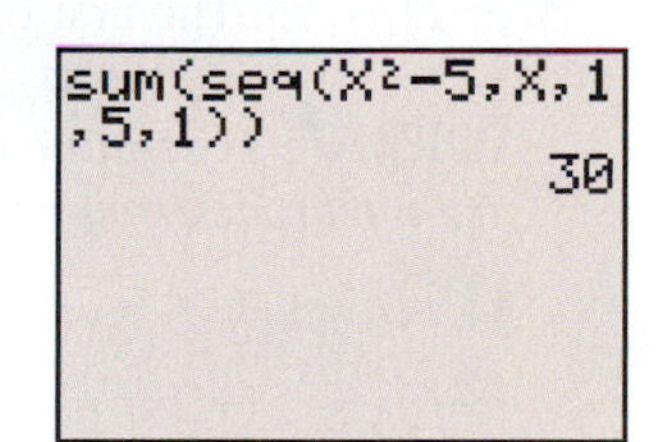


*In Problems 75–82, use a graphing calculator to find the sum.*

**75.** $\sum_{i=1}^{4}(5i + 1)$ **76.** $\sum_{i=1}^{5}(3i + 2)$ **77.** $\sum_{i=1}^{5}\dfrac{i^2}{2}$ **78.** $\sum_{i=1}^{4}\dfrac{i^3}{2}$

**79.** $\sum_{k=1}^{3}2^k$ **80.** $\sum_{k=1}^{4}3^k$ **81.** $\sum_{k=1}^{5}[(-1)^{k+1} \cdot 2k]$ **82.** $\sum_{k=1}^{8}[(-1)^k \cdot k]$

# 11.2 Arithmetic Sequences

**OBJECTIVES**

1. Determine If a Sequence Is Arithmetic
2. Find a Formula for the $n$th Term an Arithmetic Sequence
3. Find the Sum of an Arithmetic Sequence

### *Preparing for Arithmetic Sequences*

*Before getting started take the following readiness quiz. If you get a problem wrong, go back to the section cited and review the material.*

**1.** Determine the slope of $y = -3x + 1$. [Section 3.2 pp. 205–206]

**2.** If $g(x) = 5x + 2$, find $g(3)$. [Section 2.3, pp. 162–164]

**3.** Solve: $\begin{cases} x - 3y = -17 \\ 2x + y = 1 \end{cases}$ [Section 4.1, pp. 259–264]

In the last section, we looked at sequences in general. Now we look at a specific type of sequence, called an *arithmetic sequence.*

## 1 Determine If a Sequence Is Arithmetic

When the difference between successive terms of a sequence is always the same number, the sequence is called an **arithmetic sequence** (sometimes called an **arithmetic progression**). For example, the sequence

$$2, 6, 10, 14, \ldots$$

is arithmetic because the constant difference between consecutive terms is 4. If we call the first term $a$ and the **common difference** between consecutive terms $d$, then the terms of an arithmetic sequence follow the pattern

$$a, a + d, a + 2d, a + 3d, \ldots$$

### EXAMPLE 1 Determining If a Sequence Is Arithmetic

Determine if the sequence $3, 9, 15, 21, \ldots$ is arithmetic. If it is, determine the first term $a$ and the common difference $d$.

#### Solution

To determine if the sequence is arithmetic, find the difference of consecutive terms. If this difference is constant, the sequence is arithmetic. The sequence $3, 9, 15, 21, \ldots$ is arithmetic because the difference between consecutive terms is 6 ($= 9 - 3$ or $15 - 9$ or $21 - 15$. The first term is $a = 3$ and the common difference is $d = 6$.

**QUICK ✓** *Determine which of the following sequences is arithmetic. If the sequence is arithmetic, determine the first term a and common difference d.*

**1.** $-3, -1, 1, 3, 5, \ldots$ **2.** $3, 9, 27, 81, \ldots$

### EXAMPLE 2 Determining If a Sequence Defined by a Function Is Arithmetic

Show that the sequence $\{s_n\} = \{2n + 7\}$ is arithmetic. Find the first term and the common difference.

#### Solution

We could list out the first few terms and demonstrate that the difference between consecutive terms is the same, but this would be more of a demonstration, not a proof. To prove the sequence is arithmetic, we must show that for *any* consecutive terms, the difference is the same number. We do this by evaluating the sequence at the $(n - 1)$st

***Preparing for...Answers*** **1.** $-3$ **2.** 17 **3.** $(-2, 5)$

term and $n$th term in the sequence and computing the difference. If it is a constant, we've shown the sequence is arithmetic.

$$s_{n-1} = 2(n-1) + 7 = 2n - 2 + 7 = 2n + 5 \quad \text{and} \quad s_n = 2n + 7$$

Now we compute $s_n - s_{n-1}$.

$$s_n - s_{n-1} = (2n + 7) - (2n + 5)$$

Distribute: $= 2n + 7 - 2n - 5$

Combine like terms: $= 2$

The difference between *any* consecutive terms is 2, so the sequence is arithmetic with common difference $d = 2$. To find the first term, we evaluate $s_1$ and find $s_1 = 2(1) + 7 = 9$, so that $a = 9$.

**EXAMPLE 3** **Determining Whether or Not a Sequence Defined by a Function Is Arithmetic**

Show that the sequence $\{b_n\} = \{n^2\}$ is not arithmetic.

### Solution

As in Example 2, we determine the $(n-1)$st term and the $n$th term. We then show that the difference between consecutive terms is not a constant.

$$b_{n-1} = (n-1)^2 = n^2 - 2n + 1 \qquad \text{and} \qquad b_n = n^2$$

Now we compute $b_n - b_{n-1}$.

$$b_n - b_{n-1} = n^2 - (n^2 - 2n + 1)$$

Distribute: $= n^2 - n^2 + 2n - 1$

Combine like terms: $= 2n - 1$

The difference between consecutive terms is not constant—its value depends upon $n$. Therefore, the sequence is not arithmetic.

**QUICK ✓** *Determine if the sequence is arithmetic. If it is, state the first term a and the common difference d.*

**3.** $\{a_n\} = \{3n - 8\}$ **4.** $\{b_n\} = \{n^2 - 1\}$ **5.** $\{c_n\} = \{5 - 2n\}$

## 2 Find a Formula for the *n*th Term of an Arithmetic Sequence

Suppose that $a$ is the first term of an arithmetic sequence whose common difference is $d$. We want to find a formula for $a_n$, the $n$th term of the sequence. To do this, we write down the first few terms of the sequence.

$$a_1 = a$$
$$a_2 = a_1 + d = a + d$$
$$a_3 = a_2 + d = (a + d) + d = a + 2d$$
$$a_4 = a_3 + d = (a + 2d) + d = a + 3d$$
$$a_5 = a_4 + d = (a + 3d) + d = a + 4d$$
$$\vdots$$
$$a_n = a_{n-1} + d = [a + (n-2)d] + d = a + (n-1)d$$

We are led to the following result.

**THE *n*TH TERM OF AN ARITHMETIC SEQUENCE**

For an arithmetic sequence $\{a_n\}$ whose first term is $a$ and whose common difference is $d$, the $n$th term is determined by the formula

$$a_n = a + (n-1)d$$

## EXAMPLE 4 Finding a Formula for the *n*th Term of an Arithmetic Sequence

**(a)** Write a formula for the $n$th term of an arithmetic sequence whose fourth term is 8 and whose common difference is $-3$.
**(b)** Find the 14th term of the sequence.

### Solution

**(a)** We wish to find a formula for the $n$th term of an arithmetic sequence. We know that $d = -3$ and that $a_4 = 8$.

$$a_n = a + (n - 1)d$$

$d = -3;\ a_4 = 8$: $\quad a_4 = 8 = a + (4 - 1)(-3)$

We solve this equation for $a$, the first term of the sequence.

$$8 = a - 9$$

Add 9 to both sides: $\quad 17 = a$

So the formula for the $n$th term is

$a_n = a + (n - 1)d$: $\quad a_n = 17 + (n - 1)(-3)$

Distribute: $\quad = 17 - 3n + 3$

Combine like terms: $\quad = -3n + 20$

**(b)** To find the fourteenth term, we let $n = 14$ in $a_n = -3n + 20$.

$$a_{14} = -3(14) + 20$$
$$= -42 + 20$$
$$= -22$$

The fourteenth term in the sequence is $-22$.

**QUICK ✓**

**6. (a)** Write a formula for the $n$th term of an arithmetic sequence whose fifth term is 25 and whose common difference is 6.

**(b)** Find the 14th term of the sequence.

---

## EXAMPLE 5 Finding a Formula for the *n*th Term of an Arithmetic Sequence

The fourth term of an arithmetic sequence is 7, and the tenth term is 31.

**(a)** Find the first term and the common difference.
**(b)** Give a formula for the $n$th term of the sequence.

### Solution

**(a)** We know that the $n$th term of an arithmetic sequence is $a_n = a + (n - 1)d$, where $a$ is the first term and $d$ is the common difference. Since $a_4 = 7$ and $a_{10} = 31$, we have

$$\begin{cases} a_4 = a + (4 - 1)d \\ a_{10} = a + (10 - 1)d \end{cases} \quad \text{or} \quad \begin{cases} 7 = a + 3d & (1) \\ 31 = a + 9d & (2) \end{cases}$$

This is a system of two linear equations containing two variables, $a$ and $d$. We can solve this system by elimination. If we subtract equation (2) from equation (1) we obtain

$$-24 = -6d$$

Divide both sides by $-6$: $\quad 4 = d$

Let $d = 4$ in equation (1) to find $a$.

$$7 = a + 3(4)$$
$$7 = a + 12$$
Subtract 12 from both sides: $$-5 = a$$

The first term is $a = -5$ and the common difference is $d = 4$.

**(b)** A formula for the $n$th term is

$$a_n = a + (n - 1)d$$
$a = -5;\ d = 4$: $$= -5 + (n - 1)(4)$$
Distribute: $$= -5 + 4n - 4$$
Combine like terms: $$= 4n - 9$$

**QUICK ✓**

**7.** The fifth term of an arithmetic sequence is 7, and the thirteenth term is 31.

**(a)** Find the first term and the common difference.

**(b)** Give a formula for the $n$th term of the sequence.

## 3 Find the Sum of an Arithmetic Sequence

The next result gives a formula for finding the sum of the first $n$ terms of an arithmetic sequence.

**SUM OF $n$ TERMS OF AN ARITHMETIC SEQUENCE**

Let $\{a_n\}$ be an arithmetic sequence with first term $a$ and common difference $d$. The sum $S_n$ of the first $n$ terms of $\{a_n\}$ is

$$S_n = \frac{n}{2}[2a + (n - 1)d] \quad \text{or} \quad S_n = \frac{n}{2}(a + a_n)$$

We show where these results come from next.

$$S_n = a_1 + a_2 + a_3 + \cdots + a_n$$
$$= \underbrace{a}_{a_1} + \underbrace{(a + d)}_{a_2} + \underbrace{(a + 2d)}_{a_3} + \cdots + \underbrace{(a + (n - 1)d)}_{a_n}$$

We can also represent $S_n$ by reversing the order in which we add the terms, so that

$$S_n = a_n + a_{n-1} + \cdots + a_1$$
$$= \underbrace{(a + (n - 1)d)}_{a_n} + \underbrace{(a + (n - 2)d)}_{a_{n-1}} + \cdots + \underbrace{a}_{a_1}$$

Add these two different representations of $S_n$ as follows:

$$\begin{array}{l} S_n = a + (a + d) + (a + 2d) + \cdots + (a + (n - 2)d) + (a + (n - 1)d) \\ S_n = (a + (n - 1)d) + (a + (n - 2)d) + (a + (n - 3)d) + \cdots + (a + d) + a \\ \hline 2S_n = 2a + (n - 1)d + 2a + (n - 1)d + 2a + (n - 1)d + \cdots + 2a + (n - 1)d + 2a + (n - 1)d \end{array}$$

On the right side of the equation, we are adding $2a + (n - 1)d$ to itself $n$ times. This sum can be represented as $n[2a + (n - 1)d]$, so we obtain

$$2S_n = n[2a + (n - 1)d]$$

Divide both sides by 2: $$S_n = \frac{n}{2}[2a + (n - 1)d] \quad \text{Formula (1)}$$

This is one of the formulas listed in the box. If we rewrite the expression $2a + (n - 1)d$ as $a + a + (n - 1)d$, we notice that $a + (n - 1)d$ is $a_n$, so that

$$S_n = \frac{n}{2}[a + a_n] \quad \text{Formula (2)}$$

So we have two ways to find the sum of the first $n$ terms of an arithmetic sequence. Notice that Formula (1) involves the first term $a$ and common difference $d$, while Formula (2) involves the first term $a$ and the last term $a_n$. You should use whichever is easier.

### EXAMPLE 6 Finding the Sum of *n* Terms of an Arithmetic Sequence

Find the sum $S_n$ of the first 50 terms of the arithmetic sequence $2, 5, 8, 11, \ldots$.

**Solution**

Because we know the first term is $a = 2$ and the common difference is $d = 5 - 2 = 3$, we use the formula $S_n = \frac{n}{2}[2a + (n - 1)d]$ to find the sum.

**Work Smart**

There are two formulas for finding the sum of the first $n$ terms of an arithmetic sequence. Use Formula (1) if you know $n$, the first term $a$ and the common difference $d$; use Formula (2) if you know $n$, the first term $a$, and the last term $a_n$.

$$S_n = \frac{n}{2}[2a + (n - 1)d]$$

$$n = 50, a = 2, d = 3: \quad S_{50} = \frac{50}{2}[2(2) + (50 - 1)(3)]$$

$$= 25[4 + 49(3)]$$

$$= 25[151]$$

$$= 3{,}775$$

The sum of the first 50 terms, $S_{50}$, of the arithmetic sequence $2, 5, 8, 11, \ldots$ is 3,775.

**QUICK ✓**

8. Find the sum $S_n$ of the first 100 terms of the arithmetic sequence whose first term is 5 and whose common difference is 2.
9. Find the sum $S_n$ of the first 70 terms of the arithmetic sequence $1, 5, 9, 13, \ldots$.

### EXAMPLE 7 Finding the Sum of *n* Terms of an Arithmetic Sequence

Find the sum $S_n$ of the first 40 terms of the arithmetic sequence $\{-2n + 50\}$.

**Solution**

We find that the first term is $a_1 = -2(1) + 50 = 48$ and the $40^{\text{th}}$ term is $a_{40} = -2(40) + 50 = -30$. Since we know the first term and the last term of the sequence we use the formula $S_n = \frac{n}{2}[a + a_n]$ to find the sum.

$$S_n = \frac{n}{2}[a + a_n]$$

$$n = 40, a = 48, a_{40} = -30: \quad S_{40} = \frac{40}{2}[48 + (-30)]$$

$$= 360$$

The sum of the first 40 terms, $S_{40}$, of the arithmetic sequence $\{-2n + 50\}$ is 360.

**QUICK ✓**

**10.** Find the sum $S_n$ of the first 50 terms of the arithmetic sequence whose first term is 4 and fiftieth term 298.

**11.** Find the sum $S_n$ of the first 75 terms of the arithmetic sequence $\{-3n + 100\}$.

### EXAMPLE 8 Creating a Floor Design

A ceramic tile floor is designed in the shape of a trapezoid 10 feet wide at the base and 5 feet wide at the top. See Figure 5. The tiles, 6 inches by 6 inches, are to be placed so that each successive row contains one fewer tile than the preceding row. How many tiles will be required?

**Figure 5**

**Solution**

Each tile is 6 inches (0.5 foot) and the bottom row is 10 feet wide, so the bottom row requires 20 tiles. Similar logic tells us the top row requires 10 tiles. Since each successive row has one fewer tile, the total number of tiles required is

$$S = 20 + 19 + 18 + \cdots + 11 + 10$$

This is the sum of an arithmetic sequence; the common difference is $-1$. The number of terms to be added is $n = 11$, with the first term $a = 20$ and the last terms $a_{11} = 10$. The sum $S$ is

$$S_n = \frac{n}{2}[a + a_n] \qquad S_{11} = \frac{11}{2}(20 + 10) = 165$$

In all, 165 tiles will be required.

**QUICK ✓**

**12.** In the corner section of a theater, the first row has 20 seats. Each subsequent row has 2 more seats, and there are a total of 30 rows. How many seats are in this section?

# 11.2 Exercises

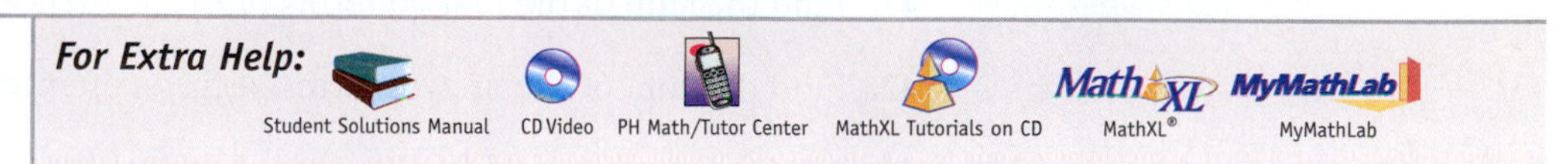


## Concepts and Vocabulary

*In Problems 1 and 2, fill in the blanks.*

**1.** In a(n) _______ sequence, the difference between consecutive terms is a constant.

**2.** For an arithmetic sequence $\{a_n\}$ whose first term is $a$ and whose common difference is $d$, the $n$th term is determined by the formula ___________.

*In Problems 3 and 4, answer True or False to each statement.*

**3.** In an arithmetic sequence, the sum of the first and last terms equals twice the sum of all the terms.

**4.** In an arithmetic sequence, the difference between the first and the last term is the common difference.

**5.** Explain how you can determine if a sequence is arithmetic.

**6.** Provide an explanation that justifies the formula for the $n$th term of an arithmetic sequence.

## Building Skills

*In Problems 7–14, an arithmetic sequence is given. Find the common difference, and write out the first four terms.*

**7.** $\{n + 5\}$ **8.** $\{n - 1\}$ **9.** $\{7n + 2\}$ **10.** $\{10n + 1\}$

**11.** $\{7 - 3n\}$ **12.** $\{5 - 2n\}$ **13.** $\left\{\frac{1}{2}n + 5\right\}$ **14.** $\left\{\frac{1}{4}n + \frac{3}{4}\right\}$

*In Problems 15–22, find a formula for the nth term of the arithmetic sequence whose first term a and common difference d are given. What is the fifth term?*

**15.** $a = 4, d = 3$ **16.** $a = 8, d = 3$ **17.** $a = 10, d = -5$ **18.** $a = 12, d = -3$

**19.** $a = 2; d = \frac{1}{3}$ **20.** $a = -3; d = \frac{1}{2}$ **21.** $a = 5; d = -\frac{1}{5}$ **22.** $a = -\frac{4}{3}; d = -\frac{2}{3}$

*In Problems 23–28, write a formula for the nth term of each arithmetic sequence. Use the formula to find the 20th term in each arithmetic sequence.*

**23.** $2, 7, 12, 17, \ldots$ **24.** $-5, -1, 3, 7, \ldots$ **25.** $12, 9, 6, 3, \ldots$

**26.** $20, 14, 8, 2, \ldots$ **27.** $1, \frac{5}{4}, \frac{3}{2}, \frac{7}{4}, \ldots$ **28.** $10, \frac{19}{2}, 9, \frac{17}{2}, \ldots$

*In Problems 29–36, find the first term and the common difference of the arithmetic sequence described. Give a formula for the nth term of the sequence.*

**29.** 3rd term is 17; 7th term is 37 **30.** 5th term is 7; 9th term is 19

**31.** 4th term is $-2$; 8th term is 26 **32.** 2nd term is $-9$; 8th term is 15

**33.** 5th term is $-1$; 12th term is $-22$ **34.** 6th term is $-8$; 12th term is $-38$

**35.** 3rd term is 3; 9th term is 0 **36.** 5th term is 5; 13th term is 7

**37.** Find the sum of the first 30 terms of the sequence $2, 8, 14, 20, \ldots$.

**38.** Find the sum of the first 40 terms of the sequence $1, 8, 15, 22, \ldots$.

**39.** Find the sum of the first 25 terms of the sequence $-8, -5, -2, 1, \ldots$.

**40.** Find the sum of the first 75 terms of the sequence $-9, -5, -1, 3, \ldots$.

**41.** Find the sum of the first 40 terms of the sequence $10, 3, -4, -11, \ldots$.

**42.** Find the sum of the first 50 terms of the sequence $12, 4, -4, -12, \ldots$.

**43.** Find the sum of the first 40 terms of the arithmetic sequence $\{4n - 3\}$.

**44.** Find the sum of the first 80 terms of the arithmetic sequence $\{2n - 13\}$.

**45.** Find the sum of the first 75 terms of the arithmetic sequence $\{-5n + 70\}$.

**46.** Find the sum of the first 35 terms of the arithmetic sequence $\{-6n + 25\}$.

**47.** Find the sum of the first 30 terms of the arithmetic sequence $\left\{5 + \frac{2}{3}n\right\}$.

**48.** Find the sum of the first 28 terms of the arithmetic sequence $\left\{7 - \frac{3}{2}n\right\}$.

## Applying the Concepts

**49.** Find $x$ so that $x + 3$, $2x + 1$, and $5x + 2$ are consecutive terms of an arithmetic sequence.

**50.** Find $x$ so that $2x$, $3x + 2$, and $5x + 3$ are consecutive terms of an arithmetic sequence.

**51. A Stack of Cans** Suppose that the bottom row in a stack of cans contains 35 cans. Each layer contains one fewer can than the layer below it. The top row has 1 can. How many cans are in the stack?

**52. A Pile of Bricks** Suppose that the bottom row in a pile of bricks contains 46 bricks. Each layer contains two fewer bricks than the layer below it. The top row has 2 bricks. How many bricks are in the stack?

**53. The Theater** An auditorium has 40 seats in the first row and 25 rows in all. Each successive row contains 2 additional seats. How many seats are in the auditorium?

**54. Mosaic** A mosaic is designed in the shape of an equilateral triangle, 20 feet on each side. Each tile in the mosaic is in the shape of an equilateral triangle, 12 inches to a side. The tiles are to alternate in color as shown in the illustration. How many tiles of each color will be required?

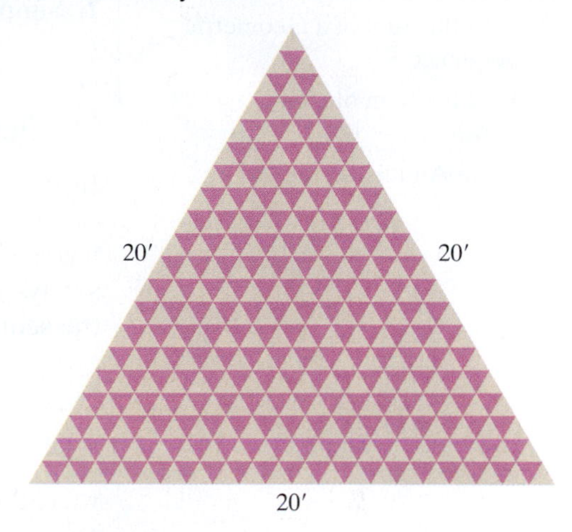


## Extending the Concepts

*In Problems 55–58, determine the number of terms that are in each arithmetic sequence.*

**55.** $-5, -2, 1, \ldots, 244$

**56.** $-9, -5, -1, \ldots, 219$

**57.** $108, 101, 94, \ldots, -326$

**58.** $99, 93, 87, \ldots, -339$

**59. Salary** Suppose you have just been hired at the starting salary of \$32,000 per year. Your contract guarantees you a \$2500 raise each year. How many years will it take before your aggregate salary is \$757,500? (**Hint:** Your aggregate salary is \$32,000 + (32,000 + \$2500) + ....)

**60. Stadium Construction** How many rows are in the corner section of a stadium containing 2030 seats if the first row has 14 seats and each successive row has 4 additional seats?

## Synthesis Review

*In Problems 61–63, (a) determine the base a of the exponential function, (b) Compute $f(1)$, $f(2)$, $f(3)$, and $f(4)$.*

**61.** $f(x) = 3^x$

**62.** $f(x) = 4^x$

**63.** $f(x) = 10\left(\frac{1}{2}\right)^x$

**64.** Compute $\frac{f(2)}{f(1)}, \frac{f(3)}{f(2)}$, and $\frac{f(4)}{f(3)}$ for each of the functions in Problems 61–63. How does the ratio in the value of the function for consecutive values of the independent variable relate to the base?

### The Graphing Calculator

*In Problems 65–68, use a graphing calculator to find the sum of each sequence.*

**65.** $\{3.45n + 4.12\}; n = 20$

**66.** $\{2.67n - 1.23\}; n = 25$

**67.** $85.9 + 83.5 + 81.1 + \cdots; n = 25$

**68.** $-11.8 + (-8.2) + (-4.6) + \cdots; n = 30$

# 11.3 Geometric Sequences and Series

**OBJECTIVES**

1. Determine If a Sequence Is Geometric
2. Find a Formula for the $n$th Term of a Geometric Sequence
3. Find the Sum of a Geometric Sequence
4. Find the Sum of a Geometric Series
5. Solve Annuity Problems

## *Preparing for Geometric Sequences and Series*

*Before getting started, take the following readiness quiz. If you get a problem wrong, go back to the section cited and review the material.*

**1.** If $g(x) = 4^x$, evaluate $g(1), g(2)$, and $g(3)$. [Section 9.1, pp. 713–714]

**2.** Simplify $\frac{x^4}{x^3}$. [Getting Ready, p. 340]

## 1 Determine If a Sequence Is Geometric

In the last section, we defined a sequence to be arithmetic if the difference between consecutive terms in the sequence was a constant. But what if the ratio of consecutive terms in the sequence is a constant? If the ratio of successive terms of a sequence is always the same number, the sequence is called a **geometric sequence.** For example, the sequence

$$2, 4, 8, 16, \ldots$$

is geometric because the constant ratio of consecutive terms is $2\left(=\frac{4}{2}=\frac{8}{4}=\frac{16}{8}\right)$. If we call the first term $a$ and the **common ratio** of consecutive terms $r$, then the terms of a geometric sequence follow the pattern

$$a, ar, ar^2, ar^3, \ldots$$

### EXAMPLE 1 Determining If a Sequence Is Geometric

Determine if the sequence 2, 6, 18, 54, 162, ... is geometric. If it is geometric, state the first term and common ratio.

#### Solution

The sequence is geometric if the ratio of consecutive terms is a constant. Because the ratio of consecutive terms is $3\left(=\frac{6}{2}=\frac{18}{6}=\frac{54}{18}=\frac{162}{54}\right)$, the sequence is geometric. The first term is $a = 2$ and the common ratio is $r = 3$.

**QUICK ✓** *Determine if the sequence is geometric. If it is, state the first term and common ratio.*

**1.** $4, 8, 16, 32, 64, \ldots$

**2.** $5, 10, 16, 23, 30, \ldots$

**3.** $9, 3, 1, \frac{1}{3}, \frac{1}{9}, \ldots$

*Preparing for...Answers*
**1.** $g(1) = 4, g(2) = 16, g(3) = 64$
**2.** $x$

## EXAMPLE 2 Determining If a Sequence Is Geometric

Determine if the sequence $\{b_n\} = \{3^{-n}\}$ is geometric. If it is geometric, find the common ratio.

### Solution

We could list out the first few terms and show that the ratio of consecutive terms is the same, but this would be more of a demonstration, not a proof. To prove the sequence is geometric, we must show that for *any* consecutive terms, the ratio is the same number. We do this by evaluating the sequence at the $(n - 1)$st term and $n$th term in the sequence and computing the ratio. If it is a constant, we've shown the sequence is geometric.

**Work Smart**

Notice the technique used in Example 2 is similar to the approach used in Section 11.2 to show a sequence in arithmetic.

$$b_{n-1} = 3^{-(n-1)} = 3^{-n+1} \quad \text{and} \quad b_n = 3^{-n}$$

Now we compute $\dfrac{b_n}{b_{n-1}}$.

$$\frac{b_n}{b_{n-1}} = \frac{3^{-n}}{3^{-n+1}}$$

$a^{m+n} = a^m \cdot a^n$: $$= \frac{3^{-n}}{3^{-n} \cdot 3^1}$$

Divide out like factors: $$= \frac{1}{3}$$

The ratio of any two consecutive terms is $\frac{1}{3}$, so the common ratio is $\frac{1}{3}$. The sequence is geometric with $r = \frac{1}{3}$.

## EXAMPLE 3 Determining If a Sequence Is Geometric

Determine if the sequence $\{a_n\} = \{n^2 + 1\}$ is geometric. If it is geometric, find the common ratio.

### Solution

We proceed as we did in Example 2 and compute $\dfrac{a_n}{a_{n-1}}$. If this ratio is a constant, then the sequence is geometric.

$$\frac{a_n}{a_{n-1}} = \frac{n^2 + 1}{(n-1)^2 + 1}$$

FOIL: $$= \frac{n^2 + 1}{n^2 - 2n + 1 + 1}$$

Combine like terms: $$= \frac{n^2 + 1}{n^2 - 2n + 2}$$

We cannot simplify the rational expression any further. Because the ratio $\dfrac{a_n}{a_{n-1}}$ depends on the value of $n$, it is not a constant. Therefore, the sequence is not geometric.

**QUICK ✓** *Determine if the sequence is geometric. If it is geometric, find the common ratio.*

**4.** $\{a_n\} = \{5^n\}$ **5.** $\{b_n\} = \{n^2\}$ **6.** $\{c_n\} = \left\{5\left(\frac{2}{3}\right)^n\right\}$

## 2 Find a Formula for the *n*th Term of a Geometric Sequence

Suppose that $a$ is the first term of a geometric sequence with common ratio $r \neq 0$. We seek a formula for the $n$th term of $a_n$. To see the pattern, we write down the first few terms.

$$\begin{aligned} a_1 &= 1a = ar^0 \\ a_2 &= ra_1 = ar^1 \\ a_3 &= ra_2 = ar^2 \\ a_4 &= ra_3 = ar^3 \\ a_5 &= ra_4 = ar^4 \\ &\vdots \\ a_n &= ra_{n-1} = r(ar^{n-2}) = ar^{n-1} \end{aligned}$$

We are led to the following result.

**THE *n*TH TERM OF A GEOMETRIC SEQUENCE**

For a geometric sequence $\{a_n\}$ whose first term is $a$ and whose common ratio is $r$, the $n$th term is determined by the formula

$$a_n = ar^{n-1}, \qquad r \neq 0$$

### EXAMPLE 4 Finding a Particular Term of a Geometric Sequence

For the geometric sequence $8, 6, \frac{9}{2}, \frac{27}{8}, \ldots$

**(a)** Find a formula for the $n$th term.

**(b)** Find the eleventh term of the sequence.

**Solution**

**(a)** We are told that the sequence is geometric. The first term is $a = 8$. To find the common ratio, we compute the ratio of consecutive terms. So $r = \frac{6}{8} = \frac{9/2}{6} = \frac{27/8}{9/2} = \frac{3}{4}$. We now substitute these values into the formula for the $n$th term of a geometric sequence.

$$\begin{aligned} a_n &= ar^{n-1} \\ &= 8\left(\frac{3}{4}\right)^{n-1} \end{aligned}$$

**(b)** The eleventh term is found by letting $n = 11$ in the formula found in part (a).

$$\begin{aligned} a_n &= 8\left(\frac{3}{4}\right)^{n-1} \\ a_{11} &= 8\left(\frac{3}{4}\right)^{11-1} \\ &= 8\left(\frac{3}{4}\right)^{10} \\ &\approx 0.45051 \end{aligned}$$

**QUICK ✓** *Find a formula for the nth term of each geometric sequence. Use this result to find the ninth term of the sequence.*

**7.** $a = 5, r = 2$

**8.** $\{50, 25, 12.5, 6.25, \ldots\}$

## 3 Find the Sum of a Geometric Sequence

The next result gives us a formula for finding the sum of the first $n$ terms of a geometric sequence.

**SUM OF $n$ TERMS OF A GEOMETRIC SEQUENCE**

Let $\{a_n\}$ be a geometric sequence with first term $a$ and common ratio $r$, where $r \neq 0, r \neq 1$. The sum $S_n$ of the first $n$ terms of $\{a_n\}$ is

$$S_n = a \cdot \frac{1 - r^n}{1 - r}, \qquad r \neq 0, r \neq 1$$

Where does this formula come from? Well, the sum $S_n$ of the first $n$ terms of $\{a_n\} = \{ar^{n-1}\}$ is

$$S_n = a + ar + ar^2 + \cdots + ar^{n-1}$$

Let's multiply both sides by $r$ to obtain

$$rS_n = ar + ar^2 + ar^3 + \cdots + ar^n$$

Now subtract $rS_n$ from $S_n$:

$$\begin{aligned} S_n &= a + ar + ar^2 + \cdots + ar^{n-1} \\ rS_n &= \phantom{a + {}} ar + ar^2 + ar^3 + \cdots + ar^n \\ \hline S_n - rS_n &= a - ar^n \end{aligned}$$

Factor $S_n$ from the expression on the left and factor $a$ from the expression on the right.

$$S_n(1 - r) = a(1 - r^n)$$

Divide both sides by $1 - r$ (since $r \neq 1$) and solve for $S_n$.

$$S_n = a \cdot \frac{1 - r^n}{1 - r}$$

### EXAMPLE 5 Finding the Sum of $n$ Terms of a Geometric Sequence

Find the sum of the first 10 terms of the sequence 2, 6, 18, 54, ....

#### Solution

This is a geometric sequence with $a = 2$ and common ratio $r = 3$ (Do you see why?). We wish to know the sum of the first 10 terms, $S_{10}$.

$$S_n = a \cdot \frac{1 - r^n}{1 - r}$$

$a = 2, r = 3, n = 10$: $$S_{10} = 2 \cdot \frac{1 - 3^{10}}{1 - 3}$$

$$= 2 \cdot \frac{-59{,}048}{-2}$$

$$= 59{,}048$$

### EXAMPLE 6 Finding the Sum of $n$ Terms of a Geometric Sequence in Summation Notation

Find the sum: $\sum_{n=1}^{8} \left[5 \cdot \left(\frac{1}{2}\right)^n\right]$

Express your answer to as many decimals as your calculator allows.

**Solution**

We can write this sum out as follows:

$$\sum_{n=1}^{8}\left[5\cdot\left(\frac{1}{2}\right)^n\right] = \frac{5}{2} + \frac{5}{4} + \cdots + \frac{5}{256}$$

We wish to find the sum of the first $n = 8$ terms of a geometric sequence with $a = \frac{5}{2}$ and common ratio $r = \frac{1}{2}$.

$$S_n = a\cdot\frac{1 - r^n}{1 - r}$$

$a = \frac{5}{2}, r = 1/2, n = 8$: $$S_8 = \frac{5}{2}\cdot\frac{1 - \left(\frac{1}{2}\right)^8}{1 - \frac{1}{2}}$$

$$= \frac{5}{2}\cdot\frac{0.99609375}{\frac{1}{2}}$$

$$= 4.98046875$$

**QUICK**  *Find the sum.*

**9.** $3 + 6 + 12 + 24 + \cdots + 3\cdot 2^{12}$

**10.** $\sum_{n=1}^{10}\left[8\cdot\left(\frac{1}{2}\right)^n\right]$

## 4 Find the Sum of a Geometric Series

**In Words**
A sequence is a list of terms. A series is the sum of an infinite number of terms.

An infinite sum of the form

$$a + ar + ar^2 + \cdots + ar^{n-1} + \cdots$$

whose first term is $a$ and common ratio is $r$, is called a **geometric series** and is denoted by

$$\sum_{n=1}^{\infty} ar^{n-1}$$

We know that the sum of the first $n$ terms of a geometric sequence is given by the formula

$$S_n = a\cdot\frac{1 - r^n}{1 - r}$$

We can write this formula as $S_n = \frac{a - ar^n}{1 - r}$ by distributing the $a$. Now if we divide $1 - r$ into each term in the numerator, we obtain

$$S_n = \frac{a}{1 - r} - \frac{ar^n}{1 - r}$$

As $n$ gets larger and larger, the expression $S_n$ will approach the value $\frac{a}{1 - r}$ because $r^n$ approaches 0 provided that $-1 < r < 1$ and we have the following result.

**SUM OF A GEOMETRIC SERIES**

If $-1 < r < 1$, the sum of the geometric series $\sum_{n=1}^{\infty} ar^{n-1}$ is

$$\sum_{n=1}^{\infty} ar^{n-1} = \frac{a}{1 - r}$$

## EXAMPLE 7 Finding the Sum of an Geometric Series

Find the sum of the geometric series: $4 + 2 + 1 + \frac{1}{2} + \cdots$.

### Solution

This is an geometric series with $a = 4$ and common ratio $r = \frac{1}{2}\left(= \frac{2}{4} = \frac{1}{2} = \frac{1/2}{1}\right)$. Since the common ratio $r$ is between $-1$ and 1, we can use the formula for the sum of a geometric series to find that

$$4 + 2 + 1 + \frac{1}{2} + \cdots = \frac{4}{1 - \frac{1}{2}} = 8$$

**QUICK ✓** *Find the sum of the geometric series.*

**11.** $10 + \frac{5}{2} + \frac{5}{8} + \frac{5}{32} + \cdots$

**12.** $\sum_{n=1}^{\infty} \left(\frac{1}{3}\right)^n$

---

**Work Smart: Study Skills**

Let's summarize the formulas for arithmetic and geometric sequences where $a$ is the first term of the sequence, $d$ is the common difference, and $r$ is the common ratio.

| | Arithmetic | Geometric |
|---|---|---|
| Find the *n*th term | $a_n = a + (n - 1)d$ | $a_n = ar^{n-1}, r \neq 0$ |
| Find the sum of the first $n$ terms | $S_n = \frac{n}{2}[2a + (n - 1)d]$ $= \frac{n}{2}(a + a_n)$ | $S_n = a \cdot \frac{1 - r^n}{1 - r}, r \neq 0, r \neq 1$ |

If you were asked to find the 5th term of the sequence 5, 20, 80, 320, ..., which formula would you use?

If you were asked to find the sum of the first 5 terms of the sequence 5, 9, 13, 17, ..., which formula would you use?

## EXAMPLE 8 Writing a Repeating Decimal as a Fraction

Express $0.\overline{1}$ as a fraction in lowest terms.

### Solution

The line over the 1 in the decimal indicates that the 1 repeats indefinitely. That is, we can write

$$\begin{aligned} 0.\overline{1} &= 0.11111\cdots \\ &= 0.1 + 0.01 + 0.001 + \cdots \end{aligned}$$

This is a geometric series with $a = 0.1$ and common ratio $r = 0.1\left(= \frac{0.01}{0.1} = \frac{0.001}{0.01}\right)$. Since the common ratio $r$ is between $-1$ and 1, we can use the formula for the sum of a geometric series to find that

$$\begin{aligned} 0.\overline{1} &= 0.1 + 0.01 + 0.001 + \cdots \\ &= \frac{0.1}{1 \quad 0.1} \\ &= \frac{0.1}{0.9} \\ \text{Multiply numerator and denominator by 10:} \quad &= \frac{1}{9} \end{aligned}$$

So $0.\overline{1} = \frac{1}{9}$.

**QUICK ✓**

**13.** Express $0.\overline{2}$ as a fraction in lowest terms.

---

### EXAMPLE 9 The Multiplier

Suppose that, throughout the United States economy, individuals spend 90% of every additional dollar they earn. Economists would say that an individual's **marginal propensity to consume** is 0.90. For example, if Roberta earns an additional dollar, she will spend $0.9(\$1) = \$0.90$ of it and save \$0.10. Whoever earns \$0.90 (from Roberta) will spend 90% of it or $0.9(\$0.90) = \$0.81$. This process of spending continues and results in a geometric series as follows:

$$\$1 + \$0.90 + \$0.81 + \$0.729 + \cdots$$

The sum of this geometric series is called the **multiplier.** Suppose that the government gives a child-tax rebate of \$500 to Roberta. What is the multiplier if individuals spend 90% of every additional dollar they earn?

#### Solution

The total impact of the \$500 tax rebate on the U.S. economy is

$$\$500 + \$500(0.9) + \$500(0.9)^2 + \$500(0.9)^3 + \cdots$$

This is a geometric series with first term $a = 500$ and common ratio $r = 0.9$. The sum of this series is

$$\begin{aligned}\$500 + \$500(0.9) + \$500(0.9)^2 + \$500(0.9)^3 + \cdots &= \frac{\$500}{1 - 0.9} \\ &= \$5000\end{aligned}$$

The United States economy will grow by \$5000 because of the child-tax credit to Roberta.

**QUICK ✓**

**14.** Redo Example 9 if the marginal propensity to consume is 95%.

---

## 5 Solve Annuity Problems

In Section 9.1, we looked at the compound interest formula, which allows us to compute the future value of a lump sum of money that is deposited in an account that pays interest compounded periodically (say, monthly). Often, though, money is invested at periodic intervals of time. An **annuity** is a sequence of equal periodic deposits. The periodic deposits may be made annually, quarterly, monthly, or daily.

When deposits are made at the same time the interest is credited, the annuity is called **ordinary.** We will only deal with ordinary annuities here. The **amount of an annuity** is the sum of all deposits made plus all interest paid.

Suppose the interest an account earns is $i$ percent per payment period (expressed as a decimal). For example, if an account pays 6% compounded monthly (12 times a year), then $i = \frac{0.06}{12} = 0.005$. To develop a formula for the amount of an annuity, suppose $\$P$ is deposited each payment period for $n$ payment periods in an account that earns $i$% per payment period. When the last deposit is made at the $n$th payment period, the first deposit has earned interest compounded for $n - 1$ payment periods, the second deposit of $\$P$ has earned interest compounded for $n - 2$ payment periods, and so on. Table 1 shows the value of each deposit after $n$ deposits have been made.

**Table 1**

| Deposit | 1 | 2 | 3 | $n-1$ | $n$ |
|---|---|---|---|---|---|
| **Future Value of Deposit of $P** | $P(1+i)^{n-1}$ | $P(1+i)^{n-2}$ | $P(1+i)^{n-3}$ | $P(1+i)$ | $P$ |

The amount $A$ of the annuity is the sum of the amounts shown in Table 1, namely,

$$\begin{aligned} A &= P(1+i)^{n-1} + P(1+i)^{n-2} + P(1+i)^{n-3} + \cdots + P(1+i) + P \\ &= P[(1+i)^{n-1} + (1+i)^{n-2} + (1+i)^{n-3} + \cdots + (1+i) + 1] \\ &= P[1 + (1+i) + \cdots + (1+i)^{n-3} + (1+i)^{n-2} + (1+i)^{n-1}] \end{aligned}$$

The expression in brackets is the sum of a geometric sequence with $n$ terms and a common ratio of $(1+i)$. As a result,

$$\begin{aligned} A &= P[1 + (1+i) + \cdots + (1+i)^{n-3} + (1+i)^{n-2} + (1+i)^{n-1}] \\ &= P \cdot \frac{1-(1+i)^n}{1-(1+i)} = P \cdot \frac{1-(1+i)^n}{-i} = P \cdot \frac{(1+i)^n - 1}{i} \end{aligned}$$

We have the following result.

**AMOUNT OF AN ANNUITY**

If $P$ represents the deposit in dollars made at each payment period for an annuity at $i$ percent interest per payment period, the amount $A$ of the annuity after $n$ payment periods is

$$A = P \cdot \frac{(1+i)^n - 1}{i}$$

## EXAMPLE 10 Determining the Amount of an Annuity

To save for retirement, Alejandro decides to place $100 into a Roth Individual Retirement Account (IRA) every month for the next 30 years. What will be the value of the IRA after 30 years assuming that his account earns 6% interest compounded monthly?

### Solution

This is an ordinary annuity with $n = 12 \cdot 30 = 360$ payments with deposits of $P = \$100$. The rate of interest per payment period is $i = \frac{0.06}{12} = 0.005$. The amount of 30 years (360 deposits) is

$$\begin{aligned} A &= \$100\left[\frac{(1+0.005)^{360} - 1}{0.005}\right] \\ &= \$100[1004.515042] \\ &= \$100{,}451.50 \end{aligned}$$

**QUICK ✓**

15. To save for retirement, Magglio decides to place $500 into a Roth Individual Retirement Account (IRA) every quarter (every three months) for the next 30 years. What will be the value of the IRA after 30 years assuming that his account earns 8% interest compounded quarterly?

# 11.3 Exercises

**For Extra Help:** Student Solutions Manual  CD Video  PH Math/Tutor Center  MathXL Tutorials on CD  MathXL®  MyMathLab

## Concepts and Vocabulary

*In Problems 1 and 2, fill in the blanks.*

**1.** In a(n) _________ sequence the ratio of successive terms is constant.

**2.** If $-1 < r < 1$, the sum of the geometric series $\sum_{n=1}^{\infty} ar^{n-1} =$ _________.

*In Problems 3 and 4, answer True or False to each statement.*

**3.** In a geometric sequence, the common ratio is always a positive number.

**4.** For a geometric sequence with first term $a$ and common ratio $r$, where $r \neq 0, r \neq 1$, the sum of the first $n$ terms is $S_n = a \cdot \frac{1 - r^n}{1 - r}$.

**5.** How do you determine if a sequence is geometric?

**6.** How do you determine if a geometric series has a sum?

## Building Skills

*In Problems 7–14, a geometric sequence is given. Find the common ratio and write out the first four terms.*

**7.** $\{4^n\}$ **8.** $\{(-2)^n\}$ **9.** $\left\{\left(\frac{2}{3}\right)^n\right\}$ **10.** $\left\{\frac{2^n}{3}\right\}$

**11.** $\{3 \cdot 2^{-n}\}$ **12.** $\left\{-10 \cdot \left(\frac{1}{2}\right)^n\right\}$ **13.** $\left\{\frac{5^{n-1}}{2^n}\right\}$ **14.** $\left\{\frac{3^{-n}}{2^{n-1}}\right\}$

*In Problems 15–26, determine whether the given sequence is arithmetic, geometric, or neither. If the sequence is arithmetic, find the common difference; if it is geometric, find the common ratio.*

**15.** $\{5n + 1\}$ **16.** $\{8 - 3n\}$ **17.** $\{2n^2\}$ **18.** $\{n^2 - 2\}$

**19.** $\left\{\frac{2^{-n}}{5}\right\}$ **20.** $\left\{\frac{2}{3^n}\right\}$ **21.** 54, 36, 24, 16, ... **22.** 100, 20, 4, $\frac{4}{5}$, ...

**23.** 2, 6, 10, 14, ... **24.** 15, 12, 9, 6, ... **25.** 1, 2, 3, 5, 8, ... **26.** 5, −2, 3, −1, 2, ...

*In Problems 27–34, (a) find a formula for the nth term of the geometric sequence whose first term and common ratio are given, and (b) use the formula to find the eighth term.*

**27.** $a = 10, r = 2$ **28.** $a = 2, r = 3$ **29.** $a = 100, r = \frac{1}{2}$

**30.** $a = 30, r = 1/3$ **31.** $a = 1, r = -3$ **32.** $a = 1, r = -4$

**33.** $a = 100, r = 1.05$ **34.** $a = 500, r = 1.04$

*In Problems 35–40, find the indicated term of each geometric sequence.*

**35.** 10th term of 3, 6, 12, 24, ... **36.** 12th term of 1, 3, 9, 27, ...

**37.** 15th term of 4, −2, 1, $-\frac{1}{2}$, ... **38.** 8th term of 10, −20, 40, −80, ...

**39.** 9th term of 0.5, 0.05, 0.005, 0.0005, ...

**40.** 10th term of 0.4, 0.04, 0.004, 0.0004, ...

*In Problems 41–48, find the sum. If necessary, express your answer to as many decimals as your calculator allows.*

**41.** $2 + 4 + 8 + \cdots + 2^{12}$ **42.** $3 + 9 + 27 + \cdots + 3^{10}$

**43.** $50 + 20 + 8 + \frac{16}{5} + \cdots + 50\left(\frac{2}{5}\right)^{10-1}$

**44.** $10 + 5 + \frac{5}{2} + \cdots + 10\left(\frac{1}{2}\right)^{12-1}$

**45.** $\sum_{n=1}^{10} [3 \cdot 2^n]$ **46.** $\sum_{n=1}^{12} [5 \cdot 2^n]$ **47.** $\sum_{n=1}^{8} \left[\frac{4}{2^{n-1}}\right]$ **48.** $\sum_{n=1}^{14} \left[10 \cdot \left(\frac{1}{2}\right)^{n-1}\right]$

*In Problems 49–58, find the sum of each geometric series. If necessary, express your answer to as many decimals as your calculator allows.*

**49.** $1 + \frac{1}{2} + \frac{1}{4} + \cdots$ **50.** $1 + \frac{1}{3} + \frac{1}{9} + \cdots$ **51.** $10 + \frac{10}{3} + \frac{10}{9} + \cdots$

**52.** $20 + 5 + \frac{5}{4} + \cdots$ **53.** $6 - 2 + \frac{2}{3} - \frac{2}{9} + \cdots$ **54.** $12 - 3 + \frac{3}{4} - \frac{3}{16} + \cdots$

**55.** $\sum_{n=1}^{\infty} \left(5 \cdot \left(\frac{1}{5}\right)^n\right)$ **56.** $\sum_{n=1}^{\infty} \left(10 \cdot \left(\frac{1}{3}\right)^n\right)$ **57.** $\sum_{n=1}^{\infty} \left(12 \cdot \left(-\frac{1}{3}\right)^{n-1}\right)$

**58.** $\sum_{n=1}^{\infty} \left(100 \cdot \left(-\frac{1}{2}\right)^{n-1}\right)$

*In Problems 59–62, express each repeating decimal as a fraction in lowest terms.*

**59.** $0.\overline{5}$ **60.** $0.\overline{3}$ **61.** $0.\overline{89}$ **62.** $0.\overline{45}$

## Applying the Concepts

**63.** Find $x$ so that $x$, $x + 2$, and $x + 3$ are consecutive terms of a geometric sequence.

**64.** Find $x$ so that $x - 1$, $x$, and $x + 2$ are consecutive terms of a geometric sequence.

**65. Salary Increases** Suppose that you have been hired at an annual salary of \$40,000 per year. You have been promised a raise of 5% for each of the next 10 years.

**(a)** What will be your salary at the beginning of your second year?
**(b)** What will be your salary at the beginning of your tenth year?
**(c)** How much will you have earned cumulatively once you have finished your tenth year?

**66. Salary Increases** Suppose that you have been hired at an annual salary of \$45,000 per year. Historically, the typical raise is 4% each year. You expect to be at the company for the next 10 years.

**(a)** What will be your salary at the beginning of your second year?
**(b)** What will be your salary at the beginning of your tenth year?
**(c)** How much will you have earned cumulatively once you have finished your tenth year?

**67. Depreciation of a Car** Suppose that you have just purchased a Honda Accord for \$20,000. Historically, the car depreciates by 8% each year, so that next year the car is worth \$20,000(0.92). What will the value of the car be after you have owned it for five years?

**68. Depreciation of a Car** Suppose that you have just purchased a Chevy Cavalier for \$16,000. Historically, the car depreciates by 10% each year, so that next year the car is worth \$16,000(0.9). What will the value of the car be after you have owned it for four years?

**69. Pendulum Swings** Initially, a pendulum swings through an arc of 3 feet. On each successive swing, the length of the arc is 0.95 of the previous length.

**(a)** What is the length of the arc after 10 swings?
**(b)** On which swing is the length of the arc less than 1 foot for the first time?
**(c)** After 10 swings, what total length will the pendulum have swung?
**(d)** When it stops, what total length will the pendulum have swung?

**70. Bouncing Balls** A ball is dropped from a height of 30 feet. Each time it strikes the ground, it bounces up to 0.8 of the previous height.

**(a)** What height will the ball bounce up to after it strikes the ground for the fourth time?

**(b)** How high will it bounce after it strikes the ground for the fifth time?

**(c)** How many times does the ball need to strike the ground before its bounce is less than 6 inches?

**(d)** What total distance does the ball travel before it stops bouncing?

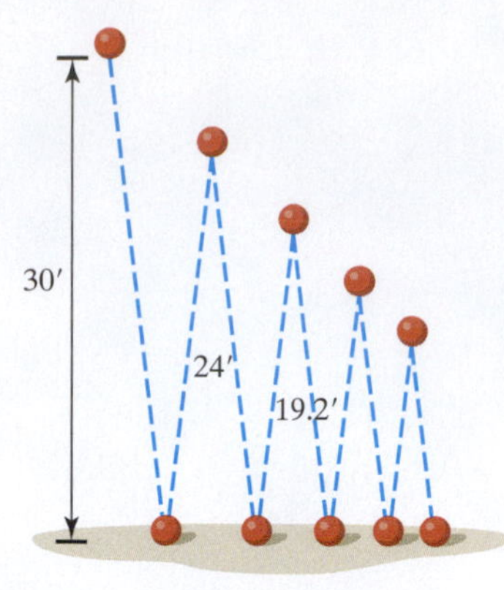


**71. A Job Offer** You are interviewing for a job and receive two offers:

*A*: \$30,000 to start, with guaranteed annual increases of 5% for the first 5 years
*B*: \$31,000 to start with guaranteed annual increases of 4% for the first 5 years

Which offer is best if your goal is to be making as much money as possible after the fifth year? Which is best if your goal is to make as much money as possible over the entire 5 years of the contract?

**72. Be an Agent** Suppose that you are an agent for a professional baseball player. Management has just offered your client a 5-year contract with a first-year salary of \$1,500,000. Beyond that, your client has three choices:

*A*. A bonus of \$75,000 each year (including the first year)
*B*. An annual increase of 4% per year beginning after the first year
*C*. An annual increase of \$80,000 per year beginning after the first year

Which option provides the most money over the 5-year period? Which the least? Which would you choose? Why?

**73. The Multiplier** Suppose the marginal propensity to consume throughout the U.S. economy is 0.98. What is the multiplier for the U.S. economy?

**74. The Multiplier** Suppose the marginal propensity to consume throughout the U.S. economy is 0.96. What is the multiplier for the U.S. economy?

**75. Stock Price** One method of pricing a stock is based upon the stream of future dividends of the stock. Suppose that a stock pays $\$P$ per year in dividends and, historically, the dividend has been increased by $i\%$ per year. If you desire an annual rate of return of $r\%$, this method of pricing a stock states that the price you should pay is the present value of an infinite stream of payments:

$$\text{Price} = P + P\cdot\frac{1+i}{1+r} + P\cdot\left(\frac{1+i}{1+r}\right)^2 + P\cdot\left(\frac{1+i}{1+r}\right)^3 + \cdots$$

The price of the stock is the sum of a geometric series. Suppose that a stock pays an annual dividend of \$2.00 and, historically, the dividend has been increased by 2% per year. You desire an annual rate of return of 9%. What is the most you should pay for the stock?

**76. Stock Price** Refer to Problem 75. Suppose that a stock pays an annual dividend of \$3.00 and, historically, the dividend has been increased by 3% per year. You desire an annual rate of return of 10%. What is the most you should pay for the stock?

**77. 401(k)** Christine contributes \$100 each month into her 401(k) retirement plan. What will be the value of Christine's 401(k) in 30 years if the per annum rate of return is assumed to be 8% compounded monthly?

**78. Saving for a Home** Jolene wants to purchase a new home. Suppose she invests \$400 a month into a money market fund. If the per annum interest rate of return on the money market is 4% compounded monthly, how much will Jolene have for a down payment in 4 years?

**79. Roth IRA** Jackson contributes \$500 each quarter into his Roth IRA. What will be the value of Jackson's IRA in 25 years if the per annum rate of return is assumed to be 6% compounded quarterly?

**80. Retirement** Raymont is planning on retiring in 15 years, so he contributes \$1500 into his IRA every 6 months (semiannually). What will be the value of the IRA when Raymont retires if the per annum interest rate is 10% compounded semiannually?

**81. What's My Payment?** Suppose that Aaliyah wants to have \$1,500,000 in her 401(k) retirement account in 35 years. How much does she need to contribute each month if the account earns 10% interest compounded monthly?

**82. What's My Payment?** Suppose that Sophia wants to have \$2,000,000 in her 401(k) retirement account in 25 years. How much does she need to contribute each quarter if the account earns 12% interest compounded quarterly?

## Extending the Concepts

**83.** Express $0.4\overline{9}$ as a fraction in lowest terms.

**84.** Express $0.85\overline{9}$ as a fraction in lowest terms.

**85.** Find the sum: $2 + 4 + 8 + \cdots + 1{,}073{,}741{,}824$

**86.** Can a sequence be both arithmetic and geometric? Give reasons for you answer.

**87.** Which yields faster growth in the terms of a sequence—an arithmetic sequence or a geometric sequence with $r > 1$? Why? Explain why Thomas Robert Malthus's conjecture that food supplies grow arithmetically while population grows geometrically provides a recipe for disaster unless population growth is curbed.

## Synthesis Review

**88.** Express $\frac{1}{3}$ as a repeating decimal. Express $\frac{2}{3}$ as a repeating decimal.

**89.** What is $\frac{1}{3} + \frac{2}{3}$? Now add the repeating decimals found in Problem 88. Conjecture the value of $0.9999999\ldots$.

**90.** Prove that $0.9999\ldots = 0.\overline{9}$ equals 1.

## The Graphing Calculator

*In Problems 91–94, use a graphing calculator to find the sum of each sequence.*

**91.** $4 + 4.8 + 5.76 + \cdots + 4(1.2)^{15-1}$

**92.** $3 + 4.8 + 7.68 + \cdots + 3(1.6)^{20-1}$

**93.** $\sum_{n=1}^{20}[1.2(1.05)^n]$

**94.** $\sum_{n=1}^{25}[1.3(0.55)^n]$

# PUTTING THE CONCEPTS TOGETHER (SECTIONS 11.1–11.3)

*These problems cover important concepts from Sections 11.1 to 11.3. We designed these problems so that you can review the chapter so far and show your mastery of the concepts. Take time to work these problems before proceeding with the next section. The answers to these problems are located at the back of the text starting on page AN-57.*

*In Problems 1–6, determine if the sequence is arithmetic, geometric, or neither. If arithmetic or geometric, determine the first term and the common difference or common ratio.*

**1.** $\frac{3}{4}, \frac{3}{16}, \frac{3}{64}, \frac{3}{256}, \ldots$

**2.** $\{2(n + 3)\}$

**3.** $\left\{\frac{7n + 2}{9}\right\}$

**4.** $1, -4, 9, -16, \ldots$ **5.** $\{3 \cdot 2^{n+1}\}$ **6.** $\{n^2 - 5\}$

**7.** Write out the sum and evaluate: $\sum_{k=1}^{6}[3k + 4]$

**8.** Express the sum using summation notation:

$$\frac{1}{2(6+1)} + \frac{1}{2(6+2)} + \frac{1}{2(6+3)} + \cdots + \frac{1}{2(6+12)}$$

*In Problems 9–12, write a formula for the nth term of the indicated sequence. Write out the first five terms of each sequence.*

**9.** arithmetic: $a = 25, d = -2$ **10.** arithmetic: $a_4 = 9, d = 11$

**11.** geometric: $a_4 = \frac{9}{25}, r = \frac{1}{5}$ **12.** geometric: $a = 150, r = 1.04$

*In Problems 13–15, find the indicated sum.*

**13.** $2 + 6 + 18 + \cdots + 2 \cdot (3)^{11-1}$

**14.** $2 + 7 + 12 + 17 + \cdots + [2 + (20 - 1) \cdot 5]$

**15.** $1000 + 100 + 10 + \cdots$

**16. Table Seating** A restaurant uses square tables in the main dining area that seat four people. For larger parties, tables can be placed together. Two tables will seat 6 people, three tables will seat 8 people, and so on (see diagram). How many tables could be needed to seat a party of 24 people?

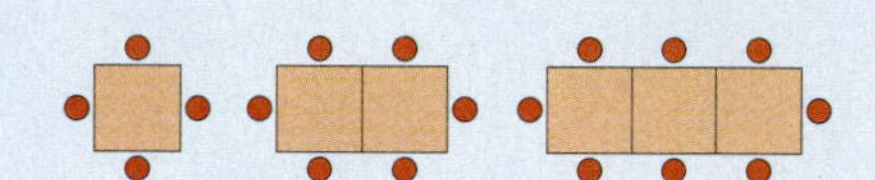

# 11.4 The Binomial Theorem

**OBJECTIVES**

1. Compute Factorials
2. Evaluate a Binomial Coefficient
3. Expand a Binomial

### *Preparing for the Binomial Theorem*

*Before getting started, take the following readiness quiz. If you get a problem wrong, go back to the section cited and review the material.*

**1.** Multiply: $(x - 5)^2$ [Section 5.2, pp. 369–372]

**2.** Multiply: $(2x + 3)^2$ [Section 5.2, pp. 369–372]

## 1 Compute Factorials

Suppose we wanted to find the product of the first 12 integers. That is, suppose we wanted to compute

$$12 \cdot 11 \cdot 10 \cdot 9 \cdot 8 \cdot 7 \cdot 6 \cdot 5 \cdot 4 \cdot 3 \cdot 2 \cdot 1$$

This product is equal to 479,001,600. Not only is this a big number, writing out the product is long. A shorthand method for writing this product is *factorial notation.*

**DEFINITION**

If $n \geq 0$ is an integer, the **factorial symbol $n!$** (read "$n$ factorial") is defined as follows:

$$0! = 1 \qquad 1! = 1$$

$$n! = n(n-1)(n-2) \cdot \cdots \cdot 3 \cdot 2 \cdot 1 \quad \text{if } n \geq 2$$

***Preparing for...Answers***
**1.** $x^2 - 10x + 25$ **2.** $4x^2 + 12x + 9$

For example, $2! = 2 \cdot 1 = 2$, $3! = 3 \cdot 2 \cdot 1 = 6$, $4! = 4 \cdot 3 \cdot 2 \cdot 1 = 24$, and so on. Table 2 lists the values of $n!$ for $0 \leq n \leq 7$.

**Table 2**

| $n$ | 0 | 1 | 2 | 3 | 4 | 5 | 6 | 7 |
|---|---|---|---|---|---|---|---|---|
| $n!$ | 1 | 1 | 2 | 6 | 24 | 120 | 720 | 5040 |

Because

$$n\underbrace{(n-1)(n-2)\cdot \cdots \cdot 3 \cdot 2 \cdot 1}_{(n-1)!}$$

we can use the formula

$$n! = n(n-1)!$$

to find successive factorials. For example, because $7! = 5040$, we have

$$8! = 8 \cdot 7! = 8(5040) = 40{,}320$$

Your calculator has a factorial key. Use it to see how fast factorials increase in value. Find the value of 69!. What happens when you try to find 70!? In fact, 70! is larger than $10^{100}$ (a **googol**), the largest number that most calculators can display.

### EXAMPLE 1 Computing Factorials

Compute the value of $\dfrac{12!}{9!}$.

**Solution**

We could directly compute 12! and then 9!, but this would be inefficient. Rather, we shall use properties of factorials. Namely,

$$\frac{12!}{9!} = \frac{12 \cdot 11 \cdot 10 \cdot 9!}{9!}$$

Divide out 9!: $= 12 \cdot 11 \cdot 10$

$$= 1320$$

**QUICK ✓** *Find the value of each factorial.*

**1.** $9!$  **2.** $\dfrac{7!}{3!}$

## 2 Evaluate a Binomial Coefficient

A formula has been given for expanding $(x + a)^n$ for $n = 2$. The *Binomial Theorem* is a formula for the expansion of $(x + a)^n$ for any positive integer $n$. If $n = 1, 2, 3$, and 4, the expansion of $(x + a)^n$ is straightforward:

$(x + a)^1 = x + a$ — Two terms, beginning with $x^1$ and ending with $a^1$

$(x + a)^2 = x^2 + 2ax + a^2$ — Three terms, beginning with $x^2$ and ending with $a^2$

$(x + a)^3 = x^3 + 3ax^2 + 3a^2x + a^3$ — Four terms, beginning with $x^3$ and ending with $a^3$

$(x + a)^4 = x^4 + 4ax^3 + 6a^2x^2 + 4a^3x + a^4$ — Five terms, beginning with $x^4$ and ending with $a^4$

Notice that each expansion of $(x + a)^n$ begins with $x^n$ and ends with $a^n$. As you read from left to right, the powers of $x$ are decreasing by 1, while the powers of $a$ are increasing by 1. Also, the number of terms that appears equals $n + 1$. Notice, too, that the degree of each monomial in the expansion equals $n$. For example, in the expansion of $(x + a)^3$, each monomial $(x^3, 3ax^2, 3a^2x, a^3)$ is of degree 3. As a result, we might conjecture that the expansion of $(x + a)^n$ would look like this:

$$(x + a)^n = x^n + \_ax^{n-1} + \_a^2x^{n-2} + \cdots + \_a^{n-1}x + a^n$$

where the blanks are numbers to be found. This is, in fact, the case, as we shall see shortly.

First, we need to introduce a symbol. We define the symbol $\binom{n}{j}$, read "$n$ taken $j$ at a time" or "$n$ choose $j$", as follows:

**Work Smart**

Do not write $\binom{n}{j}$ as $\left(\frac{n}{j}\right)$.

**DEFINITION**

If $j$ and $n$ are integers with $0 \le j \le n$, the symbol $\binom{n}{j}$ is defined as

$$\binom{n}{j} = \frac{n!}{j!(n-j)!}$$

## EXAMPLE 2 Evaluating $\binom{n}{j}$

Find:

**(a)** $\binom{4}{1}$ **(b)** $\binom{6}{2}$ **(c)** $\binom{5}{4}$

### Solution

**(a)** Here, we have $n = 4$ and $j = 1$, so

$$\binom{4}{1} = \frac{4!}{1!(4-1)!} = \frac{4!}{1!3!} = \frac{4 \cdot 3 \cdot 2 \cdot 1}{1 \cdot 3 \cdot 2 \cdot 1} = \frac{4 \cdot \cancel{3} \cdot \cancel{2} \cdot 1}{1 \cdot \cancel{3} \cdot \cancel{2} \cdot 1} = 4$$

**(b)** $$\binom{6}{2} = \frac{6!}{2!(6-2)!} = \frac{6!}{2! \cdot 4!} = \frac{6 \cdot 5 \cdot 4!}{2 \cdot 1 \cdot 4!} = \frac{6 \cdot 5 \cdot \cancel{4!}}{2 \cdot 1 \cdot \cancel{4!}} = \frac{30}{2} = 15$$

$6! = 6 \cdot 5 \cdot 4!$

**(c)** $$\binom{5}{4} = \frac{5!}{4!(5-4)!} = \frac{5!}{4! \cdot 1!} = \frac{5 \cdot 4!}{4! \cdot 1} = \frac{5 \cdot \cancel{4!}}{\cancel{4!} \cdot 1} = 5$$

**QUICK ✓** *Evaluate each expression.*

**3.** $\binom{7}{1}$ **4.** $\binom{6}{3}$

---

Four useful formulas involving the symbol $\binom{n}{j}$ are

$$\binom{n}{0} = 1 \quad \binom{n}{1} = n \quad \binom{n}{n-1} = n \quad \binom{n}{n} = 1$$

Suppose that we arrange the various values of the symbol $\binom{n}{j}$ in a triangular display, as shown next and in Figure 6.

$$\binom{0}{0}$$
$$\binom{1}{0} \quad \binom{1}{1}$$
$$\binom{2}{0} \quad \binom{2}{1} \quad \binom{2}{2}$$
$$\binom{3}{0} \quad \binom{3}{1} \quad \binom{3}{2} \quad \binom{3}{3}$$
$$\binom{4}{0} \quad \binom{4}{1} \quad \binom{4}{2} \quad \binom{4}{3} \quad \binom{4}{4}$$
$$\binom{5}{0} \quad \binom{5}{1} \quad \binom{5}{2} \quad \binom{5}{3} \quad \binom{5}{4} \quad \binom{5}{5}$$

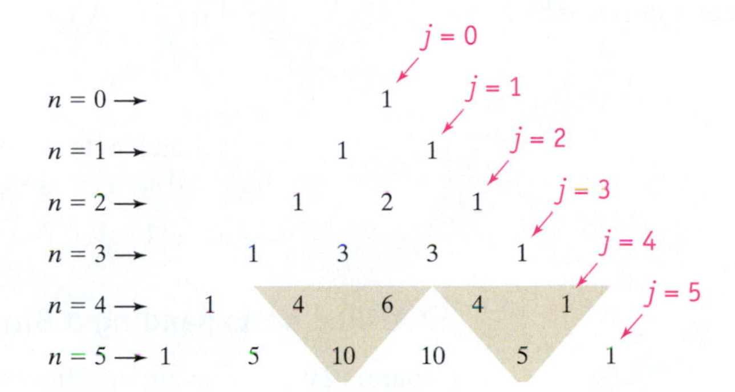


**Figure 6**
Pascal's Triangle

This display is called **Pascal's triangle,** named after Blaise Pascal (1623–1662), a French mathematician.

The Pascal's triangle has 1s down the sides. To get any other entry, add the two nearest entries in the row above it. The shaded triangles in Figure 6 illustrate this feature of the Pascal's triangle. Based on this feature, the row corresponding to $n = 6$ is found as follows:

$$\begin{array}{l} n = 5 \rightarrow \quad 1 \quad 5 \quad 10 \quad 10 \quad 5 \quad 1 \\ n = 6 \rightarrow \; 1 \quad 6 \quad 15 \quad 20 \quad 15 \quad 6 \quad 1 \end{array}$$

Although the Pascal's triangle provides an interesting and organized display of the symbol $\binom{n}{j}$, in practice it is not all that helpful. For example, if you wanted to know the value of $\binom{12}{8}$, you would need to produce 12 rows of the triangle before seeing the answer. It is much faster to use the definition of $\binom{n}{j}$.

## 3 Expand a Binomial

Now we are ready to state the **Binomial Theorem.**

**THE BINOMIAL THEOREM**

Let $x$ and $a$ be real numbers. For any positive integer $n$, we have

$$(x + a)^n = \binom{n}{0}x^n + \binom{n}{1}ax^{n-1} + \binom{n}{2}a^2x^{n-2} + \cdots + \binom{n}{j}a^jx^{n-j} + \cdots + \binom{n}{n}a^n$$

You should now see why we needed to discuss the symbol $\binom{n}{j}$; these symbols are the numerical coefficients that appear in the expansion of $(x + a)^n$. Because of this, the symbol $\binom{n}{j}$ is called the **binomial coefficient.**

### EXAMPLE 3 Expanding a Binomial

Use the Binomial Theorem to expand $(x + 3)^4$.

**Work Smart**

In the expansion, notice how the exponent on $x$ **decreases** by one for each term as we move to the right, while the exponent on 3 **increases** by 1.

**Solution**

In the Binomial Theorem, let $a = 3$ and $n = 4$. Then

$$(x + 3)^4 = \binom{4}{0}x^4 + \binom{4}{1}3^1 \cdot x^{4-1} + \binom{4}{2}3^2 \cdot x^{4-2} + \binom{4}{3}3^3 \cdot x^{4-3} + \binom{4}{4}3^4$$

Use row 4 of the Pascal's triangle in Figure 6 or use $\binom{n}{j} = \frac{n!}{j!(n-j)!}$ to evaluate the binomial coefficients.

$$= 1 \cdot x^4 + 4 \cdot 3 \cdot x^3 + 6 \cdot 9 \cdot x^2 + 4 \cdot 27 \cdot x + 1 \cdot 81$$
$$= x^4 + 12x^3 + 54x^2 + 108x + 81$$

### EXAMPLE 4 Expanding a Binomial

Expand $(2y - 3)^5$ using the Binomial Theorem.

**Solution**

First, we rewrite the expression $(2y - 3)^5$ as $[2y + (-3)]^5$. Now we use the Binomial Theorem with $n = 5$, $x = 2y$, and $a = -3$.

$$[2y + (-3)]^5 = \binom{5}{0}(2y)^5 + \binom{5}{1}(-3)^1(2y)^{5-1} + \binom{5}{2}(-3)^2(2y)^{5-2} + \binom{5}{3}(-3)^3(2y)^{5-3} + \binom{5}{4}(-3)^4(2y)^{5-4} + \binom{5}{5}(-3)^5$$

Use row 5 of Pascal's triangle or use the formula $\binom{n}{j} = \frac{n!}{j!(n-j)!}$ to evaluate binomial coefficients.

$$= (2y)^5 + 5(-3)^1(2y)^4 + 10(-3)^2(2y)^3 + 10(-3)^3(2y)^2 + 5(-3)^4(2y)^1 + (-3)^5$$
$$= 32y^5 + 5 \cdot -3 \cdot 16y^4 + 10 \cdot 9 \cdot 8y^3 + 10 \cdot -27 \cdot 4y^2 + 5 \cdot 81 \cdot 2y - 243$$
$$= 32y^5 - 240y^4 + 720y^3 - 1080y^2 + 810y - 243$$

**QUICK ✓** *Expand each binomial using the Binomial Theorem.*

**5.** $(x + 2)^4$ **6.** $(2p - 1)^5$

# 11.4 Exercises

**For Extra Help:**

## Concepts and Vocabulary

*In Problems 1–3, fill in the blanks.*

**1.** If $n \geq 2$ is an integer, then $n! =$ ________.

**2.** $0! =$ ________; $1! =$ ________; $10! =$ ________.

**3.** ________ ________ is a triangular display of the binomial coefficients.

*In Problems 4 and 5, answer True or False to each statement.*

**4.** $\binom{n}{j} = \dfrac{j!}{n!(n-j)!}$ **5.** $\binom{n}{0} = 0$

**6.** Write down the first four rows of the Pascal's triangle.

**7.** Describe the pattern of exponents of $x$ in the expansion of $(x + a)^n$.

**8.** Describe the pattern of exponents of $a$ in the expansion of $(x + a)^n$.

**9.** What is true about the degree of each monomial in the expansion of $(x + a)^n$?

**10.** Explain how you might find a particular term in a binomial expansion. For example, how might you find the fifth term in the expansion of $(x + 3)^8$?

## Building Skills

*In Problems 11–26, expand each expression using the Binomial Theorem.*

**11.** $(x + 1)^5$ **12.** $(x - 1)^4$ **13.** $(x - 4)^4$ **14.** $(x + 5)^5$

**15.** $(3p + 2)^4$ **16.** $(2q + 3)^4$ **17.** $(2z - 3)^5$ **18.** $(3w - 4)^4$

**19.** $(x^2 + 2)^4$ **20.** $(y^2 - 3)^4$ **21.** $(2p^3 + 1)^5$ **22.** $(3b^2 + 2)^5$

**23.** $(x + 2)^6$ **24.** $(p - 3)^6$ **25.** $(2p^2 - q^2)^4$ **26.** $(3x^2 + y^3)^4$

## Applying the Concepts

**27.** Use the Binomial Theorem to find the numerical value of $(1.001)^4$ correct to five decimal places. [**Hint:** $(1.001)^4 = (1 + 10^{-3})^4$.]

**28.** Use the Binomial Theorem to find the numerical value of $(1.001)^5$ correct to five decimal places. [**Hint:** $(1.001)^5 = (1 + 10^{-3})^5$.]

**29.** Use the Binomial Theorem to find the numerical value of $(0.998)^5$ correct to five decimal places.

**30.** Use the Binomial Theorem to find the numerical value of $(0.997)^5$ correct to five decimal places.

## Extending the Concepts

*Notice in the formula for expanding a binomial $(x + a)^n$, the first term is $\binom{n}{0}a^0 \cdot x^n$, the second term is $\binom{n}{1}a^1 \cdot x^{n-1}$, the third term is $\binom{n}{2}a^2 \cdot x^{n-2}$, and so on. In general, the jth term in a binomial expansion of $(x + a)^n$ is $\binom{n}{j-1}a^{j-1}x^{n-j+1}$. Use this result to find the indicated term in Problems 31–34.*

**31.** The third term in the expansion of $(x + 2)^7$

**32.** The fourth term in the expansion of $(x - 1)^{10}$

**33.** The sixth term in the expansion of $(2p - 3)^8$

**34.** The seventh term in the expansion of $(3p + 1)^9$

**35.** Show that $\binom{n}{n-1} = n$ and $\binom{n}{n} = 1$.

**36.** Show that if $n$ and $j$ are integers with $0 \le j \le n$, then $\binom{n}{j} = \binom{n}{n-j}$.

## Synthesis Review

**37.** If $f(x) = x^4$, find $f(a - 2)$ with the aid of the Binomial Formula.

**38.** If $g(x) = x^5 + 3$, find $g(z + 1)$ with the aid of the Binomial Formula.

**39.** If $H(x) = x^5 - 4x^4$, find $H(p + 1)$ with the aid of the Binomial Formula.

**40.** If $h(x) = 2x^5 + 5x^4$, find $h(a + 3)$ with the aid of the Binomial Formula.

## CHAPTER 11 ACTIVITY: PASS TO THE RIGHT

**Focus:** Review of objectives for arithmetic and geometric sequences and Binomial Theorem

**Time:** 30–35 minutes

**Group size:** 2–4

As a group, discover the relationship between Column A and Column B (i.e. <, >, or =). Be sure to discuss any differences in outcomes.

| | Column A | Column B |
|---|---|---|
| 1 | The sum of the first five terms of $a_n = \frac{3}{5}n + 1$ | The sum of the first five terms of $a_n = -\frac{1}{4}(n - 1) + 4$ |
| 2 | $\sum_{i=1}^{4}(i^2 + 2i)$ | $\sum_{i=1}^{4} i^2 + \sum_{i=1}^{4} 2i$ |
| 3 | The 8th term of an arithmetic sequence when $a_1 = 3$ and $d = \frac{3}{2}$ | The 27th term of an arithmetic sequence when $a_1 = \frac{5}{3}$ and $d = \frac{1}{3}$ |
| 4 | The 10th term of a geometric sequence with $a_1 = 3$ and $r = \sqrt{2}$ | The 9th term of a geometric sequence with $a_1 = 5$ and $r = \sqrt{3}$ |
| 5 | Find the coefficient of the 4th term of $(x + y)^{10}$ | Find the coefficient of the 3rd term of $(8x - y)^4$ |

# CHAPTER 11 REVIEW

## Section 11.1 Sequences

**KEY TERMS**

| | | |
|---|---|---|
| sequence | finite sequence | summation notation |
| terms | general term | index |
| ellipsis | alternate | partial sum |
| infinite sequence | | |

| YOU SHOULD BE ABLE TO ... | EXAMPLE | REVIEW EXERCISES |
|---|---|---|
| 1 Write the first several terms of a sequence (p. 834) | Examples 1 and 2 | 1–6 |
| 2 Find a formula for the $n$th term of a sequence (p. 836) | Example 3 | 7–12 |
| 3 Use summation notation (p. 836) | Examples 4 and 5 | 13–20 |

*In Problems 1–6, write down the first five terms of each sequence.*

**1.** $\{-3n + 2\}$

**2.** $\left\{\frac{n - 2}{n + 4}\right\}$

**3.** $\{5^n + 1\}$

**4.** $\{(-1)^{n-1} \cdot 3n\}$

**5.** $\left\{\frac{n^2}{n + 1}\right\}$

**6.** $\left\{\frac{\pi^n}{n}\right\}$

*In Problems 7–12, the given pattern continues. Write down the nth term of each sequence suggested by the pattern.*

**7.** $-3, -6, -9, -12, -15, \ldots$

**8.** $\frac{1}{3}, \frac{2}{3}, 1, \frac{4}{3}, \frac{5}{3}, \ldots$

**9.** 5, 10, 20, 40, 80, . . .

**10.** $-\frac{1}{2}, 1, -\frac{3}{2}, 2, \ldots$

**11.** 6, 9, 14, 21, 30, . . .

**12.** $0, \frac{1}{3}, \frac{1}{2}, \frac{3}{5}, \ldots$

*In Problems 13–16, write out each sum and determine its value.*

**13.** $\sum_{k=1}^{5}(5k - 2)$ **14.** $\sum_{k=1}^{6}\left(\frac{k+2}{2}\right)$ **15.** $\sum_{i=1}^{5}(-2i)$ **16.** $\sum_{i=1}^{4}\frac{i^2 - 1}{3}$

*In Problems 17–20, express each sum using summation notation.*

**17.** $(4 + 3\cdot 1) + (4 + 3\cdot 2) + (4 + 3\cdot 3) + \cdots + (4 + 3\cdot 15)$

**18.** $\frac{1}{3^1} + \frac{1}{3^2} + \frac{1}{3^3} + \cdots + \frac{1}{3^8}$

**19.** $\frac{1^3 + 1}{1 + 1} + \frac{2^3 + 1}{2 + 1} + \frac{3^3 + 1}{3 + 1} + \cdots + \frac{10^3 + 1}{10 + 1}$

**20.** $(-1)^{1-1}\cdot 1^2 + (-1)^{2-1}\cdot 2^2 + (-1)^{3-1}\cdot 3^2 + \cdots + (-1)^{7-1}\cdot 7^2$

## Section 11.2 Arithmetic Sequences

| KEY CONCEPTS | KEY TERMS |
|---|---|
| • **The $n$th Term of an Arithmetic Sequence**<br>For an arithmetic sequence whose first term is $a$ and whose common difference is $d$, the $n$th term is $a_n = a + (n - 1)d$<br>• **Sum of $n$ Terms of an Arithmetic Sequence**<br>For an arithmetic sequence whose first term is $a$ and whose common difference is $d$, the sum of the first $n$ terms is $S_n = \frac{n}{2}[2a + (n - 1)d]$ or $S_n = \frac{n}{2}(a + a_n)$. | Arithmetic Sequence<br>Common difference |

| YOU SHOULD BE ABLE TO . . . | EXAMPLE | REVIEW EXERCISES |
|---|---|---|
| 1 Determine if a sequence is arithmetic (p. 842) | Examples 1 through 3 | 21–26 |
| 2 Find a formula for the $n$th term of an arithmetic sequence (p. 843) | Examples 4 and 5 | 27–32, 37 |
| 3 Find the sum of an arithmetic sequence (p. 845) | Examples 6 through 8 | 33–36, 38 |

*In Problems 21–26, determine if the sequence is arithmetic. If so, find the common difference.*

**21.** 4, 10, 16, 22, . . .

**22.** $-1, \frac{1}{2}, 2, \frac{7}{2}, \ldots$

**23.** −2, −5, −9, −14, . . .

**24.** −1, 3, −5, 7, . . .

**25.** $\{4n + 7\}$

**26.** $\left\{\frac{n+1}{2n}\right\}$

*In Problems 27–32, find a formula for the nth term of the arithmetic sequence. Use the formula to find the 25th term of the sequence.*

**27.** $a = 3; d = 8$ **28.** $a = -4; d = -3$ **29.** $7, \frac{20}{3}, \frac{19}{3}, 6, \ldots$ **30.** 11, 17, 23, 29, . . .

**31.** 3rd term is 7 and the 8th term is 25.

**32.** 4th term is −20 and the 7th term is −32.

**33.** Find the sum of the first 30 terms of the sequence −1, 9, 19, 29, . . . .

**34.** Find the sum of the first 40 terms of the sequence 5, 2, −1, −4, . . . .

**35.** Find the sum of the first 60 terms of the sequence $\{-2n - 7\}$.

**36.** Find the sum of the first 50 terms of the sequence $\left\{\frac{1}{4}n + 3\right\}$.

**37. Cicadas** Seventeen-year cicadas emerge every 17 years to mate, lay eggs, and start the next 17-year cycle. In 2004, the Brood X cicada (the largest brood of the 17-year cicada) emerged in Maryland, Kentucky, Tennessee, and parts of surrounding states. Determine when the Brood X cicada will first appear in the 22nd century.

**38. Wind Sprints** At a certain football practice, players would run wind sprints for exercise. Starting at the goal line, players would sprint to the 10-yard line and back to the goal line. The players would then sprint to the 20-yard line and back to the goal line. This continues for the 30-yard line, 40-yard line, and 50-yard line. Determine the total distance a player would run during the wind sprints.

## Section 11.3 Geometric Sequences and Series

| KEY CONCEPTS | KEY TERMS |
|---|---|
| • **The *n*th Term of a Geometric Sequence** For a geometric sequence whose first term is $a$ and whose common ratio is $r$, the $n$th term is $a_n = ar^{n-1}$.<br>• **Sum of *n* Terms of a Geometric Sequence** For a geometric sequence whose first term is $a$ and whose common ratio is $r$, the sum of the first $n$ terms is $S_n = a \cdot \frac{1 - r^n}{1 - r}$.<br>• **Sum of a Geometric Series** If $-1 < r < 1$, the sum of the geometric series $\sum_{n=1}^{\infty} ar^{n-1} = \frac{a}{1 - r}$. | Geometric sequence<br>Common ratio<br>Geometric series<br>Marginal propensity to consume<br>Mulitplier |

| YOU SHOULD BE ABLE TO . . . | EXAMPLE | REVIEW EXERCISES |
|---|---|---|
| 1 Determine if a sequence is geometric (p. 850) | Examples 1 through 3 | 39–44 |
| 2 Find a formula for the $n$th term of a geometric sequence (p. 852) | Example 4 | 45–48, 57 |
| 3 Find the sum of a geometric sequence (p. 853) | Examples 5 and 6 | 49–52, 58 |
| 4 Find the sum of a geometric series (p. 854) | Examples 7 through 9 | 53–56 |
| 5 Solve annuity problems (p. 856) | Example 10 | 59–62 |

*In Problems 39–44, determine if the given sequence is geometric. If so, determine the common ratio.*

**39.** $\frac{1}{3}, 2, 12, 72, \ldots$ **40.** $-1, 3, -9, 27, \ldots$ **41.** $1, 1, 2, 6, \ldots$

**42.** $6, 4, \frac{8}{3}, \frac{16}{9}, \ldots$ **43.** $\{5 \cdot (-2)^n\}$ **44.** $\{3n - 14\}$

*In Problems 45–48, find a formula for the nth term of the geometric sequence. Use the formula to find the 10th term of the sequence.*

**45.** $a = 4, r = 3$ **46.** $a = 8, r = \frac{1}{4}$ **47.** $a = 5, r = -2$ **48.** $a = 1000, r = 1.08$

*In Problems 49–52, find the sum. If necessary, express your answer to as many decimal places as your calculator allows.*

**49.** $2 + 4 + 8 + \cdots + 2^{15}$ **50.** $40 + 5 + \frac{5}{8} + \cdots + 40\left(\frac{1}{8}\right)^{13-1}$

**51.** $\sum_{n=1}^{12}\left[\frac{3}{4}\cdot(2)^{n-1}\right]$

**52.** $\sum_{n=1}^{16}[-4\cdot(3^n)]$

*In Problems 53–56, find the sum of each geometric series. If necessary, express your answer to as many decimal places as your calculator allows.*

**53.** $\sum_{n=1}^{\infty}\left[20\cdot\left(\frac{1}{4}\right)^n\right]$

**54.** $\sum_{n=1}^{\infty}\left[50\cdot\left(-\frac{1}{2}\right)^{n-1}\right]$

**55.** $1+\frac{1}{5}+\frac{1}{25}+\cdots$

**56.** $0.8+0.08+0.008+0.0008+\cdots$

**57. Radioactive Decay** The radioactive isotope Tritium has a half-life of about 12 years. If there were 200 grams of the isotope initially, use a geometric sequence to determine how much would remain after 72 years.

**58. Computer Virus** In January 2004, the Mydoom e-mail worm was declared the worst e-mail worm incident in virus history accounting for roughly 20–30% of worldwide e-mail traffic. Suppose the virus was initially sent to 5 e-mail addresses and that, upon receipt, sends itself out to 5 e-mail addresses from the address book of the infected computer. If each cycle of e-mails, including the initial sending, takes 1 minute to complete, how many total e-mails will have been sent after 15 minutes?

**59. 403(b)** Scott contributes $900 each quarter into a 403(b) plan at work. His employer agrees to match half of employee contributions up to $600 per quarter. What will be the value of Scott's 403(b) in 25 years if the per annum rate of return is assumed to be 7% compounded quarterly?

**60. Lottery Payment** The winner of a state lottery has the option of receiving about $2 million per year for 26 years (after taxes), or a lump sum payment of about $28 million (after taxes). Assuming all winnings will be invested at a per annum rate of return of 6.5% compounded annually, which option yields the most money after 26 years?

**61. What's My Payment?** Sheri starts her career when she is 22 years old and wants to have $2,500,000 in her 401(k) retirement account when she retires in 40 years. How much does she need to contribute each month if the account earns 9% interest compounded monthly?

**62. College Savings Plan** On Samantha's 8th birthday, her parents open a 529 college savings plan for her and plan to contribute $400 per month until she turns 18. The per annum rate of return is assumed to be 5.25% compounded monthly and the cost per credit hour at a private 4-year university is locked-in at a rate of $340 per hour. What is the value of the plan when Samantha turns 18, and how many credit hours will the plan cover?

## Section 11.4 The Binomial Theorem

| KEY CONCEPTS | KEY TERMS |
|---|---|
| • **Factorial symbol $n!$**<br>If $n \geq 0$ is an integer, the factorial symbol $n!$ is defined as<br>$0! = 1 \quad 1! = 1 \quad n! = n(n-1)(n-2)\cdot\cdots\cdot 3\cdot 2\cdot 1 \quad \text{if } n \geq 2$<br>• **The symbol $\binom{n}{j}$**<br>If $j$ and $n$ are integers with $0 \leq j \leq n$, then $\binom{n}{j} = \frac{n!}{j!(n-j)!}$ | Factorial symbol<br>Googol<br>Pascal's triangle<br>Binomial coefficient |

*(continued)*

- **The Binomial Theorem**
  Let $x$ and $a$ be real numbers. For any positive integer $n$, we have

$$(x+a)^n = \binom{n}{0}x^n + \binom{n}{1}ax^{n-1} + \binom{n}{2}a^2x^{n-2} + \cdots + \binom{n}{j}a^jx^{n-j} + \cdots + \binom{n}{n}a^n$$

| YOU SHOULD BE ABLE TO . . . | EXAMPLE | REVIEW EXERCISES |
|---|---|---|
| 1 Compute factorials (p. 862) | Example 1 | 63–66 |
| 2 Evaluate a binomial coefficient (p. 863) | Example 2 | 67–70 |
| 3 Expand a binomial (p. 865) | Examples 3 and 4 | 71–78 |

*In Problems 63–66, evaluate the expression.*

**63.** $5!$ **64.** $\frac{11!}{7!}$ **65.** $\frac{10!}{6!}$ **66.** $\frac{13!}{6!7!}$

*In Problems 67–70, evaluate each binomial coefficient.*

**67.** $\binom{7}{3}$ **68.** $\binom{10}{5}$ **69.** $\binom{8}{8}$ **70.** $\binom{6}{0}$

*In Problems 71–76, expand each expression using the Binomial Theorem.*

**71.** $(z+1)^4$ **72.** $(y-3)^5$ **73.** $(3y+4)^6$

**74.** $(2x^2-3)^4$ **75.** $(3p-2q)^4$ **76.** $(a^3+3b)^5$

**77.** Find the fourth term in the expansion of $(x-2)^8$.

**78.** Find the seventh term in the expansion of $(2x+1)^{11}$.

# CHAPTER 11 TEST

*Remember to use your Chapter Test Prep Video CD to see fully worked-out solutions to any of these problems you would like to review.*

*In Problems 1–6, determine if the sequence is arithmetic, geometric, or neither. If arithmetic or geometric, determine the first term and the common difference or common ratio.*

**1.** $-15, -7, 1, 9, \ldots$ **2.** $\{(-4)^n\}$ **3.** $\left\{\frac{4}{n!}\right\}$

**4.** $\left\{\frac{2n-3}{5}\right\}$ **5.** $-3, 2, 0, 5, 3, \ldots$ **6.** $\{7 \cdot 3^n\}$

**7.** Write out the sum and evaluate: $\sum_{i=1}^{5}\left[\frac{3}{i^2}+2\right]$

**8.** Express the sum using summation notation: $\frac{3}{5}+\frac{2}{3}+\frac{5}{7}+\frac{3}{4}+\cdots+\frac{5}{6}$

*In Problems 9–12, write a formula for the nth term of the indicated sequence. Write out the first five terms of each sequence.*

**9.** arithmetic: $a=6, d=10$ **10.** arithmetic: $a=0, d=-4$

**11.** geometric: $a=10, r=2$ **12.** geometric: $a_3=9, r=-3$

*In Problems 13–15, find the indicated sum.*

**13.** $-2+2+6+\cdots+[4\cdot(20-1)-2]$

**14.** $\frac{1}{9}-\frac{1}{3}+1-3+\cdots+\frac{1}{9}\cdot(-3)^{12-1}$

**15.** $216 + 72 + 24 + 8 + \cdots$ **16.** Evaluate $\dfrac{15!}{8!7!}$. **17.** Evaluate $\dbinom{12}{5}$.

**18.** Expand $(5m - 2)^4$ using the Binomial Theorem.

**19. Tuition Increase** The average tuition and fees for in-state students at public four-year colleges and universities for the 2003–2004 academic year was \$4694. This represented an increase of about 14% from the previous year. If this percent increase continues each year, what will the average tuition and fees for in-state students be in the 2023–2024 academic year?

**20. Transit of Venus** On June 8, 2004, the planet Venus passed between the sun and the Earth creating a rare type of solar eclipse. Such Venus transits continually recur at intervals of 8, 121.5, 8, and 105.5 years. Since the invention of the telescope, Venus transits have been recorded in the years 1631, 1639, 1761, 1769, 1874, 1882, and 2004. Use this information to determine when the next three Venus transits will occur.

## CUMULATIVE REVIEW CHAPTERS R–11

**1.** Solve $\frac{1}{2}(x + 2) = \frac{5}{4}(x - 3y)$ for $y$.

**2.** Evaluate $f(x) = x^2 - x + 7$ for $x = 2$ and $x = -3$.

*In Problems 3–8, find all solutions to the indicated equation.*

**3.** $\frac{1}{2}x - 2 = \frac{1}{3}(x + 1) + 3$ **4.** $5x^2 - 3x = 2$

**5.** $3x^2 + 7x - 2 = 0$ **6.** $\sqrt{2x + 1} - 3 = 8$

**7.** $4^{x+1} = 8^{2x-3}$ **8.** $x^2(2x + 1) + 40 = (x^2 - 8)(x - 5)$

*In Problems 9 and 10, solve the indicated inequality. Write your answer in interval notation.*

**9.** $\frac{2}{3}x + 1 > \frac{1}{4}x - \frac{3}{2}$ **10.** $3x^2 - 2x \leq 3 - 10x$

*In Problems 11 and 12, factor the expression completely.*

**11.** $2x^2 - 5x - 18$ **12.** $6x^3 - 3x^2 + 4x - 2$

*In Problems 13–15, perform the indicated operation and simplify. Write complex numbers in standard form.*

**13.** $(5x - 3)(4x^2 - 2x + 1)$ **14.** $\dfrac{x}{x + 4} - \dfrac{3}{x - 1}$ **15.** $\dfrac{3 - i}{2 + i}$

**16.** Find the domain of the function $f(x) = \sqrt{x - 15} + \sqrt{2x - 5}$.

**17.** Find the equation of the line that passes through the points $(2, -3)$ and $(1, 4)$.

**18.** Solve the system: $\begin{cases} 2x + 3y = 5 \\ x - 2y - 6 \end{cases}$

**19.** Graph the quadratic function $f(x) = 2x^2 - 8x - 3$. Label the vertex and axis of symmetry.

**20.** Write the standard form of the equation of the circle whose center is $(4, -3)$ and whose radius is $r = 6$ units. Graph the circle.

**21.** Sketch the graph of the ellipse given by the equation $4x^2 + y^2 = 64$.

**22.** Find the sum of the first 20 terms of the arithmetic sequence whose first term is $a = -47$ and whose common difference is $d = 12$.

**23.** Find the sum of the infinite geometric series: $2, \frac{3}{2}, \frac{9}{8}, \frac{27}{32}, \ldots$.

**24. Mowing Lawns** The Robomower® automatic lawnmower can mow a 7500-square-foot lot in 5 hours. It takes the Mowbot® automatic lawnmower 6 hours to cut the same lot. How long would it take both machines to cut the lot if they work together?

**25. Aluminum Alloy** The most commonly used aluminum alloy is aluminum 3003 which is often used to make rain gutters. A manufacturer of rain gutters has 100 metric tons of an aluminum alloy that is 2.5% manganese, but this percent is too high. How much pure aluminum must be added to the 100 metric tons in order to obtain a desired alloy that is 1.2% manganese?

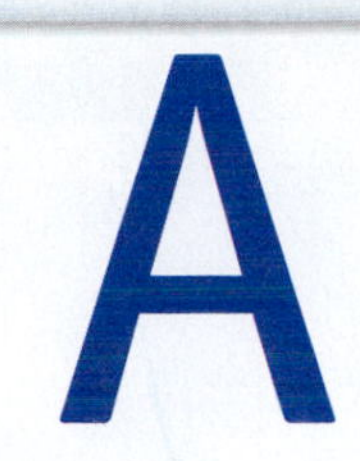

APPENDIX

# A The Library of Functions

## OBJECTIVE

1 Graph functions in the Library of Functions

### Preparing for the Library of Functions

*Before getting started, take the following readiness quiz. If you get a problem wrong, go back to the section cited and review the material.*

1. Graph $y = x^2$ by point plotting. [Section 2.1, p. 141]
2. Graph $y = x^3$ by point plotting. [Section 2.1, pp. 140–142]
3. Graph $x = y^2$ by point plotting. [Section 2.1, p. 142]

## 1 The Library of Functions

In Table 1, we review a number of the functions that we have studied in the book. Pay attention to the properties listed and the shape of the graph.

**Table 1**

| Function | Properties | Graph |
|---|---|---|
| **Linear Function** $f(x) = mx + b$ $m$ and $b$ are real numbers | • Domain and range are all real numbers. • Graph is nonvertical line with slope $= m$ $y$ intercept $= b$. | $f(x) = mx + b, m > 0$ (0, b) |
| **Identity Function** (special type of linear function) $f(x) = x$ | • Domain and range are all real numbers. • Graph is a line with slope of $m = 1$ $y$-intercept $= 0$. • The line consists of all points for which the $x$-coordinate equals the $y$-coordinate. | 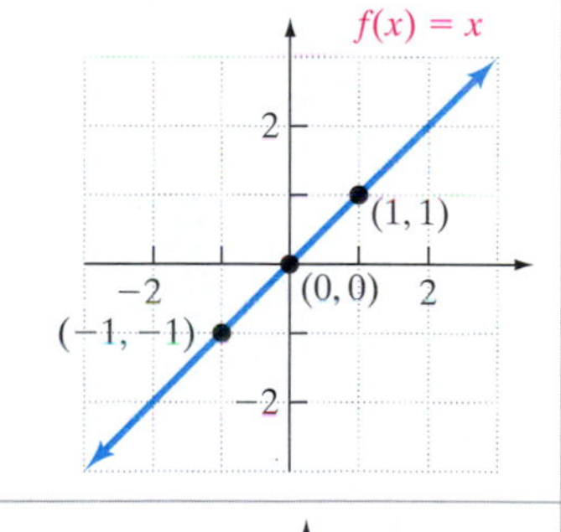  |
| **Constant Function** (special type of linear function) $f(x) = b$ $b$ is a real number | • Domain is the set of all real numbers and range is the set consisting of a single number $b$. • Graph is a horizontal line with slope $m = 0$ $y$-intercept of $b$. | 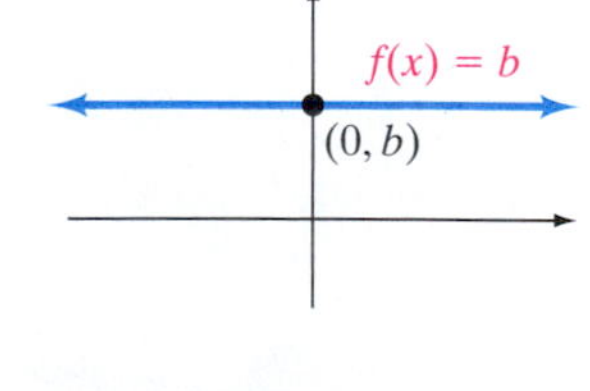  |
| **Square Function** $f(x) = x^2$ | • Domain is the set of all real numbers; its range is the set of nonnegative real numbers. • The graph is a parabola whose intercept is (0, 0). | 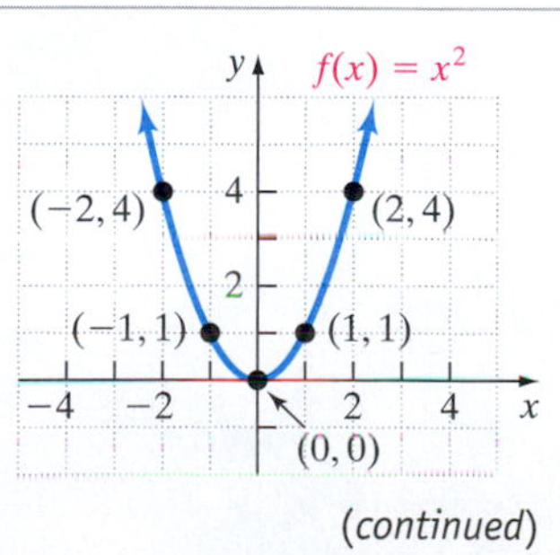  |

*(continued)*

*Preparing for...Answers*

1\.

(−4, 16) (4, 16) (−3, 9) (3, 9) (−2, 4) (2, 4) (−1, 1) (1, 1) (0, 0)

2\.

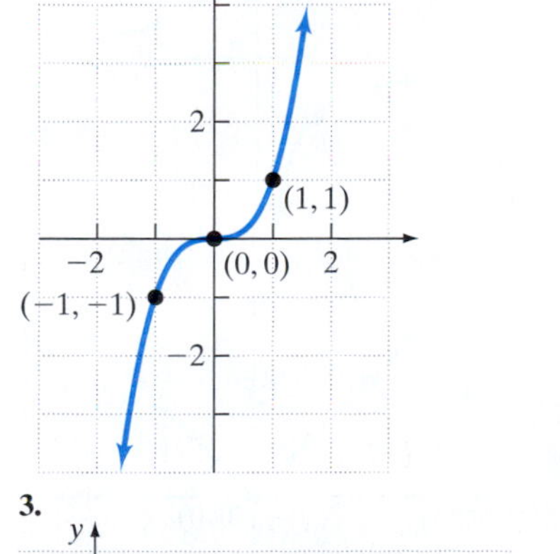


3\.

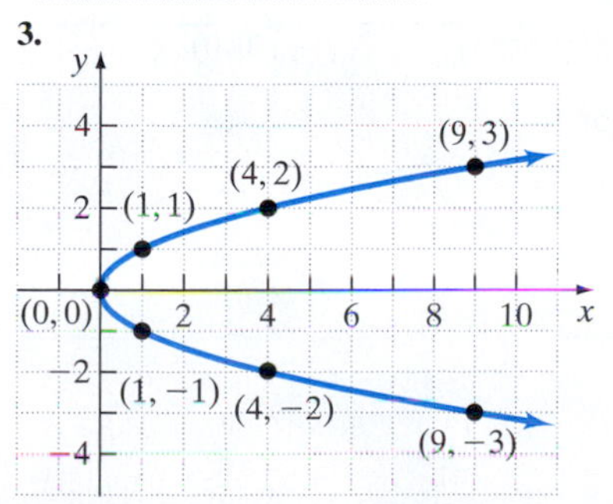

| Function | Properties | Graph |
|---|---|---|
| **Cube Function** $f(x) = x^3$ | • Domain and the range are the set of all real numbers.<br>• The intercept of the graph is at (0, 0). | 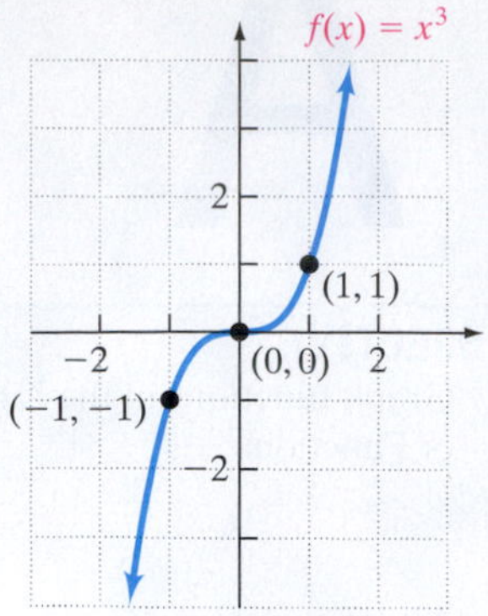 $f(x) = x^3$; (1, 1); (0, 0); (−1, −1); 2; −2 |
| **Square Root Function** $f(x) = \sqrt{x}$ | • Domain and the range are the set of nonnegative real numbers.<br>• The intercept of the graph is at (0, 0). | $f(x) = \sqrt{x}$; (4, 2); (1, 1); (0, 0); 2; 4; x; y |
| **Cube Root Function** $f(x) = \sqrt[3]{x}$ | • Domain and the range are the set of real numbers.<br>• The intercept of the graph is at (0, 0). | $f(x) = \sqrt[3]{x}$; $\left(\frac{1}{8}, \frac{1}{2}\right)$; (1, 1); $(2, \sqrt[3]{2})$; (0, 0); $\left(-\frac{1}{8}, -\frac{1}{2}\right)$; $(-2, -\sqrt[3]{2})$; (−1, −1); −4; −2; 2; 4; x; y |
| **Reciprocal Function** $f(x) = \dfrac{1}{x}$ | • Domain and the range are the set of all nonzero real numbers.<br>• The graph has no intercepts. | 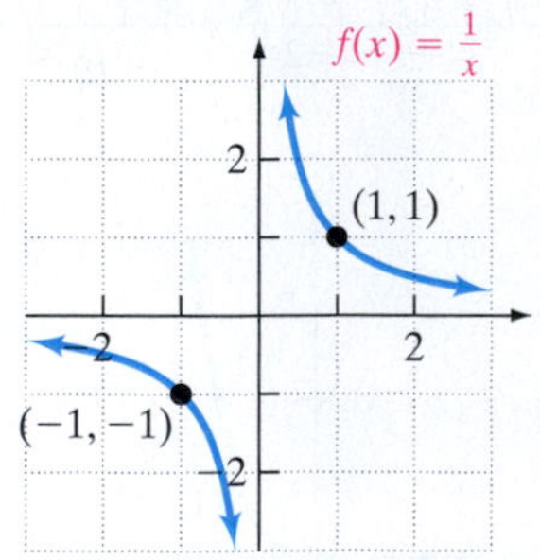 $f(x) = \frac{1}{x}$; (1, 1); (−1, −1); 2; −2 |
| **Absolute Value Function** $f(x) = \|x\|$ | • Domain of the is the set of all real numbers; its range is the set of nonnegative real numbers.<br>• The intercept of the graph is (0, 0).<br>• If $x \geq 0$, then $f(x) = x$, and the graph of $f$ is part of the line $y = x$; if $x < 0$, then $f(x) = -x$, and the graph of $f$ is part of the line $y = -x$. | 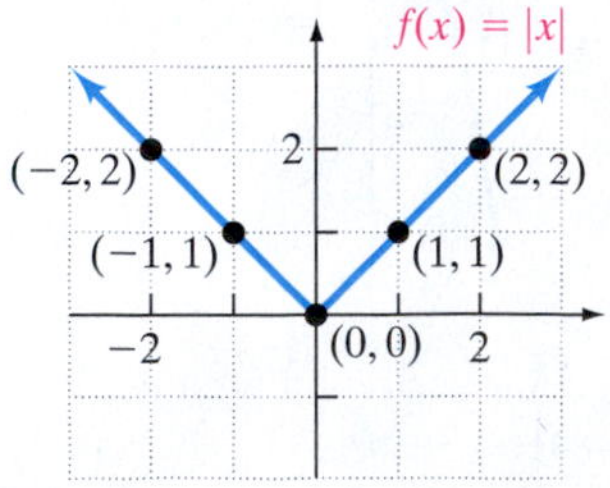 $f(x) = \|x\|$; (−2, 2); (2, 2); (−1, 1); (1, 1); (0, 0); 2; −2 |

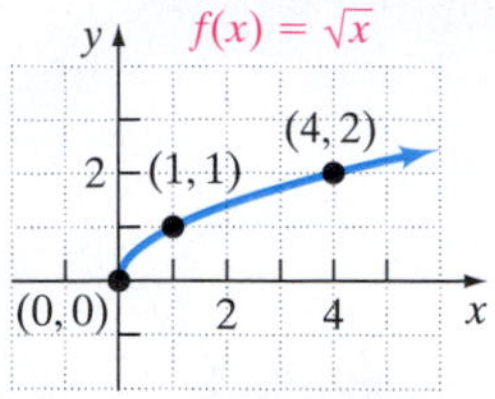

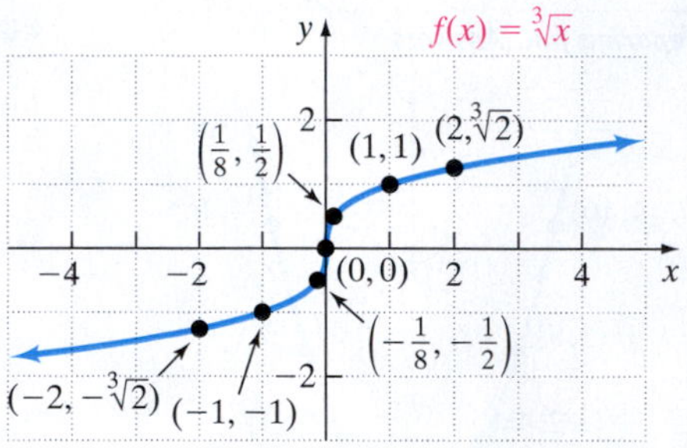

# A Exercises

**For Extra Help:**

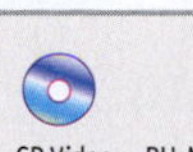
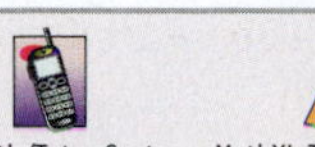
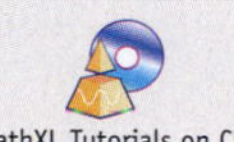

Student Solutions Manual | CD Video | PH Math/Tutor Center | MathXL Tutorials on CD | MathXL® | MyMathLab

## Concepts and Vocabulary

*In Problems 1 and 2, fill in the blanks.*

**1.** The range of the ________ ________ is the real number $b$.

**2.** The domain of the ________ ________ is the set of all nonzero real numbers.

*In Problems 3 and 4, answer True or False to each statement.*

**3.** The domain and the range of the square function is the set of all real numbers.

**4.** The domain and the range of the cube function is the set of all real numbers.

## Building Skills

*In Problems 5–12, match each graph to the function listed whose graph most resembles the one given.*

**(a)** Constant function **(b)** Linear function **(c)** Square function
**(d)** Cube function **(e)** Square root function **(f)** Reciprocal function
**(g)** Absolute value function **(h)** Cube root function

**5.** 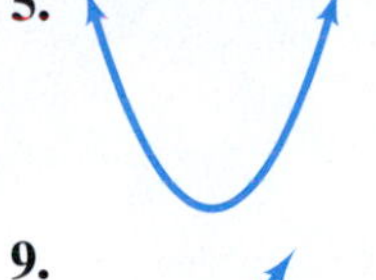

**6.**

**7.**

**8.** 

**9.**

**10.** 

**11.** 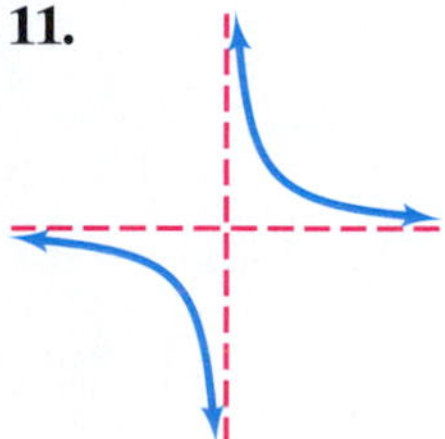

**12.**

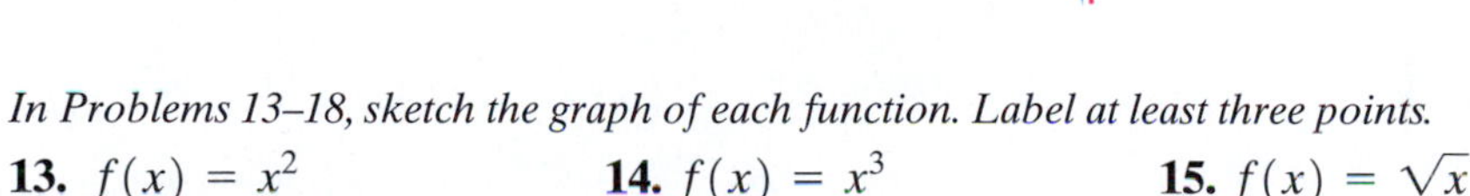

*In Problems 13–18, sketch the graph of each function. Label at least three points.*

**13.** $f(x) = x^2$

**14.** $f(x) = x^3$

**15.** $f(x) = \sqrt{x}$

**16.** $f(x) = \sqrt[3]{x}$

**17.** $f(x) = \dfrac{1}{x}$

**18.** $f(x) = |x|$

# Answers to Quick ✔ Exercises

## Chapter R

### Section R.2

**1.** set-builder: $\{x|x \text{ is a digit less than } 5\}$; roster: $\{0, 1, 2, 3, 4\}$ **2.** set-builder: $\{x|x \text{ is a digit greater than or equal to } 6\}$; roster: $\{6, 7, 8, 9\}$

**3.** True **4.** False **5.** False **6.** True **7.** True **8.** False **9.** True **10.** $10, \frac{12}{4}$ **11.** $10, \frac{0}{3}, \frac{12}{4}$ **12.** $-9, 10, \frac{0}{3}, \frac{12}{4}$

**13.** $\frac{7}{3}, -9, 10, 4.\overline{56}, \frac{0}{3}, -\frac{4}{7}, \frac{12}{4}$ **14.** $\pi$ and 5.7377377737777... **15.** All except $\sqrt{-4}$ **16. (a)** 5.694 **(b)** 5.694

**17. (a)** −4.93 **(b)** −4.94 **18.** 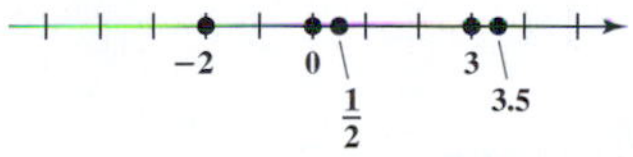 −2 0 $\frac{1}{2}$ 3 3.5 **19.** $<$ **20.** $<$ **21.** $>$ **22.** $>$ **23.** $=$

### Section R.3

**1.** 6 **2.** 10 **3.** 12 **4.** −11 **5.** −6.6 **6.** −2.2 **7.** 0 **8.** −5 **9.** $-\frac{4}{5}$ **10.** 12 **11.** $\frac{5}{3}$ **12.** 4 **13.** −9 **14.** −11

**15.** 9.1 **16.** −5.1 **17.** −16.1 **18.** −48 **19.** −60 **20.** 56 **21.** 105 **22.** 5.13 **23.** $\frac{1}{10}$ **24.** $-\frac{1}{8}$ **25.** $\frac{5}{2}$ **26.** −5 **27.** $\frac{6}{5}$

**28.** $\frac{9}{2}$ **29.** $\frac{5}{3}$ **30.** $-\frac{6}{5}$ **31.** $-\frac{5}{6}$ **32.** $\frac{4}{5}$ **33.** $\frac{1}{2}$ **34.** 8 **35.** $\frac{5}{11}$ **36.** $-\frac{1}{3}$ **37.** $\frac{11}{7}$ **38.** 1 **39.** LCD = 60; $\frac{3}{20} = \frac{9}{60}, \frac{2}{15} = \frac{8}{60}$

**40.** LCD = 90; $\frac{5}{18} = \frac{25}{90}, -\frac{1}{45} = -\frac{2}{90}$ **41.** $\frac{17}{60}$ **42.** $-\frac{1}{6}$ **43.** $-\frac{59}{150}$ **44.** $-\frac{3}{10}$ **45.** $5x + 15$ **46.** $-6x - 6$ **47.** $-4z + 32$

**48.** $2x + 3$

### Section R.4

**1.** 64 **2.** 49 **3.** −1000 **4.** $\frac{8}{27}$ **5.** −64 **6.** 125 **7.** 16 **8.** 36 **9.** 32 **10.** 48 **11.** 40 **12.** 26 **13.** $\frac{10}{13}$ **14.** 18

**15.** 50 **16.** $\frac{4}{5}$ **17.** 16 **18.** 48 **19.** $\frac{10}{9}$ **20.** $\frac{6}{7}$ **21.** 14 **22.** 7 **23.** 42 **24.** −9 **25.** 20 **26.** −18 **27.** 5 **28.** −15

### Section R.5

**1.** $3 + 11$ **2.** $6 \cdot 7$ **3.** $\frac{y}{4}$ **4.** $3 - z$ **5.** $2(x - 3)$ **6.** $5 + \frac{z}{2}$ **7.** −7 **8.** 41 **9.** 1 **10.** 3

**11.** 11,800 Yen; 118,000 Yen; 1,180,000 Yen **12.** 0°C, 30°C, 100°C **13.** $-5x$ **14.** $11x^2$ **15.** $-8x + 3$ **16.** $-4x + 8y$ **17.** $15y - 1$

**18.** $2.3x^2 + 0.9$ **19.** $-6z + 3$ **20.** $4x - 6$ **21.** $-5y + 11$ **22.** $-3z + 10$ **23.** $-6x + 4$ **24.** 1 **25.** $\frac{25x + 25}{6} = \frac{25(x + 1)}{6}$

**26. (a)** Yes **(b)** Yes **(c)** No **(d)** Yes **27. (a)** Yes **(b)** Yes **(c)** Yes **(d)** No **28. (a)** No **(b)** Yes **(c)** Yes **(d)** No

## Chapter 1

### Section 1.1

**1.** $x = 1$ **2.** $x = -7$ **3.** $z = -1$ **4.** $\{3\}$ **5.** $\{-2\}$ **6.** $\{1/5\}$ **7.** $\{2\}$ **8.** $\{-1\}$ **9.** $\{3/2\}$ **10.** $\{4\}$ **11.** $\{-4\}$
**12.** $\{3/4\}$ **13.** $\{1/2\}$ **14.** $\{2\}$ **15.** $\{-5\}$ **16.** $\{-4\}$ **17.** $\{-3/5\}$ **18.** $\{-3\}$ **19.** $\{10\}$ **20.** $\{32\}$
**21.** $\emptyset$ or { }; Contradiction **22.** $\{x|x \text{ is any real number}\}$; Identity **23.** $\{0\}$; Conditional **24.** $\{x|x \text{ is any real number}\}$; Identity

### Section 1.2

**1.** $3y = 21$ **2.** $2(3 + x) = 5x$ **3.** $x - 10 = x/2$ **4.** $y - 3 = 5y$ **5.** 18, 20, 22 **6.** 25, 26, 27 **7.** \$15 per hour **8.** \$12 per hour
**9.** 150 miles **10.** 240 minutes **11.** 40 **12.** 160 **13.** 75% **14.** \$30 **15.** \$1.20 **16.** \$32.50 **17.** \$10.50; \$1,410.50
**18.** \$67,500 in Aaa-rated bonds; \$22,500 in B-rated bonds **19.** \$5,000 in CD; \$20,000 in Corporate Bond **20.** 6 pounds of Tea A and 4 pounds of Tea B. **21.** 10 pounds of cashews; 20 pounds of peanuts **22.** After 4 hours; 240 miles **23.** After 5 hours; 450 miles

### Section 1.3

**1.** $A = \pi r^2$ **2.** $V = \pi r^2 h$ **3.** $C = 175x + 7000$ **4.** $s = \frac{1}{2}gt^2$ **5. (a)** $h = \frac{2A}{b}$ **(b)** 5 inches **6. (a)** $b = \frac{P - 2a}{2}$ **(b)** 10 cm

**7.** $P = \frac{I}{rt}$ **8.** $y = \frac{C - Ax}{B}$ **9.** $h = \frac{4x - 3}{2x - 3}$ **10.** $n = \frac{S + d}{a + d}$ **11.** width: 40 feet; length: 50 feet **12.** height: 72 inches; width: 40 inches
**13.** 4.00 inches **14.** 5.50 inches

### Section 1.4

**1.** $[-3, 2]$ 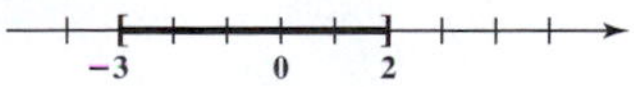 −3 0 2 **2.** $[3, 6)$ 0 3 6 **3.** $(-\infty, 3]$ 0 3

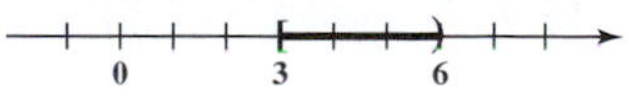

**4.** $\left(\frac{1}{2}, \frac{7}{2}\right)$ 0 $\frac{1}{2}$ $\frac{7}{2}$ 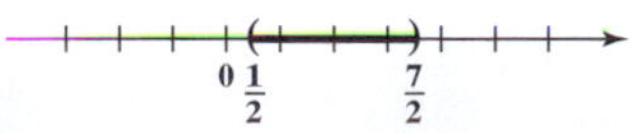 **5.** $0 < x \leq 5$ 0 5 **6.** $-6 < x < 0$ −6 0

**7.** $x > 5$ 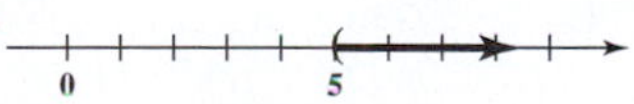 0 5 **8.** $x \leq \frac{8}{3}$ 0 $\frac{8}{3}$ 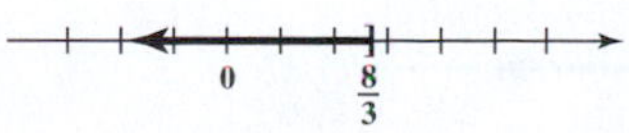 **9.** $\{x|x > 2\}$; $(2, \infty)$ 0 2

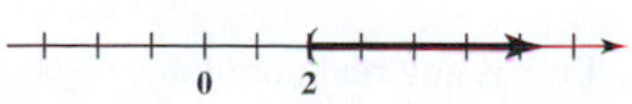

**10.** $\{x|x \leq 6\}$; $(-\infty, 6]$ 0 6

**11.** $\{x|x < 4\}$; $(-\infty, 4)$ 0 4

**12.** $\{x|x > -3\}$; $(-3, \infty)$ −3 0

**13.** $\{x|x \geq -2\}$; $[-2, \infty)$ −2 0

**14.** $\left\{x \middle| x > \frac{5}{2}\right\}$; $\left(\frac{5}{2}, \infty\right)$ 0 $\frac{5}{2}$

**15.** $\{x|x < 4\}$; $(-\infty, 4)$ 0 4

**16.** $\left\{x \middle| x \leq -\frac{7}{3}\right\}$; $\left(-\infty, -\frac{7}{3}\right]$ −4 −3 −2 −1 0

**17.** $\{x|x \geq 4\}$; $[4, \infty)$ 0 1 2 3 4 5

**18.** $\{x|x \geq 3\}$; $[3, \infty)$ 0 3

**19.** $\{x|x < 1/2\}$; $(-\infty, 1/2)$ 0 $\frac{1}{2}$

**20.** $\{x|x > -17\}$; $(-17, \infty)$ −17 −5 0 5

**21.** Any balance over $500

**22.** For any more than 24 boxes, revenue exceeds cost.

## Section 1.5

**1.** $\{1, 3, 5\}$ **2.** $\{2, 4, 6\}$ **3.** $\{1, 2, 3, 4, 5, 6, 7\}$ **4.** $\{1, 2, 3, 4, 5, 6, 8\}$ **5.** $\emptyset$ or $\{\ \}$ **6.** $\{1, 2, 3, 4, 5, 6, 7, 8\}$

**7.** $\{x|2 < x < 7\}$; $(2, 7)$ 0 2 7

**8.** $\{x|x \leq -3 \text{ or } x > 2\}$; $(-\infty, -3] \cup (2, \infty)$ −3 0 2

**9.** $\{x|x \geq 2\}$; $[2, \infty)$ 0 2

**10.** $\{x|-3 < x < 3\}$; $(-3, 3)$ −3 0 3

**11.** $\{x|1 < x < 3\}$; $(1, 3)$ 0 1 2 3 4

**12.** $\{\ \}$ or $\emptyset$

**13.** $\{1\}$ 0 1 2 3

**14.** $\{x|-1 < x < 3\}$; $(-1, 3)$ −3 0 3

**15.** $\left\{x \middle| \frac{5}{4} < x \leq 2\right\}$; $\left(\frac{5}{4}, 2\right]$ 0 $\frac{5}{4}$ 2

**16.** $\{x|-6 \leq x \leq -2\}$; $[-6, -2]$ −6 −2 0

**17.** $\{x|x < -2 \text{ or } x > 5\}$; $(-\infty, -2) \cup (5, \infty)$ −2 0 5

**18.** $\{x|x \leq 2 \text{ or } x > 6\}$; $(-\infty, 2] \cup (6, \infty)$ 0 2 6

**19.** $\{x|x \geq 1\}$; $[1, \infty)$ 0 1

**20.** $\{x|x < 4 \text{ or } x > 9\}$; $(-\infty, 4) \cup (9, \infty)$ 0 4 9

**21.** $\{x|x \text{ is any real number}\}$; $(-\infty, \infty)$ 0 4

**22.** $\{x|x \geq -1\}$; $[-1, \infty)$ −1 0

**23.** To be in the 25% tax bracket, an individual must have income between $29,700 and $71,950.

**24.** Sophia's monthly minutes for the year were between 115 minutes and 255 minutes.

## Section 1.6

**1.** $\{-7, 7\}$ **2.** $\{-1, 1\}$ **3.** $\{-2, 5\}$ **4.** $\{-5/3, 3\}$ **5.** $\{-1, 9/5\}$ **6.** $\{-5, 1\}$ **7.** $\{\ \}$ or $\emptyset$ **8.** $\{\ \}$ or $\emptyset$ **9.** $\{-1\}$

**10.** $\{-8, -2/3\}$ **11.** $\{-2, 3\}$ **12.** $\{-3, 0\}$ **13.** $\{2\}$ **14.** $\{x|-5 \leq x \leq 5\}$; $[-5, 5]$ −5 0 5

**15.** $\{x|-3/2 < x < 3/2\}$; $(-3/2, 3/2)$ $-\frac{3}{2}$ 0 $\frac{3}{2}$

**16.** $\{x|-8 < x < 2\}$; $(-8, 2)$ −8 0 2

**17.** $\{x|-2 \leq x \leq 5\}$; $[-2, 5]$ −2 0 5

**18.** $\{\ \}$ or $\emptyset$

**19.** $\{x|-2 < x < 2\}$; $(-2, 2)$ −2 0 2

**20.** $\{x|-1 \leq x \leq 7\}$; $[-1, 7]$ −1 0 7

**21.** $\{x|-2 \leq x \leq 1\}$; $[-2, 1]$ −2 0 1

**22.** $\{x|-7/3 < x < 3\}$; $(-7/3, 3)$ $-\frac{7}{3}$ 0 3

**23.** $\{x|x \leq -6 \text{ or } x \geq 6\}$; $(-\infty, -6] \cup [6, \infty)$ −6 0 6

**24.** $\{x|x < -5/2 \text{ or } x > 5/2\}$; $(-\infty, -5/2) \cup (5/2, \infty)$ $-\frac{5}{2}$ 0 $\frac{5}{2}$

**25.** $\{x|x < -7 \text{ or } x > 1\}$; $(-\infty, -7) \cup (1, \infty)$ −7 0 1

**26.** $\{x|x \leq -1/2 \text{ or } x \geq 2\}$; $(-\infty, -1/2] \cup [2, \infty)$ $-\frac{1}{2}$ 0 2

**27.** $\{x|x < -5/3 \text{ or } x > 3\}$; $(-\infty, -5/3) \cup (3, \infty)$ $-\frac{5}{3}$ 0 3

**28.** $\{x|x \neq -5/2\}$; $(-\infty, -5/2) \cup (-5/2, \infty)$ $-\frac{5}{2}$ 0

**29.** $\{x|x \text{ is any real number}\}$; $(-\infty, \infty)$ 0

**30.** $\{x|x \text{ is any real number}\}$; $(-\infty, \infty)$ 0

**31.** The acceptable belt width is between 127/32 inches and 129/32 inches.

**32.** The percentage of people that have been shot at is between 7.3 percent and 10.7 percent.

# Chapter 2

## Section 2.1

**1.** $A$: Quadrant I $B$: Quadrant IV $C$: $y$-axis $D$: Quadrant III **2.** $A$: Quadrant II $B$: $x$-axis $C$: Quadrant IV $D$: Quadrant I **3. (a)** No **(b)** Yes **(c)** Yes

**4. (a)** Yes **(b)** No **(c)** Yes **5.** **6.** **7.** **8.** **9.**

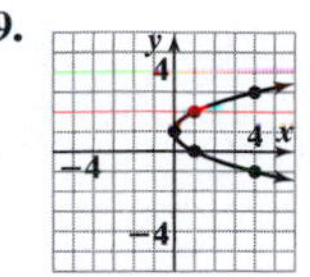

**10.** Intercepts: $(-5, 0)$, $(0, -0.9)$, $(1, 0)$, $(6.7, 0)$; $x$-intercepts: $-5, 1, 6.7$; $y$-intercept: $-0.9$ **11. (a)** \$200 thousand **(b)** \$350 thousand **(c)** The capacity of the refinery is 700 thousand gallons of gasoline per hour. **(d)** The intercept is $(0, 100)$. The cost of \$100 thousand for producing 0 gallons of gasoline can be thought of as fixed costs.

## Section 2.2

**1.** {(Max, November 8), (Alesia, January 20), (Trent, March 3), (Yolanda, November 8), (Wanda, July 6), (Elvis, January 8)}

**2.** Domain / Range: 1 → 3; 5 → 4; 8 → 4; 10 → 13

**3.** Domain: {Max, Alesia, Trent, Yolanda, Wanda, Elvis}; Range: {January 20, March 3, July 6, November 8, January 8}
**4.** Domain: {1, 5, 8, 10}; Range: {3, 4, 13} **5.** Domain: $\{-2, -1, 2, 3, 4\}$; Range: $\{-3, -2, 0, 2, 3\}$
**6.** Domain: $\{x \mid -2 \le x \le 4\}$ or $[-2, 4]$; Range: $\{y \mid -2 \le y \le 2\}$ or $[-2, 2]$
**7.** Domain: $\{x \mid x \text{ is a real number}\}$ or $(-\infty, \infty)$; Range: $\{y \mid y \text{ is a real number}\}$ or $(-\infty, \infty)$

**8.**

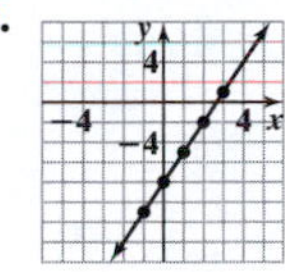

Domain: $\{x \mid x \text{ is a real number}\}$ or $(-\infty, \infty)$
Range: $\{y \mid y \text{ is a real number}\}$ or $(-\infty, \infty)$

**9.**

Domain: $\{x \mid x \text{ is a real number}\}$ or $(-\infty, \infty)$
Range: $\{y \mid y \ge -8\}$ or $[-8, \infty)$

**10.**

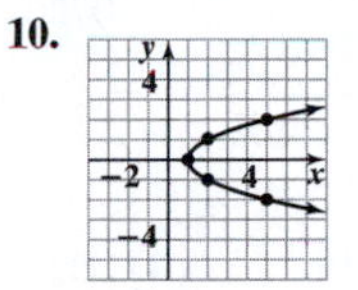

Domain: $\{x \mid x \ge 1\}$ or $[1, \infty)$
Range: $\{y \mid y \text{ is a real number}\}$ or $(-\infty, \infty)$

## Section 2.3

**1.** Function; Domain: {Max, Alesia, Trent, Yolanda, Wanda, Elvis}; Range: {January 20, March 3, July 6, November 8, January 8} **2.** Not a function **3.** Function; Domain: $\{-3, -2, -1, 0, 1\}$, Range: $\{0, 1, 2, 3\}$ **4.** Not a function **5.** Function **6.** Not a function **7.** Function **8.** Not a function **9.** Function **10.** Not a function **11.** 14 **12.** $-13$ **13.** $3x - 4$ **14.** $3x - 6$

**15.** **16.**

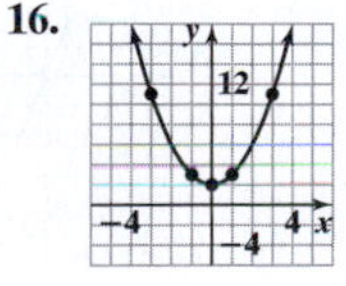

**17.**

**18. (a)** Independent variable: $t$; dependent variable: $A$
**(b)** $A(30) \approx 706.86$ square miles. After 30 days, the area contaminated with oil will be a circle covering about 706.86 square miles.

## Section 2.4

**1.** $\{x \mid x \text{ is a real number}\}$; $(-\infty, \infty)$ **2.** $\{x \mid x \ne 3\}$; $(-\infty, 3) \cup (3, \infty)$ **3.** $\{r \mid r > 0\}$; $(0, \infty)$
**4. (a)** Domain: $\{x \mid x \text{ is a real number}\}$; $(-\infty, \infty)$; Range: $\{y \mid y \le 2\}$; $(-\infty, 2]$
**(b)** $(-2, 0)$, $(0, 2)$, $(2, 0)$; $x$-intercepts: $-2$ and $2$; $y$-intercept: 2
**5. (a)** $f(-3) = -15$; $f(1) = -3$ **(b)** Domain: $\{x \mid x \text{ is a real number}\}$ or $(-\infty, \infty)$
**(c)** Range: $\{y \mid y \text{ is a real number}\}$ or $(-\infty, \infty)$
**(d)** $(-2, 0)$, $(0, 0)$, $(2, 0)$; $x$-intercepts: $-2$, $0$, and $2$; $y$-intercept: 0 **(e)** $\{3\}$
**6. (a)** No **(b)** $f(3) = -2$; $(3, -2)$ is on the graph
**(c)** $x = 5$; $(5, -8)$ is on the graph
**7.**

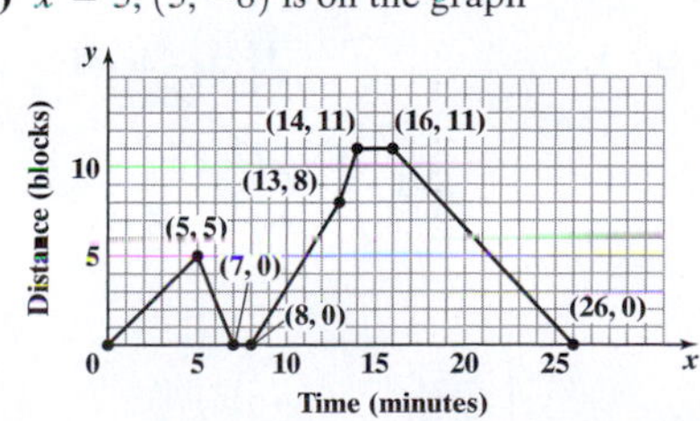

# Chapter 3

## Section 3.1

**1.** 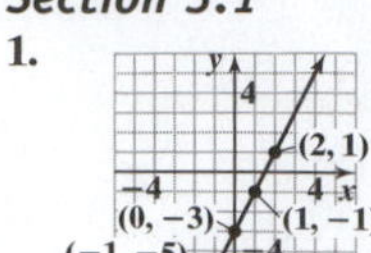

**2.** 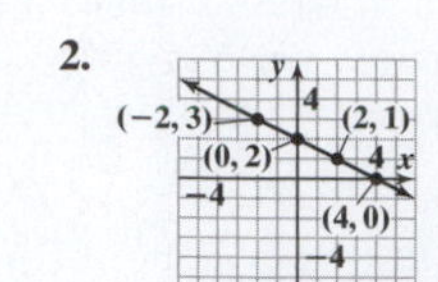

**3.** 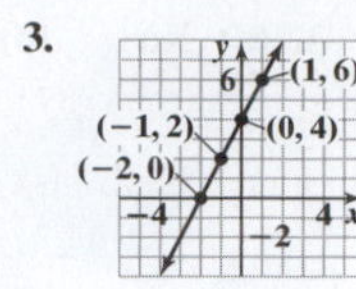

**4.** 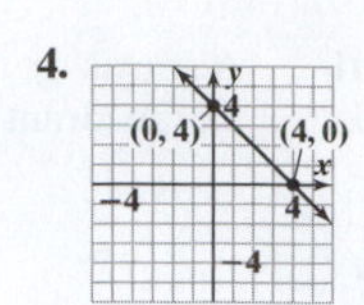

**5.** 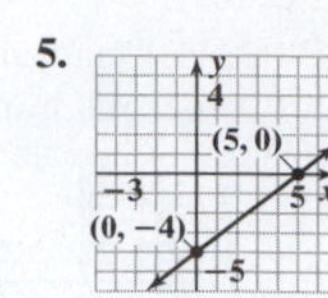

**6.** 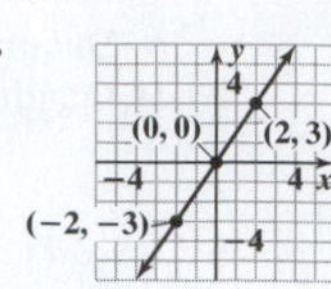

**7.** 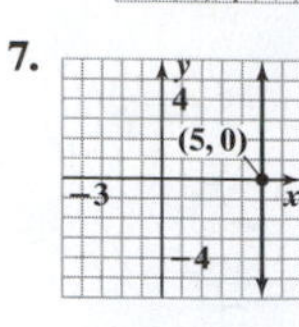

**8.** 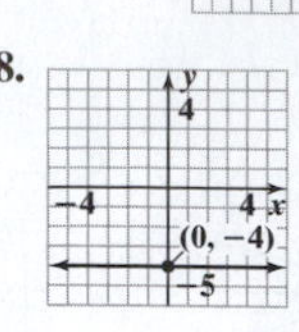

**9. (a)** $\{x \mid x \geq 0\}$ or $[0, \infty)$ **(b)** $(0, 40)$ **(c)** \$68 **(d)**

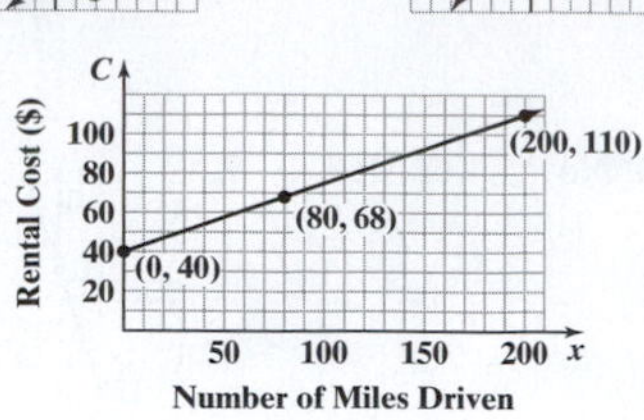


**(e)** 130 miles

## Section 3.2

**1.** 3; For every 1-unit increase in $x$, $y$ will increase by 3 units. **2.** $-\frac{7}{4}$; For every 4-unit increase in $x$, $y$ will decrease by 7 units.

**3.** 0; For every 1-unit increase in $x$, there is no change in $y$; Horizontal line **4.** Undefined; Vertical line **5.** $L_1$: $m = \frac{1}{5}$; $L_2$: $m$ is undefined; $L_3$: $m = -1$; $L_4$: $m = 0$

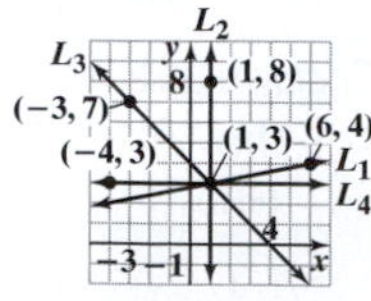

**6. (a)**

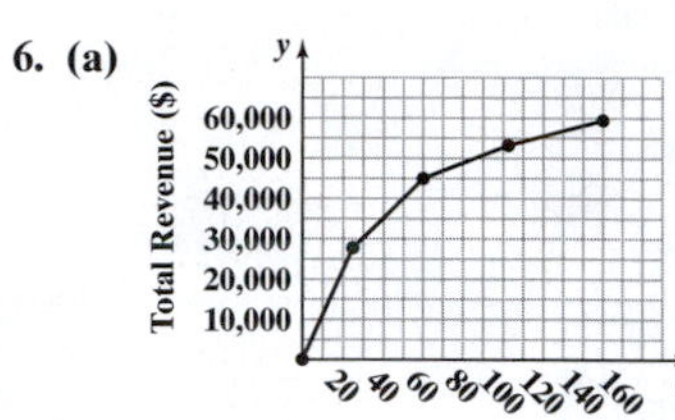


**(b)** \$1120 per bicycle. For each bicycle sold, total revenue increased by \$1120 when between 0 and 25 bicycles were sold.

**(c)** \$120 per bicycle. For each bicycle sold, total revenue increased by \$120 per bicycle when between 102 and 150 bicycles were sold.

**(d)** No, because the average rate of change (slope) is not constant.

**7. (a)** 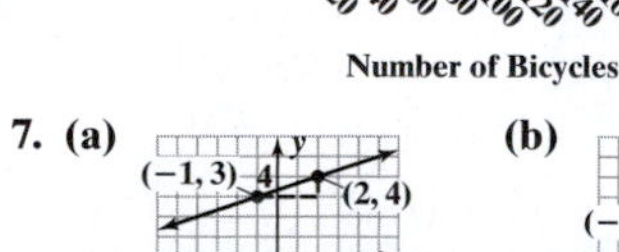

**(b)** 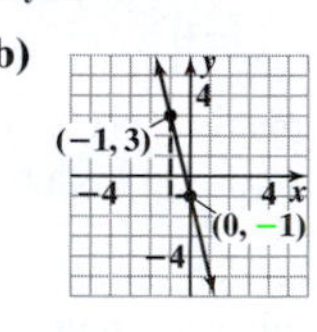

**(c)** 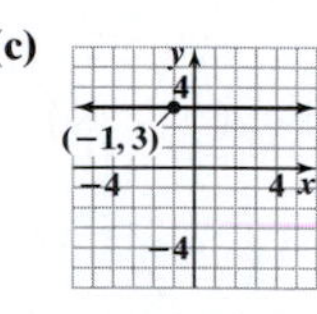

**8.** $y - 5 = 2(x - 3)$

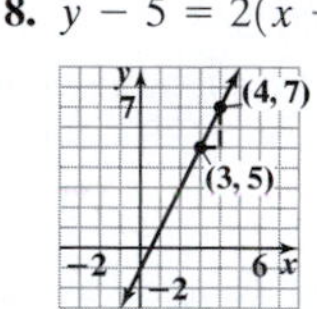

**9.** $y - 3 = -4(x + 2)$

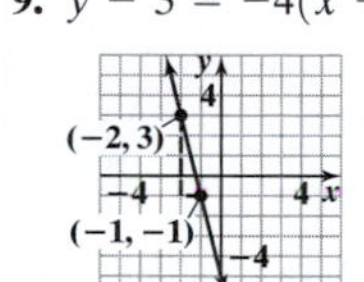

**10.** $y + 4 = \frac{1}{3}(x - 3)$

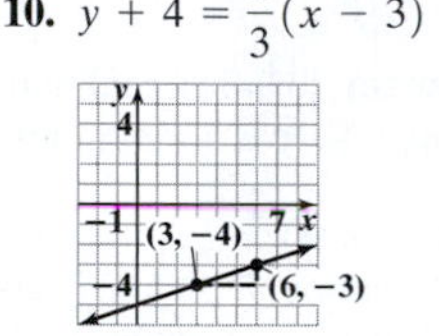

**11.** $y = -2$

**12.** $m = 3$; $y$-intercept: $-2$

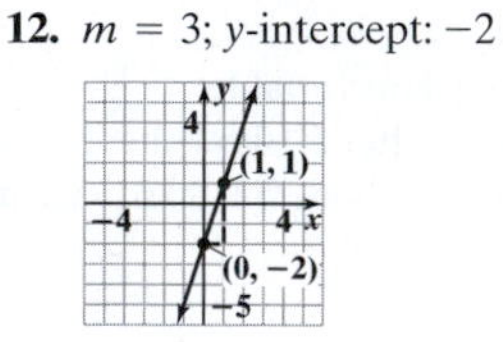

**13.** $m = -3$; $y$-intercept: 4

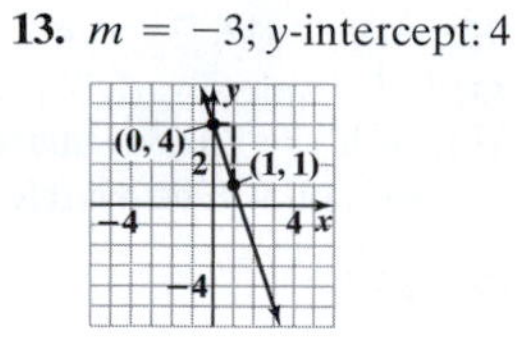

**14.** $m = \frac{3}{2}$; $y$-intercept: $-\frac{7}{2}$

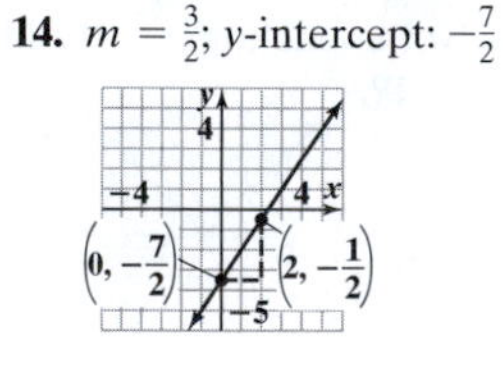

**15.** $m = -\frac{7}{3}$; $y$-intercept: 0

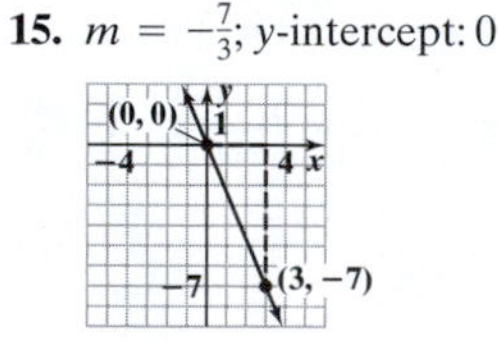

**16.** $y = 2x + 1$

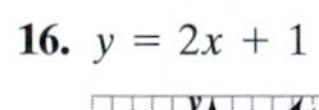

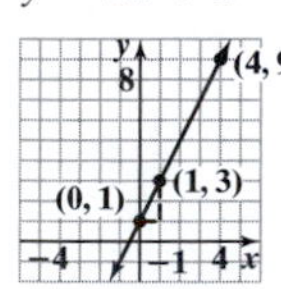

**17.** $y = -\frac{1}{2}x + 3$

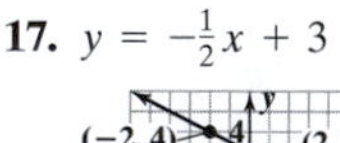

**18.** $x = 3$

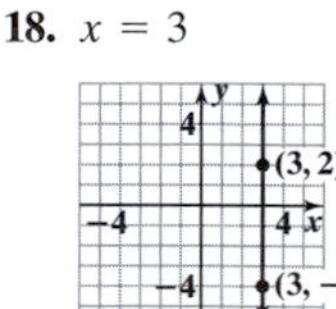

**19. (a)** $W(a) = 9a + 3082$

**(b)** 3352 g

**(c)** If the mother's age increases by 1 year, the average birth weight increases by 9 grams.

**(d)** About 25 years old

## Section 3.3

**1.** Not parallel **2.** Parallel **3.** Not parallel

**4.** $y = 3x - 7$

**5.** $y = -\frac{3}{2}x + 1$

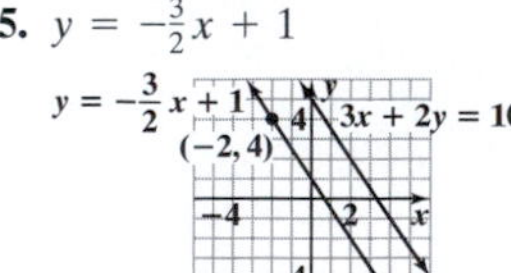

**6.** $\frac{1}{3}$

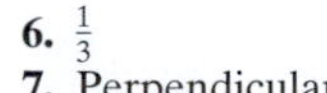

**7.** Perpendicular

**8.** Not perpendicular

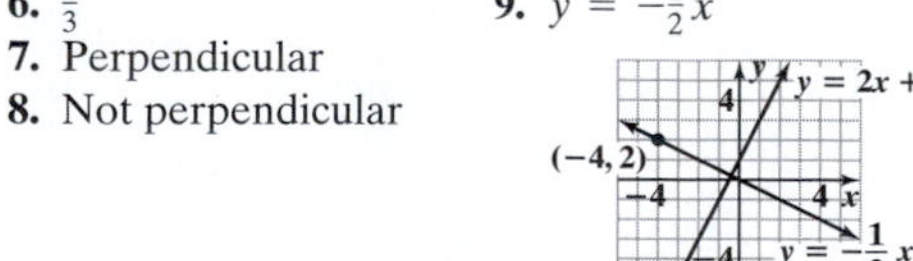

**9.** $y = -\frac{1}{2}x$

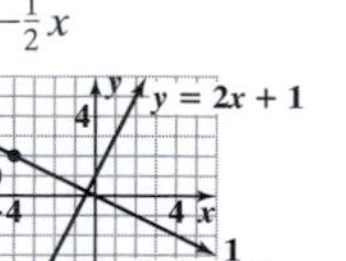

**10.** $y = -\frac{4}{3}x - 8$

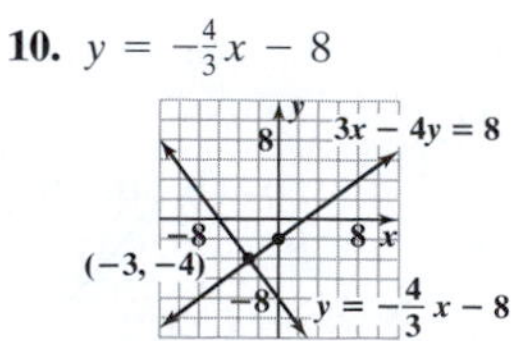

## Section 3.4

**1. (a)** No **(b)** Yes **(c)** Yes **(d)** Yes

**2.**

**3.** 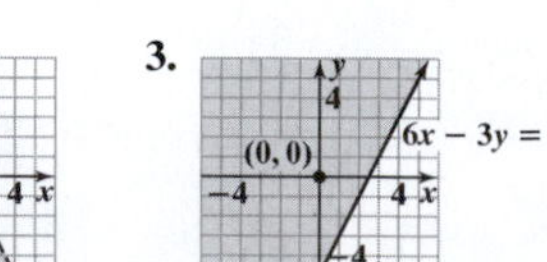

**4.** 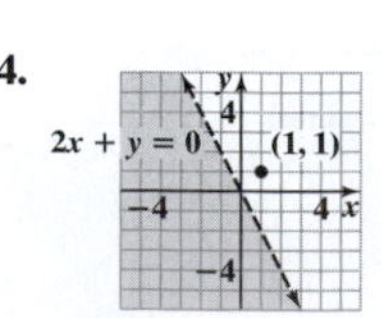

**5. (a)** $430x + 330y \leq 800$ **(b)** Yes **(c)** No

## Section 3.5

**1. (a)** $C(x) = 81x + 2000$
**(b)**

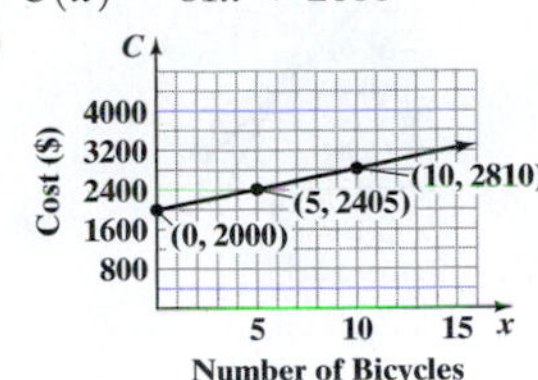


**(c)** \$3458
**(d)** 25

**2. (a)** $C(x) = 0.18x + 250$
**(b)** $[0, \infty)$
**(c)** \$307.60
**(d)**

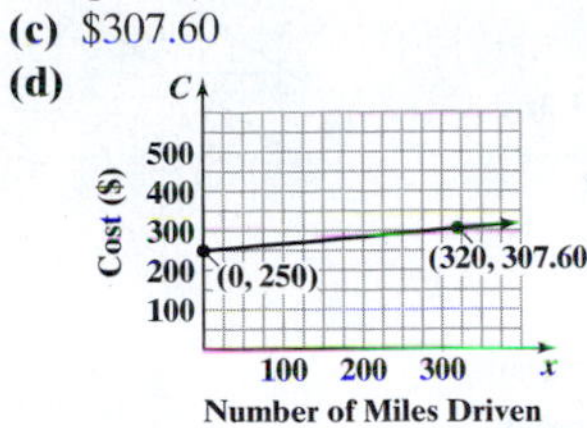


**(e)** 180 miles

**3. (a)** $C(g) = 2.79875g$
**(b)** \$15.39
**(c)**

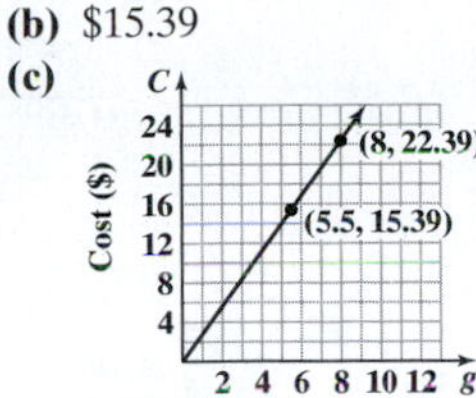


**4. (a)**

**(b)** As age increases, total cholesterol also increases.

**5.** Nonlinear **6.** Linear with positive slope
**7. (a)** Answers will vary. Using (25, 180) and (65, 269): $y = f(x) = 2.225x + 124.375$
**(b)** Answers will vary. **(c)** 211 **(d)** Each year, a male's total cholesterol increases by 2.225 mg/dL; No

# Chapter 4

## Section 4.1

**1. (a)** No **(b)** Yes **(c)** No **2.** (3, 1) **3.** (−2, 3) **4.** (−4, 7) **5.** (−6, 10) **6.** (−5, −6) **7.** (3, 6) **8.** ∅ or { }; inconsistent
**9.** $\{(x, y) | -3x + 2y = 8\}$

## Section 4.2

**1.** Cheeseburger: \$1.60; shake: \$1.85 **2.** Width: 60 yards; length: 120 yards **3.** $x = 15$, $y = 30$ **4.** \$96,000 in Aaa-rated bonds; \$24,000 in B-rated bonds **5.** 10 pounds of cashews; 20 pounds of peanuts **6.** Airspeed: 350 mph; wind resistance: 50 mph
**7. (a)** $R(x) = 230x$ **(b)** $C(x) = 160x + 2100$ **(c)** **(d)** 30 trees; \$6900

## Section 4.3

**1. (a)** No **(b)** Yes **2.** (−3, 1, −1) **3.** $\left(-\frac{5}{2}, \frac{5}{4}, \frac{1}{2}\right)$ **4.** ∅ or { }; the system is inconsistent
**5.** $\{(x, y, z) | x = -z + 1, y = 2z - 1, z \text{ is any real number}\}$ **6.** Fourteen 21-inch, eleven 24-inch, and 5 riders

## Section 4.4

**1.** $\left[\begin{array}{cc|c} 3 & -1 & -10 \\ -5 & 2 & 0 \end{array}\right]$ **2.** $\left[\begin{array}{ccc|c} 1 & 2 & -2 & 11 \\ -1 & 0 & -2 & 4 \\ 4 & -1 & 1 & 3 \end{array}\right]$ **3.** $\begin{cases} x - 3y = 7 \\ -2x + 5y = -3 \end{cases}$ **4.** $\begin{cases} x - 3y + 2z = 4 \\ 3x \qquad\quad - z = -1 \\ -x + 4y \qquad = 0 \end{cases}$ **5.** $\left[\begin{array}{cc|c} 1 & -2 & 5 \\ 0 & -3 & 9 \end{array}\right]$

**6.** $R_1 = -5r_2 + r_1$; $\left[\begin{array}{cc|c} 1 & 0 & 3 \\ 0 & 1 & 2 \end{array}\right]$ **7.** (6, −2) **8.** (3, −2, 1) **9.** $\{(x, y, z) | x = -z + 5, y = 4z + 3, z \text{ is any real number}\}$.
**10.** ∅ or { }; inconsistent

## Section 4.5

**1.** 18 **2.** −9 **3.** (−3, 5) **4.** Dependent or inconsistent system **5.** −91 **6.** (−1, 4, 1)

### Section 4.6

**1. (a)** not a solution **(b)** solution

**2.**

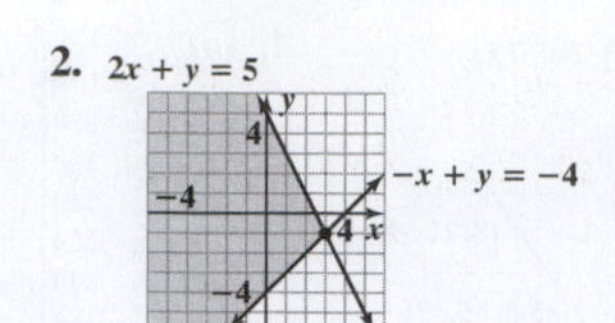


**3.**

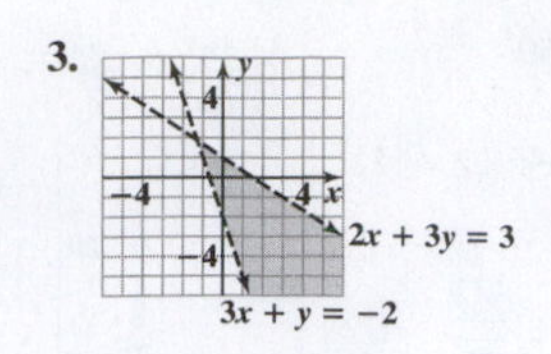


**4.**

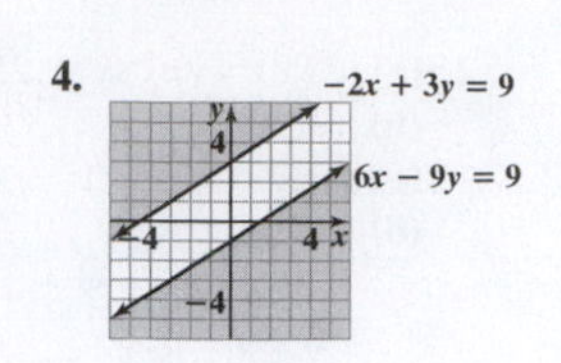


**5.**

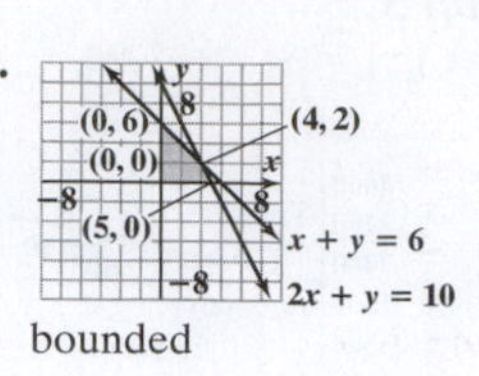


bounded

**6.** $\begin{cases} x + y \le 25{,}000 \\ x \ge 10{,}000 \\ y \le 15{,}000 \\ y \ge 0 \end{cases}$ where $x$ is amount in Treasury notes and $y$ is amount in corporate bonds

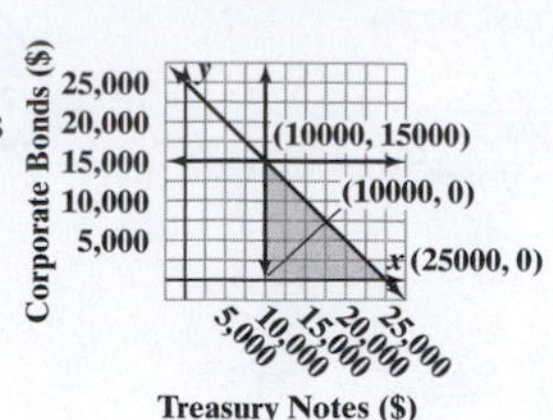


## Getting Ready for Chapter 5

**1.** 125 **2.** −243 **3.** $y^7$ **4.** $-10x^7$ **5.** $-6y^5$ **6.** 25 **7.** $y^2$ **8.** $\frac{8}{5}a$ **9.** $-\frac{3}{2}b^2$ **10.** $\frac{1}{125}$ **11.** $\frac{5}{z^7}$ **12.** $x^4$ **13.** $5y^3$
**14.** −1 **15.** 1 **16.** $\frac{9}{16}$ **17.** −64 **18.** $\frac{x^2}{9}$ **19.** 20 **20.** $\frac{1}{36}$ **21.** 100 **22.** $\frac{20x^3}{y}$ **23.** $\frac{2}{3}ab^4$ **24.** $-\frac{3}{2}b^8$ **25.** $\frac{10t^5}{3s^3}$
**26.** 64 **27.** 1 **28.** 4096 **29.** $a^{15}$ **30.** $\frac{1}{z^{18}}$ **31.** $s^{21}$ **32.** $125y^3$ **33.** 1 **34.** $81x^8$ **35.** $\frac{1}{16a^6}$ **36.** $\frac{z^4}{81}$ **37.** $\frac{32}{x^5}$ **38.** $\frac{x^8}{y^{12}}$
**39.** $\frac{27}{a^6b^{12}}$ **40.** $\frac{3x^3y^4}{4}$ **41.** $\frac{3b^5}{4a}$ **42.** $\frac{x^6}{y^{10}}$ **43.** $5.32 \times 10^2$ **44.** $-1.23 \times 10^6$ **45.** $3.4 \times 10^{-2}$ **46.** $-8.45 \times 10^{-5}$ **47.** 500
**48.** 910,000 **49.** 0.00018 **50.** 0.000001 **51.** $6 \times 10^8$ **52.** $8 \times 10^{-11}$ **53.** $2.4 \times 10^4$ **54.** $2 \times 10^2$ **55.** $2 \times 10^{-3}$ **56.** $5 \times 10^3$
**57.** $3.75 \times 10^{-13}$ **58.** 240,000,000,000 **59.** 0.003 **60.** 750 **61.** 0.000000000003

## Chapter 5

### Section 5.1

**1.** Monomial; 8; 5 **2.** Not a monomial **3.** Monomial; 12; 0 **4.** Not a monomial **5.** Coefficient: 3, Degree: 7
**6.** Coefficient: −2, Degree: 4 **7.** Not a monomial **8.** Coefficient: −1, Degree: 2 **9.** Polynomial; 3 **10.** Not a polynomial
**11.** Not a polynomial **12.** Polynomial; 2 **13.** Polynomial; 4 **14.** $6x^2 + 2x - 2$ **15.** $3w^4 - 2w^3 - 7w^2 + w - 5$
**16.** $5x^2y + 7x^2y^2 - 4xy^2$ **17.** $x^3 - 16x^2 + 7x + 2$ **18.** $11y^3 - 5y^2 - 3y - 7$ **19.** $11x^2y - 3x^2y^2 - 10xy^2$
**20. (a)** 1 **(b)** −1 **(c)** 34 **21. (a)** We estimate the fertility rate for women 15–19 years of age in 1994 was 58.93 births per 1,000 women. **(b)** We estimate the fertility rate for women 15–19 years of age in 2008 was 49.69 births per 1,000 women. **22. (a)** $2x^2 + 4x - 5$ **(b)** $4x^2 - 6x + 7$ **(c)** 1 **(d)** 35 **23. (a)** $4x - 1250$ **(b)** \$1950. If the company manufactures and sells 800 calculators, its profit will be \$1950.

### Section 5.2

**1.** $6x^7$ **2.** $-21a^4b^6$ **3.** $\frac{5}{4}x^5$ **4.** $-3x - 6$ **5.** $5x^3 + 15x^2 + 10x$ **6.** $6x^3y - 10x^2y^2 + 4xy^3$ **7.** $y^4 + \frac{1}{6}y^3 + 4y^2$ **8.** $x^2 + 5x + 4$
**9.** $6v^2 + v - 15$ **10.** $2a^2 + 9ab - 5b^2$ **11.** $2y^3 + 5y^2 - 2y - 15$ **12.** $2z^4 - 5z^3 + 7z^2 - 16z + 12$ **13.** $25y^2 - 4$ **14.** $49y^2 - 4z^6$
**15.** $z^2 - 16z + 64$ **16.** $36p^2 + 60p + 25$ **17.** $16a^2 - 24ab + 9b^2$ **18. (a)** 77 **(b)** $5x^3 + 12x^2 - 4x - 3$ **(c)** 77
**19. (a)** $x^2 - 8x + 15$ **(b)** $2xh + h^2 - 2h$

### Section 5.3

**1.** $3p^3 - 4p^2 + p$ **2.** $7m^2 - 5m + 1$ **3.** $\frac{xy^3}{4} + \frac{2y}{x} - \frac{1}{x^2}$ **4.** $x^2 + 7x - 3 + \frac{9}{x - 4}$ **5.** $x^2 + 5x + 4$ **6.** $x^3 - 5x^2 + 2 + \frac{6}{x^2 - 2}$
**7.** $2x^2 + 5x + 3 - \frac{7}{x - 2}$ **8.** $x^3 + 5x^2 - 2$ **9. (a)** $3x^2 - 4x + 3 + \frac{2x + 1}{x^2 - 2}$ **(b)** 19 **10. (a)** 42 **(b)** 0
**11. (a)** $f(-2) = -35$; $x + 2$ is not a factor **(b)** $f(5) = 0$; $f(x) = (x - 5)(2x^2 + x - 1)$

### Section 5.4

**1.** 5 **2.** $2z$ **3.** $3xy^3$ **4.** $7z(z - 2)$ **5.** $2y(3y^2 - 7y + 5)$ **6.** $2m^2n^2(m^2 + 4mn^2 - 3n^3)$ **7.** $-5y(y - 2)$ **8.** $-3a(a^2 - 2a + 4)$
**9.** $(a - 3)(4a + 3)$ **10.** $3(w - 5)(w + 1)$ **11.** $(x + y)(5 + b)$ **12.** $(w^2 + 4)(w - 3)$ **13.** $(2z + 3)(3z + 1)$ **14.** $(2x + 1)(x - 5)$

### Section 5.5

**1.** $(y + 6)(y + 3)$ **2.** $(p + 2)(p + 12)$ **3.** $(q - 4)(q - 2)$ **4.** $(x - 2)(x - 6)$ **5.** $(w - 7)(w + 3)$ **6.** $(q - 12)(q + 3)$
**7.** Prime **8.** $2x(x - 9)(x + 3)$ **9.** $-3(z + 2)(z + 5)$ **10.** $(x + 3y)(x + 5y)$ **11.** $(m + 5n)(m - 4n)$ **12.** $(2x + 3)(3x + 1)$
**13.** $(2b - 3)(b + 5)$ **14.** $(2x + 1)(4x + 5)$ **15.** $(6y - 5)(2y + 7)$ **16.** $(6x - y)(5x + 2y)$ **17.** $-1(6y + 1)(y - 4)$
**18.** $-1(3x + 2y)(3x + 5y)$ **19.** $(y^2 + 4)(y^2 - 6)$ **20.** $(x - 1)(4x - 15)$

### Section 5.6

**1.** $(x-9)^2$ **2.** $(2x+5y)^2$ **3.** $2(3p^2-7)^2$ **4.** $(z-4)(z+4)$ **5.** $(4m-9n)(4m+9n)$ **6.** $(2a-3b^2)(2a+3b^2)$
**7.** $3(b+2)(b-2)(b^2+4)$ **8.** $(p-4-q)(p-4+q)$ **9.** $(z+4)(z^2-4z+16)$ **10.** $(5p-6q^2)(25p^2+30pq^2+36q^4)$
**11.** $4(2m+5n^2)(4m^2-10mn^2+25n^4)$ **12.** $(-2x+1)(13x^2+5x+1)$

### Section 5.7

**1.** $2q(p+5q)(p-9q)$ **2.** $-3y(3x+1)(5x-9)$ **3.** $(9x-10y)(9x+10y)$ **4.** $-3n(m-7)(m+7)$ **5.** $(p-8q)^2$ **6.** $5(2x+3)^2$
**7.** $(4y-5)(16y^2+20y+25)$ **8.** $-2(2m+n)(4m^2-2mn+n^2)$ **9.** $5(2z^2-3z+7)$ **10.** $3x(2y^2+27x^2)$ **11.** $(x^2+2)(2x+5)$
**12.** $3(3x+1)(x-1)(x+1)$ **13.** $(2x+y-9)(2x+y+9)$ **14.** $(4-m-4n)(4+m+4n)$

### Section 5.8

**1.** $\{-7, 0\}$ **2.** $\left\{-\frac{3}{4}, 3\right\}$ **3.** $\{2, 3\}$ **4.** $\left\{-\frac{1}{3}, 5\right\}$ **5.** $\left\{-2, -\frac{2}{3}\right\}$ **6.** $\{-2, -1\}$ **7.** $\{-6, 4\}$ **8.** $\{-3, 1, 3\}$
**9. (a)** $-1$ and 9; $(-1, 12)$ and $(9, 12)$ are points on the graph of $g$. **(b)** 1 and 7; $(1, -4)$ and $(7, -4)$ are points on the graph of $g$.
**10.** Zeros: $-4$ and $\frac{5}{2}$; $x$-intercepts: $-4$ and $\frac{5}{2}$ **11.** 9 miles by 15 miles **12.** 60 boxes **13. (a)** After 4 seconds and after 6 seconds
**(b)** After 10 seconds

## Getting Ready for Chapter 6

**1.** $\frac{5}{6}$ **2.** $\frac{20}{3}$ **3.** $-\frac{3}{2}$ **4.** $\frac{1}{2}$ **5.** $\frac{5}{3}$ **6.** $-\frac{3}{5}$ **7.** $\frac{4}{3}$ **8.** $-\frac{5}{9}$ **9.** LCD $= 75$; $\frac{3}{25} = \frac{9}{75}, \frac{2}{15} = \frac{10}{75}$
**10.** LCD $= 126$; $\frac{5}{18} = \frac{35}{126}, -\frac{1}{63} = -\frac{2}{126}$ **11.** $\frac{19}{20}$ **12.** $\frac{17}{60}$ **13.** $-\frac{1}{6}$

## Chapter 6

### Section 6.1

**1.** $\{x \mid x \neq -6\}$ **2.** $\{z \mid z \neq 4, z \neq -7\}$ **3.** $\frac{x-4}{x+7}$ **4.** $\frac{z^2+4z+16}{2z+5}$ **5.** $-(w+5)$ **6.** $\frac{(3p+2)(p-1)}{2(p+2)}$ **7.** $-1$ **8.** $-(m+n)$
**9.** $9a^2b^3$ **10.** $\frac{(m-5)(m+1)}{2}$ **11.** $\{x \mid x \neq -6, x \neq 5\}$ **12. (a)** $R(x) = x+1$; $\left\{x \mid x \neq -3, x \neq \frac{5}{3}, x \neq 5\right\}$
**(b)** $H(x) = \frac{3x(x-5)}{4x+3}$; $\left\{x \mid x \neq -1, x \neq -\frac{3}{4}, x \neq 0, x \neq \frac{5}{3}\right\}$

### Section 6.2

**1.** $2x-1$ **2.** $\frac{3(x+3)}{x+5}$ **3.** $\frac{4x-3}{x-5}$ **4.** $24x^2y^3$ **5.** $(x+2)^2(x-7)$ **6.** $\frac{9a+8}{30a^2}$ **7.** $\frac{4x^2+10x-5}{(x-1)(x+2)}$
**8.** $\frac{(x-1)(3x+7)}{(x+2)(x+4)(2x+3)}$ **9.** $\frac{x^2+15x+5}{(2x-3)(x+2)^2}$ **10.** $\frac{-4}{x-2}$

### Section 6.3

**1.** $\frac{4(z-4)}{z}$ **2.** $\frac{1}{4}$ **3.** $\frac{4(z-4)}{z}$ **4.** $\frac{-4(x+2)}{x(x+5)}$ **5.** $\frac{ab}{3b-a}$

### Section 6.4

**1.** $\{-2\}$ **2.** $\{-2\}$ **3.** $\{8\}$ **4.** $\{\ \}$ or $\emptyset$ **5.** $\{-5\}$ **6.** $\left\{-1, \frac{3}{2}\right\}$; the points $(-1, 1)$ and $\left(\frac{3}{2}, 1\right)$ are on the graph.
**7.** The concentration of the drug will be 4 milligrams per liter after $\frac{1}{2}$ hour and after 12 hours.

### Section 6.5

**1.** $\{x \mid x < -3 \text{ or } x \geq 7\}$; $(-\infty, -3) \cup [7, \infty)$ 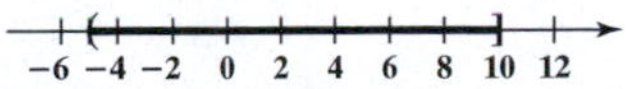

**2.** $\{x \mid -2 < x < 1\}$; $(-2, 1)$ 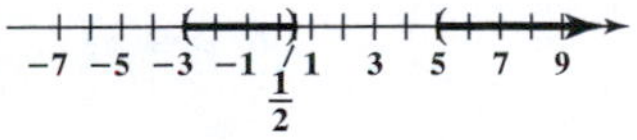


### Section 6.6

**1. (a)** $b = \frac{Y-G}{Y}$ **(b)** 0.9 **2.** $AB = 12$; $DF = 4$ **3.** 299,250,000 people **4.** $\frac{40}{3}$ hours or 13 hours, 20 minutes
**5.** 33.33 minutes or 33 minutes, 20 seconds **6.** 4 miles per hour **7. (a)** $V(l) = \frac{15{,}000}{l}$ **(b)** 300 oscillations per second **8.** 1750 joules
**9.** 1.44 ohms

## Getting Ready for Chapter 7

**1.** 9 **2.** 30 **3.** $\frac{1}{2}$ **4.** 0.4 **5.** 13 **6.** 15 **7.** 10 **8.** 14 **9.** 7 **10.** rational; 20 **11.** irrational; $\approx 6.32$
**12.** not a real number **13.** rational; $-14$ **14.** 14 **15.** $|z|$ **16.** $|2x+3|$ **17.** $|p-6|$

# Chapter 7

### Section 7.1

**1.** 5 **2.** 3 **3.** −6 **4.** not a real number **5.** $\frac{1}{2}$ **6.** 3.68 **7.** 2.99 **8.** 2.09 **9.** 5 **10.** $|z|$ **11.** $3x - 2$ **12.** 2 **13.** $-\frac{2}{3}$ **14.** 5 **15.** −3 **16.** −8 **17.** not a real number **18.** $\sqrt{b}$ **19.** $(8b)^{\frac{1}{5}}$ **20.** $\left(\frac{mn^5}{3}\right)^{\frac{1}{8}}$ **21.** 64 **22.** 9 **23.** −8 **24.** 16 **25.** not a real number **26.** 13.57 **27.** 1.74 **28.** $a^{\frac{3}{8}}$ **29.** $(12ab^3)^{\frac{9}{4}}$ **30.** $\frac{1}{9}$ **31.** 4 **32.** $\frac{1}{(13x)^{3/2}}$

### Section 7.2

**1.** $5^{11/12}$ **2.** 8 **3.** 10 **4.** $a \cdot b^{5/6}$ **5.** $\frac{4x^{1/2}}{y^{2/3}}$ **6.** $\frac{5x^{5/8}}{y^{1/8}}$ **7.** $\frac{200a^{\frac{1}{2}}}{b^{\frac{2}{3}}}$ **8.** 6 **9.** $2a^2b^3$ **10.** $\sqrt[12]{x^5}$ **11.** $\sqrt[6]{a}$ **12.** $x^{1/2}(20x + 9)$ **13.** $\frac{12x + 1}{x^{2/3}}$

### Section 7.3

**1.** $\sqrt{77}$ **2.** $\sqrt[4]{42}$ **3.** $\sqrt{x^2 - 25}$ **4.** $\sqrt[7]{20p^4}$ **5.** $4\sqrt{3}$ **6.** $12\sqrt[3]{2}$ **7.** $10|a|\sqrt{2}$ **8.** Fully simplified **9.** $2 + \sqrt{5}$ **10.** $\frac{-1 + 2\sqrt{2}}{2}$ or $-\frac{1}{2} + \sqrt{2}$ **11.** $5a^3\sqrt{3}$ **12.** $3a^2\sqrt{2a}$ **13.** $4x^2y^3\sqrt[3]{2y}$ **14.** $2ab^2\sqrt[4]{ab^3}$ **15.** $4\sqrt{3}$ **16.** $2a^2\sqrt[3]{15}$ **17.** $8ab^3\sqrt[3]{6a}$ **18.** $\frac{\sqrt{13}}{7}$ **19.** $\frac{3p}{2}$ **20.** $\frac{q\sqrt[4]{3}}{2}$ **21.** $2a^2$ **22.** $-2x$ **23.** $\frac{5a\sqrt[3]{a}}{b}$ **24.** $\sqrt[12]{10{,}125}$ **25.** $2\sqrt[6]{2250}$

### Section 7.4

**1.** $13\sqrt{13y}$ **2.** $7\sqrt[4]{5}$ **3.** $6\sqrt{2}$ **4.** $-8x\sqrt[3]{2x}$ **5.** Cannot be simplified **6.** $(8z + 5)\sqrt[3]{z}$ **7.** $2\sqrt{m}$ **8.** $3(\sqrt{6} - 10)$ **9.** $3\sqrt[3]{12} - 2\sqrt[3]{3}$ **10.** $-74 - 27\sqrt{3}$ **11.** $53 + 10\sqrt{6}$ **12.** $25 - 6\sqrt{14}$ **13.** 1

### Section 7.5

**1.** $\frac{\sqrt{3}}{3}$ **2.** $\frac{\sqrt{10}}{4}$ **3.** $\frac{\sqrt{10x}}{2x}$ **4.** $\frac{4\sqrt[3]{9}}{3}$ **5.** $\frac{\sqrt[3]{150}}{10}$ **6.** $\frac{3\sqrt[4]{p^3}}{p}$ **7.** $2(\sqrt{3} - 1)$ **8.** $\frac{\sqrt{3} + 1}{2}$ **9.** $\frac{5 + \sqrt{10} + 4\sqrt{5} + 4\sqrt{2}}{3}$

### Section 7.6

**1. (a)** 4 **(b)** $2\sqrt{7}$ **2. (a)** −1 **(b)** 3 **3.** $\{x|x \geq -6\}$ or $[-6, \infty)$ **4.** $\{t|t$ is any real number$\}$ or $(-\infty, \infty)$ **5.** $\{m|m \leq 2\}$ or $(-\infty, 2]$ **6. (a)** $\{x|x \geq -3\}$ or $[-3, \infty)$ **7. (a)** $\{x|x$ is any real number$\}$ or $(-\infty, \infty)$

**6. (b)**

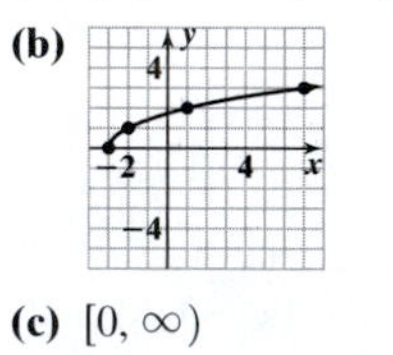

**(c)** $[0, \infty)$

**7. (b)** **(c)** $(-\infty, \infty)$

### Section 7.7

**1.** {5} **2.** {10} **3.** No real solution **4.** {4} **5.** {4} **6.** {15} **7.** {−3, 1} **8.** {12} **9. (a)** $L = \frac{8T^2}{\pi^2}$ **(b)** 32 feet

### Section 7.8

**1.** $6i$ **2.** $\sqrt{5}i$ **3.** $2\sqrt{3}i$ **4.** $4 + 10i$ **5.** $-2 - 2\sqrt{2}i$ **6.** $2 - 2\sqrt{2}i$ **7.** $1 + 11i$ **8.** $6 - 9i$ **9.** $-3 + i$ **10.** $12 + 15i$ **11.** $2 + 24i$ **12.** −18 **13.** $38 + 14i$ **14.** 73 **15.** 29 **16.** $\frac{1}{3} + \frac{4}{3}i$ **17.** $-\frac{1}{2} + \frac{3}{2}i$ **18.** $-i$ **19.** −1

# Chapter 8

### Section 8.1

**1.** $\{-4\sqrt{3}, 4\sqrt{3}\}$ **2.** {−5, 5} **3.** {−9, 9} **4.** $\{-6\sqrt{2}i, 6\sqrt{2}i\}$ **5.** $\{-3i, 3i\}$ **6.** {−13, 7} **7.** $\{5 - 4i, 5 + 4i\}$ **8.** 49; $(p + 7)^2$ **9.** $\frac{9}{4}$; $\left(w + \frac{3}{2}\right)^2$ **10.** {−4, 2} **11.** $\left\{\frac{-3 - \sqrt{11}}{2}, \frac{-3 + \sqrt{11}}{2}\right\}$ **12.** $c = 5$ **13.** $c = 6\sqrt{2}$ **14.** Approximately 17.32 miles.

### Section 8.2

**1.** $\left\{-\frac{3}{2}, 3\right\}$ **2.** $\left\{-4, \frac{1}{2}\right\}$ **3.** $\left\{1 - \frac{\sqrt{3}}{2}, 1 + \frac{\sqrt{3}}{2}\right\}$ **4.** $\left\{\frac{5}{2}\right\}$ **5.** $\{-1 - 5i, -1 + 5i\}$ **6.** Two complex solutions that are not real **7.** One repeated real solution **8.** Two irrational solutions **9.** {−3, 3} **10.** $\left\{\frac{5 - \sqrt{23}i}{4}, \frac{5 + \sqrt{23}i}{4}\right\}$ **11.** $\left\{-\frac{5}{3}, 1\right\}$ **12. (a)** 200 or 600 DVDs **(b)** 400 DVDs **13.** 16 meters by 30 meters

### Section 8.3

**1.** $\{-3, -2, 2, 3\}$ **2.** $\{-3, 3, \sqrt{2}i, -\sqrt{2}i\}$ **3.** $\{-3, -2, 2, 3\}$ **4.** $\left\{-\frac{\sqrt{3}}{2}, \frac{\sqrt{3}}{2}, -i, i\right\}$ **5.** $\left\{\frac{4}{9}, 16\right\}$ **6.** $\{25\}$ **7.** $\left\{-\frac{5}{2}, -\frac{1}{2}\right\}$
**8.** $\{-1, 125\}$

### Section 8.4

**1.**
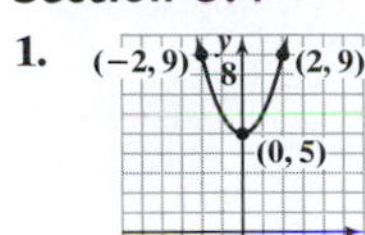


**2.**
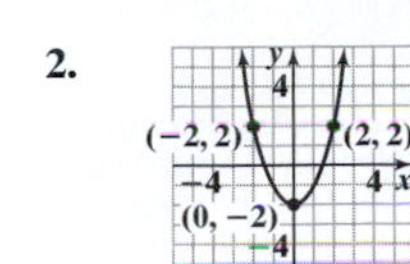


**3.**
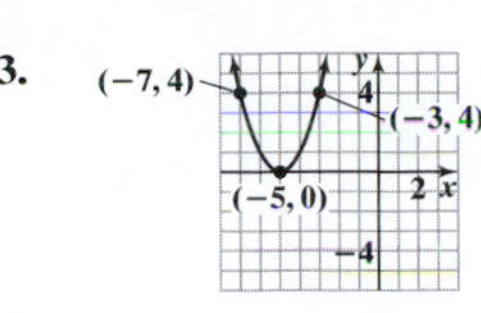


**4.**
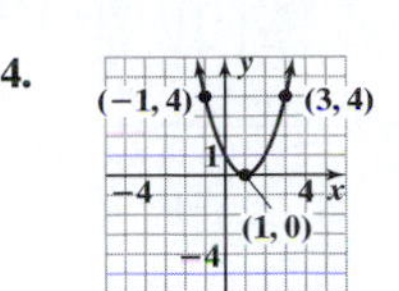


**5.**
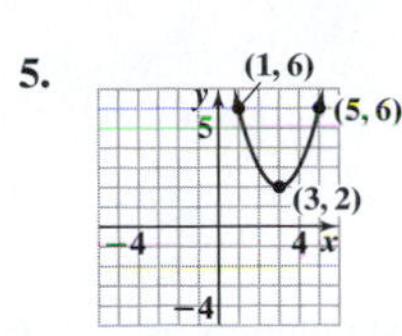


**6.**
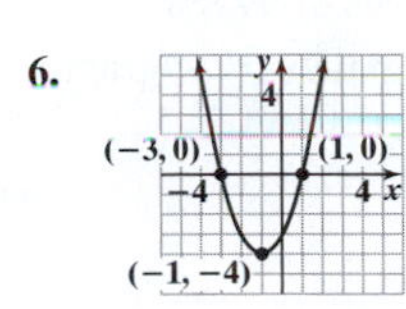


**7.**
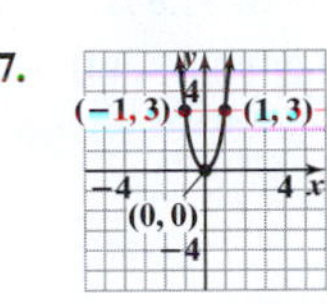


**8.**
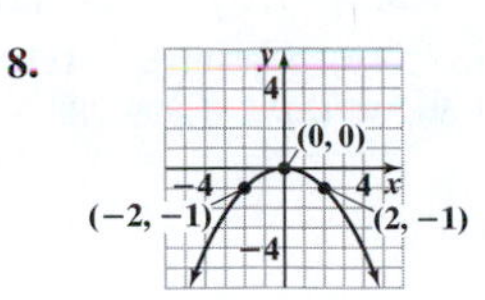


**9.**
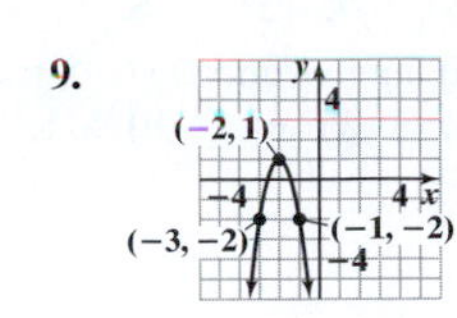


Domain: $\{x|x$ is any real number$\}$ or $(-\infty, \infty)$
Range: $\{y|y \leq 1\}$ or $(-\infty, 1]$

**10.**
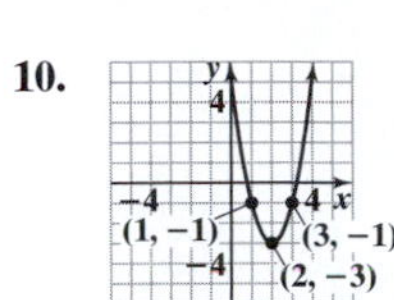


Domain: $\{x|x$ is any real number$\}$ or $(-\infty, \infty)$
Range: $\{y|y \geq -3\}$ or $[-3, \infty)$

**11.** $f(x) = -(x + 1)^2 + 2$
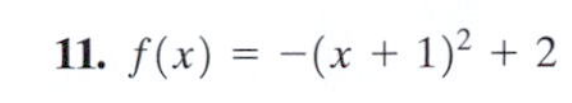

### Section 8.5

**1.**
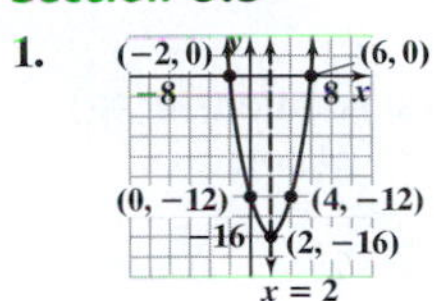


**2.** (0.71, 0) (2, 5) (3.29, 0) (0, −7) (4, −7) $x = 2$

**3.**
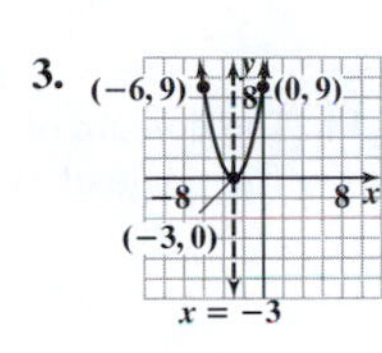


**4.**
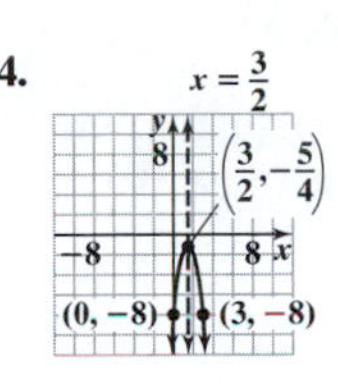


$x = \frac{3}{2}$; $\left(\frac{3}{2}, -\frac{5}{4}\right)$

**5.** Minimum; −7 **6.** Maximum; 33 **7. (a)** Revenue will be maximized when the price is \$75. **(b)** The maximum daily revenue is \$2812.50.
**8.** The maximum area that can be enclosed is 62,500 feet. The dimensions of the enclosed area are 250 feet by 250 feet.
**9.** There should be 65 boxes of CDs sold to maximize revenue. The maximum revenue would be \$4225.

### Section 8.6

**1.** $\{x|x \leq -5$ or $x \geq 2\}$; $(-\infty, -5] \cup [2, \infty)$

**2.** $\{x|x \leq -5$ or $x \geq 2\}$; $(-\infty, -5] \cup [2, \infty)$

**3.** $\{x|-6 < x < 4\}$; $(-6, 4)$

**4.** $\left\{x \middle| x < \frac{-1 - \sqrt{61}}{6} \text{ or } x > \frac{-1 + \sqrt{61}}{6}\right\}$; $\left(-\infty, \frac{-1 - \sqrt{61}}{6}\right) \cup \left(\frac{-1 + \sqrt{61}}{6}, \infty\right)$

# Chapter 9

### Section 9.1

**1. (a)** 17 **(b)** 26 **(c)** −63 **2. (a)** $9x^2 + 3x - 1$ **(b)** $3x^2 - 9x + 5$ **(c)** 29 **3.** not one-to-one **4.** one-to-one
**5.** not one-to-one **6.** one-to-one

**7.**
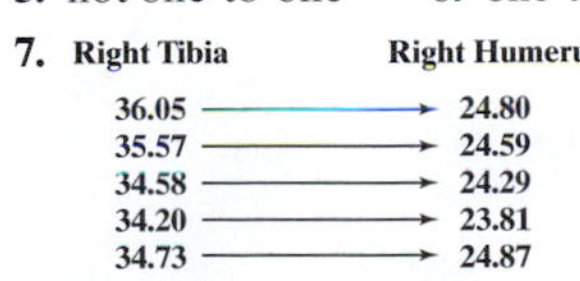


The domain of the inverse function is $\{36.05, 35.57, 34.58, 34.20, 34.73\}$. The range of the inverse function is $\{24.80, 24.59, 24.29, 23.81, 24.87\}$.

**8.** $\{(3, -3), (2, -2), (1, -1), (0, 0), (-1, 1)\}$. The domain of the inverse function is $\{3, 2, 1, 0, -1\}$. The range of the inverse function is $\{-3, -2, -1, 0, 1\}$.

**9.**
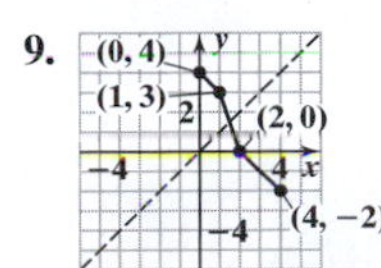


**10.** $g^{-1}(x) = \frac{x + 1}{5}$ **11.** $f^{-1}(x) = \sqrt[5]{x - 3}$

### Section 9.2

**1. (a)** 3.249009585 **(b)** 3.317278183 **(c)** 3.321880096 **(d)** 3.32211036 **(e)** 3.321997085

**2.** The domain of $f$ is all real numbers or, using interval notation, $(-\infty, \infty)$. The range of $f$ is $\{y|y > 0\}$ or, using interval notation, $(0, \infty)$.

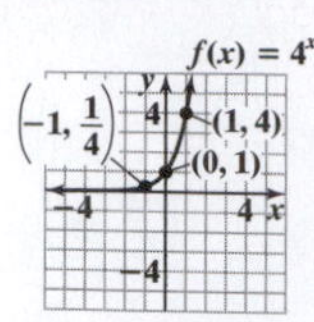


**3.** The domain of $f$ is all real numbers or, using interval notation, $(-\infty, \infty)$. The range of $f$ is $\{y|y > 0\}$ or, using interval notation, $(0, \infty)$.

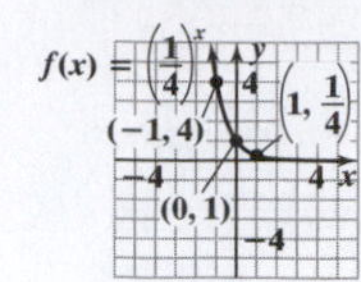


**4.** The domain of $f$ is all real numbers or, using interval notation, $(-\infty, \infty)$. The range of $f$ is $\{y|y > 0\}$ or, using interval notation, $(0, \infty)$.

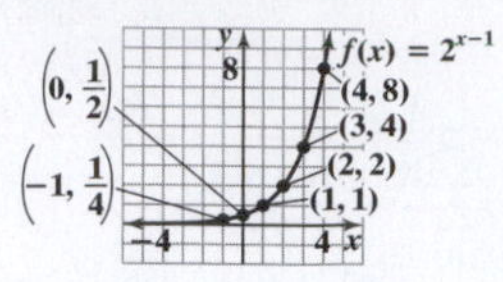


**5.** The domain of $f$ is all real numbers or, using interval notation, $(-\infty, \infty)$. The range of $f$ is $\{y|y > 1\}$ or, using interval notation, $(1, \infty)$.

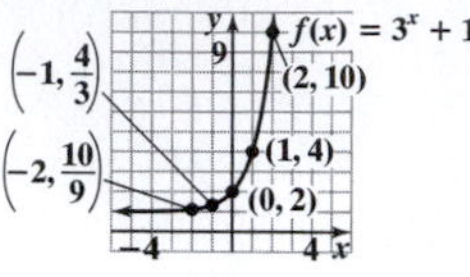


**6.** $\{3\}$ **7.** $\{2\}$ **8.** $\{0, 5\}$ **9.** $\{-1, 3\}$ **10. (a)** 0.918 or 91.8% **(b)** 0.998 or 99.8%
**11. (a)** approximately 6.91 grams **(b)** 5 grams **(c)** 0.625 gram **(d)** approximately 0.247 gram
**12. (a)** \$2,253.65 **(b)** \$11,991.60 **(c)** \$71,899.28

## Section 9.3

**1.** $3 = \log_4 w$ **2.** $-2 = \log_p 8$ **3.** $b = \log_5 125$ **4.** $2^y = 16$ **5.** $a^5 = 20$ **6.** $5^{-3} = z$ **7.** 2 **8.** −3 **9.** 2 **10.** −1

**11.** $\{x|x > -3\}$ or $(-3, \infty)$ **12.** $\left\{x \middle| x < \frac{5}{2}\right\}$ or $\left(-\infty, \frac{5}{2}\right)$

**13.** The domain of $f$ is $\{x|x > 0\}$ or, using interval notation, $(0, \infty)$. The range of $f$ is all real numbers or, using interval notation, $(-\infty, \infty)$.

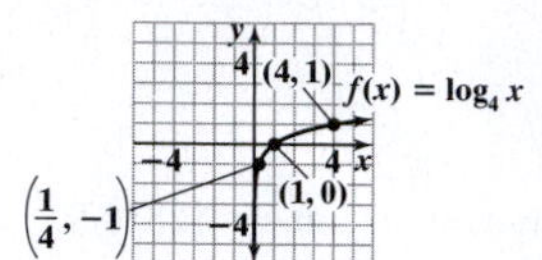


**14.** The domain of $f$ is $\{x|x > 0\}$ or, using interval notation, $(0, \infty)$. The range of $f$ is all real numbers or, using interval notation, $(-\infty, \infty)$.

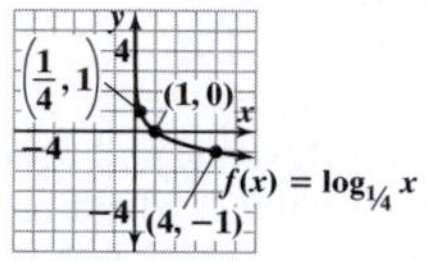


**15.** 3.146 **16.** 1.569 **17.** −0.523 **18.** $\{16\}$ **19.** $\{4\}$ **20.** $\{e^{-2}\}$ **21.** $\{10{,}020\}$ **22.** 100 decibels

## Section 9.4

**1.** 0 **2.** 0 **3.** 1 **4.** 1 **5.** $\sqrt{2}$ **6.** 0.2 **7.** 1.2 **8.** −4 **9.** $\log_4 9 + \log_4 5$ **10.** $\log 5 + \log w$ **11.** $\log_7 9 - \log_7 5$
**12.** $\ln p - \ln 3$ **13.** $\log_2 3 + \log_2 m - \log_2 n$ **14.** $\ln q - \ln 3 - \ln p$ **15.** $1.6 \log_2 5$ **16.** $5 \log b$ **17.** $2 \log_4 a + \log_4 b$
**18.** $2 + 4\log_3 m - \frac{1}{3}\log_3 n$ **19.** 2 **20.** $\log_3\left(\frac{x+4}{x-1}\right)$ **21.** $\log_5 \frac{x}{8}$ **22.** $\log_2 \frac{x^2 + 3x + 2}{x^2}$ **23.** 3.155 **24.** 2.807

## Section 9.5

**1.** $\{3\}$ **2.** $\{8\}$ **3.** $\left\{\frac{\ln 11}{\ln 2}\right\}$ or $\left\{\frac{\log 11}{\log 2}\right\}$; $\approx\{3.459\}$ **4.** $\left\{\frac{\ln 3}{2 \ln 5}\right\}$ or $\left\{\frac{\log 3}{2 \log 5}\right\}$; $\approx\{0.341\}$ **5.** $\left\{\frac{\ln 5}{2}\right\}$; $\approx\{0.805\}$

**6.** $\left\{-\frac{\ln(20/3)}{4}\right\}$; $\approx\{-0.474\}$ **7. (a)** approximately 2.85 days **(b)** approximately 32.52 days **8. (a)** approximately 6.77 years
**(b)** approximately 11.58 years

# Chapter 10

## Section 10.1

**1.** 5 **2.** $6\sqrt{5} \approx 13.42$ **3. (a)**

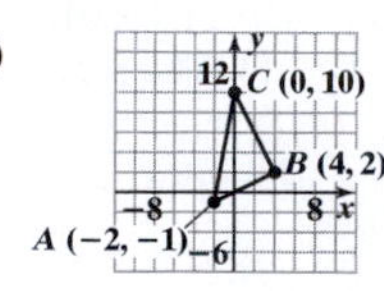


**(b)** $d(A, B) = 3\sqrt{5}$; $d(A, C) = 5\sqrt{5}$; $d(B, C) = 4\sqrt{5}$
**(c)** $[d(A, C)]^2 = [d(A, B)]^2 + [d(B, C)]^2$ **(d)** 30 square units
**4.** $\left(\frac{3}{2}, 6\right)$ **5.** $\left(1, \frac{5}{2}\right)$

## Section 10.2

**1.** $(x - 2)^2 + (y - 4)^2 = 25$ **2.** $(x + 2)^2 + y^2 = 2$

**3.** $(x - 3)^2 + (y - 1)^2 = 4$

**4.** $(x + 5)^2 + y^2 = 16$

**5.** $x^2 + y^2 - 6x - 4y + 4 = 0$

**6.** $2x^2 + 2y^2 - 16x + 4y - 38 = 0$

### Section 10.3

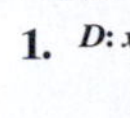
**1.**
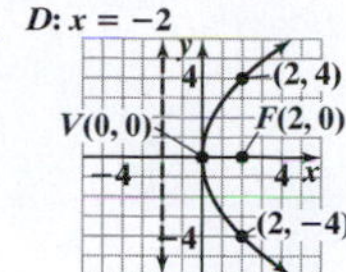

**2.**
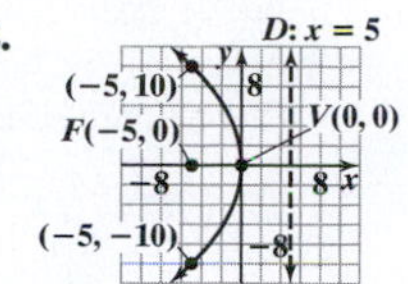


**3.**
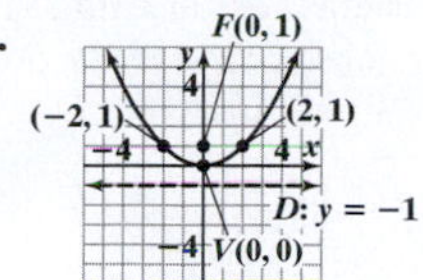


**4.**
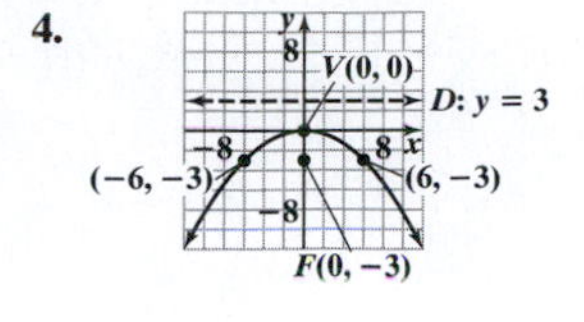

**5.**
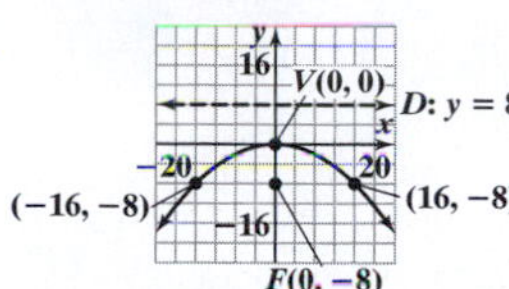

**6.**
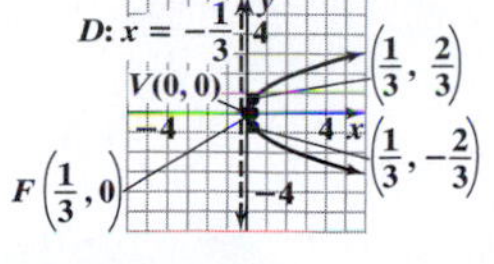


**7.** D: $x = -6$; $V(-3, 2)$; $(0, 8)$; $F(0, 2)$; $(0, -4)$

**8.** The receiver should be located 2 feet from the base of the dish along its axis of symmetry.

### Section 10.4

**1.**
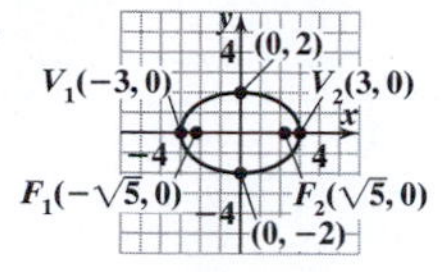


**2.**
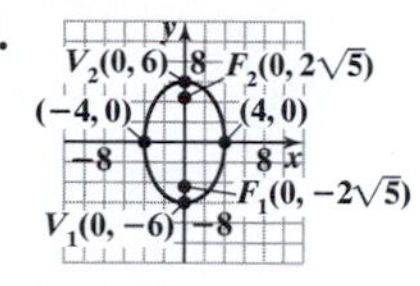


**3.** $\dfrac{x^2}{40} + \dfrac{y^2}{49} = 1$

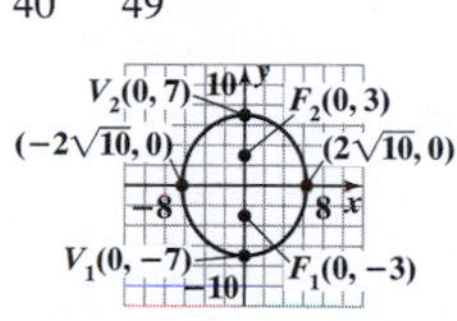


**4.**
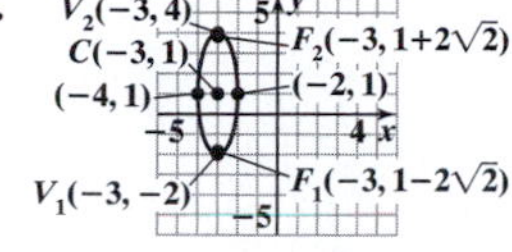


**5.** $\dfrac{x^2}{2500} + \dfrac{y^2}{1600} = 1$; 40 feet

### Section 10.5

**1.**
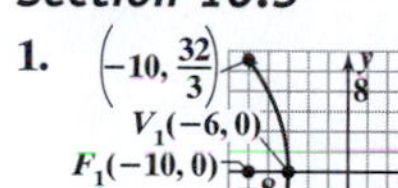
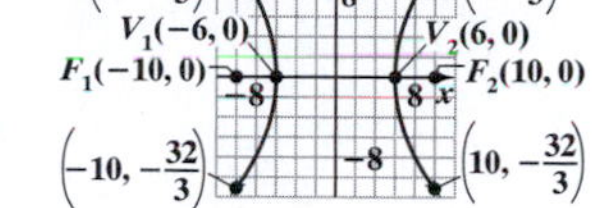


**2.**
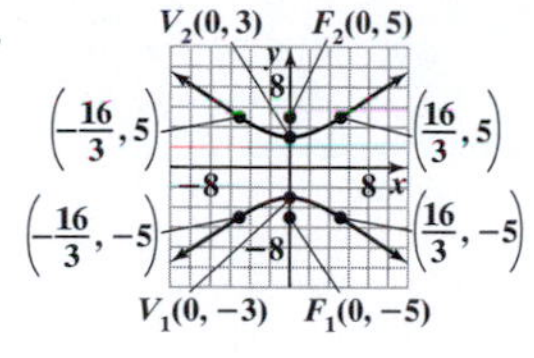


**3.** $\dfrac{x^2}{16} - \dfrac{y^2}{20} = 1$

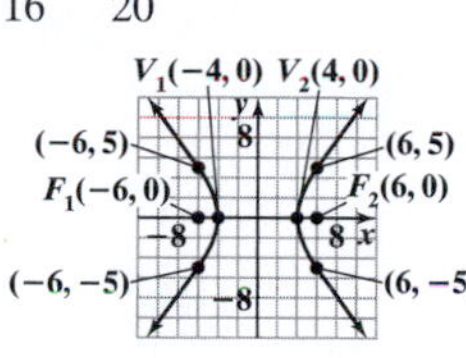


**4.**
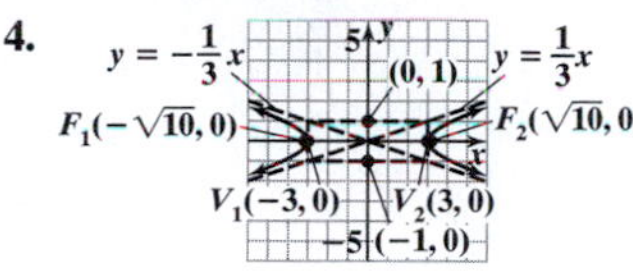


**5.**
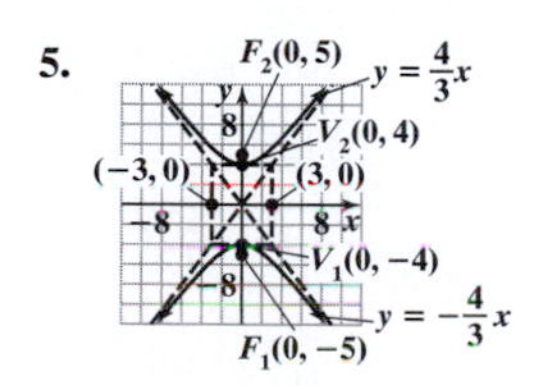


### Section 10.6

**1.** $(-3, 5)$ and $(1, -3)$ **2.** $\left(\frac{11}{5}, -\frac{22}{5}\right)$ and $(1, -2)$ **3.** $(0, -4), \left(-2\sqrt{3}, 2\right), \left(2\sqrt{3}, 2\right)$ **4.** $\varnothing$ or $\{\ \}$

## Chapter 11

### Section 11.1

**1.** $-1, 1, 3, 5, 7$ **2.** $-4, 8, -12, 16, -20$ **3.** $a_n = 2n + 3$ **4.** $b_n = \dfrac{(-1)^{n+1}}{n+1}$ **5.** $3 + 7 + 11 = 21$ **6.** $2 + 9 + 28 + 65 + 126 = 230$ **7.** $\sum_{k=1}^{12} k^2$ **8.** $\sum_{n=1}^{6} \dfrac{1}{2^{n-1}}$

### Section 11.2

**1.** Arithmetic; $a = -3, d = 2$ **2.** Not arithmetic **3.** Arithmetic; $a = -5, d = 3$ **4.** Not arithmetic **5.** Arithmetic; $a = 3, d = -2$ **6. (a)** $a_n = 6n - 5$ **(b)** 79 **7. (a)** $a = -5, d = 3$ **(b)** $a_n = 3n - 8$ **8.** 10,400 **9.** 9,730 **10.** 7,550 **11.** −1,050 **12.** 1,470 seats

### Section 11.3

**1.** Geometric; $a = 4, r = 2$ **2.** Not geometric **3.** Geometric; $a = 9, r = \dfrac{1}{3}$ **4.** Geometric; $r = 5$ **5.** Not geometric **6.** Geometric; $r = \dfrac{2}{3}$ **7.** $a_n = 5 \cdot 2^{n-1}$; $a_9 = 1280$ **8.** $a_n = 50 \cdot \left(\dfrac{1}{2}\right)^{n-1}$; $a_9 = 0.1953125$ or $a_9 = \dfrac{25}{128}$ **9.** 24,573 **10.** 7.9921875 **11.** $\dfrac{40}{3}$ **12.** $\dfrac{1}{2}$ **13.** $\dfrac{2}{9}$ **14.** The U.S. economy will increase by $10,000. **15.** $244,129.08

### Section 11.4

**1.** 362,880 **2.** 840 **3.** 7 **4.** 20 **5.** $x^4 + 8x^3 + 24x^2 + 32x + 16$ **6.** $32p^5 - 80p^4 + 80p^3 - 40p^2 + 10p - 1$

# Answers to Selected Exercises

## Chapter R

### Section R.1

Answers will vary.

### Section R.2

**1.** set **3.** rational **5.** True **7.** False **9.** True **11.** No; if a number is rational, it cannot be irrational and vice versa. No; every real number, when written in decimal form, will either terminate (rational), have a repeating block (rational), or neither terminate nor repeat (irrational). **13.** Answers will vary. **15.** If $A \subseteq B$, then all elements in $A$ are also in $B$. Plus, $A$ could equal $B$. If $A \subset B$, then all elements in $A$ are also in $B$, but $A \neq B$. **17.** If the digit *after* the specified final digit is 4 or less, then truncating and rounding will yield the same decimal approximation. **19.** $\{0, 1, 2, 3, 4, 5\}$ **21.** $\{-2, -1, 0, 1, 2, 3, 4\}$ **23.** $\emptyset$ or $\{\ \}$ **25.** True **27.** False **29.** True **31.** True **33.** $\notin$ **35.** $\in$ **37. (a)** $\{4\}$ **(b)** $\{-5, 4\}$ **(c)** $\left\{-5, 4, \frac{4}{3}, -\frac{7}{5}, 5.\overline{1}\right\}$ **(d)** $\{\pi\}$ **(e)** $\left\{-5, 4, \frac{4}{3}, -\frac{7}{5}, 5.\overline{1}, \pi\right\}$ **39. (a)** $\{100\}$ **(b)** $\{100, -64\}$ **(c)** $\left\{100, -5.423, \frac{8}{7}, -64\right\}$ **(d)** $\{\sqrt{2} + 4\}$ **(e)** $\left\{100, -5.423, \frac{8}{7}, \sqrt{2} + 4, -64\right\}$ **41. (a)** 19.9348 **(b)** 19.9348 **43. (a)** 0.0 **(b)** 0.1 **45.** $-3 \quad -\frac{3}{2} \quad 0 \quad 2 \quad 1.5$ **47.** $<$ **49.** $=$ **51.** $>$ **53.** $-282$ feet **55.** $-\$0.04$ **57.** $-6$ **59.** Answers will vary.

**61.** Answers will vary. **63.** Rounded: $\frac{8}{7} \approx 1.143$; Truncated: $\frac{8}{7} \approx 1.142$

### Section R.3

**1.** factors **3.** negative, *or* opposite; reciprocal **5.** True **7.** 0 **9.** Multiplication is distributive on a sum, but not on a product.

**11.** No, the order of subtraction is important. **13.** No, the order of division is important. **15. (a)** $-\frac{9}{5}$ **(b)** $\frac{5}{9}$ **(c)** $\frac{5}{9}$

**17. (a)** 15 **(b)** $-\frac{1}{15}$ **(c)** $-\frac{1}{15}$ **19. (a)** $-\frac{4}{3}$ **(b)** $\frac{3}{4}$ **(c)** $\frac{3}{4}$ **21. (a)** $-8$ **(b)** $\frac{1}{8}$ **(c)** $\frac{1}{8}$ **23. (a)** 4 **(b)** $-\frac{1}{4}$ **(c)** $-\frac{1}{4}$

**25. (a)** $-\frac{2}{5}$ **(b)** $\frac{5}{2}$ **(c)** $\frac{5}{2}$ **27.** $2x + 8$ **29.** $3z + 6$ **31.** $3x - 30$ **33.** $6x - \frac{15}{2}$ **35.** $\frac{z}{5}$ **37.** $\frac{6}{5}$ **39.** 15 **41.** 7 **43.** 3

**45.** $-2.5$ **47.** $-32$ **49.** 84 **51.** $\frac{5}{3}$ **53.** $-\frac{3}{7}$ **55.** $\frac{2}{3}$ **57.** $\frac{5}{4}$ **59.** $\frac{1}{6}$ **61.** $\frac{247}{210}$ **63.** $\frac{157}{210}$ **65.** $\frac{4}{3}$ **67.** 3.3 **69.** $-32$

**71.** $\frac{13}{10}$ **73.** $\frac{17}{15}$ **75.** $-\frac{139}{32}$ **77.** $\frac{1}{9}$ **79.** Commutative Property of Multiplication **81.** Multiplicative Inverse Property **83.** Reduction Property **85.** Associative Property of Addition **87.** The difference in age of the oldest and youngest president at the time of inauguration is 27 years. **89.** The Bears gained a total of 9 yards in the first 3 plays. Since 9 is less than 10, they did not obtain a first down. **91.** The difference between the highest and lowest elevation is 20,602 feet. **93.** Answers will vary.

**95. (a)** $\frac{1}{1} = 1, \frac{2}{1} = 2, \frac{3}{2} = 1.5, \frac{5}{3} \approx 1.666667, \frac{8}{5} = 1.6, \frac{13}{8} = 1.625, \frac{21}{13} \approx 1.615385, \frac{34}{21} \approx 1.619048, \frac{55}{34} \approx 1.617647, \frac{89}{55} \approx 1.618182, \frac{144}{89} \approx 1.617978, \ldots$

**(b)** 1.618 **(c)** Answers will vary. **97.** $d(P, Q) = 8$ $\quad -2 \quad 0 \quad 2 \quad 6$

**99.** $d(P, Q) = 10.9$ $\quad -10 \quad -9.3 \quad -4 \quad 0 \quad 1.6 \quad 4$ **101.** $d(P, Q) = \frac{37}{5}$ $\quad -2 \quad -7/5 \quad 0 \quad 2 \quad 6$

**103.** $\left\{-6, -3, -2, -\frac{3}{2}, -1, -\frac{1}{2}, \frac{1}{2}, 1, \frac{3}{2}, 2, 3, 6\right\}$ **105.** $-3.8$ **107.** $-19.2$ **109.** $\frac{1}{30}$ **111.** 14.4 **113.** $\frac{9}{2}$

### Section R.4

**1.** base; exponent; power **3.** False **5.** Answers will vary. **7.** Answers will vary. **9.** 7 **11.** 15 **13.** 12 **15.** $-28$ **17.** 3 **19.** 14 **21.** $-2$ **23.** 60 **25.** 180 **27.** 9 **29.** 8 **31.** $\frac{5}{2}$ **33.** 2 **35.** $\frac{6}{5}$ **37.** $\frac{2}{3}$ **39.** $-\frac{7}{4}$ **41.** $\frac{3}{2}$ **43.** $\frac{53}{14}$

**45.** $3 \cdot (7 - 2) = 15$ **47.** $3 + 5 \cdot (6 - 3) = 18$ **49.** The surface area of the cylinder is about 534.07 square inches.

**51.** After 3 seconds, the height of the ball is 6 feet. **53.** Answers will vary. **55.** Answers will vary. **57.** 175 **59.** $\frac{16}{45}$ **61.** $\frac{17}{13}$ **63.** $-212.96$ **65.** $-534.53$

### Section R.5

**1.** variable **3.** term **5.** True **7.** Answers will vary. **9.** 11 **11.** $-9$ **13.** $-21$ **15.** $\frac{3}{8}$ **17.** $\frac{6}{7}$ **19.** 29 **21.** 1 **23.** $\frac{3}{2}$

**25.** $x$ **27.** $-6z + 3$ **29.** $-z - 5$ **31.** $\frac{11}{12}x$ **33.** $4x^2 - 3x$ **35.** $1.6x$ **37.** $-3x + 1$ **39.** $-14x + 7$ **41.** $-z + 10$ **43.** $4x - 3$

**45.** $12v - 11$ **47.** $2x + \frac{9}{10}$ **49.** $\frac{13}{36}x$ **51.** $14.46x - 15.49$ **53.** $-11.08x - 5.44$ **55.** $5 + x$ **57.** $4z$ **59.** $y - 7$ **61.** $2(t + 4)$

**63.** $5x - 3$ **65.** $\frac{y}{3} + 6x$ **67. (a)** No **(b)** Yes **(c)** Yes **(d)** Yes **69. (a)** Yes **(b)** Yes **(c)** Yes **(d)** Yes

**71. (a)** No **(b)** Yes **(c)** Yes **(d)** No **73.** 1 in.$^3$; 8 in.$^3$; 27 in.$^3$; 64 in.$^3$ **75. (a)** 0 ft; 59 ft; 86 ft; 81 ft; 44 ft **(b)** The ball begins on the ground. When it is hit, it rises in the air (for somewhere around 2 seconds) and then begins to fall back towards the ground.

**77.** Let $x$ = Bob's age in years; Tony's age $= x + 5$; when Bob is 13 years old, Tony is 18 years old. **79.** Let $p$ = the original price in dollars; discounted price $= \frac{1}{2}p$; when the original price is \$900, the discount price is \$450. **81.** $\frac{4}{3}$

**83.** Answers will vary. One possible answer is, "*Twice a number z decreased by 5.*"

**85.** Answers will vary. One possible answer is, "*Twice the difference of a number z and 5.*"

**87.** Answers will vary. One possible answer is, "*One-half the sum of a number z and 3.*" **89. (a)** 3 **(b)** 15 **91. (a)** 63 **(b)** 35

**93. (a)** $-\frac{7}{5}$ **(b)** $\frac{23}{65}$ **95. (a)** 67 **(b)** 32 **97.** The calculator displays an error message because $x = 5$ makes the denominator equal to 0.

# Chapter 1

## Section 1.1

**1.** equivalent equations **3.** conditional equation **5.** False **7.** Answers will vary. **9.** $x = 2$ **11.** $m = 1$ **13.** $x = 5$ **15.** 2 **17.** 2 **19.** $-\frac{1}{4}$ **21.** 9 **23.** $-9$ **25.** $-\frac{2}{3}$ **27.** $-4$ **29.** $\{\ \}$ or $\emptyset$ **31.** $\{\ \}$ or $\emptyset$ **33.** $\{y \mid y$ is any real number$\}$ or $\mathbb{R}$ **35.** 2 **37.** $-3$ **39.** $\{\ \}$ or $\emptyset$ **41.** $-\frac{3}{5}$ **43.** $\frac{5}{2}$ **45.** 3 **47.** $-5$ **49.** $\{z \mid z$ is any real number$\}$ or $\mathbb{R}$ **51.** $-\frac{7}{2}$ **53.** $\frac{1}{7}$ **55.** $\frac{7}{2}$ **57.** $\{p \mid p$ is any real number$\}$ or $\mathbb{R}$ **59.** $\{\ \}$ or $\emptyset$ **61.** $-14$ **63.** $-3$ **65.** $-1.6$ **67.** 2 **69.** $a = -4$ **71.** $a = 3$ **73.** $x = -\frac{1}{2}$ **75.** $x = \frac{3}{4}$ **77.** $x = 1$ **79.** The card's annual interest rate is 0.15 or 15%. **81.** You earned \$18,000 in 2005. **83.** Answers will vary.

## Section 1.2

**1.** mathematical modeling **3.** True **5.** Answers will vary. **7.** Categories: Direct Translation, Mixture, Geometry, Uniform Motion, and Work Problems. Two types of mixture problems are Interest and Blends. **9.** 10 **11.** 40 **13.** 37.5% **15.** $x + 12 = 20$; 8 **17.** $2 \cdot (y + 3) = 16$; 5 **19.** $w - 22 = 3w$; $-11$ **21.** $4x = 2x + 14$; 7 **23.** $0.8x = x + 5$; $-25$ **25.** 13 and 26 **27.** 24, 25, and 26 **29.** Kendra needs an 83 on her final exam to have an average of 80. **31.** Jacob would need to print 2500 pages for the cost to be the same for the two printers. **33.** Connor: \$400,000; Olivia: \$300,000; Avery: \$100,000. **35.** The final bill will be \$616.37. **37.** The dealer's cost is about \$22,434.78. **39.** The memory cards originally cost \$17. **41.** The Nissan Altima weighs 3320 pounds, the Mazda 6s weighs 3340 pounds, and the Honda Accord EX weighs 3390 pounds. **43.** The Raiders rushed for 235 yards. **45.** Adam will get \$8500 and Krissy will get \$11,500. **47.** You should invest \$15,000 in stocks and \$9000 in bonds. **49.** The interest charge after one month will be \$29.17. **51.** The bank loaned \$225,000 at 6% interest. **53.** Pedro should invest \$9375 in the 5% bond and \$15,625 in the 9% stock fund. **55.** $50 - x$ pounds of coffee B **57.** Bobby has 15 dimes and 32 quarters saved. **59.** 36 grams of pure gold should be mixed with 36 grams of 12-karat gold. **61.** The race consists of running for 12 miles and biking for 50 miles. **63.** The slow car is traveling at 60 mph while the faster car travels at 70 mph. **65.** The boats will be 155 miles apart after 2.5 hours. **67.** One person is walking at a rate of 4 mph and the other is walking at a rate of 6 mph. **69.** Written answers will vary. The average speed of the trip to Florida and back is roughly 54.55 miles per hour. **71.** Answers will vary. **72.** Answers will vary. **73.** The train is 0.3 mile or 1584 feet long.

## Section 1.3

**1.** formula **3.** golden rectangles **5.** True **7.** $F = m \cdot a$ **9.** $V = \frac{4}{3}\pi r^3$ **11.** $r = \frac{d}{t}$ **13.** $m = \frac{y - y_1}{x - x_1}$ **15.** $x = \mu + \sigma Z$ **17.** $m_1 = \frac{r^2 F}{G m_2}$ **19.** $P = \frac{A}{1 + rt}$ **21.** $F = \frac{9}{5}C + 32$ **23.** $y = -2x + 13$ **25.** $y = 3x - 5$ **27.** $y = -\frac{4}{3}x + \frac{13}{3}$ **29.** $y = -3x + 12$ **31. (a)** $h = \frac{V}{\pi r^2}$ **(b)** The height of the cylinder is 8 inches. **33. (a)** $A = \frac{206.3 - M}{0.711}$ **(b)** An individual whose maximum heart rate is 160 should be about 65. **35. (a)** $P = \frac{A}{(1 + r)^t}$ **(b)** Roughly \$4109.64 should be deposited today to have \$5000 in 5 years in an account that pays 4% annual interest. **37.** The smaller angle measures 75° and its supplement measures 105°. **39.** The smaller angle measures 20° and its complement measures 70°. **41.** The window is 5 feet long and 8 feet wide. **43.** The area of the circle is $25\pi$ square inches (roughly 78.54 square inches). **45.** The first angle measures 40°, the second measures 55°, and the third measures 85°. **47. (a)** The patio is 17.5 ft wide and 22.5 ft long. **(b)** You would need to purchase 131.25 cubic feet of cement. **49. (a)** The deck has an area of $84\pi$ square feet (roughly 264 square feet). **(b)** It would require roughly 97.39 feet of fencing to encircle the pool. **(c)** The fence would cost about \$2434.73.

## Section 1.4

**1.** closed interval **3.** left endpoint; right endpoint **5.** True **7.** Answers will vary.

**9.** $\{x \mid x \le 6\}$; $(-\infty, 6]$

**11.** $\{x \mid x < 4\}$; $(-\infty, 4)$

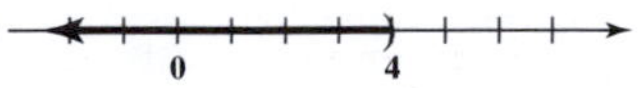

**13.** $\{x \mid x > -3\}$; $(-3, \infty)$

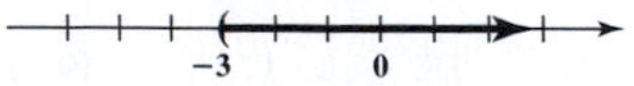

**15.** $\{x \mid x > 6\}$; $(6, \infty)$

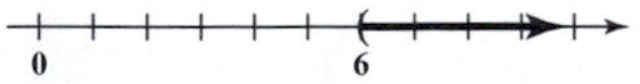

**17.** $\{x \mid x > 3\}$; $(3, \infty)$

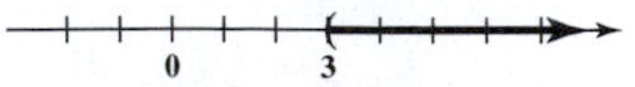

**19.** $\{x \mid x < -4\}$; $(-\infty, -4)$

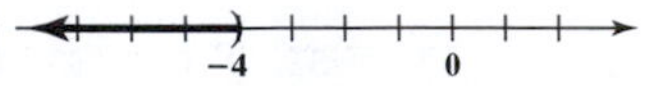

**21.** $\{x \mid x \le -1\}$; $(-\infty, -1]$

**23.** $\{x \mid x > -2\}$; $(-2, \infty)$

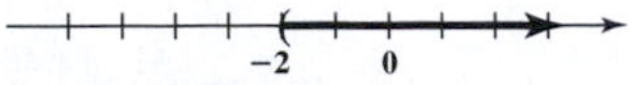

**25.** $\{x \mid x < 17\}$; $(-\infty, 17)$

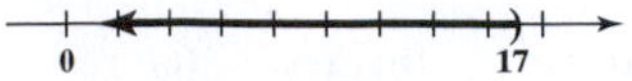

**27.** $\{x \mid x \le 4\}$; $(-\infty, 4]$

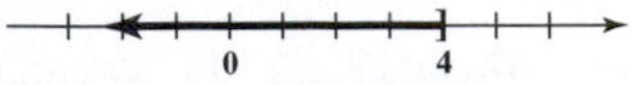

**29.** $\{x \mid x \le -30\}$; $(-\infty, -30]$

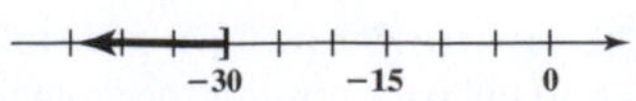

**31.** $\left\{x \mid x < \frac{1}{3}\right\}$; $\left(-\infty, \frac{1}{3}\right)$

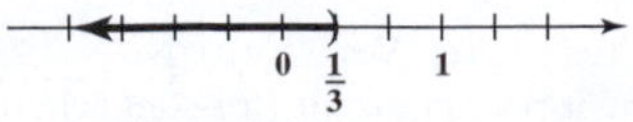

**33.** $\left\{x \middle| x < -\frac{11}{4}\right\}; \left(-\infty, -\frac{11}{4}\right)$ **35.** $\left\{x \middle| x < -\frac{16}{15}\right\}; \left(-\infty, -\frac{16}{15}\right)$ **37.** $\left\{x \middle| x \le \frac{15}{2}\right\}; \left(-\infty, \frac{15}{2}\right]$ **39.** $\left\{x \middle| x \ge \frac{21}{16}\right\}; \left[\frac{21}{16}, \infty\right)$ **41.** $\left\{x \middle| x > \frac{1}{10}\right\}; \left(\frac{1}{10}, \infty\right)$ **43.** $\{x|x < 3\}; (-\infty, 3)$ **45.** $\left\{x \middle| x < \frac{9}{2}\right\}; \left(-\infty, \frac{9}{2}\right)$ **47.** $\{y|y > -15\}; (-15, \infty)$ **49.** $\{a|a \ge 7\}; [7, \infty)$ **51.** $\left\{x \middle| x > -\frac{1}{3}\right\}; \left(-\frac{1}{3}, \infty\right)$ **53.** $\{x|x \le -14\}; (-\infty, -14]$ **55.** $\left\{x \middle| x < \frac{3}{2}\right\}; \left(-\infty, \frac{3}{2}\right)$ **57.** $\{x|x \ge -3\}; [-3, \infty)$ **59.** $\left\{x \middle| x \ge -\frac{3}{4}\right\}; \left[-\frac{3}{4}, \infty\right)$

**61.** $\{x|x \ge 4\}; [4, \infty)$ **63.** $\{z|z \le 3\}; (-\infty, 3]$ **65.** Jackie must earn at least 182 points on the final exam to earn an $A$ in Mr. Ruffatto's class. **67.** You can order no more than 3 hamburgers to keep the fat content to no more than 69 grams. **69.** The plane can carry up to 18,836 pounds of luggage and cargo. **71.** The monthly benefit will exceed \$1000 in 2012. **73.** Susan will need to sell roughly \$5,500,000 in computer systems to earn \$100,000. **75.** Supply will exceed demand when the price is greater than \$40. **77.** All real numbers are solutions; $\mathbb{R}$ **79.** When we multiply both sides of an inequality by a negative number. Or, if the sides of the inequality are interchanged.

### *Chapter 1 Putting the Concepts Together*

**1. (a)** $x = -3$ is not a solution. **(b)** $x = 1$ is a solution. **2.** $-5$ **3.** 0 **4.** identity **5.** $x - 3 = \frac{1}{2}x + 2$ **6.** $\frac{x}{2} < x + 5$ **7.** The chemist needs to mix 4 liters of the 20% solution with 12 liters of the 40% solution. **8.** After 3.4 hours, the two cars will be 255 miles apart. **9.** $y = \frac{3}{2}x - 2$ **10.** $r = \frac{A - P}{Pt}$ **11. (a)** $h = \frac{V}{\pi r^2}$ **(b)** 6 in. **12. (a)** Interval: $(-3, \infty)$ Graph: **(b)** Interval: $(2, 5]$ Graph: **13. (a)** Inequality: $x \le -1.5$ Graph: **(b)** Inequality: $-3 < x \le 1$ Graph: **14.** $[6, \infty)$ **15.** $(-\infty, 1)$ **16.** $(-\infty, 5]$ **17.** Logan can invite at most 9 children to the party.

### *Section 1.5*

**1.** intersection **3.** compound inequality **5.** False **7.** There is no real number $x$ such that $x > 4$ and $x < 2$. **9.** Yes **11.** $\{1, 4, 5, 6, 7, 8, 9\}$ **13.** $\{5, 7, 9\}$ **15.** $\emptyset$ **17. (a)** $A \cap B = \{x|-2 < x \le 5\}$ **(b)** $A \cup B = \{x|x \text{ is any real number}\}$ **19. (a)** $E \cap F = \emptyset$ **(b)** $E \cup F = \{x|x < -1 \text{ or } x > 3\}$ **21.** $\{x|-2 \le x < 3\}; [-2, 3)$ **23.** $\{x|x < -2 \text{ or } x > 3\}; (-\infty, -2) \cup (3, \infty)$ **25.** $\emptyset$ or $\{\ \}$ **27.** $\{x|x < -2\}; (-\infty, -2)$ **29.** $\emptyset$ or $\{\ \}$ **31.** $\{x|x < -2 \text{ or } x > 5\}; (-\infty, -2) \cup (5, \infty)$ **33.** $\{x|x \text{ is any real number}\}; (-\infty, \infty)$ **35.** $\{x|x < -1 \text{ or } x > 4\}; (-\infty, -1) \cup (4, \infty)$ **37.** $\{x|-1 \le x < 3\}; [-1, 3)$ **39.** $\left\{x \middle| -\frac{2}{3} \le x \le \frac{3}{2}\right\}; \left[-\frac{2}{3}, \frac{3}{2}\right]$ **41.** $\{x|x \le -3 \text{ or } x > 6\}; (-\infty, -3] \cup (6, \infty)$

**43.** $\left\{x \middle| -1 < x \le \frac{4}{5}\right\}; \left(-1, \frac{4}{5}\right]$ **45.** $\{x|0 \le x \le 8\}; [0, 8]$

**47.** $\{x|-6 \le x \le -2\}; [-6, -2]$ **49.** $\{x|x \text{ is any real number}\}$; Interval: $(-\infty, \infty)$ **51.** $\emptyset$ or $\{\ \}$ **53.** $\left\{x \middle| -\frac{5}{3} < x \le 5\right\}; \left(-\frac{5}{3}, 5\right]$

**55.** $\{x|-4 < x \le 3\}$; $(-4, 3]$ −4 0 3

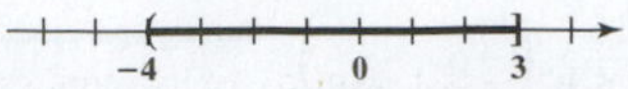

**57.** $\left\{x \middle| x < 0 \text{ or } x > \frac{5}{2}\right\}$; $(-\infty, 0) \cup (5/2, \infty)$ 0 $\frac{5}{2}$ **59.** $\{a|-3 \le a < 0\}$; $[-3, 0)$ −3 0 3

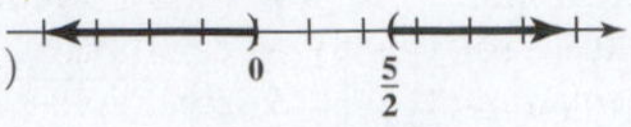

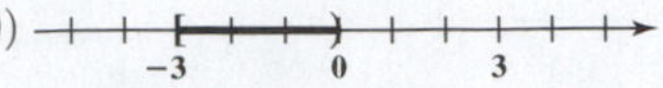

**61.** $\{x|x$ is any real number$\}$; $(-\infty, \infty)$ 0 **63.** $\left\{x \middle| -2 \le x \le \frac{8}{3}\right\}$; $\left[-2, \frac{8}{3}\right]$ −2 0 $\frac{8}{3}$

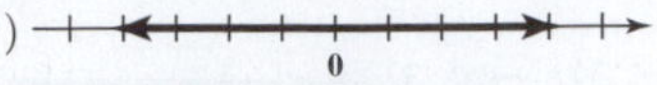

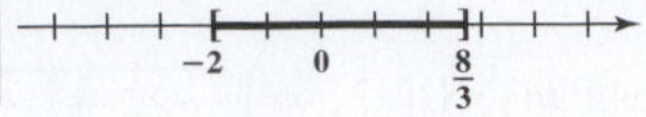

**65.** $\{x|x < -10 \text{ or } x > 2\}$; $(-\infty, -10) \cup (2, \infty)$ −10 0 2 **67.** $\{x|2 < x < 5\}$; $(2, 5)$ 0 2 5

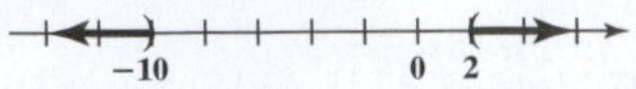

**69.** $\left\{x \middle| x \le -3 \text{ or } x > \frac{15}{4}\right\}$; $(-\infty, -3] \cup \left(\frac{15}{4}, \infty\right)$ −3 0 $\frac{15}{4}$

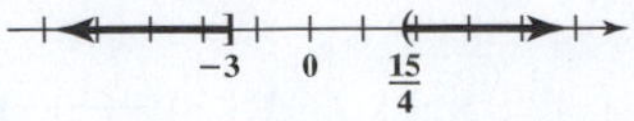

**71.** $\left\{x \middle| -5 < x \le \frac{1}{2}\right\}$; $\left(-5, \frac{1}{2}\right]$ −5 0 $\frac{1}{2}$ **73.** $a = 1$ and $b = 8$ **75.** $a = 12$ and $b = 30$ **77.** $a = -1$ and $b = 23$

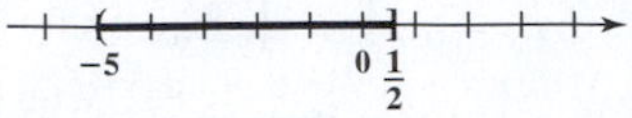

**79.** $90 < x < 140$ **81.** Joanna needs to score at least a 77 on the final. That is, $77 \le x \le 100$ (assuming 100 is the max score, otherwise $77 \le x \le 104$). **83.** The amount withheld ranges between $92.84 and $120.84, inclusive. **85.** The gas usage ranged from 150 to 165 therms.

**87. Step 1:**

$$a < b$$
$$a + a < a + b$$
$$2a < a + b$$
$$\frac{2a}{2} < \frac{a+b}{2}$$
$$a < \frac{a+b}{2}$$

**Step 2:**

$$a < b$$
$$a + b < b + b$$
$$a + b < 2b$$
$$\frac{a+b}{2} < \frac{2b}{2}$$
$$\frac{a+b}{2} < b$$

**Step 3:**

Since $a < \frac{a+b}{2}$ and $\frac{a+b}{2} < b$, it follows that $a < \frac{a+b}{2} < b$.

**89.** { } or ∅ **91.** This is a contradiction. There is no solution. If, during simplification, the variable terms all cancel out and a contradiction results, then there is no solution to the inequality. **93.** If $x < 2$ then $x - 2 < 2 - 2 \Rightarrow x - 2 < 0$. When multiplying both sides of the inequality by $x - 2$ in the second step, the direction of the inequality must switch.

### *Section 1.6*

**1.** $u = a$ or $u = -a$ **3.** $a < 0$ **5.** False **7.** $\{-10, 10\}$ **9.** $\{-1, 7\}$ **11.** $\left\{-1, \frac{13}{3}\right\}$ **13.** $\{-5, 5\}$ **15.** $\left\{-\frac{11}{2}, \frac{5}{2}\right\}$ **17.** $\{-4, 10\}$

**19.** $\{0\}$ **21.** $\left\{-\frac{7}{3}, 3\right\}$ **23.** $\left\{-7, \frac{3}{5}\right\}$ **25.** $\{1, 3\}$ **27.** $\{2\}$ **29.** $\{x|-9 < x < 9\}$; $(-9, 9)$ −9 0 9

**31.** $\{x|-3 \le x \le 11\}$; $[-3, 11]$ −3 0 11 **33.** $\left\{x \middle| -3 < x < \frac{7}{3}\right\}$; $\left(-3, \frac{7}{3}\right)$ −3 0 $\frac{7}{3}$

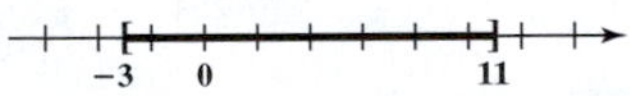

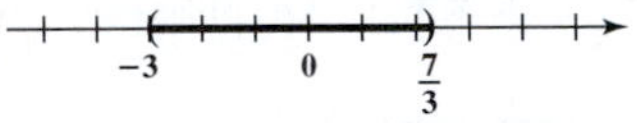

**35.** ∅ or { } **37.** $\{y|y < 3 \text{ or } y > 7\}$; $(-\infty, 3) \cup (7, \infty)$ 0 3 7

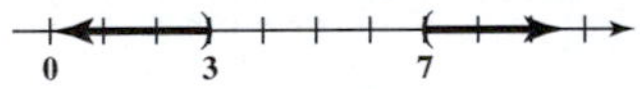

**39.** $\left\{x \middle| x \le -2 \text{ or } x \ge \frac{1}{2}\right\}$; $(-\infty, -2] \cup \left[\frac{1}{2}, \infty\right)$ −2 0 $\frac{1}{2}$

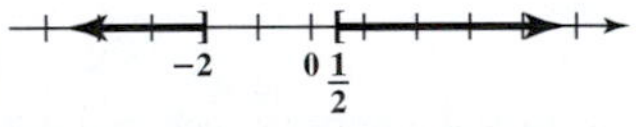

**41.** $\{y|y$ is any real number$\}$$(-\infty, \infty)$ 0 **43.** $\{x|0 < x < 6\}$; $(0, 6)$ 0 6

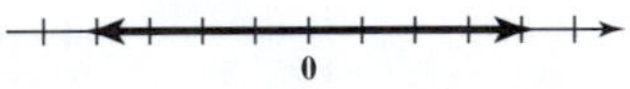

**45.** $\{x|x$ is any real number$\}$; $(-\infty, \infty)$ 0 **47.** $\left\{x \middle| -1 < x < \frac{9}{5}\right\}$; $\left(-1, \frac{9}{5}\right)$ −1 0 $\frac{9}{5}$

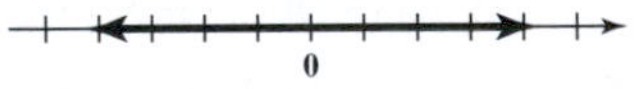

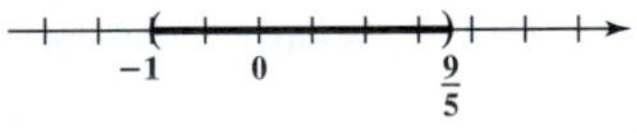

**49.** $\{x|x < 0 \text{ or } x > 1\}$; $(-\infty, 0) \cup (1, \infty)$ −1 0 1

**51.** $\{x|x \le -2 \text{ or } x \ge 3\}$; $(-\infty, -2] \cup [3, \infty)$ −2 0 3

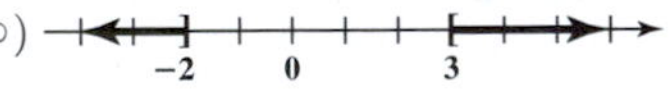

**53.** $\{x|1.995 < x < 2.005\}$; $(1.995, 2.005)$ 1.995 2 2.005

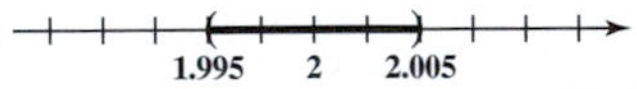

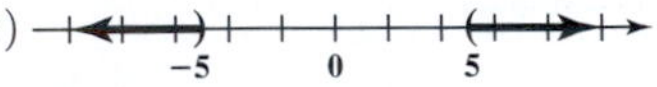

**55.** $\{x|x < -5 \text{ or } x > 5\}$; $(-\infty, -5) \cup (5, \infty)$ −5 0 5 **57.** $\{-4, -1\}$ **59.** $\{-5, 5\}$

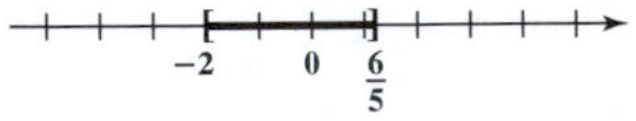

**61.** $\left\{x \middle| -2 \le x \le \frac{6}{5}\right\}$; $\left[-2, \frac{6}{5}\right]$ −2 0 $\frac{6}{5}$ **63.** ∅ or { }

**65.** $\left\{x \middle| x \le -\frac{7}{3} \text{ or } x \ge 1\right\}$; $\left(-\infty, -\frac{7}{3}\right] \cup [1, \infty)$ $-\frac{7}{3}$ 0 1

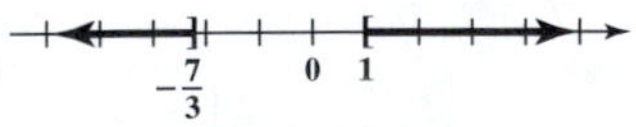

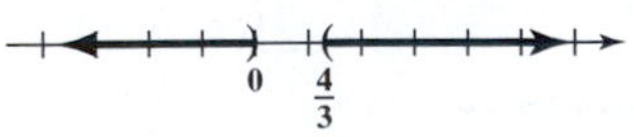

**67.** $\left\{x \middle| x < 0 \text{ or } x > \frac{4}{3}\right\}$; $(-\infty, 0) \cup \left(\frac{4}{3}, \infty\right)$ 0 $\frac{4}{3}$ **69.** $\{-1, 1\}$ **71.** ∅ or { } **73.** $\left\{-8, \frac{4}{7}\right\}$

**75.** $|5 - x| < 3$ $\{x|2 < x < 8\}$; $(2, 8)$ **77.** $|2x - (-6)| > 3$ $\left\{x \middle| x < -\frac{9}{2} \text{ or } x > -\frac{3}{2}\right\}$; $\left(-\infty, -\frac{9}{2}\right) \cup \left(-\frac{3}{2}, \infty\right)$ **79.** The acceptable rod lengths are between 5.6995 inches and 5.7005 inches, inclusive. **81.** An unusual IQ score would be less than 70.6 or greater than 129.4.

**83.** The absolute value, when isolated, is equal to a negative number which is not possible. **85.** The absolute value, when isolated, is less than −3. Since absolute values are always nonnegative, this is not possible. **87.** $\left\{-\frac{5}{2}\right\}$ **89.** $\{2\}$ **91.** ∅ or { } **93.** $\{x|x \le -5\}$; $(-\infty, -5]$

## *Chapter 1 Review*

**1.** $x = 5$ is a solution to the equation. $x = 6$ is **not** a solution to the equation. **2.** $x = -2$ is a solution to the equation. $x = -1$ is a solution to the equation. **3.** $y = -2$ is **not** a solution to the equation. $y = 0$ is a solution to the equation. **4.** $w = -14$ is **not** a solution to the equation. $w = 7$ is **not** a solution to the equation. **5.** conditional equation; $\{3\}$ **6.** conditional; $\{4\}$ **7.** conditional; $\{3\}$ **8.** conditional; $\{-8\}$ **9.** identity; $\{x|x \text{ is any real number}\}$ **10.** contradiction; $\{\ \}$ or $\emptyset$ **11.** contradiction; $\{\ \}$ or $\emptyset$

**12.** conditional; $\{-2\}$ **13.** conditional; $\{-47\}$ **14.** identity; $\{w|w \text{ is any real number}\}$ **15.** $x = -\frac{3}{2}$ must be excluded from the domain. **16.** $x = \frac{1}{2}$ must be excluded from the domain. **17.** Her Missouri taxable income was \$43,250. **18.** The regular club price for a DVD is \$21.95. **19.** $3x + 7 = 22$ **20.** $x - 3 = \frac{x}{2}$ **21.** $0.2x = x - 12$ **22.** $6x = 2x - 4$ **23.** Payton is 5 years old and Shawn is 13 years old. **24.** The five odd integers are 21, 23, 25, 27, and 29. **25.** Logan needs to get a score of 76.5 on the final exam to have an average of 80. **26.** After 1 month, Cherie will accrue about \$11.33 in interest. **27.** The original price of the sleeping bag was \$135.00. **28.** The federal minimum wage was \$5.15 (per hour). **29.** The store should mix 4 pounds of chocolate covered blueberries with 8 pounds of chocolate covered strawberries. **30.** The store should mix 7.5 pounds of baseball gumballs with 2.5 pounds of soccer gumballs. **31.** Angie should invest \$4800 at 8% and \$3200 at 18% to achieve a 12% return. **32.** About 2.89 liters would need to be drained and replaced with pure antifreeze. **33.** Josh drove 90 miles at 60 miles per hour and 210 miles at 70 miles per hour. **34.** The F14 is traveling at a speed of 1220 miles per hour and the F15 is traveling at a speed of 1420 miles per hour. **35.** $x = \frac{k}{y}$ **36.** $C = \frac{5}{9}(F - 32)$ **37.** $W = \frac{P - 2L}{2}$ **38.** $m_2 = \frac{\rho - m_1v_1}{v_2}$ **39.** $T = \frac{PV}{nR}$ **40.** $W = \frac{S - 2LH}{2L + 2H}$

**41.** $y = -\frac{3}{4}x + \frac{1}{2}$ **42.** $y = \frac{5}{4}x + \frac{5}{2}$ **43.** $y = 4x - 5$ **44.** $y = -\frac{6}{5}x + 24$ **45.** The melting point of platinum is 1772°C. **46.** The angles measure 70°, 70°, and 40°. **47.** The window measures 15 feet by 23 feet. **48. (a)** $x = 25C - 73.75$ **(b)** Debbie can talk for 426 minutes in one month and not spend more than \$20 on long distance. **49.** The patio will be $\frac{10}{27}$ of a foot thick (i.e. about 4.44 inches). **50. (a)** $r = \frac{A - \pi Rs}{\pi s}$ **(b)** The radius of the top of the frustum is 2 feet. **51. (a)** $x = \frac{C - 5.28}{0.05947}$ **(b)** Approximately 1850 kwh were used. **52.** The angle measures 60°. **53.** $(2, 7]$

**54.** $(-2, \infty)$ 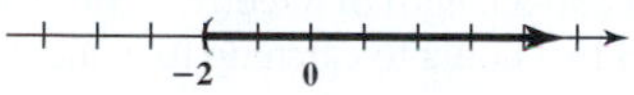 **55.** $x \le 4$

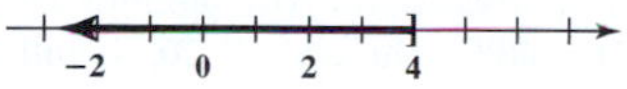

**56.** $-1 \le x < 3$ **57.** $a = 7$ and $b = 15$. **58.** $a = -1$ and $b = 5$. **59.** Solution set: $\{x|x \le -4\}$ Interval: $(-\infty, -4]$ Graph: **60.** Solution set: $\{x|x < -\frac{1}{3}\}$ Interval: $(-\infty, -\frac{1}{3})$

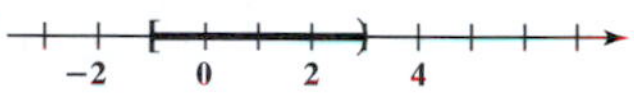

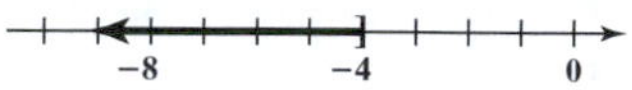

Graph: 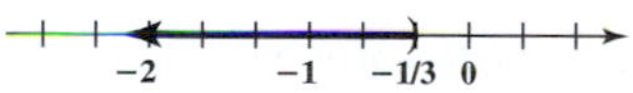 **61.** Solution set: $\{h|h \ge -\frac{2}{3}\}$ Interval: $[-\frac{2}{3}, \infty)$ Graph:

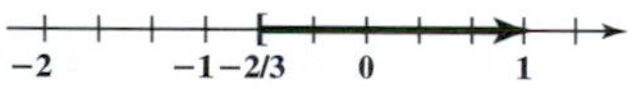

**62.** Solution set: $\{x|x > 2\}$ Interval: $(2, \infty)$ Graph:  **63.** Solution set: $\{p|p > 2\}$ Interval: $(2, \infty)$

Graph:  **64.** Solution set: $\{x|x \text{ is any real number}\}$ Interval: $(-\infty, \infty)$

Graph:  **65.** $\{\ \}$ or $\emptyset$ **66.** Solution set: $\{x|x > 4.2\}$ Interval: $(4.2, \infty)$

Graph: 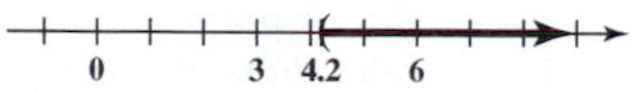 **67.** Solution set: $\{w|w > 1\}$ Interval: $(1, \infty)$ Graph:

**68.** Solution set: $\{y|y < -120\}$ Interval: $(-\infty, -120)$ Graph:

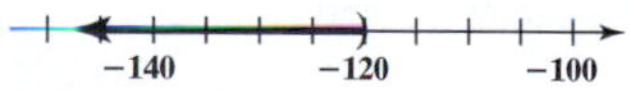

**69.** To stay within budget, no more than 60 people can attend the banquet. **70.** To stay within budget, you can drive an average of 216 miles per day. **71.** The band must sell more than 125 candy bars to make a profit. **72.** You can purchase up to 5 DVDs and still be within budget. **73.** $\{-1, 0, 1, 2, 3, 4, 6, 8\}$ **74.** $\{2, 4\}$ **75.** $\{1, 2, 3, 4\}$ **76.** $\{1, 2, 3, 4, 6, 8\}$ **77. (a)** $\{x|2 < x \le 4\}$

**(b)** $\{x|x \text{ is any real number}\}$  **78. (a)** $\{\ \}$ or $\emptyset$ **(b)** $\{x|x < -2 \text{ or } x \ge 3\}$

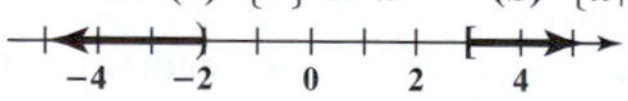

**79.** $\{x|-1 < x < 4\}$; $(-1, 4)$ Graph: **80.** $\{x|-5 < x < -1\}$; $(-5, -1)$

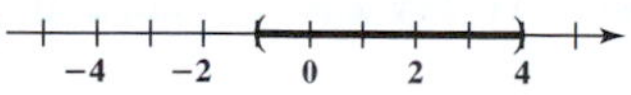

Graph: 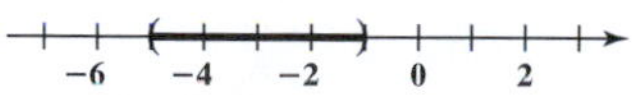 **81.** $\{x|x < -2 \text{ or } x > 2\}$; $(-\infty, -2) \cup (2, \infty)$ Graph:

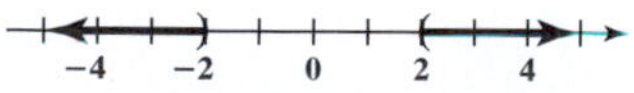

**82.** $\{x|x \le 0 \text{ or } x \ge 4\}$; $(-\infty, 0] \cup [4, \infty)$ Graph: **83.** $\{\ \}$ or $\emptyset$ **84.** $\{x|-2 \le x < 4\}$; $[-2, 4)$

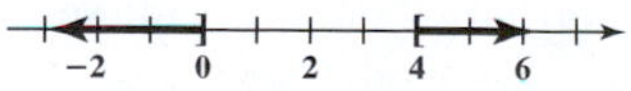

Graph: 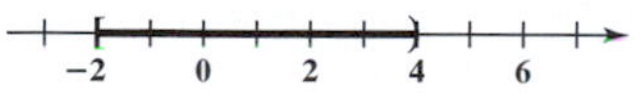 **85.** $\{x|x \le -2 \text{ or } x > 3\}$; $(-\infty, -2] \cup (3, \infty)$ Graph:

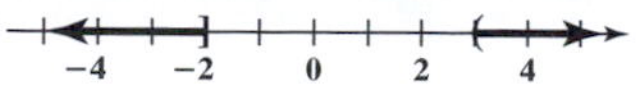

**86.** $\{x|x \text{ is any real number}\}$; $(-\infty, \infty)$ Graph:  **87.** $\{x|x < -10 \text{ or } x > 6\}$; $(-\infty, -10) \cup (6, \infty)$

Graph: 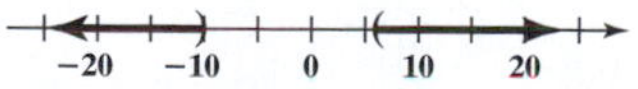 **88.** $\{x|-\frac{3}{2} \le x < \frac{5}{8}\}$; $[-\frac{3}{2}, \frac{5}{8})$ Graph: **89.** $70 \le x \le 75$

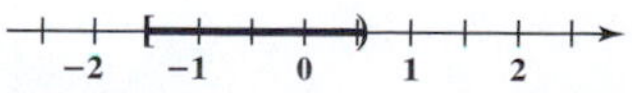

**90.** The electric usage varied from roughly 756.7 kilowatt hours up to roughly 1953.1 kilowatt hours. (recall, $x$ is the number *above* 300). **91.** $\{-4, 4\}$ **92.** $\{\frac{1}{3}, 3\}$ **93.** $\{-5, 13\}$ **94.** $\{-3, -1\}$ **95.** $\{\ \}$ or $\emptyset$ **96.** $\{-\frac{1}{2}, 2\}$ **97.** $\{x|-2 < x < 2\}$; $(-2, 2)$

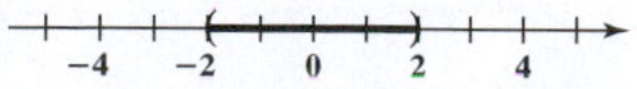

**98.** $\{x|x \le -\frac{7}{2} \text{ or } x \ge \frac{7}{2}\}$; $(-\infty, -\frac{7}{2}] \cup [\frac{7}{2}, \infty)$ **99.** $\{x|-5 \le x \le 1\}$; $[-5, 1]$

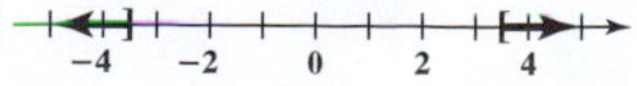

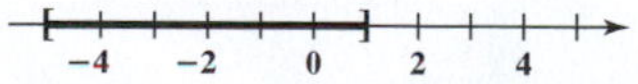

**100.** $\{x|x \le \frac{1}{2} \text{ or } x \ge 1\}; (-\infty, \frac{1}{2}] \cup [1, \infty)$

**101.** $\{x|x \text{ is a real number}\}; (-\infty, \infty)$ **102.** { } or Ø

**103.** $\{x|4.99 \le x \le 5.01\}; [4.99, 5.01]$

**104.** $\{x|x < -\frac{1}{2} \text{ or } x > \frac{7}{2}\}; (-\infty, -\frac{1}{2}) \cup (\frac{7}{2}, \infty)$

**105.** The acceptable diameters of the bearing are between 0.502 inches and 0.504 inches, inclusive. **106.** Tensile strengths below 36.08 lb/in.$^2$ or above 43.92 lb/in.$^2$ would be considered unusual.

## *Chapter 1 Test*

**1.** **(a)** $x = 6$ is a solution to the equation. **(b)** $x = -2$ is **not** a solution to the equation.

**2.** **(a)** Interval: $(-4, \infty)$ **(b)** Interval: $(3, 7]$ **3.** $3x - 8 = x + 4$

**4.** $\frac{2}{3}x + 2(x - 5) > 7$ **5.** {2} This is a conditional equation. **6.** {−4, −1} This is a conditional equation.

**7.** contradiction; { } or Ø **8.** $\{x|x \ge 3\}; [3, \infty)$

**9.** $\left\{x \middle| x > -\frac{1}{5}\right\}; \left(-\frac{1}{5}, \infty\right)$ **10.** $\left\{x \middle| x \ge \frac{1}{2}\right\}; \left[\frac{1}{2}, \infty\right)$

**11.** $\{x|-2 \le x < 6\}; [-2, 6)$

**12.** $\left\{x \middle| x < -\frac{3}{2} \text{ or } x > 4\right\}; \left(-\infty, -\frac{3}{2}\right) \cup (4, \infty)$ **13.** $y = -\frac{7}{4}x + \frac{3}{4}$ **14.** **(a)** {1, 3, 5, 6, 7, 9, 12} **(b)** {3, 9}

**15.** Glen's weekly sales must be at least $4375 for him to earn at least $750. **16.** The Crescent Rod can fit openings with widths between 59 inches and 61 inches. **17.** There were 14 children at Payton's party. **18.** The sandbox has a width of 4 feet and a length of 6 feet. **19.** The chemist needs to mix 8 liters of the 10% solution with 4 liters of the 40% solution. **20.** It will take contestant B two hours to catch up to contestant A.

## *Cumulative Review Chapters R–1*

**1.** **(a)** **(i)** 27.235 **(ii)** 27.236 **(b)** **(i)** 1.0 **(ii)** 1.1 **2.** **3.** −14 **4.** −6 **5.** 6 **6.** 81 **7.** 74

**8.** $\frac{11}{12}$ **9.** 9 **10.** $5a^2 - 4a - 13$ **11.** **(a)** No **(b)** Yes **12.** $-5x + 18$ **13.** $x = 3$ is **not** a solution to the equation. **14.** {4}

**15.** $\{\frac{13}{2}\}$ **16.** $\{-\frac{5}{2}, \frac{9}{2}\}$ **17.** $y = \frac{2}{5}x - \frac{6}{5}$ **18.** $\{x|x \ge 7\}; [7, \infty)$

**19.** $\{x|1 \le x \le 7\}; [1, 7]$

**20.** $\{x|x < -9 \text{ or } x > -5\}; (-\infty, -9) \cup (-5, \infty)$

**21.** Shawn needs to score at least 91 on the final exam to earn an A (assuming the maximum score on the exam is 100). **22.** A person 62 inches tall would be considered obese if they weighed 160 pounds or more. **23.** The angles measure 55° and 125°. **24.** The cylinder should be about 5.96 inches tall. **25.** The three consecutive even integers are 24, 26, and 28.

# Chapter 2

## *Section 2.1*

**1.** origin **3.** intercepts **5.** False **7.** Answers will vary. **9.** Answers will vary. **11.** $A$: (2, 3); I $B$: (−5, 2); II $C$: (0, −2); $y$-axis $D$: (−4, −3); III $E$: (3, −4); IV $F$: (4, 0); $x$-axis **13.**

$A$: quadrant I; $B$: quadrant III; $C$: $x$-axis; $D$: quadrant IV; $E$: $y$-axis; $F$: quadrant II

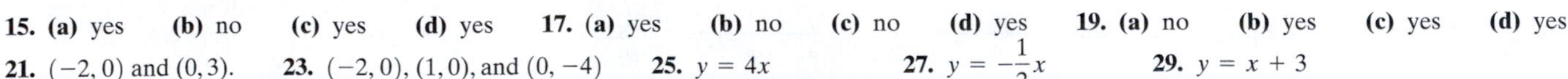

**15.** **(a)** yes **(b)** no **(c)** yes **(d)** yes **17.** **(a)** yes **(b)** no **(c)** no **(d)** yes **19.** **(a)** no **(b)** yes **(c)** yes **(d)** yes

**21.** (−2, 0) and (0, 3). **23.** (−2, 0), (1, 0), and (0, −4) **25.** $y = 4x$ **27.** $y = -\frac{1}{2}x$ **29.** $y = x + 3$

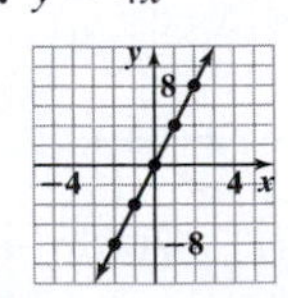

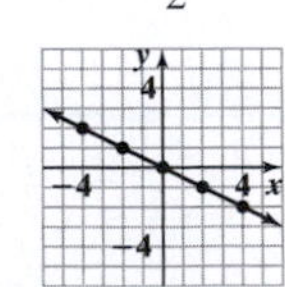

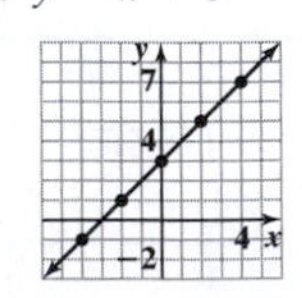

**31.** $y = -3x + 1$ **33.** 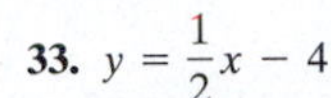 $y = \frac{1}{2}x - 4$ **35.** 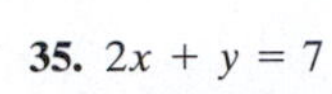 $2x + y = 7$ **37.** 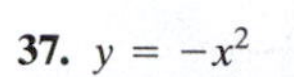 $y = -x^2$ **39.** $y = 2x^2 - 8$ **41.** $y = |x|$

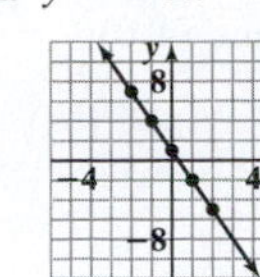
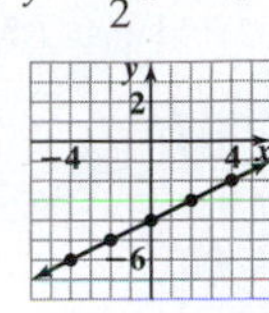
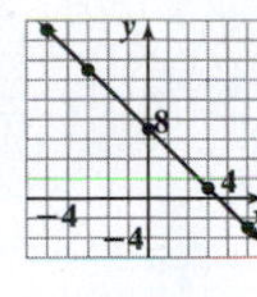
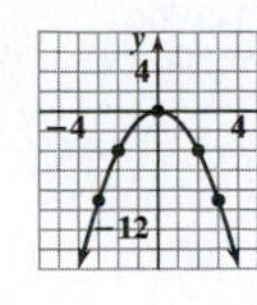
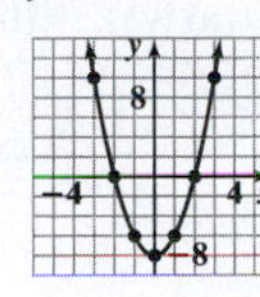
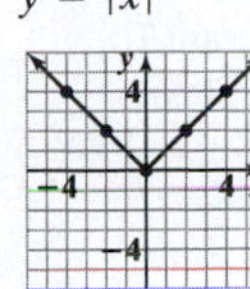

**43.** $y = |x - 1|$ **45.** $y = x^3$ **47.** $y = x^3 + 1$ **49.** $x^2 - y = 4$ **51.** $x = y^2 - 1$

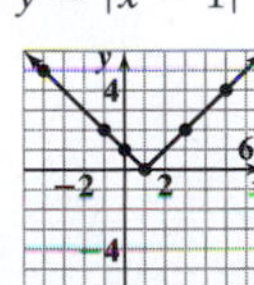
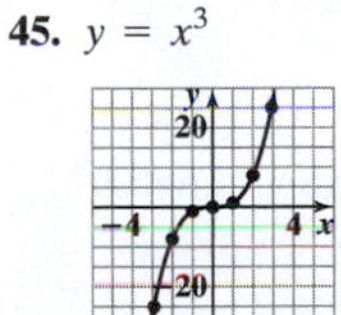
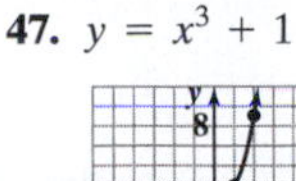
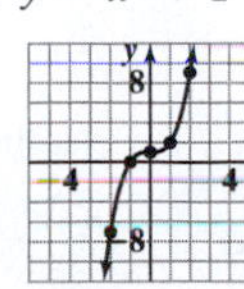
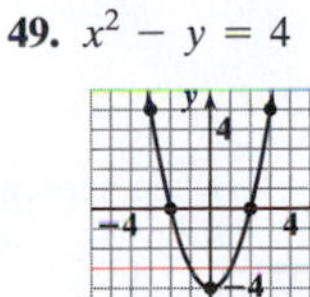
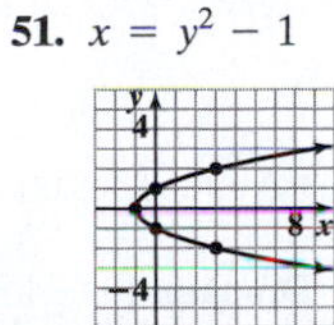

**53.** $a = \frac{7}{4}$ **55.** $b = 4$ **57.** **(a)** 400 ft$^2$ **(b)** 25 feet; 625 ft$^2$ **(c)** The $x$-intercepts are $x = 0$ and $x = 50$. These values form the bounds for the width of the opening. The $y$-intercept is $y = 0$. The area of the opening will be 0 ft$^2$ when the width is 0 feet. **59.** **(a)** \$40; \$40 **(b)** \$800 **(c)** 40; the monthly cost will be \$40 if no minutes are used.

**61.** Vertical line with an $x$-intercept of 4. **63.** Answers will vary. One possible graph is below. **65.** Answers will vary. One possibility: $y = 0$

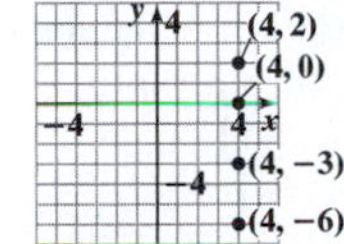
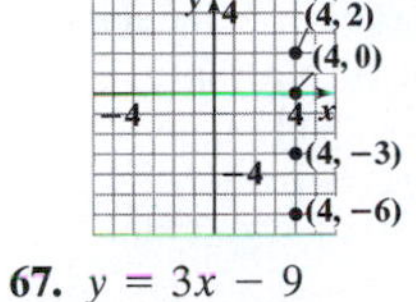

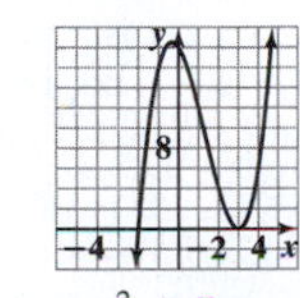
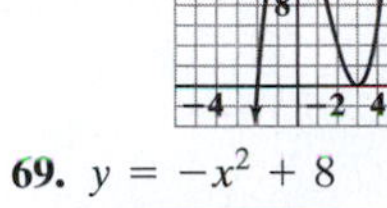

**67.** $y = 3x - 9$ **69.** $y = -x^2 + 8$ **71.** $y + 2x^2 = 13$ **73.** $y = x^3 - 6x + 1$

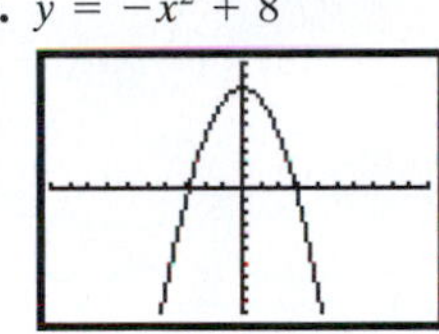
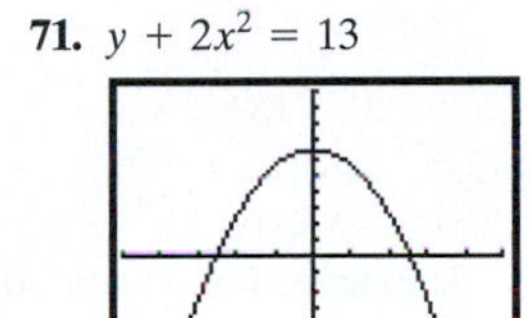
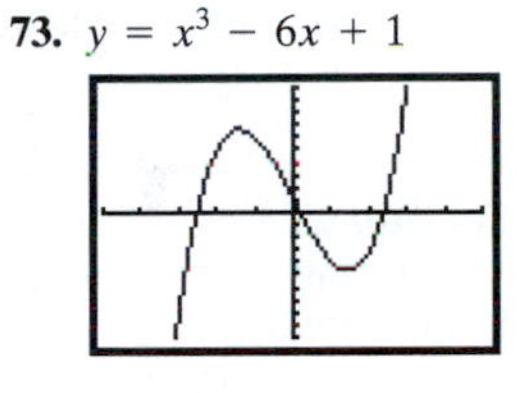

## Section 2.2

**1.** corresponds; depends on **3.** True **5.** Answers will vary. **7.** {(*USA Today*, 2.1), (*Wall Street Journal*, 1.8), (*New York Times*, 1.1), (*Los Angeles Times*, 0.9), (*Washington Post*, 0.8)}; Domain: {*USA Today, Wall Street Journal, New York Times, Los Angeles Times, Washington Post*}; Range: {0.8, 0.9, 1.1, 1.8, 2.1} **9.** {(Less than 9th Grade, \$17261), (9th–12th Grade, no diploma, \$21737), (High School Graduate, \$35744), (Associate's Degree, \$49279), (Bachelor's Degree or Higher, \$69804)}; Domain: {Less than 9th Grade, 9th–12th Grade, no diploma, High School Graduate, Associate's Degree, Bachelor's Degree or Higher}; Range: {\$17261, \$21737, \$35744, \$49279, \$69804}

**11.**

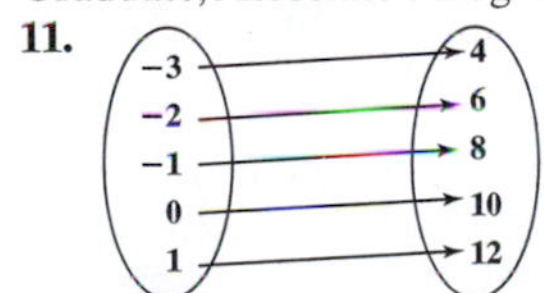

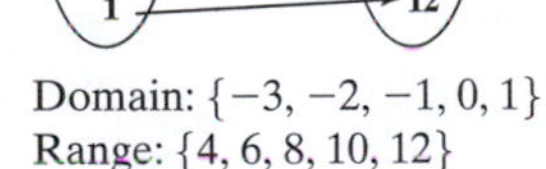

Domain: $\{-3, -2, -1, 0, 1\}$
Range: $\{4, 6, 8, 10, 12\}$

**13.**

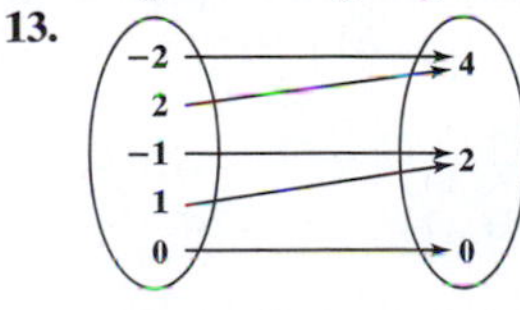

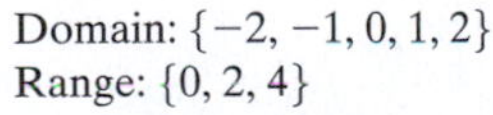

Domain: $\{-2, -1, 0, 1, 2\}$
Range: $\{0, 2, 4\}$

**15.**

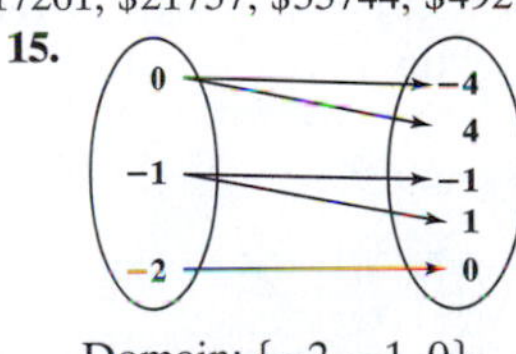


Domain: $\{-2, -1, 0\}$
Range: $\{-4, -1, 0, 1, 4\}$

**17.** Domain: $\{-3, -2, 0, 2, 3\}$
Range: $\{-3, -1, 2, 3\}$

**19.** Domain: $\{x \mid -4 \le x \le 4\}$ or $[-4, 4]$
Range: $\{y \mid -2 \le y \le 2\}$ or $[-2, 2]$

**21.** Domain: $\{x \mid -1 \le x \le 3\}$ or $[-1, 3]$
Range: $\{y \mid 0 \le y \le 4\}$ or $[0, 4]$

**23.** Domain: $\{x \mid x \text{ is a real number}\}$ or $(-\infty, \infty)$
Range: $\{y \mid y \ge -3\}$ or $[-3, \infty)$

**25.** Domain: $\{x \mid x \text{ is a real number}\}$ or $(-\infty, \infty)$
Range: $\{y \mid y \text{ is a real number}\}$ or $(-\infty, \infty)$

**27.** Domain: $\{x \mid x \text{ is a real number}\}$ or $(-\infty, \infty)$
Range: $\{y \mid y \text{ is a real number}\}$ or $(-\infty, \infty)$

**29.** Domain: $\{x \mid x \text{ is a real number}\}$ or $(-\infty, \infty)$
Range: $\{y \mid y \text{ is a real number}\}$ or $(-\infty, \infty)$

**31.** Domain: $\{x \mid x \text{ is a real number}\}$ or $(-\infty, \infty)$
Range: $\{y \mid y \text{ is a real number}\}$ or $(-\infty, \infty)$

**33.** Domain: $\{x \mid x \text{ is a real number}\}$ or $(-\infty, \infty)$
Range: $\{y \mid y \text{ is a real number}\}$ or $(-\infty, \infty)$

**35.** Domain: $\{x \mid x \text{ is a real number}\}$ or $(-\infty, \infty)$
Range: $\{y \mid y \text{ is a real number}\}$ or $(-\infty, \infty)$

**37.** Domain: $\{x \mid x \text{ is a real number}\}$ or $(-\infty, \infty)$
Range: $\{y \mid y \le 0\}$ or $(-\infty, 0]$

**39.** Domain: $\{x \mid x \text{ is a real number}\}$ or $(-\infty, \infty)$
Range: $\{y \mid y \ge -8\}$ or $[-8, \infty)$

**41.** Domain: $\{x \mid x \text{ is a real number}\}$ or $(-\infty, \infty)$
Range: $\{y \mid y \ge 0\}$ or $[0, \infty)$

**43.** Domain: $\{x \mid x \text{ is a real number}\}$ or $(-\infty, \infty)$
Range: $\{y \mid y \ge 0\}$ or $[0, \infty)$

**45.** Domain: $\{x \mid x \text{ is a real number}\}$ or $(-\infty, \infty)$
Range: $\{y \mid y \text{ is a real number}\}$ or $(-\infty, \infty)$

**47.** Domain: $\{x \mid x \text{ is a real number}\}$ or $(-\infty, \infty)$
Range: $\{y \mid y \text{ is a real number}\}$ or $(-\infty, \infty)$

**49.** Domain: $\{x \mid x \text{ is a real number}\}$ or $(-\infty, \infty)$
Range: $\{y \mid y \ge -4\}$ or $[-4, \infty)$

**51.** Domain: $\{x \mid x \ge -1\}$ or $[-1, \infty)$
Range: $\{y \mid y \text{ is a real number}\}$ or $(-\infty, \infty)$

**53.** **(a)** Domain: $\{x \mid 0 \le x \le 50\}$ or $[0, 50]$
Range: $\{y \mid 0 \le y \le 625\}$ or $[0, 625]$
**(b)** Answers may vary.

**55.** **(a)** Domain: $\{m \mid 0 \le m \le 15{,}120\}$ or $[0, 15120]$
Range: $\{c \mid 40 \le c \le 5888\}$ or $[40, 5888]$
**(b)** Answers will vary.

**57.** Actual graphs will vary but all should be horizontal lines.

## *Chapter 2 Putting the Concepts Together*

**1.** $A$: $x$-axis; $B$: quadrant II; $C$: quadrant I; $D$: $y$-axis; $E$: quadrant III; $F$: quadrant IV.

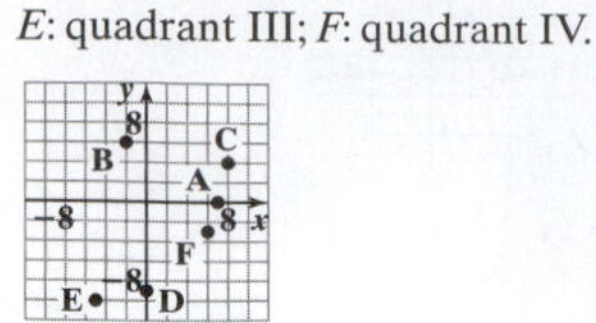

**2. (a)** yes **(b)** yes **(c)** no **3.** $y = |x| + 3$

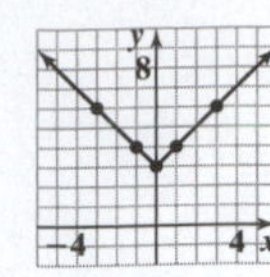

**4.** $y = \frac{1}{2}x^2 - 1$ **5.** $(0, -5)$ and $(4, 0)$

**6. (a)** About \$230,000. **(b)** July of 2003; About \$248,000. **(c)** Between September 2002 and October 2002; About \$16,000

**7.** $\{(-2, -1), (-1, 0), (0, 1), (1, 2), (2, 3)\}$ **8. (a)** Domain: $\{x|-4 \le x \le 5\}$ or $[-4, 5]$ Range: $\{y|-6 \le y \le -1\}$ or $[-6, -1]$ **(b)** Domain: $\{-4, -1, 0, 3, 6\}$ Range: $\{-3, -2, 2, 6\}$

**9. (a)** $y = |x - 2| - 3$

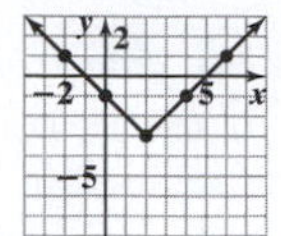

Domain: $\{x|x$ is a real number$\}$ or $(-\infty, \infty)$
Range: $\{y|y \ge -3\}$ or $[-3, \infty)$

**(b)** $y = \frac{1}{2}x^2 + 1$

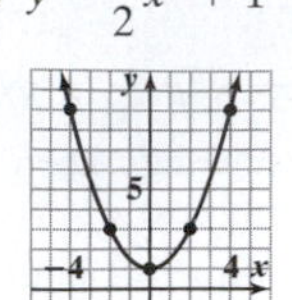

Domain: $\{x|x$ is a real number$\}$ or $(-\infty, \infty)$
Range: $\{y|y \ge 1\}$ or $[1, \infty)$

**10.** Domain: $\{t|0 \le t \le 3.8\}$
Range: $\{h|0 \le h \le 105\}$

## *Section 2.3*

**1.** function **3.** argument **5.** True **7.** map; ordered pairs; equation; graph
**9.** Function. Domain: {Virginia, Nevada, New Mexico, Tennessee, Texas} Range: $\{3, 9, 11, 32\}$ **11.** Not a function. Domain: $\{150, 174, 180\}$ Range: $\{118, 130, 140\}$ **13.** Function. Domain: $\{0, 1, 2, 3\}$ Range: $\{3, 4, 5, 6\}$
**15.** Function. Domain: $\{-3, 1, 4, 7\}$ Range: $\{5\}$ **17.** Not a function. Domain: $\{-10, -5, 0\}$ Range: $\{1, 2, 3, 4\}$ **19.** Function **21.** Function **23.** Not a function **25.** Function
**27.** Not a function **29.** Function **31.** Not a function **33.** Function **35.** Function **37. (a)** $f(0) = 3$ **(b)** $f(3) = 9$ **(c)** $f(-2) = -1$ **(d)** $f(-x) = -2x + 3$ **(e)** $-f(x) = -2x - 3$ **(f)** $f(x + 2) = 2x + 7$ **(g)** $f(2x) = 4x + 3$ **(h)** $f(x + h) = 2x + 2h + 3$ **39. (a)** $f(0) = 2$ **(b)** $f(3) = -13$ **(c)** $f(-2) = 12$ **(d)** $f(-x) = 5x + 2$ **(e)** $-f(x) = 5x - 2$ **(f)** $f(x + 2) = -5x - 8$ **(g)** $f(2x) = -10x + 2$ **(h)** $f(x + h) = -5x - 5h + 2$ **41.** $f(2) = 7$ **43.** $s(-2) = 16$ **45.** $F(-3) = 5$ **47.** $F(4) = -6$
**49.** $f(x) = 4x - 6$ **51.** $h(x) = x^2 - 2$

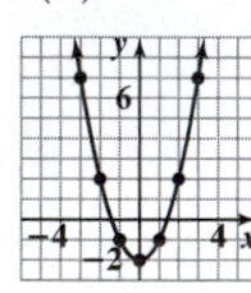

**53.** $G(x) = |x - 1|$ **55.** $g(x) = x^3$

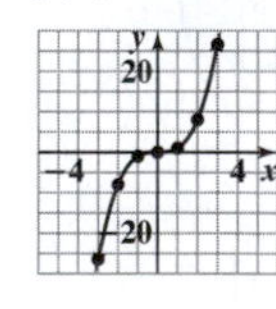

**57.** $-6 = C$ **59.** $A = 5$ **61.** $A(r) = \pi r^2$; 50.27 in.$^2$ **63.** $G(h) = 15h$; \$375

**65. (a)** The dependent variable is the population, $P$, and the independent variable is the age, $a$.
**(b)** $P(20) = 197.857$; The population of Americans that were 20 years of age or older in 2005 was roughly 198 million.
**(c)** $P(0) = 300.517$; $P(0)$ represents the entire population of the U.S. since every member of the population is at least 0 years of age. The population of the U.S. in 2005 was roughly 301 million.
**67. (a)** The dependent variable is revenue, $R$, and the independent variable is price, $p$.
**(b)** $R(50) = 7500$; Selling PDAs for \$50 will yield a daily revenue of \$7500 for the company.
**(c)** $R(120) = 9600$; Selling PDAs for \$120 will yield a daily revenue of \$9600 for the company.
**69.** Answers will vary. **71.** $f(2) = 7$ **73.** $F(-3) = 5$ **75.** $H(7) = 5$ **77.** $F(4) = -6$

## *Section 2.4*

**1.** $f(3) = 8$ **3.** domain **5.** True **7.** Answers will vary. **9.** $\{x|x$ is a real number$\}$ or $(-\infty, \infty)$ **11.** $\{z|z \ne 5\}$ or $(-\infty, 5) \cup (5, \infty)$
**13.** $\{x|x$ is a real number$\}$ or $(-\infty, \infty)$ **15.** $\left\{x \middle| x \ne -\frac{1}{3}\right\}$ or $\left(-\infty, -\frac{1}{3}\right) \cup \left(-\frac{1}{3}, \infty\right)$ **17.** $\{x|x$ is a real number $\}$ or $(-\infty, \infty)$
**19. (a)** Domain: $\{x|x$ is a real number$\}$ or $(-\infty, \infty)$ Range: $\{y|y$ is a real number$\}$ or $(-\infty, \infty)$ **(b)** $(0, 2)$ and $(1, 0)$ **21. (a)** Domain: $\{x|x$ is a real number$\}$ or $(-\infty, \infty)$ Range: $\{y|y \ge -2.25\}$ or $[-2.25, \infty)$ **(b)** $(-2, 0)$, $(4, 0)$, and $(0, -2)$

**23. (a)** Domain: $\{x|x \text{ is a real number}\}$ or $(-\infty, \infty)$
Range: $\{y|y \text{ is a real number}\}$ or $(-\infty, \infty)$
**(b)** $(-3, 0), (-1, 0), (2, 0)$, and $(0, -3)$

**25. (a)** Domain: $\{x|x \text{ is a real number}\}$ or $(-\infty, \infty)$
Range: $\{y|y \geq 0\}$ or $[0, \infty)$
**(b)** $(-3, 0), (3, 0)$, and $(0, 9)$

**27. (a)** Domain: $\{x|x \leq 4\}$ or $(-\infty, 4]$
Range: $\{y|y \leq 3\}$ or $(-\infty, 3]$
**(b)** $(-2, 0)$ and $(0, 2)$

**29. (a)** $f(-7) = -2$ **(b)** $f(-3) = 3$ **(c)** $f(6) = 2$ **(d)** negative **(e)** $\{-6, -1, 4\}$ **(f)** $\{x|-7 \leq x \leq 6\}$ or $[-7, 6]$ **(g)** $\{y|-2 \leq y \leq 3\}$ or $[-2, 3]$ **(h)** $-6, -1$, and $4$ **(i)** $-1$ **(j)** $\{-7, 2\}$ **(k)** $x = -3$

**31. (a)** $F(-2) = 3$ **(b)** $F(3) = -6$ **(c)** $x = -1$ **(d)** $x = -4$ **(e)** $y = 2$ **33. (a)** no **(b)** $f(3) = 3; (3, 3)$ **(c)** $x = 4; (4, 7)$

**35. (a)** yes **(b)** $g(6) = 1; (6, 1)$ **(c)** $x = -12; (-12, 10)$ **37.** $\{r|r \geq 0\}$ or $[0, \infty)$ **39.** $\{h|0 \leq h \leq 60\}$ or $[0, 60]$

**41.** $\{p|0 \leq p \leq 120\}$ or $[0, 120]$ **43. (a)** III **(b)** I **(c)** IV **(d)** V **(e)** II

**45.** Answers will vary.

Pulse rate (beats per min.) 160, 120, 80, 40; Time (in minutes) 0, 15, 30

**47.** Answers will vary.

Initial height of the swing

**49.** Answers will vary.

**51.** Answers will vary. One possibility.

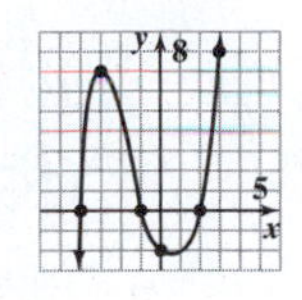

## Chapter 2 Review

**1.** $A$: quadrant IV; $B$: quadrant III; $C$: $y$-axis; $D$: quadrant II; $E$: $x$-axis; $F$: quadrant I.

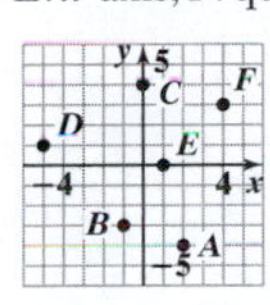

**2.** $A$: $x$-axis; $B$: quadrant I; $C$: quadrant III; $D$: quadrant II; $E$: quadrant IV; $F$: $y$-axis.

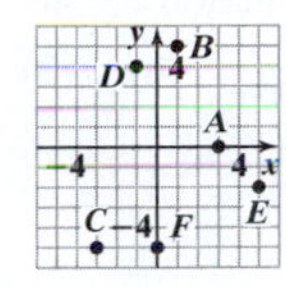

**3. (a)** yes **(b)** no **(c)** no **(d)** yes
**4. (a)** no **(b)** yes **(c)** yes **(d)** no

**5.** $y = x + 2$

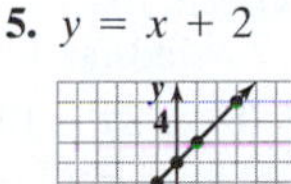

**6.** $2x + y = 3$

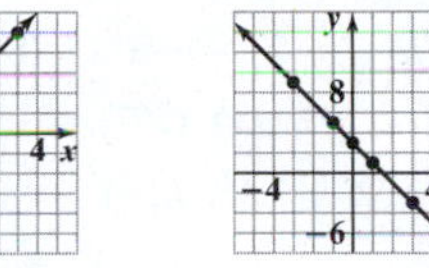

**7.** $y = 2x^2 - 3$

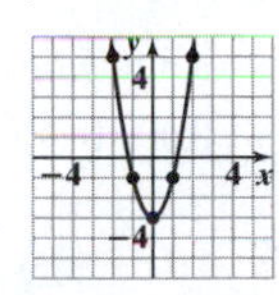

**8.** $y = -x^2 + 4$

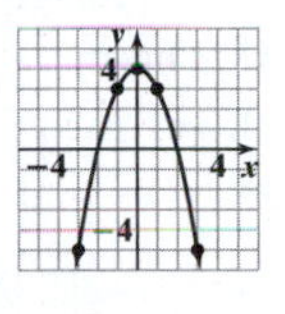

**9.** $y = -|x| - 2$ **10.** $y = |x + 2| - 1$ **11.** $y = x^3 + 2$ **12.** $y = -x^3 + 1$ **13.** $x = y^2 + 1$

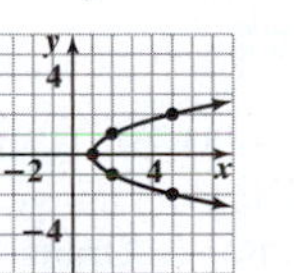

**14.** $y = \dfrac{1}{x - 2}$

**15.** $(-3, 0), (0, -1), (0, 3)$ **16.** $(-2, 0), (0, 0), (2, 0)$ **17. (a)** \$40 **(b)** About \$500 **18. (a)** 2:03 or 123 seconds **(b)** 2001; 2:00 or 120 seconds

**19.** {(Cent, 2.500), (Nickel, 5.000), (Dime, 2.268), (Quarter, 5.670), (Half Dollar, 11.340), (Dollar, 8.100)}
Domain: {Cent, Nickel, Dime, Quarter, Half Dollar, Dollar}
Range: {2.268, 2.500, 5.000, 5.670, 8.100, 11.340}

**20.** {(70, \$6.99), (90, \$9.99), (120, \$9.99), (128, \$12.99), (446, \$49.99)}
Domain: {70, 90, 120, 128, 446}
Range: {\$6.99, \$9.99, \$12.99, \$49.99}

**21.** Domain: $\{-4, -2, 2, 3, 6\}$
Range: $\{-9, -1, 5, 7, 8\}$

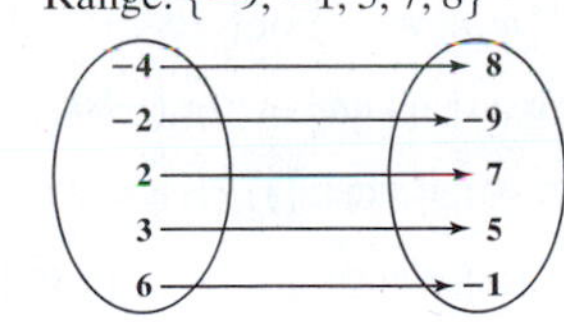

**22.** Domain: $\{-2, 1, 3, 5\}$
Range: $\{1, 4, 7, 8\}$

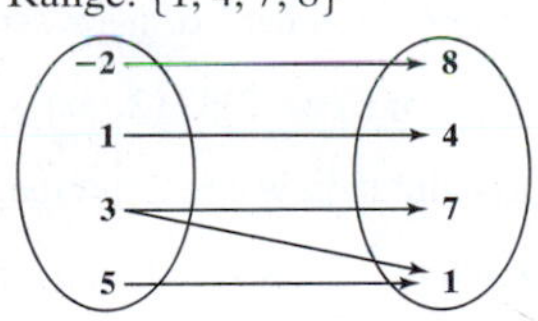

**23.** Domain: $\{x|x \text{ is a real number}\}$ or $(-\infty, \infty)$
Range: $\{y|y \text{ is a real number}\}$ or $(-\infty, \infty)$

**24.** Domain: $\{x|-6 \leq x \leq 4\}$ or $[-6, 4]$
Range: $\{y|-4 \leq y \leq 6\}$ or $[-4, 6]$

**25.** Domain: $\{2\}$
Range: $\{y|y \text{ is a real number}\}$ or $(-\infty, \infty)$

**26.** Domain: $\{x|x \geq -1\}$ or $[-1, \infty)$
Range: $\{y|y \geq -2\}$ or $[-2, \infty)$

**27.** $y = x + 2$

Domain: $\{x|x \text{ is a real number}\}$ or $(-\infty, \infty)$
Range: $\{y|y \text{ is a real number}\}$ or $(-\infty, \infty)$

**28.** $2x + y = 3$

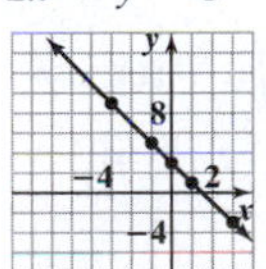

Domain: $\{x|x \text{ is a real number}\}$ or $(-\infty, \infty)$
Range: $\{y|y \text{ is a real number}\}$ or $(-\infty, \infty)$

**29.** $y = 2x^2 - 3$

Domain: $\{x|x \text{ is a real number}\}$ or $(-\infty, \infty)$
Range: $\{y|y \geq -3\}$ or $[-3, \infty)$

**30.** $y = -x^2 + 4$

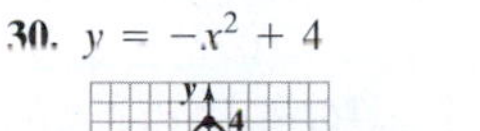

Domain: $\{x|x \text{ is a real number}\}$ or $(-\infty, \infty)$
Range: $\{y|y \leq 4\}$ or $(-\infty, 4]$

**31.** $y = -|x| - 2$

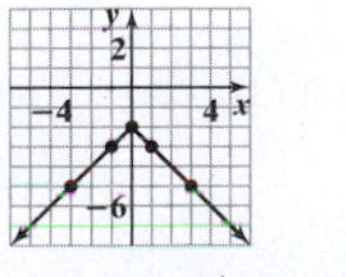

Domain: $\{x|x \text{ is a real number}\}$ or $(-\infty, \infty)$
Range: $\{y|y \leq -2\}$ or $(-\infty, -2]$

**32.** $y = |x + 2| - 1$

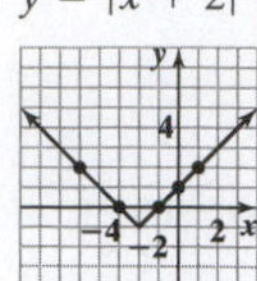

Domain: $\{x|x \text{ is a real number}\}$ or $(-\infty, \infty)$
Range: $\{y|y \text{ is a real number}\}$ or $[-1, \infty)$

**33.** $y = x^3 + 2$

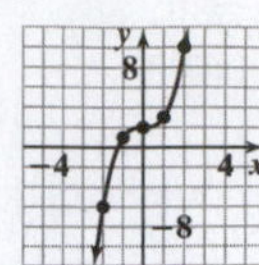

Domain: $\{x|x \text{ is a real number}\}$ or $(-\infty, \infty)$
Range: $\{y|y \text{ is a real number}\}$ or $(-\infty, \infty)$

**34.** $y = -x^3 + 1$

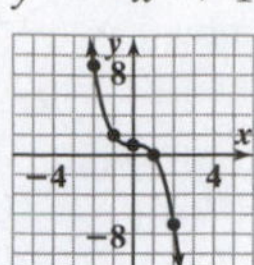

Domain: $\{x|x \text{ is a real number}\}$ or $(-\infty, \infty)$
Range: $\{y|y \text{ is a real number}\}$ or $(-\infty, \infty)$

**35.** $x = y^2 + 1$

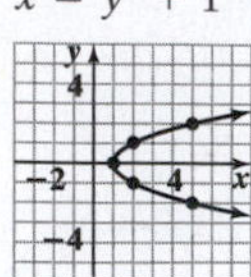

Domain: $\{x|x \geq 1\}$ or $[1, \infty)$
Range: $\{y|y \text{ is a real number}\}$ or $(-\infty, \infty)$

**36.** $y = \dfrac{1}{x - 2}$

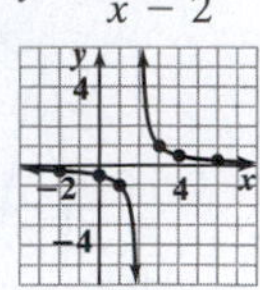

Domain: $\{x|x \neq 2\}$ or $(-\infty, 2) \cup (2, \infty)$
Range: $\{y|y \neq 0\}$ or $(-\infty, 0) \cup (0, \infty)$

**37. (a)** Domain: $\{x|0 \leq x \leq 44.64\}$ or $[0, 44.64]$
Range: $\{y|40 \leq y \leq 2122\}$ or $[40, 2122]$
**(b)** Answers may vary.

**38.** Domain: $\{t|0 \leq t \leq 4\}$ or $[0, 4]$
Range: $\{y|0 \leq y \leq 121\}$ or $[0, 121]$

**39. (a)** Not a function.
Domain: $\{-1, 5, 7, 9\}$
Range: $\{-2, 0, 2, 3, 4\}$
**(b)** Function.
Domain: {Camel, Macaw, Deer, Fox, Tiger, Crocodile}
Range: $\{14, 22, 35, 45, 50\}$

**40. (a)** Function
Domain: $\{-3, -2, 2, 4, 5\}$
Range: $\{-1, 3, 4, 7\}$
**(b)** Not a function
Domain: {Red, Blue, Green, Black}
Range: {Camry, Taurus, Windstar, Durango}

**41.** Function **42.** Not a function **43.** Not a function **44.** Function **45.** Not a function **46.** Function

**47.** Function **48.** Not a function **49. (a)** $f(-2) = -5$ **(b)** $f(3) = 10$ **50. (a)** $g(0) = -\frac{1}{3}$ **(b)** $g(2) = -5$

**51. (a)** $F(5) = -3$ **(b)** $F(-x) = 2x + 7$ **52. (a)** $G(7) = 15$ **(b)** $G(x + h) = 2x + 2h + 1$

**53.** $f(x) = 2x - 5$

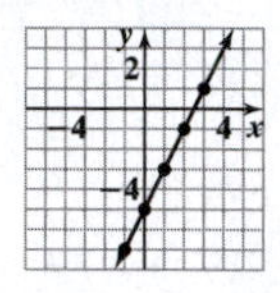

**54.** $g(x) = x^2 - 3x + 2$

**55.** $h(x) = (x - 1)^3 - 3$

**56.** $f(x) = |x + 1| - 4$

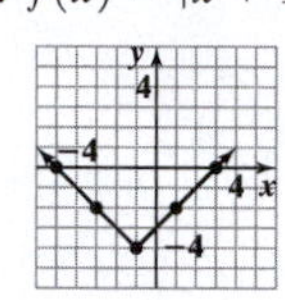

**57. (a)** The dependent variable is the population, $P$, and the independent variable is the number of years after 1900, $t$.
**(b)** $P(110) = 1119.418$; The population of Orange County will be roughly 1,119,418 in 2010.
**(c)** $P(-70) = 1272.958$; The population of Orange County was roughly 1,272,958 in 1830. This is not reasonable. (The population of the entire Florida territory was roughly 35,000 in 1830.)

**58. (a)** The dependent variable is the annual wage, $W$, and the independent variable is age, $a$.
**(b)** $W(30) = 36.261$; According to the model, a 30-year-old Wyoming resident working in the mining industry in 2000 made about \$36,261 annually on average.
**(c)** $W(16) = -2.127$; $W(16)$ represents the average annual salary of a 16-year-old Wyoming resident working in the mining industry in 2000. This result is unreasonable since annual salaries should not be negative.

**59.** $\{x|x \text{ is a real number}\}$ or $(-\infty, \infty)$ **60.** $\left\{w|w \neq -\frac{5}{2}\right\}$ or $\left(-\infty, -\frac{5}{2}\right) \cup \left(-\frac{5}{2}, \infty\right)$ **61.** $\{t|t \neq 5\}$ or $(-\infty, 5) \cup (5, \infty)$

**62.** $\{x|x \neq 2\}$ or $(-\infty, 2) \cup (2, \infty)$ **63.** $\{t|t \text{ is a real number}\}$ or $(-\infty, \infty)$ **64.** $\{x|x \text{ is a real number}\}$ or $(-\infty, \infty)$

**65. (a)** Domain: $\{x|x \text{ is a real number}\}$ or $(-\infty, \infty)$; Range: $\{y|y \text{ is a real number}\}$ or $(-\infty, \infty)$ **b.** $(0, 2)$ and $(4, 0)$

**66. (a)** Domain: $\{x|x \text{ is a real number}\}$ or $(-\infty, \infty)$; Range: $\{y|y \geq -3\}$ or $[-3, \infty)$ **(b)** $(-2, 0), (2, 0), (0, -3)$

**67. (a)** Domain: $\{x|x \text{ is a real number}\}$ or $(-\infty, \infty)$; Range: $\{y|y \text{ is a real number}\}$ or $(-\infty, \infty)$ **(b)** $(0, 0)$ and $(2, 0)$

**68. (a)** Domain: $\{x|x \geq -3\}$ or $[-3, \infty)$; Range: $\{y|y \geq 1\}$ or $[1, \infty)$ **(b)** $(0, 3)$ **69. (a)** Domain: $\{x|x \text{ is a real number}\}$ or $(-\infty, \infty)$; Range: $\{y|y \geq -4\}$ or $[-4, \infty)$ **(b)** $(-1, 0), (3, 0), (0, -2)$ **70. (a)** Domain: $\{x|x \text{ is a real number}\}$ or $(-\infty, \infty)$; Range: $\{y|y \leq 0\}$ or $(-\infty, 0]$ **(b)** $(-2, 0), (2, 0), (0, -4)$ **71. (a)** yes **(b)** $h(-2) = -11; (-2, -11)$ **(c)** $x = \frac{11}{2}; \left(\frac{11}{2}, 4\right)$

**72. (a)** no **(b)** $g(3) = \frac{29}{5}; \left(3, \frac{29}{5}\right)$ **(c)** $x = -10; (-10, -2)$

**73.**

**74.**

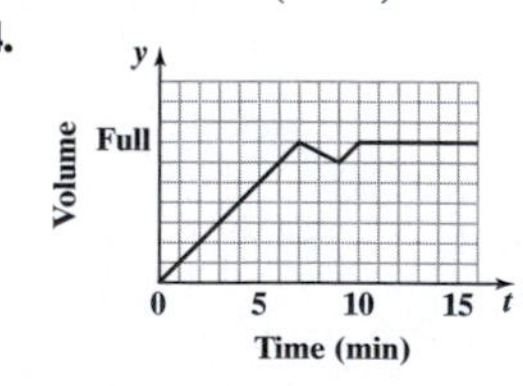


### Chapter 2 Test

**1.** $A$: quadrant IV; $B$: $y$-axis; $C$: $x$-axis; $D$: quadrant I; $E$: quadrant III; $F$: quadrant II.

**2. (a)** no **(b)** yes **(c)** yes

**3.** $y = 4x - 1$

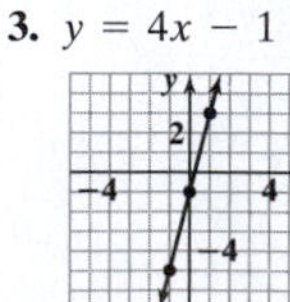

**4.** $y = 4x^2$

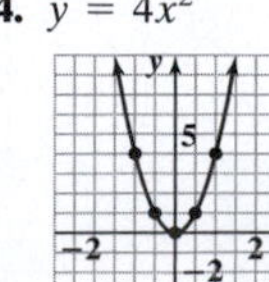

**5.** $(-3, 0), (0, 1), (0, 3)$

**6. (a)** $\approx 4.9\%$
**(b)** January; $\approx 5.9\%$
**(c)** November; $\approx 4.4\%$
**(d)** Answers will vary.

**7.** Domain: $\{-4, 2, 5, 7\}$
Range: $\{-7, -2, -1, 3, 8, 12\}$

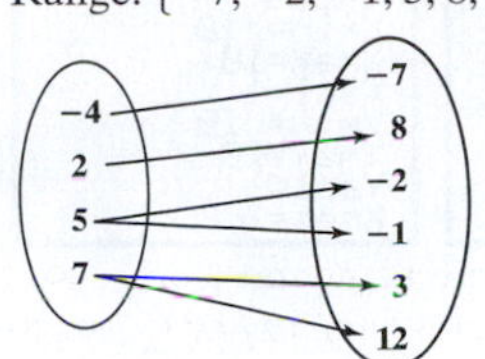


**8.** Domain: $\left\{x \middle| -\frac{5\pi}{2} \le x \le \frac{5\pi}{2}\right\}$ or $\left[-\frac{5\pi}{2}, \frac{5\pi}{2}\right]$
Range: $\{y|1 \le y \le 5\}$ or $[1, 5]$

**9.** $y = x^2 - 3$

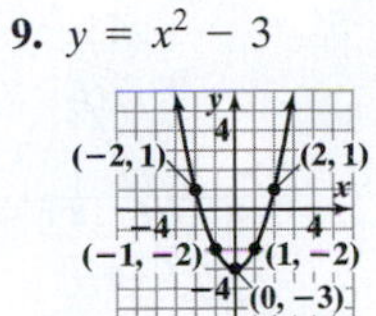


Domain: $\{x|x \text{ is a real number}\}$ or $(-\infty, \infty)$
Range: $\{y|y \ge -3\}$ or $[-3, \infty)$

**10.** Function.
Domain: $\{-5, -3, 0, 2\}$
Range: $\{3, 7\}$

**11.** Not a function.
Domain: $\{x|x \le 3\}$ or $(-\infty, 3]$
Range: $\{y|y \text{ is a real number}\}$ or $(-\infty, \infty)$

**12.** No **13.** $f(x + h) = -3x - 3h + 11$
**14. (a)** $g(-2) = 5$ **(b)** $g(0) = -1$ **(c)** $g(3) = 20$

**15.** $f(x) = x^2 + 3$

**16. (a)** The dependent variable is the ticket price, $P$, and the independent variable is the number of years after 1989, $x$.
**(b)** $P(15) = 5.71$; According to the model, the average ticket price in 2004 ($x = 15$) was \$5.71.

**17. (a)** The dependent variable is the number of registered climbers, $N$, and the independent variable is the number of years after 1960, $x$.
**(b)** $N(43) \approx 12{,}507$; According to the model, there were about 12,507 registered climbers on Mt. Rainier in 2003 ($x = 43$).
**(c)** Answers may vary.

**18.** $\{x|x \ne -2\}$ or $(-\infty, -2) \cup (-2, \infty)$

**19. (a)** yes
**(b)** $h(3) = -3$; $(3, -3)$
**(c)** $x = \frac{12}{5}$; $\left(\frac{12}{5}, 0\right)$

**20. (a)** The car stops accelerating when the speed stops increasing. Thus, the car stops accelerating after 6 seconds.
**(b)** The car has a constant speed when the graph is horizontal. Thus, the car maintains a constant speed for 18 seconds.

# Chapter 3

## Section 3.1

**1.** linear equation **3.** $x = a$; $a$ **5.** True **7.** Answers may vary.

**9.**

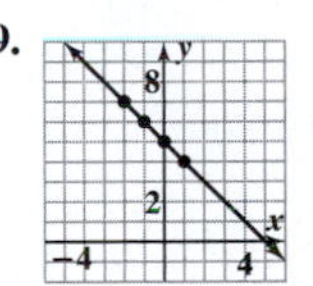

**11.**

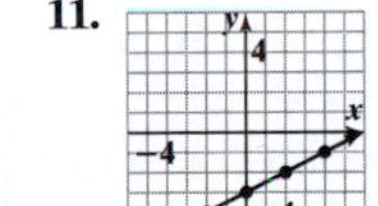

**13.**

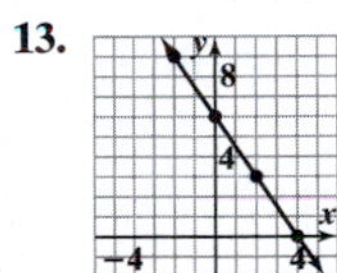

**15.**

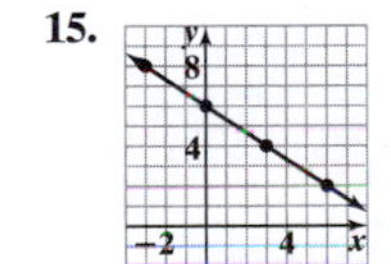

**17.**

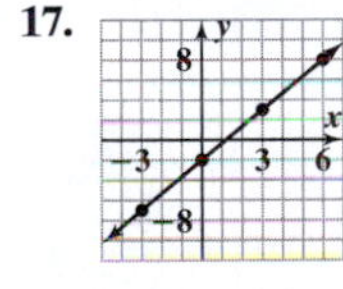

**19.**

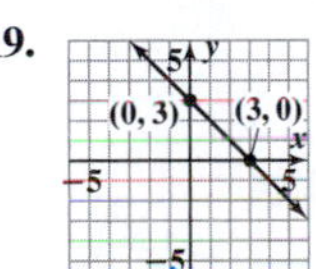


**21.**

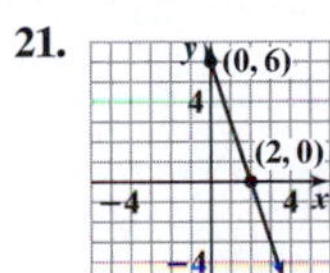


**23.**

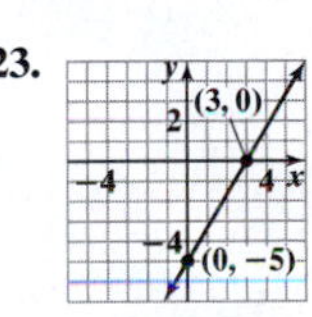


**25.**

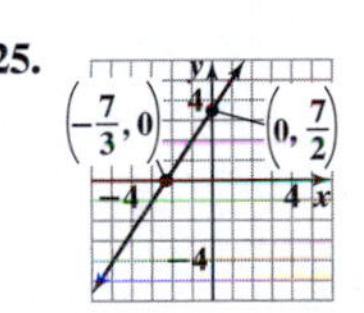


**27.**

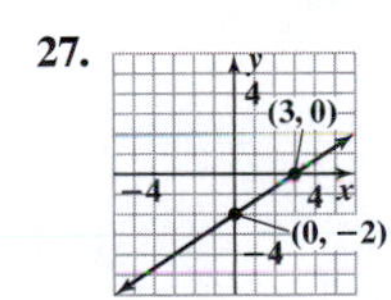


**29.**

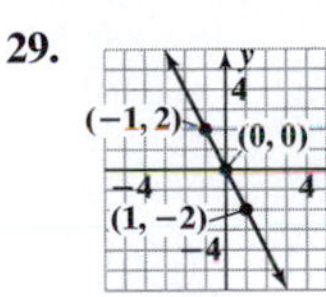


**31.**

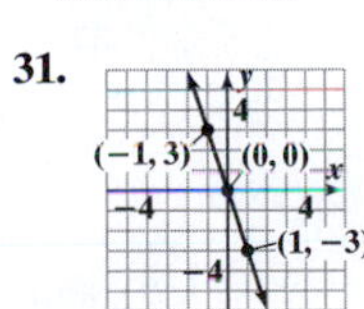


**33.**

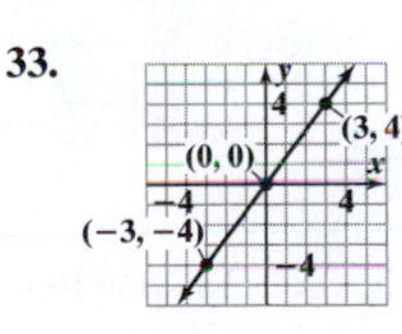


**35.**

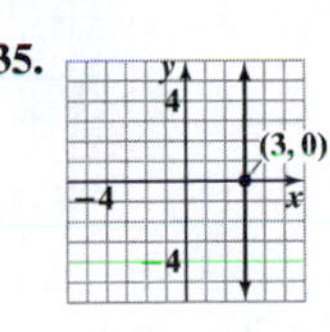


**37.**

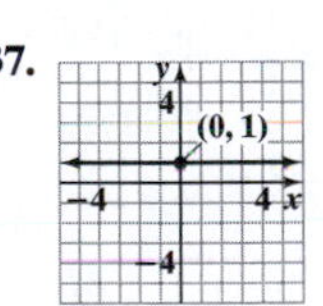


**39.**

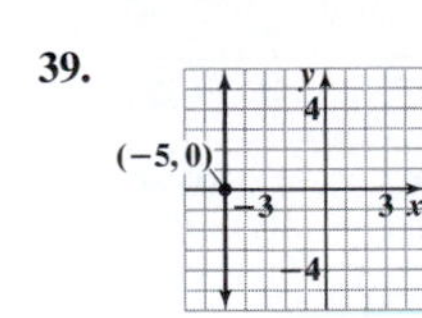


**41.**

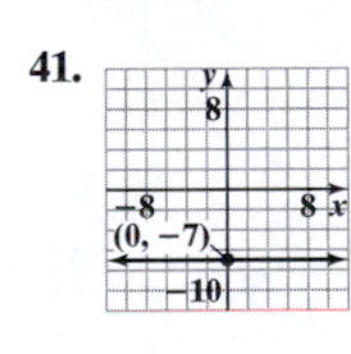


**43. (a)** $\{x|7300 \le x \le 29{,}700\}$ or $[7300, 29700]$ **(b)** \$2635
**(c)** The independent variable is $x$; the dependent variable is $T$.
**(d)**

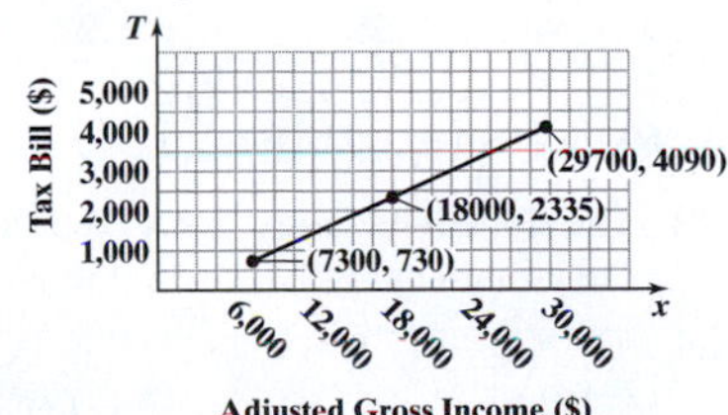


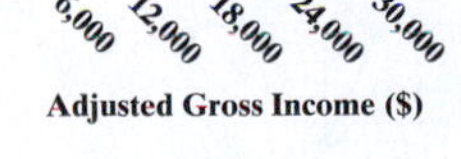


**(e)** \$25,000

**45. (a)** $\{m|m \ge 0\}$ or $[0, \infty)$ **(b)** 2; The base fare is \$2.00 before any distance is driven. **(c)** \$9.50.
**(d)**
**(e)** A person can travel 7.5 miles in a cab for \$13.25.

**47. (a)** The independent variable is $a$; the dependent variable is $H$.
**(b)** $\{a|15 \le a \le 90\}$ or $[15, 90]$ **(c)** \$566.50
**(d)**

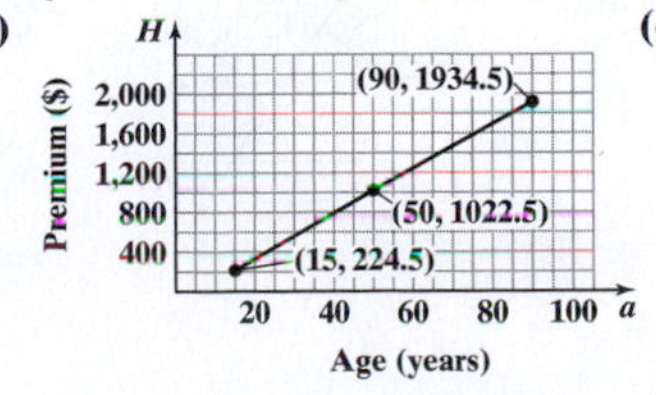


**(e)** 48 years

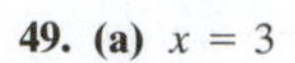
**49.** **(a)** $x = 3$ **(b)** $x = -1$
**(c)** $x = 4$
**(d)** The $y$-intercept is $-2$ and the $x$-intercept is 2.
**(e)** 2

**51.**

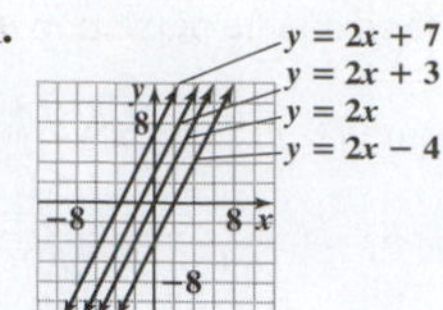


**53.**

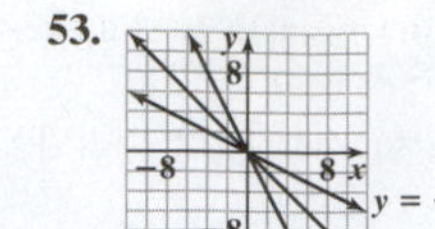


**55.** $y = -x + 3$

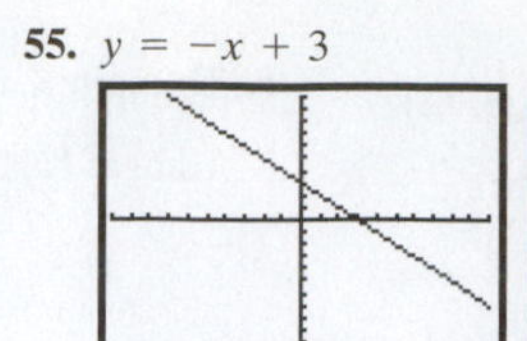


**57.** $y = -3x + 6$

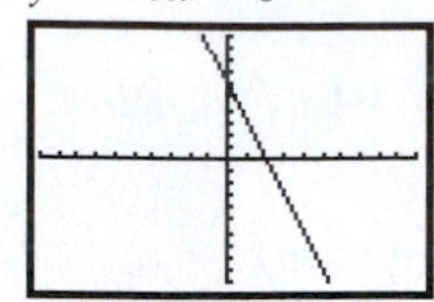


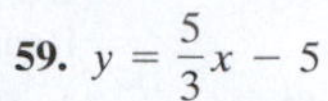
**59.** $y = \frac{5}{3}x - 5$

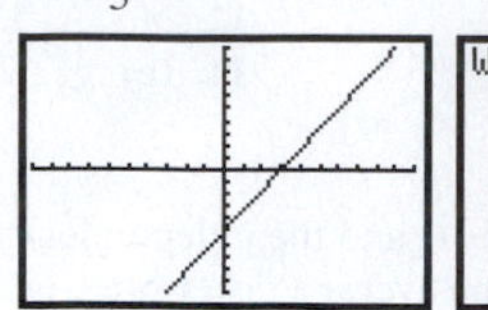


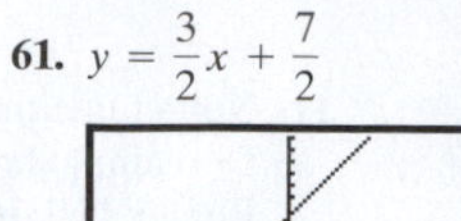
**61.** $y = \frac{3}{2}x + \frac{7}{2}$

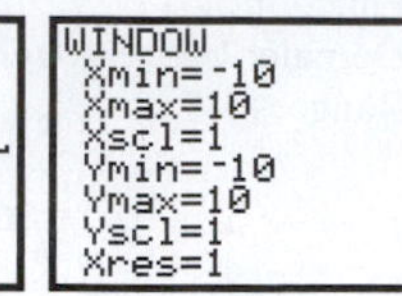


**63.** $y = \frac{2}{3}x - 2$

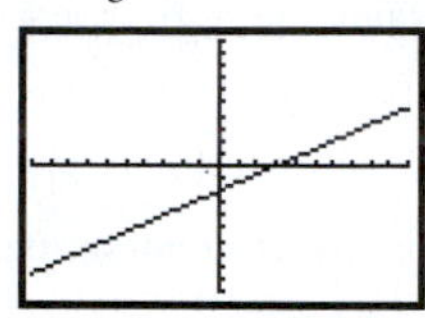


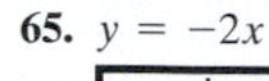
**65.** $y = -2x$

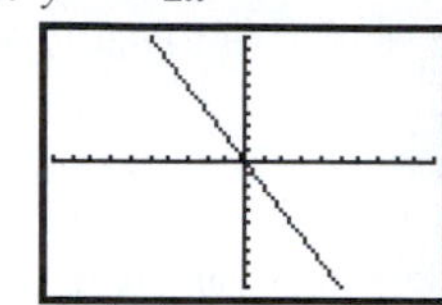


## Section 3.2

**1.** undefined **3.** 5; 1 **5.** True **7.** The only linear equation that is not a function is $x = a$, where $a$ can be any constant. **9.** A vertical line **11.** Any line that is neither vertical nor horizontal. **13.** **(a)** $\frac{4}{3}$ **(b)** For every 3-unit increase in $x$, $y$ will increase by 4 units. **15.** **(a)** $-\frac{8}{3}$ **(b)** For every 3-unit increase in $x$, $y$ will decrease by 8 units. For every 3-unit decrease in $x$, $y$ will increase by 8 units.

**17.** $m = 5$

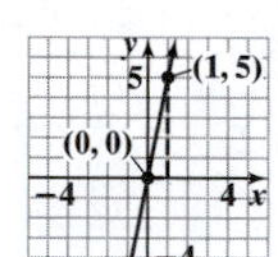


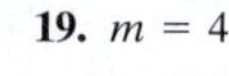
**19.** $m = 4$



**21.** $m = -3$



**23.** $m = \frac{4}{5}$

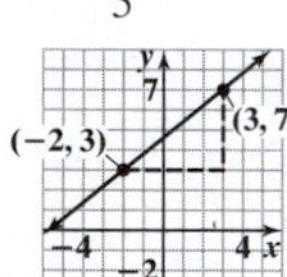


**25.** $m = 0$

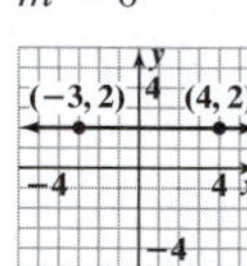


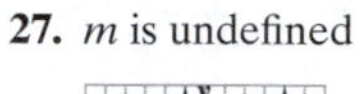
**27.** $m$ is undefined

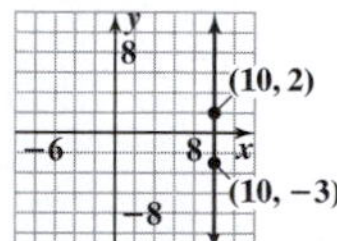


**29.** $m = \frac{2}{21}$

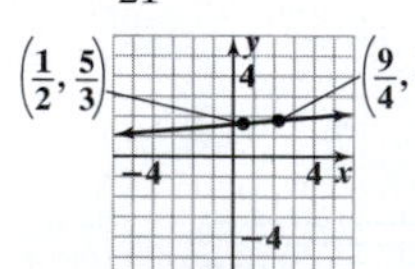


**31.** $m = -\frac{610}{87}$

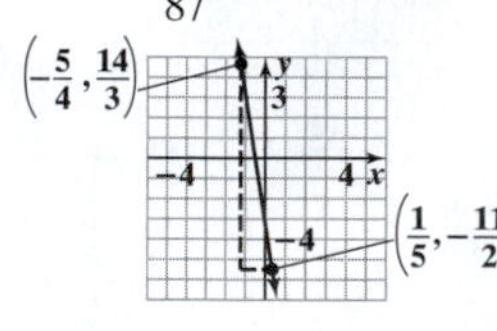


**33.**

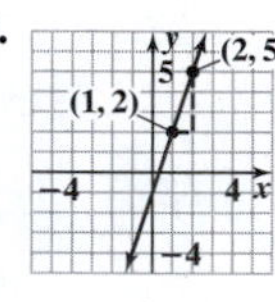


**35.**

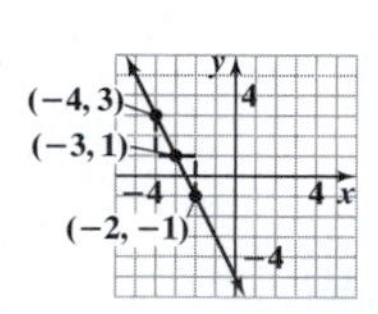


**37.**

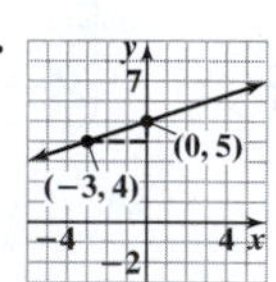


**39.**

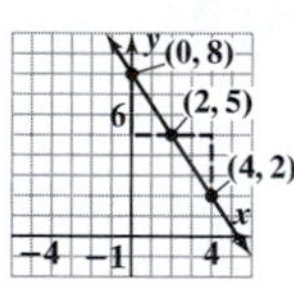


**41.**

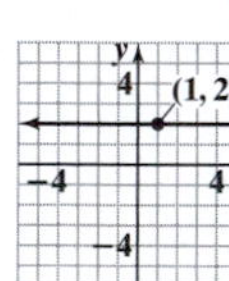


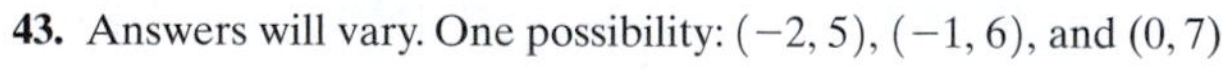
**43.** Answers will vary. One possibility: $(-2, 5)$, $(-1, 6)$, and $(0, 7)$

**45.** Answers will vary. One possibility: $(0, 8)$, $(2, 13)$, and $(4, 18)$ **47.** $y = 4x$

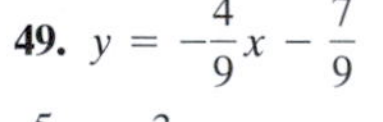
**49.** $y = -\frac{4}{9}x - \frac{7}{9}$

**51.** $y = 3$ **53.** $y = 2x$ **55.** $y = -3x - 2$ **57.** $y = \frac{4}{3}x - 2$ **59.** $y = -\frac{5}{4}x + \frac{3}{2}$ **61.** $x = 6$ **63.** $y = \frac{7}{5}x$ **65.** $y = 5x - 13$ **67.** $y = -\frac{3}{7}x + \frac{1}{7}$ **69.** $x = -1$ **71.** $y = \frac{5}{2}x + \frac{1}{2}$ **73.** $y = 4$ **75.** $y = \frac{10}{7}x - \frac{26}{21}$

**77.** The slope is 2 and the $y$-intercept is $-1$.

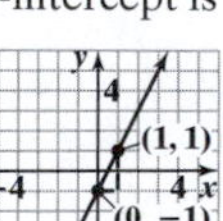


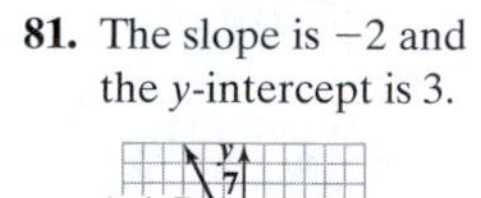
**79.** The slope is $-4$ and the $y$-intercept is 0.



**81.** The slope is $-2$ and the $y$-intercept is 3.



**83.** The slope is $-2$ and the $y$-intercept is 4.

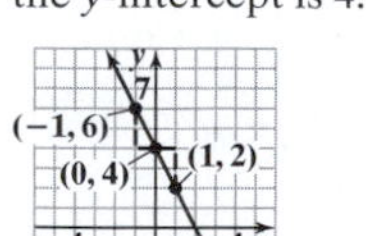


**85.** The slope is $\frac{1}{4}$ and the $y$-intercept is $-\frac{1}{2}$.



**87.** The slope is undefined and there is no $y$-intercept.

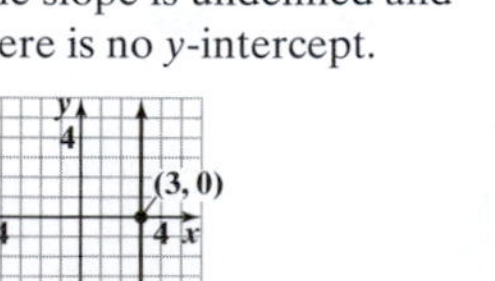


**89.**

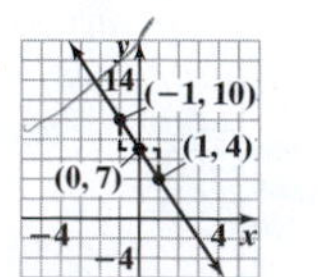


**91.**

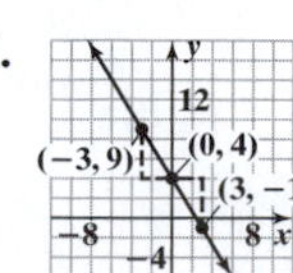


**93.** $y = 0$

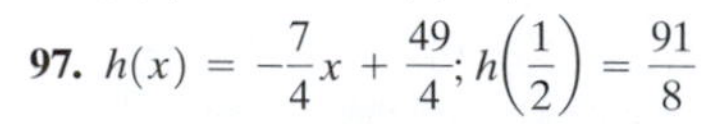
**95.** $f(x) = 2x + 2$; $f(-2) = -2$

**97.** $h(x) = -\frac{7}{4}x + \frac{49}{4}$; $h\left(\frac{1}{2}\right) = \frac{91}{8}$

**99. (a)**

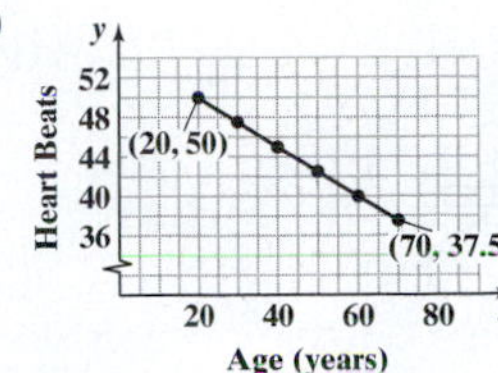


**(b)** −0.25 heartbeats per year; Between ages 20 and 30, the maximum number of heartbeats decreases at a rate of 0.25 heartbeats per year.
**(c)** −0.25 heartbeats per year; Between ages 50 and 60, the maximum number of heartbeats decreases at a rate of 0.25 heartbeats per year.
**(d)** Yes. The maximum number of heartbeats appears to be linearly related to age. The average rate of change (slope) is constant for the data provided.

**101. (a)**

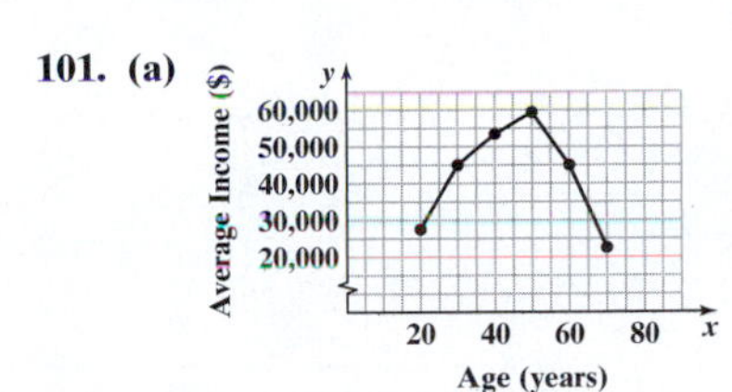


**(b)** $1676.60 per year; Between ages 20 and 30, the average individual's income increases at a rate of $1676.60 per year.
**(c)** −$1322.40 per year; Between ages 50 and 60, the average individual's income decreases at a rate of $1.332.40 per year.
**(d)** No. The average income is not linearly related to age. The average rate of change (slope) is not constant.
**103. (a)** $L(x) = -0.56x + 10.23$ **(b)** 3.51 million pounds **(c)** Swordfish landings are decreasing at a rate of 0.56 million pounds per year. **(d)** 2002

**105. (a)** $C(x) = 8350x - 2302$ **(b)** $4127.50 **(c)** The cost of diamonds increases at a rate of $8350 per carat. **(d)** 0.91 Carat
**107. (a)** $C(x) = 1.06x - 719.716$ **(b)** $7142.6 billion **(c)** The personal consumption expenditures are increasing at a rate of approximately $1.06 per $1.00 of personal disposable income. **(d)** approximately $7480.9 billion **109.** (c)

**111.**

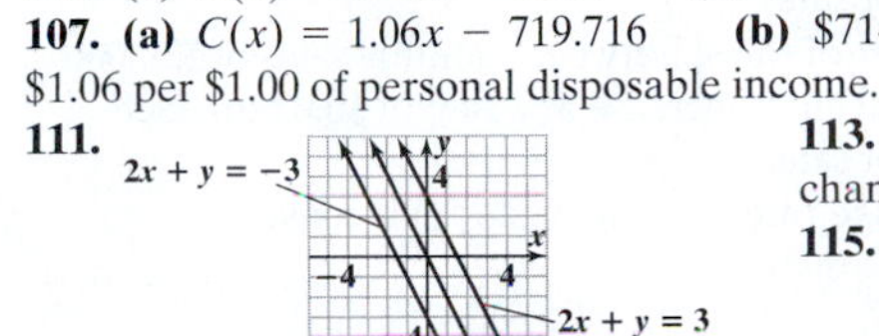


**113.** According to the ADA codes, for every one unit of vertical change in the access ramp, the horizontal change must be twelve units or more.

**115.**

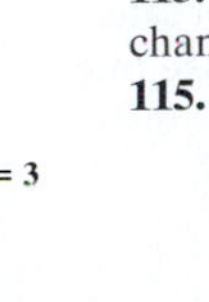

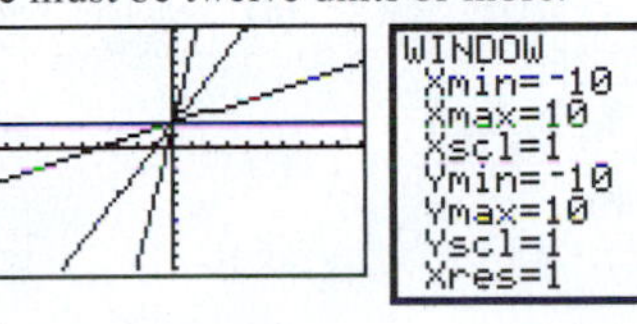


Answers may vary.

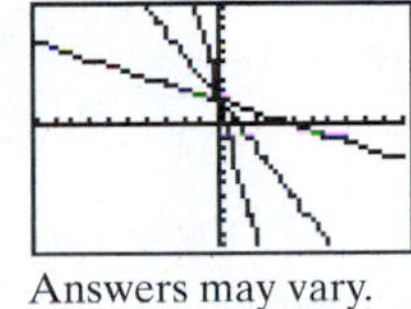

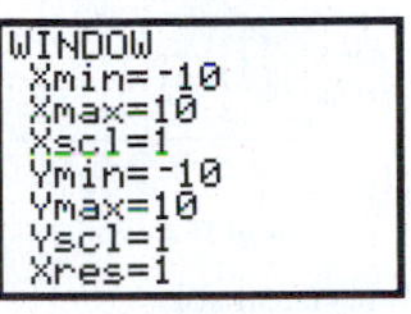


Answers may vary.

## Section 3.3

**1.** slope; $y$-intercepts **3.** False **5.** No. If the two nonvertical lines have the same $x$-intercept but different $y$-intercepts, then their slopes cannot be equal. Therefore, they cannot be parallel lines. **7. (a)** $m = 5$ **(b)** $m = -1/5$ **9. (a)** $m = -5/6$ **(b)** $m = 6/5$ **11.** Parallel.
**13.** Neither **15.** Perpendicular **17.** Parallel

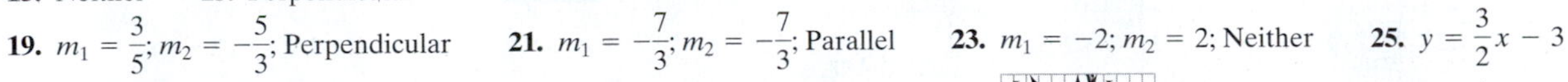

**19.** $m_1 = \frac{3}{5}$; $m_2 = -\frac{5}{3}$; Perpendicular **21.** $m_1 = -\frac{7}{3}$; $m_2 = -\frac{7}{3}$; Parallel **23.** $m_1 = -2$; $m_2 = 2$; Neither

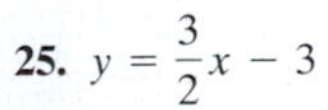

**25.** $y = \frac{3}{2}x - 3$

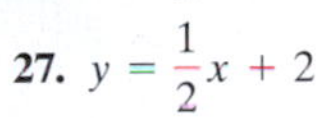

**27.** $y = \frac{1}{2}x + 2$

**29.** $y = 2$

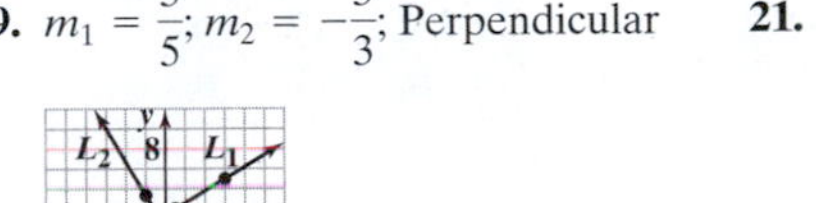

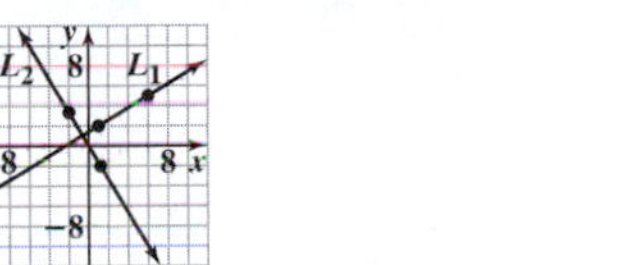

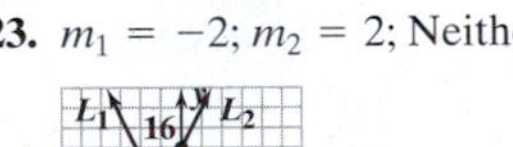

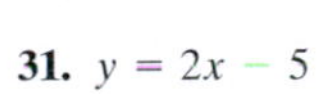

**31.** $y = 2x - 5$

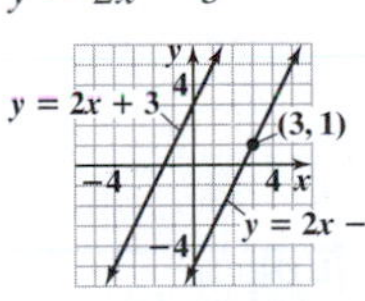


**33.** $y = \frac{1}{2}x + 2$

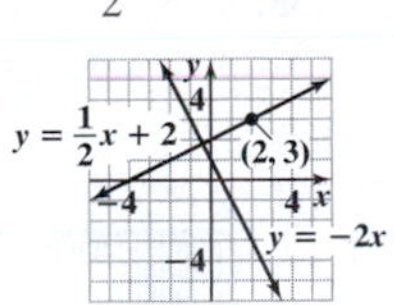


**35.** $y = -3$

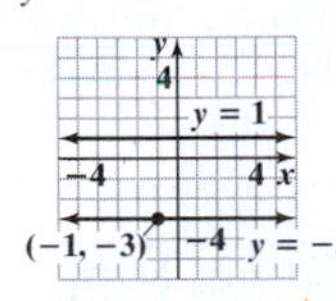


**37.** $y = 3$

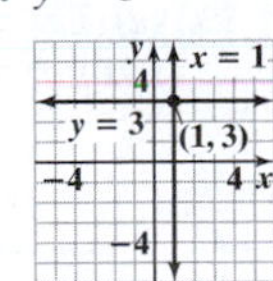


**39.** $y = 3x + 2$

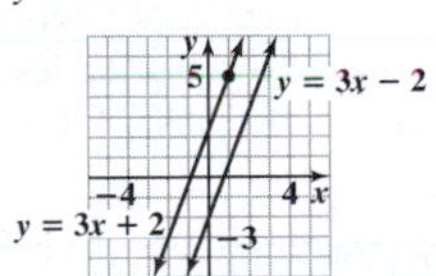


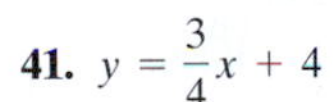

**41.** $y = \frac{3}{4}x + 4$

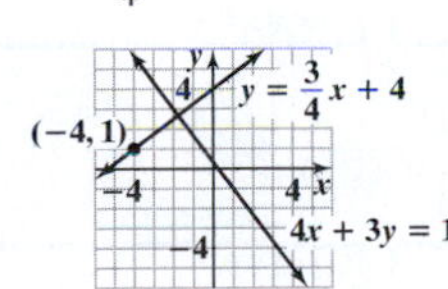


**43.** $y = -\frac{5}{2}x - 8$

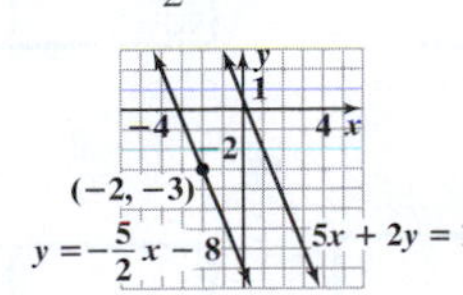


**45. (a)**

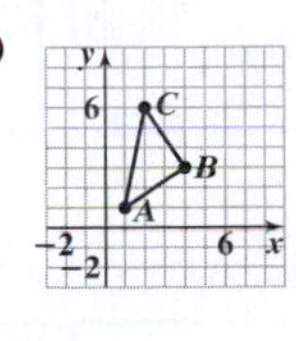


**(b)** Slope of $\overline{AB} = \frac{2}{3}$; Slope of $\overline{BC} = -\frac{3}{2}$

Because the slopes are negative reciprocals, segments $\overline{AB}$ and $\overline{BC}$ are perpendicular. Thus, triangle $ABC$ is a right triangle.

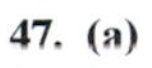

**47. (a)**

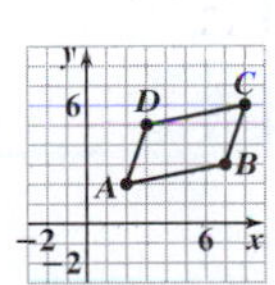


**(b)** Slope of $\overline{AB} = \frac{1}{5}$ Slope of $\overline{CD} = \frac{1}{5}$
Slope of $\overline{BC} = 3$ Slope of $\overline{DA} = 3$

Because the slopes of $\overline{AB}$ and $\overline{CD}$ are equal, $\overline{AB}$ and $\overline{CD}$ are parallel. Because the slopes of $\overline{BC}$ and $\overline{DA}$ are equal, $\overline{BC}$ and $\overline{DA}$ are parallel. Thus, quadrilateral $ABCD$ is a parallelogram.

**49.** $A = -1$ **51.** (c)

## *Putting the Concepts Together*

**1.** 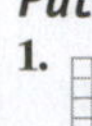 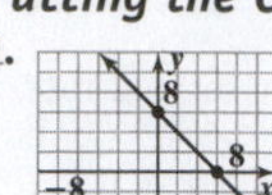


**2.** 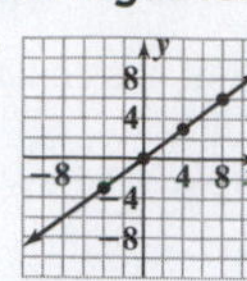


**3.** 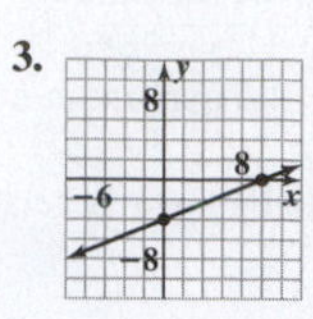


**4.** 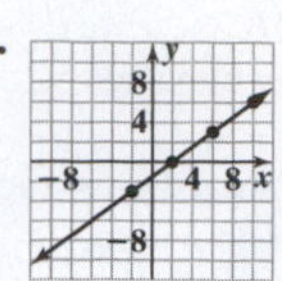


**5.** 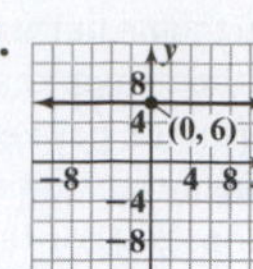


**6.** 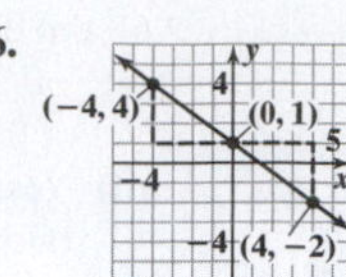


**7.** $m = \frac{1}{2}$; For every 2-unit increase in $x$, $y$ will increase by 1 unit. **8.** The slope is $\frac{4}{5}$, and the $y$-intercept is $-4$.

**9.** 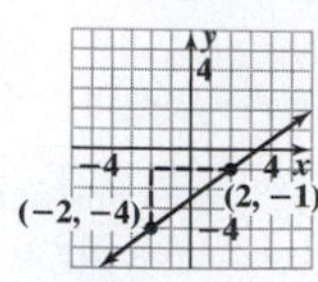


**10.** Neither **11.** $y = \frac{3}{2}x - 13$ **12.** $y = -2x + 5$ **13.** $y = \frac{2}{3}x + 6$ **14.** $y = -\frac{3}{4}x + 2$

**15. (a)** $W(x) = 40x + 5700$
**(b)** $\{x \mid x \geq 0\}$ or $[0, \infty)$
**(c)** 8180 pounds
**(d)**

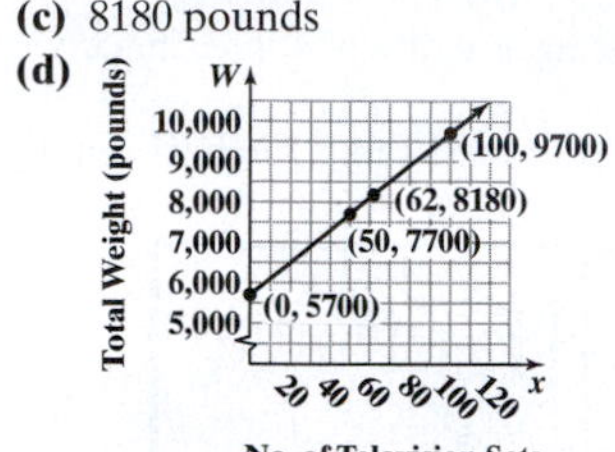


**(e)** 72 television sets

**16. (a)** 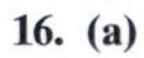

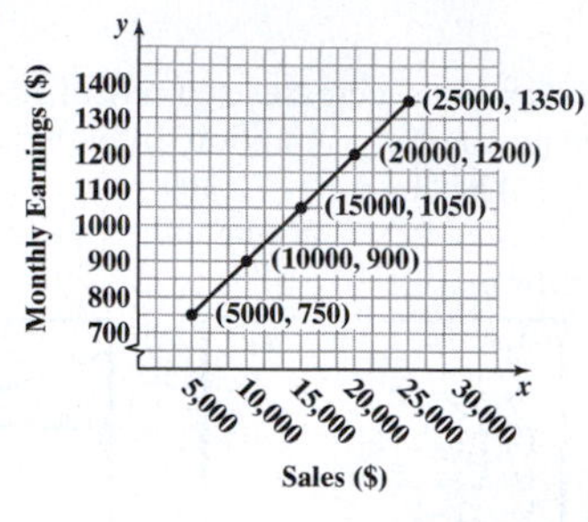


**(b)** \$0.03 per dollar of sales; Between monthly sales of \$5,000 and \$10,000, the earnings increase at a rate of \$0.03 for each additional dollar of sales.
**(c)** \$0.03 per dollar of sales; Between monthly sales of \$20,000 and \$25,000, the earnings increase at a rate of \$0.03 for each additional dollar of sales.
**(d)** Yes. The average rate of change (slope) is constant.

## *Section 3.4*

**1.** half-planes **3.** False **5.** Answers may vary. **7. (a)** yes **(b)** no **(c)** yes **9. (a)** yes **(b)** no **(c)** yes

**11.** 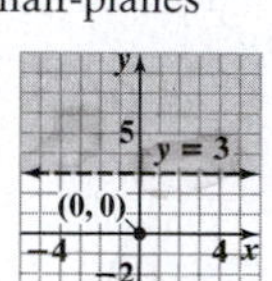


**13.** 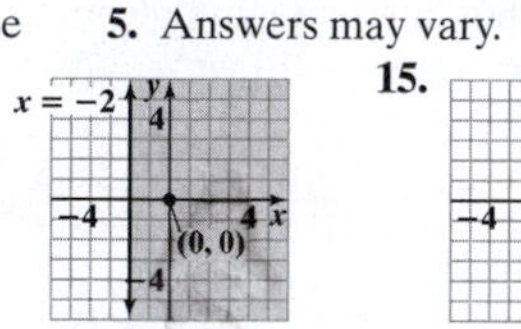


**15.** 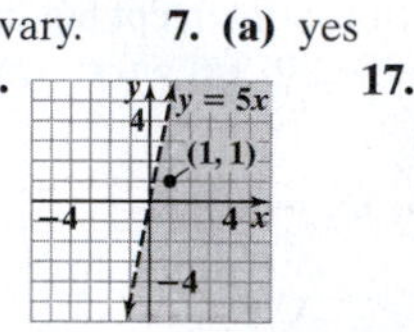


**17.** 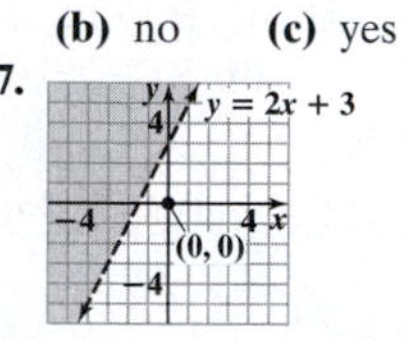


**19.** 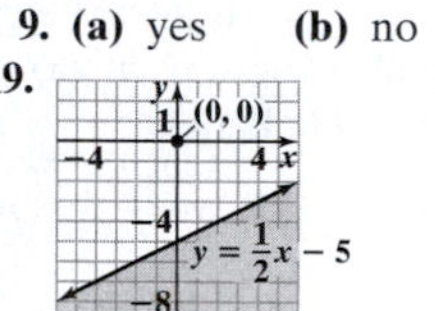


**21.** 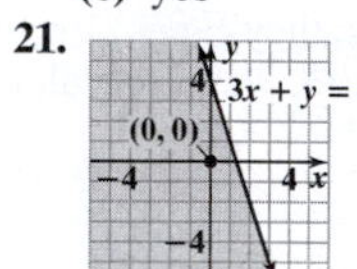


**23.** 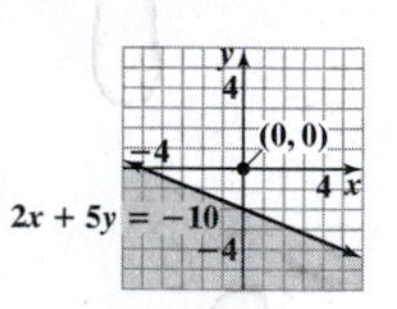


**25.** 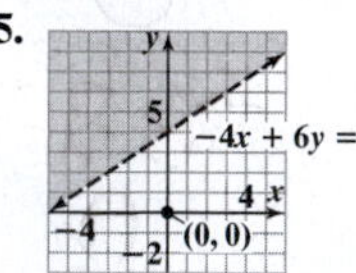


**27.** 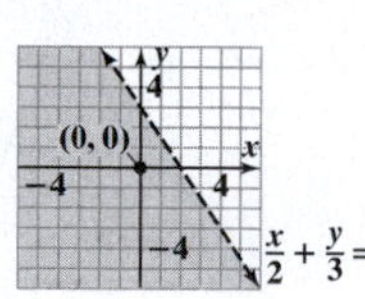


**29.** 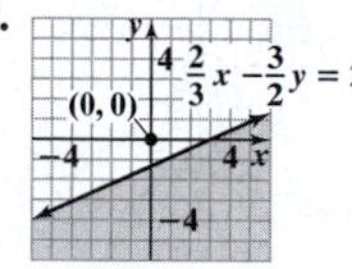


**31. (a)** Let $x$ = the number of Filet-o-Fish. Let $y$ = the number of orders of fries. $45x + 40y \leq 150$ **(b)** yes **(c)** no

**33. (a)** Let $x$ = the number Switch A assemblies. Let $y$ = the number Switch B assemblies. $2x + 1.5y \leq 80$ **(b)** no **(c)** no

**35.** $y > 2x - 2$
**37.** $y \leq -x + 1$

**39.** $y > 3$

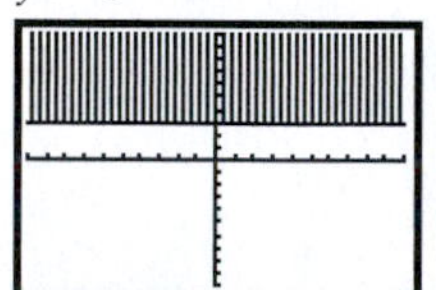

```
WINDOW
 Xmin=-10
 Xmax=10
 Xscl=1
 Ymin=-10
 Ymax=10
 Yscl=1
 Xres=1
```

**41.** $y < 5x$

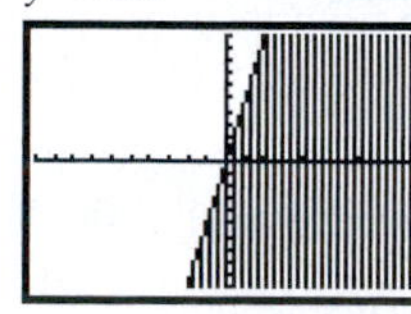

**43.** $y > 2x + 3$

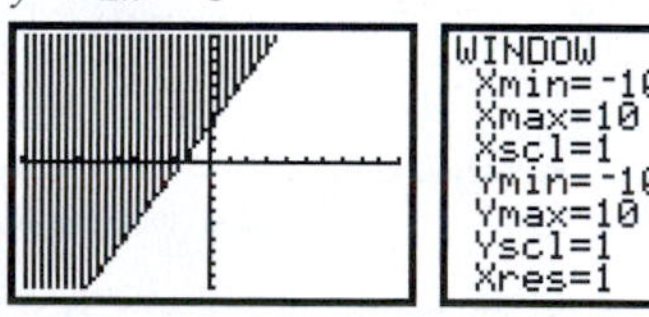


**45.** $y \leq \frac{1}{2}x - 5$

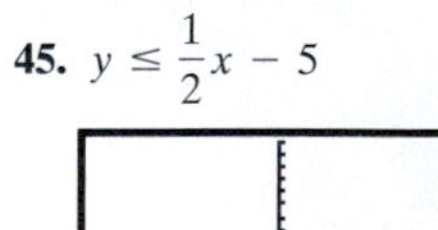

```
WINDOW
 Xmin=-10
 Xmax=10
 Xscl=1
 Ymin=-10
 Ymax=10
 Yscl=1
 Xres=1
```

**47.** $3x + y \leq 4$
$y \leq -3x + 4$

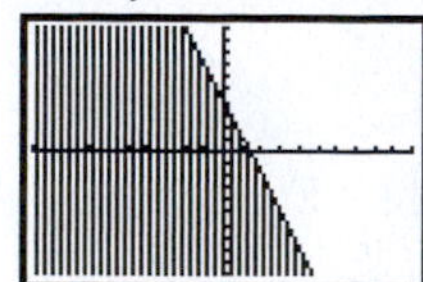

```
WINDOW
 Xmin=-10
 Xmax=10
 Xscl=1
 Ymin=-10
 Ymax=10
 Yscl=1
 Xres=1
```

**49.** $y \leq -\frac{2}{5}x - 2$

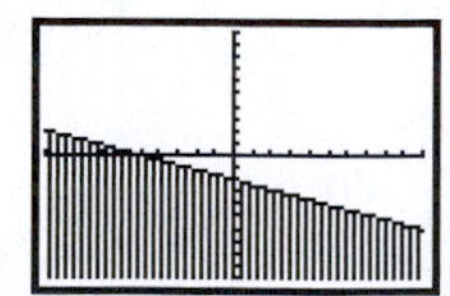

```
WINDOW
 Xmin=-10
 Xmax=10
 Xscl=1
 Ymin=-10
 Ymax=10
 Yscl=1
 Xres=1
```

## *Section 3.5*

**1.** scatter diagram **3.** $y = kx$ **5.** Answers will vary. **7. (a)** $k = 6$ **(b)** $y = 6x$ **(c)** $y = 42$

**9. (a)** $k = \frac{3}{7}$ **(b)** $y = \frac{3}{7}x$ **(c)** $y = 12$ **11. (a)** $k = \frac{1}{2}$ **(b)** $y = \frac{1}{2}x$ **(c)** $y = 15$ **13.** Nonlinear **15.** Linear with positive slope

**17. (a)**

**(b)** Answers will vary. Using the points (4, 1.8) and (9, 2.6), the equation is $y = 0.16x + 1.16$.

**(c)**

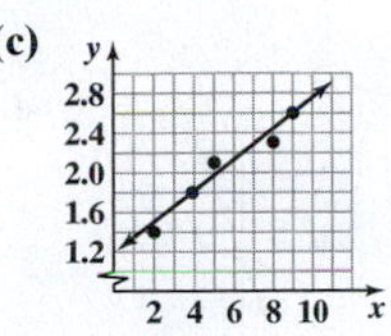

**19. (a)**

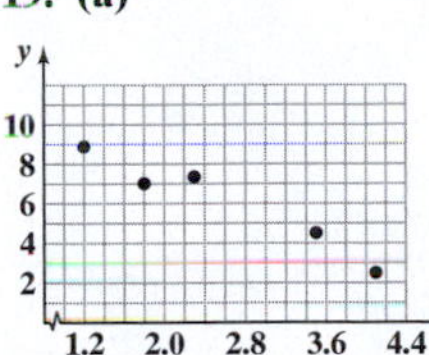

**(b)** Answers will vary. Using the points (1.2, 8.4) and (4.1, 2.4), the equation is $y = -2.1x + 10.92$.

**(c)**

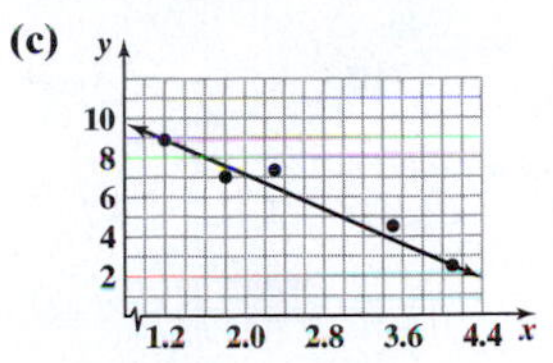

**21. (a)** $B(m) = 0.05m + 8.95$
**(b)** The independent variable is $m$; the dependent variable is $B$.
**(c)** $\{m \mid m \geq 0\}$ or $[0, \infty)$
**(d)** $23.95
**(e)** 240 minutes
**(f)**

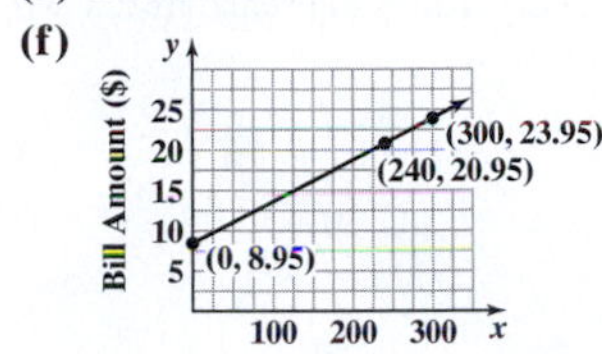


**23. (a)** $V(x) = -900x + 2700$
**(b)** $\{x \mid 0 \leq x \leq 3\}$ or $[0, 3]$
**(c)** $1800
**(d)** The $y$-intercept is 2700 and the $x$-intercept is 3.
**(e)** After two years
**(f)**

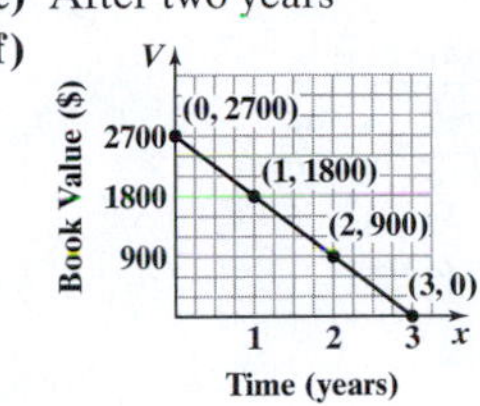


**25. (a)** $p(b) = 0.0058358b$.
**(b)** $817.01
**(c)**

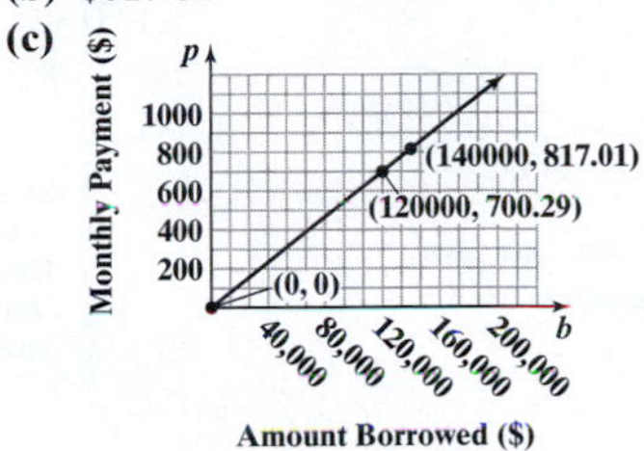


**27. (a)** $C(w) = 5.6w$
**(b)** $19.60
**(c)**

**29.** 96 feet per second

**31. (a)**

**(b)** Linear
**(c)** Answers will vary. Using the points (2300, 4070) and (3390, 5220), the equation is $y = 1.06x + 1632$.

**(d)**

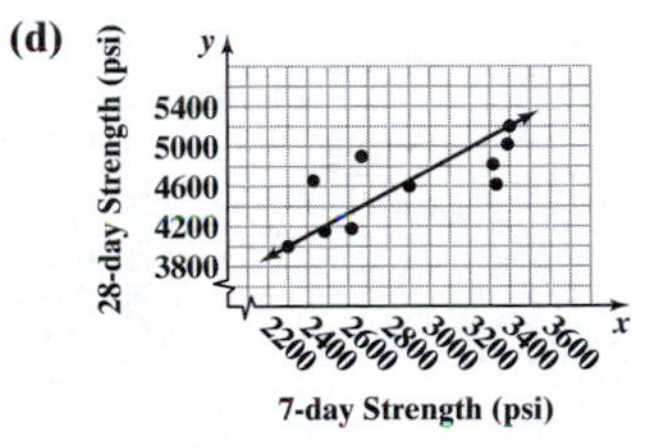


**(e)** 4812 psi
**(f)** If the 7-day strength is increased by 1 psi, then the 28-day strength will increase by 1.06 psi.

**33. (a)** No
**(b)**

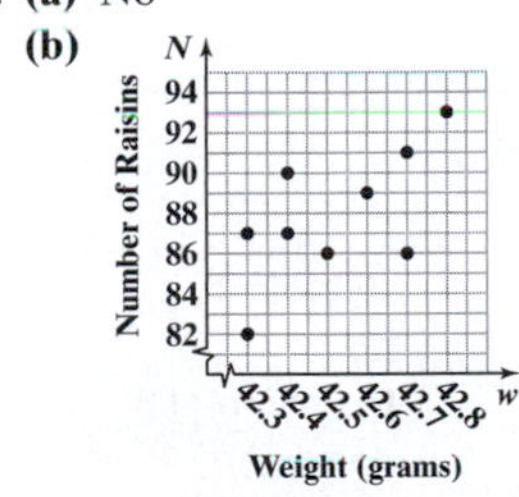


**(c)** Answers will vary. Using the points (42.3, 82) and (42.8, 93), the equation is $N = 22w - 848.6$.
**(d)**

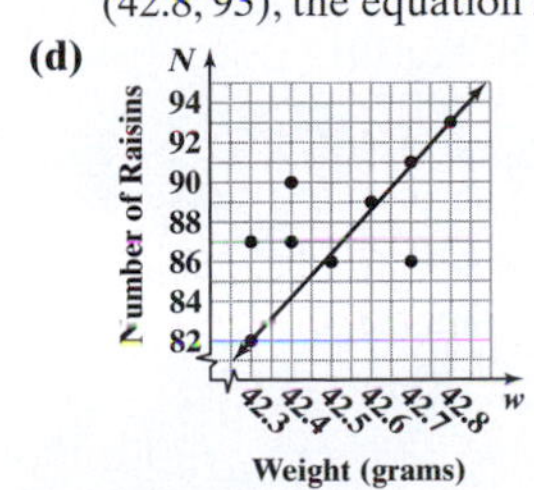


**(e)** $N(w) = 22w - 848.6$
**(f)** approximately 86 raisins
**(g)** If the weight increases by one gram, then the number of raisins increases by 22 raisins.

**35.** Answers will vary.

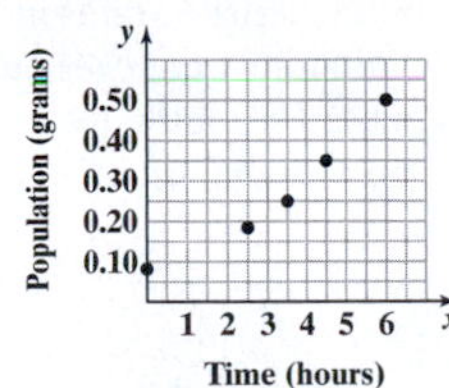


**37. (a)**

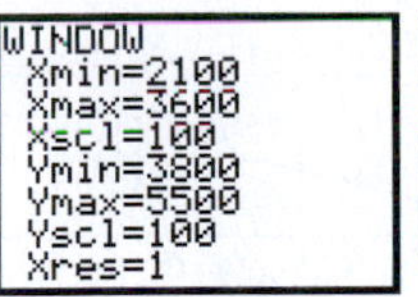


**(b)** The line of best fit is approximately $y = 0.676x + 2675.562$.

**39. (a)**

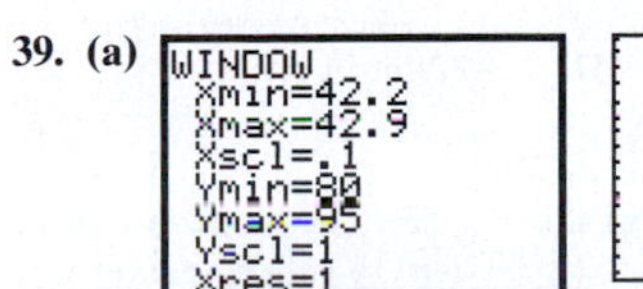


**(b)** The line of best fit is approximately $y = 11.449x - 399.123$.

## Chapter 3 Review

**1.** 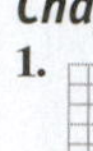 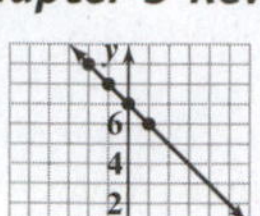

**2.** 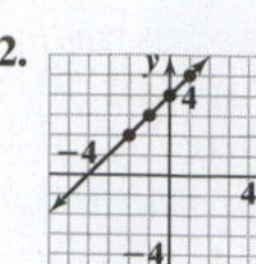

**3.** 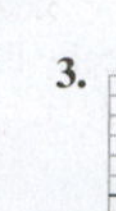

**4.** 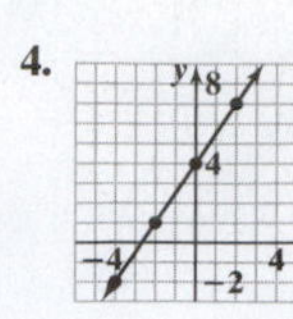

**5.** 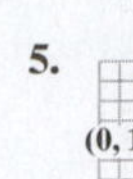


**6.** 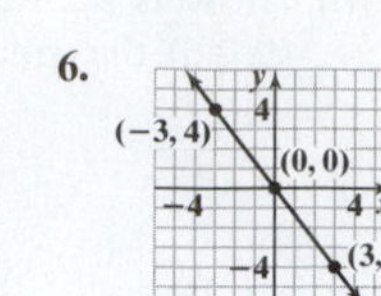


**7.** 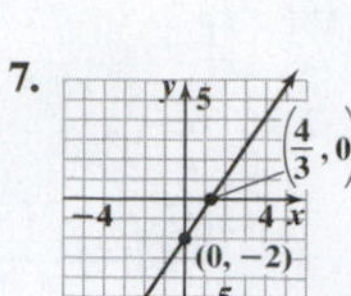


**8.** 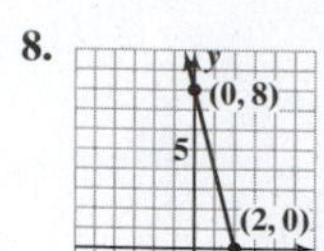


**9.** 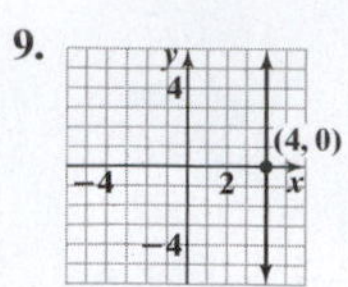


**10.** 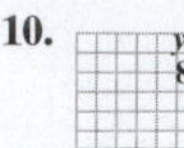


**11.** 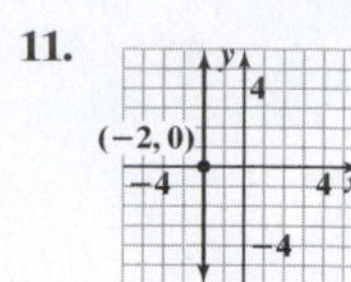


**12.** 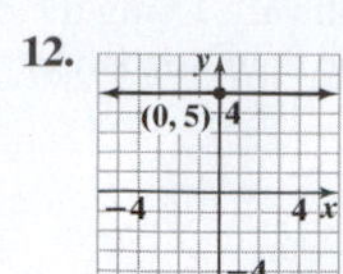


**13.** **(a)** $\{x \mid x \geq 0\}$ or $[0, \infty)$
**(b)** \$21.45
**(c)**

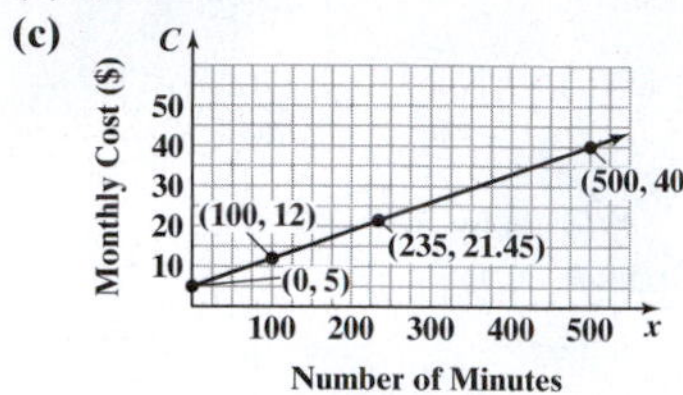


**(d)** 1000 minutes

**14.** **(a)** The independent variable is $x$; the dependent variable is $V$.
**(b)** $\{x \mid 0 \leq x \leq 5\}$ or $[0, 5]$
**(c)** \$1800
**(d)** \$1080
**(e)**

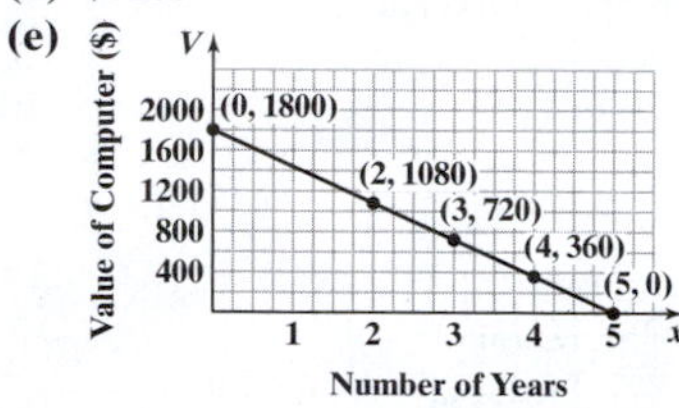


**(f)** After 5 years

**15.** **(a)** $m = \frac{5}{4}$
**(b)** For every 4-unit increase in $x$, $y$ will increase by 5 units.

**16.** **(a)** $m = -\frac{1}{4}$
**(b)** For every 4-unit increase in $x$, $y$ will decrease by 1 unit. For every 4-unit decrease in $x$, $y$ will increase by 1 unit.

**17.** $m = -2$

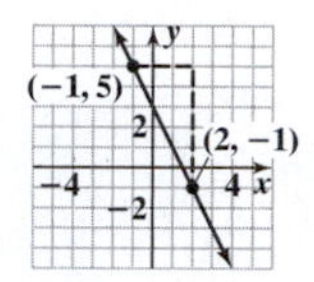


**18.** $m = \frac{3}{2}$

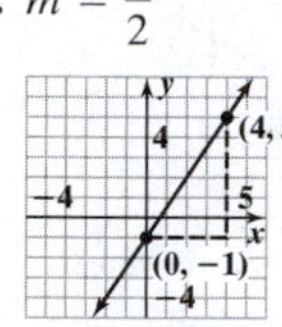


**19.** **(a)**

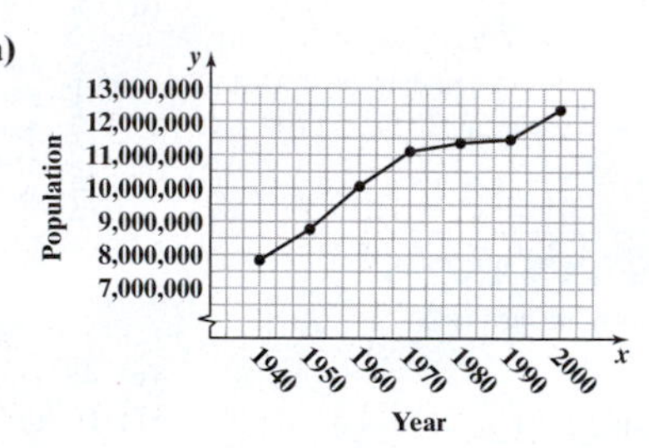


**(b)** 81,493.5 people per year; Between 1940 and 1950, the population of Illinois increased at a rate of 81,493.5 people per year.
**(c)** 319.3 people per year; Between 1980 and 1990, the population of Illinois increased at a rate of 319.3 people per year.
**(d)** 98,869.1 people per year; Between 1990 and 2000, the population of Illinois increased at a rate of 98,869.1 people per year.
**(e)** No. The average rate of change (slope) is not constant.

**20.** **(a)**

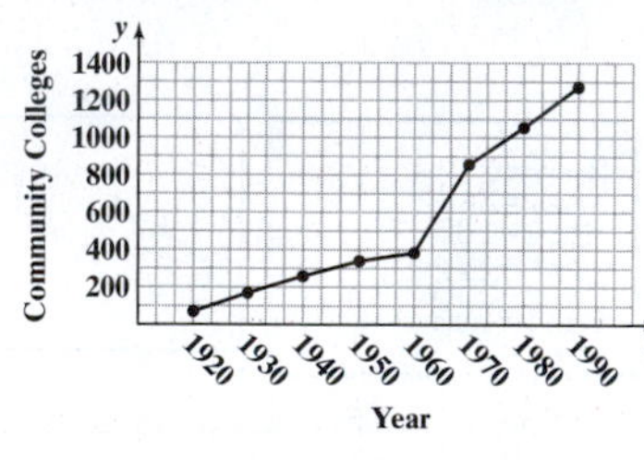


**(b)** 10.8 community colleges per year; Between 1920 and 1930, the number of public community colleges in the United States increased at a rate of 10.8 colleges per year.
**(c)** 45.7 community colleges per year; Between 1960 and 1970, the number of public community colleges in the United States increased at a rate of 45.7 colleges per year.
**(d)** 23.3 community colleges per year; Between 1980 and 1990, the number of public community colleges in the United States increased at a rate of 23.3 colleges per year.
**(e)** No. The average rate of change (slope) is not constant.

**21.** 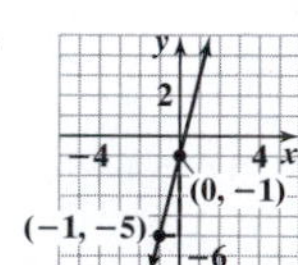


**22.** 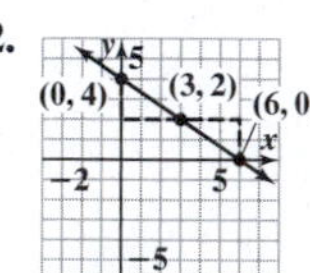


**23.** 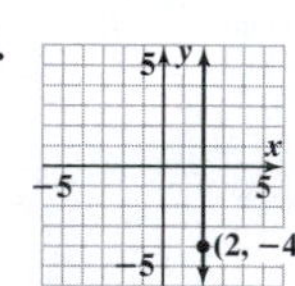


**24.** 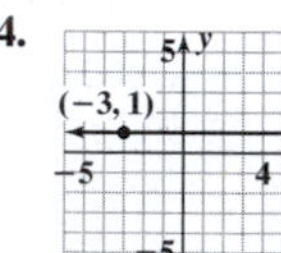


**25.** $y = -\frac{1}{2}x + 2$ or $x + 2y = 4$
**26.** $y = 3x$ or $3x - y = 0$
**27.** $y = -x + 5$ or $x + y = 5$
**28.** $y = \frac{3}{5}x + 2$ or $3x - 5y = -10$

**29.** $y = -\frac{1}{3}x + 4$ or $x + 3y = 12$ **30.** $y = 3$ **31.** $y = 2x - 9$ or $2x - y = 9$ **32.** $y = -\frac{1}{3}x + \frac{5}{3}$ or $x + 3y = 5$

**33.** The slope is 4 and the $y$-intercept is $-6$.

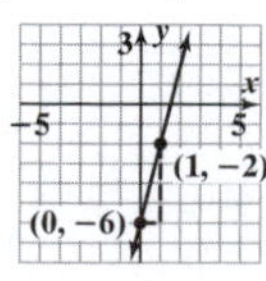


**34.** The slope is $-\frac{2}{3}$ and the $y$-intercept is 4.

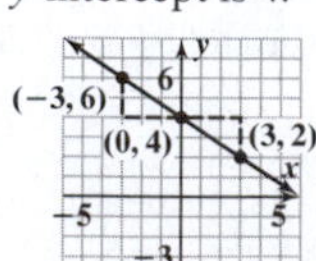


**35.** The slope is 1 and the $y$-intercept is $-4$.

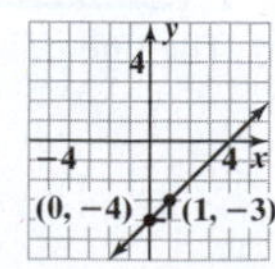


**36.** The slope is $-\frac{1}{3}$ and the $y$-intercept is $\frac{1}{2}$.

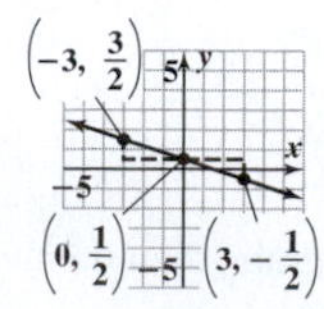

**37.** 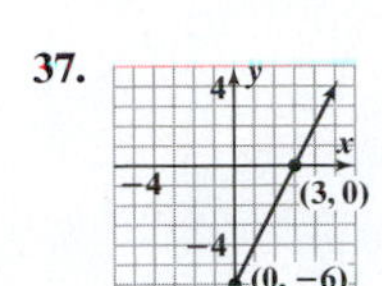 **38.** 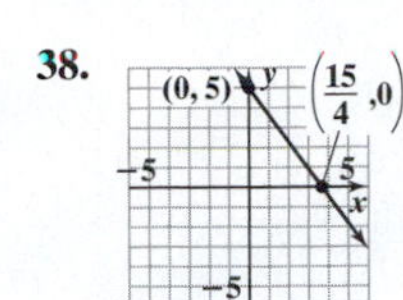 **39.** 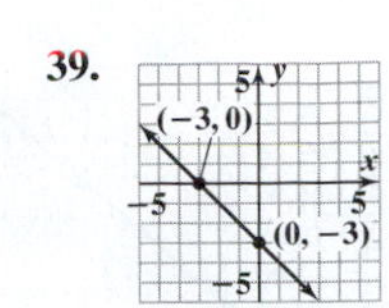 **40.** 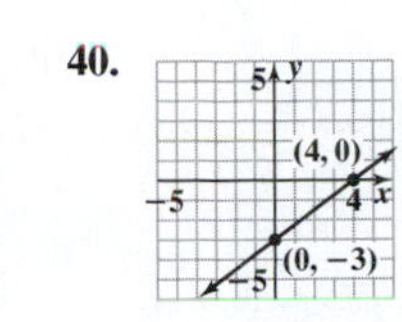

**41. (a)** $E(x) = 3.62x + 12.6$ **(b)** 41.56% **(c)** Electronically filed tax returns are increasing at a rate of 3.62% per year. **(d)** 2006

**42. (a)** $H(x) = -x + 220$ **(b)** 175 beats per minute **(c)** The maximum recommended heart rate for men under stress decreases at a rate of 1 beat per minute per year. **(d)** 52 years

**43.** $-\frac{3}{8}$ **44.** $\frac{8}{3}$ **45.** Perpendicular **46.** Neither **47.** Parallel **48.** Perpendicular

**49.** $y = -2x + 4$ **50.** $y = \frac{5}{2}x - 7$ **51.** $x = 1$ **52.** $y = -\frac{1}{3}x + 4$ **53.** $y = \frac{4}{3}x + 2$

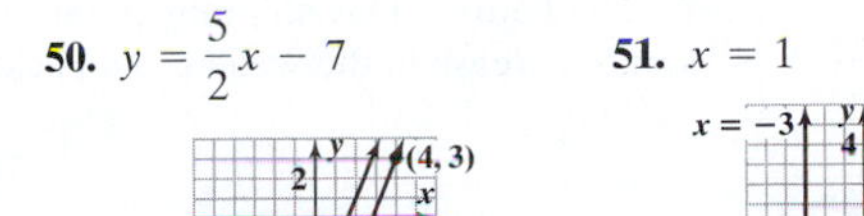
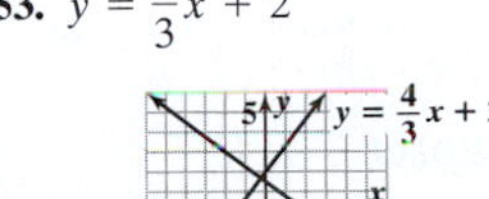
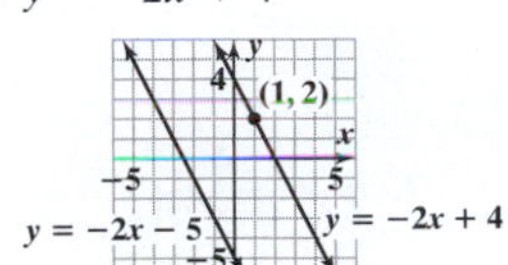
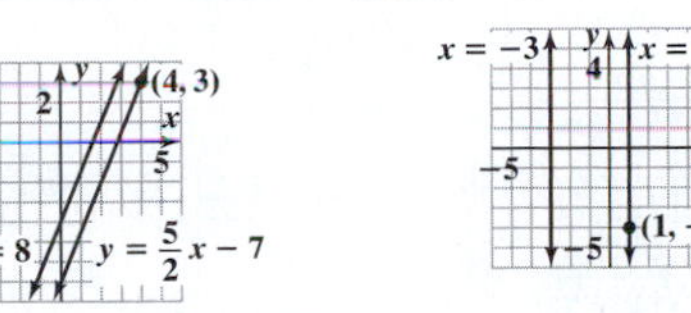
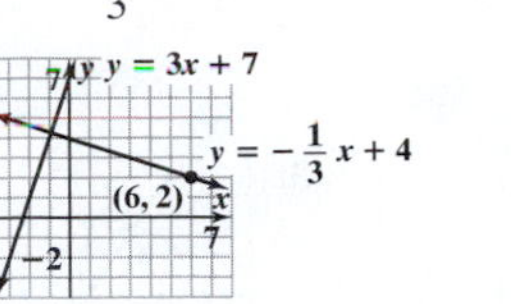

**54.** $y = -4$

**55. (a)** yes **(b)** yes **(c)** no

**56. (a)** no **(b)** yes **(c)** no

**57.** **58.** **59.** **60.**

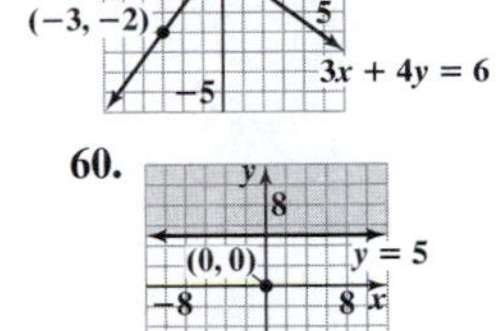
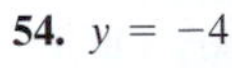
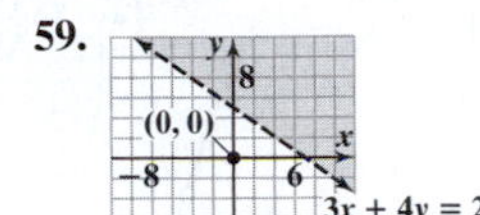
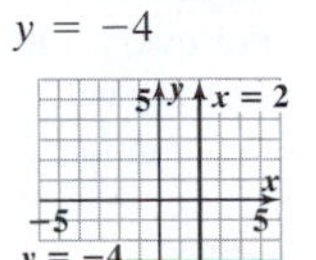
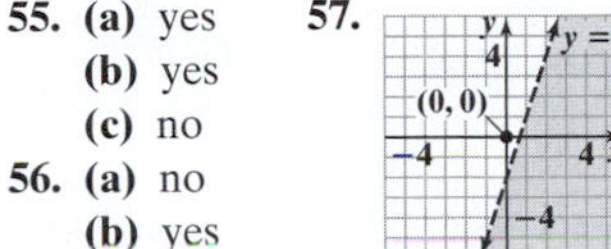
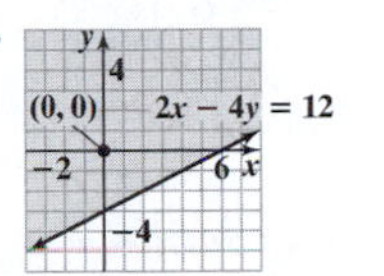

**61.** **62.**

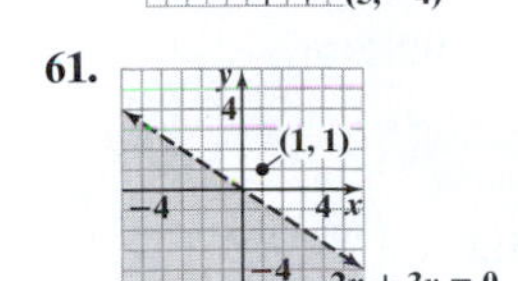
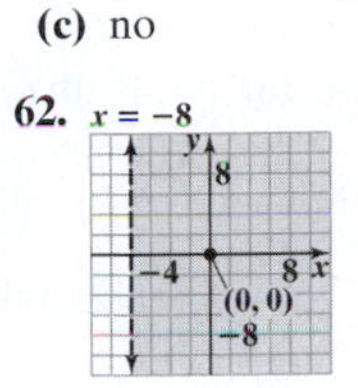

**63. (a)** Let $x$ = the number of movie tickets. Let $y$ = the number of music CDs. $7.50x + 15y \leq 60$ **(b)** no **(c)** yes

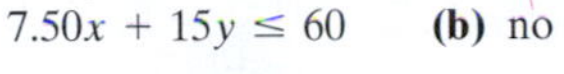
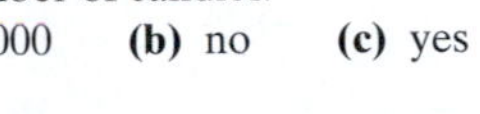

**64. (a)** Let $x$ = the number of candy bars. Let $y$ = the number of candles. $0.50x + 2y \geq 1000$ **(b)** no **(c)** yes

**65. (a)** $C(m) = 0.12m + 35$ **(b)** The independent variable is $m$; the dependent variable is $C$. **(c)** $\{m|m \geq 0\}$ or $[0, \infty)$ **(d)** \$49.88 **(e)** 268 miles were driven **(f)**

**66. (a)** $B(x) = 3.50x + 33.99$ **(b)** The independent variable is $x$; the dependent variable is $B$. **(c)** $\{x|x \geq 0\}$ or $[0, \infty)$ **(d)** \$51.49 **(e)** 7 pay-per-view movies **(f)**

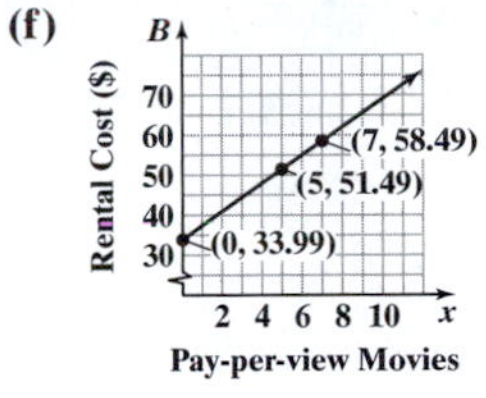

**67. (a)** $V(x) = -7050x + 84{,}600$ **(b)** $\{x|0 \leq x \leq 12\}$ or $[0, 12]$ **(c)** \$49,350 **(d)** The $y$-intercept is 84,600 and the $x$-intercept is 12. **(e)** After eight years **(f)**

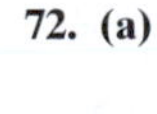

**68. (a)** $k = 12$ **(b)** $y = 12x$ **(c)** $y = 48$

**69. (a)** $k = \frac{11}{6}$ **(b)** $y = \frac{11}{6}x$ **(c)** $y = 44$

**70.** 2628 milligrams **71.** 24 feet

**72. (a)**

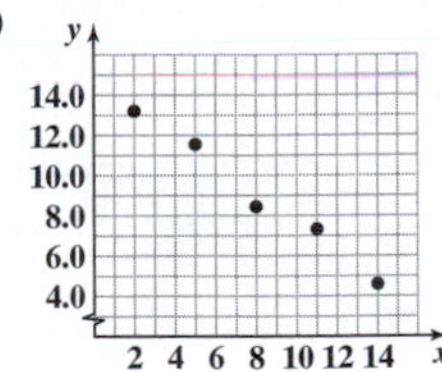

**(b)** Answers will vary. Using the points (2, 13.3) and (14, 4.6), the equation is $y = -0.725x + 14.75$.

**(c)**

**73. (a)**

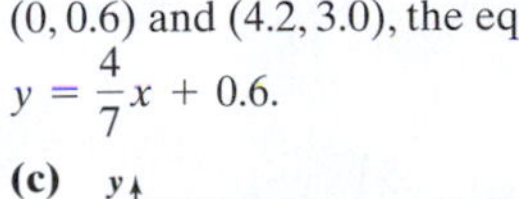

**(b)** Answers will vary. Using the points (0, 0.6) and (4.2, 3.0), the equation is $y = \frac{4}{7}x + 0.6$.

**(c)**

**74. (a)**

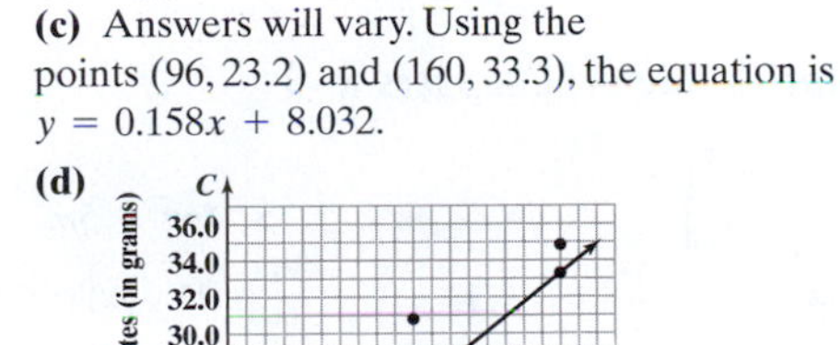

**(b)** Approximately linear.

**(c)** Answers will vary. Using the points (96, 23.2) and (160, 33.3), the equation is $y = 0.158x + 8.032$.

**(d)**

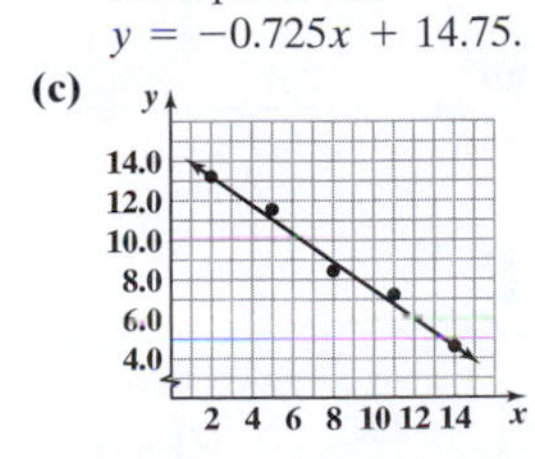
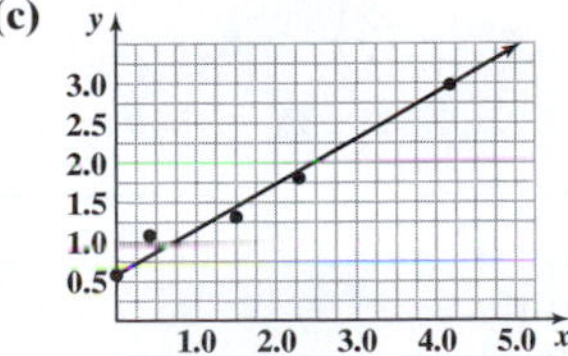
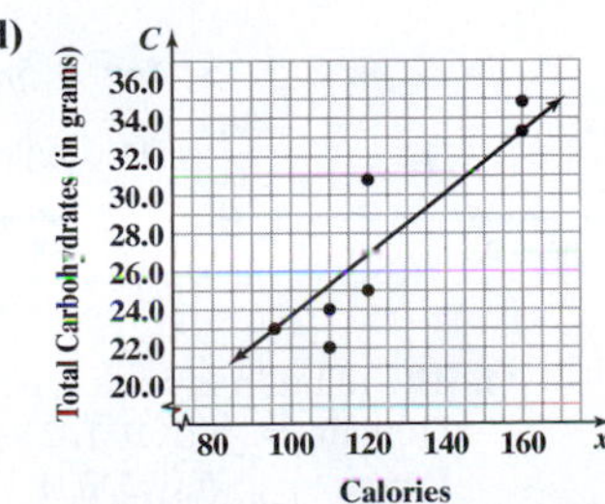

**(e)** 30.2 grams **(f)** In a one-cup serving of cereal, total carbohydrates will increase by 0.158 grams for each one-calorie increase.

**75. (a)**

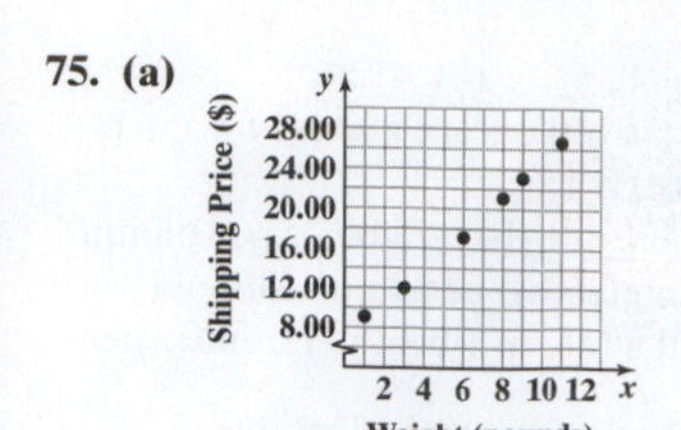


**(b)** Approximately linear.
**(c)** Answers will vary. Using the points (1, 9.25) and (11, 26.50), the equation is $y = 1.725x + 7.525$.

**(d)**

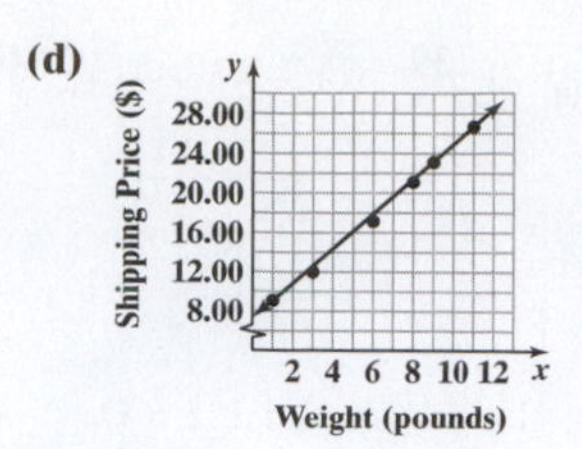


**(e)** \$16.15
**(f)** The FedEx 2Day shipping price increases by \$1.725 for each one-pound increase in the weight of a package.

## Chapter 3 Test

**1.**

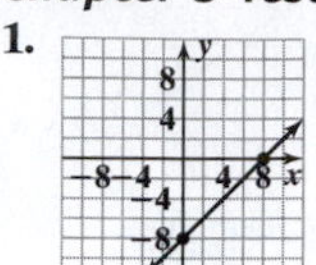

**2.**

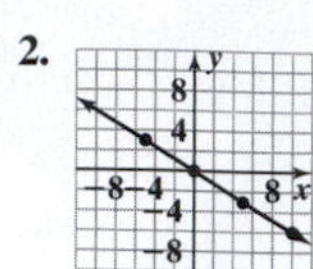

**3.**

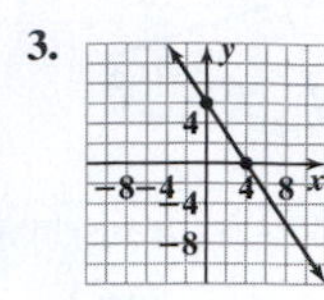

**4.**

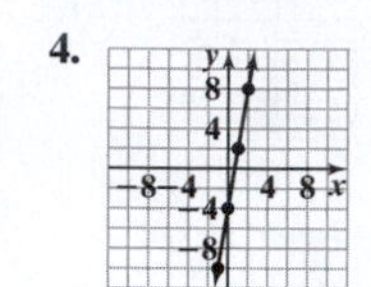

**5.**

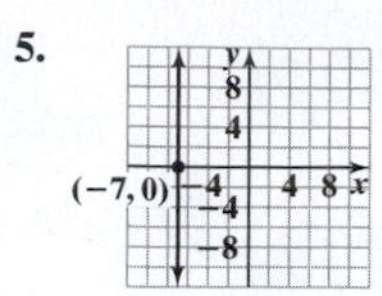


**6.**

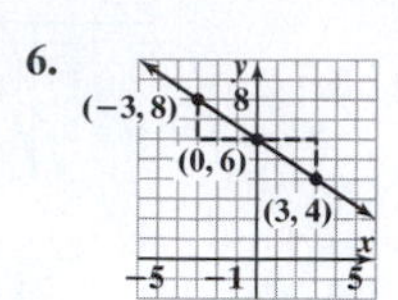


**7.** $m = -\frac{4}{3}$

For every 3-unit increase in $x$, $y$ will decrease by 4 units. For every 3-unit decrease in $x$, $y$ will increase by 4 units.

**8.**

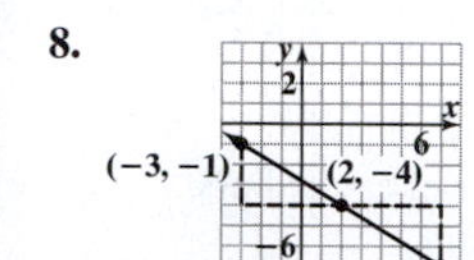


**9.** Perpendicular **10.** $y = 4x + 13$ or $4x - y = -13$ **11.** $y = -\frac{2}{3}x + 5$ or $2x + 3y = 15$

**12.** $y = \frac{1}{5}x - 3$ or $x - 5y = 15$ **13.** $y = -\frac{1}{3}x + 4$ or $x + 3y = 12$ **14. (a)** no **(b)** no **(c)** yes

**15.**

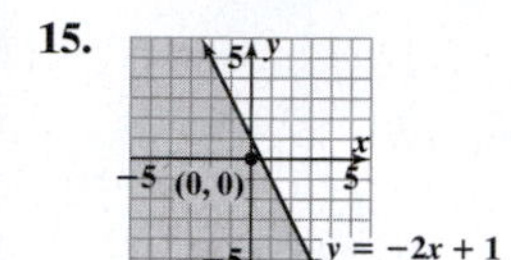


**16.**

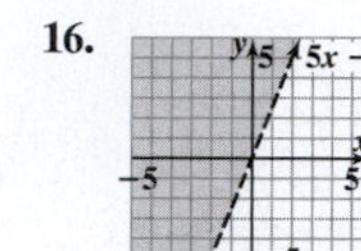


**17. (a)** $P(x) = 18x - 100$ **(b)** $\{x|x \geq 0\}$ or $[0, \infty)$ **(c)** \$512
**(d)**
**(e)** 48 shelves

**18. (a)**

**(b)** 3.14 square feet per foot; Between diameters of 1 foot and 3 feet, the area of the circle increases at a rate of 3.14 square feet per foot.
**(c)** Approximately 18.06 square feet per foot; Between diameters of 10 feet and 13 feet, the area of the circle increases at a rate of approximately 18.06 square feet per foot.
**(d)** No. The average rate of change (slope) is not constant.
**19.** \$123.50

**20. (a)**

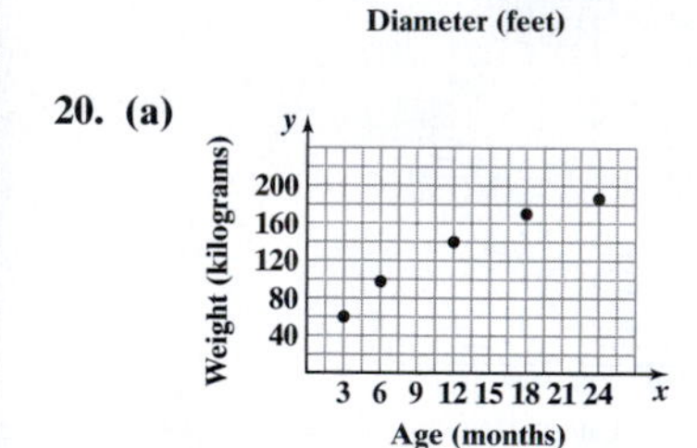


**(b)** Approximately linear.
**(c)** Answers will vary. Using the points (6, 95) and (18, 170), the equation is $y = 6.25x + 57.5$.

**(d)**

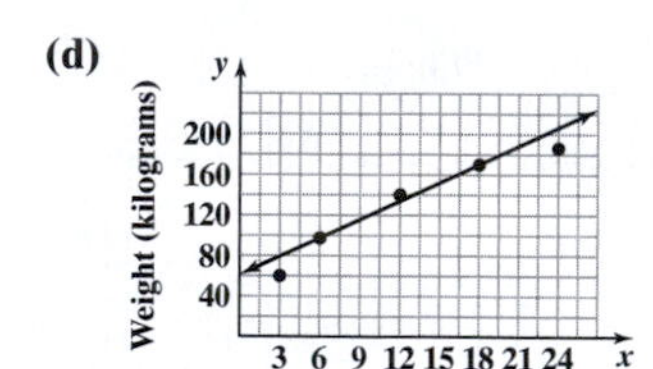


**(e)** 113.75 kilograms
**(f)** A Shetland pony's weight will increase by 6.25 kilograms for each one-month increase in age.

## Cumulative Review Chapters R–3

**1.** −16 **2.** $\frac{1}{4}$ **3.** −4 **4.** 8 **5.** $4m^2 + 5m - 9$ **6.** $\{-7\}$ **7.** $\left\{-\frac{25}{12}\right\}$ **8.** $\{-2, 7\}$ **9.** $B = \frac{2A - hb}{h}$ or $B = \frac{2A}{h} - b$

**10.** $\{x|x > -4\}$ or $(-4, \infty)$

**11.** $\{x|x \leq 5\}$ or $(-\infty, 5]$

**12.** $\{x|-7 < x < 2\}$ or $(-7, 2)$

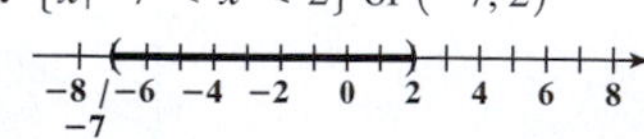

**13.**

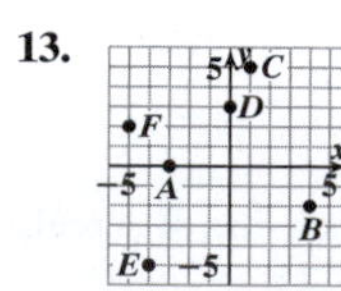


**14.** The relation is not a function.
Domain = $\{-1, 0, 1, 2\}$
Range = $\{-5, -2, 3, 4, 6\}$

**15.** The relation is not a function.
Domain = $\{x|1 \leq x \leq 5\} = [1, 5]$
Range = $\{y|-2 \leq y \leq 5\} = [-2, 5]$

**16.** The relation is a function.
Domain = $\{x \mid x \text{ is any real number}\} = (-\infty, \infty)$
Range = $\{y \mid y \text{ is any real number}\} = (-\infty, \infty)$

**17. (a)** −8 **(b)** $-\frac{13}{3}$

**18. (a)** −11 **(b)** $2x + 2h - 3$

**19.**

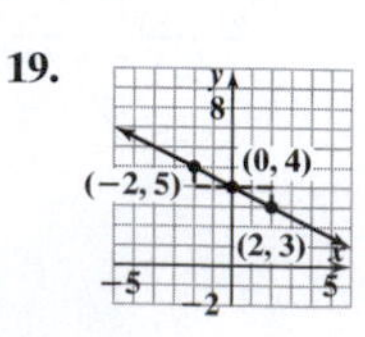

**20.** 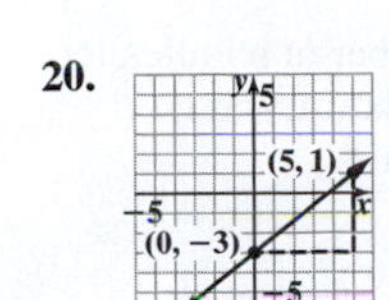

**21.** $y = -\frac{4}{3}x + 2$ or $4x + 3y = 6$ **22.** $y = -3x - 8$ or $3x + y = -8$ **23.** 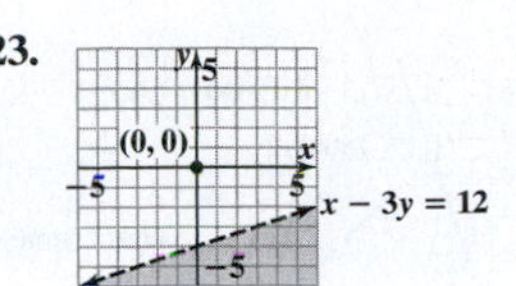


**24. (a)** The independent variable is $t$; the dependent variable is $C$. **(b)** $C(5) = 177.74$; $C(5)$ represents the projected number of cell phones that will be in use during the year 2005. That is, the function projects that 177.74 million cell phones will be in use in 2005. **(c)** $C(-8) = -7.77$; $C(-8)$ would represent the projected number of cell phones that was in use during the year 1992. That is, the function projects that $-7.77$ million cell phones were in use in 1992. This result is not reasonable. The number of cell phones in use could not be negative. **25.** 56 grams

# Chapter 4

## Section 4.1

**1.** system of linear equations **3.** consistent; dependent **5.** True **7.** Yes; answers may vary **9.** at the point $(3, -2)$ **11. (a)** no **(b)** yes **13. (a)** yes **(b)** yes **15.** $(2, -1)$ **17.** no solution **19.** $(1, 3)$ **21.** $\left\{(x, y) \middle| y = \frac{1}{2}x + 1\right\}$ **23.** $(3, -4)$

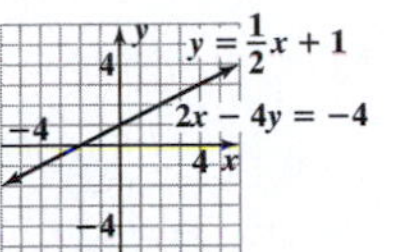

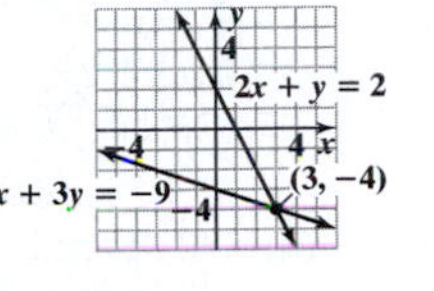


**25.** $(6, -2)$ **27.** $(-2, -3)$ **29.** no solution **31.** $\left(\frac{1}{2}, -\frac{1}{4}\right)$ **33.** $\{(x, y) | x + 3y = 6\}$ **35.** $(2500, 7500)$ **37.** $(-8, 3)$ **39.** $(11, -8)$ **41.** no solution **43.** $\left(\frac{1}{2}, -\frac{4}{5}\right)$ **45.** $\left\{(x,y) \middle| \frac{1}{3}x - 2y = 6\right\}$ **47.** $(25, 40)$ **49.** $(-9, 3)$ **51.** $\{(x,y) | x = 5y - 3\}$ **53.** $\left(\frac{39}{11}, -\frac{30}{11}\right)$ **55.** no solution **57.** $y = -2x - 5$; $y = -\frac{5}{3}x + \frac{1}{3}$; exactly one solution **59.** $y = \frac{3}{2}x + 1$; $y = \frac{3}{2}x + 1$; infinite number of solutions

**61. (a)** $y = -\frac{1}{2}x + \frac{5}{2}$; $y = x + 1$ **(b)** $(1, 2)$ **63.** (c) and (f) **65.** $A = \frac{7}{6}$ and $B = -\frac{1}{2}$

**67.** Answers will vary. One possibility follows: $\begin{cases} x + y = 3 \\ x - y = -5 \end{cases}$

**69.** $(1.2, 2.6)$

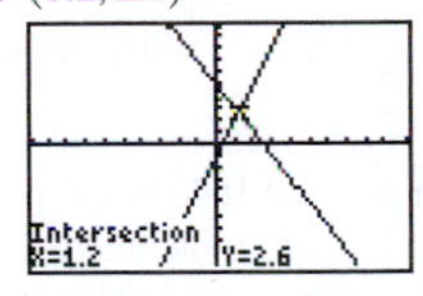


WINDOW Xmin=-10 Xmax=10 Xscl=1 Ymin=-10 Ymax=10 Yscl=1 Xres=1

**71.** $(4, 13)$

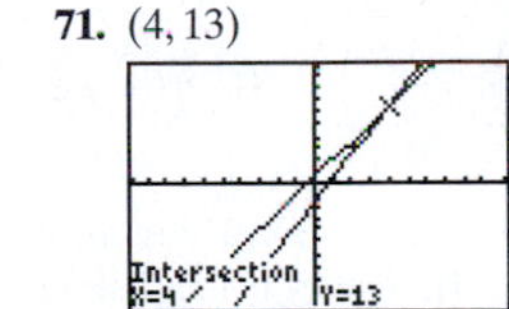


WINDOW Xmin=-10 Xmax=10 Xscl=1 Ymin=-20 Ymax=20 Yscl=2 Xres=1

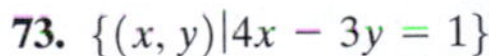
**73.** $\{(x, y) | 4x - 3y = 1\}$

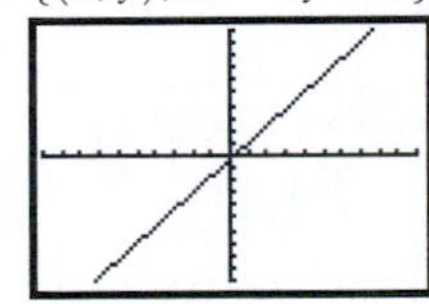

WINDOW Xmin=-10 Xmax=10 Xscl=1 Ymin=-10 Ymax=10 Yscl=1 Xres=1

**75.** approximately $(0.35, -3.76)$

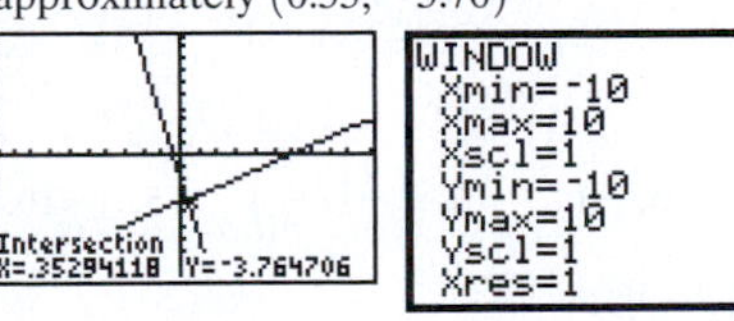


## Section 4.2

**1.** Answers will vary. **3.** 8 and 10 **5.** The first number is 12 and the second number is 23. **7.** The first number is 12 and the second number is 19.

**9. (a)**

R(x) = 12x; C(x) = 5.5x + 9880; (1520, 18240); Revenue, Cost ($); Number of Units

**(b)** $(1520, 18240)$

**11.** $60; $35 **13.** 280 calories; 210 calories **15.** 40 meters by 20 meters **17.** 15 cm, 15 cm, and 5 cm **19.** $x = 4$ and $y = 14$ **21.** $x = 4$ and $y = 8$ **23.** $22,500 in stocks and $13,500 in bonds **25.** $250,000 **27.** 60 pounds of Arabica beans and 40 pounds of African Robustas beans **29.** 50 milligrams of the first liquid and 60 milligrams of the second liquid must be mixed **31.** 8 miles per hour in still water **33.** The Lincoln traveled 400 miles and the Infiniti traveled 500 miles. The time for both trips was 10 hours each.

**35. (a)**

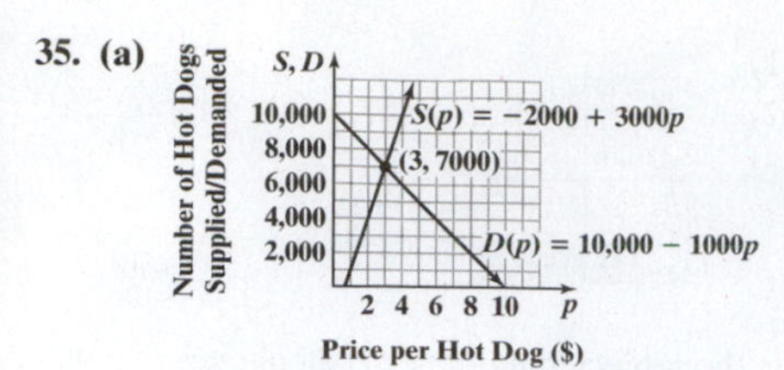


**(b)** $3; 7000 hot dogs

**37. (a)**

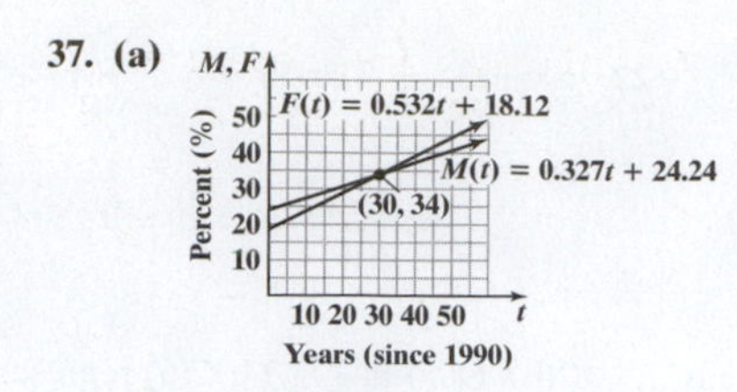


**(b)** approximately 2020; approximately 34%

**39. (a)** Let $m$ represent the number of minutes, let $A$ represent the cost of plan $A$, and let $B$ represent the cost of plan $B$.

$$\begin{cases} A(m) = 8.95 + 0.05m \\ B(m) = 5.95 + 0.07m \end{cases}$$

**(b)**

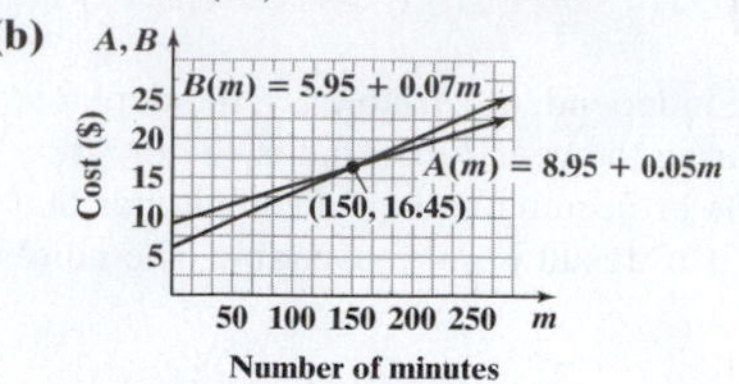


**(c)** 150 minutes; $16.45

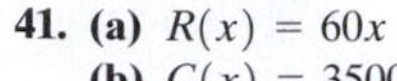

**41. (a)** $R(x) = 60x$
**(b)** $C(x) = 3500 + 35x$
**(c)**

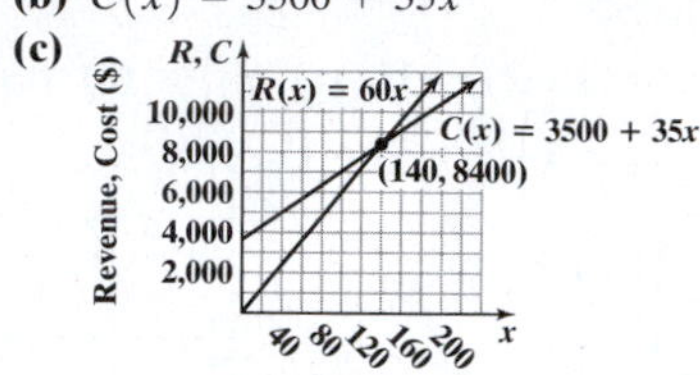


**(d)** 140 desks; The cost and revenue will both be $8400.

**43. (a) and (b)** Let $x$ represent the number of years since 1968.

**(c)** Answers may vary. Using the points (4, 20.00) and (36, 19.79), the equation is $y = -0.0065625x + 20.02625$. See graph on part (d).

**(d)** Answers may vary. Using the points (0, 22.50) and (32, 21.84), the equation is $y = -0.020625x + 22.50$.

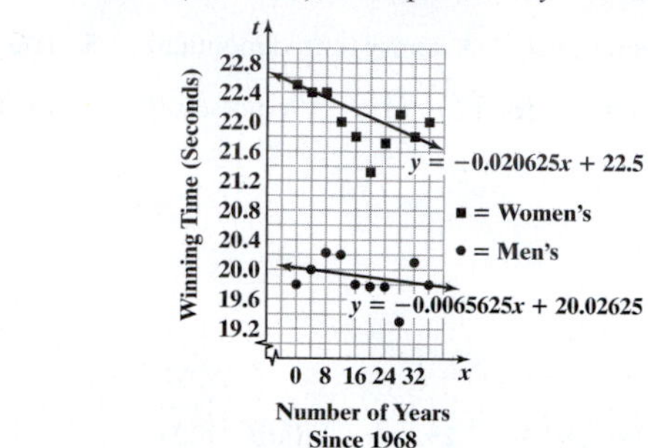


**(e)** In the year 2146; approximately 18.87 seconds
**(f)** Answers will vary.

## Section 4.3

**1.** inconsistent **3.** solution **5.** True **7.** The three planes formed by the system will intersect at the point (3, 2, 5). **9. (a)** no **(b)** yes **11.** $(3, -2, 4)$ **13.** $(3, -4, -2)$ **15.** $\left(0, 2, \frac{3}{2}\right)$ **17.** $\left(\frac{1}{3}, -\frac{2}{3}, 2\right)$ **19.** $(3, -3, 1)$ **21.** $(-5, 0, 3)$ **23.** no solution **25.** $\left(-\frac{7}{2}, -\frac{7}{2}, \frac{3}{2}\right)$ **27.** $\{(x, y, z) \mid x = 5z - 5, y = -2z + 3, z \text{ is any real number}\}$ **29.** $\left(\frac{11}{2}, 0, -\frac{1}{2}\right)$ **31.** $\left\{(x, y, z) \mid x = \frac{1}{3}z + \frac{2}{3}, y = -\frac{4}{3}z + \frac{7}{3}, z \text{ is any real number}\right\}$ **33.** Answers will vary. One possibility follows.

$$\begin{cases} x + y + z = 4 \\ x - y + z = 6 \\ x + y - z = -2 \end{cases}$$

**35. (a)** $a - b + c = -6$; $4a + 2b + c = 3$ **(b)** $a = -2, b = 5, c = 1$; $f(x) = -2x^2 + 5x + 1$ **37.** $i_1 = 1, i_2 = 4$, and $i_3 = 3$ **39.** There are 1490 box seats, 970 reserve seats, and 1640 lawn seats in the stadium. **41.** Nancy needs 1 serving of Chex® cereal, 2 servings of 2% milk, and 1.5 servings of orange juice. **43.** $12,000 in Treasury bills, $8000 in municipal bonds, and $5000 in corporate bonds. **45.** $\overline{AM} = 4$, $\overline{BN} = 2$, and $\overline{OC} = 10$ **47.** $(10, -4, 6)$ **49.** $(-2, 1, 0, 4)$

## Putting the Concepts Together

**1.** $(-3, 5)$ **2.** $(3, -2)$

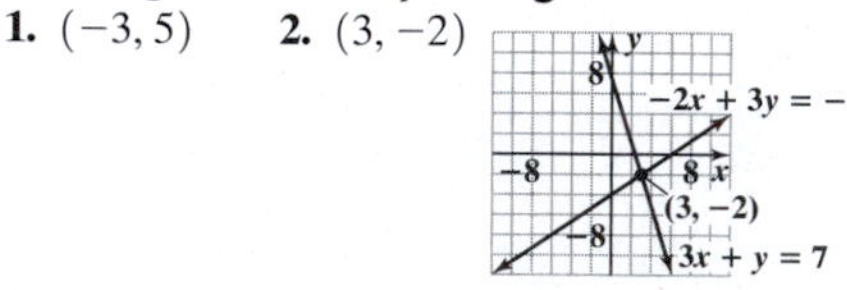


**3.** $(5, -1)$ **4.** $\left(\frac{4}{5}, -\frac{7}{5}\right)$ **5.** $(-10, 6)$ **6.** $\{(x, y) \mid 8x - 4y = 12\}$ **7.** $(-1, -3, 5)$ **8.** no solution **9.** 450 adult tickets and 375 youth tickets **10.** 290 orchestra seats, 380 mezzanine seats, and 530 balcony seats

## Section 4.4

**1.** matrix **3.** 4; 3 **5.** True **7.** Answers will vary. **9.** Multiply each entry in row 2 by $\frac{1}{5}$ (or divide each entry of row 2 by 5).

**11.** $\left[\begin{array}{cc|c} 1 & -3 & 2 \\ 2 & 5 & 1 \end{array}\right]$ **13.** $\left[\begin{array}{ccc|c} 1 & 1 & 1 & 3 \\ 2 & -1 & 3 & 1 \\ -4 & 2 & -5 & -3 \end{array}\right]$ **15.** $\left[\begin{array}{cc|c} -1 & 1 & 2 \\ 5 & 1 & -5 \end{array}\right]$ **17.** $\left[\begin{array}{ccc|c} 1 & 0 & 1 & 2 \\ 2 & 1 & 0 & 13 \\ 1 & -1 & 4 & -4 \end{array}\right]$ **19. (a)** $\left[\begin{array}{cc|c} 1 & -3 & 2 \\ 0 & -1 & 5 \end{array}\right]$

**(b)** $\left[\begin{array}{cc|c} 1 & -3 & 2 \\ 0 & 1 & -5 \end{array}\right]$ **21. (a)** $\left[\begin{array}{ccc|c} 1 & 1 & -1 & 4 \\ 0 & 3 & 5 & -11 \\ -1 & -3 & 2 & 1 \end{array}\right]$ **(b)** $\left[\begin{array}{ccc|c} 1 & 1 & -1 & 4 \\ 0 & 3 & 5 & -11 \\ 0 & -2 & 1 & 5 \end{array}\right]$ **23. (a)** $\left[\begin{array}{ccc|c} 1 & 1 & 1 & 4 \\ 0 & 1 & 5 & 5 \\ 0 & -4 & 2 & 8 \end{array}\right]$ **(b)** $\left[\begin{array}{ccc|c} 1 & 1 & 1 & 4 \\ 0 & 1 & 5 & 5 \\ 0 & -2 & 1 & 4 \end{array}\right]$

**25.** $\begin{cases} x + 4y = -5 & (1) \\ \quad\;\; y = -2 & (2) \end{cases}$ consistent and independent; $(3, -2)$ **27.** $\begin{cases} x + 3y - 2z = 6 & (1) \\ \quad\; y + 5z = -2 & (2) \\ \qquad\quad 0 = 4 & (3) \end{cases}$ inconsistent; $\varnothing$ or { } **29.** $\begin{cases} x - 2y - z = 3 & (1) \\ \quad\; y - 2z = -8 & (2) \\ \qquad\quad z = 5 & (3) \end{cases}$ consistent and independent; $(12, 2, 5)$ **31.** $(3, -5)$ **33.** no solution

**35.** $\left(\frac{1}{2}, -\frac{5}{4}\right)$ **37.** $\{(x, y) \mid 4x - y = 8\}$ **39.** $(3, -4, 1)$ **41.** $(4, 0, -5)$ **43.** $\{(x, y, z) \mid x = -2z - 0.2, y = -z - 1.4, z \text{ is any real number}\}$

**45.** no solution **47.** $\left(\frac{3}{10}, \frac{1}{10}, -\frac{1}{2}\right)$ **49.** $\left(\frac{5}{3}, \frac{2}{5}, -\frac{1}{2}\right)$ **51.** $(-2, 1, -5)$ **53.** **(a)** $a + b + c = 0$; $4a + 2b + c = 3$ **(b)** $a = 2, b = -3, c = 1$; $f(x) = 2x^2 - 3x + 1$ **55.** \$8000 in Treasury bills, \$7000 in municipal bonds, and \$5000 in corporate bonds **57.** $(-3, 7)$ **59.** $(2, 5, -4)$ **61.** $(5, -3)$ **63.** $(4, -2, 7)$

## Section 4.5

**1.** determinants **3.** square **5.** False **7.** $(-1, 2, 0)$ **9.** 10 **11.** $-2$ **13.** $-9$ **15.** $-163$ **17.** 0 **19.** $(-8, 4)$ **21.** $(3, -1)$ **23.** $\left(-\frac{1}{6}, \frac{3}{8}\right)$ **25.** $\left(\frac{3}{4}, -\frac{7}{4}\right)$ **27.** $(-2, 1, -1)$ **29.** $(3, -1, 2)$ **31.** $\left\{(x, y, z) \mid x = \frac{4}{5}z + \frac{14}{5}, y = -\frac{3}{5}z - \frac{8}{5}, z \text{ is any real number}\right\}$ **33.** $\left(\frac{3}{2}, \frac{9}{4}, -\frac{7}{4}\right)$ **35.** $\left\{(x, y, z) \mid x = -\frac{1}{2}z - 4, y = \frac{1}{4}z + \frac{3}{2}, z \text{ is any real number}\right\}$ **37.** $\left(\frac{7}{5}, -\frac{5}{3}, \frac{11}{3}\right)$ **39.** $x = 5$ **41.** $x = -2$

**43. (a)**

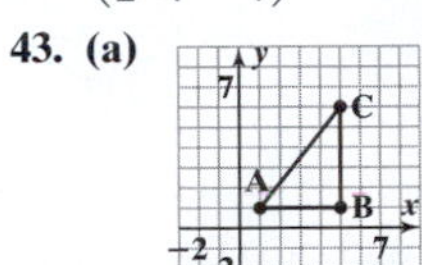


**(b)** The area of triangle $ABC$ is 10.

**45. (a)**

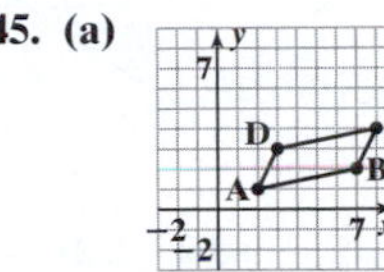


**(b)**

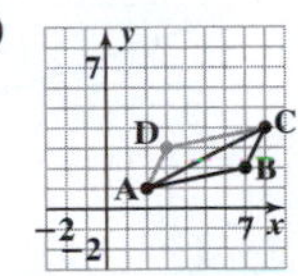


The area of triangle $ABC$ is 4.5.

**(c)** The area of triangle $ADC$ is 4.5. **(d)** 9

**47. (a)** $x + 2y = 7$ **(b)** $x + 2y = 7$ **49.** 14; $-14$; answers may vary. **51.** $(-8, 4)$ **53.** $(3, -1)$ **55.** $(-2, 1, -1)$

## Section 4.6

**1.** satisfies **3.** True **5.** Answers may vary. **7.** No. The solution set of one inequality alone is a half-plane. The other inequalities in the system would only restrict the solution set further. **9. (a)** no **(b)** yes

**11. (a)** no **(b)** no **13. (a)** no **(b)** yes

**15.** $3x + y = 10$; $x + y = 4$

**17.** $-2x + 3y = 12$; $3x + 2y = 12$

**19.**

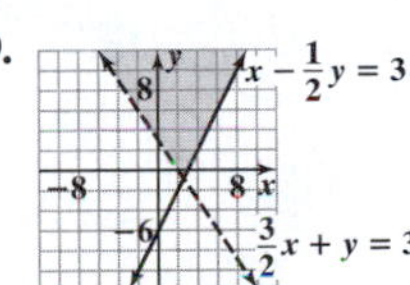


**21.**

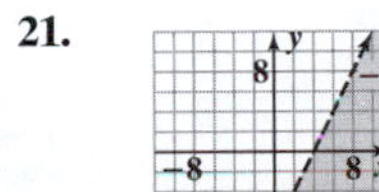


**23.**

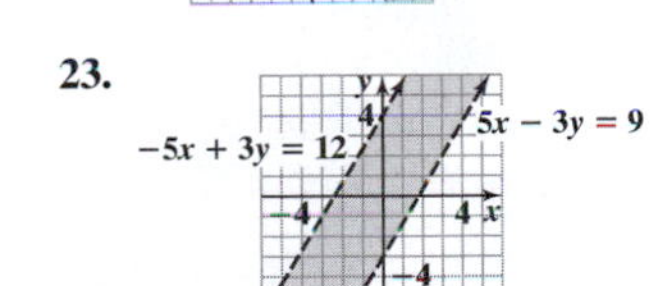


**25.**

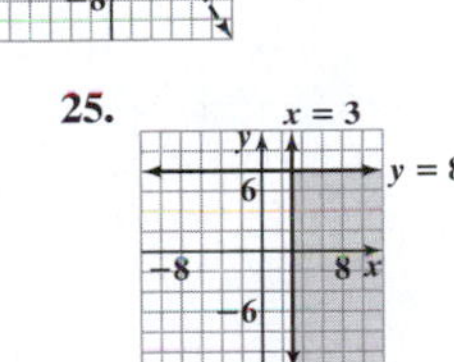


**27.**

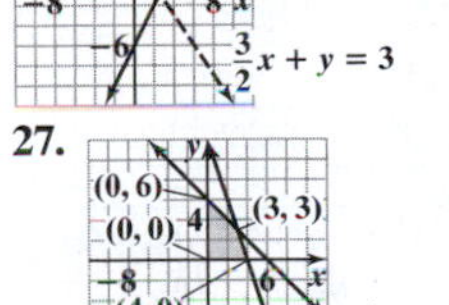


The graph is bounded.

**29.**

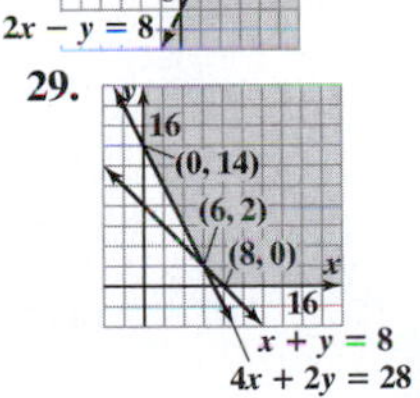


The graph is unbounded.

**31.**

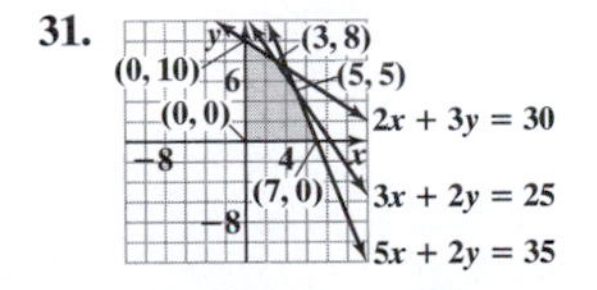


The graph is bounded.

**33.**

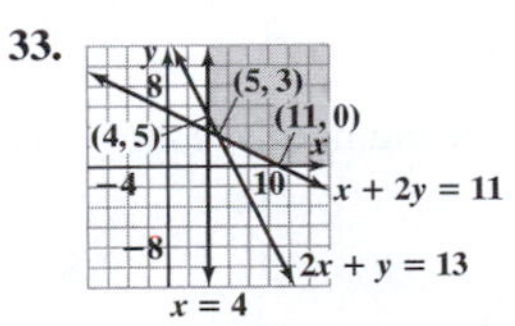


The graph is unbounded.

**35. (a)**

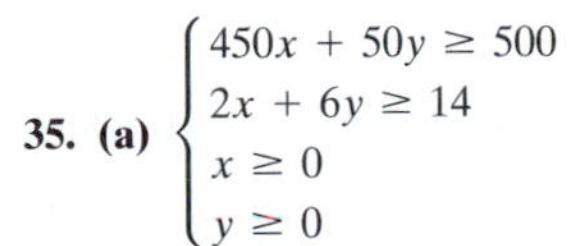

$\begin{cases} 450x + 50y \geq 500 \\ 2x + 6y \geq 14 \\ x \geq 0 \\ y \geq 0 \end{cases}$

**(b)**

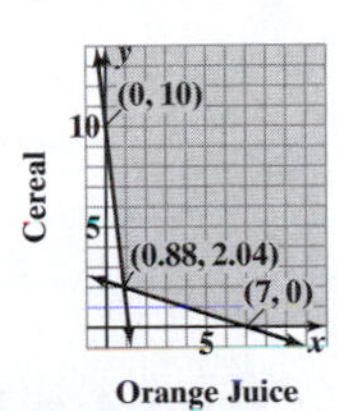


**37. (a)**

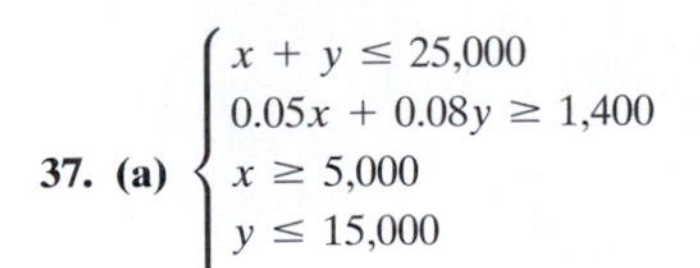

$\begin{cases} x + y \leq 25{,}000 \\ 0.05x + 0.08y \geq 1{,}400 \\ x \geq 5{,}000 \\ y \leq 15{,}000 \\ y \geq 0 \end{cases}$

**(b)**

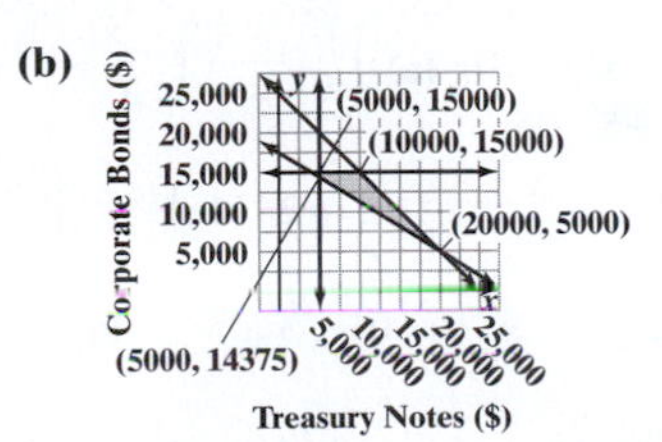


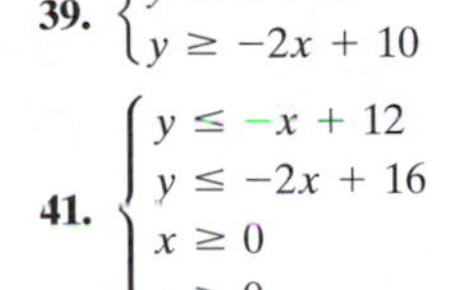

**39.** $\begin{cases} y \geq x + 4 \\ y \geq -2x + 10 \end{cases}$

**41.** $\begin{cases} y \leq -x + 12 \\ y \leq -2x + 16 \\ x \geq 0 \\ y \geq 0 \end{cases}$

**43.**

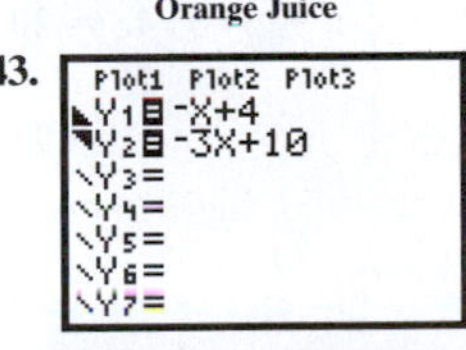

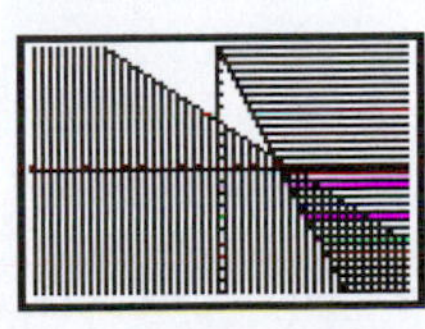

**45.** 

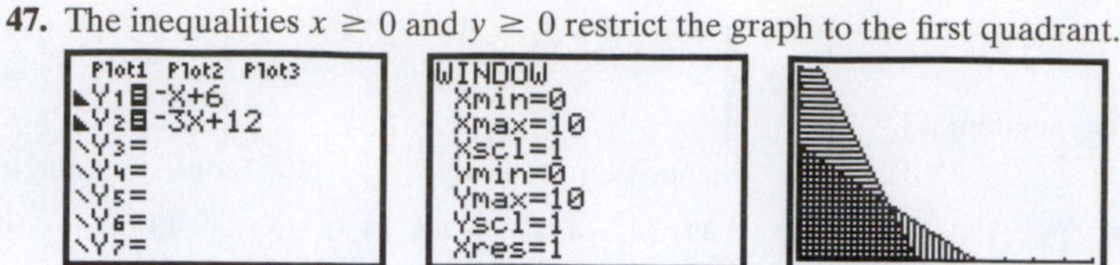


**47.** The inequalities $x \geq 0$ and $y \geq 0$ restrict the graph to the first quadrant.

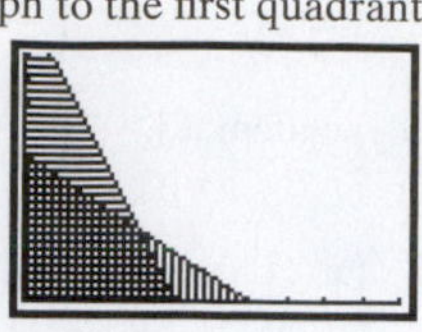

## *Chapter 4 Review*

**1. (a)** no **(b)** yes **2. (a)** yes **(b)** no **3. (a)** yes **(b)** no **4. (a)** no **(b)** yes **5.** $(4, 2)$ **6.** $(2, -1)$
**7.** $(2, -5)$ **8.** $(3, -1)$ **9.** no solution **10.** $(-3, -2)$

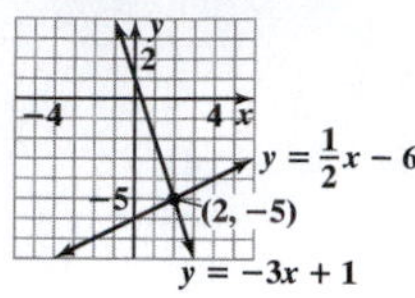


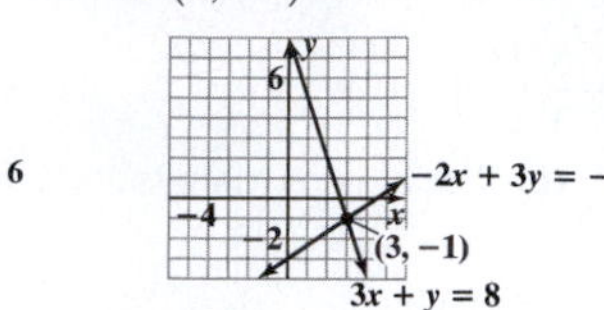


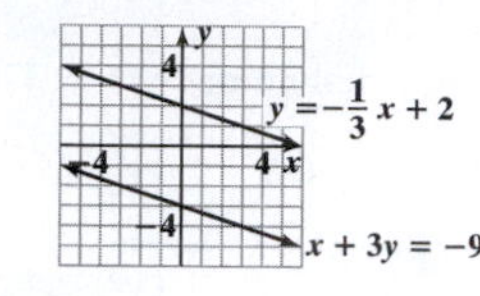


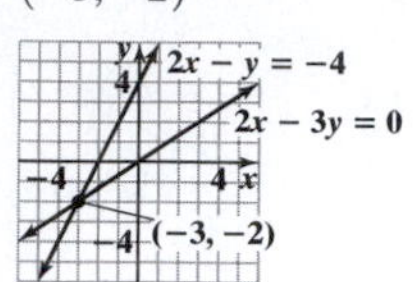


**11.** $(8, 0)$ **12.** $\left\{(x, y) \mid y = -\frac{3}{4}x + 2\right\}$ **13.** $(2, -3)$ **14.** $(-3, -5)$ **15.** $(4, -1)$ **16.** $(2, 2)$ **17.** $\{(x, y) \mid 2x - 4y = 8\}$

**18.** $(-5, -1)$ **19.** $(0, -4)$ **20.** $(-2, -4)$ **21.** $(1, -1)$ **22.** $\left(\frac{1}{2}, \frac{3}{4}\right)$ **23.** no solution **24.** $(-7, 4)$ **25.** 35 and 21

**26.** 31 males and 42 females **27.** 300 calories in a slice of pepperoni pizza and 340 calories in a slice of Italian sausage pizza
**28.** 21 inches by 13 inches **29.** angle $x = 30°$ and angle $y = 60°$ **30.** $x = 15$ and $y = 5$ **31.** 28 nickels and 12 quarters
**32.** 8 liters of the 25%-hydrochloric-acid solution and 4 liters of the 40%-hydrochloric-acid solution
**33.** \$7000 in stocks and \$3000 in bonds **34.** The plane's speed is 136 miles per hour and the wind's speed is 24 miles per hour.
**35.** $\frac{5}{6}$ hours, or 50 minutes; $66\frac{2}{3}$ miles **36.** The speed of the boat in still water is 25 miles per hour and the speed of the current is 5 miles per hour.

**37. (a)**

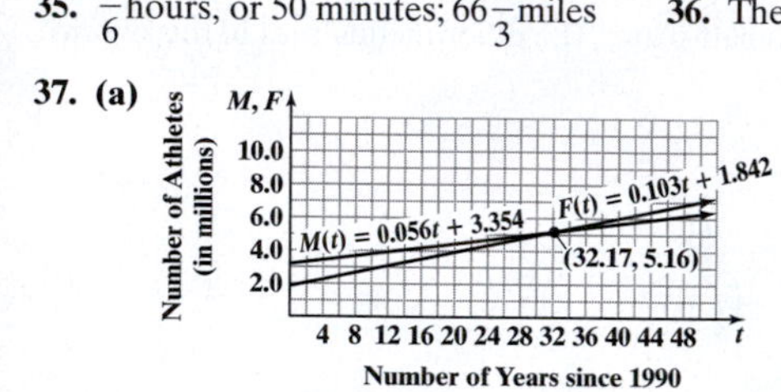


**(b)** approximately 2022

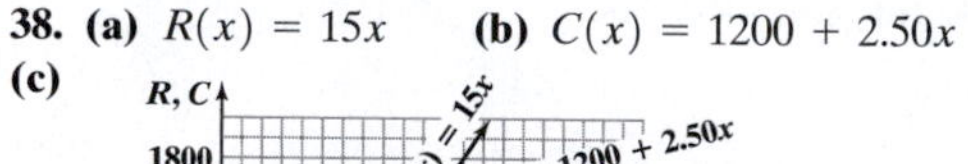

**38. (a)** $R(x) = 15x$ **(b)** $C(x) = 1200 + 2.50x$
**(c)**

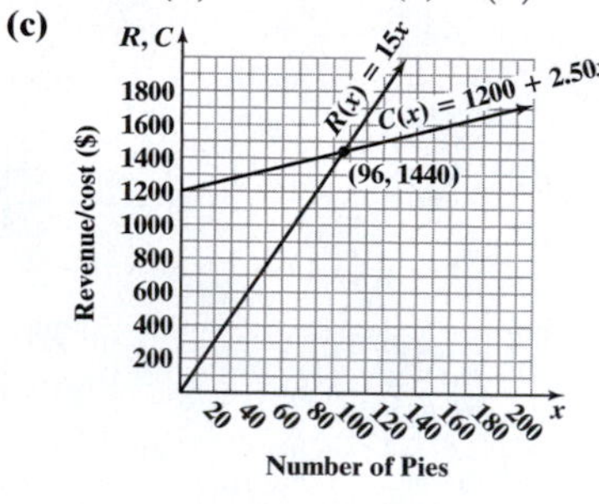


**(d)** 96 pies; \$1440

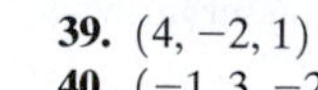

**39.** $(4, -2, 1)$
**40.** $(-1, 3, -2)$
**41.** $(5, -3, 4)$
**42.** no solution

**43.** $\left\{(x, y, z) \mid x = \frac{1}{2}z + 4, y = \frac{1}{2}z - 6, z \text{ is any real number}\right\}$ **44.** $\left(-\frac{2}{3}, \frac{1}{2}, \frac{3}{4}\right)$ **45.** $(1, 0, -6)$ **46.** $(-3, 4, 2)$

**47.** angle $x = 21°$, angle $y = 53°$, and angle $z = 106°$
**48.** A cheeseburger contains 360 calories, a medium order of fries contains 400 calories, and a medium Coke contains 280 calories.

**49.** $\left[\begin{array}{cc|c} 3 & 1 & 7 \\ 2 & 5 & 9 \end{array}\right]$ **50.** $\left[\begin{array}{cc|c} 1 & -5 & 14 \\ -1 & 1 & -3 \end{array}\right]$ **51.** $\left[\begin{array}{ccc|c} 5 & -1 & 4 & 6 \\ -3 & 0 & -3 & -1 \\ 1 & -2 & 0 & 0 \end{array}\right]$ **52.** $\left[\begin{array}{ccc|c} 8 & -1 & 3 & 14 \\ -3 & 5 & -6 & -18 \\ 7 & -4 & 5 & 21 \end{array}\right]$ **53.** $\begin{cases} x + 2y = 12 \\ 3y = 15 \end{cases}$

**54.** $\begin{cases} 3x - 4y = -5 \\ -x + 2y = 7 \end{cases}$ **55.** $\begin{cases} x + 3y + 4z = 20 \\ y - 2z = -16 \\ z = 7 \end{cases}$ **56.** $\begin{cases} -3x + 7y + 9z = 1 \\ 4x + 10y + 7z = 5 \\ 2x - 5y - 6z = -8 \end{cases}$ **57. (a)** $\left[\begin{array}{cc|c} 1 & -5 & 22 \\ 0 & -1 & 4 \end{array}\right]$ **(b)** $\left[\begin{array}{cc|c} 1 & -5 & 22 \\ 0 & 1 & -4 \end{array}\right]$

**58. (a)** $\left[\begin{array}{cc|c} 1 & -4 & 7 \\ 0 & 5 & -15 \end{array}\right]$ **(b)** $\left[\begin{array}{cc|c} 1 & -4 & 7 \\ 0 & 1 & -3 \end{array}\right]$ **59. (a)** $\left[\begin{array}{ccc|c} -1 & 2 & 1 & 1 \\ 0 & 3 & 5 & -1 \\ -1 & 5 & 6 & 2 \end{array}\right]$ **(b)** $\left[\begin{array}{ccc|c} -1 & 2 & 1 & 1 \\ 0 & 3 & 5 & -1 \\ 0 & 3 & 5 & 1 \end{array}\right]$ **60. (a)** $\left[\begin{array}{ccc|c} 1 & 3 & 4 & 4 \\ 0 & 1 & 2 & -3 \\ 0 & -4 & -7 & 7 \end{array}\right]$

**(b)** $\left[\begin{array}{ccc|c} 1 & 3 & 4 & 4 \\ 0 & 1 & 2 & -3 \\ 0 & 0 & 1 & -5 \end{array}\right]$ **61.** $(7, -3)$ **62.** $\left(\frac{1}{2}, 5\right)$ **63.** $(-4, 1)$ **64.** no solution **65.** $\{(x, y) \mid 4x - 2y = 6\}$ **66.** $\left(\frac{5}{2}, \frac{3}{8}\right)$

**67.** $(-5, 1, 4)$ **68.** no solution **69.** $\left\{(x, y, z) \mid x = -\frac{5}{6}z + 1, y = \frac{4}{3}z + 1, z \text{ is any real number}\right\}$ **70.** $(-3, 1, 3)$ **71.** 10 **72.** 0

**73.** $-2$ **74.** 9 **75.** 111 **76.** 0 **77.** $-31$ **78.** 78 **79.** $(5, -3)$ **80.** $\left(-4, \frac{1}{3}\right)$ **81.** $(-1, 2)$ **82.** $\left(\frac{7}{3}, \frac{1}{3}\right)$

**83.** no solution **84.** $\left(\frac{3}{10}, \frac{12}{5}\right)$ **85.** $(5, -2, 1)$ **86.** $(1, 7, -5)$ **87.** $(1, 2, -2)$ **88.** $\left(-\frac{1}{8}, 2, \frac{1}{2}\right)$

**89.** $\{(x, y, z) | x = -9z - 17, y = -13z - 24, z \text{ is any real number}\}$ **90.** $(2, -5, -9)$ **91. (a)** yes **(b)** no **92. (a)** no **(b)** yes

**93. (a)** yes **(b)** no **94. (a)** no **(b)** yes

**95.** **96.**

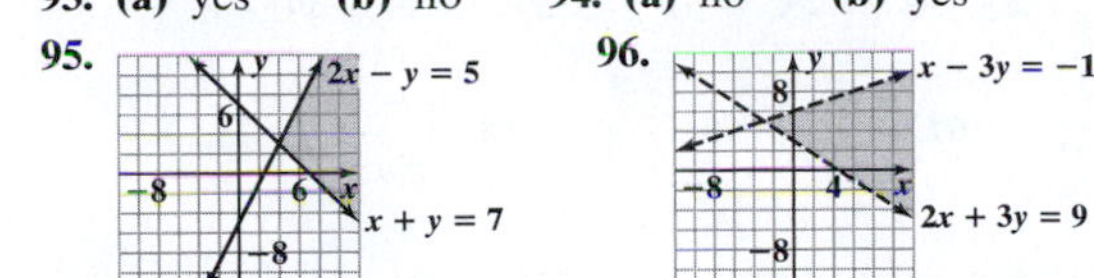


**97.**

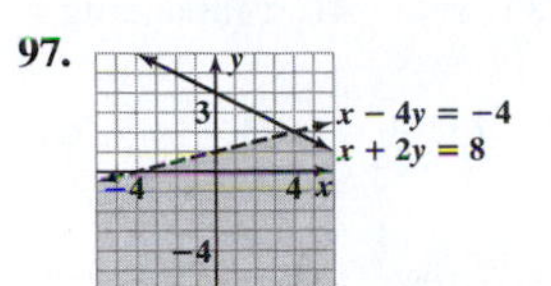


**98.**

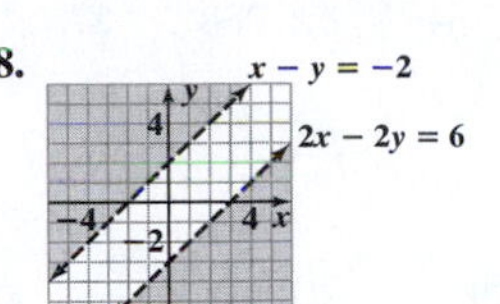


**99.**

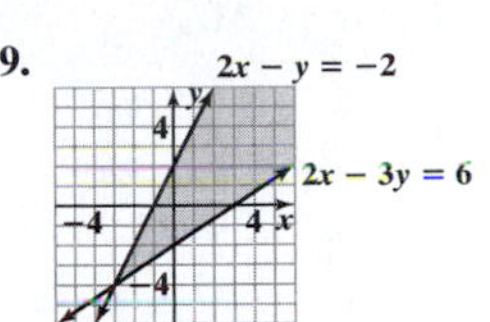


**100.**

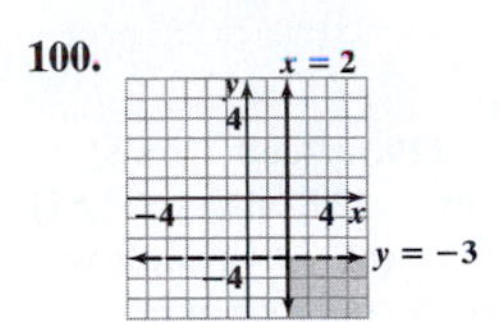


**101.**

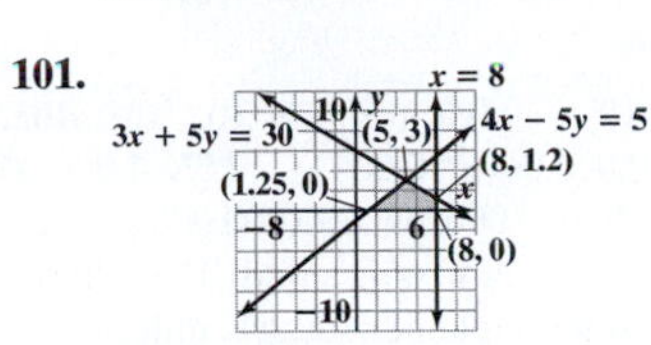


The graph is bounded.

**102.**

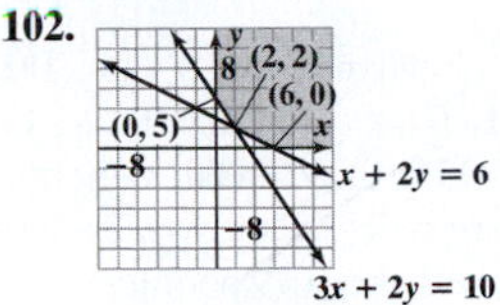


The graph is unbounded.

**103. (a)** $\begin{cases} x + y \le 4000 \\ 0.06x + 0.08y \ge 275 \\ x \ge 500 \\ y \le 2500 \\ y \ge 0 \end{cases}$

**(b)**

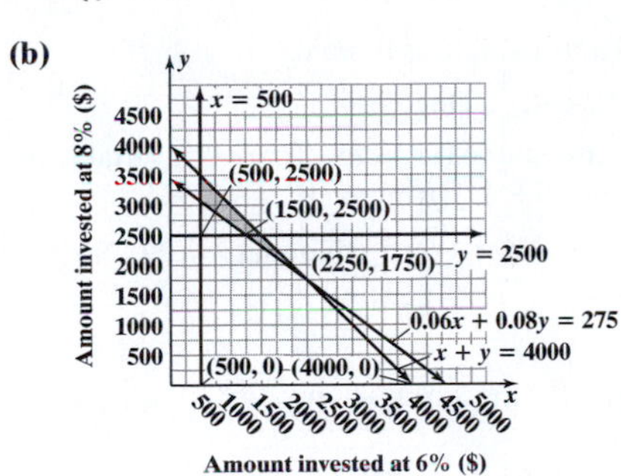


**104. (a)**

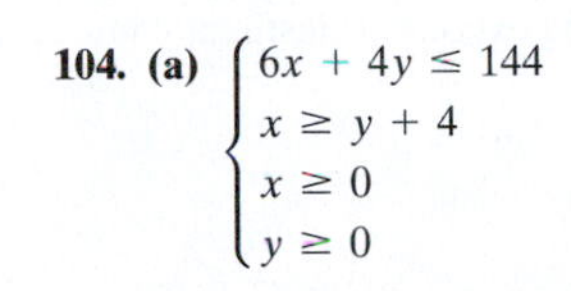

$\begin{cases} 6x + 4y \le 144 \\ x \ge y + 4 \\ x \ge 0 \\ y \ge 0 \end{cases}$

**(b)**

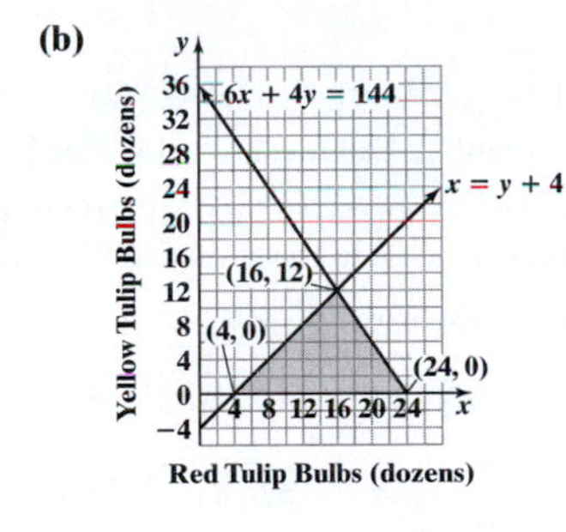


## *Chapter 4 Test*

**1.** $(-2, -4)$ **2.** $(1, -4)$ **3.** $\left(\frac{2}{3}, -\frac{5}{3}\right)$ **4.** no solution **5.** $(-6, 8)$ **6.** $\{(x, y, z) | x = z + 5, y = 2z + 2, z \text{ is any real number}\}$

**7.** $(5, 1, -3)$ **8. (a)** $\left[\begin{array}{cc|c} 1 & -3 & -2 \\ 0 & 2 & 12 \end{array}\right]$ **(b)** $\left[\begin{array}{cc|c} 1 & -3 & -2 \\ 0 & 1 & 6 \end{array}\right]$ **9. (a)** $\left[\begin{array}{ccc|c} 1 & -2 & 1 & -2 \\ 0 & 1 & -1 & 7 \\ 0 & -4 & 5 & -32 \end{array}\right]$ **(b)** $\left[\begin{array}{ccc|c} 1 & -2 & 1 & -2 \\ 0 & 1 & -1 & 7 \\ 0 & 0 & 1 & -4 \end{array}\right]$

**10.** $\left[\begin{array}{cc|c} 1 & -5 & 2 \\ 2 & 1 & 4 \end{array}\right]$; $(2, 0)$ **11.** $\left[\begin{array}{ccc|c} 1 & 2 & 1 & 3 \\ 0 & 4 & 3 & 5 \\ 2 & 3 & 0 & 1 \end{array}\right]$; $(2, -1, 3)$ **12.** $-4$ **13.** 14 **14.** $\left(-\frac{7}{4}, \frac{1}{4}\right)$ **15.** $(-5, 4, -1)$

**16.** **17.**

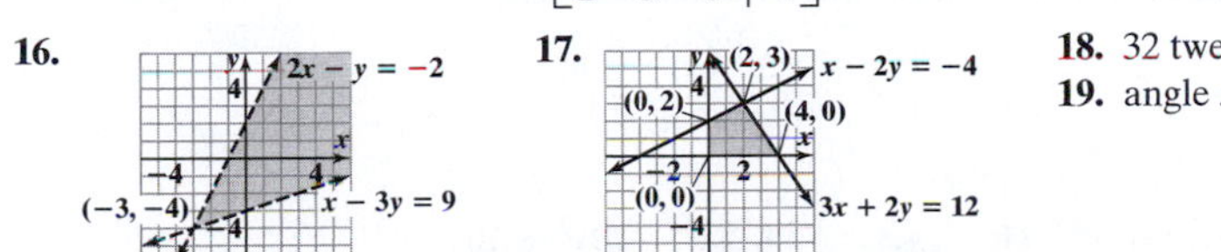


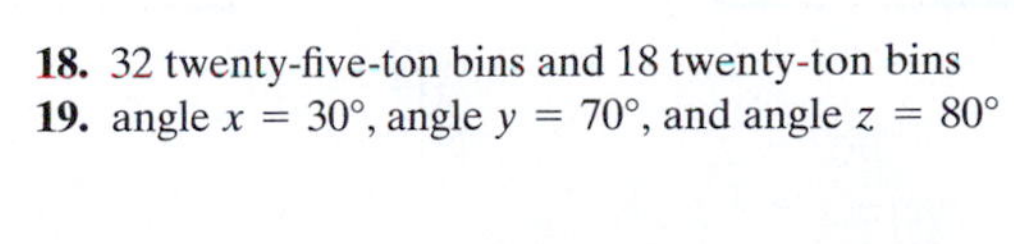

**18.** 32 twenty-five-ton bins and 18 twenty-ton bins

**19.** angle $x = 30°$, angle $y = 70°$, and angle $z = 80°$

**20. (a)**

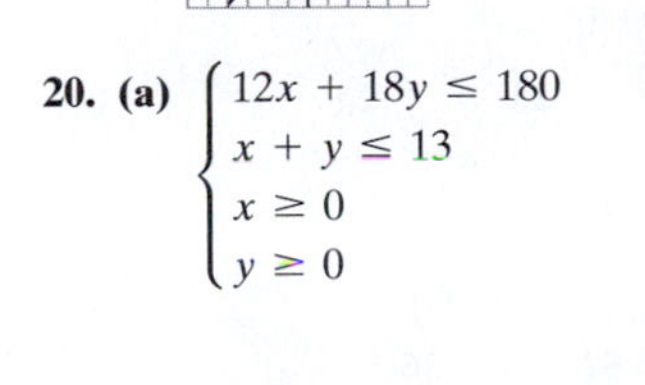

$\begin{cases} 12x + 18y \le 180 \\ x + y \le 13 \\ x \ge 0 \\ y \ge 0 \end{cases}$

**(b)**

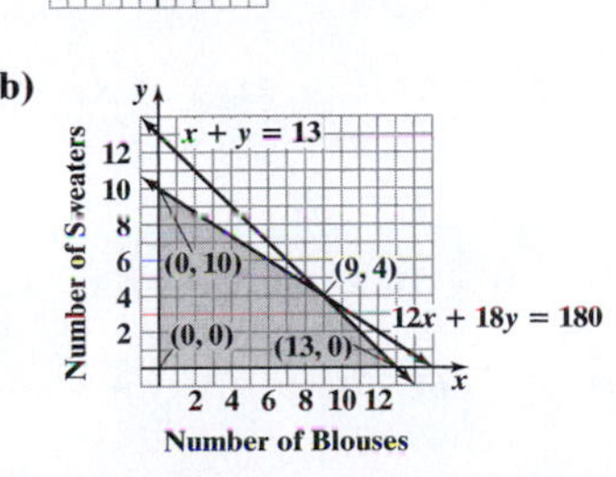

## Getting Ready for Chapter 5

**1.** base; exponent or power **3.** 0 **5.** False **7.** False **9.** Answers may vary. **11.** Answers may vary. **13.** Answers may vary. See objective 1. **15.** Answers may vary. **17.** Answers may vary. See objective 7. **19.** 25 **21.** −25 **23.** $-\frac{1}{25}$ **25.** $\frac{1}{8}$ **27.** $\frac{81}{16}$ **29.** −8 **31.** $-\frac{1}{27}$ **33.** −64 **35.** 288 **37.** $z^4$ **39.** $y^2$ **41.** 1 (assuming $x \neq 0$) **43.** 36 **45.** $-10t^5$ **47.** $\frac{5x^2}{y}$ **49.** $\frac{8}{5}a$ **51.** $4x^2y$ **53.** $\frac{3b^3}{2a}$ **55.** $\frac{3y^2}{x}$ **57.** $\frac{1}{x^8}$ **59.** $27x^6y^3$ **61.** $\frac{z^3}{64}$ **63.** $\frac{a^6}{9}$ **65.** $\frac{1}{16a^8b^{12}}$ **67.** $\frac{xy^5}{z^5}$ **69.** $\frac{2x^5}{3y^3}$ **71.** $\frac{b^4}{25a^{12}}$ **73.** 1 **75.** $-y^7$ **77.** $-x^2y^4z$ **79.** $\frac{9y^8z}{x^5}$ **81.** $\frac{3b^{16}}{2a^9}$ **83.** $4.5 \times 10^6$ **85.** $-2.3 \times 10^5$ **87.** $3.4 \times 10^{-4}$ **89.** $-1 \times 10^{-7}$ **91.** $6.8 \times 10^{-3}$ **93.** $-1.484 \times 10^{-6}$ **95.** $3 \times 10^5$ **97.** $6.4 \times 10^{19}$ **99.** $6.25 \times 10^{-3}$ **101.** $8 \times 10^6$ **103.** $-1.6 \times 10^{-13}$ **105.** 1 **107.** 12,000,000,000,000 **109.** 12,000,000,000 **111.** 0.04 **113.** 0.04 **115.** −30,000,000,000,000 **117.** 4,000,000 **119.** $x^6$ cubic units **121.** $1.276 \times 10^{-5}$ meters **123.** $1.276 \times 10^7$ meters **125.** 0.000000001 meters **127.** 40,000,000,000,000,000,000 possible states **129.** about $35,571 **131.** $5 \times 10^2$ seconds **133. (a)** 70.3 people per square mile; on average, each square mile of land contained about 70 people. **(b)** 79.4 people per square mile; on average, each square mile of land contained about 79 people. **(c)** about 9 people per square mile; on average, the number of people living on each square mile of land increased by 9, thereby making living space more crowded. **135.** $2^{x+1}$ **137.** $3^{10x+3}$ **139.** 625 **141.** $\frac{1}{2401}$ **143.** No, your friend is incorrect. He added the exponents instead of multiplying.

## Chapter 5

### Section 5.1

**1.** like terms **3.** degree **5.** True **7.** Answers may vary. **9.** A binomial has two terms and a trinomial has three terms. **11.** Answers will vary. **13.** Coefficient: 3; Degree: 2 **15.** Coefficient: −8; Degree: 5 **17.** Coefficient: $\frac{4}{3}$; Degree: 6 **19.** Coefficient: 2; Degree: 0 **21.** The exponent in the first term is not an integer that is greater than or equal to 0. **23.** There is a variable in the denominator. The expression cannot be written in the standard form for a polynomial. **25.** Yes; $5x^2 - 9x + 1$; 2; trinomial **27.** No **29.** No **31.** Yes; $\frac{5}{8}$; 0; monomial **33.** Yes; $2y^2 - 8y + 5$; 2; trinomial **35.** No **37.** Yes; $2xy^4 + 3x^2y^2 + 4$; 5; trinomial **39.** No **41.** $13z^3$ **43.** $4x^2 + 3x - 2$ **45.** $8p^3 - p^2 - 4p - 1$ **47.** $2x^2 + 4x + 3$ **49.** $3s^2t^3 - 4st^2 - 5t - 1$ **51.** $-4x^2 - 2x + 1$ **53.** $3y^3 - y^2 - 2y + 8$ **55.** $\frac{3}{4}x^2 + \frac{5}{4}x + 1$ **57.** $8x^2y^2 - 7x^2y - 3xy^2$ **59.** $x^2y + 11xy^2 + 2xy$ **61. (a)** 1 **(b)** −3 **(c)** 22 **63. (a)** 3 **(b)** 5 **(c)** −30 **65. (a)** 3 **(b)** 3 **(c)** 63 **67. (a)** $-3x + 6$ **(b)** $7x + 4$ **(c)** 0 **(d)** 11 **69. (a)** $3x^2 - 5x + 6$ **(b)** $-x^2 - 5x$ **(c)** 8 **(d)** −6 **71. (a)** $2x^3 + 6x^2 + 12x - 6$ **(b)** $6x^2 + 12x + 10$ **(c)** 58 **(d)** 28 **73.** $x^3 + x^2 - 7x + 4$ **75.** $-2b^3 + 6b^2 - 4b + 4$ **77. (a)** 4 square units **(b)** 4 square units **(c)** Since the rectangular region is in quadrant I and has one corner at the origin, the coordinates of the vertex give us the length and width of the rectangle. Since the area of a rectangle can be found by multiplying the length and width, we can just multiply the coordinates together to get the area of the region. **79. (a)** approximately $97,826 **(b)** approximately $252,790 **81. (a)** $P(x) = -0.05x^3 + 0.8x^2 + 155x - 500$ **(b)** $1836.25 **(c)** P(100) = −27,000 If 100 cell phones are sold, there would be a loss of $27,000. **83.** $a = 2$ **85. (a)** 3 **(b)** −2 **(c)** −1 **(d)** 1 **87.** $S(x) = T(x) - F(x)$

**89.**

```
Plot1 Plot2 Plot3
\Y1■4X²-7X+1
\Y2=
\Y3=
\Y4=
\Y5=
\Y6=
\Y7=
```

```
Y1(4)
            37
Y1(-2)
            31
Y1(6)
           103
```

(a)–(c)

$f(4) = 37$ $f(-2) = 31$ $f(6) = 103$

**91.** (a)–(c)

```
Plot1 Plot2 Plot3
\Y1■2X^3-5X²+X+5
\Y2=
\Y3=
\Y4=
\Y5=
\Y6=
```

```
Y1(4)
            57
Y1(-2)
           -33
Y1(6)
           263
```

$f(4) = 57$ $f(-2) = -33$ $f(6) = 263$

### Section 5.2

**1.** $A^2 - B^2$ **3.** $f(x) \cdot g(x)$ **5.** True **7.** Answers may vary. **9.** $-15x^3y^5$ **11.** $\frac{5}{3}y^4z^5$ **13.** $5x^3 + 20x^2 + 10x$ **15.** $-12a^4b - 8a^3b^2 + 4a^2b^3$ **17.** $\frac{1}{2}a^3b^2 - \frac{3}{4}a^2b^4 + 4a^2b^2$ **19.** $0.48x^4 - 0.32x^3 + 0.6x^2$ **21.** $x^2 + 8x + 15$ **23.** $a^2 + 2a - 15$ **25.** $12a^2 + 5a - 3$ **27.** $-6x^2 - 19x + 7$ **29.** $-10x^2 - 7x + 12$ **31.** $\frac{1}{3}x^2 - \frac{5}{3}x - 8$ **33.** $4a^2 - 17ab - 15b^2$ **35.** $x^3 + 5x^2 + 6x + 2$ **37.** $6a^3 - a^2 - 17a + 10$ **39.** $20z^3 + 27z^2 + 17z + 6$ **41.** $x^4 - 7x^3 + 14x^2 - 13x + 21$ **43.** $2y^3 + 13y^2 + 17y - 12$ **45.** $2w^4 + w^3 - 3w^2 - w + 1$ **47.** $3x^3 + x^2y + 2xy^2 + 4y^3$ **49.** $8a^3b - 4a^2b^2 + 2ab^3 + 20a^2 - 10ab + 5b^2$ **51.** $x^2 - 36$ **53.** $a^2 + 16a + 64$ **55.** $9y^2 - 6y + 1$ **57.** $25a^2 - 9b^2$ **59.** $64z^2 + 16yz + y^2$ **61.** $100x^2 - 20xy + y^2$ **63.** $a^6 - 4b^2$

**65.** $9x^2 - y^2 - 2y - 1$ **67.** $4a^2 + 4ab - 12a + b^2 - 6b + 9$ **69. (a)** $x^2 + 3x - 4$ **(b)** 14 **71. (a)** $8x^2 + 14x - 15$ **(b)** 99 **73. (a)** $x^3 + 3x^2 - 13x + 6$ **(b)** 21 **75. (a)** $x^2 + 4x + 5$ **(b)** $2xh + h^2$ **77. (a)** $x^2 + 9x + 12$ **(b)** $2xh + 5h + h^2$ **79. (a)** $3x^2 + 11x + 11$ **(b)** $6xh - h + 3h^2$ **81.** $5a^3b - 10a^2b^2 + 5ab^3$ **83.** $25y^2 - 1$ **85.** $24z^2 + z - 3$ **87.** $7m^2 - 6mn - 7n^2$ **89.** $x^3 + 27$ **91.** $4x^2 - 2x + \frac{1}{4}$ **93.** $p^3 + 6p^2 + 12p + 8$ **95.** $21x^2 - 29xy + 13x + 10y^2 - 9y + 2$ **97.** $3p^2 + 5p - 12$ **99.** $x^4 - x^3 - 19x^2 + 3x + 84$ **101.** $x^2 + 6x$ **103.** $4x^2 + 29x + 30$ **105.** $4x^2 + 11x + 6$ **107. (a)** $A_1 = a^2; A_2 = ab; A_3 = ab; A_4 = b^2$ **(b)** $a^2 + 2ab + b^2$ **(c)** The length of the region is $a + b$ and the width is also $a + b$. The area of the region would be $A = (a + b)(a + b) = (a + b)^2$. The result from part (b) is obtained by multiplying this expression out. **109.** $2^{2x} - 2^x - 12$ **111.** $5^{2y} - 2(5^y) + 1$

## Section 5.3

**1.** $0; (x + 1)\cdot(3x - 1)$ **3.** synthetic division **5.** True **7.** The dividend is the polynomial $f$ and has degree $n$. Since the remainder must be 0 or a polynomial that has a lower degree than $f$, the degree $n$ must be obtained from the product of the divisor and the quotient. The divisor is $x + 4$, which is of degree 1, so the quotient must be of degree $n - 1$. **9.** Yes; $(3x + 4)(2x^2 - 3x + 1)$ **11.** $2x + 3$ **13.** $\frac{2}{5}a^2 - 3a + 2$ **15.** $\frac{y}{2} + \frac{3}{2y}$ **17.** $1 + \frac{3}{2n} - \frac{9}{2m}$ **19.** $x + 3$ **21.** $2x + 5 + \frac{6}{x - 2}$ **23.** $w + 6 - \frac{7}{2w - 7}$ **25.** $x^2 + 5x - 14$ **27.** $w^2 - 4w - 5$ **29.** $3x^2 - 14x - 5$ **31.** $x - 7$ **33.** $z + 7 + \frac{4z - 5}{3z^2 + 1}$ **35.** $2x^2 - 15x + 28 + \frac{-3x + 4}{x^2 + 2x + 5}$ **37.** $x + 2$ **39.** $2x + 3$ **41.** $x + 3 + \frac{4}{x - 6}$ **43.** $x^2 + 5x + 6 + \frac{15}{x - 5}$ **45.** $3x^3 + 4x^2 - 9x - 10 - \frac{5}{x - 3}$ **47.** $x^3 - 6x^2 - 4x + 24 - \frac{35}{x + 6}$ **49.** $2x^2 + 8x + 6$ **51. (a)** $\left(\frac{f}{g}\right)(x) = x^2 - 2x + 3$ **(b)** 3 **53. (a)** $\left(\frac{f}{g}\right)(x) = x + 3$ **(b)** 5 **55. (a)** $\left(\frac{f}{g}\right)(x) = 2x - 1 + \frac{2}{x + 3}$ **(b)** $\frac{17}{5}$ **57. (a)** $\left(\frac{f}{g}\right)(x) = x^2 + 3x - 4$ **(b)** 6 **59. (a)** $\left(\frac{f}{g}\right)(x) = x - \frac{4x + 12}{x^2 - 9}$ **(b)** 6 **61.** $-5$ **63.** $-119$ **65.** 231 **67.** 2 **69.** $x - 2$ is a factor; $f(x) = (x - 2)(x - 1)$ **71.** $x + 2$ is a factor; $f(x) = (x + 2)(2x + 1)$ **73.** $x - 3$ is not a factor **75.** $x + 1$ is a factor; $f(x) = (x + 1)(4x^2 - 11x + 6)$ **77.** $a^2b - 3a + 6$ **79.** $y + 3$ **81.** $x + 6 + \frac{3x + 2}{x^2 + 5}$ **83.** $\frac{2}{3}x + \frac{1}{2x}$ **85.** $x^2 + 3x - 10 - \frac{6}{x + 4}$ **87.** $x^2 - 8$ **89.** $4x^2 + 6x + 9$ **91.** $(5x + 2)$ ft **93.** $(2x + 5)$ cm **95. (a)** $\overline{C}(x) = 0.01x^2 - 0.4x + 13 + \frac{400}{x}$ **(b)** \$26 **97.** $f(x) = 3x^2 - 10x - 25$ **99.** $f(x) = x^2 + 5x - 20$ **101.** $a = 2, b = -7, c = -12$, and $d = -13$, thus, $a + b + c + d = -30$

## Putting the Concepts Together

**1.** $5m^4 - 2m^3 + 3m + 8$ Degree: 4 **2.** $-4a^3 + 9a^2 + a - 8$ **3.** $-\frac{19}{5}y^2 + 3y - 8$ **4.** 29 **5.** $-25$ **6.** $x^2 + 4x + 10$ **7.** $2m^3n^4 - 8m^2n^4 + 12mn^3$ **8.** $9a^2 - 30ab + 25b^2$ **9.** $49n^4 - 9$ **10.** $18a^3 + 6a^2b - ab^2 + 2b^3$ **11.** $x^3 - 2x^2 + 3x + 22$ **12.** $5z^2 + 3z - 7$ **13.** $2x^2 + 7x - 1 + \frac{3}{x + 9}$ **14.** $x^2 + 3x - 1 + \frac{4}{x - 1}$ **15.** yes: $f(x) = (x + 5)(3x^2 - 7x + 12)$

## Section 5.4

**1.** factors **3.** greatest common factor **5.** False **7.** Answers may vary. **9.** $2x$ **11.** $5(a + 7)$ **13.** $-3(y - 7)$ **15.** $7x(2x - 3)$ **17.** $3z(z^2 - 2z + 6)$ **19.** $-5p^2(p^2 - 2p + 5)$ **21.** $7mn(7m^2 + 12n^2 - 5m^3n)$ **23.** $-2z(9z^2 - 7z - 2)$ **25.** $(3c - 2)(5c - 3)$ **27.** $2(a - 3)(3a - 2)$ **29.** $(x + y)(5 + a)$ **31.** $(z + 5)(2z^2 - 5)$ **33.** $(w - 5)(w + 3)$ **35.** $2(x - 4)(x - 2)$ **37.** $3x(x + 5)(x - 4)$ **39.** $(x - y)(2a - b)$ **41.** $5(w - 3)$ **43.** $(2y + 5)(y - 2)$ **45.** $3x^2y(2xy^2 - 7xy + 3)$ **47.** $(x + 1)(x^2 + 3)$ **49.** $2(x - 2)(2x - 3)$ **51.** $x^2(4 - \pi)$ square units **53.** $2\pi r(r + 4)$ square inches **55. (a)** $1.4x$ **(b)** $1.4x - 0.4(1.4x)$ **(c)** $0.84x$ **(d)** No **57. (a)** $x + 0.15x = 1.15x$ **(b)** $(1.15x) + 0.1(1.15x)$ **(c)** $1.265x$ **(d)** \$25.30 **59.** $\frac{1}{4}(x - 7)$ **61.** $\frac{1}{25}b(5b^2 + 8)$

## Section 5.5

**1.** $m \cdot n; m + n$ **3.** True **5.** The trial and error method would be better when $a \cdot c$ gets large and there are lots of factors of the product whose sums must be determined. **7.** The problem was not completely factored. The first binomial has a common factor of 3 than can be pulled out. **9.** $(x + 3)(x + 5)$ **11.** $(p + 6)(p - 3)$ **13.** $(r + 5)^2$ **15.** $(s + 12)(s - 5)$ **17.** $(x - 7)(x - 8)$ **19.** $-(w + 6)(w - 4)$ **21.** $(x + 3y)(x + 4y)$ **23.** $(p - 4q)(p + 6q)$ **25.** $2(x - 3)(x + 9)$ **27.** $-3(r - 5)(r - 8)$ **29.** $(2p + 1)(p - 8)$ **31.** $(y - 2)(4y - 3)$ **33.** $(2s - 1)(4s + 3)$ **35.** $(4z - 3)(4z + 5)$ **37.** $(2y + 1)(9y + 4)$ **39.** $2(2m - 5)(4m + 7)$ **41.** $4(4z + 1)(3z + 7)$ **43.** $(x + 7y)(2x - 3y)$ **45.** $(r - 5s)(4r - 3s)$ **47.** $(3r + 4s)(8r - 3s)$ **49.** $3x(x - 10)(x + 8)$ **51.** $4xy^2(x - 7)(2x - 5)$ **53.** $2r^2s(5r - 4)(7r + 2)$ **55.** $(x - 1)(x + 1)(x^2 - 2)$ **57.** $(mn + 7)(mn - 2)$ **59.** $(x - 7)(x + 3)$ **61.** $3(r - 2)(3r - 5)$ **63.** $(y + 2)(2y - 3)$ **65.** $(5w + 3)(2w + 7)$ **67.** $2y(8y + 5)$ **69.** $(3x + 2y)(4x + 5y)$ **71.** prime **73.** $(x^2 + 4)(x^2 + 2)$ **75.** $(z^3 + 5)(z^3 + 4)$ **77.** $(r - 4s)(r - 8s)$ **79.** $(2z + 3)(4z + 3)$ **81.** prime

**83.** **(a)** $V(3) = 1296$; when 3-inch square corners are cut from the piece of cardboard, the resulting box will have a volume of 1296 cubic inches. **(b)** $V(x) = 4x(x - 15)(x - 12)$ **(c)** $V(3) = 4(3)(3 - 15)(3 - 12) = 1296$ **(d)** Answers may vary. **85.** $2x - 5$ **87.** $\frac{1}{2}(x + 4)(x + 2)$ **89.** $\frac{1}{3}(p - 3)(p + 1)$ **91.** $\frac{4}{3}(a - 6)(a + 4)$

## Section 5.6

**1.** perfect square trinomial **3.** False **5.** Answers may vary. **7.** $(x + 2)^2$ **9.** $(w + 6)^2$ **11.** $(2x + 1)^2$ **13.** $(3p - 5)^2$ **15.** $(5a + 9)^2$ **17.** $(3x + 4y)^2$ **19.** $3(w - 5)^2$ **21.** $-5(t + 7)^2$ **23.** $2(4a - 5b)^2$ **25.** $(z^2 - 3)^2$ **27.** $(x - 3)(x + 3)$ **29.** $(2 - y)(2 + y)$ **31.** $(2z - 3)(2z + 3)$ **33.** $(10m - 9n)(10m + 9n)$ **35.** $(m^2 - 6n)(m^2 + 6n)$ **37.** $2(2p - 3q)(2p + 3q)$ **39.** $5r(4p - 7b)(4p + 7b)$ **41.** $(x + y - 3)(x + y + 3)$ **43.** $(x - 2)(x^2 + 2x + 4)$ **45.** $(m + 5)(m^2 - 5m + 25)$ **47.** $(x^2 - 4y)(x^4 + 4x^2y + 16y^2)$ **49.** $3(2x - 5y)(4x^2 + 10xy + 25y^2)$ **51.** $(p - 2)(p^2 + 5p + 13)$ **53.** $(5y + 1)(7y^2 + 4y + 1)$ **55.** $18(x^2 + 3)$ **57.** $(y^2 + z^3)(y^4 - y^2z^3 + z^6)$ **59.** $(y - 1)(y^2 + y + 1)(y^6 + y^3 + 1)$ **61.** $(5x - y)(5x + y)$ **63.** $(2x + 3)(4x^2 - 6x + 9)$ **65.** $(z - 4)^2$ **67.** $5x(x - 2y)(x^2 + 2xy + 4y^2)$ **69.** $(7m - 3n)^2$ **71.** $(y - 2)^2(y + 2)^2$ **73.** $(2ab + 3)^2$ **75.** $(x - y - 2)(x + y - 2)$ **77.** $2(n - m + 10)(n + m - 10)$ **79.** $3(y - 2)(y^2 + 2y + 4)$ **81.** $(4x + 3y - 10)(4x + 3y + 10)$ **83.** $(x - 3)(x + 3)$ square units **85.** $(x - 4)(x + 4)$ square units **87.** $\pi(R - r)(R + r)$ square units **89.** $10a(a - b)(a + b)$ cubic units **91.** $A = a \cdot a + a \cdot b + a \cdot b + b \cdot b = a^2 + ab + ab + b^2 = a^2 + 2ab + b^2 = (a + b)^2$ **93.** $b = \pm 36$; the middle term must equal $\pm$ twice the product of the quantities that are squared to get the first and last terms. **95.** $b = 9$, we need to add $9^2 = 81$ to make a perfect square trinomial. **97.** $(b - 0.2)^2$ **99.** $\left(3b - \frac{1}{5}\right)\left(3b + \frac{1}{5}\right)$ **101.** $\left(\frac{x}{3} - \frac{y}{5}\right)\left(\frac{x}{3} + \frac{y}{5}\right)$ **103.** $\left(\frac{x}{2} - \frac{y}{3}\right)\left(\frac{x^2}{4} + \frac{xy}{6} + \frac{y^2}{9}\right)$

## Section 5.7

**1.** Answers may vary. **3.** $2(x - 12)(x + 6)$ **5.** $-3(y - 3)(y + 3)$ **7.** $(2b + 5)^2$ **9.** $2(2w + y^2)(4w^2 - 2wy^2 + y^4)$ **11.** $-3(z^2 - 4z + 6)$ **13.** $(4y + 3)(5y - 6)$ **15.** $(x - 4)(x^2 + 5)$ **17.** $2(100x^2 + 9y^2)$ **19.** $(x - 3)(x + 3)(x^2 + 9)$ **21.** prime **23.** $4q(3q + 1)^2$ **25.** $3mn(4m + 3)(2m - 7)$ **27.** $3r^2(r - 2s)(r^2 + 2rs + 4s^2)$ **29.** $2(x + 4)(x - 3)(x + 3)$ **31.** $(3x^2 - 1)(3x^2 + 1)$ **33.** $(w^2 + 3)(3w^2 - 5)$ **35.** $2y(2y + 1)$ **37.** $(p - 6q - 5)(p + 6q - 5)$ **39.** $(y + 2)(y^2 - 2y + 4)(y^3 - 2)$ **41.** $(p - 1)(p + 1)(p^2 + p + 1)(p^2 - p + 1)$ **43.** $-3(x + 5)(x - 3)(x + 3)$ **45.** $3a(1 - 3a)(1 + 3a)$ **47.** $2t(t^2 + 4)(2t - 3)(2t + 3)$ **49.** $2xy(x + 5)(x - 3)(x + 3)$ **51.** $(x + 3)(3x + 7)$ square units **53.** $(x + y)(x - y)$ square units **55.** $(3x - 4)(9x^2 + 12x + 16)$ cubic units **57.** Answers may vary. The sum of two squares does not factor over the integers (or the real numbers). **59.** $x^{\frac{1}{2}}(x - 3)(x + 3)$ **61.** $x^{-2}(x + 4)(x + 2)$

## Section 5.8

**1.** polynomial equation **3.** third **5.** False **7.** The degree of a polynomial equation is the degree of the polynomial expression in the equation. **9.** $\{-1, 3\}$ **11.** $\left\{-\frac{4}{3}, 0\right\}$ **13.** $\{-3, 0, 5\}$ **15.** $\{0, 4\}$ **17.** $\{0, 8\}$ **19.** $\{-5, 3\}$ **21.** $\{4, 9\}$ **23.** $\{3\}$ **25.** $\left\{-\frac{3}{5}, 1\right\}$ **27.** $\left\{\frac{5}{6}, 3\right\}$ **29.** $\{-8, 5\}$ **31.** $\left\{-6, \frac{5}{2}\right\}$ **33.** $\{-6, 2\}$ **35.** $\left\{-\frac{7}{2}, 2\right\}$ **37.** $\{-11, 3\}$ **39.** $\{-5, -1, 2\}$ **41.** $\left\{-\frac{1}{2}, 0, 3\right\}$ **43.** $\left\{\frac{-5}{2}, -2, 2\right\}$ **45.** $\left\{-\frac{3}{5}, -\frac{2}{3}, 0\right\}$ **47.** $\left\{\frac{4}{3}, 1\right\}$ **49.** **(a)** $x = -5$ or $x = -2$ **(b)** $x = -8$ or $x = 1$ $(-5, 2), (-2, 2), (-8, 20)$, and $(1, 20)$ are on the graph of $f$. **51.** **(a)** $x = 4$ or $x = -1$ **(b)** $x = 5$ or $x = -2$ $(-1, 3), (4, 3), (-2, 15)$, and $(5, 15)$ are on the graph of $g$. **53.** **(a)** $x = 0$ or $x = 4$ **(b)** $x = 5$ or $x = -1$ $(0, 5), (4, 5), (-1, -10)$, and $(5, -10)$ are on the graph of $F$. **55.** zeros: $-7$ and $-2$; $x$-intercepts: $-7$ and $2$ **57.** zeros: $-\frac{1}{3}$ and $\frac{9}{2}$; $x$-intercepts: $-\frac{1}{3}, \frac{9}{2}$ **59.** zeros: $-5, 0$, and $4$; $x$-intercepts: $-5, 0$, and $4$ **61.** $x = 6$ or $x = -4$ **63.** $q = \frac{-7}{2}$ or $q = 2$ **65.** $b = 0$ or $b = 7$ **67.** $x = -2$ or $x = -3$ **69.** $x = -5$ or $x = 2$ or $x = -2$ **71.** The width is 16 cm and the length is 8 cm. **73.** The base is 10 ft and the height is 22 ft. **75.** 8 sides **77.** 40 meters by 20 meters (along the river) or 10 meters by 80 meters (along the river). **79.** 3 ft **81.** The width is 11 in. and the length is 16 in. **83.** **(a)** \$300 **(b)** 20 bicycles **(c)** 15 bicycles **85.** **(a)** after 2.5 seconds (on the way up), and again after 5 seconds (on the way down) **(b)** after 7.5 seconds **87.** $x = \frac{1}{2}$ or $x = -\frac{1}{2}$ or $x = 2$ or $x = -2$ **89.** $a = 0$ or $a = -1$ **91.** all real numbers except $\pm 2$, or $\{x \mid x \neq \pm 2\}$ **93.** all real numbers except $\frac{1}{2}$ and 1, or $\left\{x \mid x \neq \frac{1}{2}, 1\right\}$ **95.** $\{-0.61, 4.11\}$ **97.** $\{0.60, 24.90\}$ **99.** $\{-1.15, 1.75\}$

## Chapter 5 Review

**1.** Coefficient: $-7$; Degree: 4 **2.** Coefficient: $\frac{1}{9}$; Degree: 3 **3.** $7x^3 - 2x^2 + x - 8$; Degree: 3 **4.** $y^4 - 3y^2 + 2y + 3$; Degree: 4 **5.** $4x^2 + x - 11$ **6.** $-x^4 + 4x^3 - 5x^2 + 8x - 6$ **7.** $\frac{1}{4}x^2 - \frac{9}{2}x + \frac{1}{6}$ **8.** $\frac{5}{6}x^2 - x + \frac{13}{20}$ **9.** $10x^2y^2$ **10.** $-a^2b - 6ab^2 - 4$ **11.** **(a)** $-24$ **(b)** $-8$ **(c)** $-29$ **12.** **(a)** $-82$ **(b)** $-1$ **(c)** $-7$ 13. **(a)** $x^2 + 7x - 1$ **(b)** 29 **14.** **(a)** $2x^3 - 2x^2 + x - 12$ **(b)** $-2$ **15.** **(a)** $P(x) = -2.5x^2 + 280x - 3290$ **(b)** \$2147.50 **16.** **(a)** 6 square units **(b)** 4 square units **17.** $-12x^4y^3$ **18.** $6m^4n^7$ **19.** $-10a^3b^2 + 5a^2b^3 - 15a^2b^2$ **20.** $0.85c^3 + 2.15c^2 + 4.45c$ **21.** $x^2 - 7x - 18$ **22.** $-6x^2 + 26x - 8$ **23.** $2m^2 - 7mn - 4n^2$ **24.** $-2a^2 - 9a + 45$ **25.** $3x^3 + x^2 - 9x + 2$ **26.** $w^3 - 3w^2 - 12w + 32$ **27.** $2m^4 + m^3 - 11m^2 + 29m - 21$

**28.** $2p^3 + 11p^2q - 29pq^2 + 12q^3$ **29.** $9w^2 - 1$ **30.** $4x^2 - 25y^2$ **31.** $36k^2 - 60k + 25$ **32.** $9a^2 + 12ab + 4b^2$ **33.** $x^3 + 8$ **34.** $8x^3 - 27$ **35. (a)** $18x^2 - 27x - 35$ **(b)** 91 **36. (a)** $3x^3 + 5x^2 - x + 2$ **(b)** 270 **37.** $5x^2 - 30x + 53$ **38.** $-2xh + 3h - h^2$ **39.** $4x^2 - 2x$ **40.** $3w^4 - w^2 + 5w + 2$ **41.** $\frac{7}{2}y^2 + 6y - 3$ **42.** $\frac{m}{2n} + \frac{2}{n} - \frac{7}{2m}$ **43.** $3x + 4$ **44.** $-2x + 7 + \frac{5}{x+5}$ **45.** $3z^2 + 2 - \frac{12}{2z+3}$ **46.** $4k^2 + k - 2$ **47.** $8x^3 + 12x^2 + 18x + 27$ **48.** $2x^2 - 5x + 12 + \frac{2x+7}{x^2 - 3x + 4}$ **49.** $5x + 1 + \frac{6}{x+2}$ **50.** $9a + 4$ **51.** $3m^2 + 2m - 11$ **52.** $n^2 + 6n - 15 + \frac{7}{n-4}$ **53.** $x^3 - x^2 + 7x - 7$ **54.** $2x^2 - 4x + 13 - \frac{34}{x+2}$ **55. (a)** $x^2 + 5x - 3$ **(b)** 11 **56. (a)** $3x + 20 + \frac{9}{3x-2}$ **(b)** $\frac{112}{11}$ **57. (a)** $2x^2 + 4x - 7$ **(b)** $-7$ **58. (a)** $3x^2 - 11x + 5 + \frac{2x-3}{x^2 - x + 5}$ **(b)** $\frac{158}{17}$ **59.** 59 **60.** $-45$ **61.** $x - 2$ is a factor; $f(x) = (x-2)(3x+7)$ **62.** $x + 4$ is not a factor **63.** $(5x + 1)$ m **64.** $(2x^2 + 5x + 3)$ square centimeters **65.** $4(z + 6)$ **66.** $-7y(y - 13)$ **67.** $2xy(7x^2y - 4x + y)$ **68.** $5a^2b(6a^2b^2 + 3a - 5b)$ **69.** $(x + 5)(3x - 4)$ **70.** $(2c + 9)(3 - 4c)$ **71.** $(x - 5y)(6x + 5)$ **72.** $(a + 7)(2a - b - 1)$ **73.** $(x + 6)(x - 3)$ **74.** $(c + 2)(c - 5)$ **75.** $(7z + 8)(2z - 3)$ **76.** $(3w - 4)(7w + 2)$ **77.** $2x(x + 1)(x - 9)$ **78.** $5a^2(2a + 3)(a + 7)$ **79. (a)** $\frac{1}{2}n(n + 1)$ **(b)** 528 **80.** $R(x) = 2x(2600 - x^2)$ **81.** $(w + 2)(w - 13)$ **82.** prime **83.** $-1(t - 12)(t + 6)$ **84.** $(m + 3)(m + 7)$ **85.** $(x + 20y)(x - 16y)$ **86.** $(r - 2s)(r - 3s)$ **87.** $(x + 3)(5x - 2)$ **88.** $(2m + 11)(3m + 4)$ **89.** prime **90.** $2(t + 3)(4t - 1)$ **91.** $(2x - 1)(3x - 5)$ **92.** $(7r + 2s)(3r - s)$ **93.** $(5x - 3y)(4x - 9y)$ **94.** $-2(s - 7)(s + 1)$ **95.** $(x^2 + 1)(x^2 - 11)$ **96.** $(xy + 4)(10xy + 1)$ **97.** $(a + 7)(a - 8)$ **98.** $w(2w + 7)$ **99.** $(x + 11)^2$ **100.** $(w - 17)^2$ **101.** $(12 - c)^2$ **102.** $(x - 4)^2$ **103.** $(8y + 5)^2$ **104.** $12(z + 2)^2$ **105.** $(x - 14)(x + 14)$ **106.** $(7 - y)(7 + y)$ **107.** $(t - 15)(t + 15)$ **108.** $(2w - 9)(2w + 9)$ **109.** $(6x^2 - 5y)(6x^2 + 5y)$ **110.** $20m(2n - 1)(2n + 1)$ **111.** $(x - 7)(x^2 + 7x + 49)$ **112.** $(9 - y)(y^2 + 9y + 81)$ **113.** $(3x - 5y)(9x^2 + 15xy + 25y^2)$ **114.** $(2m^2 + 3n)(4m^4 - 6m^2n + 9n^2)$ **115.** $2(a - b)(a + b)(a^2 + ab + b^2)(a^2 - ab + b^2)$ **116.** $(y + 3)(y^2 - 6y + 21)$ **117.** $(x + 1)(x + 6)$ **118.** $(c - 12)^2$ **119.** $(z + 7)(z - 16)$ **120.** $-4xy^3(2x - 3)$ **121.** $7x(x^2 - 4x + 9)$ **122.** $3(x - 3)(x + 2)$ **123.** $(2z - 15)^2$ **124.** $(3x + 7)(4x - 7)$ **125.** $-3(x - 5)(x + 3)$ **126.** $(2n - 7)(5n + 1)$ **127.** $-(y - 2)(y + 4)$ **128.** $2(x - 5)(x^2 + 3)$ **129.** $(w - 4z - 2)(w - 2z + 2)$ **130.** $9(h + 2)(3h^2 + 4)$ **131.** $5p(pq - 4)(pq + 4)$ **132.** $(4a - 3)(9a - 5)$ **133.** $(m - 1)(m + 1)(m - 2)(m + 2)$ **134.** $-2(2m^2 - 7)(4m^4 + 14m^2 + 49)$ **135.** $(h - 1)(h + 1)(h + 2)$ **136.** $4(3x + y)(9x^2 - 3xy + y^2)$ **137.** $(x - 5)(x + 5)$ square units **138.** $(4 - \pi)x^2$ square units **139.** $\{-5, 13\}$ **140.** $\left\{0, -\frac{1}{2}, \frac{5}{3}\right\}$ **141.** $\{0, -4\}$ **142.** $\{-5, 3\}$ **143.** $\{-18, -3\}$ **144.** $\left\{\frac{2}{5}, -\frac{7}{3}\right\}$ **145.** $\{-11, 10\}$ **146.** $\{-6, -4\}$ **147.** $\{4, -2\}$ **148.** $\{-5, -2\}$ **149. (a)** $x = -8$ or $x = 3$ **(b)** $x = -7$ or $x = 2$ $(-8, 6)$, $(3, 6)$, $(-7, -4)$, and $(2, -4)$ are on the graph. **150. (a)** $x = 0$ or $x = \frac{4}{5}$ **(b)** $x = -\frac{1}{5}$ or $x = 1$ $(0, 3)$, $\left(\frac{4}{5}, 3\right)$, $\left(-\frac{1}{5}, 4\right)$, and $(1, 4)$ are on the graph. **151.** zeros: $-4$, $-2$, and 0; $x$-intercepts: $-4$, $-2$, and 0 **152.** zeros: $-\frac{3}{2}$ and 7; $x$-intercepts: $-\frac{3}{2}$ and 7 **153.** after 7 seconds **154. (a)** 0.80 **(b)** 0.90

## Chapter 5 Test

**1.** $-5x^7 + x^4 + 7x^2 - x + 1$; Degree: 7 **2.** $-\frac{5}{3}a^3b^2 + 9a^2b - 5ab - 4$ **3.** 7 **4.** $7x^3 - 4x^2 - 3x + 1$ **5.** $2a^3b^3 - 3a^3b^2 + 4a^2b$ **6.** $12x^2 + 47x - 17$ **7.** $4m^2 - 4mn + n^2$ **8.** $3z - 7 + \frac{-2z + 11}{2z^2 + 1}$ **9.** $5x + 3$ **10.** 2 **11.** 22 **12.** $4ab^2(3a^2 + 2a - 4b)$ **13.** $(2c + 7)(3c - 2)$ **14.** $(x - 16)(x + 3)$ **15.** $(2p + 3)(2 - 7p)$ **16.** $(z + 3)(5z - 8)$ **17.** $-2(7x - 4)^2$ **18.** $4((2x - 7)(2x + 7))$ **19.** $\left\{\frac{7}{3}, 1\right\}$ **20.** 9 meters by 12 meters

## Cumulative Review Chapters R–5

**1.** 5 **2.** 14 **3.** $6x + 14$ **4.** $\{-4\}$ **5.** $\{4, 6\}$ **6.** $y = \frac{4}{5}x - 6$ **7.** $\{x \mid x \geq 2\}$ or $[2, \infty)$

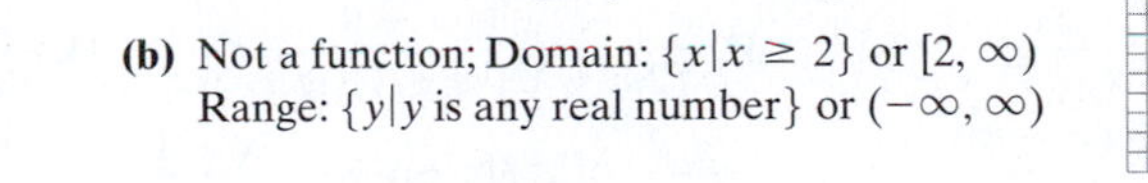

**8.** 4 liters **9.** $(-1, 0)$, $(2, 0)$, and $(0, -2)$ **10.**

Domain: $\{x \mid x$ is any real number$\}$ or $(-\infty, \infty)$
Range: $\{y \mid y \geq -3\}$ or $[-3, \infty)$

**11. (a)** Function; Domain: $\{-2, 3, 5, 7, 10\}$; Range: $\{-3, 1, 4, 13\}$ **(b)** Not a function; Domain: $\{x \mid x \geq 2\}$ or $[2, \infty)$ Range: $\{y \mid y$ is any real number$\}$ or $(-\infty, \infty)$ **12.**

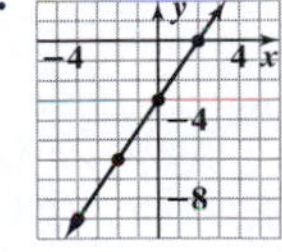

**13.** $y = \frac{2}{5}x - 8$ or $2x - 5y = 40$ **14. (a)** For roughly the first four months shown, the average price per gallon steadily decreased. Beginning in mid-December, the price began to steadily increase for the last two months. **(b)** around December 14 **(c)** roughly $1.49 **15.**

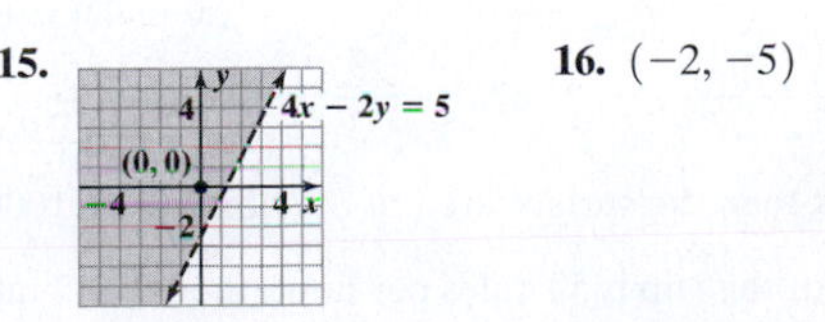


**16.** $(-2, -5)$

**17.** $(2, -5)$ **18.** \$140,000 per year **19.** Cheese pizzas cost \$6, sausage pizzas cost \$7.50, and pepperoni pizzas cost \$8.
**20. (a)** cost: $C(x) = 0.22x + 50$; revenue: $R(x) = 0.75x$ **(b)** at least 95 candy bars **(c)** at least 2729 candy bars **21.** $10x^3 + 5x^2 + x - 7$
**22.** $2x^3 - 5x^2 + 7x + 5$ **23.** $3x^2 - 5x + 2 - \frac{9}{x + 5}$ **24. (a)** $\frac{1}{3}(x + 3)^2$ **(b)** $(w - 12)(w + 5)$ **(c)** $32a^2(a - 2)(a + 2)$
**25.** $-\frac{5}{2}$ and 3

## Getting Ready for Chapter 6

**1.** lowest terms **3.** Answers may vary. **5.** Answers may vary. **7.** $\frac{1}{3}$ **9.** $-\frac{3}{7}$ **11.** 5 **13.** $\frac{5}{3}$ **15.** $-3$ **17.** $\frac{1}{12}$ **19.** $\frac{2}{5}$
**21.** $-\frac{9}{4}$ **23.** 6 **25.** 2 **27.** $\frac{5}{3}$ **29.** $\frac{19}{20}$ **31.** $\frac{13}{12}$ **33.** $-\frac{23}{30}$ **35.** $-\frac{41}{90}$ **37.** $-\frac{23}{40}$ **39.** $\frac{17}{120}$ **41.** $-\frac{41}{45}$ **43.** $-\frac{59}{150}$

# Chapter 6

### Section 6.1

**1.** rational expression **3.** True **5.** Answers will vary. **7.** Answers will vary. **9.** $\{x|x \neq -5\}$ **11.** $\{x|x \neq -2, x \neq 8\}$
**13.** $\left\{p\middle|p \neq -\frac{5}{2}, p \neq 2\right\}$ **15.** $\{x|x$ is any real number$\}$ or $(-\infty, \infty)$ **17.** $\{x|x \neq 1\}$ **19.** $\frac{2}{x - 4}$ **21.** $p + 3$ **23.** $\frac{5}{x^2}$ **25.** $\frac{q + 3}{q - 2}$
**27.** $\frac{y - 4}{y + 5}$ **29.** $-\frac{x + 3}{x + 5}$ **31.** $\frac{x - 2}{2x(x - 4)}$ **33.** $\frac{x - 3y}{x - 2y}$ **35.** $\frac{x^2 + 3}{x - 5}$ **37.** $\frac{x^2 - 2x + 4}{x - 7}$ **39.** $\frac{1}{4x(x + 3)}$ **41.** $\frac{x - 2}{x - 1}$
**43.** $\frac{(x + 3)^2}{(x + 5)(x + 7)}$ **45.** $-\frac{(2q + 1)(q + 2)}{q + 6}$ **47.** $x - 3$ **49.** 1 **51.** $\frac{9(x + 3)}{8x(x - 4)}$ **53.** $\frac{2}{ab}$ **55.** $\frac{(p - 5)(p + 3)}{p(2p + 1)}$ **57.** $\frac{x^2 + 3x + 1}{3(x + 1)}$
**59.** $\{x|x \neq 1\}$ **61.** $\left\{x\middle|x \neq -\frac{1}{2}, x \neq 4\right\}$ **63.** $\{x|x \neq -5, x \neq -1\}$ **65.** $\left\{x\middle|x \neq 2, x \neq \frac{5}{2}\right\}$ **67.** $\{x|x$ is any real number$\}$
**69. (a)** $\frac{(x + 3)(x - 1)}{2x + 3}$; $\left\{x\middle|x \neq -6, x \neq -\frac{3}{2}, x \neq 5\right\}$ **(b)** $(x - 5)(3x - 1)$; $\left\{x\middle|x \neq -6, x \neq -3, x \neq \frac{1}{3}\right\}$
**71. (a)** $\frac{(3x + 5)(x + 6)}{(x^2 + x + 1)(2x + 7)}$; $\left\{x\middle|x \neq -\frac{7}{2}, x \neq 1, x \neq 2\right\}$ **(b)** $\frac{x - 2}{x^2 + x + 1}$; $\left\{x\middle|x \neq -\frac{5}{3}, x \neq -1, x \neq 1\right\}$
**73.** $\frac{z + 3}{2}$ **75.** $\frac{m - 2n}{2(m - n)}$ **77.** $\frac{4}{w}$ **79.** $\frac{2x}{(x + 1)(x - 1)}$ **81.** Answers will vary. One possible rational expression is $\frac{x}{x - 3}$.
**83.** Answers will vary. One possible rational expression $\frac{7}{(x + 4)(x - 5)} = \frac{7}{x^2 - x - 20}$. **85.** $R(x) = \frac{4}{x^2 + x - 2}$
**87. (a)** approximately 9.8208 m/sec² **(b)** approximately 9.8159 m/sec² **(c)** approximately 9.7936 m/sec² **89. (a)** $\{x|x \neq 2\}$

**(b)**

| $x$ | 3 | 2.5 | 2.1 | 2.01 | 2.001 | 2.0001 |
|---|---|---|---|---|---|---|
| $f(x)$ | 1 | 2 | 10 | 100 | 1000 | 10,000 |

$f$ gets larger in the positive direction

**(c)**

| $x$ | 1 | 1.5 | 1.9 | 1.99 | 1.999 | 1.9999 |
|---|---|---|---|---|---|---|
| $f(x)$ | −1 | −2 | −10 | −100 | −1000 | −10,000 |

$f$ gets larger in the negative direction

**(d)** $f$ gets larger in the positive direction; $f$ gets larger in the negative direction; these results are the same as those in parts (b) and (c).

**91. (a)** $\frac{x + 5}{x + 6}$ **(b)** $-5$ **93.** $\frac{1}{x(x + 1)}$ **95.**

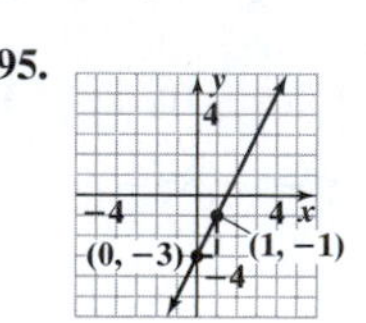


**97.**

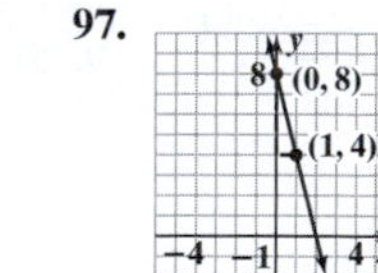


**99.**

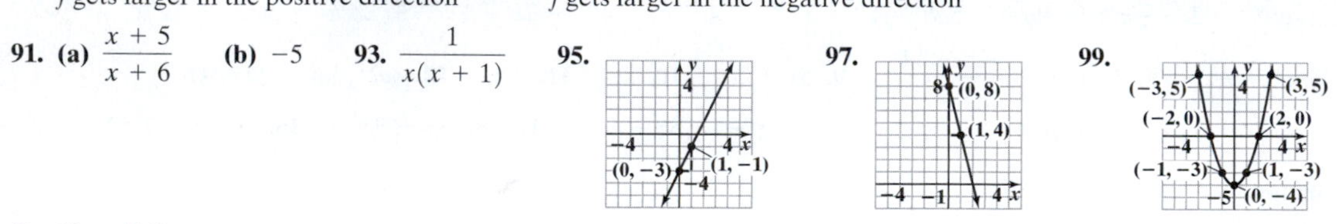


**101.** Answers will vary.

### Section 6.2

**1.** least common denominator **3.** True **5.** Answers may vary. **7.** $\frac{3x + 5}{x + 1}$ **9.** $\frac{2x - 1}{2x + 5}$ **11.** $\frac{1}{x - 5}$ **13.** $\frac{x - 3}{x + 3}$ **15.** $\frac{3x - 1}{x - 5}$
**17.** $\frac{3x^2 - 7x + 3}{x - 3}$ **19.** LCD $= 8x^3$ **21.** LCD $= 90x^3y^2$ **23.** LCD $= (x - 4)(x + 2)$ **25.** LCD $= (x - 4)(x + 3)(x - 5)$
**27.** LCD $= p^2(p + 2)(2p - 1)$ **29.** $\frac{5x + 6}{8x^2}$ **31.** $\frac{25b - 16a}{60a^2b^2}$ **33.** $\frac{14(y - 1)}{(y - 5)(y + 3)}$ **35.** $\frac{a + 4}{a + 2}$ **37.** $\frac{-2(x - 11)}{(x + 3)(x - 2)(x + 4)}$
**39.** $\frac{2x^2 - 5x + 5}{(x + 2)(x + 1)(x - 2)}$ **41.** $\frac{-11w + 9}{(2w + 1)(w + 1)(w - 3)}$ **43.** $\frac{2x^2 + xy - 5y^2}{(x - 3y)^2(x + y)}$ **45.** $\frac{1}{(x + 1)(x - 1)}$ **47.** 0 **49.** $\frac{6}{w(w - 2)}$
**51.** $\frac{(2p + 1)(p - 8)}{(p + 3)^2(2p - 5)}$ **53.** $\frac{x + 2}{x(x - 1)^2}$ **55. (a)** $\frac{5x - 1}{(x - 2)(x + 1)}$ **(b)** $\{x|x \neq -1, x \neq 2\}$ **57. (a)** $\frac{2x + 7}{(x - 4)(x + 3)}$
**(b)** $\{x|x \neq -3, x \neq -1, x \neq 4\}$ **(c)** $\frac{-1}{(x - 4)(x + 3)}$ **(d)** $\{x|x \neq -3, x \neq -1, x \neq 4\}$ **59.** $\frac{3}{x^2 + 3x + 9}$ **61.** $x - 1$ **63.** $\frac{7x + 15}{x + 3}$
**65.** $\frac{-1}{(b - 2)(b + 4)}$ **67.** $\frac{y - 3}{y + 3}$ **69.** $\frac{7x + 9}{(x - 2)(x - 3)(x + 1)}$ **71. (a)** $S(x) = \frac{2x^3 + 8000}{x}$ **(b)** $S(10) = 1000$; if the length of the base is 10 inches, then the surface area of the box will be 1000 square inches. **73. (a)** $\frac{200s + 500}{s(s + 10)}$ **(b)** $T(50) = 3.5$; if the average speed for the first 50 miles of the trip is 50 miles per hour, then it will take a total time of 3.5 hours to get to the neighboring university.
**75.** $\frac{x + y}{xy}$ **77.** $4a^2 - 12a$ **79.** $p^2 - 9$ **81.** $w^3 - 8$

## Section 6.3

**1.** complex rational expression **3.** Answers will vary. **5.** $\frac{x+1}{x-1}$ **7.** $\frac{(w-1)(w+1)}{w^2+1}$ **9.** $\frac{-2a}{(a-2)^2}$ **11.** $\frac{1}{(x-1)(x+3)}$ **13.** $\frac{x+2}{(x-2)(x-7)}$ **15.** $\frac{1}{y+3}$ **17.** $2(n-m)$ **19.** $\frac{x+3}{x(x-3)}$ **21.** $-x$ **23.** $\frac{1}{(x-1)(x-3)}$ **25.** $\frac{2(x+1)}{x+4}$ **27.** $\frac{z^2+4}{4z}$ **29.** $-b$ **31.** $\frac{3xy}{y-x}$ **33.** $\frac{mn}{(m+n)^2}$ **35.** $\frac{1}{4(b+a)}$ **37. (a)** $\frac{R_1R_2}{R_1+R_2}$ **(b)** approximately 2.857 ohms **39. (a)** $\frac{Pi\cdot(1+i)^5}{(1+i)^5-1}$ **(b)** approximately \$230.97 **41. (a)** 49.92 miles per hour **(b)** approximately 4.1145 Kb/s **43.** $x = 3$ **45.** $w = 4$ **47.** $y = 10$ or $y = -5$

### Putting the Concepts Together

**1.** $\left\{x \middle| x \neq -\frac{1}{3}, x \neq 6\right\}$ **2.** $\frac{-4n}{2n+3}$ **3.** $\frac{2p-5q}{3p-4q}$ **4.** $\frac{a(a-5)}{2(a+6)}$ **5.** $\frac{(x+2)(x-7)}{(3x+1)^2}$ **6.** $\frac{x-5}{x-2}$ **7.** $\frac{n}{(n-2)(n-3)}$ **8.** $\frac{3y^2-3y-29}{(y-3)(y-4)(y+8)}$ **9.** $\frac{3}{(x-7)(2x+1)}$ **10.** $\frac{3x^2-14x-1}{(x-4)(x-7)}$ **11.** $\frac{n+m}{mn}$ **12.** $\frac{z+3}{z-5}$

## Section 6.4

**1.** Extraneous solutions **3.** True **5.** Answers will vary. **7.** $\{-4\}$ **9.** $\left\{\frac{2}{3}\right\}$ **11.** $\{-16\}$ **13.** no solution; $\emptyset$ or $\{\ \}$ **15.** $\{-1, 5\}$ **17.** $\left\{-\frac{1}{3}, \frac{3}{2}\right\}$ **19.** $\{-3, -2\}$ **21.** $\{1\}$ **23.** $\{4\}$ **25.** $\{-5, 6\}$ **27.** $\{4\}$ **29.** $\left\{-\frac{5}{3}\right\}$ **31.** $\{-8\}$ **33.** $\left\{\frac{1}{4}\right\}$ **35.** $\{-1\}$ **37.** $x = 9$ or $x = 1$; $(1, 10)$ and $(9, 10)$ **39.** $x = -\frac{1}{2}$ or $x = -4$; $\left(-\frac{1}{2}, -9\right)$ and $(-4, -9)$ **41.** $x = 6$; $\left(6, \frac{9}{2}\right)$ **43.** $x = -5$ or $x = 3$; $(-5, 3)$ and $\left(3, \frac{1}{3}\right)$ **45.** $z = \frac{1}{4}$ **47.** $x = 3$ **49.** $z = -19$ **51.** $x = 2$ **53.** $y = 0$ or $y = -\frac{3}{5}$ **55.** either 50 or 100 bicycles **57. (a)** 80% **(b)** 90% **59.** 650 walks **61.** Answers will vary. One possibility follows: $\frac{2}{x-1} = \frac{3}{x+1}$ **63.** $x = -\frac{5}{3}$ or $x = -1$ **65.** $a = -\frac{1}{2}$ or $a = 3$ **67.** $\frac{-2x^2+6x+5}{(x-2)(x+1)}$ **69.** $x = 10$ or $x = -1$ **71.** $\{12\}$ **73.** $\{-13\}$ **75.** The calculator indicates that the solution is $-4$, but this cannot be a solution since it is not in the domain of the left side of the equation. Thus, there is no solution.

## Section 6.5

**1.** rational **3.** The statement $(-1, 4)$ indicates that neither endpoint is included in the solution set. The statement $\{x|-1 \le x \le 4\}$ indicates that both endpoints are included in the solution set. In fact, the solution of the inequality $\frac{x-4}{x+1} \le 0$ should include the endpoint 4, but it should not include the endpoint $-1$. Therefore, the statement of the solution set should be $\{x|-1 < x \le 4\}$ using set-builder notation or $(-1, 4]$ using interval notation.

**5.** $\{x|x < -1 \text{ or } x > 4\}$ or $(-\infty, -1) \cup (4, \infty)$ **7.** $\{x|-9 < x < 3\}$ or $(-9, 3)$ **9.** $\{x|x \le -10 \text{ or } x > 4\}$ or $(-\infty, -10] \cup (4, \infty)$ **11.** $\{x|-7 \le x < 8\}$ or $[-7, 8)$ **13.** $\left\{x \middle| -3 < x < \frac{1}{2} \text{ or } x > 5\right\}$ or $\left(-3, \frac{1}{2}\right) \cup (5, \infty)$ **15.** $\left\{x \middle| x \le -8 \text{ or } -\frac{5}{3} \le x < 2\right\}$ or $(-\infty, -8] \cup \left[-\frac{5}{3}, 2\right)$ **17.** $\{x|x > -1\}$ or $(-1, \infty)$ **19.** $\{x|x < -4 \text{ or } x \ge 9\}$ or $(-\infty, -4) \cup [9, \infty)$ **21.** $\left\{x \middle| \frac{3}{2} < x < 3\right\}$ or $\left(\frac{3}{2}, 3\right)$ **23.** $\{x|0 < x \le 1 \text{ or } x > 4\}$ or $(0, 1] \cup (4, \infty)$ **25.** $\{x|-5 < x < 2 \text{ or } x \ge 23\}$ or $(-5, 2) \cup [23, \infty)$ **27.** $\{x|-1 < x \le 6\}$ or $(-1, 6]$ **29.** $\left\{x \middle| -2 < x < \frac{5}{2}\right\}$ or $\left(-2, \frac{5}{2}\right)$ **31.** 100 or more bicycles **33.** Answers may vary. One possibility follows: $\frac{10}{x-2} > 0$ **35.** 2 **37.** $-\frac{7}{2}$ and 2 **39.** $\frac{2}{3}$ **41.** $\{x|x \le -4 \text{ or } x > -1\}$ or $(-\infty, -4] \cup (-1, \infty)$ **43.** $\{x|7 < x < 26\}$ or $(7, 26)$

## Section 6.6

**1.** proportion **3.** True **5.** A variable $Q$ varying jointly with $x$ and $y$ means that $Q$ is proportional to the product of $x$ and $y$. Combined variation means that direct and inverse variation are occurring at the same time. **7.** $P_1 = \frac{V_2P_2}{V_1}$ **9.** $t = \frac{R - r}{R}$ **11.** $x = \frac{y - y_1 + mx_1}{m}$ **13.** $v = \frac{\omega(I + mr^2)}{rm}$ **15.** $m = \frac{MV}{v - V}$ **17. (a)** $k = 20$ **(b)** $y = \frac{20}{x}$ **(c)** $y = 4$ **19. (a)** $k = 21$ **(b)** $y = \frac{21}{x}$ **(c)** $y = \frac{3}{4}$ **21. (a)** $k = \frac{1}{4}$ **(b)** $y = \frac{1}{4}xz$ **(c)** $y = 27$ **23. (a)** $k = \frac{13}{10}$ **(b)** $Q = \frac{13x}{10y}$ **(c)** $Q = \frac{117}{40}$ **25.** $AB = 16$ and $DF = 5$ **27.** 259,757,764 people **29.** approximately $4.10 **31.** 12.5 pounds **33.** approximately 34.3 minutes **35.** 6 hours **37.** 10 hours **39.** 35 miles per hour **41.** 1.8 feet per second **43.** Shockey will be on his own 35-yard line when Urlacher catches up to him. **45.** 8 miles per hour **47.** 45 miles per hour **49. (a)** $D(p) = \frac{375}{p}$ **(b)** 125 bags of candy **51.** 450 cc **53.** 119.8 pounds **55.** 2250 newtons **57.** $1.4007 \times 10^{-7}$ newtons **59.** 360 pounds **61. (a)** 314.16 **(b)** approximately 942.48 meters per minute; approximately 15.708 meters per second **(c)** $k \approx 0.056$ **(d)** approximately 7.86 newtons **63.** $a^{15}$ **65.** $\frac{a^4}{b^6}$ **67.** $\frac{n^{12}}{9m^4}$

## Chapter 6 Review

**1.** $\left\{x \middle| x \neq \frac{2}{3}\right\}$ **2.** $\{a|a \neq -4, a \neq 7\}$ **3.** $\{m|m$ is any real number$\}$ or $(-\infty, \infty)$ **4.** $\{n|n \neq -2, n \neq 4\}$ **5.** $\frac{6}{x - 5}$ **6.** $\frac{2}{y^3}$ **7.** $\frac{w - 7}{w + 4}$ **8.** $\frac{3a + b}{5a + 2b}$ **9.** $\frac{-1}{3m + 1}$ **10.** $\frac{n^2 + 3}{n - 4}$ **11.** $\frac{p}{2(p - 6)}$ **12.** $\frac{2q}{3(q - 5)}$ **13.** $\frac{(x - 4)(x + 6)}{(x + 2)^2}$ **14.** $\frac{2(y + 5)}{y(y - 5)}$ **15.** $\frac{3a + b}{2a - b}$ **16.** $\frac{3}{m - 2}$ **17.** $\frac{9c}{2d^3}$ **18.** $\frac{6(z - 3)}{7}$ **19.** $\frac{x + 7}{x - 3}$ **20.** $\frac{m + 4n}{m + 5n}$ **21.** $\frac{2(p - q)}{p + 3q}$ **22.** $\frac{2a + 5}{9a^2 + 6a + 4}$ **23.** $(x + 2)^2$; $\left\{x \middle| x \neq \frac{1}{2}, x \neq 5\right\}$ **24.** $\frac{x - 5}{x + 7}$; $\left\{x \middle| x \neq -7, x \neq -2, x \neq \frac{1}{2}\right\}$ **25.** $\frac{(x - 5)^2}{(2x - 1)^2}$; $\left\{x \middle| x \neq -2, x \neq \frac{1}{2}, x \neq 5\right\}$ **26.** $\frac{(x + 2)^2(x + 7)}{x - 5}$; $\left\{x \middle| x \neq -7, x \neq -2, x \neq \frac{1}{2}, x \neq 5\right\}$ **27.** $\frac{4x + 3}{x - 5}$ **28.** 4 **29.** $\frac{a + 6}{a - 2}$ **30.** $\frac{b + 5}{2b + 3}$ **31.** 3c **32.** $\frac{d + 3}{d + 1}$ **33.** LCD $= 36x^4$ **34.** LCD $= (y + 2)(y - 9)$ **35.** LCD $= p^2(2p + 5)(p - 4)$ **36.** LCD $= (q + 5)(q - 1)(q - 3)$ **37.** $\frac{m^2 + 4n^2}{m^3n^4}$ **38.** $\frac{9x - 7y^2}{6x^2y^3}$ **39.** $\frac{p^2 + q^2}{(p - q)(p + q)}$ **40.** $\frac{20x + 17}{(x - 7)(x - 3)(x + 4)}$ **41.** $\frac{y + 8}{(y - 1)^2(y + 2)}$ **42.** $\frac{11a + 7b}{(2a - 3b)(2a + 3b)}$ **43.** $\frac{6x}{x + 3}$ **44.** $\frac{5}{n^2 - 5n + 25}$ **45.** $\frac{2n}{m - 7n}$ **46.** $\frac{1}{z - 3}$ **47.** $\frac{3}{y + 2}$ **48.** $\frac{a - 3}{(a + 4)(a - 2)}$ **49. (a)** $\frac{x^2 + x + 10}{(x - 4)(x + 2)}$ **(b)** $\{x|x \neq -2, x \neq 4\}$ **50. (a)** $\frac{x + 8}{(2x - 5)(2x + 1)}$ **(b)** $\left\{x \middle| x \neq -3, x \neq -\frac{1}{2}, x \neq \frac{5}{2}\right\}$ **51.** $x + 1$ **52.** $\frac{xy}{x + y}$ **53.** $\frac{a - b}{a + b}$ **54.** $\frac{-a}{a + 3}$ **55.** $\frac{3t + 4}{5t^2 + 1}$ **56.** $\frac{1}{a + b}$ **57.** $\frac{z + 1}{z - 1}$ **58.** $\frac{x - 1}{x + 1}$ **59.** $\frac{x + y}{x - y}$ **60.** 1 **61.** $\frac{2}{x(x - 2)}$ **62.** $\frac{z - 5}{z + 5}$ **63.** $\frac{n}{m}$ **64.** $\frac{2}{(x + 1)^2(x - 2)}$ **65.** $\frac{3y^2 - 3x^2}{xy}$ **66.** $\frac{12d - 2c}{c}$ **67.** $\{10\}$ **68.** $\left\{\frac{4}{3}\right\}$ **69.** $\{-2, 7\}$ **70.** $\{9\}$ **71.** $\{2\}$ **72.** $\{3\}$ **73.** no solution; $\emptyset$ or $\{\ \}$ **74.** $\{-3\}$ **75.** $\{6\}$ **76.** $\left\{-\frac{4}{3}\right\}$ **77.** $\{5\}$ **78.** no solution; $\emptyset$ or $\{\ \}$ **79.** $x = 5$; $(5, 2)$ **80.** $x = -3$ or $x = 7$; $(-3, 4)$ and $(7, 4)$ **81.** $\{x|x < -2$ or $x \geq 4\}$ or $(-\infty, -2) \cup [4, \infty)$

−6 −4 −2 0 2 4 6 8

**82.** $\{y|-4 < y < 5\}$ or $(-4, 5)$

−6 −4 −2 0 2 4 5 6 8

**83.** $\{z|-3 < z < 3\}$ or $(-3, 3)$

−7 −5 −3 −1 0 1 3 5 7

**84.** $\{w|w < -7$ or $2 < w < 4\}$ or $(-\infty, -7) \cup (2, 4)$

−10 −8 −7 −6 −4 −2 0 2 4

**85.** $\{m|-5 < m < 2$ or $m \geq 5\}$ or $(-5, 2) \cup [5, \infty)$

−7 −5 −3 −1 1 2 3 5 7 9

**86.** $\{n|0 \leq n < 2\}$ or $[0, 2)$

−6 −4 −2 0 2 4 6 8

**87.** $\left\{a \middle| 2 < a < \frac{7}{2}\right\}$ or $\left(2, \frac{7}{2}\right)$

−6 −4 −2 0 2 $\frac{7}{2}$ 4 6 8

**88.** $\{c|c < -6$ or $0 < c < 2\}$ or $(-\infty, -6) \cup (0, 2)$

−8 −6 −4 −2 0 2 4 6

**89.** $\left\{x \middle| -\frac{3}{2} < x < 4\right\}$ or $\left(-\frac{3}{2}, 4\right)$

−6 −4 −2 $-\frac{3}{2}$ 0 2 4 6 8

**90.** $\{x|x \leq -5$ or $x > -1\}$ or $(-\infty, -5] \cup (-1, \infty)$

−11 −9 −7 −5 −3 −1 1 3 5

**91.** $C = \dfrac{C_1C_2}{C_2 + C_1}$ **92.** $T_2 = \dfrac{T_1P_2V_2}{P_1V_1}$ **93.** $G = \dfrac{4\pi^2a^2}{MT}$ **94.** $x = z\cdot\sigma + \mu$ **95.** $AB = 24$ and $DF = 10$ **96.** 18.75 feet **97.** 104 grams **98.** \$78.00 **99.** 28.8 minutes (or 28 minutes and 48 seconds) **100.** 2.8 hours (or 2 hours and 48 minutes) **101.** 21 hours; 28 hours **102.** 3 minutes **103.** 60 miles per hour **104.** 15 miles per hour **105.** Todd's average walking speed is 3 miles per hour and his average running speed is 12 miles per hour. **106.** 60 miles per hour **107. (a)** $k = 60$ **(b)** $y = \dfrac{60}{x}$ **(c)** $y = 12$ **108. (a)** $k = \dfrac{3}{4}$ **(b)** $y = \dfrac{3}{4}xz$ **(c)** $y = 42$ **109. (a)** $k = 72$ **(b)** $s = \dfrac{72}{t^2}$ **(c)** $s = 8$ **110. (a)** $k = \dfrac{8}{5}$ **(b)** $w = \dfrac{8x}{5z}$ **(c)** $w = \dfrac{9}{10}$ **111.** 1200 kilohertz **112.** 12 ohms **113.** 704 cubic centimeters **114.** 375 cubic inches

### *Chapter 6 Test*

**1.** $\left\{x \mid x \neq -\dfrac{1}{2}, x \neq 7\right\}$ **2.** $\dfrac{2m-3}{3m-1}$ **3.** $-\dfrac{1}{a+4b}$ **4.** $\dfrac{2x+5}{2x(x-4)}$ **5.** $\dfrac{y+4}{3y+1}$ **6.** $\dfrac{2q}{p-q}$ **7.** $\dfrac{3c-4}{(c-2)(c-1)}$ **8.** $\dfrac{x-4}{3(x-2)(x-5)}$; $\{x \mid x \neq -2, x \neq 0, x \neq 2, x \neq 4, x \neq 5\}$ **9.** $\dfrac{3(x^2+2x-4)}{x(x+2)(x-2)}$; $\{x \mid x \neq -2, x \neq 0, x \neq 2\}$ **10.** $\dfrac{a}{a+1}$ **11.** $\dfrac{4(d-3)}{-3(d+6)}$ **12.** $x = -1$ **13.** no solution; $\emptyset$ or { } **14.** $\left\{x \mid 2 < x \leq \dfrac{11}{2}\right\}$ or $\left(2, \dfrac{11}{2}\right]$ 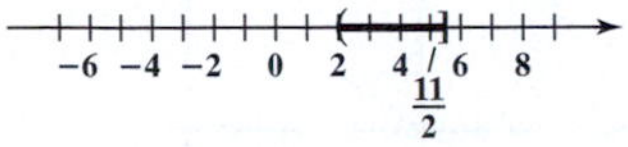

**15.** $k = \dfrac{FD^2}{q_1q_2}$ **16.** 120 seconds (or 2 minutes) **17.** 2.4 hours (or 2 hours and 24 minutes) **18.** 3 miles per hour **19.** 20 pounds **20.** 792 square centimeters

## Getting Ready for Chapter 7

**1.** radical sign **3.** $|a|$ **5.** True **7.** Perfect squares: 4, 9, 16, 25 **9.** 1 **11.** −10 **13.** $\dfrac{1}{2}$ **15.** 0.6 **17.** 1.6 **19.** not a real number **21.** rational; 8 **23.** rational; $\dfrac{1}{4}$ **25.** irrational; 6.63 **27.** irrational; 7.07 **29.** not a real number **31.** 8 **33.** 19 **35.** $|r|$ **37.** $|x+4|$ **39.** $|4x-3|$ **41.** $|2y+3|$ **43.** 13 **45.** 17 **47.** not a real number **49.** 15 **51.** −8 **53.** 6 **55.** not a real number **53.** 6 **55.** not a real number **57.** −4 **59.** 13 **61.** The square roots of 36 are −6 and 6; $\sqrt{36} = 6$ **63.** $\dfrac{4\sqrt{13}}{3}$; 4.81

## Chapter 7

### *Section 7.1*

**1.** $\sqrt[n]{a}$ **3.** index **5.** False **7.** Answers will vary. **9.** 5 **11.** −3 **13.** −5 **15.** $-\dfrac{1}{2}$ **17.** 3 **19.** 5 **21.** $|m|$ **23.** $x - 3$ **25.** $-|3p+1|$ **27.** 2 **29.** −6 **31.** 2 **33.** −2 **35.** $\dfrac{2}{5}$ **37.** −5 **39.** not a real number **41.** 32 **43.** −64 **45.** 16 **47.** 16 **49.** 8 **51.** $\dfrac{1}{12}$ **53.** 125 **55.** 32 **57.** $(3x)^{1/3}$ **59.** $\left(\dfrac{x}{3}\right)^{1/4}$ **61.** $x^{3/4}$ **63.** $(3x)^{2/5}$ **65.** $\left(\dfrac{5x}{y}\right)^{3/2}$ **67.** $(9ab)^{4/3}$ **69.** 2.92 **71.** 1.86 **73.** 4.47 **75.** 10.08 **77.** 1.26 **79.** 8 **81.** 243 **83.** not a real number **85.** $\dfrac{1}{12}$ **87.** 0.2 **89.** 127 **91.** not a real number **93.** 10; 10 **95. (a)** about 21.25°F **(b)** about 17.36°F **(c)** about −15.93°F **97. (a)** $\sqrt{\dfrac{8r\rho_w g}{3C\rho}}$ m/s **(b)** about 7.38 m/s **99.** 8 **101.** 16 **103.** $x^6y^9$ **105.** $(x-1)^{3/2}(x+3)$ **107.** $\dfrac{z-6}{z-1}$

### *Section 7.2*

**1.** $a^rb^r$ **3.** 25 **5.** 8 **7.** $\dfrac{1}{2^{7/6}}$ **9.** $\dfrac{1}{x^{7/12}}$ **11.** 2 **13.** $\dfrac{125}{8}$ **15.** $x^{1/2}y^{2/9}$ **17.** $\dfrac{x^{1/6}}{y^{1/3}}$ **19.** $\dfrac{2a}{b^{3/4}}$ **21.** $\dfrac{x^{1/18}}{2y^{4/9}}$ **23.** $8x^{1/8}y^{1/2}$ **25.** $x^2 - 2x^{1/2}$ **27.** $\dfrac{2}{y^{1/3}} + 6y^{2/3}$ **29.** $4z^3 - 32 = 4(z-2)(z^2+2z+4)$ **31.** $x^4$ **33.** 2 **35.** $2ab^4$ **37.** $\sqrt[4]{x}$ **39.** $\sqrt[6]{x^5}$ **41.** $\sqrt[8]{x^3}$ **43.** $\sqrt[6]{3^7}$ **45.** 1 **47.** $5x^{1/2}(x+3)$ **49.** $8(x+2)^{2/3}(3x+1)$ **51.** $\dfrac{6x+5}{x^{1/2}}$ **53.** $\dfrac{2(10x-27)}{(x-4)^{1/3}}$ **55.** $5(x^2+4)^{1/2}(x^2+3x+4)$ **57.** 2 **59.** 4 **61.** 10 **63.** 5 **65.** 0 **67.** $\dfrac{1}{48}$ **69.** 5 **71.** 3 **73.** $\sqrt[24]{x}$ **75.** 36 **77.** $\{x \mid x > -1\}$ or $(-1, \infty)$ **79.** $2(2a^2 - 4a - 3)$ **81.** −6

### *Section 7.3*

**1.** $\sqrt[n]{a\cdot b}$ **3.** False **5.** 0, 1, 4, 9, 16, 25 **7.** Answers may vary. **9.** $\sqrt{14}$ **11.** $\sqrt[3]{60}$ **13.** $\sqrt{15ab}$ if $a,b \geq 0$ **15.** $\sqrt{x^2-49}$ if $|x| \geq 7$ **17.** $\sqrt{5}$ if $x > 0$ **19.** $5\sqrt{2}$ **21.** $3\sqrt[3]{2}$ **23.** $4|x|\sqrt{3}$ **25.** $-3x$ **27.** $2|m|\sqrt[4]{2}$ **29.** $2|p|\sqrt{3q}$ **31.** $6\sqrt{6a}$ **33.** $9m^2\sqrt{2}$ **35.** $y^6\sqrt{y}$ **37.** $c^2\sqrt[3]{c^2}$ **39.** $m^2n\sqrt{mn}$ **41.** $5pq^2\sqrt{5p}$ **43.** $-2x^3\sqrt[3]{2}$ **45.** $pq^2\sqrt{pqr}$ **47.** $-m\sqrt[5]{16m^3n^2}$ **49.** $(x-y)\sqrt[4]{x-y}$ **51.** $2\sqrt[3]{x^3-y^3}$ **53.** 5 **55.** $3+\sqrt{2}$ **57.** $\dfrac{-4-9\sqrt{2}}{6}$ **59.** $\dfrac{1-\sqrt{2}}{2}$ **61.** 5 **63.** 4 **65.** 2 **67.** $5x\sqrt{3}$ **69.** $2b\sqrt[3]{3b}$

**71.** $18ab^2\sqrt{10}$ **73.** $3pq\sqrt[4]{4p}$ **75.** $-2ab\sqrt[5]{3a}$ **77.** $2(x-y)\sqrt[4]{3(x-y)}$ **79.** $\frac{\sqrt{3}}{4}$ **81.** $\frac{11}{10}$ **83.** $\frac{x\sqrt[4]{5}}{2}$ **85.** $\frac{3y}{5x}$ **87.** $-\frac{3x^3}{4y^4}$

**89.** 2 **91.** 4 **93.** $2a\sqrt{2}$ **95.** $\frac{2a^2\sqrt{2}}{b}$ **97.** $\frac{16a^3}{3b}$ **99.** $a^2\sqrt[3]{26}$ **101.** $\frac{3x^3\sqrt{5}}{y}$ **103.** $\sqrt[6]{432}$ **105.** $\sqrt[6]{12}$ **107.** $3\sqrt[6]{12}$

**109.** $\sqrt[3]{18}$ **111.** $\frac{\sqrt[3]{5x}}{2}$ **113.** $\sqrt[3]{45a^2}$ **115.** $6a^2\sqrt{2}$ **117.** $3a\sqrt[3]{2b^2}$ **119.** $\frac{-2\sqrt[3]{2}}{a}$ **121.** $-10m\sqrt[3]{4}$ **123.** $3ab^2\sqrt[3]{3ab}$ **125.** 6

**127.** **(a)**

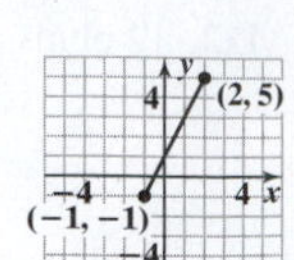


**(b)** $3\sqrt{5}$ units **129.** **(a)** roughly $1,587,000 **(b)** roughly $2,381,000 **131.** **(a)** $4x^2$ **(b)** 36

**133.** $x = -3 - \sqrt{6}$ or $x = -3 + \sqrt{6}$ **135.** $x = \frac{-2-\sqrt{7}}{3}$ or $x = \frac{-2+\sqrt{7}}{3}$ **137.** $x = -\frac{5}{7}$ or $x = 1$ **139.** $-5 \le x \le 3$

**141.** **(a)**

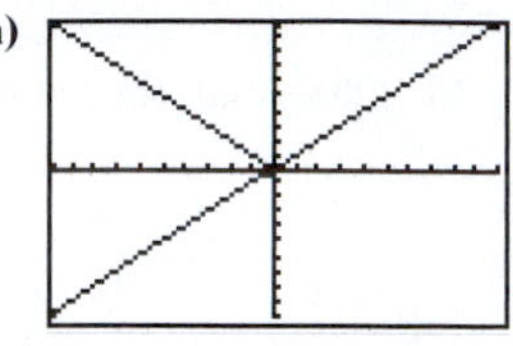

The graphs of $y = \sqrt{x^2}$ and $y = x$ are not the same so the expressions cannot be equal.

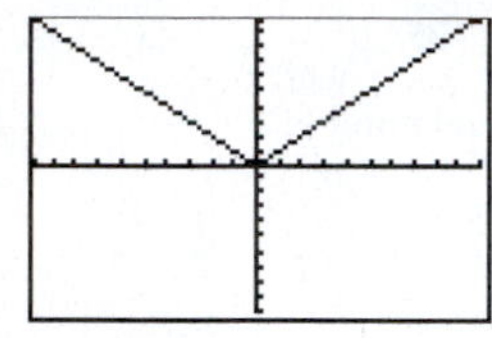

The graphs of $y = \sqrt{x^2}$ and $y = |x|$ appear to be the same. $\sqrt{x^2} = |x|$

**(b)**

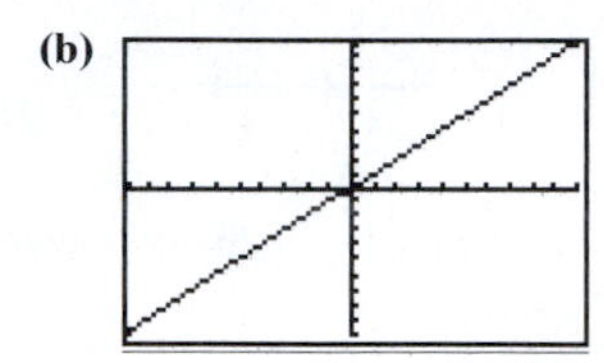

The graphs of $y = \sqrt[3]{x^3}$ and $y = x$ appear to be the same. $\sqrt[3]{x^3} = x$

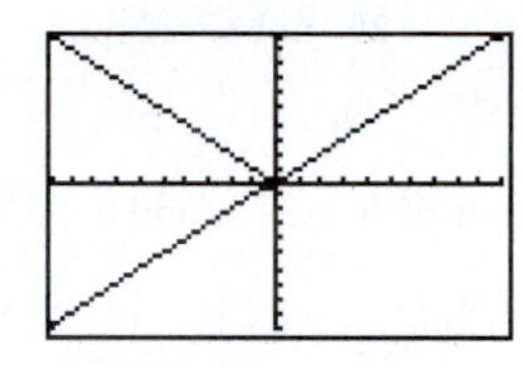

The graph of $y = \sqrt[3]{x^3}$ and $y = |x|$ are not the same so the expressions cannot be equal.

**(c)**

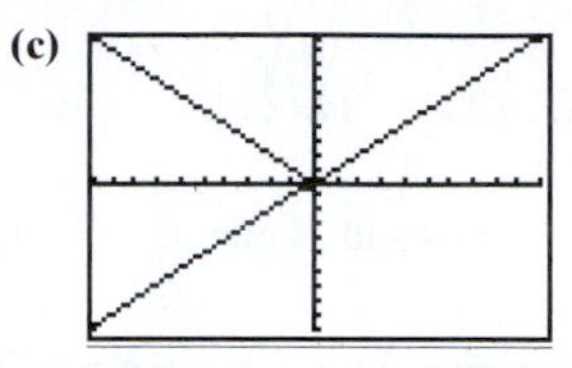

The graphs of $y = \sqrt[4]{x^4}$ and $y = x$ are not the same so the expressions cannot be equal.

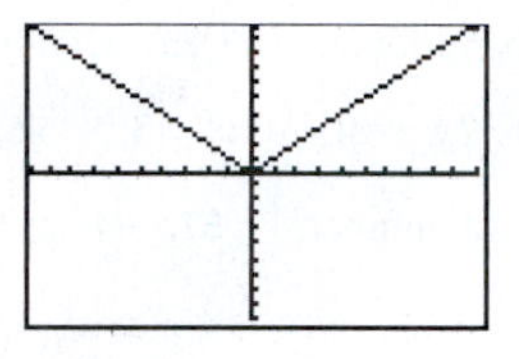

The graphs of $y = \sqrt[4]{x^4}$ and $y = |x|$ appear to be the same. $\sqrt[4]{x^4} = |x|$

**(d)** Answers will vary.

## Section 7.4

**1.** like radicals **3.** True **5.** Answers may vary. **7.** $10\sqrt{2}$ **9.** $2\sqrt[3]{x}$ **11.** $14\sqrt{5x}$ **13.** $11\sqrt[3]{5} - 11\sqrt{5}$ **15.** $8\sqrt{2}$ **17.** $-2\sqrt[3]{3}$
**19.** $-25\sqrt[3]{2}$ **21.** $9\sqrt{6x}$ **23.** $4\sqrt{2} + 3\sqrt{10}$ **25.** $32x\sqrt{3x}$ **27.** $2x\sqrt{3} - 11x\sqrt{2}$ **29.** $(3x-8)\sqrt[3]{2}$ **31.** $5\sqrt{x-1}$ **33.** $3\sqrt{x}$
**35.** $5\sqrt[3]{x}$ **37.** $2\sqrt{3} - 3\sqrt{6}$ **39.** $\sqrt{6} + 3\sqrt{2}$ **41.** $\sqrt[3]{12} - 2\sqrt[3]{3}$ **43.** $3\sqrt{2x} - 2x\sqrt{5}$ **45.** $12 + 3\sqrt{3} + 4\sqrt{2} + \sqrt{6}$
**47.** $12 - 6\sqrt{7} + 2\sqrt{3} - \sqrt{21}$ **49.** $6\sqrt{7} - 30$ **51.** $\sqrt{6} - 12\sqrt{3} + 9\sqrt{2} - 4$ **53.** $31 - 2\sqrt{15}$ **55.** $4 + 2\sqrt{3}$ **57.** $7 - 2\sqrt{10}$
**59.** $x - 2\sqrt{2x} + 2$ **61.** 1 **63.** $-11$ **65.** $2x - 3y$ **67.** $\sqrt[3]{x^2} + \sqrt[3]{x} - 12$ **69.** $\sqrt[3]{4a^2} - 25$ **71.** $\sqrt{15} + 5\sqrt{2}$ **73.** $26x^2\sqrt{7x}$

**75.** $-13$ **77.** $2\sqrt[3]{7} + \sqrt[3]{28}$ **79.** $-4 + 12\sqrt{2}$ **81.** $20\sqrt{2}$ **83.** $8 - 2\sqrt{15}$ **85.** $(3x+2)\sqrt[3]{5y}$ **87.** $2x - 7y$ **89.** $\frac{11\sqrt{5}}{25}$
**91.** The area is 108 square units. The perimeter is $30\sqrt{2}$ units. **93.** $12\sqrt{6}$ square units **95.** **(a)** $3\sqrt{3x}$ **(b)** $6\sqrt{3}$ **(c)** $6x$

**97.** Check $x = -2 + \sqrt{5}$:
$0 \stackrel{?}{=} x^2 + 4x - 1$
$0 \stackrel{?}{=} (-2 + \sqrt{5})^2 + 4(-2 + \sqrt{5}) - 1$
$0 \stackrel{?}{=} (-2)^2 + 2(-2)(\sqrt{5}) + (\sqrt{5})^2 - 8 + 4\sqrt{5} - 1$
$0 \stackrel{?}{=} 4 - 4\sqrt{5} + \sqrt{25} - 8 + 4\sqrt{5} - 1$
$0 \stackrel{?}{=} 9 - 9$
$0 = 0$ true
The value is a solution.

Check $x = -2 - \sqrt{5}$:
$0 \stackrel{?}{=} x^2 + 4x - 1$
$0 \stackrel{?}{=} (-2 - \sqrt{5})^2 + 4(-2 - \sqrt{5}) - 1$
$0 \stackrel{?}{=} (-2)^2 - 2(-2)(\sqrt{5}) + (\sqrt{5})^2 - 8 - 4\sqrt{5} - 1$
$0 \stackrel{?}{=} 4 + 4\sqrt{5} + \sqrt{25} - 8 - 4\sqrt{5} - 1$
$0 \stackrel{?}{=} 9 - 9$
$0 = 0$ true
The value is a solution.

**99.** $6p^2 + 7p - 3$ **101.** $m^2 - 16$ **103.** $x - 4$

## Section 7.5

**1.** rationalizing the denominator **3.** False **5.** $\frac{\sqrt{2}}{2}$ **7.** $-\frac{2\sqrt{3}}{5}$ **9.** $\frac{\sqrt{3}}{2}$ **11.** $\frac{\sqrt{3}}{3}$ **13.** $\frac{\sqrt{2p}}{p}$ **15.** $\frac{2\sqrt{2y}}{y^2}$ **17.** $\sqrt[3]{4}$

**19.** $\frac{\sqrt[3]{7q^2}}{q}$ **21.** $-\frac{\sqrt[3]{60}}{10}$ **23.** $\frac{\sqrt[3]{50y^2}}{5y}$ **25.** $-\frac{4\sqrt[4]{27x}}{3x}$ **27.** $\frac{12\sqrt[5]{m^2n^3}}{mn}$ **29.** $2(\sqrt{6} + 2)$ **31.** $5(\sqrt{5} - 2)$ **33.** $2(\sqrt{7} + \sqrt{3})$

**35.** $\frac{\sqrt{5} + \sqrt{3}}{2}$ **37.** $\frac{p - \sqrt{pq}}{p - q}$ **39.** $-3(2\sqrt{3} - 3\sqrt{2})$ or $3(3\sqrt{2} - 2\sqrt{3})$ **41.** $-8 - 3\sqrt{7}$ or $-3\sqrt{7} - 8$ **43.** $\frac{13\sqrt{6} - 46}{38}$

**45.** $\dfrac{p + 4\sqrt{p} + 4}{p - 4}$ **47.** $\dfrac{2 - 3\sqrt{2}}{2}$ **49.** $\dfrac{4\sqrt{3}}{3}$ **51.** $\dfrac{\sqrt{10} - \sqrt{2}}{2}$ **53.** $\dfrac{22\sqrt{3}}{3}$ **55.** 0 **57.** $\dfrac{1}{2}$ **59.** $\dfrac{\sqrt{2}}{4}$ **61.** $\dfrac{2\sqrt{3}}{3}$

**63.** $\sqrt{3} - 2$ **65.** $2(\sqrt{5} - 2)$ **67.** 2 **69.** $\dfrac{\sqrt{3}}{3}$ **71.** $\dfrac{\sqrt[3]{18}}{6}$ **73.** $\dfrac{5 - \sqrt{3}}{22}$ **75.** $\dfrac{\sqrt{6} - \sqrt{2}}{4}$ **77.** $\dfrac{1}{3(\sqrt{2} - 1)}$

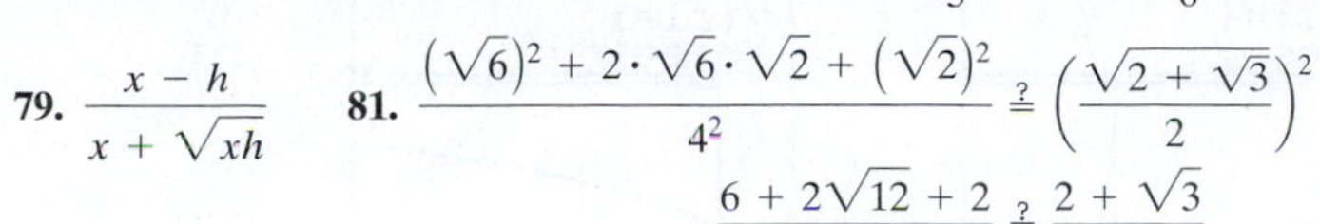

**79.** $\dfrac{x - h}{x + \sqrt{xh}}$ **81.** $\dfrac{(\sqrt{6})^2 + 2\cdot\sqrt{6}\cdot\sqrt{2} + (\sqrt{2})^2}{4^2} \stackrel{?}{=} \left(\dfrac{\sqrt{2} + \sqrt{3}}{2}\right)^2$

$$\frac{6 + 2\sqrt{12} + 2}{16} \stackrel{?}{=} \frac{2 + \sqrt{3}}{4}$$
$$\frac{8 + 2\cdot 2\sqrt{3}}{16} \stackrel{?}{=} \frac{2 + \sqrt{3}}{4}$$
$$\frac{8 + 4\sqrt{3}}{16} \stackrel{?}{=} \frac{2 + \sqrt{3}}{4}$$
$$\frac{2 + \sqrt{3}}{4} = \frac{2 + \sqrt{3}}{4}$$

**83.** (graph through $(0, -3)$, $(1, 2)$, $(2, 7)$)

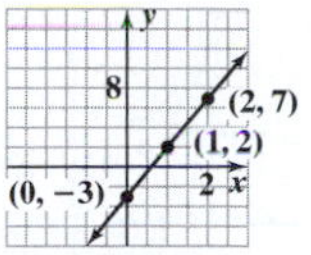

**85.**

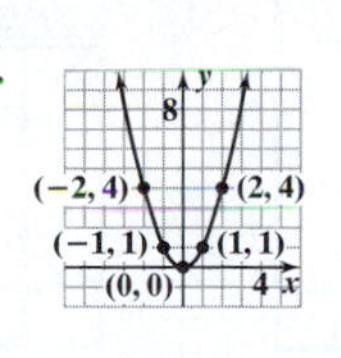


### *Putting the Concepts Together*

**1.** −5 **2.** $\dfrac{1}{16}$ **3.** $(3x^3)^{1/4}$ **4.** $7\sqrt[5]{z^4}$ **5.** $2x^{1/2}$ or $2\sqrt{x}$ **6.** $c^2 + c^3$ **7.** $\dfrac{a^2}{b^2}$ **8.** $x^{5/8}$ or $\sqrt[8]{x^5}$ **9.** $\dfrac{x^6}{y}$ **10.** $\sqrt{30ab}$ **11.** $10m^2n\sqrt{2n}$

**12.** $-2xy$ **13.** $\sqrt{3}$ **14.** $11b\sqrt{2b}$ **15.** $y\sqrt[3]{2y}$ **16.** $12x$ **17.** $7\sqrt{x}$ **18.** $14 - 28\sqrt{2} = 14(1 - 2\sqrt{2})$ **19.** $41 - 24\sqrt{2}$

**20.** $\dfrac{3\sqrt{2}}{16}$ **21.** $-\dfrac{4(\sqrt{3} + 8)}{61}$

### *Section 7.6*

**1.** even **3.** True **5. (a)** 3 **(b)** $\sqrt{14}$ **(c)** 2 **7. (a)** −5 **(b)** −1 **(c)** $-\dfrac{\sqrt{13}}{2}$ **9. (a)** 4 **(b)** $4\sqrt{6}$ **(c)** $\sqrt{6}$

**11. (a)** 2 **(b)** −2 **(c)** $-2\sqrt[3]{2}$ **13. (a)** $\dfrac{\sqrt{5}}{3}$ **(b)** $\dfrac{\sqrt{2}}{2}$ **(c)** $\dfrac{\sqrt{6}}{3}$ **15. (a)** 1 **(b)** $\sqrt[3]{4}$ **(c)** $\sqrt[3]{3}$

**17.** $\{x|x \ge 7\}$ or $[7, \infty)$ **19.** $\left\{x\middle|x \ge -\dfrac{7}{2}\right\}$ or $\left[-\dfrac{7}{2}, \infty\right)$ **21.** $\left\{x\middle|x \le \dfrac{4}{3}\right\}$ or $\left(-\infty, \dfrac{4}{3}\right]$

**23.** $\{z|z$ is any real number$\}$ or $(-\infty, \infty)$ **25.** $\left\{p\middle|p \ge \dfrac{2}{7}\right\}$ or $\left[\dfrac{2}{7}, \infty\right)$ **27.** $\{x|x$ is any real number$\}$ or $(-\infty, \infty)$

**29.** $\{x|x > -5\}$ or $(-5, \infty)$ **31.** $\{x|x \le -3$ or $x > 3\}$ or $(-\infty, -3] \cup (3, \infty)$

**33. (a)** $\{x|x \ge 4\}$ or $[4, \infty)$
**(b)**

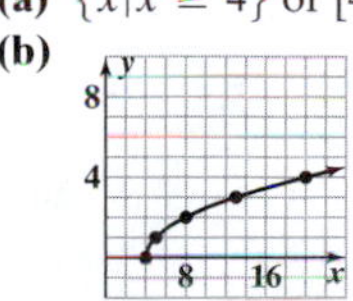

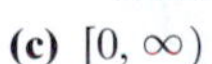

**(c)** $[0, \infty)$

**35. (a)** $\{x|x \ge -2\}$ or $[-2, \infty)$
**(b)**

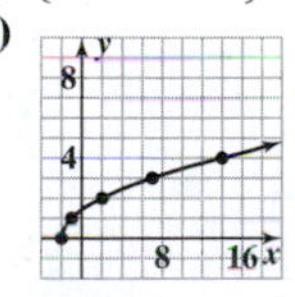

**(c)** $[0, \infty)$

**37. (a)** $\{x|x \le 2\}$ or $(-\infty, 2]$
**(b)**

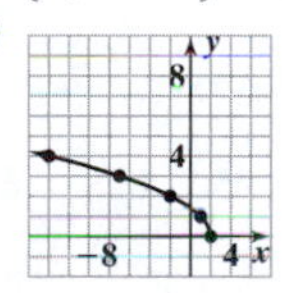

**(c)** $[0, \infty)$

**39. (a)** $\{x|x \ge 0\}$ or $[0, \infty)$
**(b)** 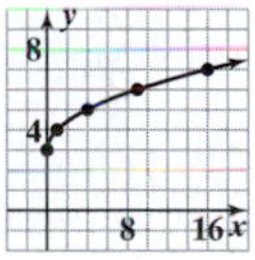
**(c)** $[3, \infty)$

**41. (a)** $\{x|x \ge 0\}$ or $[0, \infty)$
**(b)**

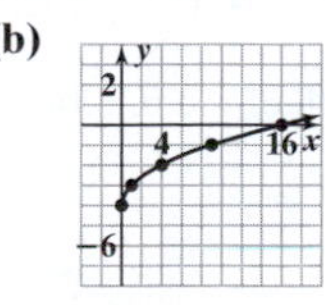

**(c)** $[-4, \infty)$

**43. (a)** $\{x|x \ge 0\}$ or $[0, \infty)$
**(b)**

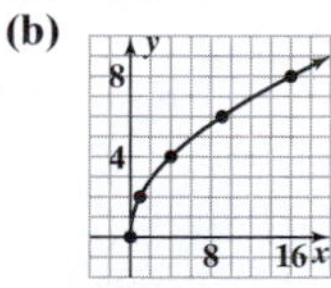

**(c)** $[0, \infty)$

**45. (a)** $\{x|x \ge 0\}$ or $[0, \infty)$
**(b)**

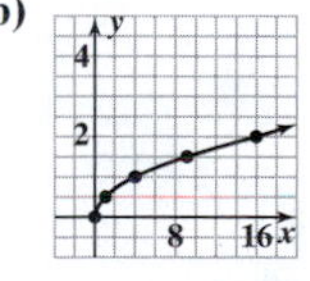

**(c)** $[0, \infty)$

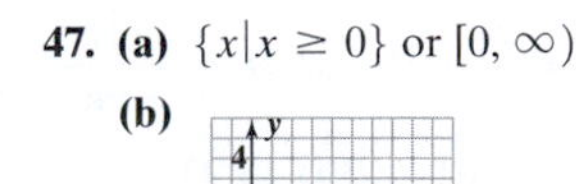

**47. (a)** $\{x|x \ge 0\}$ or $[0, \infty)$
**(b)** 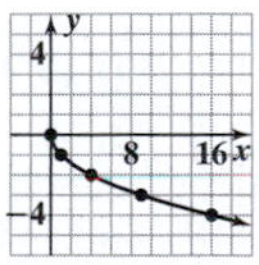
**(c)** $(-\infty, 0]$

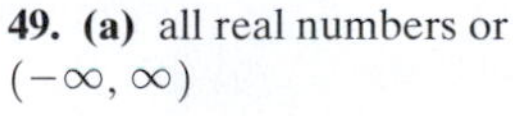

**49. (a)** all real numbers or $(-\infty, \infty)$
**(b)**

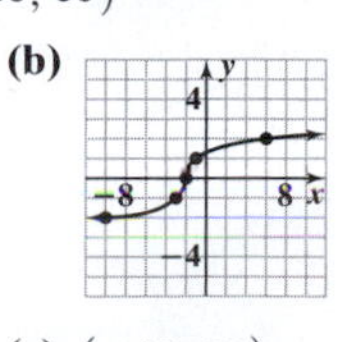

**(c)** $(-\infty, \infty)$

**51. (a)** all real numbers or $(-\infty, \infty)$
**(b)**

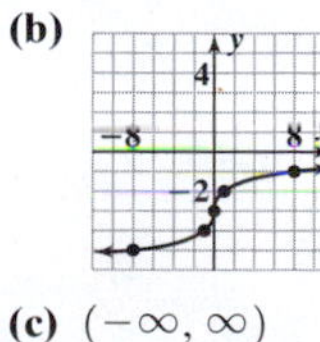

**(c)** $(-\infty, \infty)$

**53. (a)** all real numbers or $(-\infty, \infty)$
**(b)**

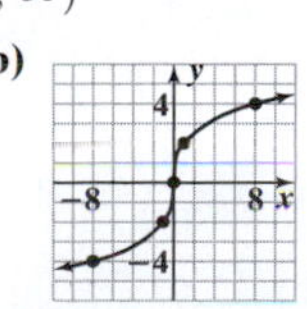

**(c)** $(-\infty, \infty)$

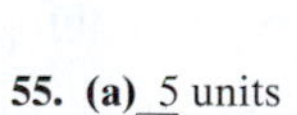

**55. (a)** 5 units
**(b)** $\sqrt{17} \approx 4.123$ units
**(c)** $5\sqrt{17} \approx 20.616$ units
**57. (a)** $4\sqrt{2} \approx 5.657$ square units
**(b)** $4\sqrt{5} \approx 8.944$ square units
**(c)** $2\sqrt{14} \approx 7.483$ square units

**59.** Shift the graph of $f(x)$ $c$ units to the right (if $c < 0$) or left (if $c > 0$). **61.** $\dfrac{5}{6}$ **63.** $\dfrac{4x + 1}{x(x + 1)}$ **65.** $\dfrac{7x + 1}{(x - 1)(x + 1)}$ or $\dfrac{7x + 1}{x^2 - 1}$

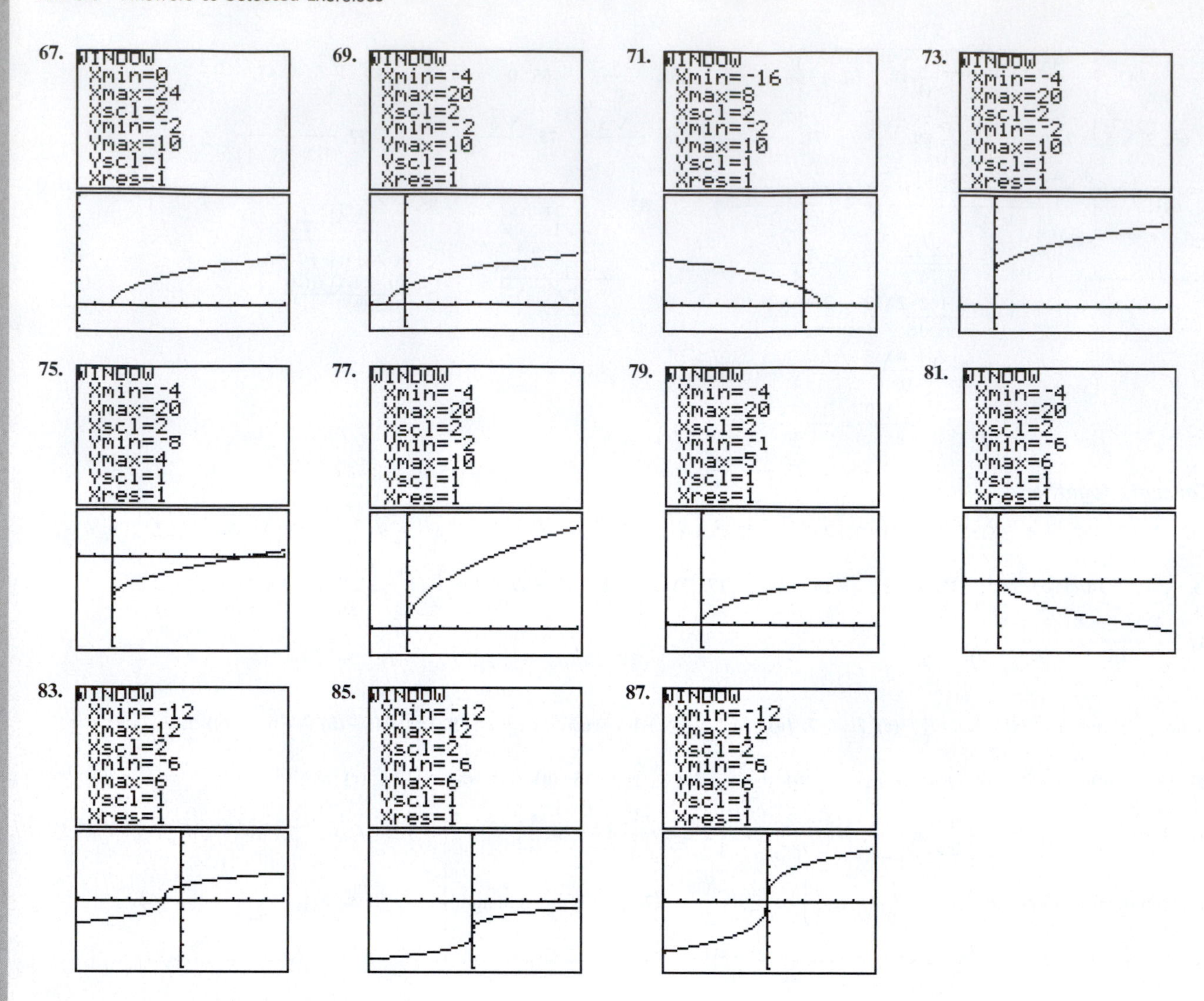


## Section 7.7

**1.** radical equation **3.** False **5.** Answers will vary. **7.** $\{16\}$ **9.** $\{7\}$ **11.** $\{11\}$ **13.** no real solution; $\emptyset$ or $\{\ \}$ **15.** $\{2\}$ **17.** $\{12\}$ **19.** $\{25\}$ **21.** $\{11\}$ **23.** $\{5\}$ **25.** $\{1\}$ **27.** $\{-5\}$ **29.** $\left\{0, \frac{1}{4}\right\}$ **31.** $\{3\}$ **33.** $\{9\}$ **35.** $\{4\}$ **37.** $\{6\}$ **39.** $\{4\}$ **41.** $\{-3\}$ **43.** $\{-1, 10\}$ **45.** $\{0, 4\}$ **47.** $\{3\}$ **49.** $\{5\}$ **51.** $\{2, 10\}$ **53.** $\{3\}$ **55.** $\{4\}$ **57.** $\{6\}$ **59.** $r = \dfrac{A^2 - P^2}{P^2}$ **61.** $V = \frac{4}{3}\pi r^3$ **63.** $F = \dfrac{q_1 q_2 r^2}{4\pi\varepsilon_0}$ **65.** no real solution; $\emptyset$ or $\{\}$ **67.** $\{4\}$ **69.** $\{2\}$ **71.** $\{9\}$ **73.** $\{-3\}$ **75.** $\{4\}$ **77.** $\{9\}$ **79.** $\{5\}$

**81.** **(a)** $(2, 0)$ **(b)** $(3, 1)$ **(c)** $(6, 2)$ **(d)** **(e)** The equation $f(x) = -1$ has no solution because the graph of the function does not go below the $x$-axis.

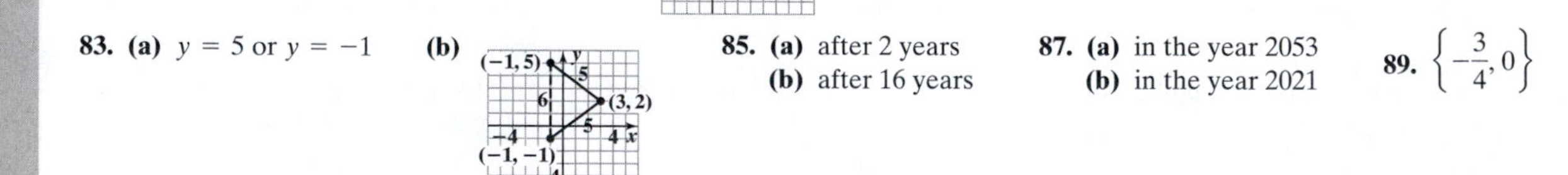

**83.** **(a)** $y = 5$ or $y = -1$ **(b)** **85.** **(a)** after 2 years **(b)** after 16 years **87.** **(a)** in the year 2053 **(b)** in the year 2021 **89.** $\left\{-\frac{3}{4}, 0\right\}$

**91.** Raising both sides of the equation to an even power. **93.** $0, -4, 12, \frac{2}{3}$, and $1.\overline{56}$ **95.** $0, -4, 12, \frac{2}{3}, 1.\overline{56}, \sqrt{2}, \pi$, and $\sqrt[3]{-4}$

**97.**

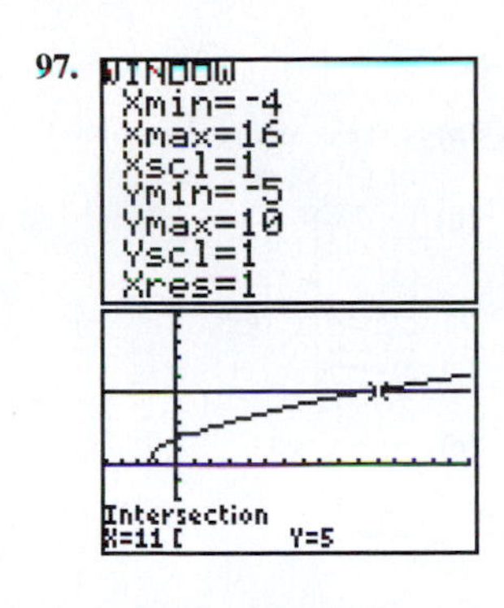


**99.**

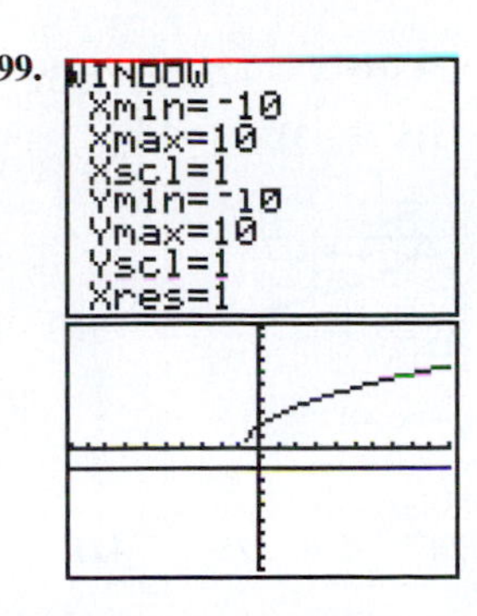


The two graphs do not intersect. Therefore, the equation has no real solution.

## Section 7.8

**1.** imaginary unit **3.** conjugate **5.** False **7.** Answers will vary. **9.** Answers will vary. **11.** $2i$ **13.** $-9i$ **15.** $3\sqrt{5}i$ **17.** $10\sqrt{3}i$ **19.** $\sqrt{7}i$ **21.** $5 + 7i$ **23.** $-2 - 2\sqrt{7}i$ **25.** $2 + i$ **27.** $\frac{1}{3} + \frac{\sqrt{2}}{6}i$ **29.** $6 - 2i$ **31.** $-4 + 6i$ **33.** $2 - 5i$ **35.** $3 - 2\sqrt{2}i$ **37.** $24 + 12i$ **39.** $-5 - 2i$ **41.** $5 + 10i$ **43.** $14 - 22i$ **45.** $34i$ **47.** $-4 + 5\sqrt{2}i$ **49.** $\frac{25}{48} + \frac{5}{24}i$ **51.** $5 + 12i$ **53.** $-9 + 40i$ **55.** $-6$ **57.** $-4\sqrt{5}$ **59.** $84 - 47i$ **61. (a)** $3 - 5i$ **(b)** 34 **63. (a)** $2 + 7i$ **(b)** 53 **65. (a)** $-7 - 2i$ **(b)** 53 **67.** $\frac{1}{3} - \frac{1}{3}i$ **69.** $\frac{2}{5} + i$ **71.** $\frac{6}{5} - \frac{3}{5}i$ **73.** $\frac{3}{29} - \frac{7}{29}i$ **75.** $i$ **77.** $1 + 3i$ **79.** $-\frac{1}{5} - \frac{7}{5}i$ **81.** $i$ **83.** $-i$ **85.** $i$ **87.** $-i$ **89.** $-15 - 8i$ **91.** $5 + 12i$ **93.** $\frac{2}{3} + i$ **95.** $\frac{1}{2} - \frac{1}{2}i$ **97.** 12 **99.** $-15 - 20i$ **101.** $-5\sqrt{6}$ **103.** $-\frac{1}{5}i$ **105.** $\frac{2}{5} + \frac{1}{5}i$ **107.** $-\frac{4}{41} - \frac{5}{41}i$ **109. (a)** $-1$ **(b)** $2i$ **111. (a)** $-7 + 6i$ **(b)** $4 - 4i$ **113. (a)** $10 - i$ ohms **(b)** 10 ohms **(c)** $-1$ ohm **115. (a)** 0 **(b)** 0 **117. (a)** 0 **(b)** 0 **(c)** 0 **119.** For a polynomial with real coefficients, the zeros will be real numbers or will occur in conjugate pairs. If the complex number $a + bi$ is a complex zero of the polynomial, then its conjugate $a - bi$ is also a complex conjugate. **121.** $y^2 - 8y + 16$ **123.** $7 - 24i$

**125.**

(1.3-4.3i)+(-5.3+0.7i)
-4-3.6i

**127.**

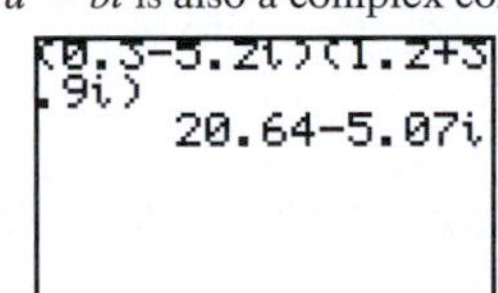


**129.**

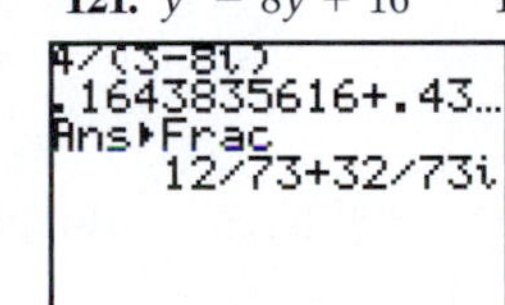


**131.**

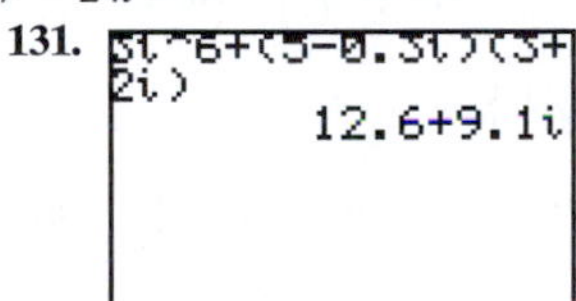


## Chapter 7 Review

**1.** 7 **2.** $-5$ **3.** $\frac{2}{3}$ **4.** 3 **5.** 3 **6.** 10 **7.** $z$ **8.** $|5p - 3|$ **9.** 9 **10.** not a real number **11.** $-2$ **12.** 9 **13.** 128 **14.** $-9$ **15.** $-1331$ **16.** 6 **17.** $-4.02$ **18.** 2.30 **19.** 4.64 **20.** 1.78 **21.** $(5a)^{1/3}$ **22.** $p^{7/5}$ **23.** $(10z)^{3/4}$ **24.** $(2ab)^{5/6}$ **25.** 64 **26.** $\frac{1}{k^{1/4}}$ **27.** $p^2 \cdot q^6$ or $(p \cdot q^3)^2$ **28.** $\frac{2b^{1/20}}{a^{3/10}}$ or $2\left(\frac{b}{a^6}\right)^{1/20}$ **29.** $10m^{1/3} + \frac{5}{m}$ **30.** $\frac{1}{2x^{1/3}}$ **31.** $\sqrt[4]{x^3}$ **32.** $11x^2y^5$ **33.** $m^2\sqrt[6]{m}$ **34.** $\frac{1}{\sqrt[3]{c}}$ **35.** $(3m - 1)^{1/4}(3m^2 - 22m + 9)$ **36.** $\frac{(3x + 5)(x - 3)}{(x^2 - 5)^{2/3}}$ **37.** $\sqrt{105}$ **38.** $\sqrt[4]{12a^3b^3}$ **39.** $4\sqrt{5}$ **40.** $-5\sqrt[3]{4}$ **41.** $3m^2n\sqrt[3]{6n}$ **42.** $p^2|q|\sqrt[4]{50}$ **43.** $8|x^3|\sqrt{y}$ as long as $y \geq 0$ **44.** $(2x + 1)\sqrt{2x + 1}$ as long as $2x + 1 \geq 0$ **45.** $wz\sqrt{w}$ **46.** $3x^2z\sqrt{5yz}$ **47.** $2a^4b\sqrt[3]{2b^2}$ **48.** $2(x + 1)$ **49.** $3\sqrt{30}$ **50.** $2\sqrt[3]{75}$ **51.** $-2x^2y^3\sqrt[3]{9x}$ **52.** $30xy\sqrt{3xy}$ **53.** $\frac{11}{5}$ **54.** $\frac{a^2\sqrt{5}}{8b}$ **55.** $\frac{\sqrt[3]{6}}{k}$ **56.** $\frac{-2w^5\sqrt[3]{20}}{7}$ **57.** $2h$ **58.** $\frac{5b^3}{2a}$ **59.** $\frac{-2x^2}{3y}$ **60.** $\frac{2n\sqrt{n}}{m}$ **61.** $\sqrt[6]{500}$ **62.** $2\sqrt[12]{2}$ **63.** $8\sqrt[4]{x}$ **64.** $6\sqrt[3]{4y}$ **65.** $5\sqrt{2} - 4\sqrt{3}$ **66.** $13\sqrt{2}$ **67.** $\sqrt[3]{2z}$ **68.** $17\sqrt[3]{x^2}$ **69.** $7\sqrt{a}$ **70.** $9x\sqrt{3}$ **71.** $4m\sqrt[3]{4m^2y^2}$ **72.** $(y - 1)\sqrt{y - 4}$ **73.** $\sqrt{15} - 3\sqrt{5}$ **74.** $3\sqrt[3]{5} + \sqrt[3]{20}$ **75.** $7 + \sqrt{5}$ **76.** $42 + 7\sqrt{2} + 6\sqrt{3} + \sqrt{6}$ **77.** $-44$ **78.** $9\sqrt[3]{x^2} + 5\sqrt[3]{x} - 4$ **79.** $x - 2\sqrt{5x} + 5$ **80.** $247 + 22\sqrt{10}$ **81.** $2a - b^2$ **82.** $\sqrt[3]{36s^2} - 5\sqrt[3]{6s} - 14$ **83.** $\frac{\sqrt{6}}{3}$ **84.** $2\sqrt{3}$ **85.** $\frac{4\sqrt{3p}}{p^2}$ **86.** $\frac{5\sqrt{2a}}{2a}$ **87.** $-\frac{\sqrt{6y}}{3y^2}$ **88.** $\frac{3\sqrt[3]{25}}{5}$ **89.** $-\frac{\sqrt[3]{300}}{15}$ **90.** $\frac{27\sqrt[5]{4p^2q}}{2pq}$ **91.** $\frac{42 + 6\sqrt{6}}{43}$ **92.** $-\frac{\sqrt{3} + 9}{26}$ **93.** $\frac{\sqrt{3}(3 - \sqrt{2})}{7}$ or $\frac{3\sqrt{3} - \sqrt{6}}{7}$ **94.** $\frac{k + \sqrt{km}}{k - m}$ **95.** $\frac{7 + 2\sqrt{10}}{3}$ **96.** $\frac{9 - 6\sqrt{y} + y}{9 - y}$ or $\frac{y - 6\sqrt{y} + 9}{9 - y}$ **97.** $\frac{10\sqrt{2} - 4\sqrt{3}}{19}$ **98.** $\frac{8\sqrt{2} - 3\sqrt{15}}{7}$ **99.** $\frac{25\sqrt{7}}{21}$ **100.** $\frac{4 + \sqrt{7}}{9}$ **101. (a)** 1 **(b)** 2 **(c)** 3 **102. (a)** 0 **(b)** 2 **(c)** 4 **103. (a)** 1 **(b)** $-1$ **(c)** 2 **104. (a)** 0 **(b)** 2 **(c)** $\frac{1}{2}$ **105.** $\left\{x \middle| x \geq \frac{5}{3}\right\}$ or $\left[\frac{5}{3}, \infty\right)$ **106.** $\{x|x \text{ is any real number}\}$ or $(-\infty, \infty)$ **107.** $\left\{x \middle| x \geq -\frac{1}{6}\right\}$ or $\left[-\frac{1}{6}, \infty\right)$

**108.** $\{x|x \text{ is any real number}\}$ or $(-\infty, \infty)$ **109.** $\{x|x > 2\}$ or $(2, \infty)$ **110.** $\{x|x < 0 \text{ or } x \geq 3\}$ or $(-\infty, 0) \cup [3, \infty)$

**111. (a)** $\{x|x \leq 1\}$ or $(-\infty, 1]$ **(b)** **(c)** $[0, \infty)$

**112. (a)** $\{x|x \geq -1\}$ or $[-1, \infty)$ **(b)** **(c)** $[-2, \infty)$

**113. (a)** $\{x|x \geq -3\}$ or $[-3, \infty)$ **(b)** **(c)** $(-\infty, 0]$

**114. (a)** $\{x|x \text{ is any real number}\}$ or $(-\infty, \infty)$ **(b)**

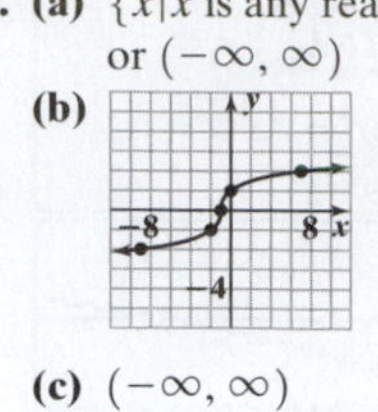

**(c)** $(-\infty, \infty)$

**115.** $\{169\}$ **116.** $\{-3\}$ **117.** $\left\{\frac{89}{3}\right\}$ **118.** no real solution; $\varnothing$ or $\{\ \}$ **119.** $\{0\}$ **120.** $\{25\}$ **121.** $\{-512\}$ **122.** $\{2\}$

**123.** $\{5\}$ **124.** $\{2\}$ **125.** $\{11\}$ **126.** $\{6\}$ **127.** $\{9\}$ **128.** no real solution; $\varnothing$ or $\{\ \}$ **129.** $\left\{\frac{15}{2}\right\}$ **130.** $\{-5, 5\}$

**131.** $h = \frac{3V}{\pi r^2}$ **132.** $v = \frac{30}{f_s^3}$ **133.** $\sqrt{29}i$ **134.** $3\sqrt{6}i$ **135.** $14 - 9\sqrt{2}i$ **136.** $2 + \sqrt{5}i$ **137.** $1 - 2i$ **138.** $-5 + 10i$

**139.** $5 - 7\sqrt{5}i$ **140.** $-5 + 7i$ **141.** $47 + 13i$ **142.** $8 - \frac{11}{6}i$ **143.** $-9$ **144.** $67 - 42i$ **145.** 145 **146.** $27 + 38i$

**147.** $\frac{6}{17} - \frac{10}{17}i$ **148.** $-\frac{21}{53} - \frac{6}{53}i$ **149.** $\frac{4}{29} - \frac{19}{29}i$ **150.** $\frac{1}{2} + \frac{7}{2}i$ **151.** $-i$ **152.** $i$

### Chapter 7 Test

**1.** $\frac{1}{7}$ **2.** $6y\sqrt[12]{x^8y^3}$ **3.** $2a^5b^4\sqrt[5]{4a^3b}$ **4.** $\sqrt{39mn}$ **5.** $4x^3y^2\sqrt{2x}$ **6.** $\frac{3a}{2b^2}$ **7.** $(x + 6)\sqrt{5x}$ **8.** $a\sqrt{b}$ **9.** $33 - 5\sqrt{x} - 2x$

**10.** $\frac{-\sqrt{2}}{18}$ **11.** $5 - 2\sqrt{5}$ **12. (a)** 1 **(b)** 3 **13.** $\left\{x \middle| x \leq \frac{5}{3}\right\}$ or $\left(-\infty, \frac{5}{3}\right]$ **14. (a)** $\{x|x \geq 0\}$ or $[0, \infty)$ **(b)**

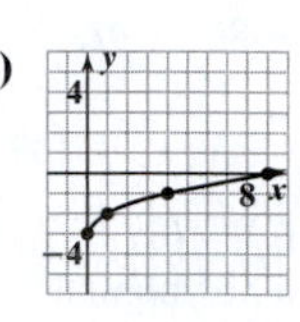

**(c)** $[-3, \infty)$ **15.** $\{13\}$ **16.** $\{3\}$ **17.** $\{2\}$ **18.** $17 - 13i$ **19.** $29 - 2i$ **20.** $\frac{73}{265} - \frac{89}{265}i$

### Cumulative Review R–7

**1.** $\frac{9}{2}$ **2.** $x + 7y + 6$ **3.** $\{4\}$ **4.** $(-\infty, 2]$ **5. (a)** 19 **(b)** 29 **6.** $\{x|x \neq 4, -2\}$ **7. (a)** 3500 games **(b)** \$120

**8.** $y = -2x + 4$ **9.** **10.** $(4, -1)$ **11.** 4 pounds of dried fruit and 6 pounds of nuts **12.** $-55$

**13.** $8x^3 - 2x^2 - 3x + 10$ **14.** $8x^3 - 20x + 9$ **15.** $3x^2 + 2x - 6 + \frac{7}{2x^2 + 3x - 5}$ **16.** $4(2x + 3)(x - 7)$

**17.** $\frac{x^2 + 6x + 3}{(x - 3)(x + 2)}$ **18.** $\{-1, 4\}$ **19.** $\{x|x < 4 \text{ or } x \geq 10\}$, or $(-\infty, 4) \cup [10, \infty)$

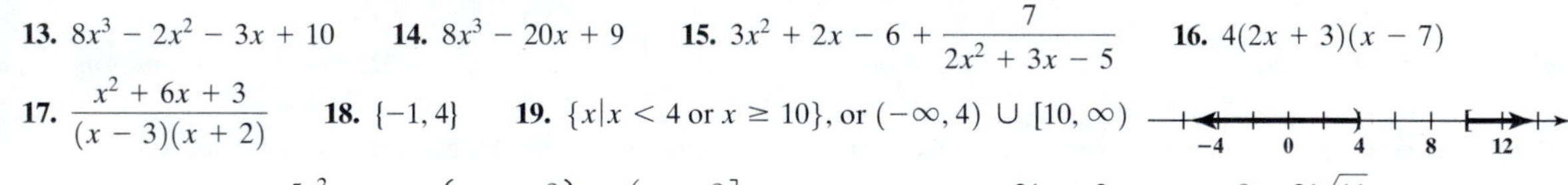

**20.** 2.4 hours **21.** $\frac{5a^2}{b}$ **22.** $\left\{x \middle| x \leq \frac{8}{3}\right\}$ or $\left(-\infty, \frac{8}{3}\right]$ **23.** $\{2\}$ **24.** $-\frac{21}{50} + \frac{3}{50}i$ **25.** $\frac{8 + 2\sqrt{11}}{5}$

# Chapter 8

### Section 8.1

**1.** $\sqrt{p}; -\sqrt{p}$ **3.** hypotenuse; legs **5.** False **7.** Answers will vary. **9.** $\{-10, 10\}$ **11.** $\{-5\sqrt{2}, 5\sqrt{2}\}$ **13.** $\{-5i, 5i\}$

**15.** $\left\{-\frac{\sqrt{5}}{2}, \frac{\sqrt{5}}{2}\right\}$ **17.** $\{-2\sqrt{2}, 2\sqrt{2}\}$ **19.** $\{-4, 4\}$ **21.** $\left\{-\frac{2\sqrt{6}}{3}, \frac{2\sqrt{6}}{3}\right\}$ **23.** $\{-2i, 2i\}$ **25.** $\{-11, 5\}$

**27.** $\{1 - 3\sqrt{2}i, 1 + 3\sqrt{2}i\}$ **29.** $\{-5 - \sqrt{3}, -5 + \sqrt{3}\}$ **31.** $\left\{-\frac{4}{3}, \frac{2}{3}\right\}$ **33.** $\left\{\frac{2}{3} - \frac{\sqrt{5}}{3}, \frac{2}{3} + \frac{\sqrt{5}}{3}\right\}$ **35.** $\{-13, 5\}$

**37.** $x^2 + 10x + 25; (x + 5)^2$ **39.** $z^2 - 18z + 81; (z - 9)^2$ **41.** $y^2 + 7y + \frac{49}{4}; \left(y + \frac{7}{2}\right)^2$ **43.** $w^2 + \frac{1}{2}w + \frac{1}{16}; \left(w + \frac{1}{4}\right)^2$

**45.** $q^2 - \frac{3}{5}q + \frac{9}{100}; \left(q - \frac{3}{10}\right)^2$ **47.** $\{-2, 7\}$ **49.** $\{-6, 2\}$ **51.** $\{2 - \sqrt{3}, 2 + \sqrt{3}\}$ **53.** $\{-4 - \sqrt{7}, -4 + \sqrt{7}\}$ **55.** $\{2 - i, 2 + i\}$

**57.** $\left\{-\frac{5}{2} - \frac{\sqrt{33}}{2}, -\frac{5}{2} + \frac{\sqrt{33}}{2}\right\}$ **59.** $\left\{\frac{3}{2} - \frac{\sqrt{17}}{2}, \frac{3}{2} + \frac{\sqrt{17}}{2}\right\}$ **61.** $\{4 - \sqrt{19}, 4 + \sqrt{19}\}$ **63.** $\left\{\frac{1}{2} - \frac{\sqrt{11}i}{2}, \frac{1}{2} + \frac{\sqrt{11}i}{2}\right\}$

**65.** $\left\{-\frac{3}{2}, 4\right\}$ **67.** $\left\{1 - \frac{\sqrt{3}}{3}, 1 + \frac{\sqrt{3}}{3}\right\}$ **69.** $\left\{\frac{5}{4} - \frac{\sqrt{17}}{4}, \frac{5}{4} + \frac{\sqrt{17}}{4}\right\}$ **71.** $\left\{-1 - \frac{\sqrt{6}i}{2}, -1 + \frac{\sqrt{6}i}{2}\right\}$

**73.** $\left\{-\frac{1}{6}-\frac{\sqrt{61}}{6}, -\frac{1}{6}+\frac{\sqrt{61}}{6}\right\}$ **75.** 10 **77.** 20 **79.** $5\sqrt{2}$; 7.07 **81.** 2 **83.** $2\sqrt{34}$; 11.66 **85.** $b = 4\sqrt{3} \approx 6.93$
**87.** $a = 4\sqrt{5} \approx 8.94$ **89.** $x = -3$ or $x = 9$ **91.** $x = -2 \pm 3\sqrt{2}$ **93.** approximately 8.944 units **95.** approximately 104.403 yards

**97.** approximately 31.623 feet **99.** **(a)** approximately 22.913 feet **(b)** 15 feet **101.** **(a)** 1 second **(b)** approximately 1.732 seconds **(c)** 2 seconds **103.** approximately 9.54% **105.** The triangle is a right triangle; the hypotenuse is 17. **107.** The triangle is not a right triangle.

**109.** $c^2 = (m^2 + n^2)^2 = m^4 + 2m^2n^2 + n^4$

$$\begin{aligned} a^2 + b^2 &= (m^2 - n^2)^2 + (2mn)^2 \\ &= m^4 - 2m^2n^2 + n^4 + 4m^2n^2 \\ &= m^4 + 2m^2n^2 + n^4 \end{aligned}$$

Because $c^2$ and $a^2 + b^2$ result in the same expression, $a$, $b$, and $c$ are the lengths of the sides of a right triangle.

**111.** $\{-4, 9\}$ **113.** $\left\{-1, \frac{1}{2}\right\}$

**115.** In both cases, the simpler equations are linear.

## Section 8.2

**1.** $\dfrac{-b \pm \sqrt{b^2 - 4ac}}{2a}$ **3.** negative **5.** False **7.** Answers may vary.

**9.** The solutions to the equation $ax^2 + bx + c = 0$, $a \neq 0$, are given by $x = \dfrac{-b \pm \sqrt{b^2 - 4ac}}{2a}$. **11.** $\{-2, 6\}$ **13.** $\left\{-\frac{3}{2}, \frac{5}{3}\right\}$

**15.** $\left\{1 - \frac{\sqrt{3}}{2}, 1 + \frac{\sqrt{3}}{2}\right\}$ **17.** $\left\{1 - \frac{2\sqrt{3}}{3}, 1 + \frac{2\sqrt{3}}{3}\right\}$ **19.** $\left\{-\frac{1}{3} - \frac{\sqrt{13}}{3}, -\frac{1}{3} + \frac{\sqrt{13}}{3}\right\}$ **21.** $\{1 - \sqrt{6}i, 1 + \sqrt{6}i\}$

**23.** $\left\{\frac{1}{2} - \frac{\sqrt{13}}{2}i, \frac{1}{2} + \frac{\sqrt{13}}{2}i\right\}$ **25.** $\left\{\frac{1}{4} - \frac{\sqrt{5}}{4}, \frac{1}{4} + \frac{\sqrt{5}}{4}\right\}$ **27.** $\left\{-\frac{2}{3} - \frac{\sqrt{7}}{3}, -\frac{2}{3} + \frac{\sqrt{7}}{3}\right\}$ **29.** 21; two irrational solutions

**31.** $-56$; two complex solutions that are not real **33.** 0; one repeated real solution **35.** $-8$; two complex solutions that are not real

**37.** 44; two irrational solutions **39.** $\left\{\frac{5}{2} - \frac{\sqrt{5}}{2}, \frac{5}{2} + \frac{\sqrt{5}}{2}\right\}$ **41.** $\left\{-\frac{8}{3}, 1\right\}$ **43.** $\left\{-\frac{7}{2}, 5\right\}$ **45.** $\{-1 - \sqrt{7}i, -1 + \sqrt{7}i\}$

**47.** $\left\{-\frac{3}{5}, 1\right\}$ **49.** $\left\{\frac{1}{7} - \frac{\sqrt{29}}{7}, \frac{1}{7} + \frac{\sqrt{29}}{7}\right\}$ **51.** $\{-4, 4\}$ **53.** $\left\{\frac{1}{4} - \frac{1}{4}i, \frac{1}{4} + \frac{1}{4}i\right\}$ **55.** $\left\{-\frac{2}{3}\right\}$ **57.** $\left\{-\frac{1}{3} - \frac{2\sqrt{7}}{3}, -\frac{1}{3} + \frac{2\sqrt{7}}{3}\right\}$

**59.** $\{2 - \sqrt{13}, 2 + \sqrt{13}\}$ **61.** $\{1 - \sqrt{5}, 1 + \sqrt{5}\}$ **63.** $\left\{\frac{1}{4} - \frac{3\sqrt{7}}{4}i, \frac{1}{4} + \frac{3\sqrt{7}}{4}i\right\}$ **65.** **(a)** $x = -7$ or $x = 3$

**(b)** $x = -4$ or $x = 0$ **67.** **(a)** $n = -1 \pm \frac{\sqrt{6}}{2}$ **(b)** $x = -1 \pm \frac{\sqrt{2}}{2}$ **69.** **(a)** $x = -\frac{1}{3} \pm \frac{\sqrt{5}}{3}i$ **(b)** $x = -\frac{1}{3} \pm \frac{\sqrt{7}}{3}$

**71.** $x = 3$; the three sides measure 3, 4, and 5 units **73.** Either $x = 1$ and the three sides measure 3, 4, and 5 units, or $x = 5$ and the three sides measure 7, 24, and 25 units.

**75.** $-2 + 2\sqrt{11}$ inches by $2 + 2\sqrt{11}$ inches, which is approximately 4.633 inches by 8.633 inches.

**77.** The base is $\frac{3}{2} + \frac{\sqrt{209}}{2}$ inches, which is approximately 8.728 inches; the height is $-\frac{3}{2} + \frac{\sqrt{209}}{2}$ inches, which is approximately 5.728 inches.

**79.** **(a)** $R(17) = 1161.1$; if 17 pairs of sunglasses are sold per week, then the company's revenue will be \$1161.10. $R(25) = 1687.5$; if 25 pairs of sunglasses are sold per week, then the company's revenue will be \$1687.50.
**(b)** either 200 or 500 pairs of sunglasses
**(c)** 350 pairs of sunglasses

**81.** **(a)** after approximately 0.6 seconds and after approximately 3.8 seconds
**(b)** after approximately 1.3 seconds and after approximately 3.0 seconds
**(c)** No; the solutions to the equation are complex solutions that are not real.

**83.** 12 inches **85.** **(a)** ages 25 and 68 **(b)** ages 30 and 63 **87.** approximately 4.3 miles per hour **89.** approximately 4.6 hours

**91.** By the quadratic formula, the solutions of the equation $ax^2 + bx + c = 0$ are $x = \dfrac{-b - \sqrt{b^2 - 4ac}}{2a}$ and $x = \dfrac{-b + \sqrt{b^2 - 4ac}}{2a}$.

The sum of these two solutions is

$$\frac{-b - \sqrt{b^2 - 4ac}}{2a} + \frac{-b + \sqrt{b^2 - 4ac}}{2a} = \frac{-2b}{2a} = -\frac{b}{a}.$$

**93.** The solutions of $ax^2 + bx + c = 0$ are $x = \dfrac{-b \pm \sqrt{b^2 - 4ac}}{2a}$.

The solutions of $ax^2 - bx + c = 0$ are

$$x = \frac{-(-b) \pm \sqrt{(-b)^2 - 4ac}}{2a} = \frac{b \pm \sqrt{b^2 - 4ac}}{2a}.$$

Now, the negatives of the solutions to $ax^2 - bx + c = 0$

are $-\left(\dfrac{b \pm \sqrt{b^2 - 4ac}}{2a}\right) = \dfrac{-b \mp \sqrt{b^2 - 4ac}}{2a} = \dfrac{-b \pm \sqrt{b^2 - 4ac}}{2a}$

which are the solutions to $ax^2 + bx + c = 0$.

**95. (a)**

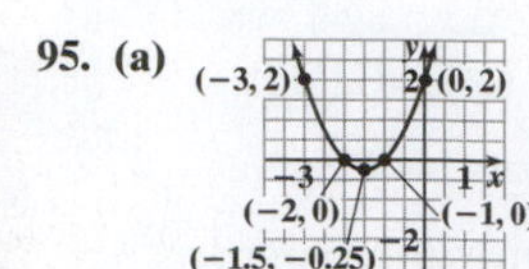


**(b)** $x = -1$ or $x = -2$

**(c)** The $x$-intercepts of the function $f(x) = x^2 + 3x + 2$ are $-2$ and $-1$, which are the same as the solutions of the equation $x^2 + 3x + 2 = 0$.

**97. (a)**

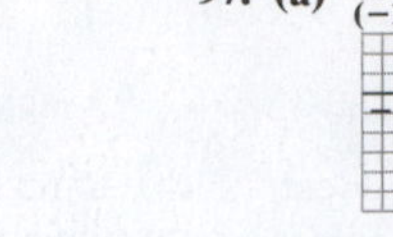


**(b)** $x = 1$

**(c)** The $x$-intercept of the function $g(x) = x^2 - 2x + 1$ is 1, which is the same as the solution of the equation $x^2 - 2x + 1 = 0$.

**99.** The discriminant is 37; the equation has two irrational solutions. This conclusion based on the discriminant is apparent in the graph because the graph has two $x$-intercepts.

**101.** The discriminant is $-7$; the equation has two complex solutions that are not real. This conclusion based on the discriminant is apparent in the graph because the graph has no $x$-intercept.

**103. (a)** $x = -3$ or $x = 8$
**(b)** The $x$-intercepts are $-3$ and 8.

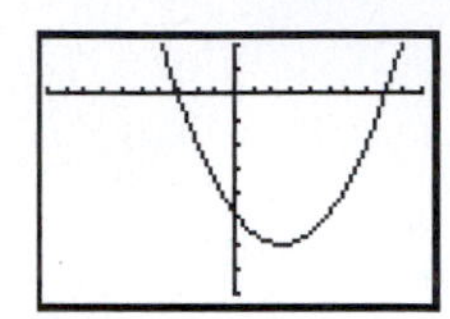

The $x$-intercepts of $y = x^2 - 5x - 24$ are the same as the solutions of $x^2 - 5x - 24 = 0$.

**105. (a)** $x = 3$
**(b)** The $x$-intercept is 3.

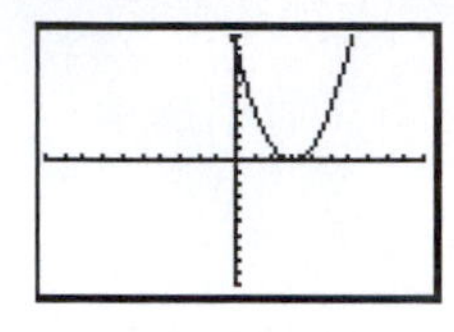

The $x$-intercept of $y = x^2 - 6x + 9$ is the same as the solution of $x^2 - 6x + 9 = 0$.

**107. (a)** $x = -\frac{5}{2} \pm \frac{\sqrt{7}}{2}i$

**(b)** The graph has no $x$-intercepts.

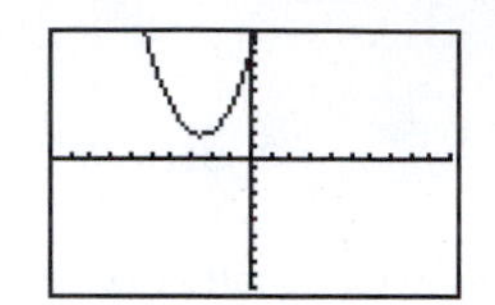

$y = x^2 + 5x + 8$ has no $x$-intercepts, and the solutions of $x^2 + 5x + 8 = 0$ are not real.

## Section 8.3

**1.** equation quadratic in form **3.** True **5.** Answers will vary. **7.** $3x - 2$ **9.** $\{-2, -1, 1, 2\}$ **11.** $\{-3i, -2i, 2i, 3i\}$

**13.** $\left\{-\frac{1}{2}, -2, \frac{1}{2}, 2\right\}$ **15.** $\{-\sqrt{2}, -\sqrt{3}, \sqrt{2}, \sqrt{3}\}$ **17.** $\{10, 2\}$ **19.** $\{-2, -3, 2, 3\}$ **21.** $\{-2i, -\sqrt{7}i, 2i, \sqrt{7}i\}$

**23.** $\{16\}$ **25.** no solution; $\emptyset$ or $\{\ \}$ **27.** $\left\{\frac{1}{4}\right\}$ **29.** $\left\{-\frac{1}{7}, \frac{1}{4}\right\}$ **31.** $\left\{-\frac{2}{3}, \frac{5}{2}\right\}$ **33.** $\{-64, 1\}$ **35.** $\{-1, 8\}$ **37.** $\{25\}$ **39.** $\left\{\frac{1}{3}, \frac{1}{2}\right\}$

**41.** $\left\{-\frac{11}{5}, -1\right\}$ **43.** $\left\{1, 3, -\frac{1}{2} - \frac{\sqrt{3}}{2}i, -\frac{3}{2} - \frac{3\sqrt{3}}{2}i, -\frac{1}{2} + \frac{\sqrt{3}}{2}i, -\frac{3}{2} + \frac{3\sqrt{3}}{2}i\right\}$ **45.** $\{-2, 4\}$ **47.** $\{-2i, -2\sqrt{2}, 2i, 2\sqrt{2}\}$ **49.** $\{16\}$

**51.** $\{-3, -2i, 3, 2i\}$ **53.** $\left\{-\frac{10}{3}, -2\right\}$ **55.** $\{9, 16\}$ **57.** $\left\{\frac{1}{2}, 5\right\}$ **59. (a)** $0, -\sqrt{7}i, \sqrt{7}i$ **(b)** $-\sqrt{6}i, \sqrt{6}i, -i, i$ **61. (a)** $0, -\sqrt{3}, \sqrt{3}$

**(b)** $-\sqrt{2}i, \sqrt{2}i, -\sqrt{5}, \sqrt{5}$ **63. (a)** $-1, \frac{1}{6}$ **(b)** $-\frac{1}{2}, \frac{1}{7}$ **65.** $-\sqrt{7}i, \sqrt{7}i, -\sqrt{2}i$ and $\sqrt{2}i$ **67.** $\frac{81}{4}$ **69.** $-\frac{8}{3}, -2$

**71. (a)** $x = 2$ or $x = 3$ **(b)** $x = 5$ or $x = 6$; comparing these solutions to those in part (a), we note that $5 = 2 + 3$ and $6 = 3 + 3$.
**(c)** $x = 0$ or $x = 1$; comparing these solutions to those in part (a), we note that $0 = 2 - 2$ and $1 = 3 - 2$.
**(d)** $x = 7$ or $x = 8$; comparing these solutions to those in part (a), we note that $7 = 2 + 5$ and $8 = 3 + 5$.
**(e)** The solution set of the equation $(x - a)^2 - 5(x - a) + 6 = 0$ is $\{2 + a, 3 + a\}$.

**73. (a)** $x = \frac{1}{2}$ or $x = 1$ **(b)** $x = \frac{5}{2}$ or $x = 3$; comparing these solutions to those in part (a), we note that $\frac{5}{2} = \frac{1}{2} + 2$ and $3 = 1 + 2$.

**(c)** $x = \frac{11}{2}$ or $x = 6$; comparing these solutions to those in part (a), we note that $\frac{11}{2} = \frac{1}{2} + 5$ and $6 = 1 + 5$.

**(d)** For $f(x) = 2x^2 - 3x + 1$, the zeros of $f(x - a)$ are $\frac{1}{2} + a$ and $1 + a$

**75. (a)** $R(1990) = 3000$; the revenue in 1990 was \$3,000 thousand (or \$3,000,000)

**(b)** $x = 2000$; in the year 2000, revenue was \$3,065 thousand (or \$3,065,000) **(c)** 2015 **77.** $x = \pm\frac{\sqrt{10 + 2\sqrt{17}}}{2}i$ or $x = \pm\frac{\sqrt{10 - 2\sqrt{17}}}{2}i$

**79.** $x = \pm\frac{3\sqrt{2}}{2}$ **81.** $x^2 - 2x + 4$ **83.** $p^{-2} + 4p^{-1} + 9$ **85.** $5\sqrt[3]{2a} - 4a\sqrt[3]{2a} = (5 - 4a)\sqrt[3]{2a}$

**87.** Let $Y_1 = x^4 + 5x^2 - 14$.

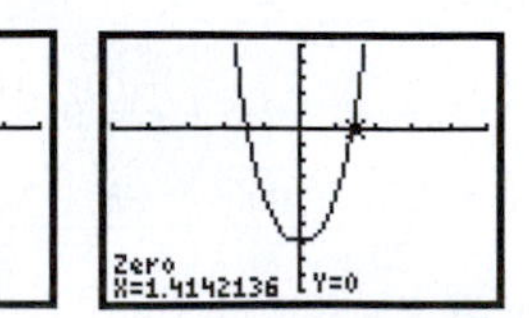


The solution set is approximately $\{-1.41, 1.41\}$.

**89.** Let $Y_1 = 2(x - 2)^2$ and $Y_2 = 5(x - 2) + 1$.

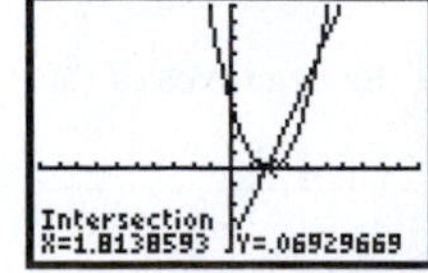


The solution set is approximately $\{1.81, 4.69\}$.

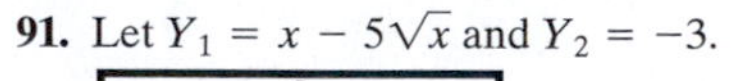

**91.** Let $Y_1 = x - 5\sqrt{x}$ and $Y_2 = -3$.

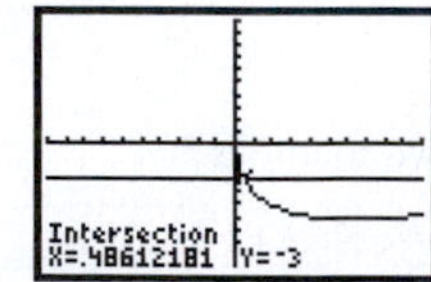


The solution set is approximately $\{0.49\}$.

**93. (a)** $Y_1 = x^2 - 5x - 6$

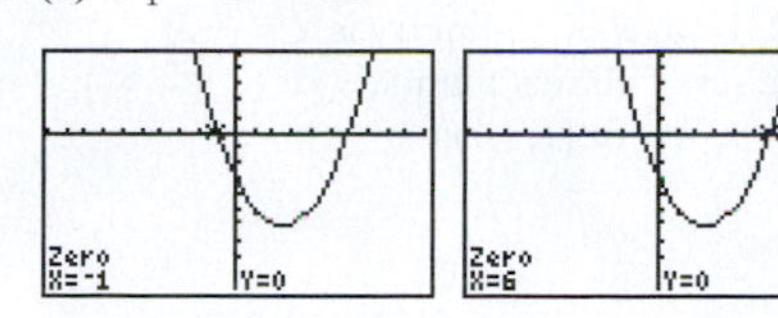


The $x$-intercepts are $-1$ and 6.

**(b)** $Y_1 = (x + 2)^2 - 5(x + 2) - 6$

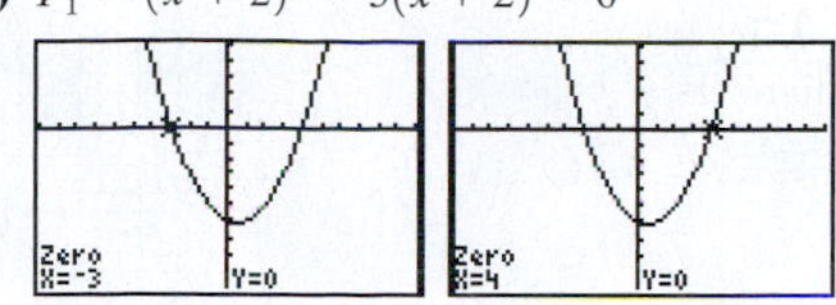


The $x$-intercepts are $-3$ and 4.

**(c)** $Y_1 = (x + 5)^2 - 5(x + 5) - 6$

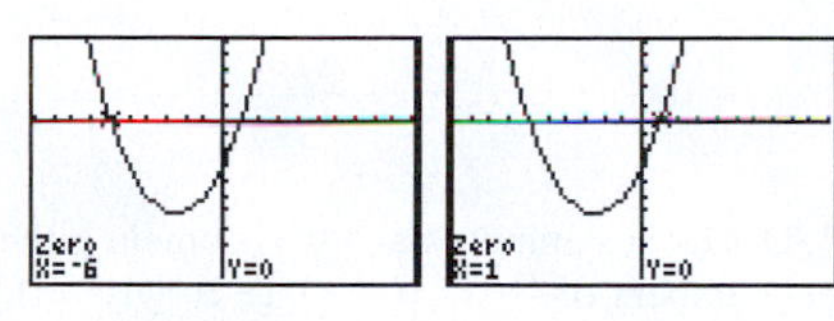


The $x$-intercepts are $-6$ and 1.

**(d)** The $x$-intercepts of the graph of $y = f(x) = x^2 - 5x - 6$ are $-1$ and 6. The $x$-intercepts of the graph of $y = f(x + a) = (x + a)^2 - 5(x + a) - 6$ are $-1 - a$ and $6 - a$.

## Putting the Concepts Together

**1.** $z^2 + 10z + 25 = (z + 5)^2$ **2.** $x^2 + 7x + \frac{49}{4} = \left(x + \frac{7}{2}\right)^2$ **3.** $n^2 - \frac{1}{4}n + \frac{1}{64} = \left(n - \frac{1}{8}\right)^2$ **4.** $\left\{\frac{1}{2}, \frac{5}{2}\right\}$ **5.** $\{4 - 2\sqrt{3}, 4 + 2\sqrt{3}\}$
**6.** $\{3 - \sqrt{2}, 3 + \sqrt{2}\}$ **7.** $\left\{-\frac{4\sqrt{5}}{7}, \frac{4\sqrt{5}}{7}\right\}$ **8.** $\{4 - \sqrt{10}, 4 + \sqrt{10}\}$ **9.** $\left\{-1 - \frac{\sqrt{3}}{3}i, -1 + \frac{\sqrt{3}}{3}i\right\}$ **10.** $\left\{-2 - \frac{\sqrt{42}}{3}, -2 + \frac{\sqrt{42}}{3}\right\}$
**11.** $b^2 - 4ac = 0$; the quadratic equation will have one repeated real solution.
**12.** $b^2 - 4ac = 60$; the quadratic equation will have two irrational solutions.
**13.** $b^2 - 4ac = -4$; the quadratic equation will have two complex solutions that are not real.
**14.** $c = \sqrt{116} = 2\sqrt{29}$ **15.** $\left\{\frac{9}{4}\right\}$ **16.** $\left\{\frac{1}{6}, -\frac{1}{3}\right\}$ **17.** Revenue will be \$12,000 when either 150 microwaves or 200 microwaves are sold.
**18.** The speed of the wind was approximately 52.9 miles per hour.

## Section 8.4

**1.** quadratic function **3.** $y$; $a$; vertically stretched; vertically compressed **5.** False
**7.** the vertex; if $a > 0$, the graph opens up and the vertex is the low point, if $a < 0$, the graph opens down and the vertex is the high point.
**9.** (I) (D) (II) (A) (III) (C) (IV) (B) **11.** shift 10 units to the left **13.** shift 12 units up
**15.** shift 5 units to the right and vertically stretch by a factor of 2 (multiply the $y$-coordinates by 2)
**17.** shift 5 units to the left, multiply the $y$-coordinates by $-3$ (which means it opens down and is stretched vertically by a factor of 3), and shift up 8 units

**19.**
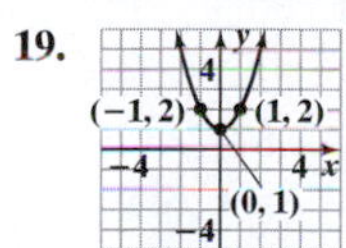


**21.**
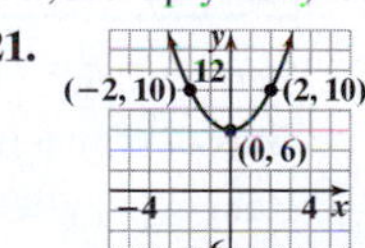


**23.**
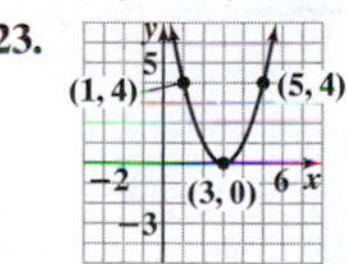


**25.**
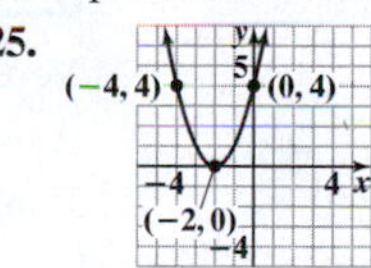


**27.**
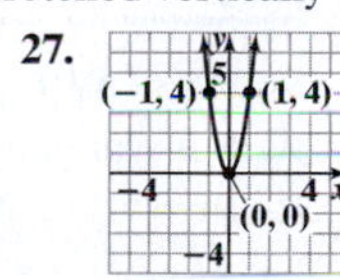


**29.**
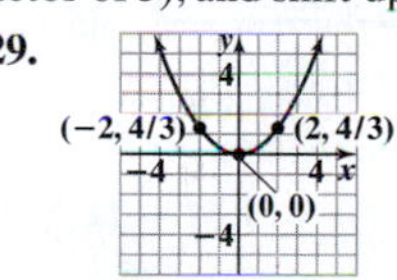


**31.**
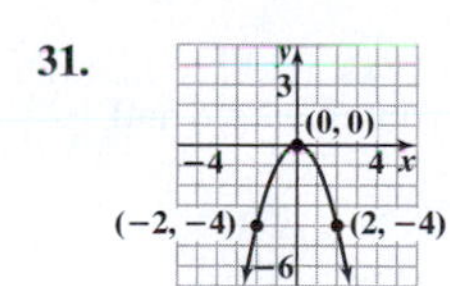


**33.**
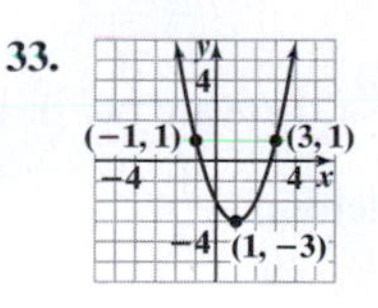


**35.**
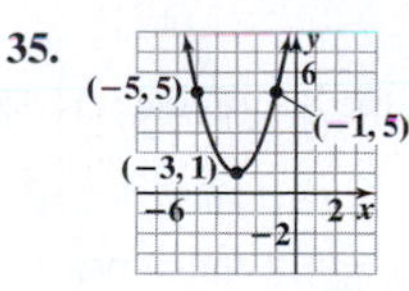


**37.**
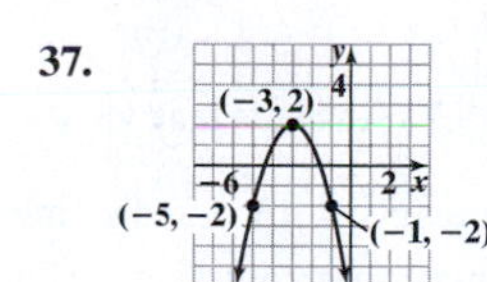


**39.**
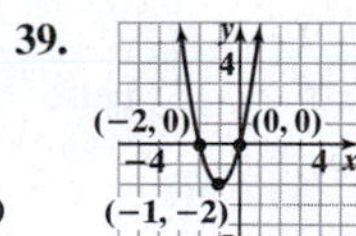


**41.**
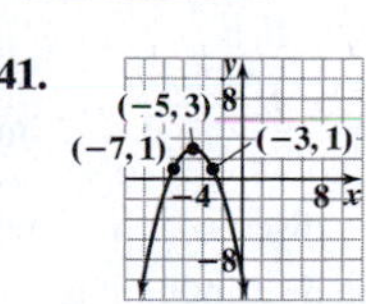


**43.** $f(x) = (x + 1)^2 - 5$

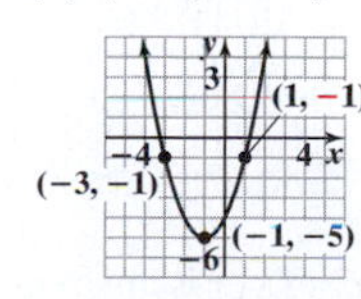


vertex is $(-1, -5)$; axis of symmetry is $x = -1$; domain is the set of all real numbers, $(-\infty, \infty)$; range is $\{y \mid y \geq -5\}$, or $[-5, \infty)$

**45.** $g(x) = (x - 2)^2 + 4$

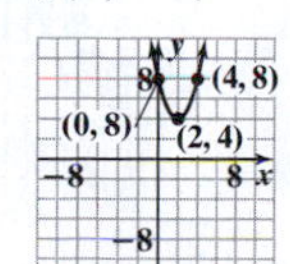


vertex is $(2, 4)$; axis of symmetry is $x = 2$; domain is the set of all real numbers, $(-\infty, \infty)$; range is $\{y \mid y \geq 4\}$, $[4, \infty)$

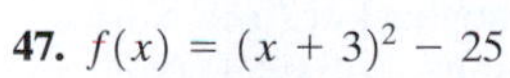

**47.** $f(x) = (x + 3)^2 - 25$

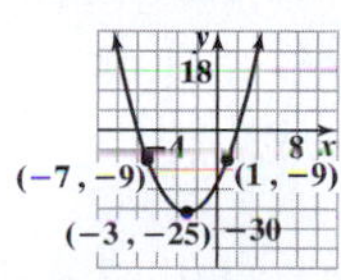


vertex is $(-3, -25)$; axis of symmetry is $x = -3$; domain is the set of all real numbers $(-\infty, \infty)$; range is $\{y \mid y \geq -25\}$, or $[-25, \infty)$

**49.** $F(x) = \left(x + \frac{1}{2}\right)^2 - \frac{49}{4}$

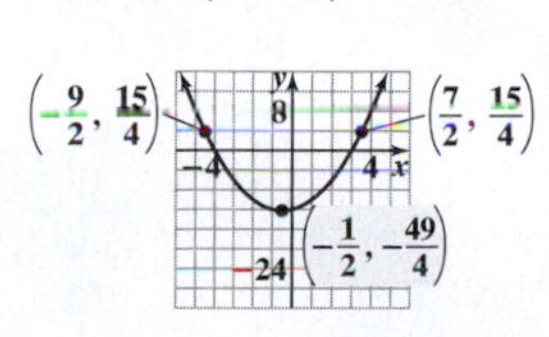


vertex is $\left(-\frac{1}{2}, -\frac{49}{4}\right)$; axis of symmetry is $x = -\frac{1}{2}$ domain is the set of all real numbers, or $(-\infty, \infty)$; range is $\left\{y \mid y \geq -\frac{49}{4}\right\}$, or $\left[-\frac{49}{4}, \infty\right)$

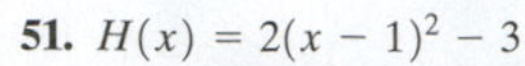

**51.** $H(x) = 2(x-1)^2 - 3$

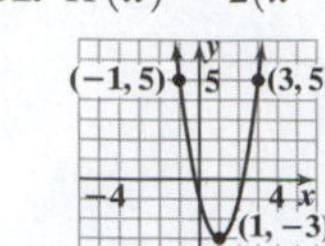


vertex is $(1, -3)$; axis of symmetry is $x = 1$; domain is the set of all real numbers, $(-\infty, \infty)$; range is $\{y \mid y \geq -3\}$, or $[-3, \infty)$

**53.** $P(x) = 3(x+2)^2 + 1$

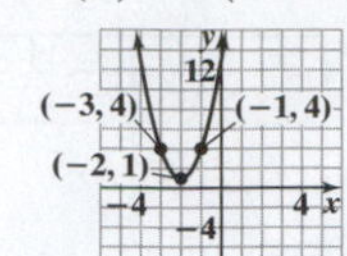


vertex is $(-2, 1)$; axis of symmetry is $x = -2$; domain is the set of all real numbers, or $(-\infty, \infty)$; range is $\{y \mid y \geq 1\}$, or $[1, \infty)$

**55.** $F(x) = -(x+5)^2 + 4$

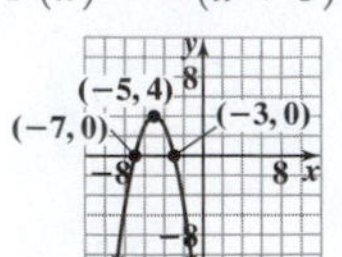


vertex is $(-5, 4)$; axis of symmetry is $x = -5$; domain is the set of all real numbers, or $(-\infty, \infty)$; range is $\{y \mid y \leq 4\}$, or $(-\infty, 4]$

**57.** $g(x) = -(x-3)^2 + 8$

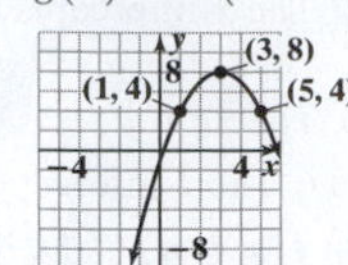


vertex is $(3, 8)$; axis of symmetry is $x = 3$; domain is the set of all real numbers, or $(-\infty, \infty)$; range is $\{y \mid y \leq 8\}$, or $(-\infty, 8]$

**59.** $H(x) = -2(x-2)^2 + 4$

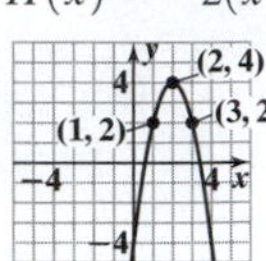


vertex is $(2, 4)$; axis of symmetry is $x = 2$; domain is the set of all real numbers, or $(-\infty, \infty)$; range is $\{y \mid y \leq 4\}$, or $(-\infty, 4]$

**61.** $f(x) = \frac{1}{3}(x-3)^2 + 1$

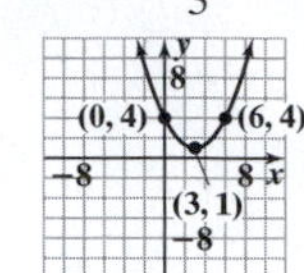


vertex is $(3, 1)$; axis of symmetry is $x = 1$; domain is the set of all real numbers, or $(-\infty, \infty)$; range is $\{y \mid y \geq 1\}$, or $[1, \infty)$

**63.** $G(x) = -12\left(x + \frac{1}{2}\right)^2 + 4$

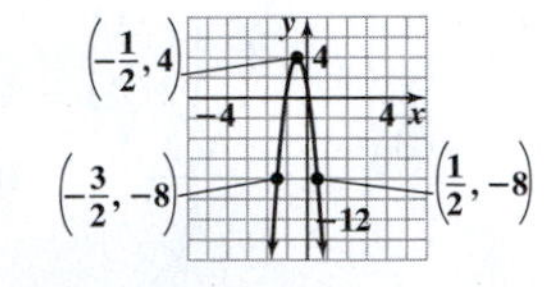


vertex is $\left(-\frac{1}{2}, 4\right)$; axis of symmetry is $x = -\frac{1}{2}$; domain is the set of all real numbers, or $(-\infty, \infty)$; range is $\{y \mid y \leq 4\}$, or $(-\infty, 4]$

**65.** Answers may vary. One possibility: $y = f(x) = (x-3)^2$ **67.** Answers may vary. One possibility: $y = f(x) = (x+3)^2 + 1$
**69.** Answers may vary. One possibility: $y = f(x) = -(x-5)^2 - 1$ **71.** $y = f(x) = 4(x-9)^2 - 6$
**73.** $y = f(x) = -\frac{1}{3}x^2 + 6$ **75.** $f(x) = (x+1)^2 - 3$ **77.** $f(x) = -2(x-3)^2 + 7$ **79.** $f(x) = (x+4)^2$ **81.** No; explanations may vary.
**83.** $29 + \frac{1}{12}$ **85.** $x^3 - 5x^2 + 4x - 2 + \frac{-2}{2x-1}$

**87.**

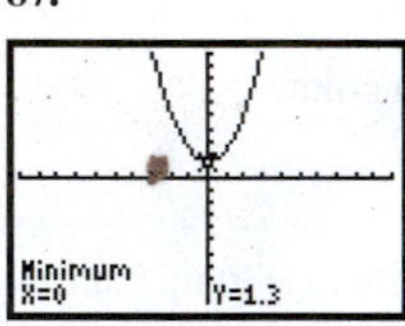


Vertex: $(0, 1.3)$
Axis of symmetry: $x = 0$
Range: $\{y \mid y \geq 1.3\} = [1.3, \infty)$

**89.**

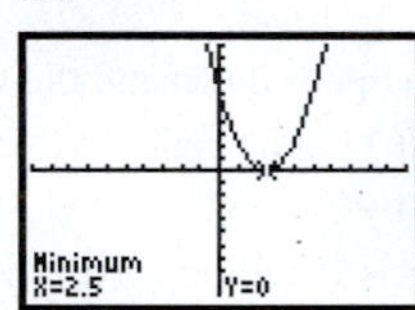


Vertex: $(2.5, 0)$
Axis of symmetry: $x = 2.5$
Range: $\{y \mid y \geq 0\} = [0, \infty)$

**91.**

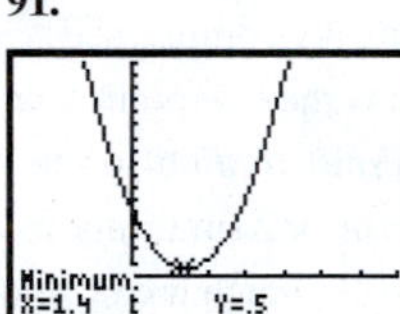


Vertex: $(1.4, 0.5)$
Axis of symmetry: $x = 1.4$
Range: $\{y \mid y \geq 0.5\} = [0.5, \infty)$

**93.**

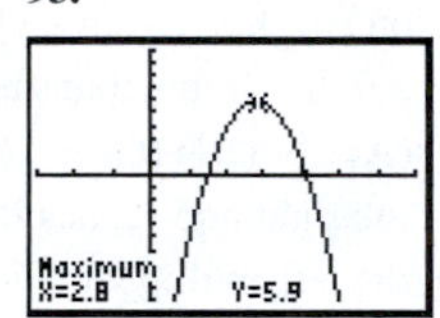


Vertex: $(2.8, 5.9)$
Axis of symmetry: $x = 2.8$
Range: $\{y \mid y \leq 5.9\} = (-\infty, 5.9]$

## Section 8.5

**1.** $x = -\frac{b}{2a}$ **3.** minimum value **5.** False **7.** Answers may vary. **9.** the discriminant is positive; there are two distinct $x$-intercepts: 8 and $-2$
**11.** the discriminant is negative; there are no $x$-intercepts **13.** the discriminant is zero; there is one $x$-intercept: $-\frac{1}{2}$
**15.** the discriminant is positive; there are two distinct $x$-intercepts: approximately $-0.39$ and approximately 0.64

**17.**

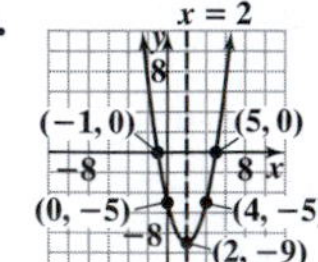


Domain: $\{x \mid x$ is any real number$\}$ or $(-\infty, \infty)$
Range: $\{y \mid y \geq -9\}$ or $[-9, \infty)$

**19.**

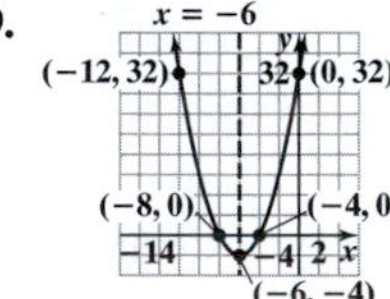


Domain: $\{x \mid x$ is any real number$\}$ or $(-\infty, \infty)$
Range: $\{y \mid y \geq -4\}$ or $[-4, \infty)$

**21.**

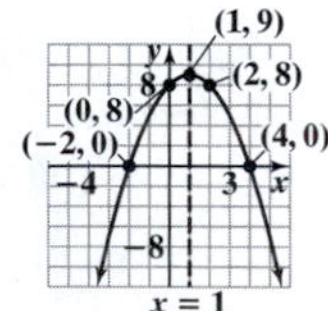


Domain: $\{x \mid x$ is any real number$\}$ or $(-\infty, \infty)$
Range: $\{y \mid y \leq 9\}$ or $(-\infty, 9]$

**23.**

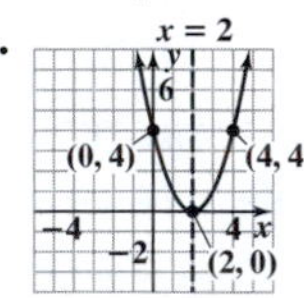


Domain: $\{x \mid x$ is any real number$\}$ or $(-\infty, \infty)$
Range: $\{y \mid y \geq 0\}$ or $[0, \infty)$

**25.**

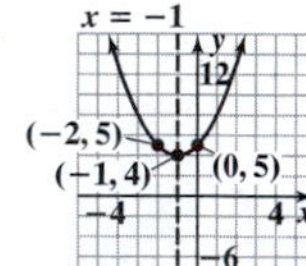
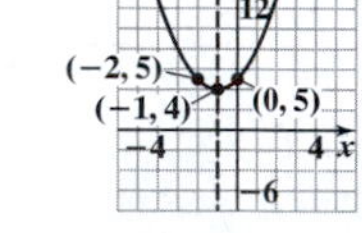


Domain: $\{x \mid x$ is any real number$\}$ or $(-\infty, \infty)$
Range: $\{y \mid y \geq 4\}$ or $[4, \infty)$

**27.**

Domain: $\{x \mid x$ is any real number$\}$ or $(-\infty, \infty)$
Range: $\{y \mid y \leq 0\}$ or $(-\infty, 0]$

**29.**

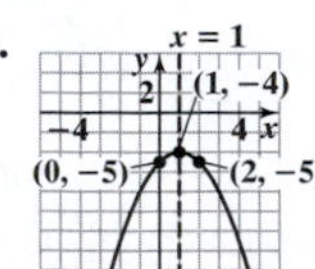


Domain: $\{x \mid x$ is any real number$\}$ or $(-\infty, \infty)$
Range: $\{y \mid y \leq -4\}$ or $(-\infty, -4]$

**31.**

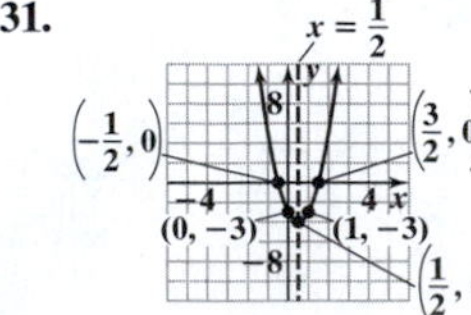


Domain: $\{x \mid x$ is any real number$\}$ or $(-\infty, \infty)$
Range: $\{y \mid y \geq -4\}$ or $[-4, \infty)$

**33.**

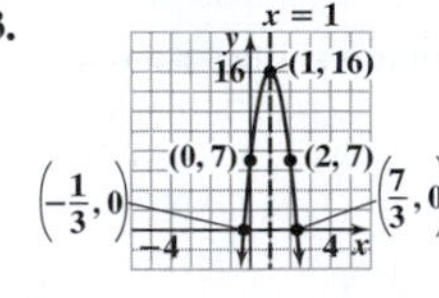


Domain: $\{x \mid x$ is any real number$\}$ or $(-\infty, \infty)$
Range: $\{y \mid y \leq 16\}$ or $(-\infty, 16]$

**35.**

Domain: $\{x \mid x$ is any real number$\}$ or $(-\infty, \infty)$
Range: $\{y \mid y \geq 0\}$ or $[0, \infty)$

**37.**
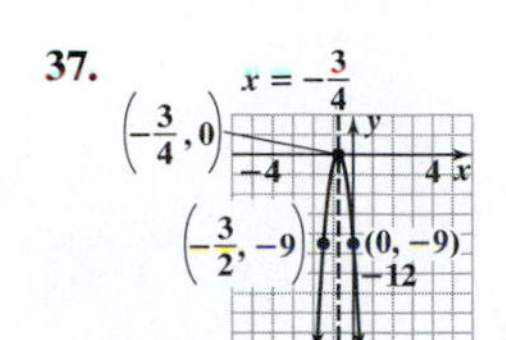

Domain: $\{x \mid x$ is any real number$\}$ or $(-\infty, \infty)$
Range: $\{y \mid y \leq 0\}$ or $(-\infty, 0]$

**39.**
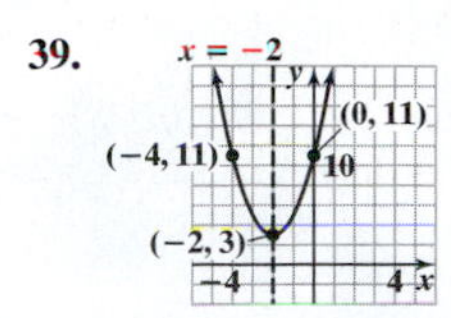

Domain: $\{x \mid x$ is any real number$\}$ or $(-\infty, \infty)$
Range: $\{y \mid y \geq 3\}$ or $[3, \infty)$

**41.**
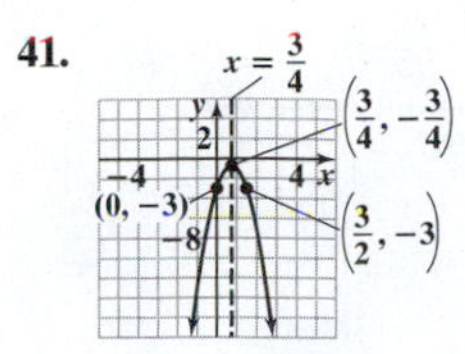

Domain: $\{x \mid x$ is any real number$\}$ or $(-\infty, \infty)$
Range: $\left\{y \mid y \leq -\frac{3}{4}\right\}$ or $\left(-\infty, -\frac{3}{4}\right]$

**43.**
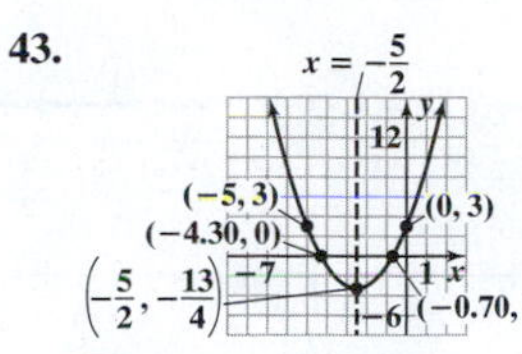

Domain: $\{x \mid x$ is any real number$\}$ or $(-\infty, \infty)$
Range: $\left\{y \mid y \geq -\frac{13}{4}\right\}$ or $\left[-\frac{13}{4}, \infty\right)$

**45.**
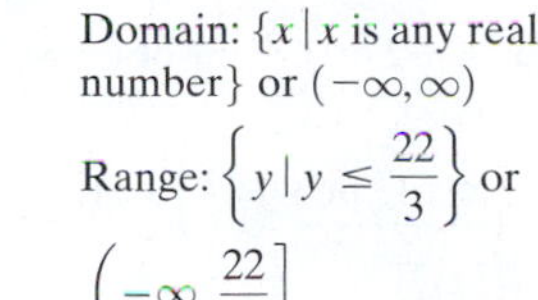

Domain: $\{x \mid x$ is any real number$\}$ or $(-\infty, \infty)$
Range: $\left\{y \mid y \leq \frac{22}{3}\right\}$ or $\left(-\infty, \frac{22}{3}\right]$

**47.**
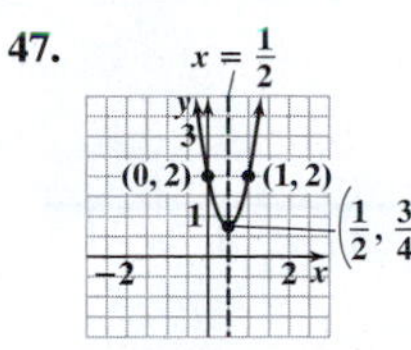

Domain: $\{x \mid x$ is any real number$\}$ or $(-\infty, \infty)$
Range: $\left\{y \mid y \geq \frac{3}{4}\right\}$ or
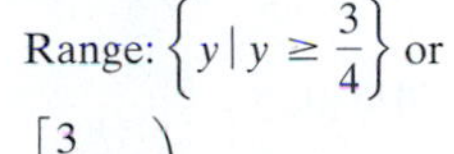
$\left[\frac{3}{4}, \infty\right)$

**49.**
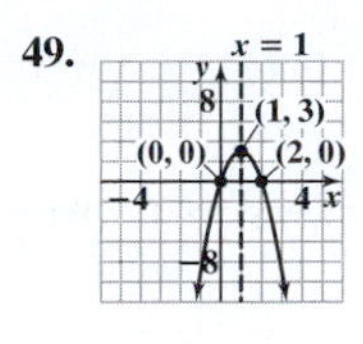

Domain: $\{x \mid x$ is any real number$\}$ or $(-\infty, \infty)$
Range: $\{y \mid y \leq 3\}$ or $(-\infty, 3]$

**51.**
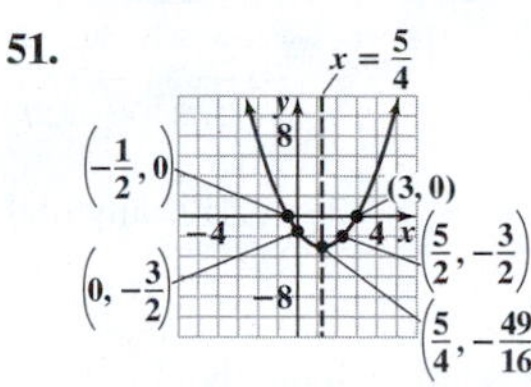

Domain: $\{x \mid x$ is any real number$\}$ or $(-\infty, \infty)$
Range: $\left\{y \mid y \geq -\frac{49}{16}\right\}$ or
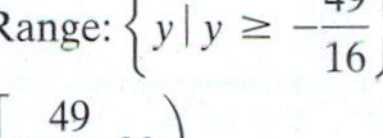
$\left[-\frac{49}{16}, \infty\right)$

**53.**
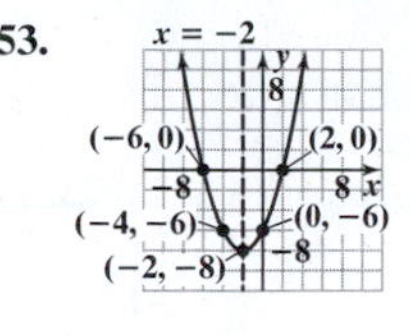

Domain: $\{x \mid x$ is any real number$\}$ or $(-\infty, \infty)$
Range: $\{y \mid y \geq -8\}$ or $[-8, \infty)$

**55.**
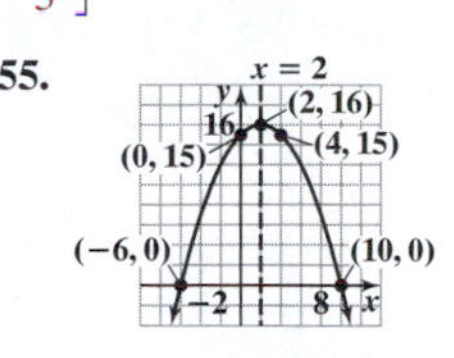

Domain: $\{x \mid x$ is any real number$\}$ or $(-\infty, \infty)$
Range: $\{y \mid y \leq 16\}$ or $(-\infty, 16]$

**57.** minimum; −3 **59.** maximum; 28 **61.** maximum; 23 **63.** minimum; −19 **65.** minimum; $-\frac{17}{8}$ **67.** maximum; $\frac{7}{3}$
**69. (a)** \$120 **(b)** \$36,000 **71.** 60; \$35 **73. (a)** after 7.5 seconds **(b)** 910 feet **(c)** about 15.042 seconds
**75. (a)** about 1753.52 feet from the cannon **(b)** about 886.76 feet **(c)** about 3517 feet from the cannon **(d)** The two answers are close. Explanations may vary. **77. (a)** about 46.5 years **(b)** \$64,661.75 **79.** 18 and 18 **81.** −9 and 9
**83.** 15,625 square yards; 125 yards × 125 yards **85.** 500,000 square meters; 500 m × 1000 m and the long side is parallel to the river
**87.** 5 inches **89. (a)** $R = -p^2 + 110p$ **(b)** \$55; \$3025 **(c)** 55 pairs **91. (a)** $f(x) = x^2 - 8x + 12$; $f(x) = 2x^2 - 16x + 24$; $f(x) = -2x^2 + 16x - 24$ **(b)** The value of $a$ has no effect on the $x$-intercepts. **(c)** The value of $a$ has no effect on the axis of symmetry. **(d)** The $x$-coordinate of the vertex is 4, which does not depend on $a$. However, the $y$-coordinate is $-4a$, which does depend on $a$.
**93.** A revenue of \$0 when the price charged is some positive number is an indication that the price was too high to keep any consumer demand. Regardless of the price charged, if no items are sold, then no revenue can be generated.

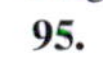
**95.**
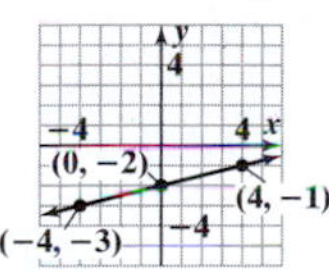


**97.**
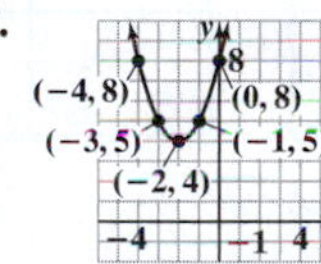


**99.** Vertex: (3.5, −9.25)
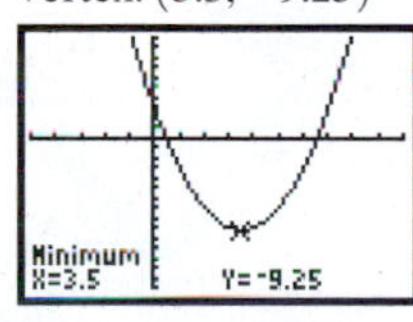


**101.** Vertex: (3.5, 37.5)
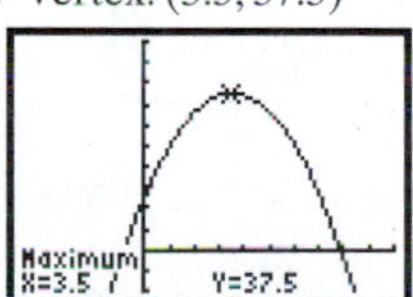


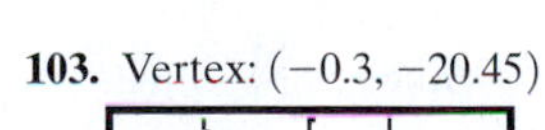
**103.** Vertex: (−0.3, −20.45)
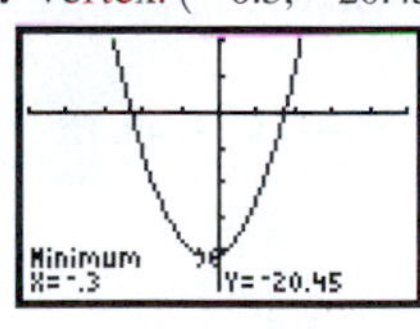


**105.** Vertex: (0.67, 4.78)
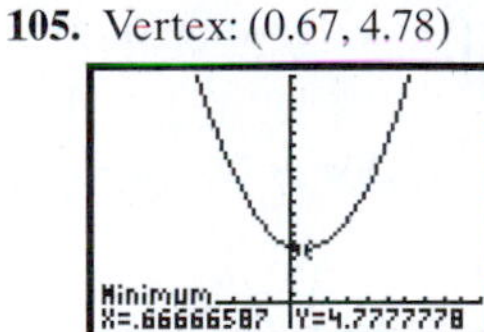


**107.** $c$ is the $y$-intercept
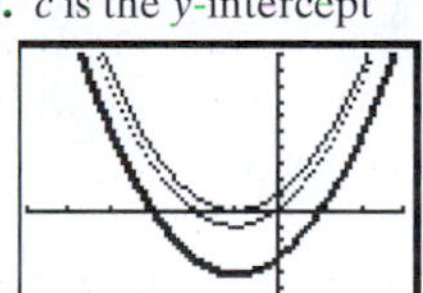

## *Section 8.6*

**1.** quadratic **3.** False **5.** Answers may vary. **7.** Answers may vary. **9. (a)** $\{x \mid x < -6 \text{ or } x > 5\}$ or $(-\infty, -6) \cup (5, \infty)$
**(b)** $\{x \mid -6 \leq x \leq 5\}$ or $[-6, 5]$ **11. (a)** $\left\{x \mid -6 \leq x \leq \frac{5}{2}\right\}$ or $\left[-6, \frac{5}{2}\right]$ **(b)** $\left\{x \mid x < -6 \text{ or } x > \frac{5}{2}\right\}$ or $(-\infty, -6) \cup \left(\frac{5}{2}, \infty\right)$

**13.** $\{x \mid x \leq -2 \text{ or } x \geq 5\}$ or $(-\infty, -2] \cup [5, \infty)$

**15.** $\{x \mid -7 < x < -3\}$ or $(-7, -3)$

**17.** $\{x \mid x < -5 \text{ or } x > 7\}$ or $(-\infty, -5) \cup (7, \infty)$

**19.** $\{n \mid 3 - \sqrt{17} \leq n \leq 3 + \sqrt{17}\}$ or $[3 - \sqrt{17}, 3 + \sqrt{17}]$ ($3 - \sqrt{17} \approx -1.12$, $3 + \sqrt{17} \approx 7.12$)

**21.** $\{m \mid m \leq -7 \text{ or } m \geq 2\}$ or $(-\infty, -7] \cup [2, \infty)$

**23.** $\left\{q \mid q \le -\frac{5}{2} \text{ or } q \ge 3\right\}$ or $\left(-\infty, -\frac{5}{2}\right] \cup [3, \infty)$

**25.** $\{x \mid -1 \le x \le 4\}$ or $[-1, 4]$

**27.** $\{x \mid x < -2 \text{ or } x > 5\}$ or $(-\infty, -2) \cup (5, \infty)$

**29.** $\left\{x \mid x \le -\frac{2}{3} \text{ or } x \ge 4\right\}$ or $\left(-\infty, -\frac{2}{3}\right] \cup [4, \infty)$

**31.** $\{x \mid -2 - \sqrt{3} < x < -2 + \sqrt{3}\}$ or $(-2 - \sqrt{3}, -2 + \sqrt{3})$; $-2 - \sqrt{3} \approx -3.73$, $-2 + \sqrt{3} \approx -0.27$

**33.** $\left\{a \mid -\frac{1}{2} \le a \le 4\right\}$ or $\left[-\frac{1}{2}, 4\right]$ **35.** $\{z \mid z \text{ is any real number}\}$ or $(-\infty, \infty)$

**37.** no solution **39.** $\{x \mid x \text{ is any real number}\}$ or $(-\infty, \infty)$

**41.** $\{x \mid 0 < x < 5\}$ or $(0, 5)$ **43.** $\{x \mid x \le -4 \text{ or } x \ge 7\}$ or $(-\infty, -4] \cup [7, \infty)$

**45.** $\left\{x \mid x < -\frac{5}{2} \text{ or } x > 2\right\}$ or $\left(-\infty, -\frac{5}{2}\right) \cup (2, \infty)$ **47.** $\{x \mid x \le -8 \text{ or } x \ge 0\}$ or $(-\infty, -8] \cup [0, \infty)$

**49.** $\{x \mid x \le -5 \text{ or } x \ge 6\}$ or $(-\infty, -5] \cup [6, \infty)$ **51.** between 2 and 3 seconds after the ball is thrown **53.** between \$110 and \$130 **55.** $x = -3$; a perfect square cannot be negative. Therefore, the only solution will be where the perfect square expression equals zero, which is $-3$. **57.** all real numbers; a perfect square must always be zero or greater. Therefore, it must always be larger than $-2$. Thus, all values of $x$ will make the inequality true. **59.** Answers may vary. One possibility follows: $x^2 + x - 6 \le 0$ **61.** Answer will vary. One possibility follows: The inequalities have the same solution set because they are equivalent. **63.** $\{x \mid x < -3 \text{ or } 1 < x < 3\}$ or $(-\infty, -3) \cup (1, 3)$ **65.** $\left\{x \,\middle|\, -\frac{4}{3} \le x \le 2 \text{ or } x \ge 6\right\}$ or $\left[-\frac{4}{3}, 2\right] \cup [6, \infty)$ **67.** $\{x \mid x < -2 \text{ or } -1 < x < 5\}$ or $(-\infty, -2) \cup (-1, 5)$ **69.** $\frac{-8m^5}{n^2}$ **71.** $\frac{b^{\frac{1}{4}}}{9a^{\frac{7}{9}}}$

**73.**

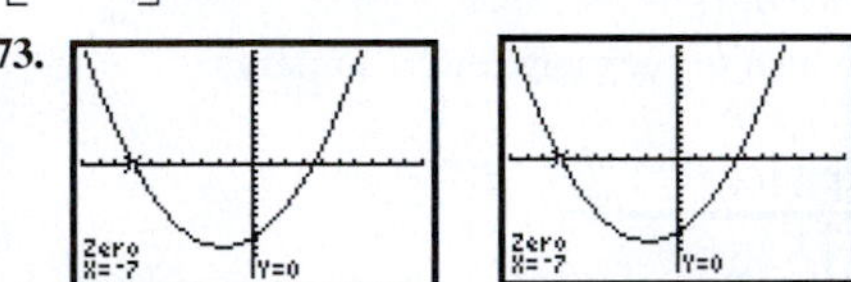


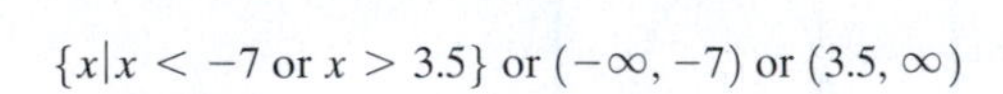

$\{x \mid x < -7 \text{ or } x > 3.5\}$ or $(-\infty, -7)$ or $(3.5, \infty)$

**75.**

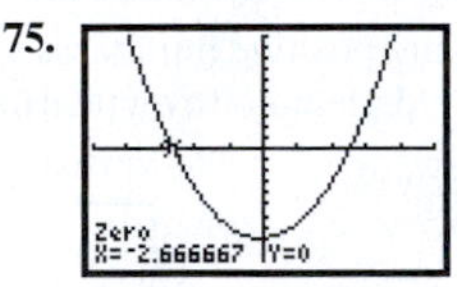


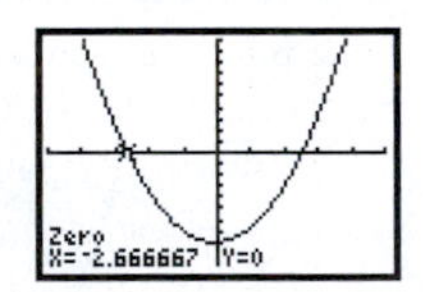


$\left\{x \,\middle|\, -\frac{8}{3} \le x \le \frac{5}{2}\right\}$ or $\left[-\frac{8}{3}, \frac{5}{2}\right]$

## Chapter 8 Review

**1.** $\{-13, 13\}$ **2.** $\{-5\sqrt{3}, 5\sqrt{3}\}$ **3.** $\{-4i, 4i\}$ **4.** $\left\{-\frac{2\sqrt{2}}{3}, \frac{2\sqrt{2}}{3}\right\}$ **5.** $\{-1, 17\}$ **6.** $\{2 - 5\sqrt{6}, 2 + 5\sqrt{6}\}$ **7.** $\left\{-5, \frac{5}{3}\right\}$ **8.** $\left\{-\frac{3\sqrt{14}}{7}, \frac{3\sqrt{14}}{7}\right\}$ **9.** $\{-4\sqrt{5}i, 4\sqrt{5}i\}$ **10.** $\left\{-\frac{3}{4} - \frac{\sqrt{13}}{4}, -\frac{3}{4} + \frac{\sqrt{13}}{4}\right\}$ **11.** $a^2 + 30a + 225; (a + 15)^2$ **12.** $b^2 - 14b + 49; (b - 7)^2$ **13.** $c^2 - 11c + \frac{121}{4}; \left(c - \frac{11}{2}\right)^2$ **14.** $d^2 + 9d + \frac{81}{4}; \left(d + \frac{9}{2}\right)^2$ **15.** $m^2 - \frac{1}{4}m + \frac{1}{64}; \left(m - \frac{1}{8}\right)^2$ **16.** $n^2 + \frac{6}{7}n + \frac{9}{49}; \left(n + \frac{3}{7}\right)^2$ **17.** $\{2, 8\}$ **18.** $\{-4, 7\}$ **19.** $\{3 - 2\sqrt{3}, 3 + 2\sqrt{3}\}$ **20.** $\left\{\frac{5}{2} - \frac{\sqrt{53}}{2}, \frac{5}{2} + \frac{\sqrt{53}}{2}\right\}$ **21.** $\left\{-\frac{1}{2} - \frac{3\sqrt{3}}{2}i, -\frac{1}{2} + \frac{3\sqrt{3}}{2}i\right\}$ **22.** $\{3 - 2\sqrt{2}i, 3 + 2\sqrt{2}i\}$ **23.** $\left\{\frac{1}{2}, 3\right\}$ **24.** $\left\{-\frac{1}{2} - \frac{3}{2}i, -\frac{1}{2} + \frac{3}{2}i\right\}$ **25.** $\left\{\frac{3}{2} - \frac{\sqrt{15}}{6}i, \frac{3}{2} + \frac{\sqrt{15}}{6}i\right\}$ **26.** $\left\{-\frac{2}{3} - \frac{\sqrt{10}}{3}, -\frac{2}{3} + \frac{\sqrt{10}}{3}\right\}$ **27.** $c = 15$ **28.** $c = 8\sqrt{2}$ **29.** $c = 3\sqrt{5}$ **30.** $c = 26$ **31.** $c = 6$ **32.** $c = 7$ **33.** $b = 3\sqrt{7}$ **34.** $a = 5\sqrt{3}$ **35.** $a = \sqrt{253}$ **36.** approximately 127.3 feet **37.** $\{-4, 5\}$ **38.** $\left\{-\frac{3}{2}, \frac{7}{2}\right\}$ **39.** $\left\{-\frac{4}{3} - \frac{\sqrt{7}}{3}, -\frac{4}{3} + \frac{\sqrt{7}}{3}\right\}$ **40.** $\left\{1 - \frac{\sqrt{10}}{2}, 1 + \frac{\sqrt{10}}{2}\right\}$ **41.** $\left\{-\frac{1}{6} - \frac{\sqrt{35}}{6}i, -\frac{1}{6} + \frac{\sqrt{35}}{6}i\right\}$ **42.** $\left\{\frac{4}{3}\right\}$ **43.** $\left\{2 - \sqrt{2}, 2 + \sqrt{2}\right\}$ **44.** $\left\{-\frac{2}{5} - \frac{1}{5}i, -\frac{2}{5} + \frac{1}{5}i\right\}$ **45.** $\left\{-\frac{5}{2} - \frac{3\sqrt{3}}{2}i, -\frac{5}{2} + \frac{3\sqrt{3}}{2}i\right\}$ **46.** $\left\{-\frac{3}{2} - \frac{\sqrt{5}}{2}i, -\frac{3}{2} + \frac{\sqrt{5}}{2}i\right\}$ **47.** 57; two irrational solutions **48.** 0; one repeated real solution **49.** $-47$; two complex solutions that are not real **50.** $-20$; two complex solutions that are not real **51.** 0; one repeated real solution

**52.** 25; two rational solutions **53.** $\{-9, 1\}$ **54.** $\left\{-\frac{5}{2}, \frac{1}{3}\right\}$ **55.** $\{-2 - 3i, -2 + 3i\}$ **56.** $\{-2\sqrt{3}, 2\sqrt{3}\}$ **57.** $\left\{1 - \frac{\sqrt{10}}{2}, 1 + \frac{\sqrt{10}}{2}\right\}$
**58.** $\{-4 - 2i, -4 + 2i\}$ **59.** $\{-3, 5\}$ **60.** $\{1 - \sqrt{2}, 1 + \sqrt{2}\}$ **61.** $\left\{-\frac{4}{3}, \frac{4}{3}\right\}$ **62.** $\{-1 - \sqrt{3}i, -1 + \sqrt{3}i\}$
**63.** $x = 10$; the three sides measure 5, 12, and 13 **64.** 12 centimeters by 9 centimeters **65. (a)** either 300 or 600 cellular phones **(b)** 450 cellular phones **66. (a)** after approximately 0.5 second and after approximately 2.7 seconds **(b)** after approximately 4.3 seconds **(c)** No; the solutions to the equation are complex solutions that are not real **67.** approximately 10.8 miles per hour
**68.** approximately 67.8 minutes **69.** $\{-4i, 4i, -3, 3\}$ **70.** $\left\{-\frac{\sqrt{3}}{2}, \frac{\sqrt{3}}{2}, -\sqrt{2}i, \sqrt{2}i\right\}$ **71.** $\left\{-\frac{10}{3}, -1\right\}$
**72.** $\{-4, 4, -2\sqrt{2}, 2\sqrt{2}\}$ **73.** $\{16, 81\}$ **74.** $\left\{\frac{9}{25}\right\}$ **75.** $\left\{-\frac{1}{3}, \frac{1}{7}\right\}$ **76.** $\left\{-343, \frac{1}{8}\right\}$ **77.** $\{16\}$
**78.** $\left\{-\frac{36}{7}, -\frac{19}{4}\right\}$ **79.** $\left\{\frac{9}{4}, \frac{49}{4}\right\}$ **80.** $\{-2\sqrt{3}, -\sqrt{5}, \sqrt{5}, 2\sqrt{3}\}$

**81.** **82.** **83.** **84.** **85.**

**86.** **87.** **88.** **89.** **90.**

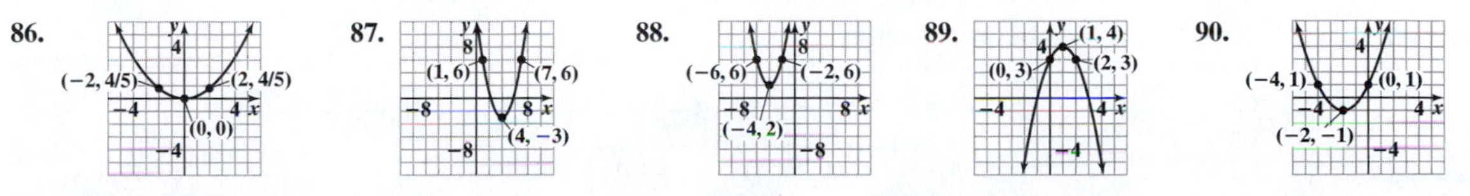

**91.** vertex is (3, 1); axis of symmetry is $x = 3$ Domain: $\{x \mid x \text{ is any real number}\}$ or $(-\infty, \infty)$ Range: $\{y \mid y \geq 1\}$ or $[1, \infty)$
**92.** vertex is $(-4, -5)$; axis of symmetry is $x = -4$
**93.** $H(x) = 2(x - 1)^2 - 5$ vertex is $(1, -5)$; axis of symmetry is $x = 1$ Domain: $\{x \mid x \text{ is any real number}\}$ or $(-\infty, \infty)$ Range: $\{y \mid y \geq -5\}$ or $[-5, \infty)$

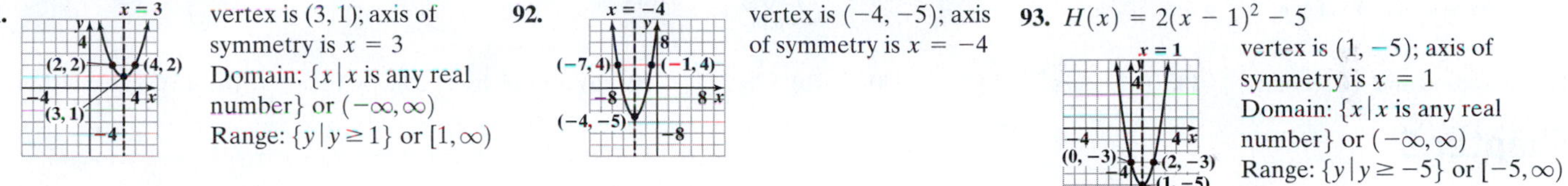

**94.** vertex is $(-3, -1)$; axis of symmetry is $x = -3$
**95.** vertex is (2, 4); axis of symmetry is $x = 2$ Domain: $\{x \mid x \text{ is any real number}\}$ or $(-\infty, \infty)$ Range: $\{y \mid y \leq 4\}$ or $(-\infty, 4]$
**96.** vertex is (2, 3); axis of symmetry is $x = 2$

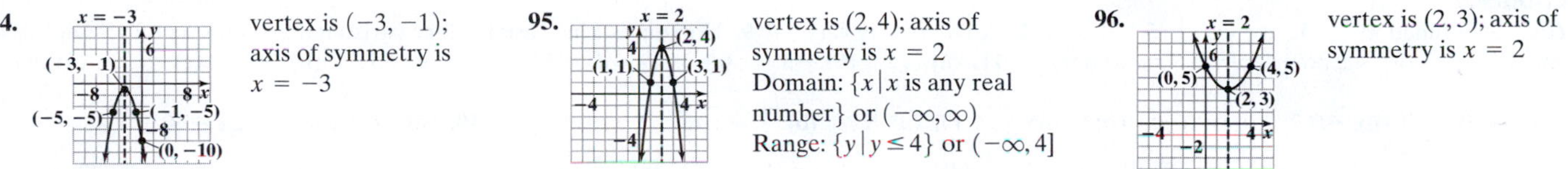

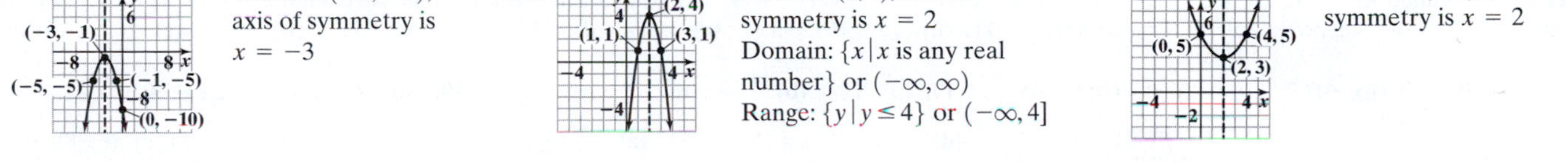

**97.** $f(x) = 2(x - 2)^2 - 4$ or $f(x) = 2x^2 - 8x + 4$ **98.** $f(x) = -(x - 4)^2 + 3$ or $f(x) = -x^2 + 8x - 13$

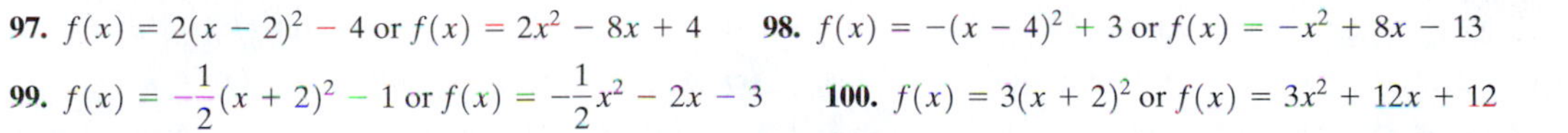

**99.** $f(x) = -\frac{1}{2}(x + 2)^2 - 1$ or $f(x) = -\frac{1}{2}x^2 - 2x - 3$ **100.** $f(x) = 3(x + 2)^2$ or $f(x) = 3x^2 + 12x + 12$

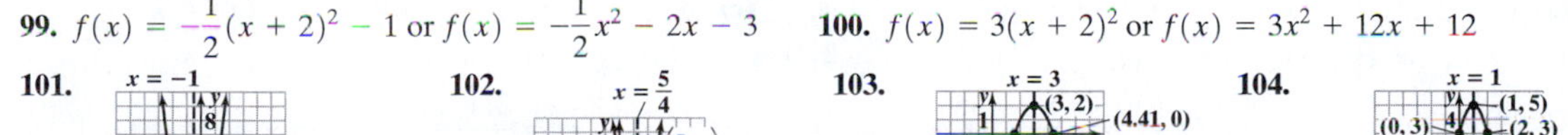

**101.** **102.** **103.** **104.** **105.**

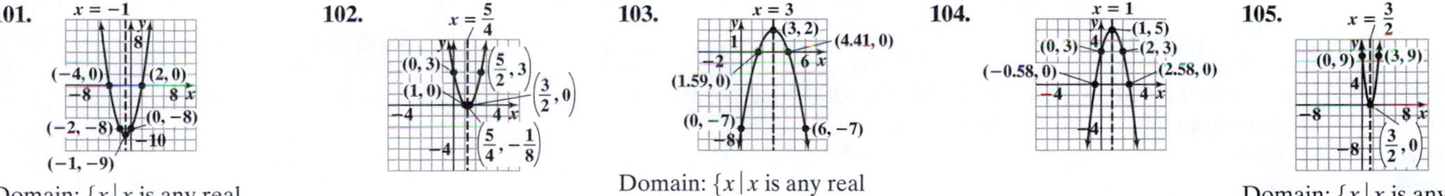

**101.** Domain: $\{x \mid x \text{ is any real number}\}$ or $(-\infty, \infty)$ Range: $\{y \mid y \geq -9\}$ or $[-9, \infty)$

**103.** Domain: $\{x \mid x \text{ is any real number}\}$ or $(-\infty, \infty)$ Range: $\{y \mid y \leq 2\}$ or $(-\infty, 2]$

**105.** Domain: $\{x \mid x \text{ is any real number}\}$ or $(-\infty, \infty)$ Range: $\{y \mid y \geq 0\}$ or $[0, \infty)$

**106.** **107.** **108.**

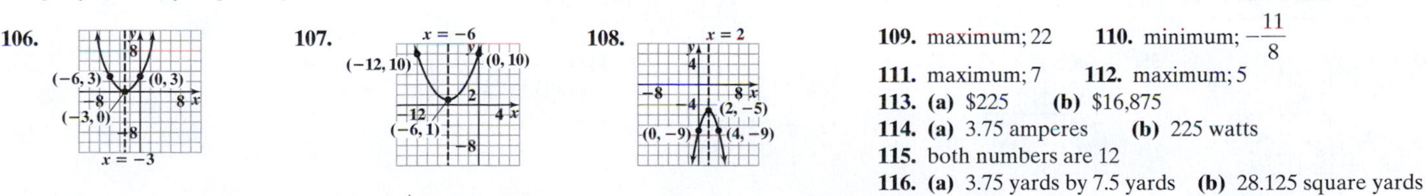

**107.** Domain: $\{x \mid x \text{ is any real number}\}$ or $(-\infty, \infty)$ Range: $\{y \mid y \geq 1\}$ or $[1, \infty)$

**109.** maximum; 22 **110.** minimum; $-\frac{11}{8}$
**111.** maximum; 7 **112.** maximum; 5
**113. (a)** \$225 **(b)** \$16,875
**114. (a)** 3.75 amperes **(b)** 225 watts
**115.** both numbers are 12
**116. (a)** 3.75 yards by 7.5 yards **(b)** 28.125 square yards
**117. (a)** 100 feet **(b)** 50 feet **(c)** 200 feet
**118. (a)** $R = -0.002p^2 + 60p$ **(b)** \$15,000; \$450,000 **(c)** 30 automobiles per month

**119. (a)** $\{x \mid x < -2 \text{ or } x > 3\}$ or $(-\infty, -2) \cup (3, \infty)$ **(b)** $\{x \mid -2 < x < 3\}$ or $(-2, 3)$
**120. (a)** $\left\{x \middle| -\frac{7}{2} \leq x \leq 1\right\}$ or $\left[-\frac{7}{2}, 1\right]$ **(b)** $\left\{x \middle| x \leq -\frac{7}{2} \text{ or } x \geq 1\right\}$ or $\left(-\infty, -\frac{7}{2}\right] \cup [1, \infty)$
**121.** $\{x \mid -4 \leq x \leq 6\}$ or $[-4, 6]$ **122.** $\{y \mid y \leq -8 \text{ or } y \geq 1\}$ or, $(-\infty, -8] \cup [1, \infty)$ **123.** $\left\{z \middle| z < \frac{4}{3} \text{ or } z > 5\right\}$ or $\left(-\infty, \frac{4}{3}\right) \cup (5, \infty)$

**124.** $\{p|-2-\sqrt{6}<p<-2+\sqrt{6}\}$ or $(-2-\sqrt{6},-2+\sqrt{6})$

**125.** $\{m|m$ is any real number$\}$ or $(-\infty,\infty)$

**126.** $\left\{w\middle|-\frac{1}{3}\le w\le\frac{7}{2}\right\}$ or $\left[-\frac{1}{3},\frac{7}{2}\right]$

### Chapter 8 Test

**1.** $x^2-3x+\frac{9}{4}$; $\left(x-\frac{3}{2}\right)^2$ **2.** $m^2+\frac{2}{5}m+\frac{1}{25}$; $\left(m+\frac{1}{5}\right)^2$ **3.** $\left\{-\frac{5}{3},-1\right\}$ **4.** $\{3-\sqrt{5},3+\sqrt{5}\}$ **5.** $\left\{1-\frac{\sqrt{2}}{2}i,1+\frac{\sqrt{2}}{2}i\right\}$

**6.** $\left\{\frac{3}{2}-\frac{\sqrt{3}}{6}i,\frac{3}{2}+\frac{\sqrt{3}}{6}i\right\}$ **7.** 57; two irrational solutions **8.** $a=6\sqrt{2}$ **9.** $\{-3,3,-2i,2i\}$ **10.** $\left\{\frac{1}{81}\right\}$

**11.**

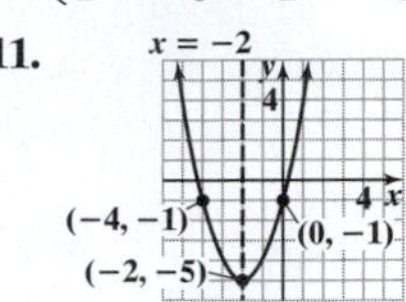

vertex is $(-2,-5)$; axis of symmetry is $x=-2$
Domain: $\{x|x$ is any real number$\}$ or $(-\infty,\infty)$
Range: $\{y|y\ge-5\}$ or $[-5,\infty)$

**12.**

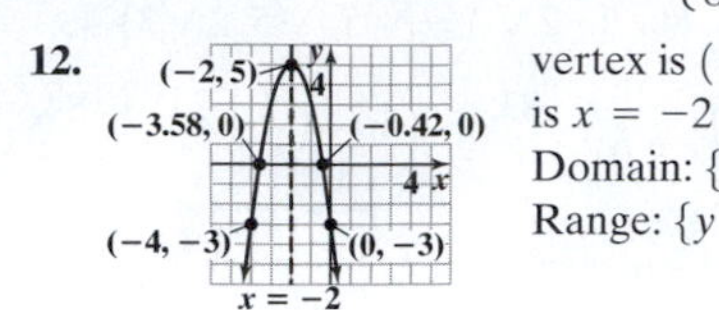

vertex is $(-2,5)$; axis of symmetry is $x=-2$
Domain: $\{x|x$ is any real number$\}$ or $(-\infty,\infty)$
Range: $\{y|y\le5\}$ or $(-\infty,5]$

**13.** $f(x)=\frac{1}{3}(x+3)^2-5$ or $f(x)=\frac{1}{3}x^2+2x-2$ **14.** maximum; 6

**15.** $\left\{m\middle|m<-3 \text{ or } m>\frac{5}{2}\right\}$ or $(-\infty,-3)\cup\left(\frac{5}{2},\infty\right)$

**16.** $\{z|-3-\sqrt{10}\le z\le-3+\sqrt{10}\}$ or $[-3-\sqrt{10},-3+\sqrt{10}]$

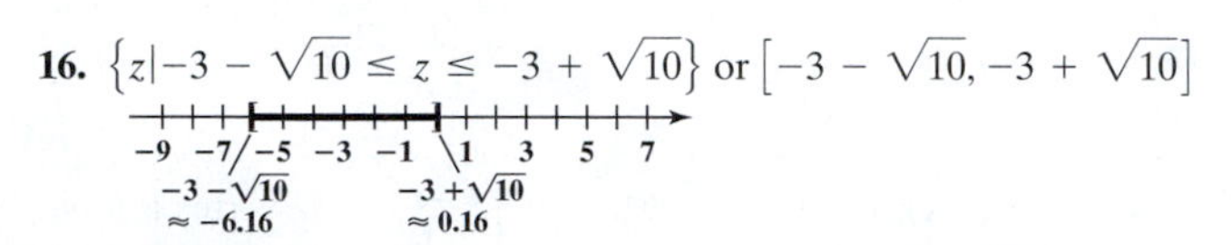

**17.** 0.4 second and 4.6 seconds **18.** 34.1 hours **19. (a)** \$340 **(b)** \$28,900 **20. (a)** 12.5 in. by 12.5 in. by 12 in. **(b)** 1875 cubic inches

# Chapter 9

### Section 9.1

**1.** composite function **3.** inverse **5.** false **7.** Answers may vary. **9.** Horizontal Line Test: If every horizontal line intersects the graph of a function $f$ in at most one point, then $f$ is one-to-one. **11. (a)** 3 **(b)** $-3$ **(c)** 19 **(d)** $-12$ **13. (a)** 85 **(b)** 19 **(c)** 29 **(d)** $-7$

**15. (a)** $-4394$ **(b)** $-507$ **(c)** 16 **(d)** $-1453$ **17. (a)** 8 **(b)** $\frac{4}{5}$ **(c)** 1 **(d)** $-\frac{4}{3}$ **19. (a)** $(f\circ g)(x)=2x+1$ **(b)** $(g\circ f)(x)=2x+2$ **(c)** $(f\circ f)(x)=x+2$ **(d)** $(g\circ g)(x)=4x$ **21. (a)** $(f\circ g)(x)=-8x+17$ **(b)** $(g\circ f)(x)=-8x-23$ **(c)** $(f\circ f)(x)=4x+21$ **(d)** $(g\circ g)(x)=16x-15$ **23. (a)** $(f\circ g)(x)=x^2-6x+9$ **(b)** $(g\circ f)(x)=x^2-3$ **(c)** $(f\circ f)(x)=x^4$ **(d)** $(g\circ g)(x)=x-6$ **25. (a)** $(f\circ g)(x)=\sqrt{x+4}$ **(b)** $(g\circ f)(x)=\sqrt{x}+4$ **(c)** $(f\circ f)(x)=\sqrt[4]{x}$ **(d)** $(g\circ g)(x)=x+8$ **27. (a)** $(f\circ g)(x)=x^2$ **(b)** $(g\circ f)(x)=x^2+8x+12$ **(c)** $(f\circ f)(x)=||x+4|+4|$ **(d)** $(g\circ g)(x)=x^4-8x^2+12$

**29. (a)** $(f\circ g)(x)=\frac{2x}{x+1}$, where $x\ne-1,0$ **(b)** $(g\circ f)(x)=\frac{x+1}{2}$, where $x\ne-1$ **(c)** $(f\circ f)(x)=\frac{2(x+1)}{x+3}$, where $x\ne-1,-3$ **(d)** $(g\circ g)(x)=x$, where $x\ne0$ **31.** one-to-one **33.** not one-to-one **35.** one-to-one **37.** not one-to-one **39.** one-to-one **41.** one-to-one **43.** not one-to-one **45.** one-to-one

**47.**

| Weight (g) | U.S. Coin |
|---|---|
| 2.500 | Cent |
| 5.000 | Nickel |
| 2.268 | Dime |
| 5.670 | Quarter |
| 11.340 | Half Dollar |
| 8.100 | Dollar |

**49.** $\{(3,0),(4,1),(5,2),(6,3)\}$ **51.** $\{(3,-2),(1,-2),(-3,-2),(9,-2)\}$

**53.** **55.** **57.**

**59.** $f(g(x))=(x-5)+5=x$
$g(f(x))=(x+5)-5=x$

**61.** $f(g(x))=5\left(\frac{x-7}{5}\right)+7=x-7+7=x$

$g(f(x))=\frac{(5x+7)-7}{5}=\frac{5x}{5}=x$

**63.** $f(g(x))=10\left(\frac{x+1}{10}\right)-1=x+1-1=x$

$g(f(x))=\frac{(10x-1)+1}{10}=\frac{10x}{10}=x$

**65.** $f(g(x))=\frac{3}{\left(\frac{3}{x}+1\right)-1}=\frac{3}{\frac{3}{x}}=3\cdot\frac{x}{3}=x$

$g(f(x))=\frac{3}{\frac{3}{x-1}}+1=3\cdot\frac{x-1}{3}+1=x-1+1=x$

**67.** $f(g(x))=\sqrt[3]{(x^3-4)+4}=\sqrt[3]{x^3}=x$

$g(f(x))=(\sqrt[3]{x+4})^3-4=x+4-4=x$

**69.** $f^{-1}(x) = \frac{x}{6}$ **71.** $f^{-1}(x) = x - 4$ **73.** $h^{-1}(x) = \frac{x + 7}{2}$ **75.** $G^{-1}(x) = \frac{2 - x}{5}$ **77.** $g^{-1}(x) = \sqrt[3]{x - 3}$ **79.** $p^{-1}(x) = \frac{1}{x} - 3$ **81.** $F^{-1}(x) = 2 - \frac{5}{x}$ **83.** $f^{-1}(x) = x^3 + 2$ **85.** $R^{-1}(x) = \frac{2x}{1 - x}$ **87.** $f^{-1}(x) = (x - 4)^3 + 1$ **89.** $A(t) = 400\pi t^2$; $3600\pi \approx 11{,}309.73$ sq ft **91. (a)** $C(x) = 2x$ **(b)** \$630 **93.** $f^{-1}(12) = 4$ **95.** Domain of $f^{-1}$: $[-5, \infty)$ Range of $f^{-1}$: $[0, \infty)$ **97.** Domain of $g^{-1}$: $(-6, 12)$ Range of $g^{-1}$: $[-4, 10]$ **99.** $x(T) = \frac{T + 300}{0.15}$ for $600 \le T \le 3960$ **101.** $\left\{-\frac{5}{2}, 3\right\}$ **103. (a)** 3 **(b)** $-3$ **(c)** 19 **(d)** $-12$ **105. (a)** 85 **(b)** 19 **(c)** 29 **(d)** $-7$ **107. (a)** $-4394$ **(b)** $-507$ **(c)** 16 **(d)** $-1453$ **109. (a)** 8 **(b)** $\frac{4}{5}$ **(c)** 1 **(d)** $-\frac{4}{3}$

**111.** $f(x) = x + 5$; $g(x) = x - 5$

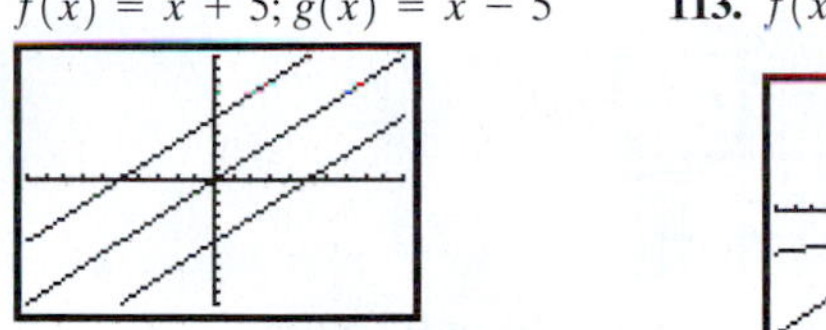

**113.** $f(x) = 5x + 7$; $g(x) = \frac{x - 7}{5}$

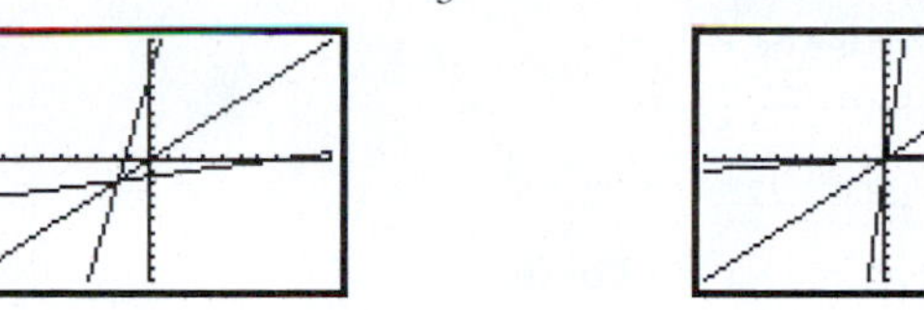

**115.** $f(x) = 10x - 1$; $g(x) = \frac{x + 1}{10}$

## Section 9.2

**1.** $>$; $\neq$ **3.** $u = v$ **5.** false **7.** As the base $a$ increases, the steeper the graph of $f(x) = a^x (a > 1)$ is for $x > 0$ and the closer the graph is to the $x$-axis for $x < 0$. **9.** Answers may vary. **11. (a)** 11.212 **(b)** 11.587 **(c)** 11.664 **(d)** 11.665 **(e)** 11.665 **13. (a)** 73.517 **(b)** 77.708 **(c)** 77.924 **(d)** 77.881 **(e)** 77.880 **15. (a)** 21.217 **(b)** 22.472 **(c)** 22.460 **(d)** 22.460 **(e)** 22.459 **17.** 7.389 **19.** 0.135 **21.** 9.974 **23.** g **25.** e **27.** f **29.** h

**31.** $f(x) = 5^x$ Domain: all real numbers or $(-\infty, \infty)$ Range: $\{y | y > 0\}$ or $(0, \infty)$

**33.** $g(x) = 10^x$ Domain: all real numbers or $(-\infty, \infty)$ Range: $\{y | y > 0\}$ or $(0, \infty)$

**35.** $F(x) = \left(\frac{1}{5}\right)^x$ Domain: all real numbers or $(-\infty, \infty)$ Range: $\{y | y > 0\}$ or $(0, \infty)$

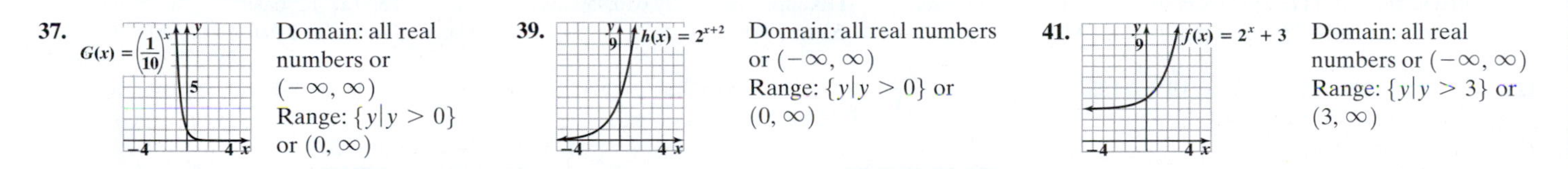

**37.** $G(x) = \left(\frac{1}{10}\right)^x$ Domain: all real numbers or $(-\infty, \infty)$ Range: $\{y | y > 0\}$ or $(0, \infty)$

**39.** $h(x) = 2^{x+2}$ Domain: all real numbers or $(-\infty, \infty)$ Range: $\{y | y > 0\}$ or $(0, \infty)$

**41.** $f(x) = 2^x + 3$ Domain: all real numbers or $(-\infty, \infty)$ Range: $\{y | y > 3\}$ or $(3, \infty)$

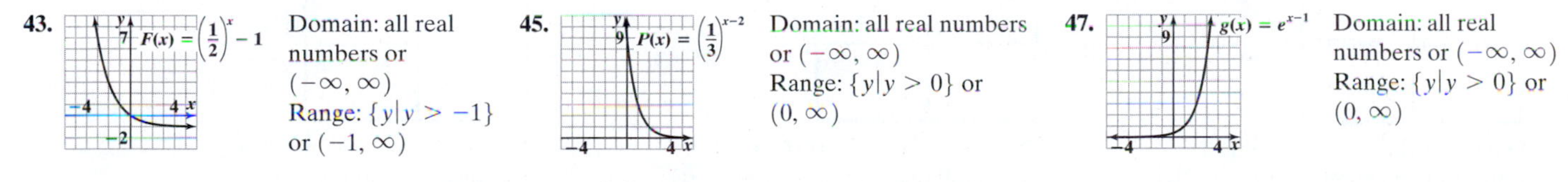

**43.** $F(x) = \left(\frac{1}{2}\right)^x - 1$ Domain: all real numbers or $(-\infty, \infty)$ Range: $\{y | y > -1\}$ or $(-1, \infty)$

**45.** $P(x) = \left(\frac{1}{3}\right)^{x-2}$ Domain: all real numbers or $(-\infty, \infty)$ Range: $\{y | y > 0\}$ or $(0, \infty)$

**47.** $g(x) = e^{x-1}$ Domain: all real numbers or $(-\infty, \infty)$ Range: $\{y | y > 0\}$ or $(0, \infty)$

**49.** $\{5\}$ **51.** $\{-4\}$ **53.** $\{5\}$ **55.** $\{5\}$ **57.** $\left\{\frac{3}{2}\right\}$ **59.** $\{1\}$ **61.** $\{-1, 4\}$ **63.** $\{-4, 2\}$ **65.** $\{9\}$ **67.** $\{-2\}$ **69.** $\{-2, 2\}$ **71.** $\{-2\}$ **73.** $\{2\}$ **75. (a)** $f(3) = 8$; $(3, 8)$ **(b)** $x = -3$; $\left(-3, \frac{1}{8}\right)$ **77. (a)** $g(-1) = -\frac{3}{4}$; $\left(-1, -\frac{3}{4}\right)$ **(b)** $x = 2$; $(2, 15)$ **79. (a)** $H(-3) = 24$; $(-3, 24)$ **(b)** $x = 2$; $\left(2, \frac{3}{4}\right)$ **81. (a)** approximately 306 million people **(b)** approximately 444 million people **(c)** The U.S. Census Bureau's prediction is 44 million people fewer than that of the model. Reasons given may vary. **83. (a)** \$5308.39 **(b)** \$5983.40 **(c)** \$6744.25 **85. (a)** \$2318.55 **(b)** \$2322.37 **(c)** \$2323.23 **(d)** \$2323.65 **(e)** Answers may vary. **87. (a)** \$14,512 **(b)** \$9757.87 **(c)** \$5380.18 **89. (a)** approximately 95.105 grams **(b)** 50 grams **(c)** 25 grams **(d)** approximately 0.661 gram **91. (a)** approximately 300.233°F **(b)** approximately 230.628°F **(c)** yes **93. (a)** approximately 29 words **(b)** approximately 38 words **95. (a)** approximately 0.238 ampere **(b)** approximately 0.475 ampere **97.** $y = 3^x$ **99. (a)** 15 **(b)** $-5$ **101. (a)** undefined **(b)** $\frac{4}{5}$ **103. (a)** 3 **(b)** $3\sqrt{3}$

**105.** $f(x) = 1.5^x$

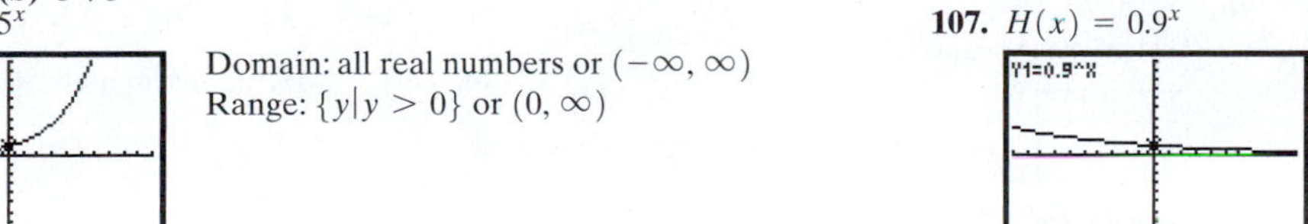


Domain: all real numbers or $(-\infty, \infty)$ Range: $\{y | y > 0\}$ or $(0, \infty)$

**107.** $H(x) = 0.9^x$

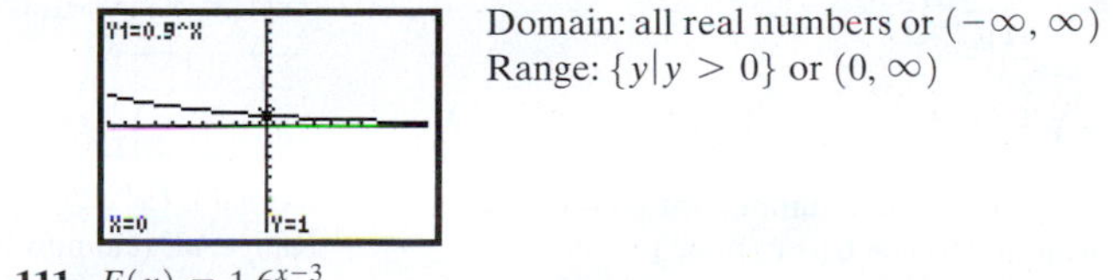


Domain: all real numbers or $(-\infty, \infty)$ Range: $\{y | y > 0\}$ or $(0, \infty)$

**109.** $g(x) = 2.5^x + 3$

Domain: all real numbers or $(-\infty, \infty)$ Range: $\{y | y > 3\}$ or $(3, \infty)$

**111.** $F(x) = 1.6^{x-3}$

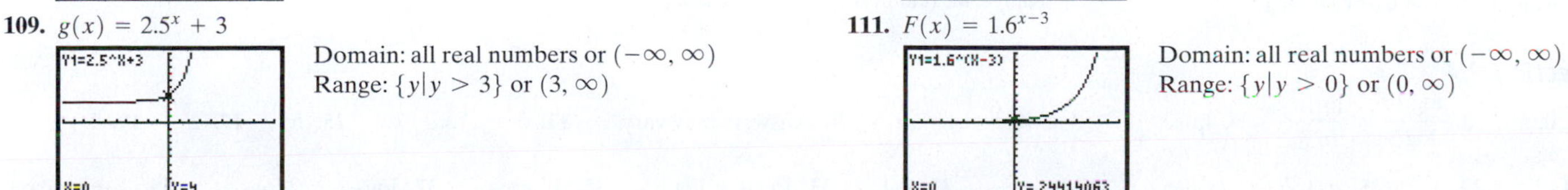


Domain: all real numbers or $(-\infty, \infty)$ Range: $\{y | y > 0\}$ or $(0, \infty)$

## Section 9.3

**1.** $x = a^y$; $>$; $\neq$ **3.** $\left(\frac{1}{a}, -1\right)$, $(1, 0)$, $(a, 1)$ **5.** false **7.** Answers may vary. **9.** $2 = \log_5 25$ **11.** $3 = \log_4 64$ **13.** $-3 = \log_2\left(\frac{1}{8}\right)$

**15.** $\ln 12 = x$ **17.** $\log_a 19 = 3$ **19.** $\log_5 c = -6$ **21.** $2^4 = 16$ **23.** $3^{-2} = \frac{1}{9}$ **25.** $e^4 = x$ **27.** $5^{-3} = a$ **29.** $a^2 = 4$

**31.** $\left(\frac{1}{2}\right)^y = 12$ **33.** 0 **35.** 3 **37.** $-2$ **39.** 4 **41.** 4 **43.** $\frac{1}{2}$ **45.** $-1$ **47.** 3 **49.** $\{x \mid x > 4\}$ or $(4, \infty)$

**51.** $\{x \mid x > -6\}$ or $(-6, \infty)$ **53.** $\{x \mid x > 0\}$ or $(0, \infty)$ **55.** $\left\{x \middle| x > \frac{2}{3}\right\}$ or $\left(\frac{2}{3}, \infty\right)$ **57.** $\left\{x \middle| x > -\frac{1}{2}\right\}$ or $\left(-\frac{1}{2}, \infty\right)$

**59.** $\left\{x \middle| x < \frac{1}{4}\right\}$ or $\left(-\infty, \frac{1}{4}\right)$

**61.**

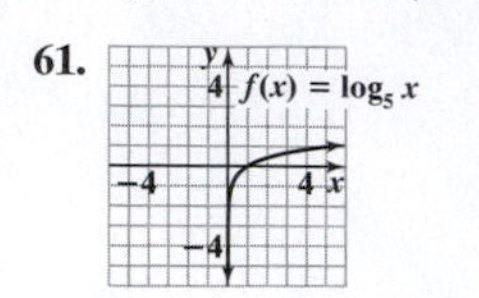


Domain: $\{x \mid x > 0\}$ or $(0, \infty)$
Range: all real numbers or $(-\infty, \infty)$

**63.**

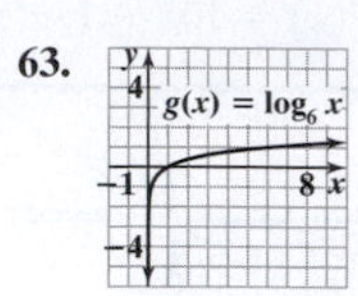


Domain: $\{x \mid x > 0\}$ or $(0, \infty)$
Range: all real numbers or $(-\infty, \infty)$

**65.**

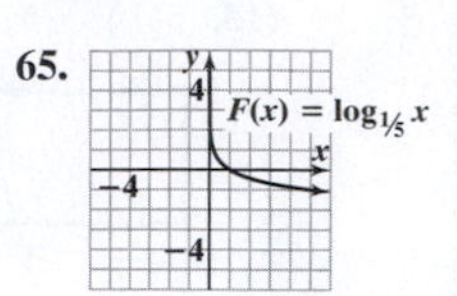


Domain: $\{x \mid x > 0\}$ or $(0, \infty)$
Range: all real numbers or $(-\infty, \infty)$

**67.**

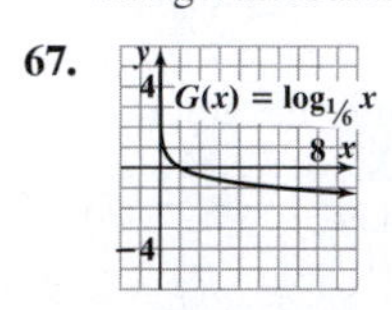


Domain: $\{x \mid x > 0\}$ or $(0, \infty)$
Range: all real numbers or $(-\infty, \infty)$

**69.**

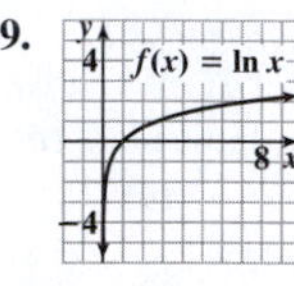


Domain: $\{x \mid x > 0\}$ or $(0, \infty)$
Range: all real numbers or $(-\infty, \infty)$

**71.** 1.826 **73.** 1.686 **75.** $-0.456$ **77.** $-1.609$ **79.** 0.097 **81.** $-0.981$ **83.** $\{4\}$ **85.** $\left\{\frac{13}{2}\right\}$ **87.** $\{6\}$ **89.** $\{3\sqrt{2}\}$ **91.** $\{10\}$ **93.** $\{e^5\}$ **95.** $\left\{\frac{11}{20}\right\}$ **97.** $\{-3\}$ **99.** $\{4\}$

**101.** $\{-3, 3\}$ **103.** **(a)** $f(16) = 4$; $(16, 4)$ **(b)** $x = \frac{1}{8}$; $\left(\frac{1}{8}, -3\right)$ **105.** **(a)** $G(7) = \frac{3}{2}$; $\left(7, \frac{3}{2}\right)$ **(b)** $x = 15$; $(15, 2)$ **107.** 20 decibels

**109.** 130 decibels **111.** approximately 7.8 on the Richter scale **113.** approximately 630,957 millimeters **115.** **(a)** 12; basic **(b)** 5; acidic **(c)** 2; acidic **(d)** $10^{-7.4}$ moles per liter **117.** $\{x \mid x < -2 \text{ or } x > 5\}$ or $(-\infty, -2) \cup (5, \infty)$ **119.** $\{x \mid x < -1 \text{ or } x > 3\}$ or $(-\infty, -1) \cup (3, \infty)$

**121.** $a = 4$ **123.** $-3x^2 - 7x + 10$ **125.** $\frac{4x^2 + 2x + 3}{(x + 1)(x - 1)(x + 2)}$ **127.** $5x\sqrt{2x}$

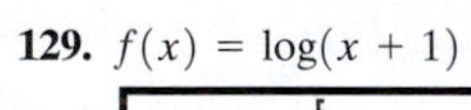

**129.** $f(x) = \log(x + 1)$

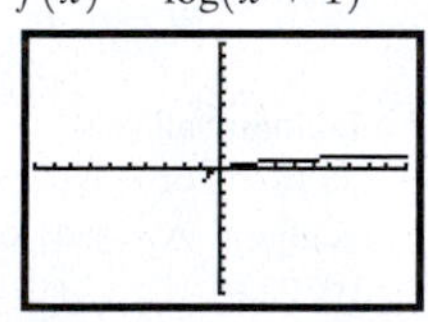

Domain: $\{x \mid x > -1\}$ or $(-1, \infty)$
Range: all real number or $(-\infty, \infty)$

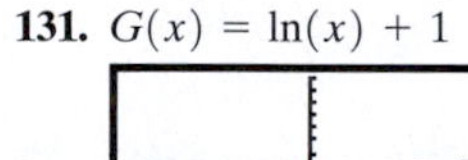

**131.** $G(x) = \ln(x) + 1$

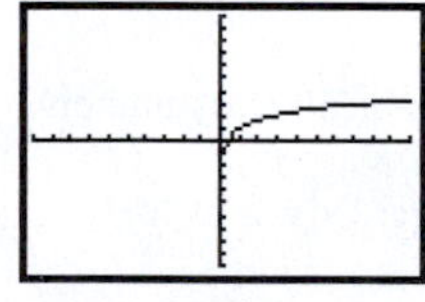

Domain: $\{x \mid x > 0\}$ or $(0, \infty)$
Range: all real number or $(-\infty, \infty)$

**133.** $f(x) = 2\log(x - 3) + 1$

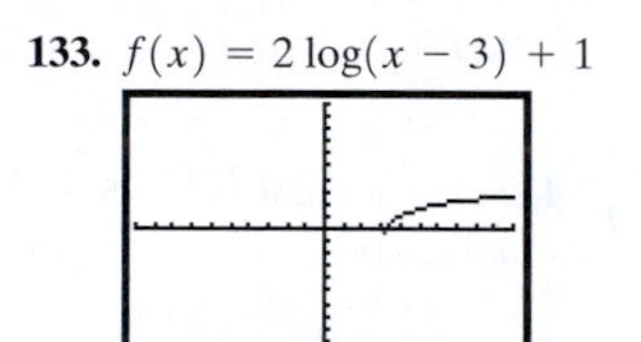

Domain: $\{x \mid x > 3\}$ or $(3, \infty)$
Range: all real number or $(-\infty, \infty)$

## Chapter 9 Putting the Concepts Together

**1.** **(a)** $(f \circ g)(x) = 4x^2 - 8x + 3$ **(b)** $(g \circ f)(x) = 8x^2 + 16x + 6$ **(c)** $(f \circ g)(3) = 15$ **(d)** $(g \circ f)(-2) = 6$ **(e)** $(f \circ f)(1) = 13$

**2.** **(a)** $f^{-1}(x) = \frac{x - 4}{3}$ **(b)** $g^{-1}(x) = \sqrt[3]{x + 4}$

**3.**

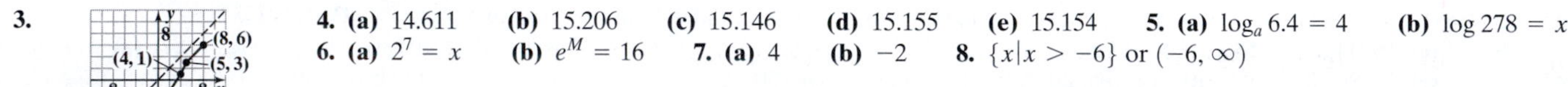


**4.** **(a)** 14.611 **(b)** 15.206 **(c)** 15.146 **(d)** 15.155 **(e)** 15.154 **5.** **(a)** $\log_a 6.4 = 4$ **(b)** $\log 278 = x$ **6.** **(a)** $2^7 = x$ **(b)** $e^M = 16$ **7.** **(a)** 4 **(b)** $-2$ **8.** $\{x \mid x > -6\}$ or $(-6, \infty)$

**9.**

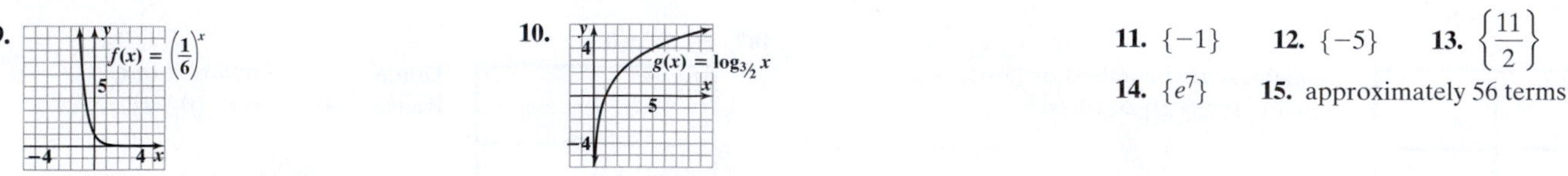


Domain: all real numbers or $(-\infty, \infty)$
Range: $\{y \mid y > 0\}$ or $(0, \infty)$

**10.** Domain: $\{x \mid x > 0\}$ or $(0, \infty)$
Range: all real numbers or $(-\infty, \infty)$

**11.** $\{-1\}$ **12.** $\{-5\}$ **13.** $\left\{\frac{11}{2}\right\}$ **14.** $\{e^7\}$ **15.** approximately 56 terms

## Section 9.4

**1.** 0; 1 **3.** $\frac{\log 10}{\log 3} = \frac{\ln 10}{\ln 3}$ **5.** false **7.** Answers may vary. **9.** Answers may vary. **11.** 3 **13.** $-7$ **15.** 5 **17.** 2 **19.** 1

**21.** 1 **23.** 4 **25.** $a + b$ **27.** $2b$ **29.** $2a + b$ **31.** $\frac{1}{2}a$ **33.** $\log a + \log b$ **35.** $4\log_5 x$ **37.** $\log_2 x + 2\log_2 y$ **39.** $2 + \log_5 x$

**41.** $2 - \log_7 y$ **43.** $2 + \ln x$ **45.** $3 + \frac{1}{2}\log_3 x$ **47.** $2\log_5 x + \frac{1}{2}\log_5(x^2+1)$ **49.** $4\log x - \frac{1}{3}\log(x-1)$

**51.** $\frac{1}{2}\log_7(x+1) - \frac{1}{2}\log_7 x$ **53.** $\log_2 x + 2\log_2(x-1) - \frac{1}{2}\log_2(x+1)$ **55.** 2 **57.** $\log(3x)$ **59.** 2 **61.** $\log_3 x^3$ **63.** $\log_4\left(\frac{x+1}{x}\right)$

**65.** $\ln(x^2y^3)$ **67.** $\log_3\left[\sqrt{x}(x-1)^3\right]$ **69.** $\log(x^2)$ **71.** $\log\left(x\sqrt{xy}\right)$ **73.** $\log_8(x-1)$ **75.** $\log\left(\frac{x^{12}}{10}\right)$ **77.** 3.322

**79.** 0.528 **81.** −2.680 **83.** 4.644 **85.** 3 **87.** 1

**89.**
$$\begin{aligned}&\log_a\left(x+\sqrt{x^2-1}\right)+\log_a\left(x-\sqrt{x^2-1}\right)\\&=\log_a\left[\left(x+\sqrt{x^2-1}\right)\left(x-\sqrt{x^2-1}\right)\right]\\&=\log_a\left[x^2-x\sqrt{x^2-1}+x\sqrt{x^2-1}-(x^2-1)\right]\\&=\log_a(x^2-x^2+1)\\&=\log_a 1\\&=0\end{aligned}$$

**91.** If $f(x) = \log_a x$, then
$$\begin{aligned}f(AB)&=\log_a(AB)\\&=\log_a A+\log_a B\\&=f(A)+f(B)\end{aligned}$$

**93.** $\left\{\frac{5}{2}\right\}$ **95.** $\{-2-\sqrt{2}, -2+\sqrt{2}\}$ **97.** {47}

**99.** $f(x) = \log_3 x = \dfrac{\log x}{\log 3}$

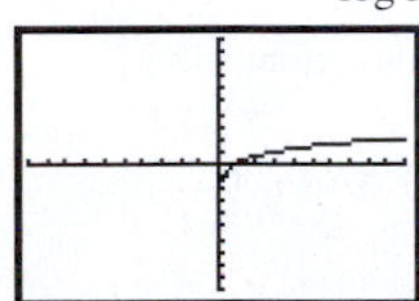

Domain: $\{x \mid x > 0\}$ or $(0, \infty)$
Range: all real numbers or $(-\infty, \infty)$

**101.** $F(x) = \log_{1/2} x = \dfrac{\log x}{\log\left(\frac{1}{2}\right)}$

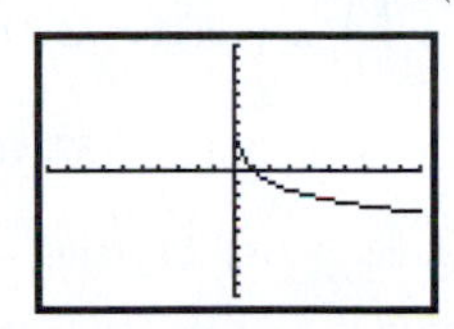

Domain: $\{x \mid x > 0\}$ or $(0, \infty)$
Range: all real numbers or $(-\infty, \infty)$

## Section 9.5

**1.** $M = N$ **3.** True (so long as the logarithms have the same base) **5.** Answers may vary. **7.** {7} **9.** {9} **11.** {3} **13.** {81}

**15.** {1} **17.** {3} **19.** $\left\{\frac{1}{3}\right\}$ **21.** $\left\{\frac{7}{5}\right\}$ **23.** $\left\{\frac{3+\sqrt{93}}{2}\right\} \approx \{6.322\}$ **25.** $\left\{\frac{1}{\log 2}\right\} \approx \{3.322\}$ or $\left\{\frac{\ln 10}{\ln 2}\right\} \approx \{3.322\}$

**27.** $\left\{\frac{\log 20}{\log 5}\right\} \approx \{1.861\}$ or $\left\{\frac{\ln 20}{\ln 5}\right\} \approx \{1.861\}$ **29.** $\left\{\frac{\log 7}{\log\left(\frac{1}{2}\right)}\right\} \approx \{-2.807\}$ or $\left\{\frac{\ln 7}{\ln\left(\frac{1}{2}\right)}\right\} \approx \{-2.807\}$ **31.** $\{\ln 5\} \approx \{1.609\}$

**33.** $\{\log 5\} \approx \{0.699\}$ **35.** $\left\{\frac{\log 13}{2\log 3}\right\} \approx \{1.167\}$ or $\left\{\frac{\ln 13}{2\ln 3}\right\} \approx \{1.167\}$ **37.** $\left\{\frac{\log 3}{4\log\left(\frac{1}{2}\right)}\right\} \approx \{-0.396\}$ or $\left\{\frac{\ln 3}{4\ln\left(\frac{1}{2}\right)}\right\} \approx \{-0.396\}$

**39.** $\left\{\frac{\log\left(\frac{5}{4}\right)}{\log 2}\right\} \approx \{0.322\}$ or $\left\{\frac{\ln\left(\frac{5}{4}\right)}{\ln 2}\right\} \approx \{0.322\}$ **41.** $\{\ln 6\} \approx \{1.792\}$ **43.** $\left\{\frac{\log 0.2}{\log 3 - \log 0.2}\right\} \approx \{-0.594\}$ or $\left\{\frac{\ln 0.2}{\ln 3 - \ln 0.2}\right\} \approx \{-0.594\}$

**45.** {8} **47.** $\left\{\frac{\log 7}{3\log 5}\right\} \approx \{0.403\}$ or $\left\{\frac{\ln 7}{3\ln 5}\right\} \approx \{0.403\}$ **49.** {2} **51.** $\{\ln 15\} \approx \{2.708\}$ **53.** $\left\{-\frac{2}{5}\right\}$ **55.** {16} **57.** {2}

**59. (a)** about the year 2021 **(b)** about the year 2050 **61. (a)** approximately 5.6 years **(b)** approximately 11.6 years
**63. (a)** approximately 1.876 years **(b)** approximately 5.369 years **(c)** approximately 13.479 years **65. (a)** approximately 2.099 seconds
**(b)** 27.62 seconds **(c)** approximately 45.876 seconds **67. (a)** approximately 5 minutes **(b)** approximately 11 minutes
**69. (a)** approximately 82 minutes **(b)** approximately 396 minutes (or 6.6 hours) **71. (a)** approximately 9 years **(b)** $t = \dfrac{\log 2}{n\log\left(1+\frac{r}{n}\right)}$
**(c)** approximately 8.693 years, which is about the same as the result from the Rule of 72 **73.** 29 minutes **75. (a)** 1 **(b)** 11 **(c)** 7
**77. (a)** $-\frac{3}{2}$ **(b)** $\frac{2}{7}$ **(c)** 0 **79. (a)** 8 **(b)** $\frac{1}{4}$ **(c)** 1 **81.** approximately {1.06} **83.** approximately {−0.70}
**85.** approximately {0.05, 1.48} **87.** approximately {0.45, 1}

## Chapter 9 Review

**1. (a)** 32 **(b)** −9 **(c)** 2 **(d)** 13 **2. (a)** 24 **(b)** −28 **(c)** −8 **(d)** 112 **3. (a)** 201 **(b)** 24 **(c)** 163 **(d)** 14
**4. (a)** 23 **(b)** 37 **(c)** −8 **(d)** 290 **5. (a)** $(f \circ g)(x) = 5x + 1$ **(b)** $(g \circ f)(x) = 5x + 5$ **(c)** $(f \circ f)(x) = x + 2$
**(d)** $(g \circ g)(x) = 25x$ **6. (a)** $(f \circ g)(x) = 2x + 9$ **(b)** $(g \circ f)(x) = 2x + 3$ **(c)** $(f \circ f)(x) = 4x - 9$ **(d)** $(g \circ g)(x) = x + 12$
**7. (a)** $(f \circ g)(x) = 4x^2 + 4x + 2$ **(b)** $(g \circ f)(x) = 2x^2 + 3$ **(c)** $(f \circ f)(x) = x^4 + 2x^2 + 2$ **(d)** $(g \circ g)(x) = 4x + 3$
**8. (a)** $(f \circ g)(x) = \dfrac{2x}{x+1}$, where $x \neq -1, 0$ **(b)** $(g \circ f)(x) = \dfrac{x+1}{2}$, where $x \neq -1$ **(c)** $(f \circ f)(x) = \dfrac{2(x+1)}{x+3}$, where $x \neq -1, -3$
**(d)** $(g \circ g)(x) = x$, where $x \neq 0$ **9.** not one-to-one **10.** one-to-one **11.** one-to-one **12.** not one-to-one

**13.** **14.**

**15.** $\{(3, -5), (1, -3), (-3, 1), (9, 2)\}$ **16.** $\{(1, -20), (4, -15), (3, 5), (2, 25)\}$

**17.** 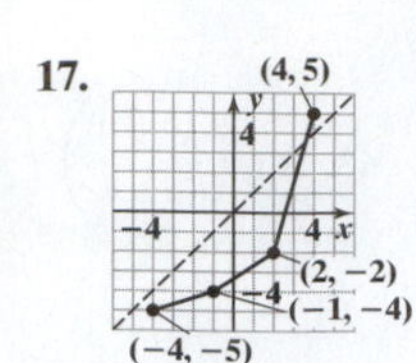


**18.** 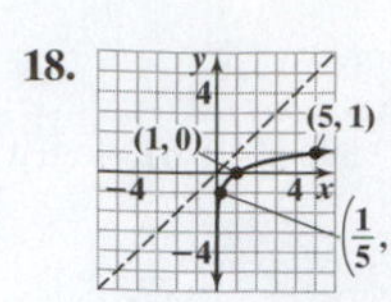


**19.** $f^{-1}(x) = \frac{x}{5}$ **20.** $H^{-1}(x) = \frac{x - 7}{2}$ **21.** $P^{-1}(x) = \frac{4}{x} - 2$ **22.** $g^{-1}(x) = \sqrt[3]{\frac{x + 1}{2}}$ **23. (a)** 27.332 **(b)** 28.975 **(c)** 29.088 **(d)** 29.093 **(e)** 29.091 **24. (a)** 1258.925 **(b)** 1380.384 **(c)** 1386.756 **(d)** 1385.479 **(e)** 1385.456 **25. (a)** 1.649 **(b)** 0.368 **(c)** 4.482 **(d)** 0.449 **(e)** 5.885

**26.** Domain: all real numbers or $(-\infty, \infty)$ Range: $\{y|y > 0\}$ or $(0, \infty)$

**27.** 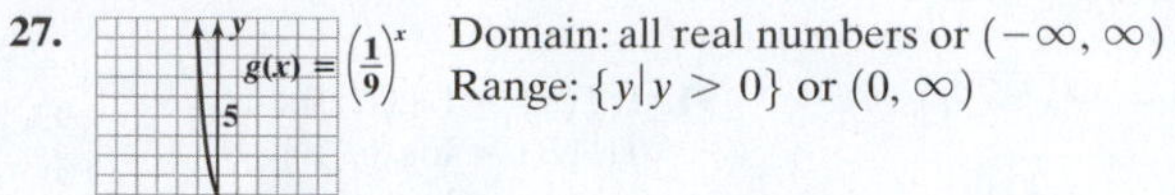 Domain: all real numbers or $(-\infty, \infty)$ Range: $\{y|y > 0\}$ or $(0, \infty)$

**28.** 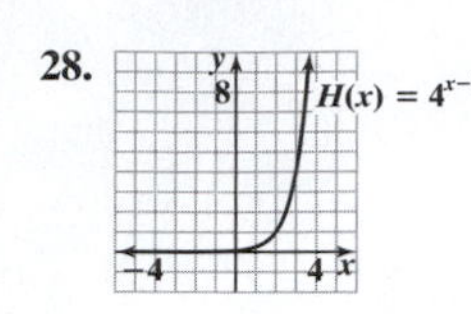


Domain: all real numbers or $(-\infty, \infty)$ Range: $\{y|y > 0\}$ or $(0, \infty)$

**29.** Domain: all real numbers or $(-\infty, \infty)$ Range: $\{y|y > -2\}$ or $(-2, \infty)$

**30.** The number $e$ is defined as the number that the expression $\left(1 + \frac{1}{n}\right)^n$ approaches as $n$ becomes unbounded in the positive direction.

**31.** $\{6\}$ **32.** $\left\{\frac{7}{2}\right\}$ **33.** $\{-4, 1\}$ **34.** $\{-2\}$ **35.** $\{9\}$ **36.** $\{-3, 3\}$ **37. (a)** \$7513.59 **(b)** \$7652.33 **(c)** \$7684.36 **(d)** \$7700.01 **38. (a)** approximately 82.034 grams **(b)** 50 grams **(c)** 25 grams **(d)** approximately 0.263 gram **39. (a)** approximately 2.708 million people **(b)** approximately 3.317 million people **40. (a)** approximately 151.449°F **(b)** approximately 94.706°F **41.** $\log_3 81 = 4$ **42.** $\log_4\left(\frac{1}{64}\right) = -3$ **43.** $\log_b 5 = 3$ **44.** $\log x = 3.74$ **45.** $8^{1/3} = 2$ **46.** $5^r = 18$ **47.** $e^2 = x + 3$ **48.** $10^{-4} = x$ **49.** $\frac{7}{3}$ **50.** 0 **51.** $-2$ **52.** $\frac{3}{2}$ **53.** $\{x|x > -5\}$ or $(-5, \infty)$ **54.** $\left\{x|x < \frac{7}{3}\right\}$ or $\left(-\infty, \frac{7}{3}\right)$ **55.** $\{x|x > 0\}$ or $(0, \infty)$ **56.** $\left\{x|x > -\frac{5}{2}\right\}$ or $\left(-\frac{5}{2}, \infty\right)$

**57.** **58.** **59.** 3.178 **60.** $-0.182$ **61.** 2.410 **62.** $-0.907$ **63.** $\{17\}$ **64.** $\{-9, 1\}$ **65.** $\left\{\frac{3}{2}\right\}$ **66.** $\{6\}$ **67.** $\{-142\}$ **68.** $\{5\sqrt{3}\}$ **69.** 80 decibels **70.** 100,000 millimeters **71.** 21 **72.** 9.34 **73.** 1 **74.** 0 **75.** 1 **76.** 16 **77.** $\log_7 x + \log_7 y - \log_7 z$ **78.** $4 - 2\log_3 x$ **79.** $3 + 4\log r$

**80.** $\frac{1}{2}\ln(x - 1) - \frac{1}{2}\ln x$ **81.** $\log_3(x^4y^2)$ **82.** $\ln\left(\frac{7\sqrt[4]{x}}{9}\right)$ **83.** $-1$ **84.** $\log_6(x - 4)$ **85.** 2.183 **86.** 0.606 **87.** $-4.419$ **88.** 3.723 **89.** $\{10\}$ **90.** $\{3\}$ **91.** $\{\sqrt{2}\} \approx \{1.414\}$ **92.** $\{64\}$ **93.** $\left\{\frac{\log 15}{\log 2}\right\} \approx \{3.907\}$ or $\left\{\frac{\ln 15}{\ln 2}\right\} \approx \{3.907\}$ **94.** $\left\{\frac{\log 27}{3}\right\} \approx \{0.477\}$ **95.** $\left\{\frac{\ln 39}{7}\right\} \approx \{0.523\}$ **96.** $\left\{\frac{\log 2}{\log 3 - \log 2}\right\} \approx \{1.710\}$ or $\left\{\frac{\ln 2}{\ln 3 - \ln 2}\right\} \approx \{1.710\}$ **97. (a)** approximately 1.453 days **(b)** approximately 23.253 days **98. (a)** about 2008 **(b)** about 2016

## Chapter 9 Test

**1.** not one-to-one **2.** $f^{-1}(x) = \frac{x + 3}{4}$ **3. (a)** 33.360 **(b)** 36.338 **(c)** 36.494 **(d)** 36.463 **(e)** 36.462 **4.** $\log_4 19 = x$ **5.** $b^y = x$ **6. (a)** $-3$ **(b)** 4 **7.** $\left\{x|x < \frac{7}{4}\right\}$ or $\left(-\infty, \frac{7}{4}\right)$

**8.** Domain: all real numbers or $(-\infty, \infty)$ Range: $\{y|y > 0\}$ or $(0, \infty)$ **9.** Domain: $\{x|x > 0\}$ or $(0, \infty)$ Range: all real numbers or $(-\infty, \infty)$

**10. (a)** 10 **(b)** 15 **11.** $\frac{1}{2}\log_4 x - 3\log_4 y$ **12.** $\log(M^4N^3)$ **13.** $-8.004$ **14.** $\{1\}$ **15.** $\{1, 3\}$ **16.** $\{4\}$ **17.** $\{-17, 17\}$ **18.** $\{9\}$ **19.** $\left\{\frac{\log 17 + \log 3}{\log 3}\right\} \approx \{3.579\}$ or $\left\{\frac{\ln 17 + \ln 3}{\ln 3}\right\} \approx \{3.579\}$ **20.** $\{2\sqrt{26}\} \approx \{10.198\}$ **21. (a)** approximately 34 million people **(b)** about 2058 **22.** 10 decibels

### *Cumulative Review R–9*

**1.** $\{5\}$ **2.** $\{x \mid -2 \le x \le 6\}$ or $[-2, 6]$ **3.** $\left\{x \,\middle|\, x \ne -3 \text{ and } x \ne \frac{7}{2}\right\}$ **4.**

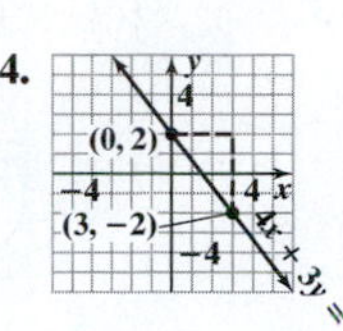


**5.** $y = -\frac{7}{5}x + 3$ or $7x + 5y = 15$

**6.** $\begin{cases} x - 2y + z = 4 & (1) \\ y - z = -2 & (2) \\ 2z = -6 & (3) \end{cases}$

Solution: $(-3, -5, -3)$ **7.** $-30$

**8.**

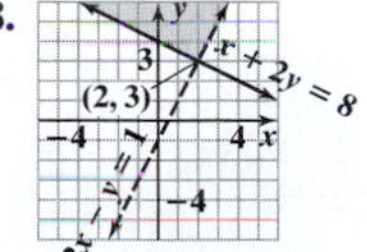


**9.** $4m^2 - 3m + 7$ **10.** $2n^3 - 5n^2 + 18$ **11.** $(4a + b)^2$ **12.** $(3y - 7)(2y - 1)$ **13.** $\frac{3x - 4}{x - 7}$

**14.** $\frac{6(p + 1)}{(p - 5)(p - 1)(p + 2)}$ **15.** $\{4\}$ **16.** $7\sqrt{6}$ **17.** $2 + \sqrt{5}$ **18.** $\{9\}$ **19.** $\left\{\frac{2 - \sqrt{22}}{3}, \frac{2 + \sqrt{22}}{3}\right\}$

**20.** $\left\{\frac{9}{4}, 4\right\}$ **21.**

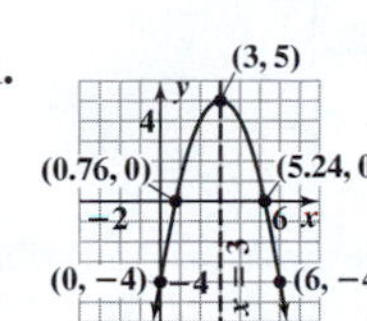


**22.** $\left\{x \,\middle|\, -2 < x < \frac{4}{3}\right\}$ or $\left(-2, \frac{4}{3}\right)$

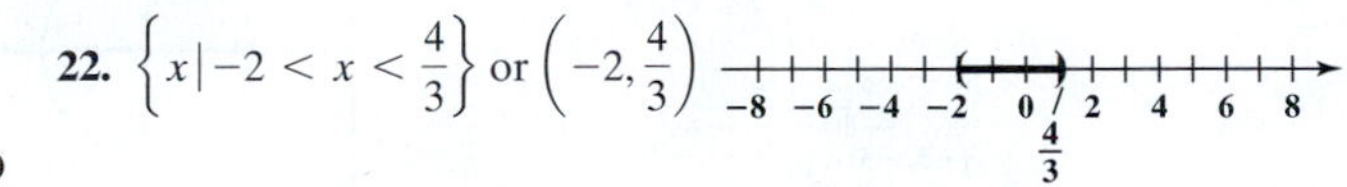

**23.**

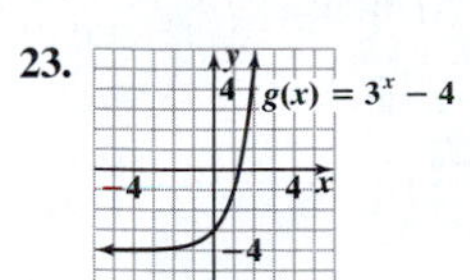


**24.** $-\frac{3}{2}$

## Chapter 10

### *Section 10.1*

**1.** $\sqrt{(x_2 - x_1)^2 + (y_2 - y_1)^2}$ **3.** false **5.** Answers may vary. **7.** 5 **9.** $4\sqrt{5} \approx 8.94$ **11.** 5 **13.** 13 **15.** 6 **17.** $3\sqrt{5} \approx 6.71$

**19.** $3\sqrt{7} \approx 7.94$ **21.** $\sqrt{12.56} \approx 3.54$ **23.** $(4, 3)$ **25.** $(3, -1)$ **27.** $\left(-1, \frac{7}{2}\right)$ **29.** $\left(-\frac{3}{2}, 0\right)$ **31.** $\left(\frac{7\sqrt{2}}{2}, \frac{5\sqrt{5}}{2}\right)$

**33.** $(0.8, -1.6)$ **35. (a)**

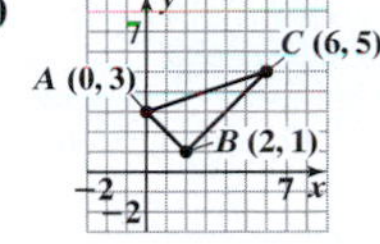


**(b)** $d(A, B) = 2\sqrt{2} \approx 2.83$; $d(B, C) = 4\sqrt{2} \approx 5.66$; $d(A, C) = 2\sqrt{10} \approx 6.32$

**(c)** $[d(A, B)]^2 + [d(B, C)]^2 \stackrel{?}{=} [d(A, C)]^2$

$(2\sqrt{2})^2 + (4\sqrt{2})^2 \stackrel{?}{=} (2\sqrt{10})^2$

$4 \cdot 2 + 16 \cdot 2 \stackrel{?}{=} 4 \cdot 10$

$8 + 32 \stackrel{?}{=} 40$

$40 = 40 \leftarrow$ True

Therefore, triangle $ABC$ is a right triangle.

**(d)** 8 square units

**37. (a)** A (−2, −4), B (3, 1), C (15, −11)

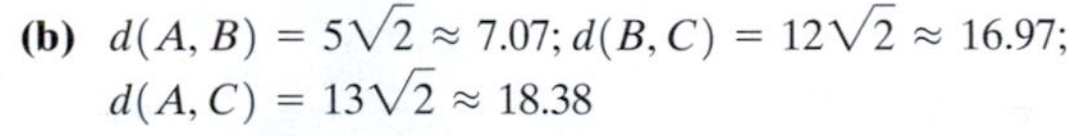

**(b)** $d(A, B) = 5\sqrt{2} \approx 7.07$; $d(B, C) = 12\sqrt{2} \approx 16.97$; $d(A, C) = 13\sqrt{2} \approx 18.38$

**(c)** $[d(A, B)]^2 + [d(B, C)]^2 \stackrel{?}{=} [d(A, C)]^2$

$(5\sqrt{2})^2 + (12\sqrt{2})^2 \stackrel{?}{=} (13\sqrt{2})^2$

$25 \cdot 2 + 144 \cdot 2 \stackrel{?}{=} 169 \cdot 2$

$50 + 288 \stackrel{?}{=} 338$

$338 = 338 \leftarrow$ True

Therefore, triangle $ABC$ is a right triangle.

**(d)** 60 square units

**39.** $(2, -3), (2, 5)$ **41.** $(-6, -3), (10, -3)$ **43. (a)** approximately 37.36 blocks **(b)** approximately 35.13 blocks **(c)** approximately 71.34 blocks **45. (a)** approximately 1.85 seconds **(b)** No. The ball will reach second base (2.65 seconds) before the runner (3.33 seconds). **47.** 9; 3 **49.** 81; 3 **51.** If $n$ is a positive integer and $a^n = b$, then $\sqrt[n]{b} = \begin{cases} |a|, & \text{if } n \text{ is even} \\ a, & \text{if } n \text{ is odd} \end{cases}$

### *Section 10.2*

**1.** circle **3.** false **5.** Answers may vary. **7.** Yes; Center: $(0, 0)$; $r = 6$ **9.** $(x - 1)^2 + (y - 2)^2 = 4$ **11.** $(x - 2)^2 + (y + 1)^2 = 16$

**13.** $x^2 + y^2 = 9$ **15.** $(x - 1)^2 + (y - 4)^2 = 4$ **17.** $(x + 2)^2 + (y - 4)^2 = 36$ **19.** $x^2 + (y - 3)^2 = 16$ **21.** $(x - 5)^2 + (y + 5)^2 = 25$

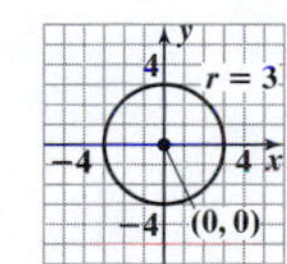


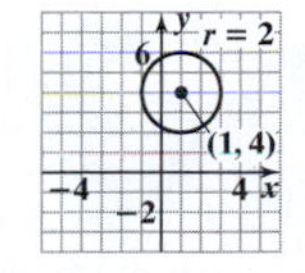


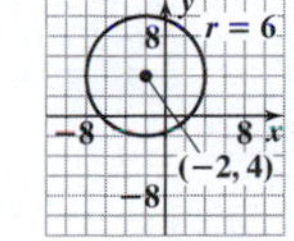


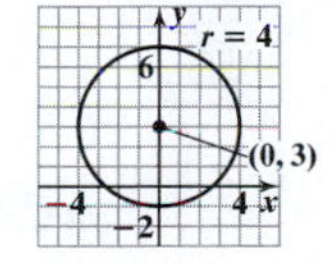


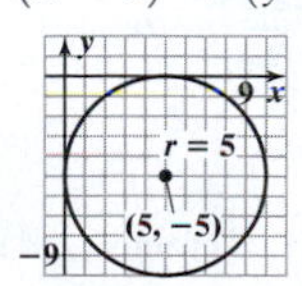


**23.** $(x - 1)^2 + (y - 2)^2 = 5$

**25.** $C = (0, 0), r = 6$ **27.** $C = (4, 1), r = 5$ **29.** $C = (-3, 2), r = 9$

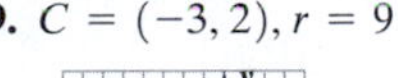

**31.** $C = (0, 3), r = 8$

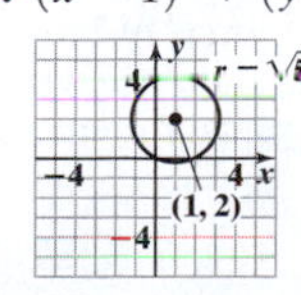


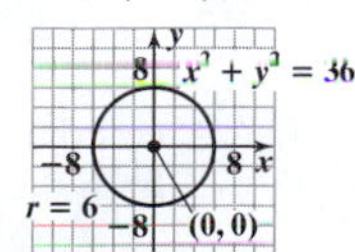


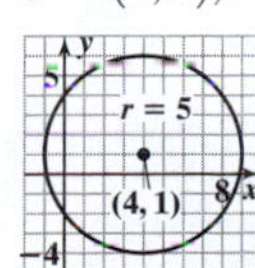


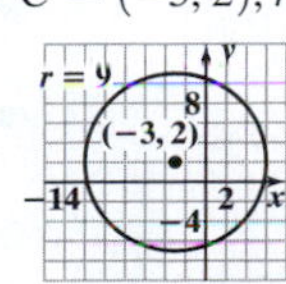


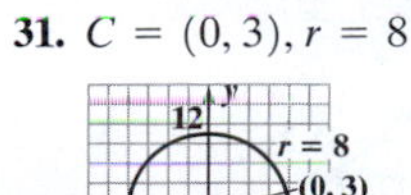

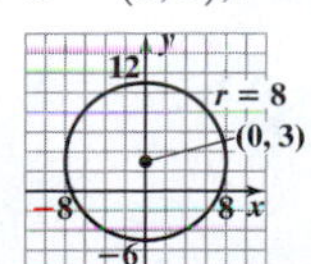

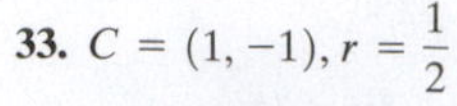

**33.** $C = (1, -1), r = \frac{1}{2}$ **35.** $C = (3, -1), r = 3$ **37.** $C = (-5, -2), r = 5$ **39.** $C = (3, -6), r = 9$

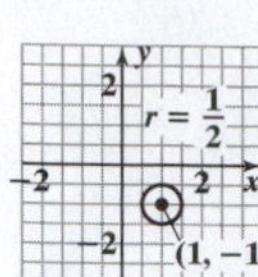


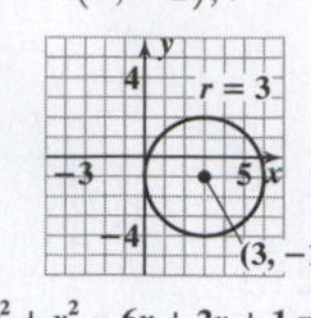


$x^2 + y^2 - 6x + 2y + 1 = 0$

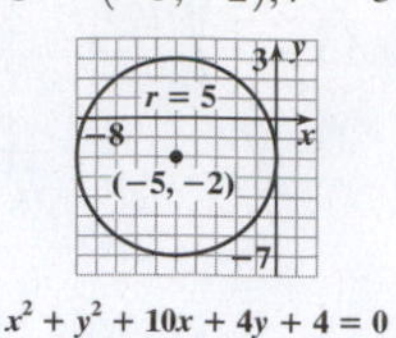


$x^2 + y^2 + 10x + 4y + 4 = 0$

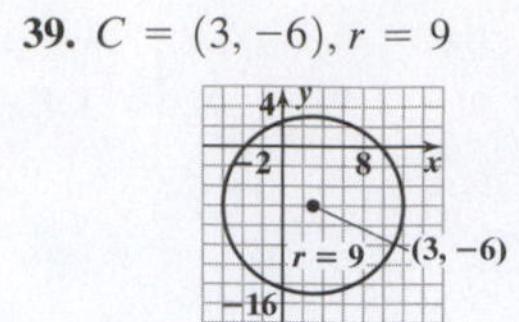


$2x^2 + 2y^2 - 12x + 24y - 72 = 0$

**41.** $x^2 + y^2 = 20$ **43.** $(x + 3)^2 + (y - 2)^2 = 9$ **45.** $(x + 1)^2 + (y + 1)^2 = 25$ **47.** $A = 64\pi$ square units; $C = 16\pi$ units
**49.** 32 square units **51.** $(x - 2)^2 + y^2 = 1$; $x^2 + y^2 - 8x + 7 = 0$

**53.** $f(x) = 4x - 3$ **55.** $g(x) = x^2 - 4x - 5$ **57.** $G(x) = -2(x + 3)^2 - 5$ **59.**

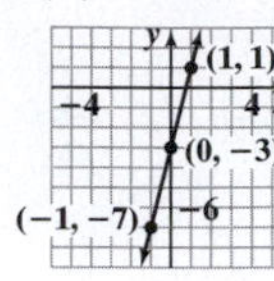


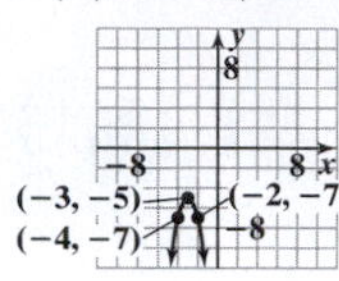


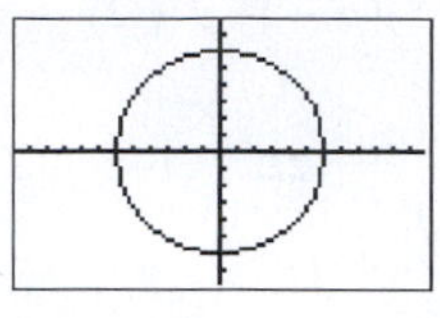

The graph here agrees with that in Problem 25.

**61.**

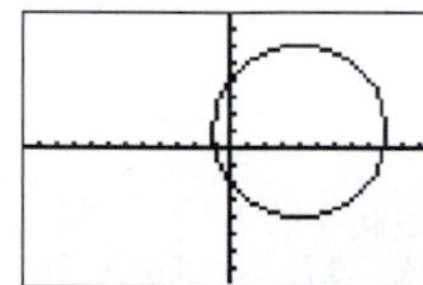

The graph here agrees with that in Problem 27.

**63.**

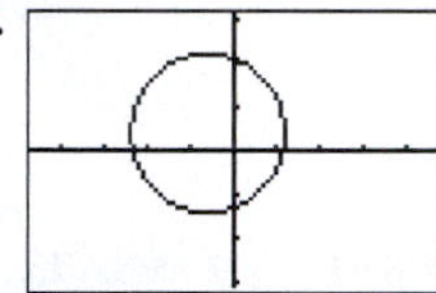

The graph here agrees with that in Problem 29.

**65.**

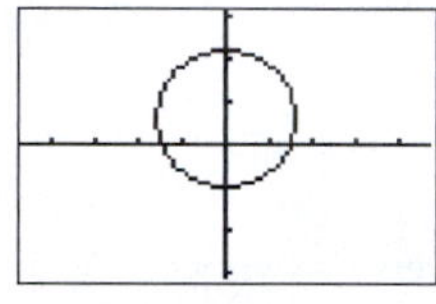

The graph here agrees with that in Problem 31.

**67.**

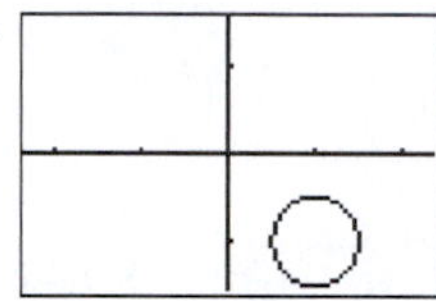

The graph here agrees with that in Problem 33.

## Section 10.3

**1.** Conics **3.** vertex **5.** false **7.** 8 units **9.** Answers may vary. **11.** c **13.** a **15.** b **17.** e

**19.** $y^2 = 20x$

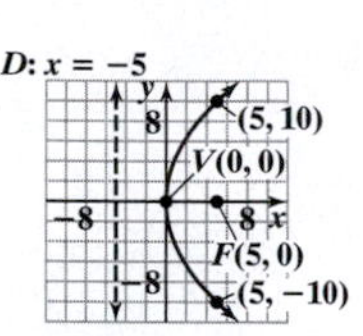


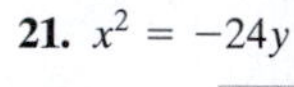

**21.** $x^2 = -24y$ **23.** $x^2 = 6y$ **25.** $x^2 = -12y$

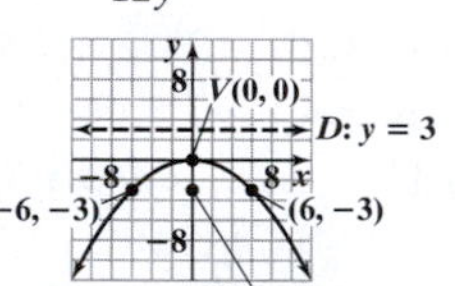


**27.** $y^2 = -12x$

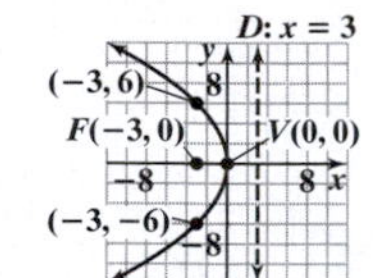


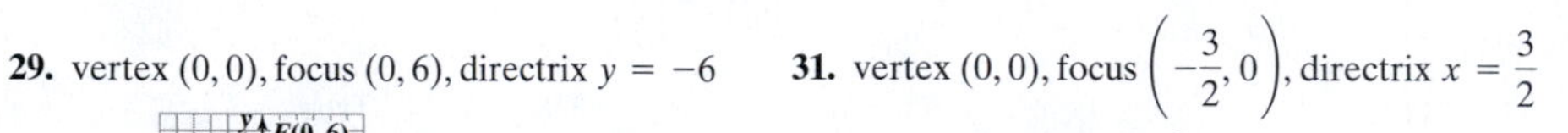

**29.** vertex (0, 0), focus (0, 6), directrix $y = -6$ **31.** vertex (0, 0), focus $\left(-\frac{3}{2}, 0\right)$, directrix $x = \frac{3}{2}$ **33.** vertex (0, 0), focus (0, −2), directrix $y = 2$

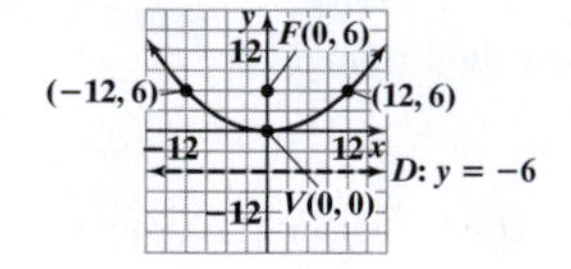


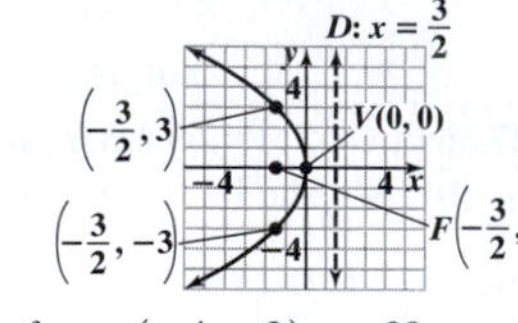


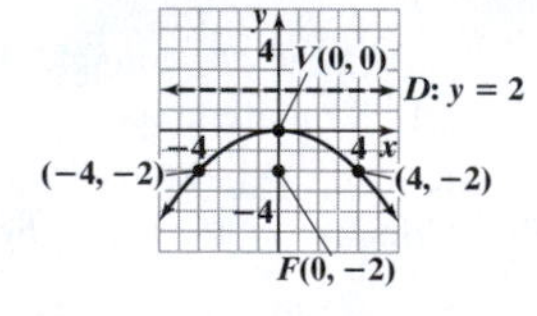


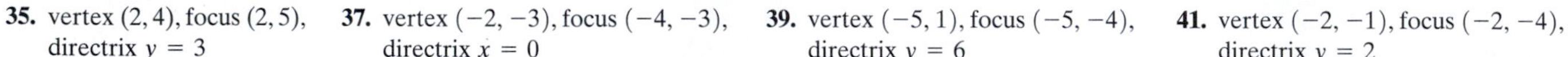

**35.** vertex (2, 4), focus (2, 5), directrix $y = 3$ **37.** vertex (−2, −3), focus (−4, −3), directrix $x = 0$ **39.** vertex (−5, 1), focus (−5, −4), directrix $y = 6$ **41.** vertex (−2, −1), focus (−2, −4), directrix $y = 2$

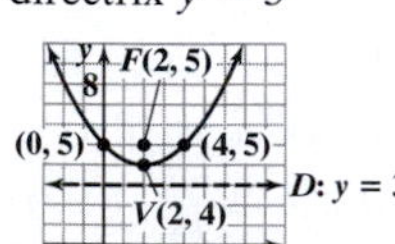


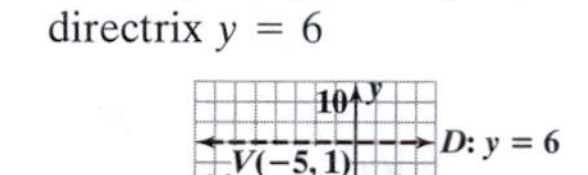

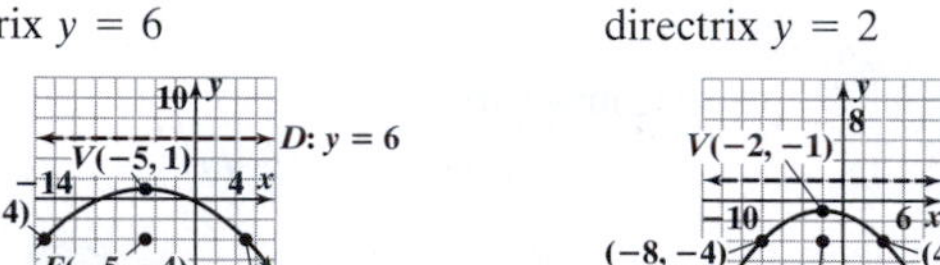


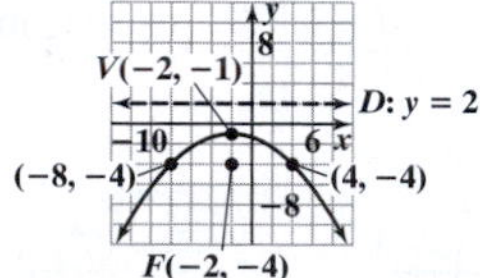


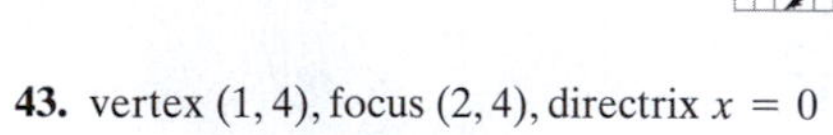

**43.** vertex (1, 4), focus (2, 4), directrix $x = 0$

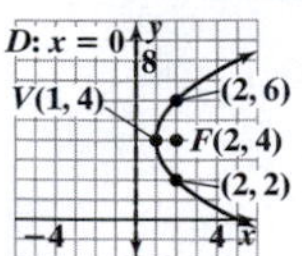


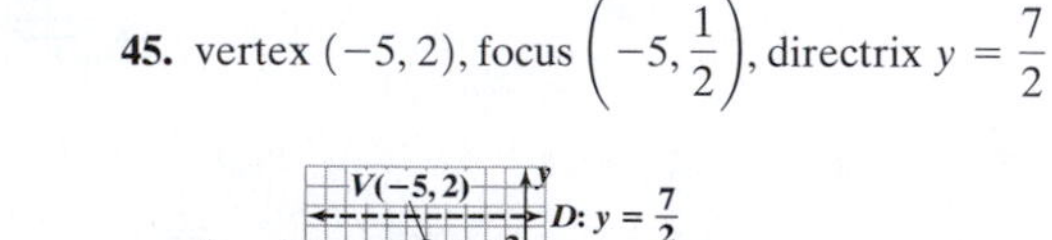

**45.** vertex (−5, 2), focus $\left(-5, \frac{1}{2}\right)$, directrix $y = \frac{7}{2}$

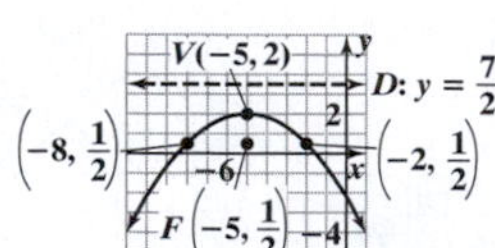


**47.** $y^2 = x$ **49.** 1 inch above the vertex along its axis of symmetry **51.** 21.6 feet
**53.** The height of the bridge is 28.8 feet at a distance of 10 feet from the center, 19.2 feet at a distance of 30 feet from the center, and 0 feet (i.e., ground level) at a distance of 50 feet from the center.

**55.** $(x - 3)^2 = 4(y + 2)$ **57.** $(y - 3)^2 = -4(x - 2)$

**59.** **(a)** Let $x = 4$ and $y = 2$:

$4^2 \stackrel{?}{=} 8 \cdot 2$

$16 = 16 \leftarrow$ True

Thus, $(4, 2)$ in on the parabola.

**(b)** The focus of the parabola is $F(0, 2)$, and the directrix is $D$: $y = -2$.

$d(F, P) = \sqrt{(0-4)^2 + (2-2)^2} = \sqrt{16} = 4$,

$d(P, D) = 2 - (-2) = 4$.

Thus, $d(F, P) = d(P, D) = 4$.

**61.** $y = (x + 3)^2$

**63.** $y = (x + 3)^2$

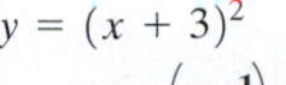

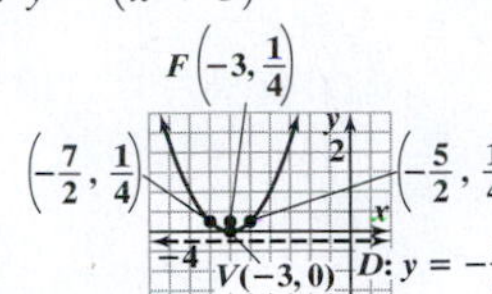


**65.** $4(y + 2) = (x - 2)^2$

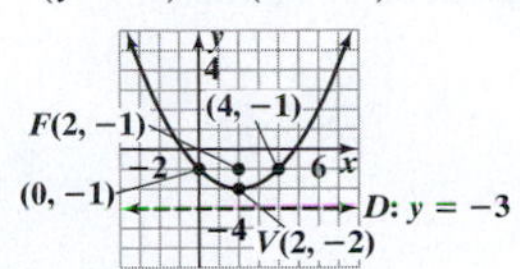


**67.**

**69.**

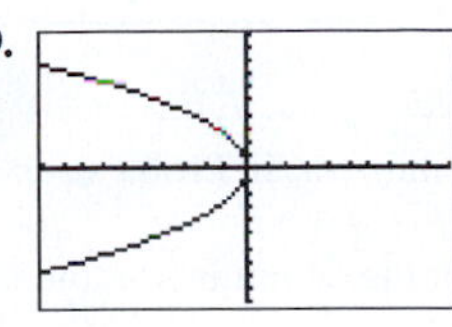

**71.**

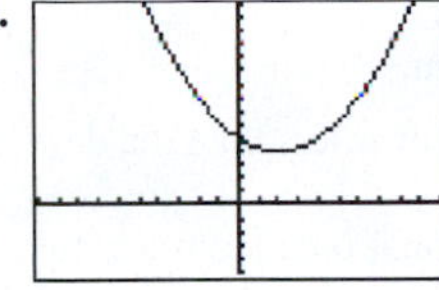

**73.**

**75.**

**77.**

## Section 10.4

**1.** ellipse, foci **3.** vertices **5.** true

**7.**

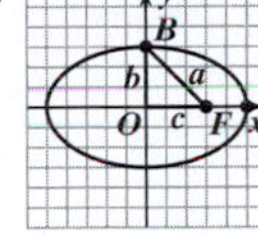


**9.** c **11.** d **13.** Foci: $(-3, 0)$ and $(3, 0)$; Vertices: $(-5, 0)$ and $(5, 0)$ **15.** Foci: $(0, -8)$ and $(0, 8)$; Vertices: $(0, -10)$, and $(0, 10)$

**17.** Foci: $(-3\sqrt{5}, 0)$ and $(3\sqrt{5}, 0)$; Vertices: $(-7, 0)$ and $(7, 0)$

**19.** Foci: $(0, -4\sqrt{3})$ and $(0, 4\sqrt{3})$; Vertices: $(0, -7)$ and $(0, 7)$

**21.** Foci: $(0, -2\sqrt{3})$ and $(0, 2\sqrt{3})$; Vertices: $(0, -4)$ and $(0, 4)$

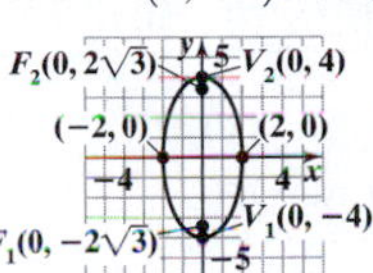


**23.** $\dfrac{x^2}{36} + \dfrac{y^2}{20} = 1$

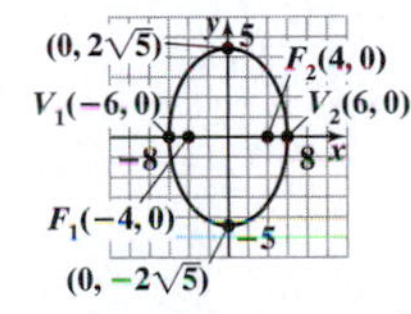


**25.** $\dfrac{x^2}{33} + \dfrac{y^2}{49} = 1$

**27.** $\dfrac{x^2}{100} + \dfrac{y^2}{64} = 1$

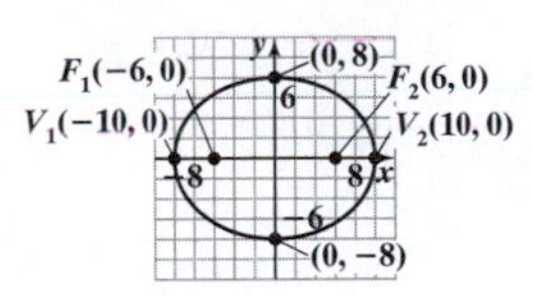


**29.** $\dfrac{x^2}{39} + \dfrac{y^2}{64} = 1$

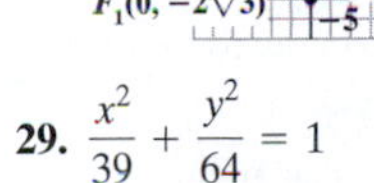

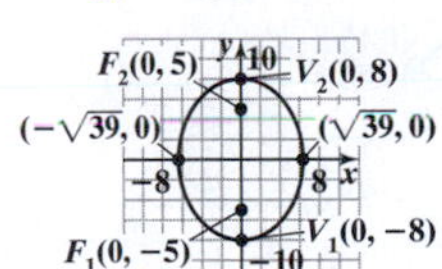


**31.** $\dfrac{(x-3)^2}{9} + \dfrac{(y+2)^2}{25} = 1$

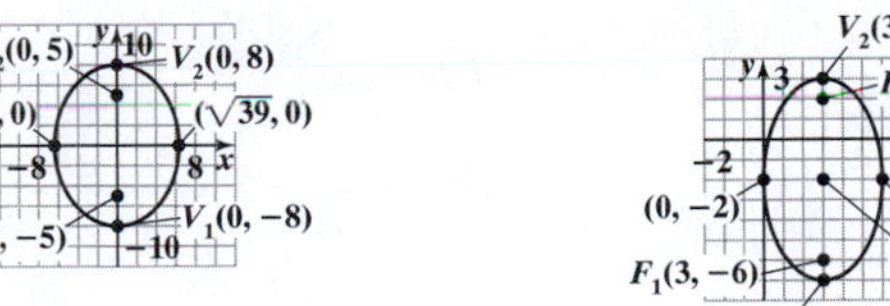


**33.** $\dfrac{(x+2)^2}{16} + \dfrac{(y-5)^2}{4} = 1$

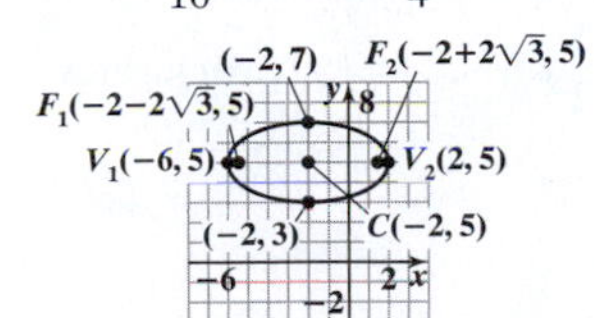


**35.** $(x-5)^2 + \dfrac{(y+1)^2}{49} = 1$

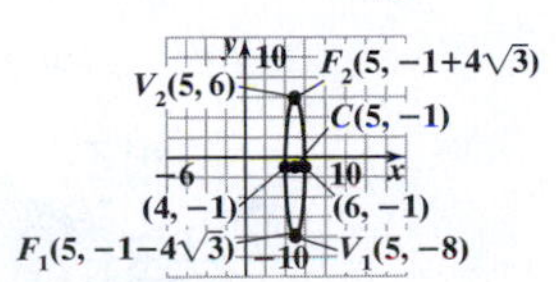


**37.** $4(x+2)^2 + 16(y-1)^2 = 64$

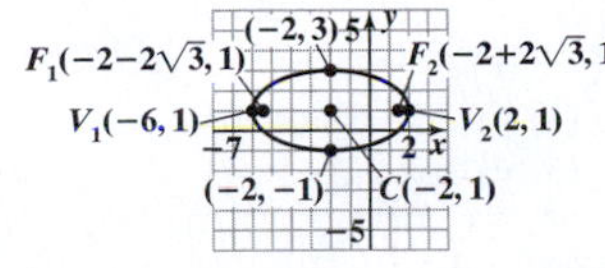


**39.** $4x^2 + y^2 - 24x + 2y - 63 = 0$

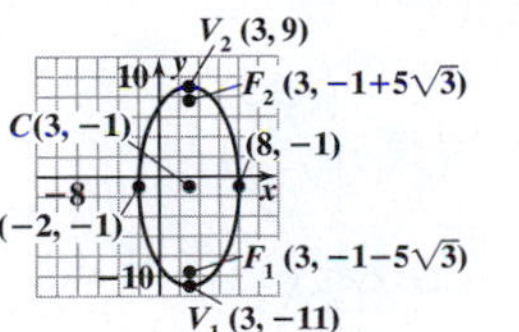


**41.** **(a)** $\dfrac{x^2}{225} + \dfrac{y^2}{100} = 1$ **(b)** Yes **(c)** No **43.** Perihelion = 91.5 million miles; $\dfrac{x^2}{8649} + \dfrac{y^2}{8646.75} = 1$ **45.** Perihelion = 460.6 million miles; Mean distance = 483.8 million miles; $\dfrac{x^2}{234{,}062.44} + \dfrac{y^2}{233{,}524.2} = 1$ **47.** $\dfrac{(x-1)^2}{16} + \dfrac{(y-2)^2}{9} = 1$ **49.** $\dfrac{(x-2)^2}{4} + \dfrac{y^2}{16} = 1$

**51.** Let $a = b$, then

$$\frac{x^2}{a^2} + \frac{y^2}{b^2} = 1$$

$$\frac{x^2}{a^2} + \frac{y^2}{a^2} = 1$$

$$a^2\left(\frac{x^2}{a^2} + \frac{y^2}{a^2}\right) = a^2(1)$$

$$x^2 + y^2 = a^2$$

which is the equation of a circle with center $(0, 0)$ and radius $a$. $c = 0$; The foci are located at the center point.

**53.**

| $x$ | 5 | 10 | 100 | 1000 |
|---|---|---|---|---|
| $f(x)$ | 0.71429 | 0.41667 | 0.04902 | 0.00499 |

**55.**

| $x$ | 5 | 10 | 100 | 1000 |
|---|---|---|---|---|
| $f(x)$ | 5.5 | 3 | 2.07216 | 2.00702 |

**57.**

| $x$ | 5 | 10 | 100 | 1000 |
|---|---|---|---|---|
| $f(x)$ | 6.83333 | 11.90909 | 101.99010 | 1001.99900 |
| $g(x)$ | 7 | 12 | 102 | 1002 |

**59.** In Problems 53 and 54, the degree of the numerator is less than the degree of the denominator. In Problems 55 and 56, the degree of the numerator and denominator are the same.
*Conjecture 1:* If the degree of the numerator of a rational function is less than the degree of the denominator, then as $x$ increases, the value of the function will approach zero (0).
*Conjecture 2:* If the degree of the numerator of a rational function equals the degree of the denominator, then as $x$ increases, the value of the function will approach the ratio of the leading coefficients of the numerator and denominator.

**61.**
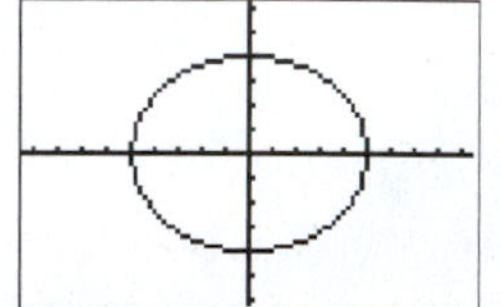

**63.**
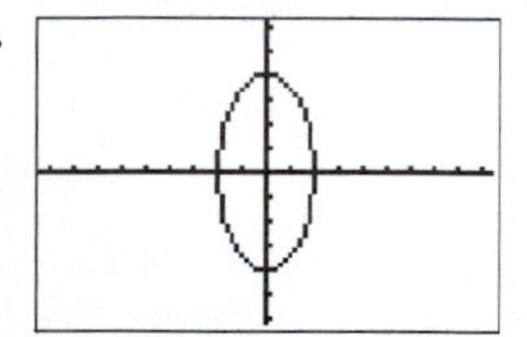

**65.**
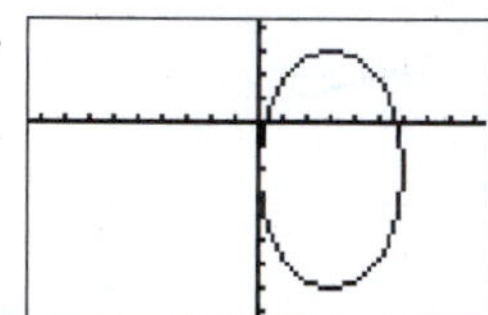

**67.**
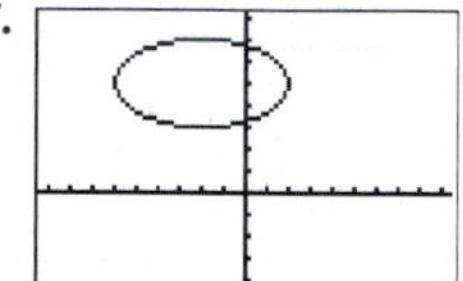

## Section 10.5

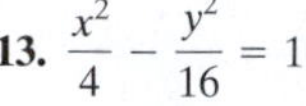

**1.** hyperbola **3.** $y = -\frac{b}{a}x$; $y = \frac{b}{a}x$ **5.** True **7.** Answers may vary. **9.** b **11.** a

**13.** $\frac{x^2}{4} - \frac{y^2}{16} = 1$

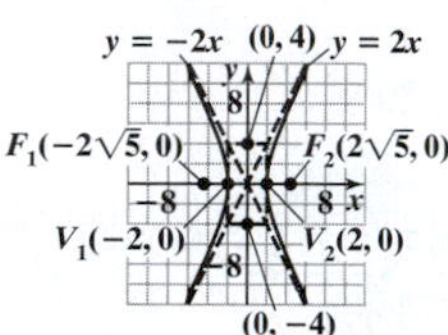


**15.** $\frac{y^2}{25} - \frac{x^2}{36} = 1$

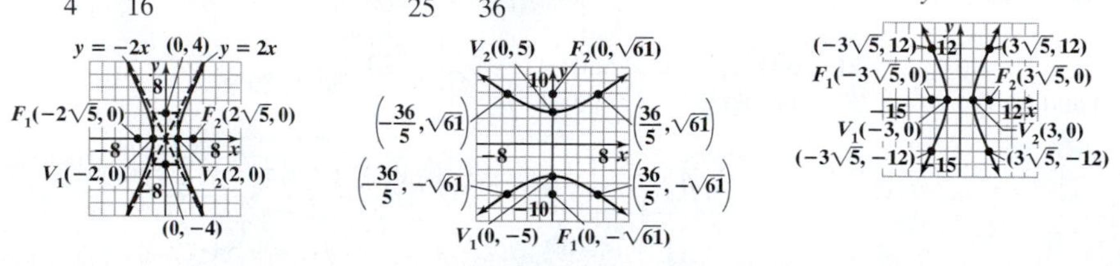


**17.** $4x^2 - y^2 = 36$

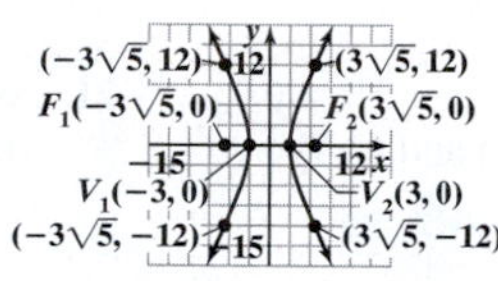


**19.** $25y^2 - x^2 = 100$

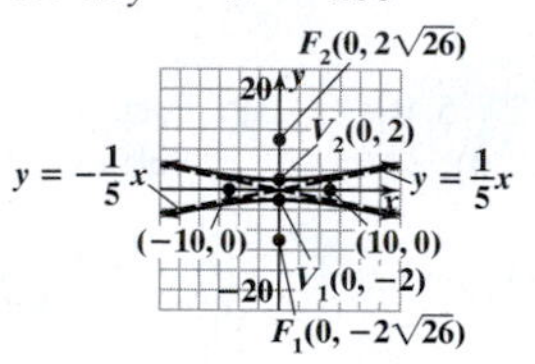


**21.** $\frac{x^2}{4} - \frac{y^2}{5} = 1$

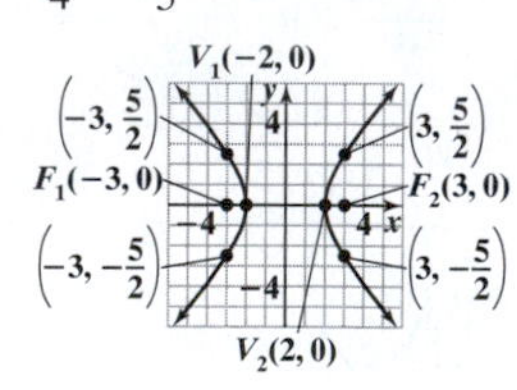


**23.** $\frac{y^2}{25} - \frac{x^2}{24} = 1$

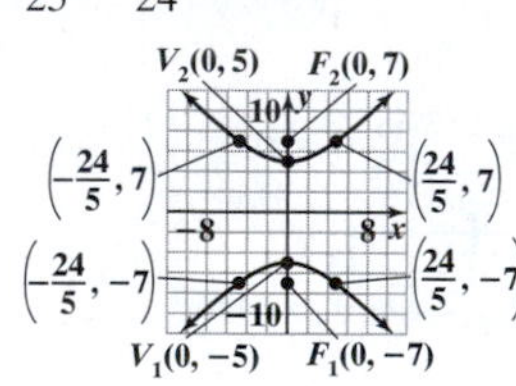


**25.** $\frac{x^2}{49} - \frac{y^2}{51} = 1$

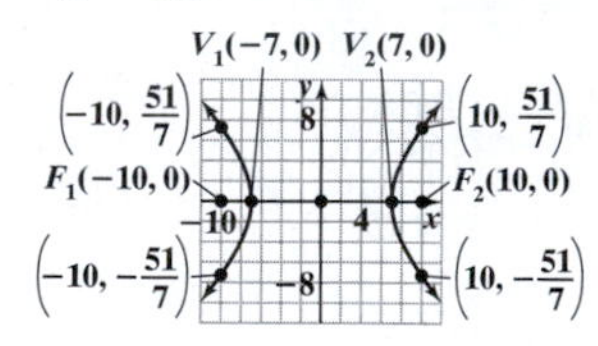


**27.** $\frac{y^2}{64} - \frac{x^2}{16} = 1$

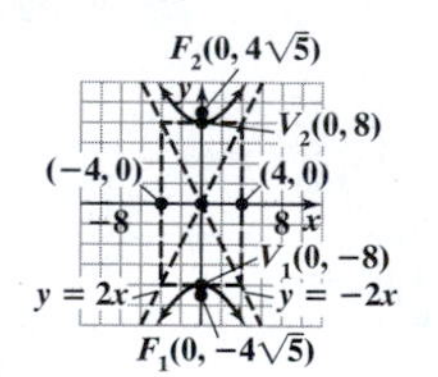


**29.** $\frac{x^2}{4.5} - \frac{y^2}{4.5} = 1$

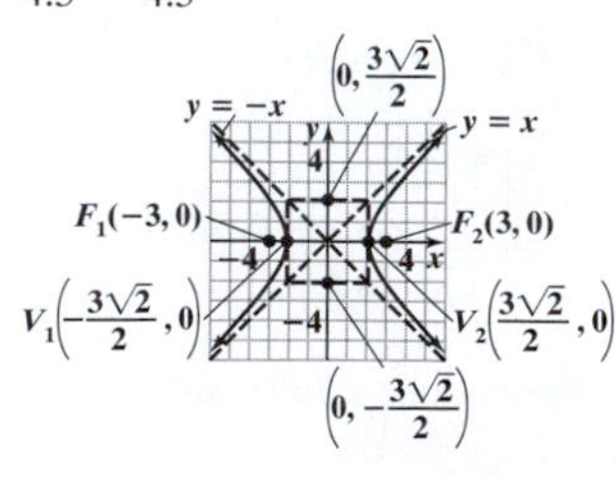


**31.** $x^2 - y^2 = 1$ **33.** $\frac{y^2}{36} - \frac{x^2}{9} = 1$

**35.** The asymptotes of both hyperbolas are $y = -\frac{1}{2}x$ and $y = \frac{1}{2}x$. Thus, they are conjugates.

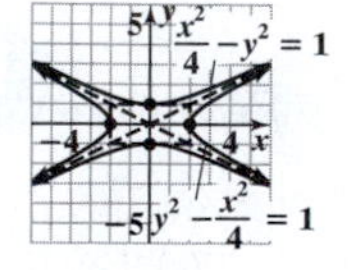


**37.** $(-3, 1)$
**39.** $\{(x, y) | 2x - 3y = 6\}$
**41.** $\{(x, y) | 6x + 3y = 4\}$
**43.**
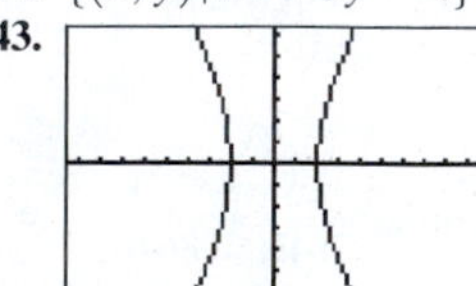

**45.**
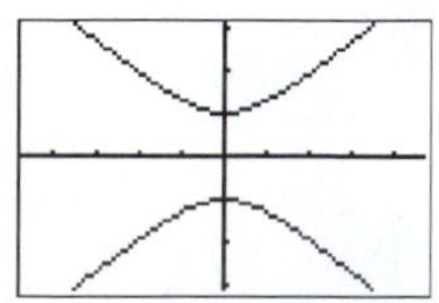

**47.**
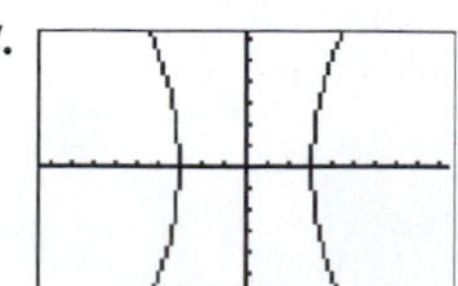

**49.**
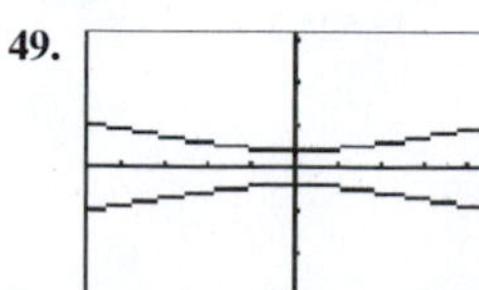

## *Putting the Concepts Together*

**1.** $3\sqrt{13}$ **2.** $(1, -3)$

**3.** $C = (-2, 8), r = 6$

$(x + 2)^2 + (y - 8)^2 = 36$

**4.** $C = (-3, 2), r = 4$

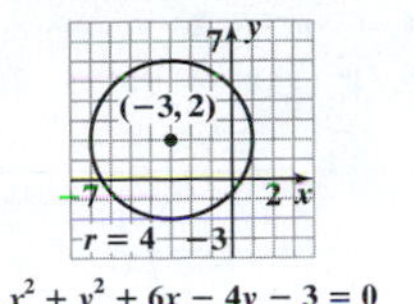


$x^2 + y^2 + 6x - 4y - 3 = 0$

**5.** $x^2 + y^2 = 169$ **6.** $(x - 2)^2 + (y - 1)^2 = 25$ **7.** Vertex: $(-2, 4)$; focus: $(-2, 3)$; directrix: $y = 5$

**8.** Vertex: $(3, -1)$; focus: $(5, -1)$; directrix: $x = 1$

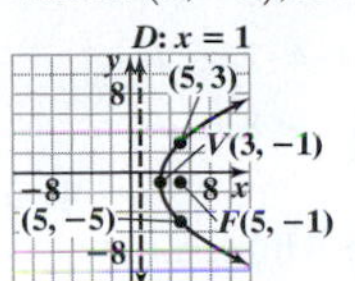


**9.** $(x + 1)^2 = -12(y + 2)$

**10.** $(y - 3)^2 = 8(x + 3)$

**11.** Foci: $(-6\sqrt{2}, 0)$ and $(6\sqrt{2}, 0)$; vertices: $(-9, 0)$ and $(9, 0)$

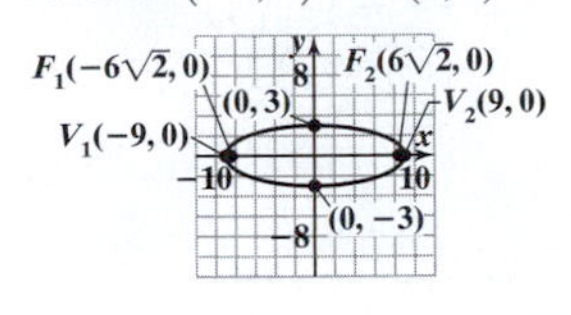


**12.** Center: $(-1, 2)$; vertices: $(-1, -5)$ and $(-1, 9)$; foci: $(-1, 2 - \sqrt{13})$ and $(-1, 2 + \sqrt{13})$

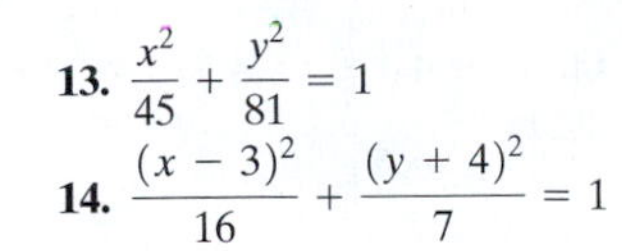

**13.** $\dfrac{x^2}{45} + \dfrac{y^2}{81} = 1$

**14.** $\dfrac{(x - 3)^2}{16} + \dfrac{(y + 4)^2}{7} = 1$

**15.** Vertices: $(0, -9)$ and $(0, 9)$; foci: $(0, -3\sqrt{10})$ and $(0, 3\sqrt{10})$; asymptotes: $y = 3x$ and $y = -3x$

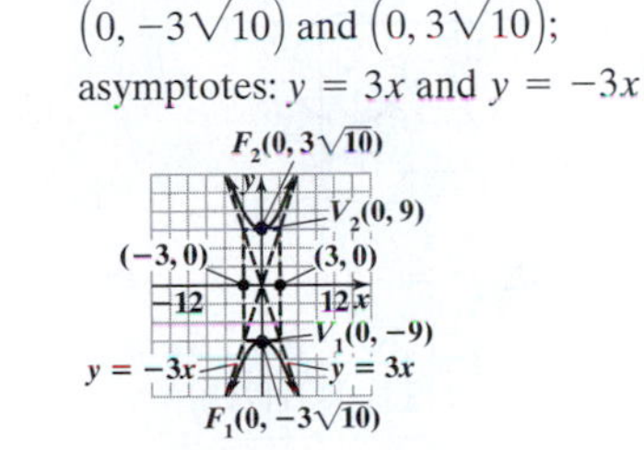


**16.** Vertices: $(-1, 0)$ and $(1, 0)$; foci: $(-\sqrt{26}, 0)$ and $(\sqrt{26}, 0)$; asymptotes: $y = -5x$ and $y = 5x$

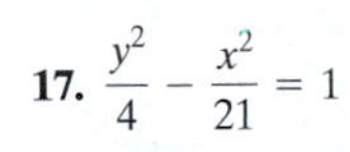

**17.** $\dfrac{y^2}{4} - \dfrac{x^2}{21} = 1$

**18.** 6.75 inches above the vertex along its axis of symmetry.

## *Section 10.6*

**1.** substitution, elimination **3.** In a system of nonlinear equations, at least one of its equations is not linear. **5.** $(0, 4)$ and $(1, 5)$ **7.** $(3, 4)$ and $(4, 3)$ **9.** $(0, -2), (-\sqrt{3}, 1)$, and $(\sqrt{3}, 1)$ **11.** $(-2, -2)$ and $(2, 2)$ **13.** $(0, -2), (0, 2), (-1, -\sqrt{3})$, and $(-1, \sqrt{3})$ **15.** $\varnothing$ **17.** $(0, 0), (-3, 3)$, and $(3, 3)$ **19.** $(-3, 7)$ and $(2, -8)$ **21.** $(-1, 11)$ and $(2, -4)$ **23.** $(-4, 0)$ and $(4, 0)$ **25.** $(0, 3)$ and $(1, 4)$ **27.** $(0, -2 - \sqrt{3}), (0, -2 + \sqrt{3}), (1, -4)$, and $(1, 0)$ **29.** $\varnothing$ **31.** $(5, 2)$ **33.** $\left(-\dfrac{8}{3}, -\dfrac{2\sqrt{10}}{3}\right), \left(-\dfrac{8}{3}, \dfrac{2\sqrt{10}}{3}\right), \left(\dfrac{8}{3}, -\dfrac{2\sqrt{10}}{3}\right)$, and $\left(\dfrac{8}{3}, \dfrac{2\sqrt{10}}{3}\right)$

**35.** $(0, -5), (3, 4), (4, 3)$, and $(5, 0)$ **37.** Either $-5$ and $-3$, or 3 and 5 **39.** 14 feet by 10 feet **41.** 19 cm by 10 cm **43.** $(0, -2), (0, 1)$, and $(2, -1)$ **45.** $(81, 3)$

**47.** If $r_1 = \dfrac{-b + \sqrt{b^2 - 4ac}}{2a}$, then $r_2 = \dfrac{-b - \sqrt{b^2 - 4ac}}{2a}$; if $r_1 = \dfrac{-b - \sqrt{b^2 - 4ac}}{2a}$, then $r_2 = \dfrac{-b + \sqrt{b^2 - 4ac}}{2a}$.

**49.** (a) $f(1) = 7$ (b) $g(1) = 2$ **51.** (a) $f(3) = 13$ (b) $g(3) = 8$ **53.** (a) $f(5) = 19$ (b) $g(5) = 32$

**55.**

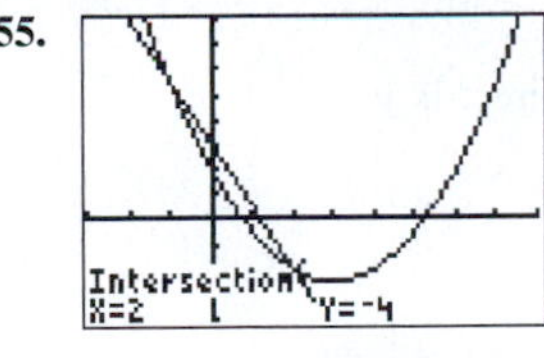


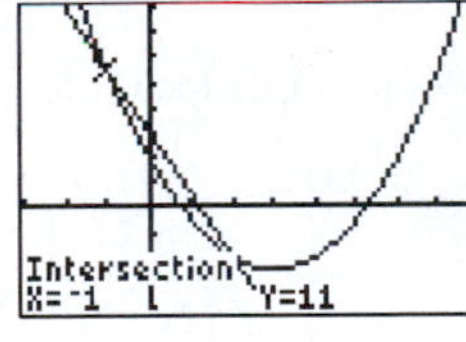


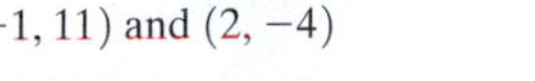

$(-1, 11)$ and $(2, -4)$

**57.**

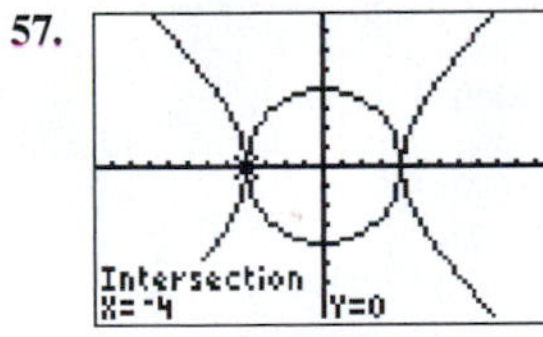


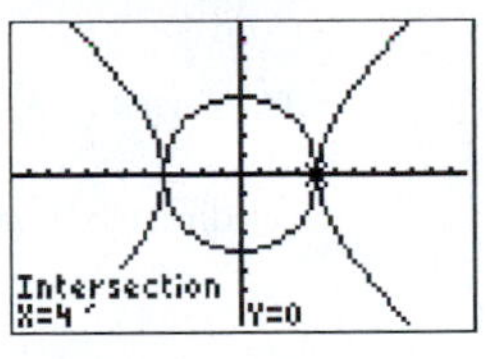


$(-4, 0)$ and $(4, 0)$

**59.**

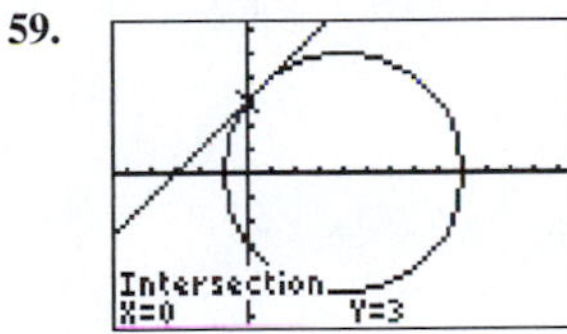


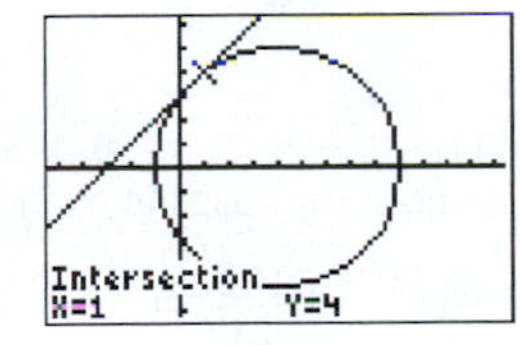


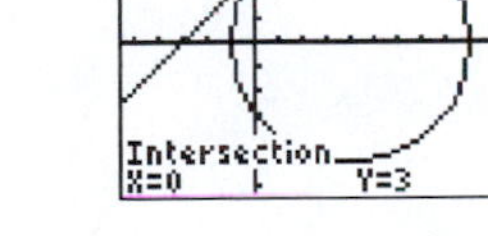

$(0, 3)$ and $(1, 4)$

**61.**

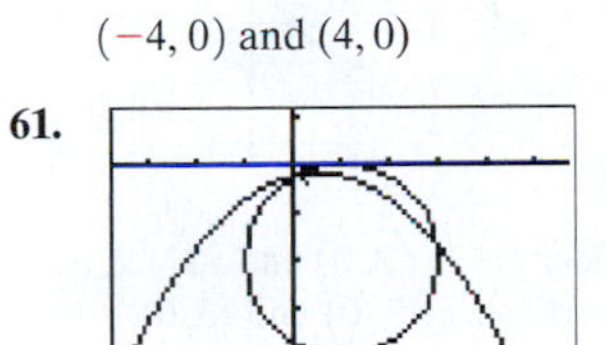


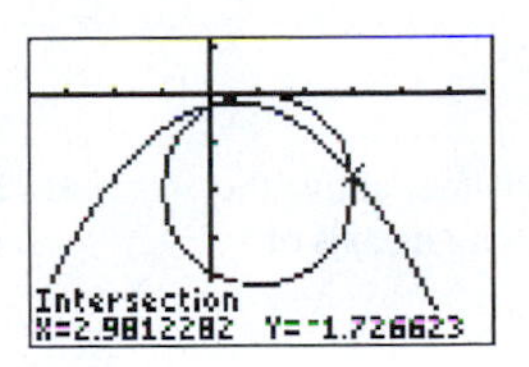


approximately $(0.056, -0.237)$ and $(2.981, -1.727)$

**63.**

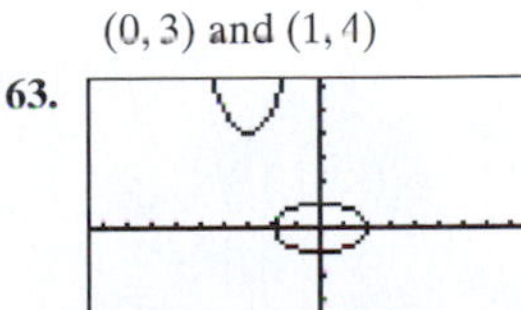

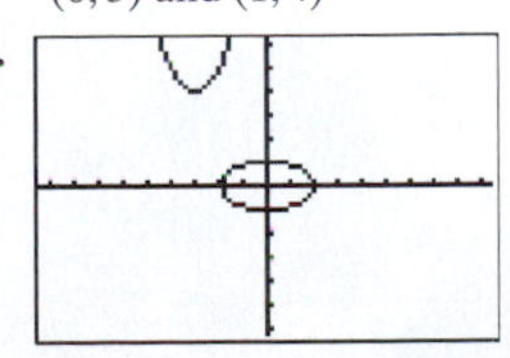

$\varnothing$

## Chapter 10 Review

**1.** 5 **2.** 10 **3.** $2\sqrt{10} \approx 6.32$ **4.** 6 **5.** $3\sqrt{19} \approx 13.08$ **6.** 2.5 **7.** $(-2, 5)$ **8.** $(6, -2)$ **9.** $(-4\sqrt{3}, -3\sqrt{6})$ **10.** $\left(\frac{5}{2}, \frac{1}{2}\right)$

**11.** $\left(\frac{3}{4}, \frac{1}{2}\right)$ **12.** **(a)**

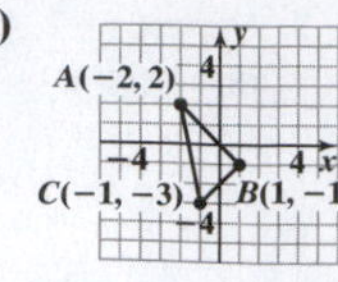


**(b)** $d(A, B) = 3\sqrt{2} \approx 4.24$; $d(B, C) = 2\sqrt{2} \approx 2.83$; $d(A, C) = \sqrt{26} \approx 5.10$

**(c)** $[d(A, B)]^2 + [d(B, C)]^2 \stackrel{?}{=} [d(A, C)]^2$

$$(3\sqrt{2})^2 + (2\sqrt{2})^2 \stackrel{?}{=} (\sqrt{26})^2$$
$$9 \cdot 2 + 4 \cdot 2 \stackrel{?}{=} 26$$
$$18 + 8 \stackrel{?}{=} 26$$
$$26 = 26 \leftarrow \text{True}$$

Therefore, triangle $ABC$ is a right triangle.

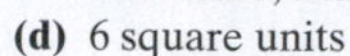
**(d)** 6 square units

**13.** $C = (-2, 1)$; $r = 4$; $(x + 2)^2 + (y - 1)^2 = 16$
**14.** $C = (5, 3)$; $r = 3$; $(x - 5)^2 + (y - 3)^2 = 9$
**15.** $x^2 + y^2 = 16$ **16.** $(x + 3)^2 + (y - 1)^2 = 9$

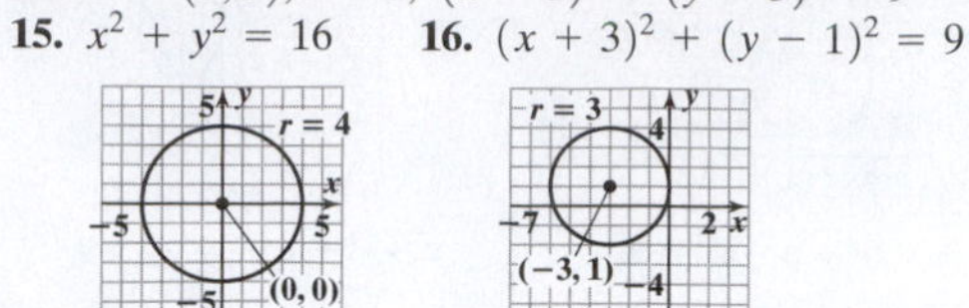
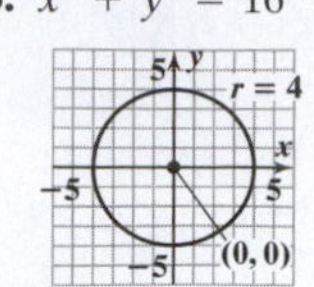


**17.** $(x - 5)^2 + (y + 2)^2 = 1$ **18.** $(x - 4)^2 + y^2 = 7$ **19.** $(x - 2)^2 + (y + 1)^2 = 25$ **20.** $(x + 1)^2 + (y - 3)^2 = 20$ **21.** $C = (0, 0), r = 5$ **22.** $C = (1, 2), r = 2$

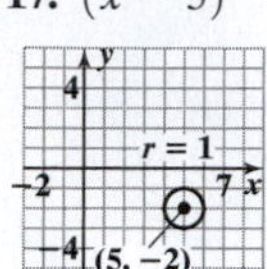


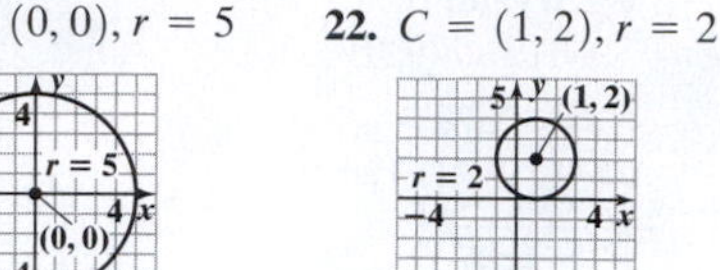


**23.** $C = (0, 4), r = 4$ **24.** $C = (-1, -6), r = 7$ **25.** $C = \left(-2, \frac{3}{2}\right), r = \frac{1}{2}$ **26.** $C = (-3, -3), r = 2$

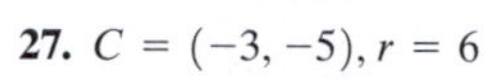
**27.** $C = (-3, -5), r = 6$

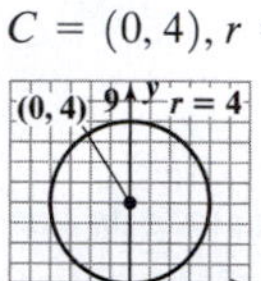


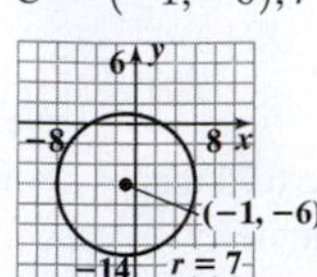


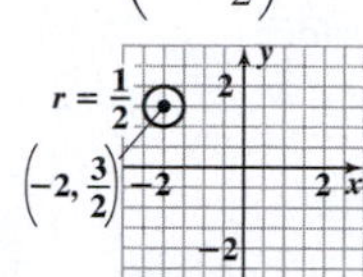

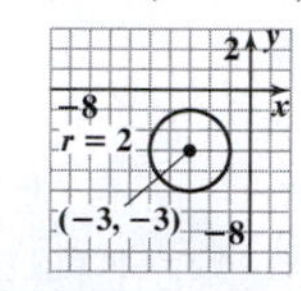


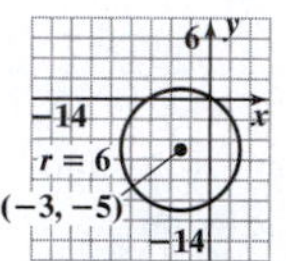


**28.** $C = (4, -2), r = 2$

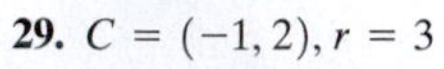
**29.** $C = (-1, 2), r = 3$

**30.** $C = (5, 1), r = 3$ **31.** $x^2 = -12y$ **32.** $y^2 = -16x$

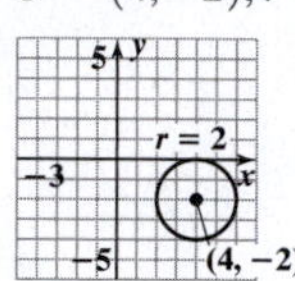


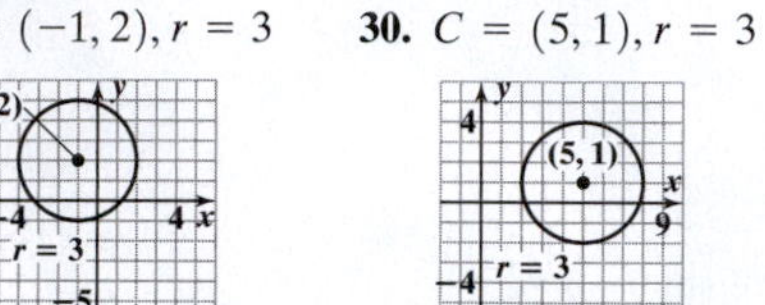


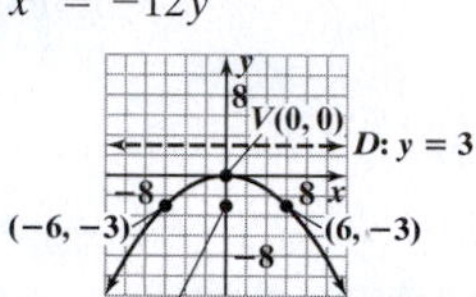


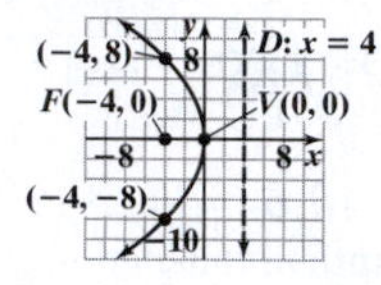


**33.** $y^2 = \frac{1}{2}x$ **34.** $x^2 = 8y$

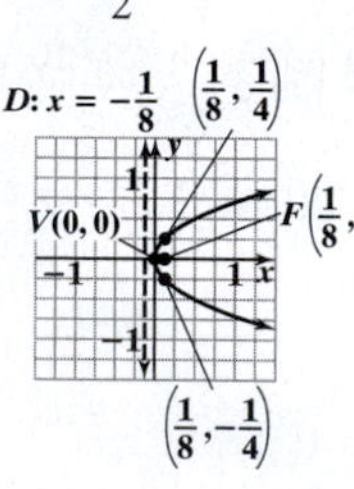

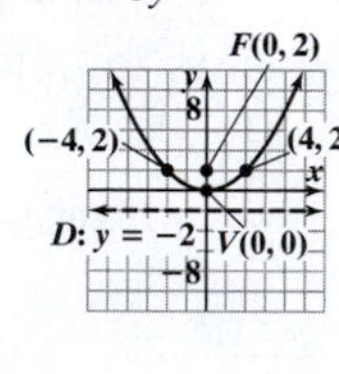


**35.** Vertex $(0, 0)$; focus $\left(0, \frac{1}{2}\right)$;

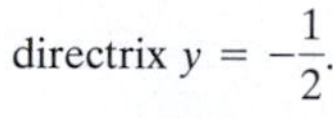
directrix $y = -\frac{1}{2}$.

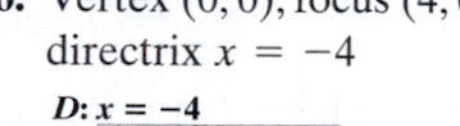
**36.** Vertex $(0, 0)$; focus $(4, 0)$; directrix $x = -4$

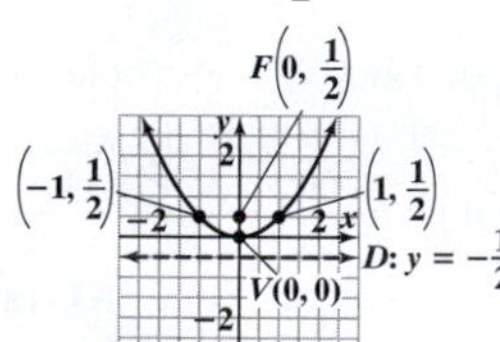


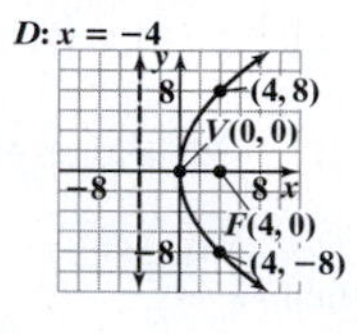


**37.** Vertex $(-1, 3)$; focus $(-1, 5)$; directrix $y = 1$

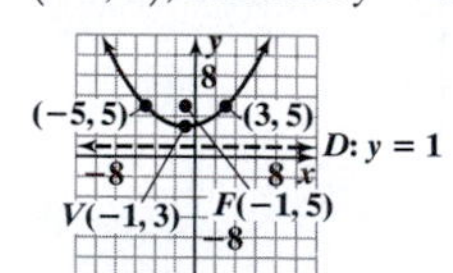


**38.** Vertex $(-3, 4)$; focus $\left(-\frac{7}{2}, 4\right)$; directrix $x = -\frac{5}{2}$

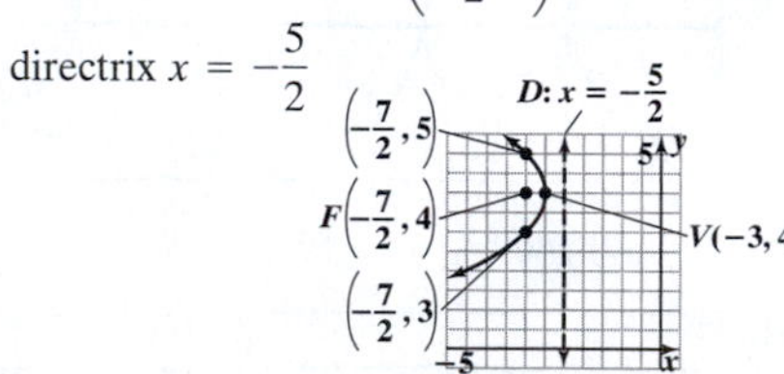


**39.** Vertex $(5, 2)$; focus $\left(5, \frac{5}{4}\right)$; directrix $y = \frac{11}{4}$

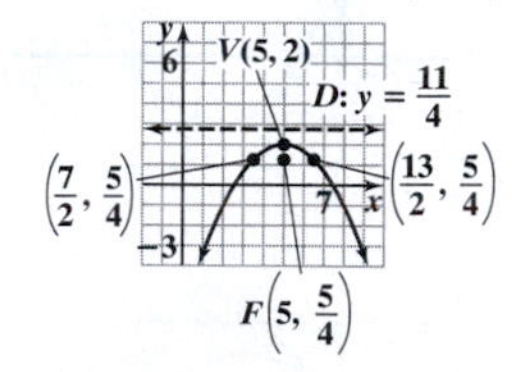


**40.** approximately 127.84 feet above the center of the dish along its axis of symmetry

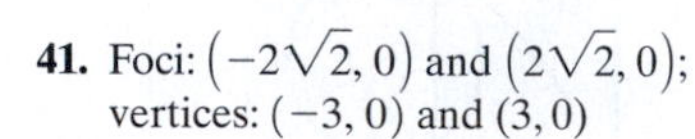
**41.** Foci: $(-2\sqrt{2}, 0)$ and $(2\sqrt{2}, 0)$; vertices: $(-3, 0)$ and $(3, 0)$

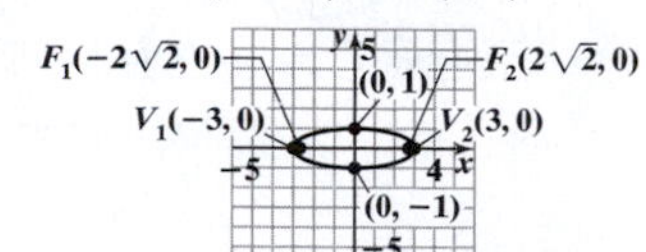


**42.** Foci: $(0, -\sqrt{5})$ and $(0, \sqrt{5})$; vertices: $(0, -3)$, and $(0, 3)$

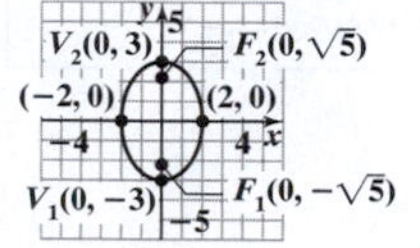

**43.** 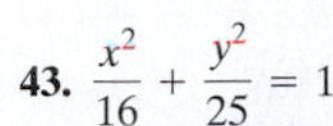
$\dfrac{x^2}{16} + \dfrac{y^2}{25} = 1$

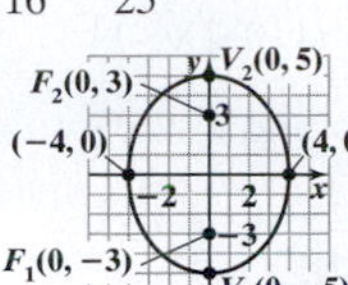


**44.** $\dfrac{x^2}{36} + \dfrac{y^2}{32} = 1$

**45.** $\dfrac{x^2}{100} + \dfrac{y^2}{36} = 1$

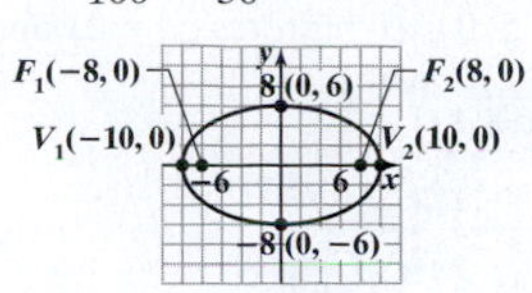


**46.** Vertices: $(-6, -2)$ and $(8, -2)$; foci: $(1 - 2\sqrt{6}, -2)$ and $(1 + 2\sqrt{6}, -2)$

**47.** Vertices: $(-3, -1)$ and $(-3, 9)$; foci: $(-3, 0)$ and $(-3, 8)$

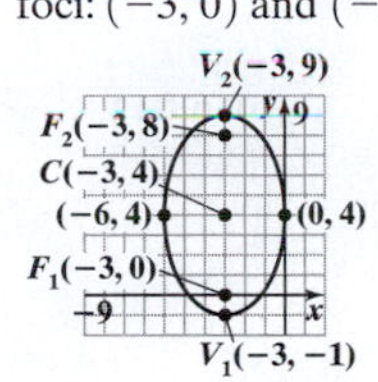


**48.** **(a)** $\dfrac{x^2}{900} + \dfrac{y^2}{256} = 1$ **(b)** Yes

**49.** Vertices: $(-2, 0)$ and $(2, 0)$; foci: $(-\sqrt{13}, 0)$ and $(\sqrt{13}, 0)$

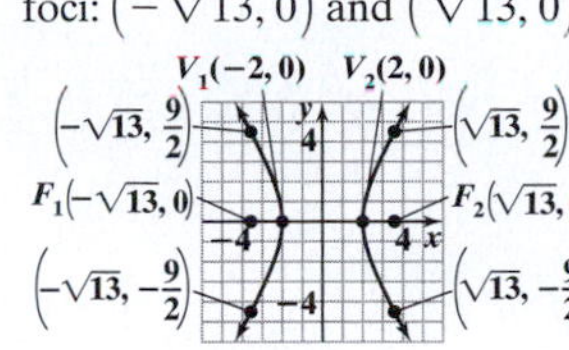


**50.** Vertices: $(0, -5)$ and $(0, 5)$; foci: $(0, -\sqrt{74})$ and $(0, \sqrt{74})$

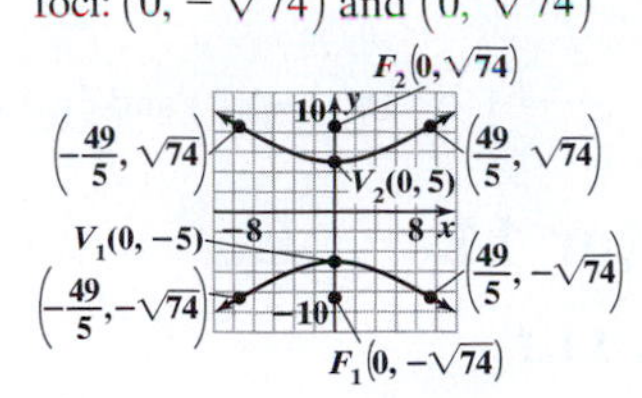


**51.** Vertices: $(0, -5)$ and $(0, 5)$; foci: $(0, -\sqrt{41})$ and $(0, \sqrt{41})$

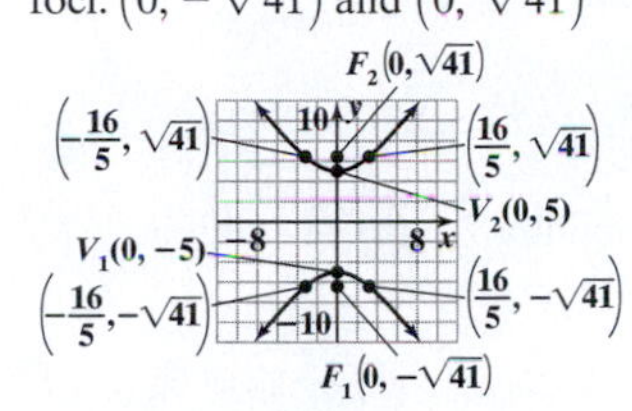


**52.** Asymptotes: $y = x$ and $y = -x$

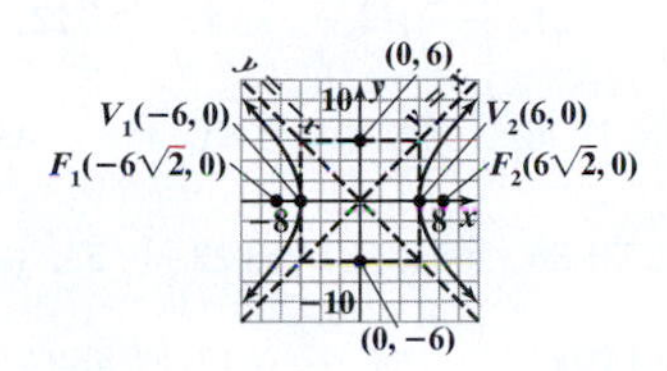


**53.** Asymptotes: $y = \dfrac{5}{2}x$ and $y = -\dfrac{5}{2}x$

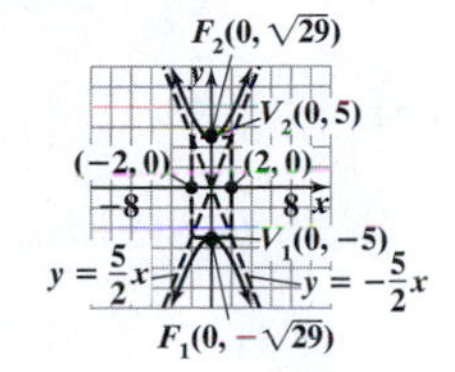


**54.** $\dfrac{x^2}{9} - \dfrac{y^2}{7} = 1$

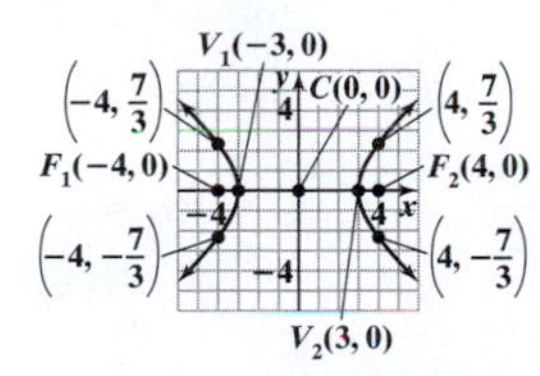


**55.** $\dfrac{y^2}{9} - \dfrac{x^2}{16} = 1$

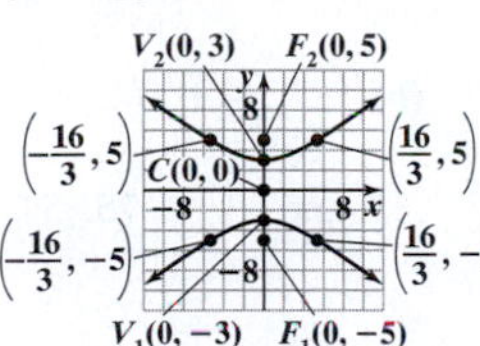


**56.** $\dfrac{y^2}{16} - \dfrac{x^2}{9} = 1$

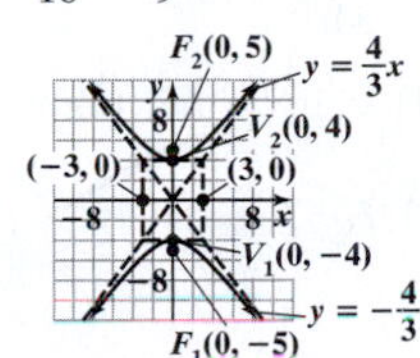


**57.** $(\sqrt{2}, \sqrt{2})$ and $(-\sqrt{2}, -\sqrt{2})$

**58.** $\left(-\dfrac{1}{2}, \dfrac{3}{2}\right)$ and $(1, 3)$ **59.** $\left(-\dfrac{1}{6}, -6\right)$ and $(1, 1)$

**60.** $(-5, -1)$, $(-5, 1)$, $(5, -1)$, and $(5, 1)$

**61.** $(2, -2\sqrt{2})$ and $(2, 2\sqrt{2})$

**62.** $(-1, 3)$ and $(1, 3)$

**63.** $(-\sqrt{3}, -\sqrt{5})$, $(-\sqrt{3}, \sqrt{5})$, $(\sqrt{3}, -\sqrt{5})$, and $(\sqrt{3}, \sqrt{5})$ **64.** $(0, 0)$, $(5, -\sqrt{15})$, and $(5, \sqrt{15})$ **65.** $(2, 4)$ and $(-1, 1)$

**66.** $(-7, -20)$ and $\left(2, \dfrac{5}{2}\right)$ **67.** $(0, 6)$ and $(-6, 0)$ **68.** $(4, 4)$ and $(1, -2)$ **69.** $\varnothing$ **70.** $(-1, -2)$, $(-1, 2)$, $(1, -2)$, and $(1, 2)$

**71.** $(-4, 0)$, $(4, 0)$, and $(0, 4)$ **72.** $(4, 0)$ and $(0, -2)$ **73.** 7 and 5 **74.** 12 cm by 5 cm **75.** 72 inches by 30 inches **76.** 12 inches and 9 inches

## Chapter 10 Test

**1.** $4\sqrt{5}$ **2.** $(-1, 2)$

**3.** $C = (4, -1)$, $r = 3$

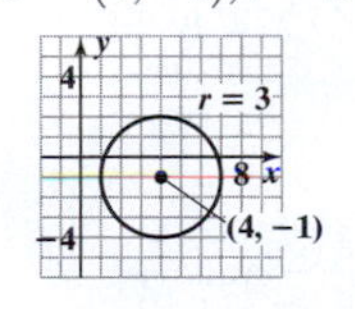


**4.** $C = (-5, 2)$, $r = 4$

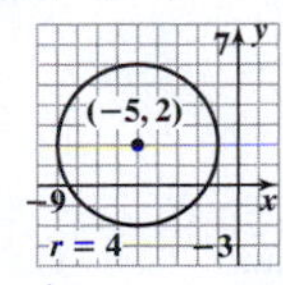


**5.** $(x + 3)^2 + (y - 7)^2 = 36$

**6.** $(x + 5)^2 + (y - 8)^2 = 100$

**7.** Vertex $(1, -2)$; focus $(2, -2)$; directrix $x = 0$

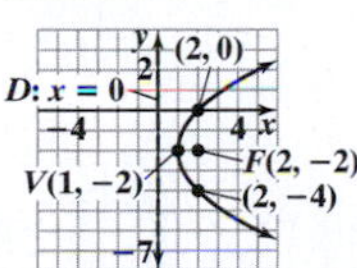


**8.** Vertex $(2, 4)$; focus $\left(2, \dfrac{13}{4}\right)$; directrix $y = \dfrac{19}{4}$

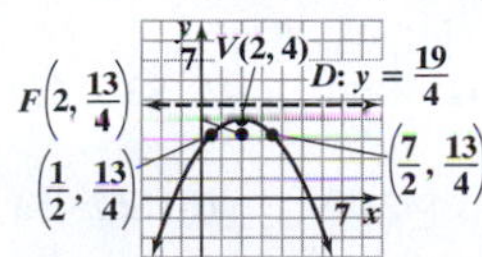


**9.** $x^2 = -16y$ **10.** $(y - 4)^2 = 8(x - 1)$

**12.** Vertices: $(2, -8)$ and $(2, 0)$; foci: $(2, -4 - \sqrt{7})$ and $(2, -4 + \sqrt{7})$

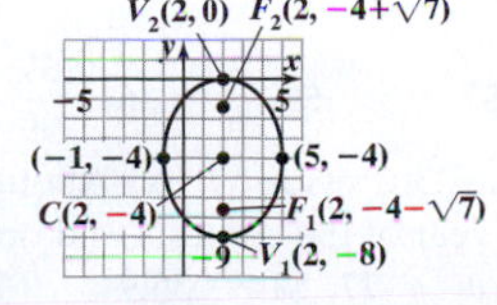


**11.** Foci: $(-4, 0)$ and $(4, 0)$; vertices: $(-5, 0)$ and $(5, 0)$

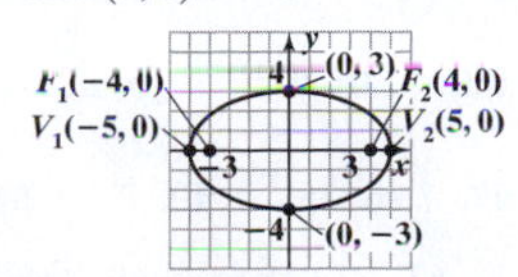

**13.** $\frac{x^2}{9} + \frac{y^2}{25} = 1$ **14.** $\frac{(x+1)^2}{16} + \frac{(y-2)^2}{25} = 1$ **15.** Vertices: $(-1, 0)$ and $(1, 0)$; foci: $(-\sqrt{5}, 0)$ and $(\sqrt{5}, 0)$; asymptotes: $y = 2x$ and $y = -2x$ **16.** Vertices: $(0, -10)$ and $(0, 10)$; foci: $(0, -2\sqrt{41})$ and $(0, 2\sqrt{41})$; asymptotes: $y = \frac{5}{4}x$ and $y = -\frac{5}{4}x$

**17.** $\frac{x^2}{9} - \frac{y^2}{55} = 1$ **18.** $(-4, 1)$ and $(1, -4)$ **19.** $(-\sqrt{5}, -2), (-\sqrt{5}, 2), (\sqrt{5}, -2)$, and $(\sqrt{5}, 2)$ **20.** **(a)** $\frac{x^2}{225} + \frac{y^2}{100} = 1$ **(b)** 6 feet

# Chapter 11

## Section 11.1

**1.** sequence **3.** partial sum **5.** false **7.** Answers will vary. **9.** Answers will vary. **11.** 8, 11, 14, 17, and 20 **13.** $\frac{1}{3}, \frac{1}{2}, \frac{3}{5}, \frac{2}{3}$, and $\frac{5}{7}$ **15.** $-1, 2, -3, 4$, and $-5$ **17.** 3, 5, 9, 17, and 33 **19.** $1, 1, \frac{3}{4}, \frac{1}{2}$, and $\frac{5}{16}$ **21.** $\frac{1}{e}, \frac{2}{e^2}, \frac{3}{e^3}, \frac{4}{e^4}$, and $\frac{5}{e^5}$ **23.** $a_n = 2n$ **25.** $a_n = \frac{n}{n+1}$ **27.** $a_n = n^2 + 2$ **29.** $a_n = (-1)^n n^2$ **31.** 54 **33.** $\frac{55}{2}$ **35.** 14 **37.** 6 **39.** 50 **41.** 45 **43.** $\sum_{k=1}^{15} k$ **45.** $\sum_{i=1}^{12} \frac{1}{i}$ **47.** $\sum_{i=1}^{9} (-1)^{i+1}\left(\frac{1}{3^{i-1}}\right)$ **49.** $\sum_{k=1}^{11} (2k + 3)$ **51.** **(a)** \$12,180 **(b)** \$12,736.36 **(c)** \$21,768.22 **53.** **(a)** 313 million **(b)** 484 million **55.** 1, 1, 2, 3, 5, 8, 13, 21, 34, and 55 **57.** 10, 10.5, 11.025, 11.57625, and 12.1550625 **59.** 8, 10, 13, 17, and 22 **61.** **(a)** 1; 2; 1.5; $1.\overline{6}$; 1.6; 1.625; $\frac{21}{13} \approx 1.615385$; $\frac{34}{21} \approx 1.619048$; $\frac{55}{34} \approx 1.617647$; $\frac{89}{55} \approx 1.618182$ **(b)** around 1.618 **(c)** 1; 0.5; $\frac{2}{3} = 0.\overline{6}$; 0.6; 0.625; $\frac{8}{13} \approx 0.615385$; $\frac{13}{21} \approx 0.619048$; $\frac{21}{34} \approx 0.617647$; $\frac{34}{55} \approx 0.618182$; $\frac{55}{89} \approx 0.617978$ **(d)** around 0.618 **63.** **(a)** $m = 4$ **(b)** $f(1) = -2; f(2) = 2; f(3) = 6; f(4) = 10$ **65.** **(a)** $m = -5$ **(b)** $f(1) = 3; f(2) = -2; f(3) = -7; f(4) = -12$ **67.** 8, 11, 14, 17, and 20 **69.** $\frac{1}{3}, \frac{1}{2}, \frac{3}{5}, \frac{2}{3}$, and $\frac{5}{7}$ **71.** $-1, 2, -3, 4$, and $-5$ **73.** 3, 5, 9, 17, and 33 **75.** 54 **77.** $\frac{55}{2}$ **79.** 14 **81.** 6

## Section 11.2

**1.** arithmetic **3.** false **5.** Answers will vary. **7.** $d = 1$; 6, 7, 8, and 9 **9.** $d = 7$; 9, 16, 23, and 30 **11.** $d = -3$; 4, 1, $-2$, and $-5$ **13.** $d = \frac{1}{2}$; $\frac{11}{2}$, 6, $\frac{13}{2}$, and 7 **15.** $a_n = 3n + 1$; $a_5 = 16$ **17.** $a_n = -5n + 15$; $a_5 = -10$ **19.** $a_n = \frac{1}{3}n + \frac{5}{3}$; $a_5 = \frac{10}{3}$ **21.** $a_n = -\frac{1}{5}n + \frac{26}{5}$; $a_5 = \frac{21}{5}$ **23.** $a_n = 5n - 3$; $a_{20} = 97$ **25.** $a_n = -3n + 15$; $a_{20} = -45$ **27.** $a_n = \frac{1}{4}n + \frac{3}{4}$; $a_{20} = \frac{23}{4}$ **29.** $a = 7$; $d = 5$; $a_n = 5n + 2$ **31.** $a = -23$; $d = 7$; $a_n = 7n - 30$ **33.** $a = 11$; $d = -3$; $a_n = -3n + 14$ **35.** $a = 4$; $d = -\frac{1}{2}$; $a_n = -\frac{1}{2}n + \frac{9}{2}$ **37.** $S_{30} = 2670$ **39.** $S_{25} = 700$ **41.** $S_{40} = -5060$ **43.** $S_{40} = 3160$ **45.** $S_{75} = -9000$ **47.** $S_{30} = 460$ **49.** $x = -\frac{3}{2}$ **51.** There are 630 cans in the stack. **53.** There are 1600 seats in the auditorium. **55.** There are 84 terms in the sequence. **57.** There are 63 terms in the sequence. **59.** It will take about 15.22 years. **61.** **(a)** 3 **(b)** 3; 9; 27; 81 **63.** **(a)** $\frac{1}{2}$ **(b)** 5; $\frac{5}{2}$; $\frac{5}{4}$; $\frac{5}{8}$ **65.** 806.9 **67.** 1427.5

## Section 11.3

**1.** geometric **3.** false **5.** Answers will vary. **7.** $r = 4$; 4, 16, 64, and 256 **9.** $r = \frac{2}{3}$; $\frac{2}{3}, \frac{4}{9}, \frac{8}{27}$, and $\frac{16}{81}$ **11.** $r = \frac{1}{2}$; $\frac{3}{2}, \frac{3}{4}, \frac{3}{8}$, and $\frac{3}{16}$ **13.** $r = \frac{5}{2}$; $\frac{1}{2}, \frac{5}{4}, \frac{25}{8}$, and $\frac{125}{16}$ **15.** arithmetic; $d = 5$ **17.** Neither **19.** geometric; $r = \frac{1}{2}$ **21.** geometric; $r = \frac{2}{3}$ **23.** arithmetic; $d = 4$ **25.** Neither **27.** **(a)** $a_n = 10 \cdot 2^{n-1}$ **(b)** $a_8 = 1280$ **29.** **(a)** $a_n = 100 \cdot \left(\frac{1}{2}\right)^{n-1}$ **(b)** $a_8 = \frac{25}{32}$ **31.** **(a)** $a_n = (-3)^{n-1}$ **(b)** $a_8 = -2187$ **33.** **(a)** $a_n = 100 \cdot (1.05)^{n-1}$ **(b)** $a_8 = 100 \cdot (1.05)^7$ **35.** $a_{10} = 1536$ **37.** $a_{15} = \frac{1}{4096}$ **39.** $a_9 = 0.000000005$ **41.** 8190 **43.** 83.3245952 **45.** 6138 **47.** 7.96875 **49.** 2 **51.** 15 **53.** $\frac{9}{2}$ **55.** $\frac{5}{4}$ **57.** 9 **59.** $\frac{5}{9}$ **61.** $\frac{89}{99}$ **63.** $x = -4$ **65.** **(a)** \$42,000 **(b)** \$62,053 **(c)** \$503,116 **67.** \$13,182 **69.** **(a)** About 1.891 feet **(b)** On the 23rd swing **(c)** About 24.08 feet **(d)** The pendulum will swing a total of 60 feet. **71.** Option A will yield the larger annual salary in the final year of the contract and option B will yield the larger cumulative salary over the life of the contract. **73.** The multiplier is 50. **75.** \$31.14 per share **77.** \$149,035.94 **79.** \$114,401.52 **81.** \$395.09, or about \$395 **83.** $0.4\overline{9} = \frac{1}{2}$ **85.** 2,147,483,646 **87.** Answers will vary. **89.** 1 **91.** approximately 288.1404315 **93.** 41.66310217

### *Putting the Concepts Together*

**1.** Geometric with $a = \frac{3}{4}$ and common ratio $r = \frac{1}{4}$ **2.** Arithmetic with $a = 8$ and $d = 2$ **3.** Arithmetic with $a = 1$ and $d = \frac{7}{9}$ **4.** Neither arithmetic nor geometric **5.** Geometric with $a = 12$ and $r = 2$ **6.** Neither arithmetic nor geometric **7.** 87 **8.** $\sum_{i=1}^{12} \frac{1}{2(6+i)}$ **9.** $a_n = 27 - 2n$; 25, 23, 21, 19, and 17 **10.** $a_n = 11n - 35$; $-24, -13, -2, 9$, and 20 **11.** $a_n = 45 \cdot \left(\frac{1}{5}\right)^{n-1}$; $45, 9, \frac{9}{5}, \frac{9}{25}$, and $\frac{9}{125}$ **12.** $a_n = 150 \cdot (1.04)^{n-1}$; 150, 156, 162.24, 168.7296, and 175.478784 **13.** $S_{11} = 177{,}146$ **14.** $S_{20} = 990$ **15.** $\frac{10{,}000}{9}$ **16.** A party of 24 people would require 11 tables.

### *Section 11.4*

**1.** $n(n-1)(n-2)\cdot \cdots \cdot 3 \cdot 2 \cdot 1$ **3.** Pascal's triangle **5.** false **7.** Beginning with $n$, the exponents of $x$ decrease by 1 in subsequent terms until the exponent is 0. **9.** The degree of each monomial is equal to $n$. **11.** $x^5 + 5x^4 + 10x^3 + 10x^2 + 5x + 1$ **13.** $x^4 - 16x^3 + 96x^2 - 256x + 256$ **15.** $81p^4 + 216p^3 + 216p^2 + 96p + 16$ **17.** $32z^5 - 240z^4 + 720z^3 - 1080z^2 + 810z - 243$ **19.** $x^8 + 8x^6 + 24x^4 + 32x^2 + 16$ **21.** $32p^{15} + 80p^{12} + 80p^9 + 40p^6 + 10p^3 + 1$ **23.** $x^6 + 12x^5 + 60x^4 + 160x^3 + 240x^2 + 192x + 64$ **25.** $16p^8 - 32p^6q^2 + 24p^4q^4 - 8p^2q^6 + q^8$ **27.** 1.00401 **29.** 0.99004 **31.** $84x^5$ **33.** $-108{,}864p^3$

**35.** $\binom{n}{n-1} = \frac{n!}{(n-1)!(n-(n-1))!} = \frac{n!}{(n-1)!1!} = \frac{n \cdot (n-1)!}{(n-1)!} = n$ $\quad \binom{n}{n} = \frac{n!}{n!(n-n)!} = \frac{n!}{n!0!} = \frac{n!}{n!} = 1$

**37.** $a^4 - 8a^3 + 24a^2 - 32a + 16$ **39.** $p^5 + p^4 - 6p^3 - 14p^2 - 11p - 3$

### *Chapter 11 Review*

**1.** $-1, -4, -7, -10$, and $-13$ **2.** $-\frac{1}{5}, 0, \frac{1}{7}, \frac{1}{4}$, and $\frac{1}{3}$ **3.** 6, 26, 126, 626, and 3126 **4.** $3, -6, 9, -12$, and 15 **5.** $\frac{1}{2}, \frac{4}{3}, \frac{9}{4}, \frac{16}{5}$, and $\frac{25}{6}$ **6.** $\pi, \frac{\pi^2}{2}, \frac{\pi^3}{3}, \frac{\pi^4}{4}$, and $\frac{\pi^5}{5}$ **7.** $a_n = -3n$ **8.** $a_n = \frac{n}{3}$ **9.** $a_n = 5 \cdot 2^{n-1}$ **10.** $a_n = (-1)^n \cdot \frac{n}{2}$ **11.** $a_n = n^2 + 5$ **12.** $a_n = \frac{n-1}{n+1}$ **13.** 65 **14.** $\frac{33}{2}$ **15.** $-30$ **16.** $\frac{26}{3}$ **17.** $\sum_{i=1}^{15}(4+3i)$ **18.** $\sum_{i=1}^{8}\frac{1}{3^i}$ **19.** $\sum_{i=1}^{10}\frac{i^3+1}{i+1}$ **20.** $\sum_{i=1}^{7}[(-1)^{i-1} \cdot i^2]$ **21.** arithmetic with $d = 6$ **22.** arithmetic with $d = \frac{3}{2}$ **23.** not arithmetic **24.** not arithmetic **25.** arithmetic with $d = 4$ **26.** not arithmetic **27.** $a_n = 8n - 5$; $a_{25} = 195$ **28.** $a_n = -3n - 1$; $a_{25} = -76$ **29.** $a_n = -\frac{1}{3}n + \frac{22}{3}$; $a_{25} = -1$ **30.** $a_n = 6n + 5$; $a_{25} = 155$ **31.** $a_n = \frac{18}{5}n - \frac{19}{5}$; $a_{25} = \frac{431}{5}$ **32.** $a_n = -4n - 4$; $a_{25} = -104$ **33.** 4320 **34.** $-2140$ **35.** $-4080$ **36.** $\frac{1875}{4}$ or 468.75 **37.** 2106 **38.** 300 yards **39.** geometric with $r = 6$ **40.** geometric with $r = -3$ **41.** not geometric **42.** geometric with $r = \frac{2}{3}$ **43.** geometric with $r = -2$ **44.** not geometric **45.** $a_n = 4 \cdot 3^{n-1}$; $a_{10} = 78{,}732$ **46.** $a_n = 8 \cdot \left(\frac{1}{4}\right)^{n-1}$; $a_{10} = \frac{1}{32{,}768}$ **47.** $a_n = 5 \cdot (-2)^{n-1}$; $a_{10} = -2560$ **48.** $a_n = 1000 \cdot (1.08)^{n-1}$; $a_{10} \approx 1999.005$ **49.** 65,534 **50.** $\approx 45.71428571$ **51.** $\frac{12{,}285}{4}$ or 3071.25 **52.** $-258{,}280{,}320$ **53.** $\frac{20}{3}$ **54.** $\frac{100}{3}$ **55.** $\frac{5}{4}$ **56.** $\frac{8}{9}$ **57.** After 72 years, there will be 3.125 grams of the Tritium remaining. **58.** After 15 minutes a total of about 38.15 billion e-mails will have been sent. **59.** After 25 years, Scott's 403(b) will be worth \$360,114.89. **60.** The lump sum option would yield more money after 26 years. **61.** Sheri would need to contribute \$534.04, or about \$534, each month to reach her goal. **62.** When Samantha turns 18, the plan will be worth \$62,950.79 and will cover about 185 credit hours. **63.** 120 **64.** 7920 **65.** 5040 **66.** 1716 **67.** 35 **68.** 252 **69.** 1 **70.** 1 **71.** $z^4 + 4z^3 + 6z^2 + 4z + 1$ **72.** $y^5 - 15y^4 + 90y^3 - 270y^2 + 405y - 243$ **73.** $729y^6 + 5832y^5 + 19{,}440y^4 + 34{,}560y^3 + 34{,}560y^2 + 18{,}432y + 4096$ **74.** $16x^8 - 96x^6 + 216x^4 - 216x^2 + 81$ **75.** $81p^4 - 216p^3q + 216p^2q^2 - 96pq^3 + 16q^4$ **76.** $a^{15} + 15a^{12}b + 90a^9b^2 + 270a^6b^3 + 405a^3b^4 + 243b^5$ **77.** $-448x^5$ **78.** $14{,}784x^5$

### *Chapter 11 Test*

**1.** arithmetic with $a = -15$ and $d = 8$ **2.** geometric with $a = -4$ and $r = -4$ **3.** neither arithmetic nor geometric **4.** arithmetic with $a = -\frac{1}{5}$ and $d = \frac{2}{5}$ **5.** neither arithmetic nor geometric **6.** geometric with $a = 21$ and $r = 3$ **7.** $\frac{17269}{1200}$ **8.** $\sum_{i=1}^{8}\frac{i+2}{i+4}$ **9.** $a_n = 10n - 4$; 6, 16, 26, 36, and 46 **10.** $a_n = 4 - 4n$; $0, -4, -8, -12$, and $-16$ **11.** $a_n = 10 \cdot 2^{n-1}$; 10, 20, 40, 80, and 160 **12.** $a_n = (-3)^{n-1}$; $1, -3, 9, -27$, and 81 **13.** 720 **14.** $-\frac{132{,}860}{9}$ **15.** 324 **16.** 6435 **17.** 792 **18.** $625m^4 - 1000m^3 + 600m^2 - 160m + 16$ **19.** about \$64,512 **20.** 2012, 2117, and 2125

### *Cumulative Review Chapters R–11*

**1.** $y = \frac{3x-4}{15}$ or $y = \frac{1}{5}x - \frac{4}{15}$ **2.** $f(2) = 9; f(-3) = 19$ **3.** $\{32\}$ **4.** $\left\{-\frac{2}{5}, 1\right\}$ **5.** $\left\{\frac{-7-\sqrt{73}}{6}, \frac{-7+\sqrt{73}}{6}\right\}$ **6.** $\{60\}$
**7.** $\left\{\frac{11}{4}\right\}$ **8.** $\{-4, -2, 0\}$ **9.** $(-6, \infty)$ **10.** $\left[-3, \frac{1}{3}\right]$ **11.** $(x+2)(2x-9)$ **12.** $(2x-1)(3x^2+2)$ **13.** $20x^3 - 22x^2 + 11x - 3$
**14.** $\frac{(x-6)(x+2)}{(x+4)(x-1)}$ **15.** $1 - i$ **16.** $\{x \mid x \geq 15\}$ or $[15, \infty)$ **17.** $y = -7x + 11$ **18.** $(4, -1)$
**19.** **20.** $(x-4)^2 + (y+3)^2 = 36$ **21.** **22.** 1340 **23.** 8

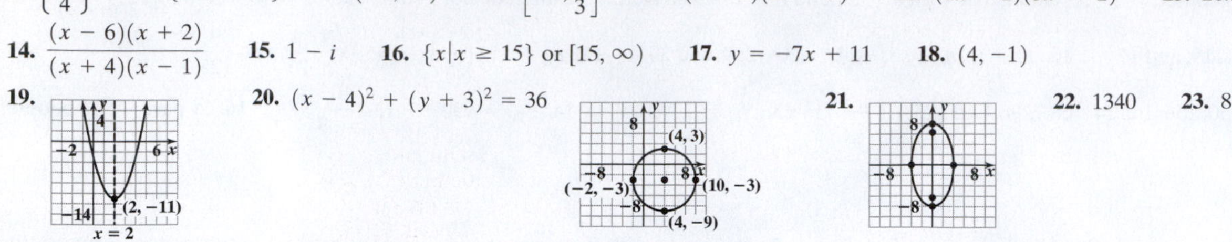


**24.** It would take about 2.73 hours to cut the lot if both machines worked together. **25.** $\frac{325}{3}$ metric tons of pure aluminum must be added.

# Appendix A

**1.** constant function **3.** false **5.** (c) **7.** (e) **9.** (b) **11.** (f)
**13.** **15.** **17.**

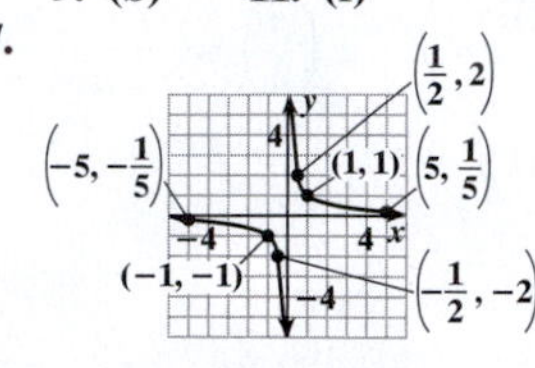

# Applications Index

# Subject Index

# Photo Credits

| | |
|---|---|
| **Chapter R** | Page 1, Edward S. May |
| **Chapter 1** | Page 51, Internal Revenue Service; Page 84 (left), Adam Crowley/Getty Images, Inc.-Photodisc; Page 84 (right), Dagli Orti/Picture Desk, Inc./Kobal Collection |
| **Chapter 2** | Page 135, Courtesy of Florence Nightingale Museum, London; Page 171, Joel Rogers |
| **Chapter 3** | Pages 189, 214, Katie Deits/Index Stock Imagery, Inc. |
| **Chapter 4** | Pages 256, 277, Bill Brooks/Masterfile Corporation |
| **Chapter 5** | Page 347 (top), AP Wide World Photos; Page 347 (bottom), Rick Fischer/Masterfile Corporation; Pages 355, 430, Jane Moriarty/Chicago from the Lake |
| **Chapter 6** | Page 455, Chase Jarvis/CORBIS-NY |
| **Chapter 7** | Page 533, The Isamu Noguchi Foundation, Inc. |
| **Chapter 8** | Pages 610, 674, AP Wide World Photos; Page 618, Jill Davis |
| **Chapter 9** | Page 697, David Weintraub/Photo Researchers, Inc. |
| **Chapter 10** | Pages 772, 804 (left) Victor Habbick Visions/Photo Researchers, Inc.; Page 793, Alan Becker/Getty Images, Inc.-Image Bank; Page 804 (right) Scott McDonald/Hedrich Blessing |
| **Chapter 11** | Page 833, Ambient Images; Page 840, Getty Images, Inc.-Image Bank |

**INTERMEDIATE ALGEBRA**
**CHAPTER TEST PREP VIDEO CD**
**Michael Sullivan, III & Katherine R. Struve**

ISBN 0-13-134607-5
CD License Agreement

Pearson Prentice Hall
Pearson Education, Inc.
Upper Saddle River, NJ 07458


**READ THIS LICENSE CAREFULLY BEFORE OPENING THIS PACKAGE. BY OPENING THIS PACKAGE, YOU ARE AGREEING TO THE TERMS AND CONDITIONS OF THIS LICENSE. IF YOU DO NOT AGREE, DO NOT OPEN THE PACKAGE. PROMPTLY RETURN THE UNOPENED PACKAGE AND ALL ACCOMPANYING ITEMS TO THE PLACE YOU OBTAINED THEM. THESE TERMS APPLY TO ALL LICENSED SOFTWARE ON THE DISK EXCEPT THAT THE TERMS FOR USE OF ANY SHAREWARE OR FREEWARE ON THE DISKETTES ARE AS SET FORTH IN THE ELECTRONIC LICENSE LOCATED ON THE DISK:**

**1. GRANT OF LICENSE and OWNERSHIP:** The enclosed CD-ROM ("Software") is licensed, not sold, to you by Pearson Education, Inc. publishing as Pearson Prentice Hall ("We" or the "Company") in consideration of your adoption of the accompanying Company textbooks and/or other materials, and your agreement to these terms. You own only the disk(s) but we and/or our licensors own the Software itself. This license allows instructors and students enrolled in the course using the Company textbook that accompanies this Software (the "Course") to use and display the enclosed copy of the Software on up to one computer of an educational institution, for academic use only, so long as you comply with the terms of this Agreement. You may make one copy for back up only. We reserve any rights not granted to you.

**2. USE RESTRICTIONS:** You may not sell or license copies of the Software or the Documentation to others. You may not transfer, distribute or make available the Software or the Documentation, except to instructors and students in your school who are users of the adopted Company textbook that accompanies this Software in connection with the course for which the textbook was adopted. You may not reverse engineer, disassemble, decompile, modify, adapt, translate or create derivative works based on the Software or the Documentation. You may be held legally responsible for any copying or copyright infringement that is caused by your failure to abide by the terms of these restrictions.

**3. TERMINATION:** This license is effective until terminated. This license will terminate automatically without notice from the Company if you fail to comply with any provisions or limitations of this license. Upon termination, you shall destroy the Documentation and all copies of the Software. All provisions of this Agreement as to limitation and disclaimer of warranties, limitation of liability, remedies or damages, and our ownership rights shall survive termination.

**4. DISCLAIMER OF WARRANTY: THE COMPANY AND ITS LICENSORS MAKE NO WARRANTIES ABOUT THE SOFTWARE, WHICH IS PROVIDED "AS-IS." IF THE DISK IS DEFECTIVE IN MATERIALS OR WORKMANSHIP, YOUR ONLY REMEDY IS TO RETURN IT TO THE COMPANY WITHIN 30 DAYS FOR REPLACEMENT UNLESS THE COMPANY DETERMINES IN GOOD FAITH THAT THE DISK HAS BEEN MISUSED OR IMPROPERLY INSTALLED, REPAIRED, ALTERED OR DAMAGED. THE COMPANY DISCLAIMS ALL WARRANTIES, EXPRESS OR IMPLIED, INCLUDING WITHOUT LIMITATION, THE IMPLIED WARRANTIES OF MERCHANTABILITY AND FITNESS FOR A PARTICULAR PURPOSE. THE COMPANY DOES NOT WARRANT, GUARANTEE OR MAKE ANY REPRESENTATION REGARDING THE ACCURACY, RELIABILITY, CURRENTNESS, USE, OR RESULTS OF USE, OF THE SOFTWARE.**

**5. LIMITATION OF REMEDIES AND DAMAGES: IN NO EVENT, SHALL THE COMPANY OR ITS EMPLOYEES, AGENTS, LICENSORS OR CONTRACTORS BE LIABLE FOR ANY INCIDENTAL, INDIRECT, SPECIAL OR CONSEQUENTIAL DAMAGES ARISING OUT OF OR IN CONNECTION WITH THIS LICENSE OR THE SOFTWARE, INCLUDING, WITHOUT LIMITATION, LOSS OF USE, LOSS OF DATA, LOSS OF INCOME OR PROFIT, OR OTHER LOSSES SUSTAINED AS A RESULT OF INJURY TO ANY PERSON, OR LOSS OF OR DAMAGE TO PROPERTY, OR CLAIMS OF THIRD PARTIES, EVEN IF THE COMPANY OR AN AUTHORIZED REPRESENTATIVE OF THE COMPANY HAS BEEN ADVISED OF THE POSSIBILITY OF SUCH DAMAGES.** SOME JURISDICTIONS DO NOT ALLOW THE LIMITATION OF DAMAGES IN CERTAIN CIRCUMSTANCES, SO THE ABOVE LIMITATIONS MAY NOT ALWAYS APPLY.

**6. GENERAL:** THIS AGREEMENT SHALL BE CONSTRUED IN ACCORDANCE WITH THE LAWS OF THE UNITED STATES OF AMERICA AND THE STATE OF NEW YORK, APPLICABLE TO CONTRACTS MADE IN NEW YORK, EXCLUDING THE STATE'S LAWS AND POLICIES ON CONFLICTS OF LAW, AND SHALL BENEFIT THE COMPANY, ITS AFFILIATES AND ASSIGNEES. This Agreement is the complete and exclusive statement of the agreement between you and the Company and supersedes all proposals, prior agreements, oral or written, and any other communications between you and the company or any of its representatives relating to the subject matter. If you are a U.S. Government user, this Software is licensed with "restricted rights" as set forth in subparagraphs (a)–(d) of the Commercial Computer-Restricted Rights clause at FAR 52.227-19 or in subparagraphs (c)(1)(ii) of the Rights in Technical Data and Computer Software clause at DFARS 252.227-7013, and similar clauses, as applicable. Should you have any questions concerning this agreement or if you wish to contact the Company for any reason, please contact in writing: Pearson Education, Inc., One Lake Street, Upper Saddle River, New Jersey 07458.

**System Requirements**

- Windows

Pentium II 300 MHz processor
Windows 98, NT, 2000, ME, or XP
64 MB RAM (128 MB RAM required for Windows XP)
4.3 MB available hard drive space (optional—for minimum QuickTime installation)
800 × 600 resolution
8x or faster CD-ROM drive
QuickTime 6.x
Sound card

- Macintosh

PowerPC G3 233 MHz or better
Mac OS 9.x or 10.x
64 MB RAM
10 MB available hard drive space for Mac OS 9, 19 MB on OS X (optional—if QuickTime installation is needed)
800 × 600 resolution
8x or faster CD-ROM drive
QuickTime 6.x

**Support Information**

If you are having problems with this software, call (800) 677-6337 between 8:00 a.m. and 8:00 p.m. EST, Monday through Friday, and 5:00 p.m. through Midnight EST on Sundays. You can also get support by filling out the web form located at: http://247.prenhall.com/mediaform

Our technical staff will need to know certain things about your system in order to help us solve your problems more quickly and efficiently. If possible, please be at your computer when you call for support. You should have the following information ready:

- Textbook ISBN
- CD-ROM ISBN
- corresponding product and title
- computer make and model
- Operating System (Windows or Macintosh) and Version
- RAM available
- hard disk space available
- Sound card? Yes or No
- printer make and model
- network connection
- detailed description of the problem, including the exact wording of any error messages.

NOTE: Pearson does not support and/or assist with the following:

- third-party software (i.e. Microsoft including Microsoft Office suite, Apple, Borland, etc.)
- homework assistance
- Textbooks and CD-ROMs purchased used are not supported and are non-replaceable. To purchase a new CD-ROM, contact Pearson Individual Order Copies at 1-800-282-0693.

## Working with Radicals (Getting Ready for Chapter 7, Chapter 7)

**Simplifying $\sqrt[n]{a^n}$**

- If $n \geq 2$ is a positive integer and $a$ is a real number, then
  $\sqrt[n]{a^n} = a$ if $n \geq 3$ is odd
  $\sqrt[n]{a^n} = |a|$ if $n \geq 2$ is even
- If $a$ is a real number and $n \geq 2$ is an integer, then $a^{1/n} = \sqrt[n]{a}$ provided that $\sqrt[n]{a}$ exists.
- If $a$ is a real number, $m/n$ is a rational number in lowest terms with $n \geq 2$, then $a^{m/n} = \sqrt[n]{a^m} = \left(\sqrt[n]{a}\right)^m$ provided that $\sqrt[n]{a}$ exists.
- If $m/n$ is a rational number and if $a$ is a nonzero real number, then $a^{-m/n} = \dfrac{1}{a^{m/n}}$ or $\dfrac{1}{a^{-m/n}} = a^{m/n}$

**Product Property of Radicals**

If $\sqrt[n]{a}$ and $\sqrt[n]{b}$ are real numbers and $n \geq 2$ is an integer, then $\sqrt[n]{a} \cdot \sqrt[n]{b} = \sqrt[n]{ab}$

**Quotient Property of Radicals**

If $\sqrt[n]{a}$ and $\sqrt[n]{b}$ are real numbers, $b \neq 0$ and $n \geq 2$ is an integer, then $\dfrac{\sqrt[n]{a}}{\sqrt[n]{b}} = \sqrt[n]{\dfrac{a}{b}}$

## Quadratic Equations and Quadratic Functions (Chapter 8)

- **Square Root Property**
  If $x^2 = p$, then $x = \sqrt{p}$ or $x = -\sqrt{p}$
- **Pythagorean Theorem**
  In a right triangle, the square of the length of the hypotenuse is equal to the sum of the squares of the lengths of the legs. That is, $\text{leg}^2 + \text{leg}^2 = \text{hypotenuse}^2$.
- **The Quadratic Formula**
  The solutions to the equation $ax^2 + bx + c = 0$, $a \neq 0$, are given by $x = \dfrac{-b \pm \sqrt{b^2 - 4ac}}{2a}$
- **Discriminant**
  For the quadratic equation $ax^2 + bx + c = 0$, $a \neq 0$:
  - If $b^2 - 4ac > 0$, the equation has two unequal real solutions.
    - If $b^2 - 4ac$ is a perfect square, the equation has two rational solutions.
    - If $b^2 - 4ac$ is not a perfect square, the equation has two irrational solutions.
  - If $b^2 - 4ac = 0$, the equation has a repeated real solution.
  - If $b^2 - 4ac < 0$, the equation has two complex solutions that are not real.
- **Vertex of a Parabola**
  Any quadratic function of the form $f(x) = ax^2 + bx + c$, $a \neq 0$, will have vertex $\left(-\dfrac{b}{2a}, f\left(-\dfrac{b}{2a}\right)\right)$
- **The $x$-Intercepts of the Graph of a Quadratic Function**
  1. If $b^2 - 4ac > 0$, the graph of $f(x) = ax^2 + bx + c$ has two different $x$-intercepts.
  2. If $b^2 - 4ac = 0$, the graph of $f(x) = ax^2 + bx + c$ has one $x$-intercept.
  3. If $b^2 - 4ac < 0$, the graph of $f(x) = ax^2 + bx + c$ has no $x$-intercepts.

## Properties of Logarithms (Chapter 9)

$a^{\log_a M} = M$ $\quad\quad \log_a a^r = r$

$\log_a(MN) = \log_a M + \log_a N$

$\log_a\left(\dfrac{M}{N}\right) = \log_a M - \log_a N$

$\log_a M^r = r \log_a M$

$\log_a M = \dfrac{\log_b M}{\log_b a} = \dfrac{\log M}{\log a} = \dfrac{\ln M}{\ln a}$

## Formulas from Chapter 10

**The Distance Formula**

$d(P_1, P_2) = \sqrt{(x_2 - x_1)^2 + (y_2 - y_1)^2}$

**The Midpoint Formula**

$M = \left(\dfrac{x_1 + x_2}{2}, \dfrac{y_1 + y_2}{2}\right)$

**Standard Form of an Equation of a Circle**

$(x - h)^2 + (y - k)^2 = r^2$ with radius $r$ and center $(h, k)$

**General Form of the Equation of a Circle**

$x^2 + y^2 + ax + by + c = 0$ when the graph exists

## Functions (Chapter 2)

- A **function** is a special type of relation where any given input, $x$, corresponds to only one output $y$. Functions can be represented through maps, sets of ordered pairs, equations, or graphs.
- **Vertical Line Test:** A set of points in the $xy$-plane is the graph of a function if and only if every vertical line intersects the graph in at most one point.
- The graph of a function, $f$, is the set of all ordered pairs $(x, f(x))$.
- When only an equation of a function is given, the **domain** of the function is the largest set of real numbers for which $f(x)$ is a real number.
- The **range** of a function is the set of all outputs of the function.

## Formulas for Lines and Slope (Chapter 3)

| | |
|---|---|
| Standard form of a line | $Ax + By = C$ |
| Equation of a vertical line | $x = a$ where $a$ is the $x$-intercept |
| Equation of a horizontal line | $y = b$ where $b$ is the $y$-intercept |
| Slope of a line | $m = \dfrac{y_2 - y_1}{x_2 - x_1}, x_1 \neq x_2$<br>Slope undefined if $x_1 = x_2$ |
| Point-slope form of a line | $y - y_1 = m(x - x_1)$ |
| Slope-intercept form of a line | $y = f(x) = mx + b$ |

## Steps for Factoring (Chapter 5)

**Step 1:** Factor out the Greatest Common Factor (GCF), if any exists.

**Step 2:** Count the number of terms.

**Step 3:** **(a)** 2 terms

- Is it the difference of two squares? If so,
  $A^2 - B^2 = (A - B)(A + B)$
- Is it the difference of two cubes? If so,
  $A^3 - B^3 = (A - B)(A^2 + AB + B^2)$
- Is it the sum of two cubes? If so,
  $A^3 + B^3 = (A + B)(A^2 - AB + B^2)$

**(b)** 3 terms

- Is it a perfect square trinomial? If so,
  $A^2 + 2AB + B^2 = (A + B)^2$ or
  $A^2 - 2AB + B^2 = (A - B)^2$
- Is the coefficient of the square term 1? If so, $x^2 + bx + c = (x + m)(x + n)$ where $mn = c$ and $m + n = b$
- Is the coefficient of the square term different from 1? If so,
  **a.** Use factoring by grouping
  **b.** Use trial and error

**(c)** 4 terms

- Use factoring by grouping

**Step 4:** Check your work by multiplying out the factored form.

## The Rules of Exponents (Getting Ready for Chapter 5, Chapter 7)

If $a$ and $b$ are real numbers and if $r$ and $s$ are rational numbers, then assuming the expression is defined,

| | | |
|---|---|---|
| Zero Exponent Rule: | $a^0 = 1$ | if $a \neq 0$ |
| Negative Exponent Rule: | $a^{-r} = \dfrac{1}{a^r}$ | if $a \neq 0$ |
| Product Rule: | $a^r \cdot a^s = a^{r+s}$ | |
| Quotient Rule: | $\dfrac{a^r}{a^s} = a^{r-s} = \dfrac{1}{a^{s-r}}$ | if $a \neq 0$ |
| Power Rule: | $(a^r)^s = a^{r \cdot s}$ | |
| Product to Power Rule: | $(a \cdot b)^r = a^r \cdot b^r$ | |
| Quotient to Power Rule: | $\left(\dfrac{a}{b}\right)^r = \dfrac{a^r}{b^r}$ | if $b \neq 0$ |
| Quotient to a Negative Power Rule: | $\left(\dfrac{a}{b}\right)^{-r} = \left(\dfrac{b}{a}\right)^r$ | if $a \neq 0$, $b \neq 0$ |

## Working with Rational Expressions (Chapter 6)

| | | |
|---|---|---|
| Multiplying Rational Expressions | $\dfrac{a}{b} \cdot \dfrac{c}{d} = \dfrac{ac}{bd}$ | $b \neq 0, d \neq 0$ |
| Adding Rational Expressions | $\dfrac{a}{c} + \dfrac{b}{c} = \dfrac{a + b}{c}$ | $c \neq 0$ |
| Subtracting Rational Expressions | $\dfrac{a}{c} - \dfrac{b}{c} = \dfrac{a - b}{c}$ | $c \neq 0$ |
| Dividing Rational Expressions | $\dfrac{a}{b} \div \dfrac{c}{d} = \dfrac{\frac{a}{b}}{\frac{c}{d}} = \dfrac{a}{b} \cdot \dfrac{d}{c} = \dfrac{ad}{bc}$ | $b \neq 0, c \neq 0, d \neq 0$ |